CONTENTS—1992 HVAC SYSTEMS AND EQUIPMENT

AIR-CONDITIONING AND HEATING SYSTEMS

AIR-HANDLING EQUIPMENT

HEATING EQUIPMENT

GENERAL COMPONENTS

UNITARY EQUIPMENT

CONTENTS—1991 HVAC APPLICATIONS

COMFORT AIR CONDITIONING AND HEATING

INDUSTRIAL AND SPECIAL AIR CONDITIONING AND VENTILATION

ENERGY SOURCES

BUILDING OPERATION AND MAINTENANCE

GENERAL APPLICATIONS

1994 ASHRAE HANDBOOK

REFRIGERATION

I-P Edition

American Society of Heating, Refrigerating and Air-Conditioning Engineers, Inc.
1791 Tullie Circle, N.E., Atlanta, GA 30329
404-636-8400

DEDICATED

TO THE ADVANCEMENT OF

THE PROFESSION

AND ITS ALLIED INDUSTRIES

Comments, criticisms, and suggestions regarding the subject matter are invited. Any errors or omissions in the data should be brought to the attention of the Editor. If required, an errata sheet will be issued at approximately the same time as the next Handbook. Notice of any significant errors found after that time will be published in the ASHRAE *Journal.*

ISBN 1-883413-15-X
ISSN 1041-2344

CONTENTS

INDUSTRIAL APPLICATIONS

LOW TEMPERATURE APPLICATIONS

REFRIGERATION EQUIPMENT

UNITARY REFRIGERATION EQUIPMENT

ADDITIONS AND CORRECTIONS

INDEX

CONTRIBUTORS

In addition to the Technical Committees, the following individuals contributed significantly to this volume. The appropriate chapter numbers follow each contributor's name.

Larry A. Rolison (1)
Tri Com Refrigeration

James D. Gaines (2)
Hussman Corporation

Robert A. Jones (2)
Nordyne Inc.

William Barbier (2)
Concept Technology

George C. Briley (3, 4, 20)
Technicold Services Inc.

Jon M. Edmonds (3)
St. Onge Ruff & Associates, Inc.

William J. Buck (3)
Engineered Refrigeration Systems

John J. Sluga (3)
Hansen Technologies Corporation

Martin L. Timm (4, 32, 50)
Henry Vogt Machine Company

Raymond H.P. Thomas (5)
Allied Signal

Robert G. Doerr (5)
The Trane Company

Danny M. Halel (5, 6)
Parker-Hannifin

Hans O. Spauschus (5)
Spauschus Associates Inc.

William H. Mapes (5)
G.E. Appliances

Alan Cohen (6)
UOP

James R. Bouril (6)
Sporlan Valve Company

John E. Hoffman (6)
Sporlan Valve Company

Thomas W. Dekleva (6)
ICI General Chemicals

Robert Yost (6)
ICI General Chemicals

Calvin J. Grisafe (6)
Witco Corporation

Kenneth W. Manz (6)
Robinair Division

Ward D. Wells (6)
E.I. DuPont de Nemours Co., Inc.

Christopher M. Powers (6)
Robinair Division

Shelvin Rosen (6)
Consultant

Carl F. Speich (7)
The Trane Company

Glenn Short (7)
CPI Engineering Services Inc.

Jose L. Reyes-Gavilon (7)
Witco Corporation

Frederick A. Hamm (7)
Texaco Inc.

Ted Mead (7)
Texaco Inc.

William H. Stein (7)
Texaco Inc.

Gordon Follette (8, 19, 20)
Follette Engineering

Kevin S.E. Wilson (8, 11, 12)
Fletcher Mechanical Ltd.

Terry D. Barber (8)
York Food Systems

Godan Nambudiripad (9)
The Pillsbury Company

William H. Sperber (9)
Grand Metropolitan-Food Sector

Katharine M.J. Swanson (9)
Grand Metropolitan-Food Sector

Bryan R. Becker (10, 15, 16, 17, 21)
University of Missouri

Ronald P. Vallort (11)
The Haskell Company

William J. Stadelman (12)
Purdue University

Benjamin C. Smith (13)
National Sea Products, Ltd.

Dennis M. Nugent (14)
TGK Consulting

Joseph Bene (18)
Bene Engineering Company

Harold R. Heath (18)
Coca Cola Foods

H.L. Shaw (18)
Shaw Engineering Inc.

Brian C. Stebleton (19)
Con Agra Frozen Foods

Evans J. Lizardos (22)
Lizardos Engineering Associates, P.C.

Gideon Zeidler (23)
University of California at Riverside

Ralph A. Ernst (23)
University of California at Davis

James F. Thompson (23)
University of California at Davis

Don Bell (23)
University of California at Riverside

Annie King (23)
University of California at Davis

Tom Leighty (24)
Refrigeration Systems

Dennis Nugent (25)
EPRI

George R. Smith (26)
HCR Incorporated

James J. Bushnell (27, 28, 29, 30)
Allied-Signal Aerospace Company

Sung Lim Kwon (27, 28)
Thermo King Corporation

Clifford A. Woodbury (28)
LTR Engineering Services

John Burgers (28)
Long Manufacturing Ltd.

William F. Anderson (29)
General Marine

A. Bruce Badger (29)
Industrial Process Equipment, Inc.

Robert W. Parkes (31, 46, 47)
Parkes Associates

Paul Adams (31, 46, 47)
Paul Adams Inc.

Carl Laverrenz (31, 46, 47)
Tyler Refrigeration

Charles Thomas (31, 46, 47)
Hussman Corporation

Ronald Babcock (32)
Northstar Ice Equipment Corporation

Paul C. Menster (33)

Russell W. Blades (33)
Basai Limited

Ted Martin (33)
Cimco Refrigeration

Reinhold Kittler (33)
Dectron, Inc.

John Vivian (33)
University of Michigan

James J. Shepherd (34)
Lewis Energy Systems, Inc.

Earl M. Clark (35)
E.I. DuPont

Brent J. Allardyce (35)
Startec Refrigeration

Richard Evans (36)
Westinghouse Hanford Company

Ram K. Shukla (37, 38)
Westinghouse Electric Company

Steve Mowrer (37)
Liebert Corporation

Kenneth R. Diller (39)
University of Texas at Austin

Donald K. Miller (40, 43)
MDK Engineering Corporation

Fred Bawel (40)
Phillips Engineering (Consultant)

CONTRIBUTORS (*Concluded*)

Keith E. Herold (40)
University of Maryland

William J. Plzak (40)
The Trane Company

Bob Reimann (40)
Carrier/United Technologies

William H. Wilkinson (40)
Battelle Columbus Labs

Brian J. Webb (41)
Environthermics Inc.

Rudy Stegmann (42, 45)
The Enthalpy Exchange

John H. Roberts (42)
The Trane Company

David W. Reid (42)
Carrier Corporation

William E. Dietrich (42)
McQuay/Snyder General

David E. Stouppe (42)
Hartford Steam Boiler

Tom O'Donnell (43)
Coors Brewing

Arnett S. Smiley (43)

Richard J. Buck (44)
Sporlan Valve Co.

Larry D. Cummings (44)
Harrison Division (GMC)

Piotr Domanski (44)
National Institute of Standards and Technology

David P. Hargraves (44)
ALCO Controls Division

Thomas R. Barratt (45)
Frick Company

Jing Zheng (45)
Allied Signal

Terry Broccard (46, 47)
Hussman Corporation

Dick Bienvenu (46, 47)
Hussman Corporation

Clark W. Bullard (48)
University of Illinois at Urbana-Champaign

Donald T. MacClellan, Jr. (48)
Columbia Gas of Kentucky

David L. Bien (48)
Sub-Zero Freezer

Richard Topping (48)
Arthur D. Little Inc.

Chinmoy Banerjee (48)
Admiral Company

Robert L. Cushman (48)
Amana Refrigeration Inc.

Ronald I. Greenwald (49)
EBCO Manufacturing Company

ASHRAE HANDBOOK COMMITTEE

Thomas B. Romine, Jr., Chair

1994 Refrigeration Volume Subcommittee: **Donald K. Miller,** Chair
Robert S. Burdick **Warren B. Johnson** **Giustino N. Mastro** **Hsien-Sheng Jason Pei** **Martin L. Timm**

ASHRAE HANDBOOK STAFF

Robert A. Parsons, Editor **Adele J. Brandstrom,** Associate Editor
Laura M. Tracy, Assistant Editor

Ron Baker, Production Manager
Nancy F. Thysell, Kellie Frady, Typography
Susan M. Boughadou, Tamara Slyker, Graphics

Frank M. Coda, Publisher
W. Stephen Comstock, Director
Communications and Publications

ASHRAE TECHNICAL COMMITTEES AND TASK GROUPS

SECTION 1.0—FUNDAMENTALS AND GENERAL

1.1 Thermodynamics and Psychrometrics
1.2 Instruments and Measurements
1.3 Heat Transfer and Fluid Flow
1.4 Control Theory and Application
1.5 Computer Applications
1.6 Terminology
1.7 Operation and Maintenance Management
1.8 Owning and Operating Costs
1.9 Electrical Systems
1.10 Energy Resources

SECTION 2.0—ENVIRONMENTAL QUALITY

2.1 Physiology and Human Environment
2.2 Plant and Animal Environment
2.3 Gaseous Air Contaminants and Gas Contaminant Removal Equipment
2.4 Particulate Air Contaminants and Particulate Contaminant Removal Equipment
2.5 Air Flow Around Buildings
2.6 Sound and Vibration Control
TG Global Climate Change
TG Halocarbon Emissions
TG Safety
TG Seismic Restraint Design

SECTION 3.0—MATERIALS AND PROCESSES

3.1 Refrigerants and Brines
3.2 Refrigerant System Chemistry
3.3 Contaminant Control in Refrigerating Systems
3.4 Lubrication
3.5 Desiccant and Sorption Technology
3.6 Corrosion and Water Treatment
3.7 Fuels and Combustion

SECTION 4.0—LOAD CALCULATIONS AND ENERGY REQUIREMENTS

4.1 Load Calculation Data and Procedures
4.2 Weather Information
4.3 Ventilation Requirements and Infiltration
4.4 Thermal Insulation and Moisture Retarders
4.5 Fenestration
4.6 Building Operation Dynamics
4.7 Energy Calculations
4.9 Building Envelope Systems
4.10 Indoor Environmental Modeling
TG Cold Climate Design

SECTION 5.0—VENTILATION AND AIR DISTRIBUTION

5.1 Fans
5.2 Duct Design
5.3 Room Air Distribution
5.4 Industrial Process Air Cleaning (Air Pollution Control)
5.5 Air-to-Air Energy Recovery
5.6 Control of Fire and Smoke
5.7 Evaporative Cooling
5.8 Industrial Ventilation
5.9 Enclosed Vehicular Facilities
TG Kitchen Ventilation

SECTION 6.0—HEATING EQUIPMENT, HEATING AND COOLING SYSTEMS AND APPLICATIONS

6.1 Hydronic and Steam Equipment and Systems
6.2 District Heating and Cooling
6.3 Central Forced Air Heating and Cooling Systems
6.4 In Space Convection Heating
6.5 Radiant Space Heating and Cooling
6.6 Service Water Heating
6.7 Solar Energy Utilization
6.8 Geothermal Energy Utilization
6.9 Thermal Storage

SECTION 7.0—PACKAGED AIR-CONDITIONING AND REFRIGERATION EQUIPMENT

7.1 Residential Refrigerators and Food Freezers
7.4 Unitary Combustion-Engine-Driven Heat Pumps
7.5 Room Air Conditioners and Dehumidifiers
7.6 Unitary Air Conditioners and Heat Pumps

SECTION 8.0—AIR-CONDITIONING AND REFRIGERATION SYSTEM COMPONENTS

8.1 Positive Displacement Compressors
8.2 Centrifugal Machines
8.3 Absorption and Heat Operated Machines
8.4 Air-to-Refrigerant Heat Transfer Equipment
8.5 Liquid-to-Refrigerant Heat Exchangers
8.6 Cooling Towers and Evaporative Condensers
8.7 Humidifying Equipment
8.8 Refrigerant System Controls and Accessories
8.10 Pumps and Hydronic Piping
8.11 Electric Motors—Open and Hermetic

SECTION 9.0—AIR-CONDITIONING SYSTEMS AND APPLICATIONS

9.1 Large Building Air-Conditioning Systems
9.2 Industrial Air Conditioning
9.3 Transportation Air Conditioning
9.4 Applied Heat Pump/Heat Recovery Systems
9.5 Cogeneration Systems
9.6 Systems Energy Utilization
9.7 Testing and Balancing
9.8 Large Building Air-Conditioning Applications
9.9 Building Commissioning
9.10 Laboratory Systems
TG Clean Spaces
TG Tall Buildings

SECTION 10.0—REFRIGERATION SYSTEMS

10.1 Custom Engineered Refrigeration Systems
10.2 Automatic Icemaking Plants and Skating Rinks
10.3 Refrigerant Piping, Controls, and Accessories
10.4 Ultra-Low Temperature Systems and Cryogenics
10.5 Refrigerated Distribution and Storage Facilities
10.6 Transport Refrigeration
10.7 Commercial Food and Beverage Cooling, Display and Storage
10.8 Refrigeration Load Calculations

SECTION 11.0—REFRIGERATED FOOD TECHNOLOGY AND PROCESSING

11.2 Foods and Beverages
11.5 Fruits, Vegetables and Other Products

PREFACE

The Refrigeration Handbook covers the equipment and systems used for applications other than human comfort. This book includes information on storing a variety of foods, industrial applications of refrigeration, and low temperature refrigeration. While the Refrigeration Handbook is primarily a reference for the practicing engineer, it is also a useful reference for anyone involved in the storage of food products.

Many of the chapters from the 1990 Refrigeration Handbook were revised for this volume. In addition, eleven refrigeration-related chapters from the 1988 Equipment Handbook were updated and moved to this book. Some of the revisions to the systems and equipment-related chapters include the following:

- Chapter 2 includes new line sizing tables for Refrigerants 134a and 502.
- Chapter 6 now has information on recycling and recovering refrigerants.
- Chapter 33 has the most current design information for ice rinks.
- Chapter 39 on biomedical applications has been completely rewritten by an expert in the field.
- Chapter 40 has been rewritten in response to renewed interest in absorption refrigeration
- Chapter 43 on component balancing has been rewritten to reflect new balancing techniques

Many food refrigeration-related chapters were revised to include recent data from research sponsored by ASHRAE and others. Some of those revisions include the following topics and chapters:

- Chapter 9 has been updated to cover basic information on microbiology of foods.
- Chapter 11 has new information on beef chilling.
- Chapter 22 has updated information on brewery production.
- Chapter 23 has been completely rewritten and gives a comprehensive overview of egg production, processing, and storage. Information on nutritive value of eggs is also included.
- Commodity storage data in Chapter 25 has been revised to reflect current research.

This book could not have been updated without the help of many volunteers, working with the society's Handbook Committee and Technical Committees. These dedicated individuals, including members of ASHRAE and non-members, spent countless hours reviewing and writing to ensure that the information was accurate and current.

This Handbook is published in two editions. One edition contains inch-pound (I-P) units of measurement and the other contains the International System of Units (SI).

A section before the index in this volume lists additions and corrections to the 1991, 1992, and 1993 volumes of this Handbook series. Any changes to this volume will be reported in the 1995 ASHRAE *Handbook—HVAC Applications* and on ASHRAE's electronic bulletin board.

If you have suggestions on improving a chapter or would like more information on how you can help revise a chapter, write Handbook Editor, ASHRAE, 1791 Tullie Circle, Atlanta, Georgia 30329.

Robert A. Parsons
Handbook Editor

CHAPTER 1

LIQUID OVERFEED SYSTEMS

OVERFEED systems are those in which excess liquid is forced, mechanically or by gas pressure, through organized-flow evaporators, separated from the vapor, and returned to the evaporators. Terms commonly used with overfeed systems include:

Low-Pressure Receiver. Sometimes referred to as an accumulator, this vessel acts as the separator for the mixture of vapor and liquid returning from the evaporators. The pumping unit is located below the low-pressure receiver. A constant refrigerant level is usually maintained by conventional control devices.

Pumping Unit. Consists of one or more mechanical pumps or gas-operated liquid circulators arranged to pump the overfeed liquid to the evaporators.

Wet Returns. The connections between the evaporator outlets and low-pressure receiver through which the mixture of vapor and overfeed liquid is drawn.

Liquid Feeds. The connections between the pumping unit outlet and the evaporator inlets.

Flow Control Regulators. Devices used to regulate the overfeed flow into the evaporators. They may be needle-type valves, fixed orifices, calibrated manual regulating valves, or automatic valves designed to provide a fixed liquid rate.

The main advantages of liquid overfeed systems are high system efficiency and reduced operating expenses. These systems have lower energy cost and fewer operating hours because:

1. The evaporator surface is used efficiently through good refrigerant distribution and completely wetted internal tube surfaces.
2. The compressors are protected. Liquid slugs resulting from fluctuating loads or malfunctioning controls are separated from suction gas in the low-pressure receiver.
3. Low-suction superheats are achieved where the suction lines between the low-pressure receiver and the compressors are short. This condition causes a minimum discharge temperature, preventing lubrication breakdown and minimizing condenser fouling.
4. With simple controls, evaporators can be hot-gas defrosted with little disturbance to the system.
5. Refrigerant feed to evaporators is unaffected by fluctuating ambient and condensing conditions. The flow control regulators do not need to be adjusted after the initial setting, since the overfeed rates are not generally critical.
6. Flash gas, resulting from refrigerant throttling losses, is removed at the low-pressure receiver before entering the evaporators. This gas is drawn directly to the compressors and is eliminated as a factor in the design of the system low side. It does not contribute to increased pressure drops in the evaporators or overfeed lines.
7. Refrigerant level controls, level indicators, refrigerant pumps, and oil drains are generally located in the equipment rooms, which are under operator surveillance.
8. Because of ideal entering suction gas conditions, compressors last longer. There is less maintenance and fewer breakdowns. The oil circulation rate to the evaporators is reduced as a result of the low compressor discharge superheat (Scotland 1963).
9. Overfeed systems have convenient automatic operation.

Possible disadvantages are:

1. In some cases, refrigerant charges are greater than those used in other systems.
2. Higher refrigerant flow rates to and from evaporators cause the size of the liquid feed and wet return lines to be larger than the high-pressure liquid and suction lines for other systems.
3. Piping insulation is costly and is generally required on all feed and return lines to prevent moisture or frost formation.
4. The installed cost may be greater, particularly for small systems or those with fewer than three evaporators.
5. The operation of the pumping unit requires added expenses that are offset by the increased efficiency of the overall system.
6. The pumping units may require more maintenance.

Generally, the more evaporators used, the more favorable are the initial costs for the liquid overfeed compared to a gravity recirculated or flooded system (Scotland 1970). Liquid overfeed systems also compare favorably with thermostatic valve feed systems for the same reason. For small systems, the initial cost for liquid overfeed may be higher than for direct expansion.

Easy operation and less maintenance are attractive features for even small ammonia systems. However, for ammonia systems operating below 0 °F evaporating temperatures, some manufacturers do not supply direct expansion evaporators because of unsatisfactory refrigerant distribution and control problems.

OPERATION OF MECHANICAL PUMP SYSTEM

Figure 1 shows the basic system in which a constant liquid level is maintained in a low-pressure receiver. A mechanical pump circulates liquid through the evaporator(s). The two-phase return mixture is separated in the low-pressure receiver. The vapor is directed to the compressor(s). The makeup refrigerant enters the low-pressure receiver by means of a refrigerant metering device.

Figure 2 shows a horizontal low-pressure receiver with a minimum pump pressure, service valves in place, and a strainer on the suction side of the pump. The strainer protects hermetic pumps when oil is miscible with the refrigerant. It should have a free area twice the transverse cross-sectional area of the line in which it is installed. A suction strainer should not be used with ammonia, and open drive pumps do not require strainers.

The preparation of this chapter is assigned to TC 10.1, Custom Engineered Refrigeration Systems.

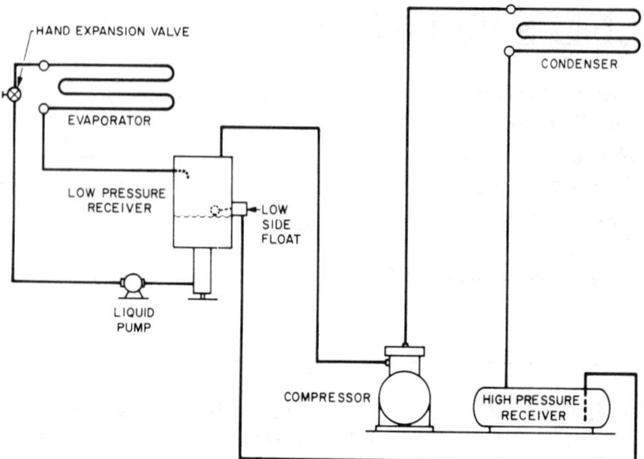

Fig. 1 Liquid Overfeed with Mechanical Pump

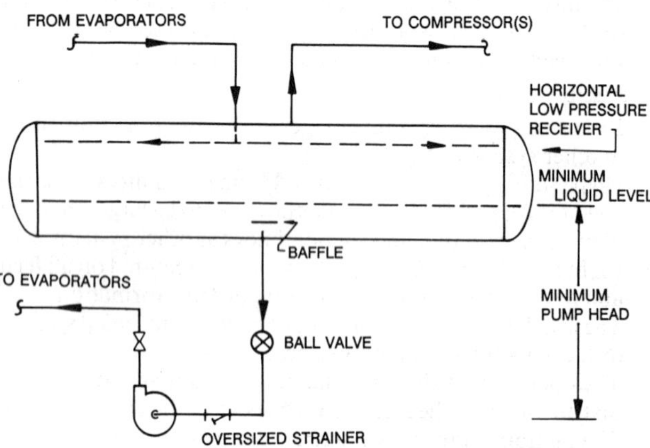

Fig. 2 Pump Circulation, Horizontal Separator

The minimum pump pressure should be at least double the net positive suction pressure to avoid cavitation. The liquid velocity to the pump should not exceed 3 ft/s.

OPERATION OF GAS PUMP SYSTEM

Figure 3 shows a liquid overfeed system in which the pumping power is supplied by gas at condenser pressure. In this system, a level control maintains the liquid level in the low-pressure receiver. There are two pumper drums; one is filled by the low-pressure receiver, while the other is drained as hot gas pushes liquid from the pumper drum to the evaporator. Pumper drum B drains when hot gas enters the drum through Valve B. To function properly, the pumper drums must be correctly vented so they can fill during the fill cycle.

Another common arrangement is shown in Figure 4. In this system, the high-pressure liquid is flashed into a controlled-pressure receiver that maintains constant liquid pressure at the evaporator inlets, resulting in continuous liquid feed at constant pressure. The flash gas is drawn into the low-pressure receiver through a receiver pressure regulator. Excess liquid drains into a liquid-pump trap from the low-pressure receiver. Check valves and a three-way equalizing valve transfer the liquid into the controlled-pressure receiver during the dump cycle. Refinements of this system are used for multistage systems.

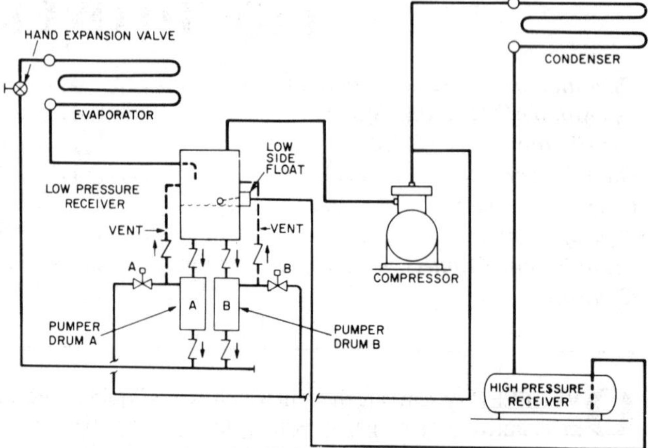

Fig. 3 Double Pumper Drum System

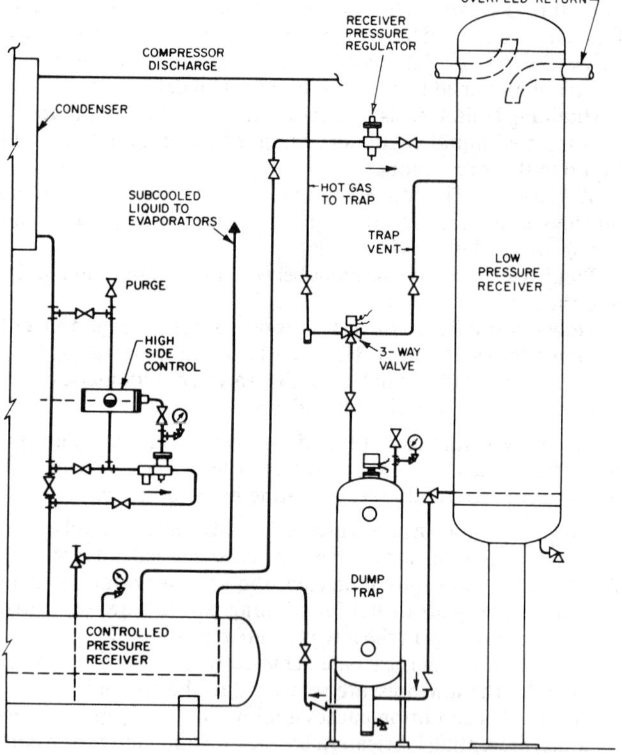

Fig. 4 Constant-Pressure Liquid Overfeed System

DISTRIBUTION

To prevent underfeeding and excessive overfeeding of refrigerants, metering devices regulate the liquid feed to each evaporator and/or evaporator circuit. An automatic regulating device continually controls refrigerant feed to the design value. Other devices used are hand expansion valves, calibrated regulating valves, orifices, and distributors.

It is time-consuming to adjust hand expansion valves to achieve ideal flow conditions. However, they have been used with some success in many installations prior to the availability of more sophisticated controls. One factor to consider is that standard hand expansion valves are designed to regulate flows caused by the relatively high pressure differences between condensing and evaporating pressure. In overfeed systems, large differences do not

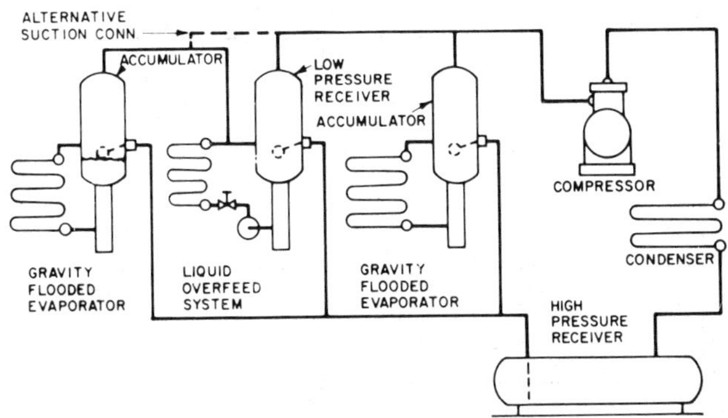

Fig. 5 Liquid Overfeed System Connected on Common System with Gravity-Flooded Evaporators

exist, so valves with larger orifices may be needed to cope with the combination of the increased quantity of refrigerant and the relatively small pressure differences—approximately 10 to 15 psi.

Calibrated, manually operated regulating valves reduce some of the uncertainties involved in using conventional hand expansion valves. To be effective, the valves should be adjusted to the manufacturer's recommendations. Because the refrigerant in the liquid feed lines is above saturation pressure, the lines should not contain flash gas. However, liquid flashing can occur if excessive heat gains by the refrigerant and/or high pressure drops build up in the feed lines.

Orifices should be carefully designed and selected. These generally are used only for top- and horizontal-feed multicircuit evaporators. Foreign matter and congealed oil globules can cause flow restriction; a minimum orifice of 0.1 in. is recommended. With the small liquid volume of ammonia normally circulated, the rate of circulation may need to be increased beyond that needed for the minimum orifice size. Pumps and feed and return lines larger than minimum may be needed. This does not apply to halocarbons because of the greater liquid volume circulated as a result of fluid characteristics.

Conventional multiple outlet distributors with capillary tubes of the type usually paired with thermostatic expansion valves have been used successfully in liquid overfeed systems. Capillary tubes may be installed downstream of a distributor with oversized orifices to achieve the required pressure reduction and efficient distribution.

Existing gravity-flooded evaporators with accumulators can be connected to liquid overfeed systems. Changes may be needed only for the feed to the accumulator, with suction lines from the accumulator connected to the system wet return lines. An acceptable arrangement is shown in Figure 5. Generally, gravity-flooded evaporators have different circuiting arrangements from overfeed evaporators. In many cases, the circulating rates developed by thermal-syphon action are greater than the circulating rates used in conventional overfeed systems.

Example 1. Find the orifice diameter of an ammonia overfeed system with a refrigeration load per circuit of 1.27 tons and a circulating rate of 7. The evaporating temperature is −30°F and the pressure drop across the orifice is 8 psi. The circulation per circuit is 0.528 gpm.

Solution: Orifice diameter may be calculated by Equation (1) as:

$$d = \left[\frac{Q}{a\,C_d\sqrt{p/S}}\right]^{0.5} \qquad (1)$$

where

a = units conversion, 29.81
Q = discharge through orifice, 0.528 gpm
p = pressure drop through orifice, 8 psi
S = specific gravity of fluid relative to water at −30°F = 5.701/8.336 = 0.6839
C_d = Coefficient of discharge for orifice = 0.61

$$d = \left[\frac{0.528}{29.81 \times 0.61\sqrt{8/0.6839}}\right]^{0.5} = 0.0922 \text{ in.}$$

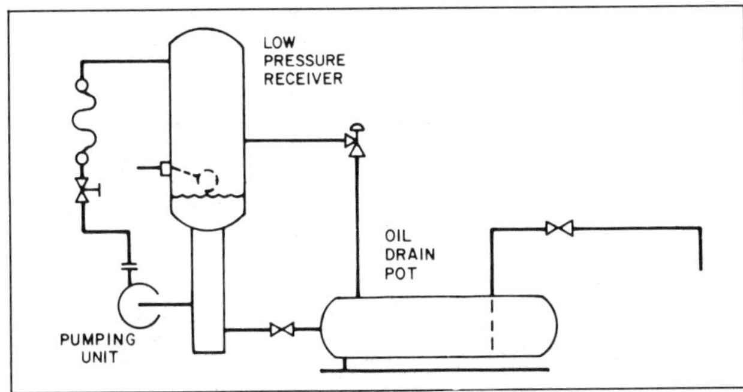

Fig. 6 Oil Drain Pot Connected to Low-Pressure Receiver

OIL IN SYSTEM

In spite of reasonably efficient compressor discharge oil separators, oil finds its way into the system low-pressure sides. In the case of ammonia overfeed systems, the bulk of this oil can be drained from the low-pressure receivers with suitable oil drainage facilities. In low-temperature systems, a separate valved and pressure-protected noninsulated oil drain pot can be placed in a warm space at the accumulator. The oil/ammonia mixture will flow into the pot and the refrigerant will evaporate. This arrangement is shown in Figure 6. Because of the low solubility of oil in liquid ammonia, thick oil globules circulate with the liquid and can restrict flow through strainers, orifices, and regulators. To maintain high efficiency, oil should be removed from the system by regular draining.

Because halocarbons are miscible with oil, positive oil return to the compressor must be assured. Many methods are used, including oil stills using both electric heat and heat exchange from high-pressure liquid or vapor. Some arrangements are discussed in Chapter 2.

Excessive oil must not be allowed to build up in evaporators, because efficiency will rapidly decrease. This is particularly critical in evaporators with high heat transfer rates associated with low volumes, such as flake-type icemakers, ice cream freezers, and scraped-surface heat exchangers. Because high refrigerant flow rate occurs through such evaporators, excessive oil can accumulate and rapidly deteriorate efficiency.

CIRCULATING RATE

In a liquid overfeed system, the mass ratio of liquid pumped to the amount of vaporized liquid is the circulating number or rate. The amount of liquid vaporized is based on the latent heat for the refrigerant at the evaporator temperature. Overfeed rate is the ratio of liquid to vapor returning to the low-pressure receiver. When vapor leaves an evaporator at saturated vapor conditions with no excess liquid, the circulating rate is 1 and the overfeed rate is 0. With a circulating rate of 4, the overfeed rate at full load is 3; at no load, it is 4. Most systems are designed for steady flow conditions. With few exceptions, the load conditions may vary, causing fluctuating temperatures outside and within the evaporator. Evaporator capacities vary considerably; with constant refrigerant flow to the evaporator, the overfeed rate fluctuates.

For each evaporator, there is an ideal circulating rate for every loading condition that will result in the minimum temperature

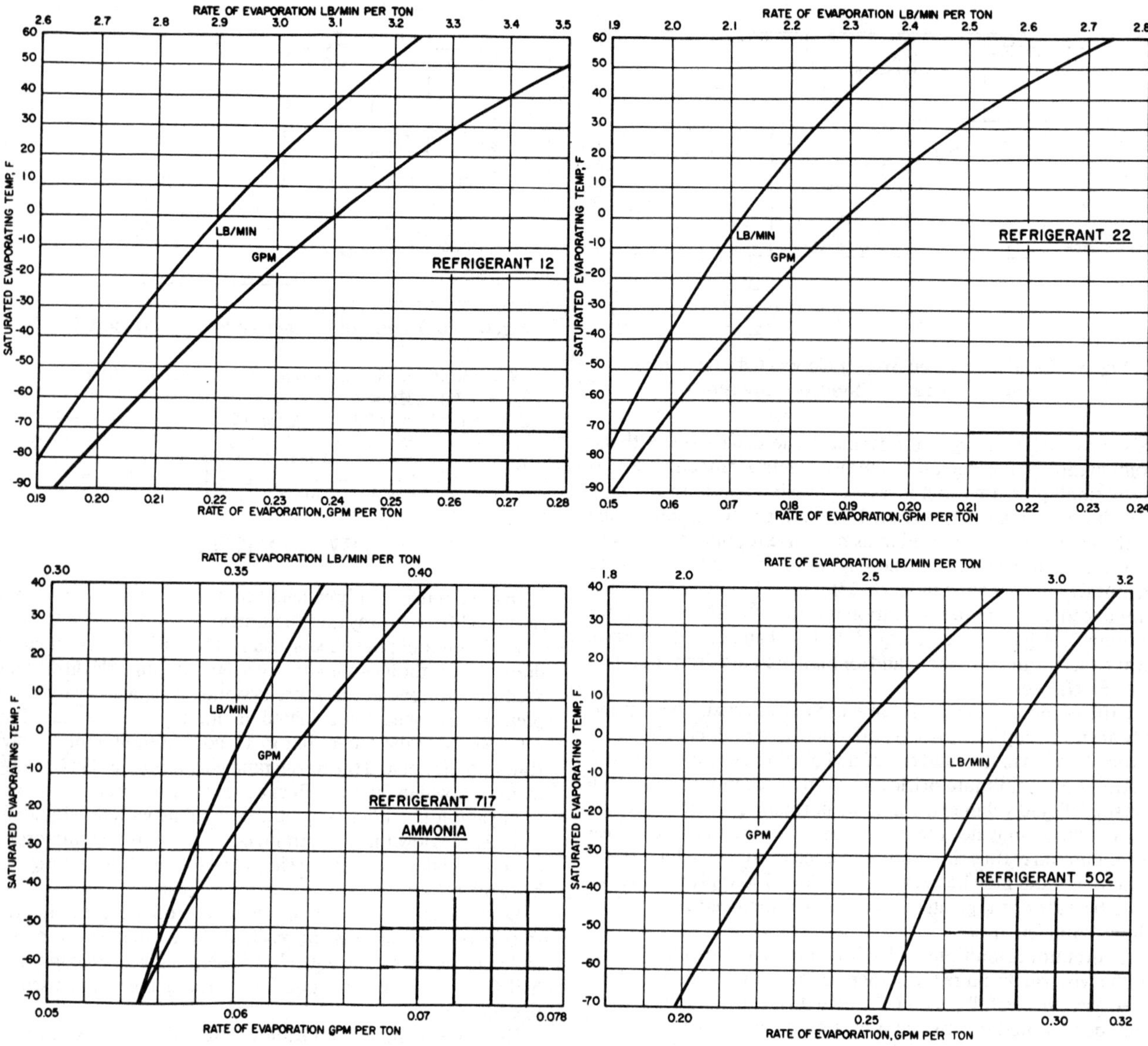

Fig. 7 Charts for Determining Rate of Refrigerant Feed (No Flash Gas)

difference and the best evaporator efficiency (Lorentzen 1968, Lorentzen and Gronnerud 1967). With few exceptions, it is impossible to predict ideal circulating rates or to design a plant for automatic adjustment of the rates to suit fluctuating loads. The optimum rate can vary with heat load, pipe diameter, circuit length, and number of parallel circuits to achieve the best performance. High circulation rates can cause excessively high pressure drops through evaporators and wet return lines. The sizing of these return lines, discussed in a separate section, can have a bearing on the ideal rates. Many evaporator manufacturers specify recommended circulating rates for their equipment. The rates shown in Table 1 agree with these recommendations.

Table 1 Recommended Minimum Circulating Rate

Refrigerant	Circulating Rate
Ammonia (R-717)	
Downfeed (large diameter tubes)	6 to 7
Upfeed (small diameter tubes)	2 to 4
R-502—upfeed	2
R-22—upfeed	3

Because of distribution considerations, higher circulating rates are common with top feed evaporators. In multicircuit systems, the refrigerant distribution must be adjusted to provide the best possible results. Incorrect distribution can cause excessive overfeed in some circuits, while others may be starved. Manual or automatic regulating valves can be used to control flow for the optimum or design value.

Halocarbon densities are about twice that of ammonia. If halocarbons R-12, R-22, and R-502 are circulated at the same rate as ammonia, the halocarbons will require 6 to 8.3 times more energy for pumping to the same height as the less dense ammonia. Since this pumping energy must be added to the system load, halocarbon circulating rates are usually lower than those for ammonia. Because ammonia has a relatively high latent heat of vaporization, for equal heat removal, much less ammonia mass must be circulated compared to halocarbons.

Even though halocarbons circulate at lower rates than ammonia, the wetting process in the evaporators is still efficient because of the liquid and vapor volume ratios. For example, at −40 °F evaporating temperatures, with constant flow conditions in the wet return connections, similar ratios of liquid and vapor are experienced with a circulating rate of 4 for ammonia, 2.5 for R-22 and R-502, and 2 for R-12. With halocarbons, some additional wetting is also experienced because of the solubility of the oil in these refrigerants.

When bottom feed is used for multicircuit coils, a minimum feed rate per circuit is not necessary since orifices or other distribution devices are not required. The circulating rate for top feed and horizontal feed coils may be determined by the minimum rates from the orifices or other distributors in use.

Figure 7 provides a method for determining the liquid refrigerant flow (Niederer 1963). The charts indicate the amount of refrigerant vaporized in a 1-ton system with circulated operation having no flash gas in the liquid feed line. The value obtained from the chart may be multiplied by the desired circulating rate and by the total refrigeration to determine total flow.

The pressure drop through the flow control regulators is usually 10 to 50% of the available feed pressure. The pressure at the outlet of the flow regulators is higher than the vapor pressure at the low-pressure receiver by an amount representing the total pressure drop of the two-phase mixture through the evaporator, any evaporator pressure regulator, and wet return lines. This outlet pressure could be 5 psi in a typical system. When using recommended liquid feed sizing practices, assuming a single-story building, the frictional pressure drop from the pump discharge to the evaporators is about 10 psi. Therefore, a pump for 20 to 25 psi should be satisfactory in this case, depending on the lengths and sizes of feed lines, the quantity and types of fittings, and the vertical lift involved.

TYPE OF PUMP

Mechanical pumps, gas pressure pumping systems, and injector systems are available for liquid overfeed systems.

Types of mechanical pumps include open, semihermetic, magnetic clutch, and hermetic rotor arrangements with positive rotary, centrifugal, or turbine vane construction. Positive rotary and gear-type pumps are generally operated at slow speeds from 90 to 300 rpm. Whatever type of pump is used, care should be taken to prevent flashing at the pump suction and/or within the pump itself.

Centrifugal pumps are preferred for larger volumes, while semihermetic pumps are best suited for halocarbons at or below atmospheric refrigerant saturated pressure.

Open-type pumps are fitted with a wide variety of packing or seals. For continuous duty, a mechanical seal with an oil reservoir or a liquid refrigerant supply to cool, wash, and lubricate the seals is commonly used. Experience with the particular application or the recommendations of an experienced pump supplier are the best guides for selecting the packing or seal. A small immersion-type electric heater within the oil reservoir can be used with low-temperature systems to assure that the oil will remain fluid. Motors are selected that have a service factor that compensates for drag on the pump if the oil is cold or stiff.

Application considerations should include ambient temperatures, heat leakage, fluctuating system pressures from compressor cycling, internal bypass of liquid to pump suction, friction heat, and motor heat conduction. They should also include dynamic conditions, cycling of automatic evaporator liquid and suction stop valves, action of regulators, gas entrance with liquid, and loss of subcooling by pressure drop. Other factors to consider are the time lag caused by the heat capacity of pump suction, cavitation, and NPSH factors (Lorentzen 1963).

The motor and stator of hermetic pumps are separated from the refrigerant by a thin nonmagnetic membrane. The metal membrane should be strong enough to withstand system design pressures. Normally, the motors are cooled and the bearings lubricated by liquid refrigerant bypassed from the pump discharge. It is good practice to use two pumps, one operating and one standby.

INSTALLING AND CONNECTING MECHANICAL PUMPS

Because of the sensitive suction conditions of mechanical pumps operating on overfeed systems, the manufacturer's application and installation specifications must be followed closely. Suction connections should be as short as possible, without restrictions, valves, or elbows. Angle or fullflow ball valves should be used. Using valves with horizontal valve spindles eliminates possible traps. Gas binding is more likely with high evaporating pressures.

Installing discharge check valves prevents back flow. Relief valves should be used, particularly for positive displacement pumps. Strainers are not usually installed in ammonia pump suction lines because they plug with oil. Strainers, although a poor substitute for a clean installation, protect halocarbon pumps from damage by dirt or pipe scale.

Pump suction connections to liquid legs (vertical drop legs from low-pressure receivers) should be made above the bottom of the legs to allow collection space for solids and sludge. Vortex eliminators should be considered, particularly when submersion of the suction inlet is insufficient to prevent the intake of gas bubbles. Lorentzen (1963, 1965) gives more complete information.

Sizing the pump suction line is important. The general velocity should be about 3 ft/s. Small lines cause restrictions, while oversized lines can cause bubble formation during decreasing evaporator temperature conditions because of the heat capacity of the liquid and piping. Oversized lines also impose increased heat gain from the ambient spaces. Oil heaters for the seal lubrication system keep the oil fluid, particularly during subzero operation. Thermally insulating all cold surfaces of pumps, lines, and receivers increases efficiency.

CONTROLS

The liquid level in the low-pressure receiver can be controlled by conventional devices such as low-pressure float valves, combinations of float switch and solenoid valve with manual regulator, thermostatic level controls, or other proven automatic devices. High level float switches are useful to stop compressors and/or operate alarms. Solenoid valves should be installed on liquid lines (minimum sized) feeding low-pressure receivers so that positive shutoff is automatically achieved with system shutdown. This prevents excessive refrigerant from collecting in low-pressure receivers, which can cause spill-over at start-up.

To prevent pumps from operating without liquid, low level float switches can be fitted on liquid legs. An alternative device, a differential pressure switch connected across pump discharge and suction connections, causes the pump to stop without interrupting liquid flow. In extreme cases, cavitation can also cause this control to operate. When hand expansion valves are used to control the circulation rate to evaporators, the orifice should be sized for operation between system high and low pressures. Occasionally, with reduced inlet pressure conditions, these valves can starve the circuit. Calibrated, manually-adjusted regulators are available to meter the flow according to the design conditions. An automatic-flow regulating valve is available specifically for overfeed systems.

Liquid and suction solenoid valves are selected for refrigerant flow rates by mass or volume, not by refrigeration ratings from capacity tables. Evaporator pressure regulators should be sized according to the manufacturer's ratings for overfeed systems. When ordering valves, tell the manufacturer they are for overfeed application, since slight modifications may be required. When evaporator pressure regulators are used on overfeed systems for controlling air-defrosting of cooling units (particularly when fed with very low-temperature liquid), the refrigerant heat gain may be achieved by sensible effect, not by latent effect. In such cases, investigate other defrosting methods. The possibility of connecting the units directly to high-pressure liquid should be considered, especially if the loads are minor.

When a check valve and a solenoid valve are paired on an overfeed system liquid line, the check valve should be downstream from the solenoid valve. When the solenoid valve is closed, dangerous hydraulic pressure can build up from the expansion of the trapped liquid as it is heated. When evaporator pressure regulators are used, the pressure of the entering liquid should be high enough to cause flow into the evaporator.

Multicircuit systems have a bypass relief valve in the pump discharge. When some of the circuits are closed, the excess liquid is bypassed into the low-pressure receiver, rather than forced through the evaporators still in operation. This prevents higher evaporating temperatures from pressurizing evaporators and reducing capacities of operating units. Where low-temperature liquid feeds can be isolated manually or automatically, relief valves can be installed to prevent damage from excessive hydraulic pressure.

EVAPORATOR DESIGN

There is an ideal refrigerant feed and flow system for each evaporator design and arrangement. An evaporator designed for gravity-flooded operation cannot always be converted to an overfeed arrangement, or vice versa, nor can systems always be designed to circulate the optimum flow rate. When top feed is used to ensure good distribution, a minimum quantity per circuit must be circulated, generally about 0.5 gpm. Distribution in bottom feed evaporators is less critical than for top or horizontal feed, since each circuit fills with liquid to equal the pressure loss in other parallel circuits.

Circuit length in evaporators is determined by allowable pressure drop, load per circuit, tubing diameter, overfeed rate, type of refrigerant, and heat transfer coefficients. The most efficient circuiting is resolved in most cases through laboratory tests conducted by the evaporator manufacturers. Their recommendations should be followed when designing systems.

TOP FEED/BOTTOM FEED

System design must determine whether evaporators are to be top or bottom fed, although both feed types can be installed in a single system. Each feed type has advantages; no best arrangement is common to all systems. Top feed advantages include (1) smaller refrigerant charge, (2) possibly smaller low-pressure receiver, (3) possible absence of static pressure penalty, (4) better oil return and (5) quicker, simpler defrost arrangements. For halocarbon systems with greater fluid densities, the refrigerant charge, oil return, and static pressure are very important.

Bottom feed is advantageous in that (1) distribution considerations are less critical, (2) relative locations of evaporators and low-pressure receivers are less important, and (3) the system design and layout are simpler. The top feed system is limited by the relative location of components. Because this system sometimes requires more refrigerant circulation than bottom feed systems, it has greater pumping load, possibly larger feed and return lines, and increased line pressure drop penalties. In bottom feed evaporators, multiple headers with individual inlets and outlets can be installed to reduce static head penalties. For high lift of return overfeed lines from the evaporators, dual-suction risers eliminate static pressure penalties (Miller 1974, 1979).

Distribution must be considered when a vertical refrigerant feed is used because of the static head variations in the feed and return header circuits. For example, for equal circuit loadings in a horizontal airflow unit cooler, using gradually smaller orifices for the bottom feed circuits than for the upper circuits can compensate for pressure differences.

When the top feed free-draining arrangement is used for air cooling units, liquid solenoid control valves can be used during the defrost cycle. This applies in particular to air, water, or electric defrost units. Any remaining liquid in the coils will rapidly evaporate or drain to the low-pressure receiver. Defrost is faster than in bottom feed evaporators.

REFRIGERANT CHARGE

Overfeed systems need more refrigerant than dry expansion systems. Top feed arrangements have smaller charges than bottom feed systems. The amount of charge also depends on the evaporator volume, the circulating rate, the sizes of flow and return lines, the operating temperature differences, and the heat transfer coefficients. Generally, top feed evaporators operate with the refrigerant charge occupying about 25 to 40% of the evaporator volume. The refrigerant charge for the bottom feed arrangement occupies about 60 to 75% of the evaporator volume with corresponding variations in the wet returns. Under certain no-load conditions in up-feed evaporators, the charge may occupy 100% of the evaporator volume. In this case, the liquid surge volume from full to no-load condition must be considered in sizing the low-pressure receiver (Miller 1971, 1974).

Evaporators with high heat flux, such as flake icemakers and scraped-surface heat exchangers, have small charges because of small evaporator volumes. The amount of refrigerant in the low side has a major effect on the size of the low-pressure receiver, especially in horizontal vessels. The cross-sectional area for vapor flow in horizontal vessels is reduced with increasing liquid level. It is important to ascertain the evaporator refrigerant charge with fluctuating loads for correct vessel design, particularly for a low-pressure receiver that does not have a constant level control but is fed through a high-pressure control.

START-UP AND OPERATION

All control devices should be checked prior to start-up. If mechanical pumps are used, the direction of operation must be correct. System evacuation and charging procedures are similar to those for other systems. The system must be operating under normal conditions to determine the total required refrigerant charge. Liquid height is established by liquid level indicators in the low-pressure receivers.

Calibrated, manually-operated regulators should be set for the design conditions and adjusted for better performance when necessary. When hand expansion valves are used, start the system by opening the valves about one-quarter to one-half turn. When balancing is necessary, cut back the regulators on those circuits not starved of liquid to force the liquid through the underfed circuits. The outlet temperature of the return line from each evaporator should be the same as the saturation temperature of the main return line, allowing for pressure drops. Starved circuits are indicated by higher temperatures than those for adequately fed circuits. Excessive feed to a circuit increases the evaporator temperature because of excessive pressure drop.

The relief bypass from the liquid line to the low-pressure receiver should be adjusted and checked to ensure that it is functioning. During operation, follow the pump manufacturer's recommendations regarding lubrication and maintenance. Regular oil draining procedures should be established for ammonia systems; a comparison should be made between the oil quantities added to and drained from each system. This comparison will determine if oil is accumulating in systems. Oil should not be drained in halocarbon systems, but drain valves should be opened, particularly after start-up, to blow out any foreign matter that may accumulate. Due to the miscibility of oil with halocarbons, it may be necessary to add oil to the system until an operating balance is achieved (Stoecker 1960, Soling 1971).

OPERATING COSTS

Operating costs for overfeed systems generally are less than for other systems. Operating costs may not be less in all cases because of the variety of inefficiencies that exist from system to system and from plant to plant. However, in cases where existing dry expansion plants were converted to liquid overfeed, the operating hours, power, and maintenance costs were reduced. The efficiencies of the early gas pump systems have been improved by using flash gas to circulate the overfeed liquid. One flash gas system is indicated in the controlled pressure system shown in Figure 4. Refinements of the double pumper drum arrangement (shown in Figure 3) have also been developed.

Gas-pumped systems, which use refrigerant gas to pump liquid to the evaporators or to the controlled-pressure receiver, require additional compressor volume, from which no useful refrigeration is obtained. These systems consume 4 to 10% or more of the compressor power to maintain the gas flow.

If the condensing pressure is reduced as much as 10 psig, the compressor power per unit of refrigeration will drop by about 7%. Where outdoor dry- and wet-bulb conditions allow, a mechanical pump can be used to pump the gas with no effect on evaporator performance. Gas-operated systems must, however, maintain the condensing pressure within a much smaller range to pump the liquid and maintain the required overfeed rate.

LINE SIZING

The liquid feed line to the evaporator and the wet return line to the low-pressure receiver cannot be sized by the method described in Chapter 33 of the 1993 ASHRAE *Handbook—Fundamentals*. Figure 7 can be used to size liquid feed lines. The circulating rate is multiplied by the evaporating rate. For example, an evaporator forming vapor at a rate of 5 lb/min and with a circulating rate of 4 needs a feed line sized for $4 \times 5 = 20$ lb/min. Alternative methods that may be used to design wet returns include:

1. Use one pipe size larger than calculated for vapor flow alone.
2. Use a velocity selected for dry expansion reduced by the factor $(1/\text{Circulating Rate})^{0.5}$. This method suggests that the wet return velocity for a circulating rate of 4 is $(1/4)^{0.5} = 0.5$ that of the acceptable dry vapor velocity.
3. Use the design method described by Chaddock *et al.* (1972). The report includes tables of flow capacities at 2°F drop per 100 ft of horizontal lines for R-717 (ammonia), R-12, R-22, and R-502.

When sizing refrigerant lines, the following design precautions should be taken:

1. Carefully size overfeed return lines with vertical risers, since more liquid will be held in risers than in horizontal pipe. This holdup increases with reduced vapor flow and increases pressure loss because of gravity and two-phase pressure drop.
2. Use double risers with halocarbons to maintain velocity at partial loads and to reduce liquid static pressure loss (Miller 1979).
3. Add the equivalent of a 100% liquid static height penalty to the pressure drop allowance to compensate for liquid holdup in ammonia systems that have unavoidable vertical risers.
4. As alternatives in severe cases, provide traps and a means of pumping liquids, or use dual-duct risers.
5. Use low loss coefficient valves installed so the stems are horizontal or nearly so (Chisolm 1971).

LOW-PRESSURE RECEIVER SIZING

Low-pressure receivers are also called liquid separators, suction traps, accumulators, liquid-vapor separators, flash-type coolers, gas and liquid coolers, intercoolers, surge drums, knock-out drums, slop tanks, or low-side pressure vessels, depending on their function and the preferred term of the user.

The sizing of low-pressure receivers is determined by the required liquid holdup volume and the allowable gas velocity. The volume must accommodate the fluctuations of liquid in the evaporators and overfeed return lines as a result of load changes. It must also handle the swelling and foaming of the liquid charge in the receiver, caused by boiling during temperature or pressure reduction. At the same time, a liquid seal must be maintained on the supply line for continuous circulation devices. A separating space must be provided for gas velocity low enough to cause a minimum entrainment of liquid drops into the suction outlet. Space limitations and design requirements result in a wide variety of configurations (Miller 1971; Stoecker 1960; Lorentzen 1966; Niemeyer 1961; Scheiman 1963, 1964; Sonders and Brown 1934; Younger 1955).

In selecting a gas and liquid separator, adequate volume for the liquid supply and a vapor space above the minimum liquid height for liquid surge must be provided. This requires an analysis of operating load variations. This, in turn, determines the Maximum Operating Liquid Level. Figures 8 and 9 identify these levels and the important parameters of vertical and horizontal gravity separators.

Vertical separators maintain the same separating area with level variations, while separating areas in horizontal separators change with level variations. Horizontal separators should have inlets and outlets separated horizontally by at least the vertical separating distance. A useful arrangement in horizontal separators distributes the inlet flow into two or more connections to reduce turbulence and the horizontal velocity without reducing the residence time of the gas flow within the shell (Miller 1971). In horizontal separators, as the horizontal separating distance is increased beyond the vertical separating distance, the residence time of the vapor passing through is longer so that higher velocities than allowed in vertical separators can be tolerated.

As the separating distance is reduced, the amount of liquid entrainment from gravity separators increases. Table 2 shows the gravity separation velocities. For surging loads or pulsating flow associated with large step changes in capacity, reduce the maximum steady flow velocity to a value achieved by a suitable multiplier such as 0.75.

The gas and liquid separator may be designed with baffles or eliminators to separate liquid from the suction gas returning from the top of the shell to the compressor. More often, sufficient separation space is allowed above the liquid level for this purpose. Such a design usually is of the vertical type, with a separation height above the liquid level of from 24 to 36 in. The shell diameter is sized to keep the suction gas velocity at a value low enough to allow the liquid droplets to separate and not be entrained with the returning suction gas off the top of the shell.

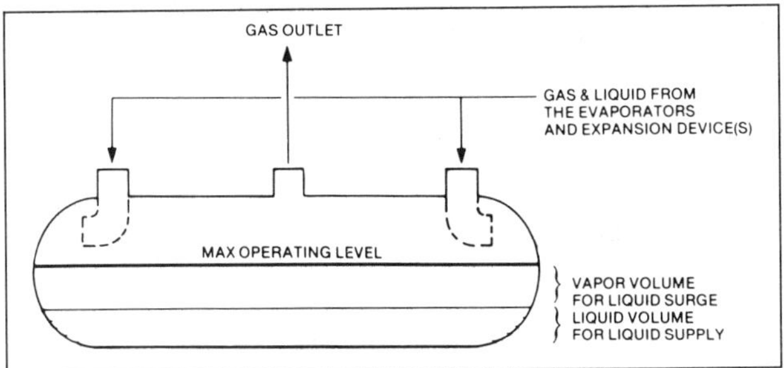

Fig. 8 Basic Horizontal Gas and Liquid Separator

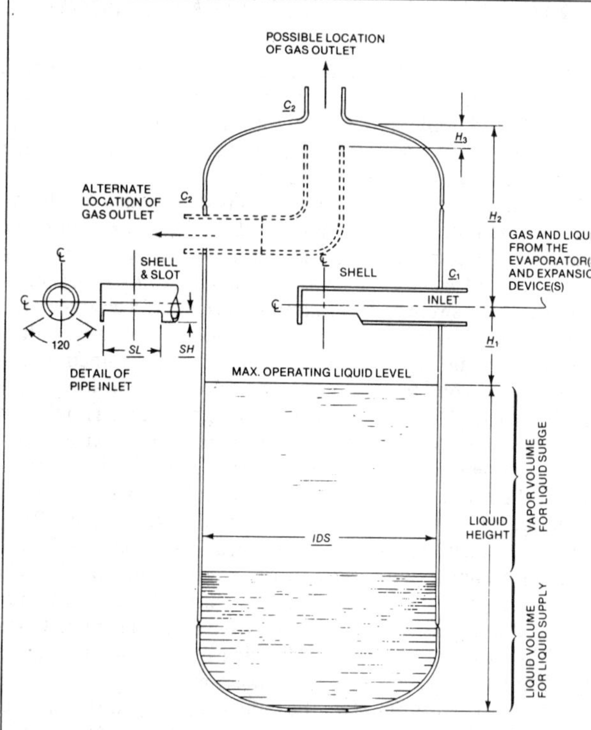

C_1 = inlet pipe diameter, OD, in.
C_2 = outlet pipe diameter, OD, in.
SH = slot height = $C_1/4$, in.
SL = slot length = $3\,C_1$, in.
H_1 = height of C_1 above maximum liquid level, in.; for pseudo SD = 24 in.
H_1 = $\sqrt{7.33\text{ cfm}/SV}$, in. (at SD = 24 in.)
cfm = maximum gas flow in the shell at maximum sustained operating conditions
SV = separation velocity, fpm
H_2 = location of C_1 from inside top of shell, in.
H_2 = SD + 0.5 depth of curved portion of head = $C_1/4$
SD = vertical separation distance, in. actual
H_3 = location of gas exit point for alternate location of C_2 measured from inside top of shell, in.
H_3 = 0.5 depth of curved portion of shell or 2 in., whichever is greater
IDS = internal diameter of shell, in.
IDS = $\sqrt{183\text{ cfm}/SV}$, in.

Fig. 9 Basic Vertical Gravity Gas and Liquid Separator

Table 2 Maximum Effective Separation Velocities for R-717, R-22, R-12, and R-502, with Steady Flow Conditions

Temp., °F	Vertical Separation Distance, in.	Maximum Steady Flow Velocity, fpm			
		R-717	R-22	R-12	R-502
+50	10	29	13	16	11
	24	125	62	70	50
	36	139	77	85	62
+20	10	42	19	22	15
	24	172	86	96	69
	36	195	102	115	83
−10	10	61	27	32	22
	24	253	120	135	97
	36	281	141	159	116
−40	10	95	41	47	33
	24	392	173	198	140
	36	428	205	230	165
−70	10	158	65	72	50
	24	649	267	303	212
	36	697	310	351	247

Adapted from Miller (1971)

Although separators are made with length to diameter (L/D) ratios of 1/1 increasing to 10/1, the least expensive separators usually have L/D ratios between 3/1 and 5/1. Vertical separators are normally used for systems with reciprocating compressors. Horizontal separators may be preferable where vertical height is critical and/or where large volume space for liquid is required. The procedures for designing vertical and horizontal separators are different.

The vertical gas-and-liquid separator is shown in Figure 9. The end of the inlet pipe C_1 is capped so that flow dispersion is directed downward toward the liquid level. The opening is four times the transverse internal area of the pipe. The height H_1 with a 120° dispersion of the flow reaches to approximately 70% of the internal diameter of the shell.

An alternate inlet pipe with a downturned elbow or mitered bend can be used. However, the jet effect of entering fluid must be considered to avoid undue splashing. The outlet of the pipe must be a minimum distance of $IDS/5$ above the maximum liquid level in the shell. H_2 is measured from the outlet to the inside top of the shell. It equals SD + 0.5, the depth of the curved portion of the head.

For the alternate location of C_2, determine IDS from Equation (2):

$$IDS = \sqrt{(183\text{ cfm}/SV) + C_2^2} \qquad (2)$$

The maximum liquid height in the separator is a function of the type of system in which the separator is being used. With some systems this can be estimated, while in others, previous experience provides the only means of selecting the proper liquid height. The accumulated liquid must be returned to the system by a suitable means at a rate comparable to the rate at which it is being collected.

When using a horizontal separator, the vertical separation distance used is an average value. The top part of the horizontal shell restricts the gas flow so that the maximum vertical separation distance cannot be used. If H_t represents the maximum vertical distance from the liquid level to the inside top of the shell, the average separation distance as a fraction of IDS is as follows:

H_t/IDS	0.1	0.2	0.3	0.4	0.5
SD/IDS	0.068	0.140	0.215	0.298	0.392
H_t/IDS	0.6	0.7	0.8	0.9	1.0
SD/IDS	0.492	0.592	0.693	0.793	0.893

The suction connection(s) for refrigerant gas leaving the horizontal shell must be located at or above the location established by the average distance for separation. The maximum cross-flow velocity of gas establishes the residence time for the gas and any entrained liquid droplets in the shell. The most effective removal of entrainment occurs when the residence time is at a maximum practical value. Regardless of the number of gas outlet connections for uniform distribution of gas flow, the cross-sectional area of the gas space is:

$$A_x = \frac{288\ SD \cdot \text{cfm}}{SV \cdot L} \qquad (3)$$

where

A_x = minimum transverse net cross-sectional area or gas space, in.^{2}
SD = average vertical separation distance, in.
cfm = total quantity of gas leaving the vessel
L = inside length of shell, in.
SV = separation velocity for the separation distance used, fpm

For nonuniform distribution of gas flow in the horizontal shell, determine the minimum horizontal distance for gas flow from point of entry to point of exit as follows:

$$RTL = \frac{144\ \text{cfm} \cdot SD}{SV \cdot A_x} \tag{4}$$

where

RTL = residence time length, in.
cfm = maximum flow for that portion of the shell

All connections must be sized for the flow rates and pressure drops permissible and must be positioned to minimize liquid splashing. Internal baffles or mist eliminators can reduce the diameter of vessels; however, test correlations are necessary for a given configuration and placement of these devices.

An alternate formula for determining separation velocities that can be applied to separators is:

$$v = k[(\rho_l - \rho_v)/\rho_v]^{0.5} \tag{5}$$

where

v = velocity of vapor, fps
ρ_l = density of liquid, lb/ft^3
ρ_v = density of vapor, lb/ft^3
k = factor based on experience without regard to vertical separation distance and surface tension for gravity separators

In gravity liquid/vapor separators that must separate heavy entrainment from vapors, use a k of 0.1. This gives velocities equivalent to those used for 12 to 14 in. vertical separating distance (*SD*) for R-717, and 14 to 16 in. vertical *SD* for halocarbons. In knockout drums that separate light entrainment, use a k of 0.2. This gives velocities equivalent to those used for 36 in. vertical *SD* for R-717 and for halocarbons.

REFERENCES

Chaddock, J.S., H. Lau, and E. Skuchas. 1976. Two-phase pressure drop in refrigerant liquid overfeed systems—Experimental measurements. ASHRAE *Transactions* 82(2):134-50.

Chaddock, J.B., D.P. Werner, and C.G. Papachristou. 1972. Pressure drop in the suction lines of refrigerant circulation systems. ASHRAE *Transactions* 78(2):114-23.

Chisholm, D. 1971. Prediction of pressure drop at pipe fittings during two-phase flow. *Proceedings* I.I.R., Washington, D.C.

Lorentzen, G. 1963. Conditions of cavitation in liquid pumps for refrigerant circulation. *Progress Refrigeration Science Technology* I:497.

Lorentzen, G. 1965. How to design piping for liquid recirculation. *Heating, Piping & Air Conditioning* (June):139.

Lorentzen, G. 1966. On the dimensioning of liquid separators for refrigeration systems. *Kaltetechnik* 18:89.

Lorentzen, G. and R. Gronnerud. 1967. On the design of recirculation type evaporators. *Kulde* 21(4):55.

Miller, D.K. 1971. Recent methods for sizing liquid overfeed piping and suction accumulator-receivers. *Proceedings* I.I.R., Washington, D.C.

Miller D.K. 1974. Refrigeration problems of a VCM carrying tanker. ASHRAE *Journal* 11.

Miller, D.K. 1979. Sizing dual suction risers in liquid overfeed refrigeration systems. *Chemical Engineering* 9.

Niederer, D.H. 1964. Liquid recirculation systems—What rate of feed is recommended. *The Air Conditioning & Refrigeration Business* (December).

Niemeyer, E.R. 1961. Check these points when designing knockout drums. *Hydrocarbon Processing and Petroleum Refiner*, June.

Scheiman, A.D. 1964. Horizontal vapor-liquid separators. *Hydrocarbon Processing and Petroleum Refiner* (May).

Scheiman, A.D. 1963. Size vapor-liquid separators quicker by nomograph. *Hydrocarbon Processing and Petroleum Refiner*, October.

Scotland, W.B. 1963. Discharge temperature considerations with multicylinder ammonia compressors. *Modern Refrigeration* (February).

Scotland, W.B. 1970. Advantages, disadvantages and economics of liquid overfeed systems. ASHRAE *Symposium Bulletin* KC-70-3, Liquid overfeed systems.

Soling, S.P. 1971. Oil recovery from low temperature pump recirculating hydrocarbon systems. ASHRAE *Symposium Bulletin* PH-71-2, Effect of oil on the refrigeration system.

Sonders, M. and G.G. Brown. 1934. Design of fractionating columns, entrainment and capacity. *Industrial & Engineering Chemistry* (January).

Stoeker, W.F. 1960. How to design and operate flooded evaporators for cooling air and liquids. *Heating, Piping & Air Conditioning* (December).

Younger, A.H. 1955. How to size future process vessels. *Chemical Engineering* (May).

BIBLIOGRAPHY

Chaddock, J.B. 1976. Two-phase pressure drop in refrigerant liquid overfeed systems—Design tables. ASHRAE *Transactions* 82(2):107-33.

Geltz, R.W. 1967. Pump overfeed evaporator refrigeration systems. *Air Conditioning, Heating & Refrigeration News* (January 30, February 6, March 6, March 13, March 20, March 27).

Lorentzen, G. and A.O. Baglo. 1969. An investigation of a gas pump recirculation system. *Proceedings* of the Xth International Congress of Refrigeration, p. 215. International Institute of Refrigeration, Paris.

Lorentzen, G. 1968. Evaporator design and liquid feed regulation. *Journal of Refrigeration* (November-December): 160.

Richards, W.V. 1959. Liquid ammonia recirculation systems. *Industrial Refrigeration* (June):139.

Richards, W.V. 1970. Pumps and piping in liquid overfeed systems. ASHRAE *Symposium Bulletin* KC-70-3, Liquid overfeed systems.

Slipcevic, B. 1964. The calculation of the refrigerant charge in refrigerating systems with circulation pumps. *Kaltetechnik* 4:111.

Thompson, R.B. 1970. Control of evaporators in liquid overfeed systems. ASHRAE *Symposium Bulletin* KC-70-3, Liquid overfeed systems.

Watkins, J.E. 1956. Improving refrigeration systems by applying established principles. *Industrial Refrigeration* (June).

CHAPTER 2

SYSTEM PRACTICES FOR HALOCARBON REFRIGERANTS

REFRIGERATION is the process of moving heat from one location to another by use of refrigerant in a closed refrigeration cycle. Oil management, gas and liquid separation, subcooling, superheating, and piping of refrigerant liquid, gas, and two-phase flow are all part of the refrigeration discipline. Refrigeration applications include air conditioning, commercial refrigeration, and industrial refrigeration.

Typical considerations for a refrigeration system may include:

- Year-round operation, regardless of outdoor ambient conditions
- Possible wide load variations (0 to 100% capacity) during short periods without seriously disrupting the required temperature levels
- Frost control for continuous performance applications
- Oil management for different refrigerants under varying load and temperature conditions
- A wide choice of heat exchange methods, *e.g.*, dry expansion, liquid overfeed, or flooded feed of the refrigerants, and the use of secondary coolants such as salt brine, alcohol, and glycol
- System efficiency, maintainability, and operating simplicity
- Operating pressures and pressure ratios that might require multistaging, cascading, and so forth

A successful refrigeration system depends on good piping design and an understanding of the accessories required to perform its functions. The fundamentals of piping and accessories applications in halocarbon refrigerant systems are discussed in this chapter. Hydrocarbon refrigerant pipe friction data can be found in petroleum industry handbooks. Use the refrigerant properties and information in Chapters 2, 16, and 17 of the 1993 ASHRAE *Handbook—Fundamentals* to calculate friction losses.

REFRIGERATION LOAD

An analysis of the refrigeration load is vital to proper equipment selection and satisfactory system performance. All load components and, often, the time profile of the load components must be considered.

Typical factors to consider are as follows:

- Heat leakage of either latent or sensible heat into the space or from the product. This flow is a function of the temperature difference and insulating efficiency of the container walls, including insulation and any heat introduced by ventilation or infiltration.
- Product load, *i.e.*, the heat that must be removed to bring the product to the desired state. This load includes both sensible and latent heat associated with a change of state.
- Internal sensible load can include heat from motors (including evaporator fan motors), lights, and other heat sources.

A plot of load versus time is important in selecting equipment that will function well at minimum as well as at peak load. Minimum load conditions are often overlooked in design so that severe operating problems such as inefficiency, short-cycling, oil migration, shortage of hot gas for defrost, or other adverse conditions arise. Often small and large capacity components must be selected and properly controlled to match operating conditions. Industrial processes are particularly prone to abrupt interruptions, going from 100 to 0% load and then back to 100% over a short time span. Improperly designed systems do not function well under these conditions.

Piping Basic Principles

Refrigerant piping systems are designed and operated to:

- Assure proper refrigerant feed to evaporators.
- Provide practical refrigerant line sizes without excessive pressure drop.
- Prevent excessive amounts of lubricating oil from being trapped in any part of the system.
- Protect the compressor at all times from loss of lubricating oil.
- Prevent liquid refrigerant or oil slugs from entering the compressor during operating and idle time.
- Maintain a clean and dry system.

REFRIGERANT FLOW

Refrigerant Line Velocities

Economics, pressure drop, noise, and oil entrainment establish feasible design velocities in refrigerant lines (see Table 1).

Higher gas velocities are sometimes found in relatively short suction lines on comfort air-conditioning or other applications where the operating time is only 2000 to 4000 h per year and where low initial cost of the system may be more significant than low operating cost. Industrial or commercial refrigeration applications, where equipment runs almost continuously, should be designed with low refrigerant velocities for most efficient compressor capacity and low equipment operating costs. An owning and operating cost analysis will reveal the best choice of line sizes. Liquid lines from condensers to receivers should be sized for 100 fpm or less to assure positive gravity flow without incurring backup of liquid flow. Liquid from receiver to evaporator lines

Table 1 Gas Line Velocities for R-22, R-134a, and R-502

Suction line	900 to 4000 ft/min
Discharge line	2000 to 3500 ft/min

The preparation of this chapter is assigned to TC 10.3, Refrigerant Piping, Controls, and Accessories.

should be sized to maintain velocities below 300 fpm, thus minimizing or preventing liquid hammering when solenoids or other electrically operated valves are used.

Refrigerant Flow Rates

Refrigerant flow rates are indicated in Figures 1 through 3 for R-22, R-134a, and R-502. To obtain the total system flow rate, select the proper rate value and multiply by the system capacity. Enter curves using saturated evaporating temperature at the evaporator outlet and actual liquid temperature entering the liquid feed device (including subcooling in condensers and liquid-suction interchanger, if used).

Since Figures 1 through 3 are based on a saturated evaporator temperature, they may indicate slightly higher refrigerant flow rates than are actually in effect when the suction vapor is superheated in excess of the conditions just mentioned. Refrigerant flow rates may be reduced approximately 3% for each 10°F increase in superheat in the evaporator.

Suction line superheating downstream of the evaporator, due to line heat gain from external sources, should not be used to reduce evaluated mass flow. This heat gain and resultant suction line superheating increases volume flow rate and line velocity per unit of evaporator capacity, but not mass flow rate. It should be considered when evaluating a suction line size for satisfactory oil return up risers.

Suction gas superheating from the use of a liquid-suction heat exchanger has an effect on oil return similar to suction line superheating. The liquid cooling that results from the heat exchange reduces mass flow rate per ton of refrigeration. This can be seen in Figures 1 through 3, because the reduced liquid temperature supplied to the evaporator feed valve has been taken into account.

Superheat resulting from heat in a space not intended to be cooled is always detrimental because the volume rate increases with no compensating gain in refrigerating effect.

REFRIGERANT LINE SIZING

In sizing refrigerant lines, cost considerations favor keeping line sizes as small as possible. However, suction and discharge line pressure drops cause loss of compressor capacity and increased power. Excessive liquid line pressure drops can cause flashing of the liquid refrigerant, resulting in faulty expansion valve operation. Refrigeration systems are designed so that friction pressure losses do not exceed a pressure differential equivalent to a corresponding change in the saturation boiling temperature. The primary measure for determining pressure drops is a given change in saturation temperature.

Pressure Drop Considerations

Pressure drop in refrigeration lines causes a reduction in the system efficiency. The correct selection must be based on cost and minimizing efficiency reduction. Table 2 indicates the approximate effect of refrigerant pressure drop on an R-22 system operating at a 40°F saturated evaporator with a 100°F saturated condensing temperature.

Pressure drop calculations are determined as normal pressure loss associated with a change in saturation temperature of the refrigerant. Typically, the refrigeration system will be sized for pressure losses of 2°F or less for each segment of the discharge, suction, and liquid lines.

Liquid Lines. Pressure drop should not be so large as to cause gas formation in the liquid line, insufficient liquid pressure at the liquid feed device, or both. Systems are normally designed so that the pressure drop in the liquid line, due to friction, is not greater than that corresponding to about a 1 to 2°F change in saturation temperature. See Tables 3 through 6 for liquid line sizing information. Liquid pressure losses for a change of 1°F saturation at 100°F condensing pressure are approximately:

Table 2 Approximate Effect of Gas Line Pressure Drops on R-22 Compressor Capacity and Horsepower*

Line Loss, °F	Capacity, %	Energy, %
Suction Line		
0	100	100
2	96.4	104.8
4	92.9	108.1
Discharge Line		
0	100	100
2	99.1	103.0
4	98.2	106.3

*Energy percentage rated at hp/ton

Refrigerant	Change, psi
R-22	2.9
R-134a	2.1
R-502	3.1

Liquid subcooling is the only method of overcoming the liquid line pressure loss to guarantee liquid at the expansion device in the evaporator. If the subcooling is insufficient, then flashing will occur within the liquid line and degrade the efficiency of the refrigeration system.

Friction pressure drops in the liquid line are caused by accessories, such as solenoid valves, filter driers, and hand valves, as well as by the actual pipe and fittings, from the receiver outlet to the refrigerant feed device at the evaporator.

Liquid line risers are a source of pressure loss and add to the total losses of the liquid line. The losses due to risers are approximately 0.5 psi per foot of liquid lift. The total losses are the sum of all the friction losses plus the pressure loss from liquid risers.

The following example illustrates the process of determining the liquid line size and checking for total subcooling required.

Example 1. An R-22 refrigeration system operating at 40°F evaporator and 105°F condensing. Capacity is 5 tons, and the liquid line is 100 ft equivalent length and a riser of 20 ft.

Solution: From Table 3, the size of the liquid line at 1°F drop is 5/8 in. OD. Use the equation in Note 3, Table 3 to compute actual temperature drop. The estimated actual friction pressure loss at 5 tons is

Actual t_{drop}	$= 1.0(5.0/6.7)^{1.8}$	= 0.59°F or about 1.8 psi
The loss for the riser	$= 20 \times 0.5$	= 10 psi
Total pressure losses	$= 10.0 + 1.8$	= 11.8 psi
The saturation pressure at 105°F condensing		= 210.8 psig
Initial pressure at beginning of liquid line		= 210.8 psig
Total liquid line losses		= 11.8 psi
Net pressure at expansion device		= 199.00 psig

The saturation temperature at 199 psig is 101.06 °F.

Required subcooling to overcome the liquid losses = (105.0 – 101.06) or 3.94°F

Refrigeration systems that have no liquid risers and where the evaporator is below the condenser/receiver benefit from a gain in pressure due to liquid weight and can tolerate larger friction losses without flashing. Regardless of the routing of the liquid lines when flashing takes place, the overall efficiency is reduced, and the system may malfunction from the flashing.

The velocity of liquid leaving a partially filled vessel (such as a receiver or shell-and-tube condenser) is limited by the height of the liquid above the point at which the liquid line leaves the vessel, whether or not the liquid at the surface is subcooled. Since the liquid in the vessel has a very low (or zero) velocity, the velocity V is the liquid line (usually at the ven contracta) is $V^2 = 2gh$,

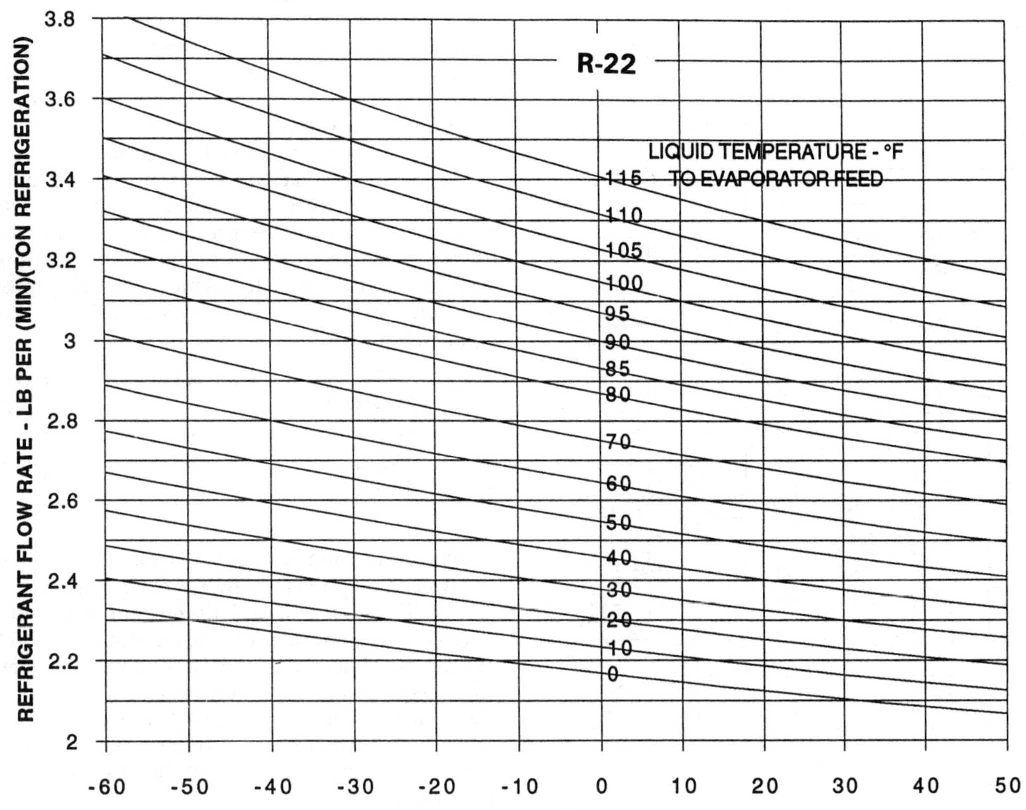

Fig. 1 Flow Rate per Ton of Refrigeration for Refrigerant 22

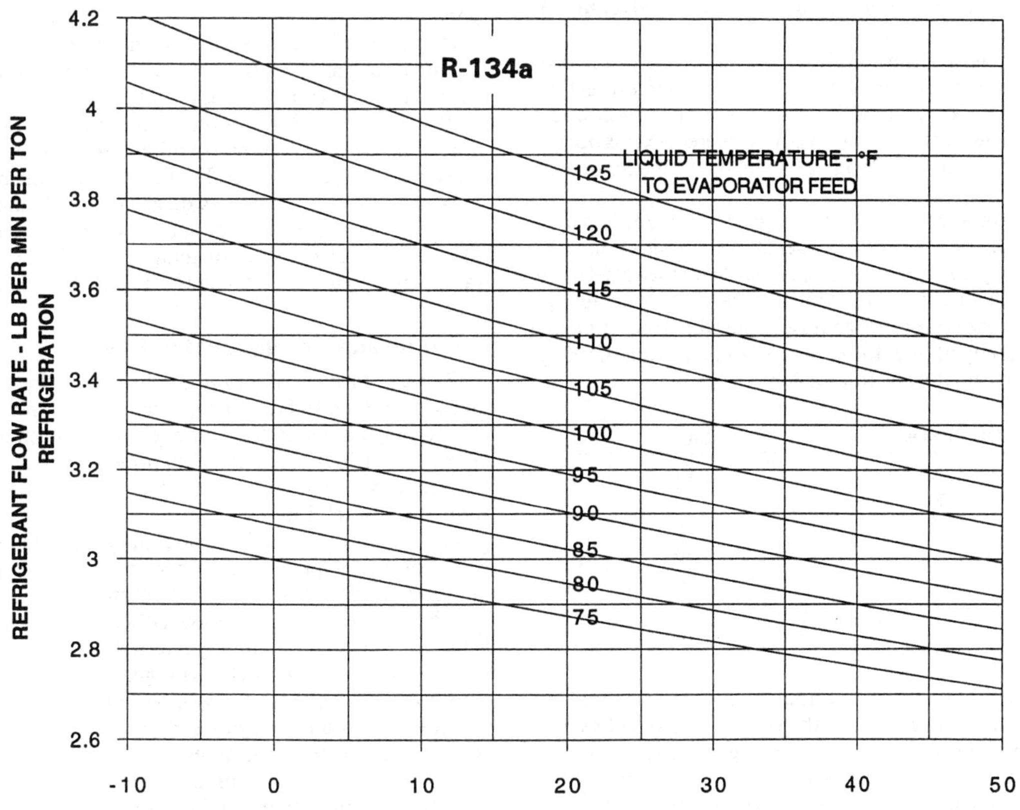

Fig. 2 Flow Rate per Ton of Refrigeration for Refrigerant 134a

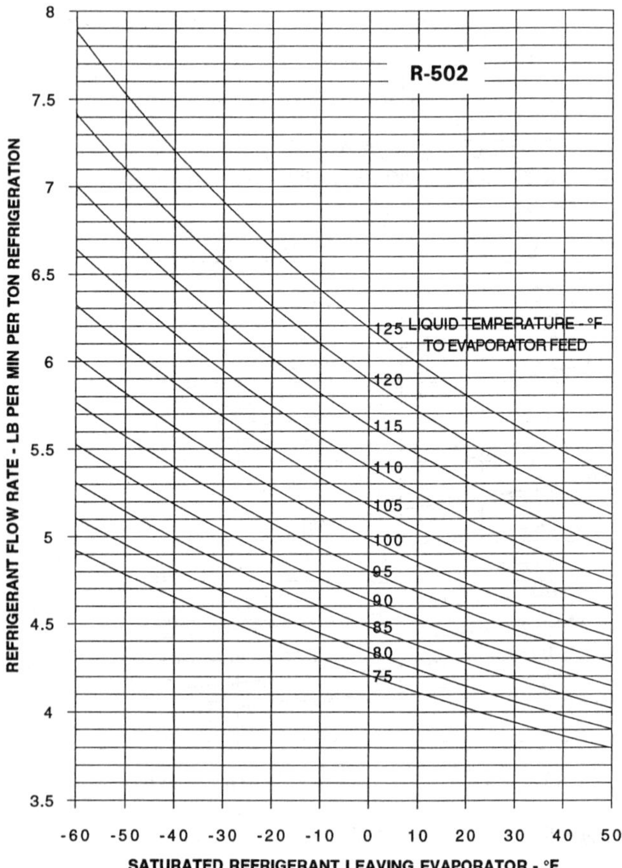

Fig. 3 Flow Rate per Ton of Refrigeration for Refrigerant 502

where h is the height of the liquid in the vessel. Gas pressure does not add to the velocity without gas flowing in the same direction. As a result, both gas and liquid flow through the line, limiting the rate of liquid flow. If this factor is not considered, excess operating charges in receivers and flooding of shell-and-tube condensers may result.

No specific data are available to precisely size a line leaving a vessel. If the height of the liquid level above the vena contracta produces the desired velocity, the liquid will leave the vessel at the expected rate. Thus, if the level in the vessel falls to one pipe diameter above the bottom of the vessel from which the liquid line leaves, the capacity of lines for R-22 at 3 lb/min per ton of refrigeration is approximately as follows for copper:

OD, Inches	Tons
1-1/8	14
1-3/8	25
1-5/8	40
2-1/8	80
2-5/8	130
3-1/8	195
4-1/8	410

The whole liquid line need not be as large as the leaving connection. After the vena contracta, the velocity is about 40% less. If the line continues down from the receiver, the value of h increases. For a 200-ton capacity with R-22, the line from the bottom of the receiver should be about 3-1/8 in. After a drop of 1 ft, a reduction to 2-5/8 in. is satisfactory.

Suction Lines. Suction lines are more critical than liquid and discharge lines from a design and construction standpoint. Refrigerant lines should be sized to (1) provide a minimum pressure drop at full load, (2) return oil from the evaporator to the compressor under minimum load conditions, and (3) prevent oil from draining from an active evaporator into an idle one. A pressure drop in the suction line reduces a system's capacity because it forces the compressor to operate at a lower suction pressure to maintain a desired evaporating temperature in the coil. The suction line is normally sized to have a pressure drop from friction no greater than the equivalent of about a 2 °F change in saturation temperature. See Tables 3 through 9 for suction line sizing information. The equivalent pressure loss at 40 °F saturated suction temperature is approximately:

Refrigerant	Suction Loss, °F	Pressure Loss, psi
R-22	2	2.91
R-134a	2	1.93
R-502	2	3.18

At suction temperatures lower than 40 °F, the pressure drop equivalent to a given temperature change decreases. For example, at −40 °F suction with R-22, the pressure drop equivalent to a 2 °F change in saturation temperature is about 0.8 psi. Therefore, low-temperature lines must be sized for a very low-pressure drop, or higher equivalent temperature losses, with resultant loss in equipment capacity, must be accepted. For very low-pressure drops, any suction or hot-gas risers must be sized properly to assure oil entrainment up the riser so that the oil is always returned to the compressor.

Where pipe size must be reduced to provide sufficient gas velocity to entrain oil up vertical risers at partial loads, greater pressure drops are imposed at full load. These can usually be compensated for by oversizing the horizontal and down run lines and components.

Discharge Lines. Pressure loss in hot-gas lines increases the required compressor power per unit of refrigeration and decreases the compressor capacity. Table 2 illustrates the power losses for an R-22 system at 40 °F evaporator and 100 °F condensing temperature. Pressure drop is kept to a minimum by generously sizing the lines for low friction losses, but still maintaining refrigerant line velocities to entrain and carry oil along at all loading conditions. Pressure drop is normally designed not to exceed the equivalent of a 2 °F change in saturation temperature. Recommended sizing tables are based on a 1 °F change in saturation temperature.

Location and Arrangement of Piping

Refrigerant lines should be as short and direct as possible to minimize tubing and refrigerant requirements and pressure drops. Plan piping for a minimum number of joints using as few elbows and other fittings as possible, but provide sufficient flexibility to absorb compressor vibration and stresses due to thermal expansion and contraction.

Arrange refrigerant piping so that normal inspection and servicing of the compressor and other equipment is not hindered. Do not obstruct the view of the oil level sight glass or run piping so that it interferes with the removal of compressor cylinder heads, end bells, access plates, or any internal parts. Suction line piping to the compressor should be arranged so that it will not interfere with removal of the compressor for servicing.

Provide adequate clearance between pipe and adjacent walls and hangers or between pipes for insulation installation. Use sleeves through floors, walls, or ceilings, sized to permit installation of both pipe and insulation. Set these sleeves prior to pouring of concrete or erection of brickwork.

Run piping so that it does not interfere with passages or obstruct headroom, windows, and doors. Refer to ASHRAE *Standard* 15, Safety Code for Mechanical Refrigeration, and other governing local codes for restrictions that may apply.

Protection against Damage to Piping

Protection against damage is necessary, particularly for small lines, since these have a false appearance of strength. Where traffic is heavy, provide protection against impact from carelessly handled hand trucks, overhanging loads, ladders, and fork trucks.

Piping Insulation

All piping joints and fittings should be thoroughly leak tested before insulation is sealed. Suction lines should be insulated to prevent sweating and heat gain. Insulation covering lines on which moisture can condense or lines subjected to outside conditions must be vapor sealed to prevent any moisture travel through the insulation or condensation in the insulation. Many commercially available types are provided with an integral waterproof jacket for this purpose. While the liquid line ordinarily does not require insulation, on installations where the suction and liquid lines are clamped together, the two lines can be insulated as a unit. When it passes through an area of higher temperature, the liquid line should be insulated to minimize heat gain. Hot-gas discharge lines usually are not insulated; however, they should be insulated if the heat dissipated is objectionable or to prevent injury from high-temperature surfaces. In the latter case, it is not essential to provide insulation with a tight vapor seal since moisture condensation is not a problem unless located outside. Hot-gas defrost lines are customarily insulated to minimize heat loss and condensation of gas inside the piping.

While all joints and fittings should be covered, it is not advisable to do so until the system has been thoroughly leak tested.

Vibration and Noise in Piping

Vibration transmitted through or generated in refrigerant piping and the resulting objectionable noise can be eliminated or minimized by proper piping design and support.

Two undesirable effects of vibration of refrigerant piping are (1) physical damage to the piping, which results in the breaking of brazed joints and, consequently, loss of charge; and (2) transmission of noise through the piping itself and through building construction with which the piping may come into direct physical contact.

In refrigeration applications, piping vibration can be caused by the rigid connection of the refrigerant piping to a reciprocating

Table 3 Suction, Discharge, and Liquid Line Capacities for Refrigerant 22 (Single- or High-Stage Applications)

Line Size Type L Copper, OD	Suction Lines (Δt = 2°F), Saturated Suction Temperature, °F: −40, Δp=0.79	−20, Δp=1.15	0, Δp=1.6	20, Δp=2.22	40, Δp=2.91	Discharge Lines (Δt = 1°F, Δp = 3.05 psi), Saturated Suction Temp., °F: −40	40	Line Size Type L Copper, OD	Liquid Lines: Vel. = 100 fpm[a]	Δt = 1°F, Δp = 3.05 psi[b]
1/2	—	—	—	0.40	0.6	0.75	0.85	1/2	2.3	3.6
5/8	—	0.32	0.51	0.76	1.1	1.4	1.6	5/8	3.7	6.7
7/8	0.52	0.86	1.3	2.0	2.9	3.7	4.2	7/8	7.8	18.2
1-1/8	1.1	1.7	2.7	4.0	5.8	7.5	8.5	1-1/8	13.2	37.0
1-3/8	1.9	3.1	4.7	7.0	10.1	13.1	14.8	1-3/8	20.2	64.7
1-5/8	3.0	4.8	7.5	11.1	16.0	20.7	23.4	1-5/8	28.5	102.5
2-1/8	6.2	10.0	15.6	23.1	33.1	42.8	48.5	2-1/8	49.6	213.0
2-5/8	10.9	17.8	27.5	40.8	58.3	75.4	85.4	2-5/8	76.5	376.9
3-1/8	17.5	28.4	44.0	65.0	92.9	120.2	136.2	3-1/8	109.2	601.5
3-5/8	26.0	42.3	65.4	96.6	137.8	178.4	202.1	3-5/8	147.8	895.7
4-1/8	36.8	59.6	92.2	136.3	194.3	251.1	284.4	4-1/8	192.1	1263.2
Steel IPS SCH								**Steel IPS SCH**		
1/2 40	—	0.38	0.58	0.85	1.2	1.5	1.7	1/2 80	3.8	5.7
3/4 40	0.50	0.8	1.2	1.8	2.5	3.3	3.7	3/4 80	6.9	12.8
1 40	0.95	1.5	2.3	3.4	4.8	6.1	6.9	1 80	11.5	25.2
1-1/4 40	2.0	3.2	4.8	7.0	9.9	12.6	14.3	1-1/4 80	20.6	54.1
1-1/2 40	3.0	4.7	7.2	10.5	14.8	19.0	21.5	1-1/2 80	28.3	82.6
2 40	5.7	9.1	13.9	20.2	28.5	36.6	41.4	2 40	53.8	192.0
2-1/2 40	9.2	14.6	22.1	32.2	45.4	58.1	65.9	2-1/2 40	76.7	305.8
3 40	16.2	25.7	39.0	56.8	80.1	102.8	116.4	3 40	118.5	540.3
4 40	33.1	52.5	79.5	115.9	163.2	209.5	237.3	4 40	204.2	1101.2

Notes:

1. Table capacities are in tons of refrigeration.
   Δp = pressure drop due to line friction, psi per 100 ft of equivalent line length
   Δt = corresponding change in saturation temperature per 100 ft, °F
2. Line capacity for other saturation temperatures Δt and equivalent lengths L_e

$$\text{Line capacity} = \text{Table capacity}\left(\frac{\text{Table } L_e}{\text{Actual } L_e} \times \frac{\text{Actual } \Delta t}{\text{Table } \Delta t}\right)^{0.55}$$

3. Saturation temperature Δt for other capacities and equivalent lengths L_e

$$\Delta t = \text{Table } \Delta t \left(\frac{\text{Actual } L_e}{\text{Table } L_e}\right)\left(\frac{\text{Actual capacity}}{\text{Table capacity}}\right)^{1.8}$$

4. Values in the table are based on 105°F condensing temperature. Multiply table capacities by the following factors for other condensing temperatures.

Condensing Temperature, °F	Suction Line	Discharge Line
80	1.11	0.79
90	1.07	0.88
100	1.03	0.95
110	0.97	1.04
120	0.90	1.10
130	0.86	1.18
140	0.80	1.26

[a] The sizing shown is recommended where any gas generated in the receiver must return up the condensate line to the condenser without restricting condensate flow. Water-cooled condensers, where the receiver ambient temperature may be higher than the refrigerant condensing temperature, fall in this category.

[b] The line pressure drop Δp is conservative; if subcooling is substantial or the line is short, a smaller size line may be used. Applications with very little subcooling or very long lines may require a larger line.

compressor. Vibration effects are evident in all lines directly connected to the compressor or condensing unit. It is thus impossible to eliminate vibration in piping; it is only possible to mitigate its effects.

Flexible metal hose is sometimes used to absorb vibration transmission along smaller pipe sizes. For maximum effectiveness, it should be installed parallel to the crank shaft. In some cases, two isolators may be required, one in the horizontal line and the other in the vertical line at the compressor. A rigid brace on the end of the flexible hose away from the compressor is required to prevent vibration of the hot-gas line beyond the hose.

Flexible metal hose is not as efficient in absorbing vibration on larger sizes of pipe because it is not actually flexible unless the ratio of length to diameter is relatively great. In practice, the length is often limited, and flexibility is reduced in larger sizes. This problem is best solved by using flexible piping and isolation-type hangers where the piping is secured to the structure.

When piping passes through walls, floors, or inside furring, it must not touch any part of the building and must be supported only by the hangers (provided to avoid transmitting vibration to the building), eliminating the possibility of walls or ceilings acting as sounding boards or diaphragms. When piping is erected where access is difficult after installation, it should be supported by isolation hangers.

Vibration and noise from a piping system can also be caused by gas pulsations from the compressor operation or from turbulence in the gas, which increases at high velocities. It is usually more apparent in the discharge line than in other parts of the system.

When gas pulsations caused by the compressor create vibration and noise, they have a characteristic frequency that is a function of the number of gas discharges by the compressor on each revolution. This frequency is not necessarily equal to the number of cylinders, since on some compressors two pistons operate together. It is also varied by the angular displacement of the cylinders, such as in the V-type compressors. Noise resulting from gas pulsations is usually objectionable only when the piping characteristics of the system amplify the pulsation by resonance. On single-compressor systems, this can be reduced by changing the size or length of the resonating line, or by installing a properly sized hot-gas muffler

Table 4 Suction, Discharge, and Liquid Line Capacities in Tons for Refrigerant 134a (Single- or High-Stage Applications)

Line Size Type L Copper, OD	Suction Lines (Δt = 2°F) Saturated Suction Temperature, °F					Discharge Lines (Δt = 1°F, Δp = 2.2 psi) Saturated Suction Temp., °F			Line Size Type L Copper, OD	Liquid Lines See notes a and b	
	0 Δp=1.00	10 Δp=1.19	20 Δp=1.41	30 Δp=1.66	40 Δp=1.93	0	20	40		Vel. = 100 fpm	Δt = 1°F Δp = 2.2
1/2	0.14	0.18	0.23	0.29	0.35	0.54	0.57	0.59	1/2	2.13	2.79
5/8	0.27	0.34	0.43	0.54	0.66	1.01	1.07	1.12	5/8	3.42	5.27
7/8	0.71	0.91	1.14	1.42	1.75	2.67	2.81	2.94	7/8	7.09	14.00
1-1/8	1.45	1.84	2.32	2.88	3.54	5.40	5.68	5.95	1-1/8	12.10	28.40
1-3/8	2.53	3.22	4.04	5.02	6.17	9.42	9.91	10.40	1-3/8	18.40	50.00
1-5/8	4.02	5.10	6.39	7.94	9.77	14.90	15.70	16.40	1-5/8	26.10	78.60
2-1/8	8.34	10.60	13.30	16.50	20.20	30.80	32.40	34.00	2-1/8	45.30	163.00
2-5/8	14.80	18.80	23.50	29.10	35.80	54.40	57.20	59.90	2-5/8	69.90	290.00
3-1/8	23.70	30.00	37.50	46.40	57.10	86.70	91.20	95.50	3-1/8	100.00	462.00
3-5/8	35.10	44.60	55.80	69.10	84.80	129.00	135.00	142.00	3-5/8	135.00	688.00
4-1/8	49.60	62.90	78.70	97.40	119.43	181.00	191.00	200.00	4-1/8	175.00	971.00
5-1/8	88.90	113.00	141.00	174.00	213.00	323.00	340.00	356.00			
6-1/8	143.00	181.00	226.00	280.00	342.00	518.00	545.00	571.00			
Steel IPS SCH									**Steel** IPS SCH		
1/2 80	0.22	0.28	0.35	0.43	0.53	0.79	0.84	0.88	1/2 80	3.43	4.38
3/4 80	0.51	0.64	0.79	0.98	1.19	1.79	1.88	1.97	3/4 80	6.34	9.91
1 80	1.00	1.25	1.56	1.92	2.33	3.51	3.69	3.86	1 80	10.50	19.50
1-1/4 40	2.62	3.30	4.09	5.03	6.12	9.20	9.68	10.10	1-1/4 80	18.80	41.80
1-1/2 40	3.94	4.95	6.14	7.54	9.18	13.80	14.50	15.20	1-1/2 80	25.90	63.70
2 40	7.60	9.56	11.90	14.60	17.70	26.60	28.00	29.30	2 40	49.20	148.00
2-1/2 40	12.10	15.20	18.90	23.10	28.20	42.40	44.60	46.70	2-1/2 40	70.10	236.00
3 40	21.40	26.90	33.40	41.00	49.80	75.00	78.80	82.50	3 40	108.00	419.00
4 40	43.80	54.90	68.00	83.50	101.60	153.00	160.00	168.00	4 40	187.00	853.00

Notes:

1. Table capacities are in tons of refrigeration.
   Δp = pressure drop due to line friction, psi per 100 ft of equivalent line length
   Δt = corresponding change in saturation temperature per 100 ft, °F
2. Line capacity for other saturation temperatures Δt and equivalent lengths L_e

$$\text{Line capacity} = \text{Table capacity}\left(\frac{\text{Table } L_e}{\text{Actual } L_e} \times \frac{\text{Actual } \Delta t}{\text{Table } \Delta t}\right)^{0.55}$$

3. Saturation temperature Δt for other capacities and equivalent lengths L_e

$$\Delta t = \text{Table } \Delta t \left(\frac{\text{Actual } L_e}{\text{Table } L_e}\right)\left(\frac{\text{Actual capacity}}{\text{Table capacity}}\right)^{1.8}$$

4. Values in the table are based on 105°F condensing temperature. Multiply table capacities by the following factors for other condensing temperatures.

Condensing Temperature, °F	Suction Line	Discharge Line
80	1.158	0.804
90	1.095	0.882
100	1.032	0.961
110	0.968	1.026
120	0.902	1.078
130	0.834	1.156

[a] The sizing shown is recommended where any gas generated in the receiver must return up the condensate line to the condenser without restricting condensate flow. Water-cooled condensers, where the receiver ambient temperature may be higher than the refrigerant condensing temperature, fall in this category.

[b] The line pressure drop Δp is conservative; if subcooling is substantial or the line is short, a smaller size line may be used. Applications with very little subcooling or very long lines may require a larger line.

Table 5 Suction, Discharge, and Liquid Line Capacities in Tons for Refrigerant 502 (Single- or High-Stage Applications)

Line Size Type L Copper, OD	Suction Lines (Δt = 2°F) Saturated Suction Temperature, °F					Discharge Lines (Δt = 1°F, Δp = 1.9 psi) Saturated Suction Temp., °F			Line Size Type L Copper, OD	Liquid Lines	
	−40 Δp = 0.92	−20 Δp = 0.133	0 Δp = 1.84	20 Δp = 2.45	40 Δp = 3.18	−40	0	40		Vel. = 100 fpm[a]	Δt=1°F Δp = 3.15[b]
1/2	0.08	0.14	0.22	0.33	0.49	0.56	0.63	0.70	1/2	1.5	2.4
5/8	0.16	0.27	0.42	0.63	0.91	1.0	1.2	1.3	5/8	2.3	4.5
7/8	0.43	0.70	1.1	1.7	2.4	2.7	3.1	3.4	7/8	4.9	11.8
1-1/8	0.87	1.4	2.2	3.4	4.8	5.5	6.3	7.0	1-5/8	8.3	24.1
1-3/8	1.5	2.5	3.9	5.8	8.4	9.6	10.9	12.1	1-3/8	12.6	42.0
1-5/8	2.4	4.0	6.2	9.2	13.3	15.2	17.2	19.1	1-5/8	17.9	66.4
2-1/8	5.0	8.2	12.8	19.1	27.5	31.4	35.6	39.5	2-1/8	31.1	138.0
2-5/8	8.8	14.5	22.6	33.7	48.4	55.3	62.8	69.5	2-5/8	48.0	243.7
3-1/8	14.1	23.2	36.0	53.7	77.0	87.9	99.8	110.5	3-1/8	68.4	389.3
3-5/8	21.0	34.4	53.5	79.7	114.3	130.5	148.1	164.0	3-5/8	92.6	579.0
4-1/8	29.7	48.5	75.4	112.3	161.0	183.7	208.4	230.9	4-1/8	120.3	816.9
5-1/8	53.2	86.7	134.6	200.3	287.1	327.3	371.3	411.3	—	—	—
6-1/8	85.6	139.5	216.2	321.3	460.6	525.2	595.9	660.1	—	—	—

Notes:

1. Table capacities are in tons of refrigeration.
   Δp = pressure drop due to line friction, psi per 100 ft of equivalent line length
   Δt = corresponding change in saturation temperature per 100 ft, °F
2. Line capacity for other saturation temperatures Δt and equivalent lengths L_e

$$\text{Line capacity} = \text{Table capacity}\left(\frac{\text{Table } L_e}{\text{Actual } L_e} \times \frac{\text{Actual } \Delta t}{\text{Table } \Delta t}\right)^{0.55}$$

3. Saturation temperature Δt for other capacities and equivalent lengths L_e

$$\Delta t = \text{Table } \Delta t \left(\frac{\text{Actual } L_e}{\text{Table } L_e}\right)\left(\frac{\text{Actual capacity}}{\text{Table capacity}}\right)^{1.8}$$

4. Values in the table are based on 105°F condensing temperature. Multiply table capacities by the following factors for other condensing temperatures.

Condensing Temperature, °F	Suction Line	Discharge Line
80	1.20	0.83
90	1.12	0.91
100	1.04	0.97
110	0.96	1.02
120	0.88	1.08
130	0.80	1.16

[a] The sizing shown is recommended where any gas generated in the receiver must return up the condensate line to the condenser without restricting condensate flow. Water-cooled condensers, where the receiver ambient temperature may be higher than the refrigerant condensing temperature, fall in this category.

[b] The line pressure drop Δp is conservative; if subcooling is substantial or the line is short, a smaller size line may be used. Applications with very little subcooling or very long lines may require a larger line.

Table 6 Suction, Discharge, and Liquid Line Capacities in Tons for Intermediate- or Low-Stage Duty for Refrigerant 22

Refrigerant and Δt Equivalent of Friction Drop[a]	Line Size Type L Copper OD	Suction Lines Saturated Suction Temperature, °F							Discharge Lines[a]	Line Size Type L Copper OD	Liquid Lines
		−90	−80	−70	−60	−50	−40	−30			
Refrigerant 22	5/8								0.7	5/8	
	7/8	0.18	0.25	0.34	0.46	0.61	0.79	1.0	1.9	7/8	
	1-1/8	0.36	0.51	0.70	0.94	1.2	1.6	2.1	3.8	1-1/8	
	1-3/8	0.6	0.9	1.2	1.6	2.2	2.8	3.6	6.6	1-3/8	
	1-5/8	1.0	1.4	1.9	2.6	3.4	4.5	5.7	10.5	1-5/8	
	2-1/8	2.1	3.0	4.1	5.5	7.2	9.3	11.9	21.7	2-1/8	
2°F Δt per 100 ft	2-5/8	3.8	5.3	7.2	9.7	12.7	16.5	21.1	38.4	2-5/8	
equivalent length	3-1/8	6.1	8.5	11.6	15.5	20.4	26.4	33.8	61.4	3-1/8	See Table 3
	3-5/8	9.1	12.7	17.3	23.1	30.4	39.4	50.2	91.2	3-5/8	
	4-1/8	12.9	18.0	24.5	32.7	43.0	55.6	70.9	128.6	4-1/8	
	5-1/8	23.2	32.3	43.9	58.7	77.1	99.8	126.9	229.5	5-1/8	
	6-1/8	37.5	52.1	71.0	94.6	124.2	160.5	204.2	369.4	6-1/8	

Notes:

1. Table capacities are in tons of refrigeration.
   Δp = pressure drop due to line friction, psi per 100 ft of equivalent line length
   Δt = corresponding change in saturation temperature per 100 ft, °F
2. Line capacity for other saturation temperatures Δt and equivalent lengths L_e

$$\text{Line capacity} = \text{Table capacity}\left(\frac{\text{Table } L_e}{\text{Actual } L_e} \times \frac{\text{Actual } \Delta t}{\text{Table } \Delta t}\right)^{0.55}$$

3. Saturation temperature Δt for other capacities and equivalent lengths L_e

$$\Delta t = \text{Table } \Delta t \left(\frac{\text{Actual } L_e}{\text{Table } L_e}\right)\left(\frac{\text{Actual capacity}}{\text{Table capacity}}\right)^{1.8}$$

4. Refer to the refrigerant thermodynamic property tables (Chapter 17, 1993 ASHRAE *Handbook—Fundamentals*) for the pressure drop corresponding to Δt.
5. Values in the table are based on 0°F condensing temperature. Multiply table capacities by the following factors for other condensing temperatures. Flow rates for discharge lines are based on −50°F evaporating temperature.

Condensing Temp., °F	Refrigerant 22 Suction	Refrigerant 22 Discharge
−30	1.09	0.58
−20	1.06	0.71
−10	1.03	0.85
0	1.00	1.00
10	0.97	1.20
20	0.94	1.45
30	0.90	1.80

[a]See section Pressure Drop Considerations.

in the discharge line immediately after the compressor discharge valve. On a paralleled compressor system, a harmonic frequency from the different speeds of multiple compressors may be apparent. This noise can sometimes be reduced by installing mufflers.

When noise is a result of turbulence, and isolating the line is not effective enough, the installation of a larger pipe diameter to reduce the gas velocity is sometimes helpful. Also, changing to a line of heavier wall or from copper to steel to change the pipe natural frequency may help.

Refrigerant Line Capacity Tables

Tables 3 to 6 show capacities for R-22, R-134a, and R-502 at specific pressure drops. The capacities shown in the tables are based on the refrigerant flow that develops a friction loss, per 100 ft of equivalent pipe length, corresponding to a 1 °F change in the saturation temperature (Δt) for discharge and liquid lines. Suction lines are based on a 2 °F change. Tables 7, 8, and 9 show suction line capacities for a 0.5 and 1 °F change of the saturation suction temperature. Pressure drops are given in degrees because this pipe sizing method is convenient and accepted throughout the industry. Corresponding pressure drops are also shown.

The refrigerant line sizing capacity tables are based on the Darcy-Weisbach relation and friction factors as computed by the Colebrook function (1938, 1939). Tubing roughness height is 0.000005 ft for copper and 0.00015 ft for steel pipe. Viscosity extrapolations and adjustments for pressures other than 1 atm were based on correlation techniques as presented by Keating and Matula (1969). Discharge gas superheat was 80 °F for R-134a and R-502, and 105 °F for R-22.

The refrigerant cycle for determining capacity is based on saturated gas leaving the evaporator. The calculations neglect the presence of oil and assume nonpulsating flow.

For additional charts and discussion of line sizing refer to Timm (1991), Wile (1977), and Atwood (1990).

Equivalent Lengths of Valves and Fittings

Refrigerant line capacity tables are based on unit pressure drop per 100 ft length of straight pipe; or a combination of straight pipe, fittings, and valves with friction drop equivalent to a 100 ft length of straight pipe.

Generally, pressure drop through valves and fittings is determined by establishing the equivalent straight length of pipe of the same size with the same friction drop. Line sizing tables can then be used directly. Tables 10, 11, and 12 give equivalent lengths of straight pipe for various fittings and valves, based on nominal pipe sizes.

The following example illustrates the use of various tables and charts to size refrigerant lines.

Example 2. Determine the line size and pressure drop equivalent (in degrees) for the suction line of a 30-ton R-22 system, operating at 40 °F suction and 100 °F condensing temperatures. The suction line is copper tubing, with 50 ft of straight pipe and six long radius elbows.

Solution: Add 50% to the straight length of pipe to establish a trial equivalent length. Trial equivalent length is 50 × 1.5 = 75 ft. From Table 3 (for 40 °F suction, 105 °F condensing), 33.1 tons capacity in 2-1/8 in. OD results in a 2 °F loss per 100 ft equivalent length. Referring to Note 4, Table 3, the capacity at 40 °F evaporator and 100 °F condensing temperature would be 1.03 × 33.1 = 34.1 tons. This trial size is used to evaluate actual equivalent length.

Straight pipe length	= 50.0 ft
Six 2 in. long radius elbows at 3 ft each (Table 10)	= 19.8 ft
Total equivalent length	= 69.8 ft

$$\Delta t = 2\,(69.8/100)(30/34.1)^{1.8} = 1.1\text{°F or } 1.6 \text{ psi}$$

Table 7 Suction Line Capacities in Tons for Refrigeration 134a (Single- or High-Stage Applications)

Line Size Type L Copper, OD	Saturated Suction Temperature, °F									
	0		10		20		30		40	
	Δt = 1°F	Δt = 0.5°F	Δt = 1°F	Δt = 0.5°F	Δt = 1°F	Δt = 0.5°F	Δt = 1°F	Δt = 0.5°F	Δt = 1°F	Δt = 0.5°F
	Δp = 0.50	Δp = 0.25	Δp = 0.60	Δp = 0.30	Δp = 0.71	Δp = 0.35	Δp = 0.83	Δp = 0.42	Δp = 0.97	Δp = 0.48
1/2	0.10	0.07	0.12	0.08	0.16	0.11	0.19	0.13	0.24	0.16
5/8	0.18	0.12	0.23	0.16	0.29	0.20	0.37	0.25	0.45	0.31
7/8	0.48	0.33	0.62	0.42	0.78	0.53	0.97	0.66	1.20	0.82
1-1/8	0.99	0.67	1.26	0.86	1.59	1.08	1.97	1.35	2.43	1.66
1-3/8	1.73	1.18	2.21	1.51	2.77	1.89	3.45	2.36	4.25	2.91
1-5/8	2.75	1.88	3.50	2.40	4.40	3.01	5.46	3.75	6.72	4.61
2-1/8	5.73	3.92	7.29	5.00	9.14	6.27	11.40	7.79	14.00	9.59
2-5/8	10.20	6.97	12.90	8.87	16.20	11.10	20.00	13.80	24.70	17.00
3-1/8	16.20	11.10	20.60	14.20	25.90	17.80	32.10	22.10	39.40	27.20
3-5/8	24.20	16.60	30.80	21.20	38.50	26.50	47.70	32.90	58.70	40.40
4-1/8	34.20	23.50	43.40	29.90	54.30	37.40	67.30	46.50	82.60	57.10
5-1/8	61.30	42.20	77.70	53.60	97.20	67.10	121.00	83.20	148.00	102.00
6-1/8	98.80	68.00	125.00	86.30	157.00	108.00	194.00	134.00	237.00	165.00
Steel										
IPS SCH										
1/2 80	0.16	0.11	0.20	0.14	0.25	0.17	0.30	0.21	0.37	0.26
3/4 80	0.36	0.25	0.45	0.31	0.56	0.39	0.69	0.48	0.84	0.59
1 80	0.70	0.49	0.88	0.61	1.09	0.77	1.34	0.94	1.64	1.15
1-1/4 40	1.84	1.29	2.31	1.62	2.87	2.02	3.54	2.48	4.31	3.03
1-1/2 40	2.77	1.94	3.48	2.44	4.32	3.03	5.30	3.73	6.47	4.55
2 40	5.35	3.75	6.72	4.72	8.33	5.86	10.30	7.20	12.50	8.78
2-1/2 40	8.53	5.99	10.70	7.53	13.30	9.35	16.30	11.50	19.90	14.00
3 40	15.10	10.60	18.90	13.30	23.50	16.50	28.90	20.30	35.20	24.80
4 40	30.80	21.70	38.70	27.20	48.00	33.80	58.80	41.50	71.60	50.50
5 40	55.60	39.20	69.80	49.10	86.50	60.93	106.00	74.95	129.00	91.00
6 40	89.90	63.40	113.00	79.60	140.00	98.50	172.00	121.00	209.00	148.00

Δp = pressure drop due to line friction, psi per 100 ft equivalent length of line
Δt = corresponding change in saturation temperature with the pressure drop, °F/100 ft

Table 8 Suction Line Capacities in Tons for Refrigerant 22 (Single- or High-Stage Applications)

Line Size Type L Copper, OD		Saturated Suction Temperature, °F									
		−40		−20		0		20		40	
		Δt = 1°F Δp = 0.393	Δt = 0.5°F Δp = 0.197	Δt = 1°F Δp = 0.577	Δt = 0.5°F Δp = 0.289	Δt = 1°F Δp = 0.813	Δt = 0.5°F Δp = 0.406	Δt = 1°F Δp = 1.104	Δt = 0.5°F Δp = 0.552	Δt = 1°F Δp = 1.455	Δt = 0.5°F Δp = 0.727
1/2		0.07	0.05	0.12	0.08	0.18	0.12	0.27	0.19	0.40	0.27
5/8		0.13	0.09	0.22	0.15	0.34	0.23	0.52	0.35	0.75	0.51
3/4		0.22	0.15	0.37	0.25	0.58	0.39	0.86	0.59	1.24	0.85
7/8		0.35	0.24	0.58	0.40	0.91	0.62	1.37	0.93	1.97	1.35
1-1/8		0.72	0.49	1.19	0.81	1.86	1.27	2.77	1.90	3.99	2.74
1-3/8		1.27	0.86	2.09	1.42	3.25	2.22	4.84	3.32	6.96	4.78
1-5/8		2.02	1.38	3.31	2.26	5.16	3.53	7.67	5.26	11.00	7.57
2-1/8		4.21	2.88	6.90	4.73	10.71	7.35	15.92	10.96	22.81	15.73
2-5/8		7.48	5.13	12.23	8.39	18.97	13.04	28.19	19.40	40.38	27.84
3-1/8		11.99	8.22	19.55	13.43	30.31	20.85	44.93	31.00	64.30	44.44
3-5/8		17.89	12.26	29.13	20.00	45.09	31.03	66.81	46.11	95.68	66.09
4-1/8		25.29	17.36	41.17	28.26	63.71	43.85	94.25	65.12	134.81	93.22
Steel											
IPS	SCH										
3/8	80	0.06	0.04	0.10	0.07	0.15	0.10	0.21	0.15	0.30	0.21
1/2	80	0.12	0.08	0.19	0.13	0.29	0.20	0.42	0.30	0.60	0.42
3/4	80	0.27	0.18	0.43	0.30	0.65	0.46	0.95	0.67	1.35	0.95
1	80	0.52	0.36	0.84	0.59	1.28	0.89	1.87	1.31	2.64	1.86
1-1/4	40	1.38	0.96	2.21	1.55	3.37	2.36	4.91	3.45	6.93	4.88
1-1/2	40	2.08	1.45	3.32	2.33	5.05	3.55	7.38	5.19	10.42	7.33
2	40	4.03	2.81	6.41	4.51	9.74	6.85	14.22	10.01	20.07	14.14
2-1/2	40	6.43	4.49	10.23	7.19	15.56	10.93	22.65	15.95	31.99	22.53
3	40	11.38	7.97	18.11	12.74	27.47	19.34	40.10	28.23	56.52	39.79
4	40	23.24	16.30	36.98	26.02	56.12	39.49	81.73	57.53	115.24	81.21
5	40	42.04	29.50	66.73	47.05	101.16	71.27	147.36	103.82	207.59	146.38
6	40	68.04	47.86	108.14	76.15	163.77	115.21	238.29	168.07	335.71	236.70
8	40	139.48	98.06	221.17	155.78	334.94	236.21	488.05	344.19	686.71	484.74
10	40	252.38	177.75	400.53	282.05	606.74	427.75	881.59	622.51	1243.64	876.79
12	ID	403.63	284.69	639.74	451.09	969.02	683.22	1410.30	995.80	1987.29	1402.63

Table 9 Suction Line Capacities in Tons for Refrigerant 502 (Single- or High-Stage Applications)

Line Size Type L Copper, OD		Saturated Suction Temperature, °F											
		−60		−40		−20		0		20		40	
		Δt = 1°F Δp=0.307	Δt = 0.5°F Δp=0.153	Δt = 1°F Δp=0.462	Δt = 0.5°F Δp=0.231	Δt = 1°F Δp=0.666	Δt = 0.5°F Δp=0.333	Δt = 1°F Δp=0.919	Δt = 0.5°F Δp=0.460	Δt = 1°F Δp=1.227	Δt = 0.5°F Δp=0.614	Δt = 1°F Δp=1.590	Δt = 0.5°F Δp=0.795
1/2		0.03	0.02	0.06	0.04	0.10	0.06	0.15	0.10	0.23	0.16	0.33	0.23
5/8		0.06	0.04	0.11	0.07	0.18	0.12	0.29	0.19	0.43	0.29	0.63	0.43
3/4		0.10	0.07	0.18	0.12	0.30	0.21	0.48	0.32	0.72	0.49	1.04	0.71
7/8		0.17	0.11	0.29	0.20	0.48	0.33	0.76	0.52	1.14	0.78	1.65	1.13
1-1/8		0.34	0.23	0.59	0.40	0.98	0.67	1.54	1.05	2.30	1.58	3.34	2.29
1-3/8		0.59	0.40	1.04	0.71	1.71	1.17	2.68	1.84	4.02	2.76	5.81	4.00
1-5/8		0.94	0.64	1.64	1.12	2.71	1.86	4.25	2.91	6.36	4.38	9.19	6.33
2-1/8		1.96	1.34	3.43	2.34	5.64	3.87	8.81	6.06	13.19	9.09	19.00	13.14
2-5/8		3.48	2.39	6.08	4.17	9.98	6.85	15.39	10.74	23.30	16.07	33.55	23.21
3-1/8		5.59	3.83	9.72	6.67	15.94	10.97	24.90	17.17	37.16	25.69	53.48	37.03
3-5/8		8.32	5.71	14.48	9.93	23.72	16.33	37.02	25.55	55.20	38.21	79.46	55.06
4-1/8		11.77	8.08	20.45	14.06	33.49	23.07	52.22	36.08	77.93	53.89	111.96	77.70
Steel													
IPS	SCH												
3/8	80	0.03	0.02	0.05	0.03	0.08	0.05	0.12	0.08	0.18	0.12	0.25	0.18
1/2	80	0.05	0.04	0.09	0.07	0.15	0.11	0.23	0.16	0.35	0.24	0.49	0.35
3/4	80	0.12	0.09	0.21	0.15	0.34	0.24	0.53	0.37	0.78	0.55	1.11	0.78
1	80	0.24	0.17	0.42	0.29	0.67	0.47	1.04	0.73	1.53	1.07	2.17	1.53
1-1/4	40	0.65	0.45	1.10	0.77	1.77	1.24	2.72	1.91	4.01	2.82	5.70	4.02
1-1/2	40	0.97	0.68	1.66	1.16	2.67	1.87	4.09	2.87	6.03	4.23	8.56	6.03
2	40	1.88	1.81	3.20	2.24	5.14	3.62	7.88	5.55	11.60	8.17	16.49	11.61
2-1/2	40	3.01	2.10	5.10	3.58	8.19	5.77	12.55	8.84	18.49	13.02	26.25	18.50
3	40	5.33	3.73	9.04	6.34	14.50	10.20	22.22	15.64	32.67	23.00	46.39	32.69
4	40	10.81	7.64	18.43	12.98	29.59	20.84	45.28	31.87	66.61	46.94	94.49	66.64
5	40	19.66	13.80	33.30	23.42	53.33	37.65	81.66	57.52	119.95	84.62	170.60	120.29
6	40	31.81	22.36	53.84	37.95	86.35	60.82	132.02	93.12	193.98	136.82	275.83	194.52
8	40	65.18	45.92	110.37	77.79	176.81	124.69	270.40	190.70	396.78	280.20	563.70	397.90
10	40	118.01	83.13	200.13	140.34	319.78	225.48	488.43	344.91	718.87	506.82	1019.75	719.75
12	ID	189.00	133.11	319.25	225.57	511.53	360.66	781.35	551.73	1148.74	810.78	1629.08	1151.46

Δp = pressure drop due to line friction, psi per 100 ft equivalent length of line
Δt = corresponding change in saturation temperature with the pressure drop, °F/100 ft

Table 10 Fitting Losses in Equivalent Feet of Pipe
(Screwed, Welded Flanged, Flared, and Brazed Connections)

Nominal Pipe or Tube Size, in.	Smooth Bend Elbows						Smooth Bend Tees			
	90° Std[a]	90° Long Rad.[b]	90° Street[a]	45° Std[a]	45° Street[a]	180° Std[a]	Flow Through Branch	Straight-Through Flow: No Reduction	Straight-Through Flow: Reduced 1/4	Straight-Through Flow: Reduced 1/2
3/8	1.4	0.9	2.3	0.7	1.1	2.3	2.7	0.9	1.2	1.4
1/2	1.6	1.0	2.5	0.8	1.3	2.5	3.0	1.0	1.4	1.6
3/4	2.0	1.4	3.2	0.9	1.6	3.2	4.0	1.4	1.9	2.0
1	2.6	1.7	4.1	1.3	2.1	4.1	5.0	1.7	2.2	2.6
1-1/4	3.3	2.3	5.6	1.7	3.0	5.6	7.0	2.3	3.1	3.3
1-1/2	4.0	2.6	6.3	2.1	3.4	6.3	8.0	2.6	3.7	4.0
2	5.0	3.3	8.2	2.6	4.5	8.2	10.0	3.3	4.7	5.0
2-1/2	6.0	4.1	10.0	3.2	5.2	10.0	12.0	4.1	5.6	6.0
3	7.5	5.0	12.0	4.0	6.4	12.0	15.0	5.0	7.0	7.5
3-1/2	9.0	5.9	15.0	4.7	7.3	15.0	18.0	5.9	8.0	9.0
4	10.0	6.7	17.0	5.2	8.5	17.0	21.0	6.7	9.0	10.0
5	13.0	8.2	21.0	6.5	11.0	21.0	25.0	8.2	12.0	13.0
6	16.0	10.0	25.0	7.9	13.0	25.0	30.0	10.0	14.0	16.0
8	10.0	13.0	—	10.0	—	33.0	40.0	13.0	18.0	20.0
10	25.0	16.0	—	13.0	—	42.0	50.0	16.0	23.0	25.0
12	30.0	19.0	—	16.0	—	50.0	60.0	19.0	26.0	30.0
14	34.0	23.0	—	18.0	—	55.0	68.0	23.0	30.0	34.0
16	38.0	26.0	—	20.0	—	62.0	78.0	26.0	35.0	38.0
18	42.0	29.0	—	23.0	—	70.0	85.0	29.0	40.0	42.0
20	50.0	33.0	—	26.0	—	81.0	100.0	33.0	44.0	50.0
24	60.0	40.0	—	30.0	—	94.0	115.0	40.0	50.0	60.0

[a] *R/D* approximately equal to 1.
[b] *R/D* approximately equal to 1.5.

Table 11 Special Fitting Losses in Equivalent Feet of Pipe

Nominal Pipe or Tube Size, in.	Sudden Enlargement, d/D			Sudden Contraction, d/D			Sharp Edge		Pipe Projection	
	1/4	1/2	3/4	1/4	1/2	3/4	Entrance	Exit	Entrance	Exit
3/8	1.4	0.8	0.3	0.7	0.5	0.3	1.5	0.8	1.5	1.1
1/2	1.8	1.1	0.4	0.9	0.7	0.4	1.8	1.0	1.8	1.5
3/4	2.5	1.5	0.5	1.2	1.0	0.5	2.8	1.4	2.8	2.2
1	3.2	2.0	0.7	1.6	1.2	0.7	3.7	1.8	3.7	2.7
1-1/4	4.7	3.0	1.0	2.3	1.8	1.0	5.3	2.6	5.3	4.2
1-1/2	5.8	3.6	1.2	2.9	2.2	1.2	6.6	3.3	6.6	5.0
2	8.0	4.8	1.6	4.0	3.0	1.6	9.0	4.4	9.0	6.8
2-1/2	10.0	6.1	2.0	5.0	3.8	2.0	12.0	5.6	12.0	8.7
3	13.0	8.0	2.6	6.5	4.9	2.6	14.0	7.2	14.0	11.0
3-1/2	15.0	9.2	3.0	7.7	6.0	3.0	17.0	8.5	17.0	13.0
4	17.0	11.0	3.8	9.0	6.8	3.8	20.0	10.0	20.0	16.0
5	24.0	15.0	5.0	12.0	9.0	5.0	27.0	14.0	27.0	20.0
6	29.0	22.0	6.0	15.0	11.0	6.0	33.0	19.0	33.0	25.0
8	—	25.0	8.5	—	15.0	8.5	47.0	24.0	47.0	35.0
10	—	32.0	11.0	—	20.0	11.0	60.0	29.0	60.0	46.0
12	—	41.0	13.0	—	25.0	13.0	73.0	37.0	73.0	57.0
14	—	—	16.0	—	—	16.0	86.0	45.0	86.0	66.0
16	—	—	18.0	—	—	18.0	96.0	50.0	96.0	77.0
18	—	—	20.0	—	—	20.0	115.0	58.0	115.0	90.0
20	—	—	—	—	—	—	142.0	70.0	142.0	108.0
24	—	—	—	—	—	—	163.0	83.0	163.0	130.0

Note: Enter table for losses at smallest diameter *d*.

Table 12 Valve Losses in Equivalent Feet of Pipe

Nominal Pipe or Tube Size, in.	Globe[a]	60° − Y	45° − Y	Angle[a]	Gate[b]	Swing Check[c]	Lift Check
3/8	17	8	6	6	0.6	5	Globe
1/2	18	9	7	7	0.7	6	and
3/4	22	11	9	9	0.9	8	vertical
1	29	15	12	12	1.0	10	lift
1-1/4	38	20	15	15	1.5	14	same as
1-1/2	43	24	18	18	1.8	16	globe
2	55	30	24	24	2.3	20	valve[d]
2-1/2	69	35	29	29	2.8	25	
3	84	43	35	35	3.2	30	
3-1/2	100	50	41	41	4.0	35	
4	120	58	47	47	4.5	40	
5	140	71	58	58	6.0	50	
6	170	88	70	70	7.0	60	
8	220	115	85	85	9.0	80	Angle
10	280	145	105	105	12.0	100	lift
12	320	165	130	130	13.0	120	same as
14	360	185	155	155	15.0	135	angle
16	410	210	180	180	17.0	150	valve
18	460	240	200	200	19.0	165	
20	520	275	235	235	22.0	200	
24	610	320	265	265	25.0	240	

Note: Losses are for valves in fully open position and with screwed, welded, flanged, or flared connections.

[a] These losses do not apply to valves with needlepoint seats.

[b] Regular and short pattern plug cock valves, when fully open, have same loss as gate valve. For valve losses of short pattern plug cocks above 6 in., check with manufacturer.

[c] Losses also apply to the in-line, ball-type check valve.

[d] For *Y* pattern globe lift check valve with seat approximately equal to the nominal pipe diameter, use values of 60° − *Y* valve for loss.

Oil Management in Refrigerant Lines

Oil Circulation. Some lubricating oil is lost from all compressors during normal operation. Since oil inevitably leaves the compressor with the discharge gas, systems using halocarbon refrigerants must return this oil at the same rate at which it leaves (Cooper 1971).

Oil that leaves the compressor or oil separator reaches the condenser and dissolves in the liquid refrigerant, enabling it to pass readily through the liquid line to the evaporator. In the evaporator, the refrigerant evaporates and the liquid phase becomes enriched in oil. The concentration of refrigerant in the oil depends on the evaporator temperature and types of refrigerant and oil used. The viscosity of the oil/refrigerant solution is determined by the system parameters. Oil separated in the evaporator is returned to the compressor by gravity or by the drag forces of the returning gas. The effect of oil on pressure drop is large, increasing the pressure drop by as much as a factor of ten in some cases (Alofs *et al.* 1990).

One of the most difficult problems in low-temperature refrigeration systems using halocarbon refrigerants is returning lubrication oil from the evaporator to the compressors. With the exception of most centrifugal compressors, and with rarely used nonlubricated compressors, refrigerant continuously carries oil into the discharge line from the compressor. Most of this oil can be removed from the stream by an oil separator and returned to the compressor. Coalescing oil separators are far better than separators using only mist pads or baffles; however, they are not 100% effective. The oil that finds its way into the system must be managed.

Oil mixes well with halocarbon refrigerants at higher temperatures. As the temperature lowers, the miscibility is reduced, and some of the oil separates to form an oil-rich layer near the top of the liquid level in a flooded evaporator. If the temperature is very low, the oil becomes a gummy mass that prevents refrigerant controls from functioning, blocks flow passages, and fouls the heat transfer surfaces. Proper oil management is often the key to a properly functioning system.

In general, direct-expansion and liquid overfeed system evaporators have fewer oil return problems than do flooded system evaporators since refrigerant flows continuously at good velocities to sweep oil from the evaporator. Low-temperature systems using hot gas defrost can also be designed to sweep oil out of the circuit each time the system defrosts. This reduces the possibility of oil coating the evaporator surface and hindering heat transfer.

Flooded evaporators can promote oil contamination of the evaporator charge since they may only return dry refrigerant vapor back to the system. Skimming systems must sample the oil-rich layer floating in the drum, a heat source must distill the refrigerant, and the oil must be returned to the compressor. Because flooded halocarbon systems can be elaborate, some designers avoid them.

System Capacity Reduction. The use of automatic capacity control on compressors requires careful analysis and design. The compressor is capable of loading and unloading as it modulates with the system load requirements through a considerable range of capacity variation. A single compressor can unload down to 25% of full-load capacity, while multiple compressors connected in parallel can unload to a system capacity of 12.5% or lower. System piping must be designed to return oil at the lowest loading, yet not impose excessive pressure drops in the piping and equipment at full load.

Oil Return Up Suction Risers. Many refrigeration piping systems contain a suction riser because the evaporator is at a lower level than the compressor. Oil circulating in the system can return up gas risers only by being transported by the returning gas or by auxiliary means such as a trap and a pump. The minimum conditions for oil transport correlate with buoyancy forces, *i.e.*, the density difference between the liquid and the vapor, and the momentum flux of the vapor (Jacobs *et al.* 1976).

The principal criteria determining the transport of oil are gas velocity, gas density, and pipe inside diameter. The density of the oil-refrigerant mixture plays a somewhat lesser role, since it is almost constant over a wide range. In addition, at temperatures somewhat lower than −40 °F, oil viscosity may be significant. Greater gas velocities are required as the temperature drops and the gas becomes less dense. Higher velocities are also necessary if the pipe diameter increases. Table 13 translates these criteria to minimum refrigeration capacity requirements for oil transport. Suction risers must be sized for minimum system capacity. Oil must be returned to the compressor at the operating condition corresponding to the minimum displacement and minimum suction temperature at which the compressor will operate. When suction or evaporator pressure regulators are used, suction risers must be sized for actual gas conditions in the riser.

For a single compressor with capacity control, the minimum capacity is the lowest capacity at which the unit can operate. For multiple compressors with capacity control, the minimum capacity is the lowest at which the last operating compressor can run.

Riser Sizing. The following example describes the use of Table 13 in establishing maximum riser sizes for satisfactory oil transport down to minimum partial loading.

Example 3. Determine the maximum size suction riser that will transport oil at the minimum loading, using R-22 with a 40-ton compressor with a capacity in steps of 25, 50, 75, and 100%. Assume the minimum system loading is 10 tons at 40 °F suction and 105 °F condensing temperatures with 15 °F superheat.

Solution: From Table 13, a 2.125 in. OD pipe at 40 °F suction and 90 °F condensing temperature has a minimum capacity of 7.5 tons. When corrected to 105 °F condensing, the minimum capacity becomes 7.2 tons. Therefore, the 2.125 in. OD pipe is suitable.

Based on Table 13, the next smaller line size should be used for marginal suction risers. When vertical riser sizes are reduced to provide satisfactory minimum gas velocities, the pressure drop at full load increases considerably; horizontal lines should be sized to keep the total pressure drop within practical limits. As long as the horizontal lines are level or pitched in the direction of the compressor, oil can be transported with normal design velocities.

Because most compressors have multiple capacity reduction features, gas velocities required to return oil up through vertical suction risers under all load conditions are difficult to maintain. When the suction riser is sized to permit oil return at the minimum operating capacity of the system, the pressure drop in this portion

Table 13 Minimum Refrigeration Capacity in Tons for Oil Entrainment up Suction Risers Type L Copper Tubing

Refrigerant	Saturated Temp., °F	Suction Gas Temp., °F	Pipe OD, in. 0.500	0.625	0.750	0.875	1.123	1.375	1.625	2.125	2.625	3.125	3.625	4.125
		Area, in²	0.146	0.233	0.348	0.484	0.825	1.256	1.780	3.094	4.770	6.812	9.213	11.970
134a	0.0	10.0	0.089	0.161	0.259	0.400	0.78	1.32	2.03	4.06	7.0	10.9	15.9	22.1
		30.0	0.075	0.135	0.218	0.336	0.66	1.11	1.71	3.42	5.9	9.2	13.4	18.5
		50.0	0.072	0.130	0.209	0.323	0.63	1.07	1.64	3.28	5.6	8.8	12.8	17.8
	10.0	20.0	0.101	0.182	0.294	0.453	0.88	1.49	2.31	4.61	7.9	12.4	18.0	25.0
		40.0	0.084	0.152	0.246	0.379	0.74	1.25	1.93	3.86	6.6	10.3	15.1	20.9
		60.0	0.081	0.147	0.237	0.366	0.71	1.21	1.87	3.73	6.4	10.0	14.6	20.2
	20.0	30.0	0.113	0.205	0.331	0.510	0.99	1.68	2.60	5.19	8.9	13.9	20.3	28.2
		50.0	0.095	0.172	0.277	0.427	0.83	1.41	2.17	4.34	7.5	11.6	17.0	23.6
		70.0	0.092	0.166	0.268	0.413	0.81	1.36	2.10	4.20	7.2	11.3	16.4	22.8
	30.0	40.0	0.115	0.207	0.335	0.517	1.01	1.70	2.63	5.25	9.0	14.1	20.5	28.5
		60.0	0.107	0.193	0.311	0.480	0.94	1.58	2.44	4.88	8.4	13.1	19.1	26.5
		80.0	0.103	0.187	0.301	0.465	0.91	1.53	2.37	4.72	8.1	12.7	18.5	25.6
	40.0	50.0	0.128	0.232	0.374	0.577	1.12	1.90	2.94	5.87	10.1	15.7	22.9	31.8
		70.0	0.117	0.212	0.342	0.528	1.03	1.74	2.69	5.37	9.2	14.4	21.0	29.1
		90.0	0.114	0.206	0.332	0.512	1.00	1.69	2.61	5.21	8.9	14.0	20.4	28.3
22	−40.0	−30.0	0.067	0.119	0.197	0.298	0.580	0.981	1.52	3.03	5.20	8.12	11.8	16.4
		−10.0	0.065	0.117	0.194	0.292	0.570	0.963	1.49	2.97	5.11	7.97	11.6	16.1
		10.0	0.066	0.118	0.195	0.295	0.575	0.972	1.50	3.00	5.15	8.04	11.7	16.3
	−20.0	−10.0	0.087	0.156	0.258	0.389	0.758	1.28	1.98	3.96	6.80	10.6	15.5	21.5
		10.0	0.085	0.153	0.253	0.362	0.744	1.26	1.95	3.88	6.67	10.4	15.2	21.1
		30.0	0.086	0.154	0.254	0.383	0.747	1.26	1.95	3.90	6.69	10.4	15.2	21.1
	0.0	10.0	0.111	0.199	0.328	0.496	0.986	1.63	2.53	5.04	8.66	13.5	19.7	27.4
		30.0	0.108	0.194	0.320	0.484	0.942	1.59	2.46	4.92	8.45	13.2	19.2	26.7
		50.0	0.109	0.195	0.322	0.486	0.946	1.60	2.47	4.94	8.48	13.2	19.3	26.8
	20.0	30.0	0.136	0.244	0.403	0.608	1.18	2.00	3.10	6.18	10.6	16.6	24.2	33.5
		50.0	0.135	0.242	0.399	0.603	1.17	1.99	3.07	6.13	10.5	16.4	24.0	33.3
		70.0	0.135	0.242	0.400	0.605	1.18	1.99	3.08	6.15	10.6	16.5	24.0	33.3
	40.0	50.0	0.167	0.300	0.495	0.748	1.46	2.46	3.81	7.60	13.1	20.4	29.7	41.3
		70.0	0.165	0.296	0.488	0.737	1.44	2.43	3.75	7.49	12.9	20.1	29.3	40.7
		90.0	0.165	0.296	0.488	0.738	1.44	2.43	3.76	7.50	12.9	20.1	29.3	40.7
502	−40.0	−30.0	0.051	0.092	0.152	0.230	0.447	0.756	1.17	2.33	4.01	6.26	9.13	12.7
		−10.0	0.053	0.095	0.157	0.237	0.461	0.779	1.21	2.41	4.13	6.45	9.41	13.1
		10.0	0.055	0.098	0.163	0.246	0.476	0.809	1.25	2.50	4.29	6.39	9.76	13.5
	−20.0	−10.0	0.068	0.122	0.201	0.303	0.591	0.999	1.54	3.08	5.30	8.27	12.1	16.7
		10.0	0.070	0.125	0.207	0.312	0.608	1.03	1.59	3.17	5.45	8.51	12.4	17.2
		30.0	0.072	0.129	0.213	0.322	0.627	1.06	1.64	3.27	5.62	8.78	12.8	17.8
	0.0	10.0	0.087	0.157	0.259	0.391	0.761	1.29	1.99	3.97	6.82	10.6	15.5	21.5
		30.0	0.089	0.160	0.264	0.399	0.777	1.31	2.03	4.05	6.96	10.9	15.9	22.0
		50.0	0.092	0.165	0.273	0.412	0.802	1.36	2.10	4.19	7.19	11.2	16.4	22.7
	20.0	30.0	0.110	0.197	0.325	0.491	0.957	1.62	2.50	4.99	8.58	13.4	19.5	27.1
		50.0	0.112	0.201	0.331	0.501	0.975	1.65	2.55	5.09	8.74	13.6	19.9	27.6
		70.0	0.115	0.207	0.342	0.516	1.01	1.70	2.63	5.25	9.02	14.1	20.5	28.5
	40.0	50.0	0.136	0.243	0.401	0.606	1.18	2.00	3.09	6.16	10.6	16.5	24.1	33.4
		70.0	0.138	0.247	0.408	0.616	1.20	2.03	3.14	6.28	10.8	16.8	24.5	34.0
		90.0	0.142	0.254	0.420	0.634	1.23	2.09	3.23	6.44	11.1	17.3	25.2	35.0

Notes:

1. The capacity in tons is based on 90 °F liquid temperature and superheat as indicated by the listed temperature. For other liquid line temperatures, use correction factors in the table below.
2. These tables have been computed using an ISO 32 mineral oil for R-22 and R-502. R-134a has been computed using an ISO 32 ester-based oil.

Refrigerant	Liquid Temperature, °F 50	60	70	80	100	110	120	130	140
134a	1.26	1.20	1.13	1.07	0.94	0.87	0.80	0.74	0.67
22	1.17	1.14	1.10	1.06	0.98	0.94	0.89	0.85	0.80
502	1.24	1.18	1.12	1.06	0.94	0.87	0.81	0.74	0.67

of the line may be too great when operating at full load. If a correctly sized suction riser imposes too great a pressure drop at full load, a double-suction riser should be used.

Oil Return Up Suction Risers—Multistage Systems. The movement of oil in the suction lines of multistage systems requires the same design approach as that for single-stage systems.

When refrigerants other than those listed in Tables 13 and 14 are used, follow the recommendations listed in Table 15. For oil to flow up along a pipe wall, a certain minimum drag of the gas flow is required. This can be represented by the friction gradient. Table 15 shows values for minimum friction gradients.

Double-Suction Risers. Figure 4 shows two methods of double-suction riser construction. While oil return in this arrangement is accomplished at minimum loads, it does not cause excessive pressure drops at full load. The sizing and operation of a double-suction riser is described as follows:

1. Riser A is sized to return oil at the minimum load possible.
2. Riser B is sized for satisfactory pressure drop through both risers at full load. The usual method is to size riser B so that the combined cross-sectional area of A and B is equal to or slightly greater than the cross-sectional area of a single pipe, which would be sized for an acceptable pressure drop at full load without regard for oil return at minimum load. The combined cross-sectional area, however, should not be greater than the cross-sectional area of a single pipe that would return oil in an upflow riser under maximum load conditions.
3. A trap is introduced between the two risers, as shown in both methods. During part-load operation, the gas velocity is not sufficient to return oil through both risers, and the trap gradually fills up with oil until the second riser B is sealed off. The gas then travels up riser A only and now has enough velocity to carry oil along with it back into the horizontal suction main.

The oil holding capacity of the trap is limited to a minimum by close-coupling the fittings at the bottom of the risers. If this is not done, the trap can accumulate enough oil on part-load operation to lower the compressor crankcase oil level. Note in Figure 4 that riser lines A and B form an inverted loop and enter the horizontal suction line from the top. This prevents oil drainage into this riser, which may be idle during partial load operation. The same purpose can be served by going horizontally into the main, providing it is larger in size than either riser.

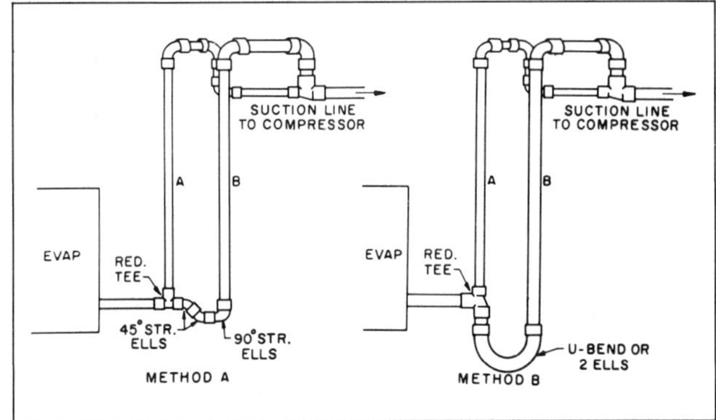

Fig. 4 Double-Suction Riser Construction

Often, double-suction risers are essential on low-temperature systems that can tolerate very little pressure drop. Any system using these risers should include a suction trap (accumulator) and a means of returning oil gradually.

For systems operating at higher suction temperatures, such as for comfort air conditioning, single-suction risers can be sized for oil return at minimum load. Where single compressors are used with capacity control, minimum capacity will usually be 25 or 33% of maximum displacement. With this low ratio, pressure drop in single-suction risers designed for oil return at minimum load are rarely serious at full load.

When multiple compressors are used, one or more may shut down while another continues to operate, and the maximum to minimum ratio becomes much larger. This may make a double-suction riser necessary.

The remaining portions of the suction line are sized to permit a practical pressure drop between the evaporators and compressors, since oil is carried along in horizontal lines at relatively low gas velocities. It is good practice to give some pitch to these lines toward the compressor. Traps should be avoided, but when that is impossible, the risers from them are treated the same as those leading from the evaporators.

Preventing Oil Trapping in Idle Evaporators. Suction lines should be designed so that oil from an active evaporator does not drain into an idle one. Figure 5A shows multiple evaporators on

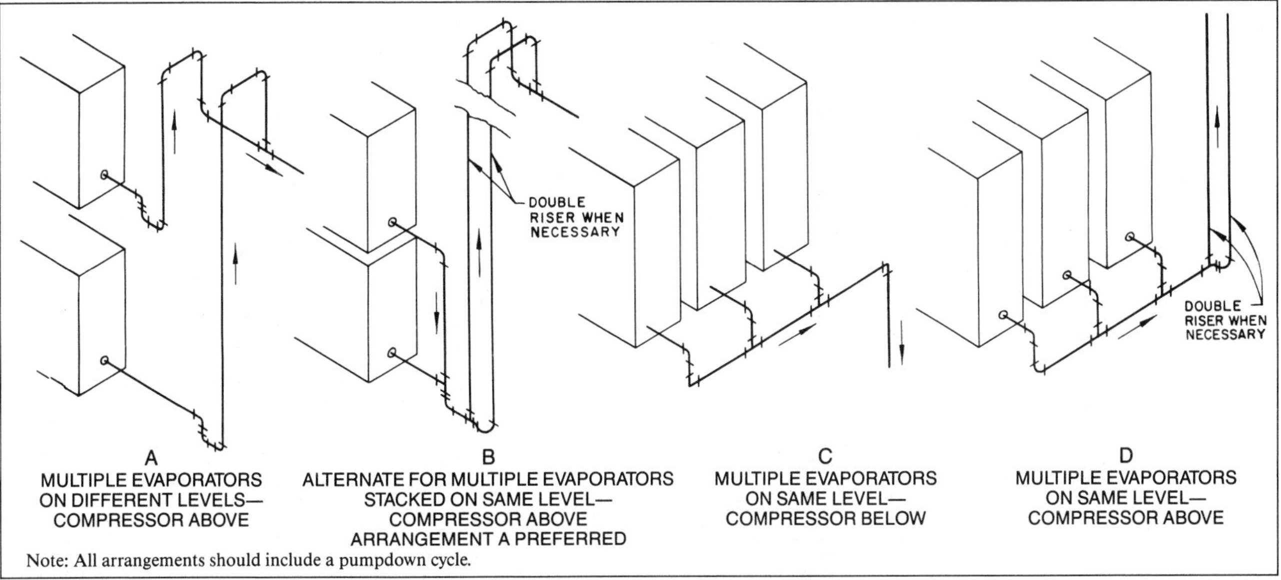

Fig. 5 Suction Line Piping at Evaporator Coils

Table 14 Minimum Refrigeration Capacity for Oil Entrainment up Hot-Gas Risers Type L Copper Tubing

Refrig-erant	Saturated Temp., °F	Discharge Gas Temp., °F	Pipe OD, in.											
			0.500	0.625	0.750	0.875	1.123	1.375	1.625	2.125	2.625	3.125	3.625	4.125
			Area, in^2											
			0.146	0.233	0.348	0.484	0.825	1.256	1.780	3.094	4.770	6.812	9.213	11.970
134a	80.0	110.0	0.199	0.360	0.581	0.897	1.75	2.96	4.56	9.12	15.7	24.4	35.7	49.5
		140.0	0.183	0.331	0.535	0.825	1.61	2.72	4.20	8.39	14.4	22.5	32.8	45.6
		170.0	0.176	0.318	0.512	0.791	1.54	2.61	4.02	8.04	13.8	21.6	31.4	43.6
	90.0	120.0	0.201	0.364	0.587	0.906	1.76	2.99	4.61	9.21	15.8	24.7	36.0	50.0
		150.0	0.184	0.333	0.538	0.830	1.62	2.74	4.22	8.44	14.5	22.6	33.0	45.8
		180.0	0.177	0.320	0.516	0.796	1.55	2.62	4.05	8.09	13.9	21.7	31.6	43.9
	100.0	130.0	0.206	0.372	0.600	0.926	1.80	3.05	4.71	9.42	16.2	25.2	36.8	51.1
		160.0	0.188	0.340	0.549	0.848	1.65	2.79	4.31	8.62	14.8	23.1	33.7	46.8
		190.0	0.180	0.326	0.526	0.811	1.58	2.67	4.13	8.25	14.2	22.1	32.2	44.8
	110.0	140.0	0.209	0.378	0.610	0.942	1.83	3.10	4.79	9.57	16.5	25.7	37.4	52.0
		170.0	0.191	0.346	0.558	0.861	1.68	2.84	4.38	8.76	15.0	23.5	34.2	47.5
		200.0	0.183	0.331	0.534	0.824	1.61	2.72	4.19	8.38	14.4	22.5	32.8	45.5
	120.0	150.0	0.212	0.383	0.618	0.953	1.86	3.14	4.85	9.69	16.7	26.0	37.9	52.6
		180.0	0.194	0.351	0.566	0.873	1.70	2.88	4.44	8.88	15.3	23.8	34.7	48.2
		210.0	0.184	0.334	0.538	0.830	1.62	2.74	4.23	8.44	14.5	22.6	33.0	45.8
22	80.0	110.0	0.235	0.421	0.695	1.05	2.03	3.46	5.35	10.7	18.3	28.6	41.8	57.9
		140.0	0.223	0.399	0.659	0.996	1.94	3.28	5.07	10.1	17.4	27.1	39.6	54.9
		170.0	0.215	0.385	0.635	0.960	1.87	3.16	4.89	9.76	16.8	26.2	38.2	52.9
	90.0	120.0	0.242	0.433	0.716	1.06	2.11	3.56	5.50	11.0	18.9	29.5	43.0	59.6
		150.0	0.226	0.406	0.671	1.01	1.97	3.34	5.16	10.3	17.7	27.6	40.3	55.9
		180.0	0.216	0.387	0.540	0.956	1.88	3.18	4.92	9.82	16.9	26.3	38.4	53.3
	100.0	130.0	0.247	0.442	0.730	1.10	2.15	3.83	5.62	11.2	19.3	30.1	43.9	60.8
		160.0	0.231	0.414	0.884	1.03	2.01	3.40	5.26	10.5	18.0	28.2	41.1	57.0
		190.0	0.220	0.394	0.650	0.982	1.91	3.24	3.00	9.96	17.2	26.8	39.1	54.2
	110.0	140.0	0.251	0.451	0.744	1.12	2.19	3.70	5.73	11.4	19.6	30.6	44.7	62.0
		170.0	0.235	0.421	0.693	1.05	2.05	3.46	3.35	10.7	18.3	28.6	41.8	57.9
		200.0	0.222	0.399	0.658	0.994	1.94	3.28	5.06	10.1	17.4	27.1	39.5	54.8
	120.0	150.0	0.257	0.460	0.760	1.15	2.24	3.78	5.85	11.7	20.0	31.3	45.7	63.3
		180.0	0.239	0.428	0.707	1.07	2.08	3.51	5.44	10.8	18.6	29.1	42.4	58.9
		210.0	0.225	0.404	0.666	1.01	1.96	3.31	5.12	10.2	17.6	27.4	40.0	55.5
502	80.0	110.0	0.192	0.344	0.567	0.857	1.67	2.82	4.36	8.71	15.0	23.4	34.1	47.3
		140.0	0.180	0.323	0.534	0.806	1.57	2.66	4.11	8.20	14.1	22.0	32.1	44.5
		170.0	0.173	0.310	0.512	0.773	1.50	2.54	3.94	7.85	13.5	21.1	30.7	42.8
	90.0	120.0	0.194	0.348	0.574	0.867	1.69	2.85	4.41	8.81	15.1	23.6	34.5	47.8
		150.0	0.182	0.326	0.538	0.813	1.58	2.68	4.14	8.26	14.2	22.2	32.3	44.8
		180.0	0.169	0.303	0.501	0.756	1.47	2.49	3.85	7.69	13.2	20.6	30.1	41.7
	100.0	130.0	0.194	0.348	0.575	0.869	1.69	2.86	4.42	8.83	15.2	23.7	34.5	47.9
		160.0	0.182	0.326	0.539	0.813	1.58	2.68	4.14	8.27	14.2	22.2	32.3	44.9
		190.0	0.170	0.304	0.503	0.739	1.48	2.50	3.87	7.71	13.3	20.7	30.2	41.9
	110.0	140.0	0.170	0.305	0.504	0.761	1.48	2.51	3.87	7.73	13.3	20.7	30.2	42.0
		170.0	0.162	0.291	0.481	0.726	1.41	2.39	3.70	7.38	12.7	19.8	28.9	40.1
		200.0	0.152	0.273	0.450	0.680	1.33	2.24	3.46	6.92	11.9	18.5	27.0	37.5
	120.0	150.0	0.170	0.305	0.503	0.760	1.48	2.50	3.87	7.73	13.3	20.7	30.2	41.9
		180.0	0.153	0.275	0.453	0.683	1.33	2.26	3.49	6.96	12.0	18.7	27.2	37.8
		210.0	0.149	0.267	0.440	0.665	1.30	2.19	3.39	6.76	11.6	18.1	26.4	36.7

Notes:

1. The capacity in tons is based on a saturated suction temperature of 20°F with 15°F superheat at the indicated saturated condensing temperature with 15°F subcooling. For other saturated suction temperatures with 15°F superheat, use the following correction factors:

Refrigerant	Saturated Suction Temperature, °F			
	−40	−20	0	+40
22	0.92	0.95	0.97	1.02
502	0.85	0.91	0.95	1.04
134a	—	—	0.96	1.04

2. These tables have been computed using an ISO 32 mineral oil for R-22 and R-502. R-134a has been computed using an ISO 32 ester-based oil.

Table 15 Sizing Data for Oil Return in Discharge or Suction Lines with Flow Vertically Upward

Saturation Temperature, °F	Line Size, 2 in. or less	Line Size, above 2 in.
0	0.35 psi/100 ft	0.20 psi/100 ft
−50	0.45 psi/100 ft	0.25 psi/100 ft

different floor levels and the compressor above. Each suction line is brought upward and looped into the top of the common suction line to prevent oil from draining into inactive coils.

Figure 5B shows multiple evaporators stacked with the compressor above. Oil cannot drain into the lowest evaporator because the common suction line drops below the outlet of the lowest evaporator before entering the suction riser.

Figure 5C shows multiple evaporators on the same level, with the compressor located below. The suction line from each evaporator drops down into the common suction line so that oil cannot drain into an idle evaporator. An alternate arrangement is shown in Figure 5D for cases where the compressor is above the evaporators.

Figure 6 illustrates typical piping for evaporators above and below a common suction line. All horizontal runs should be level or pitched toward the compressor to assure oil return.

The traps shown in the suction lines after the evaporator suction outlet are recommended by various thermal expansion valve manufacturers to prevent erratic operation of the thermal expansion valve. The expansion valve bulbs are located on the suction lines between the evaporator and these traps. The traps serve as drains and help prevent liquid from accumulating under the expansion valve bulbs during compressor off cycles. They are useful only where straight runs or risers are encountered in the suction line leaving the evaporator outlet.

DISCHARGE (HOT-GAS) LINES

Hot-gas lines should be designed to accomplish the following objectives:

- Avoid trapping oil at part-load operation
- Prevent condensed refrigerant and oil in the line from draining back to the head of the compressor
- Carefully select connections from a common line to multiple compressors
- Avoid developing excessive noise or vibration from hot-gas pulsations, compressor vibration, or both

Oil Transport Up Risers at Normal Loads. Even though a low-pressure drop is desired, oversized hot-gas lines can reduce gas velocities to a point where the refrigerant will not transport oil. Therefore, when using multiple compressors with capacity control, hot-gas risers must transport oil at all possible loadings.

Minimum Gas Velocities for Oil Transport in Risers. Minimum capacities for oil entrainment in hot-gas line risers are shown in Table 14. On multiple- compressor installations, the lowest possible system loading should be calculated and a riser size selected to give at least the minimum capacity indicated in the table for successful oil transport.

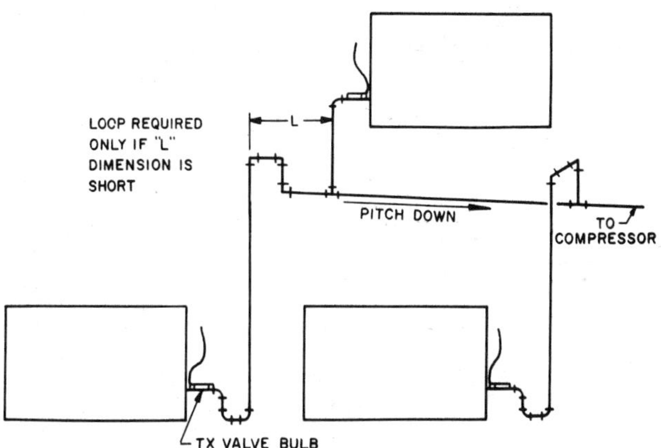

Fig. 6 Typical Piping from Evaporators Located Above and Below Common Suction Line

In some installations with multiple compressors and with capacity control, a vertical hot-gas line, sized to transport oil at minimum load, has excessive pressure drop at maximum load. When this problem exists, either a double riser or a single riser and an oil separator can be used.

Double Hot-Gas Risers. A double hot-gas riser can be used in the same way as it is used in a suction line. Figure 7 shows the double riser principle applied to a hot-gas line. Its operating principle and sizing technique is described in the section on Double-Suction Risers.

Single Riser and Oil Separator. As an alternative, an oil separator located in the discharge line just before the riser permits sizing the riser for a low-pressure drop. Any oil draining back down the riser accumulates in the oil separator. With large multiple compressors, the capacity of the separator may dictate the use of individual units for each compressor located between the discharge line and the main discharge header. Horizontal lines should be level or pitched downward in the direction of gas flow to facilitate travel of oil through the system and back to the compressor.

Piping to Prevent Liquid and Oil from Draining to Compressor Head. Whenever the condenser is located above the compressor, the hot-gas line should be trapped near the compressor before rising to the condenser, especially when the hot-gas riser is long. This minimizes the possibility that refrigerant, condensed in the line during off cycles, will drain back to the head of the compres-

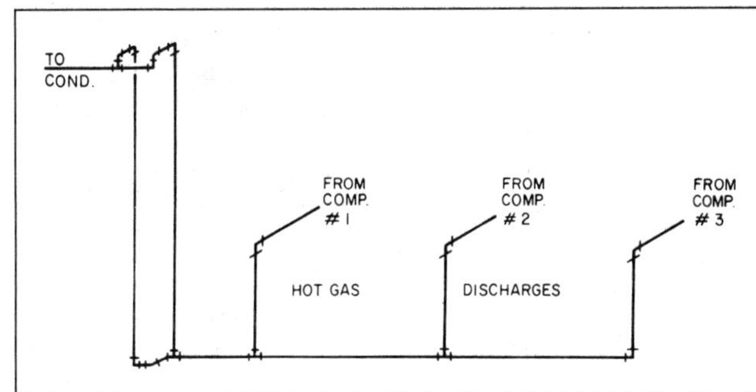

Fig. 7 Double Hot-Gas Riser

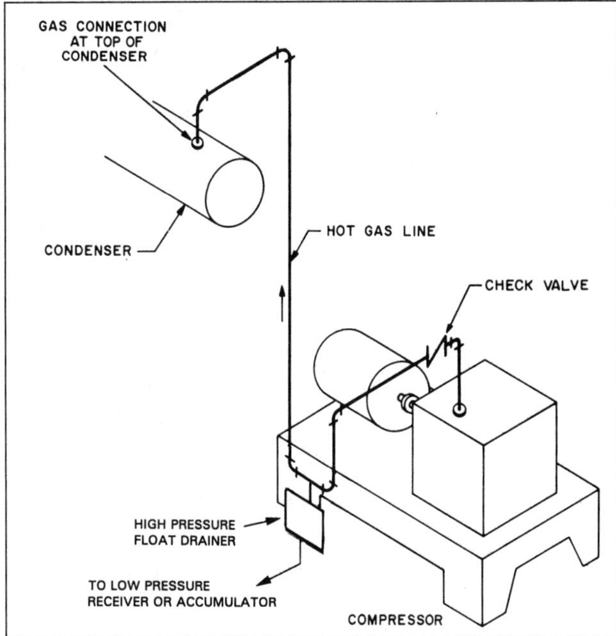

Fig. 8 Hot-Gas Loop

sor. Any oil traveling up the pipe wall also will not drain back to the compressor head.

The loop in the hot-gas line (Figure 8) serves as a reservoir and traps liquid resulting from condensation in the line during shutdown, thus preventing gravity drainage of liquid and oil back to the compressor head. A small high-pressure float drainer should be installed at the bottom of the trap to drain any significant amount of refrigerant condensate to a low-side component such as a suction accumulator or low-pressure receiver. This float will prevent an excessive buildup of liquid in the trap and possible liquid hammer when the compressor is restarted.

For multiple-compressor arrangements, each discharge line should have a check valve to prevent gas from active compressors from condensing on the heads of the idle compressors.

For single-compressor applications, a tightly closing check valve should be installed in the hot-gas line of the compressor whenever the condenser and the receiver ambient temperature are higher than that of the compressor. The check valve prevents refrigerant from boiling off in the condenser or receiver and condensing on the compressor heads during off cycles.

This check valve should be a piston type, which will close by gravity when the compressor stops running. The use of a spring-loaded check may incur chatter (vibration), particularly on slow-speed reciprocating compressors.

For compressors equipped with water-cooled oil coolers, a water solenoid and water-regulating valve should be installed in the water line so that the regulating valve maintains adequate cooling during operation, and the solenoid stops flow during the off cycle to prevent localized condensing of the refrigerant.

Hot-Gas (Discharge) Mufflers. Mufflers can be installed in hot-gas lines to dampen the discharge gas pulsations, reducing vibration and noise. Mufflers should be installed in a horizontal or downflow portion of the hot-gas line immediately after it leaves the compressor.

Because gas velocities through the muffler are substantially lower than they are through the hot-gas line, the muffler may form an oil trap. The muffler should be installed to allow oil to flow through it and not be trapped.

DEFROST GAS SUPPLY LINES

Sizing of refrigeration lines to supply defrost gas to one or more evaporators has not been an exact science. The parameters associated with sizing the defrost gas line are related to allowable pressure drop during defrost and refrigerant flow rate during the defrost.

Engineers have used an estimated two times the evaporator load for effective refrigerant flow rate to determine line sizing requirements. The pressure drop is not as critical during the defrost cycle, and many engineers have used velocity as the criterion for determining line size. The effective condensing temperature and average temperature of the gas needs to be determined. The velocity determined at saturated conditions will result in a conservative line size.

Some controlled testing (Stoecker 1984) has shown that in small coils with R-22, the defrost flow rate tends to be higher as the condensing temperature is increased. The flow rate is in the order of two to three times the normal evaporator flow rate, which supports the estimated two times used by practicing engineers.

Table 16 (R-134a, R-22, and R-502) provides guidance on selecting defrost gas supply lines based on velocity at a saturated condensing temperature of 70°F. It is recommended that initial sizing

Table 16 Refrigerant Flow Capacity Data for Defrost Lines

Pipe Size	R-134a Mass Flow Data, lb/h			R-22 Mass Flow Data, lb/h			R-502 Mass Flow Data, lb/h		
OD Copper	Velocity, 1000 fpm	Velocity, 2000 fpm	Velocity, 3000 fpm	Velocity, 1000 fpm	Velocity, 2000 fpm	Velocity, 3000 fpm	Velocity, 1000 fpm	Velocity, 2000 fpm	Velocity, 3000 fpm
1/2	110	220	330	150	300	450	230	450	680
5/8	170	350	520	240	480	720	360	730	1,090
3/4	260	510	770	350	710	1,060	530	1,070	1,600
7/8	360	720	1,090	500	1,000	1,500	750	1,510	2,260
1-1/8	620	1,230	1,850	850	1,700	2,550	1,280	2,570	3,850
1-3/8	940	1,880	2,820	1,300	2,590	3,890	1,960	3,910	5,870
1-5/8	1,330	2,660	3,990	1,840	3,670	5,510	2,770	5,540	8,310
2-1/8	2,310	4,630	6,940	3,190	6,390	9,580	4,820	9,630	14,500
2-5/8	3,570	7,140	10,700	4,930	9,850	14,800	7,430	14,900	22,300
3-1/8	5,100	10,200	15,300	7,030	14,100	21,100	10,600	21,200	31,800
3-5/8	6,900	13,800	20,700	9,510	19,000	28.500	14,300	28,700	43,000
4-1/8	9,000	17,900	26,900	12,400	24,700	37,100	18,600	37,300	55,900
5-1/8	14,000	27,900	41,900	19,300	38,500	57,800	29,100	58,100	87,200
6-1/8	20,100	40,100	60,200	27,700	55,400	83,100	41,800	83,500	125,300
8-1/8	35,100	70,100	105,200	48,400	96,700	145,100	73,000	145,900	218,900
Steel									
IPS SCH									
3/8 80	110	210	320	150	290	440	220	440	660
1/2 80	180	350	530	240	480	720	360	730	1,090
3/4 80	320	650	970	450	890	1,340	670	1,350	2,020
1 80	540	1,080	1,610	740	1,480	2,230	1,120	2,240	3,360
1-1/4 40	1,120	2,240	3,360	1,540	3,090	4,630	2,330	4,660	6,980
1-1/2 40	1,520	3,050	4,570	2,100	4,200	6,300	3,170	6,340	9,510
2 40	2,510	5,020	7,530	3,460	6,930	10,400	5,220	10,400	15,700
2-1/2 40	3,580	7,160	10,700	4,940	9,870	14,800	7,450	14,900	22,300
3 40	5,530	11,100	16,600	7,620	15,200	22,900	11,500	23,000	34,500
4 40	9,520	19,000	28,600	13,100	26,300	39,400	19,800	39,600	59,400
5 40	15,000	29,900	44,900	20,600	41,300	61,900	31,100	62,200	93,300
6 40	21,600	43,200	64,800	29,800	59,600	89,400	45,000	89,900	134,900
8 40	37,400	74,800	112,300	51,600	103,300	154,900	77,900	155,700	233,600
10 40	59,000	118,00	176,900	81,400	162,800	244,100	122,700	245,500	368,200
12 ID	84,600	169,200	253,800	116,700	233,400	350,200	176,000	352,100	528,100

Note: Refrigerant flow data based on saturated condensing temperature of 70°F.

be based on twice the evaporator flow rate and that velocities from 1000 to 2000 fpm be used for determining the defrost gas supply line size.

Gas defrost lines must be designed to continuously drain any condensed liquid.

RECEIVERS

Refrigerant receivers are vessels used to store excess refrigerant circulated throughout the system. Receivers perform the following functions:

1. Provide pumpdown storage capacity when another part of the system must be serviced or the system must be shut down for an extended time. In some water-cooled condenser systems, the condenser also serves as a receiver if the total refrigerant charge does not exceed its storage capacity.
2. Handle the excess refrigerant charge that occurs with air-cooled condensers using the flooding-type condensing pressure control (see the section on Head Pressure Control for Refrigerant Condensers).
3. Accommodate a fluctuating charge in the low side and drain the condenser of liquid to prevent reducing the effective condensing surface on systems where the operating charge in the evaporator and/ or condenser varies for different loading conditions. When an evaporator is fed with a thermal expansion valve, hand expansion valve, or low-pressure float, the operating charge in the evaporator varies considerably depending on the loading. During low load, the evaporator requires a larger charge since the boiling is not as intense. When the load increases, the operating charge in the evaporator decreases, and the receiver must store excess refrigerant.
4. On systems with multicircuit evaporators that shut off the liquid supply to one or more circuits during reduced load and pump out the idle circuit, receivers are used to hold the full charge of the idle circuit.

Connections for Through-Type Receiver. When a through-type receiver is used, the liquid must always flow from the condenser to the receiver. The pressure in the receiver must be lower than that in the condenser outlet. The receiver and its associated piping provide free flow of liquid from the condenser to the receiver by equalizing the pressures between the two so that the receiver cannot build up a higher pressure than the condenser.

If a vent is not used, the piping between condenser and receiver (condensate line) is sized so that liquid flows in one direction and gas flows in the opposite direction. Sizing the condensate line for 100 fpm liquid velocity is usually adequate to attain this flow. Piping should slope at least 0.25 in/ft and eliminate any natural liquid traps. See Figure 9 for this configuration.

The piping between the condenser and the receiver can be equipped with a separate vent (equalizer) line to allow receiver and condenser pressures to equalize. This external vent line can be piped either with or without a check valve in the vent line (see Figures 11 and 12). When piping without a check valve in the vent line, prevent the discharge gas from discharging directly into the vent line; this should prevent introducing a gas velocity pressure component on top of the liquid in the receiver. When the piping configuration is unknown, install a check valve in the vent (with the direction of flow toward the condenser). The check valve should be selected for minimum opening pressure, *i.e.*, approximately 0.5 psi. When determining the condensate drop leg height, allowance must be made to overcome both the pressure drop across this check valve as well as the refrigerant pressure drop through the condenser. This will ensure that there will be no liquid backup into an operating condenser on a multiple condenser application where one or more of the condensers are idle. The condensate line should be sized so that the velocity does not exceed 150 fpm.

The vent line flow is from receiver to condenser when the receiver temperature is warmer than the condensing temperature. Flow is from condenser to receiver when the air temperature around the receiver is below the condensing temperature. The rate of flow depends on this temperature difference as well as on the amount of receiver surface. Vent size can be calculated from this flow rate.

Connections for Surge-Type Receiver. The purpose of a surge-type receiver is to allow liquid to flow to the expansion valve without exposure to refrigerant in the receiver (*i.e.*, it can remain subcooled). The receiver volume is available for liquid that is to be removed from the system. Figure 10 shows an example of connections for a surge-type receiver. The height *h* must be adequate for a liquid pressure at least as large as the pressure loss through the condenser, liquid line, and vent line at the maximum temperature difference between the receiver ambient and the condensing temperature. The condenser pressure drop at the greatest expected heat rejection should be obtained from the manufacturer. The minimum value of *h* can then be calculated and a decision made as to whether or not the available height will permit the surge-type receiver.

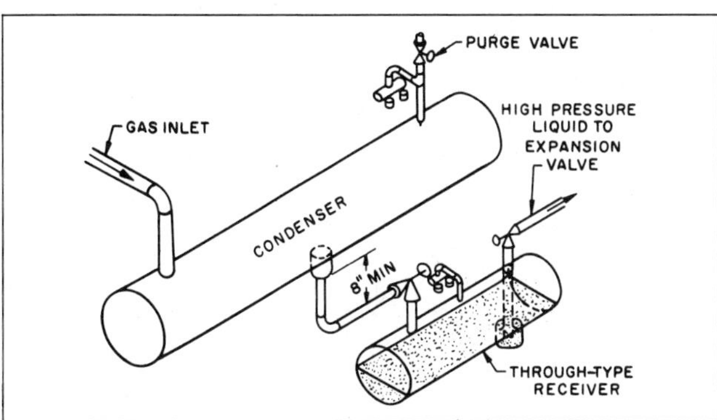

Fig. 9 Shell-and-Tube Condenser to Receiver Piping (Through-Type Receiver)

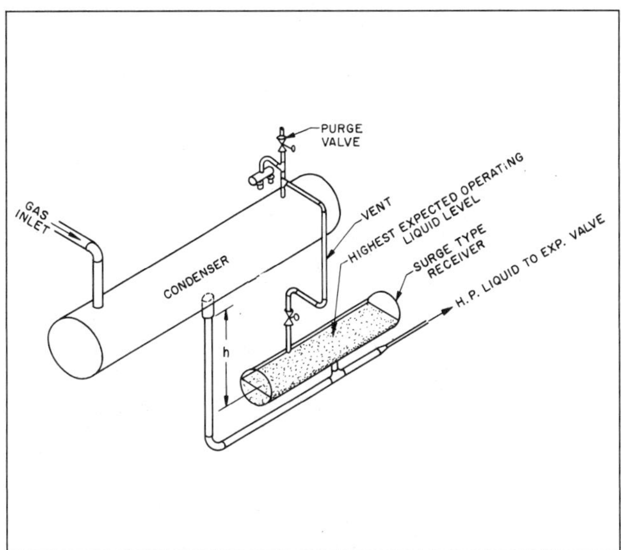

Fig. 10 Shell-and-Tube Condenser to Receiver Piping (Surge-Type Receiver)

Multiple Condensers. Two or more condensers connected in series or in parallel can be used in a single refrigeration system. When connected in series, the pressure losses through each condenser must be added. Condensers are more often arranged in parallel. The pressure loss through any one of the parallel circuits is always equal to that through any of the others, even if it results in filling much of one circuit with liquid while gas passes through another.

Figure 11 shows a basic arrangement for multiple condensers with a through-type receiver. The condensate drop legs must be long enough to allow liquid levels to adjust in them to equalize pressure losses between condensers at all operating conditions. The drop legs should be 6 to 12 in. higher than calculated to assure that liquid outlets remain free-draining. This height will provide a liquid pressure to offset the largest condenser pressure loss. The liquid seal will prevent gas blow-by between condensers.

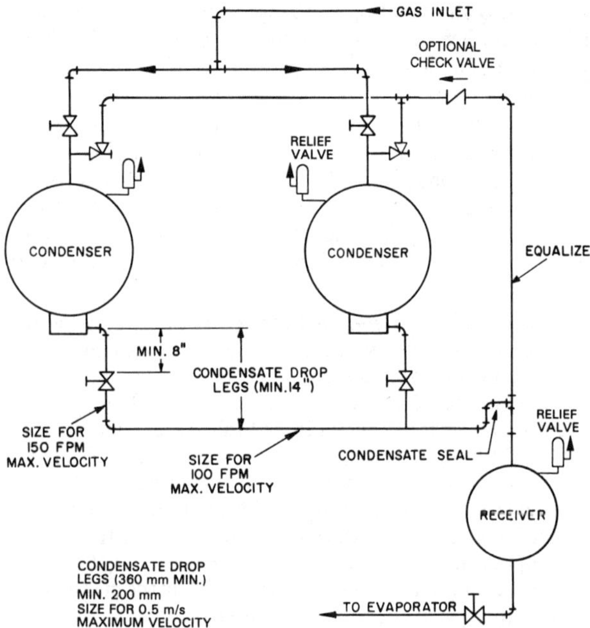

Fig. 11 Parallel Condensers with Through-Type Receiver

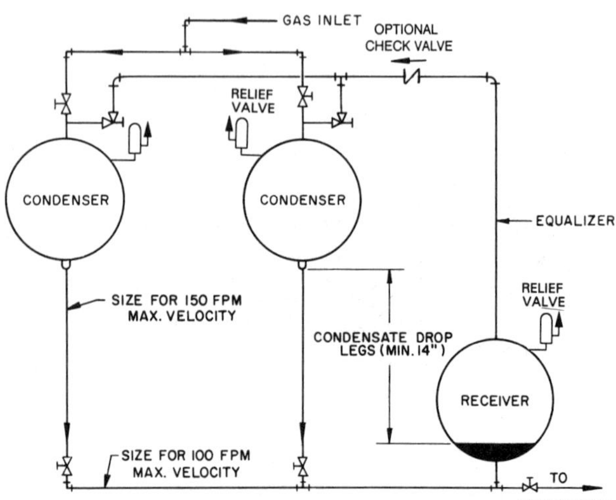

Fig. 12 Parallel Condensers with Surge-Type Receiver

Large single condensers with multiple coil circuits should be piped as if the independent circuits were parallel condensers. For example, assume the left condenser in Figure 11 has 2 psi more pressure drop than the right condenser. The liquid level on the left side will be about 4 ft higher than that on the right. If the condensate lines do not have enough vertical height for this level difference, the liquid will back up into the condenser until the pressure drop is the same through both circuits. Enough surface may be covered to reduce the condenser capacity significantly.

The condensate drop legs should be sized based on 150 fpm velocity. The main condensate lines should be based on 100 fpm. Depending on prevailing local and/or national safety codes, a relief device may have to be installed in the discharge piping.

Figure 12 shows a piping arrangement for multiple condensers with a surge-type receiver. When the system is operating at reduced load, the flow paths may not be symmetrical through each circuit. Small pressure differences would not be unusual; therefore, the liquid line junction should be about 2 or 3 ft below the bottom of the condensers. The exact amount can be calculated from pressure loss through each path at all possible operating conditions.

When condensers are water-cooled, a single automatic water valve for the condensers in one refrigeration system should be used. Individual valves for each condenser in a single system would not be able to maintain the same pressure and corresponding pressure drops.

With evaporative condensers (Figure 13), the pressure loss may be high. If parallel condensers are alike and all are operated, the differences may be small, and the height of the condenser outlets above the liquid line junction need not be more than 2 or 3 ft. If the fans of one condenser are not operated while the fans on another condenser are, then the liquid level in the one condenser must be high enough to compensate for the pressure drop through the operating condenser.

When the available level difference between condenser outlets and the liquid line junction is sufficient, the receiver may be vented to the condenser inlets (see Figure 14). In this case, the surge-type receiver can be used. The level difference must then be at least equal to the greatest loss through any condenser circuit plus the greatest vent line loss when the receiver ambient is greater than the condensing temperature.

AIR-COOLED CONDENSERS

The refrigerant pressure drop through air-cooled condensers must be obtained from the supplier for the particular unit at the

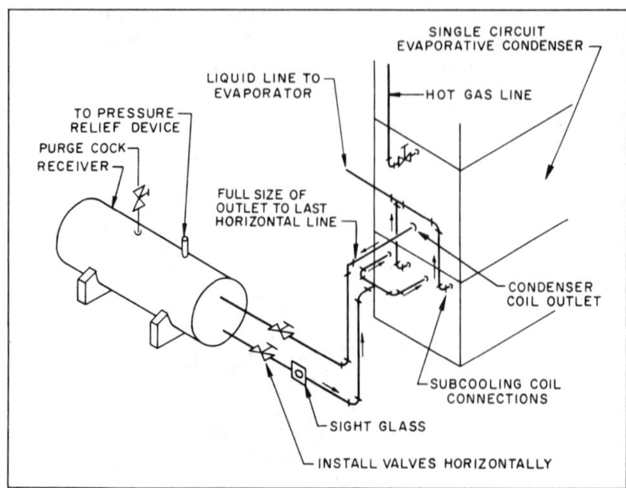

Fig. 13 Single-Circuit Evaporative Condenser with Receiver and Liquid Subcooling Coil

specified load. If the refrigerant pressure drop is low enough and it is practical to so arrange the equipment, parallel condensers can be connected to allow for capacity reduction to zero on one condenser without causing liquid backup in active condensers (Figure 15). Multiple condensers with high-pressure drops can be connected as shown in Figure 15, provided that (1) the receiver is located in an ambient equal to or lower than the inlet air temperature to the condenser; (2) capacity control affects all units equally; (3) all units operate when one operates, unless valved off at both inlet and outlet; and (4) all units are of equal size.

A single condenser with any pressure drop can be connected to a receiver without an equalizer and without trapping height if the condenser outlet and the line from it to the receiver can be sized for sewer flow without a trap or restriction, using a maximum velocity of 100 fpm. A single condenser can also be connected with an equalizer line to the hot-gas inlet if the vertical drop leg is sufficient to balance the refrigerant pressure drop through the condenser and the liquid line to the receiver.

If unit sizes are unequal, additional liquid height *H*, equivalent to the difference in full-load pressure drop, is required. Usually, condensers of equal size are used in parallel applications.

If the receiver cannot be located in an ambient temperature below the inlet air temperature for all operating conditions, sufficient extra height of drop leg *H* is required to overcome the equivalent differences in saturation pressure of the receiver and the condenser. The subcooling formed by the liquid leg tends to condense vapor in the receiver to reach a balance of rate of condensation, at an intermediate saturation pressure, with heat gain from ambient to the receiver. A relatively large liquid leg is required to balance a small temperature difference; therefore, this method is probably limited to marginal cases. The liquid leaving the receiver will nonetheless be saturated, and any subcooling to prevent flashing in the liquid line must be obtained downstream of the receiver. If the temperature of the receiver ambient is above the condensing pressure only at part-load conditions, it may be acceptable to

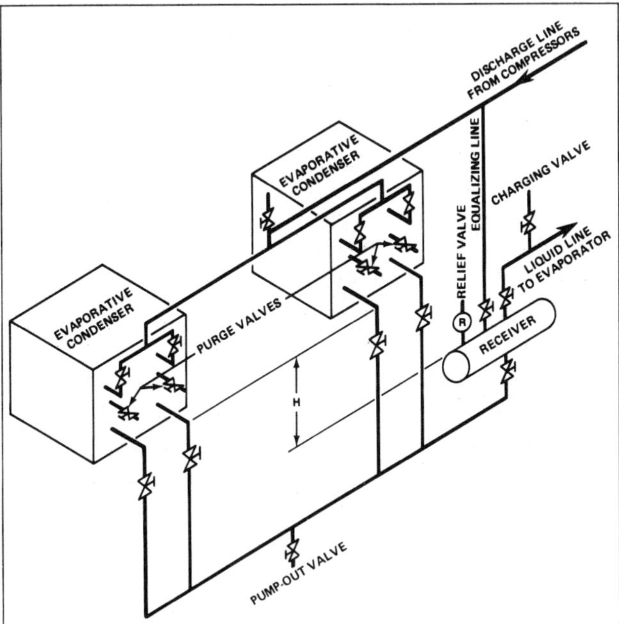

All outlets must be trapped. The height of the trap leg must be such that when one or more units are idle (fan or pump stopped), liquid may rise in the leg of the operating unit so that the static head equals the pressure drop in the operating unit at all conditions. The trap height should be 6 to 12 in. greater than the calculated height to assure that the liquid outlets remain free-draining.

Fig. 14 Multiple Evaporative Condensers with Equalization to Condenser Inlets

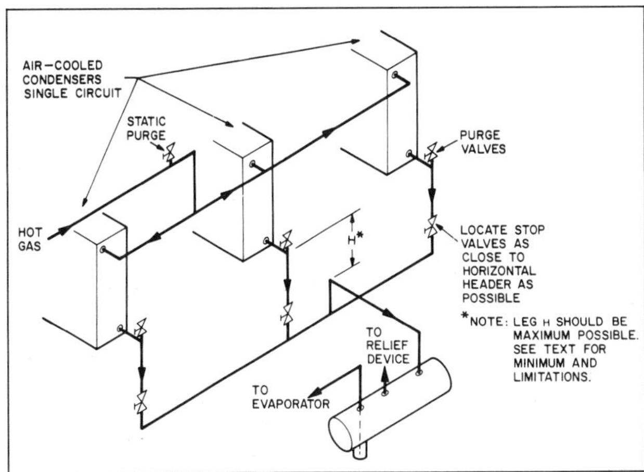

Fig. 15 Multiple Air-Cooled Condensers

back liquid into the condensing surface, sacrificing the operating economy of lower part-load head pressure for a lower liquid leg requirement. The receiver must be adequately sized to contain a minimum of the backed-up liquid so that the condenser can be fully drained when full load is required. If a low ambient control system of backing liquid into the condenser is used, consult the system supplier for proper piping.

PIPING AT MULTIPLE COMPRESSORS

Multiple compressors operating in parallel must be carefully piped to assure proper operation.

Suction Piping

Suction piping should be designed so that all compressors run at the same suction pressure and so that oil is returned in equal proportions. All suction lines should be brought into a common suction header in order to return the oil to each crankcase as uniformly as possible. Depending on the type and size of compressors, oil may be returned by designing the piping in one or more of the following schemes.

1. Oil returned with the suction gas to each compressor
2. Oil contained with a suction trap (accumulator) and returned to the compressors with a controlled means
3. Oil trapped in a discharge line separator and returned to the compressors with a controlled means (see Discharge Piping)

The suction header is a means of distributing the suction gas equally to each compressor. The design of the header can be to freely pass the suction gas and oil mixture or to provide a suction trap for the oil. The header should be run above the level of the compressor suction inlets so that oil can drain into the compressors by gravity.

Figure 16 shows a pyramidal or yoke-type suction header to maximize pressure and flow equalization at each of three compressor suction inlets piped in parallel. This type of construction is recommended for applications of three or more compressors in parallel. For two compressors in parallel, a single feed between the two compressor takeoffs is acceptable. Although not as good with regard to equalizing flow and pressure drops to all compressors, one alternative is to have the suction line from the evaporators enter at one end of the header instead of using the yoke arrangement. Then the suction header may have to be increased in size to minimize pressure drop and flow turbulence.

Suction headers designed to freely pass the gas and oil mixture should have the branch suction lines to the compressors connected to the side of the header. The return mains from the evaporators

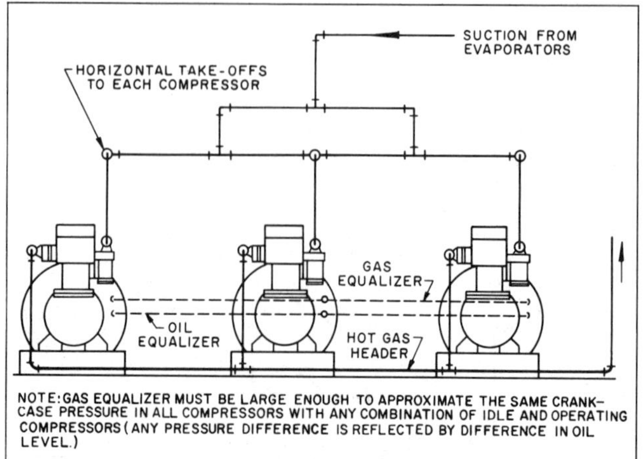

Fig. 16 Suction and Hot-Gas Headers for Multiple Compressors

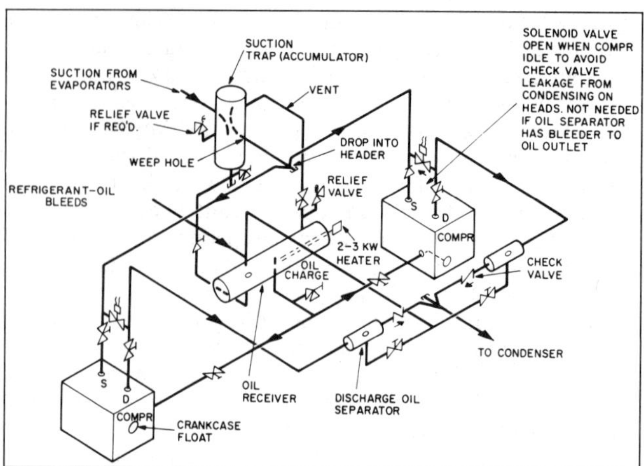

Fig. 17 Parallel Compressors with Gravity Oil Flow

should not be connected into the suction header to form crosses with the branch suction lines to the compressors. The size of the header should be full size based on the largest mass flow of the suction line returning to the compressors. The takeoffs to the compressors should be the same size as the suction header or constructed in a manner that the oil will not trap within the suction header. The branch suction lines to the compressors should not be reduced until the vertical drop is reached.

Suction traps are recommended wherever the following are used: (1) parallel compressors, (2) flooded evaporators, (3) double-suction risers, (4) long suction lines, (5) multiple expansion valves, (6) hot-gas defrost, (7) reverse cycle operation, and (8) suction pressure regulators.

Depending on the size of the system, the suction header may be designed to function as a suction trap. The suction header should be large enough to provide a region of low velocity within the header to allow for the suction gas and oil to separate. Refer to low-pressure receiver sizing in Chapter 1 to arrive at recommended velocities for separation. The suction gas flow for individual compressors should be taken off the top of the suction header. The oil can be returned to the compressor directly or through a vessel equipped with a heater to boil off the refrigerant and then allow the oil to drain to the compressors or other devices used to feed oil to the compressors.

The suction trap must be sized for effective gas and liquid separation. Adequate liquid volume and a means of disposing of it must be provided. A liquid transfer pump or heater may be used. Chapter 1 has further information on separation and liquid transfer pumps.

An oil receiver equipped with a heater effectively evaporates liquid refrigerant accumulated in the suction trap. It also assumes that each compressor receives its share of oil. Either crankcase float valves or external float switches and solenoid valves can be used to control the oil flow to each compressor.

A gravity feed oil receiver should be elevated to overcome the pressure drop between it and the crankcase. The oil receiver should be sized so that a malfunction of the oil control mechanism cannot overfill an idle compressor.

Figure 17 shows a recommended hookup of multiple compressors, suction trap (accumulator), oil receiver, and discharge line oil separators. The oil receiver also provides a reserve supply of oil for the compressors where the oil in the system external to the compressor varies with system loading. The heater mechanism should always be submerged.

Discharge Piping

The piping arrangement shown in Figure 16 is suggested for discharge piping. The piping must be arranged to prevent refrigerant liquid and oil from draining back into the heads of idle compressors. A check valve in the discharge line may be necessary to prevent refrigerant and oil from entering the compressor heads by migration. It is recommended that, after leaving the compressor head, the piping be routed to a lower elevation so that a trap is formed to allow for drainback of refrigerant and oil from the discharge line when flow rates are reduced or the compressors are off. If an oil separator is used in the discharge line, it may suffice as the trap for drainback for the discharge line.

A bullheaded tee at the junction of two compressor branches and the main discharge header causes increased turbulence, increased pressure drop, and possible hammering in the line; it should be avoided.

When an oil separator is used on multiple compressor arrangements, the oil must be piped to return to the compressors. This can be done in a variety of methods depending on the oil management system design. The oil may be returned to an oil receiver that is the supply for control devices feeding oil back to the compressors.

Interconnection of Crankcases

When two or more compressors are to be interconnected, a method must be provided to equalize the crankcases. Some compressor designs do not operate correctly with simple equalization of the crankcases. For these systems, it may be necessary to design a positive oil float control system for each compressor crankcase. A typical system allows the oil to collect in a oil receiver which, in turn, supplies oil to a device that meters oil back into the compressor crankcase to maintain a proper oil level (Figure 17).

Compressor systems that can be equalized should be placed on foundations so that all oil equalizer tapping locations are exactly level. If crankcase floats (as shown in Figure 17) are not used, an oil equalization line should connect all of the crankcases to maintain uniform oil levels. The oil equalizer may be run level with the tapping, or, for convenient access to the compressors, it may be run at the floor (Figure 18). It should never be run at a level higher than that of the tapping.

For the oil equalizer line to work properly, equalize the crankcase pressures by installing a gas equalizer line above the oil level. This line may be run to provide head room (Figure 18) or run level with the tapping on the compressors. It should be piped so that oil or liquid refrigerant will not be trapped.

Both lines should be the same size as the tapping on the largest compressor and should be valved so that any one machine can be taken out for repair. The piping should be arranged to absorb vibration.

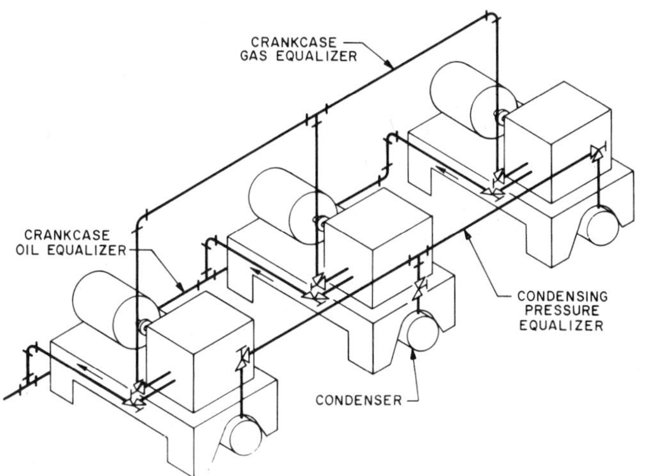

Fig. 18 Interconnecting Piping for Multiple Condensing Units

PIPING AT VARIOUS SYSTEM COMPONENTS

Flooded Fluid Coolers

For a description of flooded fluid coolers, see Chapter 38 of the 1992 ASHRAE *Handbook—Systems and Equipment.*

Shell-and-tube flooded coolers designed to minimize liquid entrainment in the suction gas require a continuous liquid bleed line (Figure 19) installed at some point in the cooler shell below the liquid level to remove trapped oil. This continuous bleed of refrigerant liquid and oil prevents the oil concentration in the cooler from getting too high. The location of the liquid bleed connection on the shell depends on the refrigerant and oil used. Refrigerants that are highly miscible with the refrigeration oil, can use a connection anywhere below the liquid level.

Refrigerants 22 and 502 can have a separate oil-rich phase floating on a refrigerant-rich layer. This becomes more pronounced as the evaporating temperature drops. When R-22 and R-502 are used with mineral oil, the bleed line usually is taken off the shell just slightly below the liquid level, or there may be more than one valved bleed connection at slightly different levels so that the optimum point can be selected during operation. With alkyl benzene lubricants, miscibility between the oil and R-22 and R-502 may be sufficient so that the oil bleed connection can be anywhere below the liquid level. The solubility charts in Chapter 7 give specific information.

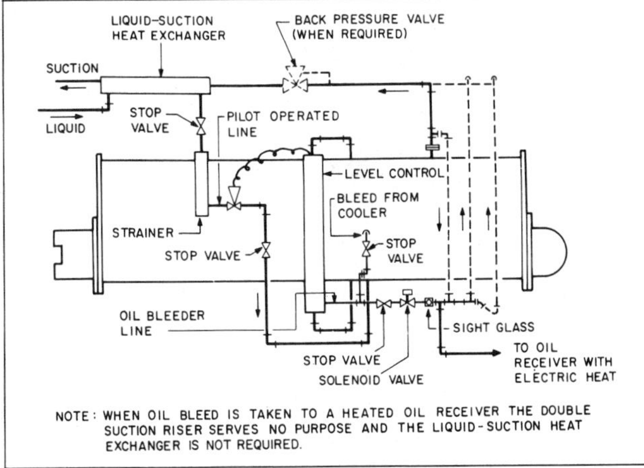

Fig. 19 Typical Piping at Flooded Fluid Cooler

Where the flooded cooler design requires an external surge drum to separate the liquid carryover from the suction gas off the tube bundle, the richest oil concentration may or may not be in the cooler. In some cases, the surge drum will have the highest concentration of oil. Here, the refrigerant and oil bleed connection is taken from the surge drum. The refrigerant and oil bleed from the cooler by gravity. The bleed sometimes drains into the suction line so that the oil can be returned to the compressor with the suction gas after the accompanying liquid refrigerant is vaporized in a liquid-suction heat interchanger. A better method drains the refrigerant-oil bleed into a heated receiver that boils the refrigerant off to the suction line and drains the oil back to the compressor.

Refrigerant Feed Devices

For further information on refrigerant feed devices, see Chapter 44. The pilot-operated low-side float control (Figure 19) is sometimes selected for flooded systems using halocarbon refrigerants. Except for small capacities, direct-acting low-side float valves are impractical for these refrigerants. The displacer float controlling a pneumatic valve works well for low-side liquid level control; it allows the cooler level to be adjusted within the instrument without disturbing the piping.

High-side float valves are practical only in systems having one evaporator, because distribution problems result when multiple evaporators are used.

Float chambers should be located as near to the liquid connection on the cooler as possible, since a long length of liquid line, even though insulated, can pick up room heat and give an artificial liquid level in the float chamber. Equalizer lines to the float chamber must be amply sized to minimize the effect of heat transmission, creating false liquid levels in the chamber. The float chamber and its equalizing lines must be insulated.

Each flooded cooler system must have a way of keeping oil concentration in the evaporator low, both to minimize the bleed-off needed to keep oil concentration in the cooler low and to reduce system losses from large stills. A highly efficient discharge gas/oil separator can be used for this purpose.

At low temperatures, periodic warm-up of the evaporator permits recovery of oil accumulation in the chiller. If continuous operation is required, dual chillers may be needed to permit deoiling an oil-laden evaporator, or an oil-free compressor may be used.

Direct-Expansion Fluid Chillers

For further information on this chiller type, see Chapter 42. Figure 20 shows typical piping connections for a multicircuit direct-expansion chiller. Each circuit contains its own thermostatic expansion and solenoid valves. One of the solenoid valves can be wired to close at reduced system capacity. The thermostatic

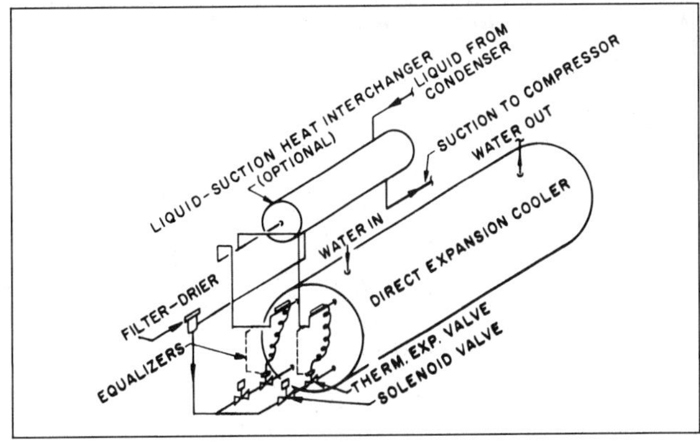

Fig. 20 Two-Circuit Direct-Expansion Cooler Connections (for Single-Compressor System)

expansion valve bulbs should be located between the cooler and the liquid-suction interchanger, if used. Locating the bulb downstream from the interchanger can cause excessive cycling of the thermostatic expansion valve, since the high-pressure liquid flow through the interchanger ceases when the thermostatic expansion valve closes; consequently, no heat source is available from the high-pressure liquid, and the cooler must starve itself to obtain the superheat necessary to open the valve. When the valve does open, excessive superheat causes it to overfeed until the bulb senses liquid downstream from the interchanger. Therefore, position the remote bulb between the cooler and the interchanger.

Figure 21 shows a typical piping arrangement that has been successful in packaged water chillers having direct-expansion coolers. With this arrangement, automatic recycling pumpdown is needed on the lag compressor to prevent leakage through compressor valves, permitting migration to the cold evaporator circuit. Liquid will slug the compressor at startup if this is not done.

On larger systems, the limited size of thermostatic expansion valves may require use of a pilot-operated liquid valve controlled by a small thermostatic expansion valve (Figure 22). The small thermostatic expansion valve pilots the main liquid control valve. The equalizing connection and the bulb of the pilot thermostatic expansion valve should be treated as a direct-acting thermal expansion valve. A small solenoid valve in the pilot line will shut off the high side from the low during shutdown. However, the main liquid valve will not open and close instantaneously.

Direct-Expansion Air Coils

For further information on this type of coil, see Chapter 21 of the 1992 ASHRAE *Handbook—Systems and Equipment*. The most common ways of arranging direct-expansion coils are shown in Figures 23 and 24. The method shown in Figure 24 provides the necessary superheat to operate the thermostatic expansion valve and is effective for heat transfer because the leaving air contacts the coldest evaporator surface. This arrangement is advantageous on low-temperature applications, where the coil pressure drop represents an appreciable change in evaporating temperature.

Direct-expansion air coils can be located in any position provided proper refrigerant distribution and continuous oil removal facilities are provided.

Figure 23 shows top-feed, free-draining piping with a vertical up airflow coil. In Figure 24, which illustrates a horizontal airflow coil, the suction is taken off the bottom header connection, providing free oil draining. Many coils are supplied with connections at each end of the suction header so that a free- draining connection can be used regardless of which side of the coil is up; the other end is then capped.

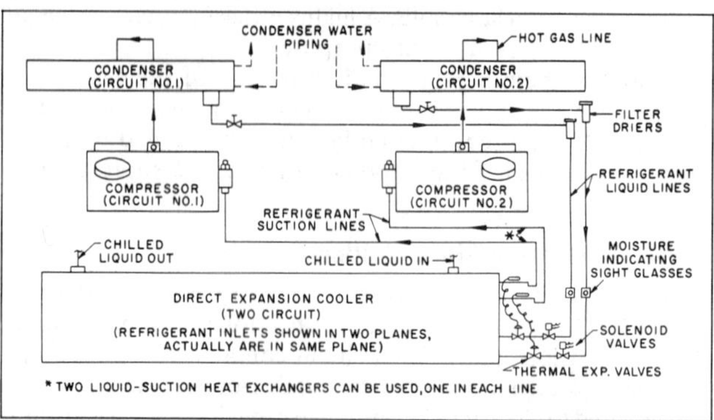

Fig. 21 Typical Refrigerant Piping in Liquid Chilling Package with Two Completely Separate Circuits

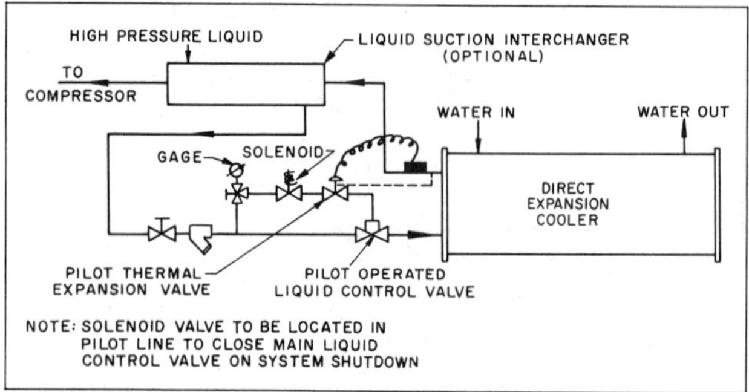

Fig. 22 Direct-Expansion Cooler with Pilot-Operated Control Valve

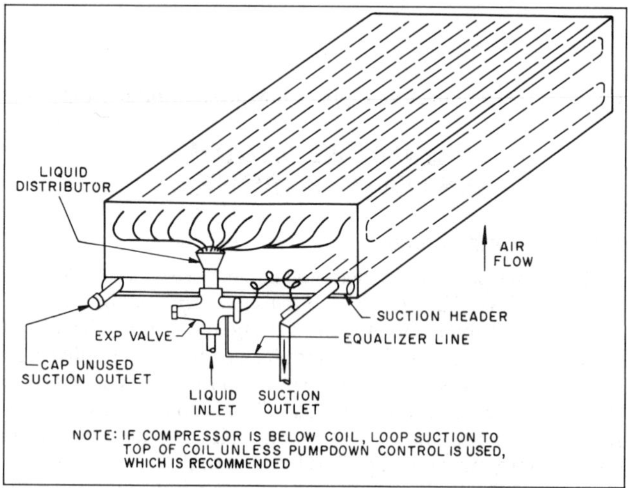

Fig. 23 Direct-Expansion Evaporator (Top-Feed, Free-Draining)

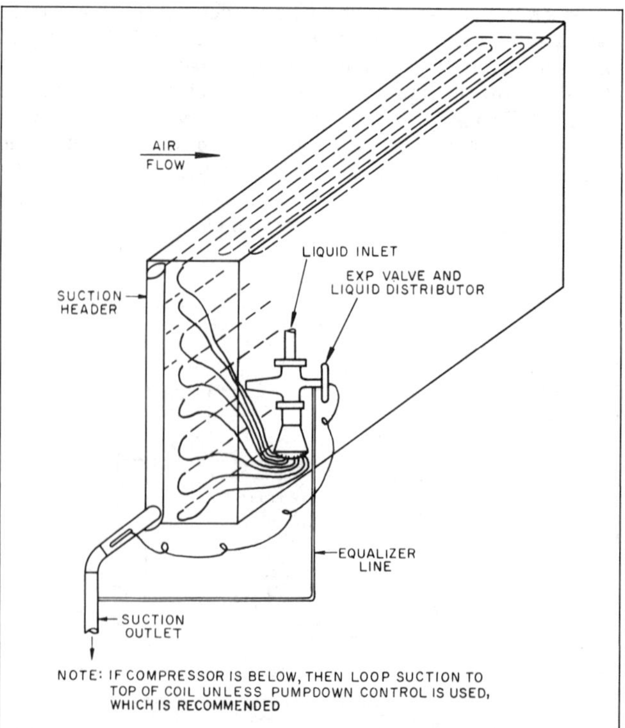

Fig. 24 Direct-Expansion Evaporator (Horizontal Airflow)

In Figure 25, a refrigerant upfeed coil is used with a vertical downflow air arrangement. Here, the coil design must provide for sufficient gas velocity to entrain oil at lowest loadings and to carry it into the suction line.

Pumpdown compressor control is desirable on all systems using downfeed as well as upfeed evaporators to protect the compressor against a liquid slugback in cases where liquid can accumulate in the suction header and/or the coil on system off cycles. Pumpdown compressor control is described later.

Thermostatic expansion valve operation and application are described in Chapter 44. Thermostatic expansion valves should be sized carefully to avoid undersizing at full load and oversizing at partial load. The refrigerant pressure drops through the system (distributor, coil, condenser, and refrigerant lines, including liquid lifts) must be properly evaluated to determine the correct pressure drop available across the valve on which to base the selection. Variations in condensing pressure greatly affect the pressure available across the valve, and hence its capacity.

Oversized thermostatic expansion valves result in a cycling condition that alternates flooding and starving the coil. This occurs because the valve attempts to throttle at a capacity below its capability, which causes periodic flooding of the liquid back to the compressor and wide temperature variations in the air leaving the coil. Reduced compressor capacity further aggravates this problem. Systems having multiple coils can use a solenoid valve located in the liquid line feeding each evaporator or group of evaporators to close them off individually as compressor capacity is reduced.

In food freezing or storage, the effect of frost accumulation on heat transfer surface efficiency must be considered and sufficient defrost time allocated. If exact conditions have to be continuously maintained, dual evaporators may have to be used and alternated to permit defrosting. When selecting evaporators for air cooling (as in cold storage warehouses), the defrost system should be based on the following tabulation of cooling hours:

Defrost Method	Maximum Hours of Operation per day
Air (above 34 °F)	16 to 18
Water or brine	20 to 22
Hot gas	20 to 22
Electrical	20 to 22

This table is based on average frost conditions. For higher frost conditions and lower evaporating temperatures, more time must be allowed for defrosting. During defrost, heat is absorbed by the frost, the metal of the air-cooling unit, and the room air. When the system returns to cooling, all the heat added to the room (except that in the condensate that was drained away) must be added back to the load because it must also be removed. The designer and the owner must agree on an economical balance of system capacity versus load.

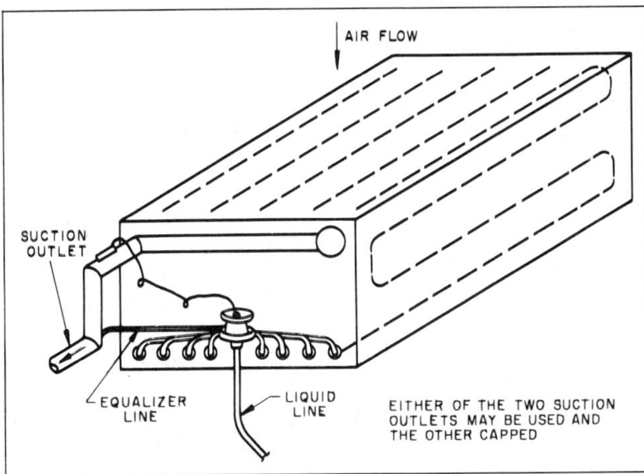

Fig. 25 Direct-Expansion Evaporator (Bottom-Feed)

Flooded Evaporators

Flooded evaporators may be desirable when a small temperature differential is required between the refrigerant and the medium being cooled, an advantageous feature in low-temperature applications.

In a flooded evaporator, the coil is kept full of refrigerant when cooling is required. The refrigerant level is generally controlled by using a high- or low-side float control. Figure 26 represents a typical arrangement showing a low-side float control, oil return line, and heat interchanger.

Circulation of refrigerant through the evaporator depends on gravity and a thermosiphon effect. A mixture of liquid refrigerant and vapor returns to the surge tank, and the vapor flows into the suction line. A baffle installed in the surge tank helps prevent foam and liquid from entering the suction line. A liquid refrigerant circulating pump (Figure 27) provides a more positive way of obtaining a high rate of circulation.

Where the suction line is taken off the top of the surge tank, difficulties arise if no special provisions are made for oil return.

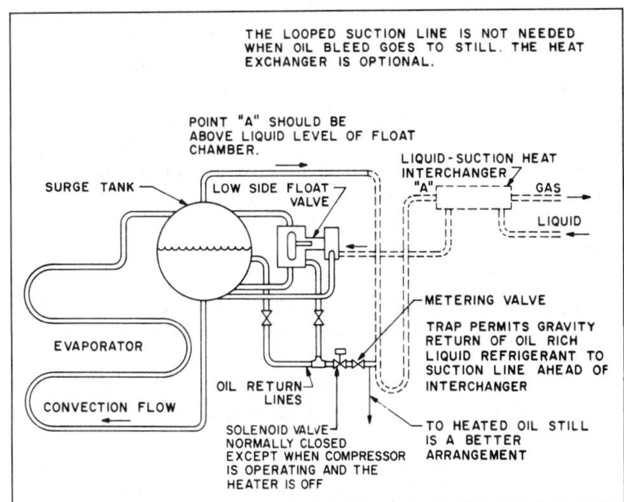

Fig. 26 Flooded Evaporator (Gravity Circulation)

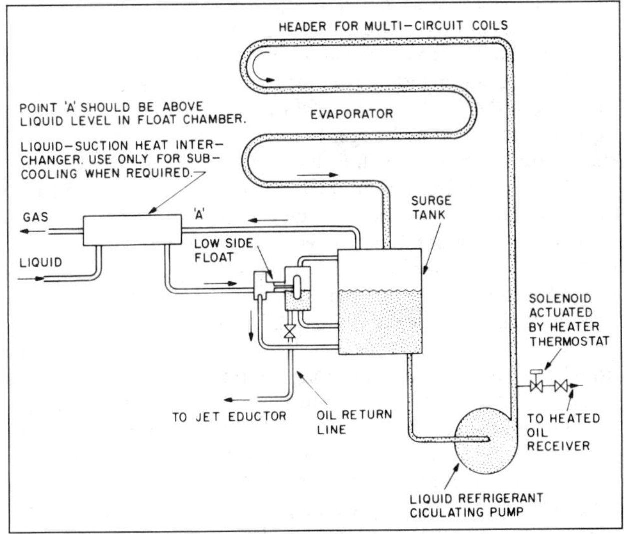

Fig. 27 Flooded Evaporator (Forced Circulation)

For this reason, the oil return lines shown in Figure 26 should be installed. These lines are connected near the bottom of the float chamber and also just below the liquid level in the surge tank (where an oil-rich liquid refrigerant exists). They extend to a lower point on the suction line to allow gravity flow. Included in this oil return line is a solenoid valve that is open only while the compressor is running and a metering valve that is adjusted to allow a constant but small volume return to the suction line. A liquid line sight glass may be installed downstream from the metering valve to serve as a convenient check on the liquid being returned.

Oil can be returned satisfactorily by taking a bleed of refrigerant and oil from the pump discharge (Figure 27) and feeding it to the heated oil receiver. If a low-side float is used, a jet ejector can be used to remove oil from the quiescent float chamber.

REFRIGERATION ACCESSORIES

Liquid-Suction Heat Exchangers

Generally, liquid-suction heat exchangers subcool the liquid refrigerant and superheat the suction gas. They are used for one or more of the following functions:

- *Increasing the efficiency of the refrigeration cycle.* The efficiency of the thermodynamic cycle of certain halocarbon refrigerants can be increased when the suction gas is superheated by removing heat from the liquid. This increased efficiency must be evaluated against the effect of pressure drop through the suction side of the exchanger, which forces the compressor to operate at a lower suction pressure. Liquid-suction heat exchangers are most beneficial at low suction temperatures. The increase in cycle efficiency for systems operating in the air-conditioning range (down to about 30°F evaporating temperature) usually does not justify their use. The heat exchanger can be located wherever convenient.
- *Subcooling the liquid refrigerant to prevent flash gas at the expansion valve.* The heat exchanger should be located near the condenser or receiver to achieve subcooling before pressure drop occurs.
- *Evaporating small amounts of expected liquid refrigerant returning from evaporators in certain applications.* Many heat pumps incorporating reversals of the refrigerant cycle include a suction line accumulator and liquid-suction heat exchanger arrangement to trap liquid floodbacks and vaporize them slowly between cycle reversals.

If the design of an evaporator makes a deliberate slight overfeed of refrigerant necessary, either to improve evaporator performance or to return oil out of the evaporator, a liquid-suction heat exchanger is needed to evaporate the refrigerant.

A flooded water cooler usually incorporates an oil-rich liquid bleed from the shell into the suction line for returning oil. The liquid-suction heat exchanger boils the liquid refrigerant out of the mixture in the suction line. Exchangers used for this purpose should be placed in a horizontal run near the evaporator. Several types of liquid-suction heat exchangers are used.

Liquid and Suction Line Soldered Together. The simplest form of heat exchanger is obtained by strapping or soldering the suction and liquid lines together to obtain counterflow and then insulating the lines as a unit. To obtain the greatest capacity, the liquid line should always be on the bottom of the suction line, since liquid in a suction line runs along the bottom (Figure 28). This arrangement is limited by the amount of suction line available.

Shell-and-Coil or Shell-and-Tube Heat Exchangers (Figure 29). These units are usually installed so that the suction outlet drains the shell. When used to evaporate liquid refrigerant returning in the suction line, the free-draining arrangement is not recommended. Liquid refrigerant can run along the bottom of the heat exchanger shell, having little contact with the warm liquid coil, and

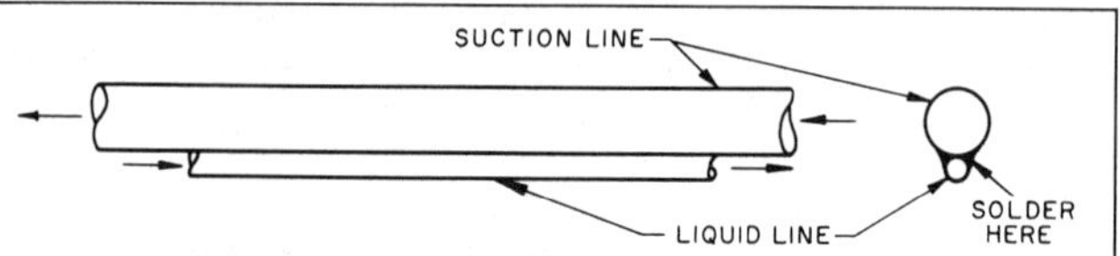

Fig. 28 Soldered Tube Heat Exchanger

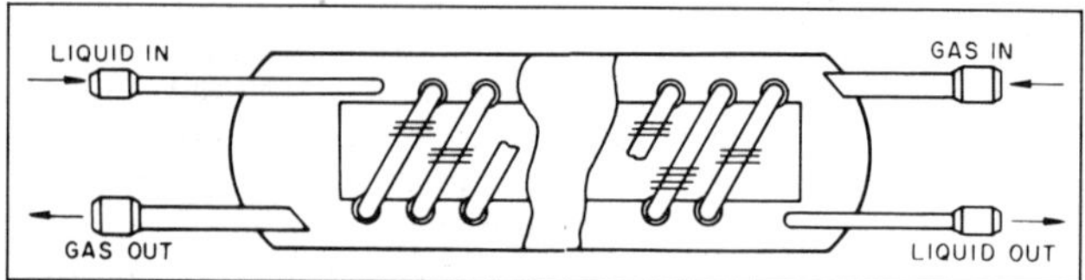

Fig. 29 Shell-and-Finned-Coil Heat Exchanger

drain into the compressor. By installing the heat exchanger at a slight angle with the horizontal (Figure 30), and with the gas entering at the bottom and leaving at the top, any liquid returning in the line is trapped in the shell and held in contact with the warm liquid coil, where most of it is vaporized. An oil return line, with a metering valve and solenoid valve (open only when the compressor is running), is required to return oil that collects in the trapped shell.

Concentric Tube-in-Tube Heat Exchangers. The tube-in-tube heat exchanger is not as efficient as the shell-and-finned-coil type. It is, however, quite suitable for cleaning up small amounts of excessive liquid refrigerant returning in the suction line. Figure 31 shows typical construction with available pipe and fittings.

Plate-Type Heat Exchangers. The plate-type heat exchanger has a high-efficiency heat transfer performance. They are very compact, have low-pressure drop, and are lightweight devices. They are good for use as liquid subcoolers.

For air-conditioning applications, heat exchangers are recommended for liquid subcooling or for clearing up excess liquid in the suction line. For refrigeration applications, heat exchangers are recommended to increase cycle efficiency, as well as for liquid subcooling and removing small amounts of excess liquid in the suction line. Excessive superheating of the suction gas should be avoided.

TWO-STAGE SUBCOOLERS

To take full advantage of the two-stage system, the refrigerant liquid should be cooled to a temperature near the interstage temperature to reduce the amount of flash gas handled by the low-

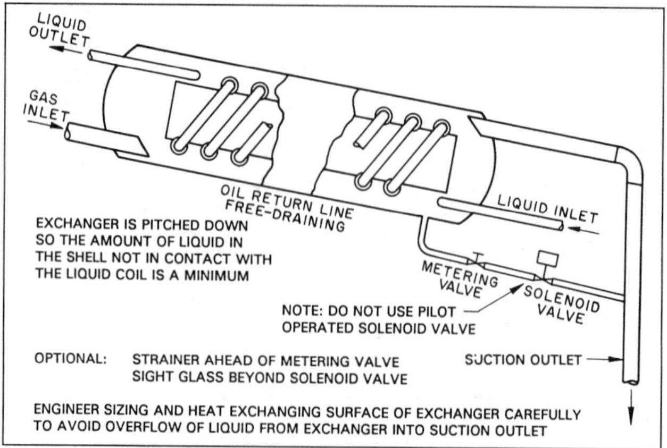

Fig. 30 Shell-and-Finned-Coil Exchanger Installed to Prevent Liquid Floodback

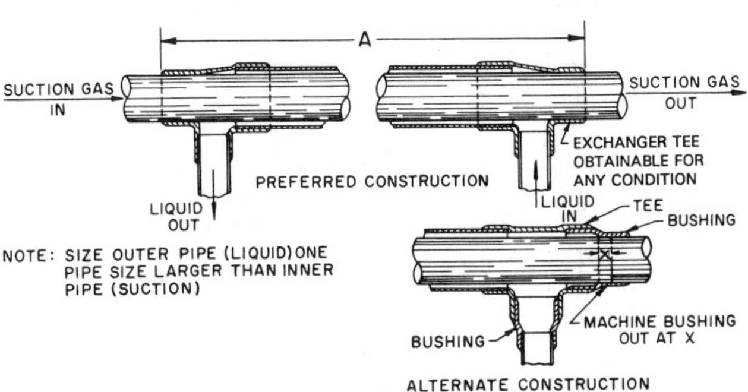

Fig. 31 Tube-in-Tube Heat Exchanger

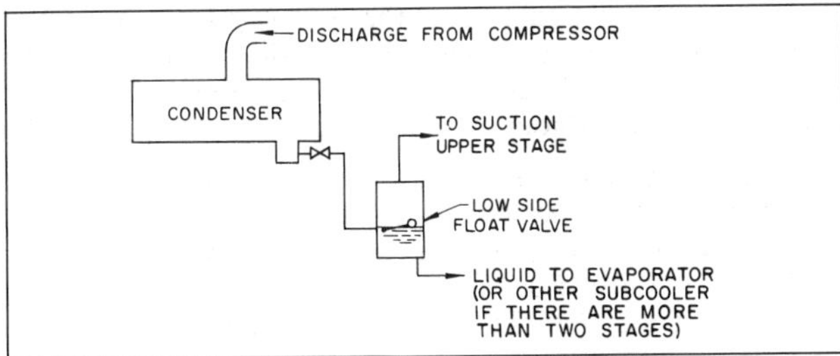

Fig. 32 Flash-Type Cooler

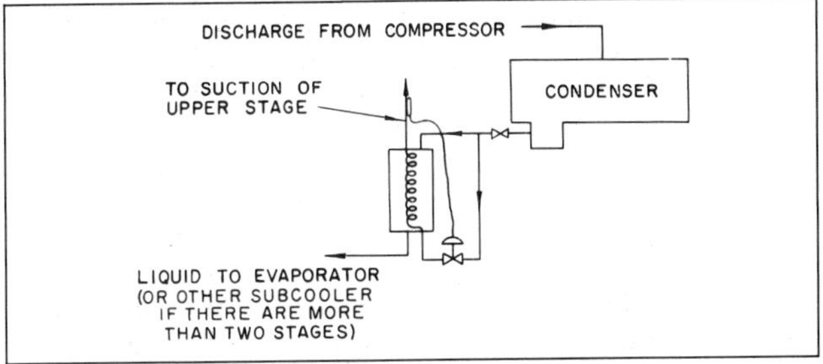

Fig. 33 Exchange-Type Subcooler

stage compressor. The net result is a reduction in total system power requirements. The amount of gain from cooling to near interstage conditions varies among refrigerants.

Figure 32 illustrates an open or flash-type cooler. This is the simplest and least costly type, which has the advantage of cooling liquid to the saturation temperature of the interstage pressure. One disadvantage is that the pressure of the cooled liquid is reduced to interstage pressure leaving less pressure available for liquid transport. Although the liquid temperature is reduced, the pressure is correspondingly reduced, and the expansion device controlling flow to the cooler must be large enough to pass all the liquid refrigerant flow. Failure of this valve could allow a large flow of liquid to the upper-stage compressor suction, which could seriously damage the compressor.

Liquid from a flash cooler is saturated, and liquid from a cascade condenser usually has little subcooling. In both cases, the liquid temperature is usually lower than the temperature of the surroundings. Thus, it is important to avoid heat input and pressure losses that would cause flash gas to form in the liquid line to the expansion device or to recirculating pumps. The cold liquid lines should be insulated, as expansion devices are usually designed to feed liquid, not vapor.

Figure 33 shows the closed or heat exchanger type of subcooler. It should have sufficient heat transfer surface to transfer heat from the liquid to the evaporating refrigerant with a small final temperature difference. The pressure drop should be small, so that full pressure is available for feeding liquid to the expansion device at the low-temperature evaporator. The subcooler liquid control valve should be sized to supply only the quantity of refrigerant required for the subcooling. This prevents a tremendous quantity of liquid from flowing to the upper-stage suction in the event of a valve failure.

Discharge Line Oil Separators

Oil is always in circulation in systems using halocarbon refrigerants. Refrigerant piping is designed to ensure that this oil passes through the entire system and returns to the compressor as fast as it leaves. Although well-designed piping systems can handle the oil in most cases, an oil separator can have certain advantages in some applications (see Chapter 44). Some of the applications where discharge line oil separators can be useful include:

- In systems where it is impossible to prevent substantial absorption of refrigerant in the crankcase oil during shutdown periods. When the compressor starts up with a violent foaming action, oil will be thrown out at an accelerated rate, and the separator will immediately return a large portion of this oil to the crankcase. Normally, the system should be designed with pumpdown control or crankcase heaters to minimize liquid absorption in the crankcase.
- In systems using flooded evaporators, where refrigerant bleedoff is necessary to remove oil from the evaporator. Oil separators reduce the amount of bleedoff from the flooded cooler needed for operation.
- In direct-expansion systems using coils or tube bundles that require bottom feed for good liquid distribution and where refrigerant carryover from the top of the evaporator is essential for proper oil removal.
- In low-temperature systems, where it is advantageous to have as little oil as possible going through the low side.
- In screw-type compressor systems, where an oil separator is necessary for proper operation. The oil separator is usually supplied with the compressor unit assembly directly from the compressor manufacturer.
- In multiple compressors operating in parallel. The oil separator can be an integral part of the total system oil management system.

In applying oil separators in refrigeration systems, the following potential hazards must be considered:

- Oil separators are not 100% efficient, and they do not eliminate the need to design the complete system for oil return to the compressor.
- Oil separators tend to condense out liquid refrigerant during compressor off cycles and on compressor start-up. This is true if the condenser is in a warm location, such as an evaporative condenser and receiver on a roof. During the off cycle, the oil separator cools down and acts as a condenser for refrigerant that evaporates in warmer parts of the system. A cool oil separator may condense discharge gas and, on compressor start-up, automatically drain it into the compressor crankcase. To minimize this possibility, the drain connection from the oil separator can be connected into the suction line. This line should be equipped with a shutoff valve, a fine filter, hand throttling and solenoid valves, and a sight glass. The throttling valve should be adjusted so that the flow through this line is only a little greater than would normally be expected to return oil through the suction line.
- The float valve is a mechanical device that may stick open or closed. If it sticks open, hot gas will be continuously bypassed to the compressor crankcase. If the valve sticks closed, no oil is returned to the compressor. To minimize this problem, the separator can be supplied without an internal float valve. A

separate external float trap can then be located in the oil drain line from the separator preceded by a filter. Shutoff valves should isolate the filter and trap. The filter and traps are also easy to service without stopping the system.

The discharge line pipe size into and out of the oil separator should be the full size determined for the discharge line. For separators that have internal oil float mechanisms, allow enough room for removal of the oil float assembly for servicing requirements.

Depending on system design, the oil return line from the separator may feed to one of the following locations:

- Directly to the compressor crankcase
- Directly into the suction line ahead of the compressor
- Into an oil reservoir or device used to collect oil. This reservoir is used as a source of oil supply for a specifically designed oil management system.

When a solenoid valve is used in the oil return line, the valve should be wired so that it is open when the compressor is running. To minimize condensed refrigerant from entering the low side, a thermostat may be installed and wired to control the solenoid in the oil return line from the separator. The thermostat sensing element should be located on the oil separator shell below the oil level and set high enough so that the solenoid valve will not open until the separator temperature is higher than the condensing temperature. A superheat-controlled expansion valve can perform the same function. If a discharge line check valve is used, it should be downstream of the oil separator.

Surge Drums or Accumulators

A surge drum is required on the suction side of almost all flooded evaporators to prevent liquid slopover to the compressor. Exceptions include shell-and-tube coolers and similar shell-type evaporators, which provide ample surge space above the liquid level or contain eliminators to separate gas and liquid. A horizontal surge drum is sometimes used where headroom is limited.

The drum can be designed with baffles or eliminators to separate liquid from the suction gas. More often, sufficient separation space is allowed above the liquid level for this purpose. Usually, the design is vertical, with a separation height above the liquid level of from 24 to 30 in. and with the shell diameter sized to keep the suction gas velocity at a value low enough to allow the liquid droplets to separate. Since these vessels are also oil traps, it is necessary to provide oil bleed.

Although separators may be fabricated with length-to-diameter (L/D) ratios of 1/1 increasing to 10/1, the lowest cost separators will usually be for L/D ratios between 3/1 and 5/1.

Compressor Floodback Protection

Certain systems periodically flood the compressor with excessive amounts of liquid refrigerant. When periodic floodback through the suction line cannot be controlled, the compressor must be protected against it.

The most satisfactory method appears to be a trap arrangement that catches the liquid floodback and may (1) meter it slowly into the suction line where the floodback is cleared up with a liquid-suction heat interchanger; (2) evaporate the liquid 100% in the trap itself by using a liquid coil or electric heater, and then automatically return oil to the suction line, or (3) return it to the receiver or to one of the evaporators. Figure 30 illustrates an arrangement that handles moderate liquid floodbacks, getting rid of the liquid by a combination of boiling off in the exchanger and a limited bleedoff into the suction line. This device, however, would not have sufficient trapping volume for most heat pump jobs or hot-gas defrost systems employing reversal of the refrigerant cycle.

For heavier floodbacks, a larger volume is required in the trap. The arrangement shown in Figure 34 has been applied successfully in reverse-cycle heat pump jobs using halocarbon refrigerants. It consists of a suction line accumulator with sufficient volume to hold the maximum expected floodback and of large enough diameter to separate liquid from suction gas. The trapped liquid is slowly bled off through a properly sized and controlled drain line into the suction line, where it is boiled off in a liquid-suction heat exchanger between cycle reversals.

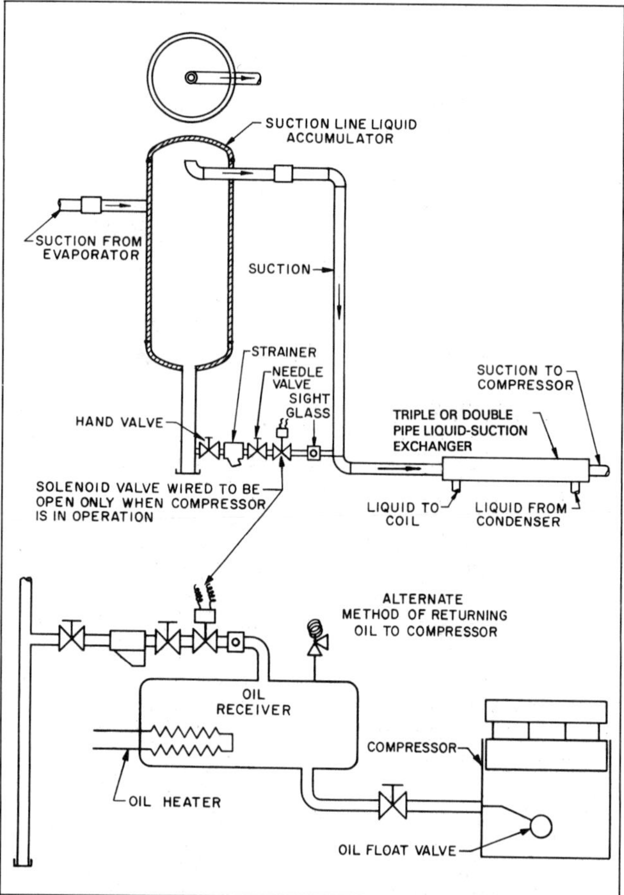

Fig. 34 Compressor Floodback Protection Using Accumulator with Controlled Bleed

With the alternate arrangement shown, the liquid-oil mixture is heated to evaporate the refrigerant, and the remaining oil is drained into the crankcase or the suction line.

Refrigerant Driers and Moisture Indicators

The effect of moisture in refrigeration systems is discussed in Chapters 5 and 6. Using a permanent refrigerant drier is recommended on all systems and with all refrigerants. It is especially important on low-temperature systems to prevent ice from forming at expansion devices. A *full-flow* drier is always recommended in hermetic compressor systems to keep the system dry and to prevent the products of decomposition from getting into the evaporator in the event of a motor burnout.

Replaceable element filter-driers are preferred for large systems, since the drying element can be replaced without breaking any refrigerant connections. The drier is usually located in the liquid line near the liquid receiver. It may be mounted horizontally or vertically with the flange at the bottom, but it should never be mounted vertically with the flange on top, because any loose material would then fall into the line when the drying element is removed.

A three-valve bypass is usually used, as shown in Figure 35, to provide a means for isolating the drier for servicing. The

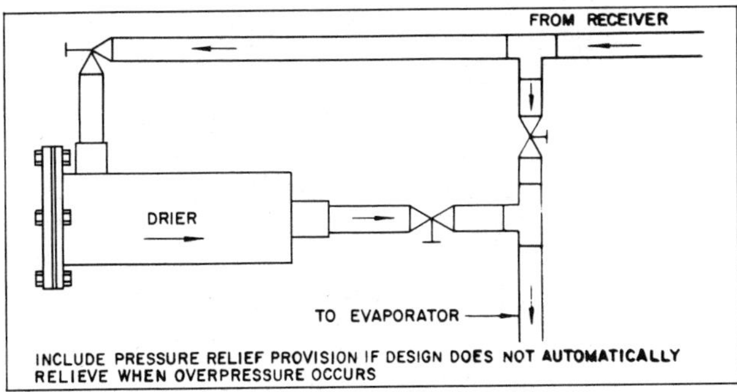

Fig. 35 Drier with Piping Connections

refrigerant charging connection should be located between the receiver outlet valve and the liquid line drier so that all refrigerant added to the system will pass through the drier.

Reliable moisture indicators can be installed in refrigerant liquid lines to provide a positive indication of when the drier cartridge should be replaced.

Strainers

Strainers should be used in both liquid and suction lines to protect automatic valves and the compressor from foreign material, such as pipe welding scale, rust, and metal chips. The strainer should be mounted in a horizontal line with the orientation so that the screen can be replaced without loose particles falling into the system.

A liquid line strainer should be installed before each automatic valve to prevent particles from lodging on the valve seats. Where multiple expansion valves with internal strainers are used at one location, a single main liquid line strainer will protect all of these. The liquid line strainer can be located anywhere in the line between the condenser (or receiver) and the automatic valves, preferably near the valves for maximum protection. Strainers should trap the particle size that could affect valve operation. In pilot-operated valves, a very fine strainer should be installed in the pilot line ahead of the valve.

Filter driers perform the dual function of refrigerant drying and filtering out particles far smaller than those trapped by mesh strainers. No other strainer is needed in the liquid line if a good filter drier is used.

Refrigeration compressors are usually equipped with a built-in suction strainer, which is adequate for the usual system with copper piping. The suction line should be piped at the compressor so that the built-in strainer is accessible for servicing.

Both liquid and suction line strainers should be adequately sized to assure sufficient foreign material storage capacity without excessive pressure drop. In steel piping systems, an external suction line strainer is recommended in addition to the compressor strainer.

Liquid Indicators

Every refrigeration system should have a way to check for sufficient refrigerant charge. The common devices used are liquid line sight glass, mechanical or electronic indicators, and an external gage glass with equalizing connections and shutoff valves. A properly installed sight glass will show bubbling when the charge is insufficient.

Liquid indicators should be located in the liquid line as close as possible to the receiver outlet or condenser outlet if no receiver is used (Figure 36). The sight glass is best installed in a vertical section of the line, a sufficient distance downstream from any valve so that the resulting disturbance does not appear in the glass. If it is installed too far away from the receiver, the line pressure drop may be sufficient to cause flashing and bubbles in the glass, even though the charge is sufficient for a liquid seal at the receiver outlet.

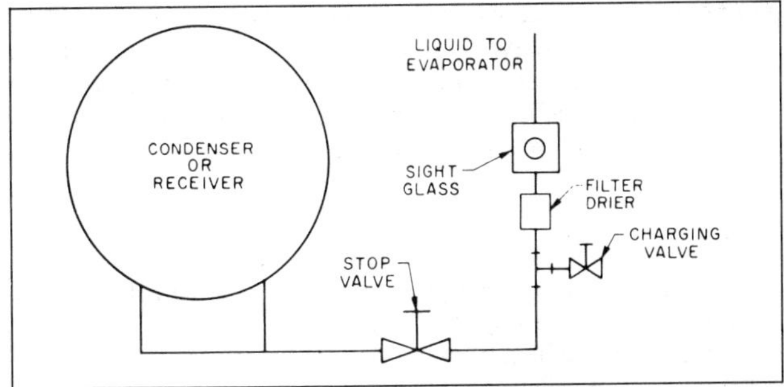

Fig. 36 Sight Glass and Charging Valve Locations

When sight glasses are installed near the evaporator, often no amount of system overcharging will give a solid liquid condition at the sight glass because of pressure drop in the liquid line or lift. Subcooling is required here. An additional sight glass near the evaporator may be needed to check on the condition of the refrigerant at that point.

Sight glasses should be installed full size in the main liquid line. In very large liquid lines, this may not be possible, and the glass can then be installed in a bypass or saddle mount that is arranged so that any gas in the liquid line will tend to move to it. A sight glass with double ports (for back lighting) and seal caps, which provide added protection against leakage, is preferred. Moisture-liquid indicators large enough to be installed directly in the liquid line serve the dual purpose of liquid line sight glass and moisture indicator.

Oil Receivers

Oil receivers serve as reservoirs for replenishing crankcase oil pumped by the compressors and provide the means to remove refrigerant dissolved in the oil. They are selected when:

- Flooded or semiflooded evaporators with large refrigerant charges are used
- Two or more compressors are operated in parallel
- Long suction and discharge lines are installed
- Double-suction line risers are installed

A typical hookup is shown in Figure 34. Outlets are arranged to prevent oil from draining below the heater level to avoid heater burnout and to prevent scale and dirt from being returned to the compressor.

Purging

Noncondensable gas separation using a purge unit is useful on most large refrigeration systems where suction pressure may fall below atmospheric pressure (see Figure 11 of Chapter 3).

HEAD PRESSURE CONTROL FOR REFRIGERANT CONDENSERS

For further information regarding head pressure control, see Chapter 36 of the 1992 ASHRAE *Handbook—Systems and Equipment*.

Water-Cooled Condensers

With water-cooled condensers, head pressure controls are used both for maintaining the condensing pressure and conserving water. On cooling tower applications, they are used only where it is necessary to maintain condensing temperatures.

Condenser Water Regulating Valves

The shutoff pressure of the valve must be set slightly higher than the saturation pressure of the refrigerant at the highest ambient temperature expected when the system is not in operation. This ensures that the valve will not pass water during off cycles. These valves are usually sized to pass the design quantity of water at about a 25 to 30 psi difference between design condensing pressure and valve shutoff pressure. Chapter 44 has further information.

Water Bypass

In cooling tower applications, a simple bypass can also be used to maintain condensing pressure, using a manual or automatic valve responsive to head pressure change. Figure 37 shows an automatic three-way valve arrangement. The valve divides the water flow between the condenser and the bypass line to maintain the desired condensing pressure. This maintains balanced flow of water on the tower and pump.

Evaporative Condensers

Among the methods used for condensing pressure control with evaporative condensers are (1) cycling the spray pump motor; (2) cycling both fan and spray pump motors; (3) throttling the spray water; (4) bypassing air around duct and dampers; (5) throttling air via dampers, either on inlet or discharge; and (6) combinations of these methods. For further information, see Chapter 36 of the 1992 ASHRAE *Handbook—Systems and Equipment.*

In water pump cycling, a pressure control at the gas inlet starts and stops the pump in response to head pressure changes. The pump sprays water over the condenser coils. As the head pressure drops, the pump stops and the unit becomes an air-cooled condenser.

Constant pressure is difficult to maintain with coils of prime surface tubing, because as soon as the pump stops, the pressure goes up and the pump starts again. This occurs because this type of coil has insufficient capacity when operating as an air-cooled condenser. The problem is not as acute with extended surface coils. Short-cycling results in excessive deposits of mineral and scale on the tubes, decreasing the life of the water pump.

One method of controlling head pressure is using cycle fans and pumps. This minimizes water-side scaling. In northern climates, an indoor water sump with a remote spray pump(s) is required. The fan cycling sequence follows:

Upon dropping head pressure:

- Stop fans.
- If pressure continues to fall, stop pumps.

Upon rising head pressure:

- Start fans.
- If pressure continues to rise, start pumps.

Damper control (see Figure 38) may be incorporated when systems require more constant head pressures. Such systems could

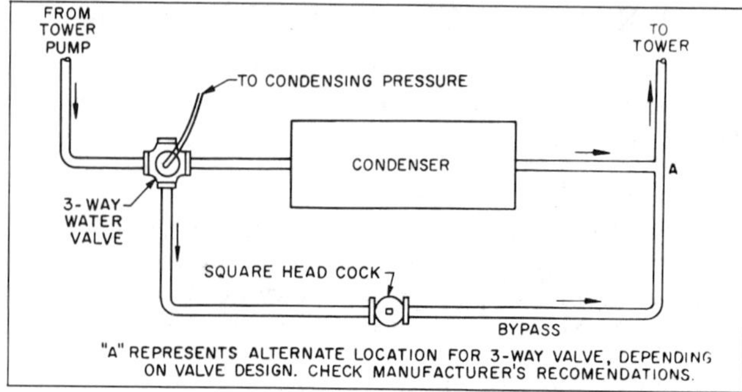

Fig. 37 Head Pressure Control for Condensers Used with Cooling Towers (Water Bypass Modulation)

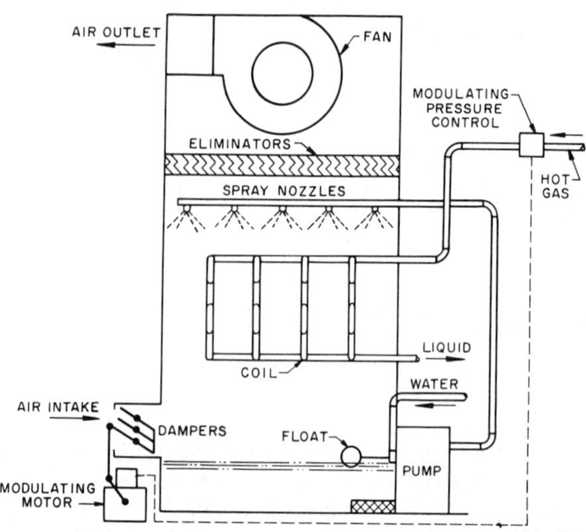

Fig. 38 Head Pressure Control for Evaporative Condenser (Air Intake Modulation)

be those using thermostatic expansion valves. One drawback of dampers is the formation of ice on the dampers and linkages.

Figure 39 incorporates an air bypass arrangement for controlling head pressure. A modulating motor, acting in response to a modulating pressure control, positions dampers so that the mixture of recirculated and cold inlet air maintains the desired pressure. In extremely cold weather, most of the air is recirculated.

Air-Cooled Condensers

Methods used for condensing pressure control with air-cooled condensers include (1) cycling fan motor, (2) air throttling or bypassing, (3) coil flooding, and (4) fan motor speed control.

The first two methods are described in the section on evaporative condensers. The third method holds the condensing pressure up by backing liquid refrigerant up in the coil to cut down on effective condensing surface.

When the head pressure drops below the setting of the modulating control valve, it opens, allowing discharge gas to enter the liquid drain line. This restricts liquid refrigerant drainage and causes the condenser to flood enough to maintain the condenser and receiver pressure at the control valve setting. A pressure difference must be available across the valve to open it. Although the condenser would impose sufficient pressure drop at full load, this may practically disappear at partial loading. Therefore, a positive restriction must be placed parallel with condenser and the control valve. Systems using this type of control require extra refrigerant charge.

In multiple fan air-cooled condensers, it is common to cycle fans off down to one fan, then to apply air throttling to that section or modulate the fan motor speed. Consult the manufacturer before using this method, as not all condensers are properly circuited for it.

Using ambient temperature change (rather than condensing pressure) to modulate air-cooled condenser capacity prevents rapid cycling of condenser capacity. A disadvantage of this method is that the condensing pressure is not closely controlled.

KEEPING LIQUID FROM CRANKCASE DURING OFF CYCLES

The control of reciprocating compressors should prevent excessive accumulation of liquid refrigerant in the crankcase during off cycles. Any one of the following control methods will accomplish this.

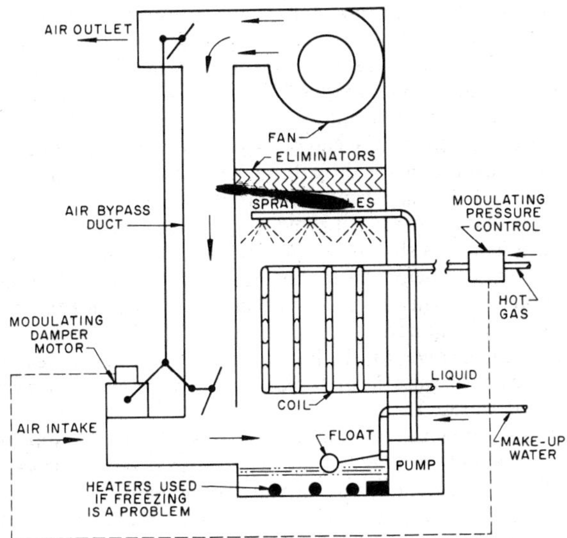

Fig. 39 Pressure for Evaporative Condenser (Air Bypass Modulation)

Automatic Pumpdown Control (Direct-Expansion Air-Cooling Systems)

The most effective way of keeping liquid out of the crankcase during system shutdown is to operate the compressor on automatic pumpdown control. The recommended arrangement involves the following devices and provisions:

- A liquid line solenoid valve in the main liquid line or in the branch to each evaporator
- Compressor operation through a low-pressure cutout providing for pumpdown whenever this device closes, regardless of whether or not the balance of the system is operating
- Electrical interlock of the liquid solenoid valve with the evaporator fan, so that the refrigerant flow will be stopped when the fan is out of operation
- Electrical interlock of the refrigerant solenoid valve with the safety devices (such as the high-pressure cutout, oil safety switch, and motor overloads), so that the refrigerant solenoid valve closes when the compressor stops, through the action of these safety devices
- Low-pressure control settings such that the cut-in point will correspond to a saturated refrigerant temperature lower than any expected compressor ambient air temperature. If the cut-in setting of the low-pressure switch is higher than this, liquid refrigerant can accumulate and condense in the crankcase at a pressure corresponding to the ambient temperature. In this case, the crankcase pressure would not rise high enough to reach the cut-in point and effective automatic pumpdown would not be obtained.

Crankcase Oil Heater (Direct-Expansion Systems)

A crankcase oil heater with or without single (nonrecycling) pumpout at the end of each operating cycle does not keep liquid refrigerant out of the crankcase as effectively as automatic pumpdown control, but many compressors equalize too fast after stopping automatic pumpdown control. Crankcase oil heaters maintain the crankcase oil at a temperature higher than that of other parts of the system, minimizing the absorption of the refrigerant by the oil.

Operation with the single pumpout arrangement is as follows. Whenever the temperature control device opens the circuit, or the manual control switch is opened for shutdown purposes, the crankcase heater is energized, and the compressor keeps running until it cuts off on the low-pressure switch. Because the crankcase heater remains energized during the complete off cycle, it is important that a continuous live circuit be available to the heater during the off time. The compressor cannot start again until the temperature control device or manual control switch closes, regardless of the position of the low-pressure switch.

This control method requires:

- A liquid line solenoid valve in the main liquid line or in the branch to each evaporator
- Using a relay or the maintained contact of the compressor motor auxiliary switch to obtain a single pumpout operation before stopping the compressor
- A relay or auxiliary starter contact to energize the crankcase heater during the compressor off cycle and deenergize it during the compressor on cycle
- Electrical interlock of the refrigerant solenoid valve with the evaporator fan so that the refrigerant flow will be stopped when the fan is out of operation
- Electrical interlock of the refrigerant solenoid valve with the safety devices (such as the high-pressure cutout, oil safety switch, and motor overloads) so that the refrigerant flow valve will close when the compressor stops, through the action of one of these safety devices

Control for Direct-Expansion Water Chillers

Automatic pumpdown control is not desirable for direct-expansion water chillers because freezing is possible if excessive cycling occurs. A crankcase heater is the best solution, with a solenoid valve in the liquid line that closes when the compressor stops.

Effect of Short Operating Cycle

With reciprocating compressors, oil will leave the crankcase at an accelerated rate immediately after starting. Therefore, each start should be followed by a sufficiently long operating period to permit the regain of the oil level. Controllers used for compressors should not produce short-cycling of the compressor. Refer to the compressor manufacturer's literature for guidelines on maximum or minimum cycles for a specified period.

HOT-GAS BYPASS ARRANGEMENTS

Most large reciprocating compressors are equipped with unloaders that allow the compressor to start with most of its cylinders unloaded. However, it may be necessary to further unload the compressor to (1) reduce starting torque requirements so that the compressor can be started both with low starting torque prime movers and on low current taps of reduced voltage starters and (2) permit capacity control down to 0% load conditions without stopping the compressor.

Full (100%) Unloading for Starting

Starting the compressor without load can be done with a manual or automatic valve in a bypass line between the hot-gas and suction lines at the compressor.

To prevent overheating, this valve is open only during the starting period and closed after the compressor is up to full speed and full voltage is applied to the motor terminals.

In the control sequence, the unloading bypass valve is energized on demand of the control calling for compressor operation, equalizing pressures across the compressor. After an adequate time delay, a timing relay closes a pair of normally open contacts to start the compressor. After a further time delay, a pair of normally closed timing relay contacts open, deenergizing the bypass valve.

Full (100%) Unloading for Capacity Control

Where full unloading is required for capacity control, hot-gas bypass arrangements can be used in ways that will not overheat the compressor. In using these arrangements, hot gas should not be bypassed until after the last unloading step.

A hot-gas bypass should (1) give acceptable regulation throughout the range of loads, (2) not cause excessive superheating of the suction gas, (3) not cause any refrigerant overfeed to the compressor, and (4) maintain an oil return to the compressor.

A hot-gas bypass for capacity control is an artifical loading device that maintains a minimum evaporating pressure during continuous compressor operation, regardless of the evaporator load. This is usually done by an automatic or manual pressure-reducing valve that establishes a constant pressure on the downstream side.

Four of the more common methods of using hot-gas bypass are shown in Figure 40. Figure 40A illustrates the simplest type of hot-gas bypass. It will dangerously overheat the compressor if used for protracted periods of time. Figure 40B shows the use of hot-gas bypass to the exit of the evaporator. The expansion valve bulb should be placed at least 5 ft downstream from the bypass point of entrance, and preferably further, to assure good mixing.

In Figure 40D, the hot-gas bypass enters after the evaporator thermostatic expansion valve bulb. Another thermostatic expansion valve supplies liquid directly to the bypass line for desuperheating purposes. It is always important to install the hot-gas bypass far enough back in the system to maintain sufficient gas velocities in suction risers and other components to assure oil return at any evaporator loading.

Figure 40C shows the most satisfactory hot-gas bypass arrangement. Here the bypass is connected into the low side between the expansion valve and the entrance to the evaporator. If a distributor is used, the gas enters between the expansion valve and distributor. Refrigerant distributors are commercially available with side inlet connections that can be used for hot-gas bypass duty to a certain extent. Pressure drop through the distributor tubes must be evaluated to determine how much gas bypassing can be effected. This arrangement provides good oil return.

Solenoid valves should be used before the constant pressure bypass valve and before the termal expansion valve used for liquid injection desuperheating, so that these devices cannot function until they are required.

The control valves for the hot gas should be close to the main discharge line because the line preceding the valve usually fills with liquid when closed.

The hot-gas bypass line should be sized so that its pressure loss is only a small percentage of the pressure drop across the valve. Usually, it is the same size as the valve connections. When sizing the valve, consult a control valve manufacturer to determine the minimum compressor capacity that must be offset, the refrigerant used, the condensing pressure, and the suction pressure.

When unloading (Figure 40C), the head pressure control requirements increase considerably, because the only heat delivered to the condenser is that caused by the motor power delivered to the compressor. The discharge pressure should be kept high

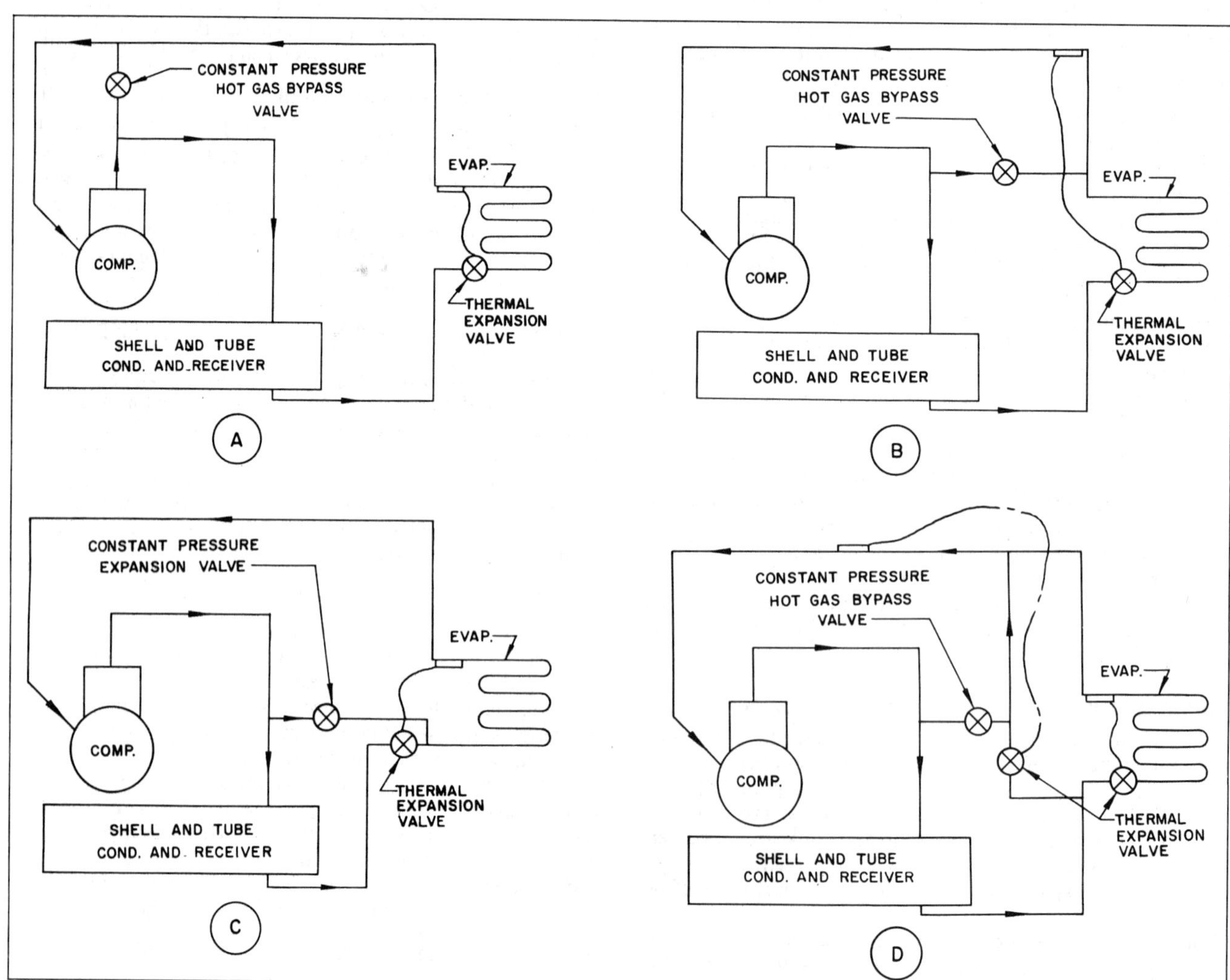

Fig. 40 Hot-Gas Bypass Arrangements

enough so that the hot gas bypass valve can deliver gas at the required rate. The condenser head pressure control must be capable of meeting this condition.

Safety Requirements

ASHRAE *Standard* 15-1992, Safety Code for Mechanical Refrigeration, and ASME B31.5 (1987 and 1991 Addendum), Refrigeration Piping, should be used as guides for safe practice, since they are the basis of most municipal and state codes. However, some ordinances require heavier piping and other features. The designer should know the specific requirements of the installation site. Only A106 Grade A or Grade B or A53 Grade A or B should be considered for steel refrigerant piping.

REFERENCES

Alofs, D.J., M.M. Hasan, and H.J. Sauer, Jr. 1990. Influence of oil on pressure drop in refrigerant compressor suction lines. ASHRAE *Transactions* 96:1.

ASHRAE.1992. Safety code for mechanical refrigeration. *Standard* 15.

ASME. 1987 and 1991 Addendum. *Standard* B31.5. Refrigeration Piping. American Society of Mechanical Engineers, New York.

Atwood, T. 1990. Pipe sizing and pressure drop calculations for HFC-134a. ASHRAE *Journal* 32(4):62-66.

Colebrook, D.F. 1938, 1939. Turbulent flow in pipes. *Journal of the Institute of Engineers* 11.

Cooper, W.D. 1971. Influence of oil-refrigerant relationships on oil return. ASHRAE Symposium Bulletin PH71(2):6-10.

Jacobs, M.L., F.C. Scheideman, F.C. Kazem, and N.A. Macken. 1976. Oil transport by refrigerant vapor. ASHRAE *Transactions* 81(2):318-29.

Keating, E.L. and R.A. Matula. 1969. Correlation and prediction of viscosity and thermal conductivity of vapor refrigerants. ASHRAE *Transactions* 75(1).

Stoecker, W.F. 1984. Selecting the size of pipes carrying hot gas to defrosted evaporators. *International Journal of Refrigeration* 7(4):225-228.

Stoecker, W.F., J.J. Lux, Jr., and R.J. Kooy. 1983. Energy considerations in hot gas defrosting of industrial refrigeration coils. ASHRAE *Transactions* 89(2):549-73.

Timm, M.L. 1991. An improved method for calculating refrigerant line pressure drops. ASHRAE *Transactions* 97(1):194-203.

Wile, D.D. 1977. *Refrigerant line sizing*. ASHRAE.

CHAPTER 3

SYSTEM PRACTICES FOR AMMONIA REFRIGERANT

REFRIGERATION is the process of moving heat from one location to another by use of refrigerant in a closed refrigeration cycle. Oil management, gas and liquid separation, subcooling, superheating, and piping of refrigerant liquid, gas, and two-phase flow are all part of refrigeration. Ammonia refrigeration applies to air conditioning, commercial refrigeration, and industrial refrigeration.

Engineered refrigeration systems require engineering analysis of all components to select balanced elements to accomplish the desired results.

Human comfort air-conditioning systems usually work over a small, well-defined range of evaporating temperatures, typically 35 to 55 °F. Because of these limited temperatures and the large market for air-conditioning equipment, manufacturers can standardize their equipment and package it for volume production at lower cost. This is particularly true in water chilling equipment.

Conversely, custom-engineered ammonia refrigeration systems often have design conditions that span a wide range of evaporating and condensing temperatures, *e.g.*, (1) a food freezing plant operating at temperatures from +50 to −50 °F; (2) a candy storage requiring 60 °F dry bulb with precise humidity control; (3) a beef chill room at 28 to 30 °F with high humidity; (4) a distribution warehouse requiring multiple temperatures for storage of ice cream, frozen food, meat, produce, docks; or (5) a chemical process requiring multiple temperatures ranging from +60 to −60 °F. Ammonia (R-717) is the choice for many industrial refrigeration systems. Typical refrigeration systems include some of the following:

- Year-round operation regardless of outdoor ambient conditions
- Possible wide load variations (0 to 100% capacity) during short periods without seriously disrupting the required temperature levels
- Frost control for continuous performance applications
- Oil management
- A wide choice of heat exchange methods, *e.g.*, direct expansion, liquid overfeed, or flooded feed of the refrigerants, and the use of secondary coolants (see Chapter 4)
- System efficiency, maintainability, and operating simplicity
- Operating pressures and pressure ratios that might require multistaging
- Special electrical requirements

For safety and minimum design criteria for ammonia as the refrigerant, refer to the *Safety Code for Mechanical Refrigeration* (ANSI/ASHRAE 15-1992); IIAR *Bulletin* 109, Minimum Safety Criteria for a Safe Ammonia Refrigeration System (1988); IIAR/ANSI 2-1984, American National Standard for Equipment, Design, and Installation of Ammonia Mechanical Refrigeration Systems; and applicable state and local codes.

REFRIGERATION LOAD

An analysis of the refrigeration load is vital to proper equipment selection and satisfactory system performance. All load components and often the time profile of the load components must be considered.

Typical factors to consider include:

1. Heat leakage of either latent or sensible heat into the space or product. This flow is a function of the temperature difference and insulating efficiency of the container walls, including insulation and any heat introduced by ventilation or infiltration.
2. Product load, which is the heat that must be removed to bring the product to the desired state. This load includes both sensible and latent heat associated with a change of state.
3. Internal sensible load can include heat from motors (including evaporator fan motors), lights, and other heat sources.

A plot of load versus time is important in selecting equipment that will function well at minimum load as well as at peak load. Minimum load conditions are often overlooked in design so that severe operating problems such as inefficiency, short-cycling, insufficient hot gas for defrost, or other adverse conditions may arise. Often small- and large-capacity components must be selected and properly controlled to match operating conditions. Industrial processes are particularly prone to abrupt interruptions, going from 100 to 0% load and then back to 100% over a short time. Improperly designed systems will not function well under these conditions.

Maximum and minimum ambient temperatures during varying loads must be considered in the design of a system with capacity reduction features, particularly when selecting a condenser.

In food freezing or storage, the effect of frost accumulation on heat transfer surface efficiency must be considered and sufficient defrost time allocated. If exact conditions have to be continuously maintained, liquid absorption coils, (see Chapter 22 of the 1992 ASHRAE *Handbook—Systems and Equipment*), multiple evaporators may have to be used and alternated to permit defrosting. When selecting evaporators for air cooling (as in cold storage warehouses), the defrost system should be based on the following tabulation of cooling hours:

Defrost Method	Maximum Hours of Operation per Day
Air (above 34 °F)	16 to 18
Water or brine	20 to 22
Hot gas	20 to 22
Electrical	20 to 22

The preparation of this chapter is assigned to TC 10.3, Refrigerant Piping, Controls, and Accessories.

This table is based on average frost conditions. For higher frost conditions and lower evaporating temperatures, more time must be allowed for defrosting.

During defrost, heat is absorbed by the frost, the metal of the air-cooling unit, and the room air. When the system returns to cooling, all of the heat added to the room (except that in the condensate that was drained away) must be added to the load because it must also be removed.

SYSTEM SELECTION

In selecting an engineered ammonia (R-717) refrigeration system, several design decisions must be considered, including whether to use (1) single-stage compression, (2) economized compression, (3) compound compression, (4) direct expansion feed, (5) flooded feed, (6) liquid recirculation feed, and (7) secondary coolants.

Single-Stage Systems

The basic single-stage system consists of evaporator(s), a compressor, a condenser, a refrigerant receiver (if used), and a refrigerant control device (expansion valve, float, etc.). Chapter 1 of the 1993 ASHRAE *Handbook—Fundamentals* discusses the compression refrigeration cycle.

Economized Systems

Economized systems are frequently used with rotary screw compressors. Figure 1 shows an arrangement of the basic components. Subcooling the liquid refrigerant before it reaches the evaporator reduces its enthalpy, resulting in a higher net refrigerating effect. Economizing is beneficial since the vapor generated during the subcooling is injected into the compressor partway through its compression cycle and must be compressed only from the economizer port pressure (which is higher than suction pressure) to the discharge pressure. This produces additional refrigerating capacity with less increase in unit energy input. Economizing is most beneficial at high pressure ratios. Under most conditions, economizing can provide operating efficiencies that approach that of two-staged compound systems, but with much less complexity and simpler maintenance.

Economized systems for variable loads should be selected carefully. At approximately 75% capacity, most screw compressors revert to single-stage performance as the slide valve moves such that the economizer port is open to the compressor suction area.

A flash-type economizer, which is somewhat more efficient, may often be used instead of the shell-and-coil economizer (Figure 1). However, ammonia liquid delivery pressure is reduced to economizer pressure.

Compound Systems

Compound compression systems compress the gas from the evaporator to the condenser in several stages. They are used to produce temperatures of −15 °F and below. This is not economical with single-stage compression.

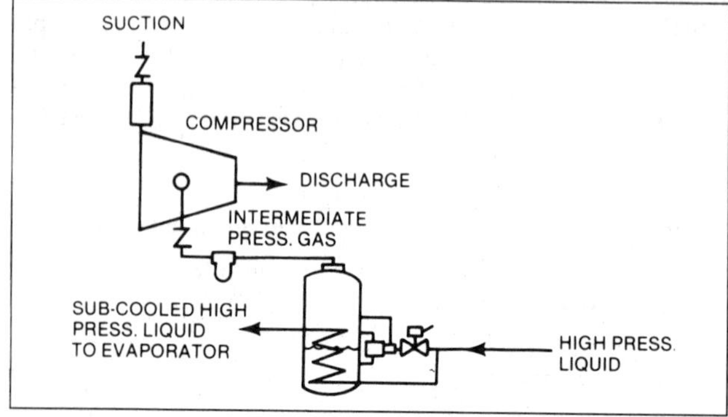

Fig. 1 Shell-and-Coil Economizer Arrangement

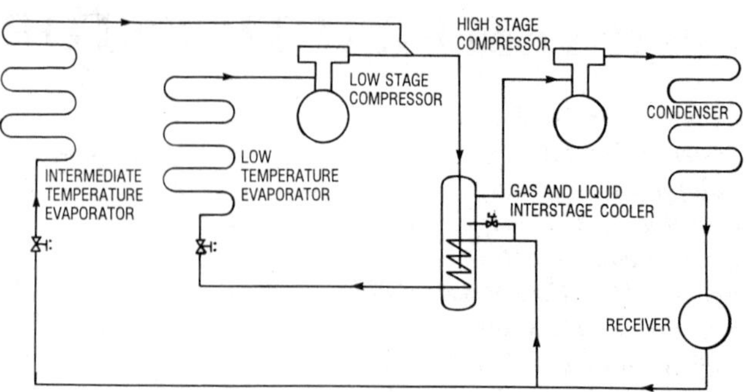

Fig. 2 Two-Stage System with High- and Low-Temperature Loads

Single-stage reciprocating compression systems are generally limited to between 5 and 10 psig suction pressure. With oil-injected economized rotary screw compressors, where the discharge temperatures are lower because of the oil cooling, the low-suction temperature limit is about −40 °F, but efficiency is very low. Two-stage systems are used down to about −70 or −80 °F evaporator temperatures. Below this temperature, three-stage systems should be considered.

Two-stage systems consist of one or more compressors that operate at low suction pressure and discharge at intermediate pressure and have one or more compressors that operate at intermediate pressure and discharge to the condenser (Figure 2).

Where either single- or two-stage compression systems can be used, two-stage systems require less power and have lower operating costs, but can have a higher initial equipment cost.

EQUIPMENT

Compressors

Various compressors available for single- and multistage applications include the following:

Reciprocating
  Single-stage (low-stage or high-stage)
  Internally compounded
Rotary vane
Rotary screw (low-stage or high-stage, with or without economizing)

The reciprocating compressor is the most common compressor used in small, 100 hp or less, single-staged or multistaged systems. The screw compressor is the predominant compressor above 100 hp, both in single- and multistaged systems. Various combinations of compressors may be used in multistage systems. Rotary vane and screw compressors are frequently used for the low-pressure stage, where large volumes of gas must be moved. The high-pressure stage may be a reciprocating or screw compressor.

Many factors must be considered in selecting a compressor, including:

- System size and capacity requirements.
- Location, such as indoor or outdoor installation at ground level or on the roof.
- Equipment noise.
- Part- or full-load operation.
- Winter and summer operation.
- Pulldown time required to reduce the temperature to desired conditions for either initial or normal operation. The temperature must be pulled down frequently for some applications for a process load, while a large cold storage warehouse may require pulldown only once in its lifetime.

Oil Cooling. When a reciprocating compressor requires oil cooling, an external heat exchanger using a refrigerant or secondary cooling is usually added.

Oil for screw compressors is usually cooled by one of three methods: (1) direct injection of the liquid refrigerant into the compressor to cool the discharge gas; (2) external cooling by a heat exchanger, or (3) external cooling using refrigerant (thermosyphon). Thermosyphon oil cooling uses refrigerant from the condenser in a flooded shell-and-tube heat exchanger to cool the oil. This is an efficient oil cooling method as it imposes no system penalties. Direct liquid refrigerant injection may shorten compressor life by thinning out the oil; it also reduces compressor efficiency. Water-cooled oil coolers are subject to corrosion and freezing where cooling tower water or water from an evaporative condenser is used. Screw compressor oil cooling can also be provided by dedicating a section of the evaporative condenser coil for cooling an inhibited glycol/water solution. This effective method requires minimal service.

Compressor Drivers. The correct electric motor(s) size(s) for a multistaged system is determined by the pulldown load. When the final low-stage operating level is −100 °F, the pulldown load can be three times the operating load. Positive displacement reciprocating compressor motors are usually selected for about 150% of operating power requirements for 100% load. The compressor's unloading mechanism can be used to prevent motor overload. Electric motors should not be overloaded, even when a service factor is indicated, to avoid overloading in low-voltage conditions. For screw compressor applications, motors should be sized by adding 10% to the operating power. Screw compressors have built-in unloading mechanisms to prevent motor overload. On any compressor application, care should be taken. The motor should not be oversized because an oversized motor has a lower power factor and lower efficiency at design and reduced loads.

Gasoline, natural gas, propane, or diesel internal combustion engines, or steam turbines, are used when electricity is unavailable, or if the selected energy source is cheaper. Sometimes they are used in combination with electricity to reduce peak demands. The power output of a given size engine can vary as much as 15% depending on the fuel selected.

The standard power rating of an engine is the absolute maximum, not the recommended power available for continuous use. Also, torque characteristics of internal combustion engines and electric motors differ greatly. The proper engine selection is at 75% of its maximum power rating. For longer life, the full-load speed should be at least 10% below maximum engine speed.

Internal combustion engines, in some cases, can reduce operating cost below that for electric motors. Disadvantages include (1) higher initial cost of the engine, (2) additional safety and starting controls, (3) higher noise levels, (4) larger space requirements, (5) air pollution, (6) requirement for heat dissipation, (7) higher maintenance costs, and (8) higher levels of vibration than with electric motors. A torsional analysis must be made to determine the proper coupling if engine drives are chosen.

Steam turbine drives for refrigerant compressors are usually limited to very large installations where steam is already available at moderate to high pressure. In all cases, torsional analysis is required to determine what coupling must be used to dampen out any pulsations transmitted from the compressor. For optimum efficiency, a turbine should operate at a high speed that must be geared down for reciprocating and possibly screw compressors. Since the gear reducer as well as the turbine cannot tolerate a pulsating backlash from the driven end, torsional analysis and special couplings are essential.

Advantages of turbines include variable speed for capacity control and low operating and maintenance costs. Disadvantages include higher initial costs and possible high noise levels. The turbine must be started manually to bring the turbine housing up to temperature slowly and to prevent excess condensate from entering the turbine. Chapter 41 of the 1992 ASHRAE *Handbook—Systems and Equipment* has further information on compressor drives.

Condensers

Condensers should be selected on the basis of total heat rejection at maximum load. Often the heat rejected at the start of pulldown is several times the amount rejected at normal, low-temperature operating conditions. Some means, such as compressor unloading, can be used to limit the maximum amount of heat rejected during pulldown. If the condenser is not sized for pulldown conditions and compressor capacity cannot be limited during this period, condensing pressure might increase enough to shut down the system.

Evaporators

Several types of evaporators are used in ammonia refrigeration systems. Fan-coil, direct-expansion evaporators can be used on ammonia systems, but they are not generally recommended unless the suction temperature is 0 °F or higher. This is due in part to the relative inefficiency of the direct expansion coil, but more importantly, the low mass flow rate of ammonia is difficult to feed uniformly as a liquid to the coil. Instead, ammonia fan coil units designed for recirculation (overfeed) systems are preferred. Typically, in this type of system, high-pressure ammonia from the system high stage flashes into a large vessel at the evaporator pressure from which it is pumped to the evaporators at an overfeed rate of 2.5 to 1 to 4 to 1. This type of system is a standard and is very efficient. See Chapter 1.

Flooded shell-and-tube evaporators are often used in ammonia systems where indirect or secondary cooling fluids such as water, brine, or glycol must be cooled.

Some problems that can become more acute in low-temperature systems than in high-temperature systems include oil transport properties, loss of capacity caused by static head from the depth of the pool of liquid refrigerant in the evaporator, deterioration of refrigerant boiling heat transfer coefficients due to oil logging, and higher specific volumes for the vapor.

The effect of pressure losses in the evaporator and suction piping is more acute in low-temperature systems because of the large change in saturation temperatures and specific volume in relation to pressure changes at these conditions. Systems that operate near zero absolute pressure are particularly affected by pressure loss.

The depth of the pool of boiling refrigerant in a flooded evaporator causes a liquid pressure that is exerted on the lower part of the heat transfer surface. Therefore, the saturation temperature at this surface is higher than the pressure in the suction line, which is not affected by the pressure. This temperature gradient must be considered when designing the evaporator.

Spray-type, shell-and-tube evaporators, while not commonly used, offer certain advantages. In this design, the liquid depth penalty for the evaporator can be eliminated since the pool of liquid is below the heat transfer surface. A refrigerant pump sprays the liquid over the surface. The pump energy is an additional heat load to the system, and more refrigerant must be used to provide the net positive suction head (NPSH) required by the pump. The pump is also an additional item that must be maintained. This evaporator design also has the design benefit of reducing the refrigerant charge requirement as compared to a flooded design (see Chapter 1).

Vessels

High Pressure Receivers. Industrial systems generally incorporate a central high-pressure refrigerant receiver, which serves as the primary source of refrigerant storage in the system. It handles the refrigerant volume variations between the condenser and the system's low side during operation and pumpdowns for repairs or defrost. Ideally, the receiver should be large enough to hold the entire system charge, but this is not generally economical. An analysis of the system should be made to determine the optimum receiver size. Receivers are commonly equalized to the condenser inlet and operate at the same pressure as the condenser. In some systems, the receiver is operated at a pressure between the condensing and highest suction pressure to allow for variations in condensing pressure without affecting the system's feed pressure.

If additional receiver capacity is needed during normal operation, extreme caution should be exercised in the design. Designers usually remove the inadequate receiver and replace it with a larger one rather than install an additional receiver in parallel. This procedure is best because even slight differences in piping pressure or temperature can cause the refrigerant to migrate to one receiver and not to the other.

Smaller auxiliary receivers can be incorporated to serve as sources of high-pressure liquid for compressor injection or thermosyphon, oil cooling, high-temperature evaporators, and so forth.

Intercoolers (Gas and Liquid). An intercooler (subcooler/desuperheater) is the intermediate vessel between the high stage and low stage in a compound system. One purpose of the intercooler is to cool the discharge gas of the low-stage compressor to prevent overheating the high-stage compressor. This can be done by bubbling the discharge gas from the low-stage compressor through a bath of liquid refrigerant or by mixing liquid normally entering the intermediate vessel with the discharge gas as it enters above the liquid level. The heat removed from the discharge gas is absorbed by the evaporation of part of the liquid and eventually passes through the high-stage compressor to the condenser. Disbursing the discharge gas below a level of liquid refrigerant will separate out any lubricant carryover from the low-stage compressor. If the volume of liquid in the intercooler is to be used for other purposes, such as liquid makeup or feed to the low stage, periodic lubricant removal is important.

Another purpose of the intercooler is to lower the temperature of the liquid being used in the system low stage. Lowering the refrigerant temperature increases the refrigeration effect and reduces the low-stage compressor's required displacement, thus reducing its operating cost.

Two types of intercoolers for compound (two-stage) compression systems are the shell-and-coil and flash-type intercoolers. Figure 3 depicts a shell-and-coil intercooler incorporating an internal pipe coil for subcooling the high-pressure liquid prior to being fed to the low stage of the system. Typically, the coil will subcool the liquid to within 10°F intermediate temperature.

Vertical shell-and-coil intercoolers with float valve feed perform well in many applications using ammonia as the refrigerant. The vessel must be sized properly to separate the liquid from the vapor that is returning to the high-stage compressor. The superheated gas inlet pipe should extend below the liquid level and have perforations or slots to distribute the gas evenly in small bubbles. The addition of a perforated baffle across the area of the vessel slightly below the liquid level protects against violent surging. A float switch that shuts down the high-stage compressor when the liquid level gets too high should always be used with this type of intercooler in case the feed valve fails to control properly.

The flash-type intercooler is similar in design to the shell-and-coil intercooler with the exception of the coil. The high-pressure liquid is flash cooled to the intermediate temperature. Caution should be exercised in selecting a flash type intercooler, since all the high-pressure liquid is flashed to intermediate pressure. While colder than that of the shell-and-coil intercooler, the liquid in the flash-type intercooler is not subcooled and is susceptible to flashing due to system pressure drop. Two-phase liquid feed to control valves may cause premature failure due to the wire drawing effect of the liquid/vapor.

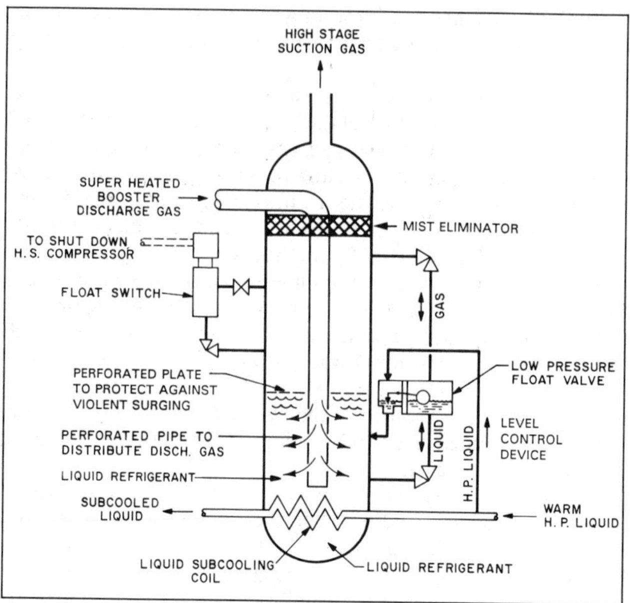

Fig. 3 Intercooler

Figure 4 shows a vertical shell-and-coil intercooler as piped into the system. The liquid level is maintained in the intercooler by a float that controls the solenoid valve feeding liquid into the shell side of the intercooler. Gas from the first-stage compressor enters the lower section of the intercooler, is distributed by a perforated plate, and is then cooled to the saturation temperature corresponding to intermediate pressure.

When sizing any intercooler, the designer must consider (1) the low-stage compressor capacity, (2) the vapor desuperheating, liquid makeup requirements for the subcooling coil load, or vapor cooling load associated with the flash-type intercooler, and (3) any high-stage side loading. Also, the volume required for normal liquid levels, liquid surging from high-stage evaporators, feed valve malfunctions, and vapor/liquid must be analyzed.

Accessories necessary are the liquid level control device and the high-level float switch. While not absolutely necessary, an auxiliary oil pot should also be considered.

Suction Accumulator. A suction accumulator (also known as a knockout drum, suction trap, pump receiver, recirculator, etc.) prevents any liquid from entering the suction of the compressor, whether on the high stage or low stage of the system. Both vertical and horizontal vessels can be incorporated. Baffling and mist eliminator pads can enhance liquid separation.

Suction accumulators, especially those not intentionally maintaining a level of liquid, should have a means of removing any buildup of ammonia liquid. Gas boil-out coils or electric heating elements are costly and inefficient.

A liquid boil-out coil (Figure 5), while one of the more common and simplest means of liquid removal, has some drawbacks. Generally, the warm liquid flowing through the coil is the source of liquid that is being boiled off. Liquid transfer pumps, gas-powered transfer systems, or basic pressure differentials are a more positive means of removing the liquid (Figures 6 and 7).

Accessories should include a high-level float switch for compressor protection along with additional pump or transfer system controls.

Vertical Suction Trap and Pump. Figure 8 shows the piping of a vertical suction trap that uses a high-head ammonia pump to transfer liquid from the system's low-pressure side to the high-pressure receiver. Float switches piped on a float column on the side of the trap can start and stop the liquid ammonia pump, sound an alarm in case of excess liquid and, sometimes, stop the compressors.

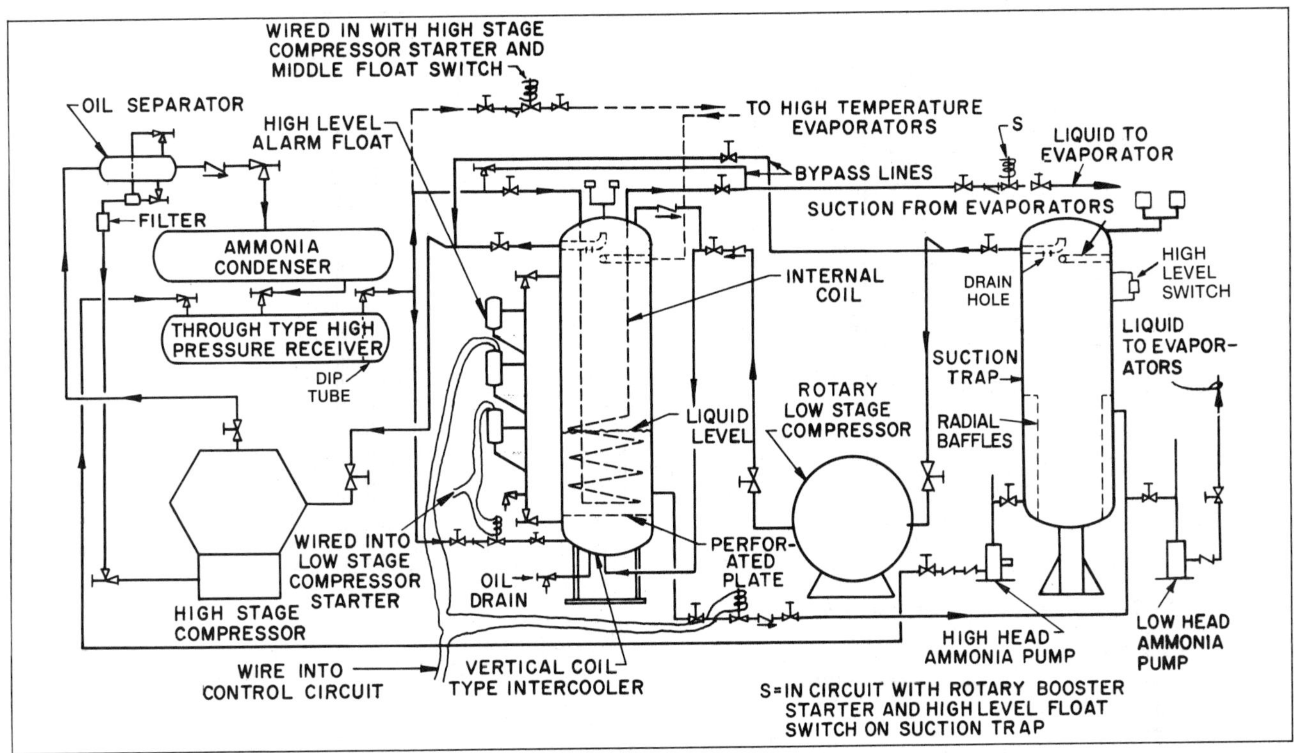

Fig. 4 Arrangement for Compound System with Vertical Intercooler and Suction Trap

When the liquid level in the suction trap reaches the setting of the middle float switch, the liquid ammonia pump starts and reduces the liquid level to the setting of the lower float switch, which stops the liquid ammonia pump. A check valve in the discharge line of the ammonia pump prevents gas and liquid from flowing backward through the pump when it is not in operation. Depending on the type of check valve used, some installations have two valves in a series as an extra precaution against pump "back spin."

Compressor controls adequately designed for starting, stopping, and capacity reduction result in minimal agitation, which aids in separating the vapor and liquid in the suction trap. Increasing the compressor capacity slowly and in small increments reduces boiling liquid in the trap, caused by the refrigeration load of cooling the refrigerant and metal mass of the trap. If another compressor is started when plant suction pressure increases, it should be brought on line slowly to prevent a sudden pressure change in the suction trap.

A high level of liquid in a suction trap should activate an alarm or stop the compressors. Although eliminating the cause is the most effective way to reduce a high level of excess surging liquid, a more immediate solution is to stop part of the compression system and raise the plant suction pressure slightly. Continuing high levels indicate insufficient pump capacity or suction trap volume.

Liquid Level Indicators. Liquid level can indicate either visual or electronic sensors or a combination of the two. Visual methods include individual circular reflex level indicators (bull's-eyes) mounted on a pipe column or stand-alone linear reflex glass assemblies (Figure 9). When operating at temperatures below the frost point, transparent plastic frost shields covering the reflex surfaces are necessary. Also, the pipe column must be insulated, especially when control devices are attached.

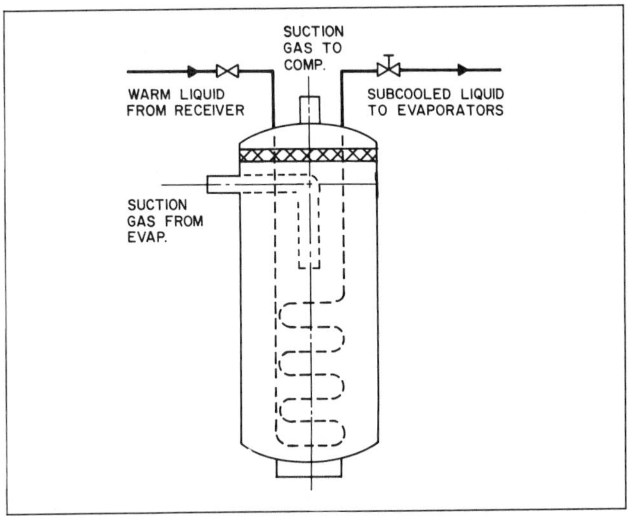

Fig. 5 Suction Accumulator with Warm Liquid Coil

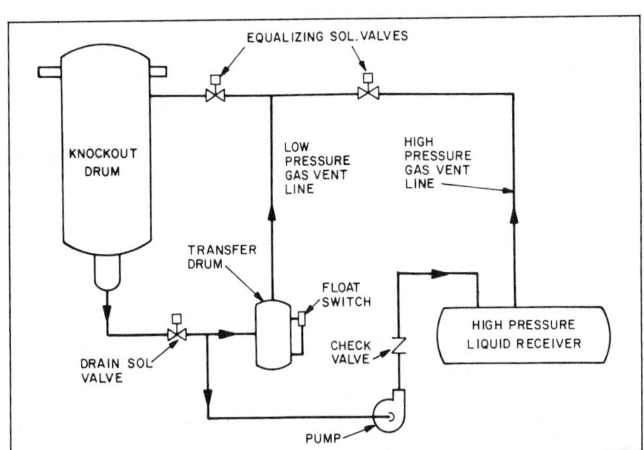

Fig. 6 Equalized Pressure Pump Transfer System

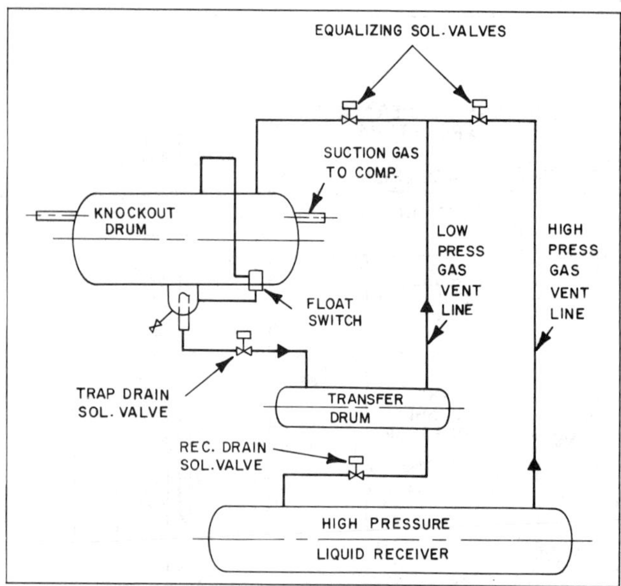

Fig. 7 Gravity Transfer System

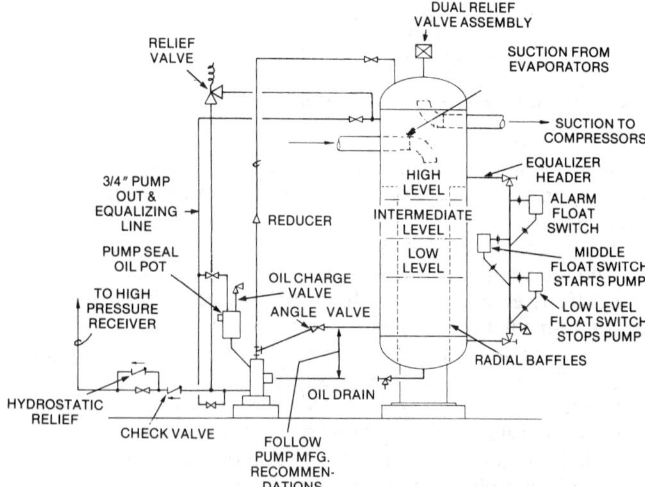

Fig. 8 Piping for Vertical Suction Trap and High-Head Pump

Electronic level sensors can continuously monitor the liquid level. Digital or graphic displays of the liquid level can be locally or remotely monitored (Figure 10).

Adequate isolation valves for the level indicators should be employed. High-temperature glass tube-type indicators should incorporate stop check or excess flow valves for isolation and safety.

Purge Units. A noncondensable gas separator (purge unit) is useful in most plants, especially when suction pressure is below atmospheric pressure. Purge units on ammonia systems are piped to carry noncondensables (air) from the receiver and condenser to the purger, as shown in Figure 11. High-pressure liquid expands through a coil in the purge unit, providing a cold spot in the purge drum. The suction from the coil should be taken to one of the low-temperature suction mains. Ammonia vapor and noncondensable gas are drawn into the purge drum, and the ammonia condenses on the cold surface. When the drum fills with air and other noncondensables, a float valve within the purger opens and permits them to leave the drum and pass into the open water bottle. Purge units are available for automatic operation.

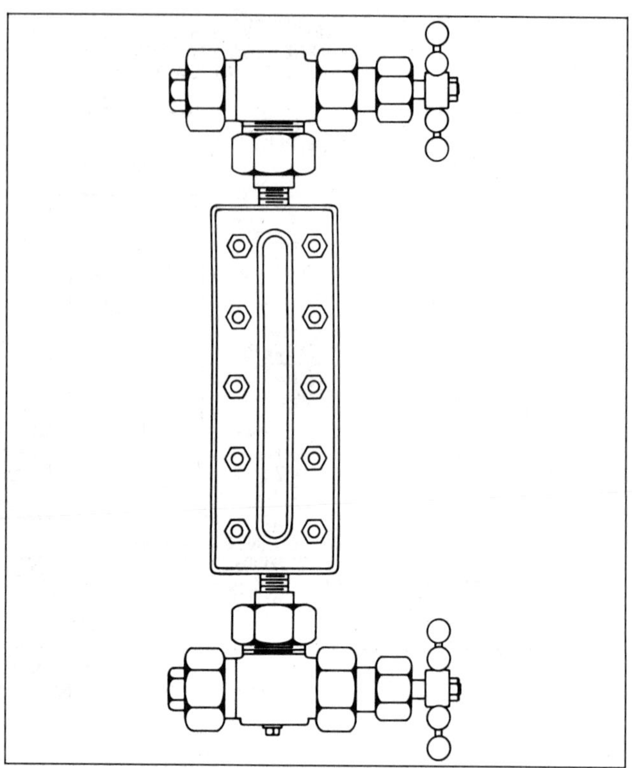

Fig. 9 Gage Glass Assembly for Ammonia

Lubricant Management

Most lubricants are immiscible in ammonia (R-717) and separate out of the liquid easily when flow velocity is low or when temperatures are lowered. Lubricants can normally be easily drained from the system. If the temperature is very low and the lubricant is not properly selected, it becomes a gummy mass that prevents refrigerant controls from functioning, blocks flow passages, and fouls the heat transfer surfaces. Proper lubricant selection and management is often the key to a properly functioning system.

In two-stage systems, proper design usually calls for lubricant separators on both the high and low-stage compressors. A properly designed coalescing separator can remove almost all the lubricant that is in droplet or aerosol form. Oil that reaches its saturation vapor pressure and becomes a vapor cannot be removed by separator. Separators equipped with some means of cooling the

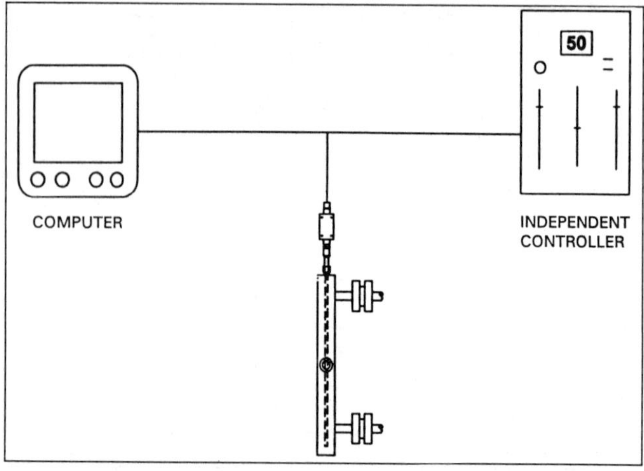

Fig. 10 Electronic Liquid Level Control

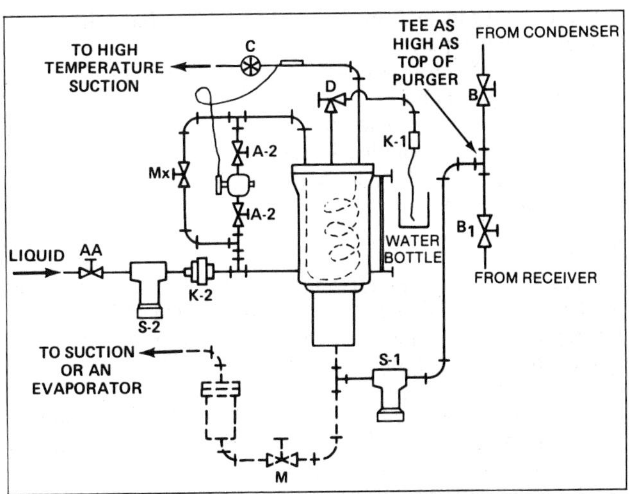

Fig. 11 Purge Unit and Piping for Noncondensable Gas

discharge gas will condense much of the vapor for consequent separation. Selection of lubricants that have very low vapor pressures below 180°F can minimize carryover to 2 or 3 parts per million (ppm). Care must be exercised, however, to be sure refrigerant is not condensed and fed back into the compressor or separator where it can lower lubricity and cause compressor damage.

In general, direct expansion and liquid overfeed system evaporators have fewer lubricant return problems than do flooded system evaporators since refrigerant flows continuously at good velocities to sweep lubricant from the evaporator. Low-temperature systems using hot-gas defrost can also be designed to sweep lubricant out of the circuit each time the system defrosts. This reduces the possibility of coating the evaporator surface and hindering heat transfer.

Flooded evaporators can promote lubricant buildup in the evaporator charge since they may only return refrigerant vapor back to the system. In ammonia systems, the lubricant is simply drained from the surge drum. At low temperatures, this procedure is difficult if the lubricant is not selected with a pour point below the evaporator temperature.

Lubricant Removal from Ammonia Systems. Most lubricants are miscible with liquid ammonia only in very small proportions. The proportion decreases with the temperature, and the lubricant separates with this decrease. The evaporation of ammonia increases the lubricant ratio, causing more lubricant to separate. The increased density causes the lubricant (saturated with ammonia at the existing pressure) to form a separate layer below the ammonia liquid.

Unless lubricant is removed periodically or continuously from the point where it collects, it can cover the heat transfer surface in the evaporator, reducing performance. If gage lines or branches to level controls are taken from low points (or lubricant is allowed to accumulate), these lines will contain lubricant. The higher lubricant density will be at a lower level than the ammonia liquid. Draining lubricant from a properly located collection point is not difficult unless the temperature is so low that the lubricant does not flow readily. In this case, maintaining the lubricant receiver at a higher temperature may be beneficial. As an alternative, a lubricant with a lower pour point can be selected.

Lubricant in the system is saturated with ammonia at the existing pressure. When the pressure is reduced, the ammonia vapor separates, causing foaming.

Draining lubricant from ammonia systems requires special care. Ammonia in the lubricant foam normally starts to evaporate and the smell of ammonia will be present. Operators should be made aware of this. On systems where lubricant is drained from a still, a spring-loaded drain valve should be installed. This type of valve will close if the valve handle is released.

CONTROLS

Refrigerant flow controls are discussed in Chapter 44. The following precautions are necessary in the application of certain controls in low-temperature systems.

Liquid Feed Control

Many controls available for single-staged, high-temperature systems may be used with some discretion on low-temperature systems. If the liquid level is controlled by a low-side float valve with the float in the chamber where the level is controlled, low pressure and temperatures have no appreciable effect on their operation. External float chambers, however, must be thoroughly insulated to prevent heat influx that might cause boiling and an unstable level affecting the float response. Equalizing lines to external float chambers, particularly the upper line, must be sized generously so that liquid can reach the float chamber, and gas resulting from any evaporation may be returned to the vessel without appreciable pressure loss.

The superheat-controlled expansion (thermostatic) valve is generally used in direct-expansion evaporators. This valve operates on the pressure difference between the bulb, which is responsive to the suction temperature, and the pressure below the diaphragm, which is the actual suction pressure.

The thermostatic expansion valve is designed to maintain a preset superheat in the suction gas. Although the pressure-sensing part of the system responds almost immediately to a change in conditions, the temperature-sensing bulb must overcome thermal inertia before its effect is felt on the power element of the valve. For this reason, when compressor capacity decreases suddenly, the expansion valve may overfeed before the bulb senses the presence of liquid in the suction line and reduces the feed. Therefore, a suction accumulator should be installed on direct-expansion low-temperature systems with multiple expansion valves.

Controlling Load during Pulldown

System transients during pulldown can be managed by controlling compressor capacity. Proper load control reduces the compressor capacity so that the energy requirements stay within the capacities of the motor and the condenser. On larger systems using screw compressors, a current sensing device reads motor amperage and adjusts the capacity control device appropriately. Cylinders can be unloaded on reciprocating compressors for similar control.

Alternatively, a downstream or outlet or crankcase pressure regulator can be installed in the suction line to throttle the suction flow should the pressure exceed a preset limit. This regulator limits the compressor's suction pressure during pulldown. The disadvantage of this device is the extra pressure drop it causes when the system is at the desired operating conditions. To overcome some of this pressure drop, the designer can use external forces to drive the valve, causing it to be held fully open when the pressure is below the maximum allowable. Systems incorporating downstream pressure regulators and compressor unloading must be carefully designed so that the two controls complement each other.

Operation at Varying Loads and Temperatures

Compressor and evaporator capacity controls are similar for multi- and single-stage systems. Control methods include compressor capacity control, hot gas bypass, or evaporator pressure regulators. Low pressure can affect control systems by significantly increasing specific volume of the refrigerant gas and pressure drop. A small pressure reduction can cause a large percentage capacity reduction.

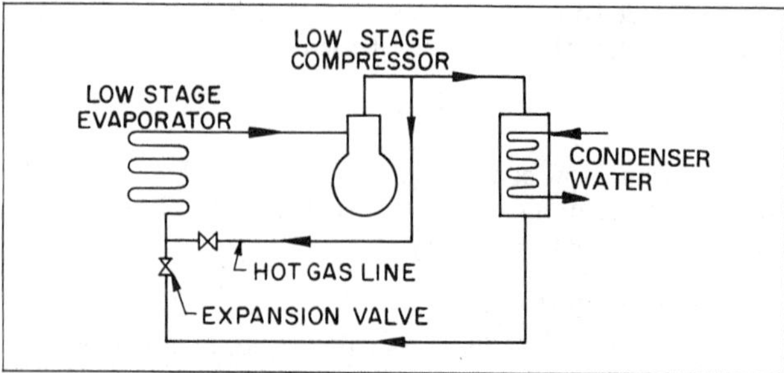

Fig. 12 Hot-Gas Injection Evaporator for Operations at Low Load

System load usually cannot be reduced to near zero, since this would result in little or no flow of gas through the compressor and consequent overheating. Additionally, high pressure ratios would be detrimental to the compressor if it is required to run at very low loads. If the compressor cannot be allowed to cycle off during low load, an acceptable alternative is a hot gas bypass. The high-pressure gas is fed into the low-pressure side of the system through a downstream pressure regulator. The gas should be desuperheated by injecting it at a point in the system where it will be in contact with expanding liquid such as immediately downstream of the liquid feed to the evaporator. Otherwise, extremely high compressor discharge temperatures can result. The artificial load supplied by the high-pressure gas can fill the gap between the actual load and the lowest stable compressor operating capacity. Figure 12 shows such an arrangement.

Electrical and Electronic Control

Microprocessor and computer-based control systems are becoming the norm for control systems on individual compressors as well as for entire system control. Almost all screw compressors use microprocessor control systems to monitor all safety functions and operating conditions. These machines are frequently linked together with a programmable controller or computer for sequencing multiple compressors so that they load and unload in response to system fluctuations in the most economical manner. Programmable controllers are also used to replace multiple defrost time clocks on larger systems for more accurate and economical defrosting. Communications and data logging permit systems to operate at optimum conditions under transient load conditions even when operatiors are not in attendance.

PIPING REQUIREMENTS

The following recommendations are given for ammonia piping. Local codes or ordinances governing ammonia mains should also be complied with.

Recommended Material

Since copper and copper-bearing materials are attacked by ammonia, they are not used in ammonia piping systems. Steel piping, fittings, and valves of the proper pressure rating are suitable for ammonia gas and liquid.

Ammonia piping should conform to ANSI/ASME, B31.5 Code for Pressure Piping, and ANSI/IIAR 2-1992, Equipment Design and Installation of Ammonia Mechanical Refrigeration Systems, which state:

1. Liquid lines 1.5 in. and smaller shall be not less than schedule 80 carbon steel pipe.
2. Liquid lines 2 in. through 6 in. shall be not less than schedule 40 carbon steel pipe.
3. Vapor lines 6 in. and smaller shall be not less than schedule 40 carbon steel pipe.
4. Vapor lines 8 through 12 in. should not be less than schedule 20 carbon steel pipe.
5. Carbon steel piping should be ASTM A-53, type E (electric resistance welded) or type S (seamless), grade A or B; or A-106 (seamless), grade A or B. A-53 grade F is not permitted for ammonia piping.

Fittings

Couplings, elbows, and tees for threaded pipe are for a minimum of 3000 psi design pressure and constructed of forged steel. Fittings for welded pipe should match the type of pipe used; *i.e.*, full weight fittings for full weight pipe and extra heavy fittings for extra heavy pipe.

Tongue and groove or ANSI flanges should be used in ammonia piping. Welded flanges for low-side piping can have a minimum 150 psi design pressure rating. On systems located in high ambients, low-side piping and vessels should be designed for 200 to 225 psig. The high side should be 250 psig if the system is water cooled or evaporative cooled condensing. Use 300 psig minimum for air-cooled designs.

Pipe Joints

Joints between lengths of pipe or between pipe and fittings can be threaded if the pipe size is 1.25 in. or smaller. Sizes of 1.5 in. or larger should be welded. An all-welded piping system is superior.

Threaded Joints. Many sealants and compounds are available for sealing threaded joints. The manufacturer's instructions cover compatibility and application method. Do not use excessive amounts or apply on female threads, because any excess can contaminate the system.

Welded Joints. Pipe should be cut and beveled before welding. Use pipe alignment guides to align the pipe and provide a proper gap between pipe ends so that a full penetration weld is obtained. The weld should be made by a qualified welder, using proper procedures such as the Standard Procedure Specifications for Welding of Pipe, Fittings and Flanges, prepared by the Mechanical Contractors Association of America.

Gasketed Joints. When using flanges, a 0.063-in. fiber gasket should be used. Before tightening flange bolts to valves, controls, or flange unions, properly align the pipe and bolt holes. If flanges are used to straighten pipe, they will put stress on adjacent valves, compressors, and controls, causing the operating mechanism to bind. To prevent leaks, flange bolts are drawn up evenly when connecting the flanges. Flanges at compressors and other system components must not move or indicate stress when all bolts are loosened.

Union Joints. Steel (3000 psi) ground joint unions are used for gage and pressure control lines with screwed valves and for joints up to 0.75 in. When tightening this type of joint, the two pipes must be axially aligned. To be effective, the two parts of the union must match perfectly. Ground joint unions should be avoided if at all possible.

Pipe Location

Piping should be at least 7.5 ft above the floor. Locate pipes carefully in relation to other piping and structural members, especially when the lines are to be insulated. The distance between insulated lines should be at least three times the thickness of the insulation for screwed fittings, and four times for flange fittings. The space between the pipe and adjacent surfaces should be three-fourths of these amounts.

Hangers located close to the vertical risers to and from compressors keep the piping weight off the compressor. Pipe hangers should be placed not more than 8 to 10 ft apart and within 2 ft of a change in direction of the piping. Hangers should be designed to bear on the outside of insulated lines. Sheet metal sleeves on the lower half of the insulation are usually sufficient. Where piping penetrates a wall, a sleeve should be installed and where the pipe penetrating the wall is insulated, it must be adequately sealed.

Piping to and from compressors and to other components must provide for expansion and contraction. Sufficient flange or union joints should be located in the piping so that components can be assembled easily during initial installation and also disassembled for servicing.

Pipe Sizing

Table 1 presents *practical* suction line sizing data based on 0.25 °F and 0.50 °F differential pressure drop equivalent per 100 ft total equivalent length of pipe. For data on equivalent lengths of valves and fittings, refer to Tables 10, 11, and 12 in Chapter 2. Table 2 lists data for sizing suction and discharge lines at 1 °F differential pressure drop equivalent per 100 ft equivalent length of pipe, and for sizing liquid lines at 100 fpm. Charts prepared by Wile (1977) present pressure drops in saturation temperature equivalents. For a complete discussion of the basis of these line sizing charts, see Timm (1991). Table 3 presents line sizing information for pumped liquid lines, high-pressure liquid lines, hot gas defrost lines, equalizing lines, and thermosyphon oil cooling ammonia lines.

Valves

Stop Valves. These valves should be placed in the inlet and outlet lines to all condensers, vessels, evaporators, and long lengths of pipe so that they can be isolated in case of leaks and to facilitate "pumping out" by evacuation. Sections of liquid piping that can be valved off and isolated must be protected with a relief device.

Installing globe-type stop valves with the valve stems horizontal lessens the chance for dirt or scale to lodge on the valve seat or disk and cause it to leak, or for liquid or oil to pocket in the area below the seat. Wet suction return lines (recirculation system) should employ angle valves only to reduce the possibility of liquid pockets and also to reduce pressure drop.

Welded flanged or weld-in-line valves are desirable for all line sizes; however, screwed valves may be used for 1.25 in. and smaller sizes. Ammonia globe and angle valves should have the following features:

- Soft seating surfaces for positive shutoff (no copper or copper alloy)
- Back seating to permit repacking the valve stem while in service
- Arrangement that allows packing to be tightened easily
- All-steel construction is preferable
- Bolted bonnets above 1 in., threaded bonnets for 1 in. and smaller

Table 1 Suction Line Capacities (Tons) for Ammonia with Pressure Drops of 0.25 and 0.50 °F/100 ft Equivalent

Steel Line Size		Saturated Suction Temperature, °F					
		−60		−40		−20	
IPS	SCH	Δt = 0.25 °F Δp = 0.046	Δt = 0.50 °F Δp = 0.092	Δt = 0.25 °F Δp = 0.077	Δt = 0.50 °F Δp = 0.155	Δt = 0.25 °F Δp = 0.123	Δt = 0.50 °F Δp = 0.245
3/8	80	0.03	0.05	0.06	0.09	0.11	0.16
1/2	80	0.06	0.10	0.12	0.18	0.22	0.32
3/4	80	0.15	0.22	0.28	0.42	0.50	0.73
1	80	0.30	0.45	0.57	0.84	0.99	1.44
1-1/4	40	0.82	1.21	1.53	2.24	2.65	3.84
1-1/2	40	1.25	1.83	2.32	3.38	4.00	5.80
2	40	2.43	3.57	4.54	6.59	7.79	11.26
2-1/2	40	3.94	5.78	7.23	10.56	12.50	18.03
3	40	7.10	10.30	13.00	18.81	22.23	32.09
4	40	14.77	21.21	26.81	38.62	45.66	65.81
5	40	26.66	38.65	48.68	70.07	82.70	119.60
6	40	43.48	62.83	79.18	114.26	134.37	193.44
8	40	90.07	129.79	163.48	235.38	277.80	397.55
10	40	164.26	236.39	297.51	427.71	504.98	721.08
12	ID	264.07	379.88	477.55	686.10	808.93	1157.59

Steel Line Size		Saturated Suction Temperature, °F					
		0		20		40	
IPS	SCH	Δt = 0.25 °F Δp = 0.184	Δt = 0.50 °F Δp = 0.368	Δt = 0.25 °F Δp = 0.265	Δt = 0.50 °F Δp = 0.530	Δt = 0.25 °F Δp = 0.366	Δt = 0.50 °F Δp = 0.582
3/8	80	0.18	0.26	0.28	0.40	0.41	0.53
1/2	80	0.36	0.52	0.55	0.80	0.82	1.05
3/4	80	0.82	1.18	1.26	1.83	1.87	2.38
1	40	1.62	2.34	2.50	3.60	3.68	4.69
1-1/4	40	4.30	6.21	6.63	9.52	9.76	12.42
1-1/2	40	6.49	9.34	9.98	14.34	14.68	18.64
2	40	12.57	18.12	19.35	27.74	28.45	36.08
2-1/2	40	20.19	28.94	30.98	44.30	45.37	57.51
3	40	35.87	51.35	54.98	78.50	80.40	101.93
4	40	73.56	105.17	112.34	160.57	164.44	208.34
5	40	133.12	190.55	203.53	289.97	296.88	376.18
6	40	216.05	308.62	329.59	469.07	480.96	609.57
8	40	444.56	633.82	676.99	962.47	985.55	1250.34
10	40	806.47	1148.72	1226.96	1744.84	1786.55	2263.99
12	ID	1290.92	1839.28	1964.56	2790.37	2862.23	3613.23

Note: Capacities are in tons of refrigeration resulting in a line friction loss (Δp in psi) per 100 ft equivalent pipe length, with corresponding change (Δt in °F per 100 ft) in saturation temperature.

Table 2 Refrigerant Suction and Discharge Line Capacities (Tons) for Ammonia (Single- or High-Stage Applications) for Pressure Drop of 1°F Equivalent

Steel Line Size		Suction Lines Δt = 1°F Saturated Suction Temperature, °F					Discharge Lines	Liquid Lines Steel Line Size			
IPS	SCH	−40 Δp = 0.31	−20 Δp = 0.49	0 Δp = 0.73	20 Δp = 1.06	40 Δp = 1.46	Δt = 1°F Δp = 2.95	IPS	SCH	Velocity/1°F 100 fpm	Δp = 2.0 psi Δt = 0.7°F
3/8	80	—	—	—	—	—	—	3/8	80	8.6	12.1
1/2	80	—	—	—	—	—	3.1	1/2	80	14.2	24.0
3/4	80	—	—	—	2.6	3.8	7.1	3/4	80	26.3	54.2
1	80	—	2.1	3.4	5.2	7.6	13.9	1	80	43.8	106.4
1 1/4	40	3.2	5.6	8.9	13.6	19.9	36.5	1 1/4	80	78.1	228.6
1 1/2	40	4.9	8.4	13.4	20.5	29.9	54.8	1 1/2	80	107.5	349.2
2	40	9.5	16.2	26.0	39.6	57.8	105.7	2	40	204.2	811.4
2 1/2	40	15.3	25.9	41.5	63.2	92.1	168.5	2 1/2	40	291.1	1292.6
3	40	27.1	46.1	73.5	111.9	163.0	297.6	3	40	449.6	2287.8
4	40	55.7	94.2	150.1	228.7	333.0	606.2	4	40	774.7	4662.1
5	40	101.1	170.4	271.1	412.4	600.9	1095.2	5	40	—	—
6	40	164.0	276.4	439.2	667.5	971.6	1771.2	6	40	—	—
8	40	337.2	566.8	901.1	1366.6	1989.4	3623.0	8	40	—	—
10	40	611.6	1027.2	1634.3	2474.5	3598.0	—	10	40	—	—
12	ID	981.6	1644.5	2612.4	3963.5	5764.6	—	12	ID	—	—

Notes:

1. Table capacities are in tons of refrigeration resulting in a line friction loss per 100 ft equivalent pipe length, with corresponding change in saturation temperature, *where*

   Δp = pressure drop due to line friction, psi per 100 ft of line

   Δt = change in saturation temperature, °F

2. Line capacity for other saturation temperatures Δt and equivalent lengths L_e

$$\text{Line capacity} = \text{Table capacity}\left(\frac{\text{Table } L_e}{\text{Actual } L_e} \times \frac{\text{Actual } \Delta t}{\text{Table } \Delta t}\right)^{0.55}$$

3. Saturation temperature Δt for other capacities and equivalent lengths L_e

$$\Delta t = \text{Table } \Delta t \; \frac{\text{Actual } L_e}{\text{Table } L_e}\left(\frac{\text{Actual capacity}}{\text{Table capacity}}\right)^{1.8}$$

4. Values in the table are based on 90°F condensing temperature. Multiply table capacities by the following factors for other condensing temperatures.

Condensing Temperature, °F	Suction Lines	Discharge Lines
70	1.05	0.78
80	1.02	0.89
90	1.00	1.00
100	0.98	1.11

5. Discharge and liquid line capacities are based on 20°F suction. Evaporator temperature is 0°F. The capacity is affected less than 3% when applied from −40 to +40°F extremes.

Table 3 Liquid Ammonia Line Capacities
(Capacity in tons of refrigeration, except as noted)

Nominal Size, in.	Pumped Liquid 3:1	4:1	5:1	High Pressure Liquid at 3 psi[a]	Hot Gas Defrost[a]	Equalizer High Side[b]	Thermosyphon Oil Cooling Lines Gravity Flow[c], 1000 Btu/h Supply	Return	Vent
1/2	10	7.5	6	30	—	—	—	—	—
3/4	22	16.5	13	69	4	50	—	—	—
1	43	32.5	26	134	8	100	—	—	—
1 1/4	93.5	70	56	286	20	150	—	—	—
1 1/2	146	110	87.5	439	30	225	200	120	203
2	334	250	200	1016	50	300	470	300	362
2 1/2	533	400	320	1616	92	500	850	530	638
3	768	576	461	2886	162	1000	1312	870	1102
4	1365	1024	819	—	328	2000	2261	1410	2000
5	—	—	—	—	594	—	3550	2214	3624
6	—	—	—	—	970	—	5130	3200	6378
8	—	—	—	—	—	—	8874	5533	11596

[a]From *Refrigerant Line Sizing* by D.D. Wile, 1977, which is available from ASHRAE (Publication Code 90250). Hot gas line sizes are based on 1.5 psi pressure drop per 100 ft of equivalent length at 100 psig discharge pressure and 3 times the evaporator refrigeration capacity.

[b]Line sizes are based on experience using total system evaporator tons.

[c]From *Thermosyphon Oil Cooling* by Frick Co. Values for line sizes above 4 in. are extrapolated.

- Consider seal cap valves in refrigerated areas and for all ammonia piping
- To keep pressure drop to a minimum, consider angle valves (as opposed to globe valves)

Control Valves. Pressure regulators, solenoid valves, and thermostatic expansion valves should be flanged for easy assembly and removal. Valves 1.5 in. and larger should have welding companion flanges. Smaller valves can have threaded companion flanges.

A strainer should be used in front of self-contained control valves to protect them from pipe construction material and dirt. A ceramic filter, installed in the pilot line to the power piston, will protect the close tolerances from foreign material when pilot-operated control valves are used.

Solenoid Valves. Solenoid valve stems should be located upright and their coils protected from moisture. They should have flexible conduit connections, where allowed by codes, and have an electric pilot light, wired in parallel, to indicate when the coil is energized. A manual opening stem is useful for emergencies.

Solenoid valves for high-pressure liquid feed to evaporators should have soft seats for positive shutoff.

Solenoid valves for other applications, such as in suction lines, hot gas lines, or gravity feed lines, should be selected for the pressure and temperature of the fluid flowing and for the pressure drop available.

Relief Valves

Safety valves must be provided in conformance with the Safety Code for Mechanical Refrigeration (ANSI/ASHRAE 15-1992) and Section VIII, Division I, of the ASME Boiler and Pressure Vessel Code. For ammonia systems, *Bulletin* No. 109-1988, *IIAR, Minimum Safety Criteria for a Safe Ammonia Refrigeration System*, published by the International Institute of Ammonia Refrigeration, also addresses the subject of safety valves.

Dual relief valve arrangements enable testing of the relief valves (Figure 13). The three-way stop valve is constructed so that it is always open to one of the relief valves if the other is removed to be checked or repaired.

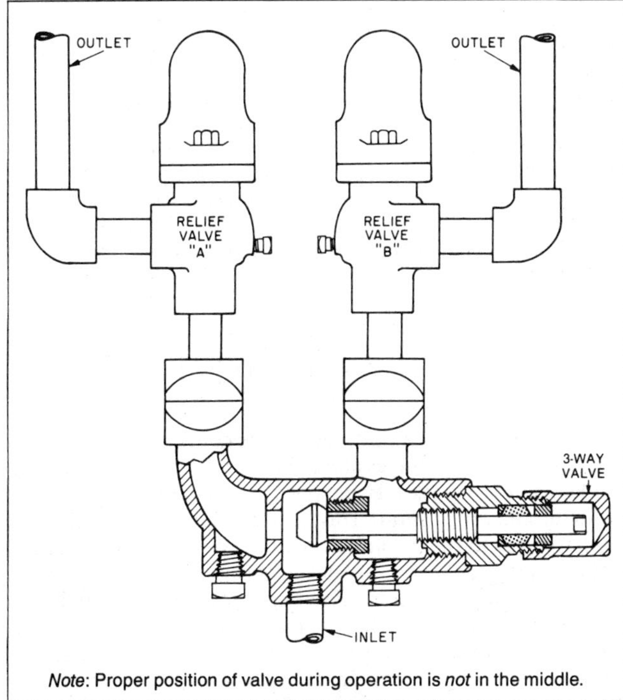

Fig. 13 Dual Relief Valve Fitting for Ammonia

Isolated Line Sections

Sections of piping that can be isolated between hand valves or check valves can be subjected to extreme hydraulic pressures if cold liquid refrigerant is trapped in them and subsequently warmed. Additional safety valves for such piping must be provided.

Insulation and Vapor Barriers

Chapters 20 and 21 of the 1993 ASHRAE *Handbook—Fundamentals* cover insulation and vapor barriers. Insulation and effective vapor barriers on low-temperature systems are very important. At low temperatures, the smallest leak in the vapor barrier can allow ice to form inside the insulation, which can totally destroy the integrity of the entire insulation system. The result can cause a significant load and power increase.

RECIPROCATING COMPRESSOR PIPING

Figure 14 shows a typical piping arrangement for two compressors operating in parallel off the same suction main. Suction mains should be laid out with the objective of returning only clean, dry gas to the compressor. This usually requires a suction trap sized adequately for gravity gas and liquid separation based on permissible gas velocities for specific temperatures. A dead-end trap can usually only trap scale and oil. As an alternative to the dead-end trap, a shell-and-coil accumulator with a warm liquid coil may be considered. Suction mains running to and from the suction trap or accumulator should be pitched toward the trap for liquid drainage. Pitch should be 1/8 in. per foot of length.

In sizing the suction mains and the takeoffs from the mains to the compressors, consider how the pressure drop in the selected piping affects the compressor size required. First costs and operating costs for compressor and piping selections should be optimized.

Good suction line systems have a total friction drop of 1 to 3 °F pressure drop equivalent. Practical suction line friction losses should not exceed 0.5 °F equivalent per 100 ft equivalent length.

A well-designed discharge main has a total friction loss of 1 to 2 psi. Generally, a slightly oversized discharge line is desirable to hold down discharge pressure and, consequently, discharge temperature, and energy costs. Where possible, discharge mains should be pitched (1/8 in. per foot) toward the condenser, without incurring a liquid trap; otherwise, pitch toward the discharge line separator.

High- and low-pressure cutouts and gages and lubricant pressure failure cutout are installed on the compressor side of the stop valves to protect the compressor.

Lubricant Separators

Lubricant separators are located in the discharge line of each compressor (Figure 14A). A high-pressure float valve drains the lubricant back into the compressor crankcase or lubricant receiver. Place the separator as far from the compressor as possible so that the extra pipe length can be used to cool the discharge gas before it enters the separator. This reduces the temperature of the ammonia vapor and makes the separator more effective.

Liquid ammonia must not reach the crankcase. Often, a valve (preferably automatic) is installed in the drain from the lubricant separator, open only when the temperature at the bottom of the separator is higher than the condensing temperature. Some manufacturers install a small electric heater at the bottom of a vertical lubricant trap instead. The heater is actuated when the compressor is not operating. Separators installed in cold conditions must be insulated to prevent ammonia condensation.

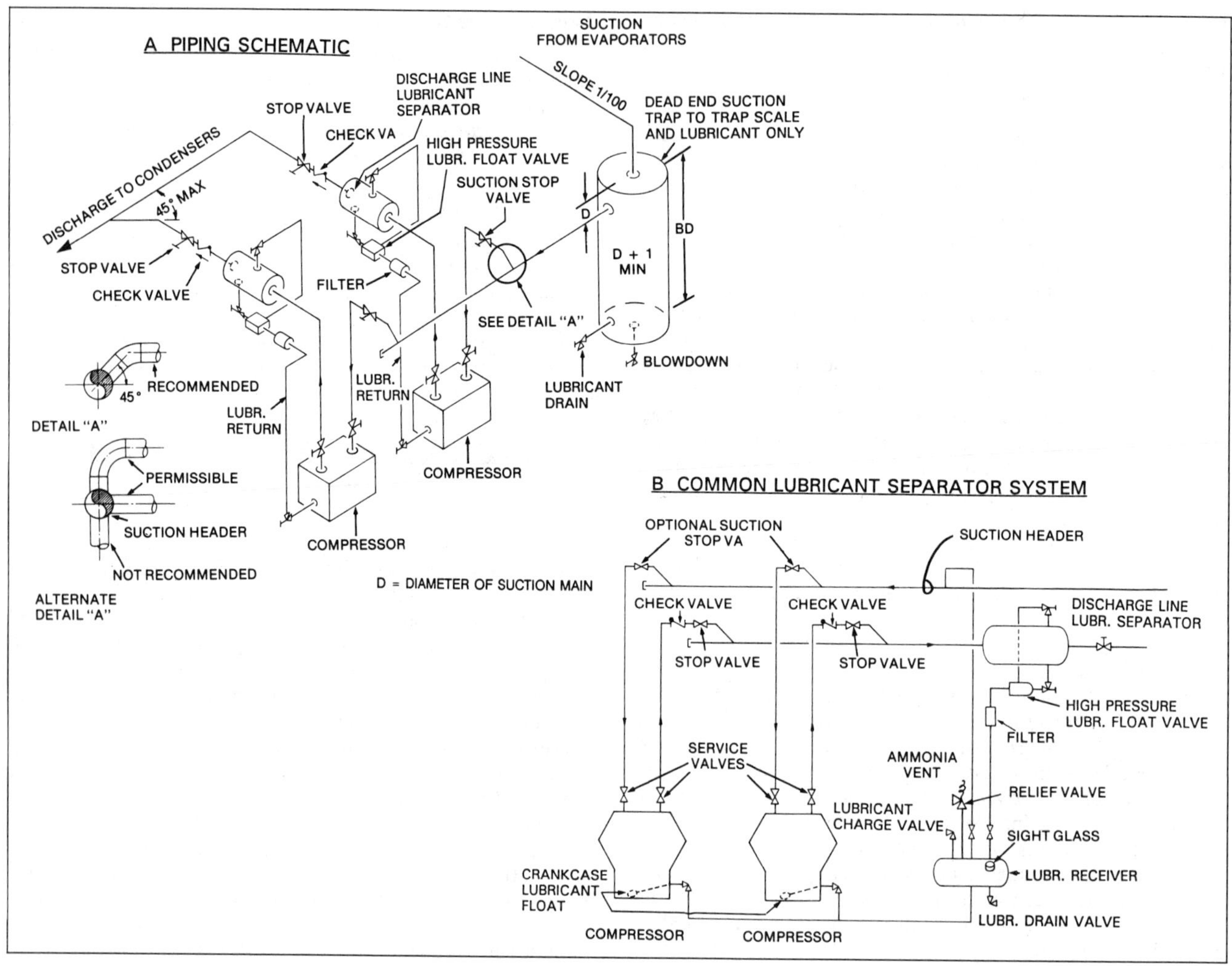

Fig. 14 Schematic of Reciprocating Compressors Operating in Parallel

A filter is recommended in the drain line on the downstream side of the high-pressure float valve.

Lubricant Receivers

Figure 14B illustrates two compressors on the same suction line with one discharge line lubricant separator. The separator float drains into a lubricant receiver, which maintains a reserve supply of lubricant for the compressors. Compressors should be equipped with crankcase floats to regulate the lubricant flow to the crankcase.

Discharge Check Valves and Discharge Lines

Discharge check valves on the downstream side of each lubricant separator prevent high-pressure gas from flowing into an inactive compressor and causing condensation (Figure 14A).

The discharge line from each compressor should enter the discharge main at a 45° maximum angle in the horizontal plane so that the gas flows smoothly.

Unloaded Starting

Unloaded starting is frequently needed to stay within the torque or current limitations of the motor. Most compressors are unloaded either by holding the suction valve open or by external bypassing. Control can be manual or automatic.

Suction Gas Conditioning

Suction main piping should be insulated, complete with vapor barrier to minimize thermal losses, to prevent sweating, and/or ice build-up on the piping, and to limit superheat at the compressor. Additional superheat results in increased discharge temperatures and reduces compressor capacity. Low discharge temperatures in ammonia plants are important to reduce lubricant carryover and because the compressor lubricant can carbonize at higher temperatures, which can cause cylinder wall scoring and lubricant sludge throughout the system. Discharge temperatures above 250°F should be avoided at all times. Lubricants should have flash point temperatures above the maximum expected compressor discharge temperature.

RECIPROCATING COMPRESSOR COOLING

Generally, ammonia compressors are constructed with internally cast cooling passages along the cylinders and/or in the top heads. These passages provide space for circulating a heat transfer medium, which minimizes heat conduction from the hot discharge gas to the incoming suction gas and lubricant in the compressor's crankcase. An external lubricant cooler is supplied on most reciprocating ammonia compressors. Water is usually the medium circulated through these passages, or *water jackets*, and through the lubricant cooler at a rate of about 0.1 gpm/ton of refrigeration. The lubricant in the crankcase (depending on type of construction) is about 120°F. Temperatures above this level reduce the lubricant's lubricating properties.

For compressors operating in ambients above 32 °F, water flow is sometimes controlled entirely by hand valves, although a solenoid valve in the inlet line is desirable to make the system automatic. Water flow *must* be stopped when the compressor stops to keep the residual gas from condensing and to conserve water. A water-regulating valve, installed in the water supply line with the sensing bulb in the water return line, is also recommended.This type of cooling is shown in Figure 15.

The thermostat in the water line leaving the jacket serves as a safety cutout to stop the compressor if the temperature becomes too high.

For compressors installed where ambient temperatures below 32 °F may exist, provide a means for draining the jacket on shutdown to prevent freeze-up. One method is shown in Figure 16. Water flow is through the normally closed solenoid valve, which is energized when the compressor starts. The water then circulates through the lubricant cooler, the jacket, and out through the water return line. When the compressor stops, the solenoid valve in the water inlet line is deenergized and stops the water flow to the compressor. At the same time, the solenoid valve opens to drain the water out of the low point to waste water treatment. The check valves in the air vent lines open when pressure is relieved and allows the jacket and cooler to be drained. Each flapper check valve is installed so that water pressure will close it, but absence of water pressure allows it to swing open.

When compressors are installed in spaces below 32 °F or where water quality is very poor, cooling is best handled by using an inhibited glycol solution or other suitable fluid in the jackets and lubricant cooler and cooling with a secondary heat exchanger. This method for cooling reciprocating ammonia compressors elimates the fouling of the lubricant cooler and jacket normally associated with city water or cooling tower water.

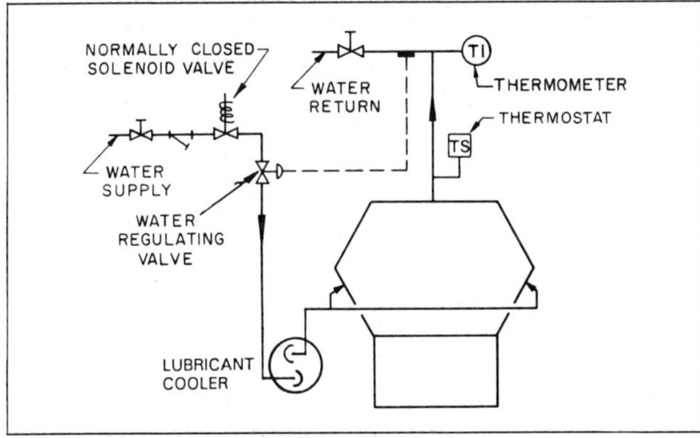

Fig. 15 Jacket Water Cooling for Ambient Temperatures above Freezing

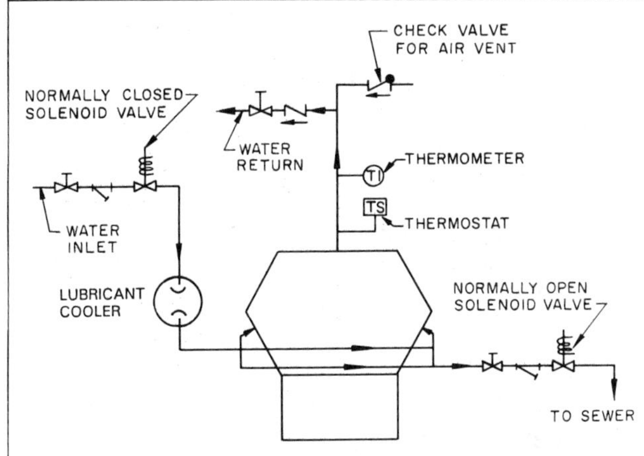

Fig. 16 Jacket Water Cooling for Ambient Temperatures below Freezing

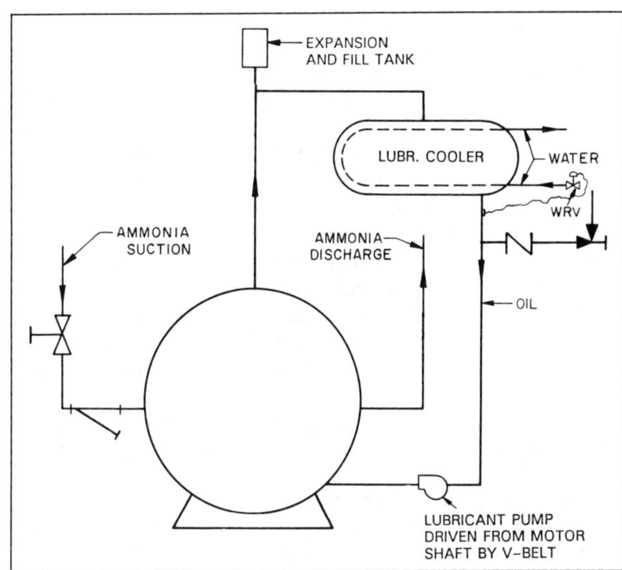

Fig. 17 Rotary Vane Booster Compressor Cooling with Lubricant

ROTARY VANE, LOW-STAGE COMPRESSOR PIPING

Rotary vane compressors have been used extensively as low-stage compressors in ammonia refrigeration systems. Now, however, the screw compressor has largely replaced the rotary vane compressor for ammonia low-stage compressor applications. Piping requirements for rotary vane compressors are the same as those for reciprocating compressors. Most rotary vane compressors are lubricated by injectors since they have no crankcase. In some designs, a lubricant separator, lubricant receiver, and cooler are required on the discharge of these compressors. A pump recirculates the lubricant to the compressor for both cooling and lubrication. In other rotary vane compressor designs, a discharge lubricant separator is not used, and the lubricant collects in the high-stage suction accumulator or intercooler from which it may be drained. Lubricant for the injectors must periodically be added to a reservoir.

The compressor jacket is cooled by circulating a cooling fluid, such as water or lubricant. Lubricant is recommended, since it will not freeze and can be the same lubricant as is used for lubrication (see Figure 17).

SCREW COMPRESSOR PIPING

Helical screw compressors are the choice for most industrial refrigeration systems. There are three types of screw compressors:

1. Fixed volume ratio (V_i) with slide valve
2. Variable volume ratio (V_i) with slide valve and slide stop
3. Fixed volume ratio (V_i) with bypass points in lieu of a slide valve

All helical screw compressors have a constant volume (displacement) design. The term V_i (volume index) refers to the internal volume ratio of the compressor. When V_i is fixed, the compressor functions most efficiently as a certain absolute compression ratio (CR). In selecting a fixed V_i compressor, consider the average CR and not the maximum CR. A guide to proper compressor selection is based on the equation V_i^k = CR, where k = 1.4 for ammonia. For example, for a screw compressor at 10°F (38.5 psia) and 95°F (195.8 psia) with 5.09 CR; $V_i^{1.4}$ = 5.09 and V_i = 3.20. Thus, a compressor with a 3.6 V_i might be the best choice. If the ambient conditions are such that the average condensing temperature is 75°F (140.5 psia), then the CR is 3.65 and the ideal V_i is 2.52. Thus, a compressor with a 2.4 V_i is the proper selection to optimize efficiency.

The compressors with fixed V_i with bypass ports in lieu of a slide valve are normally applied as a booster compressor, which normally has a V_i requirement of less than 2.9.

A variable V_i compressor makes compressor selection simpler because it can vary its V_i from 2.0 to 5.0; thus, it can automatically match the internal pressure ratio within the compressor with the external pressure ratio.

Typical flow diagrams for screw compressor packages are shown in Figures 18 and 19. Figure 18 is for indirect cooling, and Figure 19 is for direct cooling with refrigerant liquid injection. Figure 20 illustrates a variable V_i compressor which does not require a full-time lube pump but rather a pump to prelube the bearings. Full-time lube pumps are required when either fixed or variable V_i compressors are used as low-stage compressors. Lubrication systems require at least a 75 psi pressure differential for proper operation.

Screw Compressor Lubricant Cooling

The lubricant in screw compressors may be cooled three ways:

1. Liquid refrigerant injection
2. Indirect cooling with glycol or water in a heat exchanger
3. Indirect cooling with boiling high-pressure refrigerant used as the coolant in a thermosyphon process

Refrigerant injection cooling is shown schematically in Figures 19 and 21. Depending on the application, this cooling method usually decreases compressor efficiency and capacity but lowers equipment cost. Most screw compressor manufacturers publish a derating curve for this type of cooling. Injection cooling for low-stage compression has little or no penalty on compressor efficiency or capacity. However, the system efficiency can be increased by using an indirectly cooled lubricant cooler. With this configuration, the heat from the lubricant cooler is removed by the evaporative condenser or cooling tower and is not transmitted to the high-stage compressors.

The refrigerant liquid for liquid injection oil cooling must come from a dedicated source of supply. This coolant may come from the system receiver or from a separate receiver; a 5-min uninterrupted supply of refrigerant liquid is usually adequate.

An indirect or thermosyphon lubricant cooling system for low-stage screw compressors rejects the oil cooling load to the condenser or auxiliary cooling system—this load is not transferred to the high-stage compressor, which improves the system efficiency. Indirect lubricant cooling systems using glycol or water reject the lube cooling load to a section of an evaporative condenser, a separate evaporative cooler, or a cooling tower. A three-way lubricant control valve should be used to control lubricant temperature.

Thermosyphon lubricant cooling is the industry standard. In this system, high-pressure refrigerant liquid from the condenser, which boils at condensing temperature/pressure (usually 90 to 95°F design), cools the lubricant in a tubular heat exchanger. Typical thermosyphon lubricant cooling arrangements are shown in Figures 18, 20, 22, 23, and 24. Note on all figures that the refrigerant liquid supply to the lube cooler receives priority over the feed to the system low side. It is important that the gas equalizing line (vent) off the top of the thermosyphon receiver be adequately sized to match the lubricant cooler load to prevent the thermosyphon receiver from becoming gas bound.

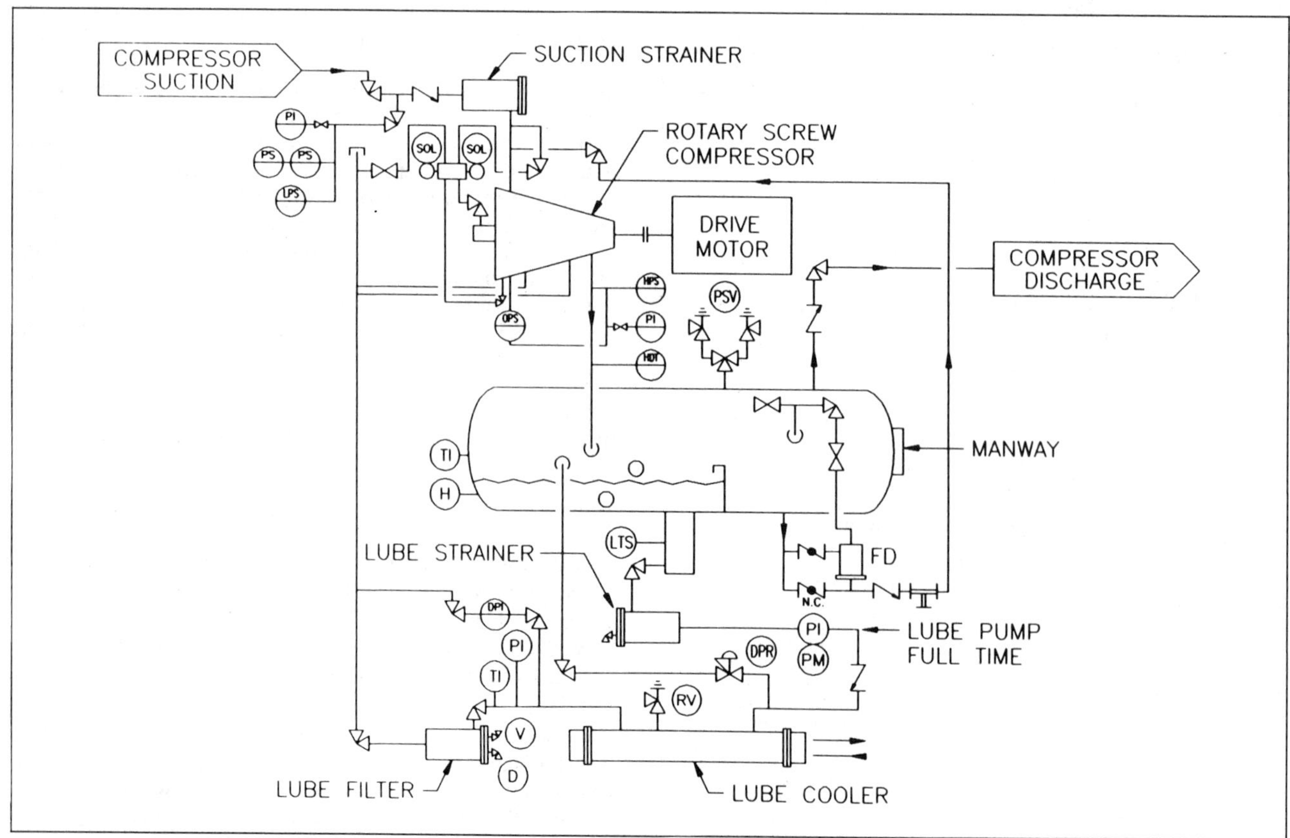

Fig. 18 Fixed V_i Screw Compressor Flow Diagram with Indirect Lubricant Cooling

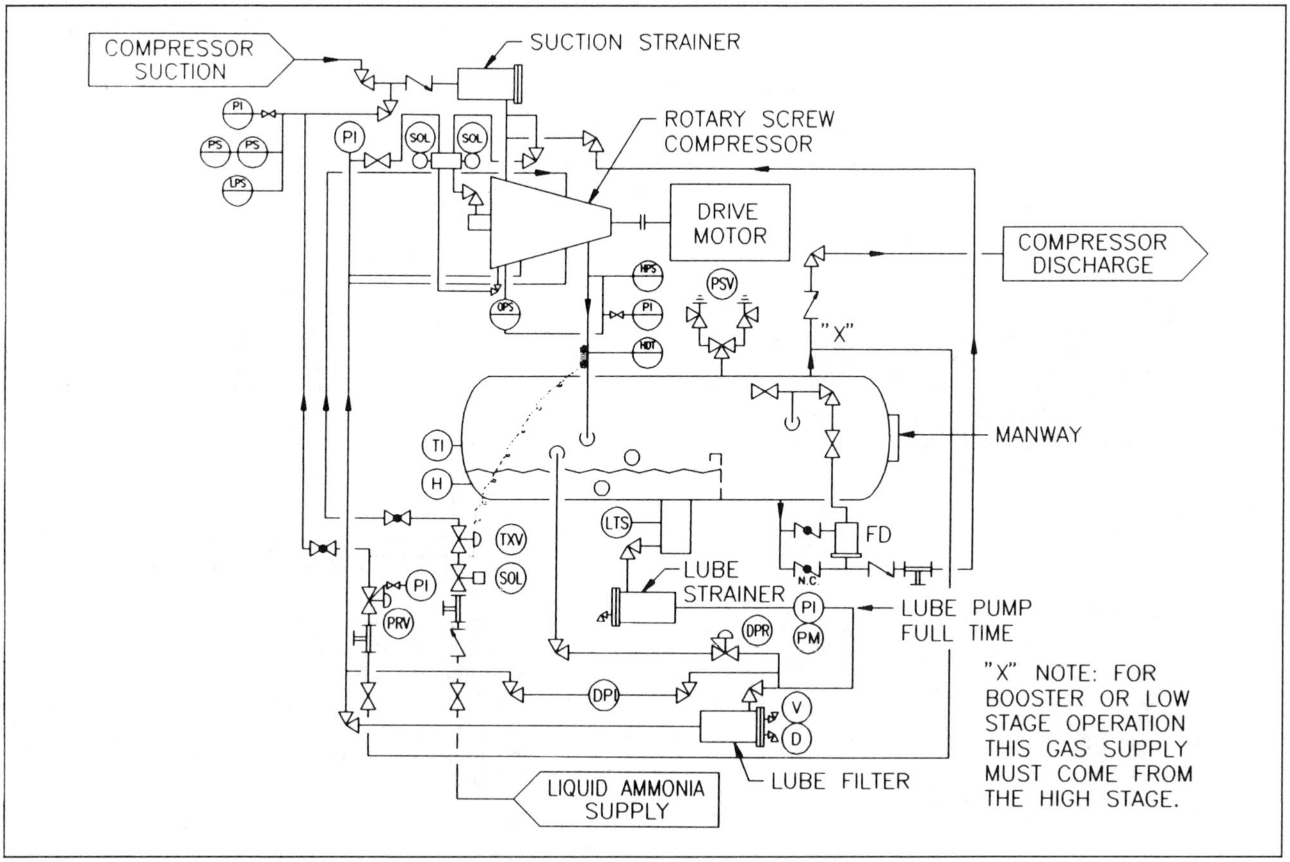

Fig. 19 Fixed V_i Screw Compressor Flow Diagram with Liquid Injection Cooling

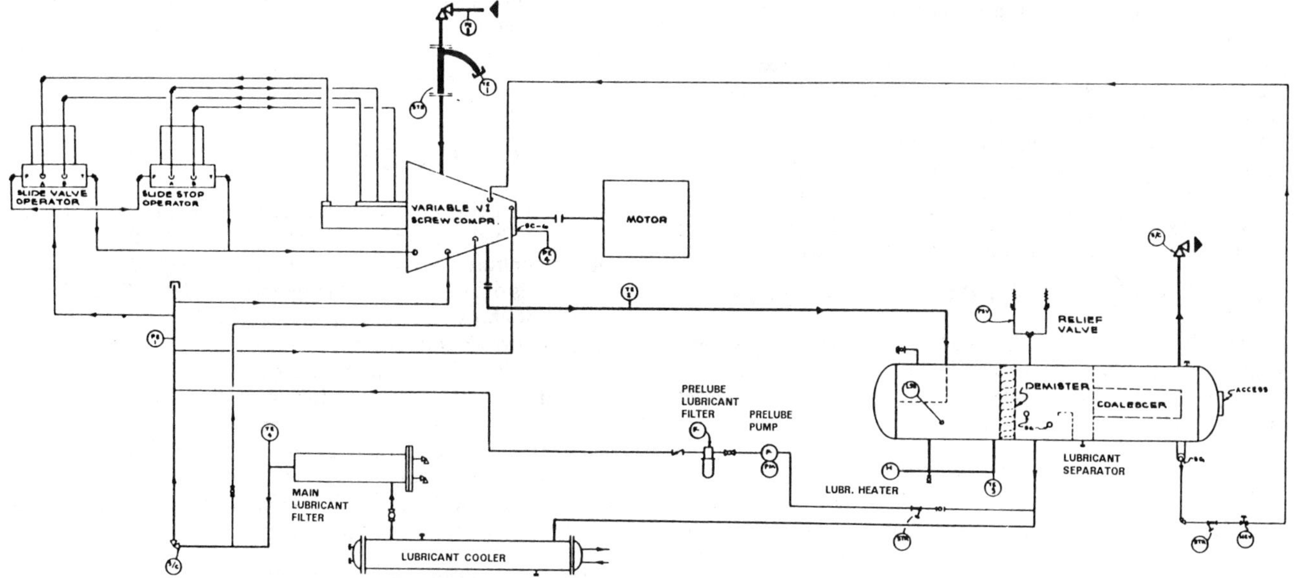

Fig. 20 Flow Diagram for Variable V_i Screw Compressor High-Stage Only

Figure 25 shows a typical capacity control system for a fixed V_i screw compressor. The four-way valve controls the slide valve position, and in turn, the compressor capacity from typically 100 to 10% with a signal from an electric, electronic, or microprocessor controller. The slide valve unloads the compressor by bypassing vapor back to the suction of the compressor.

Figure 26 shows a typical capacity and V_i control system in which two four-way control valves take their signals from a computer controller. One four-way valve controls the capacity by positioning the slide valve in accordance with the load, and the other four-way valve positions the slide stop to adjust the compressor internal pressure ratio to match the system suction and discharge

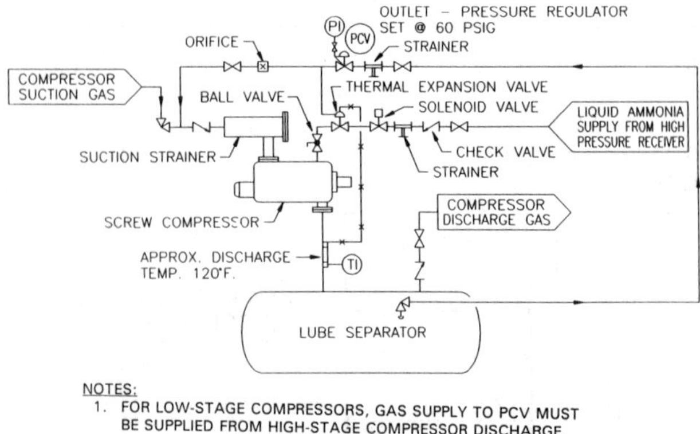

Fig. 21 Flow Diagram for Screw Compressors with Refrigerant Injection Cooling

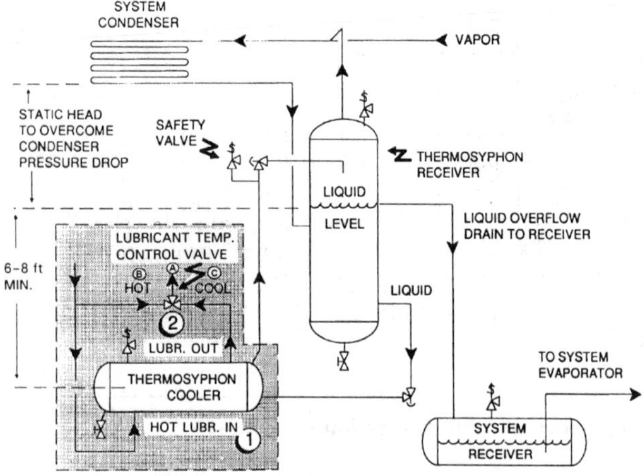

Fig. 22 Thermosyphon Lubricant Cooling System

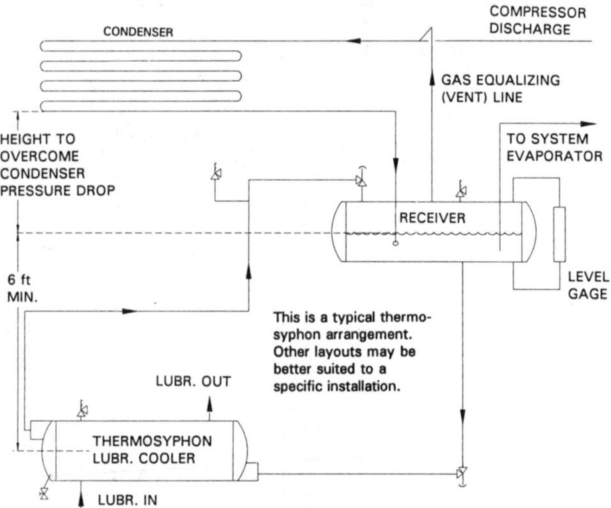

Fig. 23 Typical Thermosyphon System Receiver Mounted above Thermosyphon Oil Cooler

pressure. The slide valve works the same as that on fixed V_i compressors. The variable V_i is controlled by adjusting the slide stop on the discharge end of the compressor.

Screw compressor piping should generally be installed in the same manner as reciprocating compressors. Even though screw compressors can ingest some liquid refrigerant, they should be protected against liquid carryover. Screw compressors are furnished with both suction and discharge check valves.

CONDENSER AND RECEIVER PIPING

Properly designed piping around the condensers and receivers keeps the condensing surface at its highest efficiency by draining the ammonia out of the condenser immediately as it condenses and keeping air and other noncondensables purged.

Horizontal Shell-and-Tube Condenser and Through Receiver

Figure 27 shows a horizontal water-cooled condenser draining into a through-type (top inlet) receiver. Ammonia plants do not always require controlled water flow to maintain pressure. Usually, pressure is adequate to force the ammonia to the various evaporators without water regulation. Each situation should be evaluated by comparing water costs with input power cost savings at lower head pressures.

Water piping should be arranged so that condenser tubes are always filled with water. Air vents should be provided on condenser heads and should have hand valves for manual purging.

Receivers must be below the condenser so that the condensing surface is not flooded with ammonia. The piping should provide (1) free drainage from the condenser and (2) static height of ammonia above the first valve out of the condenser greater than the pressure drop through the valve.

The drain line from condenser to receiver is designed on the basis of 100 fpm maximum velocity to allow gas equalization between condenser and receiver. Refer to Chapter 33 of the 1993 ASHRAE *Handbook—Fundamentals* for sizing criteria.

Parallel Horizontal Shell-and-Tube Condensers

Figure 28 shows two condensers operating in parallel with one through-type (top inlet) receiver. The length of horizontal liquid drain lines to the receiver should be minimized and with no traps permitted. Equalization between the shells is achieved by keeping the liquid velocity in the drain line less than 100 fpm. The drain line can be sized from Chapter 33 of the 1993 ASHRAE *Handbook—Fundamentals.*

EVAPORATIVE CONDENSERS

Evaporative condensers are selected based on the wet-bulb temperature in which they operate. The 1% design wet bulb is that wet-bulb temperature that will be equalled or exceeded 1% of the summer months of June through September, or 29.3 h. Thus, for the majority of industrial plants that operate at least at part load all year, 99.6% of the operating time the wet-bulb temperature will be below design. The resultant condensing pressure will only equal or exceed the design condition during 0.4% of the time; but only if the design wet-bulb temperature and peak design refrigeration load occur coincidentally. This condition is more a function of how the load is calculated, what load diversity factor exists or is used in the calculation, and what safety factor is used in the calculations, than the size of the condenser.

Evaporative Condenser Location

If an evaporative condenser is located with insufficient space for air movement, the effect is the same as that imposed by an inlet damper and the fan may not deliver enough air. In addition, evaporative condenser discharge air may recirculate, which adds to the problem. The high inlet velocity causes a low-pressure region to develop around the fan inlet, inducing flow of discharge air into that region. If the obstruction is from a second condenser, the problem can be even more severe because discharge air from the second condenser flows into the air intake of the first.

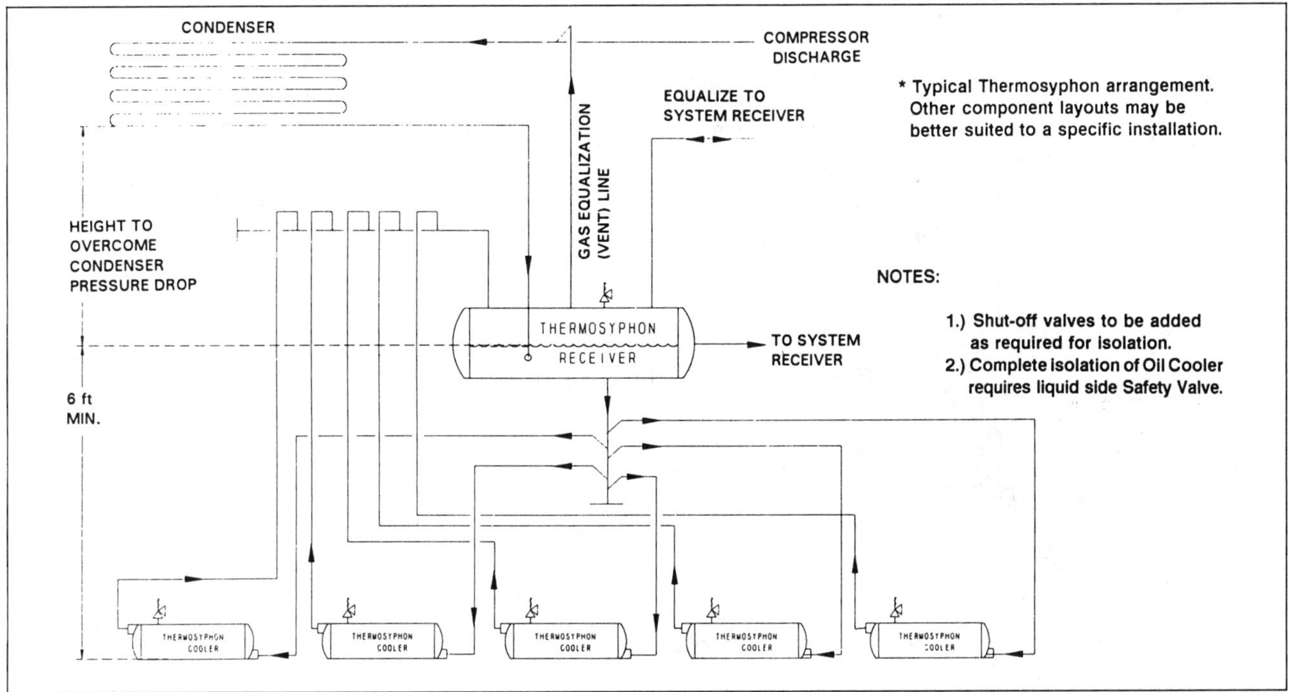

Fig. 24 Typical Thermosyphon System with Multiple Oil Coolers

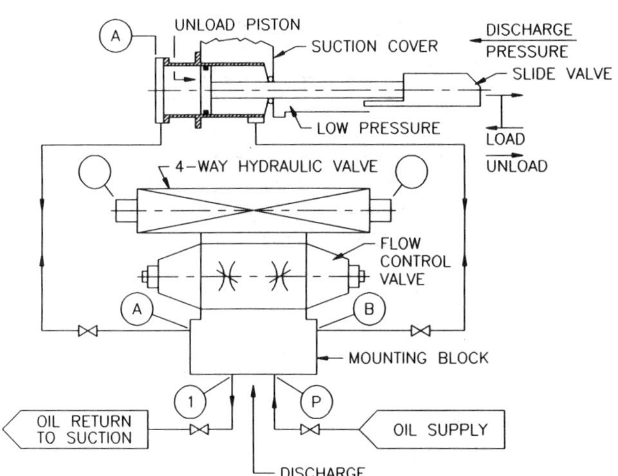

Fig. 25 Typical Hydraulic System for Slide Valve Capacity Control for Screw Compressor with Fixed V_i

Prevailing winds can also contribute to recirculation. In many areas, the winds shift with the seasons; so wind direction during the peak high humidity season must be considered.

The tops of the condensers should always be higher than any adjacent structure to eliminate downdrafts that might induce recirculation. Where this is impractical, discharge hoods can be used to discharge air far enough away from the fan intakes to avoid recirculation. However, the additional static pressure imposed by a discharge hood must be added to the fan system. The fan speed can be increased slightly to obtain proper air volume.

Evaporative Condenser Installation

A single evaporative condenser used with a through-type (top inlet) receiver can be connected as shown in Figure 29. The receiver must always be at a lower pressure than the condensing pressure. As indicated in the section on Evaporators, the receiver must be cooler than the condensing temperature.

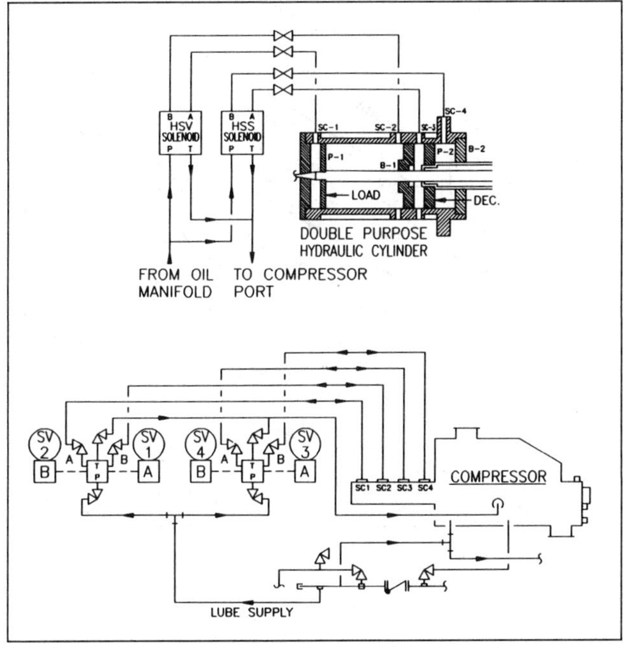

Fig. 26 Typical Positioning System for Slide Valve and Slide Stop for Variable V_i Screw Compressor

Installation in Freezing Areas. In areas having ambient temperatures below 32°F in the evaporative condenser drain pan and water circuit, the water must be kept from freezing at light plant loads. When the temperature is at freezing, the evaporative condenser can operate as a dry-coil unit, and the water pump(s) and piping can be drained and secured for the season.

Another method of preventing the water from freezing is to place the water tank inside and install it as illustrated in Figure 30. When the outdoor temperature drops, the condensing pressure drops, and a pressure switch, with its sensing element in the discharge pressure line, stops the water pump; the water is then

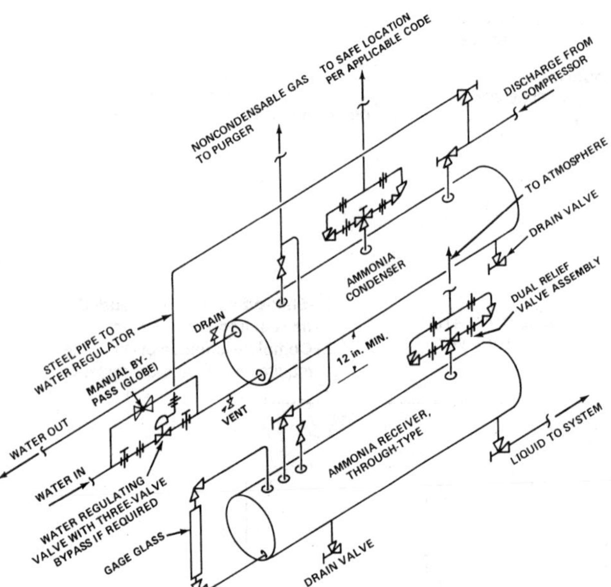

Fig. 27 Horizontal Condenser and Top Inlet Receiver Piping

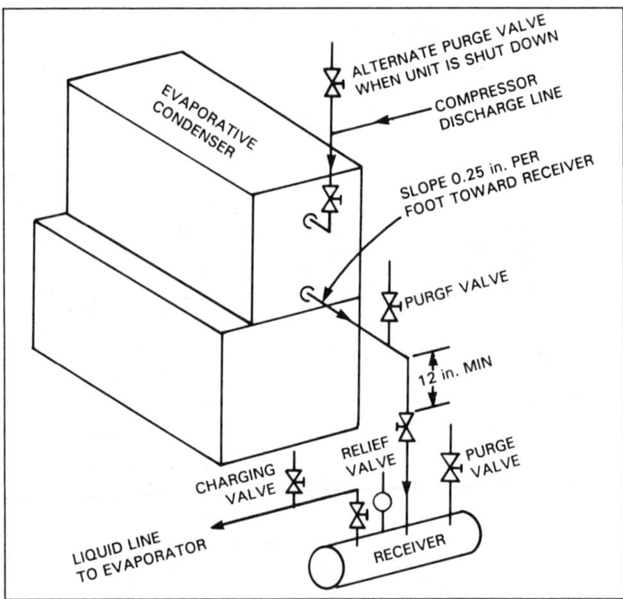

Fig. 29 Single Evaporative Condenser with Top Inlet Receiver

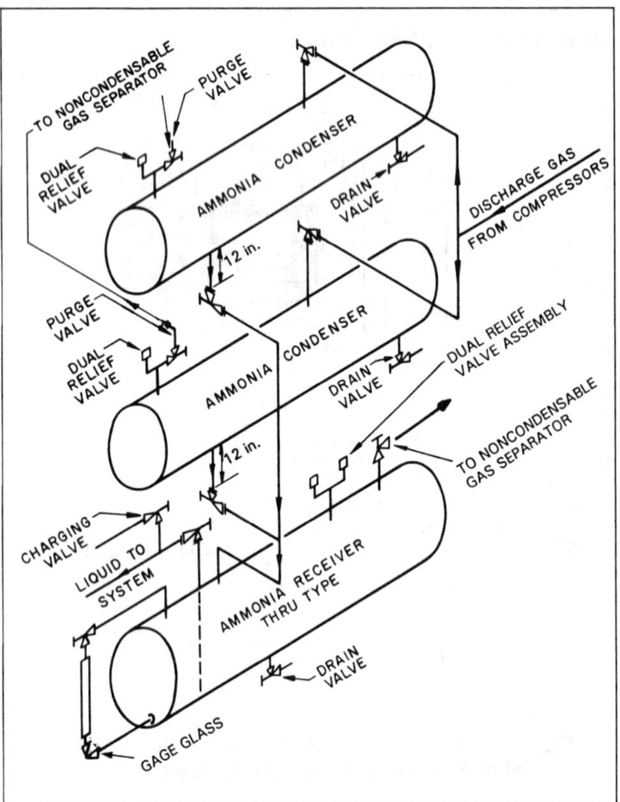

Fig. 28 Parallel Condensers with Top Inlet Receiver

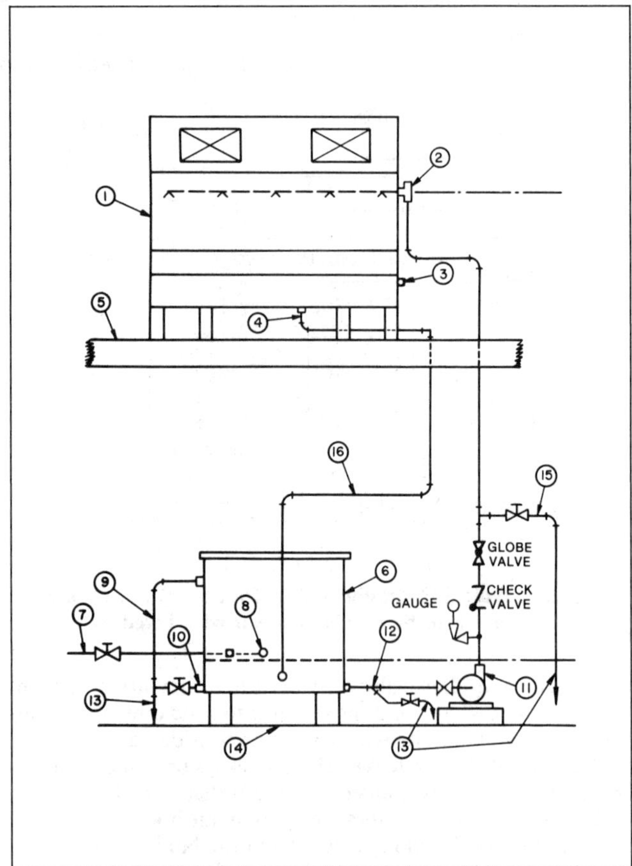

Fig. 30 Evaporative Condenser with Inside Water Tank

drained into the tank. Another method is to use a thermostat that senses the water temperature or outdoor ambient temperature and stops the pump at low temperatures. The exposed piping and any trapped water headers in the evaporative condenser should be drained into the indoor water tank.

Air volume capacity control methods include inlet, outlet, or bypass dampers; two-speed fan motors; or fan cycling in response to pressure controls.

Liquid Traps. Because all evaporative condensers have a substantial pressure drop in the ammonia circuit, liquid traps are needed at the outlets when two or more condensers or condenser coils are installed (Figure 31). Also, an equalizer line is necessary to maintain a stable pressure in the receiver to ensure free drainage from the condensers. For example, assume a 1 psi pressure drop in the operating condenser, which is producing a lower pressure

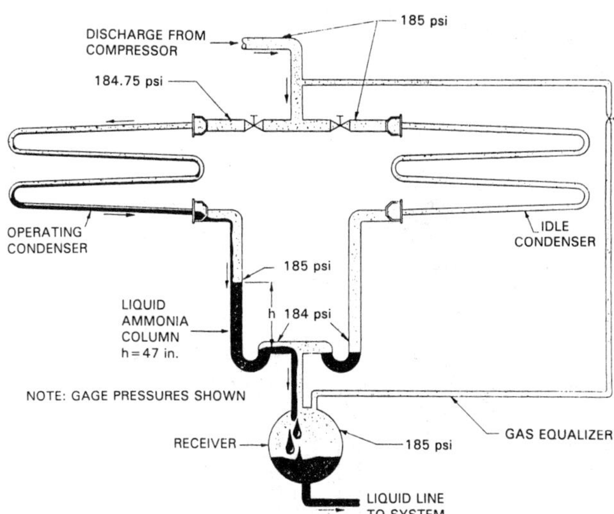

Fig. 31 Two Evaporative Condensers with Trapped Piping to Receiver

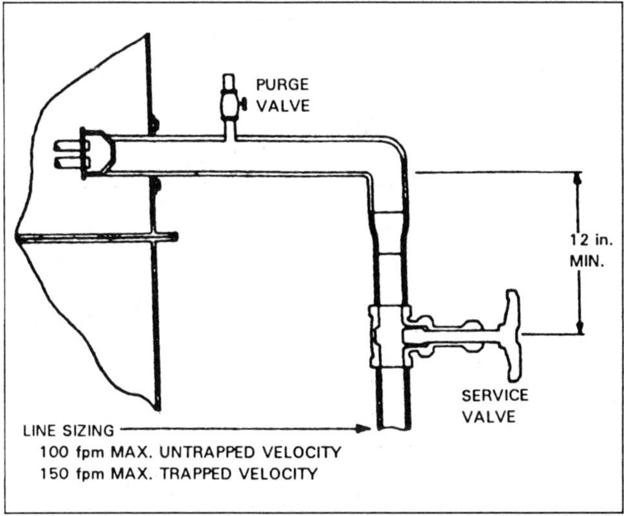

Fig. 32 Method of Reducing Condenser Outlet Sizes

(184 psig) at its outlet as compared to the idle condenser (185 psig) and the receiver (185 psig). The trap creates a liquid seal so that a liquid height *h* of 47 in. equivalent to 1 psi builds up in the vertical drop leg and not in the condenser coil.

The trap must have enough height above the vertical liquid leg to accommodate a liquid height equal to the maximum pressure drop that will be encountered in the condenser. The example illustrates the extreme case of one unit on and one off; however, the same phenomenon occurs to a lesser degree between two condensers of differing pressure drops when both are in full operation. Substantial differences in pressure drop can also occur between two different brands of the same size condenser or even different models produced by the same manufacturer.

The minimum recommended height of the vertical leg is 5 ft for ammonia. This vertical dimension *h* is shown in all evaporative condenser piping diagrams. This height is satisfactory for operation within reasonable ranges around normal design conditions and is based on the maximum condensing pressure drop of the coil. If service valves are installed at the coil inlets and/or outlets, the pressure drops imposed by these valves must be accounted for by increasing the minimum 5-ft drop-leg height by an amount

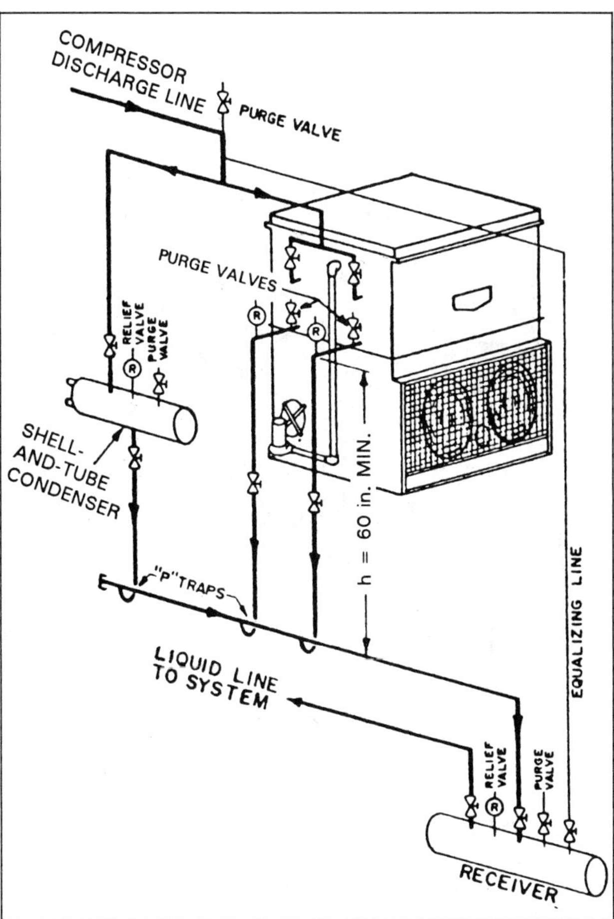

Fig. 33 Piping for Shell-and-Tube and Evaporative Condensers with Top Inlet Receiver

equal to the valve pressure drop in height of liquid refrigerant (Figure 32).

Figures 33, 34, and 35 illustrate various piping arrangements for evaporative condensers.

EVAPORATOR PIPING

Proper evaporator piping and control are necessary to keep the cooled space at the desired temperature and also to adequately protect the compressor from surges of liquid ammonia out of the evaporator. The evaporators illustrated in this section show some methods used to accomplish these objectives. In some cases, combinations of details shown on several illustrations have been used.

Figure 36 illustrates a thermostatic expansion valve on a unit cooler using hot gas for automatic defrosting.

When using hot gas or electric heat for defrosting, the drain pan and drain line must be heated to prevent the condensate from refreezing. When hot gas is used for defrosting, a heating coil is imbedded in the drain pan. The hot gas first flows through this coil and then into the evaporator coil. When using electric heat for defrosting, an electric heating coil is used under the drain pan.

Wrap-around or internal electric heating cables are used on the condensate drain line when the room temperature is below 32°F.

Since this is an automatic defrosting arrangement, hot gas must always be available at the hot-gas solenoid valve near the unit. The system must contain multiple evaporators so that the compressor will be running when the evaporator to be defrosted is shut down. The hot-gas header must be kept in a space where ammonia will

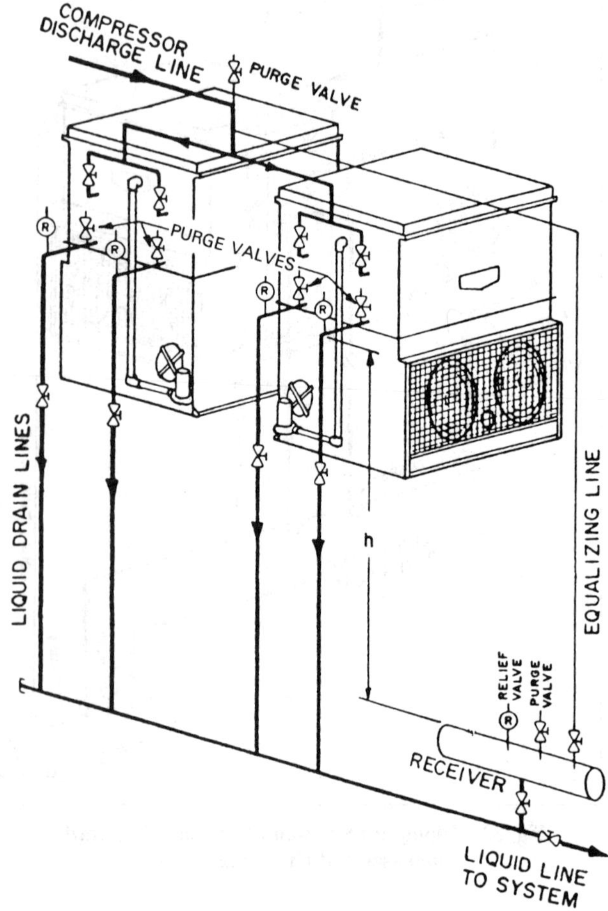

Fig. 34 Piping for Parallel Condensers with Surge-Type Receiver

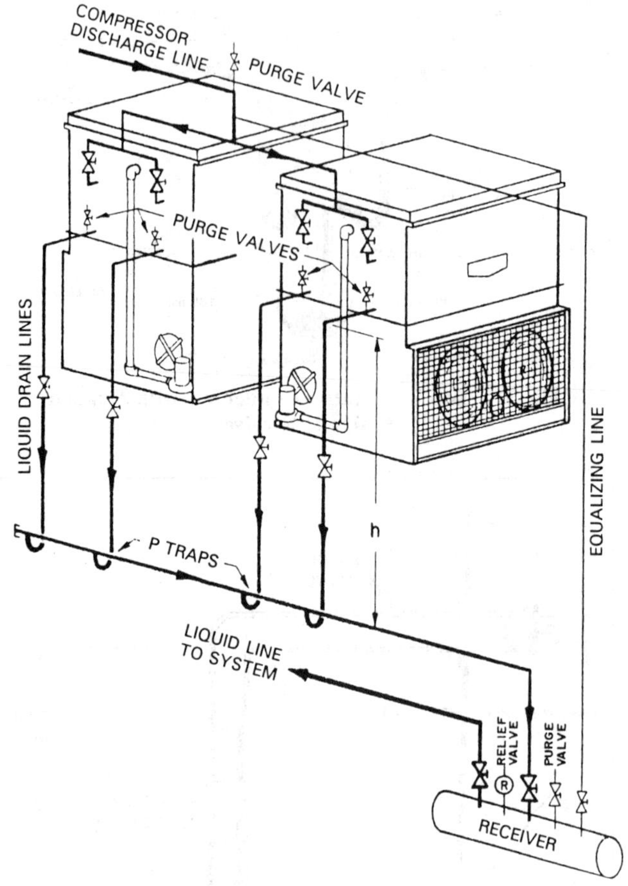

Fig. 35 Piping for Parallel Condensers with Top Inlet Receiver

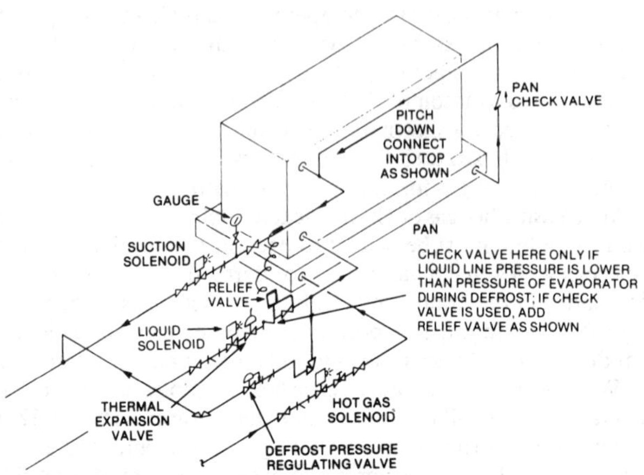

Fig. 36 Piping for Thermostatic Expansion Valve Application for Automatic Defrost on Unit Cooler

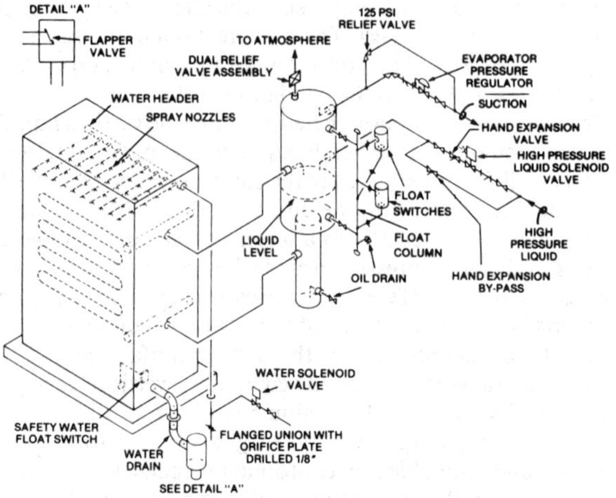

Fig. 37 Arrangement for Automatic Defrost of Air Blower with Flooded Coil

not condense in the pipe. Otherwise, the coil receives liquid ammonia at the start of defrosting and is unable to take full advantage of the latent heat of hot-gas condensation entering the coil. This can also lead to severe hydraulic shock loads. If this cannot be avoided, the insulated hot-gas main must be drained to the suction trap by a high-pressure float.

The liquid line and suction line solenoid valves are open during normal operation only and are closed during the defrost cycle. When the defrost cycle starts, the hot-gas solenoid valve is opened. Refer to IIAR *Bulletin* 116 on possible hydraulic shock when the hot-gas defrost valve is opened after a defrost.

A defrost pressure regulator maintains a gage pressure of about 70 to 80 psi in the coil.

Unit Cooler—Flooded Operation

Figure 37 illustrates a flooded evaporator with a close coupled low-pressure vessel for feeding ammonia into the coil and automatic water defrost.

The lower float switch on the float column at the vessel controls the opening and closing of the liquid line solenoid valve, regulating ammonia into the unit to maintain a liquid level. The hand expansion valve downstream of the solenoid valve should be adjusted so that it will not feed ammonia into the vessel at a rate higher than the vessel can accommodate without raising the suction pressure of gas from the vessel more than 1 or 2 psig.

The static height of liquid in the vessel should be sufficient to flood the coil with liquid under normal loads. The higher float switch should be wired into an alarm circuit and possibly a compressor shut-down circuit when the liquid level in the vessel is too high. With flooded coils having horizontal headers, distribution between the multiple circuits is accomplished without distributing orifices.

A combination evaporator pressure regulator and stop valve is used in the suction line from the vessel. During operation, the regulator maintains a nearly constant back pressure in the vessel. A solenoid coil in the regulator mechanism closes it during the defrost cycle. The liquid solenoid valve should also be closed at this time. One of the best methods to control room temperature is a room thermostat that controls the effective setting of the evaporator pressure regulator.

A spring-loaded relief valve is used around the suction pressure regulator and is set so that the vessel is kept below 125 psig.

A solenoid valve unaffected by downstream pressure is used in the water line to the defrost header. The defrost header is constructed so that it drains at the end of the defrost cycle and the downstream side of the solenoid valve drains through a fixed orifice.

Unless the room is maintained above 32 °F, the drain line from the unit should be wrapped with a heater cable or provided with another heat source and then insulated to prevent the defrost water from refreezing in the line.

The length of the water line within the space leading up to the header and the length of the drain line in the cooled space should be kept to a minimum. A flapper or pipe trap on the end of the drain line prevents warm air from flowing up the drain pipe and into the unit.

An air outlet damper may be closed during defrosting to prevent thermal circulation of air through the unit during the defrost cycle, which otherwise would affect the temperature of the cooled space. The fan is stopped during defrost.

This type of defrosting requires a drain pan float switch for safety control. If the drain pan fills with water, the switch overrides the time clock to stop the flow into the unit by closing the water solenoid valve.

There should be a 5-min delay at the end of the water spray part of the defrosting cycle so that the water can drain from the coil and pan. This limits the ice buildup in the drain pan and on the coils after the cycle is completed.

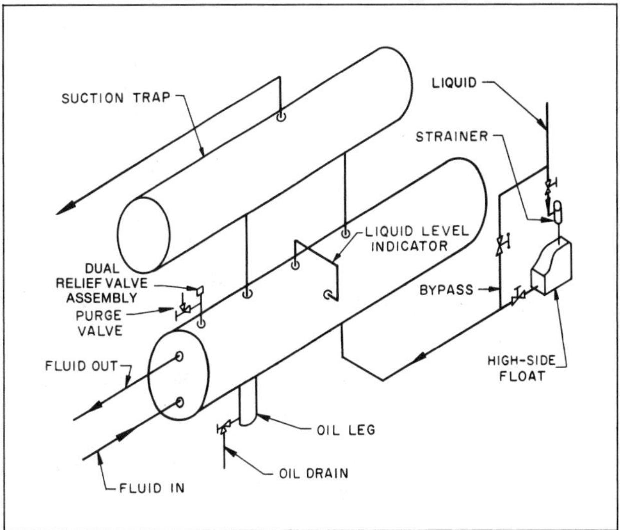

Fig. 38 Arrangement for Horizontal Liquid Cooler and High-Side Float

On completion of the cycle, the pressure in the low-pressure vessel may be about 75 psig. When the unit is opened to the suction main at a much lower pressure, some liquid surges out into the main; therefore, it may be necessary to gradually bleed off this pressure before fully opening the suction valve, to prevent thermal shock. Generally, a suction trap in the engine room removes this liquid before the gas stream enters the compressors.

The type of refrigerant control shown in Figure 37 can be used on "brine spray" type units where brine is sprayed over the coil at all times to pick up the condensed water vapor from the airstream. The brine concentration is reconcentrated continually to remove the water absorbed from the airstream.

High-Side Float Control

When a system has only one evaporator, a high-pressure float control can be used to keep the condenser drained and to provide a liquid seal between the high side and the low side. Figure 38 illustrates a brine or water cooler with this type of control. The high-side float should be located near the evaporator to avoid insulating the liquid line.

The amount of ammonia in this type of system is critical, because the charge must be limited so that, under the highest loading in the evaporator, liquid will not surge into the suction line. Some type of suction trap should be used. One method is to place a horizontal shell above the cooler, with the suction gas piped into the bottom and out of the top. The reduction of gas velocity in this shell causes the liquid to separate from the gas and draw back into the chiller.

Coolers should include a liquid indicator. Use a reflex glass lens with a large liquid chamber and vapor connections for boiling liquids and a plastic frost shield to determine the actual level. A refrigeration thermostat measuring the temperature of the chilled fluid as it exits the cooler should be wired into the compressor starting circuit to prevent freezing.

A flow switch or differential pressure switch should prove flow before the compressor starts. The fluid to be cooled should be piped into the lower portion of the tube bundle and out of the top portion.

Low-Side Float Control

For multiple evaporator systems, low-side float valves are used to control the refrigerant level in flooded evaporators. The low-pressure float shown in Figure 39 has an equalizer line from the

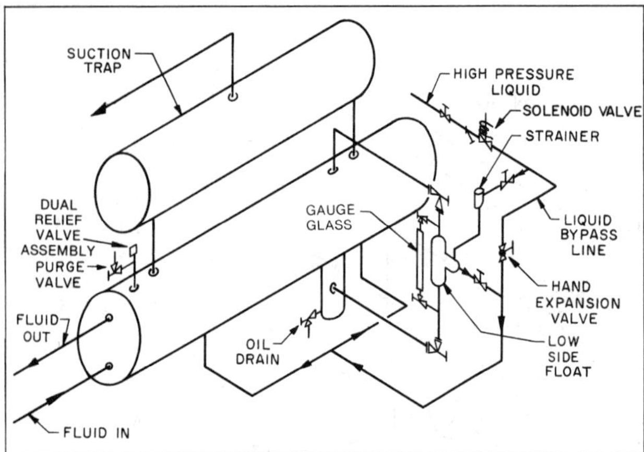

Fig. 39 Piping for Evaporator and Low-Side Float with Horizontal Liquid Cooler

top of the float chamber to the space above the tube bundle and an equalizer line out of the lower side of the float chamber to the lower side of the tube bundle.

For positive shutoff of liquid feed when the system stops, a solenoid valve in the liquid line is wired so that it is only energized when the brine or water pump motor is operating and the compressor is running.

Use a reflex glass lens with large liquid chamber and vapor connections for boiling liquids and with a plastic frost shield to determine the actual level with front extensions as required.

Usually a high-level float switch is installed above the operating level of the float to sheet the liquid solenoid valve if the float should overfeed.

AMMONIA LIQUID RECIRCULATION SYSTEMS

The following discussion gives an overview of liquid recirculating (liquid overfeed) systems. See Chapter 1 for more complete information. For additional engineering details on liquid overfeed systems, refer to Stoecker (1988).

In a liquid ammonia recirculating system, a pump circulates the ammonia from a low-pressure receiver to the evaporators. The low-pressure receiver is a shell for storing refrigerant at low-pressure and is used to supply evaporators with refrigerant, either by gravity or by a low-head pump. It also takes the suction from the evaporators and separates the gas from the liquid. Since the amount of liquid fed into the evaporator is usually several times the amount that actually evaporates there, liquid is always present in the suction return to the low-pressure receiver. Frequently, three times the evaporated amount is circulated through the evaporator (see Chapter 1).

Generally, the liquid ammonia pump is sized by the flow rate required and a pressure differential of about 25 psi. This is satisfactory for most single-story installations. If there is a static lift on the pump discharge, the differential is increased accordingly.

Size the low-pressure receiver by the cross-sectional area required to separate liquid and gas, and by the volume between the normal liquid level and the alarm liquid level in the low-pressure receiver. This volume should be sufficient to contain the maximum fluctuation in liquid from the various load conditions (see Chapter 1).

The liquid at the pump discharge is in the subcooled region. A total pressure drop of about 5 psi in the piping can be tolerated.

The remaining pressure is expended through the control valve and coil. The pressure drop and heat pickup in the liquid supply

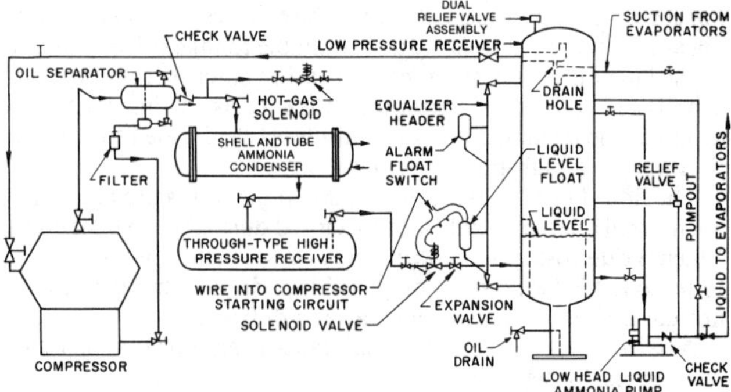

Fig. 40 Piping for Single-Stage System with Low-Pressure Receiver and Liquid Ammonia Recirculation

line should be low enough to prevent any flashing in the liquid supply line.

Provisions for liquid relief are required from the liquid main back to the low-pressure receiver, so that when the liquid line solenoid valves at the various evaporators are closed, either for defrosting or for temperature control, the excess liquid can be relieved back to the low-pressure receiver. Generally, relief valves used for this purpose are set at about 40 psi differential when positive displacement pumps are used. When centrifugal pumps are used, a hand expansion valve or a minimum flow orifice is acceptable to ensure that the pump is not dead-headed.

The suction header between the evaporators and the low-pressure receiver should be pitched 1% to permit excess liquid flow back to the low-pressure receiver. The header should be designed to avoid traps.

Liquid Recirculation in Single-Stage System. Figure 40 shows the piping of a typical single-stage system with a low-pressure receiver and liquid ammonia recirculating feed.

HOT GAS DEFROST SYSTEMS FOR AMMONIA RECIRCULATION SYSTEMS

Several methods are used for defrosting coils in areas that operate below 35 °F room temperature. These include:

1. Hot refrigerant gas, which is the predominant method
2. Water
3. Air
4. Combinations of 1, 2, and 3

The evaporator (air unit) in a liquid recirculation system is circuited so that the refrigerant flow provides maximum cooling efficiency. The evaporator can also work as a condenser if the necessary piping and flow modifications are made. When operated as a condenser, and when the fans are shut down, the hot refrigerant vapor raises the surface temperature of the coil high enough to melt any ice and/or frost on the surface so that it drains off. While the method is effective, it can be troublesome and inefficient if not properly designed.

Even while the fans are not operating, up to 50% or more of the heat given up by the refrigerant is lost to the space. To keep this heat loss to a minimum, the temperature difference between the coil surface and the room air, which is a function of the refrigerant temperature/pressure, should be kept low.

Another reason to maintain the lowest possible defrost temperature/pressure, particularly in freezers, is to keep the coil from steaming. Steam increases the refrigeration load, and the resulting icicle or frost formation has to be dealt with. Icicles increase maintenance during cleanup and ice formed during defrost tends to collect at the fan rings, which sometimes restricts fan operation.

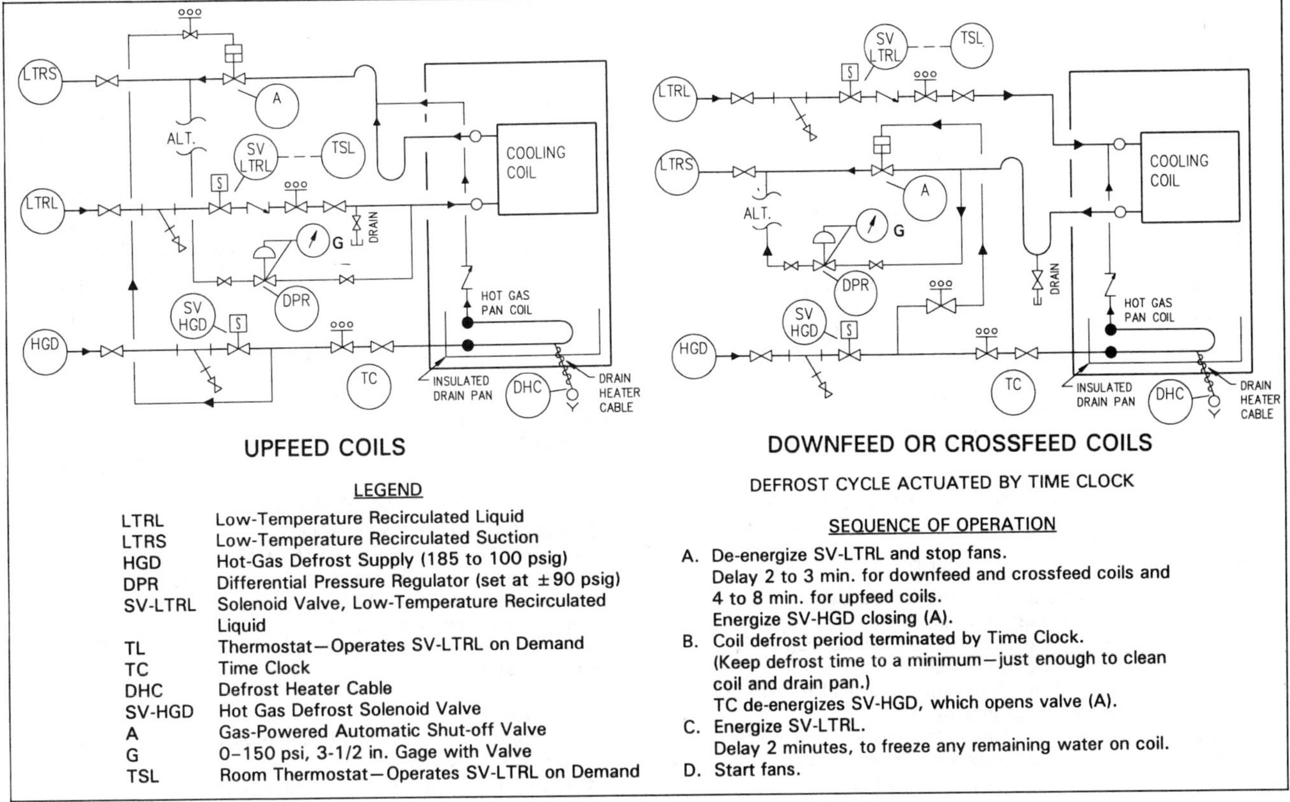

Fig. 41 Conventional Hot Gas Defrost Cycle
(For coils with 15 tons refrigeration capacity or below)

At lower defrost pressures, the defrosting takes slightly longer than at higher pressures. Obviously the less time allowed to add heat to the space, the more efficient the defrost. However, with slightly extended defrost times at lower temperatures, the overall defrosting efficiency is much greater than at higher temperature/pressure because refrigeration requirements are reduced.

Another loss during defrost can occur when hot, or uncondensed, gas blows through the coil and relief regulator and vents back to the compressor. Some of this gas load cannot be contained, and it must be vented to the compressor through the wet return line. It is most energy-efficient to vent this hot gas to the highest suction possible. Thus, an evaporator defrost relief should be vented to the intermediate or high-stage compressor, if the system is two-stage. See Figure 41 for a conventional hot gas defrost system for evaporator coils of 15 tons of refrigeration and below. Note that the wet return is above the evaporator and that a single riser is employed.

Demand Defrost

Since the efficiency of defrosting is low, it should not be initiated any more than is necessary to keep the coils clean. Less defrosting is required during the winter than during hotter, more humid periods. An effective energy-saving measure is to reset defrost schedules in the winter.

Several methods are used to initiate the defrost cycle. Demand defrost, actuated by a pressure device that measures the air pressure drop across the coil is a good way of minimizing total daily defrost time. Using this method, the coil is defrosted automatically only when necessary. This initiation system, together with the addition of a float drainer to dump the liquid formed during defrost to an intermediate vessel, comprises the most efficient defrost system available. See Figure 42 for details.

The most common defrost control method, however, is time initiated and time terminated; it includes an adjustable time of defrost duration and an adjustable number of defrost cycles per 24-h period. This control is commonly provided by a defrost timer.

Estimates indicate that the load placed on a refrigeration system by a coil during defrost is up to three times the design load during the operating cycle. Thus, it is important to properly engineer hot gas defrost systems.

Designing Hot Gas Defrost Systems

Several approaches are followed in designing hot gas defrost systems. Figure 42 shows a typical demand defrost system for both upfeed and downfeed coils. This design returns the defrost liquid to the system's intermediate pressure. An alternate way of doing this is to direct the defrost liquid into the wet suction. A float drainer or thermostatic trap with a hot gas regulator installed at the hot gas inlet to the coil is much better than the relief regulator, see Figure 42.

Most defrost systems installed today (Figure 41) use a time clock to initiate defrost; a demand defrost system is shown in Figure 42. The system uses a photo helix pressure switch to sense the air pressure drop across the coil and actuate the defrost. A thermostat terminates the defrost cycle. A timer is used as a backup to make sure the defrost terminates.

Sizing and Designing Hot Gas Piping

Hot gas is supplied to the evaporators in two ways:

1. The preferred method is to install a pressure regulator in the equipment room at the hot gas takeoff, set at approximately 100 psig, and size the piping accordingly.
2. Install a pressure regulator at each evaporator or group of evaporators and size the piping for minimum design condensing pressure, which should be 75 to 85 psig.

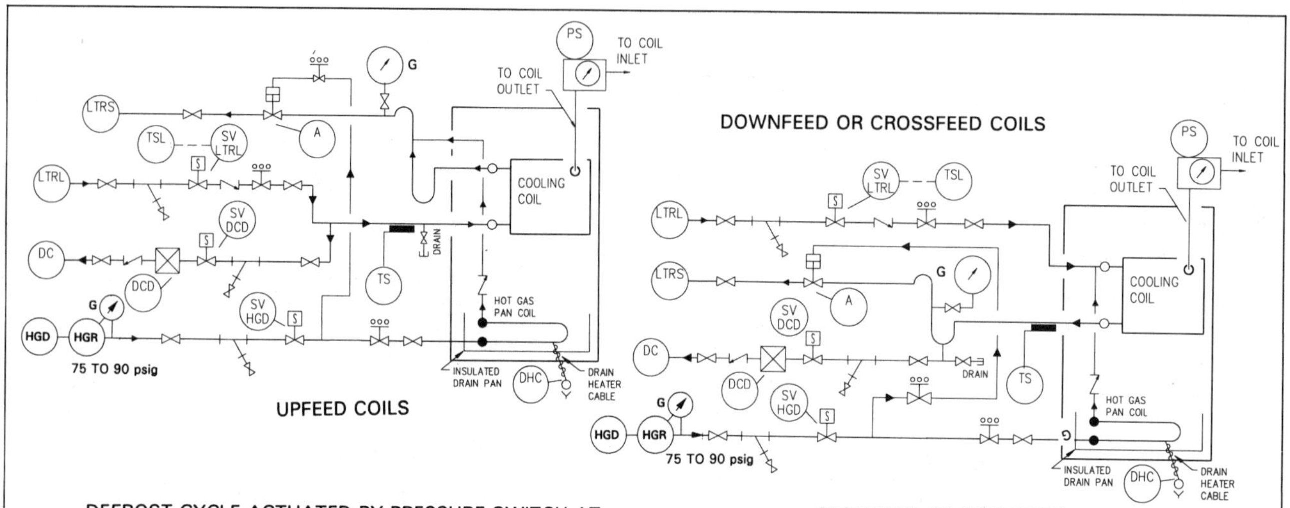

UPFEED COILS

DOWNFEED OR CROSSFEED COILS

DEFROST CYCLE ACTUATED BY PRESSURE SWITCH AT PREDETERMINED SETTING (approx. 1 in. water gage)

LEGEND

DCD	Defrost Condensate Return Drainer
LTRL	Low-Temperature Recirculated Liquid
LTRS	Low-Temperature Recirculated Suction
HGD	Hot-Gas Defrost Supply
TS	Thermostat—Terminates Hot-Gas Defrost Cycle
PS	Air Side Pressure Switch
TSL	Room Thermostat—Operates SV-LTRL on Demand
SV-LTRL	Recirculated Liquid Line Solenoid Valve
SV-HGD	Hot Gas Defrost Solenoid Valve
A	Gas-Powered Automatic Shut-off Valve
G	0–150 psi, 3-1/2 in. Gage with Valve
DHC	Defrost Heater Cable
DC	Return Liquid Defrost Condensate
HGR	Hot Gas Regulator Valve

SEQUENCE OF OPERATION

A. De-energize SV-LTRL and stop fans.
   Delay energize SV-HGD closing (A). Coil will defrost.
B. Coil defrost period terminated by TS set at 40°F.
   Use override timer after 30 min. to force SV-HGD to de-energize if TS malfunctions.
C. De-energize SV-HGD via thermostat TS opening valve (A).
D. Energize SV-LTRL.
   Delay 2 to 4 min.
E. Start fans.
F. To save energy, return liquid to intercooler of intermediate temperature recirculator.
   Alternate: Return liquid to wet suction downstream of stop valve.

Note: This defrost control method assumes a hot-gas outlet pressure regulator is in the equipment room. If high-pressure hot gas is used, an outlet pressure regulator with electric shut-off should be substituted for SV-HGD and set at 75 to 90 psig.

Fig. 42 Demand Defrost Cycle
(For coils with 15 tons refrigeration capacity or below)

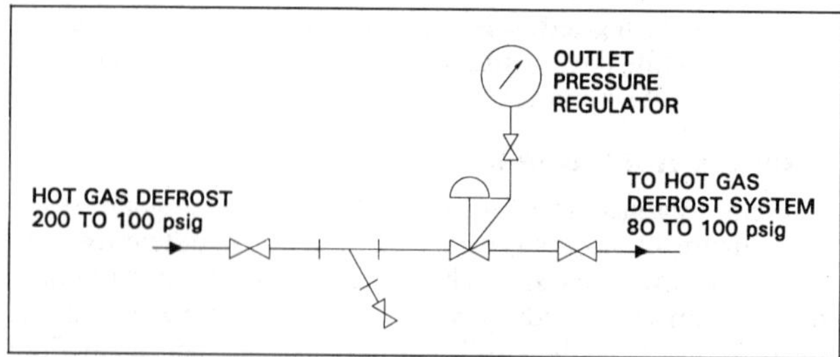

Fig. 43 Equipment Room Hot Gas Pressure Control System

A maximum of one-third of the coils in a system should be defrosted at one time. If a system has 300 tons of refrigeration coils requiring defrost, then only 100 tons of refrigeration can be defrosted at one time. Thus, the main hot gas supply pipe could be sized for 100 tons of refrigeration × 3 = 300 tons of refrigeration. The outlet pressure-regulating valve should be sized in accordance with the manufacturers' engineering data.

Reducing the defrost hot gas pressure in the equipment room has several advantages, notably less liquid will condense in the hot gas line as the condensing temperature is reduced to 52 to 64°F. A typical equipment room hot gas pressure control system is shown in Figure 43. Where hot gas lines in the system are trapped, a condensate drainer must be installed at each trap and at the low point in the hot gas line (see Figure 44). Defrost condensate (DC) liquid return piping from coils where a float or thermostatic valve

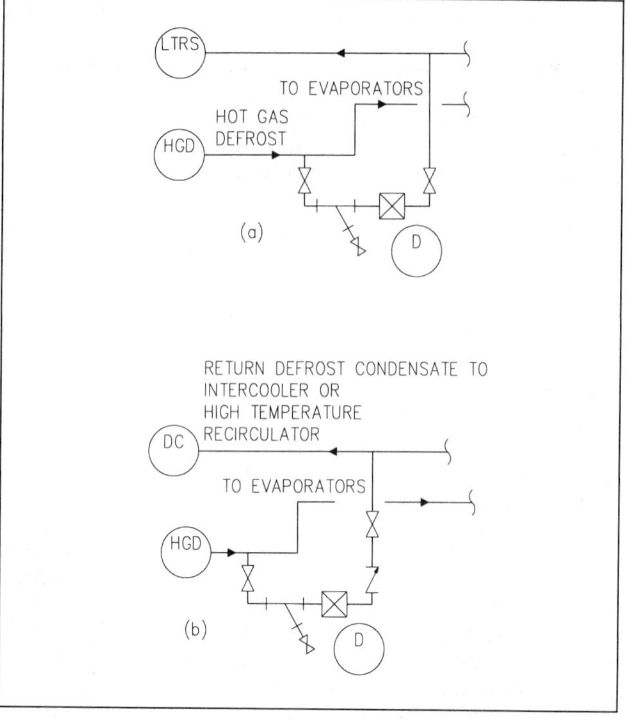

Fig. 44 Hot Gas Condensate Return Drainer

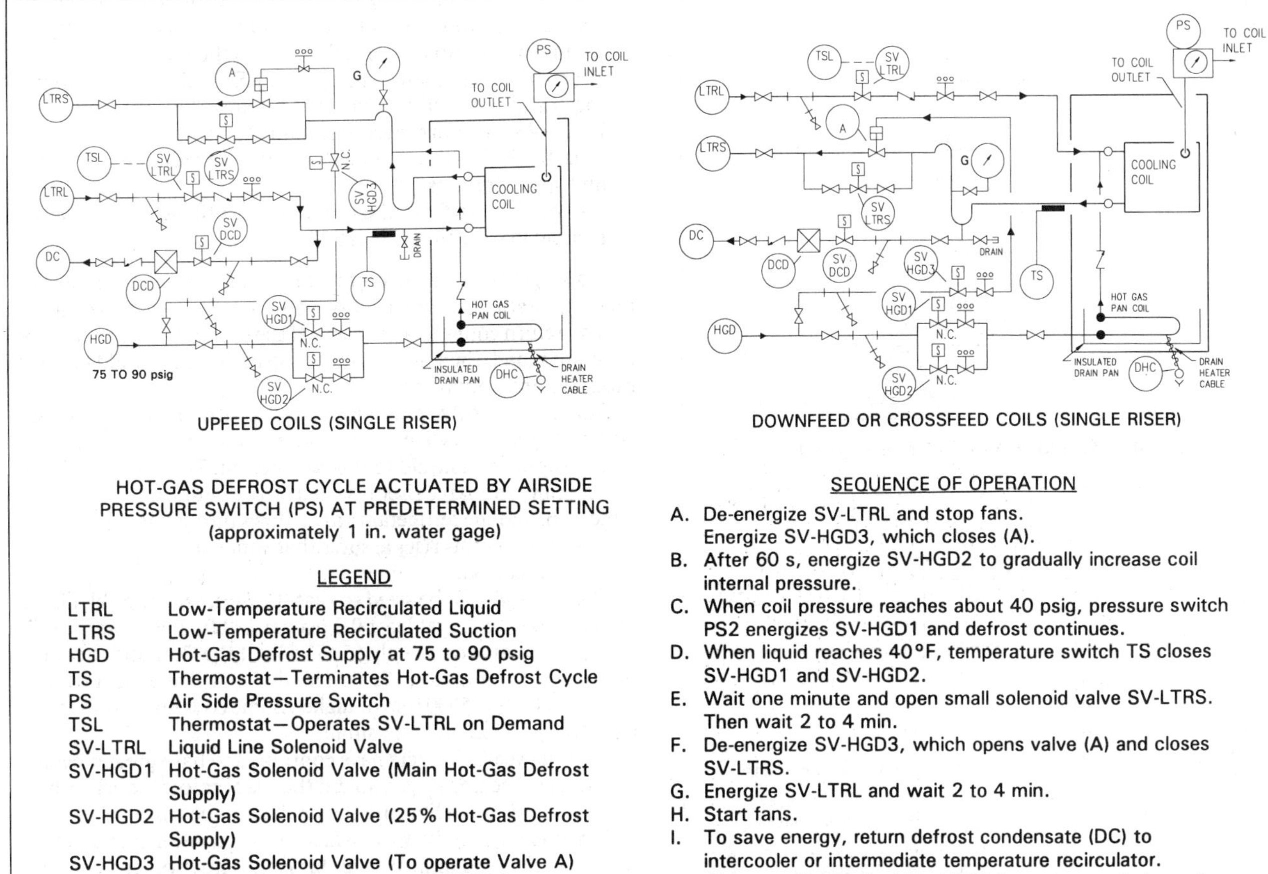

HOT-GAS DEFROST CYCLE ACTUATED BY AIRSIDE PRESSURE SWITCH (PS) AT PREDETERMINED SETTING (approximately 1 in. water gage)

LEGEND

LTRL	Low-Temperature Recirculated Liquid
LTRS	Low-Temperature Recirculated Suction
HGD	Hot-Gas Defrost Supply at 75 to 90 psig
TS	Thermostat—Terminates Hot-Gas Defrost Cycle
PS	Air Side Pressure Switch
TSL	Thermostat—Operates SV-LTRL on Demand
SV-LTRL	Liquid Line Solenoid Valve
SV-HGD1	Hot-Gas Solenoid Valve (Main Hot-Gas Defrost Supply)
SV-HGD2	Hot-Gas Solenoid Valve (25% Hot-Gas Defrost Supply)
SV-HGD3	Hot-Gas Solenoid Valve (To operate Valve A)
PS2	Pressure Switch to Operate SV-HGD1
SVLTRS	Solenoid Valve Bypass to LTRS
A	Gas-Powered Automatic Shut-off Valve LTRS
G	0–150 psi, 3-1/2 in. Gage with Valve
DHC	Drain Heater Cable
DC	Defrost Condensate Return Liquid
D	Defrost Condensate Drainer

SEQUENCE OF OPERATION

A. De-energize SV-LTRL and stop fans. Energize SV-HGD3, which closes (A).
B. After 60 s, energize SV-HGD2 to gradually increase coil internal pressure.
C. When coil pressure reaches about 40 psig, pressure switch PS2 energizes SV-HGD1 and defrost continues.
D. When liquid reaches 40°F, temperature switch TS closes SV-HGD1 and SV-HGD2.
E. Wait one minute and open small solenoid valve SV-LTRS. Then wait 2 to 4 min.
F. De-energize SV-HGD3, which opens valve (A) and closes SV-LTRS.
G. Energize SV-LTRL and wait 2 to 4 min.
H. Start fans.
I. To save energy, return defrost condensate (DC) to intercooler or intermediate temperature recirculator. *Alternate*: Return liquid to LTRS downstream of stop valve.

Note 1: This defrost method assumes the outlet pressure regulator is in the equipment room. If high-pressure hot gas is used, an outlet pressure regulator is installed at the HGD inlet.

Note 2: The soft hot-gas defrost system may be manually initiated also.

Fig. 45 Soft Hot Gas Defrost Cycle
(For coils with 15 tons refrigeration capacity or below)

is employed should be sized one size larger than the liquid feed piping to the coil.

Advantages and Features of Demand Defrost

1. Demand defrost uses the least amount of energy for defrost.
2. Demand defrost increases total system efficiency because coils are off-line for a minimum time.
3. Demand defrost imposes less stress on the piping system as there are fewer defrost cycles.
4. By regulating the hot gas to approximately 100 psig in the equipment room, the hot gas has less chance of condensing in the supply piping. Liquid in hot gas systems may cause problems due to the hydraulic shock created when the liquid is forced into an evaporator (coil). Coils in hot gas pans may rupture as a result.
5. Draining the liquid formed during defrost with a float or thermostatic drainer eliminates hot gas blow-by normally associated with pressure-regulating valves installed around the wet suction return line pilot check valve.
6. Returning the ammonia liquid to the intercooler or high-stage recirculator saves considerable energy in a system. A 20-ton refrigeration coil defrosting for 12 min will condense up to 24 lb/min of ammonia or 288 lb total. The enthalpy difference between returning to the low-stage recirculator (−40°F) and the intermediate recirculator (+20°F) is 64 Btu/lb or 18,432 Btu total or 7.68 tons of refrigeration removed from the −40°F booster for 12 min; this assumes that only liquid is drained. Note that this is the saving when liquid is drained to the intermediate point, not the total cost to defrost. If a pressure-reducing valve is used around the pilot check valve, this rate could double or triple as hot gas flows through these valves in unmeasurable quantities.

Soft Hot Gas Defrost System

The soft hot gas defrost system (SDS) is particularly well-suited to large evaporators and should be used on all coils of 15 tons of refrigeration or over. The SDS system eliminates the valve clatter, pipe movements, and noise associated with large coils during hot gas defrost. The SDS system can be used for upfeed or downfeed coils; however, the piping systems differ (see Figure 45). Coils operated in the horizontal plane must be orificed. Vertical coils that usually are cross-fed are also orificed.

The SDS system is designed to increase coil pressure gradually as defrost is initiated. This is accomplished by using a small hot gas feed with a capacity of about 25 to 30% of the estimated duty with a solenoid and a hand expansion valve adjusted to bring the

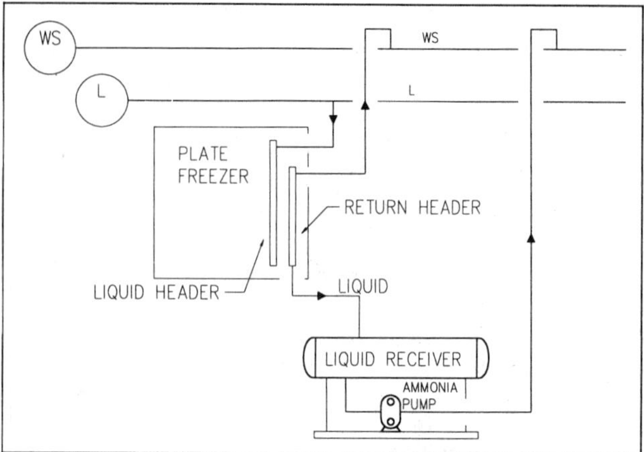

Fig. 46 Overfeed Liquid Return System

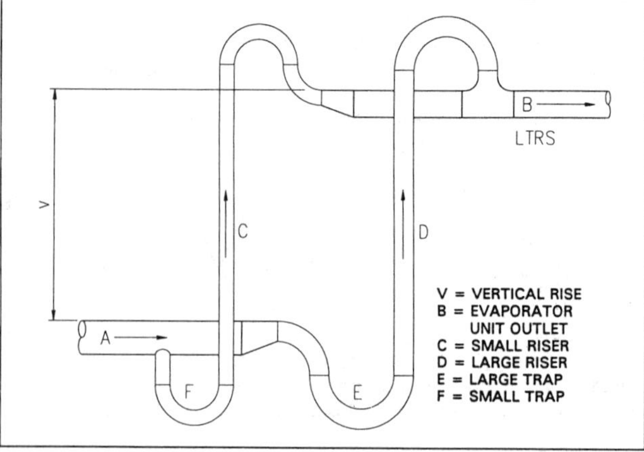

Fig. 47 Dual Low-Temperature Suction Risers

pressure up to about 40 psig in 3 to 5 min. (Refer to Sequence of Operation in Figure 45). After defrost, a small suction line solenoid is opened so that the coil can be brought down to operation pressure gradually before liquid is introduced and the fans started. The SDS can be initiated by a pressure switch just as with the demand defrost system; however, for large coils in spiral or individual quick freezing systems, manual initiation is preferred.

The soft hot gas defrost system (1) eliminates check valve chatter, and (2) eliminates most, if not all liquid hammer, *i.e.*, hydraulic problems in the piping. In addition, features 4, 5, and 6 listed in the Demand Defrost System section apply to the SDS system.

Double Riser Designs for Large Evaporator Coils

Static pressure penalty is the pressure/temperature loss associated with a refrigerant vapor stream bubbling through a liquid bath. If the velocity in the riser is high enough, it will carry over a certain amount of liquid, thus reducing the penalty. For example, at −40°F ammonia has a density of 43.08 lb/ft^3, which is equivalent to a pressure of 43.08/144 = 0.30 psi per foot of depth. Thus, a 16-ft riser has a column of liquid that exerts 16 × 0.30 = 4.8 psi. At −40°F, ammonia has a saturation pressure of 10.4 psia. At the bottom of the riser then, the pressure is 4.8 + 10.4 = 15.2 psia, which is the saturation pressure of ammonia at −27°F. This 13°F difference amounts to an 0.81°F penalty per foot of riser. If a riser was oversized to a point that the vapor did not carry the liquid to the wet return, the evaporator would be −27°F instead of −40°F. This problem can be solved in several ways:

1. Install the recirculated suction line (LTRS) below the evaporator. This method is very effective for downfeed evaporators. The suction from the coil should not be trapped. This piping arrangement also assures oil return to the recirculator.
2. Where the recirculated suction (LTRS) is above the evaporator, install a liquid return system below the evaporator (see Figure 46). This arrangement eliminates static penalty, which is particularly advantageous for plate, individual quick freeze, and spiral freezers.
3. Use dual risers from the evaporator to the low-temperature recirculated suction (Figure 47).

If a single riser was sized for minimum pressure drop at full load, the static pressure penalty would be excessive at part load; also oil return could be a problem. If the single riser were sized for minimum load, then the pressure drop in the riser would become excessive and counterproductive.

Double risers solve problems inherent with single risers. Referring to Figure 47, when the maximum load occurs, both risers return vapors and liquid to the wet suction. When the load is at a minimum, this large riser is sealed by the liquid ammonia in the large trap, and the refrigerant vapor flows through the small riser. A small trap on this riser assures that some oil and liquid return to the wet suction.

The risers should be sized so that the pressure drop calculated on a dry gas basis is at least 0.3 psi per 100 ft. The larger riser is designed for approximately 65 to 75% of the flow and the small one for 100% minus 65 to 75%. This design will result in a velocity of approximately 5000 fpm or higher. Some coils may require three risers (large, medium, and small).

Over the years, freezers have continued to have greater capacity. As they became larger, so did the evaporators (coils). Where these freezers are in line and the product to be frozen is wet, the defrost cycle can be every 4 or 8 h. Many production lines limit the defrost cycle to 30 min. Where this is the situation and the coils are large (some coils have a refrigeration capacity of 200 to 300 tons), it is difficult to design a hot gas defrost system that can complete a safe defrost in 30 min. Sequential defrost systems where the coils are defrosted alternately during production are feasible but require special treatment.

COMPOUND SYSTEMS

As pressure ratios increase, single-staged ammonia systems encounter problems including (1) high discharge temperatures on reciprocating compressors causing the oil to deteriorate, (2) loss of volumetric efficiency as high pressure leaks back to the low-pressure side through compressor clearances, and (3) excessive stresses on compressor moving parts. Thus, manufacturers usually limit the maximum pressure ratios for multicylinder reciprocating machines to approximately 7 to 9. For screw compressors, compression ratio is not a limitation, but efficiency deteriorates at high ratios. (Screw compressors incorporate oil cooling.) When the overall system pressure ratio (absolute discharge pressure divided by absolute suction pressure) begins to exceed these limits for a particular system, the pressure ratio occurring across the compressor must be reduced. This is usually accomplished by employing a two-stage system. A properly designed two-staged system exposes each of the two compressors to a pressure ratio equal to approximately the square root of the overall pressure ratio. In a three-staged system, each compressor is exposed to a pressure ratio equal to approximately the cube root of the overall ratio. When screw compressors are employed, this calculation does not always assure the most efficient system.

Another advantage to multistaging is that successively subcooling the liquid at each stage of compression increases the overall system operating efficiency. Additionally, multistaging can be used to accommodate multiple loads at different suction pressures and

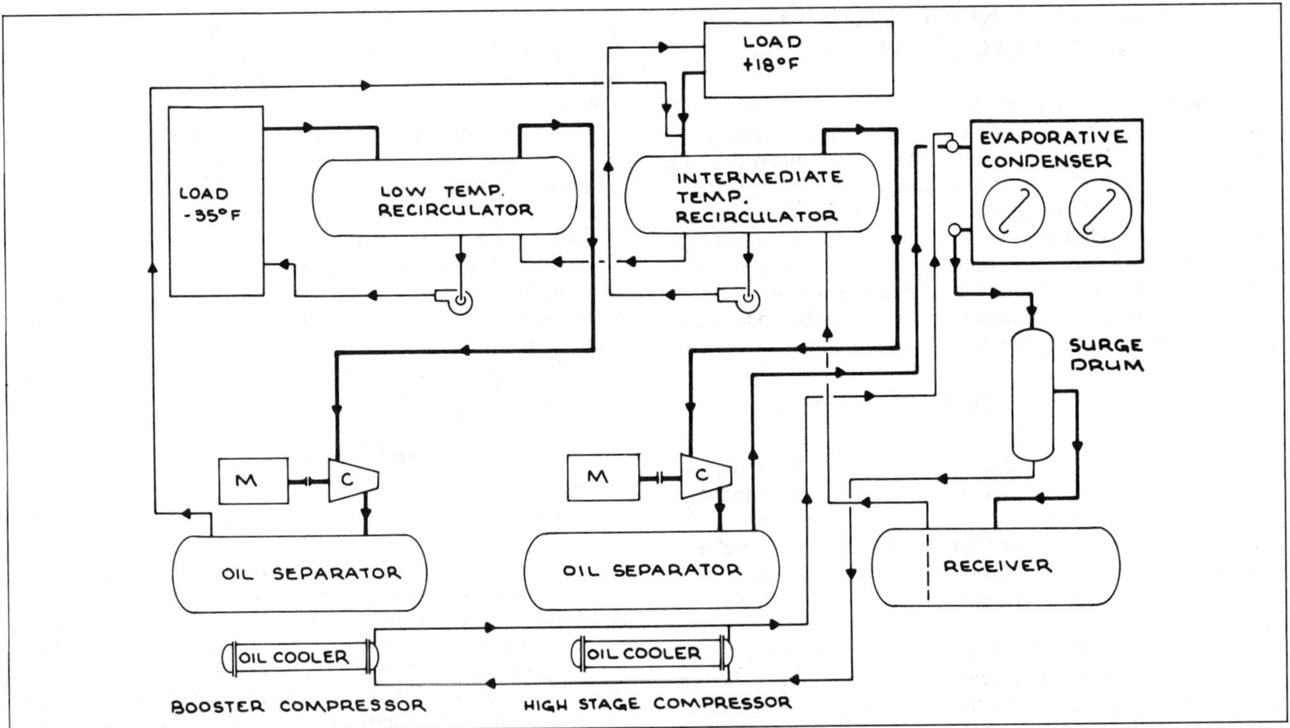

Fig. 48 Compound Ammonia System with Screw Compressor Thermosyphon Cooled

temperatures in the same refrigeration system. In some cases, two stages of compression can be contained in a single compressor such as an internally compounded reciprocating compressor. In these units, one or more cylinders are isolated from the others so that they can act as independent stages of compression. Internally compounded compressors are economical for small systems that require low temperature.

Two-Stage (Compound) Screw Compressor System

A typical two-stage, two-temperature system using screw compressors provides refrigeration for high- and low-temperature loads. For example, the high-temperature level will supply refrigerant to all process areas operating between 28 and 50°F. An 18°F intermediate suction temperature is selected. The low-temperature level requires a −35°F suction temperature for blast freezers and continuous or spiral freezers.

This type of system requires a two-stage refrigeration system (see Figure 48). It employs a flash-type intercooler that doubles as a recirculator for the +18°F load. It is the most efficient system available, if the screw compressor employs indirect lubricant cooling. If refrigerant injection cooling is used, system efficiency is decreased. This system is efficient for several reasons:

1. Approximately 50% of the booster (low-stage) motor heat is removed from the high-stage compressor load by using the refrigerant-cooled indirect lubricant cooler.

   Note: In any system, thermosyphon lubricant cooling for booster and high-stage compressors is about 10% more efficient than injection cooling. Also, plants with a piggyback, two-stage screw compressor system with no intercooling and injection cooling can be converted to a compound system with indirect cooling to increase system efficiency approximately 15%.
2. Flash-type intercoolers are more efficient than shell-and-coil intercoolers by several percent.
3. Indirect lubricant cooling of the high-stage screw compressor provides the highest efficiency available. Note that installing indirect cooling in plants with liquid injection cooling of screw compressors can increase the compressor efficiency by 3 to 4%.
4. Indirect cooling saves 20 to 30% in electric energy during the low-temperature months. During low outside air temperature conditions, the condensing pressure can be decreased to a 90 to 100 psig minimum in most ammonia systems. With liquid injection cooling, the condensing pressure can only be reduced to approximately 125 to 130 psig.
5. Variable V_i compressors with microprocessor control require less total energy when employed as high stage compressors. This controller will track the compressor operating conditions to take advantage of the ambient conditions as well as the variations in load.

Converting Single-Stage into Two-Stage Systems

When plant refrigeration capacity must be increased and the system is operating below about 10 psig suction pressure, it is usually more economical to increase capacity by adding a compressor to operate as the low-stage compressor of a two-stage system. The existing single-stage compressor then becomes the high-stage compressor of the two-stage system. Some items to consider when converting are:

- The motor on the existing single-stage compressor may have to be increased in size when used at a higher suction pressure.
- The suction trap should be checked for sizing at the increased gas flow rate.
- An intercooler should be added to cool the low-stage compressor discharge gas and to cool high-pressure liquid.
- A condenser may have to be added to handle the increased condensing load.
- A means to purge air should be added if plant suction gage pressure is below zero.
- A means to automatically reduce compressor capacity should be added so that the system will operate satisfactorily at reduced system capacity points.

AMMONIA REFRIGERANT FOR HVAC SYSTEMS

The use of ammonia for HVAC systems has received renewed interest due, in part, to the scheduled phaseout and increasing costs of CFC and HCFC refrigerants. While ammonia will not be used (or allowed) in direct refrigeration systems (ammonia in the air unit coils) for HVAC applications, ammonia secondary systems that circulate chilled water or other secondary refrigerant are a viable alternative to halocarbon systems. Ammonia package chilling units are available for HVAC applications. As with the installation of any air-conditioning unit, all applicable codes, standards, and insurance requirements must be followed.

SAFETY CONSIDERATIONS

Ammonia (R-717) is an economical choice for industrial systems. While ammonia has superior thermodynamic properties, at low concentration levels of 35 to 50 ppm, it is considered toxic. Large quantities of ammonia should not be vented to enclosed areas near open flames or heavy sparks. A proportion of 16 to 25% by volume in air in the presence of an open flame burns and can explode.

The importance of ammonia piping is sometimes minimized when the main emphasis is on selecting major equipment pieces. Mains should be sized carefully to provide low pressure drop and avoid capacity or power penalties caused by inadequate piping.

Rusting pipes and vessels in older systems containing ammonia can create a safety hazard. Oblique X-ray photographs of welded pipe joints and ultrasonic inspection of vessels may be used to disclose defects. Only vendor-certified parts for pipe, valving, and pressure-containing components according to designated assembly drawings should be used to reduce hazards. Cold liquid refrigerant should not be confined between closed valves in a pipe where the liquid can warm and expand to burst piping components. Rapid multiple pulsations of ammonia liquid in piping components, such as those developed by cavitation forces or hydraulic hammering from compressor pulsations with massive slugs of liquid carryover to the compressor, must be avoided to prevent equipment and piping damage and injury to personnel.

Most service problems are caused by inadequate precautions during design, construction, and installation (ANSI/IIAR *Standard* 2-1984, ANSI/ASHRAE *Standard* 15-1992). Ammonia is a powerful solvent that removes dirt, scale, sand, or moisture remaining in the pipes, valves, and fittings during installation. These substances are swept along with the suction gas to the compressor, becoming a menace to the bearings, pistons, cylinder walls, valves, and lubricating oil. Most compressors are equipped with suction strainers and/or additional disposable strainer liners for the large quantity of debris that can be present at initial start-up.

Moving parts are often scored when a compressor is run for the first time. Damage starts with minor scratches, which increase progressively until they seriously affect the operation of the compressor or render it inoperative.

A system that has been carefully and properly installed with no foreign matter or liquid entering the compressor will operate satisfactorily for a long time. As piping is installed, it should be power rotary wire brushed and blown out with compressed air. Then the piping system should be blown out with compressed air or nitrogen before evacuation and charging. See ANSI/ASHRAE 15-1992 for system piping test pressure.

BIBLIOGRAPHY

ANSI/ASHRAE. 1992. Safety code for mechanical refrigeration. *Standard* 15-1992.

ANSI/ASME. 1987. Code for pressure piping, B31.5-1987. American Society of Mechanical Engineers, New York.

ANSI/IIAR. 1984. Equipment, design, and installation of ammonia mechanical refrigeration systems. ANSI/IIAR 2-1984. International Institute of Ammonia Refrigeration, Chicago.

Bradley, W.E. 1984. Piping evaporative condensers. In Proceedings of IIAR meeting. International Institute of Ammonia Refrigeration, Chicago.

Briley, G.C. and T.A. Lyons. 1992. Hot gas defrost systems for large evaporators in ammonia liquid overfeed systems. IIAR Technical Pager 163.

Cole, R.A. 1986. Avoiding refrigeration condenser problems. *Heating/Piping/and Air-Conditioning,* Parts I and II, 58(7, 8).

Evaporative condenser engineering manual. 1983. Baltimore Aircoil Company, Inc., Baltimore, MD.

IIAR. 1992. Guidelines for avoiding component failure in industrial refrigeration systems caused by abnormal pressure or shock. *Bulletin* No. 116.

Loyko, L. 1989. Hydraulic shock in ammonia systems. IIAR Technical Paper T-125.

Miller, D.K. 1979. Sizing dual-suction risers in liquid overfeed refrigeration systems. *Chemical Engineering* (September 24).

Nuckolls, A.H. The comparative life, fire, and explosion hazards of common refrigerants. Miscellaneous Hazard No. 2375. Underwriters Laboratory, Northbrook, IL.

Stoecker, W.F. 1988. Chapters 8 and 9 in *Industrial refrigeration.* Business News Publishing Company, Troy, MI.

Strong, A.P. 1984. Hot gas defrost—A-one-a-more-a-time. IIAR Technical Paper T-53.

Wile, D.D. 1977. *Refrigerant line sizing.* ASHRAE.

CHAPTER 4

SECONDARY COOLANTS IN REFRIGERATION SYSTEMS

A SECONDARY coolant is a liquid that is used as a heat transfer fluid and that changes temperature as it gains or loses heat energy without changing into another phase. For the lower temperatures of refrigeration, this requires a coolant with a freezing point below that of water. In this chapter, the design considerations for components, system performance requirements, and applications for secondary coolants are discussed. Related information can be found in Chapters 2, 3, 17, 18, and 33 of the 1993 ASHRAE *Handbook—Fundamentals.*

COOLANT SELECTION

A secondary coolant must be compatible with the other materials in the system at the pressures and temperatures encountered for maximum component reliability and operating life. The coolant should also be compatible with the environment and the applicable safety regulations, and it should be economical to use and replace.

The coolant should have a minimum freezing point of 5 °F below and preferably 15 °F below the lowest temperature to which it will be exposed. When subjected to the lowest temperature in the system, the viscosity of the coolant should be low enough to allow satisfactory heat transfer and reasonable pressure drop.

The vapor pressure of the coolant should not exceed that allowed at the maximum temperature encountered. To avoid a vacuum in a low vapor pressure secondary coolant system, the coolant can be pressurized with pressure-regulated dry nitrogen in the expansion tank. However, recognize that some special secondary coolants such as those used for computer circuit cooling have a high solubility for nitrogen and must therefore be isolated from the nitrogen with a suitable diaphragm.

Load Versus Flow Rate

The secondary coolant pump is usually in the return line upstream of the chiller. Therefore, to be accurate, the pumping rate in gallons per minute is based on the density at the return temperature. The mass flow rate for a given heat load is based on the desired temperature range and required coefficient of heat transfer at the average bulk temperature.

To determine heat transfer and pressure drop, the specific gravity, specific heat, viscosity, and thermal conductivity are based on the average bulk temperature of the coolant in the heat exchanger, noting that film temperature corrections are based on the average film temperature. Trial solutions of the secondary coolant side coefficient compared to the overall coefficient and the total LMTD (log mean temperature difference) determine the average film temperature. Where the secondary coolant is cooled, the more viscous film reduces the heat transfer rate and raises the pressure drop compared to what can be expected at the bulk temperature. Where the secondary coolant is heated, the less viscous film approaches the heat transfer rate and pressure drop expected at the bulk temperature.

The greater the amount of turbulence and mixing of the bulk and film, the better the heat transfer and the higher the pressure drop. Where secondary coolant velocity in the tubes of a heat transfer device results in laminar flow, the heat transfer can be improved by inserting spiral tapes or spring turbulators that promote mixing the bulk and film. This usually increases pressure drop. The inside surface can also be spirally grooved or augmented by other devices. Since the state of the art of heat transfer is constantly improving, use the most cost-effective heat exchanger to provide optimum heat transfer and pressure drop. Energy costs for pumping the secondary coolant must be considered when selecting the fluid to be used and the heat exchangers to be installed.

Pumping Cost

Pumping costs are a function of the secondary coolant selected, the load and temperature range where energy is transferred, the pump head required by the system pressure drop (including that of the chiller), the mechanical efficiencies of the pump and driver, and the electrical efficiency and power factor where the driver is an electric motor. Small centrifugal pumps, operating in the range of approximately 50 gpm at 80 ft head to 150 gpm at 70 ft head, for 60 Hz applications, typically have 45 to 65% efficiency, respectively. Larger pumps, operating in the range of 500 gpm at 80 ft of head to 1500 gpm at 70 ft of head, for 60 Hz applications, typically have 75 to 85% efficiency, respectively.

A pump should operate near its peak operating efficiency for the flow rate and head that usually exist. The secondary coolant temperature increases slightly from the energy expended at the pump shaft. If a semihermetic electric motor is used as the driver, the motor inefficiency is added as heat to the secondary coolant, and the total kilowatt input to the motor must be considered in establishing load and temperatures.

Performance Comparisons

Assuming that the total refrigeration load at the evaporator includes the pump motor input and brine line insulation heat gains, as well as the delivered beneficial cooling, tabulating typical secondary coolant performance values assists in the coolant selection. A 1.06-in. ID smooth steel tube evaluated for pressure drop and internal heat transfer coefficient at the average bulk temperature of 20 °F and a temperature range of 10 °F for 7 fps tubeside velocity provides comparative data (see Table 1) for some typical coolants. Table 2 ranks the same coolants comparatively, using data from Table 1.

For a given evaporator configuration, load, and temperature range, select a secondary coolant that gives satisfactory velocities, heat transfer, and pressure drop. At the 20 °F level, hydrocarbon and halocarbon secondary coolants must be pumped at a rate of 2.3 to 3.0 times the rate of water-based secondary coolants for the same temperature range.

Higher pumping rates require larger coolant lines to keep the head and brake horsepower requirement for the pump within reasonable limits. Table 3 lists approximate ratios of pump power for secondary coolants. The heat transferred by a given secondary

---

The preparation of this chapter is assigned to TC 10.1, Custom-Engineered Refrigeration Systems.

coolant affects the cost and perhaps the configuration and pressure drop of a chiller and other heat exchangers in the system; therefore, Tables 2 and 3 are only guides of the relative merits of each coolant.

Other Considerations

Corrosion must be considered when selecting the coolant, an inhibitor, and the system components. The effect of secondary coolant and inhibitor toxicity on the health and safety of plant personnel or consumers of food and beverages must be considered. The flash point and explosive limits of secondary coolant vapors must also be evaluated.

Examine the secondary coolant stability for anticipated moisture, air, and contaminants at the temperature limits of materials used in the system. The skin temperatures of the hottest elements determine the secondary coolant stability.

If defoaming additives are necessary, their effect on the thermal stability and toxic properties of the coolant must be considered for the application.

DESIGN CONSIDERATIONS

The secondary coolant vapor pressure at the lowest operating temperature determines whether a vacuum could exist in the secondary coolant system. To keep air and moisture out of the system, pressure-controlled dry nitrogen can be applied to the top level of secondary coolant (*e.g.*, in the expansion tank or a storage tank). The gas pressure over the coolant plus the pressure created at the lowest point in the system by the maximum vertical height of coolant determine the minimum internal pressure for design purposes. The coincident highest pressure and lowest secondary coolant temperature dictate the design working pressure (DWP) and material specifications for the components.

To select proper relief valve(s) with settings based on the system DWP, the highest temperatures to which the secondary coolant could be subjected should be considered. This temperature would occur in case of heat radiation from a fire in the area or the normal warming of the valved-off sections. Normally, a valved-off section is relieved to an unconstrained portion of the system and the secondary coolant can expand freely without loss to the environment.

Safety considerations for the system are found in ASHRAE *Standard* 15-1992, Safety Code for Mechanical Refrigeration. The design standards for pressure piping can be found in ASME *Standard* B31.1, and the design standards for pressure vessels can be found in ASME Pressure Vessel Code.

Piping and Control Valves

Piping should be sized for reasonable pressure drop using the calculation methods in Chapters 2 and 33 of the 1993 ASHRAE *Handbook—Fundamentals*. Balancing valves or orifices in each of the multiple feed lines help distribute the secondary coolant. A reverse-return piping arrangement balances the flow. Control valves that vary the flow are sized for 20 to 80% of the total friction pressure drop through the system for proper response and stable operation. Valves sized for pressure drops smaller than 20% may respond too slowly to a control signal for a flow change. Valves sized for pressure drops in excess of 80% can be too sensitive, causing control cycling and instability.

Storage Tanks

Storage tanks can shave peak loads for brief periods, limit the size of the refrigeration equipment, and reduce energy costs. In off-peak hours, a relatively small refrigeration plant cools a secondary coolant stored for later use. A separate circulating pump sized for the maximum flow needed by the peak load is started to satisfy the peak load. Energy cost savings are enhanced if the refrigeration equipment is used to cool secondary coolant at night, when the cooling medium for heat rejection is generally at the lowest temperature.

Table 1 Secondary Coolant Performance Comparisons

Secondary Coolant	Concentration (by Weight), %	Freeze Point °F	gpm/ton[c]	Pressure Drop,[a] psi	Heat Transfer Coefficient[b] h_i, Btu/h·ft²·°F
Propylene glycol	39	−5.1	2.56	2.91	205
Ethylene glycol	38	−6.9	2.76	2.38	406
Methanol	26	−5.3	2.61	2.05	473
Sodium chloride	23	−5.1	2.56	2.30	558
Calcium chloride	22	−7.8	2.79	2.42	566
Aqua-ammonia	14	−7.0	2.48	2.44	541
Trichlorethylene	100	−123	7.44	2.11	432
d-Limonene	100	−142	6.47	1.48	321
Methylene chloride	100	−142	6.39	1.86	585
R-11	100	−168	7.61	2.08	428

[a]Based on one length of 16-ft tube with 1.06-in. ID and use of Moody Chart (1944) for 7 fps velocity. I/O losses equal one Vel. H_D for 7 fps velocity. Evaluations are at a bulk temperature of 20°F and a temperature range of 10°F.

[b]Based on a curve fit equation for Kern's adaptation (1950) of Sieder & Tate heat transfer equation (1936) using a 16-ft tube for L/D = 181 and a film temperature of 5°F lower than average bulk temperature with 7 fps velocity.

[c]Based on inlet secondary coolant temperature at the pump of 25°F.

Table 2 Comparative Ranking of Heat Transfer Factors at 7 fps[a]

Secondary Coolant	Heat Transfer Factor
Propylene glycol	1.000
d-Limonene	1.566
Ethylene glycol	1.981
R-11	2.088
Trichlorethylene	2.107
Methanol	2.307
Aqua-ammonia	2.639
Sodium chloride	2.722
Calcium chloride	2.761
Methylene chloride	2.854

[a]Based on Table 1 values using 1.06-in. ID tube 16 ft long. The actual ID and length vary according to the specific loading and refrigerant applied with each secondary coolant, tube material, and surface augmentation.

Table 3 Relative Pumping Energy Required[a]

Secondary Coolant	Energy Factor
Aqua-ammonia	1.000
Methanol	1.078
Propylene glycol	1.142
Ethylene glycol	1.250
Sodium chloride	1.295
Calcium chloride	1.447
d-Limonene	2.406
Methylene chloride	3.735
Trichlorethylene	4.787
R-11	5.022

[a]Based on the same pump head, refrigeration load, 20°F average temperature, 10°F range, and the freezing point (for water-based secondary coolants) 20 to 23°F below the lowest secondary coolant temperature.

The load profile over 24 h and the temperature range of the secondary coolant determine the minimum net capacity required for the refrigeration plant, the sizes of the pumps, and the minimum amount of secondary coolant to be stored. For maximum use of the storage tank volume at the expected temperatures, choose inlet velocities and locate the connections and the tank for maximum stratification. Note, however, that maximum use will probably never exceed 90% and, in some cases, may equal only 75% of the tank volume.

Example 1. Figure 1 depicts the load profile and Figure 2 shows the arrangement of a refrigeration plant with storage of a 23% (by weight) sodium chloride secondary coolant at a nominal 20 °F. During the peak load of 50 tons, a range of 8 °F is required. At an average temperature of 24 °F, with a range of 8 °F, the specific heat of the coolant c_p is 0.791 Btu/lb · °F. At 28 °F, the weight per unit volume of coolant at the pump (ρ_L) is 1.183 [(62.4 lb/ft³)/(7.48 gal/ft³)]; at 20 °F, the ρ_L is 1.185 [(62.4 lb/ft³)/(7.48 gal/ft³)]. Determine the minimum size storage tank for 90% use, the minimum capacity required for the chiller, and the sizes of the two pumps. The chiller and the chiller pump run continuously. The secondary coolant storage pump runs only during the peak load. A control valve to the load source diverts all coolant to the storage tank during a zero load condition, so that the initial temperature of 20 °F is restored in the tank. During the low load condition, only the required flow rate for a range of 8 °F at the load source is used; the balance returns to the tank and restores the temperature to 20 °F.

Solution: If x is the minimum capacity of the chiller, determine the energy balance in each segment by subtracting the load in each segment from x. Then multiply the result by the time length of the respective segments, and add as follows:

$$(x - 0) \times 6 + (x - 50) \times 4 + (x - 9) \times 14 = 0$$
$$6x + 4x - 200 + 14x - 126 = 0$$
$$24x = 326$$
$$x = 13.58 \text{ tons}$$

Calculate the secondary coolant flow rate (W) at peak load:

$$W = (50 \times 200)/(0.791 \times 8) = 1580.3 \text{ lb/min}$$

For the chiller at 15 tons, the secondary coolant flow rate is:

$$W = (15 \times 200)/(0.791 \times 8) = 474.1 \text{ lb/min}$$

Therefore, the coolant flow rate to the storage tank pump is 1580.3 − 474.1 = 1106.2 lb/min. The chiller pump size is determined by:

$$474.1/[(1.183 \times 62.4)/7.48] = 48 \text{ gpm}$$

Calculate the storage tank pump size as follows:

$$1106.2/[(1.185 \times 62.4)/7.48] = 111.9 \text{ gpm}$$

Using the concept of stratification in the storage tank, the interface between warm return and cold stored secondary coolant falls at the rate pumped from the tank. Since the time segments fix the total amount pumped and the storage tank pump operates only in segment 2 (see Figure 1), the minimum tank volume (V) at 90% use is determined as follows:

$$\text{Total mass} = [(1106.2 \text{ lb/min}) (60 \text{ min/h}) (4 \text{ h})]/0.90 = 295{,}000 \text{ lb}$$

and

$$V = 294{,}987/[(1.185 \times 62.4)/7.48] = 29{,}840 \text{ gal}$$

A larger tank (*e.g.*, 50,000 gal) provides flexibility for longer segments at peak load and accommodates potential mixing. It may be desirable to insulate and limit heat gains to 8000 Btu/h for the tank and lines. Energy use for pumping can be limited by designing for 46-ft head. With the smaller pump operating at 51% efficiency and the larger pump at 52.5% efficiency, the pump heat added to the secondary coolant would be 3300 Btu/h and 7478 Btu/h, respectively.

For cases with various time segments and their respective loads, the maximum load for segment 1 or 3 with the smaller pump operating cannot exceed the net capacity of the chiller minus insulation and pump heat gain to the secondary coolant. For various combinations of segment time lengths and cooling loads, the recovery or restoration rate of the storage tank to the lowest temperature required for satisfactory operation should be considered.

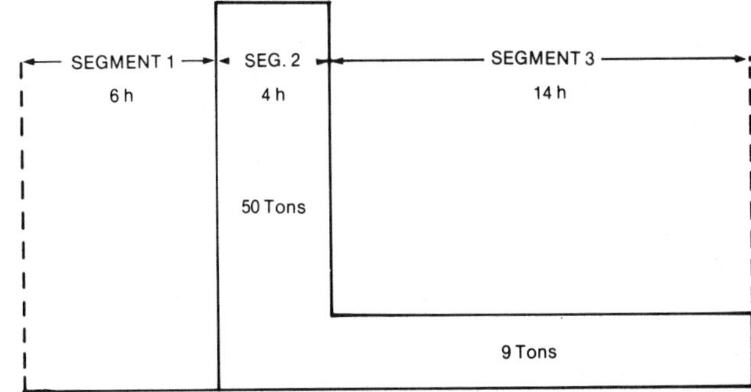

Fig. 1 Load Profile of a Refrigeration Plant where Secondary Coolant Storage Can Save Energy

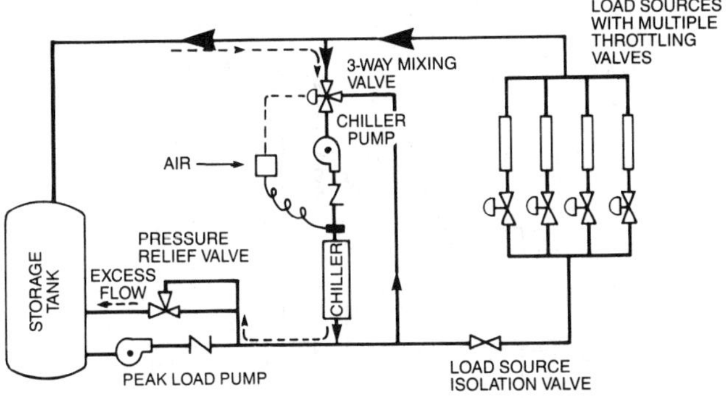

Fig. 2 Arrangement of a System with Secondary Coolant Storage

Figure 2 depicts a system arrangement with secondary coolant storage as described in Example 1.

As load source circuits shut off, the excess flow is bypassed back to the storage tank. The temperature setting of the 3-way valve is the normal return temperature for full flow through the load sources.

When only the storage tank requires cooling, the flow is as shown by the dotted lines with the load source isolation valve closed. When the storage tank temperature is at the desired level, the load isolation valve can be opened to allow cooling of the piping loops to and from the load sources for full restoration of storage cooling capacity.

Expansion Tanks

Figure 3 shows a typical closed secondary coolant system without a storage tank; it also illustrates different control strategies. The reverse-return piping assists flow balance. Figure 4 shows a secondary coolant strengthening unit for salt brines. The secondary coolant expansion tank volume is determined by considering the total coolant inventory and the differences in coolant density at the lowest temperature of coolant pumped to the load location (t_1) and the maximum temperature. The expansion tank is sized to accommodate a residual volume with the system coolant at t_1, plus an expansion volume and vapor space above the coolant. A vapor space equal to 20% of the expansion tank volume should be adequate. A level indicator, used to prevent overcharging, is calibrated at the residual volume level versus lowest system secondary coolant temperature.

Example 2. Assume a 50,000 gal charge of 23% sodium chloride secondary coolant at t_1 of 20 °F in the system. If 100 °F is the maximum temperature, determine the size of the expansion tank required. Assume that the residual volume is 10% of the total tank volume and that the vapor space at the highest temperature is 20% of the total tank volume.

$$\text{ETV} = \frac{V_S[(SG_1/SG_2) - 1]}{1 - (R_F + V_F)}$$

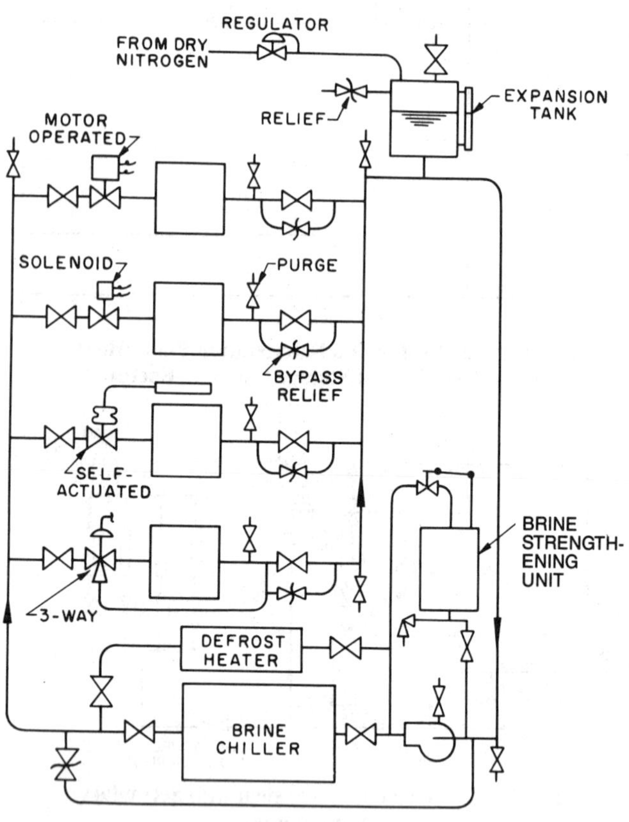

Fig. 3 Typical Closed Salt Brine System

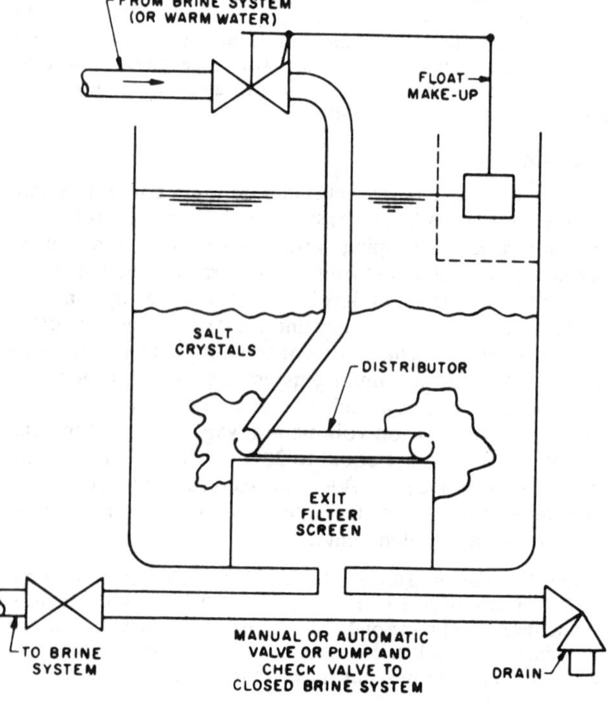

Fig. 4 Brine Strengthening Unit for Salt Brines Used as Secondary Coolants

where

ETV = expansion tank volume
V_s = system secondary coolant volume at t_1 temperature
SG_1 = specific gravity at t_1
SG_2 = specific gravity at maximum temperature
R_F = residual volume of tank liquid (low level) at t_1, expressed as a fraction
V_F = volume of vapor space at highest temperature, expressed as a fraction

If the specific gravity of the secondary coolant is 1.185 at 20 °F and 1.155 at 100 °F, the tank volume is:

$$\text{ETV} = \frac{50{,}000\,[(1.185/1.155) - 1]}{1 - (0.10 + 0.20)} = 1855 \text{ gal}$$

Pulldown Time

Example 1 is based on a static situation of secondary coolant temperature at two different loads—normal and peak. The length of time for pulldown from 100 °F to the final 20 °F may need to be calculated. For a graphical solution, required heat extraction versus secondary coolant temperature is plotted. Then, by iteration, the pulldown time is solved by finding the net refrigeration capacity for each increment of coolant temperature change. A mathematical method may also be used.

The 15-ton refrigeration system cited in the examples has a 30.03-ton capacity at a maximum of 50 °F saturated suction temperature (STP). For pulldown, a compressor suction pressure regulator (holdback valve) is sometimes used in the system. The maximum secondary coolant temperature must be determined when the holdback valve is wide open and the STP is at 50 °F. For the above example, this is at 70 °F coolant temperature. As the coolant temperature is further reduced with a constant 48.1 gpm, the capacity of the refrigeration system is gradually reduced until a 15-ton capacity is reached with 26 °F coolant in the tank. Further cooling to 20 °F will be at reduced capacity.

Temperatures of the secondary coolant mass, storage tanks, piping, cooler, pump, and insulation must all be reduced. In Example 1, as the coolant is reduced in temperature from 100 to 20 °F, the total heat removed from these items is as follows:

Brine Temperature, °F	Total Heat Removed, Million Btu
100	31.54
80	23.62
70	19.67
60	15.73
40	7.85
20	0

From a secondary coolant temperature of 100 to 70 °F, the refrigeration system capacity is fixed at 30.03 tons, and the time for pulldown is essentially linear (system net tons for pulldown is less than the compressor capacity because of heat gain through insulation and added pump heat). In Example 1, the pump heat was not considered. When recognizing the variable heat gain for a 95° ambient, and the pump heat as the secondary coolant temperature is reduced, the following net capacity is available for pulldown at the various secondary coolant temperatures:

Brine Temperature, °F	Net Capacity, Tons
100	29.86
80	29.58
70	29.44
60	25.28
40	17.80
26	14.10
20	12.70

A curve fit shows capacity is a straight line between the values for 100 and 70 °F. Therefore, the pulldown time for this interval is:

$$\theta = \frac{[(31.54 \times 10^6) - (19.67 \times 10^6)]}{12{,}000[(29.86 + 29.44)/2]} = 33.4 \text{ h}$$

From 70 to 20 °F, the capacity curve fits a second-degree polynomial equation as follows:

$$q = 9.514809086 + 0.1089883647t + 0.002524039t^2$$

where t = secondary coolant temperature, °F, and q = tons capacity for pulldown.

Rearranging to another form:

$$q = 0.002524039\,(t^2 + 43.1801429t + 3769.675938)$$
$$= 0.002524039\,[(t + 21.59007145)^2 + (57.4764713)^2]$$

Using the arithmetic average pulldown net capacity from 70 to 20 °F, the time interval would be:

$$\theta = \frac{19.67 \times 10^6}{12{,}000\,[(29.44 + 12.7)/2]} = 77.8 \text{ h}$$

If the logarithmic (base e) mean average net capacity for this temperature interval is used, the time is:

$$\theta = \frac{19.67 \times 10^6}{(19.91 \times 12{,}000)} = 82.4 \text{ h}$$

This is a difference of over 4.5 h and neither solution is correct. A more exact calculation uses a graphical analysis or calculus. When using a mathematical approach, determine the heat removed per degree of secondary coolant temperature change per ton capacity. Since the coolant's heat capacity and the heat leakage change as the temperature is reduced, this is best determined by a curve fit of the data for total heat removed versus secondary coolant temperature, using a series of iterations for secondary coolant temperature ±1 °F as the temperature is reduced.

A curve fit of the total heat removed from a given secondary coolant temperature to 20 °F gives the following:

$$Q = -7832831.847 + 391110.316t + 26.45203369t^2$$

where Q = total heat removed in Btu.

By a series of iterations for 1 °F temperature change, the heat removed per hour for one degree change in secondary coolant temperature per ton is developed from the following relationship:

$$\frac{\Delta 1\,°\text{F}}{\text{h}\cdot\text{ton}} = 0.0306818755 - 4.1498 \times 10^{-6}\,t$$

The differential equation for the differential change in coolant temperature versus time in hours is (use minus sign since coolant temperature is being reduced):

$$\frac{dt}{d\theta} = \left(\frac{-\Delta 1\,°\text{F}}{\text{h}\cdot\text{tons}}\right)\text{tons}$$

Solving for $d\theta$ and integrating determines time θ in hours:

$$\int_{70°}^{20°} d\theta = \theta = \frac{-1}{(\Delta 1\,°\text{F/h}\cdot\text{ton})}\int_{70°}^{20°}\frac{dt}{(\text{tons})}$$

The polynomial equations are then substituted for the parenthesized terms (rounding to four significant numbers):

$$\theta = \frac{-1}{(0.03068 - 0.00000415t)} \times \int_{70°}^{20°}\frac{dt}{0.002524[(t + 21.59)^2 + (57.48)^2]}$$

These are in the general form of:

$$\theta = \text{constant} \times \int\frac{dv}{v^2 + a^2} = \frac{\text{constant}}{a} \times \tan^{-1}\frac{v}{a} + C$$

where

$dv = dt$
$v = t + 21.59$
$a = 57.48$
v/a = radians

Substituting:

$$\theta = \frac{-1}{[0.002524(0.03068 - 0.000004157t)]} \times \frac{1}{57.48}\tan^{-1}\frac{(t + 21.59)}{57.48}\Bigg|_{70°}^{20°} + C$$

Subtracting values for the two temperatures, C drops out:

$$\theta_{20} - \theta_{70} = -141.1 - (-229.2) = 88.1 \text{ h}$$

The correct answer is 88.1 h, which is 7% greater than the logarithmic mean average capacity and 13% greater than the arithmetic average capacity over the temperature range.

Therefore, total time for temperature pulldown from 100 to 20 °F is:

$$\theta = 33.4 + 88.1 = 121.5 \text{ h}$$

Averaging methods would be 5.7 to 10.3 h in error on total pulldown time. Graphical methods have less error than averaging methods, but calculus produces the greatest accuracy.

Note that the general form of the equation may also develop as:

$$\theta = \text{constant}\int\frac{dv}{v^2 - a^2}$$

Therefore:

$$\theta = \frac{1}{2a}\ln\left|\frac{v - a}{v + a}\right| + C$$

The polynomial equations may be solved by computer or a handheld calculator and a suitable program or spreadsheet. The time for pulldown will be less if supplemental refrigeration is available for pulldown or if less secondary coolant is stored.

System Costs

Various alternatives may be evaluated to justify a new project or system modification. Means (1988) lists the installed cost of various projects. Park and Jackson (1984) and NBS (1978) discuss engineering and life-cycle cost analysis. Using the various time value of money formulas, the payback for storage tank handling of peak loads compared to large refrigeration equipment and higher energy costs can be evaluated. The trade-offs in these costs—initial, maintenance, insurance, increased secondary coolant, loss of space, and energy escalation—all must be considered.

Corrosion Prevention

Corrosion prevention requires choosing proper materials and inhibitors, routine testing for pH, and eliminating contaminants. Because potentially corrosive calcium chloride and sodium chloride salt brine secondary coolant systems are widely used, test and adjust the brine solution monthly. To replenish salt brines in a system, a concentrated solution may be better than a crystalline form, because it is easier to handle and mix.

A refrigerating brine should not be allowed to change from an alkaline to an acid condition. Acids rapidly corrode the metals ordinarily used in refrigeration and ice-making systems. Calcium chloride usually contains sufficient alkali to render the freshly prepared brine slightly alkaline. When any brine is exposed to air, it gradually absorbs carbon dioxide and oxygen, which eventually make the brine slightly acid. Dilute brines dissolve oxygen more readily and generally are more corrosive than concentrated brines. One of the best preventive measures is to make a closed rather than open system, using a regulated inert gas over the surface of a closed expansion tank (see Figure 2). However, many systems, such as icemaking tanks, brine spray unit coolers, and brine spray-type carcass chill rooms, cannot be closed.

A brine pH of 7.5 for a sodium or calcium chloride system is ideal, since it is safer to have a slightly alkaline rather than a slightly acid brine. Brine system operators should check pH regularly.

If a brine is acid, the pH can be raised by adding caustic soda dissolved in warm water. If a brine is alkaline (indicating ammonia leakage into the brine), carbonic gas or chromic, acetic, or hydrochloric acid should be added. Ammonia leakage must be stopped immediately so that the brine can be neutralized.

In addition to controlling the pH, an inhibitor should be used. Generally, sodium dichromate is the most effective and economical for salt brine systems. The dichromate has a bright orange color, a granular form, and readily dissolves in warm water. Since it dissolves very slowly in cold brine, it should be dissolved in warm water and added to the brine far enough ahead of the pump so that only a dilute solution reaches the pump. The quantities recommended are: 125 lb/1000 ft^3 of calcium chloride brine, and 200 lb/1000 ft^3 of sodium chloride brine.

Adding sodium dichromate to the salt brine does not make it noncorrosive immediately. The process is affected by many factors, including water quality, specific gravity of the brine, amount of surface and kind of material exposed in the system, age, and temperature. Corrosion stops only when protective chromate film has built up on the surface of the zinc and other electrically positive metals exposed to the brine. No simple test is available to determine the chromate concentration. Since the protection afforded by the sodium dichromate treatment depends greatly on maintaining the proper chromate concentration in the brine, brine samples should be analyzed anually. The proper concentration for calcium chloride brine is 7.58 gr/gal (as $Na_2Cr_2O_7 \cdot 2H_2O$); for sodium chloride brine, it is 12.128 gr/gal (as $Na_2Cr_2O_7 \cdot 2H_2O$).

Since crystals and concentrated solutions of sodium dichromate can cause severe skin rash, avoid contact. If contact does occur, wash the skin immediately. *Warning: Sodium dichromate should not be used for brine spray decks, spray units, or immersion tanks where food or personnel may come in contact with the spray mist or the brine itself.*

Polyphosphate-silicate and orthophosphate-boron mixtures in water-treating compounds are useful for sodium chloride brines in open systems. However, where the rate of spray loss and dilution is very high, any treatment other than density and pH control is not economical. For the best protection of spray unit coolers, housings and fans should be of a high quality, hot-dipped galvanized construction. Stainless steel fan shafts and wheels, scrolls, and eliminators are desirable.

While the nonsalt secondary coolants described in this chapter are generally noncorrosive when used in systems for long periods, recommended inhibitors should be used, and a pH check should be performed occasionally.

Steel, iron, or copper piping should not be used to carry the salt brines. Use copper nickel or suitable plastic. Use all-steel and iron tanks if the pH is not ideal. Similarly, calcium chloride systems usually have all-iron and steel pumps and valves to prevent electrolysis in the presence of acidity. Sodium chloride systems usually have all-iron or all-bronze pumps. When the pH can be controlled in a system, brass valves and bronze fitted pumps may be satisfactory. A stainless steel pump shaft is desirable. Consider salt brine composition and temperature to select the proper rotary seal or, for dirtier systems, the proper stuffing box.

APPLICATIONS

Applications for secondary coolant systems are extensive (see Chapters 8 through 35). A glycol coolant prevents freezing in solar collectors and outdoor piping. Secondary coolants heated by solar collectors or by other means can be used to heat absorption-cooling equipment, to melt a product such as ice or snow, or to heat a building. Process heat exchangers can use a number of secondary coolants to transfer heat between locations at various temperature levels. Using secondary coolant storage tanks for the application increases the availability of cooling and heating and reduces peak demands for energy.

Each supplier of refrigeration equipment that uses secondary coolant flow has specific ratings. Flooded and direct-expansion coolers, dairy plate heat exchangers, food processing, and other air, liquid, and solid chilling devices come in various shapes and sizes. Refrigerated secondary coolant spray wetted-surface cooling and humidity control equipment has an open system that absorbs moisture while cooling and then continuously regenerates the secondary coolant with a concentrator. Although this assists the cooling, dehumidifying, and defrosting process, it is not strictly a secondary coolant flow application for refrigeration, unless the secondary coolant also is used in the coil. Heat transfer coefficients can be determined from vendor rating data or by methods described in Chapter 3 of the 1993 ASHRAE *Handbook—Fundamentals* and appropriate texts.

A primary refrigerant may be used as a secondary coolant in a system by being pumped at a flow rate and pressure high enough that the primary heat exchange occurs without evaporation. But the refrigerant is then subsequently flashed at a low pressure, with the resulting flash gas being drawn off to a compressor in the conventional manner.

BIBLIOGRAPHY

ASHRAE. 1992. Safety code for mechanical refrigeration. ASHRAE *Standard* 15-1992.

ASME. 1992. Refrigeration piping. *Standard* B31.5. American Society of Mechanical Engineers, New York.

ASME. Boiler and pressure vessel code, Section VIII, Recommended Guidelines for the care of power boilers (non-interfiled). American Society of Mechanical Engineers, New York.

Means. Updated annually. *Means mechanical cost data,* 11th ed. Section 2, U.C.I. Division 15.5. Robert Snow Means Co., Inc., Kingston, MA.

Park, W.R. and D.E. Jackson. 1984. *Cost engineering analysis,* 2nd ed. John Wiley & Sons, New York.

NBS. 1978. National Bureau of Standards Building Science Series 113, Life cycle costing. SD Catalog Stock No. 003-003-01980-1, U.S. Government Printing Office, Washington, D.C.

CHAPTER 5

REFRIGERANT SYSTEM CHEMISTRY

A GOOD understanding of the chemical interaction of the materials in a refrigeration system helps in designing systems that have great reliability and long service life. This chapter covers the chemical aspects of both the old and new, environmentally acceptable refrigerants. Physical aspects such as measurement and contaminant control (including moisture) are discussed in Chapter 6.

EVALUATING CHEMICAL PROBLEMS

Usually, unexpected chemical problems can be attributed to inadequate testing of a new material, improper application of a previously tested material, or inadvertent introduction of contaminants into the system. Three techniques are used to chemically evaluate materials: (1) material tests in sealed tubes, (2) component tests, and (3) accelerated life and system tests.

Sealed Tube Tests

The glass sealed tube test, as described by ASHRAE *Standard* 97-1989, is widely used to assess the stability of refrigerant system materials. It is also used as a tool to identify chemical reactions that are likely to occur in operating units.

Generally, glass tubes are charged with refrigerant, oil, metal strips, and other materials to be tested. The tubes are then sealed and aged at elevated temperatures for a specified time. The tubes are inspected for color and appearance and compared to a control tube that is processed identically to the specimen tubes, but might contain a reference material rather than the test material. The contents of the tubes can also be analyzed for changes in the test materials by using methods such as gas chromatography, ion chromatography, liquid chromatography, and specific ion electrode. Wet methods, like total acid number analysis, are also used.

Originally, this test was designed to compare lubricants, but the technique is effective in testing other materials as well. For example, Huttenlocher (1972) evaluated zinc die castings, Guy *et al.* (1992) reported on the compatibilities of motor insulation materials and elastomers, and Mays (1962) studied the decomposition of R-22 in the presence of 4A-type molecular sieve desiccants. Although the sealed tube is very useful, it has some disadvantages. Since the sealed tube greatly magnifies chemical reactions likely to occur in a refrigeration system, results can be misinterpreted. Also, reactions in which mechanical energy plays a role, for example in a failing bearing, are not easily studied in a static sealed tube.

The sealed tube test, in spite of its proven utility, is only a screening tool and not a simulation of a refrigeration system. Sealed tube tests alone should not be used to predict field behavior. Material selection for refrigerant systems requires follow-up with component or system tests or both.

Component Tests

Component tests carry material evaluations a step beyond sealed tube tests because materials are not only tested in the proper environment, but also under dynamic conditions. Motorette (enameled wire, ground insulation, and other motor materials assembled into a simulated motor) tests used to evaluate hermetic motor insulation are described in UL *Standard* 984 and are a good example of this type of test. Component tests are conducted in large pressure vessels or autoclaves in the presence of a lubricant and refrigerant. Unlike sealed glass tubes, where temperature and pressure are the only means of accelerating the aging process, the autoclave tests can include external stresses that can accelerate the phenomena likely to occur in an operating system. These stresses can include mechanical vibration, on-off electrical voltages, and refrigerant liquid floodback.

System Tests

System tests can be divided into two major categories:

1. Test a sufficient number of systems under a broad spectrum of operating conditions to obtain a good, statistical reference base. The failure rates of two populations of units can be compared: (a) those containing the new materials and (b) those containing previously proven materials.
2. Test under well-controlled conditions. Temperatures, pressures, and other operating conditions are continuously monitored. A chemical analysis of the refrigerant and lubricant is done before, during, and after the test is completed.

In most cases, the tests are conducted under severe operating conditions to obtain results quickly. Analyzing the lubricant and refrigerant samples during the test and inspecting the components after teardown can yield information on (1) the nature and rate of chemical reactions taking place in the system, (2) the products formed by the reactions, and (3) the possible effect on the system life and performance. Accurate interpretation of these data determines the operating limits for the system such that chemical reactions are kept at an acceptable level.

REFRIGERANTS

Environmental Acceptability. The ozone depletion potential (ODP) of a material is a measure of its ability to destroy stratospheric ozone. The halocarbon global warming potential (HGWP) is "the ratio of calculated warming for each unit mass of gas emitted...relative to the calculated warming for a mass unit of reference gas CFC-11" (Fisher *et al.* 1989). Appliances based on a given refrigerant also have indirect (energy-related) emissions that contribute to the greenhouse effect. Environmentally acceptable materials are those (1) with low or zero ozone depletion, (2) with low global warming potentials, and (3) that provide acceptable system energy efficiencies. Materials with short atmospheric lifetimes generally have lower ODP and HGWP values. Hydrogen-containing compounds such as R-22 will have shorter atmospheric lifetimes.

Table 1 shows the boiling points, atmospheric life, ODP, HGWP, and flammability of new refrigerants and the refrigerants being replaced.

The preparation of this chapter is assigned to TC 3.2, Refrigerant System Chemistry.

Table 1 Refrigerant Properites

Refrigerant	Structure	Boiling Point °F	Atmos. Life Years	ODP	HGWP	Flam-mable
R-32	CH_2F_2	−62.0	7.3	0	0.11	Yes
R-125	CF_3-CHF_2	−54.0	40.5	0	0.84	No
R-143a	CF_3-CH_3	−54.0	64.2	0	1.1	No
R-502	R-22/115	−49.8	15.8/400	0.22	3.7	No
R-22	$CHClF_2$	−41.5	15.8	0.05	0.34	No
E-125	CF_3-O-CHF_2	−30.3	21	0		No
R-12	CCl_2F_2	−21.6	130	1.0	3.1	No
R-134a	CF_3-CH_2F	−15.7	15.6	0	0.28	No
R-152a	CHF_2-CH_3	−13.0	1.8	0	0.03	Yes
E-143a	CF_3-O-CH_3	−11.4	3.4	0		Yes
R-134	CHF_2-CHF_2	−4.0		0		No
R-227ea	CF_3-CHF-CF_3	−0.9	30*	0		No
R-227ca	CF_3-CF_2-CHF_2	2.7	15*	0		No
E-134	CHF_2-O-CHF_2	40.5	2.8			
R-143	CHF_2-CH_2F	41.0	41	0	1.1	Yes
R-236ea	CF_3-CHF-CHF_2	43.7		0		No
R-11	CCl_3F	74.5	55	1.0	1.0	No
R-245ca	CHF_2-CF_2-CH_2F	77.0	6.4	0		Yes
R-123	CF_3-$CHCl_2$	82.2	1.8	0.016	0.02	No
E-143	CHF_2-O-CH_2F	84.5		0		Yes
R-152	CH_2F-CH_2F	87.2		0		Yes

* = estimated
ODP = ozone depletion potential
HGWP = halocarbon global warming potential

Chlorofluorocarbons. Refrigerants such as CFC-12, CFC-11, CFC-114 and CFC-115 have been used extensively in the air-conditioning and refrigeration industry. Because of their chlorine content, these materials have high ODP values. The Montreal Protocol, which governs the elimination of ozone depletion substances, was strengthened in a Copenhagen meeting in 1992. By this international agreement, production of chlorofluorocarbons will be reduced by 75% by January 1, 1994, and totally phased out by January 1, 1996.

Hydrochlorofluorocarbons. These refrigerants have shorter atmospheric lifetimes (and lower ODP values) than the chlorofluorocarbons. Nevertheless, the international treaty calls for limiting the production of hydrochlorofluorocarbons by January 1, 1996. This limit will be based on 1989 production. It will be followed by a 35% reduction by January 1, 2004, a 65% reduction by January 1, 2010, a 90% reduction by January 1, 2015, a 99.5% reduction by January 1, 2020, and finally a total phaseout by January 1, 2030 (Fischer *et al.* 1991).

HCFC-22 is the most widely used hydrochlorofluorocarbon. HFC alternates for this material are now being developed. HCFC-123 is now commercially used as a replacement for CFC-11. HCFC-124 is now a leading candidate as a replacement for R-114 (Reed and Spauschus 1991).

Hydrofluorocarbons. These refrigerants contain no chlorine atoms; therefore their ODP is zero. HFC methanes, ethanes, and propanes have been extensively considered for use in air conditioning and refrigeration.

Fluoromethanes. Mixtures that include R-32 (difluoromethane CH_2F_2) are being considered as a replacement for R-22 and R-502. For very low-temperature applications, R-23 is being considered as a replacement for R-13 and R-503 (Atwood and Zheng 1991).

Fluoroethanes. R-134a (CF_3-CH_2F) of the fluoroethane series is used extensively as a direct replacement for R-12 and as a replacement for R-22 in higher temperature applications. R-125 and R-143a are used in azeotropes or zeotropic blends with R-32 and/or R-134a as replacements for R-22 or R-502. R-152a is flammable and less efficient than R-134a in applications using suction line heat exchangers (Sandvordenker 1992), but it is still being considered as a candidate for R-12 replacement. Further, it is being considered as a component, with R-22 and R-124, in zeotropic blends (Bateman *et al.* 1990, Bivens *et al.* 1989) that can be R-12 and R-500 alternatives. However, R-152 is toxic.

Fluoropropanes. Desmarteau *et al.* (1991) identified a number of fluoropropanes as potential refrigerants. R-245ca is being considered as a chlorine free replacement for R-11. Modeling work by Hughes (1992), Beyerlein *et al.* (1991), and Sand *et al.* (1991) showed R-245ca to have a COP slightly less than that of R-123. Evaluation by Doerr *et al.* (1992) showed that R-245ca was stable and compatible with key components of the hermetic system. However, Smith *et al.* (1993) demonstrated that R-245ca is slightly flammable in humid air at room temperature.

R-227ea and R-227ca blended with R-32 are under consideration as high glide zeotropic replacements for R-22. R-236ea is currently under investigation as a replacement of R-114 in naval centrifugal chillers.

Fluoroethers. Booth (1937), Eiseman (1968), Kopko (1989), O'Neill and Holdsworth (1990), O'Niell (1992), and Wang *et al.* (1991) proposed these compounds as refrigerants. The fluoroethers are usually more reactive than the fluorinated hydrocarbons both physiologically and chemically. Fluorinated ethers have found use as anesthetics and convulsants (Krantz and Rudo 1966, Terrell *et al.* 1971ab). Reactivity with glass is also characteristic of some fluoroethers (Doerr *et al.* 1993, Simons *et al.* 1977, Gross 1990).

Ammonia. Used extensively in large, open-type compressors for industrial and commercial applications, ammonia has high refrigerating capacity per unit displacement, low pressure losses in connecting piping, and low reactivity with refrigeration oil (mineral oils). (See Chapter 3 for detailed information.)

The toxicity and flammability of ammonia offset its advantages. Ammonia is such a strong irritant to the human nose (detectable below 5 mg/kg) that people automatically avoid exposure to it. Ammonia is considered toxic at 35 to 50 mg/kg. Ammonia-air mixtures are flammable, but only within a narrow range of 16 to 25% by volume. These mixtures can explode but are difficult to ignite since they require an ignition source of at least 1200°F.

The thermal stability of ammonia is often a subject of confusion. At atmospheric pressure, ammonia starts to dissociate into nitrogen and hydrogen at about 570°F in the presence of active catalysts such as nickel and iron. However, since these high temperatures are unlikely to occur in open-type compression systems, thermal stability is not a problem. Ammonia attacks copper in the presence of even small amounts of moisture; therefore, except for some specialty bronzes, copper-bearing materials must be excluded in ammonia systems. A corollary to this design practice is that copper plating, as discussed further in this chapter, is nonexistent in ammonia systems.

Flammability and Combustibility

Fedorko *et al.* (1987) studied the R-22/air flammability envelope as a function of pressure (up to 200 psia) and fuel (R-22) to oxygen ratio. They found that R-22 was nonflammable under 75 psia. In addition, the flammable compositions between 30% and 45% generated maximum heats of reaction. Their results were in general agreement with those of Sand and Andjeski (1982), who found that pressurized mixtures of R-22 and air containing at least 50% air are combustible. R-11 and R-12 did not ignite under similar conditions.

Reed and Rizzo (1991) and Lindley (1992), using different experimental arrangements, studied the combustibility of R-134a at high temperature and pressure. Lindley notes that the results depend on the nature of the equipment used in the tests. Reed and Rizzo showed R-134a is combustible at pressures above 15 psig at room temperature and air concentrations greater than 80% by volume. At 350 °F, combustibility was observed at pressures above 5 psig and air concentrations above 60% by volume. Lindley found flammability limits of 8 to 22% by volume in air at 340 °F and 100 psia. Both researchers found R-134a to be nonflammable at ambient conditions and under the likely operating conditions of air-conditioning and refrigeration equipment. Blends of R-22/152a/114 showed combustion at temperatures above 180 °F at atmospheric pressure and above, with air concentrations above 80% by volume (Reed and Rizzo 1991).

Richard and Shankland (1991) followed the ASTM E-681 method to study the flammability of R-32, R-141b, R-142b, R-152a, R-152, R-143, R-161, methylene chloride, 1,1,1-trichloroethane, propane, pentane, dimethyl ether, and ammonia. They used several ignition methods including the electrically activated match ignition source as specified in ASHRAE *Standard* 34. They also reported on the critical flammability ratio of mixtures such as R-32/125, R-143a/134a, R-152a/125, propane/R-125, R-152a/22, R-152a/124, and R-152a/134a. The critical flammability ratio is the maximum amount of flammable component that a mixture can contain and still be nonflammable, regardless of the amount of air. These data are important since mixtures containing flammable components are being considered as alternates.

Zhigang *et al.* (1992) published data on the flammability of R-152a/22 mixtures. Their measured lower flammability limit in air of R-152a is 11.4% by volume. They state that values reported in the literature range from 4.7 to 16.8% by volume. Richard and Shankland (1991) reported an average flammable range of 4.1 to 20.2% (mass) for R-152a. Zhigang *et al.* (1992) also provide data on the length of the flame as function of R-22 concentration. They found that the flame no longer existed somewhere between 17 and 40% (mass) R-22 in the mixture. This is in apparent disagreement with the data of Richard and Shankland (1991) who found a critical flammability ratio of 57.1% (mass) R-22. Comparison of results is difficult because results dependent on the exact nature of the apparatus and methods used. Grob (1991) reported on the flammabilities of R-152a, R-141b, and R-142b. He describes R-152a as having "the lowest flammable mixture percentage, highest explosive pressure and highest potential for ignition of the refrigerants studied."

COMPATIBILITY OF MATERIALS

Electrical Insulation

The insulation on electric motors is affected by the refrigerant and/or the lubricant through two primary modes, extraction of insulation polymer into the refrigerant or absorption of refrigerant by the polymer. Extraction of insulation material results in embrittlement, delamination, and general degradation of the material. In addition, extracted material can deposit out and cause components to stick or passages such as capillary tubes to clog.

Absorption of refrigerant can change the dielectric strength or physical integrity of the material through softening or swelling. Rapid desorption (off-gassing) of the refrigerant by internal heating can be more serious than the effect of refrigerant that is absorbed. Rapid desorption results in high internal pressures that cause blistering or voids within the insulation. A decrease in dielectric or physical strength is the result.

In compatibility studies of 10 refrigerants and 7 lubricants with 24 motor materials in various combinations, Doerr and Kujak (1993) showed that R-123 was absorbed to the greatest extent, but R-22 caused more damage because of more rapid desorption and higher internal pressures. They also observed damage to insulation after desorption of R-32, R-134, and R-152a in a 300 °F oven but not to the extent exhibited by R-22.

Magnet Wire Insulation. The magnet wire is coated with enamels, which are cured by heating. The most common insulation is a polyester base coat followed by a polyamide-imide overcoat. A polyester-imide base coat is also used. Acrylic and polyvinyl formal enamels are found on older motors. An enameled wire with an outer layer of polyester-glass is used in larger hermetic motors for greater wire separation and thermal stability.

The magnet wire insulation is the primary source of electrical insulation and the most critical in regard to compatibility with refrigerants. Most electrical tests (NEMA 1987) are conducted in air and may not be valid for hermetic motors. For example, wire enamels absorb R-22 up to 15 to 30% by mass (Hurtgen 1971) and at different rates depending on their chemical structure, degree of cure, and the conditions of exposure to the refrigerant. Refrigerant permeation is shown by changes in electrical, mechanical, and physical properties of the wire enamels. Fellows *et al.* (1991) measured the dielectric strength, Paschen curve minimum, dielectric constant, conductivity, and resistivity for 19 fluorocarbons.

Wire enamels in refrigerant vapor typically exhibit dielectric loss with increasing temperature, as shown in Figure 1. Depending on the atmosphere and degree of cure, each wire enamel or enamel/varnish combination exhibits a characteristic temperature t_{max} above which dielectric losses increase sharply. Table 2 shows values of t_{max} for several hermetic enamels. Continued heating above t_{max} causes aging, evidenced by the irreversible alteration of dielectric properties and an increase in the conductance of the insulating material.

Spauschus and Sellers (1969) show that the change rate in conductance is a quantitative measure of aging in a refrigerant environment. They give the aging rate for varnished and unvarnished enamels at two levels of R-22 pressure typical of high- and low-side hermetic motor operation.

Apart from the effects on long-term aging, R-22 can also affect the short-term insulating properties of certain wire enamels. Beacham and Divers (1955) demonstrated that the resistance of polyvinyl formal drops drastically when the material is submerged in liquid R-22. A parallel experiment using R-12 showed a much smaller drop, followed by quick recovery to the original

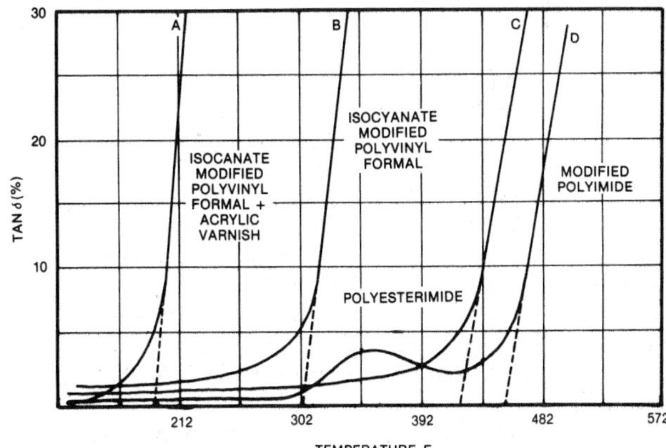

Fig. 1 Loss Curves of Various Insulating Materials
(Spauschus and Sellers 1969)

Table 2 Maximum Temperature t_{max} for Hermetic Wire Enamels in R-22 at 65 psia

Enamel Type	t_{max}, °F
Acrylic	226
Polyvinyl formal	277
Isocyanate-modified polyvinyl formal	304
Polyamide-imide	361
Polyester-imide	419
Polyimide	450

Table 3 Effect of Liquid R-22 on Abrasion Resistance

Magnet Wire Insulation	Average Cycles to Failure			
		After Time in Liquid R-22		
	As Received	7 - 10 Days	One Month	Three Months
Urethane-polyvinyl formal batch 1	40	3	2	2
Urethane-polyvinyl formal batch 2	42	2	2	7
Polyester-imide batch 1	44	15	18	6
Polyester-imide batch 2	24	10	5	6
Dual coat amide-imide top coat, polyester base	79	35	23	11
Dual coat, polyester	35	5	5	9
Polyimide	26	25	23	21

From Sandvordenker and Larime (1971).

resistance. The relatively rapid permeation of R-22 into polyvinyl formal, coupled with the low-volume resistivity of liquid R-22 and other electrical properties of the two refrigerants, explains the phenomenon.

Softening of the wire coating, which can cause the insulation to fail, occurs with certain combinations of coatings and refrigerants. Table 3 shows data on softening measured in terms of abrasion resistance for a number of wire enamels exposed to R-22. At the end of the shortest soaking period, the urethane-modified polyvinyl formal had lost all its abrasion resistance. All the other insulations, except polyimide, lost their abrasion resistance at a slower rate, which over a 3-month period approached the rate of the urethane-polyvinyl formal. Only the polyimide showed a minimal effect even though its abrasion resistance was lower originally than most of the others.

Because of the time dependency on softening, which is related to the permeation rate of R-22 into the enamel, Sanvordenker and Larime (1971) proposed that comparative tests on magnet wire be made only after the enamel is completely saturated with refrigerant, so that the effect of long-term exposure to R-22 on the enamel properties can be evaluated.

The second consequence of R-22 permeation is blistering, caused by the rapid change in pressure and temperature after a wire enamel is exposed to R-22. Heating greatly increases the internal pressure as the dissolved R-22 expands; and, since the polymer film has already been softened, portions of the enamel lift up in the form of blisters. While blistered wire has a poor appearance, field experience indicates that mild blistering is not a cause for concern, as long as the blisters do not break and the enamel film remains flexible. Currently used wire enamels have the characteristics mentioned previously and maintain dielectric strength even after blistering. However, hermetic wire enamel with strong resistance is preferred.

Varnishes. After the stator of an electric motor is wound, it is usually treated with a varnish by a vapor-pressure impregnation process for form-wound, high-voltage motors or a dip-and-bake process for low-voltage, random wound motors. The varnished motor is cured in a 275 to 350°F oven. The varnish holds the windings together in the magnetic field and acts as a secondary source of electrical insulation. The windings have a tendency to move, and independent movement of the wires causes abrasion and wear of the insulation. High-voltage motors contain form-wound coils wrapped with a porous fiberglass, which is saturated with varnish and cured as a secondary source of insulation over the enameled wire or bars.

Many different chemicals are used as motor varnishes. The most common are epoxies, polyesters, phenolics, and modified polyimides. Characteristics important to a varnish are good adhesion and bond strength to the wire enamel, flexibility and strength under both hot and cold conditions, thermal stability, good dielectric properties, and chemical compatibility with the wire enamel, sheet insulation, and refrigerant-lubricant.

The compatibility of a varnish is determined by exposing the cured varnish (in the form of a section of a thin disc and varnished magnet wire in the form of single strands, helical coils, and twisted pairs) to a refrigerant at elevated temperatures. The properties of the varnish are then compared to samples not exposed to refrigerant and to other exposed samples placed in a hot oven to rapidly remove absorbed refrigerant. The disc sections are evaluated for absorption, extraction, degradation, and changes in flexibility. The single strands are wound around a mandrel, and the varnish is examined for flexibility and effect on the wire enamel. In many cases, the varnish will not flex as well as the enamel; if bound tightly to the enamel, the varnish will remove the enamel from the copper wire. The helical coils are evaluated for bond strength (ASTM *Standard* D2519) before and after exposure to a refrigerant/lubricant mixture. The twisted pair is tested for dielectric breakdown voltage, or burnout time, while being subjected to resistance heating (ASTM *Standard* D1676).

During the compatibility testing of motor materials with alternative refrigerants, researchers observed that the varnish could absorb considerable amounts of refrigerant, especially R-123. Doerr (1992) studied the effects of time and temperature on the rate of absorption and desorption of R-123 and R-11 by epoxy motor varnishes. Absorption of refrigerant was faster at higher temperatures. Desorption of refrigerant at temperatures as high as 250°F was slow. The equilibrium absorption value for R-123 was linearly dependent on temperatures with higher absorption at lower temperatures. Absorption of R-11 remained the same at all test temperatures.

Ground Insulation. Sheet insulation material is used in slot liners, phase insulation, wedges, and for tie point insulation in hermetic motors. The sheet material is usually a polyethylene terephthalate (PET) film or an aramid mat. These can be used singly or laminated together. PET or aramid films possess excellent dielectric properties and are characterized by good chemical resistance to refrigerants and oils.

The PET film selected must contain little of the polymers with low relative molecular mass that exhibit a temperature-dependent solubility in mineral lubricants and tend to precipitate as noncohesive granules at temperatures lower than those of the motor. Another limitation of this film is that, like most polyesters, it is susceptible to degradation by hydrolysis; but the quantity of water required is more than that generally found in refrigerant systems. Sundaresan *et al.* (1991) discuss the effect of synthetic lubricants on PET films.

Elastomers

Refrigerants, oils, or mixtures of both can, at times, extract enough filler or plasticizer from an elastomer to change its physical or chemical properties. This extracted material can harm the refrigeration system by increasing its chemical reactivity or by clogging screens and expansion devices. Many elastomers are unsuitable for use with refrigerants because of exessive swell or shrinkage. Some neoprenes tend to shrink in HFC refrigerants. Nitriles swell in R-123. Hamed and Seiple (1993) determined swell data on 95 elastomers in 10 refrigerants and 7 lubricants. Data on swelling of elastomers in most of the common halocarbon refrigerants are presented in Table 12 of Chapter 16 in the 1993 ASHRAE *Handbook—Fundamentals.*

Plastics

The effect of refrigerants on plastics usually decreases as the amount of fluorine in the molecule increases. For example, R-12 has less effect than R-11, while R-13 is almost entirely inert. Cavestri (1993) determined the compatibility of 23 engineering plastics with alternative refrigerants and lubricants.

Each type of plastic material should be tested for compatibility with the refrigerant before it is used. Two samples of the same type of plastic might be affected differently by the refrigerant because of differences in polymer structure, relative molecular mass, and plasticizer. Chapter 16 of the 1993 ASHRAE *Handbook—Fundamentals* discusses the effect of refrigerants on plastic materials.

CHEMICAL REACTIONS

Thermal Stability in the Presence of Metals. All common halocarbon refrigerants have excellent thermal stability, as shown in Table 4. Bier *et al.* (1990) studied R-12, R-134a, and R-152a. For R-134a in contact with metals, traces of hydrogen fluoride were detected after 10 days at 392°F. This decomposition did not increase much with time. R-152a showed traces of HF at 356°F after five days in a steel container. Bier *et al.* suggested that vinyl fluoride is formed during the thermal decomposition of R-152a. Vinyl fluoride can then react with water to form acetaldehyde. Hansen and Finsen (1992) conducted lifetime tests on small hermetic compressors with a ternary mixture of R-22/152a/124 and an alkyl benzene lubricant. In agreement with Bier *et al.*, they found that vinyl fluoride and acetaldehyde were formed in the compressor. Aluminum, copper, and brass and solder joints lower the temperature at which decomposition begins. Decomposition also increases with time.

Under extreme conditions, such as above red heat or with molten metal temperatures, refrigerants react exothermically to produce metal halides and carbon. Extreme temperatures may occur in applications such as centrifugal compressors, if the impeller rubs against the housing when the system malfunctions. Using R-12 as the test refrigerant, Eiseman (1963) found that aluminum was most reactive, followed by iron and stainless steel. Copper is relatively unreactive. Using aluminum as the reactive metal, Eiseman reported that R-14 causes the most vigorous reaction, followed by R-22, R-12, R-114, R-11, and R-113.

Ammonia

Reactions involving ammonia, oxygen, oil degradation acids, and moisture are common factors in the formation of ammonia compressor deposits. Sedgwick (1966) suggested that ammonia or ammonium hydroxide reacts with organic acids produced by oxidation of the compressor oil to form ammonium salts (soaps), which can decompose further to form amides (sludge) and water. The reaction may be described as follows:

$$NH_3 + RCOOH \rightleftharpoons RCOONH_4 \rightleftharpoons RCONH_2 + H_2O$$

Water may be consumed or released during the reaction, depending on system temperature, metallic catalysts, and chemical state (acidic or basic condition). Compressor deposits can be minimized by keeping the system clean and dry, preventing entry of air, and maintaining proper compressor temperatures.

Reactions with Water. The halogenated refrigerants are susceptible to reaction with water (hydrolysis), but the rates of reaction are so slow that they are negligible (see Table 5). Desiccants, which are discussed in Chapter 6, are used to keep refrigeration systems dry. Cohen (1993) investigated the compatibility of desiccants with R-134a and refrigerant blends.

Lubricants

Lubricants now in use and being considered for the new refrigerants are mineral oils, alkyl benzenes, polyol esters, polyalkylene glycols, and modified polyalkylene glycols. Gunderson and Hart (1962) give an excellent introduction to synthetic lubricants including polyglycols and esters. Commercial esters (Jolley 1991) are manufactured from four types of alcohols. They are neopentylglycol (NPG) with two OH reaction sites, glycerine (GLY) with three OH sites, trimethylolpropane (TMP) with three OH sites, and pentaerythritol (PER) with four OH sites. The formulas for the four alcohol types are shown in Figure 2. The viscosities of the esters formed from the reaction of a given acid with each of the four alcohol types are given in Table 6. The extent of branching of the acid influences the miscibility of the lubricant in refrigerants.

Table 4 Inherent Stability of Refrigerants

Refrigerant	Formula	Decomposition Rated at 400°F in Steel, % per yr[a]	Temperature Where Decomposition Is Readily Observed in Laboratory,[b] °F	Temperature Where 1%/Year Decomposes in Absence of Active Materials, °F	Major Gaseous Decomposition Products[c]
22	$CHClF_2$	—	800	480	CF_2CF_2,[d] HCl
11	CCl_3F	2	1100	570[e]	R-12, Cl_2
114	$CClF_2CClF_2$	1	1100	710	R-12
115	$CClF_2CF_3$	—	1160	740	R-13
12	CCl_2F_2	Less than 1	1400	930	R-13, Cl_2
13	$CClF_3$	—	1550	1000[f]	R-14, Cl_2, R-116

Data from Norton (1957), Du Pont (1959, 1969), and Borchardt (1975).
[a]Data from Underwriters Laboratories, *Standard* 207.
[b]Decomposition rate is a bout 1% per min.
[c]Data from Borchardt (1975).
[d]A variety of side products are also produced, here and with the other refrigerants, some of which may be quite toxic.
[e]Conditions were not found where this reaction proceeds homogeneously.
[f]Rate behavior too complex to permit extrapolation to 1% per year.

Table 5 Rate of Hydrolysis in Water Grams per Litre of Water per Year

Refrigerant	Formula	1 Atm. Pressure 86°F Water Alone	1 Atm. Pressure 86°F With Steel	Saturation Pressure at 122°F With Steel
113	CCl_2FCClF_2	< 0.005	50	40
11	CCl_3F	< 0.005	10	28
12	CCl_2F_2	< 0.005	1	10
21	$CHCl_2F$	< 0.01	5	9
114	$CClF_2$-$CClF_2$	< 0.005	1	3
22	$CHClF_2$	< 0.01	0.1	—

Data from Du Pont (1959, 1969).

Table 6 Influence of Type of Alcohol on Ester-Based Fluids

Type of Alcohol	Ester Viscosity at 40°C, mm^2/s
NPG	13.3
GLY	31.9
TMP	51.7
PER	115

Note: Ester derived using the same carboxylic acid.

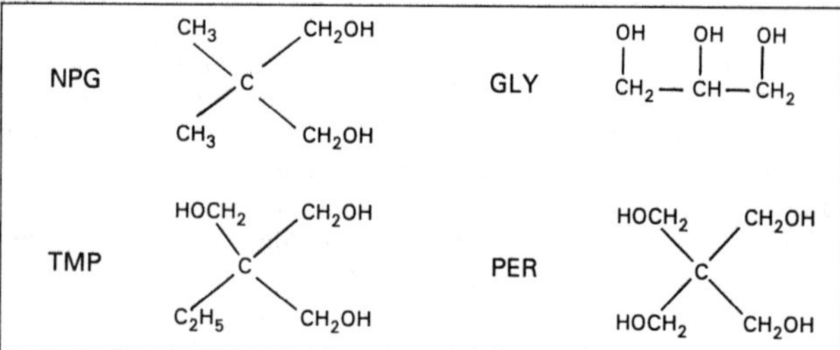

Fig. 2 Types of Alcohols Used for Ester Synthesis

Polyalkylene glycols (PAGs) are of the general formula RO—[CH_2—CHR′—O]—R′. They are now being used as lubricants in automotive applications that use HFC-134a. Linear PAGs can have one or two terminal hydroxyl groups. Modified PAGs have molecules with both ends capped by various groups. Sundaresan and Finkenstadt (1990) discuss the use of PAGs and modified PAGs in refrigeration compressors. Short and Cavestri (1992) present data on PAGs.

All of these lubricants and their associated additive packages may (1) oxidize, (2) degrade thermally, (3) react with system contaminants such as water, and/or (4) react with the refrigerant or system materials such as polyester films.

Oxidation is usually not a problem in hermetic systems using hydrocarbon oils since no oxygen is available to react with the lubricant. However if a system is not adequately evacuated or if air is allowed to leak into the system, organic acids and sludges can be formed. Lockwood and Klaus (1981) and Clark *et al.* (1985) found that iron and copper catalyze the oxidative degradation of esters. These reaction products are detrimental to the refrigeration system and can cause failure. Some have suggested that the oxidative breakdown products of PAG lubricants (Komatsuzaki *et al.* 1991) and perhaps of esters are volatile, while those of mineral oils are more likely to include sludges.

Sanvordenker (1991) studied the thermal stability of PAG and ester lubricants and found that above 400 °F, one of the decomposition products of both esters (in the presence of steel) and PAG lubricants is water. He recommends that polyol esters be used with metal passivators to enhance their stability when in contact with the metallic bearing surfaces, which can experience 400 °F temperatures. He presents data on the kinetics of the thermal decomposition of polyol esters and PAGs. These reactions are catalyzed by metal surfaces in the order low carbon steel > aluminum > copper (Naidu *et al.* 1988).

Hydrolysis of Esters

When an alcohol and an organic acid react, an organic ester and water are produced. This reaction is called esterification. The reaction is reversible. The reverse reaction of the ester and water to produce the alcohol and the organic acid is called hydrolysis.

$$\underset{\text{Ester}}{RCOOR} + \underset{\text{Water}}{HOH} \rightleftharpoons \underset{\text{Acid}}{RCOOH} + \underset{\text{Alcohol}}{ROH}$$

Jolley and others have used variations of ASHRAE *Standard* 97 sealed tube tests along with an acid number based on ASTM *Standard* D664 as a measure of hydrolysis. Hansen and Snitkjær (1991) demonstrated the hydrolysis of esters in compressor life tests run without desiccants and in sealed tubes. They detected hydrolysis by measuring the total acid number and showed that desiccants can reduce the extent of hydrolysis in a compressor. They concluded that with filter dryers, refrigeration systems using esters and R-134a can have very good reliability. Grieg (1992) also ran the Thermal and Oxidation Stability Test (TOST) by heating an oil/water emulsion to 203 °F and bubbled oxygen through the emulsion in the presence of steel and copper. Grieg shows that the hydrolysis of esters can be suppressed by appropriate additives. While agreeing that esters can be successfully used in refrigeration, Jolley points out that some types of additives are themselves subject to hydrolysis. Cottington and Ravner (1969) and Jones *et al.* (1969) studied the effect of tricresyl phosphate, a common antiwear agent, on the decomposition of esters.

System Reactions

Average bond strengths for carbon-chlorine, carbon-hydrogen, and carbon-fluorine (Pauling 1960) are 78, 93, and 100 kcal/mole, respectively. The relative stabilities of refrigerants that contain chlorine, hydrogen, and fluorine bonded to carbon can be largely understood by considering these bond strengths. The CFCs have characteristic reactions that depend largely on the presence of the C-Cl bond. Spauschus and Doderer (1961) showed that R-12 can react with a hydrocarbon oil by exchanging a chlorine for a hydrogen. This reaction is characteristic of chlorine-containing refrigerants. In the reaction, R-12 forms the reduction product R-22, R-22 forms R-32 (Spauschus and Doderer 1964), and R-115 forms R-125 (Parmelee 1965). For R-123, Carrier (1989) demonstrated that the reduction product R-133a is formed at high temperatures. The reaction of R-12 with oil to produce R-22 and chlorinated oil is as follows:

$$CCl_2F_2 + R{-}CH_2{-}CH_3 \rightleftharpoons CHClF_2 + R{-}CHCl{-}CH_3$$

Factor and Miranda (1991) studied the reaction between a chlorinated refrigerant and oil. They concluded that it can proceed by a predominantly Friedel-Crafts mechanism in which Fe^{3+} is a key catalyst. They also discovered that oil sludge can be formed by a pathway that does not generate R-22. They suggest that the free-radical mechanism that results in R-22 plays a minor role.

Huttenlocher (1992) tested 23 refrigerant-lubricant combinations for stability in sealed glass tubes. HFC refrigerants were very stable.

Fluoroethers are now being studied as alternative refrigerants. E-245 is a potential replacement for R-11. Sealed glass tube and Parr bomb stability tests with E-245 (CF_3-CH_2-O-CHF_2) showed evidence (Doerr *et al.* 1993) of an autocatylic reaction with glass, which proceeds until either the glass or the fluoroether is consumed. High pressures (about 2000 psi) usually cause the sealed glass tubes to explode.

The breakdown of CFCs and HCFCs can usually be tracked by observing the concentration of the reduction products formed. Alternately, both the amount of fluoride and chloride formed in the system can be observed. For HFCs, no chloride will be formed and reduction products are highly unlikely since the C-F bond is strong. Decomposition of HFCs is usually tracked by measuring the fluoride ion concentration in the system (Thomas and Pham 1989, Spauschus 1991, Thomas *et al.* 1993). This test showed that R-125, R-32, R-143a, R-152a, and R-134a are quite stable. In addition, Spauschus has suggested analysis for the breakdown of PAGs and esters by size exclusion chromatography.

The possibility that hydrogen fluoride released by the breakdown of the refrigerants being studied will react with glass of the sealed tube has been a concern. Sanvordenker (1985) confirmed this possibility with R-12. Spauschus (1992) found no evidence of fluoride on the glass surface of sealed tubes with R-134a.

Figures 3 and 4 show sealed tube test data for reaction rates of R-22 and R-12 with oil in the presence of copper and mild steel. The formation of chloride ion was taken as a measure of decomposition. These figures show the extent to which temperature accelerates reactions and also that R-22 is much less reactive than R-12. The data only illustrate the chemical reactivities involved and do not represent actual rates in refrigeration systems.

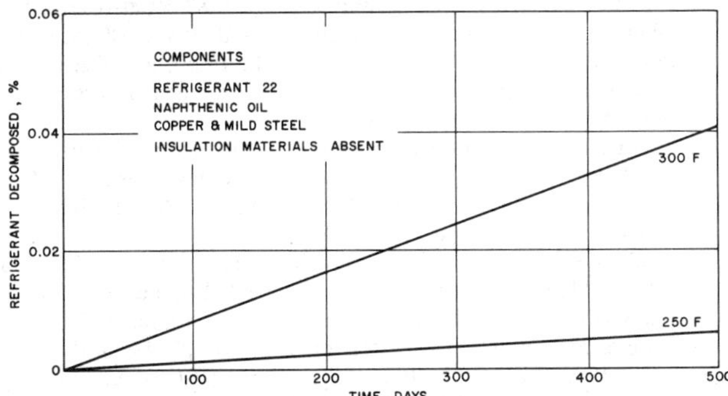

Fig. 3 Stability of Refrigerant 22 Control System
(Kvalnes and Parmelee 1957)

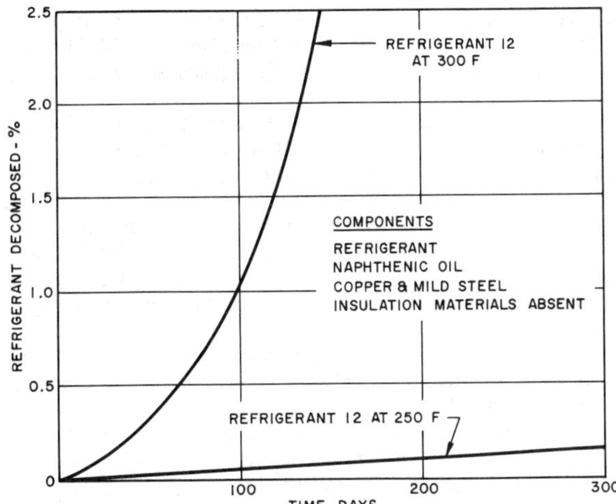

Fig. 4 Stability of Refrigerant 12 Control Systems
(Kvalnes and Parmelee 1957)

The chemistry that occurs in CFC systems that are retrofitted to use HFC refrigerants and their lubricants is an area of growing interest. Corr *et al.* (1992) point out that a major problem is the effect of chlorinated residues in the new system. Komatsuzaki *et al.* (1991) showed that R-12 and R-113 degrade PAG lubricants. Powers and Rosen (1992) ran sealed tube tests and concluded that the threshold of reactivity for R-12 in R-134a and PAG lubricant is between 1 and 3%.

Copper Plating

Copper plating is the formation of a copper film on steel surfaces in refrigeration and air-conditioning compressors. A blush of copper is often discernible on compressor bearing and valve surfaces when machines are cut apart. After several hours of exposure to air, this thin film becomes invisible, probably because metallic copper is converted to copper oxide. In more severe cases, the copper deposit, can build up to substantial thickness and interfere with compressor operation. Extreme copper plating can cause compressor failure.

Thomas and Pham (1989) compared R-12/mineral oil and R-134a/PAG copper plating. They showed that R-134a/PAG systems produced much less total copper (in solution and as precipitate) than R-12/mineral systems. They also showed that water did not significantly affect the amount of copper produced. They found that in the R-134a/PAG system, the copper was largely precipitated. In the R-12/mineral oil system, the copper was found in solution when dry and precipitated when wet. Walker *et al.* (1960) found that water below the saturation level had no significant effect on copper plating for R-12/mineral oil systems. Reyes-Gavilan (1993) observed copper plating in reciprocating compressors charged with R-134a and PAG lubricants.

Contaminant Generation by High Temperature

Hermetic motors can overheat well beyond the design levels under adverse conditions. Some of these conditions include line voltage fluctuations, "brown outs," or inadequate airflow over the condenser coils. Under these conditions, motor winding temperatures can exceed 300°F. Prolonged exposure to these thermal excursions can damage the motor insulation, depending on the thermal stability of the insulation materials, their reactivity with the refrigerant and oil, and the temperature levels encountered.

Another potential for high temperatures is in the bearings. Oil film temperatures in hydrodynamically lubricated journal bearings are usually not much higher than the bulk oil temperature; but in elastohydrodynamic films in bearings with a high slide/roll ratio, the temperature can be several hundred degrees above the bulk oil temperature (Keping and Shizhu 1991). Local hot spots in boundary lubrication can reach very high temperatures. Fortunately, the amount of material exposed to these temperatures is usually very small. The appearance of methane or other small hydrocarbon molecules in the refrigerant indicates cracking of the lubricant due to high bearing temperatures.

Thermal decomposition of organic insulation materials and some types of lubricants produces noncondensable gases such as carbon dioxide and carbon monoxide. These gases circulate with the refrigerant and increase the discharge pressure and lower unit efficiency. At the same time, the compressor temperature and the rate of deterioration of the insulation or lubricant increases. Liquid decomposition products circulate with the lubricating oil either in solution or as colloidal suspensions. Dissolved and suspended decomposition products circulate throughout the refrigeration system where they clog oil passages, interfere with the operation of expansion, suction, and discharge valves or plug capillary tubes.

Appropriate control mechanisms in the refrigeration system minimize exposure to such high temperatures. Identifying potential reactions, performing adequate laboratory tests to qualify the materials before using them in the field, and finding means to remove the contaminants generated by high-temperature excursions are equally important (see Chapter 6).

REFERENCES

ASHRAE. 1992. Number designation and safety classification of refrigerants. *Standard* 34-1992.

ASHRAE. 1989. Sealed glass tube method to test the chemical stability of material for use within refrigerant systems. *Standard* 97-1989.

ASTM. 1992. Standard test methods for film-insulated magnet wire. *Standard* D1676-92. American Society for Testing and Materials, Philadelphia.

ASTM. 1991. Standard test method for bond strength of electrical insulating varnishes by the helical coil test. *Standard* D2519-91.

ASTM. 1989. Standard test method for acid number of petroleum products by potentiometric titration. *Standard* D664-89.

ASTM. 1985. Standard test method for concentration limits of flammability of chemicals. *Standard* E681-85.

Atwood, T. and J. Zheng. 1991. Cascade refrigeration systems: The HFC-23 solution. International CFC and Halon Alternatives Conference, sponsored by The Alliance for Responsible CFC Policy, Arlington, VA.

Bateman, D.J., D.B. Bivens, R.A. Gorski, W.D. Wells, R.A. Lindstrom, R.A. Morse, and R.L. Shimon. 1990. Refrigerant blends for the automotive air conditioning aftermarket. SAE Technical Paper Series 900216. Society of Automotive Engineers, Warrendale, PA.

Beacham, E.A. and R.T. Divers. 1955. Some practical aspects of the dielectric properties of refrigerants. *Refrigerating Engineering* (July):33.

Beyerlein, A.L., D.D. Desmarteau, S.H. Hwang, N.D. Smith, and P. Joynere. 1991. Physical property data on fluorinated propanes and butanes as CFC and HCFC alternatives. International CFC and Halon Alternatives Conference. See Atwood and Zheng 1991.

Bier, K., M. Crone, M. Tuerk, W. Leuckel, M. Christill, and B. Leisenheimer. 1990. Studies of the thermal stability and ignition behavior and combustion properties of the refrigerants R-152a and R-134a. *DKV-Tagungsber* 17:169-91.

Bivens, D.B., R.A. Gorski, W.D. Wells, A. Yokozeki, R.A. Lindstrom, and R.L. Shimon. 1989. Evaluation of fluorocarbon blends as automotive air conditioning refrigerants. SAE Technical Paper Series 890306. Society of Automotive Engineers, Warrendale, PA.

Booth, H.S. 1937. Halogenated Methyl Ethers. U.S. Patent 2,066,905.

Borchardt, H.J. 1975. *Du Pont innovation*, Vol. G, No. 2 (Spring).

Carrier. 1989. Decomposition rates of R-11 and R-123. Available from ARTI Database. US DOE grant no. DE-F-G02-91 CE23810. Air-Conditioning and Refrigeration Technology Institute Database, Great Falls, VA. Carrier Corp. (Syracuse, NY).

Cavestri, R.C. 1993. Compatibility of refrigerants and lubricants with plastics. Final Report No. DOE/CE 23810-15. ARI Database (December). Air-Conditioning and Refrigeration Institute, Arlington, VA.

Clark, D.B., E.E. Klaus, and S.M. Hsu. 1985. The role of iron and copper in the degradation of lubricating oils. *Journal of ASLE* (May):80-87. American Society of Lubrication Engineers.

Cohen, A.P. 1993. Test methods for the compatibility of desiccants with alternative refrigerants. ASHRAE *Transactions* 99(1):408-12.

Corr, S., R.D. Gregson, G. Tompsett, A.L. Savage, and J.A. Schukraft. 1992. Retrofitting large refrigeration systems with R-134a. International Refrigeration Conference—Energy Efficiency and New Refrigerants Proceedings 1:221-30. Purdue University, West Lafayette, IN.

Cottington, R.L. and H. Ravner. 1969. Interactions in neopentyl polyol ester-tricresyl phosphate-iron systems at 500°F. ASLE *Transactions* 12:280-86. Amercian Society of Lubrication Engineers.

Desmarteau, D., A. Beyerlien, S. Hwang, Y. Shen, S. Li, R. Mendonca, K. Naik, N.D. Smith, and P. Joyner. 1991. Selection and synthesis of fluorinated propanes and butanes as CFC and HCFC alternatives. International CFC and Halon Alternatives Conference. See Atwood and Zheng 1991.

Doerr, R.G. 1992. Absorption of HCFC-123 and CFC-11 by epoxy motor varnish. ASHRAE *Transactions* 98(2):227-34.

Doerr, R.G. and S.A. Kujak. 1993. Compatibility of refrigerants and lubricants with motor materials. Final Report No. DOE/CE 23810-13. ARTI Database (May). Air-Conditioning and Refrigeration Technology Institute, Arlington, VA.

Doerr, R.G. and S.A. Kujak. 1993. Compatibility of refrigerants and lubricants with motor materials. ASHRAE *Journal* 35(8):42-7.

Doerr, R.G., D. Lambert, R. Schafer, and D. Steinke. 1992. Stability and compatibility studies of R-245ca, CHF_2-CF_2-CH_2F, a potential low-pressure refrigerant. International CFC and Halon Alternatives Conference. See Atwood and Zheng 1991.

Doerr, R.G., D. Lambert, R. Schafer, and D. Steinke. 1993. Stability studies of E-245 fluoroether, CF_3-CH_2-O-CHF_2. ASHRAE *Transactions* 99(2):1137-40.

Du Pont. 1959. Properties and application of the "Freon" fluorinated hydrocarbons. *Technical Bulletin* B-2, Freon Products Division, E.I. du Pont de Nemours & Co., Inc.

Du Pont. 1969. Stability of several "Freon" compounds at high temperatures. *Technical Bulletin* XIA, Freon Products Division, E.I. du Pont de Nemours & Co., Inc.

Eiseman, B.J. 1968. Chemical Processes. U.S. patent 3,362,180.

Eiseman, B.J, Jr. 1963. Reactions of chlorofluoro-hydrocarbons with metals. ASHRAE *Journal* 5(5):63.

Factor, A. and P.M. Miranda. 1991. An investigation of the mechanism of the R-12-oil-steel reaction. *Wear* 150:41-58.

Fedorko, G., G. Fredrick, and J.G. Hansel. 1987. Flammability characteristics of chlorodifluoromethane (R-22)-oxygen-nitrogen mixtures. ASHRAE *Transactions* 93(2):716-24.

Fellows, B.R., R.C. Richard, and I.R. Shankland. 1991. Electrical characterization of alternative refrigerants. XVIIIth International Congress of Refrigeration, Montreal, 45:398.

Fischer, S.K., P.J. Hughes, P.D. Fairchild, C.L. Kusik, J.T. Dieckmann, E.M. McMahon, and N. Hobday. 1991. Energy and global warming impacts of CFC alternative technologies. Sponsored by the Alternative Fluorocarbons Environmental Acceptability Study (AFEAS), Washington, D.C., and U.S. Department of Energy, Washington, D.C.

Fisher, D.A., C.H. Hales, W.C. Ko, and N.D. Sze. 1989. Relative effects on global warming of halogenated methanes and ethanes of social and industrial interest. Scientific Assessment of Stratospheric Ozone, Vol. II, Appendix: Alternative Fluorocarbons Environmental Acceptability Study (AFEAS) Report, Washington, D.C.

Grieg, B.D. 1992. Formulated polyol ester lubricants for use with HFC-134a. The role of additives and conversion of existing CFC-12 plant to HFC-134a. International CFC and Halon Alternatives Conference. See Atwood and Zheng 1991.

Grob, D.P. 1991. Summary of flammability characteristics of R-152a, R-141b, R-142b and analysis of effects in potential applications. Household Refrigerators. 42nd Annual International Appliance Technical Conference, organized by its steering committee, Batavia, IL.

Gross, T.P. 1990. Sealed Tube Tests—Grace Ether (E-134). ARTI Refrigerant Database RDB0904.

Gunderson, R.C. and A.W. Hart. 1962. *Synthetic lubricants.* Reinhold Publishing Corporation, New York.

Guy, P.D., G. Tompsett, and T.W. Dekleva. 1992. Compatibilities of non-metallic materials with R-134a and alternative lubricants in refrigeration systems. ASHRAE *Transactions* 98(1):804-16.

Hamed, G.R. and R.H. Seiple. 1993. Compatibility of refrigerants and lubricants with elastomers. Final Report No. DOE/CE 23810-14. ARI Refrigerant Database. Air-Conditioning and Refrigerating Institute, Arlington, VA.

Hamed, G.R. and R.H. Seiple. 1993. Compatibility of refrigerants and lubricants with elastomers. ASHRAE *Journal* 35:173-6.

Hansen, P.E. and L. Finsen. 1992. Lifetime and reliability of small hermetic compressors using a ternary blend HCFC-22/HFC-152a/HCFC-124. International Refrigeration Conference—Energy Efficiency and New Refrigerants. D. Tree, ed. Purdue University, West Lafeyette, IN.

Hansen, P.E. and L. Snitkjær. 1991. Development of small hermetic compressors for R-134a. XVIIIth International Congress of Refrigeration, Montreal, 223:1146.

Hughes, H.M. 1992. Non-ozone depleting refrigerants for centrifugal compressors. International CFC and Halon Alternatives Conference. See Atwood and Zheng 1991.

Hurtgen, J.R. 1971. R-22 blister testing of magnet wire. pp. 183-185. Proceedings of the 10th Electrical Insulation Conference, Chicago.

Huttenlocher, D.F. 1972. Accelerated sealed-tube test procedure for Refrigerant 22 reactions. Proceedings of the 1972 Purdue Compressor Technology Conference.

Huttenlocher, D.F. 1992. Chemical and thermal stability of refrigerant-lubricant mixtures with metals. Final Report No. DOE/CE 23810-5. ARTI Database. Air-Conditioning and Refrigeration Technology Institute, Arlington, VA.

Jolley, S.T. 1991. The performance of synthetic ester lubricants in mobile air conditioning systems. Available from ARTI Database. US DOE grant no. DE-F-G02-91 CE 23810. Air-Conditioning and Refrigeration Technology Institute Database, Great Falls, VA.

Jones, R.L., H.L. Ravner, and R.L. Cottington. 1969. Inhibition of iron-catalyzed neopentyl polyol ester thermal degredation through passivation of the active metal surface by tricresyl phosphate. Paper presented at ASLE/ASME Lubrication Conference, Houston, TX.

Keping, H. and W. Shizhu. 1991. Analysis of maximum temperature for thermo elastohydrodynamic lubrication in point contacts. *Wear* 150:1-10.

Komatsuzaki, S., Y. Homma, K. Kawashima, and Y. Itoh. 1991. Polyalkylene glycol as lubricant for HCFC-134a compressors. *Lubrication Engineering* (Dec.):1018-25.

Kopko, W.L. 1989. Extending the search for new refrigerants. Proceedings of CFC Technology Conference, Gaithersburg, MD.

Krantz, J.C. and F.G. Rudo. 1966. The fluorinated anesthetics in *Handbook of Experimental Pharamacology*, Vol XX/1. Eichler, O., H. Herken, and A.D. Welch, eds. Springer. Berlin, 501-64.

Kvalnes, D.E. and H.M. Parmelee. 1957. Behavior of Freon-12 and Freon-22 in sealed-tube tests. *Refrigerating Engineering* (November):40.

Lindley, A.A. 1992. KLEA 134a flammability characteristics at high temperatures and pressures. ARI Database. Air-Conditioning and Refrigeration Institute, Arlington, VA.

Lockwood, F. and E.E. Klaus. 1981. Ester oxidation—the effect of an iron surface. ASLE *Transactions* 25(2):236-44.

Mays, R.L. 1962. Molecular sieve and gel-type desiccants for refrigerants. ASHRAE *Transactions* 68:330.

Naidu, S.K., B.E. Klaus, and J.L. Duda. 1988. Thermal stability of esters under simulated boundary lubrication conditions. *Wear* 121:211-22.

NEMA. 1987. Magnet wire. *Standard* MW 1000-87. National Electrical Manufacturer's Association, Washington, D.C.

Norton, F.J. 1957. Rates of thermal decomposition of $CHClF_2$ and Cl_2F_2. *Refrigerating Engineering* (September):33.

O'Neill, G.J. 1992. Synthesis of fluorinated dimethyl ethers. U.K. Patent Application 2,248,617.

O'Neill, G.J. and R.S. Holdsworth. 1990. Bis(difluoromethyl) ether refrigerant. U.S. Patent 4,961,321.

Parmelee, H.M. 1965. Sealed-tube stability tests on refrigerant materials. ASHRAE *Transactions* 71(1):154.

Pauling, L. 1960. The nature of the chemical bond and the structure of molecules and crystals, In *An introduction to modern structural chemistry.* Cornell University Press, Ithaca, NY.

Powers, S. and S. Rosen. 1992. Compatibility testing of various percentages of R-12 in R-134a and PAG lubricant. International CFC and Halon Alternatives Conference, See Atwood and Zheng 1991.

Reed, P.R. and H.O. Spauschus. 1991. HCFC-124: Applications, properties and comparison with CFC-114. ASHRAE *Journal* 40(2).

Reed, P.R. and J.J. Rizzo. 1991. Combustibility and stability studies of CFC substitutes with simulated motor failure in hermetic refrigeration equipment. Proceedings of the XVIIIth International Congress of Refrigeration, Montreal, Quebec, Canada, 2:888-91

Reyes-Gavilan, J.L. 1993. Performance evaluation of naphthenic and synthetic oils in reciprocating compressors employing R-134a as the refrigerant. ASHRAE *Transactions* 99(1):349.

Richard, R.G. and I.R. Shankland. 1991. Flammability of alternative refrigerants. Proceedings of the XVIIth Congress of Refrigeration, 384.

Sand, J.R. and D.L. Andrjeski. 1982. Combustibility of chlorodifluoromethane. ASHRAE *Journal* (May):38-40.

Sand, J.R., S.K. Fischer, and P.A. Joyner. 1991. Modeled performance of non-chlorinated substitutes for CFC-11 and CFC-12 in centrifugal chillers. International CFC and Halon Alternatives Conference. See Atwood and Zheng 1991.

Sandvordenker, K.S. 1992. R-152a versus R-134a in a domestic refrigerator-freezer, energy advantage or energy penalty. International Refrigeration Conference—Energy Efficiency and New Refrigerants Proceedings, Purdue University, West Lafayette, IN.

Sanvordenker, K.S. 1991. Durability of HFC-134a compressors—The role of the lubricant. Proceedings of the 42nd Annual International Appliance Technical Conference, University of Wisconsin-Madison.

Sanvordenker, K.S. 1985. Mechanism of oil-R-12 reactions—The role of iron catalyst in glass sealed tubes. ASHRAE *Transactions* 91(1A):356.

Sandvordenker, K.S. and M.W. Larime. 1971. Screening tests for hermetic magnet wire insulation. Proceedings of the 10th Electrical Insulation Conference.

Sedgwick, N.V. 1966. *The organic chemistry of nitrogen*, 3rd ed. Clarendon press, Oxford, England.

Short, G.D. and R.C. Cavestri. 1992. High viscosity ester lubricants for alternative refrigerants. ASHRAE *Transactions* 98(1):789-95.

Simons, G.W., G.J. O'Neill, and J.A. Gribens. 1977. New aerosol propellants for personal products. U.S. Patent 4,041,148.

Smith, N.D., K. Ratanaphruks, M.W. Tufts, and A.S. Ng. 1993. R-245ca: A potential far-term alternative for R-11. ASHRAE *Journal* 35(2):19-23.

Spauschus, H.O., G. Freeman, and T.L. Starr. 1992. Surface analysis of glass from sealed tubes after aging with HFC-134a. Available from ARTI Database. RDB2729. Air-Conditioning and Refrigeration Technology Institute Database, Great Falls, VA.

Spauschus, H.O. 1991. Stability requirements of lubricants for alternative refrigerants. XVIIIth International Congress of Refrigeration, Paper No. 148.

Spauschus, H.O. and G.C. Doderer. 1964. Chemical reactions of Refrigerant 22. ASHRAE *Journal* 6(10).

Spauschus, H.O. and G.C. Doderer. 1961. Reaction of refrigerant 12 with petroleum oils. ASHRAE *Journal* 3(2):65.

Spauschus, H.O. and R.A. Sellers. 1969. Aging of hermetic motor insulation. IEEE *Transactions* E1-4(4):90.

Sundaresan, S.G. and W.R. Finkenstadt. 1991. Degredation of polyethylene terephthalate films in the presence of lubricants for HFC-134a: A critical issue for hermetic motor insulation systems. *International Journal of Refrigeration* 14:317.

Sundaresan, S.G. and W.R. Finkenstadt. 1990. Status report on polyalkylene glycol lubricants for use with HFC-134a in refrigeration compressors. Proceedings of the 1990 USNC/IIR-Purdue Conference, ASHRAE-Purdue Conference, 138-44.

Terrell, R.C., L. Spears, A.J. Szur, J. Treadwell, and T.R. Ucciardi. 1971a. General anesthetics. 1. Halogenated methyl ethers as anesthetics agents. *Journal of Medical Chemistry* 14:517.

Terrell, R.C., L. Spears, T. Szur, T. Ucciardi, and J.F. Vitcha. 1971b. General anesthetics. 3. Fluorinated methyl ethyl ethers as anesthetic agents. *Journal of Medical Chemistry* 14:604.

Thomas, R.H. and H.T. Pham. 1989. Evaluation of environmentally acceptable refrigerant/lubricant mixtures for refrigeration and air conditioning. SAE Passenger Car Meeting and Exposition. Society of Automotive Engineers, Warrendale, PA.

Thomas, R.H., W.T. Wu, and R.H. Chen. 1993. The stability of R-32/125 and R-125/143a. ASHRAE *Transactions* 99(2).

UL. 1991. Standard for safety hermetic refrigerant motor-compressors, 6th ed. *Standard* 984-91. Underwriters Laboratories, Inc., Northbrook, IL.

UL. 1993. Standard for safety refrigerant-containing components and accessories, nonelectrical, 6th ed. *Standard* 207-93. Northbrook, IL.

Walker, W.O., S. Rosen, and S.L. Levy. 1960. A study of the factors influencing the stability of mixtures of Refrigerant 22 and refrigerating oils. ASHRAE *Transactions* 66:445.

Wang, B., J.L. Adcock, S.B. Mathur, and W.A. Van Hook. 1991. Vapor pressures, liquid molar volumes, vapor non-idealities and critical properties of some fluorinated ethers: CF_3-O-CF_2-O-CF_3, CF_3-O-CF_2-CF_2H, c-CF_2-CF_2-CF_2-O-, CF_3-O-CF_2H, and CF_3-OCH_3 and of CCl_3F and CF_2ClH. *Journal of Chemical Thermodynamics* 23:699-710.

Zhigang, L., L. Xianding, Y. Jianmin, T. Xhoufang, J. Pingkun, C. Zhehua, L. Dairu, R. Mingzhi, Z. Fan, and W. Hong. 1992. Application of HFC-152a/HCFC-22 blends in domestic refrigerators.

CHAPTER 6

MOISTURE AND OTHER CONTAMINANT CONTROL IN REFRIGERANT SYSTEMS

THIS chapter covers the following topics: (1) control of moisture, which is an important and universal contaminant in refrigeration systems; (2) control of other contaminants in refrigeration systems; (3) recovery, recycling, and reclamation of refrigerants; and (4) service procedures typically used in cleaning a refrigeration system following a hermetic motor burnout.

MOISTURE

The amount of moisture in a refrigerant system must be kept below an allowable maximum to provide satisfactory operation. Moisture must be removed from refrigerant system components during manufacture and assembly to guard against moisture entering the system during installation or servicing. Any moisture that enters the refrigerant system should be removed as quickly as is practicable.

Sources of Moisture

Moisture in a refrigerant system results from:

1. Faulty equipment drying in factories and service operations
2. Introduction during installation or service operations in the field
3. Low-side leaks, resulting in entrance of moisture-laden air
4. Leakage of water-cooled heat exchangers
5. Oxidation of certain hydrocarbons of oil to produce moisture
6. Wet lubricant, refrigerant, or both
7. Decomposing cellulose insulation in hermetically sealed units
8. Moisture entering a nonhermetic refrigerant system via permeation of nonmetallic hoses and seals

Drying equipment in the factory is discussed in Chapter 45. Proper installation and service procedures as given in ASHRAE *Guideline* 3-1990 minimize the sources listed in items 2, 3, and 4. Lubricants are discussed in Chapter 7 of this volume. If purchased refrigerants and lubricants meet specifications and are properly handled, the moisture content generally remains satisfactory. See the section on hermetic motor insulation in Chapter 5 and the section on motor burnouts in this chapter.

Effects of Moisture

Excess moisture in a refrigerating system can cause one or all of the following undesirable effects:

1. Ice formation in expansion valves, capillary tubes, or evaporators
2. Corrosion of metals
3. Copper plating
4. Chemical damage to insulation in hermetic compressors or other system materials

The preparation of this chapter is assigned to TC 3.3, Contaminant Control in Refrigerating Systems.

Ice or solid hydrate separates from refrigerants if the water concentration is high enough and the temperature low enough. Solid hydrate, a complex molecule of refrigerant and water, can form at temperatures higher than those required to separate ice. Liquid water forms at temperatures above those required to separate ice or solid hydrate. Ice forms during refrigerant evaporation when the relative saturation of vapor reaches 100% at temperatures of 32 °F or below.

The separation of water as ice or liquid also is related to the solubility of water in a refrigerant. This solubility varies for different refrigerants and with temperature. Data for some refrigerants are presented in Table 1. Various investigators have obtained different results on water solubility in R-134a and R-123. The data presented here are the best available. The greater the solubility of water in a refrigerant, the less the possibility that ice or liquid water will separate in a refrigerating system. The solubility of water in ammonia, carbon dioxide, and sulfur dioxide is so high that ice or liquid water separation is not a problem.

Table 1 Solubility of Water in the Liquid Phase of Certain Refrigerants

Temp., °F	Solubility, ppm by weight								
	R-11	R-12	R-13	R-22	R-113	R-114	R-123	R-134a	R-502
160	460	700	—	4100	460	450	2600	4200	1780
150	400	560	—	3600	400	380	2300	3600	1580
140	340	440	—	3150	344	320	2000	3200	1400
130	290	350	—	2750	290	270	1800	2800	1220
120	240	270	—	2400	240	220	1600	2400	1080
110	200	210	—	2100	200	180	1400	2000	930
100	168	165	—	1800	168	148	1200	1800	810
90	140	128	—	1580	140	120	1000	1500	690
80	113	98	—	1350	113	95	900	1300	580
70	90	76	—	1140	90	74	770	1100	490
60	70	58	44	970	70	57	660	880	400
50	55	44	—	830	55	44	560	730	335
40	44	32	26	690	44	33	470	600	278
30	34	23.3	—	573	34	25	400	490	225
20	26	16.6	14	472	26	18	330	390	180
10	20	11.8	—	384	20	13	270	320	146
0	15	8.3	7	308	15	10	220	250	115
−10	11	5.7	—	244	11	7	180	200	90
−20	8	3.8	3	195	8	5	140	150	69
−30	6	2.5	—	152	6	3	110	120	53
−40	4	1.7	1	120	—	2	90	89	40
−50	3	1.1	—	91	—	1.5	70	66	30
−60	2	0.7	—	68	—	1	53	49	22
−70	1	0.4	—	50	—	0.6	40	35	16
−80	0.8	0.3	—	37	—	0.4	30	25	11
−90	0.5	0.1	—	27	—	0.2	22	18	8
−100	0.3	0.1	—	19	—	0.1	16	12	5

Data on R-134a adapted from Thrasher *et al.* (1993) and Allied-Signal Corporation. Data on R-123 adapted from Thrasher *et al.* (1993) and DuPont Company. Remaining data adapted from DuPont Company and Allied-Signal Corporation. Used by permission.

Table 2 Distribution of Water Between the Vapor and Liquid Phases of Certain Refrigerants

Temperature, °F	Water in Vapor/Water in Liquid, mass %/mass %					
	R-11	R-12	R-22	R-113	R-114	R-502
100	30.1	5.5	0.400	50.2	13.3	0.63
90	—	6.1	0.397	54.6	14.4	0.64
80	34.8	6.3	0.405	61.4	16.0	0.65
70	—	7.5	0.404	68.6	17.6	0.66
60	43.1	8.2	0.401	79.0	19.7	0.66
50	—	9.0	0.391	89.3	21.7	0.66
40	50.5	9.89	0.390	99.3	24.7	0.66
30	—	11.2	0.378	111.4	26.9	0.65
20	57.9	11.9	0.351	120.7	29.5	0.61
0	62.4	13.1	0.301	140.9	32.1	0.53
−20	71.0	15.3	0.251	172.6	37.4	0.47
−40	—	17.1	0.203	—	52.2	0.40

Data adapted from E.I. duPont de Nemours & Company, Inc. Used by permission.

Elsey and Flowers (1949) recognized that the concentration of water by mass at equilibrium is greater in the gas phase than in the liquid phase of Refrigerant 12. The opposite is true for Refrigerants 22 and 502. The ratio of mass concentrations differs for each refrigerant; it also varies with temperature. Table 2 shows the distribution ratios of water in the vapor phase to water in the liquid phase for common refrigerants. It can be used to calculate the equilibrium water concentration of the liquid phase refrigerant if the gas phase concentration is known, and vice versa. The water content in the vapor phase is determined by:

$$W = [P_w/(P_w)^0][(d_w)^0/(d_R)^0] \quad (1)$$

where

W = mass water/mass refrigerant
P_w = partial pressure of water vapor
$(P_w)^0$ = partial pressure of water vapor at saturation
$(d_w)^0$ = density of water vapor at saturation
$(d_R)^0$ = density of refrigerant vapor at saturation

Freezing at expansion valves or capillary tubes can occur when excessive moisture is present in a refrigerating system. Formation of ice or hydrate in evaporators can partially insulate the evaporator. Walker *et al.* (1962) showed that excess moisture can cause corrosion and enhance copper plating. Other factors affecting copper plating are discussed in Chapter 5.

The moisture required for freeze up is a function of the amount of flash gas formed during expansion and the distribution of water between the liquid and gas phases downstream of the expansion device. For example, in an R-12 system with a 110°F liquid temperature and a −20°F evaporator temperature, the refrigerant after expansion is 41.3% vapor and 58.7% liquid by mass. The percentage of vapor formed is determined by:

$$\%\ \text{Vapor} = 100\,\frac{h_{L(\text{liquid})} - h_{L(\text{evap})}}{h_{fg(\text{evap})}} \quad (2)$$

where

$h_{L(\text{liquid})}$ = saturated liquid enthalpy for refrigerant at liquid temperature
$h_{L(\text{evap})}$ = saturated liquid enthalpy for refrigerant at evaporating temperature
$h_{fg(\text{evap})}$ = latent heat of vaporization of refrigerant at evaporating temperature

Table 1 lists the saturated water content of the liquid phase at −20°F as 3.8 ppm. Table 2 is used to determine the saturated vapor phase water content as:

$$3.8 \text{ ppm} \times 15.3 = 58 \text{ ppm}$$

When the vapor contains more than the saturation quantity (100% rh), free water will be present as a third phase. If the temperature is below 32°F, ice will form. Using the saturated moisture values and the liquid-vapor ratios, the critical water content of the circulating refrigerant can be calculated as:

$$\begin{aligned} 3.8 \times 0.587 &= 2.2 \\ 58.0 \times 0.413 &= \underline{24.0} \\ & 26.2 \text{ ppm} \end{aligned}$$

Maintaining moisture levels below critical value keeps free water from the low side of the system.

The above analysis can be applied to all refrigerants and applications. An R-22 system with 110°F liquid and −20°F evaporating temperatures reaches saturation when the moisture circulating is 139 ppm. Note that this value is less than the liquid solubility, 195 ppm at −20°F.

Excess moisture causes paper or polyester motor insulation to become brittle, which can cause premature motor failure. However, not all motor insulations are affected adversely by moisture. The amount of water present in a refrigerant system must be small enough to avoid ice separation, corrosion, and insulation breakdown.

Exact experimental data on the maximum permissible moisture level in refrigerant systems are not known because so many factors are involved. Table 3 shows the results of moisture analysis of refrigerants taken from equipment in normal service from 1 to 16 years (ARI 1991a).

Table 3 Data on Moisture in Refrigeration Systems

Application	Range of Moisture Content Observed in Refrigerant, ppm by mass			
	R-11	R-12	R-22	R-502
Centrifugal chillers	0 - 30	0 - 25	—	—
Reciprocating and screw chillers	—	—	0 - 56	—
Refrigeration systems	—	0 - 2	1 - 122	3 - 33
Large unitary systems ≥ 10 ton	—	—	0 - 75	—
Small unitary systems < 10 ton	—	—	3 - 127	—

Moisture Indicators

Moisture-sensitive elements that change color according to moisture content can gage the moisture level in the system; the color changes at a low enough level to be safe. Manufacturer's instructions must be followed since the color change point is also affected by the liquid line temperature and the refrigerant used.

Drying Methods

Factory dehydration of refrigeration and air-conditioning units is discussed in Chapter 45 and in ASHRAE *Guideline* 3-1990. Field systems are dried by evacuation and driers. Prior to opening a system for service, refrigerant should be isolated in the system or recovered into an external storage container. After installation or service, noncondensables (air) should be removed by evacuation with a vacuum pump to lower the aboslute pressure of the system to 1000 μm of mercury or less. This will reduce the internal pressure of the system below the boiling point of water at ambient temperature. External heat may be required to vaporize water in the system. Even with these procedures, small amounts of moisture trapped under an oil film, adsorbed by the motor windings, or located far from the vacuum pump will be difficult to remove.

It is good practice to install a drier. On larger systems a drier with a replaceable element is frequently used, and may need to be changed several times before the proper degree of dryness is obtained. A moisture indicator in the liquid line can indicate when the system has been dried satisfactorily.

Special techniques are required to remove free water in a refrigeration or air-conditioning system due to a burst tube or water chiller leak. The refrigerant should be transferred to a pumpdown receiver or recovered in a separate storage tank. Parts of the system may have to be disassembled and the water drained from system low points. If the system contains a semihermetic compressor, it should be cleaned frequently by disassembling and handwiping the various parts. After reassembly, it should be dried further by passing dry nitrogen through the system and by heating and evacuation. Drying may take an extended period and require frequent changes of the vacuum pump oil. The liquid line driers should be replaced and temporary suction line driers installed. During the initial operating period, driers will need to be changed often.

If the refrigerant in the pumpdown receiver is to reused, it must be thoroughly dried before being reintroduced into the system. One way to dry it is to draw a liquid refrigerant sample and record the ambient temperature. If a chemical analysis of the sample reveals a moisture content at or near the water solubility of Table 1 at the recorded temperature, then free water is probably present. In that case, a recovery unit with a suction filter-drier and/or a moisture/oil trap must be used to transfer the bulk of the refrigerant from the receiver liquid port to a separate tank. When the free water reaches the tank liquid port, most of the remaining refrigerant can be recovered through the receiver vapor port. The water can then be drained from the pumpdown receiver.

Moisture Measurement

Techniques for measuring the amount of moisture in a compressor, or in an entire system, are discussed in Chapter 45. The following methods are used to measure the moisture content of various halocarbon refrigerants. The moisture content to be measured is generally in the ppm range, and the procedures require special laboratory equipment and techniques.

The *gravimetric method* for measuring the moisture content of refrigerants is described in ASHRAE *Standards* 63.1-1988 and 35-1992. It is not widely used in the industry. In this method, a measured amount of refrigerant vapor is passed through two tubes in series, each containing phosphorous pentoxide (P_2O_5). Any moisture present in the refrigerant reacts chemically with the P_2O_5 and appears as an increase in mass in the first tube. The second tube is used as a tare. This method is satisfactory when the refrigerant is pure, but the presence of oil will produce inaccurate results, since it will be weighed as moisture. Approximately 200 g of refrigerant are required for accurate results. Since the refrigerant must pass slowly through the tube, analysis requires many hours of elapsed time. In spite of its limitations, this method is considered the primary standard for laboratory refrigerant moisture analysis.

The *Karl Fischer Method* is suitable for measuring the moisture content of a refrigerant, even if it contains mineral oil. Although different firms have slightly different ways of performing this test and get somewhat varying results, the method remains the common industry practice for determining moisture content in refrigerants. The refrigerant sample is bubbled through predried methyl alcohol in a special sealed glass flask; any water present remains with the alcohol. Karl Fischer reagent is added, and the solution is immediately titrated to a "dead stop" electrometric end point. The Karl Fischer reagent reacts with any moisture present. The amount of water in the sample can be calculated from a previous calibration of the Karl Fischer reagent. This method, considered among the most accurate, is also suitable for measuring the moisture content of pure oil or other liquids. Special instruments designed for this particular analysis are available from laboratory supply companies. Haagen-Smit *et al.* (1970) describe improvements in the equipment and technique that significantly reduce analysis time.

DeGeiso and Stalzer (1969) discuss the *electrolytic moisture analyzer.* Refrigerant vapor is passed over a thin hygroscopic film of phosphoric acid on an electrometric probe. A meter measures the electrical conductivity of this film, which varies according to the moisture content. Oil must not coat the detector; if the detector is contaminated with oil, it must be carefully cleaned and recoated. The instrument requires at least 0.0176 ft^3 of vapor for an analysis. The analysis with this instrument is so rapid that it can be used to monitor moisture changes within a system. Brisken (1955) used this method in a study of moisture migration in hermetic systems. His work describes the sampling technique and running time necessary to obtain reproducible samples from an operating refrigeration system.

Another method, the infrared spectrophotometer, is used for moisture analysis, but requires a large sample for precise results and is subject to interference if oil is present in the refrigerant. In applications where a simple indication of "wet" or "dry" is satisfactory, commercial moisture indicators work satisfactorily.

Desiccants

Desiccants used in refrigeration systems adsorb or react chemically with the moisture contained in a liquid or gaseous refrigerant-oil mixture. Solid desiccants, used widely as dehydrating agents in refrigerant systems, remove moisture from both new or field-installed equipment. The desiccant is contained in a device called a drier (also spelled dryer) or filter-drier and can be installed in either the liquid or the suction line of a refrigeration system.

Desiccants must remove water and not react unfavorably with any other materials in the system. Activated alumina, silica gel, and molecular sieves are the most widely used desiccants acceptable for refrigerant drying. These materials remove moisture from refrigeration systems by physical adsorption. They are available in granular, bead, and block forms. The manufacturer should be consulted in selecting any desiccant.

Combinations of desiccants can be used in a single drier and may have certain advantages over a single desiccant because they can adsorb a greater variety of refrigeration contaminants. Two combinations are activated alumina with molecular sieves and silica gel with molecular sieves. Activated carbon is also used in some mixtures.

Solid core desiccants, or block forms, consist of desiccant granules held together by a binder (Walker 1963). The binder is usually a nondesiccant material. Suitable filtering action, adequate contact of the desiccant with the refrigerant, and low pressure drop are obtained by properly sizing the desiccant particles used to make up the core, and by the proper geometry of the core with respect to the flowing refrigerant.

Desiccants that take up water by chemical reaction are not recommended. Calcium chloride reacts with water to form a corrosive liquid. Barium oxide is known to cause explosions. Magnesium perchlorate and barium perchlorate are powerful oxidizing agents, which are potential explosion hazards in the presence of oil. Phosphorous pentoxide is an excellent desiccant, but its fine powdery form makes it difficult to handle and produces a high resistance to gas and liquid flow. A mixture of calcium oxide and

Table 4 Reactivation of Desiccants

Desiccant	Temperature, °F
Activated alumina	400 to 600
Silica gel	350 to 600
Molecular sieves	500 to 660

sodium hydroxide has limited use, but is not recommended as a desiccant.

Desiccants readily adsorb moisture and must be protected against it at all times until ready for use. If a desiccant has picked up moisture, it can be reactivated by heating for about 4 h at a suitable temperature, preferably with a dry air purge or in a vacuum oven (see Table 4). Only adsorbed water is driven off at the temperatures listed, and the desiccant is returned to its initial activated state. Care must be taken against repeated reactivation and excessive temperatures during reactivation as this may damage the desiccant. The desiccant in a refrigerating equipment drier should not be reactivated for reuse, because of oil and other contaminants in the drier as well as possible damage from overheating the drier shell.

Equilibrium Conditions of Desiccants. Desiccants in refrigeration and air-conditioning systems function on the equilibrium principle. If an activated desiccant contacts a moisture-laden refrigerant, the water is adsorbed from the refrigerant-water mixture onto the desiccant surface until the vapor pressures of the adsorbed water (*i.e.*, at the desiccant surface) and the water remaining in the refrigerant are equal. Conversely, if the vapor pressure of the water on the desiccant surface is higher than that in the refrigerant, water is released into the refrigerant-water mixture, and a new equilibrium point is established.

Adsorbent desiccants function by holding (adsorbing) moisture on their internal surfaces. The amount of water adsorbed from a refrigerant by an adsorbent at equilibrium is determined by pore volume, pore size, surface characteristics, and the temperature and moisture content of and solubility of water in the refrigerant.

Equilibrium curves for the various adsorbent desiccants with R-12 and R-22 are presented in Figures 1, 2, and 3. These curves (adsorption isotherms) are based on the technique developed by Gully, Tooke, and Bartlett (1954), as modified by ASHRAE *Standard* 35-1992. ASHRAE *Standards* 35-1992 and 63.1-1988 define the moisture content of the refrigerant as Equilibrium Point Dryness (EPD), and the moisture held by the desiccant as water capacity. The curves show that for any specified amount of water in a particular refrigerant, the desiccant holds a corresponding specific quantity of water.

Figure 1 shows moisture equilibrium curves for three common adsorbent desiccants in drying R-12 at 75 °F. Figure 2 presents data for drying R-22 with these desiccants also at 75 °F. As shown, the capacity of a desiccant can vary widely for different refrigerants when the same EPD is required. Generally, a refrigerant in which moisture is more soluble requires more desiccant for adequate drying than one with less solubility requires.

Figure 3 shows the effect of temperature on moisture equilibrium capacities of activated alumina and R-12. Much higher water capacities are obtained at lower temperatures, demonstrating the advantage of locating the alumina driers at relatively cool spots in the system. When using molecular sieves, the effect of temperature on water capacity is much less. ARI *Standard* 710-86 requires determining the water capacity for R-12 at an EPD of 15 ppm, and for R-22 at 60 ppm. Each determination must be made at 75 °F (see Figures 1 and 2) and 125 °F.

Figure 4 gives the water capacity of molecular sieve in liquid R-134a at 125 °F. This data was obtained using the Karl Fischer method similar to that described in Dunne and Clancy (1984).

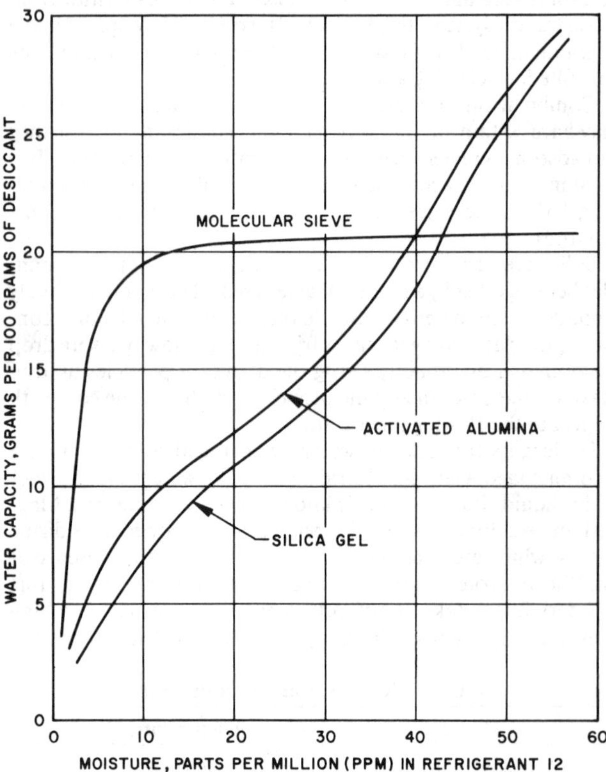

Fig. 1 Moisture Equilibrium Curves for R-12 and Three Common Desiccants at 75 °F

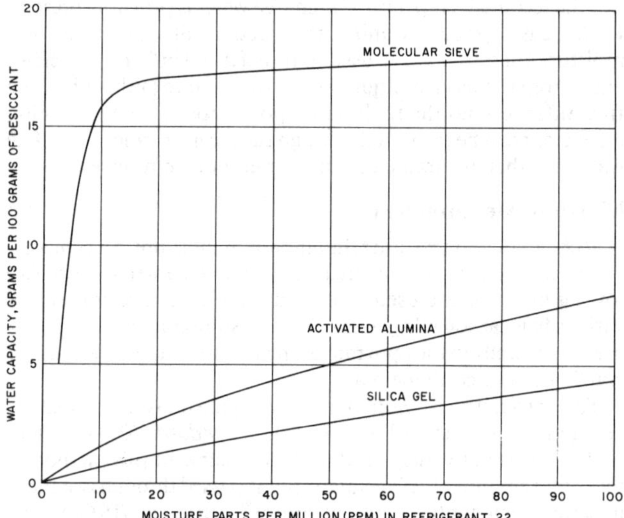

Fig. 2 Moisture Equilibrium Curves for R-22 and Three Common Desiccants at 75 °F

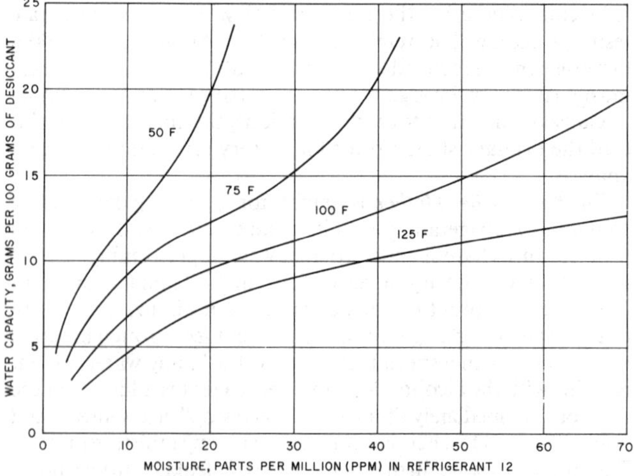

Fig. 3 Moisture Equilibrium Curves for Activated Alumina at Various Temperatures in R-12

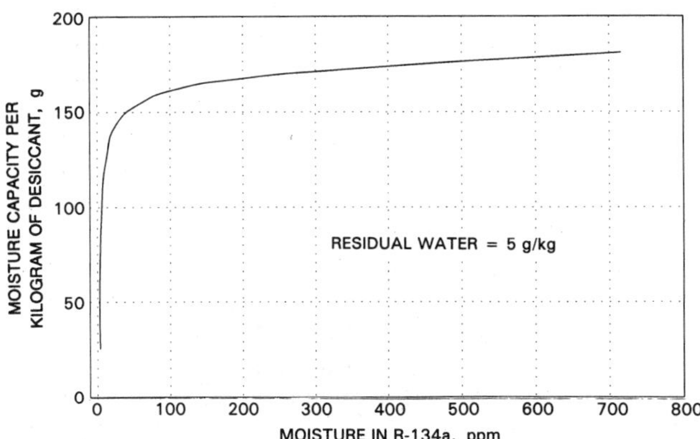

Fig. 4 Moisture Equilibrium Curve for Molecular Sieve in R-134a at 125°F
(Courtesy UOP. Reprinted with permission.)

Although the figures show that molecular sieves have higher capacities than activated alumina or silica gel at the indicated EPD, all three desiccants are suitable if sufficient quantities are used. Cost, operating temperature, other contaminants present, and equilibrium capacity at the desired EPD must be considered when choosing a desiccant for refrigerant drying. The desiccant manufacturer has information and equilibrium curves for specific desiccant-refrigerant systems.

Desiccant Applications

In addition to removing water, desiccants may be capable of adsorbing or reacting with acids, dyes, chemical additives, and refrigerant oil reaction products.

Acids. The quantity of acid that a refrigerant system tolerates depends in part on the size, the mechanical design and construction of the system, the type of motor insulation, the type of acid, and the quantity of water in the system. Generally, acids can harm refrigerant systems. Desiccants remove acids by adsorption and/or chemical reaction.

The acid capacity of desiccants in a refrigeration system is difficult to determine because the environment is complex. Refrigerant systems can have different refrigerants, oils, and concentrations of refrigerant and oil circulating through the drier, all of which affect the rate and capacity of desiccants to remove acids. In addition, acids formed in these systems can be inorganic, such as HCl and HF, or a mixture of organic acids. All factors must be considered to establish acid capacities of desiccants.

Hoffman and Lange (1962) and Mays (1962) showed that acids are removed from refrigerants and oils by adsorption and/or chemical reaction. Hoffman also showed that the concentration of water in the desiccant, the type of desiccant, and the type of acid play a major role in a desiccant's ability to remove acids from refrigerant systems.

Colors. Colored materials frequently are adsorbed by activated alumina and silica gel and occasionally by calcium sulfate and molecular sieves. Leak detector dyes may lose their effectiveness when used in systems containing desiccants. The interaction of the dye and drier should be evaluated before putting a dye in the system.

Oil Deterioration Products. Oils in refrigerant systems can react chemically to produce substances that are adsorbed by desiccants. Some of these are hydrophobic and, when adsorbed by the desiccant, reduce the rate at which it can adsorb liquid water. The rate and capacity of the desiccant to remove water dissolved in the refrigerant are not significantly impaired, however (Walker *et al.* 1955). Frequently, the reaction products are sludges or powders that can be filtered out mechanically by the drier.

Chemicals. Refrigerants that can be adsorbed by desiccants cause the drier temperature to rise considerably when the refrigerant is first admitted. This temperature rise is not the result of moisture in the refrigerant, but the adsorption heat of the refrigerant. Oil additives may be adsorbed by silica gel and activated alumina. Methanol is coadsorbed with moisture and competes with it for adsorption capacity in desiccants. Because of small pore size, molecular sieves generally do not adsorb additives or the oil.

Driers

A drier is a device containing a desiccant. It collects and holds moisture, but it also acts as a filter and adsorber of acids and other contaminants.

To prevent moisture from freezing in the expansion valve or capillary tube, a drier is installed in the liquid line close to these devices. Hot locations should be avoided. Driers can function on the low-pressure side of expansion devices, but this is not the preferred location (Jones 1969).

Moisture is reduced to a low level by one pass of the liquid refrigerant through a drier. The moisture is usually distributed throughout the entire system, and time is required for the circulating oil-refrigerant mixture to carry the moisture to the drier. Krause *et al.* (1960) showed that considerable time is required to produce moisture equilibrium in a refrigeration unit.

Driers are also used effectively to clean up systems severely contaminated due to hermetic motor burnouts and mechanical failures (see the System Cleanup Procedure after Hermetic Motor Burnout section).

Drier Selection

The drier manufacturer selection chart lists the amount of desiccants, flow capacity, filter area, water capacity, and a specific recommendation covering the type and refrigeration capacity of the drier for various applications.

The equipment manufacturer must consider the following factors when selecting a drier:

1. The *desiccant* is the heart of the drier and its selection is most important. The section on desiccants has further information.
2. The drier's *water capacity* is measured by methods described in ARI *Standard* 710-86. The reference points are set arbitrarily to prevent confusion arising from determinations made at other points. The specific refrigerant, the amount of desiccant, and the effect of temperature are all considered in the statement of water capacity.
3. The *liquid line flow capacity* is listed at 2 psi pressure drop across the drier by the official procedures of ASHRAE *Standard* 63.1-1988 and ARI *Standard* 710-86. Jones (1964) developed a gravity flow method for determining flow capacities that uses R-113 and converts the results to R-12 and R-22. Rosen *et al.* (1965) described a closed-loop method for evaluating filtration and flow characteristics of liquid line refrigerant driers. The flow capacity of suction line filters and filter-driers is determined according to ASHRAE *Standard* 78-1985 and ARI *Standard* 730-86. The latter standard gives recommended pressure drops for selecting suction line filter-driers for permanent and temporary installations. Flow capacity may be reduced quickly when critical quantities of solids and semisolids are filtered out by the drier. Whenever flow capacity drops below the machine's requirements, the drier should be replaced.
4. Although limits for particle size vary with refrigerant system size and design, and with the geometry and hardness of the particles, manufacturers publish *filtration capabilities* for comparison.

Testing and Rating

Desiccants and driers are tested according to the procedures of ASHRAE *Standards* 35-1992 and 63.1-1988, respectively. Driers are rated under ARI *Standard* 710-86. Minimum standards for listing of refrigerant driers can be found in nonelectrical UL 207, *Standard for Refrigerant Containing Components*. Filtration ratings and test standards have not been developed.

OTHER CONTAMINANTS

Refrigerant filter-driers are the principal devices used to remove contaminants from refrigeration systems. The filter-drier is not a substitute for poor workmanship or design, but a maintenance tool necessary for continued and proper system performance. Contaminants removed by filter-driers include moisture, acids, high-molecular-weight hydrocarbons, oil decomposition products, and insoluble material, such as metallic particles and copper oxide.

Metallic Contaminants and Dirt

Small contaminant particles frequently left in refrigerating systems during manufacture or servicing include chips of copper, steel, or aluminum; copper or iron oxide; copper or iron chloride; welding scale; brazing or soldering flux; sand; and other dirt. Some of these contaminants, such as copper chloride, develop from normal wear or chemical breakdown during system operation. Solid contaminants vary widely in size, shape, and density.

Solid contaminants create problems by:

1. Scoring cylinder walls and bearings
2. Lodging in the motor insulation of a hermetic system, where they act as conductors between individual motor windings or abrade the wire coating when flexing of the windings occurs
3. Depositing on terminal blocks and serving as a conductor
4. Plugging expansion valve screen or capillary tubing
5. Depositing on suction or discharge valve seats, significantly reducing compressor efficiency
6. Plugging oil holes in compressor parts, leading to improper lubrication
7. Increasing the rate of chemical breakdown in the system. At elevated temperatures, R-12 and R-22 decompose more readily when in contact with iron powder, iron oxide, or copper oxide (Norton 1957).

Liquid line filter-driers, suction filters, and strainers isolate contaminants from the compressor and expansion valve. Filters minimize the return of particulate matter to the compressor and expansion valve; but the capacity of permanently installed liquid and/or suction filters must accommodate this particulate matter without causing excessive, energy-consuming pressure losses in the refrigeration system. Equipment manufacturers should use the following procedures to ensure proper operation during the system's design life:

1. Develop system cleanliness specifications for production systems that include a practical value for maximum residual matter. Some manufacturers specify allowable quantities in terms of internal surface area.
2. Multiply the factory contaminant level by a factor of five to allow for solid contaminants that will be added during installation. The safety factor depends on the type of system and the previous experience of the installers, among other considerations.
3. Determine a value for maximum pressure drop to be incurred by the suction or liquid filter when loaded with the quantity of solid matter calculated in Step 2.
4. Conduct pressure drop tests by adding contaminants. The chosen test fluid and its flow rates should be predictable in terms of typical system flow rates and refrigerant. Contaminants in the test system should be graded for particle size and distribution and held constant from test to test.
5. Select driers for each system according to its capacity requirements and test data. In addition to contaminant removal capacity, tests can evaluate filter efficiency, maximum escaped particle size, and average escaped particle size.

Very small particles passing through filters tend to accumulate in the crankcase. Most compressors tolerate a small quantity of these particles without allowing them into the oil pump inlet, where they can damage running surfaces. If the compressor will not tolerate these particles, or if they become excessive, the particles can be reduced by changing the oil.

Organic Contaminants—Sludge, Wax, and Tars

Organic contaminants in a refrigerating system can appear when organic materials, such as oil, insulation, varnish, gaskets, and adhesives decompose. As opposed to inorganic contaminants, these materials are mostly carbon, hydrogen, and oxygen. Organic materials may be partially soluble in the refrigerant-oil mixture or may become so through the action of heat. They then circulate in the refrigerating system and can plug small orifices.

Some organic contaminants remain in a new refrigerating system during manufacture or assembly. For example, excessive solder paste introduces a wax-like contaminant into the refrigerant stream. Certain cutting oils, corrosion inhibitors, or drawing compounds frequently contain wax-like material. Normal fabricating methods can leave very small amounts of these oils deposited on the copper tubing. These contaminants are a mixture of substances that usually have no definite melting point. Organic contamination also results during the normal method of fabricating return bends. The die used during forming is lubricated with oil or other organic materials, and afterwards the return bend is brazed to the tubes to form a condenser. During brazing, residual lubricant inside the tubing and bends is baked to a resinous deposit.

If organic materials are handled improperly, certain contaminants remain in a system. Resins used in varnishes, wire coating, or casting sealers may not be cured properly and can dissolve in the refrigerant-oil mixture. Solvents used in washing stators may be adsorbed by the wire film and later, during compressor operation, carry chemically reactive organic extractables. Chips of varnish, insulation, or fibers can detach and circulate in the system. Portions of improperly selected or cured rubber parts or gaskets can dissolve in the refrigerant.

Refrigeration grade oil decomposes under adverse conditions to form a resinous liquid or a solid frequently found on refrigeration filter-driers. These oils decompose noticeably when exposed for as little as 2 h to temperatures as low as 250 °F in an atmosphere of air or oxygen. The compressor manufacturer should perform all high-temperature dehydrating operations on the machines prior to adding the oil charge. In addition, equipment manufacturers should not expose compressors to processes requiring high temperatures unless the compressors contain refrigerant or inert gas.

The result of organic contamination is frequently noticed at the expansion device. Resinous or wax-like material dissolved in the refrigerant-oil mixture, under liquid line conditions, may precipitate out at the lower temperature in the expansion device, resulting in restricted or plugged capillary tubes or sticky expansion valves (Zahorsky 1967). Wax-like material, with a mass less than a gram with a volume less than that of a few grains of rice, can render a system inoperative. These materials have physical properties that range from a fluffy powder to a solid resin entraining inorganic dirt. If the contaminant is dissolved in the refrigerant-oil mixture in the liquid line, it usually will not be removed by a filter-drier.

Chemical identification of these organic contaminants is very difficult. Infrared spectroscopy can characterize the type of organic groups present in contaminants. Materials found in actual systems vary from wax-like aliphatic hydrocarbons to resin-like materials containing double bonds, carbonyl groups and carboxyl groups. In some cases, organic compounds of copper or iron have been identified.

These contaminants can be eliminated by carefully selecting refrigeration system materials and strictly controlling cleanliness during manufacture and assembly. Because heat degrades most organic materials and enhances chemical reactions, operating conditions with excessively high discharge temperatures must be avoided to prevent formation of degradation products.

Residual Cleaning Agents

Solvents used for cleaning compressor parts are likely contaminants if left in refrigerating systems. Solvents in this category are considered pure liquids without additives. If additives are present, they are reactive materials and should not be in a refrigerating system. Some solvents are relatively harmless to the chemical stability of the refrigerating system, while others initiate or accelerate degradation reactions. For example, the common mineral spirits solvents, such as Stoddard solvent, are considered harmless to the system. Other common compounds react rapidly with hydrocarbon lubricating oils (Elsey *et al.* 1952).

Antifreeze Agents (Methyl Alcohol)

Antifreeze agents are sometimes added to refrigerating systems to act as cosolvents for the small quantities of water present. They prevent ice formation at the expansion device but add to the contaminants already present in the system.

Methyl alcohol has shown no adverse effects when used in the proportion of 3 cm^3 per pound of R-12. Larger concentrations of methyl alcohol can cause corrosion problems, especially where aluminum is present, because it will be attacked and hydrogen gas will form.

If similar systems, one dry and one with 0.05 to 1% methyl alcohol, are test-run side by side, the alcohol system will always have more stain, corrosion, copper plating, and debris on fine filter screens than the dry system. However, these systems still perform as well as those containing no methyl alcohol.

Noncondensable Gases

Gases, other than the refrigerant, are a contaminant frequently found in refrigerating systems. These gases result: (1) from incomplete evacuation, (2) when functional materials release sorbed gases or decompose to form gases at an elevated temperature during system operation, (3) through low-side leaks, and (4) from chemical reactions during system operation. Chemically reactive gases, such as hydrogen chloride, attack other components in the refrigerating system; in extreme cases, the refrigerating unit fails.

Chemically inert gases in the system, which do not liquefy in the condenser, reduce the cooling efficiency. The quantity of inert, noncondensable gas that is harmful depends on the design and size of the refrigerating system and the nature of the refrigerant. Its presence contributes to higher than normal head pressures and resultant higher discharge temperatures. Higher temperatures speed up undesirable chemical reactions.

Gases found in hermetic refrigeration units include nitrogen, oxygen, carbon dioxide, carbon monoxide, methane, and hydrogen. The first three gases listed originate from incomplete air evacuation or a low-side leak in the system. Carbon dioxide and carbon monoxide usually form when organic insulation materials are overheated. Hydrogen has been detected when a compressor is experiencing serious bearing wear. Only trace amounts of these gases are present in well-designed, properly functioning equipment.

Spauschus and Olsen (1959), Doderer and Spauschus (1966), and Gustafsson (1977) developed sampling and analytical techniques for establishing the quantities of contaminant gases present in refrigerating systems. Parmelee (1965), Spauschus and Doderer (1961, 1964), and Kvalnes (1965) applied gas analysis techniques to sealed tube tests to yield information on stability limitations of refrigerants, in conjunction with other materials used in hermetic systems.

Motor Burnouts

Motor burnout is the final result of hermetic motor insulation failure. During burnout, high temperatures and arc discharges can severely deteriorate the insulation, producing large amounts of carbonaceous sludge, acid, water, and other contaminants. In addition, a burnout can chemically alter the system's lubricating oil, and/or thermally decompose refrigerant in the vicinity of the burn. The products of burnout escape into the system, causing severe cleanup problems. If decomposition products are not removed from the system, replacement motors fail with increasing frequency.

While RSES (1988) has chosen to differentiate between mild and severe burnouts, many compressor manufacturer's service bulletins treat all burnouts alike. A rapid burn from a spot failure in the motor winding results in a mild burnout with little oil discoloration and no carbon deposits. A severe burnout occurs when the compressor remains on line and burns over a longer period, resulting in highly discolored oil, carbon deposits, and acid formation.

Since the condition of the oil can be used to indicate the amount of system contamination, the oil should be examined during the cleanup process. Wojtkowski (1964) has shown that acid in the oil should not exceed 0.05 acid number (milligrams KOH per kilogram refrigerant). Commercial acid test kits can be used for this analysis.

Various methods are recommended for cleaning a system after hermetic motor burnout (RSES 1988). However, the suction line filter-drier method is commonly used (see the System Cleanup Procedure after Hermetic Motor Burnout section).

Field Assembly

Proper field assembly and maintenance are essential for contaminant control in refrigerating systems and to prevent undesirable refrigerant emissions to the atmosphere. Refrigeration components that are adequate for intended design limits can be misapplied or mismatched. Connecting piping may be too large or too small to carry the necessary gas and still circulate oil properly. Driers may be too small or carelessly handled so that drying capacity is lost. Improper tube-joint soldering is a major source of water, flux, and oxide scale contamination. Copper oxide scale from improper brazing is one of the most frequently found contaminants. Careless tube cutting and handling can introduce excessive quantities of dirt and metal chips.

Because dehydration of the assembled system is not easy in the field, oversized driers are recommended. Even if manufacturers' components are delivered sealed and bone-dry, the weather and the open-time during assembly can introduce considerable quantities of moisture that must be removed. Inaccurate control settings, dirt-fouled air condensers, scaled or corroded heat exchangers, and improper evacuation can cause any system to break down.

REFRIGERANT RECOVERY, RECYCLING, AND RECLAMATION

Introduction

Many scientists have determined that chlorofluorocarbon (CFC) and hydrochlorofluorocarbon (HCFC) refrigerants deplete ozone in the stratosphere when released to the atmosphere. These chemicals and replacement hydrofluorocarbon (HFC) refrigerants may also contribute to global warming of the atmosphere. The

Montreal Protocol and subsequent revisions require actions to contain and properly handle refrigerants. The International Organization for Standardization (ISO) and others are developing standards or guidelines for refrigerant recovery.

The procedures involved in removing contaminants during recycling of refrigerants are similar to those discussed earlier in this chapter. Service techniques, proper handling and storage, and possible mixing of different refrigerants are of concern. Building owners, equipment manufacturers, and contractors are particularly concerned about reintroducing refrigerants with higher contaminant levels than a new refrigerant into systems.

Installation and Service Practices

Proper installation and service procedures, including proper evacuation and leak checking, are essential to minimize major equipment repairs. Service lines should be made of hose material with low permeability and should include shutoff valves. Larger systems should include isolation valves and pumpdown receivers. ASHRAE *Guideline* 3-1990 gives further detail on equipment, installation, and service requirements.

Recovering refrigerant to an external storage container and then returning the refrigerant for cleanup inside the refrigeration system is similar to the procedure described in item D in the section on System Cleanup Procedure after Hermetic Motor Burnout (Manz 1988). Some additional air and moisture contamination may be introduced in the service procedure. In general, the refrigeration system must be cleaned whether the refrigerant is isolated in the receiver, recovered into a storage container, recycled, reclaimed, or charged with new refrigerant. This cleanup is necessary because contaminants are distributed throughout the system. The advantage of new, reclaimed, or recycled refrigerant is that a properly cleaned system is not recontaminated by the addition of impure refrigerant.

Refrigerant Recovery

Recovery means to remove refrigerant in any condition from a system and to store it in an external container without necessarily testing or processing it in any way. Recovery reduces refrigerant emissions to the atmosphere and is a necessary first or concurrent step to either recycling or reclamation. The largest potential for service-related emissions of refrigerant occurs during recovery. These emissions consist of refrigerant left in the system (recovery efficiency) and losses due to service connections (Manz 1991a).

The key to reduction of emissions is proper recovery equipment and techniques. The recovery equipment manufacturer and the technician must share this responsibility to minimize refrigerant loss to the atmosphere. Training in handling halocarbon refrigerants is required to learn the proper techniques (RSES 1991). *Important: Recover refrigerants into a suitable container and keep containers for different refrigerants separate. Do not overfill containers as liquid expansion with rising temperature could cause loss of refrigerant through the pressure relief valve or even rupture of the container.*

High-pressure refrigerants (R-12) are most commonly recovered using a compressor-based recovery unit to pump the refrigerant directly into a storage container. Such a system is shown in Figure 5. Functions include the processes of evaporating, compressing, condensing, storing, and controlling. As a variation, a refrigeration unit may be used to cool the storage container to directly transfer the refrigerant. For low pressure refrigerants (R-11), a compressor or vacuum pump may be used to lower the pressure in the storage container and raise the pressure in the vapor space of the refrigeration system so that the liquid refrigerant will flow without evaporation. An alternative is to use a liquid pump to transfer the refrigerant. A pumpdown unit such as a condensing unit may be required to remove the remaining vapor refrigerant after liquid

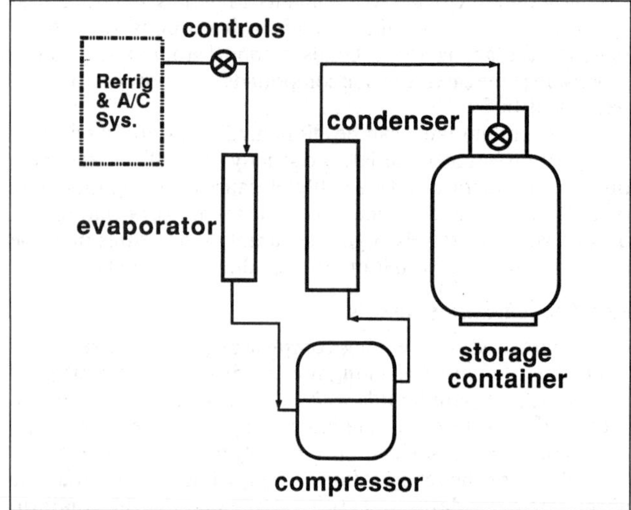

Fig. 5 Recovery Functions

removal is complete. Recovery systems for use at the factory for charging or leak testing operations will probably be larger in size and of specialized construction to meet the specific needs of the manufacturer (Parker 1988).

Components in which liquid could be trapped, such as a suction accumulator, may need to be gently heated to remove all the refrigerant. Good practice requires watching for a pressure rise after recovery is completed to determine if the recovery unit needs to be restarted to remove all refrigerant. Where visual inspection is possible, such components would be identified by frosting on external surfaces to the level of the liquid refrigerant inside.

Contaminants

The contaminants encountered in recovered refrigerants are covered in earlier sections of this chapter as they pertain to the refrigeration system operation. The main contaminants are moisture, acid, noncondensables, particulates, high boiling residue (oil and sludge), and other condensable gases (Manz 1991a).

1. Moisture is contained in the refrigerant and in the lubricant. It can be removed by lowering the temperature (dew point), and by passing the refrigerant through a desiccant (filter-drier). Some moisture is also removed by oil separation.
2. Acid consists of organic and inorganic types. Organic acids are normally contained in the oil and are removed in the oil separator and in the filter-drier. Inorganic acids, like hydrochloric acid, are removed by the noncondensable purge process, reaction with metal surfaces, and by the filter-drier.
3. Noncondensable gases consist primarily of air. These can come from the refrigeration equipment or can be introduced during servicing procedures. The control method consists of minimizing introduction through proper equipment construction and installation per ASHRAE *Guideline* 3-1990. Proper service equipment construction, connection techniques, and maintenance procedures (such as during filter-drier change) also reduce the introduction of air. Typically, a vapor purge is used to remove air.
4. Particulates are removed by suction filters, oil separation, and filter-driers.
5. High boiling residues consist primarily of refrigerant oil and sludge. Since different refrigeration systems use different oils and since the oil is a collection point for other contaminants, the oil is considered a contaminant. High boiling residues are removed by separators designed to extract the oil from the vapor phase refrigerant, or by a distillation process.
6. Other condensable gases consist mainly of other refrigerants.

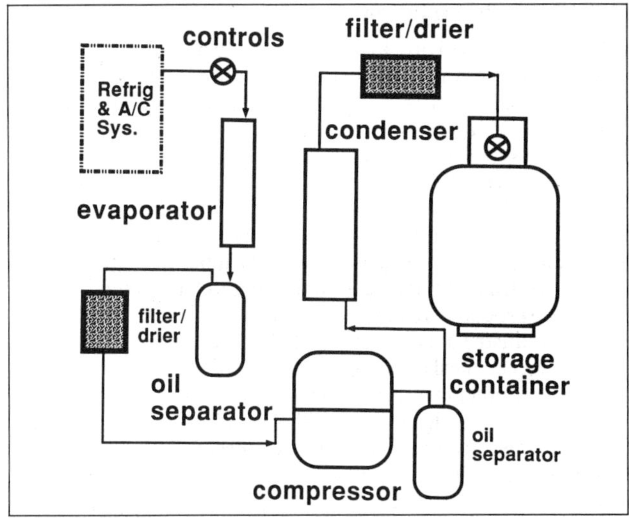

Fig. 6 Single Pass Recycling

They can be generated in small quantities by high-temperature operation or during a burnout. In rare cases, intentional mixing of refrigerants for performance or to top off with substitutes may be encountered. In general, separation of other condensable gases, if possible, can only be done at a fully equipped reclamation center.

Refrigerant Recycling

Recycle means to clean refrigerant for reuse by oil separation, noncondensable removal, and single or multiple passes through devices such as replaceable filter-driers, which reduce moisture, acidity, and particulate matter. This term usually applies to procedures implemented at the field job site or at a local service shop.

Recycling conserves limited supplies of regulated refrigerants (*e.g.*, R-12). Recycling typically consists of the three steps of oil separation, filter-drying, and noncondensable purge (Manz 1991a). A single pass recycling schematic is shown in Figure 6. In the single pass recycling unit, refrigerant is processed by oil separation and filter-drying in the recovery path. Typically, air and noncondensables are not removed during the recovery process and are handled at a later time.

In a multiple pass recycling unit, as shown in Figure 7, the refrigerant is typically processed through an oil separation process during recovery. The filter-drier may be placed in the compressor suction line or in a bypass recycling loop or both. During a continuous recycling loop, refrigerant is withdrawn from the storage tank and processed through a filter-drier and returned to the storage tank. Noncondensable purge is accomplished during this recycling loop.

The primary function of the filter-drier is to remove moisture while the secondary function is to remove acid, particulate, sludge, and varnish. The capability of the filter-drier to remove moisture and acid from used refrigerants is improved if the oil is separated prior to passing through the filter-drier (Kauffman 1992). Moisture indicators are typically used to indicate when a filter-drier change is required. For some refrigerants, these devices cannot indicate a moisture level as low as the purity level (10 ppm required by ARI *Standard* 700-88). Development work on better detectors is ongoing. *Important: Service technicians must change recovery/recycle unit filter-driers at frequent intervals as directed by the indicators and manufacturer's instructions.*

The primary advantage of recycling is performing this operation at the job site or at a local service shop and avoiding transportation costs. The likelihood of mixing refrigerants is reduced if recycling is done at the service shop compared to consolidating

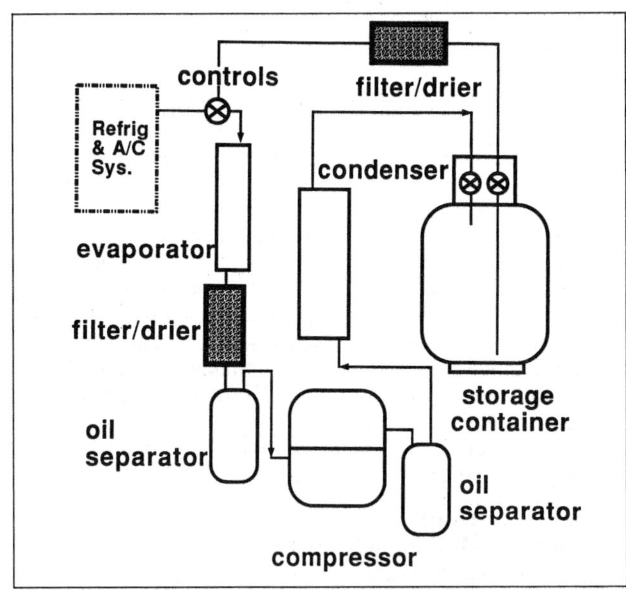

Fig. 7 Multiple Pass Recycling

refrigerant batches for shipment to a reclamation facility.

Equipment Standards

Commercial refrigerant recovery/recycling equipment exhibits considerable variation in contaminant removal capabilities, recovery performance, and portability. ARI *Standard* 740-91 (ARI 1991b) is one guide for the industry. This standard establishes methods of testing for rating performance of equipment for purity, contaminant levels, capacity, speed, and purge loss. Currently, refrigerants R-11, R-12, R-13, R-22, R-113, R-114, R-500, R-502, and R-503 are covered. ARI 740 defines a standard contaminated refrigerant sample for each type. It describes the test apparatus and method, and includes detailed sampling and chemical analysis procedures (Manz 1991a).

The manufacturer's certified ratings, sampled and verified by ARI, are printed in an ARI certification directory. Measured performance parameters include liquid recovery rate, vapor recovery rate, recycling rate, and recovery efficiency. Measured contaminant levels of recycled refrigerant without judgment of acceptance include moisture content (ppm by mass), chloride ions (pass), acidity (equivalent mg KOH/kg), high boiling residue (percent by volume), particulates/solids (pass), and noncondensables (percent by volume). A refrigerant is sampled for contaminant analysis at the time of the first filter-drier change when levels are expected to be highest.

ARI 740 provides a flexible means of rating the equipment. Manufacturer's producing equipment for "recovery" only, may place "N/A" beside any or all contaminant levels. Manufacturer's producing "recycling" or "recovery/recycling" equipment must rate their equipment for all contaminants. The standard also allows rating liquid and/or vapor feed. For example, a manufacturer of equipment for recovering liquid refrigerant only, would rate the vapor recovery rate N/A. All "recycle" equipment must meet requirements concerning refrigerant loss due to air purge and internal hose construction.

The basic distinction between recycling and reclamation is best illustrated by associating recycling with certification (ARI 740) and reclamation with analysis (ARI 700). ARI 740 covers certification testing of the recycling equipment using a "standard contaminated refrigerant sample" in lieu of chemical analysis of each batch. It provides a means of comparing equipment performance under controlled test conditions. By contrast, as described later, ARI 700 is based on chemical analysis of a refrigerant sample from each batch after contaminant removal. ARI 700 provides a means of analyzing refrigerant and comparing it to standard acceptable levels of contaminants.

ARI 740 applies to single refrigerant systems and their normal contaminants. It does not apply to refrigerant systems or storage containers with a mixture of refrigerants. No attempt has been made to rate the equipment's ability to remove different refrigerants and other condensable gases from recovered refrigerant. The standard places responsibility on the equipment operator to identify those situations and to treat them accordingly. One of the difficulties associated with recycled refrigerants is describing the purity levels when offering the refrigerant for resale. The purity of recycled refrigerant is uncertain because appropriate field measurement techniques do not exist for all contaminants listed in ARI 700. The refrigerant may be sampled and analyzed by a fully equipped laboratory.

For mobile air-conditioning equipment, refer to Society of Automotive Engineers (SAE) standards.

Special Considerations and Equipment for Handling Multiple Refrigerants

The need to keep different refrigerant types separate is paramount. Storage containers should meet applicable standards for transportation and use with that refrigerant as specified in ARI *Guideline* K (ARI 1990). Disposable cylinders (DOT specification 39) shall not be used (RSES 1991). Containers should be filled per ARI *Guideline* K and marked with the refrigerant type. Container colors for recovered refrigerants and for new and reclaimed refrigerants are specified in ARI *Guidelines* K and N (ARI 1990 and 1992), respectively.

For economic reasons, recovery/recycling (R/R) equipment should be capable of handling more than one refrigerant. The R/R equipment should only be used for labelled refrigerants. At the time the service technician desires to switch refrigerants, a significant amount of the previous refrigerant will be contained in the R/R equipment, particularly in the condenser section (Manz 1991b). This must be removed either by transferring to an evacuated or cooled cylinder or by using isolation/bypass valves to connect the condenser section to the compressor suction and to connect the compressor discharge directly to the storage container. After the bulk of the refrigerant has been removed, the system should be evacuated before changing to the appropriate storage container for the new refrigerant. This procedure should also include all lines and connecting hoses and may include replacement of the filter-driers.

The need for purging noncondensables is determined by comparing the refrigerant pressure to the saturation pressure of pure refrigerant at the same temperature. Circulation to achieve thermal equilibrium may be required to eliminate the effect of temperature difference on the pressures. A sealed bulb is often used to determine the saturation pressure for a single refrigerant system. When purging air from R/R equipment capable of handling multiple refrigerants, the difference in saturation pressures between refrigerants far exceeds any allowable partial pressure due to noncondensables. Special equipment and/or techniques are required (Manz 1991b).

Refrigerant to be recovered may be in vapor or liquid states. To optimize the recovery, R/R equipment must have the ability to handle each of these states. For some equipment, this may involve one hookup or piece of R/R equipment for liquid and a separate hookup or second piece of R/R equipment to recover vapor. In general, a single hookup is desired. When handling multiple refrigerants, traditional liquid flow control devices such as capillary tubes or expansion valves either involve compromise performance or simply will not work. Possible solutions include (1) the operator watching a sight glass for liquid flow and switching a valve; (2) multiple flow control devices with a refrigerant selection switch; and (3) a two bulb expansion valve, which controls temperature differential across the evaporator (Manz 1991b).

Refrigerant Reclamation

Reclaim means to process a refrigerant to new product specifications by means that may include distillation. Chemical analysis of the refrigerant is required to determine that appropriate product specifications are met. This term usually implies the use of processes or procedures available only at a reprocessing or manufacturing facility.

Many HVAC&R equipment warranties were drafted with the expectation that refrigerants meeting ARI *Standard* 700-88 for refrigerant purity would be used. ARI 700 applies to both new and reclaimed refrigerant. Refrigerant reclamation involves a complex distillation process to remove contaminants and return used refrigerant to industry specifications. This process is generally performed by a chemical manufacturer or reprocessor (McCain 1991).

Some equipment warranties, especially those for smaller consumer appliances, may not permit the use of refrigerants that have been reclaimed to the levels of purity specified in ARI 700-88. Manufacturers' literature should be consulted before charging with reclaimed refrigerants.

Reclamation has been used for over 30 years and has traditionally been used for systems containing 100 lb or more of refrigerant (O'Meara 1988). Assistance is often provided by the reclaimer in furnishing shipping containers and labelling instructions. Many reclaimers use air-conditioning and refrigeration wholesalers as collection points for refrigerant. Mixing of refrigerants at the consolidation points is possible. If the refrigerant is contaminated beyond limits, the price paid for the refrigerant may be reduced or the shipment may be refused. One of the advantages associated with reclaimed refrigerants is describing the purity levels when offering the refrigerant for resale.

Purity Standards

ARI *Standard* 700-88 covers the same refrigerants, regardless of source, as ARI 740 and defines acceptable levels of contaminants, which are the same as Federal Specifications for "Fluorocarbon Refrigerants" BB-F-1421B. It specifies laboratory analysis methods for each contaminant. Only fully equipped laboratories with trained personnel are currently capable of performing the analysis.

Since ARI 700 is based on chemical analysis of a sample from each batch after contaminant removal, it is not concerned with the level of contaminants before contaminant removal (Manz 1991a). This disassociation from the "standard contaminated refrigerant sample" required in ARI 740 is the basic distinction between analysis/reclamation and certification/recycling.

The standard "does not apply where refrigerant captured from a particular system is returned on site to the same system." This does not imply that recovered refrigerant returned to the same system will have equivalent performance or will provide equivalent equipment life as compared to reclaimed refrigerant. For reference, SAE *Standards* J1991 and J2099 contain recycled refrigerant purity levels for mobile air-conditioning systems using R-12 and R-134a, respectively.

SYSTEM CLEANUP PROCEDURE AFTER HERMETIC MOTOR BURNOUT

Introduction

A. This procedure is limited to positive-displacement (reciprocating or rotary) hermetic compressors. Centrifugal compressor systems are highly specialized and are frequently designed for a particular application. A centrifugal system should be cleaned according to the manufacturer's recommendations.

B. After a hermetic motor burnout, the system must be cleaned thoroughly to remove all contaminants. Otherwise a repeat burnout will *likely* occur. Failure to follow these minimum cleanup recommendations as quickly as possible increases the potential for repeat burnout.

C. Flushing the system with R-11 or similar refrigerants has been used to some extent. This method has many limitations and is no longer recommended.

Procedure

A. *Make sure a burnout has occurred.* Although a motor that will not start appears to be a motor failure, the problem may be improper voltage, starter malfunction, or a compressor mechanical fault (RSES 1988).
   1. Check for proper voltage.
   2. Check that the compressor is cool to the touch. An open internal overload could prevent the compressor from starting.
   3. Check the compressor motor for improper grounding using a megohmmeter or a precision ohmmeter.
   4. Check the external leads and starter components.
   5. Purge a small quantity of refrigerant gas from the compressor and smell it cautiously. A motor burnout is usually indicated by a characteristic burned odor.

B. *Safety.* In addition to electrical hazards, service personnel should be aware of the hazard of acid burns.
   1. When testing for odor, release a small amount of gas and smell it cautiously to avoid inhaling toxic decomposition products.
   2. If the oil or sludge in a burned-out compressor must be touched, wear rubber gloves to avoid a possible acid burn.

C. *Determine the severity of the burnout.* Burnouts have traditionally been classified as mild or severe. While the guidelines for determining mild or severe burnouts are included, no distinction based on severity is made in the recommended cleanup procedure.
   1. Obtain a small sample of the oil from the burned-out compressor and analyze it using an acid test kit. Excessive acidity (over 0.05 acid number) in the oil indicates a severe burnout. Discoloration of the oil may also indicate a severe burnout.
   2. Release a small amount of refrigerant and smell it. A characteristic burned odor indicates a severe burnout.
   3. Inspect the suction line at the compressor and the liquid line drier. Carbon deposits indicate a severe burnout.
   4. If none of the above indications of severe contamination are found, the burnout can be classified as mild.

D. *Cleanup after a burnout.* Just as proper installation and service procedures are essential to prevent compressor and system failures, proper system cleanup and installation procedures when installing the replacement compressor are also essential to prevent repeat failures. Key elements of the recommended procedures are:
   1. System refrigerant should be isolated within the system or recovered into an external storage container to avoid discharge into the atmosphere. Prior to opening any portion of the system for inspection or repairs, refrigerant should be recovered from that portion until the vapor pressure has been reduced to less than 12.7 psia (4 inches Hg vacuum) for high-pressure refrigerants (boiling point at 1 atmosphere lower than 30°F) and to less than 2.4 psia (25 inches of mercury vacuum) for low-pressure refrigerants (boiling point at 1 atmosphere higher than 30°F).
   2. Remove the burned-out compressor and install the replacement. Save a sample of the new compressor oil that has not been exposed to refrigerant and store in a sealed glass bottle. This will be used later for comparison.
   3. Inspect all system controls such as expansion valves, solenoid valves, check valves, etc. Clean or replace if necessary.
   4. Install an oversized filter-drier in the suction line to protect the replacement compressor from any contaminants remaining in the system. Install a pressure tap upstream of the filter-drier. This tap permits measuring the pressure drop from the tap to the service valve during the first hours of operation to determine if the suction line filter-drier needs to be replaced.
   5. Remove the old liquid line filter-drier, if one exists, and install a replacement drier of the next larger capacity than is normal for this system. Install a good quality moisture indicator if the system does not have one.
   6. Evacuate and leak check the system or portion opened to the atmosphere according to the manufacturer's recommendations.
   7. Recharge the system and begin operations according to the manufacturer's startup instructions. Typically:
      a. Observe pressure drop across the suction line filter-drier for the first 4 hours. Compare to pressure drop curve in Figure 8 and replace filter-driers as required.
      b. After 24 to 48 hours, check pressure drop and replace filter-driers as required. Take an oil sample and check with an acid test kit. Compare the oil sample to the initial sample saved at the time the replacement compressor was installed. Cautiously smell the oil sample. Replace oil if acidity persists or if color or odor indicates.
      c. After 7 to 10 days or as required, repeat step b.

E. *Additional suggestions*
   1. If sludge or carbon has backed up into the suction line, swab it out or replace that section of the line.
   2. If a change in the suction line filter-drier is required, change the oil in the compressor each time the cores are changed, if the compressor design permits.
   3. Remove the suction line filter-drier after several weeks of system operation to avoid excessive pressure drop in the suction line. This problem is particularly significant on commercial refrigeration systems.
   4. In some cases, noncondensable gases are produced during the burnout. Compare the measured head pressure with the saturation pressure at the condensing temperature. Purge if the head pressure is excessive.

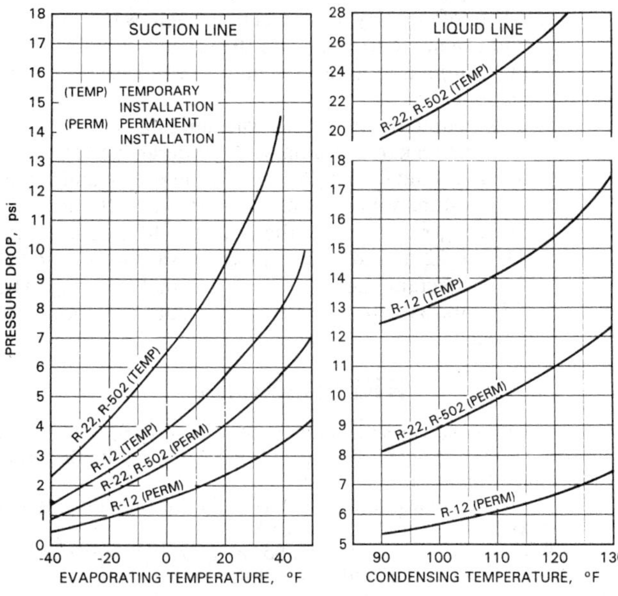

Fig. 8 Maximum Recommended Filter Drier Pressure Drop

Special System Characteristics and Procedures

Because of unique system characteristics, the procedures described here may require adaptations.

A. If an oil sample cannot be obtained from the new compressor, determine another method to get a sample from the system.
   1. Install a tee and a trap in the suction line. An access valve at the bottom of the trap permits easy oil drainage. Only 0.5 oz of oil is required for an acid analysis. Be certain the oil sample represents oil circulating in the system. It may be necessary to drain the trap and discard the first amount of oil collected, before collecting the sample to be analyzed.
   2. Make a trap from 1 3/8-in. copper tubing and valves. Attach this trap to the suction and discharge gage port connections with a charging hose. By blowing discharge gas through the trap and into the suction valve, enough oil will be collected in the trap for analysis. This trap becomes a tool that can be used repeatedly on any system that has suction and discharge service valves.

B. On semihermetic compressors, remove the cylinder head to determine the severity of burnout. Dismantle the compressor for solvent cleaning and hand wiping to remove contaminants. Consult the manufacturer's recommendations on compressor rebuilding and motor replacement.

C. In rare instances on a close-coupled system, where it is not feasible to install a suction line filter-drier, the system can be cleaned by repeated changes of the cores in the liquid line filter-drier and repeated oil changes.

D. On heat pump systems, the four-way valve and the compressor should be carefully inspected after a burnout. In cleaning a heat pump after a motor burnout, it is essential to remove any drier originally installed in the liquid line. These driers may be replaced for cleanup, or a reversible filter-drier may be installed in the common reversing liquid line.

E. Systems with a critical charge require a particular effort for proper operation after cleanup. If an oversized liquid line filter-drier is installed, an additional charge must be added. However, no additional charge is required for the suction line filter-drier that may be added to the system.

F. The new compressor should not be used to pull a vacuum. Refer to the manufacturer's recommendations for evacuation. Normally one of two methods is used, after determining that there are no refrigerant leaks in the system.
   1. Pull a high vacuum to an absolute pressure of less than 500 microns of mercury for several hours. Allow the system to stand several hours to be sure the vacuum is maintained. This requires a good vacuum pump and an accurate high-vacuum gage.
   2. Use a vacuum pump to pull a vacuum of at least 27 in. of mercury. Break this vacuum with refrigerant and allow the system to stand for 1 h. Reevacuate the system and repeat the procedure two more times for a triple evacuation. If a CFC refrigerant is used in the system, a different medium such as R-22 should be used prior to evacuation, and the CFC refrigerant should be used for final charging. The refrigerant used in intermediate evacuations should be recovered.

REFERENCES

ARI. 1986. Flow-capacity rating and application of suction-line filters and filter-driers. *Standard* 730-86. Air-Conditioning and Refrigeration Institute, Arlington, VA.

ARI. 1986. Liquid-line driers. *Standard* 710-86. Air-Conditioning and Refrigeration Institute, Arlington, VA.

ARI. 1988. Specifications for fluorocarbon refrigerants. *Standard* 700-88. Air-Conditioning and Refrigeration Institute, Arlington, VA.

ARI. 1990. Containers for recovered fluorocarbon refrigerants. *Guideline* K. Air-Conditioning and Refrigeration Institute, Arlington, VA.

ARI. 1991a. ARI refrigerant contaminant study—final report. Air-Conditioning and Refrigeration Institute, Arlington, VA.

ARI. 1991b. Performance of refrigerant recovery, recycling and/or reclaim equipment. *Standard* 740-91. Air-Conditioning and Refrigeration Institute, Arlington, VA.

ARI. 1992. Container colors for existing, new, and reclaimed refrigerants. *Guideline* N. Air-Conditioning and Refrigeration Institute, Arlington, VA.

ASHRAE. 1985. Method of testing flow capacity of suction line filter driers. ASHRAE *Standard* 78-1985 (RA 90).

ASHRAE. 1988. Method of testing liquid line refrigerant driers. ASHRAE *Standard* 63.1-1988.

ASHRAE. 1990. Reducing emission of fully halogenated chlorofluorocarbon (CFC) refrigerants in refrigeration and air-conditioning equipment and applications. ASHRAE *Guideline* 3-1990.

ASHRAE. 1992. Method of testing desiccants for refrigerant drying. ASHRAE *Standard* 35-1992.

Brisken, W.R. 1955. Moisture migration in hermetic refrigeration systems as measured under various operating conditions. *Refrigerating Engineering* (July):42.

DeGeiso, R.C. and R.F. Stalzer. 1969. Comparison of methods of moisture determination in refrigerants. ASHRAE *Journal* (April).

Doderer, G.C. and H.O. Spauschus. 1966. A sealed tube-gas chromatograph method for measuring reaction of Refrigerant 12 with oil. ASHRAE *Transactions* 72(2):IV, 4.1.

Dunne, S.R. and T.J. Clancy. 1984. Methods of testing desiccant for refrigeration drying. ASHRAE *Transactions* 90(1A):164.

Elsey, H.M. and L.C. Flowers. 1949. Equilibria in Freon-12—Water systems. *Refrigeraing Engineering* (February):153.

Elsey, H.M., L.C. Flowers, and J.B. Kelley. 1952. A method of evaluating refrigerator oils. *Refrigerating Engineering* (July):737.

Gully, A.J., H.A. Tooke, and L.H. Bartlett. 1954. Desiccant-refrigerant moisture equilibria. *Refrigerating Engineering* (April):62.

Gustafsson, V. 1977. Determining the air content in small refrigeration systems. Purdue Compressor Technology Conference.

Haagen-Smit, I.W., P. King, T. Johns, and E.A. Berry. 1970. Chemical design and performance of an improved Karl Fischer titrator. *American Laboratory* (December).

Hoffman, J.E. and B.L. Lange. 1962. Acid removal by various desiccants. ASHRAE *Journal* (February):61.

Jones, E. 1964. Determining pressure drop and refrigerant flow capacities of liquid line driers. ASHRAE *Journal* (February):70.

Jones, E. 1969. Liquid or suction line drying? *Air Conditioning and Refrigeration Business* (September).

Kauffman, R.E. 1992. Chemical analysis and recycling of used refrigerant from field systems. ASHRAE *Transactions* 98(1).

Krause, W.O., A.B. Guise, and E.A. Beacham. 1960. Time factors in the removal of moisture from refrigerating systems with desiccant type driers. ASHRAE *Transactions* 66:465.

Kvalnes, D.E. 1965. The sealed tube test for refrigeration oils. ASHRAE *Transactions* 71(1):138.

Manz, K.W. 1988. Recovery of CFC refrigerants during service and recycling by the filtration method. ASHRAE *Transactions* 94(2).

Manz, K.W. 1991a. The challenge of handling refrigerants. *Air Conditioning, Heating, & Refrigeration News.* November 4, 11, and 18 issues.

Manz, K.W. 1991b. How to handle multiple refrigerants in recovery and recycling equipment. ASHRAE *Journal* 33(4).

Mays, R.L. 1962. Molecular sieve and gel-type desiccants for Refrigerants 12 and 22. ASHRAE *Journal* (August):73.

McCain, C.A. 1991. Refrigerant reclamation protects HVAC equipment investment. ASHRAE *Journal* 33(4).

Norton, F.J. 1957. Rates of thermal decomposition of $CHClF_2$ and CF_2Cl_2. *Refrigerating Engineering* (September):33.

O'Meara, D.R. 1988. Operating experiences of a refrigerant recovery services company. ASHRAE *Transactions* 94(2).

Parker, R.W. 1988. Reclaiming refrigerant in OEM plants. ASHRAE *Transactions* 94(2).

Parmelee, H.M. 1965. Sealed tube stability tests on refrigeration materials. ASHRAE *Transactions* 71(1):154.

Rosen, S., A.A. Sakhnovsky, R.B. Tilney, and W.O. Walker. 1965. A method of evaluating filtration and flow characteristics of liquid line driers. ASHRAE *Transactions* 71(1):200.

RSES. 1988. Standard procedure for replacement of components in a sealed refrigerant system (compressor motor burnout). Refrigeration Service Engineers Society, Des Plaines, IL.

RSES. 1991. Refrigerant service for the 90's, first edition. Refrigeration Service Engineers Society, Des Plaines, IL.

SAE. 1989. Standard of purity for use in mobile air-conditioning systems. *Standard* 1991-89. Society of Automotive Engineers, Warrendale, PA.

SAE. 1991. Standard of purity for recycle HFC-134a for use in mobile air-conditioning systems. *Standard* J2099-91. Society of Automotive Engineers, Warrendale, PA.

Spauschus, H.O. and G.C. Doderer. 1961. Reaction of Refrigerant 12 with petroleum oils. ASHRAE *Journal* (February):65.

Spauschus, H.O. and G.C. Doderer. 1964. Chemical reactions of Refrigerant 22. ASHRAE *Journal* (October):54.

Spauschus, H.O. and R.S. Olsen. 1959. Gas analysis—A new tool for determining the chemical stability of hermetic systems. *Refrigerating Engineering* (February):25.

Thrasher, J.S., R. Timkovich, H.P.S. Kumar, and S.L. Hathcock. 1993. Moisture solubility in Refrigerant 123 and Refrigerant 134a. ASHRAE *Transactions* 100(1).

UL. 1986. Standard for refrigerant-containing components and accessories, nonelectrical, fifth edition. UL 207. Underwriters Laboratories, Northbrook, IL.

Walker, W.O. 1963. Latest ideas in use of desiccants and driers. *Refrigerating Service & Contracting* (August):24.

Walker, W.O., J.M. Malcolm, and H.C. Lynn. 1955. Hydrophobic behavior of certain desiccants. *Refrigerating Engineering* (April):50.

Walker, W.O., S. Rosen, and S.L. Levy. 1962. Stability of mixtures of refrigerants and refrigerating oils. ASHRAE *Journal* (August):59.

Wojtkowski, E.F. 1964. System contamination and cleanup. ASHRAE *Journal* (June):49.

Zahorsky, L.A. 1967. Field and laboratory studies of wax-like contaminants in commercial refrigeration equipment. ASHRAE *Transactions* 73(1):II, 1.1.

BIBLIOGRAPHY

Boing, J. 1973. Desiccants and driers. RSES Service Manual, Section 5, 620-16B. Refrigeration Service Engineers Society, Des Plaines, IL.

Burgel, J., N. Knaup, and H. Lotz. 1988. Reduction of CFC-12 emission from refrigerators in the FRG. *International Journal of Refrigeration* 11(4).

Cohen, A.P. 1993. Test methods for the compatibility of desiccants with alternative refrigerants. ASHRAE *Transactions* 99(1).

Cohen, A.P. and S.R. Dunne. 1987. Review of automotive air-conditioning drydown rate studies—The kinetics of drying Refrigerant 12. ASHRAE *Transactions* 93(2).

Du Pont. 1976. Mutual solubilities of water with fluorocarbons and fluorocarbon-hydrate formation. DuPont Co., Wilmington, DE.

Guy, P.D., G. Tompsett, T.W. Dekleva. 1992. Compatibilities of nonmetallic materials with R-134a and alternative lubricants in refrigeration systems. ASHRAE *Transactions* 98(1).

Kauffman, R.E. 1992. Sealed tube tests of refrigerants from field systems before and after recycling. ASHRAE *Transactions* 99(2).

Kitamura, K., T. Ohara, S. Honda, and H. SakaKibara. 1993. A new refrigerant-drying method in the automotive air conditioning system using HFC-134a. ASHRAE *Transactions* 99(1).

Sundaresan, S.G. 1989. Standards for acceptable levels of contaminants in refrigerants. *CFCs—Time of Transition*, pp. 220-23. ASHRAE.

Walker, W.O. 1960. Contaminating gases in refrigerating systems. RSES Service Manual, Section 5, 620-15. Refrigeration Service Engineers Society, Des Plaines, IL.

Walker, W.O. 1985. Methyl alcohol in refrigeration. RSES Service Manual, Section 5, 620-17A. Refrigeration Service Engineers Society, Des Plaines, IL.

Zhukoborshy, S.L. 1984. Application of natural zeolites in refrigeration industry. Proceedings of the International Symposium on Zeolites, Portoroz, Yugoslavia (September).

CHAPTER 7

LUBRICANTS IN REFRIGERANT SYSTEMS

THE primary function of a lubricant is to reduce friction and minimize wear. A lubricant achieves this by interposing a film between moving surfaces that reduces direct solid-to-solid contact or lowers the coefficient of friction.

Understanding the role of a lubricant requires an analysis of the surfaces to be lubricated. While bearing surfaces and other machined parts may appear and feel smooth, close examination reveals microscopic peaks (asperities) and valleys. With a sufficient quantity of lubricant, a layer is provided that has a thickness greater than the maximum height of the mating asperities, so that moving parts ride on a lubricant cushion.

Ideal conditions are not always easily attained. For example, when the shaft of a horizontal journal bearing is at rest, the static loads squeeze out the lubricant, producing a discontinuous film with metal-to-metal contact at the bottom of the shaft. When the shaft begins to turn, there is no layer of liquid lubricant separating the surfaces. As the shaft picks up speed, the lubricating fluid is drawn into the converging clearance between the bearing and the shaft, generating a hydrodynamic pressure that eventually can support the load on an uninterrupted fluid film.

Various regimes or conditions of lubrication can exist when surfaces are in motion with respect to one another. Regimes of lubrication are defined as follows:

Full fluid film or hydrodynamic. Mating surfaces are completely separated by the lubricant film.

Boundary. Gross surface-to-surface contact occurs because the bulk lubricant film is too thin to separate the mating surfaces.

Mixed fluid film or quasi-hydrodynamic. Occasional or random surface contact occurs.

A wide variety of materials can be used to separate and lubricate contacting surfaces. Separation can be a boundary layer on a metal surface, a fluid film, or a combination of both. The function of a lubricant extends beyond preventing surface contact. It also removes heat, provides a seal to keep out contaminants or to retain pressures, prevents corrosion, and disposes of debris created by wear. Lubricating oils are best suited to meet these various requirements.

Viscosity is the most important factor to consider in choosing a lubricant under full fluid film conditions. Under boundary conditions, the asperities are the contact points and support much, if not all, of the load. The contact pressures are usually sufficient to cause welding and surface deformation. However, even under these conditions, wear can be controlled effectively with nonfluid, multimolecular films formed on the surface. These films must be strong enough to resist rupturing, yet have acceptable frictional and shear characteristics. These films reduce surface fatigue, adhesion, abrasion, and corrosion, which are the four major sources (either singularly or together) of rapid wear under boundary conditions. The slightly active constituents left in commercially refined mineral oils give them their natural film-forming properties.

Additives have also been developed to improve lubrication under boundary conditions. These materials are characterized by terms such as oiliness agents, lubricity improvers, and antiwear additives. They form a film on the metal surface through polar attraction or chemical action. These films or coatings have lower coefficients of friction under the loads imposed during boundary conditions. In chemical action, the temperature increase brought about by friction-generated heat brings about a reaction between the additive and the metal surface. Films such as iron sulfide and iron phosphate can be formed depending on the additives and the energy available for the reaction. In some instances, organic phosphates and phosphites are used in refrigeration oils to improve boundary lubrication. The nature of the metal and the condition of the metal surfaces are important. Refrigeration compressor designers often treat ferrous pistons, shafts, and wrist pins with phosphating processes that impart a crystalline, discontinuous film of metal phosphate to the surface. This film aids boundary lubrication during the break-in period.

TESTS FOR BOUNDARY LUBRICATION

Film strength or load carrying ability are terms often used to describe lubricant lubricity characteristics under boundary conditions. Laboratory tests that measure the degree of scoring, welding, or wear have been developed to evaluate lubricants. However, bench-type tests cannot be expected to accurately simulate actual field performance in a given compressor and are, therefore, merely screening devices. Some of these tests have been standardized by the American Society of Testing and Materials (ASTM) and other organizations. The following tests evaluate lubricant performance.

In the *four-ball extreme-pressure method* (D 2783), the antiwear property is determined from the average scar diameter on the stationary balls and is stated in terms of a load-wear index. The smaller the scar, the better the load-wear index. The maximum load carrying capability is defined in terms of a weld point, *i.e.*, the load at which welding by frictional heat occurs. The *Falex* method (D 2670) allows measurement of wear during the test itself, and the scar width on the V-blocks and/or the mass loss of the pin can be used as a measure of the antiwear properties. The load carrying capability is determined from a failure, which can be caused by excess wear or extreme frictional resistance. The *Timken* method (D 2782) determines the load at which rupture of the lubricant film occurs, and the *Alpha* LFW-1 machine (D 2714) measures frictional force and wear.

However, since all these machines operate in air, available data may not apply to a refrigerant environment. Divers (1958) questioned the validity of tests in air because several of the oils that

The preparation of this chapter is assigned to TC 3.4, Lubrication.

performed poorly in Falex testing have always been used successfully in refrigerant systems. Murray *et al.* (1956) suggest that halocarbon refrigerants can aid in boundary lubrication. Refrigerant 12, for example, when run hot in the absence of oil, reacted with steel surfaces to form a lubricating film. These studies emphasize the need for laboratory testing in a simulated refrigerant environment.

In Huttenlocher's (1969) method of simulation, refrigerant vapor is bubbled through the lubricant reservoir before the test to displace the dissolved air. It is continued during the test to maintain a blanket of refrigerant on the lubricant surface. Using the Falex tester, Huttenlocher has shown the beneficial effect of Refrigerant 22 on the load carrying capability of the same lubricant compared with air or nitrogen. Sanvordenker and Gram (1974) describe a further modification of the Falex test using a sealed sample system. Both R-12 and R-22 atmospheres had beneficial effects on a lubricant's boundary lubrication characteristics when compared with tests in air.

Refrigerant 134a, which is chlorine-free and relatively stable, is a likely candidate to replace the chlorofluorocarbon R-12. Currently, wear studies are underway to determine the effect of nonchlorinated refrigerants, such as R-134a. Preliminary data suggest that these types of refrigerants also have some beneficial effect on boundary lubrication when compared to tests conducted in air. However, nonchlorinated refrigerants are unlikely to be as effective as chlorinated refrigerants.

Test parameters must simulate as closely as possible the system conditions in the base material from which the test specimens are made; namely, their surface condition, the processing methods, and the operating temperature. There are several bearings or rubbing surfaces in a refrigerant compressor, each of which may use different materials and may operate under different conditions. A different test may be required for each bearing. Moreover, hermetic compressors are precision devices with internal and bearing clearances often controlled to within several ten-thousandths of an inch. Permissible bearing wear is minimal because wear debris remains in the system and can cause other problems even if the clearances stay within working limits. Compressor system mechanics must be understood to perform and interpret simulated tests.

Some aspects of compressor lubrication are not suitable for laboratory simulation. One aspect, the return of liquid refrigerant to the compressor, can cause the lubricant to dilute or wash away from the bearings, creating conditions of boundary lubrication. Tests using operating refrigerant compressors have also been considered, and one such wear test has been proposed as a German Standard (DIN 8978). The test is functional for a given compressor system and may permit comparison of lubricants within that class of compressors. However, it is not designed to be a generalized test for the boundary lubricating capability of a lubricant. Other tests using radioactive tracers in refrigerant systems have given useful results (Rembold and Lo 1966).

REFRIGERATION LUBRICANT REQUIREMENTS

Refrigeration compressors are classified as continuous or dynamic and positive displacement. Dynamic types such as the centrifugal compressor depend on energy transfer from a rotating set of blades to the gas. Momentum imparted to the gas is converted to useful pressure by decelerating the gas. Positive-displacement compressors can be either reciprocating or rotary. Both designs confine discrete volumes of gas within a closed space and then elevate pressure by reducing the volume.

Refrigeration systems require lubricant to do more than lubricate. Oil seals compressed gas between the suction and discharge sides, and acts as a coolant to remove heat from the bearings and to transfer heat from the crankcase to the compressor exterior. Oil also reduces noise generated by moving parts inside the compressor.

Although the compression components of centrifugal compressors require no internal lubrication, rotating shaft bearings, seals, and couplings must be adequately lubricated. Turbine or other types of lubricants can be used when the lubricant is not in contact or circulated with refrigerant gas. Chapter 35 of the 1992 ASHRAE *Handbook—Systems and Equipment* describes how reciprocating and rotary compressors are lubricated.

Hermetic systems, where the motor is exposed to the lubricant, require a lubricant with electrical insulating properties. The refrigerant gas carries some lubricant with it into the condenser and evaporator. This lubricant must return to the compressor within a reasonable time and must have adequate fluidity at low temperatures. The lubricant should remain miscible with the refrigerant for good heat transfer in the evaporator and for good lubricant return. The lubricant must be free of suspended matter or components such as wax that might clog the expansion tube or deposit in the evaporator and interfere with heat transfer. In hermetic refrigeration systems, the lubricant is charged only once, so it must function for the lifetime of the compressor. The chemical stability required of the lubricant in the presence of refrigerants, metals, motor insulation, and extraneous contaminants is perhaps the most important characteristic distinguishing refrigeration oils from those used for all other applications (see Chapter 5).

As expected, an ideal lubricant does not exist; a compromise must be made to balance the requirements. A high-viscosity lubricant seals the gas pressure best, but may offer more frictional resistance. Slight foaming can reduce noise, but excessive foaming can carry too much lubricant into the cylinder and cause structural damage. Oils that are most stable chemically are not necessarily good lubricants. The oils should not be considered alone, since the functions of the lubricant are performed by lubricant-refrigerant mixtures, not by pure oils.

While the behavior of a lubricant in a refrigeration system must depend on its processing and composition, the precise relationship between composition and performance is not well defined. Standard ASTM bench tests can, however, provide such information as (1) viscosity, (2) viscosity index, (3) color, (4) specific gravity, (5) refractive index, (6) pour point, (7) aniline point, (8) oxidation resistance, (9) dielectric breakdown voltage, (10) foaming tendency in air, (11) moisture content, (12) wax separation, and (13) volatility. Other properties, particularly those involving interactions with refrigerants, must be determined by special tests described in the refrigeration literature. Among these nonstandard properties are (1) mutual solubility with various refrigerants, (2) chemical stability in the presence of refrigerants and metals, (3) chemical effects of contaminants or additives that may be in the oils, (4) boundary film-forming ability, and (5) solubility of air.

MINERAL OIL COMPOSITION

For typical applications, the numerous compounds in refrigeration oils of mineral origin can be grouped into the following structures: (1) paraffins, (2) naphthenes (cycloparaffins), (3) aromatics, and (4) nonhydrocarbons. *Paraffins* consist of all straight chain and branched carbon chain saturated hydrocarbons. N-Pentane and isopentane are examples of paraffinic hydrocarbons. *Naphthenes* are also completely saturated but consist of cyclic or ring structures; cyclopentane is a typical example. *Aromatics* are unsaturated cyclic hydrocarbons containing one or more rings characterized by alternate double bonds; benzene is a typical example. The *nonhydrocarbons* are molecules containing atoms such as sulfur, nitrogen, or oxygen in addition to carbon and hydrogen.

The preceding structural components do not necessarily exist in pure states. In fact, a paraffinic chain frequently is attached to a naphthenic or aromatic structure. Similarly, a naphthenic ring to which a paraffinic chain is attached may in turn be attached to an aromatic molecule. Because of such complications, mineral oil composition is usually described by carbon-type and molecular analysis.

In carbon-type analysis, the number of carbon atoms on the paraffinic chains, naphthenic structures, and aromatic rings is determined and represented as a percentage of the total. Thus, % C_P, the percentage of carbon atoms having a paraffinic configuration, includes not only the free paraffins but also those paraffinic chains attached to naphthenic or to aromatic rings.

Similarly, % C_N includes the carbon atoms on the free naphthenes as well as those on the naphthenic rings attached to the aromatic rings, and % C_A represents the carbon atoms on the aromatic rings. Carbon analysis describes a lubricant in its fundamental structure and correlates and predicts many physical properties of the lubricant. However, direct methods of determining carbon composition are laborious. Therefore, common practice uses a correlative method, such as the one based on the refractive index-density-relative molecular mass (n-d-m) (Van Nes and Weston 1951) or one standardized by ASTM D 2140 or ASTM D 3288. Other methods are ASTM D 2008, which uses ultraviolet absorbency and a rapid method using infrared spectrophotometry and calibration from known oils.

Molecular analysis is based on methods of separating the structural molecules. For refrigeration oils, important structural molecules are (1) saturates or nonaromatics, (2) aromatics, and (3) nonhydrocarbons. All the free paraffins and naphthenes (cycloparaffins), as well as mixed molecules of paraffins and naphthenes are included in the saturates. However, any paraffinic and naphthenic molecules attached to an aromatic ring are classified as aromatics. This representation of lubricant composition is less fundamental than carbon analysis. However, many properties of the lubricant relevant to refrigeration can be explained with this analysis, and the chromatographic methods of analysis are fairly simple (ASTM D 2549, ASTM D 2007, Mosle and Wolf 1963, Sanvordenker 1968).

The traditional classification of oils as paraffinic or naphthenic refers to the amount of paraffinic or naphthenic molecules in the refined lubricant. Paraffinic crudes contain a higher proportion of paraffin wax, and, thus, have a higher viscosity index and pour point relative to naphthenic crudes.

COMPONENT CHARACTERISTICS

Saturates have excellent chemical stability, but poor solubility with polar refrigerants, such as R-22; they are also poor boundary lubricants. Aromatics are somewhat more reactive but have very good solubility with refrigerants and good boundary lubricating properties. Nonhydrocarbons are the most reactive but are beneficial for boundary lubrication, although the amounts needed for that purpose are small. The reactivity, solubility, and boundary lubricating properties of a refrigeration lubricant are affected by the relative amounts of these components in the lubricant.

The saturate and aromatic fractions separated from a lubricant do not have the same viscosity as the parent lubricant. The saturate fraction is much less viscous, while the aromatic fraction is much more viscous than the parent lubricant. Both fractions have the same boiling range. Thus, for this range, the aromatics are more viscous than the saturates. For the same viscosity, the aromatics have a higher volatility than the saturates. Also, the saturate fraction has a lower density and a lower refractive index, but a higher viscosity index and molecular mass than the aromatic fraction of the same lubricant.

Among the saturates, the straight chain paraffins are undesirable for refrigeration applications because they precipitate as wax crystals when the lubricant is cooled to its pour point and tend to form flocs in certain refrigerant solutions (see the section on Wax Separation). The branched chain paraffins and naphthenes are less viscous at low temperatures and have extremely low pour points.

Nonhydrocarbons are mostly removed during the refining of refrigeration oils. Those that remain are expected to have little effect on the physical properties of the lubricant, except, perhaps, on its color, stability and lubricity. Since all the nonhydrocarbons

Table 1 Typical Properties of Refrigerant Lubricants

		Mineral Lubricants				Synthenic Lubricants				
Property	**ASTM**	**Naphthenic**			**Paraffinic**	**Alkyl-benzene**	**Ester**		**Glycol**	
Viscosity, cSt (SSU) at 100°F	D 445	33.1 (155)	61.9 (287)	68.6 (318)	34.2 (160)	31.7 (149)	30 (142)	100 (463)	29.9 (141)	90 (417)
Viscosity index	D 2270	0	0	46	95	27	111	98	210	135
Specific gravity	D 1298	0.913	0.917	0.9	0.862	0.872	0.995	0.972	0.99	1.007
Color	D 1500	0.5	1	1	0.5					
Refractive index	D 1747	1.5015	1.5057	1.4918	1.4752					
Relative molecular mass	D 2503	300	321	345	378	320	570	840	750	1200
Pour point, °F	D 97	−45	−40	−35	0	−50	−54	−22	−51	−40
Floc point, °F	ASHRAE 86	−68	−60	−60	−31	−100				
Flash point, °F	D 92	340	360	400	395	350	453	496	399	334
Fire point, °F	D 92	390	400	450	450	365				
Composition							Branched	Branched	PP monol	PP diol di-
Carbon-type	n-d-m						acid	acid	mono-func-	functional
$\%C_A$	Van Nes and	14	16	7	3	24	penta-	penta-	tional poly-	polypro-
$\%C_N$	Weston (1951)	43	42	46	32	None	erythritol	erythritol	propylene	pylene
$\%C_P$		43	42	47	65	76			glycol	glycol
Molecular composition	D 2549									
% Saturates		62	59	78	87	None				
% Aromatics		38	41	22	13	100				
Aniline point, °F	D 611	160	165	197	220	125				
Critical solution temperature with R-22, °F	—	25	35	74	81	−100				

(*e.g.*, sulfur compounds) are not dark, even a colorless lubricant does not necessarily guarantee the absence of nonhydrocarbons. Kartzmark *et al.* (1967) and Mills and Melchoire (1967) found indications that nitrogen-bearing compounds cause or act as catalysts toward the deterioration of oils. The sulfur and oxygen compounds are thought to be less reactive, with some types considered to be natural inhibitors and lubricity enhancers.

Solvent refining, hydrofinishing, or acid treatment followed by a separation of the acid tar formed are often used to remove more thermally unstable aromatic and unsaturated compounds from the base stock. These methods also produce refrigeration oils that are free from carcinogenic materials sometimes found in crude oil stocks.

The properties of the components naturally are reflected in the parent oil. An oil with a very high saturate content, as is frequently the case with paraffinic oils, also has a high viscosity index, low specific gravity, high relative molecular mass, low refractive index, and low volatility. In addition, it would have a high aniline point and would be less miscible with polar refrigerants. The reverse is true of naphthenic oils. Table 1 lists typical properties of several mineral-based refrigeration oils.

SYNTHETIC LUBRICANTS

The limited solubility of mineral oils with R-22 and R-502 originally led to the investigation of synthetic lubricants for refrigeration use. In more recent times, the lack of solubility of mineral oils in nonchlorinated fluorocarbon refrigerants, such as R-134a and R-32, has led to the commercial use of some synthetic lubricants. Gunderson and Hart (1962) describe a number of commercially available synthetic lubricants such as synthetic paraffins, polyglycols, dibasic acid esters, neopentyl esters, silicones, silicate esters, and fluorinated compounds. Many have properties suited to refrigeration purposes. Sanvordenker and Larime (1972) describe the properties of these synthetic lubricants, alkylbenzenes, and phosphate esters in regard to refrigeration applications using chlorinated fluorocarbon refrigerants. The phosphate esters are unsuitable for refrigeration use because of their poor thermal stability. Although very stable and compatible with refrigerants, the fluorocarbon lubricants are expensive. Among others, only the synthetic paraffins have poor miscibility relations with R-22. Dibasic acid esters, neopentyl esters, silicate esters, and polyglycols all have excellent viscosity temperature relations and remain miscible with R-22 and R-502 to very low temperatures. At this time, the three synthetic lubricants seeing the greatest use are alkylbenzene for R-22 and R-502 service and polyglycols and polyol esters for use with R-134a and refrigerant blends using R-32.

There are two basic types of alkylbenzenes, namely, branched and linear. The products are synthesized by reacting an olefin or chlorinated paraffin with benzene in the presence of a catalyst. Catalysts commonly used for this reaction are aluminum chloride and hydrofluoric acid. After the catalyst is removed, the product is distilled into fractions. The relative size of these fractions can be changed by adjusting the relative molecular mass of the side chain (olefin or chlorinated paraffin) and by changing other variables. The quality of alkylbenzene refrigeration lubricant varies, depending on the type (branched or linear) and manufacturing scheme. In addition to good solubility with refrigerants, such as R-22 and R-502, these lubricants have better high-temperature and oxidation stability than mineral oil-based refrigeration oils. Typical properties for a branched alkylbenzene are shown in Table 1.

Polyalkylene glycols (PAGs) are derived from ethylene oxide or propylene oxide. The polymerization is usually initiated with either an alcohol, such as butyl alcohol, or by water. Initiation by an alcohol results in a monol while initiation by water results in a diol. PAGs have seen commercial service with R-12 and R-22, as well as in automotive air-conditioning systems using R-134a. PAGs have excellent lubricity, low pour points, good low-temperature fluidity, and good compatibility with most elastomers. Major concerns are that these oils are somewhat hygroscopic, are immiscible with mineral oils, and require additives for good chemical and thermal stability (Short 1990).

Polyalphaolefins (PAOs) are normally manufactured using a two-step process from linear α-olefins. The first step is the synthesis of a mixture of oligomers in the presence of a $BF_3 \cdot ROH$ catalyst. Several parameters can be varied to control the distribution of the oligomers so formed. The second step involves a hydrogenation processing of the unsaturated oligomers in the presence of a metal catalyst (Shubkin 1993). PAOs have good miscibility with R-12 and R-114. Some R-22 applications have been tried but are limited by the low miscibility of the fluid in R-22. PAOs are immiscible in R-134a (Short 1990).

Neopentyl esters (polyol esters) are derived from a reaction between an alcohol and a normal or branched carboxylic acid in the presence of a catalyst. The most common alcohols used are pentaerythritol, trimethylolpropane, neopentyl glycol, and glycerin. For higher viscosities, a dipentaerythritol is often used. The acids are usually selected to give the correct viscosity and fluidity at low temperatures. Complex neopentyl esters are derived by a sequential reaction of the polyol with a dibasic acid followed by a reaction with mixed monoacids (Short 1990). This results in a lubricant with a higher relative molecular mass, high viscosity indices, and higher ISO viscosity grades. Lubricants of this type are in commercial service in rotary screw chillers using R-134a.

LUBRICANT ADDITIVES

Additives can enhance certain lubricant properties or impart new characteristics depending on their chemical composition and the nature of the base lubricant. They generally fall into three groups, namely, polar compounds, polymers, and compounds containing active elements such as sulfur. Additive types include (1) pour point depressants, (2) floc point depressants, (3) viscosity index improvers, (4) thermal stability improvers, (5) extreme pressure and antiwear additives, (6) rust inhibitors, (7) antifoam agents, (8) metal deactivators, (9) dispersants, and (10) oxidation inhibitors. Some additives offer performance advantages in one area but are detrimental in another. Some additives work best when combined with other additives. They must be compatible with system materials and be present in the optimum concentration; too little may be ineffective, while too much can be detrimental or offer no incremental improvement.

In general, additive-type oils are not required to lubricate a refrigerant compressor. However, oils that contain additives give highly satisfactory service and some additives, such as antiwear additives, offer some performance advantages over straight mineral oils. Their use is justifiable as long as the user knows of their presence and provided the additives do not significantly degrade with use.

An additive-type lubricant should only be used after thorough testing to determine whether the additive material (1) is removed by system dryers, (2) is inert to system components, (3) is soluble in refrigerants at low temperatures so as not to cause deposits in capillary tubes or expansion valves, and (4) is stable at high temperatures to avoid adverse chemical reactions such as harmful deposits. This can best be done by sealed tube and compressor testing using the actual additive/base lubricant combination intended for field use.

LUBRICANT PROPERTIES

Viscosity and Viscosity Grades

Viscosity defines a fluid's resistance to flow. It can be expressed as absolute or dynamic viscosity (centipoises, cP), kinematic viscosity (centistokes, cSt), or Saybolt Seconds Universal viscosity (abbreviated SSU or SUS). ASTM D 2161 contains tables to convert SSU to kinematic viscosity. The density must be known to convert kinematic viscosity to absolute viscosity; that is, absolute or dynamic viscosity (cP) equals density (g/cm^3) times kinematic viscosity (cSt). Refrigeration oils are sold in viscosity grades, and ASTM has proposed a system of standardized viscosity grades for industry-wide use (D 2422).

In selecting the proper viscosity grade, the environment to which the lubricant will be exposed should be considered. The viscosity of the lubricating fluid decreases if temperatures rise or if the refrigerant dissolves appreciably in the lubricant.

A large reduction in the viscosity of the lubricating fluid may affect the lubricity and, more likely, the sealing function of the lubricant, depending on the nature of the machinery. The design of some types of hermetically sealed units, such as the single-vane, rotary units, requires the lubricating fluid to act as an efficient sealing agent. In reciprocating compressors, the lubricant film is spread over the entire area of contact between the piston and the cylinder wall, providing a very large area to resist leakage from the high- to the low-pressure side. In a single-vane rotary type, however, the critical sealing area is a line contact between the vane and a roller. In this case, viscosity reduction is serious.

The lubricant with the lowest viscosity that gives the necessary sealing properties with the refrigerant used for the entire range of temperatures and pressures encountered should be chosen. A practical method for determining the minimum safe viscosity is to calculate the total volumetric efficiency of a given compressor system using several oils of widely varying viscosities. The lubricant of lowest viscosity that gives satisfactory volumetric efficiency should be selected. Tests should be run at a number of ambient temperatures, for example, 70, 90, and 110°F. As a guideline, Table 2 lists the viscosity ranges recommended for various refrigeration systems.

The International Organization for Standardization (ISO) has established a series of definite viscosity levels as a standard for specifying or selecting lubricant for industrial applications. This system, covered in the United States by ASTM D 2422, is designed to eliminate intermediate or unnecessary viscosity grades while providing enough viscosity grades for operating equipment. The system reference point is kinematic viscosity at 40°C, and each viscosity grade with suitable tolerances is identified by the kinematic viscosity at this temperature. Therefore an ISO VG 32 grade lubricant would identify a lubricant with a viscosity grade of 32 cSt at 40°C. Table 3 is a chart of the various standardized viscosity grades of lubricants.

Table 2 Recommended Viscosity Ranges

Small Systems

Refrigerant	Type of Compressor	Lubricant Viscosities at 100°F SSU	cSt
Ammonia	Screw	280-300	60-65
Ammonia	Reciprocating	150-300	32-65
Carbon dioxide	Reciprocating	280-300[a]	60-65[a]
Refrigerant 11	Centrifugal	280-300	60-65
Refrigerant 123	Centrifugal	280-300	60-65
Refrigerant 12	Centrifugal	280-300	60-65[b]
Refrigerant 12	Reciprocating	150-300	32-65[b]
Refrigerant 12	Rotary	280-300	60-65[b]
Refrigerant 134a	Centrifugal	280-300	60-65
Refrigerant 134a	Screw	280-300	60-65
Refrigerant 22	Centrifugal	280-400	60-86
Refrigerant 22	Reciprocating	150-300	32-65
Refrigerant 22	Scroll	280-300	60-65
Refrigerant 22	Screw	280-800	60-173
Halogenated refrigerants	Screw	150-4000	32-800

Industrial Refrigeration[c]

Type of Compressor	Lubricant Viscosities at 100°F SSU	cSt
Where lubricant may enter refrigeration system or compressor cylinders	150-300	32-65
Where lubricant is prevented from entering system or cylinders:		
in force-feed or gravity systems	500-600	108-129
in splash systems	150-160	32-34
Steam-driven compressor cylinders when condensate is reclaimed for ice-making	High viscosity lubricant with 140-165 SSU at 210°F	30-35 at 210°F

[a]Some applications may require lighter lubricants of 75-85 SSU (14-17 cSt); others, heavier lubricants of 500-600 SSU (108-129 cSt).
[b]Automotive applications using R-12 may require 450-500 SSU (97-107 cSt) lubricant.
[c]Ammonia and carbon dioxide compressors with splash, force-feed, or gravity circulating systems.

Table 3 Viscosity System for Industrial Fluid Lubricants (ASTM D 2422)

Viscosity System Grade Identification	Midpoint Viscosity cSt at 40°C	Kinematic Viscosity Limits cSt at 40°C Minimum	Maximum	Approximate Equivalents, SSU Units
ISO VG 2	2.2	1.98	2.42	32
ISO VG 3	3.2	2.88	3.52	a
ISO VG 5	4.6	4.14	5.06	40
ISO VG 7	6.8	6.12	7.48	a
ISO VG 10	10	9.00	11.00	60
ISO VG 15	15	13.50	16.50	75
ISO VG 22	22	19.80	24.20	105
ISO VG 32	32	28.80	35.20	150
ISO VG 46	46	41.40	50.60	215
ISO VG 68	68	61.20	74.80	315
ISO VG 100	100	90	110	465
ISO VG 150	150	135	165	700
ISO VG 220	220	198	242	1000
ISO VG 320	320	288	352	1500
ISO VG 460	460	414	506	2150
ISO VG 680	680	612	748	3150
ISO VG 1000	1000	900	1100	4650
ISO VG 1500	1500	1350	1650	7000

[a]The 36 and 50 SSU grades are not currently considered standardized grades in the United States.

Viscosity Index

Mineral oil viscosity decreases as the temperature increases and increases as the temperature decreases. The relationship between temperature and kinematic viscosity is represented by

$$\log\log(\nu + 0.7) = A + B\log T$$

where

ν = kinematic viscosity, cSt
T = thermodynamic temperature, K or °R
A, B = constants for each lubricant

This relationship is the basis for the viscosity-temperature charts published by ASTM and permits a straight line plot of viscosity over a wide temperature range. Figure 1 shows a plot for two different lubricants, such as a naphthenic mineral oil (LVI) and a synthetic lubricant (HVI). This plot is applicable over the temperature range in which the oils are homogenous liquids.

The slope of the viscosity-temperature lines is different for different lubricants. The viscosity-temperature relationship of a lubricant is described by an empirical number called the viscosity

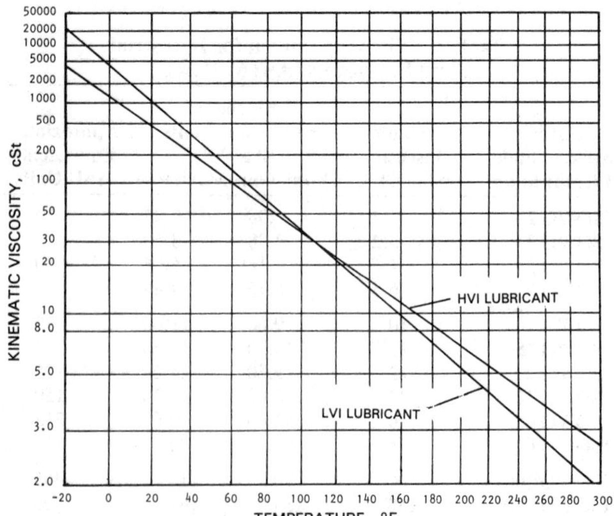

Fig. 1 Viscosity-Temperature Chart for 108 cSt (500 SSU) HVI and LVI Lubricants

index (VI) (ASTM D 2270). A lubricant with a high viscosity index (HVI) shows less change in viscosity over a given temperature range than a lubricant with a low viscosity index (LVI). In the example shown in Figure 1, both oils possess equal viscosities [32 cSt (151 SSU)] at 104°F. However, the viscosity of the LVI lubricant (a naphthenic mineral oil), as shown by the steeper slope of the line, increases to 520 cSt (2400 SSU) at 32°F, whereas the HVI lubricant (a synthetic lubricant) viscosity increases only to 280 cSt (1280 SSU) at the same temperature.

The viscosity index is related to the composition of the mineral oil. Generally, an increase in cyclic structure, aromatic and naphthenic, decreases the viscosity index. Paraffinic oils usually have a high viscosity index and low aromatic content. Naphthenic oils, on the other hand, have a lower viscosity index and are usually higher in aromatics. For the same base lubricant, the viscosity index decreases with increasing aromatic content. Generally, among common synthetic lubricants, polyalphaolefins, polyalkylene glycols, and polyol esters will have high viscosity indices. As shown in Table 1, alkylbenzenes have lower viscosity indices.

Density

The density of the lubricating oil must be known to convert kinematic to absolute viscosity. Figure 2 shows published values for pure lubricant densities over a range of temperatures. These density-temperature curves all have approximately the same slope and appear merely to be displaced from one another. If the density of a particular lubricant is known at one temperature but not over a range of temperatures, a reasonable estimate at other temperatures can be obtained by drawing a line paralleling those in Figure 2.

Density indicates the composition of a lubricant for a given viscosity. Naphthenic oils are usually more dense than paraffinic oils. Also, the higher the aromatic content, the higher the density. For equivalent compositions, higher viscosity oils have higher densities, but the change in density with aromatic content is greater than it is with viscosity.

Relative Molecular Mass

In refrigeration applications, the relative molecular mass of a lubricant is often needed. Albright and Lawyer (1959) showed that, on a molar basis, Refrigerants 22, 115, 13, and 13B1 have about the same viscosity-reducing effects on a paraffinic lubricant.

For most mineral oils, a reasonable estimate of the average molecular mass can be obtained by the standard ASTM test D 2502, based on kinematic viscosities at 100 and 210°F; or from

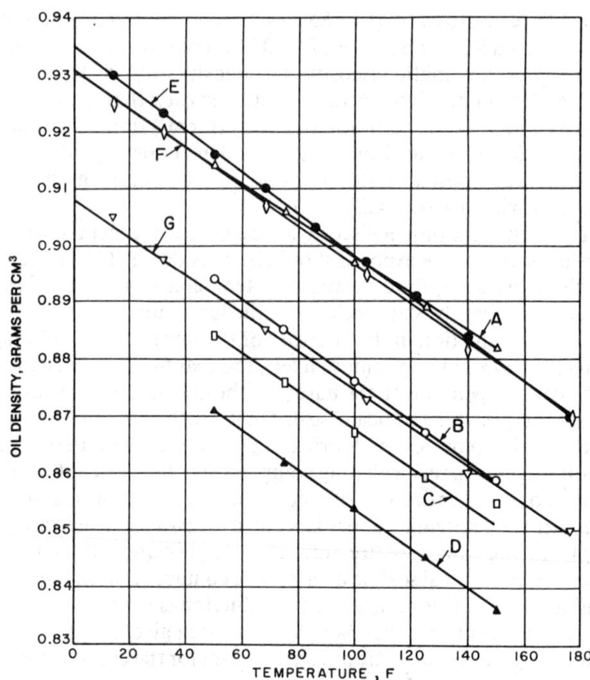

Oil	Base	Viscosity cSt 100 °F	Viscosity SSU 100 °F	Composition, % C_A	C_N	C_P	Ref.
A	Naphthene	64.7	300	—	—	—	1
B	Naphthene	15.7	80	—	—	—	1
C	Paraffin	64.7	300	—	—	—	1
D	Paraffin	32.0	150	—	—	—	1
E	Paraffin	51.7	240[a]	—	—	—	2
F	Naphthene	33.1	155[a]	12	44	44	3
G	Naphthene	45.2	210[a]	4	44	52	3

References:
1. Albright and Lawyer (1959)
2. Bambach (1955)
3. Loffler (1959)

[a]German oils, base as stated by the author; viscosities converted to SSU and interpolated at 100°F.

Fig. 2 Variation of Refrigeration Oil Densities with Temperature

viscosity-gravity correlations of Mills *et al.* (1946). Direct methods (ASTM D 2503) can also be used when greater precision is needed or when the correlative methods are not applicable.

Pour Point

Any lubricant intended for low-temperature service should be able to flow at the lowest temperature that it will encounter in the system. This requirement is usually met by specifying a suitable low pour point. The pour point of a lubricant is defined as the lowest temperature at which it will pour or flow, when tested according to the standard method prescribed in ASTM D 97.

The loss of fluidity at the pour point may manifest itself in two ways. Naphthenic oils and synthetic lubricants usually approach the pour point by a steady increase in viscosity. Paraffinic oils, unless heavily dewaxed, tend to separate out a rigid network of wax crystals, which may prevent the flow while still retaining unfrozen liquid in the interstices. Pour points can be lowered by adding chemicals known as pour point depressants. These chemicals are believed to modify the wax structure, possibly by depositing a film on the surface of each wax crystal, so that the crystals no longer adhere to form a matrix, and do not interfere with the lubricant's ability to flow. Pour point depressants are not suitable for use with halogenated refrigerants.

Standard pour test values are significant in the selection of oils for ammonia and carbon dioxide systems, and any other system in which refrigerant and lubricant are almost totally immiscible. In such a system, any lubricant that gets into the low side is essentially refrigerant free; therefore, the pour point of the lubricant itself determines whether loss of fluidity, congealment, or wax deposition will occur at low-side temperatures.

Since oil in the low side of halogenated refrigerant systems contains significant amounts of dissolved refrigerant, the pour point test, which is conducted on pure oils and in air, is of little significance. The viscosity of the lubricant-refrigerant solutions at the low-side conditions and the wax separation (or floc test) are important considerations.

Volatility—Flash and Fire Points

Since the boiling ranges and vapor pressure data on lubricating oils are not readily available, an indication of the volatility of a lubricant is obtained from the flash and fire points (ASTM D 92). These properties are normally not significant in refrigerant systems. However, some refrigerants, such as sulfur dioxide, ammonia, and methyl chloride, have a high ratio of specific heats (c_p/c_v) and consequently have a high adiabatic compression temperature. These refrigerants frequently produce carbonization of oils with low flash and fire points when operating in high ambient temperatures. Carbonization of the lubricant can also occur in some applications that use halogenated refrigerants and require high compression ratios (such as domestic refrigerator-freezers operating in high ambient temperatures). Because such carbonization or coking of the valves is not necessarily accompanied by the general lubricant deterioration, the tendency of a lubricant to carbonize is referred to as thermal instability as opposed to chemical instability. Some manufacturers circumvent such problems by using paraffinic oils, which in comparison to naphthenic oils, have higher flash and fire points. Others prevent them through appropriate system design.

Vapor Pressure

Vapor pressure is the pressure at which the vapor phase of a substance is in equilibrium with the liquid phase at a specified temperature. The composition of the vapor and liquid phases (when not pure) influences the equilibrium pressure. With refrigeration oils, the type, boiling range, and viscosity are also factors influencing vapor pressure; naphthenic oils of a specific viscosity grade will generally show higher vapor pressures than paraffinic oils.

The vapor pressure of a lubricant increases with increasing temperature, as shown in Table 4. In practice, the vapor pressure of a refrigeration lubricant at an elevated temperature is negligible compared with the refrigerant at that temperature. The vapor pressure of narrow boiling petroleum fractions can be plotted as straight line functions. If the lubricant's boiling range and its type are known, standard tables may be used to determine the lubricant's vapor pressure up to 760 mm Hg at any given temperature (API 1970).

Table 4 Increase in Vapor Pressure and Temperature

	Vapor Pressure 32 cSt (150 SSU) Oil	
Temperature, °F	Alkylbenzene, mm Hg	Naphthene Base, mm Hg
300	0.72	0.93
325	1.58	1.92
350	3.36	3.78
375	6.70	7.15
400	13.0	13.1
425	24.3	23.0
450	43.8	39.4

Aniline Point

Aniline, an aromatic compound, is more soluble in oils containing a greater quantity of similar compounds. The temperature at which a lubricant and aniline are mutually soluble is the lubricant's aniline point (ASTM D 611). Therefore, the relative aromaticity of a mineral oil can be determined by its solubility in aniline. In comparing mineral oils, lower aniline points correspond to the presence of more naphthenic and/or aromatic molecules.

Aniline point can also predict a mineral oil's effect on elastomer seal materials. Generally, a highly naphthenic lubricant swells a specific elastomer material more than a paraffinic lubricant. This is caused by the greater solvency of aromatic and naphthenic compounds present in a naphthenic-type lubricant. However, aniline point gives only a general indication of lubricant-elastomer compatibility. Within a given class of elastomer material, lubricant resistance varies widely because of differences in compounding practiced by the elastomer manufacturer. Finally, in some retrofit applications, a high aniline point mineral oil may cause elastomer shrinkage and possible seal leakage.

The behavior of elastomers in synthetic lubricants, such as alkylbenzenes, polyalkylene glycols, and polyol esters will be different than for mineral oils. For example, an alkylbenzene will have an aniline point lower than that of a mineral oil of the same viscosity grade. However, the amount of swell in a chloroneoprene type "O" ring will generally be less than that found with the mineral oil. For these reasons, lubricant-elastomer compatibility needs to be tested under conditions anticipated in actual service.

Solubility of Refrigerants in Oils

All gases are soluble to some extent in mineral oils, and many of the refrigerant gases are highly soluble. The amount dissolved depends on the pressure of the gas, and the temperature of the lubricant, on the nature of the gas, and on the nature of the lubricant. Since refrigerants are much less viscous than oils, any appreciable amount in solution causes a marked reduction in viscosity.

Two refrigerants usually regarded as poorly soluble in mineral oil are ammonia and carbon dioxide. Data showing the slight absorption of these gases by mineral oil are given in Table 5. The amount absorbed increases with increasing pressure and decreases with increasing temperature. In ammonia systems, where pressures are moderate, the 1% or less refrigerant that dissolves in the lubricant should have little, if any, effect on lubricant viscosity. However, operating pressures in CO_2 systems tend to be much higher (not shown in Table 5), and in that case, the quantity of gas dissolving in the lubricant may be enough to substantially reduce viscosity. At 390 psig, for example, Beerbower and Greene (1961) observed a 69% reduction when a 32 cSt (150 SSU) lubricant (HVI) was tested under CO_2 pressure at 80°F.

Table 5 Absorption of Low Solubility Refrigerant Gases in Oil

	Ammonia[a] (Percent by Mass)				
	Temperature, °F				
Absolute Pressure, psi	32	68	149	212	302
14.2	0.246	0.180	0.105	0.072	0.054
28.4	0.500	0.360	0.198	0.144	0.108
42.7	0.800	0.540	0.304	0.228	0.166
57.0	—	0.720	0.398	0.300	0.222
142.0	—	—	1.050	0.720	0.545

	Carbon Dioxide[b] (Percent by Mass)			
	Temperature, °F			
Absolute Pressure, psi	32	68	149	212
14.7	0.26	0.19	0.13	0.10

[a]Type of oil: Not given (Steinle 1950)
[b]Type of oil: HVI oil, 34.8 cSt (163 SSU) at 100°F (Baldwin and Daniel 1953)

LUBRICANT-REFRIGERANT SOLUTIONS

Many of the halogenated refrigerants used today are highly soluble in mineral oils at any temperature likely to be encountered. R-11 and R-12 are examples of such refrigerants. The only limit to the amount of these refrigerants that the lubricant can dissolve is established by the refrigerant pressure at a given temperature. Other halogenated refrigerants such as R-22 and R-114 may show limited solubilities with lubricant at evaporator temperatures (exhibited in the form of phase separation) and unlimited solubilities in the higher temperature regions of a refrigerant system. In some systems using the newer refrigerants, such as R-134a, a second, distinct two-phase region may occur at high temperatures. For such refrigerants, solubility studies must, therefore, be carried out over an extended temperature range.

As a result of the high solubilities of halogenated refrigerants, the lubricating fluid can no longer be treated as a pure lubricant, but rather as a lubricant-refrigerant solution whose properties are markedly different from those of pure lubricant. The amount of refrigerant dissolved in a lubricant depends on the pressure and temperature. Therefore, the composition of the lubricating fluid is different in different sections of a refrigeration system operating at steady state and changes from the time of start-up until the system attains the steady state. The most pronounced effect is on viscosity.

The crankcase of a compressor can be used as an example. If the system at start-up was at 77 °F, the viscosity of a pure 32 cSt (150 SSU) naphthenic lubricant would be about 66.9 cSt (310 SSU). Under operating conditions, it is not unusual for the lubricant temperature to be 130 °F and the viscosity of the pure lubricant to be about 16.9 cSt (85 SSU). If the system is operating with R-22 as the refrigerant, and the pressure in the crankcase is 94.7 psia, according to Little (1952), the viscosity of the lubricating fluid at start-up would be 15.7 cSt (80 SSU) rather than 66.9 cSt, and this would decrease to 9.7 cSt (58 SSU) at 130 °F.

If only lubricant properties are considered, an erroneous picture of the system is obtained. When the lubricant circulates through the system and returns from the evaporator to the compressor, a similar situation exists. The highest viscosity does not occur at the lowest temperature, because the lubricant contains a large amount of dissolved refrigerant. As the temperature increases, the lubricant loses some of the refrigerant and the viscosity reaches a maximum at a point away from the coldest spot in the system.

Similar to the lubricating fluid, the properties of the working fluid (the refrigerant) are also affected. The vapor pressure of a lubricant-refrigerant solution, for example, is markedly lower than that of the pure refrigerant. One result of this property is that the evaporator temperature is higher than if the refrigerant is pure. Another result is the so-called flooded start-up. When the crankcase and the evaporator are at about the same temperature, the fluid in the evaporator, which is mostly refrigerant, has a higher vapor pressure than the fluid in the crankcase, which is mostly lubricant. This difference in vapor pressures acts as the force driving the refrigerant to the crankcase to be absorbed in the lubricant until the pressures are equalized. At times, the moving parts in the crankcase may be completely immersed in this lubricant-refrigerant solution. At start-up, the change in pressure and the turbulence can cause excessive amounts of liquid to enter the cylinders, causing damage to the valves and creating lubricant-starvation in the crankcase. The use of crankcase heaters to prevent such problems caused by highly soluble refrigerants is discussed in Chapter 2 and by Neubauer (1958).

The occurrence of the solutions of oils and refrigerants results in a somewhat different set of rules for the lubricating fluid and for the working fluid, requiring a detailed discussion. Much of the remaining chapter deals with the properties of lubricant-refrigerant solutions.

Density

For a rough estimate of the density or specific gravity of a lubricant-refrigerant solution, assume that the solution is ideal so

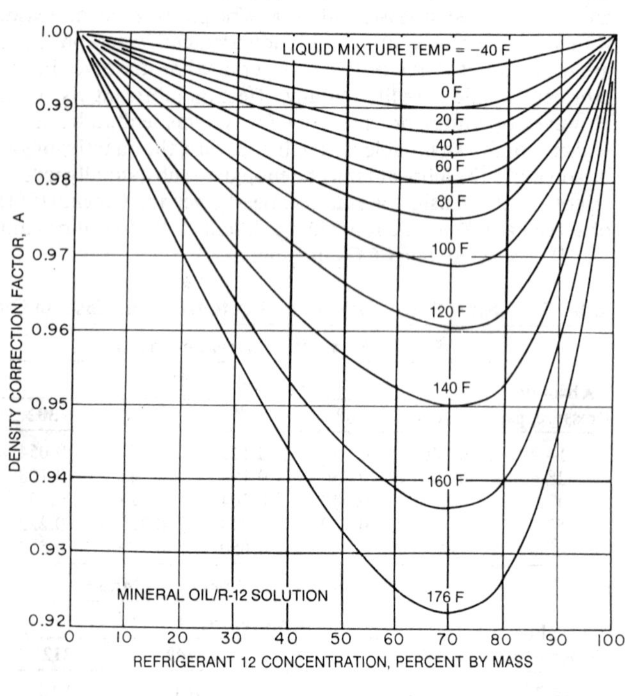

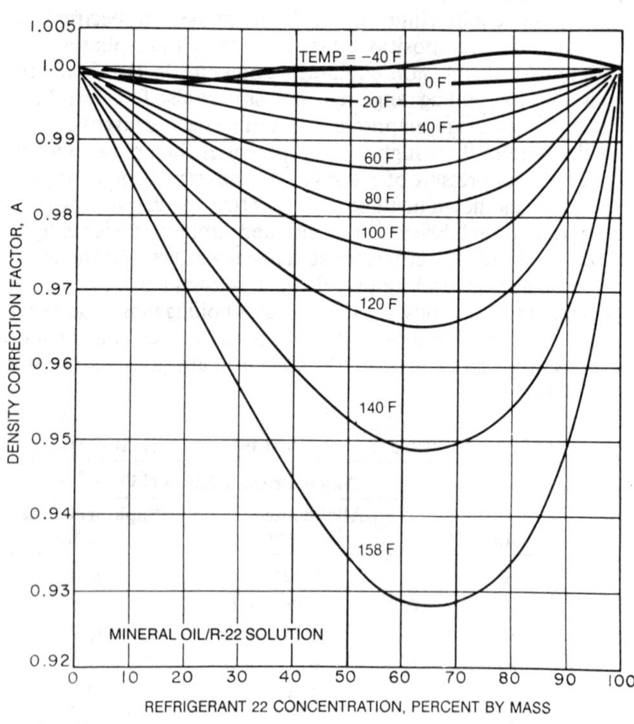

Fig. 3 Density Correction Factors
(Loffler 1959)

that the specific volumes of the components are additive. The formula for calculating the ideal density (ρ_{id}) is:

$$\rho_{id} = \rho_o/[1 + W(\rho_o/\rho_R - 1)]$$

where

ρ_o = density of pure lubricant at the solution temperature
ρ_R = density of refrigerant liquid at the solution temperature
W = mass fraction of refrigerant in solution

Depending on the refrigerant, the actual density of a lubricant-refrigerant solution may deviate from the ideal by as much as 8%. The solutions are usually more dense than calculated, but sometimes are less. For example, R-11 forms ideal solutions with oils, whereas R-12 and R-22 show significant deviations. Density correction factors for R-12 and R-22 solutions are depicted in Figure 3. The corrected densities can be obtained from the relation:

$$\text{Mixture Density} = \rho_m = \rho_{id}/A$$

where A is the density correction factor read from Figure 3 at the desired temperature and refrigerant concentration.

Similar figures are not presently available for other refrigerant/lubricant pairs. However, Van Gaalen *et al.* (1990, 1991ab) provide values of density for four refrigerant/lubricant pairs, namely, R-22/mineral oil, R-22/alkylbenzene, R-502/mineral oil, and R-502/alkylbenzene. Figures 4, 5, 6, and 7 provide data on the variation of density with temperature over a range of −25 to 250°F for R-134a in combination with a 32 ISO VG polyol ester, a 100 ISO VG polyol ester, a 32 ISO VG polyalkylene glycol, and an 80 ISO VG polyalkylene glycol, respectively (Cavestri 1993).

Thermodynamics and Transport Phenomena

Dissolving lubricant in liquid refrigerant affects the thermodynamic properties of the working fluid. The vapor pressures of refrigerant-lubricant solutions at a given temperature are always less than the vapor pressure of pure refrigerant at that temperature. Therefore, dissolved lubricant in an evaporator leads to lower

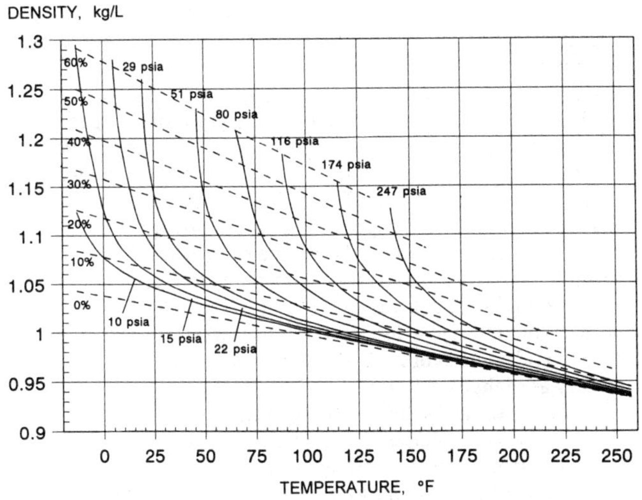

Fig. 4 Density as Function of Temperature and Pressure for Mixture of R-134a and 32 ISO VG Branched Acid Polyol Ester Lubricant

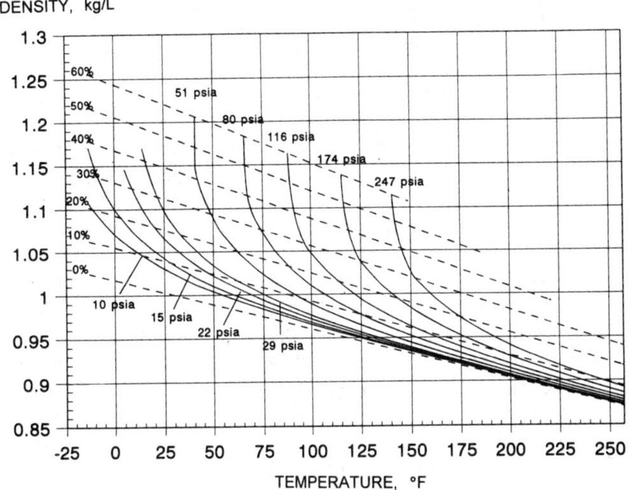

Fig. 5 Density as Function of Temperature and Pressure for Mixture of R-134a and 100 ISO VG Branched Acid Polyol Ester Lubricant

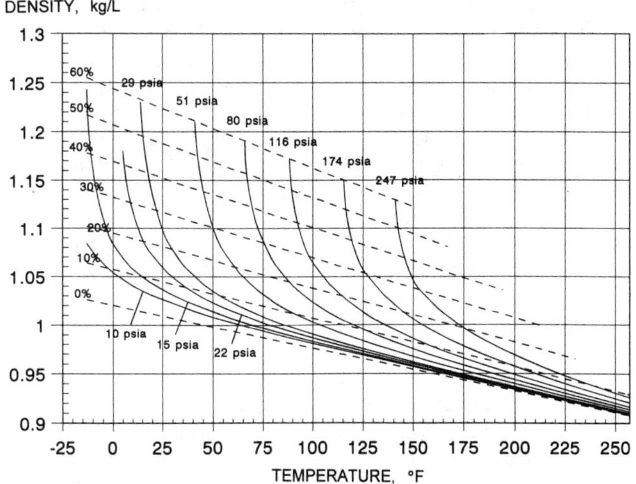

Fig. 6 Density as Function of Temperature and Pressure for Mixture of R-134a and 32 ISO VG Polypropylene Glycol Butyl Ether Lubricant

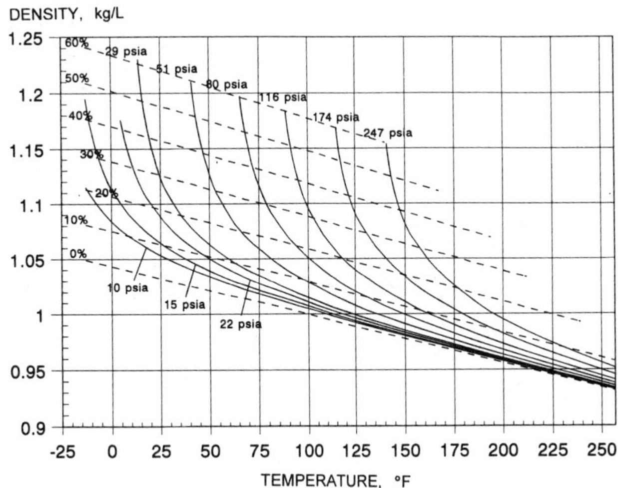

Fig. 7 Density as Function of Temperature and Pressure for Mixture of R-134a and 80 ISO VG Polyoxypropylene Diol Lubricant

suction pressures and higher evaporator temperatures than expected from pure refrigerant tables. Bambach (1955) gives an enthalpy diagram for R-12/lubricant solutions over the range of compositions from 0 to 100% lubricant and temperatures from −40 to 240 °F. Spauschus (1963) has developed general equations for calculating thermodynamic functions of refrigerant-lubricant solutions and has applied them to the special case of R-12/mineral oil solutions.

Pressure-Temperature-Solubility Relations

When a refrigerant is in equilibrium with a lubricant, a fixed and definite amount of the refrigerant is present in the lubricant at a given temperature and pressure. This is evident if the phase rule is applied to the system, which is basically a two-phase, two-component system. The lubricant, although a mixture of several compounds, may be considered one component, and the refrigerant the other. The two phases are the liquid phase and the vapor phase. The phase rule defines this system as having two degrees of freedom. Normally, the variables involved are pressure, temperature, and the compositions of the liquid and of the vapor phase. Since the vapor pressure of the lubricant is negligible compared with that of the refrigerant, the vapor phase is essentially pure refrigerant, and only the composition of the liquid phase needs to be considered. If the pressure and temperature are defined, the system is invariant, *i.e.*, the liquid phase can have only one composition. This is a different but more precise way of stating that a lubricant-refrigerant mixture having a known composition exerts a certain vapor pressure at a certain temperature. If the temperature is changed, the vapor pressure will also change.

Pressure-temperature-solubility relations are usually presented in the form shown in Figure 8. On this graph, P_1^0 and P_2^0 represent the saturation pressures of the pure refrigerant at temperatures t_1 and t_2, respectively. Point E_1 represents an equilibrium condition, where one and only one composition of the liquid, represented by W_1, is possible at the pressure P_1. If the temperature of this system is increased to t_2, some of the liquid refrigerant evaporates, and the equilibrium point shifts to E_2, corresponding to a new pressure P_2. In either case, the lubricant-refrigerant solution exerts a vapor pressure less than that of the pure refrigerant at the same temperature.

Mutual Solubility

In dealing with the lubricating and sealing problems in a compressor, the lubricating fluid is a solution of refrigerant dissolved

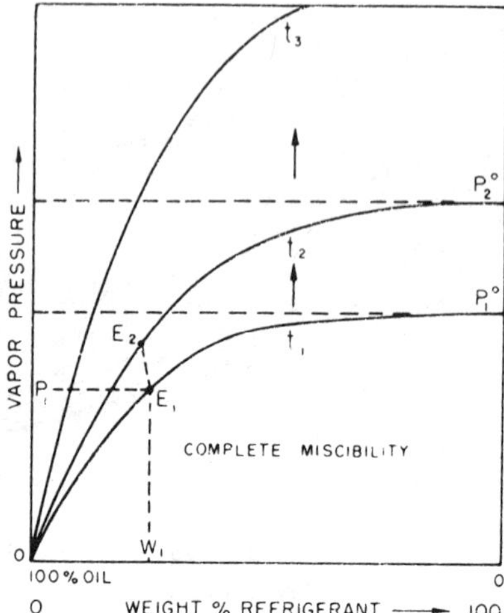

Fig. 8 P-T-S Diagram for Completely Miscible Refrigerant-Lubricant Solutions

Table 6 Mutual Solubility of Refrigerants and Mineral Oil

Completely Miscible	Partially Miscible: High Miscibility	Partially Miscible: Intermediate Miscibility	Partially Miscible: Low Miscibility	Immiscible
R-11	R-13B1	R-22	R-13	Ammonia
R-12	R-501	R-114	R-14	CO_2
R-21	R-123		R-115	R-134a
R-113			R-152a	
R-500			R-C318	
			R-502	

in lubricant. In other parts of the refrigerant system, the problem may involve a solution of lubricant in liquid refrigerant. In both instances, either the lubricant or the refrigerant could exist alone as a liquid if the other were not present; therefore, any distinction between the dissolving and dissolved component merely reflects a point of view. Either of the liquids can be considered as dissolving the other. This relationship is termed mutual solubility.

Refrigerants are classified as completely miscible, partially miscible, or immiscible, according to their mutual solubility relations with mineral oils. Since several commercially important refrigerants are partially miscible, further designation as having high, intermediate, or low miscibility is shown in Table 6.

Completely miscible refrigerants and oils are mutually soluble in all proportions at any temperature. This type of mixture always forms a single liquid phase under equilibrium conditions, no matter how much refrigerant or lubricant is present. R-11 and R-12 with mineral oil are examples.

Partially miscible refrigerant-lubricant systems are mutually soluble to a limited extent. Above the critical solution temperature (CST) or consolute temperature, many refrigerant/lubricant mixtures in this class are completely miscible, and their behavior is identical to that just described. As has already been mentioned, R-134a and some synthetic lubricants exhibit a region of immiscibility at higher temperatures.

Below the critical solution temperature, however, the liquid may separate into two phases. Such phase separation does not mean that the lubricant and the refrigerant are insoluble in each other. Each liquid phase is a solution; one is lubricant-rich and the other refrigerant-rich, depending on the predominant component. Each phase may contain substantial amounts of the leaner component in the mutual solution, and these two solutions are themselves immiscible with each other.

The importance of this concept is best illustrated by R-502, which is considered a low miscibility refrigerant exhibiting a high CST as well as a broad immiscibility range. However, even at 0 °F, the lubricant-rich phase contains about 20 mass % of dissolved refrigerant (see Figure 11). Examples of partially miscible systems, in addition to R-502, are R-22, R-114, and R-13, with mineral oils.

The basic properties of the immiscible region can be recognized by applying the phase rule to the system. With three phases (two liquid and one vapor) and two components, there can be only one degree of freedom. Therefore, either the temperature or the pressure automatically determines the composition of both liquid phases. If the system pressure is changed, the temperature of the system changes and the two liquid phases assume somewhat different compositions determined by the new equilibrium conditions.

Figure 9 illustrates the behavior of partially miscible mixtures. Point C on the graph represents the critical solution temperature t_3. There are three separate regions below this temperature on the diagram. Reading from left to right, a family of the smooth solid curves represents a region of completely miscible lubricant-rich solutions. These are followed by a wide break representing a region of partial miscibility in which there are two immiscible liquid phases. On the right side, the partially miscible region disappears

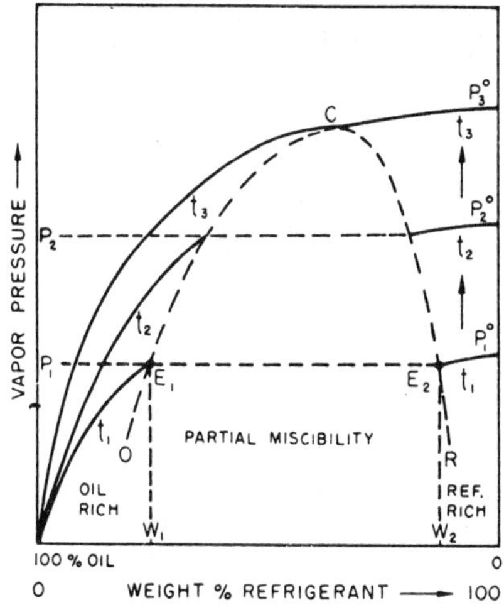

Fig. 9 P-T-S Diagram for Partially Miscible Refrigerant-Oil Solutions

into a second completely miscible region of refrigerant-rich solutions. A dome-shaped envelope (broken line curve OCR) encloses the partially miscible region; everywhere outside this dome the refrigerant and lubricant are completely miscible. In a sense, Figure 9 is a variant of Figure 8 in which the partial miscibility dome (OCR) blots out a substantial portion of the continuous solubility curves. Under the dome, *i.e.*, in the immiscible region, if one were to follow the line at temperature t_1, the two points E_1 and E_2 represent the two phases coexisting in equilibrium.

These two phases differ considerably in composition (W_1 and W_2) but have the same refrigerant pressure P_1. The solution pressure P_1 lies not far below the saturation pressure of pure refrigerant P_1^0. It is not uncommon for refrigerant-lubricant solutions near the partial miscibility limit to show less reduction in refrigerant pressure than observed at the same lubricant concentration with completely miscible refrigerants.

Totally immiscible lubricant-refrigerant systems are defined in this chapter as only very slightly miscible. In such mixtures, the immiscible range is so broad that mutual solubility effects can be ignored. Critical solution temperatures are seldom found in mixtures of the totally immiscible type. Examples are ammonia and lubricant, and carbon dioxide and mineral oil.

Effects of Partial Miscibility in Refrigerant Systems

Evaporator. The evaporator is the coldest part of the system and the most likely part in which immiscibility or phase separation will occur. If the evaporator temperature is below the critical solution temperature, phase separation is likely to occur in some part of the evaporator. The fluid entering the evaporator is mostly liquid refrigerant containing a small fraction of lubricant, whereas the liquid leaving the evaporator is mostly lubricant, since the refrigerant is in vapor form. No matter how little lubricant the entering refrigerant carries, the liquid phase, as it progresses through the evaporator, passes through the critical composition which usually lies in 15 to 20% lubricant in the total liquid phase.

Phase separation in the evaporator can sometimes cause problems. In a dry-type evaporator, there is usually enough turbulence to cause the phases to emulsify. In this case, the heat transfer characteristics of the evaporator may not be significantly affected. In flooded-type evaporators, however, the working fluid may separate into layers, and the lubricant-rich phase may float on top of the boiling liquid. In addition to the heat transfer, partial miscibility may also affect the lubricant's return from the evaporator to the crankcase. Usually the lubricant is moved by high velocity suction gas transferring momentum to the droplets of lubricant on the return line walls.

If a lubricant-rich layer separates at the evaporator temperatures, this viscous, nonvolatile liquid can migrate and collect in pockets or blind passages not easily reached by the high velocity suction gas. The lubricant return problem may be magnified and, in some cases, an oil-logging can occur. The design of the system should take into account all these possibilities, and evaporators should be designed to promote entrainment (see Chapter 2). Oil separators are frequently required in the discharge line to minimize lubricant circulation when refrigerants of poor solvent power are used or in systems involving very low evaporator temperatures (Soling 1971).

Crankcase. With certain refrigerant systems, such as R-502 and mineral oil, or even with R-22 in applications such as heat pumps, phase separation sometimes occurs in the crankcase when the system is shut down. When this happens, the refrigerant-rich layer settles to the bottom, often completely immersing the pistons, bearings, and other moving parts. At start-up, the fluid that lubricates these moving parts is mostly refrigerant with little lubricity, and severe bearing damage may result. The turbulence at start-up may cause the liquid refrigerant to enter the cylinders, carrying large amounts of lubricant with it. Precautions in design will prevent such problems in partially miscible systems.

Condenser. Partial miscibility is not a problem in the condenser, since the liquid flow lies in the turbulent region and the temperatures are relatively high. Even if phase separation occurs, there is little danger of separation of layers, the main obstacle to efficient heat transfer.

Solubility Curves and Miscibility Diagrams

Figure 10 shows mutual solubility relations of partially miscible refrigerant-lubricant mixtures. More than one curve of this type can be plotted on a miscibility diagram. Each single dome then represents the immiscible ranges for one lubricant and one refrigerant. Miscibility curves for R-13, R-13B1, R-502 (Parmelee 1964), R-22, and mixtures of R-12 and R-22 (Walker *et al.* 1957) are shown in Figure 11. Miscibility curves for R-13, R-22, R-502, and R-503 in an alkylbenzene refrigeration lubricant are shown

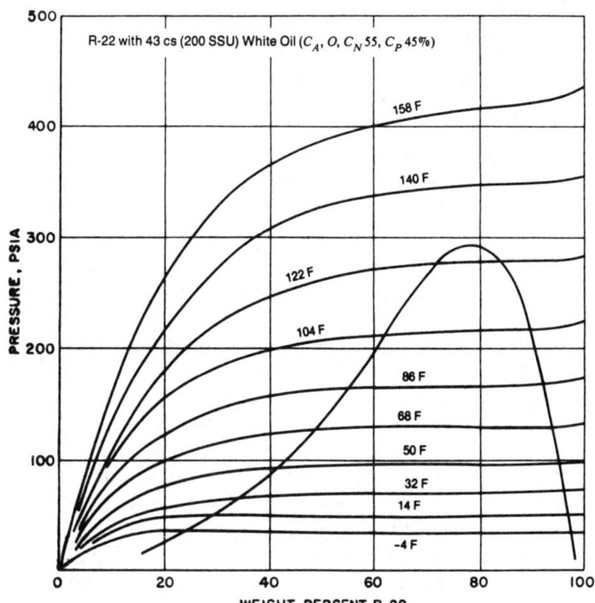

Fig. 10 P-T-S Relations of R-22 with White Oil
(Spauschus 1964)

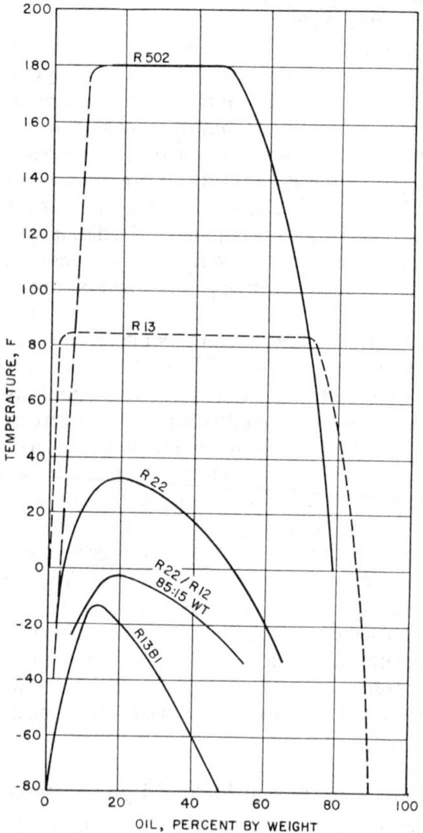

Fig. 11 Critical Solubilities of Refrigerants with 32 cSt (150 SSU) Naphthenic Lubricant (C_A 12, C_N N 44, C_P 44%)

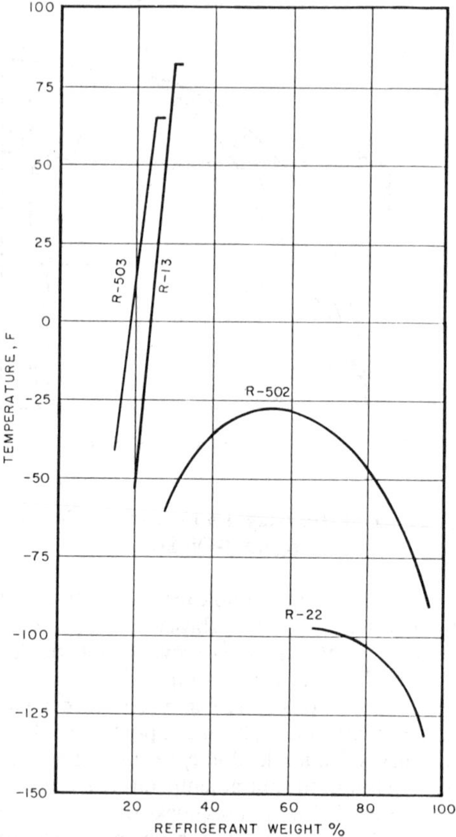

Fig. 12 Critical Solubilities of Refrigerants with 32 cSt (150 SSU) Alkylbenzene Lubricant

in Figure 12. A comparison with Figure 11 illustrates the greater solubility of refrigerants in this type of lubricant.

Effect of Lubricant Type on Solubility and Miscibility

When compared on a mass basis, low viscosity oils absorb more refrigerant than high viscosity oils. Also, naphthenic oils absorb more than paraffinic oils. However, when compared on a mol basis, some confusion arises. Paraffinic oils absorb more refrigerant than naphthenic oils, *i.e.*, reversal of the mass basis, and there is little difference between a 15.7 cSt (80 SSU) and a 64.7 cSt (300 SSU) naphthenic lubricant (Albright and Lawyer 1959, Albright and Mandelbaum 1956). The differences on either basis are small, *i.e.*, within 20% of each other. Comparisons of oils by their carbon-type analyses are not available, but in view of the data on naphthenic and paraffinic types, differences between oils with different carbon-type analyses, except perhaps for extreme compositions, are unlikely.

The effect of lubricant type and composition on miscibility is better defined than solubility. When the critical solution temperature (CST) is used as the criterion of miscibility, oils with higher aromatic contents show a lower CST. Higher viscosity grade oils show a higher CST than lower viscosity grade oils, and paraffinic oils show a higher CST than naphthenic oils (see Figures 13 and 25). When the entire dome of immiscibility is considered, a similar result is noticeable. Oils that exhibit a lower CST, usually show a narrowing of the immiscibility range, *i.e.*, the mutual solubility is greater at any given temperature.

Miscibility of R-22 With Lubricants

Parmelee (1964) showed that polybutyl silicate improves miscibility with R-22 (and also R-13) at low temperatures. Alkylbenzenes, by themselves or mixed with mineral oils, also have better miscibility with R-22 than do mineral oils alone (Seemann and Shellard 1963).

Among the mineral oils, several reports deal with miscibility differences resulting from different types and viscosity grades. Walker *et al.* (1962) provide detailed miscibility diagrams of 12 brand name oils commonly used for refrigeration systems. Walker's data show that in every case, higher viscosity lubricant of the same base and type has a higher critical solution temperature.

Loffler (1957) provides complete miscibility diagrams of R-22 and 18 oils. A portion of the properties of the oils used and the critical solution temperatures are summarized in Table 7. Although precise correlations are not evident in the table, certain trends are clear. For the same viscosity grade and base, the effect of aromatic carbon content is seen in oils 2, 3, 7, and 8 and between oils 4 and 6. Similarly, for the same viscosity grades, the effect of paraffinic structure (with essentially the same % C_A) is noticeable between oils 6 and 17 and between oils 8 and 18.

According to Loffler, the most pronounced effect on the critical solution temperature is exerted by the aromatic content of the lubricant; the table indicates that the paraffinic structure reduces the miscibility compared with naphthenic structures. Sanvordenker (1968) reported the miscibility relations of the saturated fractions and the aromatic fractions of mineral oils as a function of their physical properties. The critical solution temperatures

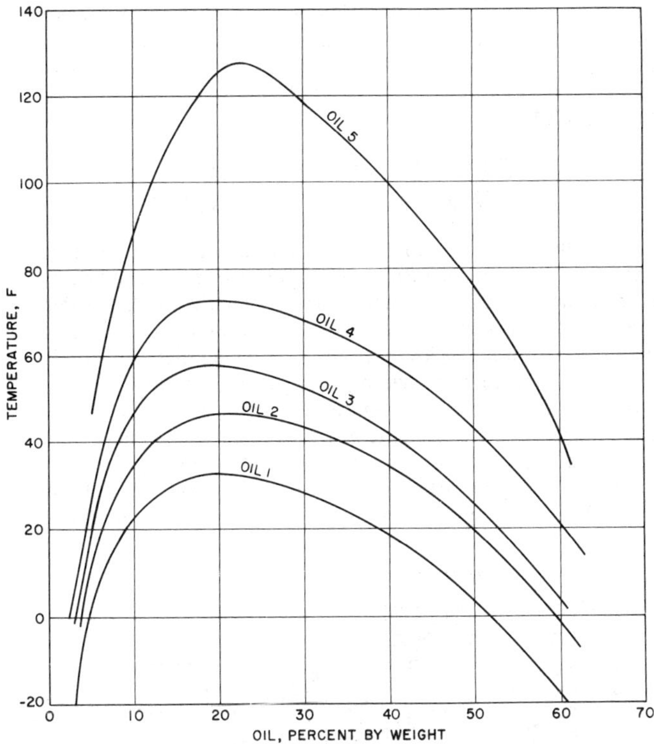

Oil No.	Viscosity cSt	Viscosity SSU (100 °F)	Compositions, % C_A	C_N	C_P	Ref.
1	34.0	159	12	44	44[a]	1
2	33.5	157	7	46	47[a]	1
3	63.0	292	12	44	44[a]	1
4	67.7	314	7	46	47[a]	1
5	41.3	192	0	55	45	2

References: 1. Walker *et al.* (1957) 2. Spauschus (1964)
[a]Estimated composition, not in original reference.

Fig. 13 Effect of Oil Properties on Miscibility with R-22

with R-22 increase with increasing viscosities for the saturates, as well as for the aromatics. For equivalent viscosities, aromatic fractions with naphthenic linkages show lower critical solution temperatures than aromatics with only paraffinic linkages.

Generally, adding R-12 to R-22 systems improves the miscibility by reducing the critical solution temperature and narrowing the immiscible range. The former effect appears as a drop of about 40 or 50°F when the mixture contains 85% R-22 and 15% R-12 (Walker *et al.* 1957, Loffler 1960). The narrowing of the immiscible range also improves the fluidity of the lubricant-rich phase, so that the minimum temperature at which the lubricant-rich phase can still flow drops by about 25°F (Loffler 1960).

Solubilities and Viscosities of Lubricant-Refrigerant Systems

Although the differences are small on a mass basis, naphthenic oils are better solvents than paraffinic oils. When considering the viscosity of lubricant-refrigerant mixtures, naphthenic oils show greater viscosity reduction than paraffinic oils for the same mass percent of dissolved refrigerant. When the two effects are compounded, under the same conditions of temperature and pressure, a naphthenic lubricant in equilibrium with a given refrigerant shows a significantly lower viscosity than a paraffinic lubricant.

Refrigerants also differ in their viscosity-reducing effects when the solution concentration is measured in mass percent. However, when the solubility is plotted in terms of mol percent, the reduction in viscosity is approximately the same, at least for Refrigerants 13, 13B1, 22, and 115 (Figure 14).

Spauschus (1964) reports numerical vapor pressure data on a R-22/white oil system; solubility-viscosity graphs on naphthenic and paraffinic oils have been published by Loffler (1960), Little (1952), and Albright and Mandelbaum (1956). Some discrepancies, particularly at high R-22 contents, have been shown in data on viscosities, which apparently could not be attributed to the properties of the lubricant and remain unexplained. However, generalized plots reported by the aforementioned authors are satisfactory for engineering and design purposes.

Spauschus and Speaker (1987) have compiled references of solubility and viscosity data. Selected solubility-viscosity data are summarized in Figures 10 and 15 through 27.

Table 7 Critical Miscibility Values of R-22 with Different Oils

Oil No.	Oil Base Type[a]	Approximate Viscosity Grade SSU	cSt	Viscosity at 122 °F Converted to SSU	to cSt	Carbon-Type Composition $\%C_A$	$\%C_N$	$\%C_P$	Critical Solution Temperature, °F
2	N	75	15	63	11.2	23	34	43	−35
3	N	75	15	60	10.2	2.5	48.5	49	21
1	N	150	32	92	18.5	13	43	44	27
8	N	200	46	118	24.6	0.6	45	55	75
7	N	200	46	127	26.7	2.8	44	54	68
5	N	200+	46+	132	27.9	22	30	47	3
4	N	250	46	135	28.6	26	28	46	−4
6	N	250	46	140	29.7	4	45	51	62
13	N	500	100	253	54.5	1.9	41	56	None[b]
12	N	500	100	282	60.7	4	41	55	None[b]
11	N	700	150	320	69.1	7	40	53	None[b]
10	N	1000	220	434	93.2	21	27	52	61
9	N	1200	220	502	109.0	27	24	50	48
18	P	200+	46+	138	29.3	0.5	33	67	None[b]
17	P	250	46	148	31.6	3.5	34	63	None[b]
16	P	300	68	164	35.2	6.4	30	63	None[b]
15	P	350	68	210	45.2	14.3	25	61	None[b]
14	P	400	100	232	50.0	18.1	22	60	111[c]

[a]P = Paraffinic, N = Naphthenic
[b]Never completely miscible at any temperature
[c]A second (inverted) miscibility dome was observed above 136°F. Above this temperature, the oil/R-22 mixture again separated into two immiscible solutions.

Wherever possible, solubilities have been converted to mass percent to provide consistency among the various charts. Figures 10 and 15 through 19 contain data on R-22 and oils, Figure 20 on R-502, Figures 21 and 22 on R-11, Figures 23 and 24 on R-12, and Figures 25 and 26 on R-114. Figure 27 contains data on the solubility of various refrigerants in alkylbenzene lubricant. Viscosity-solubility characteristics of mixtures of R-13B1 and lubricating oils have been investigated by Albright and Lawyer (1959). Similar studies on R-13 and R-115 are covered by Albright and Mandelbaum (1956).

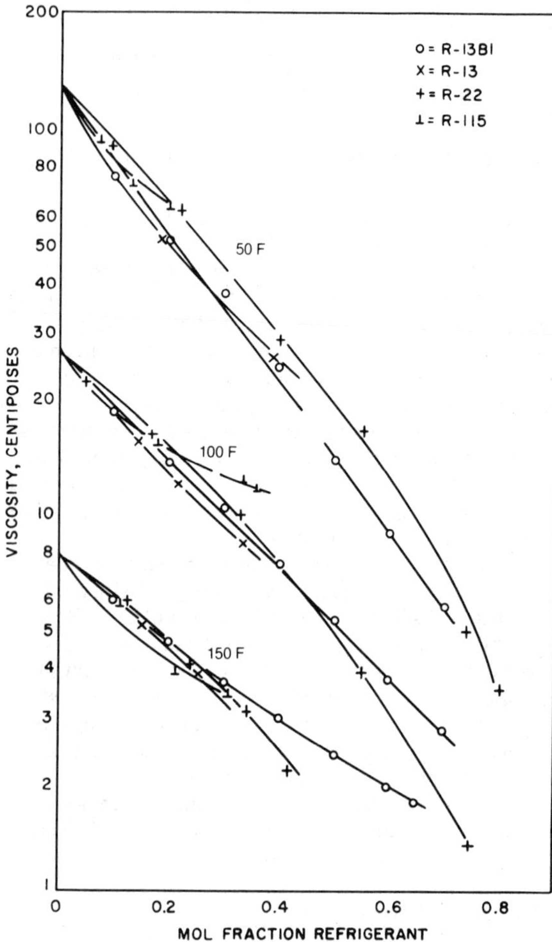

Fig. 14 Viscosity of Mixtures of Various Refrigerants and 32 cSt (150 SSU) Paraffinic Oil
(Albright and Lawyer 1959)

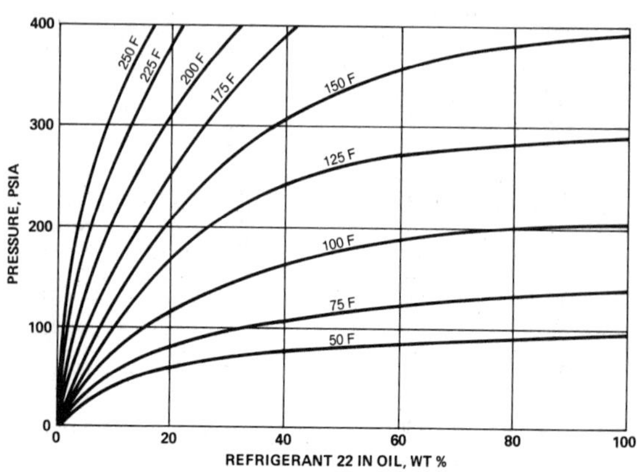

Fig. 15 Solubility of R-22 in 32 cSt (150 SSU) Naphthenic Oil
(Witco)

LUBRICANT RETURN FROM EVAPORATORS

Regardless of the miscibility relations of lubricant with refrigerants, for a refrigeration system to function properly, the lubricant must return adequately from the evaporator to the crankcase. Parmelee (1964) showed that the viscosity of the lubricant, saturated with refrigerant under low-pressure and low-temperature conditions, is important in providing good lubricant return. The viscosity of the lubricant-rich liquid that accompanies the suction gas changes as it sees rising temperatures on its way back to the compressor. Two opposing factors then come into play. First, the increasing temperature tends to decrease the viscosity of the fluid. Second, since the pressure remains unchanged, the increasing

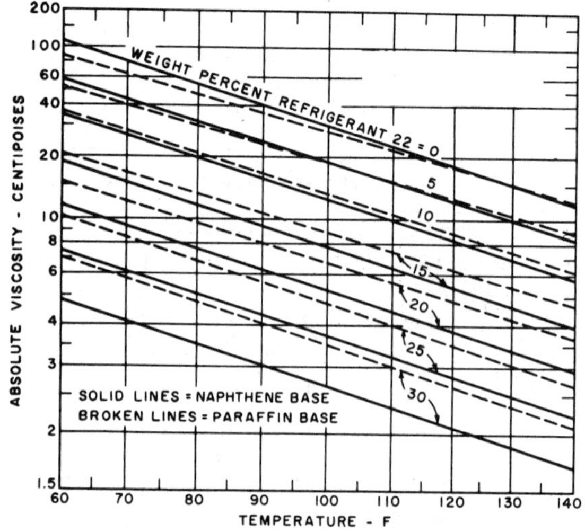

Fig. 16 Viscosity-Temperature Chart for Solutions of R-22 in 32 cSt (150 SSU) Naphthene and Paraffin Base Oils

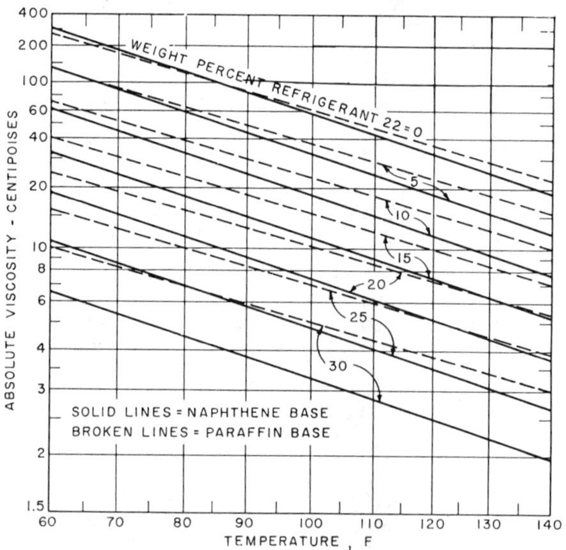

Fig. 17 Viscosity-Temperature Chart for Solutions of R-22 in 65 cSt (300 SSU) Naphthene and Paraffin Base Oils

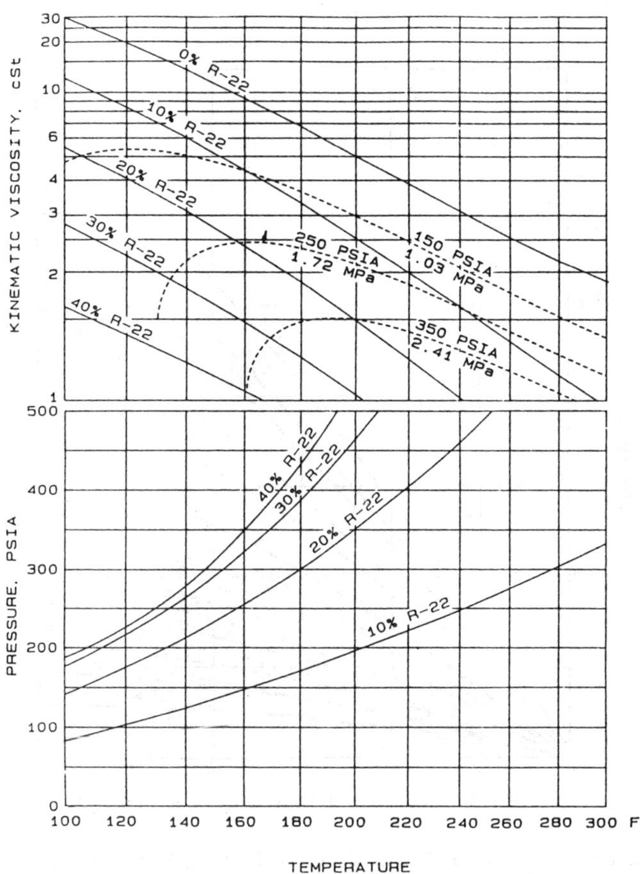

Fig. 18 Viscosity-Temperature-Pressure Chart for Solutions of R-22 in 150 SSU Naphthenic Oil
(Van Gaalen, *et al.* 1990 and 1991a)

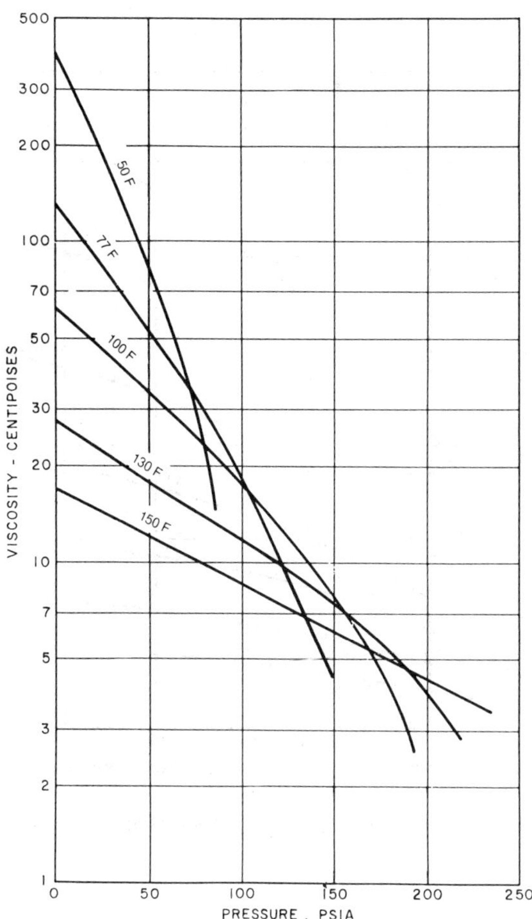

Fig. 19 Viscosity of Mixtures of 65 cSt (300 SSU) Paraffin Base Oil and R-22
(Albright and Mandelbaum 1956)

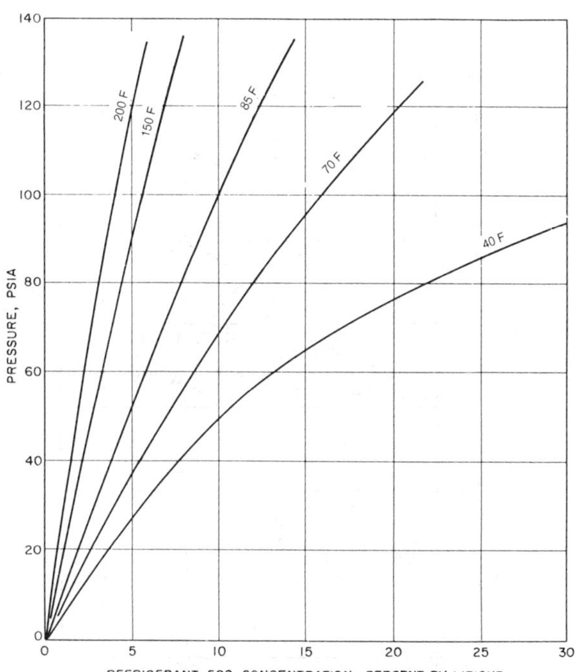

Fig. 20 Solubility of R-502 in 32 cSt (150 SSU) Naphthenic Oil (C_A 12, C_N 44, C_P 44%)

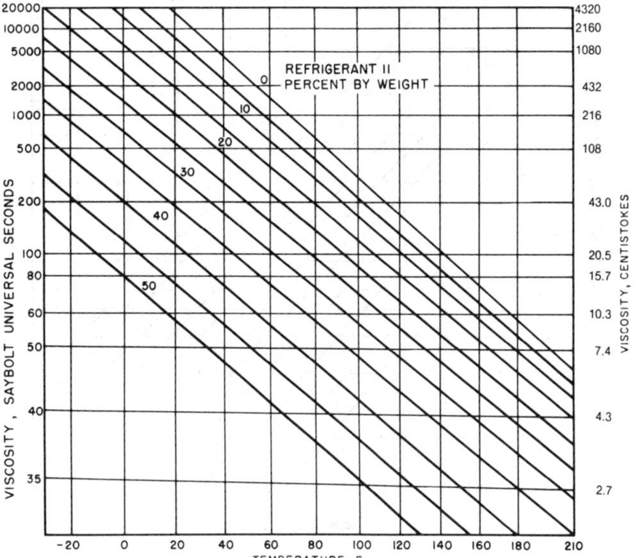

Fig. 21 Viscosity-Temperature Curves for Solutions of R-11 in 65 cSt (300 SSU) Naphthene Base Oil

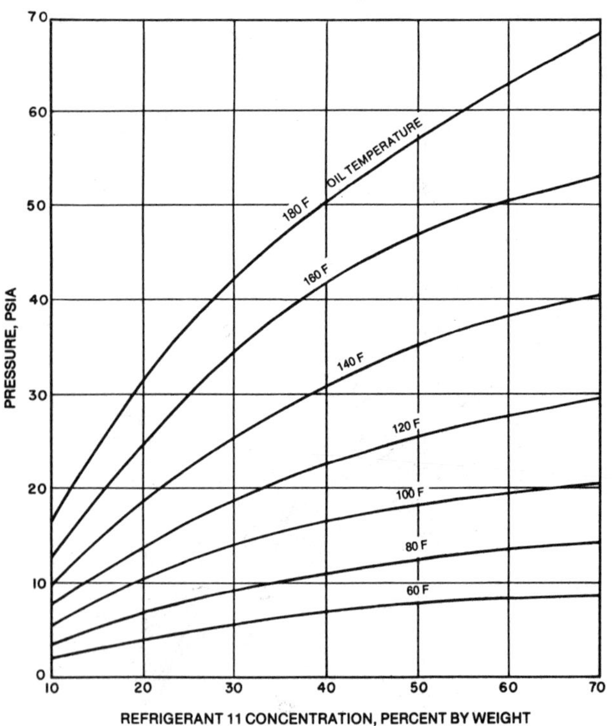

Fig. 22 Solubility of R-11 in 65 cSt (300 SSU) Oil

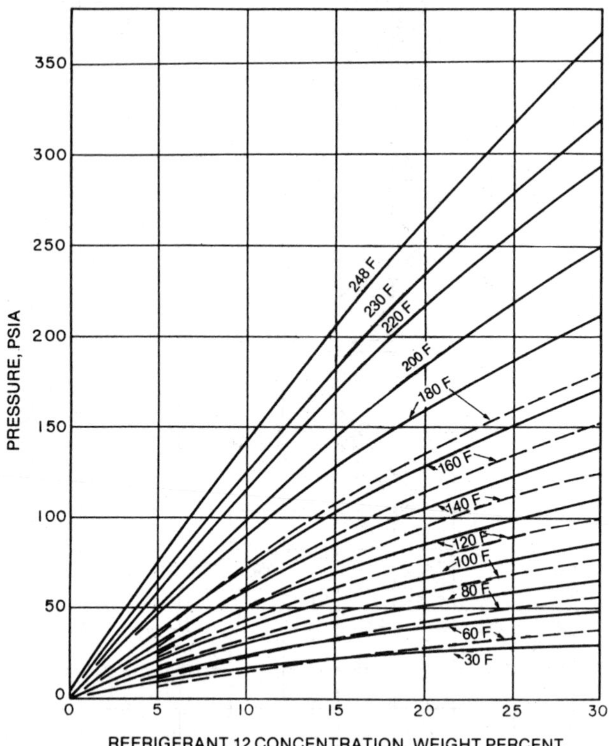

Solid lines: 32 cSt or 150 SSU naphthene base oil. Also 52 cSt or 240 SSU German base oil (Bambach 1955).
Broken lines: 32 cSt or 150 SSU and 70 cSt or 325 SSU mixed base oils.

Fig. 23 Solubility of R-12 in Refrigeration Oils

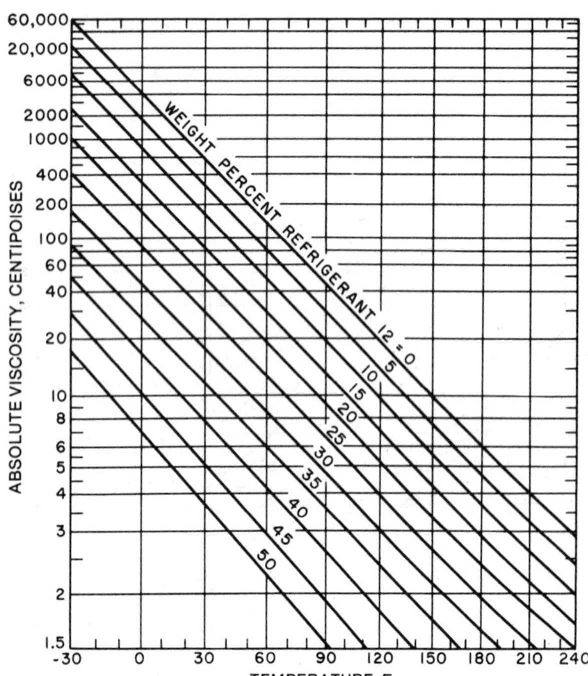

Fig. 24 Viscosity-Temperature Chart for Solutions of R-12 in 32 cSt (150 SSU) Naphthene Base Oil
(Loffler 1960)

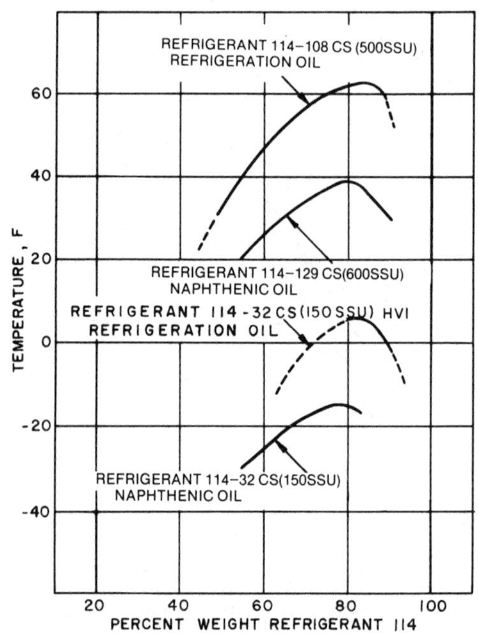

Fig. 25 Critical Solution Temperatures of R-114/Oil Mixtures

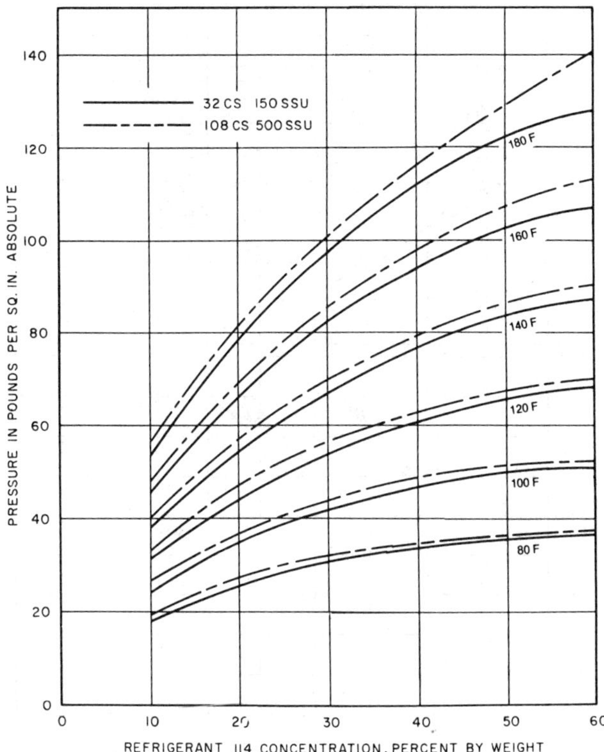

Fig. 26 Solubility of R-114 in HVI Oils

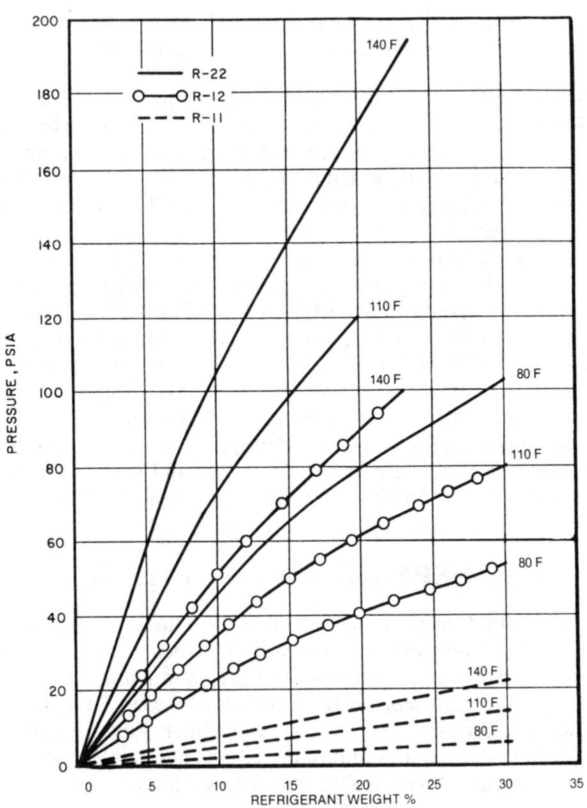

Fig. 27 Solubility of Refrigerants in 32 cSt (150 SSU) Alkylbenzene Oil

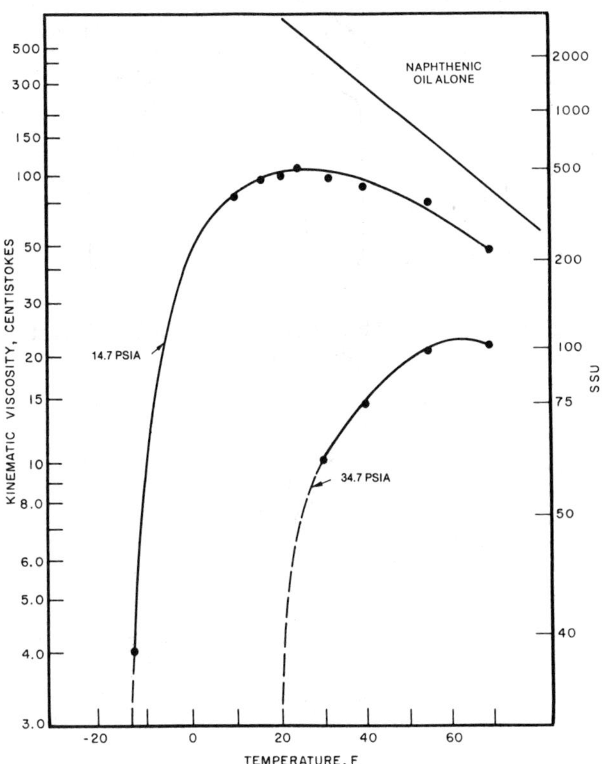

Fig. 28 Viscosity of R-12/Oil Solutions at Low-Side Conditions
(Parmelee 1964)

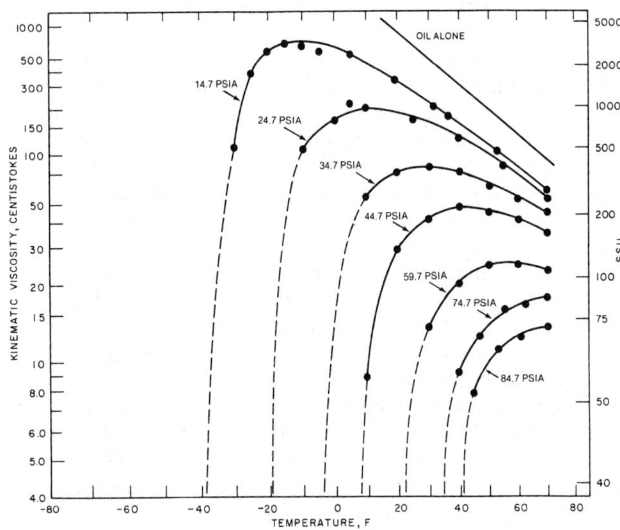

Fig. 29 Viscosity of R-22/Naphthenic Oil Solutions at Low-Side Conditions
(Parmelee 1964)

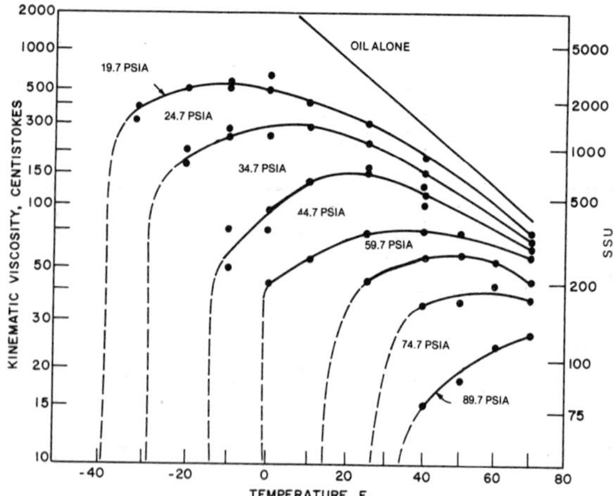

Fig. 30 Viscosity of R-502/Naphthenic Oil Solutions at Low-Side Conditions

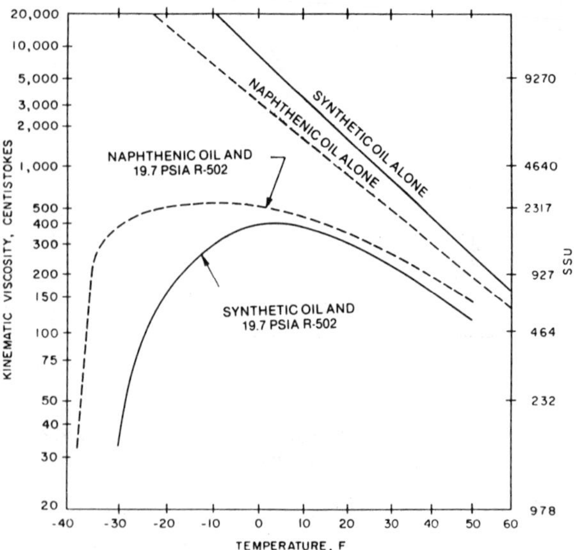

Fig. 31 Viscosities of Solutions of R-502 with 32 cSt (150 SSU) Naphthenic Oil (C_A 12, C_N 44, C_P 44%) and 32 cSt (150 SSU) Synthetic Alkylbenzene Oil

temperature also tends to drive off some of the dissolved refrigerant from the solution, thereby increasing its viscosity.

Figures 28 through 30 show the variation in viscosity with temperature and pressure for three lubricant-refrigerant solutions ranging from −40 to 70°F. In all cases, the viscosities of the solutions passed through maximum values as the temperature was changed at constant pressure, a finding that was also consistent with previous data obtained by Bambach (1955) and Loffler (1960). According to Parmelee, the existence of a viscosity maximum is significant, because the lubricant-rich solution becomes most viscous, not in the coldest regions in the evaporator, but at some intermediate point where much of the refrigerant has escaped from the lubricant. This is possibly in the suction line. The velocity of the return vapor, which might be high enough to move the lubricant-refrigerant solution in the colder part of the evaporator, might be too low to achieve the same result at the point of maximum viscosity. The designer must consider this factor to minimize any lubricant return problems. Chapters 2 and 3 have further information on velocities in return lines.

Another aspect of viscosity data at the evaporator conditions is shown in Figure 31, which compares a synthetic alkylbenzene lubricant with a naphthenic mineral oil. The two oils are the same

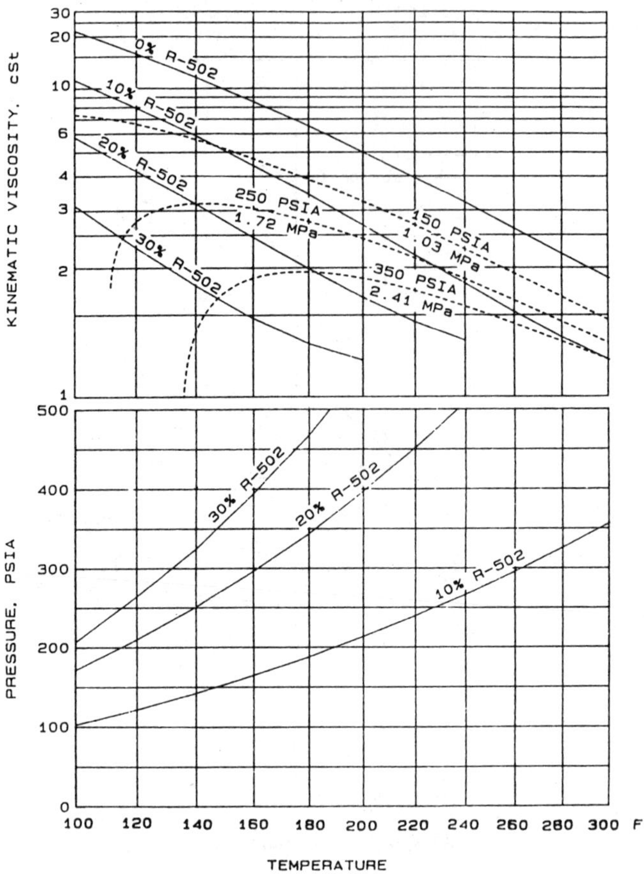

Fig. 32 Viscosity-Temperature-Pressure Chart for Solutions of R-502 in 32 cSt (150 SSU) Naphthenic Oil

viscosity grade, but the highly aromatic alkylbenzene lubricant has a much lower viscosity index in the pure state and shows a higher viscosity at low temperatures. However, at 19.7 psia or approximately −40°F evaporator temperature, the viscosity of the lubricant/R-502 mixture is considerably lower for alkylbenzene than for naphthenic lubricant. In spite of the lower viscosity index, alkylbenzene returns more easily than naphthenic lubricant.

Estimated viscosity-temperature-pressure relationships for a naphthenic lubricant with R-502 is shown in Figure 32. Figures 33 and 34 show viscosity-temperature-pressure plots of alkylbenzene and R-22 and R-502, respectively. These data are based on experimental data from Van Gaalen *et al.* (1991ab). Figures 35 and 36 show viscosity-temperature-pressure data for mixtures of R-134a and a 32 ISO VG polyalkylene glycol and an 80 ISO VG polyalkylene glycol, respectively. Figures 37 and 38 show similar data for R-134a and a 32 ISO VG polyol ester and a 100 ISO VG polyol ester, respectively (Cavestri 1993).

WAX SEPARATION (FLOC TESTS)

Petroleum-derived lubricating oils are mixtures of large numbers of chemically distinct hydrocarbon molecules. At low temperatures in the low-pressure side of refrigeration units, some of the larger molecules separate from the bulk of the lubricant, forming wax-like deposits. This wax can clog capillary tubes and cause expansion valves to stick and is, therefore, undesirable in refrigeration systems. Bosworth (1952) describes other wax separation problems in various systems.

In selecting a lubricant to use with completely miscible refrigerants such as R-11 or R-12, the wax-forming tendency of the

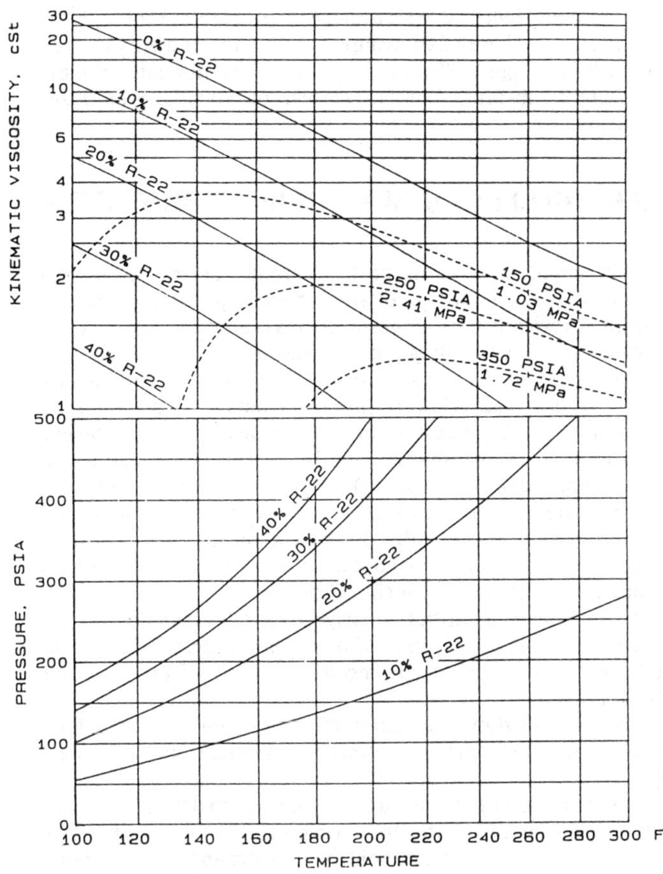

Fig. 33 Viscosity-Temperature-Pressure Chart for Solutions of R-22 in 32 cSt (150 SSU) Alkylbenzene Oil

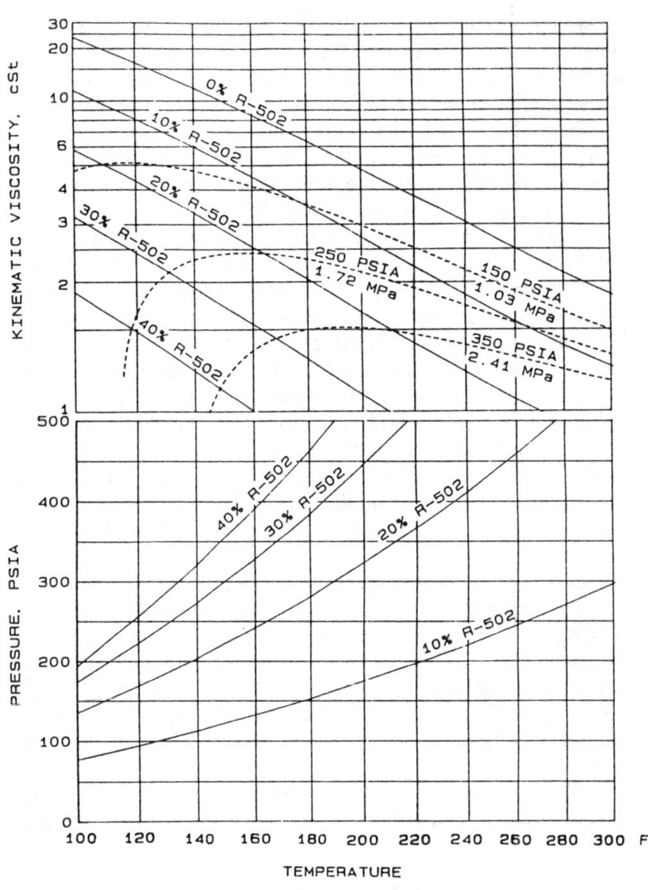

Fig. 34 Viscosity-Temperature-Pressure Chart for Solutions of R-502 in 32 cSt (150 SSU) Alkylbenzene Oil

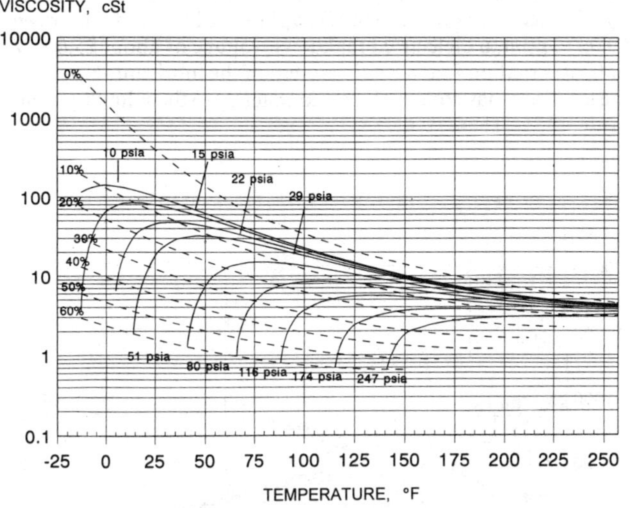

Fig. 35 Viscosity-Temperature-Pressure Plot for 32 ISO VG Polypropylene Glycol Butyl Mono Ether with R-134a

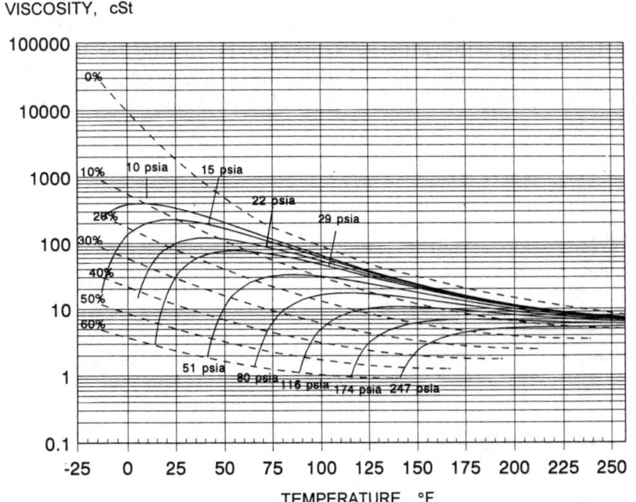

Fig. 36 Viscosity-Temperature-Pressure Plot for 80 ISO VG Polyoxypropylene Diol with R-134a

lubricant can be determined by the floc test. The test evaluates the wax precipitation tendency of a mixture that is 90% by volume refrigerant and 10% by volume lubricant. The floc point is the highest temperature at which wax-like materials or other solid substances precipitate when a mixture of 10% lubricant and 90% Refrigerant 12 is cooled under specific conditions. The test can be used on other oils as well. Since different refrigerant and lubricant concentrations are encountered in actual systems, test results cannot be used directly to predict actual system performance. The lubricant concentration in the expansion devices of most refrigeration and air-conditioning systems is considerably less than 10%, resulting in significantly lower temperatures at which wax separates from lubricant-refrigerant mixture. ASHRAE *Standard* 86-1983 describes a standard method of determining the floc characteristics of refrigeration oils in the presence of R-12.

Attempts to develop a test for the floc point of partially miscible refrigerants for use with R-22 have not been successful. The solutions being cooled often separate into two liquid phases. Once phase separation occurs, the components of the lubricant distribute themselves into the lubricant-rich phase and the refrigerant-rich phase in such a way that the highly soluble aromatics concentrate into the refrigerant phase, while the less soluble saturates concentrate into the lubricant phase. The waxy materials stay dissolved in the refrigerant-rich phase only to the extent of their solubility limit. On further cooling, any wax that separates out from the refrigerant-rich phase migrates into the lubricant-rich phase. Therefore, a significant floc point cannot be obtained with partially miscible refrigerants once phase separation has occurred. However, lack of flocculation does not mean lack of wax separation. Wax may separate in the lubricant-rich phase causing it to congeal. Parmelee (1964) has reported such phenomena with a paraffinic lubricant and R-22 system.

Floc point might not be reliable when applied to used oils. A part of the original wax may already have been deposited, and the used lubricant may contain extraneous material from the operating system.

Good design practice suggests selecting oils that do not deposit wax on the low-pressure side of a refrigeration system, regardless of single-phase or two-phase refrigerant-lubricant solutions. Mechanical design affects how susceptible equipment is to wax deposition. Wax deposits at sharp bends and the suspended wax particles build up on the tubing walls by impingement. Careful design avoids bends and materially reduces the tendency to deposit wax.

Wax separation properties are of little importance with synthetic lubricants since they do not contain wax or wax-like molecules.

SOLUBILITY OF HYDROCARBON GASES

Hydrocarbon gases such as propane (R-290) and ethylene (R-1150) are miscible with the compressor lubricating oil and are absorbed by the lubricant. The lower the boiling point or critical temperature, the less soluble the gas, all other values being equal. Gas solubility increases with decreasing temperature and increasing pressure. These relationships for propane and ethylene are illustrated in Figures 39 and 40. As with other lubricant miscible refrigerants, absorption of the hydrocarbon gas reduces lubricant viscosity.

SOLUBILITY OF WATER IN LUBRICANTS

Refrigerant systems must be dry internally for trouble-free performance (see Chapter 6). As with other components, the refrigeration lubricant must contain the least amount of moisture. Normal refinery handling practices result in a moisture content for refrigeration oils of about 30 ppm. However, this amount may increase between the time of shipment from the refinery and the time of actual use, unless proper preventive measures are taken. Small containers are usually sealed, but it is not common practice to pressure seal bulk tank cars. During transit, changes in ambient temperatures cause the lubricant to expand and contract and draw in humid air from outside. Depending on the extent of such cycling, the lubricant may have a significantly higher moisture content than at the time of shipment.

Users of large quantities of refrigeration oils frequently dry the lubricant before use. Chapter 45 discusses the methods of drying oils. Normally, removing any moisture present will also deaerate the lubricant.

Spot checks show that water solubility data taken by Clark (1940) on transformer oils apply to refrigeration oils as well (Figure 41).

A simple method, widely used in industry to detect free water in refrigeration oils, is the dielectric breakdown voltage (ASTM D 877), which is designed to control moisture and other contaminants in electrical insulating oils. According to Clark (1940), the dielectric breakdown voltage decreases with increasing moisture content at the same test temperature and increases with temperature for the same moisture content. At 80°F, when the solubility of water in a 32 cSt (150 SSU) naphthenic lubricant is between 50 to 70 ppm, a dielectric breakdown voltage of about 25,000 V indicates that no free water is present in the lubricant. However, the lubricant may contain dissolved water up to the solubility limit.

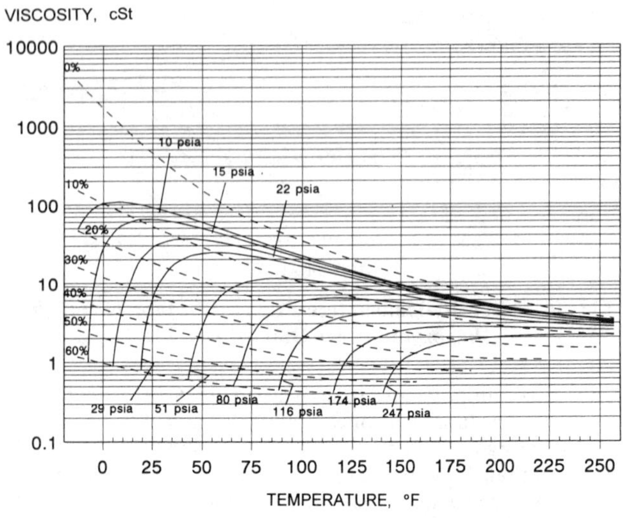

Fig. 37 Viscosity-Temperature-Pressure Plot for 32 ISO VG Branched Acid Polyol Ester with R-134a

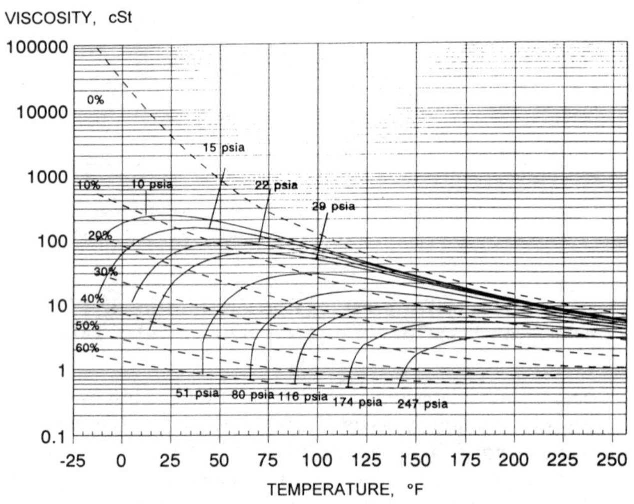

Fig. 38 Viscosity-Temperature-Pressure Plot for 100 ISO VG Branched Acid Polyol Ester with R-134a

Therefore, a dielectric breakdown voltage of 35,000 V is commonly specified to indicate that the moisture content is well below saturation. The ASTM D 877 test is not sensitive below about 60% saturation. General practice has been to measure directly the total moisture content by procedures such as the Karl Fischer (ASTM D 1533) method for low moisture levels.

SOLUBILITY OF AIR IN LUBRICANTS

Refrigerant systems should not contain excessive amounts of air or other noncondensable gases. In the case of air, the oxygen present can react with the lubricant forming lubricant oxidation products. More importantly, the air's 79% nitrogen content (which does not react with lubricant) constitutes a noncondensable gas that can interfere with refrigerating machine performance. In some systems, the tolerable volume of noncondensables is very low, and if the lubricant is added after the system is evacuated, it must not contain an excessive amount of dissolved air or any other noncondensable gas. As mentioned in the preceding section, dissolved air is removed when a vacuum process is used to dry the lubricant. However, if the deaerated lubricant is subsequently stored under dry air pressure, it will reabsorb air in proportion to the pressure. Figure 42 illustrates the volume of air under standard conditions that can be absorbed (dissolved) in mineral oil over a range of pressures. The data in Figure 42 were determined at 68 °F. However, the figure applies to other temperatures as well since the variation in air/lubricant solubility is slight.

FOAMING AND ANTIFOAM AGENTS

Excessive foaming of the lubricant is undesirable in refrigeration systems. Brewer (1951) suggests that abnormal refrigerant foaming reduces the lubricant's effectiveness in cooling the motor windings and removing heat from the compressor. Too much foaming also can cause too much lubricant to pass through the pump and enter the low side. Foaming in a pressure oiling system can result in starved lubrication under some conditions.

However, moderate foaming is beneficial in refrigeration systems, particularly for noise suppression. A foamy layer on top of the lubricant level dampens the noise created by the moving parts of the compressor. There is no general agreement on what constitutes excessive foaming or how it should be prevented. Some manufacturers add small amounts of an antifoam agent, such as silicone fluid, to refrigerator oils. Others believe that foaming difficulties are more easily corrected by equipment design.

OXIDATION RESISTANCE

Refrigeration oils used in hermetic systems are seldom exposed to oxidizing conditions. Handling and manufacturing practices include elaborate care to protect refrigeration oils against air, moisture, or any other contaminant. Oxidation resistance by itself is rarely included in refrigeration lubricant specifications.

Nevertheless, oxidation tests are justified, since oxidation reactions are generally free radical in nature and chemically similar to the reactions between oils and refrigerants. An oxygen test, using power factor as the measure, correlates with established sealed-tube tests. However, such oxidation resistance tests are not used as primary criteria of chemical reactivity, but rather to support the claims of chemical stability determined by sealed-tube and other tests.

Oxidation resistance may become a prime requirement during manufacture. The small amount of lubricant used during compressor assembly and testing is not always completely removed before the system is dehydrated. If the subsequent dehydration process is carried out in a stream of hot dry air, as is frequently the case, the hot oxidizing conditions can cause the residual lubricant to become gummy, leading to stuck bearings, overheated motors, and other operating difficulties. For these purposes, the lubricant should have high oxidation resistance. However, the lubricant used under such extreme conditions should be classed as a specialty process lubricant rather than a refrigeration lubricant. Any harmful effects occurring because of such processing is a contamination of refrigeration oils.

Once a refrigerant system is sealed against air and moisture, the oxidation resistance of a lubricant is not significant unless it reflects the chemical stability.

CHEMICAL STABILITY

Refrigeration oils must have excellent chemical stability. Within the enclosed refrigeration environment, the lubricating oil must resist chemical attack by the refrigerant in the presence of all the materials encountered, including the various metals, motor insulation, and any unavoidable contaminants trapped in the system. A lubricating oil reacts with the refrigerant at elevated temperatures, and the reaction is catalyzed by metals. Methods for evaluating the chemical stability of lubricant-refrigerant mixtures are covered in Chapter 5.

Various phenomena in an operating system, such as sludge formation, carbon deposits on valves, gumming, and copper plating of bearing surfaces have been attributed to lubricant decomposition in the presence of the refrigerant. In addition to the direct reactions of the lubricant and refrigerant, the lubricating oil may also act as a medium for reactions between the refrigerant and the motor insulation, particularly when the refrigerant extracts the lighter components of the insulation. Factors affecting the stability of various components such as wire insulation materials in hermetic systems are also covered in Chapter 5.

Copper plating is one visible sign of reactions between oils and refrigerants in an operating system. Often, after an accelerated life test and teardown of a compressor, films of copper are seen on the ferrous parts such as bearings, pistons, and valves, which operate at temperatures higher than the rest of the system and are in contact with oil. Usually, these films are too thin to affect system operation. In more severe cases, the buildup can be substantial and may interfere with the compressor's functioning. The mechanism of dissolution and copper plating in refrigerant-lubricant systems, as well as methods to minimize the phenomena, are discussed in Chapter 5.

Effect of Lubricant Type

Mineral oils differ in their ability to withstand chemical attack by a given refrigerant. In an extensive laboratory sealed-tube test program, Walker *et al.* (1960, 1962) showed that color darkening, corrosion of metals, deposits, and copper plating occur less in paraffinic oils than in naphthenic oils. Using gas analysis, Spauschus and Doderer (1961) and Doderer and Spauschus (1965) show that a white oil containing only saturates and no aromatics is considerably more stable in the presence of R-12 and R-22 than a medium-refined lubricant. Steinle (1950) has reported the effect of oleoresin (nonhydrocarbons) and sulfur content on the reactivity of the lubricant, using the Philipp test. A decrease in the oleoresin content, accompanied by a decrease in sulfur and aromatic content, showed an improvement in the chemical stability with R-12, while the oil's lubricating properties became poorer. Schwing's (1968) study on a synthetic polyisobutyl benzene lubricant reports that it is not only chemically stable but also has good lubricating properties.

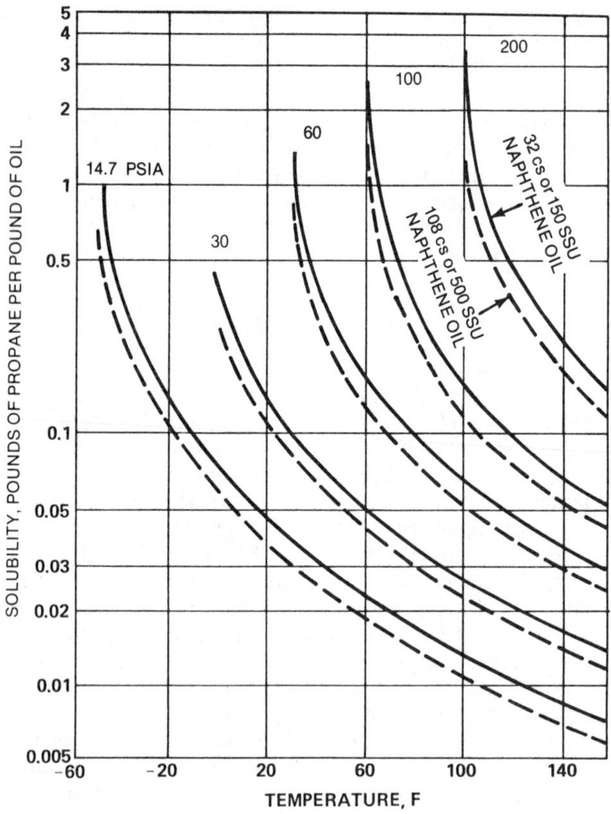

Fig. 39 Solubility of Propane in Oil

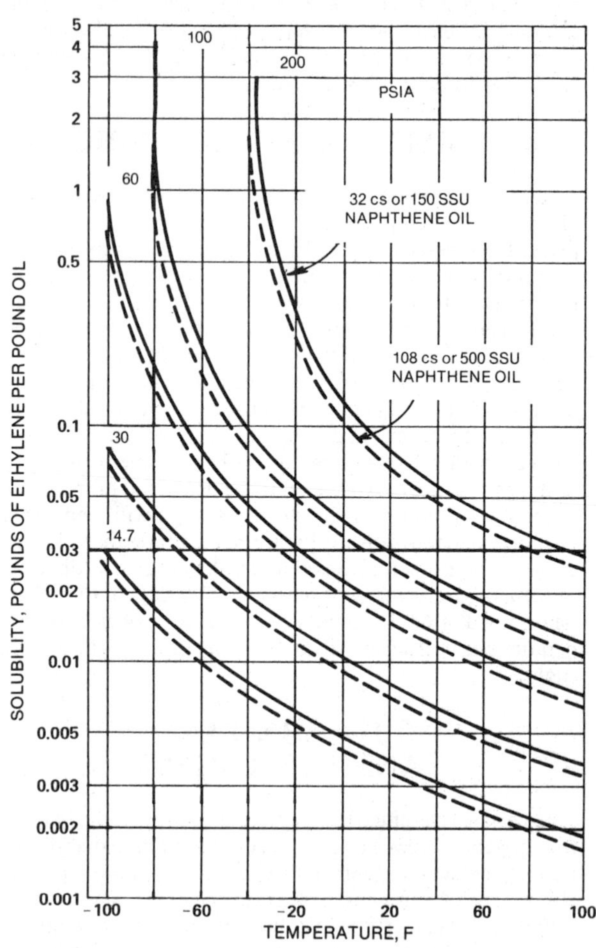

Fig. 40 Solubility of Ethylene in Oil
(Witco)

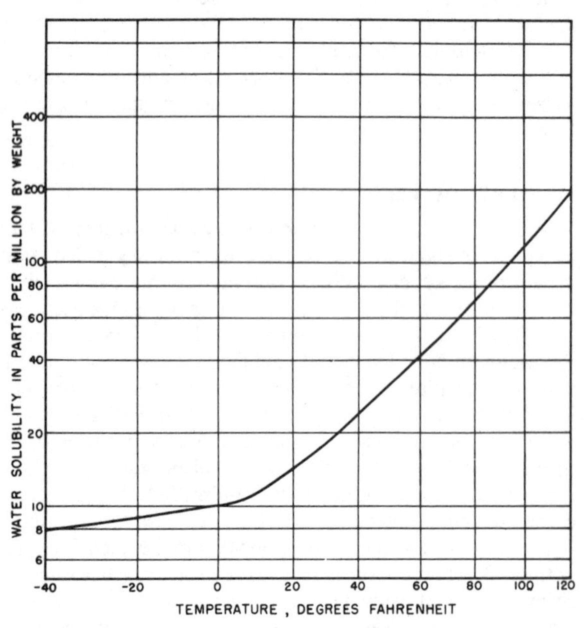

Fig. 41 Solubility of Water in Mineral Oil

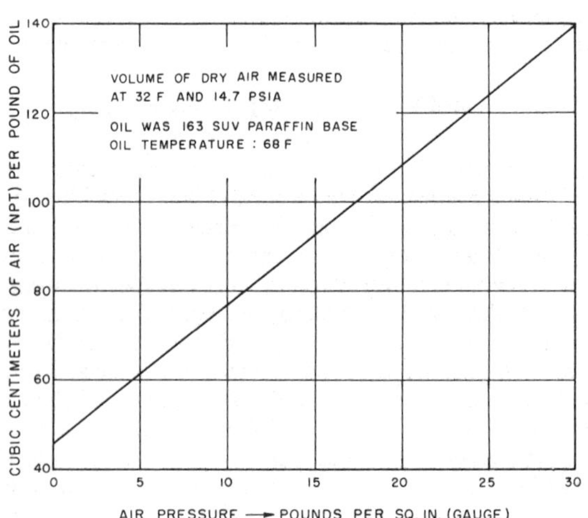

Fig. 42 Effect of Air Pressure on Solubility of Air in Mineral Oil
(Baldwin and Daniel 1953)

REFERENCES

Albright, L.F. and J.D. Lawyer. 1959. Viscosity-solubility characteristics of mixtures of Refrigerant 13B1 and lubricating oils. ASHRAE *Journal* (April):67.

Albright, L.F. and A.S. Mandelbaum. 1956. Solubility and viscosity characteristics of mixtures of lubricating oils and "Freon-13 or -115." *Refrigerating Engineering* (October):37.

API 1970. *Technical data book-Petroleum refining*, 2nd ed. American Petroleum Institute, Washington, DC.

Baldwin, R.R. and S.G. Daniel. 1953. *Journal of the Institute of Petroleum* 39:105.

Bambach, G. 1955. The behavior of mineral oil-F12 mixtures in refrigerating machines. *Abhandlungen des Deutschen Kältetechnischen Vereins*, No. 9. (Translated by Carl Demrick.) Also see abridgement in *Kältetechnik* 7(7):187.

Beerbower, A. and D.F. Greene. 1961. The behavior of lubricating oils in inert gas atmospheres. ASLE *Transactions* 4(1):87.

Bosworth, C.M. 1952. Predicting the behavior of oils in refrigeration systems. *Refrigerating Engineering* (June):617.

Brewer, A.F. 1951. Good compressor performance demands the right lubricating oil. *Refrigerating Engineering* (October):965.

Cavestri, R.C. 1993. Measurement of the solubility, viscosity and density of synthetic lubricants with HFC -134a. Final Report for ASHRAE 716-RP.

Clark, F.M. 1940. Water solution in high-voltage dielectric liquids. *Electrical Engineering Transactions* 59(8):433.

Divers, R.T. 1958. Better standards are needed for refrigeration lubricants. *Refrigeration Engineering* (October):40.

Doderer, G.C. and H.O. Spauschus. 1965. Chemical reactions of R-22. ASHRAE *Transactions* 71(I):162.

Gunderson, R.C. and A.W. Hart. 1962. *Synthetic lubricants*. Reinhold Publishing Corp., New York.

Huttenlocher, D.F. 1969. A bench scale test procedure for hermetic compressor lubricants. ASHRAE *Journal* (June):85.

Kartzmark, R., J.B. Gilbert, and L.W. Sproule. 1967. Hydrogen processing of lube stocks. *Journal of the Institute of Petroleum* 53:317.

Little, J.L. 1952. Viscosity of lubricating oil-Freon-22 mixtures. *Refrigerating Engineering* (Nov.):1191.

Loffler, H.J. 1957. The effect of physical properties of mineral oils on their miscibility with R-22. *Kältetechnik* 9(9):282.

Loffler, H.J. 1959. Density of oil-refrigerant mixtures. *Kältetechnik* 11(3):70.

Loffler, H.J. 1960. Viscosity of oil-refrigerant mixtures. *Kältetechnik* 12(3):71.

Mills, I.W., *et al.* 1946. Molecular weight-physical property correlations for petroleum fractions. *Industrial and Engineering Chemistry* 38:442.

Mills, I.W. and J.J. Melchoire. 1967. Effect of aromatics and selected additives on oxidation stability of transformer oils. *Industrial and Engineering Chemistry* (Product Research and Development) 6:40.

Mosle, H. and W. Wolf. 1963. *Kältetechnik 15:11.*

Murray, S.F., R.L. Johnson, and M.A. Swikert. 1956. Difluoro-dichloromethane as a boundary lubricant for steel and other metals. *Mechanical Engineering* 78(3):233.

Neubauer, E.T. 1958. Compressor crankcase heaters reduce oil foaming. *Refrigerating Engineering* (June):52.

Parmelee, H.M. 1964. Viscosity of refrigerant-oil mixtures at evaporator conditions. ASHRAE *Transactions* 70:173.

Rembold, U. and R.K. Lo. 1966. Determination of wear of rotary compressors using the isotope tracer technique. ASHRAE *Transactions* 72:VI.1.1.

Sanvordenker, K.S. 1968. Separation of refrigeration oil into structural components and their miscibility with R-22. ASHRAE *Transactions* 74(I):III.2.1.

Sanvordenker, K.S. and W.J. Gram. 1974. Laboratory testing under controlled environment using a falex machine. Compressor Technology Conference, Purdue University.

Sanvordenker, K.S. and M.W. Larime. 1972. A review of synthetic oils for refrigeration use. Paper presented at ASHRAE Symposium, Lubricants, Refrigerants and Systems—Some Interactions.

Schwing, R.C. 1968. Polyisobutyl benzenes and refrigeration lubricants. ASHRAE *Transactions* 74(1):III.1.1.

Seeman, W.P. and Shellard, A.D. 1963. Lubrication of Refrigerant 22 machines. IX International Congress of Refrigeration, Paper III-7, Munich, Germany.

Short, G.D. 1990. Synthetic lubricants and their refrigeration applications. *Lubrication Engineering* 46(4):239.

Shubkin, R.L. 1993. Polyalphaolefins, in *Synthetic lubricants and high performance functional fluids*. Marcel Dekker, Inc., New York.

Soling, S.P. 1971. Oil recovery from low temperature pump recirculating halocarbon systems. ASHRAE Symposium *Bulletin* PH-71-2.

Spauschus, H.O. and G.C. Doderer. 1961. Reaction of Refrigerant 12 with petroleum oils. ASHRAE *Journal* (February):65.

Spauschus, H.O. 1963. Thermodynamic properties of refrigerant-oil solutions. ASHRAE *Journal* (April):47; (October):63.

Spauschus, H.O. 1964. Vapor pressures, volumes and miscibility limits of R-22-oil solutions. ASHRAE *Transactions* 70:306.

Spauschus, H.O. and L.M. Speaker. 1987. A review of viscosity data for oil-refrigerant solutions. ASHRAE *Transactions* 93(2):667.

Steinle, H. 1950. *Kaltemaschinenole.* Springer-Verlag, Berlin, Germany, 81.

Van Gaalen, N.A., M.B. Pate, and S.C. Zoz. 1990. The measurement of solubility and viscosity of oil/refrigerant mixtures at high pressures and temperatures: Test facility and initial results for R-22/naphthenic oil mixtures. ASHRAE *Transactions* 96(2):183.

Van Gaalen, N.A., S.C. Zoz, and M.B. Pate. 1991a. The solubility and viscosity of solutions of HCFC-22 in naphthenic oil and in alkylbenzene at high pressures and temperatures. ASHRAE *Transactions* 97(1):100.

Van Gaalen, N.A., S.C. Zoz, and M.B. Pate. 1991b. The solubility and viscosity of solutions of R-502 in naphthenic oil and in an alkylbenzene at high pressures and temperatures. ASHRAE *Transactions* 97(2):285.

Van Nes, K. and H.A. Weston. 1951. *Aspects of the constitution of mineral oils*. Elsevier Publishing Co., Inc., New York.

Walker, W.O., A.A. Sakhanovsky, and S. Rosen. 1957. Behavior of refrigerant oils and Genetron-141. *Refrigerating Engineering* (March):38

Walker, W.O., S. Rosen, and S.L. Levy. 1960. A study of the factors influencing the stability of the mixtures of Refrigerant 22 and refrigerating oils. ASHRAE *Transactions* 66:445.

Walker, W.O., S. Rosen, and S.L. Levy. 1962. Stability of mixtures of refrigerants and refrigerating oils. ASHRAE *Transactions* 68:360.

Witco. Sonneborn Division, *Bulletin* 8846.

CHAPTER 8

COMMERCIAL FREEZING METHODS

FREEZING is a widely used method of food preservation that slows the physical changes and the chemical and microbiological activity causing deterioration in food products. The process of reducing temperature slows the molecular and microbial activity in food, thus extending its useful storage life. Although every product has separate and distinct ideal storage temperatures, most frozen food products are stored at temperatures of 0 to −30°F. (See Chapter 25 for frozen storage temperatures for specific products.)

The process of freezing reduces the temperature of a product from ambient level to storage level and changes most of the water in the product to ice.

The specific freezing process used has three phases: (1) cooling, which removes the product's sensible heat, reducing its temperature to the freezing point; (2) removing the product's latent heat of fusion, thus changing the water to ice crystals; and (3) continued cooling of the product to remove the sensible heat below the freezing point and reduce the temperature to the desired or optimum frozen storage temperature. Values for specific heats, freezing points, and latent heat of fusion for various products are covered in Chapter 25.

Freezing is a time-temperature related process for any typical food product (Figure 1). The longest part of the process is typically the *latent heat of fusion* as water turns to ice. Many food products are sensitive to the rate of freezing, which affects yield (dehydration), quality, nutritional value, and sensory properties. The freezing method and system selected can have substantial economic impact.

The following factors should be considered in the selection of freezing methods and systems for specific products: special handling requirements, capacity, freezing times, quality consideration, yield, appearance, first cost, operating costs, automation, space availability, and upstream/downstream processes.

This chapter covers general freezing methods and systems. Additional information for freezing of specific products is covered in Chapters 11 through 14, 18 through 21, and 24. In addition, related information can be obtained in Chapters 29 and 30 of the 1993 ASHRAE *Handbook—Fundamentals*, which cover cooling and freezing times of foods as well as their thermal properties.

FREEZING METHODS

Freezing systems can be grouped by the basic method for extracting heat from food products.

Airblast Freezing (convection). Cold air at high velocities is circulated over the product. The air removes heat from the product and releases it to an air/refrigerant heat exchanger before being recirculated.

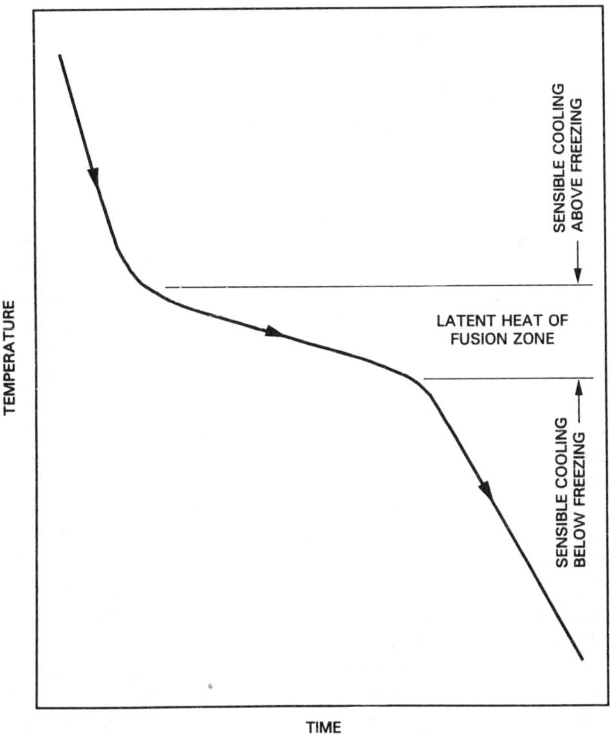

Fig. 1 Typical Freezing Curve

Contact Freezing (conduction). Food, packaged or unpackaged, is placed on or between cold metal surfaces. Heat is extracted by direct conduction through surfaces that are cooled by a circulating refrigerated medium.

Cryogenic Freezing (convection and/or conduction). Food is exposed to an environment below −76°F, which is achieved by spraying liquid nitrogen or liquid carbon dioxide into the freezing chamber.

Cryomechanical Freezing (convection and/or conduction). Food is first exposed to a cryogenic freezing treatment and then finish frozen via a mechanically refrigerated process.

Other special freezing methods, such as immersion of poultry in chilled brine, are covered under the specific product chapters.

AIRBLAST FREEZERS

Airblast freezers use air as a heat transfer medium and depend on the contact between product and air. Various levels of sophistication in airflow and conveying techniques exist, from crude blast freezing chambers to carefully controlled impingement-style freezers.

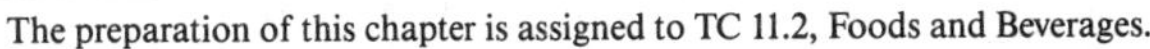

The preparation of this chapter is assigned to TC 11.2, Foods and Beverages.

The earliest airblast freezers consisted of cold storage rooms with extra fans and a surplus of refrigeration. Improvements in the form of airflow control and mechanized conveying techniques have made the heat transfer rates more efficient and the product flow more automated and less labor intensive.

While batch-type freezing is still widely used and appropriate for certain production lines, more sophisticated freezers have led to an integration of the freezing process into the production line. This integration is referred to as in-line freezing and has become essential for large volume, high quality, economic operations.

Air is the most commonly used freezing medium, and a wide range of airblast freezer systems are available:

Batch

- Cold storage room
- Stationary blast cell freezing tunnel
- Push-through trolley freezer

Continuous/In-Line

- Straight belt freezer (single, multiple)
- Fluidized bed freezer
- Spiral belt freezer
- Carton (carrier) freezer

Cold Storage Room. Although cold storage is not considered to be a freezing system, it is sometimes used for this purpose. Because the storage room was never intended to be a freezer, it lacks the necessary design characteristics and, therefore, should only be used for freezing in exceptional cases. Freezing is so slow that the quality of most products suffers. The quality of the frozen products already being stored in the room is jeopardized because flavors may be transferred from the warm products and the temperature of the frozen products may rise considerably due to the excess refrigeration load.

Stationary Blast Cell. The stationary blast cell (Figure 2) is the simplest freezer that can be expected to produce satisfactory results for most products. It is constructed of an insulated enclosure equipped with refrigeration coils and fans that circulate the air over the products in a controlled way. Products are placed on trays that are then placed into a rack in such a way that an air space is left between every layer of trays. The racks are moved in and out of the tunnel manually. It is important that the racks be placed so that air bypass is minimized. The stationary blast cell is a universal freezer—practically all products can be frozen using this method. Vegetables and other products, such as spinach, bakery products, meat patties, fish fillets, and prepared foods may be frozen in cartons or unpacked, spread in a layer on trays. The flexibility of this system makes it suitable for small quantities of varied products; however, labor requirements are relatively high and product flow is cumbersome.

Push-Through Trolley Freezer. A moderate degree of mechanization is achieved in a push-through trolley freezer (Figure 3), in which the racks are fitted with wheels. The racks, or trolleys, are usually moved on rails by a pushing mechanism that can be hydraulically powered. This type of freezer has similar characteristics to the stationary blast cell, except that labor costs and product handling are decreased by the mechanization.

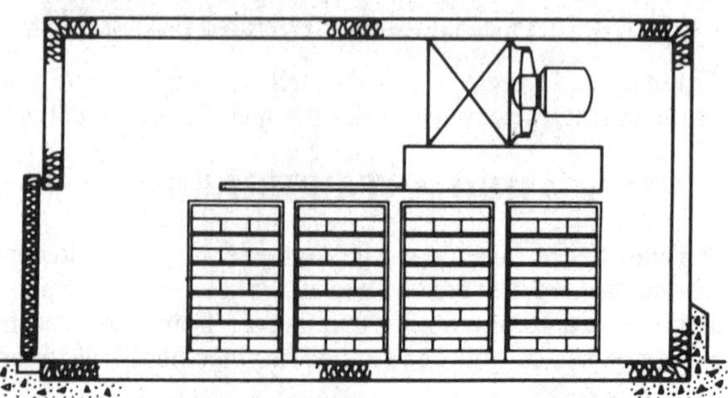

Fig. 2 Stationary Blast Cell

Straight Belt Freezer. The first mechanized airblast freezers consisted of a wire mesh belt conveyor in a blast room, which satisfied the need for continuous product flow. A disadvantage to these early systems was the poorly controlled airflow and resulting inefficient heat transfer. Current versions use a controlled vertical airflow, which forces cold air up through the product layer, thereby creating good contact with the product particles. They are generally used with fruits, vegetables, and French fried potatoes.

The principal current design is the two-stage belt freezer (Figure 4), which consists of two mesh conveyor belts in series. The first belt initially cools or crust freezes an outer layer or crust to condition the product before transferring it to the second belt for freezing and sensible cooling to 0 °F or below. Uniform product distribution over the entire belt is required to achieve uniform product contact and effective freezing. Two-stage freezers are generally operated from 15 to 25 °F refrigerant temperature in the precool section and from −25 to −40 °F in the freezing section. Capacities range from 1 to 20 tons of product per hour with freezing times from 3 to 50 min.

When products to be frozen are hot (*e.g.*, French fried potatoes from the fryer at 180 to 200 °F), an additional cooling section is used in series ahead of the normal precool section. This section supplies either refrigerated air at approximately 50 °F or filtered ambient air to cool the product and congeal the fat, thereby reducing energy use. Refrigerated air is preferred, as filtered ambient air may pose potential contamination problems and may have greater temperature variations.

Fluidized Bed Freezers. Another proprietary freezer uses air both as the medium of heat transfer and for transport, enabling the product to flow through the freezer on a cushion of cold air (Figure 5). This design is well-suited for smaller, uniform-sized

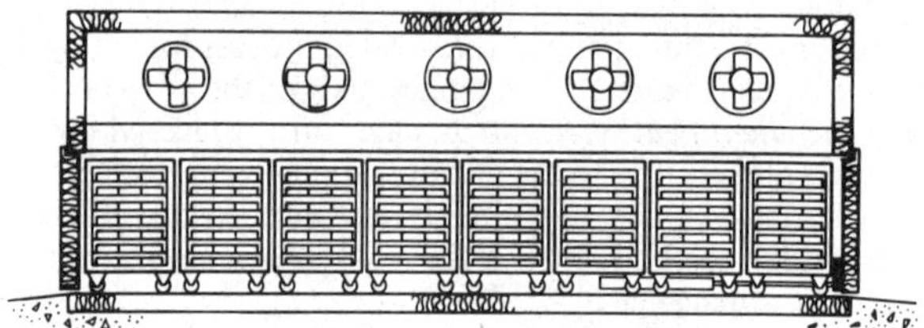

Fig. 3 Push-Through Trolley Freezer

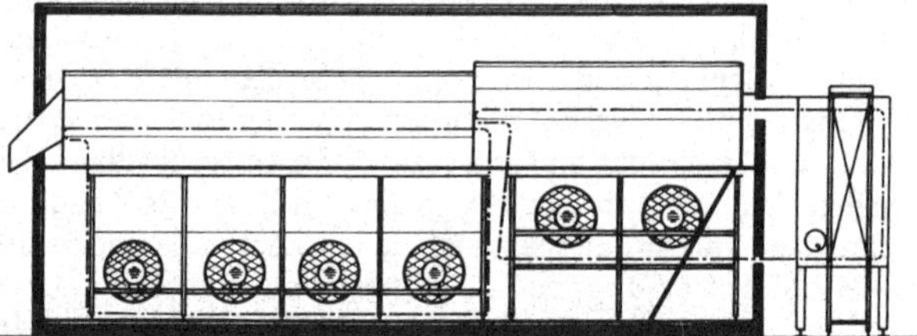

Fig. 4 Two-Stage Belt Freezer

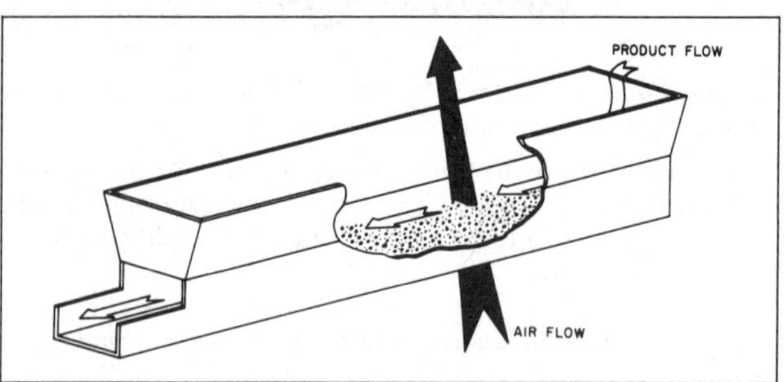

Fig. 5 Fluidized Bed Freezer

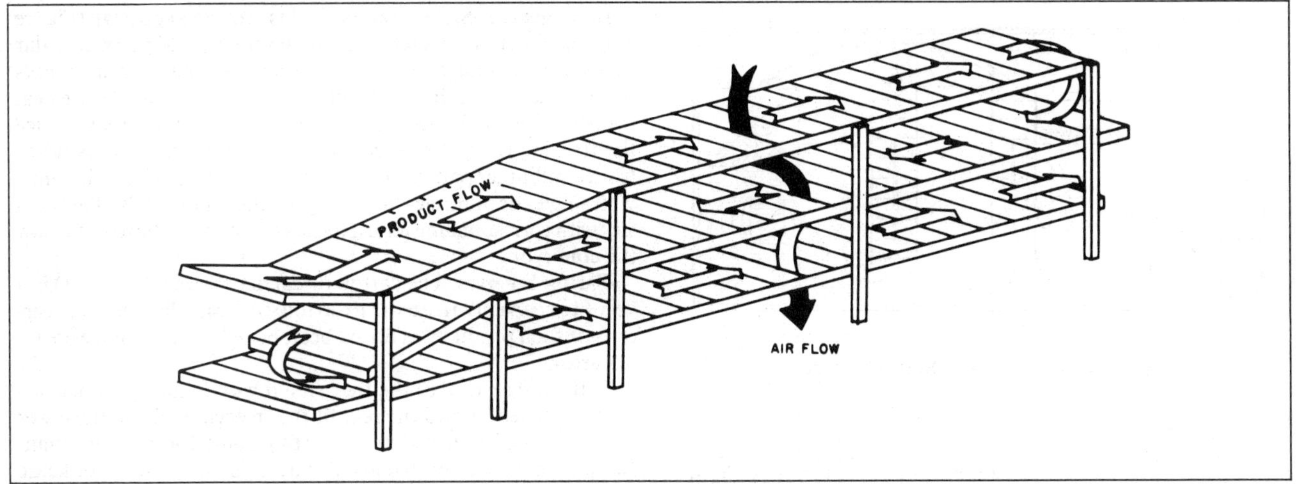

Fig. 6 Multipass, Straight Belt Freezer

particulate products such as peas, sliced and diced carrots, and shredded potatoes or cheese.

The high degree of fluidization improves the heat transfer rate and results in good use of floor space. The technique is limited to well-dewatered products of uniform size that can be readily fluidized and transported through the freezing zone. Because the principle depends on rapidly crust freezing the product, the operating refrigerant temperature must be −40°F. Fluidized bed freezers are normally manufactured as packaged, factory-assembled units with capacity ranges of 1 to 5 tons/h. The particulate products generally have a freezing time of 3 to 11 min.

Fluidized Belt Freezers. A hybrid of the two-stage belt freezer and the fluidized bed freezer, a two-stage belt freezer has a fluidizing section in the first-belt stage. An increased air resistance is designed under the first belt to provide the fluidizing conditions for wet incoming product, but at the same time, the belt is there to assist in the transport of heavier, less uniform products that do not fluidize fully. Once crust-frozen, the product can be loaded deeper for greater efficiency on the second belt. Two-stage fluidized belt freezers operate at −30 to −35°F refrigerant temperature and in capacity ranges from 1 to 15 tons/h.

Multipass, Straight Belt Freezer. For larger products with longer freezing times (up to 60 min) and higher throughputs (0.5 to 6 tons/h), a single straight-through belt freezer would require a very large floor space. Floor space can be reduced by stacking belts above each other, thus forming either a single-feed/single-discharge multipass system (usually three passes) or multiple, single-pass systems (multiple infeeds and discharges) stacked one on top of the other. The multipass (triple pass) arrangement (Figure 6) provides another benefit in that the product, after being surface frozen on the first belt, may be stacked in a rather deep bed on the lower belts. Thus, the total belt area required is reduced, as is the overall size of the freezer. This system has the potential for product damage and product jams at the belt transfers.

Spiral Belt Freezer. For products with a long freezing time (generally 10 min to 3 h), and for products that require gentle handling during freezing, it is advantageous to have an endless belt wrapped cylindrically one tier above the other, thus requiring a minimal floor space for a relatively long belt length. The spiral freezer uses a conveyor belt that can be bent laterally. The original spiral belt principle uses a spiraling rail system to carry the belt, while a more recent design uses a proprietary self-stacking belt requiring less overhead clearance. The number of tiers in the spiral can be varied to accommodate different capacities. In addition, one or more spiral towers can be used in series for products with long freezing times. Depending on the upstream process and capacity required, spiral freezers are available in a range of belt widths and are manufactured as packaged, modular, and field-erected models.

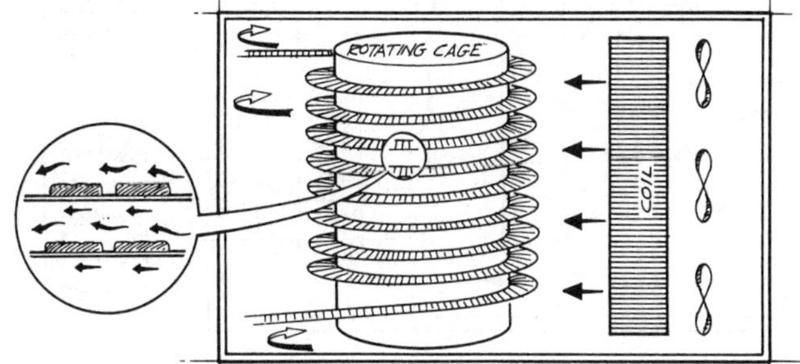

Fig. 7 Horizontal Airflow Spiral Freezer

A distinguishing feature of spiral freezer designs worthy of consideration is the configuration. The level of airflow sophistication varies from little more than an open, unbaffled spiral conveyor to very tightly controlled airflow accomplished through extensive baffling with the use of pressure-producing fans.

Horizontal airflow is applied to spiral freezers (Figure 7) with axial fans mounted along one side that blow air horizontally across the spiral conveyor with minimal baffling limited to two portions of the spiral circumference. Because the airflow is only loosely controlled, and because air will travel the path of least resistance (through the unrestricted space above each tier of belt), the air-to-product contact is haphazard and the heat transfer relatively less efficient than other more controlled designs.

Several proprietary designs have been developed to control airflow. One design (Figure 8) has a mezzanine floor that separates the freezer into two pressure zones. Baffles around the outside and inside of the belt form an air duct so that air flows down around the product and the conveyor moves the product up. The controlled airflow reduces freezing times for some products.

Another design (Figure 9) splits the airflow so the coldest air contacts the product both as it enters and leaves the freezer. The coldest air introduced on the incoming, warm product may increase surface heat transfer and more rapidly freeze the surface, which may reduce product dehydration.

Typical products frozen in spiral belt freezers include raw and cooked meat patties, fish fillets, chicken portions, pizza, and a variety of packaged products. Spiral freezers are available in a wide range of capacities, from 1000 to 12,000 lb/h. They dominate today's frozen food industry and account for the majority of unpackaged nonparticulate frozen food production.

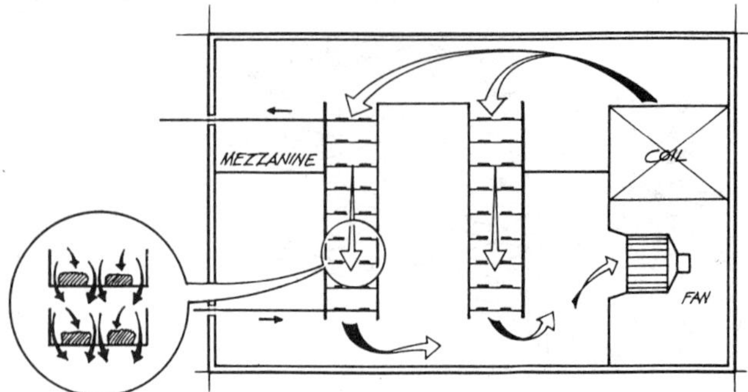

Fig. 8 Vertical Airflow Spiral Freezer

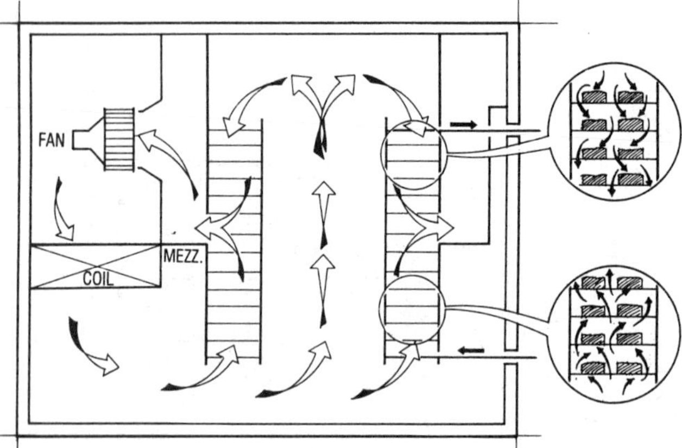

Fig. 9 Split Airflow Spiral Freezer

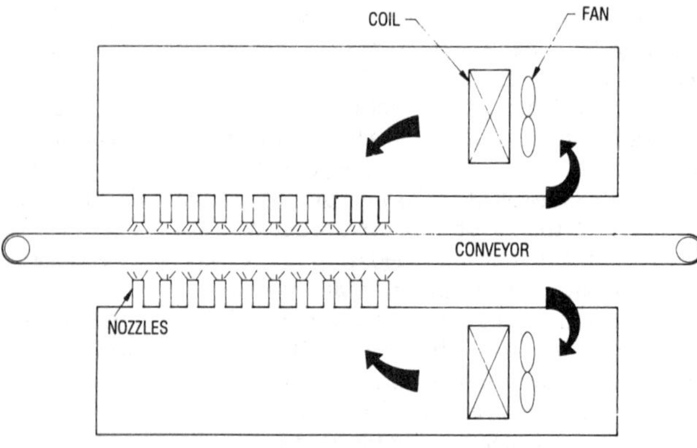

Fig. 10 Impingement-Style Freezer

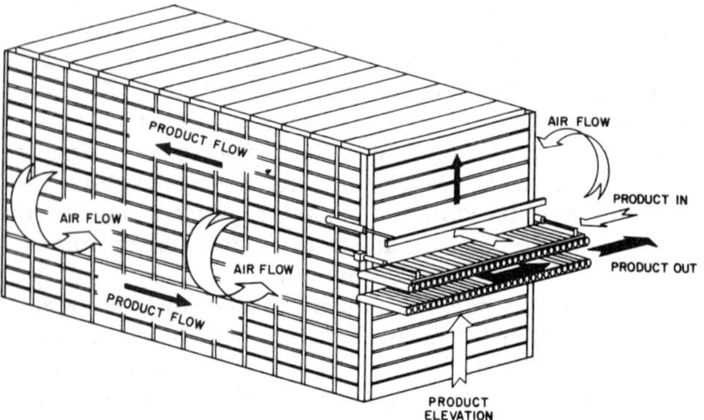

Fig. 11 Carton (Carrier) Freezer

Impingement-Style Freezers. In this proprietary design (Figure 10), cold air flows at a relatively high velocity and perpendicular to the product's largest surfaces. The airflow constantly interrupts the boundary layer that surrounds the product, so the surface heat transfer rate is enhanced. As a result, the technique may reduce the freezing time of products with large surface-to-mass ratios (thin hamburger patties, for example). Such freezers are designed with single- or multiple-pass straight belts. Thousands of air nozzles with corresponding return ducts are mounted above and below the conveyors.

Carton Freezers. The carton or carrier freezer (Figure 11) is a very high capacity freezer (5 to 20 ton/h) for medium to large cartons of such products as boxed beef, boxed poultry, or ice cream in cartons.

In the top section of the freezer, a row of loaded product carriers is pushed toward the rear of the freezer, while on the lower section they are returned to the starting point. Elevating mechanisms are located at both ends. A carrier is similar to a bookcase with shelves. When it is indexed in the loading/discharge end of the freezer, the already-frozen product is pushed off each shelf one row at a time onto a discharge conveyor. When the carrier is indexed up, this shelf aligns with the loading station, where new products are continuously pushed onto the carrier before it is moved once again to the rear of the freezer. Refrigerated air is circulated over the cartons by forced convection.

CONTACT FREEZERS

In contrast to airblast-type freezers, a contact freezer's primary means of heat transfer is by conduction; the product or package is placed in direct contact with a refrigerated surface. Contact-type freezers can be distinguished as follows:

Batch

- Manual horizontal
- Manual vertical

In-Line

- Automatic plate
- Contact belt (solid stainless steel)
- Specialized design

The most common type of contact freezer is the contact plate freezer, in which the product is pressed between metal plates. Refrigerant is circulated inside channels in the plates, which ensures an efficient heat transfer. This is reflected in short freezing times, provided the product is a good conductor of heat, as in the case of fish fillets, chopped spinach, or meat offal. However, packages or cavities should be well-filled and, if metal trays are used, they should not be distorted.

Manual and Automatic Plate Freezers. Contact plate freezers (Figure 12) are available in horizontal or vertical arrangements, using a manual loading/unloading operation. Horizontal plate freezers are also available in an automatic loading/unloading version, which generally accommodates higher capacities and provides for continuous operation. The advantage of good heat transfer in contact plate freezers is gradually reduced with increasing thickness of the product. For this reason, thickness is often limited to a maximum of 2 to 3 in. Contact plate freezers offer the advantage of efficient operation, since no fans move air through the freezer, and they are very compact. Another advantage with packaged products is that pressure from the plates minimizes any bulging that may occur. Thus, packages are even and square within close tolerances. Automatic plate freezers accommodate line capacities up to 200 packages per minute with freezing times from 10 to 150 min. When greater capacities are required, they are placed in series with associated conveyor systems to handle the loading and unloading of packages.

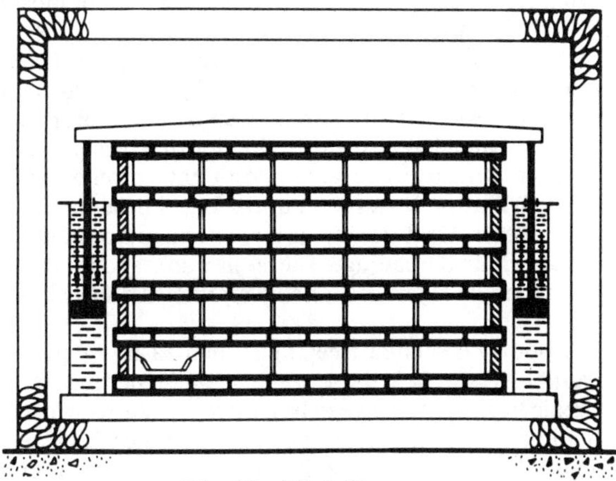

Fig. 12 Plate Freezer

Specialized Contact Freezers. For wet fish fillets and similar soft, wet products with relatively large, flat surfaces, an endless, solid stainless steel belt and a combination of air and contact freezing is used. The belts are typically 4 to 6 ft wide and may be 100 ft long. Product is loaded onto the stainless steel belt at one end and then travels in a fixed position through the freezing zone to the discharge end. Freezing is usually accomplished both by conduction through the solid stainless steel belt to a cooling medium below the belt, and by convection with controlled airflow above the belt. Frozen product with good eye appeal is produced using this freezer design, but a drawback is the physical size of the freezer. Capacities for typical products are generally limited to 1 to 2.5 tons per hour, with a freezing time of less than 30 min.

CRYOGENIC FREEZING

Cryogenic, or gas, freezing is often an alternative for (1) production on a small scale, (2) launching new products, (3) handling overload situations, or (4) using with seasonal products. Cryogenic freezers use liquid nitrogen or liquid carbon dioxide as a refrigeration medium, and the freezers may be designed as (1) batch-type cabinets, (2) straight belt freezers, (3) spiral conveyors, or (4) liquid immersion freezers.

Liquid Nitrogen Freezers. The most common type of liquid nitrogen freezer is a straight-through, single-belt, in-line tunnel. Liquid nitrogen at −320°F is introduced at the outfeed end of the freezer directly onto the product, and as it vaporizes, those cold vapors are circulated toward the infeed end where they are used for precooling and initial freezing of the products. The "warmed" vapors (typically −50°F) are then discharged to the atmosphere. The low temperature of the liquid nitrogen and the nitrogen vapor provides rapid freezing, which can be advantageous for some products from a quality and dehydration standpoint. On the downside, the freezing cost is high and the surface of products having a high water content may crack if some precautions are not taken.

Consumption of liquid nitrogen is in the range of 0.9 to 2.0 lb of nitrogen per pound of product, depending on the water content and temperature. While this results in relatively high operating costs, the small initial investment makes this freezer economical for some applications.

Carbon Dioxide Freezing. A similar cryogenic freezing method uses boiling (subliming) carbon dioxide in direct contact with foods frozen in a straight belt freezer or spiral freezer. Carbon dioxide boils at approximately −110°F, and the system operates like a liquid nitrogen freezing system, consuming cryogenic liquid as it freezes. A common application for CO_2 is freezing individual quick frozen (IQF) diced poultry in an auger or tumble freezer.

COMBINATION CRYOMECHANICAL FREEZERS

Although the technique is not new, cryomechanical (combination of cryogenic and airblast freezing) freezing applications have increased. High value, sticky products, such as IQF shrimp, and wet, delicate products, such as IQF strawberries and IQF cane berries, are a common application for these systems.

A typical cryomechanical freezer uses an initial immersion step where the product flows through a bath of liquid nitrogen to set the product surface. This rapid surface freezing reduces dehydration and improves the handling characteristics of the product, thus minimizing sticking and clumping. The cryogenically crust-frozen product is then transferred directly into a mechanical freezer, where the remainder of the heat is removed and the product temperature reduced to 0°F or lower. The cryogenic step is sometimes retrofitted to existing mechanical freezers to increase their capacity. The use of the mechanical freezing step reduces operating costs compared with cryogenic-only freezing.

OTHER FREEZER SELECTION CRITERIA

Reliability. The freezing system is probably the most vulnerable equipment in a processing line due to the harsh operating conditions. In addition, a process line usually incorporates only one freezer, which makes reliability a major concern.

Reliable equipment is essential to provide continuous, economical operation in long-term use. Freezing systems need to be designed and constructed with adequate safety factors for electrical/mechanical components and with materials that will withstand harsh environments and rugged usage for normal equipment life expectancy.

Hygiene. Cleanability and sanitary design are as important as reliability. Freezing systems should have minimal locations for the product to hang up; be constructed of noncorrosive, safe materials; and be equipped with manual and/or automatic sanitation systems for wash-down and cleanup. Product contamination can result if the equipment is not or cannot be properly cleaned and sanitized. These features are particularly important for chilled, partially cooked, and fully cooked products that may not be fully reheated or properly prepared prior to consumption.

Quality. The quality of processed food products is affected by physical changes and the speed of microbiological activity and chemical reactions. Each of these is influenced by the rate of temperature change. The freezing process physically changes the food product. The rate of physical change or freezing time determines the size of ice crystals produced. At a slow freezing rate, initially formed ice crystals can grow to a relatively large size; fast freezing forces more crystals to be seeded with a smaller average size. In practice, however, different sized ice crystals are formed because the product's surface freezes faster than its inner parts. Large ice crystals may puncture the cell walls of the product, usually increasing loss of juices during thawing. Depending on the product, this can greatly affect the texture and flavor of the remaining product tissue.

The influence of freezing time is more apparent on some products than on others. For strawberries, the drip loss is reduced from more than 20% to less than 5% with a shorter freezing time. A difference in drip loss from 20% for strawberries frozen in 12 h to 8% for strawberries frozen in 15 min is significant. Cryogenic systems perform the same freezing functions in 8 min or less, reducing the drip loss to less than 5%.

A well-applied mechanical freezer or a cryogenic freezer can crust-freeze products rapidly, minimizing loss of natural juices, aromatics, and flavor essences, and maintaining a higher quality that makes the product more marketable. Quality is also affected by storage temperature and temperature fluctuation. Lower storage temperature and fewer, less severe temperature fluctuations tend to preserve quality better.

Economics. Ironically, freezing equipment is considered both the most expensive and the least expensive link in the modern processing chain. Although the freezer frequently represents the single largest investment in a line, its operating costs are usually only 3 to 5% of the total. Packaging costs may vary widely, but generally are several times greater than total freezing cost.

One essential factor to consider when choosing freezing equipment is the loss in product mass that occurs during freezing. This loss may be about the same as the operating cost of the freezer; this applies to inexpensive products such as peas, and is even more significant for expensive products such as seafood.

Weight loss during freezing may be caused by mechanical losses, downgrading, and dehydration. Mechanical losses refer to products dropping to the floor, sticking to conveyor belts, or juice dripping, all of which are specific for each plant. A modern freezer should have almost no losses in this category. Downgrading losses refer to damaged product, breakage, and similar occurrences that render the product unsalable at the top quality price. A modern freezing system should also incur minimal losses for most products from damage and breakage.

Dehydration losses occur in any freezing system. The evaporation of water vapor from unpackaged products during freezing becomes evident as frost builds up on evaporator surfaces. Frost is also caused by excessive infiltration of warm, moist air into the freezer. Still air inside a diffusion-tight carton often creates larger dehydration losses than unpackaged products frozen in an IQF freezer. Heat transfer is poor because no circulation of air occurs within the package. The result is an evaporation of moisture that can be significant; however, the frost stays inside the carton.

A poorly designed freezing system easily operates with dehydration losses of 3 to 4%, while well-designed mechanical or cryogenic freezing systems can be built to operate with losses near zero. Liquid nitrogen tunnels normally operate with a dehydration loss of about 0 to 1.25%. This loss occurs when the nitrogen gas is circulated over the product at the infeed end of the freezer. Infeed circulation is sometimes necessary to temper the product and to use the heat capacity in the nitrogen most efficiently. Nitrogen immersion freezers have lower dehydration losses but use more liquid nitrogen. A CO_2 freezer using jet impingement operates with a dehydration loss of about 0 to 1.25%.

CHAPTER 9

FOOD MICROBIOLOGY AND ENGINEERING

REFRIGERATION'S largest overall application is the prevention or retardation of microbial, physiological, and chemical changes in foods. Even at temperatures near the freezing point, foods may deteriorate through the growth of microorganisms, through changes caused by enzymes, or through chemical reactions. Holding foods at low temperatures merely reduces the rate at which these changes take place.

In addition to controlling the deterioration of foods, refrigeration plays a major role in maintaining a safe food supply. Overall, the leading factor causing food-borne illness is improper holding temperatures for food. Another important factor is equipment that is not properly sanitized. Engineering directly impacts the safety and stability of the food supply in design of cleanable equipment and facilities, as well as maintenance of environmental conditions that inhibit microbial growth. This chapter briefly discusses the microbiology of foods and the impact of design decisions on the production of safe and wholesome foods. Methods of applying refrigeration to specific foods may be found in other chapters.

BASIC MICROBIOLOGY

Microorganisms play several roles in a food production facility. They can contribute to food spoilage, producing off-odors and flavors, or altering product texture or appearance through slime production and pigment formation. Certain organisms cause disease; others are beneficial and are required to produce foods such as cheese, wine, and sauerkraut through fermentation.

Microorganisms fall into four categories—bacteria, yeasts, molds, and viruses. Bacteria are the most common food-borne pathogens. Bacterial growth rates, under optimum conditions, are generally faster than those of yeasts and molds, making bacteria a prime cause of spoilage, especially in refrigerated, moist foods. Bacteria have many shapes, including spheres (cocci), rods, or spirals, that are usually between 0.3 and 5 to 10 μm in size. Bacteria as a group are capable of growth over a wide range of environmental conditions. Some bacteria, notably *Clostridium* and *Bacillus* spp., form endospores, *i.e.*, resting states with extensive temperature, desiccation, and chemical resistance.

Yeasts and molds become important in situations that restrict the growth of bacteria, such as in acidic or dry products. Yeasts can cause gas formation in juices and slime formation on fermented products. Mildew (black mold) on humid surfaces and mold formation on spoiled foods are also common.

Viruses are obligate, intracellular parasites, which are specific to an individual host. Human viruses, such as Hepatitis A, cannot multiply outside the human body. Design features must include facilities for good employee handwashing and sanitation practices to minimize potential for product contamination. Bacterial viruses (phage), however, may contribute to starter culture failure in bacterial fermentations if proper isolation, ventilation, and sanitation procedures are not followed. The use of commercial concentrated cultures, selected for phage resistance, has greatly reduced this problem.

The preparation of this chapter is assigned to TC 11.2, Foods and Beverages.

Sources of Microorganisms

Bacteria, yeasts, and molds are widely distributed in water, soil, air, plant materials, and the skin and intestinal tracts of humans and animals. Practically all unprocessed foods will be contaminated with a variety of spoilage and, sometimes, pathogenic microorganisms. Food processing environments that contain residual food material will naturally select the microorganisms that are most likely to spoil the particular product.

Microbial Growth

All microbial populations follow a generalized growth curve (Figure 1). An initial lag phase occurs as organisms start to grow and adapt to new environmental conditions. The lag phase is very important because the maximum extension of shelf life and length of production runs are directly related to the length of the lag phase. Once adaptation has occurred, the culture enters into the maximum (logarithmic) growth rate, and control of microbial growth is not possible without major sanitation or other drastic measures. Numbers can double as fast as every 20 to 30 min under optimum conditions. Toxin production and spore maturation, if possible, usually occur at the end of the exponential phase as the culture enters into a stationary phase. At this time, essential nutrients are depleted and/or inhibitory by-products are accumulated. Eventually the culture dies, the rate depending on the organism, the medium, and other environmental characteristics.

Critical Growth Requirements

Factors that influence microbial growth can be divided into two categories: (1) intrinsic factors that are a function of the food itself and (2) extrinsic factors that are a function of the environment in which a food is held.

Intrinsic Factors

Intrinsic factors affecting microbial growth include nutrients, inhibitors, biological sturctures, water activity, pH, and presence

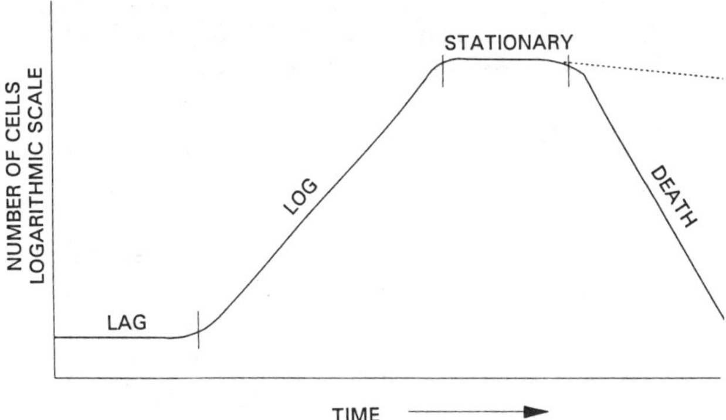

Fig. 1 Typical Microbial Growth Curve

of competing microorganisms in a food. While engineering practices have little impact on these parameters, an understanding of how intrinsic factors influence growth is useful in predicting the types of microorganisms that may be present in the production environment.

Nutrients. Like other living organisms, microorganisms require food to grow. Carbon and energy sources are usually supplied in the form of sugars and starches. Nitrogen requirements are met by the presence of protein. Vitamins and minerals are also necessary. Lactic acid bacteria have rather exacting nutritional requirements, while many aerobic spore formers have tremendous enzymatic capabilities and are capable of growth on a wide variety of substrates. Cleanable systems facilitate the removal of residual food material and deprive microorganisms of the nutrients required for growth, thus preventing a buildup of organisms in the environment.

Inhibitors. Either naturally occurring or added as preservatives, inhibitors may be present in food. Preservatives are not substitutes for hygienic practices and, with time, microorganisms may develop resistance. A cleanable processing system is still essential in preventing the development of a resistant population.

Competing Microorganisms. The presence of other microorganisms also affects the organisms in foods. Some organisms produce inhibiting compounds or have faster generation times; others are better able to use the available nutrients in a food matrix.

Water Activity. All life-forms require water for growth. Water activity (a_w) refers to the availability of water within a food system and is defined at a given temperature as:

$$a_w = \frac{\text{Vapor pressure of solution (food)}}{\text{Vapor pressure of solute (water)}}$$

The minimum water activities for growth of a variety of microorganisms, along with representative foods, are listed in Table 1. These a_w minima are also factors in environmental humidity control discussed in the Extrinsic Factors section.

In many refrigerated foods, such as fresh meats, fruits, and vegetables, microbial contamination occurs on external surfaces. These biological structures have a protective effect and are less prone to microbial growth than are cut surfaces. When enclosed in airtight packaging or in a chamber with limited air circulation, an equilibrium a_w is achieved that is equal to the a_w of the food. In these situations, the a_w of the food determines which organism will be capable of growth. If the same foods are exposed to reduced environmental relative humidity, such as meat carcasses hanging in a controlled aging room or vegetables displayed in an open case, surface dehydration acts as an inhibitor to microbial growth. Likewise, if a dry product, such as bread, is exposed to a moist environment, mold growth may occur on the surface as moisture is absorbed. Environmental relative humidity thus has a significant impact on product shelf life.

Table 1 Approximate Minimum Water Activity for Growth of Microorganisms

Organism	a_w	Foods
Pseudomonads	0.98	Fresh fruits, vegetables, meats
Salmonella spp., *E. coli*	0.95	Many processed foods
Listeria monocytogenes	0.93	
Bacillus cereus	0.92	Salted butter, fermented sausage
Staphylococcus aureus	0.86	
Molds	0.84	Soft, moist pet food
	0.80	Pancake syrup, jam
	0.70	Corn syrup
Xerotrophic molds	0.65	Caramels
Osmophilic yeasts	0.62	
Limit of microbial growth	0.60	Wheat flour
	0.40	Nonfat dry milk

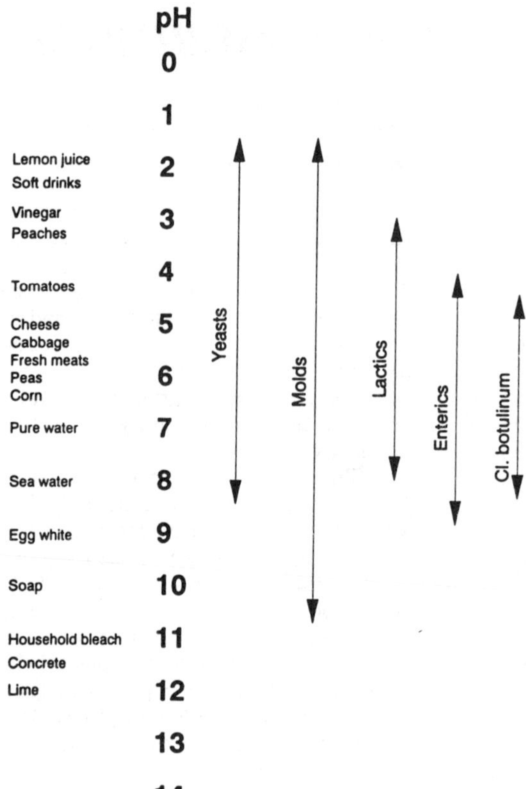

Fig. 2 pH Ranges for Microbial Growth and Representative Examples

pH. The pH, or acidity of a product, like a_w, has a major influence on microbial growth. For most microorganisms, optimal growth occurs at neutral pH, or 7.0. Few organisms grow under alkaline conditions, while some organisms, such as yeasts, molds, and lactic acid bacteria, are acid tolerant. Figure 2 depicts pH values of a variety of foods and limiting pH values for microorganisms.

Extrinsic Factors

Extrinsic factors that influence the growth of microorganisms include temperature, environmental relative humidity, and oxygen levels. Refrigeration and ventilation systems play a major role in controlling these factors.

Temperature. Microorganisms are capable of growth over a wide range of temperatures. Minimum growth temperatures for a variety of spoilage and pathogenic bacteria of significance in foods are summarized in Table 2. Previously, 45 °F was thought to be sufficient to control the growth of pathogenic organisms. However, the emergence of psychrotrophic pathogens, such as *Listeria monocytogenes*, has demonstrated the need for lower temperatures. In the United States, 40 °F is now recognized as the upper limit for safe refrigeration temperature, while 34 °F or lower may be more appropriate. Foods that will support the growth of pathogenic microorganisms may not be held between 40 and 140 °F for more than 4 h.

Temperature is used to categorize microorganisms. Those capable of growth above 113 °F, with optimum growth at 131 to 149 °F, are *thermophiles*. Thermophilic growth can be extremely rapid, with generation times of 10 to 20 min. Thermophiles can become a problem in blanchers and other equipment that maintains food

Table 2 Minimum Growth Temperatures for Some Bacteria in Foods

Organism	Possible Significance	Approximate Minimum Growth Temperature, °F
Staphylococcus aureus	Food-borne disease	50
Salmonella spp.	Food-borne disease	42
Clostridium botulinum,	Food-borne disease	
proteolytic		50
nonproteolytic		38
Lactobacillus and *Leuconostoc*	Spoilage of cooked sausage	38
Listeria monocytogenes	Food-borne disease	34
Acinetobacter spp.	Spoilage of precooked foods	31
Pseudomonads	Spoilage of raw fish, meats, poultry, and dairy products	31

at elevated temperatures for extended periods. These organisms die or do not grow at refrigeration temperatures.

Mesophiles are organisms that grow best between 68 and 113 °F. Most pathogens are in this group, with optimum growth temperatures around 98.6 °F, *i.e.*, body temperature. Mesophiles also include a number of spoilage organisms. Growth of mesophiles is quite rapid, with typical generation times of 20 to 30 min. Because mesophiles grow so rapidly, perishable foods must be cooled as fast as possible to prevent spoilage or potential unsafe conditions. Also, slower cooling rates cause mesophiles to adapt and grow at lower temperatures. With mild temperature abuse, prolific growth leading to spoilage or a potential health hazard can occur.

Psychrotrophs are organisms capable of growth at 41 °F; some are able to grow at temperatures as low as 23 °F and are a primary cause of spoilage of perishable foods. Psychrotrophic growth is slow in comparison to mesophilic and thermophilic growth, with maximum growth rates of 1 to 2 h or longer. However, control of psychrotrophic growth is a major requirement in products with extended shelf life. Because many psychrotrophs have optimum temperatures in the mesophilic range, what may seem to be an insignificant increase in temperature can have a major impact on the growth rate of spoilage organisms. Growth is roughly twice as fast with each 5 °F increase in temperature.

For all the critical growth factors, the range over which growth can occur is characteristic for a given organism. The range for growth is narrower than that for survival. For example, the maximum temperature for growth is slightly above the optimum, and death usually occurs just slightly above the maximum. This is not the case at the lower end of the temperature range. Survival of psychrotrophic and most mesophilic microorganisms is enhanced by storage at low temperatures. Freezing of microorganisms is not an effective lethal process. Some organisms, notably gram-negative bacteria, are damaged by freezing and may die slowly, but others are extremely resistant. In fact, freezing is used as an effective means of preserving microbial cultures at extremely low temperatures, *e.g.*, −112 °F.

Environmental Relative Humidity. Water, previously discussed as an essential intrinsic growth factor, is also a major extrinsic factor. Environmental water acts as a vector for transmission of microorganisms from one location to another through foot traffic or aerosols. Refrigeration drain pans and drip coils have been identified as significant contributors of *L. monocytogenes* contamination in food processing environments. Aerosols have also transmitted the agent that causes Legionnaires' disease. High relative humidity in cold rooms is a particular problem after sanitation and leads to black mold buildup on walls and ceilings as well as growth of organisms in drains and other reservoirs of water. Condensation that forms on ceilings supports microbial growth and can subsequently drip onto product contact surfaces. Inadequately drained equipment collects stagnant water and supports microbial growth that is easily transported throughout a production facility when people walk through puddles. It is extremely important to control environmental relative humidity in food production environments. Control measures are discussed further in the Regulations and Standards section.

Oxygen. Microorganisms are frequently classified by their oxygen requirement. Strictly aerobic microorganisms, such as molds and pseudomonads, require oxygen for growth. Conversely, strict anaerobes, such as *Clostridium* spp., cannot grow in the presence of oxygen. Facultatively anaerobic microorganisms (*e.g.*, coliforms) grow with or without oxygen present, and microaerophiles, such as lactobacilli, grow best in conditions with reduced oxygen levels. Controlled atmospheric chambers for fruit storage use lower oxygen levels to prolong storage life by retarding growth of spoilage organisms in addition to influencing ripening processes. Vacuum packing of foods also uses this extrinsic growth factor by inhibiting the growth of strict aerobes, such as molds and pseudomonads.

Design for Control of Microorganisms

Microorganisms can be controlled by one of three mechanisms—prevention of contamination, prevention of growth, or destruction of the organisms themselves. Design of refrigeration and ventilation systems can impact all these areas, sometimes in combination.

Prevention of Contamination

To prevent the entry of microorganisms into food production areas, ventilation systems must provide adequately clean air. Because bacteria are generally transported through air on dust particles, 95% filters (as defined in ASHRAE *Standard* 52.1-1992, Gravimetric and Dust-Spot Procedures for Testing Air-Cleaning Devices Used in General Ventilation for Removing Particulate Matter) are sufficient to remove most microorganisms. High-efficiency particulate air (HEPA) filters provide sterile air and are used for cleanrooms.

It is essential that air-filtering materials remain dry. Wet filters in ventilation systems support microbial growth, and organisms are transported throughout the production facility in the air. All ventilation systems must also be protected from water and the formation of condensate to prevent mold growth. This may require increased airflow or dehumidification systems. Positive pressure in the production environment prevents the entry of airborne contamination from sources other than ventilation ducts. Air intakes for production areas must not be positioned toward areas that are prone to contamination, such as puddles on roofs, nesting sites for birds, and so forth. Microbial contamination from airborne sources has been recorded, but quantitative assessments are not available to specify design parameters.

Refrigeration drip pans are a significant source of *L. monocytogenes* contamination. Condensation drip pans should be plumbed directly to drain to prevent contamination of floors and subsequent transport of organisms throughout a production facility. Drip pans must be easily accessible to allow scheduled cleaning, thus preventing organism growth. Air defrost should be avoided in critical areas. Continuous glycol-sprayed evaporative coils offer advantages, since glycol has been found to trap and kill microorganisms. Being hygroscopic, glycol depresses the dew point of the air providing a drier environment.

Traffic flow through a production facility should be planned to minimize contact between raw and cooked products. This is mandated through USDA regulations in plants that cook meat products. Straight line flow of a raw product from one end of a facility to the other prevents cross contamination. Walls separating raw product from cooked (or dirty from clean), with positive pressure in the cooked area, should be considered, as this provides the best means of protection. Provide adequate storage facilities to allow separate storage of raw ingredients from processed products, especially in facilities that handle meat products, which are a significant source of *Salmonella*. Raw meat must not be stored with cooked meats and/or vegetables or dairy products.

Prevention of Growth

Water control is one of the most effective and most frequently overlooked means of inhibiting microbial growth. All ventilation systems, piping, equipment, and floors must be designed to drain completely. Residual standing water supports rapid microbial growth, and foot and forklift traffic transports organisms from puddles throughout the production facility.

Condensation on ceilings and chilled pipes also supports microbial growth and may drip onto product contact surfaces that are not adequately protected. Efforts to prevent condensation are essential to prevent contamination. Insulation of pipe and/or dehumidifying systems may be necessary, particularly in chilled rooms. Increased airflow may also be useful in removing residual moisture. Maintenance of 70% rh prevents the growth of all but the most osmotolerant microorganisms; less than 60% rh prevents all microbial growth on facility surfaces (Table 1).

Sanitation procedures use much water and leave the facility and surfaces wet. Adequate dehumidification should be provided for removal of moisture during and after sanitation.

Controlling relative humidity is not always possible. For example, the aging of meat carcasses requires relative humidities of 90 to 95% to prevent excessive drying of the tissue. In these cases, temperatures just above the product freezing point should be used to inhibit microbial deterioration. Temperatures below 40°F inhibit the most common organisms that cause food-borne illness; however, 38°F is required to inhibit *L. monocytogenes*. Airflow, relative humidity, and temperature must be finely balanced to achieve maximum shelf life with limited deterioration of quality.

Freezing is also an effective means of microbial control. Limited death may occur during freezing, especially during slow freezing of gram-negative bacteria. However, freezing is not a reliable means of inactivating microorganisms. Since no microbial growth occurs in frozen foods, as long as a product remains below its freezing point, microbial safety issues are nonexistent. Frozen foods must be stored below 0°F for legal and quality reasons.

Destruction of Organisms

High temperature is an effective means of inactivating microorganisms and is used extensively in blanching, pasteurization, and canning. Moist heat is far more effective than dry heat. High temperatures (170°F) may also be used for sanitation when chemicals are not used. While hot water sanitation is effective against vegetative forms of bacteria, spores are not affected.

In addition to heat, gases such as ethylene oxide and methyl bromide, irradiation, ultraviolet light, and sanitation chemicals are effective in destroying microorganisms.

SANITATION

Cleaning and sanitation are key elements that incorporate all three strategies for control of microorganisms. The cleaning phase controls microbial growth by removing the residual food material required for proliferation. Sanitizing kills most of the bacteria that remain on surfaces. This prevents subsequent contamination of food being produced. Most microbial issues that occur in food processing environments are caused by unclean equipment, sometimes due to design. Therefore, equipment and facilities designed with cleaning and sanitizing in mind maximize the effectiveness of control.

Products that are frozen prior to packaging are particularly vulnerable to contamination. Many freezing tunnels in food processing facilities are difficult or impossible to clean because of limited access and poor drainage. Although freezing temperatures control microbial growth, proliferation of organisms does occur during downtime, such as on weekends. The following points should be considered during design phases to minimize potential problems:

Table 3 Common Cleaning and Sanitizing Chemicals

Cleaning Compounds	Sanitizers
Caustic	Chlorine
Chlorinated alkaline detergents	Iodophors
Acid cleaners	Quaternary ammonium compounds
	Acid sanitizers

- Provide good access for the cleaning crew to facilitate cleaning and adequate lighting (50 foot candles) to allow inspection of all surfaces.
- Eliminate inaccessible parts and features that permit product accumulation.
- Design equipment that is easy to dismantle using few tools, especially for areas that are difficult to clean. Design air-handling ducts for ease of cleaning. Provide removable spools or access doors. Washable fabric ducts designed and approved for such use are another option.
- Use smooth and nonporous construction materials to prevent product accumulation. Materials must tolerate common cleaning and sanitizing chemicals listed in Table 3. Consult sanitation personnel to determine chemicals likely to be used. Give special attention to insulation materials, many of which are porous. Insulation must be protected from water to avoid saturation and resultant microbial growth. An effective method is a well-sealed PVC or stainless steel cover. Avoid using fiberglass batts in food processing plants.
- All equipment must drain completely.
- Consult references and regulations on sanitary design principles.

Innovation is needed in the area of drying after cleaning is complete. Provision of adequately sloped surfaces and sufficient drains to handle water is important. Dehumidification systems and/or increased airflow in new and existing systems could greatly reduce problems associated with water, especially in the cleaning of chilled production environments.

Standard water washing procedures are not appropriate for certain food production facilities such as dry mix, chocolate, peanut butter, or flour milling operations. Refrigeration or ventilation systems for plants of this type must be made to facilitate dry cleaning, reduce condensation, and restrict water to a very confined area if it is absolutely necessary.

REGULATIONS AND STANDARDS

Facilities and equipment should be designed and installed for minimizing microbial growth and maximizing the ease of sanitation. Care should be taken to employ material that can withstand moisture and chemicals.

The food industry has developed several equipment installation standards. In the United States, examples are International Association of Milk, Food and Environmental Sanitarians (IAMFES, Ames, IA) 3-A Dairy standards, Baking Industry Sanitation Standards Committee (BISSC, Chicago, IL) bakery standards, and a select group of U.S. Department of Agriculture (USDA) standards for the meat industry. The USDA enforces the Federal Meat Inspection Act and Federal Poultry Inspection Act and requires approval of building and equipment plans.

Chapter VII, Section 701(A) of the Federal Food, Drug and Cosmetic Act as amended establishes current Good Manufacturing Practices (GMPs) in manufacturing, processing, packaging, or holding human food. These GMPs are listed in Section 21 of the Code of Federal Regulations (21 CFR), Part 110.

BIBLIOGRAPHY

Banwart, G.J. 1979. *Basic food microbiology.* AVI Publishing Company, Inc., Westport, CT.

British Association of Chemical Specialties. 1989. *The control of Legionellae by safe and effective operation of cooling systems. A code of practice.*

Chartered Institute of Building Services Engineers. 1987. Technical Memorandum TM13.

Graham, D.J. 1991-1992. *Sanitary design—A mind set.* A nine-part serial in *Dairy, Food and Environmental Sanitation.*

ICMSF. 1986. *Microorganisms in foods 2. Sampling for microbiological analysis: Principles and specific applications,* 2nd ed. University of Toronto Press, Toronto.

Imholte, T.J. 1984. *Engineering for food safety and sanitation.* Technical Institute for Food Safety, Crystal, MN.

IIR. 1979. *Recommendations for chilled storage of perishable produce.* International Institute of Refrigeration, Paris.

IIR. 1986. *Recommendations for the processing and handling of frozen foods,* 3rd ed. International Institute of Refrigeration, Paris.

Jay, J.M. 1992. *Modern food microbiology,* 4th ed. Van Nostrand Reinhold, New York.

Marriott, N.G. 1989. *Principles of food sanitation,* 2nd ed. Van Nostrand Reinhold, New York.

NAS. 1985. *An evaluation of the role of microbiological criteria for foods and food ingredients.* National Academy Press, Washington, D.C.

NFPA. 1989. Guidelines for the development, production, distribution and handling of refrigerated foods. National Food Processors Association, Washington, D.C.

Todd, E. 1990. Epidemiology of food-borne illness: North America. *The Lancet* 336:788-93.

BIBLIOGRAPHY

CHAPTER 10

METHODS OF PRECOOLING FRUITS, VEGETABLES, AND CUT FLOWERS

COOLING is generally considered the removal of field heat from freshly harvested products in time to inhibit spoilage and to maintain preharvest freshness and flavor. The term precooling implies the removal of heat before the product is shipped to a distant market, processed, or stored. Some products are room cooled in the same room in which they are stored. In this case, cooling is accomplished over a day or more. Precooling is generally done in a separate facility within a few hours or even minutes. Therefore, room cooling is not considered precooling.

PRODUCT REQUIREMENTS

Product physiology, in relation to harvest maturity and ambient temperature at harvest time, largely determines cooling requirements and methods. Some products are highly perishable and must be cooled as soon as possible after harvest. Vegetables in this category include: asparagus, snap beans, broccoli, cauliflower, sweet corn, cantaloupes, summer squash, vine-ripened tomatoes, leafy vegetables, globe artichokes, brussels sprouts, cabbage, celery, carrots, snow peas, and radishes. Vegetables such as white potatoes, sweet potatoes, winter squash, pumpkins, and mature green tomatoes need to be cured or ripened at some temperature higher than desirable for holding more perishable produce. Cooling of these products is not as important; however, some cooling is necessary if ambient temperature is high during harvest. Vegetables not listed may or may not be cooled because of lack of economic importance or susceptibility to chilling injury—for example, cucumbers.

Commercially important fruits that need to be precooled immediately after harvest include: apricots, avocados, all of the berries except cranberries, tart cherries, peaches and nectarines, plums and prunes, and tropical and subtropical fruits such as guavas, mangos, papayas, and pineapples. The tropical and subtropical fruits of this group are susceptible to chilling injury and thus need to be cooled according to individual temperature requirements. Sweet cherries, grapes, pears, and citrus fruit have a longer postharvest life than the former fruits, yet prompt cooling is essential to high quality retention during the holding period. Bananas require special ripening treatment and therefore are not precooled. Because of their keeping quality, apples generally do not need to be precooled. Early varieties, harvested when the ambient temperature is high, are normally more perishable and may benefit by precooling. Other varieties, particularly those that are stored for several months, may benefit by precooling when the apples cannot be cooled to 32°F in storage within seven days after harvest.

The preparation of this chapter is assigned to TC 11.5, Fruits, Vegetables, and Other Products.

CALCULATION METHODS

Heat Load

The refrigeration capacity needed for precooling is much greater than that required for holding a product at a constant temperature or for slow cooling of a product. Therefore, the heat load on a precooling system should be determined as accurately as possible. While it is imperative to have an adequate amount of refrigeration for effective precooling, it is uneconomical to have more refrigerating capacity available than is normally needed.

On jobs where refrigeration is needed only during a regular 8- to 10-h workday, ice builders make it possible to reduce the amount of refrigeration equipment. Equipment size can be cut in half, or more, depending on the hours of off time in relation to the hours of precooler operation.

The total heat load comes from the product, surroundings, air infiltration, containers, and heat-producing devices such as motors, lights, fans, and pumps. Product heat accounts for the major portion of total heat load on a precooling system. Product heat load depends on product temperature and cooling rate, amount of product cooled in a given time, and specific heat of the product. Heat from respiration is part of the product heat load, but only when the cooling time exceeds a few hours.

Mass-Average Temperature. An accurate determination of product temperature is essential for accurate heat load calculations. Due to the rapid heat transfer, a temperature gradient develops within the product, with faster cooling causing larger gradients. This gradient is a function of product properties, surface heat transfer parameters, and cooling rate. For example, initially hydrocooling removes heat from the exterior of a product. Consequently, temperature in this area changes rapidly while the center temperature may not change at all. Most of the product mass is in the outer portion. Thus, it is possible, based on center temperature, to calculate a small or negligible heat load while, in fact, substantial heat has been extracted. For this reason, the product mass-average temperature must be used for product heat load calculations (Smith and Bennett 1965). A mass-average temperature denotes the single value from the transient temperature distribution that would become the uniform product temperature when held for a period under adiabatic conditions.

Figure 1 can be used to determine the mass-average temperature t_{ma} of peaches during hydrocooling. Subsequently, the product cooling load can be calculated by Equation 1.

$$q = mc_p(t_i - t_{ma}) \tag{1}$$

The nomograph can be applied to products other than peaches, when temperature ratio with respect to cooling time and product size is known. Figure 2 illustrates the comparative relationship of fractional temperature difference, or temperature ratio, Y, to

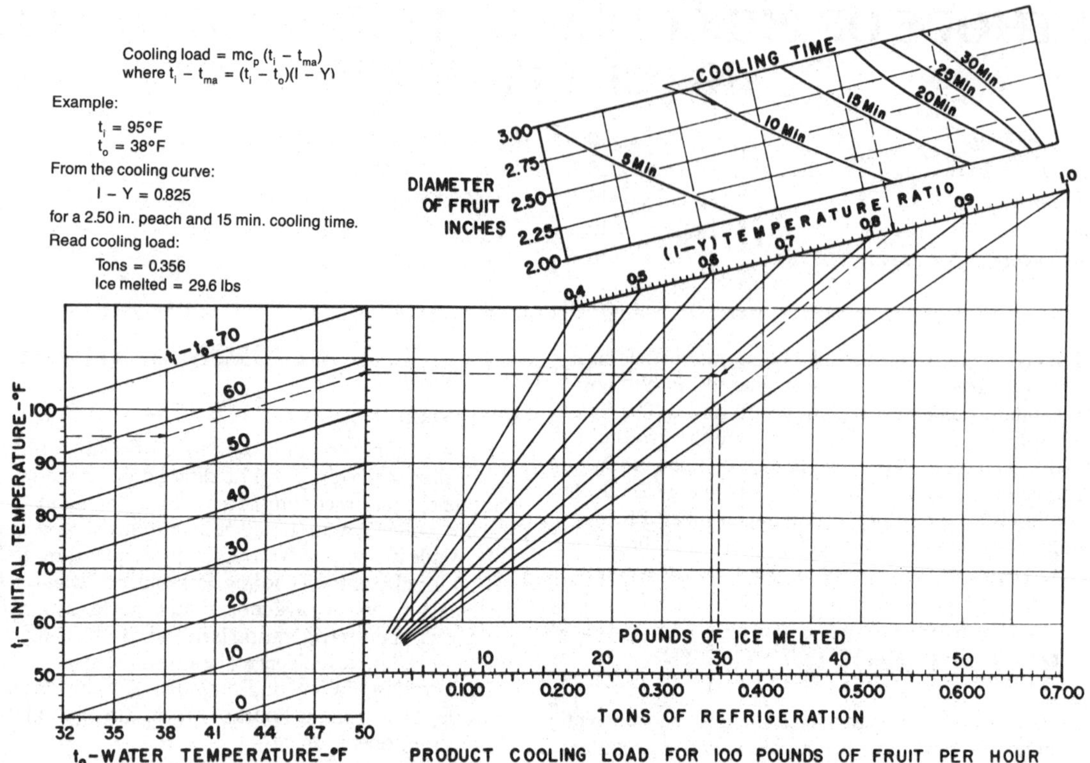

Fig. 1 Nomograph to Determine Product Heat Load of Hydrocooled Peaches

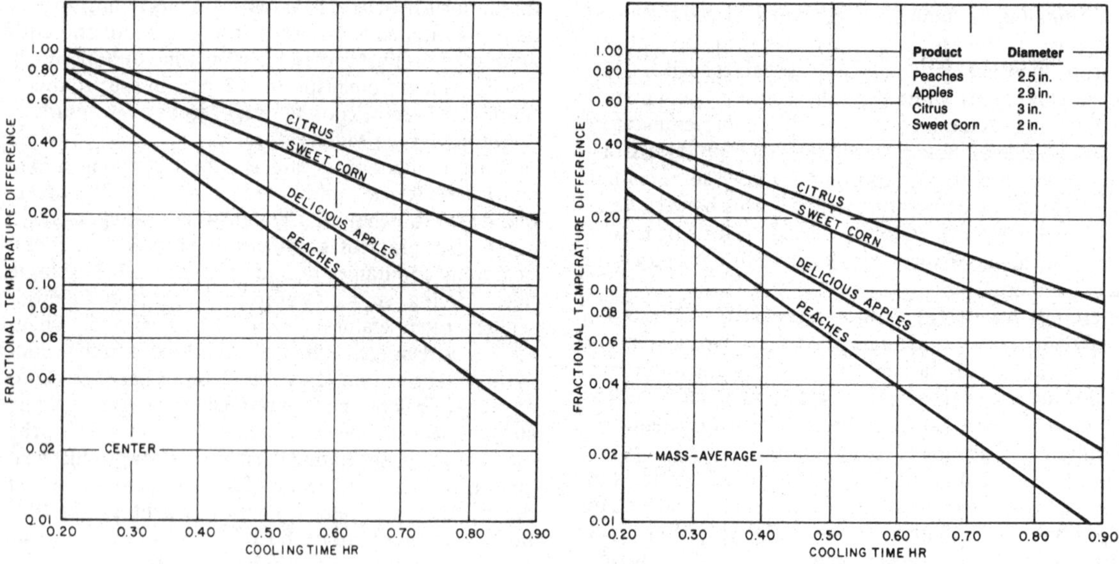

Fig. 2 Time-Temperature Response for Hydrocooled Produce

cooling time for 2.5-in. diameter peaches, 2.88-in. diameter apples and citrus fruit, and 2-in. diameter sweet corn when hydrocooled under ideal conditions of negligible surface resistance to heat transfer.

Cooling Rate

The terms *cooling coefficient, C,* and *half-cooling time, Z,* are perhaps the most meaningful in presenting cooling rate data. The temperature ratio Y is the unaccomplished temperature change at any time in relation to the total temperature change possible for the cooling condition. It is calculated by the equation,

$$Y = (t - t_o)/(t_i - t_o) \tag{2}$$

The inverted temperature ratio denotes the percent accomplished temperature change. The value of the cooling coefficient is the same in either case.

The cooling coefficient denotes the change in product temperature per unit change of cooling time for each degree temperature difference between the product and its surroundings. In most cases, a logarithmic transformation of the temperature ratio values provides a fairly accurate straight line fit to the data points. If Newtonian heat transfer occurs (a negligible temperature gradient within the product during cooling) and the straight line asymptote intercepts the temperature ratio at unity for zero time, the cooling coefficient may be calculated by the simplified formula

$$C = \ln Y/\theta \tag{3}$$

When produce is rapidly cooled, the conditions for Newton's law are seldom satisfied. A considerable temperature gradient often develops within the product, depending on its properties and rate of surface heat transfer. Also, the zero asymptote intercepts the temperature ratio at some value either greater or less than unity. In this case, the constant cooling coefficient is calculated by the equation

$$C = (\ln Y_1 - \ln Y_2)/(\theta_1 - \theta_2) \tag{4}$$

where the subscripts 1 and 2 denote points along the straight line asymptote to the cooling curve.

The half cooling time is the time required to reduce the temperature difference between the product and its surroundings by one-half. Mathematically, it is expressed as

$$Z = \ln(0.5)/C \tag{5}$$

where C is a negative value.

If certain properties and surface parameters are known, cooling coefficients may be predicted by the equation

$$C = \alpha M_1^2/l^2 \tag{6}$$

where l is the characteristic length.

$M_1^2 = G\pi^2$ if there is no resistance to heat transfer at the surface interface. Smith *et al.* (1967) derived Equation (6) and the geometry index G. When there is resistance to heat transfer at the surface, usually the case in air cooling, the transcendental function M_1^2 becomes dependent on the Biot number in relation to the product geometry. Smith *et al.* (1968) present a nomograph to obtain M_1^2 as a function of the geometry index and the Biot number.

Cooling coefficients predicted by means of Equation (6) have physical meaning only if the value of the intercept is known or can be estimated. Employing the approach of Ball and Olson (1957), Pflug *et al.* (1965) outlined a procedure for developing time-temperature response charts for objects that can be approximated by a sphere, an infinite plate, or a cylinder, and gave examples for the corresponding solution of practical problems involving use of the intercept and Biot number as functions of the surface heat transfer characteristics. The geometry analysis developed by Smith *et al.* (1967) permits extension of these solutions to anomalous shapes.

These solutions treat the product as an individual unit in a specified environment and isolated from the influence of surrounding material, which is rarely the case. Often, products are cooled in bulk or in packages of several layers, where the surrounding material influences the cooling response, particularly in air-cooling systems. The individual approach usually gives poor estimates for cooling of bulk loads.

Baird and Gaffney (1976) developed a numerical technique for calculating cooling rates in bulk loads of fruits or vegetables. This procedure, applicable to spherical products, uses finite difference solutions of the differential equations that describe heat transfer both within individual fruits and at different levels in bulk loads. Eshleman *et al.* (1976) extended this development to provide solutions for heat transfer in irregularly shaped objects. These models can be used to calculate temperatures at various points within individual products and at different depths in bulk loads at regular time intervals during cooling. Required inputs to the models include product size and shape, thermal properties of the product, the convective coefficient, and flow rate of the cooling medium.

COOLING METHODS

The principal methods of precooling are hydrocooling, forced-air cooling, package icing, and vacuum cooling. Most cooling is done at the packinghouse or in central cooling facilities. Some products can be cooled by any of these methods without suffering any adverse effects. For these products, the cooling method chosen is often determined more by such factors as economy, convenience, relation of the cooling equipment to the total packing operation, and personal preference.

HYDROCOOLING

Because of its simplicity, economy, and effectiveness, hydrocooling is a popular precooling method. The rate of heat transfer Q is directly proportional to the surface heat transfer coefficient h, the total surface area A, and the difference in temperature between the surface and its surroundings Δt. When a film of cold water flows briskly and uniformly over the surface of a warm substance, the surface temperature of the substance becomes essentially equal to that of the water. Rate of internal cooling is limited by the size and shape (volume in relation to surface area) and thermal properties of the substance being cooled.

If the fluid velocity is sufficient, as in gravity flow or forced convection over fruit and vegetable products, the resistance to heat transfer at the surface is negligible. In this case, the heat is removed as fast as it comes to the surface, and the temperature difference across the surface boundary layer is small (1°F or less). In ideal hydrocooling, the optimum average film coefficient of heat transfer is roughly 120 Btu/(h·ft²·°F); the average temperature difference across the boundary layer is about 0.8°F. On this basis, the rate of heat transfer per unit surface area is $h = 120 \times 0.8 = 96$ Btu/h per square foot of product area.

The value of h varies substantially, depending on many conditions. However, because of limitations in water temperature and product thermal properties and similarity among products, this is about as fast as any fruit or vegetable can be cooled in water. The ideal (maximum h) time-temperature response curves for hydrocooling select fruit and vegetable products are shown in Figure 2, based on both center and mass-average temperatures.

Commercial Hydrocooling

Freestone peaches, clingstone peaches hauled more than 100 miles for canning, tart cherries, and cantaloupes are hydrocooled. Few apples and citrus fruits are hydrocooled. Hydrocooling is not popular for citrus fruit because it has a long marketing season and good postharvest holding ability. Citrus fruits are also susceptible to increased peel injury and to decay and loss of quality and vitality after hydrocooling. Apples are usually cooled in the storage rooms.

Many vegetables are successfully hydrocooled. The more important of these are sweet corn, celery, radishes, and carrots. Cooling rate data for hydrocooling are given by Grierson (1957), Ford (1956), and Perry and Perkins (1968).

Types. Hydrocooling is accomplished by flooding, spraying, or immersing the product in an agitated bath of chilled water.

Commercial hydrocoolers for freestone peaches are *flood-type* and *bulk-type*. The flood-type hydrocooler cools the packaged product by flooding as it is conveyed through a cooling tunnel. Adaptations consist of conveying the product through the cooling tunnel in loose bulk or in bulk bins. The bulk-type cooler uses combined immersion and flood cooling. Loose fruit, dumped into cold water, remains immersed for half of its travel through the cooling tunnel. An inclined conveyor gradually lifts the fruit out of the water and moves it through an overhead shower. The bulk-type cooler permits greater packaging flexibility than the flood-type.

Most vegetables are moved by fork truck in unit loads of as many as 40 packages. In one system, chilled water sprays on stacks of packed vegetables placed in a refrigerated room. Water is collected in floor drains leading to a sump and recirculated over the product. One nozzle spraying approximately 90 gpm of water is located over each stack, which may contain from 30 to 40 crates (Grizzell and Bennett 1966).

Other unit load systems convey the stacks through cooling tunnels. Either spray nozzles or a flood pan may be used to deliver up to 400 gpm of chilled water to each stack. Most commercial hydrocoolers provide overhead showering at a rate of 10 to 15 gpm per square foot of top face area. This type of hydrocooler requires less space than the batch type, but is not as flexible. With the batch system, chilled water can be sprayed on the product indefinitely, depending on the season and the incoming product temperature. Also, the crates can be left in place after cooling until they are shipped.

Hydraircooling uses a mixture of refrigerated air and water in a fine mist spray that is circulated around and through the stack by forced convection (Henry *et al.* 1976). It has the advantage of reduced water requirements, the potential for improved sanitation, and the capability of adapting to fiberboard containers of the type that cannot be used in conventional hydrocooling systems. Cooling rates equal to, and in some cases better than, those obtained in conventional unit load hydrocoolers are possible.

Design and Operation. Flooded ammonia systems are used to cool the water for hydrocoolers in large packinghouses. Some use a secondary coolant. The cooling coils are contained in a tank through which water is rapidly circulated. Refrigerant temperature inside the cooling coils is approximately 28 °F, except in systems with ice building coils where it may be slightly lower.

To determine refrigeration requirements, the product mass-average temperature should be used. Because the mass-average temperature denotes the amount of heat within a product (above some reference level) no matter what the thermal gradient is, it is the only temperature value that will yield an accurate heat load computation.

Insolation, radiation from hot surfaces, convection from ambient air, or conduction from surroundings can affect the load. Protecting the hydrocooler from these sources of heat gain will enhance efficiency. Refrigeration capacity in excess of needs and cooling to a temperature below that required also reduces efficiency.

Cooling rate decreases in proportion to cooling time, provided water temperature is constant. Cooling time should be governed according to initial and final product temperature, product size, and thermal characteristics of the product with respect to surface parameters. Showalter and Grierson (1970) showed a substantial difference in half-cooling time for sizes 36 and 45 cantaloupes. A weighted average of temperatures taken at different depths showed that 20 min was required to half-cool size 36 melons and only 10 min for size 45.

When hydrocooler water is recirculated, as is usually the case in mechanically refrigerated units, decay-producing organisms accumulate. Mild disinfectants such as chlorine or approved phenol compounds will reduce buildup of bacteria and fungus spores, but they will not kill infections already in the products or sterilize either the water or the product surfaces. Brown rot and rhizopus decay spores on the surface and under the skin of peaches can be destroyed by soaking the fruit in water at 125 °F for 2 to 3 min. (Smith and Redit 1968), or by treating the fruit with a chemical fungicide, which is usually put into the hydrocooling water. The principal problem with chemicals is maintenance of uniform concentrations, particularly in ice-refrigerated equipment because of the constant dilution from melting ice. In addition, hydrocooler shower pans and/or trash screens need regular (daily) cleaning to provide maximum efficiency.

FORCED-AIR COOLING

Air cooling at a rate comparable to hydrocooling can be theoretically obtained by providing certain conditions of product exposure and air temperature.

Considering the classic heat transfer correlation of Nusselt-Reynolds-Prandtl numbers,

$$\frac{2hl}{k} = C\left(\frac{2lG'}{\mu}\right)^m \left(\frac{\mu c_p}{k}\right)^n \qquad (7)$$

for a specified air temperature and product characteristics length, the average product film coefficient of heat transfer h is a function of the interstitial mass velocity and may be written in the form:

$$h = f(G') \qquad (8)$$

The value of h, therefore, depends on the volume rate of airflow and the physical characteristics of the product.

In air cooling, the optimum value of h is considerably smaller than when cooling with water. However, Pflug *et al.* (1965) showed that apples moving through a cooling tunnel on a conveyer belt cool faster with air at 20 °F approaching the fruit at a velocity of 600 fpm than they would in a water spray at 35 °F. For this condition of air cooling, they calculated an average film coefficient of heat transfer of 7.3 Btu/(h·ft^2·°F). They note that the advantage of air is its lower temperature and that if the water were reduced to 34 °F, the time for water-cooling would be less. In tests to evaluate film coefficients of heat transfer for anomalous shapes, Smith *et al.* (1970) obtained an experimental value of 6.66 Btu/(h·ft^2·°F) for a single Red Delicious apple in a cooling tunnel with air approaching at 1570 fpm. At this rate of air flow, the logarithmic mean surface temperature of a single apple cooled for 0.5 h in air at 20 °F is approximately 35 °F. The average temperature difference across the surface boundary layer is, therefore, 15 °F and the rate of heat transfer per square foot of surface area is:

$$q/A = 6.66 \times 15 = 100 \text{ Btu/(h·ft}^2)$$

For these conditions, the cooling rate compares favorably with that obtained in ideal hydrocooling. However, these coefficients are based on single specimens isolated from surrounding fruit. Had the fruit been in a packed bed at equivalent flow rates, the values would have been less because less surface area would have been exposed to the cooling fluid. Also, the rate of evaporation from the product surface significantly affects the cooling rate.

Because of physical characteristics, mostly geometry, various fruit and vegetable products respond differently to similar treatments of airflow and air temperature. Perry *et al.* (1968) found that peaches cool faster than potatoes when they are cooled in a packed bed under similar conditions of airflow and air temperature, for example.

Surface coefficients of heat transfer are sensitive to the physical conditions involved among objects and their surroundings. Experimental surface coefficients ranging from 9 to 12 Btu/(h·ft^2·°F) were obtained by Soule *et al.* (1966) for bulk lots of Hamlin oranges and Orlando tangelos with air approaching the mass of fruit at velocities ranging from 225 to 350 fpm. Bulk bins containing 1000 lb of 2.85-in. diameter Hamlin oranges were cooled from 80°F to a final mass-average temperature of 46.5°F in 1 h with air approaching the fruit at 330 fpm (Bennett *et al.* 1966). Surface heat transfer coefficients for these tests averaged slightly above 11 Btu/(h·ft²·°F). On the basis of a log mean air temperature of 44°F, the calculated half-cooling time was 0.27 h.

By correlating data from experiments on cooling 2.8-in. diameter oranges in bulk lots with results of a mathematical model, Baird and Gaffney (1976) found surface heat transfer coefficients of 1.5 and 9 Btu/(h·ft²·°F) for approach velocities of 11 and 412 fpm, respectively. A Nusselt-Reynolds heat transfer correlation representing data from six experiments on air cooling of 2.8-in. diameter oranges and seven experiments on 4.2-in. diameter grapefruit, with approach air velocities ranging from 5 to 412 fpm, gave the relationship $Nu = 1.17\ Re^{0.529}$, with a correlation coefficient of 0.996.

Ishibashi *et al.* (1969) constructed a stage-type forced-air cooler that exposed bulk fruit to air at a progressively declining temperature (50, 32, and 14°F) as the fruit was conveyed through the cooling tunnel. Air approached the fruit at 700 fpm. With this system, 2.5-in. diameter citrus fruit cooled from 77 to 41°F in 1 h. Their half-cooling time of 0.32 h compares favorably with a half-cooling time of 0.30 h for similarly cooled Red Delicious apples at an approach air velocity of 400 fpm (Bennett *et al.* 1969). Perry *et al.* (1968) obtained a half-cooling time of 0.5 h for potatoes in a bulk bin with air approaching at 250 fpm, as compared to 0.4 h for similarly treated peaches and 0.38 h for apples. Optimum approach velocity for this type of cooling is in the range of 300 to 400 fpm, depending on conditions and circumstances.

Commercial Methods

Produce can be satisfactorily cooled (1) with air circulated in refrigerated rooms adapted for that purpose, (2) in rail cars or highway vans using special portable cooling equipment that cools the load before it is transported, (3) with air forced through the voids of bulk products moving through a cooling tunnel on continuous conveyors, (4) on continuous conveyors in wind tunnels, or (5) by the forced-air method of passing air through the containers by pressure differential. Each of these methods is used commercially, and each is suitable for certain commodities when properly applied.

In circumstances where air cannot be forced directly through the voids of products in bulk, a container type and a load pattern that permits air to circulate through the container and reach a substantial part of the product surface is beneficial. Examples of this are (1) small products such as grapes and strawberries that offer appreciable resistance to airflow through the voids in bulk lots, (2) delicate products that cannot be handled in bulk, and (3) products that are packed in the shipping containers before they are precooled.

Forced-air or pressure cooling involves definite stacking patterns and the baffling of stacks so that the cooling air is forced through, rather than around, the individual containers. Successful use of the method requires a container with vent holes placed in the direction in which the air will move and a minimum of packaging materials that would interfere with free movement of air through the containers. Under these conditions, a relatively small pressure differential between the two sides of the containers results in good air movement and excellent heat transfer. Differential pressures in use are about 0.25 to 3 in. of water, with airflows ranging from 1.0 to 3.0 cfm/lb of product.

Because the cooling air comes in direct contact with the product being cooled, cooling is much faster than with conventional room cooling. This gives the advantage of rapid product movement through the cooling plant, and the size of the plant is one-third to one-fourth that of an equivalent cold room type of plant.

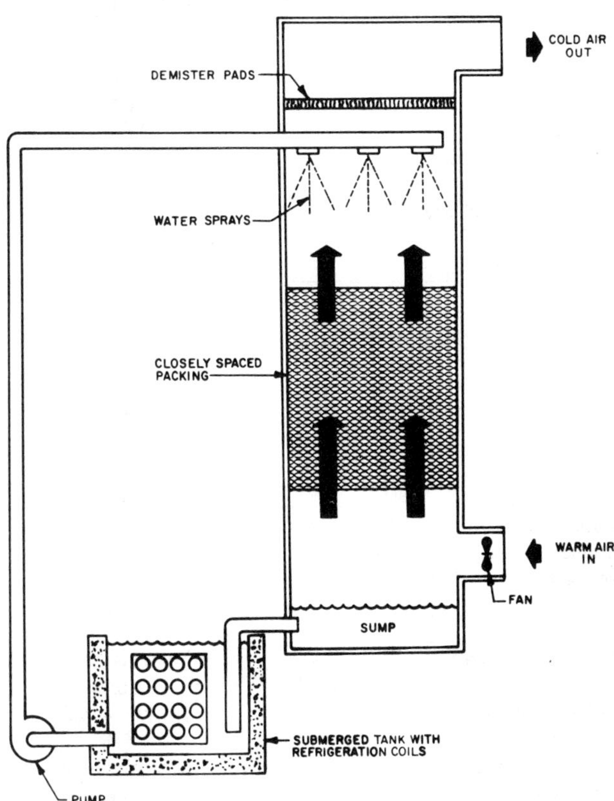

Fig. 3 Packed Tower for Forced-Air Cooling System

Mitchell *et al.* (1972) observed that forced-air cooling usually cools in one-fourth to one-tenth the time needed for conventional room cooling but that it still takes two to three times longer than hydrocooling or vacuum cooling. Guillou (1963) reported a half-cooling time of 1 h for forced-air pressure cooling of cup-packed peaches in lidded lugs as compared to 6 h for similarly packed peaches in conventional room cooling.

A proprietary direct-contact heat exchanger cools air and maintains high humidities using chilled water as a secondary coolant and a continuously wound polypropylene monofilament packing. It contains about 2000 linear feet of filament per cubic foot of packing section. Air is forced up through the unit while chilled water flows downward.

The dew-point temperature of the air leaving the unit equals the entering water temperature. Chilled water can be supplied from coils submerged in a tank. Buildup of ice on the coils provides an extra cooling effect during peak loads. This design also allows an operator to add commercial ice during long periods of mechanical equipment outage. A general diagram of the system is shown in Figure 3.

Effects of Containers and Stacking Patterns

The accessibility of the product to the cooling medium, essential to rapid cooling, may involve both access to the product within the container and to the individual container in a stack.

This effect is evident in the cooling rate data of various commodities in various types of containers reported by Mitchell *et al.* (1972). Parsons *et al.* (1972) developed a corrugated paperboard container venting pattern for palletized unit loads that produced cooling rates equal to those from conventional register stacked patterns. Fisher (1960) demonstrated that spacing apple containers on pallets reduced cooling time by 50% as compared with pallet loads stacked solidly.

Table 1 Input Values to an Engineering/Economic Model

Fixed Values	
Density	62.4 lb/ft^3
Porosity	0.39
Bulk density	38.1 lb/ft^3
Thermal conductivity	0.30 Btu/h · ft · °F
Specific heat	0.90 Btu/lb · °F
Transpiration coefficient	0.005 lb/h · ft^2 · psi
Nominal Values	
Product diameter	2 in.
Air approach velocity (Cartons)	50 fpm
(Bulk)	150 fpm
Cooling air temperature	35 °F
Product depth	4 ft
Ambient temperature	90 °F
Refrigerant	R-12
Condenser Δt	20 °F
Evaporator Δt	10 °F
Product throughput	20,000 lb/h
Annual operating time	800 h
Power cost	$0.10/kwh
Building cost	$50/ft^2
Heat exchanger cost	$0.50/*UA*(Btu/h · °F)
Compressor and accessories cost	$750/hp
Air handling system cost	$150/hp + $100
Interest rate	10%
Carton configuration	4 cartons deep
Carton vent area	4%
Final product temperature	40 °F
Initial product temperature	80 °F
Labor cost	$10/h
Labor required	1 man-hour/20,000 lb

Output results in Figure 4 are based on these input values.

Computer Solution

Baird *et al.* (1988) developed an engineering economic model for designing forced-air cooling systems. Figure 4 is an example of information that can be obtained from the model. By selecting a set of input conditions (which varies with each application), as indicated in Table 1, and varying the approach air velocity, the entering air temperature, or some other variable, the optimum (minimum cost) value can be determined. The selection of air velocity for cartons is critical, whereas the selection of entering air temperature is not as critical until the desired final product temperature of 40 °F is approached. As indicated in Table 1, the results are for four cartons deep with a 4% vent area in the direction of airflow, and they would be quite different if the carton vent area was changed. Other design parameters that can be optimized using this program are the depth of product in direction of airflow and the size of evaporator and condenser.

Table 2 gives additional information on the economics of forced-air cooling, based on the same set of input conditions in Table 1. Each unit cost is varied from zero to twice the nominal (considered to be average cost) value with the unit cooling cost given for each product. For these specific inputs, no one unit cost dominates; in fact, they all have about the same effect.

PACKAGE ICING

Finely crushed ice placed in shipping containers can effectively cool products that are not harmed by contact with ice. Spinach, collards, kale, brussels sprouts, broccoli, radishes, carrots, and onions are commonly packaged with ice (Hardenburg *et al.* 1986). Cooling a product from 95 to 35 °F requires melting ice equal to 38% of the product's mass. Additional ice must melt to remove heat leaking into the packages and to remove heat from the container. In addition to removing field heat, package ice can keep the product cool during transit.

Top icing, or placing ice on top of packed containers, is used occasionally to supplement another cooling method. Because

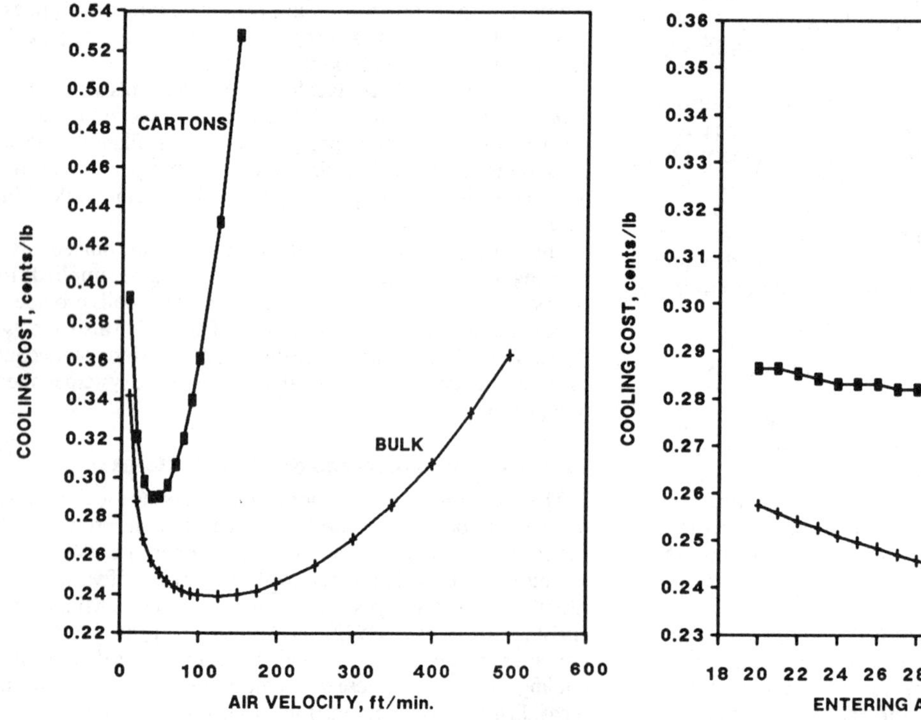

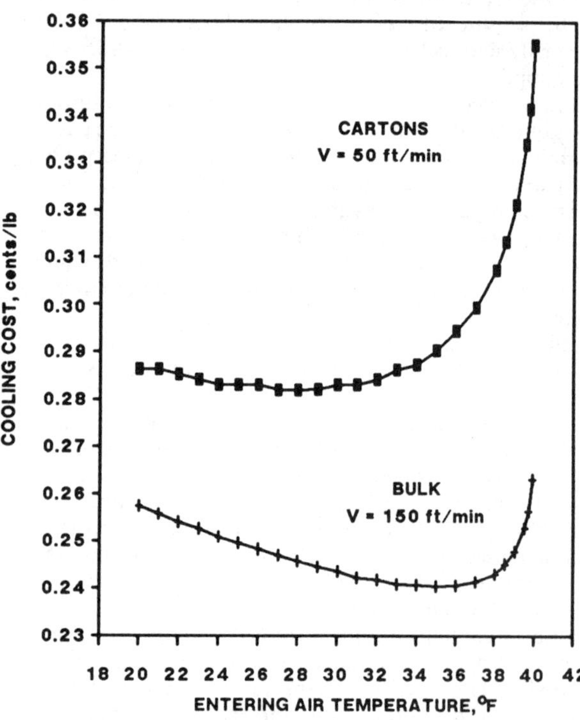

Fig. 4 Engineering-Economic Model Output for a Forced-Air Cooler
(Based on input values in Table 1)

Table 2 Relationships Among Unit Costs and Cost per Unit of Product Cooled
(Other parameters are shown in Table 1.)

	Unit Cost for Carton Cooling				
	−100%	−50%	Nominal	+50%	+100%
Building, $/ft²	0	25	50	75	100
(Cooling cost, ¢/lb)	(0.236)	(0.263)	(0.290)	(0.317)	(0.344)
Heat exchangers, $ per Btu/h·°F	0	0.25	0.50	0.75	1.00
(Cooling cost, ¢/lb)	(0.233)	(0.261)	(0.290)	(0.319)	(0.347)
Compressor, $/hp	0	375	750	1125	1500
(Cooling cost, ¢/lb)	(0.223)	(0.256)	(0.290)	(0.323)	(0.357)
Fan, $/hp	0	75	150	225	300
(Cooling cost, ¢/lb)	(0.286)	(0.288)	(0.290)	(0.292)	(0.294)
Labor, $/h	0	5	10	15	20
(Cooling cost, ¢/lb)	(0.240)	(0.265)	(0.290)	(0.315)	(0.340)
Interest rate, %	0	5	10	15	20
(Cooling cost, ¢/lb)	(0.201)	(0.241)	(0.290)	(0.345)	(0.404)
Power cost, $/kWh	0	0.05	0.10	0.15	0.20
(Cooling cost, ¢/lb)	(0.233)	(0.261)	(0.290)	(0.319)	(0.347)

corrugated containers have largely replaced wooden crates, the use of top ice has decreased in favor of forced-air and hydrocooling. Wax-impregnated corrugated containers, however, have allowed the use of icing and hydrocooling of products after packaging.

Pumping *slush ice* or *liquid ice* into the shipping container through a hose and special nozzle that connect to the package is another method used for cooling some products. Some systems can ice an entire pallet at one time.

VACUUM COOLING

Principle

Vacuum cooling of fresh produce by the rapid evaporation of water from the product works best with vegetables having a high ratio of surface area to volume. In vacuum refrigeration, water, as the primary refrigerant, vaporizes in a flash chamber under low pressure. The pressure in the chamber is lowered to the saturation point corresponding to the lowest required temperature of the water.

Vacuum cooling is a batch process. The product to be cooled is loaded into the flash chamber, the system is put into operation, and the product is cooled by reducing the pressure to the corresponding saturation temperature desired. The system is then shut down, the product removed, and the process repeated. Since the product is normally at ambient temperature before it is cooled, vacuum cooling can be thought of as a series of intermittent operations of a vacuum refrigeration system in which the water in the flash chamber is allowed to come to ambient temperature before each start. The functional relationships for determining refrigerating capacity are the same in each case.

Cooling is achieved by boiling water, mostly off the surface of the product to be cooled. The heat of vaporization required to boil the water is furnished by the product, which is cooled accordingly. As the pressure is further reduced, cooling continues to the desired temperature level. The saturation pressure for water at 212°F is 760 mm Hg. At 32°F, the saturation pressure is 4.58 mm Hg. Commercial vacuum coolers normally operate in this range.

Although the cooling rate of lettuce could be increased without danger of freezing, by reducing the pressure to 3.8 mm Hg, corresponding to a saturation temperature of 27°F, most operators do not reduce the pressure below the freezing temperature of water because of the extra work involved and the freezing potential.

Pressure, Volume, and Temperature

In a vacuum cooling operation, the thermodynamic process is assumed to take place in two phases. In the first phase, the product is assumed to be loaded into the flash chamber at ambient temperature, and the temperature in the flash chamber remains constant until saturation pressure is reached. At the onset of boiling, the small remaining amount of air in the chamber is replaced by the water vapor, the first phase ends, and the second phase begins simultaneously. The second phase continues at saturation until the product has cooled to the desired temperature.

If the ideal gas law is applied for an approximate solution in a commercial vacuum cooler, the pressure-volume relationships are:

$$\text{Phase 1} \quad pv = 29{,}318 \text{ ft·lb/lb}$$

$$\text{Phase 2} \quad pv^{1.056} = 66{,}370 \text{ ft·lb/lb}$$

where

p = absolute pressure, lb/ft²
v = specific volume, ft³/lb

The pressure-temperature relationship is determined by the value of ambient and product temperature. Based on 90°F for this value, the temperature in the flash chamber theoretically remains constant at 90°F as the pressure is reduced from atmospheric to saturation, after which it declines progressively along the saturation line. These relationships are illustrated in Figure 5. The product temperature would respond similarly but would vary depending on where temperature is measured in the product, the physical characteristics of the product, and the amount of product surface water available. While it is possible for some vaporization to occur within the intercellular spaces beneath the product surface, most of the water is vaporized off the surface. The heat required to vaporize this water is also taken off the product surface where it flows by conduction under the thermal gradient produced. Thus, the rate of cooling depends on the relation of surface area to volume of the product and the rate at which the vacuum is drawn in the flash chamber.

Since water is the sole refrigerant, the amount of heat removed from the product depends on the amount of water W_v vaporized and its latent heat of vaporization L. Assuming an ideal condition, with no heat gain from surroundings, total heat Q removed from the product is:

$$Q = W_N L_v L \qquad (9)$$

The amount of moisture removed from the product during vacuum cooling, then, is directly related to the specific heat of the product and the amount of temperature reduction accomplished. A product with a specific heat capacity of 0.95 Btu/(lb·°F) would

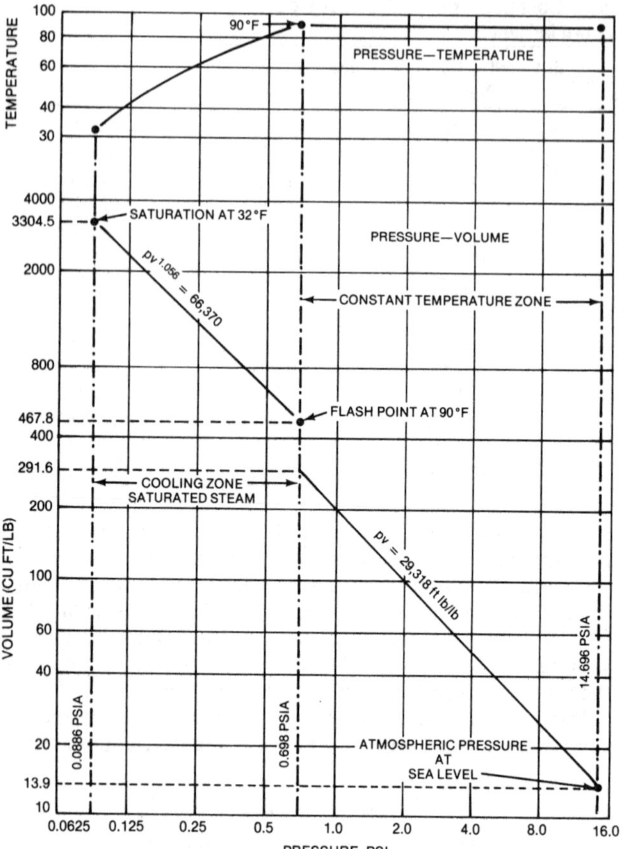

Fig. 5 Pressure, Volume, and Temperature in a Vacuum Cooler Cooling Product from 90 to 32 °F

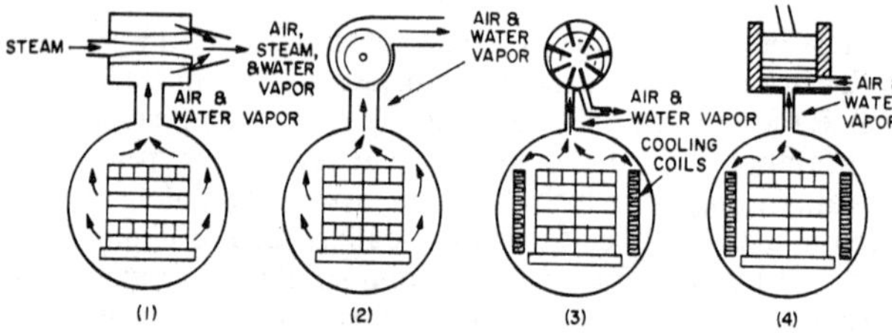

Fig. 6 Schematic Cross Sections of Vacuum-Producing Mechanisms

theoretically lose 1% moisture for each 11 °F reduction in temperature. In a study of vacuum cooling of 16 different vegetables, Barger (1963) showed that cooling of all products was proportional to the amount of moisture evaporated from the product. Temperature reductions averaged 9 to 10 °F for each 1% of weight loss, regardless of the product cooled.

Commercial Systems

The four types of vacuum refrigeration systems that use water as the refrigerant are: (1) steam ejector, (2) centrifugal, (3) rotary, and (4) reciprocating. A schematic of the vacuum-producing mechanism of each is illustrated in Figure 6.

Of these, the steam ejector type is best suited for displacing the extremely high volumes of water vapor encountered at the low pressures needed in vacuum cooling. It also has the advantage of having few moving parts, thus requiring no compressor to condense the water vapor. High-pressure steam is expanded through a series of jets or ejectors arranged in series and condensed in barometric condensers mounted below the ejectors. Cooling water for condensing is accomplished by means of an induced-draft cooling tower. In spite of these advantages, few steam ejector vacuum coolers are used today, due to the inconvenience of using steam and the lack of portability. Instead, vacuum coolers mounted on semitrailers are used to follow the seasonal crops.

The centrifugal compressor is also a high-volume pump and can be adapted to water vapor refrigeration. However, its use in vacuum cooling is limited because of inherent mechanical difficulties at the high rotative speeds required to produce the low pressures needed.

Both rotary and reciprocal vacuum pumps are capable of producing the low pressures needed for vacuum cooling, and they also have the advantage of portability. Being positive-displacement pumps, however, they have low volumetric capacity; therefore, vacuum coolers using rotary or reciprocating pumps have separate refrigeration systems to condense much of the water vapor that evaporates off the product, thus substantially reducing the volume of water vapor passing through the pump. Ideally, when it can be assumed that all of the water vapor is condensed, the required refrigeration capacity is equal to the amount of heat removed from the product during cooling.

The condenser must contain adequate surface to condense the large amount of vapor removed from the produce in a few minutes. Refrigeration is furnished from cold brine or a direct-expansion system. A very large peak load occurs from rapid condensing of so much vapor. Best results are obtained if the refrigeration plant is equipped with a large brine or icemaking tank having enough stored refrigeration to smooth out the load. A standard three-tube plant, with capacity to handle three cars per hour, will have a peak refrigeration load of at least 250 tons.

To increase cooling effectiveness and reduce product moisture loss, the product is sometimes wetted before cooling begins. However, lettuce is rarely prewetted. A propietary modification of vacuum cooling continuously circulates chilled water over the product throughout the vacuum cooling process. Among the chief advantages are increased cooling rates and residual refrigeration that is stored in the chilled water following each vacuum process. It also prevents water loss from products that show objectionable wilting after conventional vacuum cooling.

Applications

Because vacuum cooling is generally more expensive, particularly in first cost, than other cooling methods, its use is primarily restricted to products for which vacuum cooling is much faster or more convenient. Lettuce is ideally adapted to vacuum cooling. The numerous individual leaves provide a large surface area and the tissues release moisture readily. It is possible to freeze lettuce in a vacuum chamber if pressure and condenser temperatures are not carefully controlled. However, even lettuce does not cool entirely uniformly. The fleshy core, or butt, releases moisture more slowly than the leaves. Temperatures as high as 43 °F have been recorded in core tissue when leaf temperatures were down to 33 °F (Barger 1961).

Other leafy vegetables such as spinach, endive, escarole, and parsley are also suitable for vacuum cooling. Vegetables that are less suitable but adaptable by wetting are asparagus, snap beans, broccoli, brussels sprouts, cabbage, cauliflower, celery, green peas, sweet corn, leeks, and mushrooms. Of these vegetables, only cauliflower, celery, cabbage, and mushrooms are commercially vacuum cooled in California. Fruits are generally not suitable, except some of the berries, especially strawberries. Cucumbers, cantaloupes, tomatoes, dry onions, and potatoes cool very little because of their low surface-to-mass ratio and relatively impervious surface. The final temperatures of various vegetables when vacuum cooled under similar conditions are illustrated in Figure 7.

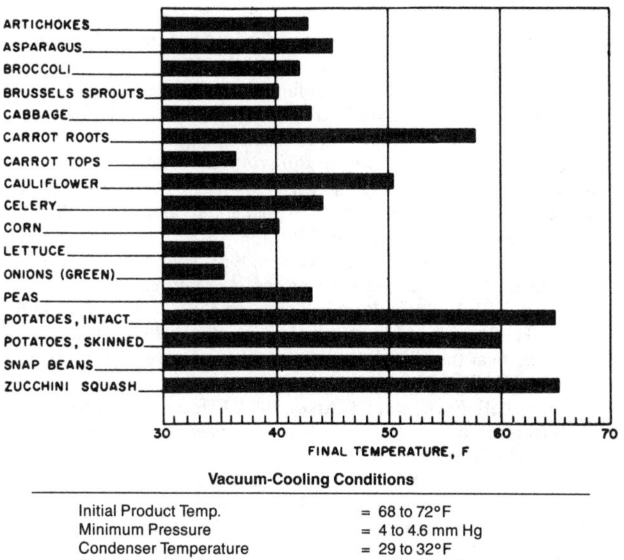

Vacuum-Cooling Conditions

Initial Product Temp.	= 68 to 72°F
Minimum Pressure	= 4 to 4.6 mm Hg
Condenser Temperature	= 29 to 32°F
Time in Vacuum Tank	= 0.42 to 0.5 h

Fig. 7 Comparative Cooling of Vegetables Under Similar Vacuum Conditions

The rate of cooling and the final temperature attained by vacuum cooling are largely affected by the ratio of the surface area of the commodity to its mass and the ease with which the product gives up water from its tissues. Consequently, the adaptability of fruits and vegetables varies tremendously for this method of precooling. For products that have a low surface-to-mass ratio, high temperature gradients occur. To prevent the surface from freezing before the desired mass-average temperature is reached, a procedure referred to as "bouncing" is practiced. This is accomplished by switching the vacuum pump off and on to keep the saturation temperature above freezing.

Mechanical vacuum coolers have been designed in several sizes. Most installations use cylindrical or rectangular retorts, sized to hold either a half- or a quarter-car of produce. However, a few are large enough to hold an entire refrigerator carload.

SELECTING A COOLING METHOD

Packinghouse size and operating procedures, response of product to the cooling method, and market demands largely dictate the type of cooling method used. Other factors considered are whether or not the product is packaged in the field or in a packinghouse, the product mix being cooled, length of cooling season, and comparative costs of dry versus water-repellant cartons. In some cases, there is little question about the type of cooling to be used. For example, vacuum cooling is most effective on lettuce and other similar vegetables. Peach packers in the southeastern United States and some vegetable and citrus packers are satisfied with hydrocooling. Air (room) cooling is used for apples, pears, peaches, plums, nectarines, sweet cherries, strawberries, and apricots. In other cases, choice of cooling method is not so clearly defined. Celery and sweet corn are usually hydrocooled, but they may be vacuum cooled as effectively. Cantaloupes may be satisfactorily cooled by several methods.

When more than one method can be used, cost becomes a major consideration. While rapid, forced-air cooling is more costly than hydrocooling, if the product does not require rapid cooling, a forced-air system can operate almost as economically as hydrocooling. In a study to evaluate costs of hypothetical precooling systems for citrus fruit, Gaffney and Bowman (1970) found that the cost for forced-air cooling in bulk lots was 20% more than that for hydrocooling in bulk and that forced-air cooling in cartons costs 45% more than hydrocooling in bulk.

COOLING CUT FLOWERS

Because of their high rates of respiration and low tolerance to heat, deterioration in cut flowers is rapid at field temperatures. Refrigerated highway vans do not have the capacity to remove the field heat in sufficient time to prevent some deterioration from occurring (Farnham *et al.* 1979). Forced-air cooling is commonly used by the flower industry. As with most fruits and vegetables, the cooling rate of cut flowers varies substantially among the various types. Rij *et al.* (1979) found that the half-cooling time for packed boxes of gypsophila was about 3 min compared to about 20 min for chrysanthemums at airflows ranging from 80 to 260 cfm per box. Within this range, cooling time was proportional to the reciprocal of airflow but varied less with airflow than with flower type.

SYMBOLS

Y = temperature ratio $(t - t_o)/(t_i - t_o)$
t = temperature of any point in product, °F
t_i = initial uniform product temperature, °F
t_o = surrounding temperature, °F
t_{ma} = mass-average temperature, °F
θ = cooling time, h
h = surface heat transfer coefficient, Btu/(h·ft^2·°F)
A = product surface area, ft^2
q = cooling load, Btu/h
Q = total heat, Btu
W_v = weight, lb
L = heat of vaporization, Btu/lb
C = cooling coefficient, reciprocal of hours
Z = half-cooling time, h
α = thermal diffusivity, ft^2/h
M_1 = first root of transcendental function
l = characteristic length, ft
G = geometry index
k = thermal conductivity, Btu/[h·ft^2 (°F/ft)]
μ = dynamic viscosity, lb/(h·ft)
G' = mass rate of airflow, lb/(h·ft^2)
c_p = specific heat, Btu/(lb·°F)
p = pressure, lb/ft^2
v = volume, ft^3

REFERENCES

Baird, C.D. and J.J. Gaffney. 1976. A numerical procedure for calculating heat transfer in bulk loads of fruits or vegetables. ASHRAE *Transactions* 82(2):525.

Baird, C.D., J.J. Gaffney, and M.T. Talbot. 1988. Design criteria for efficient and cost effective forced air cooling systems for fruits and vegetables. ASHRAE *Transactions* 94(1):1434.

Ball, C.O. and F.C.W. Olson. 1957. *Sterilization in food technology.* McGraw Hill Book Company, New York.

Barger, W.R. 1961. Factors affecting temperature reduction and weight loss of vacuum-cooled lettuce. USDA, *Marketing Research Report.*

Barger, W.R. 1963. Vacuum precooling—A comparison of cooling of different vegetables. USDA, *Marketing Research Report* No. 600.

Bennett, A.H., J. Soule, and G.E. Yost. 1966. Temperature response of citrus to forced-air precooling. ASHRAE *Journal* 8(4):48.

Bennett, A.H., J. Soule, and G.E. Yost. 1969. Forced-air precooling for Red Delicious apples. USDA, Agricultural Research Service, ARS 52-41.

Eshleman, W.D., C.D. Baird, and J.J. Gaffney. 1976. A numerical simulation of transient heat flow in irregular shaped foods. ASAE Paper No. 76-6504.

Farnham, D.S., *et al.* 1979. Comparison of conditioning, precooling, transit method, and use of a floral preservative on cut flower quality. Proceedings, *Journal of American Society of Horticultural Science* 104(4):483.

Fisher, D.V. 1960. Cooling rates of apples packed in different bushel containers and stacked at different spacing in cold storage. ASHRAE *Journal* (July):53.

Ford, K.E. 1956. Hydrocooling cantaloupes. *Proceedings of the Florida State Horticultural Society* 69:138.

Gaffney, J.J. and E.K. Bowman. 1970. An economic evaluation of different concepts for precooling citrus fruits. ASHRAE Symposium on Precooling Fruits and Vegetables, San Francisco (January).

Grierson, W. 1957. Preliminary studies for cooling Florida oranges prior to packing. *Proceedings of the Florida State Horticultural Society* 70:264.

Grizell, W.G. and A.H. Bennett. 1966. *Hydrocooling stacked crates of celery and sweet corn.* USDA, Agricultural Research Service, ARS 52-12.

Hardenburg, R.E., A.E. Watada, and C.Y. Wang. 1986. The commercial storage of fruits, vegetables, and florist and nursery stocks. USDA *Agricultural Handbook* No. 66.

Henry, F.E. and A.H. Bennett. 1973. Hydraircooling vegetable products in unit ads. *Transactions* of the ASAE 16(4):731.

Henry, F.E. A.H. Bennett, and R.H. Segall. 1976. Hydraircooling—A new concept for precooling pallet loads of vegetables. ASHRAE *Transactions* 82(2):541.

Ishibashi, S., R. Kojima, and T. Kaneko. 1969. Studies on the forced-air cooler. JSAM, Japan 31(2), September.

Mitchell, F.G., R. Guillou, and R.A. Parsons. 1972. *Commercial cooling of fruits and vegetables.* Manual 43, Division of Agricultural Sciences, University of California, Berkeley.

Parsons, R.A., F.G. Mitchell, and G. Mayer. 1972. Forced-air cooling of palletized fresh fruit. *Transactions* of the ASAE 15(4):729.

Perry, J.S., A.H. Bennett, and T.V. Minh. 1968. Experiments with a prototype commercial forced-air precooler on peaches, potatoes, apples, and strawberries. Unpublished data, University of Georgia.

Perry, R.L., and R.M. Perkins. 1968. Hydrocooling sweet corn. ASAE Paper No. 68-880 (December), Chicago.

Pflug, I.J., J.L. Blaisdell, and I.J. Kopelman. 1965. Developing temperature-time curves for objects that can be approximated by a sphere, infinite plate, or infinite cylinder. ASHRAE *Transactions* 71(1):238.

Rij, R.E., J.F. Thompson, and D.S. Farnham. 1979. *Handling, precooling, and temperature management of cut flower crops for truck transportation.* USDA-SEA Western Series No. 5 (June).

Smith, R.E. and A.H. Bennett. 1965. Mass-average temperature of fruits and vegetables during transient cooling. *Transactions* of the ASAE 8(2):249.

Smith, R.E., A.H. Bennett, and A.A. Vacinek. 1970. Convection film coefficients related to geometry for anomalous shapes. *Transactions* of the ASAE, Paper No. 69-373.

Smith, R.E., G.L. Nelson, and R.L. Henrickson. 1967. Analyses on transient heat transfer from anomalous shapes. *Transactions* of the ASAE 10(2):236.

Smith, R.E., G.L. Nelson, and R.L. Henrickson. 1968. Applications of geometry analysis of anomalous shapes to problems in transient heat transfer. *Transactions* of the ASAE 11(2):296.

Smith, W.L. and W.H. Redit. 1968. Postharvest decay of peaches as affected by hot-water treatments, cooling methods, and sanitation. USDA, *Marketing Research Report* No. 807.

Soule, J., G.E. Yost, and A.H. Bennett. 1966. Certain heat characteristics of oranges, grapefruit and tangelos during forced-air precooling. *Transactions* of the ASAE 9(3):355.

BIBLIOGRAPHY

Ansari, F.A. and A. Afaq. 1986. Precooling of cylindrical food products. *International Journal of Refrigeration* 9(3):161-63.

Arifin, B.B. and K.V. Chau. 1988. Cooling of strawberries in cartons with new vent hole designs. ASHRAE *Transactions* 94(1):1415-26.

Bennett, A.H. 1962. Thermal characteristics of peaches as related to hydrocooling. USDA, *Technical Bulletin* No. 1292.

Bennett, A.H., W.G. Chace, Jr., and R.H. Cubbedge. 1969. Heat transfer properties and characteristics of Appalachian area Red Delicious apples. ASHRAE *Transactions* 75(2):133.

Bennett, A.H., R.E. Smith, and J.C. Fortson. 1965. Hydrocooling peaches —A practical guide for determining cooling requirements and cooling times. USDA, *Agriculture Information Bulletin* No. 298 (June).

Bennett, A.H., W.G. Chace, Jr., and R.H. Cubbedge. 1970. Thermal properties and heat transfer characteristics of marsh grapefruit. USDA, *Technical Bulletin* No. 1413.

Burton, K.S., C.E. Frost, and P.T. Atkey. 1987. Effect of vacuum cooling on mushroom browning. *International Journal of Food Science & Technology* 22(6):599-606.

Chau, K.V., C.D. Baird, P.C. Talasila, and S.A. Sargent. 1992. Development of time-temperature-humidity relations for fresh fruits and vegetables. Final report for ASHRAE Research Project 678-RP.

Freeman, C.D. 1984. Cost reducing technologies in cooling fresh vegetables. ASAE Paper No. 841074.

Gaffney, J.J. and C.D. Baird. 1977. Forced-air cooling of bell peppers in bulk. ASAE *Transactions* 20(6):1174-80.

Gaffney, J.J., C.D. Baird, and W.D. Eshleman. 1976. Temperature response of avocados during cooling with chilled water. ASAE Paper No. 76-6017.

Gariepy, Y., G.S.V. Raghavan, and R. Theriault. 1987. Cooling characteristics of cabbage. *Canadian Agricultural Engineering* 29(1):45-50.

Hackert, J.M., R.V. Morey, and D.R. Thompson. 1987. Precooling of fresh market broccoli. *Transactions* of the ASAE 30(5): 1489-93.

Hayakawa, K. 1978. Computerized simulation for heat transfer and moisture loss from an idealized fresh produce. *Transactions* of the ASAE 21(5):1015-24.

Hayakawa, K. and J. Succar. 1982. Heat transfer and moisture loss of spherical fresh produce. *Journal of Food Science* 47(2):596-605.

Isenberg, F.M.R., R.F. Kasmire, and J.E. Parson. Vacuum cooling vegetables. *Information Bulletin* 186, Cornell Cooperative Extensive Service.

Jiang, H., D.R. Thompson, and R.V. Morey. 1987. Finite element model of temperature distribution in broccoli stalks during forced-air precooling. *Transactions* of the ASAE 30(5):1473-77.

Lentz, C.D. and L. van den Berg. 1977. Cabbage precooling study. *J. Inst. Can. Sci., Technol. Aliment.* 10(4):265.

Misener, G.C. and G.C. Shove. 1976. Simulated cooling of potatoes. *Transactions* of the ASAE 19(5):954.

Morey, R.V., S.A. Sargent, C.D. Baird, and M.R. Talbot. 1988. ASAE Paper No. 88-7539.

Rohrbach, R.P., R. Ferrell, E.O. Beasley, J.R. Fowler. 1984. Precooling blueberries and muscadine grapes with liquid carbon dioxide. *Transactions* of the ASAE 27(6):1950-55.

Shaw, J. and C. Kuo. 1987. Vacuum precooling green onion and celery. ASAE Paper No. 87-5522.

Stewart, J.K. and H.M. Couey. 1963. Hydrocooling vegetables—A practical guide to predicting final temperatures and cooling times. USDA, *Marketing Research Report* No. 637.

Thompson, J.F. and Y.L. Chen. 1986. Energy use in hydrocooling stone fruit. ASAE Paper No. 866556.

Thompson, J.F. and R.F. Kasmire. 1979. Evaporative cooling of chilling sensitive vegetable crops. ASAE Paper No. 79-6516.

Thompson, J.F., Y.L. Chen, and T.R. Rumsey. 1987. Energy use in vacuum coolers for fresh market vegetables. *Applied Engineering in Agriculture* 3(2):196-99, American Society of Agricultural Engineers, St. Joseph, MI.

CHAPTER 11

MEAT PRODUCTS

SOUND sanitary practices should be used at all stages of food processing, not only to protect the public but to meet aesthetic requirements. In this respect, meat processing plants are no different from other food plants. The same principles apply regarding sanitation of buildings and equipment; provision of sanitary water supplies and wash facilities; disposal of waste materials; insect and pest control; and proper use of sanitizers, germicides, and fungicides. All U.S. meat plants operate under regulations set forth in inspection service orders. For detailed sanitation guidelines to be followed in all plants producing meat under federal inspection, refer to *Agriculture Handbook* No. 570, available from the U.S. Department of Agriculture, FSIS.

Proper safeguards and good manufacturing practices should mimimize bacterial contamination and growth. This involves using clean raw materials, clean water and air, sanitary handling throughout, good temperature control (particularly in coolers and freezers), and scrupulous between-shift cleaning of all surfaces in contact with the product.

Precooked products present additional problems because favorable conditions for bacterial growth exist after the product has cooled to below 130°F. In addition, potential pathogens may experience enhanced growth because their competitor organisms were destroyed during cooking. Any delay in processing at this stage allows surviving microorganisms to multiply, especially when the cooked and cooled meat is handled and packed into containers prior to processing and freezing. Creamed products afford especially favorable conditions for bacterial growth. Filled packages should be removed immediately on filling and quickly chilled. Fast chilling not only reduces the time for growth, but can also reduce the number of bacteria.

It is even more important during processing to avoid any opportunity for the growth of pathogenic bacteria (such as *Salmonellae, Clostridium perfringens, Staphylococci,* or *Streptococci*) that may have entered the product (Thatcher and Clark 1968). While these organisms do not grow as quickly at temperatures below 40°F, they can survive freezing and prolonged frozen storage.

Storage at a temperature of about 25°F will permit the growth of psychrophilic spoilage bacteria, but at 14°F these, as well as all other bacteria, are dormant. Even though some cells of all bacteria types die during storage, activity of the survivors is quickly renewed with rising temperature. The processor should recommend safe preparation practices to the consumer. The best procedure is to provide instructions for cooking the food without preliminary thawing. In the freezer, sanitation is confined to keeping physical cleanliness and order and preventing access of foreign odors.

CARCASS CHILLING AND HOLDING

A hot carcass cooler removes live animal heat as rapidly as possible. Side effects such as cold shortening, which can reduce tenderness, must be considered. Electrical stimulation can minimize cold shortening. Rapid temperature reduction is important in reducing the growth rate of microorganisms that may exist on carcass surfaces. Conditions of temperature, humidity, and air motion must be considered to attain desired meat temperatures within the time limit and to prevent excessive shrinkage, bone taint, sour rounds, surface slime, mold, or discoloration. The carcass must be delivered with a bright, fresh appearance.

Spray Chilling Beef

Spraying cold water intermittently on beef carcasses for 3 to 8 h during chilling is currently the normal procedure in commercial beef slaughter plants (Johnson *et al.* 1988). Basically, this practice reduces evaporative losses and speeds chilling. Regulations do not allow the chilled carcass to exceed the prewashed hot carcass weight. The carcass is chilled to a large extent by evaporative cooling. As the carcass surface tissue dries, moisture migrates toward the surface, where it evaporates. Eventually, an equilibrium is reached when the temperature differential narrows and reduces the evaporative loss (Locker *et al.* 1975, Heitter 1975).

When carcasses were shrouded, a once frequently used method for reducing weight loss (shrink), typical evaporative losses ranged from 0.75 to 2.0% for an overnight chill (Kastner 1981). Heitter (1975) reported that the Chlor-chill system reduced shrink by 0.5 to 1.25%. Allen *et al.* (1987) found that spray-chilled beef sides lost 0.3% compared with 1.5% for nonspray chilled sides. Those authors stated that although variation in carcass shrink of spray-chilled sides was influenced by carcass spacing, other factors, especially those affecting the dynamics of surface tissue moisture, may be involved. Carcass washing, length of spray cycle, and carcass fatness also influence the variation in shrink. With sufficient care, however, carcass cooler shrink can be nearly eliminated.

Loin eye muscle color and shear force are not affected by spray chilling, but fat color can be lighter in spray-chilled compared to nonspray chilled sides. Over a 4-day period, color changes and drip losses in retail packs for rib steaks and round roasts were not related to spray chilling (Jones and Robertson 1989). Those authors also concluded that spray-chilling could provide a moderate reduction in carcass shrinkage during cooling without having a detrimental influence on muscle quality.

Vacuum-packaged inside rounds from spray-chilled sides had significantly more purge, *i.e.*, air removed, (0.4 kg or 0.26%) than those from conventionally chilled sides. Spacing treatments where foreshanks were aligned in opposite directions and where they were aligned in the same direction but with 6 in. between sides both result in less shrink ($P < 0.05$) during a 24-h spray-chill period than the treatment where foreshanks were aligned in the same direction but with all sides tightly crowded together (Allen *et al.* 1987). Some studies with both beef (Hamby *et al.* 1987) and pork (Greer and Dilts 1988) indicated that bacterial populations of conventionally and spray-chilled carcasses were not affected by chilling method (Dickson 1991). However, Acuff (1991) and others showed

The general responsibility for this chapter is assigned to TC 11.2, Foods and Beverages.

that use of a sanitizer (chlorine, 200 ppm, or organic acid, 1 to 3%) significantly reduces carcass bacterial counts.

Chilling Time

Although certain basic principles are identical, beef and hog carcass chilling differs substantially. The massive beef carcass is only partially chilled (although shippable) at the end of the standard overnight period; the average hog carcass may be fully chilled (but not ready for cutting) in 8 to 12 h, while the balance of the period accomplishes only temperature equalization.

The beef carcass surface retains a large amount of wash water, which provides much evaporative cooling in addition to that derived from actual shrinkage; but evaporative cooling of the hog carcass, which retains little wash water, occurs only through actual shrinkage. A beef carcass, without skin and destined largely for sale as fresh cuts, must be chilled in air temperatures sufficiently high to avoid freezing and damage to appearance. Although it must subsequently be well tempered for cutting and scheduled for in-plant processing, a hog carcass, including the skin, can tolerate a certain amount of surface freezing. Beef carcasses can be chilled with an overnight shrinkage of 0.5%, whereas equally good practice on hog carcasses will result in 1.25 to 2% shrinkage.

The bulk (16 to 20 h) of beef chilling is done overnight in high humidity chilling rooms with a large refrigeration and air circulation capacity. The balance of the chilling and temperature equalization occurs during a subsequent holding or storage period that averages one day, but can extend to 2 or 3 days, usually in a separate holding room with a low refrigeration and air circulation capacity.

Some packers load for shipment the day after slaughter, since some refrigerated transport vehicles have ample capacity to remove the balance of the internal heat in round or chuck beef during the first two days in transit. This practice is most important in rapid delivery of fresh meat to the marketplace. Carcass beef that is not shipped the day after slaughter should be kept in a beef-holding cooler at temperatures of 34 to 36°F with minimum air circulation to avoid excessive color change and weight loss.

Refrigeration Systems for Coolers

Refrigeration systems commonly used in carcass chilling and holding rooms are operated with ammonia as the primary refrigerant and are of three general types: dry coils, chilled brine spray, and sprayed coil.

Dry Coil Refrigeration. Dry coil systems comprise most chilling and holding room installations. Dry coil systems usually include unit coolers equipped with coils, defrosting equipment, and fans for air-vapor circulation. Because the coils are operated without continuous brine spray, eliminators are not required. Coils are usually finned, with fins limited to 3 or 4 per inch or with variable fin spacing to avoid icing difficulties. The units may be mounted on the floor, overhead on the rail beams, or overhead on converted brine spray decks.

Dry coil systems operated at surface temperatures below 32°F build up a coating of frost or ice, which ultimately reduces the airflow and cooling capacity. Coils must therefore be defrosted periodically, normally every 4 to 24 h for coils with 3 or 4 fins per inch, to maintain capacity. The rate of buildup, and hence the defrosting frequency, decreases with large coil capacity and high evaporating pressure.

Defrosting may be done either manually or automatically by the following:

- *Hot-gas defrost* is accomplished, with the fans off, by introducing hot gas directly from the system compressors into the evaporator coils. The evaporator suction is throttled to maintain a coil pressure of about 60 to 75 psig (at approximately 40 to 50°F). The coils then act as condensers and supply the heat for melting the ice coating. Other evaporators in the system must supply the load for the compressors during this period. Hot-gas defrost is rapid, normally requiring 10 to 30 min for completion. See Chapter 3 for hot-gas defrost piping and control systems.
- *Coil spray defrost* is accomplished (with the fans turned off) by spraying the coil surfaces with water, which supplies the heat required to melt the ice coating. Suction and feed lines are closed off, with pressure relief from the coil to the suction line to minimize the refrigeration effect. Enough water at 50 to 75°F must be used to avoid freezing on the coils, and care must be taken that freezeup does not occur in the drain lines. The sprayed water tends to produce some fog in the refrigerated space. Coil spray defrost may be more rapid than hot-gas defrost.
- *Room air defrost* (for rooms 35°F or higher) is accomplished with the fans running while suction and feed lines are closed off (with pressure relief from coil to suction line), to permit buildup of coil pressure and melting of the ice coating by transfer of heat out of the air flowing across the coils. Refrigeration therefore continues during the defrosting period, but at a drastically reduced rate. Room air defrost is slow; the time required may vary from 30 min to several hours if the coils are undersized for dry coil operation.
- *Electric defrost* is accomplished with electric heaters with fans either on or off. During defrost, refrigerant flow is interrupted.

Unit coolers may be defrosted by any one or combinations of the first three methods. All methods involve a reduction in chilling capacity, which varies with time loss and heat input. Hot-gas and coil spray defrost interrupt the chilling only for short periods, but they introduce some heat into the space. Room air defrost severely reduces the chilling rate for long periods, but the heat required to vaporize the ice is obtained entirely from the room air.

Evaporator controls customarily employed in carcass chilling and holding rooms include refrigerant feed controls, evaporator pressure controls, and air circulation control.

Refrigerant feed controls are designed to maintain, under varying loads, as high a liquid level in the coil as can be carried without excessive liquid spillover into the suction line. This is accomplished by using an expansion valve that throttles the liquid from supply pressure (typically 150 psig) to evaporating pressure (usually 20 psig or higher). The throttling of the liquid flashes some of it to gas, which chills the remaining liquid to saturation temperature at the lower pressure. If it does not bypass the coil, the flashed gas tends to reduce flooding of the interior coil surface, thus lowering coil efficiency.

The valve used may be a hand-controlled expansion valve supervised by operator judgment alone, a thermal expansion valve governed by the degree of superheat of the suction gas, or a float valve (or solenoid valve operated by a float switch) governed by the level of feed liquid in a surge drum placed in the coil suction line. This surge drum suction trap permits the ammonia flashed to gas in the throttling process to flow directly to the suction line, bypassing the coil. The trap may be small and placed just high enough so that its level governs that in the coils by gravity transfer. Or, as in the ammonia recirculation system, it may be placed below coil level so that the liquid is pumped mechanically through the coils in much greater quantity than is required for evaporation. In the latter case, the trap is sized large enough to carry its normal operating level plus all the liquid flowing through the coils, thus effectively preventing liquid spillover to the compressors. Nevertheless, it is necessary in all cases to provide further protection at the compressors' liquid return.

Present practice strongly favors liquid ammonia recirculation systems, mainly because of the greater coil heat transfer rates with the resultant greater refrigerating capacity over other systems (see Chapter 3). Some have coils mounted above the rail beams with 4 to 6 ft of ceiling head space. Air is forced through the coils, sometimes using two-speed fans.

Manual and thermal expansion valves do not provide good coil flooding under varying loads and do not bypass the flashed feed gas around the coils. As a result, evaporators so controlled are usually rated 15 to 25% less in capacity than those controlled by float valve or ammonia recirculation.

Evaporator pressure controls regulate coil temperature, and thereby the rate of refrigeration, by varying evaporating pressure within the coil. This is accomplished by using a throttling valve in the evaporator suction line downstream from the surge drum suction trap. All such valves impose a definite loss on the refrigeration system, and the amount of the loss varies directly with the pressure drop through the valve. This increases the work of compression for a given refrigeration effect.

The valve used to control evaporating pressure may be a manual suction valve set solely by operator judgment, a back pressure valve actuated by coil pressure or temperature, or a back pressure valve actuated by a temperature-sensing element somewhere in the room. Manual suction valves require excessive attention when loads fluctuate. The coil-controlled back pressure valve seeks to hold a constant coil temperature but does not control room temperature unless the load is constant. Only the room-controlled compensated back pressure valve responds to room temperature.

Air circulation control is frequently used when an evaporator must handle separate load conditions differing greatly in magnitude, such as the load in chilling rooms that are also used as holding rooms or for the negligible load on weekends. The use of two-speed fan motors (operated at reduced speed during the periods of light load) or turning the fans off and on can control air circulation to a degree.

Chilled Brine Spray Systems. These systems are generally being abandoned in favor of other systems due to such a system's large required building space, inherent low capacity, brine carryover tendencies, and difficulty of control.

Sprayed Coil Systems. These consist of unit coolers equipped with coils, brine spray banks, eliminators to prevent brine carryover, and fans for air-vapor circulation. The units are usually mounted (without ductwork) either on the floor or overhead on converted brine spray decks. Refrigeration is supplied by the primary refrigerant in the coils. Chilled or nonchilled recirculated brine is continually sprayed over the coils, thus eliminating ice formation and the need for periodic defrosting.

The brine predominantly used is sodium chloride, with caustic soda or another additive for controlling pH. Because sodium chloride brine is corrosive, bare-pipe coils (without fins) generally see service. The brine is also highly corrosive to the rail system and other cooler equipment.

Propylene glycol with added inhibitor complexes is another coil spray solution used in place of sodium chloride. As with sodium chloride brine, propylene glycol is constantly diluted by moisture condensed out of the spaces being refrigerated and must be concentrated by evaporating water from it. The reconcentration process requires special equipment designed to minimize glycol losses. Sludging in the concentrator may become an operating problem; to avoid it, additives must be selected and pH closely controlled. Finned coils are usually used with propylene glycol.

Because of its noncorrosiveness in comparison to sodium chloride, propylene glycol greatly reduces the cost of unit cooler construction as well as maintenance of space equipment.

Other Systems. Considerable attention is being directed to system designs that will reduce the amount of evaporative cooling at the time of entrance into the cooler and eliminate ceiling rail and beam condensation and drip. Good results have been achieved by using low-temperature blast chill tunnels before entrance into the chill room. The volume of ceiling condensate is reduced because the rate of evaporative cooling is reduced in proportion to the degree of surface cooling. Room condensation has been reduced by the addition of heat above carcasses (out of the main air stream), fans, minimized hot water usage during cleanup, better dry cleanup, timing of cleanup, and using wood rail supports.

Grade and yield sorting, with its simultaneous filling of several rooms, has shortened the chilling time available if refrigeration is kept off during the filling cycle. Its effect has to be offset by more chill rooms and more installed refrigeration capacity. If full refrigeration is kept at the start of filling, the peak load is reduced to the rooms being filled. Hot carcass cutting has been started with only a short prechilling time. Cryogenic chilling has also been tested for hot carcass chilling.

Beef Cooler Layout and Capacity

Carcass halves or sides are supported by hooks suspended from one-wheel trolleys running on overhead rails. The trolleys are generally pushed from the dressing floor to the chilling room by powered conveyor chains equipped with fingers that engage the trolleys, which are then distributed manually over the chilling and holding room rail system. Chilling and holding room rails are commonly placed on 3 to 4 ft centers in the holding rooms, with pullout or sorting rails between them. The rails must be placed a minimum of 2 ft from the nearest obstruction, such as a wall or building column, and the tops of the rails must be at least 11 ft above the floor. The supporting beams should be placed a minimum of 6 ft below the ceiling for optimum air distribution. Applicable to new construction in plants engaged in interstate commerce, regulations for some of these dimensions are issued by the Meat Inspection Division of the FSIS.

To assure effective air circulation, carcass sides should be placed on the rails in both chilling and holding coolers so that they do not touch each other. Required spacing varies with the size of the carcass and averages 2.5 ft per two sides of beef. In practice, however, sides are often more crowded.

A chilling room should be of such size that the last carcass loaded into it does not materially retard the chilling of the first carcass. While size is not as critical as in the case of the hog carcass chill room (because of the slower chill), to better control shrinkage and condensation, it is desirable to limit chill cooler size to hold not more than 4 h of the daily kill. Holding coolers may be as large as desired because of their ability to maintain more uniform temperature and humidity.

While overall plant chilling and holding room capacities vary widely, chilling coolers generally require a capacity equal to the daily kill; holding coolers require 1 to 2 times the daily kill.

Beef Carcasses. Dressed beef carcasses, each split into two sides, range in weight from approximately 300 to 1000 lb, averaging about 700 lb per head. Specific heats of carcass components range from 0.50 Btu/(lb · °F) for fat to 0.8 Btu/(lb · °F) or more for lean muscle, averaging about 0.75 Btu/(lb · °F) for the carcass as a whole.

The body temperature of an animal at slaughter is about 102 °F. After slaughter, physiological changes occur that generate heat and tend to increase carcass temperature, while heat loss from the surface tends to lower it.

The largest part of the carcass is the round, and at any given stage of the chilling cycle its center has the highest temperature of all carcass parts. This *deep round* temperature (about 105 °F when the carcass enters the chilling cooler) is therefore universally used as a measure of chilling progress. If it is to be significant, the temperature must be taken accurately. Incorrect techniques will show temperatures as much as 10 °F lower than actual deep round temperature. An accurate technique that yields consistent results is shown in Figure 1. The technique applies a fast-reacting, easily read stem dial thermometer, calibrated before and after tests, inserted upward to the full depth through the hole in the aitchbone.

At the time of slaughter, the water content of beef muscle is approximately 75% of the total weight. Thereafter, a gradual drying of the surface takes place, resulting in weight loss or shrink-

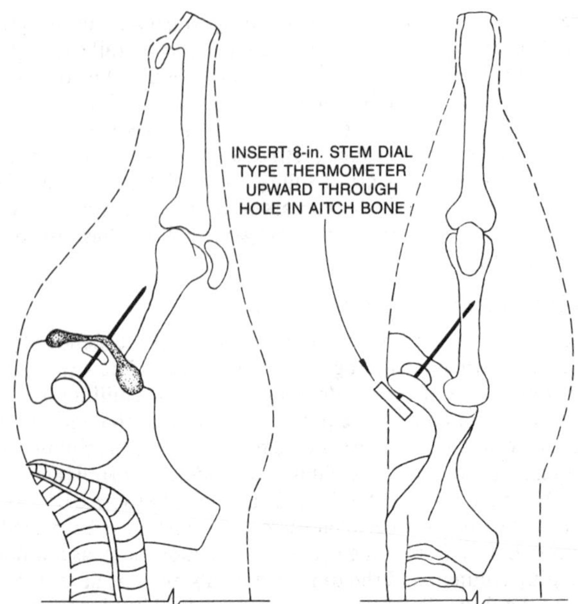

Fig. 1 Deep Round Temperature Measurement in Beef Carcass

age. Shrinkage and its measurement are greatly affected by the final operations of the dressing process: weighing and washing. Weighing must be done prior to washing if the weights are to reflect actual product shrinkage.

A beef carcass retains large amounts of wash water on its surface, which it carries into the cooler. The loss of this water, occurring in the form of vapor, does not constitute actual product loss. However, it must be considered when estimating the system capacity since the vapor must be condensed on the coils, thus constituting an important part of the refrigeration load.

The amount of wash water retained by the carcass depends on its condition and on washing techniques. A carcass typically retains 8 lb, part of which is lost by drip and part by evaporation. Water pressures used in washing vary from 50 to 300 psig, and temperatures from 60 to 115 °F.

To minimize spoilage, a carcass should be reduced to a uniform temperature of about 35 °F as rapidly as possible. In practice, deep round temperatures of 60 °F (measured as in Figure 1), with surface temperatures of 35 to 45 °F, are common at the end of the first day's chill period.

To prevent formation of surface slime most carcasses are cut, vacuum packaged, and boxed within 24 to 72 h. Otherwise, a carcass surface needs to be a certain dryness during storage. Exposed beef muscle chilled to an actual temperature of 36 °F will not slime readily if dried at the surface to a water content of 90% of dry weight (47.4% of total weight). Such a surface is in vapor pressure equilibrium with a surrounding atmosphere at the same temperature (36 °F) and 96% rh. In practice, a room atmosphere at 32 to 34 °F and approximately 90% rh will maintain a well-chilled carcass in nonsliming condition (Thatcher and Clark 1968).

Chilling-Drying Process. Curves of carcass temperature during a chilling-holding cycle are shown in Figure 2. Note that some heat loss occurs before a carcass enters the chilling cooler. The evaporative cooling of surface water dominates in the initial stages of hot carcass chilling. As chilling progresses, the rate of losses by evaporative surface cooling diminishes and the sensible transfer of heat from the carcass surface increases. Note that the time-temperature rates of change are subject to variations between summer and winter ambient conditions, which influence system capacity.

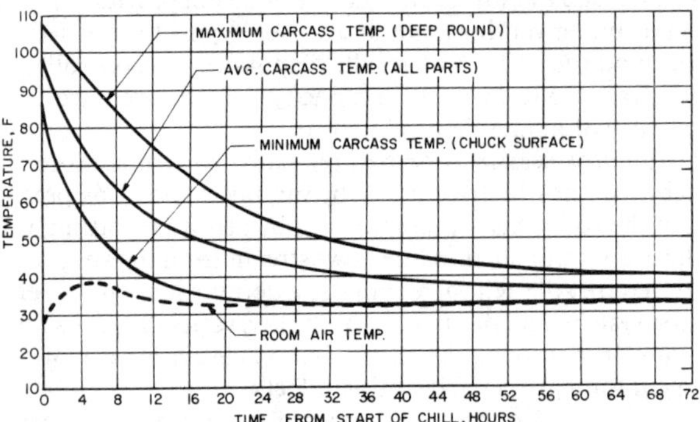

Fig. 2 Beef Carcass Chill Curves

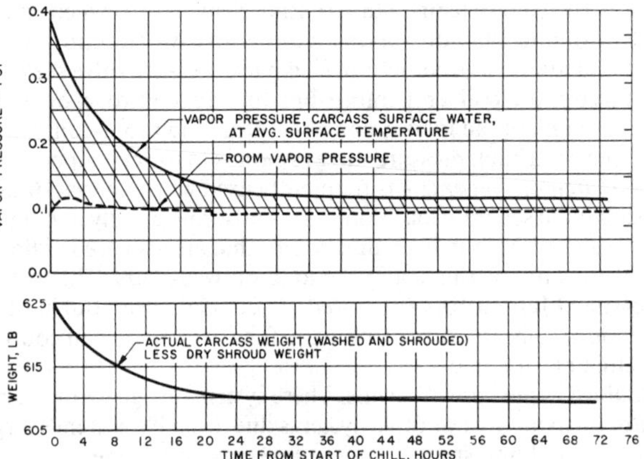

Fig. 3 Beef Carcass Shrinkage Rate Curves

The rate of transfer is increased both by more rapid circulation of air and lower air temperature, but these are limited by the necessity of avoiding surface freezing.

Estimated differences in vapor pressure between surface water (at average surface temperature) and atmospheric vapor during a typical chilling-holding cycle, and the corresponding shrinkage curve for an average carcass, are shown in Figure 3. Note the tremendous vapor pressure differences during the early part of the chill cycle when the carcass is warm. The evaporative loss could be reduced by beginning the chill with room temperature high, then lowering it slowly to minimize the pressure difference between carcass surface water and room vapor at all times. However, this slows the chill and prolongs the period of rapid evaporation. The quick chill practice is favored; but the cold shortening effect and bacterial growth must be considered in carcass quality and keeping time.

Evaporation from the warm carcass in cool air is nearly independent of room relative humidity because the warm carcass surface generates a much higher vapor pressure than the cooler vapor surrounding the carcass. If the space surrounding a warm carcass is saturated, evaporation forms fog, which can be observed at the beginning of any chill.

Evaporation from the well-chilled carcass with surface temperature at or near room temperature is different. The spread between surface and room vapor pressures approaches zero when room air is near saturation. Evaporation proceeds slowly, without forming fog. Evaporation does not cease when a room is saturated; it ceases only if the carcass is chilled through to room temperature, and no heat transfer is taking place.

The ultimate disposition of water condensed on the coils depends on the temperature of the coil surface and the method of coil operation. In continuous defrost (sprayed coil) operation,

Table 1 Weight Changes in Beef Carcass

Chilling Cooler	Weight, lb
Initial dry weight	615
Wash water pickup	8
Initial wet weight	623
Spray chill water use	16
Drip (not evaporated)	10
Weight at maximum (8 h postmortem)	633
Weight loss 8-24 h postmortem	18
Net weight loss (ideal)	0
Holding Cooler	
Weight loss/day	3
Final weight (48 h postmortem)	612

condensed and trapped water dilute the solution sprayed over the coil. In nonfrosting dry-coil operation, condensed water falls to the evaporator pan and drains to the sewer. Water frozen on the coil is lost to the sewer if removed by hot-gas or coil spray defrost. Periodic room air defrost, however, vaporizes part of the ice and returns it to room atmosphere, while losing the remainder to drain. This method of defrost is not normally used in beef chill or holding coolers because temperatures are not suitable, and thus the defrost period is excessive, resulting in abnormal room temperature variations. Most chill and holding evaporators are automatically defrosted with water or hot gas on preselected time cycles. The weight changes that take place in an average beef carcass are given in Table 1.

Chilling of the beef carcass is not completed in the chill cooler but continues at a reduced rate in the holding cooler. A carcass well chilled when it enters the holding cooler shows minimum holding shrinkage; a poorly chilled one shows high holding shrinkage.

If shrinkage values are to have any significance, they must be carefully derived. Actual product loss must be determined by first weighing the dry carcass prior to washing and then weighing it out of the cooler. In-motion weights are not sufficiently precise; carcasses must be weighed at rest. Scales must be accurate, and, if possible, the same scale should be used for both weighings. If shrinkage is to have any comparison value, it must be measured on carcasses chilled to the same temperature, since chilling occurs largely by evaporative weight loss.

Design Conditions and Refrigeration Load. Equipment selection should be based on conditions at peak load, when product loss is greatest. Room losses, equipment heat, and carcass heat add up to a total load that varies greatly—not only in magnitude but in proportion of sensible to total heat (sensible heat ratio)—throughout the chill. As the chill progresses, the vapor load decreases and the sensible load becomes more predominant.

Under peak chilling load, excess moisture condenses into fog—enough to warm the air-vapor-fog mixture to the sensible heat ratio of the heat removal process. The heat removal process of the coil therefore underestimates the actual rate of water removal by the amount of vapor condensed to fog (Table 2).

Fog does not generally form under later chilling room loads and all holding room loads, although it may form locally and then vaporize. Sensible heat ratios of air vapor heat gain and air vapor heat removal are then equal (Table 3).

Beef chilling rooms generally have evaporator capacity sufficient to hold room temperature under load approximately as shown in Figure 2. This results in an increase in room temperature to 35 to 40 °F, with gradual reduction to 32 to 34 °F. However, many installations provide greater capacity, particularly dry coil systems, which thereby avoid excessive coil frosting. In batch-loaded coolers, room temperature may be as low as 25 °F under peak load, provided it is raised to 30 °F as the chill progresses, without surface freezing of the beef. The shrinkage improvement affected by

Table 2 Load Calculations for Beef Chilling

Cooler size, ft: 62 × 74 × 18.5
Cooler capacity: 476 carcasses
Average carcass weight: 625 lb
Assumed chill rate: 50 °F in 20 h
Assumed air circulation: 167,400 cfm
Loading time: 3.3-h maximum
Assumed air to coil: 33 °F, 100% rh
Assumed fan motive power: 36 hp
Specific heat of beef: 0.75 Btu/(lb · °F)

	Loads, Btu/h		
Heat Gain—Room Load	**Sensible Heat**	**Latent Heat**	**Total Heat**
1. Transmission, infiltration, personnel, fan motor, lights, and equipment heat	162,000	4100	166,100
2. Product heat (average, first 4 h):			
a. 476 × 625 × 0.75 × 50 × 0.1	1,115,700		
b. 476 × 14.3 × 0.13 × 1070[a]	−947,000	947,000	1,115,700
3. Total heat gain (room load), kW (Items 1 + 2a + 2b)	332,700	951,100	1,283,800
Heat Removal—Coil Load			
4. Air circulation, dry air, lb/h 167,400 × 0.08 × 60 = 803,500	—	—	—
5. Heat removed per lb of dry air, total heat (Item 3)/(Item 4) = 1.6 Btu/lb	—	—	—
6. Air-vapor enthalpy, Btu/lb dry air:			
a. Air to coil, 33 °F, 100% rh	7.927	4.242	12.17
b. Btu removed, temperature drop 3.7 °F	−0.927	−0.672	−1.60
c. Air from coil, 29.3 °F, 100% rh	7.000	3.570	10.57
7. Coil air-vapor heat removal, Btu (Item 4)(Item 6b)	744,000	539,800	1,283,800
8. Room vapor condensed to fog (Item 7) − (Item 3)	411,300	−411,300	
9. Water (ice) removed by coil 476 × 14.3 × 0.13 × 144[b]		128,000	128,000
10. Total heat removal (coil load), Btu/h (Items 3 + 8 + 9)	744,000	667,800	1,411,800

[a]Heat of vaporization
[b]Heat of fusion

these lower temperatures, however, tends to be less than expected (in beef chilling) because of the relatively small part played by sensible transfer of heat.

It is standard practice in the holding room to provide evaporator capacity to keep the room temperature at 32 to 34 °F at all times. Holding room coils sized at peak load, low air vapor circulation rate, and a coil temperature 10 °F below room temperature tend to maintain the approximately 90% rh that avoids excessive shrinkage and prevents surface sliming.

From the average temperature curve shown in Figure 2 and the shrinkage curve in Figure 3, certain generalizations useful in calculating carcass chilling load may be made. In the chilling cooler, the average carcass temperature is reduced approximately 50 °F, from about 97 °F to about 47 °F, in 20 h. Simultaneously, about 14.3 lb of water is vaporized for each 625 lb carcass entering the chill; only 4.3 lb of this is actual shrinkage. The losses of sensible heat and water occur at about the same rate. In the sample load calculations, this is calculated at an average of 10% for the first 4 h of chill for sensible heat and 13% for the evaporation of moisture, which roughly agrees with the curves of Figures 2 and 3. This is the maximum rate of chill and is used for sizing the refrigeration equipment and piping.

In the holding cooler, the average carcass temperature is reduced from 47 to 39.5 °F in 24 h. Simultaneously, about 1.8 lb of water is vaporized per carcass. Where spray chilling is employed, this shrinkage approaches zero. Here also, the losses of sensible heat and water occur at about the same rate. The sample load calculations are figured at a 5% average for the first 4 h for sensible heat and 6% for latent heat.

Table 3 Load Calculations for Beef Holding

Cooler size, ft: 100 × 136 × 18.5
Cooler capacity, one day's kill: 1120 carcasses
Average carcass weight: 610 lb
Assumed chill rate: 7.5 °F in 24 h
Assumed air circulation: 91,200 cfm
Assumed air to coil: 34 °F, 96% rh
Assumed fan motive power: 24 hp
Specific heat of beef: 0.75 Btu/(lb · °F)

	Loads, Btu/h		
Heat Gain—Room Load	**Sensible Heat**	**Latent Heat**	**Total Heat**
1. Transmission, infiltration, personnel, fan motor, lights, and equipment heat	245,000	10,000	255,000
2. Product heat, Btu/h:			
a. 1120 × 610 × 0.75 × 7.5 × 0.5	192,000		
b. 1120 × 1.8 × 0.06 × 1070[a]	−131,000	131,000	192,000
3. Total heat gain (room load), Btu/h (Items 1 + 2a + 2b)	306,000	141,000	447,000
Heat Removal—Coil Load			
4. Air circulation, lb/h dry air 91,200 × 0.08 × 60 = 437,800	—	—	—
5. Heat removed per lb of dry air, Btu (Item 3)/(Item 4) = 1.023	—	—	—
6. Air-vapor enthalpy, Btu/lb dry air:			
a. Air to coil, 34 °F, 96% rh	8.167	4.227	12.394
b. Btu removed, temperature drop 2.9 °F	−0.700	−0.323	−1.027
c. Air from coil, 31.1 °F, 100% rh	7.467	3.904	11.367
7. Coil air-vapor heat removal, Btu/h	306,000	141,00	447,000
8. Room vapor condensed to fog	—	—	—
9. Water (ice) removed by coil 1120 × 1.8 × 0.06 × 144[b]	—	17,400	17,400
10. Total heat removal (coil load), Btu/h (Items 3 + 8 + 9)	306,000	158,400	464,400

[a]Heat of vaporization
[b]Heat of fusion

Under peak chilling and holding room conditions, water trapped and condensed out by the coils imposes a further latent load on the evaporators. This occurs in the form of heat extracted to freeze condensed water into ice or of heat removed to chill the returning warmed and strengthened spray solution. In the absence of a more complex evaluation, this load may be considered equal to the latent heat of fusion (144 Btu/lb) of the water removed.

Based on the data just mentioned, cooler loads may be calculated as illustrated in Tables 2 and 3. Transmission, infiltration, personnel, and equipment loads are estimated by standard methods.

The complete calculation is made to illustrate the heat removal process associated with the chilling-drying of the carcass. In particular, it illustrates that the sensible heat ratio of the heat transfer in the coil cannot be used to measure the amount of water removed from the space when fog is involved.

Evaporator Selection. Evaporator selection is a procedure of approximation only, because of the inaccuracies of load determination on the one hand and of predicting sustained field performance of coils on the other. Furthermore, there is rarely complete freedom of specification; for example, the air vapor circulation rate for a given coil may be limited to avoid spray solution carryover or excessive fan power.

Sprayed and dry coil systems perform equally well with respect to shrinkage, provided compressor capacity is adequate and the evaporators are correct for the system selected. Evaporator requirements vary widely with the type of system. Comparative evaporator data on a typical, successful flooded coil installation in the chilling cooler is presented in Table 4.

Table 4 Sample Evaporator Installations for Beef Chilling[a]

Cooler size, ft: 62 × 74 × 18.5
Cooler capacity: 476 carcasses
Deep-round chill: to 50 °F in 20 h
Design load: 1,412,000 Btu/h
Coil operation: liquid recirculation
Loading time: 3.3 h
Average carcass weight: 625 lb
Assumed air to coil: 33 °F, 100% rh
Sensible heat ratio: 53%

	Dry Coil
Coil Description:	
Type of coil	Finned
Fin spacing, fins/in.	4
Coil depth, number of pipe rows	8
Coil face area, ft^2	20.4
Coil surface area, ft^2 total	2238
Fan Description, Airflow:	
Type of fan	Centrifugal
Flow through coil, cfm	9300
Flow, cfm/ft^2 coil face area	455
Fan motive power, hp	2
Unit Rating[b] (Total Heat):	
TD for capacity rating, °F[c]	10
Chilling capacity, Btu/h	81,000
Temperature drop, air through coil, °F	3.7
Equipment for 520 Carcasses:	
Number of units required	18
Total motive power, fans and pumps, hp	36
Coil surface per carcass, ft^2	88
Airflow per carcass, cfm	350

[a]While data describe actual successful installations, other successful installations may be different.
[b]Ratings shown are estimated from performance of actual systems. Dry coil ratings are at average frost conditions, with airflow reduced by frost obstruction. While this example describes actual installations, it is not to be interpreted as an accepted standard. Other installations, employing both more and less equipment, are also successful.
[c]TD is temperature difference between refrigerant and air.

The coil U-value (overall heat transfer coefficient) and airflows shown describe sustained field performance under actual chilling conditions and loads; they should not be confused with clean coil test ratings. The U-value varies greatly with the character of the coil and its operation and is influenced by such variables as the ratio of extended-to-prime surface, which may, for example, range from 7-to-1 to 21-to-1 in standard dry coils; coil depth, which typically ranges from 8 to 12 rows in sprayed coils and from 4 to 10 rows in dry coils; fin spacing, which may be 3 or 4 per inch in typical dry coils; condition of the surface, either continuously defrosted or generally coated with frost; and airflow, which may vary from 250 to 750 cfm/ft^2 coil face area.

Greater temperature differences (TD) than those shown are sometimes used, but a higher TD is valid only at high room temperatures. The lower TD (10 °F) shown for dry coils is desirable to limit frosting. Many dry coil evaporators have higher ratios of extended-to-prime surface and higher airflows per unit face area than shown.

The difficulties involved in obtaining accurate shrinkage figures on carcasses chilled to a specified degree cause wide differences of opinion as to the coil capacity required for good chilling. While data describe actual successful installations, other successful installations may differ.

Boxed Beef

The majority of the output of beef slaughterhouses is in the form of prefabricated sections of the carcass, vacuum-packed in plastic bags and shipped in corrugated boxes. Standard cuts can be sold at cost savings to the market. The shipping density is much greater, with easier material handling, and the bones and fat are removed where their value as a byproduct is greater. Customers purchase only the sections they need, and the trim loss at final processing to primal cuts is minimized.

Vacuum-packaging with either CO_2 or N_2 has the following advantages:

- Creates anaerobic conditions, preventing the growth of mold (which is aerobic and requires the presence of oxygen for growth)
- Provides more sanitary conditions for carcass breaking
- Retains moisture, retards shrinkage
- Excludes bacteria entry, extends shelf life
- Retards bloom until opened

After normal chilling, a carcass is broken into primal cuts, vacuum packed, and boxed for shipment. Temperatures are usually held at 28 °F to prevent the development of pathogenic organisms. Aging of the beef continues after vacuum-packaging and during shipment, because the exclusion of oxygen does not slow enzymatic action in the muscle.

Freezing Times of Boneless Meat. The cooling of boneless meat from 50 to 10 °F requires the removal of about 133 Btu/lb of lean meat (74% water), most of which is latent heat liberated when the liquid water in the meat changes into ice. Most of the latent heat is produced as the meat is cooled from 30 to 25 °F. Accordingly, most of the time needed to freeze meat is spent in cooling through this range.

For boneless meat in cartons, the rate of freezing depends on the temperature and velocity of the surrounding air and on the thickness and thermal properties of the carton and the meat itself.

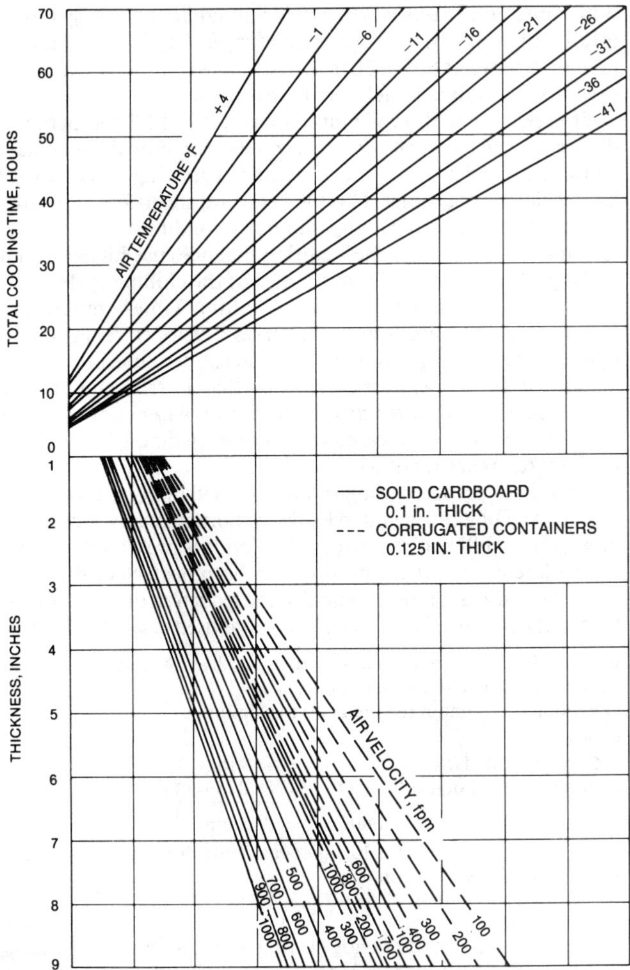

Fig. 4 Freezing Times of Boneless Meat

Figure 4 shows the effects of the first three factors on cooling times for lean meat in two carton types.

For example, the chart shows a total cooling time of 30.75 h for solid cardboard cartons 5-in. thick at an air temperature of −18 °F and a velocity of 400 fpm.

The corresponding air temperatures and cooling times may be found for any specified thickness of carton and air velocity. Conversely, the chart can be used to find combinations of air velocity and temperature needed to freeze cartons of a particular thickness in a specified time (see Figure 5).

Accuracy of the estimated freezing times is about 3% for air velocities greater than 400 fpm. Calculations are based on Plank's Equation as modified by Earle. A latent heat of 107 Btu/lb and an average freezing point of 28 °F are assumed for lean meat.

Increasing the fat content of meat reduces the water content and hence the latent heat load. Thermal conductivity of the meat is reduced at the same time, but the overall effect is for freezing times to drop as the percentage of fat rises. Actual cooling times for mixtures of lean and fatty tissue should therefore be somewhat less than the times obtained from the chart. For meat with 15% fat, the reduction is about 17%.

Hog Chilling and Tempering

The internal temperature of hog carcasses entering the chill coolers from the killing floor varies from 100 to 106 °F. The specific heat shown in the 1993 ASHRAE *Handbook—Fundamentals* is 0.62 Btu/(lb · °F), but in practice 0.7 to 0.75 Btu/(lb · °F) is used because changed feeding techniques have created leaner hogs. The dressed weight varies from 90 to 450 lb approximately, the average being near 180 lb. Present practice requires dressed hogs to be chilled and tempered to an internal ham temperature of 37 to 39 °F on an overnight basis. This limits the chilling and tempering time to 12 to 18 h.

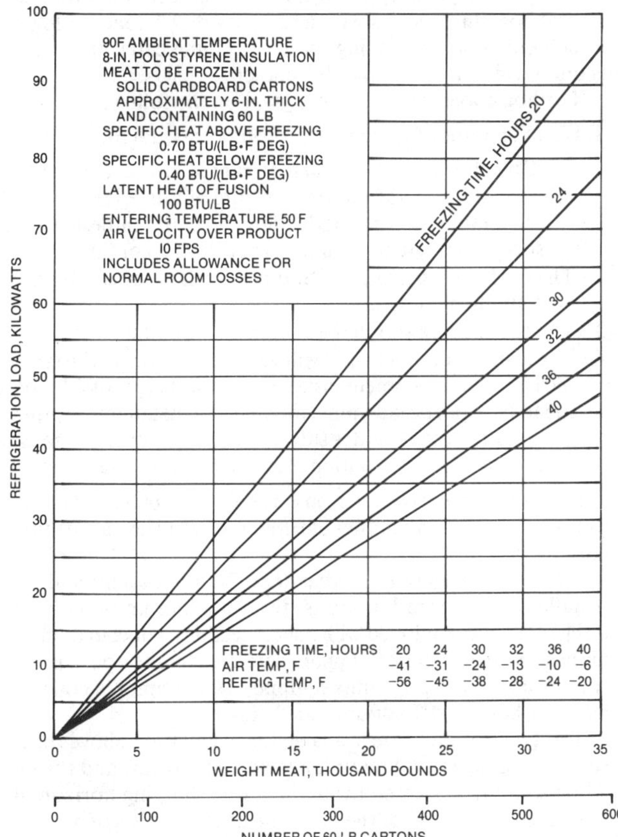

Fig. 5 Blast Freezer Loads

Cooler and refrigeration equipment must be designed to chill the hogs thoroughly with no frozen parts at the time the carcasses are moved to the cutting floor. Carcass crowding, reducing exposure to circulated chilled air, and excessively high peak temperatures are all detrimental to proper chilling.

The following hog cooler design details will provide:

- Sufficiently quick chilling to retard bacterial development and prevent deterioration.
- A cooler shrinkage from 0.1 to 0.2%.
- Firm carcasses that are dry and bright without frozen surface or internal frost, suitable for efficient cutting.

Hog Cooler Design. The capacity of hog coolers is set by the dressing rate of hogs and the planned hours of operation. However, on a one-shift basis, it is economically sound to provide cooler hanging capacity for 10-h dressing in order to properly handle the chilling of large sows that require more than 24-h exposure in the chill room; handle increased dressing volumes when market conditions warrant overtime operation; and have some flexibility in unloading and loading the cooler during normal operations. On a two-shift basis, extra cooler capacity for overtime operation is not necessary.

The rail height should be 9 ft to provide both good air circulation and adequate clearance between the floor and the largest dressed hog. (In the United States, this is a requirement of the FSIS and most state regulations.) Rails should be spaced at a minimum of 30 in. on centers to provide sufficient clearance for the hanging hogs and to prevent contact between carcasses.

The spacing of hogs on the rail varies according to the size of the hog. Hogs should be spaced so that there is at least 1.5 to 2 in. between the flank of one carcass and the back of the carcass immediately in front of it. The rail spacing of 13 in. on centers is normal for 180-lb dressed hogs.

Many meat-packing refrigeration engineers maintain that several hog chill coolers with a capacity of 2-h loading for 300 to 600 kill/h or 4-h loading for lower killing capacities is more economical than one large chill cooler.

The hog cooler should be designed on the following basis:

1. Total amount of hanging rail should be equal to:
   a. (One-shift operation) (10 h × rate of kill × 13 per 12 ft).
   b. (Two-shift operation) (16 h × rate of kill × 13 per 12 ft).
   c. For combination carcass loading, hogs and cattle, calves, or sheep, the figures should be modified accordingly.
2. The rail height should be 9 ft. It may be 11 ft for combination beef and hog coolers.
3. The rail spacing should be a minimum of 30 in.
4. The inside building height will vary depending on the type of refrigeration equipment installed. A clear height of 6 ft above the rail support is adequate for space to install units, piping, and controls; it provides sufficient plenum over the rails to ensure even air distribution over the hog carcasses.

Prechilled, intensive batch, in-line chilling is practiced in some plants, resulting in smaller shrinkages with the variations in chilling systems.

Selecting Refrigeration Equipment. Both floor units and units installed above the rail supports are used. Floor units, with a top discharge outlet equipped with a short section of duct to discharge air in a space over the rail supports, are used by a few pork processors. A few brine spray units equipped with ammonia coils, and water or hot-gas defrost units are being used.

Two types of units are available for installation above the rail supports. One uses a blower fan to force air below and through a horizontally placed coil with the air discharging horizontally from the front or top of the unit. The other consists of a vertical coil with axial fans or blowers to force the air through the unit. Both units are designed for various types of liquid feed control.

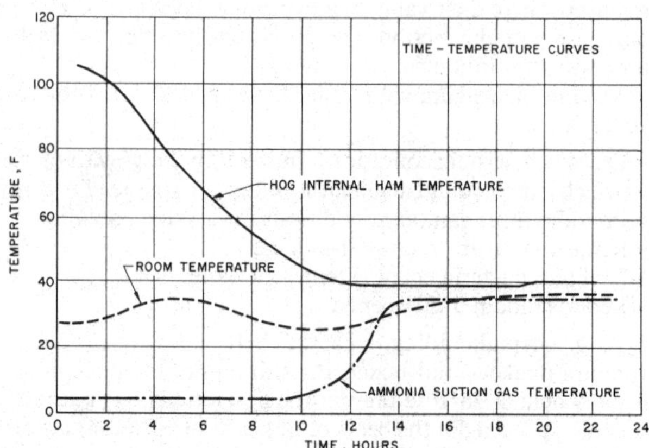

Fig. 6 Composite Hog Chilling Time-Temperature Curves

The horizontal units are equipped for both hot-gas and water defrosting. All have finned surfaces designed for hog chill cooler operation. Evaporator controls should be provided as described in the previous section on beef carcass coolers.

Careful selection of units and use of automatic controls, including liquid recirculation, provides an air circulation, temperature, and humidity balance that chills hogs with minimum shrinkage in the quickest time.

The temperature control should be set to provide an opening room temperature of 26 to 28 °F. As the cooler is loaded, the suction temperature decreases to provide the additional refrigeration effect required to handle increased refrigeration load and maintain room temperature at 30 to 32 °F. Ample compressor and unit capacity is required to achieve these results.

The selection of dry coil units based on 10 to 12 °F temperature difference (refrigerant to air) at peak operation provides adequate coil surface and a TD of 1 to 5 °F prior to opening and about 10 h after closing the cooler. This low-temperature difference results in maximum economical high humidity conditions throughout the entire chilling cycle. New cutting practices use higher initial chilling TD with lower TD at the end of the chilling cycle.

Sample Calculation. The Hog Chilling Time-Temperature Curves (Figure 6) are composite curves developed from several operation tests. The relation of the room temperature and ammonia suction gas temperature curves show that the refrigeration load decreased about 9 h after closing the cooler. After about 9 h, the room temperature is increased and the hog is tempered to an internal ham temperature of 37 to 39 °F.

Table 5 was prepared using empirical calculations to coordinate product and unit refrigeration loads. A shortcut method for determining hog chill cooler refrigeration loads is presented in Table 6. The latent heat of the product has been neglected, since the latent heat of evaporation is equal to the reduction of sensible heat load of the product. Total sensible heat was used in all calculations.

Example 1: Select cooling units for 600 hogs/h at 180 lb average dressed weight using 2-h loading time cooler.
Four coolers minimum requirement; five desirable.

Each cooler:
Capacity 1200 Hogs = 11.3 tons (Table 6)
Product Peak Load = 6 × 14 = 84.0 tons (Table 5)
Total = 95.3 tons

Select 18 units of 5.3 tons at 10.3 °F temperature difference per cooler
Approximately 198,000 cfm
Air changes per minute = 198,000/68,000 = 2.9

The refrigeration of the hog cutting room, where the carcass is cut up into its primal parts, is an important factor in maintaining the quality of the product. A maximum dry-bulb temperature of

Table 5 Product Refrigeration Load, Tons

Hours	Cooler Loading Time, h			
	1	2	4	8
1	7.20[a]	7.20	7.20	7.20
2	6.80	14.00[a]	14.00	14.00
3	6.49	13.29	20.49	20.49
4	6.19	12.68	26.88[a]	26.68
5	5.94	12.13	25.42	32.62
6	5.71	11.65	24.33	38.33
7	5.56	11.27	23.40	43.89
8	5.44	11.00	22.65	49.33[a]
9	5.38	10.82	22.09	47.51
10	—	5.38	16.38	39.71
11	—	—	10.82	34.22
12	—	—	5.38	28.03
13	—	—	—	22.09
14	—	—	—	16.38
15	—	—	—	10.82
16	—	—	—	5.38

[a]Values are for peak load:
100 hogs/h at 180 lb dressed weight average.
Chilled from 102 to 38 °F in 16 h.
Based on operating test, not laboratory standards.

Table 6 Average Chill Cooler Loads Exclusive of Product

Cooler Capacity	Room Dimensions, ft	Floor Area, ft²	Room Volume, ft³	Refrigeration tons[a]
1200 Hogs	40 × 100 × 17	4000	68,000	11.3
2400 Hogs	80 × 100 × 17	8000	136,000	22.6
3600 Hogs	120 × 100 × 17	12,000	204,000	33.9
4800 Hogs	100 × 160 × 17	16,000	272,000	45.2
6000 Hogs	100 × 200 × 17	20,000	340,000	56.5

[a]Based on 6000 ft³ room volume/ton (or use detailed calculations for building heat gains, infiltration, people, lights, and average unit cooler motors from Chapter 26 in the 1993 ASHRAE *Handbook—Fundamentals*).

50 °F should be maintained. This level is low enough to prevent excessive rise in temperature of the product during its relatively short stay in the cutting room and also complies with USDA-FSIS requirements.

Chilled carcasses entering the room may have surface temperatures as low as 30 °F. Unless the dew point of the air in the cutting room is maintained at a temperature below 30 °F, moisture will condense on the surface of the product, providing an excellent medium for bacterial growth.

Floor, walls, and all machinery on the cutting floor must be thoroughly cleaned at the end of each day's operation. Cleaning releases a large amount of vapor in the room that, unless quickly removed, will condense on the walls, ceiling, floor, and machinery surfaces. When the outside dew-point temperature is less than the room temperature, vapor removal can be accomplished by installing fans to continuously exhaust the room during cleanup.

These fans should operate only during the cleanup period and as long as required to remove the vapor produced by cleaning. Exhaust facilities equivalent to five air changes per hour should be satisfactory. When the room temperature is lower than the outside dew-point temperature, this method cannot be used. The water vapor must then be condensed out on the evaporator surfaces.

Many people work in the hog cutting room. Attention should be given to the heat given off by personnel, normally 1000 Btu/h per person. Heat given off by electric motors must also be included in the refrigeration load.

Special consideration should also be given to the latent heat load caused by knife boxes, wash water, workers, and infiltration. If the sensible heat load is not sufficient, this high latent heat load must be offset by reheat at the refrigeration units in order to maintain the desired low relative humidity.

The quantity of air circulated is influenced by the amount of sensible heat to be picked up and the relative humidity to be maintained, usually between 7 and 12 complete air changes each hour. The air distribution pattern requires careful attention to prevent drafts on the workers.

Forced air units are satisfactory for refrigerating cutting floors. Ceiling height must be sufficient to accommodate the units. A wide selection of forced air units may be applied to these rooms. They can be floor or ceiling mounted, with either dry-coil or wetted-surface units arranged for flooded, recirculated, or direct-expansion refrigerant systems.

Suction pressure regulators should be provided with both the flooded and the direct-expansion units. Automatic dry- and wet-bulb controls are essential for best operation.

Pork Trimmings

Pork trimmings come from the chilled hog carcass, principally from the primal cuts: belly, plate, fatback, shoulder, and ham. Trimmings per hog average 4 to 8 lb. Only trimmings used in sausage or canning operations are discussed here.

In the cutting or trimming room, trimmings are usually at a temperature between 38 and 45 °F; an engineer must design for the higher temperature. The product requires only moderate chilling to be in proper condition for grinding, if it is to be used locally in sausage or canning operations. If it is to be stored or shipped elsewhere, hard chilling is required. Satisfactory final temperature for local processing is 28 °F. This is the average temperature after tempering and should not be confused with surface temperature immediately after chilling. It is possible for the trimmings to have a much lower surface temperature than internal temperature immediately after chilling, especially if they have been quick-chilled.

Good operating practice requires rapid chilling of pork trimmings as soon as possible after removal from the primal cuts. This retards enzymatic action and microbial growth, which are responsible for poor flavor, rancidity, loss of color, and excessive shrinkage.

The choice of chilling method depends largely on local conditions and consists of a variation of air temperature, air velocity, and method of achieving contact between air and meat. Continuous belt equipment using low-temperature air fluids or one of the cryogenic fluids to obtain lower shrinkage is available.

Truck Chilling. Economic conditions may require existing chilling or freezer rooms to be used. Some may require an overnight chill, others less than an hour. The following methods make use of existing facilities:

- Trimmings are often chilled with CO_2 snow prior to grinding to prevent excessive temperature rise during grinding and blending. Additionally, finished products, such as sausage or hamburger in chubs, may be crusted or frozen in a glycol/brine liquid contact chiller.
- Trimmings are put on truck pans to a depth of 2 to 4 in. and held in a suitable cooler kept at 30 °F. This method requires a short chill time and results in a near uniform temperature (30 °F) of the trimmings.
- Trimmings are spread 4 or 5 in. deep on truck pans in a 0 °F freezer and held for 5 to 6 h, or until the meat is well stiffened with frost. Use of temperatures lower than 0 °F (with or without fans) will expedite chilling if time is limited. After trimmings are hard-chilled, they are removed from the metal pans and tightly packed into suitable containers. They are held in a 26 to 28 °F room until shipped or used. Average shrinkage using this system is 0.5% up to the time they are put in the containers.
- Trimmings from the cutting or trimming room are put in a meat truck and held in a cooler with a temperature from 28 to 30 °F. This method usually requires an overnight chill and is not likely to reach a temperature of 32 °F in the center of the load.
- If trimmings are not to be used within one week, they should be frozen immediately and held at −10 °F or lower.

Fresh Pork Holding

Fresh pork cuts are usually packed on the cutting floor. If they are not shipped the same day that they are cut and packed, they should be held in a cooler with a temperature of 20 to 28 °F.

Forced-ir cooling units are frequently used for holding room service because they provide better air circulation and more uniform temperatures throughout the room, minimize ceiling condensation caused by air entering doorways from adjacent warmer areas (because of traffic), and eliminate the necessity of coil scraping or drip troughs if hot-gas defrost is used.

Cooling units may be the dry type with hot-gas defrost, or wetted surface with brine spray. Units should have air diffusers to prevent direct air blast on the products. Unless the room shape is unusually odd, discharge ductwork should not be necessary.

Since the product is boxed and wrapped and the holding period is short, humidity control is not too important. Various methods of automatic control may be used. CO_2 has been applied within boxes of pork cuts. Care must be taken to maintain the ratio of pounds of CO_2 to pounds of meat for the retention period. The enclosures must be relieved and ventilated in the interest of life safety.

Calf and Lamb Chilling

Dry coils are typically used for calf and lamb chilling. These can be either the between-the-rail type, the suspended type above the rail, or floor units. The same type of refrigerating units used for pork may be used for lamb, with some modifications. For example, in the chilling of lambs and calves, it is desirable to use reduced air changes over the carcass. This is accomplished by two-speed motors, using the higher speed for the initial chill and reducing the rate of air circulation when the carcass temperatures are reduced, approximately 4 to 6 h after the cooler is loaded.

Lambs usually weigh 40 to 80 lb, with an approximate average dressed carcass weight of 50 lb. Sheep weigh up to an average of approximately 125 lb and readily take refrigeration. Adequate coil surface should be installed to maintain a room temperature below 30 °F and a relative humidity of 90 to 95% in the loading period. The evaporating capacity should be based on an average 10 °F temperature differential between refrigerant and room air temperature, with an opening room temperature of 32 °F.

The use of compensating back pressure regulating valves, which vary the evaporator pressure as the room temperature changes, is recommended. As the room and carcass temperatures drop, the temperature differential is reduced, thus holding a high relative humidity. At the end of a 4- to 6-h chill period, the air over the carcass may be reduced to help keep the bloom and color of the product.

Carcasses should not touch each other. They enter the cooler at 98 to 102 °F, with the carcass temperature taken at the center of the heavy section of the rear leg. The specific heat of a carcass is 0.7 Btu/(lb · °F). Air circulation for the first 4 to 6 h should be approximately 50 to 60 changes per hour, reduced to 10 to 12 changes per hour. The carcass should reach 34 to 36 °F internally in about 12 to 14 h and should be held at that point with 85 to 90% rh room air until shipped or otherwise processed. This gives the least possible shrinkage and prevents excess surface moisture.

In calf chilling coolers, approximately the same procedure is acceptable, with carcasses hung on 12 to 15 in. centers. The dressed weight varies at different locations, with an approximate 85 to 90 lb average in dairy country and a 200 to 350 lb (sometimes heavier) average in beef-producing localities. The same time and temperature relationship and air velocities previously given for chilling lambs are used for chilling calves, except when calves are chilled with the hide on. Also, the time may be extended for air circulation. Air circulation need not be curtailed in hide-on chilling in this application because rapid cooling gives better color to these carcasses after they are skinned.

The refrigerating capacity for lamb and calf chill coolers is calculated the same as for other coolers, but additional capacity should be added to permit reduced air circulation and maintain close temperature differential between room air and refrigerant.

Chilling and Freezing Variety Meats

The temperature of variety meats must be lowered rapidly to 28 to 30 °F to reduce spoilage. Large boxes are particularly difficult to cool. For example, a 5-in. deep box containing 70 lb of hot variety meat can still have a core temperature as high as 60 °F 24 h after it enters a −20 °F freezer. Variety meats in boxes or packages more than 3-in. thick may be chilled very effectively during freezing by adding dry ice to the center of the box.

For design calculations, variety meat has an average density and an initial temperature of about 100 °F. Specific heats of variety meats vary with the percentage of fat and moisture in each. For design purposes, a specific heat of 0.75 Btu/(lb · °F) should be used.

Quick Chilling. A better and more widely used method consists of quick chilling at lower temperatures and higher air velocities, using the same type of truck equipment as in the overnight chilling method. This method is also used for chilling trimmings. Care must be exercised in the design of the quick chilling cabinet or room to provide for the refrigeration load imposed by the hot product. One industry survey shows that approximately 50% of large establishments are using quick-chill in their variety meat operations.

The quick-chilling cabinet or room should be designed to operate at an air temperature of approximately −20 °F with air velocities over the product of 500 to 1000 fpm. During initial loading, the air temperature may rise to 0 °F. In the design of quick-chilling units, refrigeration coils are used with axial flow fans for air circulation.

The recommended defrosting method is with water and/or hot gas, except where units with continuous defrost are used. The product is chilled to the point where the outside is frosted or frozen and a temperature of 28 to 30 °F is obtained when the product later reaches an even temperature throughout in the packing or tempering room. The time required to chill the product by this method depends on the depth of product in the pans, the size of individual pieces, air temperature, and velocity. Normally, 0.5 to 4 h is a satisfactory chill period to attain the required 28 to 30 °F temperature. In addition to the obvious savings in time and space, an important advantage of this method is the low total shrinkage, averaging only 0.5 to 1%. These values were obtained in the same survey as those in Table 7.

Table 7 Storage Life of Meat Products

Product	Months			
	Temperature, °F			
	10	0	−10	−20
Beef	4 to 12	6 to 18	12 to 24	12+
Lamb	3 to 8	6 to 16	12 to 18	12+
Veal	3 to 4	4 to 14	8	12
Pork	2 to 6	4 to 12	8 to 15	10
Chopped beef	3 to 4	4 to 6	8	10
Pork sausage	1 to 2	2 to 6	3	4
Smoked ham and bacon	1 to 3	2 to 4	3	4
Uncured ham and bacon	2	4	6	6
Beef liver	2 to 3	2 to 4		
Cooked foods	2 to 3	2 to 4		

Another method of handling variety meats involves packaging the product before chilling as near as possible to the killing floor. The packed containers are placed on platforms and frozen in a freezer. Separators should permit air circulation between packages.

This method is used in preparing products for frozen shipment or freezer storage. The internal temperature of the product should reach 25 °F for prompt transfer to a storage freezer. For immediate shipment, the internal temperature of the product must be reduced to 0 °F; this may be done by longer retention in the quick freezer. Here package material and size, particularly package thickness, largely determine the rate of freezing. For example, a 5-in. thick box will take from 16 h upward to freeze, depending on the type of product, package material, size, and loading method.

The dry-bulb air temperature in these freezers is maintained at −40 to −20 °F, with air velocities over the product at 500 to 1000 fpm. The time required to reach the desired internal temperature is a combination of refrigeration capacity, size of largest package, insulating properties of package material, and so forth. A generous safety factor should be used in sizing evaporator coils for this type of service. These freezers are best incorporated within refrigerated rooms. Defrosting is by the water or hot-gas method, except where units with continuous defrost are used. Shrinkage varies in the range of only 0.5 to 1%.

The initial freezing equipment cost and design load can be reduced if carbon dioxide is included in the packaging phase as part of the operational plan. Another efficient cooling method uses plate freezers to form blocks of product that can be loaded on pallets with minimal packaging.

Packaging and Storage

Packages for variety meats do not yet have any standard sizes or dimensions. Present requirements are a package that will stand shipping, with sizes to suit individual establishments. The importance of the package is brought more into focus in the case of the hot pack freezing method. Standardization of sizes and package materials promotes faster chilling and more economical handling.

Storage of variety meats depends on its end use. For short storage (under one week) and local use, 28 to 30 °F is considered a good internal product temperature. If stored for shipping, the internal temperature of the product should be kept at 0 °F or below. Recommended length of storage is controversial; type of package, freezer temperature and relative humidity, amount of moisture removed in original chill, and the variations of the products themselves all affect storage life.

Packers' recommendations regarding storage time vary from 2 to 6 months and longer, since variety meats pick up rancidity on the surface and soft muscle tissue dehydrates while freezing. More rapid freezing and vapor-proof packaging are important in increasing storage life.

Refrigeration Load Computations

The average evaporator refrigerating load for a typical chilling process above freezing may be computed by:

$$H_r = W_m s_m (t_1 - t_2) + W_t s_t (t_1 - t_2) + H_w + H_i + H_m \qquad (1)$$

where

H_r = refrigeration load, Btu/h
W_m = weight of meat, lb/h
W_t = weight of trucks, lb/h
s_m = specific heat of meat, Btu/(lb · °F)
s_t = specific heat of truck, containers, or platforms (0.12 for steel), Btu/(lb · °F)
t_1 = average initial temperature, °F
t_2 = average final temperature, °F
H_w = heat gain through room surfaces, Btu/h
H_i = heat gain from infiltration, Btu/h
H_m = heat gain from equipment and lighting, Btu/h

The following example illustrates the method of computing the refrigeration load for a quick-chill operation.

Example 2. Find the refrigeration load for chilling six trucks of offal from a maximum temperature of 100 to 34 °F in 2 h. Each truck weighs 400 lb empty and holds 720 lb of offal. The specific heat is 0.12 for the truck and 0.75 for the offal. The room temperature is to be held at 0 °F, with an outdoor temperature of 40 °F and a rh of 70%. The walls, ceiling, and floor gain 72 Btu/ft² in 24 h and have an area of 947 ft². The volume of the room is 1881 ft³ and 12 air changes in 24 h are assumed.

Solution: Values for substitution in Equation 1 are obtained as follows:

$W_m = 6 \times 720/2 = 2160$ lb/h
$s_m = 0.75$ Btu/(lb · °F)
$t_1 - t_2 = 100 - 34 = 66$ °F
$W_t = 6 \times 400/2 = 1200$ lb/h
$s_t = 0.12$ Btu/(lb · °F)
$H_w = 72 \times 947/24 = 2841$ Btu/h
$H_i = (1.12 \times 1881 \times 12)/24 = 1053$ Btu/h where
1.12 = heat removed per ft³ of air entering, Btu
$H_m = (10 \times 2545) + (200 \times 3.4) = 26{,}140$ Btu/h
10 = assumed horsepower
2545 = Btu per horsepower hour
200 = lights, W
3.4 = Btu/W · h

Substituting in Equation (1):

$$H_r = (2160 \times 0.75 \times 66) + (1200 \times 0.12 \times 66) + 2841 + 1053 + 26{,}140 = 146{,}534 \text{ Btu/h} = 12.2 \text{ tons}$$

It is good practice to add 10 to 25% to the computed refrigeration load.

PROCESSED MEATS

Prompt chilling, handling, and storage under controlled temperatures help in the production of mild and rapidly cured and smoked meats. The product is usually transferred directly from the smokehouse to a refrigerated room, but sometimes a drop in temperature of 10 to 30 °F can occur if the transfer time is appreciable.

Since the day's production is not usually removed from the smokehouses at one time, the refrigeration load is spread over a large portion of the 24 h. Table 8 outlines temperatures, relative humidities, and time required in refrigerated rooms used in handling smoked meats.

Smoked meat prechilling results in a reduction of drips of moisture and fat, thus increasing the yield. This can be done at higher temperatures, with air velocities of up to 500 fpm (Table 8). At

Table 8 Room Temperatures and Relative Humidities for Smoking Meats

	Room Conditions			
	°F Dry-Bulb	% Relative Humidity	Final Product Temp, °F	Time h
Prechill method				
Hams, picnics, etc.				
High temperature	38 to 40	80	60	8 to 10
Low temperature	26 to 28	80	60	2 to 3
Derind bacon				
Normal	26 to 28	80	28	8 to 10
Blast	0 to 10	80	26	2 to 3
Hanging or tempering				
Ham, picnics, etc.	45 to 50	70	50 to 55	
Derind bacon	26 to 28	70	26 to 28	
Wrapping or packaging				
Hams, picnics, etc.	45 to 50	70		
Storage	28 to 45	70		

lower temperatures, air velocities of 1000 fpm and higher are used. Chilling in the hanging room or in the wrapping and packaging rooms results in slow chilling and high temperatures when packing. This is not desirable for a product that is to be stored or shipped a considerable distance.

Meats handled through smoke and into refrigerated rooms are hung or racked on cages that are moved on an overhead track or mounted on wheels. Sometimes the product is transferred from suspended cages to wheel-mounted cages between smoking and subsequent handling.

Smoked hams and picnic meats must be chilled as rapidly as possible through the incubation temperature range of 105 to 50 °F. A product requiring cooking before eating is brought to a minimum internal temperature of 140 °F to destroy possible live trichinae, while a product not requiring cooking before eating is brought to a minimum internal temperature of 155 °F.

A maximum storage room temperature of 40 °F dry bulb should be used when delivery from the plant to retail outlets is made within a short time. A room dry-bulb temperature of 28 to 32 °F is desirable when delivery is to points considerably removed from the plant and transfer is made through controlled low humidity rooms, docks, cars, or trucks, keeping the dew point below that of the product.

Bacon usually reaches a maximum temperature of 125 °F in the smokehouse. Since most smoked bacon in a sliced form is packaged, it may be transferred directly to the chill room if it has been skinned before smoking. If the bacon is to be skinned after smoking, it is usually allowed to hang in the smokehouse vestibule for 2 to 4 h, until it drops to a temperature of 90 °F before skinning.

Bacon is usually molded and sliced at temperatures just below 28 °F. Chill rooms are usually designed to reduce the internal temperature of the bacon to 26 °F in 24 h or less, requiring a room dry-bulb temperature of 18 to 20 °F. A tempering room (which also serves as storage for stock reserve), held at the exact temperature at which bacon is sliced, is often used.

Bacon molding can be done either after tempering, in which case it is moved directly to the slicing machines, or after the initial hardening, and then be transferred to the tempering room. In the latter case, care should be taken that none of the slabs is below 24 °F so that the product will not crack during molding. Bacon cured by the pickle injection process generally shows less pickle pockets if it is molded after hardening, placed no more than eight slabs high on pallets, and held in the tempering room.

In any of the rooms mentioned, air distribution must be uniform. To minimize shrinkage, the air supply from floor-mounted unit coolers should be delivered through slotted ducts or by means of closed ducts supplying properly spaced diffusers directed so that no high velocity airstreams impinge on the product itself. The exception would be in blast chill rooms where high air velocities are needed, but in reality the product is subjected to the condition for only a short time.

Refrigeration may be supplied by floor or ceiling-mounted dry or wet coil units. If the latter are selected, water, hot-gas, or electric defrost must also be used.

In recent years, many processors have begun using three different methods of chilling smoked meats: rapid blast chilling, brine chilling by direct contact spraying, and cryogenic chilling. Direct contact spraying is especially emphasized, because it affords minimization of shrinkage, longer shelf life, and more uniform chilling. This method is usually carried out in enclosures specially designed to combat the detrimental effects of salt brine. Color and salt taste may need close monitoring in contact spraying.

The product should enter the slicing room chilled to a uniform internal temperature not to exceed 50 °F for beef rounds and 40 to 45 °F for other fresh carcass parts, depending on the individual packer's temperature standard. Internal temperatures below 26 °F tend to cause shattering of products such as bacon during slicing and slow fabrication. For that type product, temperatures above 32 °F cause improper shingling from the slicing machine.

The slicing and packaging room temperature and air movement are usually the result of a compromise between the physical comfort demands of the operating personnel and the requirements of the product. The design room dry-bulb temperature should be below 50 °F, according to USDA-FSIS regulations.

An objectionable amount of condensation on the product may occur. To guard against this, the coil temperature should be maintained below the temperature of the bacon entering the room, thus keeping the dew point of the room air below the product temperature. The product should be exposed to the room air for the shortest possible time.

Bacon Slicing and Packaging Room

Exhaust ventilation should remove smoke and fumes from the sealing and packaging equipment and comply with OSHA occupancy regulations. Again, heat exchangers should be considered for reducing the resulting increased refrigeration load.

Refrigeration for this room may be supplied by forced-air units—floor or ceiling mounted, dry or wet coil—or finned-tube ceiling coils.

Dry coils should have defrost facilities if coil temperatures are to be kept at more than several degrees below freezing. Air discharge and return should be evenly distributed, using ductwork if necessary. To avoid drafts on personnel, air velocities in the occupied zone should be in the range of 25 to 35 fpm. The temperature differential between primary air and room air should not exceed 10 °F to assure personnel comfort. To provide optimum comfort conditions and dew-point control, reheat coils are necessary.

Where ceiling heights are adequate, multiple ceiling units can be used to minimize the amount of ductwork. Automatic temperature and humidity controls are desirable in cooling units.

To provide draft-free conditions, drip troughs with suitable drainage should be added to finned-type ceiling coils. However, it is difficult, if not impossible, to maintain a room air dew point low enough to approach the product temperature. Some installations operating with relative humidities of 60 to 70% do not have product condensation problems.

One control method consists of individual coil banks connected to common liquid and suction headers. Each bank is equipped with a thermal expansion valve. The suction header has an automatically operated back-pressure valve. The thermostatically controlled dual back-pressure regulator and liquid-header solenoid are both controlled by a single thermostat. This arrangement provides a simple automatic defrosting cycle.

Another system uses fin coils with glycol sprays. Humidity is controlled by varying the concentration strength of the glycol and the refrigerant temperature.

Sausage Dry Rooms

Refrigeration or air conditioning is an integral part of year-round sausage dry rooms. The purpose of these systems is to produce and control the air conditions for proper removal of moisture from the sausage.

A variety of dry sausage is manufactured, for the most part uncooked. Its keeping qualities depend on curing ingredients, spices, and removal of moisture from the product by drying. This sausage is generally of two distinct types—smoked and not smoked.

FSIS regulates the minimum temperature and amount of time that the product must be held after stuffing and prior to release, depending on the method of production. The dry room temperature shall not be lower than 45 °F and the length of time the product shall be held in the dry room prior to release is dependent on the diameter of the sausage after stuffing and the method of preparation used.

After stuffing, the sausages are held at a temperature of 60 to 75 °F and 75 to 95% rh in the sausage greenroom to develop the cure. The sausages are suspended from sticks at the time of stuffing and may be held on the trucks or railed cages or be transferred to racks in the greenroom. Sausages in 3.5- to 4-in. diameter casings are generally spaced about 6 in. on centers on the sausage sticks. The length of time the sausages are held in the greenroom depends on the method of preparation, the type and dimensions of the sausage, the operator, and the judgment of the sausage maker regarding proper flavor, pH, and other characteristics.

Those sausage varieties that are not smoked are then transferred to the sausage dry room; those that are to be smoked are transferred from the greenroom to the smokehouse and then to the dry room.

In the dry room, approximately 30% of the moisture is removed from the sausage, to a point at which the sausage will keep for a long time, virtually without refrigeration. The drying period required depends on the amount of moisture to be removed to suit trade demand, type of sausage, and type of casing. The moisture transmission characteristics of synthetic casings vary widely and greatly influence the rate of drying. The diameter of the sausage is probably the most important factor influencing the drying rate.

Small-diameter sausages, such as pepperoni, have more surface in proportion to the weight of material than do large-diameter sausages. Furthermore, the moisture from the sausage interior has to travel a much shorter distance to reach the surface, where it can evaporate. Thus, the drying time for small-diameter sausages is much shorter than that for large-diameter sausages.

Typical conditions in the dry room are approximately 45 to 55 °F and 60 to 75% rh. Some sausage makers favor the lower range of temperatures for unsmoked varieties of dry sausage and the higher range for smoked varieties.

In processing dry sausage, moisture can only be removed from the product at the rate at which the moisture comes to the surface of the casing. Any attempt to speed the drying rate results in overdrying the surface of the sausage, a condition known as case hardening. This condition is identified by a dark ring inside the casing, close to the surface of the sausage, which precludes any further attempt to remove moisture from the interior of the sausage. On the other hand, if sausage is dried at too slow a rate, excessive mold occurs on the surface of the casing, sometimes leading to an unsatisfactory appearance. An exception is the Hungarian Salami, which requires a high humidity so that a prolific mold growth can occur and flourish.

As with any other cool or refrigerated space, sausage dry rooms should be properly insulated to prevent temperatures in adjoining spaces from influencing the temperature in the dry room. Ample insulation is especially important in the case of dry rooms located adjacent to rooms of much lower temperature or rooms on the top floor, where the ceiling may be exposed to relatively high temperatures in summer and low temperatures in winter.

Insulation should be adequate to prevent the inner surface of the walls, floor, and ceiling of the dry room from differing more than a degree or two from the average temperature in the room. Otherwise, there is the possibility of condensation due to the high relative humidity in these rooms, which leads to mold growth on the surfaces themselves and, in some cases, on the sausages as well.

The sticks of sausage are generally supported on permanent racks built into the dry room. In the past, these were frequently made of wood; however, sanitary requirements have virtually outlawed the use of wood for this purpose in new construction. The uprights and rails for the racks are now made of either galvanized pipe, hot dip galvanized steel, or stainless steel. The rails for supporting the sausage sticks should be spaced vertically at a distance that leaves ample room for air circulation below the bottom row and between the top row and the ceiling. Generally, a spacing of not less than 1 or 2 ft between rails depends on the length of sausage stick used by the individual manufacturer.

The horizontal spacing between sausages should be such that they do not touch at any point, to prevent mold formation or improper development of color. Generally, with the large 4-in. diameter sausages, a spacing of 6 in. on centers is adequate.

Dry-Room Equipment. In general, two types of refrigeration equipment are currently used for attaining the required conditions in a dry room. The most common is a refrigeration-reheat system, in which the room air is circulated either through a brine spray or over a refrigerated coil and sufficiently cooled to reduce the dew point to the temperature required in the room. The other type involves the use of a hygroscopic liquid sprayed over a refrigerating coil in the dehumidifier, thus condensing the moisture from the air without the severe overcooling usually required by the refrigeration-reheat type of system. The chief advantage of this arrangement is that refrigeration and heating loads are greatly reduced.

The use of any type of liquid, brine or hygroscopic, requires periodic tests and adjusting the pH to minimize corrosion of the equipment. Although most systems depend on a type of liquid spray to prevent frost buildup on the refrigerating coils, some successful rooms use dry coils with hot-gas or water defrost.

The air for conditioning the dry room is normally drawn through the refrigerating and dehumidifying systems by a suitable blower fan (or fans) and discharged into the distribution ductwork.

Rooms used exclusively for small-diameter products with a rapid drying rate may actually have air leaving the room to return to the conditioning unit at a lower dry-bulb temperature and greater density than the point at which it is introduced. A dry-room designer needs to know what the room will be used for to determine the natural circulation of the room air. Supply and return ducts can then be arranged to take advantage of and accelerate this natural circulation to provide thorough mixing of incoming dry air with the air in the room.

Regardless of the location of the supply and return ducts, care should be taken to prevent strong drafts or high velocity airstreams from impinging on the product; this will result in local overdrying and unsatisfactory products.

A study of air circulation within the product racks will show that as air passes over the sausages and moisture evaporates from them, this air becomes cooler and heavier, and thus has a tendency to drop toward the bottom of the room, creating a vertical downward air movement within the sausage racks. This natural tendency must be considered in designing the duct installation if uniform conditions are to be achieved.

An example of the calculation involved in designing a sausage dry room follows. These calculations apply to a room used for an assortment of sausages, with an average drying time of approximately 30 days. They would not be directly applicable to a room used primarily for very large salami (which has a much longer drying period) or a room used primarily for small-diameter sausage.

In the latter case, use of the air-circulating rate shown in this example will allow the air to absorb so much moisture in passing through the room that it will be difficult to obtain uniform conditions throughout the space. Furthermore, the amount of refrigeration required to lower the air temperature enough to produce the required low inlet air dew point would be excessive. An air circulation rate of 12 air changes per hour should therefore be considered average for use in average rooms. The actual circulating rate should be adjusted to obtain the best compromise of refrigeration load and air uniformity for the particular type of product handled.

Example 3. Air conditioning for sausage drying room

Room Dimensions:

40 ft, 2 in. × 33 ft, 6 in. × 11 ft, 6 in.
Floor space: 1350 ft^2
Volume: 15,600 ft^3
Outdoor wall area: 980 ft^2
Partition wall area: 770 ft^2

Hanging Capacity:

Number of racks: 12
Length of racks: 27 ft
Number of rails high: 5
Spacing of sticks: 2/ft of rail
Number of pieces of sausage per stick: 7
Average weight per sausage: 4 lb
Total weight: 12 × 27 × 5 × 2 × 7 × 4 = 90,720 lb of product
Assume 90,000 lb green weight hanging capacity
Loading per day: 1500 lb

Assumed Outdoor Conditions (Summer):

95°F db; 74.5°F wb; 39% rh; h = 37.8 Btu/lb; 66°F dp; 96 gr/lb

Dry-Room Conditions Desired:

55°F db; 50°F wb; 70% rh; h = 20.2 Btu/lb; 46°F dp; 46 gr/lb

Sensible Heat Calculations:

Walls (2 in. insulation):		
980 (95 − 55) × 0.10	=	3920 Btu/h
Partition (4 in. insulation):		
770 (95 − 55) × 0.067	=	2060 Btu/h
Floor and ceiling:		
2700 (55 − 55) × 0.10	=	none
Infiltration: 0.5 × 15,600 (95 − 55) × 0.243 × 0.075	=	5700 Btu/h
Lights: 600 W × 3.415	=	2050 Btu/h
Motors: 5 × 1 hp × 2546	=	12,730 Btu/h
Product: 1500 × 0.8 (95 − 55)/24	=	2000 Btu/h
Total sensible heat	=	28,460 Btu/h

Latent Heat Calculations:

Product		
90,000 × 0.30 × 1000/(60 × 24)	=	18,720 Btu/h
(18,720 × 7000)/(60 × 1000)	=	2184 gr/min
Infiltration		
(0.5 × 15,600 × 0.075 (37.8 − 20.2)	=	10,300 Btu/h
(10,300 − 5700) 7000/(60 × 1000)	=	537 gr/min
Total grains of moisture = 2184 + 537	=	2721 gr/min

Assume 12 air changes per hour with an empty room volume of 15,600 ft^3 = (15,600)(12/60)(0.76) = 237 lb of air per minute. Then each pound of air must absorb 2721/237 = 11.5 gr of moisture. Since air at the desired room condition carries 46 gr/lb, the air must enter the room with only 46 − 11.5 = 34.5 gr/lb, corresponding to 41.8°F db and 40°F wb (h = 15.3 Btu/lb).

Temperature rise due to sensible heat gain (air specific heat = 0.243 Btu/lb·°F):

28,460/(237 × 0.243 × 60) = 8.2°F db.

Temperature drop due to evaporative cooling from latent heat of product only:

18,720/(237 × 0.243 × 60) = 5.4°F

Net temperature rise in the room = 8.2 − 5.4 = 2.8°F, or the air entering the room must be 55 − 2.8 = 52.2°F db at 38.5°F dp (34.5 gr/lb).

Refrigerating load = 237(20.2 − 15.3) × 60 = 69,700 Btu/h
Reheat load = (17.8 − 15.3) × 60 = 35,600 Btu/h
Room load = 237(20.2 − 17.8) × 60 = 34,100 Btu/h

Lard Chilling

In federally inspected plants, the USDA FSIS designates the types of pork fats which, when rendered, are classified as *lard.* Other pork fats, when rendered, are designated as *rendered pork fats.* The rendering process requires considerable heat, and the subsequent temperature of the lard at which refrigeration is to be applied may be as high as 120°F. The following data for refrigeration requirements may be used for either product type.

The fundamental requirement of the FSIS is good sanitation through all phases of handling. The use of copper or copper-bearing alloys that come in contact with lard should be avoided, because minute traces of copper lower the stability of the product.

Lard has the following properties:

Specific gravity at	0°F	= 0.99
	70°F	= 0.93
	160°F	= 0.88
Heat of solidification		= 48 Btu/lb
Melting begins at −35 to −40°F.		
Melting ends at 110 to 115°F.		
Point of half fusion is around 40°F.		
Specific heat in solid state		
	−110°F	= 0.28 Btu/lb·°F
	−40°F	= 0.34 Btu/lb·°F
Specific heat in liquid state		
	110°F	= 0.50 Btu/lb·°F
	212°F	= 0.52 Btu/lb·°F

In the production of lard, refrigeration is applied so that the final product will have enough texture and a firm consistency. The finest possible crystal structure is desired.

Calculations for chilling 1000 lb of lard per hour are:

Initial temperature: 120°F
Final temperature: 80°F
Heat of solidification: 48 Btu/lb
Specific heat: 0.50 Btu/(lb·°F)

$$S_f = 100\frac{t_e - t_f}{t_e - t_b} = 100\frac{115 - 80}{115 - (-40)} = 100(35/155) = 22.6\%$$

where

S_f = percent solidification at final temperature
t_e = temperature at which melting ends
t_f = final temperature
t_b = temperature at which melting begins

Latent heat of solidification:

48(22.6/100) = 10.8 Btu/lb

Sensible heat removed:

0.50(120 − 80) = 20 Btu/lb

Total heat removed:

1000(10.8 + 20) = 30,800 Btu/h = 2.57 tons refrigeration

Assuming a 15% loss because of radiation, for example, in the process, the required refrigeration for application to chill 1000 lb lard per hour would be 1.15 × 2.57 = 2.96 tons.

Filtered lard at 120°F can be chilled and plasticized in compact internal swept surface chilling units, which use either ammonia or halogenated hydrocarbons. A refrigerating capacity of about 36,000 Btu/h per 1000 lb of lard handled per hour for the product only should be provided. Additional refrigeration for the requirements of heat equivalent to the work done by the internal swept surface chilling equipment will also be needed.

When operating this type of equipment, it is essential to keep the refrigerant free of oil and other impurities so that the heat transfer surface will not have a film of oil on it to act as insulation and cut down the unit's capacity. Some installations have oil traps connected to the liquid refrigerant leg on the floor below to provide an oil accumulation drainage space.

The safety requirements for this type of chilling equipment are described in ASHRAE *Standard* 15-1992, *Safety Code for Mechanical Refrigeration.* Note that such units are pressure vessels and, as such, require properly installed and maintained safety valves.

The recommended storage temperature for packaged refined lard is 31 to 33°F. The storage temperature required for prime steam lard in metal containers is 40°F or below for up to a 6-month storage period. Lard stored for a year or more should be kept at 0°F.

Blast and Storage Freezers

The standard method of sharp-freezing a product destined for storage freezers comprises freezing the product directly from the cutting floor in a blast freezer until its internal temperature reaches the holding room temperature. The product is then transferred to holding or storage freezers.

The product to be sharp-frozen may be bagged, wrapped, or boxed in cartons. Individual loads are usually placed on pallets, dead skids, or in wire basket containers. In general, the larger the ratio of surface exposed to blast air to the volume of either the individual piece or the product's container, the greater the rate of freezing. Product loads should be placed in a blast freezer to assure that each load is well exposed to the blast air and to minimize possible short circuiting of the airflow. Each layer on a load should be separated by 2-in. spacers to give the individual pieces as much exposure to the blast air as possible.

The most popular types of blast equipment are the self-contained air handling or cooling units that consist of a fan, evaporator, and other elements in one package. These units are usually used in multiples and placed in the blast freezer to provide the optimum blast air coverage. The unit fans should be capable of high air velocity and volumetric flow; two air changes per minute is the accepted minimum.

The coils of the evaporator may have either a wet or a dry surface. Dry coils can be defrosted with hot gas or ethylene glycol, which is used as a wetting agent on wetted coil surfaces. When a wetting agent is used, concentrators must be installed to remove moisture and maintain concentration of the wetting agent. Evaporator coils are best operated flooded. Temperature differential between coil and room air should be in the range of 10 to 15°F.

Blast chill design temperatures vary throughout the industry. Most designs are within a −20 to −40°F range. For low temperatures, booster compressors that discharge through a desuperheater into the general plant suction system are used.

Blast freezers require sufficient insulation and good vapor barriers. If possible, a blast freezer should be located so that temperature differentials between it and adjacent areas are minimized in order to decrease insulation costs and refrigeration losses.

Blast freezer entrance doors should be power operated. Suitable vestibules should also be provided as air locks to decrease infiltration of outside air.

Besides normal losses, heat calculations for a blast freezer should include the loads imposed by such material handling equipment as electric trucks, skids, and spacers, together with the packaging materials for the product.

Storage freezers are usually maintained at 0 to −15°F. If the plant operates with several high and low suction pressures, the evaporators can be tied to a suitable plant suction system. The evaporators can also be tied to a booster compressor system; if the booster system is operated intermittently, provisions must be made to switch to a suitable plant suction system when the booster system is down. Storage freezer coils can be defrosted by hot gas, electricity, or water. Emphasis should be placed on not defrosting too fast with hot gas (because of pipe expansion) and on providing well-insulated, sloped, heat traced drains and drain pans to prevent freezeups.

Direct Contact Meat Chilling

Continuous processes for smoked and cooked wieners use direct sodium chloride brine tanks or deluge tunnels to chill the meat as soon as it comes out of the cooker. Everyday, the brine is prepared fresh in 2 to 13% solutions depending on the chilling temperature and the salt content of the meat.

Cooling is usually done on sanitary stainless steel surface coolers, which are either refrigerated coils or plates in cabinets. Using this type of unit allows coolant temperatures near the freezing point without damaging the cooler; damage may occur when brine is confined in a tubular cooler. The brine is circulated in quantities necessary to fully wet the surface cooler and fill the distribution troughs of the deluge.

Another type of continuous process uses a brine tank into which the wieners drop out of a conveyor that has carried them through the smoking and cooking process. The pumped brine moves the product to the end of the tank, where it is removed by hand and inserted into peeling and packaging lines.

FROZEN MEAT PRODUCTS

The handling and selling of consumer portions of frozen meats has many potential advantages compared with merchandising fresh meat. The preparation and packaging can be done at the packinghouse, allowing economies of mass production, byproduct savings, lower transportation costs, and flexibility in meeting market demands. At the retail level, frozen meat products reduce space and investment requirements and labor costs.

Prefreezing Quality of Meat

After an animal is slaughtered, physiological and biochemical reactions continue in the muscle until the complex system supplying energy for work has run down and the muscle goes into rigor. These changes continue for up to 32 h postmortem in major beef muscles. Hot boning with electrical stimulation renders meat tender on a continuous basis without conventional chilling. Freezing meat or cutting carcasses for freezing before the completion of these changes cause cold-shortening and thaw-shortening, which render meat tough. The best time to freeze meat is either after rigor has passed or later, when natural tenderization is more or less complete. Natural tenderization is completed during 7 days of aging in most major beef muscles. Where flavor is concerned, freezing as soon as tenderization is complete is desirable.

For frozen pork, the age of the meat before freezing is even more critical than it is for beef. Pork loins aged 7 days before freezing deteriorate more rapidly in frozen storage than loins aged 1 to 3 days. In tests, a difference could be detected between 1- and 3-day-old loins, favoring those only 1-day old. With frozen pork loin roasts from carcasses chilled for 1 to 7 days, the flavor of lean and fat in the roasts was progressively poorer with longer holding time after slaughter.

Effect of Freezing on Quality

Freezing affects the quality—including color, tenderness, and amount of drip—of meat.

Color. The color of frozen meat depends on the rate of freezing. Tests in which prepackaged, steak-size cuts of beef were frozen by immersion in liquid or exposure to an airblast at between −20 and −40°F revealed that airblast freezing at −20°F produced a color most similar to that of the unfrozen product. An initial meat temperature of 32°F was necessary for best results (Lentz 1971).

Flavor and Tenderness. Flavor does not appear to be affected by freezing per se, but tenderness may be affected depending on the condition of the meat and the rate and end-temperature of freezing (Jul 1969). Faster freezing to lower temperatures was found to increase tenderness; however, consensus on this effect has not been reached.

Drip. The rate of freezing generally affects the amount of drip, and meat nutrients, such as vitamins, are lost from cut surfaces after thawing. Faster freezing tends to reduce the amount of drip, although many other factors, such as the pH of meat, also have an effect on drip.

Changes in Fat. Fat of pork changes significantly in 112 days at −5°F, whereas beef shows no change within 260 days at this temperature. At −20 and −30°F, no measurable change occurs in either meat in one year.

The relationship of fat rancidity and oxidation flavor has not been clearly established for frozen meat, and the usefulness of antioxidants in reducing flavor changes during frozen storage is doubtful.

Storage and Handling

Pork remains acceptable for a shorter storage period than beef, lamb, and veal due to the differences in fatty acid chain length and saturation in the different species. Storage life is also related to storage temperature. Because animals within a species vary greatly in nutritional and physiological backgrounds, their tissues differ in susceptibility to change when stored. As a result of the differences between meat animals, packaging methods, and acceptability criteria, a wide range of storage lives is reported for any one type of meat (Table 7). Jul (1969) discusses the storage life of frozen meats and the factors that affect them.

Appreciable changes in the color and flavor of frozen beef occur at storage temperatures down to −40°F in 1 to 90 days (depending on temperature) for samples held in the dark. Changes were much more rapid (1 to 7 days) for samples exposed to light. Color changes were less pronounced after thawing than when frozen (Lentz 1971).

Reports on the effect of different storage temperatures on fat oxidation and palatability of frozen meats indicate that a temperature of 0°F or lower is desirable. Cuts of pork back held at 20, 10, 0, and −10°F show increases in peroxide value; free fatty acid is most pronounced at the two higher temperatures. For a storage period of 48 weeks, 0°F or lower is essential to avoid fat changes. Pork rib roasts of 0°F showed little or no flavor change up to 8 months, whereas at 10°F, fat was in the early stages of rancidity in 4 months. Ground beef and ground pork patties stored at 10, 0, and −10°F indicate that meats must be stored at 0°F or lower to retain good quality for 5 to 8 months. For longer storage, −20°F is desirable.

The desirable flavor in pork loin roasts stored at −6 to −8°F, with maximum fluctuations of 5 to 8 degrees, decreased slightly, apparently without significant difference between treatments. Fluctuations from 0 to 10°F did not harm quality.

Storage temperature is perhaps more critical with meat in frozen meals because of the differing stability of the various individual dishes included. Frozen meals show marked deterioration of most of the foods after 3 months at 13 to 15°F.

Storage and Handling Practices. In view of the effect of temperature on frozen foods, surveys of practices in the industry indicate why a certain amount of product reaches the consumer in poor condition. One unpublished survey indicated that 10% of frozen foods may be found to be at 6°F or higher in warehouses, 16°F or higher in assembly rooms, 21°F or higher during delivery, and 17°F or higher in display cases. All these temperatures should be maintained at 0°F for complete protection of the product.

Packaging

At the time of freezing, a package or packaging material serves to hold the product and prevent it from losing moisture. Other functions of the wrapper or box become important as soon as the storage period begins. Ideal packaging material in direct contact with meat should have: low moisture vapor transmission rate; low gas transmission rate; high wet strength; grease proofness; flexibility over a temperature range including subfreezing; freedom from odor, flavor, and any toxic substance; easy handling and application characteristics adaptable to hand or machine use; easy strippability; and reasonable price.

Individually or collectively, these properties are desired for good appearance of the package, protection against handling, preventing dehydration, which is unsightly and damages the product, and keeping oxygen out of the package.

Desiccation through use of unsuitable packaging material is one of the major problems with frozen foods. Another problem is that of distorted or damaged containers due either to lack of expansion space for the product in freezing or to selection of low-strength box material.

Whenever free space is present in a container of frozen food, ice sublimes and condenses on the film or package. Temperature fluctuation increases the severity of frost deposition.

SHIPPING DOCKS

A refrigerated shipping dock can eliminate the need for assembling orders on the nonrefrigerated dock or other area, or using a more valuable storage space for this purpose. This is especially true in the case of freezer operations. Some businesses do not really need a refrigerated order assembly area. One example is a packing plant that ships out whole carcasses or sides in bulk quantities and does not need a large area in which to assemble orders. Many such plants are constructed without any dock at all, simply having the load-out doors lead directly into the carcass-holding cooler, which is satisfactory, provided adequate refrigerating capacity is installed in the vicinity of the shipping doors to prevent undue temperature rise in the coolers during shipping.

A refrigerated shipping dock can perform a second function of reducing the refrigeration load, which is most important in the case of freezers; although it serves almost as valuable a function with coolers. Even with cooler operations, the installation of a refrigerated dock greatly reduces the load on the cooler's refrigerating units and assures a more stable temperature within the cooler. At the same time, it is possible to only provide refrigeration to maintain dock temperatures on the order of 40 to 45°F, so that the refrigerating units can be designed to operate with a wet coil. In this way, frost buildup on the units is avoided and the capacity of the units themselves substantially increased, making it unnecessary to install as many or as large units in this area.

In the case of freezers, the units should be designed and selected to maintain a dock temperature slightly above freezing, usually about 35°F. With this dock temperature, orders may be assembled and held prior to shipment without the risk of defrosting the frozen product, and the workers can assemble orders in a much more comfortable space than the freezer. The design temperature should be low enough that the dew point of the dock atmosphere is below the product temperature. Condensation on the surface of the products is one step in developing off-condition product.

With a dock temperature of 35°F, the temperature difference between the freezer itself and the outdoor summer condition is split roughly in half. Since the airflow through the loading doors or other openings is proportional to the square root of the temperature difference, this results in an approximate 30% reduction in airflow through the doors—both the doors into the dock itself and the doors from the dock into the freezer. At the same time, by cooling the outdoor air to approximately 35°F, in most cases about 50% of the total heat in the outdoor air is removed by the refrigerating units on the dock.

Since using a refrigerated dock reduces the airflow through the door into the freezer by approximately 30% and 50% of the heat in the air that does pass through this door is removed, the net effect is to reduce the infiltration load on the units within the freezer itself by about 65%. This is not a net gain; since an equal number of these units operate at a much higher temperature, the power required to remove the heat on the dock is substantially lower than it would be if the heat were allowed to enter the freezer.

The infiltration load from the shipping door, whether it opens directly into a cooler or freezer or into a refrigerated dock, is extremely high. Even with well-maintained foam or inflatable door seals, a great deal of warm air leaks through the doors whenever they are open. Such air infiltration may be calculated approximately by:

$$V = 1.4HW\theta(H)^{0.5}(t_1 - t_2)^{0.5}$$

where

V = air volume, ft^3/h at the higher temperature condition
1.4 = an empirical constant selected to convert the airflow into ft^3/h, after allowing for the contraction of the airstream as it passes through the door and for the obstruction created by a truck parked at the door with only nominal sealing
H = height of the door, ft
W = width of the door, ft
θ = time the door is open, min/h
t_1 = outdoor air temperature or the air at the higher temperature, °F
t_2 = temperature of air in the dock or cooler, °F

Time, θ, is estimated, based on the time the door is assumed to be obstructed or partially obstructed. If the doors are equipped with good seals and these are well-maintained so that the average truck will be tightly sealed to the building, this time would be assumed as only the time necessary to spot the truck at the door and complete the air seal.

The unit cooler providing refrigeration for the dock area should be ceiling suspended with a horizontal air discharge. Each unit should be aimed toward the outer wall and above each of the truck loading doors, if possible, so that cold air strikes the wall and is deflected downward across the door. This downward airflow just inside the door tends to oppose the natural airflow of the entering warm air, thus helping reduce the total amount of infiltration.

In general, a between-the-rails unit cooler has proved most successful for this purpose, since it distributes the air over a fairly wide area and at low outlet velocity. Such an airflow pattern does not create severe drafts in the working area and is more acceptable to employees working in the refrigerated space. The above comments and equation for determining air infiltration also apply to shipping doors, which open directly into storage or shipping coolers.

ENERGY CONSERVATION

Water, a utility previously considered free, frequently has the most rapid rate increases. Coupled with high sewer rates, it is the largest single-cost item in some plants. If fuel charges are added to the hot water portion of water usage, water is definitely the most costly utility. Costs can be reduced by better dry cleanup, use of heat exchangers, use of filters and/or settling basins to collect solids and greases, use of towers and/or evaporative condensers, elimination of water as a means of product transport, and an active conservation program.

Air is needed for combustion in steam generators, sewage aeration, air coolers or evaporative condensers, and blowing product through lines. Used properly in conjunction with heat exchangers, air can reduce other utility costs—fuel, sewage, water, and electricity. Nearly all plants need close monitoring of valves either leaking through or left open in product conveying. Low-pressure blowers are frequently used in place of high-pressure air, reducing initial investment and operating costs of driving equipment.

Steam generation is a source of large savings through efficient boiler operation (fuel and water sides). Reduced use of hot water and sterilizer boxes, and proper use of equipment in plants with electric and steam drives, should be promoted. Sizable reductions can be realized by scavenging heat from process-side steam and hot water and by systematically checking steam traps. In some plants, excess hot water and low energy heat can be recovered using heat exchangers and better heat balances.

Electrical energy needs can be reduced by:

- properly sizing, spacing, and selecting light fixtures and an energy program of keeping lights off (lights comprise 25 to 33% of an electric bill)
- monitoring and controlling the demand portion of electric usage
- checking and sizing motors to their actual loads for operation within the more efficient ranges of their curves
- adjusting the power factor to reduce initial costs in transformers, switchgear, and wiring
- lubricating properly to cut power demands

Although refrigeration is not a direct utility, it involves all or some of the factors just mentioned. Energy use in refrigeration systems can be reduced by:

- operating with lower condenser and higher compressor suction pressures
- properly removing oil from the system
- purging noncondensable gases from the system
- adequately insulating floors, ceilings, walls, and hot and cold lines
- using energy exchangers on exhaust and air makeup
- keeping doors closed to cut humidity or prevent an infusion of warmer air
- installing high-efficiency motors
- maintaining compressors at peak efficiency
- keeping condensers free of scale and dirt
- using proper water treatment in the condensing system
- operating with a microprocessor based management system

Utility savings are also possible when usage is considered with product line flows and storage space. A strong energy conservation program not only saves total energy but frequently results in greater product yields and product quality improvements, and therefore increased profits. Prerigor or hot processing of pork and beef products is being used to greatly reduce the energy required for postmortem chilling. Removal of waste fat and bone prior to chilling reduces the amount of chilling space by 30 to 35% per beef carcass.

REFERENCES

Acuff, G.R. 1991. Acid decontamination of beef carcasses for increased shelf life and microbiological safety. *Proceedings of the Meat Industry Resources Conference*, Chicago.

Allen, D.M., M.C. Hunt, A.L. Filho, R.J. Danler, and S.J. Goll. 1987. Effects of spray chilling and carcass spacing on beef carcass cooler shrink and grade factors. *Journal of Animal Science* 64:165.

ASHRAE. 1992. Safety code for mechanical refrigeration. *Standard* 15-1992.

Dickson, J.S. 1991. Control of *Salmonella typhimurium, Listeria monocytogenes, and Escherichia coli* O157:H7 on beef in a model spray chilling system. *Journal of Food Science* 56:191.

Earle, R.L. Physical aspects of the freezing of cartoned meat. Bulletin 2, Meat Industry Research Institute of New Zealand.

Greer, G.G. and B.D. Dilts. 1988. Bacteriology and retail case life of spray-chilled pork. *Canadian Institute of Food Science Technology Journal* 21:295.

Hamby, P.L., J.W. Savell, G.R. Acuff, C. Vanderzant, and H.R. Cross. 1987. Spray-chilling and carcass decontamination systems using lactic and acetic acid. *Meat Science* 21:1.

Heitter, E.F. 1975. Chlor-chill. *Proceedings of the Meat Industry Resources Conference AMIF*, Arlington, VA. 31-32.

Johnson, R.D., M.C. Hunt, D.M. Allen, C.L. Kastner, R.J. Danler, and C.C. Schrock. 1988. Moisture uptake during washing and spray chilling of Holstein and beef-type carcasses. *Journal of Animal Science* 66:2180.

Jones, S.M. and W.M. Robertson. 1989. The effects of spray-chilling carcasses on the shrinkage and quality of beef. *Meat Science* 24:177-88.

Jul, M. 1969. Quality and stability of frozen meats. In *Quality and Stability in Frozen Foods*, Chapter 8. Wiley-Interscience, New York.

Kastner, C.L. 1981. Livestock and meat: Carcasses, primal and subprimals. In *CRC Handbook of Transportation and Marketing in Agriculture*, edited by E.E. Finney, Jr. CRC Press, Inc., Boca Raton, FL. 239-58.

Lentz, C.P. 1971. Effect of light and temperature on color and flavor of prepackaged frozen beef. *Canadian Institute of Food Technology Journal* 4:166.

Locker, R.H., C.L. Davey, P.M. Nottingham, D.P. Haughey, and N.H. Law. 1975. In *Advances in Food Research*, edited by C.O. Chichester. Academic Press, New York. 158-217.

Thatcher, F.S. and D.S. Clark. 1968. *Microorganisms in foods: Their significance and methods of enumeration.* University of Toronto Press, Toronto, Canada.

CHAPTER 12

POULTRY PRODUCTS

POULTRY products may be ice-chilled (33 to 35 °F), deep chilled (around 28 °F), or frozen (0 °F or lower). Means of refrigeration include ice, mechanically cooled water or air, dry ice (carbon dioxide sprays), and liquid nitrogen sprays. Continuous chilling and freezing systems, with various means for conveying the product, are common. This chapter discusses the temperature-sensitive qualities of poultry products and the refrigeration and related requirements needed to maintain these qualities.

SLAUGHTERING, FEATHER REMOVAL, AND EVISCERATION

Slaughtering on a continuous conveyer line involves automatic electrical stunning to prevent struggling and to facilitate cutting and bleeding the birds with mechanical devices (Ingling 1978). Additionally, stunning meets the USDA requirement for humane slaughter. After 90-to 120-s bleeding periods, birds are scalded for periods up to 120 s to loosen the feathers, usually by immersion in a tank of warm water. However, steam-spray scalders may be used to improve sanitation.

Scalding water temperatures can range from 120 to 160 °F and higher. In commercial practice, two temperature ranges are used—semiscalding (124 to 130 °F) and subscalding (138 to 140 °F). Semiscalding removes the feathers without loss of the outer pigmented skin layer (cuticle) and is used for most broilers marketed in the chilled state. Subscalding provides complete feather removal, but the cuticle is removed during the feather picking process. Subscalding is used for all turkeys and mature chickens, and for some broilers that are to be further processed and frozen. Steam scalders have been developed successfully for subscalding, but have had only limited success for semiscalding.

Compared to a semiscalded skin surface, the exposed surface resulting from subscalding is darker and more susceptible to dehydration and further darkening during chilling and storage. Therefore, this product must be chilled at high relative humidity and all subsequent operations must be performed in a manner that prevents moisture loss from the surface. Tight, moistureproof packaging is recommended.

Feather-picking machines vary in design, depending on the speed of the picking line and the specific type of feathers to be picked. All machines remove feathers through the beating action of long rubber fingers mounted on rotating double drums or the rubbing action of rotating short stiff rubber fingers mounted in a series of discs along the length of the picker. Proper clearance between rubber fingers and bird prevents excessive skin abrasion and beating of the carcass and muscles, which may toughen the meat.

The technology of evisceration and related steps essential to the establishment of a ready-to-cook carcass have changed from hand to completely automatic machine operation. Thorough washing of the eviscerated carcass, inside and out, prior to chilling, is essential. The extent to which the body temperature of the bird is lowered when it reaches the chiller depends on the temperature of ambient air and processing water, and on the extent of washing. Internal temperatures may range from 75 to 95 °F.

During processing, sanitary practices in design, use, and cleaning of equipment are essential. Chlorinated water for processing plant use is desirable.

CHILLING

Chilling retards deteriorative changes, principally those caused by microbial growth. According to USDA regulations (1972), poultry carcasses weighing less than 4 lb should be chilled to 40 °F or below in less than 4 h, carcasses of 4 to 8 lb in less than 6 h, and carcasses of more than 8 lb in less than 8 h.

Chilling procedures and equipment have evolved to meet the changing needs of the poultry processing industry. Slow air chilling was considered adequate for semiscalded uneviscerated poultry in the past. But with the transformation to eviscerated, ready-to-cook poultry, sometimes subscalded, air chilling was replaced by chilling in tanks of slush ice. Immersion chilling is more rapid than air chilling, and more important, it prevents dehydration and affects a net absorption of water. Slush ice tank chilling, a batch process, was replaced in turn by mechanical continuous immersion slush ice chillers, which are fed automatically from the end of the evisceration conveyer line. In general, tanks are only used to hold chilled carcasses in an iced condition prior to cutting up, or to age prior to freezing.

Continuous chillers vary greatly in their mechanical design. They include: *continuous drag chillers,* in which suspended carcasses are pulled through troughs containing agitated cool water and ice slush; *slush ice chillers,* through which the carcasses are pushed by a continuous series of power-driven rakes; *concurrent tumble systems,* which involve passage of the free-floating carcasses through horizontally rotating drums suspended in tanks of successively cool water and ice slush (the movement of the carcasses is regulated by the flow rate of recirculated water in each tank); *counterflow tumble systems,* in which the carcasses are carried through tanks of cool water and ice slush by horizontally rotating drums with helical flights on the inner surface of the drums; and *rocker vat systems,* in which carcasses are conveyed by the recirculating water flow, and agitation is accomplished by an oscillating, longitudinally oriented paddle. Carcasses are removed automatically from the tanks by continuous elevators. These chillers are capable of reducing the internal temperature of broilers from 90 to 40 °F in 20 to 40 min, at processing speeds of 5000 to 10,000 birds/h. USDA regulations require a minimum overflow of 2 quarts of water per broiler from continuous immersion chillers and 1 gal per turkey. Chilling water must not be higher than 65 °F in the warmest part of the chilling system.

Adjuncts and replacements for continuous immersion chilling

The preparation of this chapter is assigned to TC 11.2, Foods and Beverages. This chapter last received a major revision in 1974.

should be used if available because immersion chilling is believed to be a major cause of bacterial contamination. Water spray chilling, air blast chilling, and use of carbon dioxide snow or liquid nitrogen spray are alternatives but are limited, as the following comparisons indicate.

Liquid water has a much higher heat transfer coefficient than any gas at the same temperature of cooling medium, so water immersion chilling is more rapid and efficient than gas chilling. Water spray chilling, without recirculation, requires much greater amounts of water than immersion chilling. Product appearance should be equivalent for water immersion or spray chilling, but inferior for air blast or carbon dioxide or nitrogen chilling, due to dehydration. Gas chilling without packaging could cause a 1 to 2% loss of moisture, while water immersion chilling permits from 4 to 15% moisture uptake, and water spray chilling up to 4% moisture uptake. Carpenter *et al.* (1979) also reported on the effect of salt brine chilling on dripless ice-packed broiler carcasses.

With adequately washed carcasses and with adequate chiller overflow in a direction counterflow to the carcasses, the bacterial count on carcasses should be reduced by continuous water immersion chilling. However, incidence of a particular low level contaminant, such as *Salmonella*, may increase during continuous water immersion chilling (Morris and Wells 1970, Surkiewicz *et al.* 1969). Such transfer of microbes can be controlled by chlorinating the chill water (Mast and MacNeil 1972, Lillard 1980). Spray chilling without recirculation has reduced bacterial surface counts 85 to 90% (Peric *et al.* 1971). Microbe transfer by spray chilling is unlikely. Chilling with air, carbon dioxide, or nitrogen presents no obvious microbiological hazards, although good sanitary practices are essential. If crust freezing is accomplished as a part of the chilling process, a reduction in bacterial load, possibly as much as 90%, can be expected.

Many chilling methods involving low-temperature air blast, carbon dioxide snow, or liquid nitrogen spray are used. Cryogenic gases are generally used in long insulated tunnels through which the product is conveyed on an endless belt. Some freezing of the outer layer (crust freezing) usually occurs, and an equilibration of temperature in outer and inner portions is allowed to occur to reach the final, intended chill temperature. In some plants, a combination of continuous water immersion chilling to reach 35 to 40 °F and a cryogenic gas tunnel to finally reach 28 °F is used. The water-chilled poultry, either whole or cut up, is generally packaged before gas chilling, to prevent dehydration.

Water absorption by the carcass in water immersion or spray chilling, and possible loss of water during gas chilling, are important economic and quality factors. USDA regulations set graded maximum levels of water absorption and retention varying with weight and type of product and specify testing procedures and basis for rejection or retention of product. The set values range from 4.3% weight increase for 27 lb and over consumer-packaged whole turkeys to 12% for ice-packed poultry at the end of the drip line. Of the water absorbed, very little appears in the muscle, most being held in the connective tissue in the skin and between skin and flesh (Klose *et al.* 1960a, Osner and Shrimpton 1966).

For example, increases in percentage of moisture due to water immersion chilling were reported from 0.5 to 2% for muscle and 10 to 12% for skin. Kotula *et al.* (1960) compared moisture uptake and drainage loss for air-agitated slush ice tank chilling and for four mechanical continuous immersion slush ice chillers. Immediately after chilling, the total moisture uptake including 2 to 3% by prechill washing ranged from 7% for 2 to 4 h of tank chilling to 17% for a counterflow tumble continuous immersion water-ice chiller. After draining, shipping, and a final 30-min drain, all uptakes ranged from 5 to 7%. Initial uptakes were much greater in a counterflow tumble system for carcasses with thigh skin area cut compared to uncut, *i.e.*, 18 versus 12%. After shipping and draining, the residual uptakes were 7 and 5%.

Although soluble solids are lost from the carcass to the chill water in immersion chilling, no evidence exists that, under accepted chilling practices, any appreciable loss of flavor or other desirable qualities occurs (Pippen and Klose 1955, Janky *et al.* 1978).

Ice requirements per bird for continuous immersion chilling depend on entering carcass temperatures and weight, entering water temperature, and exit water and carcass temperature. For a counterflow system, 60 °F entering water and 65 °F exit water, 0.25 lb of ice per pound of carcass is a reasonable estimate. This may be compared to a requirement of 0.5 to 1 lb of ice per pound of poultry for static ice slush chilling in tanks. For continuous counterflow water immersion chillers, if the plant water temperature is considerably above 65 °F, it may be economical to use a heat exchanger between incoming plant water and exiting (overflow) chill water.

Ice production for chilling is usually a complete in-plant operation, with large piping and pumps to convey the small crystalline ice or ice slush to the point of use. To reduce the use of ice, some immersion chillers are double-walled and depend on circulating refrigerant to chill the water within the chiller. The chiller has an ammonia or refrigerant oil between the outer and inner jacket with the inner jacket serving as the heat transfer medium. Agitation or a defrost cycle must be provided during periods of slack production to prevent the chiller from freezing up.

Chilling and holding to about 28 °F, the point of incipient freezing, gives the product a much longer shelf life compared with a product held at ice-pack temperatures (Stadelman 1970).

TENDERNESS CONTROL

Tenderness in cooked poultry meat is a prerequisite to acceptability. Relative tenderness decreases as birds mature, and this toughness has always been considered in the recommendations for cooking birds of various ages. Sekoguchi *et al.* (1979) studied the effects of cysteamine in relation to the tenderness of meat obtained from mature chickens. However, another type of toughness depends primarily on the length of time that the carcass is held in an unfrozen state before cooking. Birds cooked before they have time to pass through rigor are very tough. Normal tenderization after slaughter is arrested by freezing. For birds held at 40 °F, complete tenderization occurs for all muscles within 24 h and for many muscles in a much shorter time.

Other factors that interfere with normal tenderization are immersion in 140 °F water and cutting into the muscle. Formerly, birds were held unfrozen for sufficient time in the normal channels of processing and utilization to permit adequate tenderization. Shorter chilling periods, more rapid freezing, and cooking without a preliminary thawing period have shortened the period during which tenderization can occur to such an extent that toughness has become a potential consumer complaint. Hanson *et al.* (1942) observed a rapid increase in tenderness within the first 3 h of holding and a gradual increase thereafter.

Shannon *et al.* (1957), working with hand-picked stewing hens, demonstrated an increased toughness because of increased scalding temperature or increased scalding time in the ranges of 120 to 195 °F and 5 to 160 s. However, the differences in toughness that occurred within the limits of temperature and time, necessary or practical in commercial plants, were quite small.

Klose *et al.* (1959 a,b, 1961) and Pool *et al.* (1959) showed that, in addition to the beneficial effect of increasing holding time above freezing, tenderness is increased by lower scalding temperatures, shorter scalding time, and particularly by reducing the extent of beating received by the birds during picking operations. Additionally, Turkey fryers should be held at least 12 h above freezing to develop optimum tenderness. Holding fryers at 0 °F for 6 months and longer has no tenderizing effect, but holding in a thawed state (35 °F) after frozen storage has as much tenderizing effect as an equal period of chilling before freezing. Turkeys frozen 1 h after slaughter were adequately tenderized by holding for 3 days at 28 °F, a temperature at which the carcass is firm and no important quality loss occurs for the period involved. Behnke *et al.* (1973) confirmed this effect for Leghorn hens.

Machine feather picking may result in meat twice as tough as that obtained for a theoretical, hand-picked control. Toughening induced by the beating action of the mechanical feather pickers is not resolved appreciably by otherwise adequate chilling periods. This toughening action is cumulative in that increasing the number of picking machines to which the bird is exposed increases the degree of toughness. Thus, optimum tenderness can be attained through careful attention to chilling (aging) periods, feather picking conditions, and scalding practices. Exposure to unusually high temperatures during the aging period may also lead to irreversible toughening.

Overall processing efficiency could be improved by cutting up the carcass directly from the end of the eviscerating line, packaging the parts, and then chilling the still-warm packaged product in a low-temperature air blast or cryogenic gas tunnel. Several evaluations have been made of the relative tenderness of such a hot-cut product. Webb and Brunson (1972) reported that cutting the breast muscle and removing a wing at the shoulder joint before chilling significantly decreased tenderness of treated muscles, though cut carcasses were aged in ice slush before cooking. Klose *et al.* (1972) found that under commercial plant conditions, making an eight-piece hot-cut before chilling and aging significantly reduced tenderness of breast and thigh muscles, compared to cutting after chilling.

No difference in shear score was found between the outer 0.6 in. slice and the second 0.6 in. slice of breast muscle for the chill-cut birds, but a higher shear value was observed for the second slice from hot-cut birds (Treat and Goodwin 1973). Cutting the muscle from the sternum on one breast piece resulted in a higher shear value than was recorded for the other side of the breast that was left intact. Only in hot-cut birds was this found to be significant. Aging for 24 h eliminated the difference in shear value between hot-cut and chill-cut broilers (Wyche and Goodwin 1974).

Smith *et al.* (1966) indicated that too-rapid chilling of poultry might have a toughening effect, similar to cold shortening observed in red meats.

Cladfelter and Webb (1987) (U.S. Patent No. 4,675,947) describe successful electrical stimulation of broiler chickens shortly after slaughter in order to tenderize the meat. To meet the goals described by Ingling (1978), Webb *et al.* (1989) (U.S. Patent No. 4,860,403) modified the system using electrical stunning. The detailed steps in this minimum time process system (MTPS) are:

1. Live birds are hung, stunned, and killed in the normal manner. Prior to stunning, the birds' breast areas are sprayed with tap water to wet them for good electrical contact with stimulation bars.
2. The birds proceed from the neck cutter into the scalder room, which is separated from the rest of the plant by locked doors and a permanent partition to prevent employee entry without cutting electrical stimulation (ES) supply to the area. In no instance is any person to enter the ES area without following established plant guidelines for safety, including locking out the ES energy supply.
3. After entering the scalder/ES room, the birds' breast areas will rub along the energized stimulation bar as they proceed on the shackle line to the scalder. Repeated contraction of wings should be noticeable with each pulse of ES. This ES is applied during bleedout, which is a minimum of 90 s.
4. Electrical stimulation is supplied by an AC electrical source providing a pulsed current of 0.3 s on and 0.3 s off. Voltage should be 40 volts RMS for each bird (power supply must be adequate to provide this).
5. Scalding should be done in the normal manner. ES will stop just before scalding and resume just after.
6. After scalding, the birds proceed under ES to an EST (electrical stimulation treatment) area which is maintained under moist conditions to provide good electrical contact and prevent carcass feathers from drying out. ES continues for 3 min after scalding.
7. After ES, the birds exit the locked scald/ES area and proceed on the shackle line into the picking room. Sufficient shackle line is provided for at least 11 min of conditioning time between the end of ES and the beginning of picking.
8. Picking and all following operations are done in the normal manner.

An evaluation of breast muscle tenderness from the MTPS indicates that tenderness levels achieved in 40 min post mortem are equivalent to those achieved in about 6 h of normal processing. The MTPS not only shortens meat aging time, but it allows straight-through processing by eliminating the need for chillers. MTPS broilers can be processed, cut up, cooked, packaged, and frozen as a continuous line operation.

RAW POULTRY PROCESSING AND PACKAGING

Chickens and turkeys, for both chilled and frozen distribution, can be cut up in the processing plant. A popular cut for broilers is a nine-piece cut, usually two drumsticks, two thighs, two wings, two breast halves, and back. However, there are at least eight different cutting patterns, involving greater subdivision of breast and leg portion. Backs and necks are often mechanically deboned, giving a comminuted slurry which is frozen in rectangular flat cartons containing about 60 lb. Turkey breasts, legs, and drumsticks are available as separate film-packaged parts, and turkey thigh meat is marketed as a ground product resembling high quality hamburger. Partial cooking and breading and battering of broiler parts is done in some poultry processing plants.

Most packaged poultry is now tray packed, either for frozen or chilled distribution. All-plastic packages have been developed, and automated packaging lines using plastic film have been engineered. Changes in packaging methods and materials are so rapid that the best sources of information on this subject are the companies that fabricate films and packages and distribute the materials. They are listed in the most recent Encyclopedia Issue of *Modern Packaging*.

Many available packaging films are satisfactory concerning relative moisture and gas permeability, but they sometimes lack sufficient strength to withstand the rough handling encountered in normal marketing operations. Trays used in the packaging of cut-up chilled poultry need to contribute a certain amount of rigidity and form to the package and, with a blotter liner, provide absorbency to pick up the drip which exudes from the meat during storage. Plastic film overwraps should be tight fitting and reasonably moisture- and airproof. Wells *et al.* (1958) reported that partial evacuation of the air from packages wrapped in impermeable films tended to inhibit bacterial growth by reducing oxygen tension.

Ideally, packages for frozen poultry should possess low moisture-vapor permeability and should be strong, protective, attractive, and suitable for rapid freezing and prolonged storage of the product. Packages for precooked frozen poultry products must also be adaptable for reheating the product. In general, packaging requirements increase from whole to cut-up to precooked forms.

Packages for frozen, whole, and ready-to-cook poultry consist principally of plastic film bags that are tough and reasonably impermeable to moisture vapor and air. The commonly used polyvinylidene chloride, polyethylene, and polyester films are sufficient barriers to water vapor and air to give adequate protection for normal commercial times and temperatures. Turkeys, ducks, and geese are packaged mostly in the whole, ready-to-cook form, while frozen chickens appear whole and in packaged, cut-up form.

Large fiberboard cartons or containers for holding and shipping from 2 to 12 individually packaged birds should be rectangular in shape to facilitate palletizing and should be strong enough to support 16-ft high stacked loads common in refrigerated warehouses. If freezing is to be accomplished for material such as fryer turkeys that need to be frozen rapidly, holes or cutaway sections in the sides and ends are needed to permit a rapid airflow across the poultry surfaces in the air blast freezer.

DISTRIBUTION AND RETAIL HOLDING REFRIGERATION

Chilled poultry handled under proper conditions is an excellent product. However, there are limitations in its marketability because of the relatively short shelf life caused by bacterial deterioration. Bacterial growth on poultry flesh, as on other meats, has a high temperature coefficient. Studies based on total bacterial counts have shown that birds held at 36 °F for 14 days are equivalent to those held at 50 °F for 5 days or 75 °F for 1 day. Spencer and Stadelman (1955) found that birds at 31 °F had 8 days additional shelf life over those at 38 °F. The magnitude of this temperature coefficient is illustrated by Farrell and Barnes (1964) on generation times (time for population to double) for a species of *Pseudomonas*: 36 h at 29 °F, 14 h at 32 °F, 7 h at 41 °F, and less than 1 h at 77 °F. This indicates the advantage of deep chill over ice chill.

The generation time of psychrophilic organisms isolated from chickens was 10 to 35 h at 32 °F, depending on the species studied (Ingraham 1958). Raising the temperature to 36 °F reduced generation time to 8 to 14 h, again depending on the species.

The bacterial flora in poultry meat include the genera *Pseudomonas, Achromobacter, Micrococcus,* and *Flavobacterium*. Nagel *et al.* (1960) found that the distribution pattern of bacterial genera (mostly *Pseudomonas* and *Achromobacter-Alcaligenes*) was unaffected by geographical location of processing or cutting operations, or by the use of antibiotics. Ayres *et al.* (1950) observed that the growth and coalescence of colonies of the above microorganisms on the surface of chilled poultry led to a characteristic off-odor, accompanied by the development of bacterial slime. The shelf life, or time before off-odor occurred, depended not only on the holding temperature but also on the initial amount of bacterial contamination. This emphasizes the importance of adequate sanitation in processing, which minimizes initial bacterial contamination and thus prolongs shelf life. Frequent cleaning of processing equipment, as well as thorough washing of the eviscerated carcasses, is necessary. Goresline *et al.* (1951) reported a substantial decrease in bacterial contamination and an increase in shelf life by the use of 20 ppm of chlorine in processing and chilling water.

Since shelf life is limited considerably by bacterial growth (slime formation) on the skin layer, it is reasonable to assume that drastic changes in the skin surface, such as the removal of the epidermal layer by high temperature scalding, might appreciably affect shelf life. Ziegler and Stadelman (1955) reported approximately 1 day more chilled shelf life for 128 °F scalded birds than for 140 °F scalded ones.

Chickens, principally broilers, are sold as whole ready-to-cook or cut-up ready-to-cook. Poultry is shipped in wax-coated corrugated fiberboard boxes with crushed ice, or fiberboard boxes with dry-ice snow or solid particulate dry ice (carbon dioxide). Comparisons of dry-ice and water-ice packing in several types of boxes were made by Risse and Thomson (1971). While overall costs for packaging materials, refrigerant, and labor did not differ much between alternative systems, dry ice costs significantly less than water ice packing because of the greater shipping weight of 20 lb of water ice compared to 2 lb of dry-ice snow.

The limited number of warehouse receivers and retail receivers involved in this study preferred dry-ice packed poultry over water-ice packed poultry because the boxes were lighter in weight, had less drainage, and the surface appearance was superior for the dry-ice packed product. However, in the six commercial-type shipments of this study, the percentage weight loss of poultry from processing plant to receiver's warehouse was significantly higher for dry-ice packed (6.3%) than for water-ice packed (5.5%), and the warehouse arrival temperature was significantly higher for dry-ice packed (37 °F) than for water-packed poultry (33 °F). This suggests the desirability of having 28 to 30 °F truck and warehouse temperatures if dry-ice packing is used.

A number of precooked poultry meat products are being sold in wholesale and retail markets as refrigerated, nonfrozen products. Such items are usually vacuum packaged or packaged in either a carbon dioxide or nitrogen gas atmosphere. The desired temperature for such products is also 28 to 30 °F.

FREEZING

Effect on Product Quality

The behavior of muscle tissue during freezing represents the combined effects of the soluble constituents that produce a freezing point lowering and a freezing range, the size and location of the ice crystals formed, and the translocation of cellular constituents during freezing. Koonz and Ramsbottom (1939) pointed out that very slowly frozen poultry appears dark, with very large ice crystals of the extra-fiber type. When poultry is frozen at a temperature that assures a natural bloom, the ice crystals are smaller and more numerous, but are still principally of the extra-fiber type. Ice crystals are small and of the intra-fiber type when freezing is rapid enough to assure a white appearance. Ice crystals become progressively larger in poultry as the time between slaughter and freezing is increased.

Rapid freezing is essential to obtain satisfactorily light appearance in certain types of frozen poultry (van den Berg and Lentz 1958). Birds scalded at 140 °F and above completely lose their outer layer of skin during normal machine picking and become particularly susceptible to the development of a dark frozen appearance if not frozen rapidly. Also, immature fryer-roaster turkeys, which have a thin, practically fat-free skin, are naturally dark and require a rapid rate of freezing to assure a light, pleasing frozen appearance.

Increased rates of freezing were provided by use of air blast tunnels operating at air temperatures ranging from −20 to −40 °F and at air velocities of 500 fpm or more. In many cases, packaged birds freeze better on open shelves in the air blast rather than after packing the packaged birds in cartons. In an evaluation of various factors contributing to freezing rate and frozen appearance, Klose and Pool (1956) found that freezing on open shelves versus freezing in boxes, lower air blast temperature, and higher air blast velocity were important, in that order. At an air blast temperature of −20 °F or below, increasing air blast velocity beyond 600 fpm had little beneficial effect. Also at 1300 fpm, decreasing air-blast temperature from −20 to −30 °F produced almost no additional improvements in frozen appearance. Birds placed in the blast freezer immediately after evisceration and packaging and while still warm did not develop an appreciably darker frozen appearance than those chilled in ice slush before packaging and blast freezing. However, this procedure has the serious disadvantage that the duration of holding above freezing temperatures may be reduced to an amount inadequate for optimum tenderization. This concern is eliminated using the MTPS described in the

Tenderness Control section. The factor of finish, or amount of fat in the skin layer, can exert a much greater beneficial effect on frozen appearance than any possible changes in processing or freezing practices.

Poultry meat to be sold to U.S. government agencies must be fully frozen at a temperature less than 0 °F within 72 h of going into the freezer. The freezing rate of diced cooked chicken meat does affect the quality of the frozen meat. Hamre and Stadelman (1967a) reported that cryogenic freezing procedures were desirable as the resulting color was lighter, but too rapid a freezing rate resulted in a shattering of the meat cubes. The freeze-drying rates for rapidly frozen material were slower than for products frozen by slower methods. Hamre and Stadelman (1967b) indicated that tenderness of freeze-dried chicken after rehydration was affected by freezing rate prior to drying. Liquid nitrogen spray or carbon dioxide snow freezing were selected as preferred methods for overall quality of diced cooked chicken meat to be freeze-dried.

Freezing Methods

Poultry may be frozen between refrigerated double-walled plates, in a blast of refrigerated air, by immersion in a refrigerated liquid, or by a shower of liquefied gas such as nitrogen. Rectangularly packaged, cut-up poultry can be frozen in multiplate freezers within several hours. A detailed description of the available commercial equipment may be found in Chapter 8.

The major proportion of whole, ready-to-cook birds is frozen in air blast tunnel freezers, with air temperatures ranging from −10 to −40 °F and air velocities of 300 to 1000 fpm and up. It is desirable to have air temperatures at −30 °F or below during operation and air velocities over the product surfaces of at least 600 fpm. To obtain high air velocity over the product, the blast tunnel should be completely loaded across its cross section, with proper spacing of the units of the product to assure airflow around all sides

Table 1 Thermal Properties of Ready-to-Cook Poultry

Property	Value	Reference
Specific heat, above freezing	0.70 Btu/(lb · °F)	Pflug (1957)
Specific heat, below freezing	0.37 Btu/(lb · °F)	Pflug (1957)
Latent heat of fusion	106 Btu/lb	Pflug (1957)
Freezing point	27 °F	Pflug (1957)
Average Density		
Poultry muscle	67 lb/ft^3	
Poultry skin	64 lb/ft^3	
Thermal conductivity [Btu/(h · ft · °F)]		
Broiler breast muscle[b]	0.24 at 80 °F	Walters and May (1963)
Broiler breast muscle[a]	0.29 at 68 °F	Sweat *et al.* (1973)
Broiler breast muscle[a]	0.80 at −4 °F	Sweat *et al.* (1973)
Broiler breast muscle[a]	0.87 at −40 °F	Sweat *et al.* (1973)
Broiler dark muscle[a]	0.90 at −40 °F	Sweat *et al.* (1973)
Turkey breast muscle[a]	0.73 at −4 °F	Sweat *et al.* (1973)
Turkey breast muscle[b]	0.93 at −4 °F	Sweat *et al.* (1973)
Turkey leg muscle[a]	0.83 at −4 °F	Lentz (1961)

[a] and [b] indicate heat flow perpendicular and parallel, respectively, to the direction of the muscle fibers.

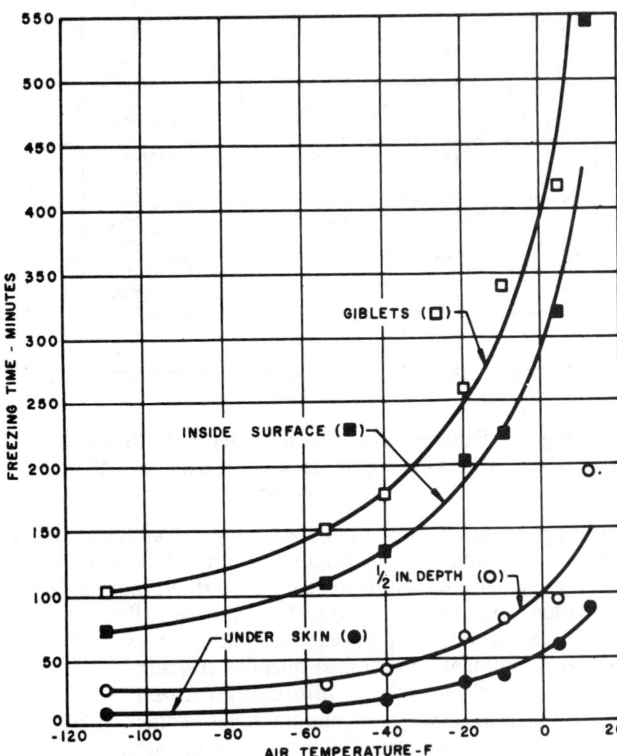

Note: Freezing time is the time required for temperature to fall from 32 to 25 °F. The values are for 5- to 8-lb chickens with initial temperature of 32 to 35 °F and with air velocity of 450 to 550 fpm.

Fig. 1 Relation between Poultry Freezing Time and Air Temperature
(van den Berg and Lentz 1958)

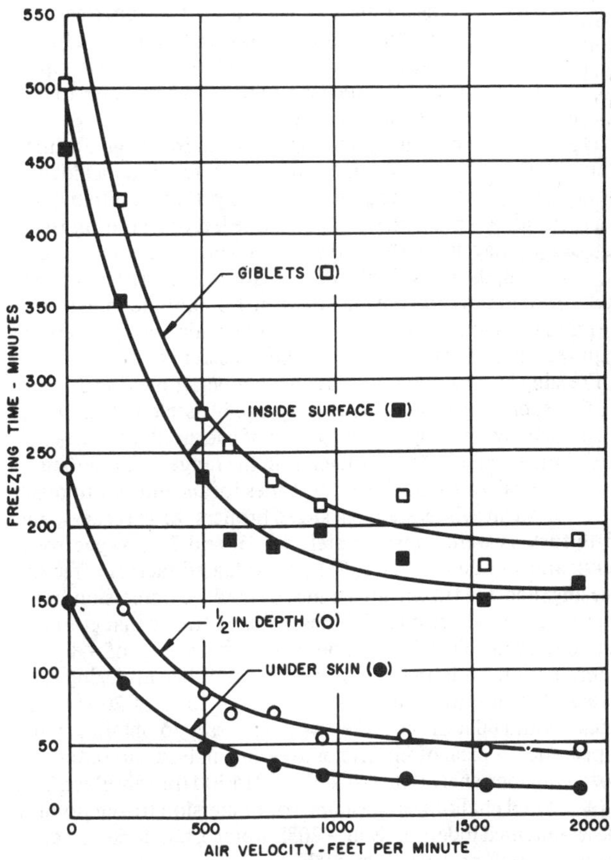

Note: Freezing time is the time required for temperature to fall from 32 to 25 °F. The values are for 5- to 8-lb chickens with initial temperature of 32 to 35 °F and with air temperature of −20 °F.

Fig. 2 Relation between Freezing Time and Air Velocity
(van den Berg and Lentz 1958)

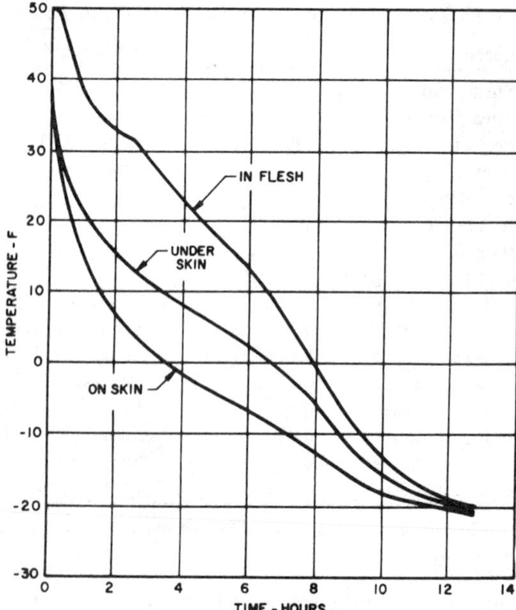

Note: For 21-lb, bronze tom turkeys on shelves in air blast.

Fig. 3 Temperature during Freezing of Packaged, Ready-to-Cook Turkeys
(Klose and Pool 1956)

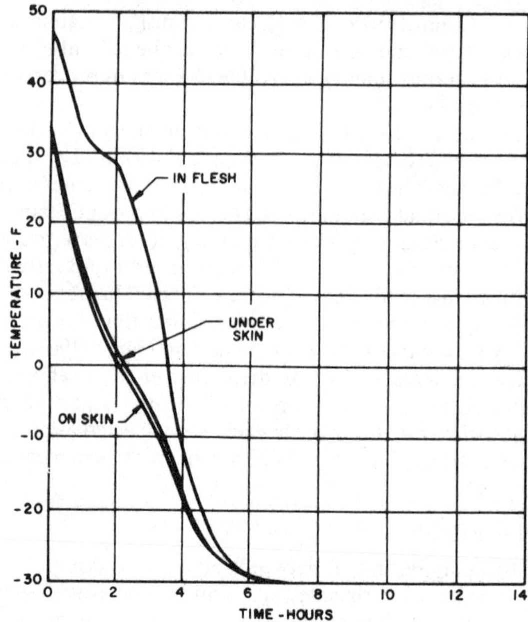

Note: For 7-lb, bronze tom turkeys on shelves in air blast.

Fig. 4 Temperature during Freezing of Packaged, Ready-to-Cook Turkeys
(Klose and Pool 1956)

and no large openings to permit bypassing of the airstream. In some cases, the whole bird may be packed into cartons or boxes, and the cartons stacked on pallets in the blast tunnel, with spacing between layers and between cartons in the same layer to permit adequate airflow and freezing rate.

However, as mentioned, freezing rates required to give a sufficiently light frozen appearance are so high for birds scalded at elevated temperatures and immature turkeys that modified freezing methods are necessary. Boxes may be left open during freezing, or they may be constructed with holes or cutouts on the ends, or, best of all, the individual, bagged birds may be frozen on open racks or shelves to provide maximum airflow and freezing rate. If freezing rates in excess of those possible in air blast tunnels are required, immersion freezing may be considered.

Freezing by direct contact with low-temperature brine or other liquids such as glycols is frequently used. Esselen *et al.* (1954) found that the freezing time per pound of packaged ready-to-cook birds immersed in −20 °F brine was in the range of 20 to 30 min. In −20 °F calcium chloride brine, times for the interior to reach 15 °F for warm eviscerated, packaged broilers, 12-lb turkeys, and 25-lb turkeys were approximately 1.5, 5, and 7 h, respectively. Lentz and van den Berg (1957) have evaluated factors affecting freezing time and frozen appearance of poultry. Immersion freezing at −20 °F produced a lightness of appearance that could be matched by air blast freezing only at a temperature of −100 °F. Combinations of immersion freezing and air blast freezing were tested. A minimum immersion time at −20 °F of 20 min for chickens and 40 min for turkeys was necessary to obtain a white appearance, typical of immersion freezing, in birds subsequently frozen on open shelves in a −20 °F, 300 to 500 fpm air blast. The light-colored chalk-like appearance of immersion frozen poultry was maintained indefinitely in −20 °F storage, but darkened after 6 weeks at 0 °F or 2 weeks at 20 °F.

Comparison of various liquids (methanol, salt brines, and glycols) that might be used in immersion freezing revealed that freezing times depended mainly on viscosity of the liquids and were maximum in glycerol and minimum in calcium chloride brine. Increased agitation decreased freezing times in the surface

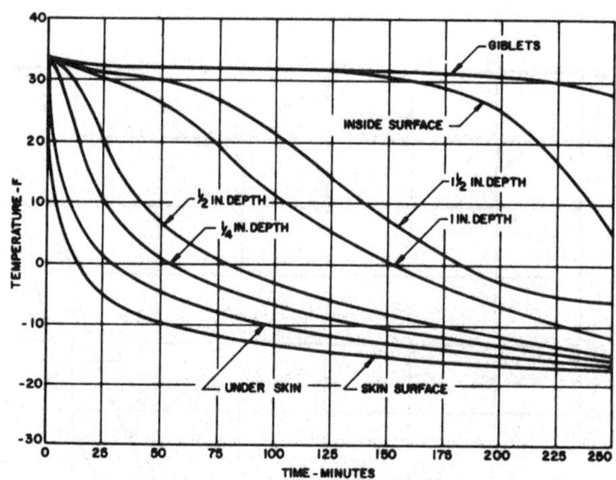

Fig. 5 Temperatures at Various Depths in Breast of 15-lb Turkeys during Immersion Freezing at −20 °F
(Lentz and van den Berg 1957)

layers by as much as 50%. Spraying the liquid was as effective in reducing freezing times as the highest rate of agitation used. Advantages of immersion freezing are lighter, more pleasing appearance; high rate of heat transfer that makes a line operation feasible; lower initial and upkeep costs than air blast; and possible use in combination with subsequent air freezing. Possible disadvantages include lack of uniformity of frozen appearance, development of leaks due to pinpoint holes or cracks in packaging film, the question of acceptance of the chalk-white birds, and lack of versatility for freezing all types of material.

Typical freezing rate curves are shown in Figures 1 to 5, which illustrate the effect of bird size and depth below skin surface, temperature and velocity of the cooling medium, and type of medium (air versus liquid). Pflug (1957) reviews the relative efficiencies and requirements for air and liquid immersion freezing and the critical importance of the conductance of the coolant-product interfacial film, citing a film heat transfer coefficient of 3 Btu/(h·ft²·°F) for air at −20°F and 500 fpm, compared to 30 Btu/(h·ft²·°F) for sodium chloride brine at 0°F and a velocity of 5 fpm. Table 1 lists physical properties of poultry related to refrigeration.

A liquid nitrogen shower system has been used to freeze the outer shell of products to −100°F or below, after which the entire product may be equilibrated in a holding room to a temperature around −10°F. Temperature in the freezer box where the liquid nitrogen is released in a shower may be as low as −250°F. IQF (Individually Quick Frozen) freezing with CO_2 is also used, particularly in preparing for freeze drying.

The freezing rate of precooked chicken affects the quality of product. Breaded precooked drumsticks frozen with liquid nitrogen are susceptible to cracking and separation of the meat from the bone. Precooked chicken that is lightly breaded (7 to 10% breading) and frozen by cryogenic procedures is susceptible to developing small white freezer burn areas next to the surface. These areas may or may not show up immediately after freezing, but will show up almost immediately after reconstituting with hot oil. In some instances, the white areas will cover the entire surface of the piece of chicken (Yingst and Goodwin 1971). Altering the fat content of thighs does not reduce the severity of this problem.

Automated units may be designed to handle packages, cartons of birds, or unwrapped pieces of chicken or turkey. Product may be transported through the freezing chamber on belts or trays. One such system is adaptable to all sizes of whole birds, packages, or cartons. The system automates the freezing operation from the point birds are placed in cartons until they are frozen and ready for a carton top to be put on. A typical design handles nearly 1 bird/s with about 150,000 lb total capacity. Refrigeration coils and fans are located at the side of the machine to give a high-velocity two-pass airflow that applies the coldest air to the warmest product. Frost or ice buildup is minimized since the shelves never come outside the freezer.

A tray system is available that automatically loads trays from a moving belt, conveys them through the air blast freezing chamber, empties the trays into a holding bin within the freezing chamber, and conveys the product onto a belt, which carries it out of the freezer to a packaging station. This system is particularly useful for cut-up chicken parts. Belt systems, using refrigerated air blast or cryogenic gases, are used for small parts, packages, and particle size poultry such as precooked diced chicken.

PRESERVING QUALITY IN STORAGE AND MARKETING

Important qualities of frozen poultry include appearance, flavor, and tenderness. If these are not optimum, the convenience and flexibility of use of the frozen product will not be sufficient inducement to the consumer. Optimum quality requires care in every phase of the marketing sequence, from the frozen storage warehouse, through transportation facilities, wholesaler, retailer, and finally to the frozen food case or refrigerator in the home.

Darkening of the bones is a condition that occurs in immature chickens and has become more prevalent as broilers are marketed at younger and younger ages. During chilled storage or during the freezing and defrosting process, some of the heme pigment normally contained in the interior of the bones of particularly young chickens leaches out through spongy areas and discolors the adjacent tissues. This in no way affects the palatability of the product, because the pigment is a natural tissue pigment, which, after being cooked, is known as hematin and is responsible for the cooked appearance of all meats. The freezing rate has no marked influence in preventing this discoloration. Also, the time between slaughter and freezing has no significant effect on the color of poultry bones or on the displacement of pigment from the interior of the bones to adjacent tissues.

Brant and Stewart (1950) found that development of dark bones was greatly reduced by a combination of freezing and storage at −30°F and immediate cooking after rapid thawing. Aside from this combination, freezing rate, temperature and length of storage, and temperature fluctuations during storage were not found to have a significant effect. Further research suggested that freezing and thawing not only liberated hemoglobin from the bone marrow cells but modified the bone structure to permit penetration by the released pigment. Roasting pieces of chicken 0.5 h prior to freezing reduced discoloration of the bone. Ellis and Woodroof (1959) found that heating legs and thighs to 180°F before freezing effectively controlled meat darkening. Methods of preheating, in order of preference, include microwave oven, steam, radiant heat oven, and deep fat frying.

During storage, poultry may become dehydrated, yielding a condition known as *freezer burn.* Dehydration can be controlled by humidification, lowering of storage temperatures, or by packaging the product adequately. Aside from adversely affecting the appearance of the product, dehydration, unless severe, does not impair quality. When freezer burn is extensive, quality is decreased because of toughening and development of oxidative rancidity of the affected area. If storage temperatures of 0°F or lower are maintained, freezer burn is usually the factor limiting the length of time that poultry can be held in storage without adequate packaging.

Wills *et al.* (1948) found that the appearance of poultry suffered greatly when stored at 20°F. The most serious defects were microbiological changes, desiccation, and development of a stale, rancid or storage odor. Serious changes in flavor and juiciness occurred in poultry that had been frozen 3 to 9 months at 10°F. Swanson and Sloan (1953) studied chemical changes in the proteins of ready-to-cook fowl held at −5°F over a period of 40 weeks. Proteolysis was indicated during storage by increases in soluble nitrogen and nonprotein nitrogen of leg and breast muscle.

The effect of variations in storage temperature on the quality of frozen turkeys was studied by Klose *et al.* (1955) for moisture losses, chemical changes, and palatability. Evaluations after 6-, 12-, and 18-month storage times indicated that, under average commercial conditions of frozen storage (moisture impermeable package and temperature range of −10 to 10°F), the only factor for which a periodic temperature fluctuation is inferior to the mean temperature is the accumulation of frost in the package. Frost formation, which influences appearance but not eating quality, increased with storage temperature, and for the −10 to 10°F fluctuation was considerably greater than for the highest (10°F) constant temperature. Results after 12-month storage indicated a definite superiority to −10°F storage over 10 and 0°F but no detectable organoleptic superiority of −30 over −10°F.

Poultry fat becomes rancid during very long storage periods or at extremely high storage temperatures. Rancidity in frozen, eviscerated whole poultry stored for 12 months is not a serious problem if the bird is packaged in essentially impermeable film and held at 0°F or below. Danger of rancidification is greatly increased when poultry is cut up before freezing and storage, because of the increased surface exposed to atmospheric oxygen.

Quality losses in frozen, packaged, and cut-up frying chickens were studied by Klose *et al.* (1959b) over temperatures of −30 to 20°F and storage periods from 1 month to 2 years. All commercial-type samples examined were acceptable after a storage period at 0°F of at least 6 months, and some were stable for more than a year. In a comparison of a superior (moisture-vapor-proof) commercial package with a fair commercial package, increased adequacy of packaging resulted in as much extension in storage life as a decrease in storage temperature of about 20°F. The results indicate that no statement on storage life can have general value unless the packaging condition is accurately specified.

Frozen storage tests by Klose *et al.* (1960b) on commercial packs of ready-to-cook ducklings and ready-to-cook geese established that these products have frozen storage lives similar to other commercial forms of poultry. Ducks and geese should be stored at 0°F or below to maintain their original high quality for 8 to 12 months.

Incorporation of polyphosphates into poultry meat by adding it to the chilling water has been shown to increase shelf life in frozen or refrigerated storage and to control loss of moisture in refrigerated storage and during thawing and cooking.

Storage studies by Hanson *et al.* (1959) on frozen fried chicken indicated that precooking produces a product much less stable than a raw product. Rancidity development is the limiting factor and, surprisingly enough, it is detected in the meat slightly sooner than in the skin and fatty coating of the fried product. The marked beneficial effect of oxygen (air)-free packaging was demonstrated in tests in which detectable off-flavors were observed at 0°F in air-packed samples after 2 months, while nitrogen-packed samples developed no off-flavors for periods exceeding 12 months.

Cooling the precooked parts in ice water prior to breading was found to reduce the TBA (thiobarbituric acid) values of precooked parts (Webb and Goodwin 1970). In this study, no difference in rancidity was noted for chicken stored 6, 8, or 10 months. By removing the skin from precooked broilers, the TBA values were lower, but yield and tenderness were reduced (Wyche *et al.* 1972). No difference was detected in the TBA values of the thighs frozen in liquid refrigerant with or without skin (Surkiewicz *et al.* 1969). Chicken parts that were blast frozen without skin were less rancid than those frozen with skin. Precooked frozen chicken parts browned for 120 s at 400°F were less rancid than those parts browned at 300°F (Love and Goodwin 1974).

The susceptibility of frozen fried chicken to storage deterioration indicates the need for low temperature (0 or −10°F) and short storage times (6 months or less) unless it is packed in an oxygen-free atmosphere. To prevent cracking and peeling of the batter coating from frozen fried chicken, Hansen and Fletcher (1963) shrunk the parts by cooking before dipping in batter. Cooling the parts to a temperature of 35°F before dredging improved the texture and adhesion of the breading (Hale and Goodwin 1968).

In contrast to a loosely packed product, such as the frozen fried chicken mentioned above, Hanson and Fletcher (1958) reported that a solid-pack product such as chicken and turkey pot pies, in which the cooked poultry is surrounded by sauce or gravy, with consequent exclusion of air, had a storage life at 0°F of at least 1 year. As is the case with raw poultry, turkey products have less fat stability than chicken products, but the stability can be increased by substituting more stable fats in the sauces or by using antioxidants. A quality defect found in precooked frozen products containing a sauce or gravy is a liquid separation and curdled appearance of the sauce or gravy when thawed for use. This separation is extremely sensitive to storage temperature. Sauces can be stored at least five times as long at 0°F as at 10°F before separation takes place. Hanson *et al.* (1951) established that the flour in the sauce was the cause of the separation, and found among a large number of alternative thickening agents that waxy rice flour produced superior stability. Sauces and gravies prepared with waxy rice flour are completely stable for about a year at 0°F.

Since precooked frozen foods are not apt to be sterilized in the reheating process in the home, the processor has an added responsibility to keep bacterial counts in the product well below hazardous levels. Extra precautions should be taken in general plant sanitation, in rapid chilling and freezing of the cooked products, and in seeing that the products do not reach a temperature that will permit bacterial growth at any time during storage or distribution.

THAWING AND USE

Under ordinary conditions, poultry should be kept frozen until shortly before its consumption. The general procedure is to defrost in air or in water. Hoffert *et al.* (1952) found no significant differences in palatability between thawing in oven, refrigerator, room, or water.

For turkeys that have been scalded at high temperatures and fast frozen to give a light appearance, the temperature in retail storage and display must be kept as low as possible (0°F is reasonable) to prevent darkening. Thawing in the package will minimize darkening.

Detailed instructions for thawing turkeys should be given on the bag used for turkey packaging. The safest procedure is to hold the turkey in the refrigerator (35 to 40°F) for 2 to 4 days depending on the size of the bird. Other methods are immersion in cool water in the bag for 4 to 6 h or holding in a paper bag or styrofoam chest for 12 to 36 h at room temperature. When using these nonrefrigerated thawing techniques, care must be taken to keep the bird's surface cool to inhibit microbiological growth (Cunningham *et al.* 1979).

Some retail stores allow frozen poultry to start thawing in the chilled (33 to 38°F) section of the meat display case where poultry is for sale on the particular day. This is an advantage to the consumer who wants to cook the poultry that night. However, this is a safe practice only if careful, constant control is maintained over the chilled inventory so that the product is not held beyond its overall shelf life in store and home. Freezing and thawing in itself does not reduce the refrigerated shelf life of the product. Elliott and Straka (1964) found that frozen-thawed chicken had a shelf life at 36°F about equal to unfrozen counterparts at 36°F, as measured by total counts of psychrophilic bacteria and by odor tests. Sauter *et al.* (1978) also reported a comparison of the microflora of fresh and thawed frozen fryers.

Ready-to-cook turkeys in a frozen, prestuffed raw form have been marketed. Extreme care should be exercised in producing and consuming this type of product to assure that the original bacterial count in the birds and stuffing is at a minimum and that, in roasting, the internal temperature reaches a value high enough to provide a safe product.

REFERENCES

Ayres, J.C., W.S. Ogilvy, and G.F. Stewart. 1950. Postmortem changes in stored meats. Part I, Microorganisms associated with development of slime on eviscerated and cut-up poultry. *Food Technology* (4):199.

Behnke, J.R., O. Fennema, and R.W. Haller. 1973. Quality changes in prerigor poultry at −3°C *Journal of Food Science* (38):275.

Brant, A.W. and G.F. Stewart. 1950. Bone darkening in frozen poultry. *Food Technology* (4):168.

Carpenter, M.D., D.M. Janky, A.S. Arafa, J.L. Oblinger, and J.A. Koburger. 1979. The effect of salt brine chilling on drip loss of icepacked broiler carcasses. *Poultry Science* (58):369.

Clatfelter, K.A. and J.E. Webb. 1987. Method of eliminating aging step in poultry processing. U.S. Patent No. 4,675,947, June 30.

Cunningham, F.E., D.R. Suderman, and M.H. Wu. 1979. Composition of drainage from thawed poultry carcasses. *Poultry Science* (58):365.

Elliott, R.P. and R.P. Straka. 1964. Rate of microbial deterioration of chicken meat at °C after freezing and thawing. *Poultry Science* (43):81.

Ellis, C. and J.G. Woodroof. 1959. Prevention of darkening in frozen broilers. *Food Technology* (13):533.

Esselen, W.B., A.F. Lexine, I.J. Pflug, and L.L. Davis. 1954. Brine immersion cooling and freezing of packaged ready-to-cook poultry. *Refrigerating Engineering* 62(7):61.

Farrell, A.J. and E.M. Barnes. 1964. Bacteriology of chilling procedures in poultry processing plants. *British Poultry Science* (5):89.

Goresline, H.E., M.A. Howe, E.R. Baush, and M.F. Gunderson. 1951. Inplant chlorination does a 3-way job. *U.S. Egg and Poultry Magazine* (4):12.

Hale, K.K., Jr. and T.L. Goodwin. 1968. Breaded fried chicken. Effects of precooking, batter composition, and temperature of parts before breading. *Poultry Science* (47):739.

Hamre, M.L. and W.J. Stadelman. 1967a. Effect of various freezing methods on frozen diced chicken. *Quick Frozen Foods* 29(4):78.

Hamre, M.L. and W.J. Stadelman. 1967b. The effect of the freezing method on tenderness of frozen and freeze dried chicken meat. *Quick Frozen Foods* 30(8):50.

Hanson, H.L. and L.R. Fletcher. 1958. Time-temperature tolerance of frozen foods. Part XII, Turkey dinners and turkey pies. *Food Technology* (12):40.

Hanson, H.L. and L.R. Fletcher. 1963. Adhesion of coatings on frozen fried chicken. *Food Technology* (17):115.

Hanson, H.L., A. Campbell, and H. Lineweaver. 1951. Preparation of stable frozen sauces and gravies. *Food Technology* (5):432.

Hanson, H.L., L.R. Fletcher, and H. Lineweaver. 1959. Time-temperature tolerance of frozen foods. Part XVII, Frozen fried chicken. *Food Technology* (13):221.

Hanson, H.L., G.F. Stewart, and B. Lowe. 1942. Palatability and histological changes occurring in New York dressed broilers held at 1.7°C (35°F). *Food Research* (7):148.

Hoffert, E., A.R. Plagge, B. Lowe, and G.F. Stewart. 1952. The defrosting method and palatability of poultry. *Food Technology* (6):337.

Ingling, A.L. 1978. Electrical terminology, measurements and units associated with the stunning technique in poultry processing plants. *Poultry Science* (57):127.

Ingraham, J.L. 1958. Growth of psychrophilic bacteria. *Journal of Bacteriology* (6):75.

Janky, D.M., A.S. Arafa, J.L. Oblinger, J.A. Koburger, and D.L. Fletcher. 1978. Sensory, physical, and microbiological comparison of brine-chilled, water-chilled, and hot-packaged (no chill) broilers. *Poultry Science* (57):417.

Klose, A.A., M.F. Pool, M.B. Wiele, H.L. Hanson, and H. Lineweaver. 1959a. Poultry tenderness. Part I, Influence of processing on tenderness of turkeys. *Food Technology* (13):20.

Klose, A.A., M.F. Pool, A.A. Campbell, and H.L. Hanson. 1959b. Time-temperature tolerance of frozen foods. Part XIX, Ready-to-cook cut-up chicken. *Food Technology* (13):477

Klose, A.A., M.F. Pool, D. de Fremery, A.A. Campbell, and H.L. Hanson. 1960a. Effect of laboratory scale agitated chilling of poultry on quality. *Poultry Science* (39):1193.

Klose, A.A., A.A. Campbell, M.F. Pool, and H.L. Hanson. 1961. Turkey tenderness in relation to holding in and rate of passage through thawing range of temperature. *Poultry Science* (40):1633.

Klose, A.A., R.N. Sayre, D. de Fremery, and M.F. Pool. 1972. Effect of hot cutting and related factors in commercial broiler processing on tenderness. *Poultry Science* (51):634.

Klose, A.A. and M.F. Pool. 1956. Effect of freezing conditions on appearance of frozen turkeys. *Food Technology* (10):34.

Klose, A.A., A.A. Campbell, and H.L. Hanson. 1960b. Stability of frozen ready-to-cook ducks and geese. *Poultry Science* (39):1136.

Klose, A.A., M.F. Pool, and H. Lineweaver. 1955. Effect of fluctuating temperatures on frozen turkeys. *Food Technology* (9):372.

Koonz, C.H. and J.M. Ramsbottom. 1939. A method for studying the histological structure of frozen products, Part I. *Poultry Food Research* (March-April):117.

Kotula, A.W., J.E. Thomson, and J.A. Kinner. 1960. Water absorption by eviscerated broilers during washing and chilling. USDA, Agricultural Marketing Service, *Marketing Research Report* No. 438 (October).

Lentz, C.P. 1961. Thermal conductivity of meats, fats, gelatin, gels, and ice. *Food Technology* (15):243.

Lentz, C.P. and L. van den Berg. 1957. Liquid immersion freezing of poultry. *Food Technology* (11):247.

Lillard, H.S. 1980. Effect on broiler carcasses and water of treating chiller water with chlorine or chlorine dioxide. *Poultry Science* (59):1761.

Love, B.E. and T.L. Goodwin. 1974. Effects of cooking methods and browning temperatures on yields of poultry parts. *Poultry Science* (53):1391.

Mast, M.G. and J.H. MacNeil. 1972. Use of glutaraldehyde as a disinfectant in immersion. *Poultry Science* (51):681.

Morris, G.K. and J.G. Wells. 1970. Salmonella contamination in a poultry processing plant. *Applied Microbiology* (19):795.

Nagel, W.C., K.L. Simpson, H. Ng, R.H. Vaugn, and G.F. Stewart. 1960. Microorganisms associated with spoilage of refrigerated poultry. *Food Technology* (14):21.

Osner, R.L. and D.H. Shrimpton. 1966. Influence of chilling and thawing procedures on the origin of constituents of fluids lost from the chicken carcasses. *British Poultry Science* 7(4):301.

Peric, M., E. Rossmanith, and L. Leistner. 1971. Verbesserung der microbiologischen qualitat von schlachthanchen durch die sprunhkuhling. *Die Fleischwirtschaft* (April):574.

Pflug, I.J. 1957. Immersion freezing found to improve poultry appearance. *Frosted Food Field* (June):17.

Pippen, E.L. and A.A. Klose. 1955. Effects of ice water chilling on flavor of chicken. *Poultry Science* (34):1139.

Pool, M.F., D. de Fremery, A.A. Campbell, and A.A. Klose. 1959. Poultry tenderness. Part II, Influence of processing on tenderness of chickens. *Food Technology* (13):25.

Risse, L.A. and J.E. Thomson. 1971. Comparative performance and costs of dry ice and water ice in shipping fresh poultry. *Marketing Research Report*, No. 906 (February). Agricultural Research Service, USDA.

Sauter, E.A., C.F. Peterson, and J.F. Parkinson. 1978. Microfloral comparison of fresh and thawed frozen fryers. *Poultry Science* (57):422.

Sekoguchi, S., R. Nakamura, and Y. Sato. 1979. Cysteamine induced changes in the properties of intramuscular collagen and its relation to the tenderness of meat obtained from mature chickens. *Poultry Science* (58):1213.

Shannon, W.G., W.W. Marion, and W.J. Stadelman. 1957. Effect of temperature and time of scalding on the tenderness of breast meat of chicken. *Food Technology* (11):284.

Smith, M.C., Jr., M.D. Judge, and W.J. Stadelman. 1966. A cold shortening effect in avian muscle *Journal of Food Science* (31):450.

Spencer, J.V. and W.J. Stadelman. 1955. Effect of certain holding conditions on shelf life of fresh poultry meat. *Food Technology* (9):358.

Stadelman, W.J. 1970. 28 to 32°F temperature is ideal for preservation, storage and transportation of poultry. ASHRAE *Journal* 12(3):61.

Surkiewicz, B.F., R.W. Johnston, A.B. Moran, and G.W. Krumm. 1969. A bacteriological survey of chicken eviscerating plants. *Food Technology* 23(8):80.

Swanson, M.H. and H.J. Sloan. 1953. Some protein changes in stored frozen poultry. *Poultry Science* (32):643.

Sweat, V.E., C.G. Haugh, and W.J. Stadelman. 1973. Thermal conductivity of chicken meat at temperatures between -75 and 20°C. *Journal of Food Science* (38):158.

Treat, D.W. and T.L. Goodwin. 1973. Effects of sex, size and time of cutting on processing yields and tenderness of broilers. *Poultry Science* (52):1348.

USDA/FSIS. 1990. Poultry products inspection regulations. Chapter 3, Sub-Chapter C, Part 381. Washington, D.C.

van den Berg, L. and C.P. Lentz. 1958. Factors affecting freezing rate and appearance of eviscerated poultry frozen in air. *Food Technology* (12):183.

Walters, R.E. and K.N. May. 1963. Thermal conductivity and density of chicken breast, muscle and skin. *Food Technology* (17):808.

Webb, J.E. and C.C. Brunson. 1972. Effects of eviscerating line trimming on tenderness of broiler breast meat. *Poultry Science* (51):200.

Webb, J.E., R.L. Dake, and R.E. Wolfe. 1989. Method of eliminating aging step in poultry processing. U.S. Patent No. 4,860,403. August 29.

Webb, J.E. and T.L. Goodwin. 1970. Precooked chicken. Effect of cooking methods and batter formula on yields and storage conditions on 2-thiobarbituric acid values. *British Poultry Science* (11):171.

Wells, F.E., J.V. Spencer, and W.J. Stadelman. 1958. Effect of packaging materials and techniques on shelf life of fresh poultry meat. *Food Technology* (12):425.

Willis, R., B. Lowe, and G.F. Stewart. 1948. Poultry storage at subfreezing temperatures—Comparisons at −10 and +10°F. *Refrigerating Engineering* (56):237.

Wyche, R.C. and T.L. Goodwin. 1974. Hot-cutting of broilers and its influence on tenderness and yield. *Poultry Science* (53):1668.

Wyche, R.C., B.E. Love, and T.L. Goodwin. 1972. Effect of skin removal, storage time and freezing methods on tenderness and rancidity of broilers. *Poultry Science* (51):655.

Yingst, L.D. and T.L. Goodwin. 1971. Freezing methods influence on fat and moisture composition of precooked thighs. *Poultry Science* (50):957.

Ziegler, F. and W.J. Stadelman. 1955. The effect of different scald water temperatures on the shelf life of fresh, non-frozen fryers. *Poultry Science* (34):237.

CHAPTER 13

FISHERY PRODUCTS

THE major types of fish and shellfish harvested from North American waters and used for food include the following:

1. Groundfish (haddock, cod, whiting, flounder, and ocean perch), lobster, clams, scallops, snow crab, shrimp, capelin, herring, and sardines from New England and Atlantic Canada.
2. Oysters, clams, scallops, striped bass, and blue crab from the Middle and South Atlantic.
3. Shrimp, oysters, red snapper, clams, and mullet from along the Gulf Coast.
4. Lake herring, chubs, carp, buffalofish, catfish, yellow perch, and yellow pike from the Mississippi Valley and the Great Lakes region.
5. Alaska pollock, Pacific pollock, tuna, halibut, salmon, Pacific cod, various species of flatfish, king and Dungeness crab, scallops, shrimp, and oysters from the Pacific Coast and Alaska.
6. Catfish, salmon, trout, oysters, and mussels from aquaculture operations in various parts of North America.

The major industrial fish used for fish meal and oil is menhaden from the Atlantic and Gulf coasts. In addition, the parts of fish not used for human consumption are often used to manufacture fish meal and oil.

Fish meal and oil are the principal components of the feed used in the aquaculture of trout and salmon. Meal also is a component of the diets of poultry and pigs. Fish oil is used in margarine, paints, and in the tanning industry. It is also being refined for pharmaceutical purposes.

This chapter deals with the preservation and processing of fresh and frozen fishery products, the care of fresh fish aboard vessels and ashore, the technology of freezing fish, and present commercial trends in the freezing, frozen storage, and distribution of seafood.

See Chapter 19 for additional information regarding fishery products for precooked and prepared foods.

FRESH FISHERY PRODUCTS

CARE ABOARD VESSELS

After fish are brought aboard a vessel, they must be promptly and properly cared for to assure maximum quality. Trawl-caught fish on the New England and Canadian Atlantic coasts, such as haddock and cod, are usually eviscerated, washed, and then iced down in the pens of the vessel's hold. The offshore Canadian fleet and the fleets of Iceland, the United Kingdom, and other European countries have been icing the fish in boxes for optimum quality. Because of their small size, other groundfish, such as ocean perch, whiting, and flounder, are not eviscerated and are not always washed. Instead, they are iced down directly in the hold of the vessel.

Crustaceans, such as lobsters and many species of crabs, are usually kept alive on the vessel without refrigeration. Warm-water shrimp are beheaded, washed, and stored in ice in the hold of the vessel; on some vessels, however, the catch is frozen either in refrigerated brine or in plate freezers. Coldwater shrimp are stored whole in ice or in chilled sea water, or they may be cooked in brine, chilled, and stored in containers surrounded with ice.

Freshwater fish in the Great Lakes and Mississippi River areas are caught in trap nets, haul seines, or gill nets and sorted according to species into 50- or 100-lb boxes, which are kept on the deck of the vessel. In most cases, fishermen carry ice aboard their vessels, and the fish are landed the day they are caught.

Freshwater fish in the lakes of Canada are iced down in the summertime and stored at collecting stations on the lakes, where they are picked up by a collecting boat with a refrigerated hold. Winter-caught Canadian freshwater fish and saltwater fish caught in Arctic water are usually weather frozen on the ice immediately after catching and are marketed as frozen fish.

Line-caught fish of the Pacific Northwest, such as halibut caught largely by bottom long-line gear and salmon caught by trolling gear, are eviscerated, washed, and iced in the pens of the vessel. Pacific salmon caught by seines and gill nets for cannery use are usually stored whole for several days, either aboard vessels or ashore in tanks of seawater refrigerated to 30°F. A small but significant volume of halibut is held similarly in refrigerated seawater aboard vessels. Tuna caught offshore by seiners or clipper vessels are usually brine-frozen at sea. However, tuna caught inshore by the smaller trollers or seiners are often iced in the round or refrigerated with a brine spray.

Fish raised by aquaculture farms are usually harvested and sold as required by the fresh fish market. They are usually shipped in containers in which they are surrounded by ice.

Icing of Fish

Fish lose quality because of bacterial or enzymatic activity or both. Reduction of storage temperature retards these activities significantly, thus delaying spoilage and autolytic deterioration.

Low temperatures are particularly effective in delaying growth of psychrophilic bacteria, which are primarily responsible for the spoilage of nonfatty fish. The shelf life of species such as haddock and cod is doubled for each 7 to 10°F decrease of storage temperature within the range of 60 to 30°F.

To be effective, ice must be clean when used aboard a vessel. Bacteriological tests on ice in the hold of a fishing vessel showed bacterial counts as high as 5 billion per gram of ice. These results indicate that chlorinated or potable water should be used in making the ice at the ice plant, ice should be stored under sanitary conditions, and unused ice should be discarded from the vessel at the end of each trip.

Both flake ice and crushed block ice are used aboard fishing vessels, although flake ice is more common because it is cheaper to produce and easier to handle mechanically.

The preparation of this chapter is assigned to TC 11.2, Foods and Beverages.

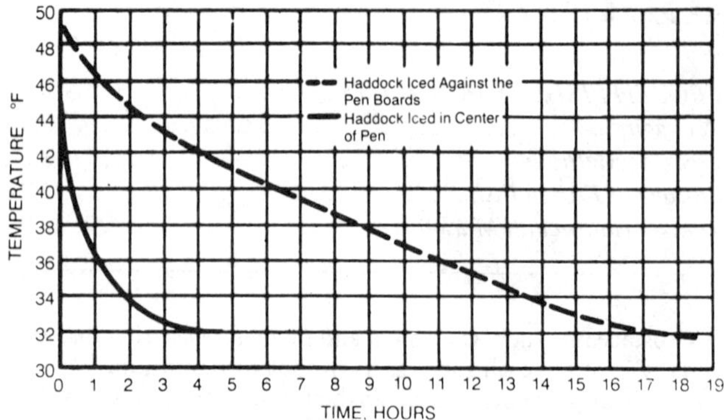

Fig. 1 Cooling Rate of Properly and Improperly Iced Haddock

The amount of ice used aboard vessels varies with the particular fishery and vessel; however, it is essential to provide sufficient ice around the fish to obtain a proper cooling rate (see Figure 1). A common ratio of ice to fish used in bulk icing on New England trawlers is one part ice to three parts fish. Experiments on English trawlers in boxing fish at sea with one part ice to two parts fish demonstrated improved quality in the landed fish, and, as ice has become more plentiful and less costly relative to the value of fish, the ratio of ice to fish continues to increase. Mechanical refrigeration is employed in some vessels to retard melting of ice en route to the fishing grounds; however, the hold temperature must be controlled after fish are taken to allow melting of the ice for effective cooling of the fish.

Saltwater Icing

Iced fish storage temperatures must be maintained close to the freezing point of fish. To obtain lower ice temperatures, the freezing point may be depressed by adding salt to the water from which the ice is made. Adequate amounts of ice made from a 3% solution of sodium chloride brine will maintain a storage environment of about 30°F. Tests conducted on the storage of haddock in saltwater ice aboard a fishing vessel showed that, under parallel conditions, fish iced with saltwater ice cooled faster and to a lower temperature than fish iced in plain ice. However, the saltwater ice melted faster than the plain ice because of its lower latent heat and the greater temperature differential of the saltwater ice. Therefore, once the saltwater ice melted, the fish stored in this ice rose to a higher temperature than those stored in plain ice. Since it is not always possible to renew ice on fish at sea, sufficient quantities of saltwater ice must be used initially to make up for the faster melting rate.

In making ice from water containing a preservative, rapid freezing or the use of a stabilizing dispersant, or both, is essential to prevent migration of the additive to the center of the ice block. This problem is not encountered in flake ice, because flake ice machines are designed to freeze water rapidly into thin layers of ice, thus fixing additive within the ice flakes. Chapter 32 describes the manufacture of flake ice in more detail.

Use of Preservatives

In the United States and Canada, the use of antibiotics in ice or in dips for treatment of whole or gutted fish, shucked scallops, and unpeeled shrimp is not permitted by regulation.

Storage of Fish in Refrigerated Sea Water

Refrigerated seawater (RSW) is used commercially for preserving fish. On the Pacific Coast, substantial quantities of net-caught salmon are stored in RSW aboard barges and cannery tenders for delivery to the canneries. On the East and Gulf coasts, RSW installations on fishing vessels are used for chilling and holding menhaden and industrial species needed for production of meal, oil, and pet food. On the east and west coasts of Canada, RSW installations are used for chilling and holding herring and capelin, which are processed on shore for their roe. Other more limited applications of RSW include holding Pacific halibut and Gulf shrimp aboard a vessel; chilling and holding Maine sardines in shore tanks for canning, and short-term holding of Pacific groundfish in shore tanks for later filleting.

With groundfish and shrimp, RSW works well for short-term storage (2 to 4 days), but it is not suitable for longer periods because of the excessive salt uptake, accelerated rancidity, poorer texture, and increased bacterial spoilage that may result. These problems can be partially overcome by introducing CO_2 gas into the RSW; the storage life of some species of fish can be increased by about one week by holding in RSW saturated with CO_2. Additional benefits of RSW are the reduction of handling that results from the bulk storage of the fish and the reduction of pressure on the fish as a result of buoyancy, faster cooling, and lower storage temperature.

In many RSW systems, the refrigeration effect is provided by ammonia flowing through external chillers or pipe coils located within the tanks. Best results have been achieved with external chillers.

Boxing at Sea

There are many advantages to using containers or boxes instead of bulk storage aboard fishing vessels. Known as *boxing at sea*, the use of containers reduces the pressure effects on the fish while they are stowed in a vessel's hold. Because significant reductions in handling during and subsequent to unloading are possible, mechanical damage and product temperature rise may be virtually eliminated and handling costs reduced. Fish can be sorted into boxes by size and species as soon as they are caught. They lend themselves more readily to mechanized handling such as machine filleting, because they are generally firmer and of more nearly uniform shape; fillet yields are generally better than they are with bulk-stored fish.

Boxing at sea is not generally practiced in the United States, except by some inshore vessels. The principal problems associated with converting a fishing vessel from bulk storage to boxed storage are the increased labor required by the crew for handling the boxes, reduced hold capacity, and a relatively large investment for boxes. Many fisheries have difficulties working out the logistics for assuring the prompt return of properly cleaned boxes to the vessel. Most of these problems have been solved in European fleets, the Canadian offshore fleet, and South American hake fishing fleets. The use of nonreturnable containers for boxing at sea simplifies logistics and reduces initial capital outlay; it has proved justifiable in some U.S. fisheries.

Reusable containers for boxing at sea are usually made of plastic. Careful icing is necessary to minimize the surface area of fish in contact with the box. Plastic boxes provide more heat transfer resistance than aluminum boxes in vessels with uninsulated fish holds and for in-plant storage prior to processing.

All fish boxes must be equipped with drains, preferably directed outside the boxes on the bottom of a stack.

SHORE PLANT PROCEDURE AND MARKETING

Proper use of ice and adherence to good sanitary practices ensure maintenance of iced fish freshness during unloading from the vessel, at the shore plant, during processing, and throughout the distribution chain. Fish landed in good quality will spoil rapidly if these practices are not carried out.

Fish unloaded from the vessel usually are graded by the buyer for species, size, and minimum quality specification. A price is based in part on the quality in relation to market requirements.

Table 1 Organoleptic Criteria of Quality Fish

Factor	Good Quality	Poor Quality
Eyes	Bright, transparent, often protruding	Cloudy, often pink, sunken
Odor	Sweet, fishy, similar to seaweed	Stale, sour, presence of sulfides, amines
Color	Bright, characteristic of species, sometimes pearlescent at correct light angles	Faded, dull
Texture	Firm, may be in rigor, elastic to finger pressure	Soft, flabby, little resilience, presence of fluid
Belly	Walls intact, vent pink, normal shape	Often ruptured, bloated, vent brown, protruding
Organs (including gills)	Intact, bright, easily recognizable	Soft to liquid, gray homogeneous mass
Muscle tissue	White or characteristic of species and type	White flesh pink to gray, spreading of blood color around backbone

Fish also may be inspected by local and federal regulatory agenices for wholesomeness and sanitary condition. Organoleptic criteria are most important for evaluating quality; however, there is a growing acceptance, particularly in Canada and some European countries, of objective chemical and physical tests as indexes of quality loss or spoilage. Organoleptic quality criteria vary somewhat among species, but the information in Table 1 can be used as a general guide in judging the quality of whole fish.

In New England and the Canadian Atlantic Provinces, groundfish unloaded from the vessel may be placed in boxes and trucked to the shore plant or conveyed directly from the hold or deck to the shore plant. Single- or double-wall insulated boxes are normally used for transporting fish. Wooden boxes are rarely used because they are a source of microbiological contamination. Ice should be applied generously to each box of fish, even if the period prior to processing is only a few hours. Fish in the plant awaiting processing for longer than those few hours should be iced heavily and stored in insulated containers or in single-wall boxes in a chill room refrigerated to 35 °F. If refrigerated facilities are not available, the boxes of fish should be kept in a cool section of the plant that is clean, sanitary, and has adequate drainage.

Large boxes of resin-coated plywood, or reinforced fiberglass that hold up to 1000 lb of fish and ice, are used by some plants in preference to icing fish overnight on the floor. These *tote* boxes are moved and stacked by forklift, can be used for trucking fish to other plants, and make better use of plant floor space. Generally, fish awaiting processing should not be kept longer than overnight.

Fresh fish are marketed in different forms: fillets, whole fish, dressed-head on, dressed-headed (head removed), or, in some instances, steaks. The method of preparing fish for marketing depends largely on the species of fish and on consumer preference. For example, groundfish such as cod and haddock usually are marketed as fillets or as dressed-headed fish. Freshwater fish such as catfish and bullheads are usually dressed and skinned; lake trout are not skinned, but are merely dressed; and lake herring are marketed in dressed, round, or filleted form.

PACKAGING FRESH FISH

Most fresh fish is packaged in institutional containers of 5- to 35-lb capacity at the point of processing. Polyethylene trays, steel cans, aluminum trays, plastic-coated solid boxes, wax-impregnated corrugated fiberboard boxes, foamed polystyrene boxes, and polyethylene bags are used.

Fresh fish is often packaged when it still contains process heat from wash water. In these cases, it is advantageous to use a packaging material that is a good heat conductor. The fresh fish industry makes little use of controlled prechilling equipment in packaging systems. As a result, product temperatures may never reach the optimum level subsequent to packaging. Traditionally, institutional fresh fish travels packed in wet ice; in this case, it may cool to the proper level in transit even if process heat is initially present. However, there is a trend toward the use of leaktight shipping containers for fresh fish, because modern transportation equipment is not designed to handle wet shipments. Also, some customers want to avoid the cost of transporting ice and demand a product that is uniformly chilled to within 32 to 36 °F when it reaches their door. Shippers who make use of leaktight shipping containers will have to upgrade their product temperature control systems to assure that the fish reaches ice temperature prior to packaging. Rapid prechilling systems that result in crust freezing can be applied to some fresh seafood products, but this practice must be used with discretion since partial freezing produces deleterious effects on quality.

Some general requirements for institutional containers that hold products such as fillets, steaks, and shucked shellfish are: (1) sufficient rigidity to prevent pressure exerted on the product, even when containers are stacked or heavily covered with ice; and (2) measures to prevent ice-melt water from contaminating the product. Some containers have drains permitting the drip associated with the fish itself to run off. Others are sealed and may be gastight, which increases shelf life. One problem associated with sealed containers is the emission of a strong odor when the package is first opened. Although this odor may be foul, it soon dissipates and has no adverse effect on quality. Dressed or whole fish may be placed in direct contact with ice in a gastight container.

Leaktight shipping containers are used with nonrefrigerated transportation systems, such as air freight, and consequently require insulation. Foamed polystyrene is particularly suited to this application. For typical air freight shipments, the most economical thickness of insulation is between 1 and 2 in. To maintain product temperature in transit, shippers use either dry ice, packaged wet ice, packaged gel refrigerant, or wet ice with absorbent padding in the bottom of the container. Foamed polystyrene containers may be of molded construction or of the composite type in which foam inserts and a plastic liner are used with a corrugated fiberboard box.

At the retail level, fresh fish may be handled in two ways. Stores with service counters display fish in unpackaged form. In some markets that do not have service counters, however, fish must be packaged prior to displaying for sale. Both types of outlets receive the product in institutional containers. If the fish is prepackaged at the market, high labor and packaging costs may be incurred, and the temperature of the product is likely to rise. Often, relatively warm fish is placed in a foam tray, wrapped, and displayed in a meat case, the temperature of which may be 40 °F or more. This drastically reduces the shelf life of the fish. Centralized prepackaging at the point of initial processing appears to have many important advantages over the present system. A number of retail chains have their suppliers prepackage the product under controlled temperature and sanitary conditions.

FRESH FISH STORAGE

The maximum storage life of fish varies with the species. In general, the storage life of East and West Coast fish properly iced and stored in refrigerated rooms at 35 °F is 10 to 15 days, with a maximum of 15 days. This depends on the condition of the fish when it is unloaded from the boat. Generally, freshwater fish properly iced in boxes and stored in refrigerated rooms may be held for only seven days. Both figures are from the time the fish is landed and processed to the time of consumption.

Cold storage facilities for fresh fish should be maintained at about 35 °F and over 90% relative humidity. Air velocity should be limited to control ice loss. Temperatures less than 32 °F retard ice melting and can result in excessive fish temperatures. This is particularly important when storing round fish such as herring, which generate heat from autolytic processes.

Floors should have adequate drainage with ample slope toward drains. All inside surfaces of the cold storage room should be easy to clean and have the capability to withstand the corrosive effects of frequent washings with antimicrobial compounds.

Radiopasteurization of Fresh Seafood

Ionizing radiation can double or triple the normal shelf life of refrigerated, unfrozen fish and shellfish stored at 33 °F (see Table 2). No off-odors, adverse nutritional effects, or other changes are imparted to the product as a result of the radiation treatment. However, irradiation of fish is still not common and is not permitted in some jurisdictions.

Table 2 Optimal Radiation Dose Levels and Shelf Life at 33 °F for Some Species of Fish and Shellfish

Species	Optimal Irradiation Dose, Rads Air Packed	Shelf Life, Weeks
Oysters—shucked, raw	200,000	3-4
Shrimp	150,000	4
Smoked chub	100,000	6
Yellow perch	300,000	4
Petrale sole	200,000	2-3 (4-5 when vac pac)
Pacific halibut	200,000	2 (4 when vac pac)
King crabmeat	200,000	4-6
Dungeness crabmeat	200,000	3-6
English sole	200,000-300,000	4-5
Soft-shell clam meat	450,000	4
Haddock	150,000-250,000	3-4
Pollock	150,000	4
Cod	150,000	4-5
Ocean perch	250,000	4
Mackerel	250,000	4-5
Lobster meat	150,000	4

Modified Atmosphere Packaging

A product environment with modified levels of nitrogen, carbon dioxide, and oxygen can curtail the growth of bacteria and extend shelf life of fresh fish. For example, whole haddock stored in a 25% carbon dioxide atmosphere from the time it is caught keeps about twice as long as in air. However, a modified atmosphere does not inhibit all microbes, and spoilage bacteria, because of their great number, usually restrict the growth of the few pathogenic bacteria present. Therefore, obvious spoilage is a safeguard against eating fish that may have dangerous levels of pathogenic bacteria.

Because modified atmosphere packaging can be a safety hazard, it is being introduced slowly in several countries, under close monitoring by regulatory agencies. This type of packaging requires a complete knowledge of regulations and a good control system that maintains proper temperature and sanitation levels.

FROZEN FISHERY PRODUCTS

The production of frozen fishery products varies with geographical location and includes primarily the production of groundfish fillets, scallops, breaded precooked fish sticks, breaded raw fish portions, fish roe, and bait and animal food in the northeastern states and in Atlantic Canada; round or dressed halibut and salmon, halibut and salmon steaks, groundfish fillets, surimi, herring roe, and bait and animal food in the northwestern states and in British Columbia; halibut, groundfish fillets, crab, salmon, and surimi in Alaska; shrimp, oysters, crabs, and other shellfish and crustacea in the Gulf of Mexico and South Atlantic states; and round or dressed fish in the areas bordering on the Great Lakes.

The fish obtained from these areas differ considerably in both physical and chemical composition. For example, cod or haddock are readily adaptable to freezing and have a comparatively long storage life, while other fatty species, such as mackerel, tend to become rancid during frozen storage and, therefore, have a relatively short storage life. The differences in composition of many species of fish and in marketing requirements necessitate consideration of the specific product with regard to quality maintenance and methods of packaging, freezing, cold storage, and handling.

Temperature is the most important factor limiting the storage life of frozen fish. At temperatures below freezing, bacterial activity as a cause of spoilage is limited. However, even fish frozen within a few hours of catching and stored at −20 °F will deteriorate very slowly until it becomes unattractive and unpleasant to eat.

Fish proteins are permanently altered during freezing and cold storage. This denaturation occurs quickly at temperatures not far below freezing, and even at 0 °F fish deteriorates rapidly. Badly stored fish is easily recognized; the thawed product is opaque, white, and dull, and juice can easily be squeezed from it. While the properly stored product is firm and elastic, poorly stored fish is spongy and, in very bad cases, the flesh may break up. Instead of the succulent curdiness of cooked fresh fish, cooked samples at first have a wet and sloppy consistency, and, on further chewing, become dry and fibrous.

Among other factors that determine how quickly quality deteriorates in cold storage are the initial quality and composition of the fish, protection of the fish from dehydration, the freezing method, and the environment during storage and transport. These factors are reflected in four principal phases of frozen fish production and handling—packaging, freezing, cold storage, and transportation.

Today, many species are brought from warm and tropical waters where parasites and toxins could infect them. In addition, food dishes that use raw seafood, such as sushi and sashimi, have gained wide popularity making these a potential health risk. Parasites are not life threatening but can cause pain and inconvenience. They are easily destroyed by cooking or by deep freezing (−40 °F). Marine toxins could be deadly and are not affected by temperature. Susceptible species should not be eaten during periods when toxins could be developed.

PACKAGING

Materials for packaging frozen fish are similar to those for other frozen foods. A package should: (1) be attractive and appeal to the consumer, (2) protect the product, (3) allow rapid, efficient freezing and ease of handling, and (4) be cost effective.

Package Considerations in Freezing

Refrigeration equipment and packaging materials are frequently purchased without considering the effect of the package size on freezing rate and efficiency. For example, a thin consumer package results in a faster rate of product freezing, lower total freezing costs, higher handling costs, and higher packaging material costs; a thicker institutional-type package results in a slower rate of product freezing, higher freezing costs, lower handling costs, and lower packaging material costs.

Tests indicate that the time required to freeze packaged fish fillets in a plate freezer is directly proportional to the square of the package thickness. Thus, if it takes 3 h to freeze packaged fish fillets 2 in. thick, it will take about 4.7 h to freeze packaged fish fillets 2.5 in. thick. The insulating effect of the packaging material, the fit of the product in the package, and the total surface area of the package must be considered. A packing material with low moisture-vapor permeability has an insulating effect, which increases freezing time and cost.

The rate of heat transfer through a packaging material is inversely proportional to its thickness; therefore, a packaging material should be used which is (1) thin enough to produce rapid freezing and adequate moisture-vapor barrier in frozen storage, and (2) thick enough to withstand heavy abuse. Aluminum foil cartons and packages offer an advantage in this regard.

Proper fit of package to product is essential; otherwise, the insulating effect of the airspace formed will reduce the freezing rate of the product and increase freezing cost. The surface area of the package is also important because of its relation to the size of the freezer shelves or plates. Maximum use of freezer space can be obtained by designing the package so that it fits the freezer properly. However, these factors often cannot be changed and still meet customer requirements for a specific package.

Package Considerations for Frozen Storage

Fish products lose considerable moisture and become tough and fibrous during frozen storage unless a package with low moisture-vapor permeability is specified. The package in contact with the product must also be resistant to oils or moisture exuded from the product, or rancidity of the oils and softening of the package material will occur. The package must fit the product tightly to minimize airspaces and thereby reduce moisture migration from the product to the inside surfaces of the package.

Unless temperatures are very low or special packaging is used, the oils in fish will oxidize in frozen storage, producing off flavor. One effective type of packaging is to replace the air surrounding the frozen fish with pure nitrogen and seal the fish in a leak-proof bag made of a material that is a barrier to the passage of oxygen.

Types of Packages

Packaging materials consist of paperboard cartons coated with various waterproofing materials, or cartons laminated with moisture-vapor-resistant films and heat-sealable overwrapping materials with a low moisture-vapor permeability. Paperboard cartons are usually made of a bleached kraft stock, coated with a suitable fortified wax, polyethylene, or other plastic material.

Overwrapping materials should be highly resistant to moisture transmission, inexpensive, heat sealable, adaptable to machinery application, and attractive in appearance. Various types of hot melt coated waxed paper, cellophane, polyethylene, and aluminum foil are available in different forms and laminate combinations that make possible the selection of an overwrapping material best suited to each product.

Consumer Packages. These usually hold less than 1 lb and are generally printed, bleached paperboard, coated with wax or polyethylene, and closed with adhesive. Fish sticks and portions, shrimp, scallops, crabmeat, precooked dinners, and entrees are packaged in this way. In the case of dinners and entrees, rigid plastic, pressboard, or aluminum trays are used inside the printed paperboard package. Rigid plastic or pressboard packages are becoming more common because they are better for microwave cooking. The packaging of these products is normally mechanized.

Materials such as polyethylene combined with cellophane, polyvinylidene chloride, or polyester and combinations of other plastic materials are used with high-speed automatic packaging machines to package shrimp, dressed fish, fish fillets, fish portions, and fish steaks prior to freezing. In some instances, tearing of the wrapping material by fins protruding from the fish has been a problem. Otherwise, this method of packaging is satisfactory and affords the product considerable protection against dehydration and rancidity at a comparatively low cost. This packaging method has also created new markets for merchandising frozen fish products. Boil-in-bag type pouches made of polyester-polyethylene and combinations of foil, polyethylene, and paper are used for packaging shrimp, fish fillets, and entrees. These packages are also suitable for microwave cooking.

Institutional Packages. The 5-lb and larger cartons used in the institutional trade are almost exclusively constructed of bleached paperboard that has been waxed or polyethylene coated. Folding cartons with self-locking covers, full-telescoping covers, or glued closures are used. Often the cartons are packaged inside a corrugated master carton or are shrink-wrapped in polyethylene film.

Products such as fish fillets and steaks are individually wrapped in cellophane or other moisture-vapor-resistant film, then packed in the carton. Fish, such as headed and dressed whiting and scallop meats, are packed into the carton and covered with a sheet of cellophane. The cover is then put on and the package is frozen upside down in the freezer. Raw, unbreaded products, such as shrimp, scallops, fillets, and steaks, are sometimes individually quick frozen (IQF) prior to packaging. When IQF, they can be glazed to enhance moisture retention. This method is preferred over freezing after packaging, because it leads to a product that is more convenient to handle and sometimes obviates the need to thaw the fish prior to cooking. For all institutional frozen fish, the trend is toward printed paperboard folding cartons coated with moisture-vapor-resistant materials instead of waxed paper or cellophane overwrap. Some frozen fish products and seafood entrees destined for institutional markets are packaged in aluminum trays or in rigid plastic trays so they may be heated within the package.

FREEZING METHODS

Product characteristics, such as size and shape, freezing method, and the rate of freezing, affect the quality, appearance, and cost of production.

Quick freezing of fish offers the following advantages:

- Chills the product rapidly, preventing bacterial spoilage
- Facilitates rapid handling of large quantities of product
- Makes use of conveyors and automatic devices practical, thus materially reducing handling costs
- Promotes maximum use of the space occupied by the freezer
- Produces a packaged product of uniform appearance, with a minimum of voids or bulges

See also Chapters 29 and 30 of the 1993 ASHRAE *Handbook—Fundamentals* and Chapter 8 of this volume.

Blast Freezing of Fish

Blast freezers for fishery products are generally small rooms or tunnels in which cold air is circulated by one or more fans over an evaporator and around the product to be frozen contained on racks or shelves. A refrigerant such as ammonia, a halocarbon, or brine flowing through a pipe coil evaporator furnishes the necessary refrigeration effect.

Static pressure in these rooms is considerable, and air velocities average between 500 and 1500 fpm, with 1200 fpm being common. Air velocities between 500 and 1000 fpm give the most economical freezing. Lower air velocities result in slow product freezing, and higher velocities increase unit freezing costs considerably.

Some factories have continuous blast freezers, in which conveyors move fish continuously through a blast room or tunnel. These freezers are built in a number of configurations, including a single pass through the tunnel, multiple passes, spiral belts, and moving trays or carpets. The configuration and type of conveyor belt or freezing surface depend on the type and quantity of the product to be frozen, the space available to install the equipment, and the capital and operating costs of the freezer.

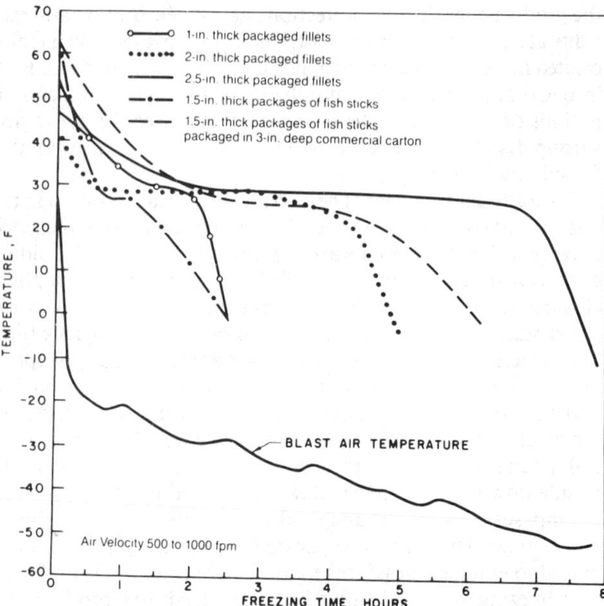

Fig. 2 Freezing Time of Fish Fillets and Fish Sticks in a Tunnel-Type Blast Freezer (Air Velocity 500 to 1000 fpm)

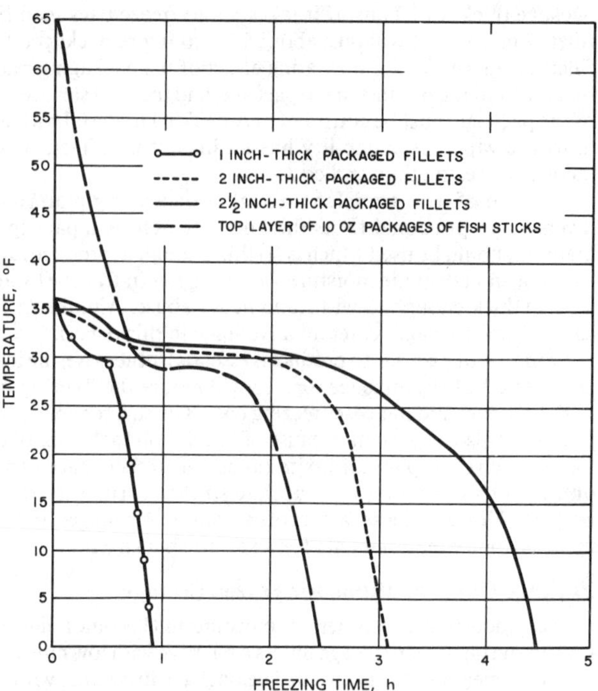

Fig. 3 Freezing Time of Fish Fillets and Fish Sticks in a Plate Freezer

Batch loaded blast freezers are used for freezing shrimp, fish fillets, steaks, scallops, or breaded precooked products packed in institutional packages; round, dressed, or panned fish; and shrimp, clams, or oysters packed in metal cans.

Conveyor-type blast freezers are widely used to freeze products prior to packaging. These products include all types of breaded, precooked seafoods; IQF fillets, loins, tails, and steaks, scallops, and shrimp; and raw, breaded fish portions. In the case of portions, which are sliced or sawed from blocks, the function of the blast freezer is to harden the batter and breading prior to packaging and to lower the temperature of the frozen fish again to storage temperatures if it has been tempered for slicing.

Dehydration of product or freezer burn may occur in freezing unpackaged whole or dressed fish in blast freezers unless the velocity of air is kept to about 500 fpm and the period of exposure to the air is controlled. Consumer packages of fish fillets or fish-fillet blocks requiring close dimensional tolerances undergo bulging and distortion during freezing unless restrained. In blast rooms or tunnels, where the product is frozen on trucks, the use of specially designed freezer trucks enabling distribution of pressure on the surfaces of the package will remedy this condition. It is difficult to control the expansion of the product on conveyor installations. The freezing times for various sizes of packaged fishery products are shown in Figure 2.

Plate Freezing of Fish

In the multiplate freezer, freezing is accomplished by refrigerant flowing through connected passageways in the horizontal movable plates stacked vertically within an insulated cabinet or in an insulated room. The plate freezer is used extensively in the freezing of fishery products packaged in consumer cartons and in 5- and 10-lb institutional-type cartons. Fish to be plate frozen should be properly packaged to keep airspaces in the package to an absolute minimum. Spacers should be used between the plates during freezing to prevent crushing or bulging of the package. For most products, the thickness of the spacers should be about 0.03 to 0.06 in. less than that of the package.

Where very close package tolerances are required, as in the manufacture of fish fillet blocks, a metal frame or tray is used to hold the packages of fish during freezing. The frame or tray is generally the same width as the package and the length of one or two blocks. It must be rigid enough to prevent bulging and to hold the fish block to close dimensional tolerances. This is sometimes accomplished with rigid spacers in order to limit the weight and cost of the tray.

Fish blocks are available in two common sizes: (1) 16.5 lb (19 by 10 by 2.5 in.) and (2) 18.5 lb (19 by 11.5 by 2.5 in.). Other blocks are sized for special applications. The fish may be packed in the block with the long dimension of the fillets generally running the length of the frame (long-pack) or the width of the frame (cross-pack). The orientation depends on the eventual cutting pattern and type of cutting used to convert the block into a finished product.

A tray is not necessary for other packaged seafoods such as shrimp, fillets, fish sticks, or scallops, where close package tolerances are not as essential. Therefore, an automatic continuous plate freezer with properly sized spacers can be satisfactorily used for these products.

The plate freezer provides rapid and efficient freezing of packaged fish products. The freezing time and energy required for freezing packaged fish sticks is greater than that for freezing packaged fish fillets, because heat transfer is slowed by the airspace within the package. The ratio of energy requirements to freezing unit mass of product increases with thickness. The freezing times of consumer and institutional size packages of fish fillets and fish sticks are shown in Figure 3.

Immersion Freezing of Fish

Immersion in low-temperature brine was one of the first methods used for quick-freezing fishery products. A number of direct immersion freezing machines were developed for freezing whole or panned fish. These machines were generally unsuitable for freezing packaged fish products, which make up the bulk of frozen fish production and have been replaced by methods employing air cooling, contact with refrigerated plates or shelves, or combinations of these methods.

Immersion freezing is used primarily for the freezing of tuna at sea and, to a lesser extent, for freezing shrimp, salmon, and Dungeness crab. Extensive research has been conducted on brine freezing of groundfish aboard the vessel, but this method is not in commercial use.

An important consideration in immersion freezing of fish is selection of a suitable freezing medium. The medium should be nontoxic, acceptable to public health regulatory agencies, easy to renew, inexpensive, and have a low freezing temperature and viscosity. It is difficult to obtain a freezing medium that meets all these requirements. Sodium chloride brine and a mixture of glucose and salt in water are acceptable media. The glucose reduces salt penetration into the fish and provides a protective glaze.

Liquid nitrogen spray and CO_2 are coming into wider use for IQF seafood products such as shrimp. The quality of fish frozen by these methods is good and, although the cost per unit mass is high, there is virtually no weight loss from dehydration, and there are space and equipment savings. In freezing fish in liquid nitrogen, the fish should not be immersed directly in the liquid nitrogen since this will cause the flesh to shatter and rupture.

Immersion Freezing of Tuna. Most tuna harvested by the United States fleet is brine-frozen aboard the fishing vessel. Freezing at sea enables the vessel to make extended voyages and return to port with a full payload of high quality fish.

Tuna are frozen in brine wells, which are lined with galvanized pipe coils on the inside. Direct expansion of ammonia into the evaporator coils provides the necessary refrigeration effect. The wells are so designed that tuna can first be precooled and washed with refrigerated seawater and then frozen in an added sodium chloride brine. After the fish are frozen, the brine is pumped overboard and the tuna kept in 10 °F dry storage. Prior to unloading, the fish are thawed in 30 °F brine on the vessel. In some cases, the fish are thawed in tanks at the cannery. Therefore, if the fish are thawed ashore, thawing on the vessel is not required beyond the stage needed to separate those fused together in the vessel's wells.

Sometimes tuna are held in the wells for a long period prior to freezing or are frozen at a very slow rate because of high well temperatures caused by overloading, insufficient refrigeration capacity, or inadequate brine circulation. These practices have a detrimental effect on product quality, especially on the smaller fish, which are more subject to salt penetration and quality changes. Tuna that is not promptly and properly frozen may undergo excessive changes, absorb excessive quantities of salt, and may even be bacteriologically spoiled when landed. Some freezing times for tuna of various sizes are shown in Figure 4.

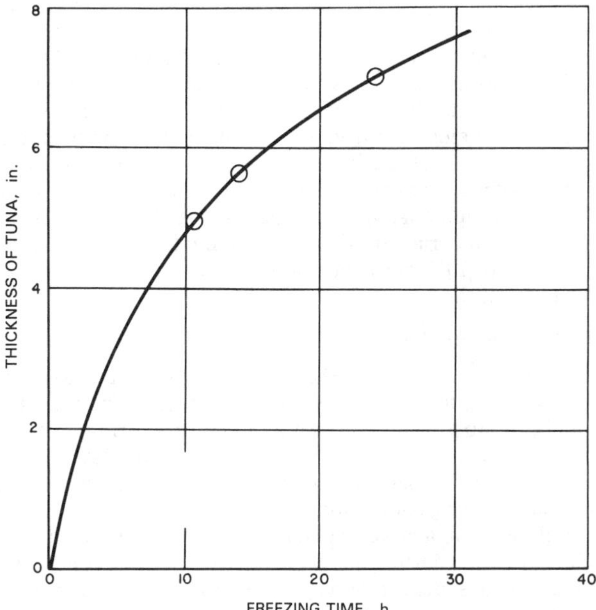

Fig. 4 Freezing Time for Tuna Immersed in Brine

Specialized Contact Freezers. Fish frozen by this method are placed on a slowly moving solid stainless steel belt. This belt conveys the fish fillets through a tunnel where they are frozen not only by an air blast, but also by direct contact between the conveyor belt and a thin layer of glycol which is pumped through the plates that support the belt. A refrigerant, such as ammonia or a halocarbon, also flows through separate channels in the plates. This provides the refrigeration effect with minimal temperature difference between the evaporating refrigerant and the product.

Freezing Fish at Sea

Freezing fish at sea has found increasing commercial application in leading fishery nations such as Japan, Russia, the United Kingdom, Norway, Spain, Portugal, Poland, Iceland, and the United States. Including freezer trawlers, factory ships, and refrigerated transports in fisheries, hundreds of large freezer vessels are operating throughout the world. United States factory freezer trawlers, factory surimi trawlers, and floating factory ships supplied by catcher vessels operate off Alaska. These vessels process mainly Alaskan pollock, cod, and flounder, although they do process other species.

Freezing groundfish at sea has not developed in the northeastern United States largely because fresh fish commands a better price than does frozen fish. For the same reason, East Coast United States producers avoid putting their product into frozen packs if they can sell it fresh. Hence, much of the frozen fish used in the United States, with the exception of Alaskan fish, is imported from other countries.

Where used, the factory vessel is equipped to catch, process, and freeze the fish at sea and to use the waste material in the manufacture of fish meal and oil. A large European factory vessel measures 280 ft in length, displaces 3700 tons, and is equipped to stay at sea for about 80 days without being refueled. About 65 to 100 people are required to operate the vessel and to process and handle the fish. On most vessels of this type, contact-plate freezers are used. The freezers can freeze about 30 tons of fish per day, and the total capacity of the frozen fish hold may be as high as 750 tons.

Because the factory trawler stays at sea for long periods, it can fully use its space for storing fish. However, because there is a limited amount of labor available on these vessels, the packs are generally restricted to the less labor-intensive types.

The freezer trawler was designed to resolve the disadvantages associated with factory freezer vessels. It is a smaller vessel equipped to freeze fish in bulk for later thawing and processing ashore. Freezer trawlers use vertical plate freezers to freeze dressed fish in blocks of about 100 lb.

Some countries use freezer trawlers to supplement the raw material for shore-based processing plants producing frozen fish products. This allows the trawlers to fill their holds in distant waters and transport the fish to home base, where the fish becomes frozen raw material that is held in storage until required for processing. In some cases, trawlers have been designed as dual fisheries, *i.e.,* fishing and freezing groundfish blocks during part of the year and catching, processing, and freezing Northern shrimp for the rest of the year.

STORAGE OF FROZEN FISH

Fishery products may undergo undesirable changes in flavor, odor, appearance, and texture during frozen storage. These changes are attributable to dehydration (moisture loss) of the fish, oxidation of the oils or pigments, and enzyme activity in the flesh. The rate at which these changes occur depends on: (1) the composition of the species of fish, (2) the level and constancy of storage room temperature and humidity, and (3) the protection afforded the product through the use of suitable packaging materials and glazing compounds.

Composition

The composition of a particular species of fish affects its frozen storage life considerably. Fish having a high oil content, such as some species of salmon, tuna, mackerel, and herring, have a comparatively short frozen storage life because of the development of rancidity as a result of the oxidation of the oils and pigments in the flesh. Certain fish, such as sablefish, are quite resistant to oxidative deterioration in frozen storage, despite their high oil content. The development of rancidity is less pronounced in fish with a low oil content. Therefore, lean fish such as haddock and cod, if handled properly, can be kept in frozen storage for many months without serious loss of quality. The relative susceptibility of various species of fish to oxidative changes during frozen storage is shown in Table 3.

Table 3 Relative Susceptibility of Representative Species of Fish to Oxidative Changes in Frozen Storage

Severe	Moderate	Minor	Very Slight
Pink salmon	Chum salmon	Cod	Yellow pike
Rockfish	Coho salmon	Haddock	Yellow perch
Lake chub	King salmon	Flounder	Crab
Whiting	Halibut	and sole	Lobster
	Ocean perch	Sablefish	
	Herring	Oysters	
	Mackerel		
	Tuna		
	Lake herring		
	Sheepshead		
	Lake trout		

Temperature

The loss of quality of frozen fish due to storage depends primarily on storage temperature and the length of time in storage. Fish stored at −20 °F will have a shelf life of more than a year. In Canada, the Department of Fisheries recommends a storage temperature of −15 °F or lower. Storage above −10 °F, even for a short period, will result in rapid loss of quality. Time-temperature tolerance studies show that frozen seafoods have memory; that is, each time they are subjected to high temperatures or poor handling practices, the loss in quality is recorded. When the product is finally thawed, the total effect of each exposure to high temperatures or other mistreatments is reflected in the quality of the product at the consumer level. Continuous storage at temperatures lower than −15 °F reduces oxidation, dehydration, and enzymatic changes, resulting in longer product shelf life. Frozen seafoods should be kept at temperatures as close to −15 °F as possible from the time they are frozen until they reach the consumer. The shelf life of frozen fish products stored at different temperatures is given in Table 4. Note the increase in shelf life at the lowest temperatures.

For many years, it was thought too costly to operate refrigerated warehouses at temperatures lower than −10 °F. However, improvements in the design and operation of refrigeration equipment have made operation of refrigerated warehouses at this temperature or lower economically possible. The production of surimi by West Coast-based factory ships has resulted in the construction of ultracold rooms for its storage. Japanese standards call for this product to be kept at −22 °F.

Table 4 Review of Effect of Storage Temperature on Shelf Life of Frozen Fishery Products

Product	Temp., °F	Shelf Life, Months	Product	Temp., °F	Shelf Life, Months
Packaged haddock fillets	10	4 to 5	Glazed whole halibut	10	3
	0	11 to 12		0	6
	−20	Longer than 12		−10	9
				−20	12
Packaged cod fillets	10	5	Whole blue fin tuna	10	4
	0	6		0 to −5	8
	−10	10 to 11		−20	12
Packaged[a] pollock fillets	20	1	Glazed whole herring	0	6
	10	2		−17	9
	0	8			
	−10	11			
	−20	24			
Packaged ocean perch fillets	15	1½ to 2	Packaged mackerel fillets	15	2
	10	3½ to 4		0	3
	0	6 to 8		−10	3 to 5
	−10	9 to 10			
Packaged striped bass fillets	15	4			
	0	9			

[a]Prepared from 1-day-old iced fish.

Humidity

A high relative humidity in the cold storage room tends to reduce the evaporation of moisture from the product. The relative humidity of air in the refrigerated room is directly affected by the temperature difference between room cooling coils and room temperature. A large temperature differential results in decreased relative humidity and an accelerated rate of moisture withdrawal from the frozen product. A small temperature difference between the air and evaporator cooling coils results in high relative humidity and reduced moisture loss from the product.

The relative humidity in commercial cold storages is 10 to 20% higher than that of an empty cold storage because of constant evaporation of moisture from the product. In a cold storage operating at 0 °F, with a 70% rh and pipe coil temperature of −10 °F, the moisture-vapor pressure of the air within the package and in direct contact with the frozen fish would be 0.0185 psia. The air in the cold storage would have a vapor pressure of 0.0132 psia, and the moisture-vapor pressure at the coils would be 0.0108 psia. These differences in moisture-vapor pressure will result in considerable moisture loss from the product unless it is adequately protected by suitable packaging materials or glazing compounds. The evaporator coils in the freezer should be sized properly so that the desired high relative humidities can be obtained. However, because of material costs and space limitations, a temperature difference of 10 °F between evaporator coils and room air is the most practical.

Packaging and Glazing

Adequate packaging of fishery products is important to prevent product dehydration and consequent quality loss. The packaging, which, in most instances, occurs prior to freezing, has been described. Individual fish, frozen in the round or dressed, cannot usually be suitably packaged; therefore, they must be protected by a suitable glazing compound.

A glaze acts as a protective coating against the two main causes of deterioration during storage, *i.e.*, dehydration and oxidation. It protects against dehydration by preventing moisture from leaving the product, and against oxidation by mechanically preventing air contact with the product. It may also act by chemical means to minimize these changes if the glaze carries a suitable antioxidant.

Table 5 Storage Conditions and Storage Life for Frozen Fish

Fish	Recommended Protection[a]	Storage Life (0 °F), Months
Chub, pink salmon	Ice glazing and packaging	4-6
Mackerel, sea herring, pollock, chub, smelts	Ice glazing and packaging	5-9
Pacific sardines, tuna	Packaging	4-6
Buffalofish, flounder, halibut, ocean perch, rockfish, sablefish, red, sockeye, silver or coho salmon, whiting, shrimp	Packaging	7-12
Haddock, blue pike, cod, hake, lingcod	Packaging	Over 12

[a]All packaging should be with moisture-resistant films.

Maximum storage life of fishery products can be obtained by employing the following procedures:

- Select only high quality fish for freezing.
- Use moisture-vapor-resistant packaging materials and fit package tightly around product or use a modified atmosphere and oxygen-barrier package.
- Freeze fish immediately after processing or packaging.
- Glaze frozen fish prior to packaging.
- Glaze round, unpackaged fish prior to cold storage.
- Put fish in frozen storage immediately after freezing and glazing, if required.
- Store frozen fish at temperatures of −15 °F or lower.
- Renew glaze on round, unpackaged fish as required during frozen storage.

The recommended protection and expected storage life for various species of fish at 0 °F are shown in Table 5.

Space Requirements

Packaged products such as fillets and steaks are usually packed in cardboard master cartons for storage and shipment. These master cartons are stacked on pallets and transferred to various areas of the cold storage room by forklift. The master cartons are strong enough to support one or two pallet loads placed on the shelf of each rack in the cold storage. In cold storages without racks, cartons should only be stacked to a height that does not cause crushing of the bottom cartons. Cartons for products in packages that contain a lot of air, such as IQF fillets, must be stronger than those for solid packages of fish to resist crushing during storage.

Whole or dressed fish frozen in blocks in metal pans, such as mackerel, chub, or whiting, are removed from the pans after freezing, glazed, and then packaged in wooden boxes lined with wax-impregnated paper or in cardboard cartons.

Round fish stored in wooden boxes can be easily reglazed at periodic intervals during frozen storage. The space requirements for the storage of fishery products are shown in Table 6.

Thawing of frozen fish. Frozen fishery products can be thawed by using circulating air or water. In thawing, the fish should not be allowed to rise above refrigerated temperatures; otherwise, rapid deterioration may occur. Thawing is a slower and more difficult process than freezing when done to ensure that quality is maintained. Each application should be carefully designed.

TRANSPORTATION AND MARKETING

Conditions of temperature and humidity recommended for frozen storage should also be applied during transportation and marketing to minimize product quality loss. Shipment in non-refrigerated or improperly refrigerated carriers, exposure to high ambient temperatures during transfer from one environment to another, improper loading of common carriers or display cases, equipment failure, and other poor practices lead to increased product temperature and consequently, quality loss.

Frozen fish are transported under mechanical refrigeration in trucks, railroad cars, or ships. Most of these vehicles are capable of maintaining temperatures of 0 °F or lower. Additional information on equipment used in the transportation and marketing of frozen fish and other foods is given in Chapters 24 to 31.

To minimize quality loss during transportation and marketing, the following procedures should be adhered to:

1. Transport frozen fish in refrigerated carriers (mechanical or dry ice systems) with ample capacity to maintain 0 °F over long distances.
2. Precool refrigerated carriers to at least 10 °F before loading.
3. Remove frozen products from the warehouse only when the carrier is ready to be loaded. Load directly into the refrigerated carrier and do not allow the product to sit on the dock.
4. Check the frozen fish temperature with a thermometer before loading.
5. Do not stack frozen fish directly against floors or walls of the carrier. Provide floor and wall racks or strips to permit air circulation around the entire load.
6. Continuously record the temperature of the refrigerated carrier during transit. Use an alarm to warn of equipment failure.
7. Measure the temperature of the product when it is removed from the common carrier at its destination.
8. If products are shipped in an insulated container, apply sufficient dry ice to maintain 0 °F or lower temperatures for the duration of the trip.

Table 6 Space Requirements for Frozen Fishery Products

Commodity	Product Package	Container for Storage	Space Required, lb/ft^3
Fish sticks, breaded shrimp, breaded scallops	8 or 10 oz	Corrugated master containers	25-30
Fish fillets, fish steaks, small dressed fish	1, 5, or 10 lb	Corrugated master containers	50-60
Shrimp	2.5 and 5 lb	Corrugated master containers	35
Panned, frozen fish (mackerel, herring, chub)	None	Wooden or fiberboard boxes	35
Round halibut	None	Wooden box	30-35
		Stacked loose	38
Round groundfish (cod, etc.)	None	Stacked loose	32
Round salmon	None	Stacked loose	33-35

9. Maintain food delivery or breakup rooms at 0 to 10°F. Do not hold products in breakup rooms any longer than necessary.
10. When received at the retail store, place the product in a 0°F storage room immediately.
11. Hold display cases in retail stores at 0°F or lower.
12. Do not overload display cases, especially above the frost line.
13. Record the temperature of the display cases. Provide an alarm to warn of excessive rise in temperature.
14. Because of the accelerated deterioration of frozen fish products in the distribution and retail chain, hold products in these areas for as short a period as possible.

BIBLIOGRAPHY

Barnett, H.J, R.W. Nelson, P.J. Hunter, S. Bauer, and H. Groninger. 1971. Studies on the use of carbon dioxide dissoved in refrigerated brine for the preservation of whole fish. *Fishery Bulletin* 69(2).

Charm, S.E. and P. Moody. 1966. Bound water in haddock muscle ASHRAE *Journal* 8(4):39.

Dassow, J.A. and D.T. Miyauchi. 1965. Radiation preservation of fish and shellfish of the Northeast Pacific and Gulf of Mexico; Ronsivalli, L.J., M.A. Steinberg, and H.L. Seagran. Radiation preservation of foods. National Academy of Science Publication No. 1273. Washington, D.C.

Feiger, E.A. and C.W. du Bois. 1952. Conditions affecting the quality of frozen shrimp. *Refrigerating Engineering* (September):225.

Holston, J. and S.R. Pottinger. 1954. Some factors affecting the sodium chloride content of haddock during brine freezing and water thawing. *Food Technology* 8(9):409.

Nelson, R.W. 1963. Storage life of individually frozen Pacific oyster meats glazed with plain water or with solutions of ascorbic acid or corn syrup solids. *Commercial Fisheries Review* 25(4):1.

Peters, J.A. 1964. Time-temperature tolerance of frozen seafood. ASHRAE *Journal* 6(8):72.

Peters, J.A., E.H. Cohen, and F.J. King. 1963. Effect of chilled storage on the frozen storage life of whiting *Food Technology* 17(6):109.

Peters, J.A. and J.W. Slavin: Comparative keeping quality, cooling rates, and storage temperatures of haddock held in fresh water ice and salt water ice. *Commercial Fisheries Review* 20(1):6.

Ronsivalli, L.J. and J.W. Slavin. 1965. Pasteurization of fishery products with gamma rays from a cobalt 60 source. *Commercial Fishery Review* 27(10):1.

Stansby, M.E., ed. 1976. *Industrial Fishery Technology*, 2nd ed. Robert E. Krieger Publishing Co., Huntington, NY.

Tressler, D.K, W.B. van Arsdel, and M.J. Copley, eds. 1968. *The freezing preservation of foods*, 4th ed. Avi Publishing Co., Westport, CT.

Wagner, R.L, A.F. Bezanson, J.A. Peters. Fresh fish shipments in the BCF insulated leakproof container. *Commercial Fisheries Review* 31(8 and 9):41.

CHAPTER 14

DAIRY PRODUCTS

RAW milk is either processed for beverage milks, creams, and related milk products for marketing, or is used for the manufacture of dairy products. Milk is defined in the United States Code of Federal Regulations, Title 21, Section 131.110. This definition includes goat's milk as defined in the Grade A Pasteurized Milk Ordinance, Part I, Section 1, A.1 of the 1991 revision. Milk products are defined in the Code of Federal Regulations, Title 21, Parts 131 through 135. Public Law 519 defines butter. Note that there are many nonstandard dairy-based products that may be processed and manufactured by the equipment described in this chapter. Dairy plant operations include receiving of raw milk; purchase of equipment, supplies, and services; processing of milk and milk products; manufacture of frozen dairy desserts, butter, cheeses, and cultured products; packaging; maintenance of equipment and other facilities; quality control; sales and distribution; engineering; and research.

Farm cooling tanks and most dairy processing equipment manufactured in the United States meet the requirements of the 3-A Sanitary Standards (IAMFES). These standards set forth the minimum design criteria acceptable for composition and surface finishes of materials in contact with the product, construction features such as minimum inside radii, accessibility for inspection and manual cleaning, criteria for mechanical (CIP) cleaning, insulation of nonrefrigerated holding and transport tanks, and other factors which may adversely affect the quality and safety of the product or the ease of cleaning and sanitizing the equipment. Also available are *3-A Accepted Practices*, which deal with construction, installation, operation, and testing of certain systems rather than individual items of equipment.

The *3-A Sanitary Standards* and *Accepted Practices* are developed by the 3-A Standards Committees which are composed of conferees representing state and local sanitarians, the U.S. Public Health Service, dairy processors, and equipment manufacturers. Compliance with the *3-A Sanitary Standards* is voluntary, but a manufacturer who complies and has authorization from the 3-A Symbol Council may affix to his equipment a plate bearing the 3-A Symbol, which indicates to regulatory inspectors and purchasers that the equipment meets the pertinent sanitary standards.

MILK PRODUCTION AND PROCESSING

Handling Milk at the Dairy

Most dairy farms have bulk milk tanks to receive, cool, and hold the milk. Tank capacity ranges from 200 to 5000 gal, with a few larger tanks. As the cows are mechanically milked, the milk flows through sanitary pipelines to an insulated stainless steel bulk tank. An electric agitator stirs the milk, and mechanical refrigeration begins to cool it even during milking.

When operated with a condensing unit of the minimum capacity given on the nameplate, a tank must have sufficient refrigerated surface at the first milking to cool 50% of its capacity in an everyday pickup tank, or 25% of its capacity in a tank for every-other-day pickup, from 90 to 50 °F within the first hour, and from 50 to 40 °F within the next hour. During the second and subsequent milking, there must be sufficient refrigerating capacity to prevent the temperature of the blended milk from rising above 45 °F. The nameplate must state the maximum rate at which the milk may be added and still meet the cooling requirements of the *3-A Sanitary Standards.*

Automatic controls maintain the desired temperature within a preset range in conjunction with the agitation. Some dairies continuously record the milk temperature in the tank, which is required in some states. Since the milk is picked up from the farm tank daily or every other day, the milk from the additional milkings generally flows into the reservoir cooled from the previous one. Some large dairy farms may use a plate or tubular heat exchanger to rapidly cool the milk. Cooled milk may be stored in an insulated silo tank (a vertical cylinder 10 ft or more in height).

Milk in the farm tank is pumped into a stainless steel tank on a truck for delivery to the dairy plant or receiving station. The tanks are well insulated to alleviate the need for refrigeration of the milk during transportation. Temperature rise when testing the tank full of water should not be more than 2 °F in 18 h, when the average temperature difference between the water and the atmosphere surrounding the tank is 30 °F.

The most common grades of raw milk are Grade A and Manufacturing Grade. The former is that used for market milk and related products such as cream. Surplus Grade A milk is used for ice cream and/or manufactured products. To produce Grade A milk the dairy farmer must meet state and federal standards. In addition to the state requirements, a few municipal governments also have raw milk regulations.

For raw milk produced under the provisions of the Grade A Pasteurized Milk Ordinance recommended by the U.S. Public Health Service, the dairy farmer must have healthy cows, adequate facilities (barn, milkhouse, and equipment), maintain satisfactory sanitation of these facilities, and have milk with a bacteria count of less than 100,000 per mL for individual producers. Commingled raw milk can not have more than 300,000 counts per mL. The milk should not contain pesticides, antibiotics, sanitizers, and so forth. However, current methods detect even minute traces, and total purity is difficult. Current regulators require no positive results on drug residue. Milk should be free of objectionable flavors and odors.

Receiving and Storage of Milk

A milk plant receives, standardizes, processes, packages, and merchandizes milk products that are safe and nutritious for human consumption. Most dairy plants either receive raw milk in bulk from a producer organization or arrange for pickup

The preparation of this chapter is assigned to TC 11.2, Foods and Beverages.

directly from dairy farms. The milk level in a farm tank is measured with a dip stick or a direct-reading gage, and the volume is converted to weight. Fat test and weight are common measures used to base payment to the farmer. A few organizations and the state of California include the percent of nonfat solids and protein content.

Some plants determine the amount of milk received by weighing the tanker, metering the milk as it is pumped from the tanker to a storage tank, and using load cells on the storage tank or other methods associated with the amount in the storage tank.

Milk is generally received more rapidly than it is processed, so ample storage capacity is needed. A holdover supply of raw milk at the plant may be needed for startup before arrival of the first tankers in the morning. Storage may be required for nonprocessing days and emergencies. Storage tanks vary in size from 1000 to 60,000 gal. The tanks have a stainless steel lining and are well insulated.

The *3-A Sanitary Standards* for silo-type storage tanks specify that the insulating material should be of a nature and an amount sufficient to prevent freezing, or an average 18-h temperature change of no more than 3 °F in the tank filled with water when the average temperature differential between the water and the surrounding air is 30 °F. Insulation material on a tank should have a total R-value at least approximately 10 Btu/h · ft² · °F, while tanks installed partially or wholly outside of a building should have insulation equivalent to the total R-value 15 Btu/h · ft² · °F. For horizontal storage tanks, the allowable temperature change under the same conditions is 2 °F.

Agitation is essential to maintain uniform milkfat distribution. Milk held in such large tanks as the silo type is continuously agitated with a slow-speed propeller driven by a gearhead electric motor or with filtered compressed air. The tank may or may not have refrigeration, depending on the temperature of the milk flowing into it and the maximum holding time. Some plants pass the milk through a plate cooler to maintain 40 °F or less on all milk directed into the storage tanks.

If cooling is provided for milk in a storage tank, it may be by refrigeration of the surface around the lining. This cooling surface may be an annular space from a plate welded to the outside of the lining for direct refrigerant cooling or circulation of chilled water or glycol solution. Another system provides a distributing pipe at the top for the chilled liquid to flow down the lining and drain from the bottom. Direct refrigerant cooling must be carefully applied to prevent milk from freezing on the lining. This limits the evaporator temperature to approximately 25 to 28 °F.

Separation and Clarification

Before pasteurizing, milk and cream are standardized and blended to control the milkfat content with legal and practical limits. Nonfat solids may also need to be adjusted for some products; some states require added nonfat solids, especially for lowfat milk such as 2% (fat) milk. Table 1 shows the approximate legal milkfat and nonfat solids requirements for milks and creams in the United States.

One means of obtaining the desired fat standard is by separating a portion of the milk. The required amount of cream or skim milk is returned to the milk. Milk with an excessive fat content may be processed through a standardizer-clarifier that removes fat to a predetermined percentage and clarifies it at the same time. This machine can be adjusted to remove 0.1 to 2.0% of the fat in milk. To increase the nonfat solids, condensed skim milk or low-heat nonfat dry milk may be added.

Milk separators are enclosed and fed with a pump. Separators designed to separate cold milk, usually not below 40 °F, have increased capacity and efficiency as the milk temperature is increased. Capacity of a separator is doubled as milk temperature is raised from 40 to 90 °F. The efficiency of fat removal with a cold milk separator decreases as temperature decreases below 40 °F. The maximum efficiency for fat removal is attained at approximately 45 to 50 °F or above. Milk is usually separated at 70 to 90 °F, but not above 100 °F in warm milk separators. If raw, warmed milk or cream is to be held for more than 20 min before pasteurizing, it should be immediately recooled to 40 °F or below after separation. The pump supplying milk to the separator should be adjusted to pump the milk at the desired rate without causing a partial churning action.

At an early stage between receiving and before pasteurizing, the milk or the resulting skim milk and cream should be filtered or clarified. An optimum time to effectively filter is during the transfer from the pickup tanker into the plant equipment. A clarifier removes extraneous matter and leucocytes, thus improving the appearance of homogenized milks.

Pasteurization and Homogenization

There are two systems of pasteurization—batch and continuous. The minimum feasible continuous operation is about 2000 lb/h. Therefore, batch pasteurization is used for relatively small quantities of liquid milk products. The product is heated in a stainless steel-lined vat to not less than 145 °F and held at that temperature or above for not less than 30 min. The Grade A Pasteurized Milk Ordinance requires that means be provided and used in batch or vat pasteurizers to keep the atmosphere above the product at a temperature not less than 5 °F higher than the minimum required temperature of pasteurization during the holding period. Whole milk, low fat milk, half-and-half, and coffee cream are cooled, usually in the vat, to 130 °F and then homogenized. Cooling is continued in a heat exchanger (*e.g.*, a plate or tubular unit) to 40 °F or lower and then packaged.

Plate coolers may have two sections, one using plant water and the second using chilled water. The temperature of the product leaving the cooler depends on the flow rates and temperature of the cooling medium. Pasteurizing vats are heated with hot water or with steam vapor in contact with the outer surface of the lining. One heating method consists of spraying the heated water

Table 1 U.S. Requirements for Milkfat and Nonfat Solids in Milks and Creams

	Legal Minimum Milkfat, %			Legal Minimum Nonfat Solids, %		
Product	**Federal**	**Range**	**Most Often**	**Federal**	**Range**	**Most Often**
Whole milk	3.25	3.0–3.8	3.25	8.25	8.0–8.7	8.25
Lowfat milk	0.5	0.5–1.9	2.0	8.25	8.25–10.0	8.25
Skim milk	0.5[a]	0.1–0.5	0.5[a]	—	8.25–9.0	8.25
Flavored milk	—	2.8–3.8	3.25	—	7.5–10.0	8.25
Half-and-half	10.5	10.0–18.0[a]	10.5	—	—	—
Light (coffee) cream	18.0	16.0–30.0[a]	18.0	—	—	—
Light whipping cream	30.0	30.0–36.0[a]	30.0	—	—	—
Heavy cream	36.0	36.0–40.0	36.0	—	—	—
Sour cream	18.0	16.0–20.0	18.0	—	—	—

[a]Maximum

around the top of the lining. It flows to the bottom where it drains into a sump, is reheated by steam injection, and again is pumped through the spray distributor. Steam-regulating valves control the temperature of the hot water.

Most pasteurizing vats are constructed and installed so that the plant's cold water is used for initial cooling of the product after pasteurization. For final vat cooling, refrigerated water is recirculated through the jacket of the vat to attain a product temperature of 40 °F or less. Cooling time to 40 °F should be less than 1 h.

Brine is rarely used in milk plants as the final cooling medium; it corrodes milk equipment and presents the danger of freezing the milk, especially with plate heat exchangers. Flow of the chilled water for final cooling should be counter-current to the product flow in heat exchangers. The rate of flow should be adjusted so that the temperature increase is not more than 10 °F during one pass through the equipment. A ratio of chilled water to product is usually about 4:1.

High temperature, short time pasteurization (HTST) is a continuous process in which the milk is heated to a temperature of not less than 161 °F and held at this temperature for at least 15 s. The complete pasteurizing system usually consists of a series of heat exchange plates contained in a press, a milk balance tank, one or more milk pumps, a holder tube, flow diversion valve, automatic controls, and sources of hot and chilled water for heating and cooling the milk. Homogenizers are used in many HTST systems as timing pumps used to process Grade A products. The heat exchanger plates are arranged so that milk to be heated or cooled flows between two plates, and the heat exchange medium flows in the opposite direction between alternate pairs of plates.

Ports in the plates are arranged to direct the flow where desired, and gaskets are arranged so that any leakage will be to the outside of the press. Terminal plates are inserted to divide the press into three sections (heating, regenerating, and cooling) and arranged with ports for inlet and outlet of milk, hot water, or steam for heating, and chilled water for cooling. To provide a sufficient heat-exchange surface for the temperature change desired in a section, the milk flow is arranged for several passes through each section. The capacity of the pasteurizer can be increased by arranging several streams for each pass made by the milk. The capacity range of a complete HTST pasteurizer is 100 to about 100,000 lb/h. A few shell-and-tube and triple-tube HTST units are in use, but the plate type is by far the most prevalent.

Figure 1 shows one example of a flow diagram for an HTST plate pasteurizing system. Raw product is first introduced into a constant level (or balance) tank from a storage tank or receiving line by either gravity or a pump. A uniform level is maintained in this tank by means of a float-operated valve or similar device. A booster pump is often used to direct the flow through the regeneration section. The product may be clarified and/or homogenized or simply directly pumped to the heating section by means of a timing pump. From the heating section, the product continues through a holding tube to the flow diversion valve. If the product is at or above the preset temperature, it passes back through the opposite sides of the plates in the regeneration section and then through the final cooling section. The flow diversion valve is set at 161 °F or above; if the product is below this minimum temperature, it is diverted back into the balance tank for repasteurization. The exchange of heat in the regeneration section causes the cold raw milk to be heated by the heated pasteurized milk going downstream from the heater section. According to the P.M.O., the pasteurized milk pressure must be maintained at least 2 psig above the raw. The flow rate of both products is the same, and the temperature change is about the same.

Most HTST heat exchangers have 80 to 90% regeneration. The cost of additional equipment to achieve more than 90% regeneration should be compared with savings in the increased regeneration to determine feasibility. The percentage of regeneration may be calculated as follows:

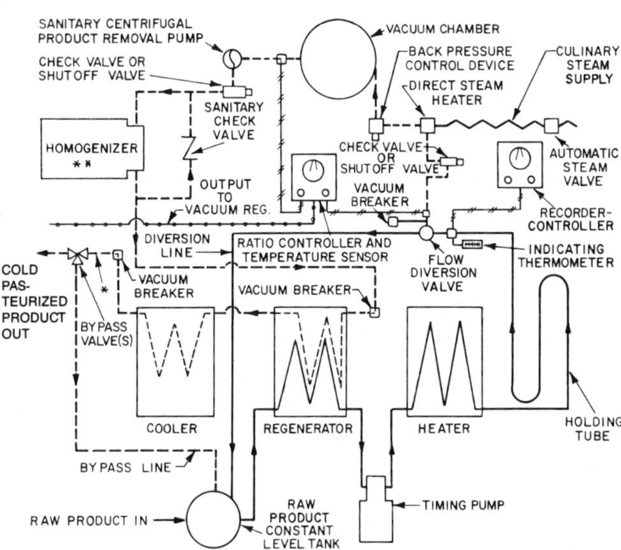

Fig. 1 Flow Diagram of Plate HTST Pasteurizer with Vacuum Chamber

$$\frac{138\,°\text{F (regeneration)} - 40\,°\text{F (raw product)}}{161\,°\text{F (pasteurization)} - 40\,°\text{F (raw product)}} = \frac{98}{121} = 81\%$$

The temperature of a product going into the cooling section can be calculated if the percent regeneration is known and the raw product and pasteurizing temperatures are determined. If they are 80%, 45 °F, and 161 °F, respectively:

$$(161 - 45) \times 0.80 = 92.8\,°\text{F}$$

$$161 - 92.8 = 68.2\,°\text{F}$$

The product should be cooled to at least 40 °F, preferably even lower, to compensate for the increase while in the sanitary pipelines or package (including filling, sealing, casing, and transfer into cold storage). The average temperature increase of milk between the time of discharge from the cooling section of the HTST unit and arrival at the cold storage in various containers is as follows: glass bottles, 8 °F; preformed paperboard cartons, 6 °F; formed paperboard, 5 °F; and semirigid plastic, 4 °F.

Many plate-type pasteurizing systems are equipped with a cooling section containing propylene glycol solution to cool the milk or milk product to temperatures lower than are practical by circulating only chilled water. This requires an additional section in the plate heat exchanger, a glycol chiller, a pump for circulating the glycol solution, and a product temperature actuated control to regulate the flow of glycol solution and prevent freezing of the product. Milk is usually cooled this way to approximately 34 °F, then packaged. The lower temperature allows the milk to absorb heat from the containers and still maintain a low enough temperature for excellent shelf life. Milk should not be cooled to temperatures between 33.5 °F and freezing because of the tendency toward increased foaming in this range. The propylene glycol is usually chilled to approximately 28 °F for circulation through the milk cooling section.

The product flow rate through the pasteurizer may be more or less than the filling rate of the packaging equipment. Pasteurized product storage tanks are generally used to hold the product until it is packaged.

The number of plates in the pasteurizing unit is determined by the volume of product needed per unit of time, the desired percentage of regeneration, and the temperature differentials between the product and the heating and cooling media. The heating section will usually have ample surface so that the temperature of the hot water entering the section will be not more than 2 to 6°F warmer than the pasteurizing, or outlet, temperature of the product. On larger units, steam may be used for the heater section instead of hot water. The cooling section is usually sized so that the temperature of the pasteurized product leaving the section is about 5°F warmer than the entering temperature of the chilled water.

The holder tube size and length is selected so that not less than 15 s will elapse for the product to flow from one end of the tube to the other. An automatic, power-actuated, flow-diversion valve, controlled by a temperature recorder-controller, is located at the outlet end of the holder tube and diverts the flow back to the raw product constant level tank as long as the product is below the minimum set pasteurizing temperature. The product timing pump is a variable speed, positive displacement, rotary type that can be sealed by the local government milk plant inspector at a maximum speed and volume. This assures a product hold of not less than 15 s in the holder tube.

As a means of reducing undesirable flavors and odors in milk (usually caused by specific types of dairy cattle feed), some plants use a vacuum process in addition to the usual pasteurization. In this process, milk from the flow diversion valve passes through a direct steam injector or steam infusion chamber where the milk is heated with culinary steam to a temperature of 180 to 200°F. The milk is then immediately sprayed into a vacuum chamber, where it is cooled by evaporation to the pasteurizing temperature and promptly pumped to the regeneration section of the pasteurizing unit. The vacuum in the evaporating chamber is automatically controlled so that the same amount of moisture will be removed as was added by steam condensate. Noncondensable gases are removed by the vacuum pump, and vapor from the vacuum chamber is condensed in a heat exchanger which is cooled by the plant water.

The vacuum chamber can be installed with any type of HTST pasteurizer. In some plants, after preheating in the HTST system, the product is further heated by direct steam infusion or injection. It then is deaerated in the vacuum chamber. The product is pumped from the chamber by a timing pump through final heating, holder, flow diversion valve, and regenerative and cooling sections. Homogenization may occur either immediately after preheating for pasteurization or after the product passes through the flow diversion valve. If the product is heated by direct steam injection and deaerated, the preferred practice is to homogenize after deaeration.

Where volatile weed and feed taints in the milk are mild, some processors use only a vacuum treatment for reduction of the off-flavor. The main objection to vacuum treatment alone is that, to be effective, the vacuum must be low enough to cause some evaporation, and the moisture so removed constitutes a loss of product. With this type of treatment, the vacuum chamber may be installed immediately after preheating, where it effectively deaerates the milk prior to heating or immediately after the flow diversion valve where it is more effective in removing the volatile taints.

Nearly all milk processed in the United States is homogenized to improve stability of the milkfat emulsion, thus preventing creaming (concentration of the buoyant milkfat in the upper portion of the containerized milk) during normal shelf life. The homogenizer is a high-pressure, reciprocating pump with 3 to 7 pistons, fitted with a special homogenizing valve. There are several types of homogenizing valves in use, all of which cause the melted fat globules in the stream of milk to be subjected to enough shear to be divided into several smaller globules. Homogenizing valves may either be single or two in series.

For effective homogenization of whole milk, the fat globules should be 2 μm or less in diameter. The usual temperature range is from 130 to 180°F, and the higher the temperature within this range, the lower the pressure required for satisfactory homogenization. The homogenizing pressure for a single-stage homogenizing valve ranges from about 1200 to 2500 psi for milk; for a two-stage valve, from 1200 to 2000 psi on the first stage plus 300 to 700 psi on the second, depending on the design of the valve and the product temperature and composition. To conserve energy, use the lowest homogenizing pressure consistent with satisfactory homogenization; the higher the pressure, the greater the power requirements.

Packaging of Milk Products

Cold product from the pasteurizer cooling section flows to the packaging machine and/or a surge tank 1000 to 10,000 gal or larger. These tanks are stainless steel, well insulated, and have agitation and usually refrigeration.

Milk and related products are packaged for distribution in paperboard, plastic, or glass containers in various sizes. Fillers vary in design. Gravity flow is used, but positive piston displacement is used on paper machines. Filling speeds range from roughly 16 to 250 units/min, but vary with container size. Some fillers handle only one size, while others may be adjusted to automatically fill and seal several size containers. Paperboard cartons are usually formed ahead of filling, but may be preformed prior to delivery to the plant. Semirigid plastic containers may be blow-molded in-plant ahead of the filler or preformed. Plastic pouches (called bags) arrive at the plant ready for filling and sealing. The filling of dispenser cans and bags is a semimanual operation.

The paperboard carton for milk consists of a 16-mil thick kraft paperboard from virgin paper with a 1-mil polyethylene film laminated onto the inside and a 0.75-mil film onto the outside. Gas or electric heaters supply heat for sealing while pressure is applied.

Blow-molded plastic milk containers are fabricated from high density polyethylene resin. The resin temperature for blow-forming varies from 340 to 425°F. The molded gallon weighs approximately 60 to 70 g, and the one-half gallon, about 45 g. Most equipment uses direct expansion water chillers or another cooling medium to cool the mold head and clutch. The refrigeration demand is sufficiently large to require the cooling load to be included in planning a plastic blow-molded operation. The blow-molded equipment manufacturer should be contacted for the refrigeration requirements of a specific machine.

Packages containing the product may be placed into cases mechanically. Stackers place the cases 5 or 6 high, and conveyors transfer the stacks into the cold storage area.

Most milk processing activities from receiving to storage can be automated. This necessitates the installation and operation of control panels and a digital computer. Automation uses a meter-based system that controls the separation, fat and/or nonfat solids content, and ingredient addition for a variety of common products. If the initial fat tests fed into the computer are correct, the accuracy of the fat content of the standardized product is ±0.01%. Added ingredients (*e.g.*, flavoring and sweetening products) must be in liquid form for a computerized operation.

Several systems to automatically clean the equipment in place (CIP) are used in milk processing plants. These may involve the holding and reuse of the detergent solution or the preparation of a fresh solution (single-use) each day. The means of programming the automatic control of each cleaning and sanitizing step also varies. Tanks, vats, and other large equipment can be cleaned by using spray balls and similar devices that assure complete coverage of soiled surfaces. Tubing, HTST units, and equipment with relatively low volume may be cleaned by the full-flood system. The solutions should have a velocity of not less than 5 ft/s and must be in contact with all soiled surfaces. Surfaces used for heating milk products, such as in batch or HTST pasteurization, are more difficult to clean than the other equipment surfaces. Other surfaces difficult to clean are those in contact with high-fat products, products containing added solids and/or sweeteners, and highly viscous products. The usual cleaning steps for this equipment are a warm water rinse, hot acid solution wash, rinse, hot alkali solution wash, and rinse. The variables of time, temperature, concentration, and velocity may need to be adjusted for effective cleaning. Just before use, the product surfaces should be sanitized with chemical solution, hot water, or steam.

Milk Storage and Distribution

Cases containing packaged products are conveyed into a cold storage room or directly to delivery trucks for wholesale or retail distribution. The temperature of the storage area should be 33 to 40 °F, and for improved keeping quality, the product temperature in the container on arrival in storage should be 40 °F or less.

The refrigeration calculation for the cold storage area should include losses through insulation, conveyor passes, and doors; heat added due to temperature of products, containers, cases, lights, conveyor drive motors, washdown water, and personnel; and lift truck operation. Relative humidity in a storage area is generally high and should be considered when selecting coils or diffuser units for maintaining the cold temperature. Frost collects rapidly on the coils, so automatic defrosting during the off-period is common. Hot gas pumped from the compressor discharge into the coils for a short time causes coils to warm and melt frost or ice. Defrosting via electric resistance heaters is also common.

Warming the room air to melt frost is generally considered too slow; in addition, the product temperature may rise.

The floor space required for cold storage depends on product volume, height of stacked cases, kind of package (glass requires more space than paperboard), whether mechanized or manual handling, and the number of processing days per week. A 5-day processing week requires a capacity for holding product supply for 2 days. A very general estimate is that 100 lb of milk product in paperboard cartons can be stored per square foot of area. Approximately one-third more area should be allowed for aisles.

Milk product may be transferred by conveyor from storage room to dock for loading onto delivery trucks. In-floor drag-chain conveyors are commonly used, especially for retail trucks. Refrigeration losses are reduced if the load-out doorway has the protrusion of cushioning material to contact the doorway frame of the truck as it is backed to the dock.

Distribution trucks need refrigeration to protect quality and extend the keeping quality of milk products. Refrigeration capacity must be sufficient to maintain Grade A products at 45 °F or less. Many plants use insulated truck bodies with integral refrigerating systems powered by the truck engine or one that can be plugged into a remote electric power source when it is parked. In some facilities, cold plates in the truck body are connected to a coolant source in the parking space. These refrigerated trucks can also be loaded when convenient and held over at the connecting station until the next morning.

Water Chilling Equipment

A large portion of the refrigeration capacity in a milk plant is required for water chilling. Chilled water is usually used for cooling pasteurized milk and other dairy products and may be used for cooling incoming milk and holding or lowering the temperature in milk storage tanks. Chilled water is refrigerated at a central source and recirculated by pumps through the various pieces of dairy equipment. Refrigeration of the water may be by flash cooling equipment or by circulation through an ice bank. Some plants use both systems.

The flash-type water chiller may be arranged as: a bank of coils having the circulated water sprayed over the coils; banks of coils or plates arranged so water flows down over the coils by gravity from distributing pipes or troughs; or refrigerated coils submerged in a water tank with rapid circulation of the water around the coils. A closed shell-and-tube type or shell-and-coil type of chiller should not be used because of the low water temperature required and the danger of freezing the water tubes.

Flash-type water chilling equipment may be used for cooling pasteurized milk and other uniform continuous cooling loads required at the same time. Also, high peak cooling loads can be handled by an ice-bank type of chilling system. To maintain a uniform chilled water supply at about 34 °F in a flash system, sufficient water should be circulated so the temperature rise of the water returning will not be greater than 10 °F. Refrigerant temperatures in the flash chiller depend on unit design (usually about 28 °F).

The ice-bank type of water chiller should be evaluated against the flash-type water chiller for each application, considering both initial capital and operating costs. The ice-bank system permits the use of a refrigerating system having considerably less capacity than is required for the peak cooling load by building ice on the refrigerated surface during the time that chilled water is not required. When chilled water is required, the melting ice adds cooling capacity to that supplied by the refrigerating system.

The ice-bank water chiller consists of an insulated tank containing refrigerated coils or plates submerged in the water and spaced so that ice can build up on the refrigerated surface to a thickness of about 1.5 in. Agitators and baffles keep water moving over the surface of the ice. The agitator may contain a motor-driven propeller to circulate water around the baffles and over the coils at a rate of five times the amount of water being pumped for product cooling, or agitation may be by means of compressed air. For air agitation, perforated pipes along the bottom of the tank extend under each stand of pipe coils so the air bubbles rising cause the water to move over the surface of the ice.

An ice-bank water chiller with a 1.5-in. thick ice buildup on 1.25-in. steel pipe (6 lb of ice per foot of pipe) has 900 Btu of cooling capacity per foot of pipe available from the melting ice. With good agitation, the water for process cooling leaves the tank at a temperature of 33 to 34 °F as long as ice remains on the refrigerated surface. Warm water returning to the tank should discharge near the tank bottom or at the inlet to the agitator. Water for process cooling should be taken from near the top of the tank. The refrigerant temperature in the ice-bank system varies from 12 to 25 °F; it is usually controlled by a dual-pressure back pressure regulator. The lower setting (12 to 25 °F) is maintained during ice building, while an automatic ice thickness control switches the regulator to the higher setting (40 °F or so) to terminate further ice building when maximum ice has been built.

In selecting or designing an ice-bank water chilling system, sufficient refrigerated surface and ice buildup capacity should be provided to handle the entire chilled water requirements for a full day. The refrigerating effect supplied by the refrigerating system while chilled water is being supplied to processing equipment should not be included in calculating capacity. This selection provides a surplus in capacity to take care of losses in insulation and circulating piping, allows for unusual peak operating conditions, and provides for some expansion.

Half-and-Half and Cream

Half-and-half is standardized to 10.5 to 12% milkfat and in most states to about the same percent nonfat milk solids. Coffee cream should be standardized to 18 to 20% milkfat. Both are pasteurized, homogenized, cooled, and packaged similarly to milk. Milkfat content of whipping cream is adjusted to 30 to 35%. Care must be taken during processing to preserve the whipping properties; this includes the omission of homogenization.

Buttermilk, Yogurt, and Sour Cream

Retail buttermilk is not from the butter churn but is rather a cultured product. To reduce the microorganisms to a low level and improve the body of the resulting buttermilk, skim milk is pasteurized at 180 °F or higher for 0.5 to 1 h and cooled to 70 to 72 °F. One percent of a lactic acid culture (starter) specifically for buttermilk is added and the mixture incubated until firmly coagulated by the correct lactic acid production (pH 4.5). The product is cooled to 40 °F or less with gentle agitation to inhibit serum separation subsequent to packaging and distribution. Salt and/or milkfat (0.5 to 1.0%) in the form of cream or small fat granules may be added. Package equipment and containers are the same as for milk. Pasteurizing, setting, incubating, and cooling are usually accomplished in the same vat. Rapid cooling is necessary, so chilled water is used. If a 500-gal vat is used, as much as 25 to 30 tons of refrigeration may be needed. Some plants have been able to cool buttermilk with a plate heat exchanger without causing a serum separation problem (wheying off).

Cultured half-and-half and cultured sour cream are manufactured similarly to the method for cultured buttermilk. Rennet may be added at a rate of 0.5 mL (diluted in water) per 10 gal cream. Care must be exercised to use an active lactic culture and to prevent postpasteurization contamination by bacteriophage, bacteria, yeast, or molds. An alternate method consists of packaging the inoculated cream, incubating, and then cooling by placing packages in a refrigerated room.

Skim milk may be used, or milkfat standardized to within the range of 1 to 5%, and a 0.1 to 0.2% stabilizer may be added to yogurt. Either vat pasteurization at 150 to 200 °F for 0.5 to 1 h or HTST at 185 to 285 °F for 15 to 30 s can be used. For yogurt to have optimum body, the milk homogenization is at 130 to 150 °F and 500 to 2000 psi. After cooling to between 110 and 100 °F, the product is inoculated with a yogurt culture. Incubation for 1.5 to 2.0 h is necessary; the product is then cooled to about 90 °F, packaged, incubated 2 to 3 h (acidity 0.80 to 0.85%), and chilled to 40 °F or below in the package. Varying yogurt cultures and yogurt manufacturing procedures should be selected on the basis of consumer preferences. Numerous flavorings are used (fruit is quite common), and sugar is usually added. The flavoring material may be added at the same time as the culture, after incubation, or ahead of packaging. In some dairy plants, a fruit (or sauce) is placed into the package before filling with yogurt.

BUTTER MANUFACTURE

Much of the butter production is in combination butter-powder plants. These plants get the excess milk production after current market needs for milk products, frozen dairy desserts, and, to some extent, cheeses are met. Consequently, seasonal variation in the volume of butter manufactured is large; spring is the period of highest volume, fall the lowest.

Separation and Pasteurization

After separation of the milk, the cream with 30 to 40% fat is either pumped to the pasteurizer or cooled to 45 °F and held for later pasteurization. Cream from cold milk separation does not need to be recooled except for extended storage. Cream is received, weighed, sampled, and, in some plants, graded according to flavor and acidity. It is pumped to a refrigerated storage vat and cooled to 45 °F if held for a short period or overnight. Cream with developed acidity is warmed to 80 to 90 °F, and neutralized to 0.12 to 0.15% titratable acidity just prior to pasteurization. If the acidity is above 0.40%, it is neutralized with a soda-type compound in aqueous solution to about 0.30% and then to the final acidity with aqueous lime solution. Sodium neutralizers include $NaHCO_3$, Na_2CO_3, and NaOH. Limes are $Ca(OH)_2$, MgO, and CaO.

Batch pasteurization is usually at 155 to 175 °F for 0.5 h, depending on intended storage temperature and time. HTST continuous pasteurization is at 185 to 250 °F for at least 15 s. HTST systems may be plate or tubular. After pasteurization, the cream is immediately cooled. The temperature depends on the time that the cream will be held before churning, whether or not it is ripened, the season (higher in winter due to fat composition), and the churning method. The range is 40 to 55 °F. The ripening process consists of adding a flavor-producing lactic starter to tempered cream and holding until acidity has developed to 0.25 to 0.30%. The cream is cooled to prevent further acid development and warmed to the churning temperature just before churning. Ripening cream is not a common practice in the United Sates, but is customary in some European countries such as Denmark. First, tap water is used to reduce the temperature to between 80 and 100 °F. Refrigerated water or brine is then used to reduce the temperature to the desired level. The cream may be cooled by passing the cooling medium through a revolving coil in the vat or through the vat jacket, or by using a plate or tubular cooler.

If the temperature of 1000 lb of cream is to be reduced by refrigerated water from 104 to 39 °F and the specific heat is 0.85 Btu/(lb · °F), the heat units to be removed would be:

$$1000\ (104 - 39)\ 0.85 = 55{,}250\ \text{Btu}$$

This heat can be removed by 55,250/144 = 384 lb of ice at 32 °F plus 10% for mechanical loss.

The temperature of the refrigerated water commonly used for cooling cream is 33 to 34 °F. The ice-bank system is efficient for this purpose. Brine is not currently used. About 265 gal of cream can be cooled from 100 to 40 °F in a vat using refrigerated water in an hour.

After a vat of cream has been cooled to the desired temperature, the temperature increases during the following 3 h. It may increase several degrees depending on the rapidity with which the cream was cooled, the temperature to which it was cooled, the richness of the cream, and the properties of the fat. The rise in temperature is due to liberation of heat when fat is changed from a liquid to a crystal form.

Rishoi (1951) presented data in Figure 2 that show the thermal behavior of cream heated to 167 °F followed by rapid cooling to 86 °F and to 50.7 °F, as compared with cream heated to 122 °F and cooled rapidly to 88.5 °F and to 53.6 °F. The curves indicate that when cream is cooled to a temperature at which the fat remains liquid, the cooling rate is normal, but when the cream is cooled to a temperature at which some fractions of the fat have crystallized, a spontaneous temperature rise takes place after cooling.

Rishoi also determined the amount of heat liberated by the part of the milkfat that crystallizes in the temperature range of 85 to 33 °F. The results are shown in Table 2 and Figure 3.

Table 2 shows that, at a temperature below 50 °F, about one-half of the liberated heat evolved in less than 15 s. The heat liberated during fat crystallization constitutes a considerable portion of the refrigeration load required to cool fat-rich cream. Rishoi states:

"If we assume an operation of cooling cream containing 40% fat from about 150 to 40 °F, heat of crystallization evolved represents about 14% of the total heat to be removed. In plastic cream containing 80% fat it represents about 30% and in pure milkfat oil about 40%."

Churning

To maintain the yellow color of butter from cream that came from cows on green pasture in spring and early summer, yellow coloring is added to the cream in the amount needed to produce the color that is obtained naturally during other periods of the year. After cooling, pasteurized cream should be held a minimum

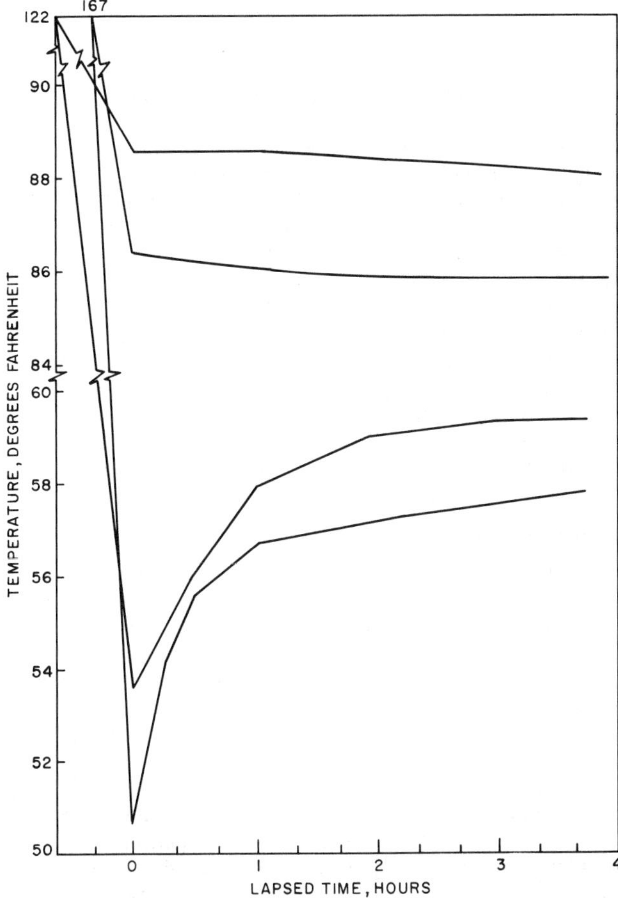

Fig. 2 Thermal Behavior of Cream Heated to 167 °F Followed by Rapid Cooling to 86 °F and to 50.7 °F; Comparison with Cream Heated to 122 °F, then Rapid Cooling to 88.5 °F and to 53.6 °F

Table 2 Heat Liberated from Fat in Cream Cooled Rapidly from about 86 °F to Various Temperatures

Calculated temperature for zero time, °F:	33.4	39.6	53.1	58.0	63.4	80.4	85.6[a]
First observed temperature:	36.2	43.7	54.5	58.6	63.8	80.4	85.6
Final equilibrium temperature:	39.3	46.1	58.0	60.7	65.3	82.6	85.6
Lapsed time, min	**Btu Liberated per Pound**						
0.25	18.3	16.7	7.75	2.9	2.3	0	0
15	23.6	20.0	12.8	7.2	5.2	0.35	0
30	25.5	23.9	18.0	12.2	9.0	2.1	0
60	30.4	27.5	21.4	14.0	10.2	2.7	0
120	32.4	29.3	25.0	14.7	10.6	2.9	0
180	32.9	30.2	26.6	16.0	10.6	4.3	0
240	34.0	30.9	27.2	17.5	10.6	4.9	
300	33.6	31.8	27.9	18.4		5.0	
360	33.6	32.5	27.2	18.8		5.0	

Percent heat liberated at zero time compared with that at equilibrium: 54.5, 51.3, 27.7, 20.7, 21.7.

Percent total heat liberated compared with that liberated at about 32 °F: 100.0, 95.7, 82.0, 55.0, 31.0, 12.5, 0.

Iodine values of three samples of butter produced while these tests were in progress were: 28.00, 28.55, and 28.24.

[a]Cooled in an ice water bath.

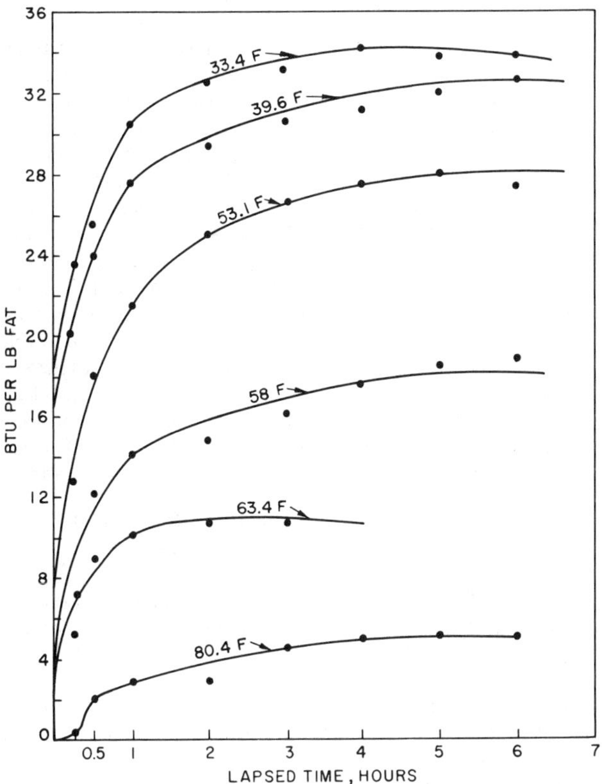

Fig. 3 Heat Liberated from Fat in Cream Cooled Rapidly from Approximately 86 °F to Various Temperatures
(Rishoi 1951)

of 2 h and preferably overnight. It is tempered to the desired batch churning temperature, which varies with the season and feed of the cows but ranges from 45 °F in early summer to 56 °F in winter, to maintain a churning time 0.5 to 0.75 h. Lower churning time results in soft butter that is more difficult (or impossible) to work into a uniform composition.

Today most butter is churned by continuous churns, but batch units remain in use, especially in smaller butter factories. Batch churns are usually made of stainless steel, although a few aluminum ones are still in use. They are cylinder, cube, cone, or double cone in shape. The inside surface of metal churns is sandblasted during fabrication to reduce or prevent butter from sticking to the surface. Metal churns may have accessories to draw a partial vacuum or introduce an inert gas, *e.g.*, nitrogen, under pressure. Working the butter under a partial vacuum reduces the air in the butter. Churns have two or more speeds, with the faster rate for churning. The speed should provide maximum agitation of the cream, usually within 0.25 to 0.5 r/s.

When churning, temperature is adjusted and the churn is filled 40 to 48% of capacity. The churn is revolved until the granules break out and attain a diameter of 5 mm or slightly larger. The buttermilk is drained and should have no more than 1% milkfat. The butter may or may not be washed. The purpose of washing is to remove buttermilk and temper the butter granules if they are too soft for adequate working. Wash water temperature is adjusted to 0 to 10 °F below churning temperature. The preferred procedure is to spray wash water over granules until it appears clear from the churn drain vent. The vent is then closed, and water is added to the churn until the volume of butter and water is approximately equal to the amount of the former cream. The churn is revolved slowly 12 to 15 times and drained or held for an additional 5 to 15 min for tempering so granules will work into a mass of butter without becoming greasy.

The butter is worked at a slow speed until free moisture is no longer extruded. The free water is drained, and the butter is analyzed for moisture content. The amount of water needed to obtain the desired content (usually 16.0 to 18.0%) is calculated and added. Salt may be added to the butter. The percent is standardized between 1.0 and 2.5% according to customer demand.

Salt may be added in dry form either to a trench formed in the butter or spread over the top of the butter. It also may be added in moistened form using the water required for standardizing the composition to not less than 80.0% fat. Working continues until the granules are completely compacted and the salt and moisture droplets are uniformly incorporated. The moisture droplets should become invisible by normal vision with adequate working. Most churns have ribs or vanes which cause tumbling and folding of the butter as the churn revolves. The butter passes between the narrow slit of shelves attached to the shell and the roll. A leaky butter is inadequately worked, possibly leading to economic losses due to reduction of weight and shorter keeping quality. The average composition of U.S. butter on the market has these ranges:

Fat	80.0 to 81.2	Moisture	16.0 to 18.0
Salt	1.0 to 2.5	Curd, etc.	0.5 to 1.5

Cultured skim milk is added to unsalted butter as part of the moisture and thoroughly mixed in during working. Cultured skim milk increases acid flavor and the diacetyl content associated with butter flavor.

Butter may be removed manually from small churns, but it is usually emptied by a mechanized procedure. One method is to dump the butter from the churn directly into a stainless steel boat on casters or a tray that has been pushed under the churn with the door removed. Butter in boats may be augered to the hopper for printing (forming the butter into retail sizes) or pumping into cartons 60 to 68 lb in size. The bulk cartons are held cold before printing or shipment. Butter may be stored in the boats or trays and tempered until printing. A hydraulic lift may be used for hoisting the trays and dumping the butter into the hopper. Cone-shaped churns with a special pump can be emptied by pumping butter from churn to hopper.

Continuous Churning

The basic steps in two of the continuous buttermaking processes developed in the United States are: fat emulsion in the cream is destabilized and the serum separated from the milkfat; the butter mix is prepared by thoroughly mixing together the correct amount of milkfat, water, salt, and cultured skim milk; this mixture is worked and chilled at the same time; butter is extruded at 38 to 50 °F with a smooth body and texture.

Several European continuous churns consist of a single machine that directly converts the cream to butter granules, drains off the buttermilk, and washes and works the butter, incorporating the salt in continuous flow. Each brand of continuous churn may vary in equipment design and specific operation details for obtaining the optimum composition and quality control of the finished product. Figure 4 shows a flow diagram of a continuous churn.

In one system, milk is heated to 110 °F and separated to cream with 35 to 50% fat and skim milk. The cream is pasteurized at 203 °F for 16 s, cooled to a churning temperature of 46 to 55 °F, and held for 6 h. The cream enters the balance tank and is pumped to the churning cylinder where it is converted to granules and serum in less than 2 s by vigorous agitation. Buttermilk is drained off and the granules are sprayed with tempered wash water while being agitated.

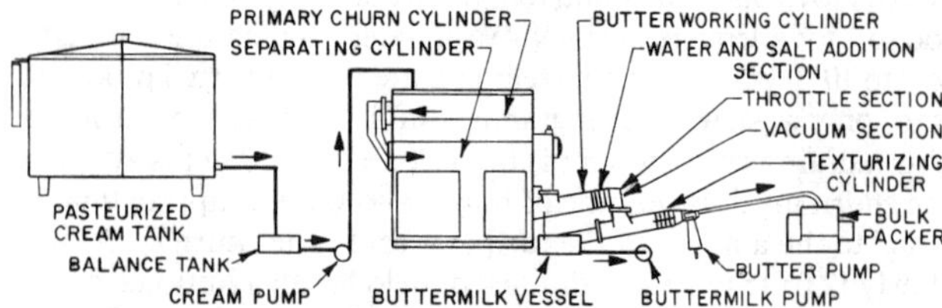

Fig. 4 Flow Diagram of Continuous Butter Manufacture

Next, salt, in the form of 50% brine prepared from microcrystalline sodium chloride, is fed into the product cylinder by a proportioning pump. If needed, yellow coloring may be added to the brine. The high speed agitators work the salt and moisture into the butter in the texturizer section and then extrude it to the hopper for packaging into bulk cartons or retail packages. The cylinders on some designs have a cooling system to maintain the desired temperature of the butter from churning to extrusion. The butterfat content is adjusted by fat test of the cream, churning temperature of the cream, and flow rate of product.

Continuous churns are designed for cleaning-in-place (CIP). The system may be automated or a cream tank may be used to prepare the detergent solution prior to circulation through the churn after the initial rinsing.

Packaging Butter

"Printing" refers to the process of forming (or cutting) butter into retail sizes. Each is then wrapped with parchment or parchment-coated foil. The wrapped prints may be inserted in paperboard cartons or overwrapped in cellophane, glassine, and so forth, and heat-sealed. For institutional uses, butter may be extruded into slabs. These are cut into patties, embossed, and each slab of patties wrapped in parchment paper. Most common numbers of patties are 48 to 72 per lb.

Butter keeps better if stored in bulk. If the butter is intended to be stored for several months, the temperature should not be above 0 °F, and preferably below −20 °F. For short periods, 32 to 40 °F is satisfactory for bulk or printed butter. Butter should be well protected to prevent absorption of off-odors during storage and weight loss due to evaporation, and to minimize surface oxidation of the fat.

The specific heat of butter and other dairy products at temperatures varying from 32 to 140 °F is given in Table 3. The temperature of the butter when removed from the churn ranges from 56 to 62 °F. Assuming a temperature of 60 °F of the packed butter, the heat that must be removed from 1000 lb to reduce the temperature to 32 °F is:

$$1000\,(60 - 32)\,0.52 = 14{,}560 \text{ Btu}$$

It is assumed that the average specific heat at the given range of temperatures is 0.52 Btu/(lb·°F). Heat units that are to be removed from the butter containers and packaging material should be added.

Table 3 Specific Heats of Milk and Milk Derivatives Btu/(lb·°F)

	32 °F	59 °F	104 °F	140 °F
Whey	0.978	0.976	0.974	0.972
Skim milk	0.940	0.943	0.952	0.963
Whole milk	0.920	0.938	0.930	0.918
15% Cream	0.750	0.923	0.899	0.900
20% Cream	0.723	0.940	0.880	0.886
30% Cream	0.673	0.983	0.852	0.860
45% Cream	0.606	1.016	0.787	0.793
60% Cream	0.560	1.053	0.721	0.737
Butter	(0.512)[a]	(0.527)[a]	0.556	0.580
Milkfat	(0.445)[a]	(0.467)[a]	0.500	0.530

[a]For butter and milkfat, values in parentheses were obtained by extrapolation, assuming that the specific heat is about the same in the solid and liquid states.

Deterioration of Butter in Storage

The development of an undesirable flavor in butter during storage may be caused by: (1) growth of microorganisms (proteolytic organisms cause putrid and bitter off-flavors); (2) absorption of odors from the atmosphere; (3) fat oxidation; (4) catalytic action by metallic salts; (5) activity of enzymes, principally from microorganisms; and (6) low pH (high acid) of salted butter.

Normally, microorganisms do not grow below 32 °F; if salt-tolerant bacteria are present, their growth will be slow below 32 °F. Growth of microorganisms does not take place at 0 °F or below, but some may survive in the butter held at this temperature. It is important to store butter in a room where the atmosphere is free of odors. Butter will readily absorb odors from the atmosphere or from odoriferous materials with which it comes into contact.

Oxidation causes a stale, tallowy flavor. Chemical changes take place slowly in butter held in cold storage, but are hastened by the presence of metals or metallic oxides.

With almost 100% replacement of tinned copper equipment with stainless steel equipment, a tallowy flavor is not as common as in the past. Factors that favor oxidation are light, high acid, high pH, and metal.

Enzymes present in raw cream are inactivated by current pasteurization temperatures and holding times. The only enzymes that may cause deterioration of butter are those produced by microorganisms that gain entrance to the pasteurized cream and butter or survive pasteurization. The chemical changes caused by enzymes present in butter are retarded by a lowering of the storage temperature.

A fishy flavor may develop in salted butter during cold storage. The development of the defect is favored by high acidity (low pH) of the cream at the time of churning and by metallic salts. With the use of stainless steel equipment and the proper control of the butter's pH, this defect now occurs very rarely. For salted butter to be stored for several months, even at −10 °F, it is advisable to use good quality cream, avoid exposing the milk or cream to strong light, copper, or iron, and to adjust any acidity developed in the cream so that the butter serum will have a pH of 6.8 to 7.0.

Total Refrigeration Load

Some dairy plants that manufacture butter also process and manufacture other products such as ice cream, fluid milk, and cottage cheese. A common refrigeration system is used. The method of determining the refrigeration load is illustrated by the following example:

Example 1. Determine the total refrigeration load for a plant manufacturing butter from 12,600 lb of 30% cream per day in three churnings.

Solution: The refrigeration requirement is obtained in the following steps A through E.

A. If the cream is separated in the plant rather than on the farm, it would have to be cooled down from 90 °F separating temperature to 40 °F for holding until it is processed.

$$\frac{12{,}600\ (90 - 40)\ 0.85}{144} = 3720 \text{ lb ice}$$

B. After pasteurization, the temperature of the cream is reduced to approximately 100 °F with city water, then down to 40 °F with refrigeration.

$$\frac{12{,}600\ (100 - 40)\ 0.85}{144} = 4460 \text{ lb ice}$$

C. After churning, the butter wash water (city water) is usually cooled to 45 °F, then used to wash the butter granules. A volume of water equal to the volume of cream churned may be used.

$$\frac{12{,}600\ (60 - 45)\ 1.00}{144} = 1310 \text{ lb ice}$$

Total ice load	9490 lb ice
Plus 10% mechanical loss	950 lb ice
Total ice required	10,440 lb ice

D. Approximately 4725 lb of butter would be obtained (12,600 lb cream × 30% fat = 3780 lb of fat). If butter contains approximately 80% fat, 3780 divided by 80% equals approximately 4725 lb of butter. The butter temperature going into the refrigerated storage room is usually about 62 °F and must be cooled down to 40 °F during the following 16 h. (For long-term storage, the butter is held at −10 to 0 °F.)

4725 (62 − 40) 0.55	= 57,170 Btu
300 lb (metal container) × (75 − 40) 0.12	= 1260 Btu
Total/24 h	58,430 Btu

E. After 24 h or longer, the butter is removed from the cooler to be cut and wrapped in 1 lb or smaller units. During this process, the butter temperature will rise to approximately 55 °F, which constitutes another product load in the cooler when it goes back for storage.

4725 (55 − 40) 0.55	= 38,980 Btu
200 lb (paper container) × (75 − 40) 0.33	= 2310 Btu
Total/24 h	41,290 Btu

Total of Steps D and E, Product Load in Cooler

$$\frac{41{,}290 + 58{,}430}{16 \text{ h}} = 6230 \text{ Btu/h}$$

On this basis, the following calculations develop:

The cream would be cooled in steps A and B. The butter would then have to be cooled through steps C, D, and E. Refrigerated water is normally used as a cooling medium in steps A, B, and C. The ice-bank system is used to produce 34 °F water, and the load should be expressed in pounds of ice required that would have to be melted to handle steps A, B, and C. This load would be added to the refrigerated water load from the various other products such as milk, cottage cheese, and so forth, in sizing the ice-bank unit.

Whipped Butter

To whip butter by the batch method, the butter is tempered to 62 to 70 °F, depending on such factors as the season, type of whipper, and so forth. The butter is cut into slabs for placing into the whipping bowl. The whipping mechanism is activated, and air is incorporated until the desired overrun is obtained, usually between 50 and 100%. The whipped butter is packaged mechanically or manually into semirigid plastic containers.

With one continuous system, butter directly from cooler storage is cut into pieces and augered until soft. However, it can be tempered and the augering step omitted. The butter is then pumped into a cylindrical continuous whipper that uses the same principles as those for incorporating air in ice cream. Air or nitrogen is incorporated until the desired overrun is obtained. Another continuous method (used less commercially) is to melt butter or standardize butter oil to the composition of butter with moisture and salt. The fluid product is pumped through a chiller-whipper. Metered air or nitrogen provides overrun control. Whipped butter is pumped in a soft state to the hopper of the filler and packaged in rigid or semirigid containers, such as plastic. It is chilled and held in storage at 32 to 40 °F.

CHEESE MANUFACTURE

Approximately 800 cheeses have been named, but there are only 18 distinctly different types. A few of the more popular types in the United States are cheddar, cottage, Roquefort or blue, cream, ricotta, mozzarella, Swiss, Edam, and Provolone. Such details of manufacture as setting (starter organisms, enzyme, milk or milk product, temperature, and time), cutting, heating (cooking), stirring, draining, pressing, salting, and curing (including temperature and humidity control) are varied to produce a characteristic variety and its optimum quality.

The production of cheddar cheese in the United States far exceeds the other cured varieties; cottage cheese production is much greater than that of the other uncured types. Another trend in the cheese industry is large factories. These plants may have sufficient curing facilities for the total production. If not, the cheese is shipped to central curing plants.

The physical shape of cured cheese varies considerably. Barrel cheese is common; it is cured in a metal barrel or similar impervious container in units of approximately 500 lb. Cheese may also be cured in rectangular metal containers holding 2000 lb.

The microbiological flora of cured cheese are important in the development of flavor and body. Heating the milk for cheese is general practice. The milk may be pasteurized at the minimum HTST conditions or be given a subpasteurization treatment that results in a positive phosphatase test. This is possible when the milk quality is good (low level of spoilage microorganisms and pathogens). Such treatments of milk give the cheese some of the characteristics of raw-milk cheese in curing, such as production of higher flavors, in a shorter time. Pasteurization to produce phosphatase-negative milk is practiced in the making of soft, unripened varieties of cheese and some of the more perishable of the ripened types such as Camembert, Limburger, and Munster.

The standards and definitions of the Federal Food and Drug Administration and of most state regulatory agencies require that cheese that is not made from phosphatase-negative milk must be cured for not less than 60 days at not less than 35 °F. Raw-milk cheese contains not only lactic acid producing organisms such as *Streptococcus lactis*, which are added to the milk during the cheese making process, but also the heterogeneous mixture of microorganisms present in the raw milk, many of which may produce gas and off-flavors in the cheese. With the advent of milk pasteurization, some control of the bacterial flora of the cheese is possible.

Freshly manufactured cheese of the cured types is rubbery in texture and has little flavor; perhaps the more characteristic flavor is slightly acid. The presence of definite flavor(s) in freshly made cheese indicates poor quality, probably resulting from off-flavored milk. On curing under proper conditions, however, the body of the cheese breaks down, and the nut-like, full-bodied flavor characteristic of aged cheese develops. These changes are accompanied by certain chemical and physical changes during the curing process. The calcium paracaseinate of cheese gradually changes into proteoses, amino acids, and ammonia. These changes are a part of the ripening process and may be controlled by time and temperature of storage. As cheese cures, varying degrees of lipolytic activity also occur. In the case of blue, or Roquefort, cheese this partial fat breakdown contributes substantially to the characteristic flavor.

During curing, the microbiological development produces changes according to the species and strains present. It is possible to predict from the microorganism data some of the usual defects in cheddar. In some cheeses (*e.g.*, Swiss), gas production accompanies the desirable flavor development.

Cheese quality is evaluated on the basis of a score card. Flavor and odor, body and texture, and color and finish are principal factors. They are influenced by milk quality, the skill of manufacturing (including starter preparation), and the control effectiveness of maintaining optimum curing conditions.

Cheddar Cheese

Manufacture. Raw or pasteurized whole milk is tempered to 86 to 88 °F. It is set by adding 0.75 to 1.25% active cheese starter and annatto yellow color, which depends on the market demand. After 15 to 30 min, 99 mL of single-strength rennet per 1000 lb milk is diluted in water 1:40 and slowly added with agitation of milk in the vat. After a quiescent period of 25 to 30 min, the curd should have developed proper firmness. The curd is cut into 0.25 to 0.40 in. cubes. After 15 to 30 min of gentle agitation, the cooking is initiated by heating water in the vat jacket by means of steam for 30 to 40 min. The curd and whey should increase 2 °F per 5 min. Then a temperature of 100 to 102 °F is maintained for approximately 45 min.

The whey is drained and the curd trenched along both sides of the vat, allowing a narrow area free of curd the length of the midsection of the vat. Slabs about 10-in long are cut and inverted at 15-min periods during the cheddaring process. When acidity of the small whey drainage is at a pH of 5.3 to 5.2, the slabs are milled (cut into small pieces) and returned to the vat for salting and stirring, or the curd goes to a machine for the automatic addition of salt and its uniform incorporation into the curd. Weighed curd goes into hoops, which are placed into a press, and 20 psi is applied. After 0.5 to 1 h, the hoops are taken out of the press, the bandage adjusted to remove wrinkles, and then the cheese is pressed overnight at 25 to 30 psi or higher. Cheese may or may not be subjected to a vacuum treatment to improve body by reducing or eliminating air pockets. After the surface is dried, the cheese is coated by dipping into melted paraffin or wrapped with one of several plastic films, or oil with a plastic film, and sealed. Yield is about 10 lb per 100 lb of milk.

As cheese factories have grown larger, a change toward faster and more mechanized methods of making cheddar cheese has evolved. The stirred curd method (whereby the cheddaring step is omitted) is being used by more cheese makers. Deep circular or oblong cheese vats with special, reversible agitators and means for cutting the curd are becoming popular. The curd is pumped from these vats to draining and matting tables with sloped bottoms and low sides, then milled, slated, and hooped. In one method the curd (except for Odenburg cheddar) is carried and drained by a draining-matting conveyor with a porous plastic belt to a second belt for cheddaring. The second belt carries the cheddared curd to the mill. The milled curd is then carried to a finishing table where it is salted, stirred, and moved out for hooping.

Another system, imported from Australia, is used in a number of cheddar cheese factories. This system requires a short method of setting. After the curd is cut and cooked, it is transferred to a series of perforated stainless steel troughs traveling on a conveyor where draining and partial fusion take place. The slabs are then transferred into buckets of a forming conveyor, transferred again to transfer buckets, and finally to compression buckets where cheddaring takes place. Cheddared slabs are discharged to a slatted conveyor which carries them to the mill and then to a final machine where the milled curd is salted, weighed, and hooped.

Curing. Curing temperature and time vary widely among cheddar plants. A temperature of 50 °F cures the cheddar more rapidly than lower temperatures. The higher the temperature above 50 °F to about 80 °F, the more rapid the curing and the more likely that off-flavors will develop. At 50 °F, 3 to 4 months are required for a mild to medium cheddar flavor. Six months or more are necessary for an aged (sharp) cheddar cheese. Relative humidity should be roughly 70%. Cheddar intended for processed cheese is cured in many plants at 70 °F because of the economy of time. Some experts suggest that the cheddar, after its coating or wrapping, should be held in cold storage at approximately 40 °F for about 30 days, then transferred to the 50 °F curing room. During cold storage, the curd particles knit together forming a close-bodied cheese. The small amount of residual lactose is slowly converted to lactic acid, along with other changes in optimum curing.

The maximum legal moisture content of cheddar is 39% and the fat must be not less than 50% of total solids. The amount of

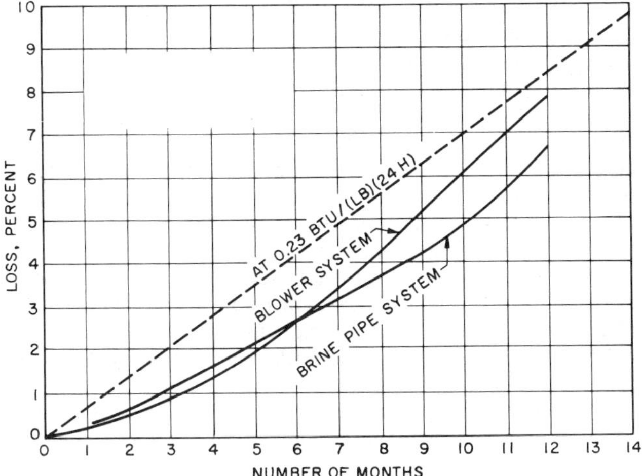

Fig. 5 Shrinkage of Cheese in Storage

moisture directly affects the curing rate to some extent within the normal range of 34 to 39%. Cheese having a loose or crumbly body and a high acidity is less likely to cure properly. For best curing, the cheese should have a sodium chloride content of 1.5 to 2.0%. A lower percentage encourages off-flavors to develop, and higher amounts retard flavor development.

Moisture Losses. The loss in weight of cheese during curing is largely attributed to moisture loss. Paraffined cheddar cheese going into cure averages approximately 37% moisture. After a 12-month cure at 40°F, paraffined cheese averages approximately 33% moisture. This loss is a real loss to the cheese manufacturer unless the cheese is sold on the basis of total solids. Control of humidity can have an important role in moisture loss. Figure 5 shows the loss from paraffined longhorns in boxes held at 38°F and 70% rh over 12 months. The conditions were well controlled but the average loss was 7%. The high loss shown on the graph was influenced by the larger surface area in 12-lb longhorns, as compared to 70-lb cheddars. Curing the cheese within a good quality sealed wrapper having a low moisture transmission (but some oxygen and carbon dioxide) largely eliminates loss of moisture.

Provolone (Pasta Filata Types)

Provolone is an Italian plastic-curd-type cheese which represents a large group of pasta-filata-type cheeses. While these cheeses vary widely in size and composition, they are all manufactured by a similar method. After the curd has been matted, like cheddar, it is cut into slabs which are worked and stretched in hot water at 150 to 180°F. The curd is kneaded and stretched in the hot water until it reaches a temperature of about 135°F. The maker then takes the amount necessary for one cheese and folds, rolls, and kneads it by hand to give the cheese its characteristic shape and smooth, closed surface. Molding machines have been developed for large scale operations to eliminate this hand labor. The warm curd of some varieties of pasta filata is placed in molds and submerged in cold water to harden into the desired shape. The hardened cheese is then salted in brine for from one to several days depending on the size and variety. Some pasta filata cheese, such as mozzarella for pizza, is packaged for shipment with wrappers to protect it for the period it is held before use. This cheese may be sealed under vacuum in plastic bags for prolonged holding.

Provolone is salted by submersion in 24% sodium chloride solution at 45°F for 1 to 3 days, depending on the size. It is hung on a light rope to dry and then transferred to the smokehouse and exposed to hickory or other hardwood smoke for 1 to 3 days. The cheese is hung in a curing room for 3 weeks at 55°F and then for 2 to 10 months at 40°F. The size varies as well as the shape, but the most common in the United States is 14 lb and pear-shaped. The moisture content ranges from 37 to 45% and the salt from 2 to 4%. Milkfat usually comprises 46 to 47% of the total solids. The yield is roughly 9.5 lb per 100 lb of milk.

Swiss Cheese

One of the distinguishing characteristics of Swiss cheese is the eye formation during curing. These eyes result from the development of CO_2. Raw or heat-treated milk is tempered to 95°F and pumped to a large kettle or vat. The starter, consisting of 27 mL of *Propioni bacterium shermanii*, 165 mL of *Streptococcus thermophilus*, and 165 mL of *Lactobacillus bulgaricus*, is added per 1100 lb of milk. After mixing, 77 mL of rennet per 1100 lb is diluted 1:40 with water and slowly added with agitation of the milk. Curd is cut when firm (after 25 to 30 min) into very small granules. After 5 min, curd and whey are agitated for 40 min, and then the steam is released into the jacket without water. Curd is heated slowly to 122 to 130°F in 30 to 45 min. Without additional steam, the cooking is continued until curd is firm and has no tendency to stick when a group of particles is squeezed together (0.5 to 1 h and whey pH 6.3). Curd is dipped into hoops (160 lb) and pressed lightly for 6 h, redressing and turning the hoops every 2 h. Pressing is continued overnight. The next step consists of soaking the cheese in brine until it has about 1.5% salt. Table 4 shows temperature and time at which curing occurs.

Table 4 Swiss Cheese Manufacturing Conditions

Processing Step	Temperature, °F	Relative Humidity, %	Time
Setting	95	—	0.4 to 0.5 h
Cooking	122 to 130	—	1.0 to 1.5 h
Pressing	80 to 85	—	12 to 15 h
Salting (brine)	50 to 52	—	2 to 3 days
Cool room hold	50 to 60	90	10 to 14 days
Warm room hold	70 to 75	80 to 85	3 to 6 weeks
Cool room hold	40 to 45	80 to 85	4 to 10 mos

Roquefort and Blue Cheese

Roquefort and blue cheese require a mold (*Penicillium roqueforti*) to develop the typical flavor. Roquefort is made from ewes' milk in France. Blue cheese in the United States is made from cow's milk. The equipment used for the manufacture and curing of blue cheese is the same as that used for cheddar with a few exceptions. The hoops are 7.5 in. in diameter and 6 in. high. They have no top or bottom covers and are thoroughly perforated with small holes. A manually or pneumatically operated device with 50 needles, which are 6 in. long and 0.125 in. in diameter, is used to punch holes in the blue curd wheels. Apparatus is also needed to feed moisture into the curing room to maintain at least 95% rh without causing a drip onto the cheese.

The milk may be raw or pasteurized and separated. The cream is bleached and may be homogenized at low pressure. Skim milk is added to the cream, and the milk is set with 2 to 3% active lactic starter. After 30 min, 3 to 4 oz of rennet per 1100 lb is diluted with water (1:35) and thoroughly mixed into the milk. When the curd is firm (after 30 min), it is cut into 5/8-in. cubes. Agitation is begun 5 min later. After whey acidity is 0.14% (1 h), the temperature is raised to 92°F and held for 20 min. The whey is drained and trenched. Approximately 2 lb of coarse salt and 1 oz of *P. roqueforti* powder is mixed into each 100 lb of curd.

The curd is transferred to stainless steel perforated cylinders (hoops). These hoops are inverted each 15 min for 2 h on a drain cloth, and curd matting is continued overnight. The hoops are removed and surfaces of the wheels covered with salt. The cheese is placed in a controlled room at 60°F and 85% rh and resalted

Table 5 Typical Blue Cheese Manufacturing Conditions

Processing Step	Temperature, °F	Relative Humidity, %	Time
Setting	85 to 86	—	1 h
Acid development	85 to 86	—	1 h
(after cutting)	92	—	120 s
Curd matting	70 to 75	80 to 90	18 to 24 h
Dry salting	60	85	5 days
Curing	50 to 55	>95	30 days
Additional curing	36 to 40	80	60 to 120 days

daily for 4 more days (total 5 days). Small holes are punched through the wheels of cheese from top to bottom of the flat surfaces. The cheese is placed in racks on its curved edge in the curing room and held at 50 to 55 °F and not less than 95% humidity. At the end of the month, the cheese surfaces are cleaned; the cheese is wrapped in foil and placed in a 36 to 40 °F cold room for 2 to 4 months (Table 5). The surfaces are again scraped clean, and the wheels are wrapped in new foil for distribution.

Originally, Roquefort and blue cheese were cured in caves with high humidity and constant cool temperature. The refrigeration of insulated blue cheese curing rooms to the optimum temperature is not difficult. However, maintaining a uniform relative humidity of not less than 95% without excessive expense seems to be an engineering challenge, at least in some plants.

Cottage Cheese

Cottage cheese is made from skim milk. It is a soft, unripened curd and generally has a cream dressing added to it. There are small and large curd types, and they may or may not have added fruits or vegetables. The cheese plant equipment may consist of receiving apparatus, storage tanks, clarifier-separator, pasteurizer, cheese vats with mechanical agitation, curd pumps, drain drum, blender, filler, conveyors, and such accessory items as refrigerated trucks, laboratory testing facilities, and whey disposal equipment. The largest vats have a 45,000-lb capacity. The basic steps are separation, pasteurization, setting, cutting, cooking, draining and washing, creaming, packaging, and distribution.

Skim milk is pasteurized at the minimum temperature and time of 161 °F for 15 s to avoid adversely affecting the curd properties. If substantially higher heat treatment is practiced, the cottage cheese manufacturing procedure must be altered to obtain good body and texture quality and reduce curd loss in the whey. Skim milk is cooled to the setting temperature, which is 86 to 90 °F for the short set (5 to 6 h) and 70 to 72 °F for the overnight set (12 to 15 h). A medium set is used in a few plants. For the short set, 5 to 8% of a good cultured skim milk (starter) and 1 to 1.5 mL of rennet diluted in water are added per 1000 lb of skim milk. For the long set, 0.25 to 1% starter and 0.5 to 1 mL of diluted rennet per 1000 lb are thoroughly mixed into the skim milk. The use of rennet is optional. The setting temperature is maintained until the curd is ready to cut. The acidity of whey at cutting time depends on the total solids content of the skim milk (0.55% for 8.7% and 0.62% for 10.5%). The pH is typically 4.80, but it may be necessary to adjust for specific make procedures.

The curd is cut into 0.5-in. cubes for large curd and 0.25 in. for small curd cottage cheese. After the cut curd sets for 10 to 15 min, heat is applied to water in the vat jacket to maintain a temperature rise in the curd and whey of 2 °F each 5 min. In very large vats, jacket heating is not practical, and superheated culinary steam in small jet streams is used directly in the vat; 20 to 30 min after cutting, very gentle agitation is applied. Heating rate may be increased to 3 or 4 °F per 5 min as the curd firms enough to resist shattering. Cooking is completed when the cubes contain no whey pockets and have the desired firmness. The final temperature of curd and whey is usually 120 to 130 °F, but some cheesemakers heat to 145 °F when making the small type curd.

After cooking is completed, the hot water in both the jacket and the whey is drained. Wash water temperature is adjusted to about 70 °F for the first washing and added gently to the vat to reduce curd temperature to 80 to 85 °F. After gentle stirring and a brief hold, the water-whey mixture is drained. The temperature of the second wash is adjusted to reduce the temperature of curd to 50 to 55 °F and to 40 °F with a third wash. Water for the last wash may have 3 to 5 ppm of added chlorine. The curd is trenched for adequate drainage. The dressing is made from low-fat cream, salt, and usually 0.1 to 0.4% stabilizer based on cream weight. Salt averages 1%, and milkfat must be 4% or more in creamed cottage cheese or 2% in low-fat cottage cheese. The dressing is cooled to 40 °F and blended into the curd.

A cheese vat can be reused sooner if the cheese pumps quickly convey the curd and whey at the completion of cooking to a special tank for drainage of the whey, washing, and blending of the dressing with the curd. Creamed cottage cheese is transferred mechanically to an automatic packaging machine. One type of filler employs an oscillating cylinder which holds a specific volume. Another type has a piston in a cylinder which discharges a definite volume. The common retail containers are 32-, 16-, 12-, and 8-oz sizes of semirigid plastic. Cottage cheese is perishable and must be stored at 40 °F or lower for prolonging the keeping quality to 2 or 3 weeks. A good yield is 15.5 lb of curd per 100 lb of skim milk with 9% total solids.

Other Cheeses

Table 6 presents data on a few additional common varieties of cheese in the United States. With the exception of the soft ripened cheeses such as Camembert and Liederkranz, freezing of cheese results in undesirable texture changes. This can be serious, as in the case of cream cheese, where a mealy, pebbly texture results. Other types, such as brick and Limburger, undergo a slight roughening of texture which is undesirable but which still might be acceptable to certain consumers. As a general rule, cheese should not be subjected to temperatures under 29 °F.

When cured cheese is held above the melting point of milkfat, it becomes greasy because of the oiling off. The oiling-off point of all types of cheese except process cheese begins at 68 to 70 °F. Consequently, storage should be substantially below the melting point (Table 7). Uncured cheese (*e.g.*, cottage, cream) is highly perishable and thus should not be stored above 45 °F and preferably at 35 °F.

Processing protects cheese from oiling off. By heating the bulk cheese to temperatures of 140 to 180 °F, and through the incorporation of emulsifying salts, a more stable emulsion is formed than in natural or nonprocessed cheese. Process cheese will not oil off even at melting temperatures. Because of the temperatures used in processing, process cheese is essentially a pasteurized product. Microorganisms causing changes in the body and flavor of the cheese during cure are largely destroyed; hence there will be practically no further flavor development. Consequently, the maximum permissible temperature for the storage of process cheese is

Table 6 Curing Temperature, Humidity, and Time of Some Cheese Varieties

Variety	Curing Temperature, °F	Relative Humidity, %	Curing Time
Brick	60 to 65	90	60 days
Romano	50 to 60	85	5 to 12 mos
Mozzarella	70	85	24 to 72 h
Edam	50 to 60	85	3 to 4 mos
Parmesan	55 to 60	85 to 90	14 mos
Limburger	50 to 60	90	2 to 3 mos

Table 7 Temperature Range of Storage, Common Types of Cheese

Cheese	Ideal Temperature, °F	Maximum Temperature, °F
Brick	30 to 34	50
Camembert	30 to 34	50
Cheddar	30 to 34	60
Cottage	30 to 34	45
Cream	32 to 34	45
Limburger	30 to 34	50
Neufchatel	32 to 34	45
Process American	40 to 45	75
Process brick	40 to 45	75
Process Limburger	40 to 45	75
Process Swiss	40 to 45	75
Roquefort	30 to 34	50
Swiss	30 to 34	60
Cheese foods	40 to 45	55

considerably higher than any of the other types. Table 7 shows the maximum temperatures of storage for cheese of various types.

Refrigeration of Cheese Rooms

Cheeses that are to be dried prior to wrapping or waxing enter the cooler at approximately room temperature. Sufficient refrigerating capacity must be provided to reduce the cheeses to drying room temperature. The product load may be taken as 1 ton for each 12,000 to 15,000 lb per day. Product load in a cheese-drying room is usually small compared to total room load. Extreme accuracy in calculating product load is not warranted.

When determining the peak refrigerating load in a cheese drying room, the factor to be remembered is that peak cheese production may coincide with periods of high ambient temperature. In addition, these rooms normally open directly into the cheese-making room, where both temperature and humidity are quite high. Also, traffic in and out of the drying room may be heavy; therefore, ample allowance for door losses should be made. Two to three air changes per hour are quite possible during the flush season.

To maintain the desired humidity, refrigerating units for the cheese-drying room should be sized to handle the peak summer load with not more than a 20°F difference between average air temperature and evaporator temperature. Units operated from a central refrigerating system should be equipped with a back pressure regulator.

Temperature control is best obtained through a room thermostat controlling a solenoid valve in the liquid supply to the unit or units. Fans should be allowed to run continuously. The modulating back pressure regulator is not a satisfactory temperature control for a cheese-drying room because it causes undesirable variations in humidity.

Air circulation should only be sufficient to assure uniform temperature and humidity throughout the room. Strong drafts or air currents should be avoided because they cause uneven drying and cracking of the cheeses. The most satisfactory refrigerating units are the ceiling-suspended between-the-rails type or the penthouse type. One unit for each 400 to 500 ft^2 of floor area usually assures uniform conditions. One unit should be placed near the door to the room to cool the warm moist air before it has a chance to spread over the ceiling. Otherwise, condensation dripping from the ceiling or mold growth will result.

Humidity control during winter presents certain problems. Since most of the refrigeration load (during the peak season) is due to insulation losses and warm air entering through the door, refrigerating units may not operate enough during cold weather to take up the moisture given up by the cheese, resulting in excess humidity and improper drying of the cheese. Within certain limits, this can be overcome by reducing the speed of the unit fans, lowering the suction pressure, and increasing the number of defrosts. If there are several units in the room, the refrigeration may be turned off on some. Fans should be left running to assure uniform conditions throughout the room. If these adjustments are not sufficient, or if automatic control of humidity is desired, it is necessary to use reheat coils in the airstream leaving the units. These may be electric heaters, steam or hot water coils, or hot gas from the refrigerating system. A heating capacity of 15 to 20% of the refrigerating capacity of the units is usually sufficient.

A humidistat may be used to operate the electric heater when the humidity rises above the desired level. The heater should be wired in series, with a second room thermostat set to shut it off if the room temperature becomes excessive. Because of variation in size and shape of drying rooms, it is impossible to generalize on air velocities and capacities. Airflow should be regulated so that the cheese feels moist for the first 24 h, after which it becomes progressively drier and firmer.

Calculation of the refrigeration load for a cheese curing room involves a simple computation of heat to be removed from the cheese at the incoming temperature in order to bring it to curing temperature, using 0.65 Btu/(lb·°F) as the specific heat of cheese. For most varieties, the heat given off during cure is negligible.

While the fermentation of lactose to lactic acid which occurs in cheese is an exothermic reaction, this process is substantially completed in the first week after cheese is made; hence further heat given off during cure is of no significance. Assuming that average conditions for American cheese curing are approximately 45°F and 70% rh, if 30 to 35°F refrigerant is used in the cooling system, a humidity of about 70% will be maintained.

FROZEN DAIRY DESSERTS

Ice cream is the most common frozen dairy dessert. Legal guidelines for the composition of frozen dairy desserts generally follow the Federal Standards. The amount of air incorporated during freezing is controlled for the prepackaged products by the standard specifying the minimum density and/or a minimum density of food solids.

The basic dairy components of frozen dairy dessert are milk, cream, and condensed or nonfat dry milk. Some plants also use butter, butter oil, buttermilk (liquid or dry), and dry or concentrated whey of the sweet type. The acid type whey (*e.g.*, from cottage cheese) can be used for sherbets.

Ice Cream

Milkfat content (called butterfat by some standards) is one of the principal factors in the legal standards for ice cream. The fat in other ingredients such as eggs, nuts, cocoa, or chocolate do not satisfy the legal minimum. Federal standards set the minimum milkfat content at 8% for bulky flavored ice-cream mixes (*e.g.*, chocolate) and 10% or above for the other flavors (*e.g.*, vanilla). Manufacturers, however, usually make two or more grades of ice cream, one being competitively priced with the minimum legal fat content, and the others richer in fat, higher in total solids, and lower in overrun for a special trade. Such ice cream may be made with a fat content of 16 or 18%, although most ice cream is made with a fat content ranging from 10 to 12%.

Serum solids content designates the nonfat solids from milk. The chief components of milk serum are lactose, milk proteins (casein, albumin, and globulin), and milk salts (sodium, potassium, calcium, and magnesium as chlorides, citrates, and phosphates). The following average composition percentages for serum solids are useful for general calculations: lactose, 54.5; milk proteins, 37.0; and milk salts, 8.5.

The serum solids in ice cream produce a smoother texture, better body, and better melting characteristics. Because serum solids are relatively inexpensive compared with fat, they are used liberally. The total solids content usually is kept below 40%.

The lower limit in the serum solids content, 6 to 7%, is found in a homemade type of ice cream, where the only dairy ingredients are milk and cream. Ice creams with an unusually high fat content are also kept near this serum solids content so that the total solids content will not be excessive. Most ice cream, however, is made with condensed or nonfat dry milk added to bring the serum solids content within the range of 10 to 11.5%. The upper extreme of 12 to 14% serum solids can be used only where rapid product sales turnover or other special means avoids sandiness.

The sugar content of ice cream is of special interest because of its effect on the freezing point of the mix and its hardening behavior. The extreme range of sugar content encountered in ice cream is from 12 to 18%, with 16% being most representative of the industry. The chief sugar used is sucrose (cane or beet sugar), either in granulated or liquid form. Many manufacturers use dextrose and corn syrup solids to replace part of the sucrose. Some manufacturers prefer sucrose in liquid form, or in a mixture with syrup, because of lower cost and easier handling in tank car lots. In some instances, 50% of the sucrose content has been replaced by other sweetening agents. A more common practice is to replace from one-fourth to one-third of the sucrose by dextrose or corn syrup solids, or a combination of the two.

Practically all ice cream is made with a stabilizer to help maintain a smooth texture, especially under the conditions that prevail in retail cabinets. Manufacturers who do not use stabilizers offset this omission by a combination of factors such as a high fat and solids content, the use of superheated condensed milk to aid in smoothing the texture and imparting body, and a sales program designed to provide rapid turnover. The stabilizing substances most commonly added are carboxymethylcellulose (CMC) and sodium alginate, a product made from giant kelp gathered off the coast of California. Gelatin is used for some ice-cream mixes that are to be batch pasteurized. Other stabilizers are locust or carob bean gum, gum arabic or acacia, gum tragacanth, gum Karaya, psyllium seed gum, and pectin. The amount of stabilizer commonly used in ice cream ranges from 0.20 to 0.35% of the mass of the mix.

Many plants now combine an emulsifier with the stabilizer to produce a smoother and richer product. The emulsifier reduces the surface tension between the water and fat phase to produce a drier-appearing product.

Egg solids in the form of fresh whole eggs, frozen eggs, or powdered whole eggs or egg yolks are used by some manufacturers. While flavor and color may motivate this choice, the most common reason for selecting them is to aid the whipping qualities of the mix. The amount required is about 0.25% egg solids, with 0.50% being about the maximum content for this purpose. To obtain the desired result, the egg yolk should be in the mix at the time it is being homogenized.

In frozen custards or parfait ice cream, the presence of eggs in liberal amounts and the resulting yellow color are identifying characteristics. Federal standards specify a minimum 1.4% egg yolk content for these products.

Milk Ices or Ice Milk

Ice milk commonly contains 3 to 4% fat and 13 to 15% serum solids; formulations with respect to sugar and stabilizers are similar to those for ice cream. The sugar content in ice milk is somewhat higher in order to build up the total solids content. The stabilizer content is also higher in proportion to the higher water content of ice milk. The overrun is approximately 70%.

Soft Ice Milk or Ice Cream

Machines that serve freshly frozen ice cream are common in roadside stands, retail ice-cream stores, and restaurants. These establishments must meet sanitation requirements and have facilities for proper cleaning of the equipment, but very few blend and process the ice-cream mix used. The mix is usually supplied either from a plant specializing in producing ice-cream mix only or from an established ice-cream plant. The mix should be cooled to about 35°F at the time of delivery, and the ice-cream outlet should have ample refrigerated space to store the mix until it is frozen. To be served in a soft condition, this ice cream mix is usually frozen stiffer than would be customary for a regular plant operation with a 30 to 50% overrun. Some mixes are prepared only for soft serve. They are 1 to 2% greater in serum solids and have 0.5% stabilizer-emulsifier to aid in producing a smooth texture. The overrun is limited to 30 to 40% during freezing to the softserve condition.

Frozen Yogurt

Hard- or soft-serve frozen yogurt is similar to low-fat ice cream in composition and processing. The significant exception is the presence of a live culture in the yogurt.

Sherbets

Sherbets are fruit (and mint) flavored frozen desserts characterized by their high sugar content and tart flavor. They must contain between 1 and 2% milkfat and not more than 5% by mass of total milk solids. While the milk solids can be supplied by milk, the general practice is to supply them by using ice-cream mix. Typically, a solution of sugars and stabilizer in water is prepared as a base for sherbets of various flavors. To 70 lb of sherbet base, 20 lb of flavoring and 10 lb of ice-cream mix are added. The sugar content of sherbets ranges from 25 to 35%, with 28 to 30% being most common. One example of a sherbet formula is: milk solids, 5%, of which 1.5 is milkfat; sugar, 13%; corn syrup solids, 22%; sherbet stabilizer, 0.3%; and flavoring, acid, and water, 59.7%.

In sherbets, and even more so in ices, a high overrun is not desirable because the resulting product will appear foamy or spongy under serving conditions. The overrun should be kept within 25 to 40%. This fact and the problem of preventing bleeding (the leakage of syrup from the frozen product) emphasize the importance of the choice of stabilizers. If gelatin is selected as the stabilizer in sherbets, the freezing conditions must be managed to avoid an excessive overrun. The gums added to ice cream are commonly used as the stabilizer in sherbets and ices.

Ices

Ices contain no milk solids, but closely resemble sherbets in other respects. To offset the lack of solids from milk, the sugar content of ices is usually slightly higher than in sherbets; it is usually 30 to 32%. A combination of sugars should be used to prevent crusty sugar crystallization, just as in the case of sherbets. The usual procedure is to make an ice-base solution of the sugars and stabilizer, from which different flavored ices may be prepared by adding the flavoring material in the same general manner as mentioned for sherbets.

Ices contain few ingredients with lubricating qualities and often cause extensive wear on scraper blades in the freezer. Frequent resharpening of the blades is necessary. Where a number of freezers are available, and the main production is ice cream, it is desirable to confine the freezing of ices and sherbets to a specific freezer or freezers, which should then receive special attention to resharpening.

Making the Ice-Cream Mix

The chosen composition for a typical ice cream would be:

Fat	12.5%	Sugar	15.0%
Serum solids	10.5%	Stabilizer	0.3%

Mixing and Pasteurizing. Generally, the liquid dairy ingredients are placed in a vat equipped with suitable means of agitation, especially to keep the sugar in suspension until it is dissolved. The dry ingredients are then added, with suitable precautions to prevent lump formation of such products as stabilizers, nonfat dry milk solids, powdered eggs, and cocoa. Gelatin should be added while the temperature is still low to allow time for the gelatin to imbibe water before its dissolving is promoted by heat. Dry ingredients that tend to form lumps may be successfully added by first mixing them with some of the dry sugar so that moisture may penetrate freely. Where vat agitation is not fully adequate, sugar may be withheld until the liquid portion of the mix is partly heated so that promptness of solution avoids settling out.

The mix is pasteurized to destroy any pathogenic organisms, to lower the bacterial count so as to enhance the keeping quality of the mix and comply with bacterial count standards, to dissolve the dry ingredients, and to provide a temperature suitable for efficient homogenization. A pasteurizing treatment of 155 °F maintained for 0.5 h is the minimum allowed. The mix should be homogenized at the pasteurizing temperature. Vat batches should be homogenized in 1 h and preferably less.

Many ice cream plants use continuous pasteurization methods that use plate-type heat exchange equipment for heating and cooling the mix. If some solid ingredients are selected, such as skim milk powder and granulated sugar, a batch is made up in a mixing tank at a temperature of 100 to 140 °F. This preheated mix is then pumped through a heating section of the plate unit where it is heated to a temperature of 175 °F or higher, and held for 25 s while passing through a holding tube. The mix is then homogenized and pumped to the precooling plate section using city, well, or cooling tower water as the cooling medium. Final cooling of the mix may be done in an additional plate section, using chilled water as the cooling medium, or the mix may be cooled through a separate mix cooler. A glycol medium is sometimes used for cooling to temperatures just above the freezing point.

Large plants generally use all liquid ingredients, especially if the production is automated and computerized. The ingredients are blended at 40 to 60 °F. The mix passes through the product-to-product regeneration section of a plate-type heat exchanger with about 70% regeneration during preheating. The mix is HTST heated to not less than 175 °F, homogenized, and held for 25 s. Greater heat treatment is common, and 220 °F for 40 s is not unusual. The final heating may be accomplished with plate equipment, a swept-surface heat exchanger, or a direct steam injector or infusor.

Steam injection and infusion equipment may be followed by vacuum chamber treatment, whereby the mix is flash cooled to 180 to 190 °F by a partial vacuum. It is further cooled through a regenerative plate section and additionally cooled indirectly to 40 °F or less with chilled water. The chief advantage of the vacuum treatment is the flavor improvement of the mix if prepared from raw materials of questionable quality.

Homogenizing the Mix. Homogenization disperses the fat in a very finely divided condition so it will not churn out during freezing. Most of the fat in milk and cream is in globules 3 to 7 μm in diameter. Some of the globules are as large as 12 μm or larger in diameter, especially if there has been some churning incidental to handling. In a properly homogenized mix, globules are seldom over 2 μm in diameter.

Cooling and Holding Mix. Methods of final cooling of ice-cream mix after pasteurization depend on the equipment used and the final mix temperature desired. The mix should be as cold as possible, to about 30 °F minimum for greater capacity and less refrigeration load on the ice cream freezers. Smaller plants generally use the vat holding system of pasteurization with either a Baudelot-type surface cooler or a plate-type heat exchanger, both with precooling and final cooling sections. The precooling may be done with city, well, or cooling tower water, and mix leaving the precooling section is about 10 °F warmer than the entering water temperature. The Baudelot-type cooler may be arranged for final cooling with chilled water, brine, or direct expansion refrigerant. A final mix temperature of 30 to 33 °F can be obtained over the surface cooler using brine or refrigerant. Final mix temperature when using chilled water will be about 40 °F.

For larger ice-cream plants, where low mix temperature is desired and where plate-type pasteurizing equipment is installed, it may be desirable to use separate equipment for the final cooling. Where the mix is preheated to about 140 °F, it will be precooled to about 10 °F warmer than the entering precooling water temperature; final cooling can be done in a remote cooler. An ammonia jacketed, scraped surface chiller is often selected. Where cold liquid mix is used through a continuous pasteurizing, high-heat vacuum system with regeneration at about 70%, the temperature of the mix to the final cooling unit will be 85 °F, assuming 40 °F original mix temperature and 190 °F temperature of mix returning through the regenerating section.

Where plants have ample ice-cream mix holding tank capacity (enabling the mix to be held overnight), part of the final mix cooling may be accomplished by means of a refrigerated surface built into the tanks. Using refrigerated mix holding tanks, the average rate of cooling may be estimated at 1 °F/h. Mixes made up with gelatin as a stabilizer should be aged 24 h to allow time for the full set of the gelatin to develop. Mixes made with sodium alginate or other vegetable-type stabilizers develop maximum viscosity on being cooled, and can be used in the freezer immediately.

Freezing Procedure

The ice-cream freezer freezes the mix to the desired consistency and whips in the desired amount of air in a finely divided condition. The aim is to conduct the freezing and later hardening to obtain the smoothest possible texture.

Freezing an ice-cream mix means, of course, freezing a mixed solution. The solutes that determine the freezing point are the lactose and soluble salts contained in the serum solids and the sugars added as sweetening agents. The other constituents of the mix affect the freezing point only indirectly, by displacing water and affecting the in-water concentration of the solutes mentioned. Leighton (1927) developed a reliable method for computing the freezing points of ice-cream mixes from their known composition. He added the lactose and sucrose content of the mix, expressed their concentration in terms of parts of sugar per 100 parts of water, and determined the freezing-point depression due to the sugars by reference to published data for sucrose. This computation is justified because lactose and sucrose have the same molecular mass.

$$\%\ \text{lactose in mix} = 0.545\ (\%\ \text{serum solids})$$

$$\frac{(\%\ \text{lactose} + \%\ \text{sucrose})\ 100}{\%\ \text{water in mix}} = \begin{array}{l}\text{parts lactose + sucrose}\\ \text{per 100 parts water}\end{array}$$

To the freezing point depression caused by these sugars, he added the depression that will be caused by the soluble milk salts. The depression caused by the salts is computed as follows:

Freezing-point depression caused by salt solids in °F
$= 4.27\ (\%\ \text{serum solids})/\%\ \text{water in mix}$

Table 8 presents the freezing points of various ice creams and a typical sherbet and an ice, as computed by Leighton's method. The freezing point represents the temperature at which freezing commences. As in the case of all solutions, the unfrozen portion becomes more concentrated as the freezing progresses, and the freezing temperature therefore decreases as the freezing progresses. In a simple solution, containing only one solute, this trend will progress until the unfrozen portion represents a saturated solution of the solute, and thereafter the temperature will remain constant until the freezing has been completed. This temperature is known as the cryohydric point of the solute. In a mixed solution such as ice cream, which contains several sugars and a number of salts, no such point can be recognized.

On the contrary, the sugars remain in solution in a supersaturated state in the unfrozen portion of the product. This is due to the fact that by the time the saturation point has been reached, the temperature is so low and the viscosity so high that essentially a glass state exists. In a mixed solution, however, the temperature required for complete freezing must be somewhat below the cryohydric point of that solute with the lowest cryohydric point. In ice cream, that solute is calcium chloride, contained as a component of serum solids. The cryohydric point of calcium chloride is −59.8 °F. Therefore, the ice cream ranges from 0 to 100% frozen within the approximate range of 27.5 to −67 °F.

Therefore, the temperature to which the ice cream has been frozen becomes a measure of the degree to which it has been frozen, as illustrated by Table 9. In the table, the freezing points of the unfrozen portions of the third ice cream listed in Table 8 have been computed when 0 to 90% of the original water has been frozen out as ice.

Refrigeration Requirements. Exact calculation of refrigeration requirements is complicated by the number of factors involved. The specific heat of the mix varies with its composition. According to Zhadan (1940), the specific heat of food products may be computed by assuming the following specific heats in Btu/(lb·°F) for the chief components: carbohydrates, 0.34; proteins, 0.37; fats, 0.40; and water, 1.00. Salts are normally not included. Where they are present in significant amounts, as in ice cream (9.5% of the serum solids), a specific heat of 0.20 is accurate. The value given by Zhadan for fats is apparently for solid fats. For milkfat in a liquid condition, Hammer and Johnson (1913) found the specific heat to be 0.52. In addition, their data clearly show that the latent heat of fusion of fats becomes involved. From their data, the latent heat of fusion of milkfat is about 35 Btu/lb.

The change from liquid to solid fat occurs over a wide temperature range, approximately 80 to 40 °F; in changing from solid to liquid fat, the range is approximately 50 to 105 °F. This wide discrepancy between solidifying and melting behavior is apparently due to the fact that milkfat is a mixture of glycerides, and mutual solubility of the glycerides is involved. In any case, the latent heat of fusion of the fat is involved in cooling the mix from the pasteurizing and homogenizing temperature down to the aging temperature of 38 to 40 °F. Instead of making detailed calculations, a specific heat of 0.80 Btu/(lb·°F) is assumed for ice-cream mix, which is high for mixes ranging from 36 to 40% total solids.

In calculating the refrigeration required for freezing and hardening, a single value of a specific heat for frozen ice cream cannot be chosen. As shown in Table 9, any change in temperature in freezing and hardening involves some latent heat of fusion of the water, as well as the sensible heat of the unfrozen mix and the ice. Near the initial freezing point, much more latent heat of fusion is involved per degree temperature change than in well-hardened ice cream, *e.g.*, at −10 to −11 °F. For this reason, instead of using an overall value of specific heat, freezing load may be computed as follows:

1. First determine the temperature to which the freezing is to be carried; then determine (by calculations such as those used to develop Table 9) how much water will be converted to ice. The heat to be removed is the product of the heat of fusion of ice and the mass of water frozen.
2. Compute the sensible heat that must be removed in the desired temperature change, by treating the product as a mix; *i.e.*, use the specific heat for ice-cream mix. The temperature change times the lb of product times 0.80 = sensible heat to be removed.

In such a calculation, the water present is treated as though it all remained in a liquid form until the desired temperature had been reached, although ice was forming progressively. Because ice has a specific heat of 0.492 instead of 1.00 as for water, this calculation will err in the direction of generous refrigeration. To offset this, the freezer agitation develops friction heat. Approximately 80% of the energy input in the motor of the freezer is converted to heat in the product. Where the product is frozen to a stiff consistency, power requirements increase, and should be added to the load calculation.

A gallon of ice-cream mix weighs from 9 lb for mixes with a high fat content, to 9.2 lb for mixes with a low fat content and a high content of serum solids and sugar. The mass of a unit volume of ice cream varies with the mix and overrun according to the following relationship:

$$\text{Percentage overrun} = 100\,\frac{(\text{Wt/gal of mix} - \text{Wt/gal of ice cream})}{\text{Wt/gal of ice cream}}$$

Freezing Ice Cream. Both batch and continuous ice-cream freezers are in general use. Both are arranged with a freezer cylinder having either an annular space or coils around the cylinder, where cooling is accomplished by direct refrigerant cooling, either in a flooded arrangement with an accumulator or controlled by

Table 8 Freezing Points of Typical Ice Creams, Sherbet, and Ice

	Composition of the Mix, %					
	Fat	Serum Solids	Sugar	Stabilizer	Water	Freezing Point, °F
Ice cream	8.5	11.5	15	0.4	64.6	27.59
	10.5	11.0	15	0.35	63.15	27.57
	12.5	10.5	15	0.30	61.7	27.55
	14.0	9.5	15	0.28	61.22	27.68
	16.0	8.5	15	0.25	60.25	27.79
	10.5	8.4	S 12, D 4	0.40	64.7	27.39
Sherbet	1.2	1.0	S 22, D 8	0.50	67.3	25.97
Ice	0	0	S 23, D 9	0.50	67.5	25.68

S = Sucrose D = Dextrose

Table 9 Freezing Behavior of a Typical Ice Cream

Water Frozen to Ice, %	Freezing Point of Unfrozen Portion, °F	Water Frozen to Ice, %	Freezing Point of Unfrozen Portion, °F
0	27.55	40	24.40
5	27.35	45	23.63
10	27.05	50	22.62
15	26.78	55	21.42
20	26.40	60	19.79
25	26.04	70	14.99
30	25.70	80	5.14
35	25.03	90	−22.29

Composition, %: fat, 12.5; serum solids, 10.5; sugar, 15; stabilizer, 0.30; water, 61.7.

a thermostatic expansion valve. The freezer cylinder has a dasher, which revolves within the cylinder. Sharp metal blades on the dasher scrape the inner surface of the cylinder to remove the frozen film of ice cream as it forms. Some batch freezers use plastic dashers and blades.

Batch freezers range in size from 2 to 40 quarts of ice cream per batch, the smaller sizes being used for retail or soft ice-cream operations, and the 40-quart size used in small commercial ice-cream plants or in large plants for running small special-order quantities. Batch freezers larger than the 40-quart size have not been used extensively since the development of the continuous freezer.

In operation, a measured quantity of mix is placed in the freezer cylinder and the required flavor, fruit, or nuts are added as freezing of the mix progresses. Freezing is continued until the desired consistency is obtained in the judgment of the operator or by the indication of a meter showing an increase in the amperage drawn by the motor as the partly frozen mix becomes stiffer. At the desired point of freezing, the refrigeration is cut off from the freezer cylinder, usually by closing the refrigerant suction valve. The dasher continues operating until enough air has been taken into the mix by the whipping action to produce the desired overrun. The overrun is checked by taking and weighing a sample from the freezer. When the desired overrun is obtained, the entire batch is discharged from the freezer cylinder into cans or cartons, and the machine is then ready for a new batch of mix.

The output of a batch-type freezer varies with the sharpness of the blades, the refrigeration supplied, and the overrun desired. The average maximum output for commerical batch freezers is eight batches per hour. This schedule allows 180 to 240 s to freeze, 120 to 180 s to whip, and about 60 s to empty the ice cream and refill with mix. For this time schedule, it is assumed the ice cream is drawn from the freezer at not over 100% overrun, at a temperature of about 24 °F and at a refrigerant temperature around the freezer cylinder of about −15 °F.

Continuous ice-cream freezers range in size from 40 to 2700 gal/h at 100% overrun. This type of machine is used almost exclusively in commercial ice-cream plants. Where large capacities are required, multiple units are installed with the ice-cream discharge from several machines connected together to supply the requirements of automatic or semiautomatic packaging or filling machines. In operation, the ice-cream mix is continuously pumped to the freezer cylinder by a positive displacement rotary pump. Air pressure within the cylinder is maintained from 20 psig to more than 100 psig, supplied by either a separate air compressor or drawn in with the mix through the mix pump. The mix entering the rear of the freezer cylinder becomes partly frozen and takes on the overrun because of air pressure and the agitation of the dasher and freezer blades as it moves to the front of the cylinder and is discharged.

The output capacity of most continuous freezers can be varied from 50 to 100% rated capacity by regulating the variable speed control supplied for the mix pump. Continuous freezers can be used for nearly every flavor of ice cream, ice milk, sherbets, and ices. Where flavors requiring nuts, whole fruits, or candy pellets are run, the base or unflavored mix is run through the continuous freezer and then passed through a fruit feeder which automatically feeds and mixes the flavor particles into the ice cream. Ice cream can be discharged from continuous freezers at temperatures of 25 °F, as required for ice-cream bar (novelty) operations, up to a very stiff consistency at 20 °F, as required for automatic filling of small packages.

Special low-temperature ice-cream freezers are available to produce very stiff ice cream for extruded shapes, stickless bars, and sandwiches. Ice-cream temperatures as low as 16 °F can be drawn with some mixes. When ample refrigerating effect is supplied, a variation of ice-cream discharge temperature can be obtained by regulating the evaporator temperature around the freezer cylinder by a suction pressure regulating valve. For filling cans and cartons, the average discharge temperature from the continuous freezer is about 22 °F, when operating with ammonia in a flooded system at about −25 °F.

To calculate accurately the refrigeration requirement for freezing the ice-cream mix in the freezer, the weight of the mix per gal and the amount of water should be known. This can be checked by weighing, knowing the percentage of water, or by calculating the weight from the mix formula, as in Example 2.

Example 2. Find the weight of mix for the following composition (by percent): milkfat, 12.0; serum solids, 10.5; sugar, 16.0; stabilizer, 0.25; egg solids, 0.25; and water, 61.0.

Solution: The specific gravity of the mix is

$$\frac{100}{\left(\frac{\%\text{ milkfat}}{0.93} + \frac{\%\text{ solids, not fat}}{1.58} + \frac{\%\text{ water}}{1.00}\right)} =$$

$$\frac{100}{(12/0.93) + (27/1.58) + (61/1.00)} = 1.099$$

$$\text{Wt per gal of mix} = \text{wt of 1 gal of water} \times \text{sp gr mix}$$

$$= 8.355 \times 1.099 = 9.18\text{ lb}$$

The overrun in ice cream varies from 60 to 100%, which affects the required refrigeration. For a continuous freezer, the required refrigeration may be calculated as in Example 3.

Example 3. Assume a typical ice-cream mix as listed in Example 2 with 100% overrun. The mix contains 61% water and goes to the freezer at a temperature of 40 °F. Freezing would start in this mix at about 27 °F, and 48% of the water would be frozen at 22 °F.

The weight of mix required to produce 100 gal of ice cream is:

$$\frac{100}{100 + \%\text{ overrun}} \times 100\text{ gal} \times \text{Wt mix per gal}$$

For the ice cream being considered, the weight of mix required for 100 gal would be:

$$\frac{100}{100 + 100} \times 100 \times 9.18 = 459\text{ lb}$$

Calculations of capacity required to freeze 100 gal/h of ice cream are as follows:

Sensible heat of mix: 459 × (40 − 27) × 0.80	=	4770 Btu/h
Latent heat: 459 × 0.61 × 0.48 × 144	=	19,350 Btu/h
Sensible heat of slush: 459 × (27 − 22) × 0.65	=	1490 Btu/h
Heat from motors: 5.5 hp × 2545	=	14,000 Btu/h
Total	=	39,610 Btu/h
5% losses from freezer and piping (estimated)	=	1980 Btu/h
Total refrigeration	=	41,590 Btu/h
	=	3.47 ton

Under the conditions given, 3.5 tons of cooling capacity per 100 gal/h of 100% overrun ice cream is required.

In continuous freezer operations, the heat gain from motors and losses from freezer and piping would remain about the same at all levels of overrun, but the necessary refrigerating effect would vary with the weight of mix required to produce 100 gal of ice cream, as shown in Table 10.

Table 10 Continuous Freezing Loads for Typical Ice-Cream Mix

Overrun, %	Ammonia Refrigeration at 3 psig Suction Pressure, Tons per 100 gal/h
60	4.04
70	3.88
80	3.74
90	3.61
100	3.50
110	3.39
120	3.30

Hardening Ice Cream. After leaving the freezer, ice cream is in a semisolid state and must be further refrigerated to become solid enough for storage and distribution. The ideal serving temperature for ice cream is about 8°F; it is considered hard at 0°F. To retain a smooth texture in hardened ice cream, the remaining water content must be frozen rapidly, so that the ice crystals formed will be small. For this reason, most hardening rooms are maintained at −20°F and some as low as −30°F. Most modern hardening rooms have forced-air circulation, either from a unit cooler or a remote bank of coils. With the ice-cream containers arranged so that air will circulate around them, the hardening time is about one-half that of rooms having overhead coils or coil shelves and gravity circulation. With forced-air circulation in the hardening room and average plant conditions, ice cream in 2.5- or 5-gal containers (or smaller packages in wire basket containers), all spaced to allow air circulation, will harden in about 10 h. Hardening rooms are usually sized to allow space for a minimum of three times the daily peak production and for a stock of all flavors, with the sizing based on 10 gal/ft^2 of floor area when stacked loose, which includes aisles.

Some larger plants use ice-cream hardening tunnels, which discharge into a low-temperature storage room. Because of the various size packages to be hardened, most tunnels are the air-blast type, operating at temperatures of −30 to −40°F and, in some cases, as low as −50°F. Containers under one-half gallon are usually hardened in these blast tunnels in about 4 h.

Contact-plate hardening machines must continuously and automatically load and unload to introduce the packages from the filler without delay. Horizontal continuous-plate freezers are described in Chapter 8. Compared to hardening tunnels, contact-plate hardeners save space and power and eliminate package bulging. They are limited to packages of uniform thickness having parallel flat sides.

Temperatures in the storage room are held at about −20°F. Space in storage rooms can be estimated at 25 gal/ft^2 when palletized and stacked solid 6 ft high, including space for aisles. Many storage rooms today use pallet storage and racking systems. Such rooms may be 30 ft tall or more, some using stacker-crane automation.

Refrigeration required to harden ice cream varies with the temperature from the freezer and the overrun. The following example calculates the refrigeration required to harden a typical ice-cream mix.

Example 4. Assume ice cream with 100% overrun enters the hardening room at a temperature of 25°F. At this temperature, approximately 30% of the water would be frozen in the ice-cream freezer with the remainder to be frozen in the hardening room. The weight of one gallon of ice cream at 100% overrun, from a mix weighing 9.18 lb/gal, is 4.59 lb. The mix is assumed to contain 61% water, and the hardening room is at −20°F.

Latent heat of hardening: 4.59 × 0.61 × 0.70 × 144	=	282 Btu
Sensible heat: 4.59 × (25 + 20) × 0.50	=	103 Btu
Total	=	385 Btu
Loss due to heat of container and exposure to outside air, assumed 10%	=	40 Btu
Total Btu per gallon to harden	=	425 Btu

Table 11 Hardening Loads for Typical Ice-Cream Mix

Overrun, %	Hardening Load, Btu/gal
60	532
70	500
80	470
90	447
100	425
110	405
120	386

Percent overrun, when calculated on the basis of the quantity of ice cream delivered by the freezer or the quantity placed in the hardening room, would affect the refrigeration required, as shown in Table 11.

Example 5. Calculate the refrigeration load in an ice-cream hardening room, assuming 1000 gal of ice cream at 100% overrun are to be hardened in 10 h in a forced-air circulation room at a temperature of −20°F. The hardening room, for three times this daily output, should have 300 ft^2 of floor area measuring approximately 15 × 20 × 9 ft high. The total insulated surface of 1230 ft^2 requires 4 to 6 in. of urethane or equivalent. For this example, the heat conductance through the insulated surface is selected at 0.04 Btu/(ft^2·h·°F). The average ambient temperature is assumed to be 90°F.

Heat leakage: 1230 × 0.04 × (90 + 20)	=	5410 Btu/h
Heat from fan motor: 2 hp × 2545/0.85	=	5990 Btu/h
Heat from lights: 600 W × 3.412	=	2050 Btu/h
Air infiltration and persons in room (approximately 20% leakage)	=	1080 Btu/h
Hardening 1000 gal ice cream × 425 Btu/gal in 10 h	=	42,500 Btu/h
Total	=	57,030 Btu/h
	=	4.75 tons

Other products such as sherbets, ices, ice milk, and novelties, usually represent a small percentage of the total output of the plant, but should be included in the total requirement of the hardening room.

Ice-Cream Bars and Other Novelties

Ice-cream plants may manufacture and merchandise a limited number of the many novelties. The most common are chocolate-coated ice-cream bars, popsicles, fudgsicles, drumsticks, ice-cream sundae cups, ice-cream sandwiches, and so forth. Small plants freeze most of these products, especially those with sticks, in metal trays each containing 24 molds, which are submerged in a special brine tank with a built-in evaporator surface and brine agitation. The product mix is prepared in a tank and cooled to 35 to 40°F. A controlled quantity of mix is poured into the tray molds or measured in with a dispenser. The tray molds are placed in the brine tank for complete freezing. Brine temperature is −30 to −36°F. The freezing rate should be rapid to result in small ice crystals, but it varies with the product and generally is 15 to 20 min. The frozen product is loosened from the molds by momentary melting of the outer layers of the product in a water bath. It is immediately removed from the molds; each is separately wrapped or put in a novelty bag and promptly placed in frozen storage for distribution.

The refrigeration calculations to freeze 100 dozen popsicles per h at 3 oz per popsicle, based on the mix containing 85% water, are:

Weight of mix estimated:

100 × 12 × 3/16 × 1.06 (sp gr)	=	239 lb/h
Cooling mix: 239 × (40 − 27) × 0.87	=	2700 Btu/h
Freezing: 239 × 0.85 × 144	=	29,250 Btu/h
Subcooling: 239 × (27 + 30) × 0.5	=	6810 Btu/h
Cooling trays (50/h) 50 × 8 lb × (60 + 30) × 0.12	=	4320 Btu/h
Heat from agitator: 1 hp × 2545	=	2550 Btu/h
Leakage through tank: 3 × 12 × 3 ft deep	=	750 Btu/h
Loss, top of tank and piping (assumed)	=	7500 Btu/h
Total refrigeration load	=	53,880 Btu/h
	=	4.5 tons

In making ice cream, ice milk, and similar kinds of bars, the mix is processed through the freezer and is extruded in a viscous form at about 22 °F. Using similar calculations, the estimated refrigeration load to freeze 100 dozen would be 2.2 tons for 3-oz ice-cream bars with 100% overrun.

The equipment to make and package novelties in large plants is available in several designs and capacities. Some are limited to the manufacture of one or a few kinds of similar novelties. Other machines have more versatility; for example, they can be used to make novelties with or without sticks, coated or uncoated, and of numerous sizes, shapes, and flavor combinations. Some of these machines include packaging in a bag or wrap, plus placement and sealing in a carton in units of 6, 8, 12, 14, 18, 24, or 48. In other plants, a separate packaging unit may be required. Some units harden the product by air at a temperature within the range of −35 to −46 °F. Brine, usually calcium chloride, with a specific gravity of 1.275 or more and a temperature of −28 to −38 °F may be the hardening medium. Capacity of novelty makers varies with the shape and size of the specific product. Common production of a novelty machine is generally within the range of 300 to 3000 dozen or more per hour. Novelty equipment in plants may be semiautomatic or automatic in performance of the necessary functions.

An example of a simple novelty machine is one that has two parallel conveyor chains on which the mold strips are fastened. The molds are conveyed through filling, stick inserting, freezing, and defrosting stages. The extractor conveyor removes the frozen product from the mold cups and carries it to packaging or through the dipping operation; it is then discharged at packaging. In the meantime, the molds go through a wash and rinse and back to be filled. The novelty is either bagged or wrapped by machine, grouped and placed into cartons, and conveyed to cold storage.

Refrigeration Equipment Operation

Countertype freezers are usually designed for use with halocarbon refrigerants and may be arranged in a self-contained cabinet with the condensing unit mounted under the freezer. Nearly all commercial ice-cream plants, particularly the larger plants, use ammonia. Some smaller plants operate continuous ice-cream freezers and refrigerate hardening rooms to acceptable temperatures with single-stage refrigerant compressors. In most cases, these plants operate their compressors at conditions above the maximum compression ratio recommended by the manufacturer.

For economy of operation, within reasonable limits of compression ratio, ice-cream plants normally use multistage compression. For freezing ice cream, producing frozen novelties, and refrigerating an ice-cream hardening room to −20 °F, one or more booster compressors may be used at the same suction pressure, discharging into second-stage compressors, which also handle the mix cooling and ingredient cold storage room loads. If a hardening tunnel is used at a temperature of −40 °F or below, at least two booster compressors should be used, one operating at the suction pressure required for the tunnel and the other operating at a higher suction pressure for the ice-cream freezers and storage room, with both units discharging into the second-stage compressor system. For plants with hardening tunnels arranged for large volumes, an analysis of operating costs may indicate savings in using three-stage compression with the low-temperature booster used for the tunnel, discharging into the second-stage booster used for freezers and storage, and then the second-stage booster discharging into the third-stage compressor system.

High-temperature loads in an ice-cream plant usually consist of refrigeration for cooling and holding cream, cooling ice-cream mix after pasteurization, cooling for mix holding tanks, and refrigeration for the ingredient cold storage room. If direct refrigerant cooling is used for the high-temperature loads, then compressor selection can be made at about 20 °F saturated suction temperature and combined with the compressor capacity required to handle the booster discharge. About the same high suction temperature can be estimated if ice-cream mix and mix holding tanks are cooled by chilled water from a flash-type water chilling system. If an ice bank chills water used for cooling pasteurized ice-cream mix, it may be desirable to provide a separate compressor to handle this ice bank refrigerating load.

Table 12 Typical Ammonia Compressor Power Requirements at 185 psig Condensing Pressure

Temperature, °F	Pressure, absolute, psi	Power, hp One-stage	Power, hp Two-stage
−25	16.0	2.8	2.2
−30	13.9	3.0	2.4
−35	12.3	3.5	2.5

Refrigeration is an important cost in an ice-cream plant, and heat transfer has a direct relationship to the effectiveness of the refrigeration system. Barriers to heat transfer are air films, frost and ice, scale, noncondensable gases, abnormal temperature differentials, clogged sprays, slow liquid circulation, poor air circulation, and foreign particles. Additional conditions that adversely affect the ice-cream freezing rate are dull scraper blades, viscous mix, low overrun percentage, high mix temperature, low ice-cream discharge temperature, and the mix composition, especially if it has a high sugar content.

Specific causes of low evaporator efficiency are rapid frost and ice development and an oil film. Automatic defrosting is one answer to the ice and frost problem. Regular oil purges help and an oil separator in the discharge line of the compressors is highly desirable. Neglected condensers lead to high head pressures and higher electrical requirements. Condenser surfaces should be kept clean of mineral deposits and other forms of scale or debris.

In compressor operation, the purging of noncondensable gases requires attention to avoid extra power use. Automatic purgers will handle this problem. The low temperatures required in ice-cream plants result in greater efficiency if two- or possibly three-stage compression systems are used. A comparison of power requirements for one- and two-stage ammonia systems at 185 psig condensing pressure is shown in Table 12.

The reduction of refrigeration losses through doors and conveyor openings and sufficient storage room insulation can contribute materially to better efficiency.

Winter Coldness Refrigerating Effect

There is some interest in using cold weather to cool products. Cold rooms and ice banks can be serviced when the ambient air temperature is 26 °F or lower.

UHT STERILIZATION AND ASEPTIC PACKAGING

Ultrahigh temperature (UHT) *sterilization* of liquid dairy products destroys viability of microorganisms with a minimum adverse effect on sensory and nutritional properties. *Aseptic packaging* (AP) containerizes the sterilized product without recontamination. Sterilization, in the true sense, is the destruction or elimination of all viable microorganisms. In industry, however, the term sterilized may refer to a product that does not deteriorate microbiologically, but in which viable organisms may have survived the sterilization process. In other words, heat treatment renders the product safe for consumption and imparts an acceptable shelf life microbiologically.

Sterilization Methods—Equipment

Retort sterilization of milk products has been a commercial practice for many years. It consists of sterilizing the product after hermetically sealing it in a metal or glass container. The heat treatment is sufficiently severe to cause a definite cooked off-flavor in milk and to decrease the heat-labile nutritional constituents of milk products. UHT-AP has the advantage of causing less cooked flavor, color change, and loss of vitamins while having the same sterilization effect as the retort method.

UHT-AP has been applied to common fluid milk products (whole milk, 2% milk, skim milk, and half-and-half), various creams, flavored milks, evaporated milk, and such frozen dessert mixes as ice cream, ice milk, milk shakes, soft-serve, and sherbets. UHT-sterilized dairy foods include eggnog, salad dressings, sauces, infant preparations, puddings, custards, and nondairy coffee whiteners and toppings.

UHT sterilization is accomplished by rapid heating of the product to the sterilizing temperature, holding the temperature for a definite number of seconds, and then rapid cooling. The methods have been classified as direct steam or indirect heating. Among the advantages of the direct methods are (1) heating is faster, (2) processing intervals between equipment cleanings are longer, and (3) the flow rate is easier to change. Among the indirect methods, advantages are: (1) regeneration potential is greater, (2) potable steam is not necessary, and (3) viscous products and those with small pieces of solids can be processed with the scraped-surface unit.

The direct steam method is subdivided into injection or infusion. In direct injection, steam is forced into the product, preferably in small streamlets, with sufficient turbulence to minimize localized overheating of the milk surfaces that the steam initially contacts. In the infusion system, the product is sprayed into a steam chamber. Advantages of infusion over injection are (1) slightly less steam pressure required (there are exceptions), (2) less localized overheating of a portion of the product, and (3) more flexibility for change of the product flow rate.

The three important indirect systems are tubular, plate, and cylinder with mechanical agitation. In the tubular type, the tube diameter must be small and the velocity of flow high to maximize heat transfer into the product.

The essential components for direct steam injection are storage or balance tank, timing pump, preheater (tubular or plate), steam injector or infuser unit, holding unit, flow-diversion valve, vacuum chamber, aseptic pump, aseptic homogenizer, plate or tubular cooler, and control instruments. The minimum items of equipment for steam infusion are the same, except that the infuser is used to heat the product from the preheat to the sterilization temperature.

The necessary equipment for the indirect systems is similar: storage or balance tank, timing pump, preheater (tubular or plate type, and preferably regenerative), homogenizer, final plate or tubular heater, holding tube or plate, flow-diversion valve, cooler (1 to 3 stages), and control instruments. The mechanically agitated heat exchanger replaces the tubes or plates in the final heating stage. Otherwise, the same items of equipment are used for this system of sterilization.

In addition to the basic equipment, many combinations of essential and supplemental items of UHT equipment are available. For example, one deviation in the indirect system is to use the pump portion of the homogenizer as a timing pump when it is installed after the balance tank. The first stage of homogenization may occur after preheating, and the second stage may occur after precooling. A vacuum chamber may be placed in the line after preheating, for precooling after sterilization, or installed in both locations. A condenser in the vacuum chamber allows the advantages of deaeration without moisture losses that otherwise would occur in the indirect system. In Europe, some indirect systems have a hold of several minutes, after preheating, to reduce the rate of solids accumulation on the final heating surfaces of the tubes or plates. A bactofuge may be included in the line after preheating to reduce a high microbiological content, especially of the bacterial spores.

Self-acting controls and other instrumentation are available to assure automatic operation in nearly every respect. Particularly important is automatic control of the temperatures for preheating, sterilizing, and precooling in the vacuum chamber, and to some extent, of the final temperature before packaging. This may include temperature-sensing elements to control heating and cooling and pressure-sensing elements for operating pneumatic valves. The cleaning cycle may be automated, beginning with a predetermined solids accumulation on specific heating surfaces. Timers regulate the various cleaning and rinsing steps.

In some systems, one or more aseptic surge tanks are installed between the UHT sterilizer and the AP equipment. Surge tanks permit continuation of the sterilizer should the operation of the AP equipment be interrupted, or the continuation of the AP equipment should sterilization be interrupted. It also makes the use of two or more AP units easier than direct flow from the UHT sterilizer to the AP machines.

When aseptic surge tanks are used, they must be constructed to withstand the steam pressure required for equipment sterilization and be provided with a sterile air venting system. Aseptic surge tanks may be unloaded by applying sterile air to force product out to the AP equipment. The pressure for air unloading can be controlled at a constant value, making uniform filling possible even when one of several AP machines is removed from service.

Aseptic surge tanks make it possible to hold bulk product, even for several days, until it is convenient to package it.

Basic Steps. After the formula is prepared and the product standardized, the processing steps are: (1) preheat to 150 to 170 °F by a plate or tubular heat exchanger; (2) heat to a sterilization temperature of 285 to 300 °F; (3) hold for 1 to 20 s at sterilizing temperature; and (4) cool to 40 to 100 °F, depending on product keeping quality needs. Cooling may be by one to three stages; generally two are used. The direct steam method requires at least two cooling stages. The first is flash cooling in a vacuum chamber to 150 to 170 °F to remove moisture equal to the steam injected during sterilization. The second stage reduces the temperature to within 50 to 100 °F. A third stage is required in most plants if the temperature is lowered additionally to 35 to 50 °F.

The products with fat are homogenized to increase stability of the fat emulsion. The direct method requires homogenization after sterilization and precooling. Homogenization may follow preheating or precooling, but usually follows preheating in the indirect method. Efficient homogenization is very important in delaying the formation of a cream layer during storage.

Sterilized plain milks (such as whole, 2%, and skim milk) are most vulnerable to having a cooked off-flavor. Consequently, the aim is to have low sterilization temperature and time consistent with satisfactory keeping quality. The total cumulative heat treatment is directly related to the intensity of the cooked off-flavor. The total processing time from preheating to cooling varies widely among systems. Most operations in the United States range from 30 to 200 s; in European UHT processes, it may be much longer.

Several factors influence the minimum sterilization temperature and time needed to control adverse effects on flavor and physical, chemical, and nutritional changes. The type of product, initial number of spores and their heat resistance, total solids of the product, and pH are the most important factors. Obviously, the relationship is direct for the number and the heat resistance of the spores. Total solids also have a direct relationship, but for an acid pH it is inverse.

Several terms are used in the designation of the influence of the UHT on the microbiological population. *Decimal reduction* refers to a reduction of 90% (*e.g.*, 100 to 10, or one log cycle). An example of a three-decimal reduction is 10,000 to 10. *Decimal reduction time*, or *D value*, is the time required to obtain a 90% decrease. *Sterilizing effect*, or *bactericidal effect*, is the number of decimal reductions obtained and expressed as a logarithmic reduction

($\log_{10}$ initial count minus $\log_{10}$ final count). A sterilizing effect of six means one organism remaining from a million per mL (10^6), and seven would be one remaining in 10 mL (a final count of 10^{-1}).

The *Z value* is the temperature increase required to reduce the D value by one log cycle (90% reduction of microorganisms with the time held constant). The *F value* (thermal death time) is the time required to reduce the number of microorganisms by a stated amount or to a specific number. For example, assuming D value of 36 s for *Bacillus substilis* spores at 250 °F and a need to reduce the spores from 10^6 per mL to < 1 per mL, the thermal treatment time would be 6×36 s = 216 s (F value).

Aseptic Packaging

Aseptic fillers are available for coated metal cans, glass bottles, plastic-paperboard-foil cartons, thermoformed plastic containers, blow-molded plastic containers, and plastic pouches. The aseptic can equipment includes a can conveyor and sterilizing compartment, filling chamber, lid sterilizing compartment, sealing unit, and instrument controls. The procedure sterilizes the cans with steam at 550 °F as they are conveyed, fills the cans by continuous flow, simultaneously sterilizes the lids, places the lids on the cans, and seals the lids onto the cans. Pressure control apparatus is not used for entry or exit of cans.

A similar system is used for glass bottles or jars. The jars are conveyed into a turret chamber; air is removed by vacuum; the jars are then sterilized for 2 s with wet steam at 60 psi and moved into the filler. The temperature of the glass equalizes to 120 °F and the filling takes place. Next, the transfer is to the capper for placement of sterile caps, which are screwed onto the jars. The filling and capping space is maintained at 500 °F.

Several aseptic blowmold forming and filling systems have been developed. Each system is different, but the basic steps using molten plastic are: (1) extruded into a parison, (2) extended to the bottom of the mold, (3) mold closed, (4) preblown with compressed air that inflates the plastic film into a bottle shape, (5) parison cut and the neck pinched, (6) final air application, (7) bottle filled and foam removed, (8) top sealed, and (9) mold opened and filled bottle ejected.

The basic steps in the manufacture of aseptic, thermoformed plastic containers are: (1) a sheet of plastic (*e.g.*, polystyrene) is drawn from a roll through the heating compartment and then multistamped into units, which constitute the containers; (2) these units are conveyed to the filler, which is located in a sterile atmosphere, and are filled; (3) a sheet of sterilized foil is heat sealed to the container tops; and (4) each container is separated by scoring and cutting.

One of the two aseptic systems for the plastic-paperboard-foil cartons draws the material from a roll through a concentrated hydrogen peroxide bath to destroy the microorganisms. The peroxide is removed by drawing the sheet between twin rolls, by exposure to ultraviolet light and hot air, or by superheated, sterilized air forced through small slits at high velocity. The packaging material is drawn downward in a vertical, sterile compartment for forming, filling by continuous flow, sealing, separation, and ejection.

In the other plastic-paperboard-foil aseptic system, the prepared, flat blanks are formed and the bottoms are heat sealed. In the next step, the inside surfaces are fogged with hydrogen peroxide. Sterilized hot air dissipates the peroxide. The cartons are conveyed into the aseptic filling and then into top-sealing compartments. The air forced into these two areas is rendered devoid of microorganisms by high efficiency filters.

Operational Problems. Aseptic operational problems are reduced by careful installation of satisfactory equipment. The equipment should comply with 3-A Standards. Milk and milk products that are processed to be commercially sterile and aseptically packaged must also meet the Grade A Pasteurized Milk Ordinance and be processed in accordance with 21 CFR Part 113, "Thermally processed low-acid foods packaged in hermetically sealed containers." Generally, the simplest system, with a minimum of equipment for product contact surfaces and processing time, is desirable. It is specifically important to have as few pumps and nonwelded unions as possible, particularly those with gaskets. The gaskets and O-rings in unions, pumps, and valves are much more difficult to clean and sterilize than are the smooth surfaces of chambers and tubing. Automatic controls, rather than manual attention, is generally more satisfactory.

Complete cleaning and sterilizing of the processing and packaging equipment are essential. Milk solids accumulate rapidly on heated surfaces; therefore, cleaning is necessary after 0.5 h of processing for the tubular or plate UHT heat exchangers, although cleaning after 3 to 4 h is more common. The cleaning practice for the sterilizer, filler, and accessory equipment usually involves the CIP method for the rinse and alkali cleaning cycle, rinse, acid cleaning cycle, and rinse. Some plants only periodically acid clean the storage tanks and packaging equipment; *e.g.*, once or twice a week. Steam sterilization just before processing is customary. At 8 to 10 psig of wet steam, 1.5 to 2.0 h (or a shorter time at higher steam pressure) may be required. Water sterilized by steam injection or the indirect method can be used for rinsing and for the cleaning solution.

Survival of spores during UHT processing, or subsequent recontamination of the product before the container is sealed, is a constant threat. Inadequate sealing of the container also may be troublesome with certain types of containers. Another source of poststerilization contamination is airborne microorganisms. These may contact the product through inadequate sterilization of air that enters the storage vat for the processed product or through air leaks into the product upstream of the sterilized product pumps or homogenizer, if a reduced pressure is created. During packaging, air may contaminate the inside of the container or the product itself during filling and sealing.

Quality Control

Poor quality of raw materials must be avoided. The higher the spore count of the product before sterilization, the larger the spore survival number at a constant sterilization temperature and time. Poor quality can also contribute to other product defects (off-flavor, short keeping quality) because of sensory, physical, or chemical changes. Heat stability of the raw product must be considered.

A good quality sterilized product has a pleasing, characteristic flavor and color that are similar to the pasteurized samples. The cooked flavor should be slight, or negligible, with no unpleasant aftertaste. The product should be free of microorganisms and adulterants such as insecticides, herbicides, and peroxide or other container residues. It should have good physical, sensory, and keeping quality.

Deterioration in storage may be evaluated by holding samples at 70, 89, 98, or 113 °F for 1 or 2 weeks. The number of samples for storage testing should be selected statistically and should include samplings of the first and last of each product packaged during the processing day. In order to identify the source of microbiological spoilage, continuous aseptic sampling into standard size containers after sterilization and/or just ahead of packaging may be practiced. Sampling rate should be set to change containers each hour.

The rate of change in storage of sterilized milk products is directly related to the temperature. Commercial practice varies with storage ranging from 35 °F to room temperature, which may go as high as 95 °F or more. In plain milks, the cooked flavor may decrease the first few days, and then remain at its optimum for 2 to 3 weeks at 70 °F before gradually declining. When the milk is held at 70 °F, a slight cream layer becomes noticeable in approximately 2 weeks and slowly continues until much of the fat has risen to the top. Thereafter, the cream layer becomes increasingly difficult to reincorporate or reemulsify.

Viscosity increases slightly the first few weeks at 70 °F and then remains fairly stable for 4 to 5 months. Thereafter, gelation gradually occurs. However, milks vary in stability to gelation depending on such factors as feeds, stage of lactation, preheat treatment, and homogenization pressure. The addition of sodium tetraphosphate to some milks causes gelatin to develop more slowly.

Occasionally, some sterilized milk products develop a sediment on the bottom of the container because of crystallization of complex salt compounds or sugars. Browning can also occur during storage. Usually, the off-flavors develop more rapidly and render the product unsalable before the off-color becomes objectionable.

Heat-Labile Nutrients

The results reported by researchers on the effects of UHT sterilization on the heat-sensitive constituents of milk products lack consistency. The variability may be attributed to the analytical methods and to the difference in total heat treatment among various UHT systems, especially in Europe. In a review, Van Eekelen and Heijne (1965) summarized the effect of UHT sterilization on milk as follows: slight or none for Vitamins A, B_2, and D, carotene, pantothenic acid, nicotinic acid, biotin, and calcium; and no decrease in biological value of the proteins. The decreases were: 3 to 10%, thiamine; 0 to 30%, B_6; 10 to 20%, B_{12}; 25 to 40%, C; 10%, folic acid; 2.4 to 66.7%, lysine; 34%, linoleic acid; and 13%, linolenic acid. Protein digestibility was decreased slightly. A substantial loss of Vitamin C, B_6, and B_{12} occurred during a 90-day storage. Brookes (1968) reported that Puschel found that babies fed sterilized milk averaged a gain of 27 g per day, compared to 20 g for the control group.

EVAPORATED, SWEETENED, CONDENSED, AND DRY MILK

Evaporated Milk

Raw milk intended for processing into evaporated milk should have a heat stability quality with little and preferably no developed acidity. As the milk is received it should be filtered and held cold in a storage tank. The milkfat is standardized to nonfat solids at the ratio of 1:2.2785. The milk is preheated to 200 to 205 °F for 10 to 20 min or 240 to 260 °F for 60 to 360 s to reduce product denaturation during the sterilizaton process. Moisture is removed by batch or (usually) continuous evaporation until the total solids have been concentrated to 2.25 times the original content.

The condensed product is pumped from the evaporator and, with or without additional heating, is homogenized at 2000 to 3000 psi and 120 to 140 °F. It is cooled to 45 °F and held in storage tanks for restandardization to not less than 7.9% milkfat and 25.9% total solids. The product is pumped to the packaging unit for filling the cans made from tin-coated sheet steel. The filled cans are conveyed continuously through a retort, whereby the product is rapidly heated with hot water and steam to 245 °F and held for 15 min to complete the sterilization. Rapid cooling with water to 80 to 90 °F follows. The evaporated milk is agitated while in the retort by the can movement. Application of labels and placement of cans in shipping cartons are done automatically.

Storage at room temperature is common, but deterioration of flavor, body, and color is decreased by lowering the storage temperature to 50 to 60 °F. Relative humidity should be less than 50% to reduce can and label deterioration. The recommended inversion of cases during storage to reduce fat separation is shown in Table 13.

Table 13 Inversion Times for Cases of Evaporated Milk in Storage

Storage Temperature, °F	Time
90	1 month initially and each 15 days
80	1 to 2 months
70	2 to 3 months
60	3 to 6 months

Sweetened Condensed Milk

Sweetened condensed milk is manufactured similarly to evaporated milk in several aspects. One important difference, however, is that added sugar replaces heat sterilization to extend storage life. Filtered cold milk is held in tanks and standardized to 1:2.2942 (fat to nonfat solids). The milk is preheated to 145 to 160 °F, homogenized at 2500 psi, and heating is continued to 180 to 200 °F for 5 to 15 min or to 240 to 300 °F for 30 s to 5 min. The milk is condensed in a vacuum pan to slightly higher than a 2:1 ratio. Liquid sugar (pasteurized) is added at the rate of 18 to 20 lb/100 lb of condensed milk.

As the mixture is pumped from the vacuum pan, it is cooled through a heat exchanger to 86 °F and held in a vat with an agitator. Nuclei for proper lactose crystalization are provided by adding finely-powdered lactose (200-mesh). The product is cooled slowly, taking an hour to reach 75 °F with agitation. Then cooling is continued more rapidly to 60 °F. Improper crystalization forms large crystals, which cause sandiness (gritty texture). The sweetened condensed milk is pumped to a packaging unit for filling into retail cans and sealing. Labeling of cans and placement in cases is mechanized, similar to the process used for evaporated milk. The product is usually stored at room temperature, but the keeping quality is improved if the storage temperature is below 70 °F.

Condensing Equipment. Both batch and continuous equipment are used to reduce the moisture content of fluid milk products. The continuous types have single, double, triple, or more evaporating effects. The improvement in efficiency with multiple effects is shown in Table 14 by the reduction in steam required to evaporate 1 lb of water.

A simple evaporator is the horizontal tube. In this design, the tubes are in the lower section of a vertical chamber. During operation, water vapor is removed from the top and the product, from the bottom of the unit. For the vertical short-tube evaporator, the chamber design may be similar to the horizontal tube. The long-tube vertical unit may be designed to operate with a rising or falling film in the tubes. The latter is common. For the falling film, the product Reynolds number should be greater than 2000 for good heat transfer. Falling-film units may have a high k factor at low temperature differentials, resulting in low steam requirements per mass of water evaporated per area of heating surface. This type (falling film) has a rapid startup and shutdown. Thermocompressing and mechanical compressing evaporators have the advantage of operating efficiently at lower temperatures, thus reducing the adverse effect on heat sensitive constituents. Vapors removed from the product are compressed and used as a source of heat for additional evaporation.

Plate-type evaporators are also in common use. They are similar to plate heat exchangers used for pasteurization in that they have a frame and a number of plates gasketed to carry the product in a passage between two plates and the heating medium in adjacent passages. They differ in that, in addition to ports for

Table 14 Typical Steam Requirements for Evaporating Water from Milk

No. of Evaporating Effects	Steam Required, lb Steam/lb Water
Single	1.30 to 1.00
Double	0.60 to 0.50
Triple	0.40 to 0.35
Quadruple	0.30 to 0.25

product, they have large ports to carry vapor to a vapor separator. Vapors flow from the separator chamber to a condenser similar to those used for other types of evaporators. Plate evaporators require less head space for installation than other types, may be enlarged or decreased in capacity by a change in the number of plates, and offer a very efficient heat exchange surface.

Equipment Operation. Positive pumps of the reciprocating type are often used to obtain 24-in. Hg vacuum in the chamber. Steam jet ejectors may be used for 25-in. Hg vacuum, for one stage; two stages permit 28-in. Hg vacuum; and three stages, 29.8-in. Hg vacuum. Condensers between stages remove the heat and may reduce the amount of vapor for the following stage. Either a centrifugal or reciprocating pump may be used to remove water from the condenser. A barometric leg may also be placed at the bottom of a 34 ft or longer condenser to remove the water by gravity.

Dry Milk and Nonfat Dry Milk

There are two important methods of drying milk—spray and drum. Each has modifications, such as the foam spray and the vacuum drum drying methods. Spray drying exceeds by far the other methods for drying milk, and the largest volume of dried dairy product is skim milk.

In the manufacture of spray-dried nonfat dry milk (NDM), cold milk is preheated to 90 °F, separated, and the skim milk for low-heat NDM is pasteurized at 161 °F for 15 s or slightly higher and/or longer. It is condensed with caution to restrict total heat denaturation of the serum protein to less than 10%. This requires using a low-temperature evaporator or operating the first effect of a regular double effect evaporator at a reduced temperature. After increasing the total solids to 40 to 45%, the condensed skim milk is continuously pumped from the evaporator through a heat exchanger to increase the temperature to 145 °F. The concentrated skim milk is filtered and enters a positive pump operating at 3000 to 4000 psi, which forces the product through a nozzle with a very small orifice, producing a mist-like spray in the drying chamber. Hot air of 290 to 400 °F or higher dries the milk spray rapidly. Nonfat dry milk with 2.5 to 4.0% moisture is conveyed from the drier by pneumatic or mechanical means, then cooled, sifted, and packaged. Packages for industrial users are 50- or 100-lb bags.

High heat nonfat dry milk is used principally for bread and other bakery products. The manufacturing procedure is the same as for low heat NDM except: (1) the pasteurization temperature is well above the minimum, *e.g.*, 175 °F for 20 s or higher; (2) after pasteurization, the skim milk is heated to 185 to 195 °F for 15 to 20 min, condensed; and (3) the concentrate is heated to 160 to 165 °F ahead of filtering and then is spray dried, similar to the process for low heat NDM. Storage of low or high heat NDM is usually at room temperature.

Dry Whole Milk. The raw whole milk in storage tanks is standardized at a ratio of fat to nonfat solids of 1:2.769. The milk is preheated to 160 °F, filtered or clarified, and homogenized at 160 °F and 3000 psi on the first stage and 750 psi on the second stage. The heating continues to 200 °F with a 180-s hold. The milk is drawn into the evaporator and the total solids are condensed to 45%. The product is continuously pumped from the evaporator, reheated in a heat exchanger to 160 °F, and spray dried to 1.5 to 2.5% moisture. Dry whole milk (DWM) is cooled (not below dew point) and sifted through a 12-mesh screen. For industrial use within 2 or 3 months, the dry whole milk is packaged in 50-lb bags and held at room temperature or, preferably, well below 70 °F.

In order to retard oxidation, the dry whole milk may be containerized in large metal drums or in customer-size cans unsealed and subjected to 28-in. Hg vacuum. Less than 2% oxygen in the head space of the package after a week of storage is a common aim. The oxygen desorption from the entrapped content in lactose is slow, and 2 vacuum treatments may be necessary with a 7- to 10-day interval between them. Warm DWM directly from the drier has a faster oxygen desorption rate than after it has cooled. Nitrogen is used to restore atmospheric pressure after each vacuum treatment. After the hold period for the first vacuum treatment, the DWM in the drums is dumped into a hopper, mechanically packaged into retail size metal cans, and given the second vacuum treatment.

Foam spray drying permits the total solids to be increased to 50 to 60% in the evaporator prior to drying. Gas, compressed air, or nitrogen is distributed, by means of a small mixing device, into the condensed product between the high pressure pump and the spray nozzle. A regulator and needle valve are used to adjust the gas flow into the product. Gas usage is approximately 0.5 ft^3/gal of concentrated product. Otherwise, the procedure is the same as for regular drying. Foam spray-dried NDM has poor sinkability but good reconstitutability in water. The density is roughly half that of regular spray-dried NDM. The additional equipment for foam spray drying is limited to a compressor, storage drum, pressure regulator, and a few accessory items. The cost is relatively small, especially if compressed air is used.

Spray driers are made in various shapes and sizes with one or many spray nozzles. The horizontal spray driers may be box shaped or a teardrop design. The vertical spray driers are usually cone or silo shaped.

Heat Transfer Calculations. The typical atomization in United States spray-drying plants is produced by a high pressure pump that forces the liquid through a small orifice in a nozzle designed to give a spreading effect as it emerges from the nozzle. Single-nozzle driers have an orifice opening diameter of 0.107 to 0.177 in. The diameter for multinozzle driers is 0.025 to 0.052 in. In Europe, the spinning disk is the most common means of atomizing in milk drying plants. Droplet sizes of 50 to 250 μm in diameter are usual. Droplet size has an inverse relationship to the rate of drying at a uniform hot air temperature. Larger droplets require a higher air drying temperature and/or longer exposure than the smaller ones.

Other essential steps in spray drying are: (1) moving, filtering, and heating the air; (2) incorporating the hot air with the product droplets; and (3) removing the moisture vapors and separating the moist air from the product particles. After passing through a rough or intermediate filter, the air is heated indirectly by steam coils or directly with a gas flame to 250 to 500 °F. During the short drying exposure time, the air temperature drops to 160 to 200 °F.

Thermal efficiency is the percentage of the total heat used to evaporate the water during the drying process. The efficiency is improved by recovery of heat from the exhaust air, decreased radiation loss, and high drying air temperature versus a low outlet air temperature. Roughly 2.2 to 3.2 lb of steam are needed to evaporate 1 lb of moisture in the drier.

$$\text{Thermal efficiency} = \frac{(1 - R/100)(t_1 - t_2)}{t_1 - t_0}$$

where

R = radiation loss, percentage of temperature decrease in drier
t_1 = inlet air temperature, °F
t_2 = outlet air temperature, °F
t_0 = ambient air temperature, °F

Most of the dried particles are separated from the drying air by gravity and fall to the bottom of the drier or the collectors. The fine particles are removed by directing the air-powder mixture through bag filters or a series of cyclone collectors. Air movement in the cyclone is designed to provide a centrifugal force for separation of the product particles. In general, several small-diameter cyclones with a fixed pressure drop will be more efficient for removal of the fines than two large units.

The drier has sensing elements to continuously record the hot air (inlet) temperature and the moist air (outlet) temperature. Adjustments of either of these temperatures during drying is done with a steam valve or gas inlet valve.

Drum Drying

Relatively little skim milk or whole milk has been drum dried in the last few years. Drum-dried products, when reconstituted, have a cooked or scorched flavor compared with the spray-dried products. The heat treatment during drying denatures the protein and results in a high insolubility index. In preparation for drying, the skim milk is separated or the whole milk is standardized to 1:2.769, (*e.g.*, 3.2 fat and 8.86 SNF). The product is filtered or clarified, homogenized after preheating, and pasteurized. If the resulting dry product is intended for bakery purposes, the milk is heated to approximately 185 °F for 10 min. The fluid product may or may not be concentrated by moisture evaporation to not more than 2 to 1. The product is then dried on the drum(s)—skim milk to not more than 4.0% and whole milk to not more than 2.5% moisture. A blade pressed against the drum scrapes off the sheet of dried product. An auger conveys the dry material to the hammer mill, where it is pulverized and sifted through an 8-mesh screen. Drum dried milks are usually packaged at the sifter into 50- to 100-lb kraft bags with a plastic liner.

A double-drum drier is more common than a single drum for drying milk. Cast iron is used more often in drum construction than stainless steel or alloy steel and chrome plate steel. The knife metal must be softer than the drum. In the double drum unit, the drums are spaced from 0.02 to 0.043 in. apart. End plates on the drums create a reservoir into which the product, at 185 °F, is sprayed the length of the drums. The steam-heated drums boil the product continuously as a thin film adheres to the revolving drums. After about 0.875 of one revolution, the film of product is dry and is scraped off. Drums normally revolve between 10 and 19 rpm. The steam pressure inside the drums is approximately 70 to 90 psi, as indicated by the pressure gage at the inlet of the condensate trap.

The steam pressure is adjusted for drying the product to the desired moisture content. Superheated steam will cause scorching of the product. Condensate inside the drums must be continuously removed, while the exterior vapors from the product are exhausted from the building with a hood and fan system. Capacity, dried product quality, and moisture content depend on many factors. Some important ones are: steam pressure in drums, rotation speed of drums, total solids of product, smoothness of drum surface and sharpness of the knives, properly adjusted gap between the two drums, liquid level in drum reservoir, and product temperature as it enters the reservoir.

REFERENCES

Brookes, H. 1968. New developments in longlife milk and dairy products. *Dairy Industries* (May).

Hammer, B.W. and A. R. Johnson. 1913. The specific heat of milk and milk derivatives. *Research Bulletin* No. 14, Iowa Agricultural Experiment Station.

IAMFES. 3-A *Sanitary Standards*. International Association of Milk, Food, and Environmental Sanitarians, Ames, IA.

Leighton, A. 1927. On the calculation of the freezing point of ice cream mixes and of the quantities of ice separated during the freezing process. *Journal of Dairy Science* 10:300.

Rishoi, A.H. 1951. *Physical characteristics of free and globular milk fat.* American Dairy Science Association, Annual Meetings (June).

Van Eeckelen, M. and J.J.I.G. Heijne. 1965. Nutritive value of sterilized milk. In *Milk sterilization*. Food and Agricultural Organization of the United Nations, Rome.

Zahadan, V.Z. 1940. *Specific heat of foodstuffs in relation to temperature.* Kholod'naia Prom. 18 (4):32. Cited from Stitt and Kennedy (Russian).

BIBLIOGRAPHY

Arbuckle, W.S. 1972. *Ice cream*, 2nd ed. AVI Publishing Co., Westport, CT.

Farrall, A.W. 1963. *Engineering for dairy and food products.* John Wiley and Sons, Inc., New York.

Griffin, R.C. and S. Sacharow. 1970. *Food packaging.* AVI Publishing Co., Westport, CT.

Hall, C.W. and T.I. Hedrick. 1971. *Drying of milk and milk products.* AVI Publishing Co., Westport, CT.

Henderson, F.L. 1971. *The fluid milk industry.* AVI Publishing Co., Westport, CT.

Judkins, H.F. and H.A. Keener. 1960. *Milk production and processing.* John Wiley and Sons, Inc., New York.

Kosikowski, F.V. 1966. *Cheese and fermented milk foods.* Published by author, Ithaca, NY.

Reed, G.H. 1970. *Refrigeration.* Hart Publishing Co., Inc., New York.

Sanders, G.P. Cheese varieties and descriptions. *Agriculture Handbook* No. 54. USDA, U.S. Government Printing Office, Washington, D.C.

Webb, B.W. and E. A. Whittier. 1970. *Byproducts from milk.* AVI Publishing Co., Westport, CT.

Wilcox, G. 1971. *Milk, cream and butter technology.* Noyes Data Corporation, Park Ridge, NJ.

Wilster, G.H. 1964. *Practical cheesemaking*, 10th ed. Oregon State University Bookstore, Corvallis, OR.

CHAPTER 15

DECIDUOUS TREE AND VINE FRUITS

THE most obvious losses from marketing fruit crops are caused by mechanical injury, decay, and aging. Losses in moisture, vitamins, sugars, and starches are less obvious, but they adversely affect quality and nutrition. Rough handling and holding at undesirably high or low temperatures increases loss. Loss can be substantially reduced by greater care in handling and by following recommended storage practices.

GENERAL PRODUCE CONSIDERATIONS

Quality and Maturity

Maximum storage life can be obtained only by storing high quality commodities soon after harvest. Different lots of fruits may vary greatly in their storage behavior due to variety, climate, soil and cultural conditions, maturity, and handling practices. When fruits are transported from a distance, grown under unfavorable conditions, or deteriorated, proper storage allowance should be made.

Fresh fruits for storage should be as free as possible from skin breaks, bruises, and decay. Much more decay will develop on bruised areas of apples than on nonbruised areas. A single severe bruise on an apple may increase the moisture loss by as much as 400%.

The amount of incipient decay infection, which influences storage potential of grapes and apples, can be predicted in the early storage period. Only lots with good storage potential should be held for late season marketing.

Maturity of the fruit at harvest time determines the refrigerated storage life and quality of the product. For any given produce there is a maturity best suited for refrigerated storage. Undermature produce will not ripen or develop good quality during or following refrigerated storage; overmature produce deteriorates quickly during storage.

Handling and Harvesting

Rising handling costs have encouraged the use of bulk handling and large storage bins for many vegetables and fruits. Moving, loading, and stacking bins by forklift trucks require care to maintain proper ventilation and refrigeration of the product. Bins should not be so deep that excessive weight damages the produce near the bottom.

Mechanical harvesters for fruits frequently cause some bruising. This damage can materially reduce the refrigerated storage life and quality of the produce.

Transportation

As in storage, losses from deterioration during distribution are affected by temperature, moisture, diseases, and mechanical damage. Gradual aging and deterioration are continuous after harvest. Time in transit may represent a large portion of the postharvest life for some commodities, such as cherries and strawberries. Thus, the environment during this period largely determines produce salability when it reaches the consumer.

To prevent undue warming and condensation of moisture, which promote decay and deterioration, most storages are built on railroad sidings. Canvas tunnels between the railcar and storage help minimize warming and condensation of moisture.

When produce is removed from storage for distribution to wholesale and retail markets, the storage operator can do little to prevent undesirable condensation. Warming the packages until they are above the dewpoint of the air would prevent it, but this takes time and space and is seldom practicable. Deterioration in flavor and condition accelerates after long periods of storage; therefore, the produce should be moved to consumers as quickly as possible.

The preparation of this chapter is assigned to TC 11.5, Fruits, Vegetables and Other Products.

STORING AND HANDLING OF FRUITS

Details on storage and handling of common fruits are given in the following sections. For more information on storage requirements and physical properties of specific commodities, see Chapter 25.

APPLES

Apples are not only the most important fruit stored on a tonnage basis, but their average storage period is considerably greater than any other fruit. The length of storage may be short for fall varieties and those going into processing, but cold storage is critical to proper handling and marketing.

Storage Temperature

For most varieties, 30 °F is the recommended storage temperature. This is 2 °F above the highest freezing point (28 °F) of most apples and should be safe for modern storage rooms. In older rooms with poor air distribution, a 30 °F temperature is unsafe, because of variable temperatures at different locations within the room.

Some apple varieties held at 30 °F develop physiological disorders that impair storage life and marketability. However, if storage temperatures are maintained at 39 °F, such disorders may not be an economic factor. Unfortunately, such elevation in temperature results in an accelerated rate of deterioration. Since the remedy may have consequences nearly as serious as the physiological disorder, a storage temperature of approximately 36 °F is often used.

Table 1 Storage Periods for Certain Apple Cultivars and Their Susceptibility to Storage Disorders

Cultivar	Months of Storage		Storage Scald Susceptibility	Other Disorders Likely to Occur in Storage[b]
	Normal	Maximum[a]		
Baldwin	4-5	6-7	Moderate	Bitter pit, brown core
Cortland	3-4	6-7	Very high	Senescent breakdown
Delicious	5-6	8-11	Moderate	Bitter pit, senescent breakdown, soft scald
Empire	4-5	8-9	Slight	—
Golden Delicious	5-6	7-11	Slight	Shriveling, bitter pit, senescent breakdown
Gravenstein	2	3	Moderate	Bitter pit, Jonathan spot
Granny Smith	5-6	7-9	High	Bitter pit, brown core, senescent breakdown
Idared	5-6	8-9	Slight	Breakdown, Jonathan spot
Jonathan	3-4	6-8	Moderate	Breakdown, Jonathan spot, soft scald
McIntosh	4-5	7-8	Moderate	McIntosh breakdown, brown core
Northern Spy	4-5	7-8	Slight	Senescent breakdown, Spy spot, bitter pit
Rome Beauty	5-6	7-8	Very high	Jonathan spot, soft scald
Spartan	5-6	7-10	Slight	Spartan breakdown, brown core
Stayman Winesap	4-5	7-8	Very high	Senescent breakdown
Winesap	5-6	8-9	High	Senescent breakdown
Yellow Newtown	5-6	8-9	High	Internal browning, senescent breakdown, bitter pit
York Imperial	4-5	6-7	Very high	Cork spot

Source: Hardenburg *et al.* (1986).
[a]For maximum storage, cultivars must be harvested at optimum maturity, stored under ideal temperature and humidity, and in most cases in the recommended controlled atmosphere. Some fruit may be stored 1-2 months longer than shown.
[b]Water core not listed for Delicious, Jonathan, Winesap, Stayman, and others, as it is present at harvest and does not develop in storage.

Storage life of apple varieties depends on harvest maturity, elapsed time, and temperature between harvest and storage, cooling rate in storage, and sometimes cultural factors. The best storage potential is usually found in apples that are mature but have not yet attained their peak of respiration when harvested. However, the grower is inclined to sacrifice storage quality for the better color often gained in red varieties by holding them longer on the tree. Even if harvesting began at the proper time, the fruit picked last may be at an advanced stage of maturity. Such late picked apples do not have good storage characteristics; neither do those harvested on the immature side, but this is seldom a problem with apples intended for storage before marketing. Harvest at proper maturity, careful handling, and prompt storage after harvest are conducive to long storage life.

Chilling injury is the term commonly applied to disorders that occur at low storage temperatures where freezing is not a factor. The exact mutual relationship of the many types of chilling injury is unknown. The principal disorders classed as chilling injuries in apples are: (1) soft scald; (2) soggy breakdown; (3) brown core; and (4) internal browning. Varieties susceptible to one or more of these disorders are Rome Beauty, Jonathan, Golden Delicious, Grimes Golden, McIntosh, Rhode Island Greening, and Yellow Newtown. In addition to variable susceptibility by variety, there are also yearly variations related to climate, fruit size, and cultural factors.

Table 1 lists the range in storage life at 30°F of several apple varieties susceptible to chilling injury at 30°F when grown under certain climatic or cultural conditions and may be more properly stored at temperatures of 36°F or higher. It also depends greatly on the expected length of storage before marketing and the availability of storage space at different temperatures.

Controlled Atmosphere Storage

Controlled atmosphere (CA) storage offers important gains in extending the market life of certain apple varieties. Chilling injury of some varieties is eliminated by elevating the storage temperature of these varieties to about 40°F and altering the composition of the atmosphere.

Only apples of good quality and high storage potential should be placed in CA storage. Harvest maturity and handling practices are crucial, and only fruits harvested at proper maturity should be considered. In any one district, this limits the apple harvest for CA storage to only a few days. Immature apples or those retained on the tree to gain better color, as is often the case with Delicious and McIntosh, are equally undesirable.

The following practices affect the condition of apples held for both conventional and CA storage:

Maturity. Because there is no reliable maturity index, growers must use personal experience of the variety, area, or orchard to decide when the crop is mature. Availability of labor, size of operation and crop, weather, storage facilities, and intended length of storage also affect the time of harvest.

Packaging. In many sections of the United States and Canada, apples are stored as harvested (*orchard-run*) in bulk bins. No sizing or sorting is done until the fruit is prepared for market. An exception is in the Pacific Northwest, where apples are sized and sorted shortly after harvest, packaged in corrugated containers that hold about 45 lb of fruit, palletized, and returned to storage. Stacking racks or supports must be used if pallets are stacked.

Handling to Storage. Apples should be cooled promptly after picking because they can deteriorate as much in one day at field temperatures as during one week held at proper storage temperatures. Normally, apples are placed in storage and cooled by the room refrigeration equipment to about 32°F in 1 to 3 days. Hydrocooling is sometimes used, but it requires careful disease control. It also interferes with scald inhibitors, which must be applied to the warm fruit.

The rooms for gas storage must be gastight, and there must be provision to remove excess CO_2 from the air. Carbon dioxide may be removed by circulating the air through washers or scrubbers filled with sodium hydroxide, ethanolamine, or plain water, or through a cabinet or room containing bags of hydrated lime.

The modified atmosphere may be obtained by filling the room with fruit, sealing it, and allowing the respiration of the fruit to provide the desired proportions. These proportions are maintained by operation of the scrubber and by ventilation.

Table 2 lists approximate temperature and atmospheric requirements for CA storage of several varieties. Since these may vary from region to region, more precise requirements should be determined from authorities within a region. For example, Jonathan,

Table 2 Requirements for Controlled Atmosphere Storage of Apples

Cultivar	Carbon Dioxide, %	Oxygen, %	Temp., °F
Cortland	5	2-3	36
	2-3	2-3	32
Delicious	1-2	1.5-2[a]	31-32
Golden Delicious	1-3	1.5-2[a]	31-32
Granny Smith	1-3	2-3	31-32
Idared	2-3	2-3	31-32
McIntosh	2-3 one month, then 5	2.5-3	36
	2-3 one month, then 5	2	38
Rome Beauty	1-3	2-3	31-32
Stayman Winesap	2-5	2-3	31-32
Yellow Newtown (Calif.)	8	3	40
(Oreg.)	5-6	3	36

Source: Hardenburg *et al.* (1986).
[a]1.5% oxygen not recommended for Delicious or Golden Delicious in New York, because less than 2% oxygen is injurious.

Rome Beauty, and Stayman Winesap are reported best at 2% CO_2 in New York, whereas a somewhat wider range is acceptable in Washington.

Disadvantages of gas storage include the difficulty in making the storage room gastight, the danger of suffocation to persons entering the room, and the impossibility of entering the room to examine the fruit without losing the desired atmosphere. Some of these disadvantages have been overcome by a method that passes the air going into the storage room through a generator, which reduces the O_2 level and raises the CO_2 content to desired levels. The process is continuous, so airtight rooms are not essential. Also, the rooms may be opened for inspection and the desired atmosphere restored quickly.

Boxes with polyethylene liners sealed after being packed with the apples have been used to create a modified atmosphere. In many instances, it is possible to closely approach the atmospheric composition of a CA room because of differential permeability of the film to O_2 and CO_2. Unfortunately, the composition of the atmospheres in different containers varies greatly, so that no dependence can be placed on improved storage life. Also, when the apples are marketed, the film liner must be opened in some way to avoid possible harmful effects of low O_2 or high CO_2 as the temperature rises. Although film liners seem impractical as a means of altering atmospheric O_2 and CO_2, unsealed film liners have proven very helpful in controlling excessive moisture loss from such varieties as Golden Delicious. Although often severe in cold storage warehouses, desiccation should not be of sufficient concern to justify film liners for fruit in CA rooms.

Storage Diseases and Deterioration

Storage problems in apples may be caused either by invading microorganisms or by the fruit's own physiological processes. Physiological disorders, although sometimes resembling rots, are related to biochemical processes within the fruit. Susceptibility to such disorders is often a variety characteristic, but it may be influenced by cultural and climatic factors and storage temperature.

Alternaria Rot. Dark brown to black, firm, fairly dry to dry storage decay centering at wounds, in skin cracks, in core area, or in scald patches—one of the blackest of storage decays. *Control*: Cultural practices that produce apples of good finish and prevent skin diseases and injuries that open way for infection.

Ammonia Gas Discoloration. Circular spots centering at lenticles; dull green on unblushed side and brown to black on blushed side. Injury may disappear from slightly affected fruits. *Control*: Ventilate as soon as possible. Examine fruit for injury at various points in the room because some sections may escape.

Bitter Pit. Many small, sunken bruise-like spots, usually on the calyx half of the fruit. Masses of brown, spongy tissue occur adjacent to surface pits or may be found deeper in the flesh. In storage, spongy tissue near surface loses moisture and tends to become hollow. New areas may appear and develop in storage. *Control*: Apply boron and calcium, as recommended, in the orchard. Follow cultural practices that promote regular bearing and stabilize moisture. Store fruit of proper maturity and cool promptly to 32 °F. Maintain humidity high enough to prevent moisture loss.

Blue Mold Rot (*Penicillium*). Spots of various sizes with decayed tissue that is soft, watery, and can be readily scooped out of the surrounding healthy flesh. Rot usually as deep as wide. Advanced stages have white tufts of mold which turn bluish green as spores are produced under moist conditions. Affected tissue has moldy or musty flavor and odor. Most prevalent type of storage decay of apples. *Control*: Handle carefully to prevent skin breaks. Cool promptly to 32 °F. Use fungicides in wash treatments. Keep picking boxes, packing house, and storage room sanitary. Whitewash walls and ceiling.

Brown Core. No external symptoms. First appears as slight browning or discoloration of core tissue between the seed cavities. Later, part or all of the flesh between the seed cavities and the core line may become brown. Serious in McIntosh and other susceptible varieties stored for long periods at 30 °F. *Control*: Pick at proper stage of maturity. Use controlled atmosphere storage at 38 °F. A disorder with similar symptoms has been reported as a result of exposure to excessive concentrations of carbon dioxide.

Freezing Injury. Watersoaked, rubbery condition of large areas or of entire apple. Vasculars (water-conducting strands) brown. Bruised areas in frozen apples large, with wrinkled gray to light brown surface. Moisture lost rapidly from affected areas. In refrigerator cars, most prevalent on floor, and at doorways; in storage rooms, most injury in bottom layer boxes, near coils, or against walls next to freezer storage. *Control*: Heat car during subfreezing weather. Prevent cold pockets in storage rooms by adequate air circulation. Minimum handling of fruit while frozen. Thaw at 40 to 50 °F. Move thawed fruit into trade channels promptly; do not allow it to become overripe.

Internal Breakdown. Mealy breakdown of internal tissue in overripe fruit. Flesh soft. Surface often duller and darker than normal. Hastened by too high storage temperature, freezing, bruising, or presence of water core which it often follows. *Control*: Pick before overmature. Cool promptly at temperatures as near 32 °F as possible for varieties that tolerate that temperature. Watch ripening rate, particularly of fruit with water core.

Internal Browning. No abnormal skin appearance. Sometimes appears only around core; the apple's outer fleshy portion remains normal in appearance. Occasionally only outer flesh is involved; but when internal browning develops in the outer fleshy portion, it is usually accompanied by browning around the core. Disease develops uniformly throughout tissue. Occurs in firm, sound apples. *Control*: Use controlled atmosphere at 38 °F for Yellow Newtowns and other susceptible varieties.

Jonathan Spot. Slate-brown to black, entirely superficial or very slightly sunken, skin-deep spots in color-bearing cells of skin. In some varieties, spots center at lenticels. *Control*: Refrigerate promptly as this disease is greatly aggravated by delayed storage. Use controlled atmosphere storage.

Lenticel Rots. Bullseye rot (*Neofabrabraea*) most common of group; of importance only from Northwest; spots fairly firm, pale centers, decay mealy, may penetrate nearly as deep as wide. Fisheye rot (*Corticum*), tough leathery spot, often follows scab, decayed tissue stringy. Side rots (*Phialophora*), spots shallow with tender skin, decayed tissue wet, slippery. *Control*: Harvest at prime maturity; store and cool promptly; use forecasting technique for Bullseye rot to determine potential keeping quality.

Scab (*Venturia*). Occasionally active scab spots on fruits at time of storage will enlarge. Fruits may be infected in orchard but show no disease at the time of storage. Disease may subsequently develop in storage as small brown or jet black spots in peel, often without breaking cuticle of fruit. *Control*: Follow recommended orchard spray schedule.

Scald. Diffuse browning and killing of skin of fruit stored for several months. Ordinarily most prevalent on immature fruit or on green portions of fruit. *Control*: Pick apples when well matured. Treat with effective scald-inhibiting chemicals. Scald develops less on fruit in controlled atmospheres.

Soft Scald. Sharply defined or slightly sunken ribbon-like areas in the skin. Affected tissue shallow and rubbery. Most severe on Jonathan, Golden Delicious, and Wealthy. *Control*: Store promptly. Use recommended controlled atmospheres, temperatures, and lengths in storage for each variety.

Soggy Breakdown. Light brown, moist, rubbery, definitely delimited areas in cortex of apple. Not visible on surface. Worst in Grimes Golden, Wealthy, and Golden Delicious. *Control*: Same as for soft scald.

Water Core. Hard, glassy, watersoaked regions in flesh of apple at core or under skin. Decreases in extent during storage but predisposes fruit to internal breakdown. *Control*: Pick as soon as mature. Watch fruit in storage and move before becoming overripe.

PEARS

Bartlett is the most important pear variety, exceeding the total of all others by a wide margin. Other Pacific Coast varieties are Hardy, Comice, Anjou, Bosc, and Winter Nelis. The eastern states have limited varieties due to the severe problem of fire blight, and grow primarily the Kieffer variety. Although most Bartlett, Hardy, and many Winter Nelis pears are canned, cold storage prior to ripening for canning is the usual procedure. A 10-day to 2-week cold storage period for Bartlett pears is commonly used by canners because it improves uniform ripening. Substantial quantities may also be stored for periods approaching maximum storage life of the variety to better use processing facilities.

Maturity at harvest has a very important bearing on subsequent storage life, as it does on the apple. However, unlike apples, pears do not ripen on the tree, nor do most varieties ripen at cold storage temperatures. If harvested too early, they are subject to excessive water loss in storage. If permitted to become overmature on the tree, their storage life is shortened and they may be highly susceptible to scald and core breakdown. Flesh firmness as measured by a pressure tester is perhaps the best measure of potential storage life of pears from any single orchard. For the Bartlett variety, a firmness of 19 to 17 lb, measured with a Magnus-Taylor pressure tester or similar device, using an 0.31 in. plunger head, indicates best storage quality. If average firmness is as low as 15 lb, storage for any prolonged period is hazardous. Pressure test information for each lot of pears going into storage may be very helpful to both the fruit owner and the cold storage operator in determining the storage program.

Careful harvesting and handling are essential to good storage quality. Bruises and skin breaks are likely sites for infection by microorganisms. Varieties such as Winter Nelis and Bosc are highly susceptible to punctures caused by stems broken in the harvesting operation. Comice is also easily damaged because of its very tender skin. Many pears are now being harvested into pallet bins holding about 1000 lb of fruit. Care in dumping fruit from a picking container is important in keeping mechanical damage to a minimum.

For best storage quality, rapid cooling after harvest is essential. Pears ripen rapidly at elevated temperatures but do not soften or change color in the early ripening stages. Therefore, a considerable part of the storage life may be used up without a visible change in the fruit. If cold storage rooms do not have adequate refrigeration and ventilation capacity for the rapid cooling of fruit, precooling in special rooms (or hydrocooling) should be considered prior to placing in the storage room. When warm fruit is placed in a room with cold fruit, the loading arrangements should be such that the temperature of the cold fruit is not elevated.

Table 3 Storage Life of Pear Varieties at 30°F

Variety	Storage Life, Months
Bartlett, Hardy, and Kieffer	2 to 3
Bosc, Comice, and Seckel	3 to 4
Anjou	6 to 7
Packham	5 to 6
Winter Nelis	7 to 8

Pears are very sensitive to temperature and should be stored at 30°F and 90 to 95% rh. Recommendations as low as 29°F have been made, but the risk of freezing injury is great unless the temperature in all parts of the room can be controlled precisely. Pears are not subject to chilling injury as are some apple varieties, so elevated storage temperatures are not required. The stacking arrangement recommended for apples in the cold storage room also applies for pears.

Since pears lose water more readily than most apple varieties, good humidity conditions in the storage room must be maintained. For long storage, 90 to 95% rh is recommended. Perforated film box liners are excellent for moisture loss control.

The approximate storage life of pears at 30°F is shown in Table 3. These values assume an additional time for transportation and marketing. If Bartlett pears for canning are harvested at the best stage of maturity and quickly cooled to 30°F, their safe storage life may be as long as 4 months, since marketing involves only ripening for processing. However, note that quality deteriorates during storage, particularly as the maximum storage life is approached.

After removal from storage, best dessert quality is attained if pears are ripened in a controlled temperature range of about 60 to 70°F. This applies to fruit for the cannery and for fresh use. Ripening at 68 to 72°F is more practical for cannery fruit than lower temperatures, because the shorter time involved reduces overhead costs with no measurable difference in quality.

The practice of controlled atmosphere storage of pears is promising. The storage life of Bartlett pears can be extended to 5 to 6 months at 30°F for fruit of desirable maturity (17 to 20 lb firmness) in an atmosphere containing 2 to 2.5% oxygen and 0.8 to 1% CO_2. Bartlett pears of advanced maturity are intolerant to elevated CO_2 and develop core and flesh browning within a few weeks. They are tolerant to low O_2, but have less storage potential than pears of desirable maturity. Chronological age and pressure test (under 16 lb firmness) is evidence of advanced maturity.

Commercial storage of pears in a controlled atmosphere has not been considered as necessary as it is in the apple industry. Since no low-temperature disorders have been recognized, there has not been a need for further extension of the storage life.

Many pears from western states, when packed for storage before marketing, use perforated polyethylene liners in the container. While such liners give excellent protection against water loss, there is no agreement as to their value in modifying the atmosphere within the container.

Storage Diseases and Deterioration

The principal storage disorders in pears are: (1) core breakdown; (2) scald and failure to ripen; and (3) fungus rots.

Core Breakdown. Often accompanies scald. Soft, brown breakdown in core area having acrid, disagreeable odor of acetaldehyde. *Control*: Do not allow pears to become overmature on tree. Cool promptly. Store at 30 °F. Ripen fruit between 65 and 75 °F.

Core breakdown is associated with overmaturity at harvest. This problem has become more important because of growth regulator sprays used to keep pears from dropping during the harvest season. Pressure test information of late harvested fruit is helpful in locating lots susceptible to core breakdown. Records of the time lapse between harvest and storage are important, since pressure test information may not be a true measure of relative storage quality where storage is delayed. Pear color, particularly in California Bartletts, is a very poor measure of potential storage life because of great variability among pears from different districts.

Scald. Often accompanies core breakdown. Brown to black softening of large areas of skin and tissues immediately beneath skin. Affected areas slough off readily. Acetaldehyde odor and flavor prominent. *Control*: Pick before overmature. Cool promptly. Store only for proper period. Cannot be controlled by oiled paper wraps.

Pear scald is associated with pears that have been stored too long and have lost their capacity to ripen. It is not related to apple scald and cannot be controlled by any supplemental treatments. The problem develops progressively earlier as the temperature is raised above 30 °F. Yellowing of the fruit is the principal storage symptom; Bartlett and Bosc are the two most susceptible varieties. Anjou and Comice may not develop scald but do lose their capacity to ripen. Periodic inspection is desirable to be sure that green pear varieties are removed from storage before yellowing progresses to the danger point. Yellow pears may show no scald in storage but may develop scald on removal to a ripening temperature. If pear scald does show in storage, the pears have been kept too long and may be worthless.

Anjou Scald. Anjou pears are often affected with a surface browning more superficial than common scald and distinct from it, resembling apple scald. Anjou scald is controlled by oiled paper wraps and effective scald inhibiting chemicals.

Gray Mold Rot (*Botrytis*). Extensive, firm, dull brown, watersoaked decay with bleached border. Dirty white to gray extensive mycelium forming nests of decayed fruits. *Control*: Wrap fruit in copper-impregnated paper. Use fungicide in spray or wax on packing line. Cool promptly to 30 °F.

Gray mold rot caused by *Botrytis cinerea* will grow at cold storage temperatures and can be a serious threat to long stored winter varieties such as Anjou and Winter Nelis. Without control measures, the disease may spread from one fruit to another by contact.

Alternaria Rot. Surface is dark brown to black. Decayed tissue is gray to black, dry in center, gelatinous at edge, easily removable as core from surrounding flesh. Found late in storage season, usually at punctures. *Control*: Prevent skin breaks. Remove from storage at first appearance of trouble.

Brown Core. Anjou and Bartlett pears stored in sealed, polyethylene-lined boxes with inadequate permeability may show various degrees of pithy brown core and desiccated air pockets. Prolonged storage in concentrations of 4% or more CO_2 often produces brown core, particularly when pears are harvested at advanced maturity or are cooled slowly after packing. *Control*: Harvest at proper maturity. Cool promptly. Store at 30 °F. Use perforated film liners to maintain CO_2 level at 1 to 3%.

Freezing Injury. Bartlett and Anjou pears exposed for 4 to 6 weeks just below their freezing point develop glassy, watersoaked external appearance with tan pithy area around core. Pears frozen sharply may break down completely or show abruptly sunken large pits where slightly bruised while frozen. *Control*: Keep transit and storage temperature above 30 °F.

GRAPES

Grapes are widely grown in the United States, but over 90% are grown in California. This state produces grapes of the *Vitis vinifera* species almost exclusively. This species can withstand the rigors of handling, transport, and storage required of table grapes for wide distribution over a long marketing period. Almost all of this fruit is precooled and much of it stored for varying periods before consumption. On the other hand, for fresh use, the fruit of the species *Vitis labrusca* (Eastern type) is largely limited to local market distribution.

Grapes grow relatively slowly, and should be mature before harvest because all of their ripening occurs on the vine. *Mature* here means that stage of physiological development when the fruit appears pleasing to the eye and can be eaten with satisfaction. However, grapes should not be overripe, as this predisposes the fruit to two serious postharvest disorders: (1) weakening of the stem attachment in some varieties, such as Thompson seedless, which causes the berries to separate from the pedicel attachment; and (2) progressively greater susceptibility to invading decay organisms. Danger of fruit decay is increased with exposure to rain or excessively damp weather before harvest (conditions favorable for the inception of field infections by *Botrytis cinerea Pers*).

Cooling and Storage

Grapes are vulnerable to the drying effect of the air because of their relatively large surface to volume ratio, especially that of the stems. Stem condition is an important quality factor and an excellent indicator of the past treatment of the fruit. Stems should be maintained in a fresh green condition not only for appearance but because they become brittle when dry and are apt to break. The stem of a grape cluster, unlike that of other fruits, is the handle by which the fruit is carried; if breakage (shatter) occurs, the fruit is lost for all practical purposes even though the shattered berries may still be in excellent condition. Therefore, careful attention should be paid to those operations that minimize moisture loss.

The rate of water loss is especially high before and during precooling because grapes are normally harvested under hot, dry conditions. Remove field heat promptly after the fruit is picked to minimize the exposure of grapes to low vapor pressure conditions. Volume and temperature of the precooling air, velocity of the air past or through the containers, and accessibility of the fruit to this air are significant factors in the rate of heat removal and are drastically influenced by the location and amount of venting of the containers, alignment of the containers (air channels), and packing materials such as curtains, cluster wraps, and pads.

Two general systems of air handling used for precooling table grapes are (1) the *velocity* or conventional system, and (2) the *pressure* system.

In the velocity system, air is forced through channels parallel to the long axis of the containers. Heat transfer is effected by conduction through the packaging materials and by penetration of the cold air to the fruit from turbulence in and around the vents. Satisfactory precooling rates can be attained if: (1) the containers are aligned so that there are no obstructed channels; (2) the velocity of the air through these channels is at least 100 fpm; and (3) air of not more than 35 °F can be supplied at the rate of at least 0.16 cfm/lb of fruit. Very humid air helps to reduce the rate of water loss from the fruit. Large cooling surfaces can be maintained, or atomized water can be added to the airstream. However, the rate of removal of field heat is the significant factor; humidifying techniques that retard precooling are likely a liability.

In the pressure system, a pressure gradient is set up so that there is a positive flow of cold air through the fruit from one vented side of the container to the other. The containers are arranged so that the air must pass *through* the containers before returning to the refrigeration surface. Precooling time may be as little as one-fifth that of the velocity system if the pressure differential across the packages is equivalent to at least 0.25 in. of water and cold air is supplied at the rate of at least 1 cfm/lb of fruit.

Recommended storage temperatures for *Vitis vinifera* (European or California type) grapes are 30°F. The relative humidity should be from 90 to 95%. Although temperatures as low as 29°F have not been injurious to well-matured fruit of some varieties, other varieties of low sugar content have been reported damaged by exposure to 31°F. Grape storage plants in California should provide uniform air circulation in the rooms. Some have precooling rooms where the grapes are cooled to about 39°F in 6 to 24 h before storage. In some plants, all of the cooling is done in the storage rooms, but only a few have sufficient air movement to cool the fruit as quickly as desired. After the fruit has been precooled, the air velocity should be reduced to a rate that will maintain uniform temperatures throughout the room (no more than 10 to 20 fpm in the channels between the lugs). Ventilation is required only to exhaust sulfur dioxide and air following fumigation.

The greatest change that takes place in grapes in storage is loss of water. The first noticeable effect is drying and browning of stems and pedicels. This effect becomes evident with a loss of only 1 to 2% of the mass of the fruit. When the loss reaches 3 to 5%, the fruit loses its turgidity and softens.

Maintaining a relative humidity of 90 to 95% in grape storage is often a problem especially at the beginning of the storage season when the rooms are being filled with dry lugs. Each lug will absorb 0.33 to 0.67 lb of water over a month's period and, unless moisture is supplied to the room, this water must come from the fruit. An effective method of supplying water to minimize shrinkage is by spray humidification. A fine spray can be obtained that will vaporize readily even at 31°F with proper balance of water and air pressure using the correct type of nozzle.

Fumigation

Vinifera grapes must be fumigated with sulfur dioxide after they are packed to prevent or retard the spread of decay. The treatment surface sterilizes the fruit, particularly wounds made during handling.

Fumigation with sulfur dioxide in storage prevents new infections of the fruit but does not control infections that have already occurred in the vineyard. Frequently, these have not developed far enough to be detected at harvest and consequently are the primary cause of decay in storage. A method of measuring field infection has been developed and used to forecast decay during storage. The forecast indicates the lots that are sound and can be safely stored and also those that are likely to decay and should be marketed early (Harvey 1955, 1984).

It has become common practice to accumulate packed fruit in the precooler during the daily packing and to fumigate the fruit in the evening. In this way, precooling is not delayed and fumigation can be done after most of the working crew has left. This initial treatment often becomes the responsibility of the refrigeration personnel.

Amount of Sulfur Dioxide. Other commodities should not be treated with the grapes or even held where the fumigant can reach them, as most of them are very easily injured by the gas. Because grapes also can be injured, they should be exposed to the minimum quantity of sulfur dioxide required, which will depend on: (1) the decay potential and condition of the fruit; (2) the amount of fruit to be treated; (3) the type of containers and packing materials; (4) the air velocity and uniformity of air distribution; (5) the size of the room; and (6) losses from leakage and sorption on walls. Under favorable conditions, a basic sulfur dioxide concentration of 0.5% by volume for 20 min. is adequate. To keep the concentration at this level, consider the absorptive capacity of the containers and fruit as well as their volume. The dosage can then be calculated from the following equation:

$$W_s = \frac{AB}{100\,E} + CD$$

where

W_s = quantity of sulfur dioxide required, lb
A = concentration of sulfur dioxide to be used, %
B = free volume of room (total volume minus 0.5 ft^3 for each container), ft^3
C = number of carloads of 28-lb lugs (1000 lugs/car)
D = quantity of sulfur dioxide absorbed by each carload, lb
E = volume occupied by 1 lb of sulfur dioxide gas at 32°F (5.5 ft^3)

For factor D, 1 lb per car is adequate when the fruit is sound. Air velocities are maintained past both sides of every container at 50 fpm or more (75 to 100 fpm if the fruit has curtains over it or the clusters are wrapped), and the room is relatively gastight with no opportunity for the fumigant to be lost on refrigeration surfaces. Conversely, a higher value of 2 lb car would be used when these factors are less favorable.

Grapes must be fumigated weekly in storage to prevent *Botrytis cinerea* from spreading from infected fruit to adjacent sound fruit. The amount of sulfur dioxide needed depends on the same factors as for the initial treatment. However, a basic concentration of 0.1% for 30 min. is adequate. Also, an absorptive factor in the range of 0.33 to 0.67 lb of sulfur dioxide per carload should be used.

Distribution Procedure. The gas must be distributed quickly and evenly to all parts of the room. This can be done by spacing special nozzles 6 ft apart along the ceiling in the room. If the outlet is placed in front of a fan, there should be one for each fan, or the air from the single fan should be distributed evenly across the room through a plenum.

The same requirements of proper container alignment, adequate fan capacity, and uniform air distribution apply here as for the initial treatment. The lugs should be oriented parallel to the airflow and channels 0.75 to 1.5 in. should be provided on both sides and kept completely unobstructed through the stacked fruit. The fruit should be stacked as near the ceiling as possible so that curtains provided over the fruit will prevent air from passing over the fruit and thus bypassing the channels. The working distance between pallets should be kept to an absolute minimum to avoid wide channels, and no holes should be left in the wall of lugs when pallets of fruit are withdrawn.

The hot-gas method of delivery may be used if the room requires 10 lb or less of gas. The steel cylinder containing the liquid sulfur dioxide is first connected to the gas inlet and the valve is then opened. The cylinder should then be placed in a pot of boiling water to vaporize the fumigant as rapidly as possible. Only about 1 lb/min can be delivered this way.

For larger quantities, the cold-gas method is usually more practical. A riser extends to the bottom of the cylinder through which the liquid sulfur dioxide rises and flows through the delivery line. Every precaution must be taken to provide enough air volume and velocity to vaporize and mix the gas thoroughly with the air before it reaches the fruit. Up to 100 lb of the material can be released in 2 to 3 min. After 30 min., the room should be purged of the gas-laden air until personnel can remain in the space without excessive discomfort.

In plants that are devoted entirely to the storage of grapes, the gas is sometimes released into the air ducts of the plant, thus using the air-cooling system for even distribution and good circulation in the rooms. With a brine spray system of refrigerating the air, a bypass around the spray chamber prevents the gas from contacting the wet metal surfaces, since it readily forms a corrosive acid in combination with water. For the same reason, sulfur dioxide should be cleared from the air of the rooms before the damper is turned and the air is circulated through the spray. It is advisable to check the acidity of the brine frequently to guard against corrosion.

Precautions. Sulfur dioxide has certain properties that demand care in its use as a fumigant in cold storage plants. The concentrations recommended for the fumigation of grapes in storage can cause respiratory spasms and death if the victim cannot escape the fumes. When working in even weak concentrations of sulfur dioxide, wear goggles to protect against injury to the eyes and a gas mask fitted with canister for acid gases (not the usual canister for ammonia gas). Concentrations as low as 30 to 40 ppm can be detected by smell. It requires several times these concentrations to cause discomfort.

Because a small segment of the population may experience severe allergic reactions to sulfites, the U.S. Environmental Protection Agency has proposed a 10 ppm tolerance for sulfite residues in table grapes (EPA 1989, Harvey *et al.* 1988). Fruit with residues exceeding the tolerance cannot be marketed.

Another precaution about sulfur dioxide which cannot be overemphasized is its injurious effects on other produce. For this reason, care must be taken that only grapes are stored in the room that is to be fumigated and that there are no leaks through wall or halls to adjacent rooms storing other produce.

Periodic inspection of the fruit is recommended to check whether the sulfur dioxide gas is reaching the center of the stacks or whether some grapes are being overtreated. If the pedicels and stems retain a yellow or green color and broken berries show no mold and appear to be dried or seared, the gas has reached the fruit in question and is having the desired effect. Serious bleaching on unbroken grapes means too high a concentration or too long an exposure, and there should be better distribution of the gas, lower concentration, or shorter fumigation periods.

Diseases

Blue Mold Rot (*Penicillium*). Watery, mushy condition. Early production of typical bluish-green spores on berries and stems. Moldy odor and flavor. *Control*: Prevent deterioration in fruit by careful handling and prompt refrigeration, preferably to 32 °F. Fumigate with sulfur dioxide in storage.

Cladosporium Rot. Black, firm, shallow decay which produces an olive-green surface mold. Common on stored grapes harvested early in the season. Infections occur on small growth cracks at the blossom end and sides of the grape. *Control*: Precool and store grapes promptly at 32 °F. After harvest, fumigate with sulfur dioxide to reduce spread.

Gray Mold Rot (*Botrytis*). Early stage, slip skin with no mold growth. Later, nest of fairly firm decay covered with abundant fine gray mold and grayish-brown, velvety spore masses. *Control*: Cull out decay when packing. Fumigate grapes with sulfur dioxide. For storage, cool grapes rapidly to 30 °F. Use forecasting technique to determine safe storage periods. Use short storage period for grapes harvested in rainy periods or after slight freezes.

Rhizopus Rot. Soft, mushy, leaky decay causing staining of lugs. Coarse extensive mycelium and black sporangia develop under moist conditions. *Control*: Prevent skin breaks. Cool promptly to below 50 °F.

Sulfur Dioxide Injury. Bleached sunken areas on berry at skin breaks or cap-stem attachment. Decolorized portions have disagreeable astringent flavor. Does not appear in full severity until cool grapes are warmed. *Control*: Apply proper concentration and distribution of gas for recommended period.

Storage Life

The normal storage life of the principal varieties of California table grapes at 30 °F is shown in Table 4. Under exceptional conditions, sound fruit will keep longer than indicated; for example, Emperor grapes have been held in good condition for 7 months, and Thompson seedless for 4 months.

Table 4 Storage Life of California Table Grapes

Variety	Storage Life, Months
Emperor, Ohanez, Ribier	3 to 5
Malaga, Red Malaga, Cornichon	2 to 3
Thompson Seedless, Tokay	1 to 2.5
Muscat, Cardinal	1 to 1.5

Table 5 Storage Life of Labrusca Grapes at 32 °F

Variety	Storage Life, Weeks
Catawba	5 to 8
Concord, Delaware	4 to 7
Niagara, Moore	3 to 6
Worden	3 to 5

The storage life of grapes is affected mostly by the attention given to selecting and preparing the fruit. Grapes should be picked at the best maturity for storage, especially Thompson seedless and Ohanez. Stems and pedicels should be well developed and the fruit should be firm and mature. Soft and weak fruit should not be stored. The display lug is a satisfactory package for storage since it can be cooled and fumigated easily.

Cooling to 40 to 45 °F is advised for grapes that are to be in transit a day or two before reaching storage. Special care should be taken during transit so that decay does not start. It is not good practice to delay fumigation until the grapes reach a distant storage plant, for in the picking and packing of grapes, many berries are injured sufficiently to permit mold to begin unless the fruit is fumigated promptly.

For labrusca (Eastern type) grapes, a storage temperature of 32 °F and humidity of 85% are recommended. Care in packing and handling the fruit, a minimum of delay before storage, and prompt cooling are important for best results with these varieties, as they are with the *vinifera* grapes. The Eastern varieties are not fumigated with sulfur dioxide due to their susceptibility to injury from it. The storage life of the important commercial varieties at 32 °F is shown in Table 5.

PLUMS

Plums are not suitable for long storage. Among the major shipping varieties, well-matured Santa Rosa, El Dorado, Nubiana, Queen Ann, Laroda, Late Santa Rosa, and Casselman are sometimes stored for short periods. The Italian Prune can be held for no more than 2 weeks before marketing begins.

Plums intended for storage should be harvested at a high soluble solids level for the variety, although doing so may delay the harvest beyond the normal picking date. Harvested fruit should be carefully graded to remove disease, defects, and injuries before packing in the shipping container.

The fruit should be thoroughly cooled before storage. Cooling may be done in the 900- to 1000-lb bulk bins that are used for harvest. The shipping containers are normally vented to aid cooling after packing. While most fruit is air cooled in conventional room coolers, some shippers use forced air to cool fruit quickly in bulk bins or shipping containers.

Plums can usually be stored for 1 month at 30 to 32 °F with 90 to 95% rh. Results of storage life tests have been variable, with some lots in certain seasons remaining in good condition even after 4 to 5 months in storage. Other lots in some seasons have not been held satisfactorily beyond 2 months. Fruit with the highest soluble solids has consistently shown the longest storage life, even when harvested several weeks after the completion of commercial harvest. Some plum varieties benefit from CA storage.

Storage Diseases and Deterioration

Plum deterioration appears as changes in appearance and flavor. A poststorage holding period should be used in judging the condition of stored fruit. Fruit that appears bright and flavorful in storage can show severe deterioration when removed to room temperature for 2 to 3 days.

Some flesh softening and a gradual loss of varietal flavor and tartness occur even at low storage temperatures. The first visual sign of deterioration is the development of translucence, first around the pit, then extending outward through the fruit. Translucence is followed by the development of progressively more severe flesh browning following the same pattern. The first noticeable loss in flavor is generally associated with the first symptoms of translucence in the tissue. It is necessary to cut through the fruit to judge condition, since fruit held under good storage conditions may appear sound from the outside while being seriously deteriorated internally. See section on Sweet Cherries for diseases. See Cold Storage and Sulfur Dioxide Injuries under Peaches.

SWEET CHERRIES

Harvesting Techniques

Sweet cherries for storage must be harvested with stems attached. Mechanical harvesting takes most of the fruit without stems and consequently the cherries must be processed or otherwise used immediately.

Cooling

Rapid cooling to 30°F is essential if this fruit is to be stored. Hydrocooling has been used successfully, and the wetting is tolerable as long as the fruit remains cold. Fungicidal postharvest sprays or dips are helpful in reducing decay during storage.

Forced-air or pressure cooling can be used to quickly cool the fruit without the problem of wetting. Moisture loss and stem drying can be minimized by rapid movement from the field to cooler, rapid cooling, and maintaining low temperatures and high humidity during cooling and storage.

Storage

When sweet cherries are stored, they are normally held in shipping containers, often with polyethylene liners. These liners permit an increase in carbon dioxide gas surrounding the fruit, which tends to reduce decay rates and increase storage time. Cherries should be stored at 30°F and may be held 2 weeks after harvest and still retain enough quality for shipment to market.

Controlled atmospheres with 20 to 25% CO_2 or 0.5 to 2% oxygen help maintain firmness and bright, full color during storage (Hardenburg *et al.* 1986). Polyethylene liners can extend the market life. The liner must be perforated when removed from storage.

Diseases

Alternaria and Cladosporium Rot. Light brown, dry, firm decay lining skin breaks that can be removed easily from surrounding healthy tissue. Mycelium on the area are fine and white above and dark green below. *Control*: Sort out cherries with cracks and other skin breaks at packing. Use fungicide in spray or sizer on packing line.

Blue Mold Rot (*Penicillium*). Circular, flat spots covering conical, soft, mushy decay that can be scooped out cleanly from surrounding healthy flesh. White fungus tufts turning to bluish green develop on surface. Musty odor and flavor. *Control*: Prevent skin breaks. Use fungicide in spray or sizer on the packing line. Market promptly. Refrigerate promptly to 32°F.

Brown Rot (*Monilinia*). See Brown Rot of Peaches. *Control*: Follow recommended orchard spray practices. Use fungicide in spray or sizer on packing line. Refrigerate promptly to 32°F. Package cherries in polyethylene bags to reduce desiccation of stem and fruit, preserve color, and reduce decay development.

Gray Mold Rot (*Botrytis*). Light brown, fairly firm, watery decay covered with extensive delicate, dirty-white mycelium. On completely decayed cherries, grayish-brown velvety spores may be found. *Control*: Handle carefully. Use fungicide in spray or sizer on packing line. Refrigerate promptly to 32°F.

Rhizopus Rot. Extensive soft, leaking decay with little change from normal color. Coarse mycelium and black spore heads are prominent under moist conditions. More prevalent in upper-layer packages in refrigerator car. *Control*: Rhizopus develops very slowly at temperatures below 50°F, so storage at recommended temperature keeps decay in check.

PEACHES AND NECTARINES

This discussion relates primarily to peaches but also applies to nectarines in many respects.

Storage Varieties

Peaches do not adapt well to prolonged storage. However, if they are sound and well matured, most freestone varieties can be stored for up to 2 weeks (some freestone and most clingstone for up to 4 weeks) without any noticeable deterioration in flavor, texture, or appearance. Storage life appears to be geared with the harvest season. Early varieties, particularly the freestone peaches now being grown in Florida and the early clingstones grown in the Southeast, have an extremely short storage life and should be used as soon as possible after harvest. However, some late season varieties can be safely stored for up to 6 weeks. In the West, the Rio Oso Gem is consistently stored for 4 to 6 weeks before being marketed.

Harvest Techniques

Peaches for fresh consumption must be in a condition to survive a postharvest holding period of several days to several weeks. The fruit must be sound and bruise-free and must be handled delicately during the harvesting and packing operations. Today, with the widespread use of bulk bins or pallet boxes, hand-picked fruit requires extra careful handling. Hydrodumpers are generally employed for dumping the pallet bins. With proper care, pallet boxes cause less bruising than small field boxes.

Cooling

Cooling peaches to 40°F soon after harvesting is essential to postharvest retention of quality and control of decay. Peaches begin to soften and decay in a few hours without proper temperature management. All peaches shipped out of the Southeast are hydrocooled. Originally, the fruit was cooled in flood-type hydrocoolers as a final operation after it was packed in containers. In the West, most fresh peaches are air cooled in pressure coolers to remove the field heat rapidly for the postharvest holding period. By using forced air or pressure cooling, peaches or nectarines in two-layer plastic tray packs with 6% side-vented corrugated containers will cool 80% in about 6 h with an airflow of 0.2 cfm/lb of fruit.

Storage

Peaches are normally stored in corrugated or tray pack shipping containers.

An environment of 31°F and 90 to 95% rh with very low air movement is best for peaches. Under these conditions, peaches can be held from 2 to 6 weeks, depending on variety.

The same storage conditions may be used for nectarines; however, they are somewhat more susceptible to shrivel than are peaches. Air velocity in the storage room should be as low as possible but still maintain proper storage temperatures. Frequent checks should be made of the fruit at the edge of alleyways, for example, to detect the first signs of shrivel.

Good experimental CA results have been obtained with peaches and nectarines held in 1% O_2 with 5% CO_2 at 32°F. Extended storage of 6 to 9 weeks is possible. The fruit ripens or softens with good flavor and is juicy on removal. Low-temperature breakdown, which is usually encountered with lengthy storage, is controlled by CA. While CA reduces decay, it does not completely control it; thus, a fungicide is needed for extended storage.

Diseases

Brown Rot (*Monolinia*). Extensive firm, brown, unsunken areas turning dark brown to black in the center and generally covered with yellowishgray spore masses. Skin clings tightly to center of old lesions. *Control*: Follow recommended field and postharvest control measures involving use of heat treatments and fungicides. Refrigerate promptly to as near 32°F as feasible.

Cold Storage Injury. Fruit loses flavor, becomes dry and mealy. Breakdown starting around pit is grayish brown, watersoaked, or mealy. *Control*: Refrigerate promptly to 32°F. Breakdown appears earlier at 38°F. Store for only 2 to 4 weeks, depending on variety.

Pustular Spot (*Coryneum*). Common on peaches from the West, occasionally on Eastern fruit. At first small purplish-red spots, later up to 0.5 in. in diameter, brown, sunken with white center. *Control*: Treat with orchard sprays. Cool harvested fruit to below 45°F.

Rhizopus Rot. Extensive, fairly firm, watery decay with uniformly brown surface color. Skin slips readily from center of lesions. Coarse mycelium; black spherical sporangia develop. *Control*: Store cannery peaches at 32°F before ripening. Prevent skin breaks. Follow recommended field and postharvest control measures. Refrigerate promptly to as near 32°F as feasible.

Sulfur Dioxide Injury. Bleached and pitted areas on fruit surface. After removal from refrigeration, injured areas of peaches are brown, dry, and collapsed. Skin may slough off. *Control*: Avoid sulfur dioxide contact of peaches (and other stone fruits) in storage or in transit with grapes.

Sour Rot. An unfamiliar postharvest disease in peaches was noticed in some packing sheds in the Southeast. First signs of the infection may be peaches that are easily skinned by the brushes and belts on the packing line. Affected peaches then develop softened and sunken brown lesions that eventually become covered with a white or creamy exudation. The infected areas generally emit a vinegar-like, sour odor. *Control*: Chlorination of dump tank water, chlorination of hydrocooling water, and careful culling of all overripe, bruised, and damaged fruit. In short, good shed sanitation and quality control are the keys to eliminating sour rot.

APRICOTS

Apricots are not stored for a prolonged time but may be held for 2 or 3 weeks if they are picked at a maturity firm enough that they will not bruise. Unfortunately, this maturity does not yeild good dessert quality fruit. Care must be used in sizing and packing the fruit going into storage, as small surface bruises can become infected with disease-producing organisms. Chapter 25 has further details.

Apricots for short-term storage are harvested in much the same way as freestone peaches, precooled, and placed in storage promptly. Storage temperature should be 32°F with a 90 to 95% rh.

Diseases and Deterioration

See Peaches and Nectarines.

BERRIES

Blackberries, raspberries, and related berries cannot be stored for more than 2 or 3 days even at 31°F with a relative humidity of 90 to 95%. An atmosphere with 20 to 40% CO_2 will increase storage life by 3 or 4 days by inhibiting fungal rots.

As they come from the field, cranberries are stored in field boxes at 36 to 40°F and 90 to 95% rh. They usually are not stored longer than 2 months. Storage at 30 to 32°F causes chilling and physiological breakdown. Modified atmospheres have not extended the storage life of fresh cranberries beyond that of conventional storage.

Diseases

Cladosporium Rot. Surfaces of berries covered with olive to olive-green mold. In raspberries, the mold is most abundant on inside or cup of berry. *Control*: Avoid bruising; pack and ship promptly. Refrigerate to 32°F.

Gray Mold Rot. Causes soft, watery rot. Fruit may be covered with dense, dusty gray growth of fungus which spreads rapidly in package, forming nests. *Control*: Avoid bruising. Refrigerate to 32°F in transit.

Anthracnose (*Gloeosporium sp.*). Berries may be completely rotted and show masses of spores glistening in salmon-colored droplets on fruit. *Control*: Refrigerate to 32°F.

Alternaria Rot. Affected berries remain firm and show gray-white woolly fungal growth from injured cap stem areas. Nesting occurs in tight clusters scattered throughout containers. *Control*: Refrigerate to 32°F.

Chilling Injury. Berries held for 4 or more weeks become tough and rubbery, surfaces are dull in appearance, red in color throughout. *Control*: Hold fruit at 37°F.

Fungus Rots (*Several Fungi*). Limited portions or entire berries are brown, soft, or collapse. Some berries turn into water bags. *Control*: Spray in field. Handle carefully. Reduce temperatures to 38°F after harvest.

STRAWBERRIES

Diseases

Gray Mold Rot (*Botrytis*). Brown, fairly firm, fairly dry decay. Dirty-gray mold and grayish-brown velvety spore masses present. Nesting common. *Control*: Apply recommended fungicides in field. Handle carefully to prevent skin breaks. Cull out all diseased berries. Cool promptly to 40°F or below.

Leather Rot (*Phytophthora*). Large, slightly discolored tough areas with indefinite purplish margins. Vascular system browned, flavor bitter. *Control*: Mulch plants to keep berries from contact with infested soil. Cool promptly to 40°F or below.

Rhizoctonia Rot. Hard dark brown decay on one side of berry, usually small quantities of soil adhering. Develops only a little after harvest. *Control*: Mulch plants to keep berries from contact with infested soil. Cull thoroughly.

Rhizopus Rot. Mushy, leaky collapse of berries associated with coarse black mycelium and sporangia. Extensive red staining of containers from leaking juice. *Control*: Reduce temperature promptly to 40°F or below. Handle carefully to prevent skin breaks.

FIGS

Diseases

Alternaria Spot. White fungal growth on surfaces which soon darkens. As fungus spots enlarge, tissue beneath becomes slightly sunken. *Control*: Cool promptly after harvest, hold at 45°F in transit.

Black Mold Rot (*Aspergillus*). Disease first appears as a dirty white to pink color of the skin and pulp. White mold growth develops within fig. Cavities formed in fruit become lined with black spore masses. *Control*: Store fresh figs at 32°F at relative humidity of 85 to 90%.

SUPPLEMENTS TO REFRIGERATION

Antiseptic Washes. Many fruits are washed before packing to remove dirt and improve appearance. In some cases, hydrocoolers are used to remove field heat. If the water is recirculated, it may become heavily contaminated with decay-producing bacteria and fungi. Chlorine can be added to the water at a level of 50 to 100 ppm to control the buildup of these organisms. Other fungicides may also be used, but they must be legally registered for the specific application.

Protective Packaging. Proper packaging protects against bruising, moisture loss, and spread of disease. Packaging materials may also contain chemicals to control spoilage. Packages must have good stacking strength for palletizing and must also perform under high humidity conditions.

Selective Marketing. The potential storage life of grapes and apples can be predicted within a few weeks after they are stored. Thus, those with a short storage life may be marketed while still in good condition and longer lived products can be stored for late season marketing. Samples taken from each lot placed in storage are kept for a few weeks at temperatures and relative humidities that favor rapid development of decay. Grapes that will not keep long can be detected in about 2 weeks and apples in about 60 days. Since both of these fruits may be stored for several months, knowing their potential storage life can significantly reduce spoilage losses.

Heat Treatment. Heat treatments to reduce decay also kill insects on the surface of the fruits and microorganisms near the surface without leaving a residue. For example, brown rot and rhizopus rot of peaches are reduced by exposing the fruit for 1.5 min. in 130°F water or for 3 min. in 120°F water.

Fungicides. Fungicides may be applied during cleaning, brushing, or waxing of some fruits. Only fungicides registered for the particular fruit and use may be used.

Irradiation. Gamma radiation has effectively controlled decay in some products. High dosages can cause discoloration, softening, or flavor loss. Commercial application of gamma radiation is limited due to the cost and size of equipment needed for the treatment and to uncertainty about the acceptability of irradiated foods to the consumer (Hardenburg *et al.* 1986).

Ultraviolet lamps are sometimes used to control bacteria and mold in refrigerated storages. While ultraviolet light kills bacteria and fungi that are sufficiently exposed to the direct rays, it does not reduce decay of packaged fruits in storage. Even ultraviolet light directed on fruits as they passed over a grader did not control decay.

REFERENCES

Environmental Protection Agency. 1989. Interim policy for sulfiting agents on grapes; Pesticide tolerance for sulfur dioxide. *Federal Register* 54 (3): 382-85.

Hardenburg, R.E., A.E. Watada, and C. Y. Wang. 1986. The commercial storage of fruits, vegetables, and florist and nursery stocks. USDA *Agriculture Handbook,* No. 66.

Harvey, J.M. 1955. A method of forecasting decay in California storage grapes. *Phytopathology* 45: 229-32.

Harvey, J.M. 1984. *Instructions for forecasting decay in table grapes for storage.* U.S. Department of Agriculture, ARS-7.

Harvey, J.M., C.M. Harris, T.A. Hanke, and P.L. Hartsell. 1988. Sulfur dioxide fumigation of table grapes: Relative sorption of SO_2 by fruit and packages, SO_2 residues, decay, and bleaching. *American Journal of Enology and Viticulture* 39: 132-36.

BIBLIOGRAPHY

Chau, K.V., C.D. Baird, P.C. Talasila, and S.A. Sargent. 1992. Development of time-temperature-humidity relations for fresh fruits and vegetables. Final report for ASHRAE Research Project 678-RP.

Ryall, A.L. and W.T. Pentzer. 1982. *Handling, transportation, and storage of fruits and vegetables*, 2nd ed. Vol. 2, Fruits and tree nuts. AVI Publishing Co., Westport, CT.

Nelson, K.E. 1979. Harvesting and handling California table grapes for market. University of California *Publication* 4095.

CHAPTER 16

CITRUS FRUITS, BANANAS, AND SUBTROPICAL FRUITS

CITRUS FRUITS

THIS chapter discusses the harvesting, handling, storage requirements, and possible disorders of fresh market citrus fruits produced in Florida, California, Texas, and Arizona.

MATURITY AND QUALITY

The degree of citrus fruit ripeness at the time of harvest is the most important factor determining eating quality. Oranges and grapefruit do not improve in palatability after harvest. They contain practically no starch, do not undergo marked composition changes after they are picked from the tree (as do apples, pears, and bananas), and their sweetness comes from natural sugars contained when they are picked.

The ripening of citrus fruits is a slow, gradual process closely related to increases in diameter and weight. Citrus fruits must be of high quality when harvested to assure quality during storage and shelf life.

Quality is often associated with the fruit rind's appearance, firmness, thickness, texture, freedom from blemishes, and color. Actually, quality determination should be based on the texture of the flesh, juiciness, content of total solids (principally sugars), total acid, aromatic constituents, and vitamin and mineral content. Age is also important, because immature fruit is usually coarse, very acid or tart, and has an internal texture that is ricey or coarse. Overripe fruit held on the tree too long may become insipid, develop off-flavors, and possess short transit, storage, and shelf life. The importance of having good quality fruit at harvest cannot be overemphasized. The main objective thereafter is to maintain quality and freshness.

HARVESTING AND PACKING

Picking. Citrus fruit is harvested in the United States throughout the year, depending on the growing area and kind of fruit. The approximate commercial shipping seasons for Florida, California-Arizona, and Texas citrus are shown in Figure 1. The picking operations are conducted by trained crews from independent packinghouses or large associations. These organizations schedule picking to meet market demands. The fruit that is not handled through cooperatives is normally sold on the tree to the shippers or processors and is picked at the latter's discretion.

The fruit is carefully removed from the trees either by hand or with special clippers and is then placed in picking bags that are emptied into field boxes. An increasing amount of fruit is handled in bulk, and the pickers put the fruit into pallet boxes or wheeled carts. In some cases, especially when fruit is picked for processing, it is loaded loose into open truck trailers. In Florida, over 90% of the oranges and slightly more than 50% of the grapefruit and specialty fruit are processed, while in California and Arizona less than 35% of all citrus fruit is processed.

At the beginning of the season, the fruit is often spot-picked; only the riper, larger, or outside fruit is harvested. Later, the trees are picked clean. In California, lemons usually are picked for size with the aid of sizing rings.

Various labor-saving devices have been tested, including mechanical platforms and positioners, tree shakers with catch frames, and air blasts for fruit removal. Mechanical harvesting, however, is limited to a very small percentage of the total crop. Because of damage incurred, fruit intended for processing only is mechanically harvested. Preharvest sprays have also been developed to improve the color and loosen the fruit to facilitate harvest.

Handling. After the fruit is received at the packinghouse, it is removed from the boxes or bulk containers by emptying it carefully to prevent damage to the fruit. It is then presized to remove fruit that is over- or undersized. Before washing, the fruit may be floated through a soak tank, which usually contains a detergent for cleaning and an antiseptic for decay control.

The washer is generally equipped with transverse brushes that revolve up to 120 rpm. If not applied at the soak tank, soap or antiseptic may be dribbled or foamed on the first series of brushes. The fruit is then rinsed by a fresh water spray.

The fruit then passes under fans that circulate warm air through the moving pieces. When dried, the fruit is polished and waxed, and then passed over roller conveyor grading tables. After grading, it is conveyed to sizing equipment to separate the pieces into the standard sizes being packed; the pieces are then dropped at stations for hand packing or conveyed to automatic or semiautomatic box-filling or bagging machines.

The packinghouse handling of California lemons for fresh market is interrupted by an extended storage period. After washing, the fruit is conveyed to a sorting table for color separation by electronic means or by human eye. Usually, four colors are recognized and are designated as dark green, light green, silver, and yellow. The dark green is a full green; the light green, a partially colored green (a green with color well broken); the silver, fully colored with a green tip (stylar end); the yellow, fully colored and mature with no green showing. Dark green fruit has a normal storage life of 4 to 6 months; and yellow, 3 to 4 weeks. These periods are approximate, as the storage (or keeping) quality of fruit varies considerably with season and grove. A light concentration of water-wax emulsion is usually applied to lemons before they are put into storage. Lemon storage is more fully discussed in a later section.

The preparation of this chapter is assigned to TC 11.5, Fruits, Vegetables, and Other Products.

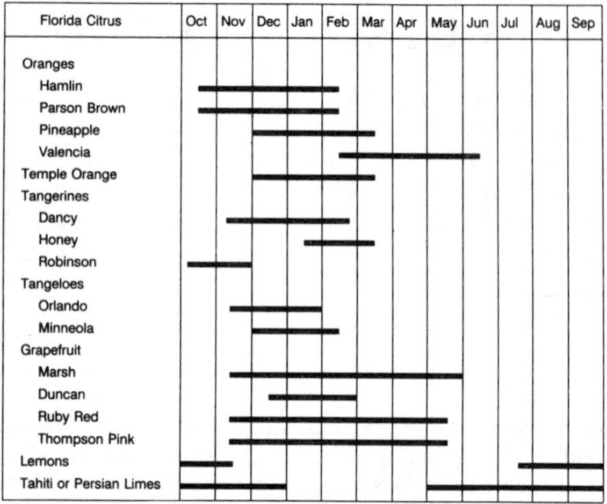

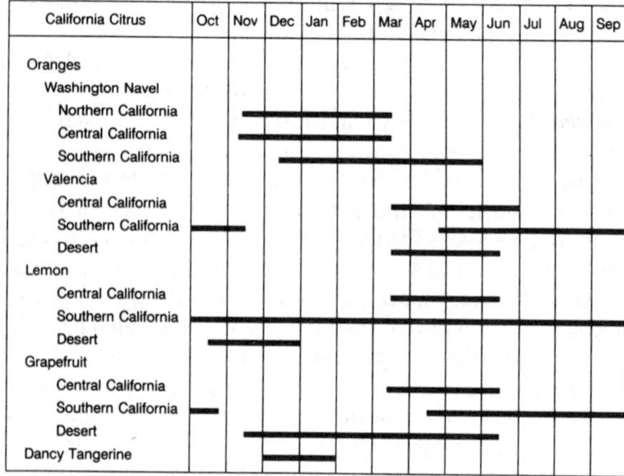

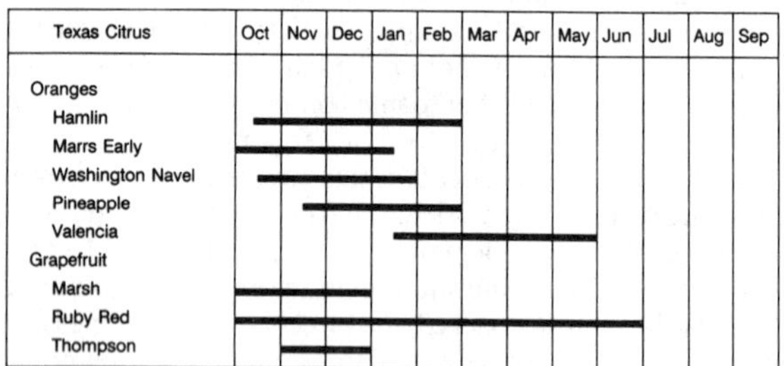

Fig. 1 Approximate Commercial Shipping Season for U.S. Citrus

After storage, lemons are waxed, then sized and packed. Post-storage washing to remove mold soilage is desirable but requires a washer incorporating very soft roller brushes.

Shipping and storage containers vary considerably for the various types of citrus fruit. The 0.8 bushel fiberboard cartons have become the standard in California and Florida. In addition, over 15% of Florida fresh fruit is consumer-packed in mesh and polyethylene bags that are normally shipped in 40-lb master cartons. After the packages are filled and closed, they are conveyed either to precooling rooms to await shipment or directly to standard refrigerator cars or trucks. The containers are stacked so that air distribution is uniform through the load.

Accelerated coloring or sweating. All varieties of citrus fruit must be mature before they are picked. Color is not always a criterion of maturity. The natural change of color in oranges from dark green to deep orange is a gradual process while the fruit remains on the tree. The fruit remains dark green from its formation until it is nearly full size and approaching maturity; then the color changes may occur very rapidly. The color change is influenced greatly by temperature variations. A few cold nights followed by warm days may completely color oranges that were previously very green. The color changes in lemons and grapefruit are similar, except that the final color is yellow. Unfavorable weather conditions may delay coloring even after maturity.

Up to a certain point, the natural color changes in Valencia oranges follow the trend described, but complete or nearly complete orange color generally develops some time before the fruit is mature. Some regreening of Valencias may occur after the fruit has reached its prime. Navel oranges in California, as well as the Florida varieties of Hamlin, Parson Brown, and Pineapple harvested in late fall and early winter, may be mature and of good eating quality, although the rind is green in color. Grapefruit, lemons, tangerines, tangelos, and other specialty fruit may also be sufficiently mature for eating before they are fully colored. Because the consumer is accustomed to fruit of characteristic color, poorly colored fruit is put through a coloring or degreening process in special rooms, bulk bins, or trailer degreening equipment.

These units are equipped to maintain temperatures and humidities at desired levels. Approximately 5 ppm ethylene in the air is maintained. The concentration of ethylene and the duration of the degreening periods depends on the variety of fruit and the amount of chlorophyl to be removed. During the operation, fresh air is introduced into the room, and a relative humidity of 88 to 92% is maintained. In Florida, temperatures of 82 to 85 °F generally are used, while in California temperatures of 65 to 70 °F are used. Lemons are usually degreened at 60 °F without added ethylene. In California, the process is called sweating instead of coloring or degreening.

Oranges, grapefruit, and specialty citrus fruits requiring ethylene treatment are frequently degreened as soon as they are delivered to the packinghouse; but they may receive a fungicide drench prior to degreening. Lemons are washed and graded or color-separated before being degreened.

A high percentage of Florida's early and midseason varieties of oranges and tangelos receive color-added treatment with a certified food dye that causes the rind of pale fruit to take on a brighter and more uniform orange color. This is usually in addition to degreening with ethylene gas. In color-added treatment, the fruit is subjected for 2 to 3 min. to the dye solution, which is maintained at about 120 °F. The color-added treatment can be given in an immersion tank filled with vegetable dye solution, or the dye can be flooded on the fruit as it passes on a roller conveyor. The color-added tank is located after the washer and before the wax applicator. Oranges with the desired color at harvest time, as well as tangerines and grapefruit, are bypassed around the dye tank, or the flow of dye may be cut off as the fruit passes over the equipment. Standards for maturity are slightly higher in Florida for oranges given color-added treatment. California oranges are not artificially colored.

Cooling. After the fruit is packed, it is cooled. The efficiency of the cooling rooms depends on the following conditions:

1. Cooling air rate per railcar load: not less than 3000 cfm
2. Relative humidity of supply air: 95% or above
3. Temperature of supply air entering room: not more than 2 °F below the selected cooling temperature

The fruit may also be cooled in a refrigerated truck trailer or container after it has been loaded.

In California, air is used to cool oranges but not lemons or grapefruit. In Florida, specialty fruits such as Temple oranges, tangerines, and tangelos may be cooled. Chapter 10 discusses cooling practices and equipment used for various commodities in more detail.

TRANSPORTATION

Fruit packed in piggybacks, trucks, ship vans, or rail cars should be stowed in appropriate modifications of the spaced bonded block to assure good air circulation, a uniform temperature, and a stable load. No dunnage is required. Such stowing provides continuous air channels through the interior of the load and improves the likelihood of sound arrival. Trailers and containers that circulate air from the bottom provide uniform temperatures throughout the load with a regular bonded-block stow.

In Florida, the present quarantine treatment for the Caribbean fruit fly, *A. suspensa*, is to subject an export load of citrus to specified temperatures for up to a 17- or 24-day period (Ismail *et al.* 1986). This treatment may be implemented in containers or in a ship's hold.

A uniform sample of 1500 fruit is withdrawn from a shipment before ship or container loading and is held at 80 °F or higher. These fruit are then examined after a 10-day incubation period. When an infestation of *A. suspensa* is found, the entire load must undergo the long treatment process. Temperature and time schedules are detailed below:

	Temperature, °F	Days
Short Treatment		
	33	10
	34	12
	35	14
	36	17
Long Treatment		
	33	14
	33.5	16
	34	17
	34.5	19
	35	20
	35.5	22
	36	24

These temperatures are required center pulp temperatures. To avoid chilling injury, a conditioning period of 7 days at 59 °F is recommended before initiation of the cold treatment process.

STORAGE

Oranges

Florida- and Texas-grown Valencia oranges can be stored successfully for 8 to 12 weeks at 32 to 34 °F with a relative humidity of 85 to 90%. The same requirements apply to Pope's Summer orange, a late-maturing Valencia-type orange. A temperature range of 40 to 44 °F for 4 to 6 weeks is suggested for California oranges. March-harvested, Arizona Valencias store best at 48 °F, but June-harvested fruits store best at 38 °F.

Oranges lose moisture rapidly, so high humidity should be maintained in the storage rooms. For storage longer than the usual transit and distribution periods, 85 to 90% relative humidity is recommended.

Florida and Texas oranges are particularly susceptible to stem end rots. Citrus fruits from all producing areas are subject to blue and green mold rot. These decays develop in the packinghouse, in transit, in storage, and in the market, but can be greatly reduced if fruit is properly treated. Proper temperature is effective in reducing decay. However, once storage fruit is removed to room temperature, decay will develop rapidly.

Storage of oranges is often complicated by the fact that prolonged holding at relatively low temperatures may induce the development of physiological rind disorders not ordinarily encountered at room temperature. Aging, pitting, and watery breakdown are the most prevalent rind disorders induced by low storage temperatures. Generally, California and Arizona oranges are more susceptible to low-temperature rind disorders than Florida oranges.

Successful long storage of oranges requires: harvest at the proper maturity, careful handling of fruit, good packinghouse methods, fungicidal treatments, and prompt storage after harvest.

The rate of respiration of citrus fruits is usually much lower than that of most stone fruits, green vegetables, and somewhat lower than that of apples. Navel oranges have the highest respiration rate, followed by Valencia oranges, grapefruit, and lemons. The heat from respiration is a relatively small part of the heat load. Table 1 shows heat generated through respiration.

Table 1 Heat of Respiration of Citrus Fruits
(Haller *et al.* 1945)

	Btu per ton of fruit per day					
	Oranges			Grapefruit		Lemons
Temp., °F	Florida	California Navels	California Valencias	Florida	Calif. Marsh	Calif. Eureka
32	700	900	400	500	500	700
40	1400	1400	1000	1100	800	1100
50	2700	3000	2600	1500	2000	2500
60	4600	5000	2800	2800	2600	3500
70	6600	6000	3900	3500	3900	5000
80	7800	8000	4600	4200	4800	5700

Grapefruit

Florida and Texas grapefruit is frequently placed in storage for 4 to 6 weeks without serious loss from decay and rind breakdown. The recommended temperature is 50 °F. A temperature range of 58 to 60 °F is recommended for the storage of California and Arizona grapefruit.

A relative humidity of 85 to 90% is usually recommended for the storage rooms in which grapefruit is held. Loss of weight and water occurs rapidly and can be avoided by maintaining the correct humidity and taking the additional precaution of a wax coating.

Decay and rind breakdown are deterrents to long storage of grapefruit and may develop in fruit during storage or following removal from storage. Proper prestorage treatments with fungicides, as discussed in the previous section on disorders and storage temperatures, will greatly reduce these problems. Also, periodic inspections of stored fruits should be made to terminate storage at the least symptom of rind pitting or excessive decay.

Export may require 10 days to 4 weeks of storage in a refrigerated hold and present problems similar to those encountered in refrigerated storage. Marsh Seedless and Ruby Red grapefruit picked before January retain appearance best when stored at 60 °F. With riper fruit, 50 to 55 °F is better for export shipments. Very ripe fruit harvested in April and May, however, develops excessive decay following storage at 50 to 60 °F.

Lemons

Most of the lemon crop is picked during the period of least consumption and stored until consumer demand justifies shipment. Lemons are generally stored near the producing areas rather than the consuming areas.

All lemons, except the relatively small percentage that are ripe when harvested, must be conditioned or cured, and degreened, before shipping. When lemons are stored prior to shipment, the curing and degreening processes proceed during storage. These lemons are usually stored at 58 to 60 °F and 86 to 88% rh. Local conditions may suggest slight modifications of these values.

Lemons picked green but intended for immediate marketing, such as most lemons grown in the desert portions of Arizona and California, are degreened and cured from 6 to 10 days at 72 to

78 °F and 88 to 90% rh. The thin-skinned Pryor strain of Lisbon lemons degreens in about 6 days, whereas the thick-skinned old-line Lisbon requires as long as 10 days.

Lemon storage rooms must have accurately controlled temperature and relative humidity; the air should be clean and uniformly circulated to all parts of the room. Ventilation should be sufficient to remove harmful metabolic products. Air-conditioning equipment is necessary to provide satisfactory storage conditions, as natural atmospheric conditions are not suitable for the necessary length of time.

A uniform storage temperature of 50 to 60 °F is important. Fluctuating or low temperatures cause lemons to develop an undesirable high color or bronzing of the rind. Temperatures 52 °F and lower cause a staining or darkening of the membranes dividing the pulp segments and may affect the flavor. Temperatures above 60 °F shorten the storage life and are favorable to the growth of decay-producing organisms.

A relative humidity of 86 to 88% is generally considered satisfactory for lemon storage, although a slightly lower humidity may be desirable in some locations. Higher humidities prevent proper curing of the lemons, encourage mold growth on walls and containers, and hasten decay of the fruit; much lower humidities cause excessive shrinkage.

Proper stacking of the fruit containers in storage rooms is important to secure uniform air circulation and temperature control. The stacks should be at least 2 in. apart and the rows, 4 in.; trucking aisles at least 6 ft wide should be provided at intervals.

Specialty Citrus Fruits

In Florida, small amounts of various specialty citrus fruits are grown commercially. These fruits, which are usually eaten out of the hand, include tangerines, tangerine hybrids (Murcott Honey oranges, Temple oranges, tangelos), the King orange, and other mandarin-type fruits.

Careful handling during picking and packing is especially necessary for these fruits. Because of their perishable nature and limited shelf life, these fruits should not be stored longer than required for orderly marketing (2 to 4 weeks). A temperature of 38 to 40 °F at 90 to 95% rh is recommended. Adequate precooling and continuous refrigeration during transit are required.

Tahiti or Persian limes, in addition to the fruits just mentioned, are grown in southern Florida. This is the only citrus fruit marketed while it is green in color. The fully ripe (yellow) fruit lacks consumer appeal and is undesirable for fresh market. Limes should be picked while still green, but after the fruit has lost the dimpled appearance around the blossom end. Good quality fruit may be stored satisfactorily for 6 to 8 weeks at temperatures of 48 to 50 °F. Mature fruit will gradually turn yellow at this temperature, however. Prevention of desiccation is very important, as is a relative humidity above 85%. Pitting occurs at temperatures below 45 °F, while temperatures above those recommended permit the development of stem end rot.

CONTROLLED ATMOSPHERE STORAGE

While some minor benefits have been obtained, tests with oranges, grapefruit, lemons, and limes using modified or controlled atmospheres have given no assurance that storage and market life can be extended. For this reason, controlled atmosphere (CA) storage is not generally recommended for citrus fruits. Atmospheres used for storage of apples and other deciduous fruits are unsatisfactory for citrus fruits and lead to rind injuries, off-flavors, and decay.

The performance of any citrus storage facility depends on: (1) the provision of sufficient capacity for peak loads; (2) evaporator and secondary refrigerating surface sufficient to permit operating at high back pressures, thus preventing low humidities and permitting economical operation; and (3) efficient air distribution, assuring velocities high enough to effect rapid initial cooling and volumes great enough to permit operation during storage with only a small temperature rise between delivery and return air. Chapters 24, 25, and 26 have further information on storage design.

STORAGE DISORDERS AND CONTROL

Postharvest Diseases

Citrus fruits often carry incipient fungus infections when they are harvested. Decay organisms may also enter minor injuries caused during harvesting and handling. The major postharvest diseases (with symptoms) encountered in storage include:

Alternaria Rot. Usually a stylar end in navel oranges as a black, dry, deeply penetrating decay. In other citrus fruits, as a slimy, leaden-brown storage decay of core starting at stem end. *Control*: Provide optimum growing conditions. Harvest oranges before over-ripe. Do not store tree-ripe lemons. Restrict storage period for other lots known to be weak. Green buttons are an indication of strong fruit. Treatment of lemons at the packinghouse with 2-4-D in wax emulsion or water improves resistance of lemons to disease.

Anthracnose (Colletotrichum). Leathery, dark brown, sunken spots or irregular areas. Internal affected tissues dark gray, fading through pink to normal color. Most serious with degreened early season tangerines, tangerine hybrids, and long-stored oranges and grapefruit. *Control*: Use recommended postharvest fungicide. Avoid long storage; move promptly.

Blue (and Green) Mold Rot (Penicillium). Soft, watery, decolorized lesions that, under moist conditions, become quickly covered with blue or olive-green powdery spores. *Control*: Prevent skin breaks. Use recommended fungicides in washes. Cool fruit to as near to 32 °F as practicable. Use biphenyl in box liners or in fruit wraps.

Brown Rot (Phytophthora). Extensive firm, brown decay having a penetrating rancid odor. Chiefly on fruit from California and Arizona. *Control*: Orchard spraying and good sanitation. Submerge fruit at packing for 2 min. in water at 114 °F.

Sour Rot (Geotrichum). Soft, watery rot with sour smell following peel injuries. Similar to early stages of mold rot, except that no powdery spores are formed. Most serious on lemons and mandarin-type fruit. *Control*: Avoid peel injuries at harvest; refrigerate at lowest practical temperature. Approved fungicides are of little or no value.

Stem End Rot (Diplodia; Phomopsis). Pliable, fairly firm, extensive, brown decay starting at stem. Sour, pungent odor. Prevalent in Florida and found occasionally in Arizona and California fruit. *Control*: Treat harvested fruit promptly in recommended fungicides and cool promptly below 50 °F.

Several chemical fungicides are approved by EPA for postharvest use on citrus fruits. These include thiabendazole (TBZ), orthophenylphenol (OPP or SOPP), and imazalil. These materials are applied after washing and before waxing or are incorporated in the wax coating. Biphenyl, a volatile fungistat, is impregnated in paper wrappers or box liners. It evaporates slowly and inhibits growth of organisms in transit and storage. Under certain conditions, it is beneficial to use a combination of these materials since all are not equally effective against the same organism. Strains of the blue and green molds (*Penicillium*), which are resistant to certain fungicides, have developed in citrus storage houses, so care must be taken in selecting a fungicide and the time of application.

Physiological Disturbances

In addition to diseases caused by fungi, some defects are caused by various physiological conditions. These defects can best be avoided by using fruit of prime maturity and by proper handling after harvest. Proper temperature and humidity levels are required during handling, storage, and transit. The physiological disorders (and symptoms) are:

Stem End Rind Breakdown. Small to large sunken, drying, discolored, firm areas in skin around stem button or on the upper part of fruit. *Control*: Pick before overmature. Avoid overheating in packinghouse treatments. Wax fruit. Store for limited period only in fairly high relative humidity, 85 to 90%. Follow storage temperatures recommended for variety and growing area.

Freezing Injury. Field freezing is found scattered through boxes. Transit and storage freezing is worse in exposed fruits in bottom-layer boxes or those nearest cooling coils. Affected fruits may show watersoaked areas in rind. The internal tissue is disorganized, watersoaked, milky, and has rind flavor. Frozen fruit loses moisture, causing drying, separation of juice vesicles, and buckling of segment walls. The freezing point of citrus fruit is about 28.5 °F.

Internal Decline. In lemons, core tissues near the stylar end break down and dry, becoming pink. *Control*: Maintain optimum moisture conditions in grove.

Pitting. This physiological disease is manifested by depressed areas of 0.1 to 0.8 in. diameter in the peel of citrus fruit. Affected tissues collapse and may appear bleached or brown. Pits occur anywhere on the fruit and may coalesce to form large irregular areas. The cause is not fully understood. In general, it is a low-temperature disorder. *Control*: Follow storage temperatures recommended for cultivar and growing area.

BANANAS

This section discusses the harvesting, transportation, processing, and storage of bananas.

HARVESTING AND TRANSPORTATION

Bananas do not ripen satisfactorily on the plant; even if they did, deterioration of ripened fruit is too rapid to allow shipping from tropical growing areas to distant markets. Bananas are harvested when the fruit is mature but unripe, with dark green peels and hard, starchy, inedible pulps. Each banana plant produces a single stem of bananas which contains from 50 to 150 individual fruits (or fingers). The stem is cut from the plant as a unit with fingers attached and transported to nearby boxing stations.

Bananas are removed from the stem, washed, and cut into consumer-sized cluster units of four or more fingers. The clusters are packed in protective fiberboard cartons that contain 40 lb of fruit. The cartons move by rail from the tropical boxing stations to port and then are loaded into the holds of refrigerated ships. On the ship, the fruit is cooled to the optimum carrying temperature, generally 56 to 58 °F, depending on variety.

Bananas are unloaded still green and unripe at seaboard and transported under refrigeration at a holding temperature of 58 °F to interior wholesale distribution centers by both truck and rail car. The objective is to maintain the product in an optimal environment and move it to its destination as quickly as possible to minimize postharvest deterioration.

DISEASES AND DETERIORATION

Bananas are subject to the following diseases and physiological disorders. Proper temperature and moisture during storage and careful handling will slow the aging and development of decay.

Anthracnose (Ripe Rot) (Gloeosporium). Shallow black spots on stems of ripening fruit. Under moist conditions, pink spore masses cover center of spots. Dark discoloration of skin may extend from stem ends over entire fruit. *Control*: Protect fruit from mechanical injury; damage is reduced if fruit is put in corrugated boxes. Schedule ripening so that fruit can be marketed and consumed before appearance of defect.

Black Rot (Ceratocystis). Transmitted from wounds via fibrovascular system of plant. Progresses into crowns and stem ends of fingers. Produces brownish-black areas in peel at fruit ends. As fruit ripens, skin becomes grayish-black in color and water soaked. Pulp rarely affected. *Control*: In the tropics, dip or spray freshly cut tips and bunches with fungicides before boxing. Avoid mechanical injury and maintain sanitation program from tropics to ripening room.

Chilling Injury. Dull gray skin color with increased tendency to darken on slight bruising. Latex in green fruit does not bleed freely and will be clear rather than cloudy. Subsurface peel tissue streaked with brown. Turning or ripe bananas are more susceptible to injury than green fruit. *Control*: Avoid temperatures below 55 °F. Moving air makes chilling more rapid.

Fungus Rots (Several Fungi). Extensive soft rot of scarred, split, or broken fruit. Affected skin and flesh moist and brown to black. Under high humidity, the surface is often covered with mold. *Control*: Handle fruit carefully to avoid bruising and mechanical injury. Cool stored fruit rapidly to 56 °F.

EXPOSURE TO EXCESSIVE TEMPERATURES

Fruit pulp temperatures only a few degrees below optimum holding temperatures, although considerably above the actual freezing point of bananas, can cause chilling injury, as described previously. The severity of chilling injury varies directly with the duration of exposure and indirectly with temperature. It is primarily a peel injury in which certain surface cells of the banana peel are killed. The contents of the dead cells eventually darken (because of oxidation) and give the fruit a dull appearance. Both green and ripe bananas are susceptible to chilling injury; severely chilled green bananas never ripen properly. Fruit pulp temperatures only a few degrees above the optimum holding temperatures can cause the fruit to ripen prematurely in transit.

Once the bananas arrive at wholesale distribution centers, they are unloaded and placed in specially equipped processing rooms for controlled ripening. As soon as the bananas have ripened to an edible state, they are rushed to retail, since ripening cannot be stopped. Even under ideal refrigeration, ripe bananas will have progressed to the point where they are too ripe to be marketed.

WHOLESALE PROCESSING FACILITIES

Wholesale banana facilities are distinguished from general wholesale produce storage facilities by special banana ripening rooms. The ripening room controls initiation and completion of fruit ripening, a natural physiological phenomenon. A typical banana room is shown in Figure 2. The ability to properly ripen bananas is so critically linked to the design of the ripening rooms that major banana importers maintain technical staffs that specialize in banana-room design. Through these technical staffs, free, nonobligatory consultation is provided to wholesalers, architects, engineers, contractors, and anyone else involved in banana ripening facility design, construction, and operation.

A typical banana processing facility consists of a bank of five or more individual ripening rooms. For design purposes, one complete turnover per week is assumed, and therefore the combined capacity of all rooms should approximately equal total weekly volume, allowing for seasonal variations.

Each load is scheduled for optimum ripeness on a particular day. Fruit shipped from this load a day ahead of schedule will be underripe; fruit shipped a day late will be overripe. Therefore, shipping for several days out of one room is not practical. There should be at least as many rooms as there are retail shipout days per week.

Because bananas cannot be processed on a continuous flow basis, individual room capacities are multiples of carlots, usually

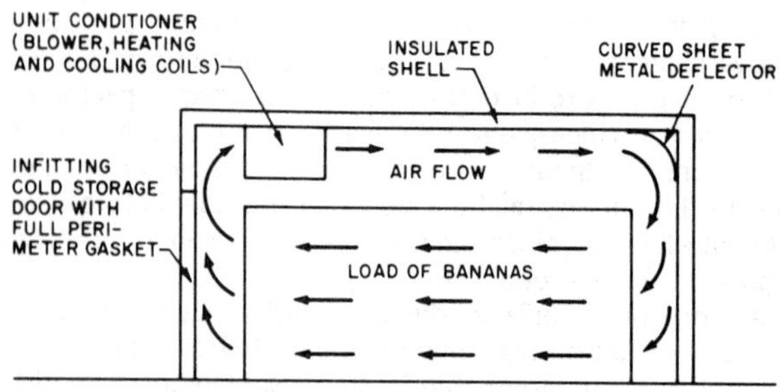

Fig. 2 Banana Room (Side View)

one half or one carlot. As the capacity of transportation equipment has increased in recent years, so has the design capacity of processing rooms. Generally, one or two large rooms would be cheaper to build than several small rooms with equivalent total capacity. However, minimizing construction cost is not the pertinent consideration; having all bananas in each particular processing room reach optimum ripeness for shipment to retail at the same time is more important.

Airtightness

Banana ripening is initiated by exposing the fruit to ethylene gas, which is introduced into the room from cylinders. The dose is 1 ft^3 of ethylene gas per 1000 ft^3 of room air space. Ethylene is explosive in air at a concentration between 2.75 and 28.6%. Many ethylene systems gas the fruit automatically over a 24-h period.

To be effective, the gas must be confined to the ripening room for 24 h, so banana rooms must be airtight. Floor drains must be individually trapped to prevent gas leakage. Special care should be taken to seal all penetrations in room walls where refrigerant piping, plumbing lines, and the like, enter rooms. Doors should have single seal gaskets all around and sweep gaskets at the floor line.

Refrigeration

A direct-expansion halocarbon system is recommended for use in banana rooms. Because of ammonia's harmful effect on bananas should leakage occur, direct-expansion ammonia systems should not be used. Malfunctioning refrigeration equipment during processing could cause heavy product losses, so, even with high initial installation costs, each ripening room should have a completely separate system.

For maintenance-free operation in the high humidity environment of processing rooms, evaporator coils should have a fin spacing of 4 fins/in. Coils should be amply sized and capacity rated at a design temperature difference of 15 °F with a refrigerant temperature of 40 °F. Air temperatures used during processing range from 45 to 65 °F. Because of the danger of banana chilling, refrigerant temperatures below 40 °F are not recommended. With programmers, suction pressure control devices or hot gas bypass systems must be installed.

Refrigeration Load Calculations

These calculations are based on the same methods used for other fresh fruits. A typical half-carlot-capacity banana room will hold approximately 432 boxes of fruit. Approximate outside dimensions for a three-tier forklift-type room shown in Figure 3 are 30 ft long by 6 ft wide by 22 ft high. Pallets are 48 in. by 40 in. and are stacked 6 pallets deep in each of 3 tiers, totaling 18 pallets per room. Boxes for use in banana rooms are approximately 10 in. high by 16 in. wide by 22 in. long and are stacked 4 boxes per each of 6 layers, totaling 24 boxes per pallet. With 18 pallets per room, 432 boxes of bananas can be stored (18 pallets by 24 boxes). Each box has a net weight of 42 lb and a gross weight of 47 lb.

Transmission load is calculated in the normal manner; the air change load is negligible. The electrical load is based on the continuous operation of multi-kilowatt fan motor(s). The peak heat of respiration is 0.5 Btu/h per pound multiplied by the total net weight of bananas in the ripening room. For product cooling, the specific heat of bananas is 0.8 Btu/(lb · °F) multiplied by the total net weight of the bananas, plus the total tare weight of the cartons multiplied by 0.4 Btu/(lb · °F), the specific heat of fiberboard. The total calculated load is thus approximately 60 Btu/h per box. A pulldown rate of 1 °F/h is assumed. Total system design capacity is calculated by assuming simultaneous peak respiration and pulldown load.

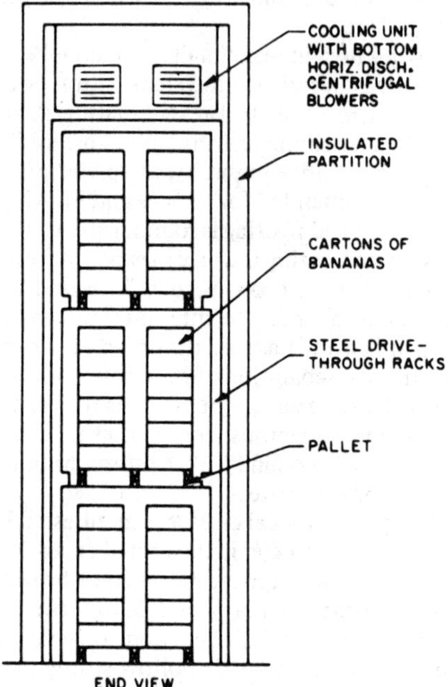

Fig. 3 Three-Tier Forklift Banana Room (End View)

Heating

Heat is not required during most ripening cycles. However, occasional loads may come in at temperatures below desired levels for treatment with ethylene gas, making heating necessary. Many banana room refrigeration units come with electrical heating elements as an integral part of the unit. If electrical heating strips are used, they should be enclosed in a corrosion-resistant sheath and have a surface temperature of not more than 800 °F in dead still air, the temperature limitation being necessary because of proximity to refrigerant coils and the inherent danger should leakage occur. Portable plug-in electric heaters are also used. Heating system capacity should be sufficient to raise load temperature at a rate of 1 °F/h. Open-flame gas heaters should never be used in banana rooms because: (1) ethylene gas used during ripening is explosive at certain concentrations; and (2) the necessary room tightness could easily result in the open-flame heaters' consuming the available oxygen within the space, thereby extinguishing the flame and permitting raw gas to enter the room.

Air Circulation

The fruit pulp temperature schedules for 4- to 8-day ripening cycles are shown in Table 2. A temperature variation of only a few degrees will considerably alter the rate of fruit ripening. For even ripening, fruit temperatures must be uniform throughout the room, so comparatively large volumes of air must be continuously circulated throughout the entire load. Centrifugal fans are necessary. They are installed for bottom horizontal discharge, so that the top boxes will not be chilled immediately in front of the unit. Fan air output should be rated at 0.62 in. water external static pres-

Table 2 Fruit Temperatures for Banana Ripening

Ripening Schedule	Temperature, °F 1st Day	2nd Day	3rd Day	4th Day	5th Day	6th Day	7th Day	8th Day
Four days	64	64	62	60				
Five days	62	62	62	62	60			
Six days	62	62	60	60	60	58		
Seven days	60	60	60	60	60	58	58	
Eight days	58	58	58	58	58	58	58	58

sure. Because of heat of respiration, heat must be continually withdrawn from the product even when it is being held at a constant temperature. Therefore, a temperature variation in the load is inevitable, with warmer fruit being downstream relative to the circulated air. Unit conditioners at the front of the room over the door discharge toward the rear of the room. This arrangement leaves riper fruit near the door to be shipped first.

For improved air distribution, a sheet metal (or other suitable material) air deflector curved to a 90° arc is mounted full width on the back room wall. This deflector reduces turbulence and directs the air downward for return through the load.

Air Volume Requirements

Circulated air volume requirements are calculated on the basis of conditions required at the end of the pulldown period. Assume a maximum allowable fruit temperature variation of 2°F, an air temperature drop through the cooling unit of 2°F, and product temperature reduction proceeding at a rate of 0.2°F/h. During the initial pulldown, the air quantity so calculated will give about a 5.5°F drop through the cooling unit. The general equation is:

$$q_t = q_r + q_p = mc_p\Delta t$$

where

q_t = total heat removed, Btu/h
q_r = heat of respiration, Btu/h
q_p = pulldown load, Btu/h
m = mass of flow rate of air, lb/h
c_p = specific heat of air, Btu/(lb·°F)
Δt = temperature change of air, °F

Using values of respiration and specific heat given in the section on load calculations, the value of q_r and q_p can be determined. q_p is calculated on the basis that at the end of the pulldown period, temperature reduction is proceeding at the rate of 0.2°F/h.

$$q_r = 0.5 \times 42 = 21 \text{ Btu/h/box}$$

$$q_p = 0.2[(0.8 \times 42) + (0.4 \times 5)] = 7.12 \text{ Btu/h/box}$$

$$q_t = 21 + 7.12 = 28.12 \text{ Btu/h/box}$$

At equilibrium, the air temperature Δt equals the fruit temperature Δt, and:

$$m = q_t/c_p\Delta t = 28.12/(0.24 \times 2) = 58.58 \text{ lb air per box}$$

$$\text{Volume} = 58.58/(0.075 \times 60) = 13.02 \text{ cfm per box}$$

For a room with 432 boxes, the airflow should be 432 × 13.02 = 5600 cfm at 0.62 in. of water external static pressure.

Humidity

A high relative humidity around the fruit is important during banana ripening. Bananas ripened under low humidity conditions are more susceptible to handling damage. When bananas were ripened on the stem, naked fruit was directly exposed to the moving airstream and automatic room humidifiers were used to prevent excessive fruit dehydration. With the advent of tropical boxing, however, banana room humidifiers are not required.

The fiberboard carton shields the fruit from the moving airstream. In addition, ample sizing of evaporator coils keeps the temperature difference across the coil to 10 to 15°F, thereby limiting dehumidification. Both natural transpiration of the fruit and airtight room design also contribute to high room humidity.

Controls

Ripening room air temperatures are varied frequently during banana processing. Temperatures should be controlled by remote bulb-type thermostats, with bulbs for heating and cooling mounted in the return airstream within the ripening room to prevent short cycling of equipment. Thermostats should be mounted on the exterior of the ripening room and have a range of 45 to 70°F, calibrated in 1°F increments, with no more than 2°F differential. The thermostats are best mounted on a control panel having a selector switch providing heating and cooling with continuous fan operation.

Automatic temperature controllers or programmers are being installed in most new facilities. Bananas produce heat continuously, but the rate of heat production varies considerably during the ripening cycle. A generalized heat-of-respiration curve is shown in Figure 4. Although more complex, the removal of heat of respiration from the load can be viewed, for the purpose of analysis, as a simple conduction process. Applying the general conduction equation: $q = kA\Delta t$, where q = heat of respiration; $kA = 1/r$; r = resistance to heat flow of packaging materials; Δt = banana minus air temperatures.

If kA is a constant, Δt must vary as q. Assuming the banana temperature is also constant, the room air temperature must be set lower as q increases during ripening. However, the exact value of q at any particular point in the ripening cycle is unknown. With the conventional, manually adjustable air-sensing thermostat-control system, fruit temperatures are taken manually with a pulp thermometer, and thermostat settings are continually adjusted to follow ripening schedules. This is essentially a trial and error procedure.

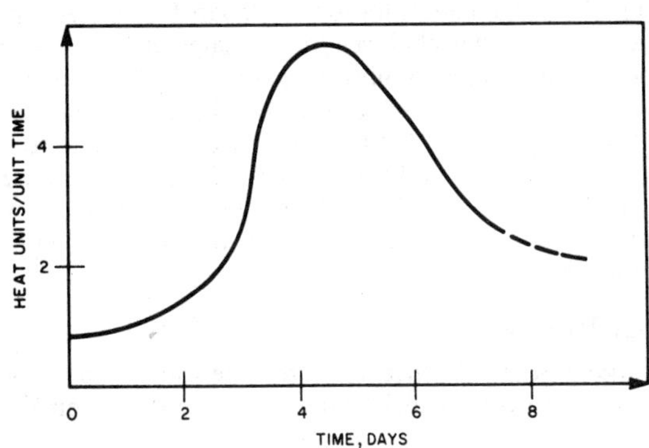

Fig. 4 Heat of Respiration during Banana Ripening

By contrast, the temperature programmer has a remote bulb, which is placed in a box of bananas in the load. Since the bulb senses fruit temperature directly, heat of respiration is compensated for during ripening. Fruit temperatures are automatically adjusted to follow preset cycles.

SUBTROPICAL FRUITS

AVOCADOS

Avocado cultivars grown in California are not grown commercially in Florida, and vice versa. Generally, Florida cultivars tend to be larger fruited than those from California. In California, the Fuerte variety accounts for 75% of the annual crop and is available from October through March. Hass (black skin) is available from April through September. In order of importance, Florida cultivars are Booth 8, Lula, Waldin, and Booth 7. Waldin appears on the market in August, followed by Booth 8 in September, and Booth 7 and Lula from October to February.

The best storage temperature for cold-tolerant Florida avocado cultivars, such as Booth 8 and Lula, is 40 °F. All Florida avocado cultivars produced during the summer, such as Waldin, are cold-intolerant and store best at 54 to 55 °F. A few cultivars, such as Fuerte, store best at 45 °F. Cold-tolerant cultivars can be held in storage a month or longer, but storage of cold-intolerant cultivars is usually limited to 2 weeks because of their susceptibility to softening and chilling injury. The best ripening temperature for avocados is 60 °F, but temperatures from 55 to 75 °F are usually satisfactory. Temperatures above 79 °F frequently cause off-flavor, skin discoloration, uneven ripening, and increased decay.

Storage Disorders

Anthracnose (Colletotrichum) are scattered black spots covering firm decayed tissue that can be removed easily from surrounding flesh. Pink spore masses form on the spots under moist conditions. Control requires preventing blemishes and other breaks in the skin.

Chilling injury is typified by small to large sunken pits in the skin, becoming brown or black in color; often accompanied by general browning of the skin and light smoky streaks in the flesh which develop independently.

MANGOES

Important early cultivars, Tommy Atkins and Irwin, mature during June and July, followed by such midseason cultivars as the Kent and Palmer, which mature during July and August. The most important cultivar produced in Florida is the large-fruited Keitt, which is late-season and matures during August and September.

The optimum storage for mangoes is 54 to 55 °F for 2 to 3 weeks, although 50 °F is adequate for some cultivars for shorter periods. Mangoes are subject to chilling injury at temperatures below 50 °F. The best ripening temperatures for mangoes are from 70 to 75 °F, but temperatures of 60 to 65 °F are also satisfactory under certain conditions. At 60 to 65 °F, the fruit develops a bright and most attractive skin color, but the flavor is usually tart and requires an additional 2 to 3 days at 70 to 75 °F to attain a sweet flavor. Mangoes ripened at 80 °F and higher frequently have a strong flavor and mottled skin.

Storage Disorders

Anthracnose (Colletotrichum) are large scattered black spots in the skin of ripening fruits. Under moist conditions, pink spore masses develop in spots. The disease is controlled on the tree by regular spraying and use of hot water treatment (131 °F for 0.12 h) after harvest.

Chilling injury causes pitting of the skin, which sometimes develops a gray cast. Fruit with chilling injury usually does not ripen uniformly. Control requires storing at proper temperatures.

PINEAPPLES

Fresh pineapples are available throughout the year, but a much larger supply is available from March through June. Only three pineapple cultivars are commercially important in the United States: the Smooth Cayenne from Hawaii, and the Red Spanish and Smooth Cayenne from Puerto Rico.

Pineapples harvested at the half-ripe stage can be held for 2 weeks at 45 to 55 °F and still have about one week's shelf life. Continuous maintenance of storage temperature is as important as the specific storage temperature. Ripe fruit should be held at 45 to 47 °F. Harvesting at the mature-green stage is not recommended because some individual fruits would be so immature that they would fail to ripen. Mature green fruit is especially susceptible to chilling injury at temperatures below 50 °F.

Storage Disorders

Black Rot (Ceratocystis). Affected tissues are extensive, soft, and leaky, ranging from normal to jet black in color. To control, treat freshly cut stem parts with benzoic acid-talc dust, prevent bruising, and cool to 50 °F.

Brown Rot (Penicillium; Fusarium) has brown, firm decay starting at eyes or cracks; it is common on overripe fruit. To control, provide good growing conditions and move before the fruit is overripe.

Chilling injury is manifested by the fruit taking on a dull hue and developing a water soaking of the flesh and a darkening of the core. To control, hold at recommended temperatures.

REFERENCES

Haller, M.H., D.H. Rose, J.M. Lutz, and P.L. Harding. 1945. Respiration of citrus fruits after harvest. *Journal of Agricultural Resources* 71:327.

Ismail, M.A., T.T. Hatton, D.J. Dezman, and W.R. Miller. 1986. In transit cold treatment of Florida grapefruit shipped to Japan in refrigerated van containers: Problems and recommendations. *Proceedings of the Florida State Horticultural Society* 99:117-121.

BIBLIOGRAPHY

Chace, W.G., Jr., J.J. Smoot, and R.H. Cubbedge. 1979. Storage and transportation of Florida citrus fruits. *Florida Citrus Industry* 51:16.

Chau, K.V., C.D. Baird, P.C. Talasila, and S.A. Sargent. 1992. Development of time-temperature-humidity relations for fresh fruits and vegetables. Final report for ASHRAE Research Project 678-RP.

Hardenburg, R.E., A.E. Watada, and C.Y. Wang. 1986. The commercial storage of fruits, vegetables, and florist and nursery stocks. USDA *Handbook* No. 66.

McCornack, A.A., W.F. Wardowski, and G.E. Brown. 1976. Postharvest decay control recommendations for Florida citrus fruit. *Florida Cooperative Extension Service Circular 359-A*.

Smoot, J.J., L.G. Houck, and H.B. Johnson. 1971. Market diseases of citrus and other subtropical fruits. USDA *Handbook*, 398.

Wardowsko, W.F., S. Nagy, and W. Grierson. 1986. *Fresh citrus fruits.* AVI Publishing, Westport, CT.

CHAPTER 17

VEGETABLES

ANNUAL losses (shrinkage) in marketing vegetables (shipping, processing, storage, and retailing) are caused, in part, by overly high temperatures during handling, storage, and transport, which increase ripening, decay, and the loss of edible quality and nutrient values. In other cases, freezing or chilling injury from overly low temperatures may be involved. Other serious losses are caused by mechanical injury from careless or rough handling and by shrinkage or wilting because of moisture loss. Many losses can be reduced substantially by following recommended handling, cooling, transport, and storage practices. Improved packaging, refrigerated transport, and awareness of the role of refrigeration in maintaining quality throughout marketing have made it possible to move vegetables to distant cities in field-fresh condition.

PRODUCT SELECTION AND QUALITY MAINTENANCE

The principal hazards to quality retention during marketing include:

1. Metabolic changes associated with respiration, ripening, and aging (composition, texture, color)
2. Moisture loss, with resultant wilting and shriveling
3. Bruising and mechanical injury
4. Parasitic diseases
5. Physiological disorders
6. Freezing and chilling injury
7. Flavor and nutritional changes
8. Growth (sprouting, rooting)

Fresh vegetables are living tissues and have a continuing need for oxygen for respiration. During respiration, stored food such as sugar is converted to heat energy, and the product loses quality and food value. Some of the refrigeration load to maintain commodity temperatures during storage or transportation can be attributed to respiration. For example, a 20,000-lb load of asparagus cooled to 39 °F can produce enough heat of respiration during a cross-country trip to melt 7900 lb of ice.

Vegetables that respire the fastest often give the greatest handling problems because they are the most perishable. Refrigeration is the best method of slowing respiration and other life processes. Chapter 25 of this volume and Chapter 30 of the 1993 ASHRAE *Handbook—Fundamentals* give more information on the respiration rates of many vegetables.

Vegetables are usually covered with microorganisms, which will cause decay given the right conditions. Deterioration because of decay is probably the greatest source of spoilage during marketing. When mechanical injuries break the skin, decay organisms will enter the produce. If it is then exposed to warm temperatures, especially under humid conditions, infection usually increases. Adequate refrigeration is the best method of controlling decay because low temperatures control the growth of most microorganisms.

Many color changes associated with ripening and aging can be delayed by refrigeration. For example, broccoli may show some yellowing in one day on a nonrefrigerated counter, while it remains green at least 3 to 4 days in a refrigerated display.

Refrigeration can retard deterioration caused by chemical and biological reactions. Sweet corn may lose 50% of its initial sugar content in a single day at 70 °F, while only about 5% will be lost in one day at 32 °F (Appleman and Arthur 1919). Also, freshly harvested asparagus will lose 50% of its vitamin C content in one day at 68 °F, whereas it takes 4 days at 50 °F or 12 days at 32 °F to lose this amount (Lipton 1968). With certain exceptions, the best temperature for retarding deterioration resulting from biological processes or from pathogens is the lowest temperature that can be maintained without freezing the commodity. This is about 2 °F above the freezing point of the vegetable.

Loss of moisture with consequent wilting and shriveling is one of the obvious ways in which freshness is lost. *Transpiration* is the loss of water in the vapor state from living tissues. Moisture losses of 3 to 6% are enough to cause a marked loss of quality for many kinds of vegetables. A few commodities may lose 10% or more in moisture and still be marketable, although some trimming may be necessary, such as for stored cabbage. For more on transpiration, see Chapter 30 of the 1993 ASHRAE *Handbook—Fundamentals*.

Postharvest Handling

Care should be exercised in stacking bulk bins in storage so that proper ventilation and refrigeration of the product is maintained. Bins should not be of such depth that excessive weight damages the product near the bottom.

The effect of rough handling of vegetables is cumulative. Several small bruises on a tomato can produce an off-flavor. Bruising also stimulates the rate of ripening of a product such as tomatoes and thereby shortens potential storage and shelf life. Mechanical damage allows increased moisture loss; skinned potatoes may lose 3 to 4 times as much weight as nonskinned ones.

After harvest, most highly perishable vegetables should be removed from the field as rapidly as possible and placed under refrigeration, or they should be graded and packaged for marketing. Since aging and deterioration of vegetables continues after harvest, the marketable life depends greatly on the temperature and care in physical handling. Quality maintenance is aided by the following procedures:

1. Harvest at optimum maturity or quality
2. Handle carefully to avoid mechanical injury
3. Handle rapidly to minimize deterioration
4. Provide protective containers and packaging
5. Use preservative chemical, heat, or modified-atmosphere treatments

The preparation of this chapter is assigned to TC 11.5, Fruits, Vegetables and Other Products.

6. Enforce good plant sanitation procedures
7. Precool to remove field heat
8. Provide high relative humidity to minimize moisture loss
9. Provide proper refrigeration throughout marketing

Cooling

Rapid cooling of a commodity after harvest, before or after packaging, and before it is stored or moved in transit, prevents deterioration of the more perishable vegetables. The faster field heat is removed after harvest, the longer produce can be maintained in good marketable condition. Cooling slows natural deterioration, including aging and ripening, slows growth of decay organisms (and thereby the development of rot), and reduces wilting, since water losses occur much more slowly at low temperatures than at high temperatures. After cooling, produce should be refrigerated continuously at recommended temperatures. If warming is allowed, much of the benefit of prompt precooling may be lost.

Types of cooling include hydrocooling, vacuum cooling, air cooling, and cooling with contact ice and top ice. These methods are discussed in detail in Chapter 10. The choice of cooling method depends on factors such as refrigeration sources and costs, volume of product shipped, and product limitations.

IN-TRANSIT PRESERVATION

Good equipment is available to transport perishable commodities to market under refrigeration by rail, trucks, piggyback trailers, and containers. A high rh of about 95% is desirable for most vegetables to prevent moisture loss and wilting. Many vegetables benefit from 95 to 100% rh. Humidity in both iced and mechanically refrigerated cars and trailers is usually high. Top ice and package ice, used most often for leafy vegetables, provide refrigeration and added moisture.

Cooling Vehicle and Product

Vehicles used to ship vegetables that require low temperature during transit should have their interiors cooled before loading. With mechanical refrigeration, the units should be operated with the doors closed until the temperature of the interior of the vehicle is reduced to the approximate transit temperature.

Generally, vegetables that require low temperature during transport should be cooled before they are loaded into transport vehicles. Cooling produce in tightly loaded refrigerator cars or trailers is a slow process, and that portion of the load exposed to the cold air discharge may be frozen when the interior of the load is still warm. It is also uneconomical to provide refrigeration capacity in vehicles for cooling.

Packaging, Loading, and Handling

Containers must protect the commodity, permit heat exchange as necessary, and serve as an appropriate merchandising unit with sufficient strength to withstand normal handling. Freight container tariffs describe approved containers and loading procedures.

Containers should be loaded to take advantage of their maximum strength and to permit adequate stripping or use of spacers to hold the load in alignment. Proper vertical alignment of containers is essential to obtain their maximum stacking strength capability, although maximum stacking frequently is incompatible with providing channels for air circulation. Channels for proper air circulation must be maintained, even at the sacrifice of some capacity and resistance of the load to shifting. Ventilation openings in containers, if any, should have the greatest possible exposure to the ventilation channels or flues.

When different types of containers are used in the same load, stacks should be separated so that one type will not damage another. If separation of stacks is not possible, containers made of lighter material, such as fiberboard, should always be loaded on top of any heavier wood containers.

Providing Refrigeration and Air Circulation

Desirable and safe transit temperatures for various vegetables and suggested temperatures to be specified for mechanically refrigerated cars and trailers are given in Table 1. For safety, the thermostat settings suggested for cool season vegetables are usually 2 to 4 °F above the freezing point. The various means for obtaining specific transit temperatures in the United States are provided under the regulations of the Perishable Protective Tariffs issued by the National Freight Committee and the Railway Express Agency.

With the many kinds of refrigeration, heating, and ventilating services now available, the shipper has only to specify the desired transport temperature. Generally, the shipper or the receiver is responsible for selecting the protective service for his commodity in transit. The various protective services are described in detail in USDA *Agriculture Handbooks* No. 669 (Ashby *et al.* 1987) (truck shipments) and No. 195 (rail shipments).

Table 1 Desirable Transit Temperatures for Various Vegetables

Vegetable	Desirable Transit Temperature, °F	Suggested Thermostat Setting[a], °F	Highest Freezing Point[b], °F
Artichokes	32	33	29.8
Asparagus	32-35	35	30.9
Beans, lima	37-41	37	31.0
Beans, snap	40-45	45	30.7
Beets, topped	32	34	30.4
Broccoli	32	34	30.9
Brussels sprouts	32	34	30.6
Cabbage	32	34	30.4
Cantaloupes	36-41	37	29.8
Carrots, topped	32	33	29.5
Cauliflower	32	34	30.6
Celery	32	34	31.1
Corn, sweet	32	34	30.9
Cucumbers	50-55	50	31.1
Eggplant	46-54	50	30.6
Endive and Escarole	32	34	31.8
Greens, leafy	32	34	—
Honeydew melon	45-50	45	30.4
Lettuce	32	34	31.6
Onions, dry	32-39	35	30.6
Onions, green	32	34	30.4
Peas, green	32	34	30.9
Peppers, sweet	45-50	46	30.7
Potatoes:			
Early crop	50-60	50	30.9
Late crop	39-50	40	30.9
For chipping:			
Early crop	64-70	64-70	30.9
Late crop	50-60	50-60	30.9
Radishes	32	34	30.7
Spinach	32	34	31.5
Squash, summer	41-50	41	31.1
Squash, winter	50-55	50	30.5
Sweet potatoes	55-61	55	29.7
Tomatoes:			
Mature green	55-70	55	30.9
Pink	46-50	50	30.6
Watermelons	50-60	50	31.3

Data from USDA *Agricultural Handbook* No. 195 (Redit 1969), USDA *Marketing Research Report* No. 196 (Whiteman 1957), and USDA *Agricultural Handbook* No. 66 (Hardenburg *et al.* 1986).

[a]For U.S. shipments of vegetables in mechanically refrigerated cars under Rule 710 and in trailers under Rule 800 of the Perishable Protective Tariff.

[b]Highest temperature at which freezing occurs.

Protection from Cold

During the winter, vegetables must be protected from freezing. Mechanically refrigerated freight cars and trucks equipped to handle the full range of both fresh and frozen commodities are designed to provide heat for cold weather protection as well as refrigeration. The heat is supplied by electric heating elements or by reverse-cycle operation of the refrigerating unit in which hot gas from the compressor is circulated in the cooling coils. The change from cooling to heating is done automatically by thermostatic controls.

Protection against freezing in transit is a major problem in moving late crop potatoes from storage to market. Cars and trailers should be warmed before loading, protected during loading with canvas door shields or loading tunnels, and loaded properly.

Checking and Cleaning Equipment

If the thermostat is out of adjustment by a few degrees, products may be damaged by freezing or by not receiving sufficient refrigeration. Thermostats should be calibrated at regular intervals to assure that the proper amount of refrigeration is furnished. Trailers and railcars should be cleaned carefully before loading. Debris from previous shipments may contaminate loads and should be swept from floors and floor racks.

Modified Atmospheres in Transit

A variety of systems provide controlled or modified atmospheres in trucks, piggyback trailers, railcars, and seavans. Though the label on each process may differ, all systems alter the levels of oxygen, carbon dioxide, and nitrogen surrounding the produce.

A frequent goal is to lower the oxygen concentration in air to a level of 1 to 5% because this level usually depresses the respiration rate. However, no single modified atmosphere can be expected to benefit more than a few commodities; crop requirements and tolerances are quite specific (Harvey 1965). Also, certain vegetables may tolerate an atmosphere at one temperature but not at another, or they may tolerate a modified atmosphere for only a limited time.

The load compartment must be fairly tight to maintain desirable atmospheres. Modified-atmosphere equipment is predominantly installed in vehicles used to transport chilled perishables in long haul movements (over five days' transit time).

Lettuce is the main vegetable shipped under a modified atmosphere. The physiological disorder known as *russet spotting* is reduced when lettuce is shipped in low oxygen atmospheres with good refrigeration (Stewart and Ceponis 1968). Lettuce is damaged by accumulated carbon dioxide (Stewart *et al.* 1970), so some fresh hydrated lime is usually placed in the cargo space to absorb carbon dioxide.

Atmosphere systems used in transport either vaporize liquid nitrogen or generate nitrogen by passing heated compressed air through hollow fibers. In either case, nitrogen is vented into the trailer to reduce oxygen levels.

Temperature is critical in determining the benefits of modified atmospheres. If temperatures are higher than those recommended for use with modified atmospheres, decay and other deterioration may be increased rather than reduced. Modified atmospheres should never be used as a substitute for good temperature management.

Use of controlled atmospheres is discussed later in this chapter. For further information on refrigerated transport by truck, railway car, ship, and air, see Chapters 27 through 30.

PRESERVATION IN WAREHOUSES

Wholesale warehouses usually do not have a whole range of controlled temperature rooms to provide optimum storage conditions for each kind of produce, and this is not necessary for short holding. About one-half the produce handled can be stored at 32 °F. Enough refrigeration capacity should be available to maintain a year-round temperature of 32 °F with 90 to 95% rh. Higher temperatures and lower humidities for more perishable vegetables accelerate quality loss and increase waste. Enough refrigerated coil surface should be provided to allow a differential of only a few degrees between the coil and air temperatures, while still providing adequate refrigeration. A difference of as little as 2 °F is desirable to permit optimum humidity conditions.

Table 2 Recommended Temperatures for Maintaining Quality of Fresh Vegetables in Wholesale Warehouses

Store at 32 °F and 90 to 95% rh		Store at 50 °F 80 to 85% rh
Artichokes	Horseradish	Beans, green
Asparagus	Kohlrabi	Cucumbers
Beans, Lima	Lettuce	Eggplants
Beets	Mushrooms	Garlic, dry
Broccoli	Onions, green	Melons
Brussels sprouts	Parsnips	Okra
Cabbage	Peas, green	Onions, dry
Carrots	Radishes	Peppers
Cauliflower	Rhubarb	Potatoes
Celery	Rutabagas	Pumpkins
Corn, sweet	Spinach	Squash, hard shell
Endive	Squash, summer	Sweet potatoes
Escarole	Turnips	Tomatoes, ripe

Adapted from Bogardus and Lutz (1961).

Desirable air temperature and humidity cannot be maintained if excessive air exchange occurs between the warehouse cold storage room and warmer areas. Operators should consider the use of air curtains or flap doors whenever doors to cold rooms must be opened often or for prolonged periods.

Some vegetables should not be stored at 32 °F because of the danger of chilling injury. Other less perishable vegetables, such as dry onions, can be held satisfactorily for short periods at warmer temperatures. For these vegetables, controlled storage conditions during wholesaling are still desirable. Table 2 shows which vegetables should be held at 32 °F and which should be held at 50 °F during wholesaling.

Mature green tomatoes, for ripening, usually need separate, controlled temperature rooms, where 55 to 70 °F temperatures with 85 to 90% rh can be maintained to delay or speed ripening as desired. Information on general aspects of cold storage design and operation can be found in Chapters 24, 25, and 26.

REFRIGERATED STORAGE

The refrigeration requirement of any storage plant must be based on peak refrigeration load. This peak usually occurs when outside temperatures are high and warm produce is being moved into the plant for cooling and storage. The peak refrigeration load depends on the amount of commodity received each day, the temperature of the commodity at the time it is placed under refrigeration, the specific heat of the commodity, and the final temperature attained.

Protective Packaging and Waxing

Vegetables should be stored in containers with adequate cushioning materials, stacking strength, and durability to protect against crushing and to withstand high humidity conditions. Bulging crates should be stacked on their sides or stripped between layers to keep weight off the commodity. Many vegetables are now stored or shipped in fiberboard or corrugated containers, but the weakening of fiberboard materials by moisture absorption at the high humidities in storage is frequently a serious problem.

Manufacturers have improved the strength and reduced the degree of moisture absorption by fiberboard materials. Special treatment of fiberboard permits the use of certain cartons during hydrocooling and with package and top ice. Cartons may be strengthened to withstand stacking by using dividers, wooden corner posts, and full telescoping covers.

Produce is consumer-packaged at production points using many types of trays, wraps, and film bags, which may present special storage problems because master containers may lack stacking strength.

Desiccation often can be minimized by using moisture retentive plastic packaging materials. Polyethylene film box liners, pallet covers, and tarpaulins may be helpful to reduce moisture loss from vegetables. Plastic films, if sealed or tightly tied, may restrict the transfer of carbon dioxide, oxygen, and water vapor, leading to harmful concentrations of these respiratory gases; and films restrict heat transfer, which retards the rate of cooling (Hardenburg 1971).

Waxes are applied to rutabagas, cucumbers, mature green tomatoes, and cantaloupes, and to a lesser extent to peppers, turnips, sweet potatoes, and certain other crops. With products such as cucumbers and root crops, waxing reduces moisture loss and thus retards shriveling. With some products, an improved glossy appearance is the main advantage. Thin wax coatings may give little if any protection against moisture loss; coatings which are too heavy may increase decay and breakdown. Waxing alone does not control decay, but waxing and fungicides combined may be beneficial. Waxing is not recommended for potatoes either before or after storage (Hardenburg *et al.* 1959).

Sprout Inhibitors

Sprout inhibitors are used when cold storage facilities are lacking or if low temperatures might injure the vegetable or affect its processing quality. Sprouting of onions, potatoes, and carrots in storage can be inhibited by spraying the plants a few weeks before harvest with a solution of maleic hydrazide. Potatoes are also sprayed or dipped in a solution that inhibits sprouting. Vaporized nonanol alcohol is circulated through ducts to suppress potato sprouting in Great Britain (Burton 1958).

Gamma irradiation suppresses sprouting of onions, sweet potatoes, and white potatoes at dosages of 0.05 to 0.15 kGy. Dosages above 0.15 kGy cause breakdown and increase decay in white potatoes (Kader *et al.* 1984).

Controlled and Modified Atmosphere Storage

Refrigeration is most effective in retarding respiration and lengthening storage life. For some products, reducing the oxygen level in the storage air and/or increasing the carbon dioxide level as a supplement to refrigeration can provide extended storage life. Careful control of the concentration of oxygen and carbon dioxide level is essential. If all of the oxygen is used, produce will suffocate and may develop an alcoholic off-flavor in a few days. Carbon dioxide given off in respiration or from dry ice may accumulate to injurious levels.

Modified or controlled atmospheres (CA) during storage of vegetables have received little application, in contrast to their wide use for apples. Many atmospheres tested on vegetables were injurious or produced only minor benefits (Dewey *et al.* 1969). If commercial use of CA for vegetables increases, it is likely to be with the use of external generators to create desired atmospheres or with the addition of nitrogen gas or dry ice, rather than by using product respiration in a gastight room.

Danish cabbage has kept better during 4 to 5 months at 32 °F in gas mixtures with 2.5 to 5% oxygen and carbon dioxide (with the balance nitrogen) than it has in air (Isenberg and Sayles 1969). Tests have indicated that mature green tomatoes may keep predominantly green for 5 to 6 weeks at 55 °F in an atmosphere at 3% oxygen with 97% nitrogen. After removal to air at 64 °F, these tomatoes ripened normally with acceptable flavor (Parsons and Anderson 1970).

Asparagus and mushrooms in refrigerated storage have kept better for short periods in atmospheres with 5 to 10% carbon dioxide than in air. The carbon dioxide inhibits soft rot of asparagus (Lipton 1965) and retards cap opening and inhibits mold in mushrooms. Some promising experimental CA results on improved quality retention during short storage also have been obtained for head lettuce, broccoli, brussels sprouts, green onions, and radishes.

Hypobaric storage, or storage at reduced atmospheric pressure, is another supplement to refrigeration that involves principles similar to those in controlled-atmosphere storage. At atmospheric pressures 0.1 or 0.2 of normal, several kinds of produce have an extended storage life. The benefits are attributed both to the low oxygen level maintained and to the continuous removal of ethylene, carbon dioxide, and possibly other metabolically active gases. Their rates of production under hypobaric ventilation are also lower.

Positive hypobaric ventilation is absolutely required to achieve the desired low ethylene concentration within and around produce to retard ripening. The continuous flow of water-saturated air at a low pressure flushes away emanated gases and prevents weight loss.

Injury

Chilling injury may be defined as an injury caused by exposure to low but nonfreezing temperatures, often in a temperature range from 32 to 50 °F. At these temperatures, vegetables become weakened because they are unable to carry on normal metabolic processes. Often, vegetables that are chilled look sound when removed from low temperatures. However, symptoms of chilling, such as pitting or other skin blemishes, internal discoloration, or failure to ripen, become evident in a few days at warmer temperatures (Morris and Platenius 1938). Vegetables that have been chilled may be particularly susceptible to decay. *Alternaria* rot is often severe on tomatoes, squash, peppers, and cantaloupes that have been chilled. Tomatoes that have been severely chilled usually ripen slowly and rot rapidly. Figure 1 shows the increasing extent of rot in mature green tomatoes held at chilling temperatures.

Both time and temperature are involved in chilling injury. Damage may occur in a short time if temperatures are considerably below the danger line, but a product may withstand a few degrees in the danger zone for a longer time. Also, the effects of chilling are cumulative. Low temperatures in the field before harvest and in transit add to the total effects of chilling that might occur in storage. A list of vegetables susceptible to chilling injury together with the symptoms and the lowest safe temperature are shown in Table 3.

Table 1 shows the *freezing points* of various vegetables. The freezing point is the highest temperature at which ice crystal formation in the tissues has been recorded experimentally. Most vegetables have a freezing point between 28 and 31 °F (Whiteman 1957). Different vegetables vary widely in their susceptibility to freezing injury.

The freezing point of the commodity is no indication of the damage to be expected from freezing. For example, tomatoes and parsnips both have freezing points of 30 to 31 °F. Parsnips can be frozen and thawed several times without apparent injury, whereas tomatoes are ruined after one freezing. Tissues injured by freezing generally appear water-soaked. Even though some vegetables are somewhat tolerant to freezing, it is desirable to avoid freezing temperatures because they shorten storage life (Parsons and Day 1970).

To minimize damage, fresh commodities should not be handled while frozen. Fast thawing damages tissues, but very slow thawing, such as at 32 to 34 °F, permits ice to remain in the tissues too long. Thawing at 39 °F is suggested (Lutz 1936).

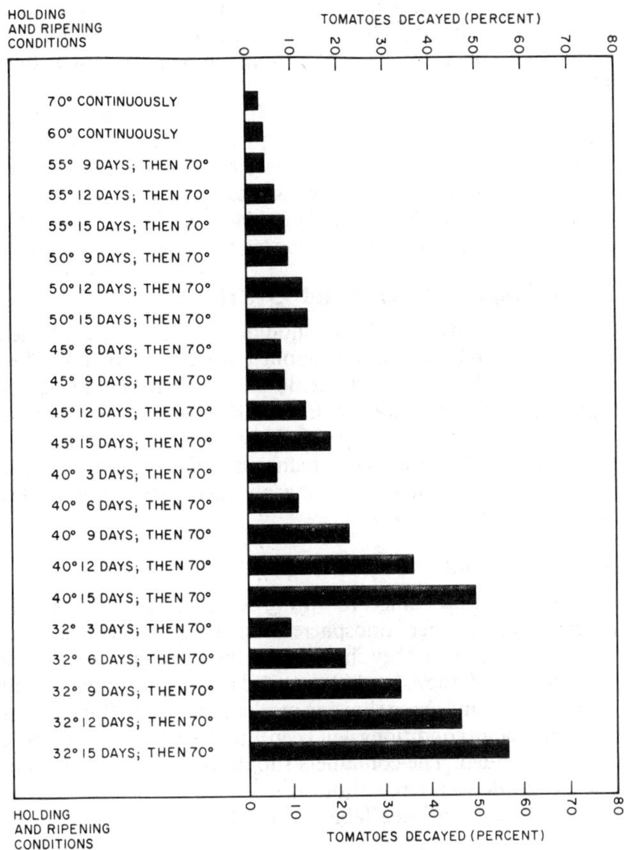

Fig. 1 Effect of Different Temperatures and Holding Periods on Rot, Principally *Alternaria*, in Tomatoes Ripened at 70 °F
(McColloch and Worthington 1954)

Table 3 Vegetables Susceptible to Chilling Injury at Moderately Low but Nonfreezing Temperatures

Commodity	Approximate Lowest Safe Temperature, °F	Character of Injury between 32 °F and Safe Temperature[a]
Beans (snap)	45	Pitting and russeting
Cucumbers	50	Pitting; watersoaked spots, decay
Eggplants	45	Surface scald; *Alternaria* rot
Melons:		
Cantaloupes	a	Pitting; surface decay
Honeydew	45 to 50	Pitting; failure to ripen
Casaba	45 to 50	Pitting; surface decay
Crenshaw and Persian	45 to 50	Pitting; surface decay
Watermelons	40	Pitting; objectionable flavor
Okra	45	Discoloration; watersoaked areas; pitting; decay
Peppers, sweet	45	Sheet pitting, *Alternaria* rot on pods and calyxes
Potatoes	37 to 39	Mahogany browning (Chippewa and Sebago); sweetening[b]
Pumpkins and Hard-Shell Squash	50	Decay, especially *Alternaria* rot
Sweet potatoes	55	Decay; pitting; internal discoloration
Tomatoes:		
Ripe	45 to 50[a]	Watersoaking and softening; decay
Mature green	55	Poor color when ripe; *Alternaria* rot

From USDA *Agricultural Handbook* No. 66 (Hardenburg *et al.* 1986).
[a]See text.
[b]Often these symptoms appear only after removal to warm temperatures, as in marketing.

STORAGE OF VARIOUS VEGETABLES

In the following sections, the temperatures and rh recommended (shown in parentheses) are the optimum for maximum storage in the fresh condition. For short storage, higher temperatures may be satisfactory for some commodities. Temperature requirements represent commodity temperature levels that should be maintained. Much of this information is taken from USDA *Agricultural Handbook* No. 66 (Hardenburg *et al.* 1986).

The quality of each lot of produce should be determined at the time of storage and regular inspections should be made during storage. Such vigilance will permit early detection of disease so that the affected commodity can be moved out of storage before serious loss occurs.

Artichokes, Globe (32 °F and 95 to 100% rh)

Globe artichokes are seldom stored, but for temporary holding a temperature of about 32 °F is recommended. A high rh of at least 95% will help prevent wilting. This product should keep for 2 weeks in storage if buds are uninjured and wilting is prevented. Perforated polyethylene liners with twenty-three ¼-in. holes/ft^2 help retard moisture loss. Hydrocooling or room cooling to 39 °F on the day of harvest reduces deterioration.

Gray Mold Rot (*Botrytis*): The most common decay of harvested artichokes. Reddish brown to dark brown firm rot (see Table 4, Note 2). *Control*: Practice sanitation in the field. Refrigerate promptly.

Asparagus (32 to 36 °F and 95 to 100% rh)

Asparagus deteriorates very rapidly at temperatures above 36 °F and especially at room temperature. It loses sweetness, tenderness, and flavor, and decay develops later. If the storage period is 10 days or less, 32 °F is recommended; asparagus is subject to chilling injury if held longer at this temperature. Asparagus is not ordinarily stored except temporarily, but at 36 °F with a high relative humidity, it can be kept in salable condition for 3 weeks. However, after a long haul to market, even under refrigeration, it cannot be expected to keep longer than 1 to 2 weeks.

Asparagus should be cooled immediately after cutting. Hydrocooling is the usual method. During transit or storage, the butts of asparagus should be placed on some moist, absorbent material to prevent loss of moisture and to maintain freshness of the spears. Sometimes asparagus bunches are set in shallow pans of water in storage.

Bacterial Soft Rot: Mushy, soft, watersoaked areas on tips and cut ends of asparagus (see Table 4, Note 1). *Control*: Avoid excessive bruising of tips; cool to 39 °.

Fusarium Rot: Watersoaked areas changing through yellow to brown, chiefly on asparagus tips; white to pink delicate mold. *Control*: Cool and ship at temperatures of 39 °F or lower; handle promptly. Keep tips dry on way to market.

Phytophthora Rot: Large, watersoaked, or brownish lesions at the side of cut asparagus stalks. Lesions later are extensively shriveled. *Control*: Cool to 39 °F. Maintain low transit temperatures. Market promptly.

Table 4 Notes on Diseases of General Occurrence

Note 1. Bacterial Soft Rot

Occurs on various vegetables as dark green, greasy or watersoaked soft spots and areas in leaves and stems. Soft, mushy, yellowish spots or soupy areas on stems, roots, and tubers of vegetables. Frequently accompanied by repulsive odor from secondary invaders. *Control*: Use sanitation practices during picking and packing to reduce contamination of harvested product. Use bactericides in postharvest wash treatments. Where possible avoid bruising and injury. Shade harvested produce in the field and reduce temperature promptly to 39 °F or lower for commodities that can withstand low temperatures.

Note 2. Gray Mold Rot (Botrytis)

Decayed tissues are fairly firm to semiwatery. Watersoaked grayish tan to brownish in color. Gray mold and grayish brown, with conspicuous velvety spore masses. *Control*: Use sanitation practices during harvesting and packing. Avoid wounds as much as possible. Use storage and transit temperatures as low as otherwise practicable because decay progresses even at 32 °F.

Note 3. Rhizopus Rot

Decayed tissues are watersoaked, leaky, and softer than those with gray mold rot or watery soft rot. Coarse mycelium and black spore heads develop under moist conditions. Nesting is common. *Control*: Insofar as possible, avoid injury and bruising. Reduce temperature promptly and maintain below 50 °F for commodities that can withstand low temperatures.

Note 4. Watery Soft Rot (Sclerotinia)

Decayed tissues are watersoaked, with slightly pinkish borders or brownish tan; very soft and watery in later stages, accompanied by development of fine white to dingy cottony mold and black to brown mustard seed-like bodies called *sclerotia*. Nesting is common. *Control*: Use sanitary practices in harvesting and packing. Cull out specimens with discolored or dead portions. Maintain temperature as low as practicable because rot progresses even at 32 °F. Do not store commodities known to have watery soft rot at harvest time.

Beans, Green or Snap (40 to 45 °F and 95% rh)

Green or snap beans are probably best stored at 39 to 45 °F, where they may keep for about 1 week. Even this recommended temperature causes some chilling injury, but is best for short storage. When stored at 40 °F or below for 3 to 6 days, surface pitting and russet discoloration may appear in a day or two following removal for marketing. The russeting will be especially noticeable in the centers of the containers where condensed moisture remains. *Contact icing is not recommended.* To prevent wilting, the rh should be maintained at 90 to 95%. Beans for processing can be stored up to 10 days at 40 °F. Containers of beans should be stacked to allow abundant air circulation; otherwise, the temperature may rise from the heat of respiration. Beans stored too long or at too high a temperature are subject to decay such as watery soft rot, slimy soft rot, and rhizopus rot.

Anthracnose (*Colletotrichum*): Circular or oval sunken spots; reddish brown around border with tan centers that frequently bear pink spore mounds. *Control*: Use resistant varieties. Plant disease-free seed. Refrigerate harvested beans promptly to 45 °F.

Bacterial Blight (*Pseudomonas*): Small, greasy-appearing, watersoaked spots in pod. Older spots show red at the center, with watersoaked area surrounding and penetrating to the seed. *Control*: Keep the field sanitary. Use disease-free seed. Reduce the transit temperature promptly to 45 °F.

Bacterial Soft Rot: See Table 4, Note 1.

Cottony Leak (*Pythium*): Pods with large, watersoaked spots, accompanied by abundant white cottony mold. *Control*: Sort out diseased pods in packing. Reduce transit temperature promptly to 45 °F.

Freezing Injury: Slight freezing results in watersoaked mottling in surface of exposed pods. Severely frozen beans become completely watersoaked, limp, and dry out rapidly. Snap beans freeze at about 31 °F.

Russeting: Chestnut brown or rusty, diffuse surface discoloration on both sides of pods. *Control*: Permit no surface moisture on warm beans. Cool promptly; avoid temperatures lower than 45 °F.

Soil Rot (*Rhizoctonia*): Large, reddish brown, sunken, decayed spots on pods. Cream-colored or brown mycelium and irregular, chocolate-colored sclerotia may develop. Nesting is common. *Control*: Maintain transit temperatures of 45 to 50 °F.

Watery Soft Rot: Presence of large black sclerotia in white mold helps to separate this from cottony leak (see Table 4, Note 4).

Beans, Lima (37 to 41 °F and 95% rh)

For best quality, lima beans should not be stored. Unshelled lima beans can be stored only about 1 week at 37 to 41 °F. They should be used promptly after removal because the pods discolor rapidly at room temperature. Even with 1 week of refrigerated storage, the pods may develop rusty brown or brown specks, spots, and larger discolored areas that reduce salability of the beans. The pod discoloration will increase sharply during an additional day at 70 °F following storage.

Beets (32 °F and 98 to 100% rh)

Topped beets are subject to wilting because of the rapid loss of water when the storage atmosphere is too dry. When stored at 32 °F with at least 95% rh, they should keep for 4 to 6 months. Before beets are stored, they should be topped and sorted to remove all diseased beets and those showing mechanical injury. Bunch beets under the same conditions will keep 1 to 2 weeks. Contact icing is recommended. The containers should be well ventilated and stacked to allow air circulation.

Baterial Soft Rot: See Table 4, Note 1.

Broccoli (32 °F and 95 to 100% rh)

Italian or sprouting broccoli is highly perishable and is usually stored for only a brief period as needed for orderly marketing. Good salable condition, fresh green color, and the vitamin C content are best maintained at 32 °F. If it is in good condition and is stored with adequate air circulation and spacing between containers to avoid heating, broccoli should keep satisfactorily 10 to 14 days at 32 °F. Longer storage is undesirable because leaves discolor, buds may drop off, and tissues soften. The respiration rate of freshly harvested broccoli is high, comparable to that of asparagus, beans, and sweet corn. This high rate of respiration should be considered when storing broccoli, especially if it is held without package ice.

Brussels Sprouts (32 °F and 95 to 100% rh)

Brussels sprouts can be stored in good condition for a maximum of 3 to 5 weeks at the recommended temperature of 32 °F. Deterioration, yellowing of the sprouts, and discoloration of the stem are rapid at temperatures of 50 °F and above. Rate of deterioration is twice as fast at 40 °F as at 32 °F. Loss of moisture through transpiration is rather high even if the relative humidity is kept at the recommended level. Film packaging is useful in preventing moisture loss. As with broccoli, sufficient air circulation and spacing between packages is desirable to allow good cooling and to prevent yellowing and decay.

Cabbage (32 °F and 98 to 100% rh)

A large percentage of the late crop of cabbage is stored and sold during the winter and early spring, or until the new crop from southern states appears on the market. If it is stored under proper conditions, late cabbage should keep for 3 to 4 months. The longest keeping varieties belong to the Danish class. Early crop cabbage, especially southern grown, has a limited storage life of 3 to 6 weeks.

Cabbage is successfully held in common storage in northern states, where a fairly uniform inside temperature of 32 to 36 °F can be maintained. An increasing quantity of cabbage is now held in mechanically refrigerated storage, but in some seasons its value does not justify the expense. Use of controlled atmospheres to supplement refrigeration aids quality retention. An atmosphere with 2.5 to 5% oxygen and 2.5 to 5% carbon dioxide can extend the storage life of late cabbage. Cabbage should not be stored with fruits emitting ethylene. Concentrations of 10 to 100 ppm of ethylene cause leaf abscission and loss of green color within 5 weeks at 32 °F.

Pallet bins are used as both field and storage containers, so the cabbage requires no handling from the time of harvest until preparation for shipment. Before the heads are stored, all loose leaves should be trimmed away; only 3 to 6 tight wrapper leaves should be left on the head. Loose leaves interfere with ventilation between heads, which is essential for successful storage. When removed from storage, the heads should be trimmed again to remove loose and damaged leaves.

Chinese cabbage can be stored for 2 to 3 months at 32 °F with 95 to 100% rh.

Alternaria Leaf Spot: Small to large spots bearing brown to black mold. This spotting opens the way for other decays. *Control*: Avoid injuries. Maintain 32 to 34 °F temperature in transit and storage. Practice sanitation in storage rooms.

Bacterial Soft Rot: This slimy decay frequently starts in the pith of a cut stem or in leaf spots caused by other organisms (see Table 4, Note 1).

Black Leaf Speck: Small, sharply sunken, brown or black specks occurring anywhere on outer leaves or in leaves throughout the head. Occurs under refrigeration in transit and storage, and in association with sharp temperature drops. *Control*: No effective control measures are known.

Freezing Injury: Heads frozen slightly may thaw without apparent injury. Freezing injury is found first as brown streaks in the stem, then as light brown watersoaking of the heart leaves and stem. *Control*: Prevent any extended exposure to temperatures below 31 °F.

Watery Soft Rot: See Table 4, Note 4.

Carrots (32 °F and 98 to 100% rh)

Carrots are best stored at 32 °F with a very high rh. Like beets, they are subject to rapid wilting if the humidity is low. For long storage, carrots should be topped and free from cuts and bruises. If they are in good condition when stored, mature carrots should keep 5 to 9 months, if promptly cooled after harvest. Carrots lose moisture readily and wilting results. Humidity should be kept high, but condensation or dripping on the carrots should be avoided, since this is conducive to decay development.

Most carrots for the fresh market are not fully mature. Immature carrots are prepackaged in polyethylene bags, either at the shipping point or in terminal markets. They are usually moved into marketing channels soon after harvest, but they can be stored for a short period to avoid a market glut. If the carrots are cooled quickly and all traces of leaf growth are removed, they can be held 4 to 6 weeks at 32 °F.

In Texas, immature carrots are often stored in clean 50-lb burlap sacks. The sacks of carrots should be stacked in such a manner that at least one surface of each sack is in contact with top ice at all times. Top ice provides some of the necessary refrigeration and prevents dehydration. Bunched carrots may be stored 10 to 14 days at 32 °F. Contact icing is recommended.

Bitterness in carrots, which may develop in storage, is due to abnormal metabolism caused by ethylene given off by apples, pears, and some other fruits and vegetables or from other sources such as internal combustion engines. Bitterness can be prevented by storing carrots away from products that give off ethylene.

Bacterial Soft Rot: See Table 4, Note 1.

Black Rot (*Stemphylium*): Fairly firm black decay at the crown, on the side, or at the tips of harvested roots. *Control*: Avoid bruising. Store at 32 °F.

Crater Rot (Rhizoctonia): Circular brown craters with white to cream-colored mold in the center. Develops under high humidity in cold storage. *Control*: Field sanitation measures. Avoid surface moisture on storage roots.

Freezing Injury: Roots are flabby, and on cutting show radial cracks in the flesh of the central part and tangential cracks in the outer part. *Control*: Prevent exposure to temperatures below 30 °F.

Gray Mold Rot: See Table 4, Note 2.

Rhizopus Rot: See Table 4, Note 3.

Watery Soft Rot: See Table 4, Note 4.

Cauliflower (32 °F and 95% rh)

Cauliflower may be stored for 3 to 4 weeks at 32 °F with about 95% rh. Successful storage depends on retarding the aging of the head or curd, preventing decay marked by spotting or watersoaking of the white curd, and preventing yellowing and dropping of the leaves. When it is necessary to hold cauliflower temporarily out of cold storage, packing in crushed ice will aid in keeping it fresh. Freezing causes a grayish brown discoloration and softening of the curd, accompanied by a watersoaked condition. Affected tissues are rapidly invaded by soft rot bacteria.

Much of the cauliflower now marketed has the leaves closely trimmed and is prepackaged in perforated cellophane overwraps and packed in fiberboard containers. Vacuum cooling is a fairly efficient method of cooling prepackaged cauliflower. In general, use of controlled atmospheres with cauliflower has not been promising. Atmospheres containing 5% carbon dioxide or higher are injurious to cauliflower, although the damage may not be apparent until after cooking.

Bacterial Soft Rot: See Table 4, Note 1.

Brown Rot (*Alternaria*): Brown or black spotting of the curd. *Control*: Use seed treatment and field spraying. Keep the curds dry. Maintain low transit temperatures. Store at 32 °F.

Celery (32 °F and 98 to 100% rh)

Celery is a relatively perishable commodity, and for storage of 1 to 2 months, it is essential that a commodity temperature of 32 °F be maintained. The rh should be high enough to prevent wilting (98 to 100%). Considerable heat is given off because of respiration, and for this reason the stacks of crates should be separated and dunnage used to allow cold air circulation under and over the crates and between the bottom crates and the floor. Forced-air circulation should be provided; otherwise there may be a 3 to 4 °F temperature differential between the top and the bottom of the room.

Celery can be cooled by forced-air cooling, by hydrocooling, or by vacuum cooling. Hydrocooling is the most common cooling method; temperatures should be brought to as near 32 °F as possible. In practice, temperatures are reduced to 40 to 45 °F. Vacuum cooling is widely used for celery packed in corrugated cartons for long-distance shipment.

Bacterial Soft Rot: See Table 4, Note 1.

Black Heart: Brown or black discoloration of tips or all of the heart leaves. Affected celery should not be stored because of rapid development of bacterial soft rot. *Control*: Good cultural practices, with special attention to available calcium. Harvest promptly after celery is mature.

Early Blight (*Cercospora*): Circular pale yellow spots on leaflets. In advanced stages, spots coalesce and become brown to ashen gray. No spots develop in storage, but affected lots lose moisture and their fresh appearance. *Control*: Control early blight in the field by spraying or dusting.

Freezing Injury: Characteristic loosening of the epidermis can best be demonstrated by twisting the leaf stem. Severe freezing causes celery to become limp and to dry out rapidly. Freezing may cause watery soft rot and bacterial soft rot. The freezing point is about 31°F.

Late Blight (*Septoria*): Small (1/8 in. or less), yellowish, indefinite spots in the leaflet and elongated spots on the leaf stalk bearing black fruiting bodies of pinpoint size on the surface and surrounding green tissue. Development of blight in storage probably is negligible, but it opens the way for storage decays. *Control*: Control late blight in the field by sanitary measures and fungicides. Store infected lots for short periods only.

Mosaic: Leaflets are mottled; the stalk shows brownish sunken streaks. Badly affected stalks shrivel. *Control*: Eradicate weeds that carry the virus. Grade out all discolored stalks at packing time.

Watery Soft Rot (*Pink Rot*): This is the principle decay of celery, often severe on field-frozen stock and on celery harvested after prolonged cool, moist weather. In early stages, it often has a pink color. *Control*: Grade out all discolored stalks at packing time. Storage at 32°F will retard but not prevent the disease.

Corn, Sweet (32°F and 95 to 98% rh)

Sweet corn is highly perishable and is seldom stored except to protect an excess supply temporarily. Corn as it usually arrives on the market should not be expected to keep even in 32°F storage for more than 4 to 8 days.

The sugar content, which so largely determines quality in corn and decreases rapidly at ordinary temperatures, decreases less rapidly if the corn is kept at about 32°F. The loss of sugar is about 4 times as rapid at 50°F as it is at 32°F.

Sweet corn should be cooled promptly after harvest. Usually, corn is hydrocooled, but vacuum cooling is also satisfactory if the corn is prewetted and top-iced after cooling. Where precooling facilities are not available, corn can be cooled with package and top ice. Sweet corn should not be handled in bulk unless it is copiously iced because of its tendency to heat throughout the pile. Sweet corn ears should be trimmed to remove most shank material before shipment to minimize moisture loss and prevent kernel denting. A loss of 2% moisture from sweet corn can result in objectionable kernel denting.

Cucumbers (50 to 55°F and 95% rh)

Cucumbers can be held only for short periods of 10 to 14 days at 50 to 55°F with a relative humidity of 95%. Cucumbers held at 45°F or below for longer periods develop surface pitting or dark-colored watery areas. These blemishes indicate chilling injury. Such areas soon become infected and decay rapidly on removal of the cucumbers to warmer temperatures. Slight chilling may develop in 2 days at 32°F and severe chilling injury within 6 days at 32°F. The susceptibility of cucumbers to chilling injury does not preclude their exposure to temperatures below 50°F for short intervals, as long as they are used immediately after removal from cold storage. Chilling symptoms develop rapidly only at higher temperatures. Thus, 2 days at 32°F or 4 days at 39°F are harmless under these conditions. Waxing is of some value in reducing weight loss and giving a brighter appearance. Shrink wrapping with polyethylene film can also prevent the loss of turgidity.

At temperatures of 50°F and above, cucumbers ripen rather rapidly, the green color changing to yellow. Ripening is accelerated if they are stored in the same room with ethylene-producing crops for more than a few hours.

Anthracnose (*Colletotrichum*): Circular, sunken, watersoaked spots that soon produce pink spore masses in the center. Later, the spots turn black. *Control*: Use disease-free seed and fungicidal applications in the field.

Bacterial Soft Rot: See Table 4, Note 1.

Bacterial Spot (*Pseudomonas*): Small, circular, watersoaked spots, later chalky or moist with gummy exudate. *Control*: If possible, avoid shipping infected cucumbers. Pack them dry and maintain temperatures as near optimum as practicable.

Black Rot (*Mycosphaerella*): Irregular, brownish, watersoaked spots of varying size, later nearly black. Black fruiting bodies are sometimes present. *Control*: Exclude infected cucumbers from the pack, if possible. Reduce carrying temperatures to about 50°F.

Chilling Injury: Numerous sunken, slightly watersoaked areas in the skin of cucumbers after removal from storage, found on cucumbers stored for longer than a week at temperatures below 45°F. *Control*: Store cucumbers at temperatures between 50 and 55°F, for not longer than 2 weeks.

Cottony Leak (*Pythium*): Large, greenish, watersoaked lesions. Luxuriant, white, cottony mold over wet decay. *Control*: Exclude infected cucumbers from the pack, if possible. Reduce carrying temperatures to about 50°F.

Freezing Injury: Large areas in cucumbers that are soft, flabby, watersoaked, and wrinkled, especially toward the stem end. *Control*: Prevent exposure to temperatures below 31°F.

Watery Soft Rot: See Table 4, Note 4.

Eggplants (46 to 54°F and 90 to 95% rh)

Eggplants are not adapted to long storage. They cannot be expected to keep satisfactorily even at the optimum temperature of 46 to 54°F for over a week and still retain good condition during retailing. Eggplants are subject to chilling injury at temperatures below 45°F. Surface scald or bronzing and pitting after sand scarring are symptoms of chilling injury. Eggplants that have been chilled are subject to decay by *Alternaria* when they are removed from storage. Exposure to ethylene for 2 or more days hastens deterioration.

Cottony Leak (*Pythium*): Decayed areas are large, bleached, discolored (tan), wrinkled, moist, and soft; later they exhibit abundant cottony mold. *Control*: Reduce carrying temperatures to about 46°F.

Fruit Rot (*Phomopsis*): Numerous, somewhat circular brown spots that later coalesce over much of the fruit with pycnidia dotting the older lesions. This is a very common decay of eggplant. *Control*: Use fungicide sprays in the field. Reduce carrying temperature to about 46°F. Move the fruit promptly if decay is evident.

Endive and Escarole (32°F and 95 to 100% rh)

Endive and escarole are leafy vegetables not adapted to long storage. Even at 32°F, which is considered the best storage temperature, they cannot be expected to keep satisfactorily for more than 2 or 3 weeks. They should keep somewhat longer if they are stored with cracked ice in or around the packages. Some desirable blanching usually occurs in endive held in storage.

Garlic, Dry (32°F and 65 to 70% rh)

Garlic should keep at 32°F for 6 to 7 months, if it is in good condition and is well cured when stored. Garlic cloves sprout most rapidly at 40 to 64°F; therefore, prolonged storage at this temperature should be avoided. In California, it is frequently put in common storage, where it can be held for 3 to 4 months or sometimes longer if the building can be kept cool, dry, and well ventilated.

Blue Mold Rot (*Penicillium*): Soft, spongy, or powdery dry decay of cloves. Affected cloves finally break down completely into gray or tan powdery masses. *Control*: Prevent bruising; keep garlic dry.

Waxy Breakdown: Yellow or amber waxy translucent breakdown of the outside cloves. *Control*: No control measures have been developed.

Greens, Leafy (32 °F and 95 to 100% rh)

Leafy greens such as collards, chard, and beet and turnip greens are very perishable and should be held as close to 32 °F as possible. At this temperature, they can be held 10 to 14 days. They are commonly shipped with package and top-ice to maintain freshness and are handled like spinach. Kale packed with polyethylene crate liners should keep at least 3 weeks at 32 °F or 1 week at 40 °F. Vitamin content and quality are retained better when wilting is prevented.

Lettuce (32 °F and 95 to 100% rh)

Lettuce is highly perishable. To minimize deterioration, it requires a temperature as close to its freezing point as possible without actually freezing it. Lettuce will keep about twice as long at 32 °F as at 37 °F. If it is in good condition when stored, lettuce should keep 2 to 3 weeks at 32 °F with a high rh. Most lettuce is packed in cartons and vacuum-cooled to about 34 to 36 °F soon after harvest. It should then be immediately loaded into refrigerated cars or trailers for shipment or placed in cold storage rooms for holding prior to shipping.

An increasing quantity of lettuce is shipped in modified atmospheres to aid quality retention. Modified atmospheres are a supplement to proper transit refrigeration, but are not a substitute for refrigeration. Lettuce is not tolerant of carbon dioxide and is injured by concentrations of 2 to 3% or higher. Romaine and leaf lettuce tolerate slightly higher carbon dioxide levels than head lettuce. Romaine is injured by 10% carbon dioxide, but not by 5% at 32 °F.

Since excess wrapper leaves are usually trimmed off before sale or use, it is suggested that lettuce be trimmed to 2 wrapper leaves before packaging, rather than the usual 5 or 6 to save space and weight. The extra wrapper leaves are not needed to maintain quality.

Bacterial Soft Rot: The most common cause of spoilage in transit and storage. Often, it starts on bruised leaves. This decay normally is the controlling factor in determining the storage life of lettuce and is much less serious at 32 °F than at higher temperatures (see Table 4, Note 1).

Brown Stain: Lesions that are typically tan, brown, or even black, and about ¼ in. wide and ½ in. long, with distinct margins that are darker than the slightly sunken centers. The margins give a halo effect. The lesions develop on head leaves just under the cap leaves. The heart and wrapper leaves are not affected. Brown stain is caused by carbon dioxide accumulation in railcars or trailers from normal product respiration. *Control*: Ventilate to keep carbon dioxide below 2% in transport vehicles by keeping one water drain open. Enclose bags of hydrated lime (in vehicles shipped under a modified atmosphere) to absorb carbon dioxide.

Gray Mold: See Table 4, Note 2.

Pink Rib: Characterized by diffuse pink discoloration near the bases of the midribs of the outer head leaves. In heads with severe symptoms, all but the youngest head leaves may be pink and discoloration may reach into large veins. The cause has not been identified, but shipment in low oxygen atmospheres at undesirably high temperatures (50 °F) can accentuate the disease. It is most common in hard to overmature lettuce. *Control*: Store and ship lettuce at recommended low temperatures.

Russett spotting: This occasionally causes serious losses. Small tan or rust-colored pitlike spots appear mostly on the midrib but possibly develop on other parts of leaves. Exposure to ethylene and to storage or transport temperatures above 37 °F are the main causes of this disorder. Hard lettuce is more susceptible to it than firm lettuce. *Control*: Avoid storing or shipping lettuce with apples, pears, or products that give off ethylene. Precool lettuce adequately to 34 to 37 °F and refrigerate it continuously. Shipment in a low oxygen atmosphere (1 to 8%) gives effective control.

Rusty Brown Discoloration: A serious market disorder of western head lettuce; a diffuse discoloration which tends to follow the veins but also spreads to adjacent tissue. The disorder starts on the outer head leaves but in severe cases may affect all leaves. *Control*: There is no known control method.

Tipburn: Dead, brown areas along the edges and tips of inner leaves. This is considered to be of field origin, but occasionally the severity of the disease increases after harvest. *Control*: Keep the affected stock well cooled and market it promptly after unloading to avoid secondary bacterial rots.

Watery Soft Rot: See Table 4, Note 4.

Melons

Cold storage is hardly used for most kinds of melons except to avoid temporarily adverse market conditions. To avoid injury by chilling, most melons are stored at 45 to 50 °F with 90 to 95% rh.

Persians should keep at this temperature range for up to 2 weeks; *honeydews* for 2 to 3 weeks; and *casabas* for 4 to 6 weeks. It is reported that these melons will be definitely injured in 8 days at temperatures as low as 32 °F. Honeydews are usually given an 18- to 24-h ethylene treatment (5000 ppm) to obtain uniform ripening. Pulp temperature should be 70 °F or above during treatment. Honeydews must be mature when harvested; immature melons fail to ripen even if treated with ethylene.

Cantaloupes harvested at the hard-ripe stage (less than full slip) can be held about 15 days at 36 to 39 °F. Lower temperatures may cause chilling injury. Full-slip hard-ripe cantaloupes can be held for a maximum of 10 to 14 days at 32 to 36 °F. They are more resistant to chilling injury. Cantaloupes are precooled by hydrocooling or forced-air cooling before loading, or by top-icing after loading in railcars or trucks.

Watermelons are best stored at 50 to 60 °F and should keep from 2 to 3 weeks. Watermelons decay less at 32 °F than at 40 °F, but they tend to become pitted and have an objectionable flavor after 1 week at 32 °F. At low temperatures, they are subject to various symptoms of chilling injury—loss of flavor and fading of red color. Watermelons should be consumed within 2 to 3 weeks after harvest, primarily because of the gradual loss of crispness.

Alternaria Rot: Irregular, circular, brownish spots, sometimes with concentric rings, later covered with black mold. Often found on melons that have been chilled. *Control*: Avoid chilling temperatures. If cold melons are to be held at room temperature, they should be so stacked that condensed moisture will evaporate readily. Market melons promptly.

Anthracnose (*Colletotrichum*): Numerous greenish, elevated spots with yellow centers, later sunken and covered with moist pink spore masses. *Control*: Apply recommended field control measures.

Chilling Injury: Honeydew and honeyball melons stored for 2 weeks or longer at temperatures of 32 to 34 °F sometimes show large, irregular, water soaked, sticky areas in the rind. *Control*: Store melons at 45 to 50 °F.

Cladosporium Rot: Small black shallow spots later covered with velvety green mold. On cantaloupes, this rot is evident on extensive shallow areas at the stem ends or at points of contact between melons and it can be rubbed off easily. *Control*: Control measures are the same as for *Alternaria rot*.

Fusarium Rot: Brown areas on white melons; white or pink mold over indefinite spots on green melons. Affected tissue is spongy and soft, with white or pink mold. *Control*: Avoid mechanical injuries; reduce carrying temperatures to 45 °F.

Phytophthora Rot: Brown, slightly sunken areas; later water-soaked and covered with a wet, appressed, whitish mold. *Control*: Cull out the affected fruits during packing. Reduce carrying temperatures to 45 °F.

Rhizopus Rot:The affected melon is soft, but not soupy and leaky as it is in similar decay on other vegetables. Coarse fungus strands may be demonstrated in decayed tissue (see Table 4, Note 3).

Stem End Rot (*Diplodia*): Brown, fairly firm decay usually starting at the stem end and affecting a large part of the watermelon. Black fruiting bodies develop later. *Control*: At the time of loading in cars, recut the stems and treat them with Bordeaux paste or another recommended fungicide.

Mushrooms (32 °F and 95% rh)

Mushrooms are usually processed or sold in a retail market within 24 to 48 h after they are harvested. They keep in good salable condition at 32 °F for 5 days, at 39 °F for 2 days, and at 50 °F or above for about 1 day. A rh of 95% is recommended during storage. While being transported or displayed for retail sale, mushrooms should be refrigerated. Deterioration is marked by brown discoloration of the surfaces, elongation of the stalks, and opening of the veils. Black stems and open veils are correlated with dehydration.

Controlled-atmosphere storage reportedly can prolong the shelf life of mushrooms held at 50 °F, if the oxygen concentration in the atmosphere is 9% or the carbon dioxide concentration is 25 to 50%.

Moisture-retentive film overwraps of caps usually are beneficial in reducing moisture loss.

Okra (45 to 50 °F and 90 to 95% rh)

Okra deteriorates rapidly and is normally stored only briefly before marketing or processing. It has a very high respiration rate at warm temperatures. Okra in good condition can be kept satisfactorily for 7 to 10 days at 45 to 50 °F. A rh of 90 to 95% is desirable to prevent wilting. At temperatures below 45 °F, okra is subject to chilling injury, which is shown by surface discoloration, pitting, and decay. Holding okra for 3 days at 32 °F may cause pitting. Contact or top-ice causes water spotting in 2 or 3 days at all temperatures.

Fresh okra bruises easily; blackening of the damaged areas occurs within a few hours. A bleaching type of injury may also develop when okra is held in hampers for more than 24 h without refrigeration.

Onions (32 °F and 65 to 70% rh)

A comparatively low rh is essential in the successful storage of dry onions. However, humidities as high as 85% and forced-air circulation have given satisfactory results. At higher humidities, at which most other vegetables keep best, onions are disposed to root growth and decay; at too high a temperature, sprouting is encouraged. Storage at 32 °F with 65 to 70% rh is recommended to keep them dormant.

Onions should be adequately cured either in the field, in open sheds, or by artificial means before or in storage. The most common method of curing in northern areas is by forced ventilation in storage. Onions are considered cured when the necks are tight and the outer scales are dried until they rustle. If not cured, onions are likely to decay in storage.

Onions are stored in 50-lb bags, in crates, in pallet boxes that hold about 1000 lb of loose onions, or in bulk bins. Bags of onions are frequently stored on pallets. Bagged onions should be stacked to allow proper air circulation.

In the northern onion-growing states, onions of the globe type are generally held in common storage because average winter temperatures are sufficiently low. They should not be held after early March unless they have been treated with maleic hydrazide in the field to reduce sprout growth.

Refrigerated storage is often used to hold onions for marketing late in the spring. Onions to be held in cold storage should be placed there immediately after curing. A temperature of 32 °F will keep onions dormant and reasonably free from decay, provided the onions are sound and well cured when stored. Sprout growth indicates too high a storage temperature, poorly cured bulbs, or immature bulbs. Root growth indicates the relative humidity is too high.

Globe onions can be held for 6 to 8 months at 32 °F. Mild or Bermuda types can usually be held at 32 °F for only 1 to 2 months. Onions of the Spanish type are often stored; if well matured, they can be held at 32 °F, at least until January or February. In California, onions of the sweet Spanish type are held at 32 °F until April or May.

Onions are damaged by freezing, which appears as watersoaking of the scales when cut after thawing. Onions that have been slightly frozen may recover with little perceptible injury, if allowed to thaw slowly and without handling. When onions are removed from storage in warm weather, they may sweat due to condensation of moisture. This may favor decay. Warming onions gradually (for example to 50 °F over 2 to 3 days) with good air movement should avoid the difficulty. Onions should not be stored with other products that tend to absorb odors.

Onion sets require practically the same temperature and humidity conditions as onions, but because they are smaller in size they tend to pack more solidly. They are handled in approximately 25-lb bags and should be stacked to allow the maximum air circulation.

Green onions (scallions) and green shallots are usually marketed promptly after harvest. They can be stored 3 to 4 weeks at 32 °F with 95% rh. Crushed ice spread over the onions will aid in supplying moisture. Packaging in polyethylene film will also aid in preventing moisture loss. Storage life of green onions at 32 °F can be extended to 8 weeks by packaging them in perforated polyethylene bags or in waxed cartons and holding them in a controlled atmosphere of 1% oxygen with 5% carbon dioxide.

Ammonia Gas Discoloration: Exposure of onions to 1% ammonia in air for 24 h causes the surface of yellow onions to turn brown, red onions to turn deep metallic black, and white onions to turn greenish yellow. *Control*: Ventilate storage rooms as soon as possible after exposure.

Bacterial Soft Rot: This decay often affects one or more scales in the interior of the bulb. Decayed tissue is more mushy than gray mold rot (see Table 4, Note 1).

Black Mold (*Aspergillus*): Black powdery spore masses on the outermost scale or between outer scales. *Control*: Store onions at just above 32 °F and at 65% rh.

Freezing Injury: A watersoaked, grayish yellow appearance of the entire outer fleshy scales results from a slight freezing injury. All scales are affected and become flabby with severe injury. Opaque areas appear in affected scales. *Control*: Prevent exposure to 30 °F temperatures and lower. Thaw frozen onions at 40 °F.

Fusarium Bulb Rot: Semiwatery to dry decay progressing up the scales from the base. Decay usually is covered with dense, low-lying white to pinkish mold. *Control*: Do not store badly affected lots. Pull out infected bulbs in slightly affected lots. Store onions at 32 °F.

Gray Mold Rot: This is the most common type of onion decay; it usually starts at the neck, affecting all scales equally. Decay often is pinkish (see Table 4, Note 2). *Control*: Cure onions thoroughly. Protect them from rain. Store them at just above 32 °F.

Smudge: Black blotches or aggregations of minute black or dark green dots on the outer drying scales of white onions. Under moist conditions, sunken yellow spots develop on fleshy scales. *Control*: Protect onions from rain after harvest. Store them just above 32 °F.

Translucent Scale: The outer 2 or 3 scales are gray and watersoaked, as in freezing. The entire scale may not be discolored; no opaque area is noticeable. Sometimes translucent scale is found in the field. *Control*: No control is known. Store onions at 32 °F after curing.

Parsley (32 °F and 95 to 100% rh)

Parsley should keep 1 to 2.5 months at 32 °F and for a somewhat shorter period at 36 to 39 °F. High humidity is essential to prevent desiccation. Package icing is often beneficial.

Bacterial Soft Rot: See Table 4, Note 1.

Watery Soft Rot: See Table 4, Note 4.

Parsnips (32 °F and 98 to 100% rh)

Topped parsnips have similar storage requirements to topped carrots and should keep for 2 to 6 months at 32 °F. Parsnips held at 32 to 34 °F for 2 weeks after harvest attain a sweetness and high quality equal to that of roots subjected to frosts for 2 months in the field. Ventilated polyethylene box or basket liners can aid in preventing moisture loss. Parsnips are not injured by slight freezing while in storage, but they should be protected from hard freezing and should be handled with great care while in a frozen condition. The main storage problems with parsnips are decay, surface browning, and their tendency to shrivel. Refrigeration and high rh will retard deterioration.

Bacterial Soft Rot: See Table 4, Note 1.

Canker (*Itersonilia sp.*): Organism enters through fine rootlets and through injuries. The surface of the infected parsnip is first brown to reddish; later, it turns black where a depressed canker is formed. *Control*: Follow the recommended field spray program. Practice crop rotation.

Gray Mold Rot: See Table 4, Note 2.

Watery Soft Rot: See Table 4, Note 4.

Peas, Green (32 °F and 95 to 98% rh)

Green peas lose part of their sugar rapidly if they are not refrigerated promptly after harvest. They should keep in salable condition 1 to 2 weeks at 32 °F. Top icing is beneficial in maintaining freshness. Peas keep better unshelled than shelled.

Bacterial Soft Rot: See Table 4, Note 1.

Gray Mold Rot: See Table 4, Note 2.

Watery Soft Rot: See Table 4, Note 4.

Peas, Southern (40 to 41 °F and 95% rh)

Freshly harvested southern peas at the mature-green stage should have a storage life of 6 to 8 days at 40 to 41 °F with high rh. Without refrigeration, they remain edible for only about 2 days, the pods yellowing in 3 days and showing extensive decay in 4 to 6 days.

Peppers, Dry Chili or Hot

Chili peppers, after drying to a moisture content of 10 to 15%, are stored in nonrefrigerated warehouses for 6 to 9 months. The moisture content is usually low enough to prevent fungus growth. A relative humidity of 60 to 70% is desirable. Polyethylene-lined bags are recommended to prevent changes in moisture content. Manufacturers of chili pepper products hold part of their supply of the raw material in cold storage at 32 to 50 °F, but they prefer to grind the peppers as soon as possible and store them in the manufactured form in airtight containers.

Peppers, Sweet (45 to 55 °F and 90 to 95% rh)

Sweet peppers can be stored for a maximum of 2 to 3 weeks at 45 to 55 °F. They are subjected to chilling injury if they are stored at temperatures below 45 °F. The symptoms of this injury are surface pitting and discoloration near the calyx, which develops in a few hours after removal from storage. At temperatures of 32 to 36 °F, peppers usually develop pitting in a few days. When stored at temperatures above 55 °F, ripening (red color) and decay development are rapid. Rapid cooling of harvested sweet peppers is essential in reducing marketing losses. It can be done by forced-air cooling, hydrocooling, or vacuum cooling. Peppers are often waxed commercially, which reduces chafing in transit and moisture loss.

Bacterial Soft Rot: See Table 4, Note 1.

Freezing Injury: The outer wall is soft, flabby, watersoaked, and dark green in color. The core and seeds turn brown with severe freezing. Sweet peppers freeze at about 31 °F.

Gray Mold Rot: See Table 4, Note 2.

Rhizopus Rot: See Table 4, Note 3.

Potatoes (Temperature, see following; 90 to 95% rh)

The proper storage environment for potatoes might be defined as that environment which will promote the most rapid healing of bruises and cuts, reduce rot penetration to a minimum, allow the least weight and other storage losses to occur, and reduce to a minimum the deleterious quality changes that might occur during storage.

Early-crop potatoes are usually not stored except during congested periods. They are more perishable and cannot be expected to keep as well or as long as late-crop tubers. Refrigerated storage at 40 °F following a curing period of 4 or 5 days at 70 °F is recommended; or they can be stored for about 2 months at 50 °F without curing. If early-crop potatoes are to be used for chipping or French frying, storage at 70 °F is recommended. Holding these potatoes in cold storage even at moderate temperatures of 50 to 55 °F for only a few days causes excessive accumulation of reducing sugars, which results in production of dark-colored chips.

Late-crop potatoes produced in the northern half of the United States are usually stored. The greater part of the crop is held in nonrefrigerated commercial and farm storages, but some potatoes are held in refrigerated storages. Potatoes in nonrefrigerated storages are usually held in bulk bins 8 to 20 ft deep. Shallower bins are used in milder climates. Some potatoes are stored in pallet boxes. In refrigerated warehouses, potatoes can be stored in sacks, pallet boxes, or bulk.

Late-crop potatoes should be cured immediately after harvest by being held at 50 to 61 °F and high relative humidity for about 10 to 14 days to permit suberization and wound periderm formation (healing of cuts and bruises). If properly cured, they should keep in sound dormant condition at 38 to 40 °F with 95% rh for 5 to 8 months. A temperature below this is not desirable, except for seed stock for late planting. For this purpose, 37 °F is best. At 37 °F or below, Irish potatoes tend to become sweet. For ordinary table use, potatoes stored at 39 °F are satisfactory, but they probably will be unsatisfactory for chipping or French frying without being desugared or conditioned at about 64 to 70 °F for 1 to 3 weeks prior to use. However, conditioning may be costly, and good results are often uncertain.

Potatoes will remain dormant at 50 °F for 2 to 4 months; and since tubers from this temperature are more desirable for both table use and processing than those from 40 °F, late-crop potatoes intended for use within 4 months should be stored at 50 °F and those for later use at 40 °F. All potatoes should be stored in the dark to prevent greening.

A storage temperature of 50 to 55 °F is recommended for most cultivars of potatoes intended for chip manufacture. At these temperatures, they usually remain in satisfactory condition if their reducing-sugar content is low enough when they are initially stored. Storage at 61 to 64 °F is less desirable because shrinkage, internal and external sprout growth, and decay are greater at these temperatures than at 50 to 55 °F. Russet Burbank potatoes for table stock or for chipping are stored at 45 °F with 95% rh.

Potatoes usually do not sprout until 2 to 3 months after harvest, even at 50 to 61 °F. However, after 2 to 3 months of storage, sprouting can be expected in potatoes stored at temperatures above 39 °F and particularly at temperatures around 61 °F. Although limited sprouting does not affect potatoes for food purposes, badly sprouted stock shrivels and is difficult to handle and market.

Certain growth-regulating chemicals have been approved by the U.S. Food and Drug Administration to control or reduce sprouting on potatoes. Potatoes treated with chemical sprout inhibitors should not be stored in the same warehouse with seed potatoes. Potatoes having the best quality and least amount of shrinkage result if 90% or slightly higher rh is maintained. Cunningham *et al.* (1971) recommend 95% or higher rh for late-crop potatoes. Condensation on the ceiling and resultant moisture drip is sometimes a problem when very high humidity is maintained.

Ventilation or air circulation in potato storage is needed to provide and maintain optimum temperature and rh throughout storage and the tubers it contains. In northern states, where average outdoor temperatures during storage are low, little circulation or ventilation is needed. Shell or perimeter circulation is extensively used in these areas for seed and table stock potatoes. Forced circulation through the potatoes is required for the higher temperature storage of processing potatoes and for table and seed stock in the warmer parts of the late-crop area. Rapid air circulation may result in lowering the rh of the air immediately surrounding the potatoes; it is conducive to drying and weight loss, which may be desirable if there are disease problems but undesirable with sound potatoes because of increased shrinkage. For late-crop Idaho potatoes, a uniform airflow, which does not have to be continuous, of 45 cfm/lb is recommended. With this ventilation, Russet Burbank potatoes stored at 45 °F with 95% rh should keep in good condition for 10 months or longer.

Potatoes should not be kept in the same room with fruits, nuts, eggs, or dairy products because of the objectionable flavor they may impart. Also, potatoes may absorb odors from cheese or from volatile chemicals.

Bacterial Ring Rot: Yellow, soft, cheesy decay of the thin layer of tissue in the vascular ring. The outer 0.25 in. of tuber and the inner part may appear normal. *Control*: Use disease-free seed; store promptly at 40 °F.

Bacterial Soft Rot (see Table 4, Note 1): This disease probably causes more loss in the early and intermediate crops than do all other potato diseases combined.

Freezing Injury: If frozen solidly, tubers become soft and cream-colored and exude moisture. Slightly frozen tubers show darkening of the vascular ring and dull gray to black areas in the flesh. Potatoes freeze at about 29 to 31 °F.

Fusarium Rot: Brown to black, spongy, and fairly dry; white or pink mold inside cavities in stored potatoes. *Control*: Avoid cutting and bruising during harvesting. After proper curing, maintain well-ventilated storage at 40 °F.

Late Blight (*Phytophthora*): Reddish brown to black granular discoloration of the outer 0.12 to 0.25 in. of tuber. The affected tissue is firm to rock hard. *Control*: Apply recommended fungicides in the field. Kill vines prior to harvesting tubers or keep tubers away from blighted vines at harvest. Keep them dry; store at 40 °F; market promptly.

Leak (*Pythium*): Large gray to black, moist, decayed area starting at bruises or the stem end of the tuber. The internal tissue is granular and cream-colored at first, turning through reddish brown to inky black. *Control*: Prevent bruising. Refrigerate tubers to 40 °F and keep them dry.

Mahogany Browning: Reddish brown patches or blotches in the flesh of tubers. Chippewa and Katahdin varieties are most susceptible. This differs from flesh discoloration caused by freezing in being reddish brown instead of gray. *Control*: Store at 40 °F or above because lower temperatures cause the discoloration.

Net Necrosis: Dark brown vascular ring and vascular netting of the flesh, most prominent at the stem end, but extends well toward the bud end; increases during storage. *Control*: Reduce storage temperature promptly to 40 °F; the infected tubers show symptoms earlier at higher temperatures.

Scald and Surface Discoloration: On early potatoes, this appears as sunken, injured areas; later, it turns black and sticky and is followed by bacterial rots. *Control*: Move potatoes promptly to market; cool to 40 °F.

Southern Bacterial Wilt: Moist, sticky exudation from the vascular ring when the tuber is cut. Sometimes there is advanced, mushy decay in the center of the tuber. *Control*: Avoid shipping infected tubers; market promptly.

Stem End Browning: Dark brown to black vascular tissue, in streaks, extending from 0.4 to 1 in. into the flesh from the stem end; develops during storage. *Control*: After curing, reduce the storage temperature promptly to 40 °F. Higher temperatures allow rapid development in susceptible lots.

Tuber Rot (*Alternaria*): Black to purplish, slightly sunken, shallow, irregularly shaped lesions, 0.25 to 1 in. in diameter, developing during storage. *Control*: Apply recommended fungicides in the field. Keep the tubers away from blighted vines as much as possible at harvest time. If the tubers are damp, inspect them frequently during storage and use forced-air ventilation to dry up excess moisture.

Pumpkins and Squash

Hard shell winter squash, such as the Hubbards, can be successfully stored for 6 months or longer at 50 to 55 °F with a relative humidity of 60 to 75%. Dry storage is needed for quality retention. All specimens should be well-matured, carefully handled, and free from injury or decay when stored. Hubbard and other dark-green skinned squashes should not be stored near apples, as the ethylene from apples may cause the skin to turn orange-yellow. Most varieties of *pumpkins* do not keep in storage for as long as the usual storage varieties of squash. Such varieties as Connecticut Field and Cushaw do not keep well and cannot be kept in good condition for more than 2 to 3 months. *Acorn squash* can be stored satisfactorily for 5 to 8 weeks at 50 °F. *Butternut squash* should keep at least 2 to 3 months at 50 °F with 50% rh.

Summer squash, such as yellow crookneck and giant straightneck, are harvested at an immature stage for best quality. The skin is tender, and these varieties are easily wounded and perishable. They should be refrigerated to about 39 °F and moved rapidly to market. They can be held for a few days at 32 to 39 °F and a relative humidity of about 90%; if they are held longer than 4 or 5 days, chilling injury causes deterioration. The storage temperature range for summer squash is 41 to 50 °F with 95% rh. A temperature of 41 °F is best for zucchini squash stored up to 2 weeks.

Black Rot (Mycosphaerella): Hard, dry, black decay, dotted with minute black pimplelike fruiting bodies, that occurs at stem ends or sides of the fruit. *Control*: Avoid skin breaks on the fruit; handle promptly.

Dry Rots (*Alternaria; Cladosporium; Fusarium*): Small, deep, dry, decayed areas. The decayed portion is easily lifted out of the surrounding healthy tissue. The surface mold is low-growing, and either greenish black or pinkish white in color. *Control*: Prevent skin breaks. Do not store hard shell squashes below 50 °F.

Rhizopus Rot: See Table 4, Note 3.

Radishes (32 °F and 95 to 100% rh)

Topped spring radishes after harvest should be precooled quickly, often by hydrocooling, to 41 °F or below. If they are then packaged in polyethylene bags, radishes can usually be held 3 to 4 weeks at 32 °F and for a somewhat shorter time at 40 °F.

Bunched radishes with tops are more perishable. They can be stored at 32 °F and a rh of 95 to 100% for 1 to 2 weeks.

Rhubarb (32 °F and 95% rh)

Fresh rhubarb stalks wilt and decay rapidly. Rhubarb in good condition can be stored 2 to 4 weeks at 32 °F with a 95% rh or above. Moisture loss during holding or storage can be minimized by using nonsealed polyethylene crate liners or by film wrapping of consumer size bunches. Removing and discarding leaf blades at harvest is desirable because it not only reduces the possibility of decay and weight loss but also reduces shipping weight and package size by one-third. Rhubarb is usually marketed with about 3/8 in. of the leaf blade attached to the petiole. Splitting of the petiole will be more serious if the entire leaf is removed.

Fresh rhubarb cut into 1-in. pieces and packaged in 1-lb perforated polyethylene bags can be held 2 to 3 weeks at 32 °F with high relative humidity. Splitting of cut ends and curling of these pieces in film bags may be a problem if marketing is at warm temperatures.

Gray Mold Rot (*Botrytis*): Grayish, smoke-colored growths and grayish brown spore masses on stalks. *Control*: Refrigerate to 32 °F.

Phytophthora Rots (*Phytophthora*): Watery, greenish brown, sunken lesions starting at the base of the leafstalk, causing brown decay. *Control*: Decay is retarded with transit and storage temperatures below 40 °F.

Rutabagas (32 °F and 98 to 100% rh)

Rutabagas lose moisture and shrivel readily if they are not stored under high humidity conditions. A hot paraffin wax coating, often given to rutabagas, is effective in preventing wilting and loss of weight and also improves appearance slightly. Too heavy a wax coating may produce severe injury from internal breakdown caused by suboxidation. Rutabagas in good condition, when stored, should keep 4 to 6 months at 32 °F.

Freezing Injury: Rare, because the commodity can stand slight freezing without injury. Severe freezing causes watersoaking and light browning of the flesh, a mustard odor, and fermentation. *Control*: Prevent repeated slight freezing or severe freezing.

Gray Mold Rot: See Table 4, Note 2.

Spinach (32 °F and 95 to 98% rh)

Spinach is very perishable, and can be stored for only short periods of 10 to 14 days at 32 °F, with 95 to 98% rh. It will deteriorate rapidly at higher temperatures. Spinach is commercially vacuum cooled and forced-air cooled. If it is thoroughly cooled, it can be held for 10 to 14 days at 32 °F, without the addition of any package ice prior to storage. When precooling facilities are not available, crushed ice should be placed in each package to provide rapid cooling and to take care of the heat of respiration. Top-ice is also beneficial.

Bacterial Soft Rot: See Table 4, Note 1.

Downy Mildew (*Peronospora*): A field disease, commonly found at the marketing stage as pale yellow irregular areas in the leaves. Downy gray mold is present on the lower surface. *Control*: Control it in the field; market promptly.

White Rust (*Albugo*): Slight yellowing of areas in the leaf above white blisterlike pustules filled with white masses of spores. *Control*: Control it in the field.

Sweet Potatoes (55 to 60 °F, 85 to 90% rh)

Most sweet potatoes are stored in nonrefrigerated commercial or farm storages. Preliminary curing at 84 °F and 90 to 95% rh for 4 to 7 days is essential for the healing of injuries received in harvesting and handling and in preventing the entrance of decay organisms. After curing, the temperature should be reduced to 55 to 61 °F, usually by ventilating the storage, and the relative humidity should be retained at 85 to 90%. Most varieties will keep satisfactorily for 4 to 7 months under these conditions. Weight loss of 2 to 6% can be expected during curing and about 2% a month during subsequent storage.

Usually, sweet potatoes will not keep satisfactorily if they have been subjected to excessively wet soil conditions just before harvest or chilled before or after harvest by exposure to temperatures of 50 °F or below. Short periods at temperatures as low as 50 °F need not cause alarm; but after a few days at lower temperatures sweet potatoes may develop discoloration of the flesh, internal breakdown, increased susceptibility to decay, and off-flavors when cooked.

Temperatures above 61 °F stimulate development of sprouting (especially at high humidities), pithiness, and internal cork (a virus disease). Refrigeration is frequently used in large sweet potato storages to extend the marketing season into warm weather when ventilation will not maintain low enough temperatures.

Sweet potatoes are usually stored in slatted crates or bushel baskets. Palletization of crates and use of pallet boxes facilitates handling. Sweet potatoes are usually washed and graded and are sometimes waxed before being shipped to market. They may be treated with a fungicide to reduce decay during marketing.

Black Rot (*Ceratocystis*): Greenish black decay, frequently fairly shallow, and sometimes circular in outline at the surface. *Control*: Follow recommended field and postharvest control measures. Heat treatment of seed roots at 106 to 109 °F for 24 h will prevent development of black rot.

Chilling Injury: Brown tinged with black discolored areas scattered or associated with vasculars. The interior becomes pithy. Chilling injury is often produced by exposure to lower temperatures for only a few days. Uncured roots are more sensitive than cured ones. *Control*: Store sweet potatoes at 55 to 60 °F.

Freezing Injury: Soft, leaky condition of the flesh. The outer layer of the potato is dark brown. *Control*: Do not subject potatoes to low temperatures; sweet potatoes may freeze at 30 °F.

Rhizopus Rot: See Table 4, Note 3. *Control*: Cure potatoes for 4 to 7 days at 84 °F before storage. Follow recommended field and postharvest control measures.

Tomatoes (Mature Green, 55 to 70 °F; Ripe, 45 to 50 °F; 90 to 95% rh)

Mature green tomatoes cannot be successfully stored at temperatures that greatly delay ripening, even at a temperature of 55 °F, which is considered to be a nonchilling temperature. Tomatoes held for 2 weeks or longer at 55 °F may develop an abnormal amount of decay and fail to reach as intense a red color as tomatoes ripened promptly at 64 to 70 °F. Temperatures of 64 to 68 °F, and a relative humidity of 90 to 95%, are probably used most extensively in commercial ripening of mature green tomatoes. At temperatures above 70 °F, decay is generally increased. A temperature range of 57 to 61 °F is probably the most desirable for slowing ripening without increasing decay problems. At this temperature, the more mature fruit will ripen enough to be packaged for retailing in 7 to 14 days. Tomatoes should be kept out of cold, wet rooms because, in addition to potential chilling injury, extended refrigeration damages the ability of fruit to develop desirable fresh tomato flavor.

Storage temperatures below 50 °F are especially harmful to mature green tomatoes; these chilling temperatures make the fruit susceptible to *Alternaria* decay during subsequent ripening. Increased decay during ripening occurs following 6 days' exposure to 32 °F, or 9 days at 39 °F (see Figure 1).

Firm ripe tomatoes may be held at 45 to 50 °F with a relative humidity of 85 to 90% overnight or over a holiday or weekend. Tomatoes showing 50 to 75% of the surface colored (the usual ripeness when packed for retailing) cannot be successfully stored for more than 1 week and be expected to have a normal shelf life during retailing. Such fruits should also be held at 45 to 50 °F and 85 to 90% rh. A storage temperature of 50 to 55 °F is recommended for pink-red to firm red tomatoes raised in greenhouses.

When it is necessary to hold firm ripe tomatoes for the longest possible time, consistent with immediate consumption on removal from storage, such as on board a ship or for an overseas military base, they can be held at 32 to 36 °F for up to 3 weeks, with some loss in quality. Mature green, turning, or pink tomatoes should be ripened before storage at this low temperature.

Ethylene gas is sometimes used to hasten and give more even ripening to mature green tomatoes. In ripening rooms, a concentration of one part ethylene per 5000 parts of air daily for 2 to 4 days will usually shorten the ripening period by about 2 days at 64 to 68 °F. Some tomatoes are gassed with ethylene in loaded railcars prior to shipping. Adding ethylene has little or no effect on tomatoes just before or after they have started to turn pink. Tomatoes themselves give off considerable ethylene as they ripen. Interest is increasing in the commercial use of low oxygen atmospheres of 3 to 5% during storage or transport to retard ripening and decay.

Alternaria Rot: Decayed area is brown to black, with or without a definite margin. Lesions are firm; rot extends into the flesh of the fruit. Dense, velvety, olive green or black spore masses frequently grow over affected surfaces. *Control*: Avoid mechanical injuries at packing time. Avoid temperatures below 50 °F in green fruit.

Bacterial Soft Rot: See Table 4, Note 1.

Cladosporium Rot: Thin, brownish blemishes or black shiny spots of shallow decay, later covered by green, velvety mold. *Control*: Take care in harvesting and packing. Ship high quality tomatoes free of field chilling injury under protective services that will provide temperatures of 55 to 68 °F.

Late Blight Rot (*Phytophthora*): Greenish brown to brown, roughened areas with a rusty tan margin. *Control*: Apply recommended field control measures. Cull tomatoes carefully before packing.

Phoma Rot: Slightly sunken, moderately penetrating, black areas at the edge of the stem scar and elsewhere on the fruit. Black pimplelike fruiting bodies develop later. Decayed tissues are firm and brown to black in color. Phoma rot is found in eastern-grown tomatoes. *Control*: Apply field control measures. Exercise care in harvesting and packing. Avoid temperatures below 55 °F.

Rhizopus Rot: See Table 4, Note 3.

Soil Rot (*Rhizoctonia*): Small circular brown spots, frequently with concentric ring markings; later, large, brown, and fairly firm lesions. In advanced stages, under warm conditions, cream-colored or brown mycelium and irregular sclerotia may develop. *Control*: Before packing, sort out tomatoes with early lesions if the disease is prevalent.

Turnips (32 °F and 95% rh)

Turnips in good condition can be expected to keep 4 to 5 months at 32 °F with 90 to 95% rh. At higher temperatures (41 °F and above), decay will develop much more rapidly than at 32 °F. Injured or bruised turnips should not be stored. Store turnips in slatted crates or bins and allow good circulation around containers.

Dipping turnips in hot melted paraffin wax gives them a glossy appearance and is of some value in reducing moisture loss during handling. However, waxing is primarily to aid in marketing and is not recommended prior to long-term storage.

Turnip greens are usually stored for only short periods (10 to 14 days). They should keep about as well as spinach at 32 °F with crushed ice in the packages.

REFERENCES

Agricultural Statistics. 1988. U.S. Department of Agriculture, Washington, D.C.

Appleman, C.O. and J.M. Arthur. 1919. Carbohydrate metabolism in green sweet corn. *Journal of Agricultural Research* 17:137.

Ashby, H.B., R.T. Hinsch, L.A. Risse, W.G. Kindya, W.L. Craig, Jr., and M.T. Turczyn. 1987. Protecting perishable foods during transport by truck. USDA *Agricultural Handbook* No. 669.

Bogardus, R.K., and J.M. Lutz. 1961. Maintaining the fresh quality in produce in wholesale warehouses. *Agricultural Marketing* 6(12):8.

Burton, W.G. 1958. Suppression of potato sprouting in buildings. *Agriculture* 65:299.

Cunningham, H.H., M.V. Zaehringer, and W.C. Sparks. 1971. Storage temperature for maintenance of internal quality in Idaho Russet Burbank potatoes. *American Potato Journal* 48:320.

Dewey, D.H., R.C. Herner, and D.R. Dilley. 1969. Controlled atmospheres for the storage and transport of horticultural crops. *Horticultural Report* No. 9, Michigan State University (July).

Hardenberg, R.E. 1971. Effect of in-package environment on keeping quality of fruits and vegetables. *HortScience* 6(3):198.

Hardenberg, R.E., H. Findlen, and H.W. Hruschka. 1959. Waxing potatoes—Its effect on weight loss, shrivelling, decay, and appearance. *American Potato Journal* 36:434.

Hardenburg, R.E., A.E. Watada, and C.Y. Wang. 1986. The commercial storage of fruits, vegetables, and florist and nursery stocks. USDA *Agriculture Handbook* No. 66.

Harvey, J.M. 1965. Nitrogen—Its strategic role in produce freshness. *Produce Marketing* 8(7):17.

Isenberg, F.M. and R.M. Sayles. 1969. Modified atmosphere storage of Danish cabbage. *Journal of the American Society for Horticultural Science* 94(4):447.

Kader, A.A. *et al.* 1984. Irradiation of plant products. Comments from CAST 1984-1. Council of Agricultural Science and Technology, Ames, IA. ISSN 0194-4096.

Lipton, W.J. 1965. Post-harvest responses of asparagus spears to high carbon dioxide and low oxygen atmospheres. *Proceedings of the American Society for Horticultural Science* 86:347.

Lipton, W.J. 1968. Effect of temperature on asparagus quality. *Proceedings*, Conference on Transportation of Perishables, Davis, CA, 147.

Lutz, J.M. 1936. The influences of rate of thawing on freezing injury of apples, potatoes and onions. *Proceedings of the American Society for Horticultural Science* 33:227.

McColloch, L.P. and J.T. Worthington. Ways to prevent chilling mature green tomatoes. *PrePack-Age* 7(6):22.

Morris, L.L. and H. Platenius. 1938. Low temperature injury to certain vegetables after harvest. *Proceedings of the American Society for Horticultural Science* 36:609.

Parsons, C.S. and R.E. Anderson. 1970. Progress on controlled-atmosphere storage of tomatoes, peaches and nectarines. *United Fresh Fruit and Vegetables Association Yearbook*, 175.

Parsons, C.S. and R.H. Day. 1970. Freezing injury of root crops—Beets, carrots, parsnips, radishes, and turnips. USDA *Marketing Research Report* No. 866.

Redit, W.H. 1969. Protection of rail shipments of fruits and vegetables. USDA *Handbook* No. 195.

Stewart, J.K. and M.J. Ceponis. 1968. Effects of transit temperatures and modified atmospheres on market quality of lettuce shipped in nitrogen-refrigerated and mechanically refrigerated trailers. USDA *Marketing Research Report* No. 832 (December).

Stewart, J.K., M.J. Ceponis, and L. Beraha. 1970. Modified atmosphere effects on the market quality of lettuce shipped by rail. USDA *Marketing Research Report* No. 863.

Watada, A.E. and L.L. Morris. 1966. Effect of chilling and nonchilling temperatures on snap bean fruits. Proceedings of the American Society for Horticultural Science 89:368.

Whiteman, T.M. 1957. Freezing points of fruits, vegetables and florist stocks. USDA *Marketing Research Report* No. 196 (December).

CHAPTER 18

FRUIT JUICE CONCENTRATES AND CHILLED JUICE PRODUCTS

THE CONSUMPTION of juices has increased significantly, largely because of (1) convenience in packaging for table use, (2) favorable comparison of the flavor of processed juice to that of fresh juice from the same fruit source, and (3) affordable pricing due to mass production and processing.

Citrus products, especially orange juice, comprise the largest percentage of the total volume of juices sold. Much of the technology used in processing noncitrus juices was developed from citrus processing.

ORANGE CONCENTRATE

Processed orange juice is sold in four principal forms:

1. Frozen concentrate (3-plus-1 concentration, in which three volumes of water are added to one volume of concentrate for reconstitution) in a variety of package sizes. These are the familiar retail products.
2. Concentrate in bulk at 65 °Brix. This is an intermediate product that is bought and sold daily as futures on the Commodity Exchange. Most of this product will ultimately be sold in one of the other forms.
3. Chilled orange juice, which is ready to drink when poured from the carton. It is made from either reconstituted concentrate or from nonconcentrated orange juice. By law, these two products must be labeled "from concentrate" or "not from concentrate."
4. Institutional or restaurant products in special packaging at 4-plus-1 or higher concentrations.

After processing, frozen citrus concentrates in retail (3-plus-1) packages must be stored at 0 °F. Highly concentrated bulk juice (65 °Brix) may be satisfactorily stored at about 15 °F. Chilled single-strength juices are stored at about 30 to 32 °F.

A schematic flow diagram of citrus processing is shown in Figure 1.

Selecting, Grading, and Handling Fresh Fruit

Fruit is selected for proper quality and maturity. Some fruit that is blemished in appearance but sound in quality and referred to as "packing house eliminations" is used. A major portion of the crop is taken directly from the grove to the processing plant. To be mature, the fruit must have the proper Brix-acid ratio, and the juice content and Brix must be above specified values. Fruit should be handled without delay because no real maturing occurs after harvesting; deterioration occurs at a rate that depends on the temperature and condition of the fruit. Citrus fruits are sufficiently rugged to withstand mechanical handling on conveyors, elevators, and belts providing the fruit is processed within a day or two after picking. Samples are taken mechanically as fruit enters the bins and records of chemical analyses are maintained. Usually fruit from two or more bins is used simultaneously to improve uniformity. The fruit passes over inspection tables both before and after temporary storage in bins, and damaged or deteriorated fruit is removed. The selection and handling of good fruit are most important.

Washing. Prior to juice extraction, the fruit is wetted by sprays. The wetting agent is dispensed and falls onto the fruit as it travels over rotating brushes. The fruit is rinsed by water sprays near the end of the washer unit. A sanitizing solution may be used to sanitize conveyors and elevators.

Juice Extraction. Citrus juices are extracted in high-speed mechanical juice extractors. Single machines handle from 300 to 700 pieces of fruit per minute. In some machines, the fruit is halved and the juice is reamed or squeezed from the half. In other machines, a tube is inserted through the middle of the fruit and pressure is applied that squeezes the juice through the fine holes into the tube, at once sieving away the seeds and large pieces of membrane. After the juice has been extracted, it passes to finishers that remove the seeds, pieces of peel, and excess cell or fruit membrane. In the past, this was a comparatively simple process involving one or two stages, but in recent years, it has become complicated and varies extensively from plant to plant. Usually, most of the pulp is separated from the juice by one or two stages of screw-type finishers.

Pulp washing, a process during which soluble solids in separated segment and cell walls are recovered by countercurrent extraction with water, is permitted in Florida, provided that the resultant extract is not used in frozen orange concentrates. It may be used in other formulated products permitted by the Federal Standard of Identity for frozen concentrated orange juice.

The juice or pulp wash liquor from the finishers may require treatment in high-speed desludging centrifuges that remove suspended matter before the juice is transferred to the evaporator. These centrifuges have peripheral discharges that open and close at intervals to discharge a thick suspension of pulp cells. This operation decreases the viscosity of the juice in the evaporator, improves the efficiency of evaporation, and improves the appearance of the final product.

Aside from this method of extraction, special means are used to classify orange pulp for inclusion in products with a high pulp content.

Heat Treatment. When frozen concentrated orange juice was first developed, heat treatment was avoided to maintain optimum flavor. Such concentrate, if prepared from good sound fruit, remains stable for a considerable time at 0 °F and for nearly a year at 5 °F. However, with large-scale production, it is not possible to assure storage below 5 °F. Concentrates originally of good quality tend to gel or clarify rapidly during storage. Heat treatment

The preparation of this chapter is assigned to TC 11.2, Foods and Beverages.

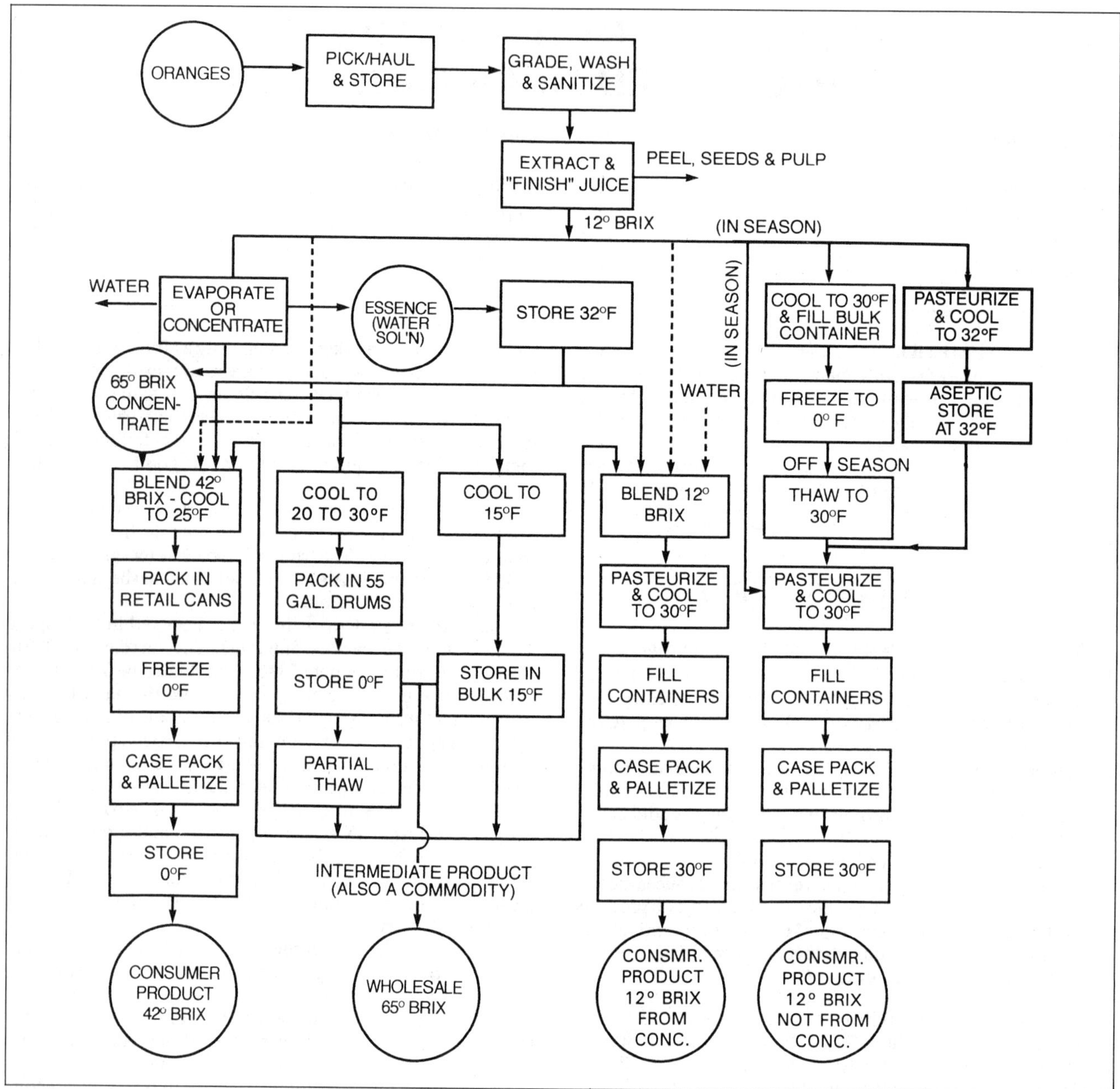

Fig. 1 Citrus Processing Schematic

is used to inactivate enzymes responsible for the development of these defects during improper storage. While earlier methods employed plate-type, steam-heated pasteurizers, heat treatment is now almost universally included as an integral part of heat conservation in the evaporation process.

Evaporation-Water Removal. Removal of water follows sequentially after juice extraction and preparation. Refer to the section on Concentration Methods for details.

Flavor Fortification. Originally, frozen concentrated orange juice was overconcentrated and fresh juice was added to reduce the concentration to the desired level, providing fresh flavor in the final product. This process is used extensively in frozen concentrates, but flavor levels cannot be standardized by this method alone. Essential oil from orange peel does not by itself supply completely balanced fresh flavor, but it is now used extensively to control flavor intensity. Some peel oil is found in the cutback juice, but little remains in the juice from the evaporator. The addition of peel oil to the finished concentrate at the rate of approximately 0.014% by volume in the reconstituted juice has become standard practice. Several variations have been introduced to supplement the use of peel oil and cutback fresh juice for flavor fortification. In most cases, the condensate from the first stage of concentration, using evaporator distillers, is used to produce an essence that is restored to the concentrate to enhance flavor. While cutback fresh juice may be used, and cold-pressed peel oil addition is practiced, most operations depend heavily on essence recovery and its incorporation into the final product.

Blending, Packaging, and Freezing. The final step in processing is the blending of concentrate with flavor-enhancing components, such as cold-pressed orange oil and liquid essence. The 65 °Brix concentrate is then reduced to about 42 °Brix, the requirement for 3-plus-1 product, which reconstitutes to about 12 °Brix when used by the consumer. The availability of cold-pressed oil and essence during that part of the year when no fresh fruit is available has led to the possibility of blending and packaging without regard for fresh juice.

Most retail product is packaged in fiber-foil cans, although aluminum and tinned steel cans may be used. Sizes range from 6 to 64 fluid ounces. Gable-top milk cartons and specialized large sizes of disposable plastic reservoirs (for dispensers) are also used. Prior to filling, the product is cooled to about 25 to 35 °F. This is usually done in the swept-surface cold-wall tank in which it is blended, but it may also be done in a bladed heat exchanger or, preferably, in a plate-type unit.

As an alternate, the 65 °Brix concentrate from the evaporator may be precooled (unblended) and packed in 55-gal open-top drums or delivered to large storage tanks for bulk storage.

The filled cans pass through air-blast freezers where the temperature is reduced to 0 °F or below. The cans are transported on link belts, and the air handlers are arranged so that air at about −25 °F is forced down through the loaded belt. From 45 to 90 min is allowed for freezing.

Thermal Properties. In calculating cooling and freezing requirements, concentrates may be considered as approximating the same heat values as sucrose solutions of the same concentration. In the range of 60 to 65 °Brix, the specific heat is about 0.68 Btu/lb·°F and for 42 °Brix, the value is about 0.74 Btu/lb·°F. Thermal conductivity in the liquid state is in the range of 0.17 to 0.18 Btu/h·ft·°F. Regarding the heat of fusion for freeze-tunnel design, in the range of 30 to 0 °F, the value should be 70 Btu/lb; however, experience has shown that a value of about 100 Btu/lb should be used. This allows for extraneous heat gains, coil defrosting, and so forth.

COLD STORAGE

Cold storage facilities for citrus processing may be generally classified into three categories, according to temperature requirements of 0, 15, and 30 °F.

Finished goods for retail and institutional markets are stored at 0 °F in insulated, refrigerated buildings. Bulk 65 °Brix product packed in drums is also stored at 0 °F, along with retail product. These buildings may range from a few thousand square feet to a few acres. Other than the usual insulation requirements, two factors are critical to the design: (1) the vapor barrier outside the insulation must be as nearly hermetic as possible, and (2) irrespective of the insulation in the floor, a method of supplying heat must be installed, beneath the insulation, to maintain a temperature at about 32 °F. Otherwise, the floor will ultimately heave from ice formation below the floor.

The 15 °F buildings are used for bulk storage of 65 °Brix concentrate. In a typical installation, such a building would house several large stainless steel tanks. Tank sizes range from a few thousand to 200,000 gal each. At the stated temperature, the product is pumpable (barely), requiring sanitary positive-displacement pumps. Since the temperature is virtually impossible to change after the product is in the tank, the product must be cooled to storage temperature before it is introduced to the tanks. Cooling is usually done in a plate-type heat exchanger.

Finally, 30 °F storage rooms are used largely for chilled single-strength juice in retail packages or for the not-from-concentrate bulk tank systems. This product is discussed in the sections Chilled Juice and Refrigeration.

CONCENTRATION METHODS

The major methods for producing concentrates may be classified as: (1) low-temperature, recirculatory, high-vacuum evaporators; (2) freeze-concentration with mechanical separation; and (3) high-temperature, single pass, multiple-effect evaporators.

TASTE Evaporator

At present, the Thermally Accelerated Short-Time Evaporator (TASTE) is the standard evaporator of the citrus industry. This unit has also been successfully applied to grape, apple, and other juices. At least 90% of all juice concentrate production in the Western Hemisphere is by TASTE evaporators. The first cost of this unit is among the lowest of all alternates, and it has excellent thermal efficiency. The flavor quality and the storage stability of juice processed in this unit compares favorably with that of juice from alternate methods of concentration.

The TASTE evaporator includes all classical methods of heat conservation. It is multiple-effect, uses concurrent vapor for staged preheating, and flashes condensate to discharge condensed vapors at the lowest possible temperature. Residence time is minimal, because all stages are single pass. Total residence times are 2 to 8 min, varying directly with evaporator capacity.

A schematic diagram of a typical TASTE evaporator is shown in Figure 2. A triple-effect unit illustrates the basic design. Typical units are offered in 4 to 7 effects (5 to 9 stages), and capacity ranges from about 10,000 to 90,000 lb/h water removal. The vapor-to-steam ratio is approximately the number of effects minus 0.5. In Figure 2, note that the juice enters each of the orifice/spray nozzles atop each stage at a temperature well above the saturation temperature at which that body operates. The resultant flashing of vapor, combined with the spray effect of the nozzle, distributes the juice on the vertical tube walls. Thus, the juice film and the water vapor flow concurrently down the tube, with the juice flowing down the wall as the vapor flows in the central area of the tube. The centrifugal pumps that transfer to the succeeding stage operate continuously near or in a cavitating condition.

As juice is transferred from stage 3 to stage 4, it must be reheated to a higher temperature (140 °F) to effect sufficient flashing as the juice enters stage 4. Reheating is typical for any effect that is divided into more than one stage. These stage divisions are necessary when, due to water removal, the remaining juice is not sufficient to wet the total required surface for that effect. Then, by flash cooling, the concentrate is quickly reduced to the pump-out temperature from 116 to 60 °F.

On a TASTE evaporator, the vapor flow is fixed and virtually unalterable. Although the juice flow shown in Figure 2 is concurrent with the vapor, in many cases the juice flow is mixed flow, in which the juice may first be introduced to an intermediate effect before being delivered to the first effect. In all cases, however, the essence-bearing material is, and must be, recovered from the vapor stream that is produced where the juice first enters the evaporator.

An essence recovery method is also shown in Figure 2. However, several different systems for essence recovery are used, some of which are proprietary. The refrigeration load for essence recovery is relatively small, usually 15 to 40 tons, according to evaporator capacity.

Freeze Concentration

In the freeze concentration system, juice is introduced and pumped rapidly through a swept-surface heat exchanger in which ice nuclei are formed at about 28 °F. This slurry is delivered to a recrystallizer where, by melting many small crystals, larger crystals are formed. The slurry of larger crystals, together with the resultant concentrate, is delivered to the wash column, where the ice rises. As it does, chilled water (melted ice) is introduced at the top to wash the ice crystals; these continue to rise and are themselves melted as they reach the top of the column. In the meantime, the concentrate (now ice-free) is drawn off the bottom of the wash column.

Centrifuges were first used to separate ice crystals from concentrate. Due to poor separation and attendant problems with crystal washing, the losses of orange soluble solids were unacceptable. However, the proprietary wash-column process has evolved,

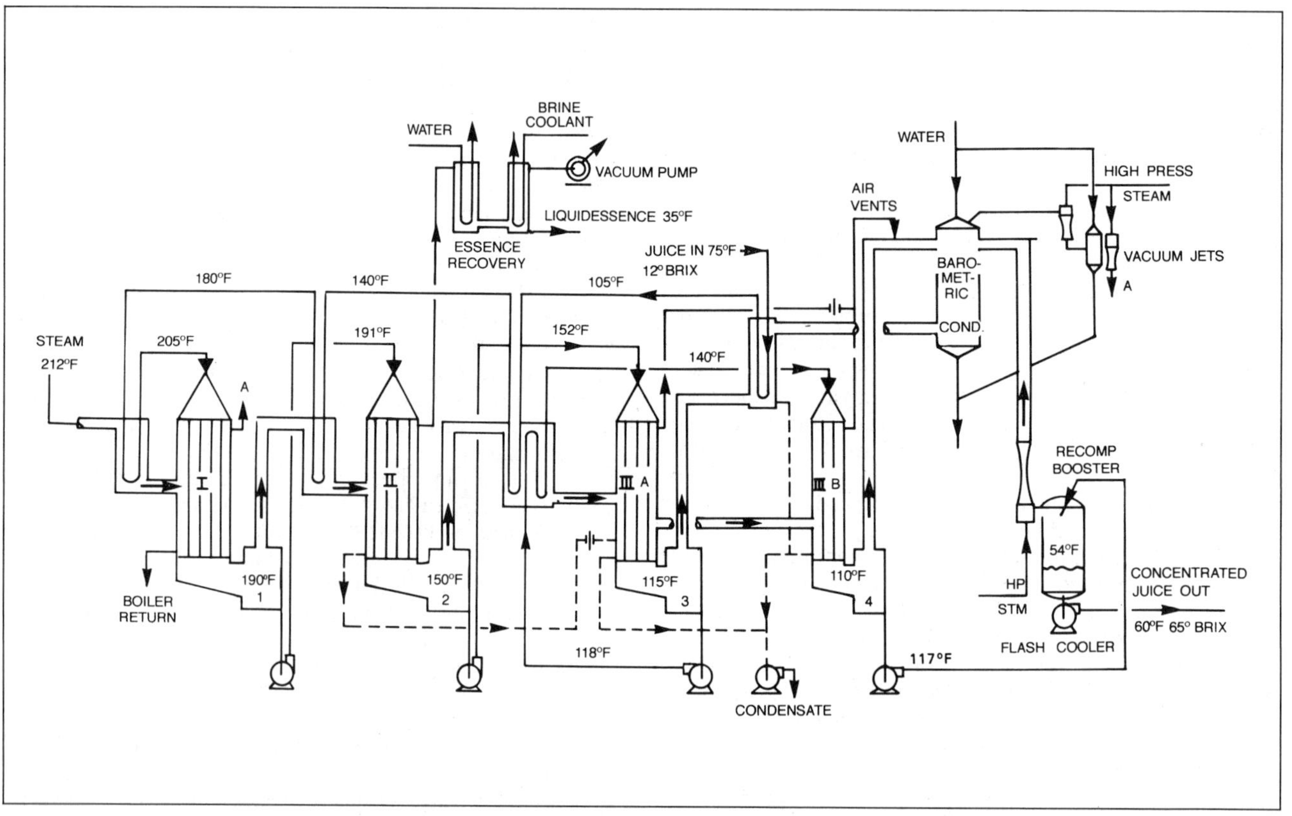

Fig. 2 Thermally Accelerated Short-Time Evaporator (TASTE) Schematic

drastically reducing these losses. The output concentration of commercial freeze-concentration equipment is presently limited to under 50 °Brix.

When first-quality oranges are used to prepare the juice, freeze concentration makes a concentrate that is indistinguishable from fresh orange juice. However, first costs for equipment and installation are relatively high, and operating costs, at best, are as much as for present-day evaporators. The cost is high despite the 144 Btu of energy required to freeze a pound of water, as opposed to about 1000 Btu for evaporation. A contributing factor is the relative cost of electricity (compressors) versus steam from fossil fuel. Another is that, by using multiple-effect evaporators, the 1000 Btu/lb can be divided by the number of effects in the evaporator, thus reducing the 1000 to the range of 150 to 250 Btu/lb, depending on the number of effects.

QUALITY CONTROL

While the most important factor in quality control is the use of good sound fruit, the quality must be checked during processing and again in the final product. Standards for the various concentrates describe the quality and serve as a basis for grading. Concentrates are checked for Brix value, Brix-acid ratio, peel oil content, and other factors. Additionally, tests may be run to see that other requirements of a particular brand are met. At periodic intervals, bacteriological samples are taken during various stages in the plant, plated on orange serum agar, and examined after incubation. While sanitation measures to be taken are known, the bacteriological samples serve as a check and indicate if the procedures have been effective. When the weather is warm, more stringent and more frequent cleanups are required than during cold weather. TASTE evaporators can run only for 6 to 12 h before substantial loss of evaporation or the release of hesperidin crystals forces a cleanup cycle. These cycles are normally spaced to coincide with other required cleaning in the juice extraction system and often take no more than 30 min of production time. The thermal conditions in a properly operated TASTE evaporator are so adverse to microorganisms and enzymes that these are not a concern in determining the cleaning cycle. Normally, low-temperature evaporators are cleaned at least every 24 h, but in some instances when arrangements are favorable, may run as long as 7 days between cleanups. Total plate counts in final product are generally maintained well below one million organisms per millilitre of reconstituted juice. Counts of a few thousand per millilitre are common and indicate a high degree of plant cleanliness. Sanitation is based on asepsis rather than antisepsis. The natural acidity and high sugar content of citrus juice concentrate normally inhibit rapid growth of organisms. Generally, counts tend to decrease in storage.

CHILLED JUICE

Chilled juice is usually packed in either fiber cartons or in plastic bottles or jugs. While the ideal storage temperature is about 30 °F, it is frequently handled in the retail trade at 40 to 45 °F and is typically stored in household refrigerators. Normal shelf life is 3 to 4 weeks.

Chilled juice is marketed in two basic forms: (1) "from concentrate" and (2) "not-from-concentrate." Due to higher overall costs, the not-from-concentrate product is higher priced. Figure 1 shows the processing steps for each of the two products. Note that the sole difference in the two products is the source of the juice.

From Concentrate Juice. Bulk concentrate, taken either from partially thawed drums or from bulk storage, is mixed in a blending tank with water, essence, and cold-pressed oil so that it is reconstituted to about 12 °Brix. This juice is then processed in a 3-stage pasteurizer. First, the juice is preheated in a regenerative section that recovers heat from juice leaving the pasteurizing section. It then flows to the pasteurizer where it is steam-heated to 180 to 190°F. Then it flows back through the regenerative section where it is partially cooled by the incoming juice. Finally, it passes through the cooling section where its temperature is reduced to about 30°F. A refrigerated glycol/water solution usually serves as the chilled brine for this purpose. These pasteurizer units are usually plate-type units, but tubular equipment is also used. From this point onward, every effort is made to maintain the 30°F juice temperature as it is packaged and placed in 30°F storage.

Not-from-Concentrate Juice. When fruit is mature and in-season, fresh juice may be taken directly from the extraction process to pasteurizing and packaging. No blending is required, and the remaining process duplicates that for the from concentrate product. However, unless special means are used in the extraction process, the peel-oil content of the juice may be excessive. It is then necessary to use a low oil extraction method or to heat and flash the juice in a special deoiler to remove excess oil. This is done prior to pasteurization.

Because chilled juice in the not-from-concentrate form is in demand year-round and mature fruit is only available a maximum of 8 months per year, two storage methods are commonly used to ensure 12-month market demands: (1) store in large bulk tanks in an aseptic environment at a temperature just above the freezing point or (2) freeze the fresh juice and store in a solid form at about 0°F.

The aseptic system usually involves a number of large stainless steel storage tanks in a cold room. The entire system must be carefully sanitized before use and must be so maintained during use. The juice is pasteurized and cooled before introduction to the tanks.

The system for freezing fresh juice predates the aseptic liquid method. The freezing method always includes cooling the juice to about 30°F prior to placing it in a container for freezing. The choice of container is related to (1) the rate at which the juice will be frozen and (2) the method used to thaw the product and remove it from the container for later processing. In some instances, the advantage in one case (freezing) may be counterbalanced by disadvantages in the other (thawing/removal). For example, juice may be cooled, placed in an open-top drum, and immediately transferred to −10°F storage where it will slowly freeze. The product quality will be satisfactory, but thawing and dumping the product from the large drum is difficult. Another method places the juice in a specially constructed storage container in which it is frozen and stored. Later, it is fully thawed and pumped from the container. Still another method encases the juice in a plastic bag. It is then quick-frozen in an airblast room. Thawing and removal methods are similar to that for drums. Some juice is also stored in open ice blocks from which it is easier to recover but more susceptible to contamination or losses.

Still another method of providing fresh juice in the off-season is to store fresh oranges under refrigeration for extraction of juice as required. This method requires large cold rooms held at about 30 to 32°F. In many cold storages, moisture should be added to keep the air at saturation. Added moisture, however, increases the refrigeration load, as all condensate from the cooling coils must be recycled to the conditioned space. Because of the cost and other limitations, fresh fruit is seldom stored for juice.

REFRIGERATION

Refrigeration Equipment

The choice of refrigerant for juice processing is almost universally R-717 (ammonia); a few small systems use R-22. Sometimes, R-22 is used in freeze-concentration systems, although ammonia has also been used for this purpose.

High-stage compressors have been of the reciprocating type, but to meet greater demands, there is a trend toward rotary screw compressors.

Most condensers are evaporatively cooled. Exposure to subtropical climate considerations make materials of construction an important consideration. In Florida, water conditioning is critical because the available make-up water has a high mineral content.

The most common refrigerant evaporators in cold rooms and air-blast freezing tunnels are finned or plain coils. In larger installations, a single, low-pressure receiver serves a multitude of coils. Liquid refrigerant is pumped through the coils at 2 to 3 times the evaporation rate, and the liquid/gas mixture is returned to the low-pressure receiver, from which the compressors take their suction. At refrigerant temperatures below 32°F, coils must be defrosted on a regular basis, generally by hot gas from the compressor discharge manifold. Some smaller ammonia installations employ air units, each with its own surge drum and controls for flooded operation. Small installations using R-22 are usually direct expansion, some of which use electric or water defrost.

To solve coil defrosting problems, some cold rooms and some freeze tunnels operate with continuous defrost, in which a strong solution of propylene glycol is sprayed continuously over the coils. The weaker glycol solution is removed and the acquired water driven off in an external concentrator. This system is more expensive to install and operate. Unless extreme care is used in designing the eliminators, the glycol solution may be entrained and be deposited on the containers. The resultant appearance of the containers is unacceptable.

Another common evaporator is a shell-and-tube unit that chills a secondary coolant, typically propylene glycol. (Ethylene glycol should not be used for food processing since it is toxic.) Large volumes of this chilled coolant are required by plate-type juice processors. While the refrigerant can be supplied directly to a gasketed plate heat exchanger, most do not satisfy the pressure requirements of the refrigerant.

Freezing tunnels have been constructed, using an alcohol-water solution, chilled in shell-and-tube equipment. The chilled solution was sprayed over the containers for quick freezing. These are now largely obsolete, having been replaced with air-blast units.

Swept-surface heat exchangers for concentrate pre-cooling, preparatory to can filling, were also used in the past. These units operated as flooded refrigerant systems, but most have been replaced by plate units, which also require glycol coolant. Today, application of swept-surface units is generally limited to freeze-concentration systems.

A typical installation has a number of small loads including:

- Essence recovery (condensing of water vapor, laden with essence). This load is typically handled with shell-and-tube equipment either by direct expansion or with a secondary coolant (glycol).
- Winterizing of cold pressed oil, for the purpose of forming precipitates of undesired materials in a quiescent storage vessel. The process goes forward at about 0 to −40°F and requires months of storage. This may be done in 55-gal drums in a cold room or, preferably, in a jacketed tank, using direct refrigerant expansion or very cold glycol.
- Tanks of essence may be stored in a 30 to 33°F cold room, or in an outdoor insulated storage tank, in which the heat gain is removed by continuous recirculation of the essence through an external plate heat exchanger cooled by glycol brine.

Refrigeration Loads

Refrigeration loads vary not only with plant throughput but also according to the relative rates at which the different products are being processed. It is apparent that for a given amount of consumable product, the volumes of chilled juice are higher than those for concentrates by a factor of 4:1 to 6:1. Thus, the refrigeration load depends largely on the mix of concentrates and chilled juices being processed.

To illustrate the magnitude of the refrigeration loads, the following fixed set of processing conditions are assumed (All percentages are expressed as a percentage of total throughput of **soluble solids.**):

- The plant has an operating evaporator capacity of 100,000 lb/h of water removal.
- 10% is packed as not-from-concentrate chilled juice. Of this, half is packaged directly for immediate sale; the remainder is frozen for storage and use during the off-season.
- 90% is processed into concentrate, which is made up of 85.5% fed to the evaporator and 4.5% used as cutback to make 42 °Brix retail packs.
- Of the concentrate from the evaporator, 30% is packed in bulk for later use as from-concentrate chilled juice; an additional 24% is similarly packed for later use as retail frozen concentrate.
- The remaining 31.5% from the evaporator is blended with the 4.5% cutback (to total 36%). This is packed into retail containers at 42 °Brix and immediately frozen.
- The chilled juice line operates for product made from concentrate. Its rate is equivalent to 30% of plant throughput solids (*i.e.*, same as quantity from the evaporator stored for this purpose).

Given the foregoing, the plant refrigeration loads would be approximately as follows:

	Refrigeration, Tons	
	Low Stage	**High Stage**
Process	250	640
Cold storage	150	280
Miscellaneous loads	—	100
Total	400	1020

Compressor Manifolding

In most cases, low-stage compressors have their own isolated high-stage machines, with an intermediate pressure of 30 to 35 psig. The remaining high-stage compressors are usually paralleled to handle all other high-stage loads at a somewhat lower suction pressure (15 to 20 psig) to optimize power requirements.

Frequently, these systems operate as central systems, with all condensers manifolded using a common receiver.

PURE FRUIT JUICE POWDERS

One manufacturer has successfully produced a vacuum dried orange juice powder. In this puff-drying process, concentrate of about 58 °Brix is introduced into a vacuum chamber where it is dried on a moving stainless steel belt. The dried powder is flavored with a locked in orange oil prepared by dispersing orange oil in a mixture of molten sugars, extruding, and cooling rapidly. Thus the oil is kept out of contact with the powdered orange juice until water is added to reconstitute the juice.

Another process know as foam-mat drying has been under investigation for citrus juices. In this process a small amount of foam stabilizer is added (0.5% of dry solids content), and the chilled concentrate of about 50 °Brix is beaten into a foam. This foam is laid out in a sheet about 1/8-in. thick on perforated trays. An air blast clears the foam from the holes and the trays are conveyed, in a stack, up a column where hot air passes up through the perforated trays and reduces the moisture to about 1.25% in 12 min. The product is then chilled to harden and scraped from the trays. In a variation of the process, a thin layer of the foam is placed on a polished stainless steel belt, dried in a stream of hot air, and finally stripped from the belt with a doctor blade in dried form. Both the puff-drying and foam-mat drying processes are the result of research by the U.S. Department of Agriculture.

OTHER CITRUS JUICES

Grapefruit Juice

In the production of frozen concentrated grapefruit juice, essentially the same equipment is used (fruit washers, finishers, and falling-film evaporators) as in the production of frozen concentrated orange juice. Some adjustments at the extractors are necessary to accommodate grapefruit. Debittering systems may also be used to improve the flavor (bittering is a defect).

Both sweetened and unsweetened frozen concentrate grapefruit juices are prepared (the sweetened product in greater quantities). The Brix of the final unsweetened product may vary from 38 to 42 °. The sweetened product must contain at least 3.47 lb of soluble grapefruit solids per gallon exclusive of added sweetening ingredients. The final Brix may vary from 38 to 48 °. In Grade A unsweetened concentrate, the Brix-acid ratio may vary from 9:1 to 14:1 and in the sweetened product from 10:1 to 13:1. All the types mentioned are reconstituted by adding three parts of water to one part of concentrate. While either seedless or seeded varieties can be used, the seeded varieties, such as Duncan, are generally preferred.

Blended Grapefruit and Orange Juice

The same procedures are used in producing a grapefruit and orange juice blend as are used in preparing the separate products. USDA grade standards recommend no less than 50% orange juice in the mixture and as much as 75% orange juice when it is light in color. Military specifications require 60 to 75% orange juice. USDA grades require 40 to 44 °Brix in unsweetened concentrates. In sweetened concentrates, the Brix must be at least 38 before sweetening and 40 to 48 ° after sweetening. For Grade A, the Brix-acid ratios in the packed concentrate may vary from 10:1 to 16:1 if unsweetened, and from 11:1 to 13:1 if sweetened.

Tangerine Juice

Differences in the nature of the tangerine require modification in the methods of handling during picking, hauling, and storage at the plant. While the grapefruit and orange are generally round, quite firm, and are able to withstand considerably rough handling, the tangerine is somewhat flat, irregular in shape, and has a loose, tender skin that is easily broken. If the skin is broken and the fruit bruised, bacteria and yeasts readily attack the fruit and undesirable enzyme actions occur. Thus, they cannot be handled in orange bins but must be handled in boxes or loose in trucks to a depth of not over 2 ft.

The processes and equipment used in manufacturing concentrated tangerine juice are, however, practically the same as those used with oranges. Since the fruit is smaller, the yield of juice from a given number of extractors is smaller, and about twice the extracting equipment is required to furnish juice to keep the evaporators operating at full capacity. Generally, the values for Brix-acid ratio, peel oil content, and concentration have followed those prescribed for the orange product. Recently, more of the pack has been sweetened, and there has been a trend to pack at a higher concentration. A Brix of 44 ° is common for a 3-plus-1 concentrate.

PINEAPPLE JUICE

Pineapple juice is prepared from small fruit and the parts of larger pineapples that are unsuitable for packing as pieces of fruit. The main sources are the cores, the layer of flesh between the shell and cylinder that is cut for the preparation of pineapple slices, and juice that drains from the crushed pineapple. The juice material amounts to about one-third of the weight of the fresh fruit. It is inspected to remove pieces of shell and spoiled flesh. The juice is extracted by passing through disintegrators and screw presses. It is then centrifuged to remove heavy foreign material and excessive insoluble solids. Processing of juice up to this point is the same for the production of single strength or concentrate.

Pineapple concentrate is produced from single-strength juice in equipment similar to that used for the production of orange and other fruit juice concentrates. The first step in the concentrating operation is to strip out the volatile flavoring materials. These are separated as about a 100-fold product and added back to the final concentrate. The concentration takes place in multiple-effect evaporators.

Pineapple concentrate is produced either as a 3-to-1 product with a Brix of about 46.5° or as a 4-1/2-to-1 product, with a Brix of about 61°. The 3-to-1 concentrate is produced in both a sterile and a frozen form. However, even the sterile product is stored and sold under refrigeration to preserve quality. The 4-1/2-to-1 concentrate is also produced in both a sterile and frozen form. It may be held for short periods without refrigeration, but it should be stored at 40 °F or less. The frozen 61 °Brix concentrate is packaged in polyethylene bags and held in 7-gal fiber containers. This product is stored under refrigeration.

Bulk pineapple concentrate is used principally as an ingredient for mixing with citrus concentrate in the production of frozen juice blends. Pineapple concentrate is also used as an ingredient in the production of many types of canned fruit drinks. A popular combination has been a pineapple-grapefruit drink.

The composition of pineapple juice varies greatly. Brix varies between 12 and 18°, with an average of about 13.5 to 14°. Acidity ranges between 0.6 and 1.1, with an average of about 0.8. Brix-acid ratio ranges from 12 to over 20 and usually averages between 16 and 17. Since pineapple concentrate is produced at a standard Brix, variation in composition shows up only in the acidity and Brix-acid ratio.

APPLE JUICE

Evaporative procedures include some method of essence (ester) recovery for incorporating the volatile components of apple flavor in the final concentrate. These procedures are designed to take advantage of the fact that in the distillation of apple juice, most of the volatile flavors are found in the first 10% of the distillate. This portion of the distillate is passed through a fractionating column to obtain the volatile flavors in concentrated form, usually about 100-fold as compared to the fresh juice. The remaining 90% of the original juice, now stripped of its volatile flavors, is then concentrated under vacuum to somewhat more than the concentration desired for the final product.

For example, by this procedure, 100 gal of apple juice prepared in a conventional manner, may be treated to yield 1 gal of 100-fold essence and 24 gal of concentrated stripped juice. On combining of these two fractions, full-flavored (4-fold) concentrate is obtained. If a higher concentrate is desired, concentration of the stripped juice is carried out to a greater extent. (In such instances, however, the juice must be depectinized to avoid excessive viscosity and gelation of the highly concentrated juice).

One report shows that 4-fold apple juice (depectinized) concentrate was essentially unchanged after one year of storage at 0 °F. Similar information is not available on concentrate prepared without depectinization of the apple juice.

GRAPE JUICE

Most of the concentrated grape juice marketed in North America is prepared from Concord grapes (*Vitis labrusca*) grown in New York, Michigan, Washington, Pennsylvania, Ohio, Arkansas, and Ontario. The grapes are harvested when the soluble solids reach a concentration of 15 to 16%. This varies with maturity and is influenced by cultural and climatic factors.

Muscadine (*Vitis rotundifolia*), typical of the southeastern United States, are processed into juice as follows: grapes are harvested in trucks to the receiving station, dumped into hoppers, crushed, potassium persulphate ($K_2S_2O_8$) added to 50 ppm free sulfite. These grapes differ from *V. labrusca* in that they do not grow in clusters but as individual berries, so that there is no need to destem them.

After crushing, the grapes are conveyed into a pneumatic press (without any heating) and pressure is exerted up to 5 bar (73 psi). Once the juice is extracted, it is quickly pasteurized by passing it through a plate heat exchanger that heats it to about 185 °F. The juice is partially cooled, and a mixture of pectinases and cellulases is added to clarify it. After 1 to 2 h, the juice is passed through an ultrafiltration (UF) tubular unit with 2.0 μm pores for filtering. The juice is then repasteurized and cooled to 45 °F in plate heat exchangers and sent to large refrigerated storage tanks for detartration and sedimentation at 28 °F. Some of the juice is concentrated to about 65 °Brix in a triple-effect evaporator with an essence recovery system. The concentrate is field-frozen and used later for bottle juice. When ready to bottle, juice is pumped through the UF filter, bottled and capped, and passed through a pasteurizer/cooler in a conveyor belt.

After washing, labrusca grapes are conveyed to the stemmer, which consists of a perforated, slowly revolving (20 rpm) horizontal drum inside of which several beaters, revolving at a much faster speed (200 rpm), knock the berries off the cluster and partially crush them before they are discharged through the drum perforations. The cluster stems are expelled from the open end of the drum. The crushed fruit is then pumped through a tubular heat exchanger, where it is heated to 140 to 145 °F for good extraction of the pigments and juice. The hot pulp then goes to hydraulic presses where the juice is removed in the same manner as apple juice. The expressed juice may be clarified in a centrifuge or filter press. If the filter press is used, from 1 to 2% of diatomaceous earth is used to maintain a high filtering rate and provide ample removal of suspended matter. Under normal operating conditions, 190 to 195 gal of juice are obtained from a ton of grapes. In some plants, screw presses are used for all or part of the crushed grapes, but this increases the suspended matter, which must be removed later.

The clarified juice is then pasteurized in tubular or plate heat exchangers to a temperature of 180 to 190 °F and cooled immediately to 30 °F before storage in tanks in refrigerated rooms maintained at 28 °F. The cooling of the juice is usually accomplished in two or more steps. In some heat exchange systems, a regeneration cycle, whereby the hot juice leaving the pasteurizer preheats entering juice, is used in the first step. Occasionally, the cooling water discharged from the heat exchangers is piped to the washers to heat the water applied to the incoming grapes.

The method of handling the cooled juice depends on the intended use of the concentrate. If it is to be used later in jelly manufacture, the juice is stored at 28 °F for 1 to 6 months to permit settling of the argols; these consist of potassium acid tartrate, tannins, and some colored materials that would give a gritty texture to the jelly or detract from its clarity. In this case, the clear juice is syphoned off the precipitate in the storage tanks and may be refiltered. Where the concentrate is to be sold as a blend formed by mixing sugar and ascorbic acid before canning and freezing, the cold storage tank merely serves as a surge tank since the juice is pumped out to the concentrator within a few hours. Here a polishing filter is used before the concentrator to minimize fouling of the evaporator tubes. The concentrate for both jelly manufacture or blended juice may be stored in tanks at 27 °F prior to processing. Whenever single-strength juice is bulk-stored at 27 °F prior to concentration, danger exists from spoilage by fermentation. To minimize yeast growth during storage, all pipe lines and equipment from the pasteurizer to the cold room should be designed for ready and frequent cleaning. The interior surfaces of storage tanks must be relatively smooth or free from crevices, and the tank should be thoroughly cleaned before use.

Concentration is carried out in two steps. In the first, the volatile flavoring materials are stripped from the juice, and the stripped juice is then concentrated to the desired density. The volatile components are removed by heating the single-strength juice to 220 to 230 °F for a few seconds in a heat exchanger, flashing a percentage of the liquid into a vapor in a jacketed tube bundle, and then discharging the liquid and vapor through an orifice tangentially into a separator. The separator should be of sufficient size to reduce the vapor velocity to 10 ft/s or less for minimal entrainment. From 20 to 30% by mass of the original juice flashes off as a vapor that is led into the base of a fractionating column filled with ceramic saddles or rings. A reflux condenser on the vapor line from the column and a reboiler section at the base of the column are used to provide the necessary reflux ratio. The vent gases from the reflux condenser are then chilled in a heat exchanger, and the condensate containing the essence is collected at a rate equivalent to 1/150 of the volume of entering flavoring material.

Methyl anthranilate, which has a boiling point of 512 °F and is only slightly soluble in water, is an important flavor component of Concord grape juice. The high boiling point and low solubility in water have caused losses of methyl anthranilate when the efficiency of the stripping column is low. These losses may be reduced by increasing the vaporizing temperature.

In a typical formulation, grape juice is concentrated to a little over 34 °Brix, and essence or fresh cutback juice is added to reduce it to this concentration. Sucrose is added to 47 °Brix and citric acid until the total acidity is 1.8% calculated at tartaric acid. When diluted with an equal quantity of water, the equivalent of sweetened single-strength juice is obtained; however, for a more palatable beverage, three parts of water are added. The product is labeled as concentrated, sweetened grape juice. The concentrate may be cooled to 20 to 30 °F in a heat exchanger or cold wall tank, filled into cans, sealed, cased, and allowed to freeze in subzero storage.

Frozen grape juice concentrate is often purchased according to its grade. The various types of concentrates and methods of rating along with objective tests of quality are listed in the USDA Standards for Grades of Frozen Concentrated Grape Juice. Similar information is available regarding the quality of concentrate purchased under government contract.

STRAWBERRY AND OTHER BERRY JUICES

Frozen strawberry juice concentrate, a seven-fold concentrate with separately packed concentrated (100-fold) essence, is used for manufacturing, especially jellies. Availability of seven-fold frozen concentrate provides a more economical way of marketing high-quality strawberry juice solids. Concentrates of red raspberry, black raspberry, and blackberry juices are also available in limited quantities.

The process for preparing strawberry and other berry juice concentrates involves essence recovery in which 12 to 20% of the juice is separated by a stripping process using a steam injection heater. Vapors containing volatile flavors are concentrated in a fractionating column to the desired degree. The juice remaining from the essence recovery step is concentrated under vacuum 3- to 7-fold by volume. A maximum temperature of 100 °F for 2-1/2 h should not be exceeded for strawberry, whereas temperatures up to 130 °F may be used in preparing boysenberry concentrate in a batch-type operation.

Preparation of juice for concentration involves chopping or coarse milling of cold, sound berries and mixing with pectic enzymes and filter aid. After several hours (4 to 5 h at room temperature), juice is expressed with a bag press or rack and clothes press. The cloudy juice is clarified in a filter press. Recovered essences are concentrated and packaged separately so that the jelly manufacturer can incorporate the essence in the jelly just before filling. This procedure greatly reduces the amount of essence lost by volatilization. The essence can also be incorporated in the concentrate to make a full-flavored product for shipping as a single unit.

Concentrated juice without essence can be packed in an enamel-lined container that need only be liquid-tight. Concentrated essence should be kept in a carefully sealed can to avoid loss of the highly volatile flavor. Both juice concentrate and essence are kept frozen for proper quality retention.

CHAPTER 19

PRECOOKED AND PREPARED FOODS

DURING the past forty years, there has been much growth in the production of precooked and prepared foods. Initially, these foods were manufactured in plants that produced major ingredients such as red meats, poultry, fish, and vegetables. As sales demands grew, facilities were constructed especially for particular categories of products. The growth in the number, variety, and amount of precooked and prepared foods is still increasing rapidly in both frozen and refrigerated categories.

Although there are many categories of precooked and prepared foods, this chapter will cover primarily main dishes, meals, vegetables, and potato production from the perspective of the refrigeration and air-conditioning concerns of facilities producing these categories of food.

MAIN DISHES, MEALS

Main dishes constitute the largest category of precooked and prepared foods. They are primarily frozen products, but many are also refrigerated. Most of these dishes can be prepared in a conventional or microwave oven. Many contain sauces and/or gravies. Consequently, sauce and/or gravy kitchens are an integral part of these production facilities. A principal characteristic of main dishes is the requirement for a substantial number of ingredients, several unit operations, an assembly-type packaging line, and subsequent cooling or freezing in individual cartons or cases. Examples of these products include:

- Soups and chowders
- Main dishes of meat, poultry, fish, and pasta items
- Complete dinners, each with a main dish usually with sauce and/or gravies, a vegetable, and dessert
- Lunches and breakfasts
- Ethnic main dishes and dinners, particularly Italian, Mexican, and Oriental styles
- Low caloric or diet versions of many of the above
- Snacks and finger foods such as pizza, fish sticks, and breaded items

General Plant Characteristics

The plant facilities for preparing, processing, packaging, cooling, freezing, casing, and storing these products vary widely. Since the variety of products is diverse, and it is beyond the scope of this chapter to cover all the operations involved in all the products, product formulations, food chemistry, and process details are not discussed. This chapter also does not cover the basic process details for preparing meat, poultry, and fishery products. For information on these subjects, refer to Chapters 11, 12, and 13, respectively.

A prepared foods plant has the following production areas and/or rooms: receiving areas; storage areas for packaging materials and supplies; storage areas for ambient, refrigerated, and frozen ingredients; rooms and/or areas for thawing and defrosting; refrigerated rooms for in-process storage; areas for mixing, cutting, chopping, and assembly; sauce and/or gravy kitchens; areas for cooking and cooling rice, other starches, and pasta; unit operations for preparing main dishes such as meat patties, ethnic foods, and poultry items; dough manufacturing for pies and pizzas; assembly, filling, and packaging areas; cooling and freezing facilities; casing and palletizing operations; finished goods storage rooms for refrigerated and frozen products; and shipping areas for outbound finished goods by truck or rail.

In addition, these operations have support and utilities areas for equipment and utensil sanitation; personnel facilities; and areas for utilities such as refrigeration, steam, water, wastewater disposal, electric power, natural gas, air, and vacuum.

Plants and equipment for the production of prepared foods should be constructed and operated to provide for minimum bacteriological contamination and ease of cleanup and sanitation. Sound sanitary practices should be adhered to at all stages of production, not only to protect the end user but also to provide wholesome, healthy products.

All meat and poultry plants in the United States, and hence most prepared food plants, operate under U.S. Department of Agriculture (USDA) regulations. For detailed information and guidelines, refer to *Agricultural Handbook* No. 570 from the USDA, Food Safety Inspection Service.

Preparation, Processing, Unit Operations

The initial steps in the production of prepared foods involve the preparation, processing, and unit manufacture of items for assembly and filling on the packaging line. These generally include scheduling ingredients; thawing or defrosting frozen ingredients, where applicable; manufacturing sauces and gravies; cooking and cooling rice, other starches, and/or pasta; unit operations for manufacturing meat patties and ethnic foods, such as burritos; mixing vegetables and/or vegetables and rice, other starches, or pasta; manufacturing dough; and cooling, storage, and transport operations prior to packaging.

The above-mentioned processes require refrigeration for controlled tempering in refrigerated rooms; plate heat exchangers or swept surface heat exchangers using chilled water or propylene glycol for preparation of sauces and gravies in processors; chilled water for pasta, other starches, or rice cooling; refrigerated preparation rooms for meat and poultry products; in-line coolers or freezers for meat patties, burritos, etc.; ice for dough manufacture; and cooler rooms for in-process storage and inventory control.

The refrigeration loads for each of these categories should be calculated individually using the methods described in Chapter 26. In addition, the loads should be tabulated by time-of-day and classified by evaporator temperature, chilled water, ice, and propylene glycol requirements. An assessment should be made whether an ice builder is appropriate for chilled water requirements.

The preparation of this chapter is assigned to TC 11.2, Foods and Beverages.

The refrigerated rooms required should be amply sized, be capable of maintaining temperatures from 31 to 49 °F, have power-operated doors equipped with infiltration reduction devices, and have evaporators that are easily sanitized. Evaporators for use in rooms where personnel are working should be equipped with gentle airflow to minimize drafts. Temperature controls for such rooms should be tamperproof.

Proper safeguards and good manufacturing practices during preparation and processing minimize bacteriological contamination and growth. This involves use of clean raw materials, clean water and air, sanitary handling of product throughout, proper temperature control, and thorough sanitizing of all product contact services during clean-up periods. Sauces, gravies, and cooked products must be cooled quickly to prevent conditions favorable for microbial growth. Refer to Chapter 9 for further information on control of microorganisms, cleaning, and sanitation.

Assembly, Filling, and Packaging

These activities cover transporting the components to packaging lines; filling or placing components into containers; placing containers into packages; closing and checking the packages; and transporting the packages to cooling and freezing equipment.

Transport of components to the filling and packaging lines depends on the plant configuration and the types of components. Items that are pumpable, such as gravies and sauces, are usually pumped to a hopper or tank adjacent to the packaging lines. Items such as free-flowing individually quick frozen (IQF) vegetables may be transported to the lines in bulk boxes via lift truck directly from cold storage facilities. For example, mixes of vegetables, rice, other starches, and/or pastas that have been prepared at the plant may be wheeled to the packaging lines in portable tanks or vats. Tempered, prefrozen meat or poultry rolls may be placed on special carts and wheeled to the packaging lines.

Packaging lines are generally tailored to accommodate specific types of prepared foods. They are usually timed, integrated systems that automate the unit operations of container fabrication or dispensing, container filling, package handling, and checking as much as is practical.

A typical packaging line for a meals product consists of a timed conveyor system with equipment for dispensing containers; a filler or fillers for components that can be filled volumetrically; volumetric timed dispensers for sauces, gravies, and desserts; net weigh filler systems for components and mixes that cannot be placed volumetrically; slicing and dispensing apparatus for placing components such as tempered meat or poultry rolls and similarly configured items; line space for personnel to place components that must be placed manually; liquid dispensers for placing spices and flavorings; and a sealing mechanism for placing a sealed plastic sheet over the containers. This system may be two or three compartments wide for high-volume production.

The timed conveyor system is followed by a single filer to align the containers in single file prior to indexing into a cartoner. The containers are automatically inserted into cartons, after which they are coded and sealed closed. Subsequent to the cartoner, the filled cartons are automatically checked for underweights and tramp metal, and then conveyed to cooling or freezing equipment. Other types of packaging lines may include some or all of the above apparatus; they are then usually specific for the prepared food items being filled and packaged.

The keys to a successful filling and packaging operation are well-prepared components, accurate volumes, weights, and placement of the items, and minimal downtime for line stoppages. Prompt action must be taken to protect and refrigerate components should significant line stoppages occur.

Previous comments regarding product and equipment safeguards, good manufacturing practices, and actions to minimize microbial growth apply to packaging activities as well.

Some of the filling and packaging lines areas are air conditioned. It is good practice to do so, particularly when these areas are subject to ambient temperatures and humidities that can affect product quality and significantly increase potential bacteriological exposure. In addition, workers in these areas are more productive with air conditioning.

Packaging area air conditioning is usually supplied through air-handling units that provide ventilation, heat, and positive pressure to the area. Some applications use separate refrigeration systems; others use chilled water available from systems installed for product chilling. Generally, no other significant refrigeration is required for the filling and packaging areas.

Cooling, Freezing, Casing

The cartons from packaging are cooled or frozen in several different ways depending on package sizes and shapes, speed of production, cooling or freezing time, inlet temperatures, plant configuration, available refrigeration systems, and labor costs relative to production requirements. Refer to Chapter 8 for additional information.

In relatively small plants, stationary air-blast or push-through trolley freezers are used when flexibility is required for a variety of products. In these cases, fully mechanized in-line freezers are not economically justified.

In larger plants where production rates are high, mechanized freezers such as automatic plate freezers, carrier freezers, and spiral belt freezers are used extensively. These freezers significantly reduce labor costs and provide for in-line freezing.

Products are refrigerated by both the above methods as well as by case cooling; in this latter method, it is particularly important that the product be rapidly reduced to storage temperatures.

The prepared foods business has many line extension additions and changes. Each product change results in component differences that may change the freezing load and/or freezing time due to inlet temperatures, latent heat of freezing, and/or package size (particularly depth). Each product should be checked to assure that production rates for the line extension or different products match the current freezing capacity.

Over time, packaging lines tend to become more efficient and capable of higher production output per unit of time due to experience as well as advances in packaging line technology. These advances often occur with existing cooling or freezing capacity, which is relatively constant. This may result in freezer exit temperatures above 0 °F if reserve freezing capacity is not available or cannot be physically added due to space limitations. For new or expanded plants, allow space and/or reserve freezer capacity. Increases in packaging line speed and efficiency of 25 to 50% are reasonable to expect. Each case should be individually evaluated.

Casing the cartoned product follows freezing with in-line production of meals and main dishes. Manual casing is used in small plants and/or with low production rates. High-speed production lines use semiautomatic and automatic casing methods to attain high productivity and to lower labor costs. Inspection at this point is necessary to ensure that freezing has been satisfactory and that the cartons are properly sealed and not disfigured. It is also important to ensure that cartons do not become defrosted due to conveyor hangups or line stoppages.

Product palletizing follows product casing. Manual palletizing is used for small plants and/or low production rates, and automatic palletizers are used for higher production rates where the capital cost can be justified with labor savings. Most palletizing is done adjacent to or near cold storage facilities and transported to the cold storage rooms via lift truck. Some manual palletizing occurs inside cold storage rooms, particularly for slow production lines, to prevent product warm-up, but it is more costly as labor rates are higher for workers in cold storage rooms.

Some plants are equipped with air-blast freezing cells to augment in-line freezing capacity. These are used primarily for cased products when existing in-line freezers are overloaded and to reduce temperatures of products that have been frozen through the latent zone but have not been sufficiently pulled down for placement in the cold storage rooms. These products are usually loosely stacked on pallets with sufficient air space around the sides of the cases to achieve rapid pull-down to 0 °F or lower. Use of spacers between tiers on pallets may be incorporated.

Finished Goods Storage and Shipping

Many plants have finished goods cold storage rooms or warehouses similar to those covered in Chapter 24. Plants that do not have their own cold storage use public cold storage warehousing that is generally located close by.

Larger prepared foods operations usually have sufficient cold storage space in which to store the necessary refrigerated and frozen ingredients and at least 72 h of finished goods production. This volume of space provides for proper inventory control, adequate scheduling of ingredients for production, and sufficient control of finished goods to assure that the product is 0 °F or lower prior to shipment and that it meets the criteria established for product quality and bacteriological counts.

The refrigeration loads for these production warehouses are calculated as suggested in Chapters 24 and 26. Special attention, however, should be given to product pull-down loads and infiltration. Liberal allowances should be made for product pull-down as freezing problems do occur, and some plants are under negative pressure at least during portions of the year due to their exhausting more air than is supplied through ventilation. This can result in infiltration, which is a serious refrigeration load (see Chapter 26) and should be corrected. This problem is not only costly in energy use, but it also makes it difficult to maintain proper storage temperatures.

Refrigerated trucks and/or rail cars are generally used for shipping. Shipping areas are usually refrigerated to 35 to 45 °F, and the truck loading doors are equipped with cushion-closure seals to reduce infiltration of outside air. Refer to Chapter 24 for additional information on refrigerated docks. Where refrigerated docks are not provided, special care should be taken to ensure that the frozen product is rapidly handled to prevent undue product warm-up.

Refrigeration Loads

Refrigeration loads cover a wide range of evaporator temperatures and different equipment types. Most plants cover these with two or three basic suction temperatures. Where two suction temperatures are provided, they are usually at −35 to −45 °F for freezing and cold storage and 10 to 20 °F for cooling loads. Where provided, a third suction temperature is usually from −20 to −10 °F for frozen product storage rooms and other medium-low temperature loads. The third suction temperature is advantageous with relatively large frozen product storage loads to reduce energy costs.

Refrigeration loads should be tabulated by time of day, season of year, evaporator temperature, and equipment type or function. This record should be made periodically for existing operations; it is essential for new and/or expanded plants. These tabulations reveal the diversity of the loadings and provide guidance for existing operations and for equipment sizing for additional capacity or new plants.

In addition, loads should be tabulated for off-shift production, weekends, and holidays to provide proper equipment sizing and for economic operation for these relatively small loads. Refer to Chapter 26 for information on load calculation procedures.

Refrigeration Systems

Most refrigeration systems for prepared foods use ammonia as the refrigerant. Two-stage compression systems are dominant, because compression ratios are high when freezing is involved and energy savings warrant the added expense and complexity. Evaporative condensers are extensively used for condensing the refrigerant. Evaporators are designed for full-flooded or liquid overfeed operation, depending on the equipment or application. Direct-expansion evaporators are seldom used.

New plant designs limit the exposure of plant employees to large quantities of ammonia. This can be accomplished by locating evaporators, such as propylene glycol chillers, water chillers, and ice builders, in or near machine rooms and away from production employees. Freezers can be located in isolated clusters near machine rooms so that the low-pressure receivers are also in or near the machine room. These practices limit exposure and provide a means for close supervision by trained, competent operators.

Some new plants use glycol chillers to circulate propylene glycol to evaporators located in production areas. This results in an energy penalty because of the secondary heat transfer, but it is deemed affordable because of less potential exposure to employees and product in the event of an ammonia spill.

Other plants have a major portion of the ammonia mains installed on the roof to reduce internal exposure to personnel and product and provide accessibility. These mains require monitoring and inspection, as an ammonia spill can still result in an injury to internal personnel and damage to product as well as injury to personnel outside the plant.

Machine room equipment should be sized not only for the full refrigeration loads imposed, but it should also be able to handle the relatively small off-shift and weekend loads without using large compressors and components at low capacity levels.

Many plants take advantage of energy savings measures, such as floating head pressure controls with oversized evaporative condensers coupled with two-speed fans, single-stage refrigeration for small areas and loads; variable-speed pumps for glycol chiller systems; ice builders to compensate for peak loads; door infiltration protection devices; added insulation; and computer-based control systems for monitoring and controlling the system. These measures should be considered for existing plants that are not so equipped and should be included in the design of plant expansions and new plants.

All refrigeration systems for prepared food plants should comply with applicable codes and standards (see Chapter 51).

Ventilation

Adequate ventilation and positive air pressure relationships are important in prepared food plants to minimize cross-contamination between raw and cooked products and to prevent condensation. Each plant has to be analyzed separately to provide proper differential pressures and to overcome air and steam exhaust quantities. Many plants are under negative pressure for at least part of the year; this condition should be corrected.

VEGETABLES

Frozen vegetables are prepared foods in that they are essentially precooked and require minimal preparation in conventional or microwave ovens prior to serving. This section covers all frozen vegetables except potato products, which are covered in the next section. Chapters 10 and 17 have further information on handling, cooling, and storing fresh vegetables.

Most vegetables to be frozen are received directly from harvest. Some are cooled and stored to smooth out production and others are processed directly. They are cleaned, washed, and graded; cut, trimmed, or chopped (if necessary); and then, blanched, cooled, and inspected prior to freezing. At this point some vegetables are filled or packed into cartons prior to freezing, whereas others are frozen and then filled into packages, bags, cases, or bulk bins.

The products cartoned prior to freezing are generally items that can only be manually packed and/or have to be check weighed prior to closing the cartons. Examples include broccoli spears, asparagus spears, leaf spinach, French cut green beans, okra, cauliflower florets, and some volumetrically machine-filled vegetables that are cartoned prior to freezing such as peas, cut corn, and cut green beans.

Products that are cartoned prior to freezing are usually frozen in manual plate freezers, automatic plate freezers, stationary air-blast tunnels, and push-through trolley freezers.

Products that are frozen prior to packaging are included in the free-flowing or individually quick frozen (IQF) categories. These include true IQF products such as peas, cut corn, cut green beans, diced carrots, and lima beans, as well as products that are more difficult to IQF such as broccoli florets, cauliflower florets, sliced carrots, sliced squash, chopped onions, etc.

The easily IQF products are usually frozen in straight belt freezers, fluidized bed freezers, or fluidized belt freezers. The more difficult products to IQF are usually frozen in fluidized bed freezers and fluidized belt freezers. Cryomechanical freezers are sometimes used where high value, sticky products are frozen. Refer to Chapter 8 for further information regarding freezers.

The products from IQF freezers are packed either directly into cartons or polyethylene bags, into cases for bulk shipment, or into tote bins for repacking at a later date or shipment to other customers for repack or use in prepared foods.

The products packed into tote bins for repacking into cartons or polyethylene bags are used for single products, products of various vegetable mixes, and products with butter or cheese sauces. They are repacked at the producing location or shipped to other plants of the same company or other companies. Repacking operations rather than direct in-season packing are preferred for most companies, as they allow the packer to produce directly for orders, which saves buying packaging material until required.

Products in tote bins for repacking are placed in tote bin dumpers in a cold storage room adjacent to the packaging lines. The products are metered onto a conveyor in proportion to the end mix required and pneumatically or mechanically conveyed to the filler hopper for volumetrically filled mixes or single products. Products and product mixes that require weighing are generally conveyed to net weigh filler systems. After packaging, the products are usually cased semiautomatically or automatically and returned to the cold storage immediately. The products are rapidly handled to minimize warm-up and clustering.

Corn-on-the-cob is prepared similarly to other vegetables, but because of its bulk and to retain quality, it is usually cooled with refrigerated water. After cooling, the product is either packaged in polyethylene bags and frozen in stationary blast cells, push-through trolleys, or manual plate freezers; or frozen bare in this same equipment or in straight belt freezers or multipass straight belt freezers, and packed into cases for institutional use or tote bins for repacking throughout the year.

Until recently, a significant quantity of corn-on-the-cob was frozen in bulk in R-12 immersion freezers. Although this rapid freezing improved quality, it has been virtually eliminated due to its high cost and the phaseout of CFCs.

Raw, whole onions for *French fried onion rings* are cleaned, sliced, and prepared prior to being coated in breading machines; after this, they are fried in an oil fryer, cooled, and frozen. Almost all production is IQF for the restaurant and food service markets. The product is frozen in various types of belt freezers. Refrigerated precoolers or precooler sections coupled with the IQF freezers are common. Product handling must be gentle, and the product should not be cooled below 5°F prior to discharge from the freezer as it can become brittle and fractured, resulting in some product downgrading.

Changes in the Prepared Vegetable Industry

Several significant changes have occurred in the prepared vegetable industry that affect refrigeration and United States production. These include a shift in production of certain vegetables to Mexico and Central and South America and an increased use of vegetables in other prepared foods and in the food service markets.

The vegetables being processed in Mexico and Central America and exported to the United States are principally those requiring large amounts of manual labor for field work and harvesting as well as for plant labor. Because labor costs are low, the cost of frozen products in these countries is considerably lower than that in the United States. Products include asparagus, broccoli, cauliflower, Brussel sprouts, okra, and fruits, such as strawberries.

The products are prepared for the United States market either as finished products in packages and polyethylene bags, or as semifinished products in cases or tote bins for repacking in the United States or for use in other prepared foods, such as meals and main dishes.

The freezing equipment used for these products is primarily operated manually as automation cannot be financially justified except for some types of belt freezers. Most of the production in the southern hemisphere is delivered in tote bins and packaged in the consuming country.

The principal products involved are those with a short season of production in the United States (*e.g.*, peas, green beans, and corn). If an entire year's estimated requirement is based on one short harvest season, it must be stored in freezer warehouses at considerable expense. On the other hand, if a portion of the estimated requirements is produced approximately six months later, the storage costs can be reduced significantly, and the requirements can be estimated more accurately.

Frozen vegetables for retail prepared foods and food service applications is most significant and ever-increasing. All the production is in bulk, either in cases or in tote bins. Some products are used frozen as a main vegetable or as part of a mix. In the case of mixes, some are used frozen and others are defrosted to mix with additional items prior to becoming a portion of a meal. Much could be done to improve product use, improve the quality of finished products, and reduce costs through cooperative relationships between vegetable processors and prepared food processors.

Refrigeration Loads and Systems

The principal refrigeration loads for vegetable operations are raw product cooling and storage, product cooling after blanching, freezing, process equipment located in freezer storage facilities, and freezer storage warehouses. Not all vegetable operations have all of these loads; each plant has unique conditions. Raw product cooling and storage are covered in Chapters 10 and 17.

Cooling after blanching is done with fresh water, with refrigerated water, and with evaporation cooling. If fresh water is used, it is usually well water at 55 to 60°F that is reused once or twice prior to blanching for product washing, cleaning, or waste product transfer. If refrigerated water is used, it is often used in combination with well or municipal water. Refrigerated water at 35 to 40°F reduces freezing loads and, for some products, enhances quality (*e.g.*, cut green beans).

Freezing loads and corresponding freezing capacity vary widely between different vegetables depending on the initial product temperature and the latent heat of fusion. Special attention must be paid to the variety and size of the particular vegetables to be frozen. A belt freezer designed to freeze 10,000 lb/h of lima beans may only freeze 7500 lb/h of 1-in. cut green beans. In addition, the freezing time will be approximately twice as long for cut green beans due to the differences in shape and bulk. Sspecific and latent heats of fusion for vegetables are listed in Chapter 25 of this volume and in Chapter 30 of the 1993 ASHRAE *Handbook—Fundamentals.*

Freezer warehouse loads are calculated as suggested in Chapters 24 and 26. Freezer storages located at vegetable processing plants have three additional potential loads to consider: (1) the extra capacity reserve needed for product pull-down during peak processing; (2) the negative pressure almost all vegetable plants are under, which can substantially increase infiltration by direct flow-through; and (3) the process machinery load (particularly pneumatic conveyors) associated with repack operations.

Almost all vegetable freezing operations use ammonia as the refrigerant. Two-stage compression systems are in general use, even in those with short peak seasons. The product cooling load is done at the intermediate suction pressure. Freezing and freezer warehouse loads use the first-stage suction. Design saturated suction temperatures vary from −32 to −40°F in the first stage to 10 to 20°F in the second stage. Design saturated condensing temperatures vary from 85 to 95°F.

A unique feature of vegetable plant refrigeration is a lack of spare equipment and redundancy for those that operate for short periods at peak capacity—usually 1500 to 2500 h/year. In these applications spare capacity cannot be justified financially. Extra care and maintenance are usually provided prior to the peak season to ensure that full capacity is available and to minimize downtime.

Flooded evaporators are often used for product cooling and cold storage facilities. Liquid overfeed systems are used for freezing apparatus and for cold storage facilities and product cooling. Direct-expansion evaporators are rarely used.

Condensing is usually done with evaporative condensers. Some older plants use shell-and-tube condensers with once-through water usage. The warmed water is reused for product washing and cleaning prior to blanching. This method provides some risk if ammonia leaks into the water stream and should not be used in new installations.

POTATO PRODUCTS

Frozen potato products constitute a large segment of the prepared foods category. The primary products include various types of French fried potato for fast food restaurant, regular restaurant, and institutional uses. They include regular, shoestring, crinkle cut, and curly fries. Potato sales for retail consumption are far less than those for institutional use. Other potato products include potato puffs, tots, and wedges which are a coproduct or made from waste-stream potatoes and rejected fries. Specialty potato products include hash browns, twice-baked potatoes, and refrigerated French fries. The refrigerated product has a relatively short shelf life and requires close control of handling, shipping, and storage.

French Fries

French fried potatoes and coproducts are processed almost year-round. In the northern United States, product is received in bulk directly from the field for 1.5 to 2 months in the fall and thereafter from storage. The storage of fresh potatoes for processing is covered in Chapter 17.

French fried potato lines are usually of large capacity, often with a raw potato capacity of 50,000 lb/h with a finished capacity of 25,000 lb/h plus 2000 lb/h of coproduct.

The bulk potatoes are metered into the processing lines, after which they are conveyed through a destoner to provide initial washing and to remove stones and other debris. They are then graded, and small potatoes are diverted to the coproduct line. Some processors also grade out large potatoes to be sold fresh as baking potatoes. The potatoes for French fries are then steam peeled and trimmed to remove blemishes. (French fries made with skins are not peeled.) Whole potatoes are cut into the desired fry shapes, and the slivers are automatically graded out and diverted to the coproduct line.

The fry shapes are then processed in one of two ways. The first method is to blanch, cool, blanch, mini-dry, and oil fry the potatoes, producing a product of approximately 28 to 30% solids. This product has a high moisture content and is more difficult to freeze because of its higher latent heat of fusion. The second method is to blanch, process the product through a two- or three-stage drier, and then oil fry the potatoes, which produces a product of approximately 34 to 36% solids, a so-called "dry fry."

After frying, the French fries are frozen on a straight belt freezer system that has three separate conveyors. In the first section, filtered ambient air or refrigerated air cools the product from 180 to 200°F to approximately 70°F to drain and/or congeal the excess fryer oil on the surface of the fries. The second and third sections, which are two mesh conveyors in series, function as a two-stage belt freezer. The first section is used for precooling and/or crust freezing to condition the fries before total freezing and sensible cooling in the third section. French fries are usually only frozen to 5 to 10°F, as they become brittle below those temperatures and may be damaged in subsequent handling.

After freezing, the fries are size graded with the longest lengths slated for institutional markets, the shorter lengths for retail, and the slivers and pieces relegated for use as cattle feed. The grading occurs in an area usually maintained at 15°F. The product from the graders is collected in tote bins for subsequent packaging, with some of the totes going directly to the packaging operation and the balance to freezer storage.

The filling of fries is accomplished with net weigh fillers. The containers used include retail cartons, retail poly bags, and institutional polyethylene bags. The institutional volume is dominant, with a typical institutional pack consisting of 6-lb polyethylene bags, six to a case. The packaging area is usually air conditioned to minimize product warm-up and clustering.

Potato Coproducts

Puffs, tots, and wedges are the coproducts manufactured from small whole potatoes that were graded out prior to peeling and slivers that were graded out after cutting the whole potatoes into shapes from the French fry line. The graded out product is about 7% of the French fried potato production. If more coproducts are desired, small whole potatoes are used.

The graded out product and/or whole potatoes that have been steam peeled or diced are inspected to remove blemishes and are then blanched. The product is then conveyed to a retrograde cooler, where the product temperature is reduced to approximately 35°F, and the product is partially dehydrated. From the cooler, the potatoes are conveyed to other equipment where they are chopped into small pieces, mixed with flavorings and condiments, and formed into the desired shapes. From the formers, the product is conveyed to the oil fryers and then to the freezer.

These products are usually frozen on spiral belt freezers, as the freezing time is relatively long due to the bulky shape and the high product inlet temperature of approximately 180°F. These products are usually not cooled in a refrigerated cooler prior to entering the spiral freezer.

The products can be packaged directly from the freezer or stored in bins for later packaging. They are distributed in both retail and institutional markets and primarily packaged in polyethylene bags.

Hash Brown Potatoes

Hash browns are manufactured from steam-peeled whole potatoes that are too small for French fries. They are then blanched and cooled conventionally or by retrograde cooling in bins, after which they are conveyed to a slicer and sliced into strips.

They are placed on a conveyor belt and formed into shapes that are scored and pulled apart to make discrete groupings of patties. They are conveyed to a straight line belt freezer and frozen to 0°F or below.

The product is generally packed into polyethylene bags for either retail or institutional distribution. A typical institutional package is a polyethylene bag containing 12 lb of hash browns.

Other Potato Products

This category includes twice-baked potatoes, potato skins, and boiled potatoes with or without skins for frozen meals or main dishes.

Refrigeration Loads and Systems

The refrigeration loads result primarily from cooling and freezing products and the associated freezer storage for in-process and finished goods storage. As noted previously, the production rates and total capacity of these plants are high, to take advantage of the economics of scale.

Belt freezer refrigeration loads are the major component, and a careful analysis should be made to ensure that adequate performance is achieved to meet the capacity requirements of the various product forms to be frozen. Refrigeration loads and capacity levels for the same freezing apparatus change significantly due to differences in latent heats of fusion, inlet and outlet temperatures, specific heats, and the size and shape of individual product pieces. Additional information regarding freezing times and refrigeration loads for specific foods may be found in Chapters 29 and 30 of the 1993 ASHRAE *Handbook—Fundamentals.*

Freezer warehouse loads are calculated as suggested in Chapters 24 and 26. Freezer storage located at potato processing plants has additional refrigeration requirements due to product pull-down loads. French fries are often discharged from belt freezers at 5 to 10°F to reduce or eliminate product breakage due to brittleness at lower temperatures. This discharge temperature coupled with subsequent packaging can result in product inlet temperature to the freezer warehouse of 15°F. The product temperature should be lowered promptly in the freezer storage to 0°F or below prior to shipping. In addition, the product quantities for a typical plant can be several hundred thousand pounds per day. Also, some freezer warehouses may be attached to plants under negative pressure, which can substantially increase infiltration. This should be corrected, as it is very difficult and costly to offset with refrigeration and infiltration reduction devices.

Potato freezing plants primarily use ammonia as a refrigerant. Two-stage compression systems are in general use, but some single-stage plants are in operation, particularly in the western United States. In two-stage systems, the product cooling is done at the intermediate pressure. Freezing and freezer warehouse loads use the first stage. Design saturated suction temperatures vary from −32 to −40°F in the first stage and 10 to 20°F in the second stage. Design saturated condensing temperatures vary from 85 to 95°F.

In single-stage systems, separate compressors are used for the 10 to 20°F product cooling and liquid refrigerant precooling. Separate compressors are used for freezing and freezer storage with design saturated suction temperatures of −28 to −32°F. The higher design suction temperature is achieved with more evaporator surface. Design saturated condensing temperatures are usually 85°F in the low wet-bulb design temperature areas where these plants are located. As the wet-bulb temperatures are even lower during the early production months in the fall, winter, and spring, the compression ratios are tolerable. Some firms find inadequate or borderline financial justification in electric power savings for the extra capital associated with two-stage systems.

Single-stage systems are simple and lhave a low first cost. Most of these plants are in rural areas, and it is easier and usually more satisfactory to train operators and mechanics for single-stage systems.

Condensing is usually done with evaporative condensers. New plants are often designed for 85°F condensing temperature with floating condensing pressures for lower wet-bulb temperatures and partial loads.

Potato processing plants have heavy refrigeration loads and operate for 6000 to 7000 h per year. Some spare machine room equipment may be needed, but it is not generally provided. Major maintenance is performed during periods of lower production as well as during downtime periods of four to six weeks per year.

The French fried potato freezers function under heavy loads and severe duty. Features include modular, rugged construction for ease of installation, evaporators of aluminum or hot-dipped galvanized tubing with variable fin spacing, axial or centrifugal fans with updraft air flow, noncorrosive materials for product contact parts, regular or sequential water defrost, belt washing apparatus, catwalks for access, and insulated panel housings with interiors constructed to withstand periodic wash-down. A few of these freezers are designed to operate continuously to provide full capacity at all times. Most are designed to be defrosted every 7 to 7.5 h, between shifts, to maintain capacity. In the latter case, the freezer should provide full capacity at the end of the shift. Spiral freezers for coproducts and straight belt freezers for hash browns are similar to those described in Chapter 8.

OTHER PREPARED FOODS

Several other types of prepared foods, including appetizers, sandwiches, breads, rolls, cakes, cookies, fruit pies, toppings, ice creams, sherbets, yogurts, and frozen novelties are covered in other chapters. Bakery products are covered in Chapter 20, and ice cream products are covered in Chapter 14.

LONG-TERM STORAGE

Most prepared foods are not produced with long-term frozen storage as an objective. Inventories are closely supervised, and production and sales are closely linked to minimize inventory. Profitability is reduced by having finished goods in storage and in the distribution chain.

One exception is some vegetables and fruits that can be processed and frozen only during the harvest season. Even here, steps are usually taken to maximize in-process storage of bulk products and to package them as required by sale orders and projections.

Some components and ingredients for prepared foods, however, must withstand long-term frozen storage if they are only produced annually or infrequently. These products require close monitoring to ensure that the quality still meets standards when used.

Regardless of the length of storage, it is important that ingredients, components, and finished goods are stored at 0°F or below with minimal temperature fluctuations.

CHAPTER 20

BAKERY PRODUCTS

THIS chapter primarily covers refrigeration and air conditioning as applied to bakery items distributed as ambient shelf-stable products. For additional information regarding bakery items distributed as refrigerated or frozen products, see Chapter 19, Precooked and Prepared Foods.

Refrigeration plays an important part in modern bakery production. It stabilizes raw material quality prior to use and helps control the temperature of the finished dough, which is one of the most important factors governing overall product quality. Refrigerated storage and freezing of finished products improve production scheduling and marketing. The availability of reliable refrigeration can determine operational procedures for large wholesale plants or small retail bakeries.

INGREDIENT STORAGE

Raw materials are generally purchased in bulk quantities except in small operations. Deliveries are made by truck or rail car and stored in bins or tanks with required temperature protection while in transit and storage. Flour is stored in bins at ambient temperatures. Smaller quantities of different types of flour such as clear, rye, and whole wheat, are usually received in bags and stored on pallets. Corn syrup, liquid sugar, lard, and vegetable oil are stored in heated tanks or in enclosed spaces where temperatures are high enough to prevent crystallization or congealing. Holding temperatures for such products are about 125 °F.

Smaller volume and specialized sugars and shortenings are received in drums, bags, or cartons and are stored at conditions that prevent mold or melting and rancidity. Many bakeries use syrups with high levulose content that may be stored at about 84 °F to maintain fluidity and pumpability. As these syrups are stored at a temperature about one-third lower than conventional syrup, less thermal input is required during storage and the refrigeration load during mixing is significantly reduced.

Yeast in cartons, bulk bags, and bulk cream should always be stored in refrigerated spaces at temperatures below 45 °F but above its freezing point, at which value severe cell mortality occurs. Cocoa, milk products, spices, and other raw materials subject to insect infestation, bacterial growth, or unstable oils subject to oxidative rancidity, should be stored under the same conditions as yeast to prevent product loss.

Total plant air conditioning is used more and more in new plant construction except in areas immediately surrounding ovens, in final proofers, and in areas where cooking vessels are located for preparing fruit fillings and hot icings. Total plant cooling was first used in plants producing Danish pastry, puff pastry, and pies and has expanded to new construction facilities for frozen dough operations and for general production. Flour dust in the air should be filtered out because it fouls air passages in the air-conditioning equipment and seriously reduces heat-transfer rates.

The preparation of this chapter is assigned to TC 11.2, Foods and Beverages.

MIXING

Bread, buns, and sweet rolls are the most important baked products from the standpoint of production volume.

Mixing is the first active process in such production. Refrigeration is required because of the generation of heat and the necessity to control the dough temperature at the end of the mix.

Metabolism of yeast is materially affected by the temperatures to which the yeast is exposed. During dough mixing, the following heat considerations are encountered: (1) *heat of friction,* by which the electrical energy input of the mixer motor is converted to heat; (2) *specific heat* of each ingredient; and (3) *heat of hydration,* evolved when a dry material absorbs water. If ice is used for temperature control, heat of fusion results. Finally, the temperature of the dough ingredients must be considered. Yeast is dormant at temperatures below 45 °F. It is extremely active in the presence of water and fermentable sugars in a temperature range of 80 to 100 °F, but the cells are killed at about 140 °F and at a lower but sustained pace below its freezing point of 26 °F. Precise temperature control is essential at all stages of storage and production, especially during mixing.

Dough may be formed by either a batch or continuous process. The batch method uses a large bowl with heavy revolving blades or agitator arms which mix flour, water, yeast, and other ingredients into a homogeneous dough.

The two principal types of batch dough mixes are the *straight-dough* process and the *sponge-dough* process. In the straight dough process, all ingredients are mixed at once. In the sponge-dough process, only part of the total amount of flour and water required are mixed with all of the yeast, yeast food, and malt. The resulting mixture, or sponge, is then fermented for a period before it is returned for a final mix with the remaining flour and water, as well as the rest of the ingredients.

The principal heat generated during mixing comes from the heat of hydration as the flour absorbs water and from the heat of friction from the mixer. To absorb this excess heat and maintain the dough at 79 to 81 °F, the ingredient water is usually supplied to the mix at a temperature of 35 to 39 °F, and the mixer is jacketed to circulate a cooling medium around the bowl.

Liquid sponge ingredients can be incorporated and fermented in special equipment either on a batch or uninterrupted basis. To render the mixture pumpable, more water is incorported than the actual amount used in a sponge to be fermented in a trough. On completion of the predetermined fermentation time, the sponge (at required pH and titratable acid levels) is chilled through heat exchange equipment from about 90 to 100 °F to about 45 to 65 °F. The cold liquid sponge is then maintained at the required temperature in a storage tank until it is weighed and pumped to the mixer, where it is combined with the remaining ingredients prior to being remixed into a dough. Regular sponges come back for remixing at about 84 °F and often need to use the refrigerated surface on the mixer jacket in conjunction with ice water or ice to achieve the required dough temperature after mixing.

The gluten matrix, which is essential to loaf volume and texture, can be developed only by properly mixing the doughs. Exact temperature control during mixing is essential. Positive dough temperature control is achieved using a cold liquid sponge at 45 to 65 °F, which limits the use of the refrigeration jacket and eliminates using ice in the doughs. In cold weather, remix dough water temperatures of 100 °F or higher are often required. A supplemental refrigeration source is used to chill the sponge to retard yeast metabolism, thereby reducing the refrigeration load at the point of final mix.

In the continuous dough mixing process, a brew, or liquid sponge is formed, using 0 to 70% of the flour, 10 to 15% of the sugar, 25 to 50% of the salt, and all of the yeast and yeast food in about 85% of the water. This mixture ferments in a series of tanks and then passes through an incorporator, where the remaining ingredients are added. The resultant thicker batter finally passes through a developer-mixer from which the finished dough, in loaf size, is extruded into the baking pans.

Properly formulated dough can also be deposited on floured belts, rounded up, and then conveyed through conventional makeup and molding equipment. Hot dog and hamburger rolls are also produced in quantity by pumping bun dough from the continuous mix developer directly to the hopper of the makeup equipment. A coating of flour on the outside dough surfaces affects gas development and retention, which in turn leads to a grain/texture more closely resembling that of sponge and dough products. External symmetry and crust characteristics are similarly changed.

Many variety breads, such as raisin and home-style loaves, cannot be made satisfactorily with the continuous mixing process. Thus, some bakeries prepare and ferment the liquid sponge and then pump it into the batch mixer along with the remaining ingredients for conventional finishing.

In continuous mixing, the liquid sponge is made up in an unjacketed tank at a temperature of 70 to 80 °F and allowed to ferment for about 1 h, during which time the temperature rises to 84 to 86 °F. The second hour of fermentation takes place in a jacketed tank where the brew is held at about 84 °F. This second stage consists of alternate transferral from the first-stage tank to a pair of tanks to maintain a continuous operation by allowing the brew to be drawn from one tank while fermentation goes on in the alternate tank.

Between final-stage tanks and an incorporator, the brew is pumped through a plate heat exchanger where it is cooled to about 70 °F. Cooling offsets the heat that will be added by mixing and by adding the remaining ingredients; thus, a predetermined final dough temperature of 79 to 82 °F can be obtained.

In many older plants, central ammonia systems are used to cool water for both ingredient use and jacket cooling. The ingredient water is cooled separately in self-contained water chillers that range in capacity from 75 to 650 gpm with 3 to 30 hp condensing units. The jacket cooling is by direct expansion of the refrigerant in the jacket at 30 °F.

For batch mixing, individual condensing units are usually located within 15 ft of each mixer. Where a battery of mixers are used in large plants, a compressor room is often located adjacent to the mixing and fermentation rooms. Table 1 lists the sizes of condensing units commonly selected for mixers.

When large batches of dough are handled in the mixers, the required cooling sometimes becomes greater than the available heat transfer surface can produce at 30 °F, and the refrigerant temperature must be lowered. A refrigerant temperature below 30 °F can, however, cause a thin film of frozen dough to form on the jacketed surface of the mixer. This frozen film effectively insulates the surface and impairs heat transfer from the dough into the refrigerant.

Formulas, batch sizes, mixing times, and almost every other part of the mixing process vary considerably from bakery to bakery and must be determined for each application. These variations usually fall within the following limits:

Table 1 Size of Condensing Units for Various Mixers

Dough Capacity, lb	Condensing Unit, hp
800	5
1000	7.5 to 10
1300	10 to 15
1600	15 to 20
2000	20

Final batch dough weights	Up to 3000 lb
Ratio of sponge to final dough weight	50 to 75%
Ratio of flour to final dough weight	50 to 65%
Ratio of water to flour	50 to 65%
Sponge mixing time	360 to 600 s
Final dough mixing time	480 to 720 s
Number of mixes per hour	2 to 5
Continuous mix production rates	Up to 7000 lb/h

The total cooling load is the sum of the heat removal required from each ingredient to bring the homogeneous mass to the desired final temperature plus the removal of the generated heat of hydration and friction. In large batch operations, the sponge and final dough are mixed in different mixers, and refrigeration requirements have to be determined separately for each process. In small operations, a single mixer is used for both sponge and final dough. The cooling load for final dough mix is greater than that for sponge mix and is used to establish refrigeration requirements.

FERMENTATION

After completion of the sponge mix, the sponge is placed in large troughs that are rolled into an enclosed conditioned space for a period of fermentation. This period lasts for 3.5 to 5 h, depending on the dough formula. The sponge comes out of the mixer at a temperature of 75 to 79 °F. During the fermentation period, the sponge temperature rises from 8 to 10 °F as a result of the heat produced by the yeast and the heat of hydration.

To equalize the temperature substantially throughout the dough mass, the room temperature is maintained at the approximate mean sponge temperature of 80 °F. Even temperature throughout the batch produces even fermentation action and a uniform product.

Water makes up a large part of the sponge and uncontrolled evaporation causes significant variations in the quality and weight of the bread. A certain amount of evaporation is required to remove excess alcohols and esters that produce too strong a flavor. The rate of evaporation from the surface of the sponge varies with the relative humidity of the ambient air and the airflow rate over the surface. The rate of movement of moisture from the inside of the sponge to the surface does not react similarly to the same external conditions, and the net result can be the drying of the surface and formation of a crust. The crust formation produces an inactive area that causes undeveloped portions. When folded into the dough mass later, these undeveloped portions produce hard, dark streaks in the finished bread.

To control the evaporation rate from the surface of the sponge, the air is maintained at 75% rh, and air movement over the surface is controlled by moving the conditioned air into, through, and out of the process room without producing crust-forming drafts.

In calculating the cooling load, the product is not considered, since the air temperature is maintained at approximately the mean of the various dough temperatures in the room. Transmission heat loss through walls, ceiling, and floor is the principal load source. Infiltration is estimated as 1.5 times the room volume per hour. Usually the light requirements for a fermentation room are about 75 W per 400 ft^2 of floor area. The conditioning units are often placed within the conditioned space, and the full motor heat load must be considered.

The only source of latent heat is the approximate 0.5% weight loss in the sponge. Under full operating load, this could account for a 1.5 °F increase in dew-point temperature. For conditions of 80 °F db and 75% rh, the dew-point temperature would be 71.5 °F, and the supply air would be introduced into the conditioned space at 72 °F dry bulb and 70 °F dew point.

In large rooms, sufficient air volume can be introduced to pick up the sensible heat load with a 8 °F rise in air temperature. In smaller rooms, a latent heat load may need to be added by spraying water directly in the room through compressed air atomizing nozzles. Water sprays have been most successful when a relatively large number of nozzles have been spaced around the periphery of the room.

Since a high removal ratio of sensible to latent heat is desirable, the condensing unit is usually specified for operation as close to 60 °F evaporator temperature as is practicable with the temperature and quantity of condensing water available.

FINAL PROOF

After it is formed into loaves and placed in pans, the dough receives its final development or proof before it is baked. It is placed in an insulated enclosure having a controlled atmosphere for 50 to 75 min. To keep the almost exhausted yeast active, the temperature is maintained at 95 to 120 °F, depending on the exact formula, the prior intensity of dough handling, and the character of the baked loaf required.

For proper crust development during the bake, the exposed surface of the dough must be kept pliable by maintaining the relative humidity of the air around 85 to 90%. There are exceptions to humidities in this range in bakeries, where they have found it necessary to compromise on a lower range because of the effect on the pan glazing, which is frequently used in place of the older method of greasing the pan.

With dew points at about 92 to 110 °F and ambient conditions going down to at least 65 °F at times, the enclosure must be adequately insulated to keep the inside surface warm enough to prevent the formation of condensation, because mold growth can be a serious problem on warm, moist surfaces. A thermal conductance of $C = 0.12$ Btu/(h · ft^2 · °F) is adequate for most conditions.

The enclosures most commonly used for the proofing process consist of a series of aisles with doors at both ends. Frequent opening of the doors to move racks of panned dough in and out makes control of the conditioned air circulation an unusually important engineering consideration. The air is recirculated at rates of 90 to 120 changes per hour and is introduced into the room to temper the infiltration air and cause a rolling turbulence throughout the enclosure. This brings the newly entered racks and pans up to room temperature as soon as possible.

The problem of air circulation becomes somewhat simpler when large automatically loaded and unloaded tray-type proofers have only minimal openings for the entrance and exit of the pans, and the load is materially reduced by elimination of the racks moving in and out.

BAKING

The fully developed loaves are baked in the oven at temperatures around 400 to 450 °F for 18 to 30 min, depending on the variety of bread. Two principal types of ovens are used: the tunnel or traveling hearth, which is loaded at one end and unloaded at the opposite end, and the tray, which is loaded and unloaded at the same end. There are three varieties of tray-type ovens. The reel type is used for small production. Single- and double-lap traveling tray ovens are used for larger production. The latter type, in which the trays travel back and forth twice, is used principally where there is an advantage in extending the oven upward to save floor space.

Heating is mostly with gas, but oil firing is more economical in some areas. Where gas interruption is a problem, combination oil and gas burners are an alternative. While both indirect- and direct-fired gas systems are used, the indirect firing is generally more popular because of its greater flexibility. Ribbon burners running across the width of the oven are operated as an atmospheric system at pressures of 6 to 8 in. of water.

The burners are located directly beneath the path of the hearth, or trays, and the flames may be adjusted along the length of the burner to equalize the heat distribution across the oven. The oven is divided into zones, with the burners on zone control so that the heat can be varied for different periods of the bake.

Controlled air circulation inside the oven assures uniform heat distribution around the product. To secure a satisfactory golden-toned crust color, the sugar on the surface is caramelized by spraying low-pressure steam on the exposed surface before the crust has begun to bake out. This is accomplished with a series of perforated steam tubes located in the first zone of baking in place of the air recirculating tubes. Heating calculations are based on 450 Btu/per pound of bread baked. Steam requirements run approximately 1 boiler horsepower per 125 lb of bread baked per hour.

BREAD COOLING

Baked loaves come out of the oven with an internal temperature equalized at 200 to 210 °F because of the evaporative cooling effect of the moisture that is driven off during baking. The crust temperature is closer to 450 °F.

The loaves are then removed from the pans and allowed to cool to an internal temperature of 90 to 95 °F. The cooling of a hygroscopic material takes place in two stages or phases, which are not distinct periods. When the bread first comes out of the oven, the vapor pressure of the moisture in the loaf is high compared to the vapor pressure of the moisture in the surrounding air. Moisture is rapidly replaced with a resultant evaporative cooling effect, which, combined with heat transmission at a relatively high temperature differential from bread to air, causes rapid cooling as indicated by the steepness of a time-temperature reduction curve in the early cooling stage (Figure 1). As the vapor pressure

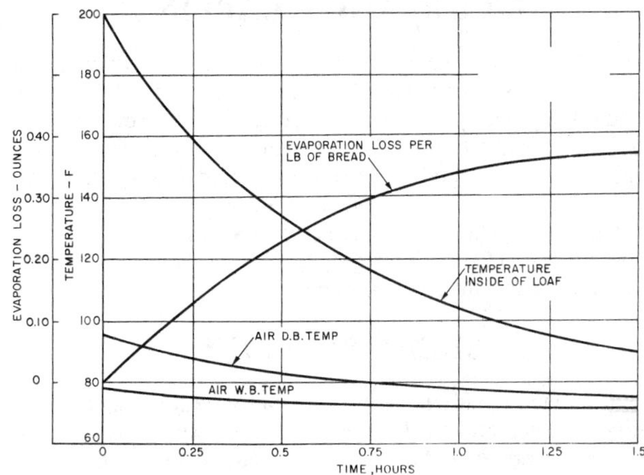

Fig. 1 Moisture Loss and Air Temperature Rise in Counterflow Bread-Cooling Tunnel

approaches equilibrium, heat transfer is mainly by transmission and the cooling curve flattens rapidly.

While small bakeries still cool bread on racks standing on the open floor for 1.5 to 3 h, depending on air conditions, many large operations cool their bread while it is moving continuously on belt-, spiral-, or tray-type conveyors. Cooling is mostly atmospheric even on these conveyors. However, to insure a uniform final product, cooling is often handled in air-conditioned enclosures with a conventional counterflow movement of air in relation to the product.

An internal temperature of 90 to 95 °F stabilizes moisture in the bread enough to reduce excessive condensation inside the wrapper, which can encourage mold development. Approximately 50 to 75 min of cooling time is required to bring the internal loaf temperature to 90 °F because of the more or less fixed time factor for movement of heat and moisture from the center of the loaf to the surface (Figure 1).

In counterflow cooling, the optimum temperature for the air introduced into the cooling tunnel is about 75 °F. The entering air at 80 to 85% rh controls the moisture loss from the bread during the latter stage of cooling and improves the keeping quality of the bread.

Attempting to shorten the cooling time by forced cooling with air temperatures below 75 °F is not satisfactory, because the rate of heat and moisture loss from the surface and the movement of the heat and moisture from the interior to the surface as a replacement becomes so unequal that the crust shrivels and cracks.

Cooling time is often shortened when mold inhibitors are used by taking the bread to the slicers at 105 to 110 °F. The product heat load is calculated assuming a sensible heat transfer based on a specific heat of 0.74 Btu/(lb · °F) for bread and a temperature reduction from 180 to 90 °F. For calculating purposes, the evaporation of moisture from the bread may be considered responsible for reducing the bread temperature to 180 °F.

The conditioning of the air can best be accomplished under normal conditions by evaporative cooling in an air washer. For periods when the outdoor wet-bulb temperature exceeds 72 °F, which is the maximum allowable based on 75 °F dry-bulb and 85% saturation, refrigeration is required.

The amount of air required is rather high because the maximum temperature rise of the air in passing over the bread is usually about 20 °F; consequently, the refrigeration load is comparatively high. In localities where wet-bulb temperatures above 72 °F are rare or of short duration, the expense of refrigeration is not considered justified, and the bread is sliced at a higher than normal temperature for this period.

SLICING AND WRAPPING

Bread from the cooler goes through the slicer where high-speed cutting blades, similar to band saw blades, cut cleanly through the bread if it has been properly cooled. If the moisture evaporation rate from the surface and the replacement rate of the surface moisture from the interior of the loaf are not kept in balance, the bread develops a soggy undercrust that fouls the blades, causing the loaf to crush during slicing. A brittle crust may also develop, which leads to excessive crumbing during the slicing. From the slicer, the bread moves automatically into the bagger.

BREAD FREEZING

The bagged bread moves normally into the shipping area for delivery to various markets by local route trucks or by long-distance haulers. Part of the production may go into a quick freezing room and then into cold storage.

Two important problems face bakers in the freezing of bread and other bakery products. The first is connected with the short work week. Most bakeries are inoperative on Saturday; thus, weekend production is much larger than for the earlier part of the week. The problem increases for bakeries on a 5-day week, with Tuesday usually being the second day off. Freezing a portion of each day's production for distribution on the days off would enable a more even daily production schedule.

A second problem is the increasing demand for variety breads and other products. The daily production run of most varieties is comparatively small, so the constant setup change is expensive and time-consuming. Running a week's supply of each variety at one time and freezing to fill daily requirements can reduce operating cost.

Both these problems are concerned with the staleness of bread, which is a perishable commodity. After baking, starch from the loaf progressively crystallizes and loses moisture. Until a critical point of moisture loss is reached, freshness can be restored by heating and reabsorbing starch crystals. A tight wrap helps keep the moisture content high over a reasonable time. The crystallization of the starch which, when complete, produces the crumbly texture of stale bread, is a spontaneous action that increases in rate, both with decreased moisture and decreased temperature. The moisture loss rate apparently increases with a temperature decrease down to the freezing point. The rate then decreases until the temperature reaches 0 °F, at which point moisture loss seems to be somewhat arrested.

Bread freezes at 16 to 20 °F (Figure 2). For freezing of all cellular structures, it is necessary to cool the bread through the freezing or latent heat removal phase as fast as possible to preserve the cell structure.

Since the moisture loss rate increases with reduced temperature, the bread should be cooled rapidly through the entire range from the initial temperature down to, and through, the freezing points. Successful freezing has been reported in room temperatures of 0, −10, −20, and −30 °F.

In USDA laboratory tests, loaves of wrapped bread placed in cold air blasts of 700 fpm were brought down from 70 to 15 °F core temperatures in the comparative times given:

Freezer Air Temperature, °F	Time, h	Temperature at End of 2 h, °F
−40	2	15
−30	2.25	16
−20	3	19
−10	3.75	21
0	5	22

Another interesting observation showed that with wrapped bread, changes in air velocity from 200 to 1300 fpm caused relatively little change in cooling the item.

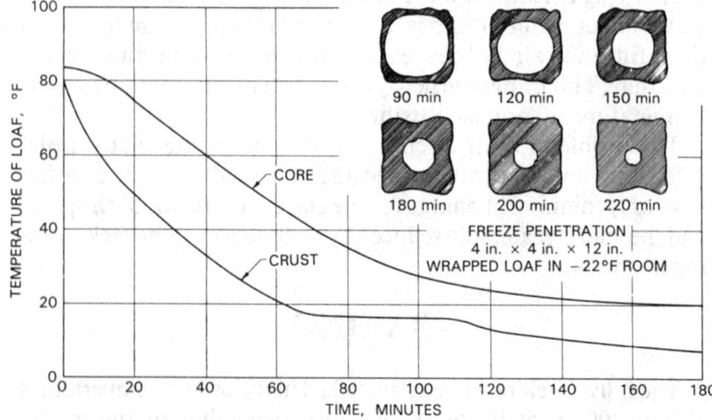

Fig. 2 Core and Crust Temperatures in Freezing Bread

Table 2 Important Heat Data for Baking Applications

Specific heat	
Baked bread (above freezing)	0.70 Btu/(lb·°F)
Baked bread (below freezing)	0.34 Btu/(lb·°F)
Butter	0.57 Btu/(lb·°F)
Dough	0.60 Btu/(lb·°F)
Flour	0.42 Btu/(lb·°F)
Ingredient mixture	0.40 Btu/(lb·°F)
Lard	0.45 Btu/(lb·°F)
Milk (liquid whole)	0.95 Btu/(lb·°F)
Liquid sponge (50% flour)	0.70 Btu/(lb·°F)
Heat of friction per horsepower of mixer motor	42.4 Btu/min
Heat of hydration of dough or sponge	6.49 Btu/lb
Latent heat of baked bread	46.90 Btu/lb
Specific heat of steel	0.12 Btu/(lb·°F)

Comparative tests between wrapped and unwrapped bread indicate a shortening of cooling time for the unwrapped bread from 10 to 30 min through the range of 0 to −20 °F, showing that the wrapper is an insulator. However, while the wrapper is a deterrent to fast freezing, its value in retaining the moisture content in the product during freezing and thawing makes freezing in the wrapped condition advisable.

Some commercial installations freeze wrapped bread in corrugated shipping cartons. The additional insulation provided by the carton increases the freezing time considerably and causes a wide variation in time between variously located loaves. One test found that a corner loaf reached 15 °F in 5.5 h and the center loaf required 9 h.

Most freezers are batch loaded rooms with the bread being placed on the wire shelves of steel racks. One of the principal disadvantages of this arrangement is that the racks must be moved in and out of the room manually. The continuous-type freezers with wire belt- or tray-type conveyors permit a steady flow and do not expose personnel to the freezer temperature. See Chapter 8 for freezer descriptions.

For air freezing temperatures of −20 °F or below, two-stage compression systems should be considered for overall economy of operation.

In addition to the primary air blowers designed to handle about 10 cfm per pound of bread frozen per hour, a series of fans are employed to assure good air turbulence in all parts of the room. Heat load calculations are based on the specific and latent heat values in Table 2.

After the quick-freeze, the bread is moved into a 0 °F holding room where the temperature evens out throughout the loaf. The bread is often placed in shipping cartons after freezing and stacked tightly on pallets.

Frozen bread must be thawed or defrosted for final use. Slow, uncontrolled defrosting requires only that the frozen item be left to stand, usually in normal atmospheric conditions. For quality control, the defrosting rate is not as critical as the freezing rate. However, some authorities suggest defrosting at the same rate as used in the freezing cycle. However, for time control, increases in relative humidity and air temperature decrease the defrosting time. Too high a relative humidity causes excessive condensation on the wrapper with some resultant susceptibiltiy to handling damage. For purposes of calculation, at 120 °F air temperature with 50% or less rh the product will defrost in about 1.75 h. Good air movement over the entire product surface at 200 fpm or higher helps to minimize the condensation and make the defrosting rate more uniform.

FREEZING OTHER BAKERY PRODUCTS

Retail bakeries freeze many products to help meet fluctuating demand. Cakes, pies, sweet yeast dough products, soft rolls, and doughnuts are all successfully frozen. A summary of tests and commercial practice indicates that these products are less sensitive to the freezing rate than are bread and rolls. Freezing at temperatures of 0 to 10 °F apparently produces just as satisfactory results as freezing at −10 to −20 °F.

Storage temperatures of 0 °F or lower will keep packaged dinner rolls and yeast-raised and cake doughnuts satisfactorily fresh for 8 weeks. Storage temperatures above 0 °F reduce the satisfactory keeping time, materially affecting their quality. Cinnamon rolls keep satisfactorily for only about 3 weeks, apparently because of the presence of raisins, which absorb moisture from the crumb of the roll. Pound, yellow layer, and chocolate layer cakes can be frozen and held at 10 °F for 3 weeks without materially affecting their quality. Sponge and angel food cakes tend to be much softer as the freezing temperature is reduced to 0 °F.

Layer cakes with icing freeze well, but condensation on exposed icing during thawing ruins the gloss so that these cakes are wrapped before freezing. Unlike cakes that can be satisfactorily frozen after baking, frozen baked pies have an unsatisfactory crust color, and the bottom crust of fruit pies becomes soggy when the pie is thawed.

The freezing of unbaked fruit pies is highly successful. Freezing time has little, if any, effect on product quality, but storage temperatures do have an effect. Frozen pies stored at temperatures above 0 °F develop badly soaked bottom crusts after 2 weeks and the fillings tend to boil out during baking, possibly due to the weakened structure of the crust.

Danish and sweet dough products are frozen baked or unbaked, depending on how quickly they will be required for sale when they are removed from the freezer. Custard and chiffon pie fillings have not had uniformly good results, but some retail bakeries have achieved satisfactory results by carefully selecting starch ingredients. Meringue does not stand up very well, but whipped cream seems to improve with freezing. Cheesecake, pizza, and cookies also freeze well.

Although some products may be of better quality if they are frozen at 0 °F while others are improved by freezing at 10 °F, variety shops must compromise on a single freezer temperature so that the various products can be placed in the same freezer. These freezers are usually maintained between 5 and 10 °F. Freezing time is not a factor because the products are kept in storage in the freezer. The freezers range from large reach-in refrigerators in retail shops to walk-in boxes in wholesale shops.

RETARDING DOUGHS AND BATTERS

Freezing is usually employed where the products are to be held from 3 days to 3 weeks. For shorter holding periods, such as might be required to have freshly baked products all day from one batch mix, a temperature only cold enough to retard fermentation action in the dough is applied. Temperatures of 32 to 40 °F slow up the leavening action sufficiently to allow holding from 3 h to 3 days.

The doughs are sometimes made up into final shape units ready for proofing and baking. Storage of cold slabs of the dough, with the units for baking being made up after thawing, are also used. This method is especially satisfactory for Danish pastry dough and other doughs with rolled-in shortening, such as pie crust. Chilling to retarded temperature seems to improve the flakiness of this type of product. Refrigeration load calculations should be made following the recommendations included in Chapter 26.

In the retarding refrigerator, about 85% rh is required to prevent the product from drying out. Condensation on the product is undesirable. This type of operation usually employs complete batch loading, so refrigeration calculations must be based on the introduction of the batch over a short period of time. The capacity of the refrigeration equipment must be sufficient to absorb the product and carrier heat load in 0.75 to 3 h, depending on cabinet size and handling technique. The products most commonly handled in this manner are Danish pastry, dough for sweet rolls and coffee cake, cookies, layer cake mixes, pie crust mixes, and bun doughs.

The required retarding temperatures are very similar to the temperatures required for storing ingredients, and the refrigerators are usually designed to handle both the ingredient storage and the retarded dough.

CHAPTER 21

CANDIES, NUTS, DRIED FRUITS, AND VEGETABLES

CANDY MANUFACTURE

AIR conditioning is essential for successful candy manufacturing. Proper atmospheric control results in increased production, lower production costs, and better and more uniform product quality.

One or more of several standardized spaces or operations are encountered in every plant. These include: hot rooms; cold rooms; cooling tunnels; coating kettles; packing, enrobing, or dipping rooms; and storage.

Sensible heat must be adsorbed by air-conditioning equipment, which includes the air-distribution system, plates, tables, and cold slabs in tunnels or similar coolers. In calculating the loads, consider such sensible heat sources as people, power, lights, sun effect, transmission losses, infiltration, steam and electric heating apparatus, and the heat of the entering product. Table 1 summarizes the optimum design conditions for refrigeration and air conditioning.

Two of the basic ingredients in candy are sucrose and corn syrup. These easily change from a crystalline form to a fluid, depending on temperature, moisture content, or combinations of the two. The surrounding temperature and humidity must be controlled to prevent moisture gain or loss, which will affect the product's texture and storage life. Temperature should be relatively low, generally below 70°F. The relative humidity should be 60% or less, depending on the type of sugar used. For chocolate coatings, temperatures of 65°F or less are desirable, with 50% rh or less.

MILK AND DARK CHOCOLATE

Cocoa butter is either the only or the principal fat in chocolate, constituting 25 to 40% or more of various types. Cocoa butter is a complex mixture of triglycerides of high molecular weight fatty acids, mostly stearic, oleic, and palmitic. Since cocoa butter is present in such large amounts in chocolate, anything affecting cocoa butter affects the chocolate product too.

Because cocoa butter is a mixture of triglycerides, it does not act as a pure compound. Its physical properties, melting point, solidification point, latent heat, and specific heat affect the mixture. Cocoa butter softens over a wide temperature range, starting at about 80°F and melting at about 94°F. It has no definite solidification point; this varies from just below its melting point to 80°F or lower, depending on the amount of cocoa butter and the time it is held at various temperatures. The presence of milk fat in milk chocolate lowers both the melting point and the solidification point of the cocoa butter. High quality milk chocolate remains fluid for easy handling at temperatures as low as 86 to 88°F. Sweet chocolate remains fluid at temperatures as low as 90 to 92°F.

Chocolate can be subcooled below its melting point without crystallization. In fact, it does not crystallize in mass but rather in successive stages, as solid solutions of a very unstable crystalline state are formed under certain conditions. The latent heat of crystallization (or fusion) is a direct function of the manner in which the chocolate has been cooled and solidified. Once crystallization has started, it will continue until completion, taking from several hours to several days, depending on its exposure to cooling, particularly to low temperatures (subcooling).

The latent heat of solidification of the grades of chocolate commonly used in candy manufacture varies from approximately 36 to 40 Btu/lb. An average value for its specific heat may be taken as 0.56 Btu/(lb·°F) before solidification and 0.30 Btu/(lb·°F) after solidification. To calculate the cooling load, start with these figures and then add a margin of safety.

Cocoa butter's cooling and solidification properties exist in five polymorphic forms: one stable form and four metastable or labile ones. Cocoa butter usually solidifies first in one of its metastable forms, depending on the rate and temperature at which it solidifies. In solidified cocoa butter, the lower melting labile forms change rapidly to the higher melting forms. The higher melting labile forms change slowly, and seldom completely, to the stable form.

Commercial chocolate blocks are cast in metal molds after the tempering process. During this process, it is desirable to cool the chocolate in the molds as quickly as possible, thus yielding the shortest possible cooling tunnel. However, over-rapid cooling (particularly in large commercial blocks, which are standard 10-lb cakes) may cause checking or cracking which, while not serious to quality, adversely affects appearance. Depositing the chocolate into metal molds at 85 to 90°F is common.

Dark chocolate should be cooled very slowly at a temperature of 90 to 92°F; milk chocolate at 86 to 88°F. Air entering the cooling tunnel, where the goods are unmolded, may be 40°F. The air may be 62°F where the goods enter the tunnel. After chocolate is deposited in the mold, it can be moved into a cooling tunnel for a continuous cooling process, or the molds can be stacked up and placed in a cooling room with forced-air circulation. In either case, temperatures of 40 to 50°F are satisfactory. The discharge room from the cooling tunnel or the room to which the molds are transferred for packing should be maintained at a low enough dew point to prevent condensation on the cooled chocolate. In load calculations for the cooling or cold room, it is necessary to account for transmission and infiltration losses, any load derived from further cooling of the molds, and the sensible and latent heat cooling loads of the chocolate itself.

The tunnel is designed to introduce a 40°F air countercurrent to the flow of chocolate, so that the coldest air enters the tunnel

The preparation of this chapter is assigned to TC 11.5, Fruits, Vegetables, and Other Products. The chapter last received a major revision in 1967.

Table 1 Optimum Design Air Conditions

Department or Process	Dry-Bulb[a] Temperature, °F	rh, %
Chocolate pan supply air	55 to 62	55 to 45
Enrober room	80 to 85	30 to 25
Chocolate cooling tunnel supply air	40 to 45	85 to 70[b]
Hand dipper	62	45
Molded goods cooling	40 to 45	85 to 70
Chocolate packing room	65	50
Chocolate finished stock storage	65	50
Centers tempering room	75 to 80	35 to 30
Marshmallow setting room	75 to 78	45 to 40
Grained marshmallow (deposited in starch) drying	110	40
Gum (deposited in starch) drying	125 to 150	25 to 15
Sanded gum drying	100	25 to 40
Gum finished stock storage	50 to 65	65
Sugar pan supply air (engrossing)	85 to 105	30 to 20
Polishing pan supply air	70 to 80	50 to 40
Pan rooms	75 to 80	35 to 30
Nonpareil pan supply air	100 to 120	20
Hard candy cooling tunnel supply air	60 to 70	55 to 40
Hard candy packing	70 to 75	40 to 35
Hard candy storage	50 to 70	40
Caramel rooms	70 to 80	40
Raw Material Storage		
Nuts (insect)	45	60 to 65
Nuts (rancidity)	34 to 38	85 to 80
Eggs	30	85 to 90
Chocolate (flats)	65	50
Butter	20	
Dates, figs, etc.	40 to 45	75 to 65
Corn syrup[c]	90 to 100	
Liquid sugar	75 to 80	40 to 30
Comfort air conditions	75 to 80	60 to 50

Conditions given in this table are intended as a guide and represent values found to be satisfactory for many installations. However, specific cases may vary widely from these values because of such factors as type of product, formulas, cooking process, method of handling, and time. Acceleration or deceleration of any of the foregoing will change the temperature, humidity, or both to some degree.

[a]Temperature and humidity ranges are given in respective order, *i.e.*, first temperature corresponds to first humidity.

[b]Optimum conditions.

[c]Depends on removal system. With higher temperatures, coloration and fluidity are greater.

as the cooled chocolate leaves the tunnel. Thus, as the tunnel air warms up on its way out, the warmest air leaves the tunnel at the point where the warmest molten chocolate enters. The leaving chocolate is markedly cooler than the entering chocolate, and the subcooling is greatly reduced. This in turn reduces the large temperature difference between the chocolate and the cooling air along the entire tunnel length.

For any particular problem, only testing will determine the length of time the chocolate should remain in the tunnel and the subsequent temperature requirements. Good cooling is generally a function of tunnel length, belt speed, and the actual time the product contacts the cooling medium.

ENROBING OR HAND DIPPING

Chocolate coatings are applied to the center material in one of two ways: either formed by hand or cast in starch or rubber molds and then dipped by hand or enrobed mechanically. The chocolate supply for hand dipping is normally kept in a pan maintained at the lowest possible temperature that will still secure sufficient fluidity for the process. Because this temperature is higher than the dipping room temperature, a heat source, such as electrically heated dipping pans with thermostatic controls, is required. The dipped candy is placed either on trays or belts while the chocolate coating sets.

Setting is controlled by conditioning the air in the dipping room. A dry-bulb temperature of 35 to 40 °F best promotes rapid setting and provides a high gloss on the finished goods. However, the temperature in the dipping room is raised for human comfort. The recommended conditions for hand-dipping rooms are 64 °F dry-bulb and a relative humidity not exceeding 50 to 55%. The principal aim is to achieve uniform air distribution without objectionable drafts. The loads for this type of installation include transmission, lights, and people, as well as the heat load from the chocolate and the heat used to warm dipping pots.

In high-speed production of bar candy, the chocolate coating is applied in an enrober machine, which consists essentially of a reservoir for the fluid chocolate that is heated and thermostatically controlled. This mass is then pumped to an upper flow pan which allows it to flow in a curtain down to the main reservoir. An open chain-type belt carries the centers through the flowing chocolate curtain where the covering is picked up. At the same time, grooved rolls pick up some chocolate and apply it to the bottom of the centers. The centers should be cooled to 75 to 80 °F to assist solidification and retention of the proper amount of coating.

The enrober tunnel sets the balance of the chocolate coating as rapidly as is consistent with high quality and good appearance of the finished goods. The discharge end of the enrober tunnel is normally in the packing room, where the finished candy is then wrapped and packed.

While not an absolute necessity, air conditioning the enrobing room is desirable. Because the coating is exposed to the room atmosphere, the atmosphere should be clean to prevent contamination of the coating with foreign material. It is advisable to maintain enrobing room conditions of 75 to 80 °F dry bulb and 50 to 55% rh, *i.e.*, low enough to prevent the centers from warming up and to assist in the setting of the chocolate after it is applied.

The coated pieces are transferred from the enrober to the bottomer slab and then pass into the enrober cooling tunnel. The function of the bottomer slab is to set the bottom coating as rapidly as possible, thus maintaining this coating and forming a firm base for the piece as it passes through the enrober tunnel. A bottomer slab is often a simple plate evaporator fed with chilled water or brine or directly supplied with refrigerant. The belt carrying the candy passes directly over this plate and heat transfer must take place from the candy through the belt to the surfaces of the bottomer slab. A bottomer slab is sometimes located *before* the enrober to obtain a good bottom prior to full coverage.

BAR CANDY

The production of bar candy calls for high-speed semiautomatic operations to keep costs to a minimum. From the kitchen, the center material is delivered to spreaders, which form layers on tables, or is cast in starch molds. Depending on the composition of the center, the hot material may be delivered at temperatures as high as 160 to 180 °F. Successive layers of different color or flavor may be placed on top of each other to build up the entire center material. These layers usually consist of nougat, caramel, marshmallow whip, or similar ingredients, to which peanuts, almonds, or other nuts may be added. Since each of the ingredients requires a different cooking process, each separate ingredient is deposited in a separate operation. Thus, a ⅛-in. layer of caramel may be first deposited and then a layer of peanuts, followed by a ¾-in. layer of nougat. Except for nuts, it is necessary to allow time for each successive layer to set before the next is applied. If the candy is spread in slabs, the slab must first be cooled and then cut with rotary knives into pieces the size of the finished center.

HARD CANDY

Manufacturing hard candy with high-speed machinery requires air conditioning to maintain temperature and humidity. The requirements of candy made of cane sugar are somewhat different from that made partly with corn syrup. For example, a dry-bulb temperature of 75 to 80°F with 40% rh is satisfactory for corn syrup (as the corn syrup percentage increases, the relative humidity must decrease), whereas the same temperature with a 50% rh is necessary for cane sugar.

Where relative humidity is to be maintained at 40% or less, standard dehydrating systems employing such chemicals as lithium chloride, silica gel, or activated alumina should be used. A combination of refrigeration and dehydration is also used.

The amount of air required is a direct function of the sensible heat of the room. Approximate rules indicate that the quantity should be between 1.5 and 2.5 cfm/ft^2 of floor area, with a minimum of 15% outdoor air or 30 cfm per person. Note that the sensible heat in hard candy is at a high temperature to keep it pliable during forming.

Where concentrations of the finished product in containers or tubs are located in the general conditioned area, the quantity of air must be increased to prevent the product sticking to the container.

Unitary air conditioners employing dry coils are satisfactory if they have a sufficient number of rows and adequate surface. A central station apparatus employing cooling and dehumidifying coils or similar design also may be used. Good filtration is essential for air purity as well as for preventing dirt accumulation on cooling coils. Reheat control is needed for constant temperature and humidity conditions. Air distribution should be designed to provide uniform conditions and eliminate the possibility of drafts.

HOT ROOMS

Such products as jellies and gums can best be dried in air-conditioned hot rooms. These products are normally cast into starch molds. The molds are contained in a tray approximately 35 by 15 by 1.5 in. with an extra 0.50 to 0.75 in. blocking at the bottom for air circulation. These trays are then racked up on trucks, with the number of trays per truck (usually 25 to 30) determined by the method of loading. The trucks are then loaded into the hot room where the actual drying is accomplished.

Normal drying temperatures average between 120 and 150°F dry bulb. While humidity is important, close humidity control is not necessary; even with the maximum outdoor temperature conditions generally encountered when this air is heated, the relative humidity will be low (from 19 to 13%).

Some operators prefer to have manual humidity control, which requires frequent inspection of actual conditions; others prefer automatic humidity control by instruments calibrated to steadily maintain desired dry-bulb temperature and relative humidity in the hot room regardless of the moisture coming out of the candy and the supply air. With full automation, the supply should provide dry air to the unit and also cool the air usually needed in hot weather to purge the hot room after completing the drying cycle.

For proper air distribution in the hot room, it is necessary to arrange for the maximum amount of air to be in contact with the product. Providing space between trays is one means of accomplishing this. The trucks within the hot room must be placed to insure a continuous airflow from truck to truck with the shortest path for the airflow. Space must also be maintained at the entering and leaving air sides to assure flow from the top to the bottom tray for each truck. A large air quantity is required to secure uniformity over the entire product zone.

One method for achieving uniformity is the ejector nozzle system. Basically, this system consists of a supply header fitted with conical ejector nozzles designed to have a tip velocity of 2000 to 5000 fpm with a static pressure behind the nozzle ranging as high as 12 in. of water. Nozzles arranged in this way will induce a flow of air about three times that actually supplied by the nozzles. This ratio gives the most economical balance between air quantity supplied and fan horsepower. By using an ejector system, the primary and induced airstreams mix over the product and space between the top tray; sufficient ceiling height must be allowed for this mixing, which rapidly decreases the necessary differential between the air supply temperature and the actual room temperature. The high airflow thus created decreases the temperature drop across the product. Since the temperature drop is proportional to the heat pickup, a greater airflow will have a lower temperature drop, so that the spread between the air temperatures entering and leaving the product zone is reduced; this promotes uniform drying.

When drying is completed, it is necessary to cool the product rapidly to facilitate unloading. This quick cooling is provided for by a second outdoor air intake which bypasses the heating coil or the air-conditioning unit. At the same time, this bypass intake is put into operation, drop dampers are opened in the bottom of the ejector header, and this air also bypasses the ejector nozzles and flushes out the heat in the room. The use of an exhaust fan is recommended to rapidly remove the heated air, which at this time rises to the ceiling.

The equipment for this type of operation consists of a fan and heating coil located outside the hot room. (No electric motor should be in the room because of the hazard of sparking.) This unit has outdoor air intakes, ejector headers, return air dampers, and dampers for the outdoor air intake. Most operators prefer to use a recording controller to maintain an accurate record of each batch. The controller simply regulates the flow of steam to the heating coil to maintain the desired room temperature. Gradual switches should be provided to control the position of the outdoor and return air dampers, since with a rise in humidity more outdoor air should be taken, and with a drop in humidity more return air should be taken. In some cases, this function can be provided automatically by a humidity control. An end position can be included on the gradual switch for the cooling down period so that when the outdoor air damper is opened wide, the exhaust fan is started.

COLD ROOMS

Many confectionary products—such as marshmallows, certain types of bar centers, and cast cream centers—require chilling and drying but cannot withstand high temperatures. Drying conditions of about 75°F and 45% rh are required, and the drying period varies from 24 to 48 h. Here an ejector-type system is used for the hot rooms, the difference being that cooling coils are provided in the unit. The load must be carefully calculated to determine the sensible and latent heat quantities from which the actual air quantity can be found, together with air supply temperature and refrigerant temperature.

When properly balanced, controlling the relative humidity becomes an inherent part of the system design. This control system is basically the same as for the hot room, except that its recording regulator must control the admission of steam to the heating coil in winter and regulate the flow of refrigerant to the cooling coil in summer. Flushing dampers and a cooling down cycle are necessary in this type of operation. One precaution must be observed in the equipment for the cold room in connection with starch or sugar dust picked up in the return airstream. During the cooling cycle, the cooling coil normally condenses moisture, and accumulated starch or sugar dust rapidly forms a paste on the cooling coil, reducing its capacity and necessitating frequent maintenance and cleaning of the equipment. Thus, filters should always be used for the outdoor and return air entering the cooling coil.

COOLING TUNNELS

Various candy plant cooling requirements can best be handled in a cooling tunnel, including the cooling of coated centers after leaving the enrober, the cooling of cast chocolate bars, and the cooling of hard candy. These operations are usually set up for a continuous flow of high rate production. The product is normally conveyed on belts either through the enrober or the casting machine and then through a cooling chamber.

A cooling tunnel consists of an insulated box placed around the conveyor so that the product may travel through it in a continuous flow. Cold air is supplied to this enclosure to cool the product. To achieve maximum heat transfer between the air and product, air should flow counter to the material flow. In general, air supply temperatures of 35 to 45 °F with air velocities up to 2500 fpm have been found satisfactory. The high velocity improves the heat transfer and creates a turbulent condition in the airflow which further improves heat-transfer efficiency.

The actual size of the tunnel is determined by the size of the conveyor belt and the air quantity. The air quantity depends on the heat load and the desired rate of cooling. The rise in air temperature through the tunnel should be limited to 15 to 20 °F maximum. One air-conditioning unit, which normally consists of a fan and coil together with the necessary duct connections to and from the unit, is generally used for each tunnel. An outdoor air intake is advisable, since cooling can be frequently accomplished with outdoor air, without operating the refrigeration plant. The tunnel should be made as tight as possible and the entrances and exits for the candy should be reduced to a minimum to limit air loss from the tunnel or air infiltration into the tunnel; in some cases, it is advisable to use a flexible canvas curtain to control airflow. As some loss is unavoidable, it is practicable to take a small amount of outdoor air or air from adjoining spaces to provide a slight excess pressure in the tunnel.

For chocolate-enrobing work, the condition of the air is the paramount factor in securing the best possible luster and most even coating. The best results are obtained with rather slow cooling, but this requires either a low production rate or excessively long tunnels. The final design is a compromise. The coating must be in the proper condition when it is poured over the centers, since improper temperature at this point will result in blushing or loss of luster. Proper temperature, however, is a function of the enrober machine and its operation; no amount of correction in the tunnel will compensate.

A variation of the standard single-pass counterflow tunnel has been used for enrobing. The tunnel is divided horizontally by an uninsulated sheet metal partition. The belt carrying the candy rides directly on this partition, and the return belt is brought back through the lower space below the partition. The cold air is supplied to the lower chamber near the enrober, progresses to the opposite end of the tunnel, is transferred to the top chamber, progresses back to a point near the enrober, and then is returned to the cooling equipment. This tunnel has two important advantages—it chills the return belt so that this belt can act as a bottomer slab to give a quick setting of the base of the coated piece, and the uninsulated partition with the coldest air below assists in this bottoming operation. In this manner, the air supply has already absorbed some of its heat load by the time it is actually introduced to the product-cooling zone. The method approaches the advantage of slow cooling but keeps the tunnel at a minimum length.

COATING KETTLES OR PANS

The application of air conditioning to revolving coating kettles or pans is satisfactory. Originally these pans were merely provided with a supply of warm air, ranging from 80 to 125 °F. This air was then exhausted from the kettle to the room, creating a severe nuisance because of sugar dust blowing out of the kettle. In addition, a portion of the energy required to produce rotation of the kettle is converted to heat in the centers being coated, which causes enough expansion to produce cracking or checking. Thus, a portion of the coating is applied, the product is withdrawn for a seasoning period of up to 24 h, and the material is then returned to the kettles for additional coating.

Some installations overcome most of the sugar-dust problem by using a conditioned supply of air to the kettles and providing positive exhaust from the kettles. The wet- and dry-bulb temperature of the air supplied to the kettles is so controlled that the rate of coating evaporation and the drying are uniform with a high production rate and reduced labor cost. Evaporation of the moisture in the coating material tends to take place at the wet-bulb temperature of the air supplied to the kettle. A large portion of the heat of crystallization entering the product is absorbed, and the centers are not overheated, thus eliminating the need for a seasoning period and permitting continuous operation. With air conditioning applied to coating kettles, the number of rejects caused by splitting, cracking, uneven coating, or doubles can be reduced considerably. Sugar dust recovery is accomplished with a cyclone-type dust collecting device.

PACKING ROOMS

Air conditioning is essential to production in a number of rooms or departments in every candy plant. The average manufacturer spends considerable time and effort in the design and application of packaging materials, since this affects product keeping quality. Important packaging considerations are moistureproof containers, antimoisture insulating for abnormally high humidity conditions, and protection against freezing or extreme heat. Manufacturers often overlook the fact that the air in the packing room surrounds the products when they are placed in one of these moistureproof containers. Controlling this air is essential for proper packing. For instance, products packed in a room at 85 °F dry bulb and 60% rh, which may easily be encountered in a normal summer, would have air surrounding the product at a 69 °F dew point. If this sealed package were subjected to temperatures below 69 °F, the air in the package would become supersaturated and moisture would condense on the surfaces of the container and the product. If the product was then subjected to a higher temperature, the moisture would reevaporate; but in this process, chocolates would lose their luster or show sugar bloom, and marshmallows would develop either a sticky or a grained surface, depending on the formula used. In practice, conditions of 65 °F dry bulb and 50% rh have been found most practicable. This could be improved if the relative humidity were reduced several points. In the case of hard candy, which is intensely hygroscopic, the relative humidity should be reduced to 35 to 40% at 70 to 75 °F.

REFRIGERATION PLANT

Large candy manufacturers often use a central refrigeration plant for cooling water or brine, for circulation throughout the plant to meet various requirements. In some smaller plants, small units are used for each requirement or each group of requirements. Thus, one compressor may be installed on a cooling tunnel, while separate compressors are used in each of the departments for packing, storing, and other operations. This arrangement permits a planned program of expansion or rehabilitation of the plant a portion at a time. In a large plant, this results in a multiplicity of small units and adds considerably to the work of the operation.

The type of equipment used depends on plant size, initial and operating costs, plans for future expansion, and the like (see Chapters 1 through 4). The *Safety Code for Mechanical Refrigeration* (ASHRAE *Standard* 15-1978) should be adhered to in the design of the plant, depending on the refrigerant used.

Some larger plants use centrifugal compressors in a central plant design. The centrifugal machine can be driven by motor or turbine. Most candy plants generate steam at high pressure for use in various cooling operations. Turbines can either use this high-pressure steam, exhausting at a back pressure for lower temperature cooking operations, or can condense exhaust steam. The centrifugal unit is extremely reliable, with low operating cost, ease of maintenance, and long life.

Brine at 28 to 32 °F is desirable for central installations. This secondary coolant temperature is low enough to obtain the necessary low dew-point temperatures for many processes, but is high enough to prevent frosting on the finned evaporators. Solutions of alcohol and water, or a weak solution of calcium chloride brine, have been used in some plants. Increased production has caused the lowering of brine temperatures, particularly to provide cooler air supplied to tunnels.

A secondary coolant distribution system makes it possible to connect all the varied services to one source of refrigeration and, by using control devices, maintain dry-bulb and dew-point temperatures more precisely. Cold slabs are supplied for caramels, nougats, and similar products with coolant at this temperature, and its use is extended to comfort cooling in nearby offices and clerical operations.

For smaller manufacturing plants, multiple condensing units using halocarbon refrigerants and direct expansion air-conditioning units are more satisfactory. Similarly, cooling tunnels are equipped with separate cold diffusers and connected to individual halocarbon refrigerant compressors (or a group of units is connected to one single or duplex compressor). These compressors vary in size from 5 to 50 hp and can be located adjacent to the spaces conditioned.

STORAGE

CANDY

Most candies are held for one week to more than a year between manufacture and consumption. Storage may be in the factory, in warehouses during shipping, or in retail outlets. It is important that the candy not lose quality during that time.

Low-temperature storage does not produce undesirable results if: (1) the candies are made of proper ingredients for refrigerated storage; (2) the packages have a moisture barrier; (3) the storage room is held at equilibrium humidity with desirable moisture conditions for preserving the candy; and (4) the candy is brought to room temperature before the packages are opened.

The storage period depends on: (1) the marketing season of the candy; (2) the stability of the candy; and (3) the storage temperature and humidity (see Table 2).

The shelf life of a candy is determined by the stability of its individual ingredients. Common candy ingredients are sugar (including sucrose, dextrose, corn syrups, corn solids, and invert syrups), dark and milk chocolates, nuts (including coconut, peanuts, pecans, almonds, walnuts, and others), fruits (including cherries, dates, raisins, figs, apricots, and strawberries), dried milk and milk products, butter, dried eggs, cream of tartar, gelatin, soybean flour, wheat flour, starch, and artificial colors.

Refrigerated storage of candy ingredients is especially advantageous for seasonal products, such as peanuts, pecans, almonds, cherries, coconut, and chocolate. Ingredients with delicate flavors and colors, such as butter, dried eggs, and dried milk, retain quality more evenly year-round if kept properly refrigerated. Otherwise, ingredients containing fats or proteins may lose considerable flavor or develop off-flavors before being used.

Table 2 Expected Storage Life for Candy

Candy	Moisture Content, %	Relative Humidity, %	Storage Life, Months — Storage Temperature, °F 68	48	32	0
Sweet chocolate	0.36	40	3	6	9	12
Milk chocolate	0.52	40	2	2	4	8
Lemon drops	0.76	40	2	4	9	12
Chocolate-covered peanuts	0.91	40 to 45	2	4	6	8
Peanut brittle	1.58	40	1	1.5	3	6
Coated nut roll	5.16	45 to 50	1.5	3	6	9
Uncoated peanut roll	5.89	45 to 50	1	2	3	6
Nougat bar	6.14	50	1.5	3	6	9
Hard creams	6.56	50	3	6	12	12
Sugar bonbons	7.53	50	3	6	12	12
Coconut squares	7.70	50	2	3	6	9
Peanut butter taffy kisses	8.20	40	2	3	5	10
Chocolate-covered creams	8.09	50	1	3	6	9
Chocolate-covered soft creams	8.22	50	1.5	3	5	9
Plain caramels	9.04	50	3	6	9	12
Fudge	10.21	65	2.5	5	12	12
Gumdrops	15.11	65	3	6	12	12
Marshmallows	16.00	65	2	3	6	9

Candies are semiperishable: the finest candies or candy ingredients may be ruined by a few weeks of improper storage. This includes many candy bars, packaged candies, and some choice bulk candies. Unless refrigeration is provided from time of manufacture through the retail outlet, the types of candies offered for sale must be greatly reduced in the summer.

Benefits from refrigerated storage of candies, especially during the summer, are: (1) insects are rendered inactive at temperatures below 48 °F; (2) the tendency to become stale or rancid is reduced; (3) candies remain firm as an assurance against sticking to the wrapper or being smashed; (4) loss of color, aroma, and flavor is reduced; and (5) candies can be manufactured year-round and accumulated for periods of heavy sales.

Color

Many colors used in hard candies, hard creams, and bonbons gradually fade when stored at room temperature, especially in the light. However, the most marked effect of storage temperature on color occurs with chocolates. In candies containing high protein and nuts, there is a gradual darkening of color, especially at higher storage temperatures.

Temperatures from 85 to 95 °F cause graying of chocolates in only a few hours and darkening of nuts within a month. Graying or fat bloom appears as *cold grease* on the surface and occurs because of crystals of fat on the surface of the chocolate coating. While this is usually associated with old candies, new candies can become gray in one day under adverse storage conditions. Chilling of chocolates following exposure to high temperatures produces graying very quickly, but chilling without previous heat exposure does not.

Sugar blooming of chocolate looks similar to graying or fat blooming and is caused by deposition of crystallized sugar on the surface from condensation of moisture, following removal of the candy from refrigerated storage without proper tempering. Experiments have shown that chocolate tempered by gradually raising the temperature to normal without opening the package did not incur sugar blooming even after storage at 0 °F or lower.

Sugar bloom may also occur because of storage in overhumid air and migration of moisture from the centers to the surface.

Flavor

Keeping candy fresh is one of the chief reasons for refrigerated storage. Most flavors added to candies, including peppermint, lemon, orange, cherry, and grape, are distinctive and stable during storage. Less pronounced flavors such as those of butter, milk, eggs, nuts, and fruits, are more sensitive to high temperatures.

Low temperatures retard the development of staleness and rancidity in fats, preserve flavors in fruit ingredients, and prevent staleness and other off-flavors in candies containing such semiperishable ingredients as milk, eggs, gelatin, nuts, and coconut. Candies containing fruits become strong in flavor when they are stored at room temperature or higher for more than a few weeks; those containing nuts become rancid. There are no specific critical temperatures at which undesirable changes occur, but the lower the temperature, the more slowly they take place.

Texture

Candy becomes increasingly soft at high temperatures and increasingly hard and brittle at low temperatures, reaching an optimum for eating at about 70°F. These changes in texture are reversible from below 0 to 80°F, enabling refrigerated candies to be returned satisfactorily to any desired temperature for eating. This is extremely important, since the texture of most candies subjected to very low temperatures (or even shipped in contact with dry ice at about −110°F) is not permanently changed.

Most candies are manufactured at controlled temperatures. Their texture (except in hard candies and hard creams) is maintained best at temperatures below 68°F.

Insects

Candies containing fruits, chocolate, nuts, or coconut are favorite hosts for insects. Since fumigation and insect repellants are seldom permissible with candies, refrigeration is used to inactivate insects in candy and candy ingredients.

Common insects become active at about 50°F and activity increases as the temperature is raised to 100°F. While common cold storage temperatures do not kill many insects, they do inactivate them at temperatures below 50°F. Both adults and eggs may exist for months at above freezing temperatures without feeding or propagating. Candies with insect eggs on either the product or wrappers may be refrigerated for long periods with no apparent damage, but when brought back to warm rooms, a serious insect infestation may develop.

Both adults and eggs are killed when infested materials are stored at or near 0°F. Lower temperatures and long storage periods are lethal to insects. However, storing candies at 0°F for a few weeks usually destroys all forms of insect life.

Storage Temperature

Regarding their effect on candies, storage temperature is difficult to separate from humidity, but the latter is more important. There are no specific critical refrigerated temperatures at which a certain type of candy must be held. In general, the lower the temperature, the longer the storage life, but the greater the problem of moisture condensation on removal.

Air conditioning (68 to 70°F). Since the storage of candies begins in the tempering room of the manufacturing plant, 68°F and 50% rh is desirable to prevent soft sticky pieces from being packaged. Under these conditions, all candies remain firm and there will be little or no graying of chocolates; their original luster can be held.

Hard candies or types containing only sugar ingredients keep in good condition for more than 6 months at 68 to 70°F. Other types become stale, lose flavor and luster, and darken in color. Under prolonged storage, candy containing nuts or chocolate becomes musty or rancid, and even the colors and flavors of some hard candies may fade. Only *summer candies* should be held for more than a few weeks at 68 to 70°F, or higher. Unless precautions are taken, the temperature of truck or rail shipments may rise to the melting point of semisoft candies or to the graying point of chocolates. Candies temporarily stored in the sunshine or in warm places in buildings may suffer severely in loss of shape, luster, and color. Some companies use refrigerated trucks and railroad cars for hauling candies. As portable refrigerators become more generally available, their use for candies should increase and thereby supply the missing link of conditioned storage for candies from manufacturer to consumer.

Cool storage (48 to 50°F). Candies stored at this temperature remain firm and retain good texture and color; only those containing nuts, butter, cream, or other fats become stale or rancid within 4 months. Candies that remain practically fresh for 4 months are fudge, caramels, sugar bonbons, gumdrops, marshmallows, lemon drops, hard creams, and semisweet chocolates; they are wrapped in aluminum foil to give added protection. Candies that become stale at this temperature are peanut butter, taffy kisses, peanut brittle, uncovered peanut rolls, chocolate-covered peanut rolls, and nougat bars.

Cold storage (32 to 34°F). Most candies can be successfully held in cold storage for at least one year, and many for much longer. Only those containing nuts, coconut, chocolate, or other fatty materials will become stale or rancid.

Freezer storage (0°F). The need and economic justification for freezing candy is the same as for any other food—better preservation for a longer time. This method of preservation is suitable for candies (1) in which high quality standards must be maintained; (2) in which a longer shelf life is desired than is accomplished from other methods of storage; (3) which are normally manufactured from 6 to 9 months in advance of consumption; and (4) which are especially suitable for retailing as frozen items. Because of their high sugar and low moisture content, little ice formation occurs in candies at 0°F.

One of the chief reasons for freezing candies is to hold them in an unchanged condition for as long as 9 months, then thaw and sell them as fresh candies. Experience shows that this is not only possible but practical if the manufacturer (1) freezes only those candies that would lose quality when held at a higher temperature; (2) eliminates the few kinds that crack during freezing; (3) packages the candies in moistureproof containers similar to those used for other frozen foods; and (4) thaws the candies in the unopened packages to avoid condensation of moisture on the surface.

Moistureproof packaging. Experiments show that candies for freezing require more protection than those for common storage because the storage period is usually longer and a greater tendency exists for condensation on removal. A single layer of moistureproof material—aluminum foil, polyethylene, saran, mylar, polypropylene, cellophane, glassine, or laminations including one of these—afford adequate protection. Candies not fully protected from desiccation become hard, grainy, and lose flavor.

Protection is provided when the moisture barrier is in contact with the candy in the form of a sealed, individual wrapper. Inner liners for the boxes protect candy, provided they are sealed (which is difficult and seldom accomplished). The usual manner of applying moisture barriers is as overwraps for the boxes, chiefly because overwraps are easiest to apply and seal by machines. Overwraps provide less protection than wraps for individual pieces of candy because of the relatively large amount of air enclosed in the box. Also, boxes with extended edges offer less protection.

Thawing. Frozen candies should be thawed in the unopened packages. While freezing itself affects only a few candies, the manner of removal from storage affects all of them, especially candies unprotected by special coatings or individual wraps.

Improvement. A few candies are improved by freezing. Candies that are improved in freshness, mellowness, or textural smoothness by freezing include those with high moisture content and without protective coatings or individual wrapping. Usually these are candies ordinarily subject to surface drying. Marshmallows, jellies, caramels, fudges, divinities, coconut macaroons, fruit loaves, coconut bonbons, panned Easter eggs, malted milk balls, and chocolate puffs are in this group.

The stability of candies after freezing is good. Candies may be held frozen for 6 months or more, carefully thawed, and then sold as fresh. Some candies are prepared especially for freezing. These are made of low melting point fat, have more flavor and softer texture than most candies, and should be eaten while they are cold.

Humidity Requirements

Sugar ingredients are stable over a wide range of storage temperatures, but they are sensitive to high or low humidity. The initial moisture content of candies largely determines the optimum relative humidity of the storage atmosphere. Candies with a moisture content of 12 to 16%—marshmallows, gumdrops, coconut sticks, jelly beans, and fudge—should be stored at 60 to 65% rh to avoid becoming sticky, runny or moldy, or hard and crusty. Candies with a moisture content of 5 to 9%—most fine candies, nougat bars, nut bars, hard and soft creams, bonbons, and caramels—should be stored at 50 to 55% rh to retain the original weight, finish, and texture. Candies with moisture contents below 2%—milk chocolate bars, chocolate covered nuts, and all kinds of hard candies—should be stored at 45% rh or lower.

The hygroscopicity of the ingredients also determines the relative humidity at which candies must be held to retain the original firmness and finish. Candies with a high proportion of invert syrups, such as taffy kisses, must be kept in a very dry place, even though their moisture content is not extremely low. Other candies containing high proportions of invert syrup, honey, or corn syrup, must be held in a drier atmosphere than the moisture content indicates.

Candies stored at low temperatures have a wider range of critical relative humidities than those are stored at high temperatures. For example, nougat bars stored at 65% rh (10% too high) will become sticky within a few days at room temperature, but at 40 °F or lower, stickiness might not show up for many weeks. Similarly, marshmallows stored in a room with 55% rh (10% too low) become dry and crusty within a few days at room temperature, but if stored at 40 °F or lower, show little change for several weeks. Thus, refrigeration retards the ill effects of storage under improper humidity conditions. Humectants, such as sorbitol, glycerine, and high conversion corn syrup, are advantageous for maintaining original moisture content of certain candies.

NUTS

Nuts commonly refrigerated include peanuts, walnuts, pecans, almonds, filberts, chestnuts, and imported cashews and Brazil nuts. The advantages of refrigerated storage of nuts are: (1) marketable life is increased as much as ten times; (2) the natural texture, color, and flavor are retained almost perfectly from one season to the next; (3) staleness, rancidity, and molding are retarded for more than two years, depending on the temperature; and (4) insect activity is arrested at temperatures below 48 °F. With optimum temperature, humidity, atmospheric conditions, and packaging, good quality nuts may be successfully stored for up to 5 years.

Temperature

Other conditions being equal, the lower the temperature, the longer the storage life of nuts. Life may be extended from 2 to 3 times with each 20 °F drop in temperature. The freezing point of nuts, depending on the moisture content, is about 23 °F for chestnuts; 14 °F for walnuts, pecans, and filberts; and 13 °F for peanuts. A normal moisture content (percentage) for stored nuts is: chestnuts, 30; peanuts, 6; walnuts, 4.5; pecans, 4; and filberts, 3.5.

Shelled nuts stored from one harvest season to the next without appreciable loss in quality must be held at 36 °F or lower; those held for 6 to 9 months must be kept at 48 °F or lower; and all nuts stored for 4 to 6 months should be held below 68 °F. The period of storage, at a given temperature, doubles if the nuts are unshelled.

Relative Humidity

While the storage temperature of nuts may range from 68 to −20 °F or lower, the relative humidity must range between 65 and 75%. This is to maintain the optimum moisture content for desired texture, color, flavor, and stability. Should the moisture content rise as much as 2% above normal, the nuts (except chestnuts) will darken, become stale, and may become moldy. If the moisture drops more than 2% below normal, the nuts become objectionably hard and brittle.

When the relative humidity is suitable, nuts that are too high or too low in moisture may be safely stored with the assurance that rapid air circulation will bring the moisture content to a safe level. In this sense, storage acts as a conditioning room.

Atmosphere

All nuts (again, except chestnuts) contain 45% or more oil and readily absorb odors and flavors from the atmosphere and surrounding products. Certain gases, particularly ammonia, react with tannin in the seed coats of nuts, causing them to turn black. Therefore, the atmosphere in the nut storage room must be free of all odors. This includes the containers, walls of the room, pallets, and other stored products.

Products that can be safely stored with nuts include dried fruits, candies, rice, and goods packaged in cans, bottles, or barrels. Commodities that should *not* be stored with nuts are onions, meats, cheese, chocolate, fresh fruits, and other products with an odor or high moisture content.

Packaging

The storage life of nuts may be greatly influenced by the choice of package. When nuts are shelled, they become bruised and oil *crawls* over the surface and onto the package in a very thin film. Unless this is retarded by a package that acts as a barrier, contains an antioxidant, or removes air by vacuum, the nuts become stale and rancid. Furthermore, some packaging materials such as polyethylene should be avoided because they impart an undesirable odor to the nuts.

DRIED FRUITS AND VEGETABLES

Dried fruits and vegetables differ from each other in that: (1) fruits contain sugars that render them more hygroscopic, harder to dry, and greater absorbers of moisture during storage; (2) the moisture in dried fruits may range from 32% with sorbate treatment to as low as 2%, while that in vegetables ranges from 7% to a low of 0.3%; (3) dried fruits are acid, more highly colored, and more stable during storage than vegetables, which are nonacid; (4) fruits are generally dried raw with active enzyme and respiration systems, while vegetables are blanched or precooked, with no active enzymes (thus dried fruits are more responsive to storage temperature and humidity conditions than are vegetables); (5) the high sugar and acid content of dried fruits provides an adverse physical environment for bacteria and thus makes their growth almost impossible even though the moisture level is higher than that in dried vegetables.

Dried fruits and vegetables maintain quality longer when stored at low temperatures. Packaging in a nitrogen or carbon dioxide atmosphere and in the presence of a desiccant to achieve further reduction in moisture content has been used to extend storage life of dehydrated vegetables. Staleness (charred- or off-flavor) and other deteriorative changes in dehydrated vegetables are inhibited by packing in nitrogen. In air-packed samples, the rate of staling is reduced at low temperature. Treatment with sulfite to retain color and extend storage life has been widely used for cut fruits and dried vegetables. A light coating of laundry starch applied to diced carrots prior to dehydration has achieved excellent results in retention of color and other quality factors during storage in cellophane at 84 °F without sulfite.

Refrigerated storage at 40 to 50 °F or lower retards and controls insect infestation. Substantial killing occurs with exposure at 32 °F for 6 months or longer, and a temperature of 0 °F kills insects within a few hours. An alternative method is fumigation, which is generally used in commercial practice.

Nonenzymatic browning (browning in products that have been scaled or blanched adequately to inactivate enzymes) is reduced in dehydrated vegetables at low temperature. Cold storage offers protection for several years.

Increasing the temperature of dried apricots accelerates oxygen consumption, carbon dioxide production, disappearance of sulfur dioxide, and darkening.

Molds and yeasts will not grow in dried fruits that have an adequate sulfur dioxide content or less than 25% moisture. At 32 °F, relative humidity is less important than at higher temperatures.

Other methods of dehydration include the following:

Dehydrofreezing. In this process, the raw, prepared product is dried to about 50% in weight, followed by freezing. This method has yielded excellent fruit and vegetable products when storage was at 0 °F or lower. Concentration of juices by low temperature and high vacuum and preservation of the concentrate by freezing (Chapter 18) is another application of the process.

Freeze-drying. In this process, the prepared product is frozen quickly to produce small ice crystals, followed by sublimation of the moisture with the use of vacuum and low heat, in a manner to preserve porosity of the product. This method is successfully used with products of high value, high protein, low fat, and low sugar content. While refrigerated storage is not necessary to prevent spoilage of freeze-dried food, it is necessary to preserve maximum flavor and natural color.

Dried Fruit Storage

Refrigeration is beneficial in augmenting drying as a means of preserving fruits. The optimum conditions for holding most dried fruits are at about 55% rh and just above each fruit's individual freezing temperature. Since the sugar content of these products is high, the freezing point varies from about 22 to 26 °F. Refrigerated storage is beneficial in retaining natural flavor, ascorbic acid, carotene, and sulfur dioxide, and in controlling browning, insects, rancidity, and molding. Other than for insect control, low humidity is more important than low temperatures for storing dried fruit. Packaged, sulfured cut fruits will keep adequately at higher humidities.

While most dried fruits are adversely affected by softening and injured by freezing, dates are held best by freezing. Before storage, dried fruits should be brought to the desired moisture content.

When practical, dried fruits should be packed in moistureproof containers made of metal or foil, which not only assure a constant moisture content in storage, but prevent injury from moisture condensation on removal from storage. The permeability of the packaging medium the dried fruits are sealed in is extremely important in relation to adverse effects of storage humidity. Dried fruit storage recommendations are:

Raisins. At 32 to 40 °F and at 50 to 60% rh, sugaring is prevented for one year, provided the moisture content of the dried fruit is not unusually high. Raisins contain 15 to 18% moisture; for extremely long storage, the lowest possible moisture content should be maintained.

Figs. These may be held for a year at 32 to 40 °F and at 50 to 60% rh. A temperature of 55 °F or lower prevents darkening for more than 5 months, and low humidity controls sugaring.

Prunes. These may be held for a year at 32 to 40 °F with a relative humidity from 50 to 60%. For storage of 4 to 5 months, a relative humidity of 75 to 80% is not detrimental.

Apples. At 32 to 40 °F with a relative humidity of 55 to 65%, apples retain excellent color and texture for more than one year. A relative humidity of 70 to 80% is not objectionable at 32 °F, but at 40 °F enough moisture may be gained to cause the fruit to mold within 8 months. Browning develops gradually at temperatures of 40 °F and above. Even at 40 °F, browning will develop but at a slower rate.

Pears. Same as for apples.

Peaches. Sun-dried freestone peaches are harder to store than most dried fruits. Therefore, the temperature should be held close to 32 °F with a rh of 55 to 65%. At 40 °F and moderate humidity, the moisture pickup will cause molding and browning to develop rapidly.

Clingstone peaches (dehydrated after steam scalding) should be stored at 32 to 40 °F and at 55 to 75% rh. Sun-dried peaches will tolerate a slightly higher humidity than dehydrated peaches.

Apricots. Dried apricots are easy to keep in refrigerated storage at 32 to 40 °F with a 55 to 65% rh. They remain in excellent condition for more than one year. At 40 °F with moderate humidity there is enough gain in moisture content to cause molding.

Dates (sucrose or hard type). For storage of 6 months or less, dates may be held at 32 °F with 70 to 75% rh, but for longer storage they should be stored at 24 to 26 °F. Usually it is more convenient to store them at 32 °F, at which temperature they can be stored for over a year.

Soft or invert sugar-type dates may be held for 6 months at 28 to 32 °F, but if stored for 9 to 12 months they should be stored from 0 to 10 °F. Uncured dates should be stored at 0 to 10 °F.

Dried Vegetable Storage

Few specific recommendations for dehydrated vegetable storage temperatures are available. Decrease in storage temperature retards deterioration, but cold storage is considered necessary only for long storage periods. Among the advantages are: (1) control of insects at 45 °F or lower; (2) preservation of natural colors; and (3) retention of initial flavors and vitamins.

Since most dried vegetables have very low moisture content, are well packaged, and are usually surrounded by an atmosphere of nitrogen or rarefied air, refrigeration is less essential than it is for undried vegetables.

CONTROLLED ATMOSPHERE

Low oxygen in the storage atmosphere suppresses the growth of insects and molds, retards rancidity and staleness, and reduces oxidative changes in flavors, odors, and colors. In small packages, oxygen can be reduced by vacuum; in storage and large shipping containers, oxygen can best be flushed out with nitrogen. Excellent results have been obtained by substituting up to 98% of the atmosphere with nitrogen. Nitrogen is preferred to carbon dioxide, ethylene, or other gases for storage of low moisture products; it greatly extends the shelf life even with refrigeration.

CHAPTER 22

BEVERAGE PROCESSES

THIS chapter discusses the processes and use of refrigeration in breweries, wineries, and carbonated beverage plants.

BREWERIES

Malt is the primary raw ingredient used in the brewing of beer. While adjuncts such as corn grits and rice contribute considerably to the composition of the extract, they do not possess the necessary enzymatic components required for the preparation of the wort. They lack nutrients (amino acids) required for yeast growth and contribute little to the flavor of beer. Malting is the initial stage in preparing raw grain to make it suitable for mashing. Traditionally, this operation was carried out in the brewery, but in the past century, this phase has become so highly specialized that it is now almost entirely the function of a separate industry.

Various grains such as wheat, oats, rye, and barley can be malted; however, barley is the predominate grain used in the preparation of mash because it has a favorable ratio of protein to starch. It has the proper enzyme systems required for conversion, and the barley hull provides an important filter bed during lautering. Also, barley is readily available in most of the world.

To malt barley, the raw grain is steeped in 40 to 65 °F water for two or three days, until its moisture content increases from about 12 to 45%. The water is changed frequently, and the grain is aerated. After two or three days, the kernels germinate, and the white tips of rootlets appear at the end of the kernels. During this stage, the water is drained from the batch, which is then spread out on a floor, where germination continues. During the growing period, the green malt is turned over continuously to assure uniform growth of the kernels. However, this floor method of malting has been displaced because it is labor intensive.

Current practice is to use slowly revolving drums that turn over the growing malt, or, in the compartment system, slow-moving, mechanically driven, plow-like agitators are used. At the desired stage in its growth, the green malt is transferred to a kiln, where further growth is checked by reducing the moisture content. The kilning is usually done in two stages. First, the moisture content of the malt is reduced to approximately 8 to 14%; then, the heat is increased until the moisture is further reduced to 4%. Using this heating procedure reduces excessive destruction of enzymes. The desired color and aroma are obtained by controlling the final degree of heat.

After kilning, the malt is cleaned to separate dried rootlets from the grain, which then is stored for future use. The finished malt differs from the original grain in several significant ways. The hard endosperm was modified and now is chalky and friable. The enzymatic activity has been greatly increased, especially alpha amylase, which is not present in unmalted barley. The moisture content is reduced, making it more suitable for storing and subsequent crushing. It now has a distinctive flavor and aroma, and the starch and diastase are readily extractable in the brewhouse.

CHEMICAL ASPECTS

Two distinct types of chemical reactions are used in brewing beer. In the first, which is carried out in the brewhouse and is called mashing, the starches in the grains are hydrolyzed into sugars and complex proteins are broken down into simpler proteins, polypeptides, and amino acids. These reactions are brought about by first crushing the malt and suspending it in warm (100 to 122 °F) water by means of agitation in the mash tun. When adjuncts are used, a portion of the malt is cooked separately with the adjunct, usually corn grits or rice. After boiling, this mixture is combined with the main mash, which has been so proportioned that a combining temperature generally in the range of 145 to 162 °F results. Within this temperature range, the alpha and beta amylases degrade the starch to mono-, di-, tri-, and higher saccharides. By suitably choosing a time and temperature regimen, the brewer controls the amount of fermentable sugars produced. The enzyme diastase (essentially a mixture of alpha and beta amylase), which induces this chemical reaction, is not consumed but acts merely as a catalyst. Some of the maltose subsequently is changed by another enzyme, maltase, into a fermentable monosaccharide, glucose.

Mashing is complete when the starches are converted to iodine-negative sugars and dextrins. At this point, the temperature of the mash is raised to a range from 167 to 170 °F, which is the "mashing-off" temperature. This stops the amylolytic action and fixes the ratio of fermentable to nonfermentable sugars. The wort is separated from the mash solids using a lauter tub, a mash filter, or proprietary equipment (MBAA 1981). Hot water (168 to 170 °F) is then "sparged" through the grain bed to recover additional extract. Wort and sparge water are added to the brew kettle and boiled with hops, which may be in the form of pellets, extract, or whole cones. After boiling, the brew is quickly cooled and transferred to the starting cellar, where yeast is added to induce fermentation. Figure 1 shows a double-gravity system with the grains stored at the top of the brewhouse. As the processing continues, the flow is downward by gravity. The hot wort from the bottom of the brewhouse is then pumped to the top of the stockhouse, where it is cooled and again proceeds by gravity through fermentation and lagering.

The preparation of this chapter is assigned to TC 11.2, Foods and Beverages.

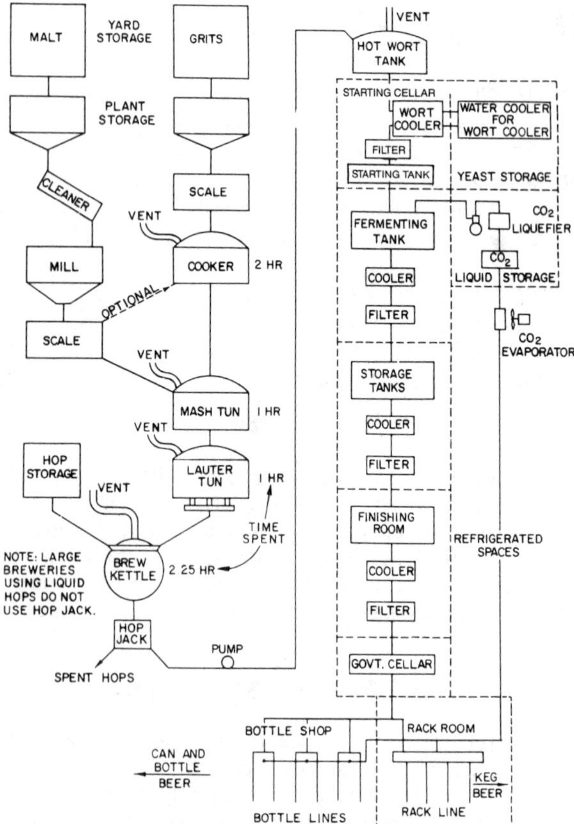

Fig. 1 Brewery Flow Diagram

After cooling the wort, yeast and sterile air are injected into it. The yeast is pumped in as a slurry at a rate of between 1 and 3 lb of slurry for each barrel of wort. The wort may be oxygenated with pure oxygen. However, normally, oil-free compressed air is filtered and treated with ultraviolet light and then added to the wort, which is nearly saturated with approximately 11 ppm of oxygen.

The fermentation process takes place in two phases. During the first phase, called the respiratory or aerobic phase, the yeast consumes the oxygen present.

$$\text{Extract} + \text{yeast} + O_2 \longrightarrow CO_2 + H_2O + \text{more yeast}$$

During this phase, which typically lasts 6 to 8 h, the yeast uses a metabolic pathway, preparing it for the anaerobic fermentation to follow. The depletion of the oxygen present signals the yeast to start anaerobically metabolizing the sugars present in the extract, releasing heat and producing CO_2 and ethanol as metabolic byproducts.

1. $$\underset{180}{C_6H_{12}O_6} \longrightarrow \underset{92}{2\,C_2H_5OH} + \underset{88}{2\,CO_2}$$

2. $$\underset{342}{C_{12}H_{22}O_{11}} + \underset{18}{H_2O} \longrightarrow \underset{184}{4\,C_2H_5OH} + \underset{176}{4\,CO_2}$$

3. $$\underset{504}{C_{18}H_{32}O_{16}} + \underset{36}{2\,H_2O} \longrightarrow \underset{276}{6\,C_2H_5OH} + \underset{264}{6\,CO_2}$$

Using these reactions, the amount of CO_2 produced can be calculated for the typical North American lager wort.

Table 1 Total Solids in Wort

Percent Solids[a]	Specific Gravity	Weight per Barrel, lb Total	Weight per Barrel, lb Solids	Specific Heat, Btu/lb·°F
0	1.0000	258.7	0.00	1.000
1	1.0039	259.7	2.60	0.993
2	1.0078	260.7	5.21	0.986
3	1.0118	261.8	7.85	0.979
4	1.0157	262.8	10.51	0.972
5	1.0197	263.8	13.19	0.965
6	1.0238	264.9	15.89	0.958
7	1.0278	265.9	18.61	0.951
8	1.0319	267.0	21.36	0.944
9	1.0360	268.0	24.12	0.937
10	1.0402	269.1	26.91	0.930
11	1.0443	270.2	29.72	0.923
12	1.0485	271.2	32.55	0.916
13	1.0528	272.4	35.41	0.909
14	1.0570	273.4	38.28	0.902
15	1.0613	274.6	41.18	0.895
16	1.0657	275.7	44.11	0.888
17	1.0700	276.7	47.06	0.881
18	1.0744	277.9	50.03	0.874
19	1.0788	279.1	53.03	0.867
20	1.0833	280.2	56.05	0.860

[a]Saccharometer readings.

For example, if the fermentable extract is 10, 43, and 12% by mass of the sugars in equations 1, 2, and 3, respectively, the theoretical yield is 0.512 lb CO_2/lb extract.

During the early fermentation phase, the yeast multiplies rapidly, then more slowly as it consumes the available sugars. Normal multiplication for the yeast is 5 to 6 times. A representative value for the heat released during fermentation is 280 Btu/lb of extract (sugar) fermented.

Wort is measured by the saccharometer (measures sugar content), which is a hydrometer calibrated to read the percentage of solids as maltose in solution with water. The standard instrument is the Plato saccharometer, and the reading is referred to as percentage of solids by saccharometer, or degrees Plato (°P). Table 1 illustrates the various data deducible from reading the saccharometer.

The same instrument is used to check the progress of the fermentation. While it still gives an accurate measure of density of the fermenting liquid, it is no longer a direct indicator of dissolved solids because the solution now contains alcohol, which is less dense than water. This saccharometer reading is called the apparent extract, which is always less than the real extract (apparent attenuation is calculated from the hydrometer reading of apparent extract and the original extract). In engineering computations, 81% of the change in apparent extract may be considered a close approximation of the change in real extract. Thus, 81% of the difference between the solids shown in Table 1 for saccharometer readings before and after fermentation would represent the mass of maltose fermented. This weight in pounds per barrel times 280 Btu/lb gives the heat of fermentation in Btu per barrel. (Each barrel has a capacity of 31 gal.) The difference between the original solids and mass of fermented solids gives the residual solids per barrel. It is assumed that there is no change in the volume because of fermentation. The specific heat of beer may be assumed to be the same as that of the original wort, but the mass per barrel is decreased according to the apparent attenuation.

Bottom fermentation-type yeast (*Saccharomyces uvarium*, formerly *carlsbergensis*) is used in fermenting lager beer. Top fermentation-type yeast (*Saccharomyces cerevisiae*) is used in making ale. They are so called because, after fermentation, one settles to the bottom and the other rises to the top. A more significant difference between the two types is that in the top fermentation type, the fermenting liquid is allowed to attain a higher temperature before a continued rise is checked. The brewing of ale may differ from the brewing of lager in the following ways:

- For ale, a more highly kilned malt is used
- Malt forms a greater proportion of the total grist (less adjunct)
- Infusion mashing is used and a wort of higher original specific gravity is generally produced
- More hops are added during the kettle boil
- A different yeast and temperature of fermentation are used

Therefore, ale may have a somewhat higher alcohol content and a fuller, more bitter flavor than lager beer. With the bottom fermentation yeasts, fermentation is generally carried out between 45 and 65 °F and most commonly between 50 and 60 °F. Ale fermentations are generally carried out at somewhat higher temperatures, often peaking in the range of 70 to 75 °F. In either type, the temperature during fermentation would continue to rise above that desired but is checked by cooling coils or attemperators through which a cooling medium such as propylene glycol, ice water, brine, or ammonia is circulated. In the past, these attemperators were manually controlled, but more recent installations are automatic.

PROCESSING

Wort Cooling

To prepare the boiling wort from the kettle for fermentation, it must first be cooled to a temperature of 45 to 55 °F. To avoid contamination with foreign organisms that would adversely affect the subsequent fermentation, this cooling must be done as quickly as possible, especially through the temperatures around 100 °F. Besides the primary function of wort cooling, other beneficial effects accrue that are essential to good fermentation, including precipitation, coagulation of proteins, and oxidation because of natural or induced aeration depending on the type of cooler used.

In the past, the Baudelot cooler was almost universally used because it is inherently easy to clean and affords the necessary aeration of the wort. However, the traditional open-type Baudelot cooler was replaced by one consisting of a series of swinging leaves encased within a removable enclosure into which sterilized air is introduced for aeration. This modified form, in turn, has virtually been replaced by the totally enclosed heat exchanger. Air for aeration is admitted under pressure into the wort steam, usually at the discharge end of the cooler. The air is first filtered and then irradiated to kill bacteria, or it can be sterilized by heating in a double-pipe heat exchanger with steam. By the injection of 0.167 ft^3 of air per barrel (bbl) of wort, which is the amount necessary to saturate the wort, a normal fermentation should result. The quantity can be increased or diminished accurately as the subsequent fermentation indicates.

The coolant section of wort coolers is usually divided into two or three sections. For the first section, a potable source of water is used. The heated effluent goes to hot water tanks where, after additional heating, it is used for succeeding masking and sparging brews. Final cooling is done in the last section, either by direct expansion of the refrigerant or by means of an intermediate coolant such as chilled water or propylene glycol. Between these two, a third section may be used from which the warm water can be recovered and stored in a wash-water tank for later use in various washing and cleaning operations around the plant.

Closed coolers save space and expensive cooler room air-conditioning equipment. They also permit a faster cooling rate and provide accurate control of the degree of aeration. The problem of cleaning plate coolers is solved by having a spare set of plates ready for weekly replacement. The cleaning between successive brews is accomplished by circulating cleaning or sterilizing solutions, or both, through the cooler.

In selecting a wort cooler, the following should be considered:

- The cooling rate should allow the contents of the kettle can to be cooled in less than 2 h.
- The heat transfer surfaces to be apportioned between the first section, using an available water supply, and the second section, using refrigeration, should be such that the most economical use is made of each of these resources. Cost of water, its temperature, and its availability should be balanced against the cost of refrigeration. Usual design practice is to cool the wort in the first section to within 10 °F of the available water.
- Usable heat should be recovered (the effluent from the first section is a good source of preheated water). After additional heating, it can be used for succeeding brews and as wash waste in other parts of the plant. At all times, the amount of heat recovered should be consistent with the overall plant heat balance.
- Meticulous sanitation and maintenance costs are important factors.

The size of the wort coolers is determined by the rate of cooling desired, the rate of water flow, and the temperature differences used. A brew, which may vary in size from 50 to 1000 bbl and over, is ordinarily cooled in less than 1 to 4 h. Open-type coolers are made in stands up to about 20 ft long. Where more length is needed, two or more stands are operated in parallel.

Open coolers are best operated with a wort flow of 10 to 11.5 bbl/h per foot of stand. As the flow is increased beyond this rate, an increasingly larger part of the wort is splashed from the top tube of the cooler and dropped directly into the collecting pan below without contacting the cooler surfaces. An increased amount of wort, which flows over the surfaces, must be subcooled to offset what has been bypassed.

In the plate-type cooler, this bypassing does not occur, and wort velocities can be increased to a point where the friction pressure through the cooler approaches the maximum design pressure of the press and gasketing. The number of passes and streams per pass afford the designer much latitude in selecting the most favorable parameters for optimum performance and economical design. This design is based on the specific heat of wort, its initial temperature and the range through which it is to be cooled, the temperature of the available water supply, and the ratio of the quantity of cooling water to wort that is to be used. Design and operating features of a typical plate cooler are as follows:

Specifications

Quantity of wort to be cooled	17,000 lb/h
Temperature of hot wort	210 °F
Temperature of cooled wort	40 °F
Temperature of available water (maximum)	70 °F
Water used, not to exceed	34,000 lb/h
Temperature of water leaving cooler	140 °F
Temperature of wort leaving first section of cooler	80 °F
Temperature of incoming recirculated chilled water	34 °F

Plate cooler (first section)

Number of plates	40
Heat transfer surface per plate	4.3 ft^2
Heat transfer surface in first section	172 ft^2
Number of passes	5
Number of streams per pass	4
Water flow rate	34,000 lb/h
Wort flow rate	17,000 lb/h

Plate cooler (second section)

Number of plates	24
Heat transfer surface per plate	4.3 ft^2
Heat transfer surface in second section	103 ft^2
Number of passes	3
Number of streams per pass	4
Chilled water flow rate	52,000 lb/h

A shell-and-tube or plate-type cooler with two stages of cooling can cool the wort efficiently. In the first (hot) stage, potable water is used counterflow to the wort, and the usual discharge temperature is about 168 °F. This hot water is then used in the following brews at various blended temperatures. Excess is used in the brewery's general operations.

The second stage of wort cooling is accomplished by a closed system of refrigerated water at about 36 °F through a closed cooler, which cools the wort to 50 °F or lower, depending on the brewer. Lower temperature water (33 °F) may be used in open units where no danger of freezing exists.

Wort cooling may be accomplished in one stage, depending on the potable water temperature available and plant refrigeration capacity. This might be a somewhat simpler and less expensive arrangement, but sewerage cost may be an important factor where water saving is mandatory. These heat exchangers have many benefits, including easier cleaning, uniform temperature control, uniform control of the sterile air injected for aeration, reduction of overall steam requirements for brewing, and a minimum of water wasted. The water consumption has been reduced from 17 to 20 barrels per barrel of beer to 6 to 10 barrels per barrel of beer.

Fermenting Cellar

After cooling, the wort is pitched with yeast and collected in a fermenting tank, where respiration and fermentation occur according to the chemical reaction previously discussed. The daily rate of fermentation varies depending on the operating procedure adopted in each plant. On the first day, a representative rate might be 2 lb of converted maltose per barrel of wort. The rise in temperature caused by fermentation and by the growth and changing physiology of the yeast increases this rate to 7 lb on the second day. By now, the maximum desired temperature has been attained, and a further rise is checked by an attemperator, so that on the third day, another 7 lb is converted. This rate continues through the fourth day. Two examples of the fermentation rate follow; one is for normal gravity brewing, and the other is for high (heavy) gravity brewing.

Example 1 Normal Gravity Brewing

Fermentation Day	°Plato	Real Extract	Extract per bbl, lb	Extract Fermented per bbl, lb
0	11	11.0	29.63	—
1	10	10.2	27.38	2.25
2	8	8.6	22.95	4.43
3	5	6.1	16.12	6.83
4	3	4.5	11.81	4.31
5	2.5	4.1	10.75	1.06
				18.88

Example 2 High Gravity Brewing

Fermentation Day	°Plato	Real Extract	Extract per bbl, lb	Extract Fermented per bbl, lb
0	16	16.0	43.97	—
1	15	15.2	41.64	2.33
2	12	12.8	34.73	6.91
3	7	8.7	23.11	11.51
4	4	6.3	16.66	6.56
5	3.5	5.9	15.58	1.08
				28.39

By now, the amount of unconverted maltose remaining in the beer is greatly diminished. Because alcohol, carbon dioxide, and other products of fermentation inhibit further yeast propagation, the action nearly stops on the fifth day, when only about 3 lb are converted. At this stage the yeast begins to flocculate (clump together) and either settles to the bottom of the fermenter (bottom yeast) or rises to the top (top yeast). Because of the reduced fermentation rate, the temperature of the beer begins to fall, either as the result of increased attemperation applied to the tank itself, heat loss from the tank to the surrounding area, or a combination of both. Many fermentation programs call for the beer to be cooled to a temperature ranging from 35 to 45 °F at this time. This period of more rapid cooling aids in settling the yeast. At the completion of this cooling period, the fermentation rate is essentially zero, and the beer is ready to be transferred off the settled yeast. Complete fermentation generally occurs in about 7 days. The introduction of new types of beers (*i.e.*, reduced calorie, reduced alcohol) and the more general use of high gravity brewing have led to the use of a variety of fermentation programs both between brewers making the same product and within the same brewery for different products.

While complete fermentation can be accomplished in less than 7 days, most brewers may take 7 to 14 days or more for the fermentation and subsequent cooling, depending on original gravity, whether a secondary fermentation is used, and available cooling capacity. Most brewers cool the beer to between 42 and 38 °F after ending fermentation or after the final days of quick cooldown in the fermenting tank. In addition, the long rest allows time for the yeast to settle. Some brewers agitate the beer in cylindrical fermenters, which enables them to ferment the beer faster and then to separate out the yeast by centrifuge. Most brewers cool the beer to the desired 29 to 45 °F temperature before it goes into storage for resting and settling between fermentation and final aging.

Fermenting Cellar Refrigeration

The agitation necessary for heat exchange between the attemperator and the beer is provided partly by convection resulting from temperature gradients in the beer. Agitation is principally by the ebullition caused by the CO_2 bubbles rising to the surface of the liquid. In estimating the heat transfer surface required, a heat transfer rate range from 15 to 30 Btu/h · ft^2 · °F is reasonable. The heat loss from tank walls and the surface of the liquid may be disregarded when calculating the attemperator coil surface requirements. However, if the room temperature is allowed to drop appreciably below 50 °F, the heat dissipated through the metal tank walls becomes important. Depending on the degree of heat dissipation, fermentation may be retarded or even inhibited. In such instances, wooden fermenting tanks or the insulation of the tank walls and bottoms are indicated so that the control of heat removal remains in the attemperator.

Refrigeration requirements are based on the maximum volume of wort being fermented, as illustrated by Example 3.

Example 3. Figure 2 illustrates the volume of wort production based on a 500 bbl/day production rate. The days are represented by the abscissa, and the pounds of solids converted per day, by the ordinate. The individual brews in fermentation on any particular day are additive. For example, on the fifth day, Brew No. 1 is finishing with a conversion rate of 3 lb for that day; Brew No. 5, which is just beginning the fermentation cycle, is fermenting at the rate of 2 lb; and Brews No. 2, 3, and 4 are each at the maximum rate of 7 lb/bbl per day. The total solids fermented on this day is 26 lb/bbl for the 2500 bbl in fermentation, totalling 13,000 lb of solids converted per day. Since the heat of fermentation is 280 Btu/lb, the refrigeration load is

$$(13,000 \times 280)/(24 \times 12,000) = 12.6 \text{ tons}$$

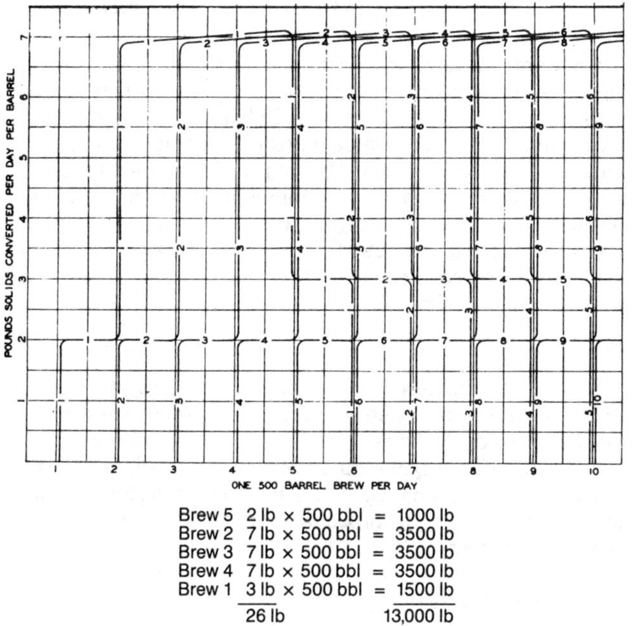

Fig. 2 Solids Conversion Rate

Calculations for sizing of attemperators must consider (1) the internal dimensions of the fermenting tank and its capacity; (2) the temperature difference between the coolant and the fermenting beer; (3) the maximum daily sugar conversion rate; and (4) the heat evolved, which is at the rate of 280 Btu/lb of fermentable sugar converted.

Assuming a square fermenting tank 13 ft per side to hold a brew of 500 bbl, and allowing 1 ft between the tank wall and the attemperator for easy cleaning, an 11 ft square attemperator can be used, giving 44 perimetrical feet of tubing.

From Figure 2, the maximum daily conversion rate is shown to be 7 lb/bbl. Calculating for 500 bbl per day at 280 Btu/lb of sugar converted

$$7 \times 500 \times 280 = 980{,}000 \text{ Btu/day or } 40{,}833 \text{ Btu/h}$$

Then, assuming a 50 °F fermenting beer temperature and a 20 °F brine temperature, a temperature difference of 30 °F, and a heat transfer rate of 15 Btu/h · ft² · °F for the attemperator, the number of square feet of surface required would be:

$$40{,}833/(15 \times 30) = 90.7 \text{ ft}^2$$

Considering 4-in. OD tubing with an external area of 1.05 ft²/ft, the number of lineal feet required would be:

$$90.7/1.05 = 86.4 \text{ ft}$$

Two attemperators each 11-ft square would give 88 ft of tubing, which is adequate for the conditions outlined.

The amount of CO_2 generated per barrel of beer fermented depends on the original gravity of the beer and the degree of fermentation. A little more than half of the generated CO_2 is collected and liquefied for later use in the brewery for carbonating; for counterpressure in transfer operations; and for the bottle, can, and keg lines.

Adequate fresh air must be provided in an open-tank fermenting cellar to safely dilute the CO_2 emanating from the fermenters. The amount permissible should be such that there is no health hazard to the employees working in these spaces. Concentrations below 0.5% are considered safe. Increasing amounts reduce a worker's efficiency, and 4% concentrations make the performance of work for protracted periods untenable. Since it is heavier than air, CO_2 tends to settle to the floor, from where it may be withdrawn by scupper connections at the floor level. Fresh air may be introduced through the ceiling, the openings being located so as not to remove or disturb the layer of protective CO_2 gas on top of the fermenting beer.

Old attemperators usually consisted of one or more rings of 3 or 4 in. copper or stainless tubing, concentric with the walls of the tank and supported at about two-thirds of the height of the liquid. Later designs consist of a partial exterior jacket at about the same height as the old coils. The jacket is a multipass of three or more baffled passes, or a dimple plate design, providing good flow and heat transfer. A glycol solution or liquid ammonia may be circulated through the cooling jackets. These tank changes, dictated by automation and economics, allow easier in-place cleaning of tanks and provide more cooling effect in fermenting.

Stock Cellar

The stock cellar is a refrigerated room containing tanks into which the cooled beer from the fermenter is transferred for the purpose of aging and settling. The period of retention of beer in the stock cellar is the brewer's discretion and may be one week to several months. The beer is stored in the presence of some yeast remaining from the original fermentation. Under these conditions, slow, subtle, but important changes take place and contribute to the flavor characteristic of the beer, including the coagulation of protein, which might produce haze in the finished product.

While the CO_2 pollution of air in the stock cellar is far less than in the fermenting room, adequate provision must be made to supply fresh air in sufficient amounts to keep the CO_2 concentration below 0.5%. Air-conditioning equipment using chemical dehumidification and refrigeration is generally used to maintain dry conditions such as 32 °F and 50% rh in storage areas. This decreases mold growth and rusting of steel girders and other steel structures. To maintain lower CO_2 concentration in tightly closed cellars and to reduce operational cost, heat exchange sinks and thermal wheels are used to cool incoming fresh air and exhaust cold stale air.

Air compressor systems commonly use air driers with refrigerated aftercoolers, 32 °F glycol coolers, desiccant drying, or a combination. This is necessary if lines pass through areas below 32 °F.

A continuous aging process used in multistory buildings, all gravity flow, is shown in Figure 3. The process is better for larger operations that principally produce one brand of beer.

Kraeusen Cellar

Instead of carbonating the beer during the finishing step, some brewers prefer to carbonate by the Kraeusen method. In this

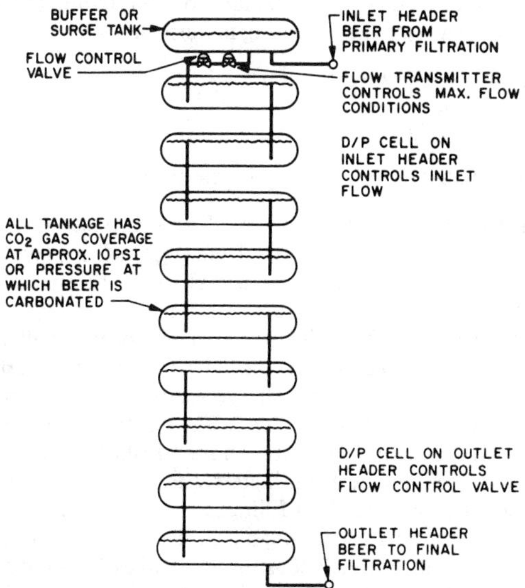

Fig. 3 Continuous Aging Gravity Flow

procedure, the fully fermented beer is moved from the fermenting tank to a tank capable of holding about 20 psig and a small percentage of actively fermenting beer is added. The tank is allowed to vent freely for 24 to 48 h, then it is closed and the CO_2 pressure is allowed to build. Since the amount of CO_2 retained in the beer is a function of temperature and pressure, the brewer can achieve the desired carbonation level by controlling either or both pressure and temperature. At the conclusion of the Kraeusen fermentation, generally a week or more, the beer may be moved to another storage tank. However, the brewer can accomplish the same effect by leaving the beer in the Kraeusen tank and cooling the beer either by space cooling, tank coils, or both.

Heat is generated by this secondary fermentation, but the temperature of the liquid does not rise as high as it did in the fermenter because the fermentable sugars are available only from the small percentage of actively fermenting beer, added as Kraeusen. Furthermore, the bulk of the liquid may have a lower starting temperature than in primary fermentation. Typically, a temperature of 40 to 50°F may be reached at the peak, after which the liquid cools to the ambient temperature of the room. This cooling can be accelerated in the tanks by means of attemperators through which a cooling liquid, such as propylene glycol, is circulated. Since heat is generated during the Kraeusen fermentation, refrigeration load calculations must include removal of this heat by transfer to the air in the cellar, by means of tank coils, or by a combination of both. Furthermore, if the tank is to be used as a storage tank, the calculation must include the necessary heat removal to reduce the beer temperature to the desired level.

Finishing Operations

After flavor maturation and clarification in the storage tanks, the beer is ready for finishing. Finishing includes the processes of carbonation, stabilization, standardization, and clarification.

Carbonation. Any of the following processes are used to raise the CO_2 concentration from 1.2 to 1.7 volumes/volume to about 2.7 volumes/volume:

- Kraeusen
- In-line
- In-tank with stones
- Saturator
- Aging train

Stabilization. The formation of colloidal haze caused by soluble proteins and tannins forming insoluble protein-tannin complexes is reduced by any of the following materials:

- Enzymes (papain)
- Tannic acid
- Tannin absorbents,
- Protein absorbents, silica gel, bentonite

Standardization. Chilled, deaerated, and carbonated water are added to adjust original gravity from high-gravity level (14 to 16°Brix) down to normal package levels (10 to 12°Brix or lower for low-calorie beer).

Clarification. In the finishing cellar, the beer is polished by filtration and is then carbonated by any of several methods. The filtering usually is done through a series of cellulose pulp filters and/or diatomaceous earth filters. The number of filters used depends on the brilliance desired in the finished product. After this final processing, beer is transferred to the government cellar and held until it is needed for filling kegs in the racking room, or bottles or cans in the packaging plant. In some breweries, initial clarification is accomplished using centrifuges. This reduces the load on the filtration system, allowing higher flow rates and longer filter runs.

OUTDOOR STORAGE TANKS

Some breweries are using vertical outdoor fermenting and holding tanks (similar to those popular with dairies). These tanks have working capacities of 2000 to 10,000 barrels. The geometry of these tanks includes a conical bottom and height-to-diameter ratios from 1:1 to 5:1. The tanks are jacketed and use propylene glycol or the direct expansion of ammonia for cooling. Insulation is usually 4 to 6 in. thick polyurethane foam with an aluminum cladding. They may be built as fermenters, as aging tanks, or in many cases, fermenting and aging are completed in the same tank with no beer transfer.

Hop Storage

Hops should be stored at a temperature of 32 to 34°F with 55 to 65% rh and with very little air motion to prevent excessive drying. Sweating of the bales should not be permitted because this would carry off the light aromatic esters and deteriorate the fine hop character. Nothing else should be stored in the hops cellar as foreign odors may be absorbed by the hops, which results in off-flavors in the beer.

Yeast Culture Room

In the yeast culture room, yeast is propagated to be used in reseeding and replacing yeast that has lost its viability. Normal fermentation of aerated wort also propagates yeast. The amount of yeast will roughly triple during fermentation depending on the degree of aeration. A portion of this yeast is repitched in later fermentation, and the balance is discarded as waste yeast (sometimes sold for other purposes). Clean yeast, usually the middle layer of the yeast deposit remaining in a fermenting tank after removal of the beer, is selected for repitching.

Repitched yeast is carefully handled to avoid contamination with bacteria and is stored in the yeast room as a liquid slurry (yeast balm) in suitable vats. If open vats are used, 80% rh is required to prevent the yeast from hardening on the vat walls. The CO_2 blanket on top of the vats should not be disturbed by excessive air motion. There is considerable variation in yeast handling and recycling practices.

PASTEURIZATION

Plate pasteurizers heat the beer to a temperature sufficient for proper pasteurization (15 s at 160°F or 10 s at 165°F) and then cool the pasteurized product with incoming cold beer. Plate pasteurizers and microfiltration are used to produce a beer that is similar to draft beer but does not require refrigeration to prevent spoilage. It is distributed in bottles and cans that can be of slightly lighter construction since they do not have to withstand the high pasteurizing pressure created in tunnel-type pasteurizers.

CARBON DIOXIDE

CO_2 Collection

Carbon dioxide gas, produced as a by-product of fermentation, can be collected from closed fermenters, compressed, and stored in pressure tanks for later use. It may be used for final carbonation, counterpressure in storage and finishing tanks, transfer, and the bottling and canning of the product. In the past, the CO_2 was stored in the gaseous state at about 250 psig. However, in most medium and large breweries, the gas is collected and, after thorough washing and purification, it is liquefied and stored. Carbon dioxide stored in the liquid state would occupy about 2% of the volume of an equal mass of gas at the same pressure at room temperature.

As an example, from each barrel of wort fermented, about 13 lb of CO_2 are generated over a period of five days, though not at a constant daily rate. Therefore, brews must be carefully scheduled to provide the necessary CO_2 gas, thereby minimizing storage requirements. As a general rule, only about 50 to 60% of the total gas generated is collected. The gas generated at the beginning and the end of the fermentation cycle is unsuitable and is discarded because of excessive air content and other impurities.

From the fermenting tank, the gas is piped through a foam trap to a gas pressure booster. Surplus gas is discharged to the outside from a water-column safety relief tank, which also protects the fermenting tank from excessive gas pressure. To compensate for friction pressure loss in the long lines to the compressor and to increase its capacity, the booster raises the pressure from as low as 1 in. of water to 5 or 6 psig.

Compressors, which in the past were of the two-stage type with water injection, are being replaced by nonlubricated compressors using carbon or nonstick fluorocarbon rings. Today, lubricated screw and reciprocating compressors are used for food and beverage-grade CO_2 production in commercial CO_2 plants. These may be single two-stage compressors or two compressors comprising individual high and low stages. A complete collection system consists of suction and foam trap; rotary boosters, where required; scrubber; deodorizer; compressor (or compressors); intercoolers and aftercoolers; dehumidifying tower; condenser (with refrigeration from plant cascade system or separate compressor); liquid storage tanks; and vaporizers, all interconnected and automatically controlled.

Liquefaction

From tables of the properties of CO_2, the condensing pressures at several temperatures are:

−20 °F	200 psig
−14 °F	225 psig
−8 °F	252 psig

The latent heat at saturation temperature is about 120 Btu/lb. The refrigerant for liquefying the compressed gas should be about −21 °F to effectively condense the CO_2. Most of the moisture must be removed from the compressed gas; this may be done by passing the gas through a horizontal-flow finned coil located in a 36 °F cellar to condense out about 80% of the moisture, *i.e.*, the condensate being drained from the system. Also, this is done effectively with refrigerant-cooled precoolers, intercoolers, and aftercoolers. Further removal is effected by passing the gas through desiccant driers. The emerging gas has a slightly higher temperature, but has a dew point around −70 °F or lower.

Under these conditions, the gas is liquefied when it comes in contact with the liquefying surfaces, which stay ice-free because of the low moisture content (−40 °F dew point) of the gas and thus assure continuous service. The driers are installed in duplicate, with automatic timing for regeneration of the desiccant material. Desiccant driers usually rely on liquid CO_2 as the refrigeration gas. An earlier method used dual sets of double-pipe driers in which moisture was frozen out and retained in the heat exchanger.

Liquefiers are vertical shell-and-tube, inclined double-pipe, or shell-and-tube types. The refrigerant side is operated fully flooded, with the refrigerant being supplied from the main plant and a booster compressor discharging into the plant suction main. Carbonating systems have undergone changes with all-closed fermenters, refrigerated condensing systems, and large liquid CO_2 holding tanks. See Figures 4 and 5 for collecting and liquefaction system flow diagrams.

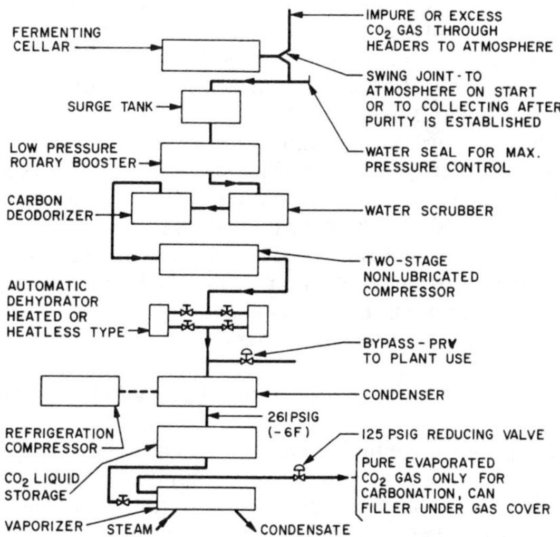

Fig. 4 Typical Arrangement of CO_2 Collecting System

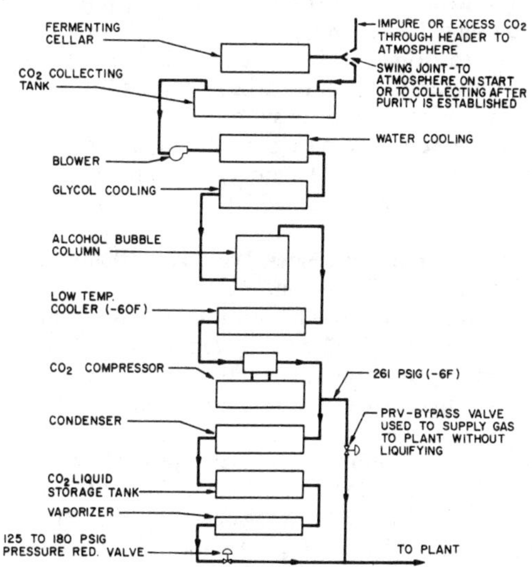

Fig. 5 Special CO_2 Collecting and Liquefaction System

CO_2 Storage and Reevaporation

The condensed CO_2 drains into a storage tank, which is usually designed for a working pressure of 300 psig and varying storage capacities of 10,000 to 120,000 lb each. The vessel is enclosed in an insulated box and is equipped with equalizing connections, safety valves, liquid-level indicator or try cocks, and electric heating units. Tests are regularly conducted for gas purity from samples withdrawn from above the liquid level.

As liquid is withdrawn from the tank, it is introduced to a steam-heated liquid evaporator, which is automatically controlled to give the desired superheat to the reevaporated gas. This type of storage tank is gradually replacing other types because of its ability to control the temperature of the reevaporated gas closely. Then the evaporated gas is stored temporarily in buffer tanks until it is needed in the brewhouse and package plant.

HEAT BALANCE

Most of the steam required for processing, heating water, and general plant heating can be obtained as a by-product. Because the manufacture of beer is a batch process with various peaks occurring at different times, the study of the best heat balance possible is difficult. In a given plant, it depends on many variables, and a comprehensive study of all factors is most important.

In brewery plant locations where electric energy costs are high, the installation of cogeneration facilities can be favorable. However, in plants that produce in excess of 3-million bbl annually, the steam turbine as a prime mover often comes into prominence. A bleeder-type steam turbine operating at 400 psig can be used to drive a refrigeration compressor, electrical generator, or both; steam bled from it can be used for process and other needs requiring low-pressure steam. In smaller plants, less favorable heat balances must be accepted in line with more economical plant investment programs. Each brewery requires individual study to procure the most economical program.

COMMON REFRIGERATION SYSTEMS

Absorption Machine for Heat Balance (especially for air conditioning and water cooling for wort cooling). The unit requires a careful heat balance study to determine if it is economical.

Halocarbon Refrigerant Cascade System. Oil-free ammonia as brine can be pumped at a 4:1 ratio. The system requires pumps, but more often is expanded on a 1:1 ratio if connected with a compression system. The ratio of motive power to refrigeration at 25 and 185 psig is 1.5 bhp/ton.

Direct Centrifugal. This is often an oil-free ammonia system with a probable 1:1 ratio, or it can be recirculated. The unit requires pumps or a pressure transfer system.

Oil-Sealed Screw Compressors. Ammonia is circulated at a 1:1 ratio in this system. The units require pumps or a pressure transfer system. The ratio of motive power to refrigeration is 1.38 bhp/ton.

Oil-Free Compressors, Screw Type. The ratio of motive power to refrigeration is 1.5 bhp/ton.

Large Balanced Opposed Horizontal Double-Acting Reciprocating Compressor. While the system is not oil-free, good oil separation equipment minimizes this problem. It can use recirculation or direct expansion. The ratio of motive power to refrigeration is 1.25 bhp/ton.

Further automation has been accomplished by programming the flow of materials in the brewhouse, as well as the entire brewing operation. The newest brewing operations are fully automated.

Where necessary, a cooling tower may be used to reduce thermal pollution or to conserve water in the pasteurizing phase. Ecology plays an important part in the brewery; stacks are monitored for particulates, effluent is checked, and heat from the kettle vents and others is recovered. Water usage is more closely regulated, and water-saving equipment, including evaporative condensers, is used in the refrigeration systems.

VINEGAR PRODUCTION

Vinegar is produced from any liquid capable of first being converted to alcohol (such as wine, cider, and malt) and syrups, glucose, molasses, and the like.

First Stage: Conversion of sugar to alcohol by yeast (anaerobic)

$$C_6H_{12}O_6 \longrightarrow 2CH_3CH_2OH + 2CO_2$$

Second Stage: Conversion of alcohol to acetic acid by bacterial action (aerobic)

$$CH_3CH_2OH + O_2 \longrightarrow CH_3COOH + H_2O$$

Bacteria are active only at the surface of the liquid where air is available. Two methods are used to increase the air-to-vinegar surface:

The *packed generator* or Frings generator is a vertical cylinder with a perforated plate and filled with oak shavings or other inert support material intended to increase column surface area. The weak alcohol and the vinegar culture are introduced, and the solution is continuously circulated through a sparger arm, with air being introduced through drilled holes in the top of the tank. A heat exchanger is used to remove the heat generated and to maintain the solution at 86°F. This is a batch process requiring 72 h.

In the *submerged fermentation process*, the air is distributed to the bacteria by continuously disbursing air bubbles through the mash in a tank that is practically filled with cooling coils to maintain 86°F. This also is a batch process requiring 39 h.

Concentration is best accomplished by removing some of the water in the form of ice, which achieves a 12 to 40% increase in the concentration of the acid. In freezing out the water, a rotator is often used. About 0 to 10°F is required on the evaporating surface to produce the best crystals; then the ice is separated in a centrifuge. The vinegar is then stored 30 days before filtering. Effective concentration can also be achieved by distillation (as is done for distilled white vinegar).

WINE MAKING

The use of refrigeration to control the rates of various physical, chemical, enzymatic, and microbiological reactions in commercial wine making is well established. Periods at elevated temperatures, followed by rapid cooling, can be used to denature oxidative enzymes and proteins in grape juices, to retain desirable volatile constituents of the grapes, to enhance the extraction of color pigments from the skins of red grapes, to modify the aroma of the juices from certain white grape cultivars, and to inactivate the fungal populations of mold-infected grapes. Reduced temperatures can slow the growth rate of natural yeast and of the enzymatic oxidation of certain phenolic compounds, assist in the natural settling of grape solids in juices, and favor the formation of certain by-products during fermentation. Also, reduced temperatures can be used to enhance the nucleation and crystallization of potassium bitartrate from wines, to slow the rate of aging reactions during storage, and to promote the precipitation of wood extractives of limited solubility from aged brandies.

The extent to which refrigeration is used in these applications depends on such factors as the climatic region in which the grapes have been grown, the grape cultivars used, the physical condition of the fruit at harvest, the styles and types of wines being produced, and the discretion of the winemaker.

Presently, the wine industry in the United States is heavily committed to the production of table wines (ethanol content less than 14% by volume), with considerable less emphasis on the production of dessert wines and brandies than in the past. Additionally, the recent growth in wine cooler popularity has significantly altered winery operations where they are produced. A variety of enological practices and winery equipment can be found between the batch emphasis of small wineries (crushing tens of tons per season) and the continuous emphasis of large wineries (crushing hundreds or thousands of tons per season).

The applications of refrigeration will be classified and considered in the following order:

1. Must cooling
2. Heat treatment of red musts
3. Juice cooling
4. Heat treatment of juices
5. Control of fermentation temperature
6. Potassium bitartrate crystallization
7. Control of storage temperatures
8. Chill-proofing of brandies

MUST COOLING

Must cooling is the cooling of crushed grapes prior to the separation of the juice from the skins and the seeds. White wine grape musts will often be cooled prior to being introduced to a juice-draining system or a skin-contacting tank; this is done to reduce the rate of oxidation of certain juice components, as well as to prevent the onset of spontaneous fermentation by wild and potentially undesirable organisms. Must cooling can be used when grapes are delivered to the winery at excessively high temperatures or when they have been heated to aid in pressing or extracting red color pigments.

In general, tube-in-tube or spiral heat exchangers are used for this application. Tubes of at least 4-in. internal diameter with detachable end sections of large-radius return bends are necessary to reduce the possibility of blockage by any stems that might be left in the must after the crushing-destemming operation. The cooling medium can be chilled water, a glycol solution, or a directly expanding refrigerant. Overall heat transfer coefficients for must cooling range between 70 and 125 Btu/h · ft^2 · °F, depending on the proportions of juice and skins, with the must side providing the controlling resistance. In small wineries, jacketed draining tanks and fermenters are often used to cool musts in a relatively inefficient batch procedure in which the overall coefficients are on the order of 1.70 to 5.30 Btu/h · ft^2 · °F because the must is stationary and, therefore, rate controlling.

HEAT TREATMENT OF RED MUSTS

Most red grapes have white or greenish-white flesh (pulp) and juice. The coloring matter or pigments (anthocyanins) reside in the skins. Color can be rapidly extracted from these varieties by heat treating the musts so that the pigment-containing cells are disrupted prior to actual fermentation. This is done in several countries throughout the world when grape skins are low in color or when color extraction during fermentation is poor. Heating the must is necessary to produce the desirable flavor profile of some varieties, most notably Concord, in the manufacture of juices, jellies, and wines. The series of operations is referred to as thermovinification. The must is heated to temperatures in the range 135 to 167 °F, generally by draining off some of the juice, condensing steam, and returning the hot juice to the skins for a given contact time, often 30 min. The complete must can be cooled prior to separation and pressing, or the colored juice can be drawn off and cooled prior to the fermentation.

Heat treatment of red grapes can also be used to inactivate the more active oxidative enzymes found in red grapes infected by the mold *Botrytis cinerea*, or to aid the action of pectic enzyme preparations added to facilitate the pressing of some cultivars. In all cases, the temperature-time pattern used is a compromise between desirable and undesirable reactions. The two most undesirable reactions are caramelization and accelerated oxidation of the juice. Condensing steam and tube-in-tube exchangers are generally used for these applications, with design coefficients similar to those given previously for must cooling.

JUICE COOLING

Juices separated from the skins of white grapes are usually cooled to between 35 and 70 °F to aid in the natural settling of suspended grape solids, to retain volatile components in the juice, and to prepare it for a cool fermentation. Tube-in-tube, shell-and-tube, and spiral exchangers and small jacketed tanks are used either with direct expansion refrigerant, propylene glycol solution, or chilled water as the cooling medium. Overall coefficients for juice cooling range between 95 and 150 Btu/h · ft^2 · °F for the exchanger and 4.40 to 8.80 Btu/h · ft^2 · °F for small jacketed tanks. The small jacketed tank values can be improved significantly by juice agitation. Transport and thermal properties of 24% (by mass) sucrose solutions can be used for grape juices. There is a general tendency for medium and large wineries to use continuous-flow juice cooling arrangements of jacketed tanks.

HEAT TREATMENT OF JUICES

Juices from sound grapes can be exposed to a high-temperature, short-time (HTST) treatment to denature grape proteins, reduce the number of unwanted microorganisms, and, some winemakers believe, enhance the varietal aroma of certain juices. The denaturation of proteins reduces the need for their removal from the finished wine by absorptive clays such as bentonite. However, turbid juices and wines can result from this treatment, presumably because of modification of the pectin and polysaccharide fraction. Clarified juices from mold-infected grapes can be treated in a similar way to denature oxidative enzymes and to inactivate the molds. Most wineries in the United States rely on "pure culture" fermentation to achieve consistently desirable results, hence the value of HTST treatment in the control of microbial populations.

A typical program would include rapidly raising the juice temperature to 194 °F, holding it for 2 s, and rapidly cooling it to 60 °F. Plate heat exchangers are used for this treatment because of their thin film paths and high overall coefficients. Grape pulp and seeds can cause problems in this equipment if they are not completely removed beforehand.

FERMENTATION TEMPERATURE CONTROL

The anaerobic conversion of grape sugars to ethanol and carbon dioxide by yeast cells is exothermic, although the yeast is capturing a significant quantity of the overall energy change in the form of high-energy phosphate bonds. Experimental values of the heat of reaction range between 79.3 and 95.3 Btu/mole, with the value of 94.4 Btu/mole being generally accepted for fermentation calculations (Bouffard 1985).

One gallon of juice at 16.37 lb/ft^3 sucrose (24 °Brix) will produce approximately 60 gal of carbon dioxide during fermentation. Allowing for the enthalpy lost by this gas, with its saturation levels of water and ethanol vapors, the corrected heat release value is 90.9 Btu/mole at 59 °F and 83.6 Btu/mole at 77 °F. The adiabatic temperature rise of the 16.37 lb/ft^3 juice would then be 87 °F, based on the 59 °F value, and 82 °F, based on the 77 °F value. Whether or not a fermentation will approach these adiabatic conditions depends on the difference between the rate of heat generation by fermentation and the rate of heat removal by the cooling system. For constant temperature fermentations, which are the most common type of temperature control practiced, these rates must be equal. Red wine fermentations are generally controlled at temperatures between 75 and 90 °F, while white wines are fermented at 50, 60, or 68 °F, depending on the cultivar and the type of wine being produced. The more rapid fermentations of red wines are used in the cooling load calculations of individual fermenters; a more involved composite calculation, allowing for both red and white fermentations staggered in time, is necessary for the overall daily fermentation loads.

At 68 °F, red wines have average fermentation rates in the range of 2.5 to 3.1 lb/ft^3 per day, which correspond to heat release rates of approximately 21 to 26 Btu/ft^3 per hour. The peak fermentation rate is generally 1.5 times the average, leading to values of 31.5 to 39.0 Btu/ft^3 per hour. This value, multiplied by the volume of must fermenting, provides the maximum rate of heat generation. The heat transfer area of the jacket or external exchanger can then be calculated from the average coolant temperature and the overall heat transfer coefficient.

The largest volume of a fermenter of given proportions, whose fermentation can be controlled by jacket cooling alone, is a function of the maximum fermentation rate and the coolant temperature. The limitation occurs because the volume (and hence the heat generation rate) increases with the diameter cubed, while the jacket area (and hence the cooling rate) only increases with the diameter squared. Similarly, the temperature rise in small fermenters cooled only by ambient air depends on the fermenter's volume and shape and the ambient air temperature (Boulton 1979a).

The development of a kinetic model for wine fermentations (Boulton 1979b) has made it possible to predict the daily or hourly cooling requirements of a winery. The many different fermentation temperatures, volumes, and starting times can now be incorporated into algorithms for the optimal control of refrigeration compressors by the prediction of future demands and the scheduling of off-peak electricity usage.

POTASSIUM BITARTRATE CRYSTALLIZATION

Freshly pressed grape juices are usually saturated solutions of potassium bitartrate. The solubility of this salt decreases as alcohol accumulates so that newly fermented grape wines are generally supersaturated. The extent of supersaturation and even the solubility of this salt depends on the type of wine. Young red wines can hold almost twice the potassium content at the same tartaric acid level as young white wines, with other effects due to pH and ethanol concentration. Since salt solubility also decreases with temperature, wines will usually be cold stabilized so that the crystallization occurs in the tank rather than in the bottle, if the finished wine is chilled. In the past, this was done by holding the wine as close to its congealing temperature (usually 25 to 21 °F for table wines) for two to three weeks. The crystallization at these temperatures can be increased dramatically by the introduction of nuclei, either potassium bitartrate powder or other neutral particles, and subsequent agitation. In modern wineries, it is particularly important to supply nuclei for crystallization, since stainless steel tanks do not offer convenient sites for rapid growth as do older wooden cooperage. The holding times can be reduced to between 1 and 4 h by these methods. Several continuous and semicontinuous processes have been developed, most incorporating an interchange of the cold exit stream with the warmer incoming wine (Riese and Boulton 1980). Dessert wines can be stabilized in the same manner, except that the congealing temperatures are usually in the range of 12 to 7 °F. It is usual for the stabilization to take place some time after the harvest period and for the suction temperatures of the refrigeration compressors to be adjusted in favor of low coolant temperatures rather than refrigeration capacity.

STORAGE TEMPERATURE CONTROL

The control of storage temperature is perhaps the most important aspect of the postfermentation handling of wines, particularly generic white wines. The transfer of wine from a fermenter to a storage vessel generally results in at least a partial saturation with oxygen. The rates at which oxidative browning reactions (and the associated development of acetaldehyde) advance depend on the wine, its pH and free sulfur dioxide level, and its storage temperature. Berg and Akiyoshi (1956) indicate that in the oxidation of white wines, for temperatures below ambient, the rate was reduced to one-fifth its value for each 18 °F reduction in temperature. Similar studies of the hydrolysis of carboxylic esters (Ramey and Ough 1980) produced during low-temperature fermentations indicate that the rate was more than halved for each 18 °F reduction in temperature. These latter data suggest that on average the esters have half-lives of 380, 600, and 940 days when wine is stored at 60, 50, and 40 °F, respectively. As a result, wines should generally be stored at temperatures between 40 and 50 °F if oxidation and ester hydrolysis are to be reduced to acceptable levels. The importance of cold bulk wine storage will likely increase as vintners strive to reduce the amount of sulfur dioxide used to control undesirable yeasts.

The cooling requirements during storage can easily be calculated from the dimensions and materials of construction of the vessels and the thickness of the insulation used.

CHILL-PROOFING BRANDIES

In the production of brandies, refrigeration is used in the chill-proofing step, just prior to bottling. When the proof is in the range of 100 to 120 proof (50 to 60% v/v ethanol), aged brandies will contain polysaccharide fractions extracted from the wood of the barrels. When the proof is reduced to 80 proof for bottling, some of these components with limited solubility will become unstable and precipitate, often as a dispersed haze. These components are removed by rapidly chilling the diluted brandy to a temperature in the 0 to 32 °F range and filtering while cold. This is generally done by using a plate heat exchanger and a pad filter. The outgoing filtered brandy is then used to precool the incoming stream in order to reduce the cooling load. Calculations can be made by using the properties of equivalent ethanol-water mixtures, with particular attention to the viscosity effects on the heat transfer coefficient.

CARBONATED BEVERAGES

Refrigeration equipment is used in many carbonated beverage plants. The refrigeration load varies with plant and production conditions; small plants may use 150 tons of refrigeration and large plants may require over 1500 tons.

Dependency on refrigeration equipment has diminished in new carbonated beverage plants using the latest deaerating, carbonating, and high-speed beverage container-filling equipment. During the past 20 years, new technology and equipment design have been introduced to improve water deaeration, as well as new production machinery approaches for container-filling operations. These developments have diminished the need for refrigeration primarily used to precool product water essential in many earlier beverage plants with vintage filling equipment.

In facilities that use refrigeration, the product water is often deaerated before cooling to aid carbonation. In addition, cooling the product at this stage of production (1) facilitates carbonation to obtain maximum stability of the carbonated beverage during filling (to reduce foaming), (2) permits reducing the pressure at which the beverage is filled into the container (minimizing glass bottle breakage at filler), and (3) reduces overall filling equipment size and investment.

Immediately before filling operations, beverage product preparation requires the use of equipment for proportioning, mixing, and carbonation so the finished beverage has the proper release of carbon dioxide gas when it is served. The equipment for these functions is frequently found as an integrated apparatus (device), frequently called a mixer-carbonator or a proportionator.

Table 2 lists the volume of carbon dioxide dissolved per volume of water at various temperatures. Table 2 indicates that at 60 °F and atmospheric pressure, a given volume of product water will absorb an equal volume of carbon dioxide gas. If the carbon dioxide gas is supplied to the product water under a pressure of approximately 15 psig (interpolating between 10 and 20 psig), it will absorb two

Table 2 Volume of CO_2 Gas Absorbed in One Volume of Water

Temperature °F	Pressure in Bottle, psig 0	10	20	30	40	50	60	70	80	90	100
32	1.71	2.9	4.0	5.2	6.3	7.4	8.6	9.7	10.9	12.2	13.4
40	1.45	2.4	3.4	4.3	5.3	6.3	7.3	8.3	9.2	10.3	11.3
50	1.19	2.0	2.8	3.6	4.4	5.2	6.0	6.8	7.6	8.5	9.5
60	1.00	1.7	2.3	3.0	3.7	4.3	5.0	5.7	6.3	7.1	7.8
70	0.85	1.4	2.0	2.5	3.1	3.7	4.2	4.8	5.4	6.1	6.6
80	0.73	1.2	1.7	2.2	2.7	3.2	3.6	4.1	4.6	5.2	5.7
90	0.63	1.0	1.5	1.9	2.3	2.7	3.2	3.6	4.0	4.5	4.9
100	0.56	0.9	1.3	1.7	2.0	2.4	2.8	3.2	3.5	3.9	4.3

volumes. For each additional 15 psig, one additional volume of gas is absorbed by the water. Reducing the temperature of the product water to 32 °F increases the absorption rate to 1.7 volumes. Therefore, at 32 °F product temperature, each increase of 15 psig in CO_2 pressure results in the absorption of an additional 1.7 volumes instead of one volume as when the product water temperature is 60 °F. Carbonated levels for different products vary from less than 2.0 volumes to around 5.0 volumes.

BEVERAGE AND WATER COOLERS

The main sanitation requirements for beverage and/or water coolers are hygiene and ease of cleaning, particularly if the beverage is cooled rather than the water. The key point is that water freezes easily, and the design of cooler equipment needs to avoid this. The early Baudelot tank solved the problem by not forming ice; however, since sanitation of such systems is a problem, their use is not recommended.

If a cooler is needed, most plants choose plate heat exchangers and careful control of temperature. Plate heat exchangers reduce the tendency of ice formation through high turbulence, which reduces thermal gradients. Furthermore, they are hygienic and easy to clean. Such heat exchange devices are normally fed by a brine (not direct refrigeration) and are protected against brine leakage, for example, by ensuring that the brine pressure is lower than the beverage product/water pressure.

In many existing beverage plants, coolers with patented direct-expansion refrigeration equipment are used to achieve system security, hygiene, ease of cleaning, etc., using a Baudelot-type system. However, this equipment is used only for water cooling (not product), making it easier to clean. This is achieved by the equipment manufacturer's long experience with a proprietary system and its attention to detailed equipment design.

When coolers are necessary, it is recommended that this component of the refrigeration system be located adjacent to or integrated with the proportioner-mixer-carbonator. Usually these devices are physically positioned next to the beverage container filler. Normally, the refrigeration plant itself should be housed separately from the product processing and filling areas, preferably located together with the other plant utilities (boilers, hot water heater, air compressors, etc.).

It is most important to keep the beverage free from contamination by foreign substances or organisms picked up from the atmosphere or metals dissolved in transit. Consequently, coolers are designed for ease of cleaning and freedom from water stagnation. The coolers and product water piping are fabricated of corrosion-resistant nontoxic metal (preferably stainless steel); however, certain plastics are usable. For example, acrylonitrile-butadiene-styrene (ABS) is used in the beverage industry for raw water piping.

Refrigeration Plant

Halogenated hydrocarbons or ammonia refrigerants are commonly used for those plants requiring beverage product and/or water coolers. Refrigeration compressors vary from the two-cylinder vertical units to the larger, multicylinder V-style compressors.

The refrigeration plant should be centralized in larger production facilities. With the multiplicity of product sizes, production speed variations, and other factors affecting the refrigeration load, an automatically controlled central plant conserves energy, reduces electrical energy costs, and improves opportunities for a preventive maintenance program.

Makeup water and electrical energy costs have encouraged careful selection and use of compressors, air-cooled condensers, evaporative condensers, and cooling towers. In some plants, increased economy has been gained by using the spent water from empty can and bottle rinsers as makeup water for the evaporative condensers and cooling towers.

As indicated earlier, the temperature to which the product must be cooled depends on the type of filling machinery used, as well as the deaerating-mixing-proportioning-carbonating equipment used. These cooling needs may be divided into three general categories: (1) those that use water at supply temperature or less, (2) those that operate with water at 45 to 55 °F, and (3) those that require water at 40 °F or lower. The exact temperature to which the product should be cooled depends on the specific requirements of the beverage product and the needs of the particular plant. These requirements are primarily based on product preparation, production equipment availability, and capital costs versus operating costs.

Refrigeration Load

The refrigeration load is determined by the amount of water being cooled per unit of time. This is derived from the maximum fluid output of the beverage filler. Most cooling units are of the instantaneous type; they must furnish the desired output of cold water continuously, without relying on storage reserve.

Knowing the water temperature from the supply source, the temperature to which the water is to be cooled, and the water demand, the refrigeration load can be determined by

$$q_R = Qc_p(t_s - t_c)/24$$

where

q_R = cooling load, tons of refrigeration
Q = water flow rate, gpm
t_s = supply water temperature, °F
t_c = cold water temperature, °F
c_p = specific heat of water = 1 Btu/lb · °F

In the computation of the refrigeration load, one of the most troublesome values to determine is the highest temperature the incoming supply water can be expected to reach. This temperature usually occurs during the hottest summer period. Moreover, allowance should be made for additional warming of the supply water flowing through the piping and water treatment equipment in the beverage plant.

SIZE OF PLANT

The output of each plant depends on the beverage-filling capacity of the plant production equipment. Small, individual filling units turn out approximately 600 cases of 24 beverage containers (approximately half-pint capacity) per hour or 240 containers per minute (240 cpm); intermediate units turn out up to 1200 cases per hour (480 cpm); and high-speed, fully automatic machines begin at approximately 1200 cases per hour and go through several increases in size up to the largest units, which approach 5000 cases per hour (2000 cpm).

Operation of these filling machines, which also determines the demand on refrigeration machinery, usually exceeds 8 h per day, especially during summer months when market demands are highest.

An arbitrary classification of beverage plants may be (1) small plants that produce under 1.25 million cases per year, (2) intermediate plants that produce about 2.5 million cases per year, and (3) large plants that produce 15 million or more cases per year. The latter require installation of multiple-filling lines.

The usual distribution area of the finished beverages is within the metropolitan area of the city in which the plant is located. Some plants have built such a reputation for their goods, that they ship to warehouses several hundred miles distant. Local distribution is made from there. A few nationally known products are shipped long distances from producing plants to specialized markets.

In the warehouse, cans and nonreturnable bottles filled with precooled beverage are commonly warmed to a temperature exceeding the dew-point temperature to prevent condensation and resulting package damage. Bottled goods should be protected against excessive temperature and direct sunlight while in storage and transportation. At the point of consumption, the carbonated beverage is often cooled to temperatures close to 32 °F.

LIQUID CARBON DIOXIDE STORAGE

Liquefied carbon dioxide, used to carbonate water and beverages, is truck delivered in bulk to the beverage plant. The liquid is piped from the trucks to large outdoor storage-converter tanks equipped with mechanical refrigeration and electrical heating. The typical tank unit is maintained at internal temperatures not exceeding 0 °F, so that the equilibrium pressure of the carbon dioxide does not exceed 300 psig and the storage tanks need not be built for excessively high pressures. Full-controlled equipment heats or refrigerates, and safety relief valves discharge sufficient carbon dioxide to relieve excess pressure.

REFERENCES

Berg, H.W. and M. Akiyoshi. 1956. Some factors involved in the browning of white wines. *American Journal of Enology* 7:1.

Bouffard, A. 1985. Determination de la chaleur degagee dans la fermentation alcoolique. Comptes rendus hebdomadaires des seances, Academie des Sciences, Paris 121:357. *Progres agricole et viticole* 24:345.

Boulton, R. 1979a. The heat transfer characteristics of wine fermentors. *American Journal of Enology and Viticulture* 30:152.

Boulton, R. 1979b. A kinetic model for the control of wine fermentations. Biotechnology and Bioengineering Symposium No. 9:167.

MBAA. 1981. *The Practical Brewer*, 2nd ed. Master Brewer's Association of the Americas, Madison, WI.

Ramey, D.R. and C.S. Ough. 1980. Volatile ester hydrolysis or formation during storage of model solutions and wine. *Journal of Agriculture and Food Chemistry* 28:928.

Riese, H. and R. Boulton. 1980. Speeding up cold stabilization. *Wines and Vines* 61(11):68.

CHAPTER 23

EGGS AND EGG PRODUCTS

ABOUT 75% of the table eggs produced in the United States are sold as shell eggs. The remainder are further processed into liquid, frozen, or dehydrated egg products that are used in food service or as an ingredient in food products. Small amounts of further processed eggs are converted to retail egg products, mainly mayonnaise and salad dressings. However, this area has developed rapidly in recent years as major food companies have entered this market. Shell egg processing includes cleaning, washing, candling for quality, sizing, and packaging. Further processed eggs require shell removal, filtering, blending, pasteurization, and possibly freezing or dehydration.

Following processing, shell eggs intended for use within several weeks are stored at 45 to 55 °F and relative humidities of 75 to 80%. These conditions reduce the evaporation of water from the egg, which reduces the egg's weight and hastens the breakdown of the albumen—an indicator of quality and grade. Shell eggs are also refrigerated during transportation, during short- and long-term storage, in retail outlets, and at the institutional and consumer levels.

Research has shown that microbial growth can be curtailed by holding eggs at less than 40 °F. The USDA has proposed that processed eggs be kept below 45 °F until they reach the consumer. This requirement is intended to prevent the growth of *Salmonella* (see October 27, 1992, United States Federal Register). The proposal is expected to include the retail trade, manufacturers, institutional users, and restaurants. If so, major changes in the way eggs are handled will have to be implemented. Storage and display areas not now refrigerated will have to be corrected. All egg storage areas will have to be able to maintain ambient temperatures at 45 °F.

SHELL EGGS

EGG STRUCTURE AND COMPOSITION

Physical Structure

The parts of an egg are shown in Figure 1 and physical properties of eggs are given in Table 1. The *shell* is about 11% of the egg weight and is deposited on the exterior of the outer shell membrane. It consists of a mammillary layer and a spongy layer. The shell contains large numbers of pores (approximately 17,000) that allow water, gases, and small particles, such as microorganisms, to move through the shell. A thin, clear film (cuticle) on the exterior of the shell covers the pores. This material is thought to retard the passage of microbes through the shell and serves to prevent moisture loss from the egg's interior. The shape and structure of the shell provide enormous resistance to pressure stress, but very little resistance to breakage caused by impact.

Two tough fibrous *membranes* surround the albumen. As the egg ages and moisture is lost, an air cell develops on the large end of the egg between these two membranes. The size of the air cell is an indirect measure of the egg's age and is used to evaluate interior quality.

The *white* (albumen) constitutes about 58% of the egg weight. The white consists of a thin, inner chalaziferous layer of firm protein containing fibers that twist into chalazae on the polar ends of the yolk. These structures anchor the yolk in the center of the egg (Figure 1). The albumen consists of inner thin, thick, and outer thin layers.

The *yolk* constitutes approximately 31% of the egg weight. It consists of a yolk (vitelline) membrane and concentric rings of six yellow layers and narrow white layers (Figure 1). In the intact egg, these layers are not visible. Most of the egg's lipids and cholesterol are bounded into a lipoprotein complex that is found more in

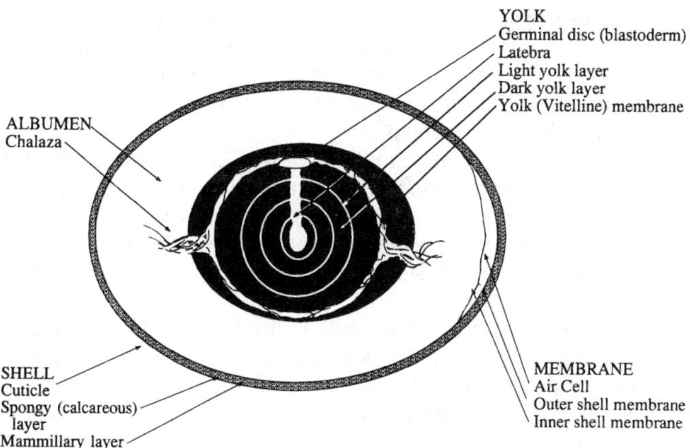

Fig. 1 Structure of an Egg

Table 1 Physical Properties of Chicken Eggs
(Burley and Vadehra 1989)

Property	Whole Egg	Albumen	Yolk
Solids, %	26.4	11.5	52.5
pH (fresh eggs)		7.6	6.0
Density, lb/ft^3	67.5	64.7	64.7
Surface tension, psi			6.38×10^{-4}
Freezing point, °F		31.2	31.0
Specific heat, Btu/lb · °F	0.772		
Viscosity, mPa · s			
Thick white		164	
Thin white		4	
Electrical conductivity, mho/cm $\times 10^{-4}$		8.25	0.07
Water activity, % relative humidity		97.8	98.1

The preparation of this chapter is assigned to TC 11.2, Foods and Beverages.

the white layers. The yolk contains the germinal disc, which consists of about 20,000 cells if the egg is fertile. However, eggs produced for human consumption are not fertile because the hens are raised without roosters.

Chemical Composition

The weight of the chicken egg varies from 35 to 80 g or more. The main factors affecting weight and size are the bird's age, breed, and strain. Nutritional adequacy of the ration and ambient temperature of the laying house also influence egg size. Size affects the composition of the egg in that the proportion of the parts changes as egg weight increases. For example, small eggs laid by young pullets just coming into production will have relatively more yolk and less albumen than eggs laid by older hens. Table 2 presents the general composition of a typical egg weighing 60 g.

The *shell* is low in water content and high in inorganic solids. The solids are mainly calcium carbonate as calcite crystals plus small amounts of phosphorus and magnesium and some trace minerals. Most of the shell's inorganic matter is protein. It is found in the matrix fibers closely associated with the calcite crystals and in the cuticle layer covering the shell surface. Protein fibers are also present in the pore canals and in the two shell membranes. The membranes contain keratin, a protein that makes the membranes tough even though they are very thin.

Egg albumen, or egg white, is a gel-like substance consisting of ovomucin fibers and globular-type proteins in an aqueous solution. Ovalbumin is the most abundant protein in egg white. When heated to about 140°F, coagulation occurs and the albumen becomes firm. Several fractions of ovoglobulins have been identified by electrophoretic and chromatographic analyses. These proteins impart excellent foaming and beating qualities to egg white when making cakes, meringues, candies, etc. Ovomucin is partly responsible for the viscous characteristic of raw albumen and also has a stabilizing effect on egg-white foams, an important property in cakes and candy.

Egg white contains a small amount of carbohydrates. About half of this is present as free glucose and half as glycoproteins containing mannose and galactose units. In dried egg products, glucose interacts with other egg components to produce off-colors and off-flavors during storage; therefore, glucose is enzymatically digested before drying.

The *yolk* comprises one third of the edible portion of the egg. Its major components are water (48 to 52%), lipids (33%), and proteins (17%). The yolk contains all of the fatty material of the egg. The lipids are very closely associated with the proteins. These very complex lipoproteins give yolk special functional properties, such as emulsifying power in mayonnaise and foaming and coagulating powers in sponge cakes and doughnuts.

Nutritive Value

Eggs are a year-round staple in the diet of nearly every culture. The composition and nutritive value of eggs differ among the various avian species. However, only the chicken egg is considered here, as it is the most widely used for human foods.

Eggs contain high-quality protein, which supplies essential amino acids that cannot be produced by the body or that cannot be synthesized at a rate sufficient to meet the body's demands. Eggs are also an important source of minerals and vitamins in the human diet. Although the white and yolk are low in calcium, they contain substantial quantities of phosphorus, iron, and trace minerals. Except for vitamin C, one or two eggs daily can supply a significant portion of the recommended daily allowance for most vitamins, particularly the vitamins A and B_{12}. Eggs are second only to fish liver oils as a natural food source of vitamin D.

The fatty acids in the egg yolk are divided into saturated and unsaturated in a ratio of 1:1.8, with the latter further subdivided into mono- and polyunsaturated fatty acids in a ratio of 1:0.3. Eggs are a source of oleic acid, a monosaturated fatty acid; they also contain polyunsaturated linoleic acid, an essential fatty acid. The fatty acid composition of eggs and the balance of saturated to unsaturated fatty acids can be changed by modifying the hen's diet. Several commercial egg products with modified lipids have been marketed.

Table 2 Composition of Whole Egg

Egg Component	Protein, %	Lipid, %	Carbohydrate, %	Ash, %	Water, %
Albumen	9.7–10.6	0.03	0.4–0.9	0.5–0.6	88.0
Yolk	15.7–16.6	31.8–35.5	0.2–1.0	1.1	51.1
Whole egg	12.8–13.4	10.5–11.8	0.3–1.0	0.8–1.0	75.5

Note: Shell is not included in above percentages.

	Percent of Egg	Calcium Carbonate	Magnesium Carbonate	Calcium Phosphate	Organic Matter
Shell	11	94.0	1.0	1.0	4.0

Source: Stadelman and Cotterill (1990).

EGG QUALITY

Quality Grades and Weight Classes

In the United States, the Egg Products Inspection Act of 1970 requires that all eggs moving in interstate commerce be graded for size and quality. Cracked or dirty eggs may be sold directly to consumers only on the farm or at an authorized processing plant. Loss eggs (inedible eggs) such as leakers (broken shell and broken membranes), blood and meat spots, rots, or eggs with developed embryos may not be used for human consumption, but may be

Table 3 United States Standards for Quality of Shell Eggs

Quality Factor	AA Quality	A Quality	B Quality
Shell	Clean Unbroken Practically normal	Clean Unbroken Practically normal	Clean to slightly stained[a] Unbroken Abnormal
Air Cell	1/8 in. or less in depth Unlimited movement and free or bubbly	3/16 in. or less in depth Unlimited movement and free or bubbly	Over 3/16 in. in depth Unlimited movement and free or bubbly
White	Clear Firm	Clear Reasonably firm	Weak and watery Small blood and meat spots present[b]
Yolk	Ouline slightly defined Practically free from defects	Outline fairly well defined Practically free from defects	Outline plainly visible Enlarged and flattened Clearly visible germ development but no blood Other serious defects

For eggs with dirty or broken shells, the standards of quality provide two additional qualities. These are:

Dirty	Check
Unbroken. Adhering dirt or foreign material, prominent stains, moderate stained areas in excess of B quality.	Broken or cracked shell but membranes intact, not leaking.[c]

[a]Moderately stained areas permitted (1/32 of surface if localized, or 1/16 if scattered).
[b]If they are small (aggregating not more than 1/8 in. in diameter).
[c]Leaker has broken or cracked shell and membranes, and contents are leaking or free to leak.

Source: *Federal Register*, CFR7, Part 56, May 1, 1991. USDA *Agriculture Handbook* No. 75, p. 18.

used in pet foods. Shell eggs for intrastate commerce are not regulated by the USDA unless they are part of the USDA's shield program, which is voluntary. Most states, however, have egg-grading laws or regulations very similar or identical to those of the USDA.

The USDA standards for quality of individual shell eggs are shown in Table 3. The quality of shell eggs begins to decline immediately after the egg is laid. Water loss from the egg causes a thinning of the albumen and an increase in the size of the air cell. Carbon dioxide migration from the egg results in an increase in albumen pH and a decrease in vitelline membrane strength.

Classes for shell eggs are shown in Table 4. The average weight of shell eggs from commercial flocks varies with age, strain, diet, and environment. Typical case weights for white Leghorn hens are shown in Table 5. Practically all eggs produced on commercial poultry farms are processed mechanically. They are washed, candled, sized, oiled, then packed. Although eggs are sold by units of 6, 12, 18, or 30 per package, the packaged eggs must maintain a minimum weight that relates to the egg size.

Quality Factors

Besides legal requirements, egg quality encompasses all the characteristics that affect an egg's acceptability to a particular user. The specific meaning of egg quality may vary. To a producer, it might mean the number of cracked or loss eggs that cannot be sold, or the percentage of undergrades on the grade-out slip. The processor associates quality with prominence of yolk shadow under the candling light and the resistance of the shell to damage on the automated grading and packing lines. The consumer looks critically at shell texture and cleanliness and the appearance of the broken-out egg.

Shell Quality. Strength, texture, porosity, shape, cleanliness, soundness, and color are factors determining shell quality. Of these, shell soundness is the most important. It is estimated that about 10% of all eggs produced are cracked or broken between oviposition and retail sale. Eggs that have shell damage can only be salvaged for their liquid content, but eggs that have both shell and shell membrane ruptured are regarded as a loss and cannot be used for human consumption. Shell strength is highly dependent on shell thickness, which is affected by genetics, nutrition, length of continuous lay, disease, and environmental factors.

Table 4 United States Egg Weight Classes for Consumer Grades

Size or Weight Class	Minimum Net Weight per Dozen, oz	Minimum Net Weight per 30-Dozen Case, lb	Minimum Weight for Individual Eggs, oz
Jumbo	30	56.0	2.42
Extra Large	27	50.5	2.17
Large	24	45.0	1.92
Medium	21	39.5	1.67
Small	18	34.0	1.42
Peewee	15	28.0	

Table 5 Effect of Flock Age on Case Weight
(30 dozen eggs/case)

Age, weeks	Mean, lb	Expected Range, lb	
		Minimum	Maximum
21	37.0	34.1	39.6
22	38.6	36.2	40.8
23	39.9	38.5	41.8
24	40.9	39.8	42.9
25	41.9	40.2	43.7
26	42.7	40.8	44.3
27	43.5	41.5	44.8
28	44.1	42.0	45.2
29	44.7	42.9	45.8
30–33	45.9	43.7	47.6
34–37	47.3	45.7	48.7
38–41	48.3	46.7	49.4
42–45	49.1	47.4	50.2
46–49	49.6	48.0	50.8
50–59	50.3	48.5	51.8
60–69	50.9	49.3	50.3
70–79	51.2	49.6	52.4
Mean	48.5		

Eggs with smooth shells are preferred over those with a sandy texture or prominent nodules that detract from the egg's appearance. Eggs with rough or thin shells or other defects are often weaker than those with smooth shells. Although shell texture and thickness deteriorate as the laying cycle progresses, the exact causes of these changes are not fully understood. Some research suggests that debris in the oviduct collects on the shell membrane surface, resulting in rough texture formation (nodules).

The number and structure of pores are factors in microbial penetration and loss of carbon dioxide and water. Eggs without a cuticle or with a damaged cuticle are not as resistant to water loss, water penetration, and microbial growth as those with this outer proteinaceous covering. External oiling of the egg shell provides additional protection.

Eggs have an oval shape with shape indexes (breadth/length × 100) ranging from 70 to 74. Eggs that deviate excessively from this norm are considered less attractive and break more readily in packaging and in transit.

Shells with visible soil or deep stains are not allowed in a high-quality pack of eggs. Furthermore, soil usually contains a heavy load of microorganisms that may find their way into egg contents and cause spoilage.

Shell color is a breed characteristic. Brown shells owe their color to a reddish-brown pigment, ooporphyrin, which is derived from hemoglobin. The highest content of the pigment is near the surface of the shell. White shells contain a small amount of ooporphyrin, too, but it is degraded soon after laying by exposure to light. Brown shelled eggs tend to vary in color.

Albumen Quality. Egg white viscosity differs in various areas of the egg. A dense layer of albumen is centered in the middle and is most visible when the egg is broken out onto a flat surface. Raw albumen has a yellowish-green cast. In high-quality eggs, the white should stand up high around the yolk with a minimum spreading of the outer thin layer of the albumen. The quality of thick albumen in the freshly laid egg is affected by genetics, duration of continuous production, and environmental factors. Albumen quality generally declines as age progresses, especially in the last part of the laying cycle. Breakdown of thick white is a continuing process in eggs held for food marketing or consumption. The rate of quality loss depends on holding conditions.

Intensity of the color is associated with the amount of riboflavin in the ration. The albumen of top-quality eggs should be free of any blood or meat spots. Incidence of non-meat spots such as blood spots and related problems has been reduced to such a low level by genetic selection that it is no longer a serious concern.

The chalazae may be very prominent in some eggs and can create a negative reaction from consumers who are unfamiliar with these structures (Figure 1). The twisted, rope-like cords are merely extensions of the chalaziferous layer surrounding the yolk and are a normal part of the egg. The chalazae stabilize the yolk in the center of the egg.

Yolk Quality. Shape and color are the principal characteristics of yolk quality. In a freshly laid egg, the yolk is nearly spherical, and when the egg is broken out onto a flat surface, the yolk stands high with little change in shape. Shell and albumen tend to decline in quality as the hen ages. However, yolk quality, as measured by shape, remains relatively constant throughout the laying cycle.

Yolk shape depends on the strength of the vitelline membrane and the chalaziferous albumen layer surrounding the yolk. After oviposition, these structures gradually undergo physical and chemical changes that decrease their ability to keep the yolk's spherical shape. These changes alter the integrity of the vitelline membrane so that water passes from the white into the yolk, increasing the yolk's size and weakening the membrane.

Color as a quality factor of yolk depends on the desires of the user. Most consumers of table eggs favor a light-to-medium yellow color, but some prefer a deeper yellowish-orange hue. Processors of liquid, frozen, and dried egg products generally desire a darker yolk color than users of table eggs because these products are used in making mayonnaise, doughnuts, noodles, pasta, and other foods that depend on eggs for their yellowish color. If laying hens are confined, yolk color is easily regulated by adjusting the number of carotenoid pigments supplied in the hen's diet. Birds having access to growing grasses and other plants usually produce deep-colored yolks of varying hues.

Yolk defects that detract from their quality include blood spots, embryonic development, and mottling. Blood on the yolk can be from (1) hemorrhages occurring in the follicle at the time of ovulation, or (2) embryonic development that has reached the blood-forming stage. The second source is a possibility only in breeding flocks where males are present.

Yolk surface mottling or discoloration can be present in the fresh egg or may develop during storage and marketing. Very light mottling, resulting from an uneven distribution of moisture under the surface of the vitelline membrane, can often be detected on close examination, but this slight defect usually passes unnoticed and is of little concern. Certain coccidiostats (nicarbazin) and wormers (piperazine citrate and dibutylin dilaurate) have been reported to cause mottled yolks and should not be used above recommended levels in layer rations. More serious are the olive-brown mottled yolks produced by rations containing cottonseed products with excessive amounts of free gossypol. This fat-soluble compound reacts with iron in the yolk to give the discoloration. Cottonseed meal may also have cyclopropanoid compounds that increase vitelline membrane permeability. When iron from the yolk passes through the membrane and reacts with the conalbumen of the white, a pink pigment is formed in the albumen. Cyclopropanoid compounds also cause yolks to have a higher proportion of saturated fats than normal, giving the yolks a pasty, custard-like consistency when they are cooled.

Flavors and Odors. When birds are confined and fed a standard ration, eggs have a uniform and mild flavor. Off-flavors can be caused by rations with poor-quality fish meal containing rancid oil or by birds having access to garlic, certain wild seeds, or other materials foreign to normal poultry rations. Off-flavors or odors from rations are frequently found in the yolk, as many compounds imparting off-flavors are fat-soluble. Once eggs acquire off-flavor during storage, their quality is unacceptable to consumers. Eggs have a great capacity to absorb odors from the surrounding atmosphere (Carter 1968). Storage should be free from odor sources such as apples, oranges, decaying vegetable matter, gasoline, and organic solvents (Stadelman and Cottermill 1990). If this cannot be avoided, odors can be controlled with charcoal absorbers or periodical ventilation.

Control and Preservation of Quality

Egg quality is evaluated by shell appearance, air-cell size, and the apparent thickness of the yolk and white. A low storage temperature and shell oiling slow down the escape of carbon dioxide and moisture and prevent shrinkage and thinning of the white. Clear white mineral oil sprayed on the shell after washing partially protects the egg, and it is used in most commercial operations. Rapid cooling will also reduce moisture loss.

Egg quality loss is slowed by maintaining egg temperatures near the freezing point. Albumen freezes at 32.8 °F and the yolk freezes at 31 °F. Stadelman *et al.* (1954) and Tarver (1964) found that eggs stored for 15 or 16 days at 45 to 50 °F had significantly better quality than eggs stored at 57 to 61 °F.

Stadelman and Cotterill (1990) recommend that storage and humidity should be controlled and maintained between 75 and 80%. As a rule, eggs lose about 1% of their weight per week in storage. When large amounts of eggs are palletized, humidity in the center of the pallet may be higher than that of the surrounding air. Therefore, air flow through the eggs is needed to remove excess humidity above 95% to prevent mold growth and decay.

Albumen quality loss is associated with carbon dioxide loss from the egg. Quality losses can be reduced by increasing carbon dioxide levels around the eggs. Controlled atmosphere storage and modified atmosphere packaging have been studied, but they are not used commercially because eggs typically do not need long-term storage. Oiling also helps retard carbon dioxide loss.

Egg Spoilage and Safety

Bacteriological Spoilage. Shell eggs deteriorate in three distinct ways: (1) decomposition by bacteria and molds, (2) changes from chemical reactions, and (3) changes because of absorbtion of flavors and odors from the environment. Dirty or improperly cleaned eggs are the most common source of bacterial spoilage. Dirty eggs are contaminated with bacteria. Improper washing caused by using water colder than the eggs or water with high iron content increases the possibility of contamination, although it removes evidence of dirt. Most improperly cleaned eggs spoil during long-term storage. Safe washing procedures for storage eggs requires extremely high sanitary standards.

Eggs contaminated with certain microorganisms spoil quickly, resulting in black-, red-, or green-rot, crusted yolks, mold, etc. However, eggs occasionally become heavily contaminated without any outward manifestations of spoilage. Clean, fresh eggs are seldom contaminated internally. However, if the shell becomes wet or sweaty, particularly under conditions of fluctuating temperatures, bacteria and/or molds may cause spoilage.

Preventing Microbial Spoilage. Egg quality can be severely jeopardized by the invasion of microorganisms that cause off-odors and off-flavors. With frequent gathering, proper cleaning, and refrigeration, sound-shell eggs that move quickly through market channels have few spoilage problems.

Sound-shell eggs have a number of defenses against microbial attack, some of which are mechanical and some chemical. Although most of the shell pores are too large to impede bacterial movement, the cuticle layer, and possibly materials within the pores, offer some protection, especially if the shell surface remains dry. Bacteria successful in penetrating the shell are next confronted by a second physical barrier, the shell membrane.

Microorganisms reaching the albumen find it unfavorable for growth. Movement is retarded by the viscosity of the egg's white. Also, most bacteria prefer a pH near neutral; but the pH of egg white, initially at 7.6 when newly laid, increases to 9.0 or more after several days, providing a deterring alkaline condition.

Conalbumen, which is believed to be the main microbial defense system of albumen, complexes with iron, zinc, and copper, thus making these elements unavailable to the bacteria and restricting their growth. The chelating potential increases with the rise in albumen pH.

Eggs are capable of warding off a limited quantity of organisms, but should be handled in a manner that minimizes contamination. Egg washing must be done with care. Proper overflow, maintenance of a minimum water temperature of 95 °F (as required by USDA regulations), and use of a sufficient quantity of approved detergent-sanitizer are important for effective cleaning. The wash water should be at least 20 °F warmer than the internal temperature of the eggs to be washed. Likewise, the rinse water should be a few degrees higher than the wash water. Under these conditions, the contents of the eggs expand to create a positive pressure, which tends to repel penetrations of the shell by microorganisms.

Regular changes of the wash water, as well as thorough daily cleaning of the washing machine, are very important. When the wash water temperature exceeds the egg temperature by more than 50 °F, an inordinate number of cracks in the shells, called thermal cracks, occur. Excessive shell damage also occurs if the washer and its brushes are not properly adjusted. Most egg processors use wash waters at temperatures of 110 to 125 °F.

Some changes that take place during storage are caused by chemical reaction and temperature effects. As the egg ages, the pH of the white increases, the thick white thins, and the yolk membrane thins. Ultimately, the white becomes quite watery, although total protein content changes very little. Some coincidental loss in flavor usually occurs, although it develops more slowly.

EFFECT OF REFRIGERATION ON EGG QUALITY AND SAFETY

Refrigeration is the most effective and practical means for preserving quality of shell eggs. It is widely used in farm holding rooms, processing plants, and in marketing channels. United States production of shell eggs seldom fluctuates more than 10% above or below the monthly average for the year and demand is fairly constant. Thus, warehousing eggs is a matter of backlogging stocks for handling and grading and for adjusting supply to current demand, thereby decreasing the need for long-term cold storage.

Refrigeration conditions for shell eggs to prevent quality loss during short- and long-term storage are as follows:

Temperature, °F	Rel. Humidity, %	Storage Period
50 to 60	75 to 80	2 to 3 weeks
45	75 to 80	2 to 4 weeks
29 to 31	85 to 92	5 to 6 months

For long-term storage, eggs should be kept just above their freezing point, 27 °F. However, low temperatures are seldom used because most eggs are consumed within a short period. Low temperatures can cause sweating, *i.e.*, condensation of moisture on the shell, which causes microbial spoilage.

Cooling Rates

Henderson (1957) showed that air rates of 105 to 600 fpm flowing past an individual egg caused it to cool within one hour to 90% of the difference between the initial egg temperature and the temperature of the cooling air. Eggs packed in filler flats required 4 to 5 h to achieve 90% of total possible temperature drop. Bell and Curley (1966) reported that 55 °F air forced around fiber flats in vented corrugated fiberboard boxes cooled eggs from 90 to 60 °F in 2 to 5 h. Unvented cartons with the same pack required more than 30 h to cool. Czarick and Savage (1992) found that it takes more than 5 days for a pallet of eggs in cases to cool from 85 or 90 °F to 45 °F in a 45 °F cold room. Henderson (1957) also showed that forced ventilation of palletized eggs produced cooling times close to that of cooling individual eggs.

Adequate air flow through a box requires that the box be vented. In a study done for fruits and vegetables, Baird *et al.* (1988) showed that cooling cost increases rapidly when carton face vent area decreased below that of 4% of the total area. Other packing materials, such as liners, wraps, flats, or cartons, must not prevent the air that enters the box from contacting the eggs.

Egg moisture loss is not increased by rapid cooling. Funk (1935) found that weight loss was the same for eggs in wire baskets cooled in 1 h with circulating air or 15 h with still air.

Egg temperature after washing and packing in in-line systems can typically reach 75 to 85 °F and in rare cases may reach nearly 100 °F. Eggs could be cooled between washing and packing just prior to being placed into cases, perhaps in a space-saving spiral belt cooler currently being used for food freezing. This would allow the use of current packaging, which is usually impervious to air flow. Cooling times would be similar to those for individual egg tests, and cooling times of 1 h or less could be expected.

Off-line and In-line Processing

Poultry farms either send eggs to a processing plant or package them themselves. On commercial farms, the hens reside in cages with sloped floors. Eggs immediately roll onto a gathering tray or conveyor where they are either gathered by hand or conveyed directly to a packing machine (in-line) operation. Machines have the ability to package both in-line and off-line eggs, thereby increasing the flexibility of the operation (Figure 2). Off-line operations have coolers both for incoming eggs and for outgoing finished product (Figure 3). An in-line operation has only one cooler for the outgoing finished product (Figure 4).

During egg packaging in both off-line (Figure 5) and on-line facilities, as eggs enter the processing area, they are oriented into a series of single rows—commonly 12 to 18. They then enter a washing unit, where a series of sprays and brushes wash the eggs with a warm detergent solution (110 to 125 °F). Following washing, the eggs are rinsed with warm water and sanitized with an approved sanitizing agent such as chlorine. The eggs are then dried by air, oiled, and placed on a conveyer, which rotates the eggs as they enter the candling booth. There an intense light under the conveyer illuminates both the egg's internal defects and shell defects. Two operators (candlers) identify the defective eggs, which are removed. The eggs are then weighed and sized automatically, and the different sizes are packaged into cartons (12 or 18 eggs) or flats (20 or 30 eggs).

One trend in packaging equipment is to increase output to 450 to 500 cases per hour. This development could be possible if automatic candling is developed, as the human eye can only efficiently respond to about 100-150 cases per hour. An automatic crack detector has been developed and can be incorporated in egg grading machines. Kuney *et al.* (1992) demonstrated the high cost of good eggs overpulled in error by candlers. Machine speed was the major factor related to overpulling. Packaging is another area that could be automated because feeding packaging materials, packaging cartons or flats into cases, and palletizing are still largely manual operations.

Cooling requirements for shell eggs obviously vary with the weight of eggs to be cooled and their initial temperature. For an in-line processing system, eggs are moved on a conveyer to the processing room immediately after being laid. Czarick and Savage (1992) reported that incoming egg temperature in an in-line system reached equilibrium with the layer house temperature. This

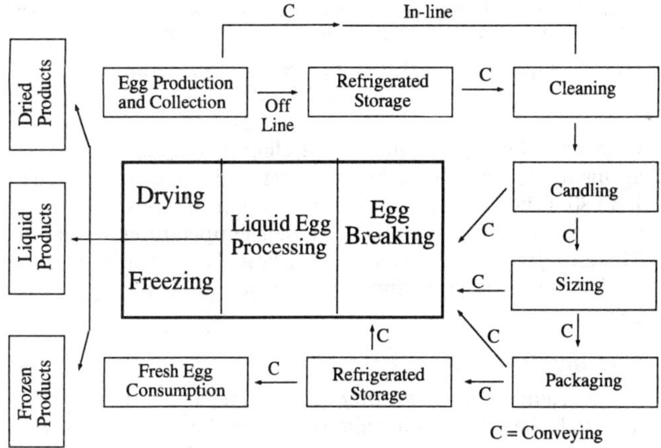

Fig. 2 Unit Operations in Off-Line and In-Line Egg Packaging

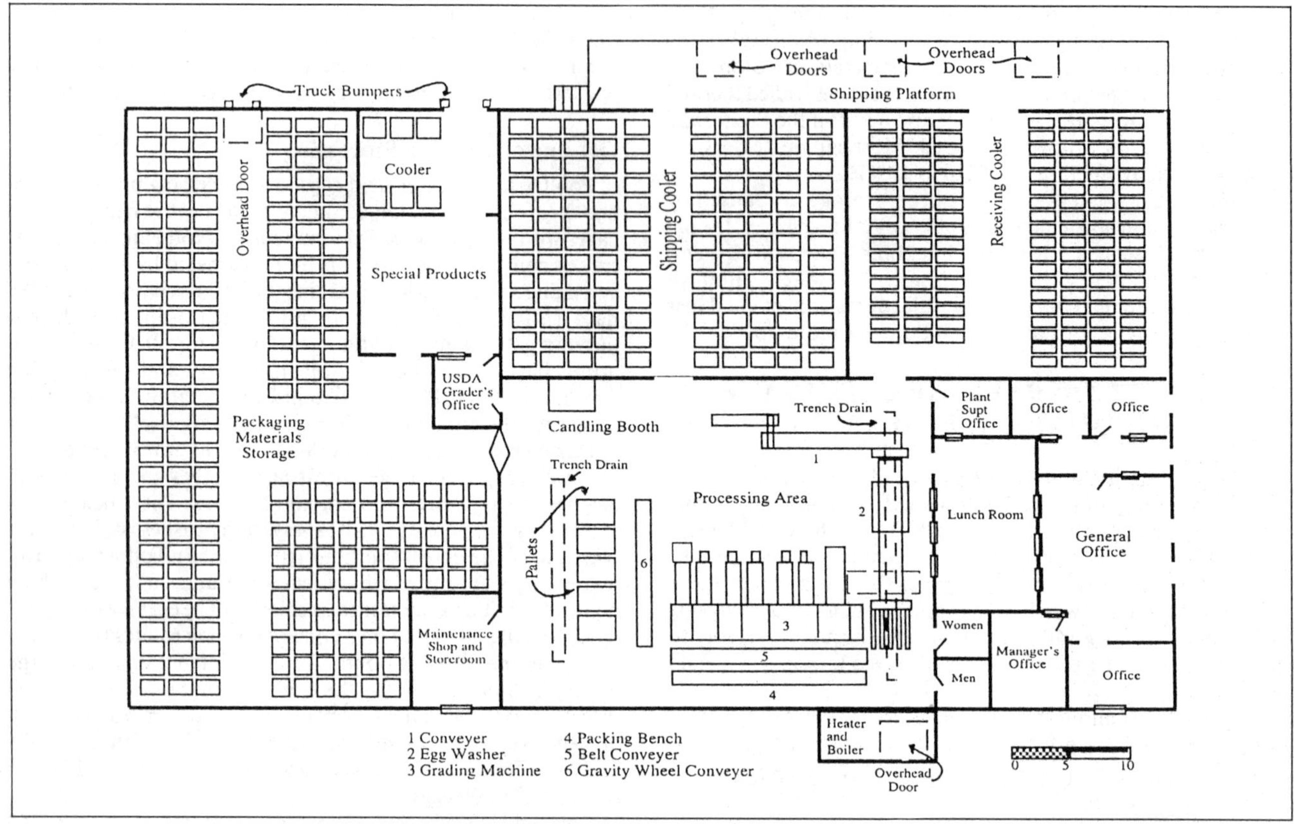

Fig. 3 Off-Line Egg Processing Operation
(Goble 1980)

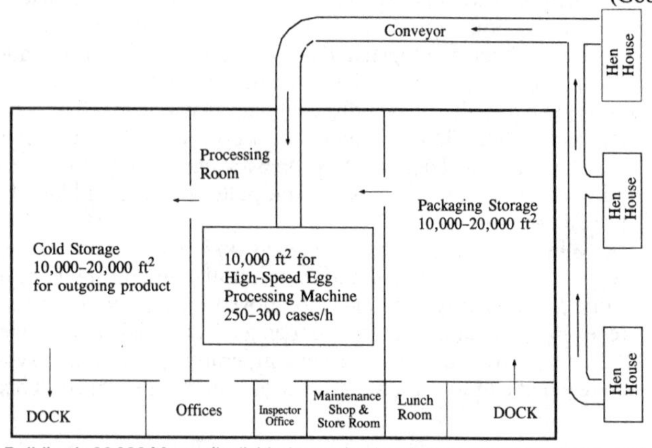

Building is 30,000 ft² equally divided to packaging storage, processing room, and cold storage with possible expansion from cold storage side to 50,000 ft².

Fig. 4 Typical In-Line Processing Operation
(Zeidler and Riley 1993)

varies considerably with house type, climate, and ventilation management. House temperatures are often maintained in the area of 75 to 80 °F, however, 90 °F sometimes occurs. Czarick and Savage (1992) reported that washing raised egg temperatures about 5 °F. Therefore, with an in-line system, cartoned eggs would enter the refrigerated storage at about 5 °F above the laying house temperature.

Cooling Problems

Experience has shown that quality defects are more readily detected when the eggs are allowed to age. Thus, in off-line processing, eggs from production units are usually stored overnight at 55 to 60 °F before processing. Eggs enter the storage room at approximately the prevailing house temperature. But with present cooling methods, eggs would need to be kept in cold storage about 48 h to cool completely. In contrast, with in-line systems, eggs are commonly processed while still warm from the house and are often packaged in a warm state.

Research has shown that with current practices of handling, packed eggs require up to one week of storage before they reach the temperature of the storage room. The packaging materials effectively insulate the eggs from the surrounding environment, especially in the center of the pallet. In addition, pallets are often stacked too near each other and may be wrapped in plastic, which further insulates the inner cases and reduces airflow. Also, most eggs are moved from storage within days of processing, so they are only partially cooled. But delaying shipment to allow the eggs to cool results in less-than-fresh products being delivered to the consumer and interior quality suffers.

Because of the inefficiency of cooling eggs in containers, it would seem best to cool them before packing. Since cold eggs are more subject to thermal cracking when exposed to hot water, the logical time to cool the eggs is before the packaging. However, existing equipment is not designed to incorporate this procedure.

Another method to improve the cooling rate in containers is to use semi-open or perforated fiber cases. Blasts of cold air directed at the sides of such cases or forcing air through the pallets would greatly hasten cooling. Cases would have to be stacked to allow air to circulate freely around the pallets.

Condensation on Eggs

Moisture often condenses on the shell surface when cold eggs are moved from the cool storage into hot and humid outside conditions or if the temperature varies widely inside the cooler. Sweating results in a wet egg, and if microbes are present on wet egg

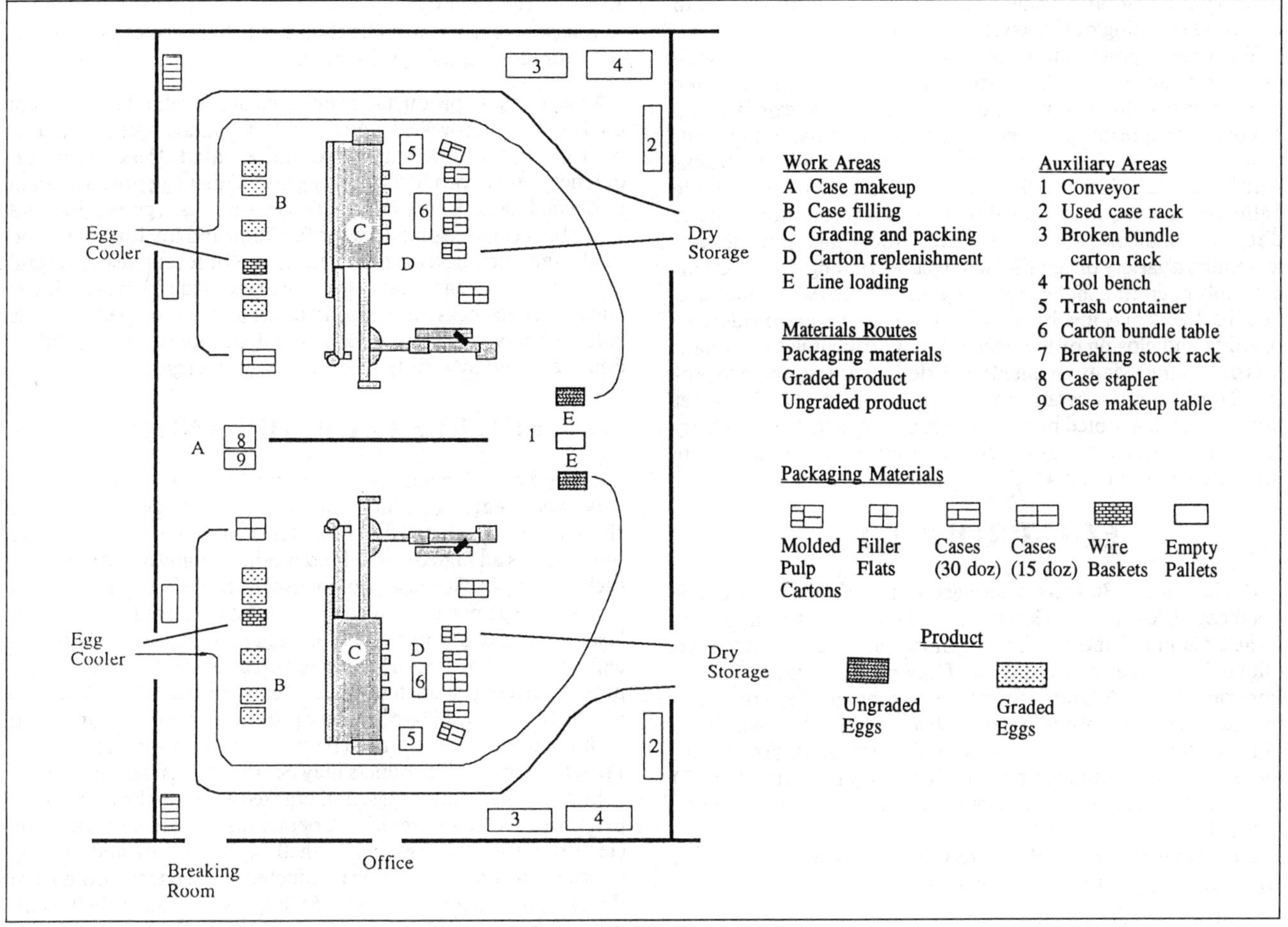

Fig. 5 Material Flow in Off-Line Operation
(Hamann *et al.* 1978)

Table 6 Ambient Conditions When Moisture Condenses on Cold Eggs

Outside Temperature, °F	Outside Relative Humidity, % Egg Temperature 45°F	55°F
55	70	—
60	60	85
65	50	70
70	40	60
75	35	50
80	30	43
85	25	36
90	21	31

shells, they can pass easily into the egg. In addition, wet eggs are more likely to become stained when handled.

Condensation or sweating can be predicted from a psychometric chart. Table 6 lists typical conditions in which sweating may occur.

Plastic wrapped around the pallets to stabilize the load for shipping can also prevent moisture loss and increase the humidity in the load, which can cause mold problems when eggs are held too long in this condition.

PACKAGING

Shell eggs are packaged for the individual consumer or the institutional user. Consumer packs are usually a one dozen carton or variations of it. The institutional user usually receives shell eggs in 30 dozen cases on twelve 30-egg filler-flats.

Consumer cartons are generally made of paper pulp or foam plastic. Some cartons have openings in the top designed for viewing the eggs, which also facilitates cooling. Cartons are generally delivered to the retailer in corrugated containers that hold 15 to 30 dozen eggs, in wire or plastic display baskets that hold 15 dozen eggs, or on rolling display carts. The wire baskets and the rolling racks allow more rapid cooling, but are also more expensive and take up more space in storage and in transport.

TRANSPORTATION

Shell eggs are transported from the off-line egg production site to the egg processing plants, and from the processing plant to local or regional users, and less frequently from one region of the country to another or overseas. Truck transport is most common and refrigerated trucks will be mandatory in the United States, with possible exemption for small quantities delivered to short-distance customers.

Cases and baskets are generally stored and transported on wooden pallets in 30-case lots (five cases high with six cases per layer). The common carrier for local and long-distance hauling is the refrigerated tractor/trailer combination. Trailers carry 24 to 26 thirty-case pallets of eggs, often of one size category. A typical load of 720 to 780 cases weighs about 44,000 lb. Some additional cases may be added when small or medium eggs are being transported. Eggs are not generally stacked above six cases high to allow the cold air to travel to the rear of the trailer and to minimize crushing of the lower level cases.

The interregional shipment of eggs is quite common with production and consumption areas often 1500 miles apart. Such shipments usually require two to three days using team driving.

Local transportation of eggs may be with similar equipment, especially when delivered to retailer warehouses. Smaller trucks with capacities of 250 to 400 cases are often used when multiple deliveries are required. Local deliveries are commonly made directly to retail or institutional outlets. Individual store deliveries require a variety of egg sizes to be placed on single pallets. This assembly operation in the processing plant is very labor-intensive. Local delivery may involve multiple short stops and considerable opening and closing of the storage compartment with resultant loss of cooling. Many patented truck designs are available to protect cargo from temperature extremes during local delivery, yet none has been adopted by the egg industry. A 1993 USDA survey found that over 80% of the trucks used to deliver eggs were unsuitable to maintain 45 °F.

EGG PRODUCTS

During the last 30 years, total egg consumption in many developed countries such as Germany, the United Kingdom, Spain, Israel, Canada, and the United States has steadily declined (Table 7), whereas consumption of egg products has increased. For example, in the United States the percentage of egg products in total egg consumption rose from 9% in 1960 to 24.4% in 1992. Other countries, such as China, Columbia, India, Mexico, Japan, and Hungary, continue to build their poultry industry (Table 7) to take advantage of the nutritional value of eggs, which provide the lowest cost animal protein.

Egg products are classified into four groups according to the American Egg Board Guidelines (1981):

1. Refrigerated egg products
2. Frozen egg products
3. Dried egg products
4. Specialty egg products (including hard-cooked eggs, omelets, scrambled eggs, egg substitutes)

Table 7 Consumption of Eggs and Egg Products

Country	g/person · day				(kg/yr)
	1983	1986	1988	1990	1990
Australia	33.8	23.9	28.4	28.3	10.3
Canada	34.4	33.2	31.4	30.9	11.3
China	7.9	12.9	15.8	17.6	6.4
Colombia	15.1	16.7	18.3	18.7	6.8
France	39.9	43.1	44.5	41.1	18.0
Germany Fed. Rep.	48.3	44.6	42.7	39.4	14.4
Hungary	51.8	50.7	56.3	59.5	21.7
India	2.6	2.9	3.0	3.6	1.3
Israel	56.1	52.1	51.9	47.6	17.3
Italy	31.8	30.7	33.2	33.6	12.3
Japan	46.6	48.2	51.2	50.9	18.6
Jordan	17.2	19.3	10.5	11.3	4.1
Malaysia	17.0	19.4	22.3	22.7	8.3
Mexico	22.2	28.9	31.0	27.0	9.9
Netherlands	38.2	36.0	32.3	31.7	11.6
Norway	31.9	33.1	32.7	31.4	11.5
Spain	45.3	45.1	44.6	41.6	15.2
Switzerland	33.7	34.6	32.9	31.4	11.5
Syria Arab Rep.	21.3	18.6	16.5	10.9	4.0
Taiwan	21.7	22.0	21.9	25.1	9.2
Turkey	11.4	11.6	15.0	16.4	6.0
UK	33.7	32.4	32.7	31.2	11.4
USA	41.8	40.8	39.4	37.4	13.7
USSR (Former)	38.9	41.1	42.2	40.4	14.8

Source: FAO

Most of these products are not seen at the retail level, but are used as further processed ingredients by the food processing industry for such products as mayonnaise, salad dressing, pasta, quiches, bakery products, and egg nog. Other egg products such as deviled eggs, Scottish eggs, frozen omelets, egg patties, and scrambled eggs are prepared for fast food and institutional food establishments, hotels, and restaurants. In recent years, several products such as egg substitutes (which are made from egg whites) and scrambled eggs have appeared. Yet to be developed are large volume items such as aseptically filled, ultrapasteurized, chilled liquid egg and low cholesterol chilled liquid eggs.

EGG BREAKING OPERATIONS

The egg breaking process transforms shell eggs into liquid products: whole egg, egg white, and yolk. Liquid egg products are chilled, frozen, or dried. These items can be used as is or are processed as an ingredient in food products. Only a few products, such as hard-cooked eggs, do not use the breaking operation system. Dried egg powder, which is the oldest processed egg product, lost ground as a proportion of total egg products, whereas chilled egg products are booming due to superior flavor, aroma, pronounced egg characteristics, and convenience. Most liquid egg products (about 44% of all egg products) must be consumed within a relatively short time because of their short storage life. Frozen or dried egg products may be stored considerably longer.

Surplus eggs, small eggs, and cracked eggs are the major supply source for egg breaking operations. Those eggs must be cleaned in the same manner as shell eggs. Washing and loading of eggs to be broken must be conducted in a separate room from the breaking operation (Figure 6). Eggs with broken shell membrane (leakers) or blood spots are not allowed to be broken for human consumption. Most egg breaking operations are close to egg production areas and in many cases are merely a separate area of a shell egg processing and packaging facility. An egg breaking operation usually receives its eggs from several processing plants in the area that do not have breaking equipment. Storage and transport of eggs, and especially of cracked eggs, reduces the quality of the end product. Therefore, the opening in 1992 of the first plant to use the entire egg production of its farms was a major improvement.

Two types of egg breaking equipment are available:

1. *Basket centrifuge.* Shell eggs are dumped into a centrifuge and a whole egg liquid is collected. Several states and some local health authorities ban this equipment for breaking eggs for human consumption due to the high risk of contamination. However, similar centrifuges are used to extract liquid egg residue from the discarded egg shells. This inedible product is used mostly for pet food purposes.
2. *Egg breaker and separator.* These machines can process up to 100 cases per hour (36,000 eggs), which is still a slow process compared to 250 to 300 cases per hour handled by modern table egg packaging equipment.

The basic unit is composed of a cracker head and a yolk-white separator. The cracker head cracks the egg at the center and pulls the two half-shells away from each other. The yolk-white separator is composed of two cups positioned one above the other. The top cup retains the yolk and lets the white slide to the bottom cup. Many of these basic breaking units are attached to a round or oval rotating table, and they can be easily detached for cleaning or removing inedible eggs. An egg loader feeds the machine. The rotation of the table provides enough time for the yolk to separate

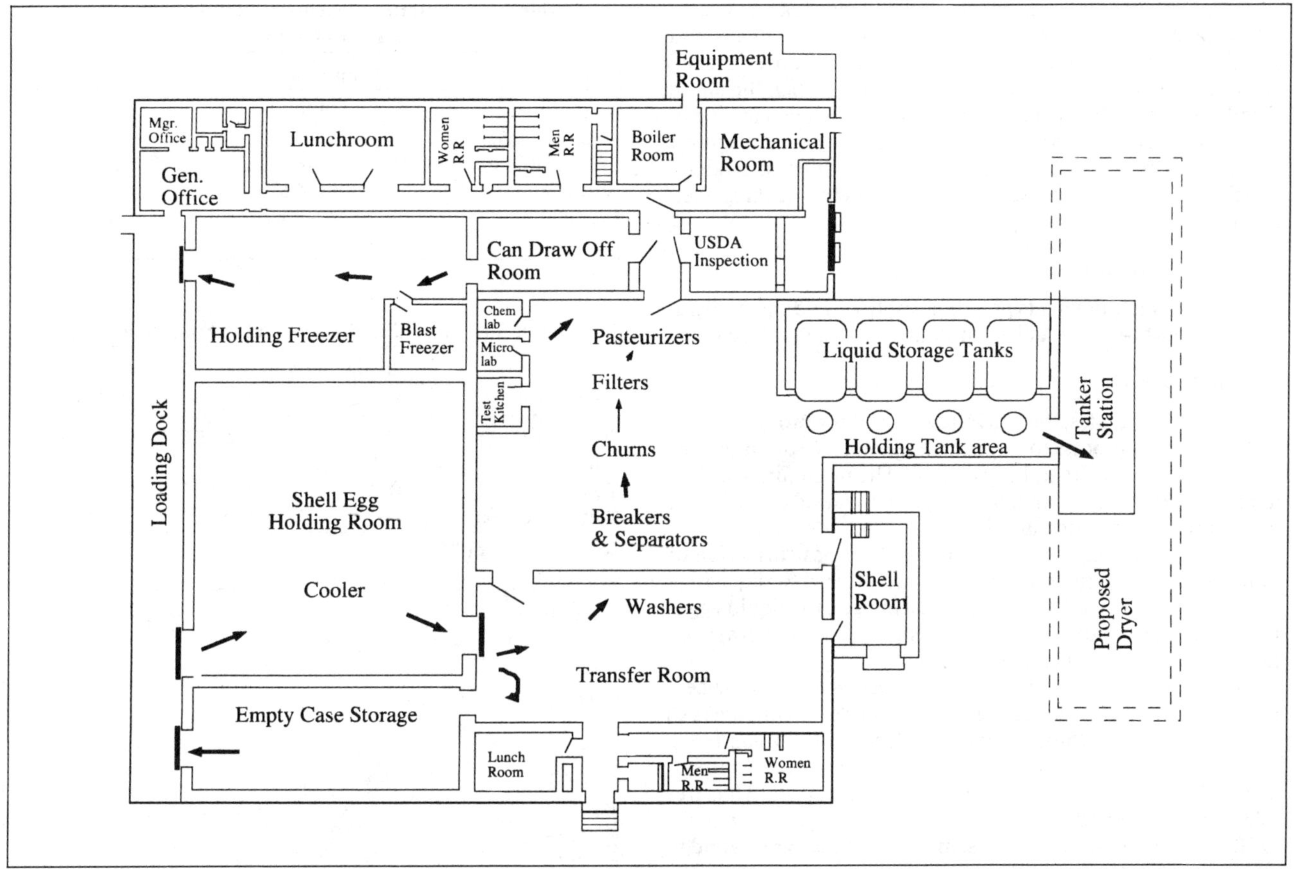

Fig. 6 Floor Plan and Material Flow in Large Egg Breaking Plant
(Courtesy of Seymour Food)

Table 8 Minimum Cooling and Temperature Requirements for Liquid Egg Products
(unpasteurized product temperature within 2 h from time of breaking)

Product	Liquid (Other than salt product) Held 8 h or less	Liquid (Other than salt product) Held in excess of 8 h	Liquid Salt Product	Temperatures within 2 h after pasteurization	Temperatures within 3 h after stabilization
Whites (not to be stabilized)	55 °F or lower	45 °F or lower	—	45 °F or lower	—
Whites (to be stabilized)	70 °F or lower	55 °F or lower	—	55 °F or lower	[a]
All other products (except product with 10% or more salt added)	45 °F or lower	40 °F or lower		If held 8 h or less, 45 °F or lower. If held in excess of 8 h, 40 °F or lower.	
Liquid egg product (10% or more salt added)	—	If to be held 30 h or less, 65 °F or lower. To be held in excess of 30 h, 45 °F or lower.	—	65 °F or lower[b]	—

Source: Regulations governing the inspection of eggs and egg products (7C Federal Register Part 59) May 1, 1991.

[a]Stabilized liquid whites shall be dried as soon as possible after the removal of glucose. The storage of stabilized liquid whites shall be limited to that necessary to provide a continuous operation.

[b]The cooling process shall be continued to assure that any salt product held in excess of 24 h is cooled and maintained at 45 °F or lower.

Table 9 Pasteurization Requirements of Various Egg Products

Liquid Egg Products	Minimum Temperature, °F	Minimum Holding Time, minutes
Albumen (without use of chemicals)	134 132	3.5 6.2
Whole egg	140	3.5
Whole egg blends (less than 2% added non-egg ingredients)	142 140	3.5 6.2
Fortified whole eggs and blends (24 to 38% egg solids, 2 to 12% non-egg ingredients)	144 142	3.5 6.2
Salt whole egg (2% salt added)	146 144	3.5 6.2
Sugar whole egg (2 to 12% sugar added)	142 140	3.5 6.2
Plain yolk	142 140	3.5 6.2
Sugar yolk (2% or more sugar added)	146 144	3.5 6.2
Salt yolk (2 to 12% salt added)	146 144	3.5 6.2

Source: Regulations governing the inspection of eggs and egg products (7C Federal Register Part 59) May 1, 1991.

from the white. If whole egg product is needed, no yolk-white separation takes place and processing becomes much faster and cheaper. The liquid products are then filtered to remove egg shell particles, membranes, and other solid particles. Then the liquid is collected in holding tanks where additives such as salt, sugar, or carbohydrates may be added to stabilize the product against freezing according to the specifications of the finished product. Holding tanks are kept in a separate room from the breaking operation room.

Holding Temperatures

The USDA requirements prepasteurization holding temperatures for out-of-shell liquid egg products are listed in Table 8.

Pasteurization

In the United States, the USDA requires all egg products made by the breaking process to be pasteurized and free of pathogens and requires all plants to be inspected. The minimum required temperatures and holding times for pasteurization of each type of egg product are listed in Table 9.

Plate heat exchangers, commonly used for pasteurization of milk and dairy products, are also commonly used for liquid egg products. Prior to entering the heat exchanger, the liquid egg is moved through a clarifier that removes solid particles such as vitelline (the yolk membrane) and shell pieces.

Egg white solids may be made *Salmonellae* negative by heat treatments. Spray-dried albumen is heated in closed containers so that the temperature throughout the product is not less than 130 °F for not less than 7 days, until it is free of *Salmonellae*. For pan-dried albumen, the requirement is 125 °F for 5 days until it is free of *Salmonellae*. For the dried whites to be labeled pasteurized, the USDA requires that each lot be sampled, cultured, and found to contain no viable *Salmonellae*.

Temperature, time, and pH affect the pasteurization of liquid eggs. Various countries specify different pasteurization time, temperature, and pH, but all specifications provide the same pasteurization effects (Table 10). Higher pH requires lower pasteurization temperature and pH 9.0 is most commonly used for egg whites (Figure 7). Various egg products demonstrate different destruction curves (Figure 8), and therefore, different pasteurization conditions were set for these products (Table 10).

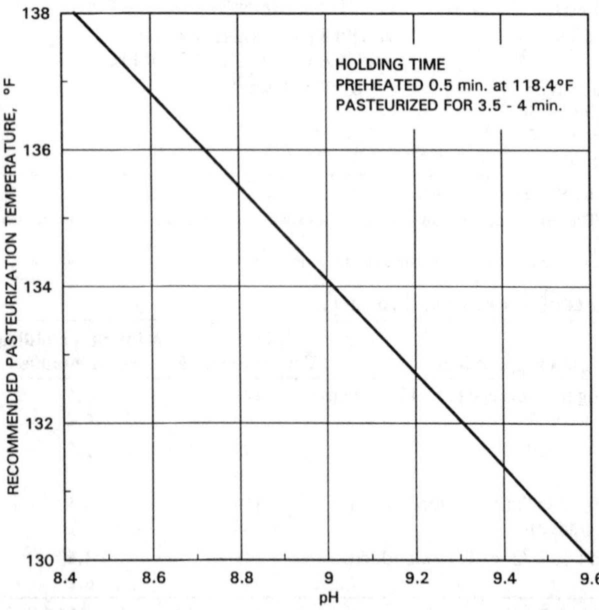

Fig. 7 Effect of pH on Pasteurization Temperature of Egg White

Table 10 Minimum Pasteurization Requirements in Various Countries

Country	Temperature, °F	Time (minutes)
Great Britain	148	2.5
Poland	151–154	3
China (PRC)	146	2.5
Australia	144.5	2.5
Denmark	149–156.5	1.5–3
USA	140	3.5

Source: Stadelman *et al.* (1988)

Table 11 Liquid and Solid Yields From Shell Eggs

	Liquid (% by weight)		
Constituent	Mean	Range	Solids (%)
Shell	10.5	7.8 to 13.6	99.0
Whites	58.5	53.1 to 68.9	11.5
Yolk	31.0	24.8 to 35.5	52.5
Edible whole egg	89.5	86.4 to 92.2	24.5

Source: Shenstone (1968)

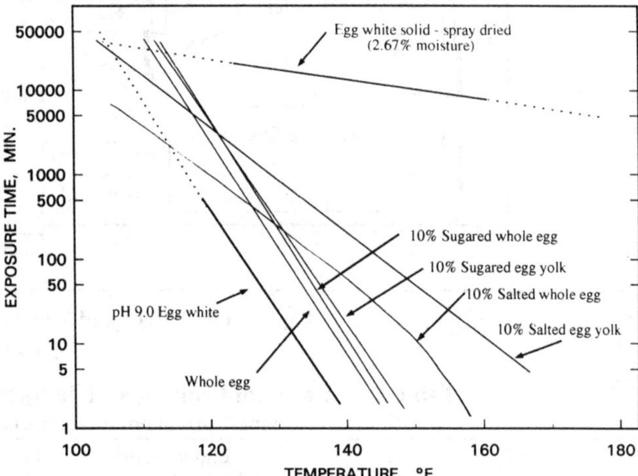

Fig. 8 Thermal Destruction Curves of Several Egg Products (Stadelman and Cotterill 1990)

Egg whites are more sensitive to higher temperatures than whole eggs or yolk, and therefore will coagulate. Thus, lactic acid is added to adjust the pH to 7.0 to allow the egg whites to withstand 141 to 143 °F. Egg whites can be pasteurized at 125 to 127 °F for 1.5 minutes if, after the heat treatment, 0.075 to 0.1% hydrogen peroxide is added for 2 min, followed by its elimination with the enzyme catalase. Liquid yolk, on the other hand requires higher temperatures for pasteurization than liquid whole eggs (144 °F for 3.5 min).

Yields

The ratios of white, yolk, and shell vary with the size of the egg. During the laying cycle, the hens lay small, medium, and large size eggs, which have different proportions of yolk and white. Therefore, the distribution of egg sizes that the breaking plant receives during the year varies with season, breed, egg prices, and surplus sizes. As a result, the processing yields of white, yolk, and shell vary accordingly (Table 11).

CHILLED LIQUID EGG PRODUCTS

Liquid egg products are extremely perishable and should be cooled immediately after production and kept cool until used. Chilled liquid egg products are convenient to use, do not need defrosting, and can be delivered in bulk tank trucks or compact refrigerated units, which reduces packaging costs. However, shelf-life at 34 to 30 °F is about 2 to 3 weeks; therefore, this product is used mostly as an ingredient in food processing.

Extending the shelf life of liquid egg products is a top priority of the egg industry. This task is complicated, as egg proteins are much more heat-sensitive than dairy proteins. As a result, eggs must be kept at lower temperatures. Microbial standards for postpasteurization products are shown in Table 12. Ball *et al.* (1987) used ultra-pasteurization and aseptic packaging to extend the shelf life of chilled whole egg to 24 weeks. However, in contrast to ultrapasteurized milk, this product still needs to be kept under refrigeration.

FROZEN EGG PRODUCTS

Egg products are usually frozen in cartons, plastic bags, 30-lb plastic cans, or 55-gal drums (for bulk shipment). Table 1 in Chapter 30 of the 1993 ASHRAE *Handbook—Fundamentals* lists thermal properties involved in freezing egg products. Freezing is usually accomplished by air blasts at temperatures ranging from −10 to −40 °F. Nonpasteurized egg products must be solidly frozen or reduced to at least 10 °F within 60 h after breaking. Pasteurized products designated for freezing must be frozen solid or cooled to a temperature of at least 10 °F within 60 h after pasteurization. Newer freezing techniques for products containing cooked white (deviled eggs or egg rolls) include individual quick freezing at very low temperatures (−4 to −240 °F).

Defrosting. Frozen eggs may be defrosted at 5 °C in approved metal tanks in 40 to 48 h. If defrosted at higher temperatures (up to 50 °F), the time cannot exceed 24 h. Running water can be used for defrosting. When the frozen mass is crushed by crushers, all sanitary precautions must be followed.

Chilled and Frozen Products

Chilled or frozen liquid whole egg, yolk, and whites are the major high volume products.

Stabilized egg products. Additives are added to yolk products before freezing to prevent coagulation during thawing. Ten percent salt is added to yolks used in mayonnaise and salad dressings, and 10% sugar is added to yolks used in baking, ice cream, and confectionary manufacturing. Whole egg products are also fortified with salt or sugar according to finished product specifications. However, egg whites are not fortified as they do not have gelation problems during defrosting. The approximate composition of several egg products is listed in Table 13.

UHT products (ultraheat treatment). UHT development was initially aimed at producing sterile milk with superior palatability by replacing conventional sterilization at 250 °F for about 12 to 20 min. with 275 °F for 2 to 5 s. UHT treatment of liquid eggs is more complicated, as egg proteins are more sensitive to heat treatment; therefore, UHT liquid eggs must be kept under refrigeration. Ultrapasteurized, aseptically filled, chilled, whole liquid egg product is now limited to institutional food establishments.

Egg Substitutes. In order to satisfy the demand for low-cholesterol egg products (when low-cholesterol eggs are unavailable), substitutes are made from egg whites, which do not contain cholesterol. The yolk is replaced with vegetable oil, food coloring, gums, and nonfat dry milk. Recent formulations have reduced the fat content to almost zero. These products are packaged in cardboard containers and sold frozen or chilled in numerous formula variations.

Low-Cholesterol Eggs. Many techniques have been developed to remove cholesterol from eggs, yet no commercial product is currently available, although market testing is currently being conducted. In 1992, two patented methods were considered for commercial manufacturing:

1. Removal of cholesterol by beta-cyclodextrin [U.S. Patent 4,980,180 (Cully and Vollenbrecht 1990)]. The cholesterol is bound to the beta-cyclodextrin, which is added to the liquid yolk, and the complex is removed by centrifugation. Residual beta-cyclodextrin is then eliminated from the yolk with alpha-amylase.

Table 12 Microbiological Specifications for Some Egg Products

Product	spc/g[a]	Coliform per g	Yeast mold per g	E. coli per g	Salmonella	Staphylococcus[b]
Pasturized frozen whole egg	<10,000	<10	<10	<10	Neg.	Neg.
Frozen sugared yolk	<10,000	<10	<10	Neg.	Neg.	Neg.
Frozen salted yolk	<10,000	<10	<10	Neg.	Neg.	Neg.
Frozen egg whites	<10,000	10 max	<10	Neg.	Neg.	Neg.
Spray-dried whole eggs	<10,000	<10	<10	—	Neg.	—
Spray-dried ablumen	<10,000	<10	—	—	—	—
Spray-dried yolk	<10,000	<10	<10	—	Neg.	—

Source: Stadelman *et al.* (1988) [a]Standard plate count per gram [b]Coagulase positive

Table 13 Approximate Composition of Selected Egg Products (per 100g)

		Composition, g				
	Calories	Water	Protein	Lipid	CHO	Ash
Product			**Frozen Eggs**			
Whole egg	158	74.57	12.14	11.15	1.20	0.94
White	49	88.07	10.14	Trace	1.23	0.56
Yolk, pure	377	48.2	16.1	34.1	—	1.69
Yolk, commercial*	323	55.04	14.52	28.65	0.36	1.43
Yolk, sugared	323	50.82	12.52	25.50	9.49	1.27
			Dehydrated Eggs			
Whole egg	594	4.14	45.83	41.81	4.77	3.45
Whole egg, stabilized	615	1.87	48.17	43.95	2.38	3.63
White, stabilized, flakes	351	14.62	76.92	0.04	4.17	4.25
White, stabilized, powder	376	8.54	82.40	0.04	4.47	4.55
Yolk	687	4.65	30.52	61.28	0.39	3.16

Source: Stadelman (1992)
*Contains approximately 17% white

2. Cholesterol can be rapidly extracted from egg yolk by high-pressure (600 to 700 psi) homogenizing of the yolk and vegetable oil (*e.g.*, soybean oil mix) followed by separation of the cholesterol-containing oil by centrifugation at 10,000 rpm. The cholesterol is then removed from the oil by steam stripping, which allows the oil to be reused (Zeidler 1992). The cholesterol can then be used to manufacture pharmaceutical products or as growth-promoting feed for aquacultured shrimp, lobster, or other seafood.

DEHYDRATED EGG PRODUCTS

Spray drying is the most commonly used method for egg dehydration. However, other methods are used for specific products such as scrambled eggs, which are made by freeze drying, and egg white products, which are usually made by pan drying to produce a flake-like product. In spray drying (Figure 9), the liquid is atomized by nozzles which operate at 500 to 600 psi. The centrifugal atomizer, in which a spinning disc or rod rotates at 3,500 to 50,000 rpm, creates a hollow cone pattern for the liquid, which enters the drying chamber. The atomized droplets meet a 250 to 450 °F hot air cyclone, which is created and driven by a fan blowing in the opposite direction. Because the surface area of the atomized liquid is so large, the moisture evaporates very rapidly. The dry product is separated from the air, cooled, and, in many cases, sifted before being packaged into fiber drums lined with vapor retarder liners. Military specifications usually call for gas-packaging in metal cans. Moisture level in dehydrated products is usually around 5%, whereas in pan dryer products moisture level is around 2%.

Spray dryers are classified as vertical or horizontal types. However, large variations exist in methods of atomizing, drying air movement, and powder separation.

Whole egg and yolk products naturally contain reduced sugar. In order to extend the shelf life of these products and to prevent color change through browning (Mylard reaction), the glucose in the egg is removed by baker's yeast, which consumes the glucose within 2 to 3 h at 86 °F. The liquid is then pasteurized in continuous heat exchangers at 142 °F for 4 min and dried. Whole egg and yolk powder have excellent emulsifying, binding, and heat coagulating properties, whereas egg white possesses whipping capabilities.

Different dry egg products are used in different baking products such as sponge cakes, layer cakes, pound cakes, doughnuts, and cookies. Numerous dry products exist because it is possible to dry eggs together with other ingredients such as milk, other dairy products, sucrose, corn syrup, and other carbohydrates.

Common Dried Products. Figure 9 shows the processing for several dried egg products. Some common dried egg products include

- Pan-dried egg whites, spray-dried egg white solids, whole egg solids, yolk solids
- Stabilized (desugared) whole egg, stabilized yolk
- Free-flowing (sodium silicoaluminate) whole egg solids, free flowing yolk solids
- Dry blends: whole egg or yolk with carbohydrates such as sucrose, corn syrup
- Dry blends with dairy products such as scrambled egg mix

EGG PRODUCT QUALITY

Criteria usually used in evaluating egg product quality are odor, yolk color, bacteria count, solids and fat content (for yolk and whole egg), yolk content (for whites), and performance. All users want a wholesome product with a normal odor that performs satisfactorily in the ways it will be used. For noodles, a high solids content and color are important. Bakers are particular about performance—whites do not perform well in angel food cake if

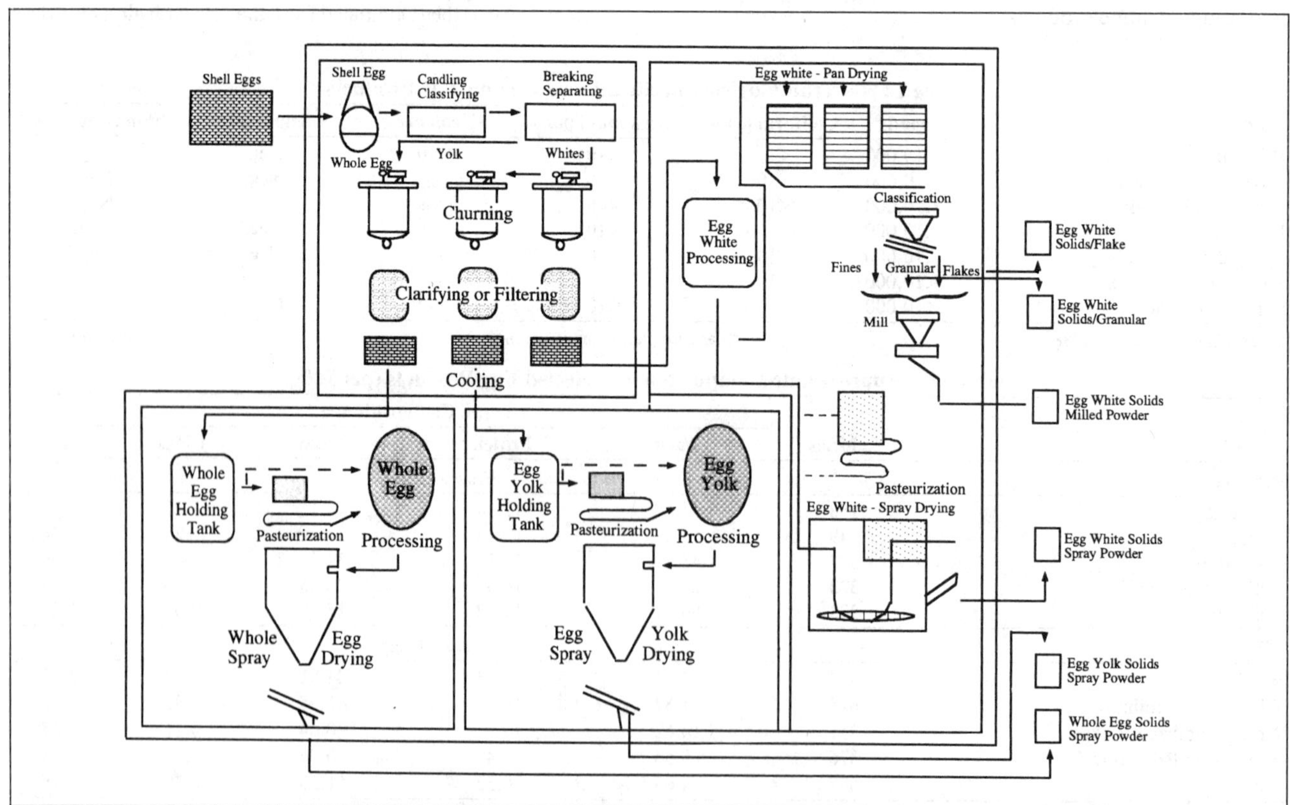

Fig. 9 Steps in Egg Product Drying

excessive yolk is present. They test the foaming performance of whites based on the height and volume of angel food cake and custard. Performance is also critical for candy (using whites). Salad dressing and mayonnaise are used to evaluate the performance of the yolk as an emulsifier, and emulsion stability is tested.

SANITARY STANDARDS AND PLANT SANITATION

In the United States, Egg 3-A Sanitary Standards and Accepted Practices are formulated by the cooperative efforts of the U.S. Public Health Service; the U.S. Department of Agriculture; the Poultry and Egg Institute of America; the Dairy Industry Committee; International Association of Milk, Food, and Environmental Sanitarians; and the Dairy Food Industries Supply Association. The Standards are published by the *Journal of Food Protection* (formerly the *Journal of Milk and Food Technology*).

Egg processing facilities and equipment require daily cleaning and sanitation. Plastic egg flats should be sanitized after each use to avoid microbial contamination of eggs. Chlorine or quaternary-based sanitizers are often used for egg washing and for cleaning equipment, egg flats, floor, walls, etc. Water for egg washing should have low iron content (below 2 ppm) to prevent bacterial growth.

Filters in forced-air egg drying equipment should be cleaned a minimum of once per week. Egg processing rooms should be well ventilated. Inlet air filters should be cleaned weekly. Egg cooling rooms should be kept clean and free from dust or molds.

REFERENCES

Ball, H.R., Jr., M. Hamid-Samirin, P.M. Foegeding, and K.R. Swartzel. 1987. Functional and microbial stability of ultrapasteurized aseptically packaged refrigerated whole egg. *J. Food Sci.* 52:1212-18.

Baird, C.D., J.J. Gafney, and M.T. Talbot. 1988. Design criteria for efficient and cost effective forced air cooling systems for fruits and vegetables. ASHRAE *Transactions* 94(1):1434-54.

Bell, D.D. and R.G. Curley. 1966. Egg cooling rates affected by containers. *California Agriculture* 20(6):2-3.

Burley, R.W. and D.V. Vadehra. 1989. *The avian egg: Chemistry and biology.* John Wiley and Sons, New York.

Carter, T.C. (ed). 1968. *Egg Quality: A study of the hen's egg.* Oliver & Boyd, Edinburgh.

Cully, J. and H.R. Vollenbrecht. 1990. Process for removal of beta-cyclodextrin from egg yolk or egg yolk plasma. U.S. Patent 4,980,180.

Czarick, M. and S. Savage. 1992. Egg cooling characteristics in commercial egg coolers. *J. Appl. Poultry Res.* 1:258-70.

Funk, E.M. 1935. The cooling of eggs. Missouri Ag. Exp. Sta. Bull. 350.

Goble, J.W. 1980. Designing egg shell grading and packaging plants. USDA Marketing Research Report No. 1105.

Hamann, J.A., R.G. Walters, E.D. Rodda, G. Serpa, and E.W. Spangler. 1978. Shell egg processing plant design. USDA/ARS Market Research Report No. 912.

Henderson, S.M. 1957. On-the-farm egg processing: Cooling. *Agricultural Engineering* 38(8):598-601, 605.

Kuney, D.R., S. Bokhari, G. Zeidler, R. Ernst, and D. Bell. 1992. Factors affecting candling errors. Proceedings of the 1992 Egg Processing, Packaging and Marketing Seminar, San Bernardino and Modesto, California.

Shenstone, F.S. 1968. The gross composition, chemistry, and physico-chemical basis of organization of the yolk and white. In: *Egg quality: A study of the hen's egg.* T.C. Carter (ed). Oliver and Boyd, Edinburgh.

Stadelman, W.J., E.L. Baum, J.G. Darroch, and H.G. Walkup. 1954. A comparison of quality in eggs marketing with and without refrigeration. *Food Technology* 8:89-102.

Stadelman, W.J. 1992. Eggs and egg products. In: *Encyclopedia of Food Science and Technology*, Vol. 2. Y.H. Hui (ed). J. Wiley & Sons, Inc., New York.

Stadelman, W.J. and O.J. Cottermill (eds). 1990. *Egg science and technology*, 3rd ed. Food Products Press, Binghamton, New York.

Stadelman, W.J., V.M. Olson, G.A. Shemwell, and S. Pasch. 1988. *Egg and poultry meat processing.* Ellis Harwood, Ltds., Chickester, England.

Tarver, F.R. 1964. The influences of rapid cooling and storage conditions on shell egg quality. *Food Technology* 18(10):1604-06.

USDA. 1990. Egg grading manual. *Agriculture Handbook* No. 75. Agricultural Marketing Service.

Zeidler, G. 1992. Decholesteralyzing of eggs with ready-to-use cholesterol as by-product. U.S. Patent pending.

Zeidler, G. and D. Riley. 1993. The role of humidity in egg refrigeration. Proceedings of the 1993 Egg Processing, Packaging, and Marketing Seminar.

BIBLIOGRAPHY

Anderson, K.E., F.T. Jones, and P.A. Curtis. 1992. Heat loss from commercially packed eggs in post-processing coolers. Extension Report in Poultry Science, North Carolina Cooperative Extension Service.

Anonymous. 1991. Salmonella risk reduction—Integrated guidelines for table egg producers. Proceedings of the 95th Annual Meeting of the U.S. Animal Health Association, San Diego, California.

Balander, R. 1992. Interior egg quality as affected by oiling temperature and storage. *Mid-state Poultry Newsletter* 3(2).

Bell, D.D., G. Johnston, M. Swanson, and R. Ernst. 1978. Egg shell damage during washing. *Progress in Poultry* 12:1-17.

Blake, J.A. 1966. Pre-process handling of eggs. Proceedings of the Rutgers Liquid Egg Processing Course, 17-35. Atlantic City, NJ, March 22-24.

Cottrill, O.J. and G.S. Giger. 1977. Egg product trend yields from shell eggs. *Poultry Science* 56:1027-31.

Cottrill, O.J. 1981 (Revised in 1990). A scientist speaks about egg products. An American Egg Board publication, Park Ridge, IL.

Dawson, L.E. and J.A. Davidson. 1951. Farm practices and egg quality: Part III. Egg-holding conditions as they affect decline in quality. Quarterly Bulletin, Michigan Agricultural Experiment Station 34(1):105-44.

Douglas, C.R., B.L. Damron, and R.D. Jacobs. 1992. Characterization of temperature patterns in transport vehicles. Southeastern International Poultry Exposition.

Harris, C. 1976. An egg grading and processing plant for high volume production. USDA/ARS Marketing Research Report No. 837.

Henderson, S.M. 1958. On-the-farm egg processing: Moisture loss. *Agricultural Engineering* 39(1):28-30, 34.

Henderson, S.M. and F.W. Lorenz. 1951. Cooling and holding eggs on the ranch. Calif. Ag. Exp. Sta. Circular 405.

Hinton, H.R. 1968. Storage of eggs. In: *Egg quality: A study of the hen's egg.* Oliver and Boyd, Edinburgh, U.K.

Mitchell, F.G., R. Guillou, and R.A. Parsons. 1972. Commercial cooling of fruits and vegetables. Calif Ag. Exp. Sta. Manual 43.

Romanoff, A.L. and A.J. Romanoff. 1949. *The avian egg.* John Wiley & Sons, Inc., New York.

Rhorer, A.R. 1991. What every producer should know about refrigeration. *Egg Industry* (May/June):16-25.

Sicer, J.W. 1966. Egg sweating problem. Poultry for profit (leaflet), Purdue University.

Tressler, D.K. and W.J. Sultan. 1975. *Foods products formulary Vol 2. Cereals, baked goods, dairy and egg products.* The AVI Publishing Co., Inc., Westport, CT.

USDA. Agricultural statistics. U.S. Government Printing Office, Washington, D.C.

Van Rest, D.J. 1967. Operations research on egg management. *Transactions of the American Society of Agricultural Engineers*, Chicago, IL. (Dec.):752-55.

Wells, R.G. and C.G. Belyavin. 1984. *Egg quality: Current problems and recent advances.* Butterworth's, London.

Zadow, J.G. 1986. Ultra heat treatment of milk. *Food Technology in Australia* 38(7):290-97.

CHAPTER 24

REFRIGERATED WAREHOUSE DESIGN

A REFRIGERATED warehouse is any building or section with controlled storage conditions, using refrigeration. Two basic storage facilities are (1) coolers that protect commodities at temperatures usually above 32 °F and (2) low-temperature rooms (freezers), operating under 32 °F to prevent spoilage or to maintain or extend product life.

The conditions within a closed refrigerator chamber must be maintained to preserve the stored product. This refers particularly to seasonal, shelf life, and long-term storage. Specific items for consideration include:

- Uniform temperatures
- Length of air blow and impingement on stored product
- Effect of relative humidity
- Effect of air movement on employees
- Controlled ventilation, if necessary
- Product entering temperature
- Expected duration of storage
- Required product outlet temperature
- Traffic in and out of storage area

The Association of Food and Drug Officials (AFDO) developed a guide for establishing standards for all phases of handling frozen foods. The code treats receiving, handling, freezing, storage, and transporting frozen foods and calls for sanitary measures, as well as temperature requirements of 0 °F and lower for holding frozen foods. These standards must be recognized in the design and operation of refrigerated storage warehouses.

Regulations of the Occupational Safety and Health Act (OSHA), Environmental Protection Agency (EPA), U.S. Department of Agriculture (USDA), and other standards must also be incorporated in warehouse operations and procedures.

Refrigerated warehouses may be operated for or by a private company for storage of their own products or as a public facility whereby storage services are offered to many concerns, or for both. Important locations for warehouses, public or private, are (1) point of processing, (2) intermediate points for general or long-term storage, and (3) final distributor or distribution point.

The five categories for the classification of refrigerated storages for preservation of food quality are:

- Controlled atmosphere for long-term storage of fruits and vegetables
- Coolers at temperatures of 32 °F and above
- High-temperature freezers at 27 to 28 °F
- Low-temperature storage rooms for general frozen products, usually maintained at −10 to −20 °F
- Low-temperature storages at −10 to −20 °F, with a surplus of refrigeration for freezing products received above 0 °F

It should be noted that due to ongoing recent research, the trend is toward lower temperatures for frozen foods. Refer to Chapters 20 and 21 in the 1993 ASHRAE *Handbook—Fundamentals* for further information.

The preparation of this chapter is assigned to TC 10.5, Refrigerated Distribution and Storage Facilities.

INITIAL BUILDING CONSIDERATIONS

Private refrigerated space is usually adjacent to, or in the same building with, other operations of the owner.

Public space should be located to serve a producing area, a transit storage point, a large consuming area, or various combinations to develop a good average occupancy. Also, it should have:

- Convenient location for producers, shippers, and distributors, considering the present tendency toward decentralization and avoidance of congested areas
- Good railroad switching facilities and service with minimum switching charges from all trunk lines to plant tracks if a railhead is necessary for the profitable operation of the business
- Easy access from main highway truck routes as well as local trucking, but avoiding location on congested streets
- Location with ample land for trucks, truck movement, and plant utility space plus future expansion
- Location with a reasonable land cost
- Adequate power and water supply
- Provisions for surface, waste, and sanitary water disposal
- Consideration to zoning limitations and fire protection
- Location that avoids residential areas where noise of outside operating equipment, such as fans and engine-driven equipment on refrigerated vehicles, would be objectionable
- External appearance may be a factor in the community
- Taxes and insurance to be investigated
- Plant security
- Favorable undersoil bearing conditions and good surface drainage

Plants are often located away from congested areas or even outside city limits where the cost of increased trucking distance is offset by better plant layout possibilities, a better road network, better or lower priced labor supply, or other economies of operation.

Size Determination

Building orientation and size of a cold storage warehouse are determined by the following factors:

- Is the receipt and shipment of goods to be primarily by rail or truck? Shipping practices will affect the platform areas and internal traffic pattern.
- What relative percentage of merchandise is for cooler and freezer storage? Products requiring specially controlled conditions, such as fresh fruits and vegetables, may justify or demand several individual rooms. Seafood, butter, and nuts also require special treatment. However, overall occupancy may be reduced because of seasonal conditions.
- What percentage is anticipated for long-term storage? Products that are stored long term can usually be stacked more densely.

- Will the product be primarily in small or large lots? The drive-through rack system or a combination of pallet racks and a mezzanine have proved effective in achieving efficient operation and effective space use. Mobile or moving rack systems are also plausible.
- How will the product be palletized? A dense product such as meat, tinned fruit, drums of concentrate, and cases of canned goods can be stacked very efficiently. Palletized containers and special pallet baskets or boxes effectively hold meat, fish, and loose products.

  The slip sheet system, which requires no pallets, eliminates the waste space of the pallet and can be used effectively for some products.

  Pallet stacking racks make it feasible to use the full height of the storage and palletize any closed or boxed merchandise.
- Will rental space be provided for tenants? Rental space usually requires special personnel and office facilities. An isolated area for the tenant operations is also desirable. These areas are usually leased on a unit area basis, and plans are worked into the main building layout.

Stacking Arrangement

Typically, height of refrigerated spaces varies between 28 and 35 ft or higher clear space between the floor and structural steel to allow fork truck operation. Pallet rack systems use the greater height. The practical height for stacking pallets without racks is 15 to 18 ft. The clear space above the pallet stacks is used for air distribution, lighting, and sprinkler lines. Overhead space is inexpensive, and since the refrigeration requirement for the extra height is insignificant, a minimum of 20 ft clear height is desirable. Greater heights are valuable if automated or mechanized material-handling equipment is contemplated.

High stacking may influence insurance rates. The floor area in a warehouse where a diversity of merchandise is to be stored can be calculated on the basis of 8 to 10 lb per gross cubic foot to allow about 40% for aisles and space above the pallet stacks.

For special purpose or production warehouses, products can be stacked with less aisle and open space, to result in an allowance factor of about 20%. Docks should be refrigerated and are an absolute necessity in humid and warm climates. The relatively high cost for doors, door cushion closures, and refrigeration influences the economic dock size and number of doors. A commercial warehouse usually requires more truck dock space than specialized storage due to the variety of products handled.

BUILDING DESIGN

Most refrigerated warehouses are single-story structures. Small columns on wide centers permit palletized storage with minimal lost space. This type of building usually provides additional highway truck unloading space. The single-story design has the following characteristics that must be considered: (1) horizontal traffic distances, which to some extent offset the vertical travel required in a multistory building; (2) difficulty of using the stacking height with many commodities or with small lot storage and movement of goods; (3) the necessity for treatment of the floor below freezers to give economical protection against possible ground heaving; and (4) high land cost for capacity of the building. A one-story design for moderate or low stacking heights has a high cubage cost because of the high ratio of construction costs and added land cost compared to product storage capacity. However, the first cost and operating cost will always be lower than that of a multistory facility.

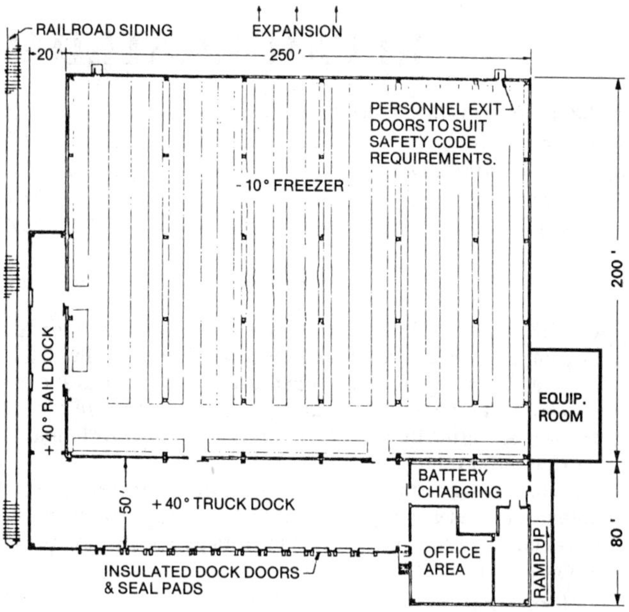

Fig. 1 Typical Plan of One-Story Warehouse

One-Story Configuration

Figure 1 shows a one-story layout incorporating facilities that comply with current practices. The following essential items and functions are considered:

- Equipment room
- Refrigerated shipping docks with seal-cushion closures on the doors
- Automatic doors
- Batten curtains or strip doors
- Low-temperature storage held at −10°F or lower
- Pallet-rack systems to facilitate the handling of small lots and to comply with regulations of First-In-First-Out, which is required for some products
- A blast freezer or separate sharp freezer room for isolation of products being frozen
- Cooler or convertible space
- Space for brokerage offices and for distributors
- Space for empty pallet storage and repair
- Space for shop and battery charging
- Automatic sprinklers in accordance with NFPA regulations

Other areas that must be included in a complete operable facility are:

- Electrical area
- Shipping office
- Administration office
- Personnel welfare facilities

A modified one-story design is sometimes used to reduce horizontal distances and land costs. An alternative is to locate nonproductive services (including offices and the mechanical room) on a second-floor level, usually over the truck platform work area to permit full use of the ground floor for production work and storage. However, potential vibration of the second floor from equipment must be considered.

One story, or modification of the one story, gives the maximum capacity per unit of investment with a minimum of overall operating expense, including cost of investment, refrigeration, and labor. Mechanization must be considered as well.

Designs that give minimum overall costs restrict office facilities and utility areas to a minimum. Dock area is of major importance to assure efficiency in the operation of loading and unloading merchandise.

Shipping and Receiving Docks

Regulations on temperature control during all steps of product handling have led to designing the trucking dock as a refrigerated anteroom to the cold storage area. Typically, loading and unloading of transport vehicles is handled by separate work crews. One crew moves the product in and out of the vehicles, and a warehouse crew moves the product in and out of the refrigerated storage. This procedure may allow the merchandise to accumulate on the shipping dock. Maintaining the dock at 35 to 45 °F offers the following advantages:

- Reduced refrigeration load in the low-temperature storage area, where energy demand per unit capacity of refrigeration is high
- Less ice or frost forms in the low-temperature storage because less warm air infiltrates into the area
- Refrigerated products held on the dock maintain a favorable temperature
- Packaging remains in good condition because it stays dry. Warehouse personnel are more comfortable because temperature differences are less
- Requires less maintenance because condensation on powered lift trucks and other equipment is reduced
- Reduces or eliminates the need for anterooms or vestibules

Utility Space

Space for a general office, locker room, and machine room is needed. A superintendent's and warehouse records office should be located near the center of operations and a checker's office should be in view of the dock and traffic arrangement. Rented space should be isolated from warehouse operations.

The machine room should include ample space for refrigerating equipment and maintenance, adequate ventilation, standby capacity for emergency ventilation, and adequate segregation from other areas. Separate exits are required by most building codes. A maintenance shop and space for parking, charging, and servicing warehouse equipment should be located adjacent to the machine room. Electrically operated materials handling equipment is used to eliminate inherent safety hazards of combustion-type equipment. Battery-charging areas should be designed with high roofs and must be ventilated. This is necessary due to potential combustible fumes that result from the charging activity.

SPECIALIZED STORAGE DESIGNS

Warehouse handling methods and storage requirements dictate design. Automated materials handling within the storage, particularly for high stack piling, may be an integral part of the structural system or require special structural treatments. Controlled atmospheres and minimal air circulation require special building designs and mechanical equipment to achieve required conditions. Drive-in and/or drive-through rack systems can improve product inventory control and can be used in combination with stacker cranes, special narrow-aisle high stacker cranes, and automatic conveyors in various phases of automation. Mobile racking systems may be considered where space is at a premium. In general, warehouses may be classified as follows:

- Public refrigerated warehouse designed to handle all commodities with several chambers to control temperatures from 35 to 60 °F with humidity control and to −20 °F without humidity control.
- Refrigerated warehouse for case and breakup distribution, automated to varying degrees, with racks arranged with the appropriate number of pallet spaces in acceptable pallet faces, to accommodate the projected usage.
- Warehouse designed for a processing operation with a block storage for frozen ingredients and rack storage for palletized outshipment of processed merchandise. An economic adaptation is to adjoin a commercial warehouse to a processing plant. The warehouse would receive inventory of frozen ingredients and deliver to the processor on order. It would also receive processed merchandise, take inventory, and make up outshipments for commercial carrier pickup and delivery. Process specialization for the manufacturer and experienced efficient operation of the warehouse staff are advantages.
- Public warehouse serving several production manufacturers, storing and inventorying products in lots and assembling outshipments for commercial carrier pickup and delivery or for delivery service offered by the public warehouse.
- Specialized mechanized warehouse with stacker cranes, racks, infeed and outfeed conveyors, and special conveyor vestibules. In the design of such a facility, which may be 60 to 100 ft high, the cooling units must be mounted in the highest internal area to pick up moisture infiltration. They must also be mounted to prevent condensation from dripping on the stored product and mechanical equipment below. A penthouse that houses the refrigerating evaporators eleminates equipment in the room and facilitates access for equipment maintenance. Frequently, added rack storage space is available in penthouse designs, and evaporator maintenance can be more convenient because access to this equipment is from the roof.

Controlled Atmosphere Storage Rooms

Controlled atmosphere (CA) storage rooms may be required for specialized storages, particularly those handling apples. In addition to refrigeration apparatus to control temperature, these storages also include special gastight seals to facilitate the maintenance of an atmosphere that is lower in oxygen and higher in nitrogen and carbon dioxide than normal atmosphere. These storage rooms require apparatus to control the CO_2 concentration and, in many cases, use nitrogen-generating or oxygen-consuming devices.

The desired atmosphere must be experimentally determined for the commodity as produced in the specific area that the storage is to serve.

Information is available for certain commodities. Commercial application of CA storage has been limited to fresh fruits and vegetables that respire in storage, consuming O_2 and producing CO_2. The storages may be classified as having either (1) product-generated atmospheres, where the room is sufficiently well sealed so that the normal oxygen consumption of the fruit can consume the O_2 that is in the storage space and that leaks into the room, or (2) externally generated atmospheres where nitrogen generators or O_2 consumers assist the normal action of the fruit. The second type of system can cope with a poorly sealed room, but the cost of operation may be quite high. However, even with the external gas generator system, a well-sealed room is desired.

In most cases, a CO_2 scrubber is required. The only exception is in fruit storages where the total of the desired O_2 and CO_2 content is 21%. In such cases, a normal balance between oxygen consumption and CO_2 exists, and no CO_2 scrubbing is required. Carbon dioxide may be removed by (1) passing the room air over dry lime that is replaced periodically; (2) passing the air through wet caustic solutions where the NaOH is replaced periodically; (3) water scrubbers where the CO_2 is absorbed from the room air by a water spray and then desorbed from the water by passing outdoor air through the water in a separate compartment; (4) monoethanolamine scrubbers that may have the solution regenerated periodically by a manual process or regenerated continuously by automatic equipment; or (5) dry adsorbents automatically regenerated on a cyclic basis.

Among the systems of room sealing are (1) lining the walls and ceiling of the room with 28-gage galvanized steel and connecting this into a floor-sealing system; (2) specially faced plywood or plywood with an impervious sealing system applied to the inside face; (3) sprayed urethane carefully applied and finished with thermal (fire) barriers.

To be sufficiently tight for use as an hermetic room, the space should meet the following test: after being pressurized to 1 in. water gage, the rate of pressure loss is observed; if, at the end of one hour, 0.1 to 0.2 in. remains and the test has been conducted under uniform temperature and barometric conditions, the room is sufficiently sealed. A satisfactory external gas generator type of room may lose pressure at double the above rate. The test prescribed for the room using product-generated atmosphere is about one air change of the empty room in a 30-day period.

Extreme care in all details of construction is required to obtain a seal that passes this test. Doors are sealed and have sills that can be bolted down and sealed. Electrical conduits and special seals around all pipe and hanger penetrations allow some movement, but the seal remains intact. Structural penetrations through the seal must be avoided, and the structure must be stable. CA rooms in multifloor frame buildings, where the structure deflects appreciably under load, are extremely difficult to seal. Most gas seals are applied in the wrong place, *i.e.*, at the cold side of the insulation, so they can be readily maintained and points of leakage can be detected. Some moisture entrapment is to be expected, and insulation materials must be carefully selected so that minimal damage will result from this incorrect placement of a vapor barrier.

In some installations, cold air with a dew point lower than the inside surface temperature is circulated through a space between the insulation and the gas seal for dryness. For additional information on CA storage, see Chapters 15 and 17.

Jacketed Storages

Mechanically cooled walls, floors, and ceilings may be economical for long-term storage duty, such as bin storage, CA storage, and for products that deteriorate with active air movement. Embedded pipe coils or airspaces through which refrigerated air is recirculated can provide the cold surface. With this method, leakage is absorbed in the jacket and prevented from entering the refrigerated space.

The following must be considered in the initial design of the storage:

- Initial cool-down of the product, which can impose short-term peak loads
- Service loads when loading product in and out
- Odor contamination from products that are not compatible
- Product heat of respiration in cooler spaces

Supplementary refrigeration or conditioning units in the refrigerated space, which would be operated only as required, can usually alleviate these problems.

Automated Warehouses

Automated warehouses are usually tall fixed rack arrangements with stacker cranes under full automatic, semiautomatic, or hand control. The control systems can be tied into a computer system to retain a complete inventory of product and location.

Some of the advantages are:

- First-In-First-Out inventory can be controlled.
- The enclosure structure is high, requiring a minimum of floor space and favorable cubage cost.
- Product damage and pilfering are minimized.
- Direct material handling costs are minimized.

Some of the disadvantages are:

- Very high first cost of the racking system and building compared to conventional designs
- Acces may be slower, depending on product flow requirements
- Cooling equipment may be difficult to access for maintenance
- Air distribution must be carefully evaluated in the design phase

CONSTRUCTION METHODS

Cold storages, more than most construction, require correct design, quality materials, good workmanship, and close supervision. Design should ensure that proper installation can be accomplished under various adverse job-site conditions. Materials must be compatible with each other. Installation must be made by careful workers directed by an experienced, well-trained superintendent. Close cooperation between the general, roofing, insulation, and other contractors increases the possibility of a successful installation.

Construction methods can be classified as either (1) insulated structural panel, (2) mechanically applied insulation, or (3) adhesive or spray-applied foam systems. These construction techniques seal the insulation within an air- and moisture-tight envelope, which must not be violated by major structural components.

Three ways are used to achieve an uninterrupted envelope. The first and simplest is total encapsulation of the structural system by an exterior vapor barrier under the floor, on the outside of the walls, and over the roof deck (Figure 2). This method offers the least number of penetrations through the vapor barrier, as well as the lowest cost. The second method is an entirely interior system, where the vapor barrier envelope is placed within the room and insulation is added to the walls, floors, and suspended ceiling (Figure 3). This technique is used where walls and ceilings must

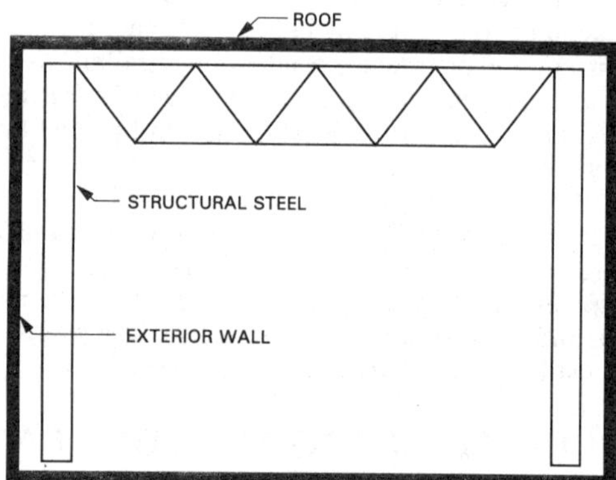

Fig. 2 Total Exterior Vapor Barrier System

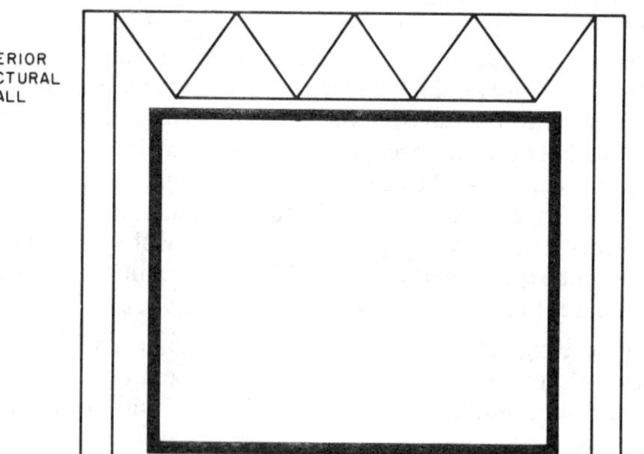

Fig. 3 Entirely Interior Vapor Barrier System

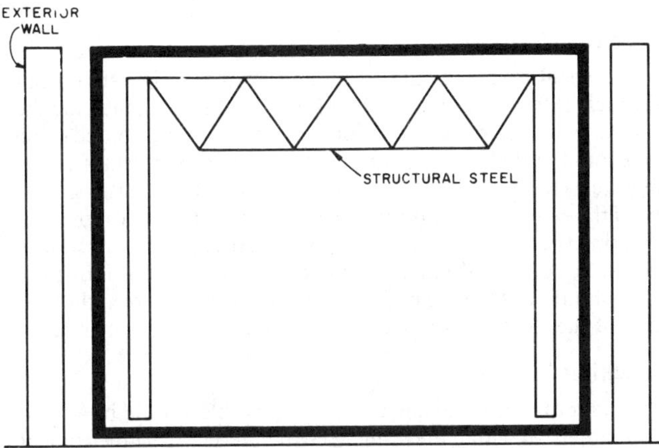

Fig. 4 Interior-Exterior Vapor Barrier System

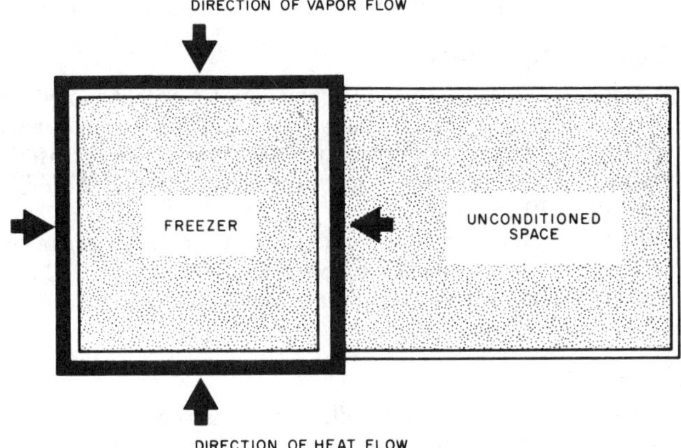

Fig. 5 Separate Vapor Barrier System for Each Area of Significantly Different Temperature

be washed, where an existing structure is converted to low-temperature space, or for smaller rooms located within large coolers or unrefrigerated facilities. The third method is inside-outside construction (Figure 4), involving an exterior curtain wall of masonry or similar material tied to an interior structural system. Adequate space allows the vapor barrier insulation system to turn up over a roof deck and be incorporated into a roofing system, which serves as the vapor barrier. This construction method is a viable alternative, although it offers more interruptions in the vapor barrier than the exterior system.

The total exterior vapor barrier system (Figure 2) is best because of the advantages of fewest penetrations and lowest cost. Each area of widely varying temperature should be divided into separate envelopes to retard heat and moisture flow between them (Figure 5).

Space Adjacent to Envelope

Condensation at the envelope is usually caused by high humidity and inadequate ventilation. Most often, this occurs within a dead air space, such as a ceiling plenum or hollow masonry unit, through-metal structure, or beam cavity. All closed air spaces should be eliminated, except those large enough to be ventilated adequately. Ceiling plenums, for instance, are best ventilated by mechanical vents, which move air above the envelope.

The insulation envelope and vapor barrier should not be penetrated if possible. Insulate and vapor-seal all steel beams, columns, and large pipes that project through the insulation with a 4-ft wrap of insulation. Conduit, small pipes, and rods should be insulated for a distance four times the regular wall insulation thickness. (Conduits and small pipes should be vapor-sealed on the inside to prevent moisture flow.) In both cases, the thickness of insulation on the projection should be half that on the regular wall or ceiling. Fill any voids within metal projections. Locate the wrap insulation on the warm side where practicable, and seal the metal projection on the warm side.

Air/Vapor Treatment at Junctures

Air and vapor leakage at wall/ceiling junctures is probably the predominant construction problem in cold storage facilities.

When a cold room of interior-exterior design (Figure 4) is lowered to operating temperature, the structural elements (roof deck and insulation) contract and can pull the ceiling away from the wall. Because of negative pressure in the space of the wall/ceiling juncture, warm moist air can leak into the room and form frost and ice. Therefore, proper design of the air/vapor seal is critical.

An air/vapor flashing sheet system is best for preventing leakage. This is a transition from the roof system/vapor barrier to the exterior wall system/vapor barrier. A good corner flashing sheet must be flexible, tough, airtight, and vaportight. Proper use of flexible insulation at overlaps, mastic adhesive, and a good mastic sealer ensure leak-free performance. To remain airtight and vaportight during the life of the facility, a properly constructed vapor barrier should:

- Be flexible enough to withstand building movements that may occur at operating temperatures
- Allow for thermal contraction of the insulation as the room is pulled down to operating temperature
- Be constructed with a minimum of penetrations that might cause leaks (Wall ties and structural steel that extends through the corner flashing sheet may eventually leak no matter how well sealed during construction; keep these to a minimum, and make them accessible for maintenance.)
- Have corner flashing sheet properly lapped and sealed with adhesive and mechanically fastened to the wall vapor barrier
- Have corner flashing sealed to roof without openings
- Have floor to exterior vapor barriers that are totally sealed

The interior-exterior design is likely to be unsuccessful because of extreme difficulty in maintaining an air- or vaportight environment.

The practices outlined for the wall/roof junction apply for other insulation junctions. The insulation manufacturer and designer must coordinate the details of the corner flashing design together.

Poor design and shoddy installation cause moist air leakage into the facility, resulting in frost and ice formation, energy loss, poor appearance, loss of useful storage space, and, eventually, expensive repairs.

Floor Construction

Refrigerated facilities held above freezing need no special underfloor treatment. A below-the-floor vapor barrier is needed in facilities held below freezing, however. In these facilities, the subsoil eventually freezes, and any moisture in this soil will also freeze and cause floor frost heaving. In moderate climates, underfloor tubes vented to ambient air are sufficient to prevent heaving. Artificial heating, either by air circulated through underfloor ducts or glycol circulated through plastic pipe is the preferred method to prevent frost heaving. Electric heating cables installed under the floor can also be used to prevent frost formation. The choice of heating method depends on energy cost, reliability, and maintenance requirements.

Future facility expansion must be considered when underfloor heating systems are being designed. Therefore, a system including artificial heating methods that do not require building exterior access is preferred.

Surface Preparation

When an adhesive is used, the surface against which the insulating material is to be applied should be smooth and dust-free. Where room temperatures are below freezing, masonry walls should be leveled and sealed with cement back plaster. Smooth poured concrete surfaces may not require back plastering.

No special surface preparation is needed for a mechanical fastening system, assuming that the surface is reasonably smooth and in good repair.

The surface must be warm and dry for a sprayed foam system. Any cracks or construction joints must be prepared to prevent projection through the sprayed insulation envelope. All loose grout and dust must be removed to assure a good bond between the sprayed foam insulation and the surface. Very smooth surfaces may require special bonding agents.

No special surface preparation is needed for panels used as a building lining or as the primary wall structure, assuming the surfaces are sound and reasonably smooth.

Finishes

Insulated structural panels with metal exterior and metal or reinforced plastic interior faces are popular for both coolers and freezers. Their use keeps moisture from the insulation and leaves only the joints between panels as potential areas of moisture penetration. They are also available with surface finishes that meet government requirements.

For sanitary washdowns, a scrubbable finish is sometimes required. Such finishes generally have low permeance, and when applied on the inside surface of the insulation, a lower permeance treatment is required on the outside of the insulation.

All insulated walls and ceilings should have an interior finish. The finish should be impervious to moisture vapor and should not serve as a vapor retarder, except for panel construction. The permeance of the in-place interior finish should be significantly greater than the permeance of the in-place vapor barrier.

To select an interior finish to meet in-use requirements of the installation, the following factors should be considered: (1) fire resistance, (2) washdown requirements, (3) mechanical damage, (4) moisture and gas permeance, and (5) government requirements. All interior walls of insulated spaces should be protected by bumpers and curbs wherever there is a possibility of damage to the finish.

Hung Ceilings

Suspended insulated ceilings perform well when they are surfaced with a vapor barrier flashed to the wall insulation vapor barrier on the top or warm side. The space above the ceiling must be ventilated a minimum of six air changes per hour to minimize the possibility of condensation. Roof-mounted exhaust fans and uniformily spaced vents around the perimeter of the plenum may be used. The mechanical ventilators should be thermostat and/or humidstat controlled to turn off when (1) the outside temperature is below 50 °F and (2) the outside humidity is below 60% rh. At these conditions, condensation rarely occurs, and the ventilation only reduces the insulating effect of the dead air space.

Hung ceilings should be designed for light foot traffic for inspection and maintenance access. Fastening systems for ceiling panels include spline U channel and camlock.

Floor Drains

Floor drains should be avoided if possible, particularly in freezers. If necessary, they should have short, squat dimensions, and be placed high enough on top of floor insulation in the room to allow the drain and piping to be installed above the insulation envelope.

Electrical Wiring

Electrical wiring should be brought into a refrigerated room through as few locations as possible (preferably one), piercing the wall vapor barrier and insulation only once. Plastic-coated cable is recommended for this service where codes permit. If codes require conduit, the last fitting on the warm side of the run should be of the explosion-proof type, sealed to prevent water vapor from entering the cold conduit. The light fixtures in the room should not be vapor-sealed but should allow free passage of moisture. Care should be taken to maintain the vapor seal between the outside of the electrical service and the cold room vapor barrier.

Tracking

Cold room meat tracking, wherever possible, should be erected and supported within the insulated structure, entirely independent of the building itself. This places all the weight of the tracking on the cold room floor, eliminates flexure of the roof structure or overhead members, and simplifies maintenance.

Cold Storage Doors

Doors should be strong and, at the same time, light enough for easy opening and closing. Hardware should be of good quality, so that it can be set to compress the gasket uniformly against the casing. All doors to rooms operating below freezing should be equipped with heaters.

In-fitting doors are not recommended for rooms operating below freezing, unless they are provided with heaters, and they should not be used at temperatures below 0 °F with or without heaters.

Hardware

All metal hardware, either within the construction or exposed to conditions that will rust or corrode the base metal, should be heavily galvanized, plated, or otherwise protected.

Refrigerated Docks

The type of warehouse, whether it is for distribution, in-transit storage, or seasonal storage, will dictate the loading dock requirements. Shipping docks and corridors should provide liberal space for (1) movement of goods to and from storage, (2) storage of pallets and idle equipment, (3) sorting, and (4) inspecting. The dock should be 30 ft or wider.

Floor heights of refrigerated vehicles vary widely but are often higher than unrefrigerated vehicles. Rail dock heights and building clearances should be verified by the railroad serving the plant. Truck dock heights must comply with the requirements of fleet owners and clients, as well as the requirements for local delivery trucks. A 54-in. dock height above the rail is typical for refrigerated rail cars. Trucks generally require a 54-in. height above the pavement, although local delivery trucks may be much lower. Some reefer trucks are up to 58 in. above grade. Adjustable ramps at some of the truck spots will partly compensate for height variation. Three to five railroad car spots per million cubic feet of storage should be planned. If dimensions permit, seven to ten truck spots per million cubic feet should be provided in a public warehouse.

Refrigerated docks maintained at temperatures of 35 to 45 °F require about 5 tons of refrigeration per 1000 ft^2 of floor area. Cushion-closure seals around the truck doorways reduce infiltration of outside air. An inflatable or telescoping enclosure can be extended to seal the space between a railcar and the dock. Insulated doors for docks must be mounted on the inside walls.

REFRIGERATION SYSTEM SELECTION

The selection of the refrigeration system for a refrigerated warehouse must be established in the early stages of planning. If it is a single-purpose, low-temperature storage building, almost any type of system can be applied. However, if commodities to be stored require different temperatures and humidities, a system must be selected that may require the use of several isolated rooms and conditions.

Using factory-built package unitary equipment may have merit for the smallest structures and also for a multiroom facility that requires a variety of storage conditions. Conversely, the central compressor room has been the accepted standard for larger installations, especially where energy conservation is important.

Direct refrigeration, either a flooded or pumped recirculation system serving fan coil units, is a dependable choice for a central compressor room. Screw compressors, programmable logic controllers, and microprocessor controls complement the central engine room refrigeration equipment.

Load Determination

The refrigeration loads of warehouses of the same capacity vary widely. Many factors, including building design, indoor and outdoor temperatures, and most important, the type and flow of goods expected, plus the daily freezing capacity, contribute to the load. Therefore, no simple design rules apply. Experience from comparable buildings and operations is valuable, but an analysis of any projected operation should be made. Compressor and room cooling equipment should be designed for maximum daily requirements, which will be well above any monthly average.

The factors to be considered include:

- Heat transmission through insulated enclosures
- Heat and vapor infiltration load from warm air passing into refrigerated space
- Heat from pumps or fans circulating refrigerated brine or air, power equipment, personnel working in refrigerated space, product moving equipment, and lights
- Heat removed from goods in reducing them from receiving to storage temperatures
- Heat produced by goods in storage
- Heat to be removed in freezing goods received unfrozen
- Other loads, such as office air conditioning, car precooling, or special operations inside the building
- Refrigerated shipping docks
- Heat released from automatic defrost units from the fan motors and defrosting, which increases overall refrigerant system requirements
- Blast freezing or process freezing extra

Areas subject to high humidity, warm temperatures, or manual product handling may dramatically effect the design, particularly the refrigerating system.

Heat leakage or transmission load can be calculated using the known overall heat transfer coefficient of various portions of the insulating envelope, the area of each portion, and the temperature difference between the lowest design cold room temperature and the highest average air temperature for three to five consecutive days at the building location. For floors on ground, the average yearly temperature should be used instead of the maximum or the average underfloor temperature to be maintained under freezer storages.

Heat infiltration load varies greatly with the following variables: size of room, number of openings to warm areas, protection on openings, traffic through openings, cold and warm air temperatures, and humidities. Basis for calculation should be on experience, remembering that most of the load usually occurs during the day-time operations. Chapter 26 presents a complete analysis of load calculation.

A summation of the average proportional effect of the load factors, as itemized previously and influenced by the type of warehouse design and usage, is shown in Table 1 as a percentage of total load.

Heat from goods received for storage can be approximated from the quantity expected daily and knowing the source. Generally, 10 to 20 °F of temperature reduction can be expected, but for some newly processed items and for fruits and vegetables direct from harvesting, 60 °F or more temperature reduction may be required. For general public cold storage, the load should run 4 to 8 tons cooling capacity per million cubic feet to allow for items received direct from harvest in a producing area.

Heat is produced by many commodities in cooler storage, principally fruits and vegetables. Heat of respiration is a sizable factor, even at 32 °F, and is a continuing load throughout the storage period. Refrigeration loads should be calculated for maximum expected occupancy of such commodities.

The refrigerating load to be provided for freezing will vary from zero for the purely distribution warehouse, to the major portion of the total load for a warehouse near a producing area. The freezing load will depend on the commodity, the temperature at which it is received, and the method of freezing. More refrigeration is required for blast freezing than for still freezing without forced-air circulation.

Air Units and Coils

Fan and Coil Units. These units may have either direct-expansion, flooded, or recirculating liquid evaporators with either primary or finned coil surfaces or a brine spray coolant. Storage temperature, packaging method, type of product, etc., must be considered when selecting a unit. The coil surface area, temperature difference between the refrigerant coil and return air, and airflow volume depend on the particular application. If air units must be located above the entrances to a refrigerated space, they will draw in warm, moist air from the adjacent space, which will increase the frequency of defrosting.

Automatic Defrosting. Properly engineered and installed systems can be automatically defrosted successfully with hot gas, water, electric, or continuously sprayed brine. The sprayed brine system has the advantage of producing the full refrigeration capacity at all times; however, it does require a supply and return pipe system with a means of boiling off the absorbed condensed moisture.

Table 1 Refrigeration Design Load Factors for Typical 100,000 ft^2 Single-Floor Freezer

	Long-Term Storage		Short-Term Storage		Distribution Operation	
	Cooling Capacity		Cooling Capacity		Cooling Capacity	
Items of Refrigeration Load	Tons	Percent	Tons	Percent	Tons	Percent
Transmission losses	98	49	98	43	98	36
Infiltration	10	5	20	9	40	15
Internal operation loads	50	25	56	24	62	22
Cooling of goods received	7	3	15	6	30	11
Other factors	35	18	41	18	45	16
Total design capacity	200	100	230	100	275	100

Water defrost systems require large supply and drain water lines, which must be designed for quick and effective draining in freezers on the defrost cycle. Hot gas and electric systems on freezer duty must be provided with insulated heated drain pan and drain lines, which are pitched to assure quick and complete drain-off of melted frost. The drain lines are usually insulated with a removable, noncombustible material.

Some unit designs are enclosed in insulated casings and have automatic dampers to recirculate the warmed defrosting air within the casing. These design features provide higher efficiency and keep defrost vapors from escaping into the refrigerated space and connecting ductwork.

Regardless of the type of air units selected, air distribution is very important. For uniform temperature and air motion, room coils widely distributed over the ceiling give good results. When using air units, the air should be distributed as evenly as possible.

The cold side of the ceiling insulation should be vapor sealed adjacent to fan coil units hung directly under the insulated ceilings. During the defrost cycle, vapor and warm air at the front and rear of the units will rise and condense out as frost or free water on the underside of the insulated ceiling. If moisture penetrates the ceiling insulation, the freeze-thaw cycle will destroy the insulation. Under these circumstances, the ceiling insulation in this area should be protected with a vaportight sheet such as aluminum or glassboard.

Multiple Installations. To distribute air without ductwork, multiple installations of coil and fan units have been used. The quantity of air circulated per unit capacity of refrigeration may be higher than that required for sprayed unit systems. Thus, proper application is important. For single-story buildings, sprayed units installed in penthouses with duct distribution of the air have been used to make full use of floor space in the storage area (Figure 6). Either prefabricated or field-erected refrigeration systems or cooling units connected to a central refrigeration plant can be incorporated in penthouse design.

Unitary cooling units are located in a penthouse, with distributing ductwork projected through the penthouse floor and under the insulated ceiling below. Return air passes up through the penthouse floor grille. This system avoids the interference of fan coil units hung below the ceiling in the refrigerated chamber. It also facilitates maintenance access.

Defrost water drain piping passes through the penthouse insulated walls and onto the main storage roof. Refrigerant mains and electrical conduit can be run over the roof on suitable supports to the central compressor room or packaged refrigeration units on the adjacent roof. Temperature control thermostats and electrical equipment can be housed in the penthouse.

A personnel access door into the penthouse is required for convenient equipment service. The inside insulated penthouse walls and ceiling must be vaportight to keep condensation from deteriorating the insulation and to maintain the integrity of the building vapor barrier. Some primary advantages are:

- Cooling units, catwalks, and piping do not interfere with product storage space and are not subjected to physical damage from stacking truck operations.
- Service to all cooling equipment and controls can be handled by a single individual from essential floor or roof deck location.
- Maintenance and service costs are minimized.
- Main piping, control devices, and block valves are not located within the refrigerated space.
- If control and block valves are located outside the penthouse, potential refrigerant leaks will occur outside the refrigerated space.

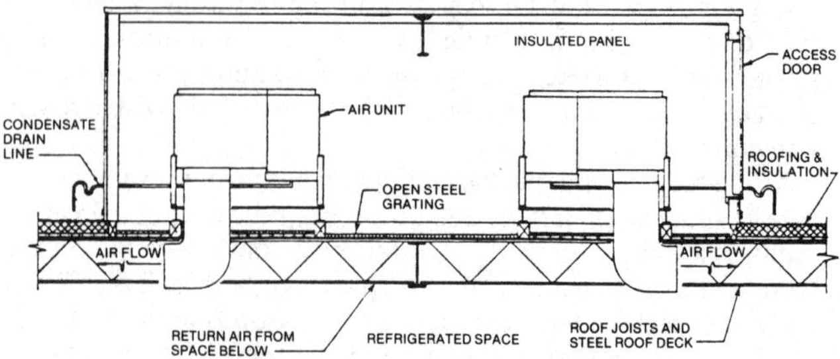

Fig. 6 Penthouse Application of Cooling Units

Freezing Facilities

The public refrigerated warehouse for unfrozen goods may be located in an area where products from processors are available for quick freezing and storage. Air-blast freezing is the usual method. The equipment must be of adequate capacity to accommodate the anticipated load. Variables in freezing characteristics of the products make using two or more freezing cells with dedicated fans and coils advisable.

These freezing cells can be constructed within the freezer storage space and oriented near the shipping platform to ensure efficient material handling. To ensure air movement through the stacked product and resultant efficient freezing, special racks or pallets with spacers for stacking the product in the blast freezer chamber should be provided.

Types of Refrigeration Systems and Equipment

Direct refrigeration systems, rather than brine systems, are used almost exclusively. The refrigerant is either flooded using surge drums controlled by float valves, or recirculated with a pump from an accumulator system. The latter system is advantageous for heavy concentrated loads, such as blast freezers, and is usually preferred for room unit coolers with forced-air circulation.

The indirect brine system has advantages in simplicity of operation, ease of control, avoidance of possible leakage of refrigerant into storage areas, and other factors that tend to offset the savings of the direct system. Brine is particularly desirable for convection-coil installation and for widespread systems. However, the high initial and operating cost has almost eliminated the use of brine systems in new construction.

The refrigerant generally used for central plant systems has been ammonia. The initial and operating cost of halocarbon equipment is slightly higher than ammonia equipment, but in some cases, this additional investment is justified to meet local conditions. However, environmental concerns and future availability should be considered.

For larger plants, the screw-type compressor may have a somewhat lower initial cost, and maintenance should be less than with reciprocating compressors.

In ammonia systems, oil accumulates in vessels normally located outside the product storage rooms. These vessels are equipped with properly valved (and sometimes heated) oil collection pots to remove oil from the system. Modern screw compressors have sophisticated oil separation systems, which dramatically reduce oil contamination problems.

Automation in the form of individual room temperature control, automatic defrosting, and safety shutdown in case of equipment malfunction is common in most warehouses. A central panel monitors mechanical and electrical equipment and announces trouble areas immediately.

Refrigeration plant equipment should be designed for maximum load of the plant. All equipment should be in multiple units to allow for economical operation at light loads and continuity of operation in case of failure of one unit.

For temperatures below −25 °F, two-stage compression is generally used. Compound compressors having capacity control on each stage may be used; but for variable loads of the refrigerated warehouse, separate high- and low-stage (or booster) compressors may be better. Each should have capacity control, or the units should be of different capacity for balancing both high- and low-stage operating conditions. If blast freezers are included, pipe connections should be arranged so that sufficient booster capacity for the blast freezers can be operated at a lower suction pressure, while the other booster capacity is at higher suction pressure for the freezer room load.

In a two-stage system, the liquid refrigerant should be precooled at the high-stage suction pressure to reduce the low-stage load by 10 to 20%. A good, automatic purger to remove air and other noncondensable gases is essential. Ammonia plants should have oil collection vessels at all points where oil may accumulate in the system. Traps should be placed on the compressor discharge, condensers, intercoolers, and evaporators so that oil may be removed periodically without loss of ammonia. Chapters 1 through 7 have further information on refrigeration systems.

The use of commercial, air-cooled, package or factory-assembled units is common, especially in smaller plants. These units have the advantages of lower initial cost, smaller space requirements, and no need for a special machine room or operating engineer. But they use more energy and have higher operating costs, maintenance costs, and a shorter life expectancy for components (usually compressors) than central refrigeration systems.

INSULATION TECHNIQUES

The two main functions of an insulation envelope are to reduce economically the refrigeration requirements for the refrigerated space and to prevent condensation on the exterior.

Vapor Retarder Systems

The primary concern in the design of a low-temperature facility is the vapor retarder system, which should be 100% effective. *The success or failure of an insulation envelope is due entirely to the vapor retarder systems used to prevent water vapor transmission into and through the insulation.*

Once water vapor passes a vapor retarder, a series of events, all of which are detrimental to the refrigerated warehouse and its operation, begins. For example, during warm and humid weather, water vapor migrating through the insulation is chilled to its dew point somewhere in the insulation and condenses to water. In the case of a cooler, the insulation remains wet, while in a freezer, ice forms. In both cases, the value of the insulation will gradually decrease, and it will eventually be destroyed.

Part of the water vapor entering the insulation, after condensing or freezing, will revaporize or sublime. This vapor may again condense as it advances from the warm side to the cold side of the insulation. Additional infiltration moisture follows, and the insulation gradually becomes less effective because of moisture buildup. Some moisture passes into the refrigerated space and collects on the refrigerating coils, but it is not usually sufficient to dry out the insulation unless the vapor barrier discontinuity is located and corrected.

In walls with insufficient insulation, the dew point of the migrating water vapor may, during certain periods, be reached at the inside wall surface and cause condensation and freezing. This can also happen to a wall that originally had adequate insulation but, through water or ice formation in the insulation, lost its insulating value. In either case, the ice deposited on the wall will gradually push the insulation and protective covering away from the wall until the insulation structure collapses.

It is extremely important to properly install vapor retarders and seal joints between the vapor barrier material to ensure the continuity of the vapor barrier from one surface to another, *i.e.*, wall-to-roof, wall-to-floor, or wall-to-ceiling. The failure of vapor retarder systems is almost entirely because of the workmanship during application of the systems for a warehouse.

Types of Insulation

Rigid Insulation. Insulation materials, such as polystyrene, polyisocyanurate, polyurethane, and phenolic material, have proven satisfactory when installed with the proper vapor barrier and finished with materials that provide protection from fire and form a sanitary surface. Consideration should be given to the selection of the proper insulation material based primarily on economics on an installed basis including the finish, sanitation, and fire protection.

Panel Insulation. The use of prefabricated insulated panels for insulated wall and roof construction is widely accepted. These panels are assembled around the building structural frame or can be installed as liner panels in an existing facility. Economics normally prohibits the use of liner panels. These panels can be insulated at the factory with either polystyrene or urethane. Other insulations do not lend themselves to panelized construction.

The basic advantage, besides economics, in using insulated panel construction is that repair and maintenance are simplified because the outer skin also serves as a vapor barrier and is conveniently available. This is of great benefit if the structure is to be enlarged in the future. Proper vapor barrier tie-ins then become practical.

Formed-in-Place Insulation. This application method is gaining acceptance as a result of the developments in polyurethane insulation and equipment for installing this insulation. Portable blending machines are available with a spray or frothing nozzle connected by hoses to pumps and material tanks or with pre-packed pressurized material tanks. The partially expanded mixture in the case of froth foaming is fed from the nozzle into the insulation wall, floor, or ceiling cavities to fill the insulation space provided for monolithic insulation construction without joints. Froth foams set in approximately 2 min. Spray foams, however, are normally from a fast-reacting urethane system, which rises and sets in about 10 to 15 s. Expansion pressures are considerably higher for sprayed urethane than for froth urethane.

INSULATION THICKNESS

The R-value of insulation required varies with the temperature held in the refrigerated space and the conditions surrounding the room. Generally, the R-values in Table 2 are recommended for different types of facilities. The range in R-values results from variation in cost of energy, insulation materials, and climatic condi-

Table 2 Recommended R-Values

Type of Facility	Temperature Range, °F	Floors R	Walls/Suspended Ceilings R	Roofs R
Cooler[a]	40 to 50	Perimeter insulation only	25	30 to 35
Chill cooler[a]	25 to 35	20	24 to 32	35 to 40
Holding freezer	−10 to −20	27 to 32	35 to 40	45 to 50
Blast freezer[b]	−40 to −50	30 to 40	45 to 50	50 to 60

Note: Because of the wide range in the cost of energy and insulation materials on the thermal performance basis, a recommended R-value is given as a guide in each of the respective areas of construction. For more exact values, consult a designer and/or insulation supplier. R-value units are $h \cdot ft^2 \cdot °F/Btu$.

[a] If a cooler has the possibility of being converted to a freezer in the future, the owner should consider insulating the facility with the higher R-values from the freezer section.

[b] R-values shown are for a blast freezer built within an unconditioned space. If built within a cooler or freezer, consult a designer and/or insulation supplier.

tions. For more exact values, consult a designer and/or insulation supplier. R-values less than those shown should not be used.

APPLYING INSULATION

The method and materials used to apply insulating materials to walls, floors, ceilings, roofs, and doors need careful consideration. Four accepted construction techniques are described in this section.

Roofs

Insulating materials should preferably be placed on the roof or floor above the refrigerated space, not under the ceiling. If this type of construction is not feasible and the insulation must be installed under a concrete ceiling or hung from a structural framework, the vapor barrier, insulation, and finish materials should be mechanically supported from the structure above, and should not rely on adhesive application only. Suspending a wood or metal deck from the roof structure and applying insulation and a vapor barrier to the top of the deck is another method of hanging ceiling insulation. Prefabricated panels may also be suspended under the roof. Methods of application and the support structure vary with type of insulation, but skill of application and attention to positive air and vapor seals are essential to continued effectiveness.

Suspended insulated ceilings, whether built-up or prefabricated, require adequate ventilation to maintain the plenum space near ambient conditions to minimize condensation and deterioration of vapor barrier materials. A minimum of six air changes per hour is adequate for most cases. The need for permanent sealing in insulating hanger rods, columns, conduit, and other penetrations is also important.

When installing insulation on top of metal decking or concrete structural slabs for a building larger than 100 ft by 100 ft, the structural designer usually includes roofing expansion joints. Since the refrigerated space is not normally subject to temperature variations, structural framing is usually designed with no expansion or contraction joints if the structural framing is entirely enclosed within the insulation envelope. Board insulation laid on metal decking should be installed in two or more layers with the seams staggered.

An examination of the coefficients of linear expansion for typical roof construction materials illustrates the need for careful attention to this phase of the building design. The data given in Table 3 should be verified for specific products.

Table 3 Coefficients of Linear Expansion

Material	Linear Thermal Expansion, $10^{-6}/°F$
Aluminum	12
Brick	3.4
Wood (pine)	3.0
Rigid glass foam	4.6
Polystyrene, bead board	15
Polystyrene, foam	40
Polyurethane, foam	60
Corkboard	80
Concrete	5.5
Steel	6.7
Tile	3.6
Asphalt, roof-grade	120
Roofing, asphalt-impregnated	
Cross	15
Length	25
Coal tar pitch	90
Expanded perlite board	
Cross	900
Length	2000

While asphalt built-up roofs have been used, loosely laid membrane roofing has become popular and has proven relatively maintenance free.

Walls

Wall construction must be designed so that as few structural members as possible penetrate the insulation envelope. Insulated panels applied to the outside of the structural frame prevent conduction through the framing. Structural framing must be independent of the exterior wall where masonry or concrete wall construction is used. The exterior wall cannot be used as a bearing wall unless a hung insulated ceiling is used instead of the envelope method of insulation construction. Usually, tie rods connect the top of the masonry wall to the interior structural steel. These tie rods require special attention to seal the vapor barrier properly.

Where interior insulated partitions are required, a double column arrangement at the partition prevents structural members from penetrating the wall insulation. Envelope construction should always be used where possible for satisfactory operation and long life of the insulation structure.

Governing codes for fire prevention and sanitation must be followed before a finish is selected for wall insulation. For conventional insulation materials other than prefabricated panels, a vapor retarding system should be selected for walls of the membrane type, *i.e.*, free from the structural wall and flexible enough to withstand building movement.

Abrasion-resistant membranes, such as 10-mil thick black polyethylene film with a minimum of joints, are suitable vapor retarders. These membranes may be mechanically attached to the wall without adhesives. Rigid insulation can then be installed dry and finished with plaster or sheet finishes, as the specific facility requires. Contraction of the interior finish is of more concern than expansion, since temperatures are usually held far below installation ambient temperatures.

Floors

Some freezer buildings have been constructed without floor insulation, and some operate without difficulty. However, the possibility of failure is so great that this practice is seldom recommended.

Underfloor ice formation with resultant heaving of floors and columns can be prevented by adding heat to the soil or fill under the insulation. Air ducts, electric heating elements, or pipes through which a liquid is recirculated are used.

The air duct system works well for smaller storages. For a larger storage, it should be supplemented with fans and a source of heat when the length of pipe is more than 100 ft. The end openings should be screened to keep out rodents and insects, as well as any material that might close off the air passages. The ducts must be sloped for drainage to remove condensed moisture.

The electrical systems are simple to install and maintain if the heating elements are run in conduit or pipe so they may be replaced; however, operating costs may be very high. Adequate insulation should be used, since this has a direct influence on the energy consumed. Perforated pipes should not be used.

The pipe grid system as shown in Figure 7 is usually best, since it can be designed and installed to warm where needed and can later be regulated to suit varying conditions. Extensions of this system can be placed in vestibules and corridors to reduce the objectionable conditions of ice and wet floors. A source of heat for this system can be obtained by a heat exchanger in the refrigeration system, steam, or gas engine exhaust. The temperature of the recirculated fluid is controlled at 50 to 70°F, depending on design requirements. Almost universally, the pipes are made of plastic.

The pipe grid system is usually placed in the base concrete slab directly under the insulation. A vapor barrier should be placed

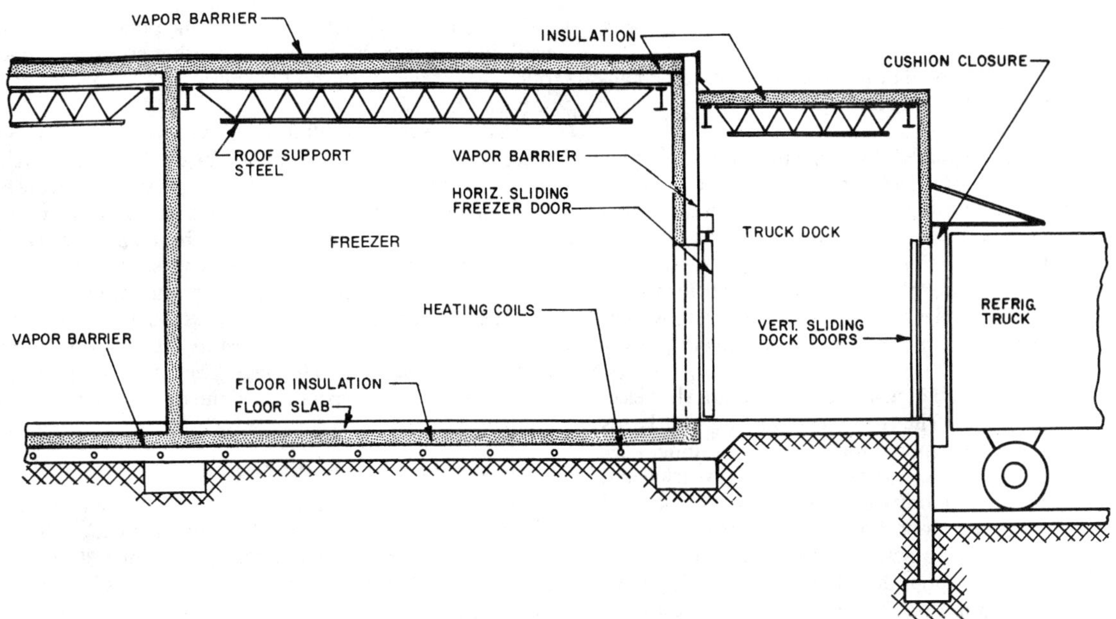

Fig. 7 Typical One-Story Construction with Underfloor Warming Pipes

below the pipe to prevent corrosion of metal pipe. The fluid should be an antifreeze solution such as glycol, with the proper inhibitor.

The amount of warming for any system can be calculated and is about the same for medium and large refrigerated spaces regardless of ambient conditions. The flow of heat from the earth serves as a safety factor and is about 1.3 Btu/h per square foot of floor. The calculated heat input requirement is the floor insulation leakage based on the temperature difference between the 40 °F underfloor earth and room temperature (*e.g.*, 50 °F temperature difference for a −10 °F storage room). The underfloor pipe grid also facilitates future expansion.

Doors

The selection and application of cold storage doors are a fundamental part of cold storage warehouse design and have a strong bearing on the overall economy of warehouse operation. The trend is to have fewer and better doors. Manufacturers offer many types of doors supplied with the proper thickness of insulation for the intended use. Four basic types of doors are swinging, horizontal sliding, vertical sliding, and double acting. Door manufacturers' catalogs give detailed illustrations of each. Doors used only for personnel cause few problems. In general, a standard swinging personnel door, 3 ft wide by 6 ft, 6 in. high, designed for the temperatures and humidity involved, is adequate.

The proper door for heavy traffic areas should provide maximum traffic capacity, minimum loss of refrigeration, and minimum maintenance.

Factors to consider when selecting cold storage doors are:

- Automatic doors are a primary requirement with lift truck and automatic conveyor material-handling systems.
- Careless lift truck operators are a hazard to door effectiveness and operation. Guards can be provided but are effective only when the door is open. Photoelectric and ultrasonic beams across the doorway or proximity loop control on both sides of the doorway can provide additional protection by monitoring objects in the door openings or approaches. These systems can also provide control to open and close the door.
- The selection of automatic door systems to suit traffic requirements and the building structure justifies experienced technical guidance.
- To ensure continuous door performance, the work area near the doors must be supervised, and the doors must have planned maintenance.
- Cooled or refrigerated shipping platforms increase door efficiency and reduce door maintenance because of lower humidity and temperature difference. Icing of the door is lessened, and fogging in trafficways is reduced.

Air curtains and plastic or rubber strip curtains, when used with doors, give varied effectiveness. Often the curtain suggests to the lift truck driver that it is a substitute for the door, so the door is left open with a concurrent loss of refrigerated air.

Swing or Slide. A door with hinges on the right side, when facing the side on which the operating hardware is mounted, is called a right-hand swing. When looking at a sliding door from the side of the wall on which it is mounted, a door sliding to the right to open is called a right-slide door.

Vertical Sliding Doors. These doors, which are hand or motor operated with counterbalanced spring or weights, are used on truck receiving and shipping openings.

Refrigerated Room Doors. Doors for pallet material handling are usually automatic horizontal sliding doors, either single-slide or biparting.

Metal or Plastic Cladding. Light metal cladding or a reinforced plastic skin protect most doors. Areas of abuse must be further protected by heavy metal, either partial or full height.

Heat. To prevent ice formation with subsequent faulty door operation, doors are available with automatic electrical heat—not only in the sides, head, and sill of a door or door frame, but also in switches and cover hoods of power-operated units. The use of such heating elements is necessary on all four edges of double-acting doors in low-temperature rooms. Safe devices that meet electrical codes must be used.

Guard Posts and Bumpers. Power-operated doors require protection from abuse. Bumpers embedded in the floor on both sides of the wall and on each side of the passageway help preserve the life of the door. Correctly placed guard posts protect sliding doors from traffic damage.

Buck and Anchorage. Effective door operation is impossible without good buck and anchorage provisions. The recommendations of the door manufacturers should be coordinated with wall construction.

Door Location. Doors should be located to suit material handling safely. Irregular aisles and blind spots in trafficways near doors should be avoided.

Door Size. A hinged insulated door opening should provide at least 1 ft clearance on both sides of a pallet. Thus, 6 ft should be the minimum size for a 4-ft pallet load. Double-acting doors should be 8 ft wide. Specific conditions at a particular doorway can require variations from this recommendation. A standard height of 10 ft accommodates all high stacking lift trucks.

Sill. A concrete sill minimizes the rise at the door sill. A thermal break should be provided in the floor slab at or near the plane of the front of the wall.

Power Doors. Horizontal sliding doors are standard when electric operation is provided. The two-leaf biparting unit keeps opening and closing time to a minimum. Also, since leading edges of both leaves have safety edges, personnel, doors, trucks, and product are protected. The door is out of the way when open and protected from damage. The door should remain open for a minimum amount of time. A combination of pull cord to open and a time-delay relay, proximity-loop control, or photoelectric cell for closing are used. Major door damage may be reduced by the proper location of pull-cord switches. Doors also must be protected from moisture and frost with heat or baffles. Automatic doors should have a preventive maintenance program to check gaskets, door alignment, electrical switches, safety edges, and heating circuits. Safety releases on locking devices are necessary to prevent entrapment of personnel.

Fire Rated Doors. Available in both swing and slide types, fire rated doors are also insulated. As refrigerated buildings have increased in size and the contents have increased in value, insurance companies and fire authorities are requiring fire walls and doors.

Vestibules. Door openings large enough to accommodate lift trucks with high masts, two-pallet-high loads on the forks, and tractor drawn trailers are of such size to cause appreciable refrigeration loss. Infiltration of moisture is objectionable, since it forms as condensate or frost on stacked merchandise and within the warehouse structure. Door heights up to 10 and 12 ft are frequently required, especially where drive-through racks are used. These conditions can be particularly serious when doors are located in opposite walls of a refrigerated space and crossflow of air is possible. It is important to reduce this infiltration by enclosed refrigerated loading docks or, in some instances, by one-way traffic vestibules.

OTHER CONSIDERATIONS

Temperature Pulldown

Because of the low temperatures within freezer facilities, contraction of structural members in this space will be substantially greater than in any surrounding ambient or cooler facilities. Therefore, contraction joints must be properly designed to prevent structural damage during facility pulldown.

The first stage of temperature reduction should be from ambient down to 35 °F at whatever rate of reduction the refrigeration system can achieve.

The room should then be held at that temperature until it is dry. Finishes are especially subject to damage when temperatures are lowered rapidly. Portland cement plaster should be fully cured before the room is refrigerated.

If there is a possibility that the room is airtight (most likely for small rooms, 20 ft by 20 ft maximum), swinging doors should be partially open during pulldown to relieve the internal vacuum caused by cooling of the air, or vents should be provided.

The concrete slab will contract during pulldown, causing slab-wall joints, contraction joints, and other construction joints to open. At the end of the holding period (*i.e.*, at 35 °F), any necessary caulking should be done.

An average time for achieving this condition is 72 h. However, there are other indicators that may be used, such as watching the rate of frost formation on the coils or measuring the rate of moisture removal by capturing the condensation during defrost.

After the refrigerated room is dry, the temperature can then be reduced again at whatever rate is achievable with the refrigeration equipment until the operating temperature is reached. Rates of 10 °F per day have been used in the past, but if care has been taken to remove all the construction moisture in the previous steps, faster rates are possible without damage. The total drying process could take up to 4 weeks, depending on the facility, size, and moisture content.

Material Handling Equipment

Frozen food public warehouses are a high-volume, year-round operation with fast-moving order pick areas backed by in-transit bulk storage. Distribution warehouses may carry 300 to 3000 items or as many as 30,000 lots. Palletized loads stored either in bulk or on racks are transported by lift trucks or high-rise S/R (storage/retrieval) machines in a 0 to −20 °F environment. Standard, battery-driven forklifts that can lift up to 25 ft can service one-deep, two-deep reach-in, drive-in, drive-through, or gravity-flow storage racks. Special forklifts can lift up to 60 ft.

Automated S/R machine storage operations make better use of storage volume, require fewer personnel, and reduce the refrigeration load because the building has less roof and floor area. This equipment operates in a height range of 23 to 100 ft to service one-deep, two-deep reach-in, two- to twelve-deep rollpin, or gravity-flow pallet storage racks. On-line computers and bar code identification allow a system to automatically control the retrieval, transfer, and delivery of products. In addition, these systems can record product location and inventory and load several delivery trucks simultaneously from one order pick conveyor and sorting device.

A refrigerated plant may have two or more material handling systems where the storage area contains fast- and slow-moving reserve storage, plus slow-moving order pick. Fast-moving items may be delivered and order picked by a conventional lift truck pallet operation with up to 30 ft stacking heights. In the fast-moving order pick section, the storage room internal height is raised to 20 ft to accommodate S/R machines, reserve pallet storage, order pick slots, multilevel palletizing, and the infeed, discharge, and order pick conveyors. Mezzanines may be considered to gain maximum access to the order pick slots. Intermediate level fire protection sprinklers may be required in the high rack or mezzanine areas above 14 ft high.

Fire Protection

Ordinary wet sprinkler systems can be applied to refrigerated spaces that are above freezing temperature. In lower temperature spaces, they cannot be applied without modification. Two types of modified systems are (1) the brine system and (2) the dry-pipe system. The brine system has been replaced with the dry-pipe system because of brine damage to stored merchandise.

In the dry-pipe system, sprinkler pipes in the refrigerated spaces are filled with dry compressed air. The water is sealed off from the compressed air by a special dry-pipe valve. The air pressure, which is controlled, keeps this valve closed as long as no sprinkler failure occurs. When the air pressure is relieved, the water pressure forces the valve open and feeds water into the sprinkler system.

If a sprinkler head malfunctions or is mechanically damaged, entering water cannot be drained and will freeze in rooms below freezing. If this occurs, the affected piping must be removed, thawed, dried, replaced, and retested. Safety precautions can be designed to prevent water from entering the system, except when heat is released by a fire. This emergency protection can be provided by installing heat-sensing devices in the sprinkler dry-valve system and a special automatic water shutoff valve, which is normally open but will instantly close whenever sprinkler system air is released without heat action of one or more of the heat-sensing devices.

The design of a dry sprinkler system operating in areas below 32 °F requires special knowledge and should not be undertaken without expert guidance. Freezer storage with rack storages 30 ft high or more may require special design, and the initial design should be shown to the insuring company.

Local regulations may require ceiling isolation smoke curtains and smoke vents near the roof in large refrigerated chambers. These features allow smoke to escape and help the fire-fighting crew locate the fire. If the building does not have a sprinkler system, central reporting or warning systems are available for hazardous areas.

Maintenance

Buildings can change in dimension due to settling, temperature change, and other factors; therefore, cold storage facilities should be inspected regularly to spot problems early so that preventive maintenance can be performed in time to avoid serious damage.

Inspection and maintenance procedures fall into two categories: *basic system*—floor, wall, and roof/ceiling systems; and *apertures*—doors, frames, and other access to cold storage rooms.

Basic System

- Stack pallets at a sufficient distance (18 in.) from walls or ceiling to permit air circulation.
- Examine walls and ceiling at random every month for frost buildup. If buildup persists, locate the break in the vapor barrier.
- For insulated ceilings below a plenum, inspect the plenum areas for possible roof leaks or condensation.
- If condensation or leaks are detected, make repairs immediately.

Apertures

- Remind personnel to close doors quickly to reduce frosting in rooms.
- Check the rollers and door travel periodically to be sure that the seal at the door edge is effective. If leaks are detected, adjust the door to restore a moisture- and airtight condition.
- Check doors and door edges to detect damage from forklifts or other traffic. Repair any damage immediately to prevent door icing or motor overload caused by excessive friction.
- Lubricate doors according to the maintenance schedule from the door manufacturer to ensure free movement and complete closure.
- In the walls or ceiling, periodically check seals around openings for ducts, piping, and wiring.

BIBLIOGRAPHY

Aldrich, D.F. and R.H. Bond. 1985. Thermal performance of rigid cellular foam insulation at subfreezing temperatures. Proceedings of the ASHRAE/DOE/BTECC Conference, ASHRAE SP 49, 739-50.

Baird, C.D., J.J. Gaffney, and M.T. Talbot. 1988. Design criteria for efficient and cost effective forced air cooling systems for fruits and vegetables. ASHRAE *Transactions* 94(1):1434-54.

Ballou, D.F. 1981. A case history of a frost heaved freezer floor. ASHRAE *Transactions* 87(2):1099-1105.

Beatty, K.O., E.B. Birch, and E.M. Schoenborn. 1951. Heat transfer from humid air to metal under frosting conditions. *Journal of the ASRE* (December):1203-7.

Cole, R.A. 1989. Refrigeration load in a freezer due to hot gas defrost and their associated costs. ASHRAE *Transactions* 95(2).

Coleman, R.V. 1983. Doors for high rise refrigerated storage. ASHRAE *Transactions* 89(1B):762-65.

Corradi, G. 1973. Air cooling units for the refrigerating industry and new equipment. Revue Generale du Froid 1(January):45-51.

Courville, G.E., J.P. Sanders, and P.W. Childs. 1985. Dynamic thermal performance of insulation metal deck roof systems. Proceedings of the ASHRAE/DOE/BTECC Conference, ASHRAE SP 49, 53-63.

Crawford, R.R., J.P. Mavec, and R.A. Cole. 1992. Literature survey on recommended procedures for the selection, placement, and type of evaporators for refrigerated warehouses. ASHRAE *Transactions* 98(1).

D'Artagnan, S. 1985. The rate of temperature pulldown. ASHRAE *Journal* 27(9):36.

Hampson, G.R. 1981. Energy conservation opportunities in cold storage warehouses. ASHRAE *Transactions* 87(2):845-49.

Hendrix, W.A., D.R. Henderson, and H.Z. Jackson. 1989. Infiltration heat gains through cold storage room doorways. ASHRAE *Transactions* 95(2).

Holske, C.F. 1953. Commercial and industrial defrosting: General principles. *Refrigerating Engineering* 61(3):261-62.

Kerschbaumer, H.G. 1971. Analysis of the influence of frost formation on evaporators and of the defrost cycles on performance and power consumption of refrigerating systems. Proceedings of the 13th International Congress of Refrigeration, 1-12.

Kurilev, E.S. and M.Z. Pechatnikov. 1966. Patterns of airflow distribution in cold storage rooms. Bulletin of the International Institute of Refrigeration, Annex 1966-1, 573-79.

Lehman, D.C. and J.E. Ferguson. 1982. A modified jacketed cold storage design. ASHRAE *Transactions* 88(2):228-334.

Lotz, H. 1967. Heat and mass transfer and pressure drop in frosting finned coils; Progress in refrigeration science and technology. Proceedings of the 12th International Congress of Refrigeration, Madrid 2:49-505.

Niederer, D.H. 1976. Frosting and defrosting effects on coil heat transfer. ASHRAE *Transactions* 82(1):467-73.

Paulson, B.A. 1988. Air distribution in freezer areas. International Institute of Ammonia Refrigeration 10th Annual Meeting, No. 10, 261-71.

Powers, G.L. 1981. Ambient gravity air flow stops freezer floor heaving. ASHRAE *Transactions* 87(2):1107-16.

Sainsbury, G.F. 1985. Reducing shrinkage through improved design and operation in refrigerated facilities. ASHRAE *Transactions* 91(1B):726-34.

Sastry, S.K. 1985. Factors affecting shrinkage of foods in refrigerated storage. ASHRAE *Transactions* 91(1B):683-89.

Shaffer, J.A. 1983. Foundations and superstructure systems for stacker crane high-rise freezers and coolers. ASHRAE *Transactions* 89(1).

Sherman, M.H. and D.T. Grimsrud. 1980. Infiltration-pressurization correlation: Simplified physical modeling. ASHRAE *Transactions* 86(2):778-807.

Soling, S.P. 1983. High rise refrigerated storage. ASHRAE *Transactions* 89(1B):737-61.

Stoecker, W.F. 1960. Frost formations on refrigeration coils. ASHRAE *Transactions* 66:91-103.

Stoecker, W.F., J.J. Lux, and R.J. Kooy. 1983. Energy considerations in hot-gas defrosting of industrial refrigeration coils. ASHRAE *Transactions* 89(2A):549-68.

Treschel, H.R., P.R. Achenbach, and J.R. Ebbets. 1985. Effect of an exterior air infiltration barrier on moisture condensation and accumulation within insulated frame wall cavities. ASHRAE *Transactions* 91(2A):545-59.

Tye, R.P., J.P. Silvers, D.C. Brownell, and S.E. Smith. 1985. New materials and concepts to reduce energy losses through structural thermal bridges. Proceedings of the ASHRAE/DOE/BTECC Conference, ASHRAE SP 49, 739-50.

Voelker, J.T. 1983. Insulating considerations for stacker crane high rise freezers and coolers. ASHRAE *Transactions* 89(1B):766-68.

Wang, I.H. and Touber. 1987. Prediction of airflow pattern in cold stores based on temperature measurements. Proceedings of Commission D, International Congress of Refrigeration, Vienna, 52-60.

CHAPTER 25

COMMODITY STORAGE REQUIREMENTS

THIS chapter presents information on the essential average storage requirements of most of the important perishable foods that enter the market on a commercial scale. Also included is a short discussion on the storage of furs and fabrics. The information is derived from scientific experimentation and the best commercial practice. The data are based on the storage of fresh quality commodities that have been properly harvested and handled and that have had the field heat removed properly. For products transported from a distance or deteriorated products, appropriate allowances should be made.

The recommended temperatures are the optimum values for long storage and are actual commodity temperatures, not air temperatures. For short storage, higher temperatures are often satisfactory. Conversely, products subject to chilling injury can sometimes be held at a lower temperature for a short time without injury. Exceptions are bananas, cranberries, cucumbers, eggplant, melons, okra, pumpkins and squash, white and sweet potatoes, and tomatoes. Adhere to minimum recommended temperatures for these products.

The storage life recommendations are based on usual commercial practice, not necessarily including special treatments, which could, in certain instances, extend storage life considerably.

The values for water content and freezing points are actual laboratory determinations, but they are only approximate because of the great variability in plant and animal tissues and the products thereof. For long-term storage of fresh products, the optimum storage temperature for many foods is just above their freezing point. Knowledge of freezing points is useful to the refrigerated storage industry in determining how to handle various commodities in storage. The highest temperature at which freezing may occur is generally given. In previous editions, the average freezing point was given, which may be somewhat lower. The highest freezing point may be a better guide for commodities damaged by freezing.

Values of the foods' water content are useful to the refrigerating engineer as a basis for calculating specific heats and the latent heat of freezing. Specific heat is usually calculated by Siebel's formula, as follows:

$$c_p = 0.008a + 0.20 \text{ (above freezing)}$$

$$c_p = 0.003a + 0.20 \text{ (below freezing)}$$

where c_p signifies the specific heat of a substance in Btu/(lb · °F) containing a, the percent of water; 0.20 is the value in Btu/(lb · °F) representing the specific heat of the solid constituents of the substance.

The composition of foods varies to some extent, depending on the water content, where the food has been grown, and other factors. In tables compiled by different editors, slightly different formulas may have been used in calculations to determine both specific and latent heat. The values given in these tables can be used for refrigeration load calculations. When in doubt, use the higher values. For more precise work, determine the specific heat and other values for the specific sample in question.

Fresh fruits and vegetables are live products, and their heat of respiration must be considered as a part of the refrigeration load in storage handling. The approximate rate of heat evolution for various commodities is given in Chapter 30 of the 1993 ASHRAE *Handbook—Fundamentals.*

The preparation of this chapter is assigned to TC 10.5, Refrigerated Distribution and Storage Facilities.

REFRIGERATED STORAGE

Cooling

Rapid removal of field heat and cooling the product to the storage temperature substantially increases the storage life. Because deterioration occurs much faster at warm than at low temperatures, the faster field heat is removed after harvest, the longer produce can be maintained in good marketable condition in storage. Chapter 10 covers cooling methods in detail.

Storage Requirements and Deterioration

Storage requirements for specific fruits and vegetables are so varied that no generalizations can be made. Table 1 presents the recommended storage requirements for various products. Some products require a curing period before storage. Other products, such as Irish potatoes, require different storage conditions depending on the intended use.

Causes for the deterioration of meat, fish, poultry, eggs, and dairy products are given in Chapter 9 and in the commodity chapters for those products.

Deterioration. The environment in which harvested produce is placed may greatly influence not only the respiration rate but other changes and products formed in related chemical reactions. In fruits, these changes are described as ripening. In many fruits, such as bananas and pears, the process of ripening is required to develop the maximum edible quality. However, as ripening continues, deterioration begins and the fruit softens, loses flavor, and eventually undergoes tissue breakdown.

In addition to deterioration after harvest by biochemical changes within the product, desiccation and diseases caused by microorganisms are also important.

Deterioration rate is greatly influenced by temperature and is reduced as temperature is lowered. The specific relationships between temperature and deterioration rate vary considerably among commodities and diseases. A generalization, assuming a deterioration rate of 1 for a fruit at 30 °F, is shown in Table 2. The best temperature to slow down deterioration is the lowest temperature that can safely be maintained without freezing the commodity, which is 1 to 2 °F above the freezing point of the fruit or vegetable.

Table 1 Storage Requirements and Properties of Perishable Products

Commodity	Storage Temperature, °F	Relative Humidity, %	Approximate Storage Life[a]	Water Content, %	Highest Freezing, °F	Specific Heat above Freezing[b], Btu/lb·°F	Specific Heat below Freezing[b], Btu/lb·°F	Latent Heat,[c] Btu/lb
Vegetables								
Artichokes								
Globe	32	95 to 100	2 weeks	84	29.9	0.87	0.45	120
Jerusalem	31 to 32	90 to 95	4 to 5 months	80	28.0	0.83	0.44	114
Asparagus	32 to 35	95 to 100	2 to 3 weeks	93	30.9	0.94	0.48	133
Beans								
Snap or Green	40 to 45	95	7 to 10 days	89	30.7	0.91	0.47	127
Lima	34 to 40	95	3 to 5 days	67	30.0	0.73	0.40	94
Dried	50	70	6 to 8 months	11		0.32	0.23	
Beets								
Roots	32	95 to 100	4 to 6 months	88	30.4	0.90	0.46	126
Bunch	32	98 to 100	10 to 14 days		31.3			
Broccoli	32	95 to 100	10 to 14 days	90	30.9	0.92	0.47	130
Brussels Sprouts	32	95 to 100	3 to 5 weeks	85	30.6	0.88	0.46	122
Cabbage, late	32	98 to 100	5 to 6 months	92	30.4	0.94	0.47	132
Carrots								
Topped-immature	32	98 to 100	4 to 6 weeks	88	29.5	0.91	0.46	126
Topped-mature	32	98 to 100	7 to 9 months	88	29.5	0.91	0.46	126
Cauliflower	32	95 to 98	3 to 4 weeks	92	30.6	0.93	0.47	132
Celeriac	32	95 to 100	6 to 8 months	88	30.4	0.91	0.46	126
Celery	32	98 to 100	2 to 3 months	94	31.1	0.95	0.48	135
Collards	32	95 to 100	10 to 14 days	87	30.6	0.90	0.46	125
Corn, Sweet	32	95 to 98	4 to 8 days	74	30.9	0.79	0.42	106
Cucumbers	45 to 50	95	10 to 14 days	96	31.1	0.97	0.49	137
Eggplant	45 to 54	90 to 95	7 to 10 days	93	30.6	0.94	0.48	133
Endive (Escarole)	32	95 to 100	2 to 3 weeks	93	31.9	0.94	0.48	133
Frozen Vegetables[h]	−10 to 0		6 to 12 months					
Garlic, dry	32	65 to 70	6 to 7 months	61	30.6	0.69	0.40	89
Greens, leafy	32	95 to 100	10 to 14 days	93	31.5	0.94	0.48	133
Horseradish	30 to 32	95 to 100	10 to 12 months	75	28.7	0.78	0.42	104
Jicama	55 to 65	65 to 70	1 to 2 months					
Kale	32	95 to 100	2 to 3 weeks	87	31.1	0.89	0.46	125
Kohlrabi	32	98 to 100	2 to 3 months	90	30.2	0.92	0.47	129
Leeks, green	32	95 to 100	2 to 3 months	85	30.7	0.88	0.46	122
Lettuce, head	32 to 34	95 to 100	2 to 3 weeks	95	31.7	0.96	0.48	136
Mushrooms	32	95	3 to 4 days	91	30.4	0.93	0.47	130
Okra	45 to 55	90 to 95	7 to 10 days	90	28.7	0.92	0.46	129
Onions								
Green	32	95 to 100	3 to 4 weeks	89	30.4	0.91	0.47	127
Dry, and onion sets	32	65 to 75	1 to 8 months	88	30.6	0.90	0.46	126
Parsley	32	95 to 100	1 to 2 months	85	30.0	0.88	0.45	122
Parsnips	32	98 to 100	4 to 6 months	79	30.4	0.84	0.44	112
Peas								
Green	32	95 to 98	1 to 2 weeks	74	30.9	0.79	0.42	106
Dried	50	70	6 to 8 months	12		0.30	0.24	
Peppers								
Dried	32 to 50	60 to 70	6 months	12		0.30	0.24	17
Sweet	45 to 50	90 to 95	2 to 3 weeks	92	30.7	0.94	0.47	132
Potatoes								
Early	38 to 40	90 to 95	4 to 5 months	81	30.9	0.85	0.44	116
Main crop	38 to 40	90 to 95	5 to 8 months	78	30.9	0.82	0.43	111
Sweet	50 to 55	85 to 90	4 to 7 months	69	29.7	0.76	0.41	99
Pumpkins	50 to 55	50 to 75	2 to 3 months	91	30.6	0.92	0.47	130
Radishes								
Spring	32	95 to 100	3 to 4 weeks	95	30.7	0.95	0.48	134
Winter	32	95 to 100	2 to 4 months	95	30.7	0.95	0.48	134
Rhubarb	32	95 to 100	2 to 4 weeks	95	30.3	0.95	0.48	134
Rutabagas	32	98 to 100	4 to 6 months	89	30.0	0.91	0.47	127
Salsify	32	98 to 100	2 to 4 months	79	30.0	0.83	0.44	113
Seed, vegetable	32 to 50	50 to 65	10 to 12 months	7 to 15		0.29	0.23	16
Spinach	32	95 to 98	10 to 14 days	93	31.5	0.94	0.48	133
Squash								
Acorn	45 to 50	70 to 75	5 to 8 weeks		30.6			
Summer	41 to 50	95	5 to 14 days	94	31.1	0.95	0.48	135
Winter	50 to 55	50 to 75	4 to 6 months	85	30.6	0.88	0.45	122
Tamarillos	37 to 40	85 to 95	10 weeks					
Tomatoes								
Mature green	55 to 60	90 to 95	1 to 3 weeks	93	31.0	0.94	0.48	133
Firm, ripe	45 to 50	90 to 95	4 to 7 days	94	31.1	0.95	0.48	134
Turnips								
Roots	32	95	4 to 5 months	92	30.0	0.93	0.47	132
Greens	32	95 to 100	10 to 14 days	90	31.6	0.92	0.47	129
Watercress	32	95 to 100	2 to 3 weeks	93	31.4	0.94	0.48	133
Yams	61	85 to 90	3 to 6 months	74		0.79	0.42	105

Table 1 Storage Requirements and Properties of Perishable Products (*Continued*)

Commodity	Storage Temperature, °F	Relative Humidity, %	Approximate Storage Life[a]	Water Content, %	Highest Freezing, °F	Specific Heat above Freezing[b], Btu/lb·°F	Specific Heat below Freezing[b], Btu/lb·°F	Latent Heat,[c] Btu/lb
Fruits and Melons								
Apples	32 to 38	90 to 95	3 to 8 months	84	29.3	0.87	0.45	121
Apples, dried	41 to 48	55 to 60	5 to 8 months	24		0.42	0.27	
Apricots	30 to 32	90 to 95	1 to 3 weeks	85	30.1	0.88	0.46	122
Avocados	40 to 55	85 to 90	2 to 8 weeks	76	31.5	0.81	0.40	94
Bananas	56 to 58	85 to 95		75	30.6	0.81	0.42	108
Blackberries	31 to 32	90 to 95	3 days	85	30.6	0.88	0.46	122
Blueberries	31 to 32	90 to 95	2 weeks	83	29.7	0.86	0.45	118
Cantaloupes	36 to 40	95	5 to 15 days	92	29.8	0.93	0.48	132
Casaba Melons	45 to 50	85 to 95	4 to 6 weeks	93	30.0	0.94	0.48	133
Cherries								
Sour	32	90 to 95	3 to 7 days	84	29.0	0.87	0.45	121
Sweet	30 to 31	90 to 95	2 to 3 weeks	80	28.8	0.84	0.44	114
Coconuts	32 to 35	80 to 85	1 to 2 months	47	30.4	0.58	0.34	67
Cranberries	36 to 40	90 to 95	2 to 4 months	87	30.4	0.90	0.46	124
Crenshaw Melons	45 to 50							
Currants	31 to 32	90 to 95	1 to 4 weeks	85	30.2	0.88	0.45	122
Dates, cured	0 to 32	75 or less	6 to 12 months	20	3.7	0.38	0.26	29
Dewberries	31 to 32	90 to 95	2 to 3 days	85	29.7	0.88	0.45	122
Elderberries	31 to 32	90 to 95	1 to 2 weeks	80		0.84	0.44	115
Figs								
Dried	32 to 40	50 to 60	9 to 12 months	23		0.39	0.27	34
Fresh	31 to 32	85 to 90	7 to 10 days	78	27.6	0.82	0.43	112
Frozen fruits	−12 to 0	90 to 95	18 to 24 months					
Gooseberries	31 to 32	90 to 95	2 to 4 weeks	89	30.0	0.91	0.46	127
Grapefruit	58 to 60	85 to 90	6 to 8 weeks	89	30.0	0.90	0.46	127
Grapes[g]								
American	31 to 32	85 to 90	2 to 8 weeks	82	29.7	0.86	0.45	118
Vinifera	30 to 31	90 to 95	3 to 6 months	82	28.1	0.86	0.45	118
Guavas	40 to 50	90	2 to 3 weeks	83		0.86	0.45	119
Honeydew Melons	41 to 50	90 to 95	2 to 3 weeks	93	30.4	0.94	0.48	133
Kiwifruit	31 to 32	90 to 95	3 to 5 months	82	29	0.86	0.45	118
Lemons	52 to 55[d]	85 to 90	1 to 4 months	89	29.4	0.91	0.46	127
Limes	45 to 48	85 to 90	6 to 8 weeks	86	29.1	0.91	0.46	123
Loganberries	31 to 32	90 to 95	2 to 3 days	83	29.7	0.86	0.45	119
Loquats	32	90	3 weeks	87		0.89	0.46	125
Lychees	35	90 to 95	3 to 5 weeks	82		0.86	0.45	118
Mangoes	50 to 55	85 to 90	2 to 3 weeks	81	30.4	0.85	0.44	117
Nectarines	31 to 32	90 to 95	2 to 4 weeks	82	30.4	0.84	0.44	118
Olives, fresh	41 to 50	85 to 90	4 to 6 weeks	75	29.4	0.80	0.42	108
Oranges, CA and AZ	38 to 48	85 to 90	3 to 6 weeks	86	29.7	0.88	0.46	124
Oranges, FL and TX	32 to 34	85 to 90	8 to 12 weeks	86	30.6	0.89	0.46	123
Papayas	45	85 to 90	1 to 3 weeks	91	30.6	0.91	0.47	130
Peaches	31 to 32	90 to 95	2 to 4 weeks	89	30.4	0.91	0.46	127
Peaches, dried	32 to 41	55 to 60	5 to 8 months	25		0.43	0.28	
Pears[g]	29 to 31	90 to 95	2 to 7 months	83	29.2	0.86	0.45	118
Persian Melons	45 to 50	90 to 95	2 weeks	93	30.6	0.94	0.48	133
Persimmons	30	90	3 to 4 months	78	28.1	0.84	0.48	112
Pineapples, ripe	45	85 to 90	2 to 4 weeks	85	30.2	0.88	0.45	122
Plums	31 to 32	90 to 95	2 to 4 weeks	86	30.6	0.88	0.45	123
Pomegranates	40	90 to 95	2 to 3 months	82	26.6	0.86	0.44	118
Prunes								
Fresh	31 to 32	90 to 95	2 to 4 weeks	86	30.5	0.88	0.45	123
Dried	32 to 41	55 to 60	5 to 8 months	28		0.46	0.28	
Quinces	31 to 32	90	2 to 3 months	85	28.4	0.88	0.45	122
Raisins				18		0.38	0.25	
Raspberries								
Black	31 to 32	90 to 95	2 to 3 days	81	30.0	0.84	0.44	117
Red	31 to 32	90 to 95	2 to 3 days	84	30.9	0.87	0.45	120
Strawberries	31 to 32	90 to 95	5 to 7 days	90	30.6	0.92	0.47	129
Tangerines	40	90 to 95	2 to 4 weeks	87	30.0	0.90	0.46	125
Watermelons	40 to 50	90	2 to 3 weeks	93	31.3	0.97	0.48	133
Seafood (Fish)								
Haddock, Cod, Perch	31 to 34	95 to 100	12 days	81	28	0.85	0.44	117
Hake, Whiting	32 to 34	95 to 100	10 days	81	28	0.85	0.44	117
Halibut	31 to 34	95 to 100	18 days	75	28	0.80	0.42	107
Herring								
Kippered	32 to 36	80 to 90	10 days	61	28	0.70	0.38	87
Smoked	32 to 36	80 to 90	10 days	64	28	0.72	0.39	92
Mackerel	32 to 34	95 to 100	6 to 8 days	65	28	0.73	0.40	93
Menhaden	34 to 41	95 to 100	4 to 5 days	62	28	0.71	0.39	89
Salmon	31 to 34	95 to 100	18 days	64	28	0.72	0.39	92
Tuna	32 to 36	95 to 100	14 days	70	28	0.77	0.40	100
Frozen fish	−20 to −4	90 to 95	6 to 12 months					

Table 1 Storage Requirements and Properties of Perishable Products (*Continued*)

Commodity	Storage Temperature, °F	Relative Humidity, %	Approximate Storage Life[a]	Water Content, %	Highest Freezing, °F	Specific Heat above Freezing[b], Btu/lb·°F	Specific Heat below Freezing[b], Btu/lb·°F	Latent Heat,[c] Btu/lb
Seafood (Shellfish)[a]								
Scallop meat	32 to 34	95 to 100	12 days	80	28	0.84	0.44	114
Shrimp	31 to 34	95 to 100	12 to 14 days	76	28	0.81	0.43	109
Lobster, American	41 to 50	In sea water	Indefinitely	79	28	0.83	0.44	113
Oysters, Clams (meat and liquid)	32 to 36	100	5 to 8 days	87	28	0.89	0.46	125
Oyster in shell	41 to 50	95 to 100	5 days	80	27	0.84	0.44	115
Frozen shellfish	−30 to −4	90 to 95	3 to 8 months					
Meat (Beef)								
Beef, fresh, average	28 to 34	88 to 95	1 week	62 to 77	28 to 29[f]	0.70 to 0.84	0.39 to 0.43	89 to 110
Beef carcass								
Choice, 60% lean	32 to 39	85 to 90	1 to 3 weeks	49	29	0.61	0.35	70
Prime, 54% lean	32 to 34	85	1 to 3 weeks	45	28	0.58	0.34	64
Sirloin cut (choice)	32 to 34	85	1 to 3 weeks	56		0.66	0.37	80
Round cut (choice)	32 to 34	85	1 to 3 weeks	67		0.50	0.40	96
Dried, chipped	50 to 59	15	6 to 8 weeks	48		0.60	0.34	69
Liver	32	90	5 days	70	29	0.77	0.41	100
Veal, lean	28 to 34	85 to 95	3 weeks	66		0.74	0.40	94
Beef, frozen	−10 to 0	90 to 95	6 to 12 months					
Meat (Pork)								
Pork, fresh, average	32 to 34	85 to 90	3 to 7 days	32 to 44	28 to 29[f]	0.48 to 0.57	0.30 to 0.33	46 to 63
Carcass, 47% lean	32 to 34	85 to 90	3 to 5 days	37		0.52	0.31	53
Bellies, 35% lean	32 to 34	85	3 to 5 days	30		0.47	0.29	43
Backfat, 100% fat	32 to 34	85	3 to 7 days	8		0.30	0.22	
Shoulder, 67% lean	32 to 34	85	3 to 5 days	49	28[f]	0.61	0.35	70
Pork, frozen	−10 to 0	90 to 95	4 to 8 months					
Ham								
74% lean	32 to 34	80 to 85	3 to 5 days	56	29[f]	0.66	0.37	80
Light cure	37 to 41	80 to 85	1 to 2 weeks	57		0.67	0.37	82
Country cure	50 to 59	65 to 70	3 to 5 months	42		0.56	0.33	60
Frozen	−10 to 0	90 to 95	6 to 8 months					
Bacon								
Medium fat class	37 to 41	80 to 85	2 to 3 weeks	19		0.38	0.26	27
Cured, farm style	61 to 64	85	4 to 6 months	13 to 20		0.34 to 0.39	0.24 to 0.26	19 to 29
Cured, packer style	34 to 39	85	2 to 6 weeks					
Frozen	−10 to 0	90 to 95	2 to 4 months					
Sausage								
Links or bulk	32 to 34	85	1 to 7 days	38		0.53	0.31	54
Country, smoked	32	85	1 to 3 weeks	50	25	0.62	0.35	72
Frankfurters, average	32	85	1 to 3 weeks	56	29	0.66	0.37	80
Polish style	32	85	1 to 3 weeks	54		0.65	0.36	77
Meat (Lamb)								
Fresh, average	28 to 34	85 to 90	3 to 4 weeks	60 to 70	28 to 29[f]	0.69 to 0.77	0.38 to 0.41	86 to 100
Choice, lean	32	85	5 to 12 days	61	28	0.70	0.38	87
Leg, choice, 83% lean	32	85	5 to 12 days	65		0.73	0.40	93
Frozen	−10 to 0	90 to 95	8 to 12 months					
Meat (Poultry)								
Poultry, fresh, average	28 to 32	95 to 100	1 to 3 weeks	74	27	0.80	0.42	106
Chicken, all classes	28 to 32	95 to 100	1 to 4 weeks	74	27	0.80	0.42	106
Turkey, all classes	28 to 32	95 to 100	1 to 4 weeks	64	27	0.72	0.39	92
Turkey breast roll	−4 to −1		6 to 12 months					
Turkey frankfurters	0 to 15		6 to 16 months					
Duck	28 to 32	95 to 100	1 to 4 weeks	69	27	0.76	0.41	99
Poultry, frozen	−10 to 0	90 to 95	12 months					
Meat (Miscellaneous)								
Rabbits, fresh	32 to 34	90 to 95	1 to 5 days	68		0.75	0.40	97
Dairy Products								
Butter	32	75 to 85	2 to 4 weeks	16	−4 to 31	0.36	0.25	23
Butter, frozen	−10	70 to 85	12 to 20 months					
Cheese, Cheddar,								
long storage	32 to 34	65	12 months	37	8	0.52	0.31	53
short storage	40	65	6 months	37	8	0.52	0.31	53
processed	40	65	12 months	39	19	0.50	0.31	56
grated	40	65	12 months	31		0.45	0.29	44
Ice cream, 10% fat	−20 to −15	90 to 95	3 to 23 months	63	21	0.70	0.39	86
Ice Cream, premium	−30 to −40	90 to 95	3 to 23 months					
Milk								
Fluid, pasteurized	39 to 43		7 days					
Grade A (3.7% fat)	32 to 34		2 to 4 months	87	31	0.92	0.46	125

Table 1 Storage Requirements and Properties of Perishable Products (*Concluded*)

Commodity	Storage Temperature, °F	Relative Humidity, %	Approximate Storage Life[a]	Water Content, %	Highest Freezing, °F	Specific Heat above Freezing[b], Btu/lb·°F	Specific Heat below Freezing[b], Btu/lb·°F	Latent Heat,[c] Btu/lb
Dairy Products (*Continued*)								
Milk (*Continued*)								
Raw	32 to 39		2 days					
Dried, whole	70	Low	6 to 9 months	2		0.26	0.21	28
Dried, nonfat	45 to 70	Low	16 months	3		0.26	0.21	4
Evaporated	40		24 months	74	29.5	0.79	0.42	106
Evaporated, unsweetened	70		12 months	74	29.5	0.79	0.42	106
Condensed, sweetened	40		15 months	27	5	0.42	0.28	40
Whey, dried	70	Low	12 months	5		0.28	0.22	7
Eggs								
Eggs								
Shell	29 to 32[e]	80 to 90	5 to 6 months	66	28[f]	0.73	0.40	96
Shell, farm cooler	50 to 55	70 to 75	2 to 3 weeks	66	28[f]	0.73	0.40	96
Frozen,								
Whole	0		1 year plus	74		0.80	0.42	106
Yolk	0		1 year plus	55		0.65	0.36	79
White	0		1 year plus	88		0.90	0.46	126
Whole egg solids	35 to 40	Low	6 to 12 months	2 to 4		0.22	0.21	4
Yolk solids	35 to 40	Low	6 to 12 months	3 to 5		0.23	0.21	6
Flake albumen solids		Low	1 year plus	12 to 16	16	0.31	0.24	20
Dry spray albumen solids		Low	1 year plus	5 to 8		0.26	0.22	11
Candy								
Milk chocolate	0 to 34	40	6 to 12 months	1		0.25	0.20	1
Peanut brittle	0 to 34	40	1.5 to 6 months	2		0.26	0.21	3
Fudge	0 to 34	65	5 to 12 months	10		0.32	0.23	14
Marshmallows	0 to 34	65	3 to 9 months	17		0.37	0.25	24
Miscellaneous								
Alfalfa meal	0	70 to 75	1 year plus					
Beer								
Keg	35 to 40		3 to 8 weeks	90	28	0.92	0.47	129
Bottles and cans	35 to 40	65 or below	3 to 6 months	90				
Bread	0		3 to 13 weeks	32 to 37	0.70	0.34	46 to 53	
Canned goods	32 to 60	70 or lower	1 year					
Cocoa	32 to 40	50 to 70	1 year plus					
Coffee, green	35 to 37	80 to 85	2 to 4 months	10 to 15		0.32 to 0.35	0.23 to 0.24	14 to 21
Fur and fabrics	34 to 40	45 to 55	Several years					
Honey	50		1 year plus	17		0.35	0.26	26
Hops	28 to 32	50 to 60	Several months					
Lard (without	45	90 to 95	4 to 8 months	0				
antioxidant)	0	90 to 95	12 to 14 months	0				
Maple syrup				33		0.48	0.31	5[illegible]
Nuts	32 to 50	65 to 75	8 to 12 months	3 to 6		0.22 to 0.25	0.21 to 0.22	4 to 8
Oil, vegetable, salad	70		1 year plus	0				
Oleomargarine	35	60 to 70	1 year plus	16		0.32	0.25	22
Orange juice	30 to 35		3 to 6 weeks	89		0.91	0.47	127
Popcorn, unpopped	32 to 40	85	4 to 6 weeks	10		0.31	0.24	19
Yeast, baker's compressed	31 to 32			71		0.77	0.41	102
Tobacco								
Hogshead	50 to 65	50 to 65	1 year					
Bales	35 to 40	70 to 85	1 to 2 years					
Cigarettes	35 to 46	50 to 55	6 months					
Cigars	35 to 50	60 to 65	2 months					

Note: The text in this chapter or the appropriate commodity chapter gives additional information on many of the commodities listed. The following individuals helped obtain data for certain commodities: R.L. Hiner, L. Feinstein, and A. Kotula, Meat and Poultry; J.W. White and C.O. Willits, Honey and Maple Syrup; E.B. Lambert, Mushrooms; H. Landani, Furs and Fabrics; M.K. Veldhuis, Orange Juice; A.L. Ryall, Plums and Prunes; L.P. McCulloch, Tomatoes; J.W. Slavin and J.A. Peters, Fish; T.I. Hedrich, Dairy Products; and L. Henrickson, Meat.

[a]Storage life is not based on maintaining nutritional value.

[b]Specific heat c_p (in Btu/lb·°F) is calculated by Siebel's formula. For temperatures above freezing, $c_p = 0.008a + 0.20$, where a is the percent water content of the commodity; for temperatures below freezing, $c_p = 0.003a + 0.20$. Siebel's formula is not very accurate in the frozen region because foods are not simple mixtures of solids, and liquids are not completely frozen, even at −20°F.

[c]The latent heat of fusion is the latent heat of fusion of water (143.4 Btu/lb) multiplied by the water content of the commodity.

[d]Lemons stored in production areas for conditioning are held at 55 to 58°F, but sometimes they are held at 32°F.

[e]Eggs with weak albumen freeze just below 30°F.

[f]Average freezing point.

[g]For a complete listing of grapes and pears, see *Recommendations for Chilled Storage of Perishable Foods*, International Institute of Refrigeration, 1979.

[h]For a complete listing of frozen food practical storage life, see *Recommendations for the Processing and Handling of Frozen Foods*, 3rd ed. International Institute of Refrigeration, 1986.

Table 2 Approximate Relationship of Temperature and Deterioration Rate in Fresh Produce

Temperature, °F	Deterioration Rate
68	8 to 10
50	4 to 5
41	3
37	2
32	1.25
30	1

Some produce will not tolerate low storage temperatures. Severe physiological disorders that develop because of exposure to low but not freezing temperatures are classed as *chilling injury.* The banana is a classic example of a fruit displaying chilling injury symptoms, and storage temperatures must be elevated accordingly. Certain apple varieties exhibit this characteristic, and prolonged storage must be held at a temperature well above that usually recommended. The degree of susceptibility of an apple variety to chilling may vary with climatic and cultural factors. Products susceptible to chilling injury, its symptoms, and the lowest safe temperature are discussed in Chapters 9, 15, 16, and 17.

Desiccation. Water loss, which causes a product to shrivel, is a physical factor related to the evaporative potential of air, and can be expressed as follows:

$$p_D = p(100 - \phi)/100 \tag{1}$$

where

p_D = vapor pressure deficit, indicating combined influence of temperature and relative humidity on evaporative potential of air
p = vapor pressure of water at given temperature
ϕ = relative humidity, percent

For example, comparing the evaporative potential of air in storage rooms at 32°F and 50°F dry bulb, with 90% rh in each room, the vapor pressure deficit at 32°F is 0.018 in. Hg, while at 50°F, it is 0.036 in. Hg. Thus, if all other factors are equal, commodities tend to lose water twice as fast at 50°F dry bulb as at 32°F at the same rh values. For equal water loss at the two temperatures, the rh has to be maintained at 95% at 50°F in comparison to 90% at 32°F. These comparisons are not precise because the water in fruits and vegetables contains a sufficient quantity of dissolved sugars and other chemical materials to cause the water to be in equilibrium with water vapor in the air at 98 to 99% rh instead of 100% rh. Lowering the vapor pressure deficit by lowering the air temperature is an excellent means of reducing water loss during storage.

Other important factors in desiccation include product size, the kind of protective surface on the product, and air movement. Of these, the storage operator can control only the last, and this control is greatly influenced by the container, kind of pack, and stacking arrangement (*i.e.*, the ability of the air to move past individual fruits and vegetables).

As a rule, shrivelling does not become a serious market problem until fruits lose about 5% of their weight, but any loss reduces the salable amount. Moisture losses of 3 to 6% are enough to cause a marked loss of quality for many kinds of produce. A few kinds may lose 10% moisture and still be marketable, although some trimming may be necessary, as for stored cabbage.

The vapor pressure deficit cannot be kept at a zero level, but it should be maintained as low as possible. A maximum of about 0.018 in. Hg, which corresponds to 90% rh at 32°F, is recommended. Some compromise is possible for short storage periods. In many instances, the refrigerated storage operator may find it desirable to add moisture, or, in special cases, the owner of the produce may find it desirable to use moisture barriers such as film liners.

REFRIGERATED STORAGE PLANT OPERATION

Checking Temperatures and Humidity

To maintain top product quality, temperature in the cold storage room must be accurately maintained. Variations of 2 to 3°F in the product temperature above or below the desired temperature are too large in most cases. Storage rooms should be equipped either with accurate thermostats or with manual controls that receive frequent attention.

In refrigerated storage rooms, thermometers are usually placed at a height of about 5 ft for convenience in reading. Temperatures should be monitored where they might be undesirably high or low—one or two aisle temperatures is not enough. A record of both product and air temperatures is necessary to determine performance of the storage plant. A thermometer or recording device of good quality is essential.

Temperature in less accessible storage locations, such as the middle of stacks, can be obtained conveniently with distant reading thermometer equipment such as thermocouples or electrical-resistance thermometers.

Storage instructions or recommendations usually specify a relative humidity within 3 to 5% of the desired levels. An ordinary sling psychrometer at temperatures of 32°F or lower cannot be read that closely. An error of 0.5°F in reading either the wet- or the dry-bulb thermometer will cause an error of 5% rh. Carefully calibrated thermometers graduated to 0.1°F with a range of 25 to 40°F are best adapted for this purpose in fruit storage. A convenient device for measuring humidity consists of a pair of these thermometers, mounted in a short length of metal casing attached to a spring or motor-operated fan that draws air past the thermometers at a speed of 3 fps or faster. The thermometers should be placed so that they will not be heated by the fan motor, and they should be read quickly to prevent warming. The advantage of this instrument over the sling psychrometer is that it can be left in the room long enough to get a true wet-bulb reading. This may require 15 min or more if ice is formed on the wet bulb. Under these conditions, a thin coating of ice is preferable to a thick one in getting accurate readings. Hair hygrometers are satisfactory if not subjected to sudden large changes in humidity and temperature and if checked regularly with a psychrometer.

Perhaps a more accurate method of determining the relative humidity is to electrically record the dew-point temperature of the air and to use a resistance thermometer of suitable sensitivity to record the ambient or dry-bulb temperature. From these temperature records, the relative humidity can be calculated.

Air Circulation

Air must be circulated to keep refrigerated storage rooms at an even temperature throughout. Commodity temperatures in a storage room may vary because the air temperature rises as the air passes through the room and absorbs heat from the commodity; also, heat leakage may vary in different parts of the storage. In a duct system, the air near the return ducts will be warmer than the air near delivery ducts. In many storages, refrigeration units are installed over the center aisle. Air circulates from the center of the rooms outward to the walls, down through the rows of produce, and back up through the center of the room.

Rapid air circulation is needed most during removal of field heat. Sometimes this is best done in separate cooling rooms that have more refrigerating and air-moving capacity than regular refrigerated storage rooms (see Chapter 10). After field heat is removed, a high air velocity is usually undesirable. Air movement is needed only to remove respiratory heat and heat entering the room through exterior surfaces and doorways. Excessive air movement can increase moisture loss from a product, with a resulting loss of both weight and quality. However, some air circulated by fans or blowers must be uniformly distributed through all parts of the room.

The nature of the container and the manner of stacking are important factors that influence cooling performance. An elaborate system for air distribution is useless if poor stacking prevents airflow. If spacing is irregular, the wider spaces will get a greater volume of air than narrower ones. If some spaces are partially blocked, dead air zones will occur with resultant higher temperatures.

Sanitation and Air Purification

The refrigerated storage and the product containers must be clean. Fruits and vegetables coming into the storage are generally contaminated with mold spores, which can enter through punctures or breaks in the skin of the product. Commodities so contaminated can foster a rapid development of the spores, which are carried by air currents throughout the storage. Removing decaying raw material from the storage and sanitizing the product container will reduce the problem. If mold contamination is excessive, the storage develops a musty or moldy odor that is quickly picked up by the fruits and vegetables, a fault of many apple varieties and other products held for several months in refrigerated storage. This problem can best be controlled by means of special cleaners, sanitizers, and deodorizers (see Chapters 15, 16, and 17 for details on product diseases).

During several months of storage, even at 31 °F, molds may grow on the surface of packages and on the walls and ceilings of rooms under high relative humidity conditions. These surface molds generally will not rot fruits and vegetables. However, because surface molds are unsightly, storage warehouses should have a thorough cleaning at least once a year. Good air circulation alone is of considerable value in minimizing growth of surface molds. If floors and walls become moldy, they can be scrubbed with a cleaner containing sodium hypochlorite or trisodium phosphate, then rinsed, and aired. Field boxes and equipment can be cleaned with 0.25% calcium hypochlorite solutions or by exposing to steam for 2 min.

All inspected plants and warehouses operate under regulations with sanitation requirements clearly set forth in inspection service orders. Plants should be constructed to prevent the entrance of insects and rodents. This involves rat-proof building construction and adequate screening. For doors frequently opened, special measures must be taken to prevent the entrance of insects.

The frequency of cleanup operations and the detergents and sanitizing agents used should be specified by the quality control leader. A representative from quality control should inspect all areas after cleanup and determine whether or not a suitable job has been done. Waste and miscellaneous trash accumulation areas must receive special attention around warehouses, as they become breeding places for rodents and insects. In any food warehouse, the successful quality control and sanitation program depends on the cooperation and vigilance of management.

Air may be purified in storage rooms where odors or volatiles may contribute to off-flavors and hasten deterioration. Air may be cleaned with trays or canisters containing 6 to 14 mesh-activated coconut shell carbon. Pinewood volatiles are removed by activated carbon air-purifying units. Some produce volatiles are also removed, but ethylene, a ripening gas, is not removed by activated carbon alone.

Air washing with water to remove volatiles does not retard fruit ripening. Air washing may increase the relative humidity and thus aid in maintaining good fruit appearance by reducing weight loss.

Removal of Produce from Storage

When produce is removed from storage, undue warming and condensation of moisture, which promote decay and deterioration, must be prevented. Because most storages are built on railroad sidings, canvas tunnels should be installed between the car and the storage through which the produce will be conveyed to minimize warming and condensation of moisture.

When produce is removed from storage for distribution to wholesale and retail markets, the storage operator can do little to prevent undesirable condensation. Warming the packages until they are above the dew point of the air would prevent it, but this takes time and space and is seldom practicable. Deterioration in flavor and condition proceeds rapidly after long storage periods. Therefore, the produce should be moved to consumers as rapidly as possible.

STORAGE OF FROZEN FOODS

Frozen foods deteriorate during the period between production and consumption. The extent of deterioration depends on many factors, such as protection afforded by the package. However, the most important factors are storage temperature and storage time.

Bacteria in frozen foods may be killed during freezing and frozen storage, but all the bacteria present are never completely destroyed. When defrosting, frozen foods are still subject to bacterial decomposition.

SPECIFIC PRODUCTS

Beer

Since beer in bottles or cans is either pasteurized or filtered to destroy or remove the living yeast cells, it does not require as low a storage temperature as keg beer. Bottled beer may be stored at ordinary room temperature 70 to 75 °F, but for convenience, it is often stored with keg beer at a lower temperature of 35 to 40 °F. The bottled product should be protected from strong light, especially direct sunlight. The storage life will vary from 3 to 6 months, depending largely on the method of processing and packaging. Keg beer, usually stored at 35 to 40 °F, has a storage life of 3 to 6 weeks.

Canned Foods

Canned foods that are heat processed in hermetically sealed containers do not benefit from refrigerated storage if they are to be stored for no longer than 2 or 3 months at temperatures that rarely exceed 75 °F. However, seasonal commodities produced to provide an inventory for an entire year or longer do benefit from reduced temperature and humidity storage, which delays the onset of considerable color, texture, and flavor changes, loss of nutrients, and container corrosion. Notable examples are canned asparagus, cherries, and catsup. Reduced temperature and humidity storage for canned goods is essential in environments where ambient temperatures and humidities regularly exceed 90 °F and 70% rh.

Dried Foods

Dried foods and feeds, particularly those expected to supply a high level of protein, such as dehydrated milk or alfalfa meal, should be protected against high temperatures and humidities. Good packaging, such as canning in vacuum, is helpful and can maintain dried food nutritive value and quality for a year or longer if ambient temperatures do not exceed 90 °F regularly. When stored in bulk or in bags that are not good water vapor barriers, the storage life at 40% rh and 70 °F should be limited to less than one year, and at 90 °F to less than 6 months. At 60% rh and 70 °F, the storage life should be limited to 6 months, and at 90 °F to not more than 3 to 4 months. The only process by which dried foods can maintain quality and nutritive values for a year and longer at elevated temperatures is by packaging in zero oxygen. Similar or

Table 3 Temperature and Time Requirements for Killing Moths in Stored Clothing

Storage Temperature, °F	All Eggs Dead After, Days	All Larvae Dead After, Days	All Adults Dead After, Days
−0.4 to 5	1	2	1
5 to 10	2	21[a]	1
10 to 15	4	—	1
15 to 19	—	—	1
20 to 25	21	67	4
25 to 30	21	125[b]	7
30 to 35	—	283[c]	—

Table adapted from USDA *Publication* AMS-57 (1955).
[a]50 to 95% of larvae may be killed in 2 days.
[b]A few larvae survived this period.
[c]Larvae survived this period.

better results can be obtained with 0 °F storage regardless of adequacy of the package or relative humidity.

Furs and Fabrics

Refrigerated storage effectively protects furs, floor coverings, garments, and other materials containing wool against insect damage. The commonly used refrigerated storage temperatures do not kill the insects, but inactivate them, preventing insect damage while the susceptible items are in storage (Table 3). However, if insects are present, the article is susceptible to damage as soon as it is removed from refrigerated storage.

Articles should be free of any possible infestation before placement in refrigerated storage. Those items that can be cleaned should be so treated. Others can be either fumigated or mothproofed.

Furs and garments should be stored at 34 to 40 °F. A temperature of 40 °F is most widely used commercially. The low temperature not only inactivates fabric insects but also preserves the vitality and luster of furs and the tensile strength of fabrics.

Continuous storage below the 34 to 40 °F range is a wasteful expense as far as protection from insect damage is concerned. Food should not be stored with fur garments. Some storage firms maintain constant temperatures in their fur vaults between 14 and 32 °F and claim excellent results. However, no research indicates that temperatures in this range are required for storing dressed furs or fabrics. Cured raw furs (but not processed) should be stored at −10 to 10 °F with 45 to 60% rh and will keep up to 2 years.

Honey

Both extracted (liquid) and comb honey can be held satisfactorily in common dry storage for about a year. The slow darkening and flavor deterioration at ordinary room temperatures becomes objectionable after this time. Although cold storage is not necessary, temperatures below 50 °F will maintain original quality for several years and retard or prevent fermentation. The range between 50 and 65 °F should be avoided, as it promotes granulation; this increases the probability of fermentation of raw (unheated) honey. As storage temperature increases in the 80 to 100 °F range, deterioration is accelerated; temperatures constantly above 85 °F are unsuitable and above 90 °F, quite damaging.

Honey for export is best kept in cold storage, since the half-life for honey diastase at 77 °F is about 17 months.

Raw honey of greater than 20% moisture is always in danger of fermentation; the likelihood is much less at or below 18.6% moisture. Below 17% moisture, raw honey will not ordinarily ferment. Granulation increases the possibility of fermentation of raw honey by increasing the moisture content of the liquid portion. Properly pasteurized honey will not ferment at any moisture content. Granulated honey can be reliquefied by warming to 120 to 140 °F.

Comb honey should not be stored above 60% rh to avoid moisture absorption through the wax, which leads to fermentation.

Finely granulated honey (honey spread, Dyce process honey, and honey cream) must not be stored above about 75 °F. Higher temperatures will, in time, cause partial liquefaction and destroy the texture. Any subsequent regranulation by lower temperatures will produce an undesirable coarse texture. For holding more than 4 months, cold storage is required.

Maple Syrup

Maple syrup keeps indefinitely at room temperatures without darkening or losing flavor, if packed hot (at or within a few degrees of its boiling point) and in clean containers, which are promptly closed airtight and laid on their sides or inverted to self-sterilize the closure, and then cooled. Refrigerated storage is not necessary. However, once opened, the syrup in a bottle, can, or drum may become contaminated by organisms in the air. Mold or yeast spores, which may be present in improperly pasteurized syrup, though unable to germinate in full-density syrup, may grow in the thin syrup on the surface caused by condensed water. Small packages not completely sterile, containing spores, can be kept free of vegetative growth by periodically inverting the containers to redisperse any thin syrup on the surface caused by condensed water. Maple syrup should never be packaged at temperatures below 180 °F. After pasteurizing, cool the syrup as quickly as possible to prevent stack burn, which darkens the syrup and causes a lowering of its grade.

Nursery Stock and Cut Flowers

The temperature and approximate storage life given in Table 4 for cut flowers allow for a reasonable shelf life after removal from storage; the storage period may be extended beyond that recommended here.

Low temperature (31 to 33 °F) and dry packaging prevent, or at least greatly retard, flower disintegration and extend the storage life. These conditions, while not widely used commercially, are recommended. Proper dry packing requires a moisture-vapor-proof container in which flowers can be sealed. No free water is added because the package prevents almost all water loss.

Flowers held in water should not be crowded in the containers and should be arranged on shelves or racks to allow good air circulation. Forced air circulation should be provided, but flowers must be kept out of a direct draft. Use clean water and clean containers.

Many kinds of nursery stock can also be stored at temperatures of 31 to 35 °F. Open packages and harden flowers before marketing if the blooms have been stored for long periods. Flowers conditioned at about 50 °F following storage regain full turgidity most rapidly. Cut or crush stem ends and then place in water or a food solution at 80 to 100 °F for 6 to 8 h.

Many kinds of cut flowers and greens are injured if stored in the same room with certain fruits, principally apples and pears, which give off gases (such as ethylene) during ripening. These gases cause premature aging of blooms and may defoliate greens. Greens should not be stored in the same room with cut flowers because the greens, acting in the same way as fruit, can hasten bloom deterioration.

Greens, bulbs, and certain nursery stock are usually packaged or crated when stored. Some bulbs and nursery stock are packed in damp moss or similar material, and low temperatures are required to keep them dormant. Polyethylene wraps or box liners are very effective for maintaining quality of strawberry plants, bare-root rose bushes, and certain cuttings and other nursery stock in storage. Strawberry plants can be stored up to 10 months in polyethylene-lined crates at 30 to 32 °F.

Table 4 Storage Conditions for Cut Flowers and Nursery Stock

Commodity	Storage Temperature, °F	Relative Humidity, %	Approximate Storage Life	Method of Holding	Highest Freezing Point, °F
Cut Flowers					
Calla Lily	40	90 to 95	1 week	Dry pack	
Camellia	45	90 to 95	3 to 6 days	Dry pack	30.6
Carnation	31 to 32	90 to 95	3 to 4 weeks	Dry pack	30.8
Chrysanthemum	31 to 32	90 to 95	3 to 4 weeks	Dry pack	30.5
Daffodil (Narcissus)	32 to 33	90 to 95	1 to 3 weeks	Dry pack	31.8
Dahlia	40	90 to 95	3 to 5 days	Dry pack	
Gardenia	32 to 34	90 to 95	2 weeks	Dry pack	31.0
Gladiolus	36 to 42	90 to 95	1 week	Dry pack	31.4
Iris, tight buds	31 to 32	90 to 95	2 weeks	Dry pack	30.6
Lily, Easter	32 to 35	90 to 95	2 to 3 weeks	Dry pack	31.1
Lily-of-the-valley	31 to 32	90 to 95	2 to 3 weeks	Dry pack	
Orchid	45 to 55	90 to 95	2 weeks	Water	31.4
Peony, tight buds	32 to 35	90 to 95	4 to 6 weeks	Dry pack	30.1
Rose, tight buds	32	90 to 95	2 weeks	Dry pack	31.2
Snapdragon	40 to 42	90 to 95	1 to 2 weeks	Dry pack	30.4
Sweet Peas	31 to 32	90 to 95	2 weeks	Dry pack	30.4
Tulips	31 to 32	90 to 95	2 to 3 weeks	Dry pack	
Greens					
Asparagus (plumosus)	35 to 40	90 to 95	2 to 3 weeks	Polylined cases	26.0
Fern, dagger and wood	30 to 32	90 to 95	2 to 3 months	Dry pack	28.9
Fern, leatherleaf	34 to 40	90 to 95	1 to 2 months	Dry pack	
Holly	32	90 to 95	4 to 5 weeks	Dry pack	27.0
Huckleberry	32	90 to 95	1 to 4 weeks	Dry pack	26.7
Laurel	32	90 to 95	2 to 4 weeks	Dry pack	27.6
Magnolia	35 to 40	90 to 95	2 to 4 weeks	Dry pack	27.0
Rhododendron	32	90 to 95	2 to 4 weeks	Dry pack	27.6
Salal	32	90 to 95	2 to 3 weeks	Dry pack	26.8
Bulbs					
Amaryllis	38 to 45	70 to 75	5 months	Dry	30.8
Caladium	70	70 to 75	2 to 4 months		29.7
Crocus	48 to 63		2 to 3 months		29.7
Dahlia	40 to 48	70 to 75	5 months	Dry	28.7
Gladiolus	45 to 50	70 to 75	5 to 8 months	Dry	28.2
Hyacinth	63 to 68		2 to 5 months		29.3
Iris, Dutch, Spanish	68 to 77	70 to 75	4 months	Dry	29.3
Gloriosa	50 to 63	70 to 75	3 to 4 months	Polyliner	
Candidum	31 to 33	70 to 75	1 to 6 months	Polyliner and peat	
Croft	31 to 33	70 to 75	1 to 6 months	Polyliner and peat	
Longiflorum	31 to 33	70 to 75	1 to 10 months	Polyliner and peat	28.9
Speciosum	31 to 33	70 to 75	1 to 6 months	Polyliner and peat	
Peony	33 to 35	70 to 75	5 months	Dry	
Tuberose	40 to 45	70 to 75	4 to 12 months	Dry	
Tulip	63	70 to 75	2 to 6 months	Dry	27.6
Nursery Stock					
Trees and shrubs	32 to 36	95	4 to 5 months	[a]	
Rose bushes	31 to 36	85 to 95	4 to 5 months	Bare rooted with polyliner	
Strawberry plants	30 to 32	80 to 85	8 to 10 months	Bare rooted with polyliner	29.9
Rooted cuttings	31 to 36	85 to 95		Polywrap	
Herbaceous perennials	27 to 28	80 to 85	4 to 8 months	[a]	
	31 to 35	80 to 85	3 to 7 months		
Christmas trees	22 to 32	80 to 85	6 to 7 weeks		

Data from USDA *Agricultural Handbook* No. 66.
a For details for various trees, shrubs, and perennials, see *Storage of Nursery Stock* (AAN 1960).

Table 5 Space, Weight, and Density Data for Commodities Stored in Refrigerated Warehouses

Commodity	Type of Package	Outside Dimensions of Package, in.	Average Gross Weight of Package, lb	Average Net Weight Merchandise, lb	Average Gross Weight Density, lb/ft³	Average Net Weight Density, lb/ft³
Apples	Wood box, Northwestern	19-1/2 × 11 × 12-3/16	50	42	33.1	27.8
	Fiber tray carton	20-1/2 × 12-1/2 × 13-1/4	46-3/4	43	23.8	21.9
	Fiber master carton	22-1/2 × 12-1/2 × 13	44-3/4	41	21.2	19.4
	Fiber bulk carton	19 × 12-1/2 × 13	44-3/4	41	25.0	22.9
	Pallet bin	47 × 47 × 30	1030	900	26.9	23.5
Beef						
Boneless	Fiber carton	28 × 18 × 6	146	140	83.4	80.0
Fores	Loose					22.2
Hinds	Loose					22.2
Celery	Wirebound crates	20-1/4 × 16 × 9-3/4	60	55	32.8	30.0
	Fiber carton	16 × 11 × 10	36	32	35.4	31.4
Cheese	Hoops	16 × 16 × 13	84	78	43.6	40.5
	Wood, export	17 × 17 × 14	87	76	37.1	32.5
Cheese, Swiss	Wheels	32-1/2 × 32-1/2 × 7		171		40.0
Chili peppers	Bags	45 × 21 × 26	234	229	16.5	16.1
Citrus fruits						
Oranges	Box	12-1/8 × 13-1/4 × 26-1/4	77	69	31.5	28.3
	Bruce box	13 × 11 × 26-1/4	88	83	40.5	38.2
	Pallet, 40 cartons	40 × 48 × 58-1/2	1690	1480	26.0	22.8
California oranges	Fiber carton	16-3/8 × 10-1/16 × 10-1/2	40	37	38.0	35.2
Florida oranges	Fiber carton	19-1/4 × 12-1/4 × 8	45	37	41.3	33.9
Lemons	Fiber carton	16-3/8 × 10-1/16 × 10-1/2	40	37	40.0	37.0
Grapefruit	Fiber carton	19-1/4 × 12-1/4 × 8	40	38	36.7	34.9
Coconut, Shredded	Bags	38 × 18-1/2 × 8	101	100	31.0	30.7
Cranberries	Fiber carton	15-3/4 × 11-1/4 × 10-1/2	26	24	24.1	22.2
Cream	Tins	12 × 12 × 14	52-3/4	50	45.2	42.9
Dried fruit	Wood box	15-1/2 × 10 × 6-1/2	26-1/2	25	45.4	42.9
Dates	Fiber carton	14 × 14 × 11	32	30	25.7	24.0
Raisins, prunes, figs, peaches	Fiber carton	15 × 11 × 7	32	30	47.9	44.9
Eggs, Shell	Wood cases	26 × 12 × 13	55	45	23.4	19.1
Eggs, Frozen	Cans	10 × 10 × 12-1/2	32	30	44.2	41.5
Frozen fishery products						
Blocks	4/13 1/2 lb carton	20-3/4 × 12-1/8 × 6-3/4	56	54	57.0	55.0
	4/16 1/2 lb carton	19-3/4 × 10-3/4 × 11-1/4	68	66	49.2	47.8
Filets	12/16 oz carton	12-3/4 × 8-5/8 × 3-13/16	13.5	12	55.8	49.6
	10/5 lb carton	14-1/2 × 10 × 14	52.25	50	44.6	42.7
	5/10 lb carton	14-1/2 × 10 × 14	52.2	50	44.5	42.7
Fish sticks	12/8 oz carton	11 × 8-3/8 × 3-7/8	6.9	6	33.6	29.3
	24/8 oz carton	16-7/16 × 8-5/16 × 4-5/8	13.8	12	37.8	32.9
Panned fish portions	None, glazed	Wooden boxes				35.0
	2, 3, 5, and 6 lb carton	Custom packing				29-33
Round ground fish	None, glazed	Stacked loose				33-35
Round Halibut	None, glazed	Wooden box, loose				30-35
		Stacked loose				38.0
Round Salmon	None, glazed	Stacked loose				33-35
Shrimp	2-1/2 and 5 lb cartons	Custom packing				35.0
Steaks	1, 5, or 10 lb packages	Custom packing				50-60
Frozen Fruits, Juices, Vegetables						
Asparagus	24/12 oz carton	13-1/2 × 11-3/4 × 8-1/4	21	18	27.7	23.8
Beans, Green	36/10 oz carton	12-1/2 × 11 × 8	25-1/2	22-1/2	40.1	35.3
Blueberries	24/12 oz carton	12 × 11-1/2 × 8	20	18	31.3	28.2
Broccoli	24/10 oz carton	12-1/2 × 11-1/2 × 8-1/2	18-1/2	15	26.2	21.2
Citrus concentrates	Fiber carton 48/6 oz	13 × 8-3/4 × 7-1/2	27	26	54.7	52.7
Peaches	24/1 lb carton	13-1/2 × 11-1/4 × 7-1/2	27	24	41.0	36.4

Table 5 Space, Weight, and Density Data for Commodities Stored in Refrigerated Warehouses (*Concluded*)

Commodity	Type of Package	Outside Dimensions of Package, in.	Average Gross Weight of Package, lb	Average Net Weight Merchandise, lb	Average Gross Weight Density, lb/ft³	Average Net Weight Density, lb/ft³
Peas	6/5 lb carton	17 × 11 × 9-1/2	32	30	31.1	28.2
	48/12 oz carton	21-1/2 × 8-1/2 × 12-1/2	38	36	28.7	27.2
Potatoes, French fries	12/16 oz carton					28.6
	24/9 oz carton					24.0
Spinach	24/14 oz carton	12-1/2 × 11 × 8-1/2	24	21	35.5	31.0
Strawberries	30 lb can	12-1/2 × 10 × 10	32	30	44.2	41.5
	24/1 lb carton	13 × 11 × 8	28	24	42.3	36.2
	450 lb barrel	35 × 25 × 25		450		35.5
Grapes, California	Wood lug box	6-1/2 × 15 × 18	31	28	32.4	29.2
Lamb, Boneless	Fiber box	20 × 15 × 5	57	53	65.7	61.0
Lard (2/28 lb)	Wood export box	18 × 13-1/4 × 7-3/4	64	56	59.8	52.5
Lettuce, Head	Fiber carton	20-1/2 × 13-1/2 × 9-1/2	37-1/2	35	24.7	
	Fiber carton	21-1/2 × 14-1/4 × 10-1/2	45-55	42-52	26.9	25.2
	Pallet, 30 cartons	42 × 50 × 66	1350	1170	16.8	14.6
Milk, Condensed	Barrels	35 × 25-1/2 × 25-1/2	670	600	50.9	45.6
Nuts						
Almonds, In shell	Sacks	24 × 15 × 33	91-1/2	90	13.3	13.1
Almonds, Shelled	Cases	6-3/4 × 23-1/2 × 11	32	28	31.7	27.7
English Walnuts, In shell	Sacks	25 × 11 × 31	103	100	20.9	20.3
English Walnuts, Shelled	Fiber carton	14 × 14 × 10	27	25	23.8	22.0
Peanuts, Shelled	Burlap bag	35 × 10 × 15	127	125	39.2	38.6
Pecans, In shell	Burlap bag	35 × 22 × 12	126-1/2	125	23.7	23.4
Pecans, Shelled	Fiber carton	13 × 13 × 11	32	30	29.8	27.9
Peaches	3/4 bushel	16-7/8 top dia	41	48	43.9	40.7
	1/2 bushel	14-1/2 top dia	28	25	45.0	40.2
	Wirebound crate	19 × 11-3/4 × 11-1/8	42	38	29.2	26.4
	Wood lug box	18-1/8 × 11-1/2 × 5-3/4	26	23	38.0	33.1
Pears	Wood box	8-1/2 × 11-1/2 × 18	52	48	51.0	47.1
Pears, Place pack	Fiber carton	18-1/2 × 12 × 10	52	46	40.5	35.6
Pork						
Bundle bellies	Bundles	23-1/2 × 10-1/2 × 7	57	57	57.0	57.0
Loins, Regular	Wood box	28 × 10 × 10	60	54	37.0	33.3
Loins, Boneless	Fiber box	20 × 15 × 5	57	52	65.7	59.9
Potatoes	Sack	33 × 17-1/2 × 11	101	100	27.5	27.2
Poultry, Fresh (eviscerated)						
Fryers, whole, 24-30 to pkg.	Wirebound crate	24 × 10 × 7	65	60	27.5	25.4
Fryer parts	Wirebound crate	17-3/4 × 10 × 12-1/2	54	50	42.1	38.9
Poultry, Frozen (eviscerated)						
Ducks, 6 to pkg.	Fiber carton	22 × 16 × 4	32-1/2	31	39.9	38.0
Fowl, 6 to pkg.	Fiber carton	20-3/4 × 18 × 5-1/2	33 1/2	31	28.2	26.1
Fryers, Cut up, 12 to pkg.	Fiber carton	17-1/4 × 15-3/4 × 4-1/4	30 1/2	28	45.4	41.7
Roasters, 8 to pkg.	Fiber carton	20-3/4 × 18 × 5-1/2	32 1/2	30	27.3	25.2
Turkeys,						
3-6 lb, 6 to pkg.	Fiber carton	21 × 17 × 6-1/2	30	27	22.5	20.1
6-10 lb, 6 to pkg.	Fiber carton	26 × 21-1/2 × 7	52-1/2	48	23.3	21.2
10-13 lb, 4 to pkg.	Fiber carton	26-1/2 × 16 × 7-1/2	50	46	27.2	25.0
13-16 lb, 4 to pkg.	Fiber carton	29 × 18-1/2 × 9	67-1/2	62	24.2	22.2
16-20 lb, 2 to pkg.	Fiber carton	17 × 16 × 9	39	36	27.7	25.4
20-24 lb, 2 to pkg.	Fiber carton	19 × 16-1/2 × 9-1/2	47-1/2	44	27.6	25.5
Tomatoes						
Florida	Fiber carton	19 × 10-7/8 × 10-3/4	43	40	33.3	31.0
	Wirebound crate	18-3/4 × 11-15/16 × 11-15/16	64	60	41.3	38.7
California	Wood lug box	17-1/2 × 14 × 7-3/4	34	30	30.9	27.3
Texas	Wood lug box	17-1/2 × 14 × 6-5/8	34	30	36.2	31.9
Veal, Boneless	Fiber carton	20 × 15 × 5	57	53	65.7	61.0

Adapted from the table *Density of Commodites Carried in Cold Storage Warehouses,* courtesy of the Missouri Valley Chapter of the National Association of Refrigerated Warehouses, and from other sources.

Popcorn

Store popcorn at 32 to 40°F and about 85% rh. This relative humidity yields the optimum popping condition and the desired moisture content of about 13.5%.

Vegetable Seeds

Seeds generally benefit from low temperatures and low humidity storage. High temperatures and high humidity favor loss of viability. Most vegetable seeds undergo no significant decrease in germination during one season when stored at 50°F and 50% rh. Full viability is retained far longer than one year as temperature and humidity are reduced. A temperature of 20°F and 15 to 25% rh are considered ideal but rarely necessary unless seed viability must be maintained for many years. Aster, pepper, tomato, and lettuce seeds stored under these conditions had equal or better viability after 13 years than did the fresh seed.

Low moisture content of the seed is important for germination. Hemp seed containing 9.5% moisture was mostly dead after 12 years storage at 50°F, but lost only 12% viability when moisture content was 5.7%. If stored at about 0°F, moisture percentage should not exceed 10% for many species. Seed with higher moisture content, when stored at 0°F, eventually equilibrates at a lower moisture level but may initially suffer frost damage. At higher temperatures, if it is impossible to keep humidity low enough, seeds must be stored in moistureproof containers.

DENSITY OF STORED COMMODITIES

Table 5 is a compilation of data giving the types of containers used for storage, their dimensions, gross weights, net weights, and density. Additional information on gross weights of packed containers and on dimensions and densities of pallet loads of produce is given in USDA *Marketing Research Report* No. 467.

BIBLIOGRAPHY

ASHRAE. 1974. Relative humidity and the storage of fresh fruits and vegetables—Recent research results and developments. ASHRAE Symposium CH-73-7.

Bogardus, R.K. 1961. Wholesale fruit and vegetable warehouses—Guides for layout and design. *Marketing Research Report* No. 467.

Dilley, D.R. 1982. Principles and effects of hypobaric storage of fruits and vegetables. ASHRAE *Transactions* 88(1):1461-78.

Gacula, M.C., Jr. and J.J. Kubala. 1975. Statistical models for shelf life failures. *Journal of Food Science.*

Hankin, L. and D. Shields. 1982. Relation of code dates to retail milk supply. *Journal of Food Protection* 45.

Hardenburg, R.E., A.E. Watada, and C.Y. Wang. 1986. The commercial storage of fruits, vegetables, and florist and nursery stocks. USDA *Handbook* No. 66. USDA, Washington, D.C.

Hunt Ashby, B. 1987. Protecting perishable foods during transport by truck. USDA *Handbook* No. 669. USDA, Washington, D.C.

IIR. 1986. General principles for the freezing, storage and thawing of foodstuffs. Recommendations for the processing and handling of frozen foods. International Institute of Refrigeration, Paris.

IIR. 1990. Principles of refrigerated preservation of perishable foodstuffs. Manual of refrigerated storage in the warmer developing countries. International Institute of Refrigeration, Paris.

ISO. 1974. Avocados—Guide for storage and transport. International Organization for Standardization, Geneva.

Mahlstede, P. and W.E. Fletcher. 1960. *Storage of nursery stock.* American Association Nurserymen, Washington, D.C.

Morris, L.L., *et al.* 1981. Rose of CA, including CO, in prolonging the storage life of tomatoes. Proceedings of the 3rd National Controlled Atmosphere Research Conference on Controlled Atmospheres for Storage and Transport of Perishable Agricultural Commodities, Corvallis, OR.

Nesbit, C.D. 1974. Changing concepts in refrigerated warehouse construction. Patterson Enterprises, Jacksonville, FL.

Post, K. and C.W. Fisher, Jr. 1952. Commercial storage of cut flowers. Cornell University Extension *Bulletin* No. 853.

RRF. 1989. *Commodity storage manual.* Refrigeration Research Foundation, Bethesda, MD.

Ryall, A.L. and W.J. Lipton. 1978. *Handling, transportation and storage of fruits and vegetables.* Vol. 1, Vegetables and melons. AVI Publishing Co., Inc., Westport, CT.

Ryall, A.L. and W.T. Pentzer. 1982. *Handling, transportation and storage of fruits and vegetables.* Vol. 2, Fruits and tree nuts. AVI Publishing Co., Inc., Westport, CT.

Scott, V.N. 1989. Interaction of factors to control microbial spoilage of refrigerated foods. *Journal of Food Protection.*

Schlimme, D.V., M.A. Smith, and L.M. Ali. 1991. Influence of freezing rate, storage temperature, and storage duration on the quality of cooked turkey breast roll. ASHRAE *Transactions* 97(1).

Schlimme, D.V., M.A. Smith, and L.M. Ali. 1991. Influence of freezing rate, storage temperature, and storage duration on the quality of turkey frankfurters. ASHRAE *Transactions* 97(1).

Tressler, D.K. and C.F. Evers. 1957. *The freezing preservation of foods,* 3rd ed. AVI Publishing Co., Inc., Westport, CT.

Ulrich, R. 1979. Fruits and vegetables. Recommendations for chilled storage of perishable produce. International Institute of Refrigeration, Paris.

USDA. 1955. Protecting stored furs from insects. USDA *Publication* AMS-57.

Valenzuela Segura, G., D.D. Delgado, and D.R. Ramirez. 1972. Handling, storage and transport systems for exported refrigerated perishable foods. Revista del Instituto de Investigaciones Technologicas, Bogota, Columbia.

Webb, B.H., A.H. Johnson, and J.A. Alford. 1973. *Fundamentals of dairy chemistry.* AVI Publishing Co., Inc., Westport, CT.

CHAPTER 26

REFRIGERATION LOAD

THE segments of total refrigeration load are (1) *transmission load*, which is heat transferred into the refrigerated space through its surface; (2) *product load*, which is heat removed from and produced by products brought into and kept in the refrigerated space; (3) *internal load*, which is heat produced by internal sources, *e.g.*, lights, electric motors, and people working in the space; (4) *infiltration air load*, which is heat gain associated with air entering the refrigerated space; and (5) *equipment-related load.*

The first four segments of load constitute the net heat load for which a refrigeration system is to be provided; the fifth segment consists of all heat gains created by the refrigerating equipment. Thus, net heat load plus equipment heat load is the total refrigeration load for which a compressor must be selected.

This chapter contains load calculating procedures and data for the first four segments and load determination recommendations for the fifth segment. Information needed for the refrigeration of specific foods can be found in Chapters 10 through 23.

TRANSMISSION LOAD

Sensible heat gain through walls, floor, and ceiling varies with type of insulation, thickness of insulation, construction, outside wall area, and temperature difference between refrigerated space and ambient air.

The thermal conductance values for several cold storage insulations are listed in Table 1. The insulation thickness that this table refers to is actual insulation thickness, not the overall wall thickness. Comparative values of various insulation materials are in Chapter 22 of the 1993 ASHRAE *Handbook—Fundamentals.*

The overall coefficient of heat transfer U of the wall, floor, or ceiling can be derived by the following equation:

$$U = \frac{1}{1/f_i + x/k + 1/f_o} \tag{1}$$

where

U = overall heat transfer coefficient, Btu/h · ft² · °F
x = wall thickness, in.
k = thermal conductivity of wall material, Btu · in/h · ft² · °F
f_i = inside film or surface conductance, Btu/h · ft² · °F
f_o = outside film or surface conductance, Btu/h · ft² · °F

A value of 1.65 Btu/h · ft² · °F for f_i and f_o is frequently used for still air. If the outer surface is exposed to 15 mph wind, f_o is increased to 6 Btu/h · ft² · °F.

The preparation of this chapter is assigned to TC 10.8, Refrigeration Load Calculations.

With thick walls and low conductivity, the resistance x/k makes U so small that $1/f_i$ and $1/f_o$ have little effect and can be omitted from the calculation. Walls are usually made of more than one material; therefore, the value x/k represents the composite resistance of the materials. The U-value for a wall with flat parallel surfaces of materials 1, 2, and 3 is given by the following equation:

$$U = \frac{1}{x_1/k_1 + x_2/k_2 + x_3/k_3} \tag{2}$$

where x values indicate thicknesses and k values indicate conductivities of the materials used. Table 1 shows possible values of x/k.

The effect on the overall thermal performance of the prefabricated or insulated panels' metal surfaces is negligible and should not be considered in calculating the U-value. In this case, the reciprocal of the conductance values in Table 1 can be substituted for x/k in Equations (1) or (2).

After establishing U, the heat gain is given by the basic equation:

$$q = UA\Delta t \tag{3}$$

where

q = heat leakage, Btu/h
A = outside area of section, ft²
Δt = difference between outside air temperature and air temperature of the refrigerated space, °F

Table 2 lists minimum insulation thicknesses based on expanded polyurethane as recommended by the refrigeration industry. Equivalent thicknesses of other insulation materials can be found by comparing C values in Table 1. Consideration should be given to conductivity increases with time (see Chapter 20 of the 1993 ASHRAE *Handbook—Fundamentals*).

Chapter 24 of the 1993 ASHRAE *Handbook—Fundamentals* gives outdoor design temperatures for major cities; values for 1% should be used.

If the refrigerated room is exposed to the sun, additional heat will be added to the heat load. For practical purposes, the temperature can be adjusted to compensate for solar effect. The values given in Table 3 apply throughout the 24-h period and are added to the ambient temperature when calculating wall heat gain.

Latent heat gain due to moisture transmission through walls, floors, and ceilings of modern-construction refrigerated facilities is negligible. Data in Chapter 22 of the 1993 ASHRAE *Handbook—Fundamentals* may be used to calculate this load if moisture-permeable materials are used.

Table 1 Insulation Conductance Values C (x/k) for Walls, Floor, and Ceiling

Insulation Thickness	Polyurethane (Expanded) $k = 0.16$[b]	Polyurethane (Board) $k = 0.18$[b]	Polystyrene (Extruded) $k = 0.20$[b]	Glass Fiber and Polystyrene (Molded Beads) $k = 0.25$[b]	Corkboard $k = 0.30$[b]
in.	Btu/h·ft²·°F	Btu/h·ft²·°F	Btu/h·ft²·°F	Btu/h·ft²·°F	Btu/h·ft²·°F
1	0.160	0.180	0.200	0.250	0.300
2	0.080	0.090	0.100	0.125	0.150
3	0.053	0.060	0.067	0.083	0.100
4	0.040	0.045	0.050	0.063	0.070
5	0.032	0.036	0.040	0.050	0.060
6	0.027	0.030	0.033	0.042	0.050
7	0.023	0.026	0.029	0.036	0.043
8	0.020	0.022	0.025	0.031	0.038
9	0.018	0.020	0.022	0.028	0.033
10	0.016	0.018	0.020	0.025	0.030

Note:
Where wood studs are used in walls, multiply above values by 1.1.

[a]Commonly used values of conductance are tabulated. See Chapters 20 and 22 in the 1993 ASHRAE *Handbook—Fundamentals* for variations due to temperature, aging, and other considerations.
[b]Thermal conductivity k is expressed in Btu·in/h·ft²·°F.

Table 2 Minimum Insulation Thickness

Storage Temperature	Expanded Polyurethane Thickness	
	Northern U.S.	Southern U.S.
°F	in.	in.
50 to 60	1	2
40 to 50	2	2
25 to 40	2	3
15 to 25	3	3
0 to 15	3	4
−15 to 0	4	4
−40 to −15	5	5

Table 3 Allowance for Sun Effect

Typical Surface Types	East Wall °F	South Wall °F	West Wall °F	Flat Roof °F
Dark colored surfaces				
Slate roofing	8	5	8	20
Tar roofing				
Black paint				
Medium colored surfaces				
Unpainted wood	6	4	6	15
Brick				
Red tile				
Dark cement				
Red, gray, or green paint				
Light colored surfaces				
White stone	4	2	4	9
Light colored cement				
White paint				

Note: Add °F to the normal temperature difference for heat leakage calculations to compensate for sun effect—do not use for air-conditioning design.

PRODUCT LOAD

The primary sources of refrigeration load from products brought into and kept in the refrigerated space are (1) heat removal required to reduce product temperature from receiving to storage temperature and (2) heat generation by products in storage, mainly fruits and vegetables.

A product placed in a refrigerated room at a higher temperature than the room temperature loses heat as it approaches room temperature. The quantity of heat to be removed can be calculated from knowledge of the product, including state on entering the refrigerated space, final state, mass, specific heat above and below freezing temperature, and latent heat. When cooling a product from one state and temperature to another state and temperature, the following equations apply:

1. Heat removal in cooling from the initial temperature to some lower temperature above freezing:

$$Q_1 = mc_1(t_1 - t_2) \tag{4}$$

2. Heat removal in cooling from the initial temperature to the freezing point of the product:

$$Q_2 = mc_1(t_1 - t_f) \tag{5}$$

3. Heat removal to freeze the product:

$$Q_3 = mh_{if} \tag{6}$$

4. Heat removal in cooling from the freezing point to the final temperature below the freezing point:

$$Q_4 = mc_2(t_f - t_3) \tag{7}$$

where

Q_1, Q_2, Q_3, Q_4 = heat removal, Btu
m = weight of the product, lb
c_1 = specific heat of the product above freezing, Btu/lb·°F
t_1 = initial temperature of the product above freezing, °F
t_2 = lower temperature of the product above freezing, °F
t_f = freezing temperature of the product, °F
h_{if} = latent heat of fusion of the product, Btu/lb
c_2 = specific heat of the product below freezing, Btu/lb·°F
t_3 = final temperature of the product below freezing, °F

Refrigeration system capacity for products brought into refrigerated spaces is determined from the time allotted for heat removal and assumes that the product is properly exposed to remove the heat in that time. The calculation is:

$$q = \frac{Q_2 + Q_3 + Q_4}{n} \tag{8}$$

where

q = product cooling load, Btu/h
n = allotted time period, h

Specific heats above and below freezing and latent heats of fusion for many products are given in Chapter 25. A product's latent heat of fusion is related to its water content and can be estimated by multiplying the product's percent of water (expressed as a decimal) by the water's latent heat of fusion, which is 144 Btu/lb. Most food products freeze in the range of 26 to 31 °F. When the exact freezing temperature is not known, assume that it is 28 °F.

Example 1. 220 lb of lean beef is to be cooled from 65 to 40 °F, then frozen and cooled to 0 °F. Specific heat of beef before freezing is 0.77 Btu/lb · °F; after freezing, 0.40 Btu/lb · °F. The latent heat of fusion is 100 Btu/lb.
Solution:

To cool from 65 to 40 °F in a chilled room:

$$220 \times 0.77 \times (65 - 40) = 4235 \text{ Btu}$$

To cool from 40 °F to freezing point in freezer:

$$220 \times 0.77 \times (40 - 28) = 2033 \text{ Btu}$$

To freeze:

$$220 \times 100 = 22{,}000 \text{ Btu}$$

To cool from freezing to storage temperature:

$$220 \times 0.40\ (28 - 0) = 2464 \text{ Btu}$$

Total:

$$4235 + 2033 + 22{,}000 + 2464 = 30{,}732 \text{ Btu}$$

Fresh fruits and vegetables respire and release heat during storage. This heat of respiration varies with product and temperature; the colder the product, the less the heat of respiration. Table 1 of Chapter 25 gives the heat evolution rates for various products.

Calculations in *Example 1* do not cover heat gained from product containers brought into the refrigerated space. When pallets, boxes, or other packing materials are a significant portion of the total mass introduced, this heat gain should be calculated.

Equations (4) through (8) are used to calculate the total heat gain to be extracted by the refrigeration equipment. Any moisture shrinkage involved in the extraction process appears as latent heat gain. The amount of moisture involved is usually provided by the end-user as a percentage of product mass and, with such information, the latent heat component of the total heat gain may be determined. Subtracting the latent heat component from the total heat gain determines the sensible heat component.

Table 4 Heat Equivalent of Electric Motors

Motor hp	Connected Load in Refrigerated Space[a]	Motor Losses Outside Refrigerated Space[b]	Connected Load Outside Refrigerated Space[c]
	Btu/hp · h	Btu/hp · h	Btu/hp · h
1/8 to 1/3	4600	2550	2100
1/2 to 3	3800	2550	1300
5 to 20	3300	2550	800

[a]For use when both useful output and motor losses are dissipated within refrigerated space; motors driving fans for forced circulation unit coolers.
[b]For use when motor losses are dissipated outside refrigerated space and useful motor work is expended within refrigerated space; pump on a circulating brine or chilled water system; fan motor outside refrigerated space driving fan circulating air within refrigerated space.
[c]For use when motor heat losses are dissipated within refrigerated space and useful work expended outside of refrigerated space; motor in refrigerated space driving pump or fan located outside of space.

Table 5 Heat Equivalent of Occupancy

Refrigerated Space Temperature, °F	Heat Equivalent/ Person, Btu/h
50	720
40	840
30	950
20	1050
10	1200
0	1300
−10	1400

Note: Heat equivalent may be estimated by $q_p = 1295 - 11.5t(°\text{F})$

INTERNAL LOAD

All electrical energy dissipated in the refrigerated space (from lights, motors, heaters, and other equipment) must be included in the internal heat load. Heat equivalents of electric motors are shown in Table 4. People release heat at varying rates depending on temperature, type of work, clothing, size, and other factors. Table 5 shows the average load caused by occupancy of the refrigerated space. When people go into the refrigerated space for short durations, they carry a considerable amount of heat over and above that listed in Table 5. If traffic load is heavy, adjustments must be made.

The latent heat component of internal load is usually very small compared to total refrigeration load and is customarily regarded as all sensible heat in total load summaries. However, the latent heat component may be great where water is involved in processing or cleaning and should be independently calculated.

INFILTRATION AIR LOAD

Infiltration air heat gain and associated equipment-related load can amount to more than half the total refrigeration load of distribution-type warehouses and similar applications.

Infiltration by Air Exchange

The most common cause of infiltration is room-to-room air-density difference (see Figures 1 and 2). For the typical case where mass inflow equals mass outflow less condensed moisture, the room must be sealed except for the opening in question. If the cold room is not sealed, direct in-flow, which is discussed later, may apply.

The equation for heat gain through doorways from air exchange is as follows:

$$q_t = qD_tD_f(1 - E) \tag{9}$$

where

q_t = average heat gain for the 24-h or other period, Btu/h
q = sensible and latent refrigeration load for fully established flow, Btu/h
D_t = doorway open-time factor
D_f = doorway flow factor
E = effectiveness of doorway protective device

Gosney and Olama (1975) provide the following air exchange equation for fully established flow:

$$q = 795.6A\ (h_i - h_r)\ \rho_r\ (1 - \rho_i/\rho_r)^{0.5}\ (gH)^{0.5}F_m \tag{10}$$

where

A = doorway area, ft²
h_i = enthalpy of infiltration air, Btu/lb
h_r = enthalpy of refrigerated air, Btu/lb
ρ_i = density of infiltration air, lb/ft³
ρ_r = density of refrigerated air, lb/ft³
g = gravitational constant = 32.174 ft/s²
H = doorway height, ft
F_m = density factor

$$F_m = \left(\frac{2}{1 + (\rho_r/\rho_i)^{1/3}}\right)^{1.5} \tag{11}$$

(Chapter 6 of the 1993 ASHRAE *Handbook—Fundamentals* and the ASHRAE Psychrometrics Chart give air enthalpy and density values.)

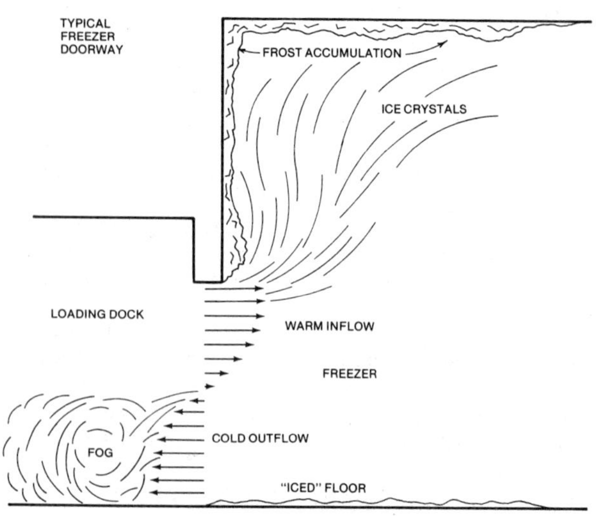

Fig. 1 Flowing Cold and Warm Air Masses that Occur for Typical Open Freezer Doors

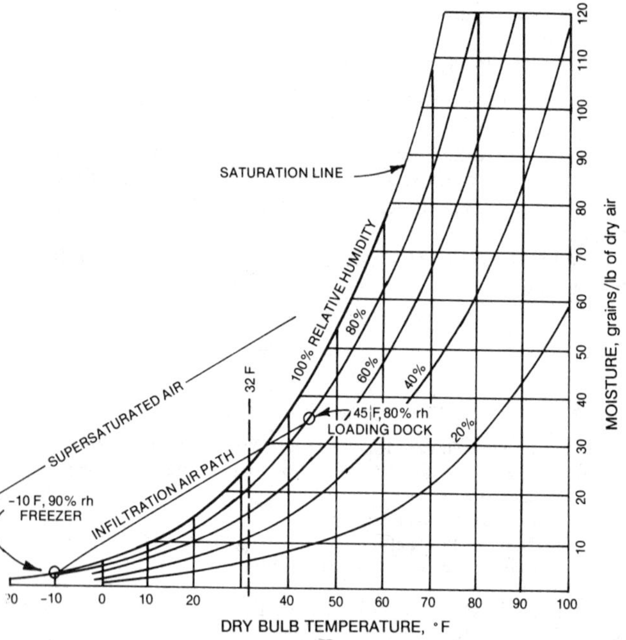

Fig. 2 Psychrometric Depiction of Air Exchange for Typical Freezer Doorway

Table 6 SHR (R_s) for Infiltration from Outdoors to Refrigerated Spaces

Outdoors Temp. °F	rh, %	Cold Space at 90% rh Dry-Bulb Temperature, °F −30	−20	−10	0	10	20	30	40	50	60
100	50	0.59	0.57	0.55	0.53	0.51	0.49	0.48	0.46	0.45	0.45
	40	0.65	0.63	0.61	0.59	0.57	0.56	0.54	0.53	0.54	0.57
	30	0.71	0.69	0.68	0.66	0.65	0.64	0.63	0.64	0.66	0.76
	20	0.79	0.77	0.76	0.75	0.74	0.74	0.75	0.78	0.87	—
95	60	0.58	0.56	0.54	0.52	0.49	0.47	0.45	0.43	0.42	0.41
	50	0.62	0.60	0.58	0.56	0.54	0.52	0.51	0.49	0.48	0.50
	40	0.67	0.66	0.64	0.62	0.60	0.59	0.57	0.57	0.58	0.64
	30	0.73	0.72	0.70	0.69	0.67	0.66	0.66	0.68	0.72	0.89
90	60	0.61	0.59	0.57	0.55	0.52	0.50	0.48	0.46	0.45	0.45
	50	0.65	0.63	0.61	0.59	0.57	0.55	0.54	0.52	0.52	0.56
	40	0.70	0.68	0.67	0.65	0.63	0.61	0.61	0.61	0.63	0.74
	30	0.76	0.74	0.73	0.71	0.70	0.69	0.70	0.72	0.80	—

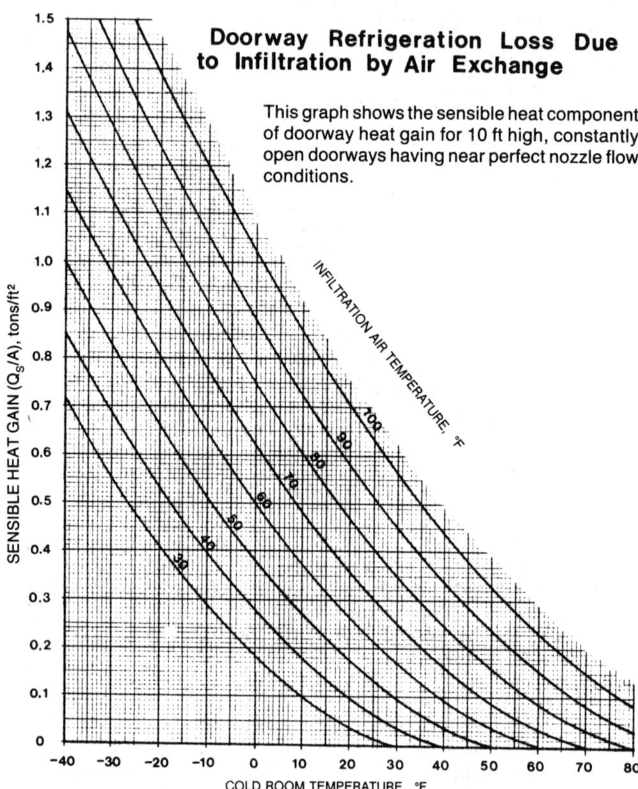

Fig. 3 Sensible Heat Gain by Air Exchange for Continuously Open Door with Fully Established Flow

Table 7 SHR (R_s) for Infiltration from Warmer to Colder Refrigerated Spaces

Warm Space Temp. °F	rh, %	Cold Space at 90% rh Dry-Bulb Temperature, °F −40	−30	−20	−10	0	10	20	30	40	50
70	100	0.60	0.58	0.56	0.53	0.50	0.47	0.44	0.41	0.37	0.34
	80	0.66	0.64	0.61	0.59	0.56	0.53	0.50	0.48	0.46	0.44
	60	0.72	0.70	0.68	0.66	0.63	0.61	0.59	0.58	0.59	0.64
	40	0.79	0.78	0.76	0.75	0.73	0.72	0.71	0.73	0.80	—
60	100	0.66	0.64	0.62	0.59	0.56	0.52	0.49	0.45	0.41	0.35
	80	0.71	0.69	0.67	0.64	0.62	0.59	0.56	0.53	0.52	0.53
	60	0.77	0.75	0.73	0.71	0.69	0.67	0.65	0.65	0.70	—
	40	0.83	0.82	0.81	0.79	0.78	0.77	0.78	0.83	—	—
50	100	0.72	0.70	0.67	0.64	0.61	0.57	0.53	0.49	0.43	—
	80	0.76	0.74	0.72	0.70	0.67	0.64	0.61	0.59	0.62	—
	60	0.81	0.80	0.78	0.76	0.74	0.72	0.71	0.75	—	—
	40	0.87	0.86	0.84	0.83	0.82	0.82	0.85	—	—	—
40	100	0.77	0.75	0.72	0.69	0.66	0.62	0.57	0.51	—	—
	80	0.81	0.79	0.77	0.74	0.72	0.69	0.66	0.67	—	—
	60	0.85	0.84	0.82	0.80	0.78	0.77	0.79	0.99	—	—
	40	0.90	0.89	0.88	0.87	0.86	0.88	0.97	—	—	—
30	100	0.82	0.80	0.77	0.74	0.70	0.66	0.59	—	—	—
	80	0.85	0.83	0.81	0.79	0.76	0.73	0.73	—	—	—
	60	0.88	0.87	0.86	0.84	0.83	0.83	0.94	—	—	—
	40	0.92	0.91	0.90	0.90	0.91	0.96	—	—	—	—
20	100	0.86	0.84	0.82	0.79	0.75	0.69	—	—	—	—
	80	0.89	0.87	0.85	0.83	0.81	0.80	—	—	—	—
	60	0.91	0.90	0.89	0.88	0.88	0.95	—	—	—	—
	40	0.94	0.94	0.93	0.94	0.97	—	—	—	—	—
10	100	0.90	0.88	0.86	0.83	0.78	—	—	—	—	—
	80	0.92	0.90	0.89	0.87	0.86	—	—	—	—	—
	60	0.94	0.93	0.92	0.92	0.96	—	—	—	—	—
	40	0.96	0.96	0.96	0.98	—	—	—	—	—	—
0	100	0.92	0.91	0.89	0.85	—	—	—	—	—	—
	80	0.94	0.93	0.92	0.91	—	—	—	—	—	—
	60	0.96	0.95	0.95	0.97	—	—	—	—	—	—
	40	0.97	0.97	0.98	—	—	—	—	—	—	—

Equation (12), when used with Figure 3, is a simplification of Equation (10):

$$q = 3790WH^{1.5}(Q_s/A)(1/R_s) \tag{12}$$

where

Q_s/A = sensible heat load of infiltration air per square foot of doorway opening as read from Figure 3, tons/ft²
W = doorway width, ft
R_s = sensible heat ratio of the infiltration air heat gain, from Tables 6 or 7 (or from a psychrometric chart)

R_s values in Tables 6 and 7 are based on 90% rh in the cold room. A small error occurs where these values are used for 80 or 100% rh. This error together with loss of accuracy due to simplification results in a maximum error for Equation (12) of approximately 4%.

For cyclical, irregular, and constant door usage, alone or in combination, the doorway open-time factor D_t can be calculated as follows:

$$D_t = \frac{(P\theta_p + 60\theta_o)}{3600\,\theta_d} \tag{13}$$

where

D_t = decimal portion of time doorway is open
P = number of doorway passages
θ_p = door open-close time, s/passage
θ_o = time door simply stands open, min
θ_d = the daily (or other) time period, h

The value of θ_p for conventional pull-cord operated doors is approximately 25 s per passage through the doorway and 10 to 15 s per passage for high-speed doors. θ_o and θ_d should be provided by the user.

The doorway flow factor D_f is the ratio of actual air exchange to fully established flow. Fully established flow occurs only in the unusual case of an unused doorway standing open to a large room or to the outdoors, where the cold outflow is not impeded by obstructions (such as stacked pallets within or adjacent to the flow path either within or outside the cold room) and can rapidly escape the vicinity of the doorway. Under these conditions, D_f is 1.0.

Hendrix *et al.* (1989) found with traffic flow equalling one entry and exit per minute through fast-operating doors that a flow factor D_f of 0.8 is conservative for a 28 °F temperature difference. Tests by Downing and Meffert (1993) at temperature differences of 12 °F and 18 °F found a flow factor of 1.1. In view of these results, the recommended flow factor for cyclically operated doors with temperature differentials less than 20 °F is 1.1, and the recommended flow factor for higher differentials is 0.8.

The effectiveness E of open-doorway protective devices is 0.95 or higher for newly installed strip doors, fast-fold doors, and other nontight-closing doors. However, depending on the traffic level and door maintenance, E may quickly drop to 0.8 on freezer doorways and to about 0.85 for other doorways. Airlock vestibules with strip doors or push-through doors have an effectiveness ranging between 0.95 and 0.85 for freezers and between 0.95 and 0.90 for other doorways. The effectiveness of air curtains range from very poor to more than 0.7.

Infiltration by Direct In-Flow

Where refrigerated spaces are equipped with constantly or frequently open doorways or other through-the-room passageways, a negative pressure created elsewhere in the building because of mechanical air exhaust without mechanical air replenishment is a common cause of heat gain from the direct in-flow of warm infiltration air. The effect is identical to that of open doorways exposed to the wind; the effect may be great in either case. Equation (14) for heat gain from infiltration by direct in-flow provides the basis for either correcting the negative pressure or adding to refrigeration capacity.

$$q_t = 60VA(h_i - h_r)\rho_r D_t \tag{14}$$

where

q_t = average refrigeration load, Btu/h
V = average air velocity, ft/min
A = opening area, ft²
h_i = enthalpy of infiltration air, Btu/lb
h_r = enthalpy of refrigerated air, Btu/lb
ρ_r = density of refrigerated air, lb/ft³
D_t = decimal portion of time doorway is open

Area A is the smaller of the inflow and outflow openings. If the smaller area has leaks around truck load-out doors in well-maintained loading docks, the leakage area can vary from 0.3 ft² to over 1.0 ft² per door. For high merchandise movement loading docks, the facility manager should estimate the time these doors are fully or partially open.

To evaluate velocity V, the magnitude of negative pressure or other flow-through force must be known. If differential pressure across a doorway can be determined, velocity can be predicted by converting static head to velocity head. However, attempting to estimate differential pressure is usually not possible; generally, the only alternative is to assume a commonly encountered velocity. The magnitude of these velocities is frequently in the range of 60 to 300 ft/min.

The effectiveness of nontight-closing devices on doorways subject to infiltration by direct flow-through is not readily determined. Depending on the flow-through pressure differential, its tendency to vary, and the ratio of inflow area to outflow area, the effectiveness of these devices can be very low.

Sensible and Latent Heat Components

When calculating q_t for infiltration air, the sensible and latent heat components may be obtained by plotting the infiltration air path on the appropriate ASHRAE psychrometric chart, determining the air sensible heat ratio R_s from the chart, and calculating as follows:

$$\text{sensible heat: } q_s = q_t R_s \tag{15}$$

$$\text{latent heat: } q_l = q_t(1 - R_s) \tag{16}$$

where

$R_s = \dfrac{\Delta h_s}{\Delta h_t}$, as shown on ASHRAE psychrometric charts

EQUIPMENT-RELATED LOAD

Equipment-related load consists essentially of fan heat where forced air circulation is used, reheat where humidity control is provided, defrosting heat gain where defrosting occurs, and moisture evaporation where the defrosting process is exposed to refrigerated air. To accurately select heat-extracting equipment, a distinction should be made between those equipment heat loads that are felt within the refrigerated space and those that are introduced directly to the refrigerating fluid.

Equipment heat gain is usually minor at space temperatures above approximately 30 °F. Where reheat or other artificial loads are not imposed, total equipment heat gain, including defrosting heat gain where it applies, is about 5% of the four preceding segments of load.

Equipment heat gain becomes major at freezer temperatures. At −20 °F for example, the theoretical contribution to total refrigeration load due to fan power and coil defrosting alone exceeds, for many systems, 15% of the four preceding loads. This percentage assumes perfect adjustment and operation during defrosting and the absence of light-density frost on the coils. The actual percentage for use may be considerably higher, particularly where room sensible heat ratio (RSHR) is more than a few points below 1.00.

SAFETY FACTOR

Generally, a 10% safety factor is applied to the calculated load to allow for possible discrepancies between the design criteria and actual operation. Safety factors should be selected in consultation with the facility user and should be applied individually to the first four heat load segments. A separate safety factor should be added to the coil-defrosting portion of the equipment load for freezer applications that use dry-surface refrigerating coils.

Little data is available to predict heat gain from coil defrosting. For this reason, the experience of existing similar facilities should be sought to obtain an appropriate defrosting safety factor. Similar facilities should have similar room sensible heat ratios.

TOTAL REFRIGERATION LOAD

A correct calculation of the total refrigeration load is necessary to properly size equipment, run a load diversification analysis, operate the system, and estimate operating costs, The load sensible heat ratio should be estimated for installations with dry-surface refrigerating coils and included with the refrigeration load calculations.

BIBLIOGRAPHY

Cole, R.A. 1987. Infiltration load calculations for refrigerated warehouses. *Heating/Piping/Air Conditioning* (April).

Cole, R.A. 1984. Infiltration: A load calculation guide. Proceedings of the International Institute of Ammonia Refrigeration, 6th annual meeting, San Francisco (February).

Cole, R.A. 1989. Refrigeration loads in a freezer due to hot gas defrost, and their associated costs. ASHRAE *Transactions* 95(2):1149-54.

Dickerson, R.W. 1972. Computing heating and cooling rates of foods. ASHRAE *Symposium Bulletin* No-72-03.

Downing, C.C. and W.A. Meffert. 1993. Effectiveness of cold-storage door infiltration protective devices. ASHRAE *Transactions* 99(2).

Ellis, W.P. 1964. Interior finishes for cold storage facilities. ASHRAE *Journal* 6(12).

Fisher, D.V. 1960. Cooling rates of apples packed in different bushel containers, stacked at different spacings in cold storage. ASHRAE *Transactions* 66.

Fultz, D.A. 1971. A rating system for evaluating low-temperature construction. ASHRAE *Symposium Bulletin* WA-71-03.

Gentry, J.P. and R. Guillou. 1966. Fog spray humidification in cold storage. ASHRAE *Journal* 8(10).

Gosney, W.B. and H.A.L. Olama. 1975. Heat and enthalpy gains through cold room doorways. Paper presented before The Institute of Refrigeration at the Faculty of Environmental Science and Technology, The Polytechnic of the South Bank, London (December).

Govan, F.A. 1969. Basic factors to be considered in insulating a refrigerated warehouse. ASHRAE *Symposium Bulletin* DV-69-05.

Hamilton, J.J., D.C. Pearce, and N.B. Hutcheon. 1959. What frost action did to a cold storage plant. ASHRAE *Journal* 1(4).

Haugh, C.G., W.J. Standelman, and V.E. Sweat. 1972. Prediction of cooling/freezing times for food products. ASHRAE *Symposium Bulletin* NO-72-03.

Heaton, E.K. and J.G. Woodroff. 1970. Humidity and weight loss in cold storage pecans. ASHRAE *Journal* 12(4).

Hendrix, W.A., D.R. Henderson, and H.Z. Jackson. 1989. Infiltration heat gains through cold storage room doorways. ASHRAE *Transactions* 95(2).

Hovanesian, J.D., H.F. Pfost, and C.W. Hall. 1960. An analysis of the necessity to insulate floors of cold storage rooms at 35 °F. ASHRAE *Transactions* 66.

Jones, B.W., B.T. Beck, and J.P. Steele. 1983. Latent loads in low humidity rooms due to moisture. ASHRAE *Transactions* 89(1).

Karam, H.J. and H. Kreiner. 1962. Reducing heat losses in a commercial refrigerator. ASHRAE *Journal* 5(1).

Kayan, C.F. and J.A. McCague. 1959. Transient refrigeration loads as related to energy-flow concepts. ASHRAE *Journal* 1(3).

Lamiman, D.R. and J.A. Mixon. 1969. Urethane insulation applied to refrigerant buildings. ASHRAE *Journal* 11(5).

Lotz, W.A. 1964. Heat and air transfer in cold storage insulation. ASHRAE *Transactions* 70.

McGarvey, A.R. 1964. Wrap your cold storage room in a vapor barrier. ASHRAE *Journal* 6(12).

Meyer, C.S. 1964. "Inside-out" design developed for low-temperature buildings. ASHRAE *Journal* 5(4).

Pham, O.T. and D.W. Oliver. 1983. Infiltration of air into cold stores. Meat Industry Research Institute of New Zealand, Inc., Hamilton, New Zealand, IIF-IIR, 16th International Congress of Refrigeration, Paris, France.

Pichel, W. 1966. Soil freezing below refrigerated warehouses. ASHRAE *Journal* 8(10).

Popper, K. and F. Nury. 1966. Depot sterilization of cold storage rooms. ASHRAE *Journal* 8(8).

Powell, R.M. 1970. Public refrigerated warehouses. ASHRAE *Journal* 12(8).

Ruff, A.W. and F.J. Webber. 1969. The panel building for refrigerated warehouse construction. ASHRAE *Symposium Bulletin* DV-69-05.

Seward, R.M. 1970. New thoughts on an old subject—Cold storage insulation. ASHRAE *Journal* 12(7).

Slopa, R.E. 1971. Selection and installation of cold storage doors. ASHRAE *Symposium Bulletin* WA-71-03.

Vevoda, R.F. 1968. Evaluating insulation systems for cold storage facilities. ASHRAE *Journal* 10(1).

Waite, H.J. 1969. Technical requirements for supporting structures. ASHRAE *Symposium Bulletin* DV-69-05.

Wilder, C.M. 1969. Doors for refrigerated warehouses. ASHRAE *Symposium Bulletin* DV-69-05.

CHAPTER 27

TRUCKS, TRAILERS, AND CONTAINERS

THE transport of perishable commodities may be as simple as a direct delivery from farm to market by a small truck, or may require travel by sea, air, rail, highway, or combinations of these modes to satisfy distant markets. For intermodal aspects of container transfer between highway and rail carriage, see Chapter 28; for marine aspects of intermodal container applications, see Chapter 29; and for air transport, see Chapter 30. The Agricultural Research Service (ARS) and the Agricultural Marketing Service (AMS) within the U.S. Department of Agriculture have published many reports on the subject of refrigerated transportion.

TYPES OF EQUIPMENT

Refrigerated transport equipment can be broadly classified by type of operation—highway and intermodal equipment, marine service equipment, or straight trucks. In the United States, each state fixes the maximum weight and dimensions of trucks and trailers that may be operated over its highways. A summary of current state size and weight limits for trucks and truck-trailers is issued by the American Trucking Association (ATA).

Highway and Intermodal Trailers

Refrigerated semitrailers for overland use are up to 55 ft long, 8.5 ft wide, and 14 ft overall height. Where permitted, double or triple trailers may be pulled by one tractor, or a trailer may be pulled by a refrigerated truck. USDA *Agricultural Handbook* 669 (Ashby *et al.* 1987) gives recommended loading methods and temperatures for many perishable products.

Highway trailers, removed from their tractors, are carried piggyback on railroad flat cars (TOFC = trailer on flat car). Trucks and trailers with tractors are driven onto ships (roll-on roll-off). Containers are carried on truck chassis, on trailer chassis, on railroad flat cars (COFC = container on flat car), above and below deck on container ships, and, in limited fashion, on cargo aircraft.

Containers differ from standard bodies primarily in the hardware required to facilitate attachment to the chassis and in structural hardware required for lifting and stacking. In a typical trailer-ship transfer operation, for example, the container is lifted from the trailer chassis and placed either on deck or in the hold of the ship. Special fittings permit these containers to be stacked six-high in ship holds (and on shoreside terminals). On-deck carriage of containers may call for stacking two-high under a third dry cargo van. The corner posts must be designed to carry the floor load and added vertical load applied by lashing gear. The American National Standards Institute (ANSI) and International Standards Organization (ISO) have developed standards for refrigerated containers for intermodal use (ANSI 1990, ISO 1988a).

The preparation of this chapter is assigned to TC 10.6, Transport Refrigeration.

Sizes are designated by the TEU (20 ft equivalent unit). For example, 2 TEU = 40 ft long.

For long-haul land transportation, either highway or rail, each vehicle body or container requires an independent means for refrigeration. Containers interchanged between marine and land use can be refrigerated with independent systems, or from the ship's central system at sea and with clip-on systems on land. Two or more containers are sometimes connected end-to-end to form a single unit for long-haul operation and separated for local delivery.

Some railroad yards transfer trailer bodies from the trailer chassis to gondola or flat cars in a manner similar to those for the trailer-to-ship operation. Various container transfer machines are used to move, stack, or load containers. Finger-lift, pincer-lift, or cable-lift systems each apply characteristic stresses that must be considered in structural design. Many truck and rail carriers apply a standard piggyback technique, *i.e.*, the trailer is driven or lifted onto a railroad flat car and attached to the car for haul by rail.

Trucks

Refrigerated trucks are used primarily for short-haul wholesale delivery within a population center or between population centers. Body styles have been developed to suit the character and distribution needs for perishable products. Trucks and trailers may contain more than one compartment held at different temperatures. Trucks are used in operations that may require more door openings than trailers. Penney and Phillips (1967) measured the air exchange and cooling load caused by door openings of typical refrigerated trucks.

Various devices such as power lift gates and conveyors, which facilitate loading and unloading, are built into the body. Trucks for retail delivery may be furnished for either walk-in or reach-in service. Wheeled racks, which can be rolled into place in the truck, are used to expedite loading.

BODY DESIGN AND CONSTRUCTION

An insulated body, when matched with a suitable refrigeration system, should provide economic temperature control for the commodity being transported or stored. The necessary features are determined by the type of service intended. Trucks may have doors on the two sides rather than at the rear, but most trailers and containers have full opening rear doors large enough to permit forklift trucks to be driven into the vehicle. Many trailers are also equipped with a curbside door. Meat rails are provided if the vehicle is to be used for hanging meat. Temporary or permanent discharge ducts from the refrigerating unit can be used to improve air movement through the cargo.

Insulation

The goal of body builders is to construct a lightweight body having sufficient strength in which the insulation will stay dry and maintain its original insulating value. Insulation should have moderate cost, low density, low thermal conductivity, low moisture permeability and retention, ease of application, uniformity, resistance to breakdown at temperature extremes, and fire resistance. The insulation should resist cracking, crumbling, shifting, and packing from the shock, vibration, and flexing of the body structure.

Bodies for low-temperature use (about 0°F) have been insulated with about 6 in. of an insulating material having a k-factor lower than 0.3 Btu·in/h·ft^2·°F. Legal requirements and other factors limit the outside width, height, and length of vehicles. Therefore, the designer must establish an insulation thickness to obtain optimum cargo space and operating performance. For example, increasing insulation thickness from 3 to 4 in. in a 40-ft long trailer decreases cargo space by 100 ft^3, or 5%. Decreased insulation thickness, however, requires an increase in refrigerating or heating capacity. In determining the optimum insulation thickness, the transport body and its refrigeration unit must be considered as one system.

Although glass fiber, expanded polystyrene, and other materials continue to be used as insulation materials, urethane foam now predominates. Such insulation may have a k-factor of 0.18 or lower, even after aging, which permits thinner construction. Loose fill material that settles during use should not be used. Batts or sheets of semirigid materials are usually compressed slightly as a precaution against settling. Chapter 22 of the 1993 ASHRAE *Handbook—Fundamentals* lists characteristics of most available materials.

The heat transferred from outside to inside the vehicle, excluding heat that enters when the door is open for loading or unloading cargo, includes heat transferred (1) through insulation, (2) through structural members, and (3) by air and moisture leakage. Air leakage can affect overall heat transfer significantly, so some buyers specify a maximum rate of air leakage with a given air pressure imposed in the cargo space.

Floor Insulation. Floor loads are frequently supported on rigid insulating material to eliminate floor beams. Most truck, trailer, and container floors must support forklift trucks. Some medium- or high-temperature trailers are built without any insulation in the floor, but engine waste heat and radiation from hot road surfaces can raise the temperature of the exposed undersurface as much as 20°F above the ambient air temperature.

Floor surfaces may be of wood or metal. If wood floors are furnished, they should be tongue-and-groove treated lumber, well-sealed against water penetration. Metal floors should be watertight, and if separate floor racks are not furnished, the floor surface should allow adequate air movement under the load. Several formed or extruded floor surfaces are available, but not all of these permit enough air circulation under the load.

Galvanized or treated steel and aluminum are used widely for floors in ice cream trucks, while aluminum is most often used for trailer floors. Corkboard, expanded polystyrene, and polyurethane foam are used for floor insulation. A metal skirt reaching at least 6 in. up the walls should be bonded to the floor so that water running down the wall or collecting on the floor will not enter the insulation. Drains, if used, must be self-closing. Plywood, aluminum, other metals and certain plastics are used for interior wall surfaces. Glass-reinforced plastic materials are being used for both interior and exterior surfaces.

Reducing Heat Transmission

Often the buyer specifies the maximum allowable heat transfer rate, usually at 100°F and 50% rh outside and 0°F inside air temperature. Bodies for low temperature (0 to −20 °F) typically have 3 to 4 in. of polyurethane insulation. For temperatures above feezing, 1 to 2.5 in. of polyurethane is applied. At the option of the customer, the insulation thickness is usually nonuniform on the different surfaces, with the front wall generally thicker for structural reasons. For multistop delivery operations, trailers and trucks are sometimes provided with roll-up insulated doors.

Moisture Penetration

All exterior surfaces of an insulated body must be made as airtight and water vaportight as possible. Water vapor will pass through any opening or nonvaporproof barrier in the outer shell and will condense in the insulation itself or in cavities in the insulation space. Some ice cream truck bodies have increased more than 500 lb in weight over a period of use, and large semitrailers in frozen food service have gained up to 1500 lb. Since the k-factor of wet insulation may be much higher than its original value, either the required low temperature may not be obtained or a much greater load is placed on the refrigerating equipment with resultant increased operating and maintenance costs. Use of closed-cell insulation in place of fibrous types minimizes moisture buildup problems.

Air Leakage

Eby and Collister (1955) pointed out the serious heat gain penalty imposed by air entering the insulation space through cracks in the front of the vehicle caused by the motion of the vehicle. They showed that this driving force is 1.21 in. of water at 50 mph, and that a 6-ft length of unsealed seam can permit 1440 ft^3/h of air to enter. At conditions of 100°F ambient temperature at 50% rh and a trailer temperature of 0°F, the heat gain caused by this amount of infiltration air could be 4600 Btu/h, assuming that this leakage air left the vehicle saturated at 0°F.

Refrigerated vehicles leak air even when they are stationary, probably because of the stack effect of the temperature difference between inside and outside. This driving force for air infiltration, for a body 8 ft high and a temperature difference of 100°F, is about 0.030 in. of water (Phillips *et al.* 1960). Openings with an aggregate area of 1 in^2 each at the top and bottom of the exterior skin would permit an air leakage of about 120 ft^3/h, if they function as thin plate orifices. For a nominal 35-ft trailer, observed leakage rates of more than 700 ft^3/h and an accumulation of nearly 1 lb/h of ice at ambient laboratory conditions of 100°F at 50% rh and 0°F trailer temperature illustrate the value of improving the vaportightness of the exterior shell.

The need for better sealing is even more significant at road speeds. For example, if a 1-in^2 opening were located where it was subjected to 1.21 in. of water ram air pressure at 50 mph, approximately 1150 ft^3/h of air could be driven into the insulation space of the vehicle. At ambient temperature and humidity conditions of 100°F and 50%, the heat gain from this leakage would be about 3800 Btu/h, if the air was cooled to the trailer temperature of 0°F.

Road tests of nominal 35-ft commercial trailers by Phillips *et al.* (1960) have shown air leakage rates as high as 1590 ft^3/h at 50 mph. Figure 1 shows the heat gain resulting from air leaking into a vehicle with an interior temperature of 0°F, at various ambient temperatures and relative humidities.

When moist air moves through the insulated space, the water vapor left behind causes trouble in several ways:

- Increased heat gain caused by the latent heat of vaporization and fusion
- Loss of payload because of gain in mass
- Corrosion
- Increased heat gain through insulation
- Coating of coils with frost resulting in loss of refrigerating effect
- More frequent defrosting
- Physical damage to insulation
- Rotting of wood members
- Odor

Additional sealing of the metal skin of a truck or trailer is required to make it leakproof. Three methods are popular: (1) use of foamed-in-place plastic insulations; (2) lining the inside of the exterior skin with a nonpermeable vapor barrier, such as aluminum foil coated with a plastic binder, which can be sealed at the

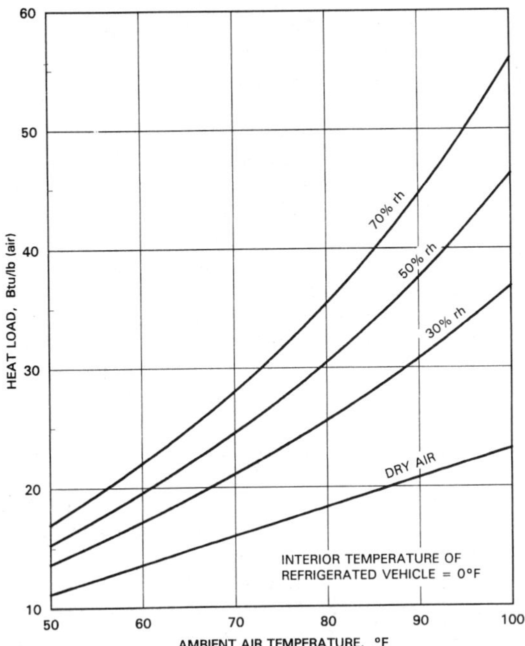

Fig. 1 Heat Load from Air Leakage

joints; or (3) coating the interior surface of the exterior skin with some type of vapor-sealing compound such as neoprene. The vapor seal must not be destroyed where wiring, piping, or frames penetrate the barrier.

Typical trailer specifications call for air leakage of less than 2 cfm at a negative pressure of 0.5 in. of water in the trailer. The interior skin need not, and probably should not, be vaportight; it must be water resistant so that cleaning does not readily soak the insulating material, however. All doors must fit properly and be gasketed to reduce air and heat leakage. Hardware for these doors must be substantial enough to withstand severe conditions.

The interior walls of trailers, which may be used in vacuum cooling of fresh produce, must be vented, preferably into the cargo space, to prevent pressure damage to the insulated spaces during the vacuum cooling operation.

Insulated bodies with one-piece molded plastic exterior shells are now being produced. Types and sizes range from small retail meat trucks to maximum legal length trailers and containers. High impact, thermal resistance, and good vapor-sealing characteristics are claimed for this type of construction.

Air Circulation

Inadequate air distribution is probably the principal cause of improper refrigeration of cargo during transport. Satisfactory product temperatures will not be maintained unless the load is surrounded by proper air or surface temperatures. Either forced or gravity circulation from plate or pipe coils maintains the product temperature. Forced-air units typically discharge air over the stacked cargo either directly from the unit or through ducts directing the air toward the rear of the cargo space.

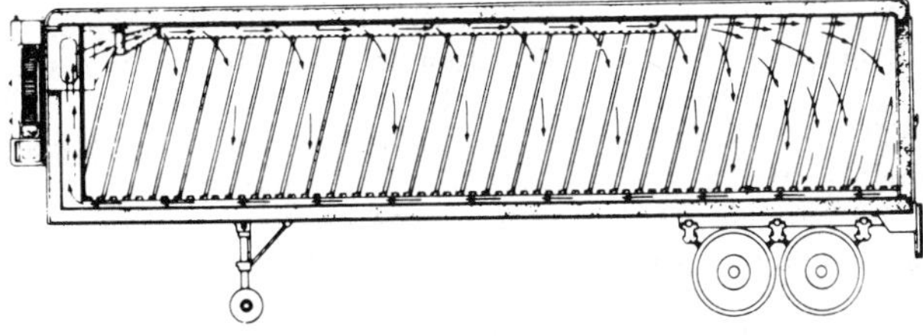

Note false bulkhead which forms return-air plenum taking air from floor.

Fig. 2 Section of Trailer Showing Air Circulation

Products that are to be cooled in transit, or loads that produce heat, must be arranged so that refrigerated air can move not only around the load but through it as well. Nonheat-producing loads, placed in the vehicle at the proper temperature, need only be blanketed with air at the points where heat could enter the product. Most systems using fans can move an adequate air supply over the top of the load. When light density loads approach the ceiling, or when cross bows interfere with the throw of air, supply ducts, either permanent or temporary, facilitate the delivery of air to the rear of the vehicle.

Air channels must be provided on the side walls and inside of the rear doors. Attention must be paid to the movement of air along the floor under the load. Many vehicles have formed metal floors, which allow air to flow the length of the vehicle. Figure 2 shows a trailer with a supply duct and a return air plenum formed by a false front bulkhead extending to the floor racks. Wall strips and floor racks provide channels for airflow.

AUXILIARY EQUIPMENT

Heaters

Most trailer and truck refrigeration units with independent power are designed with a means for heating the cargo space. An augmented hot-gas cycle, in which the evaporator fan then circulates the heated air, is frequently used. Heaters that burn alcohol, kerosene, butane, propane, or charcoal are also used. Because of potential hazards from open flame burners, such heaters must conform to applicable safety regulations.

Modified Atmosphere

Many systems have been developed to control the makeup of the interior environment of the refrigerated vehicle; most limit the amount of oxygen in produce-hauling vehicles and containers. In general, because the time for overland transport is so short, the quality of the products carried under low oxygen is not significantly different from similar products that are adequately cooled and held at proper storage temperature. Some exceptions are claimed for leafy vegetables. A high concentration of carbon dioxide is deleterious to some products, but it has proven beneficial for mold control in others, such as strawberries. Because of the longer transit time, marine shipments should benefit from controlled or modified atmosphere during shipment.

Thermometers

Refrigerated vehicles usually have an indicating thermometer to monitor cargo space temperature. More complex instruments are sometimes used to indicate, control, or perhaps record the cargo space temperature.

TYPES OF REFRIGERATING SYSTEMS

Ventilation

Control of temperature in an insulated vehicle by ventilation with outside air is the least dependable of any system. Obviously, it is limited to areas or times of the year when outdoor conditions are appropriate. Ventilation is usually accomplished by leaving small doors open at the front and rear of the vehicle. Pressure caused by forward motion of the vehicle and the adjustment of doors by the operator control the airflow. Ventilation can be used with mechanical systems to help remove field heat or reduce the concentration of ripening gases generated by a commodity.

Product Subcooling

Some refrigerated commodities are transported without providing refrigerating equipment on the truck or trailer by using the heat-absorbing capacity of the product itself. Milk, for example, has been transported in tank trucks and trailers, some as large as 5300 gal capacity. Orange juice and many hot liquids are also transported in this way.

For example, 5300 gal of milk will absorb 42,380 Btu/°F temperature rise. If an average tank has a heat leakage of 6500 Btu/h during the trip, the milk could be transported 1600 miles (40 h at an average speed of 40 mph) with a temperature rise of less than 6.5 °F and without any refrigeration other than the heat-absorbing capacity of the milk. Orange juice shipped in bulk from Florida to New York, a distance of nearly 1400 miles, has warmed less than 9 °F with no refrigeration other than the product itself. Product quality and other factors must be considered, however, before relying on this method for shipments of perishable products.

Water Ice

This means for transport cooling has largely been replaced by other methods. Small amounts of ice are still placed on top of loads to supplement mechanical refrigeration in the removal of field heat.

Dry Ice (Carbon Dioxide)

In the past, dry ice was used to maintain frozen commodities in delivery trucks. Currently it is used for shipping small quantities of specialty items or items requiring very low shipping temperatures. Dry ice sublimes at −109 °F and absorbs about 240 Btu/lb in the process.

Liquid Nitrogen or Liquid Carbon Dioxide Spray

Liquid nitrogen has a boiling point of −320 °F and, with superheating to 0 °F, has a total refrigerating effect of about 160 Btu/lb. Liquid carbon dioxide, with a boiling point of −109 °F and superheating to 0 °F, has a net refrigerating effect of about 120 Btu/lb.

For road operation, the refrigerant is carried in storage vessels located either in or out of the refrigerated space. To eliminate the need for high-pressure nitrogen tanks, the storage vessels are insulated and equipped with relief valves set for about 25 psig and vented into the cargo space. Liquid carbon dioxide must be stored at pressures in excess of 60.4 psig and at temperatures below 87.8 °F. For cooling cargo after loading, some terminals use liquid refrigerant sprays fed from dock-mounted storage tanks. In this operation, the distributing tube can be introduced through a partially open door. Caution must be taken to prevent refrigerant from being released when personnel are in the cargo space because of possible harm from freezing or the risk of suffocation. Liquid carbon dioxide systems must be designed to prevent nozzle blockage by snow formation.

Guilfoy (1973) describes a liquid nitrogen clip-on unit for land refrigeration of intermodal containers. During shipboard operation, the containers are refrigerated either by immersion in a temperature-regulated hold or by connecting ducts to a central system.

Eutectic Plates for Holdover

Station Charging. Some retail service trucks use a holdover plate arrangement to maintain low temperatures. A holdover plate consists of a coil for the primary refrigerant mounted inside a thin tank filled with a eutectic solution with a freezing temperature sufficiently low to meet the required conditions. Holdover or eutectic plates vary in size from 18 to 36 in. wide, 40 to 120 in. long, and 1 to 3 in. thick. The refrigerating capacity and weight for a given size plate varies with the thickness. For example, a 30 by 66 in. plate of 2.63 in. thickness has a refrigerating capacity of about 17,000 Btu (at −8 to −9 °F eutectic temperature) and weighs about 300 lb, while the same size plate of 1.5 in. thickness has a capacity of about 9000 Btu and weighs about 225 lb. Standard eutectic plates operating at temperatures of −59, −29, −14, −12, −9, −8, −6, 18, 23, and 27 °F are available.

Several of these plates are mounted on the walls, ceiling, or both, or are used as shelves or compartment dividers in the vehicle to provide the required refrigerating capacity while the vehicle is away from the garage or station where the plates are refrozen. When the truck is garaged, flexible connections from the stationary refrigerating plant are attached to connecting devices provided on the truck. A charging island, where several trucks can be connected at one time, can be used for fleet operation. Any refrigerant can be used—the most common being Refrigerant 717 or, more recently, Refrigerant 22. Expansion valves may be mounted on the truck or on the garage wall, and various types of automatic-closure quick connecting devices are available for attaching the flexible lines. In plants using a chilled brine circuit, the plates may be designed to circulate brine through the plates to freeze the eutectic solution.

To provide proper cooling capacity, the plates must be mounted so that air can move freely on both sides, usually 1.5 to 2 in. from walls with top edges 6 to 8 in. from the ceiling. On ceilings, plates should be sloped at least 1 in. per foot across the shortest dimension and should be at least 2 in. below the ceiling at the higher edge. Hangers must be securely fastened to walls, preferably by fastenings incorporated in the studs. Fastenings must not penetrate the outside wall surface. A drip trough may be desirable under each plate, although many vehicles are defrosted by scraping the frost without warming the plates. To protect plates from damage, guard rails should be placed at least 1 in. away from plates, if necessary. Eutectic plate systems usually circulate air by gravity but some have fans with ducts and dampers for temperature control.

Vehicle-Mounted Condensing Unit. Some trucks are equipped with condensing units powered by electric motors. These units operate on plug-in power to refreeze the eutectic plates when the vehicle is stopped. The comparative length of the time for road operation and for refreezing are factors in determining the required size of the condensing unit and the amount of eutectic solution (Guilfoy and Mongelli 1971).

Variations of truck-mounted condensing unit systems include those with a separate engine drive, those with a compressor driven by the truck engine, and those with a nose mount or chassis mount condensing unit.

Plastic Plates or Tubes. Several European manufacturers have developed holdover systems that contain the eutectic solution in either rectangular or round cross-sectional tubes. The plastic parts replace relatively heavy steel containers.

Mechanical Refrigeration

Independent Engine or Electric Motor Driven. Many styles of independent engine and/or electric motor driven mechanical refrigerating units are available and constitute the most popular application for both trucks and trailers. The one-piece, plug-type, self-contained unit mounts in an opening in the front wall of the vehicle. The condensing section on the outside and the evaporator section on the inside are separated by an insulated plug that attaches to the vehicle wall and supports the various parts of the refrigerating unit. For trailers, some of the plug units are tall and slender enough to mount between the tractor and trailer; others are short and deep and extend over the tractor roof. For intermodal containers, the most widely used refrigerating unit is electric motor-driven, usually hermetic, with or without a companion engine-driven generator. The mounting plug for the unit essentially forms the entire end wall of the container.

The refrigerating capacity of self-contained, independent engine-driven plug-type trailer units in an ambient temperature of 100 °F range as follows: 20,000 to 48,000 Btu/h at 35 °F trailer temperature; 12,000 to 34,000 Btu/h at 0 °F trailer temperature; and 6000 to 25,000 Btu/h at −20 °F trailer temperature. These units weigh from 800 to 1700 lb. Similar self-contained units, which weigh from 500 to 1000 lb and have a lower cooling capacity, are available for use on trucks.

Some units have two belt-driven compressors, others have one compressor that is either belt- or direct-driven by the auxiliary engine. These auxiliary engines may operate on gasoline from the truck tank, from a separate gasoline or diesel fuel supply on trailers, or on liquefied petroleum gas. In addition to the engine, many refrigeration units are equipped with an electric motor for standby operation. Most units can be used for heating and can be defrosted automatically by the hot-gas method. Most units are thermostatically controlled, starting and stopping or reducing speed as refrigeration need requires. Condenser and evaporator fans are driven by the unit engine, and tight-fitting dampers close off the evaporator airflow to permit defrosting without unduly warming the vehicle interior. The one-piece construction facilitates unit replacement in the event of breakdown or removal for maintenance.

Two-piece units are also used for both trucks and trailers. One style uses an engine-driven generator mounted under the trailer to provide electric power to operate one or two independent refrigerating systems contained in a plug-type unit mounted in the front wall. The plug-type unit consists of a self-contained, hermetic compressor system, and the housing under the trailer contains an engine-driven alternator. The compressor can be operated from house current during standby. Another style uses an evaporator section, usually with electric motor-driven fans, mounted in the nose of the vehicle, and an engine-driven condensing unit installed under the body. In this case, permanent piping is run between the condensing unit and evaporator, and a generator on the engine provides electric power to operate the evaporator fan. Some units are designed to cycle on thermostatic control, others operate continuously but change the speed of the engine in response to the need for refrigeration within the body. Some units are designed for automatic defrosting; others are defrosted by manual operation of hot-gas valving or electric heaters.

Power Take-Off from Vehicle Engine or Transmission. For most cases, this type of refrigerating system is limited to trucks, although an axle-driven unit has been devised for trailers. Several types of power take-off means are available. One manufacturer has an alternator, belt-driven from the engine crankshaft, which produces a regulated alternating current voltage that is rectified to drive direct current motors for the compressor and fans in the body. Temperature control is accomplished by cycling the compressor. Refrigeration is produced only when the truck engine is in operation; full refrigerating capacity is claimed for all engine speeds above 500 rpm.

One manufacturer uses an all-electric, self-contained hermetic unit to provide refrigeration for medium-temperature truck bodies. A specially regulated alternator produces an alternating current power supply of variable voltage and frequency at essentially constant current for engine speed from idle operation to 4000 rpm. The alternator is mounted on the truck engine and belt-driven from the engine crankshaft pulley. Standby operation is provided by electrical connection to regularly available alternating current power. A two-speed centrifugally operated gear shift is available to provide more suitable generator speed if the truck engine speed range is too wide.

Power take-off systems of the mechanical type sometimes have the condenser mounted in front of the truck radiator, with flexible tubing to connect to the unit in the cargo space. The most common system uses an engine-mounted compressor, which is belt-driven from the crankshaft pulley. One manufacturer has a flexible shaft drive for the unit compressor and condenser fan, with a belt-driven electric clutch at the forward end of the flexible shaft. In this system, of course, the compressor speed is a function of engine speed.

A variation of the power transmission mechanism for use with a belt-driven flexible shaft includes a standby electric motor to drive the compressor. The flexible shaft drives an electric clutch attached to a universal joint. An overriding clutch idles the electric motor when the compressor is being driven by the flexible shaft. When the electric motor operates, the electric clutch is disengaged. The flexible shaft is driven by a belt from the engine crankshaft, and the electric clutch is used for temperature control.

Hydraulic power transmission is a special case of power take-off from the truck transmission or engine. The hydraulic pump is either mounted in the truck engine compartment and belt-driven from the engine crankshaft or is direct-driven from power take-off in the truck transmission. Hydraulic drive systems may be designed to provide a near constant output speed from a variable input speed.

MECHANICAL REFRIGERATION COMPONENTS

The major components of a vapor compression transport are power source, compressor, condenser, evaporator system, and controls.

Power Source

Refrigeration systems for short trucks that require moderate cooling typically use the truck engine for power. A belt on the engine crankshaft drives the compressor, and the alternator powers the fans. An optional drive motor can be plugged into power mains. If the cooling demand is higher (as for frozen food, a large truck body, or frequent door openings), the system likely is powered by a small engine (usually diesel with a 25 to 35 in^3 displacement).

Eutectic systems with 5 to 7.5 hp motors may also meet the needs of this application.

Larger diesel engines, up to about 80 in^3, provide power for the largest refrigeration units for large trucks (up to 28 ft) having a high service load or carrying frozen products. Most truck units driven by a self-contained engine are available with an optional electric motor drive.

Trailer refrigeration units are typically driven by diesel engines with 100 to 140 in^3 displacement. Optional electric standby may be available.

Compressor

Small compressors driven by the truck engine are available from the automobile air-conditioning industry. These include in-line or V-piston, rotary vane, swash-plate piston, and scroll designs. Larger truck compressors generally have in-line or V-piston configurations with two to six cylinders. Trailer units use piston-type compressors with up to 40 in^3 displacement. Aluminum is used on some designs to reduce weight.

Condensers

Truck engine-powered condensers may be mounted (1) in front of the truck engine radiator, (2) above the truck cab with a battery powered fan, or (3) below the truck body. The condenser and fans for larger truck and trailer units are packaged with the compressor, engine, and other components. Condensers are designed for forced convection cooling and generally have aluminum fins and copper tubes.

The high-pressure side of any refrigeration system that operates when the vehicle is in motion must include the following safeguards not normally required of a stationary unit:

- Design all components to withstand shock loads and vibration. Avoid short, rigid lines, which are subject to cold working and are likely to fail.
- Individually anchor heavy objects such as driers, valves, and sight glasses.
- Consider water and dirt in the design.

- Protect all elements from corrosion, especially components for intermodal containers, which are exposed to sea water.
- Protect bearings of all exposed parts.
- Seal shaft seals, particularly against dirt or abrasive material penetration.
- Shield electric motors from splashed or air-carried water.

Evaporator Systems

The evaporator arrangement for eutectic plate systems is described elsewhere in the Eutectic Plates for Holdover section. Other transport refrigeration systems use plate-fin and tube evaporator coils; fans or blowers provide forced convection airflow. During the defrost cycle, airflow to the cargo is stopped either by stopping the fans or by closing dampers. The heat required for defrost is provided by hot refrigerant gas or electric heaters.

Control Requirements and Devices

Frozen and fresh produce must be maintained close to its optimum temperature for best retention of food value and appearance. The quality of some frozen foods is adequately maintained at 0 °F, while others, such as ice cream specialty products, may require a temperature of −20 °F. Compartmentalized trailers or trucks with two or three temperatures are often used for wholesale delivery of frozen, fresh, or dry commodities in the same vehicle. Certain pharmaceuticals, films, or electronic devices may require humidity control combined with temperature control.

Chapter 25 presents a table of preferred temperatures for long-term storage of perishable products. USDA *Agricultural Handbooks* 669 (Ashby *et al.* 1987) and 668 (McGregor 1987) discuss many aspects of maintenance of quality of fruits, vegetables, plants, and flowers during transport. The Association of Food and Drug Officials has prepared a model code that outlines storage and transportation temperature limits for frozen foods.

The rigid requirements for maintaining product temperature dictate a suitably designed and maintained truck or trailer with provisions for perimeter airflow, a satisfactory loading pattern, an adequate supply of cooled or heated air from the unit within an acceptable temperature differential, and a precise thermostat with a proper sensor position.

Where the control sensor is located in the return airstream of a refrigerated load, all points upstream of the sensor (*i.e.*, all of the load space) are subject to temperatures lower than the sensor. In this case, the possibility exists for overcooling part of the load. Conversely, if the system is heating a load, all of the space is subject to temperatures higher than the sensor.

In general, little difficulty is experienced with the sensor located in the return air for frozen loads or precooled fresh produce; but the thermostat should be set at a higher set-point temperature to avoid overchilling products requiring cooling in transit. In all cases with return air sensing, the thermostat should be set 4 or 5 °F above the product chill or freeze point. For a product with a high specific heat cooling load, the thermostat setting should be increased to 8 to 10 °F above the chill or freeze point. If the control sensor is positioned in the supply airstream, the temperature relationship between sensor and the return air is reversed.

The use of various capacity reduction or modulation devices tends to reduce the air temperature differences in the cargo space.

Control Devices. Electronic thermostats used for load temperature control on container units operate over a wide control range with a high degree of accuracy. Controls are multistep and/or modulating for both heating and cooling modes. Auxiliary features include temperature indication, multipoint data logging, and out of range control. Recent developments plan communication by satellite to locate a load or determine the load temperature.

Other operational or safety controls include oil safety devices for engines or compressors, and controls for automatic engine starting and automatic defrost initiation and termination.

Many mechanical refrigeration units are designed to operate from −20 to 70 °F in the cargo space and to pulldown from a higher temperature every trip. Various means are used to limit the power required during the transient pulldown period. Such devices include compressor crankcase pressure-limiting controls, compressor unloading, and thermostatic expansion valves with outlet pressure control.

CALCULATION OF COOLING LOADS

Phillips and Penney (1967) describe a method of rating refrigerated trucks. In an earlier study, Phillips *et al.* (1960) describe a rating method for refrigerated trailers. The Truck-Trailer Manufacturers Association developed a method for rating heat transmission in refrigerated vehicles (TTMA 1989). The International Standards Organization has prepared thermal test standards (ISO 1988b).

Although suitable standards are not in general use for rating the performance of all types of refrigerating systems used for trucks and trailers, the Air-Conditioning and Refrigeration Institute has developed a standard (ARI 1983) dealing with speed-governed and variable-speed transport refrigeration units using forced circulation air coolers. Until standards are available for all types of systems, the comparison of the capacity of refrigerating units should be made on the basis of net refrigerating capacity at specified conditions of air temperature at the condenser and evaporator inlet, for example, at 100 °F ambient temperature and 0 °F trailer interior temperature.

The Refrigerated Transport Foundation developed a classification system based on the combined thermal performance of both trailers and refrigeration units (CGTFL 1988). The trailer thermal rating is based on a practice recommended by the Truck Trailers Manufacturers Association (TTMA 1989). The refrigeration unit cooling capacity is determined by ARI *Standard* 1110 (ARI 1983). By this classification system, the combined trailer and unit can be classified in one or more of four temperature ranges:

DF for deep frozen, −20 °F
F for frozen, 0 °F
C35 for chilled temperature, 35 °F
C65 for chilled temperature, 65 °F

To qualify for the C35 and C65 classification, sufficient heat must be available to achieve the classification at an ambient of 0 °F. The classification is voluntary for all parties—the trailer manufacturer, the unit manufacturer, and the carrier or owner. Trailers that are classified display a permanent plate or decal that lists (1) temperature range(s), (2) trailer U-factor in Btu/h · °F, (3) data on area provided for airflow around the periphery of the loading space, (4) type of bulkhead, (5) presence or not of air distribution ducts on the ceiling, (6) amount of excess unit cooling capacity beyond the steady-state requirements at the coldest range chosen while at 100 °F ambient, and (7) airflow available from the unit.

Insulated trailers range from 36 to 53 ft in length, and steady-state heat gain values (U-factor) range from 80 to 240 Btu/h · °F. The refrigeration system must have an additional cooling capacity beyond that needed for steady-state heat transfer to satisfy deterioration factors, solar radiation, air infiltration, door openings, and product loads. Extra capacity also reduces the pulldown time prior to loading, which improves equipment use.

Phillips and Penney (1967) found that solar radiation can increase the cooling load of stationary vehicles more than 20% for several hours during a sunny day. Penney and Phillips (1967) also noted that the cooling load caused by door usage can be five times the body heat transmission when a truck is used in multistop operation.

Manufacturers of refrigeration units for trucks and trailers have devised computer programs or charts to match the unit to the body for the specific operational details planned by the customer. The input for the calculations include body U-factors, losses from door openings, product data factors, ambient conditions, and refrigeration unit capabilities.

Sample Calculation for Trailer Cooling Load

Assume a trailer is loaded with 38,000 lb of peaches at an average pulp temperature of 54 °F. The load will be delivered 72 h later at an average product temperature of 34 °F. Average ambient temperature is 85 °F.

The trailer U-factor is 155 Btu/h · °F as determined either from data on the trailer or estimated from information on insulation properties, size, and condition of the trailer.

From Chapter 25, the specific heat of peaches (above freezing) is 0.91 Btu/lb · °F.

To estimate the heat of respiration, the average temperature of peaches during transit is assumed to be 40 °F. From Chapter 25, the heat of respiration of peaches at 40 °F is an average of 1700 Btu/(24 h) (ton). A more precise calculation of heat of respiration would recognize that it varies with product temperature, which decreases during the pulldown period.

Total heat load	= Product heat load + Heat of respiration + Heat transmitted
Product heat load	= (Specific heat) (Weight) (Temperature change)
	= (0.91) (38,000) (20)
	= 692,000 Btu
Heat of respiration	= (Heat of respiration factor) (Tons) (Days)
	= (1700) (19) (72/24) = 96,900 Btu
Heat transmitted	= (U-factor) (Hours) (Ambient temperature − Thermostat set temperature)
	= (155) (72) (85 − 36) = 547,000 Btu
Total heat load for 72 h	= 692,000 + 96,900 + 547,000
	= 1,336,000 Btu
Average load/h	= 18,600 Btu/h

If ice was used instead of mechanical refrigeration, the amount of ice melted would be about 1,336,000/144 = 9280 lb. Liquid nitrogen absorbs about 175 Btu/lb during evaporation and warming to 32 °F. Thus, if liquid nitrogen were used in this example, the amount of liquid nitrogen required would be 1,336,00/175 = 7630 lb.

The preceding calculation neglects initial cooling requirements for the trailer structure, air in the trailer, and cargo packing materials.

Perishables should be fully cooled prior to loading. However, some cooling is usually required in transit because the product temperature can rise during loading, and, during periods of high production, cooling facilities can be overloaded or insufficient time is permitted to fully cool the center of a product. Cooling during transit can reduce the overall time from harvest to consumption. However, airflow through the load is essential to cool the product in transit successfully. The section on Control Requirements and Devices also discusses product cooling.

REFERENCES

AFDO. Code of recommended practices for the handling of frozen foods. Association of Food and Drug Officials, York, PA.

ANSI. 1990. Requirements for road/rail closed dry van containers. ANSI *Standard*. American National Standards Institute, New York.

ARI. 1983. Mechanical transport refrigeration units. *Standard* 1110-77. Air-Conditioning and Refrigeration Institute, Arlington, VA.

Ashby, H.B., R.T. Hinsch, L.A. Risse, W.G. Kindya, W.L. Craig Jr., and M.T. Turczyn. 1987. Protecting pershable foods during transport by truck. *Agricultural Handbook* 669. U.S. Department of Agriculture, Washington, D.C.

ATA. Summary of size and weight limits. American Trucking Association, Alexandria, VA.

CGTFL. 1988. Refrigerated transportation foundation method classification of controlled temperature vehicles. Recommended practice No. 1-89. California Grape and Tree Fruit League, Fresno, CA.

Eby, C.W. and R.L. Collister. 1955. Insulation in refrigerated transportation body design. *Refrigerating Engineering* (July): 51.

Guilfoy, R.F., Jr. 1973. Refrigeration systems for transporting frozen foods. ASHRAE *Journal* (May):58.

Guilfoy, R.L., Jr. and R.C. Mongelli. 1971. A method for measuring cost and performance of refrigeration systems in local delivery vehicles. ARS *Publication* 52-64. USDA, Agricultural Research Service.

IIR. 1985. Technology advances in refrigerated storage and transport. International Institute of Refrigeration, Paris.

ISO. 1988a. Series I freight containers—external dimensions and ratings, 4th ed. *Standard* 668:1988. International Standards Organization, Geneva.

ISO. 1988b. Series I freight containers—specifications and testing—part 2: thermal containers, 3rd ed. *Standard* 1496-2:1988. International Standards Organization, Geneva.

McGregor, B. 1987. Tropical products transport handbook. *Agriculture Handbook Number* 668. USDA.

Penney, R.W. and C.W. Phillips. 1967. Refrigeration requirements for truck bodies—effects of door usage. Agricultural Research Service Technical Bulletin No. 1375. USDA.

Phillips, C.W., *et al.* 1960. A rating method for refrigerated trailer bodies hauling perishable foods. Marketing Research Report No. 433. Agricultural Marketing Service, USDA.

Phillips, C.W. and R.W. Penney. 1967. Development of a method for testing and rating refrigerated truck bodies. USDA Agricultural Research Service, Technical Bulletin No. 1376.

TTMA. 1989. Method for rating heat transmissions of refrigerated vehicles. Recommended Practice No. 38-89. Truck-Trailer Manufacturers' Association, Washington, D.C.

BIBLIOGRAPHY

ASHRAE. 1971. Refrigeration systems for perishable food delivery vehicles. Symposium Bulletin WA-71-4.

ASHRAE. 1972. Long-haul transportation of respiring perishable commodities in refrigerated containers. Symposium Bulletin NO-72-7.

IIR. 1985. Long distance refrigerated transport—land and sea. International Institute of Refrigeration, Paris.

IIR. 1986. Recommendations for the processing and handling of frozen foods. International Institute of Refrigeration, Paris.

United Nations. 1970. Agreement on the international carriage of perishable foodstuff and the special equipment to be used for such carriage (ATP). Economic Commission for Europe, Inland Transport Committee, United Nations.

CHAPTER 28

RAILROAD REFRIGERATOR CARS

THE development of refrigerator car service by the railroad industry and private refrigerator car lines during the past century has been an important factor in establishing the present nationwide system of distributing perishables. The railroad industry also transports perishables in *piggyback* refrigerated trailers and containers, which are discussed in Chapter 27.

REFRIGERATOR CARS IN THE UNITED STATES AND CANADA

Various types of railroad cars are classified, defined, and given designating letters by the Association of American Railroads (AAR). Refrigerator cars are designated as Class R cars, except those equipped for passenger train service, which fall under Class B (Passenger Equipment Cars), and those designed for transporting solid carbon dioxide, under Class I (Special Cars). The first of the designating letters indicates the car class.

With Class R cars, the second designating letter indicates the type of cooling equipment installed. The letter B indicates that no cooling equipment is provided, C indicates cars using a cryogen, P indicates mechanical refrigeration, and S indicates cars equipped with ice bunkers. The letter L is added to the designating letters of the car equipped with adjustable loading or stowing devices, B to the car designed for use in bulk loading and equipped with interior slope sheets and equipment for loading and unloading, and C to the car equipped with permanently affixed containers. The AAR definitions and letter designations for various types of refrigerator cars are as follows:

RB: Bunkerless refrigerator car with or without ventilating devices and with or without a device for attaching portable heaters. Constructed with insulation in side, ends, floor, and roof to meet maximum requirements of 250 Btu/h·°F for 50-ft cars and 300 Btu/h·°F for 60-ft cars. Effective for cars ordered new after March 1, 1984.

Note: Cars built or rebuilt prior to March 1, 1984, must have been constructed with a minimum of 3 in. insulation in the sides and ends and 3 in. in floor and roof, based on the insulation requirements given in AAR *Standard* S-2010 or a thickness reduced in proportion to the thermal conductivity of the insulation.

RBL: Car similar in construction to an RB-type car, but equipped with adjustable loading or stowing device.

Note: Cars equipped with interior side rails only, built new, rebuilt, or reclassified on and after January 1, 1966, in order to qualify for the RBL designation, shall have a minimum of 4 usable side rails on each car wall, each extending from doorway to approximately 4 ft from the end of a car.

RC: Refrigerator car similar to an RB car, using a cryogen to produce temperatures to transport frozen commodities.

RP: Mechanical refrigerator car equipped with or without means of ventilation and provided with apparatus for furnishing protection against heat and/or cold. Apparatus operated by power other than from the car axle.

RPB: Mechanical refrigerator. Similar to RP-type car, but designed for use in bulk potato or similar type loading as cars are equipped with interior slope sheets and conveyors and/or equipment for mechanical loading and unloading.

RPC: Mechanical refrigerator car similar in design to RP car but equipped with permanently affixed container(s).

RPL: Mechanical refrigerator. Similar to RP but equipped additionally with adjustable loading or stowing device.

RS: Bunker refrigerator car equipped with ice bunkers. Designed primarily for use of chunk ice and with or without means of ventilation.

PROTECTIVE SERVICES

Refrigerator cars are exposed to outdoor air conditions that vary from summer desert temperatures to low winter temperatures; however the desired transit temperature for perishable commodities varies from 0°F or slightly lower for frozen foods to as high as 65°F for certain types of bananas and pears. To handle this problem, the National Perishable Freight Committee has established different classes of protective services. These services, together with rules and regulations and charges, are published in the *Perishable Protective Tariff* (Doean 1983).

Perishable freight refers to any commodity susceptible to deterioration or decay and/or that may be protected by cooling, ventilating, or heating. A handbook giving recommended protective services and loading methods for the proper care of perishable agricultural commodities in refrigerator cars is available (Redit 1969). This publication includes a comprehensive bibliography, an explanation of the Perishable Tariff Rules, and methods of calculating the refrigeration loads involved.

Mechanical Protective Service

Cars equipped with mechanical refrigeration units offer temperature control service, with car interior settings from 0 to 70°F at any ambient temperature. These cars cool and heat to maintain the desired interior temperature (see Figure 1).

Mechanical protective service is available both at the standard tariff rate, which includes operating the mechanical system from point of origin to destination, and in modified forms whereby the unit is not started until the indicated car temperature reaches a specified high or low level, or whereby the unit is operated until the desired temperature is reached, after which it is turned off. One such starting or stopping operation is permitted for each trip, especially with commodities that do not require close temperature control, but do require protection when shipped between points with extreme differences in outdoor temperatures.

The preparation of this chapter is assigned to TC 10.6, Transport Refrigeration.

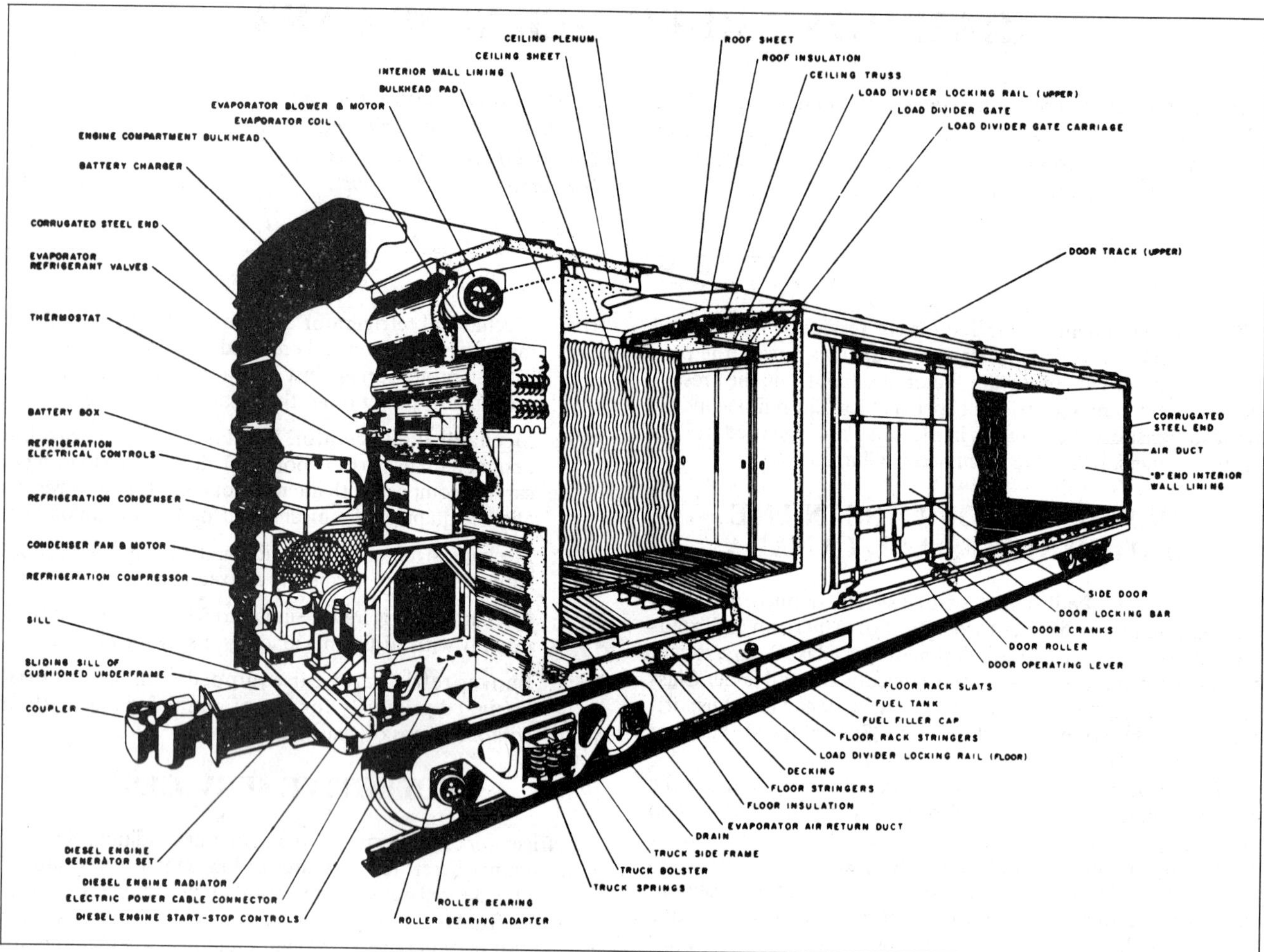

Fig. 1 Typical Refrigeration Components for Mechanical Refrigerator Cars

Refrigeration Services Using Ice

Top or *body* icing often used when shipping leafy and root vegetables, melons, keg or draught beer, and Christmas trees; crushed or snow ice is distributed throughout the body of the car (mostly on top of the load) in direct contact with the lading. Top or body icing at the shipment's origin and the placement of ice in loading containers is usually done by the shipper.

The *Perishable Protective Tariff* permits shippers to cool perishable commodities in refrigerator cars at loading stations during or after the loading process. The methods of cooling shipments in cars prior to transit cooling, including contact or top icing and vacuum cooling, are discussed in Chapter 10.

Heater Services

Mechanical refrigerator cars are equipped to provide both heat and cooling to maintain any desired temperatures from 0 to 70 °F in any ambient temperature. Therefore, no external devices are needed to provide heater service with mechanical refrigerator cars.

Several types of heater service for protection against frost, freezing, or artificial overheating of shipments during the heater season are available (which varies according to heater service and commodity). With *carrier's protective service,* the carrier furnishes whatever protection it considers necessary. With *shipper's protective service,* heaters are furnished and installed by the shipper and are serviced by the carrier as directed by the shipper. With *shipper's specified service,* heaters are furnished, installed, and serviced by carriers as directed by the shipper.

Protection furnished by the carrier after service has been changed in transit from refrigeration or ventilation is known as *modified carrier's protective service.* Protection supplied without request on a shipment originating outside regular heater territory and west of the Mississippi River is known as *voluntary heater service.* Protection using liquid fuel heaters furnished, installed, and serviced by the carrier with the thermostat set as directed by the shipper and ventilators manipulated at the temperature specified by the shipper is *special heater protective service.*

Other Special Services

In addition to the abovementioned protective services, several special services or privileges are provided by the railroad industry for the movement of perishable freight. Shipments of commodities under refrigeration may be stopped in transit and fumigated in the refrigerator car on instructions from the shipper, who may also fumigate the cargo in the cars prior to departure from the loading track. Fumigation is accomplished by the release of certain proprietary gases into the car from pressure tanks.

Sampling or inspection of a shipment in transit may be done by the owner, the owner's representative, or the prospective buyer when authorized by the shipper or owner. When so instructed, a carrier will change, in transit, the type of protective service, the destination of the shipment, or the consignee. About half of all refrigerator car shipments are reconsigned in transit. Stopover privileges are permitted to complete loading or to unload partially, and transit privileges are permitted for rehandling, transferring, or reconditioning.

SOURCES OF REFRIGERATION

A refrigerator car fleet consists of ice bunker cars using water ice as their source of refrigeration and mechanical refrigerator cars, which use mechanical compression machines as a refrigeration source. However, the number of ice bunker cars is rapidly diminishing; they are being replaced almost exclusively by mechanically refrigerated cars, most of which are owned by private companies. Those bunker cars still serviceable are being used either in non-ice bunker services, such as ventilation or body ice, or as insulated box cars.

Mechanical refrigeration was determined to be most feasible by using standard compression refrigeration system components, but with an independent electric generator set as its own source of power. This makes the refrigeration system independent of the car movement. By adopting standard commercial components using standard voltages, these systems can also be plugged into carside power outlets to make refrigeration available during loading and unloading without running the engine.

Mechanical Refrigeration

Refrigeration systems are conventional, and the air circulation systems and capacity controls are designed to provide the humidity control required for handling fresh commodities. Powered with their own fuel supply, they are independent and may be operated anywhere at any time. This offers the advantage of providing maximum cooling as soon as the car is loaded; in conventional ice bunker fan cars, the cooling rate was small until the car moved.

The conventional mechanical car is used for transporting frozen foods and fresh perishables in the range from 0 to 70 °F. The capacity is sufficient to carry frozen foods in any ambient temperature encountered in rail service; therefore, the same system has more than adequate capacity to hold fresh perishables in the 35 °F range, as well as those products at 50 °F or higher. Excess refrigeration at the higher temperature levels is controlled by unloading compressor cylinders, speed variation, or both.

The mechanical car is not effective as a precooler for removing field heat, since the close stacking and tight packaging of commodities in the car prevents the free circulation of air around the individual packages, even though the refrigeration capacity may be available. Therefore, most fresh perishables have the greater portion of their field heat removed before being loaded in these cars; the car is used to obtain the final cooling and desired storage temperature in transit.

Other Sources of Refrigeration

Nitrogen has been used to cool frozen foods to very low temperatures so that the temperature rise for the transit period would allow arrival with temperatures still below 0 °F. Initial temperature was predicated on the expected length of the trip, type of packaging, and heat leak of the car.

In other applications, liquid nitrogen is carried in the vehicle in cryogenic vessels at about 20-psig operating pressure; this pressure delivers the nitrogen to the lading compartment through overhead spray nozzles. The refrigeration rate is controlled by a thermostat and solenoid-operated flow control valve. The principal applications have been with frozen foods and meat.

Because the atmospheric boiling point of liquid nitrogen is −320 °F, overcooling fresh perishables has been a problem. Early systems depended on gas diffusion through the lading space to cool the commodity, but systems now use air-circulating fans to provide more uniform temperatures. These fans are driven by pneumatic motors using the vent gas from the nitrogen storage tanks as a power source.

The principal deterrent to further development of liquid nitrogen systems has been the high cost of the refrigerant. With some applications, such as local delivery of frozen foods, liquid nitrogen systems are less expensive than mechanical refrigeration.

Liquid carbon dioxide has also been tested, and several systems are now commercially available. These may be either the direct-release type or systems using secondary refrigerants flowing in wall or ceiling coils. Application of liquid CO_2 has generally been limited to cars in captive service, such as for meat in regular schedules, since it depends on setting up a supply and distribution system for the liquid refrigerant. Here again, local cost factors determine whether liquid CO_2 will supplant other refrigeration.

MECHANICAL REFRIGERATION EQUIPMENT

With the exception of cars having underslung units, mechanical cars have a louvered engine compartment at one end, which houses most of the mechanical refrigeration equipment. The compartment is separated from the loading space by an insulated bulkhead. The system is usually divided into major subassemblies, including an engine-generator, a compressor-condenser, evaporator coil and blower, a refrigerant control, and an electrical control panel (see Figure 1). Some direct-drive systems have all components combined into a single unit mounted in the machinery compartment on a sliding base for ease of installation, inspection, and replacement. All components are specially selected and mounted because of the vibration and impact forces in freight-car service.

The refrigeration capacity for a mechanical refrigerator car depends primarily on the type of refrigeration service for which the car is intended.

Power and Drive Equipment

Diesel engines have been used almost exclusively to drive the refrigeration systems in mechanical cars. Engines are equipped with shutdown controls to protect against high coolant temperature and loss of lubricating oil pressure.

The generators used with electric-drive systems are connected directly to the engine and provide 220 V, three-phase, 60-Hz current with a rated output from 12 to 25 kW. A majority of the mechanical refrigerator cars have a 20-kW generator connected directly to a 34-hp, two-cycle diesel.

Two-speed engine operation is used to reduce engine maintenance and fuel costs. Full engine speed is used for pulldown and, after the car temperature is within a few degrees of the temperature control setting, low-speed operation is maintained. With electric drive systems, 60-Hz operation is obtained at full speed, and 40-Hz operation at low speed. Generator characteristics are designed so that the voltage varies directly as the frequency; thus the generator, battery charging system, electric control, temperature controls, relays, starters, and so forth, all function at both 40 and 60 Hz. Being inductive devices, their impedance varies as the frequency, and they draw approximately the same current at both speeds.

Compressors range from two-cylinder, belt-driven types through six-cylinder, semihermetic types and include multiple cylinder open types and hermetic types. Many compressors are equipped with unloaders on all or some of the cylinders to obtain fully unloaded starting conditions. Cylinder unloaders are also used for refrigerant capacity control.

Condensers are of standard air-cooled design, usually constructed of copper tubes and aluminum fins. Air is drawn through louvers in the side of the car across the condenser and discharged out the other side of the car after passing over the engine. Condenser fans are the propeller type, mounted directly on the motor shaft, and delivering about 8000 cfm with a 2-hp, 220-V, three-phase induction motor.

Refrigeration System Components

Evaporators are made of copper tube and aluminum fins and have reinforced tube sheets capable of transmitting the forces from the lading into the bulkhead on which they are mounted in the event of load shifting. Electric tubular heating elements used for car heating and defrosting are placed in the coil to effect good heat distribution for proper defrosting, although some systems place the heater assembly beneath the evaporator coil for easy removal. Electric heating elements are also placed in the evaporator drain pan to assist in the removal of meltage formed during defrosting. These are energized only during the defrost cycle.

Some systems use hot gas in lieu of electric heaters for heating and defrosting. Various arrangements use either full reverse cycle operation or modified versions in which the principal refrigerant charge is trapped in the condenser and only gas flows between compressor and evaporator.

Refrigeration systems are generally of the direct-expansion type with a thermostatic expansion valve controlling refrigerant flow. In addition, the refrigerant systems contain a solenoid liquid stop valve, filter, dryer, and suction-liquid heat exchanger. The refrigerant controls are usually clustered on a panel and are either mounted on the condensing assembly or on the adjacent bulkhead wall. Connections between these assemblies and the evaporator are made with flexible hoses having union connections to facilitate unit replacements. System components are equipped with shutoff valves enabling the removal and replacement of components without complete loss of refrigerant charge.

Vibration and shock conditions are severe for refrigerant piping and require flexible vibration isolators between compressors and condensers. In addition, all piping is thoroughly secured with braces and clamps to prevent movement or contact with adjacent members. Condensing assemblies are, in turn, shock mounted to the car floor, and all parts of the system are designed for 15-g loading in the direction of travel, 5-g in the vertical direction, and 2-g laterally.

The refrigeration system starts automatically on starting the engine generator set. When the proper voltage is reached, the electrical control circuit is actuated and the condenser and evaporator fans start immediately. The time delay relay is also actuated and, after an approximate 60-s delay, starts the compressor motor. This delay permits the engine to settle down and clear itself after starting before the major load of the compressor is applied.

Electric-drive mechanical refrigeration systems are equipped with a plug-receptacle, which makes it possible to shut the engine down and operate the system from a standby electric power source. Standby operation has been used while correcting engine difficulties and also when the engine noise or exhaust fumes are objectionable during loading and unloading. Fuel for the internal combustion engine is usually carried in fuel tanks located under the car floor.

Electrical controls are in a dustproof panel usually mounted on the bulkhead adjacent to the condensing assembly. This panel contains the motor starters and other electrical devices required by the particular system, such as timers, annunciator lights, and switches. Annunciator lights are generally mounted on the side of the panel visible from the doorway at the side of the car; they indicate whether the system is in a *cooling* or *heating* mode, and, in some cases, an additional light indicates a defrost cycle.

Temperature control is accomplished using a thermostat with a sensing element located in the return air flue to the evaporator coil. The thermostat has multiple switches acting in sequence on a temperature fall to unload the compressor, reduce speed on multispeed systems, stop the compressor, and, if the temperature continues to fall, switch to heat.

Figure 2 shows a typical thermostat operation, with the switch from high cooling to low cooling at 2 °F above set point and the system shutoff just below the thermostat set point. If the temperature continues to drop, the unit will switch to heating at 2 °F below set point. If the low cooling range is insufficient to hold the car temperature, it will rise and switch the unit back to high cooling at 2 to 3 °F above the set point.

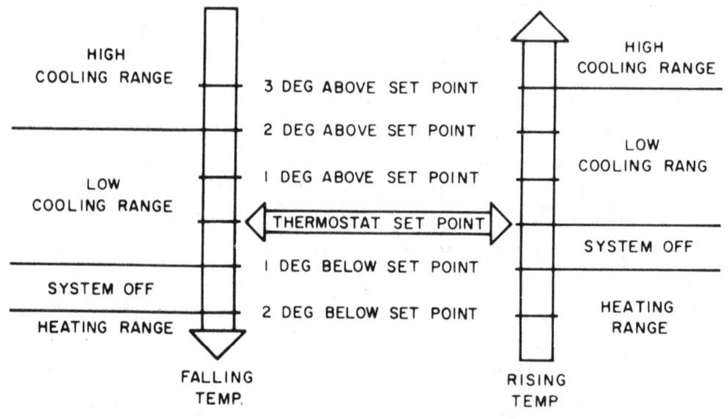

Fig. 2 Thermostat Operation

Defrosting is automatic and controlled by an electric timer in some systems. Others use a differential pressure switch to measure the pressure drop across the evaporator coil, and the switch is adjusted to initiate the defrost cycle when enough frost has formed to reduce the airflow about one-third. Once tripped, the defrost relay holds the heaters on until a thermostat on top of the coil breaks the circuit and the system returns to normal operation. The termination thermostat setting is usually about 55 °F; in addition, a limiting device set at 150 °F is installed to protect against high temperature in case of malfunction of the regular controls.

During the defrost cycle, the control circuit stops the evaporator fan to prevent any of the heat produced in the coil from reaching the lading compartment. At the termination of defrost, the evaporator fan starts again and runs continuously during cooling, heating, or the off cycle.

All mechanical refrigerator cars are equipped with air-circulating fans, which are usually mounted above the evaporator coil in the bulkhead, between the machinery compartment and the loading area (see Figure 1). Nearly all are of the backward-curved centrifugal type and draw air from the floor through the evaporator coil and discharge it into the ceiling distribution duct. The fans consist of two or three blowers mounted directly on the electric motor shaft or on a separate belt-driven shaft. Fans usually deliver about 4000 cfm and require a 1.5-hp, 220-V motor.

Temperature-Indicating Devices

Two types of temperature-indicating devices have been used in refrigerator cars. Both consist of a liquid-filled metallic bulb located inside the car, connected by a flexible metal capillary tube to a dial indicator mounted outside the car. One type uses mercury under a pressure of 300 to 600 psi, a Bourdon tube to actuate the dial pointer, and a bimetallic strip to compensate for temperature changes in the case. The other type uses toluene under a pressure of 16 to 60 psi and three brass bellows to actuate the dial pointer. One of the bellows responds to the liquid in the bulb and connecting capillary tube. Liquid in a second capillary tube, which is not connected to the bulb but runs parallel to and is the same length as the first tube, actuates a second bellows, which compensates for temperature changes in the tubing. The third bellows compensates for temperature changes in the case.

Since the Canadian railroads use tariff rules for perishable protective services that are based on inside temperature control, nearly all Canadian cars are equipped with temperature-indicating devices. The bellows type with two dial pointers (one for top and one for bottom temperatures) is standard for Canadian cars.

SPECIAL PURPOSE REFRIGERATED CARS

Several special purpose mechanically refrigerated cars have been introduced. One design for bulk shipment is based on the standard mechanical refrigerator-car arrangement (Figure 3). The engine compartment is in one end, but the addition of three subfloor hoppers in the center of the car permit the unloading of bulk products. Mechanically operated hinged endwalls can be moved into a sloped position to facilitate the unloading of bulk products over the flat floor sections. This car can be used as a conventional mechanical refrigerator car when the endwalls are in their vertical position and floor grates are placed over the hoppers.

Another design (Figure 4) is based on a covered hopper car that has three compartments with a full-length hatch and a gravity-type outlet for each. The exterior of the car is covered with a layer of spray-on polyurethane foam insulation protected by a special coating; the roof and other high-wear areas are protected with an additional fiberglass covering. The interior surfaces of the hoppers are covered with a highly abrasion-resistant coating, which has been approved for use with foodstuffs.

Three electric heating elements supply the heat usually needed. If fast preheating is desired, three additional units operate for maximum heat. Noncombustible electric heat preserves oxygen, which is important for lading protection. For cooling, a refrigeration unit provides controlled cold air automatically. The unit is mounted on the end of the car to provide access for service and maintenance and protection against vibration, moisture, and dirt.

In operation, air is drawn off the top of the load and passes through a cooling/heating section at the end of the car where it is brought to the preset temperature. A high-pressure fan forces the conditioned air through the car's side sills, into the bottom of each compartment, and upward through the cargo.

Another conventional mechanical refrigerator car with a slightly different machinery arrangement—a mechanical car with the refrigeration unit in the side door—has been developed. The engine generator that furnishes the electrical power is located beneath the car. This feature allows the unit to be removed for high-temperature applications.

Another type of refrigerating equipment circulates air that is reverse to that in the conventional unit. In this case, the blower pulls air down through the coil and discharges it under the floor. It passes up through the load and returns to the cooling unit over the top of the load.

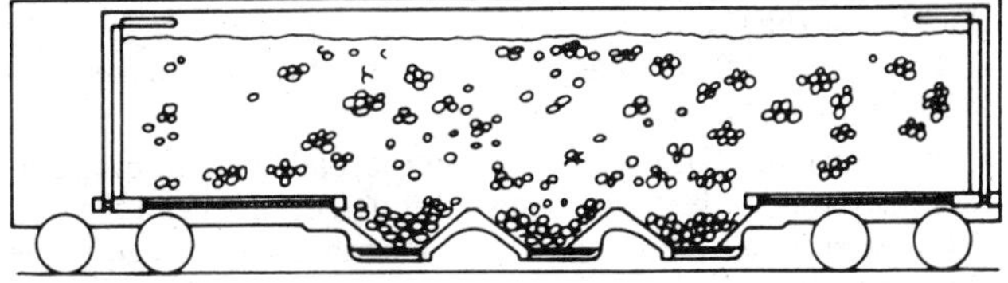

Fig. 3 Mechanical Refrigerator Car for Bulk Loading

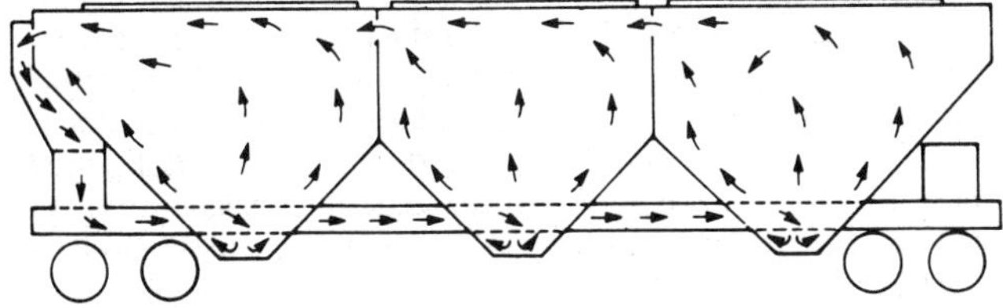

Fig. 4 Covered Hopper Car for Refrigerated Bulk Loading

CAR DESIGN AND CONSTRUCTION

The basic design and construction of refrigerator cars is very similar to that for railroad box cars because both types of cars must meet all the structural, safety, and performance requirements demanded in railroad freight service. Specifications for standard freight refrigerator cars have been established by the Association of American Railroads and are often revised to incorporate improved standards (AAR 1984).

The principal construction features of refrigerator cars that are different from those found in conventional freight cars include: thermal insulation and water vapor barriers incorporated within the structure of the car, mechanical refrigeration equipment, and various features that facilitate the circulation of air.

Refrigerator car construction features common to the standard box car include: (1) trucks (wheels, axles, bearings, springs, and supporting members); (2) steel underframe; (3) brakes and brake equipment; (4) couplers and draft gears; and (5) a superstructure consisting of structural members, exterior steel sheathing (on the sides, ends, and roof), interior wood or plastic lining and floor, side door openings, and various safety appliances.

Refrigerator cars have cushioned underframes, which absorb and dissipate impact forces through a special cushioning device built into the *center sill* (the longitudinal steel member that is the backbone of the underframe). The center sill of the car is fixed, and a sliding sill within the center sill contains the couplers at each end of the car. The two are connected by the hydraulic cushioning device built into the center sill. Other refrigerator cars have separate cushioning devices in each end of the car that perform a similar function by isolating the coupler impact forces from the main car structure. To permit palletized loading and unloading of refrigerator cars with forklift trucks, cars have door widths from 10 to 10.5 ft to facilitate turning in the doorway area.

Movable steel load dividers (or compartmentizers) brace the load and split the loading compartment into three sections. These load dividers reduce impact damage and segment loads billed to stop for partial loading or unloading.

Most cars now have at least a 4000 ft^3 capacity, with a load-carrying ability of 130,000 lb. A complete list of railway equipment is published quarterly in *The Official Railway Equipment Register.*

Insulation of Car Structure

The usual structural members of rail cars constitute high-conductivity heat paths and interfere with the placement of insulation. Various service conditions (including impact and vibration of the car, air pressure caused by the movement of the car at high speeds, body and top icing within the car, periodic washing of the car interior, and reversal of the direction of water vapor flow with winter and summer seasons) make it difficult to retain the initial resistance to heat flow. Wetting and settling are problems aggravated by these service conditions.

The AAR specifications for freight refrigerator cars establish insulation requirements for newly built cars. The cars are to be insulated by means of rigid, closed-cell polyurethane foam to obtain the desired insulation value. RP and RS series cars should be designed to obtain a total heat loss rate of 120 Btu/h · °F, based on a 50-ft-long car. RB series cars should meet a maximum UA factor requirement of 250 Btu/h · °F for 50-ft cars and 300 Btu/h · °F for 60-ft cars.

Membrane-type water vapor barriers usually serve as the covering for blanket insulation. Membrane barriers with one or two reflective surfaces have been placed adjacent to the steel sides in addition to being installed in the roof. Some mechanical cars have a layer of water vapor barrier paper above the top of the ceiling plenum and also on the back side of the inside wood lining.

The use of closed cellular plastic insulation has become standard. The insulation is practically impervious to water and water vapor, and many types possess sufficient compressive strength to withstand the entire floor load in a refrigerator car. Some high-conductivity heat paths through the car structure can be avoided because closed cellular plastic insulation provides sufficient strength to eliminate various conventional structural members such as horizontal belt rails and floor stringers.

The insulation is either board form or foamed-in-place. Mechanical refrigerator cars have either foam-in-place polyurethanes or prefabricated panels. Foam is added after the interior and exterior structure is in place, or prefabricated sandwich panels incorporating the interior lining and insulation are fastened to the car structure at final assembly.

A typical car with foamed-in-place insulation is shown in Figure 5. Horizontal belt rails are no longer used; all interior members are placed vertically in the walls or transversely in the floor and roof. This permits the injection and free rise of the foam material so that it will fill all cavities without interference. The metal posts are on the outside of the car to prevent penetration of the insulation panels.

Mechanical refrigerator cars insulated with the foamed-in-place polyurethanes show a heat leakage only half that of previous insulation and car construction methods. In addition, foam seals the structure against air leakage and prevents infiltration through the insulation spaces from the interior air circulation system, the latter contributing heavily to total refrigeration load in many cases. Mechanical refrigerator cars typically have a heat transfer rate of 90 to 100 Btu/h·°F.

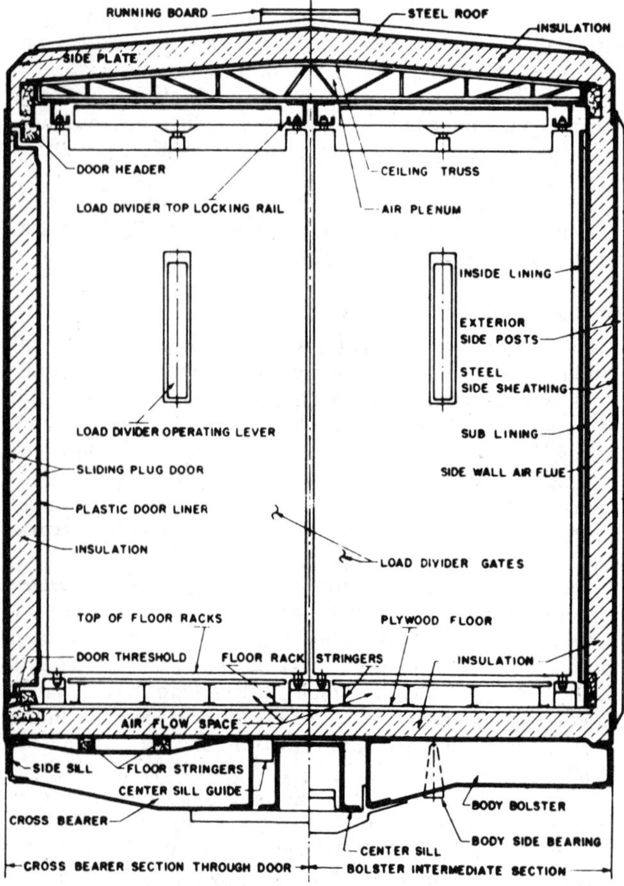

Fig. 5 Cross Section of Typical Mechanical Refrigerator Car Construction

Air Leakage and Other Losses

In addition to conductivity losses through insulation and car structure, a considerable load derives from air infiltration and solar radiation. Doors and drains are principal sources of infiltration. A large amount of air may enter through seams and joints in the car structure. Door gaskets and hardware are carefully designed and applied, as the trend to larger doors increases the exposed areas. Drains under the evaporator drip trap use drip water from the cooling coil as the sealant. Drains elsewhere in the car are self-closing flappers.

The interior car lining is also sealed against leakage, since the air circulation system maintains a constant positive pressure in the ceiling plenum and a negative pressure in the air-return flues. All interior lining joints, especially in the evaporator and blower areas, are sealed with tape or some type of mastic sealer.

By improving car sealing methods, the air leakage rate of mechanical cars is usually less than 100 ft^3/h at 0.5-in. water pressure within the car. This is especially important as more mechanical cars move under modified atmospheric conditions, where excessive infiltration destroys the beneficial effects of the atmosphere.

Cars are finished in light colors, and roofs are usually white or aluminum to obtain high reflectivity. To maintain desired reflectivity, cars are cleaned or repainted regularly because dirty surfaces increase the solar load, which forms a considerable part of the total heat load.

Air Circulation

Floor racks, side and end wall flues, and special ceiling ducts in refrigerator cars facilitate air circulation. Floor racks allow air circulation under the load. Openings in the longitudinal stringers permit side-to-side air circulation and are especially important in cars equipped with side wall flues. Floor racks are designed to carry forklift truck wheel loads of 6000 to 12,500 lb per wheel. Height of floor racks varies from about 4.8 to 7.5 in. Minimum clearance under floor rack slats is 4 in.

Refrigerator cars are equipped with vertical side wall flues that protect loads from direct contact with very warm or cold side walls. Side wall flues are formed by continuous plywood or tongue and groove lining spaced 0.5 to 1.5 in. from the regular inside wall by vertical spacers and open at the top (above the load) and at the bottom (below the floor racks). Cars equipped with mechanical refrigeration have end wall flues in addition to side wall flues. Side wall flues in mechanical cars with a ceiling duct extend from the duct to the underside of the floor rack slats; with the complete envelope system, they extend from ceiling plenum to air return duct below the floor.

The principal types of air-circulation systems used in mechanical refrigerator cars are as follows:

Open. Air is discharged directly over the load from the evaporator end of the car. Air passes down through the load and the wall flues and returns to the evaporator via the channels formed by the floor rack stringers.

Plenum. Same as the open system, except the air is discharged into a slotted or perforated ceiling supply duct that distributes the air uniformly over the top of the load.

Envelope. The air is discharged into a ceiling duct. The ceiling duct, wall ducts, and a special floor duct together form a completely closed envelope around the car. The air circulated within the envelope does not enter the loading area, nor does it escape from the envelope even when the side doors are opened. Figure 6 shows the cross section of the vehicle; some air circulates down the side wall flue and some circulates through the load itself. Both air passes join in the floor duct for return to the cooling unit.

Figure 7 shows the typical air circulation of a mechanical car. The blower draws air through the cooling coil and discharges it

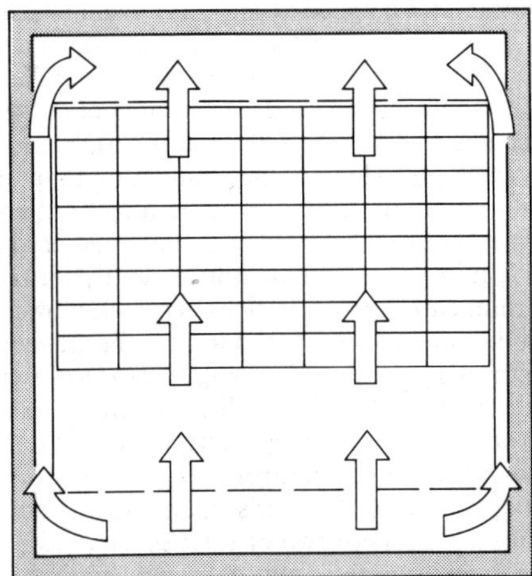

Fig. 6 Cross Section Showing Air Circulation

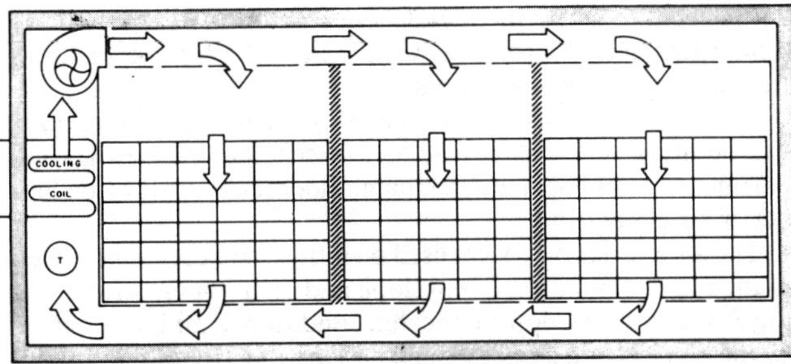

Fig. 7 Conventional Air Circulation System

into a ceiling duct or air plenum above the load. From here, it is distributed into either the lading compartment or wall flues on either side and the far end of the car. It then goes into the floor duct, which carries the air back to the cooling unit. Just prior to entering the cooling coil, the air flows over the thermostat sensing element, marked *T.* When the load divider gates of the mechanical car are in place, dividing the load into three spaces, the air circulation pattern is essentially the same because airflow is from ceiling to floor, and the load divider gates do not interfere with this circulation pattern.

As the airflow diagrams indicate, air will flow through the load from ceiling to floor. However, the amount that flows through the load, compared with that flowing in the side wall flues, depends entirely on how tight the load is in comparison to the openings in the side wall flues. Wider spaces between containers will increase air circulation through the load itself when it is necessary to remove either field heat or heat of respiration from within the load itself.

In the case of frozen foods (a heavy, dense load), air does not need to circulate through the lading space; all air circulation takes place around the periphery of the load to intercept heat from outside. In this case, little or no air should bypass through the lading itself, and all air must flow around the load to give maximum protection. Therefore, frozen loads are usually packed quite densely.

HEATING EQUIPMENT

Air-circulating fans and heaters were discussed previously in reference to mechanical refrigerator cars. Some specially equipped RB cars also offer heating protection service. These heating devices are typically powered by propane, natural gas, diesel, and gasoline fuels. They circulate heated fluid to fan coil units.

REFRIGERATOR CARS IN EUROPE

One of the major refrigerator car builders is located in Germany, and it supplies refrigerator rolling stock to many European countries. Two major types of refrigerator rolling stock are used in Europe: refrigerator sections (usually five-car sections) and autonomous refrigerator cars. These all-purpose types can carry any perishable product in the −10 to 70 °F temperature range, with ambient design temperature varying from 105 to −60 °F. Some of the cars are specialized for transporting cooled (not frozen) meat and are equipped with beef rails and hooks.

An autonomous refrigerator car is completely independent, consisting of a fully insulated lading compartment and two engine compartments (one at each end). Each engine compartment has a unitized refrigeration and heating unit mounted on top under the car ceiling. This design allows better air circulation and temperature equalization throughout the load. The two independent refrigeration systems also contribute to car reliability. If one system fails, the second unit can usually provide the necessary cooling or heating capacity to maintain interior temperature within the range allowable for the load. Each unit (running continuously) provides 75% of maximum cooling or heating capacity under extreme difference between interior and ambient temperatures.

A diesel-generator in each engine compartment powers the refrigeration/heating unit. All electrical controls and monitoring devices are mounted in dust- and waterproof panels on the engine compartment bulkhead wall.

All cars are equipped with plug-receptacles on the outside of the car, which allow the system to be operated from an outside power source during loading and unloading, or when the car is not running and engine noise and exhaust fumes might be objectionable. Each engine compartment has equipment failure-indicating lights and audio signals on the outside. Temperature control boxes provided on every car allow interior temperature readings in transit without entering the load or engine compartment, which are sealed after equipment start-up by authorized personnel. The operation is automatic after initial start-up and is controlled by interior thermostats, which are set at the desirable temperature, depending on the type of load.

A five-car refrigerator section consists of four load refrigerator cars and one diesel car in the middle of the section, which carries the central power plant and the maintenance crew.

Load cars are similar to autonomous refrigerator cars except they have no diesel generators. Instead, electric power for the refrigeration and heating equipment is provided through a central power plant, which usually consists of two diesel generators; each generator is capable of supplying electrical power for up to six fully loaded refrigeration units.

The car also has a central control panel from which all equipment operation and power distribution are controlled. Equipment failure indicators are also located on the central panel. Although refrigeration and heating equipment operation is completely automated (providing there is electrical power), automatics can be overridden by the crew if necessary.

Five-car refrigerator sections are also built in Russia. They are slightly larger than those used in the rest of Europe. Section arrangement is similar, but load cars have two refrigeration units, both located at one end of the car. Each unit also has the ability to provide a capacity of up to 75% of the maximum cooling or heating load.

Mechanical Refrigeration Equipment

Autonomous refrigerator cars and refrigerator sections built in Germany use completely unitized refrigeration units, which can be easily removed and replaced within an hour through an access door located on the car's end wall. The refrigeration unit is of the direct-expansion type with thermostatic expansion valve, and it uses a semihermetic two-stage compressor. The condenser is air cooled by two independent fans driven by separate induction electric motors. Fan operation is controlled by a thermostat, with or without fan condenser operation, depending on the load. The evaporator has two blowers which can be turned on without cooling or heating to provide air circulation and ventilation. Electric tubular heating elements, mounted in front of the evaporator are used for car load heating.

Hot gas from the compressor is used for evaporator defrosting. Two solenoid valves are energized alternatively: one on the liquid line to the evaporator, which opens during normal cooling operation and the other, on the defrost line, which opens to supply hot refrigerant to the evaporator during defrosting (the liquid solenoid valve is closed at that time). The defrost cycle is controlled automatically either by an evaporator pressure switch or by a timing device. A hot-gas circuit is also used for pressure equalization before compressor start and, in that case, is controlled by time relay.

Pressure gages mounted on the unit frame and connected to the appropriate points of the system allow vital information, such as suction and discharge pressures and compressor oil pressure, to be read during system operation. Provision is also made for reading compressor intermediate pressure. The electric control panel is mounted on the unit frame in a dustproof box.

Refrigeration units of five-car sections built in Russia are similar to the split-type system with two compressor-condenser units, using open compressors located in the car engine compartment and two evaporators, located in the load compartment behind the insulated wall. Each car has only one compartment; it accommodates both refrigeration units (one on top of the other), which can work independently and be controlled automatically depending on the cooling load. Every compressor-condenser unit is generally associated with its own evaporator; however, the piping arrangement and manual bypass valves allow both evaporators to be connected to either refrigeration unit.

Diesel generators in autonomous refrigerator cars are mounted on a sliding base together with a fuel tank, battery, and electrical control box. The engine compartment side door can be opened wide to allow easy and fast unit replacement. Plug-in electrical conduits are the only disconnections required. The generator provides 220-V, three-phase, 50-Hz (European standard) power. All motors on refrigeration units are of the three-phase induction type. A transformer provides 36-V (dc) current for control circuits.

The power plant for a five-car refrigerator section is located in the middle of the section and consists of two diesel generators, which provide 220-V, three-phase, 50-Hz electrical current transmitted to load cars through electrical conduits. Each is designed to supply at least 75% of the current required under maximum load conditions, which makes it possible to operate six out of eight fully loaded refrigeration units. Normally each diesel generator provides power to an assigned half of the section (two cars, four refrigerator units), but in case one generator fails, a bypass electrical contactor can be activated to allow the second generator to supply power through the entire section. This feature contributes to section reliability and can also save fuel, when load conditions do not require simultaneous operation of both refrigeration units in every car. Fuel tanks are located under the car.

Some five-car sections built in Germany also have a supplementary small diesel generator, similar to the one used in autonomous refrigerator cars. Such a generator can be used to operate auxiliary equipment, such as fuel pumps for pumping diesel fuel from undercar tanks to the dispensing tank under the ceiling of the engine compartment and air compressors (diesel engines on those sections use compressed air start-up). It also allows the car battery to charge while the car is idle for a long time. When the car is in motion, the battery is automatically charged by the axle generator. The small generator is sometimes used when load conditions require only air circulation or ventilation. In this case, small generator capacity is sufficient to run all blower fans in an entire section simultaneously, without using the main diesel generator. Since the diesel fuel consumption for the main generator is much greater than that of the small diesel, this also saves fuel.

The central control panel located next to the engine compartment of the car allows full control of the refrigeration/ heating equipment operation. As five-car sections are operated by a crew (which usually consists of two or three mechanics), the central control panel can be used for remote manual equipment control, overriding automatic operation, in case of the automatic or equipment failure or for reducing the fuel consumption.

The rest of the special car comprises the crew's living quarters, with sleeping compartments, kitchen, bathroom with shower, and dining area. Heat is provided by the boiler (which works on diesel fuel) and hydronic baseboards.

All refrigerator cars have passenger-type trucks for softer rides. Car bodies are completely insulated (except load car engine compartments) with polyurethane foam. Floor racks are provided for better air circulation.

The air-circulation system used on refrigerator cars built in Germany is a combination of envelope and open types. After passing through the evaporator/heater, conditioned air is discharged by the blowers into a ceiling duct; it then moves to the continuous wall ducts on both sides of the load compartment and enters the load compartment through the floor racks. After passing through the load from bottom to top, air is drawn back to the evaporator. This system, together with the unit's location at both ends of the car, contributes to equal air and temperature distribution throughout the car load.

REFERENCES

AAR. 1984. *Manual of standard and recommended practice.* Recommended Practice RP-253-88. Association of American Railroads, Mechanical Division, Washington, D.C.

AAR. *Supplement to the manual of standard and recommended practice.* Association of American Railroads, Mechanical Division, Washington, D.C.

Code of rules for handling perishable freight. Circular 40. National Perishable Freight Committee, Chicago.

Doean, J.J. 1983. Perishable protective tariff No. 619-B. National Perishable Freight Committee, Chicago.

Redit, W.H. 1969. Protection of rail shipments of fruits and vegetables. USDA *Agricultural Handbook* No. 195.

The car and locomotive cyclopedia, 5th ed. 1984. Simmon-Boardman Books, Inc., Omaha, NE.

The official railway equipment register. R.E.R. Publishing Corp., New York.

USDA. 1969. Protection of rail shipments of fruits and vegetables. USDA *Agricutural Handbook* No. 195, July.

Yuriev, Y.M. and L.B. Lavric-Karamin. 1989. *Isothermal rail cars built in Germany.* Transport Publishing, Moscow.

CHAPTER 29

MARINE REFRIGERATION

CARGO REFRIGERATION

MARINE transport is an interim operation between preshipment storage of indeterminate duration and early distribution at destination ports. Frequently, the marine transport period is equal to or even exceeds the full high-quality life of the perishables being transported. Good design, therefore, requires that shoreside criteria be applied to the floating cold storage plant.

The increased use of various sizes of cargo containers has affected savings, mostly by providing faster unloading and reloading of the vessels. Most refrigerated cargo can be containerized. The need for controlled atmospheric environments for perishable produce imposes additional limitations over those for dry cargo.

Containers will never completely supplant the ship's built-in cargo refrigerator for such services as those provided by passenger ships, logistic supply ships, all-refrigerated fruit carriers, and special service vessels; these services will continue to require the insulation and refrigeration of the principal structural compartments of a ship. The physical and mechanical aspects of both containerized services, as well as the break-bulk (or individual package) type of operation are discussed in this section.

BUILT-IN REFRIGERATORS

The location and arrangement of insulated compartments and compartment subdivisions within the hull should reflect the trade in which the vessel will serve. The volume of trade, the scheduling of ports of call, and the efficient and speedy handling of produce with minimum exposures are all factors that will influence these arrangements. Perishables should be the last cargo loaded and the first to be discharged.

Arrangement and Utility

Limitations of dimension and arrangement are due to the ship's structure, compliance with floodability compartmentation of the hull, and the fire-resistant regulations to which many ships are subject.

The refrigerators should not be designed exclusively for high-temperature cargo services unless it is certain that the vessel will always remain in that limited trade. Otherwise, all compartments should be readily convertible to any temperature service from below −20 to 60 °F, thus allowing flexibility in segregating cargo according to temperature requirements and ports of call.

In a vessel in which only a portion is refrigerated and designed for late loading and early discharge, the compartments are usually located under the topmost deck to the hull. The naval architect considers the effect on trim of the vessel in normal loading conditions. Generally, the added weights of the insulation and the perishable cargo are not as influential as the weight of dry cargoes, which are usually of greater density. Often these compartments are empty or lightly loaded during some of the voyage. Trim conditions generally require the refrigerators to be placed forward of the midlength of the vessel or grouped about the middle section, where the moment arms for trim are of lesser consequence. Almost without exception, refrigerators are arranged symmetrically about the ship's longitudinal centerline.

The rooms or areas where the refrigerating control valves are installed should be arranged and located so that the apparatus or controls are accessible to the operating personnel through trunks or passageways at all times. When the boundary bulkheads parallel hatches, they should clear the openings by 3 ft as a safety measure, providing adequate room for handling hatch beams and covers.

The greater the number of subdivisions in the refrigerated compartments, the greater the loss to the ship's revenue spaces, due to the volumes occupied by insulated partitions, cooling apparatus, insulated piping, and access areas. On this basis, the all-refrigerated ship, with only the main structural boundaries insulated, make the most efficient use of a ship's costly enclosures. On the other hand, the conventional all-refrigerated ship, with insulated hatch plugs at the weather deck, has several undesirable features, such as the difficulty of providing uniform compartment temperatures and of refrigerating hatch areas far from the refrigeration apparatus.

With large-volume rooms, the extended period in which the hatch is open to the outside atmosphere for loading and discharge is not conducive to a good environment for frozen or fresh cargoes. Seldom are frozen goods sufficiently cooled to tolerate an extended loading interval. Fresh products often are not precooled before loading, and the accumulation and delayed extraction of respiratory heat in the compartment is detrimental. During long discharge intervals in ports, atmospheric moisture causes sweating or frosting of products, some of which may suffer from this exposure. With partial delivery at a port of call, such exposure may unfavorably affect the remaining cargo. Also, harbor workers may refuse to work in spaces with the refrigerating equipment in operation. Air curtains help substantially to keep heat from entering the refrigerated area.

A more suitable arrangement of the all-refrigerated ship can be made by fitting the insulated compartments in the 'tween decks with access doors in hatch-side bulkheads, and by means of various combinations of insulated plug hatches at the lower hold and 'tween deck levels. Hatch areas may be used for general cargo or treated as separately refrigerated compartments. Many ships are fitted with side port doors at the 'tween deck level, through which cargo is handled by conveyors which are independent of the ship's overall cargo gear.

The preparation of this chapter is assigned to TC 10.6, Transport Refrigeration.

The central refrigeration machinery plant should be located in, or immediately adjacent to, the main propulsion machinery room, where the ship's responsible watch officers are in constant attendance. A central location for the refrigeration machinery usually results in economy of space and close connections for power and pumping facilities.

Insulation and Construction

Insulation. Moisture-vapor and water-resistant insulation is of particular importance on board ship because of frequent and extreme temperature cycles due to intermittent refrigeration. On termination of refrigeration at discharging ports, insulation will be at lower temperatures than the open room, and often the room surfaces stand dripping wet with atmospheric moisture, which enters through the open door or hatch. Both the warm and the cold side should be moisture-sealed equally, and cold-side breather ports are not recommended. Other common sources of water in ships' refrigerators are melting of ice (packed with vegetables) and defrosting of cool surfaces.

Severe service conditions, which subject the insulation to injury or change by mechanical damage or vibration, and intermittent refrigeration, place exacting requirements on insulation for ships' refrigerators. The ideal shipboard composite insulation should have the following characteristics:

- High insulating value
- Imperviousness to moisture from any source
- Light in weight
- Flexibility and resilience to accommodate ships' stresses and loading
- Good structural strength
- Resistance to infiltrating air
- Resistance to disintegration or deterioration
- Fire resistance or fireproof self-extinguishing requirement
- Odorless
- Not conducive to harboring rodents or vermin
- Reasonable installation cost
- Workability in construction

In the United States, the properties of the insulator and the details of construction must meet approval by the U.S. Coast Guard and the U.S. Public Health Service. For information on insulation materials and moisture barriers, see Chapters 20 and 22 of the 1993 ASHRAE *Handbook—Fundamentals.*

Construction. The three principal parts to the refrigerator boundary include the envelope (or basic structure), the insulating material, and the room lining.

The envelope is usually partly composed of the ship's hull, the watertight decks, or watertight main bulkheads whose members resist the entry of vapor from the warm side. The inboard boundaries outlining the refrigerators should have an equal ability to resist moisture. A continuous steel internal bulkhead with lap seams and welded stiffeners provides a boundary of adequate strength and tightness. Details of design may accommodate dimensioned insulators or facilitate means of fastening these materials. Doorway main bucks of steel channel provide good structure but are usually a source of sweating on low-temperature rooms because of heat gain through the metal. Wooden door bucks minimize sweating but are a retreat from efforts to eliminate concealed wooden structure.

Partitional bulkheads may be of similar detail, but airtight sealing is less important. Some installations are framed with angle-bar grids, between which the insulator is installed. In passenger vessels (over 12 passengers), Coast Guard regulations governing fire-resistant construction restrict wood assembly. Under no circumstances should wood be a part of the floor assembly since it deteriorates rapidly under the prevailing conditions.

The assembled boundary of a ship's refrigerator must withstand heavy floor loads and several wall thrusts of cargo when the vessel rolls or pitches in a heavy sea; it must be able to flex with the hull structure being stressed in any angle. The assembly must resist vibration caused by the propelling machinery, the sea, and the careless handling of cargo. The vapor seal of all surfaces must remain intact.

Only in extreme cases should voids in the insulation assembly be concealed. Filling such volumes with insulating material is cheaper and more effective than constructing internal framing. The exceptions to this rule are in the deep volumes formed by bilge brackets, deck brackets, and open box girders. Solid filling results in more insulation thickness than is needed for a heat barrier in the overhead and the ship's side where beams and frames are deep.

The third part of the boundary assembly is the room lining. This surface must be sturdy enough to withstand the impact of frequent cargo loading and handling. On passenger vessels, U.S. Coast Guard regulations require that the lining be fire-resistant. Tongue-and-groove lumber is considered obsolete for any ship, and other linings are employed. On freight vessels where wood is permissible, exterior grade plywood is sometimes applied. A few installations have been made either of laminated plastic sheets or wood fiber hardboard; both are satisfactory when properly supported. Steel linings are costly, impractical, and difficult to maintain and repair. The favored lining is the cement and fire-resistant fiber hardboard panel with aluminum sandwich lamination. When the insulator is secured with adhesives containing volatile solvents, the aluminum laminations should be of perforated metal or mesh.

The U.S. Public Health Service requires all linings to be rat-proofed. Vulnerable linings should have an underlay of 1/4 in., 16-gage galvanized wire mesh for rat-proofing.

Applying Insulation

In the application of panels with adhesives over block insulators, the butted joints should be separated sufficiently for the adhesive to extrude, provide a moisture-sealed joint, and accommodate movement of the panels by flexing of the ship's structure. The joints may be covered by cargo battens, which are also secured by adhesive and with brass screws. The walls of all refrigerators should be fitted with vertical cargo battens on 15- to 18- in. centers to hold the cargo clear of the insulated wall. This spacing permits circulation of air and prevents the contacting package from assuming the part of a room insulator.

Container vessels of cellular-type construction move the containers along guiding columns; the boundaries of the hold, the hatch coamings excepted, are not subject to mechanical damage. Here the insulation may be of the simplest, low-cost, rat-proofed form without fitted hard panels or cargo battens.

Urethane foam floor insulation requires a restricting surface to confine and distribute the expanding material. In one method, plywood is secured over foam spacer blocks, and the material is injected into the space through properly spaced holes. The membrane and the wearing surface covering are laid over the plywood. But wood in the floor is not good practice, nor is the wood a suitable base for a heavy-duty wearing surface material. An alternate method calls for laying expanded board or blocks in an approved adhesive. The use of the more resilient corkboard laid in adhesives may be good practice in some applications.

The greatest weakness in ship refrigerator construction is the floor covering. Research and practice have not brought forth a totally satisfactory covering. Unlike a warehouse, a tightly packed cargo refrigerator must have floor gratings both to assure the circulation of air and to protect the bottom tier of products from heat leakage. The grating supports carry the cargo weight to concentrated load areas of the floor. The room may be filled with warm general cargo in alternate service, and a thermoplastic covering may be punctured by the supports.

The floor covering must be flexible enough to withstand the flexing of a ship or extreme temperature fluctuations, as well as maintain a moistureproof cover over the insulation. The most satisfactory material is a mastic composed of emulsified asphalt, sand, and cement. This material is applied cold; on setting, it has good load-bearing qualities, is impervious to water, has a small degree of ductility, and may be used in less thickness than concrete. It should always be reinforced, and expansion joints should be included to accommodate shrinkage and adjustment to movements of the ship.

All rigid or semirigid floor coverings should have rubber-base composition expansion joints capable of bonding to the edges of the floor slabs or wall and not subject to shrinking from age. The expansion joints should trace the periphery of the room, the line of all underdeck girder systems, bulkhead offsets to pillars, and similar lines of anticipated ship stress.

Water from ice-packed vegetables and defrost water draining onto the floor covering requires it to be impervious to moisture in the slab as well as the joints. If the floor insulation is wetted, it will deteriorate, lose its efficiency as an insulator, and give off odors. If wet floor insulation freezes, it will lift the deck covering and destroy it. If water penetrates to the steel, the ships' structure will corrode unnoticed.

All floor coverings will crack or become damaged during the ship's life. As a secondary security against water, a membrane should be laid between the insulator and the floor covering. The membrane need not be flashed to the wall if suitable sealing expansion joints are fitted at the juncture of the wall panel and the floor covering.

When an attempt is made to seal insulation with waterproof paper, it should be applied in double thickness, and the laps and perforations should be cement-sealed. Wall and ceiling panels attached with suitable adhesives do not require waterproof paper inner linings.

Floors and Doors

The floor is the weakest point in the ship's refrigerator. The weakest element in the floor is the floor drain or scupper because of the difficulty in bonding the floor covering with the metal drain fitting. Water will often find its way between the covering and the insulation. The conventional floor drain is fitted with a perforated plate flush with the floor covering and hidden by the floor gratings. If the perforations become clogged by debris, water will accumulate at the scupper. A drain fitting near a wall or corner will, on most occasions, create a weak section in the floor covering and will develop cracks running to the wall or across the corner. Figure 1 shows a satisfactory scupper fitting. It is flush with the top of the gratings, has a liftout cover for easy cleaning, and is bonded to the floor covering with expansion joint material. Drains between decks should be omitted wherever practicable.

Refrigerator doors are generally a manufactured product. They should have generously designed steel hardware, and the door and frame should be metal sheathed and have a flat sill and double gasket. Very large or double doors should have additonal dogs to assure proper sealing when closed.

Sliding doors should be installed wherever possible to reduce interferences and conserve adjacent revenue space. When used, brackets should be installed to support portable horizontal spars inside the doors to prevent cargo from falling against the doors in a seaway. Molded glass fiber swinging doors insulated with urethane foams poured in place are available. They are strong and lightweight, are easily handled, and can be fitted with lightweight hardware.

In finishing, wooden surfaces should be varnished rather than shellacked, since the latter material has little protective penetration. Manufactured nonmetallic surfaced materials may be painted or varnished, but if they are nonhygroscopic, their original surface will usually present a good appearance for longer than a painted coating.

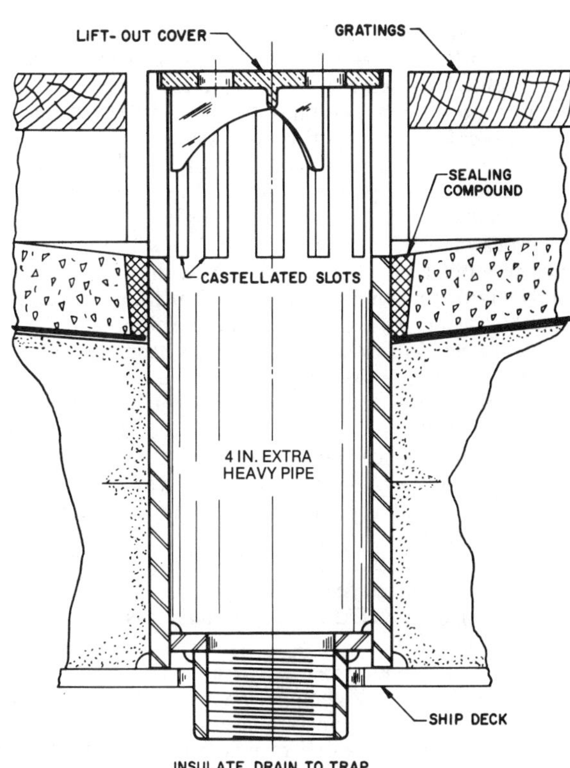

Fig. 1 Floor Drain Fitting

Low-temperature apparatus or piping should be inspected carefully during installation. All joints and surfaces should be generously sealed to keep out atmospheric moisture. Special attention should be given to pipe covering ends, valves, and bulkhead penetrations. On subzero services, special composition adhesives should be used. The smallest omission or breach of a seal will allow progressive destruction of the covering.

On shipboard, where piping systems are relatively short, the function of the insulation is more that of preventing sweating or frosting of cold surfaces than preventing heat gain.

REFRIGERATION SYSTEM

Refrigerants 12 and 22 are used in reciprocating compressors, and R-11 is used in centrifugal units. R-134a is a substitute for R-12, and R-123 for R-11. These substitutes and R-22 have little or no ozone depletion potential. Refrigerant 22 has temperature-pressure characteristics similar to ammonia, but its use in direct expansion (DX) systems requires special attention to applications issues, such as discharge temperature limits, choice of oil, crankcase heater use, and oil returns. It is suitable for close-connected brine cooling systems under all conditions. Ammonia is widely used on fishing and seafood processing vessels. This section describes halogenated hydrocarbon systems only.

Compressors

Modern ships are outfitted with refrigerators with a minimum design temperature of −20 °F. This range may be reached without excessive compression ratios with R-12 units in compound compression or brine cascading. Centrifugal, reciprocating, and screw compressors are used. Chapter 35 of the 1992 ASHRAE *Handbook—Systems and Equipment* describes compressors in detail.

Air and airborne moisture leakage into a refrigerating system is minimized when the low side is maintained at or above atmospheric pressure. However, centrifugal compressors using R-11 and operating at pressures below atmosphere have little trouble from this source because of the effectiveness of shaft seals characteristic of rotative machinery and the continuously operating purging units auxiliary to this type of compressor. The use of brine circulation is mandatory for centrifugal systems using R-11.

Reciprocating R-12 compressors should not be operated in parallel because of the oil's tendency to leave the crankcase with the refrigerant flow and return to flood one unit and starve another. Consequently, each direct expansion evaporator system should be served by its own compressor. Multicompressor units in multiple-evaporator direct-expansion plants are undesirable; parallel brine systems with fewer condensing units are better.

Capacity control of compressors may be affected by speed variations, automatic cylinder cutouts, unloaders, or bypasses. Intermittent operation or cycling should be reduced to a minimum by speed regulation or unit sizing to provide maximum uniformity in plant conditions.

Centrifugal compressors are essentially large-capacity units with capacities varied by speed controls, bypassing, throttling, or metering of refrigerant. The condenser pressure is controlled by circulating water flow.

Reciprocating units are V-belt driven or direct-connected to an electric motor. Centrifugal compressors may be driven by a step-up gear and motor or by steam turbine direct drive. A steam turbine reduces electric generator demands and often fits well in the main steam plant heat balance.

Compressors should be installed in a fore-and-aft centerline position to reduce the effect of gyroscopic bearing loads when the ship rolls. This position also favors lubrication of the unit.

Reserve compressor capacity and spare part lists are specified by the American Bureau of Shipping. This agency recommends two condensing units, one of which, running continuously, should be capable of maintaining full cargo lading in tropical waters. The rules further state that when refrigerated spaces are of 15,000 ft^3 total capacity or less, consideration should be given to the installation of a single condensing unit with adequate spare parts. An experienced ship operator would consider numerical machine reserves necessary as reserve in refrigeration capacity and would no doubt exceed these minimums. Table 1 suggests reserve capacity of various installations.

Condensers and Coolers

Halocarbon condensers are of the conventional shell-and-tube design and, since these refrigerants are noncorrosive to all commercial metals commonly used in refrigeration, a wide variety of materials is available. High heat transfer characteristics and resistance to corrosion/erosion by circulating sea water should be considered in selection and design.

The refrigerant is usually between the tubes and shell of the condenser, and sea water circulates through the tubes in a multiple-pass arrangement. Steel is usually used for shells with cupronickel tubes and cupronickel tube sheets. Bronze water boxes are preferred. If used, cast iron water boxes should have zinc anodes and be coated with a protective material such as epoxy. The tendency to substitute high velocities of circulating water for cooling surface in heat exchange should be resisted to provide long life. Water velocity of 6 fps in the tubes is considered high, and water box velocities above 2 fps are conducive to turbulence and accelerated erosion of tubes. Cooling surfaces are designed for maximum conditions.

Table 1 Operating and Reserve Capacities of Condensing Units

No. of Units, 100% Load	Additional or Reserve Unit, %	Total No. of Units
1	100	2
2	50	2
3	33⅓	4
4	25	5
5 or more	20	6 or more

Shell-and-tube condensers need double drains if installed level; vertical outlet lines long enough to ensure a liquid seal at the expected angle of list or trim are essential. A single outlet may be used if the condenser is inclined at an angle greater than the expected angle of list or trim. Horizontal receivers must be drained in a similar manner. Vertical receivers with a single outlet are also an option. Receivers may be constructed of welded steel and should be arranged to assure submergence of the liquid outlet under all sea conditions. The receiver should have sufficient capacity to hold a complete charge of the system, plus 20% reserve volume. Means should be provided for operators to observe the liquid level.

Brine cooler specifications are similar to those of the condensers except that the materials must resist the corrosive effects of brine. Steel tubes and tube sheets are generally used. In conventional design, the refrigerant circulates through the tubes to reduce the refrigerant charge to a minimum and provides maximum wetting of cooling surfaces. Large coolers may have two or more independent tube groups in parallel, each with its own expansion valve. Multiple parallel expansion results in improved flexibility, increased reliability, and better control at partial loads.

REFRIGERATION DISTRIBUTION

Refrigeration may be distributed by direct-expansion (DX) systems or by brine as the secondary refrigerant. Because the refrigerated compartments are located far from the machinery, direct-expansion systems have many shortcomings. The extended return lines require a large total pressure drop to produce high enough velocities at low loads to return the oil to the machinery. A full load on the same line will reduce the capacity of the compressor cylinders because of the resultant rarefied vapor.

When the evaporators or the return lines in DX systems are lower than the compressor suction, oil traps are possible; these may, under certain conditions, cause alternate starving and slugging of the compressors. Small dual risers (the bottom one of which forms a trap) provide a suitable oil lift when the vapor velocity in vertical return lines falls below 1500 fpm when one compressor is operating. The long low-side lines are generally concealed by insulation or are inaccessible for repair at sea or when the ship is loaded. The cargo may be seriously damaged if vibrations, the flexing of the ship, or damages cause the lines to develop leaks. In addition, the extended systems require large charges of refrigerant with many more scattered points where moisture or air could enter the system.

Brine has many advantages over DX systems. The full charge of the primary system of such a plant is small in quantity, and, in case of accident to the piping or the components, escaping refrigerant is confined to the refrigerating machinery room. The short runs of piping result in small total pressure drops and permit an efficient layout with fewer compressor units. This facilitates the return of oil to the compressors.

Brine cushions abrupt thermal changes and eliminates the unwanted sensitivity of control that is characteristic of DX latent heat systems. Even though the primary system should be automatically controlled, excellent results in cargo space conditions are obtainable with manual control of return brine flow. Brine systems provide great flexibility in plant operations by the capabilities of compounding or sharing refrigeration loads. Total centralized control is feasible.

Space Cooling

The refrigerators in ships are tightly stowed, without aisles or clearances, and the method of stowage has more influence on cooling than the design of the refrigeration equipment. Marine refrigerators are of countless variations in size, dimension, proportion, and configuration. As a result, conventional tests have little significance as guides to operations.

Because cargoes are so tightly stowed, wall coil refrigeration or gravity air circulation is obsolete. Circulating air removes heat from the cargo, but does accelerate dehydration of vulnerable cargo. A long air path requires a higher velocity airstream than a short path for the same air temperature rise.

One method of refrigeration, used often in banana and other fruit carriers, uses external coil bunkers and fan systems to circulate chilled air through ducts and ported false bulkheads. Typically, brine-cooled serpentine coils of pipe have been used. This system supplies air to one side of the compartment and fans exhaust it from the opposite side. Some installations permit reversal of the direction of the airflow. Refrigeration effectively reaches the entire volume if it is properly stowed and the ports are correctly adjusted. The usual distribution results in low air velocities and relatively low moisture pickup. European builders often employ vertical airstreams between floor and ceiling plenums, but this system has not been adopted by American operators.

Cooling Units—Diffusers

These units are composed of three sections: a bottom air inlet, the middle encasing the cooling coils, and the top housing the motor-driven fans, which are usually in multiple arrangement on a common shaft. Large units may be arranged horizontally in similar sequence. Diffusers are packaged units that are easily and cheaply installed. These units are fitted with distributing ducts that direct the air along the ceiling and down the bulkheads between the cargo battens.

The cooling coils are, in some cases, small tube DX evaporators with closely spaced fins. Each vertically arranged tube pass is fed liquid from the distributing head of a thermoexpansion valve. The outlet consists of a connecting manifold at the base of the coils. When air flows horizontally, the evaporator tubes are horizontal and the connecting manifold vertical. Other types employ prime surface tubes or coils cooled by DX or brine.

The manufacturers of the units will usually specify cooling performance as refrigeration capacity divided by the air-to-refrigerant temperature difference. The ratio provides a ready comparison of competitive designs or proposals.

Cooling units or diffusers are installed behind guards against which the cargo may be piled and behind which access is provided for maintenance. The cargo space lost to these units and their associated ducts increases proportionately as room volume becomes smaller. Fans should have output adjustment in steps to half-rated speed, and the motors should be watertight.

Defrosting Systems

Most marine refrigerator cooling surfaces operate below 32 °F, so they rapidly accumulate frost. Prime surface coils require less frequent defrosting than do extended surfaces of equivalent capacity rating. The problem becomes most severe during initial cooling. Maintaining a small temperature difference between the evaporator and room air is the principal means of providing desirable high humidities for fresh perishable cargo.

All cooling surfaces should have means to melt accumulated frost and remove the resulting water. Direct-expansion systems, if not too far from the compressors, may be defrosted by hot gas or, alternatively, with warm water spray. Because hot gas defrosting is complex, most plants defrost with heated fresh or sea water. When using heated water, its temperature should not exceed 90 °F to avoid high refrigerant pressures in the evaporator.

Brine-cooled surfaces may be defrosted by passing heated brine through the coils. A steam-heated, thermostatically controlled brine heater may be installed in a central location for the purpose.

In some long voyage chill services, such as for bananas, citrus fruits, and apples, replacement air is necessary to remove excessive quantities of carbon dioxide and other gases. Continuous airflow maintains uniform conditions within the room and is preferred over intermittent airflow. Outside air should first pass through the cooling unit before distribution.

Citrus fruits should have replacement air at the rate of 2 to 3% per minute based on the gross volume of the refrigerator; bananas require as much as 5% replacement air. When replacement ventilation is provided, accessible, efficient, and well-insulated dampers should be fitted to cut off this air supply when frozen food is carried.

Plant Layout and Piping

Plant layout aboard ship should be as simple as possible without sacrificing reliability. In addition, the machinery plant should be closely associated with the main power plant to provide short piping and power connections and facilitate close supervision.

The personnel changes of engineers and refrigeration crew members demand that strangers to the installation be able, on short notice, to trace well-labeled systems and place the plant in operation or maintain it without undue hazards to the machinery or cargo. Even at the expense of valuable revenue space, an uncrowded machinery room should be provided that will give ample space for operations, maintenance, and the repair of apparatus or the ships's structure, and proper clearances should be allowed for the insulation of low-temperature parts.

All machinery should have sturdy foundations, and all parts should be secured against vibrations set up either by themselves or the main propulsion plant. High-speed machinery should be mounted on fore-and-aft centerlines, and all gravity feeds, drains, and tanks should be designed or installed with full consideration of the effects of trim, roll, or pitch of the vessel.

Each refrigeration system must work independently of other systems in the primary refrigerant side. Table 1 lists the recommended number of units for a plant. The entire plant should be cross-connected to back up the load from any system normally assigned to a specific use.

Multiple compartmentalized refrigerator arrangements operating with DX can become a complex central plant, particularly with hot gas defrosting. Remotely located evaporators result in decentralized control with apparatus scattered over the ship.

Halocarbon return piping may be designed on the basis of a 2 °F total temperature drop between the evaporator and the compressor. Oil trapping should be avoided. Minimum vapor velocities in horizontal lines of 700 to 800 fpm are recommended for oil return flow. The liquid refrigerant should be subcooled immediately after it leaves the receiver, because flashing is easier to prevent than recondensation. This requires placing the liquid-suction heat interchanger close to the condenser and receiver.

Liquid lines should be arranged to prevent line flashing either from high temperature or from excessive pressure drop in the static head of risers. In situations that require a high lift, a refrigerant-cooled subcooler may be needed in place of a liquid-suction heat interchanger. Liquid lines should be fitted with efficient driers and strainers.

Two-Temperature Brine Plant

Brine refrigerant plants require a condensing unit for each evaporator, but the brine system may be cross connected. Most marine plants operate from subzero to ambient temperatures with any or all compartments being interchangeable. A two-temperature brine plant allows high-temperature brine to refrigerate fresh products with minimum desiccation effect, and low-temperature

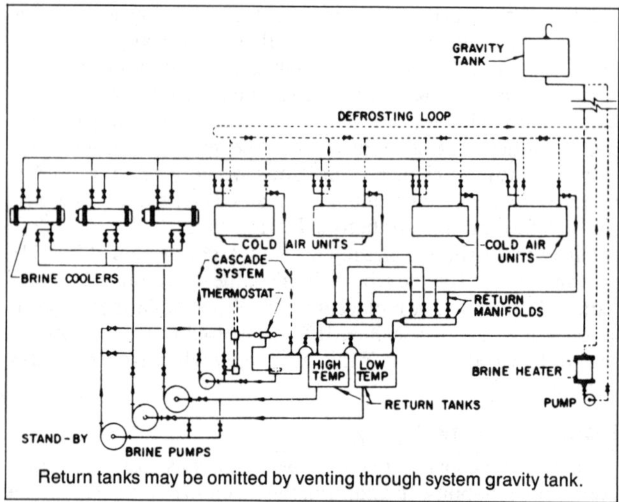

Return tanks may be omitted by venting through system gravity tank.

Fig. 2 Two-Temperature Brine System

brine to refrigerate frozen cargoes (see Figure 2). A two-temperature brine plant suggests two circulating piping systems served by two pumps and a standby pump available for either system. The desired brine can be selected by setting valves at the refrigerators.

Brine returns from each cooling unit should run to the central plant return-valve twin manifold where room temperatures may be remotely controlled by brine flow. By controlling return flow, the systems are kept under pressure, and air binding will be at a minimum. The returning brine is next diverted to its respective collector tank. Two types of systems are in use: the *open system*, in which the return brine streams are in view of the operator; and the *closed return system*, in which a static head is maintained in the system by a vented overhead gravity tank. Brine systems use a refrigerant other than R-12.

A simple two-temperature brine system can be made by including either a closed high-temperature brine loop or a cooling unit with a recirculating pump. The selective brine temperature is maintained by a thermostatically operated pilot valve which feeds low-temperature brine to the loop or cooler as required. The central plant thus becomes a one-temperature installation.

With the use of a secondary refrigerant such as brine, an additional temperature differential results between the primary evaporating refrigerant and the compartment air temperature. Cascading of an R-12 refrigerant plant is desirable for a room temperature of −20 °F. In this operation, the high-temperature brine circulates through the condenser in lieu of sea water and reduces head pressure and compression ratios over those of single-stage compression.

For defrosting with hot brine, an independent small pipe loop having its own pump and heater is suggested. In this operation, the cold brine valves are closed at the refrigerator, and the cooling coils are placed in series with the pump and heater. Only the brine content of the coil is displaced; the entire pipeline is not heated.

Thermometers

The thermometer is the principal indicator of how a refrigerator plant is functioning. Room temperature depends on the proper placement of the sensing bulb within that space. In rooms operating with wide variation below or above the freezing temperature of the product, the bulb can be placed almost anywhere, but in rooms operating only slightly above the commodity's freezing temperature, care must be taken to place the bulb where it will read the temperature accurately and avoid freezing. The bulb should not be placed in direct contact with warm-side surfaces or refrigerating equipment. Rooms with cold air units should have the bulbs located in the supply airstream, that being the point of critical temperatures for chill cargo. When 32 to 33 °F supply air is distributed, the heat of the load is removed as rapidly as possible with no danger of freezing the product.

Recording thermometers are essential to proper operation and management of cargo refrigerators. Indicating thermometers should have the dial installed outside and near the access doors in each room. All bulbs used for comparative readings should be installed adjacent to each other. Mercury thermometers will, because of their design, be near the refrigerator.

Indicating thermometers inserted in walls through the insulation are not satisfactorily accurate or representative. Indicating thermometers should have the dial installed externally and near the access doors in each room. All bulbs used for comparative readings should be installed adjacent to each other. Switch-operated, electric resistance indicating thermometers should be installed with the dial located at the log desk.

Recording thermometers are essential to proper operation and management of cargo refrigerators. Mercury actuated instruments will, because of system limitations, have the instrument near the refrigerator. Electric resistance recording instruments are best located in the machinery room. In some cases, entire cargoes are electronically monitored and controlled from a central location.

REFRIGERATION LOAD

Specifications

The refrigeration specifications should set forth the extreme operating conditions of loading, ambient and sea temperatures, and rates of pulldown. In the all-purpose installation, each compartment should be designed for the refrigeration of warm, fresh products from the field or orchard; for the overall condition, a percentage division of chill and freezer cargo with simultaneous and total loading should be stated.

Typical conditions include the following:

1. Arrangement and net volume capacities of the refrigerated compartments	
2. Thicknesses and kinds of insulation	
3. Ambient temperatures	
Weather surfaces	100 °F
Adjacent machinery spaces	100 °F
Other adjacent spaces	85 °F
Sea temperatures	85 °F
4. Overall stowage factor	70 ft³/ton
5. Percentage total loading as chill	75%
Percentage total loading as freezer	25%
6. Receiving temperature, chill cargo	80 °F
Receiving temperature, frozen cargo	25 °F
7. Carrying temperature, chill cargo	34 °F
Carrying temperature, frozen cargo	0 °F
8. Initial period of cargo heat removal (equivalent)	72 h
9. Replacement air at 85 °F db, 75 °F wb	3%

The owner should describe the kind of refrigeration system to be installed and specify the number of compressors and other auxiliary parts or apparatus together with sources of emergency pumping and water facilities. All equipment and installations must be specified as complying with the rules and regulations of the Classification Societies (American Bureau of Shipping, Lloyd's Register, and others), the U.S. Coast Guard, the U.S. Public Health Service, ASHRAE *Standard* 15-1992, Safety Code for Mechanical Refrigeration, and ASHRAE *Standard* 26-1992, Mechanical Refrigeration Installations on Shipboard.

The specification writer should specify machinery performance only, unless the writer wishes to assume full responsibility for functioning of the plant and all its parts. The specifier should, however, obtain from the vendors full descriptions, details, capacities, and specifications of the equipment proposed to permit comparative analysis.

Completion tests should be required to determine workmanship and functional performance. Performance guarantees should cover operations under loaded service conditions.

Calculations

The following method of refrigeration load calculation for a general service plant carrying heterogeneous chill cargo may appear simplified, but it is justified by the great range of conditions common to marine installations. Refrigeration loads for freezer cargo may be calculated in a similar manner using the same stowage factor, a specific heat of 0.40 Btu/lb · °F, and an equivalent pulldown period of 72 h. The loads for respiration heat, replacement air, or latent heat of fusion will not be present.

Specialized service in known ambient conditions may be calculated more precisely, but an arbitrary 10% margin should still be added to the results to compensate for aging and unforeseen heat gains.

With general calculations, the following operating conditions should be assumed:

1. *Weather ambient conditions*: Up to 100 °F.
2. *Ambient sea conditions*: Up to 85 °F.
3. *Conductivity of insulation*: According to standards given for the material, urethane foam with an installed k-value of 0.20 Btu · in/h · ft^2 · °F is suggested.
4. *Resistivity of outer boundaries, inner linings, and surface films*: These factors should be ignored because boundaries and linings are usually dense and have high conductivity values.
5. *Infiltration and open door leakage*: For cargo refrigeration installations, such losses at sea are nil. Port exposures during loading and discharge reestablish pulldown conditions.
6. *Ventilation or replacement air*: This factor is often omitted when carrying heterogeneous cargo for short to medium length voyages. For specialized service, it may be as much as 300% of the gross room volume per hour.
7. *Electrical energy conversion*: The energy load from fans and brine pumps will be on demand load rather than connected load. An arbitrary value of 3000 Btu/h per brake horsepower may be used.
8. *Product load*: This factor ranges widely for heterogeneous cargo. Its volume will vary from 40 to 120 ft^3/ton and its specific heat will vary from 0.22 to 0.95 Btu/lb · °F. The gross weight of the package and the specific heat of the product should be used. An average volume is 70 ft^3/ton.
9. *Receiving temperature of cargo*: Chill cargo will range from carrying temperature to ambient. Frozen cargo will range from −20 to 28 °F.
10. *Carrying temperature of cargo*: Ranges from 32 to 55 °F for chill cargo, and 0 to −20 °F for frozen foods.
11. *Respiratory heat of chill cargo*: Meat products, eggs, and dairy products have no respiratory heat. Chapter 30 in the 1993 ASHRAE *Handbook—Fundamentals* lists heat of respiration of many horticultural products at various storage temperatures.

CONTAINER VANS

The early programs for marine refrigerated containerization borrowed from developments in land transport, particularly the highway vehicular concept. Currently called the conventional container, the shortcomings of its precursor have been inherited along with the merits. Highway transport has developed on the premise of an interim transport of relatively short duration between source and delivery by a single carrier. For fresh horticultural products, initial transport is assumed to occur when the products are hardiest and most likely to endure poor environments. These influences are reflected in the conventional marine container. However, the use of interchangeable (or intermodal) transport of containerized products demanding excellent conditions is increasing. For the usual interval of an overseas voyage, which is often added to storage and land haul, good cold storage is essential.

Standards and Assemblies

As surface transport equipment often passes through many countries, international container standards for overall dimensions, fastenings (lashings), handling, and lifting has been published as the *U.S.A. Standard Specifications for Containers* (USASI MH 5.1), sponsored by the American Materials Handling Society and the American Society of Mechanical Engineers. The American Bureau of Shipping has issued a *Guide for Certification of Cargo Containers*. Lloyd's Register of Shipping has published rules covering refrigerated cargo containers.

Universal basic standards of dimension are 8 ft wide and 8 ft high. Except for width, which conforms to highway regulations, ship operators have disregarded basic standards, adapting different lengths and heights best suited to a particular operation. These variations make cargo transfers difficult if not impossible. A disregard of universal standards can result in rehandling, with hazardous exposures of sensitive perishable products, at interchange stations.

Most containers are 8.5 ft high, 8 ft wide, and either 20 or 40 ft long. Some 8, 9, or 9.5-ft high containers are also in use. A demand for 40-ft general service containers continues as adequate port terminal facilities are made available and ships' holds are outfitted to accommodate them. However, the problems of air circulation within a conventional refrigerated container rapidly accelerate with increasing length.

True intermodality is further frustrated by the large number of attachable refrigerating machines, all of which are made incompatible by varying characteristics and dimensions. European standards, which are based on centralized marine systems, have reduced this problem. These systems are served by a ship's refrigeration plant and external attachable refrigeration for land haul connects to the air-circulating connections used in sea transport.

Little design choice is left to the owner except for overall length and height, type of insulation, and method of applying refrigeration. Universal practice requires stacking containers six units high. Aluminum shapes, plates, and extrusions are generally used in container assemblies. However, containers are being assembled with plastic embodiment, and others are being built with metal framework and panels of glass fiber and plywood.

Container insulation is mostly urethane foam because of its low density, its closed-cell characteristics, and its high bonding properties. For containers capable of carrying frozen foods, a nominal thickness of 3 in. on the walls and floor and 4 in. in the top is suitable. Considering aging, skin densities, possible imperfections of installation, and the inevitable penetrations and heat bridges, an average k-value of 0.20 Btu · in/h · ft^2 · °F is recommended in the calculation of transmitted heat loads. Floors are of extruded aluminum or stainless steel; linings are aluminum or fiberglass. All must have cargo battens or corrugations on the walls to prevent surface contact by the cargo and to assure ventilation.

The methods of assembly, the calculation of refrigeration loads, and the testing procedures for marine containers are the same as for highway vehicles.

The Conex logistics container was initially designed for military dry cargo. Its dimensions and allowable loaded weight promote mobility in handling by forklift trucks, flatbed trucks, rail cars, and helicopter cranes. The larger container is 6 ft-3 in. wide,

6 ft-10.5 in. high, and 8 ft-6 in. long. This size container, if adapted to refrigerated service, is too small and would be needed in too great a quantity to fit each unit with individual attached refrigerating machines. However, it is feasible to fit an insulated container with communicating air ports and an internal fan to circulate the chilled air of a ship hold or warehouse through the cargo. Without auxiliary sources of refrigeration, these containers would have limits in shoreside hauling or holding. For extended shoreside service, attachable machines may be affixed to a closable aperture in the nose of the container.

Conventional Containers

The conventional marine container is fitted with a nose-mounted refrigerating machine powered by a dismountable diesel generator set for land-based operation. To accommodate on-deck stowage alignment with the dry cargo containers, a recess for the refrigerating machine is built into the container. The capacity and effect on evaporator relative humidity are of concern when applied to the long sea voyage.

With the earlier models, cold air blankets the top of the load. The air is recirculated by a fan built as part of the internally extended evaporator. The plant is thermostatically controlled and may include automatic defrosting. In most new containers, the cold air flows between the T-bars that form the deck and up through the cargo. Microprocessor-based control of the incoming air temperature and high airflow can maintain the product temperature within ± 1 °F throughout the container.

Some units have dual drive, where an internal combustion engine is used in shoreside services and electric motors drive the units on board ship. Others are all electrically driven, the current supplied by chassis-mounted generators or by the ship's electric plant.

Because of the considerable rejected heat of the machines and cargo, conventional containers are carried on the weather decks. Many container refrigeration units are equipped with both a water- and air-cooled condenser. When it is operated in the hold, the water-cooled unit is connected to a cooling water source, and the fan on the air-cooled condenser is turned off.

Developments in Heat Removal Equipment

One development in separable refrigerating units involves an integrated refrigeration plant, with or without an associated internal combustion engine power source. The unit covers the entire transverse area of the container's nose with air-circulating ports that match a ship's resident refrigerating system. The nose of the container is permanently insulated, and the refrigeration unit attaches to the container by snap-on toggles.

Liquid nitrogen is economically feasible for many services. In such arrangements, the supply tanks and control apparatus may be carried by the transporting chassis, and the attachments to the spray header and the sensing bulb are separable.

Modified Atmospheres in Transport

As supplements to refrigeration, modified atmospheres slow the metabolism of horticultural products during transport. Sea transport conditions do not allow precise control of the atmosphere. Low-oxygen atmospheres in conventional refrigeration ships' holds are not acceptable because of the hazard to personnel. Individual control of concentrations for shipments in conventional containers is not operationally feasible, but the one-shot application at the point of loading of gastight containers does have promise. Chemical carbon dioxide scrubbers should be enclosed with the cargo when long voyages are anticipated.

Criteria for Marine Carriage

Products intended for export must be of the highest quality and in prime condition. They must be packed in packages that will not be crushed during transport and will allow heat removal; these are responsibilities of the shipper and should be subject to requirements of the marine carrier. All products carried in cargo containers should be cooled to carrying temperature before original departure, and the container should be expertly loaded to allow air movement through the stack, yet remain secure during transport. Uniformity of spacing and product temperature are requisites of good refrigeration for frozen foods as well as fresh horticultural products.

Even with orderly initial stacking within the container, disarrangement may occur en route because of the ship's motion. Vibrations caused by propulsion machinery can cause improperly secured packages to creep. Shifting of packages can lock air passages and effectively retard refrigeration. The shipper's method of stowage, like the quality of the products, is concealed from the marine carrier who assumes the responsibility for good delivery at destination.

Projected Developments

The following developments are projected for cargo refrigeration:

- Increased container size to 48 ft or more
- Remote monitoring and control via powerline communication and satellite link
- Microprocessor control with automated diagnostics and calibration procedures
- Controlled atmosphere systems, which have the ability to sense and control the concentration of some gases

SHIPS' REFRIGERATED STORES

Most vessels carry enough provisions for long voyages without replenishing en route. The refrigerating equipment must operate under extreme ambient conditions. Attention is being given to such factors as the storage of frozen foods, packaging, humidity, air circulation, and the increase of space requirements.

COMMODITIES

Perishable foods can be fresh, dehydrated, canned, smoked, salted, and frozen. For some, refrigeration is necessary; for others it may be omitted if the storage period is not too long. Space aboard ship is costly and limited; many rooms at different temperatures cannot be provided. The benefits of refrigeration and other conditions can be obtained by providing conditions outlined in Table 2 and described in the following sections.

Meats and Poultry

Substantial savings in space and preparation labor and better quality can be obtained with precut, boned, frozen meat, and poultry packed in moisture-vaporproof cartons and wrappers. For this reason, increased 0 °F storage space should be anticipated. Fresh meats are less suitable because of their relatively short storage life. Also, the space required for fresh meats is two to four times more than that needed for prepackaged meats.

Fish, Ice Cream, and Bread

Good quality fish, properly prepared and packaged, will remain odorless and palatable for a long time.

Commercially prepared ice cream is nearly always available and used to a great extent for both passenger and freight vessels. Ice cream for immediate use should be kept at a slightly higher temperature in an ice cream cabinet in the galley or pantry. Ice cream-making equipment may be desired, in which case provision must be made for hardening the ice cream, ices, and sherbets.

Table 2 Classifications for Ships' Refrigeration Services

Service	Temp., °F	Passenger Vessels	Freight Vessels
Freezer rooms			
Meats/Poultry	−20	X	X
Frozen foods	−20	X	X
Ice cream	−20	X	X
Fish	−20	X	X
Ice	28	X	
Bread	0	X	
Chill rooms			
Fresh fruit/Vegetables	34	X	X
Dairy products/Eggs	32	X	X
Thaw rooms	40 to 45	X	X
Wine rooms	48	X	
Bon voyage packages	40	X	
Service boxes			
Main galley			
Cooks' boxes	40	X	X
Butchers' boxes	40	X	
Bakers' boxes	40	X	
Salad pantry refrigerator	40	X	
Coffee pantry refrigerator	40	X	
Ice cream cabinet	10	X	
Ice chest		See text	
Mess rooms or pantries	40	X	X
Deck pantries	40	X	
Wine stewards' box	40	X	
Bars/Fountains	Various	X	
Miscellaneous			
Ice cube freezers	See text	X	
Ice cream freezers	See text	X	
Biologicals	40	X	
Drinking water systems	See text	X	X
Canned foods	See text	X	X
Dehydrated foods	See text	X	X
Ventilated stores			
Hardy root vegetables	See text	X	X
Flour cereals	See text	X	X

Excellent results can be obtained by purchasing freshly baked bread, sealing it in moistureproof wrappers, and storing it at 0 °F. This supply may be supplemented by bread and other bakery items made on the ship. Frozen bread may be thawed in its wrapper in a few hours.

Fruits and Vegetables

Packaged frozen fruits, fruit juices or concentrates, and vegetables may be stored in any freezer room. All packaged frozen products may be held in a common 0 °F storage space. However, the increased usage, especially on large passenger vessels, may justify separate refrigerated spaces for some or all of the products.

In some cases, fresh-grown product is desired. These items may be stored in a common room, but some compromises with their optimum storage conditions must be expected.

Dairy Products, Ice, and Drinking Water

All dairy products may be stored in a single room, following customary shoreside practice. Strong cheeses with odors that might be adsorbed by other foods should be stored in a tightly enclosed chest or cabinet placed in the dairy refrigerator. Eggs may be processed by oil dipping or heat stabilization to make them less sensitive to unfavorable humidity conditions or odors. Large passenger vessels should be fitted with a separate egg storage room. Butter for reserve supply should come aboard frozen and be kept in a freezer. Frozen homogenized milk has been perfected to a degree that it can be carried for reasonably long periods. Aseptically canned whole milk is now available and may be stored without benefit of refrigeration, but this product has some limitations because of the detectable cooked flavor.

Flake ice machines and automatic ice cube makers are common on passenger and freight vessels. Chilled drinking water is piped to many parts of a ship. The water is cooled in closed-system scuttle-butts, and the necessary circulating lines serve living and machinery spaces where drinking fountains and carafe-filling taps are installed. Remote stations that would require unusually long insulated piping runs are better served by independent domestic-type refrigerated drinking fountains.

STORAGE AREAS

Many borderline perishables, such as potatoes and onions, are satisfactorily stored with ventilation only. Hardy root vegetables are carried on freight vessels not destined for winter zones in slatted bins on a protected weather deck. Flour and cereals must be stored in cool, well-ventilated spaces to minimize conditions conducive to the propagation of weevils and other insects.

Storage Space Requirements

Space requirements for refrigerated ships' stores can be approximated by formulas. However, catering officials and supervising stewards have specific ideas regarding the total volume and subdivisions, and these sometimes vary greatly. A freight ship in ordinary scheduled service seldom exceeds 45 days between replenishment of stores and passenger vessels, considerably less. In addition, deliveries en route are possible. Passenger vessels diverted for cruises can expand into the otherwise unused refrigerated cargo spaces.

The space provided should allow suitable working floor areas for good storekeeping. When possible, stable piling 6 ft high is good practice; and, if the clear height of the room is less than 7 ft, allowances must be made for air circulation. A storage factor of 90 ft^3/ton should suffice and allow floor working area. In the absence of a directing caterer or steward for consultation, Equation (1) may be used to estimate the total refrigerated storage space for merchant vessels. Ice storage is not included because of the various methods used in supplying it.

$$V = \left(\frac{NDP}{2000}\right)F \qquad (1)$$

where

V = total volume of refrigerated storage (not including ice), ft^3
N = number of crew and passengers
D = number of days between re-storing
P = mass of refrigerated perishables per person per day, lb
= 10 lb for freighters
= 13.5 lb for passenger vessels
F = stowage factor (approximately 90 ft^3/ton)

For example, for a freighter on a 45-day voyage with a crew of 53 and 12 passengers,

$$V = \left(\frac{65 \times 45 \times 10}{2000}\right) 90 = 1316 \text{ ft}^3$$

With 6-ft high stowage, the net floor area would be 219 ft^2. With 8-ft high ceilings, the gross volume would be 1752 ft^3.

Gross volume represents actual space available for storage of foods up to ceiling height and does not include the space occupied by cooling units, coils, gratings, or other equipment.

Stores' Arrangement and Location

Next to the arrangement of ships to meet their major purpose, the planning of ships' housekeeping facilities is most important.

Efficient operation by culinary workers requires not only well-arranged working spaces, but also convenient supply stores. Storerooms are usually located in spaces least suitable for living quarters or revenue-earning volumes and in areas adjacent to the main galley and pantry. The arrangement should provide easy access, which generally places the reserve storage refrigerators on the deck below the galley.

Aboard freight vessels, the refrigerators serve for daily issue as well as reserve storage. Aboard passenger vessels, the reserve storerooms are less frequently entered, and greater use is made of the service or work boxes.

Passenger vessels carry a corps of steward's storekeepers, who should have an issue counter and office located within sight of the exits serving this area. The storerooms should extend to the ship's sides or have passageways reaching to sideport doors through which the stores may be loaded directly into the ship. However, the arrangement of passenger ship stores will likely be compromised because of the interferences of structure, machinery or access hatches, and ventilation trunks.

In addition to the requirements for reserve storage of perishable foods, refrigerators (often referred to as working boxes) must be provided for the galley and pantry crew. On cargo vessels, a large domestic-type refrigerator will suffice. When more space is needed, a commercial walk-in box can be used. On passenger vessels, larger boxes are built-in like reserve refrigerators. The capacity of the passenger ship refrigerators will be governed by the number of passengers carried, the variety of the menu, and the arrangement of the galley and pantries.

Ice cream stored in the reserve boxes is too hard for serving, hence a dry or closed type of serving cabinet that will maintain temperatures from 5 to 10°F must be installed in the pantry. Passenger vessels may require ice cream-making machines as well as bar and soda fountain equipment, the latter being fitted with commercial, independent refrigerating units.

SHIP REFRIGERATOR DESIGN

Marine refrigeration equipment for off-shore vessels should be designed, selected, and applied so that it will function properly under extreme conditions with a minimum dependence on expert servicing.

Refrigerated Room Construction

Free water that might enter the insulation through faulty floor or wall surfaces is the most harmful element to ships' refrigerators. Room linings and floor coverings should be made of the materials and have the surface character that will give life-long resistance to the absorption of water by the insulation and the adherence of moisture on the room's interior surfaces. The construction of reserve and built-in refrigerators should follow details similar to those of the conventionally designed cargo refrigerators described in the Cargo Refrigeration section.

Adequate floor drains of the type that may be cleaned without lifting floor gratings should be provided and located so that, with the probable stowage plan, the scuppers will be accessible for cleaning without moving shelving or the excessive weights of stores.

All details must be in compliance with the regulations of the U.S. Public Health Service, which also emphasizes rat-proofing. American ships are also subject to strict fire-resistance regulations.

All doors and frames should be of sturdy construction to resist frequent slamming and should have metal sheathing or reinforced glass fiber doors. They should be large enough to facilitate the loading of stores. The locking device should permit release of its fastenings from the inside by a person accidentally locked in.

All rooms should have galvanized iron or stainless steel racks or shelves to meet storekeeping needs, and they should be easily removable from their supports to facilitate rapid and thorough cleaning. The meat room and thaw room should have a single fore-and-aft meat rail for miscellaneous uses and thawing, respectively. Floor gratings or duckboards, fitted to each room, should be of a size and weight to facilitate removal and cleaning of gratings and the room.

The refrigerators should be fitted with waterproof lighting fixtures well guarded from damage by storing operations. The wiring should be bronze basket-weave cables, surface mounted, and well secured. Lighting switches should be mounted inside each room at the door with indicating lights in the outer passageway. Each room should have an efficient audible alarm for the use of any person inadvertently locked inside.

Remote reading thermometers, from which room temperatures can be read in the outside passageway, are essential to good operations. The bulb should be located in a representative location in the room—generally the geometric center at the ceiling. A large passenger installation justifies a duplicate electrical-resistance thermometer, with the instrument located in the refrigeration machinery room.

Service boxes in the galley and pantries should be constructed with a minimum amount of wood. The linings and shelving should be made of materials and have a surface character that facilitates thorough cleaning. Service refrigerators should not have raised door sills, and the floor should present a flush surface that is easily drained and cleaned. The cooling surfaces should be totally accessible for cleaning. Small units should be mounted without floor clearance on elevated bases, or be provided with at least an 8-in. clearance to facilitate scrubbing underneath.

Methods of Refrigeration

Freighters that carry no refrigerated cargo or that normally carry perishables in one direction, will only, in most cases, use direct-expansion systems for stores. Freight ships normally carrying refrigerated cargoes round trip may be arranged to use brine refrigeration for the combined service. Passenger vessels may have brine circulating systems, individual condensing units on each fixture, or a suitable multiplex arrangement.

The location of the condensing units for ships' stores, particularly for galley service, bars, and soda fountains, often presents a plumbing problem with the circulating water. This may make air-cooled units advisable for such applications. When air-cooled units are used, an ample supply of the coolest available air must be assured, and the hot air leaving the condenser must be expelled if the units are located in small rooms. Ventilating fans are mandatory for air-cooled condenser installations. Condensing units should not be installed in hot machinery rooms. Standby units for the more important refrigerating system components are advisable to ensure against extended interruptions of service at sea.

Direct-Expansion Systems. These systems may be classed as central plant, multiplexed, and unit installations. The central plant installation uses two condensing units, one as a standby. If both low- and high-temperature refrigerators are to be served by the single operating unit, it must be selected and based on the lowest refrigerant temperature to be used in the system; this entails some sacrifice of compressor capacity. For successful operation, the smallest cooling unit must balance with the condensing unit at a saturated suction temperature above the low-pressure safety cutout switch setting. If this condition is not met, hot gas bypass and desuperheating valves will be needed.

Multiplexed Systems. In these systems, the low-temperature fixtures are grouped on one condensing unit, the medium-temperature services on another, and water cooling and high-temperature loads on another. If the refrigerant temperatures required by the various fixtures on one group differ greatly, the higher temperature evaporators should be supplied with evaporator pressure regulators for automatic control. Assuming a series of four refrigerators for ships' foods—one for dairy products, one for fruits and vegetables, one for meats, and another for frozen foods—each would have its own R-12 condensing unit with associated cooling units. This provides maximum protection of perishables in case of mechanical failure. A better arrangement provides one or more standby condensing units. With proper piping and valving, the standby unit can be connected to substitute for any unit that has failed.

Unit Installations. These systems apply a single condensing unit to each fixture—an arrangement preferred by some engineers.

Cooling Units. Packaged forced-air cooling units with finned tubes speed up and reduce the cost of the installation. Automatic defrosting may be used, where average room temperatures are about 34 °F and above. Air should be directed above the food; it should then be allowed to diffuse down and return to the cooling unit for recirculation. One or more of the units may be installed in a room after considering air distribution. If finned-tube, forced air units are used for temperatures below 34 °F, some positive means of defrosting must be provided. Hot gas and electric defrost are successful if properly applied and used.

When wall coils are used, paneled baffles should be installed over them both to prevent the produce from contacting the low-temperature surfaces and to promote air circulation. To cool products in larger spaces, ceiling-mounted fans should be installed for positive air distribution. The fan is generally located in the center of the room. In designing direct-expansion wall coils, the circuits must not be so long that excessive superheat of the expanded refrigerant results. On the other hand, a shortage of cooling surface results in excessive differences between refrigerant and room air temperature, with consequent dehydration of stored products.

Wall coils with an odd number of horizontal passes simplify series connection. Flow should be from top to bottom to minimize oil hang-up. Intermediate risers must be sized to ensure proper velocity for transport of the oil and unevaporated refrigerant leaving that coil section. Each riser may need to be preceded by a close-coupled trap to minimize oil hang-up in the bottom pass. The final riser should be sized as described in Chapter 2. Double risers may be needed to cope with reduced capacity operation.

Suction lines must be connected to overhead mains so oil cannot drain into idle or low-side equipment. Suction lines should run level and above the compressor suction inlets. If a riser must be used, a suction line accumulator may be needed. If a double suction riser is used, a suction line accumulator is essential to avoid oil slugging.

Controls. The most reliable system is the simplest one with a minimum of automatic apparatus. For marine applications, liquid controls should be of the automatic expansion type. Thermostatic valves are suitable for most applications. Rolling, pitching, trim, and list of vessels make float valve operation difficult. Expansion valves, being spring operated, are not affected by these conditions. Capillary tubes are commonly used in unit refrigerators and will function on vessels. If a float-type control is unavoidable, its vertical plane of operation should be fore-and-aft rather than athwartship. Float controls are unsuitable in vessels with propulsion machinery aft or which normally trim aft.

Temperature controls using thermostats, low-pressure controls, evaporator-regulating valves, and solenoid valves perform satisfactorily in ships' service systems. Such equipment, however, must be unaffected by vibration or ships' motion. Mercury bulb controls should not be used.

Direct-expansion controls should be located adjacent to their respective units, be installed clear of damage hazards caused by handling stores, have strong guards, and be fitted with locked covers to prevent unauthorized persons from tampering with them. Controls should be installed outside the refrigerators; on closely coupled systems using a central plant, they may be installed in the machinery room without the benefit of guards.

The calculation of refrigerant loads of a ship's service refrigerator system should follow the same procedure indicated for cargo systems in the section on Cargo Refrigeration. A larger margin for miscellaneous heat gain, such as frequently opened doors, should be allowed. An arbitrary figure of 10 to 25% of the total load may be used.

FISHING VESSELS

All vessels harvesting fish products, from the small open gill net boats used in inland lakes to the large factory processing ships on the high seas, use some form of refrigeration, whether it is ice picked up each day sufficient for the day's catch, or a version of the most advanced blast freezing system.

REFRIGERATION SYSTEM DESIGN

When designing a refrigeration system, the following issues should be considered:

- The vessel owner and the design engineer must be aware of the monetary value of a fully loaded fish hold. The money saved by selecting and using substandard equipment may be a needless and expensive gamble.
- The vessel may be several hundred miles from a qualified service technician and have very limited resources on board for emergency repair. In the event of a system failure, effective initial design may maintain temperatures longer, thus preserving the product.
- Marine refrigeration systems are subjected to severe conditions, including high engine room temperatures, low ambient temperatures, electrolysis (corrosion), impacts, and vibrations. In some cases, these conditions are compounded by little or no maintenance, or even worse, abusive maintenance.
- The system should be well laid out and designed in such a way as to allow new operators to adapt to the system quickly.
- All safety and operating controls should be employed. In the event of a component failure, a backup system should be available, or, ideally, built into the system.
- On completion, the vessel should be provided with all wiring and refrigerant flow diagrams, an operator's manual, and a supply of spare parts.

In the initial planning, the designer must know:

- For what fisheries the vessel is being equipped and in what area of the world the vessel will operate.
- In what future fisheries the vessel may be required to work. (At this point, such considerations will probably add little or no cost to the system.) Necessary alterations may be as few as increasing the spacing in the freezing racks.

Hold Preparation

On any vessel presently being refrigerated or being fitted for future refrigeration installations, 6 in. of insulating spray-on urethane is required. Special attention must be given to insulating areas of high heat such as engine rooms, bulkheads, and the underside of the main decks. High heat sources such as the hatch combing, shaft log, and fuel tanks with fuel returns from the engines, must also be insulated. The insulation must be protected to prevent moisture from destroying its insulating quality. In the northern Pacific, laid up fiberglass is commonly used because of its strength, light weight, and versatility. Pen board guides, mounting brackets, and plate racks are at times fiberglassed into the liner, and thus become a very secure part of the vessel. Fiberglass has the advantage of being easily cleaned and sanitized.

Case Study

In this case study, the owner of a 50-ft north Pacific trawler requested plate-type freezing surfaces secured to the front and rear bulkheads and under the deck to control heat gain through insulation, and a freezing rack made up of plates on which fish would be laid during freezing. All equipment uses a single compressor.

Heat Load Calculations. Depending on the sizes of the vessels, the available spaces, and budgets, most vessels employ one or more systems. U.S. law requires the product to be core frozen to −20°F and stored at or below that temperature.

1. Precalculated load from all heat gain sources (not including product load), including a 10% safety factor, is 8000 Btu/h.
2. Saturated suction temperatures: −40°F
3. Required hold temperature: −20°F
4. Temperature difference (TD): −40 − (−20) = 20°F
5. The U-factor, or the amount of heat absorbed by the plate surface, is 2.0 Btu/h · ft² · °F (below 32°F in still air). The U-factor can be increased by increasing airflow over the plates. The heat transfer per square foot of plate surface then equals:

$$TD \times U = 20 \times 2 = 40 \text{ Btu/h} \cdot \text{ft}^2$$

6. Plate heat transfer surface required: 8000/40 = 200 ft²
7. On a visit to the vessel, it was found that 22 in. by 48 in. plates would be best suited to this vessel. The number of plates required (assuming both sides of the plate provide cooling in this application) are as follows:

$$22 \times 48 \times 2 \text{ sides}/144 = 14.5 \text{ ft}^2 \text{ cooling surface per plate}$$
$$200/14.5 = 14 \text{ plates required}$$

8. The hatch combing heat load has been considered in the calculation. The hatch is heavily insulated, has a small entry port, and an insulated floor, again with a small entry port into the hold. The hatch should be kept at above 32°F for bulk fresh food storage. It is equipped with two additional plates and an EPR valve (see Figure 3).
9. A typical method of mounting under the deck plate surfaces is shown in Figure 4.

Freezer Cell Loading. A typical freezing cell (Figure 5) is usually located across the front bulkhead or in another suitable space. It is constructed with 4- to 5-in. spaces between the plates and with 2 or 3 spaces at 5 to 7 in. for salmon. Tuna requires spacing of 6 in. and greater. The freezing cell is usually constructed of noncorrosive materials, such as aluminum, with stainless bolts.

1. The example vessel expects to handle an average catch of 300 lb of salmon daily, and to load its freezing cell twice with 1500 lb per load. On average, 6 to 7 lb of fish per square foot should be used for proper freezing. The surface area required is:

$$\frac{1500 \text{ lb}}{6 \text{ lb/ft}^2} \times 1.1 \text{ safety factor} = 275 \text{ ft}^2$$

Note: Only one side of a plate can be used.

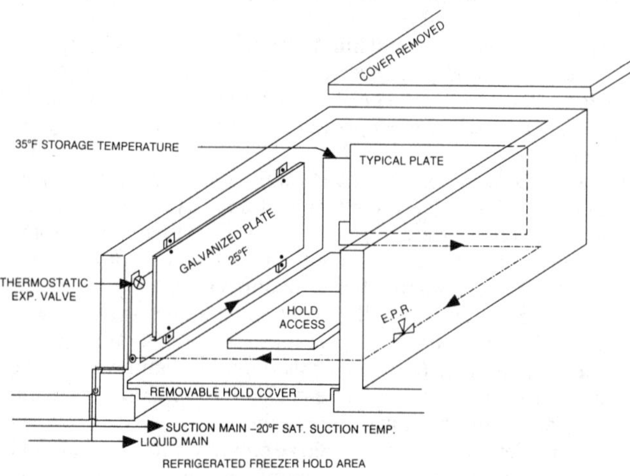

Fig. 3 Cross Section of Product Store

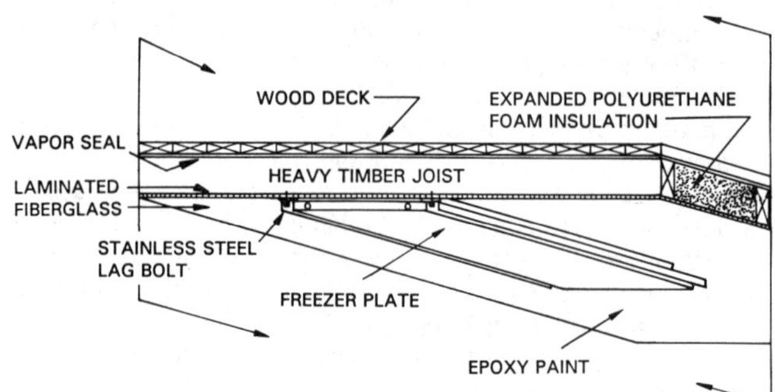

Fig. 4 Typical Under-Deck Freezer Plate Installation

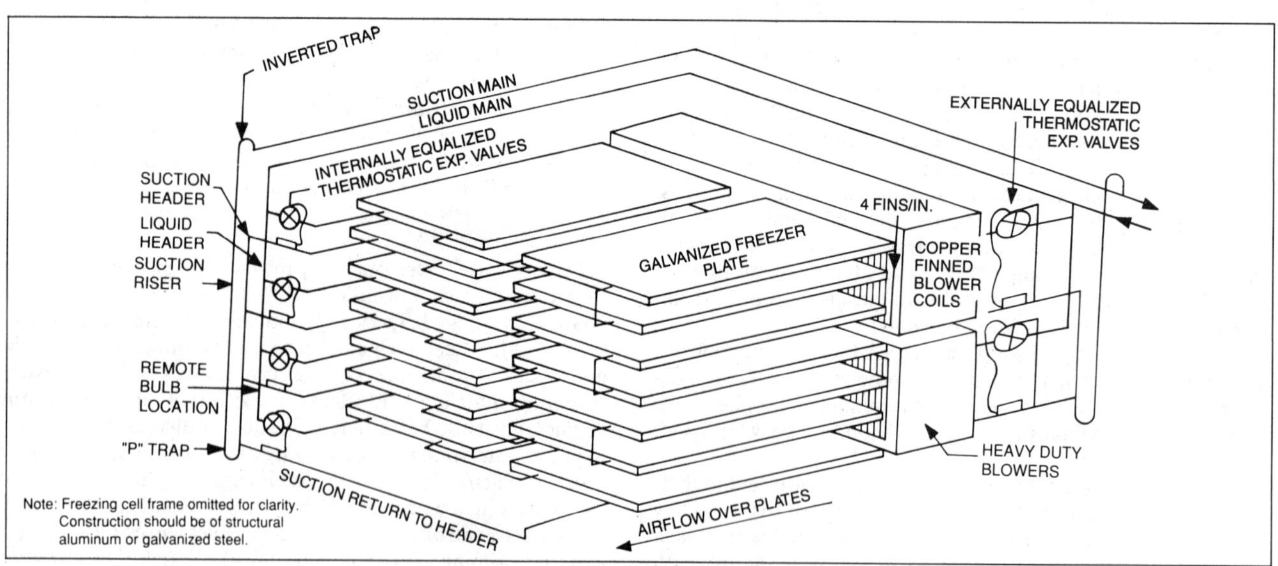

Fig. 5 Typical Marine Freezing Cell

2. A 48 in. by 48 in. freezing surface allows easy access to 3 sides and requires a minimum of deck area. It also permits the use of storage pens on both sides. The number of freezing surfaces required are as follows:

$$48 \times 48/144 = 16\ \text{ft}^2/\text{surface}$$

$$\text{Number of freezing surfaces} = 275/16 = 17.2, \text{ or } 18 \text{ freezing surfaces}$$

3. The covering plate not used for loading can be used as a heat transfer surface and must be considered in these calculations. Total heat transfer surface available is:

$$18 \times 16 \times 2 \text{ sides} = 576\ \text{ft}^2$$

$$\text{Heat absorption capacity} = UA\,(\text{TD}) = 2 \times 576 \times 20 = 23{,}000\ \text{Btu/h}$$

4. Total product load for 1500 lb of salmon frozen to −20 °F in 12 h is:

$$q = m\,[C_1\,(t_1 - t_f) + h_{if} + C_2\,(t_f - t_2)]/\theta$$

where

q = freezing load, Btu/h
m = mass of product, lb
C_1 = specific heat of product above freezing = 0.86 Btu/lb·°F for salmon
C_2 = specific heat of product below freezing = 0.39 Btu/lb·°F for salmon
t_1 = initial temperature of product = 50 °F
t_2 = final temperature of product = −20 °F
h_{if} = latent heat of product = 122.4 Btu/lb of salmon
θ = time period, h

$$q = 1500\,\{0.86\,(50 - 32) + 122.4 + 0.39\,[32 - (-20)]\}/12 = 19{,}770\ \text{Btu/h}$$

Assuming a 10% safety factor

$$q = 19{,}770 \times 1.1 = 21{,}700\ \text{Btu/h}$$

Note that the plate absorption capacity and product bond are closely correlated in still air.

Plate capacity may be increased in two ways. One way is to increase the airflow across the product, which will increase the U-factor of the plate to as high as 4 Btu/h·ft²·°F. The other way is to install a compressor that runs the system at a greater TD. This method has a higher initial cost, however. In addition, problems associated with lower suction temperatures and pressures may arise. In most cases, a fan forcing 200 to 400 cfm over the plate will increase capacity by about 20%. The fan should be turned off when freezing is no longer required.

Total Evaporator Loading. The total evaporator load, then, is the heat absorption capacity of the freezing plates and the heat gain from other sources.

$$q_e = 8{,}000 + 23{,}000 = 31{,}000\ \text{Btu/h}$$

An additional evaporator currently in use for freezing and holding, other than plates, consists of blower coils. As a general rule, these coils have to be custom made due to the size of the loads, hold configurations, and other site-related influences. They should be constructed with not more than 4 fins per inch and with no dissimilar metals to help prevent corrosion. If an all-aluminum coil is used, it should be constructed using marine grade material. Copper-aluminum construction should not be used.

The most effective blast freezing system is a plate freezing cell, sized and constructed as previously described, with hot gas defrost and equipped with a blast coil to force air over the plates. The blast coil should have about 4 fins per inch and deliver an air blast of 1000 cfm/ft² of coil face area. The air blast discharges across one-half of the plate freezing cell and returns across the other half (Figure 5).

Compressor and Drive Selection. The compressor may have two duties: (1) to hold the frozen product, and (2) to freeze the fresh product entering the hold. When not required to freeze fresh product, the evaporator surface assists in pulling the hold down to a selected temperature.

The compressor chosen for this application has a capacity rating of 40,000 Btu/h at −40 °F saturated suction temperature (SST) and +90 °F saturated discharge temperature (SDT) at 1750 rpm. At 1450 rpm, the rating is 33,000 Btu/h, which is closer to the load of this application and is preferred for longer compressor life. Manufacturers' ratings should be consulted to ensure that the compressor will operate within all specifications (*i.e.*, above minimum rpm to ensure proper lubrication and head and oil cooling, if required).

Mounting Frame. The mounting frame should prevent flexing and vibration. Special consideration should be given to diesel-driven units to prevent harmonic vibration due to their higher compression ratios. The combination of high compression ratios and the great weight of most diesel engines precludes flimsy frame construction. The material used in the frame construction of the example was 3/8 in. × 8 in. channel, well cross-braced. This base construction should be considered a minimum for all marine units.

To save space, most marine units are direct driven. The coupling must be aligned to the manufacturer's tolerances, or better, to prevent compressor damage. A good dial gage and knowledge of its use is required for this procedure.

Since a vessel has many pieces of equipment operating at various speeds and under varying load conditions, harmonic vibrations may be created. These high-frequency vibrations cannot be detected in some cases. To prevent damage, all piping and tubing connected to the compressor should have vibration eliminators. Gages mounted on the compressor should be oil-filled and equipped with antipulsation dampers. When possible, all controls, gages, switch gear, and so forth should be mounted on a remote panel away from the compressor or unit-mounting skid. A remote panel gives greater access for servicing the compressor drive system while protecting delicate components. The high noise levels in the engine room require the engineer to use ear protection devices, which makes it necessary to include an alarm system to indicate when a system shuts down. For operator convenience, the alarm, if not audible in the wheelhouse, should be equipped with a flashing indicator light. This light, along with operating pressure gages and remote reading thermometers, should be mounted in a control console within easy view of the operator.

Condenser Selection. Because condensers are exposed to the heat of the system, cold water, and the corrosive action of salt, zinc anodes need to be used as a sacrificial metal, regardless of what material the condenser is constructed.

On a small vessel, a keel cooler may be fixed to the underside of the vessel, and the refrigerant is piped directly to it. However, this system is not in wide use because, if the vessel hits a submerged object and the keel cooler is ruptured, refrigerant is lost and the system fills with sea water. Secondly, as this system is primarily used for economic reasons, they are at times undersized and do not provide enough heat transfer surface. Keel coolers are inefficient when the vessel is tied up because no water flows past the hull; they are too small to create adequate natural circulation around their surface. This situation causes a higher operating pressure and temperature and, often, a complete system shutdown.

The most widely used condensing system uses cupronickel shell-and-tube receiver condensers or tube-in-tube receiver condensers. Regardless of which condenser is selected, it must have removable heads that are accessible inside the vessel to permit inspection and cleaning. The water system should have at least two pickup points, so if one becomes plugged, the other may be of use. These intakes should be located so that the pumps will not cavitate during rough sea conditions. Easily cleaned sea water filters should be included and the operating head pressure should be controlled with a 3-way water-regulating valve.

The receiver-condenser should have sufficient capacity to hold the refrigerant charge. It should be installed in the most stable position available, with a liquid outlet at both ends. If the receiver condenser does not have sufficient holding capacity, both outlets should be piped to a vertical receiver of sufficient size to hold the additional charge. Vertical receivers with a bottom outlet always ensure that a liquid refrigerant column is available to the expansion valves to prevent flash gas regardless of the vessel's attitude. The vertical receiver should be designed into a system with a tube-in-tube condenser with no pumpdown capacity.

Another condensing system widely used in medium-sized vessels employs an oversized keel cooler to permit full operation at all times. An antifreeze (propylene glycol) and water solution is circulated through the keel cooler, using a land-based or marine condenser. Operating pressures may be controlled with a 3-way water regulator, thus bypassing excess solution around the condenser back into the keel cooler. Natural water movement allows the refrigeration system to operate at full capacity at all times, even while stationary and even over kelp beds, which may plug the water intakes of salt water cooling systems.

Oil Separators. Oil separators should be used with all low-temperature systems, especially with plate or pipe-style evaporators, because they may oil log if the suction gas velocity is restricted. Crankcase pressure regulators are required to prevent compressor and drive motors from overloading during start-up.

Suction Accumulators. Properly designed suction accumulators prevent compressor wash-outs by liquid refrigerant. The accumulator should be sized to hold at least the low-temperature evaporator operating charge. Accumulators equipped with liquid boil out coils not only help to boil off liquid refrigerant in these accumulators, but also provide some measure of liquid subcooling. Boil out coils are simply several loops of the liquid line inside the accumulator. Boil out accumulators used in conjunction with properly sized liquid suction heat exchangers increase system capacity through liquid subcooling.

Liquid Pumps. Liquid pumps, available in a full range of voltages suitable for marine refrigeration, are used on some vessels. This addition allows the system to operate at lower head pressures, while supplying a full column of liquid refrigerant to the expansion valves, thus preventing flash gas.

Mechanical Subcooling. Mechanical subcooling, which uses an additional refrigeration system to cool the liquid of the main system, provides up to 30% increase in capacity. For R-502 subcooling from 100 to 50°F, units and chiller vessels should be sized at 3500 Btu/h per ton of refrigeration capacity. For subcooling below 50°F, 4000 Btu/h per ton should be used. Where an existing system is undersized and cannot be increased for reasons such as space limitations or power limitations of the main system, subcooling becomes an important addition to any system.

Liquid Line Driers. Liquid line driers are usually of the cartridge style and are installed with bypass valves to allow quick and easy changing of the drier blocks. The drier should be placed in a visible location, preferably upstream from the liquid moisture indicator and in an area with minimum vibration. Driers should also be equipped with purge valves.

Other Considerations. Refrigerant flow to the freezing plates must be accurately controlled. There are a few methods to accomplish this. The most common practice is to use a 1-ton, internally equalized expansion valve, feeding between 30 to 50 ft^2 of plate surface area. The valve chosen should operate below −40°F.

A liquid header of 7/8 in. or larger, to which all expansion valves are attached, ensures a full column of liquid to all valves. Isolation valves should be used wherever possible to isolate various sections of the vessel's refrigerating system. Isolating areas such as the port, starboard, under-deck plates, and other sections allows partial use of the system in the event of component failures. Use of these valves should be included in the overall piping plan. All refrigerant lines should be well protected from damage and properly secured to prevent damage through vibration and various vessel movements. If clamps of dissimilar metal to the piping are to be used, the pipe must be protected from direct contact with the clamp, using a good quality rubberized gasket material. This material should be wrapped around the pipe before the clamp is secured to prevent electrolysis that would corrode the copper.

All flow control devices must be of the best possible quality, ensuring that replacement parts are available for future service. All controls, gages, and thermometers should be waterproof and capable of withstanding strenuous use. All electrical wiring must be completed by a qualified marine electrician, while all refrigeration equipment must be installed by a qualified refrigeration technician.

REFRIGERATION WITH ICE

Ice is commonly used for the preservation of groundfish, shrimp, halibut, and most other commercial species. Bin or pen boards are installed to divide the hold as desired (Figure 6). Ice is usually stored in alternate bins so that it is handy for packing around the fish as the fish is loaded into the adjacent bin. The crushed ice varies in size up to 5 in. lumps. As the fish are stowed with crushed ice, each pen is generally divided horizontally by the insertion of boards so that the bottom fish will not be crushed. The compartmentalized sections should not be more than 30 in. high if undesirable crushing and bruising are to be avoided.

The approximate amount of ice required is 1 ton for each 2 tons of fish in summer, and 1 ton to each 3 tons of fish in winter, based on a voyage of about 8 days. Less ice is needed if the ship has supplemental refrigeration.

The method of stowing the fish in ice is very important to the keeping quality. The depth of ice on the floor of the pen should be a minimum of 2 in. at the end of the voyage. This is obtained by having the initial bedding of ice 1 in. thick for each day of the voyage. In stowing, one or two layers of fish are laid on the bedding ice so that the ice is just completely covered. In no case should the layer of fish exceed 12 in. in thickness. The top covering layer of ice is about 9 in. thick, heaped up higher in the center than along the sides. This method of stowing permits the pile to adjust itself to melting and settling and results in good drainage of water and fish slime.

Most small fishing vessels are constructed of wood; the holds are not insulated. The large vessels are of steel construction with

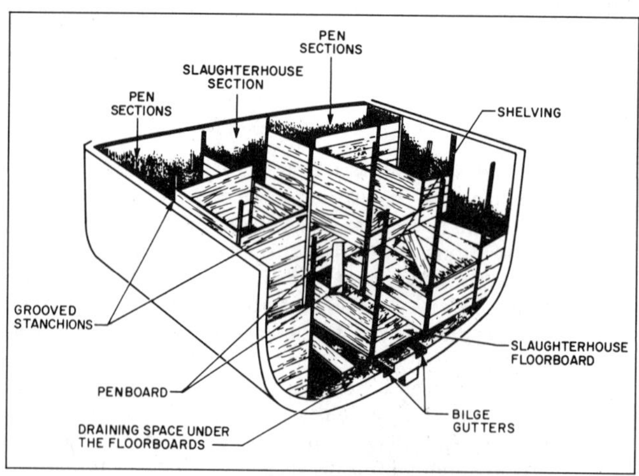

Fig. 6 Typical Layout of Pens in Hold

insulated holds lined with wood or aluminum. Mechanical refrigeration is used on some vessels to keep the ice from melting fast and to maintain lower temperatures. The most common mechanical system uses DX cooling coils under the deckhead of the hold and sometimes around the entire shell. Another system uses both a double-wall construction with cold air circulated around the entire hold between the insulated hull and deck and an inner metal lining.

REFRIGERATION WITH SEA WATER

In the United States, recirculated refrigerated sea water is commonly used instead of ice for holding industrial fish (*i.e.*, menhaden and salmon) in satisfactory condition. When used for food fish, it is most suitable for larger fish where salt penetration is not a problem. The sea water is continuously pumped either around the fish and over baffled cooling coils placed along one side of the insulated tank or through external brine coolers and then sprayed over the fish. Welded reinforced aluminum tanks and aluminum cooling coils have been used successfully. The smaller trollers require about 1 ton refrigeration capacity for each 2 tons of fish capacity, while the larger fishing vessels require about 1 ton refrigeration capacity for each 5 to 6 tons fish capacity.

Refrigerated sea water is used to preserve the highly perishable menhaden. In one commercial method of preservation, sea water is chilled at the rate of 5 gpm per ton of refrigeration from 50 to 45 °F with 37.5 °F suction at the compressor, based on cooling 200 tons of menhaden to 35 °F in 24 h at the rate of 3000 fish per ton. The calculated load of 43 tons of refrigeration includes 10% for insulation losses, so a plant of 60 tons capacity provides considerable margin for contingencies. The chilled water is sprayed over the partially filled fish hold and returned from the bottom of the hold to the chiller.

Distant water vessels are usually equipped for freezing and handling the catch at sea because they stay out for periods of 3 to 9 months, making storage with ice unfeasible. Although a large number of different vessels and freezing systems are used, the general types can be classified as either those freezing large whole fish, such as tuna and halibut, and those freezing processed and semiprocessed fishery products, such as fish blocks or groundfish in bulk lots.

The method of freezing is determined by the physical and biochemical characteristics of the fish and the desired end product. For the most part, large fish such as tuna, which are eventually canned and somewhat resistant to salt intake, are conveniently frozen in brine wells where space savings and ease of handling offer convincing benefits. Cod, haddock, hake, and similar demersal or midwater species, which are more delicate than tuna, are usually frozen rapidly either in vertical or horizontal plate freezers, as described in the previous section, or in air blast freezers.

TUNA SEINERS

Three main factors distinguish the tuna seine fishery from almost all others. These are the big hauls that may be made at any one time, the great size of the larger fish with its corresponding slow heat removal, and the high temperatures of landed tropical tunas.

A system suitable for freezing tuna must be designed to take account of these factors and also meet other general requirements. Physically, it must be reliable and permit large quantities of tuna to be moved in and out in a short time. It must be compact, accessible for service, and compatible with the other space, weight, and trim requirements of the vessel. The cost of installation and maintenance must be low enough to permit an adequate return on the investment. Finally, the system must be able to chill, hold, freeze, store, and thaw tuna under conditions that assure high quality when delivered to the cannery. Some of these factors are now considered in more detail.

For Vessels. Purse-seining is the predominant catch method used in the eastern Pacific tuna fishery. The vessel capacity may vary from 400 to 2000 tons of fish. For tuna in brine wells, the maximum storage density is approximately 50 lb/ft^3. Maximum storage density for tuna in dry refrigerated holds is approximately 36 lb/ft^3. Enough refrigeration capacity is needed to freeze fish from 70 to 0 °F in 2 to 3 days and to store the frozen fish at 0 °F in dry well.

Operational Data. Vessels can range several thousands of miles and may be required to hold fish for 12 weeks or more, although the average trip time is nearer 6 to 8 weeks. The average catch rate varies considerably. Tuna landed for canning vary in size from 5 to 130 lb. The cross-sectional dimensions of these tuna at the point of greatest girth are approximately 7 in. for 26-lb fish and 13 in. for 130-lb fish. Tropical tunas are fished in waters up to 85 °F. The temperatures of the landed fish are higher, reaching as much as 90 °F. The required rate of unloading fish from the vessel is 200 to 400 ton/24 h.

A tuna seiner is over 250 ft long, has a cruising range of 4000 nautical miles, and a hold capacity of over 2000 tons of frozen tuna. To accommodate catches of up to 2000 tons, a large seiner may have 20 brine wells (86,000 ft^3) lined with galvanized pipe coils using ammonia direct expansion refrigeration. A total of 400 tons of refrigeration is required for fish storage, ship service, cold storage room, and air conditioning. Each brine well is insulated on all sides with 6 in. foamed-in-place polyurethane. Basic refrigeration for the brine well evaporated coils is supplied by five reciprocating ammonia compressors with capacity reduction controls, horizontal oil separators, and automatic oil return.

The hold is divided into steel wells or tanks arranged on both sides of the shaft alley. The well insulation is about 5 to 6 in. thick. The wells are constructed of 5-in. wooden planking except on sides adjacent to the skin of the ship where 2-in. planking is used. The shaft alley accommodates pipelines, control valves, brine pumps, and other mechanical devices all closely arranged in a very limited space. The tanks are lined with cooling coils, and each is equipped with a brine circulating pump, sea water inlet and outlet, and connections to the brine transfer lines. Additional fuel oil is carried in some fish wells, since the permanent fuel tanks cannot carry enough fuel oil for the long trip.

The wells on large seiners are prepared for receiving fish by being filled with fresh sea water and being chilled to 29 °F with the refrigeration system. Since sea water freezes at about 28 °F, it is not practical to cool below 29 °F in the preliminary chilling operation. Before the first well is completely filled with fish, the second well is filled with fresh sea water, cooled down, and made ready for loading. With the refrigeration plant operating and the brine recirculating, the fish are chilled down to 30 °F internal temperature in 24 to 72 h. This preliminary chilling time varies with the operator, some chilling the fish as rapidly as possible and others preferring a three-day precooling period, feeling that the longer time for chilling seals the pores of the fish better and prevents excessive salt penetration, which may increase freezing time when dense brine is used in the later freezing process.

The next operation strengthens the brine so that fish may be frozen at a lower temperature. This is accomplished by dumping salt directly into the well. It requires about 100 lb of salt for each ton of fish capacity of the well. This makes dense brine of the sea water, the dense brine having a specific gravity of 1.126, a specific heat of 0.833 Btu/lb · °F, and a freezing point of 8.2 °F. A cubic foot of the dense brine and fish mixture weighs 70.5 lb, of which 50 lb is fish and 20.5 lb is brine.

Within 2 or 3 days of operation with the dense brine and additional refrigeration, the internal temperature of the fish reaches 20 °F or lower as the brine temperature is maintained at about 18 °F. The fish become rigid because about 75% or more of their water content is frozen. The dense brine solution is then pumped to another well for reuse, although in some cases it may be pumped overboard if badly contaminated with fish slime and blood. After brine freezing, the well is drained of dense brine and the temperature of the fish may be further reduced by the refrigeration coils only.

The fish are maintained in a dry frozen condition at 18°F or lower until port is almost reached. About one day from port, the circulation of sea water may be started through some of the wells so that the tuna will be sufficiently loosened from each other to start unloading. The fish must be thoroughly thawed before entering the cannery processing line. Steel buckets are lowered into the wells, and the fish are manually thrown into the buckets. Power winches lift the buckets to the wharf. The fish are unloaded into flumes, which carry them to the weighing tank and then into the cannery.

For this type of refrigeration duty, each tuna vessel generally has three or more refrigeration compressors, driven by electric motors, and three different suction lines—one for 29°F brine, one for 18°F brine, and one for holding. Each fish well and tank is connected to each of the three suction lines, and the ammonia compressors are cross-connected so that any or all of them may operate on any of the three different types of loads. Ammonia is generally used because of the desirability of parallel operation of compressors and because oil problems prevent R-12 compressors from being operated interconnected aboard ship.

For a large tuna clipper (see Figure 7), duplicate 10-in. bait pumps, each capable of delivering 2300 gpm of sea water against a 20-ft head, and one 3-in. brine transfer pump, capable of handling 300 gpm of sea water against a 50-ft head, are provided. For emergency operation, the brine transfer pump may be cross-connected on the suction and discharge sides with the ship's general service pump.

Each brine tank and each of the three bait tanks, which are also arranged for freezing fish, are provided with a 2-in. brine circulating pump with shutoff valves on inlet and outlet. Each pump circulates 200 gpm of sea water against a 20-ft head. Pumps are mounted in the shaft alley. These pumps draw from the bottoms of the tanks and discharge into the hatches above the tanks through 2.5-in. galvanized piping. This circulation of the brine improves the heat transfer of the cooling coils and makes possible the rapid, uniform cooling and subsequent freezing of the fish.

Freezer Trawler, Factory Vessels, and Mother Ships

On a worldwide basis, the largest quantity of fish frozen at sea comprises such species as cod, hake, pollock, redfish, and other demersal or midwater fish. The fish are caught by bottom or midwater trawls and may be either frozen in bulk on the vessel or processed into fish fillets or fish fillet-type blocks and then frozen. The catch rate varies with the particular fishing operation, but may be as high as 60 to 70 tons per day with averages well below 20 tons per day in most northern fisheries and 35 to 40 tons per day on an overall average. The vessels vary in size from 150 ft for the smaller bulk freezer trawlers, to over 200 ft for factory freezer vessels, and over 300 ft for the larger mother ships.

In bulk freezing aboard vessels in the United Kingdom, the fish are landed on the vessel, eviscerated, and then frozen in bulk in vertical plate freezers using R-22. A typical British bulk freezer trawler will have a length of 215 ft with a capacity of about 1800 gross tons. A vessel with this classification has a 10 to 12 station top loading, side unloading, vertical plate freezer, producing blocks of whole frozen whitefish, 42 in. long by 21 in. wide by 4 in. thick, each block weighing approximately 100 lb. The storage rate is 55 to 60 ft^3 per ton. Total freezing capacity is about 35 tons of fish per day.

In a typical operation, fish are landed on the vessel, sorted and gutted by hand, washed in a rotating cylindrical washer, and conveyed to storage bins alongside the two rows of freezers that run fore and aft on either side of the factory space. The fish are packed between the pairs of freezer plates and are reduced to a temperature of about −5°F in about 3.5 h. Hot trichlorethylene is then circulated through the evaporated plates both to help release the blocks from the freezer and to ensure that the wet fish do not stick to the plates when reloading begins. The discharged frozen blocks of fish are chuted through the insulated hatch in the factory deck to the cold storage room below, which operates at −20°F and has a capacity of about 500 tons. Trawlers of this type make voyages of 30 to 60 days, depending on the catching rate, and fish principally off Greenland, Newfoundland, and Labrador.

Factory trawlers are equipped to catch, fillet, package, and freeze the product on the vessel in a manner very similar to shoreside processing operations. A large vessel with a length of about 250 ft has refrigeration capacity sufficient to freeze 30 tons of fish fillets a day and to store 590 tons at 20°F. After being landed, the fish are stored in ice or chilled sea water and then are headed, filleted, and skinned by machine. The fillets are packed in trays and frozen in blocks, usually in horizontal plate freezers, and then transferred to the cold storage room. This pattern of operations is similar in most factory freezing vessels. Russian, Polish, and German companies have emphasized use of blast freezers, and some of the larger vessels can freeze at a rate of 100 tons a day. Some vessels may produce a combination of factory-finished products, bulk frozen fish, and iced fish that is the last day's catch.

The mother ship is a processing vessel with no catching capacity. It is the central part of a total fishing complex consisting of a fleet of fishing trawlers and fuel supply vessels that operate at considerable distances from home port. The fishing trawlers, which fish for several days, transfer their catch to the mother ship, where the fish are processed, frozen, and stored in much the same manner as on the factory vessel.

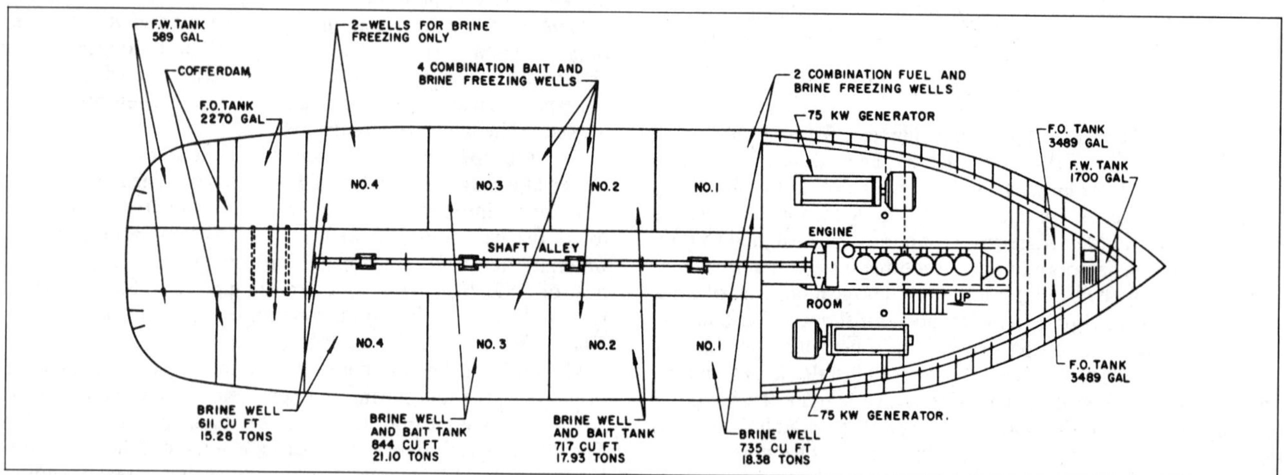

Fig. 7 Plan of Hold for Tuna Clipper

CHAPTER 30

AIR TRANSPORT

AIR freight service is provided by all-cargo carriers and passenger airlines. The latter companies also have all-cargo aircraft. Wide-body aircraft have a combination of passenger and cargo mix on the main deck, increasing cargo capacity (Figures 1 and 2). All lines maintain regularly scheduled flights so shippers may adequately plan delivery time. Special charter flights are also available from regular terminals and from airports located close to the producing areas. Payload range comparisons of wide-body jets are shown in Figure 2. Prospective shippers should contact the airlines serving their locality to obtain specific details for handling perishable shipments.

6-PALLET ARRANGEMENT

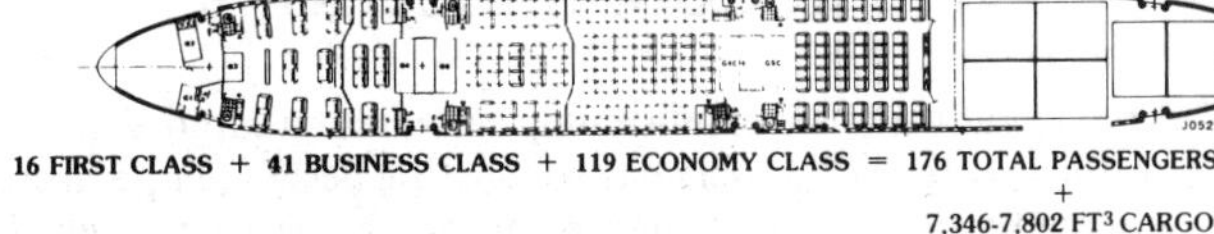

10-PALLET ARRANGEMENT

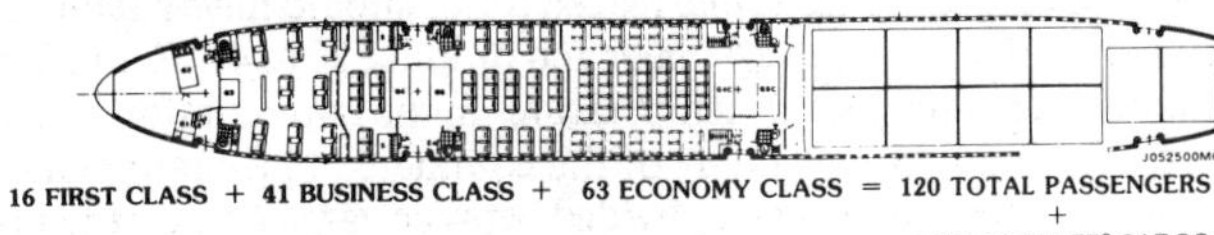

LOWER DECK ARRANGEMENT

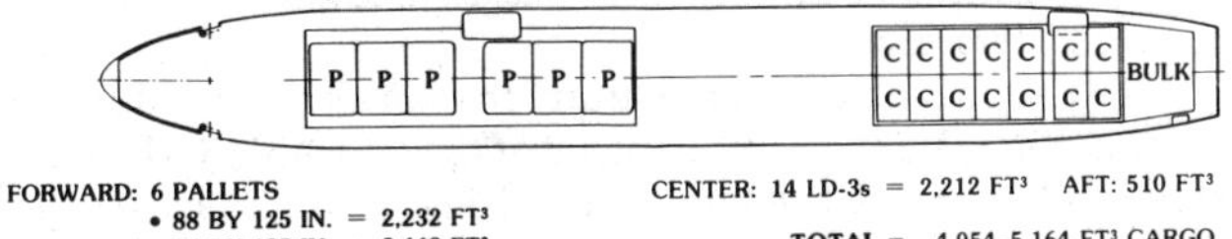

Fig. 1 Flexible Passenger/Cargo Mix

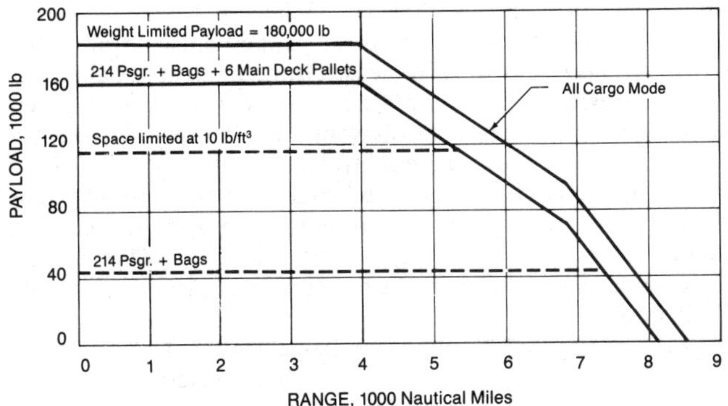

Fig. 2 Payload-Range Comparison for Wide Body Jet

The preparation of this chapter is assigned to TC 10.6, Transport Refrigeration.

PERISHABLE AIR CARGO

Some aircraft have cargo compartment temperature control with setting options from just above freezing to normal room temperature. Most compartments have a single temperature control. The control is achieved by balancing the skin heat loss with the supply of expended passenger cabin air and, when necessary, the introduction of hot jet engine bleed air through eductors. Skin heat exchangers are used to assist in maintaining the lower temperatures at high (cold) altitudes. This mode of refrigeration is not available at low altitudes or on the ground where skin temperatures can exceed the compartment temperature significantly. Refrigeration techniques for aircraft rely primarily on precooling, insulated containers, dry ice charged containers, quick handling, and short-time exposure to adverse conditions. Airports seeking to expand cargo operations are adding refrigerated warehouses internationally. The availability of refrigerated warehouses is generally the result of specific market demands and competition.

Fruits and vegetables, flowers and nursery stock, poultry and baby chicks, hatching eggs, meats, seafoods, dairy products, live animals, whole blood, body organs, and drugs (biologicals) are transported exclusively by air. Items so shipped are generally of such perishable nature that slower modes of transportation result in excessive deterioration in transit, so air movement is the only possible means of delivery. Certain early season and specialty fruits and vegetables can be flown to distant markets economically because of the high market prices when there is a short supply. Some items, such as cut flowers and papayas, arrive at distant markets in better condition than they would otherwise, so the extra transportation cost is justified. Flowers are shipped on a regular basis from Hawaii to the mainland and from California and Florida to the large midwestern and eastern cities. Air movement of strawberries has increased tremendously, including direct shipments to global destinations. Papayas are shipped from Hawaii almost exclusively by air.

When carefully handled, ice cream is shipped successfully to Japanese markets from the United States; however, some unsuccessful shipments have occurred because customs inspectors have opened containers for inspection and have taken too much time. The lowering of trade barriers has reduced the risk.

Fruits and Vegetables

All fresh fruits, vegetables, and cut flowers are alive and remain living throughout their entire salable period. Being alive, they respond to their environment and have definite limitations regarding the conditions they can tolerate. They remain alive by the process of respiration, which breaks down stored foods into energy, carbon dioxide, and water, with the uptake of atmospheric oxygen. Respiration, together with accompanying chemical changes, results in quality changes and the eventual death of the commodity. These internal changes associated with life cannot be stopped but should be retarded if a high level of quality is to be retained for a prolonged period.

Seafood

Seafood and fish also benefit from the speed of air freight. The abundance of fresh fish at restaurants and markets throughout the country is the result of air shipment.

Animals

Design of aircraft cargo compartments for animals is based on SAE *Standard* AIR 1600-85 and the Code of Federal Regulations (U.S.), Title 9. Temperature and ventilation regulations and recommendations for birds and animals of all sizes are included in these documents. Air transportation limits exposure to the extremes that would otherwise require special handling and additional cost for animal safety in accordance with the regulations.

PERISHABLE COMMODITY REQUIREMENTS

The justification for the air movement of perishable commodities is based on: (1) the time element; and (2) the delivery of a higher quality product than is possible by other modes of transportation. Better delivered quality makes possible increased returns to the shipper. This not only offsets the added transportation costs but increases consumer demand and acceptance as well. The market quality of perishable items is definitely controlled by a time and temperature relationship. Temperature cannot be ignored even for the few hours now required for transcontinental air movement. Proper temperature and humidity must be maintained at all times.

Pentzer *et al.* (1958) lists desirable transit environments for most perishable horticultural commodities. Figure 3 shows the result of a test of air shipments of strawberries from California to Chicago in a refrigerated but uninsulated container. The shipments were exposed to high ambient temperatures during ground handling at origin, resulting in fruit temperatures ranging from 50 to 60 °F instead of the desired 32 to 34 °F. These berries were compared with those shipped by rail in 4.5 days with temperatures averaging 38 °F for the transit period. Appearance and decay on delivery were about the same for both lots. Thus, the advantage of the short 22-h air movement was offset by a loss in quality arising from the unfavorable temperature.

Top quality of many of the most perishable commodities can be significantly reduced by only a few hours' exposure to unfavorably high temperatures. Many drugs (biologicals) and other items, such as whole blood, can be rendered completely ineffective or toxic if not kept at the specified low temperature.

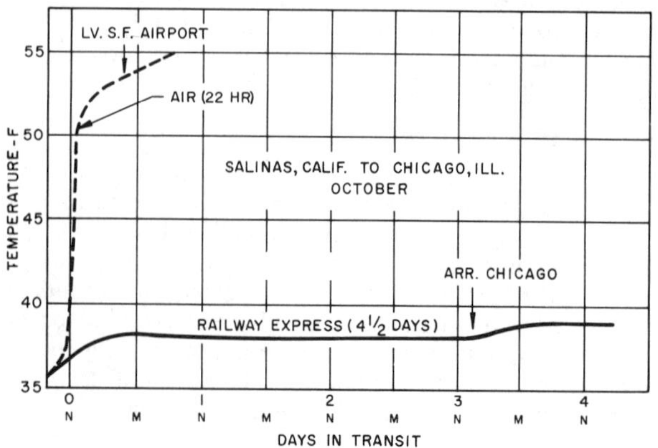

Fig. 3 Temperature of Strawberries Shipped by Air and Rail

Some flowers, fruits, and vegetables respond favorably to reduced oxygen levels, increased amounts of carbon dioxide, or both, which could be maintained by gastight packaging or containers.

The maintenance of temperatures near freezing is not desirable for all products because some are subject to chilling injury when they are exposed to temperatures well above this point. Chilling injury is most pronounced in tropical products, such as bananas, tomatoes, cucumbers, avocados, and orchids. Temperatures above 55 °F are usually safe for cold-sensitive commodities. Other items, such as cut flowers, require temperatures between 32 and 55 °F.

Certain fruits and vegetables require humidity control. Humidity should be kept between 85 and 95% to prevent wilting and general loss of water. The relative humidity in the cabin of an airplane flying at about 40,000 ft is generally less than 10%. However, the respiration of fruits and vegetables, placed in closed containers with recirculated cooling air, should produce the required humidity level with no additional water added.

Certain vegetables, such as peas, broccoli, lettuce, and sweet corn, have high respiration rates and the heat produced may amount to the equivalent of 250 lb or more of ice meltage per ton of vegetables per day at 60 °F. In designing the refrigeration systems for aircraft containers, the additional evaporator capacity required to handle the heat of respiration should be considered.

DESIGN CONSIDERATIONS

A refrigeration or air-conditioning system for a cargo airplane or airborne cargo containers has conflicting design temperature requirements, depending on the type of cargo to be carried, which makes it difficult to use one optimum refrigeration system for all kinds of cargo. For example, frozen foods should have a temperature of 0 °F or lower, fresh meat and produce 30 to 45 °F, and live animals generally require temperatures in the same comfort range as for humans. Today, many of the commercial jet cargo planes operate with the main cabin divided between cargo compartments and passenger compartments, and they are supplied by a single air system controlled to the comfort of the human occupants. In this case, perishable cargo must be packed in containers, insulated, and iced or precooled.

The design ambient temperatures that an airplane will experience in flight are given in Chapter 9 of the 1991 ASHRAE *Handbook—Applications*. A cargo jet cruising close to Mach 0.9, has an increase in skin temperature over ambient of about 50 °F. With an all-cargo load, the basic air-conditioning systems are capable of maintaining main cargo compartment temperatures on a design hot day from 40 °F at 30,000 ft to 30 °F at 40,000 ft.

The air-conditioning systems are equipped with controls to prevent freezing of moisture condensed from the air at low altitudes. With the extremely dry air prevailing at cruise altitudes, an override of these anti-icing controls would permit an even lower cabin temperature, although it is doubtful that storage temperatures of frozen goods could be met. Thus, some insulation would still be required in frozen food containers. Further, the airplanes are often required to hold at relatively low altitudes of 20,000 ft or less, because of heavy traffic at the busier airports, for periods of 30 min or more.

Permanent attachment of a mechanical refrigeration system to a cargo container may not be desirable for several reasons: increased load, reduction in usable volume, and difficulty in rejecting the condensing unit heat load overboard. These objections are particularly applicable to containers carried in the main cargo hold. On the other hand, permanently attached units would permit refrigeration of just part of the cargo load while the remainder could be held at temperatures in the normal human or animal comfort zone. Temperature control of products requiring widely differing transit and storage temperatures would be more feasible with onboard refrigeration units.

SHIPPING CONTAINERS

Fruits and vegetables are generally shipped in the same containers used for surface transportation: wooden boxes, veneer crates of various types, or fiberboard cartons. Most flower containers are constructed of either plain or corrugated fiberboard materials. Wooden cleats are used for bracing material, generally as dividers or corner braces inside the cargo box. Where lading may be exposed to very cold surfaces, external cleats may be used as spacers to prevent direct contact. Certain flowers such as gardenias and orchids may be packaged in individual cellophane-wrapped boxes or trays and placed in a master container. Any tightly sealed film wrap must be perforated by at least one small hole to permit release of air from the container during ascent to high altitudes.

Containers, built on pallets and shaped to make maximum use of the interior airplane volume, are in use. Containers presently in use with the airlines are described in the IATA Register. Containers for aircraft, except for the belly cargo holds, are not shaped to make maximum use of the interior volume of the airplane. One reason for this is that the individual packages that will fill the containers are generally rectangular in shape anyway. Another reason is to permit easier intermodal transport, *e.g.*, from motor truck to the airplane and vice versa. Because of the size of aircraft loading doors and irregular aircraft cross-sections compared to surface vehicles and vessels, containerization may require this compromise.

Containerization is a system of moving goods in sealed, reusable freight containers too large for manual handling, and which do not have wheels permanently attached. The advantages are: far less cargo damage and pilferage, lower packaging costs, minimized handling, and lower shipping rates. Presently, these containers may be loaded at the air freight terminal or loaded at the shipper's facility and transported by flatbed truck-trailer or railroad, or both, to the air terminal.

The critical condition for design of insulation and refrigeration systems (detachable plug-in type or permanently installed) for cargo containers is the time that the container is on the dock in the hot sun waiting for shipment. For this condition, an ambient temperature of 100°F db is assumed. The average outside skin temperature of an unpainted metal container is about 115°F.

Under the conditions just mentioned, the 8 ft by 8 ft by 10 ft container with 0.5 in. of high efficiency insulation (recirculating the air and considering no latent load) requires about 18,000 Btu/h of refrigeration to maintain 35°F inside and about 24,000 Btu/h to maintain 0°F inside. For quick pulldown to these temperatures of the container and fresh perishable contents (assuming prefreezing of frozen products), the capacities should increase by about 50%.

Fresh fish, shrimp, and oysters may be packed in boxes, barrels, or special containers. Proper precaution must be taken to prevent drippage from melted ice into the cargo space. Live lobsters are packed in insulated containers with salt water seaweed. Frozen foods are always packed in insulated containers. Whole blood is shipped in specially developed containers. Insulated bags are also used.

The configurations and dimensions of two insulated containers are shown in Figure 4. Insulated with closed-cell, rigid plastic foam, the containers are a fabricated sandwich structure and are sized to fit conventional pallets and materials handling systems. The heat transfer rate for the entire standard container is 28 Btu/(h·°F) and 32 Btu/(h·°F) for the commercial size. More recent

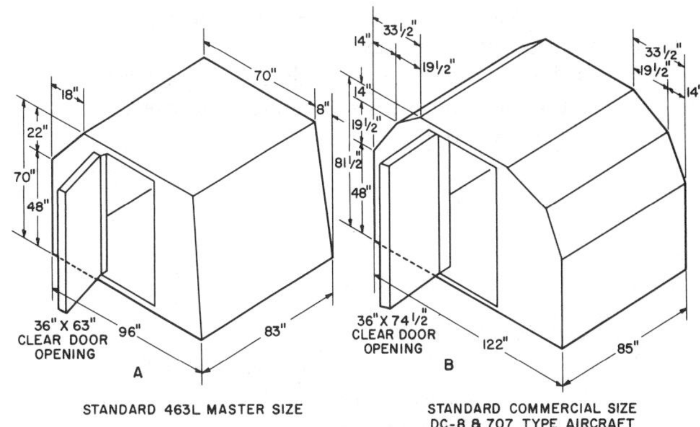

Fig. 4 Insulated Containers Designed to Fit Configuration of Cargo Aircraft

aircraft such as the MD-11 depicted in Figure 1, use pallets 125 in. wide by 64.4, 88, and 96 in. which may be loaded to 64 in. high and retained by straps. Load capacities are 6700 lb, 10,000 lb, and 11,000 lb, respectively. Containers are LD-3 half width, and LD-6 full width, the latter having twice the capacity and width. Both are 60.4 in. deep and 64 in. high. The LD-3 width is 79 in., the volume is 158 ft^3, and the capacity is 3300 lb. A plug-in portable mechanical refrigerating unit may be positioned in the doorway for standby operation. Tight construction permits controlled atmosphere application. A smaller shipping unit, insulated with a foamed plastic, has inside dimensions of 45 in. by 21 in. by 24 in., *i.e.*, a total area of 35.1 ft^2 and a capacity of 13.1 ft^3. The heat transfer rate of the entire container is 2.8 Btu/(h·°F).

TRANSIT REFRIGERATION

Many commodities must receive refrigeration in transit. In most cases, this is accomplished by a refrigerant in the package. Water ice, dry ice, and certain proprietary refrigerants are used. Because no method of transit refrigeration can economically cool a warm commodity to its desired transit temperature, all perishable items must be cooled before shipment. Flowers generally are wrapped in several layers of paper or light insulating material and kept cool by water ice. The ice (solid, chopped, or flaked) is placed in a plastic bag or wrapped in many layers of paper and tied to one of the cleats of the container. In some instances, the block of ice may be chilled to 0°F in a freezer before putting it in the package, thereby obtaining a slightly greater refrigeration capacity. Newspapers are sometimes wadded up and thoroughly wetted and then frozen to 0°F or lower. In all cases, the paper helps absorb the ice meltage water and reduces the chances for leakage into the cargo space.

Some voids should be left in the containers to permit air circulation and uniform cooling. Boxes should be sealed to prevent air exchange. Placing the ice or water for freezing in sealed plastic bags eliminates drippage, but melting ice in open containers increases humidity, which is particularly desirable for cut flowers. A packaged refrigerant with no escape of free liquid must be used with commodities that would be damaged by water. Water ice acts as refrigerant in special-type blood containers.

Dry ice is used extensively with frozen products and fresh strawberries, the amount depending on the type of container and the length of the journey. Sometimes it is placed in with water ice, not only for its own refrigerating value, but also to slow down meltage of the water ice and extend its value to the end of the transit period. Dry ice alone is seldom used for flowers because its very low temperature may cause freezing damage to adjacent blooms if not properly spaced or insulated. The use of large amounts of dry ice may cause a buildup of carbon dioxide gas in concentrations dangerous to humans and animals unless proper ventilation of the cargo compartment is provided.

When the heat transfer rate of insulated shipping containers is known, the amount of refrigeration required can be estimated with reasonable accuracy from

$$Q = HD\Delta t$$

where

Q = total heat transfer, Btu
H = heat transfer rate of entire container, Btu/(h·°F)
Δt = difference between ambient temperature and that at which product is to be carried, °F
D = duration of transit, h

For example, assume the small container previously described holds 15 standard strawberry trays, each holding 13 lb of berries (195 lb total), with the fruit cooled to and carried at 35 °F at an ambient temperature of 75 °F for a transit time of 24 h:

$$H = 2.8 \text{ Btu/(h·°F)}$$

$$\Delta t = 75 - 35 = 40 \text{°F}$$

$$D = 24 \text{ h}$$

$$Q = (2.8 \times 40 \times 24) = 2688 \text{ Btu}$$

The heat of respiration generated by the berries at 35 °F is about 4000 Btu/(ton·24 h) (see Table 2 in Chapter 30 of the 1993 ASHRAE *Handbook—Fundamentals*). For 195 lb of berries 24 h in transit:

$$4000 \times (195/2000) = 390 \text{ Btu}$$

The ice required to absorb this heat is

$$(2688 + 390)/144 = 21.4 \text{ lb}$$

The amount of dry ice required would be about 11.9 lb.

These simple calculations can be made only when the thermal conductance or heat-transfer rate of the container is known. It would therefore be of considerable value to all concerned—shipper, carrier, and receiver—to have this factor determined for all insulated shipping containers and clearly displayed. Such ratings have been made on truck-trailer bodies, as discussed in Chapter 27.

When package refrigeration is not available, rapid warmup can be retarded by insulated containers or blanket insulation over stacks or pallet loads. This method has been satisfactory with some of the less perishable fruits such as peaches. Temperatures can be maintained for several hours in flight with the proper use of these blankets. Care must be exercised in loading to ensure that insulating material is wrapped completely around the cargo and that containers are not in direct contact with hot or cold surfaces. Some cargo compartments on passenger aircraft may be cooled by the air-conditioning system, but the temperature will not be in the optimum range for most perishable commodities.

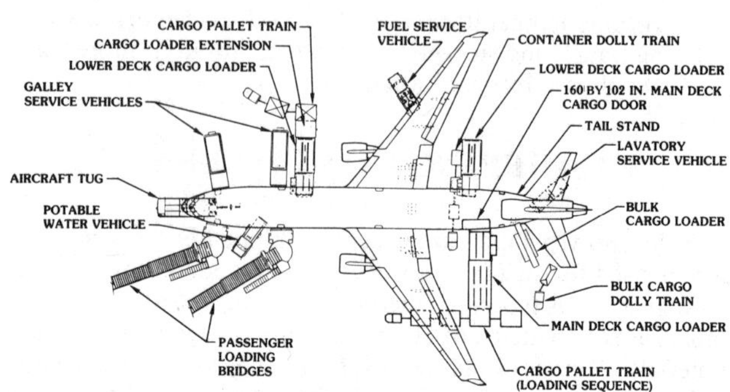

Fig. 5 Ground Service Equipment Arrangement

GROUND HANDLING

All the advantages of speed can be lost if the shipper, carrier, and receiver do not follow good handling practices that keep deterioration to a minimum.

Ground handling can amount to over 70% of the total elapsed time from shipper to receiver. To reduce this ground time, load palletization and special pallet carriers and loaders, in conjunction with improved load-handling systems aboard the aircraft, are used. Air freight terminals are now designed and built to use these new handling techniques. Combination cargo/passenger jets present unique loading techniques. A typical ground service equipment arrangement is shown in Figure 5.

Fast pickup and delivery are also essential. Because of the generally high ground temperatures at shipping point terminals and intermediate points, most perishable agricultural commodities should be cooled as soon after harvesting as possible and delivered to the air terminal in properly refrigerated vehicles, particularly if they are shipped in uninsulated containers. At the terminal, these shipments must be held at proper temperatures if prompt loading from the pickup vehicle is not possible. Holding rooms, refrigerated mechanically or by ice, should be provided. During seasonal loading peaks, refrigerated trucks or trailers may be used as temporary holding rooms. During hot weather, cargo space must be cooled before loading. Portable air-conditioning equipment, such as that used for passenger aircraft, is used.

The airlines have developed rules for handling various perishable commodities. These include the temperatures desired in transit, the amount of seasonal protection needed, loading methods for various types of containers, and other factors involved in proper handling.

REFERENCES

International Airport Association. Updated annually. IATA *Register of containers and pallets*. Montreal.

Pentzer, W.T., Jr., *et al.* 1958. Air transportation of fruits, vegetables and cut flowers: Temperature and humidity requirements and perishable nature. USDA, AMS Report No. 280.

SAE. 1985. Animal environment in cargo compartments. *Standard* AIR 1600-85. Society of Automotive Engineers, Warrendale, PA.

Taylor, W.P. 1990. In Proceedings of the 25th Intersociety Energy Conversion Engineering Conference 4:285-87. IEEE Catalog No. 90CH2942-1.

Tyree, L., Jr. 1973. Refrigerated containerized transport for "jumbo" jets. *Progress in Refrigeration Science and Technology* 4:515-25. AVI Publishing Company, Westport, CT.

USDA. 1993. *Animal Welfare*. In Chapter 1 of Code of Federal Regulations, Title 9, Animals and Animal Products.

RETAIL FOOD STORE REFRIGERATION

REFRIGERATION equipment used in self-service retail food stores may be broadly grouped into display refrigerators, storage refrigerators, processing refrigerators, mechanical refrigeration machines, and refrigeration. Chapters 46 and 47 present this equipment.

DISPLAY REFRIGERATORS

Open, self-service refrigerated display cases for medium and low temperatures are widely used in food markets. However, glass-door multideck models have rapidly gained in popularity, particularly for low-temperature applications. *Deck* is an industry term for a shelf, pan, or rack that supports the displayed product. Many operators combine single-deck and multideck models in most departments where perishables are displayed and sold. Sometimes, closed service refrigerators are used to display unwrapped fresh meat, delicatessen food, and, frequently, fish on crushed ice, instead of or supplemented by mechanical refrigeration. More complex layouts of display refrigerators have developed as new or remodeled stores strive to be distinctive and more attractive. Refrigerators are allocated in relation to expected sales volume in each department, such as meat or dairy. Thus, floor space is allocated to provide balanced stocking of merchandise and smooth flow of traffic in relation to expected peak volume periods.

The small store accommodates a wide variety of merchandise in a limited floor space. Thus, display refrigerators installed in small and medium stores tend to be the glass reach-in type, which can display 50 to 100% more quantity and variety of merchandise in the same amount of floor space as single-deck refrigerators. This concentration of large refrigeration loads in a small space makes year-round temperature and humidity control essential.

Product Temperatures

Display refrigerators are designed to merchandise food to maximum advantage, while providing short-term protection (24 to 72 h). Table 1 lists temperatures for the maintenance of the product.

Display refrigerators are not designed to cool the product; they are designed to maintain product temperature. The merchandise, when put into the case, should be at or near the proper temperature. Therefore, adequate refrigerated storage in another space is needed. Food placed directly in the refrigerator on delivery to the store should come from properly refrigerated trucks. Little or no delay in transferring perishables from storage or trucks to the display refrigerator should be permitted.

Display refrigerators should be loaded properly. The product on display should never be piled so high that it is out of the refrigerated zone or be stacked so that circulation of refrigerated air is blocked. The load line recommendations of the manufacturer must be followed to obtain good refrigeration performance. Proper refrigerator design and loading minimize energy use, maximize the efficiency of the refrigeration equipment, and minimize product loss.

The preparation of this chapter is assigned to TC 10.7, Commercial Food and Beverage Cooling, Display and Storage.

Table 1 Temperatures in Display Refrigerators

Type of Fixture	Temperatures[a] Minimum[a], °F	Maximum[b], °F
Dairy		
Multideck	36	38
Produce, Packaged		
Single-Deck	35	38
Multideck	35	38
Meat, Unwrapped (Closed Display)		
Display area	36[b]	b
Deli smoked meat		
Multideck	32	34
Meat, Wrapped (Open Display)		
Single-Deck	24	26
Multideck	24	26
Frozen food		
Single-Deck	c	−13[d]
Multideck, Open	c	−10[d]
Glass door reach-in	c	−5[d]
Ice cream		
Single-Deck	c	−24[d]
Glass door reach-in	c	−12[d]

[a]These temperatures are air temperatures, with the thermometer in the outlet of the refrigerated airstream and not in contact with the product displayed.
[b]Unwrapped fresh meat should only be displayed in a closed, service type display case. The meat should be precooled to 36°F internal temperature prior to placing on display. The case air temperature should be adjusted to keep the internal meat temperature at 36°F for minimum dehydration and optimum display life. Display case air temperature varies with manufacturer.
[c]Minimum temperatures for frozen foods and ice cream are not critical (except for energy conservation); maximum temperature is important for proper preservation of product quality.
[d]The differences in display temperatures among the three different styles of frozen food and ice cream display cases are a result of the orientation of the refrigeration air curtain and the size and style of the opening.
The single deck has a horizontal air curtain and opening of 30 in.
The Multishelf, Open has a vertical air curtain and opening of 42 to 50 in.
The glass door reach-in has a vertical air curtain protected by a multipane insulated glass door.

Store Ambient Effect

Display fixtures are affected significantly by temperature, humidity, and movement of surrounding air. These display cases are designed primarily for supermarkets, virtually all of which are either air conditioned or in cool climates.

The application engineer needs to verify that the year-round store ambient conditions are within the performance ratings of the various cases selected for the store. Usually, the satisfactory performance of display case refrigerators is seriously affected when store conditions exceed 75°F and 55% rh, which defines a dew

Table 2 Relative Refrigeration Requirements with Varying Store Ambient Conditions

	70 °F Dry-Bulb Temperature Relative Humidity, %					78 °F Dry-Bulb Temperature Relative Humidity, %		
Case Model	**30**	**40**	**55**	**60**	**70**	**50**	**55**	**65**
Multideck dairy	0.90	0.95	1.00	1.08[a]	1.18[b]	0.99	1.08[a]	1.18[b]
Multideck low temperature	0.90	0.95	1.00	1.08[a]	1.18[a]	0.99	1.08[a]	1.18[b]
Single deck low temperature	0.90	0.95	1.00	1.08[a]	1.15	0.99	1.05	1.15
Single deck red meat	0.90	0.95	1.00	1.08[a]	1.15	0.99	1.05	1.15
Multideck red meat	0.90	0.95	1.00	1.08[a]	1.18[b]	0.99	1.08[a]	1.18
Low temperature reach-in	0.90	0.95	1.00	1.05[a]	1.10[a]	0.99	1.05[a]	1.10

Note: Package warm-up may be more than indicated. Standard flood lamps are clear PAR 38 and R-40 types.
[a]More frequent defrosts required
[b]More frequent defrosts required plus internal condensation (not recommended)

point of 57.5 °F and a humidity ratio of 0.0102 lb of moisture per pound of dry air. The refrigeration load for food store display cases is normally rated at store ambient conditions of 75 °F and 55% rh. When store ambient relative humidity is different from that at which store display cases are rated, the energy requirements for case operation will vary. Howell (1993) concludes that display case energy savings at 35% store relative humidity rather than 55% ranged from 5% for glass door reach-in cases to 29% for multishelf deli cases. Table 2 lists correction factors for the effect of store relative humidity on display case refrigeration requirements when the dry-bulb temperature is 70 and 78 °F.

Additional savings can be achieved by controlling anticondensate heaters and reducing defrost frequency at relative humidities below 55%. Energy savings credit for reduced display case antisweat heaters can only be taken if the display cases are equipped with humidity-sensing controls that reduce the amount of power supplied to the heaters as the store dew point decreases. Also, defrost savings can be considered when a demand type of defrost control is used. This control reduces the frequency of defrost as the store relative humidity decreases. Individual manufacturers have specific anticondensate and defrost values for their equipment.

Since relative humidity varies throughout the year, the dew point for each period should be analyzed. The sum of these values provides the total annual energy consumption. In a store designed for a maximum relative humidity of 55%, the air-conditioning system will dehumidify only when the relative humidity exceeds 55%.

In climates where the outdoor air temperature is low in winter, infiltration of outdoor air can cause store humidity to drop below 55% rh. Separate calculations need to be done for periods during which mechanical dehumidification is used and periods when it is not required. As an example, in Boston, Massachusetts, mechanical dehumidification is required for only about 3 1/2 months of the year, while in Jacksonville, Florida, it is required for almost 7 1/2 months of the year. Also, in Boston, there are 8 1/2 months when the store relative humidity is below 40%, while Jacksonville has these conditions for only 4 1/2 months. The engineer must weigh the savings at lower relative humidity against the cost of the mechanical equipment required to maintain relative store humidity levels at, for example, below 40% instead of 55%.

Adverse Heat Sources

Lighting. Refrigeration performance and food temperatures are also adversely affected by display lighting and other forms of radiant heat. High-intensity lighting raises product temperatures several degrees, as shown in Figure 1. Light also discolors smoked, cured, and table-ready meats. No more than 70 footcandles of

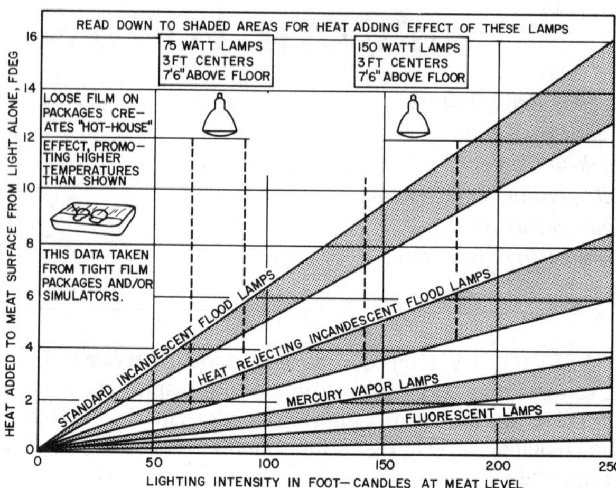

Note: Package warmup may be more than indicated. Standard flood lamps are clear PAR 38 and R-40 types. Blue, pink, or other hues in these types reduce the amount of light, but not the heat load.

Fig. 1 Heating Effect of Lighting on Packaged Meat in Open Display Refrigerators

display lighting should be used on smoked, cured, and table-ready meats unless they will be sold within 24 h. Discoloration is directly proportional to both light intensity and length of time displayed.

The heat rejecting lamp is questionable, and conventional incandescent floodlights and supplemental incandescent lighting should never be used to display retail meats.

Ceiling Temperatures. An 80 °F ceiling can raise the surface temperature of meat in a display case 3 to 5 °F. A 100 °F ceiling can raise the meat surface temperature 4 to 8 °F. Proper ceiling and light fixture design minimizes this problem.

Packaging. The surface temperature of a loosely wrapped package of meat with an air space between the film and surface may be 2 to 4 °F above the ambient temperature. This is called the greenhouse effect.

Load Lines and Loading. The following observations indicate the effect of loading:

- Voids in the display raise the surface temperature of a package in front of a void 2 to 6 °F.
- For all makes and styles of display refrigerators, keeping the top of the product 2 in. below the maximum load line improves the surface temperatures 2 to 6 °F.

If meat has been handled sanitarily until it is placed in the display case, elevated temperatures can be more tolerable. When the meat surfaces are contaminated by dirty knives, meat saws, table tops, and the like, even optimum display temperatures will not prevent premature discoloration and subsequent downgrading of the meat. The elimination of excessive heat from lighting, high ceiling temperatures, and poor display practices should be required in addition to strict sanitation practices.

MEAT PROCESSING

In a self-service meat market, the meat preparation operations of cutting, wrapping, sealing, weighing, and labeling involve precise production control and scheduling to meet varying sales demands. The faster the processing, the less critical the temperature and corresponding refrigeration demand.

The preparation room should not be too dry, but condensation on the meat (providing a medium for bacterial growth) should be avoided by maintaining a reasonable dew-point temperature. Low-velocity fan coil units are generally used. Fan coil units should be selected with a maximum of 10 °F temperature difference between the entering air and the evaporator temperature. Gravity coils are also available and have the advantage of lower room air velocities.

The meat preparation area is generally cooled to about 45 to 55 °F, which is desirable for the personnel, but not low enough for meat storage. Thus, meat should be held in that room only long enough for the cutting and packaging operation; then, as soon as is possible, it should be moved to a packaged product storage cooler held at 28 °F.

The meat preparation room may be a refrigerated room adjacent to the meat storage cooler or one compartment of a two-compartment cooler. In such a cooler, one compartment is refrigerated at about 28 to 32 °F and used as a meat storage cooler, while the second compartment is refrigerated at about 45 °F and used as a cutting and packaging room. However, best results are attained when the meat is cut and wrapped in a temperature of 28 to 32 °F.

Wrapped Meat Storage

At some point between the preparation room and the display refrigerator, refrigerated storage for the wrapped cuts of meat must be provided. Without this space, a balance cannot be maintained between the cutting-packaging rate and the selling rate for each particular cut of meat. Display refrigerators with refrigerated bottom storage compartments, equipped with racks for holding trays of meats, offer one solution to this problem. However, the amount of stored meat is not visible, and the inventory cannot be controlled at a glance.

A second method uses a pass-through, reach-in cabinet. This cabinet has both front and rear insulated glass doors and is located between the preparation room and the display refrigerators. After wrapping, the meats are passed into the cabinet for temporary storage at 28 °F and then withdrawn from the other side for restocking the display refrigerator. Since these pass-through cabinets have glass doors, the inventory of wrapped meats is visible and, therefore, controllable.

The third and most common method uses a section of the back-room walk-in meat storage cooler. The cooler is usually equipped with rolling racks into which the trays of meat can be slid. This method of storage also offers visible inventory control.

A fourth and preferred method is a completely separate packaged meat storage cooler, which has convenient access to both the preparation room and the display cases. The overriding philosophy in successful meat preparation and merchandising can be summarized as: Keep it clean, keep it cold, and keep it moving.

WALK-IN COOLERS

Walk-in coolers are required for the storage of meat, fresh produce, dairy products, frozen food, and ice cream. Medium and large stores have separate produce and dairy coolers usually in the 35 to 40 °F range. Meat coolers are used in all food stores with storage conditions between 28 and 32 °F. Meat, fish, and poultry should each be stored in separate coolers to prevent odor transfer.

Moisture conditions must be confined to a relatively narrow range because an excessive humidity level encourages bacteria and mold growth, which leads to sliming. A level that is too low leads to excessive dehydration.

Air circulation must be maintained at all times to prevent stagnation, but it should not be so rapid as to cause drying of an unwrapped product. Forced-air blasts must not be permitted to strike products; therefore, low-velocity coils are recommended.

For optimum humidity control, unit coolers should be selected with a 10 °F temperature differential (TD) between entering air temperature and evaporator temperature. Note that the published ratings of commercial unit coolers do not reflect the effect of frost accumulation on the evaporator. The unit cooler manufacturer can determine the correct frost derating factor for its published capacity ratings. From experience, a minimum correction multiplier of 0.80 is typical.

A low-temperature storage capacity equivalent to the total volume of the low-temperature display equipment in the store is satisfactory. Storage capacity requirements can be reduced by frequent deliveries.

Generally, forced-air coils are selected for low-temperature coolers where humidity is not critical for packaged products. For low-temperature coolers, gas or electric defrost is required. Off-cycle defrosts are used in produce and dairy coolers. Straight-time or time-initiated, time or temperature terminated gas or electric defrosts are generally used for meat coolers.

REFRIGERATORS AND SYSTEMS

Food stores sell all types of perishable foods that require a variety of refrigeration systems to best preserve and most dramatically display each product. The variations in refrigerators require different evaporator/pressure requirements ranging from the highest for the meat processing room to the lowest for the ice cream level. Produce and meat preparation rooms may approach the suction pressures used in air-conditioning applications. Open ice cream display cases may have suction pressures as low as −40 °F. All other cases and coolers fall between these pressures.

Temperature controls also vary greatly, from a produce preparation room (which may operate with a wet coil) requiring no defrost to the ice cream case requiring induced heat to defrost the coil periodically. Electronic sensor control provides the most accurate control and can also provide temperature alarm to prevent food loss. Various defrost methods include (1) off-time defrost, (2) gas defrost, (3) electric defrost, and (4) defrost using ambient air induced into the refrigerator.

Various Refrigeration System Design Solutions

The selection of high-side equipment used to operate display refrigerators and storage rooms for food stores involves the following considerations: (1) cost/space limitations, (2) reliability, (3) maintainability and complexity, and (4) operating efficiency. Solutions span the simplest one compressor and associated controls on one refrigerator to the complex central refrigeration plant operating all refrigerators in a store.

Condensing-unit systems, which use single compressors with multiple cases or coolers, provide a popular, suitable approach. This method places produce, dairy, meat, or frozen food refrigerators in groups, each on a single compressor, which may have its own condenser or may have an air- or water-cooled remote condenser.

Another common refrigeration technique couples two compressors in parallel. Oil distribution, temperature control at more than one suction pressure, and defrosting must be considered. Parallel units usually operate connected to one or more large condensers. The condensers are usually remote air cooled or remote evaporative cooled, but they can also be built as part of the compressor rack assembly.

Carrying the sequence one step further leads to a large assembly tying three to nine compressors in parallel. The same needs as defined for two units in parallel apply to the larger styles. However, efficiency may be sacrificed when greatly differing suction pressures are grouped in one suction pressure designed system. Split suction manifolds separate load groups on a rack. To match changing evaporator loads, rack capacity is varied by cycling compressors, varying the speed of one or more compressors, and/or unloading compressor cylinders or changing valves.

Running all the refrigeration equipment at the low-suction pressure required for ice-cream on low-temperature systems and for meat on medium-temperature systems is inefficient. In large parallel systems, the ice cream and meat refrigeration are frequently isolated. Single compressor satellites tied into the parallel compressors are often used for small ice cream or meat loads, while split suction, parallel equipment is commonly used for larger loads. In both systems, different suction pressures are obtained, but all compressors discharge into a common header. Moreover, the refrigerating system must be highly reliable because it must keep operating 24 h a day for ten or more years, protecting the large investment in highly perishable foods.

LOAD VERSUS RATINGS

Food store refrigerator manufacturers publish case load ratings to match the proper condensing unit with fixture load. For single compressor applications only, the ratings can be stated, for selection convenience, as the capacity the condensing unit must deliver at an arbitrary suction pressure (evaporator temperature). In general, manufacturers of open display refrigerators use ASHRAE *Standards* 72-1983 RA and 117-1992, which specify a standard refrigeration load requirement at a rating condition of 75 °F and 55% rh in the sales area. Display refrigerators are a specialty, so manufacturers' recommendations must be followed to achieve proper results in both efficiency and product integrity.

Multiplexed Systems

When dissimilar display case refrigerator loads are connected to the same condensing system, the manufacturer should be consulted to determine the true maximum suction pressure at the fixture refrigerant line outlet for each system and the true load that such a system adds to the total. The condensing unit must deliver the total of all the loads at a suction pressure at the machine no higher than the lowest system pressure requirement, less the suction line pressure drop. The other systems must have suction line restriction (by evaporator pressure regulators or some other sure means) to prevent higher temperature evaporators from adding unnecessarily to the load. The pressure regulator then assures that the condensing unit will pull down to the pressure necessary for the lowest evaporator temperature required.

Electronically or mechanically actuated pilot-operated evaporator pressure regulation valves control evaporator pressures in multiplexed systems. These valves cause little or no measurable pressure drop when they are in the full open position. Another design uses liquid or discharge pressure to open and close the EPR valves, so the pressure drop is negligible.

The suction gas temperature leaving display fixtures is often superheated. Particularly on low-temperature fixtures, the suction line gas temperature increase from heat gained from the store ambient can be substantial. This increase, which adversely affects condensing unit capacity and compressor discharge gas temperature, must be considered for system design; both capacity effects and the cooling needs of accessible hermetic compressors are involved.

One solution to excessively high superheat is to run the suction line and liquid line tightly together from the refrigerant line outlet, with the pair insulated together for a distance of 30 to 60 ft from the fixture outlet. This technique cannot be used with gas defrost or refrigerants requiring low suction superheat at the compressor suction (for example, R-22 low-temperature single-stage systems). Most manufacturers produce suction to liquid heat exchangers installed in the display fixture. This technique converts a loss into a gain by allowing the suction gas to pick up heat from the liquid instead of the store ambient. Under all conditions, the suction line should be insulated from the point where it leaves the display case to the suction service valve on the compressor. The insulation and its installation must be vapor resistant.

To ensure proper thermostatic expansion valve operation, the engineer should verify that the liquid entering the fixture is subcooled as planned and that this anticipated event did occur. Some case and/or system designs require liquid line insulation. This is very important when ambient outdoor air or mechanical subcooling is being used to improve system efficiency.

CONDENSING METHODS

Many commercial refrigeration installations use air-cooled condensers. Alternatively, evaporative condensers or water-cooled condensers with cooling towers may be specified. To obtain the lowest operating costs, equipment should operate at the lowest head pressure possible. The minimum allowable pressures are determined by other design and component considerations incorporated by the designer.

Techniques that permit a system to operate satisfactorily with lower condensing temperatures include (1) insulating liquid lines and/or receiver tank, (2) subcooling the liquid refrigerant by design, and (3) connecting the receiver as a surge tank with appropriate valving. Condensing pressure must still be controlled, at least to the lower limit required by the expansion valve, gas defrosting, and heat reclaim. The typical thermostatic expansion valve is capable of feeding the evaporator properly, assuming a solid liquid column of refrigerant is always supplied to the expansion valve and there is sufficient pressure drop across the valve. Balanced port thermostatic expansion valves have enhanced the opportunity for floating pressures down with varying ambient temperatures below the design point.

Air-Cooled Machine Room

The arrangement of standard air-cooled condensing units in a separate air-cooled machine room is less prevalent today, but is still used in some supermarkets. Dampers, which may be powered or gravity operated, supply air to the room; fans or blowers, controlled by room temperature at a thermostat, exhaust the air.

A complete air-cooled condensing unit located indoors requires ample, well-distributed ventilation. Ventilation requirements vary depending on maximum summer conditions and evaporator temperature, but 750 to 1000 cfm per condensing unit horsepower has given proper results. Exhaust fans should be spaced for an even distribution of air (see Figure 2).

Rooftop air intake units should be sized for 750 fpm velocity or less to keep airborne moisture from entering the room. When condensing units are stacked (as shown in Figure 2), the ambient

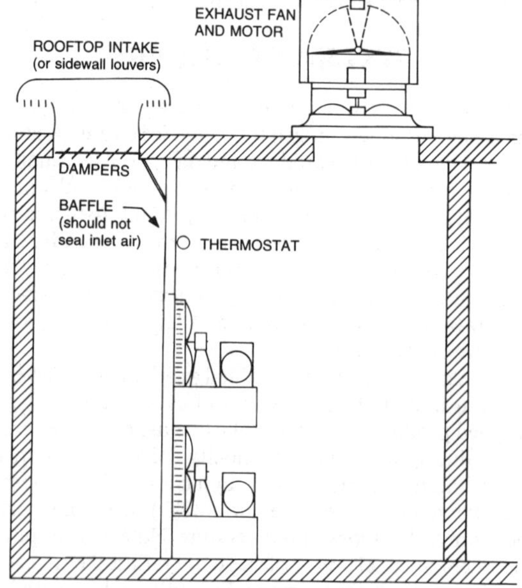

Fig. 2 Typical Air-Cooled Machine Room Layout

air design should provide upper units with adequate ventilation. Rooftop intakes are preferred because they are not as sensitive to wind as sidewall intakes, especially in winter in cold climates. Butterfly dampers installed in upblast exhaust fans, which are controlled by a thermostat in the compressor room, exhaust warm air from the space.

The air baffle helps prevent intake air from short-circuiting to the exhaust fans (Figure 2). Because air recirculation is needed around the condensers for proper winter control, the intake air should not be baffled to flow only through the condensers.

Ventilation fans for air-cooled machine rooms normally do not have a capacity equal to the total of all the individual condenser fans. Therefore, if the air is baffled to flow only through the condenser during maximum ambient temperatures, the condensers will not receive full free air volume when all or nearly all condensing units are in operation. Also, during winter operation, tight baffling of the air-cooled condenser would prevent recirculation of condenser air, which is essential to maintaining sufficiently high room temperature for proper refrigeration system performance.

Machine rooms that are part of the building need to be airtight so that air from the store is not drawn by the exhaust fans into the machine room. Additional load is placed on the store air-conditioning system if the compressor machine room, with its large circulation of outside air from the rest of the store, is not isolated.

Condensers Sizing

To minimize energy consumption, condensers should be more generously sized than they are for typical air-conditioning applications. Typical condenser selection is based on the temperature differential between the air entering the condenser and the saturated condensing temperature. Generally accepted TDs for refrigeration condenser sizing are 15 °F for medium-temperature systems and 10 °F for low-temperature systems.

Remote Air-Cooled Condenser

The remote condenser may be placed outdoors or it may be placed indoors to heat portions of the building in winter. Regardless of the arrangement of the remote condenser or the indoor-outdoor mechanical package, the following design points are relevant. The air-cooled condenser may be either a single-circuit or a multiple-circuit condenser. The manufacturer's heat rejection factors should be followed to ensure that the desired temperature difference is accommodated.

Pressure must be controlled on most outdoor condensers. Fan cycle controls work well down to 50 °F on condensers with single compressors or with parallel groups of compressors. Below 50 °F, condenser flooding (with the refrigerant) can be used alone or with fan controls. Flooding requires a larger refrigerant charge and larger liquid receivers. In contrast, split condensers with solenoid valves in the hot gas lines and bleed-off in the liquid return lines can reduce the condenser surface during cold weather. Natural subcooling can be integrated into the design to save energy.

Fans are controlled by pressure controls, liquid line thermostats, or a combination of both. Ambient control of condenser fans is commonly used; however, it may not give the degree of condensing temperature control required in systems designed for high efficiency gain. Ambient control of condenser fans is not recommeded except in mild climates down to 50 °F. Sometimes, pressure switches, in conjunction with gravity louvers, cycle the condenser fan. This system requires no refrigerant flooding charge.

The sizing of the receiver tank for all design on the high-pressure side, especially for remote condensers, must be considered. Remote condenser installations, particularly when associated with heat recovery, have substantially higher internal high-side volume than other types of systems. Much of the high side is capable of holding liquid refrigerant, particularly if runs are long and lines are large.

Roof-mounted condensers should have at least 3 ft of space between the roof deck and the bottom of the condenser slab to minimize the radiant heat load from the roof deck to the condenser surface. Also, free airflow to the condenser should not be restricted. Remote condensers should be placed at least 3 ft from any wall, parapet, or other airflow restrictor. Two side-by-side condensers should be placed at least 6 ft from each other. In Chapter 14 of the 1993 ASHRAE *Handbook—Fundamentals*, the problems of locating equipment for proper airflow are discussed in detail.

Evaporative Condenser Arrangements

Evaporative condensers are also available as single- or multiple-circuit condensers. Manufacturer conversion factors for operating at a given condensing temperature and wet-bulb temperature must be applied to determine the required size of the evaporative condenser.

In cold climates, the condenser must be installed to guard against freezing during winter. Evaporative condensers demand a regular program of maintenance and water treatment to ensure uninterrupted operation. The receiver tank should be capable of storing the extra liquid refrigerant during warm months. Line sizing must be considered to keep a reasonable tank size.

Closed water condenser/evaporative cooler systems are used frequently. In this arrangement, an evaporative condenser cools water instead of refrigerant. This water flows in a closed, chemically stabilized circuit with a regular water-cooled condenser (a two-stage heat transfer system). Heat from the condensing refrigerant transfers to the closed water loop in the regular water-cooled condenser. The warmed water then passes to the evaporative cooler.

The water-cooled condenser and evaporative cooler must be selected considering the temperature difference (1) between the refrigerant and the circulating water and (2) between the circulating water and the available wet-bulb temperature. The double temperature difference results in higher head pressures than when the refrigerant is condensed in the evaporative condenser. On the other hand, this arrangement causes no corrosion inside the refrigerant condenser itself because the water flows in a closed circuit and is chemically stabilized.

The extreme temperature of the entering discharge gas is the prime cause of evaporative condenser corrosion. The severity of corrosion can be substantially reduced by using the closed water condensing arrangement. The extent of this corrosion reduction relates directly to the temperature reduction between discharge temperatures experienced even with generously sized evaporative condensers on the one hand, versus the entering water temperature designed into the closed water circuit on the other.

Water flow in the closed water circuit can be balanced between multiple condensers on the same evaporative cooler circuit with water-regulating valves. Usually, low head pressures are prevented by temperature control of the closed water circuit. Three-way valves provide satisfactory water distribution control between condensers.

Cooling Tower Arrangements

Few supermarkets use water-cooled condensing units, since the trend is toward air cooling. Nearly all water-cooled condensing units are installed with a water-saving cooling tower because of the high cost of water and sewage disposal.

The engineering of water cooling towers for perishable foods is different than for air conditioning because (1) the hours of operation required are much greater than that for space conditioning; (2) refrigeration is required year-round; and (3) in some applications, cooling towers must survive severe winters. A thermostat must control the tower fan for year-round control of the condensing pressure. The control is usually set to turn off the fan when the water temperature drops to a temperature that produces the lowest desired condensing pressure. Water-regulating valves are sometimes used in a conventional manner. Dual-speed fan control is also used.

Some engineers use balancing valves for water flow control between condensers and rely on water temperature control to avoid low head pressure. Proper bleedoff and regular water treatment is required to ensure the satisfactory performance and full life of the cooling towers, condensers, water pumps, and piping. Water treatment specialists should be consulted because each locality has different water and atmospheric conditions. A regular program of water treatment is mandatory.

METHODS OF DEFROST

The most common defrost methods are condensing unit or system off-time, electric heat, latent heat, and air defrost.

Condensing Unit Off-Time

This method simply shuts off the unit and allows it to remain off until the evaporator reaches a temperature that permits defrosting and gives ample time for condensate drainage. Since this method obtains its heat from the air circulated in the display fixture, it is quite slow and limited to open fixtures maintaining temperatures of 34 °F or above. Defrost may be controlled by (1) suction pressure control, no time clock required; (2) time clock initiation and termination; (3) time clock initiation and suction pressure termination; and (4) time clock initiation and temperature termination.

Suction Pressure Control. This control is adjusted for a cut-in pressure high enough to allow defrosting during the off cycle. This method is usually used in fixtures maintaining temperatures from 36 to 43 °F. It provides some cooling effect during the defrost period because air is circulated over melting frost on the coil. If an excessive heat or humidity load should occur and cause the evaporator to ice, the evaporator pressure will be lowered to the cutout point of the control, thus initiating a defrost cycle to clear the evaporator.

However, condensing units and/or suction line may, at times, be subjected to low ambient temperatures below the evaporator's temperature. This prevents the buildup of suction pressure to the cut-in point and allows the condensing unit to remain off for prolonged periods. In such instances, fixture temperatures may become excessively high and displayed product temperatures will increase.

A similar situation can exist if the suction line from a fixture is installed in a trench or conduit with numerous other cold lines. The other cold lines may prevent the suction pressure from building to the cut-in point of the control.

Methods 2, 3, and 4 of controlling off-cycle defrosting use defrost time clocks to break the electrical circuit to the condensing unit initiating a defrost cycle. The difference lies in the manner in which the defrost period is terminated.

Time Initiation and Termination. A timer initiates and terminates the defrost cycle after the selected time interval. The length of the defrost cycle must be determined and the clock set accordingly.

Time Initiation and Suction Pressure Termination. This method is similar to the time initiation and termination method, except that suction pressure terminates the defrost cycle. The length of the defrost cycle is automatically adjusted to the condition of the evaporator, as far as frost and ice are concerned. However, to overcome the problem of the suction pressure not rising because of the defrost cut-in pressure previously described, the timer has a fail-safe time interval to terminate the defrost cycle after a preset time, regardless of suction pressure.

Time Initiation and Temperature Termination. This method is also similar to the time initiation and termination method, except that temperature terminates the defrost cycle. The length of the defrost cycle varies depending on the amount of frost on the evaporator or in the airstream leaving the evaporator. A temperature sensor is located on a tube of the evaporator or in the airstream leaving the evaporator. The timer also has a fail-safe setting in its circuit to terminate the defrost cycle after a preset time, regardless of the temperature.

Demand and Proportional Defrost. This system initiates defrost based on demand (need) or proportioned to humidity or dew point. Techniques vary from measuring temperature-spread change of the air entering and leaving the coil, to changing the defrost frequency based on store relative humidity. Other systems use a device that senses the frost level on the coil.

Electric Defrost

Electric defrost methods usually apply heat externally to the evaporator and require a longer defrost period than the hot gas defrost method, usually about 1.5 times longer. The heating element may be in direct contact with the evaporator, depending on conduction for defrost; or it may be located between the evaporator fans and the evaporator, depending on convection or a combination of conduction and convection for defrost. In either instance, the manufacturer generally installs a temperature-limiting device on or near the evaporator to prevent excessive temperature rise if any controlling device fails to operate.

The electric defrost method simplifies the installation of low-temperature fixtures. The controls used to automate the cycle usually include one or more of these devices (1) defrost timer, (2) solenoid valve, (3) electrical contactor, and (4) case evaporator fan delay switch.

Hot Gas Defrost

Gas defrost uses heat from the compressor's discharge gas to defrost the evaporators. To remove the coil frost, the discharge pressure gas from the compressors is directed to the case evaporator. One method tempers the hot gas impact on the cold suction line by beginning the defrost cycle with saturated gas from the top of the liquid in the receiver. Occasionally, supplemental electric heaters are added to ensure rapid and reliable defrosting. A timer generally terminates the defrost cycle, although temperature termination is sometimes used.

Air Defrost

Air defrost moves air from the store ambient into the refrigerator. A variety of systems is used; some use supplemental electric heat to ensure reliability. Heat content of store ambient air during the winter is critical for good results from this method.

REFRIGERANT LINES

Sizing refrigerant lines, both liquid and suction, is critical in the average refrigeration installation because of the typically long horizontal runs and the frequent use of vertical risers. Correct liquid line sizes are essential to ensure a full feed of liquid to the expansion valve and proper suction-line oil return to the compressor without excessive pressure drop. Oversizing of liquid lines must also be avoided to prevent system pump-down or defrost cycles from operating improperly in single-compressor systems.

Oil separates in the evaporator and moves toward the compressor at a lower velocity than the refrigerant. Unless the suction line is properly installed, the oil can accumulate at low places, causing problems such as compressor damage from liquid slugging, excessive pressure drop, and reduced system capacity. To prevent these problems, horizontal suction lines must pitch down as the gas flows toward the compressor, the bottom of all suction risers must be trapped, and the refrigerant velocity in suction risers must be maintained according to piping practices described in Chapters 2 and 3. To overcome a large pressure drop, the suction lines may be oversized on long runs, but they still must pitch down toward the compressor.

Manufacturers' recommendations and appropriate line sizing charts should be followed to avoid adding heat to either the suction or liquid lines. In large stores, both suction and liquid lines can be insulated profitably, particularly if subcooling is present.

EFFECTS OF NEW REFRIGERANTS

Current refrigerants used in the refrigeration industry are R-12, R-22, and R-502. Most countries, including the United States, have created legislation to eliminate these refrigerants over time due to their effect on ozone depletion. Display and storage equipment, as the merchandiser views and uses it, will probably not be affected. However, the refrigeration system components and operating characteristics will undoubtedly change.

At this time, the alternative refrigerants that will be chosen or the total effect on the refrigeration operation are unknown. Currently, one short-term solution is to stay with the R-22 design until firm recommendations are available from the chemical suppliers and compressor manufacturers. For the supermarket designer, the compressor is the heart of the system. Its performance with the new refrigerants and their compatibility with oil and motor insulation are extremely important. Changing the type of refrigerant in an existing system without careful planning and knowledge could result in seriously premature compressor failures.

INTERACTION BETWEEN REFRIGERATION AND AIR CONDITIONING

Open display equipment often extracts enough heat to reduce the store indoor temperature as much as 16°F below air temperatures desired in the customer aisles. The air-conditioning return duct system or fans can be used to withdraw chilled air from the floor in front of the cases and discharge it overhead or back to the store air handler.

Heat Reclamation

Heat reclaim condensers and related controls operate as alternates to or in a series with the normal refrigeration condensers. They can be used in winter to return most of the refrigeration and compressor heat to the store. They may also be used in mild spring and fall weather when some heating is needed to overcome the cooling effect of the refrigeration system itself. Another use is for cooling coil reheat when needed for humidity control in spring, summer, and fall. Excess humidity in the store must be avoided because it can increase the display case refrigeration load as much as 20% at the same dry-bulb temperature. Also, heat reclamation can be used to heat water for store use. The section on supermarkets in Chapter 2 of the 1991 ASHRAE *Handbook—Applications* has more detailed information on the interrelation of the store environment and the refrigeration equipment.

CONDENSING UNIT NOISE

Air-cooled condensing units located outdoors, either as single units with weather covers or grouped in prefabricated machine rooms, produce sounds that must be evaluated. The largest source of noise is usually moving air with propeller-type condenser fans. Other sources are noise from compressor and fan motors, high-velocity refrigerant gas noise, general vibration, and amplification of sound where vibration is transmitted to mounting structures. The last item is most critical when units are roof-mounted.

A fan speed or fan cycle control is helpful in controlling air noise because only the amount of air necessary to maintain proper head pressure is generated. Care should be taken not to restrict discharge air. Whenever possible, it should be discharged vertically upward.

Resilient mountings for fan motors and small compressors and isolation pads for larger motors and compressors are helpful in reducing noise. Discharge line mufflers are the best solution for high-velocity gas noise. Lining enclosures with sound-absorbing material is of minimal value. Isolation pads can help on roof-mounted units, but even more important is choosing the right location in regard to the supporting structure so that structural member vibration does not amplify the noise.

If sound levels are still excessive after the foregoing controls have been implemented, location becomes the greatest single factor. Distance from a sensitive area is most important in choosing a location; for each time the distance is doubled, the noise level is halved. Direction is also important. Condenser air intakes should face parking lots, open fields, or streets zoned for commercial use. In sensitive areas, avoid ground-level installation close to building walls, as the walls will reflect the sound.

When it is impossible to meet requirements by the foregoing methods, barriers can be used. While a masonry wall is an effective barrier, it may be objectionable because of cost and weight. If a barrier is used, it must be sealed at the bottom, because any opening will allow sound to escape. Barriers also must not restrict condenser entering air. Keep the open area at the top and sides at least equal to the condenser face area. When noise is a consideration, (1) purchase equipment designed to operate as quietly as possible, for example 850 rpm condenser fan motors instead of higher speed motors; (2) choose the location carefully; and (3) use barriers when the first two steps do not meet requirements.

REFERENCES

Howell, R.H. 1993a. Effects of store relative humidity on refrigerated display case performance. ASHRAE *Transactions* 99(1).

Howell, R.H. 1993b. Calculation of humidity effects on energy requirements of refrigerated display cases. ASHRAE *Transactions* 99(1).

CHAPTER 32

ICE MANUFACTURE

MOST commercial ice production is done with ice makers that produce three basic types of fragmentary ice of a type and size required for a particular application. The basic types of fragmentary ice are plate, tubular, and flake. Chapter 54 of the ASHRAE *Handbook and Product Directory—1978 Applications* includes information on block ice manufacturing. Among the many areas where manufactured ice is used are:

- Processing: Fish, meat, poultry, dairy, bakery products, and hydrocooling
- Storage and transporation: Fish, meat, poultry, and dairy products
- Manufacturing: Chemicals and pharmaceuticals
- Others: Retail consumer ice; concrete mixing and curing; and off-peak thermal storage

ICE MAKERS

Flake Ice

Flake ice is produced by applying water to the inside or outside of a refrigerated drum or to the outside of a refrigerated disk. The drum is either vertical or horizontal and may be either stationary or fixed. The disk is vertical and rotates about a horizontal axis. Ice removal devices fracture the thin layer of ice produced on the freezing surface of the ice maker, breaking it free from the freezing surface and allowing it to fall into an ice bin, which is generally located below the ice maker.

The thickness of the ice produced by flake ice machines can be varied by adjusting the speed of the rotating part of the machine, varying evaporator temperature, and regulating the water flow on the freezing surface. Production of flake ice is on a continuous basis as contrasted with tube or plate ice, which is made using an intermittent cycle or harvest type of operation. The thickness of ice produced ranges from 0.04 to 0.18 in. A continuous operation without a harvest cycle results in less refrigeration capacity required to produce a ton of ice than any other type of manufactured ice with similar makeup water and evaporating temperatures. The exact amount of refrigeration required varies by type and design of the flake ice machine. Typical flake ice machines are shown in Figures 1 and 2.

All water used by flake ice machines is converted into ice; therefore, there is no waste or spillage. Usually flake ice makers are operated at a lower evaporating temperature than tube or plate ice makers, and the ice is colder when it is removed from the ice-

The preparation of this chapter is assigned to TC 10.2, Automatic Icemaking Plants and Skating Rinks.

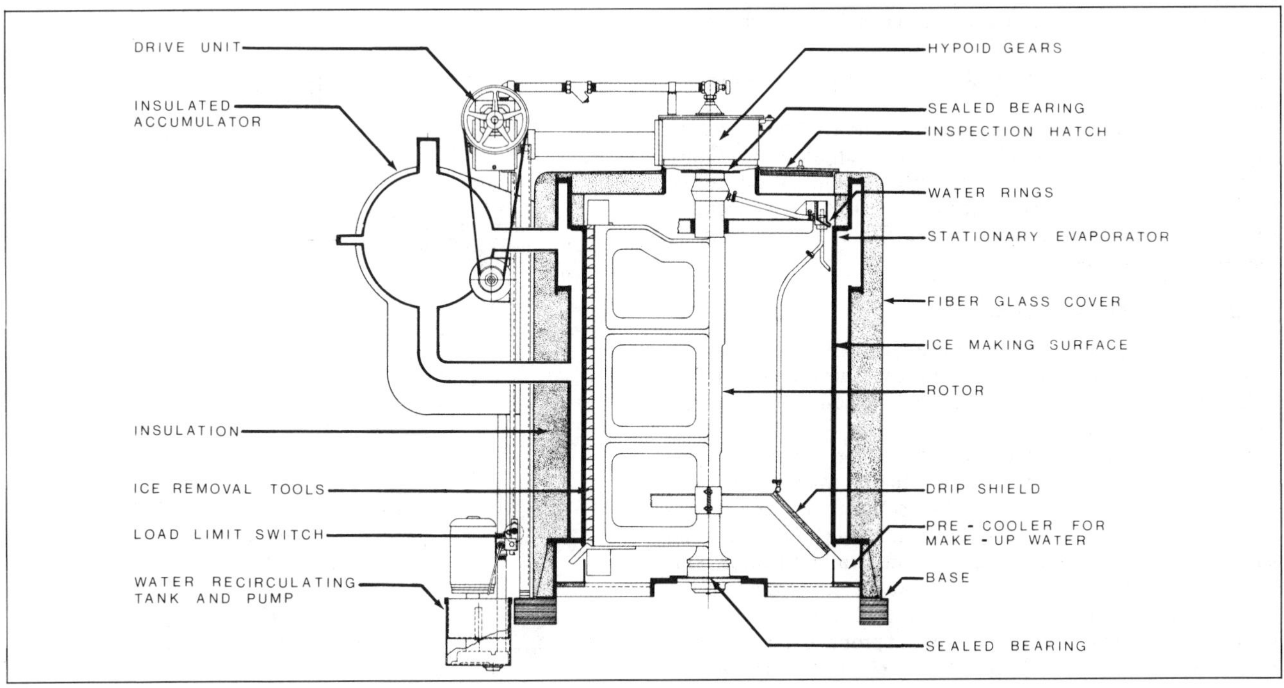

Fig. 1 Flake Ice Maker

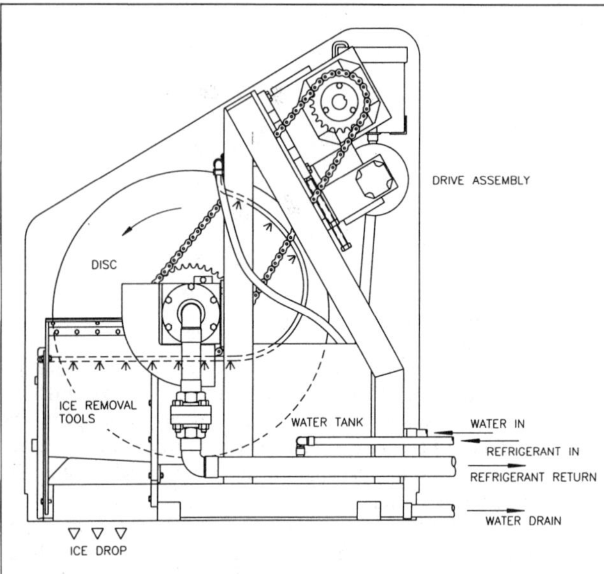

Fig. 2 Disk Flake Ice Maker

making surface. The surface of flake ice is not wetted by thawing during removal from the freezing surface, as is common with other types of ice. Since it is produced at a colder temperature, flake ice is most adaptable to automated storage, particularly when low-temperature ice is desired.

The rapidity of freezing the ice on the freezing surface results in the opaque appearance of flake ice, which is caused by entrained air. For this reason, flake ice is not commonly used for applications where a clear ice appearance is important. For some applications, such as chemical processing and concrete cooling, where rapid cooling is important, flake ice is ideal because the flakes present the maximum amount of cooling surface for a given amount of ice.

When used as ingredient ice in sausage making or other food grinding and mixing, flake ice provides rapid cooling while minimizing mechanical damage to other ingredients and wear on mixing/cutting blades.

Some flake ice machines can produce salty ice from seawater. These are particularly useful in ship-board applications. Other flake ice machines require adding trace amounts of salt to the makeup water to enhance the release of ice from the refrigerated surface. In rare cases, the presence of salt in the finished product may be objectionable.

Tubular Ice—Outside Tube

Tubular ice is produced by freezing a falling film of water either on the outside of a tube with evaporating refrigerant on the inside or by freezing water on the inside of tubes surrounded by evaporating refrigerant on the outside.

When ice is produced on the outside of a tube, the freezing cycle is normally from 8 to 15 min., with the final ice thickness from 0.2 to over 0.5 in. following the curvature of the tube. The refrigerant temperature inside the tube continually drops from an initial suction temperature of about 25 °F to the terminal suction temperature in the range of 10 to −15 °F. At the end of the freezing cycle, the circulating water is shut off. Introducing hot discharge gas results in harvest defrost. To maintain proper harvest temperatures, typical discharge gas pressure is 160 psia. This drives the liquid refrigerant in the tubes up into an accumulator and melts the inside of the tube of ice, which slides down the tube through a sizer and mechanical breaker, and then down an ice slide through an opening into storage. The defrost cycle is normally about 30 s. The unit returns to the freezing cycle by returning the liquid refrigerant to the tubes from the accumulator.

This type of ice maker operates with refrigerants R-717, R-12, and R-22. Higher capacity units of 10 tons per 24 h and larger usually use R-717. The capacity of the unit increases as the terminal suction pressure decreases. A typical unit with 70 °F makeup water and R-717 as the refrigerant will produce 19.3 tons of ice per 24 h with a terminal suction pressure of 38.5 psia and requires 35.7 tons of refrigeration. This equates to 1.85 tons of refrigeration per ton of ice. The same unit will produce 41.6 tons of ice per 24 h with a terminal suction pressure of 21 psia and requires 80 tons of refrigeration. This equates to 1.92 tons of refrigeration per ton of ice. Figure 3 shows the physical arrangement for an ice maker that makes ice on the outside of the tubes.

Tubular Ice—Inside Tube

When ice is produced inside a tube, it can be harvested as a cylinder or as crushed ice. The freezing cycle is approximately from 13 to 26 min. The tube is usually 0.9 to 2 in. in diameter, producing a cylinder that can be cut to desired lengths. The refrigerant temperature outside the tube is continually dropping, with an initial temperature of 25 °F and a terminal suction temperature ranging from 20 to −5 °F. At the end of the freezing cycle, the circulating water is shut off and the ice is harvested by introducing hot discharge gas into the refrigerant in the freezing section. To maintain gas temperature, typical discharge gas pressure is 180 psia. This releases the ice from the tube; the ice descends to a motor-driven cutter plate that can be adjusted to cut the ice cylinders to the length desired (up to 1.5 in.). At the end of the defrost cycle, the discharge gas valve is closed and the circulating water resumed.

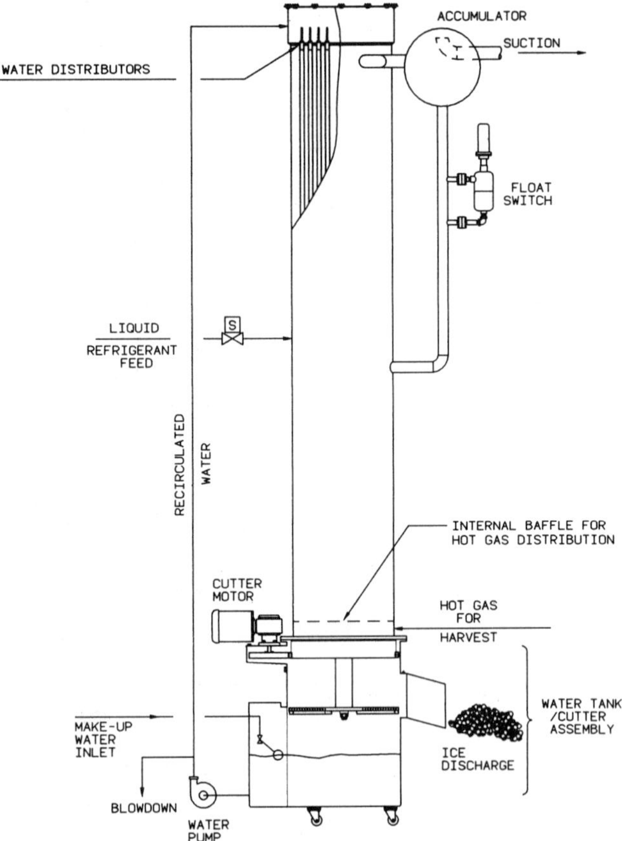

Fig. 3 Tubular Ice Maker

This type of unit can use refrigerants R-717, R-12, and R-22; the capacity again increases as the terminal suction pressure decreases. A typical unit with 70 °F makeup water and R-717 as the refrigerant will produce 43 tons of ice per 24 h with a terminal suction pressure of 40 psia and requires 74.5 tons of refrigeration. This equates to 1.73 tons of refrigeration per ton of ice. The same unit will produce 66 tons of ice per 24 h with a terminal suction pressure of 30 psia and requires 135 tons of refrigeration. This equates to 2.04 tons of refrigeration per ton of ice.

Tubular ice makers are advantageous because they produce ice at higher suction pressures than other types of ice makers. They can make a relatively thick and clear ice, the curvature of which helps prevent bridging in storage. Tubular ice makers have a greater height requirement for installation than that of plate or flake ice makers. However, tubular ice makers have a smaller footprint. Provision must be made in the refrigeration system high side to accommodate the volume of refrigerant required for the proper amount of harvest discharge gas. Ice temperatures are similar to those achieved by plate ice makers, but they are generally higher than the temperature capabilities of flake ice makers.

Supply water. Supply water temperature has a great effect on the capacity of either type of ice maker freezing ice on tubes. If the supply water temperature is reduced from 70 to 40 °F, the ice production of the unit will increase approximately 18%. In larger systems, the economics of precooling the water in a separate water cooling system with higher suction pressures should be considered.

Plate Ice

Plate ice makers are commonly defined as those that build ice on a flat vertical surface. Water is applied above freezing plates and flows by gravity over the freezing plates during the freeze cycle. Liquid refrigerant at a temperature of between −5 and 20 °F is contained in internal circuiting inside the plate. The freezing cycle time governs the thickness of ice produced. Ice thicknesses in the range of 0.25 to 0.75 in. are quite common, with freeze cycles varying from 12 to 45 min. Figure 4 shows a flow diagram of a plate ice maker using water for harvest. All plate ice makers use a sump and recirculating pump concept whereby an excess of water is applied to the freezing surface. Water not converted to ice on the plates is collected in the sump and is recirculated as precooled water for ice making.

Ice is harvested from plate ice makers by one of two methods. One method involves the application of hot gas to the refrigerant circuit to warm the plates to 40 to 50 °F, causing the ice surface touching the plate to reach its melting point and thereby release from the plate. The ice falls by gravity to the storage bin below or to a cutter bar or crusher that further reduces the ice to a more uniform size. Plate ice makers using the hot gas method of harvesting are capable of producing ice on one or two sides of the plate, depending on the design.

In the second method of harvesting ice, warm water flows on the backside of the plate. In so doing, the refrigerant inside the plate is heated above the ice melting point, and the ice is released. Ice makers using the water warming harvest principle manufacture ice on one side of the plates. Harvest water is prechilled by passing over the plates. It is then collected in the sump and recirculated to become precooled water for the next batch of ice.

For plate ice makers, the freezing time, harvest time, and the related water, pump, and refrigeration are controlled by adjustable electromechanical or electronic devices. Since a wide variety of thicknesses and freezing times is available, plate ice makers can produce clear ice at the longer and slower freezing times. Thus the plate ice maker is commonly used in applications where clear ice is desired.

Because of the harvest cycle involved, plate ice makers require more refrigeration per unit mass of ice produced than flake ice makers. This disadvantage is offset by the capability of plate ice makers to operate at higher evaporating temperatures; thus, connected motor power is usually less than for flake ice makers. During the harvest cycle, the suction pressure rises considerably, depending on the design of the ice maker. When a common refrigeration system is used for multiple refrigerated requirements, a stable suction pressure can be maintained for other refrigeration loads by using a dedicated compressor for the ice machine, or by using a dual pressure suction regulator at each ice machine to minimize the load placed on the suction main during harvest. This may occur in large processing plants, refrigerated warehouses, and so forth. Large plate ice makers can be arranged for harvesting of

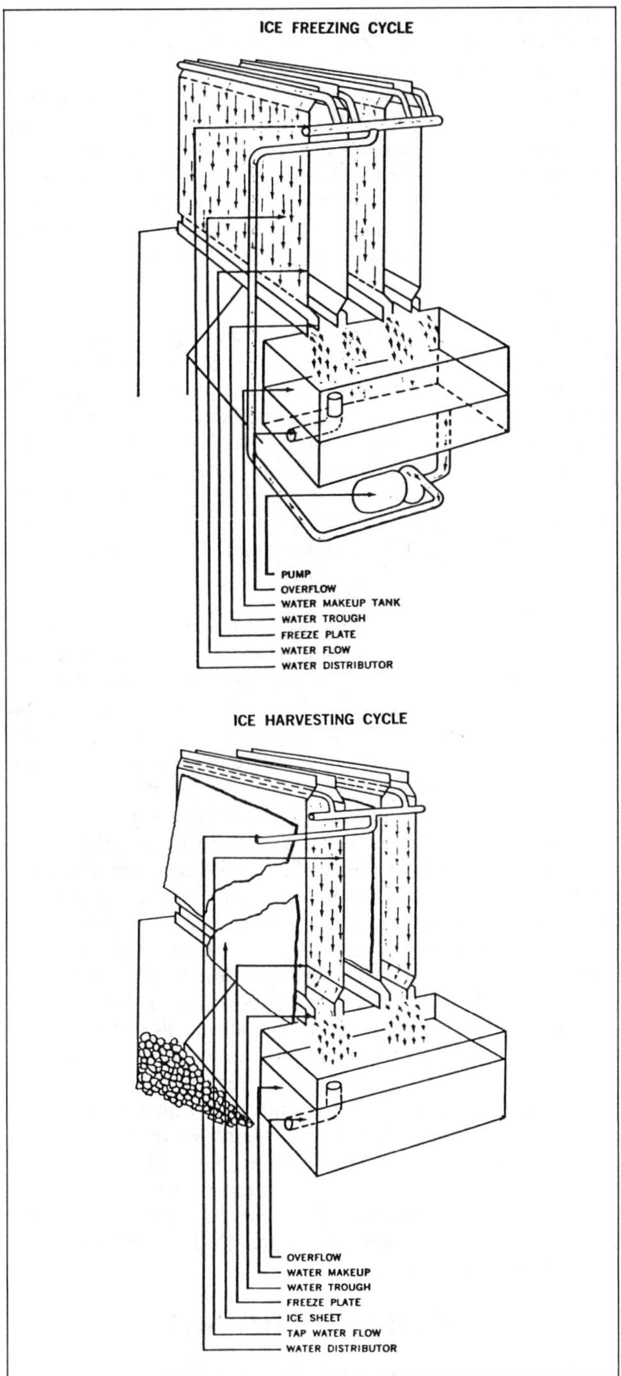

Fig. 4 Plate Ice Maker

only sections or groups of plates at one time. Properly adjusting the time spacing for harvesting each section can reduce the fluctuation in suction pressure.

Plate ice makers using the water harvest principle rely on the temperature of the water for harvesting. A minimum of 65 °F for the water is usually recommended to minimize both the harvest cycle time and harvest water consumption in excess of ice making requirements. For installations in cold water areas, or where wintertime inlet water temperatures are low, it is advisable to provide auxiliary means for warming the inlet water to 65 °F.

Ice Builders

Ice builders comprise various types of apparatus in which ice is produced on the refrigerated surfaces of coils or plates submerged in water in an insulated tank. This equipment is commonly known as an ice bank type of water chiller. The ice built on the freezing coils is not used as a manufactured ice product but rather as a means of cooling water circulating through the tank. The ice builder is most often used in applications where high peak and intermittent cooling loads requiring chilled water occur. See Chapter 39 in the 1991 ASHRAE *Handbook—Applications.*

Scale Formation

The performance of all ice makers is affected by the characteristics of the inlet water used. Impurities and excessive hardness can cause a scale to be deposited on the freezing surface of the ice maker. The deposit reduces the heat transfer capability of the freezing surface with a resultant reduction in ice making capacity. Deposited scale may also further reduce ice making capacity by causing poor ice removal from the freezing surface during the harvest process. The ice tends to stick on the freezing surface. The rated capacity of all ice makers is based on substantially releasing all the ice from the freezing surface during the removal period. Because the process of freezing water into ice tends to freeze a greater proportion of pure water on the ice maker freezing surface, impurities tend to remain in the excess or recirculated water. A blowdown, or bleedoff, whereby a portion of the recirculated water is bled off and discharged, can be installed. The bleedoff system can control the concentration of chemicals and impurities in the recirculated water. Determining the necessity of a bleedoff system and the effectiveness of this concept for controlling scale deposits depends on local water conditions. Some refrigeration system loss is experienced, since the recirculated water that is bled off to drain has been precooled. Water that is bled off may be passed through a heat exchanger to precool incoming makeup water. Water conditions, water treatment, and related water problems in ice making are covered in Chapter 43 of the 1991 ASHRAE *Handbook—Applications.*

THERMAL STORAGE

Interest in energy conservation renewed interest in the ice storage concept to provide for thermal storage of cooling capacity for air-conditioning or process applications. Using lower off-peak and weekend power rates, the ice is produced and stored. During the day, stored ice is used to provide the refrigeration of the chilled water system. The design and features of thermal storage equipment are covered in Chapter 39 of the 1991 ASHRAE *Handbook—Applications.*

ICE STORAGE

Fragmentary ice makers have the capability of producing ice either on a continuous basis or a constant number of harvest cycles per hour. The use of the ice is generally not at a constant rate but on a batch basis. Batches vary greatly based on user requirements. The ice must be stored and recovered from storage on demand. Ice storage and storage bin design therefore become important where labor savings, economics, the quantity of ice to be stored, the amount of automation desired, and user delivery requirements are concerned.

Ice makers can produce ice 24 h a day. By making ice during off-shifts and weekends, as well as during work shifts, considerable savings in total ice making and refrigeration system requirements can be achieved. In addition, by using electrical power during off-peak hours, peak loads on the power system are reduced during the day. Many power companies offer reduced rates at off-peak hours.

Ice storages vary in type from short-term to prolonged-term storage, with degrees of automation for filling and discharge ranging from manual shoveling to a completely automatic rake system.

Short-term storage generally requires provision for the ice production of one day. The ice maker is mounted over a bin, and ice falls by gravity into the bin. The bin is an insulated, airtight enclosure with one or more insulated doors for access. Ice is removed from the bin by shoveling or scooping. In such a storage, the subcooling effect of the ice generally offsets the heat loss through the insulated bin walls without excessive melting. In most situations, it is not necessary to provide refrigeration units in the ice storage bin where the ice production is being used on a daily basis and where ambient temperatures are reasonable.

Prolonged ice storage requires a refrigerated, airtight, insulated storage bin. Some designs provide for false walls and floor, which produce an envelope effect that allows cold air to circulate completely around the mass of ice in storage. If wet ice is placed in a bin refrigerated to a temperature below 32 °F, it will freeze together and may be difficult to remove.

Time and pressure affect the storage quality of fragmentary ice. Even though a bin is refrigerated to a temperature well below 32 °F, pressure can cause local melting near the bottom. Thus, there is a limit to the size and configuration of a gravity-filled storage bin. The ice falling from an ice maker forms a cone directly underneath the drop in the bin. With slight variation because of the type of ice, the angle of repose is approximately 30°. Fusion of ice under pressure limits the practical ice storage depth to 10 to 12 ft. To use the volume of the bin more efficiently, a leveling screw mounted in the overhead can be used to carry the ice away from the top of the ice cone.

There is also a practical limit to the size of a storage bin in which ice can be manually removed through refrigerator doors. The simplest device used to remove ice from the bin is a screw conveyor with a trough at floor level, which is equipped with gratings and removable sectional covers. The removable covers protect the screw from ice blockage when the conveyor is not running. The gratings are for the protection of personnel.

Ice Rake and Live Bottom Bins

The ice rake system is used for larger and fully automated storages. These storages generally have a 10- to 300-ton capacity for a single rake system. Depending on plant demands, combinations of rake systems can be developed into an integrated production, storage, and delivery system with capabilities of up to 1000 tons storage with multiple delivery systems by screw conveyor or pneumatic conveying. A range of delivery rates up to 60 tons of ice per hour can be achieved. The advantages of such rake systems are the elimination of labor for storing and transporting ice, longer and more effective distribution systems, faster delivery and termination of ice flow, less waste of ice, and the elimination of physical contamination.

Storages that incorporate rake systems are of two basic types. One type encloses the ice storage and rake system in an arrangement of steel framework and panels with the complete unit installed in a refrigerated room. The second type involves the construction of an insulated enclosure around the ice bin and rake

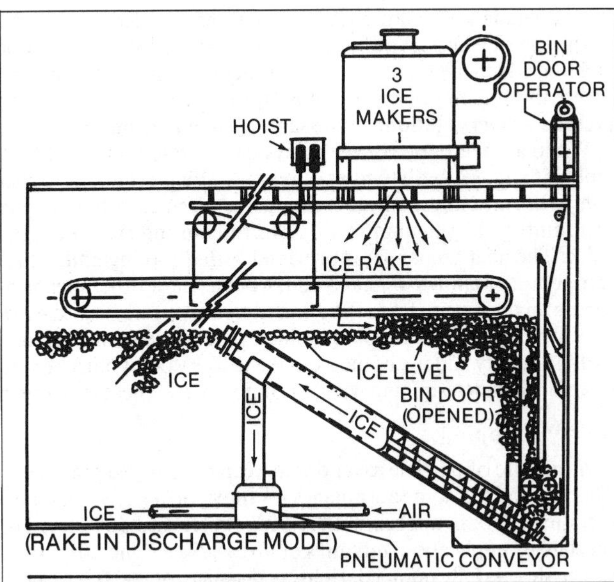

Fig. 5 Ice Rake System

system. This type can be installed in a building or outside, depending on the type of weather protection provided. For either type, the ice makers are mounted outside of the refrigerated space. Figure 5 shows the arrangement of components for a typical rake system.

Once fragmentary ice comes to rest in the storage bin, it will not flow freely, and a mechanical force is necessary to start the ice moving. The deeper the ice is stored, the greater the pressure on the ice near the bottom. The ice near the bottom tends to fuse together faster. Rake systems work from the top of the ice; they continuously level and fill the storage bin, as well as automatically remove the ice on demand. The systems operate in nonrefrigerated or refrigerated bins. Since most users of large storages also want dry ice for ease of handling, large automated storages are usually refrigerated.

The ice rake itself consists of a structural steel mechanism with drive, which operates similar to the tracks of a crawler tractor. By means of a hoist and timer, the rake is raised or lowered to automatically maintain its position suspended and in close contact with the ice level. Wide scraper-type conveyors, mounted across the tracks the full width of the bin, spread the ice out and drag it toward the back of the bin during the filling mode. To dispense the ice at delivery, the scraper conveyors reverse direction and drag the ice to the opposite end of the bin, dropping it into a screw conveyor mechanism. From this point, the ice is transferred to the external delivery system of screw conveyors.

Some rake systems have features that allow ice deliveries from the bin to be remotely controlled and volumetrically metered. Ice deliveries can be recorded on digital counters at the storage bin, remote stations, and control centers. Accuracy is in the range of ±2%. Another method of metering ice from a rake system storage has the screw conveyor deliver the ice to a weigh belt. As the ice passes along the moving belt, it is electronically weighed and the weight is recorded. Selection of the belt material carrying the ice is critical to prevent ice from sticking to the belt. The weigh belt is often installed in a refrigerated area adjacent to the ice storage.

Another type of ice storage with delivery system capabilities is the *live bottom* type with a multiplicity of screws arranged in various configurations on the bottom of the storage bin. Because of ice fusion, these bins are limited to short-term storage. The success of this type of bin depends on the type and quality of fragmentary ice being stored and the ability of the design to overcome particle fusion. Particle fusion may result in ice bridges forming over the top of the screws; then the screws will bore holes in the ice rather than empty the bin.

Primarily for the consumer bagged-ice industry, a bin and automatic storage system is used in which the entire floor moves, carrying the ice load into slowly rotating beaters. As the ice breaks loose, it drops to a screw conveyor, which feeds an ice bagger. This type of bin is located in a refrigerated room, and the ice makers are located away from the bin. The ice makers must be shut off so that no ice can flow into the storage during the discharge and bagging process.

The ice silo is used for long- or short-term storage with capacities in the range of 20 to 100 tons. The silo tank comprises a cylindrical part and a tapered, conical part leading the ice to the outlet at the bottom of the tank. From this point, the ice is transported by a screw delivery system. A rotating flexible chain arrangement is provided in the silo to assist in ice removal and to partially overcome the fusion problem. The ice maker is mounted over the top of the silo, and the ice falls into the storage. No leveling of ice is required in the bin, since the diameter of the silo is sized to be compatible with the ice maker. The larger the ice storage, the higher the silo. As a result, proper ice discharge becomes more critical in the design when considering the fusion of ice and the fact that the ice must finally pass through a relatively small opening at the bottom of a tapered zone.

DELIVERY SYSTEMS

The location of ice manufacture is rarely the location of ice usage. Usually it is necessary to move ice from the ice machine or storage bin to some other area where it will be used; thus, a conveying system is required. Most conveyor applications use screws, belts, or pneumatic-type systems. Great care must be taken in selecting the size and type of conveyor to be used, because no matter what type of fragmentary ice is being handled, problems such as fines, freezeup, and ice jams can be encountered with an improperly designed system. Fines, or snow, describe the small particles of ice that chip off the larger pieces during harvesting, crushing, or conveying operations.

Screw and Belt Conveyors

Screw conveyors are the most popular of all the conveyances used for transporting ice. Screw conveyors are manufactured in sizes of 4 in. diameter and up, as well as in various screw pitches. Most ice-conveying operations use 6-, 9-, and 12-in. diameter screws.

The sizing and drive power requirements of screw conveyors are determined by the ice delivery rate, the inclination of the conveyor, and the conveyor screw pitch. With fragmentary ice, the selection of an undersized conveyor will result in excessive conveyor speed or require that the conveyor run too full of ice. These conditions can produce excessive fines.

When screw conveyors transport ice through high ambient inside areas, or outside in the weather, such as in icing fishing vessels, it is advisable to insulate the screw conveyor trough and provide the conveyor with insulated covers. Rain is as problematic as sunshine for contributing to ice meltage and delivery difficulties. For this reason, most screw conveyors operating in the weather are provided with sectional and removable covers.

Belt conveyors are often used when excess moisture has to be removed from the ice or to minimize the fines. The mesh-type belts allow snow and excess water to fall through. Stainless steel, galvanized steel, or high-density polyethylene are commonly used for belting.

PNEUMATIC ICE CONVEYING

Pneumatic ice conveying systems have proved desirable, economical, and practical when transporting fragmentary ice distances of 100 ft or more and when multiple delivery stations must be served. A pneumatic system is advantageous when delivery stations are in different directions or at different elevations, when delivery through a pressure hose is needed, or when flexibility is required for future changes or the addition of delivery stations.

The basic principle of conveying ice by a pneumatic system involves using a rotary blower, which delivers air to a rotary airlock valve or conveying valve. Ice is fed into the conveying valve, and compressed air conveys the mixture of air and ice at high velocity through thin-walled tubing (aluminum, stainless steel, or plastic). Figure 6 shows the diagrammatic arrangement of pneumatic system components.

Delivery rates are generally between 10 to 40 tons per h, and conveying distances up to 600 ft are common. Delivery distances exceeding 600 ft can be achieved at reduced delivery rates, with the maximum distance practical being approximately 1000 ft. Conveying pressures range from 4 to 10 psig, depending on the delivery rate and the maximum distance the ice is to be conveyed. The air velocity required to keep the ice in suspension in the conveying line will vary among the different types of fragmentary ice. A pneumatic system cannot satisfactorily convey all sizes of fragmentary ice. The manufacturer of the ice-making equipment should be consulted for recommended line velocities. Sometimes, storage bins for ice plants using a pneumatic delivery system are refrigerated to assure cold, free-flowing ice with minimum moisture. Since the ice remains in the tubing a very short time, tubing insulation is seldom needed or used. However, in warm climates, shading the conveying line reduces the solar load. The tubing typically used has a diameter of 4 to 8 in. and requires minimum support, making installation easy and economical.

Multiple delivery points are served by automatic Y-type diverter valves or multiple way slide valves, either air or electrically operated. Pneumatically blown ice can be delivered under pressure out of the end of a hose. An alternate method of delivery is a cyclone receiver, which takes the ice at high velocity, dissipates the air, and drops the ice by gravity. Combinations of hose stations and cyclone delivery stations in the same system are common.

When a pneumatic conveying system is used in areas of high ambient and wet-bulb temperatures on systems requiring higher conveying pressures, a heat exchanger is often used to cool and dehumidify the pneumatic air prior to entering the conveying valve. The heat exchanger is provided with a cooling coil, either refrigerant or chilled-water cooled, a demister or other means of separating moisture from the air, and a condensate trap to expel the entrapped moisture. Geographical location, system pressure, and the quality and use of the ice at the delivery point must be considered when determining whether or not to use a heat exchanger.

Slurry Pumping

A mixture of particle ice and water can be pumped as a slurry. This method has some advantages for transporting ice. Generally, the slurry mix is approximately 50% water and 50% ice. For specialized application, mixtures of up to 80% ice and 20% water can be successfully pumped. Delivery distances of 800 ft have been achieved with delivery rates of 60 tons of slurry mix per h. This practice has been extensively used in the produce industry and has potential in concrete cooling, chemical processing, and other ice or chilled-water related applications. Ice slurry mixes are of particular interest where there is a requirement for low-temperature chilled water at or near 32 °F. In converting ice to water, the absorption of latent heat at the usage point enables more cooling to be done with a slurry mix than with straight chilled-water cooling. Pumping volumes and line sizes are minimized, and ice meltage during mixing and pumping does not normally exceed 1 to 3%. The system has a capability of automation for continuous operation.

The basic system for slurry pumping involves a mixing tank in which the ice and water are mixed. The ice is carried by any of the conventional conveying methods from the ice storage bin to the

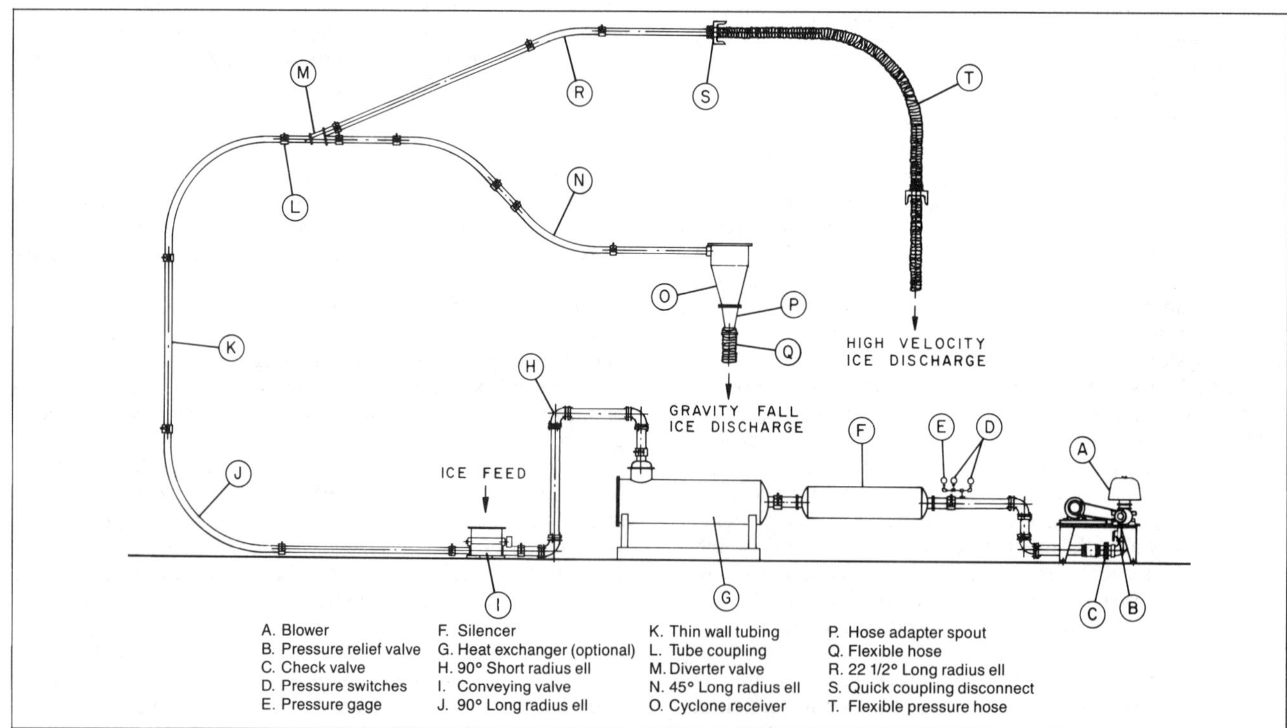

Fig. 6 Typical Flake Ice Pneumatic Conveying System

slurry mix tank. Agitators in the tank operate continuously to maintain a mixture with uniform consistency. Pumps then discharge the slurry mix through pipelines to the point of use. The pumps are of the centrifugal type, modified for pumping slurry. When the icing cycle at the usage points is intermittently demanded, a recirculating system returns unused slurry to the mixing tank. In this way, the slurry is kept moving at all times, and the possibility of ice blockage in the lines is minimized. The temperature of the slurry solution in the tank is maintained at 32 °F.

The fresh produce industry offers a unique application for slurry mixes. Body icing of the fresh produce, which is generally of nonuniform size and configuration, is achieved by applying the slurry mix to the dry packed product. Drain holes are provided in the shipping container to remove water. Because of its suspension in water, ice is carried to all parts of the container. The ice solidifies as the water drains from the container. The product is then completely surrounded with ice, the voids are filled, and potential hot spots are eliminated. The drained water can be collected and returned to the mixing tank.

COMMERCIAL ICE

Commercial ice is primarily used for human consumption. It is also called *packaged* or *consumer* ice and is used for cooling beverages and for other applications in restaurants, hotels, and similar institutions. This use requires packaging at the ice plant for storage and eventual distribution. In bagged form, commercial ice is also available for sale in grocery stores and automatic, coin-operated vending machines. When the ice is to be used in beverages, ice produced by plate or tube ice makers is preferred because of the clear appearance and the fact that it can be made in greater thicknesses. Rake systems are often used to store packaged ice and convey it into the packaging system.

A packaging system normally comprises two pieces of equipment: an ice bagger and a bag closer. These components are available from ice packaging equipment manufacturers in various types and sizes. In the bagging process, the ice is fed from the ice storage bin into the bagging machine by a screw or belt conveyor. The bagging machine meters ice into a bag placed below the discharge chute of the bagger. The amount of ice measured into the bag can be determined by weight, volume, or sight approximation, depending on the equipment used.

When packaging by weight, the ice bag is placed on a weighing table located on the bagging machine. Ice is then dispensed into the ice bag until the desired weight is in the bag. At this point, a switch mounted on the weigh scale stops the flow of ice.

The volumetric bagging machine deposits the ice in a rotating chamber, which is adjustable in volume. After a predetermined volume of ice enters the changer, the ice is discharged into the bag below. Since the shape and size of the ice is not constant, the volumetric chamber is usually set to produce a 3 to 5% overage by volume. Therefore, the proper minimum weight of ice in the bag is assured. The bag closer consists of a mechanical unit that ties and seals the top of the bag with a wire ring, wire twist tie, or plastic clip. Smaller bagging operations do not use a bag closer, and the bags are manually closed with plastic ties, wire rings, staples, and so forth.

In more elaborate systems, the bag is formed from roll stock, filled with ice, automatically removed from the bagging machine, and automatically closed before it is dropped onto a conveyor, which carries the bagged ice to the refrigerated storage room. The degree of automation for the bagging and closing operation is determined by the number of bags of ice to be produced per day, the size of the bags, and the cost benefit relationship between automated equipment and reduced labor costs.

Ice bags are made of plastic, most often polyethylene, or, very rarely, heavy moisture-resistant paper. Plastic bags are used in most modern plants.

Packaged ice requires a refrigerated warehouse or room where the ice is stored prior to distribution. The ice storage is sized to meet the daily production of the plant and the distribution requirements. Generally, a bag ice storage has a capability of storing 3 to 7 days' production. Although the ice will not melt at a storage temperature of below 32 °F, it is important that the storage be maintained at a temperature between 10 and 25 °F. The lower temperature provides for a subcooling of the ice and avoids meltage during distribution. Depending on the type and quality of the ice being used, the bagged ice can contain some water. The percentage of water can range from 0 to 5%. For this reason, provision is made in the storage room refrigeration system for the product load of refreezing the water.

ICE SOURCE HEAT PUMPS

Ice making systems can be configured to provide heating alone, or heating and cooling, for a building or process. The conversion of water to ice occurs at a relatively high evaporator temperature and coefficient of performance compared to air source heat pumps operating at low ambient temperatures. Systems can provide necessary heating, with the resulting ice disposed of by melting with low-grade heat, such as solar. In addition, it can be used for useful cooling through daily, weekly, or seasonal storage.

The concept was originally considered mainly for residential heating and cooling. Currently, installations are proving feasible for larger structures, such as office buildings. Energy consumption savings resulting from the coefficient of performance of a conventional heat pump system are achieved. Using off-peak night and weekend rates can reduce power costs, and the ice produced is used for building cooling requirements. As a result, a system can be developed that consumes less energy, and the energy is consumed at a lower rate.

Ice source heat pumps follow two basic approaches. The first involves using the ice builder principle, with coils in a large tank, as the evaporator component of a heat pump system. The second approach uses a fragmentary type of ice maker as the evaporator of the heat pump system, with ice being stored in a tank as a mixture of ice and water. Many variations, combinations, and adaptations may be developed from these basic systems. The requirement for thermal storage of a large quantity of ice dictates new planning in architectural building design. Heating and cooling system designs for the building are also influenced.

BIBLIOGRAPHY

Dorgan, C.E. 1985. Icemaker heat pumps operation and design. ASHRAE *Transactions* 91(1).

Dorgan, C.E., G.C. Nelson, and W.F. Sharp. 1982. Icemaker heat pump performance—Reedsburg Center. ASHRAE *Transactions* 88(1).

CHAPTER 33

ICE RINKS

ANY level sheet of ice made by refrigeration (the term *artificial ice* is sometimes used) is referred to in this chapter as an ice rink regardless of use and whether it is located indoors or outdoors.

The freezing of an ice sheet is usually accomplished by the circulation of a heat transfer fluid through a network of pipes or tubes located below the surface of the ice. The heat transfer fluid is predominantly a secondary coolant such as glycol, methanol, or calcium chloride (see Chapter 18 of the 1993 ASHRAE *Handbook—Fundamentals*).

R-22 and ammonia are most frequently used for chilling secondary coolants for ice rinks. R-12 and R-502 have also been used; however, due to the phaseout of the CFC refrigerants, they should no longer be considered for use. R-22 will also be phased out in the future, so for new rinks R-22 and CFC replacements should be evaluated according to status and availability when equipment is being selected.

In some rinks, R-22, and ammonia to a lesser degree, have been applied as a direct coolant for freezing. The direct refrigerant rinks operate at higher compressor suction pressures and temperatures, thus achieving an increased COP, compared to secondary coolants. However, due to emissions regulations, the projected R-22 phaseout, building codes, and fire regulations, R-22 and ammonia should not be used to freeze ice directly in rinks.

APPLICATIONS

Most ice surfaces are used for a variety of sports, although some are constructed for specific purposes and are of specific dimensions. Usual rink sizes include:

Hockey. The accepted North American hockey rink size is 85 by 200 ft. Radius corners of 28 ft are recommended by professional and amateur rules. The Olympic and international hockey rink size is 96 by 196 ft, with 20-ft radius corners. Many rinks are 85 by 185 ft, 80 by 180 ft, and 70 by 170 ft and are considered adequate. In substandard size rinks, a corner radius of not less than 20 ft should be provided to permit the use of mechanical resurfacing equipment.

Curling. Regulation surface for this sport is 14 by 146 ft; however, the width of the ice sheet is often increased to allow space for installation and dividers between the sheets, particularly at the circles. Most are laid out on ice sheets measuring 15 by 150 ft.

Figure Skating. School or compulsory figures are generally done on a patch approximately 16 by 40 ft. Freestyle and dance routines generally require an area of 60 by 120 ft or more.

Speed Skating. Indoor speed skating has traditionally been on hockey-size rinks. The Olympic-size outdoor speed skating track is a 1400 ft oval, 35 ft wide with 392 ft straightaways and curves with an inner radius of 87.5 ft. Most speed skating ovals are outdoors; however, some recently constructed speed skating rinks are full size and indoors.

Recreational Skating. Recreational skating can be done on any size or shape rink, as long as it can be efficiently resurfaced. Generally, 30 ft^2 is allowed for each person actually skating; 25 ft^2 per skater is acceptable, except where a large number of preteens are skating. An 85 by 200 ft hockey rink with 28-ft radius corners has an area of 16,327 ft^2 and will accommodate a mixed group of about 650 skaters.

Public Arenas, Auditoriums, and Coliseums

Public arenas, auditoriums, field houses, and so forth, are designed primarily for spectator events. They are alternately used for ice sports, ice shows, and recreational skating, as well as for non-ice events, such as basketball, boxing, tennis, conventions, exhibits, circuses, rodeos, and stock shows. The refrigeration system can be designed so that, with adequate manpower, the ice surface can be produced within 12 to 16 h. However, general practice is to leave the ice sheet in place and hold other events on an insulated floor placed on the ice. This approach saves significant time, labor, and energy.

REFRIGERATION REQUIREMENTS

The heat load factors considered in the following section include type of service, length of season, usage, type of enclosure, radiant load from roof and lights, and the geographic location of the rink with associated wet- and dry-bulb temperatures. In the case of outdoor rinks, the sun effect and weather conditions must also be considered.

A fairly accurate estimate of refrigeration requirements can be made based on data from a number of rink installations with the pipes covered by not more than 1 in. of sand or concrete and not more than 1.5 in. of ice—a total of 2.5 in. sand or concrete and ice.

The refrigeration load may be estimated either by: (1) calculating the refrigeration necessary to freeze the ice to required conditions in a specified time, or (2) calculating the refrigeration necessary to maintain the ice surface and temperature during the most severe usage and operating conditions that coincide with the maximum ambient environmental conditions.

In the time-to-freeze method, the quantity of ice required (rink surface area multiplied by thickness) is calculated first. Then the refrigeration is determined to: (1) reduce the water from application temperature to 32 °F, (2) freeze the water to ice, (3) reduce the ice to the required temperature, and (4) handle the heat loads and system losses during the freezing period. The total requirement is divided by system efficiency and freezing period to determine the required refrigeration.

Example 1. Calculate the refrigeration required to build 1 in. thick ice on a 16,300 ft^2 rink in 24 hours.

Assume the following material properties and conditions:

The preparation of this chapter is assigned to TC 10.2, Automatic Icemaking Plants and Skating Rinks.

Material	Specific Heat, Btu/lb·°F	Temperature, °F Initial	Temperature, °F Final	Density or Weight
6 in. concrete slab	0.16	35	20	150 lb/ft^3
Supply water	1.0	52	32	62.5 lb/ft^3
Ice	0.49	32	24	—
Ethylene glycol, 35%	0.83	40	15	32,000 lb

Latent heat of freezing water = 144 Btu/lb
Building and pumping heat load = 50 tons of refrigeration
System losses = 15%
Mass of water = 16,300 ft^2 × 1/12 ft × 62.5 lb/ft^3 = 85,000 lb
Mass of concrete = 16,300 ft^2 × 1/2 ft × 150 lb/ft^3 = 1,230,000 lb

Then:

$$Q_R = (\text{Sys. losses})(Q_F + Q_C + Q_{SR} + Q_{HL})$$

where

Q_R = refrigeration requirement
Q_F = water chilling and freezing
Q_C = concrete chilling load
Q_{SR} = refrigeration to cool secondary coolant
Q_{HL} = building and pumping heat load
Q_F = 85,000 lb[1(52 − 32) + 144 Btu/lb + 0.49(32 − 24)]/(24 h × 12,000 Btu/h·ton) = 49.6 ton
Q_C = 1,230,000 × 0.16(35 − 20)/(24 × 12,000) = 10.3 ton
Q_{SR} = 32,000 × 0.83(40 − 15)/(24 × 12,000) = 2.3 ton
Q_R = 1.15(49.6 + 10.3 + 2.3 + 50) = 129 ton

When no time restrictions apply, the estimated refrigeration is the amount needed to offset the usage loads plus the coincidental heat loads during the most severe operating conditions. Table 1 lists approximate refrigeration requirements for various rinks with controlled and uncontrolled atmospheric conditions. Table 1 should only be used to check the calculated refrigeration requirements. Table 2 shows the distribution of various load components for basic construction, and estimated potential load reductions that may be obtained when energy-conserving design and operating techniques are used.

Heat Loads

Energy and operating costs for ice rinks are very significant, and these costs should be analyzed during design. A good estimate of required refrigeration can be calculated by summing the heat load components at design operating conditions. The heat loads for ice rinks consist of conductive, convective, and radiant components. Connelly (1976) collected the performance data summarized in Tables 2 and 3. The amount of control over each load source is indicated with an approximate percentage of the maximum reduction of the load that is possible through effective design and operation.

Table 1 Range of Ice Rink Refrigeration

4 to 5 Winter Months, Above 37° Latitude	ft^2/ton (refrigeration)
Outdoors, unshaded	85 to 300
Outdoors, covered	125 to 200
Indoors, uncontrolled atmosphere	175 to 300
Indoors, controlled atmosphere	150 to 350
Curling rinks, indoors	200 to 400
Year-Round (Indoors) (Controlled Atmosphere)	**ft^2/ton (refrigeration)**
Sports arena	100 to 150
Sports arena, accelerated ice making	50 to 100
Ice recreation center	130 to 175
Figure skating clubs and studios	135 to 185
Curling rinks	150 to 225
Ice shows	75 to 130

Table 2 Ice Rink Heat Loads, Indoor Rinks

Load Source	Approx. Max. Percentage of Total Load, *	Max. Reduction through Design and Operation, %
Conductive loads:		
Ice resurfacing	12	60
System pump work	15	80
Ground heat	4	80
Header heat gain	2	40
Skaters	4	0
Convective loads:		
Rink air temperature	13	50
Rink humidity	15	40
Radiant loads:		
Ceiling radiation	28	90
Lighting radiation	7	40
Total	100	

*Load distribution for basic rink without insulation below rink floor.

Table 3 Ice Rink Heat Loads, Outdoor Rinks

Load Sources	Approx. Max. Percentage of Total Load, *	Max. Reduction through Design and Operation, %
Conductive loads:		
Ice resurfacing	9	50
System pump work	12	80
Ground heat	2	40
Header heat gain	1	30
Skaters	1	0
Convective loads:		
Air velocity	0 to 15	10
Air temperature	0 to 15	0
Humidity	0 to 15	0
Radiant loads:		
Solar load	10 to 30	60
Total	100	

*Load distribution basic rink without below rink floor insulation.

Conductive Loads. If a rink is uninsulated, *heat gain from the ground* below the rink and at the edges averages 2 to 4% of the total heat load. Also, permafrost may accumulate and frost heaving, which is detrimental to both the rink and the piping, may result. In addition, heaving makes it more difficult to maintain a usable ice surface.

The heat gain from the ground and perimeter is highest when the system is first placed in operation; however, it decreases as the temperature of the mass beneath the rink decreases and permafrost accumulates. Ground heat gain is reduced substantially with insulation. Chapter 22 of the 1993 ASHRAE *Handbook—Fundamentals* gives details on computing heat gain with insulation.

Heat gain to the piping is normally about 2 to 4% of the total refrigeration load depending on length of piping, surface area, and ambient temperatures. The ice and frost that naturally accumulate on headers reduce the heat gain. Insulation can be applied to reduce the heat gain to the piping and keep ice from accumulating. A header designed for balanced circuit flow to the freezing grid, may, with precautions and using steel headers and piping, be imbedded within the rink floor. The imbedded headers contribute to the ice freezing and eliminate the trench to rink floor piping penetrations.

A circuit loop should be placed around the rink perimeter to prevent soft ice from developing at the edges (see the section on Rink Piping and Pipe Supports).

Heat gain from coolant circulating pumps can represent up to 12% of the refrigeration load. The pumps normally operate 24 h per day. The pumping heat load is the pump horsepower times the conversion factor of 2550 Btu/hp plus adjustment for the pump and motor operating efficiency. Energy consumption from pump operation can be reduced by using pump cycling, two-speed motors, multiple pumps, or variable-speed motors with the appropriate controls. Proprietary variable motor speed controls are also available. The coolant flow should be sufficient at all times for acceptable chiller operation and to maintain a balanced flow through the piping grid.

The refrigeration system auxiliaries, such as condenser pumps, condensers, cooling tower or evaporative condensers, and condenser fans consume substantial electrical energy. Appropriate design of the system and its control and good equipment selection should keep these auxiliary electric loads reasonable.

Ice resurfacing represents a significant operating heat load. About 0.25 in. of water is flooded onto the ice surface, normally at temperatures between 130 and 180°F, to restore the ice surface condition. The heat load resulting from the flood water application may be calculated as follows:

$$Q_f = 62.5\,V_f\,[1.0(t_f - 32) + 144 + 0.49(32 - t_i)]$$

where

Q_f = heat load, Btu/flood
V_f = flood water volume = area × thickness (normally 0.25 in.), ft³
t_f = flood water temperature, °F
t_i = ice temperature, °F

The resurfacing water temperature affects the load and time required to freeze the flood water. Maintaining good water quality through proper treatment may permit the use of lower flood water temperature and less volume.

Convective Loads. The convective load from the air to the ice may represent as much as 28% or more of the total heat load to the ice (Tables 2 and 3). The convective heat load is affected by air temperature, relative humidity, and air velocity near the ice surface. Precautions should be taken to minimize the influence of air movement across the ice surface in the design of the rink heating and dehumidification air distribution system. The convection heat load may be estimated using the procedure from Appendix 5 in the publication, "Energy Conservation in Ice Skating Rinks" (DOE 1980). The estimated convective heat transfer coefficient can be calculated using the formula:

$$h = 0.6 + 0.00318V$$

where

h = convective heat transfer coefficient, Btu/h · ft² · °F
V = air velocity over the ice, ft/min

The effective heat load (including the latent heat effect of convective mass transfer) is given by the following equation:

$$Q_{cv} = h(t_a - t_i) + [K(X_a - X_i)(1226\text{ Btu/lb})(18\text{ lb/mole})]$$

where

Q_{cv} = convective heat load, Btu/h · ft²
K = mass heat transfer coefficient
t_a = air temperature, °F
t_i = ice temperature, °F
X_a = mole fraction of water vapor in air, lb mol/lb mol
X_i = mole fraction of water in saturated ice, lb mol/lb mol

When the mole fraction of air is calculated using a relative humidity of 80% and a dry bulb of 38°F, X_a is approximately 6.6×10^{-3}, and X_i for saturated ice at 100% and a temperature of 21°F is 3.6×10^{-3}. On the basis of the Chelton Colburn analogy, $K \approx 0.17$ lb/h · ft² (DOE/TIC 1980).

In locations with high ambient wet-bulb temperatures, dehumidification of the building interior should be considered. This process lowers the load on the icemaking plant and reduces condensation and fog formation. Traditional air conditioners are inappropriate because the large ice slab tends to maintain a lower than normal dry-bulb temperature.

Radiant Loads. Indoor ice rinks create a unique condition where a large, relatively cold plane (the ice sheet) is maintained beneath an equally warm plane (the ceiling). The ceiling is warmed by conductive heat flow from the outside and by normal stratification of arena air. Up to 35% of the heat load on the ice sheet comes from radiant sources. On outdoor rinks radiant sources are the sun or a warm cloud cover. Vertical hanging cloth suspended from east-west horizontal overhead wires has been used to reduce the winter sun load.

In indoor and covered rinks, lighting is the major source of radiant heat to the ice sheet. The actual quantity depends on the type of lighting and how the lighting is applied. The direct radiant heat component of the lighting can be as much as 60% of the kilowatt rating of the luminaires. A radiant heating system can be another source of radiant heat gain to the ice. If radiant heat is used to maintain the comfort level in the promenade or spectator area, the radiant heaters should be located and directed to avoid direct radiation to the ice surface. The infrared components of the lighting can be estimated from manufacturers' data.

The infrared heat gain component from the ceiling and building structure, which is warmer than the ice surface, can be calculated by applying the Stefan-Boltzmann equation as follows:

$$q_r = A f_{ci} \sigma (T_c^4 - T_i^4)$$

$$f_{ci} = \left[\frac{1}{F_{ci}} + \left(\frac{1}{\epsilon_c} - 1\right) + \frac{A_c}{A_i}\left(\frac{1}{\epsilon_i} - 1\right)\right]^{-1}$$

where

q_r = radiant heat load, Btu/h · ft²
A_c = ceiling area, ft²
A_i = ice area, ft²
ϵ = emissivity
f_{ci} = gray body configuration factor, ceiling to ice surface
F_{ci} = angle factor, ceiling to ice interface (from Figure 1)
T = temperature, °R
σ = Stefan-Boltzmann constant = 0.1714×10^{-8} Btu/(h · ft² · °R⁴)

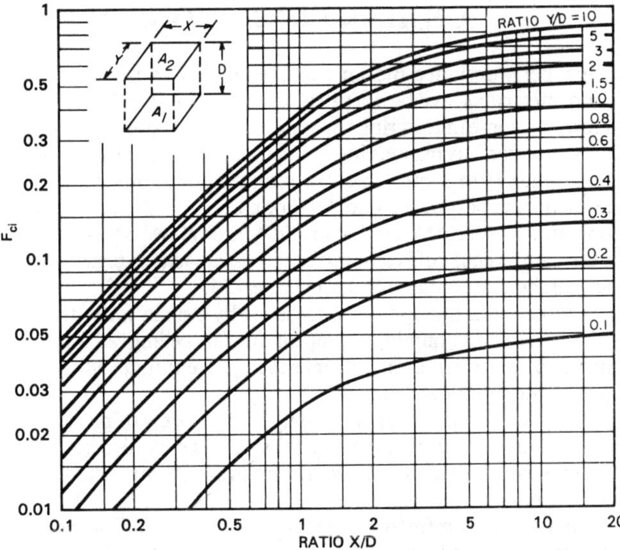

Fig. 1 Angle Factor for Radiation between Parallel Rectangles F_{ci}

Example 2. An ice rink has the following conditions:

Ice dimension: 85 ft × 200 ft = 17,000 ft^2
Ice temperature: 24 °F (484 °R), ϵ_i = 0.95
Ceiling radiating area: 90 ft × 200 ft = 18,000 ft^2
Ceiling mid-height: 25 ft
Ceiling temperature: 60 °F (520 °R), ϵ_c = 0.90

$$x/d = 90/25 = 3.5$$

$$y/d = 200/25 = 8.0$$

From Figure 1, F_{ci} = 0.68

$$f_{ci} = \left[\frac{1}{0.68} + \left(\frac{1}{0.90} - 1\right) + \frac{18{,}000}{17{,}000}\left(\frac{1}{0.95} - 1\right)\right]^{-1} = 0.611$$

Then:

$$q_r = 18{,}000 \times 0.611 \times 0.1714 \times 10^{-8}(520^4 - 484^4)$$
$$= 343{,}840 \text{ Btu/h}$$

The ceiling radiant heat load can be reduced by lowering the temperature of the ceiling, by keeping warm air away from the ceiling, by increasing the roof insulation, and, more significantly, by lowering the emissivity of the ceiling material to shield the ice from the building structure.

Ceiling and roof materials and exposed structural members have an emissivity that may be as high as 0.9. Special aluminum paint can lower the emissivity to between 0.5 and 0.2. Polished metal such as polished aluminum or aluminum foil have an emissivity of 0.05.

Also because a low-emissivity ceiling is cooled very little by radiant loss, most of the time its temperature remains above the dew point of the rink air, so condensation and dripping is substantially reduced or eliminated.

Low emissivity fabric or tiled ceilings frequently are incorporated into new and existing rinks to reduce radiation loads, decrease condensation problems, and reduce the overall lighting required.

Radiant heat gain to the ice, especially in outdoor rinks, can be further controlled by painting the ice about 1 in. below the surface with whitewash or slaked lime. Commercial paints with a low solar absorptivity, which are generally water based, are also available.

ICE RINK CONDITIONS

Properly designed indoor rinks, as well as properly designed renovated rinks, can be operated year-round without shut down. However, some indoor rinks operate from 6 to 11 months and shut down for various reasons including maintenance, rink construction, inability to control indoor conditions, or unprofitable operation during part of the year. Outdoor, uncovered rinks generally operate from early November to mid-March above 40° North latitude. However, if sufficient refrigeration capacity is provided, the ice can be maintained for a longer period.

Indoor rinks are operating successfully even in warm tropical climates. Relative humidity, temperature, and ceiling radiant losses must be controlled in these climates to prevent fog, ceiling dripping, and high operating cost.

Steel frame, brick, concrete, and various forms of plastic have been used to enclose ice skating rinks. Rinks have also been built under air-supported structures for seasonal use and usually over a multipurpose surface.

Arena heating is frequently provided for skater and/or spectator comfort and can be provided in conjunction with a dehumidification system. Heat recovery from the refrigeration system may be used for limited heating, supplementing the heating system, or dehumidification reheat. Ice rink temperatures are usually maintained between 40 and 60 °F; however, for skater or spectator comfort, higher temperatures are sometimes preferred. The relative humidity maintained in the arena depends on factors such as building construction, indoor temperature, and outdoor wet bulb.

The system should be designed to reduce fogging and ice surface condensation. Relative humidity at or below 80% at rink temperatures between 40 and 60 °F is usually sufficient to eliminate fogging; however, condensation can occur on the ceiling or roof structure due to radiation from the building structure to the ice. Low relative humidity is needed to reduce this condition when a high emissivity ceiling is exposed to the ice surface.

Ventilation should be the minimum required for the building occupancy so that the humidity introduced with outdoor air is kept as low as is feasible; but enough outdoor air must enter to maintain acceptable indoor air quality (see ASHRAE *Standard* 62-1989). Gas engine resurfacing machines should be equipped with catalytic exhaust convertors to reduce carbon monoxide emissions. The makeup air or ventilation air in humid climates should be dehumidified prior to being supplied to the arena.

Carbon monoxide and nitrogen dioxide are pollutant emissions from gasoline- or propane-fueled ice resurfacers. The concentration of these chemicals can reach dangerously high levels if they are not controlled or eliminated. In some areas, regulations require sensors to detect and alarm at unsafe chemical concentration. Check the health regulations for local requirements.

Each rink user group has its own preference for the type of ice used. Hockey players and curlers prefer hard ice; figure skaters prefer softer (*i.e.*, warmer) ice so they can clearly see the tracings of their skates; and recreational skaters prefer even softer ice, which minimizes the buildup of shavings and scrapings.

Since ice surface temperature can not be measured easily, the ice condition is customarily controlled either from a predetermined coolant average temperature or from the ice temperature measured by a themocouple embedded beneath the ice surface. At approximately 45 °F air temperature and one 1 in. ice thickness, ice at 20 to 22 °F is satisfactory for hockey, 24 to 26 °F for figure skating, and 26 to 28 °F for recreational skating. A 1 °F higher ice temperature may be feasible when water with a low mineral content is used for resurfacing. To achieve these ice temperatures, the coolant temperature is maintained about 6 to 10 °F lower than the ice temperature. The temperature of the coolant must be lowered to maintain the same ice conditions at higher wet-bulb temperatures or when abnormally high loads are experienced, such as when television lighting is used.

EQUIPMENT SELECTION

Compressors

Two or more refrigeration compressors should be used in an ice rink system. When two compressors are used, one compressor should be specified with ample capacity to maintain the ice sheet under normal load and operating conditions. When greater capacity is required during the initial ice freezing or under high heat loads, the second compressor picks up the load. In multiple compressor installations, a multistage thermostat and/or a motorized sequence control may be used to control the operation of the compressors. The use of multiple compressors serves as a backup to maintain the ice in the event of compressor failure or a service requirement.

Compressors and evaporators should operate at a suction pressure corresponding to a 10 °F mean temperature difference between the coolant and primary refrigerant in systems operating with secondary coolants, or between the ice and the refrigerant in direct refrigerant rinks.

Condensers and Heat Recovery

Wells, lakes, or rivers can be good sources of condenser cooling water, if they are available. Capacity is easy to regulate and the low coolant temperature maintains low condensing pressures, which saves energy. But, condensers require high quality water, which may need treatment to prevent scale formation, fouling, or corrosion in the condenser tubes.

Cooling towers used in conjunction with water-cooled condensers, evaporative condensers, or air-cooled condensers are alternatives. When selecting a cooling tower or evaporative condenser, not only the maximum expected wet-bulb temperature during the skating season should be considered, but also suitable controls to cover the wide range in capacities and protection against freezeup needed in cold weather. In addition, a water treatment specialist should be consulted.

Air-cooled condensers are used in northern climates, particularly where the rink is used only in the winter. They can be economically sized and require no water, so that the possibility of freezeup is eliminated. This type of condenser, however, is not economical for year-round operation, and for seasonal operation it must have wide-range capacity control. Heat rejected by the condensers can be recovered and used with water or air-cooled condensing systems.

When a cooling tower system is selected, heat from condensers can be used for such energy-saving applications as arena heating, subfloor heating, domestic water heating, and snow melting. This is generally done either by circulating the condenser cooling water through heat exchangers or with fan coil units. The circulated cooling water can also be used in conjunction with a heat pump as a heat sink/source for heating or cooling various areas within the rink building and for water heating.

Ice Temperature Control

Ice temperature may be controlled by various methods. Thermostats that sense the return coolant temperature or the differential temperature between the supply and return coolant can be used to control the refrigeration system. They may also be used in controlling operation of the coolant pump. To be effective, a differential sensor should sense a small temperature difference. The return coolant temperature can be sensed by multistage sensors that sense a larger temperature difference. Another strategy varies coolant flow by controlling the pump with a temperature sensor buried in the ice. Direct refrigerant systems can be controlled by regulating compressor operation with a sensor in the ice. This method has been used with a direct refrigerant impulse pumping system. Compressor capacity and pump operation may be controlled from the low-pressure receiver when refrigerant pumps are used to circulate the refrigerant.

Rink Piping and Pipe Supports

High flow rate secondary systems use standard mild steel pipe 0.75, 1, or 1.25 in. in diameter; thin-walled polyethylene plastic pipe 1 in. in diameter; or UHMW (ultrahigh molecular weight) polyethylene plastic pipe 1 in. in diameter. These are placed at 3.5- or 4-in. centers on the rink floor. A proprietary low flow rate secondary coolant system uses 0.25-in. tubing made of flexible plastic with tube spacing averaging 0.75 in. or one dual tube every 1.5 in. Direct refrigerant rinks generally use 0.63 to 0.88-in. steel tubing, which is placed on 3-in. centers for outdoor rinks and 4-in. centers for indoor rinks.

The pipe grid must be maintained as close to level as possible, regardless of the rink piping system used. When a pipe rink surface is of the open type with sand fill around and over the pipes, the conduit usually rests on pressure-treated sleepers set level with the subbase; however, the sleepers can be omitted in a rink that is to be operated year-round. The piping is then spaced with clips, plastic stripping, or punched metal spacers.

In permanent concrete floors, the pipe or steel tubes are supported on notched iron supports or welded chair supports. The latter must be used in the case of plastic pipe.

Headers and Expansion Tanks

Secondary coolant rinks using large-diameter pipe generally have the piping running lengthwise, with the supply and return headers across one end. Small-diameter tubing rinks generally run crosswise, with the supply and return headers along one side. Direct refrigerant rinks generally run lengthwise, with the supply header at one end and the return header at the opposite end in a balanced system. The header must be sized to assure an even distribution of coolant through every pipe. The systems are generally designed with low coolant velocities, which do not need balancing valves. If at all possible, the return header should be placed at the same elevation as the rink piping, with a minimum of two air vents to eliminate the trapping of air.

The three-pipe reversed return header and distribution arrangement (Figure 2) is commonly used. However, a properly sized two-pipe header system (Figure 3) is frequently applied and gives nearly uniform circuit flow with no discernible differences in the ice surface. To allow for thermal contraction and expansion, headers and main piping should be free to move without producing excessive stress.

Polyethylene distribution headers should only be used with proper allowances for expansion and contraction. The coefficient of thermal expansion for steel is relatively low and very close to that of concrete, while the polyethylene pipe expansion coefficient is much higher. Pipe clamp connections must remain accessible for inspection and tightening. Clamps are not considered permanent joints.

A closed secondary coolant system requires an expansion tank to safely accommodate the expansion and contraction of the coolant resulting from fluid temperature changes. The expansion tank must be installed so that it cannot be isolated from the system.

Coolant Equipment

The coolant circulating pumps must be sized for the particular type of rink and system involved. Large-diameter pipe rinks require 10 to 15 gpm per ton refrigeration to maintain the required 2 to 3 °F temperature differential between incoming and outgoing coolant. These operate at approximately 20 to 25 psig. Low flow rate dual tubing or mat rinks use about 2.4 gpm per ton. Differentials of 4 to 6 °F are normal, but 10 to 12 °F differentials can be experienced in high load conditions with no reduction in

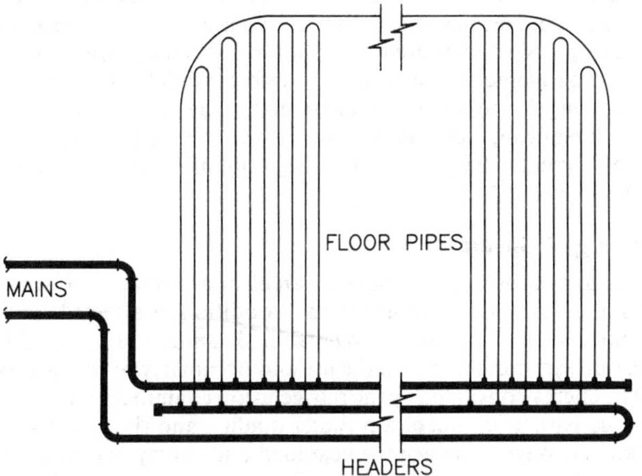

Fig. 2 Reverse Return System of Distribution

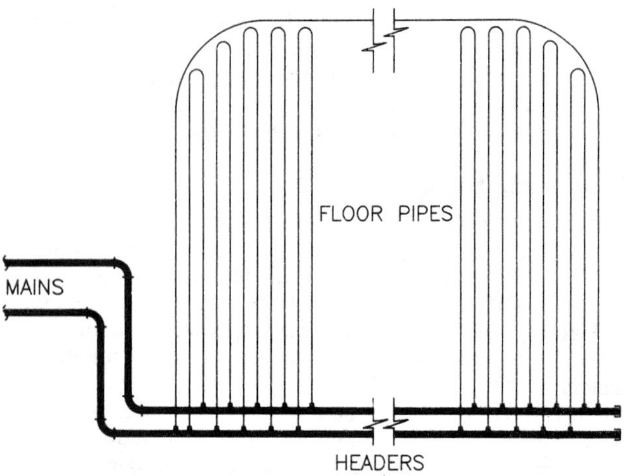

Fig. 3 Two Pipe Header and Distribution

ice quality. Uniform temperatures are achieved in mat rinks by temperature averaging between closely spaced, adjoining counterflow tubes operating at approximately 40 to 50 psig pressure.

For auditoriums and sports arenas, the rink surface should have provision for deicing in less than 4 h. In this operation, the floor is heated to about 50 °F so that the ice can be peeled off by first breaking the bond between the floor and ice and then breaking the ice and removing it with power tractors.

A standard heat exchanger can be used, with piping so arranged that all the coolant can be pumped through the heater, with the coolant flowing in the tubes and the steam or water in the shell. Approximately 350 Btu/h per square foot of rink surface is needed to heat the coolant in the system enough to warm the floor and break the ice bond.

In sports arena rinks designed for frequent deicing, ice may be required every other day, and 0.63 in. ice or more must be frozen in 12 h to allow the ice to temper before being skated on the following day. In such cases, a cold coolant accumulator keeps the size of the refrigeration equipment and connected power within reasonable limits. This accumulator is a coolant storage tank bypassed from the thawing cycle. When refrigeration on the ice surface is not required, a large volume of coolant in the accumulator may be cooled to approximately −25 °F and be ready to be pumped into the rink piping when needed. The cold coolant accumulator should store a sufficient volume to cool the entire cooling system coolant volume from 65 to 0 °F. This cold coolant tank usually holds more than three times the volume of the cooling system's charge.

With the increase in construction and operating costs, the use of accumulators has been declining. In place of accumulators, ice-making equipment is sized to handle the demand loads, and arenas are programmed to eliminate the need for quickly making and removing the ice. The use of phase change materials shows promise of reducing the accumulator size by 6 to 8 times, eliminating high demand charges, and allowing total off-peak operation at lower electric rates.

Energy Consumption

Energy consumption for an ice arena and rink facility is somewhat unique. Maintaining internal conditions is affected by the cold ice sheet. The lighting, ventilation, heating, and dehumidification systems depend on the use and occupancy of the facility. The energy consumed by the refrigeration equipment is affected by construction, operation, water quality, and the various use factors. Ways to reduce the heat load and energy consumption should be considered in both the design and operation of an ice rink. These include:

- Install low emissivity ceilings to reduce refrigeration and lighting loads and to permit compressors to operate at a higher saturated suction temperature
- Reclaim the refrigerant superheat to preheat shower water, heat the ice resurfacing water, melt ice shavings, heat the subfloor, etc.
- Select a pumping system and controls to reduce coolant flow during part load conditions
- Install an energy management system
- Insulate the subfloor and header piping
- Control the temperature and humidity in the arena to reduce sensible and latent heat gain to the ice
- Install high efficiency luminaires
- Use demineralized water or water with a very low mineral content for the ice and resurfacing

GENERAL RINK FLOOR DESIGN

Generally, five types of rink surface floors are used (Figure 4):

- Open or sand fill type, for plastic or metal piping or tubing
- Permanent, general-purpose type, with piping or tubing embedded in concrete on grade
- All-purpose type, with piping or tubing embedded in concrete with floor slab insulated on grade
- All-purpose floors, supported on piers or walls
- All-purpose floor with reheat; use this type when the water table and moisture are severe problems or when the rink is to operate for more than six months

The open sand fill floor is the least expensive type of rink floor. The cooling pipes rest on wood sleepers over a bed of crushed stone or other fill. The clean washed sand is filled in around the cooling pipes. Curling rink floors, as well as hockey and skating rinks, where first cost is a factor and the building is not intended to be used for other purposes, are usually constructed in this manner. Clay or cinders should never be used in the bed or for fill around the pipes. Tubing rinks do not need supports or sleepers; the tubes are laid on accurately leveled sand.

Rinks using 1-in. plastic pipe or the mat type are usually covered with sand to a depth of 0.5 to 1 in. to provide additional strength to the ice surface and to reduce cracking. Many portable outdoor rinks have used this arrangement for laying the plastic pipes or tubing mats on top of existing sodded areas, black top, or concrete. More permanent installations of outdoor semiportable rinks have used this same arrangement where recreational area is at a premium. Such an installation consists of steel pipes supported on notched steel sleepers, which in turn are supported on concrete piers down to solid ground.

To obtain a better return on investment, most indoor rinks that operate with an ice surface for only a portion of the year have a permanent general-purpose concrete floor with subfloor insulation and heat pipes so that the floor may be used for other purposes when the skating season is over. The floor should withstand the average street load and is usually designed with 1 or 1.25-in. steel or plastic pipe embedded in a steel-reinforced concrete slab 4 to 6 in. thick, depending on the anticipated loading and coolant pipe diameter.

In sports arenas, where the ice is removed and the floor made ready for other sports and entertainment, the ice floor must be constructed to withstand the frequent change from hot to cold, and the refrigerating machinery must be of sufficient capacity to freeze a sheet of ice 0.63 in. thick in 12 h. This type of floor is always insulated.

Subfloor insulation must be installed when quick changeovers are desired, when there is a high moisture content in the subsoil, when the floor is elevated, or when the rink is in continuous use for more than 9 months. This subfloor insulation serves to reduce the refrigeration load on ice-making equipment and to slow down,

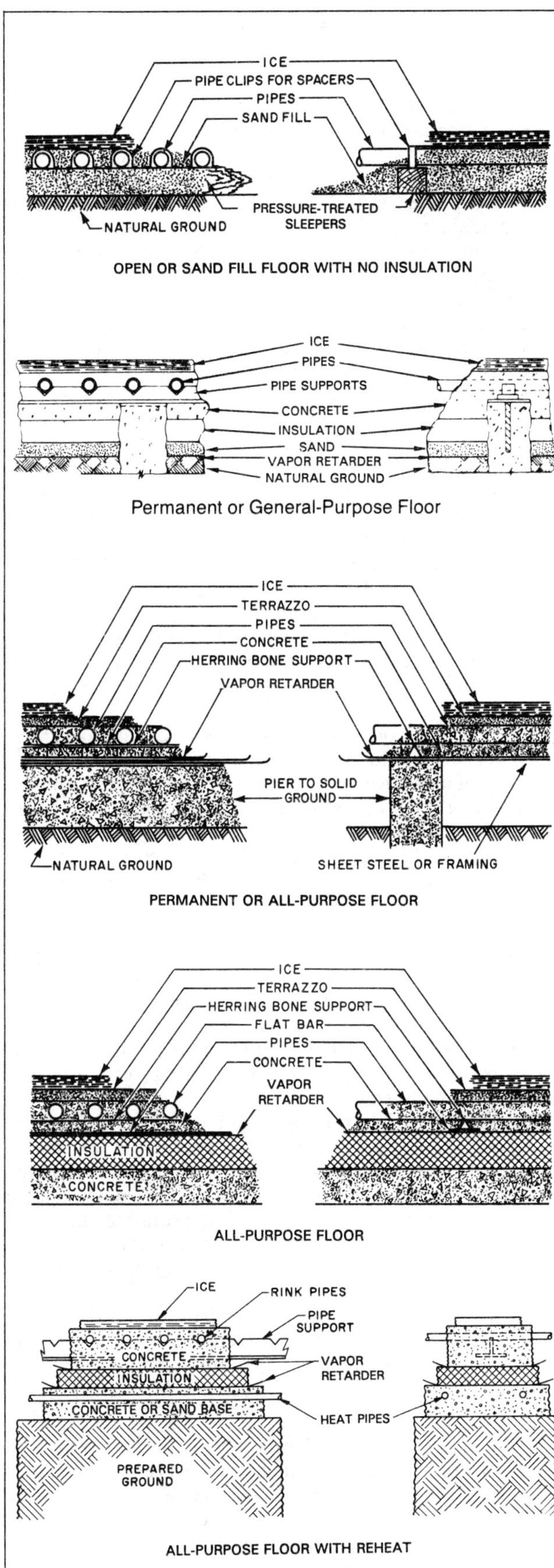

Fig. 4 Ice Rink Floors

but not eliminate, the cooling of the subsoil on surfaces installed on grade.

Drainage

The suitability of the subsoil on which an ice rink is built has a great influence on the rink's success. Complete ice surfaces have had to be rebuilt because of poor drainage and the ultimate heaving of the ice surface. Thus, skating rinks should not be built on swampy or low-lying land unless adequate drainage is provided.

Moist subsoil will freeze in the ground to a depth of 4 ft or more. The frozen water will heave the ice surface when freezing takes place at a depth of 6 in. or more. Heaving creates an uneven skating surface; moves and raises walls, piers, and header trenches; cracks walls and piping; and necessitates the eventual drainage and rebuilding of the rink floor.

Not only should there be a complete drainage system around the footings of the rink to prevent seepage, but there should also be one under the rink surface itself. This is particularly important when a sand fill floor is used; a good system will assure that the ice melted after the skating season will completely drain away and the sand will dry out as quickly as possible.

Subfloor Heating for Freeze Protection

Subfloor heating by electrical heating cables or a pipe or tubing recirculating system using a warm antifreeze solution is found in most new rinks to prevent below floor permafrost development and the resultant heaving. Pipes or tubing are on 12 to 24-in. centers located under 2 to 4 in. of insulation. They are generally installed in sand rinks, which are used year-round, although they may be poured into a concrete base slab with insulation between the base slab and the rink slab (see Figure 4).

Alternatively, the heating pipes may be laid directly in the subfoundation below the rink pipe or insulation. However, an installation not equipped with insulation requires a greater depth between the heating pipes and the ice-making pipes to prevent an increased load.

Neither water nor warm air should be used for subfloor heating. Water, if inadvertently allowed to freeze, cannot be readily melted out. In time, warm air ducts become filled with frost and ice because of high rink humidity and air duct leakage.

Usually, the same fluid used for the coolant in the ice-making system is used for subfloor heating; it can be heated to the necessary 40 to 50 °F in a heat exchanger warmed by compressor waste heat. Subfloor insulation should be of a rigid moistureproof board, such as high-density polystyrene foam, and be completely enveloped in a polyethylene vapor retardant.

Preparation for Rink Floor

When building on natural ground, regardless of whether a sand fill or a permanent general-purpose floor is intended, proper preparation of the bed is important unless the rink is built on elevated sand and gravel subsoil. If the rink is to be built on clay, part clay, or rock subsoil, water should be prevented from collecting in low areas. Either the clay or rock should be excavated or the rink level should be built up with crushed stone and gravel to a height of about 4 ft, after which it should be well rolled. Water should not be used for settling the fill.

In the case of sand fill rinks, quickly draining the melted ice at the end of the skating season will ensure rapid drying of the sand and rink piping and result in a longer life for the steel piping. Cinders should never be used as fill in open sand fill rinks because of the possibility of sulfur in the cinders which, when damp, accelerates corrosion of steel piping.

Care must be taken to assure a level surface over the entire rink with no more than ±0.13 in. in any 10 ft^2 area and ±0.25 in. overall.

Permanent General-Purpose Rink Floor

When constructing a permanent general-purpose floor, the same precaution must be taken regarding subsoil as for a sand fill rink. The concrete floor should withstand, at a minimum, the average road pavement load.

When local conditions make it advisable, the rink floor should be insulated. Insulation may be laid on a level concrete or sand base.

The concrete mixture should have a 28-day strength of 3000 to 5000 lb/ft^2 and be put in place in a quality manner (a concrete engineer is recommended to specify concrete, its placement, and curing). Suitable cross-reinforcing is necessary in addition to pipe supports.

Concrete floors with mat-type tubing are poured in two courses. A first course is poured and leveled; the mats are then rolled out and positioned. A 6 by 6 in. wire mesh is laid on top of the mats; then a second course, with grouting between it and the first, is poured on top of the first course, mats, and wire. Water pressure should be kept in the tubing to spot any leaks or cuts that may develop. Once started, the pouring of each course of the concrete floor should be continuous with interruptions not to exceed 15 min.

General-purpose rink floors should not be defrosted too frequently. When a rink constructed with a general-purpose floor is to be used during the ice season for purposes that require an ice-free floor, it is preferable to place an insulated portable-section wood floor over the ice for each occasion.

All-Purpose Floors

If a rink floor as used in sports arenas is to withstand both the expansion and contraction of frequent frosting and defrosting and thermal shock because of the circulation of very low-temperature coolant, then extra precautions must be taken in its construction, such as provisions for the free movement of the freezing slab with respect to the subfloor.

Header Trench

A well-constructed header trench of sufficient size to house the headers and connections and the subfloor heating system, if applicable, is essential unless the steel distribution headers are cast into the concrete slab as part of the rink. Provisions for movement of pipes due to thermal expansion and contraction should be incorporated into the design. This trench should be equipped with removable covers and be well-drained to facilitate drying out. The headers and piping in the trench are not usually insulated, which allows for periodic inspection and painting of the piping. However, unless a large trench is provided, consideration should be given to insulating the headers on rinks that operate year-round because of the massive buildup of frost. Provision must be made for purging air from the rink piping and header system.

Snow Pit

A snow-melting pit should be provided at a suitable point, usually at one end of the rink. It should be of sufficient size to handle the scraped off snow and the ice accumulated during planing, or it may be made large enough to accommodate the complete ice removal.

Discharge water from shell-and-tube condensers, a waste heat recovery system, or some other heat source should be provided to melt the snow and ice. An average load for a snow melting pit from a mechanical ice resurfacer is between 140,000 and 180,000 Btu/h for a 16,300 ft^2 rink. A large drain with overflow, as well as a large removable screen to filter out trash, should be provided.

BUILDING, MAINTAINING, AND PLANING ICE SURFACES

Regardless of the type of rink floor used, when the plant is first placed in operation, the equipment should be operated long enough for a sharp frost to appear on the surface. Then the entire surface should be uniformly covered with a fine spray. This process should be repeated until a 0.5 in. thickness of ice is built, or until the surface is level. After applying a layer of water base white paint, another 0.38 in. thick layer of ice is built before painting the red and blue lines. Red and blue lines are available in plasticized paper; however, they need to be covered with a minimum of 0.5 in. of ice to protect against damage. It is essential that sand floors be thoroughly wet before freezing because dry sand has poor conductivity. The surface should not be frozen any colder than required after this buildup so as to allow the ice to temper before it is used for skating and also to deter cracking.

To maintain an ice surface, it is customary to scrape off the snow after each skating session or hockey period. In all but the smallest rinks, this is done by a motorized resurfacer. On small rinks, the scraping is done manually with a wide hardened-steel scraper blade. The most satisfactory method of resurfacing the ice between sessions is to use a sprinkler tank filled with hot water, which is wheeled over the ice. The sprinkler has an adjustable valve to control the quantity of water, which is sprayed into a terry cloth bag that wipes the fine snow off the ice surface and fills the crevices cut by the skaters. In this manner, the least amount of water is added, reducing the ice buildup and refrigeration load.

By far the most common method is the use of automatic resurfacing machines. Mounted on a four-wheel drive chassis, the machines plane the ice, pick up the snow, and lay down a new ice surface using hot or cold water. Hot water generally gives harder ice, since air bubbles are removed, but high energy costs have led many rinks to alternate hot and cold water resurfacings. Rink corners should be at least a 20-ft, preferably 28-ft, radius for effective use of this equipment. Smaller equipment is available for studio and small rinks.

Because of inattentive ice making, improper sprinkling equipment, or deep cutting of the ice during public skating, the ice may become uneven and excessively thick. There may be a fairly slight variation in the ice thickness across the rink, but more serious is the resulting variation in the condition of the ice. In any case, the low spots on the ice must be built up, increasing the thickness and refrigeration requirements.

For example, under assumed conditions, where 18 °F coolant would be cold enough to hold a 1.5 in. thickness of ice, calculations show that −5 °F coolant would be required if the ice were permitted to build up to 6 in., with a corresponding decrease in effective refrigeration capacity and an increase in operating costs. In other words, every additional inch of ice thickness required from the refrigeration system increases 8 to 15%, depending on system heat load (DOE 1980).

Since ice of 0.5 to 1 in. thickness is satisfactory for skating and is the most economical thickness to freeze and hold, the ice should be periodically planed to maintain the desired thickness.

Water Quality

The quality of the water affects energy consumption and ice quality. Water contaminants, such as minerals, organic matter, and dissolved air, can affect both the freezing temperature and the ice thickness necessary to provide satisfactory ice conditions. Proprietary treatment systems for arena flood water are available. When these treatments are properly applied, they reduce or eliminate the effects of contaminants and improve ice conditions.

RINK FOG AND CEILING DRIPPING

During mild weather, particularly in early fall and late spring in the northern United States and Canada, condensation often drips from the roofs and roof supports of rinks (especially curling rinks), due to construction, internal conditions, or insufficient internal heat loads. The condensate dropping on the ice ruins the curling surface and the fog obstructs the view. These conditions cannot be solved by ventilation because the introduction of outdoor air only aggravates the problem when the weather is mild and humid. Insulating the roof also aggravates the drip during mild outside weather conditions. Low-emissivity ceilings stay warmer and thus reduce condensation and drip.

Under these conditions, to prevent condensation in the roof space and to clear the fog, a six-sheet curling rink, with 12,500 ft^2 of ice, would require the removal of 33 lb of moisture per hour, necessitating about 6 tons of refrigeration.

Units using the coolant from the rink piping as a cooling medium avoid frosting by recirculating with a small bypass pump to keep coil inlet brine above 32 °F. Reheat coils are often included and frequently use back waste heat from condenser cooling water to counter the cooling effect of the dehumidifier coil. Self-contained, air-cooled, compressor-type packaged dehumidifying units, as well as desiccant drier types with gas or electric regeneration, are available.

Various dehumidification and defogging systems should be evaluated in an owning and operating cost analysis. Unlike the normal behavior of moist air, air with high moisture content does not rise in an ice rink. Instead, the air remains near the ice surface because it is colder than the surrounding air. The air circulation system should remove the cold, moist air away from the ice with minimum draft on the ice surface. Increased air velocity near the ice surface increases the heat load to the ice and can cause surface wetness. Air circulation is also important to remove carbon monoxide from resurfacing equipment, which tends to remain below the top of the dasher boards and near the ice surface.

IMITATION ICE SKATING SURFACES

A number of different imitation ice-skating surfaces have been marketed; these use semiporous plastic panels dressed with a synthetic lubricant. The coefficient of friction of ice is approximately 0.03 at 26 °F and is even less because of the film of water produced by pressure under the skate. In considering the use of imitation surfaces, the actual friction coefficients of these surfaces—both when freshly lubricated and after a period of usage—should be investigated.

BIBLIOGRAPHY

Albern, W.F. and J.J. Seals. 1983. Heat recovery in an ice rink? They did it at Cornell University. ASHRAE *Journal* 25(9):38-39.

ASHRAE. 1989. Ventilation for acceptable indoor air quality. ANSI/ASHRAE *Standard* 62-1989.

ASHRAE. 1968. Ice skating rinks. *Symposium* at ASHRAE meeting in Columbus, OH.

Banks. N.J. 1990. Desiccant dehumidifiers in ice arenas. ASHRAE *Transactions* 96(1):1269-72.

Blades, R.W. 1992. Modernizing and retrofitting ice skating rinks. ASHRAE *Journal* 34(4):34-42.

Brauer, M., J.D. Spengler, K. Lee, and Y. Yanagisana. 1992. Air pollutant exposures inside hockey rinks: Exposure assessment and reduction strategies. *Proceedings* Second International Symposium on Safety in Ice Hockey, Pittsburgh, PA.

Canadian Electrical Associates. 1992. Potential electricity savings in ice arenas and curling rinks through improved refrigeration plant. CEA No. 9129-858 Manbek Resource Consul Book.

Connelly, J.J. 1976. ASHRAE *Seminar* on Ice Rinks (February), Dallas, TX.

DOE. 1980. Energy conservation in ice skating rinks. Prepared by B.K. Dietrich and T.J. McAvoy. U.S. Department of Energy.

Matus, S.E. *et al.* 1988. Carbon monoxide poisoning at an indoor ice skating facility. *Proceedings* ASHRAE IAQ 88 Conference, pp. 275-283.

Minnesota Department of Health. 1990. Indoor air quality unit: Regulating air quality in ice arenas.

Rein, R.G. and C.M. Burrows. 1981. Basic concepts of frost heaving. ASHRAE *Transactions* 87(2):1087-97.

CONCRETE DAMS AND SUBSURFACE SOILS

REFRIGERATION is one of the more important tools of the heavy construction industry, particularly in the temperature control of large concrete dams. It is also used to stabilize both water-bearing and permanently frozen soil. This chapter briefly describes some of the cooling practices that have been used for these purposes.

CONCRETE DAMS

Without the application of mechanical refrigeration during construction of massive concrete dams, much smaller construction blocks or monoliths would have to be used, which would slow construction. By removing unwanted heat, refrigeration can speed construction, improve the quality of the concrete, and lower the overall cost.

METHODS OF TEMPERATURE CONTROL

Temperature control of massive concrete structures can be achieved by (1) selecting the type of cement, (2) replacing part of the cement with pozzolanic materials, (3) using embedded cooling coils, or (4) precooling the materials. The measures used depend on the size and type of structure and on the time permitted for its construction.

Cement Selection and Pozzolanic Admixtures

The temperature rise that occurs after concrete is placed is due principally to the cementing materials' heat of hydration. This temperature rise varies directly with cement content per unit volume and, more significantly, with the type of cement. Ordinary Portland cement (Type I) releases about 180 Btu/lb, half of which is typically generated in the first day after the concrete is placed. Depending on specifications, Type II cement may generate slightly less heat. Type IV is a low heat cement that generates less heat at a slower rate.

Pozzolanic admixtures may be used in lieu of part of the cement. Such pozzolans include fly ash, calcined clays and shales, diatomaceous earths, and volcanic tuffs and pumicites. The heat-generating characteristics of these pozzolans vary, but are generally about one-half that of cement.

When determining the system refrigeration load, heat release data for the cement being used should be obtained from the manufacturer.

Cooling with Embedded Coils

In the early to mid 1900s, the heat of curing on large concrete structures was removed by embedded cooling coils for glycol or water recirculation. Further, they lowered the temperature of the structure to its final state during construction. This is desirable where volumetric shrinkage of a large mass is necessary during construction, for example, to allow the contraction joint grouting of intermediate abutting structures to be completed.

In the past, thin wall tubing was placed as a grid-like coil on top of each 5 or 7.5 ft lift of concrete in the monoliths. Chilled water was then pumped through the tubing, using a closed loop system to remove the heat. A typical system used 1-in. OD tubing with a flow of about 4 gpm through each embedded coil. While the number of coils in operation at any time varies with the size of the structure, 150 coils is not uncommon in larger dams. Initially, the temperature rise in each coil can be as much as 8 to 10 °F, but it later becomes 3 to 4 °F. An average temperature rise of 6 °F is normal. When sizing the refrigeration equipment, the heat gain of all the circuits is added to the heat gain through the headers and connecting piping.

For a typical system with 150 coils based on a design temperature rise of 6 °F in the embedded coils and a total heat loss of 6 °F elsewhere, the size of the refrigeration plant would be about 300 tons. Figure 1 shows a flow diagram of a typical embedded coil system.

Cooling with Chilled Water and Ice

The actual temperature of the mix at the time of placement has a greater effect on the overall temperature changes and subse-

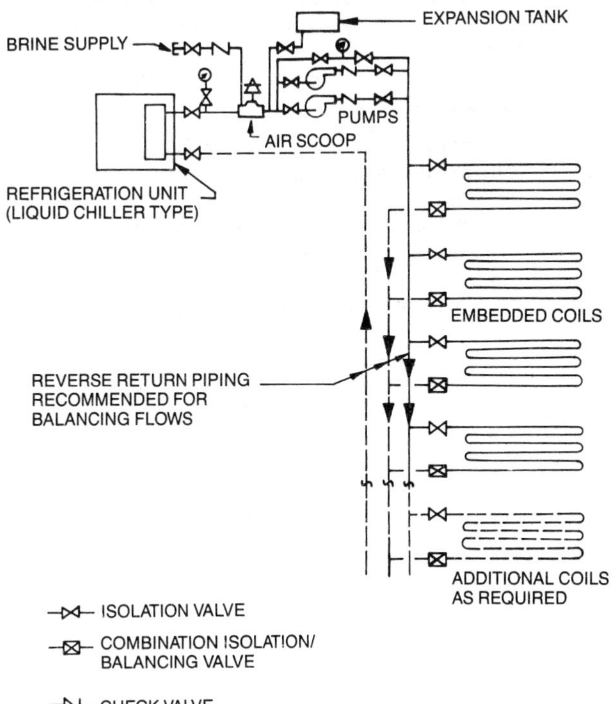

Fig. 1 Flow Diagram of Typical Embedded Coil System

The preparation of this chapter is assigned to TC 10.1, Custom Engineered Refrigeration Systems.

quent contraction of the concrete than any change caused solely by varying the heat-generating characteristics of the cementing materials. Further, placing the concrete at a lower temperature normally results in a smaller overall temperature change than that obtained with embedded coil cooling. Because of these inherent advantages, precooling measures have been applied to most concrete dams.

Glen Canyon Dam illustrates the installation required. The concrete was placed at a maximum placing temperature of 50 °F during summer months when aggregate temperature was about 87 °F, cement temperature was as high as 150 °F, and the river water temperature was about 85 °F. Maximum air temperatures averaged over 100 °F during the summer months. The selected system included cooling aggregates with 35 °F water jets on the way to the storage bins, adding refrigerated mix water at 35 °F, and adding flaked ice for part of the cold water mix. Subsequent cooling of the concrete to temperatures varying from 40 °F at the base of the dam to 55 °F at the top was also required. The total connected brake power of the ammonia compressors in the plant was 6200 hp, with a refrigeration capacity equivalent to making 6000 tons of ice per day.

The maximum amount of chilled water that may be added to the concrete mix is determined by subtracting the amount of surface water from the total mix water, which is free water. Frequently, if a chemical admixture is specified, some water must be added to dissolve the admixture—usually about 20% of the total free water. This limits the amount of ice that can be added to the remaining 80% of free water available. After the amount of ice that is needed for cooling is determined, the size of the ice-making equipment can be fixed. When determining the capacity of the equipment, allowances should also be made for cleaning, service time, and storage for the ice during nonproductive times.

When calculating the actual heat removal, the ice is considered to be 32 °F when introduced into the mixer. The chilled water is assumed to be 40 °F entering the mixer, even though the chiller may supply the water at a lower temperature.

Cooling by Inundation

The temperatures specified today cannot be achieved solely by adding ice to the mix. In fact, it is not possible on heavy construction of this type (in view of low cement content and low water-cement ratio specified) to put a sufficient amount of ice in the mix to obtain the specified temperatures. As a result, inundation (deluging or overflowing) of aggregates in refrigerated water was developed and was one of the first uses of refrigeration in dams.

Table 1 Temperature of Various Size Aggregates Cooled by Inundation

	Aggregate Size, in.				
Time, min.	**6**	**3**	**1.5**	**0.75**	**0.375**
1	85 °F	69 °F	49 °F	39 °F	38 °F
2	77	59	41	38 (42)	
5	66	46 (45)	38 (42)		
10	56	40 (42)			
15	50	38			
20	46				
25	44				
30	42				
40	40				
50	39				

Source: Peugh, V.L. and I. Tyler. 1934. "Mathematical theory of cooling concrete aggregates." In *The inundation method of cooling concrete aggregates at Bull Shoals Dam* (House 1949). Numbers in parentheses are from tests by R. McShea.

Note: The temperatures listed are at the center of a cobble with an assumed thermal diffusivity of 0.07 ft^2/h. The aggregate initial temperature is 90 °F and the cooling water temperature is 35 °F.

When aggregates are cooled by inundation with water, generally the three largest sizes are placed in large cylindrical tanks. Normally two tanks are used for each of the three aggregate sizes to provide backup capacity and a constant flow of materials. The cooling tanks, loaders, unloaders, chutes, screens, and conveyor systems from the tanks into the concrete plant should be enclosed and cooled from 45 to 40 °F by refrigeration units, with blowers placed at appropriate points in the housing around the tanks and conveyors.

Peugh and Tyler (House 1949) determined the inundation (or soaking) time required by calculations and actual tests. Pilot tests corroborated their computations of aggregates' cooling times as indicated in Table 1.

This study indicated that an immersion period of about 40 min will bring the aggregate down to an average temperature of 40 °F. Theoretically, smaller sizes can be brought to the desired temperature in less time. Considerations, such as the rate at which cooling water can be pumped, make it unlikely that a cooling period less than 30 min should be considered. Any excess cooling will provide the needed safety factor. However, the limiting factor on the overall cycle is the cooling time for the largest aggregate, which is nearly 45 min, plus about 15 min for loading and unloading. Backup capacity should be considered for maintaining a constant flow of materials.

Air Blast Cooling

Following the inundation method's development, air-blast cooling was developed. This method of cooling aggregates does not require particular changes to handling the materials or additional tanks for inundation; the aggregate is cooled by blowing cold air through the aggregate in the batching bins above the concrete mixers. Also, the air cycle used in cooling can be used in heating aggregates during cold weather. The aggregate is cooled during the final stage of handling; this does not increase the moisture content.

The compartmented bins where air-blast cooling is usually accomplished are generally sized so that if any supply breakdowns occur, the mixing plant will not have to shut down before a particular pour can be completed. On this basis, the average concrete octagonal bin above the mixing plant on a large job holds at least 600 yd^3. The size is usually more than adequate to allow time for air cooling. However, certain minimum requirements must be considered. If possible, based on a one-hour loading and cooling schedule, at least 2 h of storage volume should be provided for each size aggregate that air will cool. The minimum volume should be 1.5 h plus cycling time, based on the tables shown for cooling 6-in. aggregate.

The bin compartment analysis shown in Table 2 may be used as a starting point in determining the air refrigeration loads and static pressures. This type of analysis should give approximately equal storage periods for each aggregate size used. In practice, after air cycle cooling is calculated, more air volume is needed in

Table 2 Bin Compartment Analysis for Determining Refrigeration Loads and Static Pressures

Material Size, in.	lb/yd³	% of Total	No. of Bin Compartments	Bin %
Stones				
6 to 3	800	21.33	4	25.00
3 to 1.5	700	18.67	3	18.75
1.5 to 0.75	650	17.33	2	12.50
0.75 to 0.25	600	16.00	3	18.75
Sand	1000	26.67	4	25.00
Total	3750	100	16	100

Note: In practice, these relative sizes will vary from time to time; the amounts shown are for an assumed design mix of the principal classes of concrete. On any given job, several design mixes requiring different amounts for each size are needed.

Table 3 Resistance Pressure

Aggregate Size, in.	Velocity, fpm	Resistance Pressure, in. water/ft of ht.
6 to 3	300	0.29
3 to 1.5	200	0.32
1.5 to 0.75	100	0.27
0.75 to 0.25	60	0.30

the smaller aggregate compartments, and less in the sand and aggregate, to obtain maximum cooling. This is because of the higher air resistance in the smaller aggregate sections and the fact that air does not cool the sand compartment as effectively.

Computing Air-Blast Cooling Loads. To calculate the required cooling, several assumptions must be made:

Assumption 1. Normally, the lowest temperature of air leaving the cooling coils is 38 to 40 °F. While lower air temperatures may be achieved, these should not be trusted because a temperature lower than 35 °F usually causes rapid frosting of the coils.

Assumption 2. The heat transfer between the aggregate and air will be only 80 to 90% effective. To allow for air temperature rise in the ducts, heat leakage, pressure drop, etc., an empirical factor of 85% may be used. Thus, with 80 °F aggregate and 40 °F cooling air, the effective temperature differential would be 0.85 (80 − 40) = 34 °F.

Assumption 3. The rise in temperature of air passing through the aggregate compartment normally will not exceed 80% of the difference between the entering aggregate and the entering air temperatures. Thus with 80 °F aggregate and 40 °F air, the temperature rise of the air is 0.80 (80 − 40) = 32 °F. The maximum temperature of the return air is 40 + 32 = 72 °F.

Assumption 4. An allowance should be included for heat leakage into the air ducts on the aggregate bin sides—normally about 2% of a total air-blast cooling load.

Another important consideration is the static pressure against air flowing through a body of aggregates. The resistance to air flowing through a bin varies as the square of the air volume or velocity; this is summarized in Table 3. The resistance pressure listed is for a unit height of aggregate. To use the values shown in Table 3, the cross-sectional area of the aggregate compartment and the height of the aggregate column must be known. The manufacturers of concrete mixing bins and mixing equipment can supply this information.

Other Cooling Methods

As specifications continued to require lower and lower placing temperatures, direct methods of cooling sand and cement were initiated to obtain the overall heat removal required. Sand, but not cement, can be cooled by inundation and spray methods. But the free moisture content of the sand when batched causes trouble in correctly proportioning the mix.

Attempts have been made to cool sand using hollow flight screw conveyors through which chilled water flows. Since cooling either of these components below the dew point could result in condensation and serious handling problems, this method shows little promise. A vacuum system, which evaporates the surface moisture to reduce the aggregate temperature, has also been used.

For concrete pours less than 200,000 cubic yards, liquid nitrogen is sometimes used to reduce the temperature to the final pour temperature. Nitrogen is used because it costs less than a mechanical system.

SYSTEM SELECTION PARAMETERS

For larger installations, plant selection depends on a number of factors, including the following:

1. Normal pouring rate, yd^3/h (contractor or contract specified).
2. Maximum pouring rate, yd^3/h (contractor or contract specified).
3. Total allowable mixing water, lb/yd^3 (usually contract specified).
4. Required concrete placement temperature (usually contract specified).
5. Concrete temperature when coming from the mixer (to be determined considering materials handling to placement site, time in transit, and storage at placement site).
6. Average ambient temperature and aggregate temperature during period of maximum placement. The average ambient temperature of the aggregates is assumed to be the mean ambient temperature—including night and day—during the period of storage, which is determined by the amount of storage capacity provided by the contractor and the rate of concrete placement. If the minimum live storage is, for example, 100,000 tons of aggregates and the pouring rate is 2000 yd^3/day, the consumption would be approximately 3800 tons of aggregates per day. If the job is working a 6-day week, this would provide 26 days of storage. Unless weather conditions are unusual, the temperature of the rock delivered to the reclaiming tunnel is assumed to equal the average ambient temperature for the 26 days preceding the delivery.
7. Specific heat of materials. The specific heat of the aggregates and sand may vary with project location. Typical values include: sand = 0.106 Btu/lb·°F; aggregates = 0.12 Btu/lb·°F; water = 1.0 Btu/lb·°F; cement = 0.12 Btu/lb·°F.
8. Heat release rate for cement (material specifications).

Where the aggregate cooling range from initial to final temperature of the mix is relatively small (15 to 20 °F), or where the required pour temperature is relatively high (65 °F or more), chilled water plus ice in the mix or chilled water plus air blast on larger aggregates may handle the entire cooling load. When the overall temperature reduction is greater than this, or when lower pour temperatures are specified, such as 50 °F or less, a combination of all three types of cooling will probably be required because only a limited amount of heat removal can be obtained by one of these methods alone.

Cooling by air blast alone is limited by the entering air temperature, which must be maintained high enough to prevent coil frosting. Cooling by inundation, although it requires large inundation tanks, offers the most positive and sure method of cooling. Ice can also be added to the mix to remove the remainder of the heat. The result is a very satisfactory blending of the aggregates and exact control of the amount of water in the mix.

CONTROL OF SUBSURFACE WATER FLOW

Refrigeration has been used successfully since 1880 to freeze moisture in unstable and water-bearing soil and to stop underground flows of water in pervious material or gravelly stream beds. Other common methods of stopping the flow of water include sheet piling, cement grout, chemicals, and well points. In many cases, freezing has been the last resort after other methods are unsuccessful. In a number of cases, a combination of well points, grouting, and freezing have solved the problem.

Using refrigeration, the common practice is to lay a series of concentric pipes in a line or arch pattern in the path of the subsurface water flow. A wall of earth is frozen by pumping cold brine down the inside pipe and letting it flow back through the annular space between the inside and outside pipe. The growth of frozen soil on the outside of the concentric pipe proceeds until it connects with the frozen cylinder formed on the adjacent pipe.

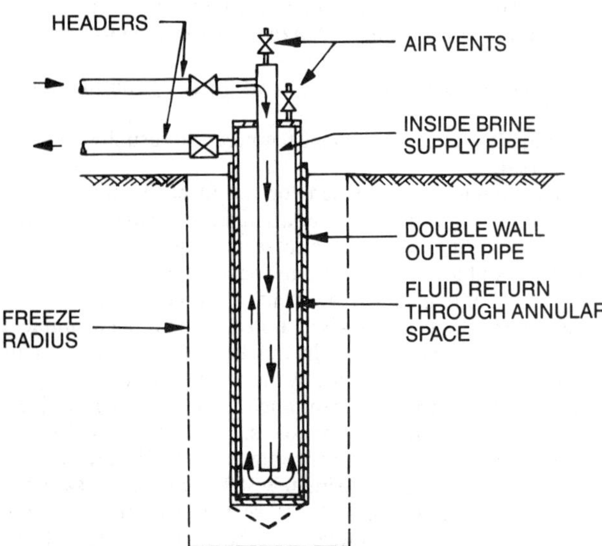

Fig. 2 Typical Freezing Point

Spacing between pipes can vary, depending on the time available to complete the wall of ice, but spacings of 2 to 4 ft on center have been used successfully. Freezing pipes have been successfully used to control subsurface water in both dam and mining projects. Figure 2 shows a typical freezing pipe and system diagram.

The brine or refrigerant and method of containment should be carefully selected in system design. If a leak develops in the system, groundwater contamination could result or the soil saturated with the refrigerant may have to be excavated and cleaned, depending on the type of refrigerant used. Double-wall piping, or an environmentally safe refrigerant that will vaporize when exposed to the air, should be considered.

Ammonia is generally used as the basic refrigerating medium. Then brine, chilled by the ammonia, is circulated through the freezing pipes. The brine is commonly calcium chloride ($CaCl_2$), but magnesium chloride ($MgCl_2$) is recommended as it has less tendency to precipitate and clog the piping at low temperatures. To monitor the progress of freezing, thermocouples are normally located throughout the area to be frozen. Brine temperatures may range from +15 to −20°F depending on the state of the freezing and the amount of time available to complete the task.

Liquid nitrogen has been used for small projects.

SOIL STABILIZATION

In northern latitudes where areas of permanently frozen earth, or permafrost, prevail, methods of soil stabilization for building and equipment foundations are required. These methods range from providing a nonfrost-susceptible gravel pad to rigid below-grade insulation plus an active or passive refrigeration system to freeze the soil and keep it frozen. Since many of these systems are in remote areas, simplicity and reliability become major factors in system design.

THERMAL DESIGN

Piling Design

Frost heave has long been a problem for the designers of piles and buildings in the arctic or subarctic. Uplift forces as high as 13.5 tons/ft of perimeter must be taken into account when designing nonthermal piling (Long and Yarmak 1982). Designers have used sleeves, greases, waxes, and plastics to reduce the adfreeze bond in the active layer and lower frost heave forces. Almost all the methods for reducing heave forces have proven to be only temporary (Long and Yarmak 1982). Increased pile embedment remains the only sure method of preventing frost heaving of conventional piling. The use of thermopiles can effectively eliminate frost heaving forces (Long and Yarmak 1982). The thermopile freezes the active layer radially from the pile. As a result, active layer temperatures adjacent to the pile remain at the same temperature as the rest of the pile.

Slab-on-Grade Buildings, Outdoor Slabs, and Equipment Pads

When a slab-on-grade building, outdoor slab, or equipment pad is constructed on a permafrost area, the resulting soil thawing or thaw bulb must be considered. Thermopiles, thermoprobes, or an active refrigerated foundation system may be required to stabilize the structure, slab, or pad.

Design Considerations

Soil properties have a pronounced effect on the capacity requirements for passive or active refrigeration systems. The soil within the radius of influence is an integral part of the refrigeration system. Highly conductive soils will increase the radius of influence of the system and allow more heat to be pumped out of the subsoils. Conversely, poorly conductive soils will cause the radius of influence to be small; the thermal lag through the soils will be high, and the heat transfer rate, low. Soil moisture contents and soil classifications are valuable for estimating thermal conductivities. Backup capabilities for active and passive systems should be considered to avoid foundation failure.

PASSIVE COOLING

The three processes used by passive systems for heat removal are air convection, liquid convection, and two-phase liquid/vapor convection. All passive refrigeration systems rely on the temperature differential between the soil and the winter air to operate. When the temperature of the soil in contact with the refrigeration system is lower than the air temperature, the system is dormant.

Air Convection Systems

Air convection has been used to provide subgrade cooling below on-grade and pile-supported structures. For a passive air convection system to work, two criteria must be met in the design—the air outlet must be higher than the inlet to promote convection, and the air distribution system must be designed so that the friction imposed by the distribution system is low enough to allow convection to begin. Air convection systems are usually designed to take advantage of the prevailing winds at a specific site. The wind can push air through a distribution system at greater velocities than convection would allow; however, the wind in the arctic frequently carries large quantities of snow. Distribution ducts can be blocked by snow and ice causing the system to fail (Long and Yarmak 1982). Ducting fans are built into many air convection systems for active refrigeration backup.

Liquid Convection Systems

Liquid convection has been used to provide subgrade cooling below on-grade structures. Radiator and heat absorber portions of the system are normally connected by either a double or single pipe with a flow splitter to decrease frictional losses in the system and to provide maximum cooling of the working fluid. Examples of some working fluids are trichloroethylene, kerosene, and methanol and water. Frictional losses within the liquid system are high and limit heat transfer rates. Large-diameter pipes may be used to overcome frictional losses within liquid systems so that high heat transfer rates can be achieved. Some liquid convection systems allow an option for mechanical circulation of the working fluid as an active refrigeration backup. Liquid systems must be sealed to avoid leakage into the subsoils. Introducing the working fluid into the permafrost subsoils could depress the soil freezing point and increase the probability of foundation failure.

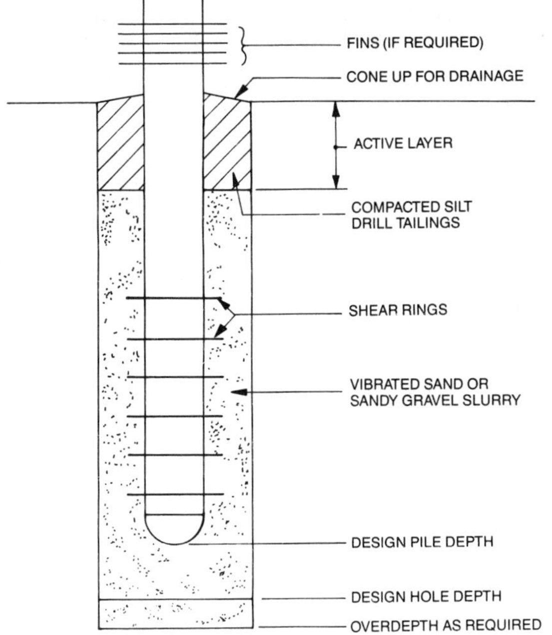

Fig. 3 Thermo Ring Pile Placement

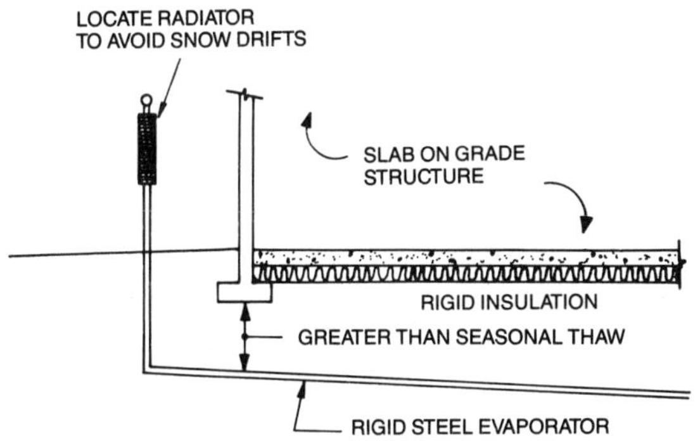

Fig. 4 Typical Thermo-Probe Installation

Two-Phase Systems (Heat Pipes)

Two-phase liquid/vapor convection systems are the most widely used passive refrigeration systems for permafrost foundations and earth stabilization. A typical two-phase unit is constructed of pipe enclosed at both ends and charged with a passive refrigerant gas. The radiator (aboveground condenser) portion of the unit can have a bare or finned surface, depending on heat transfer requirements (Figure 3). The evaporator portion of the unit can have any configuration as long as a slope remains between the evaporator and the radiator (Figure 4).

Refrigeration of the subgrade occurs when the radiator has a lower temperature than the soil in contact with the bottom of the evaporator, where the liquid portion of the refrigerant is pooled (Yarmak and Long 1982). Condensation occurs in the radiator, initiating evaporation of the refrigerant in the evaporator. The condensate wets the walls of the unit and flows down to the evaporator. Reevaporation of refrigerant condensate with subsequent cooling occurs where the soil in contact with the evaporator is warmer than the soil adjacent to the liquid pool of refrigerant at the bottom of the unit. Then the entire evaporator unit is reduced in temperature, cooling the surrounding soil.

A two-phase system will start with a temperature differential of as little as 0.01 °F between the radiator and the evaporator. Liquid and air convection systems may require temperature differentials of 4 °F to 15 °F before they start.

Propane, butane, halocarbons, anhydrous ammonia, and carbon dioxide have been used as the refrigerant gas in two-phase systems. Choice of refrigerant gas depends primarily on the allowable internal pressure capabilities of the vessel containing the gas, the quality of available gases, the molecular stability of the gas, and the preference of either the customer or manufacturer of the system. Relatively low-pressure systems using propane, halocarbons, or anhydrous ammonia have been known to gas lock, that is, gases other than the refrigerant gas accumulate in the radiator portion of the unit and prevent the refrigerant gas from condensing. Purging or venting of noncondensable gases may be required on a system following start-up.

ACTIVE SYSTEMS

Active ground refrigeration systems are used to keep building or equipment foundations stable when a passive system will not maintain the required stability. The system is normally a network of underground ductwork or piping through which a cooling fluid flows. Heat is removed using a heat exchanger or cooling coil, with refrigeration provided by medium- or low-temperature refrigeration units. System design can include provisions to bypass the refrigeration units during cold weather. Outside air should not be used directly in systems that use air as a cooling fluid due to the moisture introduced into the system in the form of snow and

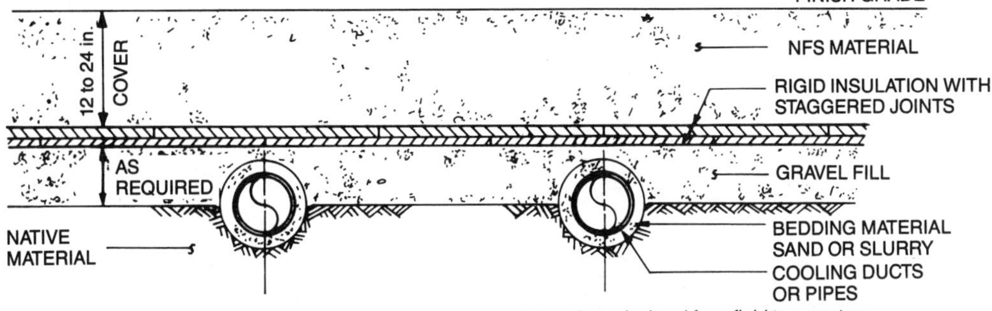

Fig. 5 Active Ground Stabilization System

humidity. A defrost system for the cooling coil in a circulating air system should be considered. Figure 5 shows a typical system installation.

REFERENCES

House, R.F. 1949. The inundation method of cooling concrete aggregates at Bull Shoals Dam. USA Corps of Engineers, Little Rock, AR.

Long, E.L. and Edward Yarmack, Jr. 1982. Permafrost foundations maintained by passive refrigeration. Petroleum Division, Ocean Engineering Division—ASME (March).

Yarmak, E., Jr. and E.L. Long. 1982. Some considerations regarding the design of two-phase liquid/vapor convection type passive refrigeration systems.

BIBLIOGRAPHY

Kinley, F.B. 1955. Refrigeration for cooling concrete mix. *Air Conditioning, Heating, and Ventilating* (March): 94.

Long, E.L. 1963. The long thermopile, permafrost. Proceedings of International Conference, National Academy of Sciences, Washington, D.C., 487-91.

Townsend, C.L. Control of cracking in mass concrete structures. *Engineering Monograph* No. 34. U.S. Department of Interior, Bureau of Reclamation.

CHAPTER 35

REFRIGERATION IN THE CHEMICAL INDUSTRY

CHEMICAL industry refrigeration systems range in capacity from one ton of refrigeration or less to thousands of tons. Temperature levels range from those associated with chilled water through the cryogenic range. The degree of sophistication and interrelation with the chemical process varies from that associated with comfort air conditioning of laboratories or offices to that where the production of reliable refrigeration is vital to product quality or to the safety of the operation.

However, two significant characteristics identify most chemical industry refrigeration systems. Almost exclusively, they are engineered one-of-a-kind systems, and equipment used for normal commercial application may be unacceptable for chemical plant service.

This chapter gives guidance to refrigeration engineers in working with chemical plant designers so they can design an optimum refrigeration system. Refrigeration engineers must be familiar with the chemical process for which the refrigeration facilities are being designed. An understanding of the overall process is also desirable. Computer programs are also available that can calculate cooling loads based on the gas chromatographic analysis of a process fluid. These programs accurately define not only the thermodynamic performance of the fluid to be chilled, but also the required heat transfer characteristics of the chiller.

Occasionally, because the process is secret, refrigeration engineers may have limited access to process information. In such cases, chemical plant design engineers must be aware of the restrictions this may place on providing a satisfactory refrigeration system.

FLOW SHEETS AND SPECIFICATIONS

The starting point in attaining a sound knowledge of the chemical process is the flow sheet. Flow sheets serve as a road map to the unit being designed. They include such information as heat and material balances around the major system components and pressures, temperatures, and composition of the various streams within the system. Flow sheets also include refrigeration loads, the temperature level at which refrigeration is to be provided, and the manner in which refrigeration is to be provided to the process (such as via a primary refrigerant or via a secondary coolant). They indicate the nature of the chemicals and processes to be anticipated in the vicinity in which the refrigeration system is to be installed. This information should indicate the need for special safety considerations in the design of the refrigeration system or for construction materials that resist corrosion by process materials or process fumes.

Different portions of a process flow sheet may be developed by different process engineers; consequently, the temperature levels at which refrigeration is specified may vary by only a few degrees. A study of such a situation might reveal that a single temperature is satisfactory for several or even all of the users of refrigeration, which could reduce project cost by eliminating multilevel refrigeration facilities.

Most process flow sheets indicate the design maximum refrigeration load required. The refrigeration engineer should also know the minimum design load. Process loads in the chemical industry tend to fluctuate through a wide range, creating potential operational problems.

Flow sheets also indicate the significance of the refrigeration system to the overall process and the desirability of providing redundant systems, interlocking systems, and so forth. In some cases, refrigeration is mandatory to ensure safe control of a process chemical reaction or to achieve satisfactory product quality control. In other cases, loss or malfunction of the refrigeration system has much less significance.

Other sources of information are also valuable. A properly prepared set of specifications and process data expands on flow sheet information. These generally cover the proposed process design in much more detail than the flow sheets and may also detail the mechanical systems. Information regarding the design philosophy, including continuity of operation, safety hazards, degree of automation, and special startup requirements, is generally found within the specifications. Equipment capacities, design pressures and temperatures, and materials of construction may be included. Specifications for piping, insulation, instrumentation, electrical, pressure vessels and heat exchangers, painting, and so forth are normally issued as part of the design package available to a refrigeration engineer.

It is imperative that refrigeration engineers establish effective communication with chemical process engineers. The refrigeration engineer must know what information to request and what information to give to the chemical process engineer for design optimization. The following sections outline some of the significant characteristics of chemical industry refrigeration systems. A full understanding of these peculiarities is of value in achieving effective communication with chemical plant designers.

REFRIGERATION—SERVICE OR UTILITY

Refrigeration engineers unfamiliar with the chemical industry must understand that unless the chemical process is cryogenic in nature, chemical plant designers probably consider refrigeration merely as a service or utility of the same nature as steam, cooling water, compressed air, and the like. Chemical engineers expect the reliability of the refrigeration to be of the same quality as other services. When a steam valve is opened, chemical engineers expect steam to be available instantly, in whatever quantities demanded. When steam is no longer required, the engineer expects to be able to shut off the steam supply at any time without adversely affecting any other steam user or the source of steam. The same response from the refrigeration system will be expected. This high degree of reliability is usually so strongly implied that no specific mention of it may be made in specifications.

The preparation of this chapter is assigned to TC 10.1, Custom Engineered Refrigeration Systems. This chapter last received a major revision in 1971.

Because refrigeration is frequently considered a service, process designers spend insufficient time analyzing temperature levels, potential load combinations, energy recovery potentials, and the like. The potential for minimizing the size of the refrigeration system, the total plant investment, or both, by providing refrigeration at a minimum number of temperature levels, is frequently not investigated by process engineers. Likewise, the potential for power recovery is frequently overlooked.

Part of the reason for this attitude is that refrigeration facilities represent only a minor part of the total plant investment. The entire utilities installation for the chemical industry usually falls in the range of 5 to 15% of the total plant investment, with the refrigeration system only a small portion of the utilities investment. Process requirements may be overruling, but process engineers must recognize legitimate process necessities and avoid unnecessary and costly restrictions on the refrigeration system design.

LOAD CHARACTERISTICS

Flow sheet values generally indicate direct process refrigeration requirements and do not include heat gains from the equipment or piping. Flow sheet peaks and average loads generally do not allow for unusual startup conditions or off-normal process operation that may impose unusual refrigeration loads. This information must be gained by a thorough understanding of the process and by discussing the potential effect on refrigeration system design for off-normal process conditions with the process engineer.

Once the true peak loads are established, the duration and frequency of the peaks must be considered. For some simple processes this information is fairly straightforward; however, if the plant is designed for both batch operation and multiproduct manufacture, this can be an enormous task. Computer simulation of such a combination of processes has led to optimization of not only the refrigeration equipment but also the process equipment. Computer simulation assures that the refrigeration machine, secondary coolant storage tanks, and circulating pumps are properly sized to handle both average and peak loads with a minimum investment. Few applications require computer simulation, but a thorough understanding of the relationship of peak loads to average loads and their influence on refrigeration system component sizing is vital to good design.

Sometimes unusually light load conditions must be met. If the process is cyclic and an on-off operation is undesirable, the refrigeration system may run for significant periods under a very light load or even a no-load condition. Light loads often require special design of the system controls, such as multistep unloading and hot gas bypass with reciprocating compressors, a combination of suction throttling and hot gas bypass with centrifugal compressors, and slide valve unloading and hot gas bypass with screw compressors. A secondary coolant system might require a bypass arrangement.

The investment in refrigeration equipment can be kept to a minimum when only a few levels of refrigeration are required. Checking the specified temperatures to be sure they are based on some process requirement may show that fewer temperature levels than shown on the flow sheet are necessary. If multiple levels of refrigeration are required, a compound system should be evaluated. The evaluation must consider the limited ability of a compound system to provide the precise temperatures required of some processes.

Production Philosophy

The flow sheet and specifications generally indicate whether a chemical process is in continuous or batch operation, but further research may be required to understand the required continuity of service for the refrigeration facilities. The chemical industry frequently requires a high degree of continuity; plant production rates are often based on 8000 h or more per year. In general, the refrigeration equipment is worked extremely hard, with no off-peak period because of seasonal changes, and any unscheduled interruption of refrigeration service may create large production losses. In most cases, scheduled maintenance shutdowns are not only infrequent but also highly vulnerable to cancellation or delays because of the press of production requirements.

As a result, reliability is key in the design of chemical industry systems. Equipment that is satisfactory in commercial or light industrial service is frequently unfit where high service rates and minimum availability for maintenance are the rule. In some cases, duplicate systems are justified. More often, multiple part-capacity units are installed so that a refrigeration system breakdown will not create a total process production loss. Major equipment and hardware items that require minimal maintenance or that permit maintenance with the refrigeration system in operation should be selected (for instance, dual lube oil filters or a bypass line that permits temporary reversal of condenser water flow for cleaning). Particular attention should be paid to equipment layout, so that adequate access, tube pullout space, and laydown space are available to minimize refrigeration system maintenance time. In some cases, overhead steel supports for rigging heavy equipment, or permanent monorails, are justifiable.

Flexibility Requirements

The chemical industry constantly develops new processes; consequently, the usual chemical plant undergoes constant modification. On occasion, total processes are rendered obsolete and scrapped before design production rates are ever reached. Thus flexibility should be designed into the refrigeration system so it may be adaptable to some process modification. Designing for optimum flexibility is difficult; however, a study of the potentials or probabilities of process modifications and of the expected life of the process facility will help in making design decisions.

SAFETY REQUIREMENTS

Most chemical processes require special design to ensure safe operation. Many raw materials, intermediates, or finished products are themselves corrosive or toxic or are potential fire or explosion hazards. Frequently, the chemical reactions involved in the process generate extremes of pressure or temperature that must be properly contained for safe operation. Refrigeration engineers must be aware of these potential hazards as well as any abnormal hazards that may develop during startup, unscheduled shutdown, or other upset within the chemical process. When designing modifications or expansions for an existing facility, the possibility that certain construction or maintenance techniques may be safety hazards must be considered.

Corrosion

The shell, tubes, tube sheets, gaskets, packing, O-rings, seal materials, and components of instrument or control hardware must be properly specified. The potential hazards of leakage between the normal process side and the refrigeration side of heat exchange equipment must be investigated, because an undesirable chemical reaction may occur between the process material and the refrigerant.

An additional corrosive hazard may result from leaks, spills, or upsets within the process area. Safe chemical plant design must anticipate the unusual as well as the usual hazards. For example, if refrigerant piping will run adjacent to a flanged piping system containing a highly corrosive material, special materials for the piping and insulation systems may be justified.

Toxicity

If the refrigeration system indirectly contacts a toxic material via heat exchange equipment, flanges and such elements as gaskets, packing, and seals in direct contact with the toxic material must be designed carefully. The possibility of a leak in equipment that might allow refrigerants or secondary coolants to mix with process chemicals and cause a toxic or otherwise dangerous reaction must also be considered. In some cases, the potential for toxic leaks may be so high that a special ventilation system may be required.

Even though containment and ventilation can handle toxic materials under normal process conditions, toxic material may need to be vented in abnormal situations to avoid the hazards of fire or explosion. In such cases, the toxic material is frequently vented or diluted through a tall stack so that ground or operating level concentrations do not reach toxic limits. Refrigeration engineers should evaluate the desirability of locating the refrigeration equipment itself or its controls outside the operating areas. An alternate solution is to ventilate either the refrigeration system or its controls with a system that has a remote air intake.

The hazards of toxicity are not confined to the chemical process itself. Some common refrigeration chemicals are toxic in varying degrees. Because of the chemical industry's interest in safety, refrigeration engineers are required to treat some of the common refrigerants and secondary coolants with much more caution than required for the usual commercial or industrial system.

Fire and Explosion

Plant specifications normally define the area classifications, which determine the need for special enclosures for electrical equipment. For areas designated as being in an explosion-proof environment, standard components are normally grouped in a single large explosion-proof cabinet. Intrinsically safe control systems, which eliminate the need for explosion-proof enclosures at all end devices, are also used. Intrinsically safe shutdown systems eliminate arcing at the end device by using low-voltage signals controlled by a microprocessor. Control equipment can also be mounted in standard or weatherproof cabinets equipped with an inert gas purge system.

Items other than the electrical system may require special consideration in areas of high fire or explosion hazard. The use of flammable materials should be carefully reviewed. Stainless steel instrument tubing should be chosen rather than unprotected plastic tubing, for example. Both insulation materials and insulation finish systems should minimize flame spread in the event of a fire. In some extremely hazardous areas, nonflammable refrigerants or secondary coolants may be required, rather than flammable materials that would provide otherwise superior performance.

In areas of high explosion potential, modification of the usual refrigeration system design may be required. Design pressures for refrigeration vessels or piping may be determined by process considerations as well as refrigeration system requirements. Special pressure relief systems, such as rupture disk in series with relief valves, dual full-sized relief valves, or a parallel relief valve and rupture disc with transfer valves between them, are often required.

Refrigeration System Malfunction

A malfunction can itself be a significant safety hazard, whether the upset is caused by an internal failure of the system or by fire, explosion, or some other catastrophe. Most process areas designated as being in a hazardous environment are protected by automatic gas and flame detection systems that shut down the refrigeration system when explosive mixtures or fire are detected in an area. Loss of refrigeration may permit a process reaction to run out of control and cause loss of product, fire, explosion, or the release of toxic material in an area remote from the original source of trouble. Thus, one or several degrees of redundancy may be needed to minimize the consequence of refrigeration system malfunction.

Storage of cold coolant, ice, or cold eutectic, or an alternate emergency supply of cooling water may be necessary to meet emergency peaks. Alternate sources of electric power to the refrigeration system may be desirable. Dual drive capability by either electric motor or steam turbine might even be justified. Uninterruptible power systems may also be used for control systems.

Frequently, special facilities are used to protect against the extreme consequences of refrigeration system malfunction. One example is the quenching of the process reaction via an inhibiting or neutralizing chemical introduced to the process in the event of an unusual pressure or temperature rise. Another simpler and more common example requires closing one or more process valves in the event of a loss of refrigeration.

Maintenance

In an operating chemical plant, because of hazards frequently encountered, the normal maintenance procedures may not be permitted. Because welding, burning, or the use of an open flame is often prohibited throughout large areas of a chemical plant, maintenance flanges or screwed connections are used to permit replacement of piping and equipment. Sometimes extra access space or handling facilities (such as monorails) are provided to permit efficient removal of machinery to an area where welding, burning, and the like, is permitted.

EQUIPMENT CHARACTERISTICS

Automation

In chemical plant operations, instrumentation represents a significant percentage of the plant investment. For this reason, most chemical plant designers insist on standardization of instrumentation throughout the plant. These requirements may not include familiar refrigeration components and may create problems in the refrigeration design. Therefore, it is vital that the refrigeration engineer and the chemical or instrumentation engineer must agree on the instrument requirements early in the design phase.

A second concept frequently adopted is the control and monitoring of all plant operation from a single central control room. It is not unusual to operate a multimillion dollar processing facility with two central control room operators and perhaps a single roving operator. This concept influences the refrigeration system in several ways. Refrigeration system controllers and alarm and shutdown lights are mounted on the central control room (CCR) panel, as are recorders or indicators that display refrigeration system temperatures, pressures, flows, and so forth.

Even with central control room operation, a local panel is required for the display or recording of additional information that can aid in troubleshooting an emergency shutdown. Frequently, startup control is available only at this local panel, to assure that startup is not attempted unless an operator is present to witness the operation of major items of refrigeration equipment. The CCR-mounted hardware would certainly conform to the standards of the process instrumentation to minimize operator confusion either in reading informative devices or in operating control devices. Most plants now use centralized computer or microprocessor controls. In some cases, the refrigeration unit is controlled from the main computer or microprocessor.

When applying a CCR concept, it is important to determine exactly how much information and control are to be provided at the control room and how much are to be provided locally. Transmission of unnecessary information to the CCR can be costly, but, if sufficient information is not available, serious process upsets are inevitable. Most process operators are not trained to understand the intricacies of refrigeration machinery.

The CCR system must permit monitoring of the refrigeration system performance and control of that performance to suit process needs.

Operators require alarms to indicate abnormal conditions for which they can make corrections, either at the CCR or in the field, and to indicate a system breakdown or shutdown. They also require a locally mounted manual shutdown station in the event of an emergency. Devices such as sequencing alarms and lube oil or bearing temperature recorders, which are either troubleshooting aids or can be checked or logged by the roving operator, are best installed locally.

The concept of designing for a minimum of operator attention with personnel that are not refrigeration specialists is yet another reason that a high degree of reliability is required for a chemical industry refrigeration system.

Outdoor Construction

Another chemical industry characteristic is outdoor construction. The chemical industry installs sophisticated process and auxiliary facilities outdoors all over the world. Whether the problems are those imposed by low temperatures, heavy snows and freezing rain, dust storms, baking heat, or hurricane force winds with salt-laden rains, they are generally unfamiliar to the uninitiated refrigeration engineer. For example, explosion-proof electrical construction is not necessarily weather resistant. Lube oil heaters and a prestartup circulation of heated lube oil may be required for compressors or other rotating machinery. In areas with high winds, special attention must be paid to the detailed installation instructions for insulation jacketing applied to pipe or vessels.

Winter operation of cooling towers may require multiple-cell construction, two-speed or reverse rotation cooling-tower fans, or even facilities for steam heating the cooling-tower basin. Instrument air for transmission of signals or power to pneumatic operators should be dried to a dew point (under pressure conditions) lower than anticipated ambient conditions to avoid condensate or ice from forming in the instrument air lines or in the instruments themselves. In fact, it has become almost standard that the instrument air provided is both oil-free and dried to a low dew point. The effect of ambient temperature, as well as radiation from the sun, should be considered in determining system design pressures, especially when equipment may be idle. Ambient temperatures up to 120 °F and vessel skin temperatures of 165 °F can be experienced in hot climates. To determine whether or not purchased equipment meets the requirements for outdoor installation, a detailed check of vendors' drawings, specifications, descriptive literature, and vendor-procured components is required.

Energy Recovery

Both the installed cost and the operating costs for the refrigeration system of some chemical processes can be reduced by the intelligent use of energy recovery techniques. If the process requires large quantities of low-pressure steam, the use of back pressure turbines to drive the refrigeration compressors could significantly reduce refrigeration system energy costs. On the other hand, if the process generates an excess of low-pressure steam, an absorption system may provide an overall saving. For processes with an excess of low-pressure steam only during the summer months when heating requirements are at a minimum, a condensing turbine may be economical if it is sized to operate at the low steam pressure when the excess is available and at a higher pressure during the heating season.

Other energy imbalances occurring within a chemical process can be advantageous. A waste heat boiler installed in a high-temperature gas stream may provide a source for low-cost steam. If the gas stream is at a moderate pressure as well as a high temperature, the possibility of a gas-driven power recovery turbine should be considered.

Another means by which operating costs frequently can be reduced is the reuse of once-through cooling water. Frequently, turbine-driven centrifugal refrigeration machines can use cooling water from the refrigerant condenser to condense the turbine exhaust steam, either in a shell-and-tube or a low-level jet condenser. Another possibility is the reuse of refrigeration system cooling water in process heat exchangers. Again, a thorough understanding of the process is a prerequisite for understanding the energy recovery potential, and a flow sheet should be helpful.

Performance Testing

Frequently, a requirement for performance testing of the refrigeration facilities is included within the contract for a chemical industry processing unit. Agreement should be reached as early as possible between the owner and the contractor regarding the exact procedure to be used for testing. If the test is to be run at some condition other than design conditions, both parties must agree on the methods of converting the test results to design conditions. Approximation techniques, such as those outlined in ARI *Standard* 550, are usually unacceptable in the chemical industry. The refrigeration engineer must be assured that adequate facilities for an equitable test are designed into the refrigeration system. This may require additional flowmetering devices or more accurate temperature-measuring devices than are required for normal plant operation.

Insulation Requirements

The service conditions imposed by the chemical industry on both piping and equipment insulation are frequently more exacting than those experienced in the usual commercial or industrial installation. Not only must the initial integrity be as near perfect as possible, but it must also resist the high degree of both physical and chemical abuse that it is likely to incur during its lifetime. To achieve both a minimum permeability and a maximum resistance to abuse, multicomponent finish systems may be required. In some cases, a vapor barrier mastic coating system (which may include reinforcing cloth) is covered with aluminum, stainless steel, or an epoxy-coated carbon steel jacket to protect against physical and chemical abuse. Piping and equipment insulated under ideal shop working conditions must be designed to withstand loading and unloading and erection into position on the job site without damaging the vapor barrier.

Because the fire hazard is generally high and the potential loss of personnel and investment resulting from a fire is prohibitive, a strict limit is usually placed on insulation finish systems having a high flame-spread rating, particularly in indoor construction.

The frequent use of stainless steel in the chemical industry for piping and equipment creates another problem with regard to insulation system design. Many stainless steels fail when they are exposed to chlorides. Stress corrosion cracking can occur in a matter of hours. Consequently, chloride-bearing insulation materials must not contact stainless steels even in minute quantities and should not be used anywhere in the system unless a valid vapor retarder is interposed. Chapter 21 of the 1993 ASHRAE *Handbook—Fundamentals* covers the general subject of thermal insulation and water vapor barriers.

Design Standards and Codes

Relatively few suppliers of refrigeration equipment regularly manufacture to meet codes or standards that apply to the chemical industry. Another variation from commercial or industrial design practice is the use of company standards. For the usual commercial or industrial plant, the client will at best provide performance specifications and a statement of what is to be done, leaving the preparation of detailed specifications for equipment, piping, ducting, insulation, and painting to the designer. However, many such items are covered by company standards, which, though established primarily for use in the process cycles, can yield corporate benefits if they are also used in design of the refrigeration systems. A request for all applicable company standards at the start of the design and an effort to use them will avoid costly rework of the design following a review by the client.

STARTUP AND SHUTDOWN

Processes are most hazardous during startup and shutdown. Though present in batch or discontinuous processing units, the problem is usually more severe in continuously operating units for the following reasons:

1. Instrumentation and control must be designed for the normal condition, and the cost of features intended for use only during startup or shutdown often cannot be justified. Frequently, conditions at startup or shutdown fall outside the range of the operating instruments and control, so that manual control is necessary.
2. The same argument holds for much of the process equipment, so that extraordinary measures, such as severely throttled flow, minimum-flow bypasses, and recycling may be needed.
3. The operators go through these conditions only infrequently and may have forgotten the techniques of operation at the time they are most needed.
4. The process conditions at startup and shutdown are usually not recorded on flow sheets or in descriptions because they occur so infrequently and usually vary continually as the units are brought on and off stream. As a consequence, designers have a tendency to overlook them and to concentrate on the conditions in the operating range.

Refrigeration engineers must inquire whether startup or shutdown is likely to impose any special conditions on the refrigeration system. Startup and shutdown is a special burden in this respect, since operators are particularly busy with the processing equipment and cycle during these times and generally cannot monitor or adjust the operation of what they regard as a service system. Therefore, process engineers must be made thoroughly aware of the precise limitations that startup or shutdown of the refrigeration system may impose on process operation.

REFRIGERANTS

Such factors as flammability, toxicity, and compatibility with proposed construction materials may influence the final selection of a refrigerant more than in other applications. Special attention should be paid to the consequences of leakage between the process materials and the refrigerant or the secondary coolant. Chapters 16 and 18 of the 1993 ASHRAE *Handbook—Fundamentals* discuss refrigerants and secondary coolants in detail.

Because of their nontoxic, nonflammable properties, halogenated hydrocarbons have been used predominantly, but recent environmental concerns have reduced their application throughout the chemical industry. Hydrocarbons such as methane, propane, ethane, propylene, ethylene, or ammonia are used in many cases where the process stream involves them as constituents. In the petrochemical industry, the use of these materials within the process is common.

Of the secondary coolants, calcium and sodium chloride brines have been used most often, although glycols and such halocarbons as methylene chloride, trichloroethylene, R-11, and R-12 also have been frequent choices. Again, environmental concerns predicate against the use of R-11 and R-12 in new facilities.

Many of the same factors that influence refrigerant selection must be considered in choosing a secondary coolant. Corrosivity, toxicity, and stability are of special significance in determining suitability for chemical plant service.

Refrigeration Systems

An indirect system, in which brine or chilled water is circulated to air washers, cooling coils, and process heat exchangers from a central refrigeration plant, is much more prevalent in the chemical industry than in the food industry or in residential or light commercial comfort air conditioning. This is particularly true where large capacities or low temperature levels are involved. An indirect system permits centralization of the refrigeration equipment and associated auxiliaries, which may offer significant advantages in operation and maintenance, particularly if remote location of the refrigeration equipment permits design, operation, and maintenance in a nonhazardous location. It also may permit the installation of a minimum number of large units rather than many small units located in remote areas. For low-temperature systems of significant capacity, an indirect brine cooling system installed in the process area close to the process users is common.

Where the number of process heat exchangers requiring cooling and the length of piping can be kept to a minimum, a direct system, which uses the refrigerant in the process heat exchange equipment, often is the optimum design, particularly for small or medium loads. Since an indirect heat exchanger is not required in this case, a higher operating suction pressure and consequent lower operating and investment costs may be possible. Direct systems are also used when a refrigerant is involved in the manufacturing process stream, as in the production of ammonia or many petrochemicals. Here, the length of refrigerant lines, with possible high refrigerant losses because of leakage, is a less significant factor in system selection. Direct systems are usually of the dry expansion or flooded evaporator design; flooded coil systems of the gravity feed or pumped liquid overfeed design are relatively uncommon.

Direct systems for larger capacity, low-temperature, multiple user service have advantages and disadvantages that must be considered for proper system selection. Some of the disadvantages are:

1. Maintaining an extensive refrigeration piping system free of leaks is difficult. Leakage from piping for secondary coolants is frequently less objectionable than the refrigerant gases. Checking for refrigerant leaks or repairing them in certain high explosion hazard process areas can be a problem because halide and electronic leak detectors may not be permitted and burning or welding may not be possible without a plant shutdown. If air or moisture leaks into a system operating at vacuum conditions, extensive icing and corrosion problems can result.
2. Higher piping costs are often involved when all items are considered, including large and expensive vapor and liquid-control valves at individual heat exchangers, particularly for the halocarbon refrigerants. Generous refrigerant knockout separators are necessary at each stage. Although both refrigerant and secondary coolant lines require insulation to prevent capacity losses, sweating, and icing, refrigerant lines, especially on the vapor return to the compressor, can greatly exceed the size of those required for brine recirculation.
3. No system reserve capacity is available as is the case with a secondary coolant, particularly if the latter is designed as a storage system. Process upsets can directly and suddenly increase the load on the refrigeration unit, causing cycling of the equipment and possible damage. For some processes, meeting short, sharp load peaks is of paramount importance to avoid off-standard production or unsafe operating conditions.
4. Constant temperature control is often more difficult or more costly to maintain with direct refrigeration than with a secondary coolant.
5. Initial testing of an extensive direct refrigeration system may be a significant problem. Testing must be done pneumatically rather than hydrostatically to prevent problems associated with water left in the refrigerant system. Pneumatic testing is a haz-

ardous operation and is forbidden in some chemical plants. The alternative of extensive posttesting dehydration is usually both expensive and time consuming.
6. The initial cost for refrigerants is usually much higher in an extensive direct refrigeration system than in a secondary coolant system operating at temperatures most frequently encountered in the chemical industry. In the case of system leaks, the costs of makeup coolant are generally less than the costs of makeup primary refrigerant.

Some of the advantages offered by direct refrigeration systems include the following:

1. Careful control of corrosion inhibitors may be necessary to keep secondary coolants stable so that they do not cause extensive equipment damage.
2. Less equipment and maintenance may be required; secondary coolant circulation and control or coolant mixing and makeup facilities are not needed.
3. Power costs are generally lower because of higher suction pressures and, in some designs, because pumps are not required.
4. Damage because of equipment freezing is not likely. Such damage can occur in a secondary coolant system if the coolant condition or the refrigeration plant is not properly operated.

Thus, the broad scope of refrigeration applications within the chemical industry permits the use of virtually any refrigeration system under the proper process conditions.

REFRIGERATION EQUIPMENT

For the most part, the refrigeration equipment used in the chemical industry is identical to, or closely parallels, the equipment used in other industries. The chemical industry is unique, however, in the wide variety of applications, the large temperature ranges covered by these applications, the diversity of equipment usage, and the variation of mechanical specifications required. Where possible, the chemical industry uses standard equipment, but this is frequently impossible because of the particularly rigorous demands of chemical plant service. Therefore, this section only briefly describes the application and modification of refrigeration equipment for chemical plant service.

Compressors

Refrigeration engineers may find difficulty in applying conventional refrigeration compressors to chemical plant service. Most process engineers are familiar with heavy duty, forged steel, high pressure, single or double throw reciprocating gas compressors; they are uncomfortable with the high speed, cast iron, or steel compressors that are standard to the refrigeration industry. Another difference between commercial and chemical plant usage is the greater use of open-drive equipment in the chemical plant.

The large capacities and low temperatures frequently encountered in chemical plant duty have led to wide use of either centrifugal compressors or high capacity rotary or screw compressors. These large machines vary from standard commercial equipment principally in the amount and complexity of controls or other auxiliaries provided. Load control devices such as multistep unloaders or hot-gas bypass systems are often required to permit a compressor turndown to 10% of full load or, in some cases, to permit no-load operation without either compressor surge problems or on-off operation. Most systems with large multistage centrifugal compressors use economizers to minimize power and suction volume requirements. Compressor lube oil systems are often provided with auxiliary oil pumps, dual oil filters, dual oil coolers, and the like, to permit routine maintenance without shutdown and to minimize shutdown frequency. Compressor control and alarm systems are frequently tied into central control room panel boards to permit monitoring and/or control of compressors.

Compressors for hydrocarbon gas refrigerants find their greatest use within the chemical industry, particularly in the field of petrochemicals. The relatively low cost and ready availability of pure hydrocarbons and hydrocarbon mixtures frequently dictate their use. Many offer the additional advantage of positive pressure operation throughout the entire refrigeration cycle.

Since refrigeration systems within the chemical industry are often required to operate for a year or more without shutdown, standby compression equipment is frequently installed. Even the larger refrigeration loads sometimes require 100% standby protection. Special controls may be required to provide rapid and automatic startup of the standby equipment. The main drive is commonly an electric motor and the standby drive may be either a steam turbine or an internal combustion engine. Provisions must be made via nonelectric drivers or emergency generating equipment to keep all necessary auxiliaries and controls operative during the electrical outage. Oversized crankcase heaters may be required, as well as electric or steam tracing of various lubricant system components.

High in-service requirements, plant standardization, explosion hazards, and corrosive atmospheres all require special controls. Often, the copper instrument tubing normally used on commercial equipment must be replaced with steel or stainless steel tubing more suitable to the proposed plant atmosphere. Lubricant piping must be stainless steel pipe with nonferrous valves, coolers, and filters. This requirement is primarily to minimize expensive delays in initial plant startup that may result when rust or scale within lubricant systems causes damage to bearings or seals in high-speed centrifugal or screw compressors.

Absorption Equipment

Chapter 1 of the 1993 ASHRAE *Handbook—Fundamentals* and Chapter 40 of this volume discuss absorption equipment in detail. Absorption equipment has seen little use in chemical plants, even though plant waste heat is available to operate it.

By special design and reselection of materials, hot streams of many fluids can be used as the energy source instead of using hot water or steam. Direct fired units are available. Hot condensable vapors can also be used as the energy source.

Commercial absorption machine equipment must be modified to permit outdoor operation. Manufacturers' recommendations should be followed regarding changes necessary to prevent freezing on the water side and solution crystallization on the absorption side of the equipment, particularly during shutdown.

Condensers

Water-Cooled Condensers. Units for chemical plant service require relatively minor design changes from those provided for commercial installations. Since cooling water is frequently of low quality, special materials of construction may be required throughout the tube side. An example is the necessity to switch from copper to a cupronickel when cooling water comes from a tidal source that is high in chlorides. If the cooling water is high in mud or silt content, it is sometimes justifiable to install piping and valving that will permit backflushing the condenser without requiring a refrigeration machine shutdown. Chemical plant requirements normally dictate shell-and-tube condensers of the replaceable tube type. Process engineers may insist on conservative tube side velocities (8 ft/s or less as a maximum for copper tubes) and replaceable tube bundles. The long hours of operation without opportunity for cleaning and the types of cooling water used frequently require that a higher water-side fouling factor be assumed than on commercial installations.

Air-Cooled Condensers. With increasing restrictions on the use of water for condensing, air-cooled condensing systems have been used in several instances, even in larger centrifugal-type plants. These have usually been installed in warmer locations where the increase in condensing pressure (temperature) over that from the use of cooling tower water or once-through water systems is minimal.

Air-cooled condensers for chemical plant service are normally fabricated to one of the API standards for forced convection coolers. Care must be taken when specifying these coolers so that the manufacturer understands the type of duty associated with a condensing refrigerant. The service required of an air-cooled condenser in a chemical plant atmosphere dictates either the use of more expensive alloys in the tube construction or conventional materials of greater wall thickness, to give acceptable service life. Air coolers may be more difficult to locate because recirculation of hot discharge air or fouling by hot process exhaust gases must be avoided.

Evaporative Condensers. Evaporative condensers, particularly for smaller refrigerating loads, are used extensively in the chemical industry, and they should become more prevalent as more emphasis is placed on the reduction of thermal contamination of rivers, lakes, and streams. In a few larger installations, the combination of an air cooler and an evaporative condenser operating in series satisfies the condensing requirements.

In most cases, the commercial evaporative condenser is totally unsuitable for chemical plant service, but satisfactory results can be obtained if this equipment is carefully specified. The major items of concern are the atmospheric conditions to which such equipment may be exposed and the long in-service requirements of the chemical plant. The chemical plant atmosphere, which may abound in vapors or dusts that are corrosive in themselves, can be an even more serious problem when these vapors and dusts are passed over surfaces that are constantly being wetted. Another problem is that dusts from nearby raw material storages or grinding operations may infiltrate the water recirculating system and plug the spray nozzles. The problems of water treatment and winter freeze protection are usually much more severe in chemical plant service because of the lower quality water that is frequently available and the demand for both year-round operation and a high turndown ratio. Light load operation in freezing weather calls for extreme care in design to avoid freezeup.

Two other areas of commercial evaporative condenser design that must frequently be strengthened for chemical plant duty are the electrical equipment, which must be satisfactory for the plant environment, and the fans, dampers, and recirculating pumps, which must be suitable for long-life low-maintenance service.

Evaporators

The general familiarity of chemical plant design personnel with heat exchanger design and application may sometimes lead them to suggest that refrigeration evaporators for the chemical plant should be designed similarly to evaporators in nonrefrigeration service. While the general laws of heat transfer apply in either case, there are special requirements for evaporators in refrigeration service which are not always present in other types of heat exchanger design. Refrigeration engineers must coordinate the process engineers' experience with the special requirements of a refrigeration evaporator. To do so, the standards of the Tubular Exchangers Association (TEMA) should be consulted to ensure that the end product is familiar to the plant engineer while still performing efficiently as a refrigeration chiller.

Paramount in these special requirements are the proper treatment of oil circulation in the refrigeration evaporator and proper evaluation of liquid submergence as it may affect low-temperature evaporator performance. When the evaporator in chemical plant service is being used with reciprocating and rotary screw compression equipment, continuous oil return from the evaporator must normally be provided. If continuous oil return is not possible, an adequate oil reservoir for the compression equipment, with periodic transfer of oil from the low side of the system, may be needed. On evaporators used with centrifugal compression equipment, continuous oil return from the evaporators is not necessary. In general, centrifugal compressors pump very little oil, so oil contamination of the low side of the system is not as serious as with positive displacement equipment. However, even with centrifugal equipment, the low side evaporators eventually become contaminated with oil, which must be removed. Most centrifugal systems operate for several years before oil accumulation in the evaporator adversely affects evaporator performance. Newer tube surfaces with porous coatings may be more sensitive to the presence of oil in the refrigerant than would be conventional finned surfaces. For newer surfaces, a continuous oil return system may be essential for centrifugal systems.

Flooded shell refrigeration evaporators operating at extremely low temperatures and low suction pressures may build up an excessive liquid head, which can create higher evaporating pressures and temperatures at the bottom of the evaporator than at the top. Spray-type evaporators with pump recirculation of refrigerant eliminate this static head penalty.

Special materials for evaporator tubes and shells of particularly heavy wall thickness are frequently dictated to cool process streams of a highly corrosive nature. Corrosion allowances in evaporator design, which are seldom a factor in the commercial refrigeration field, are often required in chemical service. Ranges of permissible velocities are frequently specified to prevent sludge deposits or erosion at tube ends.

Process side construction suitable for high pressures seldom encountered in usual refrigeration applications is frequently necessary. Choice of process side scale factors must also be made carefully without overstatement.

Differences between process inlet and outlet temperatures of 100 °F or more are not uncommon. For this reason, special consideration must be given to thermal stresses within the refrigerant evaporator. U-tube or floating tube sheet construction is frequently specified in chemical plant service, but minor process side modifications may permit the use of less expensive standard fixed tube sheet design. The refrigerant side of the evaporator may be required to withstand pressures resulting from maximum process temperature or the evaporator must be able to bypass the process stream under certain high-temperature conditions, *e.g.*, in a refrigeration system failure.

Relief devices and safety precautions common to the refrigeration field normally meet chemical plant needs but should be reviewed against individual plant standards and local statutory requirements. Forged steel relief valves are becoming more common as they meet the applicable refinery piping codes. In hazardous service, relief valves are sized for emergency discharge in the event of fire. The effect of chemical vapors on the downstream (outlet) internal parts of relief valves may call for special materials or trapped outlet piping with isolating liquid seals.

Process requirements frequently call for sudden or unexpected load changes on the refrigeration evaporator. Possible thermal shocks, with attendant stresses, must be evaluated, and the evaporator must be designed to meet any such conditions.

Evaporators in chemical plant duty normally require inspection and cleaning on an annual basis. For this reason, they should be located for accessibility and ease of tube replacement. Possible contamination of the process stream or the refrigerant side, because of leakage, should be evaluated. Special means of leak detection from one side of the evaporator to the other may be justified on occasion.

Low-temperature refrigeration in the chemical industry often creates extremely high viscosities on the process side of the equipment. Special evaporator designs may be needed to minimize pressure drops on the process side and to maintain optimum heat-transfer performance. Small tube diameters may not be compatible with the process stream because certain processes may call for extra large tubes. For extremely low temperature and high viscosity duties, evaporators are sometimes provided with rotating internal wall scrapers to assure flow of high viscosity fluids through the evaporators. Similarly, jacketed process vessels are used to cool highly viscous materials, while rotary scrapers keep the vessel walls clean.

For proper process flow, evaporators usually are remote from the other refrigeration equipment to minimize piping and pumping costs. Because remotely located evaporators place special emphasis on proper refrigerant piping practices, secondary coolant systems may be used. Chemical plants frequently use flooded refrigeration systems, which pump refrigerant from the central compressor station to remote evaporators. The use of these systems often reduces the design difficulties in assuring adequate oil return, and special provisions must be made at the central refrigeration station to protect the compressor against liquid carryover in the suction gas. The system must have an adequate accumulator to assure dry gas to the compressor.

Standard air-side evaporators may require modification, mainly to solve special corrosion problems in handling air or process gases that attack standard coil materials. Occasionally, process requirements demand coil designs that do not match standard commercial air-side pressure drops, air-side design temperature range, or both. Coils of special depth and special finning may be required and coil casings and fan casings of alloy steel are common.

Instrumentation and Controls

Since the heart of the chemical plant is its instrument control system, it follows that the instrumentation and control is much more advanced in the chemical industry than in commercial or the usual industrial refrigeration applications. As previously discussed, chemical industry refrigeration instrumentation hardware is much more sophisticated in design, particularly in regard to providing increased safety, reliability, and compatibility with process instrumentation devices. This sophistication extends to the application and design of individual hardware items. The chemical industry seldom settles for integral control devices such as self-contained pressure regulators or capillary-actuated thermal-control valves. The usual chemical industry control loop consists of a sensing device, a transmitter, a recorder/controller, a positioner, and an operator, all pneumatically or electrically interconnected. Many plants use central computer and microprocessor controls. Interfacing between the refrigeration system and control system may be necessary.

Cooling Towers and Spray Ponds

In a refrigeration system that uses water-cooled rather than air-cooled or evaporative condensers, heat may be rejected to once-through cooling water, spray ponds, or cooling towers. The chemical industry uses mechanical draft towers almost exclusively. These are generally of the induced draft design and are about evenly divided between crossflow and counterflow operation. Although a familiarity with these items is necessary, chemical plant engineers are usually responsible for their design.

Miscellaneous Equipment

Pumps. Refrigeration system pumps are usually of a high quality centrifugal design, the primary exception being small positive displacement pumps for compressor lube oil systems. In the past, heavy duty design was the rule rather than the exception, and secondary coolant and chilled-water units were usually of a horizontal split-case design, patterned after boiler-house or water plant construction. Chemical process designers have advocated standard chemical plant pump designs, which usually have a vertically split case and an end suction. If the selection is made carefully, this design is successful in many applications, and the resultant savings in pump costs, space requirements, and spare parts stocking requirements make it economically attractive. For pumped materials difficult to contain, such as most refrigerants and many secondary coolants, mechanical shaft seals of various designs are frequently used. As a result of their highly successful use in pumping difficult process fluids, canned or sealless pumps are used in such applications as liquid overfeed systems using halocarbons. Because the pumping of difficult fluids is a common problem, chemical process designers can be of invaluable assistance.

Piping. As a consequence of several factors, including low fluid temperatures, large pipe sizes, congested pipe alley space, and the industry's reluctance to use expansion joints for high duty service and in corrosive atmospheres, piping flexibility problems are much more complex. Expansion joints are frequently prohibited, which increases space requirements dramatically. Secondly, piping and valve standards that apply to both process and service facilities are frequently established by the process designer. The engineer who is accustomed to using carbon-steel piping systems with tongue and groove flanging and valves may find that plant standards call for a welded nickel-steel system with raised face flanges, spirally-wound stainless steel gaskets, and cast-steel valving.

Most piping construction problems resulting from the difference between expectations of the process engineer and the experience of the refrigeration engineer can be resolved by constructing the system to meet ANSI/ASME *Standard* B31.3, Chemical Plant and Petroleum Refinery Piping. The ANSI/ASME B16 series of standards that defines the flanges and fittings of the process industry should also be followed. Chapters 1 through 4 also discuss piping sizing for various refrigeration systems.

Tanks. Chemical plants use storage tanks for both refrigerants and secondary coolants more frequently than most commercial or industrial plants. In chilled water or brine circulation systems, storage tanks often serve a dual purpose: (1) to store secondary coolants during operation to provide a reserve capacity and thus smooth out short-term peak requirements and (2) to store secondary coolants during a maintenance shutdown of process evaporators. In some cases, brine mix and storage facilities are provided, so that any brine lost due to leakage or unusual maintenance demands can be quickly replaced, thus minimizing unscheduled process outages. In many cases, refrigerant pumpout compressors and storage receivers can minimize loss of the refrigerant and unscheduled outage time because of refrigeration system failures on the refrigerant side.

The chemical industry designs all pressure vessels in accordance with the ASME boiler and pressure vessel codes, in particular Section VIII, Division I, for unfired pressure vessels, regardless of local government regulations requiring such design. In most plants, standards are established regarding such items as pressure relief devices, manhole design, insulation supports, and tank supports. A thorough knowledge of the plant standards to be applied should be gained before specifications and design details are established for refrigeration system tankage.

CHAPTER 36

ENVIRONMENTAL TEST FACILITIES

ENVIRONMENTAL test facilities are used to simulate an environment or combination of environments under laboratory controlled conditions that duplicate or exaggerate the effects found in actual service. They assist the engineer and scientist in exploring the effects of equipment and in developing equipment for resistance to the many environmental forces.

The acceptance of and demand for environmental simulation facilities result from the following factors: (1) parallel and reproducible tests can be made; (2) equipment being tested can usually be observed and analyzed during testing; and (3) supporting equipment requirements are reduced to a minimum. Field testing and product development costs are reduced, lead time required for completion of product development is shortened, and most desirable reliability features can be incorporated in the original manufacture of the product.

Environmental test facilities are used not only to determine the performance of mechanical and electrical equipment, but for certain tests on personnel as well. Personnel testing includes: (1) checking protective equipment and clothing; (2) altitude and space procedures indoctrination; and (3) studying physiological and psychological effects on the human body and mind.

Environmental testing is usually divided into two general classifications—climatic and dynamic. The climatic tests of primary interest include the following:

Temperature Chambers. These are used for (1) temperature soaks at high and low extremes; (2) temperature shock testing in which the part is subjected to rapid high and low temperature cycling; and (3) programmed cycling in which the parts are subjected to repetitive expansion and contraction stresses and breathing.

Humidity Chambers. These may involve simply exposing the equipment to a constant humidity level, or *cycling*, wherein the temperature and relative humidity are varied. Cycle testing induces breathing and condensation within the parts tested. Subcooling may also be used to produce icing conditions.

Salt Spray Chambers. These are used for study of the corrosive action of salt vapor on components, usually at constant high humidity test conditions.

Fungus Chambers. These are used to stimulate the growth of fungus on equipment, primarily electrical, being studied to assure protection against tropical climates.

Miscellaneous Climatic Chambers. These include chambers for the simulation of desert sand and dust with high velocity air movement, sunshine, snow, and rain, as well as chambers for proof testing of explosion-resistant wiring devices.

Altitude and Space Chambers. Altitude chambers have been used for some time in testing aircraft equipment. In recent years, space chambers have been developed for missiles, rockets, and space vehicle development work.

Combined Environment Testing. This type of test facility combines two or more of the above environments in one system, with all the complexity implied. Current thinking is that this type of testing is the only satisfactory means of proving that a piece of equipment will stand up in actual service.

Dynamic or nonclimatic tests include vibration, physical shock, acceleration, mechanical stress, nuclear radiation, cosmic radiation, micrometeorite bombardment, and many other types of stress.

The air-conditioning and refrigerating engineer may be directly concerned with design of test chambers for the simulation of most climatic environments. Some suggested approaches to design of test equipment for the production and control of various environments will be given; the dynamic tests, however, are beyond the scope of this text. Because of the wide variety of tests that must be produced, detailed descriptions are not possible within this chapter.

Many techniques used in the design of environmental test equipment are the same as those employed for low-temperature metallurgy (Chapter 38), space simulators (Chapter 37), biomedical applications (Chapter 39), and cryogenic equipment (Chapter 37).

Many existing federal specifications outline basic environmental test specifications and the design approach for some of the chambers to be used in this work. The rapid pace of missile and space vehicle development has led to a series of informal special test criteria developed by project contractors. Environmental test equipment designers are extending the state of their art to develop equipment to meet these needs.

COOLING SYSTEMS

Temperature reduction in test chambers can be accomplished by both mechanical refrigeration systems and the use of expendable refrigerants. Engineered refrigeration systems are discussed in the first part of Chapter 3. Both approaches can be used directly or in conjunction with secondary heat-transfer fluids.

Primary Refrigerants

Any refrigerant suitable for mechanical refrigeration can be used as the primary refrigerant. Table 1 lists characteristics of several refrigerants at low temperatures. Selection requires an evaluation of equipment size, cooling load, type of evaporator, temperatures to be produced, method of condensing, hazards, lubrication, and special operating requirements. However, the use of chlorofluorcarbons is being eliminated.

Table 1 Characteristics of Selected Low-Temperature Primary Refrigerants

	Saturation Temperature, °F, at pressure indicated		
Refrigerant	**10 psia**	**14.7 psia**	**200 psia**
22	−55.6	−41.4	96.3
502	−64.3	−50.1	89.8
13	−127	−114.6	−7.8
503	−139.8	−127.6	−12.1
14	−207.7	−198.3	−105.6

The preparation of this chapter is assigned to TC 9.2, Industrial Air Conditioning.

The halocarbon refrigerants are commonly selected because they are not toxic or flammable, and they are stable at elevated temperatures. R-22 or R-502 is used in single-stage systems and in the high stage of two-stage cascasde systems; either R-13 or R-503 is used in the low stage. R-14 is used in the low stage of three-stage cascade systems. R-502 is commonly used in a single stage to an evaporating temperature of −40 °F, and R-502 with R-13 or R-503 refrigerants in a cascade system to −120 °F evaporating temperature. R-14 is used in the low stage of a three-stage cascade system for lower evaporating temperatures. These choices of refrigerant are common in practice and do not mean that single-stage systems using R-22 or R-502 for temperatures below −40 °F are not used or that −120 °F is the lowest evaporating temperature possible of a two-stage cascade system. Although single-stage and cascade systems are the most common designs presently used, a two- or three-stage compound system may be selected for certain load applications.

Test chambers are frequently required to operate at high as well as low temperatures. When using primary refrigerants in evaporators, which may be subjected to temperatures above 300 °F, thermal isolation and cooling of the evaporator are necessary to prevent oil decomposition or other possible deterioration of the refrigerant circuit. Refrigerant 13 is more stable than R-22. Refrigerant filter cartridges should be changed frequently in systems where the evaporator is subjected to high temperatures. A secondary refrigerant heat exchange fluid, which is pumped out of the cooling coil at some safe, predetermined temperature, is another method for protection against overheating of a primary refrigerant coil. However, it is more common to use expendable refrigerants for high-temperature equipment, even with the disadvantage of higher operating costs at low temperature.

Expendable Refrigerants

Expendable refrigerants, such as dry ice, liquid carbon dioxide, liquid nitrogen, liquid helium, and others discussed in Chapter 37, are suitable for producing low temperatures in environmental chambers. Sublimation of dry ice within the chamber, chilling brine for circulation through heat exchangers, or direct expansion of the liquids within the test space are common expendable refrigerant methods. Liquid nitrogen, liquid helium, and other cryogenic liquids work particularly well below the temperature range of mechanical refrigeration systems.

The advantages of expendable refrigerants in environmental test equipment are reduced initial cost, basic simplicity, reduced weight and size of the chamber, and the ability to produce very rapid pull-down rates to low temperatures. Disadvantages include higher operating costs, the need for a reliable source of expendable refrigerants, the potential personnel hazard resulting from the absence of oxygen when the air is displaced in the test space, and the possible detrimental effect of submerging the tested product in the gas or liquid of an expendable refrigerant which is expanded directly into the test space. Indirect brine systems frequently approach the first cost of mechanical refrigeration systems. Direct expansion of expendable refrigerants in altitude chambers is impracticable.

Chambers using dry ice as the expendable refrigerant have, in most cases, been replaced with direct injection of liquid carbon dioxide or liquid nitrogen, which eliminates the need for a dry ice compartment. Also, the chamber is smaller and the temperature more easily controlled. Only on-off cycling is possible for control of direct injection liquid CO_2 systems because of the triple point of CO_2. Solenoid actuated two-way valves and pneumatically or electrically actuated ball valves have been used with much success. If reduced capacity operation is desirable, the control valve should be time-pulsed or more than one valve should be used with small and large expansion orifices. Solenoid valves with built-in orifices designed for this specific application are available.

Liquid CO_2 is used at two pressures; low-pressure bulk liquid is stored in refrigerated receivers at approximately 0 °F and 300 psig. Its latent heat is approximately 120 Btu/lb. High-pressure liquid is stored at room temperature at pressures ranging from 750 to 1000 psig, depending on ambient temperatures. Its latent heat varies from approximately 50 to 75 Btu/lb, depending on initial temperatures. High-pressure liquid is used only for very small or infrequent cooling loads because the latent heat is low, the liquid fills only 60 to 70% of a high-pressure cylinder, and the cost is considerably higher than for low-pressure liquid. For increased efficiency when high-pressure liquid CO_2 is selected, an economizer that precools incoming liquid by heat exchange to the exhaust cold vapors can be used to reduce the CO_2 requirement by 15 to 30%.

Liquid nitrogen can be applied either directly or indirectly to produce temperatures cheaply and easily down to its approximate −321 °F boiling point (at 1 atm). Control is accomplished either by on-off cycling of special solenoid actuated valves or by specially designed flow-metering valves. The liquid is stored in Dewar flasks or vacuum-insulated tanks. Transfer is accomplished either by self-pressurizing the storage vessel or by using dry air or nitrogen, properly pressure regulated, to pressurize the flask and force the liquid from a discharge tube.

Piping, valves, heat exchangers, and so forth should be fabricated of nonferrous metals, stainless steel, or high nickel steel. Few plastics are suitable seals in this temperature range. Although its latent heat of evaporation at 1 atm pressure is only 85 Btu/lb, another 56 Btu/lb is available from the superheating of gas to −100 °F. Therefore, it has a greater capacity of heat absorption than does low-pressure liquid CO_2 when operating at a −100 °F or higher control point. Suitable precautions should be taken against the low-temperature hazards to operating personnel. Oxygen monitoring systems should be located in the test cells to warn personnel if oxygen depletion occurs. Carbon dioxide and nitrogen gases should be vented to the outside atmosphere after they have been used in the test chamber.

Other cryogenic liquids such as liquid helium and liquid hydrogen provide test temperatures close to absolute zero. These are too expensive for routine work at temperatures which can be obtained by more economical means. Special storage and control valves are required. More detailed information on cryogenic liquids, their handling and piping, may be found in Chapter 37.

Secondary Coolants

Low-temperature heat transfer fluids are used where: (1) control flexibility is better accomplished by their use; (2) large central cooling systems have been chosen; (3) the high temperatures experienced will rule out primary refrigerants because of complex mechanical design problems; (4) expendable refrigerants are used and direct cooling of air or a product is not possible; (5) the wall of a temperature chamber or the shroud of a space simulator is to be conditioned and a close temperature gradient is required; or (6) thermal shock by alternately immersing the test device in liquids of different temperatures is required.

Halocarbons, liquid hydrocarbons, alcohol, some primary refrigerants, silicone fluids, and aqueous glycol solutions are used as secondary coolants. Thermal considerations include viscosity, specific heat, specific gravity, thermal conductivity, freezing and boiling points, and coefficient of expansion. Other important considerations are flammability, toxicity, corrosiveness, vapor pressure, and water solubility (see Table 2).

To minimize evaporation losses, the more volatile fluids are used in closed systems with suitable expansion tanks or accumulators and are frequently pressurized by an inert gas such as nitrogen. However, some secondary refrigerants dissolve in nitrogen. In that case, other methods must be used to pressurize the system such as an expansion tank with a diaphragm separator. Also, certain

Table 2 Selected Characteristics of Wide-Range Heat-Transfer Fluids

Property	Santa Barbara Chemical Lexsol 408M	3M Fluorinert FC-77	Liquid Carbonic LQ-1575	Florida Chemical D-Limonene	R-1120** Trichloroethylene
Boiling point, °F	320	207	350	310	187
Freezing point, °F	< −100	−166*	−140	−142	−144
Density at 68 °F, lb/ft³	48.6	111.1	53	53.3	90.5
Specific heat at 68 °F, Btu/lb·°F	0.5	0.25	0.8	0.45	0.23
Viscosity, centipoise					
68 °F	1	1.8	0.9	0.66	0.58
−40 °F	3.5	2.2	2.0	1.1	1.17
−60 °F	7	3.6	—	1.2	1.4
−80 °F	—	7.5	—	1.3	1.6
Thermal conductivity at 68 °F, Btu/h·ft·°F	0.08	0.038	0.07	0.10	0.06

*Pour point
**A hazardous chemical

fluids must be kept refrigerated to prevent boiling off under standby conditions.

Secondary refrigerants are cooled in insulated sublimation tanks by spraying the liquid over dry ice or directly injecting liquid CO_2 or liquid N_2 into the secondary refrigerant. Such systems require a pump for recirculation of the secondary refrigerant. Common operating difficulties in sublimation systems include both cavitation in pumps because of the release of dissolved CO_2 or liquid N_2 gas and foaming in the sublimation tank with resultant carry-over of the secondary heat-transfer liquid with the vented gases. Positive displacement pumps and positive suction pressures are recommended.

In a system with a wide temperature range, the pump and shaft seal must be carefully selected. Magnetic drive and canned pumps have no external seals, which eliminates the shaft seal problem.

Secondary coolants can also be cooled by a mechanical refrigeration system or by a heat exchanger cooled by liquid nitrogen. These systems avoid carryover, cavitation, and moisture problems. The evaporator or heat exchanger must be designed to avoid freeze up. In Chapter 18 of the 1993 ASHRAE *Handbook—Fundamentals*, more detail is included and additional coolants and their properties are discussed.

HEATING SYSTEMS

Electric heat is most commonly used in environmental chambers. Prime or extended surface, tubular, or strip heaters are suitable for circulating airstream systems. This method is a conduction or convection type of heating used to duplicate conditions in storage, in transportation, and when a protective housing is provided around equipment. With proper precautions, open nichrome, strip, or coil wire heaters can be used; they offer the advantages of a rapid response because of their low thermal mass. All types of electric heaters require proper insulation and protection against moisture. Generally speaking, temperature test chambers (dry-bulb control) may use open wire resistance heaters, although condensation at low temperatures may produce excessive moisture and corrosion on the heaters.

Some specifications prohibit direct radiation from chamber heaters to the test object. In this case, heater baffling may be required.

Salt-spray chambers are heated indirectly, usually by circulating heated water or air around the outside of the chamber shell outside the salt-fog atmosphere. Heaters for sand and dust chambers must use sheathed heaters that have good resistance against the erosion of the high-velocity sand and dust. Explosion-proof testing chambers contain an explosive fuel and air mixture. Heaters must have temperature-limiting devices on the sheath with terminals extended to the outside to prevent ignition of the mixture.

Altitude and space chambers present special heat problems. Arcing between terminals under vacuum conditions must be guarded against. General practice brings the terminals to the outside of the vacuum space through a vacuum-tight fitting. The reduced convective heat transfer of a heater under vacuum must be considered in the design to prevent burnout. Thermal and mechanical bonding of sheathed heaters to the exterior wall of the vacuum structure has been successfully used to avoid some of these problems. Internal forced convection heating is usually suitable up to about 50,000 ft of altitude, approximately 0.1 atm.

Indirect heating is also suitable for environmental chambers. Hot water, steam, brines, and oils can be used in coils in various configurations. These heating systems are particularly applicable where close tolerance must be provided within the system because complete modulation is possible. Modulation control is also possible in electrical resistance heating systems by using power proportioning equipment. Hot water recirculating systems and steam systems can be used with relatively standard design approaches. Complete drainage of the coils and piping within the chamber must be assured if the chamber will be operated below freezing. Proper water treatment should be provided.

Many brines and heat-exchange fluids that are not limited to the temperature ranges of water or steam are available. No one fluid for the commonly required wide ranges of environmental test equipment, such as −100 to 300 °F, is available at present. Most chemical brines present some problems with viscosity, toxicity, flammability, chemical reaction with the piping, and other limitations. Any heat-exchange fluid must be carefully evaluated for limitations.

Another type of heating system is the radiant type, which is required to simulate such situations as sun exposure where both the heat-flux density and the wavelength are important; more commonly, radiant sources are used in equipment exposed to radiation from some high-temperature source. Radiant sources in environmental chambers vary from exposed nichrome wire heaters to specialized equipment such as arc lamps and mercury Xenon lamps used to simulate sunshine.

Since the wavelength of the source is a function of its temperature, the conditions to be simulated must be understood before a suitable radiant heat source can be designed. Where the problem is simply to produce high watt density on the product to be tested, bare nichrome wires, sheathed heaters, carbon and silicon carbide rod heating elements, and similar devices are available. Tubular quartz heat lamps are often suitable for very rapid heat-up conditions. Sunshine at sea level is usually simulated by the use of mercury vapor lamps or combinations of tungsten filament heat lamps to produce the proper wavelength and watt density. Radiant heating systems may be required to produce heat densities in excess of 125 kW/ft^2. Careful design of the chamber structure, proper support of the heating elements, and correct location of the heating elements are required to prevent mechanical and thermal stress problems resulting from large temperature differentials and rapid heat-up and cool-down cycles.

Under altitude simulation conditions, removal of heat from a test object is possible by convection up to about 100,000 ft (8.3 mm Hg, absolute). To accomplish this, the air mover should be of the largest practical size so that the residual air in the chamber is passed at high velocity through the cooling coils and directed on and around the test object. An envelope of cooled, rarified air, within which temperature gradients will be reasonably small, is thus produced around the test object.

Large gradients may occur throughout the rest of the chamber but will not affect the test results. Airflow should be reduced at low temperatures to prevent fan motor overload and minimize the addition to the cooling load. Cold wall construction reduces the heat load to be transferred to the cooling coils within the vacuum space, provides surface for radiant transfer, and overcomes some of the losses through the structure. Use a pumped secondary refrigerant to prevent high wall temperature gradients.

Evaporators should be maintained at the lowest practical temperature to provide the maximum temperature difference for both radiant and convective heat transfer. In simulated altitudes over 100,000 ft, a radiation heat-transfer system should be designed into the chamber to produce the desired temperature-control conditions on the test part. Treating the walls of the chamber to produce high emissivity is necessary in this case; close control of the entire wall surface is generally advisable. The limitation of radiant heat transfer in the low-temperature range must be considered because the amount of heat transferred from one surface to another is a function of the ratio of the square of the absolute temperatures. With temperatures in the −100 °F range, the heat-transfer rate is small. The detailed calculation of this effect is described in Chapter 3 of the 1993 ASHRAE *Handbook—Fundamentals*, where the Stefan-Boltzmann equation is discussed.

Air Movement

All types of fans and air movers are used in test chambers. Propeller, axial flow, and centrifugal fans are generally selected. Positive displacement or multistage turbine-type blowers provide the necessary total pressure for ram air simulation under altitude conditions. Drives for internally mounted fans must be located externally when extreme conditions are encountered in the airstream. To eliminate corrosion and minimize heat transmission, stainless steel shafts are commonly used. In almost all instances, a vapor seal is required to minimize or eliminate transmission to or from the ambient air. Internal bearings must withstand the full range of environmental conditions with a reasonable life expectancy. Altitude chamber fan shafts must be equipped with vacuum-tight shaft seals.

Air distribution within the work space of the test equipment is important in producing both uniform gradient conditions and system response. Frequently, close temperature gradients specified for air circulation rate and distribution will complicate the control problem. Air densities may vary almost three-to-one in temperature test chambers and more than fifty-to-one in altitude test chambers. Therefore, motor sizing and speed control of fans require careful attention. The airflow for a specific gradient is directly proportional to the net heat gain or loss in the work space and inversely proportional to the permissible temperature gradient and density of the air.

Usually the air volume must be increased at elevated temperatures because of the lower density. The circuit pressure drop will be approximately proportional to the air density. Therefore, the system balance points should be determined from the fan characteristic curves for various operating range conditions in the chamber.

Humidification

Common means of raising the relative humidity within a test chamber include directly introducing low-pressure steam, vaporizing water by electric immersion heaters in an open evaporation pan, and directly atomizing water sprays alone or in combination with electric heaters to aid vaporization of the spray.

Steam and vapor generators are best for increasing humidity as well as temperatures, since considerable sensible heat is introduced. Control of steam may be modulating or on-off. With the latter, control set point anticipation is suggested to minimize overrun because of too rapid a rise in moisture content. When self-contained steam generators are used, they should include a low water cut-off device, suitable pressure relief valves, and other safety devices. Makeup water should be provided from a distilled or demineralized source.

Vapor generators operating at atmospheric pressure are commonly selected when it is desirable to produce test conditions of 95 ± 5% rh at temperatures between 100 and 160 °F. Protection against immersion heater burnout is mandatory for good design.

For production of high humidities at temperatures near and below ambient, some form of atomized water spray is recommended because of the adiabatic cooling effect on evaporation. Although water spray systems do not have as rapid a response as steam, this is no disadvantage at lower temperatures where the humidity ratio is low. By controlling the temperature of the sprayed water, both humidification and dehumidification can be achieved with a single system. Heating and cooling means may be located either inside the conditioning spray plenum or externally in the recirculating water circuit. Close control of humidities is possible with such a design approach. Distilled or demineralized water is suggested, since the hardness of tap water may be corrosive or cause contamination. Recirculating spray systems should be periodically flushed out because of possible airstream contamination of equipment or test parts. For the system to operate properly, careful consideration of the construction material must be exercised.

Dehumidification

Low humidity conditions with controlled dew points above freezing are usually produced by mechanical refrigeration cooling and electric reheating because most environmental chambers require mechanical refrigeration for extended dry-bulb temperature range control. For the control of humidity conditions requiring dew points below freezing and continuous operation, either alternately defrosted dual evaporators or automatically regenerating desiccant dehumidifiers are used. If an enclosure is sufficiently vaportight and there is no internal latent load, an evaporator sufficiently large to handle the frost buildup because of initial latent load may be an acceptable solution.

Ram air coolers that simulate the cooling air supply for electronic airborne equipment at controlled temperature, humidity, and altitude conditions, present special dehumidification problems. If 100% outdoor air is required, it is best handled by a wet coil which removes a large part of the moisture just above freezing, followed by a set of parallel alternately defrosted evaporators. For ram air simulators in a closed-loop arrangement, simpler systems are possible if the cooling of air does not involve dehumidification.

When using certain liquid or solid desiccants in very high or low dry-bulb ranges, mechanical refrigeration cooling may be required for precooling or aftercooling. In such systems, a portion of the chamber air is continuously withdrawn by an auxiliary blower and passed through the drying agent and any necessary precooling or aftercooling coils.

Some tests require the conditioned air dew point to remain sufficiently low to prevent condensation within the test article during dry-bulb control temperature cycling. This requirement can be a problem during rapid heat pull-up and may require a desiccant dehumidifier.

SYSTEM CONTROL AND INSTRUMENTATION

Control Tolerances

Environmental test facilities generally require minimal control tolerances over a wider range than that required for ordinary HVAC systems. Control of temperatures as close as ±0.5 to 1.0 °F are commonly specified. Relative humidity control tolerance is at most ±5% and more often ±3% or even less. To meet these requirements, the control and instrumentation system must have an even better accuracy since operating tolerances often exceed control tolerances.

When, in addition, the test chamber is designed to provide a wide range of test conditions, the HVAC equipment must be oversized for most of those conditions. The control system is then required to modulate equipment output to a small percentage of capacity, with all of the difficulties inherent in controlling oversized equipment.

For these reasons, the control devices used with test chambers must be of the highest quality. In general this will be "industrial quality," as used in process control applications. Most systems operate in a modulating mode, using proportional plus integral (PI) or, occasionally, proportional plus integral plus derivative (PID). Proportional Only control is not acceptable due to its inherent deviation from set point. For discussions of control theory, operating modes and typical systems, see Chapter 41 of the 1991 ASHRAE *Handbook—Applications* and Haines (1987).

Temperature Controls

Systems for temperature control are of the types discussed in Chapter 41 of the 1991 ASHRAE *Handbook—Applications* and Haines (1987). Controllers should operate in PI mode, although PID may be needed when rapid cycling over a wide range of set points is required.

Temperature sensors may be thermocouples, thermisters, or RTDs. Thermisters and RTDs should be specified to have an absolute accuracy of ± 0.5 °F and a sensitivity of at least 0.1 °F. Most good RTD and thermister sensors can meet these requirements. Thermisters must be specified with factory certified performance and a guarantee of not more than 1.0 °F drift per year so that they can be readily replaced. Note that thermisters do drift and must, therefore, be recalibrated or replaced at frequent intervals, depending on the tolerance specifications. Wound wire RTDs may be specified with no drift over time. The best quality wound wire RTDs are platinum; these may be obtained with certified accuracies, traceable to industry standards. Platinum RTDs using solid state deposition techniques are also available. These are accurate and have a fast response but are subject to drift over time. Sensitivity and accuracy may not be suitable for close tolerances when bulb and capillary-type sensors are used with pneumatic controllers. RTD sensors used with pneumatic controllers can greatly improve accuracy.

Sensors must be suitable for the temperature ranges required. Consult the sensor manufacturer for this information. Where wide ranges are required, two or more instruments may provide better accuracy over the entire range.

Recorders may be digital or of the strip or circular chart type. They may use the same sensors as are used by the controllers. If separate sensors are used they must be calibrated together. For accurate calibration, laboratory-type mercury thermometers should be used, depending on the range.

Humidity Controls

Accurate humidity sensing is very difficult. Materials that vary dimensionally with changes in relative humidity are not suitable for close control because of hysteresis and drift. Systems using electrolytes and wet-bulb/dry-bulb sensors require essentially continuous maintenance to ensure accuracy. For any system, accuracies at extremely high or below freezing temperatures may be questionable. For reasonably close control (± 5% rh), solid-state deposition-type sensors of the capacitance or resistance type are satisfactory and have a fast response. They must be calibrated regularly—the best have a drift of about 1% per year. For very close control, for calibration, and for continuous accurate sensing, the chilled-mirror dew-point sensor should be used. This device is available in several packaging arrangements and, if needed, can be obtained with a certification of accuracy traceable to an agency standard. Its response is somewhat slow for use with rapidly changing conditions. In some cases, it has been used to back up and check the solid-state instruments which are actually used for control.

Controllers must operate in the PI or PID mode and recorders may use sensors in common with the controllers.

Pressure Controls

An altitude measurement and control system may consist of a simple manometer for indicating pressure and a hand-throttling or bleed valve to control the level; or the system may be completely automatic, operated by a mechanical or electronic sensing device that actuates modulating bleed controls to automatically regulate the vacuum. The U-tube manometer shows relative pressure between the site barometric pressure and the internal vacuum of the chamber. Absolute vacuum measuring manometers are recommended for altitude chambers to avoid the need to correct for local barometric pressure variations.

Opposed-bellows absolute pressure controllers and electrical instruments using strain gage pressure transducers with bridge circuit electronic controllers are available. These instruments are suitable for altitudes to 200,000 ft. They provide on-off control or proportioning output signals and can control various functions of the vacuum system. Electronic vacuum-sensing devices measure the change in interference across an air gap between two electronic devices because of the reduction in gas molecules in the space. Coupled with amplifiers, this effect initiates action in control circuits at a set point.

One instrument system has an alpha emitter and measures the amount of energy reaching the grid to determine the vacuum level. Another system uses ionization of a filament. A thermocouple measures the change in the heat-exchange rate as a function of vacuum. Most electronic vacuum instruments are available with microamp or millivolt output that may be coupled to a standard potentiometer to provide both a record and an output signal for controlling the vacuum system.

The most common way of automatically controlling a vacuum system is to bleed air into the chamber through a modulating valve. This floods the pump to the desired capacity so that good control can be maintained with the proper size bleed valve. Frequently, two or more valves are required to control the pressure in the chamber over a wide range.

Altitude chambers can be controlled by cycling a solenoid valve in the pump suction line or simply turning the pump on and off. The latter measure is not recommended because of the excessive strain on the motor and the drive mechanism. These two methods are not practical for diffusion pumping systems.

Valves and Dampers

Haines (1987) discusses the factors to be addressed in selecting control valves and dampers. In test chambers where a wide range of conditions must be simulated, the system gains are aggravated by the need to size the HVAC equipment for the most severe conditions, which means that it is oversized for every other condition, sometimes extremely so. Oversizing the valve or damper inevitably leads to poor control. It follows that valves and dampers should be undersized rather than oversized if close control is to be obtained. Undersizing penalizes the air or water system hydraulics, but may be necessary to obtain the desired results. Use of multiple small dampers should be considered.

Computers as Controllers

Typically, DDC (Direct Digital Control) with computers is used. The computer simply takes over the controller function. It can be programmed to any kind of operating and timed sequence desired. It can also provide visual and printed output of system status, either in real time or as history, including graphic and statistical presentations. The engineer should select the functions needed to satisfy the design parameters.

Accessories

To provide satisfactory operation, a control system must include other accessories besides sensing elements and controllers. A pneumatic system should include a pressure-reducing valve, a filter-drier, and a surge tank; all are standard with such systems. For systems operating below freezing, nitrogen or dried compressed air should be considered as the pneumatic medium rather than compressed air if the system is to run within the chamber.

The selection of safety and other interlock relays and switches should be carefully considered not only to provide safe operation but to control the system and reduce the number of switch settings to be made by the operator. Vacuum switches are frequently used to cut out the heaters or switch them to series wiring whenever an altitude chamber is under vacuum, thus preventing burnout because of low convection transfer. Often there is a temperature limit controller to prevent damage that may occur because of excessive heating or cooling of the test module.

In addition to the conventional control systems for temperature, humidity, and vacuum, the designer may encounter requirements for measuring or controlling many other variables. These variables include: hydrogen-ion concentration (pH); fluid velocity and pressure, dust, and smoke density; explosive fuel and air mixtures; ozone, oxygen, and carbon dioxide levels; and others.

DESIGN CALCULATIONS

Cooling and Heating Loads

Because the rate of temperature change is usually important in the operation of test chambers, both the steady-state and transient load must be calculated to size the refrigeration and heating system. The steady-state cooling load includes heat transferred through the insulation, framing, windows, sleeves and penetrations, fan shafts and other conducting materials, and the internal live heat loads of lighting, product, fans, and personnel. The sum of these loads that occur simultaneously at the lowest air temperature in the chamber is the minimum design capacity. The equipment needs to be sized to handle this design capacity, plus a factor of safety, at the refrigerant evaporating temperature required and as determined by the evaporator size.

Transmission losses must also be determined to calculate the total steady-state heating load. The live loads, such as blowers and lights, will decrease the heating load, but usually they are not deducted from the transmission losses. The heaters are then sized with the load at the high temperature and derated for low voltage and operating temperature.

For transient conditions, the thermal mass of all items, including chamber air, liner and bracing, insulation and framing, evaporators, heat exchangers, heaters, blower, brackets, shelves or other gear, and the test specimen must be determined. The product of the mean specific heat and the mass gives the average thermal mass. The product of the time rate of temperature change and the thermal mass gives the instantaneous load due to temperature transients. The total load may include transmission and live loads as well.

Under transient conditions, a gradient will exist between the room ambient and the heat source or sink. For small chambers, conservative design assumes that the temperature of all items directly contacting the chamber air changes at the same rate as the air temperature. In cases with very rapid temperature changes and high mass loads, the surface areas, air velocity, and point of temperature control must be considered in designing the necessary refrigerating or heating system. A larger system cooling requirement may be calculated by this method rather than by one based on the total load changing with the air temperature— especially if a linear rate of change is required and the final temperature is low on the capacity curve of the refrigeration equipment. This larger capacity is also required for heating, except electric heating, where the capacity is essentially uniform at all temperatures.

Relative Humidity

Equipment selection for relative humidity control is similar to normal air-conditioning applications, only more complex in that widely varied control conditions must be produced. A typical control range is 35 to 185 °F dry-bulb temperature and from 20 to 98% rh above 35 °F dew point. Very low dew points may be required for special test programs. Sensible cooling loads may be calculated in a manner similar to that for low-temperature loads, but must also account for such additional factors as the sensible heat of steam when used for humidification.

Since dehumidification is normally accomplished by mechanically cooling and reheating the air, the most common approach is to take a portion of the total recirculating air, cool it to the required dew point, then reheat it as required to keep it at the desired dry-bulb temperature level. If the total air volume were cooled and then reheated, rather large dehumidification and reheat capacities would result. When long-term tests are required at controlled dew points near freezing, heat transfer cooling surfaces must be amply sized to operate without freezing.

Where large internal heat gains are present and high humidities are desired, spray-wetted surfaces producing adiabatic saturation may be used. By atomizing the spray and controlling the temperature of the recirculating spray water, systems may be used for simultaneous cooling and humidification.

The sizing of heaters for increasing dry-bulb temperature is the same as for normal chamber heating. Atmospheric pressure vapor generators and direct steam injection are alternate methods of humidification. The amount of water that must be added to achieve the desired increase in absolute humidity for the known volume of air within the chamber determines the rate of moisture addition. Both of these methods add sensible heat, as well as latent heat, which must be added to the cooling load to obtain the total load.

CHAMBER CONSTRUCTION

Inner liners of test chambers for high- and low-temperature and humidity are most commonly made of type 302 or 304, heliarc welded stainless steel. The outside liner is made of aluminum or mild steel, welded or sealed to provide a vapor-tight structure for the insulation. For walk-in rooms, prefabricated, urethane insulated, foamed-in-place modular panels are popular. Special care must be taken in the sealing of the panel joints.

A means must be provided to prevent an excessive pressure differential between the inside and outside of the chamber during rapid changes of temperature. Even opening and closing a door, when test space conditions are different from ambient, may change the air temperature and consequent pressure enough to buckle the liner of a well-sealed chamber if it is not properly vented.

The structure must be designed to provide for the considerable expansion and contraction that take place during wide-range temperature cycling. Doors are a particularly difficult problem. Any fan baffles, coil mountings, or other accessories must be carefully planned so that moving parts are not interfered with during temperature cycling runs. Shock loads induced by liquid nitrogen and other refrigerants can cause violent stress concentrations. Humidity chambers must be designed so that any condensate formed on the coils, walls, or floor will drain rapidly from the chamber. This feature is important in programmed humidity chambers in which operation ranges from high- to low-humidity conditions.

Altitude Chambers

Generally, a chamber that simulates altitude to 250,000 ft (0.016 torr) is considered an altitude chamber; above that altitude, the chamber is considered a space simulator. Interior liners of chambers that simulate altitude are usually constructed of aluminum, mild steel, or stainless steel, with exterior structural steel reinforcing designed to withstand a vacuum or 15 psig.

Depending on the desired test space configuration, chamber liners may be rectangular or cylindrical. Design of the vessel to withstand the exterior pressure should be in accordance with the latest edition of the American Society of Mechanical Engineers' Pressure Vessel Code.

Altitude chambers that control temperature or humidity or both, in combination with altitude simulation, are built with the pressure liner on the inside or outside of the insulation space. The inside liner, whether or not it is the pressure liner, should be made of a corrosion resistant material such as stainless steel. The outside liner may be mild steel. If the outside liner is the pressure liner, it must be of sufficient thickness or reinforced to withstand a vacuum or 15 psi pressure differential. It must also be treated for corrosion resistance due to equalization with the test space, which will introduce various combinations of temperature and (high) humidity into the liner surfaces and insulation.

If the inner liner is the pressure liner, it must be stressed for the temperature range involved as well as the pressure. The inner liner should be the pressure liner if high altitude use is a major consideration, because the insulation space is not exposed to the vacuum. However, because of the liner and reinforcing mass, the cooling and heating load during temperature transition is greater than with an outside pressure liner. With an outside pressure liner, the insulation space must be equalized with the test space to prevent inner liner collapse. Achieving high vacuum on the insulation and insulation space is very difficult unless it is dry; therefore, this fact must be considered during temperature or temperature-humidity runs, or when diving the chamber to atmospheric pressure after an altitude test.

Seam welds of the vacuum vessel should be on the vacuum side and should be of one continuous pass with skip welding, if required, on the outside. This method of welding prevents virtual leaks and allows for leak checking.

Insulation

Chapters 20, 21, and 22 of the 1993 ASHRAE *Handbook—Fundamentals* discuss insulation fundamentals and applications in detail.

Doors

Both overlap and plug-type doors are used, the latter most frequently on both high-humidity chambers (because of internal condensation) and high-temperature chambers, to prevent or minimize warping. They must be rugged and well fitted, particularly for altitude chambers. Vapor sealing with multiple gaskets is important, particularly on smaller chambers, since defrosting means are seldom provided. Suitable gasket materials would be natural rubber, synthetic rubber, and silicone. Plastics or light-gage stainless steel are recommended for thermal breaker strips; pressed wood hardboard materials generally have insufficient resistance to physical damage, high temperatures, or high humidities. Low wattage heater cables, installed under the breaker strip, prevent larger doors from freezing closed. Safety measures are essential for walk-in freezers, including doors easily opened from inside.

For altitude chambers, the door assembly must register against a gasket that will easily vacuum seal. Vacuum gaskets must be properly retained to prevent their being drawn into the chamber.

Windows

Special hermetically sealed multi-light window assemblies are manufactured for special conditions. They are suitable for wide-range temperature and humidity tests and, if provided with an inner pane of sufficient strength, for altitude tests as well. Tempered plate is required for the inner panes so that temperature shocks will not cause failure. Because of the pressures that build up, sealed assemblies for higher temperatures require careful design. For temperatures up to 1000 °F, windows of tempered glass are required and hermetic sealing is difficult, but research indicates a possible range of −300 to 1000 °F. The number of panes and design for a given task are usually specified by the window manufacturer. A typical −100 to 300 °F window contains six lights of glass enclosing dry gas spaces.

Accessories

Interior lighting is mostly incandescent, in vaportight and sometimes explosionproof fixtures. Ample illumination will be given usually by 6 to 10 W/ft^2 of floor space. Lamps are not recommended for use inside chambers above 600 °F.

Power leads may handle normal alternating current or special aircraft voltages and frequencies.

Thermocouples in almost any number may be called for. Bare iron-constantan should not be installed in a high-humidity chamber. For the operator's convenience, thermocouples are frequently connected to plugs and outlets.

Turning shafts, if small and manually operated, may be coupled to adjustable devices on the item being tested. These shafts should be stainless steel tubes. O-rings on the outside of the chamber may be used for pressure sealing.

Power shafts, provided for driving rotating equipment, must have vacuumtight shaft seals for vacuum chambers.

Sleeves and plugs are required for general purpose testing. Holes of various sizes may be specified. On altitude chambers, threaded or flanged caps must be provided.

Special lines for pneumatically or hydraulically operated equipment may require pressure pipes with special connections at the ends.

Protective panels are used to prevent damage to the chamber if an internal explosion or sudden release of pressure occurs.

Window wipers are installed on windows to remove condensation if vision is required when temperature adjustments are made within the chamber.

Reach-in ports, with or without gauntlets, may be provided for minor work within a test space without opening doors or disturbing internal conditions.

BIBLIOGRAPHY

Haines, R.W. 1987. *Control systems for heating, ventilating and air conditioning*, 4th ed. Van Nostrand Reinhold, New York.

Holladay, W.L. 1950. Low temperature test chamber design. *Refrigerating Engineering* (July):656.

Missimer, D.J. 1956. Cascade refrigeration systems for ultra low temperatures. *Refrigerating Engineering* (February).

Missimer, D.J. 1972. Mechanical system can reach −140 °C. *Research/Development* (July):40.

Missimer, D.J. 1973. Ultra low-temp systems—A practical summary. *Refrigeration Service and Contracting* (December):18.

Missimer, D.J. and W.L. Holladay. 1967. Cascade refrigerating systems—State-of-the-art. ASHRAE *Journal* (April):70.

U.S. Government Printing Office. 1990. Environmental test methods and engineering guidelines. MIL-Std-810E. Washington, D.C. (February).

CHAPTER 37

CRYOGENICS

CRYOGENIC is a Greek word meaning "creation or production by means of cold." At present, cryogenic technology is used to produce appropriate changes in gases, in the form of a liquid or solid, to achieve a desired objective. Figure 1 illustrates this temperature range, which has an upper limit arbitrarily defined as −150°F (−250°F by some) and a lower limit of absolute zero. These limits separate it from the temperature range generally used in refrigerating engineering.

One important application of cryogenics is the separation and purification of air into its various components (oxygen, nitrogen, argon, and the rare gases). Other important developments have been the large-scale production of liquid hydrogen; helium extraction from natural gas; storage and transport of liquefied gases such as oxygen, argon, nitrogen, helium, neon, xenon, and hydrogen; liquefaction of natural gas for ocean transport and peak shaving; and many new types of cryogenic refrigeration devices.

Cryogenic processes generally range from ambient conditions to the boiling point of the cryogenic fluid. Cryogenic cycles also incorporate two or more pressure levels. These properties must also cover the vapor, vapor-liquid, liquid, and sometimes the solid regions. Therefore, the physical properties of fluids over a great range of temperatures and pressures must be known.

Solubility of contaminants must be known to design for their removal. The main physical properties needed for design are those used in unit operations, such as fluid flow, heat transfer, and the like. In addition, those properties directly related to the Joule-Thomson effect and expansion work are needed. Properties such as density, viscosity, thermal conductivity, heat capacity, enthalpy, entropy, vapor pressure, and vapor-liquid equilibriums are generally obtained in graphical, tabular, or equation form, as a function of temperature and pressure.

Selection of materials for low-temperature application depends on the properties of these materials at the desired temperatures, especially regarding brittleness, elasticity, thermal conductivity, thermal expansion, etc. Chapter 38 briefly discusses materials for low-temperature application.

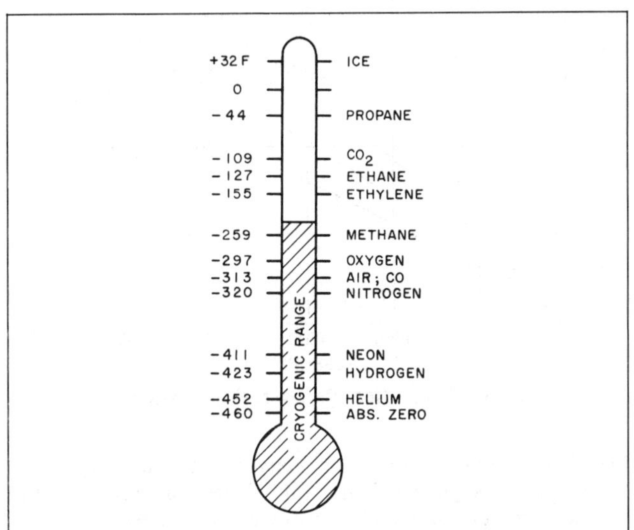

Fig. 1 Cryogenic Thermometer Showing Normal Boiling Temperatures at Atmospheric Pressure

The preparation of this chapter is assigned to TC 10.4, Ultralow Temperature Systems and Cryogenics. The chapter last received a major revision in 1971 and a slight revision in 1993.

PRODUCING LOW TEMPERATURES

Figure 2 shows the minimum Carnot cycle work for refrigeration at various temperature levels and the rapid increase in the theoretical work with decreasing temperature. Theoretical work becomes infinite at absolute zero. The actual work for low-temperature refrigeration increases at a more rapid rate than that indicated by the theoretical curve, since the thermodynamic efficiency decreases with decreasing temperatures.

The decrease in efficiency results from the greater heat leak at lower temperatures and from other irreversible processes associated with the refrigerator or liquefier. To attain cryogenic temperatures, correct refrigerating methods and equipment must be selected.

Of the many basic processes by which low temperatures are attained, the most commonly used are the Joule-Thomson expansion process (including the cascade) and the expander or work process. The vapor compression process, so widely applied in conventional refrigeration, is not found in cryogenic applications because the critical temperatures of all cryogenic fluids are below normal ambient temperature. Other processes, such as adiabatic demagnetization and adiabatic desorption, have been successfully applied in specialized laboratory studies.

Joule-Thomson Refrigeration

In the Joule-Thomson process, a gas at high pressure is expanded through a restriction to a low pressure, with a resulting change in gas temperature. If the high-pressure gas is initially below its inversion temperature, the gas temperature decreases as a result of the throttling. By passing such a gas through an effective countercurrent heat exchanger prior to or during the expansion process, it is possible to obtain extremely low temperatures and to partially liquefy the gas.

A typical Joule-Thomson refrigerator is shown in Figure 3. High-pressure gas at pressure p_1 and temperature t_1 enters the heat exchanger and is cooled by a countercurrent stream of low-pressure gas. At a low temperature, the gas is expanded to a low pressure through a restriction, shown schematically as a Joule-Thomson

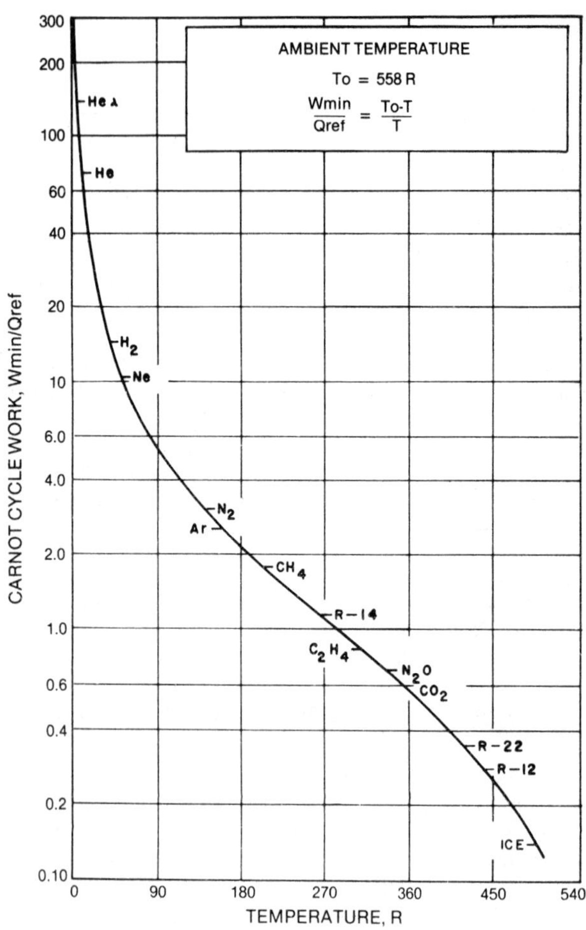

Fig. 2 Effect of Carnot Cycle Work

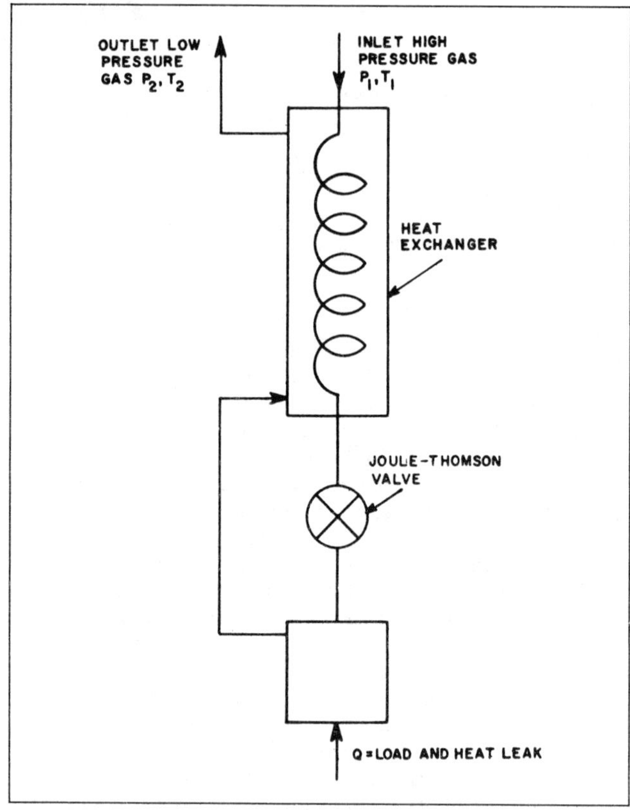

Fig. 3 Simple Joule-Thomson Refrigerator

valve, and a portion is liquefied. The liquid is vaporized at constant temperature by the heat leak and load Q. The saturated gas at low pressure returns through the heat exchanger, cooling the incoming high-pressure stream, and leaves the exchanger at temperature t_2 and pressure p_2.

An energy balance around the refrigerator shows that the sum of the heat leak and load, Q, equals the enthalpy difference between the low- and high-pressure gas streams as:

$$Q = H(p_2, t_2) - H(p_1, t_1) \quad (1)$$

Introduction of the enthalpy of the low-pressure gas stream at the temperature t (the enthalpy of the gas stream if it is warmed completely to the temperature of the inlet gas) yields:

$$Q = [H(p_2, t_1) - H(p_1, t_1)] - [H(p_2, t_1) - H(p_2, t_2)] \quad (2)$$

The first term is the isothermal enthalpy change for a Joule-Thomson expansion process. It equals the maximum or theoretical refrigeration available for such a process. The second is an energy term caused by the temperature difference at the warm end of the heat exchanger. Since this term can be expressed in terms of the specific heat of the gas and the temperature difference, Equation (2) may be rewritten as:

$$Q = \text{(Theoretical refrigeration)} - c_p \Delta t \quad (3)$$

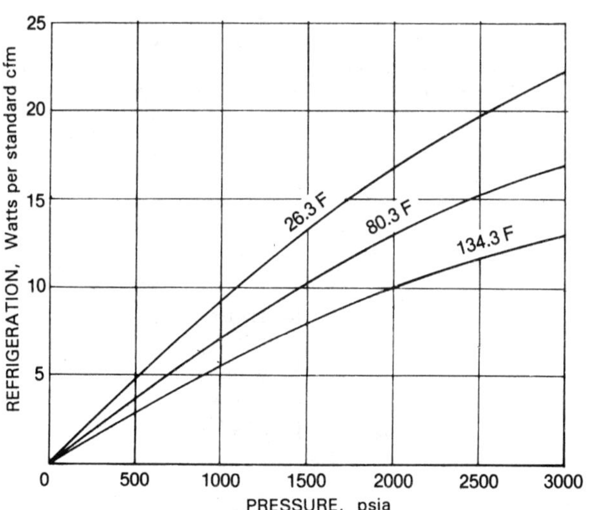

Fig. 4 Theoretical Refrigeration—Nitrogen

in which Δt equals the difference between temperature t_1 and t_2. Since t_1 must be greater than t_2 for a finite heat exchanger, the second term in Equation (3) is positive, thus representing a loss in refrigeration as a result of the heat exchanger inefficiency. Since this loss is proportional to the temperature difference, the most efficient exchangers obtain the maximum amount of refrigeration. Figure 4 shows the theoretical refrigeration of nitrogen as a function of several inlet temperatures and various operating pressures for $p_2 = 1$ atm. These curves show trends typical of gases that exhibit a Joule-Thomson cooling effect; the theoretical refrigeration increases as the pressure increases or the inlet temperature decreases.

The loss in nitrogen refrigeration caused by the warm-end temperature difference of the exchanger equals approximately 1.1 Btu/(h·cfm·°F). For a refrigerator operating at 1000 psi and 80°F, for example, the theoretical refrigeration per unit of gas flow is approximately 14.8 Btu/(h·cfm·°F). From Equation (3), the exchanger temperature difference must be less than 25°F to obtain useful refrigeration or low temperatures. Since the coldest temperature obtained in the heat exchanger is about −321°F, the normal boiling point of nitrogen, the exchanger efficiency, as normally defined, must be greater than

$$\frac{(80 - 25) - (-321)}{80 - (-321)} \times 100 = 94\%$$

Hence, heat exchangers must be extremely efficient to realize useful amounts of refrigeration. In practice, exchanger efficiencies of 98% (warm temperature difference of 8°F for nitrogen systems) or higher are needed.

The Joule-Thomson throttling process is irreversible, and its improper use in a refrigeration process can cause poor process efficiency. A temperature-entropy thermodynamic chart shows that the entropy increase in a throttling process decreases with decreasing temperature and is the least when in the throttling liquid phase. For maximum process efficiency, therefore, the Joule-Thomson throttling valve is always located at the lowest possible temperature.

Gases such as the fluorinated hydrocarbons, methane, argon, and nitrogen, exhibit cooling Joule-Thomson effects when at normal ambient temperatures and may be used to provide refrigeration at temperatures as low as about −325°F. However, to reach lower temperatures, the basic Joule-Thomson system must be modified. Neon and hydrogen boil in the −424 to −406°F range. However, these gases exhibit an inverse Joule-Thomson effect at normal ambient temperatures: *i.e.*, the gases will warm during an expansion from a high to a low pressure. Since the desired cooling effect is obtained when the gas is cooled to lower temperatures (to the boiling points of nitrogen or argon), as shown in Figure 5, an auxiliary Joule-Thomson circuit may be used to precool them prior to their Joule-Thomson expansion system.

A similar effect is exhibited by helium in Figure 6, which shows the theoretical refrigeration of helium. Note that extremely low

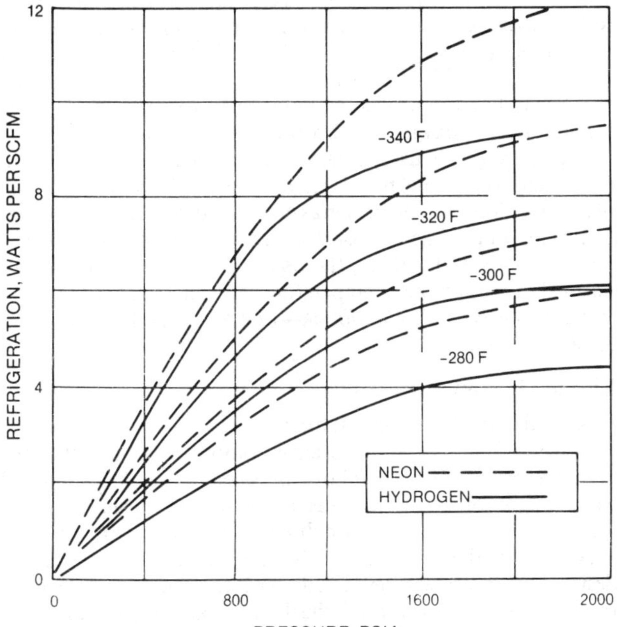

Fig. 5 Theoretical Refrigeration—Neon and Hydrogen

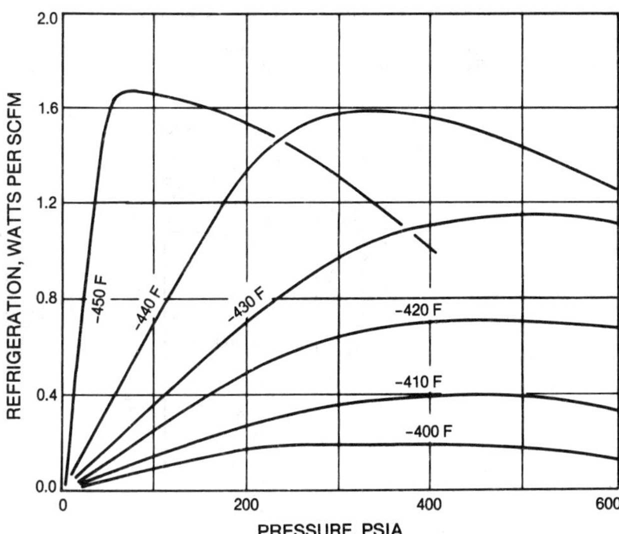

Fig. 6 Theoretical Refrigeration—Helium

temperatures must be obtained for precooling the helium to produce appreciable amounts of refrigeration. The loss in refrigeration caused by the warm-end temperature differences of the exchangers equals about 0.013 Btu/(ft^3·°F) for neon and helium and for hydrogen below about −316°F.

A cascade process is one in which gas is precooled by another Joule-Thomson process. Usually the precooling includes phase change with a high-pressure condensing stream in heat exchange with a low-pressure boiling stream. Several stages are usually cascaded together for improved performance. (Figure 14 is an example of a cascade cycle.)

Expander Refrigeration

In the Joule-Thomson process, the high pressure is significant only in that, under these conditions, the nonideality of gases at certain temperature levels may be used to produce low temperatures. A thermodynamically more efficient process results if the high-pressure gas is allowed to do work during its expansion process. In performing work, the energy content of the gas is reduced, with a resulting temperature decrease.

A simple expander or work cycle is shown in Figure 7. High-pressure gas enters the heat exchanger and is cooled by a countercurrent stream of low-pressure gas. At an intermediate temperature, some gas is withdrawn from the heat exchanger and sent to an expansion engine or turbine where it is cooled by expanding work on the machine. The withdrawal temperature is often chosen so that the gas leaving the expander is at low pressure and is a saturated vapor.

The expansion process at constant entropy is ideal. The efficiency of the expander is defined as the ΔH across the expander divided by the ΔH that would result from a constant entropy expansion. The gas not withdrawn to the expander continues to be cooled in the heat exchanger and cools to a temperature much lower than could be obtained in a comparable Joule-Thomson refrigerator. It is expanded in a Joule-Thomson valve and may be partially liquefied. The two gas streams then reunite and, after absorbing energy from the heat load, return through the exchanger.

An energy balance around the system shows that the sum of the heat load and heat leak, Q, equals the sum of the work produced in the expander and the enthalpy difference at the warm end of the exchanger:

$$Q = W + H(\text{l-p outlet}) - H(\text{h-p inlet}) \qquad (4)$$

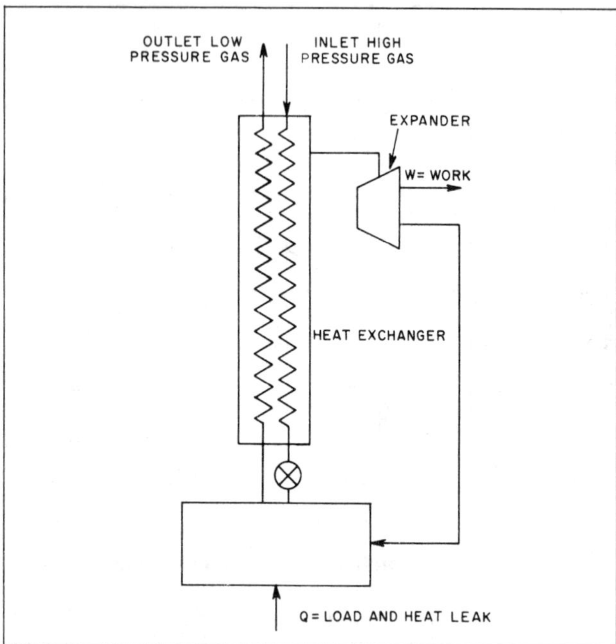

Fig. 7 Simplified Expander Refrigeration Cycle

This equation is comparable to Equation (1), derived for the Joule-Thomson process. The work *W* represents the refrigeration available through the expansion of the gas. The enthalpy difference may be composed of a Joule-Thomson effect and the warm-end temperature loss (a term that represents the heat exchanger inefficiency), as in the Joule-Thomson process. Hence, in analogy to Equation (3):

$$\text{Refrigeration} = (\text{Work}) + (\text{J-T Effect}) - c_p \Delta t \qquad (5)$$

Thus, the refrigeration obtained from such a cycle will be considerably more than that obtained for the Joule-Thomson cycle. The use of helium represents an extreme example. Helium at ambient temperature exhibits a rise in temperature when expanded from a high pressure to a low pressure by a Joule-Thomson process. Thus, according to the definition used here, the J-T effect is negative. However, the possibility of obtaining work using helium in an expander makes it possible to obtain a finite amount of cooling or refrigeration. Figure 8 shows the refrigeration and work for a helium refrigerator producing refrigeration at about −320°F (normal boiling point of nitrogen). At this warm temperature, the helium remains a gas throughout, and all the high-pressure helium is withdrawn to the expander.

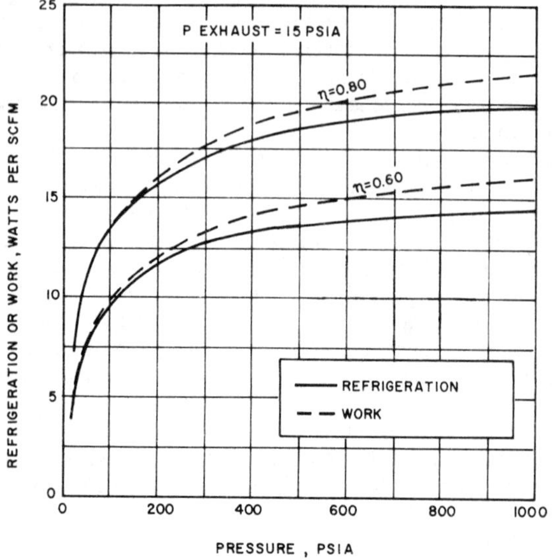

Fig. 8 Work and Refrigeration from Helium Expander Cycle

The losses contributed by the heat exchanger inefficiencies are assumed to be negligible. The higher value of expansion efficiency is obtained in larger refrigerators, and the lower value is practicable for small laboratory units. The difference between the refrigeration and work curves represents the contribution of the negative Joule-Thomson effect of the helium.

Particular attention must be given to the heat exchanger design. The effects of heat exchanger inefficiency, as manifest in the warm-end temperature differences, apply to this cycle as well as to the Joule-Thomson process. Thus, it is possible to produce only a minute amount of refrigeration, in spite of high expander efficiencies, if the exchangers are poorly designed.

Cycles

The method of adapting the previously described refrigerating means to actual refrigerating or liquefying cycles is determined by many factors, including: (1) the temperature level desired; (2) the ultimate load or product requirement; (3) the product purity; (4) the distribution of auxiliary heat leaks and loads; and (5) the availability of auxiliary refrigeration such as liquid nitrogen. Although basic cycles may be specified, each refrigerator or liquefier must be considered individually because of the varied requirements. In the following sections, the application of the fundamentals to specific requirements are discussed.

OXYGEN LIQUEFACTION

Oxygen is the second most abundant substance in air (21% by volume, 23% by mass) and is produced in large quantities by distillation of liquid air. Liquid oxygen has a characteristic blue color caused either by the presence of the polymer O_4 or by the unpaired electrons responsible for its paramagnetism. Saturated liquid oxygen at 1 atm and −297°F has a density of about 71 lb/ft^3. Oxygen gas is colorless, odorless, and tasteless. It is only slightly soluble in water and is a poor conductor of electricity and heat.

Oxygen is chemically very active, especially with hydrocarbons. Although oxygen does not burn, it supports combustion and combines directly or indirectly with all the elements except the rare gases. Because of its chemical activity, oxygen presents a safety problem in handling. Serious explosions have resulted from the combination of oxygen and hydrocarbon lubricants.

Liquid oxygen is used in large quantities as an oxidizer with fuels such as kerosene, hydrogen, and others, for rocket and missile propulsion and for steel making. It is also produced in large quantities for commercial use. Delivered in special cryogenic railway tank cars and trailers, the liquid oxygen is vaporized and warmed to near ambient temperature before it is used. Chapter 17 in the 1993 ASHRAE *Handbook—Fundamentals* lists properties of saturated oxygen.

Liquid Oxygen Generator

A condensed flow diagram for a liquid oxygen generator cycle is shown in Figure 9. Filtered air is compressed in a multistage compressor to 2000 to 3000 psi. Between the second and third stages of compression, the air is passed through a caustic scrubbing system to remove carbon dioxide from the airstream.

Air leaving the fifth stage of the compressor passes through a desiccant or molecular sieve filled vessels to remove moisture from the air. After passing through the drier, the airstream is split. One stream passes through the main heat exchangers, which are cooled by effluent gases from the distillation column; it is then expanded

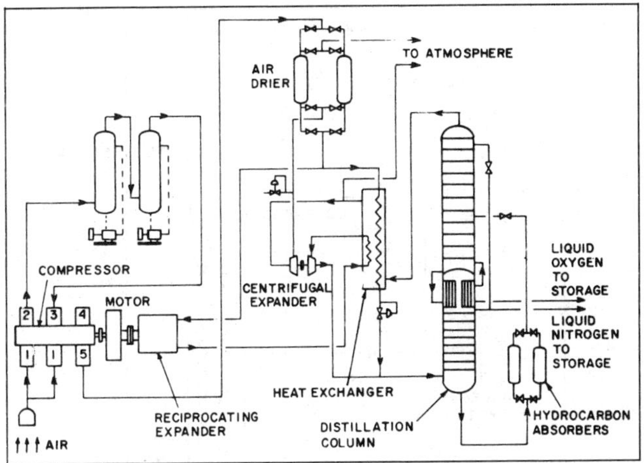

Fig. 9 Simplified Flow Diagram of Liquid Oxygen Generator

through a valve to the pressure in the high-pressure column. The other portion of the airstream is expanded through a reciprocating expander, from compressor discharge pressure to an intermediate pressure, is warmed through coils in the main heat exchanger, and is further expanded in a centrifugal expander, to the pressure of the high-pressure column.

A double rectification column is used. Crude liquid oxygen from the high-pressure column is passed through hydrocarbon adsorbers before being expanded to the low-pressure column. Liquid nitrogen product is withdrawn from the stream, which also provides liquid nitrogen reflux to the low-pressure column. Liquid oxygen is withdrawn from the bottom of the low-pressure column. Effluent nitrogen-rich gas from the low-pressure column is warmed in the main heat exchangers countercurrently with incoming air and discharged to the atmosphere. A portion of the warm effluent gas loads the centrifugal expander and is then used for reactivating the air drier, hydrocarbon adsorbers, and reciprocating expander oil filters, since dual assemblies of each are provided to ensure continuous operation of the liquid oxygen generator.

Reciprocating and Centrifugal Expanders

The dual reciprocating and centrifugal expander with intermediate reheat of the airstream increases efficiency. This scheme can be effectively applied where airflows are relatively large, since a centrifugal expander works well at lower expansion pressures. The refrigeration increase is the result of increased enthalpy drop, which is caused by the higher inlet temperature to the centrifugal expander.

The reciprocating expander is more efficient when it operates at a higher exhaust pressure because of the smaller losses in the expansion engine ports, valves, and cylinder. This is illustrated in Figure 10, by comparing the conventional arrangement of a single reciprocating engine expanding from compressor discharge pressure to high-pressure column pressure. About 14% increase in total expander refrigeration is realized by using this scheme. The reciprocating expansion engine is considerably smaller in size because of the higher density of the exhaust air from the engine.

The overall cycle efficiency of a liquid oxygen generator is about 780 kWh/ton of liquid oxygen plus liquid nitrogen. Because the use of the liquid oxygen may be cyclic with high peak requirements and periods of full storage, generators must be taken off-stream and put back on-stream with a minimum interruption of production. A warm liquid oxygen generator unit requires about 5 h to be placed in full production; if the generator unit is shut down cold, it requires approximately 30 min to return to production.

Any high-pressure cycle, liquid oxygen generating unit occasionally requires defrost of the main heat exchangers. One unit requires

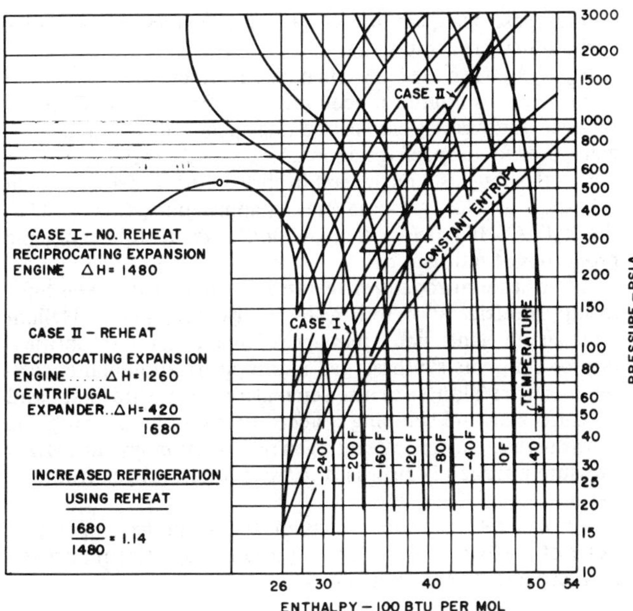

Fig. 10 Effect of Reheat on Expander Refrigeration

4 h to defrost and then cool down the main heat exchangers for normal liquid oxygen production again.

Another arrangement used to make commercial liquid oxygen incorporates an expander-type refrigeration unit which operates on the nitrogen gas produced at the top of the high-pressure column. Cold nitrogen vapor withdrawn from the column is warmed to ambient temperature where it is compressed. The aftercooled compressed nitrogen is partially recooled in countercurrent heat exchange with the warming nitrogen and split into two streams. One stream is fed to an expansion engine, where the gas is work-expanded, cooled, and then returned to the column. The other stream continues in heat exchange with the warming nitrogen and is liquefied. This liquid is returned to the column and provides the necessary refrigeration, not only for overcoming the usual process heat influxes but also for withdrawal of the liquid oxygen product from the bottom of the low-pressure column.

Although not as efficient as the preceding process, the latter operates at pressures conveniently handled by centrifugal compression and expansion machinery and offers advantages typical of this equipment, such as low investment, ease of operation, low maintenance, and freedom from contamination of the process streams with hydrocarbon lubricants. It can also be used for easy conversion of a gaseous oxygen plant to a liquid oxygen plant by adding the refrigeration equipment.

NITROGEN REFRIGERATION

Nitrogen is the major constituent of air (78% by volume or 75% by mass) and, as discussed previously, is produced commercially by distillation of liquid air. Liquid nitrogen is a clear, colorless, odorless, and tasteless fluid resembling water in appearance. It is only slightly soluble in water and is a poor conductor of electricity and heat. As the main ingredient of air, nitrogen is mostly inert and it does not produce toxic or irritating vapors.

Nitrogen at room temperature and low pressure does not react readily with other elements. It neither burns nor supports combustion. At elevated temperatures it combines with some of the more active metals such as calcium, sodium, and magnesium to form nitrites. With high pressures and temperatures or in the presence of catalysts, a wide variety of compounds can be formed, among these being ammonia, the nitrous oxides, and complex carbon-metallic compounds.

Nitrogen, when used as gas in bulk quantities, is often sold and transported as liquid for economy and convenience. Gaseous nitrogen is extensively used in inserting applications. Liquid nitrogen is used for deep refrigeration storage, food freezing, pulverizing, deflashing, as a refrigerant in shrink-fitting metal parts, in cold traps, and for various laboratory applications. Chapter 17 in the 1993 ASHRAE *Handbook—Fundamentals* lists cryogenic properties of nitrogen.

Large-scale nitrogen refrigeration systems have been built to supply refrigeration to large liquid hydrogen plants. Helium liquefaction plants, liquid nitrogen scrubbing plants for purification of ammonia synthesis gas, argon purification plants, low-temperature plants for the separation of natural gas, and food freezing and shipping systems are examples of cryogenic processes that incorporate nitrogen refrigeration systems. High-pressure nitrogen is also used to extract natural gas from the earth.

Nitrogen refrigeration systems are applied in the temperature range of −344 to −260°F. The important properties of nitrogen as a refrigerant are as follows:

Molecular weight	28.02
Normal boiling point	−320.5°F
Latent heat at boiling point	2405 Btu/lb-mol
Liquid density at boiling point	50.4 lb/ft^3
Critical temperature	−232.78°F
Critical pressure	33.5 atm
Critical density	19.4 lb/ft^3
Melting (triple) point	−345.9°F

Typical overall requirements of actual power used per quantity of refrigeration obtained range from 26 bhp/ton at −280°F to 36 bhp/ton at −321°F.

The Carnot cycle theoretical work requirement is given to be:

$$W/q = [(t_1 - t_2)/(460 + t_2)]\,[12{,}000/2545] \tag{6}$$

where

W = Carnot work, brake horsepower
q = refrigeration, tons
t_1 = cooling water temperature, °F
t_2 = refrigeration temperature level, °F

Thus, if $t_1 = 85$°F and $t_2 = -315$°F,

$$\begin{aligned} W/q &= \{[85 - (-315)]/[460 + (-315)]\}\,[12{,}000/2545] \\ &= 13 \text{ bhp/ton} \end{aligned} \tag{7}$$

Therefore, overall cycle efficiency E of a nitrogen refrigerator removing heat at −315°F is:

$$E = 100\frac{W_T/q_R}{W_A/q_R} = 100\frac{13}{36} = 36\% \tag{8}$$

Significant factors that contribute to the inefficiency are: (1) inefficient mechanical equipment (such as compressors, expanders, and motors); (2) temperature differences and pressure drop required for heat transfer; (3) heat leak from ambient into cold equipment; and (4) piping configuration.

A typical nitrogen refrigeration cycle is shown in Figure 11. This Claude (expansion engine) cycle incorporates a double flash system. Variations of this cycle include multiple refrigeration sources such as a second expansion engine or a high-temperature Joule-Thomson refrigerator to precool a split feed stream. Another variation is shown in Figure 12. This dual pressure arrangement uses a moderate pressure for the stream that passes through the Joule-Thomson valve and is liquefied, and a low pressure for the stream that is recycled through the expansion engine.

The moderate pressure is at about the critical pressure of nitrogen to minimize the adverse thermodynamic effect of the large isothermal heat load of condensing nitrogen. The pressure of the recycle stream is based on optimum performance of highly efficient centrifugal compression and expansion machinery. Although the performance of the cycle is not as efficient as the high-pressure cycles, the lower operating pressures provide advantages in lower initial investment, ease of operation, and decreased maintenance.

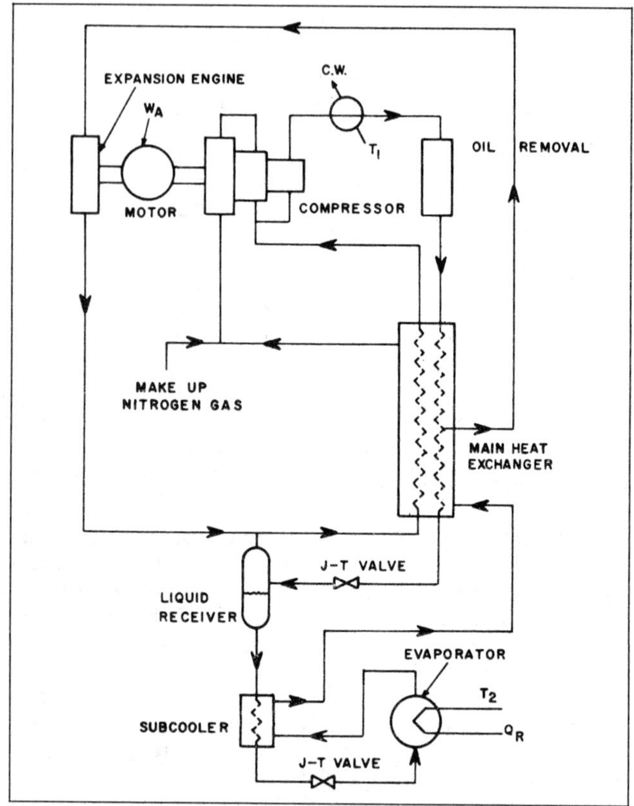

Fig. 11 Nitrogen Refrigeration System

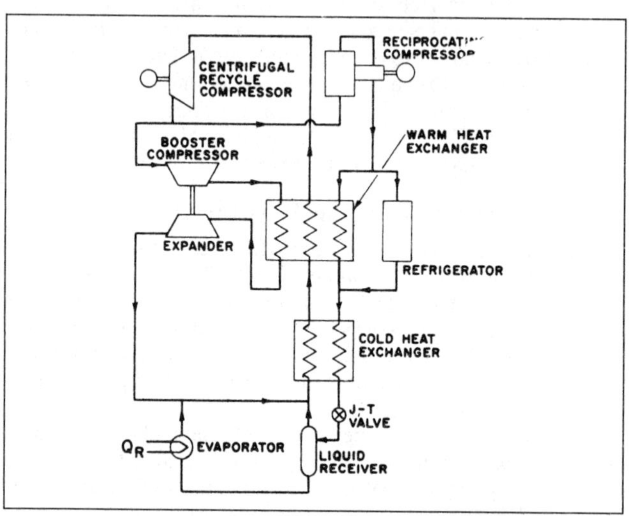

Fig. 12 Dual-Pressure Nitrogen Refrigeration System

Nitrogen required to make up losses from the system can be supplied, if an air separation plant is part of the facility, as in the case of argon purification plants or liquid nitrogen scrub units. If nitrogen from an air separation plant is not available, makeup nitrogen can be supplied by hauled-in liquid nitrogen, or hauled-in gaseous nitrogen, either in conventional cylinders or in tube trailers, or from an inert gas generator. Leakage from the system should be minimized. One of the major sources of leakage is the compressor.

Reciprocating compressors are usually selected for cycles requiring a discharge pressure between 1500 and 3000 psi. Compressor designs that minimize leakage are available. Oil-lubricated reciprocating compressors generally give the best service for compressing dry nitrogen, and highly efficient oil removal devices are necessary to prevent oil from entering the cold section of the system, where it would gradually plug the heat exchangers and prevent continuous operation.

For cycles requiring low to intermediate discharge pressures (<700 psi), centrifugal compressors are widely used, especially for large refrigerators. These machines are well adapted to handling large volumes of gas, are lower in cost than reciprocating compressors, show good operating efficiency, and eliminate oil contamination.

HYDROGEN LIQUEFACTION

Gaseous hydrogen is colorless, odorless, and tasteless; therefore, its presence cannot be detected by the human senses. It is the lightest of all elements and diffuses rapidly through porous materials and through some metals at red heat. The thermal conductivity of hydrogen gas at atmospheric temperature and pressure is about seven times greater than that of air. Hydrogen gas is flammable in air over a relatively wide range of mixtures (4 to 75%). It burns, in air, with a pale blue, almost invisible flame. When mixed in the proper proportions with air, oxygen, or other oxidizers, it forms an explosive mixture. It is nontoxic, but can cause asphyxiation by exclusion of air in confined areas.

Liquid hydrogen is transparent and odorless. Its density is about 1/14 that of water. It is not corrosive or significantly reactive. The low temperature of liquid hydrogen can solidify any gas except helium. Liquid hydrogen has a relatively high coefficient of thermal expansion, which must be considered in the design of equipment for handling the liquid. With the exception of helium, all known substances are essentially insoluble in liquid hydrogen. Hydrogen reacts with air or oxygen only in the presence of a catalyst (*e.g.*, platinum-black) or stimulus such as an arc, spark, or flame. It is hyperbolic with fluorine (requires no source of ignition).

Large-scale hydrogen liquefaction ranges in size from 6.5 to 60 ton/day. The problems encountered in the liquefaction of hydrogen are (1) purification, (2) ortho-to-para conversion, (3) precooling and liquefaction, and (4) transfer and storage. Chapter 17 in the 1993 ASHRAE *Handbook—Fundamentals* lists cryogenic properties of saturated hydrogen, and *Thermodynamic Properties of Refrigerants* (Stewart *et al.* 1986) lists properties at various pressures.

Purification

Hydrogen has been produced from electrolytic cell gas, refinery off-gas, and steam reduction or partial oxidation processes. Hydrogen cell gas has mainly been used for supplying small liquefiers up to 1100 lb per day capacity; for larger plants, other sources are preferred. A partial oxidation process is considered when oil is the only available reactant for the locality. A steam reformer process is preferred when natural gas or naphtha is available. Substantial savings in capital investment can be realized by the use of refinery off-gas.

Purification of hydrogen prior to liquefaction depends on the source. In all cases, the required drying is normally accomplished near ambient temperature, with activated alumina or molecular sieve desiccant. The alumina may be reactivated at a lower temperature and has a lower initial cost, whereas the molecular sieve has the advantages of (1) only slight capacity reduction from heavy hydrocarbon contents in the hydrogen feed, and (2) permitting removal of H_2S and CO_2 along with the moisture in a single step.

Benzene and toluene have been separated from refinery off-gas by an oil scrub system. The other hydrocarbons may be separated in the low-temperature equipment, down to the freezing point of methane (−296.5°F). By cooling the feed gas against liquid nitrogen boiling at about 4 atm, and following it by liquid-gas phase separation, the hydrocarbon concentration may be reduced to a level of about 0.9% by volume. These residual hydrocarbon impurities, along with nitrogen, carbon monoxide, argon, and oxygen, may be removed by one of three methods: (1) adsorption by activated charcoal; (2) absorption by liquid methane, liquid propane, or both; or (3) adsorption by molecular sieve. Adsorption by charcoal is adaptable to small and large plants, and impurities have been reduced to below 1 ppm in a single step. The methane-propane absorption process offers the advantage of less operator attention, whereas the charcoal adsorption method allows for greater flexibility during plant startup and shutdown. The adsorption by molecular sieve can reduce impurities to 0.001 ppm and applies to large plants.

Ortho-to-Para Conversion

Hydrogen, at ambient temperature, normally exists as 75% ortho (normal hydrogen), whereas, at the normal boiling temperature, its equilibrium concentration is 99.8% para. These forms differ by having parallel (ortho) or opposed (para) nuclear spins in the two atoms forming the hydrogen molecule. In time, the unassisted conversion will take place at the normal liquid temperature by the amount of:

$$1/O_f - 1/O_i = K\theta \tag{9}$$

where

θ = time, h
K = reaction rate constant, 0.0124/h
O_f = final ortho mol fraction (Grilly 1953)
O_i = initial ortho mol fraction (Grilly 1953)

The reaction is exothermic; 609 Btu of heat are released per pound of hydrogen converted. As shown in Table 1, the boiloff rates in storage would be substantial until a para concentration level of 95 to 99% was reached. This is the level at which the liquid is normally produced.

Catalysts increase the conversion rate during the cooling and liquefaction of the hydrogen. Hydrous iron oxide and Cr_2O_3 on

Table 1 Boiloff Loss in Ortho-to-Para Liquid Hydrogen Conversion

Para Concentration, %	Boiloff Loss, % per Day
98	0.019
95	0.119
90	0.48
80	1.9
70	4.3
60	7.6
50	11.9
40	17.2
25	26.8

an Al_2O_3 gel carrier, and NiO on Al_2O_3 gel have been used. The NiO catalyst is about 90 times as active as the others and produces one of the fastest catalytic reactions known.

One or two stages of conversion at the liquid nitrogen level, followed by staged or continuous conversions down to the liquid hydrogen level, are commonly used in large plants. The staged and continuous conversion methods offer almost equal advantages for increasing cycle efficiencies. Important considerations for the choice between staged and continuous conversion include the process to be used with it and the manner in which the two are integrated to give an overall efficient plant of minimum capital investment.

Precooling and Liquefaction

The generally accepted cycle for small capacity plants is the single or dual pressure, liquid nitrogen precooled, J-T cycle, shown in Figure 13. By staging the conversion at the liquid nitrogen level, the energy consumed is about 5 kWh, and 1 gal of liquid nitrogen is used per gallon of 95% para liquid hydrogen for a plant size in the range of 500 to 1000 lb per day capacity. The process typically uses industrial grade or electrolytic cell gas as the feed, which must be compressed to 1500 psi pressure, dried in a desiccant bed of activated alumina, and cooled to the vacuum liquid nitrogen temperature level of about −340°F. This requires a nitrogen evaporative pressure of about 2.7 psia.

Final cleanup and staged para conversion is then accomplished, followed by a final pass against the vacuum nitrogen to remove the heat of conversion and heat leakage. The precooled hydrogen is finally cooled in the J-T heat exchanger against the flashed return vapors and then reduced in pressure to near atmospheric. The flashed gases are separated, and the liquid is converted to above 95% para in the separator-converter; vapors are returned to help cool the 1500 psi stream, and the liquid is transferred from the separator into storage.

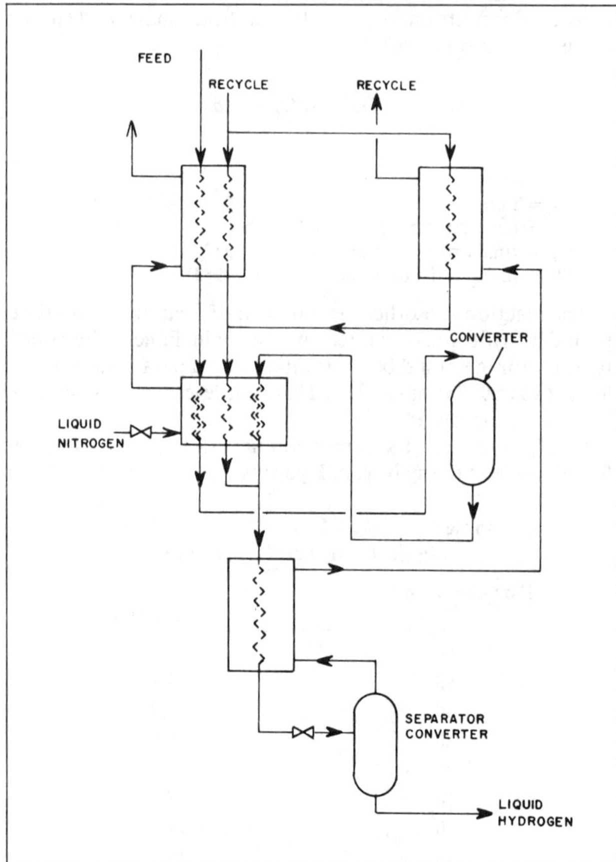

Fig. 13 Hydrogen Liquefier

High-capacity hydrogen liquefiers are commonly equipped with a single or dual pressure, nitrogen precooled, expander cycle. Variations of this type of cycle use up to a 1500 psi hydrogen recycle head pressure. The process will produce liquid hydrogen with an energy consumption of about 3 kWh per gallon into storage, including all refrigeration requirements.

For the liquefiers of small capacity (13 gal/h and less), the cold equipment is contained in one or two vacuum-jacketed Dewars to minimize heat leakage. For large units, the equipment above normal boiling temperature of nitrogen may be contained in a steel panel cold box, using perlite or rockwool insulation. Equipment below boiling nitrogen temperature is contained in vacuum-jacketed Dewars.

LIQUEFIED NATURAL GAS

Liquefied natural gas (LNG) is the liquid form of regular natural gas consisting primarily of methane, hydrogen, and a mixture of other hydrocarbons (approximately 95% CH_4, 3% C_2H_6, 1% N_2, 0.5% C_3H_8, and 0.5% $C_nH_{2(n+1)}$). The gas is nontoxic, flammable, and lighter than air. The liquid, which is produced from gas by liquefaction, is clear, clean, and odorless. It has a density of 26.5 lb/ft³ at −258°F and 1 atm.

The gas is primarily used for residential and industrial heating and to a very minor extent, for the production of electrical energy. LNG has been applied increasingly as a base load in peak shaving applications in satellite operations. In all of these applications, the end use is nearly always as a gas.

Liquefying natural gas reduces its specific volume by a ratio of 600:1, which makes handling and storage economically possible despite the added cost of liquefaction and the specially insulated transport and storage equipment. The two principal types of LNG liquefaction plants, peak shaving and base load, are characterized by the disposition of the LNG.

Peak shaving has been applied the most extensively, and the majority of liquefaction plants have been built for this purpose. Peak shaving is a method of smoothing the demand on a gas pipeline by liquefying a portion of the gas during the summer or off-peak season, storing the LNG, and subsequently vaporizing it during peak periods in winter. In this way, fuel requirements are met with a smaller pipeline than would be required without peak shaving.

The base load plant liquefies all of the gas available to it for transportation to another geographical location. These large-capacity plants are located in industrially undeveloped regions where natural gas is available in large quantities. Liquefaction makes natural gas available in areas where natural gas reserves are limited, nonexistent, or being depleted.

Feed Treatment

Natural gas must be treated, usually before liquefaction, to remove components that would freeze and stop the process equipment. The type of feed treatment depends on the source of the natural gas as well as the nature and quantity of the components it contains. Such components commonly consist of carbon dioxide, hydrogen sulfide, water with its associated hydrates, hexane and heavier saturated hydrocarbons, and cyclic compounds such as benzene and toluene.

If the natural gas feed is obtained directly from a pipeline, as in a peak shaving facility, it may already have been purified to a certain degree. Hydrogen sulfide may have been removed, and in the process, some carbon dioxide will also have been removed. The recovery of LPG removes the C_6 + hydrocarbons. Generally, all that will require removal is additional carbon dioxide and water. If the natural gas feed is obtained from the well head, removal of all the components listed must be considered. The extent of removal depends on the composition and temperature of the liquid phases that form during liquefaction. Feed impurities must be reduced in concentration to the extent that the residual amounts are completely soluble in the liquid phases at all points in the process. Typically, carbon dioxide and hydrogen sulfide are reduced to 150 ppm by volume, while water is removed to a dew point of about −100 °F, or 1 ppm at maximum.

The most popular processes to remove acid gases are amine scrubbing, hot potassium carbonate scrubbing, cold methanol scrubbing, and adsorption on molecular sieve. Water can be removed by methanol or glycol systems, but removal on a desiccant such as molecular sieve or activated alumina is most widely applied. Sometimes a single bed of molecular sieve is used to remove hydrogen sulfide, carbon dioxide, and water. The pretreatment system must be considered carefully with respect to the liquefaction system, and is often closely tied into it.

Liquefaction

Natural gas liquefaction cycles are generally cascade or expander cycles. The cascade cycle provides refrigeration by the Joule-Thomson process or isenthalpic throttling of separate refrigerant streams. The expander cycle provides refrigeration at lower temperature levels by expanding the gas in an engine producing work, ideally, at constant entropy. Figure 14 shows a simplified three-stage cascade cycle containing three independent refrigeration systems. The initial refrigeration is supplied by circulating propane, the intermediate refrigeration by ethylene, and the low-temperature refrigeration by methane.

The pressure-enthalpy chart (Figure 14) indicates the mechanism by which refrigeration is provided by the propane system. In this example, the propane is compressed from a cold section temperature of −30.5 °F at 20 psia pressure to 190 psia, ideally at constant entropy. The heat of compression and the refrigeration load that vaporizes the propane are rejected to cooling water in the compressor aftercooler at constant condensing pressure, at a saturation temperature of 100 °F. The condensed propane is throttled, at constant enthalpy, to 20 psia and cools to its saturation temperature, the initial −30.5 °F. This two-phase stream then flows to the evaporator where it absorbs heat from the other streams (process feed gas, ethylene, and methane) and vaporizes. The pressure-enthalpy chart indicates that, under these conditions, each pound of circulating propane provides 180.5 − 75.9 = 104.6 Btu of refrigeration.

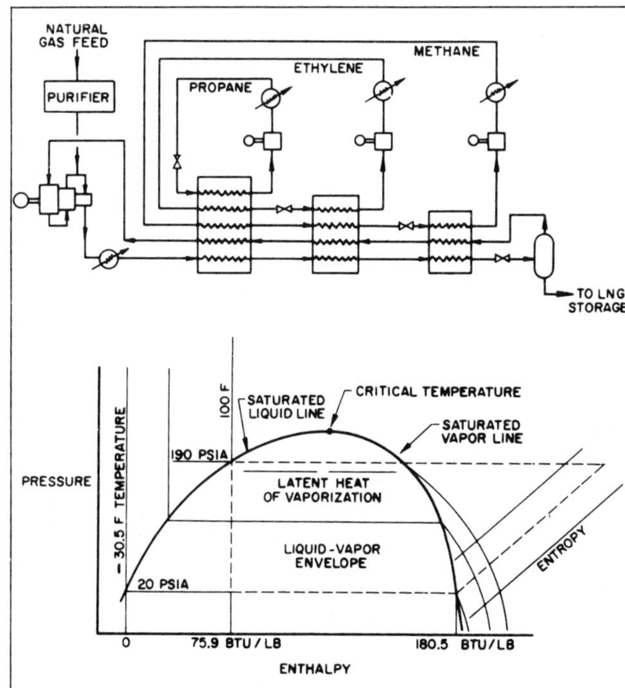

Fig. 14 Cascade Cycle

The ethylene and methane refrigeration systems operate in similar fashion. However, these streams are cooled with the higher boiling refrigerants, in addition to water. The compressor discharge pressure for ethylene is selected so that total liquefaction is obtained at the propane evaporator temperature; the condensing pressure for methane is at the ethylene evaporator temperature.

The three-stage cycle shown in Figure 14 is a simplified version of the cascade cycle. Present liquefiers incorporate from five to ten stages of cooling by evaporating each refrigerant at pressures intermediate between the final discharge pressure and the first-stage suction pressure. The overall refrigeration system is more reversible thermodynamically, because it provides refrigeration at several levels at intermediate pressures, thereby reducing the compression power significantly. Other cycle variations integrate the methane refrigerant system with the natural gas feed. One variation replaces the methane refrigeration shown in Figure 14 by an LNG flash separator stage and returns the vapor overhead at a higher pressure, which simplifies the machinery requirements for the facility.

Mixed Refrigeration Cascade Cycle. A variation of the cascade system is the mixed refrigeration or autorefrigerated cascade cycle. A simplified version of this cycle is shown in Figure 15. Instead of using several pure refrigerants in separate closed loops, this process combines all the refrigerants in a single multicomponent stream. Such a refrigerant stream might contain, for example, a mixture of nitrogen, methane, ethane, propane, and butane, with the composition to provide the necessary refrigeration at the required temperature levels.

The compressed refrigerant stream is cooled in the compressor aftercooler where most of the heavier components of the mixture are partially condensed. The liquid and vapor phases are separated and further cooled in separate passes of the first-stage heat exchanger against returning low-pressure refrigerant. The subcooled liquid is throttled into the low-pressure returning refrigerant stream to provide the necessary refrigeration to partially condense the separated vapor streams. This process is repeated in several stages of heat exchange, with each successive liquid fraction consisting of lighter hydrocarbons as the temperature decreases. The compressed natural gas is cooled and liquefied in countercurrent heat exchange with the returning refrigerant stream in a separate pass in each heat exchanger.

With the proper composition of the mixed refrigerant, this process can be made thermodynamically more efficient than a multistage cascade process. This occurs when the refrigeration required by the cooling stream is provided by the warming stream

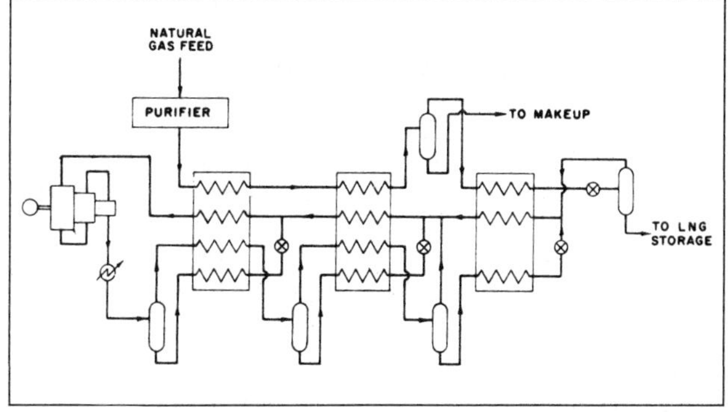

Fig. 15 Mixed Refrigeration Cascade Cycle

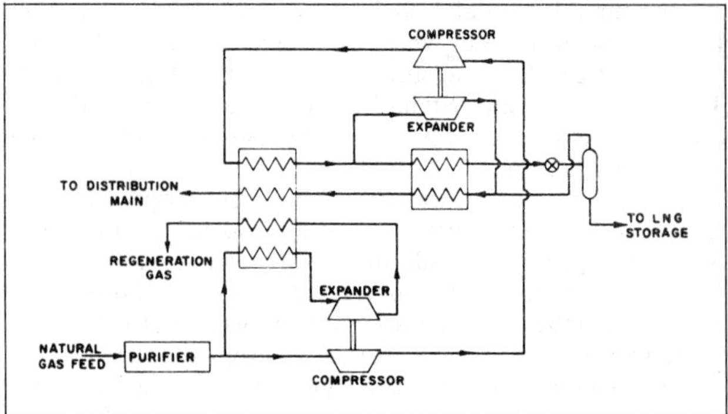

Fig. 16 Dual Expander Cycle

only to the extent required and at a temperature only slightly lower than the cooling stream. The resulting temperature-enthalpy curve for the warming stream is matched closely to the temperature-enthalpy curve for the cooling stream, which represents a condition of low thermodynamic irreversibility.

Additional advantages of the mixed refrigeration cascade cycle are simplicity in equipment and piping, including the use of a single refrigerant loop, a single compressor, reduced plant maintenance, use of the feed gas as the source of components for the refrigerant, and the elimination of facilities for the storage of refrigerant. The refrigerant mixture is usually obtained by separating the necessary components from the feed stream.

These advantages are especially important in remote locations, since the mixed refrigerant system has proved particularly useful in base load LNG plants. It is seldom used in peak shaving plants.

Expander Cycle. In the expander cycle, heat exchange and expander steps are arranged according to the pressure level at which the gas is originally available, the gas composition, and the relative cost and availability of auxiliary refrigeration. Cycle efficiency can be increased by using multiple expansion stages to produce refrigeration at several temperature levels. However, it soon reaches an optimum point, which represents balance between the cost of additional heat transfer surface and expansion engines and the savings obtained through reduced power. Figure 16 shows a dual expander cycle in which refrigeration is produced at two different temperature levels.

The natural gas feed is purified and split into two streams. One stream is precooled before being expanded to produce high temperature level refrigeration; the other is compressed in the two expander-driven booster compressors before precooling to the second expander. The higher pressure drop across the second expander produces refrigeration at a lower temperature level. Refrigeration is recovered from both expanded streams by cooling and liquefying the natural gas feed in countercurrent heat exchange.

The dual expander cycle is used to reduce the investment in plants whose pipeline pressure exceeds that required by the local distribution system. The excess pressure is used to produce the refrigeration needed to liquefy the natural gas, which eliminates the need for the refrigerant cycle compressor.

The type of cycle selected for an LNG plant depends primarily on plant capacity. Large plants invariably use a cascade cycle because the high investment cost imposed by cycle complexity is countered by the low operating cost resulting from high cycle efficiency. Intermediate and small plants usually use an expander cycle because low investment is more important. One small plant has made use of a Joule-Thomson throttling cycle to obtain minimum investment.

LNG Storage

Tanks for LNG storage are among the largest for cryogenic tankage. In general, storage tanks for a peak shaving facility are larger in relation to the capacity of a liquefaction unit than for a base load shipping or receiving facility. This is because the peak shaving plant must liquefy and store enough gas during its 200 to 275-day off-season to accommodate peak loads during the remaining season. This peak demand is five to ten times the minimum sent out during off-peak operation. The base load plant must provide no more storage than needed to accept plant liquefaction capacity for the maximum periods that ships are not in port. Tanks with a capacity of 0.5 billion cubic feet at standard conditions are common and one tank is 230 ft in diameter with a 176-ft liquid depth and holds 4 billion cubic feet at standard conditions.

Three types of tank construction are used: double-wall metal tanks, prestressed concrete, and frozen earth cavity. The double-wall metal tank consists of one container nested within the other with an annular space 4 to 5 ft thick filled with insulation, commonly perlite. The inner container holds the LNG and is usually made of 9% nickel steel, although aluminum alloy is also used. The outer tank serves as a vaportight container for the insulation and is made of carbon steel. The insulation space is purged with dry gas to exclude atmospheric moisture. Tank design often includes direct communication between the inner container and the insulation space, so that the boiloff vapors may serve as a purge. This type of tank is erected above ground on a specially prepared site which consists of load-bearing insulation over a select sand fill. Heating coils may be embedded below the tank to prevent frost heaving.

Prestressed concrete tanks are usually built when large capacity is needed. Such tanks are built in a variety of configurations featuring above and below-ground installations with single- and double-wall tanks and external and internal insulation. In-ground installation provides a measure of safety by containing spills in the event of tank leakage or rupture.

In the frozen earth cavity tank, the earth itself serves as the container. The earth is prefrozen by pipes sunk into the ground in a cylindrical pattern. Refrigerant is passed through the pipes and when a satisfactory impermeable frozen wall is obtained, the cavity is excavated, a suitable insulated concrete or metal roof is added, and the tank is ready for use. Mixed success has been attained with frozen earth cavity tanks because of uncertainties of underground water flow and the inability to prevent cracking in some frozen-earth structures.

The large capacity and thickness of insulation keep heat leakage into the tank very low. Boiloff rates as low as 0.05% of tank capacity per day are normally obtained for the doublewall metal tanks and many of prestressed concrete tanks. The frozen earth tank usually exhibits higher boiloff rates on initial fill, but boiloff slowly decreases with time. Boiloff vapors are collected, compressed, and returned to the distribution system.

Sendout Systems

A sendout system consists of pumps, vaporizers, and piping to withdraw LNG from storage and inject the vaporized natural gas into local distribution systems at the proper pressure. Equipment must be sized to deliver sendout requirements imposed by maximum peak loads. Use of sendout equipment at peak shaving units is low because it operates only during peak load seasons. However, reliability is of great importance, and sendout systems are usually provided in duplicate, with each unit capable of delivering maximum sendout. Base load units use sendout equipment almost continuously.

Commonly used vaporizers are of two types. Each relies on natural gas as the source of heat, and each uses an intermediate fluid to transfer the heat from the combustion gases to the LNG. In one type, water is warmed by submerged combustion burners, and the LNG is vaporized in tubes immersed in the water bath. In the second type, isopentane absorbs heat in a direct-fired burner and delivers it in an LNG exchanger. One vaporizer uses sea water for its heat source. The water flows by gravity over tubes in an open rack exchanger, with the LNG flowing inside the tubes.

Pumps are generally multistage, immersed centrifugal units mounted in an insulated vessel outside the main storage tank. Sealed, submerged motor-type pumps have also been used.

Using the Cold LNG

The large amounts of refrigeration available at temperatures as low as −250 °F have prompted many to examine potential uses for it. Typically, refrigeration recovery would be practiced at base load rather than at peak shaving units because the large capacity and uniform demand facilitates coordination with the refrigeration user. Obvious uses include food refrigeration and freezing and central air conditioning, although the use pattern of the latter is seasonal and out of phase with fuel demand during the heating season. The safety aspects of using a hydrocarbon stream in connection with air separation would have to be carefully considered. A variation of the air separation plant application is a system of conserving refrigeration. Nitrogen would be liquefied at the sendout site, in exchange with vaporizing LNG, and returned via the methane tanker to the LNG liquefaction plant where it would relinquish its refrigeration, thus substantially reducing the required power.

An olefins petrochemical plant has the necessary scale of operations to make effective use of all the refrigeration available from a typical base load LNG operation. By providing separation equipment, the LNG can also supply the ethane and propane for feedstock to the olefins plant. Such an integration of facilities would provide mutual benefit to both chemical plant operator and gas supplier.

HELIUM REFRIGERATION AND LIQUEFACTION

Helium, classified as a rare gas, is one of the most difficult gases to liquefy and its unusual properties have created so much interest that it has been the object of more experimental and theoretical research than any of the other cryogenic fluids. Liquid helium does not freeze under its own vapor pressure even if the temperature is reduced to absolute zero. It must be compressed to a pressure of 25 atm before it will freeze. Liquid helium has a density of 7.8 lb/ft^3 under saturated conditions of −450 °F and 1 atm. Liquid helium is odorless and colorless and is hard to see because of its low index of refraction.

Helium exists as several isotopes, the most common being helium 4. The next line, helium 3 is much rarer than helium 4 and is generally unknown outside of the research laboratory. Whenever helium is referenced without isotope designation, it can be assumed to be helium 4.

As a liquid, helium exhibits two highly different characteristics depending on temperature. Between the boiling point (4.2 K) and the lambda point at 2.18 K (the branching point of the specific heat curve resembling the Greek letter lambda) it is known as liquid helium I, and exhibits characteristics typical of other liquids. At 2.18 K the liquid undergoes a transition (the lambda transition) and becomes liquid helium II. This colder liquid exhibits the phenomenon of superfluidity (virtually zero viscosity) and attains an extremely high thermal conductivity, over 1000 times greater than that of copper.

All commercial helium is obtained from natural gas where it exists in concentrations of 0.5 to 2.5% by volume and in rare cases by as much as 8%. Chapter 17 in the 1993 ASHRAE *Handbook—Fundamentals* lists cryogenic properties of saturated helium and *Thermodynamic Properties of Refrigerants* (Stewart *et al.* 1986) lists properties at various pressures.

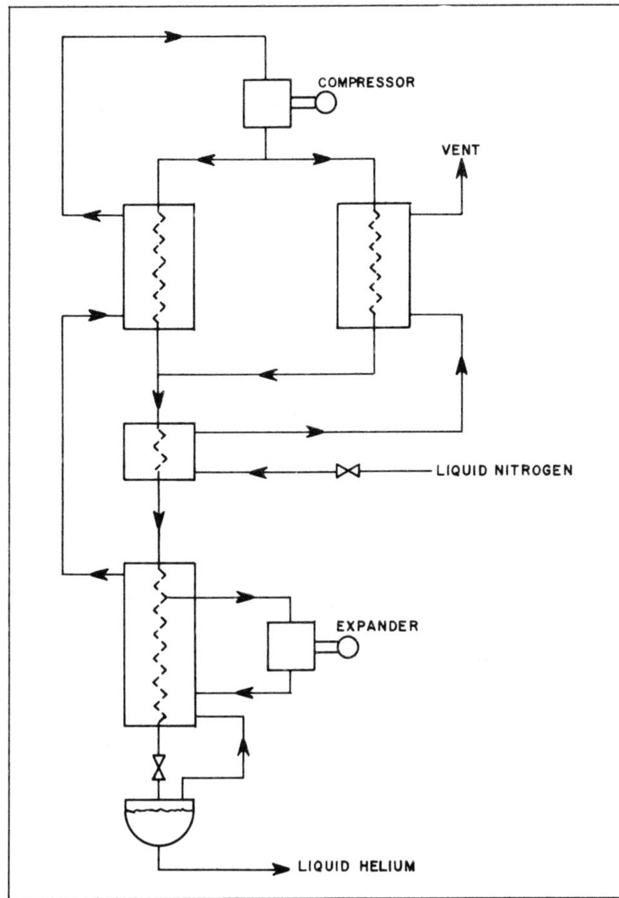

Fig. 17 Small-Scale Helium Liquefier

Helium liquefaction and refrigeration is applied in space chambers, computers, maser cooling, superconduction studies, and cryogenic research. Refrigerator capacities range from a fraction of a watt to as high as several tons of refrigeration, at levels ranging from 4.5 to 45 °R. Liquefiers, ranging in size from 2.5 to 25 gal/h of normal saturated liquid at 7.6 °R, have been constructed and operated.

Liquefier Design

Basic liquefier cycles have been (1) the nitrogen-hydrogen cascade and (2) the precooled expansion engine cycle. The latter is the more popular of the two cycles and ordinarily uses normal boiling liquid nitrogen as the precoolant. Figure 17 is a flow scheme of a typical expander cycle. The requirements for this liquefier, which has a capacity of 13 gal/h (to storage), would be 155 hp and 280 gal per day of liquid nitrogen. A second expander could be used to replace the liquid nitrogen refrigeration requirement.

For a liquefier of greater capacity (26 gal/h and more), the cycle variation as shown in Figure 18 could be used. The utility requirement would be about 1100 hp for a 190 gal/h (to storage) liquefier, plus a liquid nitrogen requirement of 4100 gal per day.

Refrigerator Design

The most common type of refrigerator used with space chambers applies the process shown in Figure 19. This cycle eliminates the need for a precoolant by sizing the expander to make up for the heat leak and the warm-end loss. This requires additional heat exchanger surface area to minimize the losses.

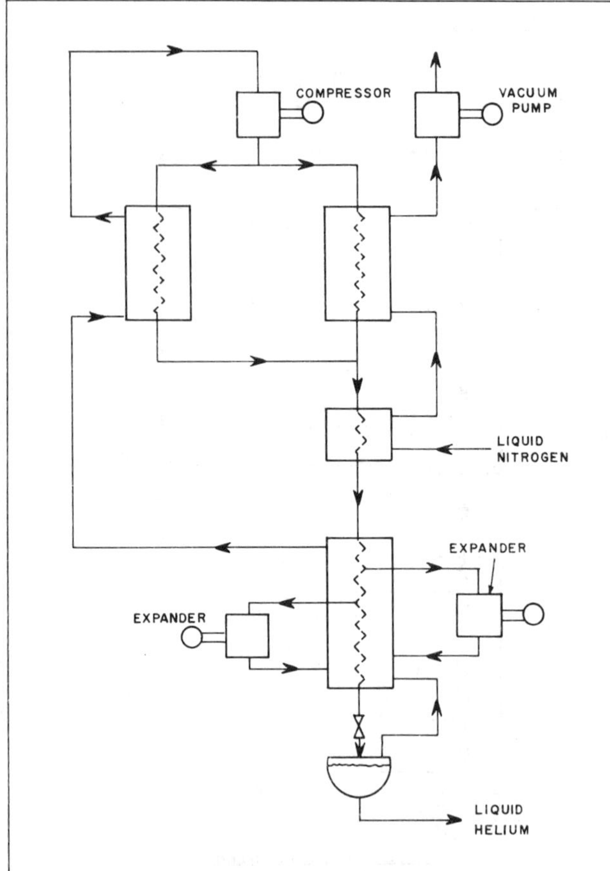

Fig. 18 Large-Scale Helium Liquefier

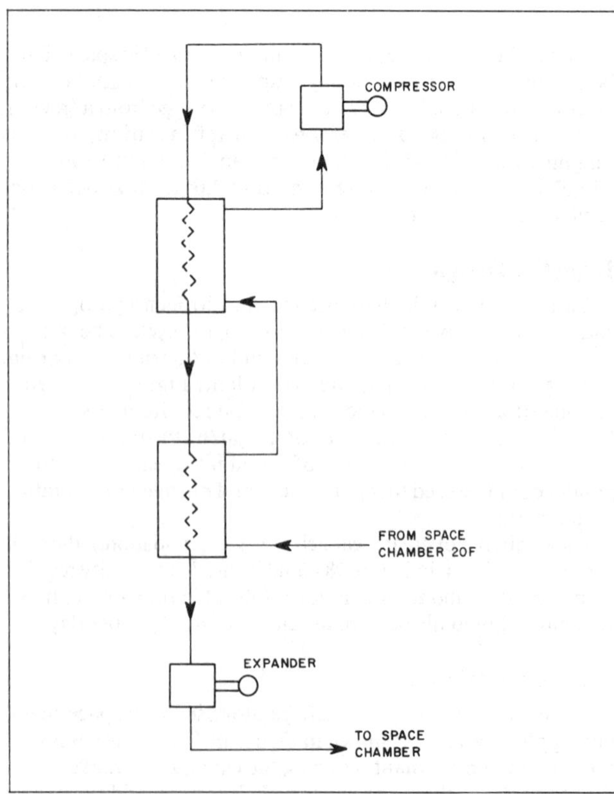

Fig. 19 Helium Refrigerator

Equipment Design

For larger capacity liquefiers that use liquid nitrogen precoolant, the equipment operating above the boiling nitrogen temperature level may be insulated in steel panel boxes or round casings filled with perlite, vermiculite, or rockwool, whereas the lower temperature equipment should be enclosed by a vacuum-jacketed Dewar, using either multilayer insulation or liquid nitrogen shielding to reduce the heat leak to an economical level.

The main factors that determine the efficiency of a liquefier, once the cycle is chosen, are (1) expander efficiency, (2) transfer losses, and (3) storage heat leak. Each percentage point of expander efficiency will influence the capacity of the liquefier by about 2%. Transfer and storage losses can also influence the plant size by 20 to 30%.

Liquefaction-grade helium, which contains less than 50 ppm by volume impurities, requires final purification before reduction below liquid nitrogen temperature. An adsorber located at the nitrogen precoolant level is normally used for this service. The adsorbent must be completely reactivated to remove moisture in order to maintain its capacity for trace impurities in the helium. Following this preliminary reactivation at 350 to 450°F, subsequent normal reactivations may be accomplished by purging the adsorber with ambient temperature helium.

ARGON EXTRACTION

Argon is present in atmospheric air in a concentration of about 1% by volume or 1.25% by mass. Since its boiling point lies between that of liquid oxygen and nitrogen, a good grade of argon (90 to 95% pure) can be obtained by adding a small auxiliary argon recovery column to an air separation plant. Argon gas is clear, colorless, odorless, and tasteless, with properties similar to those of nitrogen. Saturated liquid argon at 1 atm and −302°F has a density of 87.5 lb/ft^3.

Argon is extremely inert, and with the exception of fluorine, forms no known compounds. As such, it is primarily used as an inserting gas, especially in welding applications. See Chapter 17 in the 1993 ASHRAE *Handbook—Fundamentals* for cryogenic properties of argon.

Argon can be produced as a byproduct of the oxygen production cycle (Figure 9). Binary separation of argon (an impurity) and oxygen occurs in the section immediately below the upper column in Figure 9. The argon concentration increases on each succeeding tray until a maximum concentration of 6 to 8% is reached. Crude argon is then refined to the desired purity in a separate column.

MINIATURE CLOSED-CYCLE REFRIGERATORS

Cryogenically cooled, solid state devices (masers, infrared detectors, cryotrons, and parametric amplifiers) require miniature refrigerators. Successful miniaturization has required adapting simplified basic cycles and an appreciably reduced scale of equipment, although fundamental refrigeration principles apply.

Application of J-T Effect

A typical miniature refrigerator using the Joule-Thomson effect is shown in Figure 20 (Koontz and Lashmet 1963). In this refrigerator, neon is cascaded against a nitrogen circuit, resulting in temperature levels of approximately −406°F. This cycle has been successfully adapted to the cooling of infrared detectors, which require a refrigeration capacity of approximately 0.85 Btu/h. For comparison with the larger refrigerators described earlier, flow rates of approximately 0.33 cfm at standard conditions are used, and the entire system, including compressors, weighs approximately 50 lb.

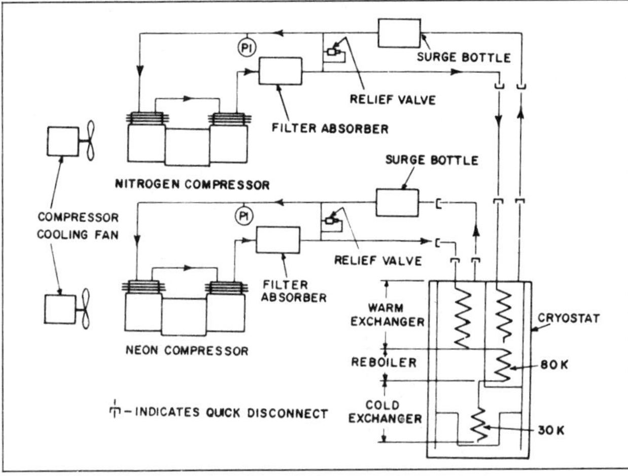

Fig. 20 Cascade Refrigerator (54 °R) for Infrared Detector Cooling

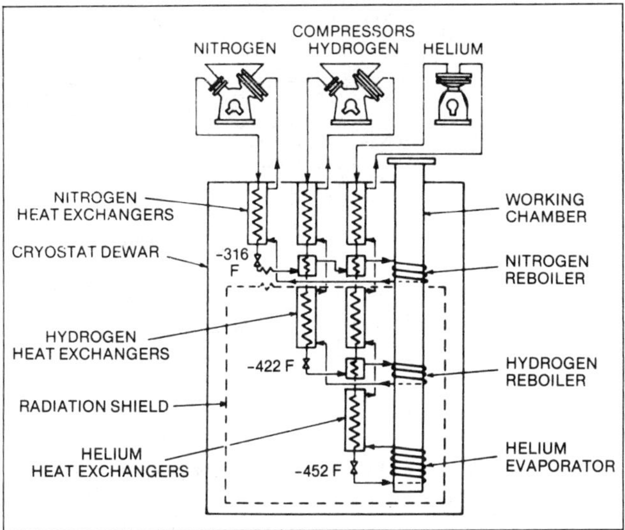

Fig. 21 Cascade Refrigerator (7.9 °R) for Maser Cooling

The extension of the Joule-Thomson principle to produce liquid helium temperatures is shown in Figure 21 (Geist and Lashmet 1963). In this system, nitrogen provides refrigeration at approximately 144 °R. Hydrogen, precooled to this temperature, transfers some of the refrigeration to the 38 °R temperature level. The final stage of refrigeration is provided by helium, precooled to about 38 °R. A typical refrigerator of this type, which is suitable for cooling masers, will develop approximately 4.3 Btu/h of useful refrigeration at 7.9 °R. The normal operating conditions for the three circuits are as follows:

Circuit	Discharge Pressure, psia	Suction Pressure, psia	Flow Rate, scfm
Nitrogen	2800 to 3000	15 to 20	3.6 to 4.0
Hydrogen	2000 to 2200	16 to 20	4.3 to 5.0
Helium	250 to 270	15 to 20	3.3 to 3.9

An example of the application to both the expansion engine principle and the Joule-Thomson effect is shown in Figure 22. This particular refrigerator, described by Zeitz and Woolfendon (1963), was designed to produce 0.7 Btu/h of useful refrigeration at 4.5 °R. It uses an expander circuit and a Joule-Thomson circuit. All refrigeration is provided by the expander circuit. During steady state,

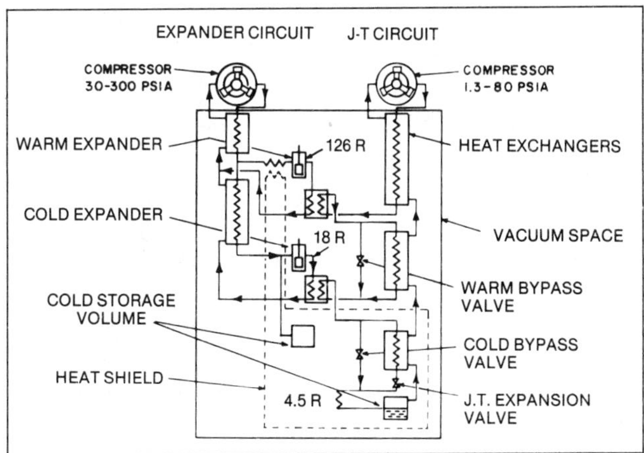

Fig. 22 Expander Refrigerator (4.5 °R)

it operates from 30 to 300 psia, with a flow of 7 cfm. This circuit uses two expansion engines, operating at approximately 126 °R and 18 °R, respectively. No Joule-Thomson valve is included.

To provide more refrigeration during cooldown, the expander circuit operates between 60 and 600 psia. The expansion engines, which normally operate at 300 rpm, are increased to 600 rpm, and available refrigeration is increased four-fold.

The Joule-Thomson circuit circulates 0.4 cfm between 1.3 and 80 psia. It picks up refrigeration from both the warm expander and the cold expander, or at roughly 126 °R and 18 °R. Final expansion down to operating temperatures is achieved through a Joule-Thomson exchanger and expansion valve.

Two independent circuits are used to operate the circuit providing the refrigeration in a supercharged condition—60 psia at cooldown and 30 psia at steady-state. If a single circuit were used, all the flow would have to be handled at 1.5 psia, requiring much larger compressors and exchangers and reducing the practicality of this type of system.

Stirling Cycle

In its ideal form, the Stirling cycle is composed of two reversible isothermal processes separated by two reversible constant volume processes. Therefore, it has the theoretical efficiency of a Carnot cycle.

In a real refrigerating machine, the process steps are carried out on a fixed gas charge by the sinusoidal motion of a pair of coupled pistons either in the same cylinder or in a pair of cylinders with a fixed spatial orientation. The pistons are connected to a motor drive through a crank mechanism which preserves a phase relationship between their motion such that the gas charge is shuttled to and from the cylinder through a regenerator. The regenerator acts alternately as a heat source and sink to preserve the temperature difference between the near constant volume compression and expansion processes.

Heat exchangers mounted close to both ends of the regenerator provide the near isothermal heat addition and extraction portions of the cycle. Two arrangements for Stirling refrigeration machines are shown in Figure 23. In the coaxial arrangement, the expansion and compression volumes are defined by the motion of the two pistons within a single cylinder. This configuration has some advantages of compactness and, as the pressures in the cold and warm working volumes are nearly balanced, the sealing problem is made easier.

Figure 24 shows an arrangement for a Stirling refrigerator to cool at 36 °R. While it is possible to produce refrigeration at this temperature by applying a single expansion cycle, a more efficient solution is to use two stages of expansion in a configuration like that shown (Prast 1965). A miniature device patterned after this

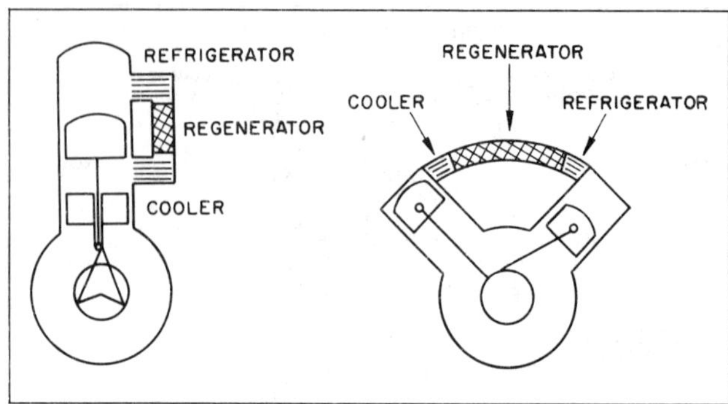

Fig. 23 Two Arrangements of Stirling Refrigerator

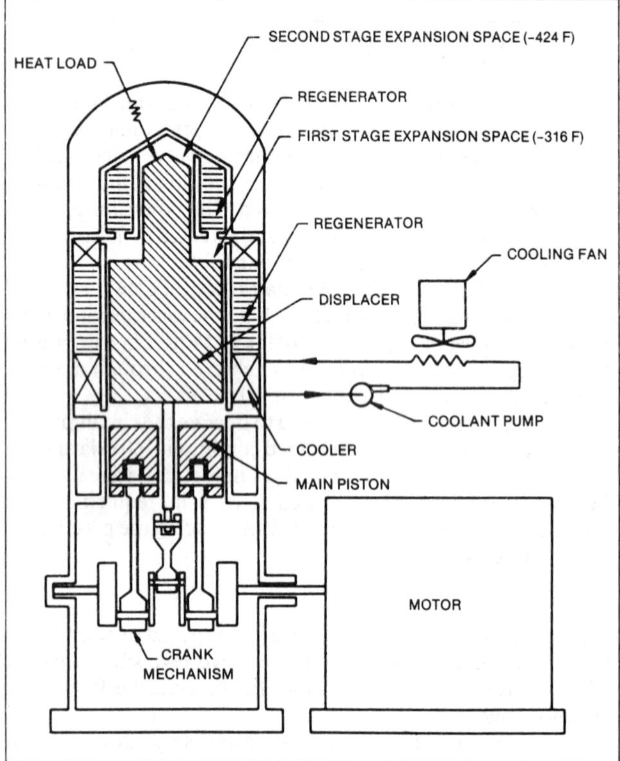

Fig. 24 Two Stage (36 °R) Stirling Refrigerator

concept, producing useful refrigeration at 36 °R and 3.4 Btu/h, at 45 °R, requires about 2000 Btu/h of input power and weighs about 25 lb. The working fluid is helium with a mean working pressure of 125 psig. A no-load cooldown time to 47 °R is about 10 min. The item to be cooled is most conveniently mounted at the cold head of the device, but remote cooling through a secondary heat transfer loop has been used.

Because of limiting low-temperature heat capacities of regenerator materials, Stirling refrigerators will not, by themselves, effectively provide refrigeration below about 22 °R. Combined with a Joule-Thomson loop, they can cool to liquid helium temperatures. Vibration, induced microphonics in direct-mounted solid state devices, performance degradation from contamination of the regenerators, and piston ring wear are some of the problems associated with these refrigerators.

The Gifford-McMahon cycle (Gifford and Hoffman 1961, Hoffman 1963), a refrigerator made to operate on a modified Stirling cycle, is shown in Figure 25. Like the Stirling device, it uses regenerators to store and release thermal energy alternately while maintaining fixed mean temperature differences between the expansion volumes and the warm-end process fluid. Refrigeration is produced by expansion processes in volumes provided by a piston-cylinder configuration. In this case, two expansion stages in series are used. The expansion processes are made to take place by the motion of the sinusoidally driven piston and by controlled actuation of valves that alternately communicate the expansion spaces to the inlet and discharge side of the compressor. In the configuration shown, the pressure at the top and base of the piston is almost balanced at all times so that the crosshead drive must provide only the forces necessary to overcome friction and pressure losses in the regenerators. Sealing between gas spaces at different temperatures is also made easier by the design feature.

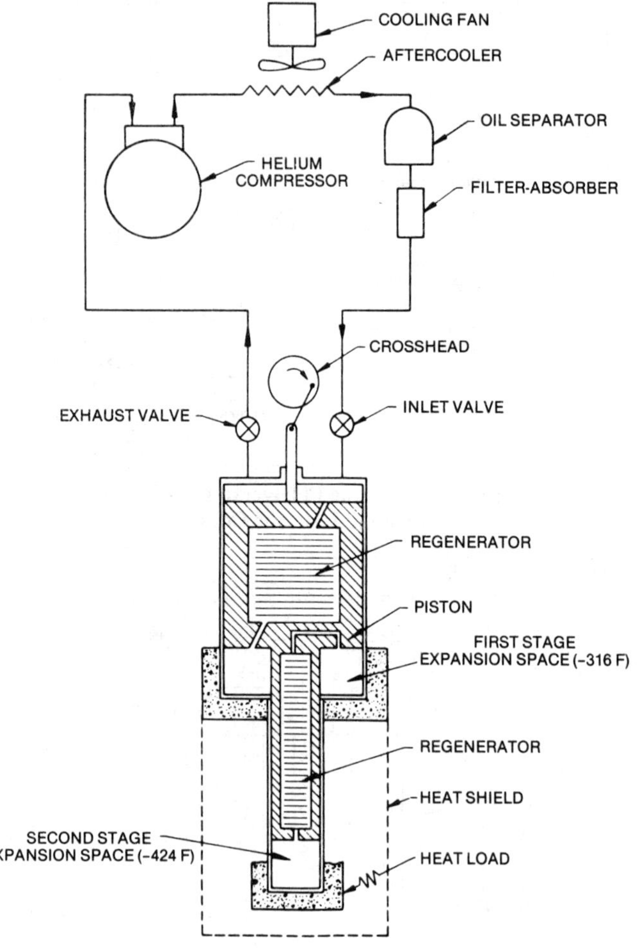

Fig. 25 Two Stage (36 °R) Modified Stirling Refrigerator

Unlike the straight Stirling refrigerator, the expansion and compression processes are completely separable in both the mechanical and thermodynamic sense. Therefore, more or less conventional compression equipment is used and the expansion engine pod can be located remotely from the compressor. The connection between compressor and engine is provided by warm pressure and return lines. The Gifford-McMahon cycle is not ideally reversible, and real refrigeration devices using this cycle do not exhibit the efficiency of the Stirling machine. However, relatively good performance is achieved by using highly effective regenerators and multistage expansion at different temperature levels.

Like the straight Stirling refrigerator, this modified version cannot effectively cool below about 22 °R and for the same reason. Similarly, to reach liquid helium temperatures, it is used as a precooler for a Joule-Thomson loop.

EQUIPMENT

The machinery generally used by the cryogenic industry can be grouped as (1) compressors, (2) expanders, and (3) pumps.

Compressors

Compressors may be classified as (1) centrifugal; (2) axial; (3) reciprocating, nonlubricated, or lubricated; (4) diaphragm; and (5) screw or lobe.

A *centrifugal compressor* is used mostly as the main air supply for separation plants. It has a horsepower range from 1200 to 10,000 and a flow capacity between 10,000 and 60,000 scfm. The discharge pressure is usually about 60 to 100 psia. These machines are generally fitted with adjustable inlet guide vanes to facilitate capacity reduction and maintain economical power requirements. Reductions in flow of up to 30% are obtainable without serious power penalties or the danger of surging.

An *axial compressor* is used for the main air supply to the largest air separation plants. Such a compressor typically has a capacity of 125,000 scfm of air at a discharge pressure of about 100 psia and requires 22,000 hp. These machines are usually driven by gas turbines or combination gas and steam turbines. The steam turbine has the additional function of starting the gas turbine. Capacity is adjusted by means of inlet guide vanes, as used with centrifugal compressors.

A *reciprocating compressor* is widely applied in the cryogenic industry. It is used to compress gases to high pressure, *i.e.*, 3000 to 4000 psi. When the flow rates required are 10,000 cfm and under, and high efficiencies are necessary, this machine should be used in preference to the centrifugal type.

The reciprocating compressor is available in lubricated and nonlubricated construction. This refers only to cylinder construction. The main bearings, connection rods, crossheads, and pins are lubricated by hydrocarbon oils in the usual manner on both machines.

Nonlubricated compressors used in oxygen compression have carbon- or bronze-filled TFE piston rings and piston rod packing. The suction and discharge valves are specially constructed for oxygen service. The distance pieces that separate the cylinders from the crankcase are purged with an inert gas, such as nitrogen, to preclude the possibility of high concentrations of oxygen in the area in the event of excessive rod packing leakage. The safety of the operator must be the prime consideration. Consequently, oxygen compressors with shutdown devices that detect high temperature and high pressures in the interstage piping are frequently employed. This type of compressor is usually limited to low-pressure ratios across each stage, so the discharge temperatures are below 325 °F.

Another popular nonlubricated reciprocating compressor has pistons that depend on labyrinths on the piston cylindrical surface rather than on piston rings to form a seal. These machines are built only in the vertical configuration and, like other nonlubricated machines, are used to compress oxygen because of their inherent safety features, or whenever an uncontaminated gas supply is required.

Another type of nonlubricated reciprocating compressor used in the cryogenic industry is the diaphragm machine. The diaphragm is actuated by a piston connected to the crankshaft by the usual crosshead and connecting rod arrangement. The piston displaces the diaphragm hydraulically. The hydraulic fluid is either a soap and water solution or one of the synthetic oils that are compatible with oxygen, if the machine is used with oxygen. The diaphragm machine is used for high-pressure, low flow rate applications, requiring an uncontaminated gas stream for such service as filling industrial gas bottles.

The screw and lobe compressor is widely applied as a main air source compressor in smaller air separation plants. The spiral rotors are so designed that the air is compressed to discharge pressure within the cavities between the rotors, prior to discharge from the machine. These machines run at rotor speeds of approximately 5000 rpm and are used where airflows are on the order of 8000 to 12,000 scfm and discharge pressures are as high as 100 psia. The lobe compressor is used on low-pressure and on low- to medium-pressure flow rates.

The lubricated reciprocating compressor, as used in the cryogenic industry, is most widely applied as a high-pressure air supply, or on nitrogen recycle service. Machines on this service usually have cylinders lubricated with a synthetic oil of the TCP group, to minimize the possibility of ignition in interstage piping, coolers, or separators. Otherwise, the construction is according to standard air machine practice. To minimize pulsations in the system, close attention is given to the design of the interstage piping. Capacity control on this type of machine is by clearance pockets, automatically operated through a selector switch on the control panel.

Expanders

Expanders may be grouped as (1) reciprocating (nonlubricated or lubricated), and (2) turbine. These are used to expand various gases efficiently from high to low pressures to obtain refrigeration. The gases may be air, nitrogen, natural gas, hydrogen, or helium, depending on the cycle. The expanders obtain high efficiencies by causing the gas to do work on a piston or turbine wheel and thereby realize temperature drops in excess of those obtainable with a J-T valve for the same pressure ratio.

The expanders in modern air separation plants reach high adiabatic efficiencies. For reciprocating types, efficiencies of 80% are normally quoted, and even higher values are quoted for high-capacity turbo machinery.

Varying the inlet valve closing point controls the capacity of *reciprocating expanders* by controlling the amount of gas admitted to the cylinder before expansion takes place. These machines are used to expand pressure gas streams from 3000 psia to pressures as low as 100 psia, depending on the cycle selected by the process designer.

Cylinder lubrication on this type of machine is usually one of the refrigeration compressor oils. Warm water in jackets around the ring-pass section of the cylinder prevents the oil from freezing in the cylinder. The discharge temperatures of an air expander will usually approach, within a few degrees, the saturation temperature at the discharge pressure, which is −275 °F at 100 psia.

Nonlubricated reciprocating expansion engines are generally applied where possible oil contamination of the process is intolerable or where extremely low operating temperatures preclude the use of cylinder lubricants. This type of expansion engine is found in hydrogen and helium liquefaction plants and in helium refrigerators. Some nonlube machines are operating in air separation plants where the inlet pressures are moderate, about 1500 psi.

The *turbo-expander* is used when flow conditions are reasonably constant, oil contamination is not permissible, and inlet pressures do not exceed 100 psia. These machines are most frequently the radial-inflow-type turbine. The adiabatic efficiency of the machine can be as high as 90% when inlet nozzles, turbine wheels, and blower wheels are accurately matched and process conditions are favorable.

Turbo-expanders are loaded by means of compressors, blowers, generators, and dynamometers to absorb the shaft power produced by expansion of the process gas through the turbine. Compressors and blowers are used either to compress a process stream or simply to compress atmospheric air and discharge back to atmosphere. The generator converts the turbine shaft work to electrical energy, which is returned to the electrical supply system. Dynamometers are usually used in small-scale operations where energy is dissipated by pumping recycled oil through a restriction.

Pumps

Pumps may be grouped as (1) centrifugal and (2) piston. Pumps used in the cryogenic industry differ from those in other industries in materials of construction, design, and mode of operation. A cryogenic centrifugal pump is used as a transfer pump on tank trucks, in in-plant transfer service, and for liquid product pressurization prior to vaporization. The pumps are usually designed for specific duties and are not usually shelf items.

Depending on the duty cycle and discharge pressure, pump casings are made from stainless steel or aluminum. Aluminum is usually selected for impellers. Nonlubricated bearings are bronze, running on a stainless steel sleeve or shaft. Both vertical and horizontal pumps are built. Multistage pumps are usually vertical, and single-stage pumps are horizontal. The reciprocating pumps used in cryogenic applications are generally used for high discharge pressures required for filling industrial gas cylinders and recharger applications. Many single and multicylinder reciprocating pumps are available for pressures up to 3000 psi.

Heat Exchangers

Heat exchangers used in cryogenics must be compact and highly efficient: *i.e.*, effective for the required heat transfer using small stream-to-stream temperature differences. Efficiency stems from the second law of thermodynamics that leads to:

$$W_{lost} = T_o \Delta S_T \tag{10}$$

where

W_{lost} = work required beyond that needed in a thermodynamically ideal process
T_o = ambient temperature
S_T = entropy change of the universe caused by the process under study

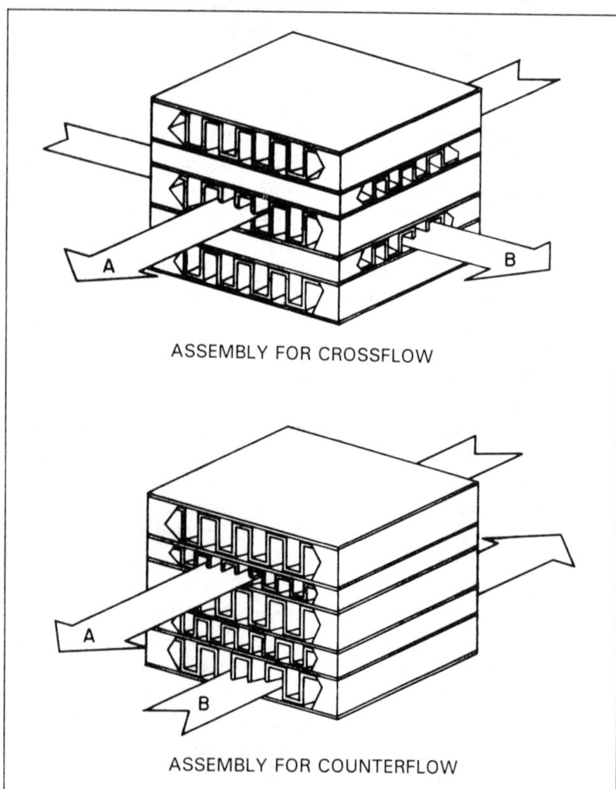

Fig. 26 Plate-and-Fin Heat Exchangers

In heat exchanger steps, Equation (10) becomes:

$$W_{lost} = \int T_o \frac{T_1 - T_2}{T_1 T_2} dH \tag{11}$$

where

$T_1 - T_2$ = stream-to-stream ΔT at any point in the heat exchanger
T_1, T_2 = hot and cold stream temperatures at any point in the heat exchanger
H = enthalpy level along the heat transfer path

The work lost is especially large where T_1 and T_2 are both small. Thus, any low-temperature ΔT becomes especially costly. The necessity for compactness stems from the need to reduce heat leakage from ambient into all cryogenic process equipment. The requirement of low ΔT also forces the cryogenic engineer to use large surface areas for heat transfer since the heat transfer coefficient cannot easily be changed. Thus, the designer requires a surface area inversely proportional to the ΔT.

Therefore, most cryogenic heat transfer processes take place in heat exchangers of three general types: (1) plate-and-fin exchangers, (2) coiled tube-in-shell exchangers, and (3) regenerators. Each of these devices offers large heat transfer surface area per unit volume; 1000 to 3500 ft^2/ft^3 are normal. Each has particular advantages and limitations.

Plate-and-fin heat exchangers are generally made of aluminum corrugated and flat sheets stacked alternately and brazed en masse by dipping in a molten salt bath. The resulting core is connected with headers to control stream flows, thus producing the final exchanger. Both crossflow and counterflow arrangements are possible, and a wide variety of corrugation designs and flow passage cross sections are available. Figure 26 shows the arrangement schematically. Typical parting-sheet to parting-sheet distances are 0.4 in., and normal heat transfer areas are about 1500 ft^2/ft^3.

Plate-and-fin exchangers have the advantages of high surface-to-volume ratios, relatively low cost, and wide flexibility in multistream flow arrangements and in passages assigned for pressure drop or high heat transfer rate. Disadvantages are that these exchangers are available in limited sizes only, thus requiring frequent manifolding and limiting use up to only moderate pressures, currently about 1000 psi.

The *coiled tube-in-shell heat exchanger* is used extensively in cryogenic systems and is most useful in high-pressure systems where the pressure difference between the tube-side stream and the shell-side stream is large.

The tubing can be either aluminum or copper. Aluminum is most suitable for pressures inside the tube of less than 600 psi. The tubes are headered into tube sheets or, in the case of high-pressure service, pipe headers. The unit is usually designed so that all the tubes in the heat exchanger are the same length, a feature necessary for obtaining uniform distribution.

The cryogenic heat exchanger often operates with a large temperature difference from its warm to its cold end. Thus, significant longitudinal heat flow can occur in the structural members of the exchanger. Generally, this is reduced by building the exchanger relatively long in proportion to its cross section, though this increases the pressure drop and may allow greater heat leakage. Another caution is the use of low-conductivity material for any large member that extends the length of the exchanger. Thus, in the coiled tube-in-shell exchanger, both the mandrel and the outside shell are usually made of stainless steel.

Coiled tube-in-shell heat exchangers are made as small as a finger for use in maser cooling and cryosurgery devices, and as large as 16 ft in diameter and 115 ft long for use in LNG base load plants. This diversity of sizes and the possibility for use at almost any processing pressure are the main advantages; the main disadvantage is relatively high cost.

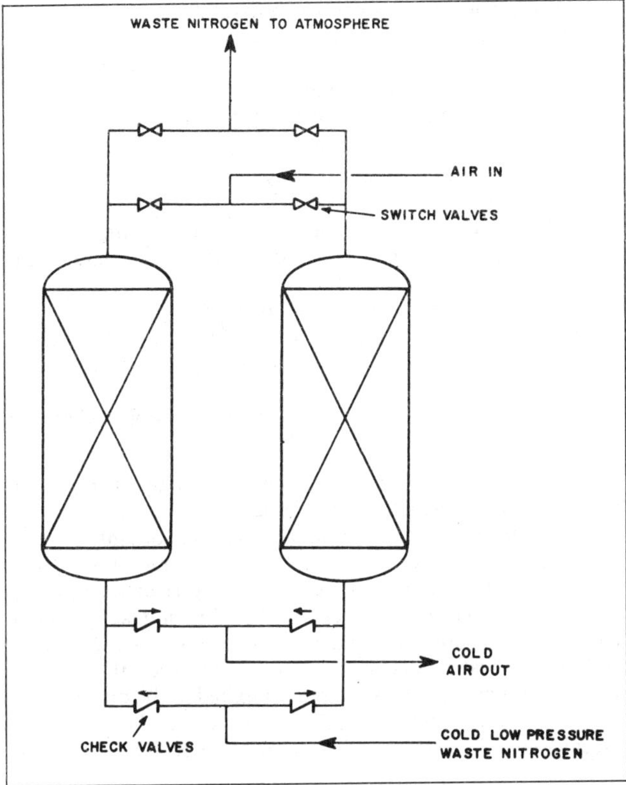

Fig. 27 Regenerator Set

Regenerators. Figure 27 shows a regenerator set, typical of those used in air separation plants. Each regenerator is a vessel packed with a material of relatively high heat capacity, such as quartz pebbles, and, as shown, two such vessels are needed to replace one heat exchanger. Each vessel is alternately traversed by each gas stream. In one period, the incoming airstream warms the packing in one vessel, while the effluent gas stream cools the packing in the other vessel. After a short interval (up to 10 min), the gas streams are switched from one vessel to the other by means of a switch and check valve arrangement.

In small cryogenic refrigerator service, such as in the Stirling cycle or the Gifford-McMahon cycle, the regenerator is usually a single vessel packed with copper wool or a similar material. In these regenerators, flow is reversed in a much more rapid cycle, perhaps with each stroke of the compressor piston.

Regenerators have the following advantages as compared to normal heat exchangers:

1. They are relatively easy to manufacture and less costly.
2. They concentrate a very large surface area into a small volume.
3. Well-designed regenerators have low pressure drop characteristics.
4. They allow for impurities, such as the carbon dioxide and water content of atmospheric air, to be deposited on the packing from the gas being cooled during one period, and then allow the same impurities to be evaporated into gas being warmed during the other period, thus eliminating the need for removal of impurities by other means.

Especially in air separation systems, regenerators also have disadvantages that have eliminated them from use in new plants. These include:

1. Their operation depends on the reliable operation of check valves and switch valves, some of which must operate at low temperature.
2. Each switch produces a severe physical blow on all the process piping and machinery.
3. Regenerator operation is inherently cyclical, so the operation of low-temperature expanders and distillation equipment is made difficult.
4. A pure product cannot be withdrawn from a regenerator since each switch contaminates the vessel with a charge of impure gas.
5. Regenerator packing tends to disintegrate in the environment of cyclical changes in temperature and pressure, thus sending dust into the rest of the plant.
6. Regenerators have a high maintenance cost due to cyclic operation.

These disadvantages are not so apparent in small refrigerator applications, so regenerators are extensively used for these processes. Regenerators are usually limited in air separation to heat exchange of relatively low-pressure gases, and where the pressure difference between the two gas streams is small. Regenerator vessels for low-temperature service are usually made of either aluminum, stainless steel, or 9% nickel steel.

Valves

The basic criterion for selecting valves for any service is the efficiency with which they accomplish shutoff or throttling. In low-temperature circuits, the additional consideration of heat leakage must be included. Most process valves are located in the cold box and are constructed of series 300 stainless steel. With these valves, extended bonnets of stainless steel are required to minimize the thermal conductance, thus keeping the packing at near ambient temperatures.

Teflon-impregnated non-asbestos packing is suitable for oxygen service and most cold applications. A thin wall column is constructed around the shaft of liquid-control valves to reduce the heat loss. To reduce leaks, the inaccessible cold valves have no end or bottom flanges. With top-opening valves, the trim can be changed without opening the insulating jacket.

In sizing cold control valves, a designer should calculate the valve capacity to give adequate control for normal cold liquid or gas flow conditions and to allow as much warm defrost flow as is practical. For cryogenic liquid expansion valves, valve sizing should include both liquid and gas calculations, since flashing occurs. It is good practice to install valve positioners on extended stem-control valves.

Vacuum-jacketed valves are used extensively at liquid hydrogen and helium temperatures. Construction considerations include extra long, thin vacuum-jacketed stems to lengthen the heat path, thus reducing the heat leak. Expansion bellows are used to isolate valves from line stress during cooldown.

CRYOGENIC INSULATION

Cryogenic insulations can generally be categorized as (1) straight (high) vacuum, (2) filled vacuum (evacuated multilayers, evacuated powders, or evacuated fibers), and (3) nonvacuum (powders, foams, or fibers).

Evacuated systems predominate in those applications concerned with long-term storage of the more volatile fluids, particularly when fluid quantities to be stored are relatively small. Nonvacuum systems predominate in those applications concerned with short-term storage and transfers of the less volatile fluids, storage of large quantities (*e.g.*, field-erected tanks), and production plants.

In general, evacuated multilayer insulations, when properly designed and fabricated, provide the highest impedance to heat flow. They typically trade off some loss of thermal impedance through the mechanism of solid conductance for considerable gain in minimizing gas conduction and radiation. Evacuated powders, typically five to ten times more conductive than evacuated multilayers, enjoy the advantages of cost and tolerance to poor vacuum (*e.g.*, 10^{-2} torr versus 10^{-4} torr for multilayer). Nonvacuum insulations enjoy a further cost advantage, but their considerably higher conductances render them useful primarily for applications in which equipment volume is no problem (*e.g.*, large stationary equipment) and thick insulation can be used. The choice of insulation depends on the particular application.

The following sections give the more important characteristics and properties of various types. Designers can use this information as a guide to selection of insulation for a given use.

Vacuum Insulation

Straight (High) Vacuum Insulation. Heat transfer through high vacuum insulation predominantly occurs by radiation, but gas conduction may contribute significantly.

The net exchange of radiant energy between two surfaces in Btu/h is:

$$q = EA_1\sigma(T_2^4 - T_1^4) \tag{12}$$

where

E = effective emissivity of the two surfaces
A_1 = inner surface area, ft^2
σ = Stefan-Boltzmann constant, 0.1714×10^{-8} Btu/h·ft²·°R⁴
T_2 = hot wall temperature, °R
T_1 = cold wall temperature, °R

Table 2 gives the values of E in terms of individual surface emissivities for three commonly used geometries. Table 3 gives the minimum recorded emissivities for a number of materials used in cryogenics. These data include normal and hemispherical emissivities and absorptivities, since their differences can usually be ignored for engineering calculations.

Table 2 Effective Emissivity E in Terms of Individual Emissivities of Cold and Hot Walls, e_1 and e_2

	Specular Reflection	Diffuse Reflection
Parallel plates	$\dfrac{e_1 e_2}{e_2 + (1 - e_2) e_1}$	$\dfrac{e_1 e_2}{e_2 + (1 - e_2) e_1}$
Long coaxial cylinders ($L >> d$)	$\dfrac{e_1 e_2}{e_2 + (1 - e_2) e_1}$	$\dfrac{e_1 e_2}{e_2 + (A_1/A_2)(1 - e_2)e_1}$
Concentric spheres	$\dfrac{e_1 e_2}{e_2 + (1 - e_2) e_1}$	$\dfrac{e_1 e_2}{e_2 + (A_1/A_2)(1 - e_2)e_1}$

Table 3 Selected Minimum Total Emissivities[a]

Surface	Surface Temperature, °R 7.2	36	139	540
Copper	0.0050		0.008	0.018
Gold			0.01	0.02
Silver	0.0044		0.008	0.02
Aluminum	0.011		0.018	0.03
Magnesium				0.07
Chromium			0.08	0.08
Nickel			0.022	0.04
Rhodium			0.078	
Lead	0.012		0.036	0.05
Tin	0.012		0.013	0.05
Zinc			0.026	0.05
Brass	0.018			0.035
Stainless steel, 18-8			0.048	0.08
50 Pb 50 Sn solder			0.032	
Glass, paints, carbon				0.9
Silver plate on copper		0.013	0.017	
Nickel plate on copper		0.027	0.033	

[a]These are actually absorptivities for radiation from a source at 540°R. Normal and hemispherical values are included indiscriminately.

Existing experimental data on low-temperature emissivity allow the following generalizations to be made:

1. Materials having the lowest emissivities also have the lowest electrical resistances.
2. Emissivity decreases with decreasing temperature.
3. The apparent emissivity of good reflectors is increased by surface contamination.
4. Alloying a metal increases its emissivity.
5. Emissivity is increased by treatments such as mechanical polishing, which result in work-hardening of the surface layer of the metal.
6. Visual appearance (brightness) is not a reliable criterion of reflecting power at long wavelengths.

The best reflecting surfaces are pure, good conductor metals, annealed and cleaned, using a method that avoids strain, such as one involving the use of acids and residue-free organic solvents.

As a rule, the objective when using high-vacuum insulation is to obtain a vacuum of such quality (10^{-6} torr or less) that the heat transfer by residual gas conduction does not seriously contribute to the overall heat transfer. Since this is not always possible, gaseous heat conduction at low pressures, where the mean free path of the gas molecules is large compared to the dimension of the concentric spheres, coaxial cylinders, or parallel plates, may be calculated by using an adaptation of the Knudsen formula.

$$Wgc = \frac{\gamma + 1}{\gamma - 1}\alpha\left(\frac{R}{8\pi MT}\right)^{0.5} p(T_2 - T_1) \tag{13}$$

$$\alpha = (\alpha_1\alpha_2)/[\alpha_2 + \alpha_1(1 - \alpha_2)(A_1/A_2)] \tag{14}$$

where

Wgc = net energy transfer per unit time per unit area of inner surface, Btu/h·ft²
γ = c_p/c_v, the specific heat ratio of the gas, assumed constant
R = molar gas constant
p = pressure, mm Hg
M = molecular mass of the gas
T = temperature at the point where p is measured, °R
α = overall accommodation coefficient
A = area, and subscripts 1 and 2 refer to the inner and outer surfaces, respectively, ft^2

This expression reduces to

$$Wgc = C\alpha p(T_2 - T_1) \tag{15}$$

where

$$C = \frac{\gamma + 1}{\gamma - 1}\left(\frac{R}{8\pi MT}\right)^{0.5} \tag{16}$$

Values of this constant for cryogenic applications have been calculated by Corruccini (1957) and are shown in Table 4. The temperature at the pressure gage is assumed to be 80°F.

Table 4 Constants for the Gas Conduction Equation

Gas	T_2 and T_1, °R	Constant
N_2	≤ 720	28.0
O_2	≤ 540	26.0
H_2	540 and 139 or 540 and 162	93.0
H_2	139 and 36	70.1
H	any	49.3
Air	≤ 720	27.5

Table 5 Approximate Accommodation Coefficients

Temperature, °R	He	H_2	Air
540	0.3	0.3	0.8–0.9
139	0.4	0.5	1
36	0.6	1	1

Table 5 gives some suggested values for individual accommodation coefficients. The experimental data are meager, and the uncertainty in these values is great. However, for most engineering calculations, these data are adequate.

Evacuated Multilayer Insulation. In principle, this insulation consists of many brightly reflective radiation shields, separated by poor conductors, in a vacuum. When properly applied, this insulation has an apparent mean thermal conductivity between 80 and −424 °F of 2×10^{-5} to 4×10^{-5} Btu/(h·ft·°R).

The radiating shields are usually supported and separated by a spacer with low solid conductance but with relative thermal transparency. In general, the spacers cause a net increase in conductance over the case in which the same number of radiating shields would be floating without spacers. The limit of performance (no spacers) could be evaluated by:

$$Q = [E\sigma A/(n + 1)](T_2^4 - T_1^4) \tag{17}$$

where

σ = Stefan-Boltzmann constant
A = area
E = effective emissivity
n = number of floating radiation shields
T_1 and T_2 = absolute temperatures of hot and cold boundaries

The spacer can cause considerable deviation from this limiting performance. Thus, for design purposes, the thermal performance of multilayer insulation is expressed by an apparent mean thermal conductivity k as follows:

$$q_T = [kA(T_2 - T_1)]/b \tag{18}$$

where

b = thickness of the insulation

Two multilayer systems in wide commercial use are:

1. Thin sheets of mylar that have been vacuum coated with aluminum on one side to produce a low-emissivity surface. In this system, the mylar itself acts as the spacer. The coated mylar is usually crinkled to reduce contact area between successive layers.
2. Thin layers of low-emissivity metal foil (usually aluminum 0.00025-in. thick), separated by a low-conductance spacer (usually glass fiber paper).

Tables 6 and 7 give typical thermal conductivity values for these two materials. Table 7 shows the undesirable effect of mechanical compression on the three aluminum-glass systems. It also illustrates the improved performance as the spacer is made thinner, allowing a larger number of shields per unit thickness of insulation. Cost of these systems rises with the use of thinner spacers.

Table 6 Apparent Mean Thermal Conductivity of Aluminized Mylar (Boundary Temperatures 540 and 139 °R)

Number of Shields per in.	Density, lb/ft^3	Conductivity, Btu/(h·ft·°R) × 10^6
61	0.0131	28.0
102	0.0219	25.0
150[a]	0.0331	24.0

[a]Optimum density.

Table 7 Apparent Mean Thermal Conductivity of Three Glass Fiber Paper-Aluminum Foil Insulations (Boundary Temperatures 540 and 139 °R)

Number of Shields per in.	Density, lb/ft^3	Conductivity, Btu/(h·ft·°R) × 10^6	Type of Insulation
12.7[a]	0.0156	101	Linde SI-10
25.4	0.0312	231	Linde SI-10
71.1[a]	0.0625	20	Linde SI-62
142.2	0.125	58	Linde SI-62
109.2[a]	0.0938	12	Linde SI-92[b]
218.4	0.188	29	Linde SI-92[b]

[a]Optimum density.
[b]Spacer is opacified with aluminum flakes (Grunert *et al.* 1969).

The added cost of the improved performance must be justified by a more critical application.

The following precautions should be taken when using multilayer insulations:

1. Use vacuum lower than 10^{-4} torr for maximum efficiency. Deteriorating effects of elevated pressure are presented by Glaser (1967).
2. Avoid high mechanical loads on the multilayer stack.
3. Use low-emissivity reflecting surfaces.
4. Minimize gaps between adjacent shields at corners or joints because they can add significantly to the total heat transport.

Evacuated Powder Insulation. The insulating value of powders is increased greatly by removing the interstitial gas. Therefore, when powders are used at low pressures (10^{-2} torr or less), gas conduction is negligible, and heat transport is chiefly by radiation and solid conduction. For some powders, radiation absorption reduces heat transfer more than solid conduction increases it, and these powders thus reduce the overall heat transport. Table 8 shows the apparent mean thermal conductivity between −323 °F and 80 °F of some selected powders. Figure 28 shows the apparent mean thermal conductivity of several powders as a function of interstitial gas pressure (Fulk 1959).

The amount of heat transport because of radiation through these dielectric powders can be reduced by adding metallic powders. Figure 29 shows the effect of adding metallic powders (Hunter *et al.* 1960).

The vacuum level required with evacuated powder insulation is much less extreme than with straight vacuum or multilayer insu-

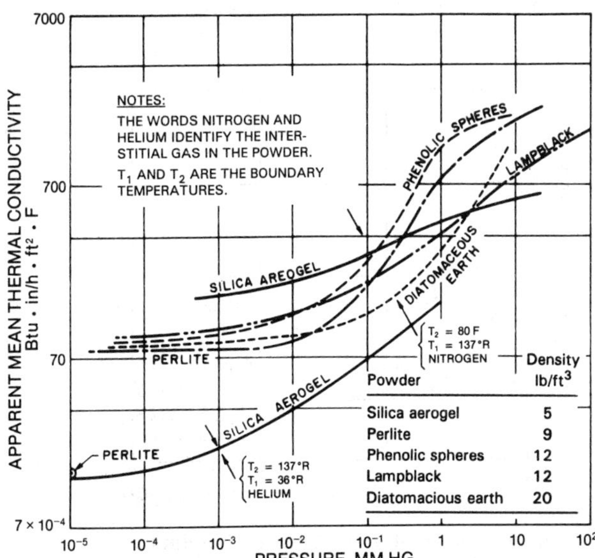

Fig. 28 Thermal Conductivities of Powders

Table 8 Apparent Mean Thermal Conductivity of Several Powders
(25.4-mm sample thickness; boundary temperatures of 540 and 137°R; and wall emissivities greater than 0.8)

Powder	Remarks	Density, lb/ft^3	Gas Pressure, mm Hg	Interstitial Gas	Conductivity Btu/(h·ft·°R)
		6.2	10^{-4}	—	12×10^{-4}[a]
	Chemically	6.2	628[e]	Nitrogen	113×10^{-4}
	prepared	6.2	628	Helium	358×10^{-4}
	250A	6.2	628	Hydrogen	462×10^{-4}
Silica aerogel	+ 10% free silicon dust (by mass)	7.0	10^{-4}	—	10.5×10^{-4}
Silica	Flame prepared	3.7	10^{-4}	—	12×10^{-4}
	150-200A	3.7	630	Nitrogen	107×10^{-4}
	+ 30 mesh	3.7	10^{-4}	—	12×10^{-4}
	+ 30 mesh	6.2	10^{-4}	—	10.5×10^{-4}
	+ 30 mesh	6.2	628	Nitrogen	192×10^{-4}
	+ 30 mesh	6.2	628	Helium	735×10^{-4}
	+ 30 mesh	6.2	628	Hydrogen	845×10^{-4}
	− 30 + 80 mesh	8.1	10^{-4}	—	7×10^{-4}
Perlite, expanded	− 30 + 80 mesh	8.1	628	Nitrogen	188×10^{-4}
	− 30 + 80 mesh	8.1	628	Helium	728×10^{-4}
	− 30 + 80 mesh	8.1	628	Hydrogen	838×10^{-4}
	− 80 mesh	8.7	10^{-4}	—	5.8×10^{-4}
	− 80 mesh	8.7	628	Nitrogen	202×10^{-4}
	− 80 mesh	8.7	628	Helium	780×10^{-4}
	− 80 mesh	8.7	628	Hydrogen	840×10^{-4}
	− 30 mesh	8.7	10^{-4}	—	5.8×10^{-4}[b,c,d]
Vermiculite	10-14 mesh		760		260×10^{-4}
		15.0	10^{-4}	—	9×10^{-4}
Diatomaceous earth	1 – 100μm	15.6	10^{-4}	—	8×10^{-4}
		18.0	10^{-4}	—	5.8×10^{-4}
Alumina, fused	− 50 + 100 mesh	125	10^{-4}	—	10.5×10^{-4}
Alumina, laminar	0.1 – 10 microns	4.4	10^{-4}	—	13.3×10^{-4}
	− 20 + 30 mesh 30%	9.4	10^{-4}	—	10.5×10^{-4}
Mica, expanded	− 30 + 15 mesh 70%	9.4	628	Nitrogen	290×10^{-4}
	− 30 + 15 mesh 70%	9.4	628	Helium	867×10^{-4}
Lampblack	—	12.5	10^{-4}	—	7.2×10^{-4}
Charcoal peach pits	20-30 mesh	30	10^{-4}	—	10.5×10^{-4}
Carbon + 7% Ash[f]	30% 44μm	12.5	10^{-4}	—	3.5×10^{-4}
Calcium silicate	0.02 μm	10.6	10^{-4}	—	4.3×10^{-4}
(synthetic)	0.02-0.07 μm	22.5	10^{-4}	—	3.3×10^{-4}
	0.02-0.07 μm	22.5	628	Nitrogen	263×10^{-4}

[a] 1.2×10^{-4}·Btu/h·ft·°R for boundary temperatures of 137 and 36°R
[b] 3.8×10^{-4}·Btu/h·ft·°R for boundary temperatures of 540 and 36°R
[c] 1.2×10^{-4}·Btu/h·ft·°R for boundary temperatures of 137 and 36°R
[d] 0.46×10^{-4}·Btu/h·ft·°R for boundary temperatures of 137 and 7.6°R
[e] Barometric pressure at Boulder, Colorado
[f] Mostly SiO_2 and Al_2O_3

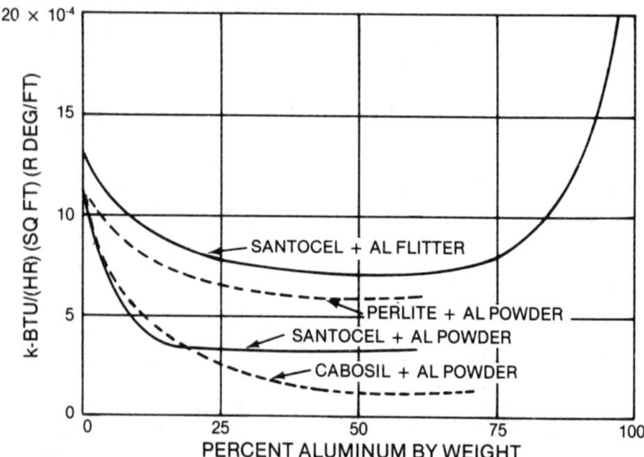

Fig. 29 Apparent Mean Thermal Conductivity between 540 and 137°R of Evacuated Powders with Added Aluminum

lations. Most powders, however, can adsorb considerable quantities of gases and water. Evacuation to even moderate pressure levels is usually difficult. The accumulation of water can be avoided by suitable powder handling techniques. Adsorbed gases must be driven off at elevated temperatures, or outgassing of noncondensable gas must be accommodated by including adsorbents or getters in the vacuum space. In addition, fine mesh filters must be used to protect the vacuum pumps from these highly abrasive particles.

Evacuated fibrous insulations have performance levels comparable to evacuated powders, but usually they are more expensive and require higher vacuum levels to attain their best performance. Both their cost and sensitivity to vacuum level are generally related to fiber size and density; better performance and higher cost are usually associated with small fiber diameter and high density. Typical performance of evacuated fibrous systems is shown in Figure 30.

Nonvacuum Insulations

Nonvacuum insulations may consist of powders, fibers, or foams. Heat conduction in such insulations is primarily because of the interstitial gas and is roughly the thermal conductivity of the gas. For example, the mean thermal conductivity in the range of −323 to 80 °F is 0.084 Btu/(h • ft • °F) for a perlite powder containing hydrogen gas at atmospheric pressure. The corresponding value for hydrogen itself is 0.072. Similarly, perlite containing helium is 0.073, and a sample of glass fiber containing helium is 0.077, while the corresponding value for helium gas is 0.063. Again, perlite containing nitrogen at atmospheric pressure is 0.019, the glass fiber containing nitrogen is 0.015, while nitrogen itself is 0.010. In the same temperature range, air-filled Styrofoam is 0.015, while air is 0.010.

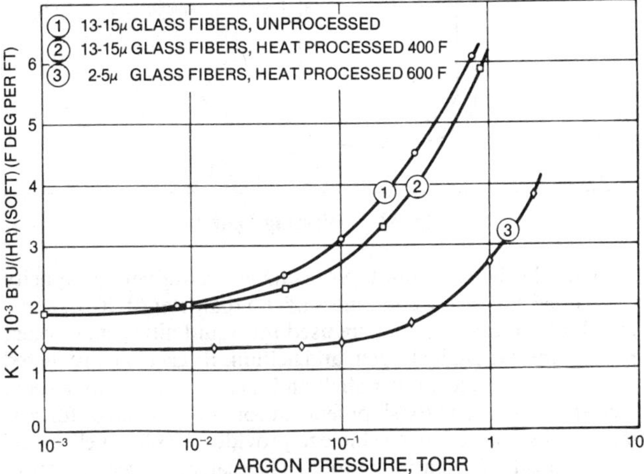

Fig. 30 Thermal Conductivity of Glass Fiber Mats under 750 mm Hg Mechanical Load

Table 9 Apparent Mean Thermal Conductivity of Selected Foams

Foam Type	Density, lb/ft³	Boundary Temperature, °R	Test Space Pressure, mm Hg	Conductivity, Btu·in/(h·ft²·°R)
Polystyrene[a]	2.4	540/137	760	0.23
	2.9	540/137	760	0.18
	2.9	137/36	10^{-5}	0.056
Epoxy resin[b]	5.0	540/137	760	0.23
	5.0	540/137	0.01	0.12
	5.0	540/137	0.004	0.090
Polyurethane[c]	5 to	540/137	760	0.23
(isocyanate)	8.5	540/137	0.001	0.083
Rubber[d]	5.0	540/137	760	0.25
Silica[e]	10.0	540/137	760	0.38
Glass[f]	9.0	540/137	760	0.24

[a]Dow Chemical Co. *Styrofoam*
[b]Debell & Richardson *Du Ra Foam*
[c]Nopco Chemical Co. *Lock Foam*
[d]U.S. Rubber
[e]Pittsburgh Corning *Foam Sil*
[f]Pittsburgh Corning *Foam Glass*

Among the powders, silica aerogel has superior insulating value when unevacuated because of its extremely small effective particle size.

Rigid Foam Insulations. Of most interest in cryogenic applications are the more or less closed-cell foams of polystyrene, epoxy, polyurethane, rubber, silica, and glass. The thermal conductivities of some selected samples are shown in Table 9. In general, the thermal conductivity of a foam is determined by the interstitial gas contained in the cells, plus internal radiation and a contribution because of solid conduction. Evacuation of most foams reduces the apparent thermal conductivity, thus showing the partial open-cell nature of most of these materials (Kropschot 1959).

Over time, many CFC-blown foams will be permeated by air, which may increase their conductivity by as much as 30%. If hydrogen or helium is allowed to permeate the cells, the conductivity can be increased by a factor of three or four. Glass and silica foams appear to be the only ones that have fully closed cells. Note that even under evacuation, foam insulations have much higher conductivities than the other types of insulation previously discussed.

STORAGE AND TRANSFER

Cryogenic fluids require more elaborate storage than fluids boiling near ambient temperature. For example, propane and LPG can be stored in carbon steel containers without insulation. At 70 °F, propane has a vapor pressure of 110 psig. More expensive materials of construction are required when the stored fluid is below −40 °F. For fluid temperatures down to about −320 °F, or liquid nitrogen at normal boiling temperature, some type of insulation is required to minimize boiloff losses. Liquid hydrogen and helium below −320 °F requires a vacuum system or its equivalent to prevent air condensation and resultant high boiloff rates.

To minimize losses of fluids boiling below −320 °F that are handled in small quantities, a liquid nitrogen shield is used on containers having high vacuum insulation. In larger sizes, multilayer insulation is a less expensive alternate to the liquid nitrogen shield for liquid hydrogen and helium containers. Table 10 shows a classification for cryogenic storage according to fluid, size, insulation, and application. This classification is only approximate, and economics must be considered in cases where overlap may occur.

Although normal criteria for line sizing apply in the transfer of cryogenic fluids, a designer must also consider cooldown and heat leak losses, the behavior of the construction materials at cryogenic temperatures, and the servicing required to maintain good performance. The selection of insulation for the transfer system is determined on the basis of fluid properties, transfer rate, line

Table 10 Classification of Cryogenic Storage

Fluid	Size, gal	Insulation	Application
Liquid oxygen or nitrogen	Up to 26	Vacuum	Use
Liquid oxygen or nitrogen	Up to 42	Multilayer	Transport and use
Liquid hydrogen, helium, or neon	Up to 40	Vacuum and liquid nitrogen shield or multilayer	Transport and use
Liquid oxygen, nitrogen, or argon	Up to 7000	Evacuated powder	Transport
Liquid oxygen or nitrogen	Up to 13,000	Evacuated powder	Production and use
Liquid oxygen or nitrogen	Greater than 13,000	Powder	Production and use
Liquid argon	Up to 50,000	Evacuated powder	Production and use
Liquid hydrogen	Up to 30,000	Multilayer	Transport
Liquid hydrogen	Up to 850,000	Evacuated powder	Production and use
Liquid helium	Up to 9000	Multilayer	Transport and use
Liquid helium	Up to 30,000	Multilayer	Production
Liquid natural gas	Up to 12,600,000	Powder with nitrogen purge	Production (above ground)
Liquid natural gas	Up to 10,500,000	Powder or foam with nitrogen purge	Sea transport

lengths, duty, and service. For example, a noninsulated line is normally used for liquid oxygen or nitrogen transfer at high rates, over relatively short distances, and for short durations. For liquid helium transfer, however, multilayer insulation and short transfer lengths are essential.

Dewars

Dewar vessels vary in size from less than a quart to several thousand gallons. The small Dewars for liquid oxygen and nitrogen are vacuum-jacketed and constructed of glass, glass fiber-reinforced epoxy, copper, or stainless steel. Figure 31 shows the basic designs of small-capacity Dewars. Part A of this figure shows the simplest design, which ranges in inside diameter from several inches to about 4 ft and in inside depth from less than a foot to more than 10 ft. In large sizes, these Dewars are made of stainless steel, and a floating radiation shield is included to decrease the heat leak. Also included in the larger sizes is a getter to maintain the sealed vacuum. This is important because of the outgassing of noncondensables, such as hydrogen, from the metal surfaces.

Dewars are useful for transferring liquid nitrogen and oxygen from larger stationary storage into test apparatus. Their design may also include a flange, which permits bolting them to a top plate enclosure. The Dewar may be used as the enclosure for a cryostat where the low-temperature refrigeration heat exchange surface or test apparatus is supported from the top plate. For the larger capacities (up to 26 gal), having lower boiloff rates, the design shown (the LOX-LIN Dewar design) in Part B of Figure 31 is used. Liquid hydrogen and helium may be stored in the design shown in Figure 32. The liquid nitrogen shielding reduces the radiation heat leak to the hydrogen or helium, which would otherwise boil off at a much higher rate.

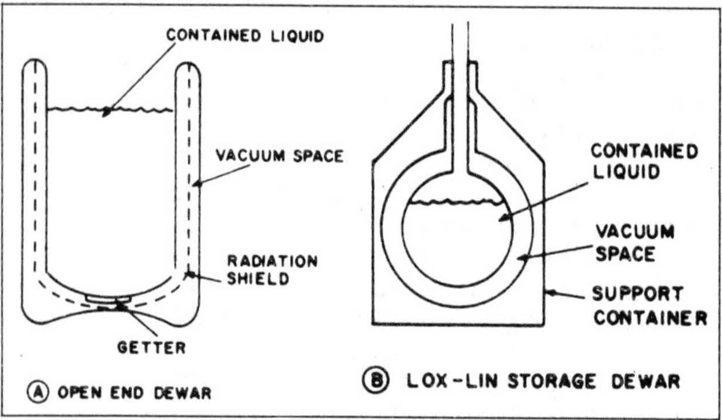

Fig. 31 Storage Dewars for Liquid Oxygen or Nitrogen

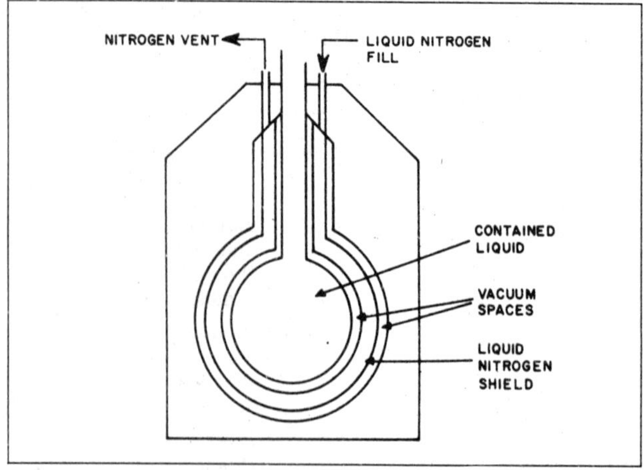

Fig. 32 Storage Dewar for Liquid Hydrogen or Helium

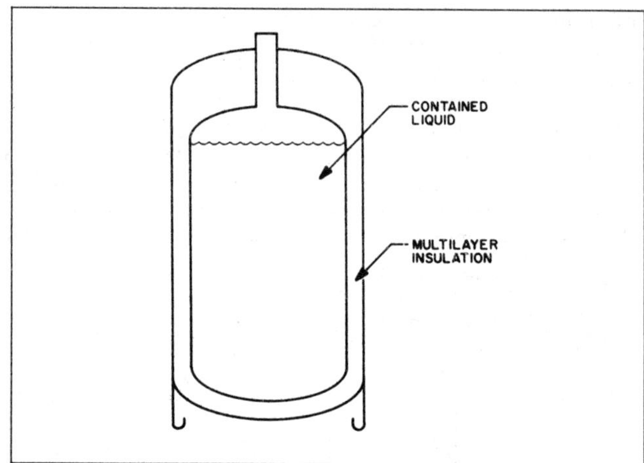

Fig. 33 Shipping Dewar

Figure 33 shows a multilayer insulated Dewar with a capacity of 26 to 66 gal. These containers have a diameter of 20 in. and a length of 60 to 80 in., and are used for liquid nitrogen, oxygen, natural gas, argon, hydrogen, and helium. Ruggedly constructed with a stainless steel inner shell and a carbon steel outer shell, Dewars are intended for shipment via common carrier or for permanent installation on a vehicle to provide LNG for fuel or LIN for in-transit refrigeration of food. They can be moved directly to a point of use via hand truck and can have a heavy-walled inner shell to supply use-point pressures up to 200 psi.

Stationary Storage

The simplest and least costly container for cryogenic liquids consists merely of an inner shell of nickel steel, stainless steel, or aluminum, enclosed by a thin carbon steel outer shell. The space between the outer and inner shells is filled with a powder or fiber-type insulation, such as perlite or rockwool, to insulate the inner shell contents. This design is commonly used for liquid oxygen and nitrogen storage containers of greater than 13,000 gal capacity and eliminates the need to construct a heavy outer shell to support loads because of an evacuated insulation space.

The lower heat leak, evacuated-powder insulated construction is preferred in smaller sizes because its higher surface-to-volume ratio yields higher rates of heat in-leak per unit of contents. Evacuated-powder insulated containers are shop-fabricated to minimize cost. They are thus limited to 13,000 gal by normal shipping clearances, but tanks to 50,000 gal are shippable with special handling.

Liquid hydrogen storage containers range in capacity to almost 1-million gal. Evacuated powder is used on tanks larger than 50,000 gal, which must be field-erected. Smaller, shop-fabricated tanks may also use evacuated powder, unless the more expensive but lower heat-leak multilayer insulation can be justified. The factors favoring multilayer insulation are the lower temperature, higher volatility, and higher cost of liquid hydrogen, compared with liquid oxygen and nitrogen. On the other hand, many liquid hydrogen storage tanks supply gas to a use point, and vent losses can usually be used.

For the storage of liquid helium, the economic justification for high vacuum or multilayer insulation plus liquid nitrogen or vent gas shielding is unquestionable. Liquid helium is not only expensive, but it is more volatile than liquid hydrogen. Only 9.2 Btu are required to vaporize one gallon of liquid helium, compared with 11.4 Btu for liquid hydrogen and 544 Btu for LIN. Since liquid helium cannot normally be shipped and stored without vent losses, multilayer insulated vessels can be designed to recover refrigeration from the vent gas by using this cold gas to chill the insulation and reduce heat leakage. Without this vent gas shielding or high vacuum insulation with liquid nitrogen shielding, liquid helium cannot be retained for more than a few hours.

Transport

A single trailer is normally adaptable for the transportation of liquid nitrogen, oxygen, or argon. Trailers with capacities up to 7000 gal have been constructed. Highway load restriction has established this as the maximum capacity. For servicing receiver tank pressures up to 250 psi, a specially designed cryogenic centrifugal or turbine pump is used. The pumps may be installed on the trailer or on the ground with each receiver tank, many of which are low pressure (15 psig or less). A simple pressure transfer is used for servicing such tanks, using a built-in self-pressurization coil on the trailer to maintain a pressure differential between the trailer and receiver.

Cryogenic trucks with capacities up to about 2000 gal are widely used to service smaller receiver tanks, using an on-board cryogenic pump. Some trucks may have a second, high-pressure pump and vaporizing system for refilling gas storage systems up to 3000 psi. Railroad tank cars with capacities of up to about 20,000 gal are also in use to transport liquid nitrogen, oxygen, and argon. Trailers, trucks, and tank cars use evacuated-powder insulation.

Liquid hydrogen is transported in trailers up to 14,000 gal capacity, and in railroad tank cars up to 30,000 gal. Multilayer insulation is almost always used. The low mass of liquid hydrogen allows the design of transports with capacity limited by size restrictions, not weight limits. Liquid hydrogen is usually transferred by pressure difference, using a self-pressurization coil. A few trailers are equipped with pumps for servicing higher pressure receiver tanks. These pumps are expensive, because of the low density and high volatility of liquid hydrogen.

Liquid helium is generally shipped in portable Dewars of up to 26 gal capacity. A small number of larger transportable containers and trailers up to 9000 gal capacity have been built and are used for domestic or overseas shipment from producing plant to distribution stations, where smaller Dewars are filled for shipment to the user. Liquid helium is transferred by differential pressure. Usually, the normal pressure buildup in the supply vessel is sufficient. If an external source of helium gas is needed for additional pressure, the liquid-gas interface should not be disturbed. Improper techniques can convert the entire liquid contents to gas in a brief period. Even the best techniques cause a substantial percentage of the transferred liquid to change to gas. When large volumes are transferred, loss recovery techniques are employed to compress gas into storage.

Transfer of Servicing

For the transfer of liquid oxygen and nitrogen, a foam glass block insulation of vacuum-jacketing has been used. For high transfer rates over short durations, uninsulated lines are usually used. The economical breakeven point between no insulation, foam insulation, and vacuum-jacketing depends on transfer rate and quantities, transfer distances, liquid cost, and cooldown and maintenance requirements.

As the multilayer line fabrication becomes less expensive, its application for liquid oxygen and nitrogen service is more accepted in cases of higher product loss costs. Transfer lines for liquid hydrogen service normally use the vacuum-jacketed design or the multilayer insulation in conjunction with the vacuum jacket. Liquid helium requires the use of either multilayer insulation or liquid nitrogen shielding and vacuum-jacketing of lines to accomplish economic transfers.

One of the larger problems with cryogenic vacuum-jacketed transfer line design concerns the contraction of the inner line on cooldown. In the past, stresses have been relieved by expansion loops every 300 ft, by outer line bellows, or by a combination of the two. The use of invar for the inner line material has decreased or eliminated the requirement for the expansion bellows. The inner line is allowed to become stressed between two anchor points, since the invar expansion coefficient is low, relative to the materials previously used.

The optimum line size for the transfer of liquid hydrogen and helium at low pressures is usually the size that gives close to the minimum storage transfer pressure. This is because of the costly storage involved and the higher losses associated with higher transfer pressures. For liquid oxygen and nitrogen, a transfer pump is generally sized to overcome the maximum differential pressure, plus a nominal head for the line pressure drop.

INSTRUMENTATION

Except for in-line devices, cryogenic plants use essentially the same instruments as most chemical processes. Cryogenic fluid in the sensing leads that directly operate controllers, transmitters, and switches is allowed to increase to atmospheric temperatures through warmup loops before entering the elements. Occasionally, when analyzing some gas mixtures from the liquid phase, a separator must be provided to assure no fractional distillation of the components when they change from the liquid to the warm gas phase.

Flow measurement of cryogenic liquids also requires special consideration. For both positive displacement meters and variable head meters, the measurement depends on exact, fixed dimensions. At low temperatures, thermal contraction presents clearance and lubrication problems with positive displacement meters; it also causes an area correction factor of less than 1.0 on variable area meters. Materials such as Monel or 304 stainless steel, which have lower expansion factors, are often selected for this service.

Pressure drop across an orifice plate can cause flashing if the liquid is not sufficiently subcooled before the orifice. Accurate temperature measurements are essential, since the density of some cryogenic fluids varies considerably with temperature. Many low-temperature primary measuring devices, *e.g.*, sonic probes and resistance thermometers, require insertion through vacuum-jacketed vessels. Special electrical connectors that provide a good vacuum seal at liquid helium temperatures are available.

Temperature Measurement

For normal commercial cryogenic temperature measurement, three types of thermometers are widely used: (1) thermocouples; (2) resistance bulbs; and (3) vapor pressure bulbs.

Thermocouples are the most frequently used temperature measuring device in the −350 to 0 °F range. They are relatively inexpensive and have low heat capacity, thus giving good response characteristics. Their small size lends itself to construction into many shapes and forms, thus allowing their use in many difficult locations. Except in galvanometer-type meters, thermocouples are usually connected to the bridge circuit of a potentiometer, which allows readings to be taken at a considerable distance from the installation.

The EMF produced by thermocouples below −350 °F becomes quite small, making them impracticable below this range, unless special precautions are taken. In addition, inhomogeneities in the wire can cause errors because of local thermocouple effects.

Copper-constantan is used in most thermocouples for moderate cryogenic temperature. With premium grade thermocouple wire, and with a −300 to −75 °F temperature range, the limit of error is ±1% of the temperature being measured. For low-temperature measurement, No. 20 B&S wire gage size is recommended. In cold boxes, thermocouples are soldered to the pipe to be measured, with an insulating piece installed to provide a heat barrier at the junction.

Thermocouples using gold, 2.1% atomic cobalt, and copper are also used. This type of thermocouple has the advantage of higher thermoelectric power, but gold tends to be less homogeneous than most other materials. Normally, thermocouples follow standard temperature versus millivolt tables. Calibration deviation can be checked with the bulb and the table.

Resistance thermometers of the older, pure metallic type (mostly platinum) are used to measure temperatures in the cryogenic range of 36 °R to ambient. Platinum is chosen particularly because of its chemical inertness, good workability, high purity, and because of the availability of test data. It is limited at the lower range because of its insensitivity and low resistivity. The sensing element is essentially a coil of fine wire wound around a frame of insulating material and carefully supported so that it will not be subject to the mechanical strain caused by differential thermal expansion.

Platinum resistance thermometers are ideally suited for differential temperature measurement and for temperature control when the span is less than 9 °R and the accuracy requirements are around 0.1 °R. The semiconductor resistance thermometer gives good results from 2 to 180 °R. Semiconductors have negative temperature coefficients below approximately 27 °R, giving them higher sensitivity than platinum, and, used with standard indicators and recorders, need only a precision power source. At low temperatures, the conduction process in semiconductors varies with different temperature ranges. This makes calibration difficult and is therefore used most frequently below the platinum range of 36 °R.

The vapor pressure thermal system consists of a pressure measurement element, connected by capillary tubing to a temperature-sensing bulb. These systems are relatively sensitive in the cryogenic ranges and function with good reproducibility. Construction of the bulb depth is critical to prevent convection. Extra care is needed in system design to assure that the lowest temperature is in the bulb. Impurities in the sensing fluid can also produce errors. In measuring liquid hydrogen, a catalyst must be inserted in the bulb to assure a conversion from ortho to para hydrogen. Vapor pressure thermometers are frequently used to check thermocouples and resistance bulbs at the normal boiling points of cryogenic liquids.

Liquid Levels

Liquid levels at low temperatures are often measured with standard instrumentation, but certain techniques must be followed to get satisfactory results. On cryogenic tank storage, the familiar oil tank system of a float connected to a steel line and then to a perforated tape, to drive a local or remote readout device, is a standard application. Suitable materials are used in the parts within the tank. Design must be such that there is no freezeup of the float system where it comes through the tank roof. Levels of storage tanks and of various vessels within the cold box use the common differential pressure meter. The principle used with cryogenic fluids is that both top and bottom meter legs are filled only with gas at the level taps to the meter, and the difference in head is the mass of the fluid.

On low boiling point fluids, the lines are run so that as soon as the liquid leaves the vessel, the heat leak quickly produces a change to the gaseous state at that elevation. With high boiling point liquids, particularly those composed of some higher boiling constituents, fractional distillation can take place at the point of heat leak and cause a buildup of liquid in the lower leg; this will produce an erroneous reading. The density of the vapor of higher boiling point liquids can be significant if the system is operated at higher pressures. Considerable care should therefore be taken when designing and installing liquid level legs. Back-purging with a warm lower boiling point gas is recommended on difficult installations.

Capacitance probes have been used for both single-point (on-off) liquid level measurement and continuous (analog) measurement. Typical construction for a continuous reading system consists of a concentric set of tubes, each of which is a plate of a capacitor. As the temperature considerably affects the dielectric constant of cryogenic fluids, it is necessary to include a dielectric constant cell below the main section of the tube that forms the reference capacitor, to make automatic correction for dielectric changes. The capacitor operates on a 400 Hz current, working on a self-balancing bridge circuit. Readout is obtained on any standard potentiometric indicator or recorder. The simpler point contact probe for on-off service requires no compensating capacitance or balancing bridge circuit.

Ultrasonic probes are basically used for point sensing of cryogenic liquids. An electrical signal from an amplifier external to the probe drives a diaphragm, which is the face of the probe. This motion in the ultrasonic range (above 20,000 Hz) is feedback to the amplifier and operates a control relay sensitive to the amount of oscillation. When the face of the probe becomes immersed in the liquid, the damping effect reduces the feedback oscillation and thus trips the control relay. This system can be used for high or low alarms or to control and measure tank filling and emptying cycles.

Displacer-type liquid level sensing devices have been used in cryogenic liquids. These instruments operate on the principle that the buoyant force acting on a cylindrical displacer reduces the apparent mass of the displacer in proportion to the amount of its submergence in the liquid phase. Standard instruments designed for use in room temperature liquids may be used without modification in heavier cryogens such as liquid nitrogen. For liquid hydrogen or liquid helium applications, the standard instrument should be modified to improve its sensitivity in these low-density fluids. The output of the instrument may be a standard pneumatic or electrical signal or a proportional control signal.

SAFETY

Processing and handling cryogenic fluids requires special safety precautions. Hazards occur because of the temperature levels, the toxic or flammable nature of the fluids, their lack of compatibility with common materials of construction, interaction between the cryogens and impurities found in cryogenic mixtures, and composition changes caused by vaporization of liquid mixtures.

Cryogenic liquids are so cold compared to the human body that contact between these liquids and the skin can cause severe cold burns, which are more painful and harder to treat than hot burns. Unprotected skin contacting very cold surfaces (such as pipelines or vessels containing cryogenic liquids) may stick fast, and the flesh may be torn in removal. When handling these liquids, protective gloves, face shields, and complete leg and arm coverings should be used. Open or porous shoes should not be used, and safety shoes are recommended. Even exposure to cold vapor cloud may be dangerous if intense enough for the body temperature to drop.

The low temperature levels also ensure that heat leaking into the liquid container will keep the cryogen boiling under whatever pressure is kept in the vessel. If the vessel is not vented, large pressures can build up and may rupture the container as a result of liquid vaporization from heat in-leakage. Figure 34 shows the pressure buildup over a closed tank of liquid nitrogen (Vance and Duke 1962). Similar pressure buildups are found for all cryogenic liquids. Since vapors escaping from a cryogenic vessel are cold, it is possible for a vent line to become clogged with ice, causing pressures to develop even in vented tanks.

Heat leakage and vaporization of cryogenic fluid trapped within valves, fittings, and in sections of piping could cause excessive pressure buildup and rupture of the equipment. For example, liquid hydrogen expands about 800 times when warmed to room temperature. The maximum should be limited through proper equipment design and pressure relief devices such as safety valves and burst disks.

Table 11 lists the toxicity and flammability characteristics of common cryogens. Even though some gases are nontoxic, none of the gases except oxygen will support life. Most cryogens are odorless and hence give no warning of their presence. Thus, vessels filled with cryogenic fluid may be hazardous because of: (1) the toxic effect of the residue vapor; (2) the inability of a worker to tolerate such vapors as nitrogen or helium; or (3) the flammability of the vapor mix (as with hydrogen or methane). Important safety considerations for handling such gases include adequate ventilation, monitoring the atmosphere, a personnel watch, and planned corrective action to be taken in the event that low oxygen levels are encountered.

Combustion in any system requires a fuel, an oxidant, and a means of ignition. In oxygen-containing systems, combustion is possible with almost any fuel, including many metallic and non-metallic materials suitable for industrial use. This is of special concern to designers and users of vessels, piping, and machinery for handling oxygen. Safety considerations in the design of oxygen systems include attempts to prevent ignition and to prevent the propagation of combustion if initiated. Metallic and nonmetallic materials vary in their susceptibility to ignition and combustion. Many common contaminants such as oil, rust, cloth, paper, tobacco, wood, paint, and metal chips are easily ignited in oxygen and may burn sufficiently to increase the temperature so as to ignite more stable materials such as low carbon steel, stainless steel, or aluminum.

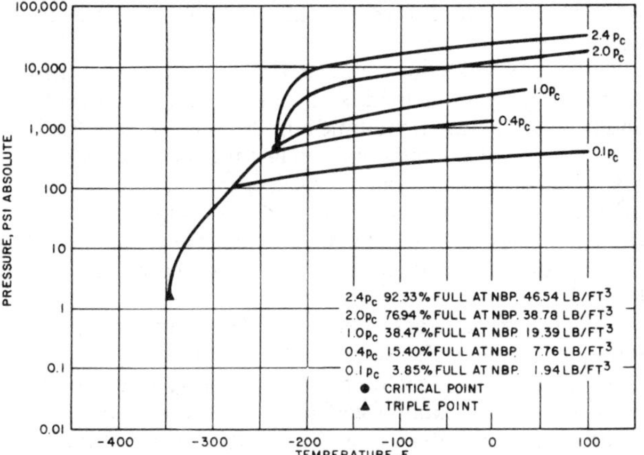

Fig. 34 Pressure Developed on Warming Confined Liquid Nitrogen
(Courtesy of Air Products and Chemicals, Inc., Allentown, PA)

Contaminants in cryogenic systems also result in functional interference with components in the system by particle deposits on moving parts, valve seats, system sensors, and controls. All cryogenic systems must be subjected to specific purging and cleaning procedures.

Prior to the loading of cryogenic vessels and working on such systems, they must be made inert. This can be accomplished by vacuum, pressure, or flowing purge. Vacuum purge is most satisfactory, because it requires fewer operations and eliminates any fuel pockets. It is accomplished by venting the systems to atmosphere, evacuating to a relatively low pressure, repressurizing the system with inert gas to a positive pressure, and again venting to atmospheric pressure. The volume of a gas is inversely proportional to the pressure, and the pressure of a mixture of gases equals the sum of its parts.

If a gaseous H_2 tank at atmospheric pressure is reduced to 10 mm Hg of pressure and then repressurized to 3 atm with an inert gas (N_2), the tank will contain:

$$\frac{10 \text{ mm Hg}}{(760 \text{ mm/atm} \times 3 \text{ atm}) - 10} \times 100 = 0.44\% \; H_2$$

For hydrogen, since the explosive limit in air ranges from 4 to 75%, the system can be assumed to be inert. Should entry into the

Table 11 Properties of Cryogenic Fluids

Gas	Boiling Point, °R	Volume Expansion to Gas[a]	Flammability Limits in Air, Mole %	Toxicity	Specific Gravity, (Air = 1)	Odor
Helium-3	5.8	470/1			0.137	
Helium-4	7.6	757/1			0.137	
Hydrogen	36.7	851/1	4.65-94		0.06952	
Deuterium	42.5	976/1	yes	Radioactive		
Tritium	45.2	—	yes	Radioactive		
Neon	48.8	1438/1			0.6964	
Nitrogen	139.3	696/1			0.6970	
Carbon monoxide	146.9	—	12.5-74.2	High	0.6978	
Fluorine	153.0	957/1		High	1.312	Sharp
Argon	157.1	842/1			1.38	
Oxygen	162.4	860/1			1.1053	
Methane	201.1	648/1	50-15.0	Slight	0.5544	
Krypton	215.6	693/1			2.818	
Refrigerant 14 (CF_4)	261.4	446/1		Moderate	1.62	
Ozone	190.3	—	yes	High	1.658	Sulfurous
Xenon	295.2	573/1			4.53	
Ethylene	304.7	487/1	2.7-40		0.978	Sweet
Boron trifluoride	310.9	—		High	2.37	Pungent
Nitrous oxide	330.5	666/1		Moderate	1.530	Sweet
Ethane	332.6	436/1	3.0-12.5	Slight	1.047	
Hydrochloric acid	338.4	—		High	1.268	Pungent
Acetylene	340.4	—	2.5-80	Slight	0.9073	Garlic
Fluoroform	340.4	—		Moderate		
1,1-Difluoroethylene	342.0	—		Slight		Faint ether
Refrigerant 13 (CF_3Cl)	344.9	—		Slight		Mild
Carbon dioxide	389.7[b]	641/1		Slight	1.5829	Slight pungent

[a]Gas at 529 °R and 1 atm pressure.
[b]Triple point.

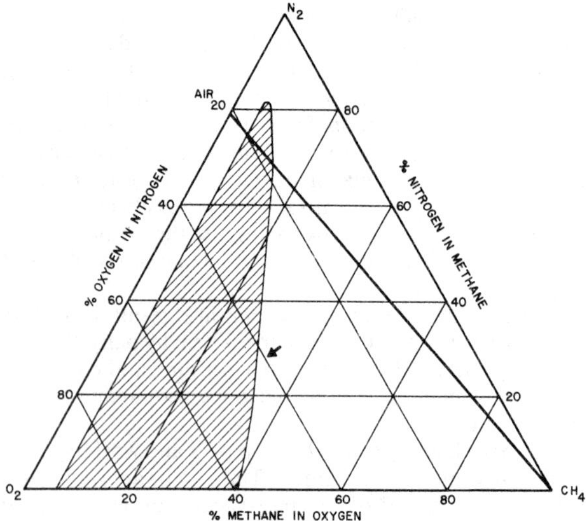

Fig. 35 Flammable Limits of CH_4—O_2—N_2
(Courtesy of Air Products and Chemicals, Inc., Allentown, PA)

vessel be required, the vessels can be thoroughly purged with air, without any combustion occurring. The O_2 content should be raised above 19.5% by volume before entering.

A pressure purge is less satisfactory than a vacuum purge. It requires alternate pressurizing and venting of the system until a safe atmosphere is reached. The method of determining flammable gas content is similar to vacuum purge. Raising the tank pressure to 3 atm with N_2 in the initial cycle gives a VH_2 of 33.3%; the second cycle would be $VH_2 = 33.3/3 = 11.1\%$; and the third, $11.1/3 = 3.7\%$. While a flowing purge is the simplest method, it is the least satisfactory because it provides no positive assurance that a completely inert atmosphere has been attained. Purging configurations should assure turbulent flow of the purging gas; flow rates should be high enough to thoroughly purge all parts of the system. The importance of using safe procedures to purge and render the system inert can be obtained from a study of the flammability limits of gas mixtures.

Figure 35 shows the flammability limits of CH_4— N_2—O_2 at or near atmospheric pressure (Vance and Duke 1962). The cross-hatched area includes all compositions in which the mixture is flammable. At other compositions, combustion does not raise the gas temperature high enough to sustain combustion. The line drawn from 21% O_2/79% N_2 binary to pure CH_4 represents the compositions that can result from mixing CH_4 with air. Note that flammability occurs from about 5 to 15% CH_4 in air. If an LNG tank were to be purged with air, the gas mixture resulting would at first be too rich in CH_4 for combustion. However, as purging continues, flammable mixtures would develop, resulting in a hazardous condition. Data on flammability limits may be found in the literature and interpreted in the light of the behavior shown here.

Organic contaminants, which may enter with a cryogenic liquid, are usually frozen and may float or sink in liquid. Thus, they can collect at low or high points in a system. If this occurs, an explosive mixture can be generated with oxygen when the equipment is warmed up and both the liquid and the deposited solids are vaporized. This has been the cause of many explosions in the cryogenic areas of air separation plants. In these units, hydrocarbons, especially acetylene, from the compressor oil or from the atmosphere can enter the system in trace quantities and must be removed by adsorption or by venting, especially if they are present in concentrations greater than their solubility in the cryogenic liquid.

The release of cryogenic liquids and vapors must be carefully controlled. Although many cryogenic liquids produce vapors of lower molecular weight than air, the lower temperatures result in a more dense fluid. The vapors travel along the ground and collect in low places. Exposure to these vapors is hazardous. Exposure to oxygen vapors could result in clothing or any equipment with oil-lubricated parts becoming oxygen enriched. Incidents of fire have occurred in many such cases. Any personnel exposed to such vapors should ventilate their clothing for at least 20 min before exposing themselves to a source of ignition. Open flames should not be permitted where oxygen enrichment exists.

Venting for cryogenic vapors should be designed to provide for sufficient dilution of the vapors, making them safe to personnel or operating equipment. Safe distance depends on the vent direction, amount of flow, liquid or gas, wind direction, terrain, and other factors. For oxygen, about 50 ft is a recommended safe distance from the vent to equipment where large quantities are vented.

The disposal of flammable cryogens is best accomplished in a burnoff system in which the liquid or gas is piped to a remote area and burned with air in multiple burner arrangements. Such systems should include pilot ignition methods, warning systems in case of flameout, and means for purging the vent line. Small quantities may be vented unburned from a single vent, which must be released at least 15 ft above a roof peak; the amount depends on the wind direction and velocity, distance from dwellings, and the like.

Should a fire occur at the vent or at other locations, the fuel supply should be shut off. No attempt should be made to put out a hydrogen or LNG fire while it is still being supplied with fuel. If a hydrogen fire is extinguished and the flow of hydrogen is not stopped, a hazardous combustible mixture may start forming at once. The mixture will almost certainly be ignited and there will be an explosion that will cause more damage and restart the fire. Water should be used to keep the metal parts cool until the fire burns itself out.

Spills with liquid hydrogen and LNG (two cryogens produced in bulk) result in a vapor blanket including zones of combustible mixtures that could ignite the entire mass. Both these fluids burn clean; H_2 gives a nearly invisible flame. Compared to a flammability limit of 2 to 9% (by volume) for jet fuel, in air, hydrogen has a flammability limit of 4 to 75% and LNG from 5 to 15%. The ignition of explosive mixtures of H_2 with oxygen or air occurs with low energy input; about one-tenth that of a gasoline-air mixture. All ignition sources should be eliminated, and all equipment and connections should be grounded. Lightning protection in the form of lightning rods, aerial cables, and ground rods, suitably connected, should be provided at all preparation, storage, and use areas for flammable cryogenic fluids.

The mechanical properties of materials at low temperatures must be known to design cryogenic equipment. The low temperatures limit the number of materials that can safely be used, since many materials become brittle as the temperature is lowered. This is especially true of some iron alloys where a transition occurs from ductile to brittle failure modes. The temperature level and range over which this transition occurs depends on the metal composition, its treatment history, and its application conditions. On the other hand, materials without this transition are likely to have greater strength at cryogenic temperatures than at ambient. However, since any system built to operate with cryogens will be warmed to ambient from time to time, the design must permit safe operation at all temperatures from ambient to the lowest temperature possible for the operation.

Seals are a particularly difficult problem where cryogens are to be used. Most materials to be used as sealants, gaskets, and packing act as fuels in an oxygen atmosphere and/or become brittle at low temperatures. Moreover, the wide operating temperature range results in large dimensional changes or stresses in the materials. Even welds cause difficulties because of the composition variations that occur at the edges of the weld and because of the problem of controlling weld heating and cooling rates.

Although plastics have served well in some cryogenic systems, they must be used with care. Some polymers become brittle at low temperatures, whereas others tend to flow when cold. In every case, the problem of relative expansion and contraction rates between the polymeric material and a metal joined to it must be solved. Finally, many materials are attacked chemically by cryogens, especially oxygen and hydrogen. Data should be obtained on the suitability of each material of construction for the temperature range and the specific cryogen to be used before it is used in a cryogenic system.

Engineers entering this field should proceed cautiously, checking the cryogenic literature for guidelines on good design and operating practice. In every case, safe operation will require detailed design and operating consideration of all the possible hazards. The bibliography lists a few sources of information.

SPACE SIMULATORS

A space simulator produces an environment as close to actual space conditions as possible. Normally, the high vacuum and thermal properties of space are the primary simulation goals. Satisfactory methods of producing zero gravity conditions (or the high energy particle and electromagnetic radiation conditions of space) have not been achieved.

A nearly perfect vacuum is required for true space simulation. Table 12, which gives the presently accepted values for pressure at various altitudes, shows that pressures below 10^{-9} mm Hg are necessary to simulate altitudes above 420 mi. The vacuum in interplanetary space is 10^{-12} and 10^{-14} mm Hg.

To achieve such pressures in a space simulation chamber, extremely high pumping speeds, accompanied by a minimization of the gas leak rate into the chamber, must be provided. Materials used in chamber construction must be chosen for low outgassing (release of adsorbed gas under vacuum), since this may provide a major portion of the vacuum pumping load.

Space is a nearly perfect heat sink, equivalent to a black body at a temperature of about 7 °R. This condition is simulated in a chamber by surrounding the test object with a cooled, absorptive sink, usually maintained near liquid nitrogen temperature. Radiation is then provided to simulate the sun's radiation, earth albedo, or other effects.

The satisfactory simulation of solar radiation, including a duplication of the solar spectrum, requires elaborate equipment; *i.e.*, carbon arcs of mercury xenon arc lamps, with complex optical systems for collimation and filtering. For many applications, the use of infrared lamps or tubular heaters placed within the chamber, close to the vehicle, is a satisfactory substitute. These can be programmed in various ways to simulate orbital conditions. When this type of heat source is used, a small reflector is placed behind them to collimate the beam partially. Correction of heat flux density is required because of change in emissivity with wavelength. Earth albedo can also be simulated with these lamps or with warm panels, which can be placed in suitable locations within the chamber.

Table 12 Altitude versus Pressure Chart

Altitude, ft	Pressure, mm Hg
240,000	0.0253
260,000	0.00892
280,000	0.00291
300,000	9.5×10^{-4}
350,000	8.5×10^{-5}
400,000	1.6×10^{-5}
500,000	3.5×10^{-6}
600,000	1.5×10^{-6}
800,000	4.0×10^{-7}
1,000,000	1.3×10^{-7}
1,200,000	5.0×10^{-8}
1,500,000	1.4×10^{-8}
2,000,000	2.3×10^{-9}
2,100,000	1.7×10^{-9}
2,200,000	1.2×10^{-9}
2,300,000	8.8×10^{-10}

Table 13 Outgassing Rate for Mild and Stainless Steel
(Giles 1965)

Material and Condition	1 h	10 h	100 h
Mild steel			
Degreased	1.7×10^{-6}	3.2×10^{-7}	6.2×10^{-8}
After 750 °F bake	—	3.9×10^{-10}	—
Stainless steel			
As received	6.2×10^{-6}	6.2×10^{-7}	—
Degreased	9.5×10^{-9}	6.2×10^{-9}	2.9×10^{-9}
After 930 °F bake	—	5.3×10^{-10}	2.6×10^{-13}

Note: The outgassing rate is in torr · ft^3/(s · ft^2) where 1 torr = 1 mm Hg. A torr · ft^3 is a unit volume of gas corrected to one torr absolute pressure, or

$$\text{Total outgassing rate of material, ft}^3 = \frac{\text{torr}\cdot\text{ft}^3}{\text{s}\cdot\text{ft}^2} \times \frac{\text{Material area, ft}^2}{\text{System pressure, torr}}$$

Although high vacuums are required to simulate space conditions closely, thermal balance studies only require pressures less than about 10^{-5} mm Hg to eliminate heat transfer because of convection. Effects that can be observed at lower pressures include volatilization of various materials, lubrication problems with moving parts, and cold welding, which occurs below 10^{-9} mm Hg.

Chamber Construction

Chambers for space simulation are of welded construction, generally stainless steel or stainless-clad plate. External stiffeners are used to reduce the plate thickness of the tank; these can be mild steel. The interior is polished to reduce the area for outgassing and to reduce the radiant heat transfer. The data in Table 13 show the outgassing rates expected from both mild and stainless steel and the advantage of using stainless steel. Conventional fabrication and welding techniques are used on these chambers, but double welding is avoided to eliminate trapped gas pockets that provide virtual leaks that cannot be detected by ordinary methods. Reliable metallic or elastomer *O* ring seals are used up to large diameters. When double *O* rings are used, the space between them is usually evacuated to provide a guard vacuum for the seal.

When pressures below 10^{-8} or 10^{-9} mm Hg are required, bakeout provisions are included in the form of strip heaters and insulation. Baking the chamber at temperatures of 400 to 750 °F removes adsorbed gases and reduces the outgassing rate of the metal walls. This, however, places certain limitations on chamber design, and generally requires cooling for elastomer seals, which are nonbakeable.

Vacuum Pumping

Vacuum pumping is accomplished by mechanical, diffusion, ion, or sublimation pumps.

Mechanical Vacuum Pumps. Rotary piston-type pumps and vane-type vacuum pumps are most commonly used to produce vacuum conditions within space simulation chambers. Single-stage units blank off at about 10^{-2} mm Hg and two-stage units at 10^{-3} mm Hg.

Systems with gas loads in the micrometre range often employ positive displacement-type blowers as boosters. These blowers, when backed by rotary or vane pumps, give a vacuum usable throughout the range from 10^{-2} to 10^{-3} mm Hg (Dowling *et al.* 1961).

Mechanical vacuum pumps are displacement machines that use a special oil as the sealant. They tend to collect any condensable gases in the crankcase, thereby limiting the ultimate pressure to the vapor pressure of the contaminants. Gas ballast is frequently used to help discharge moisture from rotary piston pumps. Crankcase heaters may also be useful. When heavy concentrations of condensables are to be pumped, cold traps (condensers) are often used.

Diffusion Pumps. These are a form of booster pump using a special oil and jet assembly. The largest size commonly used on space chambers has a pumping capacity of 1800 ft^3/s. Diffusion pumps are usually installed around the periphery of, and close to, the chamber to maximize pumping speed.

For best results and lowest ultimate pressures, a single refrigerated baffle or series of baffles must be placed between the diffusion pump and the chamber to minimize oil backstreaming. The baffles are generally cooled with water or liquid nitrogen.

Ion and Sublimation Pumps. Ion pumps, or getter-ion pumps, rely on the gas-absorbing power of nascent metal films. This is enhanced by ionization of the pumped gases. The metal film is formed continuously by evaporation or by sputtering. In some cases, a magnetic field is used to increase the path length and improve ionization efficiency. Sublimation pumps rely on an active metal film such as titanium, which reacts chemically with the gases and therefore removes them from the system. These pumps have an advantage over diffusion pumps in that they are free from oil and cannot contaminate the system. Performance improvement of sublimed films may be made by cooling the metal films to liquid nitrogen temperatures (Hunt *et al.* 1962), thereby increasing the pumping speed, particularly for hydrogen.

Thermal Shrouds

Simulation of space conditions requires that a test object be surrounded with surfaces highly absorptive to thermal radiation and emitting minimal radiation. This is best accomplished with blackened surfaces cooled to cryogenic temperatures. These surfaces are known as cryopanels, heat sinks, or thermal shrouds. Cryogenic surfaces are generally cooled with liquid nitrogen, not only because it is readily available, inexpensive, and safe to use, but because its temperature range gives an adequate simulation of space conditions.

If actual space conditions are to be represented by a temperature of about 7 °R, any shroud temperature above this will lead to an increase in the test object temperature above what it would attain in space. This increase is called the simulation error, which is calculated by:

$$\Delta T = \frac{T_s^4}{7T_o^3} + \frac{T_o}{7}\left(\frac{A_o}{A_s}\right)\left(\frac{1}{E_s} - 1\right) \qquad (19)$$

where

ΔT = temperature error, °R
T_o = test object temperature, °R
T_s = shroud temperature, °R
A_o = test object area, ft^2
A_s = shroud area, ft^2
E_s = shroud emissivity

The first term in the expression, which contains both shroud and test object temperatures, is the error caused by radiant emission of the shroud in excess of what would come from space. The second term, which does not include shroud temperature, is the error caused by reflections from the shroud or by its failure to absorb radiation from the test object. This term would be eliminated if the shroud emissivity were unity.

In a well-designed simulator, the area ratio of the shroud and test object can be made large enough so that, with normally attained values of E_s, the second term will be small with respect to the first term in the equation and can be neglected. For a vehicle temperature of 540 °R, the error for a 144 °R shroud is 1.2 °R; for a 198 °R shroud, it is 2.5 °R. Thus, shroud temperatures can rise considerably above the boiling point of liquid nitrogen before serious errors are introduced. A commonly selected maximum temperature is 198 °R, which gives an effective overall temperature of about 173 °R. After this temperature is selected, the shroud is designed to maintain temperature under the heat load anticipated in the chamber.

Shrouds are usually of sheet metal construction, with tubes containing liquid nitrogen affixed to the sheet. Tubes may be welded on or formed into the sheet by various processes. A commonly used material is *tube-in-strip*, in which the tubes are formed into the billet before rolling and are inflated by hydraulic pressure after the sheet is rolled. Another method uses a printed pattern of tubes on a metal sheet, which is then bonded to a second sheet and inflated. *Plate-coil* construction is also used to some extent (particularly for stainless steel shrouds) and consists of embossed and welded sheets containing a pattern of tubes.

Liquid Nitrogen Supply Systems

The low-temperature operation of these units limits the choice of materials: aluminum, copper, and stainless steel are the metals normally used. Aluminum and copper have a distinct advantage in thermal conductivity which permits a sizable web between tubes without significant temperature rise. Aluminum presents some fabrication problems over copper but retains its reflectivity much better; this is important in these systems.

Surface treatment of the panels is required to obtain high reflectance to the chamber walls and high absorptivity on the inside. A bright dip etch of aluminum provides a permanently bright surface with an emissivity of 0.1 or less. The other side can be black anodized or painted with certain black paints. The paints have proved more suitable than the black anodize: they have high absorptivity over a wider range of wavelengths and show less tendency to bleach out under ultraviolet radiation. Extended surfaces on the black side will improve the situation to some extent, but will greatly increase both the weight and cost of the shroud.

The shroud panels are generally fastened to a support ring or framework inside the chamber, which is suspended by long stainless rods to minimize heat leak. The fastening of the panels to the support structure, and to each other, should be as loose as possible to allow for the relatively large contractions that take place during cooldown.

Liquid nitrogen supply systems for the shrouds generally are of two types, usually described as flash systems or subcooled systems (Figure 36). The flash system pumps liquid directly from the storage tank through the load, then expands it through a valve, either directly into the tank or into a separator from which the remaining liquid is drained into the tank by gravity. The subcooled system pumps the liquid in a closed loop, using a heat exchanger to remove the heat with liquid nitrogen boiling at atmospheric pressure.

Both of these systems provide single-phase flow of liquid under pressure through the shroud. This is necessary to prevent hot spots or vapor locking in the tubes and to ensure maximum heat transfer. The flash-type system is simpler, but the tank must be elevated to provide net positive head for the pumps. Usually a separator, or specially designed tank, is also required to take care of the flash process.

The subcooler uses liquid from a standard tank, with no special provisions, and has the advantage of somewhat smaller pump losses because of the pump suction operating at tank pressure. The tank can be pressurized to 20 to 30 psig, thus reducing the amount of pressure that must be generated by the pump.

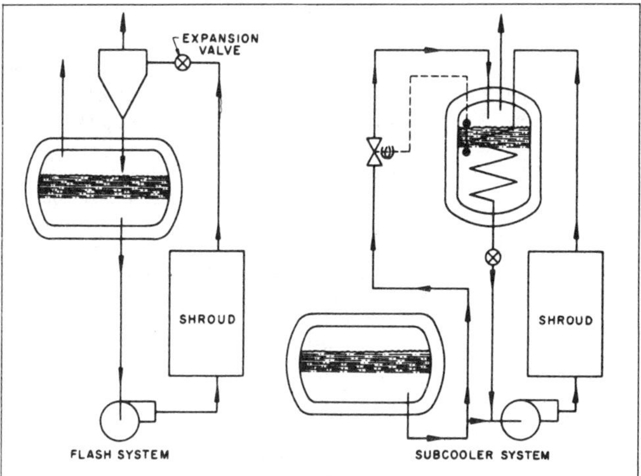

Fig. 36 Liquid Nitrogen Systems

From the pumping system, the liquid is generally fed to a valving system, which divides the liquid stream into the various zones of the shroud. This may be an insulated valve box or separate valves mounted on the chamber with pneumatic operators. In addition to flow control valves, vent valves allow each zone to be cooled individually by venting the vapor; they are then switched on stream when the proper temperature is reached.

Refrigeration and Heat Systems

When temperatures above that of liquid nitrogen are required, other types of systems may be selected. In some cases, a heat transfer fluid or brine is pumped through the shroud. This fluid is heated or cooled externally to produce the desired temperature. Cooling is usually accomplished with single or multistage refrigerators, and temperature is generally limited to about −100 °F. In some cases, carbon dioxide is used as a refrigerant for these systems. Heating can be done with electric immersion heaters or with steam in a heat exchanger. Such systems can be used to control temperature environments over a fairly wide range.

Another method for covering a wide range of temperatures is the circulation of gas, commonly nitrogen, through the shroud. To increase the mass flow, the gas is circulated in a pressurized loop, generally at 100 to 200 psig. It can then be heated or cooled externally to established controlled temperature levels in the chamber. Liquid nitrogen, which cools the gaseous nitrogen stream, permits operating at all temperatures to −300 °F. Heating with steam or electric heaters permits temperature elevation with such a system, but temperatures are generally limited to about 300 °F. Electric heating elements sealed in silicone rubber have also been used.

Cryopumping

Lower pressures are possible with cryogenic pumping, which uses a large portion of the chamber walls as a pumping surface, than provided by oil diffusion pumps. Further, large pumping surfaces minimize the possibility of molecules rebounding from the chamber walls and impinging on the object under test. In this sense, cryogenics offers a better simulation of the infinite molecular sink of space. At 36 °R, the vapor pressure of the air gases (oxygen and nitrogen) is about 10^{-12} mm Hg, so that surfaces cooled to this temperature act as pumps to pressures as low as 10^{-10} or 10^{-11} mm Hg.

Various panel arrays are used. Although shielding somewhat reduces the pumping speed, these panels are generally shielded from the inside of the chamber, so that they will not receive any direct thermal radiation. In all cases, the shielding panels are part of the cold wall of the chamber and are cooled to liquid nitrogen temperature to minimize the heat load on the panels. The panels are usually reflective and, when possible, are faced with reflective surfaces to further reduce this load.

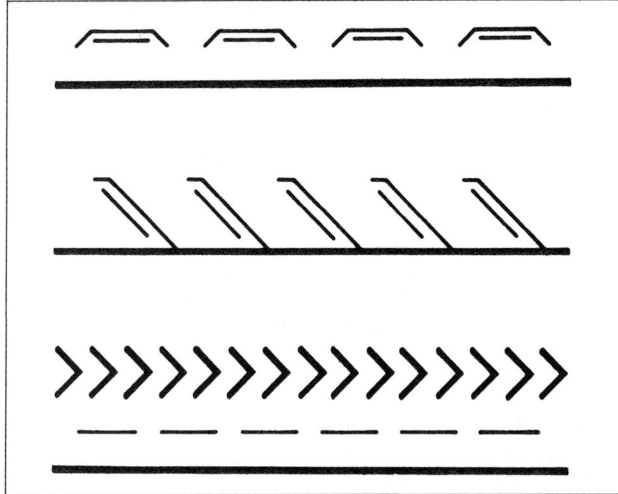

Fig. 37 Cryopumping Arrays (Arrangement of Panels)

The pumping speed of these arrays is determined by a quantity defined as the capture probability of the array, a specific property determined by the arrangement and dimensions of the panels and shields. The flat-panel array and chevron-shielded array (top and bottom in Figure 37) have capture efficiencies in the vicinity of 0.20. The angle-fin array (center of Figure 37) has an efficiency of about 0.35 (Barnes and Hood 1961, Pinson and Peck 1962, Hood 1963). The actual speed may be computed by multiplying the orifice speed of 380 ft/s by the capture efficiency. Thus, the angle fin has a speed in the vicinity of 1300 ft/s. Exact prediction of these speeds depends on other factors such as test object size, fraction of surface covered by the array, heat sink temperature, and cryopanel temperature, but the above method gives estimates accurate enough for most purposes.

Since the gas molecules and thermal radiation follow the same laws of reflection from the surfaces, the heat load on the panels of the arrays increases as the capture efficiency increases. Since refrigeration to 36 °R is quite expensive, a compromise is made between maximum speed and allowable heat load. For most purposes, the angle-fin array has proved to be quite satisfactory, combining adequate shielding with high pumping speed and relative simplicity of construction.

In most space simulation chambers with cryopumping, carbon dioxide, water vapor, and other condensable gases are pumped by the shroud, and oxygen, nitrogen, and argon are handled by the cryopump panels. The remaining hydrogen, helium, and neon are then removed by the diffusion pumps. Thus, any large loads of the latter three gases will result in large pressure increases unless the number of diffusion pumps is quite large.

The refrigeration for cryopumping is generally supplied by helium refrigerators that pump a stream of cold helium gas through the panels. The helium refrigeration cycle used is a simple compressor-expander system. The gas is cooled with a heat exchanger to about 40 °R before expansion, is expanded isentropically, and enters the load at about 18 to 27 °R.

Either reciprocating or turbine-type expanders may be used. The cold gas returns from the panels at about 36 °R and passes through the counterflow heat exchanger where it is warmed to room temperature before returning to the compressor. These systems usually operate at a pressure of about 300 psig and with pressure ratios of 5 to 10. A refrigerator of this type, capable of absorbing 3400 Btu/h of heat at the 36°R level, is sufficient to handle cryopumping arrays of 1000 ft^2 area, which would be found in chambers 30 to 40 ft in diameter.

Further improvement in a cryopumped vacuum chamber can be gained by lowering the panel temperature to 7.6°R with liquid helium. However, the large expense involved with the production and use of liquid helium has limited its use to small vacuum systems.

Cryosorption Pumping

One technique used for the evacuation of space simulation chambers is cryosorption—the physical adsorption of gases on cryogenically cooled surfaces. Materials such as charcoal, zeolite, silica gel, and alumina, when cooled to cryogenic temperatures, possess strong attractive forces for the common gases to be pumped in vacuum chambers. These forces effectively reduce the vapor pressure of a gas at the cryosurface temperature, enabling it to be pumped at its normal boiling point, *e.g.*, nitrogen on a 139°R adsorbent surface or hydrogen on a 36°R surface.

The efficiency of a cryosorption system depends on the maintenance of an adsorbent surface at low temperature. Since an adsorbent is inherently a poor thermal conductor, a thin layer (*e.g.*, 0.12 in.) is bonded in some fashion to a cryogenically cooled metal plate. Cryosorption panels using a molecular sieve as an adsorbent possess high pumping speeds and large capacities for hydrogen when cooled to 36°R, the normal boiling point of hydrogen. Likewise, cryosorption panels cooled to 7.6°R, the normal boiling point of helium, have demonstrated a high pumping speed and capacity for helium (Grenier and Stern 1966). The capacity of a cryosorption panel is extremely important in that the volumetric pumping speed of such a system decreases substantially as saturation is approached.

Regeneration of the panel can be achieved only by bakeout at elevated temperatures (480°F) for periods of up to 24 h, thus desorbing the pumped gas and, presumably, aborting any test under way in the vacuum chamber. Data indicate, however, that the capacity of a panel for hydrogen is sufficient, under normal gas loads, to provide for a test run of several months or more without regeneration (Stern *et al.* 1966).

Since 36°R cryosorption panels can handle only noncondensable gases, principally hydrogen, care must be taken to protect these panels from condensation of the condensable gases and resultant decrease of hydrogen capacity. For example, condensed nitrogen has a detrimental effect on panel operation. For this reason, 36°R baffles must be placed in front of the adsorbent panel to prevent condensation of the condensable gases on its surface.

Design for Human Testing

The emphasis in the space program on manned space flights has placed humans in a simulated space vacuum environment. Adequate safeguards to ensure the safety of an astronaut in a space chamber are essential. Locks must be provided for astronaut rescue operations, crew exchange, and repair or modification of the spacecraft while it is under space vacuum conditions. Biomedical facilities must be present in the chamber vicinity for monitoring, observation, and emergency rescue.

One problem of these facilities has been the repressurization system, installed for the protection of occupants in case of space suit failure and resultant exposure to chamber vacuum. Repressurization systems admit high-pressure air at a sufficient rate to raise the chamber pressure to that at 30,000 ft in as little as 5 s. The remainder of the repressurization to standard atmospheric pressure is then achieved at a slower rate of 25 s. Several areas, such as time, noise level, and fog, are critical for repressurization systems. Extremely dry air must be used for repressurization to prevent excessive fogging and resultant difficulty in finding an injured person.

REFERENCES

Barnes, C.B. and C.B. Hood. 1961. Correlation between pumping speed and cryoplate geometry. *Advances in Cryogenic Engineering* 7:64.

Corruccini, R.J. 1957. Gaseous heat conduction at low pressures and temperatures. *Vacuum* 7:19.

Dowling, D.L., D.B. Herrick, and W.E. Rose. 1961. Evacuating large space chambers with positive displacement blowers. *Vacuum Symposium Transactions*, 1235.

Fulk, M.M. 1959. Evacuated powder insulation for low temperatures. *Progress in Cryogenics* 1:63. Heywood and Co., London.

Geist, J.M. and P.K. Lashmet. 1963. *Advances in Cryogenic Engineering* 8:199. Plenum Press, Inc., New York.

Gifford, W.E. and T.E. Hoffman. 1961. A new refrigeration system for 4.2 K. *Advances in Cryogenic Engineering* 6:82. Plenum Press, Inc., New York.

Giles, S. 1965. Outgassing of systems (Space vacuum technology). *Test Engineering and Management* XIII (April).

Glaser, P.E., *et al.* 1967. Thermal insulation systems. NASA SP-5027.

Grenier, G. and S. Stern. 1966. Cryosorption pumping of helium at 4.2 K. *Journal of Vacuum Science and Technology* 3(6):334.

Grilly, E.R. 1953. *The review of scientific instruments* 24(1):1.

Grunert, W.E., *et al.* 1969. Opacified fibrous insulation. Proceedings of the 4th Thermophysics Conference, AIAA.

Hoffman, T.E. 1963. Reliable, continuous, closed-circuit 4 K refrigeration for a Maser application. *Advances in Cryogenic Engineering* 8:213. Plenum Press, Inc., New York.

Hood, C.B. 1963. Cryopumping speeds in large space simulation chambers. Proceedings of the Institute of Environmental Sciences, 81.

Hunt, A.L., C.D. Damm, and E.C. Popp. 1962. Gettering of residual gases and adsorption of hydrogen on evaporated molybdenum films at liquid-nitrogen temperatures. *Advances in Cryogenic Engineering* 8:110.

Hunter, B.J., *et al.* 1960. Metal powder additives in evacuated powder insulation. *Advances in Cryogenic Engineering* 5:146. Plenum Press, New York.

Koontz, J.K. and P.K. Lashmet. 1963. Paper presented at Annual Meeting of ASME, Philadelphia, November 17-22.

Kropschot, R.H. 1959. Cryogenics insulation materials and techniques. ASHRAE *Journal* (September):48.

Pinson, J.D. and A.W. Peck. 1962. Monte Carlo analysis of high speed pumping systems. *Vacuum Symposium Transactions*, 406.

Prast, G. 1965. A modified Philips-Stirling cycle for very low temperatures. *Advances in Cryogenic Engineering* 10:40. Plenum Press, Inc., New York.

Stern, S., R. Hemstreet, and D. Ruttenbur. 1966. Cryosorption pumping of hydrogen at 20 K. *Journal of Vacuum Science and Technology* 3:165.

Vance, R.W. and W.M. Duke. 1962. *Applied cryogenic engineering*. John Wiley and Sons, New York.

Zeitz, K. and B.K. Woolfendon. 1963. *Advances in Cryogenic Engineering* 8:206. Plenum Press, Inc., New York.

BIBLIOGRAPHY

Coward, H.F. and G.W. Jones. 1965. Limits of flammability of gases and vapors. Bureau of Mines *Bulletin* 503, U.S. Government Printing Office, Washington, D.C.

Hydrogen Safety Manual. 1968. NASA-Lewis Research Center, Cleveland, OH, NASA TM X52454.

Lapin, A. 1972. *Liquid and gaseous oxygen safety review*, 4 vols. NASA CR-120922, Air Products and Chemicals, Inc. (June).

Lewis, B. and G. von Elbe. *Combustion, flames and explosions of gases*, 2nd ed. Academic Press, Inc., New York.

NFPA. 1986. *Fire protection handbook*, 16th ed. National Fire Protection Association. Quincy, MA.

Product Pamphlets. Compressed Gas Association, Inc., New York.

Zabetakis, M.G. 1965. Flammability characteristics of combustible gases and vapors. Bureau of Mines *Bulletin* 627. U.S. Government Printing Office, Washington, D.C.

Zabetakis, M.G. 1967. *Safety with cryogenic fluids.* Department of the Interior, U.S. Bureau of Mines, Plenum Press, New York.

CHAPTER 38

LOW TEMPERATURE METALLURGY

THIS chapter describes the use of low temperatures for changing properties of metals, the effects of low temperatures on metallic properties, acceptability criteria for design, and the basic structure of construction materials.

MECHANICAL PROPERTIES

Mechanical and physical properties, cost, fabricability, and availability, are some of the important factors that are considered in selecting a material for a low-temperature application. Few generalizations can be made, except that a decrease in temperature increases hardness, strength, and modulus of elasticity. The effect of low temperatures on ductility and toughness varies considerably, however. The ductility of some metals increases with decrease in temperature, while others show an increase in ductility to some limiting low temperature followed by a decrease at lower temperatures. Still others show a decrease in toughness and ductility as the temperature is decreased below that normally encountered in the atmosphere.

Tensile Strength

In Figure 1, the relation of tensile strength to temperature is shown for metals generally selected for structural components. The slopes of the curves indicate that the increase in strength with decrease in temperature varies among the different metals. However, tensile strength is not a particularly good criterion for determining the suitability of a material for low-temperature service because most failures result from embrittlement. Ductility values obtained from the static tensile test may give some clue to degree of embrittlement, but the notched-bar impact test gives a better indication both of how the material performs under dynamic loading and how it reacts to complex multidirectional stress.

Ductility

Ductility, as measured by percentage of elongation in the tensile test, is shown in relation to temperature for several metals in Figure 2. The curves for copper, aluminum (5083), and lead show that ductility of these metals, which have a face-centered cubic crystal structure, increases as the temperature is lowered. However, not all face-centered cubic metals exhibit an increase in ductility at lower temperatures, as illustrated by the AISI 304 stainless steel. Most metals that have a body-centered cubic or hexagonal close packed crystal structure decrease drastically in ductility as the temperature is lowered, similar to zinc and SAE 1010 steel. These metals rupture with little plastic deformation even when tested under static loading.

Notched-Bar Impact

Some question exists as to whether values obtained on notched specimens approximately 0.4 in. square reflect behavior of full-size sections used in engineering structures. Considerable support has been given to the concept that the mode of propagation of

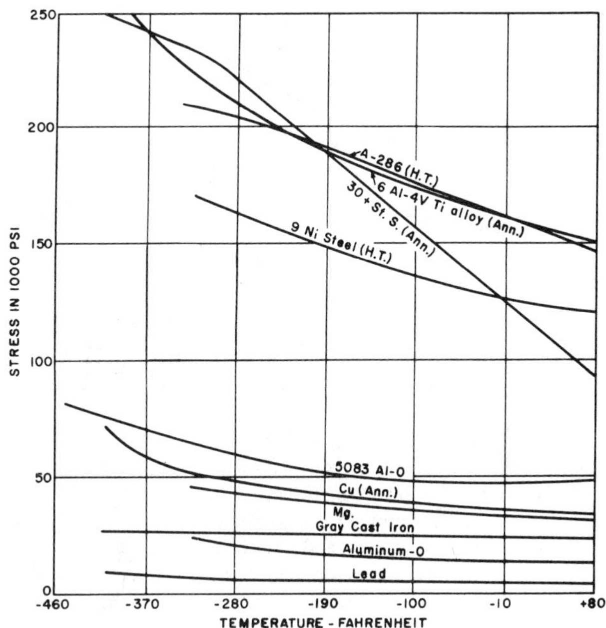

Fig. 1 Relation of Temperature to Tensile Strength of Several Metals

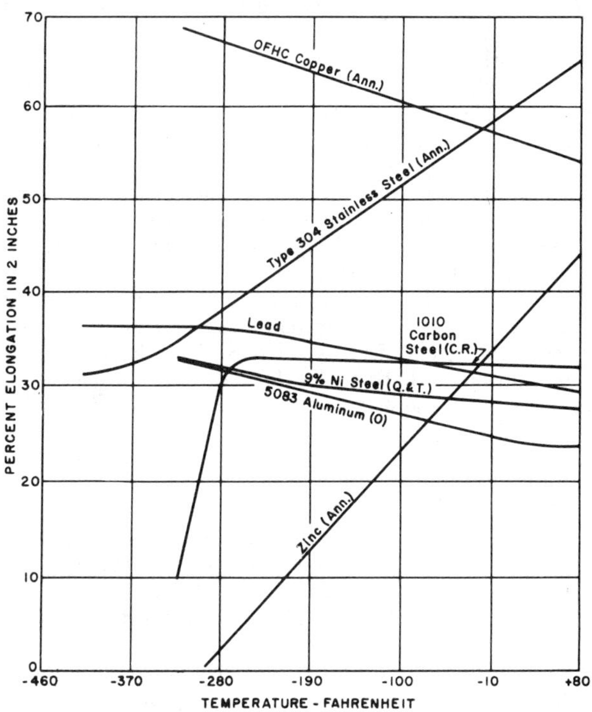

Fig. 2 Elongation Versus Temperature of Several Metals

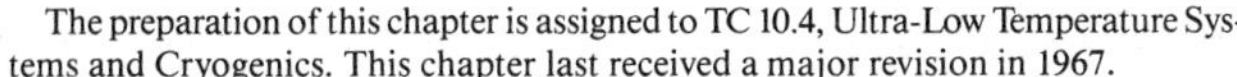

The preparation of this chapter is assigned to TC 10.4, Ultra-Low Temperature Systems and Cryogenics. This chapter last received a major revision in 1967.

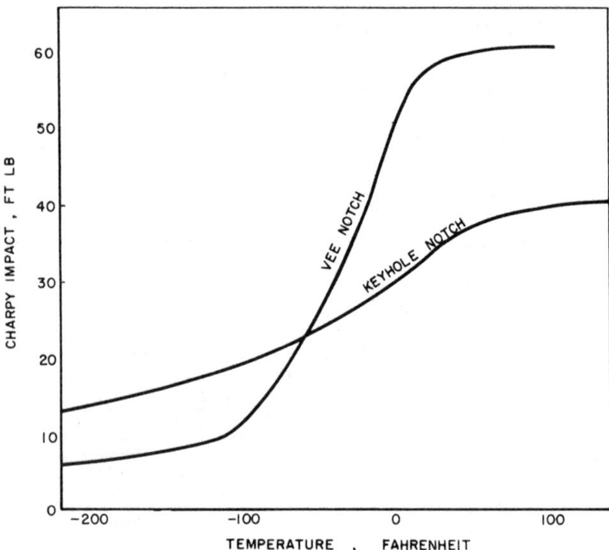

Fig. 3 Effect of Different Notches on Impact Properties of a Normalized and Tempered Cast Nickel Steel

fracture (*i.e.*, shear or cleavage) is of greater importance than total energy to rupture. Another group advocates that the amount of plastic deformation should be the criterion.

The most popular impact tests are the Charpy V and Keyhole notched specimens (see Figure 3) and the NRL drop weight test. The primary purpose of any low-temperature, service evaluation test is to define the temperature or temperature range at which a material fails in a brittle rather than a ductile manner. The temperature distinguishing between ductile and brittle behavior is the *transition temperature.*

Any test can be used to determine the transition temperature of steels, but the transition temperature determined by one test is not necessarily the same as that determined by another. In fact, the type of transition curve (see Figure 2) may change with the test method; and if the criterion is based on a condition of fracture common to all tests, such as 50% cleavage and 50% shear, transition temperature may vary as much as 50 °F or more in the same steel when tested by different methods. Temperature, the strain rate, and the degree of restraint all affect the temperature of change from ductile to brittle behavior in any particular steel; as these variables change according to different test methods, differences in transition temperature occur.

Nonferrous alloys commonly used at cryogenic temperatures do not exhibit a transition temperature, as noted in Figure 2. This does not mean that nonferrous materials all have high degrees of resistance to impact. Impact values for many nonferrous alloys are low, but there is no transition temperature below which these materials are subject to brittle failure or are extraordinarily notch-sensitive.

Notch Tensile Strength

Notch tensile testing defines the effect of a notch of particular geometrical configuration. Testing has been done of notches with varying calculated stress concentration factors, so it is sometimes difficult to compare data. However, some advantages of notch tensile testing are:

1. Sheet material can be tested that is not thick enough for accurate notched-bar impact tests.
2. Notched-unnotched ratios less than unity indicate the point at which a notch is effective in producing decreased strengths.
3. This type of test is perhaps more representative of the normal stresses applied to a structure than is the impact test.

Table 1 Typical Notch Tensile Data for Several Materials at −423 °F

	Yield Strength, ksi	Tensile Strength, ksi	Elongation, %	Notched-Unnotched Ratio[a]
A286, 0.1-in. thick, solution treated and aged	127	215	30	0.96
Ti-5A1-2.5Sn, 0.063 in., annealed	258	263	2	0.71
Type 310 Stainless Steel, 0.020 in.	261	290	1-12	1.12
Hastelloy B, 0.011 in.	240	283	16	1.09
Inconel X, 0.063 in.	134	233	30	0.85
5052-H38 Aluminum, 0.40 in.	56.5	81.2	37	0.96

[a]K_T for all tests = 6.3.

Although the notched-unnotched ratio of unity has been used in many instances as an acceptability criterion for a material, relative degrees of brittleness can be tolerated in many situations and ratios of less than unity may be acceptable. No one number separates materials into categories of good or bad for either notch tensile testing or impact testing. Examples of the results of some typical notch tensile tests are shown in Table 1.

Fracture Toughness

For fracture to occur, a crack must first be initiated, then propagated. However, in most engineering structures, microscopic cracks are already present and fracture is possible when the applied stress is sufficient to propagate the preexisting microcrack. With this in mind, various fracture toughness tests have been devised. In essence, a crack is introduced into a test specimen and the growth of the crack is recorded as a function of temperature and applied stress. Values of stress needed to fracture the specimen as a function of temperature indicate the sensitivity of the alloy to brittle fracture. The fracture toughness is expressed either as stress factor K_c (critical crack extension force) or G_c (critical stress intensity factor), both of which are adjusted values of the fracture stress. These parameters are regarded as basic material parameters, just as the ultimate stress, and may be used in fracture mechanic calculations. ASTM *Standard* E381 describes this test in detail.

MATERIAL SELECTION

Nonferrous Alloys

Aluminum alloys are characterized by increasing tensile strength, yield strength, and ductility, as temperature is lowered. The impact strength of aluminum alloys may rise, remain at a constant level, or decrease, with decreasing test temperatures. Successful use is being made of most alloys at temperatures at least as low as that of liquid hydrogen (−423 °F) and liquid helium (−453 °F). No transition temperature exists for aluminum alloys. Typical strength curves for the aluminum-magnesium 5083, 5086, and 5154 alloys are shown in Figure 4. Aluminum alloys that cannot be heat treated can be welded in heavy and light sections; limiting tensile strengths of 40,000 to 50,000 psi make these alloys desirable only for moderate stress levels.

Copper and copper alloys behave similarly to aluminum alloys as testing temperatures are decreased. Copper alloys are used in liquefaction plants. For these as well as aluminum alloys, the strength characteristics are inversely proportional to their impact resistance (high strength alloys have low impact resistance). Welding or brazing of heavy sections can present serious fabrication problems.

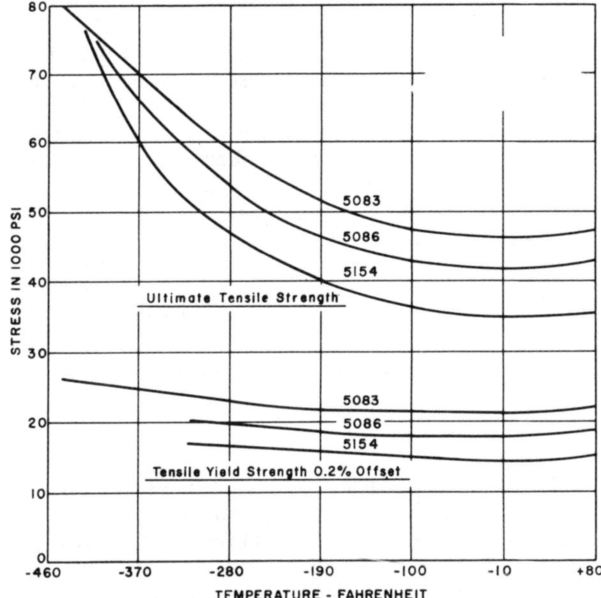

Fig. 4 Typical Tensile Properties of Alloys at Low Temperatures

Nickel and nickel alloys do not exhibit brittle failure tendencies at low temperatures, and high strength alloys such as Inconel X can be used at cryogenic temperatures. Nickel alloys can be welded successfully, but their high cost has decreased their more general use. A 36% nickel-iron alloy has been used where low expansion properties are important.

Materials used in joining, such as the *lead-tin solder alloys*, can be used at low temperatures, but increasing tin contents create transition-type impact curves. Silver-bearing brazing alloys are not generally subject to brittle failure at temperatures at least as low as that of liquid hydrogen (−423 °F).

Titanium alloys are useful as cryogenic materials, primarily for the aerospace industry, where high strength-mass ratios are important. Some titanium alloys do exhibit transition curves, but the Ti-6 Al-4V alloy has been used extensively at −320 °F and the Ti-5A1-2.5 Sn alloy is suitable at liquid hydrogen temperatures (−423 °F). High fabrication and material costs limit the usefulness of titanium alloys except where extremely high strength-mass ratios are of prime importance. Yield strengths of 100,000 psi coupled with low densities make these alloys attractive.

Ferrous Alloys

The low-temperature properties of steel are dependent on deoxidation practice, heat treatment, and alloy content. The use of aluminum for deoxidation of steel provides a fine-grain structure that is inherently more resistant to brittle failure than is a coarse-grain structure. Steels for low-temperature service are frequently specified to be made to fine-grain practice.

Quenched and tempered steels provide better impact properties than similar steels in the normalized condition; the annealed or soft condition generally provides the worst impact properties. Commercial quenching practices limit the thicknesses for which carbon steel can be fully hardened. Therefore, the use of quenched and tempered steels must be limited to those steels with an alloy content that will allow them to be fully hardened on quenching.

Transition temperatures, below which steels are generally considered to be unusable for cryogenic service, are sometimes ascribed to the transformation of austenite to martensite, with a sharp decrease in impact strength over a rather narrow temperature range. Alloying elements reduce the tendency for transformation to take place.

Nickel is the most effective alloying element for increasing the resistance of steel to low-temperature embrittlement. The temperature at which embrittlement occurs decreases as nickel content increases up to about 15% nickel. With higher percentages of nickel, the impact properties of low carbon steels become practically nonvariant at temperatures below atmospheric (see Figure 5). Manganese, within limits, is also beneficial in steel for increasing the resistance to embrittlement at low temperatures. Other alloys may slightly improve steel's resistance to low-temperature embrittlement, but their effect is generally an indirect one of modifying the form and distribution of the carbides rather than the direct one of having an alloying effect on irons, as is the case with nickel.

Quenched and tempered low alloy steels of the AISI or SAE types are frequently used in light and moderate sections of machinery parts that are to operate at temperatures down to −150 °F. A type of steel should be selected that will fully harden on quenching; if any welding is performed, it should be done prior to the quenching treatment.

The choice of materials for large pressure vessels and towers has probably received more attention than any other phase of the low-temperature problem. Because of almost universal use of welded construction for these parts, it is essential that a material be used that can be welded by an accepted commercial process and that welds have resistance to low-temperature embrittlement approaching that of the base material. Refer to ASME Section VIII, "Rules for Construction of Pressure Vessels."

In addition, the properties of the base material must not suffer deterioration from the heat of welding or, at least, not to an extent that cannot be corrected by a simple form of heat treatment such as stress relief. This requirement immediately eliminates steels that derive their resistance to low-temperature embrittlement from full heat treatment. As-rolled or normalized aluminum-treated low carbon steel, welded with AWSE 7015 electrodes, is used for comparatively thin wall vessels for temperatures down to −50 °F. Low carbon, 2.25% nickel steel, welded with AWSE 8105 nickel steel electrodes, is generally used for temperatures down to −75 °F or slightly lower, and low carbon, 3.5% nickel steel, welded with AWSE 8015 nickel steel electrodes, has been employed for temperatures down to −150 °F.

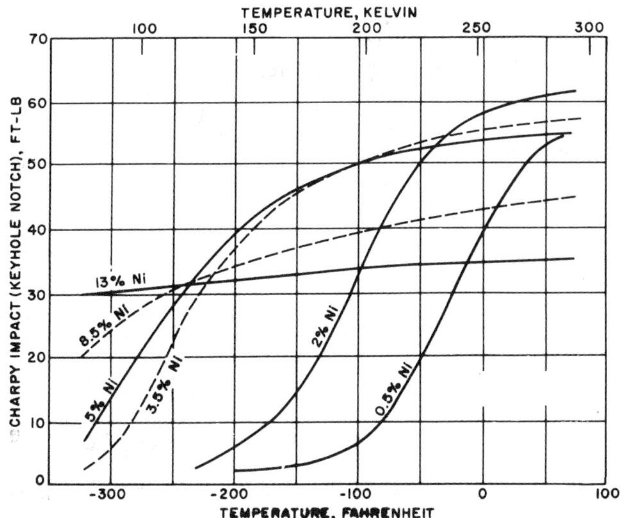

All steels contained 0.10% carbon except the nickel-free steel (0.20% carbon) and the 2% nickel steel (0.15% carbon) (Armstrong and Brophy).

Fig. 5 Effect of Nickel on the Low-Temperature Embrittlement of Normalized Low Carbon Steels

Low-carbon quenched and tempered steel plate is used for pressure vessels; but in this case, welding is performed on the heat treated steel plate, and further heat treatment is not usually applied. The limiting low temperature for such vessels is usually dependent on the properties of the weld metal. HY-80 quenched and tempered steel plate has been used in submarine pressure hulls. This alloy steel has a transition temperature below −120 °F and extremely high yield strengths, in excess of 80,000 psi. It can be welded without subsequent stress relief, but the welds are presently limited to somewhat higher transition temperatures. T1 quenched and tempered steel plate has been used in high strength applications and has acceptable impact properties to −50 °F. This alloy is generally stress relieved after welding.

Low-carbon, 9% nickel steel is used successfully at temperatures as low as that of liquid nitrogen (−320 °F). In the quenched and tempered condition, this alloy has a yield strength of 75,000 psi and can be welded with inconel-type filler metals without the necessity for additional stress relief. Use of this material has increased for cryogenic applications because of its low cost and high strength. Chemical and tensile requirements are shown in Table 2.

Chromium-nickel stainless steels have been used more than any other ferrous alloys for use at temperatures to that of liquid helium (−453 °F). No stress relief is generally required after welding, and impact strengths vary only slightly with decreasing temperature. A popular steel for low-temperature service in the chromium-nickel group is type 304, which has a yield strength in excess of 35,000 psi. Where higher strengths are required and welding can be avoided, strain hardened forms of the chromium-nickel stainless steels are available with yield strengths of 100,000 psi.

The *precipitation hardening stainless steels*, such as 17-7PH and 17-4PH, have received some attention in the aircraft and missile industry. Their ability to develop extremely high strength levels may be advantageous for cryogenic purposes down to −200 °F, but they have not been extensively used for these purposes.

Maraging steels have been developed with alloy contents of 18 to 25% nickel. These alloys exhibit high strength at temperatures as low as −320 °F, since they develop up to 250,000 psi yield strength.

Cast iron is usually considered a brittle material at normal atmospheric temperatures, but it is used successfully in many applications, even when some degree of shock resistance is required. Impact measurements on cast iron are usually made on unnotched arbitration test bars 1.2 in. in diameter, and although the impact resistance decreases as the temperature is lowered, a sharp transition temperature has not been observed. The plain cast irons have very low impact resistance at both room and low temperatures, but some of the alloyed cast irons possess a moderate degree of toughness.

The high alloy irons of the Ni-Resist type have appreciably higher toughness, but both the low alloyed and Ni-Resist types show a drop of 25 to 30% in impact values from room temperature down to −300 °F. Since the impact resistance of low alloyed high strength cast irons is approximately 25 to 30 ft lb at room temperature, and that of Ni-Resist, 80 ft lb or more, the impact values at −300 °F are about 20 ft lb for grey iron and 50 to 90 ft lb for Ni-Resist. These values cannot be compared with impact properties of steels and metals that are measured with notched specimens about 0.4 in. square, as such specimens are unsuitable for measuring differences in inherently brittle metals.

Table 2 Chemical and Tensile Requirements of Low-Carbon, 9% Nickel Steel
(From ASTM *Standard* A 353)

Chemical Requirements Element	Composition, %
Carbon, maximum*	0.13
Manganese, maximum	
Heat analysis	0.90
Product analysis	0.98
Phosphorus, maximum*	0.035
Sulfur, maximum	0.035
Silicon	
Heat analysis	0.15 to 0.40
Product analysis	0.13 to 0.45
Nickel	
Heat analysis	8.50 to 9.50
Product analysis	8.40 to 9.60
Tensile Requirements	
Tensile strength, ksi	100 to 120
Yield strength (0.2% offset), ksi	75
Elongation in 2 in., min. %	20.0

*Applies to both heat and product analysis

Analysis for Use

The choice of material for a specific use is often a compromise involving several factors. Cost is of major importance. The stress level at which the product will operate may require section sizes in certain of the lower strength materials that are impractical to manufacture or assemble. The operating temperature will further limit acceptable materials. The ability to weld or to stress relieve after welding may have a critical effect on the eventual choice of materials.

The possibilities of corrosion often must be considered. Thermal expansion characteristics have been critical in certain applications. Elastic modulus is certainly of interest in bolting and other applications. Resistance to thermal shock has sometimes been of major importance and, in these cases, alloys with high thermal conductivity are often considered. No general rules can really be established for the choice of material.

PRODUCT IMPROVEMENT

The principal areas in which cold treatment has been practiced to advantage in product improvement of ferrous alloys may be summarized as follows:

1. Increased hardness and wear resistance.
2. Dimensional stability of tools, gages, and machine parts.
3. Elimination of grinding cracks.
4. Increased cutter tool life.
5. Improved magnetic properties.
6. Ability to store aluminum alloy.

Since austenite is relatively soft and ductile, any appreciable quantity present in the surface structure of a tool or machine part will reduce hardness, which may lead to more rapid wear. Carburizing can increase hardness. Increased hardness values of up to 15 points Rockwell C have been noted in carburized cases of SAE 3310 gears after multiple cold treatment (−120 °F) retempering cycles on a case intentionally carburized to a high surface carbon content.

Where maximum dimensional stability is required in tools, gages, or machine components, cold treatment (or *cold stabilization*) may be used to convert the retained austenite and resultant expansion prior to the final dimensioning of the part.

Dimensional stability is important in gage manufacture and in close tolerance machine parts subject to subsequent low-temperature service conditions in high altitude aircraft. SAE 52100 steel is used in such applications and is particularly sensitive to austenitizing temperature. When austenitized at 1700 °F, approximately 13% retained austenite exists in the quenched structure.

Seasoning for dimensional stability is a variation of cold stabilization involving cycling several times over the temperature range found in service. Although conversion of retained austenite may occur, stress relief or equalization of stress is an important factor in producing stability in this application. Formerly, for applications requiring dimensional stability, iron castings were seasoned by storage for months before machining. Today, seasoning consists of cycling cast iron parts from −150 to 700 °F, allowing for possible contraction.

The elimination of retained austenite in hardened steel reduces the possibility of grinding cracks normally resulting from thermal stresses caused by the localized heating action of the grinding wheel. This condition is further aggravated by the expansion of retained austenite, which transforms thermally or by the mechanical shock of the wheel and causes conversion to martensite with severe surface stresses.

If high speed steel is properly hardened and then tempered 1050 °F for 2.5 h, very little retained austenite should remain in the structure. With the usual addition of two or more multiple tempering cycles at 1050 °F, insufficient retained austenite should remain in the structure to warrant cold treatment. However, cold treatment may serve to reduce the number of required multiple tempering cycles.

High-carbon, high-chrome (1.5% carbon, 12 to 14% chromium, 1% molybdenum), 1% carbon, 5% chromium, 1% molybdenum die steels respond very well to cold treatment. Multiple cycling of high-carbon, high-chrome steels between −120 and 950 °F produces greater stability than is possible by secondary hardening alone in the temper at 950 °F.

Wear resistance can be increased in the 1% carbon, 5% chromium steel by cooling to −150 °F prior to a 400 °F temper. This is particularly effective whenever the surface carbon content is increased by pack hardening in carburizing compound. This technique of carburizing certain tool steels to improve surface wear resistance is also used with the 0.5% carbon, 1 to 2% tungsten, 1% chromium punch and die steel.

Hardenable magnet steels, such as the 1% carbon types with either 3.5% chromium, 3% cobalt, or 5% tungsten, tend to retain austenite in the hardening process. This nonmagnetic austenite may be converted to magnetic martensite by cooling to −150 °F as a continuation of the quenching cycle, with a resultant increase of about 5% martensite.

Aluminum alloy rivets made from 2017 and 2024 aluminum will normally start to age at 32 °F within 16 h after the solution treatment. They may be stored for about a week at 0 °F and held indefinitely at −40 °F without hardening taking place until after the riveting operation. Cold treatment is applied to solution treated castings, forgings, and sheet metal blanks for storage prior to cold working operations. To promote dimensional stability and better machined finishes in precipitation hardenable aluminum and magnesium castings, a combination of cold treatment at −140 °F followed by overaging 50 °F above the aging temperature has been used. This allows permanent growth to take place prior to the final dimensioning of precision parts and minimizes possible service failures caused by temperature fluctuations over the same range.

Expansion fitting differs from shrink fitting in that an internal part is thermally contracted by cold treatment, inserted into a mating part, and allowed to expand to a tight fit on returning to atmospheric temperature. Shrink fitting, on the other hand, involves the heating of the external part and its subsequent shrinkage when assembled and allowed to cool around an inside part. A combination of both expansion and shrink fitting may be used when greater assembly strength is required.

Shrink disassembly of bushings, sleeves, and hollow inserts is possible if a suitable refrigerated coolant can be allowed to flow through the internal part. Cold treatment in the assembly of parts offers a definite advantage in ease of operation, with freedom from oxidation, distortion, and softening caused by tempering of hardened parts when heated for shrink fitting. A disadvantage of the process is the limited size change possible when cooling with liquid nitrogen to −320 °F compared to the thermal expansion resulting from heating through several hundred degrees.

In a proprietary process, cold treatment is applied to control the phase change between liquid and solid mercury used as replica material instead of the wax or plastic employed in the lost wax or precision investment casting process. Mercury is rapidly frozen in a master mold by cooling to −100 °F, removed as a solid replica of the part, and then held below −40 °F during the investment process.

The various means of producing subatmospheric temperatures and the temperatures obtainable are:

Salt and ice	−6 °F
Mechanical refrigeration, single stage	−50 °F
Solid CO_2 and methyl alcohol or acetone	−100 °F
Solid CO_2	−109 °F
Mechanical refrigeration, multiple stage	−150 °F
Liquid air	−297 °F
Liquid nitrogen	−320 °F
Liquid helium	−453 °F

Mechanical refrigeration is generally operated continuously at the desired temperature. Both liquid baths and circulated air chambers are used for cooling parts; however, the advantage of rapid heat transfer in the refrigerated liquid bath is somewhat offset by the greater tendency toward cracking.

CHAPTER 39

BIOMEDICAL APPLICATIONS OF CRYOGENIC REFRIGERATION

THE controlled exposure of biological materials to subzero (*i.e.*, potentially freezing) states has multiple practical applications, which have been rapidly multiplying in recent times. Primary among these applications are the long-term preservation of cells and tissues, the selective surgical destruction of tissue by freezing, the preparation of aqueous specimens for electron microscopy imaging, and the study of biochemical mechanisms used by a multitude of living species to withstand the rigors of extreme environmental cold. Some of the applications are restricted to the research laboratory, but clinical and commercial environments are increasingly frequent venues for activities in low-temperature biology. The success of much of this work depends on the design and availability of apparatus that can control temperatures and thermal histories. This apparatus can be adapted and programmed to meet the specific needs of particular applications.

This chapter briefly describes many of the principles driving the present growth and development of low-temperature biological applications. An understanding of these principles is required to optimally design practical apparatus for executing low-temperature biological processes. Although this field is growing in both breadth and sophistication, this chapter is restricted to processes that involve temperatures below which ice formation is normally encountered, *i.e.*, 0 °C, and to an overview of the state-of-the-art.

PRESERVATION APPLICATIONS

Principles of Biological Preservation

Successful cryopreservation of living cells and tissues is coupled to control the thermal history during exposure to below freezing temperatures. The objective of cryopreservation is to reduce the specimen's temperature to such an extent that the rates of chemical reactions that control processes of degeneration become very small, creating a state of effective suspended animation. An Arrhenius analysis (Benson 1982) shows that temperatures must be maintained well below freezing to reduce reaction kinetics enough to store specimens injury free for an acceptable time (usually measured in years). Consequently, one of two types of processes is typically encountered: either the specimen freezes or it undergoes a transition to a glassy state (vitrification). Although both of these phenomena may lead to irreversible injury, most of the destructive consequences of cryopreservation can be avoided.

A change in chemical composition occurs with freezing as the water segregates in the solid ice phase, leaving a residual solution that is rich in electrolytes. This process occurs progressively as the solidification process proceeds through a temperature range that defines a "mushy zone" between the ice nucleation and eutectic states (Körber 1988). If this process follows a series of equilibrium states, the liquidus line on the solid/liquid phase diagram for a system of the chemical composition of the specimen defines the relationship between the system temperature and the solute concentration. The fraction of total water that is solidified increases as the temperature is reduced, according to the function defined by applying the lever rule to the phase diagram liquidus line for the initial composition of the specimen (Prince 1966). This relationship has been worked out for a simple binary model system of water and sodium chloride and has been used to calculate the thermal history of a specimen of defined geometry during cryopreservation (Diller *et al.* 1985). As explained later, the osmotic stress on the cells with a concurrent efflux of intracellular water results from chemical changes. The critical range of states over which this process occurs corresponds closely to the temperature extremes defined by the mushy zone. At higher temperatures there is no phase change so osmotic stress does not exist. At lower temperatures the permeability of the cell plasma membrane is reduced significantly (as described via an Arrhenius function), and the membrane transport impedance is so high that no significant efflux can occur. Thus, the specimen's chemical history and osmotic response is coupled to its thermal history as defined by the phase diagram properties.

The property of a cell that dictates the response to freezing is the permeability of the plasma membrane to water and permeable solutes. The permeability determines the mass exchange between a cell and its environment when an osmotic stress develops during cryopreservation. The magnitude of the permeability decreases exponentially with the absolute temperature. Thus, the resistance to the movement of chemical species in and out of the cell becomes much larger as the temperature is reduced during a freezing process. Since the osmotic driving force also increases as the temperature decreases, in general, the balance between the osmotic force and resistance determines the extent of mass transfer that will occur during freezing. At high subfreezing temperatures (defined generally by the mushy zone), the osmotic force dominates and extensive transport occurs. At low subfreezing temperatures, the resistance dominates and the chemical species are immobilized either inside or outside the cells. The amount of mass exchanged across the membrane is a direct function of the amount of time spent in states for which the osmotic force dominates the resistance. Thus, at slow cooling rates, the cells of a sample experience extensive dehydration, and at rapid cooling rates, very little net transport occurs. The absolute magnitude of the cooling rate that defines the slow and rapid regimes for a specific cell depends on the plasma membrane permeability. A cell with a large permeability requires a rapid cooling rate to prevent extreme transport. The converse holds for cells having a small membrane permeability; they require prolonged high-temperature exposure to effect significant accumulated transport.

When very little transport occurs before low temperatures are reached during cooling, water becomes trapped within the cell in

The preparation of this chapter is assigned to TC 10.4, Ultra-Low Temperature Systems and Cryogenics.

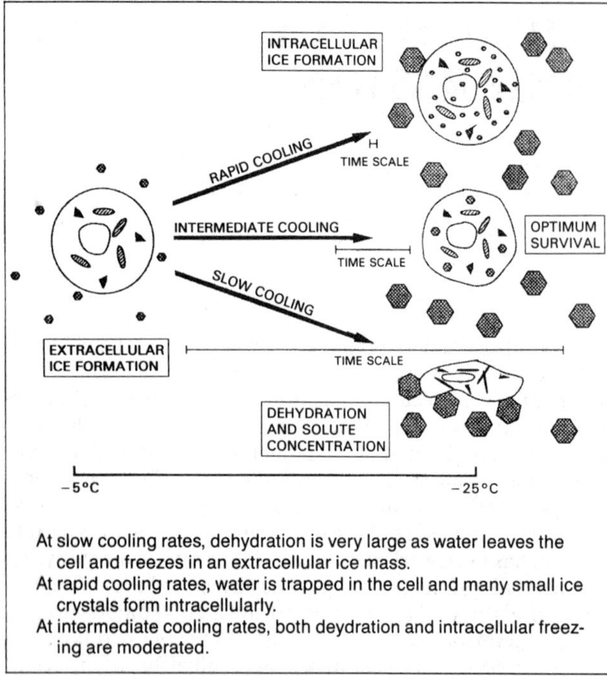

Fig. 1 Schematic of Response of Single Cell during Freezing as Function of Cooling Rate

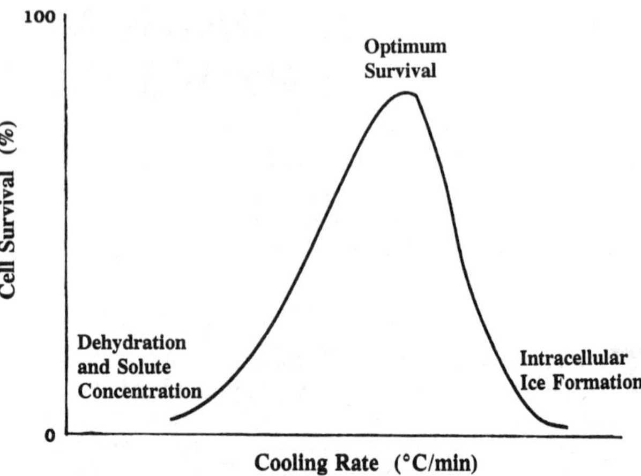

Fig. 2 Generic Survival Signature Indicating Independent Injury Mechanisms Associated with Extremes of Slow and Rapid Cooling Rates during Cell Freezing

a subcooled state. Chemical equilibration is achieved with extracellular ice by the intracellular nucleation of ice. This phenomenon is referred to as intracellular freezing. In this process, a substantial degree of liquid subcooling occurs prior to nucleation, so the resulting ice structure is dominated by numerous very small crystals. Further, at the low temperatures, the extent of subsequent recrystallization is minimal and the intracellular solid state surface energy is high.

At slow cooling rates and at high subfreezing temperatures, both extensive dehydration of cells and an extended period of exposure to concentrated electrolyte solutions occur. Clear evidence shows that some combination of the dehydration and the exposure to concentrated solutes leads to irreversible injury (Mazur 1970, Meryman *et al.* 1977). Mazur (1977) has also demonstrated that freezing at cooling rates that are rapid enough to cause intracellular ice formation causes a second mechanism of irreversible cell injury. These processes are illustrated in Figure 1, which shows that each of the extremes of the cooling process during freezing produces a potential for damaging cells. Figure 1 also implies that an intermediate cooling rate should minimize the aggregate effects of these injury processes and define the conditions at which optimum recovery from cryopreservation can be achieved.

Experimental data has been obtained for the survival of a large number of cell types for freezing and thawing as a function of the cooling rate. Nearly without exception the survival function follows an inverted V profile when plotted against cooling rate (Figure 2). This plot has been described as the survival signature of a cell. It illustrates the tradeoff between competing heat and mass transfer processes that govern the cryopreservation process. Solution concentration/osmotic effects lead to slow cooling rate injury. In this state there is adequate time for transport of water out of the cell before sufficient heat transport occurs to lower the temperature enough to drive the membrane permeability to nearly zero. Conversely, at rapid cooling rates the cell temperature is lowered so quickly that there is insufficient time for dehydration, and injury is caused by the formation of intracellular ice. The magnitude of the optimum intermediate cooling rate is a function of the magnitude of the membrane transport permeability. Higher permeabilities result in higher optimum cooling rates. Thus, the optimum thermal history for any cell type must be tailored for its unique constitutive properties.

For most cell types, the band width of cooling rates that defines the conditions for optimum cryopreservation survival is small, and the highest achievable survival is unacceptably low. Fortunately, for practical clinical applications, the spectrum of working cooling rates can be broadened and the maximum survival increased by modifying the sample prior to freezing by adding a cryoprotective agent (CPA). Although a wide range of chemicals exhibits cryoprotective properties, the most commonly used include glycerol, dimethyl sulfoxide (DMSO), and polyethylene glycol. Numerous theories have been postulated to explain the action of CPAs. In simplest terms, they modify the processes of solute concentration and/or intracellular freezing (*e.g.*, Lovelock 1954, Mazur 1970). Introducing CPAs to cell systems results in a major modification of the phase diagram for the system (Fahy 1980). In particular, the rate of electrolyte concentration with decreasing temperature may be reduced by nearly ten times, and the eutectic state depressed by as much as 60 to 80 K. These consequences greatly extend the regime of the mushy zone during solidification (Cocks *et al.* 1975, Jochem and Körber 1987).

Although phase diagrams provide much information for understanding the possible states that may occur during the cryopreservation of living tissues, their interpretation is limited due to two major factors. First, the chemical complexity of living systems is far greater than the simple binary, ternary, or quaternary mixtures that are used to model their behavior. Second, and more importantly, the thermal data used to generate phase diagrams is usually obtained for near equilibrium conditions. In contrast, most cryopreservation is executed under conditions far from the equilibrium state. For some situations, the goal is to maintain a state of disequilibrium; this domain includes vitrification methods that are applied to reach a solid glassy state in order to avoid ice crystal formation, latent heat effects, and solute concentration effects. In many cases, the degree of thermodynamic equilibrium reached for the low-temperature storage state may differ significantly between the intracellular and extracellular volumes (Mazur 1990). The equilibration can be controlled by manipulating the thermal boundary conditions of the cryopreservation protocol and by alternating the system's chemical composition prior to initiating cooling. Many of the same chemicals used for cryoprotection may be added at higher concentrations to decrease the probability of ice crystal formation at subzero temperatures and elicit vitrification (Fahy 1988).

In addition to the thermal history of the interior of a specimen, the thermodynamic relations determining the release of the latent heat of fusion as a function of temperature in the mushy zone (Hayes *et al.* 1988) must be considered. A specimen of finite dimension has a distribution of thermal histories within it during the freezing process (Meryman 1966). The pattern assumed for modeling the evolution of latent heat during freezing has a large effect on the cooling rates predicted as a function of local position in a specimen. Consequently, the anticipated spatial distribution of cell survival as a consequence of the preservation process may depend strongly on the model chosen for the thermodynamic coupling between the system's thermal and osmotic properties. Hartman *et al.* (1991) applied this principle to evaluate how to choose the optimum location for a thermal sensor to record the most representative thermal history during the freezing of a specimen of finite dimensions. Hartman *et al.*'s analysis indicated that the geometric center of a sample is a poor selection for positioning the sensor. A position approximately one third of the distance from the center to the periphery more accurately represents the integrated thermal history experienced by the mass during freezing.

Preservation of Biological Materials by Freezing

Biological materials are primarily cryopreserved by freezing them to deep below freezing temperatures. Among clinical and commercial tissue banks, freezing is the predominant method for preservation. Following the discovery of the cryoprotective properties of glycerol (Polge *et al.* 1949) and other CPAs, procedures for cryopreservation were developed for storing a variety of cells and tissues. Table 1 summarizes the types of tissue frozen by standard cryopreservation techniques as of 1993.

A typical protocol for cryopreservation consists of the following steps:

1. Place specimen in an appropriate container.
2. Add CPA by sequential increments at reduced temperatures.
3. Cool to subzero temperatures.
4. Possibly induce extracellular ice nucleation followed by a controlled period of thermal and osmotic equilibration.
5. Cool through high subfreezing temperatures with the greatest degree of thermal control invoked for the entire process then quench to storage temperature, usually in liquid nitrogen.
6. Store for extended periods.
7. Warm relatively rapidly by immersion in a heated water bath.

Table 1 Spectrum of Various Types of Living Cells and Tissues Commonly Stored by Freezing (As of 1993)

Tissue	Comments	References
Blood vessels	DMSO used for CPA; cooling rate < 1 °C/min.	Gottlob *et al.* 1982
Bone marrow stem cells	DMSO is usual CPA; widely used in cancer therapy.	McGann *et al.* 1981
Cornea	DMSO is usual CPA.	Armitage 1991
Erythrocytes	Usual CPA is glycerol; high concentrations for slow cooling and low concentrations for rapid cooling; wide spread clinical use.	Turner 1970, Valeri 1976, AATB 1985, Huggins 1985
Embryos		
Mouse	Many species of mammalian embryos have been cryopreserved successfully. The most common CPAs for these applications are glycerol and DMSO. 1,2-Propanediol is used for humans. Variations in the required thermal protocol and processing steps exist among the different species. Ice crystal nucleation is usually controlled by seeding.	CIBA 1977, Zeilmaker 1981
Rat		Whittingham *et al.* 1972, Wilmut 1972
Goat		Whittingham 1975
Sheep		Bilton and Moore 1976
Rabbit		Willadsen *et al.* 1976
Bovine		Bank and Maurer 1974
Drosophila		Wilmut and Rowson 1973
Human		Mazur *et al.* 1992
		Troundson and Mohr 1983
Heart valves	DMSO is usual CPA; cooling rate of 1-2 °C/min.	Angell *et al.* 1987
Hepatocytes	DMSO is usual CPA; cooling rate of 2.5 °C/min.	Fuller and Woods 1987
Islets	DMSO used for CPA; cooling rate < 1 °C/min.	Rajotte *et al.* 1983, Taylor and Benton 1987
Lymphocytes	DMSO is usual CPA; primary application is in clinical testing.	Knight 1980, Scheiwe *et al.* 1981
Microorganisms	DMSO and glycerol used for CPAs.	James 1987
Oocytes		
Hamster	Many species of mammalian oocytes have been cryopreserved successfully. The most common CPAS are glycerol and DMSO. Variations in the required thermal protocol and processing steps exist among the different species. Clinical applications in humans have been difficult to achieve.	Bernard 1991
Mouse		Critser *et al.* 1986
Primate		Whittingham 1977
Rabbit		DeMayo *et al.* 1985
Rat		Diedrick *et al.* 1986
Human		Kasai *et al.* 1979
		Van Uem *et al.* 1987
Parathyroid	DMSO used for CPA; cooling rate of 1 °C/min.	Wells *et al.* 1977
Periosteum	DMSO used for CPA; cooling rate of 1 °C/min.	Kreder *et al.* 1993
Plants	Selected plants are cold hardy; some germplasm is cryopreserved.	Grout 1987, Withers 1987
Platelets	Best success is with DMSO as CPA; high sensitivity to freezing and osmotic injury.	Schiffer *et al.* 1985, Sputtek and Körber 1991
Skin	Both glycerol and DMSO used as CPA.	Aggarwal *et al.* 1985
Sperm		
Animal	First mammalian cells frozen successfully. Broad applications for animals and humans using glycerol as CPA.	Polge 1980
Human		Sherman 1973

8. Serially dilute to remove CPA using nonpenetrating solutes to control the intracellular/extracellular osmotic balance.
9. Harvest the specimen for its intended application.

Details vary among individual tissue types; the references in Table 1 give sources of specific parameter values for individual tissues. The refrigeration requirements may vary considerably among different tissues, but basic principles and processes of the cryopreservation processes are generally consistent. The bibliography and the references in Table 1 identify appropriate introductory references.

Most of the initial applications of cryobiology were in clinical, research, and nonprofit (*e.g.*, the Red Cross) venues. However, the commercial sector has recently adopted cryopreservation methods (McNally and McCaa 1988). As the arsenal of practical cryopreservation methods has grown, the profit potential of freezing tissues for prolonged storage is being recognized and exploited. Thus, an added set of incentives and motives is driving the development of techniques which make use of challenging refrigeration schemes.

Preservation of Biological Materials by Freeze Drying

Freeze drying extends long-term storage at ambient temperatures without the threat of product deterioration. This process removes water from the specimen while it is frozen by sublimation. As a result, no thawing process occurs during rewarming. Thus, none of the decay processes associated with the presence of water in the liquid state are active. Freeze drying has been applied widely in the food processing industry where the product need not be rehydrated in the living state. Other applications, such as taxidermy, also avoid this stringent requirement. The list of biological materials that are frequently processed by freeze drying is extensive and encompasses various microorganisms, protein solutions, pharmaceuticals, and bone. Rowe (1970) reviewed the early state-of-the-art of the physical and engineering aspects of freeze drying; more recently, Franks (1990) has reviewed the physical and chemical principles that govern the freeze drying process from the perspective of achieving an optimal process design.

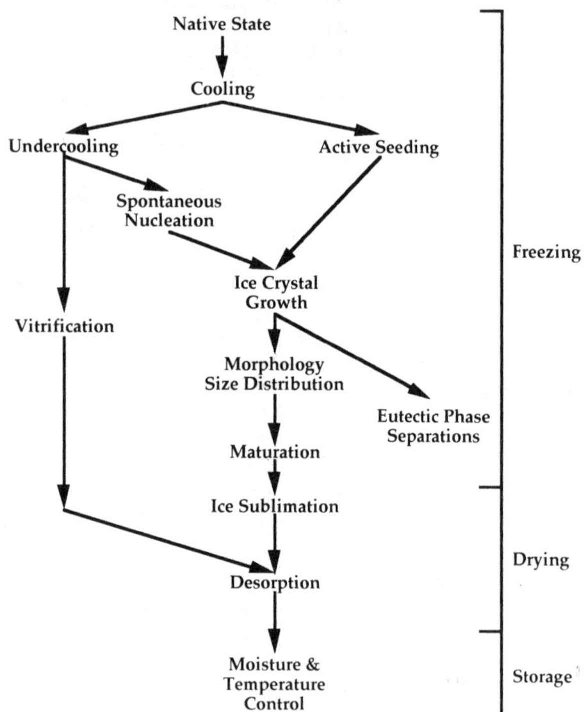

Fig. 3 Key Steps in Freeze Drying Process

Franks (1990) characterized the processing steps for freeze drying, and these are summarized in Figure 3. In the figure, note the alternate process pathways from the native to the stored state. The path can be controlled by equipment design and operator intervention. The material shown initially is in the native state from which cooling is initiated. As subfreezing temperatures are reached, either the material remains subcooled in the liquid state or ice crystals form either by spontaneous nucleation or by active seeding a substrate on which a solid phase may form. The material will vitrify if sufficiently subcooled. Ice crystals will grow in a nucleated material, with the rate of temperature change determining the structure and size distribution. Simultaneously, the solute becomes concentrated until the eutectic state is reached. At this state an additional solid phase may form, or the liquid solution may become supersaturated as the temperature is reduced further.

The material is dried, either in the crystalline or vitreous state by drawing a vacuum on the system at low temperature. Finally, the dried material may be stored at ambient temperatures, although the material is often stored at high subfreezing temperatures to minimize the probability of product deterioration by the activity of residual water. Production methods have been developed mainly by empirical experience and art. However, Franks (1990) argues for the need of a stronger scientific base to more effectively increase the process productivity and quality.

During freeze drying, many phase transitions either never occur or are precipitated at states far from equilibrium, and the slow kinetics of subsequent diffusion processes at low temperatures limit the system from moving toward equilibrium. Even if water crystallizes, the residual solutes are likely to never crystallize fully, if at all. As a supersaturated solution is cooled, the viscosity becomes so large that crystallization processes become undetectable. The so called glass transition temperature is defined by the intersection of the liquidus curve on the phase diagram and an isoviscosity curve for which the mechanical properties of the material are glass-like. These states are illustrated on a state diagram, as shown in Figure 4. By definition, the state diagram does not represent a locus of the system's equilibrium states, but it provides a map of the temperature and composition combinations of defined kinetic properties (Franks 1985).

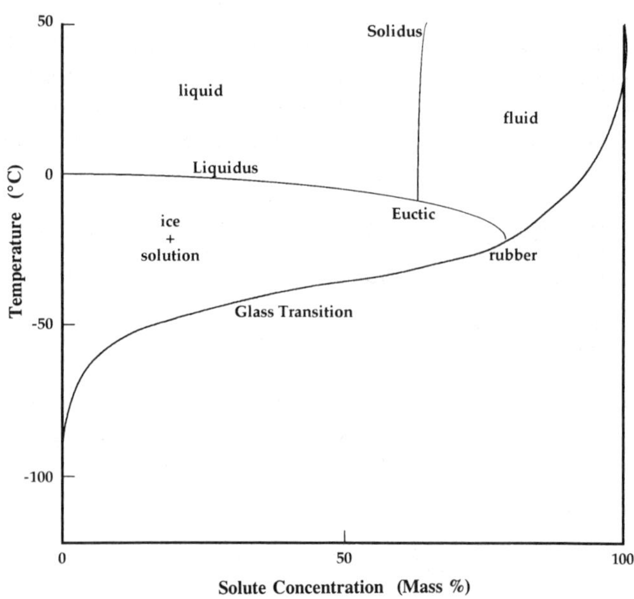

Fig. 4 Generic State Diagram for Aqueous Solution

The intersection of the liquidus curve, when extended past the eutectic temperature t_e into the supersaturated region, and the glass transition curve defines a specific glass transition temperature t_g. This property depends on the combination of the solute concentration and composition. At states above the glass transition threshold, the material behaves as a viscoelastic medium, which is unacceptable for long-term storage. An important aspect of the state diagram is that the slope of the glass transition curve is very steep at high concentrations of solute. As a consequence, the glass transition temperature t_g is well above 0 °C for a pure solute, thus providing for stable storage conditions. However, since small amounts of residual water in the system can significantly lower the glass transition temperature, it is important to check and control the moisture content of a freeze dried product. To this end, Levine and Slade (1988) provided extensive data on the glass transition temperature and unfreezable water fraction of many molecular solutions of interest in the design of freeze drying processes. It is most important that the water is removed from the material at a state temperature lower than the glass transition value at the local solute concentration value. If salts do not precipitate into a solid phase, then unfrozen water remains, which can affect the freeze-drying kinetics (Murase *et al.* 1991).

Preservation of Biological Materials by Vitrification

The vitrified state has been noted as playing an integral, albeit partial and secondary, role in the preservation of tissues by freezing and by freeze drying. Vitrification may also be used as a storage technique in its own right. Since solidification is avoided, problems associated with the freezing concentration of solutes are avoided. In addition, there are no complications due to latent heat removal from the specimen at a moving phase front. However, water cannot achieve a vitreous state simply by cooling a solution of physiological composition. Therefore, vitrification is achieved only with the prior modification of the specimen by adding high concentrations of solutes (*i.e.*, CPAs) to alter the kinetics of the crystallization process and the locus of the liquidus curve and the glass transition curve on the state diagram. Although the CPA concentration that must be achieved prior to cooling is higher for vitrification than for freezing, the vitrification process produces no subsequent solution concentration phenomenon as does freezing. Therefore, if the higher initial CPA concentrations can be tolerated without injury at above 0 °C where the addition occurs, vitrification may present a distinct benefit as an approach to long-term cryopreservation.

Fahy (1988) presents an extensive summary of various constitutive properties of candidate CPAs for vitrification, as well as empirical data for the crystallization properties of solutes in aqueous solutions. Other important sources of data for the design of vitrification processes are publications of Boutron, who has measured the thermal and glass forming properties of solutions of particular relevance to cryopreservation. For example, Boutron (1993) deals with the glass forming tendency and stability of the amorphous (glassy) state of 2,3—butanediol in physiological solutions of varying chemical complexity.

In addition to modifying a system to be cryopreserved by adding a CPA, Fahy *et al.* (1984) has explored modifying the state behavior of tissues by cooling under high pressures. Pressures of up to 100 MPa were used during cooling to reduce the melting temperature of water to about −9 °C and the homogeneous nucleation temperature to −54 °C, which is equivalent to the reduction in phase change state achieved by introducing a 3 M concentration of a common CPA. However, limiting factors associated with thermodynamic properties and design of apparatus must be solved before this technique can be considered for practical applications.

The growth of submicroscopic (light) ice crystals, primarily during warming, has been hypothesized to be injurous to vitrified cells and tissues. Several approaches have been pursued to control this process. Rapid warming through the region of sensitive temperatures where crystal nucleation and growth are most probable is used to reduce the time of exposure to these processes (*e.g.*, Marsland 1987). Problems with this technique have included insuring a homogeneous temperature throughout the tissue and matching the hardware to the impedance properties of the specimen, especially for large organs composed of heterogeneous tissues. Alternatively, Rubinsky *et al.* (1992) have used biological antifreezes from polar fishes (which adsorb to specific faces of ice crystals to inhibit crystal growth) as a CPA constituent to reduce the susceptibility of mammalian tissues to injury. Accordingly, antifreeze glycopeptides have been added to the vitrifying solution to increase the post thaw viability of vitrified porcine oocytes and embryos.

Vitrification of tissues has been most successful with physically small specimens. Freezing cryopreservation has also been most successful with specimens having this same characteristic, *e.g.*, suspensions of isolated cells and small multicellular tissues. One of the major advantages anticipated for vitrification techniques is in processing whole organs for cryopreservation. To date this potential has not been realized, due in part to difficulty in solving engineering problems associated with the processing. The specimen must cool rapidly throughout to prevent significant numbers of ice crystals from forming in any portion of the tissue volume, which could then later propagate into other areas. Unfortunately, boundary conditions and heat transfer characteristics of relatively large organs do not allow such rapid cooling. The threshold cooling rates can be altered as a function of the tissue's chemical composition by adding a CPA; that is, the most promising approach to resolving this limitation is likely to be chemical rather than thermal. Nonetheless, more effective control of the thermal boundary conditions could be beneficial.

The cooling process also produces a second problem, in direct conflict with satisfying the threshold cooling rate requirement. As progressively larger temperature gradients are created within the specimen in order to boost the cooling rate, corresponding internal thermal stresses are generated. In the glassy state the elastic strength of the vitrified tissue can easily be exceeded, causing mechanical fracture of the tissue to be preserved (Fahy 1990). This phenomenon is obviously irreversible and totally unacceptable. Thus, cooling must be designed to reduce the temperature fast enough to avoid ice nucleation but slow enough to avoid mechanical fracture. Fortunately, some possible solutions to this quandary have been tested (*i.e.*, annealing stages at appropriate thermal states) and hold promise for vitrification of large organs.

Preservation of Biological Materials by Undercooling

An option for cryopreservation in the undercooled state has found a limited range of applications. This technique avoids heterogeneous nucleation of ice crystals in subcooled water and maintains the storage temperature above the value at which homogeneous nucleation occurs (Franks 1988).

The undercooling method of storage is based on the fact that aqueous solutions can be cooled to temperatures substantially below the equilibrium phase change state without the nucleation of ice crystals. The temperature of spontaneous homogeneous nucleation for pure water is approximately −40 °C. Thus, if externally induced heterogeneous nucleation can be blocked, a substantial window of subzero temperatures can be used for storage of biological materials. This approach avoids the injurious effects of ice formation and the freeze concentration of solutes as well as the need to add and remove chemical CPAs to the specimen; although the temperature range available is not low enough to insure long-term storage without product deterioration. Because the physical basis of the undercooling process is much different from the alternate methods described previously, the strategy for developing effective storage is also substantially different.

The key to undercooled storage is the ability to control (prohibit) the nucleation of ice in the specimen. Although the homogeneous nucleation temperature is about 40 °C below the equilibrium freezing state, in practice it is difficult to reach even −20 °C owing to heterogeneous nucleation by particulate matter in the specimen. Further, the presence of just a single ice crystal nucleus is adequate to feed the growth of ice throughout a large volume of aqueous medium. However, because heterogeneous nucleation occurs in the extracellular subvolume of a cell suspension, Franks (1983) suspended the biological material in a medium of innocuous oil formed into microdroplets, thereby dispersing the bulk aqueous suspending solution. In effect, the material, such as cells, was suspended in a very thin film of aqueous solution, which dramatically depresses the ice nucleating tendency of the extracellular matrix. By this method, living cells may be undercooled to nearly the homogeneous nucleation temperature (Franks *et al.* 1983). Subsequently, many different types of cells have been undercooled in water-in-oil dispersions to −20 °C or lower without injury (Mathias *et al.* 1985).

A similar approach has been developed for the storage of biochemicals. For example, an aqueous protein solution can be dispersed in an oil carrier formulated to form a gel, thereby trapping the biological material in very small isolated droplets in the inert matrix. Each of the microdroplets is unable to communicate with any neighboring droplets, thus preventing local ice nuclei from providing a substrate for ice growth in the material. Challenges of this process involve creating microdroplet dispersions for effective storage that recover when returned to ambient temperatures. The temperature must be precisely controlled to avoid both homogeneous nucleation by becoming too cold and accelerated product deterioration by becoming too warm. Typical storage temperatures are in the range of −20 °C.

RESEARCH APPLICATIONS

Electron Microscopy Specimen Preparation

Freezing has been adopted widely as a method of preparing specimens for examination by electron microscopy. The advantages of freezing are that it need not involve chemical modification of the specimen in the active liquid state and that the physical substructure of components may be preserved. Conversely, the cooling process may cause ice crystals to form, which would alter or mask the structure to be imaged and which could concentrate the solute locally and cause internal osmotic flows that would produce image artifacts. Thus, control of the thermal history during cooling is critical in obtaining a high quality preparation for viewing on the microscope. Cooling rates of 10^5 to 10^6 °C/s or higher are desirable to minimize osmotic dehydration of cells and to avoid ice crystal nucleation and growth. Cooling removes heat from the surface of the specimen and in most cases the highest cooling rates occur at the boundary of the specimen. Thus the quality of preparation may vary significantly as a function of position, so the specimen should be mounted so that the dimension normal to the primary direction of heat transfer is as small as possible.

Bald (1987) analyzed factors that govern the cooling process during the cryopreparation of specimens. In each case, the objective is to cool the specimen as rapidly as possible. Three different approaches have been developed for cryofixation; these are classified as slamming, plunging, and spraying. Cooling by slamming is effected by mechanically driving the specimen and its mounting holder onto the surface of a cryogenically refrigerated solid block, which has a large thermal inertia in comparison with the specimen. The impact velocity of the specimen against the cold block is high to achieve as rapid a change in the thermal boundary conditions as possible. The drive mechanism is spring loaded to maintain continuous contact with the block following impact so that thermal resistance to the specimen is minimized.

The plunging technique uses a liquid rather than a solid refrigeration sink. As in slamming, the specimen is driven into a relatively large volume of cryogenic liquid. In common practice, the liquid is prepared in a subcooled or supercritical state so that heat transport from the specimen is not limited by a boiling boundary layer at the interface (Bald 1984). It is also important to eliminate a stratified layer of chilled vapor above the liquid through which the specimen would pass during plunging. Such a vapor layer would cool the specimen somewhat before contact with the liquid cryogen in the vapor medium, but because it has a relatively low convective coefficient the effective cooling rate is substantially reduced.

For the spraying method of cryofixation the specimen is held in a stationary mount, and a jet of liquid cryogen is directed onto the specimen. Heat is removed by a combination of evaporation and convection of the cryogen.

Analysis by Bald (1987) indicated that slamming is potentially the most effective method of rapid cooling for cryofixation. The velocity of the specimen during plunging must be 20 m/s or greater to reach thermal performance levels characteristic of slamming. In general, it is easier to design apparatus to achieve the velocities required for satisfactory performance by spraying than plunging. Further, high plunge velocities are more likely to damage the specimen than are equivalent spray velocities. The most effective cryogen for both plunging and spraying is subcritical ethane.

After the temperature is reduced, further preparation for viewing on the electron microscope may involve mechanical fracture of the specimen, chemical substitution of one constituent such as water (Hunt 1984), or removal of a chemical constituent such as by vacuum sublimation of water (Linner and Livesey 1988, Livesey and Linner 1988, Echlin 1992). Sectioning and fracturing techniques are used to expose internal structure and constituents of a specimen. This approach to preparation is particularly appropriate at cryogenic temperatures since biological materials become quite brittle and very little plastic deformation occurs that would alter the morphology. The exposed internal surfaces may be either imaged directly or modified by mechanical or chemical means. Echlin (1992) presented a comprehensive summary of the cryoprocessing of materials for electron microscopy.

Cryomicroscopy

Initial investigations that made use of cryomicroscopy were conducted in the early 1800s and have been pursued ever since. From its earliest adoption, cryomicroscopy made it possible to obtain useful information about the behavior of living tissues at subfreezing temperatures, but the rigor and breadth of application has been limited primarily by the difficulty in controlling the refrigeration applied to the specimen.

Diller and Cravalho (1970) designed a cryomicroscope in which independently controlled refrigerating and heating sources controlled the specimen temperature and its time rate of change during both cooling and heating. Heating was produced by applying a variable voltage across a transparent, electrically resistive thick film coating deposited on the underside of a glass plate on which the biological specimen was mounted. The local temperature was monitored via a microthermocouple positioned in direct contact with the specimen, and this signal was applied as the input to the electronic control system. By miniaturizing the thermal masses of all components of the system, much higher rates of temperature change were achieved with this system than previously possible (cooling rates approaching 10^5 °C/min). This system was cooled by circulating a chilled refrigerant fluid through a closed chamber directly beneath the plate on which the specimen was mounted.

Subsequently, the design was modified by McGrath *et al.* (1976) to eliminate the flow of refrigerant fluid from passing through the optical path of the microscope. Rather, heat was conducted away from the specimen via a thin radial plate that was chilled on its periphery by a refrigerant. This design is mechanically more satisfactory and offers a thinner working cross-section through the optical path, but the lateral temperature gradients are much higher. These two designs are known as convection and conduction cryomicroscopes, respectively (Diller 1988). The former has been adapted to allow for the simultaneous alteration of both the chemical and thermal environments of the specimen (Walcerz and Diller 1990), and the latter has been commercially marketed with a computer control system (McGrath 1987).

The operation of these cryomicroscopes is based on modulating the temperature over time at the site where the specimen is mounted on the microscope to create the desired thermal history for an experimental trial. The dimensions of the specimen are limited by the field of view of the microscope optics, since the specimen is stationary during a trial. An alternative approach has been adapted to study the control of a different set of variables. In this system a steady state temperature gradient is established across the viewing area of the microscope, and the specimen is moved in time through the gradient to produce the desired temperature history (Rubinsky 1984, Körber 1988). Advantages of this system are that specimens of macroscopic size may be frozen, as has been adapted to controlled thermal preparation of specimens for electron microscopy (Bischoff *et al.* 1990), and the cooling rate applied to a specimen can be investigated as defined by the product of the spatial temperature gradient and the velocity of advance of the phase interface (Beckmann *et al.* 1990). A similar gradient stage was built by Koroush and Diller (1983) for analysis of solidification processes. This system included feedback control of the temperatures at the ends of the gradient to view a stationary specimen. Gradient cryomicroscopes have seen little application for cryobiology owing to the limited range of cooling rates that can be achieved.

Cryomicrotome

The refrigerated microtome maintains tissue specimens at a subfreezing temperature in a mechanically rigid state so that very thin sections may be cut in preparation for viewing by electron microscopy. The degree of rigidity required is a function of the thickness of the specimen to be cut; thinner sections require greater rigidity, which is achieved by lower temperatures. Stumpf and Roth (1965) have determined that temperatures above $-30\,^{\circ}\mathrm{C}$ are adequate to obtain sections 1 μm thick while temperatures below $-70\,^{\circ}\mathrm{C}$ facilitate the cutting of sections thinner than 1 μm. Thus, the apparatus must produce both a wide range of temperatures and accurate thermal control during the processing. The apparatus must also be designed to exclude environmental moisture that could contaminate the specimen, and to isolate the refrigeration apparatus from the sectioning chamber in order to minimize mechanical vibrations that could compromise the dimensional integrity of the delicate cutting process.

CLINICAL APPLICATIONS

Hypothermia

Although accidental hypothermia is the most widely encountered clinical condition of lowered body core temperature, induced hypothermia has been developed as a method of reducing the rate of metabolism of selected organs, such as the heart and brain, during surgical procedures. This procedure is of particular benefit in neonatal patients in which the blood vessels and surgical field are too small to effectively apply standard cardiac bypass procedures for maintaining peripheral circulation during surgery. If the temperature can be reduced to a suitably low level (12 to 20 °C), then it is possible to stop the heart and to pursue surgical procedures in the absence of blood perfusion without incurring irreversible injury. The period for which the body can be subjected to the absence of perfused oxygenated blood is a function of the hypothermic temperature, and may last as long as an hour. These procedures require (1) the temperature of the organ to be within tolerances that limit tissue damage and (2) the ability to quickly lower and raise the temperature to provide the maximum fraction of the low temperature period for the surgical procedure. For example, Eberhart addressed the challenge of achieving a suitably rapid rate of cooling for the brain by perfusion through the vascular network with a chilled solution (Olsen *et al.* 1985).

The most effective approach to cooling an internal organ is to circulate the blood through a heat exchanger outside the body. The blood is then perfused through the vascular system of the organ, which acts as a physiological heat exchanger. Weinbaum and Jiji (1989) demonstrated the efficacy of thermal equilibration between various components of the vascular tree and the local embedding tissue. Earlier procedures relied primarily on surface cooling to chill internal organs, which is significantly less effective than perfusion in most applications. The results of Olsen *et al.* (1985) indicate that the brain can be very rapidly cooled to a hypothermic state by an infusion of cold arterial blood. However, when blood circulation was stopped for cardiac surgical procedures, a gradual, but significant rewarming of the brain occurred due to parasitic heat flow from surrounding structures that had not been cooled. Thus, a combination of cold perfusion through the vascular system and surface cooling seems to provide the best control of the body core temperature during hypothermic surgical procedures.

Cryosurgery

In contrast with the previous applications in which the objective is to maximize the survival of tissues exposed to freezing and thawing, cryosurgery has the goal of selective total destruction of a targeted area of tissue within the body. Cryosurgery is applied to destroy and/or excise tissue that is either dead or diseased. It is usually one of several treatment alternatives and has risen and fallen in favor as a method of treating various types of lesions. In general, it has been most effective in treating lesions for which there is direct or easy external access to allow mechanical placement of a cryoprobe or the spray of a cryogenic fluid. The most commonly accepted uses of cryosurgery include the treatment of skin, mucosal and gynecological lesions, liver cancer, and in cardiac surgery for treatment of tachyarrhythmias (Gage 1992). Other uses that have demonstrated efficacy but not such broad adoption are the treatment of hemorrhoids, oral, prostate and anorectal cancer, bone tumors, vertigo, retinal detachment, and visceral tumors (Gage 1992).

Primary advantages of cryosurgery are that (1) it provides a bloodless approach to surgery, (2) in some applications it reduces the rate of death, and (3) the extent of destruction inside the affected area can be imaged with noninvasive methods (Gilbert *et al.* 1985). This latter process makes use of a continuous ultrasonic scan of the freezing zone to monitor the interface between the solid and liquid phases as it grows into the targeted tissue. Experimental evidence indicates that a close correlation exists between the extent of phase interface propagation and the boundary of the zone of tissue destruction (Rubinsky *et al.* 1990), and these results may be explained in large part by a model for the mechanism of destruction of the freezing process (Rubinsky and Pegg 1988). The model asserts that during freezing of tissue, ice forms preferentially in the vascular network and the ice also propagates through the vessels as the solidification front advances. The cells near the vascular network dehydrate due to osmotic stress, and this water then freezes in the vascular lumina. As a result, vessels may expand by as much as a factor of two (for electron micrographs see Rubinsky *et al.* 1990) causing irreversible injury. Thus, the primary action of freezing in destroying tissue during cryosurgery may be by rendering the vascular system nonfunctional rather than by causing direct cryoinjury. Without an active microcirculatory blood flow, the thawed tissue will die rapidly.

Tissue is best destroyed by using a true cryogenic fluid, which is most commonly liquid nitrogen at −196 °C. The size of the probe and the flow rate of cryogen through it determine the volume of tissue that may be frozen. For example, a 9-mm diameter probe will produce in tissue an ice ball with a diameter as large as 25 mm (Dilley *et al.* 1993). Frequently, tumors exceed the capacity of a single probe, but at present, commercial multiprobe cryosurgery systems do not exist. As a result, multiple systems are used, which are both hardware intensive and compromise control over the freezing process (Onik and Rubinsky 1988). Thus, opportunities exist to improve cryosurgical apparatus.

Recent innovations have included operating the refrigerant system under vacuum thereby creating liquid phase heat transfer with the active heat transfer surface of the probe, which has a considerably lower thermal resistance than a boiling interface (Baust *et al.* 1992). This approach to enhancing thermal performance is similar to that used to cool specimens for electron microscopy rapidly (Bald 1987).

Other problems in the design of cryosurgical equipment remain to be solved. For example, parasitic heat leakage along the probe stem to the cold tip extends the active surface capable of causing tissue damage away from the area designed for destruction. This leakage is particularly compromising to the surgical procedure for treating malignant diseases in locations other than on the body surface (Onik and Rubinsky 1988). The simple and convenient interchangeability of probe tips having various geometries and thermal capacities would enhance the flexibility of cryosurgical apparatus. Further, the increasing incidence of sexually transmitted diseases dictates the need for a cryosurgical probe that may either be effectively sterilized (Evans 1992) or be disposable (Baust 1993).

REFRIGERATION HARDWARE FOR CRYOBIOLOGICAL APPLICATIONS

In general, two classes of refrigeration sources have been adopted successfully to biological applications: vapor compression cycle cooling and boiling of liquid cryogens. Also two types of thermal performance standards may be required of these refrigeration sources. As indicated in the previous sections, the thermal history during cooling is very often a critical factor in determining the success of a cryobiological procedure. The refrigerating apparatus must achieve a critical cooling rate within a specimen and regulate the cooling rate within specified tolerances over a designated range of temperatures. If the refrigeration apparatus is designed for general applications, these criteria will be demanded for a large variety of procedures.

A second important performance standard is the minimum specimen temperature that can be maintained in the system. Many biological applications depend on continuously holding the specimen at a temperature below a value at which significant process kinetics may occur. Of most importance are (1) control of the nucleation of ice or other solid phases in vitrified materials, and (2) limitation of recrystallization of small ice crystals that form during cooling. Many cryopreservation procedures require that the specimen be warmed from the stored state as rapidly as possible to avoid the above phenomena for which the kinetics are most favorable at higher subfreezing temperatures. For long-term storage of biological materials, temperatures below −120 °C are generally considered to be safe from the effects of devitrification and crystal growth. This state pushes the limits of refrigeration that can be produced by mechanical means, however.

An example of a generic cooling, storage, and warming protocol for cryopreservation is shown in Figure 5. The protocol is divided into seven steps. The first (a) consists of adding a cryoprotective agent at a temperature slightly above freezing. This operation is usually executed with the specimen held in a constant temperature circulating bath. The mixing and osmotic equilibration process may occur in several serial steps and last for half an hour or longer. The specimen is then immersed into a second constant temperature bath held at a high subfreezing temperature (such as −10 °C). The cooling rate during this process (b) is uncontrolled, being governed by the inherent heat transfer characteristics of the container and the refrigerant fluid. This constant temperature holding period (c) enables nucleation of ice in the specimen at a predetermined thermodynamic state and provides time for release of the latent heat of fusion and for osmotic equilibration between the intracellular and extracellular volumes. Subsequently, the specimen is placed into a controlled rate refrigerator to execute process (d) during which the temperature is reduced at a rate that maintains a balance between an acceptable osmotic state of the cells and avoids intracellular ice formation. As discussed earlier, the absolute magnitude of this cooling rate depends on the properties of the subject cell, and it may vary over several orders of magnitude for different specimen types. When the specimen reaches a temperature where kinetic rate processes approach zero (*e.g.*, −80 °C), the specimen may be plunged (e) into a liquid nitrogen bath for long-term storage (f). Finally, the specimen is warmed and thawed by removing it from the refrigerator and immersing it directly in a water bath (g).

In practice, many variations exist on the cryopreservation scheme shown in Figure 5. One of the most frequent simplifications is to eliminate one or more of the steps (b through d). Whether or not such a simplification is acceptable depends on the sensitivity of the specimen to variations in the thermal history. This sensitivity is a function of the properties of the cells, the physical geometry of the specimen and its packaging for cryopreservation, and chemical modifications performed during step (a).

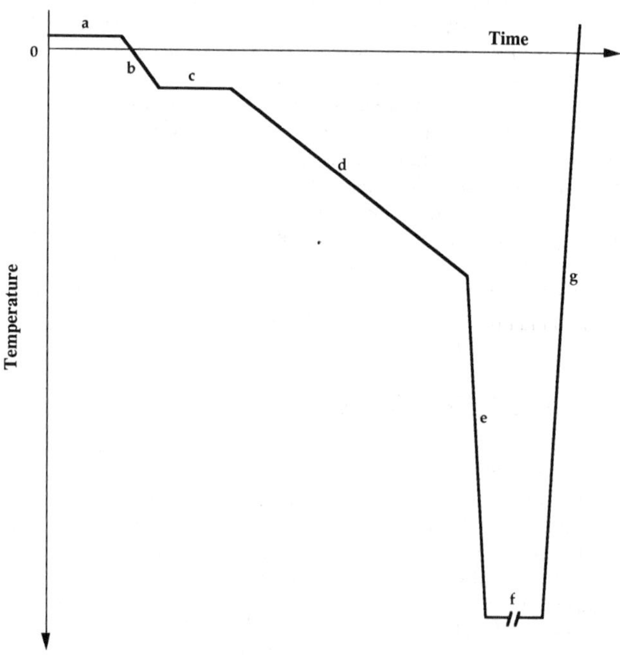

Fig. 5 Generic Thermal History for Example Cryopreservation Procedure

As the scientific basis for understanding and designing optimal protocols for processes in cryobiology has been strengthened, the specificity and sophistication of the associated refrigeration apparatus has likewise progressed. Therefore, considerable opportunity for improvements in cryobiology hardware remains. The decade of the 1980s witnessed the founding of many new commercial ventures with the objective of exploiting this potential. A common theme was an effective link to the scientific and/or medical community to insure that equipment was designed to address the needs of the customers.

REFERENCES

Aggarwal, S.J., C.R. Baxter, and K.R. Diller. 1985. Cryopreservation of skin: An assessment of current clinical applicability. *J. Burn Care Rehabil.* 6:469-76.

AABB. 1985. *Technical manual of the American Association of Blood Banks*. American Association of Blood Banks, Arlington, VA.

Angell, W.W., J.D. Angell, J.H. Oury, J.J. Lamberti, and T.M. Greld. Long-term follow-up of viable frozen aortic homografts. A viable homograft valve bank. *J. Thorac. Cardiovasc. Surg.* 93:815-22.

Armitage, W.J. 1991. Preservation of viable tissues for transplantation. In *Clinical applications of cryobiology*. B.J. Fuller and B.W.W. Grout, editors. CRC Press, Boca Raton, FL, 170-89.

Bald, W.B. 1984. The relative efficiency of cryogenic fluids used in the rapid quench cooling of biological samples. *J. Micros.* 134:261-70.

Bald, W.B. 1987. *Quantitative cryofixation*. Adam Hilger, Bristol, England.

Bank, H. and R.R. Maurer. 1974. Survival of frozen rabbit embryos. *Exp. Cell Res.* 89:188-96.

Baust, J.G. 1993. Cautions in cryosurgery. *Cryo-Letters* 14:1-2.

Baust, J.G., Z. Chang, and T.C. Hua. 1992. Emerging technology in cryosurgery. *Cryobiology* 29:777.

Beckmann, J., Ch. Körber, G. Rau, A. Hubel, and E.G. Cravalho. 1990. Redefining cooling rate in terms of ice front velocity and thermal gradient: First evidence of relevance to freezing injury of lymphocytes. *Cryobiology* 27:279-87.

Benson, S.W. 1982. *The foundation of chemical kinetics*. Robert E. Kreiger, Malabar, FL.

Bernard, A.G. 1991. Freeze preservation of mammalian reproductive cells. In *Clinical applications of cryobiology*. B.J. Fuller and B.W.W. Grout, editors. CRC Press, Boca Raton, FL, 149-68.

Bilton, F.J. and N.M. Moore. 1976. *In vitro* culture, storage and transfer of goat embryos. *Aust. J. Biol. Sci.* 29:125-29.

Bischoff, J., C.J. Hunt., B. Rubinsky, A. Burgess, and D.E. Pegg. 1990. Effects of cooling rate and glycerol concentration on the structure of the frozen kidney: Assessment by cryo-scanning electron microscopy. *Cryobiology* 27:301-10.

Boutron, P. 1993. Glass-forming tendency and stability of the amorphous state in solutions of a 2,3—butanediol containing mainly the levo and dextro isomers in water, buffer, and Euro-Collins. *Cryobiology* 30:86-97.

CIBA Foundation. 1977. *The freezing of mammalian embryos*. North Holland/Elsevier, Amsterdam, Holland.

Cocks, F.H., W.H. Hildebrandt, and M.L. Shepard. 1975. Comparison of the low temperature crystallization of glasses in the ternary systems H_2O-NaCl-dimethyl sulfoxide and H_2O-NaCl-glycerol. *J. Appl. Phys.* 46:359-72.

Critser, J.K., B.W. Arneson, D.V. Aaker, and G.D. Ball. 1986. Cryopreservation of hamster oocytes: Effects of vitrification or freezing on human sperm penetration of zona-free hamster oocytes. *Fertil. Steril.* 46:277-84.

DeMayo, F.J., R.G. Rawlins, and W.R. Dukelow. 1985. Xenogenous and *in vitro* fertilisation of frozen/thawed primate oocytes and blastomere separation of embryos. *Fertil. Steril.* 43:295-300.

Diedrick, K., S. al-Hasani, H. Van der Ven, and D. Krebs. 1986. Successful *in vitro* fertilisation of frozen thawed rabbit and human oocytes. *Journal of In Vitro Fertilization and Embryo Transplantation* 3:65.

Diller, K.R. Cryomicroscopy. In *Low temperature biotechnology: Emerging applications and engineering contributions*. J.J. McGrath and K.R. Diller, editors. ASME, New York, 347-62.

Diller, K.R. and E.G. Cravalho. 1970. A cryomicroscope for the study of freezing and thawing processes in biological cells. *Cryobiology* 7:191-99.

Diller, K.R., L.J. Hayes, and M.E. Crawford. 1985. Variation in thermal history during freezing with the pattern of latent heat evolution. *AIChE Symposium Series* 81:234-39.

Dilley, A.V., D.Y. Dy, A. Warlters, S. Copeland, A.E. Gillies, R.W. Morris, D.B. Gibb, T.A. Cook, and D.L. Morris. 1993. Laboratory and animal model evaluation of the Cryotech LCS 2000 in hepatic cryotherapy. *Cryobiology* 30:74-85.

Echlin, P. 1992. *Low-temperature microscopy and analysis*. Plenum Press, New York.

Evans, D.T.P. 1992. In search of an optimum method for the sterilization of a cryoprobe in a sexually transmittable diseases clinic. *Genitorin. Med.* 68:275-76.

Fahy, G.M. 1980. Analysis of "solution effects" injury: Equations for calculating phase diagram information for the ternary systems NaCl-dimethylsulfoxide-water and NaCl-glycerol-water. *Biophys. J.* 32:837-50.

Fahy, G.M. 1988. Vitrification. In *Low temperature biotechnology: Emerging applications and engineering contributions*. J.J. McGrath and K.R. Diller, editors. ASME, New York, 113-46.

Fahy, G.M., D.R. MacFarlane, C.A. Angell, and H.T. Meryman. 1984. Vitrification as an approach to cryopreservation. *Cryobiology* 21:407-26.

Fahy, G.M., J. Saur, and R.J. Williams. 1990. Physical problems with the vitrification of large biological systems. *Cryobiology* 27:465-71.

Franks, F. 1985. *Biophysics and biochemistry at low temperatures*. Cambridge University Press, Cambridge.

Franks, F. 1988. Storage in the undercooled state. In *Low temperature biotechnology: Emerging applications and engineering contributions*. J.J. McGrath and K.R. Diller, editors. ASME, New York, 107-12.

Franks, F. 1990. Freeze drying: From empiricism to predictability. *Cryo-Letters* 11:93-110.

Franks, F., S.F. Mathias, P. Galfre, S.D. Webster, and D. Brown. 1983. Ice nucleation and freezing in undercooled cells. *Cryobiology* 20:298-309.

Fuller, B.J. and R.J. Woods. 1987. Influence of cryopreservation on uptake of ^{99m} Tc Hida by isolated rat hepatocytes. *Cryo-Letters* 8:232-37.

Gage, A. 1992. Progress in cryosurgery. *Cryobiology* 29:300-4.

Gilbert, J.C., G.M. Onik, W.K. Hoddick, and B. Rubinsky. 1985. Real time ultrasonic monitoring of hepatic cryosurgery. *Cryobiology* 22:319-30.

Gottlob, R., L. Stockinger, and G.F. Gestring. 1982. Conservation of veins with preservation of viable endothelium. *J. Cardiovasc. Surg.* 23:109-16.

Grout, B.W.W. 1987. Higher plants at freezing temperatures. In *The effects of low temperatures on biological systems*. B.W.W. Grout and G.J. Morris, editors. Edward Arnold, London, 293-314.

Hartman, U., B. Nunner, Ch. Körber, and G. Rau. 1991. Where should the cooling rate be determined in an extended freezing sample? *Cryobiology* 28:115-30.

Hayes, L.J., K.R. Diller, H.J. Chang, and H.S. Lee. 1988. Prediction of local cooling rates and cell survival during the freezing of cylindrical specimens. *Cryobiology* 25:67-82.

Huggins, C.E. 1985. Preparation and usefulness of frozen blood. *Ann. Rev. Med.* 36:499-503.

Hunt, C.J. 1984. Studies on cellular structure and ice location in frozen organs and tissues: The use of freeze-substitution and related techniques. *Cryobiology* 21:385-402.

James, E. 1987. The preservation of organisms responsible for parasitic diseases. In *The effects of low temperatures on biological systems*. B.W.W. Grout and G.J. Morris, editors. Edward Arnold, London, 410-31.

Jochem, M. and Ch. Körber 1987. Extended phase diagrams for the ternary solutions H_2O-NaCl-glycerol and H_2O-NaCl-hydroxyethyl-starch (HES) determined by DSC. *Cryobiology* 24:513-36.

Kasai, M., A. Iritani, and B.C. Chang. 1979. Fertilisation *in vitro* of rat ovarian oocytes after freezing and thawing. *Biol. Reprod.* 21:839-44.

Körber, Ch. 1988. Phenomena at the advancing ice-liquid interface: Solutes, particles and biological cells. *Quart. Rev. Biophysics* 21:229-98.

Knight, S.C. 1980. Preservation of leukocytes. In *Low temperature preservation in medicine and biology*. M.J. Ashwood-Smith and J. Farrant, editors. University Park Press, Baltimore, 121-28.

Kreder, H.J., F.W. Keeley, and R. Salter. 1993. Cryopreservation of periosteum for transplantation. *Cryobiology* 30:107-12.

Levine, H. and L. Slade. 1988. Principles of "Cryostabilization" technology from structure/property relationships of carbohydrate/water systems. *Cryo-Letters* 9:21-63.

Lipton, J.M. 1985. Thermoregulation in pathological states. In *Heat transfer in biology and medicine: Volume 1 analysis and applications*. A. Shitzer and R.C. Eberhart, editors. Plenum Press, New York, 79-105.

Linner, J.G. and S.A. Livesey. 1988. Low temperature molecular distillation drying of cryofixed biological samples. In *Low temperature biotechnology: Emerging applications and engineering contributions*. J.J. McGrath and K.R. Diller, editors. ASME, New York, 117-58.

Livesey, S.A. and J.G. Linner. 1988. Cryofixation methods for electron microscopy. In *Low temperature biotechnology: Emerging applications and engineering contributions*. J.J. McGrath and K.R. Diller, editors. ASME, New York, 159-74.

Lovelock, J.E. 1954. The protective action by neutral solutes against haemolysis by freezing and thawing. *Biochem. J.* 56:265-70.

Marsland, T.P. 1987. The design of an electomagnetic rewarming system for cryopreserved tissue. In *The biophysics of organ cryopreservation*. D.E. Pegg and A.M. Karow, Jr., editors. Plenum Press, New York, 367-85.

Mathias, S.F., F. Franks, R.H.M. Hatley. 1985. Preservation of viable cells in the undercooled state. *Cryobiology* 22:537-46.

Mazur, P. 1963. Kinetics of water loss from cells at subzero temperature and the likelihood of intracellular freezing. *J. Gen Physiol.* 47:347-69.

Mazur, P. 1970. *Cryobiology*: The freezing of biological systems. *Science* 168:939-49.

Mazur, P. 1977. The role of intracellular freezing in the death of cells cooled at supraoptimal rates. *Cryobiology* 14:251-72.

Mazur, P. 1990. Equilibrium, quasi-equilibrium and nonequilibrium freezing of mammalian embryos. *Cell Biophysics* 17:53-92.

Mazur, P., K.W. Cole, J.W. Hall, P.D. Schreuders, and A.P. Mahowald. 1992. Cryobiological preservation of *Drosophila* embryos. *Science* 258:1932-35.

McGann, L.E., A.R. Turner, M.J. Allalunis, and J.M. Turc. 1981. Cryopreservation of human peripheral blood stem cells: Optimal cooling and warming conditions. *Cryobiology* 18:469-72.

McGrath, J.J., E.G. Cravalho, and C.E. Huggins. 1975. An experimental comparison of intracellular ice formation and freeze-thaw survival of hela S-3 cells. *Cryobiology* 12:540-50.

McGrath, J.J. 1987. Temperature-controlled cryogenic light microscopy—An introduction to cryomicroscopy. In *The effects of low temperatures on biological systems*. B.W.W. Grout and G.J. Morris, editors. Edward Arnold, London, 234-67.

McNally, R.T. and C. McCaa. 1988. Cryopreserved tissues for transplant. In *Low temperature biotechnology: Emerging applications and engineering contributions*. J.J. McGrath and K.R. Diller, editors. ASME, New York, 91-106.

Meryman, H.T. 1966. The interpretation of freezing rates in biological materials. *Cryobiology* 2:165-70.

Meryman, H.T., R.J. Williams, and M. St J. Douglas. 1977. Freezing injury from "Solution" effects and its prevention by natural or artificial cryoprotection. *Cryobiology* 14:287-302.

Murase, N., P. Echlin, and F. Franks. 1991. The structural states of freeze-concentrated and freeze-dried phosphates studied by scanning electron microscopy and differential scanning calorimetry. *Cryobiology* 28:364-75.

O'Brien, M.F., G. Stafford, M. Gardner, P. Pohlener, D. McGriffin, N. Johnston, A. Brosna, and P. Duffy. 1987. The viable cryopreserved Allograft aortic valve. *J. Cardiac. Surg.* 2:153-67.

Olson, R.W., L.J. Hayes, E.H. Wissler, H. Nikaidoh, and R.C. Eberhart. 1985. Influence of hypothermia and circulatory arrest on cerebral temperature distributions. *Trans. ASME, J. Biomech. Engr.* 107:354-60.

Onik, G. and B. Rubinsky. 1988. Cryosurgery: New developments in understanding and technique. In *Low temperature biotechnology: Emerging applications and engineering contributions*. J.J. McGrath and K.R. Diller, editors. ASME, New York, 57-80.

Polge, C. 1980. Freezing of spermatozoa. In *Low temperature preservation in medicine and biology*. M.J. Ashwood-Smith and J. Farrant, editors. University Park Press, Baltimore, 45-64.

Polge, C., A.U. Smith, and A.S. Parkes. 1949. Revival of spermatazoa after vitrification and dehydration at low temperatures. *Nature* (London) 49:666.

Prince, A. 1966. *Alloy phase equilibria*. Elsevier Publishing Company, Amsterdam, Holland.

Rajotte, R.V., G.L. Warnock, L.C. Bruch, and A.W. Procyshyn. 1983. Transplantation of cryopreserved and fresh rat islets and canine pancreatic fragments: Comparison of cryopreservation protocols. *Cryobiology* 20:169-84.

Rall, W.F. and G.M. Fahy. 1985. Ice free cryopreservation of mouse embryos at −196 °C by vitrification. *Nature* 313:573-75.

Rowe, T.W.G. 1970. Freeze-drying of biological materials: Some physical and engineering aspects. In *Current trends in cryobiology*. A.U. Smith, editor. Plenum Press, New York, 61-138.

Rubinsky, B., A. Arav, and A.L. DeVries. 1992. The cryoprotective effect of antifreeze glycopeptides from antarctic fishes. *Cryobiology* 29:69-79.

Rubinsky, B. and M. Ikeda. 1985. A cryomicroscope using directional solidification for the controlled freezing of biological material. *Cryobiology* 22:55-68.

Rubinsky, B., C.Y. Lee, L.C. Bastacky, and G. Onik. 1990. The process of freezing and the mechanism of damage during hepatic cryosurgery. *Cryobiology* 27:85-97.

Rubinsky, B. and D.E. Pegg. 1988. A mathematical model for the freezing process in biological tissue. *Proc. Royal Soc. London B* 234:343-58.

Scheiwe, M.W., Z. Pusztal-Markos, U. Essers, R. Seelis, G. Rau, Ch. Körber, K.H. Stürner, H. Jung, and B. Liedtke. 1981. Cryopreservation of human lymphocytes and stem cells (CFU-c) in large units for cancer therapy—A report based on the data of more than 400 frozen units. *Cryobiology* 18:344-56.

Schiffer, C.A., J. Aisner, and J.P. Dutcher. 1985. Platelet cryopreservation using dimethyl sulfoxide. *Ann. N.Y. Acad Sci.* 459:353-61.

Sherman, J.K. 1973. Synopsis of the use of frozen human semen since 1964: State of the art of human semen banking. *Fertil. Steril.* 24:397-412.

Sputtek, A. and Ch. Körber. 1991. Cryopreservation of red blood cells, platelets, lymphocytes, and stem cells. In *Clinical applications of cryobiology*. B.J. Fuller and B.W.W. Grout, editors. CRC Press, Boca Raton, FL, 95-147.

Stumpf, W.F. and L.J. Roth. 1965. Frozen sectioning below −60 °C with a refrigerated microtome. *Cryobiology* 1:227-32.

Taylor, M.J. and M.J. Benton. 1987. Interaction of cooling rate, warming rate and extent of permeation of cryoprotectant in determining survival of isolated rat islets of langerhans during cryopreservation. *Diabetes* 36:59-65.

Troundson, A. and L. Mohr. 1983. Human pregnancy following cryopreservation, thawing and transfer of an 8-cell embryo. *Nature* 305:707-9.

Turner, A.R. 1970. *Frozen blood—A review of the literature 1949-1968*. Gordon and Breach, London.

Valeri, C.R. 1976. *Blood banking and the use of frozen blood products*. CRC Press, Boca Raton, FL.

Van Uem, J.F.H.M., D.R. Siebzehnrueble, B. Schuh, R. Koch, S. Trotnow, and N. Lang. 1987. Birth after cryopreservation of unfertilized oocytes. *Lancet* 1:752-53.

Walcerz, D.B. and K.R. Diller. 1991. Quantitative light microscopy of combined perfusion and freezing processes. *J. Microscopy* 161:297-311.

Weinbaum, S. and L.M. Jiji. 1989. The matching of thermal fields surrounding countercurrent microvessels and the closure approximation in the Weinbaum—Jiji Equation. *Trans. ASME, J. Biomech. Engr.* 111:234-37.

Wells, S.A., J.C. Gunnells, R.A. Gutman, J.D. Shelburne, S.G. Schneider, and L.M. Sherwood. 1977. The successful transplantation of frozen parathyroid tissue in man. *Surgery* 81:86-91.

Whittingham, D.G. 1975. Survival of rat embryos after freezing and thawing. *J. Reprod. Fertil.* 43:575-78.

Whittingham, D.G. 1977. Fertilisation *in vitro* and development to term of unfertilised mouse oocytes previously stored at −196 °C. *J. Reprod. Fertil.* 49:89-94.

Whittingham, D.G., P. Mazur, and S.P. Leibo. 1972. Survival of mouse embryos frozen to −196 °C and −269 °C. *Science* 178:411-14.

Willadsen, S.M., C. Polge, L.E.A. Rowson, and R.M. Moor. 1976. Deep freezing of sheep embryos. *J. Reprod. Fertil.* 46:151-54.

Wilmut, I. 1972. The effect of cooling rate, cryoprotective agent and stage of development on survival of mouse embryos during freezing and thawing. *Life Sci* 11:1071-79.

Wilmut, I. and L.E.A. Rowson. 1973. Experiments on the low-temperature preservation of cow embryos. *Vet. Rec.* 93:686-90.

Withers, L.A. 1987. The low temperature preservation of plant cell, tissue and organ cultures and seed for genetic conservation and improved agricultural practice. In *The effects of low temperatures on biological systems.* B.W.W. Grout and G.J. Morris, editors. Edward Arnold, London, 389-409.

Zeilmaker, G. (editor) 1981. *Frozen storage of laboratory animals.* Gustav Fischer, Stuttgart.

BIBLIOGRAPHY

Primary literature: The main English language sources for general literature on cryobiology are found in two archival journals: *Cryobiology* and *Cryo-Letters.* In addition, the *Bulletin of the International Institute of Refrigeration* provides a timely listing of the world literature in low-temperature biology. Other references are distributed among a large number of journals that are either more general or are oriented toward specific physiological or applications areas.

Monographs: A number of monographs have been written on the principles and applications of low-temperature biology. In general, these have been edited works in which a number of contributing authors provide a series of expositions in focused areas of expertise. Over the last forty years, they have appeared rather consistently. A list of selected monographs follows:

Bald, W.B. 1987. *Quantitative cryofixation.* Adam Hilger, Bristol.

Diller, K.R. 1992. Modeling of bioheat transfer processes at high and low temperatures. In *Advances in heat transfer: Bioengineering heat transfer* 22. Y.I. Cho, editor. Academic Press, Boston, 157-357.

Franks, F. 1985. *Biophysics and biochemistry at low temperatures.* Cambridge University Press, Cambridge.

Fuller, B.J. and B.W.W. Grout (editors). 1991. *Clinical applications of cryobiology.* CRC Press, Boca Rotan, FL.

Grout, B.W.W. and G.J. Morris (editors). 1987. *The effects of low temperatures on biological systems.* Edward Arnold Publishers, Ltd, London.

McGrath, J.J. and K.R. Diller (editors). 1988. *Low temperature biotechnology: Emerging applications and engineering contributions.* ASME, New York.

Meryman, H.T. (editor). 1966. *Cryobiology.* Academic Press, New York.

Pegg, D.E. and A.M. Karrow (editors). 1987. *The biophysics of organ cryopreservation.* Plenum Press, New York.

Smith, A.U. 1961. *Biological effects of freezing and supercooling.* Williams and Wilkins, Baltimore.

Smith, A.U. (editor). 1970. *Current trends in cryobiology.* Plenum Press, New York.

CHAPTER 40

ABSORPTION COOLING, HEATING, AND REFRIGERATION EQUIPMENT

THIS chapter describes both the standard absorption cooling and heating units available and new concepts that are evolving. Absorption units offer several advantages. They reduce the use of CFC refrigerants and eliminate concerns about lubricants in refrigerants. In addition, particularly with direct-fired systems, they level the demand for natural gas year-round, which promotes efficient maintenance of pipeline systems and lowers costs.

Refrigeration with absorption is possible at evaporator temperatures ranging from 50°F to as low as −75°F with a variety of cycles and fluids. For heat pumps and heat transformers, evaporator fluid inlet temperatures can easily reach 212°F.

Absorption equipment, which is heat driven, can both cool and heat for human comfort and process control. Heat sources for driving an absorption unit include the following:

- Direct gas and oil firing
- Indirect steam heating from boilers
- Centrifugal compressor steam turbine exhaust
- Waste process steam
- Solar-heated or diesel or gas engine-heated hot water
- Cogeneration heat recovery steam or hot water
- Hot process fluids
- Heat recovery from process streams and flue gases

With larger absorption units, water is usually cooled by passing it through the tubes of the evaporator. This cooled water then flows through air coils to air condition spaces or through other exchangers to cool process fluids. The concept can be applied to directly cool any process fluid passed through the tubes of the absorption unit evaporator, as long as the materials of construction are compatible with the fluid and the temperatures required are attainable by the refrigerant. Smaller absorption units can be used to directly cool air or gases that are passed over finned evaporator coils.

DEFINITIONS

The following terms refer to the basic absorption cycle and equipment:

Solution. A liquid of at least two components, one soluble in the other. The combination is called the working fluid or *fluid pair* of an absorption system. Chapter 1 of the 1993 ASHRAE *Handbook—Fundamentals* discusses the characteristics of practical fluid pairs, and Chapter 17 lists thermodynamic property data for the two commonly used solutions, water-lithium bromide and ammonia-water.

Sorbent. The portion of the solution that transports the absorbed refrigerant through the processes that make up the thermochemical compressor.

The preparation of this chapter is assigned to TC 8.3, Absorption and Heat Operated Machines.

Refrigerant. A volatile substance that (1) leaves the solution at the desorber or generator, (2) performs the refrigeration process at an evaporator, and (3) is reabsorbed at the absorber to complete the thermodynamic cycle.

Performance additive. A surfactant (usually one of the octyl alcohols) that is added in minute quantities to the lithium bromide solution. It reduces surface tension and triggers a violent convection at the interface between refrigerant vapor and solution (the Marangoni Effect). The effect considerably enhances the rate of absorption of water vapor by the solution.

Coefficient of Performance (COP). For a chiller, the ratio of cooling capacity to required heat energy input. For a heat pump or heat transformer, the ratio of heat energy output capacity to the required heat energy input.

Thermochemical compressor. The portion of the absorption refrigeration cycle in which a heat source provides the energy to compress the refrigerant vapor, thus replacing the mechanically driven compressor in the vapor compression cycle.

Concentrated (strong) solution. Solution with relatively little refrigerant and a high concentration of sorbent.

Dilute (weak) solution. Solution with relatively large amounts of refrigerant and a low concentration of sorbent.

Absorption. The process in which refrigerant vapor is absorbed into a concentrated (strong) solution. The heat of condensation of the water and the heat of mixing are released into the fluid by the absorption process. The fluid must be cooled (usually with cooling tower water) to allow sufficient refrigerant to be continuously absorbed into solution while maintaining a low-pressure condition.

Desorption (generation). The process in which heat is added to a dilute (weak) solution to drive refrigerant vapor from the solution, thus concentrating the solution and permitting the refrigerant to be recycled for reuse at the evaporator.

Rectification. Thermally induced mass transfer in which vapor of a volatile sorbent is stripped from the refrigerant vapors driven from the dilute solution in the desorption (generation) process. Not all sorbents are sufficiently volatile to require rectification. For example, in a water-lithium bromide working fluid, water is the refrigerant, and lithium bromide is a nonvolatile sorbent that does not need rectification. However, in an ammonia-water working fluid, ammonia is the refrigerant, and water, the sorbent, is sufficiently volatile to require rectification.

Condensation. Liquefaction of the refrigerant vapor caused when it gives up its latent heat and any superheat (if present) to an available heat sink.

Evaporation. Vaporization of liquid refrigerant by heat supplied from a heat source, thus producing a cooling effect on the heat source.

Thermal recuperation. Heat exchange usually between strong and weak solutions within the cycle that improves the overall thermal efficiency.

Crystallization. The freezing or conversion of the pumpable liquid solution into a solid or slushy mass when it is excessively strong in sorbent at a given temperature. Not all solutions experience this problem; water-lithium bromide does but ammonia-water does not, except at very low temperatures.

P-T-x equilibrium diagram. The plot of equilibrium properties for a given solution relating pressure (P), temperature (T), and concentration (x).

Heat pump. A Type 1 heat pump is an absorption machine that uses heat from a high temperature source to elevate the quantity and temperature of heat from available sources and moves this energy to another location at an intermediate temperature. A Type 2 heat pump, which is usually called a heat transformer, is an absorption machine that substantially elevates the temperature of energy from available sources at reduced quantities, using the inherent latent heats and chemical heat of mixing for the fluids of the cycle.

BASIC CYCLES

Single-Effect Cycle for Cooling

Figure 1 shows the absorption cycle in two portions: a refrigeration portion and a thermochemical compressor portion, thus establishing a parallel with vapor compression refrigeration. A volatile liquid refrigerant, such as water or ammonia, which is suitable for the operating condition, evaporates in the evaporator vessel at a low pressure, thus producing cooling at a low temperature.

Instead of the refrigerant vapor being drawn into a compressor at low pressure and being physically compressed as in a mechanical refrigeration cycle, the refrigerant vapor is absorbed by a separate adjacent absorber into a solution that has had its vapor pressure reduced while it was being cooled. The heat of condensation and heat of mixing caused by the high chemical affinity of the solution for the refrigerant is removed and rejected to a heat sink by a second fluid such as water or ambient air to prevent the evaporator and absorber pressure from rising. Usually this heat is rejected by a water-cooled, an air-cooled, or an evaporative condenser.

The diluted sorbent solution is usually pumped to the higher temperature and pressure generator or desorber side of the system, where the solution can be reconcentrated to complete the cycle.

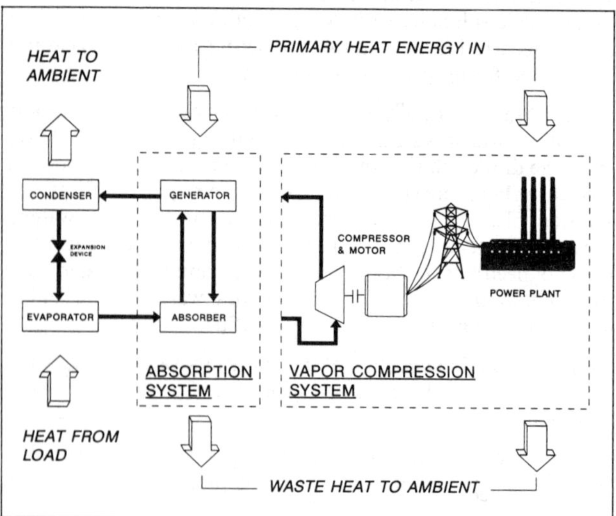

Fig. 1 Similarities between Absorption and Vapor Compression Systems

A recuperative energy exchange usually takes place between the hot concentrated solution leaving the generator and the cooler dilute solution being pumped from the absorber back to the generator.

Periodic Single-Effect Systems for Cooling

More than a century ago, the first application of absorption cooling did not have materials and pump seals capable of functioning in the continuous or circulated cycle described previously. The system used the extremely corrosive fluid pair of sulfuric acid (sorbent) and water. To avoid pumping, a batch process used only two major components: a vessel containing the solution and a heat-transfer vessel that would contain only refrigerant. The upper portions of the vessels were connected by a passage that permitted vapor to pass between them.

At first, in the activating mode of operation, only the solution vessel contained a fluid. Heat from an outside source drove refrigerant from the solution, the solution vessel acting like a generator. Vapor generation proceeded until the vapor pressure was sufficient to permit the refrigerant vapor to condense in the refrigerant vessel that was cooled by the ambient.

When sufficient condensed refrigerant collected in the refrigerant vessel, the activation mode was stopped by removing the high-temperature heat source. In the second mode, the solution vessel was recooled by the ambient, and the refrigerant vessel was connected to the cooling load. Cooling the solution vessel started an absorption process that dropped the vapor pressure so that the refrigerant liquid in the refrigerant vessel started boiling at the reduced vapor pressure to produce cooling.

Similar periodic cycles are currently being used with sorbents that cannot be pumped because they are solid. Usually more than one set of interconnected vessels is used to simulate a continuous process.

Heat Transformer

The single-effect absorption heat pump cycle is sometimes called a Type 1 heat pump, and a heat transformer is referred to as a Type 2 heat pump. The single-effect cycle involves sinks and sources at three temperature levels: the highest temperature is for the heat input to the generator, the intermediate temperature is for the heat rejection from both the condenser and the absorber (usually near ambient temperatures), and the lowest temperature is for the heat addition to the evaporator. For a cooling device, the useful result is the cooling of the load by the evaporator, and the rejection is to the ambient atmosphere.

For a Type 1 heat pump, the useful result is the rejection heat, which is transferred to the load being heated, and the atmospheric reference is the heat source for the evaporator. In both cases, the driving energy input is the heat to the generator. A COP (heating) of 1.3 to 1.8 is routinely achieved for respective temperature rises of 72 to 36°F with corresponding waste heat source temperatures of 104 to 194°F and driving heat input source temperatures of 290 to 304°F using water-lithium bromide machines.

In the Type 2 heat pump (heat transformer), the heat sink of a Type 1 system becomes a heat source. The object is to take heat at an intermediate temperature that has no value (because it is not hot enough) and generate a smaller amount of heat at a higher temperature that can be used. The pressure level of the evaporator/absorber pair is arranged to be higher than that of the condenser/generator pair. The highest temperature usable output is from the absorber, and the lowest temperature is the heat rejected to the ambient heat sink of the condenser. Heat at the intermediate temperature level is added to both the evaporator and the regenerator. In a simple heat transformer, the amount of heat produced at the higher temperature absorber is about half the heat added at the intermediate level evaporator and regenerator. Temperature rises of 45 to 90°F are easily achieved for processes with corresponding intermediate source temperatures of 175 to 200 °F using water-lithium bromide machines.

Multiple-Effect Systems for Cooling

Since the thermochemical compressor is governed by the limits of the Carnot cycle (defined by the absolute temperatures of the heat sources and the heat sinks), its efficiency may be improved by increasing the temperature of the generator. Constraints of system assemblies and the characteristics of the chosen working fluids determine the ability to make thermodynamic improvements.

In single-effect systems, the change in concentration and the difference between equilibrium temperatures of the boiling dilute and the outgoing strong solutions for the generator are small. The solution flow rates relative to the flow of refrigerant are large. Significant thermodynamic cycle losses are tied to any inefficiency of the weak solution/strong solution heat exchanger. Temperature approaches should be as small as practical.

The term *double effect* is generally restricted to systems in which a second generator/condenser pair is added to the system; its input is from the heat-driving energy source, and the heat rejection from its condenser is the heat input to the original generator. A given amount of external heat into the first-effect generator provides about 50 to 80% more cooling in a double-effect system than in a single-effect system.

The attractiveness of additional condenser/desorber stages is countered by limits of pressure, temperature, corrosion, increases in first costs, and the tendency to generate noncondensables. However, coupling into *triple-effect* systems with higher temperatures, new inhibitors, and new materials is being researched.

WATER-LITHIUM BROMIDE ABSORPTION TECHNOLOGY

Absorption equipment using water as the refrigerant and lithium bromide as the absorbent is classified by the method of heat input to the primary generator (the firing method) and whether the absorption cycle is single- or multiple-effect. Single- and double-effect absorption chillers are described in the Absorption Chillers section.

Machines using steam or hot liquids as a heat source are *indirect fired*, while those using direct combustion of fossil fuels as a heat source are *direct fired*. Machines using clean, hot waste gases as a heat source are also classified as indirect fired but are often referred to as *heat-recovery chillers*.

Components

Solution recuperative heat exchangers, also referred to as economizers, are typically shell-and-tube or plate heat exchangers. They transfer heat between hot and cold absorbent solution streams, thus recycling energy. The material of construction is mild steel or stainless steel.

Condensate subcooling heat exchangers, a variation of solution heat exchangers, are used on steam-fired double-effect machines. These heat exchangers use hot condensate from the primary generator to preheat the solution entering the generator.

Indirect-fired generators are usually of the shell-and-tube type, with the absorbent solution either flooded or sprayed outside the tubes, and the heat source (steam or hot fluid), inside the tubes. The absorbent solution boils outside the tubes, and the resulting intermediate or strong concentration absorbent solution flows from the generator through an outlet pipe. The refrigerant vapor evolved passes through a vapor/liquid separator consisting of baffles and eliminators and then flows to the condenser section. Ferrous materials are used for absorbent containment; copper, copper-nickel alloys, stainless steel, or titanium are used for the tube bundle.

Direct-fired generators consist of a fire-tube section, a flue-tube section, and a vapor/liquid separation section. The fire tube is typically a double-walled vessel with an inner cavity large enough to accommodate a radiant or open-flame fuel oil or natural gas burner. Dilute solution flows in the annulus between the inner and outer vessel walls and is heated by contact with the inner vessel wall. The flue-tube section is typically a tube- or plate-type heat exchanger connected directly to the fire tube.

Heated solution from the fire-tube section flows on one side of the heat exchanger, and flue gases flow on the other side. Hot flue gases further heat the absorbent solution and cause it to boil. The flue gases leave the generator, while the partially concentrated absorbent solution and refrigerant vapor mixture pass to a vapor and liquid separator chamber. This chamber separates the absorbent solution from the refrigerant vapor. Materials of construction are mild steel for the absorbent containment parts and stainless steel for the flue gas heat exchanger.

Secondary or second-stage generators are used only in double- or multistage machines. They are usually of the shell-and-tube type and operate similarly to indirect-fired generators of single-effect machines. The heat source, which is inside the tubes, is high-temperature refrigerant vapor from the primary generator shell. Materials of construction are mild steel for absorbent containment and usually copper or copper-nickel alloys for the tubes, although small double-effect direct-fired chiller-heaters may use mild steel tubes. Eliminators are stainless steel.

Evaporators are heat exchangers, usually of the shell-and-tube type, over which liquid refrigerant is dripped or sprayed and evaporated. The liquid to be cooled passes through the inside of the tubes. Evaporator tube bundles are usually copper or a copper-nickel alloy. Refrigerant containment parts are mild steel. Eliminators and drain pans are stainless steel.

Absorbers are tube bundles over which strong absorbent solution is sprayed or dripped in the presence of refrigerant vapor. The refrigerant vapor is absorbed into the absorbent solution, thus releasing heat of dilution and heat of condensation. This heat is removed by cooling water that flows through the tubes. Dilute absorbent solution leaves the bottom of the absorber tube bundle. Materials of construction are mild steel for the absorbent containment parts and copper or copper-nickel alloys for the tube bundle.

Condensers are tube bundles located in the refrigerant vapor space near the generator of a single-effect machine or the second-stage generator of a double-effect machine. The water-cooled tube bundle condenses refrigerant from the generator on the surface of the tubes. Materials of construction are mild steel for the refrigerant containment parts and copper for the tube bundle. For special waters, the condenser tubes can be copper-nickel, which will derate the unit.

High-stage condensers are found only in double-effect machines. This type of condenser is typically the inside of the tubes of the second-stage generator. Refrigerant vapor from the first-stage generator condenses inside the tubes, and the resulting heat is used to concentrate absorbent solution in the shell of the second-stage generator when heated by the outside surface of the tubes.

Pumps move absorbent solution and liquid refrigerant in the absorption machine. Pumps can be configured as individual (one motor, one impeller, one fluid stream) or combined (one motor, multiple impellers, multiple fluid streams). The motors and pumps are hermetic or semihermetic. Motors are cooled and bearings lubricated either by the fluid being pumped or by a filtered supply of liquid refrigerant. Impellers are typically brass, cast iron, or stainless steel; volutes are steel or impregnated cast iron, and bearings are babbitt-impregnated carbon journal bearings.

Refrigerant pumps (when used) recirculate liquid refrigerant from the refrigerant sump at the bottom of the evaporator to the evaporator tube bundle in order to effectively wet the outside surface and enhance heat transfer.

Dilute solution pumps take dilute concentration absorbent solution from the absorber sump and pump it to the generator.

Absorber spray pumps recirculate absorbent solution over the absorber tube bundle to assure adequate wetting of the absorber surfaces. These pumps are not found in all equipment designs. Some designs use a jet eductor for inducing concentrated solution flow to the absorber sprays.

Purge systems are required on lithium bromide absorption equipment to remove noncondensables (air) that leak into the machine or hydrogen (a product of corrosion) that is produced during equipment operation. Noncondensable gases, present even in small amounts, can cause reduction in chilling capacity and even lead to solution crystallization. Purge systems for larger sizes above 100 tons of refrigeration typically consist of these components:

- Vapor pickup tube(s) usually located at the bottom of large absorber tube bundles
- Noncondensable separation and storage tank(s), located in the absorber tube bundle or external to the absorber/evaporator vessel
- A vacuum pump or valving system using solution pump pressure to periodically remove noncondensables collected in the storage tank

Some variations include jet pumps (eductors), powered by pumped absorbent solution and placed downstream of the vapor pickup tubes to increase the volume of sampled vapor, and water-cooled absorbent chambers to remove water vapor from the purged gas stream.

Because of their size, smaller units have fewer leaks, which can be more easily detected during manufacture. As a result, small units may use variations of solution drip and entrapped vapor bubble pumps plus purge gas accumulator chambers.

Palladium cells, found in large direct-fired and small indirect-fired machines, continuously remove small amounts of hydrogen gas that is produced by corrosion. These devices operate on the principle that thin membranes of heated palladium are permeable to hydrogen gas only.

Corrosion inhibitors, typically lithium chromate, lithium nitrate, or lithium molybdate, protect machine internal parts from the corrosive effects of the absorbent solution in the presence of air. Each of these chemicals is used as a part of a corrosion control system. Acceptable levels of contaminants and the correct solution pH range must be present for these inhibitors to work properly. Solution pH is controlled by adding lithium hydroxide or hydrobromic acid.

Performance additives are used in most lithium bromide equipment to achieve design performance. The heat- and mass-transfer coefficients for the simultaneous absorption of water vapor and cooling of lithium bromide solution have relatively low values that must be enhanced. A typical additive is one of the octyl alcohols.

Refrigerant flow control between condensers and evaporators is typically achieved with orifices (suitable for high- or low-stage condensers) or liquid traps (suitable for low-stage condensers only).

Solution flow control between generators and absorbers is typically achieved with flow control valves (primary generator of double-effect machines), variable-speed solution pumps, or liquid traps.

Absorption Chillers

Figure 2 is a schematic diagram of a commercially available, single-effect indirect-fired liquid chiller, showing one of several configurations of the major components. During operation, heat is supplied to tubes of the *generator* in the form of a hot fluid or steam, causing dilute absorbent solution on the outside of the tubes to boil. This evolved refrigerant vapor (water vapor) flows through eliminators to the *condenser*, where it is condensed on the outside of tubes that are cooled by a flow of water from a heat sink (usually a cooling tower). Both the boiling and condensing processes take place in a vessel that has a common vapor space at a pressure of about 0.9 psia.

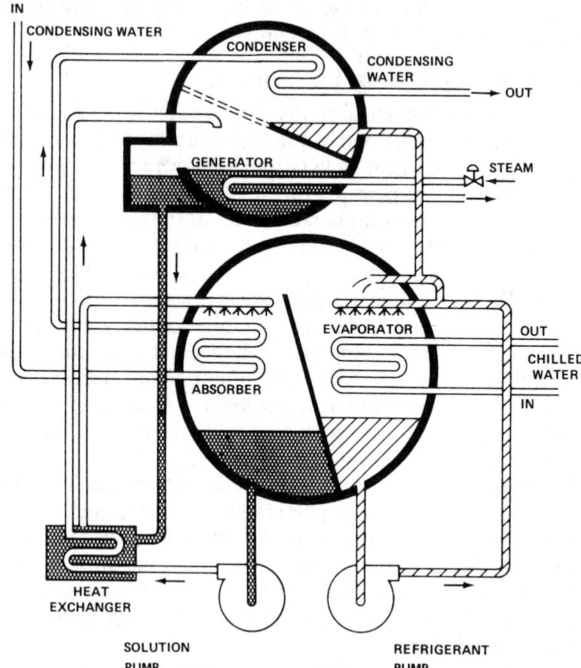

Fig. 2 Two-Shell Lithium Bromide Cycle Water Chiller

The condensed refrigerant passes through an orifice or liquid trap in the bottom of the condenser and enters the evaporator. In the *evaporator*, the liquid refrigerant boils as it contacts the outside surface of tubes that contain a flow of water from the heat load. In this process, the water in the tubes is cooled as it releases the heat required to boil the refrigerant. Refrigerant that does not boil is collected at the bottom of the evaporator, flows to a *refrigerant pump*, is pumped to a distribution system located above the evaporator tube bundle, and is sprayed over the evaporator tubes again.

The dilute absorbent solution that enters the generator increases in concentration (percentage of sorbent in the water) as it is boiled and releases water vapor. The resulting strong absorbent solution leaves the generator and flows through one side of a *solution heat exchanger* where it cools as it heats a stream of dilute absorbent solution passing through the other side of the solution heat exchanger on its way to the generator. This increases the efficiency of the machine by reducing the amount of heat that must be added to the dilute solution before it begins to boil in the generator.

The cooled, strong absorbent solution then flows (in some designs via a jet eductor) to a solution distribution system located above the *absorber tubes* and drips or is sprayed over the outside surface of the absorber tubes. The absorber and evaporator share a common vapor space. This allows refrigerant vapor, which is evolved in the evaporator, to be readily absorbed into the absorbent solution flowing over the absorber tubes. This absorption process releases heat of condensation and heat of dilution, which are removed by cooling water flowing through the absorber tubes. The resulting dilute absorbent solution flows off the absorber tubes and then to the absorber sump and *solution pump*. The pump and piping convey the dilute absorbent solution to the heat exchanger, where it accepts heat from the strong absorbent solution. From there, the dilute solution flows into the generator, thus completing the cycle. During operation, the pressure in the absorber/evaporator vapor space is about 0.1 psia.

These machines are typically fired with low-pressure steam or medium-temperature liquids. Several manufacturers have machines with capacities ranging from 50 to 1660 tons of refrigeration. Machines of capacities 5 to 10 tons are also available from international sources.

Performance characteristics	
Steam input pressure	9 to 12 psig
Steam consumption	18.3 to 18.7 lb/ton·h
Hot fluid input temperature	240 to 270°F, with as low as 190°F for some smaller machines
Heat input rate	18,100 to 18,500 Btu/ton·h, with as low as 17,100 Btu/ton·h for some smaller machines
Cooling water temperature in	85°F
Cooling water flow	3.6 gpm/ton, with up to 6.4 gpm/ton for some smaller machines
Chilled water temperature off	44°F
Chilled water flow	2.4 gpm/ton, with 2.6 gpm/ton for some smaller international machines
Electric power	0.01 to 0.04 kW/ton with a minimum of 0.004 kW/ton for some smaller machines

Physical characteristics	
Nominal capacities	50 to 1660 tons, with 5 to 10 tons for some smaller machines
Length	11 to 33 ft, with as low as 3 ft for some smaller machines
Width	5 to 10 ft, with 3 ft minimum for some smaller machines
Height	7 to 14 ft, with 6 ft for some smaller machines
Operating weight	11,000 to 115,000 lb, with 715 lb for some smaller machines

Typical COPs for large single-effect machines at standard ARI (American Refrigeration Institute) rating conditions are 0.7 to 0.8.

Figure 3 is the schematic diagram of a commercially available, double-effect indirect-fired liquid chiller. All major components are similar to the single-effect chiller except for an added generator (first-stage or primary generator), condenser, heat exchanger, and condensate subcooling heat exchanger.

Operation of the double-effect absorption machine is similar to that for the single-effect machine. The primary generator receives heat from the external heat source, which boils dilute absorbent solution. The pressure in the vapor space of the primary generator is about 15 psia. This vapor flows to the inside of tubes in the second-effect generator. It is sufficiently hot to boil and concentrate absorbent solution on the outside of these tubes, thus creating additional refrigerant vapor with no additional primary heat input.

The extra solution heat exchanger (high-temperature heat exchanger) is placed in the intermediate and dilute solution streams flowing to and from the primary generator to preheat the dilute solution. Because of the relatively large pressure difference between the vapor spaces of the primary and secondary generators, a mechanical solution flow control device is required at the outlet of the high-temperature heat exchanger to maintain a liquid seal between the two generators. A valve at the heat exchanger outlet that is controlled by the liquid level leaving the primary generator can maintain this seal.

One or more condensate heat exchangers may be used to remove additional heat from the primary heat source steam by subcooling the steam condensate. This heat is added to the dilute or inter-

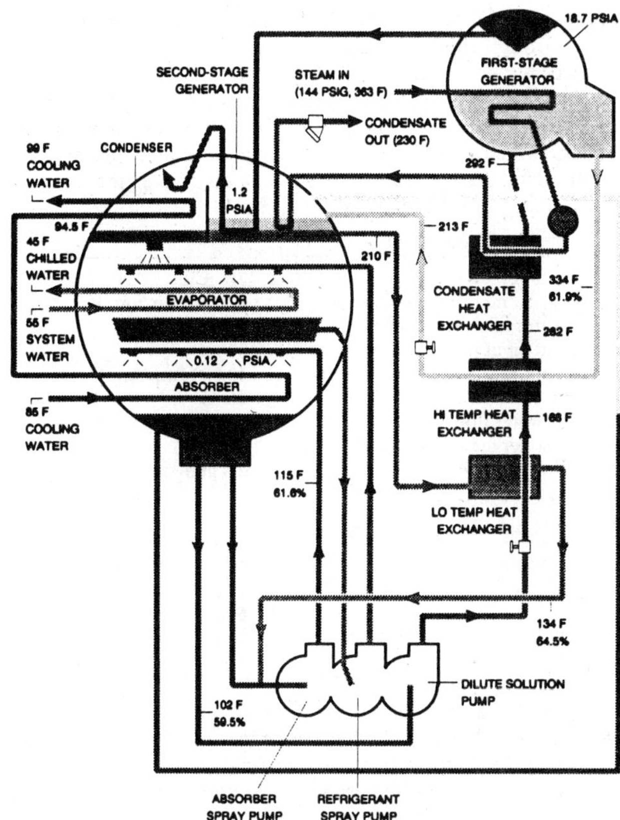

Fig. 3 Double-Effect Indirect-Fired Chiller

mediate solution flowing to one of the generators. The result is a reduction in the quantity of steam required to produce a given refrigeration effect; however, the required heat input remains the same. The COP is not improved by condensate exchange.

As with the single-effect machine, the strong absorbent solution flowing to the absorber can be mixed with dilute solution and pumped over the absorber tubes or can flow directly from the low-temperature heat exchanger to the absorber.

Also, as with the single-effect machines, the four major components can be contained in one or two vessels.

The solution flow cycle shown in Figure 3 is a series flow and is one of several solution flow cycles that are used in double-effect absorption equipment. The following solution flow cycles may be used:

Series flow. All solution leaving the absorber runs through a pump and then flows sequentially through the low-temperature heat exchanger, high-temperature heat exchanger, first-stage generator, high-temperature heat exchanger, second-stage generator, low-temperature heat exchanger, and absorber.

Parallel flow. Solution leaving the absorber is pumped through appropriate portions of the combined low- and high-temperature solution heat exchanger and is then split between the first- and second-stage generators. Both solution flow streams then return to appropriate portions of the combined solution heat exchanger, are mixed together, and flow to the absorber.

Reverse parallel flow. All solution leaving the absorber is pumped through the low-temperature heat exchanger and then to the second-stage generator. At this point, the solution stream is split, with a portion going to the low-temperature heat exchanger and on to the absorber. The remainder goes sequentially through a pump, the high-temperature heat exchanger, the first-stage generator, and the high-temperature heat exchanger. This stream then rejoins the solution from the second-stage generator; both streams flow through the low-temperature heat exchanger and to the absorber.

These machines are typically fired with medium-pressure steam of 80 to 144 psig or hot liquids of 300 to 400°F. Typical operating COPs are 1.1 to 1.2. These machines are available commercially from several manufacturers and have capacities ranging from 100 to 1700 tons of refrigeration.

Performance Characteristics	
Steam input pressure	115 psig
Steam consumption (with condensate saturated conditions)	9.7 to 10 lb/ton·h
Hot fluid input temperature	370°F
Heat input rate	10,000 Btu/ton·h
Cooling water temperature in	85°F
Cooling water flow	3.6 to 4.5 gpm/ton
Chilled water temperature off	44°F
Chilled water flow	2.4 gpm/ton
Electric power	0.01 to 0.04 kW/ton

Physical Characteristics	
Nominal Capacities	100 to 1700 tons
Length	10 to 31 ft
Width	6 to 12 ft
Height	8 to 14 ft
Operating weight	15,000 to 132,000 lb

Figure 4 is a schematic diagram of a commercially available, double-effect direct-fired liquid chiller. All major components are similar to the double-effect indirect-fired chiller except for substitution of the direct-fired primary generator for the indirect-fired primary generator and elimination of the steam condensate subcooling heat exchanger. Operation of these machines is identical to that of the double-effect indirect-fired machines.

These machines are typically fired with natural gas or fuel oil (most have dual fuel capabilities). Typical operating COPs are 0.92 to 1.0. These machines are available commercially from several manufacturers and have capacities ranging from 100 to 1500 tons. Machine capacities of 20 to 100 tons are also available from international sources.

Performance Characteristics	
Fuel consumption (high heating value of fuel)	12,000 to 13,044 Btu/ton·h
COP (high heating value)	0.92 to 1.0
Cooling water temperature in	85°F
Cooling water flow	4.4 to 4.5 gpm/ton
Chilled water temperature off	44°F
Chilled water flow	2.4 gpm/ton
Electric power	0.01 to 0.04 kW/ton

Physical Characteristics	
Nominal capacities	100 to 1500 tons
Length	10 to 34 ft, with minimum of 5 ft for some machines
Width	5 to 21.3 ft, with minimum of 4 ft for some machines
Height	7 to 12 ft
Operating weight	11,000 to 174,600 lb, with a minimum of 3300 lb for some machines

Operation

Modern water-lithium bromide chillers are trouble free and easy to operate. As with any equipment, careful attention should be

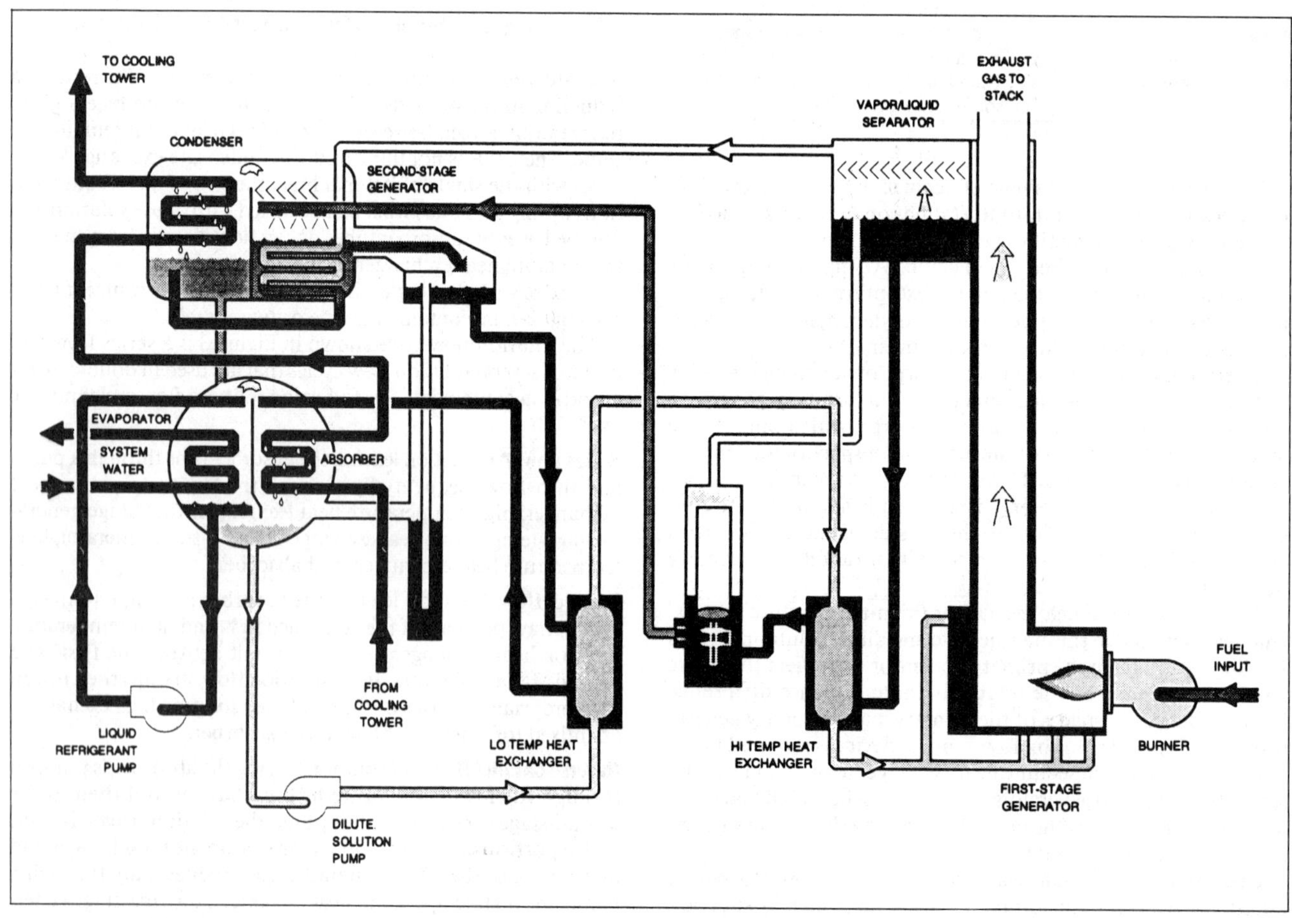

Fig. 4 Double-Effect Direct-Fired Chiller

paid to operational and maintenance procedures recommended by the manufacturer of the equipment. A discussion of some key points common to all types of lithium bromide absorption equipment follows.

Operational Limits. Chilled water temperature leaving the evaporator should normally be between 40 and 60 °F. The upper limit is set by the pump lubricant and is somewhat flexible. The lower limit exists because the refrigerant (water) freezes at 32 °F.

Cooling water temperature entering the absorber tubes is generally limited to between 45 and 110 °F, although some machines limit the entering cooling water temperatures to between 70 and 95 °F. The upper limit exists because of hydraulic and differential pressure limitations between the generator-absorber, the condenser-evaporator, or both, and to reduce absorbent concentrations and corrosion effects. The lower temperature limit exists because, at excessively low cooling water temperature, the condensing pressure drops too low and excessive vapor velocities carry over solution to the refrigerant in the condenser. Sudden lowering of cooling water temperature at high loads will also promote crystallization; therefore, some manufacturers will dilute the solution with refrigerant liquid to help prevent crystallization. The supply of refrigerant is limited, however, so this dilution is done in small steps.

Operational Controls. Modern absorption machines are equipped with electronic control systems. The primary function of the control system is to safely operate the absorption machine and modulate its capacity in order to satisfy the load requirements placed upon it.

The temperature of the chilled water leaving the evaporator is set at a desired value. Deviations from this set point indicate that the machine capacity and the load applied to it are not matched. Machine capacity is then adjusted as required by modulation of the heat input control device. Modulation of heat input results in changes to the concentration of absorbent solution supplied to the absorber if the pumped solution flow remains constant.

Some equipment uses solution flow control to the generator(s) in combination with capacity control. The solution flow may be reduced with modulating valves or solution pump speed controls as the load decreases (which reduces the required sensible heating of solution in the generator to produce a given refrigeration effect), thereby improving part-load efficiency.

Operation of lithium bromide machines with low entering cooling water temperatures or a rapid decrease in cooling water temperature during operation can cause liquid carryover from the generator to the condenser and possible crystallization of absorbent solution in the low-temperature heat exchanger. For these reasons, most machines have a control that limits heat input to the machine based on entering cooling water temperature. Since colder cooling water enhances machine efficiency, the ability of machines to use colder water, when available, is important.

Use of electronic controls with advanced control algorithms have improved part-load and variable cooling water temperature operation significantly compared to older pneumatic or electric controls. Electronic controls have also made chiller setup and operation simpler and more reliable.

These steps are involved in a typical start-run-stop sequence of an absorption chiller with chilled and cooling water flows pre-established:

1. Cooling required signal is initiated by building control device or in response to rising chilled water temperature.
2. All chiller unit and system safeties are checked.
3. Solution and refrigerant pumps are started.
4. Heat input valve is opened or burner is started.
5. Chiller begins to meet the load and controls chilled water temperature to desired set point by modulation of heat input control device.
6. During operation, all limits and safeties are continually checked. Appropriate action is taken, as required, to maintain safe chiller operation.
7. Load on chiller decreases below minimum load capabilities of chiller.
8. Heat input device is closed.
9. Solution and refrigerant pumps continue to operate for several minutes to dilute the absorbent solution.
10. Solution and refrigerant pumps are stopped.

Limit and Safety Controls. In addition to capacity controls, these chillers require several protective devices. Some controls keep the units operating within safe limits and others stop the unit before damage occurs due to a malfunction. Each limit and safety cutout function usually uses a single sensor when electronic controls are used. The following limits and safety features are normally found on absorption chillers:

Low-temperature chilled water control/cutout. Allows the user to set the desired temperature for chilled water leaving the evaporator. Control then modulates the heat input valve to maintain this set point. This control incorporates chiller start and stop by water temperature. A safety shutdown of the chiller is invoked if a low-temperature limit is reached.

Low-temperature refrigerant limit/cutout. A sensor in the evaporator monitors refrigerant temperature. As the refrigerant low-limit temperature is approached, the control limits further loading, then prevents further loading, then unloads, and finally invokes a chiller shutdown.

Chilled water, chiller cooling water, and pump motor coolant flow. Flow switches trip and invoke chiller shutdown if flow stops in any of these circuits.

Pump motor over-temperature. A temperature switch in the pump motor windings trips if safe operating temperature is exceeded and shuts down the chiller.

Pump motor overload. Current to the pump motor is monitored, and the chiller shuts down if the current limit is exceeded.

Absorbent concentration limit. Key solution and refrigerant temperatures are sensed during chiller operation and used to determine the temperature safety margin between solution temperature and solution crystallization temperature. As this safety margin is reduced, the control first limits further chiller loading, then prevents further chiller loading, then unloads the chiller, and finally invokes a chiller shutdown.

In additional to this type of control, most chiller designs incorporate a built-in overflow system between the evaporator liquid storage pan and the absorber sump. As the absorbent solution concentration increases in the generator/absorber flow loop, the refrigerant liquid level in the evaporator storage pan increases. The initial charge quantities of solution and refrigerant are set such that liquid refrigerant will begin to overflow the evaporator pan when maximum safe absorbent solution concentration has been reached in the generator/absorber flow loop. The liquid refrigerant overflow goes to the absorber sump and prevents further concentration of the absorbent solution.

Burner fault. Operation of the burner on direct-fired chillers is typically monitored by its own control system. A burner fault indication is passed on to the chiller control and generally invokes a chiller shutdown.

High-temperature limit. Direct-fired chillers typically have a temperature sensor in the liquid absorbent solution near the burner fire tube. As this temperature approaches its high limit, the control first limits further loading, then prevents further loading, then unloads, and finally invokes a chiller shutdown.

High-pressure limit. Double-effect machines typically have a pressure sensor in the vapor space above the first-stage generator. As this pressure approaches its high limit, the control first limits further loading, then prevents further loading, then unloads, and finally invokes a chiller shutdown.

Machine Setup and Maintenance

Large capacity lithium bromide absorption water chillers are generally put into operation by factory-trained technicians. Proper procedures must be followed in order to insure that the machines will function as designed and continue to function in a trouble-free manner for their intended design life (20 plus years). Steps required to set up and start a lithium bromide absorption machine include:

1. Level the unit so that internal pans and distributors can function properly.
2. Isolate the unit from foundations with pads if it is located near noise-sensitive areas.
3. Confirm that factory leaktightness has not been compromised.
4. Charge the unit with refrigerant water (distilled or deionized water is required) and lithium bromide solution.
5. Add corrosion inhibitor to the absorbent solution if required.
6. Calibrate all control sensors and check all controls for proper function.
7. Start the unit and bring it slowly to design operating condition while adding performance additive (usually one of the octyl alcohols).
8. If necessary to obtain design conditions, adjust absorbent and/or refrigerant charge levels. This procedure is known as trimming the chiller, and, if done correctly, will allow the chiller to operate safely and efficiently over its entire operating range.
9. Fine tune control settings.
10. Check purge operation.

Recommended periodic operational checks and maintenance procedures typically include:

- Purge operation and air leaks. Confirm that the purge system operates correctly and that the unit does not have chronic air leaks. Continued leakage of air into an absorption chiller will deplete the corrosion inhibitor, cause corrosion of internal parts, contaminate the absorbent solution, reduce chiller capacity and efficiency, and may cause crystallization of the absorbent solution.
- Sample absorbent and refrigerant periodically and check for contamination, pH, corrosion-inhibitor level, and performance additive level. Use these checks to adjust the levels of additives in the solution and as an indicator of internal machine malfunctions.

Mechanical systems such as the purge, solution pumps, controls, and burners all have periodic maintenance requirements recommended by the manufacturer.

Chiller Performance at Other Than Design Rating

The performance of lithium bromide absorption machines is affected by the operating conditions and the heat transfer surface chosen by the manufacturer. Manufacturers of this equipment can provide detailed performance information for their equipment at specific alternate operating conditions.

AMMONIA-WATER ABSORPTION TECHNOLOGY

Figure 5 shows a typical schematic of an ammonia-water machine, which is available as a direct-fired air-cooled liquid chiller in capacities of 3 to 5 tons. Ammonia-water equipment varies from water-lithium bromide equipment to accommodate three major differences:

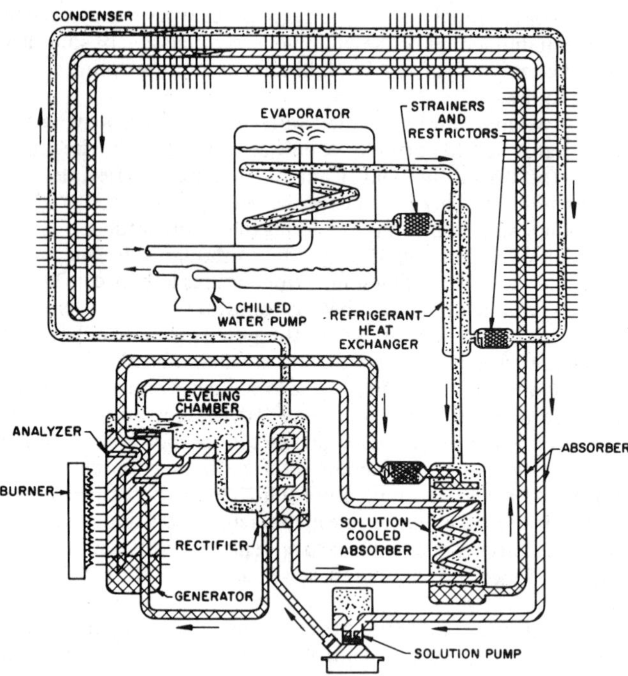

Fig. 5 Ammonia-Water Direct-Fired Air-Cooled Chiller

1. Water (the absorbent) is also volatile, so the regeneration of weak absorbent to strong absorbent is a fractional distillation process.
2. Ammonia (the refrigerant) causes the cycle to operate at condenser pressures of about 300 psia and at evaporator pressures of approximately 70 psia. As a result, vessel sizes are held to a diameter of 6 in. or less to avoid construction code requirements on small systems, and positive-displacement solution pumps are used.
3. Air cooling requires condensation and absorption to occur inside the tubes so that the outside can be finned for greater air contact.

Component Descriptions

Generator. The vertical vessel is finned on the outside to extract heat from the combustion products. Internally, a system of analyzer plates creates intimate counterflow contact between the vapor generated, which rises, and the absorbent, which descends. Atmospheric gas burners depend on the draft of the condenser air fan to sustain adequate combustion airflow to fire the generator. The exiting flue products mix with the air that has passed over the condenser and absorber.

Heat Exchangers. Heat exchange between strong and weak absorbents takes place partially within the generator-analyzer. A tube bearing strong absorbent spirals through the analyzer plates and in the solution-cooled absorber, where strong absorbent metered from the generator through the solution capillary passes over a helical coil bearing weak absorbent. In the solution-cooled absorber, the strong absorbent absorbs some of the vapor from the evaporator, thus retaining its heat of absorption within the cycle to improve its COP. The strong absorbent and unabsorbed vapor continue from the solution-cooled absorber into the air-cooled absorber, where absorption is completed and the heat of absorption is rejected to the air.

The *rectifier* is a heat exchanger that consists of a spiral coil through which weak absorbent from the solution pump passes on its way to the absorber and generator. Some type of packing is included to assist counterflow contact between condensate from the coil (which is refluxed to the generator) and the vapor (which continues on to the air-cooled condenser). The function of the rectifier is to concentrate the ammonia in the vapor from the generator by cooling and stripping out some of the water vapor.

Absorber and Condenser. These finned tubes are arranged so that most of the incoming air flows over the condenser tubes and most of the exit air flows over the absorber tubes.

Evaporator. The liquid to be chilled drips over a coil bearing evaporating ammonia, which absorbs the refrigeration load. On the chilled-water side, which is at atmospheric pressure, a pump circulates the chilled liquid to the load source. Refrigerant to the evaporator is metered from the condenser through restrictors. A tube-in-tube heat exchanger provides the maximum refrigeration effect per unit mass of refrigerant. The tube-in-tube design is particularly effective in this cycle because water present in the ammonia produces a liquid residue that evaporates at increasing temperatures as the amount of residue decreases.

Solution Pumps. The reciprocating motion of a flexible sealing diaphragm moves solution through suction and discharge valves. Hydraulic fluid pulses delivered to the opposite side of the diaphragm by a hermetic vane or piston pump at atmospheric suction pressure impart this motion.

Capacity Control. A thermostat usually cycles the machine on and off. A chilled-water switch shuts the burners off if the water temperature drops close to freezing. Units may also be underfired by 20% to derate to a lower load.

Protective Devices. Typical protective devices include (1) flame ignition and monitor control, (2) a sail switch that verifies airflow before allowing the gas to flow to the burners, (3) a pressure relief valve, and (4) a generator high-temperature switch.

Equipment Performance and Selection

Ammonia absorption equipment is built and rated to meet ANSI *Standard* Z21.40.1-81, Gas-fired Absorption Summer Air Conditioning Appliances, Sixth Edition, for outdoor installation. The rating conditions are ambient air at 95 °F dry bulb and 75 °F wet bulb and chilled water delivered at the manufacturer's specified flow at 45 °F. A COP of about 0.5 is realized, based on the higher heating value of the gas.

Although most units are piped to a single furnace, duct, or fan coil and operated as air conditioners, multiple units supplying a multicoil system for process cooling and air conditioning are common. Also, chillers are packaged with an outdoor boiler and can supply chilled or hot water as the cooling or heating load requires.

Physical Characteristics

Cooling capacities	36,000 to 60,000 Btu/h
Length	40 to 48.5 in.
Width	29.1 to 33.5 in.
Height	37.6 to 46 in.
Weight	550 to 775 lb

AMMONIA-WATER-HYDROGEN CYCLE

Domestic absorption refrigerators use a modified absorption cycle with ammonia, water, and hydrogen as working fluids. Wang and Herold (1992) reviewed the literature on this cycle. These units are popular for recreational vehicles because they can be dual-fired by gas or electric heaters. They are also popular for hotel rooms because they are silent. The refrigeration unit is hermetically sealed. All spaces within the system are open to each other and, hence, are at the same total pressure, except for minor variations caused by fluid columns used to circulate the fluids.

The key elements of the system shown in Figure 6 include a generator (1), a condenser (2), an evaporator (3), an absorber (4), a rectifier (7), a gas heat exchanger (8), a liquid heat exchanger (9), and a bubble pump (10). The following three distinct fluid circuits

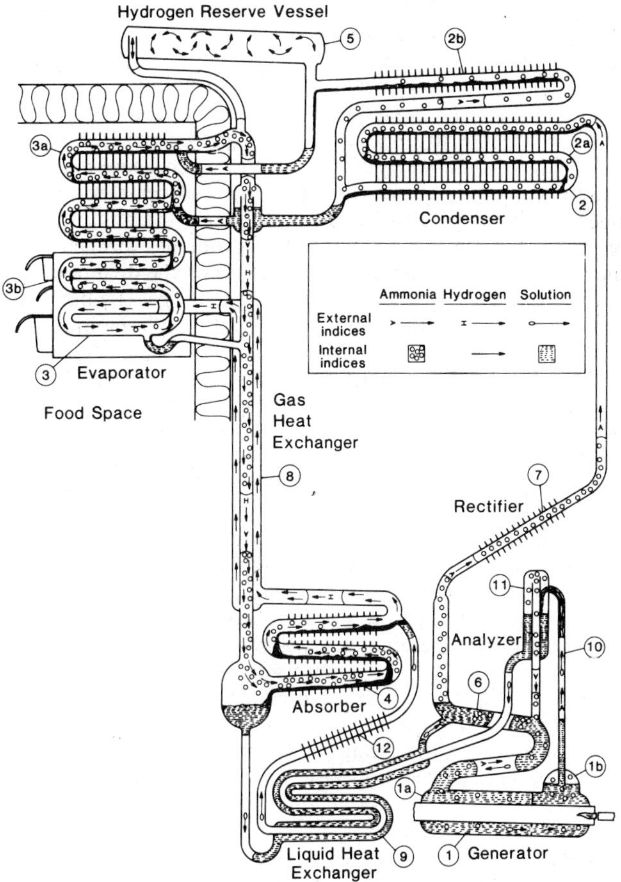

Fig. 6 Domestic Absorption Refrigeration Cycle

exist in the system: (I) an ammonia circuit, which includes the generator, condenser, evaporator, and absorber; (II) a hydrogen circuit, which includes the evaporator, absorber, and gas heat exchanger; and (III) a solution circuit, which includes the generator, absorber, and liquid heat exchanger.

Starting with the generator, a gas burner or other heat source applies heat to expel ammonia from the solution. The ammonia vapor generated then flows through an analyzer (6) and a rectifier (7) to the condenser (2). The small amount of residual water vapor in the ammonia is separated by atmospheric cooling in the rectifier and drains to the generator (1) through the analyzer (6).

The ammonia vapor passes into section (2a) of the condenser (2), where it is liquified by air cooling. Fins on the condenser increase the cooling surface. The liquified ammonia then flows into an intermediate point of the evaporator (3). A liquid trap between the condenser section (2a) and the evaporator prevents hydrogen from entering the condenser. Ammonia vapor that does not condense in the condenser section (2a) passes to the other section (2b) of the condenser and is liquified. It then flows through another trap into the top of the evaporator.

The evaporator has two sections. The upper section (3a) has fins and cools the food space directly, while the lower section (3b) cools the freezing compartment directly.

Hydrogen gas, carrying a small partial pressure of ammonia, enters the lower evaporator section (3) and, after passing through a precooler, flows upward and counterflow to the downward flowing liquid ammonia, increasing the partial pressure of the ammonia in the vapor as the liquid ammonia evaporates. While the total pressures in the evaporator and the condenser are the same, typically 20 bar, substantially pure ammonia is in the space where condensation takes place, and the vapor pressure of the ammonia essentially equals the total pressure. In contrast, the ammonia partial pressures entering and leaving the evaporator are typically 1 and 3 bars, respectively.

The gas mixture of hydrogen and ammonia leaves the top of the evaporator and passes down through the center of the gas heat exchanger (8) to the absorber (4). Here, ammonia is absorbed by liquid ammonia-water solution, and hydrogen, which is almost insoluble, passes up from the top of the absorber, through the external chamber of the gas heat exchanger (8), and into the evaporator. Some ammonia vapor passes with the hydrogen from the absorber to the evaporator. Because of the difference in molecular mass of ammonia and hydrogen, the gas circulation is maintained between the evaporator and absorber by natural convection.

Countercurrent flow in the evaporator permits placing the box cooling section of the evaporator at the top of the food space, which is the most effective location. Also, the gas leaving the lower temperature evaporator section (3b) can pick up more ammonia at the higher temperature in the box cooling evaporator section (3a), thus increasing capacity and efficiency. In addition, the liquid ammonia flowing to the lower temperature evaporator section is precooled in the upper evaporator section. The dual liquid connection between the condenser and the evaporator permits extending the condenser below the top of the evaporator to provide more surface, while maintaining gravity flow of liquid ammonia to the evaporator. The two-temperature evaporator partially segregates the freezing function from the box cooling function, thus giving better humidity control.

In the absorber, the strong absorbent flows counter to and is diluted by direct contact with the gas. From the absorber, the weak absorbent flows through the liquid heat exchanger (9) to the analyzer (6) and then to the weak absorbent chamber (1a) of the generator (1). Heat applied to this chamber causes vapor to pass up through the analyzer (6) and to the condenser. The solution passes through an aperture in the generator partition into the strong absorbent chamber (1b). Heat applied to this chamber causes vapor and liquid to pass up through the small-diameter bubble pump (10) to the separation vessel (11). While liberated ammonia vapor passes through the analyzer (6) to the condenser, the strong absorbent flows through the liquid heat exchanger (9) to the absorber. The finned air-cooled loop (12) between the liquid heat exchanger and the absorber precools the solution further. The heat of absorption is rejected to the surrounding air.

The hydrogen reserve vessel (5), which is connected between the condenser outlet and the hydrogen circuit, is a reservoir for the hydrogen gas and compensates for changes in load and environmental conditions.

Controls

Burner Ignition and Monitoring Control. These controls are either electronic or thermomechanical. Electronic controls ignite, monitor, and shut off the main burner as required by the thermostat. For thermomechanical control, a thermocouple monitors the main flame. The low-temperature thermostat then changes the input to the main burner in a two-step mode. A pilot is not required because the main burner acts as the pilot on low fire.

Low-Temperature Thermostat. This thermostat monitors the temperature in the cabinet and controls the gas input.

Safety Device. Each unit has a fuse plug to relieve pressure in the event of fire. Gas-fired installations require a flue exhausting to outside air. Nominal operating conditions are as follows:

Ambient temperature	95 °F
Freezer temperature	10 °F
Input	100 Btu/h · ft^3 of cabinet interior

SPECIAL APPLICATIONS

A single-stage indirect-fired lithium bromide absorption chiller as a bottoming cycle for an exhausting steam turbine that drives a centrifugal water chiller uses the steam twice. As a result, thermal efficiencies and costs are comparable to those of the double-effect absorption chiller. The chilled water flows in series through the absorption unit, then out of the centrifugal chiller. The cooling water can flow through the centrifugal unit condenser and then the absorption unit, or the cooling water can be piped in parallel. Approximately two-thirds of the cooling duty is carried by the absorption unit and one-third by the centrifugal unit.

The exhausting steam turbine drive for the centrifugal compressor is cheaper than the usual condensing turbine required for high efficiency, and the absorption generator becomes the main steam condenser. The steam turbine uses superheated steam at 100 psig and exhausts the steam at between 9 and 12 psig to the absorption chiller generator.

Single-stage absorption units have been used to condense hot hydrocarbon process vapors from the top of a distillation column while cooling solvents for chemical plants. They have also been installed to use heat extracted directly from hot gasoline in the tubes of the generator while directly cooling lean oil for oil refinery absorbers. These applications took considerable engineering development and are difficult to justify today; however, they show the possibilities.

Single-stage absorption units have used solar-heated and diesel or gas engine-heated hot water at 180 °F to cool buildings for comfort.

An absorption unit could be added to condense waste steam to subcool refrigerant from an existing conventional single-stage refrigeration plant and boost the available cooling capacity of the existing compressors. Also, with heat recovery features of absorption units provided by some manufacturers, hot exhaust gases supplemented by gas or oil firing could provide comfort and process cooling and heating.

EVOLVING CONCEPTS

The absorption chiller and heat pump industry has changed greatly in recent years. The following systems and cycles are close to commercial realization.

Triple-Effect Cycles

Triple-effect absorption cooling can be classified as single-loop and dual-loop cycles. Single-loop triple-effect cycles are basically double-effect cycles with an additional generator and condenser. The resulting system with three generators and three condensers operates similarly to the double-effect system. Primary heat (from a natural gas or fuel oil burner) concentrates absorbent solution in a first-stage generator at about 450 °F. The refrigerant vapor produced is then used to concentrate additional absorbent solution in a second-stage generator at about 300 °F. Finally, the refrigerant vapor produced in the second-stage generator concentrates additional absorbent solution in a third-stage generator at about 200 °F. The usual internal heat-recovery devices (solution heat exchangers) can be used to improve cycle efficiency. As with the double-effect cycles, several variations of solution flow paths through the generators are possible.

Theoretically, the COP obtainable with these triple-effect cycles is about 1.7 (not taking into account burner efficiency). Difficulties with these cycles include the following:

- High solution temperatures pose problems to solution stability, performance additive stability, and material corrosion
- High pressure in the first-stage generator vapor space requires costly pressure vessel design and high-pressure solution pump(s)

A double-loop triple-effect cycle consists of two cascaded single-effect cycles. One cycle operates at normal single-effect operating temperatures and the other at higher temperatures. The smaller high-temperature topping cycle is direct-fired with natural gas or fuel oil and has a generator temperature of about 450 °F. Heat is rejected from the high-temperature cycle at 200 °F and is used as the energy input for the conventional single-effect bottoming cycle. Both the high-and low-temperature cycles remove heat from the cooling load at about 44 °F.

Theoretically, the overall COP obtainable with this triple-effect cycle is about 1.8 (not taking into account burner efficiency).

As with the single-loop triple-effect cycle, high temperatures create problems with solution and additive stability and material corrosion. Also, the use of a second loop requires additional heat exchange vessels and additional pumps. However, both loops operate below atmospheric pressure and, therefore, do not require costly pressure vessel designs.

GAX (Generator-Absorber Heat Exchange) Cycle

The air-cooled absorption air-conditioning equipment presently available operates at gas-fired cooling COPs of just under 0.5 at ARI rating conditions. The absorber heat exchange cycle of past air conditioners had a COP of about 0.67 at the rating conditions. In recent years, several projects have been initiated around the world to develop generator-absorber heat exchange (GAX) cycle systems. The best known programs have been directed toward cycle COPs of about 1.0.

The GAX cycle is a heat-recuperating cycle in which absorber heat is used to heat the lower temperature section of the generator as well as the rich ammonia solution being pumped to the generator. This cycle, like others capable of higher COPs, is more complex and difficult to develop than the ammonia single-stage and absorber heat exchange cycles, but its potential gas-fired COPs of 0.9 in cooling mode and 1.8 in heating mode make it capable of significant energy savings on an annual basis. In addition to providing a more effective use of heat energy than the most efficient furnaces, the GAX heat pump is able to supply all the heat a house requires to outdoor temperatures below 0 °F without the use of supplemental heat.

Solid-Vapor Sorption Systems

Solid-vapor heat pump technology is being developed for zeolite, silica-gel, activated-carbon, and coordinated complex adsorbents. The cycles are periodic in that the refrigerant is transferred periodically between two primary vessels. Several concepts providing quasi-continuous refrigeration have been developed. One advantage of solid-vapor systems is that no solution pump is needed. The main challenge in designing a competitive solid-vapor heat pump is to package the adsorbent in such a way that good heat and mass transfer are obtained in a small volume. A related constraint is that good thermal performance of periodic systems requires that the thermal mass of the vessels be small to minimize cyclic heat-transfer losses. Research in the area of solid-vapor technology can be found in the proceedings of two recent conferences (IIR 1991 and 1992).

Liquid Desiccant-Absorption Systems

In efforts to reduce a building's energy consumption, designers have successfully integrated liquid desiccant equipment with standard absorption chillers. These applications have been building specific and are sometimes referred to as application hybrids. In a more general approach, the absorption chiller is modified so that rejected heat from its absorber can be used to help regenerate the liquid desiccant. Only liquid desiccants are appropriate for this integration because they can be regenerated at lower temperatures than solid desiccants.

The desiccant dehumidifier dries ventilation air sufficiently that, when it is mixed with return air, the building's latent load is satisfied. The desiccant drier is cooled by cooling tower water so that a significant amount of the cooling load is transferred directly to the cooling water. Consequently, the absorption chiller size is significantly reduced, potentially to as little as 60% of the size of the chiller in a conventional installation.

Because the air handler is restricted to sensible load, the evaporator in the absorption machine will run at higher temperatures than normal. Consequently, a machine operating at normal concentrations in its absorber will reject heat at higher temperatures. To permit convenient regeneration of the liquid desiccant, only moderate increases in solution concentration are required. These are subtle but significant modifications to a standard absorption chiller.

Such combined systems seem to work best when about one-third of the supply air comes from outside the conditioned space. These systems do not require 100% outside air for ventilation, so they should be applicable to conventional buildings as newly mandated ventilation standards are accommodated. Because they always operate in a form of economizer cycle, they are particularly effective during shoulder seasons (spring and fall). As lower cost liquid desiccant systems become available, reduced first costs may join the advantages of decreased energy use, better ventilation, and improved humidity control.

ALTERNATIVE ABSORPTION WORKING FLUIDS

The fluids chosen for liquid-vapor absorption cycles are usually water-lithium bromide or ammonia-water. These fluids currently exhibit the best combination of properties for many typical applications. However, these choices limit the machine design in several ways. Limiting characteristics of water-lithium bromide include (1) evaporator temperature must be above 32 °F to avoid freezing the water, (2) coolant temperatures must be relatively low to avoid crystallization of the lithium bromide (making air cooling difficult), and (3) corrosion is a major factor at temperatures above 350 °F. Limiting characteristics of ammonia-water include (1) the toxicity of ammonia and (2) its high vapor pressure.

Many fluid alternatives have been considered by researchers. Several systems appear to be approaching commercial use.

High-Temperature Fluid. An aqueous mixture of nitrate salts has good high-temperature properties for industrial absorption heat pumps and topping cycles. These salts have been used previously as corrosion inhibitors in water-lithium bromide systems.

A large effort is being expended to find corrosion inhibitors for lithium bromide systems that will allow high-temperature operation. In particular, a lithium bromide-based triple-effect cycle will require such inhibitors or will require metals having greater corrosion resistance.

BIBLIOGRAPHY

Alefeld, G. 1985. Multi-stage apparatus having working-fluid and absorption cycles, and method of operation thereof. U.S. Patent No. 4,531,374.

Bogart, M. 1981. *Ammonia absorption refrigeration in industrial processes.* Gulf Publishing Co., Houston.

DeVault, R.C. 1986. Triple-effect absorption chiller utilizing two refrigeration circuits. U.S. Patent No. 4,732,008.

Eisa, M.A.R., S.K. Choudhari, D.V. Paranjape, and F.A. Holland. 1986. Classified references for absorption heat pump systems from 1975 to May 1985. *Heat Recovery Systems* 6:47-61. Pergamon Press Ltd., Great Britain.

Erickson, D.C. 1986. High temperature absorbent for water vapor. U.S. Patent No. 4,563,295.

Grossman, G. and K.W. Childs. 1983. Computer simulation of a lithium bromide-water absorption heat pump for temperature boosting. ASHRAE *Transactions* 89(1).

Hanna, W.T. and W.H. Wilkinson. 1982. Absorption heat pumps and working pair developments in the U.S. since 1974, New working pairs for absorption processes, pp. 78-80. *Proceedings of Berlin Workshop* by the Swedish Council for Building Research, Stockholm, Sweden.

Huntley, W.R. 1984. Performance test results of a lithium bromide-water absorption heat pump that uses low temperature waste heat. Oak Ridge National Laboratory *Report* No. ORNL/TM9702, Oak Ridge, TN.

IIR. 1991. *Proceedings of the XVIIIth International Congress of Refrigeration* (August 10-17, 1991, Montreal, Canada), Volume III. International Institute of Refrigeration, Paris, France.

IIR. 1992. *Proceedings of Solid Sorption Refrigeration Meetings of Commission Bl* (November 18-20, 1992, Paris, France). International Institute of Refrigeration, Paris, France.

Meckler, M. 1988. Off-peak desiccant cooling and cogeneration combine to maximize gas utilization. ASHRAE *Transactions* 94, paper DA-88-1-5.

Niebergall, W. 1981. *Handbuch der Kaltetechnik*, Volume 7: Sorptionsmachinen. R. Plank, ed. Springer Verlag, Berlin.

Phillips, B.A. 1990. Development of a high-efficiency, gas-fired, absorption heat pump for residential and small-commercial applications: Phase I Final *Report*: Analysis of advanced cycles and selection of the preferred cycle. ORNL/Sub/86-24610/1, September.

Scharfe, J., F. Ziegler, and R. Radermacher. 1986. Analysis of advantages and limitations of absorber-generator heat exchange. *International Journal of Refrigeration* 9:326-33.

Stoecker, W.F. and J.W. Jones. 1982. *Refrigeration and air conditioning*. McGraw-Hill Book Company, New York.

Vliet, G.C., M.B. Lawson, and R.A. Lithgow. 1982. Water-lithium bromide double-effect cooling cycle analysis. ASHRAE *Transactions* 88(1):811-23.

Wang, L. and K.E. Herold. 1992. Diffusion-absorption heat pump. Annual *Report* to Gas Research Institute, GRI-92/0262.

Wilkinson, W. H. 1991. A simplified high efficiency DUBLSORB system. ASHRAE *Transactions* 97, paper NY-91-2-1, 1991.

CHAPTER 41

FORCED-CIRCULATION AIR COOLERS

FORCED-CIRCULATION unit coolers and product coolers are designed to operate continuously within refrigerated enclosures. A cooling coil and motor-driven fan make up the basic components of these coolers. These components provide the cooling or freezing temperatures and proper airflow to the room. Coil defrost equipment is added for low-temperature operations.

Any unit, such as a blower coil, unit cooler, product cooler, cold diffuser unit, or air-conditioning air handler is considered a forced-air cooler when operated under refrigeration conditions. Many design and construction choices are available, including: (1) various spaced fin coils; (2) electric, gas, air, or water defrosting; (3) discharge air velocity; (4) centrifugal or propeller fans, either belt- or direct-driven; (5) ducted or nonducted, and/or (6) freestanding or ceiling suspended.

TYPES OF FORCED-CIRCULATION AIR COOLERS

Sloped Front Unit Coolers

These units, which range from 5 to 10 in. high, are commonly used in back-bar and under-the-counter fixtures, as well as in vertical, self-serve, glass door reach-in enclosures. These are small unit coolers, with airflows usually not exceeding 150 cfm per fan. The sloped fronts are designed for horizontal top mounting as a single unit, or installation as a group of parallel connected units. Direct-drive fans are sloped to fit within the restricted return air sweep rising across the enclosure access doors.

Low Air Velocity Unit Coolers

These units can have a half-round appearance, although long, narrow, dual-coil units are used in meat-cutting rooms and in meat and floral walk-in coolers. The unit has an amply finned coiled surface to maintain high humidities in the room. Discharge air velocities at the coil face range from 85 to 200 fpm.

Standard Air Velocity Unit Coolers

Low silhouette units are 12 to 15 in. high. Medium or midheight units are 18 to 48 in. high or higher. Some units can be classified as high silhouette unit coolers. The air velocity at the coil face can be as high as 600 fpm, but generally is 300 to 400 fpm.

High Air Velocity Product Coolers

These units are used in blast tunnel freezing and special cooling of products that are not adversely affected by moderate dehydration during rapid cooling. They generally draw air through the cooler at discharge velocities of about 2000 fpm.

The preparation of this chapter is assigned to TC 8.4, Air-to-Refrigerant Heat-Transfer Equipment.

Forced-circulation coolers direct air over a refrigerated coil in the enclosure. The coil lowers the air temperature below its dewpoint, which causes condensate or frost to form on the coil surface. However, the normal refrigeration load is a sensible heat load, *i.e.*, dry-bulb referenced, and the coil surface is considered dry. Rapid and frequent defrosting on a timed cycle can maintain this dry-surface condition.

Sprayed Coil Product Coolers

Spray coils feature a saturated coil surface that can cool the processed air closer to the coil surface temperature than can a dry coil. In addition, the spray continuously defrosts the low-temperature coil. Unlike unit coolers, spray coolers are usually floor mounted and discharge air vertically. The unit sections include a drain pan/sump, coil with spray section, moisture eliminators, and fan with drive. The eliminators remove airborne droplets of spray to prevent their discharge into the refrigerated area. Typically, belt-driven centrifugal fans draw air through the coil at 600 fpm or less.

Water can be used as the spray medium for coil surfaces with temperatures above freezing. For coil surfaces with temperatures below freezing, a suitable material such as listed below must be added to the water to lower the freezing point to 12 °F or lower than the coil surface temperature. Some recirculating solutions include:

- **Sodium Chloride.** This solution is limited to a room temperature of 10 °F or higher. Its minimum freezing point is −6 °F.
- **Calcium Chloride.** This can be used for enclosure temperatures down to about −10 °F, but its use may be prohibited in enclosures containing food products.
- **Aqueous Glycol Solutions.** These solutions are commonly used in water and/or sprayed coil coolers operating below freezing. Food-grade propylene glycol solutions are commonly used because of their low oral toxicity, but they generally become too viscous to pump at temperatures below −13 °F. Ethylene glycol solutions may be pumped at temperatures down to −40 °F. Because of its toxicity, sprayed ethylene glycol in other than sealed tunnels or freezers (no human access allowed during process) is usually prohibited by most jurisdictions. When a glycol mix is sprayed in food storage rooms, any carryover of the spray must be maintained within the limits prescribed by the USDA, the Bureau of Animal Industries, and all applicable local codes.

All brines are hygroscopic; they absorb condensate and become progressively weaker. This dilution can be corrected by continually adding salt to the solution to maintain sufficient below-freezing temperature. Salt is extremely corrosive, so it must be contained in the sprayed coil unit with suitable corrosive-resistant materials, which must receive periodic inspection and maintenance.

Sprayed coil units are usually installed in refrigerated enclosures requiring high humidity, *e.g.*, chill coolers. Paradoxically, the same sprayed coil units can be used in special applications requiring low relative humidity. For such dehydration applications, both a high brine concentrate (near its eutectic point) and a wide difference between the process air and the refrigerant temperature are maintained. Process air reheat downstream of the sprayed coil corrects the dry-bulb temperature.

COMPONENTS

Draw-through and Blow-through Airflow

Unit fans may draw air through the cooling coil and discharge it through the fan outlet into the enclosure; or the fans may blow air through the cooling coil and discharge it from the coil face into the enclosure. Blow-through units have a slightly higher thermal efficiency because heat from the fan is removed from the forced airstream by the coil. Draw-through fan energy adds to the heat load of the refrigerated enclosure, but neither load from the fractional horsepower fan motors is significant. Selection depends more on a manufacturer's design features for the unit size required, as well as the air throw required within the particular enclosure.

Blow-through design has a lower discharge air velocity stream because the entire coil face area is usually the discharge opening (grilles and diffusers not withstanding). An air throw of 33 ft or less is common for the average standard air velocity from a blow-through unit. Greater throw, in excess of 100 ft, is normal for draw-through centrifugal fan units. The propeller fan in the high silhouette draw-through unit cooler is popular for intermediate ranges of air throw.

Fan Assemblies

Direct-drive propeller fans (motor plus blade) are popular because they are simple, economical, and can be installed in multiple assemblies in a unit cooler housing. Additionally, they require less motor power for a given airflow rate capacity.

The centrifugal fan assembly usually includes belts, bearings, sheaves, and coupler drives along with each of their inherent problems. Yet, this design is necessary for applications having high air distribution static pressure losses. These applications include enclosures with ductwork runs, tunnel conveyors, and high density stacking of products. Centrifugal fan-equipped units are also used in produce ripening rooms, where a large air blast is required during the ripening process.

Casing

Casing materials are selected for compatibility with the enclosure environment in which they are installed. Construction usually features coated aluminum or galvanized steel. Stainless steel is also used in food storage or preparation enclosures where sanitation must be maintained. On larger cooler units, internal framing is fabricated of sufficiently substantial material, such as galvanized steel, and casings are usually made with similar material. Some plastic casings are used in small unit coolers, while some large, ceiling-suspended units may feature all aluminum construction.

Coil Construction

Coil construction varies from uncoated (all) aluminum tube and fin to hot-dipped galvanized (all) steel tube and fin, depending on the type of refrigerant used and the environmental exposure to the coil surface. The most popular unit coolers have copper tube/aluminum fin construction. Ammonia refrigerant equipment is not constructed from copper tube coil. Also, sprayed coils are not constructed from aluminum fin and tube materials.

Fin spacings vary from 6 to 8 fins per inch for coils with surfaces above 32°F and between 0 and 32°F when latent loads are insignificant. Otherwise, 3 to 6 fins per inch is the accepted spacing for coil surfaces below 32°F, with a spacing of 4 fins per inch when latent loads exceed 15% of the total load.

Defrosting

Coils must be defrosted when frost accumulates on their surfaces. The frost (or ice) is usually greatest at the air entry side of the coil; therefore, the required defrost cycle is determined by the inlet surface condition. In contrast, a reduced secondary surface-to-primary surface ratio produces greater frost accumulations at the coil outlet face. In theory, the accumulation of greater frost at the coil entry air surface improves the heat transfer capacity of the coil. However, accumulated coil frost usually has two negative effects: (1) it impedes heat transfer because of its insulating effect and (2) it reduces airflow because it restricts the free air area within the coil.

Depending on the defrost method, as much as 80% of the defrost heat load could be transferred into the enclosure. This heat load is not normally included as part of the enclosure heat gain calculation, but it is usually accounted for by the factor that estimates the hours per day of refrigeration running time.

A longer time between defrost cycles can be achieved by using more coil tube rows and wider fin spacing. Ice accumulation should be avoided to reduce defrost time. For example, in low-temperature applications having high latent loads, unit coolers should not be located above freezer doors.

Controls

Electromechanical controls cycle to maintain the desired enclosure temperature. Modulating control valves such as evaporator pressure regulators are also used. In its simplest form, the temperature control is a thermostat mounted in the enclosure that either cycles the compressor on and off or a liquid line feed solenoid that opens and closes. A suction pressure switch can substitute for the wall-mounted thermostat.

An electronically based energy management system (EMS) maintains optimum energy control by equating suction pressure transducer readings with the signal from a temperature diode sensor in the enclosure. EMS controls are commonly used in large warehouses and supermarkets.

AIR MOVEMENT AND DISTRIBUTION

Air distribution and velocity are important concerns in selecting and locating a unit cooler. The direction of the air and air throw should be such that air moves where there is a heat gain; this principle applies to the enclosure walls and ceiling, as well as to the product. Unit coolers should be placed (1) so they do not discharge air at any doors or openings; (2) away from doors that do not incorporate an entrance vestibule or pass to another refrigerated enclosure to keep from inducing additional infiltration into the enclosure; and (3) away from the airstream of another unit to avoid defrost difficulties.

The velocity and relative humidity of air passing over an exposed product affect the amount of surface drying and weight loss. Air velocities of 500 fpm over the product are typical for most freezer applications. Higher velocities require additional fan power and, in some cases, only slightly decrease the cooling time. For example, air velocities in excess of 500 fpm for freezing plastic-wrapped bread reduce freezing time very little. However, increasing the air velocity from 500 to 1000 fpm over unwrapped pizza reduces the freezing time and product exposure by almost half. This variation shows that product testing is needed to design the special enclosures intended for blast freezing and/or automated food processing. Sample tests should yield the following information: ideal air temperature, air velocity, product weight loss, and dwell time. With this information, the proper unit or product coolers, as well as the supporting refrigeration equipment and controls, can be selected.

RATING UNITS

Currently, no industry standard exists for rating unit and product coolers. Part of the difficulty in developing a workable standard is that many variables are encountered. Cooler coil performance and capacities should be based on a fixed set of conditions and greatly depend on (1) air velocity, (2) refrigerant velocity, (3) temperature difference, (4) frost condition, and (5) superheating adjustment. The most significant items are refrigerant velocity, as related to refrigerant feed through the coil, and frost condition and defrosting in low-temperature applications. The following sections address some of the problems involved in arriving at a common rating.

Refrigerant Velocity

Depending on the commercial refrigerant feed method used, both the capacity ratings of the cooler and the refrigerant velocity vary. The following feed methods are used.

Dry Expansion. In this system, a thermostatically controlled, direct-expansion valve allows just enough liquid refrigerant into the cooling coil to ensure that it vaporizes at the outlet. In addition, 5 to 15% of the coil surface is used to superheat the vapor. The direct expansion (DX) coil ratings are usually the lowest of the various feed methods.

Recirculated Refrigerant. This system is similar to a dry expansion feed because it has a hand expansion valve or metering device to control the flow of the entering liquid refrigerant. The coil is intentionally overfed to eliminate superheating of the refrigerant vapor. The amount of liquid refrigerant pumped through the coil may be two to six times that of a dry expansion coil. As a result, this coil's capacity is higher than that for a dry expansion feed (see Chapter 1).

Flooded. This system has a liquid reservoir (surge drum) located adjacent to each unit or set of units. The surge drum is filled with a subcooled refrigerant and connected to the cooler coil. To ensure gravity flow of this refrigerant and a completely wet internal coil surface, the liquid level in the surge drum must be higher than the top of the coil. The capacity of gravity-recirculated feed is usually the highest attainable.

Brine. This term encompasses any liquid or solution that absorbs heat within the coil without a change in state. Ethylene glycol and water, propylene glycol and water, and R-11 are used, in addition to solutions of calcium chloride or sodium chloride and water. The capacity ratings for brine coils depend on many variables such as flow, viscosity, specific heat, and density. As a result, this rating is obtained by special request from a coil manufacturer and generally is 10 to 40% less than its flooded rating application.

Frost Condition

Frost condition and defrosting are perhaps the most indeterminate variables that affect the capacity rating of forced-air cooler coils. Any unit operating below 32 °F coil temperature accumulates frost or ice. Although a light frost accumulation slightly improves heat transfer of the coil, continuous accumulation negatively affects coil performance. Ultimately, defrosting is the only solution.

Enclosure Air Temperature Above 35 °F. Whenever the enclosure air is 35 °F or slightly warmer, it can be used to defrost the coil. However, some of the moisture on the coil surface evaporates into the air, which is undesirable for low-humidity application. The following methods of control are commonly used.

1. If the refrigeration cycle is interrupted by a defrost timer, the continually circulating air melts the coil frost and ice. The timer can operate either the compressor or a refrigerant solenoid valve.
2. An oversized unit cooler controlled by a wall thermostat defrosts during its normal *off* and *on* cycling. The thermostat can control a refrigeration solenoid in a multiple-coil system or the compressor in a unitary installation.
3. Pressure control can be used for slightly oversized unitary systems. A low-pressure switch connected to the compressor suction line is set at a cut-out point such that the design suction pressure corresponds to the saturated temperature required to handle the maximum enclosure load. The suction pressure at the compressor drops and causes the compressor motor to stop as the enclosure load fluctuates or as the oversized compressor overcomes the maximum loading.

   A thermostatic expansion valve on the unit cooler controls the liquid refrigerant flow into the coil, which varies with the load. The cut-in point, which starts the compressor motor, should be set at the suction pressure, which corresponds to the equivalent saturated temperature of the desired refrigerated enclosure air. The pressure differential between the cut-in and cut-out points corresponds to the temperature difference between the enclosure air and the coil temperatures. The pressure settings should allow for the pressure drop in the suction line.

Enclosure Air Temperature Below 35 °F. Whenever the enclosure air is below 35 °F, supplementary heat must be introduced into the enclosure to defrost the coil surface and drain pan. Unfortunately, some of this defrost heat remains in the enclosure until the unit starts operation after the defrost cycle. No two defrost methods have the same timing input.

1. Hot-gas defrost can be the fastest and most efficient method if an adequate supply of hot gas is available. The hot refrigerant discharge gas internally cleans the draw pan assembly tubes and aids in returning the lubricants back to the compressor. It can be used for small commercial units and large industrial systems, and for low-temperature applications. Hot-gas defrost can also improve capacity because it can remove some of the load from the condenser when used to defrost multiple evaporators alternately on a large, continuously operating compressor system. This method of defrost puts the least amount of heat into the enclosure air, especially when latent gas defrosting is used.
2. Electric defrost effectiveness depends on the location of the electric heating elements. The electric heating elements can be placed either in contact with the finned coil surface or inserted into the interior of the coil element. In this latter method, either special fin holes or dummy tubes are used. Electric defrost can be efficient and rapid; it is simple to operate and maintain, but it does dissipate the most heat into the enclosure.
3. Heated air may be used as a defrost method. Some unit coolers are constructed to isolate the frosted coil from the cold enclosure air. Once isolated, the air around the coil is heated by hot gas or electric heating elements and is circulated to hasten the defrost. This heated air also must heat a drain pan, which is needed in all enclosures at temperatures of 34 °F or less. Some units have specially constructed housings and ducting to run warm air from adjoining areas.
4. Water defrost is the quickest method of defrosting a unit. It is efficient and effective for rapid cleaning of the complete coil surface. Water defrost can be performed manually or on an automatic timed cycle. This method becomes less desirable as the enclosure temperature decreases much below freezing, but it has been successfully used in cases as low as −40 °F. Water defrost is used more for industrial product cooler units than for small ceiling-suspended units.

5. Hot brine can be used as a defroster, as brine-cooled coils can have a heater to heat brine for the defrost cycle. This system heats from within the coil and is as rapid as hot-gas defrost. The heat source can be steam, electric resistance elements, or condensing water.

Defrost Control. Each defrosting method is done with the fan turned off. Frequently, two methods are run simultaneously to shorten the defrost cycle. Both inadequate defrost time and overdefrosting can degrade overall performance; thus, a defrost cycle is best ended by monitoring temperature. A thermostat may be mounted within the cooler coil to sense a rise in the temperature of the finned or tubed surface. A temperature of at least 40 °F indicates the removal of frost and automatically returns the unit to the cooling cycle.

Fan operation is delayed, usually by the same thermostat, until the coil surface temperature approaches its normal operating level. This practice prevents unnecessary heating of the enclosure after defrost and also prevents drops of defrost water from being blown off the coil surface. In some applications, fan delay after defrost is essential to prevent a rapid buildup of air pressure, which could structurally damage the enclosure.

Initiation of defrost can be automated by time clocks, running time monitors, air pressure differential controls, or by monitoring the air temperature difference through the coil (which increases as the airflow is reduced by frost accumulation). Adequate supplementary heat for the drain pan and drain lines should be considered. Also, drain lines should be pitched properly and trapped outside the cold area to keep them from freezing.

Basic Cooling Capacity

Most rating tables state gross capacity and assume that the fan assembly or defrost heat is included as part of the enclosure load calculation. Some manufacturers' cooler coil ratings may appear as sensible, while some may be listed as total capacity, which includes the sensible and latent capacity. Some ratings involve reduction factors to account for frost accumulation in low-temperature applications or for some unusual condition. Some include multiplication factors for various refrigerant types.

The rating, known as the basic cooling capacity, is based on the temperature difference between the inlet air and the refrigerant within the coil; Btu/h per degree TD (temperature difference) is the dimension used. The coil inlet air temperature is considered the same as the enclosure air temperature. This practice is common for smaller enclosure applications. For larger installations, as well as heavy-use process work, manufacturer's published ratings should be applied on the average of the coil inlet-to-outlet temperatures. This is regarded as the average enclosure temperature. The refrigerant temperature is usually the temperature equivalent to the saturated pressure at the coil outlet.

The TD necessary to obtain the unit cooler capacity varies with each application. It may be as low as 8 °F for wet storage coolers and as high as 25 °F for workrooms. The TD can be related to the desired humidity requirements. The smaller the TD, the smaller the amount of dehumidification caused by the coil. The following information gives general guidance for selecting a proper TD.

Medium-temperature applications above 25 °F saturated suction:

- For a very high relative humidity (about 90%), a temperature difference of 8 to 10 °F is common.
- For a high relative humidity (approximately 80%), a temperature difference of 10 to 12 °F is recommended.
- For a medium relative humidity (approximately 75%), a temperature difference of 12 to 16 °F is recommended.
- Temperature differences beyond these limits usually result in low enclosure humidities, which dry the product. However, for packaged products and workrooms, a TD of 25 to 30 °F is not unusual. Paper storage or similar products also require a low humidity level. Here, a TD of 20 to 30 °F may be necessary.

For ow-temperature applications below 25 °F saturated suction, the temperature difference is generally kept below 15 °F because of system economics and frequency of defrosting rather than for humidity control.

INSTALLATION AND OPERATION

Whenever possible, refrigerating air-cooling units should be located away from enclosure entrance doors and passageways. This practice helps reduce coil frost accumulation, as well as fan blade icing. The cooler manufacturer's installation, start-up, and operation instructions generally give the best information. Upon installation, the unit nameplate data (model, refrigerant type, electrical data, warning notices, certification emblems, and so forth) should be recorded. This information should be compared to the job specifications, as well as the manufacturer's instructions for correctness.

GENERAL INFORMATION

Additional information on the selection, ratings, installation, and maintenance of cooler coil units is available from unit manufacturers. Also, the section on Food Refrigeration has specific product cooling information.

REFERENCES

ARI. 1989. Unit coolers for refrigeration. *Standard* 420. Air Conditioning and Refrigeration Institute, Arlington, VA.

ASHRAE. 1990. Method of testing forced and natural convection air coolers for refrigeration. *Standard* 25.

CHAPTER 42

LIQUID CHILLING SYSTEMS

A LIQUID chilling system cools water, brine, or other secondary coolant for air conditioning or refrigeration. The system may be either factory assembled and wired or shipped in sections for erection in the field. The most frequent application is water chilling for air conditioning, although both brine cooling for low-temperature refrigeration and chilling of fluids in industrial processes are also common.

The basic components of a liquid chilling system include a compressor, a liquid cooler (evaporator), a condenser, a compressor drive, a liquid refrigerant expansion or flow-control device, and a control center; the system may also include a receiver, an economizer, and/or a subcooler. In addition, certain auxiliary components may be used, such as an oil cooler, an oil separator, an oil-return device, a purge unit, an oil pump, a refrigerant transfer unit, and/or additional control valves.

GENERAL CHARACTERISTICS

PRINCIPLES OF OPERATION

Liquid (usually water) enters the cooler, where it is chilled by liquid refrigerant evaporating at a lower temperature. The refrigerant vaporizes and is drawn into the compressor, which increases the pressure and temperature of the gas so that it may be condensed at the higher temperature in the condenser. The condenser cooling medium is warmed in the process. The condensed liquid refrigerant then flows to the evaporator through a metering device. A fraction of the liquid refrigerant changes to vapor (*flashes*) as the pressure drops between the condenser and the evaporator. Flashing cools the liquid to the saturated temperature at the evaporator pressure. It produces no refrigeration effect in the cooler. The following modifications minimize the problem of flashing.

Subcooling. Condensed refrigerant may be subcooled to a temperature below its saturated condensing temperature in either the subcooler section of a water-cooled condenser or a separate heat exchanger. Subcooling reduces the amount of flashing and increases the refrigeration effect within the chiller.

Economizing can be either a direct-expansion (DX) or flash-type system. In a DX system, the main liquid is usually cooled in the shell side of a shell-and-tube heat exchanger at condensing pressure, from the saturated condensing temperature to within several degrees of the intermediate saturated temperature. Before cooling, a small portion of the liquid flashes and evaporates in the tube side of the heat exchanger to cool the main liquid flow. Although subcooled, the liquid will still be at the condensing pressure.

In a flash-type system, the entire liquid flow is expanded to the intermediate pressure in a vessel. This supplies liquid to the cooler at the saturated intermediate pressure; however, the liquid is at the intermediate pressure.

In either case, the flash gas enters the compressor at a high-pressure stage of a multistage centrifugal compressor, at the intermediate stage of an integral two-stage reciprocating compressor, at an intermediate pressure port of a screw compressor, or at the inlet of a high-pressure stage on a multistage reciprocating or screw compressor.

Both of these methods (sometimes combined for maximum effect) reduce flash gas and increase the net refrigeration effect per unit power consumption.

Liquid Injection. Condensed liquid is throttled to the intermediate pressure and injected into the second-stage suction of the compressor to prevent excessively high discharge temperatures and, in the case of centrifugal machines, to reduce noise. In the case of screw compressors, condensed liquid is injected into a port fixed at slightly below discharge pressure to provide oil cooling.

COMMON LIQUID CHILLING SYSTEMS

Basic System

The refrigeration cycle of a basic system is shown in Figure 1. Chilled water enters the cooler at 54 °F, for example, and leaves at 44 °F. Condenser water leaves a cooling tower at 85 °F, enters the condenser, and returns to the cooling tower at 95 °F. Condensers may also be cooled by air or through evaporation.

This system, with a single compressor and one refrigerant circuit employing a water-cooled condenser, is extensively used to chill water for air conditioning. It is relatively simple and compact.

These liquid chillers are available with hermetic, semihermetic, and open-type compressors with various means of starting.

The preparation of this chapter is assigned to TC 8.1, Positive Displacement Compresors, and TC 8.2, Centrifugal Machines.

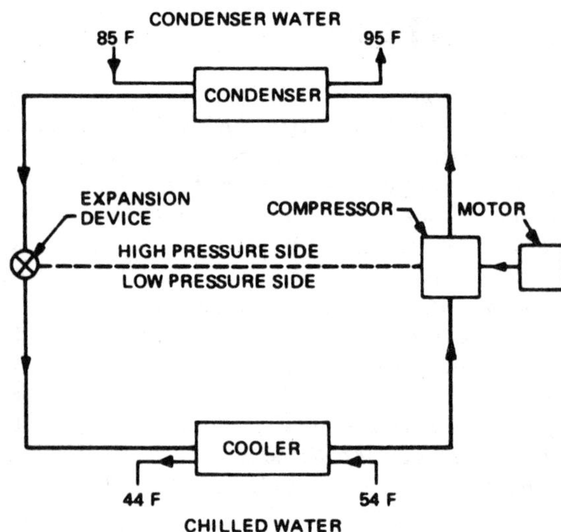

Fig. 1 Equipment Diagram for Basic Liquid Chiller

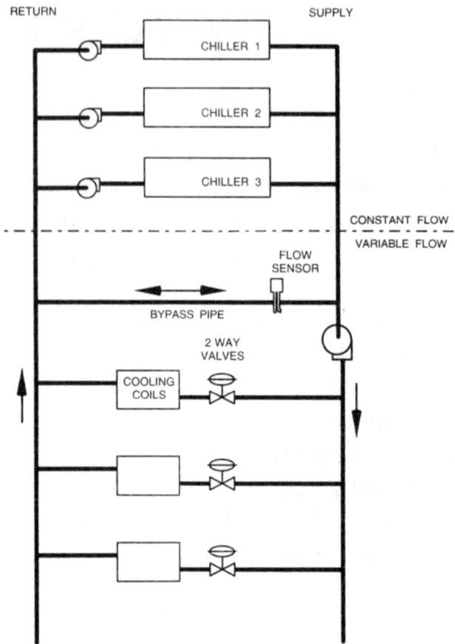

Fig. 2 Decoupled System

Multiple Chiller System

Multiple chiller systems offer some standby capacity if repair work must be done on one chilling machine. Starting in-rush current is reduced, as well as power costs at partial-load conditions. Maintenance can be scheduled for one chilling machine during part-load times, and sufficient cooling can still be provided by the remaining unit(s). These advantages require an increase in installed cost and space, however.

Constant water flow through the chillers is required for stable control of the chilled water system. Load variation is temperature-related and is easily detected by temperature controls. In contrast, with variable water flow through the chillers, load variation is flow-related. Since a temperature control system cannot sense flow variation, it is unable to adjust its response to that needed to produce stable control. However, some applications have variable water flow through the cooling coils. In this case, a decoupled system (primary/secondary) is typically used to maintain constant water flow through the chillers. This system involves separating the distribution pumping system from the production pumping system. It allows variable flow through the cooling coils but maintains constant water flow through the chillers, allowing good control of the multiple chillers.

A typical decoupled system is shown in Figure 2. The flow system uses multiple pumps. The common element is a length of bypass piping that connects the return and supply headers. From the control standpoint, the entire system is greatly simplified. Each chiller-pump combination operates as an individual and is independent from the remaining chillers. Capacity control is exactly as if each chiller was alone.

Instead of using temperature as an indicator of system demand, relative flow is the indicator. If the distribution system demands greater flow than that supplied by the number of chiller-pump combinations, return water is forced through the bypass into the supply header. This is a clear indication of the need for additional chiller capacity and another chiller-pump set is started. Overcapacity is detected by bypass flow in the opposite direction, and the chiller-pump sets are turned off.

Two basic multiple chiller systems are used: *parallel* and *series* chilled water flow. In the parallel arrangement, liquid to be chilled is divided among the liquid chillers; the multiple chilled streams are combined again in a common line after chilling. As the cooling load decreases, one unit may be shut down, but the remaining unit(s) must then provide colder-than-design chilled liquid so that when all streams combine (including one from the idle machine), design chilled liquid supply temperature is provided in the common line.

When the design chilled water temperature is above approximately 45 °F, all units should be controlled by the combined exit water temperature or by the return water temperature (RWT), since overchilling will not cause dangerously low water temperature in the operating machine(s). Chilled water temperature can be used to cycle one unit off when it drops below a value corresponding to a capacity that can be matched by the remaining units.

When the design chilled water temperature is below about 45 °F, each machine should be controlled by its own chilled water temperature, both to prevent dangerously low evaporator temperatures and to avoid frequent shutdowns by the low-temperature cutout. In this case, the temperature differential setting of the RWT must be adjusted carefully to prevent short cycling caused by the step increase in chilled water temperature when one chiller is cycled off. These control arrangements are shown in Figures 3 and 4.

In the series arrangement, the chilled liquid pressure drop may be higher if shells with fewer liquid-side passes or baffles are not available. No overchilling by either unit is ever required, and compressor power consumption is lower than it is for the parallel arrangement at partial loads. Since the evaporator temperature never drops below the design value (because no overchilling is necessary), the chances of evaporator freezeup are minimized. However, the chiller should still be protected by a low-temperature safety control.

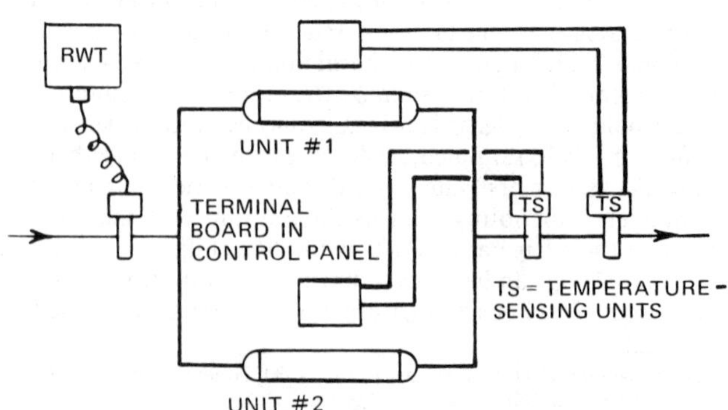

Fig. 3 Parallel Operation High Design Water Leaving Coolers (Approximately 45 °F and Above)

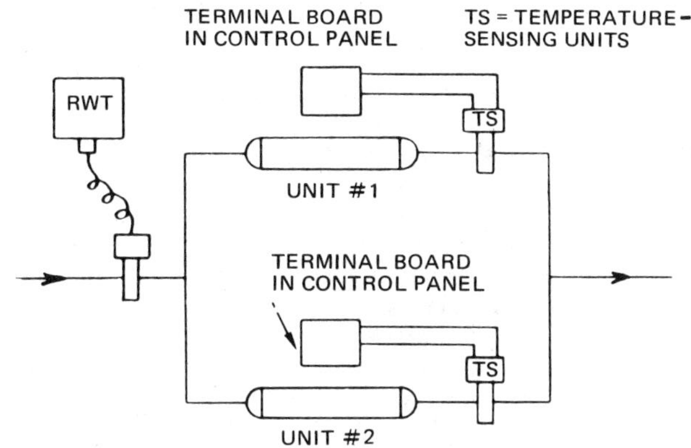

Fig. 4 Parallel Operation Low Design Water Leaving Coolers (Below Approximately 45 °F)

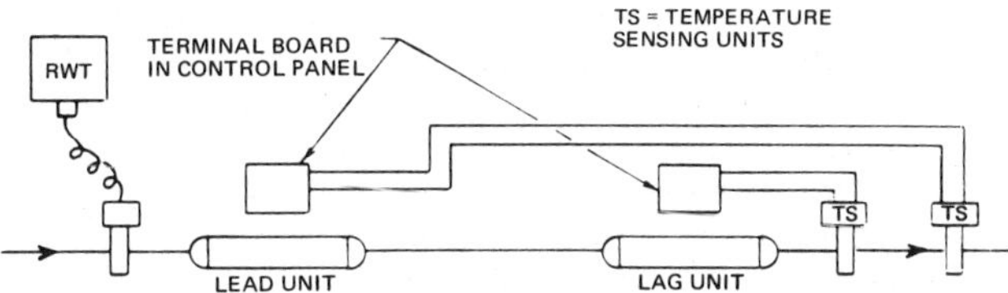

Fig. 5 Series Operation

When condensers are water-cooled, they are best piped in series counterflow so that the lead machine is provided with warmer condenser and chilled water and the lag machine is provided with colder entering condenser and chilled water. Thus, refrigerant compression for each unit is nearly the same. If about 55% of design cooling capacity is assigned to the lead machine and about 45% to the lag machine, identical units can be used. In this way, either machine can provide the same standby capacity if the other is down, and lead and lag machines may be interchanged to equalize the number of operating hours on each.

A control system for two machines in series is shown in Figure 5. (On reciprocating chillers, RWT sensing is usually used instead of leaving water sensing because it allows closer temperature control.) Both units are modulated to a certain capacity; then, one unit shuts down, leaving less than 100% load on the operating machine.

One machine should be shut down as soon as possible, with the remaining unit carrying the full load. This not only reduces the number of operating hours on a unit, but also leads to less total power consumption because the COP tends to decrease below the full load value when unit load drops much below 50%.

Heat-Recovery Systems

Any building or plant requiring the simultaneous operation of heat-producing and cooling equipment has the potential for a heat-recovery installation. Heat-recovery equipment should be considered for all new or retrofit installations. In some cases, the installed cost may be less because of the elimination or reduction of both heating equipment and the space required for it.

Heat-recovery systems extract heat from chilled liquid and reject some of that heat, plus the energy of compression, to a warm-water circuit for reheat or heating. Air-conditioned spaces thus furnish heating for other spaces in the same building. During the full-cooling season, all heat must be rejected outdoors, usually by a cooling tower. During spring or fall, some heat is required inside, while a portion of the heat extracted from the air-conditioned spaces must be rejected outside simultaneously.

Heat recovery offers the user low heating costs and reduces space requirements for mechanical equipment. A control system

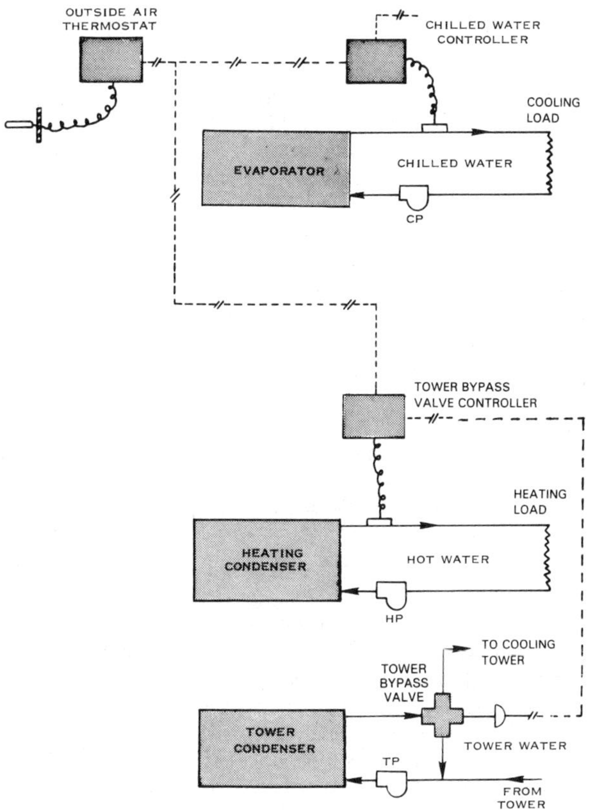

Fig. 6 Heat-Recovery Control System

must be designed carefully, however, to take the greatest advantage of the recovered heat and to maintain proper temperature and humidity in all parts of the building. Chapter 8 of the 1992 ASHRAE *Handbook—Systems and Equipment* covers balanced heat-recovery systems.

Since cooling tower water is not satisfactory for heating coils, a separate, closed warm-water circuit with another condenser bundle or auxiliary condenser, in addition to the main water chiller condenser, must be provided. In some cases, it is economically feasible to use a standard condenser and a closed-circuit water cooler.

A recommended control scheme is shown in Figure 6. The heating water temperature is controlled by a cooling tower bypass valve, which modulates the flow of condenser cooling water to the tower. An outside air thermostat resets the hot water control point upward as the outdoor temperature drops and resets the chilled water temperature control point upward on colder days. In this way, extra power is not consumed unnecessarily by the compressor in attempting to maintain summer design coil temperatures during dry, cold outdoor conditions.

EQUIPMENT SELECTION

The largest factor in total liquid chiller owning cost is the cooling load size; therefore, an accurate calculation of total needed liquid chiller capacity should be made. The practice of adding a 10 to 20% safety factor to load estimates is not only unnecessary because of the availability of accurate load estimating methods, but it also proportionately increases costs related to equipment purchase, installation, and poor efficiency resulting from wasted power. Oversized equipment can also cause operational difficulties such as frequent on-off cycling or surging of centrifugal machines at low loads. The penalty for a small underestimation of cooling load, however, is not serious. On the few design load days of the year, an increase in chilled liquid temperature is often acceptable. However, for some industrial or commercial loads, a safety factor can be added to the load estimate.

The life-cycle cost approach should be used to minimize the overall purchase and operating costs. It is discussed in Chapter 33 of the 1991 ASHRAE *Handbook—Applications*. Total owning cost is composed of the following:

1. **Purchase Price.** Each machine type and/or manufacturer's model should include all the necessary auxiliaries such as starters and vibration mounts. If these are not included, their price should be added to the base price.
2. **Installation Cost.** Factory-packaged machines are both less expensive to install and usually considerably more compact, resulting in space savings. The cost of field assembly of field-erected chillers must also be evaluated.
3. **Energy Cost.** Using an estimated load schedule and part-load power consumption curves furnished by the manufacturer, a year's energy cost should be calculated.
4. **Maintenance Cost.** Each bidder may be asked to quote on a maintenance contract on a competitive basis.
5. **Insurance and Taxes.**

For heat-recovery package chillers, system cost and performance should be compared, in addition to equipment costs. For example, the heat-recovery chiller installed cost should be compared with the installed cost of a chiller plus a separate heating system. The following factors should also be considered: (1) energy costs, (2) maintenance requirements, (3) life expectancy of the equipment, (4) standby arrangement, (5) relationship of heating to cooling loads, (6) effect of package selection on sizing, and (7) type of peripheral equipment.

Condensers and coolers are often available with either *liquid heads*, which require disconnection of water piping for tube access and maintenance, or *marine-type water boxes*, which permit tube access with water piping intact. The former is considerably lower in price. The cost of disconnecting piping must be greater than the additional cost of marine-type water boxes to justify their use. However, it is only necessary to install a union or flange connection in the piping to facilitate the removal of heads.

The above approaches are helpful in choosing between equipment types and makes. Chapter 33 of the 1991 ASHRAE *Handbook—Applications* is recommended for further information. The following guidelines can be used to determine the types of liquid chillers that are generally used for air conditioning:

Up to 25 tons	— Reciprocating
25 to 80 tons	— Screw or Reciprocating
80 to 200 tons	— Screw, Reciprocating, or Centrifugal
200 to 800 tons	— Screw or Centrifugal
Above 800 tons	— Centrifugal

For air-cooled condenser duty, brine chilling, or other high pressure applications from 80 to about 200 tons, reciprocating and screw liquid chillers are more frequently installed than centrifugals. Centrifugal liquid chillers (particularly multistage machines), however, may be applied quite satisfactorily at high pressure conditions.

Factory packages are available to about 2400 tons and field-assembled machines to about 10,000 tons.

CONTROL

Liquid Chiller Controls

The *chilled liquid temperature sensor* sends an air pressure (pneumatic control system) or electrical signal (electronic control system) to the control circuit, which then modulates compressor capacity in response to leaving or return chilled liquid temperature change from its set point.

Compressor capacity adjustment is accomplished differently on the following liquid chillers:

Reciprocating chillers use combinations of cylinder unloading and on-off compressor cycling of single or multiple compressors.

Centrifugal liquid chillers, driven by electric motors, commonly use adjustable prerotation vanes. Turbine drives and inverter-driven, variable-speed electric motors allow the use of speed control in addition to prerotation vane modulation, reducing power consumption at partial loads.

Screw compressor liquid chillers use a slide valve to adjust the length of the compression path. Inverter-driven, variable-speed electric motors and turbine and engine drives can also modulate screw compressor speed to control capacity.

In air-conditioning applications, most centrifugal and screw compressor chillers modulate from 100% to approximately 10% load. Although relatively inefficient, hot-gas bypass can be used to reduce capacity to nearly 0% with the unit in operation.

Reciprocating chillers are available with simple on-off cycling control in small capacities and with multiple steps of unloading down to 12.5% in the largest multiple compressor units. Most intermediate sizes provide unloading to 50, 33, or 25% capacity. Hot-gas bypass can reduce capacity to nearly 0%.

The *water temperature controller* is a thermostatic device that unloads or cycles the compressor(s) when the cooling load drops below minimum unit capacity. An *antirecycle timer* is sometimes used to limit starting frequency.

On centrifugal or screw compressor chillers, a *current limiter* or *demand limiter* limits compressor capacity during periods of possible high power consumption (such as pulldown) to prevent current draw from exceeding the design value; such a limiter can be set to limit demand, as described in the section on Centrifugal Liquid Chillers.

Controls That Influence the Liquid Chiller

Condenser cooling water control is required to regulate pressure. Normally, the temperature of the water leaving a cooling tower can be controlled by fans, dampers, or a water bypass around the tower. Bypass around the tower allows water velocity through the condenser tubes to be maintained, thereby preventing low-velocity fouling.

A flow-regulating valve is another common means of control. The orifice of this valve modulates in response to condenser pressure. For example, a reduction in pressure decreases the water flow, which, in turn, raises the condenser pressure to the desired level.

For air-cooled or evaporative condensers, compressor discharge pressure can be controlled by cycling fans, shutting off circuits, or flooding coils with liquid refrigerant to reduce the heat transfer.

A reciprocating chiller usually has a thermal expansion valve, which requires a restricted range of pressure to avoid starving the evaporator (at low pressure).

An expansion valve(s) usually controls a screw compressor chiller. Cooling tower water temperature can be allowed to fall with decreasing load from the design condition to the chiller manufacturer's recommended minimum limit.

Screw compressor chillers above 150 tons may use flooded-type evaporators and evaporator liquid refrigerant controls similar to those used on centrifugal chillers.

A thermal expansion valve may control a centrifugal chiller at low capacities, while higher capacity machines employ a high-pressure float, orifice(s), or even a low-side float valve to control refrigerant liquid flow to the cooler. These latter types of controls allow relatively low condenser pressures, particularly at partial loads. Also, a centrifugal machine may surge if pressure is not reduced when cooling load decreases. In addition, low pressure reduces compressor power consumption and operating noise. For these reasons, in a centrifugal installation, cooling tower water temperature should be allowed to fall naturally with decreasing load and wet-bulb temperature, except that the liquid chiller manufacturer's recommended minimum limit must be observed.

Safety Controls

Some or all of the cutouts listed below may be provided in a liquid chilling package to stop the compressor(s) automatically. Cutouts may be manual or automatic reset.

1. **High Condenser Pressure.** This pressure switch opens if the compressor discharge pressure exceeds the value prescribed in ASHRAE *Standard* 15-1992, Safety Code for Mechanical Refrigeration.
2. **Low Refrigerant Pressure (or Temperature).** This device opens when evaporator pressure (or temperature) reaches a minimum safe limit.
3. **High Oil Temperature.** This device protects the compressor if loss of oil cooling occurs or if a bearing failure causes excessive heat generation.
4. **High Motor Temperature.** If loss of motor cooling or overload because of a failure of operating controls occurs, this device shuts down the machine. It may consist of direct-operating bimetallic thermostats, thermistors, or other sensors embedded in the stator windings; it may be located in the discharge gas stream of the compressor.
5. **Motor Overload.** Some small reciprocating compressor hermetic motors may use a directly operated overload in the power wiring to the motor. Some larger motors use pilot-operated overloads. Centrifugal and screw compressor motors generally use starter overloads or current-limiting devices to protect against overcurrent.
6. **Low Oil Sump Temperature.** This switch is used either to protect against an oil heater failure or to prevent starting after a prolonged shutdown before the oil heaters have had time to drive off refrigerant dissolved in the oil.
7. **Low Oil Pressure.** To protect against clogged oil filters, blocked oil passageways, loss of oil, or an oil pump failure, a switch shuts down the compressor when oil pressure drops below a minimum safe value or if sufficient oil pressure is not developed shortly after compressor startup.
8. **Chilled Liquid Flow Interlock.** This device may not be furnished with the liquid chilling package but is needed in the external piping to protect against a cooler freezeup in the event of a liquid flow stoppage.
9. **Condenser Water Flow Interlock.** This device is sometimes used in the external piping.
10. **Low Chilled Liquid Temperature.** Sometimes called *freeze protection*, this cutout operates at a minimum safe value of leaving chilled liquid temperature to prevent cooler freezeup in the case of an operating control malfunction.
11. **Relief Valves.** In accordance with ASHRAE *Standard* 15-1992, relief valves or rupture disks, set to relieve at the shell design working pressure, must be provided on most pressure vessels. Fusible plugs may also be used in some locations. Pressure relief devices should be vented outdoors or to the low-pressure side, in accordance with the standard.

STANDARDS

ARI *Standards* 550-92, Centrifugal and Rotary Screw Water-Chilling Packages, and 590-92, Positive Displacement Compressor, provide guidelines for the rating of centrifugal and reciprocating liquid chilling machines, respectively.

The design and construction of refrigerant pressure vessels are governed by the ASME *Boiler and Pressure Vessel Code,* Section VIII, except when design working pressure is 15 psig or less (as is usually the case for R-123 liquid chilling machines). The water-side design and construction of a condenser or cooler is not within the scope of the ASME code unless the design pressure is greater than 300 psi or the design temperature is greater than 210 °F.

ASHRAE *Standard* 15-1992 applies to all liquid chillers and new refrigerants on the market. New standards for equipment rooms are included. Methods for the measurement of unit sound levels are described in ARI *Standard* 575-87, Method of Measuring Machinery Sound Within an Equipment Space.

METHODS OF TESTING

All tests of reciprocating liquid chillers for rating or verification of rating should be conducted in accordance with ASHRAE *Standard* 30-1978, Methods of Testing Liquid Chilling Packages.

Centrifugal or screw liquid chiller ratings should be derived and verified by test in accordance with ARI *Standard* 550-92.

GENERAL MAINTENANCE

Listed below are some of the general maintenance specifications that apply equally to reciprocating, centrifugal, and screw chillers. The equipment should be neither overmaintained nor neglected. A preventive maintenance schedule should be established; the items covered can vary with the nature of the application.

The following list is intended as a guide; in all cases, the manufacturer's specific recommendation should be followed.

Continual Monitoring

Condenser water treatment—treatment is determined specifically for the condenser water used.
Operating conditions—daily log sheets are recommended.

Periodic Checks

Leak check
Purge operation
System dryness
Oil level
Oil filter pressure drop
Refrigerant quantity or level
System pressures and temperatures
Water flows
Expansion valves operation

Regularly Scheduled Maintenance

Condenser and oil cooler cleaning
Calibrating pressure, temperature, and flow controls
Tightening wires and power connections
Inspection of starter contacts and action
Dielectric checking of hermetic and open motors
Oil filter and drier change
Analysis of oil and refrigerant
Seal inspection
Partial or complete valve or bearing inspection, as per manufacturer's recommendations

RECIPROCATING LIQUID CHILLERS

EQUIPMENT DESCRIPTION

Components and Their Function

The *reciprocating compressor* described in Chapter 35 of the 1992 ASHRAE *Handbook—Systems and Equipment* is a positive-displacement machine that maintains fairly constant volume flow rate over a wide range of pressure ratios. The following three types of compressors are commonly used in liquid chilling machines:

1. Welded hermetic, to about 25 tons chiller capacity
2. Semihermetic, to about 200 tons chiller capacity
3. Direct-drive open, to about 200 tons chiller capacity

Open motor-driven liquid chillers are usually more expensive than *hermetics* but are more efficient. *Hermetic motors* are generally suction gas cooled; the rotor is mounted on the compressor crankshaft.

Condensers may be evaporative, air- or water-cooled. Water-cooled versions may be either tube-in-tube, shell-and-coil, or shell-and-tube. Most shell-and-tube condensers can be repaired, while others must be replaced if a refrigerant-side leak occurs.

Air-cooled condensers are much more common than evaporative condensers. Less maintenance is needed for air-cooled heat exchangers than for the evaporative type. Remote condensers can be applied with condenserless packages. (Information on condensers can be found in Chapter 36 of the 1992 ASHRAE *Handbook—Systems and Equipment.*)

Coolers are usually direct-expansion, in which refrigerant evaporates while it is flowing inside tubes and chilled liquid is cooled as it is guided several times over the outside of the tubes by shell-side baffles. Flooded coolers are sometimes used on industrial chiller packages. Flooded coolers maintain a level of refrigerant liquid on the shell side of the cooler, while the liquid to be cooled flows through tubes inside the cooler. Tube-in-tube coolers are sometimes used with small machines; they offer low cost when repairability and required space are not important criteria. (A detailed description of coolers can be found in Chapter 38 of the 1992 ASHRAE *Handbook—Systems and Equipment.*) The thermal expansion valve, capillary, or other expansion device modulates refrigerant flow from the condenser to the cooler to maintain enough suction superheat to prevent any unevaporated refrigerant liquid from reaching the compressor. Excessively high values of superheat are avoided so that unit capacity is not reduced. (For additional information, see Chapter 44.)

Oil cooling is not usually required for air conditioning. However, if oil cooling is necessary, a refrigerant-cooled coil in the crankcase or a water-cooled oil cooler may be used. Oil coolers are often used in conjunction with low suction temperatures or high-pressure ratio applications when extra oil cooling is needed.

Capacities and Types Available

Available capacities range from about 2 to 200 tons. Multiple reciprocating compressor units have become popular for the following reasons:

1. The number of capacity increments is greater, resulting in closer liquid temperature control, lower power consumption, less current in-rush during starting, and extra standby capacity.
2. Multiple refrigerant circuits are employed, resulting in the potential for limited servicing or maintenance of some components while maintaining cooling.

Selection of Refrigerant

R-12 and R-22 have been the primary refrigerants used in chiller applications. CFC-12 is being replaced with HFC-134a, which has similar properties. However, it should be noted that R-134a requires synthetic lubricants because it is not compatible with mineral oils. R-134a is suitable for both open and hermetic compressors.

R-22 provides greater capacity than R-134a for a given compressor displacement. R-22 is the refrigerant of choice for most open and hermetic compressors, but as an HCFC, it is scheduled for phaseout in the future. R-717 (ammonia) has similar capacity characteristics to R-22, but, because of odor and toxicity, R-717 is subject to restrictions for use in public or populated areas. However, R-717 chillers are becoming more popular because of bans on CFC and HCFC refrigerants. R-717 units are open-drive compressors and are piped with steel because copper cannot be used in ammonia systems.

Alternative HFC refrigerants, azeotropes, zeotropes, and blends are being developed. For more information, refer to ASHRAE *Standard* 34-1992, Number Designation and Safety Classification of Refrigerants.

PERFORMANCE CHARACTERISTICS AND OPERATING PROBLEMS

A distinguishing feature of the reciprocating compressor is its pressure rise versus capacity characteristic. Pressure rise has only a slight influence on the volume flow rate of the compressor, and, therefore, a reciprocating liquid chiller retains nearly full cooling capacity, even on above-design wet-bulb days. It is well suited for air-cooled condenser application and low-temperature refrigeration.

A typical performance characteristic is shown in Figure 7 and compared with the centrifugal and screw compressors. Methods of capacity control are furnished by the following:

1. Unloading of compressor cylinders (one at a time or in pairs)
2. On-off cycling of compressors
3. Hot-gas bypass
4. Compressor speed control
5. A combination of the above methods

Figure 8 illustrates the relationship between system demand and performance of a compressor with three steps of unloading. As cooling load drops to the left of the fully loaded compressor line (A), compressor capacity is reduced to that represented by line (B), which produces the required refrigerant flow. Since cooling load varies continuously while machine capacity is available in fixed increments, some compressor on-off cycling or successive loading and unloading of cylinders is required to maintain fairly constant liquid temperature. In practice, a good control system minimizes the load-unload or on-off cycling frequency while maintaining satisfactory temperature control.

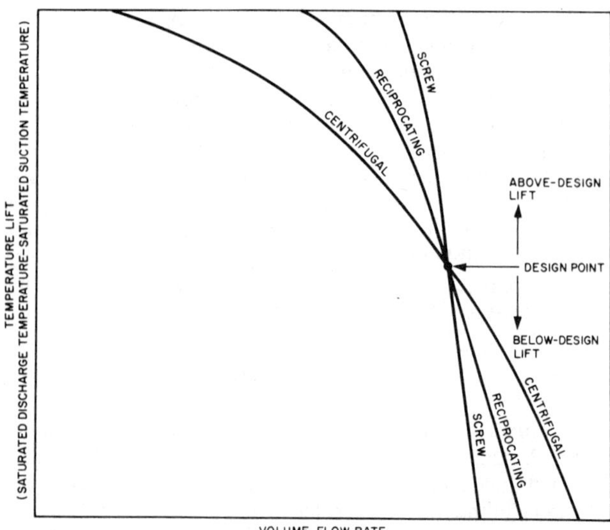

Fig. 7 Comparison of Single-Stage Centrifugal, Reciprocating, and Screw Compressor Performance

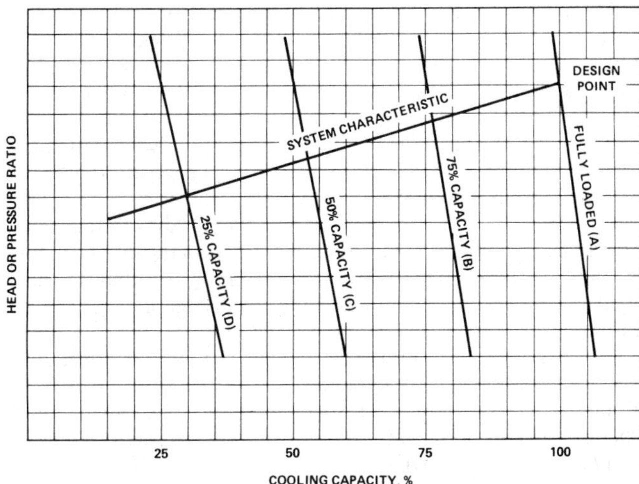

Fig. 8 Reciprocating Liquid Chiller Performance with Three Equal Steps of Unloading

METHOD OF SELECTION

Ratings

Two types of ratings are published. The first, for a packaged liquid chiller, lists values of capacity and power consumption for many combinations of leaving condenser water and chilled water temperatures (ambient dry-bulb temperatures for air-cooled models). The second type of rating shows capacity and power consumption for different condensing temperatures and chilled water temperatures. This type of rating permits selection with a remote condenser that can be evaporative, water-, or air-cooled. Sometimes the required rate of heat rejection is also listed to aid in selection of a separate condenser.

Power Consumption

With all liquid chilling systems, power consumption increases as condensing temperature rises. Therefore, the smallest package, with the lowest ratio of input to cooling capacity, can be used when condenser water temperature is low, the remote air-cooled condenser is relatively large, or when leaving chilled water temperature is high. The cost of the total system, however, may not be low when liquid chiller cost is minimized. Increases in cooling tower or fan coil cost will reduce or offset the benefits of reduced compression ratio. Life-cycle costs (initial cost plus operating expenses) should be evaluated.

Fouling

A fouling allowance of 0.00025 $ft^2 \cdot °F \cdot h/Btu$ is included in manufacturers' ratings in accordance with ARI *Standard* 590-92. However, fouling factors greater than 0.00025 should be considered in the selection if water conditions are other than ideal.

SPECIFIC CONTROL CONSIDERATIONS

A reciprocating chiller is distinguished from centrifugal and screw compressor-operated chillers by its use of increments of capacity reduction rather than continuous modulation. Therefore, unique arrangements must be used to establish precise chilled liquid temperature control while maintaining stable operation free from excessive on-off cycling of compressors or unnecessary loading and unloading of cylinders.

To help provide good temperature control, return chilled liquid temperature sensing is normally used by units with steps of capacity control. The resulting flywheel effect in the chilled liquid circuit damps out excessive cycling. Leaving chilled liquid temperature sensing has the advantage of preventing excessively low leaving chilled liquid temperatures if chilled liquid flow falls significantly below the design value. It may not provide stable operation, however, if rapid changes in load are encountered.

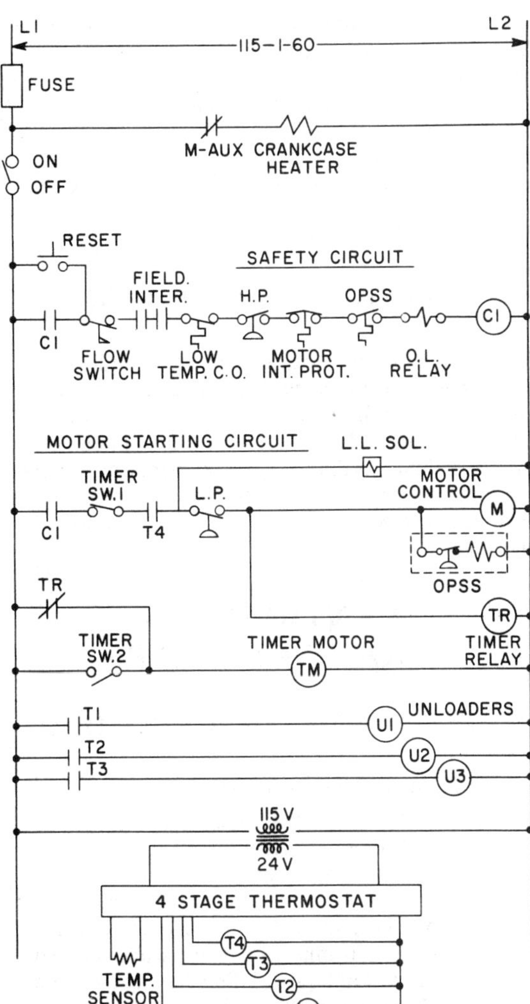

Fig. 9 Reciprocating Liquid Chiller Control System

An example of a basic control circuit for a single compressor-packaged reciprocating chiller with three steps of unloading is shown in Figure 9. The unit is started by moving the on-off switch to the *on* position. The programmed timer will start operating. Assuming that the flow switch, field interlocks, and chiller safety devices are closed, pressing the momentarily closed reset button will energize control relay C1, locking in the safety circuit and the motor starting circuit. When the timer completes its program, timer switch 1 closes and timer switch 2 opens. Timer relay TR energizes, stopping the timer motor. When timer switch 1 closes, the motor starting circuit is completed and the motor contactor holding coil is energized, starting the compressor.

The four-stage thermostat controls the capacity of the compressor in response to the demand of the system. Cylinders are loaded and unloaded by de-energizing and energizing the unloader solenoids. If the load is reduced so that the return water temperature drops to a predetermined setting, the unit shuts down until the demand for cooling increases.

Opening a device in the safety circuit will de-energize control relay C1 and shut down the compressor. The liquid line solenoid is also de-energized. Manual reset is required to restart. The crankcase heater is energized whenever the compressor is shut down.

If the automatic reset, low-pressure cutout opens, the compressor shuts down, but the liquid line solenoid remains energized. The timer relay (TR) is de-energized, causing the timer to start and complete its program before the compressor can be restarted. This prevents rapid cycling of the compressor under low-pressure conditions. A time delay low-pressure switch can also be used for this purpose with the proper circuitry.

SPECIAL APPLICATIONS

For multiple chiller applications and a 10 °F chilled liquid temperature range, the use of a parallel chilled liquid arrangement is common because of the high cooler pressure drop resulting from the series arrangement. For a large (18 °F) range, however, the series arrangement eliminates the need for overcooling during operation of one unit only. Special coolers with low water pressure drop may also be used to reduce total chilled water pressure drop in the series arrangement.

CENTRIFUGAL LIQUID CHILLERS

EQUIPMENT DESCRIPTION

Components and Their Function

The *centrifugal compressor* is described in Chapter 35 of the 1992 ASHRAE *Handbook—Systems and Equipment*. Since it is not of the constant displacement type, it offers a wide range of capacities continuously modulated over a limited range of pressure ratios. By altering built-in design items (including number of stages, compressor speed, impeller diameters, and choice of refrigerant), it can be used in liquid chillers having a wide range of design chilled liquid temperatures and design cooling fluid temperatures. Its ability to vary capacity continuously to match a wide range of load conditions with nearly proportional changes in power consumption makes it desirable for both close temperature control and energy conservation. Its ability to operate at greatly reduced capacity makes for more on-line time with infrequent starting.

The hour of the day for starting an electric-drive centrifugal liquid chiller can often be chosen by the building manager to minimize peak power demands. It has a minimum of bearing and other types of contacting surfaces that can wear; this wear is minimized by providing forced lubrication to those surfaces prior to startup and during shutdown. Bearing wear is usually more dependent on the number of startups than the actual hours of operation. Thus, by reducing the number of startups, the life of the system is extended, and maintenance costs are reduced.

Both open and hermetic compressors are selected. Open compressors may be driven by steam turbines, gas turbines or engines, or electric motors, with or without speed-changing gears. (Engine and turbine drives are covered in detail in Chapter 41 and electric motor drives in Chapter 40 of the 1992 ASHRAE *Handbook—Systems and Equipment*.)

Packaged electric-drive chillers may be of the open- or hermetic-type and use two-pole, 50- or 60-Hz polyphase electric motors, with or without speed-increasing gears. Hermetic units use only polyphase induction motors. Speed-increasing gears and their bearings, in both open- and hermetic-type packaged chillers, operate in a refrigerant atmosphere, and the lubrication of their contacting surfaces is incorporated in the compressor lubrication system.

Magnetic and SCR (silicon controlled rectifier) motor controllers are used with packaged chillers. When purchased separately, the controller must meet the specifications of the chiller manufacturer to ensure adequate equipment safety. When timed step starting methods are used, the time between steps should be long enough for the motor to overcome the relatively high inertia of the compressor and attain sufficient speed to minimize the electric current drawn immediately after transition.

Flooded coolers are commonly used, although direct-expansion coolers are employed by some manufacturers in the lower capacity ranges. The typical flooded cooler uses copper tubes that are mechanically expanded into the tube sheets, and, in some cases, into intermediate tube supports, as well.

Since refrigerant liquid flow into the compressor increases power consumption, mist eliminators or baffles are often used in flooded coolers to minimize refrigerant liquid entrainment in the suction gas. (Additional information on coolers for liquid chillers can be found in Chapter 38 of the 1992 ASHRAE *Handbook—Systems and Equipment*.)

The condenser is generally water cooled, with refrigerant condensing on the outside of copper tubes. Large condensers may have refrigerant drain baffles, which direct the condensate from within the tube bundle directly to the liquid drains, reducing the thickness of the liquid film on the lower tubes.

Air-cooled condensers can be used with units that use higher pressure refrigerants, but with considerable increase in unit energy consumption at design conditions. Operating costs should be compared with systems using cooling towers and condenser water circulating pumps.

System modifications, including subcooling and economizing as described under Principles of Operation, are often used to conserve energy. (Additional information concerning thermodynamic cycles can be found in Chapter 1 of the 1993 ASHRAE *Handbook—Fundamentals*. For information on condensers and subcoolers, see Chapter 36 of the 1992 ASHRAE *Handbook—Systems and Equipment*.)

Some units combine the condenser, cooler, and refrigerant flow control in one vessel; a subcooler may also be incorporated.

Capacities and Types Available

Centrifugal packages are currently available from about 80 to 2400 tons at nominal conditions of 44 °F leaving chilled water temperature and 95 °F leaving condenser water temperature. This upper limit is continually increasing. Field-assembled machines extend to about 10,000 tons. Single-stage and two-stage internally geared machines and two- and three-stage direct-drive machines are commonly used in packaged units. Electric motor-driven machines constitute the majority of units sold.

Units with hermetic motors, cooled by refrigerant gas or liquid, are offered from about 80 to 2000 tons. Open-drive units are not offered by all hermetic manufacturers in the same size increments but are generally available from 80 to 10,000 tons.

Selection of Refrigerant

The selection of refrigerant is an important consideration in determining equipment and operating costs. In choosing refrigerant, consider coefficient of performance, operating pressures, flow rate, heat-transfer properties, stability, toxicity, flammability, cost, and availability. Halogenated hydrocarbons are normally used as refrigerants because they offer reasonable characteristics with respect to the application. Some refrigerants are being phased out because of environmental concerns. ASHRAE *Standards* 15-1992 and 34-1992 cover the use and handling of refrigerants.

The centrifugal compressor is particularly suitable for handling relatively high flow rates of suction vapor. As the volumetric flow of suction vapor increases with higher capacities and lower suction temperatures, the higher pressure refrigerants, for example, R-134a and R-22, are used. The physical size and weight of the refrigerant piping, and, often, other components of the refrigeration system, are reduced by the use of higher pressure refrigerants. In order of decreasing volumetric flow and increasing pressures, the commonly used refrigerants are R-113, R-123, R-11, R-114, R-134a, R-12 or R-500, and R-22.

Pressure vessels for use with R-113, R-11, and R-123 usually have a design working pressure of 15 psi on the refrigerant side. The vessel shells are usually stronger than necessary for this requirement to ensure sufficient rigidity and prevent collapse under vacuum.

The thermal stability of the refrigerant and its compatibility with materials it contacts are also important. Special attention is given to the selection of elastomers and electrical insulating materials because many common materials of this sort are affected by the refrigerants. (Additional information concerning refrigerants can be found in Chapters 16 and 17 of the 1993 ASHRAE *Handbook—Fundamentals.*)

PERFORMANCE AND OPERATING CHARACTERISTICS

Figures 10 and 11 assume isentropic conditions to show reversible or specific work available from a compressor, and required for a system, as a function of volumetric flow. Figure 10 shows a compressor's performance at constant speed with various inlet guide vane settings. Figure 11 shows that compressor's performance at various speeds with open inlet guide vanes. The required system specific work can be obtained from refrigerant tables and charts for particular suction and condensing temperatures obtained from cooler and condenser performance data.

However, for comparative purposes, the *temperature lift* (condensing temperature minus evaporating temperature) closely represents the specific work when a particular refrigerant is considered. At less than full load, the percent volumetric flow will usually be a little less than the percent load, unless some vapor passes through or bypasses the refrigerant expansion device. Single or series orifices can provide some bypass vapor at reduced loads, which makes the percent volumetric flow equal to or greater than the percent load.

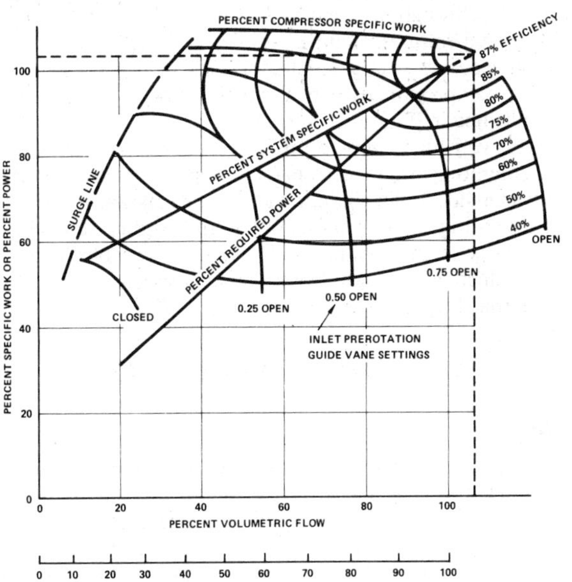

Fig. 10 Typical Centrifugal Compressor and System at Constant Speed

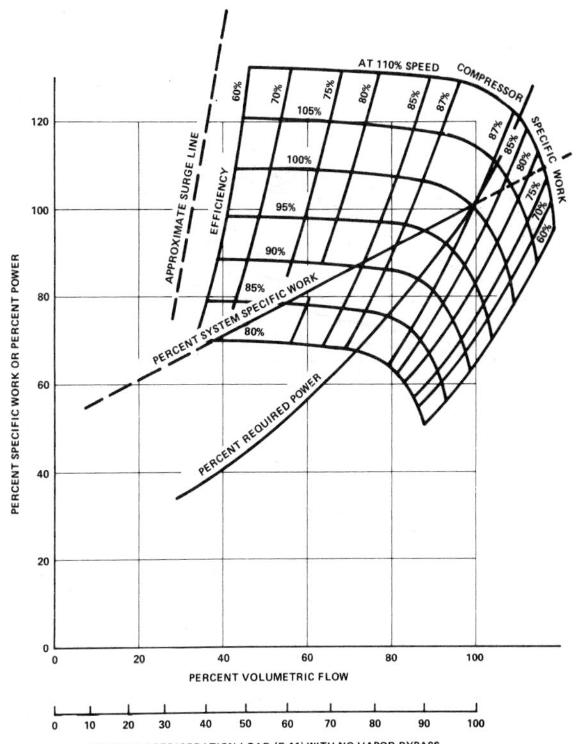

Fig. 11 Typical Centrifugal Compressor Performance at Various Speeds

A *hot-gas bypass* allows the compressor to operate down to zero load. This feature is a particular advantage for such intermittent industrial applications as the cooling of quenching tanks. Bypass vapor obtained by either method maintains the power consumption at the same level attained just prior to starting bypass, regardless of load reductions. At light loads, some bypass vapor, if introduced into the cooler below the tube bundle, may increase the evaporating temperature by agitating the liquid refrigerant and thereby more thoroughly wetting the tube surfaces.

Capacity is modulated at constant speed by automatic adjustment of prerotation vanes that whirl the refrigerant gas at the impeller eye. This effect matches system demand by shifting the compressor performance curve downward and to the left (as shown in Figure 10). Compressor efficiency, when unloaded in this manner, is superior to suction throttling. Some manufacturers also automatically reduce diffuser width or throttle the impeller outlet with decreasing load.

Speed control for a centrifugal compressor offers even lower power consumption. Down to about 50% capacity, the speed may be reduced gradually without surging. Control is transferred to the prerotation vanes for operation at lower loads. While capacity is related directly to a change in speed, the pressure produced is proportional to the square of the change in speed. Therefore, the pressure produced by reducing the speed may be less than that required by the load. Combined use of hot-gas bypass, prerotation vanes, etc., would then be necessary.

Figure 12 shows how temperature lift varies with load. A typical reduction in entering condenser water temperature of 10°F helps to reduce temperature lift at low load. Other factors producing lower lift at reduced loads are (1) the reduction in condenser cooling water range (the difference between entering and leaving temperatures, resulting from decreasing heat rejection); (2) the decrease in temperature difference between condensing refrigerant and leaving condenser water; and (3) a similar decrease between evaporating refrigerant and leaving chilled liquid temperature.

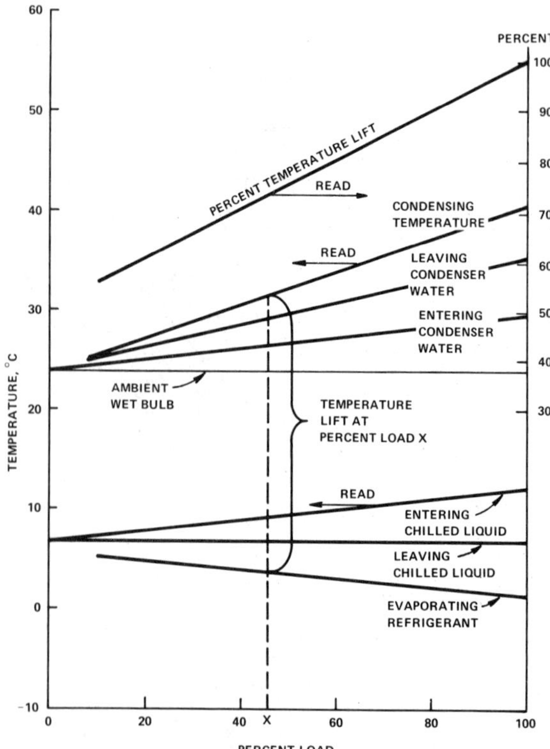

Fig. 12 Temperature Relations in a Typical Centrifugal Liquid Chiller

In many cases, the actual reduction in temperature lift is even greater because the wet-bulb temperature usually drops with the cooling load, producing a greater decrease in entering condenser water temperature.

In Figure 10, the temperature lift versus percent load of Figure 12 is translated into percent specific work versus percent volumetric suction vapor flow and superimposed on a constant speed centrifugal performance map. Below the horizontal scale for percent volumetric flow, corresponding percentages of the refrigeration load are indicated. The compressor map has curves for several inlet guide vane settings from *closed* to *open*, showing the developed specific work versus percent volumetric flow with the 100% values of both specific work and volumetric flow corresponding to design load conditions. At any other load, some compressor specific work curve at a specific fractional vane opening will intersect the system specific work curve. Note that if the system specific work curve is extended, it intersects the open compressor curve at 106% system volumetric flow and 103% system specific work. The compressor has some extra capacity.

The compressor map also has a number of contour lines of constant efficiency. At each point along the percent system specific work curve, the efficiency can be interpolated. Thus, for any specific percent load, the percent volumetric suction vapor flow, the specific work, and the compressor efficiency can be obtained from Figure 10. Using the evaporating temperature shown in Figure 12 and the refrigerant tables and/or chart, the specific volume of the suction vapor can be obtained.

Specific work, by definition, is the reversible work per unit mass. The total work per unit mass is therefore the specific work divided by the efficiency, neglecting the mechanical losses in bearings, gears, etc., because such losses are usually not included in the compressor efficiency. Power (the work done per unit time) is thus the product of the work per unit mass and the mass flow rate, the latter being the volumetric flow, divided by the specific volume, *i.e.*:

$$P_{th} = W_s Q_1 / \eta_s V_1$$

where

P_{th} = theoretical power
W_s = isentropic specific work
η_s = isentropic efficiency
Q_1 = suction volumetric flow
V_1 = specific volume of suction vapor

Figure 10, from such calculations, shows the percent power curve. Mechanical losses can be added or theoretical power may be divided by the mechanical efficiency to obtain the total power required.

The above analysis is for single-stage compressors or for multistage units without interstage cooling or introduction of vapor between stages. The specific work shown for multistage units is then the total specific work developed and the volumetric flow of the suction vapor to the first stage. Where interstage flash coolers are used, each stage must be handled separately.

Figure 11 shows the effects resulting from changes in compressor speed for a compressor with similar performance characteristics as the one shown in Figure 10. The load curve of Figure 10 is also indicated in Figure 11. Note that the compressor's design parameters, including impeller diameters, flow areas, and 100% speed, are selected to match the system specific work curve at a volumetric flow slightly greater than for obtaining peak efficiency. As the load decreases from 100% and the speed progressively lessens, the operating efficiency remains high for a considerable range in capacity. Further reductions in capacity and speed show progressively lower efficiencies and a progression toward instability as indicated by the approximate surge line. The curved nature of the percent required power curve shows the effects resulting from the changes in efficiency along the load line as the specific work and capacity vary.

As stated earlier, speed control is usually used from 100% down to about 50% load; below 50%, inlet vane control is used with performance similar to that shown in Figure 10. Power consumption is reduced when the coldest possible condenser water is used, consistent with the chiller manufacturer's recommended minimum condenser water temperature. In cooling tower applications, minimum water temperatures should be controlled by a cooling tower bypass and/or by cooling tower fan control, not by reducing the water flow through the condenser. Maintaining a high flow rate at lower temperatures minimizes fouling and power requirements.

Surging occurs when the system specific work becomes greater than the compressor developed specific work or above the surge line indicated in Figures 10 and 11.

Excessively high temperature lift and corresponding specific work commonly originate from (1) excessive condenser or evaporator water-side fouling beyond the specified allowance; (2) inadequate cooling tower performance and higher-than-design condenser water temperature; and (3) noncondensables in the condenser, which increase condenser pressure. When surging occurs, correct the above conditions that exist.

METHOD OF SELECTION

Ratings

A refrigeration machine with specified details is chosen from selection tables for given capacities and operating conditions, or through computer-generated selection or performance programs. Rating tables differ from selection tables in that they list the capacities and operating data for each refrigeration machine under various operating conditions, often with specific details for the listed conditions. The details specified for centrifugal systems include the number of passes and the water-side pressure drop in each of the heat exchangers, the required kilowatt input, electrical characteristics, and part-load performance.

The maximum number of condenser and cooler water passes should be used, without producing excessive water pressure drop. The greater the number of water-side passes, the less the power consumption. Sometimes a slight reduction in condenser water flow (and slightly higher leaving water temperature) allows a better selection (lower power consumption or smaller model) than will the choice of fewer water passes when a rigid pressure-drop limit exists.

Fouling

In accordance with ARI *Standard* 590-92, a fouling allowance of 0.00025 $ft^2 \cdot °F \cdot h/Btu$ is included in manufacturers' ratings. (Chapter 36 of the 1992 ASHRAE *Handbook—Systems and Equipment* has further information about fouling factors.)

To reduce fouling, a minimum water velocity of about 3.3 ft/s is recommended in coolers or condensers. Maximum water velocities exceeding 11 ft/s are not recommended because of potential erosion problems.

Proper water treatment and regular tube cleaning are recommended for all liquid chillers to reduce power consumption and operating problems. See Chapter 43 of the 1991 ASHRAE *Handbook—Applications* for water treatment information.

Continuous or daily monitoring of the quality of the condenser water is desirable. Checking the quality of the chilled liquid is also desirable. The intervals between checks become greater as the possibilities for fouling contamination become less—for example, an annual check should be sufficient for closed-loop water-circulating systems for air conditioning. Corrective treatment is required, and periodic, usually annual, cleaning of the condenser tubes usually keeps fouling within the specified allowance. In applications where more frequent cleaning is desirable, an on-line cleaning system may be economical.

Noise and Vibration

An important consideration in the application of large equipment is the noise generated and transmitted to adjoining areas. A standard for sound measurement and rating can be found in ARI *Standard* 575-87. Additional information on chiller noise, measurement, and control is given in Chapter 7 of the 1993 ASHRAE *Handbook—Fundamentals* and Chapter 42 of the 1991 ASHRAE *Handbook—Applications.*

The chiller manufacturer's recommendations for mounting should be followed to prevent transmission or amplification of vibration to adjacent equipment or structures. Auxiliary pumps, if not connected with flexible fittings, can induce vibration of the centrifugal unit, especially if the rotational speed of the pump is nearly the same as either the compressor prime mover or the compressor. Flexible tubing becomes less flexible when it is filled with liquid under pressure and some vibration can still be transmitted.

SPECIFIC CONTROL CONSIDERATIONS

The section on Control describes the chilled liquid temperature sensor. In centrifugal systems, it is usually placed in thermal contact with the leaving chilled water. In electrical control systems, the electrical signal is transmitted to an electronic control module, which controls the operation of an electric motor(s) positioning the capacity controlling inlet guide vanes. A current limiter is usually included on electric motor-driven machines. An electrical signal from a current transformer in the compressor motor controller is sent to the electronic control module. The module receives indications of both the leaving chilled water temperature and the compressor motor current. The portion of the electronic control module responsive to motor current is called the current limiter. It overrides the demands of the temperature sensor.

The inlet guide vanes, independent of the demands for cooling, do not open more than the position that results in the present setting of the current limiter. Pneumatic capacity controls operate in a similar manner. The chilled liquid temperature sensor provides a pneumatic signal. The controlling module receives both that signal and the motor current electrical signal and controls the operation of a pneumatic motor(s) positioning the inlet guide vanes. Both controlling systems have sensitivity adjustments.

The current limiter on most machines can limit current draw during periods of high electrical demand charges. This control can be set from about 40 to 100% of full load amperes. Whenever power consumption is limited, cooling capacity is correspondingly reduced. If cooling load is only 50% of the full value, the current (or demand) limiter can be set at 50% without loss of cooling. By setting the limiter at 50% of full current draw, any subsequent high demand charges are prevented during pulldown after startup. Even during periods of high cooling load, it may be desirable to limit electrical demand if a small increase in chiller liquid temperature is acceptable. If the temperature continues to decrease after the capacity control has reached its minimum position, a low-temperature control stops the compressor and restarts it when a rise in temperature indicates the need for cooling. Manual controls may also be provided to bypass the temperature control. Provision is included to ensure that the capacity control is at its minimum position when the compressor starts to provide an unloaded starting condition.

Additional operating controls are needed for appropriate operation of oil pumps, oil heaters, purge units, and refrigerant transfer units. An antirecycle timer should also be included to prevent frequent motor starts. Multiple unit applications require additional controls for capacity modulation and proper sequencing of units. (See the Multiple Chiller System section.)

Safety controls protect the unit under abnormal conditions. Safety cutouts that may be required are for high condenser pressure, low evaporator refrigerant temperature or pressure, low oil pressure, high oil temperature, high motor temperature, and high discharge temperature. Auxiliary safety circuits are usually provided on packaged chillers. At installation, the circuits are field wired to field-installed safety devices, including auxiliary contacts on the pump motor controllers and flow switches in the chilled water and condenser water circuits. Safety controls are usually provided in a lockout circuit, which will trip out the compressor motor controller and prevent automatic restart. The controls reset automatically, but the circuit cannot be completed until a manual reset switch is operated and the safety controls return to their safe positions.

AUXILIARIES AND SPECIAL APPLICATIONS

Auxiliaries

Purge units are required for centrifugal liquid chilling machines using R-123, R-11, R-113, or R-114 because evaporator pressure is below atmospheric pressure. If a purge unit were not used, air and moisture would accumulate in the refrigerant side over a period of time. Noncondensables collect in the condenser during operation, reducing the heat-transfer coefficient and increasing condenser pressure as a result of both their insulating effect and the partial pressure of the noncondensables. Compressor power consumption increases, capacity is reduced, and surging may occur.

Moisture may build up until saturation of refrigerant occurs and free moisture is present. Acids will then be produced from reaction with the refrigerant and internal corrosion will begin. A purge unit is designed to prevent the accumulation of noncondensables and ensure internal cleanliness of the liquid chilling machine. It is not intended, however, to reduce the need for proper leak checking and repairing of leaks, which is required maintenance for any liquid chiller. Purge units may be manual or automatic, compressor-operated, or compressorless.

ASHRAE *Standard* 15-1992 requires purge and rupture disk venting to be piped outdoors for all refrigerants. Because of environmental concerns and the increasing cost of refrigerants, new high efficiency (air to refrigerant) purges are available that reduce refrigerant losses during normal purging operations.

Oil coolers may be water cooled, using condenser water when the quality is satisfactory, or chilled water when a small loss in net cooling capacity is acceptable. Oil coolers may also be refrigerant or air cooled, eliminating the need for water piping to the cooler.

Refrigerant Transfer Unit. A refrigerant transfer unit may be provided for centrifugal liquid chillers using refrigerants with a boiling point below ambient temperature at atmospheric pressure (R-134a, R-12, R-114, R-22, R-500). The unit consists of a small reciprocating compressor with electric motor drive, a condenser (air- or water-cooled), an oil reservoir and oil separator, and valves and interconnecting piping. Refrigerant transfer has three steps:

1. *Gravity Drain.* When the receiver is at the same level as or below the cooler, some liquid refrigerant may be transferred to the receiver by opening valves in the interconnecting piping.
2. *Pressure Transfer.* By resetting valves and operating the compressor, refrigerant gas is pulled from the receiver and is used to pressurize the cooler, forcing refrigerant liquid from the cooler to the storage receiver. If the chilled liquid and condenser water pumps can be operated to establish temperature differences, the migration of refrigerant from the warmer vessel to the colder vessel can also be used to assist in the transfer of refrigerant.
3. *Pump-Out.* After the refrigerant liquid has been transferred, valve positions are changed and the compressor is operated to pump refrigerant gas from the cooler to the transfer unit condenser, which sends condensed liquid to the storage receiver. If any chilled liquid (water, brine, etc.) remains in the cooler tubes, pump-out must be stopped before cooler pressure drops below the saturation condition corresponding to the freezing point of the chilled liquid.

When recharging, if the saturation temperature corresponding to cooler pressure is below the chilled liquid freezing point, refrigerant gas from the storage receiver must be introduced until the cooler pressure is above this condition. The compressor can then be operated to pressurize the receiver and move refrigerant liquid into the cooler without danger of freezeup.

Water-cooled transfer unit condensers provide fast refrigerant transfer. Air-cooled condensers eliminate the need for water, but they are slower and more expensive.

Special Applications

Heat-Recovery System. One special application of centrifugal liquid chillers is heat recovery. Instead of rejecting all heat extracted from the chilled liquid to a cooling tower, a separate closed condenser cooling water circuit is heated by the condensing refrigerant for such purposes as comfort heating, preheating, or reheating. Factory packages are now available that include an extra condenser water circuit, either in the form of a double-bundle condenser or an auxiliary condenser.

The control requirements for a centrifugal heat-recovery package are as follows:

1. *Control of chilled liquid temperature* is accomplished by a sensor in the leaving chilled liquid line signaling the capacity control device.
2. *Control of hot water temperature* is accomplished by a sensor in the hot water line that modulates a cooling tower bypass valve. As the heating requirement increases, hot water temperature drops, opening the tower bypass slightly. Less heat is rejected to the tower, condensing temperature increases, and hot water temperature is restored as more heat is rejected to the hot water circuit.

The hot water temperature selected has a bearing on the installed cost of the centrifugal package, as well as on the power consumption while heating. Lower hot water temperatures of 95 to 105°F result in a less expensive machine that uses less power. Higher temperatures require a greater compressor motor output, perhaps higher pressure condenser shells, sometimes extra compression stages, or a cascade arrangement. Installed cost of the centrifugal heat-recovery machine is increased as a result.

Another consideration in the design of a central chilled water plant with heat-recovery centrifugals is the relative size of the cooling and heating loads seen by the liquid chilling machine. It is best to equalize the heating and cooling loads on each machine so that the compressor may operate at optimum efficiency during both the full cooling and full heating seasons. When the heating requirement is considerably smaller than the cooling requirement, multiple packages will lower operating costs and allow standard air-conditioning centrifugal packages of lower cost to be used for the remainder of the cooling requirement (ASHRAE Symposium *Bulletin* DV-69-4, Centrifugal Heat Pump Systems). In multiple packages, only one unit is designed for heat recovery and carries the full heating load.

Free Cooling. The use of centrifugal liquid chillers to cool without operating the compressor is called *free cooling.* When a supply of condenser water is available at a temperature below that of the needed chilled water temperature, the chiller can operate as a thermal siphon. Low-temperature condenser water condenses refrigerant, which is either drained by gravity or pumped into the evaporator. Higher-temperature chilled water causes the refrigerant to evaporate, and vapor flows back to the condenser because of the pressure difference between the evaporator and the condenser. Free cooling is limited to about 10 to 30% of the chiller design capacity. The actual free cooling capacity depends on the chiller design and the temperature difference between the desired chilled water temperature and the condenser water temperature.

Air-Cooled System. Two types of air-cooled centrifugal systems are prevalent. One consists of a water-cooled centrifugal package with a closed-loop condenser water circuit. The condenser water is cooled in a water/air heat exchanger. This arrangement results in higher condensing temperature and increased power consumption. In addition, winter operation requires the use of glycol in the condenser water circuit, reducing the heat-transfer coefficient.

In the other type, systems that are directly air-cooled eliminate the intermediate heat exchanger and condenser water pumps, resulting in lower power requirements. However, attention must be given to keeping the condenser and refrigerant piping leaktight. Air-cooled centrifugal packages are now on the market.

Because a centrifugal will surge if it is subjected to a pressure appreciably higher than design, the air-cooled condenser must be designed for the required heat rejection. In common practice, the selection of a reciprocating air-cooled machine is based on an outside dry-bulb temperature that will be exceeded 5% of the time. A centrifugal may be unable to operate during such times because of surging, unless the chilled water temperature is raised proportionately. Thus, the compressor impeller(s) and/or speed should be selected for the maximum dry-bulb temperature to ensure that the desired chilled water temperature will be maintained at all times. In addition, the condenser coil must be kept clean.

An air-cooled centrifugal chiller should allow the condensing temperature to fall naturally to about 70°F during colder weather. The resulting decrease in compressor power consumption is greater than that for reciprocating systems controlled by thermal expansion valves.

During winter shutdown, precautions must be taken to prevent freezing of the cooler liquid resulting from a free cooling effect from the air-cooled condenser. A thermostatically controlled heater in the cooler, in conjunction with a low refrigerant pressure switch to start the chilled liquid pumps, will protect the system.

Brine Cooling. The most frequent use of centrifugal liquid chilling units is for water chilling applications. Centrifugals are also applied for brine cooling duty with brines such as calcium chloride, methylene chloride, ethylene glycol, and propylene glycol. (For a complete description of brines, see Chapter 18 of the 1993 ASHRAE *Handbook—Fundamentals*.) Brine properties must be considered in calculating heat-transfer performance and pressure drop. Because of the greater temperature lift, higher compressor speeds and possibly more stages may be required for brine cooling duty. Compound and/or cascade systems are required for low-temperature applications.

Vapor Condensing. Many process applications require the condensation of vapors such as ammonia, chlorine, or hydrogen fluoride. Centrifugal liquid chilling units are used for these applications.

OPERATION AND MAINTENANCE

Proper operation and maintenance are essential for reliability, longevity, and safety. See Chapter 35 of the 1991 ASHRAE *Handbook—Applications* for general information on principles, procedures, and programs for effective maintenance. The manufacturer's operation and maintenance instructions should also be consulted for specific procedures.

Normal operation conditions should be established and recorded at initial startup. Changes from these conditions can be used to signal the need for maintenance. One of the most important items is to maintain a leaktight unit. Leaks on units operating at subatmospheric pressures will result in air and moisture entering the unit and increasing the condenser pressure.

While the purge unit can remove noncondensables sufficiently to prevent an increase in condenser pressure, continuous entry of air and attendant moisture into the system promotes refrigerant and oil breakdown and corrosion.

Periodic analysis of the oil and refrigerant charge can also be used to identify system contamination problems. High condenser pressure or frequent purge unit operation indicate leaks that should be corrected as soon as possible. With positive operating pressures, leaks result in loss of refrigerant and such operating problems as low evaporator pressure. A leak check should also be included in preparation for a long-term shutdown. (See Chapter 6 for a detailed discussion of the harmful effects of air and moisture.)

Normal maintenance should include periodic oil and refrigerant filter changes as recommended by the manufacturer. All safety controls should be checked periodically to ensure that the unit is protected properly.

Cleaning of inside tube surfaces may be required at various intervals, depending on the water condition. Condenser tubes may only need annual cleaning if proper water treatment is maintained. Cooler tubes need less frequent cleaning if the chilled water circuit is a closed loop.

If it is necessary to remove the refrigerant charge and open the unit for service, the unit should be leak-checked, dehydrated, and evacuated properly before recharging. (See Chapter 45 for dehydrating, charging, and testing and ASHRAE Symposium *Bulletin* LP-68-2, The Operation and Maintenance of Centrifugal Units.)

SCREW LIQUID CHILLERS

EQUIPMENT DESCRIPTION

Components and Their Function

Single- and twin-screw compressors are both positive-displacement machines with nearly constant flow performance. Compressors for liquid chillers can be both oil-injected and oil-injection-free. (Chapter 35 of the 1992 ASHRAE *Handbook—Systems and Equipment* describes screw compressors in greater detail.)

The cooler may be flooded or direct-expansion. There is no particular cost advantage to one design over the other. The flooded cooler is more sensitive to freezeup, requires more refrigerant, and requires closer evaporator pressure control, but its performance is easier to predict and it can be cleaned. The direct-expansion cooler requires closer mass flow control, is less likely to freeze, and returns oil to the oil system rapidly. The decision to use one or the other is based on the relative importance of these factors on a given application.

Screw coolers have the following characteristics: (1) high maximum working pressure, (2) continuous oil scavenging, (3) no mist eliminators (flooded coolers), and (4) distributors designed for high turndown ratios (direct-expansion coolers). A suction gas, high-pressure liquid heat exchanger is sometimes incorporated into the system to provide subcooling for increased thermal expansion valve flow and reduced power consumption. (For further information on coolers, see Chapter 38 of the 1992 ASHRAE *Handbook—Systems and Equipment*.)

Flooded coolers were once used in units with a capacity larger than about 400 tons. Direct-expansion coolers are also used in larger units up to 800 tons with a servo-operated expansion valve having an electronic controller that measures evaporating pressure, leaving brine temperature, and suction gas superheat.

The condenser may be included as part of the liquid chilling package when water-cooled, or it may be remote. Air-cooled liquid chilling packages are also available. When remote air-cooled or evaporative condensers are applied to liquid chilling packages, a liquid receiver generally replaces the water-cooled condenser on the package structure. Water-cooled condensers are the cleanable shell-and-tube type (see Chapter 36 of the 1992 ASHRAE *Handbook—Systems and Equipment*).

Oil cooler loads vary widely, depending on the refrigerant and application, but they are substantial because oil injected into the compressor absorbs a portion of the heat of compression. Oil cooling is by (1) a water-cooled oil cooler using condenser water, evaporative condenser sump water, chilled water, or a separate

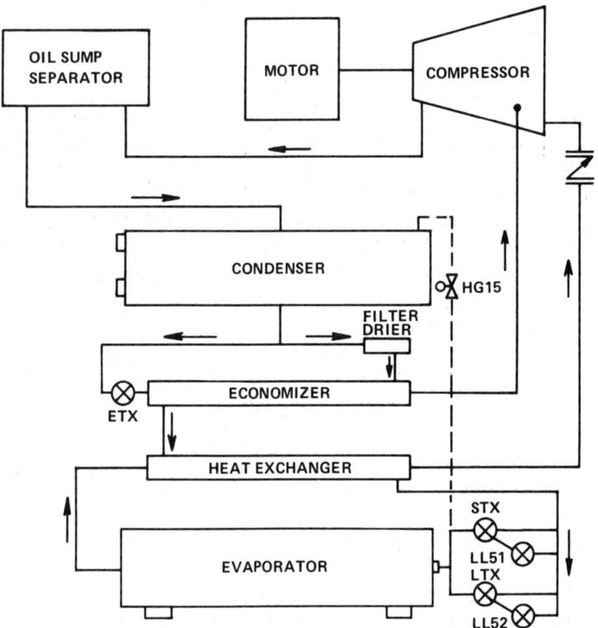

Fig. 13 Refrigeration System Schematic

water- or glycol-to-air cooling loop; (2) an air-cooled oil cooler using an oil-to-air heat exchanger; (3) a refrigerant-cooled oil cooler (where oil cooling load is low); (4) liquid injection into the compressor; or (5) condensed refrigerant liquid thermal recirculation (thermosyphon), where appropriate head pressure is available. The latter two are the most economical both in first cost and overall operating cost because cooler maintenance and special water treatment are eliminated.

Efficient oil separators are required. The types and efficiencies of these separators vary according to refrigerant and application. Field built-up systems require better separation than complete factory-built systems. Ammonia applications are most stringent because there is no appreciable oil return with the suction gas from the flooded coolers normally used in ammonia applications. However, separators are available for ammonia packages, which do not require the periodic addition of oil that is customary on other R-717 systems. The types of separators in use are centrifugal, de-mister, gravity, coalescer, and combinations of these.

Hermetic compressor units may use a centrifugal separator as an integral part of the hermetic motor while cooling the motor with discharge gas and oil simultaneously. A schematic of a typical refrigeration system is shown in Figure 13.

Capacities and Types Available

Screw compressor liquid chillers are available as factory-packaged units from about 30 to 1250 tons. Both open and hermetic styles are manufactured. Packages without water-cooled condensers, with receivers, are made for use with air-cooled or evaporative condensers. Most factory-assembled liquid chilling packages use R-22 and some are beginning to use R-134a.

Additionally, compressor units, comprised of a compressor, hermetic or open motor, oil separator, and oil system, are available from 20 to 2000 tons. These are used with remote evaporators and condensers for low, medium, and high evaporating temperature applications. Condensing units, similar to compressor units in range and capacity but with water-cooled condensers, are also built. All of the above are suitable for R-12, R-134a, R-502, and R-22. Similar open motor-drive units are available for R-717, as are booster units.

Selection of Refrigerant

The refrigerants most commonly used with screw compressors on liquid chiller applications are R-22, R-134a, and R-717. The active use of R-12 and R-500, which are CFCs, has been discontinued for new equipment.

The screw compressor operating characteristic shown in Figure 7 is compared with reciprocating and centrifugal performance. Additionally, since the screw compressor is a positive-displacement compressor, it does not surge. Since it has no clearance volume in the compression chamber, it pumps high volumetric flows at high pressure. Because of this, screw compressor chillers suffer the least capacity reduction at high condensing temperatures.

The screw compressor provides stable operation over the whole working range because it is a positive-displacement machine. The working range is wide because the discharge temperature is kept low and is not a limiting factor because of oil injection into the compression chamber. Consequently, the compressor is able to operate single-stage at high pressure ratios. An economizer system can be used to improve the capacity and lower the power consumption at full-load operation.

An example of such an economizer arrangement is shown in Figure 13, where the main refrigerant liquid flow is subcooled in a heat exchanger connected to the intermediate pressure port in the compressor. The evaporating pressure in this heat exchanger is higher than the suction pressure of the compressor. Oil separators must be sized for the size of compressor, type of system (factory- assembled or field-connected), refrigerant, and type of cooler. Direct-expansion coolers have less stringent separation requirements than do flooded coolers.

In a direct-expansion system, the refrigerant evaporates in the tubes, which means that the velocity is kept so high that the oil rapidly returns to the compressor. In a flooded evaporator, the refrigerant is outside the tubes, and some type of external oil-return device must be used to minimize the oil concentration in the cooler. Suction or discharge check valves are used to minimize spinback and oil loss during shutdown.

Since the oil system is on the high-pressure side of the unit, precautions must be taken against oil dilution. Dilution can also be caused by excessive floodback through the suction or intermediate ports; unless properly monitored, it may go unnoticed until serious operating or mechanical problems are experienced.

METHOD OF SELECTION

Ratings

Screw liquid-chiller ratings are generally presented similarly to those for centrifugal-chiller ratings. Tabular values include capacity and power consumption at various chilled water and condenser water temperatures.

In addition, ratings are given for packages without the condenser, listing capacity and power versus chilled water temperature and condensing temperature. Ratings for compressors alone are also common, showing capacity and power consumption versus suction temperature and condensing temperature for a given refrigerant.

Power Consumption

Typical part-load power consumption is shown in Figure 14. Power consumption of screw chillers benefits from reduction of condensing water temperature as the load decreases, as well as operating at the lowest practical pressure at full load. However, because direct-expansion systems require a pressure differential, the power consumption saving is not as great at part load as shown.

Fouling

A fouling allowance of 0.00025 $ft^2 \cdot °F \cdot h/Btu$ is incorporated in screw compressor chiller ratings. Excessive fouling (above the design value) increases power consumption and reduces capacity. Fouling of water-cooled oil coolers results in higher than desirable oil temperatures.

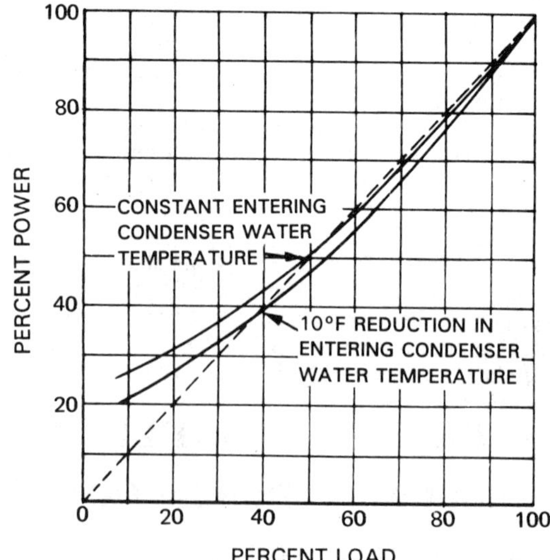

Fig. 14 Typical Screw Compressor Chiller Part-Load Power Consumption

SPECIFIC CONTROL CONSIDERATIONS

Screw chillers provide continuous capacity modulation, from 100% capacity down to 10% or less. The leaving chilled liquid temperature is sensed for capacity control. Safety controls commonly required are (1) oil failure switch, (2) high discharge pressure cutout, (3) low suction pressure switch, (4) cooler flow switch, (5) high oil and discharge temperature cutout, (6) hermetic motor inherent protection, (7) oil pump and compressor motor overloads, and (8) low oil temperature (flood back/dilution protection). The compressor is unloaded automatically (slide valve driven to minimum position) before starting. Once it starts operating, the slide valve is controlled hydraulically by a temperature-load controller that energizes the load and unload solenoid valves.

The current limit relay protects against motor overload from higher than normal condensing temperatures or low-voltage conditions and also allows a demand limit to be set, if desired. An antirecycle timer is used to prevent overly frequent recycling. Oil sump heaters are energized during the off cycle. A hot gas capacity control is optionally available and prevents automatic recycling at no-load conditions such as is often required in process liquid chilling. A suction to discharge starting bypass sometimes aids starting and allows the use of standard starting torque motors.

Some units are equipped with electronic regulators specially developed for the screw compressor characteristics. These regulators include PI-control (Proportional-Integrating) of the leaving brine temperature and such functions as automatic/manual control, capacity indication, time circuits to prevent frequent recycling and to bypass the oil pressure cutout during startup, switch for unloaded starting, etc. (Typical external connections are shown in Figure 15.)

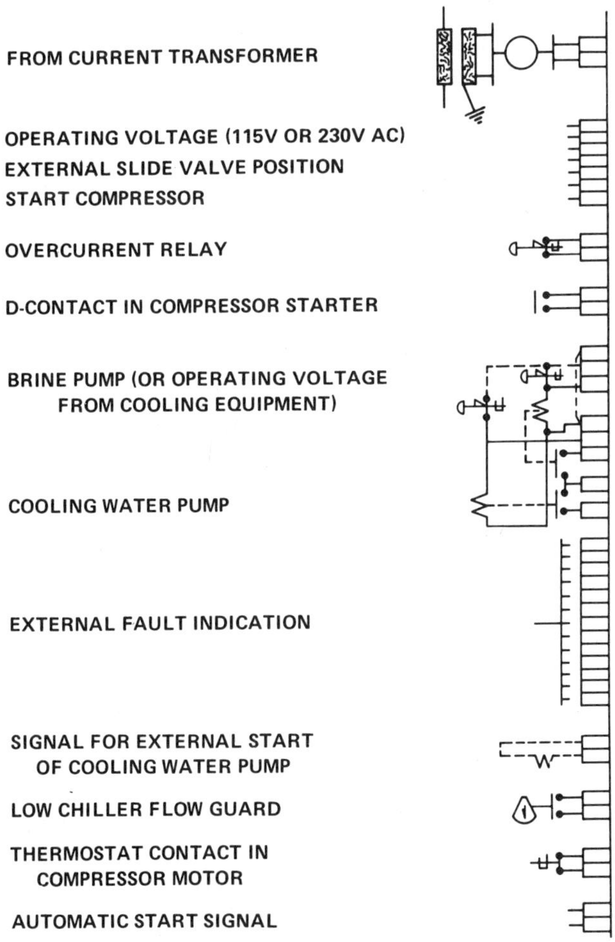

FROM CURRENT TRANSFORMER
OPERATING VOLTAGE (115V OR 230V AC)
EXTERNAL SLIDE VALVE POSITION
START COMPRESSOR
OVERCURRENT RELAY
D-CONTACT IN COMPRESSOR STARTER
BRINE PUMP (OR OPERATING VOLTAGE FROM COOLING EQUIPMENT)
COOLING WATER PUMP
EXTERNAL FAULT INDICATION
SIGNAL FOR EXTERNAL START OF COOLING WATER PUMP
LOW CHILLER FLOW GUARD
THERMOSTAT CONTACT IN COMPRESSOR MOTOR
AUTOMATIC START SIGNAL

Fig. 15 Typical External Connections for Screw Compressor Chiller

AUXILIARIES AND SPECIAL APPLICATIONS

Auxiliaries

A refrigerant transfer unit is similar to the unit described in the section on Auxiliaries under CENTRIFUGAL LIQUID CHILLERS. It is designed for R-22 operating pressure. Its flexibility is increased by including a reversible liquid pump on the unit. It is available as a portable unit or mounted on a storage receiver.

An oil-charging pump is useful for adding oil to the pressurized oil sump. Two types are used: a manual pump and an electric motor-driven positive-displacement pump.

Acoustical enclosures are available for installations that require low noise levels.

Special Applications

Because of the screw compressor's positive-displacement characteristic and oil-injected cooling, its use for high differential applications is limited only by power considerations and maximum design working pressures. Therefore, it is being used for a number of special applications because of reasonable compressor cost and no surge characteristic. Some of the fastest growing areas are listed below:

1. Heat-recovery installations
2. Air-cooled
   a. Split packages with field-installed interconnecting piping
   b. Factory-built rooftop packages
3. Low-temperature brine chillers for process cooling
4. Ice rink chillers
5. Power transmission line oil cooling

High temperature compressor and condensing units are being used increasingly for air conditioning because of the higher efficiency of direct air-to-refrigerant heat exchange resulting in higher evaporating temperatures (see Chapter 21 of the 1992 ASHRAE *Handbook—Systems and Equipment*). Many of these installations have air-cooled condensers (see Chapter 36 of the 1992 ASHRAE *Handbook—Systems and Equipment*).

MAINTENANCE

Periodic maintenance of screw liquid chillers is important; it is essential that the manufacturer's instructions be followed, especially since there are some items that differ substantially from reciprocating or centrifugal units.

Water-cooled condensers must be cleaned of scale periodically (see the General Maintenance section). If the condenser water is also used for the oil cooler, this should be considered in the treatment program. Oil coolers operate at higher temperatures and lower flows than condensers, so it is possible that the oil cooler may have to be serviced more often than the condenser.

Since large oil flows are a part of the screw compressor system, the oil filter pressure drop should be monitored carefully and the elements changed periodically. This is particularly important in the first month or so after startup of any factory-built package and is absolutely essential on field-erected systems. Since the oil and refrigeration systems merge at the compressor, much of the loose dirt and fine contaminants in the system eventually find their way to the oil sump, where they are removed by the oil filter. Similarly, the filter-drier cartridges should be monitored for pressure drop and moisture during initial start and regularly thereafter. Generally, if a system reaches an acceptable dryness level, it stays that way unless it is opened.

It is good practice to check the oil for acidity periodically, using commercially available acid test kits. Oil does not have to be changed unless it is contaminated by water, acid, or metallic particles. Also, a refrigerant sample should be analyzed yearly to determine its condition.

There are certain procedures that should be followed on a yearly basis or during a regularly scheduled shutdown. These include checking and calibrating all operation and safety controls, tightening all electrical connections, inspecting power contacts in starters, dielectric checking of hermetic and open motors, and checking the alignment of open motors.

Leak testing of the unit should be performed regularly. For a water-cooled package used for summer cooling, this should be performed yearly. A flooded unit with proportionately more refrigerant in it, used for year-round cooling, should be tested every four to six months. A process air-cooled chiller designed for year-round operation 24 hours per day should be checked every one to three months.

The screw compressor is a simple, rugged machine. The decision concerning whether to tear down, and how far to go, depends, in part, on the nature of its application, *i.e.*, process, computer cooling, air conditioning, the design of the rest of the system, whether there is standby equipment, and the cost of a breakdown.

Based on 6000 operating hours per year and depending on the above considerations, a reasonable inspection or changeout timetable is shown below:

Shaft seals	1.5 to 4 yr	Replace
Hydraulic cylinder seals	1.5 to 4 yr	Replace
Thrust bearings	4 to 6 yr	Check preload via shaft end play every 6 months and replace as required
Shaft bearings	7 to 10 yr	Inspect

CHAPTER 43

COMPONENT BALANCING IN REFRIGERATION SYSTEMS

A REFRIGERANT is a fluid used for heat transfer in a refrigerating system. The fluid absorbs heat at a low temperature and pressure and transfers heat at a higher temperature and pressure. Usually, the process involves a change of state of the fluid. Energy transfer is a function of the heat transfer coefficients, the temperature differences and the amount, type, and configuration of the heat transfer surface and, hence, the heat flux on either side of the heat transfer device.

In an evaporator, refrigerant vaporizes as its enthalpy or heat content increases. A compressor pulls the vapor from the evaporator through suction piping and compresses the refrigerant gas to a higher pressure and temperature. Compression deviates from a constant entropy line because of the adiabatic efficiency of the compressor for the pressure ratio involved and because of mechanical friction. Power input to the shaft of the compressor is added to the refrigerant, and compression increases the pressure, temperature, and enthalpy as heat of the refrigerant.

Multiple compressors in series can further compress the refrigerant gas if necessary. Some interstage liquid cooling and gas desuperheating may be used to increase the efficiency of the system and compression. At the higher pressure, the refrigerant gas can be condensed on a condenser surface and heat is rejected to a coolant of air, water, brine, or water spray.

An intermediate-temperature condenser can serve as a cascading device. A low-temperature high-pressure refrigerant condenses on one side of the cascade condenser surface by giving up heat to a low-pressure refrigerant that is boiling on the other side of the surface. The vapor of this boiling refrigerant transfers the energy to the next compressor (or compressors); heat of compression is added and, at a higher pressure, the last refrigerant is condensed on the final condenser surface.

Heat is rejected to a heat sink created by air, water, brine, or water spray. The saturation temperatures of evaporation and condensation throughout the system fix the terminal pressures that the single or multiple compressors must operate against.

Generally, the smallest differential between the saturated evaporator and the saturated condensing temperatures results in the lowest energy requirement for compression. Liquid refrigerant cooling or subcooling in multi-staged systems should be used to minimize energy consumption.

Where intermediate pressures are not specified by a specific evaporator load and temperature requirement, the compressors automatically balance at their respective suction and discharge pressures as a function of their relative displacements and compression efficiencies. This chapter covers techniques used to determine the balance points for typical components in a single-stage brine chiller system.

COMPONENTS

Evaporators are designed to fulfill specific functions and applications. Evaporators may have flooded, direct expansion, or liquid overfeed cooling coils with or without fins. Evaporators are used to cool air, gases, liquids, and solids; condense volatile substances; and freeze products.

Ice builder evaporators accumulate ice to store cooling energy for later use. Embossed plate evaporators are available (1) to cool a falling film of liquid; (2) to cool, condense, and/or freeze out volatile substances from a fluid stream; or (3) to cool or freeze a product by direct contact. Brazed plate fluid chillers are being used to reduce refrigerant charge.

Ice, wax, or food products are frozen and scraped from some freezer surfaces. Electronic circuit boards and mechanical or food products are being flash cooled by direct immersion in boiling refrigerants. These are some of the diverse applications demanding innovative configurations and materials that perform the function of an evaporator.

Compressors take the form of positive displacement, reciprocating piston, vane rotary, scroll, single and double dry and lubricant flooded screw devices, and single- or multi-stage centrifugals. They can be operated in series or parallel with each other, in which case special controls may be required.

The drivers for the compressors can be direct hermetic, semi-hermetic, or open with mechanical seals on the compressor. In hermetic and semi-hermetic drives, the motor inefficiencies are added to the refrigerant as heat. Open compressors are driven with electric motors, fuel-powered reciprocating engines, or steam or gas turbines. Intermediate gears, belts, and clutch drives may be included in the drive.

Cascade Condensers are used with high-pressure, low-temperature refrigerants (such as R-23) on the bottom cycle, and high-temperature refrigerants (such as R-22, azeotropes, and refrigerant blends or zeotropes) on the upper cycle. Condensers are manufactured in many forms, including: shell and tube, embossed plate, submerged, and direct-expansion double coils. The high-pressure refrigerant from the compressor(s) on the lower cycle condenses at a given intermediate temperature. A separate, lower-pressure refrigerant evaporates on the other side of the surface at a somewhat lower temperature. The vapor formed from the second refrigerant is compressed by the higher cycle compressor(s) until it can be condensed at an elevated temperature. The total heat is then rejected to a heat sink.

Subcoolers can be of shell-and-tube, shell-and-coil, or tube-in-tube construction. Friction losses reduce the liquid pressure that feeds refrigerant to an evaporator. Subcoolers are used to improve the efficiency of the system and to prevent refrigerant liquid from flashing due to pressure loss caused by friction and the vertical rise in lines. Refrigerant blends (zeotropes) can take advantage of temperature glide on the evaporator side with a direct-expansion-in-tube serpentine or coil configuration. In this case, the temperature glide from the bubble point to the dew point promotes increased efficiency and lower surface requirements for the subcooler. A flooded shell for the evaporating refrigerant requires use of only the higher dew-point temperature.

Lubricant coolers remove friction heat and some of the superheat of compression. Heat is usually removed by water, air, or a direct-expansion refrigerant.

The preparation of this chapter is assigned to TC 10.1, Custom Engineered Refrigeration Systems.

Condensers that reject heat from the refrigeration system are available in many standard forms. These include: water or brine cooled shell-and-tube, shell-and-coil, plate-and-frame, or tube-in-tube condensers; water cascading over or spray over plate or coil serpentine models; and air-cooled, fin-coil condensers. Special heat pump condensers are available in other forms such as tube-in-earth and submerged tube bundle, or as serpentine and cylindrical coil condensers that heat baths of boiling or single-phase fluids.

SELECTING DESIGN BALANCE POINTS

The refrigeration load at each designated evaporator pressure, the refrigerant properties, and the liquid refrigerant temperature feeding each evaporator determine the required flow rate of refrigerant in a system. The additional flow rates of refrigerant that provide refrigerant liquid cooling, desuperheating, and compressor lubricant cooling, where used, depend on the established liquid refrigerant temperatures and intermediate pressures.

For a given refrigerant and flow rate, the suction line pressure drop, suction gas temperature, pressure ratio and displacement, and volumetric efficiency determine the required size and speed of rotation for a positive displacement compressor. At low flow rates, particularly at very low temperatures and in long suction lines, heat gain through insulation can significantly raise the suction temperature. Also, at low flow rates a large, warm compressor casing and suction plenum can further heat the refrigerant before it is compressed. These heat gains increase the required displacement of a compressor. The compressor manufacturer must recommend the superheating factors to apply. The final suction gas temperature due to suction line heating is calculated by an iterative process.

Another concern is that more energy is required to compress the refrigerant to a given condenser pressure as the suction gas gains more superheat. This can be seen by examining a pressure-enthalpy diagram for a given refrigerant such as R-22, which is shown in Figure 6 of Chapter 17 in the 1993 ASHRAE *Handbook—Fundamentals*. As suction superheat increases along the horizontal axis, the slopes of the constant entropy lines of compression decrease. This means that a greater enthalpy change must occur to produce a given pressure rise. For a given flow then, the power required for compression is increased. With centrifugal compressors, pumping capacity is related to wheel diameter and speed, as well as volume flow and acoustic velocity of the refrigerant at the suction entrance. If the thermodynamic pressure requirement becomes too great for a given speed and volume flow, the centrifugal compressor will experience periodic back flow and surging.

Figure 1 shows an example of a system of curves needed to represent the maximum refrigeration capacities for a brine chilling plant. The example shows only one type of positive displacement compressor in a single-stage system operating at a steady-state condition. The figure is a graphical method of expressing the first law of thermodynamics with an energy balance applied to a refrigeration system.

One set of nearly parallel curves (A) represents the capacity of the cooler at various brine temperatures versus saturated suction temperature (a pressure condition) at the compressor, allowing for suction line pressure drops. The (B) set of curves represents the capacities of the compressor as the saturated suction temperature is varied and the saturated condenser temperature (a pressure condition) is varied. The (C) set of curves represents the heat transferred to the condenser by the compressor. It is calculated by adding the heat input at the evaporator to the energy imparted to the refrigerant by the compressor. The (D) set of curves represents the condenser performance at various saturated condenser temperatures as the inlet temperature of a fixed quantity of cooling water is varied.

The (E) set of curves represents the combined compressor and condenser performance as a "condensing unit" at various saturated suction temperatures for various cooling water temperatures. These curves were cross-plotted from the (C) and (D) curves back to the set of brine cooler curves as indicated by the dotted construction lines for the 80 and 92 °F cooling water temperatures. Another set of construction lines (not shown) would be used for the 86 °F cooling water. The number of construction lines used can be increased as necessary to adequately define curvature (usually no more than three per condensing-unit performance line).

The intersections of curves (A) and (E) represent the maximum capacities for the entire system at those conditions. For example, these curves show that the system develops 150 tons when cooling the brine to 44 °F at 36.4 °F (saturated) suction and using 80 °F cooling water. At 92 °F cooling water, the capacity drops to 134.5 tons if the required brine temperature is 42 °F and the required saturated suction temperature is 35 °F. The corresponding saturated condensing temperatures can be determined by inspection of the curves. The condenser heat rejection is apparent from the (C) curves at a given balance point.

The equation at the bottom of the figure may be used to determine the shaft horsepower (BHP) required at the compressor for any given balance point. A sixth set of curves could be drawn to indicate the power requirement as a function of capacity versus saturated suction and saturated condensing temperatures.

The same procedure can be repeated to calculate the performance of cascade systems. The rejected heat at the cascade condenser would be treated as the chiller load in making a cross-plot of the upper cycle, high-temperature refrigeration system.

For cooling air at the evaporator(s) and for condenser heat rejection to ambient air or evaporative condensers, the same procedures would be used. The performance of coils and expansion devices such as thermostatic expansion valves may also be graphed, once the basic concept of heat and mechanical energy input equivalent combinations is recognized. Chapter 1 of the 1993 ASHRAE *Handbook—Fundamentals* has further information.

This graphical plotting method finds the natural balance points of compressors operating at their maximum capacities. For multiple stage loads at several specific operating temperatures, the usual way of controlling compressor capacities is with a suction pressure control and compressor capacity control device. This control then accommodates any mismatch in the pumping capabilities of series connected compressors instead of allowing each compressor to find its natural balance point.

Computer programs could be developed to determine balance points of complex systems. However, because applications, components, and piping arrangements are so diverse, many designers use available capacity performance data from vendors and plot the balance points for chosen components. Individual computer programs may be available for specific components, which will speed the process.

Desuperheating of suction gas at intermediate pressures where multi-stage compressors balance is essential to reduce discharge temperatures of the upper-stage compressor. Desuperheating also helps reduce oil carryover and reduces energy requirements. Subcooling improves the net refrigeration effect of the refrigerant supplied to the next lower temperature evaporator and reduces the system energy requirements.

ENERGY AND MASS BALANCES

A systematic, point-to-point flow analysis of the system (including the piping) is essential to account for pressure drops and heat

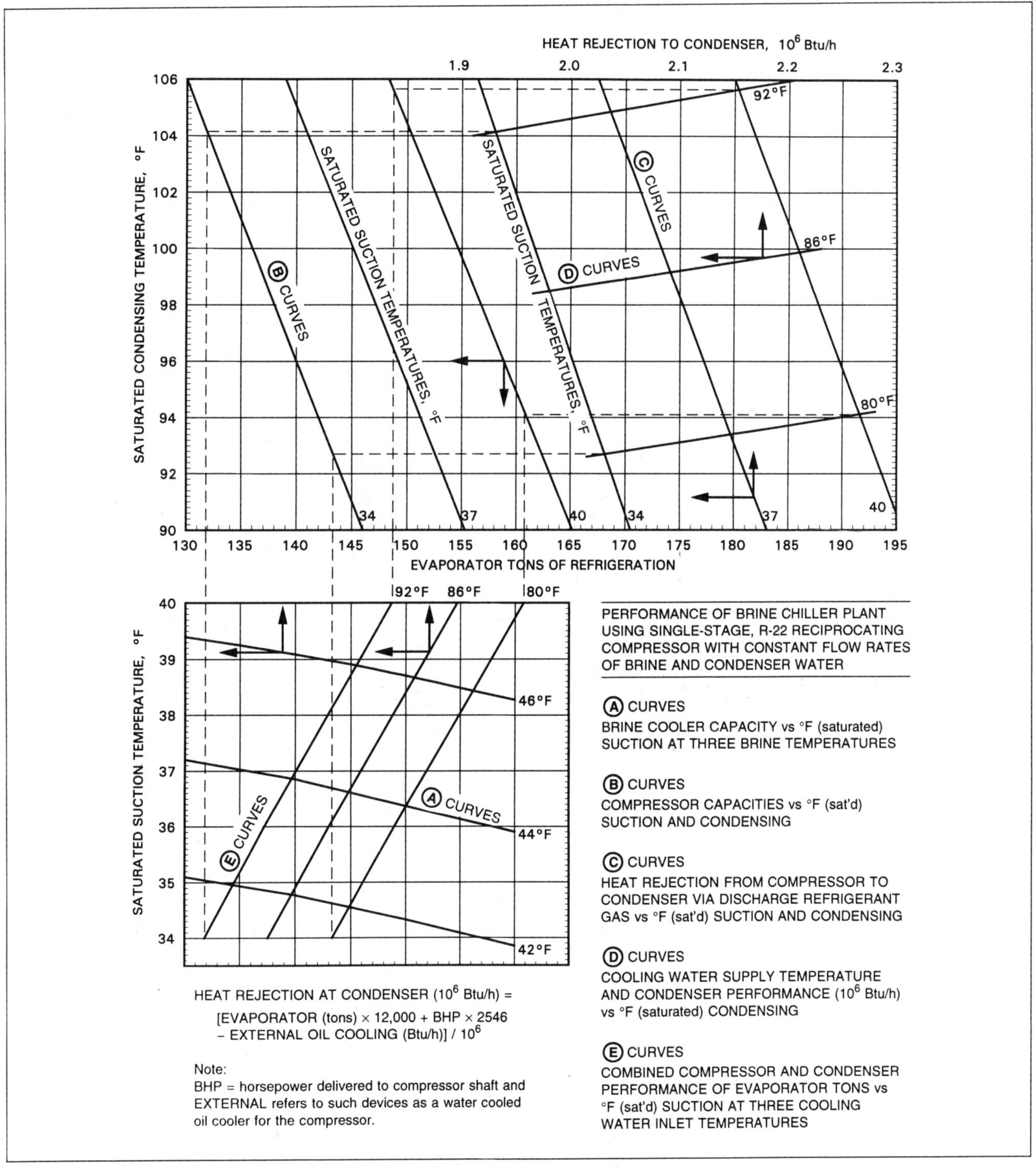

Fig. 1 Brine Chiller Balance Curve

gains, particularly in long suction lines. Air-cooled condensers, in particular, can have large pressure drops, which must be included in the analysis to estimate a realistic balance. Making a flow diagram of the system with designated pressures and temperatures, loads, enthalpies, flow rates, and energy requirements helps in identifying all important factors and components.

An overall energy and mass balance for the system is also essential to avoid mistakes. The overall system represented by the complete flow diagram should be enclosed by a dotted line envelope. Any energy inputs to or outputs from the system that directly affect the heat content of the refrigerant itself should cross the dotted envelope line and must enter the energy balance equations. Accurate estimates of the ambient heat gains through insulation and heat losses from discharge lines where they are significant improves the comprehensiveness of the energy balance and accuracy of equipment selections.

Cascade condenser loads and subcooler or desuperheating loads carried by a refrigerant are internal to the system and thus do not enter into the overall energy balance. The total energy entering the system equals the total energy leaving the system.

If calculations do not show an energy balance within reasonable tolerances for the accuracy of data used, then an omission occurred or a mistake was made and should be corrected.

The dotted envelope technique can be applied to any section of the system, but all energy transmissions must be included in the equations including the enthalpies and mass flow rates of streams that cross the dotted line.

SYSTEM PERFORMANCE

Rarely are sufficient sensors and instrumentation devices available, nor are conditions proper at a given job site to permit the calculation of a comprehensive, accurate energy balance for an operating system. Water cooled condensers and oil coolers for heat rejection and the use of electric motor drives, where motor efficiency and power factor curves are available, offer the best hope for estimating the actual performance of the individual components in a system. The evaporator heat loads can be derived from the measured heat rejection and derived mechanical or measured electrical energy inputs. A comprehensive flow diagram assists in a field survey.

Various coolant flow detection devices are available for direct measurement inside a pipe and for measurement from outside the pipe with variable degrees of accuracy. Sometimes flow rates may be estimated by simply weighing or measuring an accumulation of coolant over a brief time interval.

Temperature and pressure measurement devices should be calibrated and be of sufficient accuracy. Digital calibrated scanning devices for comprehensive simultaneous readings are best. Electrical power meters are not always available, so voltage and current at each leg of a motor power connection must be measured. Voltage drops for long power leads must be calculated when the voltage measurement points are far removed from the motor. The motor load versus efficiency and power factor curves must be used to determine motor output to the system.

Gears and belt or chain drives have friction and windage power losses that must be included in any meaningful analysis.

Stack gas flows and enthalpies for engine or gas turbine exhausts as well as air inputs and speeds must be included. In this case, heavy reliance must be placed on the performance curves issued by the vendor to estimate the energy input to the system.

Calculating steam turbine performance requires the measurements of turbine speed, steam pressures and temperatures, and condensate mass flow coupled with confidence that the vendor's performance curves truly represent the current mechanical condition. Plant personnel normally experience difficulty in obtaining operating data at specified performance values.

Heat rejection from air cooled condensers or coolers is extremely difficult to measure accurately because of changing ambient temperatures and the extent and scope of airflow measurements required. Often, one of the most important issues is the wide variation or cycling of process flows, process temperatures, and product refrigeration loads. Hot gas false loading and compressor continuous capacity modulations complicate any attempt to make a meaningful analysis.

The prediction of and measurement of the performance of systems using refrigerant blends (zeotropes) are especially challenging because of the temperature variations between bubble points and dew points.

Nevertheless, ideal conditions of nearly steady state loads and flows with a minimum of cycling sometimes occur frequently enough to permit a reasonable analysis. Several sets of nearly simultaneous data at all points over a short time enhance the accuracy of any calculation concerning the performance of a given system. In all cases, the proper purging of condensers and elimination of excessive lubricant contamination of the refrigerant at the evaporators are essential to determine system capabilities accurately.

CHAPTER 44

REFRIGERANT-CONTROL DEVICES

THE control of refrigerant flow is essential in any refrigeration system. The Control Switches section of this chapter details control switches, including (1) pressure control switches, (2) temperature control switches, (3) fluid flow-sensing swtiches, (4) differential control switches, and (5) float switches. The Control Valves section addresses the operation, selection, and application of control valves, including (1) thermostatic expansion valves, (2) electric expansion valves, (3) constant pressure expansion valves, (4) evaporator pressure and temperature regulators, (5) suction pressure regulators, (6) condenser pressure regulators, (7) high-side float valves, (8) low-side float valves, (9) solenoid valves, (10) condensing water regulators, (11) check valves, and (12) relief devices. The third section is Discharge-Line Oil Separators, the fourth section is Capillary Tubes, and the fifth section is Short Tube Restrictors. For further information on automatic control, see Chapter 41 of the 1991 ASHRAE *Handbook—Applications.*

While most examples, references, and capacity data in this chapter refer to the more common refrigerants (R-12 and R-22) currently used in the industry, new data is being developed for the new alternative refrigerants (*e.g.*, R-134a). When possible, data for the new refrigerants has been included in this revision. Additional information will be included in future revisions as it becomes available.

CONTROL SWITCHES

The *control switch* operates one or more sets of electrical contacts, which are used, for example, to open or close water or refrigerant valves, engage and disengage compressor clutch coils and relays, active timers, and thermostats. Control switches respond to a variety of physical changes such as pressure, temperature, liquid level, flow velocity, and proximity. Pressure- and temperature-responsive controls have one or more power elements, which may use bellows, diaphragms, snap discs, or bourdon tubes to produce the force needed to operate the mechanism. Level-responsive controls may use floats, mercury balance tubes, or electronic probes to operate (directly or indirectly) one or more sets of electrical contacts.

Refrigeration controls may be categorized into three basic groups: (1) operating, (2) primary, and (3) limit. Operating controls such as thermostats, turn systems on and off. Primary controls, such as floats, provide safe continuous operation. Limit controls, such as the high-pressure cut-out switch, protect a refrigeration system from unsafe operation.

PRESSURE CONTROL SWITCHES—COMMERCIAL APPLICATIONS

The refrigerant pressure is applied directly to the power element, which moves against a spring that can be adjusted to control any operation at the desired pressure. If the control is to operate in the subatmospheric (or *vacuum*) range, the bellows or diaphragm force is sometimes reversed to act in the same direction as the adjusting spring. To counteract this, an additional spring may be needed to overcome the reversed bellows force; consequently, the force of both the bellows and the vacuum spring oppose the adjusting spring. However, the controls are sometimes built without any spring and use the spring force of the bellows or diaphragm in place of the adjusting spring.

The force available for doing work in this control switch depends on the pressure in the system and on the area of the bellows or diaphragm. With proper area, a sufficient force can be produced to operate heavy switches, water valves, or refrigerant valves. In heavy-duty controls, the minimum differential is large because of the massiveness of the bellows and the opposing spring system.

PRESSURE CONTROL SWITCHES—AUTOMOTIVE APPLICATIONS

Pressure control switches in most automotive refrigeration systems are used primarily to control electrical engagement of the compressor clutch or condenser fan. These switches may be either line mounted or installed into the compressor body. The following is a list of types of pressure control switches with their corresponding functions:

The preparation of this chapter is assigned to TC 8.8, Refrigerant System Controls and Accessories.

Type	Function
High-pressure cut-out (HPCO)	Disengages compressor clutch when excessive pressure occurs
High-side low-pressure (HSLP)	Disengages compressor clutch under low ambient and/or low-pressure conditions
High-side fan cycling (HSFC)	Cycles condenser fan on and off to provide proper condenser pressure
Low-side low-pressure (LSLP)	Disengages compressor clutch when low charge or system blockage occurs
Low-side clutch cycling (LSCC)	Cycles compressor clutch on and off to provide proper evaporator pressure

The modern automotive pressure control switch incorporates one or more steel snap discs, which provide positive pressure at the electrical contacts. Use of the snap action disc construction also assures consistent differential pressure between on and off settings. (See Figure 1.) One very important benefit of this construction over the earlier direct contact or creep switch is the substantial reduction in electrical contact bounce or flutter, which can be very damaging to compressor clutch assemblies, fan relays, and electronic control modules (ECM).

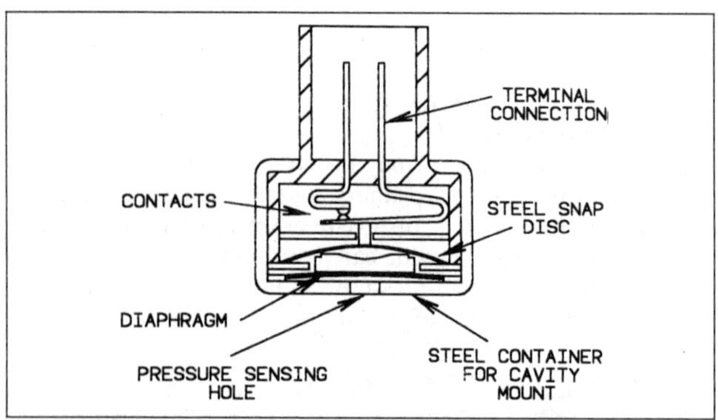

Fig. 1 Pressure Control Switch (Compressor Mounted)

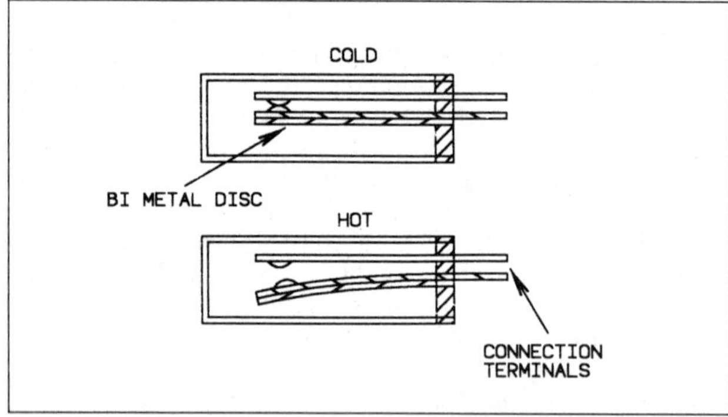

Fig. 2 Direct Temperature Control Switch

TEMPERATURE CONTROL SWITCHES

Indirect temperature control switches for refrigeration are pressure control switches in which the pressure-responsive power element is replaced by a temperature-responsive power element. The exact temperature-pressure or temperature-volume relationship of the fluid used in the power element allows the temperature of the bulb to control the switch accurately. Operation of the control switch results from these changes in pressure or volume.

Direct temperature control switches generally contain bimetallic discs that activate electrical contacts when increasing or decreasing temperature occurs. The bimetallic discs or blades are constructed of low- and high-expansion metal. As the temperature of the disc or blade is increased or decreased, distortion occurs, causing the attached electrical contacts to engage or disengage. This type of control switch is often used in thermal limit applications since on/off differentials are too coarse for use in most primary refrigerant control systems. The same principle is used in thermostat construction in which a semicoiled strip of bimetallic material is used to activate electrical contacts. (See Figure 2.)

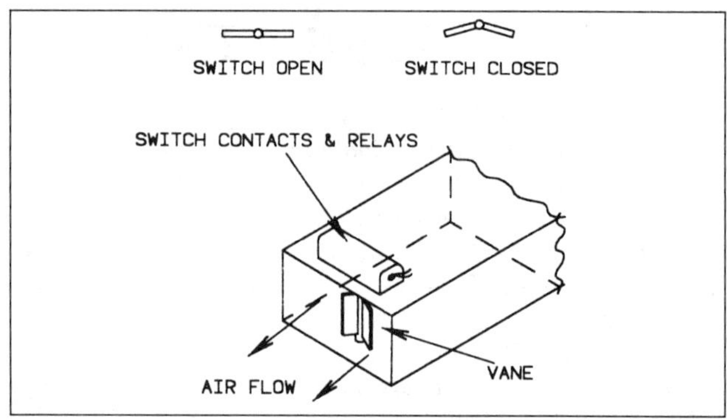

Fig. 3 Air Flow Switch

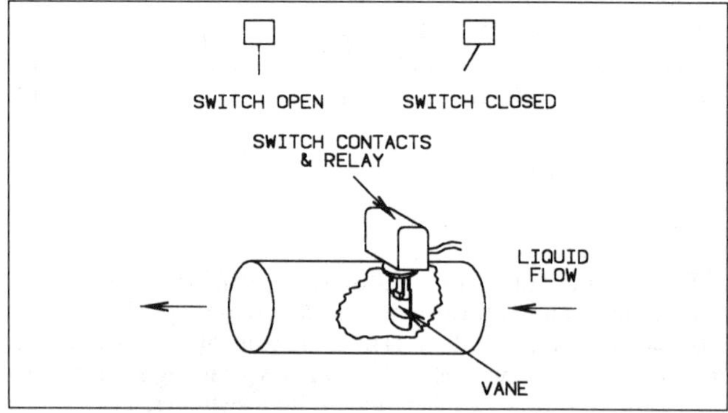

Fig. 4 Liquid Flow Switch

FLUID FLOW-SENSING SWITCHES

Refrigeration fluid flow-sensing switches are of two basic types: air flow and liquid flow.

Air flow switches (Figure 3) may be used in any application for which it is important to detect the presence of air flow, for example, refrigeration air ducts. An air flow switch may be used to shut down fan blower motors or total systems if air stoppage occurs. The spring-loaded vanes are installed into a regulated high-velocity air stream. The pressure of the air perpendicular to the vane causes it to deflect and make electrical contact. Stoppage of air flow causes the vanes to retract and break electrical contact.

Liquid flow switches work in a similar manner. (See Figure 4.) Water or refrigerant flow applied perpendicular to the vane causes it to deflect and make electrical contact. Liquid flow switches are often used to indicate sufficient water or refrigerant flow through chillers or condensers.

DIFFERENTIAL CONTROL SWITCHES

Control switches that maintain a given difference in pressure or temperature between two pipe lines or spaces are called *differential control switches.* An example of this type of control is the oil safety switch used with reciprocating compressors that have forced-feed pressure lubrication. These controls have two elements (either pressure- or temperature-sensitive) because they sense conditions in two different locations.

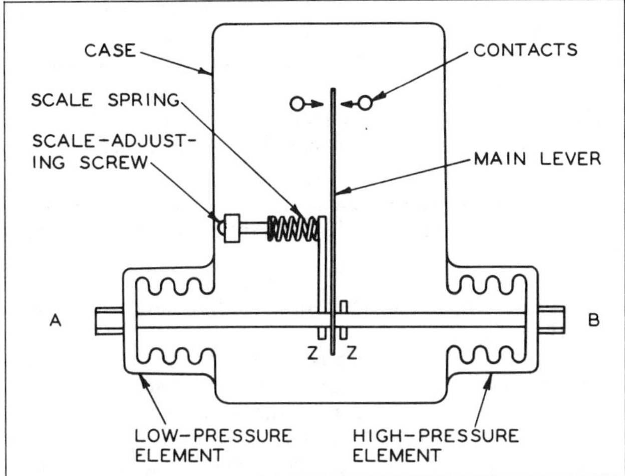

Fig. 5 Differential Pressure Control Switch

Figure 5 is a schematic diagram of a differential pressure control switch that uses bellows as power elements. As shown, the two elements are rigidly connected by a rod so that motion of one causes motion of the other. On the connecting rod, a power takeoff operates either single-pole double-throw contacts (as shown) or valves. A compression spring permits the setting of the differential pressure at which the device operates. The sum of the forces developed by the low-pressure bellows and the scale spring equals the force developed by the high-pressure bellows at the control point.

Instrument differential is the difference in pressure between the low- and the high-pressure elements for which the instrument is adjusted. In the case of a temperature element, this difference is expressed in degrees. *Operating differential* is the difference in pressure or temperature required to open or close the switch contacts. It is actually the change in instrument differential from *cut-in* to *cut-out* for any setting. Operating differential can be varied by a second spring that acts in the same direction as the first and takes effect only at the cut-in or cut-out point without affecting the other spring. A second method is the adjustment of the distance between collars *Z-Z* on the connecting rod. The greater the distance between them, the greater the operating differential.

If a constant instrument differential is required on a temperature-sensitive differential control switch throughout a large temperature range, it is usually necessary to use a different fill in one element than in the other, if both are of the vapor-charged type. The alternative is to use liquid-filled power elements.

A second type of differential-temperature control uses two sensing bulbs and capillaries connected to one bellows with a liquid fill. This is known as a *constant-volume* fill, because the operating point corresponds to a constant volume of the two bulbs, capillaries, and bellows. If the two bulbs have equal volume, a rise in the temperature of one bulb requires an equivalent fall in the temperature of the other to maintain the operating point.

FLOAT SWITCHES

A *float switch* has a float that operates one or more sets of electrical contacts through variation in the level of a liquid. It is connected by equalizing lines to the vessel in which the liquid level is to be maintained or indicated.

Operation and Selection

Some float switches (see Figure 6) operate from the movement of a magnetic armature located in the field of a permanent magnet. Others use true solid-state circuits in which a variable signal is generated by liquid contact with a probe that replaces the float. The latter methods are adapted to remote-controlled applications and are preferred for ultralow temperature applications. Switches that have mercury-tube contacts are usually not recommended for installation in an ambient temperature lower than −25 °F, since mercury freezes solid at temperatures of approximately −38 °F.

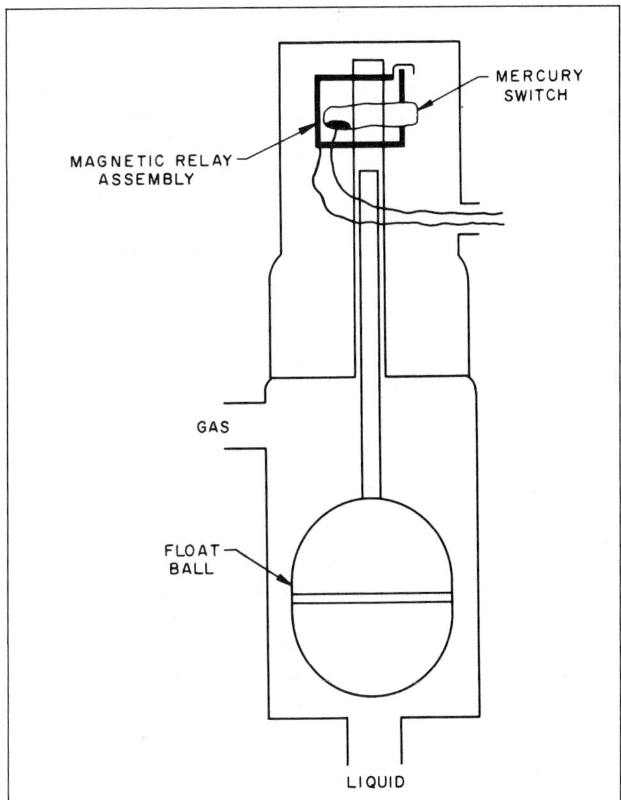

Fig. 6 Magnetic Float Switch

Application

The float switch can maintain or indicate the level of a liquid, operate an alarm, control the operation of a pump, or perform other functions. A float switch, solenoid liquid valve, and hand expansion valve combination can control the refrigerant level on the high- or low-pressure side of the refrigeration system in the same way that high- or low-side float valves are used. The hand expansion valve, located in the refrigerant liquid line immediately downstream of the solenoid valve, is initially adjusted to provide a refrigerant flow rate at maximum load to keep the solenoid liquid valve in the open position 80 to 90% of the time; it need not be adjusted thereafter. From the outlet side of the hand expansion valve, the refrigerant passes through a line and enters either the evaporator or the surge drum, depending on the unit design.

When the float switch is applied for low-side level control, proper precaution must be taken to provide a quiet liquid level that falls in response to an increase in evaporator load and rises with a decrease in evaporator load. The same recommendations for insulation of the body and liquid leg of the low-side float valve apply to the float switch when it is used for refrigerant-level control on the low-pressure side of the refrigeration system. To avoid floodback in this application, controls should be wired to prevent the opening of the solenoid liquid valve when the solenoid suction valve closes or the compressor stops.

CONTROL VALVES

Valves are used to start, stop, direct, and modulate the flow of refrigerant to satisfy system requirements in accordance with load requirements. To ensure satisfactory performance, valves should be protected adequately from foreign material, excessive moisture, and corrosion in refrigeration systems by the installation of properly sized strainers, filters, and/or filter-driers. Other valve designs and constructions are available. Either a diaphragm or a bellows can be used in various types of refrigerant flow-control valves.

THERMOSTATIC EXPANSION VALVES

The thermostatic expansion valve controls the flow of liquid refrigerant entering the evaporator in response to the superheat of the gas leaving it. Its basic function is to keep the evaporator active without permitting liquid to be returned through the suction line to the compressor. This is done by controlling the mass flow of refrigerant entering the evaporator so it equals the rate at which it can be completely vaporized in the evaporator by the absorption of heat. Since the thermostatic expansion valve is operated by superheat and is responsive to changes in superheat, a portion of the evaporator must be used to superheat the refrigerant gas.

Unlike the constant pressure-type, the thermostatic expansion valve is not limited to constant load applications. It is used successfully for controlling the refrigerant flow to all types of direct-expansion evaporators in air-conditioning and commercial, low-temperature, and ultralow-temperature refrigeration systems.

Operation

A schematic cross section of the typical thermostatic expansion valve, with the principal components identified, is shown in Figure 7. The following forces govern thermostatic expansion valve operation:

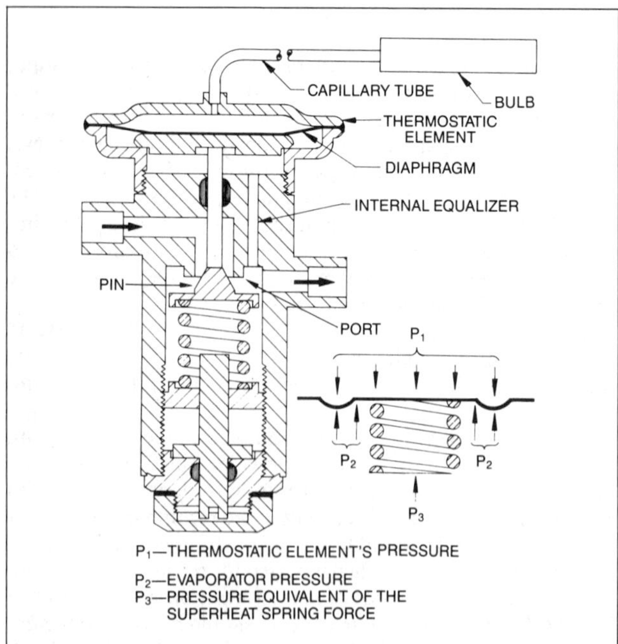

Fig. 7 Typical Thermostatic Expansion Valve

P_1 = the pressure of the thermostatic element (a function of the bulb's charge and the bulb temperature), which is applied to the top of the diaphragm and acts to open the valve

P_2 = the evaporator pressure, which is applied under the diaphragm through the equalizer passage and acts in a closing direction

P_3 = the pressure equivalent of the superheat spring force, which is applied underneath the diaphragm and is also a closing force

At any constant operating condition, these forces are balanced and $P_1 = P_2 + P_3$.

An additional force, which is small and not considered fundamental, arises from the unbalanced pressure across the valve port. To a degree, it can affect thermostatic expansion valve operation. For the configuration shown in Figure 7, the force resulting from port imbalance is the product of the pressure drop across the port and the difference in area of the port and the stem; it is an opening force. In other designs, depending on the direction of flow through the valve, the port imbalance may result in a closing force.

The principal effect of port imbalance is on the stability of the valve control. As with any modulating control, if the ratio of the diaphragm area to the port is kept large, the imbalanced port effect is minor. However, depending on this ratio or system operating conditions, valves are made with balanced port construction.

Figure 8 shows an evaporator operating at a saturation temperature of 40 °F (68.5 psig). Liquid refrigerant enters the expansion valve, is reduced in pressure and temperature at the valve port, and enters the evaporator at Point A as a mixture of saturated liquid and vapor. As flow continues through the evaporator, more of the refrigerant is evaporated. Assuming there is no pressure drop, the refrigerant temperature remains at 40 °F until the liquid is entirely evaporated at Point B. From this point, additional heat absorption increases the temperature and superheats the refrigerant gas, while the pressure remains constant at 68.5 psig, until, at Point C (the outlet of the evaporator), the refrigerant gas temperature is 50 °F. At this point, the superheat is 10 °F (50 to 40 °F).

An increase in the heat load on the evaporator increases the temperature of the refrigerant gas leaving the evaporator. The bulb of the thermostatic expansion valve senses this increase; the thermostatic charge pressure, P_1, increases and causes the valve to open wider. The increased flow rate results in a higher evaporator pressure, P_2, and a balanced control point is again established. Conversely, a decrease in the heat load on the evaporator

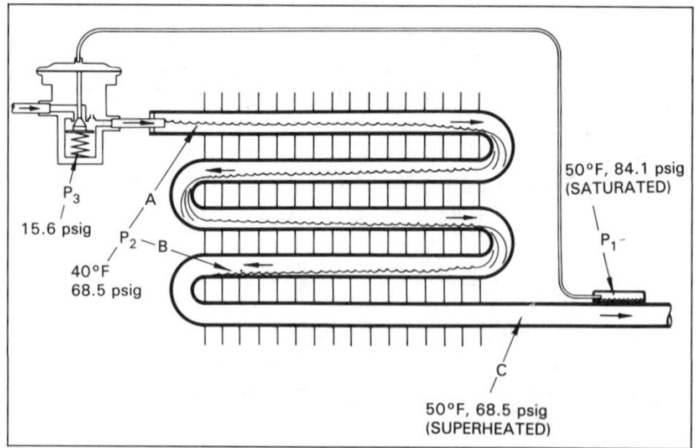

Fig. 8 Thermostatic Expansion Valve Controlling Flow of Liquid Refrigerant 22 Entering Evaporator, Assuming Refrigerant 22 Charge in Bulb

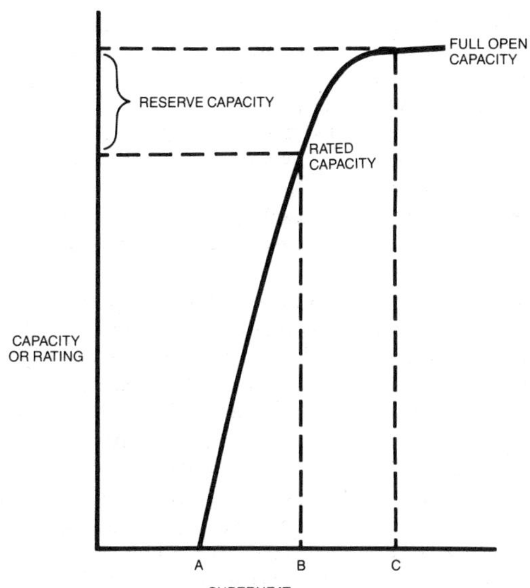

Fig. 9 Typical Gradient Curve for Thermostatic Expansion Valves

decreases the temperature of the refrigerant gas leaving the evaporator and causes the thermostatic expansion valve to move in a closing direction.

The new control point, following an increase in valve opening, is at a slightly higher operating superheat because of the spring rate of the diaphragm and superheat spring. Conversely, a decrease in load results in an operating superheat slightly lower than the original control point.

These superheat changes in response to load changes are illustrated by the gradient curve of Figure 9. Superheat at no load A, *static superheat,* ensures sufficient spring force to keep the valve closed during equipment shutdown. An increase in valve capacity or load is approximately proportional to superheat until the valve is open fully. The *opening superheat,* represented by the distance AB, may be defined as the superheat increase required to open the valve to match the load. *Operating superheat* is the sum of static superheat and opening superheat.

Capacity

The *factory superheat setting* (static superheat setting) of thermostatic expansion valves is made when the valve starts to open. Valve manufacturers establish capacity ratings on the basis of opening superheat from 4 to 8 °F, depending on valve design, valve size, and application. Full-open capacities usually exceed rated capacities by 10 to 40% to allow a reserve, represented by the distance BC in Figure 9, for manufacturing tolerances and application contingencies.

Valve selection should not be based on use of the reserve capacity of the thermostatic expansion valve, which is available only at higher operating superheat. The added superheat may have an adverse effect on performance. Because valve gradients used for rating purposes are selected to produce optimum modulation for a given valve design, manufacturers' recommendations should be followed.

Thermostatic expansion valve capacities are normally published for various evaporator temperatures and valve pressure drops. (See ASHRAE *Standard* 17-1986 and ARI *Standard* 750-1987 for testing and rating methods.) Nominal capacities apply at 40 °F evaporator temperature. Capacities are reduced at lower evaporator temperatures. These reductions in capacity are the result of the change in the refrigerant pressure-temperature relationship at lower temperatures. For example, if Refrigerant 22 is used, the change in saturated pressure between 40 and 45 °F is 7.5 psi, whereas between −20 and −15 °F the change is 3.0 psi. Although the valve responds to pressure changes, capacities are based on superheat gradients. Thus, the valve opening and, consequently, the valve capacity, is less for a given superheat change at lower evaporator temperatures.

Pressure drop across the valve port is always the net pressure drop available at the valve, rather than the difference between compressor discharge and compressor suction pressures.

Allowances must be made for the following:

1. Pressure drop through condenser, receiver, liquid lines, fittings, and liquid line accessories, such as filters, driers, solenoid valves, etc.
2. Static pressure in a vertical liquid line. If the thermostatic expansion valve is at a higher level than the receiver, there will be a pressure loss in the liquid line because of the static pressure of liquid.
3. Distributor pressure drop.
4. Evaporator pressure drop.
5. Pressure drop through suction line and accessories, such as evaporator pressure regulators, solenoid valves, accumulators etc.

Variations in valve capacity related to changes in system conditions are approximately proportional to the following relationship:

$$Q \cong C\sqrt{\rho \Delta p}\,(h_g - h_f)$$

where

Q = heat flow
C = constant related to thermostatic expansion valve design
ρ = entering liquid density
Δp = valve pressure difference
h_g = enthalpy of vapor exiting evaporator
h_f = enthalpy of liquid entering the thermostatic expansion valve

Thermostatic expansion valve ratings are based on vapor-free liquid entering the valve. If flash gas is present in the entering liquid, the valve capacity is reduced substantially because the gas must be handled along with the liquid. Flashing of the liquid refrigerant may be caused by pressure drop in the liquid line, the filter-drier, the vertical lift, or a combination of these. If the refrigerant subcooling at the receiver outlet is not adequate to prevent the formation of flash gas, additional subcooling means must be used to remove it. Liquid-to-suction heat exchange provides a moderate degree of subcooling, but for extreme requirements, a separate liquid-cooling coil may be necessary.

Thermostatic Charges

Each type of thermostatic charge has certain advantages and limitations. The principal types of thermostatic charges and their characteristics are described below.

Gas Charge. Conventional gas charges are limited liquid charges that use the same refrigerant in the thermostatic element that is used in the refrigeration system. The amount of charge is such that, at a predetermined temperature, all of the liquid has vaporized and any temperature increase above this point results in virtually no increase in element pressure. Figure 10 shows the pressure-temperature relationship of the Refrigerant 22 gas charge in the thermostatic element. Because of the characteristic pressure-limiting feature of its thermostatic element, the gas-charged valve can provide compressor motor overload protection on some systems by limiting the maximum operating suction pressure (MOP). It also helps to prevent *floodback* (the return of refrigerant liquid to the compressor through the suction line) on starting. Increasing the superheat setting lowers the maximum operating suction

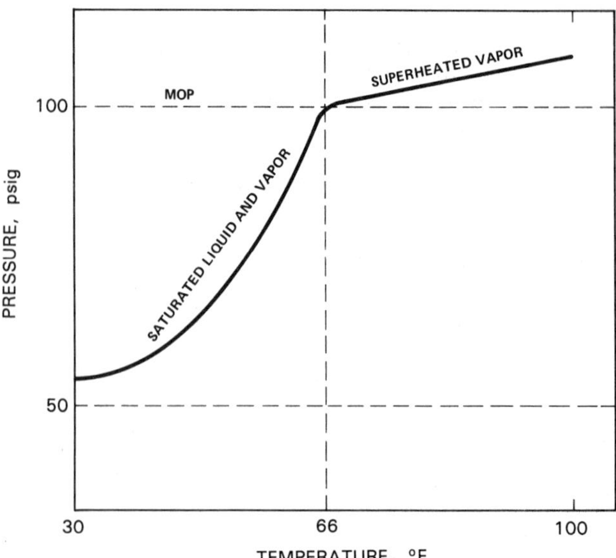

Fig. 10 Pressure-Temperature Relationship of Refrigerant 22 Gas Charge in Thermostatic Element

pressure; decreasing the superheat setting raises it because the superheat spring, together with the evaporator pressure, act to balance the element pressure through the diaphragm.

Gas-charged valves must be carefully applied to avoid loss of control from the bulb. If either the diaphragm chamber or the capillary tube becomes colder than the bulb, the small amount of charge in the bulb condenses at the coldest point. This results in the valve throttling or closing. This is detailed in the Application section.

Liquid Charges. Straight liquid charges use the same refrigerant in the thermostatic element that is used in the refrigeration system. The volumes of the bulb, capillary tubing, and diaphragm chamber are so proportioned that the bulb contains some liquid under all temperature conditions. Therefore, the bulb always controls the valve operation, even with a colder diaphragm chamber or capillary tube.

The characteristics of the straight liquid charge (see Figure 11) result in an increase in operating superheat as the evaporator temperature decreases. This usually limits the use of the straight liquid charge to moderately high evaporator temperatures. The valve

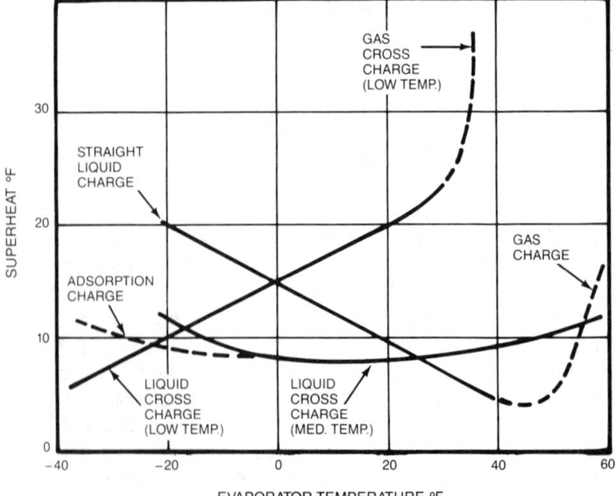

Fig. 11 Typical Superheat Characteristics of Common Thermostatic Charges

setting required for a reasonable operating superheat at a low evaporator temperature may cause floodback during pulldown from normal ambient temperatures.

Liquid Cross Charges. The liquid cross charges, unlike the conventional liquid charges, use a liquid in the thermostatic element that is different from the refrigerant in the system. Cross charges have flatter pressure-temperature curves than the system refrigerants with which they are used. Consequently, their superheat characteristics differ considerably from those of the straight liquid or gas charges.

Cross charges in the commercial temperature range generally have superheat characteristics that are nearly constant or that deviate only moderately through the evaporator temperature range. This charge, also illustrated in Figure 11, is generally used in the evaporator temperature range of 40 to 0°F or slightly below.

For evaporator temperatures substantially below 0°F, a more extreme cross charge may be used. At high evaporator temperatures, the valve controls at a high superheat. As the evaporator temperature is reduced to the normal operating range, the operating superheat is also reduced to normal. This characteristic prevents floodback on starting, reduces the load on the compressor motor after start-up, and permits a rapid pulldown of suction pressure. To avoid floodback, valves with this type of charge must be set for the optimum operating superheat at the lowest evaporator temperature expected.

Gas Cross Charges. The gas cross charges combine the features of the gas charge and the liquid cross charge. They use a limited amount of liquid, thereby providing a maximum operating pressure. The liquid used in the charge is different from the refrigerant in the system and is chosen to provide superheat characteristics similar to those of the liquid cross charges (low temperature). Consequently, they provide both the superheat characteristics of a cross charge and the maximum operating pressure of a gas charge (Figure 11). While a commercial (medium-temperature) gas cross charge is possible, its uses are limited.

Adsorption Charge. Typical adsorption charges depend on the property of an adsorbent, such as silica gel or activated carbon, that is used in an element bulb to adsorb and desorb a gas such as carbon dioxide, with accompanying changes in temperature. The amount of adsorption or desorption changes the pressure in the thermostatic element. Since adsorption charges respond primarily to the temperature of the adsorbent material, they are essentially unaffected by cross-ambient conditions. The comparatively slow response time of the adsorbent results in a charge characterized by its stability. Superheat characteristics can be varied by using different charge materials, adsorbents, and/or charge pressures. The pressure-limiting feature of the gas or gas cross charges is not available with the adsorption-type element.

Type of Equalization

Internal Equalizer. When the refrigerant pressure drop through a single-circuit evaporator is equivalent to not more than a 2°F change in evaporator temperature, a thermostatic expansion valve that has an internal equalizer may be used. Internal equalization describes valve outlet pressure transmitted through an internal passage to the underside of the diaphragm (see Figure 7).

The pressure drop in many evaporators is greater than the 2°F equivalent. When a refrigerant distributor is used, the pressure drop across the distributor will cause the pressure at the outlet of the expansion valve to be considerably higher than the pressure at the evaporator outlet. As a result, an internally equalized valve will control at an abnormally high superheat. Under these conditions, the evaporator does not perform efficiently because it is starved for refrigerant. Furthermore, the distributor pressure drop is not constant but varies with evaporator load and cannot be compensated for by adjusting the superheat setting of the valve.

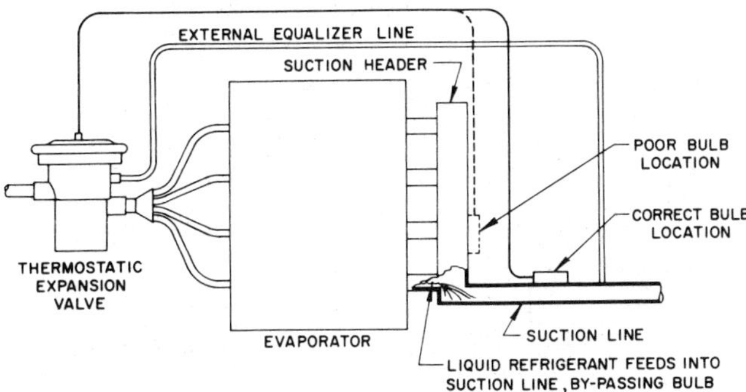

Fig. 12 Bulb Location for Thermostatic Expansion Valve

External Equalizer. Because evaporator and/or refrigerant distributor pressure drop produces poor system performance with an internal equalizer valve, a valve that has an external equalizer is used. Instead of the internal communicating passage shown in Figure 7, an external connection to the underside of the diaphragm is provided. The external equalizer line is connected either to the suction line, as shown in Figure 12, or into the evaporator at a point downstream from the major pressure drop.

Alternate Types of Construction

Pilot-operated thermostatic expansion valves are used on large systems in which the required capacity per valve is beyond the range of direct-operated valves. The pilot-operated valve consists of a piston-type pilot-operated regulator, which is used as the main expansion valve, and a low-capacity thermostatic expansion valve, which serves as an external pilot valve. The small pilot thermostatic expansion valve supplies pressure to the piston chamber or, depending on the regulator design, bleeds pressure from the piston chamber in response to a change in the operating superheat. Pilot operation permits the use of a characterized port in the main expansion valve to provide good modulation over a wide loading range. Therefore, the pilot-operated valve performs well on refrigerating systems that have some form of compressor capacity reduction, such as cylinder unloading. Figure 13 illustrates such a control valve applied to a large capacity, direct-expansion chiller.

The auxiliary pilot controls should be sized to handle only the pilot circuit flow. For example, in Figure 13, a small solenoid valve in the pilot circuit, installed ahead of the thermostatic expansion valve, converts the pilot-operated valve into a stop valve when the solenoid valve is closed.

Equalization Features. When the compressor stops, a thermostatic expansion valve usually moves to the *closed* position. This movement sustains the difference in refrigerant pressures in the evaporator and the condenser. Low-starting torque motors require that these pressures be equalized to reduce the torque needed to restart the compressor. One way to provide pressure equalization is to add, parallel to the main valve port, a small fixed auxiliary passageway, such as a slot or drilled hole in the valve seat or valve pin. This opening permits a limited fluid flow through the control, even when the valve is closed and allows the system pressures to equalize on the *off* cycle. The size of such a fixed auxiliary passageway must be limited so its flow capacity is not greater than the smallest flow that must be controlled in normal system operation.

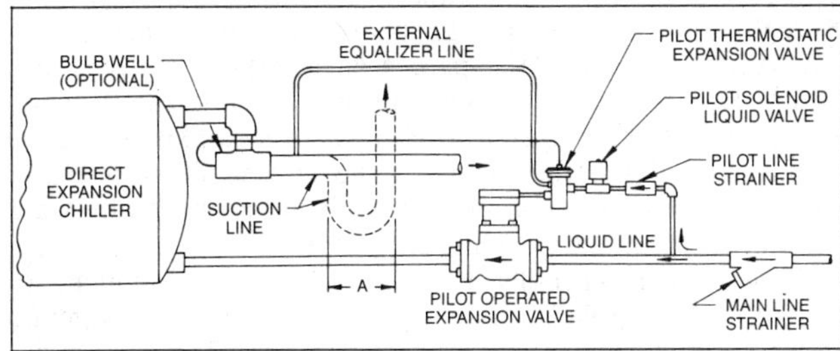

Fig. 13 Pilot-Operated Thermostatic Expansion Valve Controlling Liquid Refrigerant Flow to Direct-Expansion Chiller

Another more complex control is available for systems requiring shorter equalizing times than can be achieved with the fixed auxiliary passageway. This control incorporates an auxiliary valve port, which bypasses the primary port and is opened by the element diaphragm as it moves toward and beyond the position at which the primary valve port is closed. The flow capacity of such an auxiliary valve can be considerably larger than that of the fixed auxiliary passageway, so that pressures can equalize more rapidly.

Flooded System. Thermostatic expansion valves are seldom applied to flooded evaporators because superheat is necessary for proper valve control; only a few degrees of suction vapor superheat in a flooded evaporator incurs a substantial loss in capacity. If the bulb is installed downstream from a liquid-to-suction heat exchanger, a thermostatic expansion valve can be made to operate at this point on a higher superheat. Valve control is apt to be poor because of the variable rate of heat exchange as flow rates change (see Application section).

Expansion valves with modified thermostatic elements are available in which electrical heat is supplied to the bulb. The bulb is inserted in direct contact with refrigerant liquid in a low-side accumulator. The contact of cold refrigerant liquid with the bulb overrides the artificial heat source and throttles the expansion valve. As the liquid falls away from the bulb, the valve feed increases again. Although similar in construction to a thermostatic expansion valve, it is essentially a modulating liquid level control valve.

Desuperheating Valves. Thermostatic expansion valves with special thermostatic charges are used to reduce gas temperatures (superheat) on various air conditioning and refrigeration systems. Suction gas in a single-stage system can be desuperheated by injecting liquid directly into the suction line. This cooling may be required with or without discharge gas bypass used for compressor capacity control. The line upstream of the valve bulb location must be long enough so the injected refrigerant can mix adequately with the gas being desuperheated. On compound compression systems, liquid may be injected directly into the interstage line upstream of the valve bulb to provide intercooling.

Application

Hunting is alternate overfeeding and starving of the refrigerant feed to the evaporator, which produces sustained cyclic changes in the pressure and temperature of the refrigerant gas leaving the evaporator. Extreme hunting reduces the capacity of the refrigeration system because the mean evaporator pressure and temperature are lowered and the compressor capacity is reduced. If overfeeding of the expansion valve causes intermittent flooding of liquid into the suction line, the compressor may be damaged.

Although hunting is commonly attributed to the thermostatic expansion valve, it is seldom solely responsible. One reason for hunting is that all evaporators have a time lag. When the bulb signals for a change in refrigerant flow, the refrigerant must traverse the entire evaporator before a new signal reaches the bulb. This lag or time lapse may cause continuous overshooting of the valve both opening and closing. In addition, the thermostatic element, because of its mass, has a time lag that may be in phase with the evaporator lag and amplify the original overshooting.

It is possible to alter the response rate of the thermostatic element by either using thermal ballast or changing the mass or heat capacity of the bulb, thereby damping or even eliminating the hunting. A change in valve gradient may produce the same result.

Extremely high refrigerant velocity in the evaporator can also cause hunting. Liquid refrigerant under these conditions moves in waves, called *slugs*, that fill a portion of the evaporator tube and erupt into the suction line. These unevaporated slugs chill the bulb and temporarily reduce the feed of the valve, resulting in intermittent starving of the evaporator.

On multiple-circuit evaporators, a lightly loaded or overfed circuit will also flood into the suction line, chill the bulb, and throttle the valve. Again, the effect is intermittent; when the valve feed is reduced, the flooding ceases, and the valve reopens.

Hunting can be minimized or avoided by the following:

1. Select the proper valve size from the valve capacity ratings rather than nominal valve capacity; oversized valves aggravate hunting.
2. Change the valve adjustment. A lower superheat setting usually (but not always) increases hunting (Huelle 1972, Wedeking and Stoecker 1966, Stoecker 1966).
3. Select the correct thermostatic element charge. Cross-charged elements have inherent antihunt characteristics.
4. Design the evaporator section for even refrigerant and airflow. Uniform heat transfer from the evaporator is only possible if refrigerant is distributed by a properly selected and applied refrigerant distributor and air distribution is controlled by a properly designed housing. (Air-cooling and dehumidifying coils, including refrigerant distributors, are detailed in Chapter 21 of the 1992 ASHRAE *Handbook—Systems and Equipment*.)
5. Size and arrange suction piping correctly.
6. Locate and apply the bulb correctly.
7. Select the best location for the external equalizer line connection.

Bulb Location. Most installation requirements are met by strapping the bulb to the suction line to obtain good thermal contact between them. Normally, the bulb is attached to a horizontal line upstream of the external equalizer connection (if used) at a 3 or 9 o'clock position as close to the evaporator as possible. While the bulb is not normally placed near or after suction line traps, some system designers test and prove locations that differ from these recommendations. A good moisture-resistant insulation over the bulb and suction line eliminates any adverse effect of varying ambient temperatures at the bulb location.

Occasionally, the bulb of the thermostatic expansion valve is installed downstream from a liquid-suction heat exchanger to compensate for a capacity shortage due to an undersized evaporator. While this procedure seems to be a simple method of obtaining maximum evaporator capacity, installing the bulb downstream of the heat exchanger is undesirable from a control standpoint. As the valve modulates, the liquid flow rate through the heat exchanger changes, causing the rate of heat transfer to the suction vapor to change. An exaggerated valve response follows, resulting in hunting. It may be possible to find a bulb location downstream from the heat exchanger that reduces the hunt considerably. However, the danger of floodback to the compressor normally overshadows the need to attempt this method.

Certain installations require increased bulb sensitivity as a protection against floodback. The bulb, if located properly in a well in the suction line, has a rapid response feature because of its intimate contact with the refrigerant stream. The bulb sensitivity can be increased by the use of a bulb smaller than is normally supplied. However, the use of the smaller bulb is limited to gas-charged valves. Good piping practice also effects expansion valve performance.

Figure 14 illustrates the proper piping arrangement when the suction line runs above the evaporator. An oil trap that is as short as possible is located downstream from the bulb. The vertical riser(s) must be sized to produce a refrigerant velocity that ensures continuous return of oil to the compressor. The terminal end of the riser(s) enters the horizontal run at the top of the suction line; this avoids both interference from the overfeeding of any other expansion valve or any drainback during the *off* cycle.

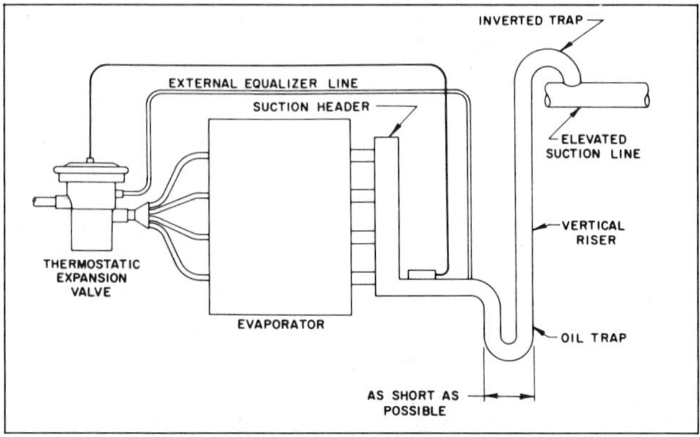

Fig. 14 Bulb Location When Suction Main Is Above Evaporator

If circulated with oil-miscible refrigerants, a heavy concentration of oil elevates the refrigerants' boiling temperatures. The response of the thermostatic charge of the expansion valve is related to the saturation pressure and the temperature of pure refrigerant. In an operating system, the false pressure-temperature signals of oil-rich refrigerants cause both floodback or operating superheats considerably lower than indicated and quite often cause erratic valve operations. To keep the oil concentration at an acceptable level, either the oil pumping rate must be reduced or an effective oil separator must be used.

The external equalizer line is ordinarily connected at the evaporator outlet, as shown in Figure 12. It may also be connected at the evaporator inlet or at any other point in the evaporator downstream of the major pressure drop. On evaporators with long refrigerant circuits that have inherent lag, hunting may be minimized by changing the connection point of the external equalizer line.

The ambient temperature at the valve does not ensure a corresponding temperature of the diaphragm chamber because a certain amount of heat is conducted to the cooler valve body. It is not unusual for the diaphragm-chamber temperature to fall below the bulb temperature. When this fall occurs with a gas-charged valve, the entire thermostatic charge can condense in the diaphragm chamber, and the refrigerant feed can then be controlled from that point. Extreme starving of the evaporator, to the point of complete cessation of feed, is characteristic of this condition. For this reason, gas-charged valves are normally used only when sufficient pressure drop exists between the outlet of the valve and the bulb location. On multiple-circuit evaporators, the refrigerant distributor serves this purpose. The pressure drop through the distributor elevates the valve body temperature above that of the evaporator and assists in maintaining control at the valve bulb. Straight liquid and liquid cross-charged valves operate properly when the temperature of the space in which they are located is temperate. However, consideration must be given to the location of the valve when the temperature of the space is extreme because of its adverse effect on the pressure in the bulb, capillary tube, and diaphragm-chamber assembly. Adsorption-charged valves are unaffected by the surrounding ambient temperature.

Direct-expansion chillers located in a heated environment are among the few applications in which gas-charged valves operate successfully without refrigerant distributors. This is possible because operation occurs at considerably lower superheats (4 to 6°F) and narrower ranges of temperatures than other types of direct-expansion evaporators. When this type of chiller is subjected to cold outdoor ambients, special considerations to prevent charge migration to the diaphragm housing of the valve element are needed. The adsorption charge can be used for these applications because it is unaffected by the colder ambients. If a pressure-limiting feature is necessary, special hydraulic elements can use the gas cross charge in such a way as to prevent charge migration.

ELECTRIC EXPANSION VALVES

Electronically controlled expansion valves have become widely available. The electric expansion valve, similar to the thermostatic expansion valve, is a liquid refrigerant flow-control device. These valves may be categorized into the following four types:

1. Heat motor operated
2. Magnetically modulated
3. Pulse width modulated; *on-off* type
4. Step motor driven

Heat motor valves may be either of two types. In the first type, one or more bimetallic elements are heated electrically, causing them to deflect. The bimetals are linked mechanically to a valve pin or poppet. In a second type of heat motor expansion valve, a volatile material charge is contained within an electrically heated chamber so that the charge temperature (and pressure) is controlled by electrical power input to the heater. The charge pressure is made to act on a diaphragm or bellows, which is balanced against either ambient air pressure or refrigeration system suction pressure. The diaphragm is linked to a pin or poppet.

In *magnetically modulated valves*, a direct current electromagnet modulates smoothly, while an armature, or plunger, compresses a spring progressively as a function of coil current. The modulating electromagnet plunger may be connected to a valve pin or poppet directly or may be used as the pilot element in a servo loop to operate a much larger valve. When the modulating plunger operates a pin or poppet directly, the valve may be of a pressure-balanced port design so that pressure differential has little or no influence on valve opening.

The *pulse width modulated valve* is an *on-off* solenoid valve with special design features that allow it to function as an expansion valve through a life of millions of cycles. Even though the valve is either fully opened or closed, it operates as an infinitely variable metering device by pulsing the valve open regularly. The duration of each opening, or pulse, is regulated by the electronics. For example, a valve may be pulsed every six seconds. If 50% flow is needed, the valve would be held open three seconds and closed for three seconds.

A *step motor* is an electronically commutated multiphase motor capable of running continuously in forward or reverse, or it can be discretely positioned in increments of a small fraction of a revolution. Step motors are used in instrument drives, plotters, and other applications where accurate positioning is required. Step motors require electronics, such as an integrated circuit, to switch windings in proper sequence. When used to drive expansion valves, a lead screw changes the rotary motion of the rotor to a linear motion suitable for stroking a valve pin or poppet. The lead screw may be driven directly from the rotor, or a reduction gearbox may be placed between the motor and lead screw. The motor may be hermetically sealed within the refrigerant environment, or the motor and gearbox can be sealed to operate in air.

Electric expansion valves may be controlled by either digital or analog electronic circuits. Electronic control gives the additional flexibility to consider control schemes that are impossible with conventional valves.

CONSTANT PRESSURE EXPANSION VALVES

The constant pressure expansion valve is operated by the evaporator or valve-outlet pressure; it regulates the mass flow rate of liquid refrigerant entering the evaporator and maintains this pressure at a constant value. Although this valve was first used as a liquid refrigerant expansion valve, other applications are also addressed.

Operation

Figure 15 shows a schematic cross section of a constant pressure expansion valve. The valve has both an adjustable spring, which exerts its force on top of the diaphragm in an opening direction, and a spring beneath the diaphragm, which exerts its force in a closing direction. Evaporator pressure is admitted beneath the diaphragm, through either the internal or external equalizer passage, and the combined forces of the evaporator pressure and the closing spring act to counterbalance the opening spring pressure.

With the valve set and refrigerant flowing at a given pressure, a small increase in the evaporator pressure forces the diaphragm up, which restricts the refrigerant flow and limits the evaporator pressure. When the evaporator pressure drops below the valve setting due to a decrease in load, the top spring pressure moves the valve pin in an open direction. As a result, the refrigerant flow increases and raises the evaporator pressure to the balanced valve setting. This valve controls evaporation of the liquid refrigerant in the evaporator at a constant pressure and temperature.

Constant pressure expansion valves automatically adjust the flow of liquid refrigerant to the evaporator to balance compressor pumping capacity. When this balance is established, evaporator pressure remains constant during the remainder of the running phase of the refrigeration cycle. This low-side pressure balancing point is selected by turning the valve adjuster to the desired pressure setting. To avoid floodback to the compressor during low heat load periods, the valve setting should be selected to use evaporator surface effectively when there is a minimum heat load on the system. When there is a maximum heat load on the system, the increase in evaporator pressure causes the valve to close partially. This prevents the pressure from rising any further and overloading the compressor.

For reasons of operation and adjustment, constant pressure expansion valves are most effective on applications with a constant heat load. When the compressor stops at the end of the running

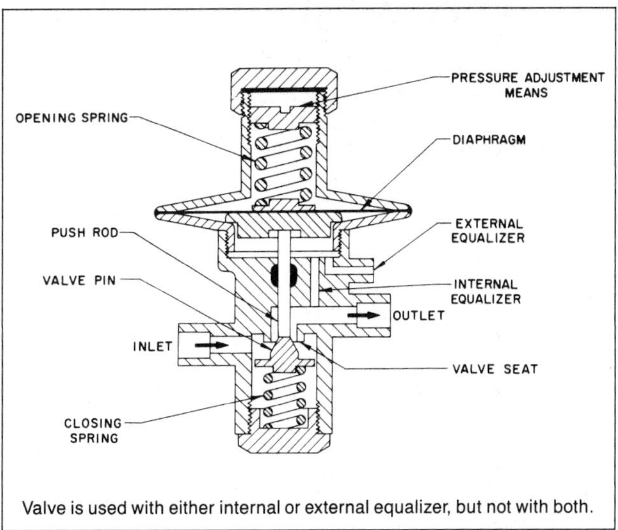

Valve is used with either internal or external equalizer, but not with both.

Fig. 15 Constant Pressure Expansion Valve

phase of the cycle, standard constant pressure expansion valves close to prevent off-cycle refrigerant flow. The rapid equalization features described in the Thermostatic Expansion Valves section can also be built into constant pressure expansion valves to provide off-cycle pressure equalization for use with low-starting torque compressor motors.

Selection

The constant pressure expansion valve for liquid refrigerant expansion service should be selected for the required capacity and system refrigerant at the lowest expected pressure drop across the valve and should have an adjustable pressure range to provide the required evaporator (valve outlet) pressure. The system designer should decide whether the valve should be a bleed-type or a standard expansion valve.

Application

The constant pressure expansion valve, when applied as a liquid refrigerant expansion valve, is suitable only for constant load applications; therefore, its use is limited. Ordinarily, only one such liquid expansion valve is used on each system. When applied to a variable load, this valve starves the evaporator at the high load and overfeeds it at the low load, with possible compressor damage resulting in the latter case. Because of the difference in the operating principles of the constant pressure expansion valve and the thermostatic expansion valve, both cannot be used in the same system as liquid-refrigerant expansion valves.

The constant pressure expansion valve is used for the pilot valve in some of the large suction pressure regulating valves.

The constant pressure expansion valve may be used as the pilot valve on a modified suction pressure regulator, or it may be used to bypass compressor discharge gas in refrigeration systems. Such applications may arise from either the requirement for a reduction in compressor capacity or the requirement for a low-limit control of either the evaporator, the suction pressure, or both. After passing through the valve, the hot gas may be introduced directly into the suction line, in which case some additional desuperheating means may be required, or it may be introduced at the evaporator inlet, where it mixes with the cold refrigerant and is desuperheated before it reaches the compressor. Some applications suitable for constant pressure expansion valves are drink dispensers, food dispensers, water coolers, ice cream freezers, and self-contained room air-conditioning units.

EVAPORATOR PRESSURE AND TEMPERATURE REGULATORS

The evaporator pressure regulator (back-pressure regulator) regulates the evaporator pressure (pressure entering the regulator) at a constant value. It is used in the evaporator outlet or suction line wherever low-limit control of the evaporator pressure or temperature is required.

Operation

As illustrated in Figure 16, the inlet pressure acts on the bottom of the seat disk and opposes the adjusting spring. The outlet pressure acts on the underside of the bellows and the top of the seating disk and, because the effective areas of the bellows and the port are equal, the two forces cancel each other, and the valve responds to inlet pressure only. When the evaporator pressure rises above the force exerted by the spring, the valve begins to open. When the evaporator pressure drops below the force exerted by the spring, the valve closes. In actual operation, the valve assumes a throttling position to balance system load.

This change in the pressure, which acts on and operates the diaphragm or bellows, is called *differential*. The *total valve differential* is the difference between the pressure at which the valve operates at rated stroke and the pressure at which the valve starts to open.

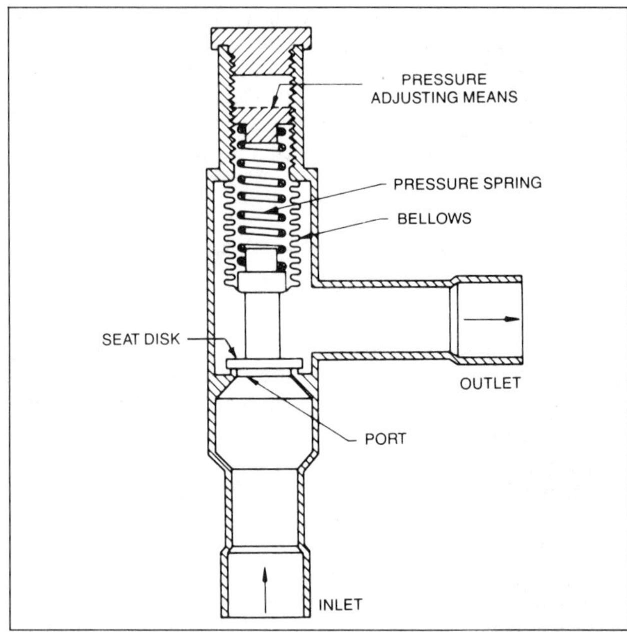

Fig. 16 Direct-Acting Evaporator Pressure Regulator

Pressure drop is the difference between the pressure at the valve inlet and the pressure at the valve outlet. Understanding the difference between differential and pressure drop when determining the proper valve size for a given set of load conditions is essential.

Pilot-operated evaporator pressure regulators are either self-contained or high-pressure driven. The self-contained regulator (Figure 17) starts to operate when the evaporator pressure rises above the pressure setting of the diaphragm spring; the diaphragm moves in an opening direction, increasing the pressure above the piston. This increase moves the piston down, thereby moving the main valve in an opening direction and causing the evaporator pressure to drop back to the pressure setting of the pilot.

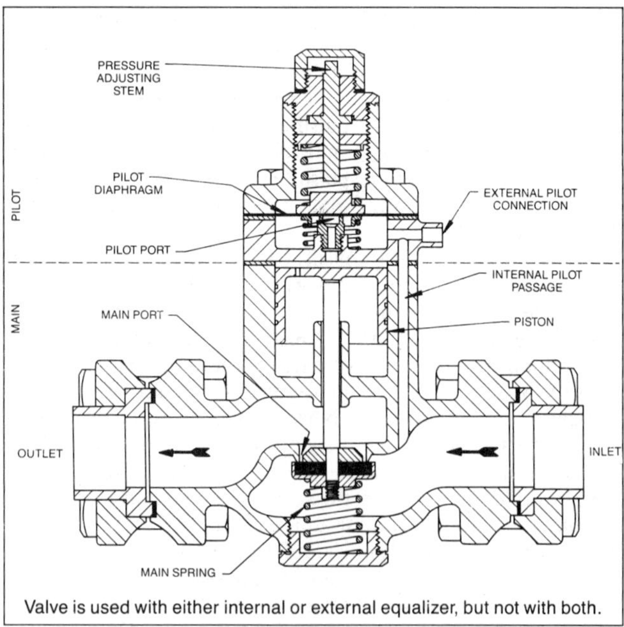

Valve is used with either internal or external equalizer, but not with both.

Fig. 17 Pilot-Operated Evaporator Pressure Regulator (Self-Contained)

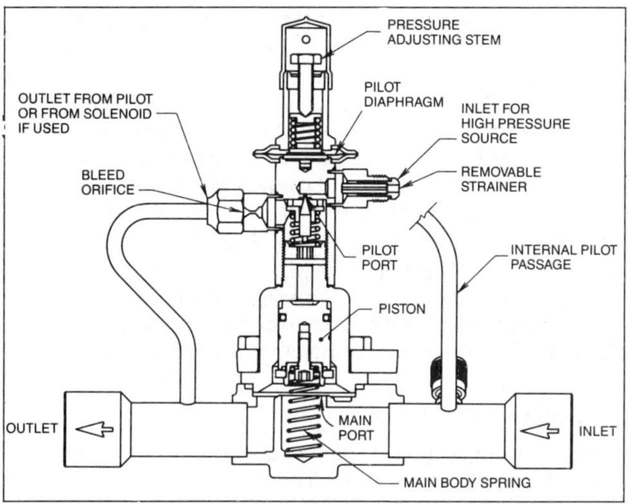

Fig. 18 Pilot-Operated Evaporator Pressure Regulator (High Pressure Driven)

When this happens, the pilot valve moves in a closing direction, allowing the pressure above the piston to decrease. Then, the main spring moves the main valve in the closing direction, preventing the evaporator pressure from falling below the pressure setting of the pilot. In operation, the pilot valve and the main valve assume either intermediate or throttling positions, depending on the load.

Many pilot-operated regulators are of a normally open design and require high-pressure liquid or gas to provide a closing force. The advantage of this regulator is that it requires no minimum operating suction pressure drop to operate. When the valve inlet pressure increases above set point, the pilot valve diaphragm lifts against the adjustment spring load (Figure 18). When the diaphragm moves up in response to the increasing inlet pressure, the pilot valve pin closes the pilot valve port. Gas or liquid from the high-pressure side of the system is throttled by the pilot, allowing the main piston spring to move the valve to a more open position. Because the top of the piston chamber bleeds to the downstream side of the valve through a bleed orifice, a continuous flow of high-pressure liquid or gas through the pilot valve drives the piston down to a closed or partially closed position. A solenoid valve may be used to drive the piston down, either by porting pressure directly to the top of the piston or by closing the bleed orifice, thus closing the valve for defrost operation.

Selection

Unless otherwise dictated by special system requirements, the evaporator pressure regulator should be selected for the required capacity and system refrigerant at the required evaporator (regulator inlet) pressure and the lowest pressure drop across the regulator that are consistent with good regulator performance and economical compressor operation.

Application

Evaporator pressure regulators are used on finned-coil evaporators where frosting should be prevented or evaporator pressure must be maintained constant and higher than suction pressure to prevent dehumidification. These regulators are used on water and brine chiller applications to prevent freeze-up of the chillers under low-load conditions.

On multiple evaporator installations, as shown in Figure 19, evaporator pressure regulators (direct-acting or pilot) can be installed to control evaporator pressure and temperature in each unit. The regulators maintain the desired evaporator temperature in the warmer units, while the compressor continues to operate to

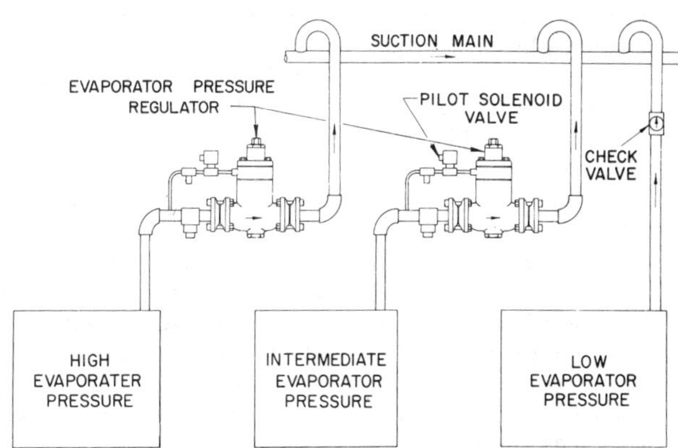

Fig. 19 Evaporator Pressure Regulators in Multiple System

satisfy the coldest unit. With such a system, the compressor may be controlled by either a low-pressure switch or by thermostats installed in the individual units.

The evaporator pressure regulator, with internal pilot passage, receives its source of pressure for pilot operation at the regulator inlet connection, whereas the regulator with the external pilot connection not only permits a choice of pressure source for pilot operation, but also permits the use of a remote pressure pilot or pilot valves for other purposes.

If the regulator inlet pressure is unsteady and adversely affects the regulator performance, a source of pressure steadier than that available at the regulator inlet may be obtained for the pilot by connecting the external pilot line to a surge drum or enlarged section of suction line upstream from the regulator.

A remote pressure pilot installed in the external pilot line facilitates adjustment of the pressure setting when the regulator must be installed in an inaccessible location. A pressure pilot on the self-contained design with a pneumatic connection permits resetting and controlling of the evaporator pressure and temperature by a pneumatic control system.

In Figure 19, a pilot solenoid valve installed in the external pilot line enables the regulator to function as a suction stop valve, as well as an evaporator pressure regulator. This feature is particularly useful on a flooded evaporator to prevent *pumping-out* of the evaporator, which occurs when the thermostat is satisfied and the compressor continues to operate to cool other evaporators.

A dual-pressure regulator may be attained by using two pressure pilots. The lower pressure pilot is piped through a pilot solenoid valve and, when energized, controls the lower pressure. The higher pressure, which occurs when the pilot solenoid valve is deenergized, is generally used for a warmer evaporator requirement or for pressure above 32 °F saturation for defrosting.

In some instances, it is desirable to control, at a constant temperature, the air or liquid entering or leaving the evaporator by modulating the evaporator pressure and temperature in accordance with the load demand so that the evaporator temperature is decreased during heavy loads and increased during lighter loads. This can be accomplished with an evaporator pressure regulator modified in one of the following ways:

1. By using a temperature-actuated pilot in the external pilot line in place of the pressure pilot and by placing the bulb of the temperature pilot where it will respond to the temperature of the air or liquid being cooled. The temperature pilot modulates the position of the regulator according to the load requirement and maintains the air or liquid at a constant temperature.
2. By using a pneumatic thermostat and a pressure pilot with a pneumatic connection so that the evaporator pressure and temperature are raised by an increase in air pressure. This increase is supplied to the pressure pilot at reduced loads and is lowered

by a decrease in air pressure supplied to the pressure pilot at increased loads, as required to maintain the temperature of the pneumatic thermostat bulb. This bulb is placed to respond to the temperature of the air or liquid being cooled. Thus, the air or liquid being cooled is maintained at a constant temperature.

3. By using a pressure pilot that is adaptable to a reversible electric motor drive, which resets both the pressure pilot and a potentiometer-type thermostat to control the reversible electric motor. The evaporator pressure and temperature are raised at reduced loads and lowered at increased loads, as required to maintain the temperature of the thermostat bulb, which is placed to respond to the temperature of the air or liquid being cooled. The air or liquid being cooled is maintained at a constant temperature.

In addition to the normal applications, these regulators, when equipped with either high-pressure pilots and suitable valve-seat material or a higher range pressure spring, such as in the direct-acting regulator, are used in several methods of application in refrigeration systems that have air-cooled condensers. The regulators are used for maintaining low-limit control during cold weather operation of the compressor discharge, the liquid line pressure, or both.

Electronically piloted, suction throttling valves have been developed to control temperature in a food merchandising refrigerator or other refrigerated space (Figure 20). This valve is a form of evaporator pressure regulator, although it responds only to temperature in the space, rather than pressure in the evaporator or suction line. The system consists of a temperature sensor, an electronic control circuit, and a suction throttling valve. A temperature setting is made by turning a calibrated potentiometer or rotary switch normally located on the control circuit. The valve responds to the difference between set point temperature and the prevailing temperature. A temperature above the set point drives the valve further open, while a temperature below set point modulates the valve in the closing direction. During defrost, the control circuit drives the valve tightly closed.

By modulating suction gas flow in response to the measured temperature, the refrigerated space may be held close to the set point regardless of variations in heat load or compressor suction pressure. An electronic regulator is particularly useful on refrigerators containing fresh meat or other products where close control or temperature is important. In some instances, an energy savings may be realized, because a temperature regulating control reduces refrigeration during low heat load conditions more effectively than an evaporator pressure regulator.

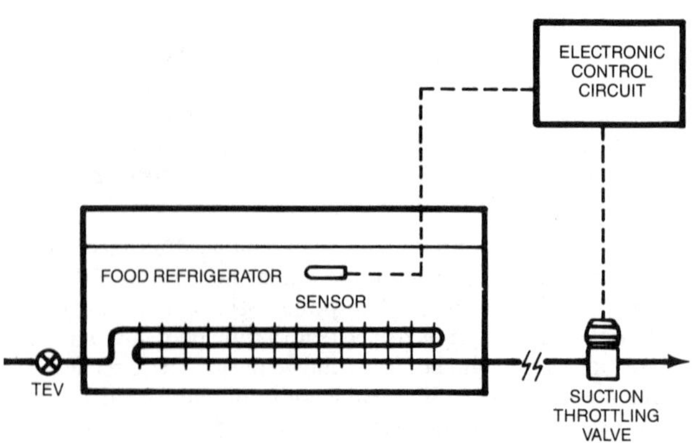

Fig. 20 Electronic Temperature Controlling Suction Throttling Regulator

SUCTION PRESSURE REGULATORS

The suction pressure regulator (holdback valve or crankcase pressure regulator) limits the compressor suction pressure (regulator outlet pressure) to a maximum value. This type of regulator should be used in the suction line at the compressor on any refrigeration installation in which the liquid expansion valve cannot limit the suction pressure and compressor motor overload would otherwise exist because of the following:

1. Excessive starting load
2. Excessive suction pressure following the defrost cycle
3. Prolonged operation at excessive suction pressure
4. Low-voltage and high-suction pressure conditions

Operation

Direct-acting suction pressure regulating valves sense only their outlet or downstream pressure (compressor crankcase or suction pressure). As illustrated in Figure 21, the inlet pressure acts on the underside of the bellows and the top of the seating disk. Since the effective areas of the bellows and port are equal, these forces cancel each other and do not affect valve operation. The valve outlet pressure acts on the bottom of the disk and exerts a closing force, which is opposed by the adjustable spring force. When the outlet pressure drops below the equivalent force exerted by the spring, the valve moves in an opening direction. If the outlet pressure rises, the valve moves in a closing direction and throttles the refrigerant flow to maintain the set point of the valve.

Externally and internally pilot-operated suction pressure regulators are available for larger system applications. Although their design is more complex, due to pilot operation, their method of operation is similar to that described above.

Selection

The suction pressure regulator should be selected for the required capacity, system refrigerant, required regulator inlet pressure, and lowest practical pressure drop across the regulator to minimize any loss in system capacity.

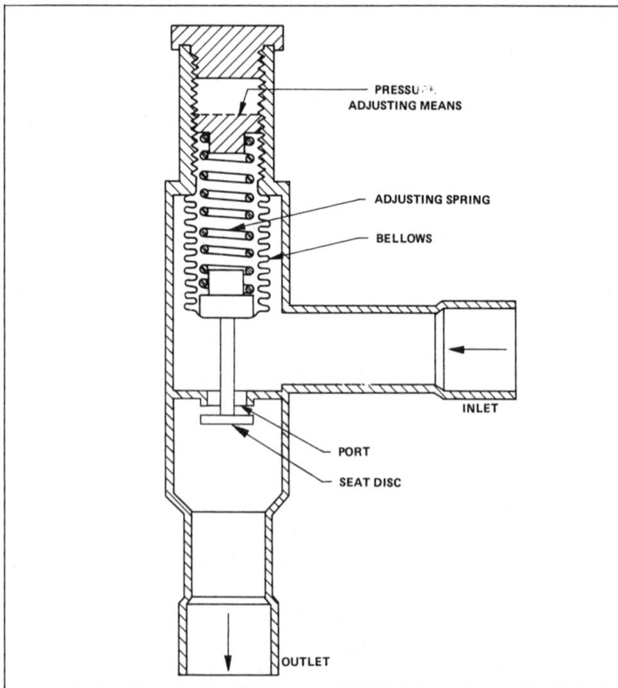

Fig. 21 Direct-Acting Suction Pressure Regulator

Application

In addition to the normal applications, these regulators—when equipped with high-pressure pilots, suitable valve-seat material, or a higher range pressure spring such as in the direct-acting regulator—are used in several methods of application in refrigeration systems with air-cooled condensers. They are used for maintaining low-limit control of the compressor discharge, liquid line pressure, or both, during cold weather operation. In addition, such modified regulators are used to bypass compressor discharge gas in refrigeration systems, as described in the Application section for the constant pressure expansion valve.

CONDENSER PRESSURE REGULATORS

Various condenser pressure-regulating valves are used to maintain sufficient condensing pressure to allow air-cooled condensers to operate properly during the winter. Both single- and two-valve arrangements have been used for this purpose. See Chapter 2 of this volume and Chapter 36 of the 1992 ASHRAE *Handbook—Systems and Equipment* for more information.

The two-valve arrangement often uses a valve that is constructed and operates similarly to the evaporator pressure-regulating valve shown in Figures 16 and 17. This control is installed either in the liquid line between the condenser and receiver or in the discharge line. It throttles when the condenser or discharge pressure falls as a result of a low ambient condition.

The second valve in the two-valve arrangement bypasses discharge gas around the condenser to the receiver to mix with cold liquid and maintain adequate high-side pressure. Several bypass valves are available, some of which are similar to the suction pressure-regulating valve shown in Figure 21. This valve responds to outlet pressure (receiver pressure). When receiver pressure decreases as a result of a decrease in ambient temperature, the valve opens and bypasses discharge bypass gas to the receiver. Figure 22 shows another device that responds to changes in pressure between its inlet and outlet. As the differential pressure increases, the valve opens. Thus, when the other valve in this two-valve arrangement throttles and restricts liquid flow, a differential is created, and this bypass device opens.

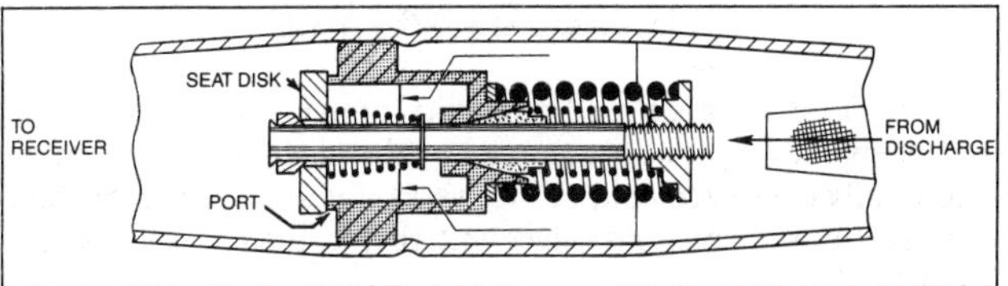

Fig. 22 Condenser Bypass Valve

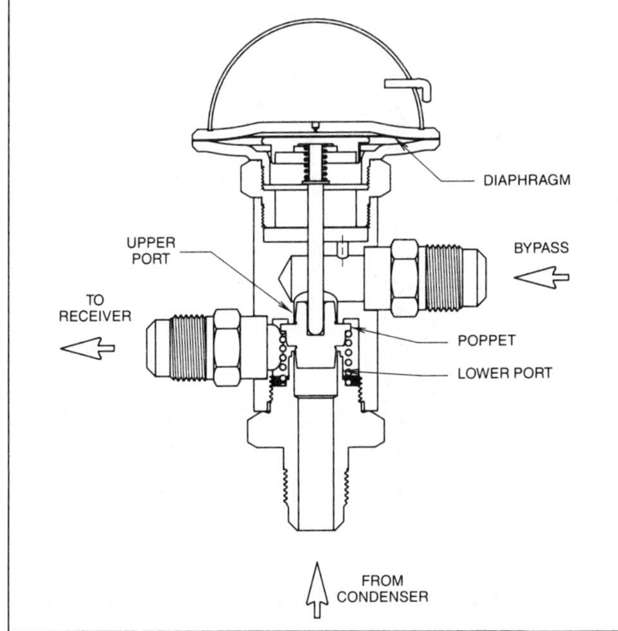

Fig. 23 Three-Way Condenser Pressure-Regulating Valve

It is sometimes an advantage to substitute a single three-way condenser pressure-regulating valve for the two-valve arrangement described previously. The three-way valve (Figure 23) simultaneously holds back liquid in the condenser and passes compressor discharge into the receiver to maintain pressure in the liquid line. The lower side of a metal diaphragm is exposed to system high-side pressure, while the upper side is exposed to a noncondensable gas charge (usually dry nitrogen). A pushrod connects the diaphragm to the valve poppet, which seats on either the upper or lower port and throttles either the discharge gas or the liquid from the condenser, respectively. During system start-up in extremely cold weather, the poppet may be tight against the lower seat, stopping all liquid flow from the condenser and bypassing discharge gas into the receiver until adequate system pressure is developed. During stable operation in cold weather, the poppet modulates at an intermediate position, with liquid flow from the condensing coil mixing with compressor discharge gas within the valve and flowing to the receiver. During warm weather, the poppet seats tightly against the upper port, allowing free flow of liquid from the condenser but preventing flow of discharge gas.

Three-way condenser pressure-regulating valves are not usually adjustable by the user. The pressure setting is established by the pressure of the gas charge placed in the dome above the diaphragm during manufacture.

HIGH-SIDE FLOAT VALVES

Operation

A *high-side float valve* controls the mass flow rate of refrigerant liquid entering the evaporator so it equals the rate at which the refrigerant gas is pumped from the evaporator by the compressor. Figure 24 shows a cross section of a typical valve. The refrigerant liquid flows from the condenser into the high-side float valve body, where it raises the float and moves the valve pin in an opening direction, permitting the liquid to pass through the valve port, expand, and flow into the evaporator. Most of the system refrigerant charge is contained in the evaporator at all times. The high-side float system is a flooded system.

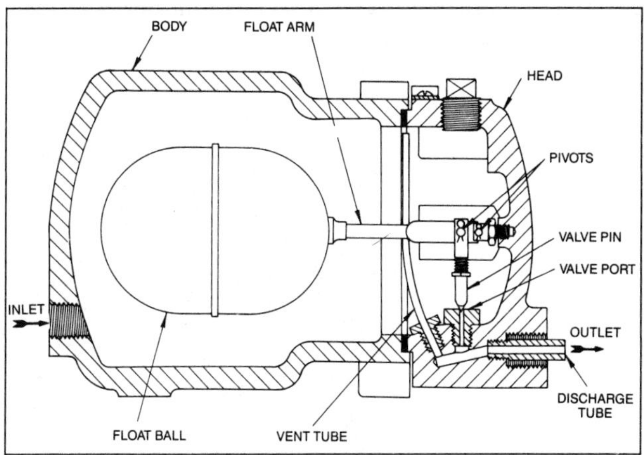

Fig. 24 High-Side Float Valve

Selection

For acceptable performance, the high-side float valve is selected for the system refrigerant and a rated capacity neither excessively large nor too small. The orifice is sized for the maximum required capacity with the minimum pressure drop across the valve. The valve operated by the float may be a pin-and-port construction (Figure 24), a butterfly valve, a balanced double-ported valve, or a sliding gate or spool valve. The internal bypass vent tube allows installation of the high-side float valve near the evaporator and above the condenser without danger of the float valve becoming gas bound. Some large-capacity valves use a high-side float valve for pilot operation of a diaphragm or piston-type spring-loaded expansion valve. This arrangement can provide improved modulation over a wide range of load and pressure-drop conditions.

Application

A refrigeration system in which a high-side float valve is used consists ordinarily of a single evaporator, compressor, and condenser. The operating receiver or a liquid sump at the condenser outlet can be quite small. A full-sized receiver is required for pumping out the flooded evaporator. Under certain conditions, the high-side float valve may be used to feed more than one evaporator in a system; consequently, additional control valves are required. One of the disadvantages of the high-side float valve system is that the amount charge is critical. The use of an excessive amount of system charge causes floodback, while an insufficient amount causes a reduction in system capacity.

LOW-SIDE FLOAT VALVES

Operation

The *low-side float valve* performs the same function as the high-side float valve, but it is connected to the low-pressure side of the system. When the evaporator liquid level drops, the float opens the valve, which allows refrigerant liquid from the liquid line to flow through the valve port and directly enter the evaporator or surge drum. In another type of valve design, the refrigerant liquid flows through the valve port, passes through a remote feedline, and enters the evaporator through a separate connection. (A typical direct-feed valve construction is shown in Figure 25.) The low-side float system is a flooded system.

Selection

Low-side float valves are selected in the same manner as the high-side float valves discussed previously.

Application

In the low-side float valve system, the refrigerant charge is not critical. The low-side float valve can be used in multiple evaporator systems in which some of the evaporators may be controlled by other low-side float valves and some by thermostatic expansion valves.

Depending on its design, the float valve is mounted either directly in the evaporator or surge drum or in an external chamber connected to the evaporator or surge chamber by equalizing lines, *i.e.*, a gas line at the top and a liquid line at the bottom. In the externally mounted type, the float valve is separated from the float chamber by a gland that maintains a quiet level of liquid in the float chamber for steady actuation of the valve.

In evaporators with high boiling rates or restricted liquid and gas passages, the boiling action of the liquid raises the refrigerant level during operation. When the compressor stops or the solenoid suction valve closes, the boiling action of the refrigerant liquid ceases, and the refrigerant level in the evaporator drops. Under these conditions, the high-pressure liquid line supplying the low-side float valve should be shut off by a solenoid liquid valve to prevent overfilling of the evaporator. Otherwise, excess refrigerant will enter the evaporator on the *off* cycle, which can cause floodback when the compressor starts or the solenoid suction valve opens.

When a low-side float valve is used, precautions must be taken that the float is in a quiet liquid level that falls properly in response to an increase in evaporator load and rises with a decrease in evaporator load. In low-temperature systems particularly, it is important that the equalizer lines between the evaporator and either the float chamber or the surge drum be generously sized to eliminate any reverse response of the refrigerant liquid level in the vicinity of the float. Where the low-side float valve is located in a nonrefrigerated room, the equalizing liquid and gas lines and the float chamber must be insulated to provide a quiet liquid level for the float.

SOLENOID VALVES

A *solenoid valve* is closed by gravity, pressure, or spring action and opened by a plunger actuated by the magnetic action of an electrically energized coil, or vice versa. Figures 26 and 27 show cross sections of solenoid valves with their principal components identified.

Because solenoid valves are actuated electrically, they may be conveniently operated in remote locations by any suitable electric

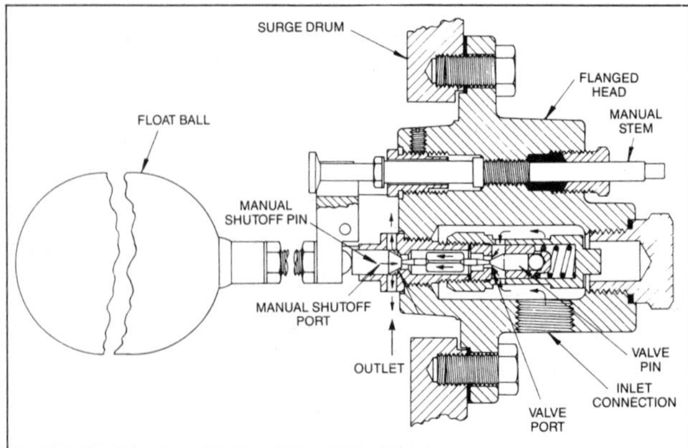

Fig. 25 Low-Side Float Valve

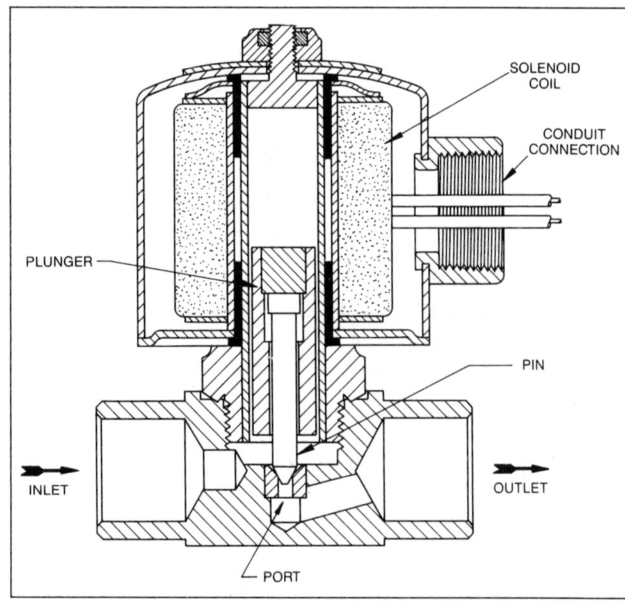

Fig. 26 Normally Closed Direct-Acting Solenoid Valve with Hammer-Blow Feature

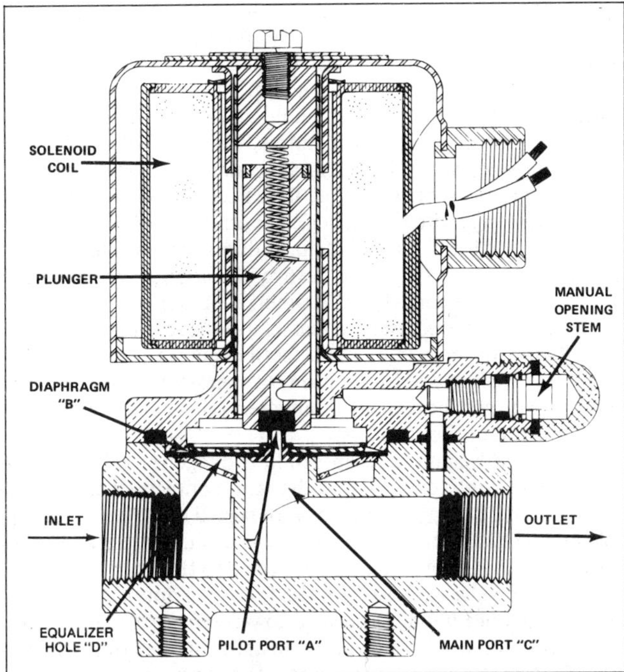

Fig. 27 Normally Closed Pilot-Operated Solenoid Valve with Direct-Lift Feature

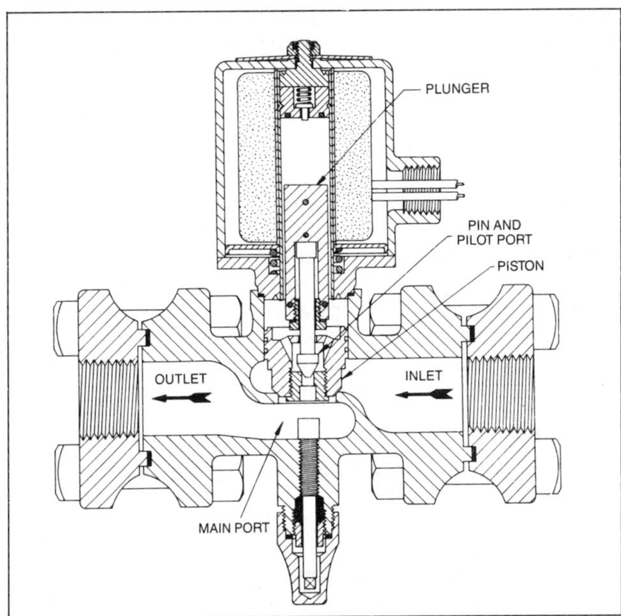

Fig. 28 Normally Closed Pilot-Operated Solenoid Valve with Hammer-Blow and Mechanically Linked Piston-Pin Plunger Features

switch. These valves are always fully open or fully closed, in contrast to motorized valves, which may operate in a modulating position. Solenoid valves may be used to control the flow of many different fluids if the pressures and temperatures involved, the viscosity of the fluid, and the suitability of the materials used in the valve construction are carefully considered.

Solenoid valves can be divided into the following general types:

1. **Normally closed solenoid valves,** in which the closure member moves away from the port to open the valve when the coil is energized, *e.g.*, a two-way solenoid valve.
2. **Normally open solenoid valves,** in which the closure member moves to the port to close the valve when the coil is energized, *e.g.*, a two-way solenoid valve.
3. **Multiaction solenoid valves,** which combine (in one body) the action of one or more normally open and one or more normally closed solenoid valves, *e.g.*, a three-way solenoid valve or a four-way solenoid valve.

Operation

While all types of solenoid valves are used, the normally closed type is used far more extensively. In the normally closed direct-acting solenoid valve shown in Figure 26, the solenoid coil, acting on the plunger, pulls the valve pin away from and off the valve port, thereby opening it directly. Because this valve depends on the power of the solenoid coil for operation, its port size for a given operating pressure differential is limited by the limitations of solenoid coil size.

Figure 27 shows a medium-sized, normally closed pilot-operated solenoid valve. In this valve, the solenoid coil, acting on the plunger, does not open the main port directly but opens the pilot Port A. Pressure trapped on top of Diaphragm B is released through the pilot port, thus creating a pressure imbalance across the diaphragm, forcing it upward and opening the main Port C. When the solenoid coil is de-energized, the plunger drops and closes pilot Port A. Then the pressures above and below the diaphragm equalize again through the equalizer Hole D, and the diaphragm drops and closes the main port. In some pilot-operated solenoid valve designs, a piston is used for the main closing member instead of a diaphragm. In medium-sized valves, the pilot port is usually located in the main closing member.

Such pilot-operated valves depend on a certain minimum pressure drop across the valve (approximately 0.5 psi or more) to hold the piston or diaphragm in the *open* position. If a valve is oversized such that inadequate pressure drop is developed, the valve may chatter or fail to open fully. Valves should be sized by the capacity tables provided by the manufacturer, rather than by pipe or tube size or port diameter. Where it is desirable to keep the valve open without this pressure drop penalty, such as on refrigeration suction lines, the piston or diaphragm may be linked mechanically to the solenoid valve pin and plunger, as shown in Figure 28. The opening and closing actions are the same as before. However, the increased pulling force of the plunger as it approaches its stop position in the coil is used to hold the piston in the open position, without requiring valve pressure drop.

To obtain the maximum operating pressure differential for a given solenoid pulling power, many valves leave the plunger free to gain momentum before it knocks the valve pin out of the valve port or pilot port with an *impact* or a *hammer-blow* effect (see Figures 26 and 28). Solenoid valves with this hammer-blow feature must have the full rated voltage, within the customary tolerance of +10 to −15%, applied instantaneously to their coils so that the valves open under rated conditions. Otherwise, they will fail. If, after the valve is in the open position, the line voltage drops below the hold-in voltage value but not to zero, the plunger drops and the valve closes. After the line voltage builds up again, the valve will not reopen, whether it is used on alternating or direct current, because the hammer-blow effect is lost. However, in the case of an alternating current valve, the coil will overheat and may burn out under this condition because the inherent high inrush current continues to flow through the coil when the plunger is not pulled all the way into the coil to close the air gap.

Direct-lift solenoid valves (see Figure 27), in which the valve pin is an integral part of the plunger, can be designed to open fully at rated voltage, +10 to −15%, whether the voltage is applied gradually or instantaneously. Normally open solenoid valves are usually held in the *open* position by gravity or a spring force. When energized, the power of the solenoid coil, acting on the plunger, pulls the valve pin on the valve port or pilot port to close it and, therefore, reverses the valve action.

Multiaction solenoid valves are available to accommodate many different flow configurations, such as (1) a common inlet three-way valve, which directs flow from a common inlet connection to one of two outlet connections; or (2) the four-way valve described in the Refrigerant-Reversing Valves section. Design compromises usually result in the ports of these direct-acting valves being smaller (or the maximum operating pressure differential lower) than the equivalent normally closed two-way solenoid valves. As a result, multiaction pilot-operated solenoid valves are used in most cases. Their use warrants careful application analysis because the pressure differences required to shift and hold the valves may not exist under all operating conditions.

Selection

When solenoid valves are selected, the following factors should be considered:

1. Basic flow configuration, such as two-way and three-way normally closed.
2. Type of fluid to be handled.
3. Temperature and pressure conditions of the entering fluid.
4. Allowable fluid flow pressure drop across the valve needed to establish the port size for the required capacity.
5. Capacity in appropriate terms; do not size for less pressure drop than is needed to open a piloted valve. Use capacity tables for sizing to ensure that pressure drop at least equals the minimum pressure drop required for operation as specified by the manufacturer. If a piloted valve is too far oversized, it may chatter or fail to open fully. If no minimum pressure drop is specified, 0.5 psi is normally safe. If the valve is advertised to have a *zero pressure drop* opening feature, minimum pressure drop need not be considered.
6. Maximum operating pressure differential under which the valve will be required: (*a*) to open, for the normally closed valve; and (*b*) to close, for the normally open valve. Three- and four-way valves require additional information on the operating conditions for which they are intended.
7. Safe working pressure. This should not be confused with the pressure difference under which the valve is required to open.
8. Type and size of line connections.
9. Electrical characteristics for the solenoid coil. Voltage and frequency must be specified for alternating current, but only voltage is specified for direct current.
10. Ambient temperature in which the valve will be located.
11. Cycling rate of valve.
12. Hazard of location, which may make explosion-proof coil housings necessary.

Application

Spring-loaded solenoid valves (see Figure 27) usually can be installed in vertical lines or any other position. The solenoid valves in Figure 26 should be installed upright in horizontal lines.

Solenoid valves must each be used with the correct individual current characteristics. Momentary overvoltage is not harmful, but sustained overvoltage of more than 10% may cause solenoid coils to burn out under unfavorable conditions. Undervoltage is harmful to alternating current-operated valves if it reduces operating power enough to prevent the valve from opening when the coil is energized. This condition may cause burnout of an alternating current coil, as discussed previously.

When the solenoid valve is energized by a control transformer of limited capacity, the transformer must be able to provide proper voltage during the inrush load. As the inrush alternating current may be several times the holding current, it is useless to check the voltage at the coil leads when only holding current is being supplied. For such applications, the inrush current in amperes, multiplied by the rated coil voltage, gives the necessary volt-ampere capacity, which must be provided by the transformer for each solenoid valve simultaneously actuated. The inrush and holding currents of a direct current solenoid valve are equal.

Fuses protecting electrical supply lines for solenoid valves should be sized according to holding current and should preferably be of the slow blowing type. To protect the coil insulation and the controlling switch, a capacitor or other device may be wired across the coil leads of a high-voltage, direct current solenoid valve when installed to absorb or destroy the counter-voltage surge generated by the coil when the circuit is broken.

Whenever it is necessary to reassemble the solenoid valve after installation, the magnetic coil sleeves (if required) must be replaced in their correct respective positions to operate the valve properly. Leaving the coil sleeves out of an alternating current valve may result in coil burnout.

To avoid valve failure because of low voltage, the solenoid valve coil should not be energized by the same contacts or at the same instant that a heavy motor load is connected to the electrical supply line. The solenoid coil can be energized immediately before or after the heavy motor load is connected to the line.

Solenoid valves are used for the following applications:

1. **Refrigerant Liquid.** To prevent flow of refrigerant liquid to the evaporator, a solenoid valve is installed in the liquid line just ahead of the expansion valve to (1) prevent flow of refrigerant liquid to the evaporator when the compressor is idle, (2) provide individual temperature control in each room of a multiple system, or (3) control the number of evaporator sections used as the load varies on a central air-conditioning installation.
2. **Refrigerant Suction Gas.** In commercial multiple systems, especially those having evaporators containing a large amount of refrigerant, solenoid valves are often provided in the suction line from each unit or room, as well as in the liquid line, to isolate each evaporator completely. Otherwise, refrigerant gas migrates from one evaporator to another through the suction line during the *off* cycle, which causes uneven performance and possible floodback when the compressor starts again.

   In applications 1 and 2, the solenoid valves are operated by thermostats. The compressor may be operated by a low-pressure switch or directly by a thermostat. The compressor may or may not be operated on a pumpdown cycle, either a continuous one or a pumpdown and lockout cycle.
3. **Refrigerant Discharge Gas.** In many hot-gas defrost applications, a solenoid valve, installed in a line connected to the discharge line between the compressor and the condenser, feeds the evaporator with hot gas, which provides heat for the defrosting operation. The solenoid valve remains closed, except during the defrosting operation. A solenoid valve installed in a bypass around one or more compressor cylinders provides compressor-capacity control. This valve may be used to bypass the entire compressor output and reduce the compressor starting load where required.
4. **Water and Other Liquids.** Solenoid valves are used to control the flow of water and many other liquids. Water is one of the more harmful liquids because it deposits solids on the internal solenoid valve surfaces, causing corrosion; therefore, solenoid valves for water service should be easy to dismantle and clean.
5. **Air.** Many air systems rely on solenoid valves to operate controls or actuators. Since rapid cycling is often required, solenoid valves for air service should be selected with consideration for endurance.
6. **Steam.** The application of solenoid valves in industrial steam systems is quite varied. Because of the continuous high steam temperatures involved, special high-temperature solenoid coils are usually required. The ambient temperatures in which the valves are located also need consideration.

Pilot Solenoid Valve Application

Pilot solenoid valves are often used in industry. A few applications pertaining to refrigerating systems follow.

Two-way, normally closed, direct-acting pilot solenoid valves are used with (1) a large, piston-type, spring-loaded expansion valve to provide refrigerant liquid shutoff service, as shown in Figure 13; (2) either an evaporator-pressure regulator to provide shutoff service or a pressure pilot selector; (3) a large piston-type, spring-loaded regulator to provide refrigerant gas shutoff service; and (4) four-way refrigerant switching valves on heat-pump systems to accomplish cooling, heating, and defrosting.

Three-way, direct-acting, pilot solenoid valves are used to operate (1) a cylinder unloading mechanism for compressor capacity reduction; and (2) three- and four-way reversing valves on heat-pump systems to accomplish cooling, heating, and defrosting.

Refrigerant-Reversing Valves. These are three- or four-way two-position valves that are usually operated by pilot solenoid valves and designed for reversing or changing the direction of refrigerant flow through certain parts of a refrigeration system. They are used on refrigerating and year-round air-conditioning (heat-pump) systems to control cooling, heating, and defrosting operations.

Operation

Valves may be operated by either a two-, three-, or four-way pilot solenoid valve, which may be an integral part of the reversing valve or a separate valve connected to the reversing valve by tubes. The reversing valves can be divided into the poppet valve and the slide valve group. Although design modifications can be found within each of these groups, the following explanations present the general principles of operation.

The *four-way reversing valve* may be connected so the refrigeration system is *fail-safe* on either the heating or cooling cycle if the solenoid valve coil fails. The valve is connected by interchanging the two lines on the reversing valve, which connect to the inside and outside heat-exchange coils. Figure 29 shows refrigeration flow through a four-way slide reversing valve in the cooling (or defrosting) cycle.

During the cooling cycle shown in Figure 29, the pilot valve is energized, which opens pilot Port A to bleed high-pressure refrigerant into Chamber C. Simultaneously, pilot Port B is connected to low pressure and bleeds refrigerant from Chamber D. The pressure difference across Piston G develops enough force on Piston G to cause Pistons G and H and the connecting structural member to move Slide F from encompassing Ports K and L to encompassing Ports M and L. When G reaches the end of its stroke, the valve has reversed, and the inside coil is connected to low-pressure cool refrigerant, while the outside coil is connected to high-pressure hot refrigerant.

To reverse to the heating cycle (Figure 30), the pilot valve is de-energized. This opens pilot Port B, which bleeds high-pressure refrigerant into Chamber D and connects pilot Port A to low pressure, which bleeds refrigerant from Chamber C. The valve slide moves in the same manner as before, and the inside coil is connected to high-pressure hot refrigerant, while the outside coil is connected to low-pressure cool refrigerant.

A similar reversing action of Pistons G and H can be produced by a three-way pilot valve. The connection J to the high-pressure tube of the main valve is omitted, and the pilot valve operates to connect either Chamber C or D to the low-pressure source. The high pressure in Chamber E acts on the Pistons G and H and moves the Slide F toward the chamber with the low pressure. In this case, a bleed hole in each piston allows refrigerant to fill Chambers C and/or D when these chambers expand in volume. Also, the piston must seal the main valve end-cap ports to prevent leakage through the pilot valve to the suction Tube S of the main valve.

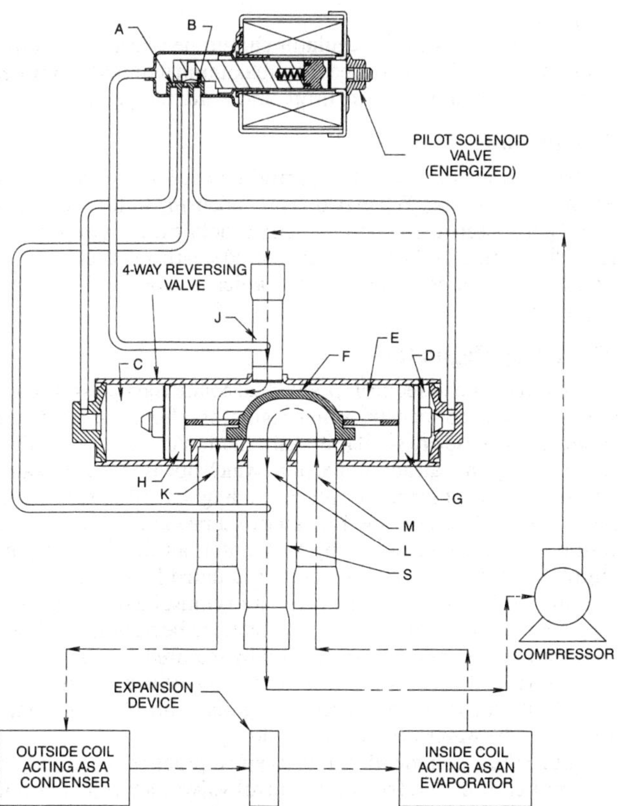

Fig. 29 Four-Way Slide-Type Refrigerant-Reversing Valve Used in Cooling (or Defrosting) Cycle of Refrigeration System

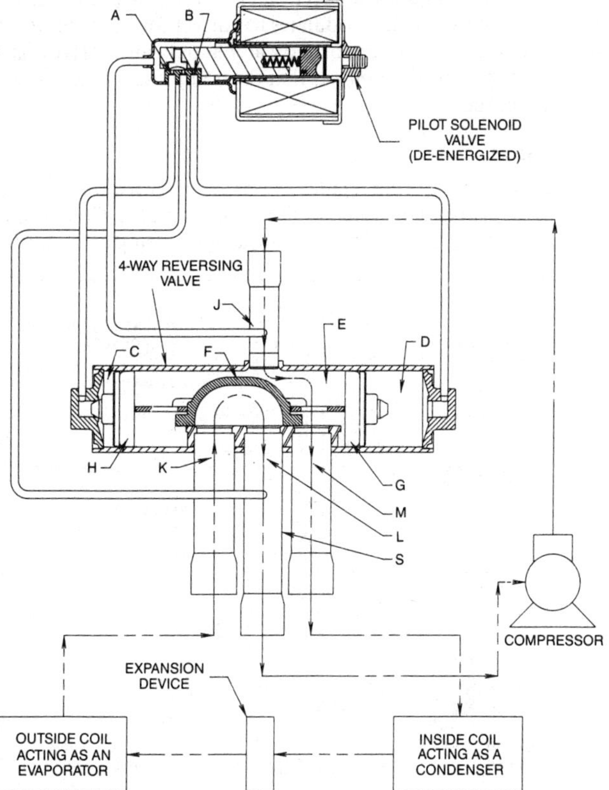

Fig. 30 Four-Way Slide-Type Refrigerant-Reversing Valve Used in Heating Cycle of Refrigeration System

Reversing valves operate well with most fluorinated refrigerants, but the type of refrigerant used affects the size of the valve selected for a given capacity. The valve should have minimum pressure drops through the valve passages in both the discharge and suction gas paths so that the compressor capacity is not reduced.

Application

In addition to their year-round air-conditioning (heat-pump) application on residential or commercial systems, four-way reversing valves are used in refrigeration systems in motor trucks, trailers, and railway refrigerator cars for the transportation of perishable cargo. The transport systems can be arranged to operate automatically and to provide cooling, heating, and defrosting, as required.

While the four-way valves are sometimes adapted for other applications, three-way reversing valves are designed specifically for commercial refrigeration systems to defrost the evaporator or for heat reclaim using an auxiliary condenser.

Regardless of the type of installation to which the reversing valve is added (*e.g.*, window unit, residential unit, commercial unit, transportation unit, or customer-built systems), the system may be made to operate automatically. If a dual-action thermostat is used, cooling automatically occurs when the temperature rises above a preset value, and heating automatically occurs when the temperature drops below a preset value.

CONDENSING WATER REGULATORS

Two-Way Regulators

The condensing water regulator modulates the quantity of water passing through a water-cooled refrigerant condenser in response to the condensing pressure. This regulator is used on a vapor-cycle refrigeration system to maintain a condensing pressure that loads but does not overload the compressor motor. The regulator automatically modulates to correct for both variations in temperature or pressure of the water supply and variations in the quantity of refrigerant gas that the compressor is sending to the condenser.

Operation

The condensing water regulator consists of a valve and an actuator that are linked together, as shown in Figure 31. The actuator consists of a metallic bellows and adjustable spring combination

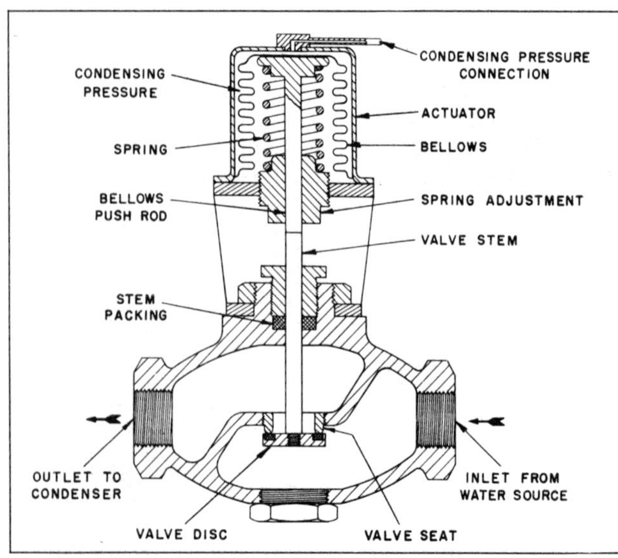

Fig. 31 Two-Way Condensing Water Regulator

connected to the system condensing pressure. For large water-flow capacities, a small condensing water regulator is used for pilot-operation of a diaphragm-type main valve.

After a compressor starts, the condensing pressure begins to rise. When the opening pressure setting of the regulator spring is reached, the bellows moves to open the valve disk gradually or slide from its seat. The regulator continues to open as the condensing pressure rises until a balance point is reached between the water flow and the heat rejection requirement, at which point the condensing pressure is stabilized. When the compressor stops, the continuing water flow through the regulator causes the condensing pressure to drop gradually, closing the regulator, which becomes fully closed when the opening pressure setting of the regulator is reached.

Selection

Selection of a condensing water regulator depends on the system refrigerant used, the water-flow rate required, and the available water-pressure drop across the regulator. While one standard bellows operator can sometimes handle any of several refrigerants, special springs or bellows may be required for very high- or low-pressure refrigerants, as in the case of Refrigerant 717 (ammonia), for which a stainless steel bellows must be used in place of a brass bellows.

The water-flow rate required depends on condenser performance, the temperature of available water, the quantity of heat that must be rejected to this water, and the allowable leaving water temperature. For a given opening of the valve seat, which corresponds to a given pressure rise above the regulator opening point, the flow rate handled by a given size water regulator is a function of the available water-pressure drop across the valve seat. Available water-pressure drop is determined for the required flow rate by deducting condenser water-pressure drop, pipeline pressure drop, and static pressure losses from the pressure of the water at its supply point.

The condensing water regulator should be selected from the manufacturer's data on the basis of maximum required flow rate, minimum available pressure drop, and water temperature.

Application

Oversizing of a regulator should be avoided because this encourages hunting. When two-way condensing water regulators (see Figure 31) are used on recirculating cooling tower systems, the throttling action of the valves during cold weather causes a reduction in pump and tower circulation, which is undesirable for that equipment.

Three-Way Regulators

On tower systems requiring individual condensing pressure control, three-way condensing water regulators should be used (see Figure 32). These are similar in construction to two-way regulators, but they have an additional port, which opens to bypass water around the condenser as the port controlling water flow to the condenser closes. Thus, the tower decking or sprays and the circulating pump receive a constant supply of water, although the water supply to individual condensers is modulated for control.

Three-way condensing water regulators must be supplemented by other means if cooling tower systems are to be operated in freezing weather. An indoor sump is usually required, and a temperature-actuated three-way water control valve is used to divert periodically all of the condenser leaving water directly to the sump whenever the water becomes too cold.

Unlike the recommended and accepted practice pertaining to the application of refrigerant-control valves, a strainer is not ordinarily used with a water regulator because a strainer usually requires more cleaning and servicing than a regulator without a strainer.

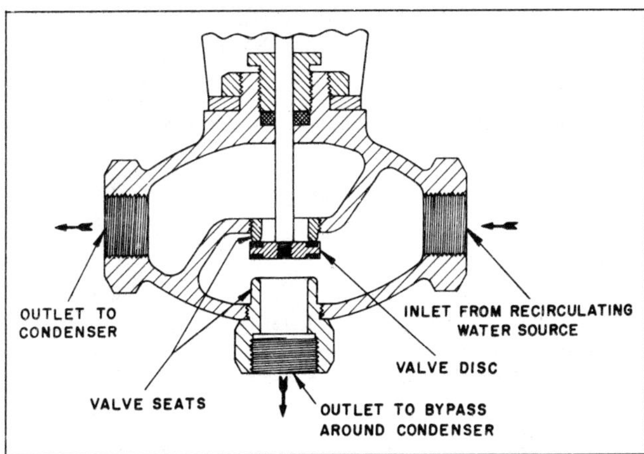

Fig. 32 Three-Way Condensing Water Regulator

CHECK VALVES

Refrigerant check valves are normally used in refrigerant lines in which pressure reversals can cause undesirable reverse flows. A check valve is usually opened by a portion of the pressure drop, which causes flow in the pipeline. Closing usually occurs either when a reversal of pressure takes place or when the pressure drop across the check valve is less than the minimum opening pressure drop in the normal flow direction.

Conventional check valve designs frequently use piston construction and use the globe pattern in large sizes, while in-line designs are commonly smaller than the 2-in. size. Either design may be used with closing springs; the heavier springs give more reliable and tighter closing but require greater pressure drop for the check valve to open. The in-line check valve design does not permit a manual opening stem. Although conventional check valves may be designed to open at less than 1-psi drop, they may not be reliable at temperatures below −25 °F because the light closing springs may not overcome the viscous oil.

Other check valve designs are used for various special functions not covered here, such as *excess flow checks,* which close only when flow exceeds the maximum desired rate; *electrically lifted checks,* which require no pressure drop to remain open; *remote pressure-operated checks,* which are normally open but close when they are supplied with a higher pressure source of refrigerant; and *pressure differential valves,* which maintain a uniform pressure differential between system components for special functions in a refrigeration system.

Seat Materials

Although precision metal seats may be manufactured nearly bubble-tight, they are not economically practical for refrigerant check valves. Seats made of synthetic rubbers provide excellent tightness at medium and high temperatures but may leak at low temperatures due to lack of resilience. Because high temperatures deteriorate most rubbers suitable for refrigerants, the use of plastic seat materials has become increasingly successful, despite the possibility of damage by large pieces of foreign matter in the systems.

Applications

In *compressor discharge lines,* check valves are used to prevent flow from the condenser to the compressor during the *off* cycle or to prevent flow from an operating compressor to an idle compressor. While a 2- to 6-psi pressure drop is tolerable, the check-valve design must resist pulsations of the compressor and the temperature of discharge gas, and it must be bubble-tight to prevent accumulation of liquid refrigerant at the compressor discharge valves or in the crankcase.

In liquid lines, check valves prevent reverse flow through the unused expansion device on heat-pump systems or prevent backup into the low-pressure liquid line of a recirculating system during a defrost period. While a 2- to 6-psi pressure drop is usually acceptable, the check-valve seat must be bubble-tight.

In the suction line of a low-temperature evaporator, a check valve may be used to prevent the transfer of refrigerant vapor to a lower temperature evaporator on the same suction main. In this case, the pressure drop must be less than 2 psi, the valve seating must be reasonably tight, and the check valve must be reliable at low temperatures.

Normally, open pressure-operated check valves are used to close suction lines, gas, or liquid legs in gravity recirculating systems during defrost.

In hot-gas defrost lines, check valves may be used in the branch hot-gas lines connecting the individual evaporators to prevent crossfeed of refrigerant during the cooling cycle when the defrost operation is not taking place. In addition, check valves are used in the hot-gas line between the hot-gas heating coil in the drain pan and the evaporator, to prevent pan coil sweating during the refrigeration cycle. Tolerable pressure drop is typically 2 to 6 psi, seating must be nearly bubble-tight, and seat materials must withstand high temperatures.

To prevent chatter or pulsation, check valves should be sized for the particular pressure drop that ensures that they are in the wide-open position at the desired flow rate.

RELIEF DEVICES

Refrigerant relief devices have either safety or functional uses. A safety relief device is designed to relieve positively at its set pressure for one crucial occasion without prior leakage. The relief may be to the atmosphere or to the low side.

A *functional* relief device is a control valve that may be called on to open, modulate, and reclose with repeatedly accurate performance. For reasons of system control, relief is usually from a portion of the system at higher pressure to a portion at lower pressure. Design refinements of the functional relief valve usually make it unsuitable or uneconomical as a safety relief device.

Safety Relief Valves

These are most commonly *pop-type designs,* which abruptly open when the inlet pressure exceeds the outlet pressure by the

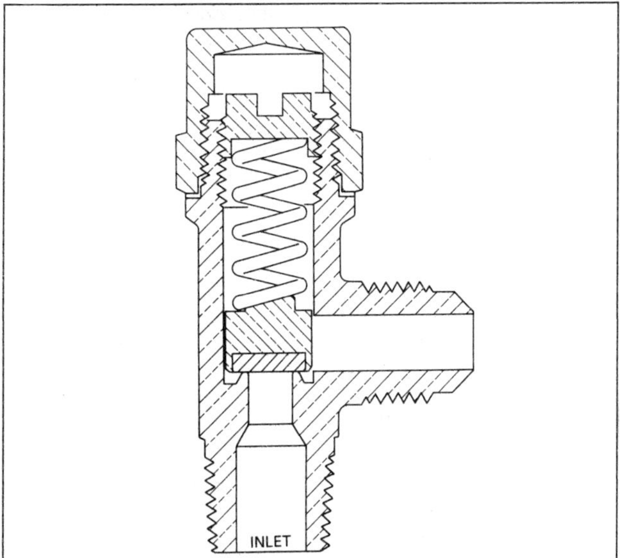

Fig. 33 Pop-Type Safety Relief Valve

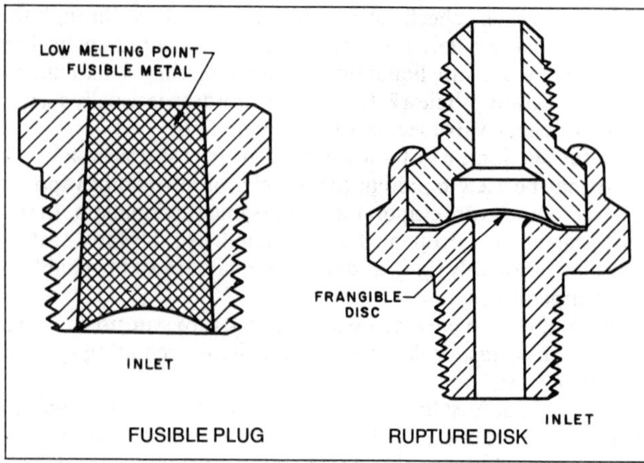

Fig. 34 Safety Relief Devices

valve setting pressure (see Figure 33). Seat configuration is such that once lift begins, the resulting increased active seat area causes the valve seat to pop wide open against the force of the setting spring. Because the flow rate is measured at a pressure of 10% above the setting, the valve must open within this 10% increase in pressure.

This relief valve operates on a fixed pressure differential from inlet to outlet. Because the valve is affected by back pressure, the installation of a rupture disk at the valve outlet is not permissible.

Relief valve seats are made of metal, plastic, lead alloy, or synthetic rubber. The last is commonly used because it has greater resilience and, consequently, probable reseating tightness. For valves that have lead-alloy seats, an emergency manual reseating stem is occasionally provided to permit reforming of the seating surface by tapping the stem lightly with a hammer. The advantages of the pop-type relief valve are simplicity of design, low initial cost, and high-discharge capacity.

Other Safety Relief Devices

Two devices performing similar safety relief operations are the *fusible plug* and the *rupture disk* (Figure 34). The former contains a fusible member that melts at a predetermined temperature corresponding to the safe saturation pressure of the refrigerant, but is limited in application to pressure vessels with internal gross volumes of 3 ft^3 or less and internal diameters of 6 in. or less. The rupture member contains a frangible disk designed to rupture at a predetermined pressure.

Discharge Capacity

The minimum required discharge capacity of the pressure relief device or fusible plug for each pressure vessel is determined by the following formula, specified by the ASHRAE *Standard* 15-1992:

$$C = fDL$$

where

C = minimum required air discharge capacity of the relief device, lb/min
D = outside diameter of vessel, ft
L = length of the vessel, ft
f = factor dependent on the refrigerant, as shown in Table 1

Capacities of pressure relief valves are determined by test in accordance with the provisions of the ASME *Boiler and Pressure Vessel Code* (1989). Relief valves approved by the National Board of Boiler and Pressure Vessel Inspectors are stamped with the code symbol, which consists of the letters UV in a clover leaf design with the letters NB stamped directly below this symbol. In addition, the pressure setting and capacity are stamped on the valve.

Table 1 Values for *f* for Discharge Capacity of Pressure Relief Devices

Refrigerant	Value of *f*
When used on the low side of a limited-charge cascade system:	
R-170, R-744, R-1150	1.0
R-13, R-13B1, R-503	2.0
R-14	2.5
Other applications:	
R-717	0.5
R-11, R-40, R-113, R-123, R-142b, R-152a, R-290, R-600, R-600a, R-611, R-764	1.0
R-12, R-22, R-114, R-134a, R-C 318, R-500, R-1270	1.6
R-115, R-502	2.5

Notes:
1. The values of *f* listed do not apply if fuels are used within 20 ft of the pressure vessel. In this case, the methods in API RP 520 shall be used to size the pressure relief device.
2. When one pressure relief device or fusible plug is used to protect more than one pressure vessel, the required capacity shall be the sum of the capacities required for each pressure vessel.
3. For refrigerants not listed, consult ASHRAE *Standard* 15-1992.

When relief valves are used on pressure vessels of 10 ft^3 internal gross volume or more, a relief system consisting of a three-way valve and two relief valves in parallel is required.

The rated discharge capacity of a rupture member or fusible plug that discharges to the atmosphere under critical flow conditions is determined by calculation, using the formula provided in the ASHRAE *Standard* 15-1992.

Pressure Setting

The maximum pressure setting for a relief device is limited by the design working pressure of the vessel to be protected. Pressure vessels normally have a safety factor of 5. Therefore, the minimum bursting pressure is five times the rated design working pressure. The relief device must have enough discharge capacity to prevent the pressure in the vessel from rising more than 10% above its design pressure. Since the capacity of a relief device is measured at 10% above its stamped setting, the setting cannot exceed the design pressure of the vessel.

To prevent loss of refrigerant through pressure relief devices during normal operating conditions, the relief device setting must be substantially higher than the system operating pressure. For rupture members, the setting should be 50% above a static system pressure and 100% above a maximum pulsating system pressure. Failure to provide this margin of safety causes fatigue of the frangible member and rupture well below the stamped setting.

For relief valves, the setting should be 25% above the maximum system pressure. This safety factor will provide spring force on the valve seat sufficient to maintain a tight seal and still allow for setting tolerances and other factors that cause settings to vary. Although relief valves are set at the factory to be close to the stamped setting, the variation may be as much as 10% after the valves have been stored or placed in service for a period of time.

Discharge Piping

The size of the discharge pipe from the pressure relief device or fusible plug should not be less than the size of the pressure relief device or fusible plug outlet. The maximum length of the discharge piping is provided in a table or may be calculated from the formula provided in the ASHRAE *Standard* 15-1992.

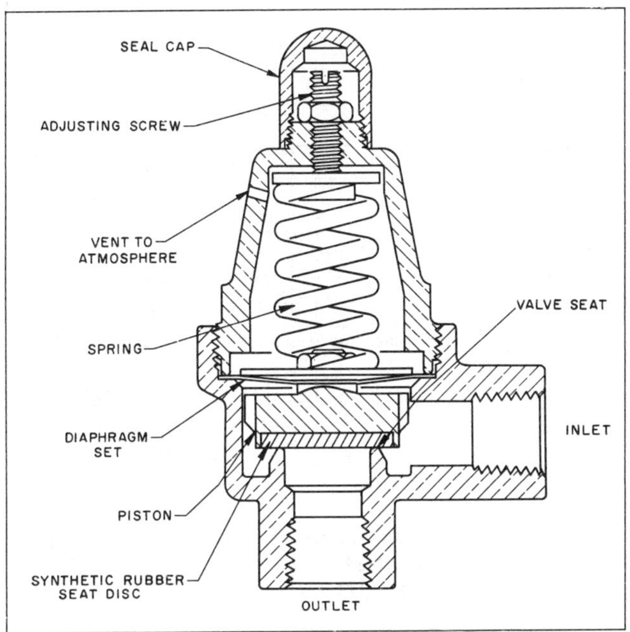

Fig. 35 Diaphragm-Type Relief Valve

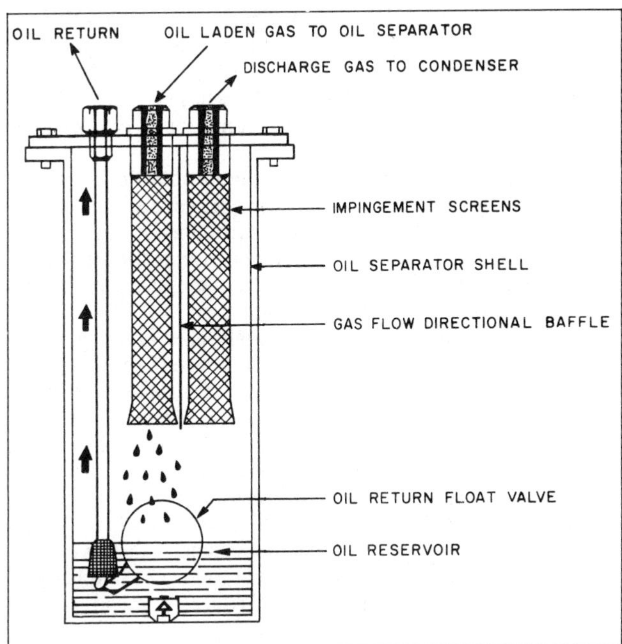

Fig. 36 Discharge-Line Oil Separator

Relief Device Summary

1. A relief device with sufficient capacity for code requirements and one suitable for the type of refrigerant used should be selected.
2. The proper size and length of discharge tube or pipe should be used.
3. The relief device should not be discharged prior to installation or when pressure testing the system.
4. For systems containing large quantities of refrigerant, a three-way valve and two relief valves should be used.
5. A pressure vessel that permits the relief valve to be set at least 25% above the maximum system pressure should be provided.

Functional Relief Valves

Functional relief valves are usually diaphragm types in which the system pressure acts on a diaphragm that lifts the valve disk from the seat (Figure 35). The other side of the diaphragm is exposed to both the adjusting spring and atmospheric pressure. The ratio of effective diaphragm area to seat area is high, so the outlet pressure has little effect on the operating point of the valve.

Because the lift of the diaphragm is not great, the diaphragm valve is frequently built as the pilot or servo of a larger piston-operated main valve, thereby providing fine sensitivity and high-flow capacity. Construction and performance are similar to the previously described pilot-operated evaporator pressure regulator, except that the diaphragm valves are constructed for higher pressures. Thus, the valves are suitable for use as defrost relief from evaporator to suction pressure, as large-capacity relief from a pressure vessel to the low side, or as a liquid refrigerant pump relief from pump discharge to the accumulator to prevent excessive pump pressures when some evaporators are valved closed.

DISCHARGE-LINE OIL SEPARATORS

The discharge-line oil separator removes oil from the discharge gas of lubricated helical rotary (screw) and reciprocating compressors. Oil is separated by (1) reducing gas velocity, (2) changing direction of flow, (3) impingement on baffles, (4) mesh pads or screens, and (5) centrifugal force. The separator reduces the amount of oil reaching the low side, helps maintain the oil charge in the compressor oil sump, and muffles the sound of the gas flow.

Figure 36 shows a small separator incorporating inlet and outlet screens and a high-side float valve. A space below the float valve provides for dirt or carbon sludge. When oil accumulates to raise the float ball, oil passes through a needle valve and returns to the low-pressure crankcase. When the level falls, the needle valve closes, preventing the release of hot gas into the crankcase. Insulation and electric heaters may be added to prevent the refrigerant from condensing when the separator is exposed to low temperatures. A wide variety of horizontal and vertical flow separators is manufactured with one or more of such elements as centrifuges, baffles, wire mesh pads, or cylindrical filters.

SELECTION

Separators are usually given system capacity ratings for several refrigerants at several suction and condensing temperatures. Another rating method for selection purposes gives the capacity in terms of the compressor displacement volume. Some separators also show a marked reduction in separation efficiency at some stated minimum capacity.

Because the compressor capacity increases when the suction pressure is raised or the condensing pressure is lowered, the system capacity at its lowest compression ratio should be the criterion for selecting the capacity of the separator.

APPLICATION

A discharge-line oil separator is best for ammonia or hydrocarbon refrigerants to reduce oil fouling in the evaporator. With oil-soluble halocarbon refrigerants, only certain flooded systems, low-temperature systems, or systems with long suction lines or other oil return problems need oil separators. (See Chapter 2 for more information about oil separator applications.)

CAPILLARY TUBES

Every refrigerating unit requires a pressure-reducing device to meter the flow of refrigerant to the low side in accordance with the demands placed on the system. The capillary tube achieved popularity, especially with the smaller unitary hermetic equipment such as household refrigerators and freezers, dehumidifiers, and room air conditioners. Capillary tube use has been extended to include larger units such as unitary air conditioners in sizes up to 10 tons capacity.

The capillary operates on the principle that liquid passes through it much more readily than gas. It consists of a small diameter line, which, when used for controlling a system's refrigerant flow, connects the outlet of the condenser to the inlet of the evaporator. It is sometimes soldered to the outer surface of the suction line for heat-exchange purposes.

A high-side liquid receiver is not normally used with a capillary; consequently, a corresponding reduction in refrigerant charge could result. In a few applications, such as household refrigerators, freezers, room air conditioners, and heat pumps, a small low-side accumulator may be used. The pressure-equalizing characteristic of a capillary makes the use of a low-starting-torque motor compressor possible. Inherently, a capillary does not operate as efficiently over a wide range of conditions as does a thermostatic expansion valve; however, due to counter-balancing factors in most applications, its performance is generally good. The simplicity of the capillary gives it the advantage of reduced cost.

THEORY

Because the capillary passes liquid much more readily than gas, it is a practical metering device. When a condenser-to-evaporator capillary is sized to permit the desired flow of refrigerant, liquid will seal its inlet. If a system imbalance occurs and some gas (uncondensed refrigerant) enters the capillary, this gas considerably reduces the mass flow of refrigerant, which increases condenser pressure. This condition causes subcooling and increases the mass flow of refrigerant. If properly sized for the application, the capillary compensates automatically for load and system variations and gives acceptable performance over a wide range of operating conditions.

A common flow condition is to have subcooled liquid at the entrance to the capillary. Bolstad and Jordan (1948) described the flow behavior from temperature and pressure measurements along the tube as follows:

> "With subcooled liquid entering the capillary tube, the pressure distribution along the tube is similar to that shown in the graph (see Figure 37). At the entrance to the tube, section 0-1, a slight pressure drop occurs, usually unreadable on the gauges. From point 1 to point 2, the pressure drop is linear. In the portion of the tube 0-1-2, the refrigerant is entirely in the liquid state, and at point 2, the first bubble of vapor forms. From point 2 to the end of the tube, the pressure drop is not linear, and the pressure drop per unit length increases as the end of the tube is approached. For this portion of the tube, both the saturated liquid and saturated vapor phases are present, with the percent and volume of vapor increasing in the direction of flow. In most of the runs, a significant pressure drop occured from the end of the tube into the evaporator space.
>
> "With a saturation temperature scale corresponding to the pressure scale superimposed along the vertical axis, the observed temperatures may be plotted in a more efficient way than if a uniform temperature scale were used. The temperature is constant for the first portion of the tube 0-1-2. At point 2, the pressure has dropped to the saturation pressure corresponding to this temperature. Further pressure drop beyond point 2 is accompanied by a corresponding drop in temperature, the temperature being the saturation temperature corresponding to the pressure. As a consequence, the pressure and temperature lines coincide from point 2 to the end of the tube."

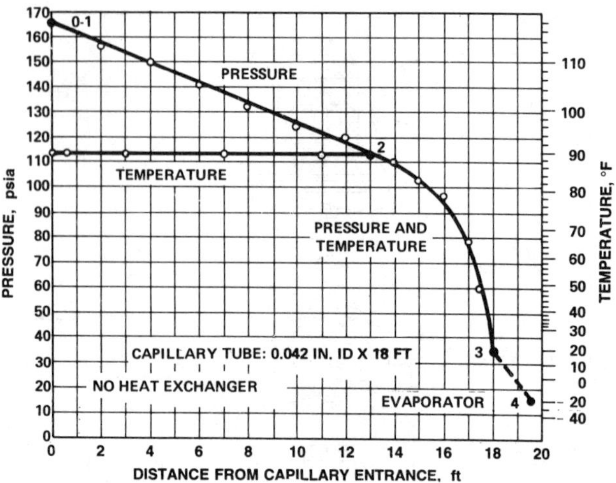

Fig. 37 Pressure and Temperature Distribution along Typical Capillary Tube
(Bolstad and Jordan 1948)

More detailed measurements performed later (for example, see Mikol 1963 or Li *et al.* 1990) showed that generation of the first vapor bubble does not occur at the point where the liquid pressure reaches the saturation pressure (Point 2 on Figure 37), but rather the refrigerant remains in the liquid phase for some limited length past Point 2, reaching a pressure below the saturation pressure. This delayed evaporation, often referred to as metastable or superheated liquid condition, has to be accounted for in analytical modeling of the capillary tube, or a significant underestimation of the mass flow rate of refrigerant will result (Kuehl and Goldschmidt 1991).

The rate of refrigerant flow through a capillary always increases with an increase in inlet pressure. Flow rate also increases with a decrease in external outlet pressure down to a certain critical value, below which the flow does not change. Figure 37 illustrates a case in which the outlet pressure inside the capillary has reached the critical value, which is higher than the external pressure. Such a condition is typical for normal operation. The point at which the first gas bubble appears, is called the *bubble point.* The preceding portion of capillary is called the *liquid length,* and that following is called the *two-phase length.*

SYSTEM DESIGN FACTORS

A capillary tube must be designed to be compatible with other system components. Compressor size, heat exchanger capacity as well as other system components have an effect on the capillary size. In general, once the compressor and heat exchangers have been selected to meet the requirement at design condition, the capillary size and system charge are determined. Detailed design considerations for systems using a capillary tube may be different for different applications (domestic refrigerator, window air conditioner, residential split heat pump).

Capillary size and system charge together determine subcooling and superheat for a given system operating at design conditions. Performance at off-design conditions should be checked for desired performance characteristics. Capillary tube systems are generally much more charge sensitive than expansive valve systems.

The high side must be designed carefully for use with a capillary. To prevent hydrostatic rupture failure in case of capillary stoppage, the high-side volume should be sufficient to contain the entire refrigerant charge. It may be necessary to provide sufficient refrigerant storage volume to protect against excessive discharge pressures during high-load conditions.

Another consideration in high-side design, where cyclic operation is involved, is unloading during the *off* period. When unit operation ceases, the capillary continues to pass refrigerant from the high side to the low side until pressures are equalized. Good drainage of the liquid into the capillary during this unloading interval should be provided. If liquid is trapped in the high side, it will evaporate there during the *off* cycle, pass to the low side as a warm gas, condense, and add latent heat to the evaporator. Liquid trapping may also increase the time required for the pressures to equalize after the compressor stops operating. If this interval is too long, the compressor may not be sufficiently unloaded to permit easy starting, especially in domestic refrigerators.

The amount of refrigerant in the evaporator is at its maximum value during the *off* cycle and its minimum during the *running* cycle. The suction piping should be arranged to reduce the adverse effects of the variable-charge distribution. A suitable liquid accumulator is sometimes necessary.

In some systems (*e.g.*, refrigerators), a capillary tube is soldered to a suction line to heat it up and avoid condensation of water vapor from the ambient air. This suction line heat exchanger may result also in efficiency improvement, depending on the refrigerant and application.

When a suction line heat exchanger is used, the excess capillary length may be coiled and located at either end of the heat exchanger. Although more heat is exchanged with the excess coiled at the evaporator, system stability is enhanced if a portion of the capillary is located at the condenser. Forming of coils and bends should be done carefully to avoid local restrictions. The effect of forming on the restriction should be considered when specifying the capillary.

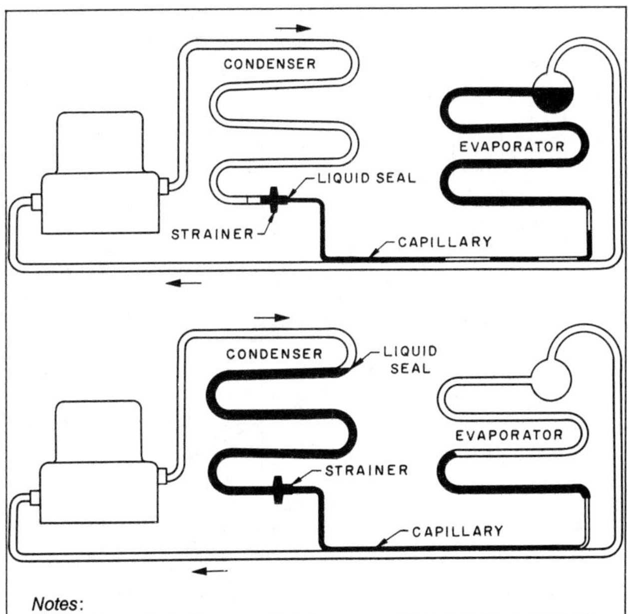

Notes:
1. Capillary selected for capacity balance conditions. Liquid seal at capillary inlet but no excess liquid in condenser. Compressor discharge and suction pressures normal. Evaporator properly charged.
2. Too much capillary resistance—liquid refrigerant backs up in condenser and causes evaporator to be undercharged. Compressor discharge pressure may be abnormally high. Suction pressure below normal. Bottom of condenser subcooled.

Fig. 38 Effect of Capillary Tube Selection on Refrigerant Distribution

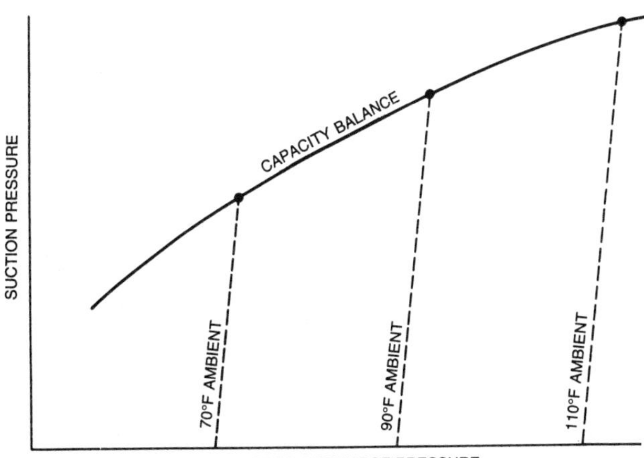

Note: Operation below this curve results in a mixture of liquid and vapor entering the capillary. Operation above the capacity balance points causes liquid to back up in the condenser and elevate its pressure.

Fig. 39 Capacity Balance Characteristic of Capillary System

CAPACITY BALANCE CHARACTERISTIC

The selection of a capillary tube depends on the application and anticipated range of operating conditions. One approach to the problem involves the concept of *capacity balance.* A refrigerating system may be operating at the condition of capacity balance when the resistance of the capillary is sufficient to maintain a liquid seal at its entrance without excess liquid accumulating in the high side of the system (see Figure 38). Only one such *capacity balance point* exists for any given compressor discharge pressure. A curve through the capacity balance points for a range of compressor discharge pressures is called the *capacity balance characteristic* of the system. Such a curve is shown in Figure 39. Ambient temperatures are drawn on the chart for a typical air-cooled system. A given set of compressor discharge and suction pressures is associated with fixed condenser and evaporator pressure drops; these pressures establish the capillary inlet and outlet pressures.

The capacity balance characteristic curve for any combination of compressor and capillary may be determined experimentally by the arrangement shown in Figure 40, which makes it possible to vary independently the suction and discharge pressures until capacity balance is obtained. The desired suction pressure may be obtained by regulating the heat input to the low side, usually by electric heaters; the desired discharge pressure may be obtained by a suitable controlled water-cooled condenser. A liquid indicator is located at the entrance to the capillary. The usual test procedure is to hold the high-side pressure constant and, with gas bubbling through the sight glass, slowly increase the suction pressure until a liquid seal forms at the capillary entrance. Repeating this procedure at various discharge pressures determines the capacity balance characteristic curve similar to that shown in Figure 39. This equipment may also be used as a calorimeter to determine simultaneously the capacity of the refrigerating system.

OPTIMUM SELECTION AND REFRIGERANT CHARGE

Whether the initial capillary selection and charge are optimum for the unit is always questioned, even in such simple applications as a condenser-to-evaporator capillary for a room air-conditioning unit. The refrigerant charge in the unit can be varied using a small refrigerant bottle (valved off and sitting on a scale) connected to the circuit. The interconnecting line must be flexible and arranged

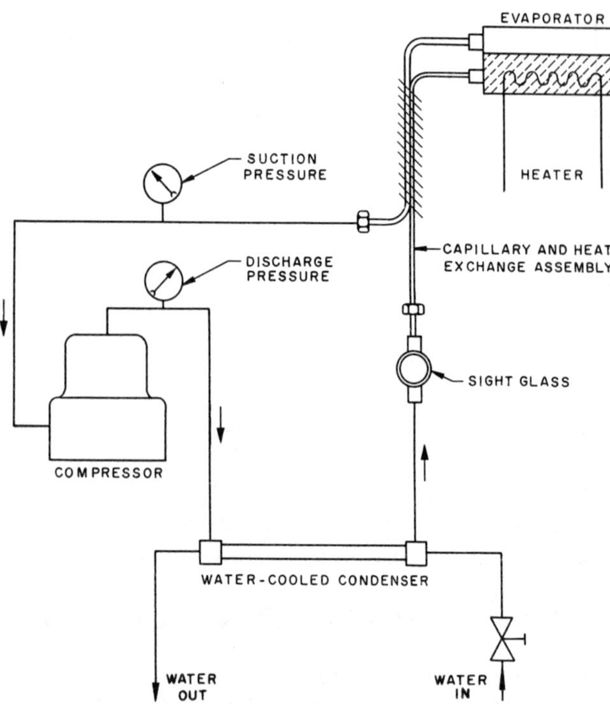

Fig. 40 Test Setup for Determining the Capacity Balance Characteristic of Given Combination of Compressor and Capillary and Heat Exchanger Assembly

so that it is filled with vapor instead of liquid. The charge is brought into the unit or removed from it by heating or cooling the bottle.

The only test for varying the capillary restriction is to remove the element, install a new selection, and determine the optimum charge, as outlined above. A method occasionally used is to pinch the capillary to determine whether or not increased resistance is needed.

It is necessary to operate the unit through the expected range of operation to determine power and cooling capacity for any given selection and charge combination.

APPLICATION

Processing and Inspection

To prevent mechanical clogging caused by foreign particles, a strainer should precede the capillary. Also, all parts of the system must be evacuated adequately to eliminate water vapor and noncondensable gases, which may cause clogging by corrosion. The oil should be free from wax separation at the minimum operating temperature.

The interior capillary surface should be smooth and uniform in diameter. Although plug-drawn copper is more common, wire-drawn or sunk tubes are also available. Life tests should be conducted at low evaporator temperatures and high condensing temperatures to check on the possibility of corrosion and plugging. Material specifications for seamless copper tube are given in ASTM *Standard* B75. Similar information on hand-drawn copper tubes can be found in ASTM *Standard* B360-92.

A procedure should be established to ensure uniform flow capacities, within reasonable tolerances, for all capillaries used in product manufacture. This procedure may be conducted as follows. The final capillary, determined from tests, is removed from the unit and given an airflow capacity rating, using the wet-test meter method described in ASHRAE *Standard* 28-1988. Master capillaries are then produced, by using the wet-test meter airflow equipment, to provide the maximum and minimum flow capacities for the particular unit. The maximum flow capillary has a flow capacity equal to that of the test capillary, plus a specified tolerance. The minimum flow capillary has a flow capacity equal to that of the test capillary, less a specified tolerance. One sample of the maximum and minimum capillaries is sent to the manufacturer of capillary tubes to be used as tolerance guides for elements supplied for a particular unit. Samples are also sent to the inspection group for quality control.

Preliminary Selection

The preliminary selection of a condenser-to-evaporator capillary for a given compressor rating may be determined by referring to Figure 41. Note that the compressor rating is based on the refrigeration per unit mass specified. If desired, the ratings may be converted to mass flow by dividing the compressor capacity by the refrigeration per unit mass given on the chart. For the preliminary selection, no distinction is made between units with and without heat exchangers. The subcooling in Figure 41 pertains to the total subcooling of the refrigerant entering the capillary

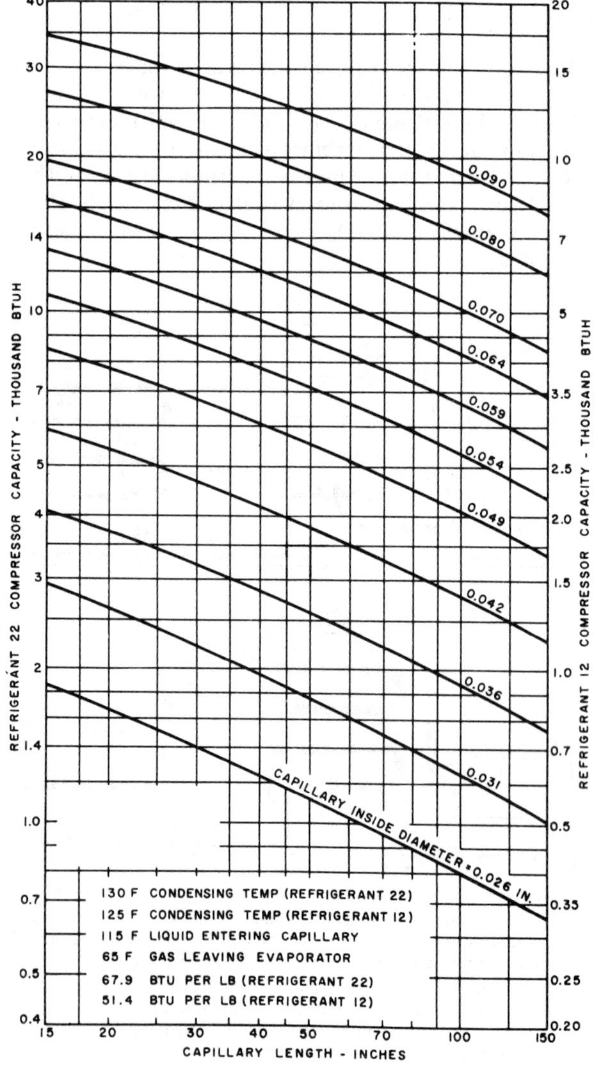

Fig. 41 Preliminary Selection Chart for Refrigerants 12 and 22 Condenser-to-Evaporator Capillary

tube, whether this is in the condenser or in the condenser and a heat exchanger.

In the selection of a capillary for a specific application, practical considerations influence the length. For example, the minimum length will be determined by such geometric considerations as the physical distance between the high side and low side and the length of capillary tube required for optimum heat exchange. It may also be dictated by considerations of exit velocity and noise and the possibility of plugging with foreign materials. The maximum length may be determined primarily by considerations of cost. It is fortunate, therefore, that the flow characteristics of a capillary can be adjusted independently by varying either its bore or its length. Thus, it is feasible to select the most convenient length independently and then (within certain limits) select a bore to give the desired flow. An alternate procedure is to select a standard bore and then adjust the length, as required.

Standard diameters and wall thickness for capillary tubes are given in ASTM *Standard* B360-1992. Many nonstandard tubes are also used, resulting in nonuniform interior surfaces and variations in flow.

Simplified Calculation Procedure

The optimum capillary size corresponding to any given set of system operating conditions can be calculated. These conditions of pressure, mass flow, and inlet subcooling or quality are a function of the unit design and a choice of the service operating conditions.

The capillary on room air conditioners is adjusted to give maximum unit capacity at ASHRAE rating conditions of 80 °F dry bulb, 67 °F wet bulb indoors and 95 °F dry bulb, 75 °F wet bulb outdoors. Optimum performance may be obtained when there is considerable subcooling in the refrigerant prior to its entrance into the capillary.

The unit must be tested to ensure the capillary performs properly under various limiting conditions. For example, the capillary of a household refrigerator, which may have been sized to give optimum performance on no-load cycles at a particular ambient temperature, should be tested to ensure it operates properly during pulldown under maximum and minimum ambient and loading conditions.

Figures 42 and 43 present general application ratings for Refrigerants 12 and 22. The ratings are quite versatile, as they can be used to calculate flow rate directly from a given capillary tube selection and flow condition or used to determine a capillary tube selection from a flow condition and flow rate. These ratings, while presented in the form shown originally by Hopkins (1950), show different values because they were recalculated from the later work of Whitesel (1957). Very limited data are available on performance of capillary tubes with new ozone-benign refrigerants. For comparison of R-12 and R-134a, see Wijaya (1991).

The ratings based only on Figures 42 and 43 do not use suction pressure as a parameter. This is not serious because Bolstad and Jordan (1948) demonstrated that drop in back pressure over a wide range causes a negligible increase in refrigerant flow rate. Hopkins (1950) showed that using the flow factor to correct tube diameter and length results in some additional inaccuracy.

The effect of the suction pressure on capillary capacity may be calculated. First, the critical pressure is determined from Figure 44 and the correction factor is obtained from Figure 45. The capacity obtained from Figures 42 and 43 is multiplied by the correction factor to obtain the final value of flow. Relatively high back pressures cause substantial reductions in flow below the figure given by the basic rating chart for critical outlet pressure.

The effect of a suction line heat exchanger can normally be considered by subtracting the liquid temperature drop through this element from the liquid temperature at capillary inlet. This method holds rigidly when the refrigerant bubble point comes after the heat exchanger. Figure 46, which was derived mathematically for the latter condition, may be used to determine the heat-exchanger liquid subcooling.

Figure 47 may be used to determine liquid length, which indicates liquid pressure drop. Although in preparing this chart the Refrigerant 12 pressure drop values were calculated to be about 5.5% less than those for Refrigerant 22 at a given mass flow, the two sets of values were averaged for simplification. Average values of densities between −40 and 140 °F were used.

The ratings given in the figures in this chapter should be used only for a preliminary selection of a capillary tube; some discrepancy with test data has been reported in the literature. For example, Wijaya (1991) showed that Figures 42 and 43 underpredict the refrigerant mass flow rate by as much as 17.4%, depending on the operating conditions. A similar discrepancy was reported by Kuehl and Goldschmidt (1991).

Calculation procedures are illustrated in Examples 1 to 6.

Example 1: Determine the flow rate of Refrigerant 22 for a restrictor of 0.054 in. ID, 111 in. long, operating without heat exchange at 100 °F saturated condensing temperature with 10 °F subcooling.

Solution: From Figure 42, at 100 °F saturated condensing temperature, or 210.6 psi, and 10 °F subcooling, the flow rate for an 80 in. length of 0.064 in. ID restrictor is found to be 97.0 lb/h.

To correct for actual diameter and length, the flow factor ϕ_1 is found from Figure 43. For 0.054 in. ID and 111 in. length, $\phi_1 = 0.55$. Corrected flow rate is $97.0 \times 0.55 = 53.4$ lb/h.

Example 2: Determine the condition of flow for the restrictor of Example 1 if the flow rate falls to 40 lb/h at 120 °F saturated condensing temperature, or 274.3 psi pressure.

Solution: The flow factor $\phi_1 = 0.55$ (from Example 1). With a flow rate of 40 lb/h, the uncorrected flow rate to be used with Figure 42 is $40/0.55 = 72.7$ lb/h.

From Figure 42, with a flow rate of 72.7 lb/h and 274.3 psi condensing pressure, the flow condition is read as 12% inlet quality at the restrictor entrance.

Example 3: Determine the length of 0.042 in. ID tube, which is equivalent to 30 in. of 0.036 in. ID tube.

Solution: From Figure 43, the flow factor for 30 in. of 0.036 in. of ID tube is read as 0.35. For this value of flow factor, it is found from Figure 43 that 73 in. of 0.042 in. ID tube are required.

Example 4: Determine the length of 0.054 in. ID capillary that will pass 50 lb/h of Refrigerant 12, with no subcooling and no gas at the inlet, when the inlet pressure outside the capillary is 150 psig.

Solution: The absolute inlet pressure is 164.7 psi. From Figure 42, the flow rate is 65 lb/h. The flow factor ϕ_1 is $50/65 = 0.77$. From Figure 43, for $\phi_1 = 0.77$ and 0.054 in. ID, the length of capillary required is 53 in.

Example 5: What will be the flow rate for either Refrigerant 12 or Refrigerant 22 for a 0.070 in. ID capillary 131 in. long with an inlet pressure of 175 psi, 20% inlet quality, and an outlet pressure of 150 psi?

Solution: From Figure 42, the basic rating is 41 lb/h. From Figure 43, $\phi_1 = 0.98$. The basic flow is then $0.98 \times 41 = 40.1$ lb/h. $L/1250\,D = 131/(1250)(0.070) = 1.5$.

Enter Figure 44 at 175 psi inlet pressure and proceed vertically to 20% inlet quality. Proceed horizontally to the $L/1250D$ line of unity. Proceed vertically to the $L/1250D$ line of 1.5. Proceed horizontally to a critical outlet pressure of 45 psi. The value of ψ for Figure 45 is $(150 - 45)/(175 - 45) = 0.808$. From Figure 45, $\phi_2 = 0.53$. The corrected flow is $40.1 \times 0.53 = 21.3$ lb/h.

Example 6: Determine the amount of subcooling obtained in a counterflow heat exchanger having a 0.43 in. ID suction tube 2 ft long. The flow rate of Refrigerant 12 is 200 lb/h, and the temperature difference between the inlet liquid and inlet gas is 60 °F.

Solution: Enter Figure 46 at 0.43 in. ID suction line and proceed vertically to the diagonal Refrigerant 12 line. Proceed horizontally from the intersection to the 200 lb/h flow rate. From this point, drop vertically to the 2 ft length curve and proceed horizontally to the 60 °F temperature difference line. The liquid subcooling is read on the horizontal scale at the bottom of Figure 46 at 12 °F.

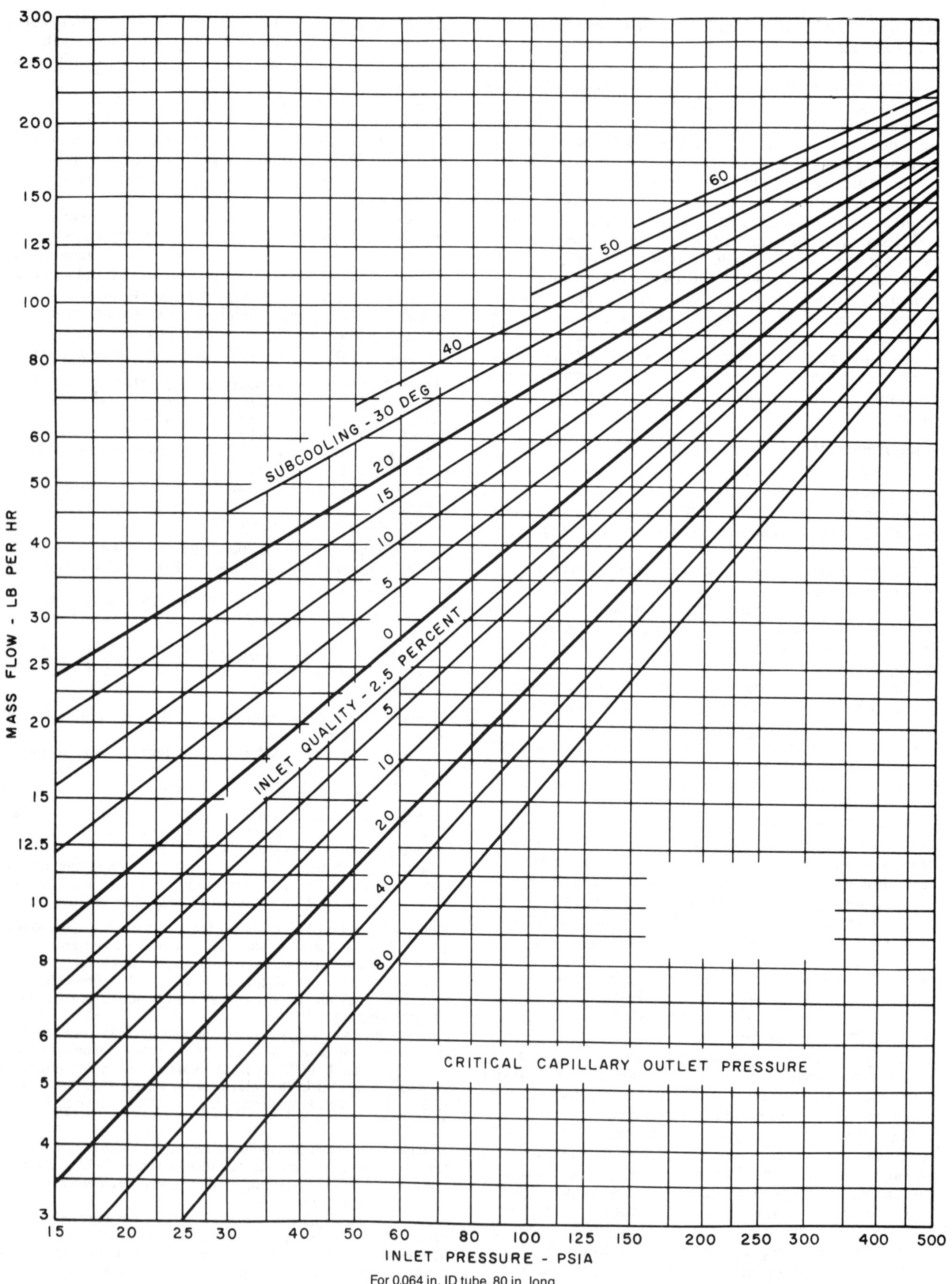

For 0.064 in. ID tube, 80 in. long

Fig. 42 Basic Rating Curves for Condenser-to-Evaporator Capillary (Refrigerants 12 and 22)

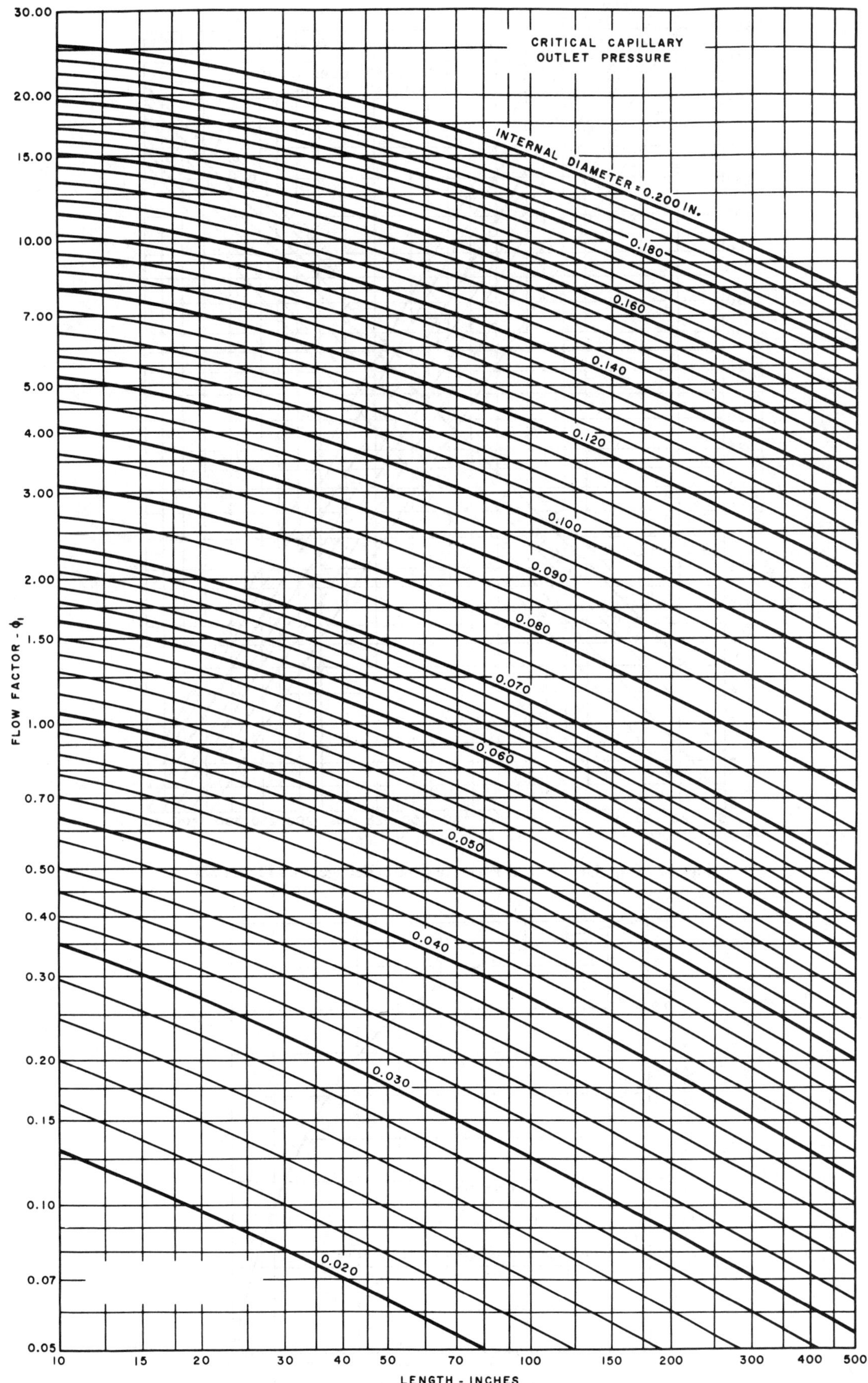

Fig. 43 Capillary Flow Factors (Refrigerants 12 and 22)

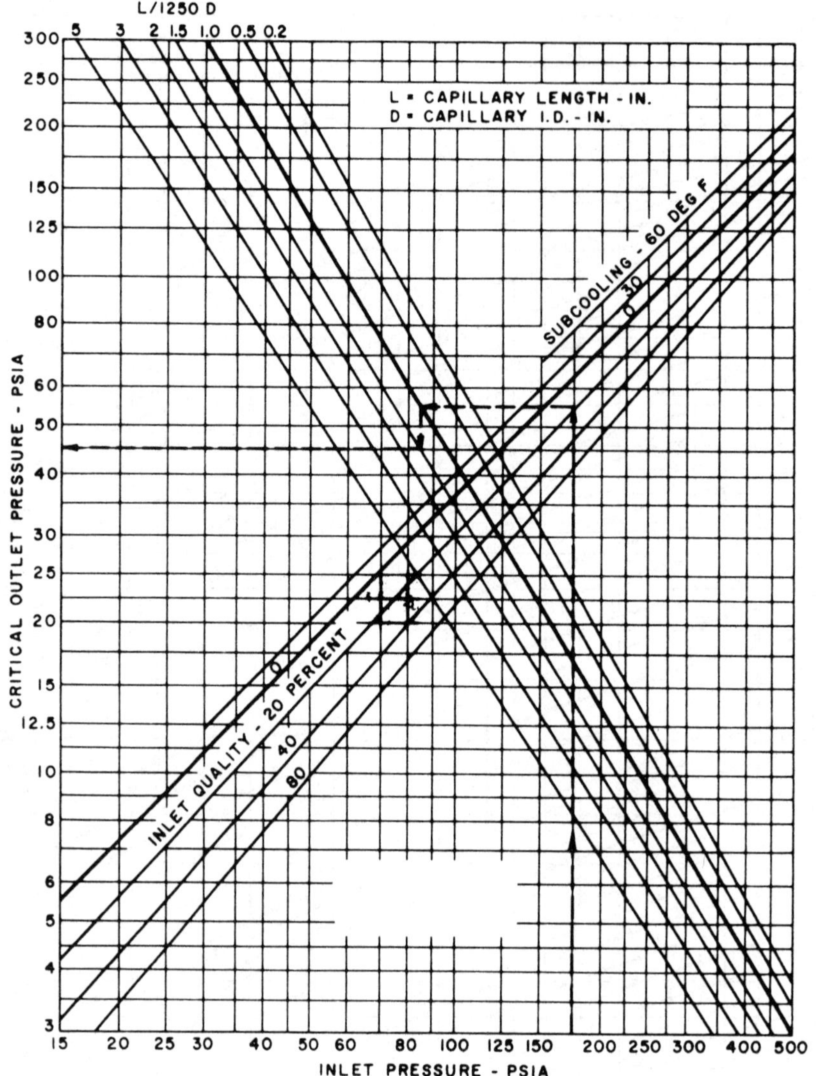

Fig. 44 Capillary Critical Pressure Chart (Refrigerants 12 and 22)

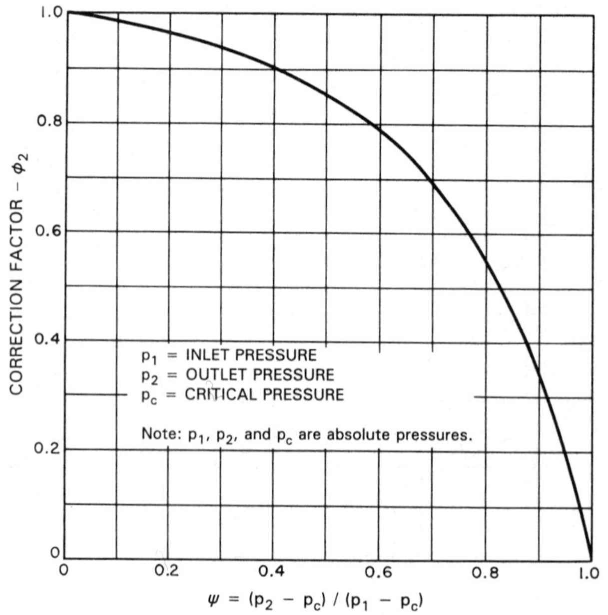

Fig. 45 Critical Correction Factor (Refrigerants 12 and 22)

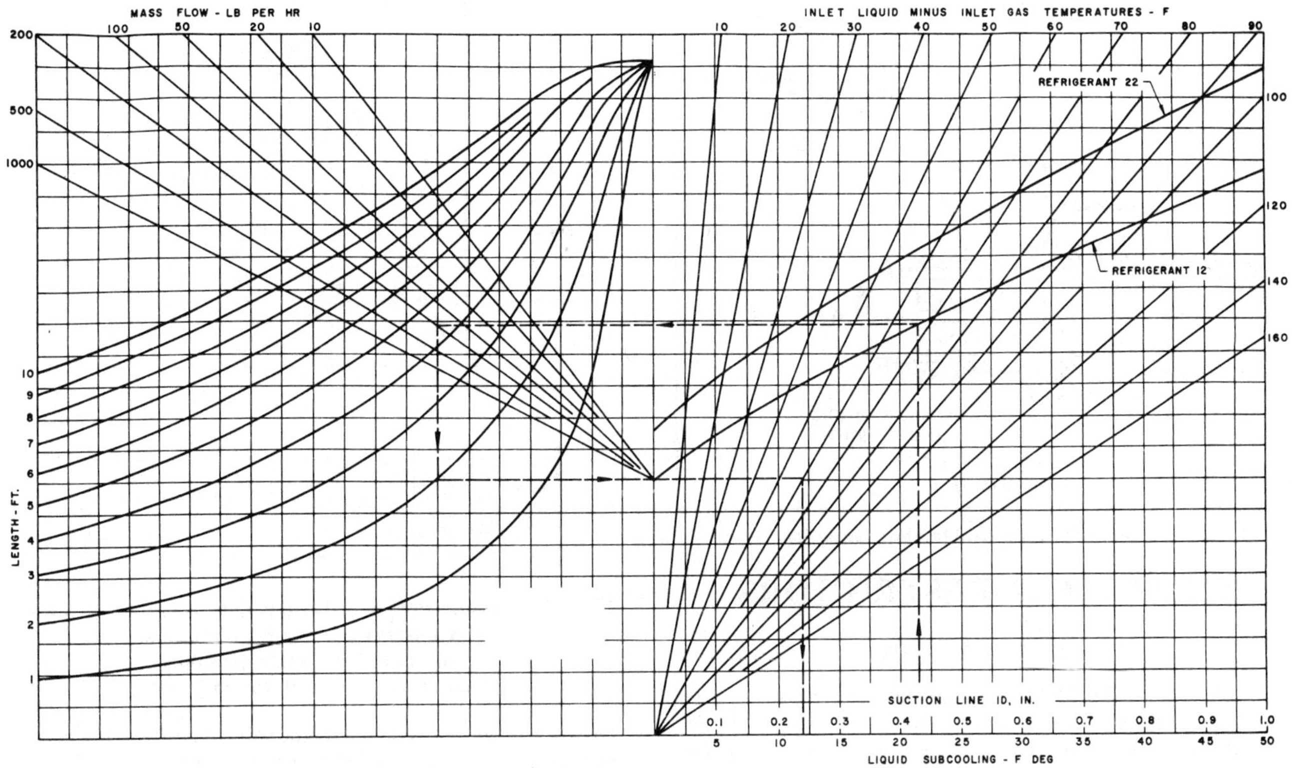

Fig. 46 Performance Chart for Suction Line and Capillary Heat Exchanger (Liquid-to-Gas Counterflow)

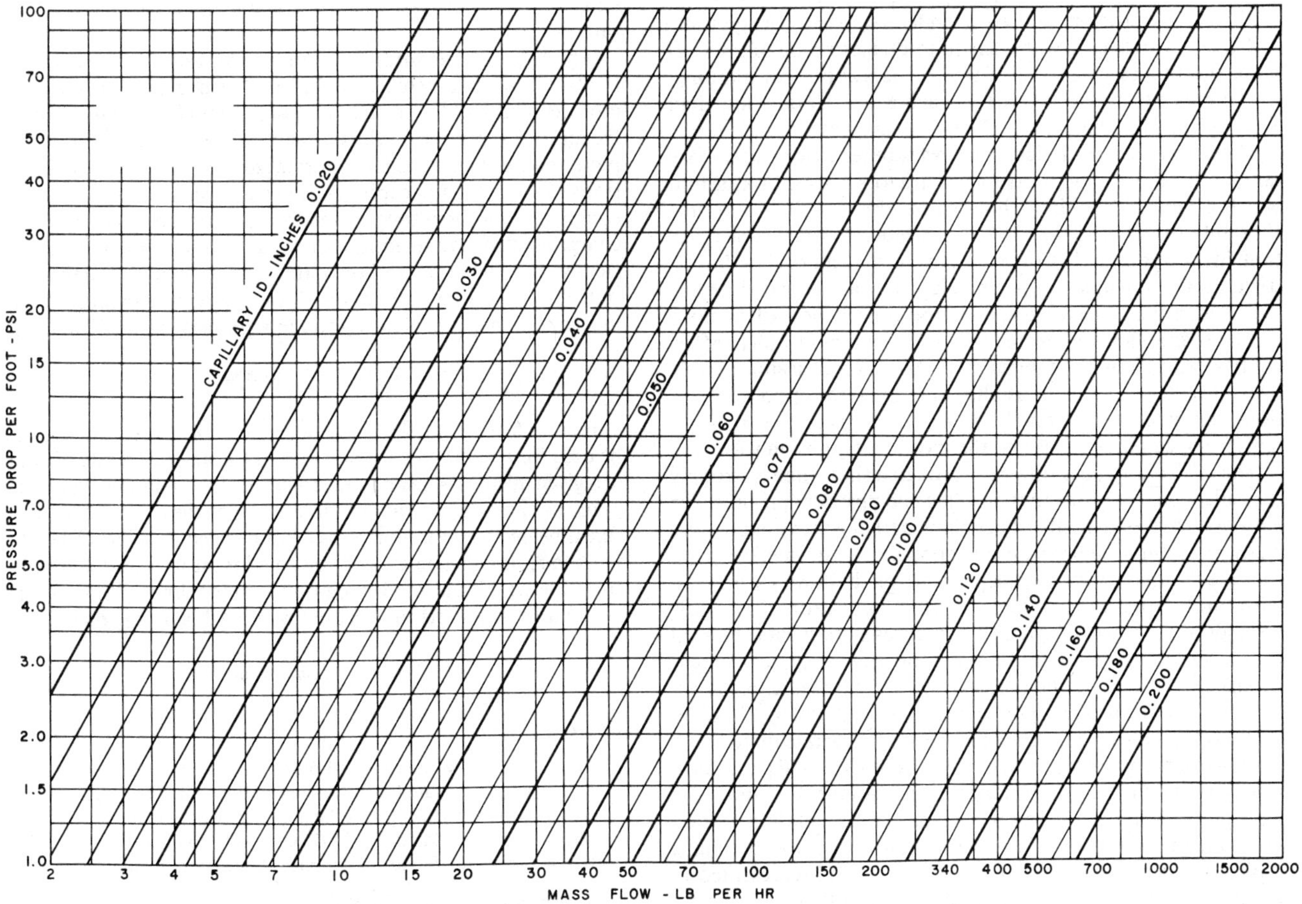

Fig. 47 Pressure-Drop Chart for Capillary Liquid (Refrigerants 12 and 22)

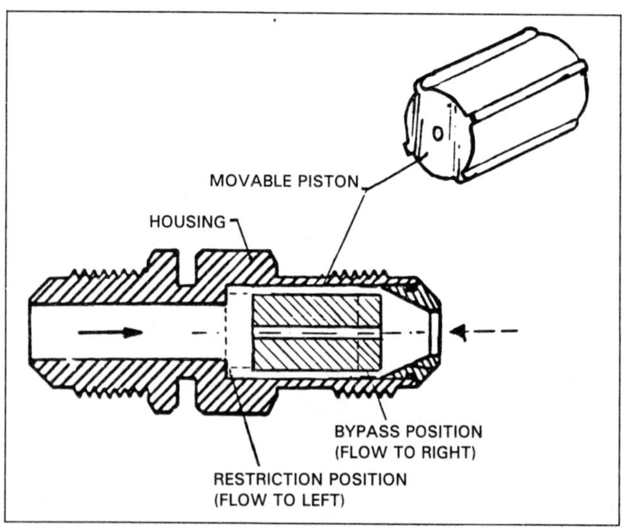

Fig. 48 Schematic of a Movable Short Tube Restrictor

SHORT TUBE RESTRICTORS

APPLICATION

In recent years, short tube restrictors have become widely used in residential air conditioners and heat pumps. They offer the advantages of low cost, high reliability, ease of inspection and replacement, and potential elimination of check valves in the design of a heat pump. Due to their pressure-equalizing characteristics, short tubes allow the use of a low-starting-torque compressor motor.

Short tube restrictors, as used in residential systems, are typically 3/8 to 1/2 in. in length, with a length-to-diameter ratio of $3 < L/D < 20$. Short tubes are also called plug orifices or orifices, although the latter is reserved for restrictors having an L/D ratio less than 3. Capillary tubes have an L/D ratio much greater than 20.

There are two basic designs for short tube restrictors: stationary and movable. Movable short tube restrictors consist of a piston, which is movable within its housing (see Figure 48). A movable short tube offers restriction for the refrigerant flowing in one direction; in the opposite flow direction, the refrigerant pushes the restrictor away from its seat, opening a larger area for the flow. While the stationary design is used in units operating in the cooling mode only, movable short tubes are used in heat pumps in which different flow restrictions are required for the cooling and heating modes. Two movable short tubes, installed in series in the opposite direction, eliminate the need for check valves, which are needed in systems using capillary tubes and thermostatic expansion valves.

The refrigerant mass flow rate through a short tube is strongly dependent on the upstream subcooling and upstream pressure. For a given inlet pressure, inlet subcooling, and downstream pressure below the saturation pressure corresponding to the inlet temperature, the flow has a very weak dependence on the downstream pressure, indicating a nearly choked-flow condition. This flow dependence is shown in Figure 49, and represents test data obtained on a 0.5 in. long, laboratory-made short tube at three different downstream pressures and the same upstream pressure. A significant drop in downstream pressure from approximately 170 to 70 psia produces a smaller increase in the mass flow rate than does a modest change of downstream pressure from 190 to 170 psia. There is only a slight pressure drop along the length of the short tube. The large pressure drop at the entrance is due to the rapid fluid acceleration and the inlet losses. The large pressure drop in the exit plane, typical for heat pump operating conditions and represented in Figure 49 by the bottom pressure line, indicates that choked-flow condition has nearly occurred.

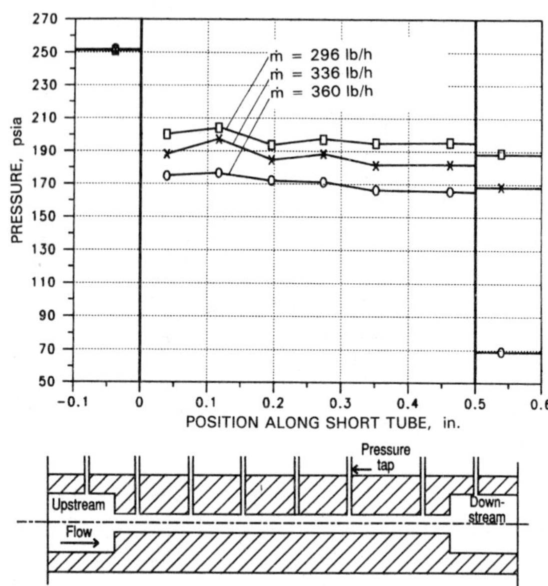

Fig. 49 Pressure Profile at Various Downstream Pressures with Constant Upstream Conditions: L = 0.5 in., D = 0.053 in., subcooling 25°F
(Adapted from Aaron and Domanski, 1990)

Among geometric parameters, the short tube diameter has the strongest influence on the mass flow rate. Chamfering the inlet of the short tube may increase the mass flow rate by as much as 25%, depending on the length-to-diameter ratio and chamfer depth. Chamfering the exit causes no appreciable change in the mass flow rate.

Although refrigerant flow inside a short tube is different than flow inside a capillary tube, the choked-flow phenomenon is a common denominator for both flows, making both types of tubes suitable for use as metering devices. Because of this phenomenon, basic performance characteristics important for air-conditioning applications are similar for capillary tubes and short tubes. Systems equipped with short tubes, as with capillary tubes, are critically charged systems. Inherently, a short tube does not operate as efficiently over a wide range of operating conditions as does a thermostatic expansion valve. However, due to counterbalancing factors, its performance is generally good in a properly charged system.

SELECTION

Figures 50 through 53, from Aaron and Domanski (1990), can be used for a preliminary evaluation of the mass flow rate of R-22 at a given inlet pressure and subcooling in air-conditioning and heat pump applications (*i.e.*, applications in which the downstream pressure is below the saturation pressure of refrigerant at the inlet). The method requires reading mass flow rate for the reference short tube from Figure 50 and modifying the reading with multipliers that account for the short tube geometry according to the equation

$$m = m_r \phi_1 \phi_2 \phi_3$$

where

m = mass flow rate for the short tube
m_r = mass flow rate for the reference short tube from Figure 50
ϕ_1 = correction factor for tube geometry from Figure 51
ϕ_2 = correction factor for L/D versus subcooling from Figure 52
ϕ_3 = correction factor for chamfered inlet from Figure 53

Aaron and Domanski (1990) also provide a correlation that includes downstream pressure as input and is more accurate than

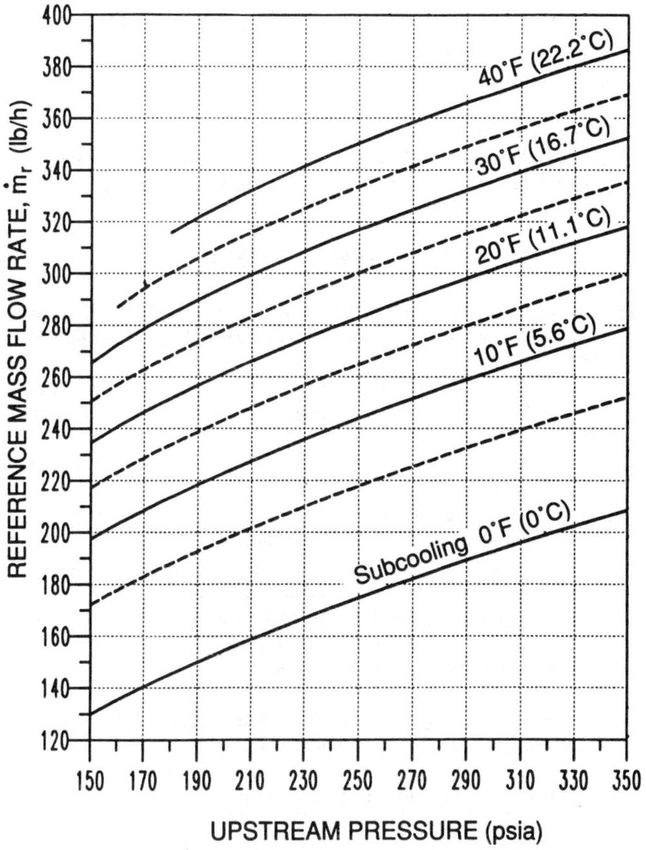

REFERENCE MASS FLOW RATE, $\dot{m}_r$ (lb/h)

40°F (22.2°C)
30°F (16.7°C)
20°F (11.1°C)
10°F (5.6°C)
Subcooling 0°F (0°C)

UPSTREAM PRESSURE (psia)

Fig. 50 Mass Flow Rate versus Condenser Pressure for Reference Short Tube. L **= 0.5 in.,** D **= 0.053 in., sharp-edged.**

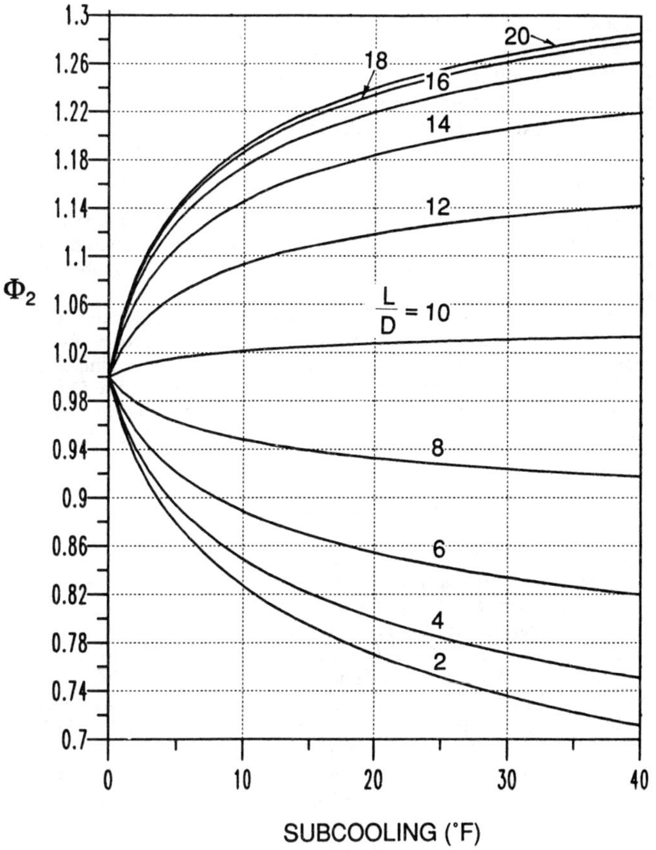

Φ_2

$\frac{L}{D}$ = 10

SUBCOOLING (°F)

Fig. 52 Correction Factor for L/D **versus Subcooling**

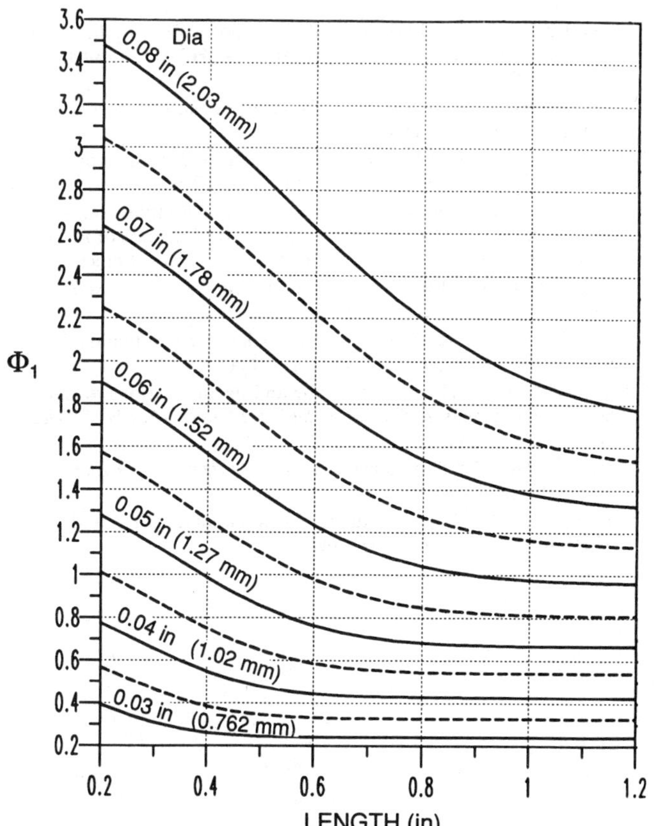

Fig. 51 Correction Factor for Short Tube Geometry

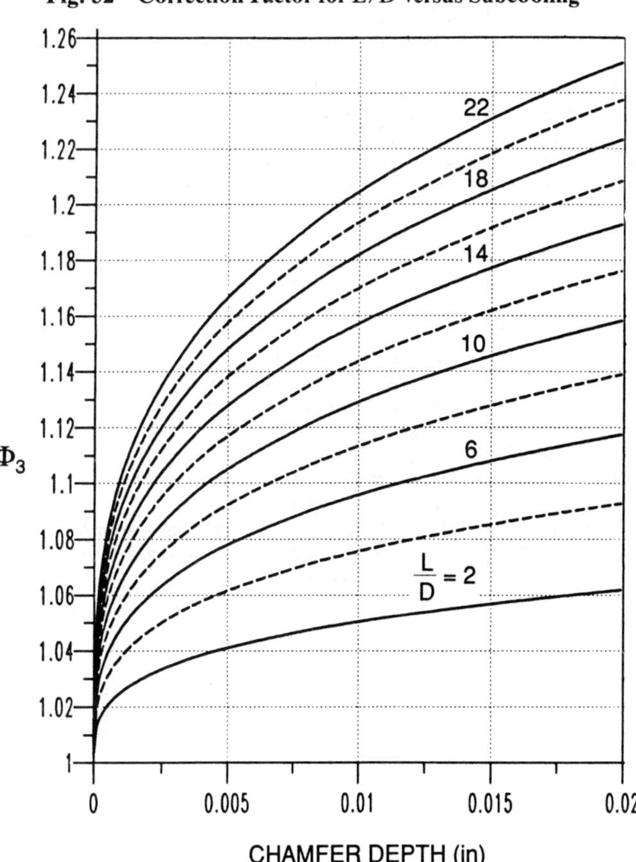

Fig. 53 Correction Factor for Inlet Chamfering

the graphical method included here. Neglecting downstream pressure on the graphs may introduce an error in the mass flow rate prediction as compared to the correlation results; however, this discrepancy should not exceed 3% due to the choked-flow condition at the tube exit. It should be noted that the lines for 0 and 40°F subcooling in Figure 50 were obtained by extrapolation beyond the test data and may carry a large error.

Example 7: Determine the mass flow rate of Refrigerant 22 through a short tube restrictor 0.375 in. long, of 0.06 in. inside diameter, and chamfered 0.01 in. deep at an angle of 45°. The inlet pressure is 240 psia, and subcooling is 10°F.

Solution: From Figure 50, for 240 psia and 10°F subcooling, the flow rate for the reference short tube, m_r, is 240 lb/h. The value for ϕ_1 from Figure 51 is 1.62. The value for ϕ_2 from Figure 52 for 10°F subcooling and $L/D = 0.375/0.06 = 6.25$ is 0.895. The value of ϕ_3 from Figure 53 for $L/D = 6.25$ and a chamfer depth of 0.01 in. is 1.098. Thus, the predicted mass flow rate through the restrictor is $240 \times 1.62 \times 0.895 \times 1.098 = 382$ lb/h.

REFERENCES

Aaron, D.A. and P.A. Domanski. 1990. Experimentation, analysis, and correlation of Refrigerant-22 flow through short tube restrictors. ASHRAE *Transactions* 96(1):729-42.

API. 1993. Sizing, selection, and installation of pressure-relieving devices in refineries. Part I—Sizing and selection. 6th Edition. *Standard* RP 520 PT I-93. American Petroleum Institute, Washington, D.C.

API. 1988. Sizing, selection, and installation of pressure-relieving devices in refineries. Part II—Installation. 3rd Edition. *Standard* RP 520 PT II-88. American Petroleum Institute, Washington, D.C.

ARI. 1987. Thermostatic refrigerant expansion valves. *Standard* 750-1987. Air-Conditioning and Refrigeration Institute, Arlington, VA.

ASHRAE. 1986. Method of Testing for capacity rating of thermostatic refrigerant expansion. *Standard* 17-1986.

ASHRAE. 1988. Method of testing flow capacity of refrigerant capillary tubes. *Standard* 28-1988.

ASHRAE. 1992. Safety code for mechanical refrigeration. *Standard* 15-1992.

ASTM. 1992. Standard specification for hand-drawn copper capillary tube for restrictor applications. *Standard* B360-92. American Society for Testing and Materials, Philadelphia, PA.

ASTM. 1992. Standard Specification for Seamless Copper Tube. *Standard* B75 Rev A-92. American Society for Testing and Materials, Philadelphia, PA.

Bolstad, M.M. and R.C. Jordan. 1948. Theory and use of the capillary tube expansion device. *Refrigerating Engineering* (December):519.

Hopkins, N.E. 1950. Rating the restrictor tube. *Refrigerating Engineering* (November):1087.

Huelle, Z.R. 1972. The MSS line—A new approach to the hunting problem. ASHRAE *Journal* (October):43.

Kuehl, S.J. and V.W. Goldschmidt. 1991. Modeling of steady flows of R-22 through capillary tubes. ASHRAE *Transactions* 97(1):139-48.

Mikol, E.P. 1963. Adiabatic single and two-phase flow in small bore tubes. ASHRAE *Journal* 5(11): 75-86.

Stoecker, W.F. 1966. Stability of an evaporator-expansion valve control loop. ASHRAE *Transactions* 72(II):IV.3.1-8.

Wedeking, G.L. and W.F. Stoecker. 1966. Transient response of the mixture-vapor transition point in horizontal evaporating flow. ASHRAE *Transactions* 72(II):IV.2.1-15.

Whitesel, H.A. 1957. Capillary two-phase flow. *Refrigerating Engineering* (April):42.

Whitesel, H.A. 1957. Capillary two-phase flow—Part II. *Refrigerating Engineering* (September):35.

Wijaya, H. 1991. An experimental evaluation of adiabatic capillary tube performance for HFC-134a and CFC-12, pp. 474-83. Proceedings of the International CFC and Halon Alternatives Conference. The Alliance of Responsible CFC Policy, Arlington, VA.

BIBLIOGRAPHY

Bolstad, M.M. and R.C. Jordan. 1949. Theory and use of the capillary tube expansion device. Part II—Nonadiabatic flow. *Refrigerating Engineering* (June):577.

Cooper, L., W.R. Brisken, and C.K. Chu. 1957. Simple selection method for capillaries derived from physical flow conditions. *Refrigerating Engineering* (July): 37.

Halter, E.J. 1975. Droplet and drop size and distribution estimation. ASHRAE *Transactions* 81(2):459-70.

Koizumi, H. and K. Yokoyama. 1980. Characteristics of refrigerant flow in a capillary tube. ASHRAE *Transactions* 86(2):19-27.

Kuehl, S.J. and V.W. Goldschmidt. 1990. Transient response of fixed area expansion devices. ASHRAE *Transactions* 96(1):743-50.

Kuijpers, L.J.M. and M.J.P. Janssen. 1983. Influence of thermal non-equilibrium on capillary tube mass flow, pp. 307-15. Proceedings of the XVIth International Congress of Refrigeration, Commission B2, Paris, France.

Lathrop, H.F. 1948. Application and characteristics of capillary tubes. *Refrigerating Engineering* (August):129.

Li, R.Y., S. Lin, and Z.H. Chen. 1990. Numerical modeling of thermodynamic nonequilibrium flow of refrigerant through capillary tubes. ASHRAE *Transactions* 96(1):542-49.

Marcy, G.P. 1949. Pressure drop with change of phase in a capillary tube. *Refrigerating Engineering* (January):53.

Pasqua, P.F. 1953. Metastable flow of freon-12. *Refrigerating Engineering* 61:1084-88.

Pate, M.B. and D.R. Tree. 1987. An analysis of choked flow conditions in a capillary tube-suction line heat exchanger. ASHRAE *Transactions* 93(1):368-80.

Prosek, J.R. 1953. A practical method of selecting capillary tubes. *Refrigerating Engineering* (June):644.

Schulz, U. 1985. State of the art: The capillary tube for, and in, vapor compression systems. ASHRAE *Transactions* 91(1):92-105.

Schulz, U. 1987. Critical two-phase flow in a capillary tube expansion device, pp. 290-306. Proceedings of the XVIIth Congress of Refrigeration, International Institute of Refrigeration, Paris, France.

Smith, F.G. 1956. Turbulent flow of air through capillary tubes. *Refrigerating Engineering* (October):48.

Staebler, L.A. 1948. Theory and use of a capillary tube for liquid refrigerant control. *Refrigerating Engineering* (January):54.

Staebler, L.A. 1950. The capillary tube and its applications to small refrigerating systems. *Refrigeration Service Engineers Society Service Manual*, Section 18.

Swart, R.H. 1946. Capillary tube exchangers. *Refrigerating Engineering* (September):221.

Sweedyk, J.M. 1981. Capillary tubes—Their standardization and use. ASHRAE *Transactions* 87(1):1069-76.

CHAPTER 45

FACTORY DEHYDRATING, CHARGING, AND TESTING

PROPER testing, dehydration, and charging help ensure the proper performance and extended life of the refrigeration system. This chapter covers the methods used to perform these functions. The chapter does not address criteria such as allowable moisture content, refrigerant quantity, and performance, which are specific to each machine.

DEHYDRATION

Factory dehydration may only be feasible for certain sizes of equipment. On large equipment, which will be open to the atmosphere when it is connected in the field, factory treatment is usually limited to purge and backfill, with an inert holding charge of nitrogen. In most instances, this equipment is only stored for short periods, so this method suffices until total system evacuation and charging can be done at the time of installation.

Excess moisture in refrigeration systems may lead to the freezeup of the capillary or expansion valve. These contaminants lead to valve breakage, motor burnout, and bearing and seal failure. Chapter 6 has more information on moisture and other contaminants in refrigerant systems.

Normally, these effects, with the exception of freezeup, are not detected by a standard factory test. Therefore, it is important to use a dehydration technique that results in a safe moisture level without adding foreign elements or solvents. In conjunction with this technique, an accurate method of moisture measurement must be established. Many factors such as the size of unit, its application, and the type of refrigerant determine the acceptable moisture content. Table 1 shows moisture limits specified by several manufacturers.

Sources of Moisture

The main sources of moisture include those that are (1) retained on the surfaces of metals; (2) produced as a product of the combustion of a gas flame; (3) contained in liquid fluxes, oil, and refrigerant; (4) absorbed in the hermetic motor insulating materials; (5) a part of the factory ambient at the point of unit assembly; and (6) provided by free water. The moisture contained in the refrigerant has no effect on the dehydration of the component or unit at the factory. However, because the refrigerant is added after the dehydration process, it must be considered in determining the overall moisture content of the completed unit. The moisture in the oil may or may not be removed in the dehydration process, depending on when the oil is added to the component or system.

Bulk oils, as received, have 20 to 30 ppm of moisture. Refrigerants have an accepted commercial tolerance on bulk shipments 10 to 15 ppm. Therefore, controls at the factory are needed to ensure the maintenance of these moisture levels in the oils and refrigerant.

The newer insulating materials in hermetic motors retain much less moisture when compared to the old rag paper and cotton-insulated motors. However, tests by several manufacturers have shown that the stator, with its insulation, is still the major source of moisture in compressors.

Dehydration by Heat, Vacuum, or Dry Air

Heat may be applied by placing components in an oven or by using infrared heaters. Oven temperatures of 180 to 340°F are usually maintained. The oven temperature should be selected carefully to prevent damage to the synthetics used and to avoid breakdown of any residual run-in oil that may be present in compressors. The air in the oven must be maintained at a low humidity level. When using heat alone, the time and escape area are critical; therefore, the size of the parts that can be economically dehydrated by this method is restricted.

The vacuum method reduces the boiling point of water below that of the ambient temperature. The moisture then changes to vapor, which is pumped out by the vacuum pump. Table 3 in Chapter 6 of the 1993 ASHRAE *Handbook—Fundamentals* shows the relationship of temperature and pressure.

Vacuum is classified according to the following:

Low Vacuum	29.92 to 1.0 in. Hg
Medium Vacuum	1.0 in. Hg to 1 μm Hg
High Vacuum	1 to 1×10^{-3} μm Hg
Very High Vacuum	1×10^{-3} to 1×10^{-6} μm Hg
Ultra High Vacuum	1×10^{-6} μm Hg and beyond

The degree of vacuum achieved and the time required to obtain the specified moisture level is a function of (1) the type and size of vacuum pump used, (2) the internal volume of the component or system, (3) the size and composition of water-holding materials in the system, (4) the initial amount of moisture in the volume, (5) piping and fitting sizes, (6) the shape of the gas passages, and (7) the maintained external temperatures. The pumping rate of the vacuum pump is critical only if the unit is not evacuated through a conductance-limiting orifice such as a purge valve.

If dry air or nitrogen is drawn or blown through the equipment for dehydration, it removes moisture by becoming totally or partially saturated. In systems with several passages or blind passages, flow may not be sufficient to dehydrate. The flow rate should obtain optimum moisture removal, and its success depends on the overall system design and temperature.

Combination Methods

Each of the following methods can be effective if controlled carefully, but a combination of two or even three of the methods is preferred because of the faster and more uniform dryness of the treated system.

Heat and Vacuum Method. The heat of this combination method drives the deeply sorbed moisture to the surfaces of materials and removes it from walls; the vacuum lowers the boiling point, making the pumping rate more effective. The heat source can be an oven, infrared lamps, or an ac (or dc) current circulating through the internal motor windings of semi-hermetic and her-

The preparation of this chapter is assigned to TC 8.1, Positive Displacement Compressors.

Table 1 Factory-Dehydrating and Moisture-Measuring Methods for Refrigeration Systems

Mfg.	Component or System	Method of Dehydration	Method of Moisture Determination	Moisture Limit
A	Evaporator coil, small	−70°F dew point dry-air sweep 240 s	P_2O_5 dry-air sweep 24 h	25 mg
	Evaporator coil, large			65 mg
	0.25 hp condensing unit	Purchased dry	P_2O_5 sample refrigerant	25 mg
	7.5 hp condensing unit			80 mg
B	Compressor	Evacuate 4 h in 250°F oven	Cold trap	180 mg
	Condenser and evaporator	Oven, hot dry-air purge	Cold trap	15 mg
	Air conditioner from above components	Vacuum at room ambient to 240 μm of mercury	P_2O_5	35 ppm
C	Evaporator and condenser	200°F air at −20°F dew point	None	None
	Compressor	240°F oven 10 h sweeping 0.2 cfm at −20°F dew point	Repeat dehydration process	1 g
D	Coils and tubing	250°F oven, sweep with −70°F dew point air	Sweep 1 ft^3/h dry nitrogen over dew point recorder	10 mg
	Compressor	300°F oven, sweep with −70°F dew point air		100 mg
E	Freezer	Purchase dry components −40°F dew point, sweep with −70°F dew point air in room ambinet	P_2O_5 test on refrigerant in system	10 ppm
	Drier	350°F oven sweeping with −40°F dew point air for 3 h		
F	Compressor	0.5-h dc winding heat to 356°F 0.25-h vacuum repeat	Cold trap 4 h at 302°F	0.2 g
	Condenser and evaporator	1 h in oven 302°F with dry-air sweep	Cold trap 4 h at 302°F	0.2 g
	Drier	6-h bake at 347°F		
	Refrigerator from above components	Vacuum, dc winding heat, oven 230°F	Cold trap 4 h at 302°F	0.2 g
G	Compressor	0.5-h vacuum, winding heat to 190°F	Cold trap 6 h at 200°F	1.2 g
	Air conditioner using above compressor	0.5-h vacuum, winding heat 3 h oven 250°F, 0.5-h vacuum	Aminco Weaver check of refrigerant in system	25 mg/kg R-22
H	Compressor	Before or after welding, pass through 280°F oven maintained at −60°F dew point, 5.5 h	Cold trap 6 h	0.2 g
I	Compressors:			
	3 to 5 ton	Dry air at 275 ±5°F for 3 h	Cold trap	0.25 g
	7.5 to 15 ton	0.5 h, evacuation at 275 ±15°F		0.75 g
	20 to 40 ton	Dry-nitrogen sweep at 275 ±15°F for 3.5 h evacuate to 200 μm of mercury	Cold trap	0.75 g
	50 to 100 ton	Evacuate at 270 ±5°F for 4 h to 1000 μm of mercury	Cold trap	1.00 g
J	Compressors:			
	1.5 to 5 ton	−100°F dew point dry air 340°F oven for 1.5 h	Moisture monitor checked daily; audited by cold trap	0.1 to 0.5 g
	2 to 40 ton semihermetics	−100°F dew point dry air 250°F oven for 3.5 h	Electrolytic moisture monitor audited by cold trap	0.1 to 1.1 g
	5 to 150 ton open compressors	1 mm of mercury vacuum in 175°F oven for 1.5 h	Audited by cold trap	0.4 to 2.7 g
K	Refrigerants 500 and 22; 3 and 4 G oils	As purchased	Karl Fischer or equivalent	10 ppm 30 ppm

metic compressors. Combinations of vacuum, heat, and then vacuum can also be used.

Heat and Dry-Air Method. The heat drives the moisture from the materials. The dry air picks up this moisture and removes it from the system or component. The dry air used should have a dew point between −40 and −100°F. The sources of heat are the same as those mentioned previously. Heat can be combined with a vacuum to accelerate the process. The heat and dry-air method is effective with open, hermetic, and semi-hermetic compressors. The heating temperature should be selected carefully to prevent damage to the compressor parts or breakdown of any residual oil that may be present.

The advantages and limitations of the various methods depend greatly on the system or component designs and the results expected. Goddard (1945) considers double evacuation with an air sweep between vacuum applications the most effective method, while Larsen and Elliot (1953) believe the dry-air method, if controlled carefully, is just as effective as the vacuum method and much less expensive, although it incorporates an evacuation of one and a half hours after the hot-air purge. Tests by one manufacturer show that a 280°F oven bake for one and a half hours, followed by a twenty-minute evacuation, effectively dehydrates compressors that use the newer insulating materials. For additional discussions on moisture measurements, see Chapter 6.

MOISTURE MEASUREMENT

Measuring the correct moisture level in a dehydrated system or part is important but not always easy. Table 1 shows measuring methods used by various manufacturers, and others are described in the literature. Few standards are available, however, and the moisture limits accepted by various manufacturers vary.

Cold-Trap Method. This common method of determining residual moisture generally monitors the production dehydration system to ensure that it produces equipment that meets the required moisture specifications. An equipment sample is selected after completion of the dehydration process, placed in an oven, and heated in the range of 150 to 275°F (depending on the limitations of the sample) for 4 to 6 h. During this time, a vacuum is drawn through a cold-trap bottle immersed in an acetone and dry-ice solution, or equivalent, which is generally held at about −100 °F. The vacuum levels are between 10 to 100 μm of mercury, with the lower levels preferred. Important factors in this method of moisture determination are leaktightness of the vacuum system and cleanliness and dryness of the cold-trap bottle.

Vacuum Leakback. The rate of vacuum leakback is another means of checking components or systems to see that no water vapor is present. This method is used primarily in conjunction with a unit or system evacuation that removes the noncondensables prior to final charging. This test allows a check of each unit, but too rapid a pressure buildup may signify a leak, as well as incomplete dehydration. The time factor may be critical in this method and must be examined carefully. Blair and Calhoun (1946) show that a small surface area in connection with a relatively large volume of water may only build up vapor pressure slowly. This method also does not give the actual condition of the charged system.

Dew Point. When dry air is used, a reasonably satisfactory check for dryness is a dew point reading of the air as it leaves the part being dried. If the airflow is relatively slow, there should be a marked difference between the dew point of air entering and air leaving the part, followed by a decrease in dew point of the leaving air until it eventually equals the dew point of the entering air. As is the case with all systems and methods described in this chapter, the values considered acceptable depend on the size, usage, and moisture limits desired. Different manufacturers use different limits.

Gravimetric Method. In this method, described by ASHRAE *Standard* 35-1992, a controlled amount of refrigerant is passed through a train of flasks containing phosphorous pentoxide (P_2O_5), and the weight increase of the chemical caused by moisture is measured. Although this method is satisfactory when the refrigerant is pure, any oil contamination produces inaccurate results. This method must be used only in a laboratory or under carefully controlled conditions. Also, it consumes considerable time and cannot be used when production quantities are high. Furthermore, the method is not effective in systems containing only small charges of refrigerant because it requires from 200 to 300 grams of refrigerant for accurate results. If it is used on systems where withdrawal of any amount of refrigerant changes the performance, recharging is required.

Aluminum Oxide Hygrometer. This sensor consists of an aluminum strip that is anodized by a special process to provide a porous oxide layer. A very thin coating of gold is evaporated over this structure. The aluminum base and the gold layer form two electrodes that essentially form an aluminum oxide capacitor.

In the sensor, the water vapor passes through the gold layer and comes to equilibrium on the pore walls in direct relation to the vapor pressure of water in the ambient surrounding the sensor. The number of water molecules absorbed in the oxide structure determines the sensor's electrical impedance, which modulates an electrical current output that is directly proportional to the water vapor pressure. This device is suitable for both gases and liquids over a temperature range of 158 to −166 °F and a pressure range of about 10 μm of mercury absolute to 5000 psig. The *Henry's Law* constant must be determined for use with each fluid. This constant is the saturation parts per million by mass of water for the fluid divided by the saturated vapor pressure of water at a constant temperature. For many fluids, this constant must be corrected for the operating temperature at the sensor.

Christensen Moisture Detector. For a quick check of uncharged components or units, the Christensen Moisture Detector is used on the production line. In this method, dry air is blown first through the dehydrated part and then over a measured amount of $CaSO_4$. The temperature of the $CaSO_4$ rises in proportion to the quantity of water absorbed by the $CaSO_4$, and desired limits can be set and monitored. One manufacturer reports that coils were checked in 10 s with this method. Moisture limits between 2 and 60 mg by this detector. Corrections must be made for variations in the desiccant grain size, quantity of air passed through the desiccant, and difference in instrument and component temperatures.

Karl Fischer Method. In systems containing refrigerant and oil, moisture may be determined by (1) measuring the dielectric strength or (2) the Karl Fischer Method (Reed 1954). In this method, a sample is condensed and cooled in a mixture of chloroform, methyl alcohol, and Karl Fischer reagent. The refrigerant is then allowed to evaporate as the solution warms to room temperature. When the refrigerant has evaporated, the remaining solution is titrated immediately to a *dead stop* electrometric end point, and the amount of moisture is determined. This method requires a sample of 50 to 60 g and takes about 1 h to perform. It is generally considered inaccurate below 15 ppm; however, as the method does not require that oil be boiled off the refrigerant, it can be used for checking complete systems. Reed points out that additives in the oil, if any, must be checked to ensure that they do not interfere with the reactions of the method. The Karl Fischer Method may also be used for determining moisture in oil alone (Reed 1954, ASTM 1985, Morton and Fuchs 1960).

Instrument Described by Taylor. Taylor (1956) describes an electrolytic water analyzer designed specifically to analyze moisture levels in a continuous process, as well as discrete samples. The device passes the refrigerant sample, in a vapor form, through a sensitive element consisting of a phosphoric acid film surrounding two platinum electrodes—with the acid film absorbing the moisture. When a dc voltage is applied across the electrodes, the water absorbed in the film is electrolyzed into hydrogen and oxygen, and the resulting dc current, in accordance with *Faraday's First Law of Electrolysis,* flows in proportion to the weight of the products electrolyzed. Liquids and vapor may be analyzed because the device has an internal vaporizer. This device handles all of the popular hydrocarbon refrigerants, but the samples must be free of oils and other contaminants. In tests on desiccants, this method is quick and accurate with R-22.

Sight-Glass Indicator. In fully charged halocarbon systems, a sight-glass indicator can be used in the refrigerant lines. This device consists of a colored chemical button visible through the sight glass, which indicates excessive moisture by a change in color. This method requires that the system be run for a reasonable time to allow moisture to circulate over the button. This method compares moisture only qualitatively to a fixed standard. It has been used on factory-dehydrated split systems to ensure that they are dry after field installation and charging and are in common use in conjunction with filter-driers to monitor moisture in operating systems.

Special Considerations. Although all the methods described in this section can effectively measure moisture, their use in the factory requires certain precautions. Operators must be trained in the use of the equipment or, if the analysis is made in the laboratory, the proper method of securing samples must be understood. Sample flasks must be dry and free of contaminants; lines must be clean, dry, and properly purged. The procedures for weighing the sample, the time during the cycle, and the location of the sample part should be clearly defined and followed carefully. Checks and calibrations of the equipment must be made on a regular basis if consistent readings are to be obtained.

CHARGING

The accuracy needed when charging refrigerant or oil into a unit depends on its size and application. Charging equipment must also be adapted to the particular conditions of the plant: manual or automatic charging devices are used. *Standard-type charging* is used where extreme accuracy is not necessary or the production rate is not high. Fully automatic charging boards check the vacuum in the units, evacuate the charging line, and meter the desired amount of oil and refrigerant into the system. These devices are accurate and suitable for high production.

Refrigerant and oil must be handled carefully during charging; the place and time to charge the oil and refrigerant have a great bearing on the life of a system. If a complete unit is charged prior to performance testing, the presence of liquid refrigerant in the crankcase can cause damage because of slugging. If the oil is added after the refrigerant is already in the crankcase, excessive foaming and oil vapor lock may cause bearing damage. Refrigerant lines must be dry and clean, and all charging lines must be kept free of moisture and noncondensable gases. Also, new containers must be connected with proper purging devices. Carelessness in observing these precautions may lead to excess moisture and noncondensables in the refrigeration system.

Oil drums must lie on their sides in storage to prevent accumulation of dirt and water on top, which might get into the oil as soon as a drum is opened. Drums of oil should be opened in the last moment before charging, even on applications requiring that the oil be degasified. Regular checks for moisture or contamination must be made at the charging station to ensure that the oil and refrigerant delivered to the unit meet the required specifications. Compressors charged with oil for storage or shipment must be charged with dry nitrogen. Compressors without oil may be charged with dry air.

TESTING FOR LEAKS

It is important to detect leaks prior to charging. Extended warranties and critical refrigerant charges add to the importance of proper leak detection.

Basically, the *allowable leakage rate* depends entirely on the system or component characteristics. Any leak on the low pressure side of a system operating below atmospheric pressure is dangerous, no matter how large the refrigerant charges. A system that has 4 to 6 oz of refrigerant and a 5-year warranty must have virtually no leak (1 oz in ten years or more), whereas in a system that has 10 to 20 lb of refrigerant, the loss of 1 oz of refrigerant in a year is not critical. Before any leak testing is done, the component or system should be strength tested at a pressure considerably higher than the leak test pressure. This test ensures safety when the unit is being tested under pressure in an exposed condition. Applicable design test pressures for high- and low-side components have been established by UL, ASME, ANSI, and ASHRAE. Units or components using composition gaskets as joint seals should have the final leak test after dehydration. Many have found that a final torquing of this type of joint after dehydration is beneficial in reducing leaks.

Leak-Detection Methods

Water Submersion Test. The most popular method of leak and strength testing used is the water submersion test. The unit or component is pressurized to the specified positive pressure and submerged in a well-lighted tank filled with clean water. A long time may be needed to obtain the leak test sensitivity desired.

Pressure Testing. The unit is sealed off under pressure or vacuum, and a decrease or rise in pressure versus time is noted. The disadvantages of this test are the time involved, the lack of sensitivity, and the inability to determine the exact location of the leak.

Soap Bubble Leak Test. A pressurized system's suspected leak areas are brushed with a soapy solution. Bubbles will form in the immediate leak zone.

Halide Leak Testing. The halide torch is used on systems charged with a halogenated refrigerant. The sensitivity of this test is approximately 1 oz per year. The gas, drawn across a faintly bluish flame, turns the flame greenish blue and varies in intensity with the size of the leak. Each joint or area can easily be probed, thus locating the leak. The sensitivity of halide torches is reduced by refrigerant contamination; therefore, testing should be done in well-ventilated areas or chambers. Large leaks, even in well-ventilated areas, may cause contamination levels so high that small leaks are not detected. **Caution:** Some of the newer halocarbon refrigerant blends may contain flammable components; the use of a halide torch may be dangerous, especially if the leak is large (see ASHRAE *Standard* 34-1992).

Electronic Leak Testing. The electronic leak detector consists of a probe that draws air over a platinum diode, whose positive ion emission is greatly increased in the presence of a halogen gas. This increased emission is translated into a visible or audible signal. Electronic leak testing shares with halide torches the disadvantages that every suspect area must be explored and that contamination makes the instrument less sensitive; however, it does have some advantages. The main advantage is increased sensitivity. With a well-maintained detector, it is possible to identify leakage at a rate of 10^{-3} mm^3/s (standard), which is roughly equivalent to the loss of 1 oz of refrigerant in 100 years. Another advantage is that the detector can be calibrated in many ways so that a leak can be measured quantitatively. The instrument also can be desensitized to the point that leaks below a predetermined rate are not found. Some models have an automatic compensating feature to accomplish this.

With increased sensitivity, the problem of contamination becomes more critical. To use this improved sensitivity, the unit under test is placed in a chamber slightly pressurized with outside air, which keeps contaminants out of the production area and carries contaminating gas from leaky units. An audible signal allows the probe operator to concentrate on probing, without having to watch a flame or dial. Equipment maintenance presents a problem because the sensitivity of the probe must be checked at short intervals. Any exposure to a large amount of refrigerant causes loss of probe sensitivity. A rough check (such as air underwater testing), prior to use of the electronic device, is frequently used to find large leaks.

Fluorescent Leak Detection. This system involves infusing a small quantity of a fluorescent additive into the oil/refrigerant charge of an operating system. Leakage is observed as a yellow-green glow under an ultraviolet (UV) lamp. This method is suitable for halocarbon systems.

Mass Spectrometer. The most sensitive leak detection method is probably the mass spectrometer. The unit to be tested is evacuated and then surrounded by a helium and air mixture. The vacuum is then sampled through a mass spectrometer, and any trace of helium indicates one or more leaks. The sensitivity of the mass spectrometer is extremely high, as leaks of 10^{-7} mm^3/s can be detected. Effective test levels in the manufacturing environment, though, are closer to 10^{-2} mm^3/s. The helium for testing is normally kept inside a chamber completely closed except at the bottom. The unit to be tested is simply raised into the lighter-than-air helium atmosphere.

This method, in addition to being extremely sensitive, has the advantage of measuring all leaks on all joints simultaneously. Therefore, a quick test is possible. However, the cost of equipment is high, helium is expensive, the instrument must be maintained carefully, and a method of locating individual leaks must be developed.

The concentration of helium needed depends on the maximum leak permissible, the configuration of the system under test, the time it can be left in the helium atmosphere, and the vacuum level in the system being tested; the lower the vacuum level, the higher the helium readings. The longer a unit is exposed to the helium atmosphere, the lower a concentration is necessary to maintain the required sensitivity. If, because of the shape of the test unit, a leak occurs at a point distant from the point of sampling, a good vacuum must be drawn, and sufficient time must be allowed for traces of helium to appear on the mass spectrometer.

In general, a helium concentration of more than 10% is costly. The inherently high diffusion rate causes it to disperse to the atmosphere, no matter how effectively the chamber is designed. As in the case of other methods described in this chapter, the best testing procedure in using the spectrometer is to locate calibrated leaks at extreme points of the test unit and to adjust exposure time and helium concentration in the most economical manner. One manufacturer of refrigeration equipment found leaks at a rate of 0.05 oz of refrigerant per year by using a 10% concentration of helium and exposing the tested system for 10 min.

Although the mass spectrometer method is extremely sensitive, the sensitivity that can be used may be limited by the characteristics of the tested system. Since only the *total leak rate* is found with this method, it is impossible to tell whether a leakage rate of, for example, 1 oz per year, is the result of one fairly large leak or a number of small leaks. If a sensitivity is desired that rejects units outside of the sensitivity range of tests listed earlier in this chapter, it is necessary to use a helium probe for the location of leaks. In this method, the component or system to be probed is evacuated fully to clear it of helium; then, while it is connected to the mass spectrometer, a fine jet of helium is sprayed over each joint or suspect area. If a large system is tested, a waiting period is necessary because some time is required for the helium to pass from the leak point to the mass spectrometer. Isolated areas, such as return bends on one end of a coil, may be hooded and sprayed with helium to determine whether the leak is in this region, thus saving time.

Special Considerations

Two general categories of leak detection may be selected: one group furnishes a leak check before refrigerant is introduced into the system, and the other group requires the use of refrigerant. The methods that do not use refrigerant have the advantage that heat applied to repair a joint has no harmful effects. Repairs requiring heat on units containing refrigerant must have the refrigerant removed and vented before attempting to weld, braze, or solder. This practice prevents breakdown of the refrigerant and pressure buildup, which would prevent the successful completion of a sound joint.

All leak-testing equipment must be calibrated frequently to ensure maximum sensitivity. The electronic leak detector or the mass spectrometer is usually calibrated with equipment furnished by the manufacturer. Mass spectrometers, for example, are usually checked by a flask containing helium. A glass orifice in the flask allows the helium to escape at a known rate, and the operator maintains the desired sensitivity by comparing the noted escape rate with a known standard.

The effectiveness of the detection system can best be checked with calibrated leaks made of glass, which can be bought commercially. These leaks can be built into a test unit and sent through the normal leak-detection cycles to evaluate the effectiveness of the detection method. Care must be taken that the test leakhole does not become closed. To check against closing, the leakage rate of the test leak must be determined before and after each system audit.

From a manufacturing standpoint, the use of any leak-detection method should be secondary to the prevention of leaks. Improper brazing and welding techniques, unclean parts, untested sealing compounds or improper fluxes and brazing materials, and poor workmanship result in leaks that occur in transit or later. Careful control and analysis of each joint or leak point make it possible to concentrate tests on areas where leaks are most likely to occur. If operators must scan hundreds of joints on each unit, the probability of finding all leaks is rather small, whereas concentration on a few suspect areas reduces field failures considerably.

PERFORMANCE TESTING

Since there are many types and designs of refrigeration systems, this section only presents specific information on reciprocating compressor testing and covers some important aspects of performance testing of other components and complete systems.

Compressor Testing

The two prime considerations in compressor testing are power and capacity. Secondary considerations are leakback rate, low-voltage starting, noise, and vibration.

Testing Without Refrigerant. A number of tests measure compressor power and capacity before the unit is exposed to refrigerant. In cases where excessive power is caused by friction of running gear, *low-voltage tests* spot defective units early in assembly. In these tests, voltage is increased from a *low* or *zero* value to the value that causes the compressor to *break away*, and this value is compared with an established standard. When valve-plates are accessible, performance can be tested by using an air pump for *leakback tests.* Air at fixed pressure is put through the unit to determine the flow rate at which proper valve opening occurs. The air pressure exerted against the closing side of the valve indicates its efficiency. This method is effective only when the valves are reasonably tight, and its use is difficult when a valve must be run in before it seats properly.

Extreme care should be taken when a compressor is used to pump air because the combination of oil, air, and high temperatures caused by compression can cause a diesel effect or result in an explosion.

In a common test using the compressor as an air pump, the discharge airflow is measured through a flow meter, orifice, or other flow-measuring device. When the volumetric efficiency of the compressor with refrigerant is known, the flow rate that can be expected with air at a given pressure may be calculated. Since this test adiabatically compresses the air, the head pressure must be low to prevent overheating of discharge lines and oil oxidation if the test lasts longer than a few minutes. (The temperature of adiabatic compression is 280°F at 35 psig, but 540°F at 125 psig.) When the compressor is run long enough to stabilize temperatures, both power and flow can be compared with established limits. Temperature readings at discharge and rpm measurements will aid in analyzing defective units. If a considerable amount of air is discharged or trapped, the air used in the test must be dry enough to prevent condensation from causing rust or corrosion on the discharge side.

Another method of determining compressor performance requires the compressor to pump from a free air inlet into a fixed volume. The time required to reach a given pressure is compared against a maximum standard acceptable value. The pressure used in this test is approximately 125 psig, so that a reasonable time spread can be obtained. The time needed for measuring the capacity of the compressor must be sufficient for accurate readings but short enough to prevent overheating. Power readings can be recorded at any time in the cycle. By shutting off the compressor, the leakback rate can be measured as an additional check. In addition to the pump-up and leakback tests noted above, a vacuum test should also be performed.

The vacuum test should be performed by closing off the suction side with the discharge open to the atmosphere. The normal vacuum obtained under these conditions is 27 in. of mercury. Abrupt closing of the suction side also allows the oil to serve as a check on the priming capabilities of the pump because of the suppression of the oil and attempt to deaerate. This test also checks for porosity and leaking gaskets. To establish reasonable pumpup times, leakback rates, and suctions, a large number of production units must be tested to allow for production variations.

In any capacity test using air, to prevent compressor contamination, only clean, dry air should be used.

The acceptance test limits described are best established by taking compressors of known capacity and power and observing their performance during the test. Precautions should be taken to prevent oil used repeatedly for the lubrication of many compressors from becoming acid or contaminated.

Testing With Refrigerant. The most common test method is the *run around* cycle. Successful variations and modifications of this cycle are described in ASHRAE *Standard* 23-93. This method requires a condenser large enough to handle the heat of compression, and an expansion device. The gas compressed by the compressor is flash-cooled until its enthalpy is the same as that at the suction conditions, and it is then expanded back to the suction state. This method eliminates the need for an evaporator and uses a condenser about one-fifth the size normally used with the compressor. On compressors of small capacity, a piece of tubing that connects discharge to suction and has a hand expansion valve can be used effectively. The measure of performance is usually the relationship of suction and discharge pressures to power. When a water-cooled condenser is used, the head pressure is usually known, and the water temperature rise and flow are used as capacity indicators.

As a further refinement, flow-measuring devices can be installed in the refrigerant lines. This system is charge-sensitive if predetermined head and suction pressures and temperatures are to be obtained. This is satisfactory when all units have the same capacity and one test point is acceptable, since the charge desired can be determined with little experimentation. When a variety of sizes is to be tested, however, or more than one test point is desired, a liquid sump or receiver after the condenser can be used for full-liquid expansion.

Refrigerant must be free of contamination, inert gases, and moisture; the tubing and all other components should be clean and sealed when they are not in use. In the case of hermetic and semi-hermetic systems, a motor burnout on the test stand makes it imperative not to use the stand until it has been thoroughly flushed and is absolutely acid-free. In all tests, oil migration must be observed carefully, and the oil must be returned to the crankcase.

The length of a compressor performance test depends on various factors. Stabilization of conditions is a prerequisite if accuracy is to be obtained. If oil-pump or oil-charging problems are inherent, the compressor should be run long enough to ensure that all defective units are detected. Most manufacturers use test periods from approximately 30 min to 1 h. A check of the test system is usually made by running a sample of the production units on a calorimeter under controlled conditions.

Testing of Complete Systems

In a factory, testing of any system may be done in a controlled ambient temperature or in an existing shop ambient temperature. In both cases, tests must be run carefully, and proper corrections must be made, if necessary. Since measuring air temperature and flow is difficult, production-line tests are usually more reliable when secondary conditions are used as capacity indicators. Measurements of water temperature and flow, power, cycle time, refrigerant pressures, and refrigerant temperatures are reliable capacity indicators.

When testing self-contained air conditioners, for example, a fixed load may be applied to the evaporator, using any air source and either a controlled ambient or shop ambient temperature. As long as the load is relatively constant, its absolute value is not important. For water-cooled units, in which water flow can be absolutely controlled, capacity is best measured by the heat rejected from the condenser. Suction and discharge pressures can be measured for the analysis.

Suction and discharge pressures can be used as a direct measure of capacity in units that have air-cooled condensers. As long as the load is relatively constant, the absolute value is not important. Air distribution, velocity, or temperature over the coil of the test unit must be kept constant during the test, and the performance of the test unit must then be correlated with the performance of a standard unit. Power measurements supplement the suction and discharge pressure readings. As a rule, suction and discharge temperatures are useful in determining unsatisfactory operation of the unit and are particularly important when the evaporator or condenser loads are not reasonably stable. In such cases, simultaneous readings of suction temperature and pressure throughout an entire cycle permit the experienced observer to judge the performance of the unit.

The length of the test run depends on the test used, but, in general, stabilization should be achieved in approximately 30 min. For units in which air flows over an evaporator, the conditions should load the test unit with a dry coil. This reduces the time necessary to balance the test system. The test requires preheating of the air temperature to a level that maintains the test unit evaporator saturated temperature above the dew point of the supplied air. For units with air-cooled condensers, the air leaving the condenser can be recirculated back to the evaporator for use as the load. This second arrangement is simple and inexpensive, but it may cause wide variation in performance if the system is not controlled carefully.

The primary function of the factory performance test is to ensure that a unit is constructed and assembled properly. Therefore, all equipment must be compared to a standard unit. This standard unit should be typical of the unit used to pass the ARI and AHAM certification programs for compressors and other units. ARI and AHAM provide rating standards with applicable maximum and minimum tolerances. Several ASHRAE standards specify applicable rating tests.

Normal causes of malfunction in a complete refrigeration system are overcharging, undercharging, presence of noncondensable gases in the system, blocked capillaries or tubes, and excessive power. To determine the validity and sensitivity of any test procedure, it is best to use a unit with known characteristics and then establish limits for deviations from the test standard. If the established limits for charging are ± 1 oz of refrigerant, for example, the test unit is charged first with the correct amount of refrigerant and then with 1 oz more and 1 oz less. If this procedure does not establish clearly defined limits, it cannot be considered satisfactory and new values must be established. This same procedure should be followed regarding all variables that influence performance and result in deviations from established limits. All equipment must be maintained carefully and calibrated if tests are to have any significance. Gages must be checked at regular intervals and protected from vibration. Capillary test lines must be kept clean and free of contamination. Power leads must be kept in good repair to eliminate high resistance connection, and electrical metres must be calibrated and protected to obtain consistent data.

In plants where component testing and manufacturing control have been so well managed that the average unit performs satisfactorily, units are tested only long enough to find major flaws. Sample lot testing is sufficient to ensure product reliability. This approach is sound and economical because complete testing taxes power and plant capacity and is not necessary.

When the evaporator load is static, as in the case of refrigerators or freezers, time, temperature, and power measurements are used to measure performance. The time elapsed between start and first compressor shutoff or the average on-and-off period during a predetermined number of cycles in a controlled or known ambient temperature determines performance. Also, concurrent suction and discharge temperatures in connection with power readings are used to establish conformity to standards. On units where the necessary connections are available, pressure readings may be taken. Such readings are usually possible only on units where refrigerant loss is not critical because some loss is caused by gages.

Units with complicated control circuits usually undergo an operational test to ensure that controls function within design specifications and operate in the proper sequence.

Testing of Components

Component testing must be based on a thorough understanding of the use and purpose of the component. Pressure switches may be calibrated and adjusted with air in a bench test and need not be checked again if there is no danger of blocked passages or pulldown tripout during the operation of the switch. However, if the switch is brazed into the final assembly, precautions are needed to prevent blocking of the switch capillary.

Capillaries for refrigeration systems are checked by air testing. When the capillary limits are known, it is relatively easy to establish a flow-rate and pressure-drop test for eliminating crimped or improperly sized tubing. When several capillaries are used in a distributor, a series of water manometers check for unbalanced flow and can find damaged or incorrectly sized tubes.

In plants with good manufacturing control, only sample testing of evaporators and condensers is necessary. Close control of coils during manufacture leads to the detection of improper expansion, poor bonding, split fins, or uneven spacing. Proper inspection eliminates the need for costly test equipment. In testing the sample, either a complete evaporator or condenser or a section of the heat-transfer surface is tested. Since liquid-to-liquid furnishes the most easily and accurately measurable heat-transfer medium, a tube or coil can be tested by passing water through it while it is immersed in a bath of water. The temperature of the bath is kept constant, and the capacity is calculated by measuring the coil flow rate and the temperature differential between water entering and leaving the coil.

REFERENCES

ASHRAE. 1992a. Number designation and safety classification of refrigerants. ANSI/ASHRAE *Standard* 34-1992.

ASHRAE. 1992b. Method of testing dessicants for refrigerant drying. ASHRAE *Standard* 35-1992.

ASHRAE. 1993. Methods of testing for rating positive displacement refrigerant compressors and condensing units. ANSI/ASHRAE *Standard* 23-93.

ASTM. 1985. Guide to test methods and specifications for electrical insulating oils of petroleum origin. ASTM *Standard* D117-85. American Society for Testing and Materials, Philadelphia.

Blair, H.A. and J. Calhoun. 1946. Evacuation and dehydration of field installations. *Refrigerating Engineering* (August):125.

Goddard, M.B. 1945. Moisture in Freon refrigerating systems. *Refrigerating Engineering* (September):215.

Larsen, L.W. and J. Elliot. 1953. Factory methods for dehydrating refrigeration compressors. *Refrigerating Engineering* (December):1325.

Morton, J.D. and L.K. Fuchs. 1960. Determination of moisture in fluorocarbons. ASHRAE *Transactions* 66:434.

Reed, F.T. 1954. Moisture determination in refrigerant oil solutions by the Karl Fischer method. *Refrigerating Engineering* (July):65.

Taylor, E.S. 1956. New instrument for moisture analysis of "Freon" fluorinated hydrocarbons. *Refrigerating Engineering* (July):41.

CHAPTER 46

RETAIL FOOD STORE REFRIGERATION EQUIPMENT

THE modern retail store is a high-volume food sales outlet with maximum inventory turnover. Almost half of retail food sales is made up of perishable or semi-perishable foods requiring refrigeration. These foods include fresh meats, dairy products, perishable produce, frozen foods, ice cream and frozen desserts, and various special items such as bakery and deli products and prepared meals. These foods are displayed in highly specialized and flexible storage, handling, and display apparatus.

These food products need to be kept under safe temperatures during storage and processing, as well as during display. The back room of a food store is both a processing plant and a warehouse distribution point. It includes specialized refrigerated rooms, which must be coordinated during construction planning because of the interaction between the store's environment and the refrigeration equipment (Figure 1). Chapter 2 of the 1991 ASHRAE *Handbook—Applications* and Chapter 31 of this volume cover, in more detail, the importance of coordination.

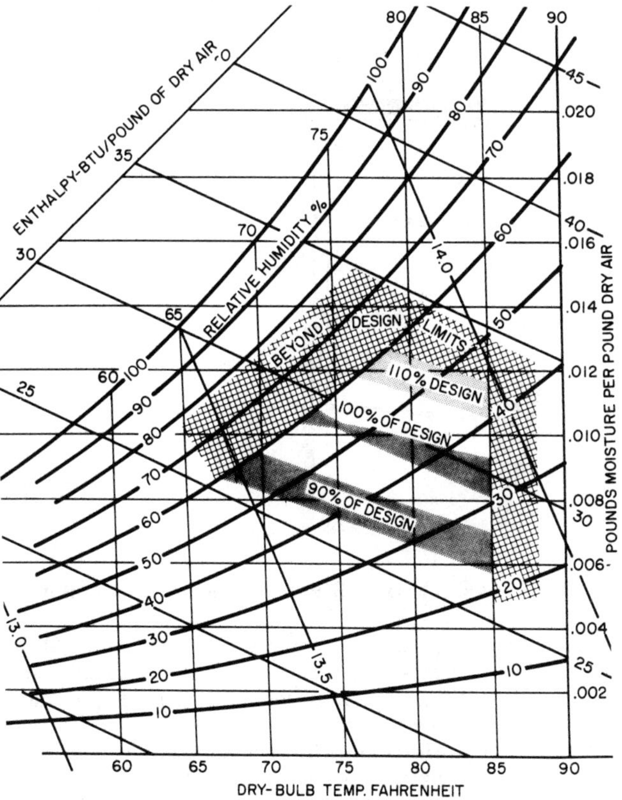

Fig. 1 Open Display Refrigerator Load Factors Superimposed on Psychrometric Chart

The preparation of this chapter is assigned to TC 10.7, Commercial Food and Beverage Cooling, Display and Storage.

DISPLAY REFRIGERATORS

Each category of perishable food has its own physical characteristics, handling logistics, and display requirements that dictate specialized display shapes and flexibility required for merchandising. Also, the same food product requires different display treatment in different locations, depending on such things as neighborhood preferences, neighborhood income level, size of store, volume of sales, and the local availability of food items by type.

Table 1 summarizes a study of ambient conditions in retail food stores. Individual store ambient readings showed that only 5% of all readings (including those when the air conditioning was not working or not turned on) exceeded 75 °F dry bulb or 0.0113 lb of moisture per pound of dry air. Based on these data, the industry has chosen 75 °F dry bulb and 64 °F wet bulb (55% rh) as *summer design conditions.*

Manufacturers sometimes publish ratings for open refrigerators at lower ambient conditions than this standard because the milder conditions may significantly reduce the heat load on the refrigerators. In addition, lower ambient conditions may permit fewer defrosts and reductions in anti-sweat heaters, which make it possible to save substantial amounts of energy on a complete store basis.

Attention should be given to the opposite condition, whereby store environments higher than the industry standard will dramatically raise the refrigeration requirements and consequently the energy demand. Higher store ambient temperatures may also significantly affect the performance of the refrigeration equipment—particularly open, self-service models.

Meat Display

Most meat is sold in packaged form that is often cut and packaged on the store premises. Control of temperature, time, and sanitation from the truck to the checkout counter is important. Surface temperature on the meat in excess of 40 °F shortens the salable life of meat products significantly and increases the rate of discoloration.

Sanitation is also important. Even if all else is kept equal, good sanitation can as much as double the salable life of meat in a display refrigerator. In this chapter, sanitation includes the control of the time during which meat is exposed to temperatures above 40 °F.

Table 1 Average Store Conditions in the United States

Season	Dry Bulb °F	Wet Bulb °F	Pounds Moisture per Pound Dry Air	rh, %
Winter	69	54	0.0054	36
Spring	70	58	0.0079	50
Summer	71	61	0.0091	56
Fall	70	58	0.0079	50

Store Conditions Survey conducted by Commercial Refrigerator Manufacturers' Association. Approximately 2000 store readings in all parts of the country, in all types of stores, during all months of the year reflected the above ambient store conditions.

The design of open meat display refrigerators, either well-type single deck or vertical multi-deck, is limited by the freezing point of meat. Ideally, the refrigerators are set to operate as cold as possible without freezing the meat. The temperatures are maintained with a minimum of fluctuations (with the exception of defrost) to ensure the coldest possible stable internal and meat surface temperatures.

Along with molds and natural chemical changes, bacteria also discolors meat. With good control of sanitation and refrigeration, experiments in stores have produced shelf life of one week and more. Bacterial population is greatest on the exposed surface of displayed meat because it becomes warmer than the interior. Although the cold airflow refrigerates each package, the surface temperature cumulatively increases (along with bacterial growth) by the following:

1. Infrared rays from lights
2. Infrared rays from the ceiling surface
3. High stacking of meat products
4. Voids in display
5. Store drafts that disturb refrigerator air

Where these factors are handled improperly, the surface temperature of the meat will often be above 50°F. It takes great care in every building and equipment detail, as well as refrigerator loading, to maintain meat surface temperature below 40°F. However, the required diligence is rewarded by excellent shelf life, improved product integrity, higher sales volume, and less scrap or spoilage.

The surface temperature will rise significantly during defrost. However, tests have compared matched samples of meat—one goes through normal defrost and the other is removed from the refrigerator during its defrosting cycles. While defrosting characteristics of refrigerators vary, such tests have shown that the effects of properly handled defrosts on shelf life are negligible. Tests for a given installation can easily be run to prove the effects of defrosting on shelf life for that specific set of conditions.

Self-Service Meat Refrigerators. The meat department planner can select from a wide variety of available meat display possibilities. Self-service meat products are displayed in packaged form.

1. Single-deck refrigerators with or without rear or front access storage doors (Figure 2). *Deck* refers to a shelf, pan, or rack that supports the product.
2. Multi-deck refrigerators with or without rear access (Figures 3 and 4).
3. Either of the above with or without glass fronts.
4. Processed meat versions of the above, often designed for somewhat higher temperatures but including special merchandising shelves or accessories.

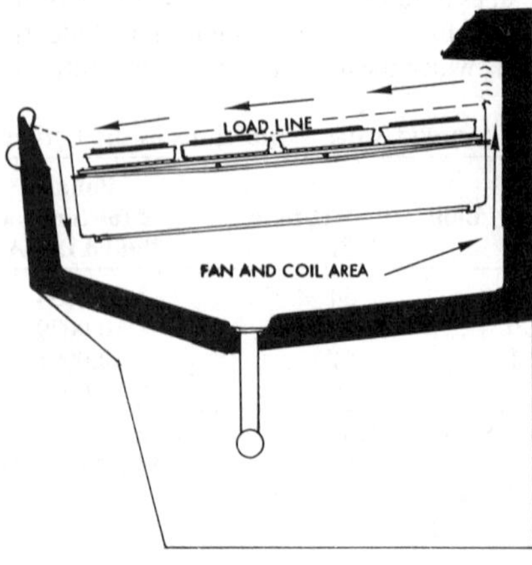

Fig. 2 Single-Deck Meat Display Refrigerator

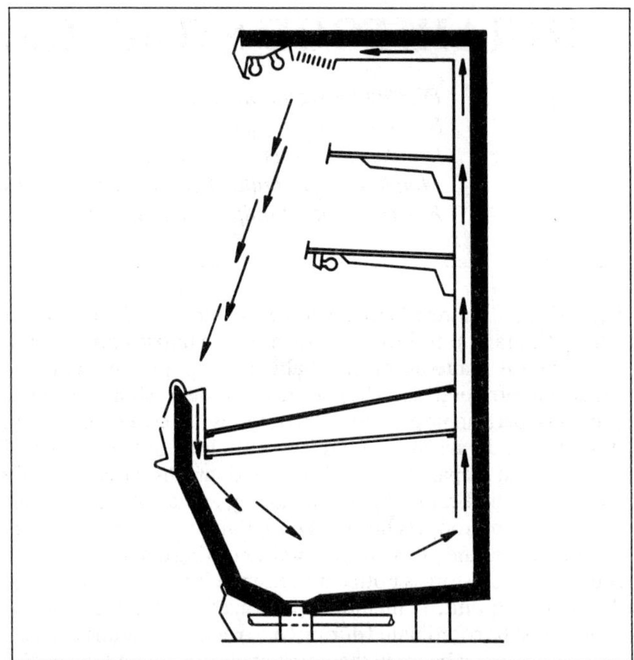

Fig. 3 Multiple-Deck Meat Refrigerator

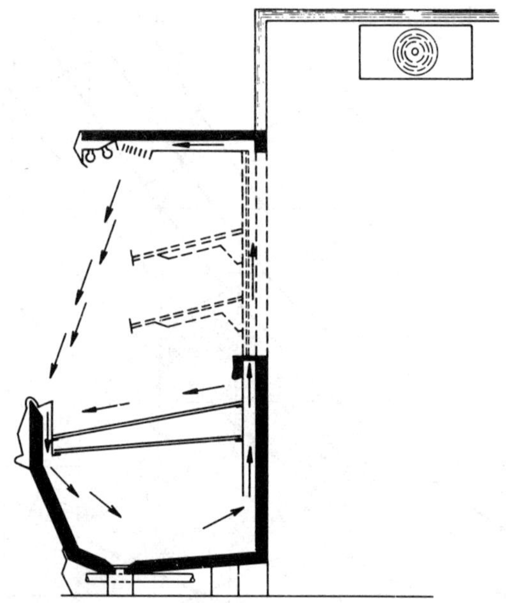

Fig. 4 Multiple-Deck Rear-Service Meat Refrigerator

All of the above refrigerators are available with a variety of lighting, superstructures, shelving, and other accessories tailored to the special merchandising needs. Storage compartments are rarely used in self-service meat refrigerators.

Closed Service Meat or Deli Refrigerators. Service meat products are generally displayed in bulk, unwrapped. Generally, closed refrigerators are definable in one of the following categories:

1. Fresh red meat, with or without storage compartment (Figure 5).
2. Delicatessen and smoked or processed meats, with or without storage.
3. Fresh fish and poultry, usually without storage but designed to display the fresh fish and/or poultry on a bed of cracked ice.

These refrigerators are offered in a variety of configurations. They are available with gravity or forced convection coils, and their

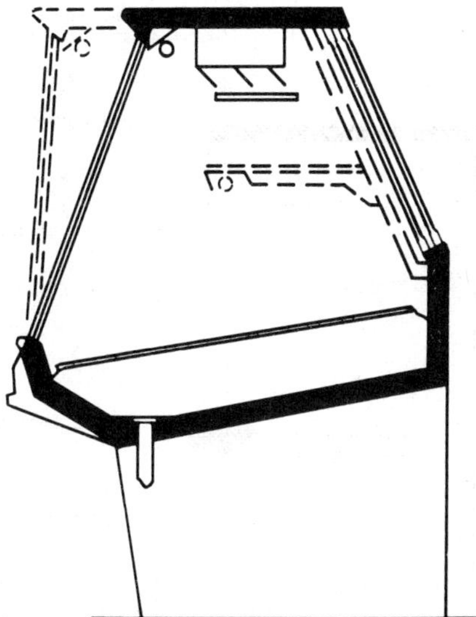

Fig. 5 Closed-Service Display Refrigerator
(Gravity coil model)

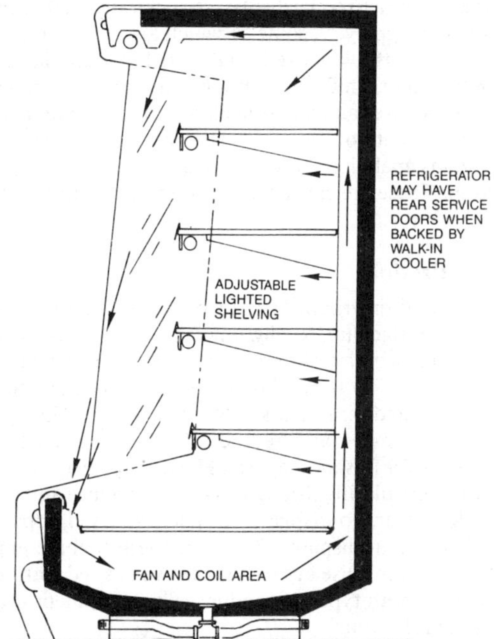

Fig. 6 Multi-Deck Dairy Refrigerator

fronts may be nearly vertical or angled as much as 20° in flat or curved glass panels, either fixed or hinged. Gravity coils are usually preferred for the more critical products; although forced-air coil models using various forms of humidification systems are also common. This equipment can be used for red meats, delicatessen items, or fish products in a variety of design combinations.

These service refrigerators typically have sliding rear-access doors, which are sometimes removed during busy periods. This practice is not recommended by manufacturers, however, because the internal product display zone temperature and humidity levels are affected. There is a conflict between the need for highlighting the product in a closed refrigerator and the need for maintaining a high moisture level in the atmosphere within the refrigerator. Even the lighting within the store is involved. Bright store lighting requires even brighter refrigerator lighting to highlight the product in the refrigerator. Lights in the refrigerator produce heat. Reheating refrigerated air increases the total refrigeration load on the fixture.

The best solution for this problem is to subdue ambient lighting around the refrigerators so that minimum lighting within the refrigerator is not dominated. Meat and deli refrigerators usually maintain a long shelf life when the air temperatures in the refrigerator are between 35 and 37 °F. At the same time, the velocity of the refrigerated air over the product should be minimized. A high relative humidity and a stable temperature in the product zone should be maintained for maximum shelf life. An evaporator pressure control is the preferred method for refrigeration control. A minimum temperature differential thermostat is used as a backup for times when the load is small. This control method keeps product dehydration to a minimum.

Dairy Display

Dairy products include such items with great sales volume as fresh milk, butter, eggs, and margarine. They also include the myriad of small items of fresh (and sometimes processed) cheeses, special above-freezing pastries, and other perishables (Figure 6). The available equipment includes the following:

1. Full-height, full-adjustable shelved displayers without doors in back for use against a wall or with doors in back for rear service or for service from the rear through a dairy cooler. The effect of rear service openings on the surrounding refrigeration must be considered (Figure 7).
2. Closed-door displays built in the wall of a walk-in cooler with adjustable shelving behind doors. Shelves are located and stocked in the cooler.

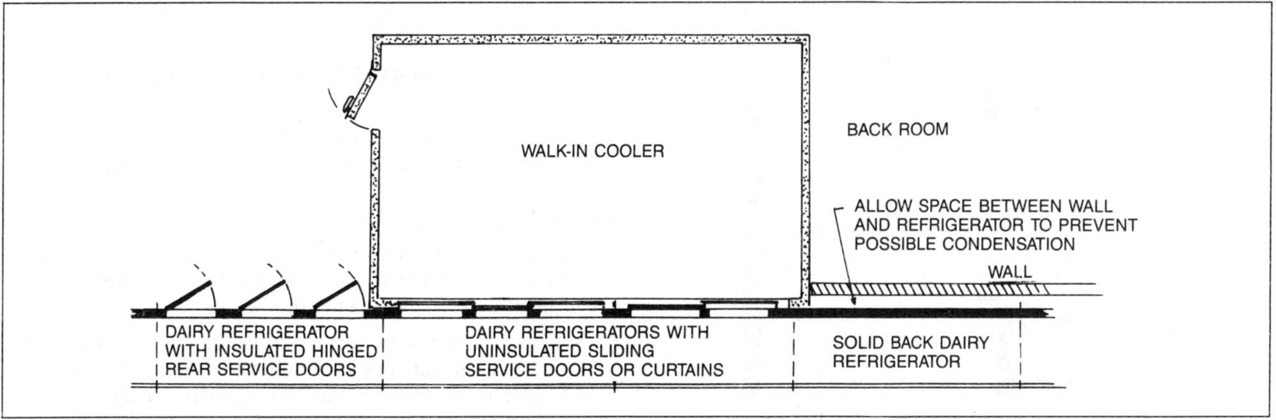

Fig. 7 Possible Dairy Display Arrangements

3. A variety of other special displayers, including single-deck and island-type displayers, some of which are self-contained and reasonably portable for seasonal, perishable specialities.
4. The refrigerator, similar to Item 1, but able to receive either conventional shelves and a base shelf and front or pre-made displays on pallets or carts. This version comes with either front-load capability only or rear-load capability only (Figures 8 and 9). These are called front roll-in or rear roll-in display cases.

Fruit and Produce

Wrapped and unwrapped produce is often intermixed in the same display refrigerator. Ideally, unwrapped produce should have low velocity refrigerated air forced up through the loose product. Water is usually also sprayed on the leafy vegetables, either by manually operated spray hoses or by automatic misting systems, to retain their crispness and freshness. Produce is often displayed on a bed of ice for freshness appeal. However, packaging prevents this air from circulating through wrapped produce and requires higher velocity air. To display both packaged and unpackaged produce in the same display refrigerator, the available equipment is usually a compromise between these two desired features and is suitable for both types of product. The equipment available includes the following:

1. Wide or narrow single-deck displayers with or without mirrored superstructures.
2. Two- or three-deck displayers, similar to the one in Figure 10, usually for multiple case lineups near the above display refrigerators.
3. Because of the nature of produce merchandising, there is a variety of non-refrigerated displayers of the same family design, which are usually designed for connection in continuous lineup with the refrigerators.
4. The refrigerator, like Item 2, but able to receive either conventional shelves and a base shelf and front or pre-made displays on pallets and carts. This version comes with either front-load capability only or rear-load capability only (Figures 8 and 9).

This produce equipment is generally available with a variety of merchandising and other accessories, including bag compartments, sprayers for wetting the produce, night covers, scale racks, sliding mirrors, and other display shelving and apparatus.

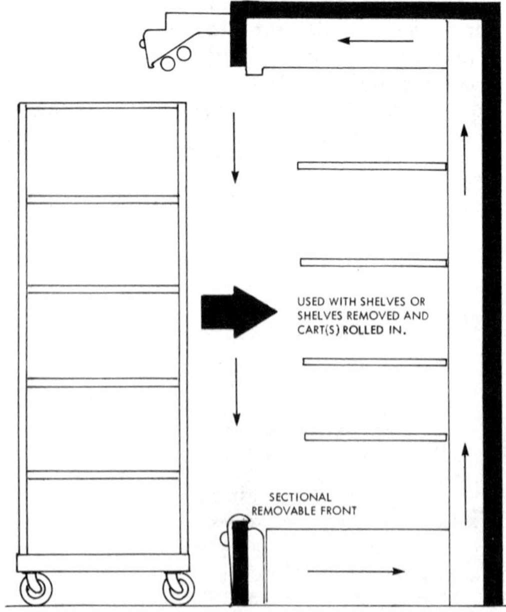

Fig. 8 Vertical Front-Load Dairy (or Produce) Refrigerator with Roll-In Capability

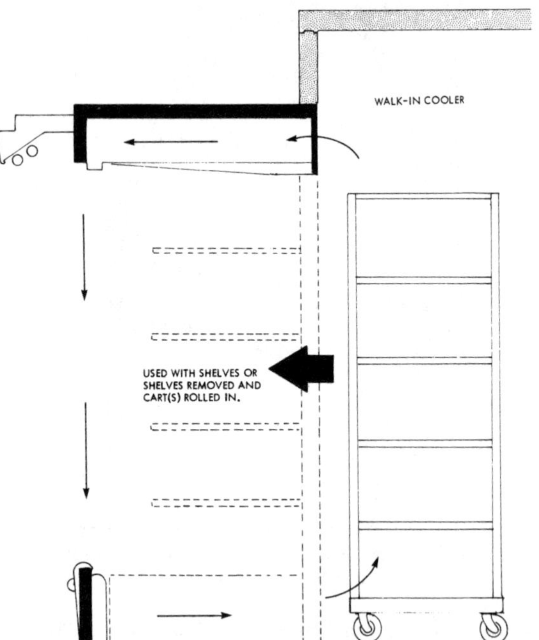

Fig. 9 Vertical Rear-Load Dairy (or Produce) Refrigerator with Roll-In Capability

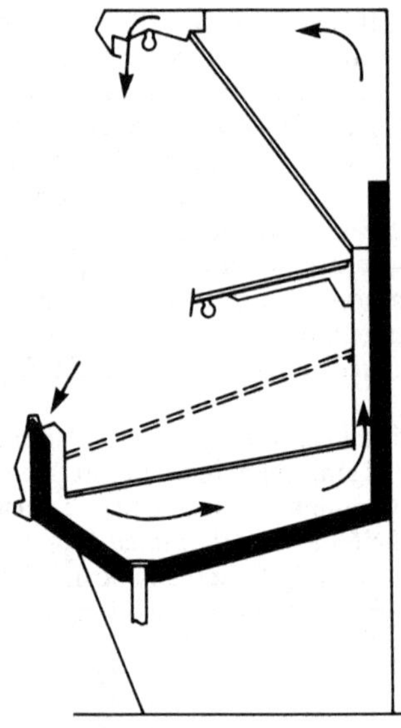

Fig. 10 Multiple-Deck Fruit and Produce Refrigerator

Frozen Food and Ice Cream

To display frozen foods most effectively (depending on varied need), many types of display refrigerators have been designed and are available. These include the following:

1. Single-deck well-type refrigerators for one-side shopping. Many types of merchandising superstructures for related non-refrigerated foods are available. Configurations of these refrigerators are designed for matching lineup with fresh meat refrigerators, and there are similar refrigerators for matching

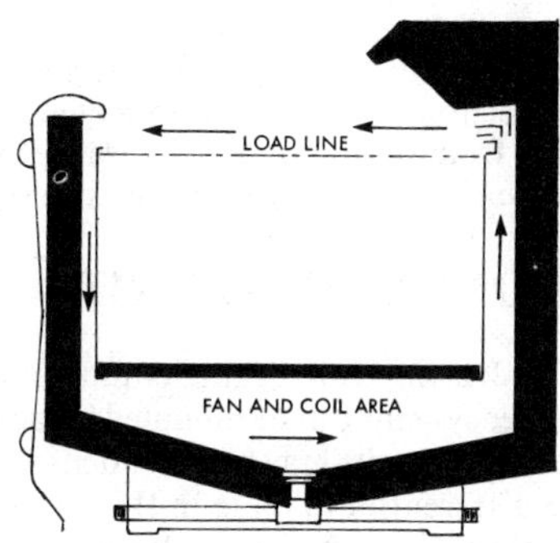

Fig. 11 Single-Deck Well-Type Frozen Food Refrigerator

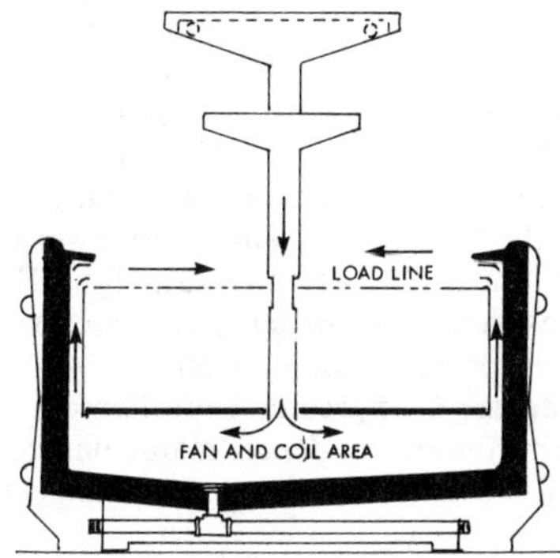

Fig. 12 Single-Deck Island-Type Frozen Food Refrigerator

lineup of ice cream refrigerators with their frozen food counterparts. These refrigerators are offered with or without glass fronts (Figure 11).

2. Single-deck island for shop-around. These are available in widths ranging from the above single-deck refrigerators to refrigerators of double width, with various sizes in between. Some across-the-end increments are available to complete the shop-around configuration. They are available with or without various merchandising superstructures for selling related nonrefrigerated food items (Figure 12).
3. Freezer shelving in two to six levels with many refrigeration system configurations (Figure 13). Multi-deck, self-service, frozen food and ice cream fixtures are generally more complex in design and construction than single-deck models. Because they have wide, vertical display compartments, they are more affected by ambient conditions in the store. Generally, open multi-deck models have two or three air curtains to maintain product temperature and shelf-life requirements.
4. Glass door, front reach-in refrigerators, usually of a continuous lineup design. This style allows for maximum inventory volume and variety in minimum floor space. These advantages must be balanced against the barrier produced by the doors and the greater labor requirement in restocking. The front-to-back dimension of these cabinets is usually about 24 in; greater attention must be given to the back product to provide the desired rotation. While many believe that these refrigerators consume less energy in their operation than open multi-deck, low-temperature refrigerators, specific comparisons should be made by models to determine original and operating costs (Figure 14).
5. Spot merchandising refrigerators, usually self-contained and sometimes arranged for quick change from the nonfreezing to freezing temperature to allow for promotional items of either type (*i.e.*, fresh asparagus or ice cream).

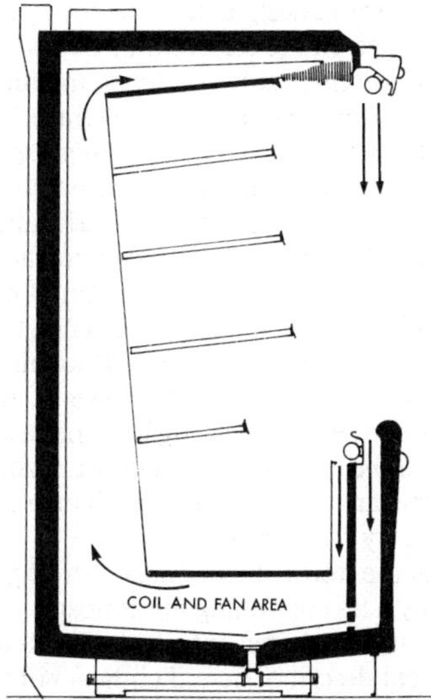

Fig. 13 Multiple-Deck Frozen Food Refrigerator

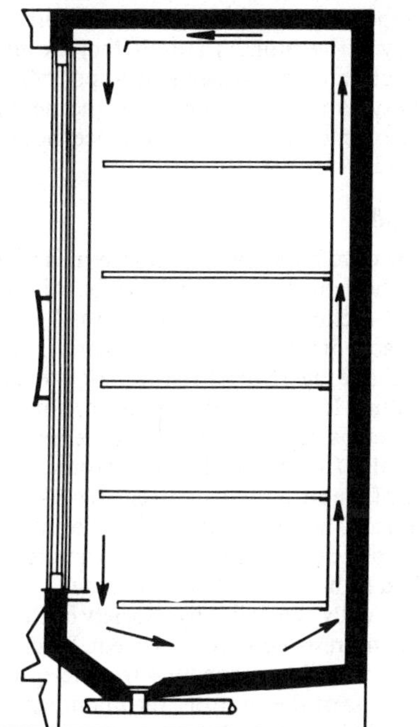

Fig. 14 Glass Door, Frozen Food Reach-In

6. Versions of most of the above items are available for ice cream and usually have modified defrost heaters and other changes necessary for the approximately 10°F colder temperature required. As display temperature lowers to below 0°F (product temperature), the problem of frost and ice accumulation in flues and product zone increases drastically. Proper product rotation and restocking frequently minimize frost accumulation.

Refrigerator Construction

Commercial refrigerators for market installations are usually of the endless construction-type, which allows a continuous display as refrigerators are joined. Separate end sections are provided for the first and last units in a continuous display. Methods of joining self-service refrigerators vary, but they are usually bolted or cam locked together.

All refrigerators are constructed with surface zones of transition between the refrigerated area and the room atmosphere.

Thermal breaks of various designs separate the zones to minimize the amount of surface on the refrigerator below the dew point. Surfaces in front of discharge air nozzles, sometimes on the nose of the shelving and sometimes at the refrigerator's front rails or center flue, may be below the dew point. In glass door reach-in freezers or medium-temperature refrigerators, the frame jambs and the glass are below the dew point. In these locations, resistance heat is used effectively to raise the exterior surface temperature above the dew point to prevent accumulation of condensation.

With the current emphasis on energy efficiency, the designer has developed means other than resistance heat to raise the temperature of surfaces above the dew point. However, when no other technique is known, resistance heating becomes a necessity. That heat may sometimes be controlled by cycling and/or proportional controllers to vary heat with store ambient changes, thereby reducing the annual energy consumption.

The designers of the stores can do a great deal to promote energy efficiency. Not only does controlling the atmosphere within a store reduce the refrigeration requirements, it also holds down the heating of the equipment described here. This heat not only consumes energy; it also places added demand on the refrigeration load.

Evaporators and air-distribution systems for display refrigerators are highly specialized and are usually fitted precisely into the particular display refrigerator. As a result, they are inherent in the fixture and are not standard independent evaporators. The design of the air-circuit system, the evaporator, and the means of defrosting are the result of extensive testing to produce the particular display results desired.

Defrost Methods

The defrosting of refrigerators is primarily an application matter, even though the defrosting mechanism is built into the refrigerator. Refer to the section "Methods of Defrost" in Chapter 31.

Defrosting is accomplished by electric heaters, cycling off the compressor, selective ingestion of store air, or latent heat-reverse cycle gas defrosting. In the defrost operation, not only must the evaporator be defrosted, particularly in the low-temperature equipment, but also frost in the flues and around the fan blades in various areas of the air-distribution system must be melted and completely drained. The design details are the result of laboratory testing; calculations can give only an approximate indication of the heat required or how it must be distributed.

Defrosting is usually controlled by a variety of clocks. They are often part of a compressor controller system. Electronic systems often have communication capabilities from telephones outside the store. Regardless of the controls, the manufacturer's recommendations should be followed.

Cleaning and Sanitizing Equipment

Since the evaporator coil is the most difficult part to clean, the operator should consider the judicious use of high-pressure, low-liquid-volume sanitizing equipment. This type of equipment enables personnel to spray cleaning and sanitizing solutions into the duct, grill, coil, and waste outlet areas with a minimum of disassembly and a maximum of effectiveness. However, this equipment must be used carefully because the high pressure stream can easily displace sealing and caulking materials. High-pressure streams should not be directed toward electrical devices as well. Hot liquid can also break the glass on models with glass fronts and closed service fixtures.

STORAGE COOLERS

Each category of food product displayed is usually backed up by storage in the back room. This storage usually consists of refrigerated rooms with sectional walls and ceilings equipped with the necessary storage racks for a particular food product.

Walk-in coolers, which serve a dual purpose of storage and display, are equipped with either sliding or hinged glass doors on the front. These door sections are often prefabricated and set into an opening within the front of the cooler. Allowance must be made in computing the refrigeration load for the extra service load.

CONDENSING UNITS

Air-Cooled Condensing Unit. A single-unit compressor with air-cooled condenser systems can be mounted in racks up to three high to save space. These units may have condensers sized so that the temperature differential (TD) is in the 10 to 25°F range. Optionally available next-larger-size condensers are often used to achieve lower TDs and higher Energy Efficiency Ratios (EER) in some supermarkets, convenience stores, and other applications. Single compressors with heated crankcases and heated insulated receivers and other suitable outdoor controls are assembled into weatherproof racks for outdoor installations. Sizes range from 0.5 to 30 horsepower.

Water-Cooled Condensing Unit. Water-cooled units range in size from 0.5 to 30 horsepower and are best for hot, dry climates. The city water-cooled condensing unit is usually no longer economical due to the high cost of water and sewer fees. Cooling towers, in which one tower cools the water for all compressors, have been used instead. Closed-water systems are also used in areas that can use an evaporative water cooler.

Remote Condenser-Compressor Unit. Remote units operate efficiently with minimum condenser maintenance. Evaporative condensers are also used in some areas. Sizes range from 0.5 to 30 horsepower.

Recommended sizing for remote air-cooled condensers is 10°F TD for low-temperature application and 15°F TD for medium-temperature application. Remote water-cooled condensers are often used in areas with abundant water.

Single-Compressor Control. Single-compressor units make up half of the supermarket compressor equipment presently used. A solid-state pressure control for single units can help control excess capacity when the ambient temperature drops. The control senses the pressure and adjusts the cutout point to eliminate short cycling, which ruins many compressors in low load conditions. This control also saves energy by maintaining a higher suction pressure than would otherwise be possible and by reducing overall running time.

Parallel Compressors. Multiple compressors may be operated in parallel with a large receiver and multi-station manifolds for liquid, suction, and discharge gas. These systems usually have a remote condenser and frequently recover heat from a secondary condenser coil in the air handler. A separate compressor for ice cream refrigerators on a low-temperature system or for meat refrigerators on a medium-temperature system can be physically mounted on the rack and piped so that only the heat removed from the lower temperature refrigerators is supplied at the less efficient rate. Parallel operation is also applied in two-stage or compound R-22 systems for low-temperature applications. Two-stage compression includes interstage gas cooling before the second stage of compression to avoid excessive discharge temperatures. Parallel compressors may be equal or unequal in size and may have unloaders or variable speed drives.

Remote Air-Cooled Condenser and Heat-Recovery Coil. Remote air-cooled condensers are popular for use with parallel compressors. The condenser coil TD is in the range of 10°F for low temperature and 15°F for medium temperature applications. For energy conservation, generous sizing of the condenser with a lower TD is recommended. Figure 15 diagrams a basic parallel system with an air-cooled condenser and heat-recovery coil.

Evaporative Condenser. Evaporative condensers have a coil in which refrigerant is condensed or closed loop circulating water is

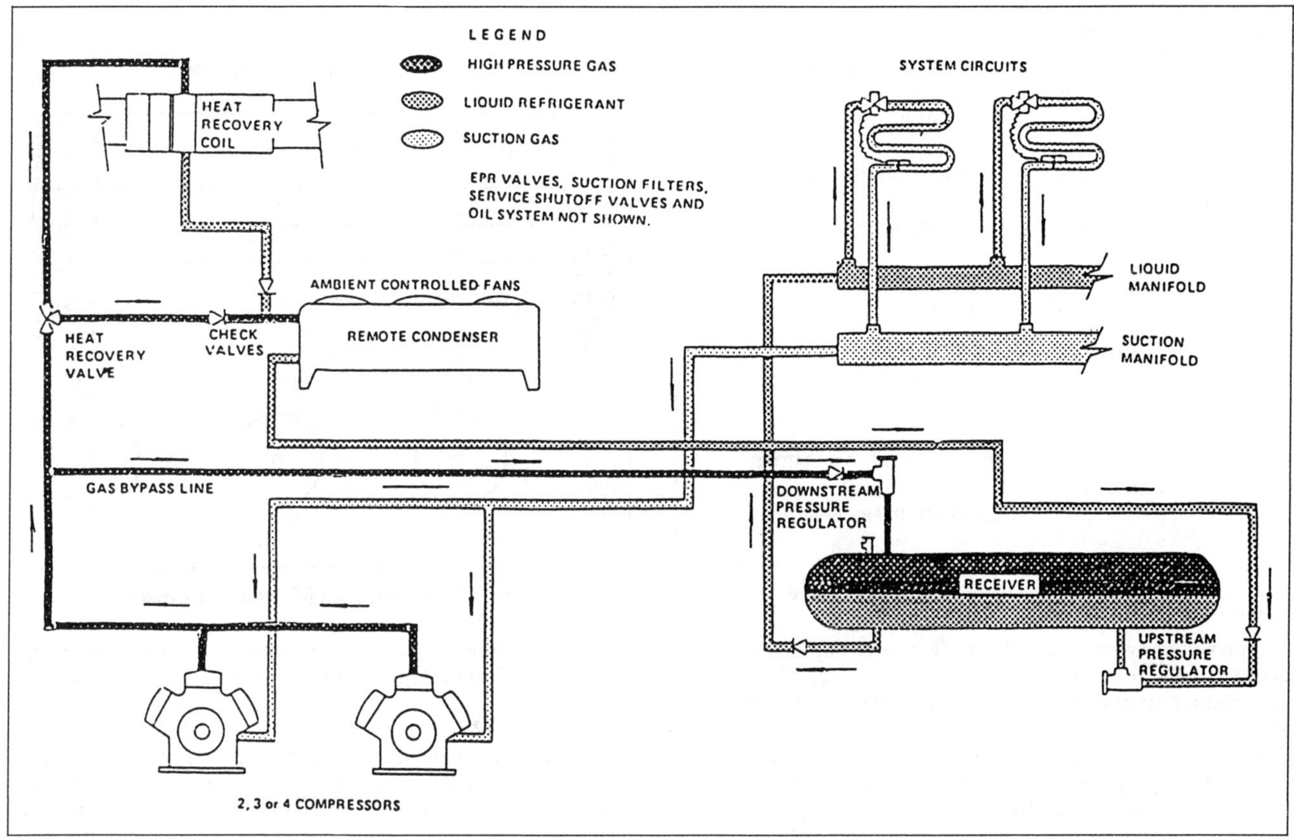

Fig. 15 Basic Parallel System with Remote Air-Cooled Condenser and Heat Recovery
(Reprinted by Permission of Tyler Refrigeration)

cooled and has a means to supply air and water over its external surfaces. Heat is transferred from the condensing refrigerant inside the coil or the water in the loop to the external wetted surface and then into the moving airstream principally by evaporation. In areas where the wet-bulb temperature is about 30°F below the dry-bulb temperature, the condensing temperature can run from 10 to 30°F above the wet-bulb temperature. This lower condensing temperature saves energy. One evaporative condenser can be installed for the entire store. Chapter 36 in the 1992 ASHRAE *Handbook—Systems and Equipment* gives more details.

MATCHING REFRIGERATOR REQUIREMENTS WITH COMPRESSOR CAPACITY

The designer matches requirements of refrigerator lineups to the capacity of a single-compressor unit or divides them into manageable circuits for parallel compressor systems. The parallel systems adapt easily to gas defrost. Gas directly from the compressor discharge, or, in some instances, from the top of the receiver at a lower temperature, flows through a manifold. Electric defrost, reverse air defrost, and off-cycle defrost can also be used on both parallel and single-unit systems. Liquid and/or suction line solenoid valves control the circuits for defrosting. Often, a suction stop is combined with the temperature controlling Evaporator Pressure Regulator (EPR). The entire system with its individual circuits is controlled by a multi-circuit time clock. Temperature control on the branch circuits is also achieved by refrigerator thermostats actuating liquid line solenoid valves. Shutoff valves isolate each circuit for service convenience. Refrigerator requirements are now often given as a refrigeration load per unit length, with a lower value sometimes allowed for parallel systems. The rationale for this lower value is that peak loads are less with programmed defrost. Refrigerator temperature recovery after a defrost period is less of a strain than it would be to a single-compressor system.

The published refrigerator load requirements allow for extra capacity for temperature pull-down after defrost, per ASHRAE *Standard* 72-1983. The industry considers a standard store condition to be 75°F and 55% rh, which should be maintained with air conditioning. Much of this air-conditioning load is carried by the open refrigerators, and credit for the heat removed by them should be considered in sizing the air-conditioning system.

Typical Parallel Compressor Operation

A typical supermarket includes one or more medium-temperature parallel systems for meat, deli, dairy, and produce refrigerators and the medium-temperature walk-in coolers. The system may have a separate compressor for the meat or deli refrigerators, or all units may have a single compressor. Energy Efficiency Ratios (EERs) typically run from 8 to 9 Btu/h per watt for the main load. Low-temperature refrigerators and coolers are grouped on one or more parallel systems with ice cream refrigerators on a satellite or on a single compressor. EERs range from 4 to 5 Btu/h per watt for the frozen foods and ice cream units as low as 3.5 to 4.0 Btu/h per watt. Cutting and preparation rooms are most economically placed on a single unit, since the refrigeration EER is at nearly 10 Btu/h per watt. Air-conditioning compressors are also separate, since their EER can range up to 11 Btu/h per watt (Figure 16).

Capacity Control of Parallel Systems

This system must be designed to maintain proper refrigerator temperatures under peak summer load. During the remainder of the year, store conditions can be easily maintained at a more ideal condition, and refrigeration load will be lower. In the past, refrig-

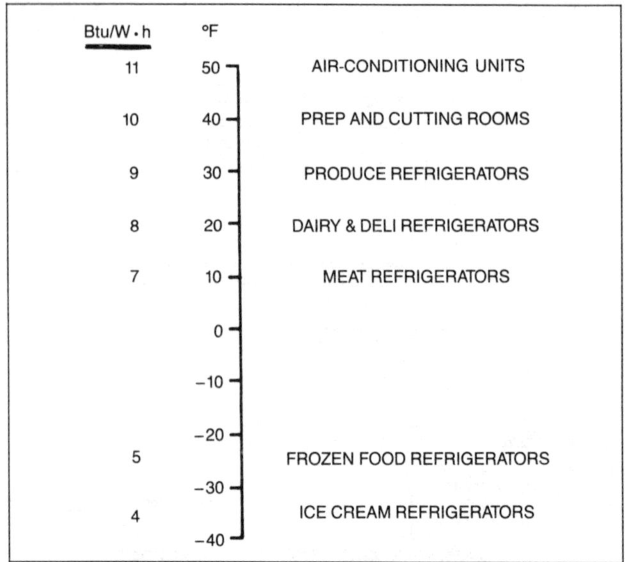

Fig. 16 Typical Compressor Efficiency

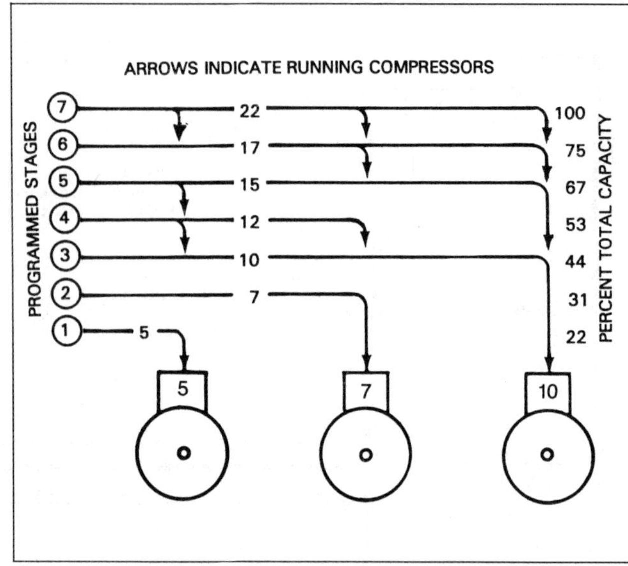

Fig. 17 Stages with Mixed Compressors

eration systems were operated at 90 °F condensing conditions or above to maintain enough high-side pressure to feed the refrigerated display fixture expansion valves properly. When outdoor ambient conditions allow, today's technology permits the condensing temperature to follow the ambient down to about 70 °F or less. When proper liquid-line piping practices are observed, the expansion valves feed the evaporators properly under these low condensing temperatures. Therefore, at partial load, the system has excessive capacity to perform adequately.

Multiple compressors may be controlled or staged based on a drop in system suction pressure. If the compressors are equal in size, a mechanical device can turn off one compressor at a time until only one is running. The suction pressure will be perhaps 5 psi or more below optimum. Newer, solid-state control devices can cycle units *on* and *off* while the system remains at one economical pressure range. The run time for each compressor motor can be equalized. Satellite compressors can also be controlled accurately with one control that also monitors other functions such as oil pressure, alarm functions, etc. Compressor cylinder unloading can also be used to vary rack capacity.

Staging Unequally Sized Compressors

Unequally sized compressors can be staged to obtain more steps of capacity than the same number of equally sized compressors. Figure 17 shows seven stages of capacity from a 5, 7, and 10 hp compressor parallel arrangement. Improved unloaders on multibank compressors promise to provide greater flexibility in the future, as does variable speed, usually on the lead compressor, for excellent capacity control.

Subcooling Liquid Refrigerant

Allowing refrigerant to *naturally subcool* in cool weather as it returns from the remote condenser also saves energy. This can be done by several means. The most common is to flood the condenser and allow the liquid refrigerant to cool close to the ambient temperature. Mechanical subcooling may also be economical in many areas. A subcooling compressor can be combined on a parallel system. Another method is to have a separate parallel system with branches to subcoolers on the other parallel and single systems in the store. The mechanical subcooling would be set to operate when the ambient temperature raises the refrigerant temperature above the desired subcooled temperature. The advantage to this method is that the mechanical subcooling compressor can run at twice the efficiency of the main system, thus saving energy through year-round liquid temperature control.

Heat Recovery

Heat recovery may be an important feature of virtually every compressor system, parallel or single. A heat-recovery coil is simply a second condenser coil placed in the air handler. If the store needs heat, this coil is energized and run in series with the regular condenser (Figure 15). The heat-recovery coil can be sized for a 30 to 50 °F TD depending on the capacity in cool weather. Lower head pressures in parallel systems permit little heat recovery. However, when heat is required in the store, simple controls create a higher head pressure for heat recovery. When compared with the cost of auxiliary gas or electric heat, the higher energy consumption may be compensated for by the value of the heat gained.

Water can be heated by a desuperheater on one large, single unit; or, commonly, water is heated by the interstage desuperheater on two-stage or compound R-22, parallel systems.

Factory-Assembled Equipment

Factory assembly of the necessary compressor systems with either direct air-cooled or any style of remote condenser is common practice. Both single and parallel systems can be housed, prepiped, and pre-wired at the factory. The complete unit is then delivered to the job site for placement on the roof or beside the store.

ENVIRONMENTAL EQUIPMENT AND CONTROL

Major components of common store environmental equipment include (1) central air handler with fresh makeup air-mixing box, (2) air-cooling coils, (3) heat-recovery coils, (4) supplemental heat equipment, (5) connecting ducts, and (6) termination units such as air diffusers and return grilles.

Environmental control is the heart of energy management. Control panels, designed for the unique heating, cooling, and humidity control requirements for food stores, provide several stages of heating (up to 8) and cooling (up to 3), plus a dehumidification stage. When high humidity exists in the store, cooling is activated to remove moisture. The controller receives input from temperature and dew point sensors that are located in the sales area.

Some controllers include night setback for cool climates and night setup for warm climates. This feature may save energy by turning the air handler off and allowing the store temperature to fluctuate several degrees, above or below the store temperature set point. However, store warm-up practices will impose an energy use penalty to the display case refrigeration systems and will have an effect on display case performance, particulary open models.

Following are rules for good air distribution in food stores:

Air Circulation. Operate 100% of the time the store is open at a volume of 0.6 to 1 cfm/ft^2 of sales area. Air supply and return grilles should be located so they do not disturb the air in open display cases and negatively affect case performance.

Fresh Air. Introduce whenever the air handler is operating. It should equal the required air change per hour or an allowance for all exhaust fans, whichever is greater.

Discharge Air. Discharge most or all of the air in areas where heat loss or gain occurs. This load normally is at the front of the store and around glass areas and doors.

Return Air. Locate return air registers as low as possible. With low registers, return air temperature may be 50 to 55 °F. This practice reduces heating and cooling requirements and temperature stratification compared to high returns. A popular practice, where store construction allows, is to return air under case ventilated bases and through trenches.

CHAPTER 47

FOOD SERVICE AND GENERAL COMMERCIAL REFRIGERATION EQUIPMENT

FOOD service requires refrigerators that meet a variety of needs. This chapter covers refrigerators available for (1) restaurants, (2) fast-food stores, (3) cafeterias, (4) commissaries, (5) hospitals, (6) schools, and (7) other specialized applications.

Many of the refrigeration products used in food service applications are self-contained, and the corresponding refrigeration systems are conventional. Some systems, however, do employ ice for fish, salad pans, or specialized preservation and/or display. (Further information on some of these products will be found in Chapters 46, 48, and 50.)

Generally, electrical and sanitary requirements of refrigerators are covered by criteria, standards, and inspections of Underwriters Laboratories, the National Sanitation Foundation, and the U.S. Public Health Service.

Frame construction is usually of metal. Occasionally, wood is used in minor amounts, but it is rarely visible in the final assembly. Nonmetallic materials are increasingly used for decoration, finish, trim, thermal break strips, and gaskets.

Nickel-chromium stainless steel is the most common sheet metal used for these refrigerators. Other sheet metals used are aluminum and carbon steel, either hot or cold rolled. Stainless steel usually has a grained or polished finish when exposed in final assembly. Aluminum generally is unfinished, although it may occasionally be lacquered or coated with some other organic finish. Carbon steel may be black; plated; porcelain enameled; aluminized; or coated with either high solids powder paint, enamel, or one of the vinyl finishes.

Insulation is usually one of the following: (1) foam polyurethane, poured in place; (2) slab polyurethane; (3) slab polystyrene; or (4) glass fiber batts. Foam polyurethane is advantageous because it conforms to various cavity shapes. Through adherence bonding to cabinet liners, it provides added structural integrity.

REACH-IN AND SPECIALTY CABINETS

The **reach-in refrigerator** is an upright, box-shaped cabinet with straight vertical front(s) and hinged or sliding doors (Figure 1). It is usually about 2.5 to 3 ft deep, about 6 ft high, and ranges in width from about 3 to 10 ft. Capacity ranges are from about 20 to 90 ft^3. These capacities and dimensions are standard from most manufacturers.

The typical reach-in cabinet (Figure 1) is available in many styles and combinations, depending on its intended application. Other shapes, sizes, and capacities are available on a custom basis from some manufacturers (Figure 2).

There are many varied adaptations of refrigerated spaces for storing perishable food items. Reach-ins, by definition, are normal or low temperature refrigerators small enough to be moved into a building. This definition also includes refrigerators and freezers built for special purposes, such as mobile cabinets or refrigerators on wheels, display refrigerators for such products as pies, cakes, and bakery goods. The latter cabinets usually have glass doors. Candy refrigerators also have specialized size, shape, and temperature.

Refrigerated vending machines satisfy the general definition of reach-ins, but they also receive coins and dispense products individually. Generally, the full product load of a vending machine

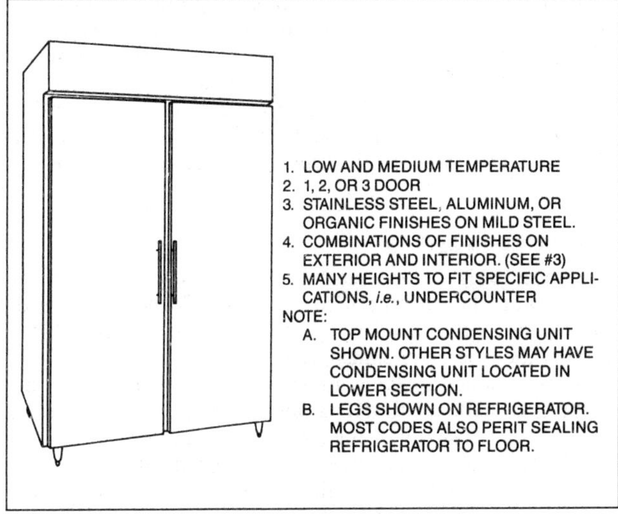

Fig. 1 Reach-In Food Storage Cabinet Features

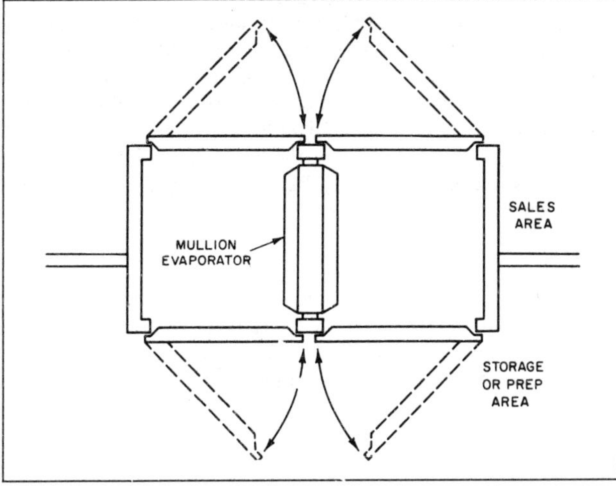

Fig. 2 Pass-Through Styles Facilitate Some Handling Situations

The preparation of this chapter is assigned to TC 10.7, Commercial Food and Beverage Cooling, Display and Storage.

is not accessible to the customer. Beverage-dispensing units dispense a measured portion into a cup, rather than in a bottle or can.

TYPES OF CONSTRUCTION

Reach-in refrigerators are available in two basic types of construction. The older style is a wood frame substructure clad with a metal interior and exterior. The newer style is a welded assembly of exterior panels with insulation and liner inserts.

In descending order of cost, the materials used on exteriors and interiors are stainless steel, porcelain enamel on steel, aluminum, and synthetic enamel on steel. The requirements are for a material that (1) matches or blends with that used on nearby equipment; (2) is easy to keep clean; (3) is not discolored or etched by commonly used cleaning materials; (4) is strong enough to resist denting, scratching, and abrasion; and (5) provides the necessary frame strength. The material chosen by an individual purchaser depends a great deal on layout and budget.

Temperature Ranges

Reach-in refrigerators are available for medium or low temperature ranges. The medium temperature range has a maximum of 45 °F and a minimum of 32 °F internal product temperature, with the most desirable average temperature being close to 38 °F. The low temperature range need not be below −10 °F and should not exceed 10 °F internal product temperature. The desirable average is 0 °F for frozen foods and −5 °F internal product temperature for ice cream. Both temperature ranges are available in cabinets of many sizes, and some cabinets combine both ranges.

Refrigeration Systems

Remote refrigeration systems are often used when cabinets are installed in a hot or otherwise unfavorable location, where the noise or heat of the condensing units would be objectionable, or under other special circumstances.

Self-contained systems, in which the condensing unit and controls are built into the refrigerator structure, are of two general types and are usually of the air-cooled style. The first has the condensing unit beneath the cabinet; in some designs it takes up the entire lower part of the refrigerator, while in others it occupies only a corner at one lower end. The second type has the condensing unit on top. There is no advantage to locating a self-contained condensing unit beneath the refrigerator; although the air near the floor is generally cooler, and thus beneficial to the condensing unit, it is usually dirtier. Putting the condensing unit on top of the cabinet allows full use of cabinet space, and, although the air passing over the condenser may be warmer, it is cleaner and more abundant. In addition, top mounting of both condensing unit and coil offers the same physical and constructional advantages. Frequently, bare-tube condensers are used to minimize dust clogging.

Styles

Reach-ins have doors on the front. Refrigerators that have doors on both front and rear are called *pass-through* or *reach-through refrigerators.* Doors are either full height (one per section) or half height (two per section). Doors may have windows or be solid, hinged, or sliding.

Interiors

Shelves are standard interior accessories and are usually furnished three or four per full height section. Generally, various types of shelf standards are used to provide vertical shelf adjustment.

Modifications and Adaptations

Food service applications often require extra shelves or tray slides, pan slides, or other interior accessories to increase food-holding capacity or make operation more efficient. With increasing use of foods prepared off-premises, specialization of on-premise storage cabinets is growing. This is developing an increasing pressure for designs that consider new food shapes, as well as in-and-out handling and storage.

Beverage service applications use standard cabinets if reach-ins are used, except when glass doors and special interior racks are used for chilled product display.

Meal factories, such as airline or central feeding commissaries, require rugged, heavy-duty equipment, often fitted for bulk in-and-out handling.

Retail bakeries also have special requirements. The dough retarder refrigerator and the bakery freezer permit the baker to spread the work load over the entire week and to offer a greater variety of products. The recommended temperature for a dough retarder is 36 to 40 °F. The relative humidity should be in excess of 80% (never lower) to prevent crusting or other undesirable effects. In the freezer, the temperature should be held at 0 °F. All cabinets or wheeled racks should be equipped with racks to hold the 18 by 26 in. bun pans, which are standard throughout the baking industry.

Retail stores use reach-ins for many different nonfood applications. Drug stores often have refrigerators with special drawers for storage of biologicals. (See section on Nonfood installations.)

Retail florists use reach-in refrigerators for displaying and storing flowers. Although a few floral refrigerator designs are considered conventional in the trade, the majority are custom built. The display refrigerator located in the sales area at the front of the shop may include a picture window display front and have one or more display-type access doors, either swinging or sliding. A variety of open refrigerators may also be used.

For the general range of flowers in a refrigerator, most retail florists have found that best results are obtained at temperatures from 40 to 45 °F. The refrigeration coil and condensing unit should be selected to maintain a high relative humidity. Some florists favor a gravity-type cooling coil because the circulating air velocity is low. Others, however, choose forced-air cooling coils, which develop a positive but gentle airflow through the refrigerator. The forced-air coil has an advantage when the in-and-out service is especially heavy because it provides quick temperature recovery during these peak conditions.

Nonfood installations use a wide range of reach-ins, some standard, except for accessory or temperature modifications, and some completely special. Examples are (1) biological and pharmaceutical cabinets; (2) blood bank refrigerators; (3) low and ultralow temperature cabinets for bone, tissue, and red-cell storage; and (4) special-shape refrigerators to hold column chromatography and other test apparatus.

Blood bank refrigerators for whole blood storage are usually standard models, ranging in size from under 20 to 45 ft^3 but with the following modifications:

1. Temperature is controlled at 37 to 41 °F.
2. Special shelves and/or racks are sometimes used.
3. A temperature recorder, usually hand-wound, with a 24-hour or seven-day chart is furnished.
4. An audible and/or visual alarm system is supplied to warn of unsafe blood temperature variation.
5. Sometimes an additional alarm system is provided to warn of power failure.

Most biological serums and vaccines require refrigeration for proper preservation and to retain highest potency. In hospitals and laboratories, the refrigerator temperatures should be 34 to 38 °F.

In other locations, temperatures may be slightly higher. The refrigerator should provide low humidity and, of course, should not freeze.

Products in these refrigerators are kept in specially designed stainless steel drawers sized for convenient storage, labeled for quick and safe identification, and perforated for proper air circulation.

Biological refrigerators, those for laboratory use, and mortuary refrigerators involve the same technology as that for food preservation. Storage in mortuary refrigerators is usually short-term, normally 12 to 24 hours at 34 to 38 °F. Refrigeration is obtained by a standard air- or water-cooled condensing unit with a forced-air cooling coil.

Mortuary refrigerators are built in various sizes and arrangements, the most common being two- and four-cadaver self-contained models. In the former cabinet are two individual storage compartments, one above the other, with the condensing unit compartment located above and indented into the upper front of the cabinet; ventilation grills are on the front and top of this section. The four-cadaver cabinet is equivalent to two two-cadaver cabinets set together, the storage compartments being two cabinets wide by two cabinets high with the compressor compartment above. Six- and eight-cadaver cabinets are built along the same lines. The two-cadaver refrigerator is approximately 38 in. wide by 94 in. deep by 77 in. high and is shipped completely assembled.

Each compartment contains a mortuary rack consisting of a carriage supporting a stainless steel tray. The carriage is telescoping, equipped with roller bearings so that it slides out through the door opening and is self-supporting even when extended. The tray is removable. Some specifications call for a thermometer to be mounted on the exterior front of the cabinet to show the inside temperature reading.

ROLL-IN CABINETS

These cabinets are very similar in style and appearance to reach-ins. *Roll-ins* (Figure 3) are usually part of a food-handling or other specific-purpose system (Figure 4). Pans, trays, or other specially sized and/or shaped receptacles are used to serve a specific system need, such as the following:

1. Schools, hospitals, cafeterias, and other food-handling institutional facilities
2. Meal manufacturing
3. Bakery processing
4. Pharmaceutical products
5. Body parts preservation (*e.g.*, blood)

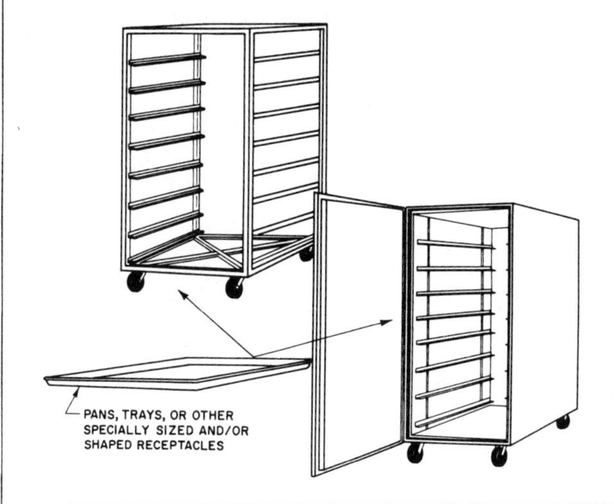

Fig. 3 Roll-In Cabinet

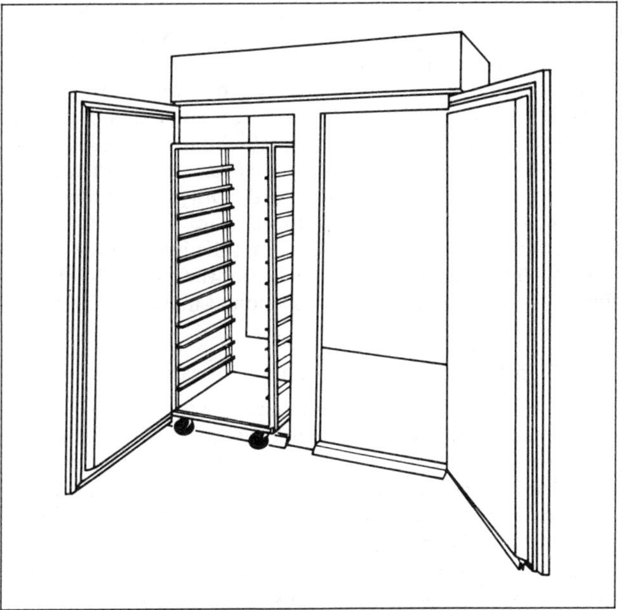

Fig. 4 Roll-Ins—Usually Part of a Food-Handling or Other Special-Purpose System

The roll-in differs from the reach-in in the following ways:

1. The inside floor is at about the same level as the surrounding room floor, so wheeled racks of product can be rolled directly from the surrounding room into the cabinet interior.
2. Cabinet doors are full height, with drag gaskets on their bottoms.
3. Cabinet interiors have no shelves or other similar accessories.

The racks that roll in and out of these cabinets are generally fitted with slides to handle 18 by 26 in. pans, although some newer systems call for either 12 by 20 in. or 12 by 18 in. steam table pans. Racks designed for special applications are available, but usually custom designed.

Manufacturers and contractors offer various methods of insulating the floor area. This problem must not be ignored, particularly if the roll-in is to hold frozen food.

FOOD FREEZERS

Chapter 8 covers the commercial freezing of food products. However, some hospitals, schools, commissaries, and other mass-feeding operations use on-premises freezing to level work loads and operate kitchens efficiently on normal schedules. Industrial freezing equipment is usually too large for these applications, so operators are using either regular frozen food storage cabinets for limited amounts of freezing or special reach-ins that are designed and refrigerated to operate as batch-type blast freezers.

WALK-IN COOLERS/FREEZERS

This type of commercial refrigerator is a factory-made, prefabricated, sectional version of the built-in, large-capacity cooling room. It closely matches the reach-in type in meeting a wide variety of applications.

Its function is to store foods and other perishable products in larger quantities and for longer periods than the reach-in refrigerator. Good refrigeration practice dictates that dissimilar foods be stored in separate rooms because they require different temperatures and moisture conditions and because odors from some foods are absorbed by others. Large food operations usually require three rooms: one for fruits and vegetables, one for meats and poultry, and one for dairy products. A fourth room, at 0 °F, may be added for frozen foods.

The sectional cooler offers flexibility over the built-in type: it is easily erected, easily moved, and, by adding standard sections, can be readily altered to meet changing requirements, uses, or layouts. Also, the sectional walk-in cooler can be erected outside a building, providing more refrigerated storage with no building costs except for footings and a low-cost roof supported by the cooler. Further advantage results from the high degree of skill applied in the design and construction of the factory-built product, which is subject to close inspection of materials and vapor-sealing techniques.

Self-Contained Sectional Walk-In Coolers

The versatility of sectional walk-in coolers was greatly increased by the introduction of self-contained models. There are various methods of application (Figure 5). These self-contained units use complete refrigeration systems, usually air-cooled, in a single compact package. The units are installed in the sectional cooler/freezer wall or ceiling panels.

Foam plastic materials in both self-contained and remote sectional coolers have further improved. Polyurethane foamed-in-place between two skins of metal suitably box-formed makes a lightweight, water-resistant panel. Additionally, the foam is a more efficient insulator, allowing slimmer panels for equal insulation value compared to most other insulations.

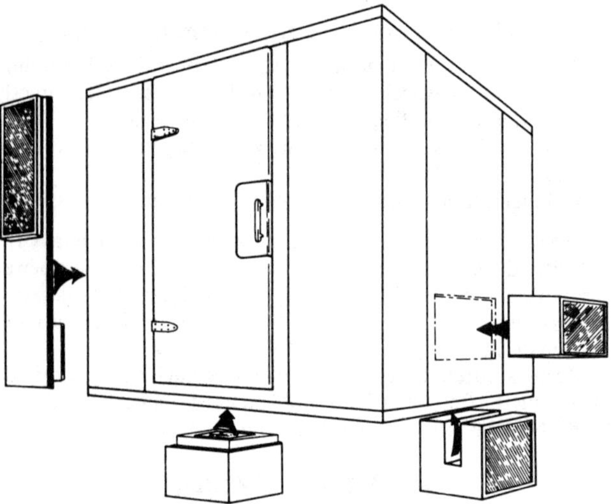

Fig. 5 Equipment Added to Make a Walk-In Cooler Self-Contained

Walk-in Floors

Sectional walk-in coolers, termed *floorless* by the supplier, are furnished with floor splines to fasten to the existing floor to form a base for the wall sections. Models with an insulated floor are also available.

A normal temperature (above freezing) cooler can be erected on an uninsulated concrete floor (on the ground) with only about 5% greater energy requirements than a cooler on an insulated floor. About 1.3 Btu/h · ft^2 enters from the ground. Generally, floor losses are considered negligible.

Level entry is becoming more important as the use of hand and electric trucks increases. The advantage and convenience of level entry afforded by a floorless cooler can also be obtained by recessing a sectional insulated floor.

Design Characteristics

The factory-made walk-in cooler consists of standard top, bottom, wall, door, and corner sections, which are shipped to the user and erected on the job. The frames are filled with insulation and are covered on the inside and outside with metal. The edges of these frames are usually of tongue-and-groove construction and either fitted with a gasket material or provided with suitable caulking material to ensure a tight vapor seal when assembled. These sections are assembled on the site with either lag screws or hooks operated from inside of the cooler.

Exterior and interior surfaces are made of one or more of the following:

- Zinc coated steel
- Aluminum
- Stainless steel
- Vinyl, fused on steel (infrequently used)
- Enamel, baked on steel (infrequently used)

At one time, coolers were used primarily to hold sides or quarters of beef, lamb carcasses, crates of vegetables, and other bulky items. Food operations now rarely, if ever, use such items. If they do, the items are broken down, trimmed, or otherwise processed before going into refrigerated storage. The modern cooler is not a storage room for large items, but a temporary place for quantities of small, partially or totally processed products. The food cooler, therefore, is likely to be equipped with sturdy, adjustable shelving about 18 in. deep and arranged in tiers, three or four high, around the inside walls. Alternatively, the cooler is often provided with rolling racks that are actually shelving on wheels. These racks are rolled directly into (and out of) the cooler.

CHAPTER 48

HOUSEHOLD REFRIGERATORS AND FREEZERS

THIS chapter covers the design and construction of full-sized household refrigerators and freezers, the most common of which are illustrated in Figure 1. Small portable and secondary refrigerators are not addressed here specifically. Some of these small refrigerators use absorption systems, special forms of compressors, and, in some cases, thermoelectric refrigeration. A major application has developed in the recreational vehicle market.

The section Refrigeration Systems only covers the vapor-compression cycle, which is almost universally used for full-sized household refrigerators and freezers. In these applications, where several hundred Btu/h are pumped through perhaps 100 °F differential from freezer to room temperature, other *electrically powered* systems compare unfavorably to the manufacturing and operating costs of vapor-compression systems. Typical operating efficiencies of the three most practical refrigeration systems are as follows for a 0 °F freezer and 90 °F ambient:

Thermoelectric—approximately 0.3 Btu/watt-hour
Absorption—approximately 1.5 Btu/watt-hour
Vapor-compression—approximately 4.5 Btu/watt-hour

An absorption system may operate from gas at a lower cost per unit of energy, but the initial cost, size, and weight of this system have made it unattractive for major appliances where electric power is available. Because of its simplicity, thermoelectric refrigeration could replace other systems if an economical thermoelectric material were developed.

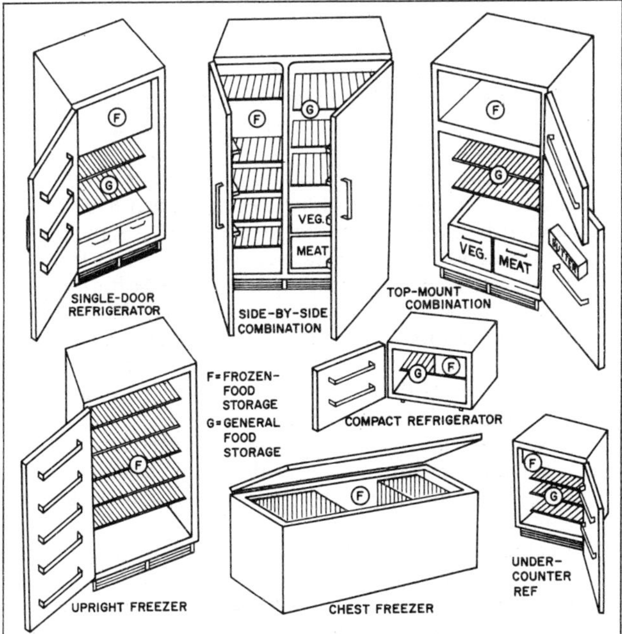

Fig. 1 Configurations of Contemporary Household Refrigerators and Freezers

The preparation of this chapter is assigned to TC 7.1, Residential Refrigerators and Food Freezers.

PRIMARY FUNCTIONS

Food storage space at reduced temperature is the primary function of a refrigerator or freezer, with ice making an essential secondary function. For the preservation of fresh food, a general storage temperature between 32 and 39 °F is desirable. Higher or lower temperatures or a humid atmosphere are more suitable for storing certain foods. A discussion of special-purpose storage compartments to provide these conditions may be found in the Cabinet section. Food freezers and combination refrigerator-freezers used for long-term storage are designed to hold temperatures near 0 °F and always below 8 °F. In single-door refrigerators, the frozen food space is usually warmer and is not intended for long-term storage. Optimum conditions for food preservation are addressed in more detail in Chapters 8 through 26.

PERFORMANCE CHARACTERISTICS

A refrigerator or freezer must maintain desired temperatures and have reserve capacity to cool to these temperatures when started on a hot summer day. Most models cool down within a reasonable time in a 110 °F ambient at rated voltage.

Overall system efficiency has become important, as rising energy costs have driven cost of operation upward and federal energy standards dictate consumption limits. The challenge for the designer to control noise and vibration has been made more complex with the need for fans for forced-air circulation and new compressors with higher efficiencies and capacities. The need for increased storage volumes and better insulation efficiency has resulted in almost universal use of foam insulation, which is less acoustically absorbent than glass fiber. Vibrations from running or stopping the compressor must be isolated to prevent mechanical transmission to the cabinet or to the floor and walls where it may cause additional vibration and noise.

SAFETY REQUIREMENTS

American manufacturers comply with Underwriters' Laboratories *Standard* UL-250, Standards for Safety, Household Refrigerators and Freezers, which protects the user from electrical shock, fire dangers, and other hazards under normal and some abnormal conditions. Specific areas addressing product safety focus on motors, hazardous moving parts, grounding and bonding, stability (cabinet tipping), door-opening force, door-hinge strength, shelf strength, component restraint (shelves and pans), glass strength, cabinet and unit leakage current, leakage current from surfaces wetted by normal cleaning, high voltage breakdown, ground continuity, testing and inspection of polymeric parts, and uninsulated live electrical parts accessible with an articulated probe.

DURABILITY AND SERVICE

Studies show that the average refrigerator and freezer will belong to its original owner for 10 to 15 years and will continue to give satisfactory service for a considerably longer period. There is a wide variation in life span, however; some refrigerators run for over 30 years. Their reliability, particularly that of the older and simpler refrigerators, has led people to expect a 15- to 20-year life of the hermetic unit, and, in many respects, the appliance must be designed to protect itself over this period. Motor overload protectors are normally incorporated, and an attempt is made to design fail-safe circuits so that the hermetic motor of the compressor will not be damaged by a failure of a minor external component, unusual voltage extremes, or voltage interruptions.

Customer-operated devices must withstand frequent use. For example, a refrigerator door may be opened and shut over 300,000 times in its lifetime. To protect the customer against the cost of premature failure, most manufacturers will replace faulty parts within one year and repair or replace a faulty hermetic system within five years or longer at no charge for materials. Beyond these warranties, the terms vary among manufacturers; they are sometimes combined with an extended service contract.

In the design of refrigerators and freezers, provisions must be made for economical and effective servicing if damage or malfunction occurs in the field.

CABINETS

A good cabinet design achieves the optimum blend of the following:

- Maximum food storage volume for the floor area occupied by the cabinet
- Maximum utility, performance, convenience, and reliability
- Minimum heat gain
- Minimum cost to the consumer

Use of Space

The fundamental factors in cabinet design are usable food-storage capacity and its external dimensions. Food storage volume has been considerably increased without a corresponding increase in the external dimensions of the cabinet. This has been accomplished mainly by using thinner but more effective insulation and by reducing the space occupied by the compressor and condensing unit. The method of computing storage volume and shelf area is described in ANSI/AHAM *Standard* HRF-1-1988.

Frozen Food Storage

The increased use of frozen foods requires refrigerators with much larger frozen food storage compartments. In single-door models, the frozen food storage volume is provided by a freezing compartment across the top of the general food storage compartment. An insulated baffle beneath the evaporator allows it to operate at the low temperatures required for short-term frozen food storage, while maintaining temperatures above freezing in the general food storage compartment. Sufficient air passes around the baffle to cool the general storage compartment.

In larger refrigerators, the frozen food space often represents a large part of the total volume, and it usually has a separate exterior door or drawer and a lower temperature capability; in this case, the model is classified as a combination refrigerator-freezer. The frozen food compartment in these combinations is most often positioned across the top, but on the largest models, there has been a trend toward side-by-side arrangements. A few bottom-mounted models have reappeared in the market. Two-door cabinets are sometimes built with two separate inner liners, housed in a single outer shell. Others use only a single liner divided into two sections by an insulated partition integral with the liner or installed as a separate piece.

Food freezers are offered in two forms: upright (vertical) and chest. Locks are often provided on the lids or doors as protection against accidental door opening or access by children. A power supply indicator light or a thermometer with an external dial may be provided to warn of high storage temperatures.

Special-Purpose Compartments

Special-purpose compartments provide a more suitable environment for storage of specific foods. For example, a warmer compartment for maintaining butter at a suitable spreading temperature is often found in the door. Some refrigerators have a meat storage compartment that can maintain storage temperatures just above freezing and may include an independent temperature adjustment feature. Some models have a special compartment for fish, which is maintained at approximately 30°F. High-humidity compartments for storage of leafy vegetables and fresh fruit are found in practically all refrigerators. These generally tight-fitting drawers, located within the food compartment, protect vulnerable foods from the drying effects of circulating dry air in the general storage compartment. The desired conditions are maintained in the special storage compartments and drawers by (1) enclosing them to prevent air exchange with the general storage area and (2) surrounding them with cold air to maintain the desired temperatures.

Ice and Water Service

Through a variety of manual and automatic means, refrigerators provide ice. Ice trays are placed into the freezing compartment in a stream of air that is substantially below 32°F or placed in contact with a directly refrigerated evaporator surface for manual operation.

Automatic ice-making equipment in household refrigerators has increased. Almost all automatic defrost refrigerators include factory-installed automatic ice makers or can accept a field installable ice maker.

The ice-maker mechanism is located in the freezer section of the refrigerator and requires an attachment to a water line. The ice-freezing rate is primarily a function of the system design. Most ice makers are in no-frost refrigerators, and the water is frozen by passing refrigerated air over the ice mold. Because the ice maker has to share the available refrigeration capacity with the freezer and food compartments, the ice production rate is usually limited by design to 4 to 6 lb per 24 h. An ice production rate of about 4 lb per 24 h, coupled with an ice storage container capacity of 7 to 10 lb, is adequate for most users.

When designing an ice maker, the various methods of automatically accomplishing the basic functions must be considered to determine if they meet the design objectives. These basic functions are as follows:

1. **Initiating** the ejection of the ice as soon as the water is frozen is necessary to obtain a satisfactory production rate. Ejection before complete freezing causes wet cubes to freeze together in the storage container or may cause the ice mold to overfill. One method is to initiate ejection in response to the temperature of a selected location in the mold when complete freezing is indicated. Another successful method is to initiate ejection based on the time required to freeze the water under normal freezer temperatures. In either method, the temperature or time required may vary in different applications, depending on the cooling air temperature, as well as the rate and direction of the airflow.

2. **Ejecting** the ice from the mold must be a reliable operation. In several designs, this is accomplished by freeing the ice from the mold with an electric heater and pushing it from the tray into an ice storage container. In other designs, water is frozen in a plastic tray by passing refrigerated air over the top so that the water freezes from the top down. The natural expansion that takes place during this freezing process causes the ice to partially *freeze free* from the tray. By twisting and rotating the tray, the ice can be completely freed and ejected into a container.
3. **Driving** the ice maker in most designs is done by a gear motor, which operates the ice ejection mechanism and may also be used to time the freezing cycle and the water-filling cycle, and to operate the stopping means.
4. **Filling** the ice mold with a constant volume of water, regardless of the variation in line water pressure, is necessary to ensure uniform-sized ice cubes and prevent overfilling. This is done by timing a solenoid flow-control valve or by using a solenoid-operated, fixed-volume slug valve.
5. **Stopping** is necessary when the ice storage container is full until some ice is used. This is accomplished by using a feeler-type ice level control or a weight control.

Ice service has become more convenient in some models by dispensing ice through the door. In one case, the ice is ejected into a storage container, which is accessible from the outside of the freezer door as a tilt-out compartment. In another design, ice is delivered through a trap door in the freezer door by an auger mechanism operating in the ice storage container. The auger motor is energized when a push-button switch is contacted by the action of placing a glass under the trap door. An additional selector switch is available to actuate an ice crusher as the cubes pass through the door on some designs. Chilled water and/or juice dispensing are provided on still other designs.

Thermal Considerations

The total heat load imposed on the refrigerating system comes from heat sources that are external and internal to the cabinet. The relative values of the basic or predictable portions of the heat load (which are independent of usage) are shown in Figure 2. A large portion of the peak heat load may result from door openings, food loading, and ice making, which are variable and unpredictable quantities dependent on customer use. As the beginning point for the thermal design of the cabinet, the significant portions of the heat load are normally calculated and then confirmed by test.

The major predictable heat load is the heat passing through the cabinet walls. Table 1 shows the insulating value of fibrous and foam insulations commonly used to insulate the cabinet; for further information, see Chapter 20 of the 1993 ASHRAE *Handbook—Fundamentals.*

External sweating can be avoided by keeping exterior surfaces warmer than the dew point of the environment. Condensation is most likely to occur around the hardware, on door mullions, along the edge of door openings, and on any cold refrigerant tubing that may be exposed outside the cabinet. In a 90 °F room, no external surface temperature on the cabinet should be more than 5 or 6 °F below the room temperature. If it is necessary to raise the exterior surface temperature to avoid sweating, this can be done by either

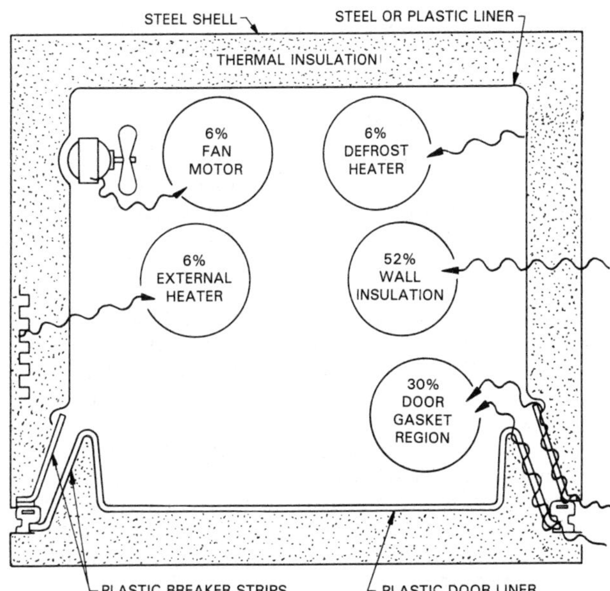

Fig. 2 Cabinet Cross Section Showing Typical Contributions to Total Heat Load

routing a loop of the condenser tubing under the front flange of the cabinet outer shell or by locating low-wattage wires or ribbon heaters behind the critical surfaces. Most refrigerators that incorporate electric heaters have power-saving electrical switches that allow the user to deenergize these electrical heaters when the environmental conditions do not require their use.

Temporary condensation on internal surfaces may occur with frequent door openings, so the interior of the general storage compartment must be designed to avoid objectionable accumulation or drippage.

Figure 2 shows the design features of the throat section where the door meets the face of the cabinet. On products with metal liners, metal-to-metal contact between inner and outer panels is avoided by using thermal breaker strips. Since the air gap between the breaker strip and the door panel provides a low-resistance heat path to the door gasket, the clearance should be kept as small as possible and the breaker strip as wide as practical. If the inner liner is made of plastic rather than steel, there is no need for separate plastic breaker strips because these would be formed as an integral part of the liner.

Cabinet heat leakage can be reduced by using door gaskets with more air cavities to reduce conduction or by using internal secondary gaskets. Care must be taken so that the maximum door opening force of 15 lb (as specified in United States *Federal Register* Vol. 38, No. 242, Tuesday, December 18, 1973) is not exceeded.

Structural supports, used to support and position the food compartment liner from the outer shell of the cabinet, are usually constructed of a combination of steel and plastics to provide adequate strength with maximum thermal insulation.

Internal heat loads, which must be overcome by the system's refrigerating capacity, are generated by fan motors used for air

Table 1 Effect of Thermal Insulation on Cabinet Wall Thickness

Insulation	Thermal Conductivity, Btu·in/h·ft²·°F	Wall Thickness, in. — For Threshold of External Sweating in 90 °F at 75% rh: 0 °F	38 °F	Common Practice: 0 °F	38 °F
Mineral or glass fiber, air filled	0.22 to 0.28	2.0 to 2.7	1.3 to 1.75	3.0 to 3.5	2.3 to 2.75
Foamed-in-place urethane foam, heavy gas filled	0.13	1.25	0.85	1.9	1.6
Foamed slab urethane foam, heavy gas filled	0.16	1.5	1.0	2.0	1.5

circulation and by heaters used to prevent undesirable internal cabinet sweating or frost buildup or to modify storage temperatures, where required.

Structure and Materials

The external shell of the cabinet is usually a single fabricated steel structure, which supports the inner food compartment liner, the door, and the refrigeration system. The space between the inner and outer walls of the cabinet is usually filled with foam slabs or foamed-in-place insulation. In general, the door and breaker strip construction is similar to that shown in Figure 2, although breaker strips and food liners formed of a single plastic sheet are also common. The doors cover the whole front of the cabinet, and plastic sheets become the inner surface for the doors, so no separate door breaker strips are required. The door liners are usually formed to provide an array of small door shelves and racks. Cracks and crevices are avoided, and edges are rounded and smooth to facilitate cleaning. Interior lighting uses incandescent lamps controlled by mechanically operated switches, actuated by the opening of the refrigerator door(s) or chest freezer lid. Table 2 summarizes the most common materials and manufacturing methods used in the construction of refrigerator and freezer cabinets.

The cabinet design must provide for the special requirements of its refrigerating system. For example, it may be desirable to refrigerate the freezer sections by attaching evaporator tubing directly to the food compartment liner. Also, it may be desirable, particularly with food freezers, to attach the condenser tubing directly to the shell of the cabinet to prevent external sweating. Both designs influence the cabinet heat leakage and the amount of insulation required.

The method of installing the refrigerating system into the cabinet is also important. Frequently, the system is installed in two or more component pieces and then assembled and processed in the cabinet. Unitary installation of a completed system directly into the cabinet allows the system to be tested and charged beforehand. The cabinet design must be compatible with the method of installation chosen. In addition, systems using forced air frequently require ductwork in the cabinet or insulation spaces.

Mechanically, the overall structure of the cabinet must be strong enough to withstand shipping, in which case it will be strong enough to withstand daily usage. Porcelain-enameled inner food liners must be designed to prevent porcelain chipping or crazing at the points of support. Plastic food liners must withstand the thermal stresses they are exposed to during shipment and usage, and they must be unaffected by the common contaminants they might encounter in a kitchen environment. Shelves must be designed not to deflect excessively under the heaviest anticipated load. Refrigerator doors and associated hardware must withstand a total of about 300,000 door openings.

Expanded-in-place foam insulation has had an important influence on cabinet design and assembly procedures. Not only is the wall thickness reduced, but the rigidity and bonding action of the foam usually eliminates the need for structural supports. The foam is normally expanded directly into the insulation space, adhering to the food compartment liner and the outer shell. Unfortunately, this does not permit disassembly of the cabinet for service or repairs. Alternative constructions that overcome this objection use prefoamed slabs or expand the foam against an inner mold, which is later withdrawn and replaced with the food compartment liner. In either case, the liner is not bonded to the foam, and it can be easily removed if required. However, the foam no longer provides the structural tie between the liner and the outer shell.

The steel outer shells of a refrigerator or freezer cabinet are typically finished with a synthetic baked enamel, which may be applied after fabrication. However, the use of prepainted steel is becoming more common, thus reducing volatile emissions that accompany the finishing process. Either finish method must provide a consistently durable finish to enhance product appearance and avoid corrosion. Steel food compartment liners are commonly finished in acid-resistant porcelain enamel, which usually consists of one ground coat and one or more finish coats, each separately fired. Organic finishes are available for use on steel and aluminum liners, and prefinished stock has been used where edges are lock seamed.

Use of Plastics. As much as 15 or 20 lb of plastic is incorporated in a typical refrigerator or freezer, and the use is increasing, largely because of:

- Wide range of physical properties
- Good bearing qualities
- Electrical insulating ability
- Moisture and chemical resistance
- Low thermal conductivity
- Ease of cleaning
- Pleasing appearance with or without an applied finish
- Potential of multifunctional design in a single part
- Transparency, opacity, and colorability
- Ease of forming and molding
- Lower cost

A few examples illustrate the versatility of plastics. High impact polystyrene and ABS plastics are used for inner door liners and food compartment liners (see Table 2). In these applications, no applied finish is necessary. These and similar thermoplastics such as polypropylene and polyethylene are also selected for evaporator doors, baffles, breaker strips, drawers, pans, and many small items. The phenolics are used for decorative door panels, terminal boards and terminal covers, and as a binder for the glass fiber insulation. The good bearing qualities of nylon and acetal are used advantageously in such applications as hinges, latches, and rollers for sliding shelves. Gaskets, both for the refrigerator and the evaporator doors, are generally made of vinyl or rubber.

Many items (such as ice cubes and butter) readily absorb odors and tastes from materials to which they are exposed. Accordingly, manufacturers take particular care to avoid using any plastics or other materials in the interior of the cabinet that will impart an odor or taste.

Table 2 Cabinet Materials and Manufacturing Methods

Component	Material	Common Thickness, in.	Manufacturing Method	Finish
Outer cabinet assembly			Welded assembly	
Wrapper and top	Low carbon cold-rolled steel	0.024–0.036	Roll form and bend	Organic finish (polyester, alkyd, acrylic, etc.)
Back	Low carbon cold-rolled steel	0.022–0.033	Draw and stamp	
Bottom	Low carbon cold-rolled steel	0.016–0.025	Draw and stamp	
Inner cabinet liner	Enameling iron	0.024–0.036	Bend and weld	Vitreous enamel
	Low carbon cold-rolled steel	0.024–0.036	Bend and weld	Organic finish
	Aluminum	0.024–0.036	Bend and weld	Anodized or organic
	Plastic	0.050–0.200	Injection molded or vacuum formed	None
Inner door liner	Plastic	0.075–0.095	Vacuum formed or injection molded	None
Breaker strips	Plastic	0.075–0.095	Extruded or injection molded	None

Moisture Sealing

To retain the original insulating qualities of the cabinet, the insulation must be kept dry. Moisture may get into the insulation through leakage of water from the food compartment liner, the defrost water disposal system, or, most commonly, through vapor leaks in the outer shell.

Outer shell construction is generally seam or spot welded and carefully sealed against vapor transmission by using mastics and hot melt asphaltic or wax compounds at all joints and seams. In addition, door gaskets, breaker strips, and other parts should provide maximum barriers for vapor flow from the room air to the insulation. When refrigerant evaporator tubing is attached directly to the food compartment liner, as is generally done in chest freezers, moisture will not migrate from the insulation space, and special efforts must be made to vapor-seal this space.

Although urethane foam insulation tends to inhibit moisture migration, it does have a tendency to trap water when migrating vapor reaches a temperature below its dew point. The foam then becomes permanently wet, and its insulation value is decreased. For this reason, a vaportight exterior cabinet is equally important with foam insulation.

Door Latching

Latching of doors is accomplished by mechanical or magnetic latches that act to compress relatively soft compression gaskets made of extruded rubber or vinyl compounds. Gaskets with magnetic materials embedded in the gasket are generally used. Chest freezers are sometimes designed so that the weight of the lid acts to compress the gasket, although most of the weight is counterbalanced by springs in the hinge mechanism.

In 1956, the Refrigerator Safety Act, Public Law 84-930, was enacted prohibiting shipment in interstate commerce of any household refrigerator that was not equipped with a device or system permitting the door to be opened from the inside. Although freezers are not included under PL 930, Underwriters' Laboratories *Standard* UL-250 has adopted the requirements of PL 930 to prevent entrapment in freezers.

In addition, since freezers are sometimes located in areas that have public access, they are often equipped with key locks to prevent pilferage. Underwriters' Laboratories requires these key locks to be the non-self-engaging type, and the key must be self-ejected from the slot when not in place to prevent entrapment caused by accidentally locking the door.

Cabinet Testing

Specific tests necessary to establish the adequacy of the cabinet as a separate entity include (1) structural tests, such as repeated twisting of the cabinet and door; (2) door slamming test; (3) tests for vapor-sealing of the cabinet insulation space; (4) odor and taste transfer tests; (5) physical and chemical tests of plastic materials; and (6) heat leakage tests. Cabinet testing is also discussed later in this chapter.

REFRIGERATING SYSTEMS

The vapor compression refrigerating systems used with modern refrigerators vary considerably in capacity and complexity, depending on the cabinet application. They are hermetically sealed and normally require no replenishment of refrigerant or oil during the useful life of the appliance. The components of the system must provide optimum overall performance and reliability at minimum cost. In addition, all safety requirements of UL *Standard* 250 must be met. The fully halogenated refrigerant, R-12, has been used in household refrigerators for many years. However, due to its strong ozone depletion property, appliance manufacturers must replace R-12 with environmentally acceptable refrigerants.

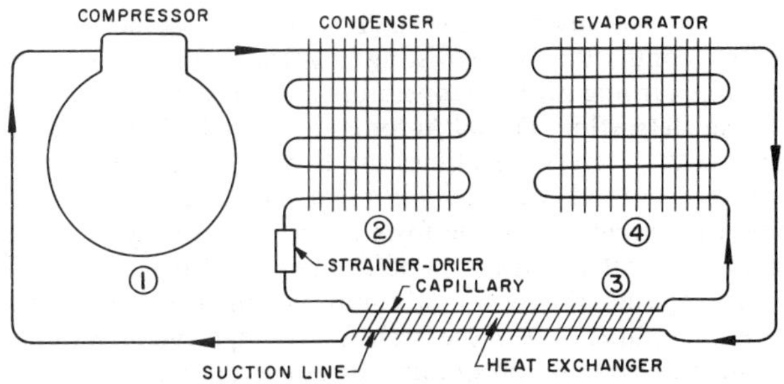

Fig. 3 Refrigeration Circuit

Design engineers are evaluating substitutes that can match the performance and reliability of existing designs.

The design of refrigerating systems for refrigerators and freezers has improved because of new refrigerants, wider use of aluminum, smaller and more efficient motors and compressors, universal use of capillary tubes, and simplified electrical components. These refinements have kept the vapor compression system in the best competitive position for household application.

Refrigerating Circuit

Figure 3 shows the refrigerant circuit for a vapor compression refrigerating system. The refrigeration cycle is as follows: (1) electrical energy supplied to the motor drives a positive displacement compressor, which draws cold, low-pressure refrigerant vapor from the evaporator and compresses it; (2) the resulting high-pressure, high-temperature discharge gas then passes through the condenser, where it is condensed to a liquid, while the heat is rejected to the ambient air; (3) the liquid refrigerant passes through a metering capillary tube to the evaporator at a reduced pressure; and (4) the low-pressure, low-temperature liquid in the evaporator absorbs heat from its surroundings, evaporating to a gas, which is again withdrawn by the compressor.

Note that energy enters the system through the evaporator (heat load) and through the compressor (electrical input). Thermal energy is rejected to the ambient by the condenser and the compressor shell. A portion of the capillary tube is usually soldered to the suction line for heat exchange. By using the effect of the cool suction gas, system capacity and efficiency are further increased.

A strainer-drier is usually placed ahead of the capillary tube to remove foreign material and moisture. Refrigerant charges of 1 lb or less of R-12 are common. A thermostat (or cold-control) cycles the compressor to provide the desired temperatures within the refrigerator. During the off cycle, the capillary tube permits the pressures to equalize throughout the system.

Materials used in refrigeration circuits are selected for (1) their mechanical properties, (2) their compatibility with the refrigerant and oil on the inside, and (3) their resistance to oxidation and galvanic corrosion on the outside. Evaporators are usually made of aluminum tubing, either with integral extruded fins or with extended surfaces mechanically attached to the tubing. Condensers are usually made of steel tubing with an extended surface of steel sheet or wire. Steel tubing is used on the high-pressure side of the system, which is normally dry, and copper is used for suction tubing, where condensation can occur. Because of its ductility, corrosion resistance, and ease of brazing, copper is used for capillary tubes and often for small connecting tubing. Wherever aluminum tubing comes in contact with copper or iron, it must be protected against moisture to avoid electrolytic corrosion.

Defrosting

A few manufacturers still use simple manual defrost in which the cooling effect is generated by gravity circulation of air over a refrigerated surface (evaporator) located at the top of the food compartment. The refrigerated surface forms some of the walls of a frozen food space, which usually extends across the width of the food compartment. Defrosting is typically accomplished by manually turning off the temperature-control switch.

Cycle Defrosting (Partial Automatic Defrost). Combination refrigerator-freezers sometimes use two separate evaporators for the fresh food and freezer compartments. The fresh-food compartment evaporator defrosts during each off cycle of the compressor, with the energy for defrosting provided mainly by the heat leakage into the fresh-food compartment. The cold control senses the temperature of the fresh-food compartment evaporator and cycles the compressor on when the evaporator surface is above 32 °F. The freezer evaporator requires infrequent manual defrosting.

No-Frost Systems (Automatic Defrost). Most combination refrigerator-freezers and some upright food freezers are often refrigerated by air that is fan-blown over a single evaporator, concealed from view. Because the evaporator is colder than the freezer compartment, it collects practically all of the frost, and there is little or no permanent frost accumulation on the frozen food or on exposed portions of the freezer compartment. The evaporator is defrosted automatically by electric heat or hot refrigerant gas, and the defrosting period is short to limit food temperature rise. The resulting water is disposed of automatically by draining to the exterior, where it is evaporated in a pan located in the warm condenser airstream or on warm condenser coils. Defrosting is usually initiated by a timer at intervals of up to 24 h. If the timer operates only when the compressor runs, the accumulated time tends to reflect the probable frost load.

Developments in electronics have allowed introduction of micropressor-based control systems to some household refrigerators. An adaptive defrost function is usually included in the software. Various parameters are monitored so that the period between defrosts varies according to actual conditions of use. Adaptive defrost tends to reduce energy consumption and improve food preservation.

Forced Heat for Defrosting. All no-frost systems add heat to the evaporator to accelerate melting during the short defrosting cycle. The most common method uses a 300 to 1000 W electric heater. The typical defrost cycle is initiated by a timer, which stops the compressor and energizes the heater.

When the evaporator has melted all the frost, a defrost termination thermostat opens the heater circuit. In most cases, the compressor is not restarted until the evaporator has drained for a few minutes and the system pressures have subsided, thus reducing the applied load for restarting the compressor. Commonly used defrost heaters include metal-sheathed heating elements in thermal contact with evaporator fins and radiative heating elements positioned to radiate heat to the evaporator.

Evaporator

The *manual defrost* evaporator usually is a box with three or four sides refrigerated. Refrigerant may be carried in tubing brazed to the walls of the box, or the walls may be constructed from double sheets of metal, which are brazed or metallurgically bonded together with integral passages for the refrigerant. These constructions usually use aluminum, and special attention is required to avoid (1) contamination of the surface with other metals that would promote galvanic corrosion and (2) configurations that may be readily punctured in use.

The *cycle defrost* evaporator for the fresh-food compartment is designed for natural defrost operation and is characterized by its low thermal capacity. It may be a lightweight vertical plate, usually made from bonded sheet with integral refrigerant passages, or a serpentine coil with or without fins. In either case, it should be located near the top of the compartment and arranged for good water drainage during the defrost cycle. In some designs, this cooling surface has been located in an air duct remote from the fresh-food space, with air circulated continuously by a small fan.

The *no-frost forced-convection* evaporator is usually a forced-air fin-and-tube arrangement designed to minimize the effect of frost accumulation, which tends to be relatively rapid in a single evaporator system. The coil is usually arranged for airflow parallel to the long dimension of the fins.

The fins may be more widely spaced at the air inlet to provide for preferential frost collection and to minimize the air restriction effects of the frost. All surfaces must be heated adequately during the defrost cycle to ensure complete defrosting, and provision must be made for draining and evaporating the defrost water outside the food-storage spaces.

Evaporators for chest freezers usually consist of tubing that is in good thermal contact with the exterior of the food compartment liner. The tubing is usually concentrated near the top of the liner with wider spacing near the bottom to take advantage of gravity circulation of the air inside. Upright food freezers usually have refrigerated shelves and a refrigerated surface at the top of the food compartment. These are commonly connected in series with an accumulator at the exit end. No-frost freezers usually incorporate a fin-and-tube evaporator and an air circulating fan as used in the no-frost combination refrigerator-freezers.

Condenser

The condenser is the main heat-rejecting component in the refrigerating system. It may be cooled by natural draft on free-standing refrigerators and freezers or fan-cooled on larger models and on models designed for built-in applications.

The *natural draft condenser* is located on the back wall of the cabinet and is cooled by natural air convection under the cabinet and up the back. The most common form of natural draft condenser consists of a flat serpentine of steel tubing with steel cross wires welded on 0.25-in. centers on one or both sides perpendicular to the tubing. Tube-on-sheet construction may also be used.

The *hot wall condenser,* another natural draft arrangement, is used principally with food freezers. It consists of condenser tubing attached to the inside surface of the cabinet shell. The shell thus acts as an extended surface for heat dissipation. With this construction, external sweating is seldom a problem.

The *forced-draft condenser* may be of fin-and-tube, folded banks of tube-and-wire or tube-and-sheet construction. Various forms of condenser construction are used to minimize clogging caused by household dust and lint. The compact, fan-cooled condensers are usually designed for low airflow rates because of noise limitations. Air ducting is often arranged to use the front of the machine compartment for the entrance and exit of air. This makes the cooling air system largely independent of the location of the refrigerator and permits built-in applications.

A portion of the condenser may be located under the defrost water evaporating pan to promote water evaporation. The condenser may also incorporate a section for *compressor cooling* from which the partially condensed refrigerant is routed to an oil-cooling loop in the compressor, where the liquid refrigerant, still at high pressure, absorbs heat and is reevaporated. The vapor is then routed through the balance of the condenser, to be condensed in the normal manner. In some designs, as noted previously, a portion of the condenser tubing is routed internally in contact with the outer case in place of antisweat heaters.

Condenser performance may be evaluated directly on calorimeter test equipment similar to that used for compressors. However, the final design of the condenser must be determined by performance tests on the refrigerator under a variety of operating conditions.

Generally, the most important design requirements for a condenser include (1) sufficient heat dissipation at peak-load conditions, (2) storage volume that adequately prevents excessive pressures during pulldown or in the event of a restricted or plugged capillary tube, (3) good refrigerant drainage to minimize the off cycle losses and the time required for equalization of system pressures, (4) external surface easily cleaned or designed to avoid dust and lint accumulation, and (5) adequate safety factor against bursting.

Capillary Tube

The refrigerant metering device generally used is the capillary tube, a small-bore tube connecting the outlet of the condenser to the inlet of the evaporator. The regulating effect of this simple control device is based on the principle that a given weight of liquid passes through a capillary more readily than the same weight of gas at the same pressure. Hence, if uncondensed refrigerant vapor enters the capillary, the mass flow will be reduced, giving the refrigerant more cooling time in the condenser. On the other hand, if liquid refrigerant tends to back up in the condenser, the condensing temperature and pressure rises, resulting in an increased mass flow of refrigerant. Under extreme conditions, the capillary either passes considerable uncondensed gas or backs liquid refrigerant well up into the condenser. Under normal operating conditions, a capillary tube gives good performance and efficiency. Figure 4 shows the typical effect of capillary refrigerant flow rate on system performance.

A capillary tube has the advantage of extreme simplicity and no moving parts. It also lends itself well to being soldered to the suction line for heat-exchange purposes. This position prevents sweating of the otherwise cold suction line, and, at the same time, increases the refrigerating capacity and efficiency. Another advantage is that the pressure equalizes throughout the system during the off cycle and reduces the starting torque required of the compressor motor. The capillary is the narrowest passage in the refrigerant system and the place where low temperature first occurs. For that reason, a combination strainer-drier is usually located directly ahead of the capillary to prevent it from being plugged by ice or any foreign material circulating through the system.

Selection. The optimum metering action can be obtained by variations in either the diameter or the length of the tube. Such factors as the physical location of the system components and the heat exchanger length (36 in. or more is desirable) may determine the selection of the length and bore of the capillary tube for any given application.

Capillary selection is covered in detail in Chapter 44. Once a preliminary selection is made, an experimental unit can be equipped with three or more different capillaries that can be activated independently. System performance can then be evaluated by using each capillary with slightly different flow characteristics in turn.

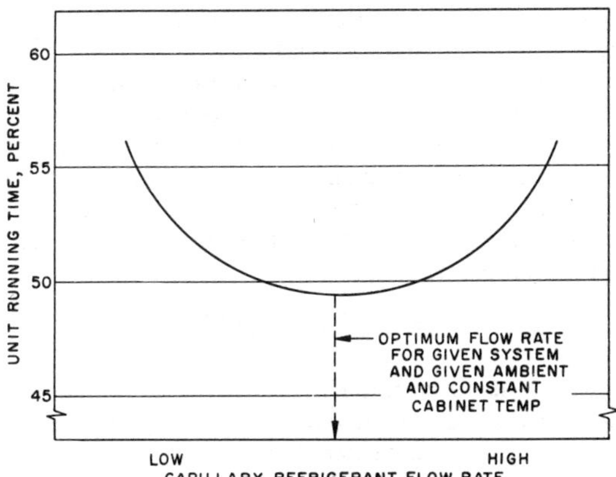

Fig. 4 Typical Effect of Capillary Tube Selection on Unit Running Time

Final selection of the capillary requires an optimization of performance under both no load and pulldown conditions, with maximum and minimum ambient and load conditions. The optimum refrigerant charge can also be determined during this process.

Compressor

While a more detailed description of compressors can be found in Chapter 35 of the 1992 ASHRAE *Handbook—Systems and Equipment*, a brief discussion of the small compressors used in household refrigerators and freezers is included here.

These products use positive displacement compressors in which the entire motor-compressor is hermetically sealed in a welded steel shell. Capacities range from about 300 Btu/h to about 2000 Btu/h when measured at the usual rating conditions of −10 °F evaporator, 130 °F condenser, 90 °F ambient, with the suction gas superheated to 90 °F and the liquid subcooled to 90 °F.

Design emphasis is placed on ease of manufacturing, reliability, low cost, quiet operation, and efficiency. Figure 5 illustrates the two types of rotary compressors and two reciprocating piston compressor mechanisms, which are used in virtually all conventional refrigerators and freezers, with no one type being much less costly than the others. While rotary compressors are somewhat more compact than reciprocating compressors, a greater number of close tolerances are involved in their manufacture.

These compressors are directly driven by squirrel cage induction motors of the two-pole, 3450 rpm type, although some four-pole, 1750 rpm motors are also used. Field windings are insulated

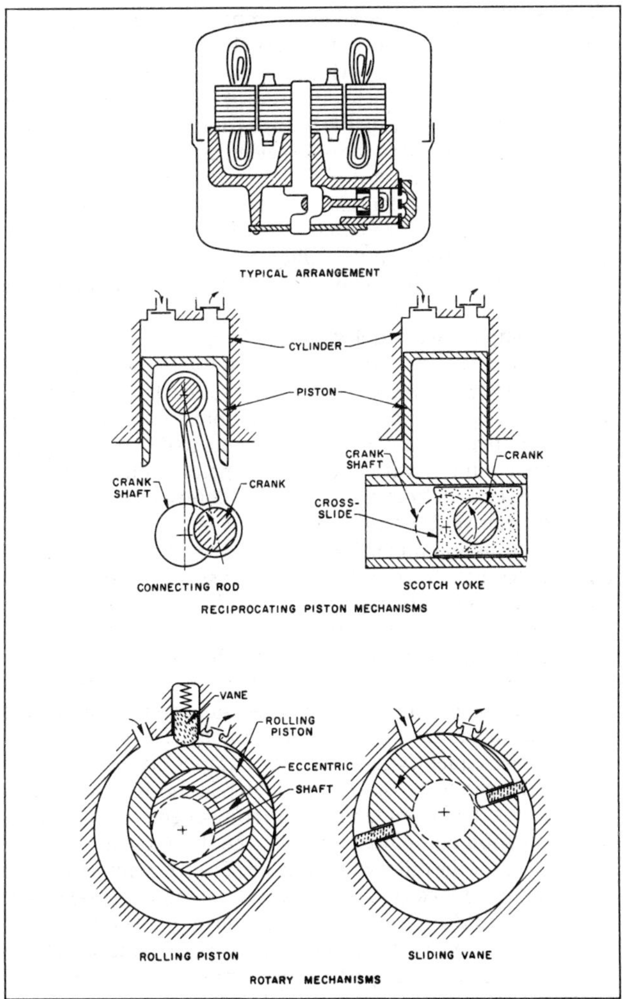

Fig. 5 Refrigerator Compressors

with special wire enamels and plastic slot and wedge insulation; all are chosen for their compatibility with the refrigerant and oil. During continuous runs at rated voltage, motor winding temperatures may be as high as 250°F for 110°F ambient temperature. In addition to maximum operating efficiency at normal running conditions, the motor must provide sufficient torque for starting and temporary peak loads because of start-up and pulldown of a warm refrigerator, and for the load conditions associated with the defrosting action. These conditions should be met at the anticipated extremes of line voltage.

Starting torque is provided by a split-phase winding circuit, which may include a starting capacitor in the larger motors. When the motor comes up to speed, an external electromagnetic relay or positive temperature coefficient (PTC) device disconnects the start winding. A run capacitor sometimes is employed for greater motor efficiency. Motor overload protection is provided by an automatic resetting switch, which is sensitive to a combination of motor current and compressor case temperature, or internal winding temperature.

The compressor is cooled by rejecting heat to the surroundings. This is easily accomplished with a fan-cooled system. However, an oil-cooling loop carrying partially condensed refrigerant may be necessary when the compressor is used with a natural-draft condenser and in some forced-draft systems above 1000 Btu/h.

Temperature-Control System

The temperature-control thermostat is generally an electromechanical switch actuated by a temperature-sensitive power element that has a condensable gas charge, which operates a bellow or diaphragm. At operating temperature, this charge is in a two-phase state, and the temperature at the gas-liquid interface determines the pressure on the bellows. To maintain temperature control at the bulb end of the power element, the bulb must be the coldest point at all times.

The thermostat must have an electrical switch rating for the inductive load of the compressor and other electrical components carried through the switch. The thermostat is usually equipped with a shaft and knob for adjustment of the operating temperature.

In the simple gravity-cooled system, the sensing bulb of the thermostat is normally clamped to the evaporator. The location of the bulb and the degree of thermal contact is selected to produce both a suitable cycling frequency for the compressor and the desired refrigerator temperature. Small refrigerators sold in Europe sometimes are equipped with a manually operated push button to prevent the control from coming on until defrost temperatures are reached; afterward, normal cycling is resumed.

In the combination refrigerator-freezer with the split air system, the location of the thermostat sensing bulb depends on whether an automatic damper control is used to regulate the airflow to the fresh-food compartment. When such an auxiliary control is used, the sensing bulb is usually located to sense the temperature of the air leaving the evaporator. In manual damper controlled systems, the sensing bulb is usually placed in the cold airstream to the fresh-food compartment. The sensing bulb location is frequently related to the damper effect on the airstream. Depending on the design of this relationship, the damper may become the freezer temperature adjustment, or it may serve the fresh-food compartment, with the thermostat being the adjustment for the other compartment. The temperature-sensing bulb should be located to provide a large enough temperature differential to drive the switch mechanism, while (1) avoiding excessive cycle length; (2) avoiding short cycling time, which can cause compressor starting problems; and (3) avoiding annoyance to the user from frequent noise level changes.

In some refrigerators, microprocessor-based control systems have replaced the electromechanical thermostat switch and, in some cases, both compartment controls use thermistor-sensing devices that relay electronic signals to the microprocessor. Electronic control systems provide a higher degree of independent temperature adjustments for the two main compartments.

System Design and Balance

A principal design consideration is the selection of components that will operate together to give the optimum system performance when the total cost is considered. Normally, a range of combinations of values for these components meet the performance requirements, and the lowest cost is only obtained through a careful analysis or a series of tests—usually both. For instance, for a given cabinet configuration, food-storage volume, and temperature, the following can be traded off against one another: (1) insulation thickness and overall shell dimensions, (2) insulation material, and (3) system capacity. Each of these variables affects the total cost, and most of them can be varied only in discrete steps.

The experimental procedure involves a series of tests. Calorimeter tests may be made on the compressor and condenser, separately or together, and on the compressor and condenser operating with the capillary tube and heat exchanger. Final selection of the components requires performance testing of the system when installed in the cabinet. These tests also determine the refrigerant charge, airflows for the forced draft condenser and evaporator, temperature control means and calibration, necessary motor protection, etc. The Evaluation section covers the final evaluation tests to be made on the complete refrigerator. The interaction between components is further addressed in Chapter 43. This method assumes a knowledge (equations or graphs) of the performance characteristics of the various components, including the heat leakage of the cabinet and the heat load imposed by the customer. The analysis may be performed manually point by point. If enough component information exists, it can be programmed into a computer simulation capable of responding to various design conditions or statistical situations. Although the available information may not always be adequate for an accurate analysis, this procedure is often useful, though it must be followed by confirming tests.

Processing and Assembly Procedures

All parts and assemblies that are to contain refrigerant are processed to avoid or remove unwanted substances from the final sealed system and to charge the system with refrigerant and oil. Each component should be thoroughly cleaned and then stored in a clean, dry condition prior to assembly. The presence of free water in stored parts produces harmful compounds such as rust or aluminum hydroxide, which are not removed by the normal final assembly process. Procedures for dehydration, charging, and testing may be found in Chapter 45.

Assembly procedures are somewhat different, depending on whether the sealed refrigerant system is completed as a unit before being assembled to the cabinet, or whether components of the system are first brought together on the cabinet assembly line. Using the unitary installation procedure, the system may be tested for its ability to refrigerate and then be stored or delivered to the cabinet assembly line.

EVALUATION

Once the unit is assembled, laboratory tests, supplemented by field testing, are necessary to determine actual performance. The following aspects are considered in this section:

- Test facilities required
- Established test procedures, published by standard, technical, and industry organizations

- Special performance testing
- Materials testing
- Life testing of components
- Field testing

Environmental Test Rooms

Controlled temperature and humidity test rooms are essential for performance testing of refrigerators and freezers. ANSI/AHAM *Standard* HRF-1-1988 describes the environmental conditions to be maintained. The rooms should be capable of providing ambient temperatures ranging from 70 to 110°F accurate to within 1°F of the desired value. The temperature gradient and the air circulation within the room should also be maintained closely. To provide more flexibility in testing, it may be desirable to have an additional test room that can also cover the range between 0 and 70°F.

At least one test room should have the capability of maintaining a desired relative humidity within ±2%, up to 85%.

All instruments should be calibrated at regular intervals. The instruments for adequate performance testing of a refrigerator or freezer are described in ANSI/AHAM *Standard* HRF-1-1988. Instrumentation should have accuracy and response capabilities of sufficient quality to measure the dynamics of the systems tested.

Thermocouple recorders, indicating and recording wattmeters, ammeters, voltmeters, and potentiometers are used in testing refrigerators and freezers. Refrigerator test laboratories have developed automated means of data acquisition (with computerized data reduction output) and the automated programming of tests.

Standard Performance Test Procedures

ANSI/AHAM *Standard* HRF-1-1988 describes tests for determining the performance of refrigerators and freezers. It specifies the standard ambient conditions, power supply, and means for selecting samples and measuring temperatures. Test procedures include the following:

No-Load Pulldown Test. This tests the ability of the refrigerator or freezer in a 110°F ambient temperature to pull down from a stabilized warm condition to design temperatures within an acceptable period.

Simulated Load Test (*Refrigerators*) or Storage Load Test (*Freezers*). This test determines the electrical energy (kWh) consumption rate per 24-h period, the percent operating time of the compressor motor, and temperatures at various locations within the cabinet at 70, 90, and 110°F ambient temperatures for a range of temperature control settings. The cabinet doors remain closed during the test. The freezer compartment is loaded with filled frozen packages. Each test point may take 8 h or more to ensure steady-state condition and accuracy of data. The data taken are usually plotted as shown in Figure 6 for a combination refrigerator-freezer with only a fresh-food temperature control. If there is a separate control for freezer temperature, these graphs can carry additional curves for high and low freezer control settings. Freezers are tested similarly, but in a 90°F ambient. Under actual operating conditions in the home, with frequent door openings and ice making, the performance may not be as favorable as that shown by this test. However, the test indicates general performance, which can serve as a basis for comparison.

Ice-Making Test. This test performed in a 90°F ambient determines the rate of making ice with the ice trays or other ice-making equipment furnished with the refrigerator.

External Surface Condensation Test. This test determines the extent of moisture condensation on the external surfaces of the cabinet in a 90°F, high-humidity ambient when the refrigerator or freezer is operated at normal cabinet temperatures. Although ANSI/AHAM *Standard* HRF-1-1988 calls for this test to be made at a relative humidity of 75 ±2%, it is customary to determine the

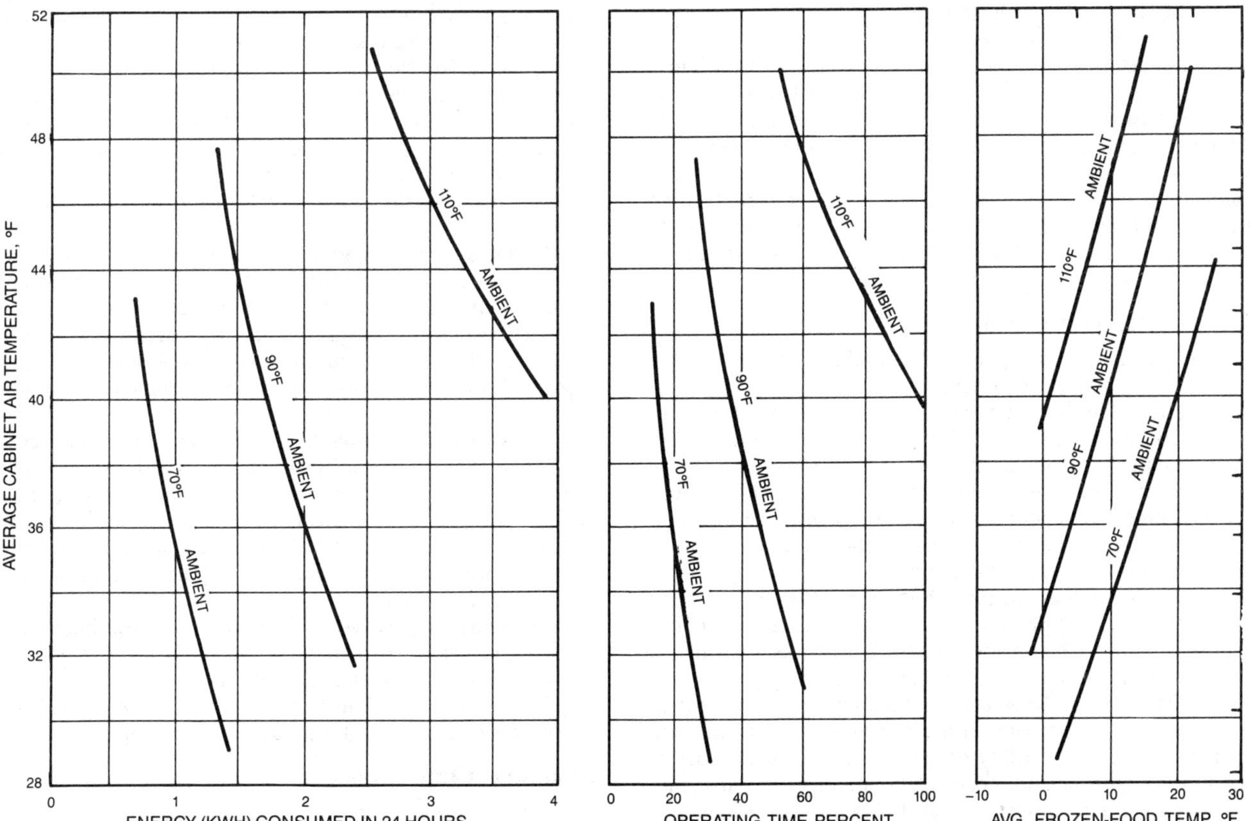

Fig. 6 Sample Plot of Test Results of Simulated Load Tests

sweating characteristics through a wide range of relative humidity up to 85%. This test also determines the need for, and the effectiveness of, anticondensation heaters in the cabinet shell and door mullions.

Internal Moisture Accumulation Test. This dual-purpose test is also run under high-temperature, high-humidity conditions. First, it determines the effectiveness of the moisture sealing of the cabinet in preventing moisture from getting into the insulation space and degrading the performance and life of the refrigerator. Secondly, it determines the rate of frost buildup on refrigerated surfaces, the expected frequency of defrosting, and the effectiveness of any automatic defrosting features, including means for defrost water disposal.

This test is made in ambient conditions of 90°F and 75% rh with the cabinet temperature control set for normal temperatures. The test period extends over a 21-day period with a rigid schedule of door openings over the first 16 h of each day. For a refrigerator, the test calls for 96 door openings per day for the general refrigerated compartment; 24 per day for the freezer compartment and food freezers.

Current Leakage Test. This test determines the electrical current leakage through the entire electrical insulating system under severe operating conditions. To eliminate the possibility of a shock hazard, a current leakage of 0.75 mA at rated line voltage, in accordance with ANSI/UL *Standard* 250-83, is considered the maximum allowable industry limit.

Handling and Storage Test. In common with most other major appliances, it is during shipping and storage that a refrigerator is exposed to its most severe impact forces, to vibration, and to the extremes of temperature. When packaged, it should withstand, without damage, a drop of several inches onto a concrete floor, the impact experienced in a freight car coupling at 10 mph, and jiggling equivalent to a trip of several thousand miles by rail or truck. The widespread use of plastic parts makes it important to select materials that also withstand the high and low temperature extremes that may be experienced.

This test determines the ability of the cabinet, when packaged for shipment, to withstand handling and storage conditions in extreme temperatures. It involves raising the crated cabinet 6 in. off the floor and suddenly releasing it on one corner. This is done for each of the four corners. This procedure is carried out at stabilized temperature conditions—first in a 140°F ambient temperature, and then in a 0°F ambient. At the conclusion of the test, the cabinet is uncrated, operated, and all accessible parts are examined for damage.

Special Performance Testing

To ensure customer acceptance, several additional performance tests are customarily performed.

Usage Test. This is similar to the internal moisture condensate test, except that additional performance data will be taken during the test period, including (a) electrical energy consumption per 24-h period; (b) percent running time of the compressor motor; and (c) cabinet temperatures. These data give an indication of the reserve capacity of the refrigerating system and the temperature recovery characteristics of the cabinet.

Low Ambient Temperature Operation. It is customary to conduct a simulated load test and an ice-making test at ambient temperatures of 55°F or below. This test determines performance under unusually low temperature conditions.

Food Preservation Tests. This test determines the food-keeping characteristics of the general refrigerated compartment and is useful for evaluating the utility of special compartments such as vegetable crispers, meat keepers, high-humidity compartments, and butter keepers. This test is made by loading the various compartments with food, as recommended by the manufacturer, and periodically observing the condition of the food.

Noise Tests. The complexity and increased size of refrigerators has made it difficult to keep the sound level within acceptable limits. Thus, sound testing is important to ensure customer acceptance.

A meaningful evaluation of the sound characteristics may require a specially constructed room with a background sound level of 30 dB or less. The wall treatment may be either reverberant, semireverberant, or anechoic, with the reverberant room construction usually favored in making an instrument analysis. A listening panel is most commonly used for the final evaluation, with most manufacturers striving to correlate instrument readings with panel judgment.

High- and Low-Voltage Tests. The ability of the compressor to start and to pull down the system after an ambient soak is tested with applied voltages that are at least 10% above and below the rated voltage. (The starting torque is reduced at low voltages; the motor tends to overheat at high voltage.)

Special Functions Tests. Refrigerators and freezers with special features and functions may require additional testing. In the absence of formal procedures for this purpose, test procedures are usually improvised, as required.

Determining Energy Consumption. This is a special 90°F ambient, no-door-opening test using the test procedure for electrical refrigerators and electric refrigerators-freezers and freezers published by the Department of Energy (DOE) in the *Federal Register.* This test procedure specifies a statistical sampling plan, which must be followed to establish the estimated annual cost of energy for labeling under the FTCs Energyguide Program, as well as conformance with DOE energy standards. The DOE test procedure presently references AHAM *Standard* HRF-1-1979 for methods of testing.

Materials Testing

The materials used in a refrigerator or freezer should meet certain test specifications. All materials in contact with foods must meet U.S. Food and Drug Administration requirements. Metals, paints, and surface finishes may be tested according to procedures specified by ASTM and others. Plastics may be tested according to procedures formulated by the Society of the Plastics Industry appliance committee. In addition, the following tests on materials, as applied in the final product, are assuming importance in the refrigeration industry (Federal Specification AA-R-00211 H[GL]):

Odor and Taste Contamination. This test determines the intensity of odors and tastes imparted by the cabinet air to uncovered, unsalted butter stored in the cabinet at operating temperatures.

Stain Resistance. The degree of staining is determined when the cabinet exterior surfaces and the surface of plastic interior parts are coated with a typical staining food (*e.g.*, prepared cream salad mustard).

Environmental Cracking Resistance Test. This tests the cracking resistance of the plastic inner door liners and breaker strips at operating temperatures when coated with a 50/50 mixture of oleic acid and cottonseed oil. The cabinet door shelves are loaded with weights, and the doors are slammed on a prescribed schedule extending over an 8-day test period. The parts are then examined for cracks and crazing.

Breaker Strip Impact Test. This test determines the impact resistance of the breaker strips at operating temperature when coated with a 50/50 mixture of oleic acid and cottonseed oil. The breaker strip is impacted by a 2-lb dart dropping from a prescribed height. The part is then examined for cracks and crazing.

Component Life Testing

Various components of a refrigerator and freezer cabinet are subject to continual use by the consumer throughout the life of the product, and must be adequately tested to ensure their durability for at least a 10-year life. Some of these items are (1) hinges, (2) latch mechanism, (3) door gasket, (4) light and fan switches, and (5) door shelves. These components may be checked by an automatic mechanism, which opens and closes the door in a prescribed manner. A total of 300,000 cycles is generally accepted as the standard for design purposes. Door shelves should be loaded as they would be for normal home usage. Several other important characteristics may be checked during the same test: (1) retention of door seal, (2) rigidity of door assembly, (3) rigidity of cabinet shell, and (4) durability of inner door panels.

Life tests on the electrical and mechanical components of the refrigerating system may be made, as required.

Field Testing

Additional information may be obtained from a program of field testing in which test models are placed in selected homes for observation. Since high temperature and humidity are the most severe conditions encountered, the Gulf Coast is a popular field test area. Laboratory testing has limitations in the complete evaluation of a refrigerator design, and field testing can provide the final assurance of customer satisfaction.

Field testing is only as good as the degree of policing and the completeness and accuracy of reporting. However, if done properly, the information is important, not only in product evaluation, but in providing criteria for more realistic and timely laboratory test procedures and acceptance standards.

REFERENCES

AHAM. 1988. Household refrigerators, combination refrigerator freezers, and household freezers. ANSI/AHAM *Standard* HRF-1-1988. Association of Home Appliance Manufacturers, Chicago.

Federal Register. 1973. Vol. 38, No 242. Dated Tuesday, December 18, 1973.

Federal Register. 1989. Vol. 54, No. 221. Dated Friday, November 17, 1989, pp. 47916–47945.

Federal Specification. 1977. Refrigerators, mechanical, household (electrical, self-contained), AA-R-00211H(GL).

Office of Business Research and Analysis. 1975. Economic significance of fluorocarbons, *U.S. Department of Commerce.*

Refrigeration Safety Act. *Public Law* 84-930.

UL. 1992. Household refrigerators and freezers, 9th ed. UL *Standard* 250-83. Underwriters' Laboratories, Northbrook, IL.

CHAPTER 49

DRINKING WATER COOLERS AND CENTRAL SYSTEMS

UNITARY COOLERS

A MECHANICALLY refrigerated drinking water cooler consists of a factory-made assembly in one structure. This cooler, by means of a complete mechanical refrigeration system, has the primary function of cooling potable water and dispensing it by integral and/or remote means.

Water coolers differ from water chillers in that water coolers dispense potable water, and water chillers are used in air-conditioning systems for residential, commercial, and industrial applications, as well as for cooling water for industrial processes.

The capacity of a water cooler is expressed in gallons per hour and is the quantity of water cooled in one hour from a specified inlet temperature to a specified dispensing temperature (see the section Ratings). Normal standard capacities of water coolers range from 1 to 30 gallons per hour (gph).

Types

Figure 1 shows the three basic types of water coolers.

A *bottle water cooler* uses a bottle or reservoir for storing the supply of water to be cooled and a faucet or similar means for filling glasses, cups, or other containers. It also includes a wastewater receptacle.

A *pressure-type water cooler* is supplied with potable water under pressure and includes a wastewater receptacle or means for disposing of water to a plumbing drainage system (Figure 2). These coolers use a faucet or similar means for filling glasses or cups, or a valve to control the flow of water as a projected stream from a bubbler so that water may be consumed without using glasses or cups.

A *remote-type cooler* is factory-assembled in one structure that uses a complete mechanical refrigeration system and functions primarily to cool potable water for delivery to a separately installed dispensing means.

In addition to the basic descriptions, coolers are also described by (1) specialized conditions of use, (2) additional functions they perform, or (3) type of installation.

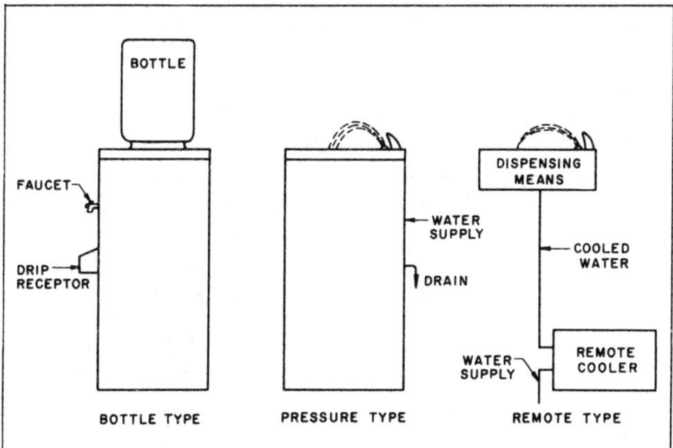

Fig. 1 Basic Drinking Water Coolers

Specialized Uses

- An explosion-proof water cooler is constructed for safe operation in hazardous locations, as classified in Article 500 of the *National Electrical Code.*
- A cafeteria cooler is one that is supplied with water under pressure from a piped system and is intended primarily for use in cafeterias and restaurants for dispensing water rapidly and conveniently into glasses or pitchers. It includes a means for disposing of wastewater to a plumbing drain system.
- A water cooler for wheelchair and handicap access is designed and installed at an appropriate level for debilitated persons.

Additional Functions

- A water cooler may also have a refrigerated compartment with or without provisions for making ice.
- A water cooler may also include a means for heating and dispensing potable water for making instant hot beverages and soups.

Type of Installation (see Figure 3)

a. Freestanding
b. Flush-to-wall
c. Wall-hung
d. Semirecessed
e. Fully recessed

Refrigeration Systems

Hermetically sealed motor compressors are commonly used for both 50- and 60-Hz AC applications. Belt-driven compressors are generally used only for DC and 25-Hz supply. Compressors are similar to those used in household refrigerators and range from 0.05 to 0.75 hp.

Forced air-cooled condensers are most commonly used. In coolers rated less than 5 gph, natural convection, air-cooled (static) condensers are sometimes included. Water-cooled condensers are used on models intended for high ambient temperatures or where lint and dust in the air make air-cooled types impractical.

Capillary tubes are used almost exclusively for refrigerant flow control in hermetically sealed systems. In belt-driven systems and some hermetically sealed systems, expansion valves are used.

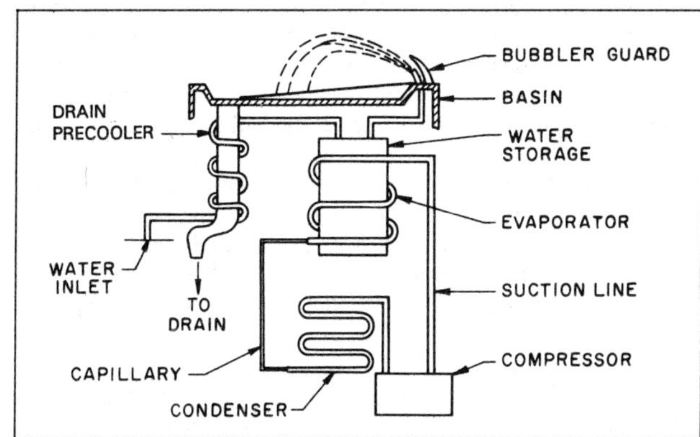

Fig. 2 Pressure Water Cooler

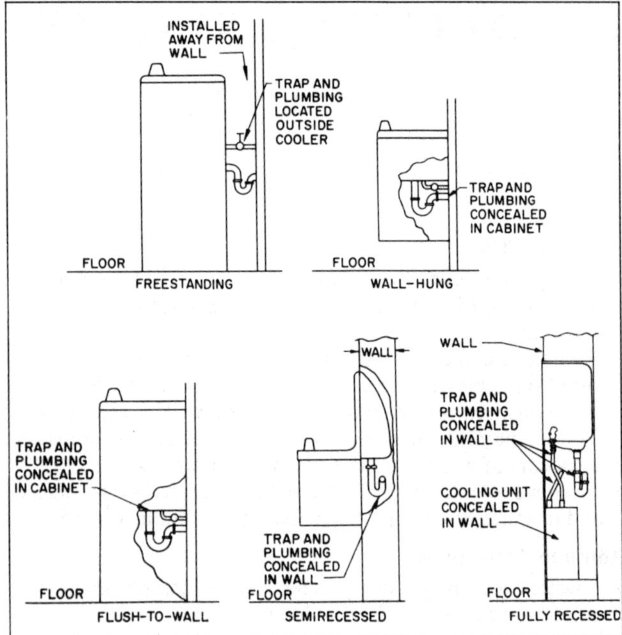

Fig. 3 Types of Installation for Drinking Water Coolers

Most water coolers have the evaporator formed by refrigerant tubing bonded to the outside of a water circuit. However, some are made with an immersed evaporator.

The water circuit is usually a tank or a coil of large tubing. Materials used in the water circuit are usually nonferrous or stainless steel. Since the coolers dispense water for human consumption, sanitary requirements are essential.

Pressure coolers are often equipped with precoolers to exchange heat from the supply water to the waste water. When drinking from a bubbler stream, the user wastes about 60% of the cold water, which runs down the drain. A precooler puts the incoming water in heat exchange relationship with the wastewater. Sometimes the cold wastewater subcools liquid refrigerant in an arrangement called a subcooler. Coolers intended only to dispense water into cups are not equipped with precoolers, since there is no appreciable quantity of wastewater.

Water coolers providing a refrigerated storage space are commonly referred to as compartment coolers. These refrigeration systems vary from a single-series system to two independent systems.

Compartment coolers use a single-series system to feed the water cooling evaporator first and then the compartment evaporator. When the compressor operates, both water and compartment cooling take place. The thermostat is usually located where it is most affected by the compartment temperature. Therefore, the water cooling is affected greatly by compartment loading and usage.

Some compartment coolers with a single compressor have two temperature controls. These controls operate the compressor in conjunction with the solenoid valve(s), which direct refrigerant flow through the desired evaporator, as required. Also, some compartment coolers use two independent refrigeration systems.

Stream Regulators

Since the principal function of a pressure water cooler is to provide a drinkable stream of cold water from a bubbler, it is usually provided with a valve to maintain a constant stream height, independent of supply pressure. A flow rate of 0.5 gpm from the bubbler is generally accepted as giving an optimum stream for drinking.

Table 1 Standard Rating Conditions
(Adapted from ARI *Standard* 1010-84)

Type of Cooler	Temperature, °F Ambient	Inlet Water	Cooled Water	Heated Potable Water[a]	Spill, %
Bottle types	90	90	50	165	None
Pressure types					
Using Precooler or Non-precooler Drain (bubbler service)	90	80	50	165	60
Not using precooler or other heat-transferring device	90	80	50	165	None
Compartment coolers	Standard rating conditions for water cooling noted above apply. During the standard rating test, no ice shall melt, nor shall the average temperature exceed 46°F in the refrigerated compartment.				

[a]This temperature shall be referred to as the Standard Rating Temperature (Heating).

Notes:

1. For water-cooled condenser water coolers, the established flow of water through the condenser shall not exceed 2.5 times the base rate capacity, and the outlet condenser water temperature shall not exceed 130°F.
2. ARI *Standard* 110-90, Air-Conditioning and Refrigerating Equipment Nameplate Voltages.

Ratings

Water coolers are rated on the basis of continuous flow capacity under specified water temperature and ambient conditions. ARI *Standard* 1010-84, Drinking-Fountains and Self-Contained Mechanically-Refrigerated Drinking-Water Coolers, gives the generally accepted rating conditions and references test methods as prescribed in ASHRAE *Standard* 18-1987, Methods of Testing for Rating Drinking-Water Coolers with Self-Contained Mechanical Refrigeration Systems. Table 1 gives the standard rating conditions.

Applications

The following guidelines are suggested for placement and selection of a water cooler:

- **General Locations.** Drinking fountains or water coolers should be available within 200 ft of any location where persons are regularly engaged in work.
- **Prohibited Locations.** Drinking fountains or water coolers should not be installed in toilet rooms or any location where the equipment is exposed to contamination from toxic or otherwise hazardous materials.
- **Ventilation.** Adequate ventilation should be provided for water coolers equipped with air-cooled condensers in accordance with the manufacturer's installation instructions.
- **Minimum Requirements.** Table 2 presents minimum requirements for the application of water coolers based on recognized industry practice.

Table 2 Water Cooler Requirements

	Persons Served per Gallon per Hour
Offices, schools, hospitals, retail stores, buildings, office building lobbies, theater lobbies, airline terminals	25
Light manufacturing	15
Heavy manufacturing	12
Hot, heavy manufacturing	10

ANSI *Standard* A117.1-1986, Buildings and Facilities—Providing Accessibility and Usability for Physically Handicapped People, provides guidelines for the placement of specialized water coolers for wheelchair/handicapped individuals.

Example 1. A manufacturing facility employs 625 people. There are 95 office personnel: 60 have access to one water cooler and 35 have access to another. There are 51 people in a hot, heavy manufacturing environment with access to one water cooler. The remaining 479 people are involved in light manufacturing operations and are equally divided among 5 water coolers. Required water capacities are calculated as follows:

$$\text{Capacity per water cooler} = \frac{\text{Persons served}}{\text{Value from Table 2} \times \text{No. of water coolers}}$$

Capacity for office water coolers

$$\frac{60}{25 \times 1} = 2.4 \text{ gph}$$

$$\frac{35}{25 \times 1} = 1.4 \text{ gph}$$

Capacity for hot, heavy manufacturing coolers

$$\frac{51}{10 \times 1} = 5.1 \text{ gph}$$

Capacity for light manufacturing coolers

$$\frac{479}{15 \times 5} = 6.4 \text{ gph}$$

Standards and Codes

In addition to ARI *Standard* 1010-84 and ASHRAE *Standard* 18-1987, Underwriters Laboratories *Standard* 399 and Canadian Standards Association *Standard* C22.2 No. 120-M91 cover safety requirements. Four commercial item descriptions (A-A-1151, A-A-1152, A-A-1153, and A-A-1154) are specified by the U.S. government.

Many local plumbing codes apply directly or indirectly to water coolers. These codes are directed primarily toward eliminating any possibility of cross-connection between the potable water system and the wastewater (or refrigerant) system. Most coolers have a double-wall construction to eliminate the possibility of conflict with any code.

CENTRAL SYSTEMS

A central chilled drinking water system may be considered in a multistory office building where drinking fountains are stacked one floor level above the other. The system should be designed to provide 50 °F water to the drinking fountains. To allow for heat gain in the distribution system, the chiller should be sized to provide 45 °F outlet water. Water system working pressures should be limited to 80 psi gage.

The components in a central chilled drinking water system can be broken down into the following categories: (1) chiller, (2) distribution piping, and (3) drinking fountains.

Chillers

The chiller may be a field-built system or a factory-assembled unit; it is common practice to use factory-assembled units. In either event, the chiller consists of the following components:

Compressor. A semihermetic or hermetic direct-driven compressor.

Condenser. A condenser that is either water-cooled or air-cooled is used. Large air-cooled condensers are often remotely located.

Evaporator. A direct expansion-type evaporator is used. It may be of the shell-and-tube type, with a separate storage tank or an immersion-type coil in a storage tank. If a separate tank is used, a circulating pump is needed to circulate the water between the evaporator and storage tank. (Check local codes for allowable construction.) Most areas require two thicknesses of metal between potable water and refrigerant circuits.

Storage. A storage capacity of approximately one-half the rated chiller capacity is a good starting point for sizing the tank. Unique and unusual situations require the storage tank size to be modified. Care must be taken in selecting tank materials to reduce the possibility of galvanic action. Refer to the section Distribution Piping System for systems designed without a storage tank.

Pumps. System water-circulating pumps are normally bronze-fitted, close-coupled, single-stage pumps with mechanical seals. Pump sizing is determined from the size of the recirculating system. The pump flow rate should be such that there is no more than a 5 °F rise through the loop.

Controls. Typical system controls include (1) high- and low-pressure cutouts, (2) freeze control, (3) water temperature control, and (4) a flow switch to ensure that there is water flowing before the compressor is permitted to operate.

Options. The following options should also be considered: (1) duplex system pumps with timers, (2) split refrigeration systems with timers, (3) tank anodes, and (4) water filtration combined with activated carbon treatment.

Distribution Piping System

The distribution piping system delivers chilled water to the drinking fountains. The piping can be galvanized steel, copper, or brass designed for a working pressure of 80 psi gage. Recommendations on the use of piping materials that minimize corrosion and galvanic action are presented in Chapter 44 of the 1991 ASHRAE *Handbook—Applications.*

The makeup cold-water lines are made of the same material as the distribution piping. When the water supply has objectionable characteristics, such as a high iron or calcium content, or contains odoriferous gases in solution, filtration and/or activated carbon treatment should be applied in the makeup water.

Some manufacturers choose not to use storage tanks in their chiller design; rather, they use large-diameter manifolds in the distribution system. The chiller manufacturer should be contacted for assistance in manifold sizing.

Insulation is necessary on all the distribution piping and the storage tanks. The insulation should be glass fiber insulation, such as that normally used on chilled water piping, with a conductivity of 0.22 $\text{Btu} \cdot \text{in}/(\text{h} \cdot \text{ft}^2 \cdot {}^\circ\text{F})$ at a 50 °F mean temperature, and with a vapor barrier jacket, or equal. All valves and piping, including the branch to the fixture, should be insulated. The waste piping from the drinking fountain, including the trap, should be insulated. This insulation is the same as that recommended for use on cold water lines.

Drinking Fountains

Any standard drinking fountain can be used on a central drinking water system. They are made of stainless steel, bronze, vitreous china, or marble. It is important, however, that the automatic volume or stream regulator provided with the fountain provide a constant stream height from the bubbler, with inlet pressures up to 80 psi gage.

System Design

The load is based on building population, activity level of the population, and the environment in which the activity takes place. Table 2 lists the appropriate factor(s). Table 3 converts the flow rate to refrigeration load. The heat gain from the distribution piping system is based on a circulating water temperature at 45 °F. Table

4 lists the heat gains for various ambient temperatures. The length of all lines must be included when calculating the heat gain in the distribution piping. Table 5 tabulates the heat input from various sizes of circulating pump motors.

The total cooling load consists of the heat removed from the makeup water, the heat gains from the piping, the heat gains from the storage tank, and the heat input from the pumps. Oversizing the chiller may be included if there is a potential for either future building expansion or a higher population. Otherwise, the cold water storage tank should compensate for any abnormal water draw requirements.

The circulating pump is sized to circulate a minimum of 3 gpm per branch or the gpm necessary to limit the temperature rise of the circulatory water to 5 °F, whichever is greater. Table 6 lists the circulating pump capacity to limit the temperature rise of the circulated water to 5 °F. If a separate pump circulates water between the evaporator and the storage tank, the energy input to this pump must be included in the heat gain.

The storage tank capacity should be at least 50% of the hourly usage. The hourly usage may be selected from Table 2.

Table 3 Refrigeration Load[a]

Water Inlet Temperature, °F	Btu per Gal Cooled to 45 °F
65	167
70	208
75	250
80	291
85	333
90	374

[a]Multiply values by flow rate in gph to calculate the refrigeration load in Btu/h. Refrigeration load = (Specific heat) (Density)(Temperature difference)

Table 4 Circulating System Line Loss

Nominal Pipe Size, in.	Heat Gain, Btu/(h·ft·°F)	Heat Gain, Btu/h per 100 ft (45 °F circulating water) Room Temperature, °F 70	80	90
1/2	0.110	280	390	500
3/4	0.119	300	420	540
1	0.139	350	490	630
1 1/4	0.155	390	550	700
1 1/2	0.174	440	610	790
2	0.200	500	700	900
2 1/2	0.228	570	800	1030
3	0.269	680	940	1210

Table 5 Circulating Pump Heat Input

Motor horsepower	1/4	1/3	1/2	3/4	1
Pump heat input, Btu/h	640	850	1270	1910	2550

Table 6 Circulating Pump Capacity

Nominal Pipe Size, in.	Room Temperature, °F 70	80	90
	(gal/h per 100 ft of pipe)		
1/2	8.0	11.1	14.3
3/4	8.4	11.8	15.2
1	9.1	12.8	16.5
1 1/4	10.4	14.6	18.7
1 1/2	11.2	15.7	20.2

Notes: Capacity includes all branch lines needed to limit temperature rise to 5 °F (water at 45 °F). Values include a 20% safety factor.

For pump pressure, figure the longest branch only. Install pump on the return line to discharge into the cooling air. Makeup connections should be between the pump and the cooling unit.

General criteria for sizing distribution piping for a central chilled drinking water system are as follows:

- Limit the maximum velocity of the water in the circulating piping to 3 ft/s to avoid giving the water a milky appearance.
- Avoid excessive friction losses. Energy required to circulate water enters the water as heat and requires additional capacity in the water chiller. Accepted practice limits the maximum friction loss to 10 ft of head per 100 ft of pipe.
- Dead-end piping, such as that from main riser to fountain, should be kept as short as possible and should not exceed 15 ft in length. Maximum diameter of such dead-end piping should not exceed 3/8 in. IPS, except on very short runs.
- Size piping on total number of gallons circulated. This includes gallons consumed plus gallons necessary for heat leakage.

General criteria for design layout of piping for a central chilled drinking water system are as follows:

- Keep pipe runs as straight as possible with a minimum of offsets.
- Use long sweep fittings wherever possible to reduce friction loss.
- In general, limit maximum pressure developed in any portion of the system to 125 psi. If the height of a building causes pressures in excess of 125 psi, divide the building into two or more systems.
- If more than one branch line is used, install balancing cocks on each branch.
- Provide pressure-relief valves and air vents at high points in chilled water loop.

Example 2. Design a central chilled drinking water system for the 10-story building shown in Figure 4. The net floor area is 14,600 ft^2 per floor, and occupancy is assumed at 100 ft^2 per person. The entering (makeup) water is 70 °F maximum, and the ambient temperature is 90 °F maximum. Applicable codes are the Uniform Plumbing Code, Uniform Building Code, National Standard Plumbing Code, and all local plumbing codes.

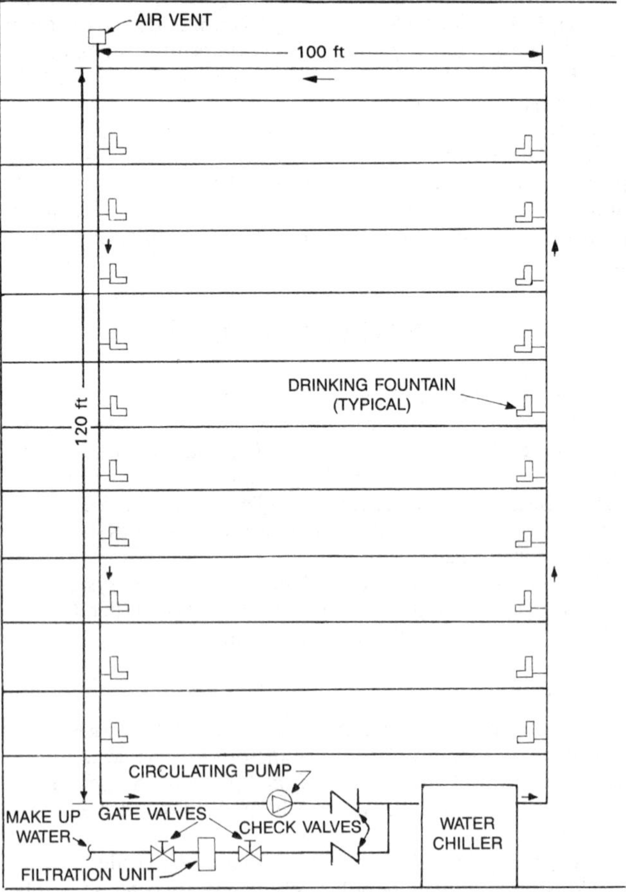

Fig. 4 Typical Central System

Solution:

1. Number of drinking fountains required:
   Occupancy = 14,600/100 = 146 people per floor
   The National Standard Plumbing Code requires one fountain for every 100 people, or 146/100 = 1.46 fountains per floor. Therefore, use 2 fountains per floor, for a total of 20 fountains.
2. Estimated usage load:
   Makeup water = 146 people per floor × 10 floors/25 persons per gal/h (from Table 2) = 58.4 gph
3. Refrigeration load to cool makeup water:
   Refrigeration load = 58.4 gph × 208 Btu per gallon (from Table 3) = 12,150 Btu/h
4. Refrigeration load resulting from piping heat gain:
   Assume a 3/4-in. diameter chilled water circuit. Then, the heat gain from the piping system of Figure 4 is (from Table 4):

Risers	120 ft × 540 Btu/h · 100 ft × 2 risers	= 1296
Distribution mains:	100 ft × 540 Btu/h · 100 ft × 2 mains	= 1080
	Total piping heat gain	= 2376 Btu/h

   The chilled water must be circulated at a minimum of 3 gpm.
5. Refrigeration load resulting from circulating pump input:
   In some chiller systems, one pump circulates water through the evaporator heat exchanger and the building plumbing system. In this case, there is no need to account for the heat gain from the pump, since it is already included in the chiller's capacity. If there are two separate pumps, the heat input from the circulating pump needs to be considered. Assume that this system uses two pumps.

   Pump selection:

   15.2 gph/100 ft (Table 6, 3/4-in. pipe at 90 °F ambient) × 440 ft = 67 gph or slightly over 1 gpm. Since this is less than the established 3 gpm minimum circulating rate, the 3 gpm figure will be used in determining the plumbing system pressure drop.

   Pressure drop calculations:

   Pipe length = 440 ft
   Increase pipe length 50% to allow for fittings. If an unusually large number of fittings is used, consider each for its actual contribution to pressure drop. Therefore:

   Equivalent pipe length = 440 (1.5) = 660 ft
   Water flow = 3 gpm
   Pipe size = 0.75 in.
   Pressure drop per 100 ft of pipe (Table 7) = 4.1 ft of head
   Total pressure = 4.1 × 660/100 = 27.1 ft of head

   Pump manufacturer's literature shows that a 1/3-hp pump motor is required. From Table 5, the heat input of the pump motor is 850 Btu/h.
6. Refrigeration load resulting from storage tank heat gain:
   Normally, a tank sized for 50% of the anticipated hourly demand is specified. In this example, 60 gal/h is the hourly demand; therefore, a 30-gal storage tank is used. This is approximately the capacity of a 16-in. diameter by 48-in. long tank. Assume 1.5-in. insulation, 45 °F water, with the tank being in a 90 °F room. Assume the insulation to have an overall heat transfer coefficient of 0.13 Btu/(h · ft^2 · °F). The surface area of the tank is about 20 ft^2. Therefore, the heat gain is:

$$20 \times 0.13\,(90 - 45) = 117 \text{ Btu/h}$$

7. Load summary:

Item	Heat gain, Btu/h
Makeup water	12,150
Piping	2,376
Circulating pump heat	850
Storage tank	117
Required chiller capacity	15,493

Codes and Regulations

Most mechanical installations are regulated by local codes, most of which are based on guide codes prepared by state or national code-writing organizations. Usually, one of these guide codes has been selected by a particular municipality or other governmental body and has been modified to suit local conditions. For this reason, it is important to refer to the actual code used in the locality. Other codes that require careful review include refrigeration codes. Many follow ASHRAE *Standard* 15-1992, Safety Code for Mechanical Refrigeration, but they may have exceptions. Electrical regulations, as they apply to control and power wiring, and ASME requirements for tanks and piping must also be followed.

Table 7 Friction of Water in Steel Pipes

Flow Rate, gpm	1/2 in. Pipe		3/4 in. Pipe		1 in. Pipe		1-1/4 in. Pipe		1-1/2 in. Pipe	
	Velocity	Pressure Drop	Velocity	Pressure Drop	Velocity	Pressure Drop	Velocity	Pressure Drop	Velocity	Pressure Drop
1	1.05	2.1	—	—	—	—	—	—	—	—
2	2.10	7.4	1.20	1.9	—	—	—	—	—	—
3	3.16	15.8	1.80	4.1	1.12	1.26	—	—	—	—
4	—	—	2.41	7.0	1.49	2.14	0.86	0.57	—	—
5	—	—	3.01	10.5	1.86	3.25	1.07	0.84	0.79	0.40
10	—	—	—	—	3.72	11.7	2.14	3.05	1.57	1.43
15	—	—	—	—	—	—	3.20	6.50	2.36	3.0
20	—	—	—	—	—	—	—	—	3.15	5.2

Note: Pressure drop is in feet of water due to friction through 100 ft of smooth, straight steel pipe. Velocity is in ft/s.

CHAPTER 50

AUTOMATIC ICE MAKERS

THIS chapter addresses commercial-size automatic ice makers—their construction, operation, and application. Specifically omitted are ice makers used in domestic refrigerators (covered in Chapter 48), large plant units requiring trained personnel (covered in Chapter 32), and ice machines used for thermal storage (see Chapter 39 in the 1991 ASHRAE *Handbook— Applications*). This chapter is concerned with machines that make small, fairly uniform pieces of ice, such as ice flakes or ice cubes, up to approximately 2 oz in mass. These units usually include a bin or ice storage facility and are arranged and controlled to maintain a certain stated amount of product in storage. Unit capacities range from production rates of 15 lb/24 h to several tons a day.

Fundamentally, ice means spot cooling. Water resulting from meltage is often used in applications where products, especially vegetables, tend to lose weight (through dehydration) and become less attractive in appearance. Today, the largest markets for ice makers are bars, restaurants, cafeterias, soft drink parlors, motels, hotels, hospitals, fish and vegetable markets, concession stands, fast-food services, and prepackaged ice for home use.

Although ice can be made from sea water (with use usually confined to fresh fish preservation) or almost any other water, ice in the automatic ice-maker industry is customarily made from potable water and is kept relatively uncontaminated. Further classifications are *clear ice* and *cloudy ice*; there are different degrees of cloudiness, depending on the ice-maker design.

Machines using a batch process make clear ice particles or chunks; machines using a continuous process either produce cloudy flakes or they compact the flakes into more or less opaque pieces of various geometric shapes. An appreciation of the characteristics and source material (water) of product ice is important not only to the design engineer but also to anyone wishing to apply, specify, or purchase ice-making equipment for specific applications.

Fundamentally, ice is crystalline in structure and follows the physical laws of crystalline material. However, water is seldom pure H_2O and, depending on its source, is generally made up of a complex mixture of layers of tiny water crystals interspersed with layers of an amorphous matrix of other chemicals. This composition accounts not only for the clarity (or lack of it), but also for a number of other qualities of ice, such as hardness or, particularly, the shearing quality.

Factors influencing the actual character of the ice formed are not only the water of which it is made but also the freezing rate and degree of washing of the interface between the already frozen ice and the water to be frozen.

The process of removing heat from a body of water to turn it into ice is normally a concentrating process that tends to freeze pure H_2O and leave the remaining liquid water with a higher percentage of extraneous chemicals than it had at the beginning of the process. This concentrating occurs at the interface between the newly formed ice and the water surrounding it. If the freezing rate is too fast, the rejected chemicals are frozen into a matrix surrounding the pure H_2O crystals, and the remaining water is consumed or converted into ice at the original concentration. If, however, freezing is slow enough or the ice interface is washed by a flow of water, these rejected chemicals can be removed and the resulting ice is significantly purer than the water from which it was made.

Continuous ice makers make ice that contains most of the chemicals in the original water, although some rejection of extraneous chemicals may result if sufficient bleeding of the makeup water mixture is continuously maintained. Units operating on a batch cycle, however, tend to produce ice that is purer than its source water.

Since water easily dissolves many substances, many impurities are always present in natural water, the most common of which are compounds of sodium, calcium, magnesium, and iron. In addition, the water may carry such suspended impurities as fine clay, sand, and fragments of vegetation, as well as entrapped air and microscopic organisms, including bacteria.

Calcium and magnesium salts make water hard and tend to come out of solution and deposit in an ice maker, eventually making the ice maker inoperative if periodic cleaning and lime removal are not performed conscientiously. Typhoid, cholera, and dysentery are primarily spread by infected water supplies, so only water that is bacteriologically and chemically safe must be used for product ice that is to be used in beverages or is to come in direct contact with food.

Practically all domestic water systems have different mixtures of the same common chemicals. Whenever the total chemical content exceeds 400 ppm, or if especially objectionable gases or compounds are present, auxiliary water treatment is indicated.

DEFINITIONS

The following terms are used in the ice-making industry:

Pounds of Corrected Products, or **144 Btu Ice** describes an imaginary product that absorbs 144 Btu/lb while melting at 32°F.

To convert the product of any particular machine to 144 Btu ice, a sample of the actual product as it leaves the evaporator is tested in a calorimeter. The reading in Btu/lb of product is then divided by 144, which gives a multiplier to be used with that machine to establish an equitable comparison of pounds of ice per 24 h among all types of machines.

Cube Ice normally refers to a fairly uniform product that is hard, solid, usually clear, and generally weighs less than 2 oz per piece—as distinguished from flake, crushed, or fragmented ice.

Flake Ice or **Flaked Ice** is made in a thin sheet anywhere from approximately 0.06 to 0.18 in. thick. The sheet may be flat or curved, but the thin ice is generally broken into random-sized flakes when harvested. The term also refers to machines that produce ice in this manner and, additionally, compress or extrude a product in larger chunks that either resemble pebbles or are in irregular shapes but fairly uniform sizes.

Crushed Ice is made in hard masses that are later crushed into a smaller size. This ice is characterized by the amount of fine or slush ice mixed in with the more uniform larger chunks.

The preparation of this chapter is assigned to TC 10.2, Automatic Icemaking Plants and Skating Rinks.

Blowdown or **Bleedoff** is the rejection of a certain amount of recirculated or ingredient water in order to control the amount of chemicals that are present because of the concentrating effect of water frozen into ice.

Harvesting or **Harvest Cycle** (sometimes called the **defrost cycle**) is the removal or separation of the manufactured ice from the evaporator.

Auger originally referred to the scraper or helical wedging device that rotated inside a cylindrical evaporator, removed the ice from the interior wall, and pushed the separated ice up and out of the evaporator section. Later, the term was applied to the ice remover (usually helical), even when the ice was made on the outside of a small, vertical, cylindrical evaporator.

Density, as used in this chapter, refers to the mass per unit volume (lb/ft^3) of a sample of the product from a given machine and not to the density of an individual particle of product. The definition is useful in determining the amount of ice that can be stored in a bin having a known usable storage volume.

ICE-MAKER CONSTRUCTION

Types of Product and Evaporators

Flaked ice is uniformly thin, randomly shaped in its perimeter, and sometimes clear but more often cloudy or actually white. Evaporators currently used to produce flaked ice are of four main forms: cylinders, flat plates, flexible belt, and disks.

The most common evaporator is a cylinder, usually brass, with refrigerated tubing wrapped on the outside and bonded to it. The evaporator may also be two concentric cylinders with water on either the outside of the outside cylinder or the inside of the inside cylinder, and with the annular space between cylinders occupied by the refrigerant. The simplest form, however, is a cylinder with the refrigerant on the inside and the water to be frozen on the outside. These evaporators require augers, helixes, or equivalent components to remove the ice.

Both cylinder types can be made with an extended auger section in which the ice flakes removed are compressed and the resultant product extruded and ejected. This compressed product differs considerably in both density and appearance from the uncompressed flaked ice and can be binned and dispensed far more readily.

The cylinder evaporator is subject to many variations. Both clear and opaque ice can be produced, depending on the means of distributing the water. The cylinder can also be stationary or rotated, as well as placed in a vertical or horizontal position. The harvesting means is also open to many variations, depending on whether ice is formed on the inside or the outside of the cylinder.

The cylinder evaporator has been produced to meet capacities from 100 lb to 20 tons/day, using the whole range of refrigerants from the halogenated fluorocarbons and methyl chloride to ammonia. Larger flexible-cylinder machines have also been refrigerated with circulated cold brine. In general, if the type of ice produced is satisfactory for the intended application, this evaporator form can be the least expensive to produce and the most efficient to operate thermodynamically. These machines are almost exclusively operated on a continuous cycle rather than in a batch process, although batch-process machines are available when clear ice is sought. Cracked ice can be produced on a flat evaporator, using a batch process and either an ice crusher or some other ice-breaking feature.

The *belt-ice* method is used mostly in industrial-size machines. Belt ice is produced from a metal belt turning over two pulleys—the belt sliding upward across a slightly-curved inclined refrigerated surface. Flowing water strikes the belt near the upper end of its length and flows downward, forming ice as it flows. The ice thus formed continues past the point of impact of the water and leaves the belt in a ribbon or sheet as the belt passes over the upper pulley.

Cube ice is produced from a great variety of evaporators. Most of these machines fall into five main types: flat plates (using either the whole surface or selected spots), multiple cells or molds, tube machines, rod machines, and channel machines.

Flat-plate evaporators may be vertical or in an inclined position; it may have water flowing on either the top or the underside of the evaporator. The main characteristics are that the ice is formed in a rectangular slab during the freezing cycle, and it is usually quite clear. During the harvest cycle, the evaporator plate is heated, and the ice drops or slides off. This slab is then cut into square or rectangular pieces, most often with a grid of either electrically heated wires or small tubes carrying a heated fluid. The slab thickness can usually be varied at will, but the cube sizes are determined by the previously selected grid pattern.

There are several other versions of the flat-plate evaporator in which grids of metal or plastic are pressed against the flat evaporator and a third movable element closes the side opposite the evaporator plate, thereby forming individual cells. Recirculated water is supplied to each cell. In these versions, the individual cubes are ejected during the harvest cycle, and no cutting is required.

Cell- or **mold-type evaporators** make the individual cubes without any additional cutting. The cups or cells may be any shape from round to square to polysided. They may be inverted, and then water is sprayed up into the cavities to produce clear ice. The cells are usually perpendicular to the vertical plate. Water then flows by gravity from the top of the evaporator and washes into the cells enroute to a recirculating water pan. During the harvest cycle, the water is shut off, and the cavities are heated in various ways to free the ice cubes. The cubes fall onto a grill that directs the ice to a chute or bin below.

Tube machines make ice inside the tubes, with the refrigerant outside the tubes. The ice is made in a long cylindrical shape, usually with a hole in the center, and is broken into short lengths as it emerges from the evaporator. The ice is kept clear by making sure that an excess of water is continuously passing through the tubes.

In one version, tube sheets within a vertical shell hold a series of tubes in place. The shell and tubes act as a flooded evaporator during the freezing process and as a condenser during the harvesting cycle. Water is pumped to the top of the shell and flows by gravity down the inside walls of the freezing tubes.

Another version is a single tube within a tube. The ice is formed on the inside of the inner tube, and the refrigerant is in the annular space between the two tubes. Water is pumped through the inner tube until the resistance to flow becomes greatly increased, at which time hot gas is introduced into the annular space. When ice is free, it is forced out by water pump pressure and cut or broken into lengths of two or three diameters.

A third cube machine of the tube type forms separate ice bodies in square, vertical, stainless steel tubes. These tubes are banded at uniformly spaced intervals with copper heat conductors, which, in turn, are refrigerated. Water flows down the inside of the tubes and turns to ice in the region of the bands, thereby producing clear ice cubes with an hourglass-shaped hole through the center. Hot-gas harvesting is used, and the cubes fall by gravity into a storage bin below.

Rod units usually employ a series of short refrigerated rods (or, in some cases, short tubes) sealed at the bottom, which drop into a tank of flowing or agitated water. Clear, thimble-like pieces of ice form, and at a predetermined time, the water tank is removed (or the fingers are raised). When the hot gas thaws the ice loose, the ice falls free and is guided into an opening that leads to an ice bin. Both the length of the ice piece and its outer diameter can be varied.

Channel-type evaporators are made by forming thin, stainless steel sheets into a series of channels. These sheets are then mounted upright so that the channels are vertical. On one side, water flows down the sides and bottom. On the opposite side and at right angles to the channels is a refrigerant-carrying serpentine tube. This tube is bonded to the stainless steel at equally spaced intervals, thus forming a series of cold spots over which the water flows. When hot gas replaces the cold refrigerant, individual pieces of ice are freed and fall into a bin below.

Almost as many types of evaporators and ice machines have been built as there are makes of automobiles. The machines that have survived have the fewest number of moving parts, are economical to manufacture, and have controls as close to fail-safe as possible. As long as the ice is compatible with its ultimate use, the acceptance of any specific design is based as much on reliability as on efficiency or even first cost.

The Complete Ice-Maker Package

Automatic ice-making equipment can be purchased in a wide variety of models, with some companies offering as many as three or four different ice shapes. Some units are made of separate modules capable of being assembled into complete units to satisfy a wide range of applications. These modules often include an automatic ice-making section for either cubes or flakes; the section shuts off when it is set on a container or bin of some kind. Crusher sections are offered that can be set to produce crushed ice from ice cubes and direct uncrushed ice into one bin section and crushed into another. Various bin sizes with different access ports are often available. Additionally, sanitary dispensing units are offered as an alternate to a simple ice bin. Other add-on accessories include water stations for filling drinking water glasses with ice and water and soda fountain dispensing units.

In addition to the modular approach, units are specifically designed for many special applications. These special units may be designed to conserve floor space or fit under counters. Hospital units must protect the product from contamination, so the units must be able to withstand thorough cleaning and sanitizing by cleaning personnel.

Coin-operated machines that make, bag, and vend ice (all automatically) have not proved to be economical because the cost and complication of the machines are too great for the yearly volume of ice sold. However, vending machines that simply vend bags of ice are gaining acceptance. In this case, the ice is made and bagged elsewhere. This operation is especially economical in connection with merchandising requiring refrigerated holding rooms for storing backup supplies.

Performance and General Operating Characteristics

ARI (1990) shows typical water use and energy input for ice cube machines tested in accordance with ASHRAE *Standard* 29-1988 as follows:

	Air Cooled	Water Cooled
Potable water used, gal/100 lb ice	13.2 to 63.1	13.2 to 75.7
Condenser water used, gal/100 lb ice	na	108 to 301
Energy input, kWh per 100 lb ice	5.4 to 22.5	4.7 to 14.2

System Design

All automatic ice-making systems can be broken down into the following subsystems:

1. Refrigeration circuit
2. Water (for ice making) circuit
3. Ice removal and/or harvest system
4. Unit electrical controls
5. Ice storage and dispensing systems, where included
6. General safety and sanitation codes as they apply to each of the above systems
7. Special application and specifications that may bear on individual system designs

Refrigeration Circuit. Any refrigeration system is a balance of the compressor capacity, the condenser capacity, and the evaporator capacity. In the previous sections, the types of evaporators in current production were reviewed. The thinner the ice produced, the more thermally efficient the production method; as a corollary, the more prime refrigerant surface per in.^{3} of ice, the more efficient the heat transfer. Also, a continuous process without a harvesting and a freezing cycle is thermally more efficient than a batch process. The determining factors in any evaporator design are (1) production of the kind of ice required for the specific end use, (2) manufacturing economics, and (3) the ease of cleaning and extent of service required to keep the evaporator running for at least eight or ten years.

The best heat-transfer metals are the most desirable for an evaporator. The surface on which ice is made, however, must be corrosion resistant, nontoxic, nonporous, and readily cleanable; most important of all, the refrigerant passages must be clean and clear of all oxides and corrosive fluxes. The cross-sectional areas must be uniform so gas velocities do not fall below those necessary to prevent oil separation or accumulation.

The high side (compressor, condenser, and receiver) is generally made up by the ice maker manufacturer from separately selected components suitable for the specific design, rather than from stock high-side units furnished by compressor manufacturers.

Both air- and water-cooled condensers have their particular application advantages. The air-cooled units are the simplest; however, proper ventilation of the installation area may not be practical, or the rejected heat may be too much of a load for the air-conditioning system if the unit is installed in an air-conditioned space. Thus, water-cooled units may be the simplest way of conducting all the rejected heat from the installation area. An alternative is a remotely located air-cooled condenser, which rejects heat away from the air-conditioned space and yet retains the economy of the air-cooled system.

Commercial high sides are usually low-temperature units for flaked-ice machines and medium-temperature units for cube-ice machines (batch process systems where hot gas is used for defrost). The compressor motors are generally of high-torque design.

Refrigerant 502 is most commonly selected, although R-22, R-12, and several refrigerant blends are also used, especially in the larger capacities.

The liquid control means may be (1) an automatic expansion valve (especially good if the evaporator load is fairly steady), (2) a fixed superheat thermal expansion valve (the easiest to apply), or (3) a capillary restrictor tube (the least expensive to provide, but the most difficult to apply).

If a thermal expansion valve is used, a liquid receiver generally is required, but if an automatic expansion valve or a capillary tube is used for liquid control, a suction line accumulator is generally used.

Unless the manufacturer has sophisticated process equipment and a reliable quality-control system is in operation, all systems should have strainer-driers in the high side. Ice-making systems are essentially low-temperature systems and are, therefore, very sensitive to excess moisture in the refrigerant circuit, particularly in systems using Refrigerant 12.

For most cubers and other batch-process units, a hot-gas solenoid valve bypasses hot compressor discharge gas around the condenser and moves it directly into the evaporator during the harvest cycle. This is normally an electrically operated closed valve, which must be large enough to (1) provide as short a harvest cycle as practical without creating such a thermal shock that the ice tends to break up and (2) prevent excessive compressor overload.

High- and low-pressure controls may or may not be used in the refrigerant circuit. Most water-cooled condensing units use automatic-reset high-pressure controls. Depending on the extent of system safety, the low-pressure cutout is most often a manual reset control; this arrangement ensures that the unit will be investigated to determine the cause when an excessively low pressure develops.

Ingredient Water Circuit. The potable water used to make ice is normally a part of the city water supply. In the majority of flaked-ice units a float valve maintains a constant water volume in the system. This mechanically operated valve is usually mounted in a separate float chamber.

A flexible line runs from this float chamber to the evaporator section. A constant and fixed water level is maintained in the evaporator because of the vertical relationship between it and the water level in the float chamber.

Where clear ice is made, a water pump usually circulates water, taken from a water tank or sump, through the evaporator, from which the water returns to the sump. A float valve normally maintains a constant water level in the sump. In this case, the float valve may be either mounted directly in the sump or have a separate float chamber.

The float valve and its enclosure must meet sanitary code requirements so that no back siphonage can occur to contaminate the primary water supply. This is usually accomplished by ensuring that the valve-outlet orifice is at least 1 in. above the float chamber or any possible flood level within the machine proper. Also, if such a flood develops (for instance, because of a failure of the float valve to shut off), the exposed electrical equipment must not be shorted out by splash or free-running water.

Unit protection from water failure with air-cooled condensers is normally limited to a low-pressure cutout in the refrigerant circuit. Not all units use water pumps. In some machines, a stirring device keeps water moving. Compressed air has also been used to agitate water adjacent to the ice.

The trend in ice makers is to have a minimum quantity of water in the water system at any one time. If the unit is of the continuous-process type and is making clear or fairly clear ice, a constant bleedoff must be used to control the mineral concentration. With this type of operation, no relationship exists between quantity of bleedoff and total system water quantity. In batch processes, where clear ice is made, either a constant bleedoff or batch dumping a certain amount of water at the end of each cycle is used. As stated earlier, keeping the water quantity to a minimum maintains a good flushing action at the end of each cycle. This dumping can be obtained by a siphon effect, initiated by stopping the water pump and allowing all the system water to flow back into the water sump, thus increasing the water level to the point where the siphon starts. Alternatively, it may be simply a matter of energizing a solenoid valve in the water sump drain line, allowing any given amount of water to escape during the harvest cycle.

In all cases, drainage must be provided at the lowest point in the water system so the unit can be drained completely after any routine cleaning operation, before shipping, and when preparing for winter storage. All materials that come into contact with the ingredient water or the ice must be nontoxic, noncorrosive, smooth, impervious, nonodor imparting, cleanable, and durable. Minimum requirements for sanitary design are specified in the National Sanitation Foundation *Standard* 12; which is almost mandatory.

Ice Removal and/or **Harvesting Systems.** Usually an auger removes the ice from flaked-ice continuous-process machines. The design of most of these machines is based on breaking the ice away from the evaporator at the interface between the ice and the metal freezing surface. However, this bond is not consistent in shear strength; this, as well as the irregular pattern of the ice leaving the freezing surface, causes a highly irregular torque requirement on the auger drive shaft. As a result, side thrusts on the auger vary considerably. As designed, cutting or breaking the bond is balanced equally between cutters approximately 180° apart. When ice breaks loose irregularly, this balancing effect is lost. Side thrusts become of major proportion; thus, if close clearances are to be maintained between cutter and evaporator, the evaporator shell and auger must be quite sturdy.

The strength of the primary bond or fracture plane between evaporator and ice is influenced by many variables. Strength can vary with the temperature of the ice at the bonding plane, the water composition, the conditions of the evaporator surface, and the angle and shape of the cutting blade. These conditions can change from day to day or season to season. Thus, the ice-harvesting design becomes crucial, and this factor alone has been the downfall of many promising evaporator designs.

Most ice-cube machines are batch-process units that use hot gas to melt the bond between the ice and the evaporator. If the ice is made in cups, the adhesion or surface tension can create a vacuum and prevent the ice from falling. Either vacuum-breaking holes or some means of exerting pressure on the ice may be required to release the ice. When ice forms on ice rods, it must reach a critical mass to overcome the adhesive forces at the interface between evaporator and ice so it will drop off. Ice is abrasive, and this fact must be considered on all ice-handling components.

Unit Electrical Controls. The control systems for continuous-process machines are relatively simple, requiring only that the unit be shut down manually or automatically when an ice bin has been filled. To protect the machine, controls must shut the unit down whenever the water supply fails, an auger overloads its drive motor, excessively high discharge pressures are experienced, or abnormally low suction pressures are encountered.

A batch ice maker, which goes through a freezing and harvesting cycle, requires a more complicated control system than one that freezes continuously. A batch cycle must terminate the freezing at the proper time to start the thawing cycle and to resume the freezing operation when harvesting is complete. Timers, pressure-operated switches, ambient-compensated thermostats, water-overflow actuators, two-element thermostats, ice thickness feelers, and combinations of these have been used. Selection depends on what is most suitable for the particular evaporator design. Additional control is needed to shut off the unit when the storage bin is full of ice; a temperature-sensing element, which can be contacted by the ice, is often used. Mechanical feelers and photoelectric devices have also been applied successfully.

All controls must be effective over a range of ambient air temperatures from 40 to 110°F and supply water temperatures from 40 to 100°F.

Numerous safety devices help prevent damage to the apparatus or injury to personnel. These include (1) high-pressure cutout, (2) low-pressure cutout, (3) motor overload protector, (4) overfreeze protection, (5) fusable plugs in receiver shells, (6) safety switches interlocking with access panels, and (7) other devices that a particular design may require.

Ice Storage and **Ice-Dispensing Systems.** Most automatic ice makers provide an ice storage bin designed to hold approximately a 10- to 12-hour ice production. A certain demand does exist for units of a medium-to-large capacity of both flaked ice and ice cubes with no integral bins. These units are ideal for applications where high demands occur less frequently than every 12 hours. For example, country clubs have high weekend demands. Also, supermarkets often use ice beds to display vegetables, meats, or fish; these beds are remade only once or twice a week. Often, a standard bin is made by the ice maker manufacturer and holds the usual 12-hour production of ice. Bins made by independent ice bin manufacturers are better for larger quantities of ice for special applications.

Ice bins are usually built to NSF construction standards, with from 1.5 to 3.0 in. of insulation on the sides and top and from

2 to 4 in. of insulation on the bottom. These bins are seldom refrigerated because the ice may refreeze together, so drainage must be provided. Drains should never be less than 0.5 in. and preferably should be 0.75 in. in internal diameter. A strainer should be included at the drain inlet. Access to the ice can be achieved at the top, bottom, or side of the storage bin.

Ice is quite impervious to outside attack. However, as air near the ice cools, moisture and any contaminants present in the air are condensed onto the ice surface. Besides unsanitary areas, the greatest source of contamination lies in scoops, shovels, or other instruments introduced into the ice and then withdrawn and left in other unsanitary areas.

Many sanitary dispensers are available, but the first cost has held sales down, except in those applications where sanitation is crucial. Such an ice dispenser (1) must protect the ice from outside contamination while it is in storage, (2) must be readily cleanable so that frequent cleaning can be done by regular housekeeping help, and (3) should cost less than the ice machine. Many current designs meet these objectives, and sanitary ice dispensers are an increasing part of the standard product line for major manufacturers. A need for storing and dispensing uncontaminated ice continues to be of major interest to hospitals, motels, restaurants, and operations where ice may come in direct contact with food or beverages.

Safety and Sanitary Standards. The following standards apply to ice-making equipment.

National Sanitation Foundation *Standard* 12, *Automatic Ice Making Equipment.*

Underwriters Laboratories, Inc. (UL) *Standard* 563-84, *Ice Makers*, covers electrical safety.

Canadian Standards Association (CSA) *Standard* 313 covers electrical safety of ice-maker products for the Canadian market.

Air-Conditioning and Refrigeration Institute (ARI) *Standard* 810-79, *Automatic Commercial Ice-Makers*, details the conditions and methods of rating. These ratings are based on ASHRAE *Standard* 29-78, *Method of Testing Automatic Ice Makers.*

Other standards, such as the ASME *Unfired Pressure Vessel Code,* may apply, as well as individual state and city plumbing and sanitary codes.

Special Concerns. Noise and vibration, while of little concern in a busy kitchen, should be considered for equipment installed in a dining room or quiet hospital area. A noise level equal to that of a window air conditioner is acceptable for most applications, although some specifications may quote specific NC sound power levels.

Condensation or sweating on the bin or ice-dispensing part of the equipment may be a concern. Generally, equipment is satisfactory for most applications if no moisture drips from the unit when it operates for 4 h in an ambient of 90 °F dry bulb, 78 °F wet bulb.

Installation, operation, and owner maintenance determines the eventual life of the unit, no matter how well it is designed and built. Therefore, the designer should make the unit as simple and easy to clean as possible. In addition, operating and maintenance instructions that are clear and positive will help the owner extend the life of the machine.

Special Applications and Specifications. Several special applications require additional nonstandard design features. Shipboard application usually requires that units continue to produce ice when they are subjected to a 15-degree pitch and a 30-degree roll; for naval vessels, however, 15-degree permanent list and high shock are additional requirements. Also, a no-radio interference requirement is often added.

Explosion-proof units are sometimes required, but this is a special feature that normally is not a production item. Usually, odd electrical current characteristics can be handled by transformers, since direct current is now rarely encountered. For the international market, many manufacturers provide specific models that meet specific local voltage and frequency requirements. Occasionally, measures are required to prevent fungus and high humidity.

APPLICATION

Many ice-maker designs, capacities, and ice shapes are available. The primary factors in selecting the best equipment for any specific application include the following considerations:

1. What type (or types) of ice best satisfies the application?
2. What storage or dispensing methods will be required?
3. Is the cost of ice of primary importance?
4. Is the cost of electricity and/or water of primary importance?
5. What is the quality of the available water supply?
6. Are there noise-level restrictions that must be met?
7. Are reliability and long life of primary importance?
8. Are service and maintenance facilities readily available?
9. What are the space limitations (if any) for the installation?
10. Are there any general environmental conditions or restrictions that must be met?

Choice of Ice Types

Differentiating among the types of ice, two general physical characteristics are apparent: (1) ice can be clear, white, or have varying degrees of opacity; and (2) the individual ice pieces can come in a myriad of sizes and shapes. Generally, the shape description is cubes, flakes, crushed, or aggregate. This last designation is not generally used, but it is meant to cover the more or less uniform ice that results from thin ice flakes that have been compressed and extruded and then broken again as the extrusion is forced against a stop of some kind. (Trade designations are *pebble ice, granular ice, nugget ice,* or *cracked ice.*) This ice differs from crushed ice in that it has relatively few fines and is neither entirely clear nor entirely white, but has various degrees of cloudiness. Also, ice that is stored for a few hours may vary greatly from ice just coming out of the ice-maker discharge chute. The stored ice, unless cubed, is said to cure. In addition to having been drained of most of its excess water, it has changed its structure so that both its density and its thermal capacity (Btu/lb) have changed. This change can be important when figuring bin capacities.

Sometimes, maximum ice size is important. For instance, it must be small enough to readily enter the opening of a patient's carafe for hospital use. It must not have so many fines that it clogs straws or glass drinking tubes.

Shape can also be important; for instance, ice flakes are suited for packing flowers because they do not bruise the petals. This shape is also ideal for fish because it tends to conform to the surface on which it lies. On the other hand, an egg shape has long-lasting qualities because it presents a minimum of outside surface to the liquid to be cooled. Also, various forms of ice react differently in bins. Some do not pack well, and some do not lend themselves to deep bins. Ice cubes generally flow more easily.

Probably the most important consideration for the selection of one ice form over another is user preference. In many applications, one type of ice is clearly superior to others, but, frequently, any form will do the job. When ice is to be delivered to specification, however, all ice-making capacities must be quoted from ASHRAE *Standard* 29-1988, which converts all product mass to 144 Btu/lb ice per 24 h at stated ambient and water conditions. Also, ice bin capacities generally are understood to mean pounds of cured ice and not 144 Btu/lb ice, unless the latter is specifically mentioned. Normally, these are not significant points, but they become important if strict specifications are involved or actual cooling ability is guaranteed.

In the beverage industry, the ice must not impart any flavor to the beverage. If the local water contains gases or solids that impart a distinctive odor or taste, a flake-ice machine may not be satisfactory, whereas a cube ice maker would eliminate most of the objectionable flavor. On the other hand, proper water pretreatment may make flake ice acceptable.

Ice Storage

Most automatic ice makers not only produce ice, they also store it and provide some means of dispensing it. Storage is expected to be an insulated container with 2 to 3 in. of insulation designed for periodic cleaning or sanitizing. (Ideally, it will meet NSF standards). Additionally, it must be designed so that ice can be discharged from the bin with a minimal hazard that the withdrawal will affect the sanitary conditions of the remaining ice. (This ideal prevents the use of scoops or shovels that touch the ice remaining in storage.)

This design also precludes any openings into the ice-containing area in which packages or bottles can be left to cool and possibly contaminate the ice in storage. Whether or not the equipment is maintained in a sanitary condition is entirely up to the owner. The equipment supplier, however, must ensure that the equipment is capable of being maintained in a sanitary condition when straightforward cleaning instructions supplied by the manufacturer are followed.

Table 1 Ice-Maker Application Data[a]

Classification	Application	Preferred Form of Ice	Use Cycle[b]	Consumption[c]
		Food and Drink Service		
Fast-food drive-in	Water, soft drinks	Crushed/flake	7 days; 150% 2 consec. days	0.25 to 0.5 lb/customer
Night clubs	Food-drinks	Cube	7 days; 200% 2 consec. days	2 to 4 lb/seat
Bar cocktail lounges	Drinks			
Carry home	Retail stores			
	Recreation areas	Cube/crushed	7 days; 200% 3 consec. days	Varies by season, available in
	Marinas, fishing piers			5 and 10 lb size bags
Caterers	Banquets	All	Varies	1 lb per meal
	Truck (mobile)	Flaked/crushed	7 days; 140% 5 consec. days	200 to 400 lb/truck
	In-plant feeding	All	7 days; 140% 5 consec. days	0.5 to 1.0 lb/meal
Auditoriums, stadiums		Crushed/flake	Varies	0.5 lb/customer
Churches	Banquets	Cube	7 days; 200% 2 consec. days	0.5 lb/meal
Theaters	Snack bar	Crushed/flake	7 days; 200% 2 consec. days	0.5 lb/snack bar patron
Hotel-motel, resorts	Banquet, meeting room	Cube	7 days; 200% 2 consec. days	1.0 lb/meal
	Dining room	Cube/flake	7 days; 200% 2 consec. days	1.0 lb/meal
	Coffee shop	Cube/flake	Daily, 100%	1.0 lb/meal
	Guest room service	Cube	Daily, 100%	5 lb/room
	Kitchen	Flake/crushed	Daily, 100%	0.5 lb/meal
	Buffets-display	Flake/crushed	Varies Daily, 100%	10 lb/ft^2 of display area
Restaurants, cafeterias	Dining room	All	7 days; 150% 2 consec. days	0.5 lb/meal
	Serving line display	Flake/crushed	7 days; 100%	10 lb/ft^2 of display
	Kitchen	Flake/crushed	7 days; 100%	0.5 lb/meal
Military bases	Mess halls	Flake/crushed	7 days; 100%	0.5 lb/meal
	Post exchange clubs	Cube/crushed	Daily, 100%	1.5 lb/customer
Prisons	Dining halls	Flake/crushed	Daily, 100%	0.33 lb/meal
Airlines, airline caterers	In-flight feeding	Cube	7 days; 150% 2 consec. days	1.0 lb/meal
Clubs, country and private	Banquet, meeting room	Cube	7 days; 200% 2 consec. days	0.5 lb/meal
	Dining room	Cube	7 days; 200% 2 consec. days	0.5 lb/meal
	Bar, cocktail lounge	Cube	7 days; 200% 2 consec. days	0.5 to 1.0 lb/customer
	Kitchen	Crushed/flake	1 day; 100%	0.5 lb/meal
	Golf course	All	1 day; 100%	1 to 4 lb/gallon of water cooler capacity
		Food Preservation		
Seafood markets	Merchandising (display)	Crushed/flake	7 days; 140% 5 consec. days	5 lb/ft^2 of display
Florists	Shipping	Flakes	Varies	2 lb/box of flowers
Groceries	Display	Crushed/flake	Daily	5 lb/ft^2 of display
		Medical		
Hospitals	Dietary	Crushed/flake	Daily, 100%	0.5 lb/meal
	Nursing service	Crushed/flake	Daily, 100%	5.0 lb/bed
Nursing homes	Dietary	Crushed/flake	Daily, 100%	0.33 lb/meal
	Nursing service	Crushed/flake	Daily, 100%	3 lb/bed
Athletic fields	Ice packs and drinking water	Crushed/flake	Daily, 100%	3 to 5 lb/person per day
Construction	Drinking water coolers	Crushed/cubes	Daily, 100%	3 to 5 lb/person per shift

[a]Additional application information other than food, beverages, or display counters may be found under general refrigeration requirements by industry or product.
[b]These values assist in balancing bin capacity and ice-making capability to obtain the most economical combinations. The length of the normal use cycle is shown first. The next figure is an approximation of how much ice might be consumed during a peak period in the use cycle; it is expressed as a percentage of the average daily production. The last figure indicates how long the high consumption might last.
[c]Ice consumption is generally cyclic. Proper storage capacity helps ensure an adequate supply of ice during high-consumption periods. The values shown are average consumption for the indicated applications.

Ice-storage capacities are commonly stated in terms of how much ice, or product, can be accommodated when the bin is full. Since ice has a natural angle of repose, the machine generally turns off before reaching the full bin capacity. This factor should be considered when selecting equipment. In any application, ice bin and machine production capacities must balance to fit the ice-demand cycle. Equipment must also be selected for maximum demand, and machine capacities at lowest production rates must be compatible with the installation environment.

Ice Costs

Ice is a relatively inexpensive commodity. If ice is used in small quantities, the cost of making it is secondary to the space the ice maker occupies, the convenience of keeping the ice as close as possible to the point of use, or the original first cost. This relative importance must be kept in mind when choosing the ice form or size and when considering reliability. Thus, two small machines are safer than one large one, and the machine that makes ice for the least amount of money is the logical choice.

Equipment Life

Unit life depends on installation, environment, and regular cleaning and maintenance. Unless these factors are considered at the time of installation, ice-making equipment must be classified as relatively short-lived. Where ice is added to food or beverages, the regular cleaning and inspection required for sanitary purposes should promote longer life. An awareness of the necessity for regular cleaning, not only of the ice-storage facilities but also of the ice-making parts of the ice maker, must be impressed on the user. Neglected equipment in areas of poor makeup water may have a life as short as three years, whereas if water treatment is used and the equipment is regularly cleaned and maintained, it could extend to a ten-year life.

Installation

Proper installation is the foremost requirement of economical, maintenance-free operation. Manufacturer's recommendations should be followed closely.

Often, the necessary drain and its pitch are overlooked when a location is chosen. The drain must take care of blowdown, ice meltage, and, in the case of water-cooled condensers, waste water. Bin drains should be insulated to prevent moisture from condensing on their cold exteriors; this condensation is often mistaken for a leaking bin.

Bad water conditions can create innumerable problems for ice-making equipment. When melted, the ice may leave a scum on water in a glass, or it may have an objectionable taste. Bad water also may have deleterious effects on the equipment. When the local water conditions are known to cause problems, water treatment must be considered at the outset.

Water treatment firms in the vicinity of the installation are the best source of water treatment recommendations. In general, demineralization offers the best overall water conditioning preparation for ice-making equipment. Softening by ion exchange (sodium cycle zeolite) eliminates most scaling problems, but when the dissolved solids content of makeup water exceeds 400 ppm, cloudy mushy ice is apt to result. Treating makeup water with polyphosphates reduces only the scaling tendencies, but this lengthens equipment life.

Ambient conditions have an important influence on performance and maintenance requirements. By avoiding locations where temperatures are extreme—either high or low—better performance may be ensured. Also, ease of accessibility for cleaning and maintenance helps prolong life. Most ice makers create some noise, so it is important to locate them where noise is not objectionable.

Table 1 lists ice-maker application data and is a digest of the experiences of many individuals and firms recommending and installing ice-making equipment in diverse areas of the country under various local conditions. Therefore, the figures are averages, and on any particular application they must be modified up or down according to local conditions.

Environmental Considerations

Many areas experience severe shortages of electric power and supply water, so it is important for ice-making systems to operate with maximum efficiency. Some measures of operating efficiency are as follows:

1. Pounds of ice per gallon of supply water
2. Pounds of ice per kWh of power consumed by the ice-making system
3. Pounds of ice per gallon of condensing water
4. Pounds of ice per square foot of floor space

BIBLIOGRAPHY

ARI. 1990. Directory of certified automatic commercial ice cube machines and ice storage bins. Air-conditioning and Refrigeration Institute, Arlington, VA.

ASHRAE. 1988. Method of testing automatic ice makers. ASHRAE *Standard* 29-1988.

CSA. 1964. Standards for safety, ice makers. CSA *Standard* 133. Canadian Standards Association.

Eddy, D.E. 1965. Manufacture, storage, handling and uses of fragmentary ice. ASHRAE *Journal* (September):66.

FAO. 1968. Ice in fisheries. *FAO Fisheries Report.* Food and Agriculture Organization of the United Nations, Rome.

Hardenberg, R.E., A.E. Watada, and C.Y. Wang. 1986. Commercial storage of fruits, vegetables and florist and nursery stocks. United States Department of Agriculture, *Agriculture Handbook* No. 66, Agricultural Research Service, Washington, D.C.

NSF. 1992. Automatic ice making equipment. NSF *Standard* 12-92. National Sanitation Foundation, Ann Arbor, MI.

Proctor, W.T. 1977. Ice vending—A growing market. *American Automatic Merchandisers* (May):36.

UL. 1992. Ice makers. UL *Standard* 563-92. Underwriters Laboratories Inc., Northbrook, IL.

CHAPTER 51

CODES AND STANDARDS

THE Codes and Standards listed in Table 1 represent practices, methods, or standards published by the organizations indicated. They are valuable guides for the practicing engineer in determining test methods, ratings, performance requirements, and limits applying to the equipment used in heating, refrigerating, ventilating, and air conditioning. *Copies can usually be obtained from the organization listed in the Publisher column.* These listings represent the most recent information available at the time of publication.

Table 1 Codes and Standards Published by Various Societies and Associations

Subject	Title	Publisher	Reference
Air Conditioners	Room Air Conditioners	CSA	C22.2 No. 117-1970 (R 1992)
Room	Room Air Conditioners	AHAM	ANSI/AHAM (RA C-1)
	Method of Testing for Rating Room Air Conditioners and Packaged Terminal Air Conditioners	ASHRAE	ANSI/ASHRAE 16-1983 (RA 88)
	Method of Testing for Rating Room Air Conditioners and Packaged Terminal Air Conditioner Heating Capacity	ASHRAE	ANSI/ASHRAE 58-1986 (RA 90)
	Methods of Testing for Rating Room Fan-Coil Air Conditioners	ASHRAE	ANSI/ASHRAE 79-1984 (RA 91)
	Heating and Cooling Equipment	CSA/UL	CAN/CSA-C22.2 No. 236.M90
	Performance Standard for Room Air Conditioners	CSA	CAN/CSA-C368.1-M90
	Room Air Conditioners (1993)	UL	UL 484
Packaged Terminal	Packaged Terminal Air Conditioners	ARI	ARI 310-90
	Packaged Terminal Heat Pumps	ARI	ARI 380-90
	Standards for Packaged Terminal Air-Conditioners and Heat Pumps	CSA	C744-93
Transport	Air Conditioning of Aircraft Cargo (1978)	SAE	SAE AIR806A
	Nomenclature, Aircraft Air-Conditioning Equipment (1978)	SAE	SAE ARP147C
Unitary	Load Calculation for Commercial Summer and Winter Air Conditioning, 4th ed. (1988)	ACCA	ACCA Manual N
	Application of Sound Rated Outdoor Unitary Equipment	ARI	ARI 275-84
	Commercial and Industrial Unitary Air-Conditioning Equipment	ARI	ANSI/ARI 360-86
	Sound Rating of Outdoor Unitary Equipment	ARI	ARI 270-84
	Unitary Air-Conditioning and Air-Source Heat Pump Equipment	ARI	ANSI/ARI 210/240-89
	Methods of Testing for Rating Heat Operated Unitary Air-Conditioning Equipment for Cooling	ASHRAE	ANSI/ASHRAE 40-1986 (RA 92)
	Methods of Testing for Rating Unitary Air-Conditioning and Heat Pump Equipment	ASHRAE	ANSI/ASHRAE 37-1988
	Methods of Testing for Seasonal Efficiency of Unitary Air Conditioners and Heat Pumps	ASHRAE	ANSI/ASHRAE 116-1983
	Performance Standard for Split-System Central Air Conditioners and Heat Pumps	CSA	CAN/CSA-C273.3-M91
	Performance Standard for Single Package Central Air Conditioners and Heat Pumps	CSA	CAN/CSA-C656-M92
	Performance Standard for Rating Large Air Conditioners and Heat Pumps	CSA	CAN/CSA-C746-93
	Air Conditioners, Central Cooling (1982)	UL	ANSI/UL 465-1984
Air Conditioning	Commercial Low Pressure, Low Velocity Duct System Design	ACCA	Manual Q
	Duct Design for Residential Buildings	ACCA	ACCA Manual D
	Load Calculation for Residential Winter and Summer Air Conditioning, 7th ed. (1986)	ACCA	ACCA Manual J
	Gas-Fired Absorption Summer Air Conditioning Appliances (with 1982 addenda)	AGA	ANSI Z21.40.1-1981; Z21.40.1a-1982
	Heating and Cooling Equipment (1990)	CSA/UL	ANSI/UL 1995-1992
	Environmental System Technology (1984)	NEBB	NEBB
	Automotive Air-Conditioning Hose (1989)	SAE	ANSI/SAE J51 MAY89
	HVAC Systems—Applications, 1st ed. (1986)	SMACNA	SMACNA
	HVAC Systems—Duct Design (1990)	SMACNA	SMACNA
	Installation Standards for Residential Heating and Air Conditioning Systems (1988)	SMACNA	SMACNA
	Requirements for Gas-Fired, Engine-Driven Air Conditioning Appliances	AGA	4-89
	Requirements for Gas-Fired Desiccant Type Dehumidifiers and Air Conditioners	AGA	9-90
	Air Conditioning Equipment, General Requirements for Subsonic Airplanes (1961)	SAE	SAE ARP85E
	General Requirements for Helicopter Air Conditioning (1970)	SAE	SAE ARP292B
	Testing of Commercial Airplane Environmental Control Systems (1973)	SAE	SAE ARP217B

Table 1 Codes and Standards Published by Various Societies and Associations (*Continued*)

Subject	Title	Publisher	Reference
Air Conditioning (Continued)	Service Hose for Automatic Air Conditioning	SAE	SAE J22196
	Standard of Purity for use in Mobile Air Conditioning Systems	SAE	SAE J1991
	Extraction and Recycle Equipment for Mobile Automotive Air Conditioning Systems	SAE	SAE J1990
	Guide to the Application and Use of Passenger Car Air Conditioning Compressor Face Seals	SAE	SAE J1954
	Rating Air Conditioner Evaporator Air Delivery and Cooling Capacities	SAE	SAE J1487
	Information Relating to Duty Cycles and Average Power Requirements of Truck and Bus Engine Accessories	SAE	SAE J1343
	Test Method for Measuring Power Consumption of Air Conditioning and Brake Compressors for Trucks and Buses	SAE	SAE J1340
	Design Guidelines for Air Conditioning Systems for Off-Road Operator Enclosures	SAE	SAE J169
	Aircraft Ground Air Conditioning Service Connection	SAE	SAE AS4262
	Recommended Practice for the Design of Tubing Installations for Aerospace Fluid Power Systems	SAE	SAE ARP994
	Control of Excess Humidity in Avionics Cooling	SAE	SAE ARP987
	Guide for Qualification Testing of Aircraft Air Valves	SAE	SAE ARP986
	Air Cycle Air Conditioning Systems for Military Air Vehicles	SAE	SAE ARP4073
	Engine Bleed Air Systems for Aircraft	SAE	SAE ARP1796
	Air Conditioning of Subsonic Aircraft at High Altitudes	SAE	SAE AIR795
	Aircraft Fuel Weight Penalty Due to Air Conditioning	SAE	SAE AIR1168/8
Unitary	Method of Rating Computer and Data Processing Room Unitary Air Conditioners	ASHRAE	ANSI/ASHRAE 127-1988
	Method of Rating Unitary Spot Air Conditioners	ASHRAE	ANSI/ASHRAE 128-1988
Air Curtains	Air Curtains for Entranceways in Food and Food Service Establishments	NSF	NSF-37
	Air Distribution Basics for Residential and Small Commercial Buildings	ACCA	ACCA Manual T
	Residential Equipment Selection	ACCA	ACCA Manual S
	Test Code for Grilles, Registers and Diffusers	ADC	ADC 1062:GRD-84
	Metric Units and Conversion Factors	AMCA	AMCA 99-0100-76
	Test Methods for Air Curtain Units	AMCA	AMCA 220-91
	Air Terminals	ARI	ANSI/ARI 880-89
	Method of Testing for Rating the Performance of Air Outlets and Inlets	ASHRAE	ANSI/ASHRAE 70-1991
	Standard Methods for Laboratory Air Flow Measurement	ASHRAE	ANSI/ASHRAE 41.2-1987 (RA 92)
	Rating the Performance of Residential Mechanical Ventilating Equipment	CSA	CAN/CSA-C260-M90
	Rating the Performance of Residential Mechanical Ventilating Equipment	CSA	CAN/CSA-C260-M90
	High Temperature Pneumatic Duct Systems for Aircraft (1981)	SAE	ANSI/SAE ARP699D
	Direct Gas-Fired Door Heaters	AGA	ANSI Z83.17-1990; Z21.17a-1991
Air Ducts and Fittings	Commercial Low Pressure, Low Velocity Duct Systems	ACCA	Manual Q
	Duct Design for Residential Winter and Summer Air Conditioning	ACCA	Manual D
	Flexible Air Duct Test Code	ADC	ADC FD-72R1-1979
	Flexible Duct Performance and Installation Standards	ADC	ADC-91
	Pipes, Ducts and Fittings for Residential Type Air Conditioning Systems	CSA	B228.1-1968
	Installation of Air Conditioning and Ventilating Systems (1993)	NFPA	ANSI/NFPA 90A-1993
	Installation of Warm Air Heating and Air-Conditioning Systems (1993)	NFPA	ANSI/NFPA 90B-1993
	Ducted Electric Heat Guide for Air Handling Systems (1971)	SMACNA	SMACNA
	HVAC Air Duct Leakage Test Manual (1985)	SMACNA	SMACNA
	HVAC Duct Construction Standards—Metal and Flexible, 1st ed. (1985)	SMACNA	SMACNA
	HVAC Duct Systems Inspection Guide (1989)	SMACNA	SMACNA
	Rectangular Industrial Duct Construction (1980)	SMACNA	SMACNA
	Round Industrial Duct Construction (1977)	SMACNA	SMACNA
	Thermoplastic Duct (PVC) Construction Manual (Rev. A, 1974)	SMACNA	SMACNA
	Closure Systems for Use with Rigid Air Ducts and Air Connectors (1991)	UL	UL 181A
	Factory-Made Air Ducts and Air Connectors (1990)	UL	UL 181
	Marine Rigid and Flexible Air Ducting (1986)	UL	ANSI/UL 1136-1986
Air Filters	Method for Measuring Performance of Portable Household Electrical Cord-Connected Room Air Cleaners	AHAM	ANSI/AHAM AC-1
	Commercial and Industrial Air Filter Equipment	ARI	ARI 850-84
	Residential Air Filter Equipment	ARI	ARI 680-86
	Gravimetric and Dust Spot Procedures for Testing Air Cleaning Devices Used in General Ventilation for Removing Particulate Matter	ASHRAE	ANSI/ASHRAE 52.1-1992
	Method for Sodium Flame Test for Air Filters	BSI	BS 3928
	Particulate Air Filters for General Ventilation—Requirements, Testing Marking	BSI	BS EN 779:1993
	Electrostatic Air Cleaners (1989)	UL	ANSI/UL 867-1988
	High-Efficiency, Particulate, Air Filter Units (1990)	UL	ANSI/UL 586-1990
	Test Performance of Air Filter Units (1987)	UL	ANSI/UL 900-1987

Table 1 Codes and Standards Published by Various Societies and Associations (*Continued*)

Subject	Title	Publisher	Reference
Air-Handling Units	Commercial Low Pressure, Low Velocity Duct Systems	ACCA	Manual Q
	Duct Design for Residential Winter and Summer Air Conditioning	ACCA	Manual D
	Central Station Air-Handling Units	ARI	ANSI/ARI 430-89
	Direct Gas-Fired Make-up Air Heaters	AGA	ANSI Z83.4-1991; Z83.4a-1992
Air Leakage	Air Leakage Performance for Detached Single-Family Residential Buildings	ASHRAE	ANSI/ASHRAE 119-1988
Boilers	A Guide to Clean and Efficient Operation of Coal Stoker-Fired Boilers	ABMA	ABMA
	Boiler Water Limits and Steam Purity Recommendations for Watertube Boilers	ABMA	ABMA
	Boiler Water Requirements and Associated Steam Purity—Commercial Boilers	ABMA	ABMA
	Fluidized Bed Combustion Guidelines	ABMA	ABMA
	Guidelines for Industrial Boiler Performance Improvement	ABMA	ABMA
	Matrix of Recommended Quality Control Requirements	ABMA	ABMA
	Operation and Maintenance Safety Manual	ABMA	ABMA
	Recommended Design Guidelines for Stoker Firing of Bituminous Coals	ABMA	ABMA
	(Selected) Summary of Codes and Standards of the Boiler Industry	ABMA	ABMA
	Thermal Shock Damage to Hot Water Boilers as a Result of Energy Conservation Measures	ABMA	ABMA
	Commercial Applications Systems and Equipment	ACCA	Manual CS
	Boiler and Pressure Vessel Code (11 sections) (1989)	ASME	ASME
	Boiler, Pressure Vessel, and Pressure Piping Code	CSA	B51-M1986
	Heating, Water Supply, and Power Boilers—Electric (1991)	UL	ANSI/UL834-1991
Cast-Iron	Ratings for Cast-Iron and Steel Boilers (1993)	HYDI	IBR
	Testing and Rating Heating Boilers (1989)	HYDI	IBR
Gas or Oil	Gas-Fired Low-Pressure Steam and Hot Water Boilers	AGA	ANSI Z21.13-1991; Z21.13a-1993
	Gas Utilization Equipment in Large Boilers (with 1972 and 1976 addenda; R-1983, 1989)	AGA	ANSI Z83.3-1971; Z83.3a-1972; Z83.3b-1976
	Requirements for High Pressure Steam Boilers	AGA	3-89
	Control and Safety Devices for Automatically Fired Boilers	ASME	ANSI/ASME CSD.1-1988
	Oil-Fired Steam and Hot-Water Boilers for Residential Use	CSA	B140.7.1-1976 (R 1991)
	Oil-Fired Steam and Hot-Water Boilers for Commercial and Industrial Use	CSA	B140.7.2-1967 (R 1991)
	Single Burner Boiler Operations	NFPA	ANSI/NFPA 8501-1992
	Prevention of Furnace Explosions/Implosions in Multiple Burner Boiler-Furnaces (1991)	NFPA	NFPA 85C-1991
	Commercial-Industrial Gas Heating Equipment (1973)	UL	UL 795
	Oil-Fired Boiler Assemblies (1990)	UL	ANSI/UL 726-1990
	Standards and Typical Specifications for Deaerators, 5th ed. (1992)	HEI	HEI
	Method and Procedure for the Determination of Dissolved Oxygen, 2nd ed. (1963)	HEI	HEI
Building Codes	ASTM Standards Used in Building Codes	ASTM	ASTM
	National Building Code, 11th ed. (1993)	BOCA	BOCA
	National Property Maintenance Code, 3rd ed. (1990)	BOCA	BOCA
	One- and Two-Family Dwelling Code (1992)	CABO	CABO
	Model Energy Code (1992)	CABO	CABO
	Uniform Building Code (1991)	ICBO	ICBO
	Uniform Building Code Standards (1991)	ICBO	ICBO
	Directory of Building Codes and Regulations (1993 ed.)	NCSBCS	NCSBCS
	Standard Building Code (1991 ed. with 1992/1993 Revisions)	SBCCI	SBCCI
Mechanical	Safety Code for Elevators and Escalators (plus two yearly supplements)	ASME	ANSI/ASME A 17.1-1990
	BOCA National Mechanical Code, 7th ed. (1993)	BOCA	BOCA
	Uniform Mechanical Code (1991) (with Uniform Mechanical Code Standards)	ICBO/ IAPMO	ICBO/IAPMO
	Standard Gas Code (1991 ed. with 1992/1993 revisions)	SBCCI	SBCCI
	Standard Mechanical Code (1991 ed. with 1992/1993 revisions)	SBCCI	SBCCI
Burners	Guidelines for Burner Adjustments of Commercial Oil-Fired Boilers	ABMA	ABMA
	Domestic Gas Conversion Burners	AGA	Z21.17-1991; Z21.17a-1993
	Installation of Domestic Gas Conversion Burners	AGA	ANSI Z21.8-1984; Z21.8a-1990
	General Requirements for Oil Burning Equipment	CSA	CAN/CSA-B140.0-M87 (R 1991)
	Installation Code for Oil Burning Equipment	CSA	CAN/CSA-B139-M91
	Oil Burners; Atomizing Type	CSA	CAN/CSA-B140.2.1-M90
	Pressure Atomizing Oil Burner Nozzles	CSA	B140.2.2-1971 (R 1991)
	Replacement Burners and Replacement Combustion Heads for Residential Oil Burners	CSA	B140.2.3-M1981 (R 1991)
	Vaporizing-Type Oil Burners	CSA	B140.1-1966 (R 1991)
	Commercial-Industrial Gas Heating Equipment (1973)	UL	UL 795
	Oil Burners (1989)	UL	ANSI/UL 296-1988

Table 1 Codes and Standards Published by Various Societies and Associations (*Continued*)

Subject	Title	Publisher	Reference
Capillary Tubes	Capillary Tubes Method of Testing Flow Capacity of Refrigerant	ASHRAE	ANSI/ASHRAE 28-1988
Chillers	Methods of Testing Liquid Chilling Packages	ASHRAE	ASHRAE 30-1978
	Absorption Water-Chilling and Water Heating Packages	ARI	ARI 560-92
	Centrifugal and Rotary Screw Water-Chilling Packages	ARI	ARI 550-92
	Positive Displacement Compressor Water-Chilling Packages	ARI	ARI 590-92
	Performance Standard for Rating Packaged Water Chillers	CSA	C743-93
Chimneys	Design and Construction of Masonry Chimneys and Fireplaces	CSA	CAN/CSA-A405-M87
	Chimneys, Fireplaces, and Vents, and Solid Fuel Burning Appliances	NFPA	ANSI/NFPA 211-1992
	Chimneys, Factory-Built, Medium Heat Appliance (1986)	UL	ANSI/UL 959-1992
	Chimneys, Factory-Built, Residential Type and Building Heating Appliance (1989)	UL	ANSI/UL 103-1988
Cleanrooms	Procedural Standards for Certified Testing of Cleanrooms (1988)	NEBB	NEBB-1988
Coils	Forced-Circulation Air-Cooling and Air-Heating Coils	ARI	ARI 410-91
	Methods of Testing Forced Circulation Air Cooling and Air Heating Coils	ASHRAE	ASHRAE 33-1978
Comfort Conditions	Thermal Environmental Conditions for Human Occupancy	ASHRAE	ANSI/ASHRAE 55-1992
Compressors	Compressors and Exhausters (reaffirmed 1986)	ASME	ANSI/ASME PTC 10-1965 (R 1986)
	Displacement Compressors, Vacuum Pumps and Blowers	ASME	ANSI/ASME PTC9-1974 (R 1992)
	Safety Standard for Air Compressor Systems	ASME	ANSI/ASME B19.1-1990
	Safety Standard for Compressors for Process Industries	ASME	ASME/ANSI B19.3-1991
	Compressed Air and Gas Handbook, 5th ed. (1988)	CAGI	CAGI
Refrigeration	Ammonia Compressor Units	ARI	ANSI/ARI 510-87
	Method for Presentation of Compressor Performance Data	ARI	ARI 540-91
	Positive Displacement Refrigerant Compressors, Compressor Units and Condensing Units	ARI	ANSI/ARI 520-90
	Method for Presentation of Compressor Performance Data	ARI	ARI 540-91
	Methods of Testing for Rating Positive Displacement Refrigerant Compressors and Condensing Units	ASHRAE	ANSI/ASHRAE 23-1993
	Hermetic Refrigerant Motor-Compressors (1991) (Harmonized with CAN/CSA-C22.2 No.140.2-M91)	CSA/UL	UL 984
Computers	Protection of Electronic Computer/Data Processing Equipment	NFPA	ANSI/NFPA 75-1992
Condensers	Commercial Applications Systems and Equipment (for equipment selection only)	ACCA	Manual CS
	Remote Mechanical Draft Air-Cooled Refrigerant Condensers	ARI	ARI 460-87
	Water-Cooled Refrigerant Condensers, Remote Type	ARI	ARI 450-87
	Methods of Testing for Rating Remote Mechanical-Draft Air-Cooled Refrigerant Condensers	ASHRAE	ASHRAE 20-1970
	Methods of Testing Remote Mechanical-Draft Evaporative Refrigerant Condensers	ASHRAE	ANSI/ASHRAE 64-1989
	Methods of Testing for Rating Water-Cooled Refrigerant Condensers	ASHRAE	ANSI/ASHRAE 22-1992
	Addendum I, Standards for Steam Surface Condensers (1989)	HEI	HEI
	Standards for Steam Surface Condensers, 8th ed. (1984)	HEI	HEI
	Standards for Direct Contact Barometric and Low Level Condensers, 5th ed. (1970)	HEI	HEI
Condensing Units	Commercial Applications Systems and Equipment	ACCA	Manual CS
	Residential Equipment Selection	ACCA	Manual S
	Commercial and Industrial Unitary Air-Conditioning Condensing Units	ARI	ANSI/ARI 365-87
	Methods of Testing for Rating Positive Displacement Refrigerant Compressors and Condensing Units	ASHRAE	ANSI/ASHRAE 23-1993
	Standard for Safety Heating and Cooling Equipment (Harmonized with CAN/CSA C22.2 No. 236-M90)	UL	UL 1995-90
	Refrigeration and Air-Conditioning Condensing and Compressor Units (1987)	UL	ANSI/UL303-1988
Contactors	Definite Purpose Contactors for Limited Duty	ARI	ANSI/ARI 790-86
	Definite Purpose Magnetic Contactors	ARI	ANSI/ARI 780-86
Controls	Quick-Disconnect Devices for Use with Gas Fuel	AGA	ANSI Z21.41-1989; Z21.41a-1990; Z21.41b-1992
	Energy Management Control Systems Instrumentation	ASHRAE	ANSI/ASHRAE 114-1986
	Temperature-Indicating and Regulating Equipment	CSA	C22.2 No. 24-1993
	Control Centers for Changing Message Type Electric Signals (1991)	UL	UL 1433
	Industrial Control Equipment (1993)	UL	UL 508
	Limit Controls (1989)	UL	ANSI/UL 353-1988
	Primary Safety Controls for Gas- and Oil-Fired Appliances (1985)	UL	ANSI/UL 372-1985

Table 1 Codes and Standards Published by Various Societies and Associations (*Continued*)

Subject	Title	Publisher	Reference
Controls (continued)	Solid State Controls for Appliances (1986)	UL	ANSI/UL244A-1987
	Temperature-Indicating and -Regulating Equipment (1988)	UL	ANSI/UL 873-1987
	Tests for Safety-Related Controls Employing Solid-State Devices (1989)	UL	UL 991
Commercial and Industrial	General Standards for Industrial Control and Systems	NEMA	NEMA ICS 1-1988
	Industrial Control Devices, Controllers and Assemblies	NEMA	NEMA ICS 2-1988
	Instructions for the Handling, Installation, Operation and Maintenance of Motor Control Centers	NEMA	NEMA ICS 2.3-1983 (R 1990)
	Maintenance of Motor Controllers after a Fault Condition	NEMA	NEMA ICS 2.2-1983 (1988)
	Preventive Maintenance of Industrial Control and Systems Equipment	NEMA	NEMA ICS 1.3-1986
Residential	Automatic Gas Ignition Systems and Components	AGA	ANSI Z21.20-1989; Z21.20a-1991; Z21.20b-1992
	Gas Appliance Pressure Regulators	AGA	ANSI Z21.18-1987; Z21.18-1989; Z21.18-1992
	Gas Appliance Thermostats	AGA	ANSI Z21.23-1989; Z21.23a-1991
	Manually Operated Gas Valves for Appliances, Appliance Connector Valves and Hose End Valves	AGA	ANSI Z21.15-1992
	Manually-Operated Piezo Electric Spark Gas Ignition Systems and Components	AGA	ANSI Z21.77-1989
	Hot Water Immersion Controls	NEMA	NEMA DC-12-1985 (R 1991)
	Line Voltage Integrally-Mounted Thermostats for Electric Heaters	NEMA	NEMA DC 13-1979 (R 1985)
	Quick Connect Terminals	NEMA	ANSI/NEMA DC 2-1982 (R 1988)
	Residential Controls—Class 2 Transformers	NEMA	NEMA DC 20-1986 (R 1992)
	Residential Controls—Surface Type Controls for Electric Storage Water Heaters	NEMA	NEMA DC 5-1989
	Safety Guidelines for the Application, Installation, and Maintenance of Solid State Controls	NEMA	NEMA ICS 1.1-1984 (R 1988)
	Temperature Limit Controls for Electric Baseboard Heaters	NEMA	NEMA DC 10-1083 (R 1989)
	Wall-Mounted Room Thermostats	NEMA	NEMA DC 3-1989
	Electrical Quick-Connect Terminals (1991)	UL	UL 310
Coolers	Refrigeration Equipment	CSA	CAN/CSA-C22.2 No. 120-M91
Air	Unit Coolers for Refrigeration	ARI	ANSI/ARI 420-89
	Methods of Testing Forced Convection and Natural Convection Air Coolers for Refrigeration	ASHRAE	ANSI/ASHRAE 25-1990
	Commercial Bulk Milk Dispensing Equipment	NSF	NSF 20
Bottled Beverage	Methods of Testing and Rating Bottled and Canned Beverage Vendors and Coolers	ASHRAE	ANSI/ASHRAE 32-1986 (RA 90)
	Refrigerated Vending Machines (1989)	UL	ANSI/UL 541-1988
Drinking Water	Application and Installation of Drinking Fountains and Drinking Water Coolers	ARI	ANSI/ARI 1020-84
	Drinking Fountains and Self-Contained, Mechanically Refrigerated Drinking Water Coolers	ARI/ANSI	ANSI/ARI 1010-84
	Methods of Testing for Rating Drinking-Water Coolers with Self-Contained Mechanical Refrigeration Systems	ASHRAE	ANSI/ASHRAE 18-1987 (RA 91)
	Drinking-Water Coolers (1993)	UL	ANSI/UL 399-1992
	Manual Food and Beverage Dispensing Equipment	NSF	ANSI/NSF 18
Liquid	Refrigerant-Cooled Liquid Coolers, Remote Type	ARI	ANSI/ARI 480-87
	Methods of Testing for Rating Liquid Coolers	ASHRAE	ANSI/ASHRAE 24-1989
Cooling Towers	Commercial Applications Systems and Equipment	ACCA	Manual CS
	Atmospheric Water Cooling Equipment	ASME	ANSI/ASME PTC 23-1986 (R 1991)
	Water-Cooling Towers	NFPA	ANSI/NFPA 214-1992
	Acceptance Test Code for Spray Cooling Systems (1985)	CTI	CTI ATC-133-1985
	Acceptance Test Code for Water Cooling Towers: Mechanical Draft, Natural Draft Fan Assisted Types, Evaluation of Results, and Thermal Testing of Wet/Dry Cooling Towers (1990)	CTI	CTI ATC-105-1990
	Certification Standard for Commercial Water Cooling Towers (1991)	CTI	CTI STD-201-1991
	Code for Measurement of Sound from Water Cooling Towers	CTI	CTI ATC-128-1981
	Fiberglass-Reinforced Plastic Panels for Application on Industrial Water Cooling Towers	CTI	CTI STD-131-1986
	Nomenclature for Industrial Water-Cooling Towers	CTI	CTI NCL-109-1983
	Recommended Practice for Airflow Testing of Cooling Towers	CTI	CTI-1993
Dehumidifiers	Commercial Applications Systems and Equipment	ACCA	Manual CS
	Dehumidifiers	AHAM	ANSI/AHAM DH 1
	Dehumidifiers	CSA	C22.2 No. 92-1971 (R 1992)
	Dehumidifiers (1987)	UL	ANSI/UL 474-1987
Desiccants	Method of Testing Desiccants for Refrigerant Drying	ASHRAE	ANSI/ASHRAE 35-1992
Driers	Method of Testing Liquid Line Refrigerant Driers	ASHRAE	ANSI/ASHRAE 63.1-1988
	Liquid Line Driers	ARI	ANSI/ARI 710-86

Table 1 Codes and Standards Published by Various Societies and Associations (*Continued*)

Subject	Title	Publisher	Reference
Electrical	Voltage Ratings for Electrical Power Systems and Equipment	ANSI	ANSI C84.1-1989
	Canadian Electrical Code, Part 1 (16th ed.)	CSA	C22.1-1990
	Application Guide for Ground Fault Interrupters	NEMA	NEMA 280-1990
	Application Guide for Ground Fault Protective Devices for Equipment	NEMA	NEMA PB 2.2-1988
	Enclosures for Electric Equipment	NEMA	NEMA 250-1991
	Enclosures for Industrial Control and Systems	NEMA	NEMA ICS 6-1988
	General Requirements for Wiring Devices	NEMA	NEMA WD 1-1983 (R 1989)
	Low Voltage Cartridge Fuses	NEMA	NEMA FU 1-1986
	Molded Case Circuit Breakers	NEMA	NEMA AB 1-1986
	Terminal Blocks for Industrial Use	NEMA	NEMA ICS 4-1983 (R 1988)
	National Electric Code (1990)	NFPA	ANSI/NFPA 70-1993
	Compatibility of Electrical Connectors and Wiring (1988)	SAE	SAE AIR1329 A
	Manufacturers' Identification of Electrical Connector Contacts, Terminals and Splices (1982)	SAE	SAE AIR1351 A
	Class T Fuses (1988)	UL	ANSI/UL 198H-1987
	Enclosures for Electrical Equipment (1992)	UL	UL 50
	Fuseholders (1993)	UL	ANSI/UL 512-1992
	High-Interrupting-Capacity Fuses, Current-Limiting Types (1986)	UL	ANSI/UL 198C-1986
	Molded-Case Circuit Breakers and Circuit-Breaker Enclosures (1991)	UL	UL 489
	Terminal Blocks (1993)	UL	UL 1059
	Thermal Cutoffs for Use in Electrical Appliances and Components (1983)	UL	ANSI/UL 1020-1986
Energy	Air Conditioning and Refrigerating Equipment Nameplate Voltages	ARI	ARI 110-90
	Energy Conservation in Existing Buildings—High Rise Residential	ASHRAE	ANSI/ASHRAE/IES 100.2-1991
	Energy Conservation in Existing Buildings—Commercial	ASHRAE	ANSI/ASHRAE/IES 100.3-1985
	Energy Conservation in Existing Facilities—Industrial	ASHRAE	ANSI/ASHRAE/IES 100.4-1984
	Energy Conservation in Existing Buildings—Institutional	ASHRAE	ANSI/ASHRAE/IES 100.5-1991
	Energy Conservation in Existing Buildings—Public Assembly	ASHRAE	ANSI/ASHRAE/IES 100.6-1991
	Energy Conservation in New Building Design—Residential only	ASHRAE	ANSI/ASHRAE/IES 90A-1980
	Energy Efficient Design of New Buildings Except Low Rise Residential Buildings	ASHRAE	ASHRAE/IES 90.1-1989
	Model Energy Code (MEC) (1992)	CABO	BOCA/ICBO/SBCCI
	Uniform Solar Energy Code (1991)	IAPMO	IAPMO
	Energy Directory (1991)	NCSBCS	NCSBCS
	Energy Management Guide for the Selection and Use of Polyphase Motors	NEMA	NEMA MG 10-1983 (R 1988)
	Energy Management Guide for the Selection and Use of Single Phase Motors	NEMA	MEA MG 11-1977 (R 1992)
	Total Energy Management Handbook, 3rd ed.	NEMA	NEMA 05101-1986
	Energy Conservation Guidelines (1984)	SMACNA	SMACNA
	Energy Recovery Equipment and Systems, Air-to-Air (1991)	SMACNA	SMACNA
	Retrofit of Building Energy Systems and Processes (1982)	SMACNA	SMACNA
	Energy Management Equipment (1984)	UL	ANSI/UL 916-1987
Exhaust Systems	Commercial Low Pressure, Low Velocity Duct Systems	ACCA	Manual Q
	Fundamentals Governing the Design and Operation of Local Exhaust Systems	ANSI	ANSI/AIHA Z9.2-1991
	Laboratory Ventilation	ANSI	ANSI/AIHA Z9.5-1992
	Open-Surface Tanks—Ventilation and Operation	ANSI	ANSI/AIHA Z9.1-1991
	Safety Code for Design, Construction, and Ventilation of Spray Finishing Operations (reaffirmed 1971)	ANSI	ANSI/AIHA Z9.3-1985
	Ventilation and Safe Practices of Abrasives Blasting Operations	ANSI	ANSI/AIHA Z9.4-1985
	Method of Testing Performance of Laboratory Fume Hoods	ASHRAE	ANSI/ASHRAE 110-1985
	Compressors and Exhausters	ASME	ANSI/ASME PTC 10-1974 (R 1986)
	Mechanical Flue-Gas Exhausters	CSA	CAN 3-B255-M81
	Exhaust Systems for Air Conveying of Materials	NFPA	ANSI/NFPA 91-1992
	Draft Equipment (1993)	UL	UL 378
Expansion Valves	Thermostatic Refrigerant Expansion Valves	ARI	ANSI/ARI 750-87
	Method of Testing for Capacity Rating of Thermostatic Refrigerant Expansion Valves	ASHRAE	ANSI/ASHRAE 17-1986
Fan Coil Units	Room Fan-Coil Air Conditioners	ARI	ARI 440-89
	Methods of Testing for Rating Room Fan-Coil Air Conditioners	ASHRAE	ANSI/ASHRAE 79-1984 (R 1991)
	Fan-Coil Units and Room Fan-Heater Units (1986)	UL	ANSI/UL 883-1986
Fans	Commercial Low Pressure, Low Velocity Duct Systems	ACCA	Manual Q
	Duct Design for Residential Winter and Summer Air Conditioning	ACCA	Manual D
	Designation for Rotation and Discharge of Centrifugal Fans	AMCA	AMCA 99-2406-83
	Drive Arrangements for Centrifugal Fans	AMCA	AMCA 99-2404-78
	Drive Arrangements for Tubular Centrifugal Fans	AMCA	AMCA 99-2410-82
	Inlet Box Positions for Centrifugal Fans	AMCA	AMCA 99-2405-83

Table 1 Codes and Standards Published by Various Societies and Associations (*Continued*)

Subject	Title	Publisher	Reference
Fans (continued)	Laboratory Methods of Testing Fans for Rating	AMCA	ANSI/AMCA 210-85
	Motor Positions for Belt or Chain Drive Centrifugal Fans	AMCA	AMCA 99-2407-66
	Site Performance Test Standard Power Plant and Industrial Fans	AMCA	AMCA 803-87
	Standards Handbook	AMCA	AMCA 99-86
	Fans and Blowers	ARI	ARI 670-90
	Laboratory Methods of Testing Fans for Rating	ASHRAE	ANSI/ASHRAE 51-1985 ANSI/AMCA 210-85
	Methods of Testing Dynamic Characteristics of Propeller Fans—Aerodynamically Excited Fan Vibrations and Critical Speeds	ASHRAE	ANSI/ASHRAE 87.1-1992
	Fans	ASME	ANSI/ASME PTC 11-1984 (R 1990)
	Fans and Ventilators	CSA	C22.2 No. 113-M1984 (R 1993)
	Rating the Performance of Residential Mechanical Ventilating Equipment	CSA	CAN/CSA C260-M90
	Electric Fans (1991)	UL	UL 507
Ceiling	AC Electric Fans and Regulators	ANSI	ANSI-IEC Pub. 385
Filters	Flow-Capacity Rating and Application of Suction-Line Filters and Filter Driers	ARI	ANSI/ARI 730-86
	Exhaust Hoods for Commercial Cooking Equipment (1990)	UL	ANSI/UL 710-1992
	Grease Filters for Exhaust Ducts (1979)	UL	UL 1046
Fire Dampers	Fire Dampers (1990)	UL	ANSI/UL 555-1989
Fireplaces	Factory-Built Fireplaces (1988)	UL	ANSI/UL 127-1992
Fire Protection	Standard Method for Fire Tests of Building Construction and Materials	ASTM	ASTM E 119-88
	Method of Test of Surface Burning Characteristics of Building Materials	ASTM/ NFPA	ASTM E 84-89a; NFPA 255-1990
	BOCA National Fire Prevention Code, 8th ed. (1990)	BOCA	BOCA
	Uniform Fire Code (1991)	IFCI	IFCI
	Uniform Fire Code Standards (1991)	IFCI	IFCI
	Interconnection Circuitry of Non-Coded Remote Station Protective Signalling Systems	NEMA	NEMA SB 3-1969 (R 1989)
	Fire Doors and Windows	NFPA	ANSI/NFPA 80-1992
	Fire Prevention Code	NFPA	ANSI/NFPA 1-1992
	Fire Protection Handbook, 17th ed.	NFPA	NFPA
	Flammable and Combustible Liquids Code	NFPA	ANSI/NFPA 30-1990
	Life Safety Code	NFPA	ANSI/NFPA 101-1991
	National Fire Codes (issued annually)	NFPA	NFPA
	Smoke Control Systems	NFPA	NFPA 92A-1988
	Smoke Management Systems	NFPA	NFPA 92B-1993
	Fire Tests of Door Assemblies	NFPA	ANSI/NFPA 252-1990
	Standard Fire Prevention Code (1991 ed. with 1992/1993 revisions)	SBCCI	SBCCI
	Fire Tests of Building Construction and Materials (1992)	UL	UL 263
	Fire Tests of Through-Penetration Firestops	UL	UL 1479-1984
	Heat Responsive Links for Fire-Protection Service (1987)	UL	ANSI/UL 33-1987
Fireplace Stoves	Fireplace Stoves (1988)	UL	ANSI/UL 737-1988
Flow Capacity	Method of Testing Flow Capacity of Suction Line Filters and Filter Driers	ASHRAE	ANSI/ASHRAE 78-1985 (RA 90)
Freezers	Refrigeration Equipment	CSA	CAN/CSA-C22.2 No. 120-M91
Household	Household Refrigerators, Combination Refrigerator-Freezers, and Household Freezers	AHAM	ANSI/AHAM; HRF 1
	Capacity Measurement and Energy Consumption Test Methods for Refrigerators, Combination Refrigerator-Freezers, and Freezers	CSA	CAN/CSA-C300-M91
	Household Refrigerators and Freezers	CSA	C22.2 No. 63-M1987
	Household Refrigerators and Freezers (1983)	UL	ANSI/UL 250-1991
Commercial	Dispensing Freezers	NSF	ANSI/NSF-6-1989
	Food Service Refrigerators and Storage Freezers	NSF	NSF-7
	Ice Cream Makers (1993)	UL	ANSI/UL 621-1992
	Commercial Refrigerators and Freezers (1992)	UL	ANSI/UL 471-1991
	Ice Makers (1992)	UL	ANSI/UL 563-1991
Furnaces	Commercial Applications Systems and Equipment	ACCA	Manual CS
	Residential Equipment Selection	ACCA	Manual S
	Direct Vent Central Furnaces	AGA	ANSI Z21.64-1990; Z21.64a-1992
	Gas-Fired Central Furnaces (except Direct Vent)	AGA	ANSI Z21.47-1990; Z21.47a-1990; Z21.47b-1992
	Gas-Fired Duct Furnaces	AGA	ANSI Z83.9-1990; Z83.9a-1992
	Gas-Fired Gravity and Fan Type Direct Vent Wall Furnaces	AGA	ANSI Z21.44-1991; Z21.44a-1992
	Gas-Fired Gravity and Fan Type Floor Furnaces	AGA	Z21.48-1992
	Gas-Fired Gravity and Fan Type Vented Wall Furnaces	AGA	Z21.49-1992

Table 1 Codes and Standards Published by Various Societies and Associations (*Continued*)

Subject	Title	Publisher	Reference
Furnaces (continued)	Methods of Testing for Annual Fuel Utilization Efficiency of Residential Central Furnaces and Boilers	ASHRAE	ANSI/ASHRAE 103-1993
	BOCA National Mechanical Code	BOCA	BOCA
	Installation Code for Solid-Fuel-Burning Appliances and Equipment	CSA	CAN/CSA-B365-M91
	Solid Fuel-Fired Central Heating Appliances	CSA	CAN/CSA-B366.1-M91
	Heating and Cooling Equipment	CSA/UL	CAN/CSA C22.2 No.236-M90
	Oil Burning Stoves and Water Heaters	CSA	B140.3-1962 (R 1991)
	Oil-Fired Warm Air Furnaces	CSA	B140.4-1974 (R 1991)
	Installation of Oil Burning Equipment	NFPA	NFPA 31-1992
	Standard Gas Code (1991 ed. with 1992/1993 revisions)	SBCCI	SBCCI
	Standard Mechanical Code (1991 ed. with 1992/1993 revisions)	SBCCI	SBCCI
	Commercial-Industrial Gas Heating Equipment (1973)	UL	UL 795
	Oil-Fired Central Furnaces (1986)	UL	UL 727-1986
	Oil-Fired Floor Furnaces (1987)	UL	ANSI/UL 729-1987
	Oil-Fired Wall Furnaces (1987)	UL	ANSI/UL 730-1986
	Residential Gas Detectors (1991)	UL	UL 1484
	Single and Multiple Station Carbon Monoxide Detectors (1992)	UL	UL 2034
	Solid-Fuel and Combination-Fuel Central and Supplementary Furnaces (1991)	UL	ANSI/UL 391-1991
Heat Exchangers	Remote Mechanical-Draft Evaporative Refrigerant Condensers	ARI	ANSI/ARI 490-89
	Method of Testing Air-to-Air Heat Exchangers	ASHRAE	ANSI/ASHRAE 84-1992
	Standard Methods of Test for Rating the Performance of Heat-Recovery Ventilators	CSA	CAN/CSA-C439-88
	Standards for Power Plant Heat Exchangers, 2nd ed. (1990)	HEI	HEI
	Standards of Tubular Exchanger Manufacturers Association, 7th ed. (1988)	TEMA	TEMA
Heat Meters	Method of Testing Thermal Energy Heat Meters for Liquid Streams in HVAC Systems	ASHRAE	ANSI/ASHRAE 125-1992
Heat Pumps	Commercial Applications Systems and Equipment	ACCA	Manual CS
	Heat Pump Systems: Principles and Applications (Commercial and Residence)	ACCA	Manual H
	Residential Equipment Selection	ACCA	Manual S
	Commercial and Industrial Unitary Heat Pump Equipment	ARI	ANSI/ARI 340-86
	Ground Water-Source Heat Pumps	ARI	ARI 325-85
	Water-Source Heat Pumps	ARI	ANSI/ARI 320-86
	Methods of Testing for Rating Unitary Air-Conditioning and Heat Pump Equipment	ASHRAE	ANSI/ASHRAE 37-1988
	Heating and Cooling Equipment	CSA/UL	CAN/CSA C22.2 No. 236-M90
	Installation Requirements for Air-to-Air Heat Pumps	CSA	C273.5-1980 (R 1991)
	Performance Standard for Split System Central Air Conditioners and Heat Pumps	CSA	CAN/CSA-C273.3-M91
	Heat Pumps (1985)	UL	ANSI/UL 559-1985
Gas-Fired	Requirements for Gas-Fired, Absorption and Adsorption Heat Pumps	AGA	10-90
Heat Recovery	Gas Turbine Heat Recovery Steam Generators	ASME	ANSI/ASME PTC 4.4-1981 (R 1992)
	Energy Recovery Equipment and Systems, Air-to-Air (1991)	SMACNA	SMACNA
	Requirements for Heat Reclaimer Devices for Use with Gas-Fired Appliances	AGA	1-80
Heaters	Direct Gas-Fired Industrial Air Heaters	AGA	ANSI Z83.18-1990; Z83.18a-1991; Z83.18b-1992
	Gas-Fired Construction Heaters	AGA	ANSI Z83.7-1990; Z83.7a-1991
	Gas-Fired Infrared Heaters	AGA	ANSI Z83.6-1990; Z83.6a-1992
	Gas-Fired Pool Heaters	AGA	ANSI Z21.56-1991
	Gas-Fired Room Heaters, Vol. I, Vented Room Heaters (with 1989 and 1990 addenda)	AGA	ANSI Z21.11.1-1991
	Gas-Fired Room Heaters, Vol. II, Unvented Room Heaters (with 1990 addenda)	AGA	ANSI Z21.11.2-1992
	Requirements for Unvented Room Heaters Equipped with Oxygen Depletion Safety Shutoff Systems	AGA	2-79
	Gas-Fired Unvented Commercial and Industrial Heaters (with 1984 and 1989 addenda)	AGA	ANSI Z83.16-1982; Z83.16a-1984; Z83.16b-1989
	Requirements for Gas-Fired Vented Catalytic Type Room Heaters	AGA	1-81
	Requirements for High Pressure LP Infrared Poultry and Livestock Heating Systems	AGA	4-87
	Requirements for Direct Gas-Fired Circulating Heaters for Agricultural Buildings	AGA	5-88
	Requirements for Residential Radiant Tube Heaters	AGA	7-89
	Requirements for Gas-Fired Infrared Patio	AGA	5-90

Table 1 Codes and Standards Published by Various Societies and Associations (*Continued*)

Subject	Title	Publisher	Reference
Heaters (continued)	Gas-Fired Unvented Catalytic Room Heaters for Use with Liquefied Petroleum (LP) Gases	AGA	ANSI Z21.76-1991; Z21.76a-1992
	Desuperheater/Water Heaters	ARI	ARI 470-87
	Air Heaters	ASME	ANSI/ASME PTC 4.3-1968 (R 1991)
	Standards for Closed Feedwater Heaters, 5th ed. (1992)	HEI	HEI
	Fuel-Fired Heaters—Air Heating—for Construction and Industrial Machinery (1989)	SAE	SAE J1024 MAY89
	Motor Vehicle Heater Test Procedure (1982)	SAE	SAE J638 JUN82
	Electric Air Heaters (1980)	UL	ANSI/UL 1025-1991
	Electric Central Air Heating Equipment (1986)	UL	ANSI/UL 1096-1985
	Electric Dry Bath Heaters (1989)	UL	ANSI/UL 875-1989
	Electric Heaters for Use in Hazardous (Classified) Locations (1991)	UL	ANSI/UL 823-1990
	Electric Heating Appliances (1987)	UL	ANSI/UL 499-1987
	Electric Oil Heaters (1991)	UL	ANSI/UL 574-1990
	Fixed and Location-Dedicated Electric Room Heaters (1992)	UL	UL 2021
	Commercial-Industrial Gas Heating Equipment (1973)	UL	UL 795
	Movable and Wall- or Ceiling-Hung Electric Room Heaters (1992)	UL	UL 1278 (1992)
	Oil-Fired Air Heaters and Direct-Fired Heaters (1975)	UL	UL 733
	Oil-Burning Stoves (1973)	UL	UL 896
	Room Heaters, Solid Fuel-Type (1988)	UL	ANSI/UL 1482-1988
	Unvented Kerosene-Fired Room Heaters and Portable Heaters (1993)	UL	UL 647
Combination	Requirements for Gas-Fired Combination Space Heating/Water Heating Appliances	AGA	11-90
Heating	Commercial Applications Systems and Equipment	ACCA	Manual CS
	Residential Equipment Selection	ACCA	Manual S
	Determining the Required Capacity of Residential Space Heating and Cooling Appliances	CSA	CAN/CSA-F280-M90
	Automatic Flue-Pipe Dampers for Use with Oil-Fired Appliances	CSA	B140.14-M1979 (R 1991)
	Electric Duct Heaters	CSA	C22.2 No. 155-M1986 (R 1992)
	Heater Elements	CSA	C22.2 No.72-M1984 (R 1992)
	Oil-Fired Service Water Heaters and Swimming Pool Heaters	CSA	B140.12-1976 (R 1991)
	Performance Requirements for Electric Heating Line-Voltage Wall Thermostats	CSA	C273.4-M1978 (R 1992)
	Portable Industrial Oil-Fired Heaters	CSA	B140.8-1967 (R 1991)
	Portable Kerosine-Fired Heaters	CSA	CAN 3-B140.9.3 M86
	Electric Air Heaters	CSA	C22.2 No.46-M1988
	Advanced Installation Guide for Hydronic Heating Systems, 1991	HYDI	IBR 250
	Comfort Conditioning Heat Loss Calculation Guide	HYDI	IBR H-21 1984, IBR H-22 (1989)
	Installation Guide for Residential Hydronic Heating Systems, 6th ed. (1988)	HYDI	IBR 200
	Radiant Floor Heating (1993)	HYDI	IBR400
	Environmental System Technology (1984)	NEBB	NEBB
	Pulverized Fuel Systems	NFPA	ANSI/NFPA 8503-1992
	Aircraft Electrical Heating Systems (1965) (reaffirmed 1983)	SAE	SAE AIR860
	Performance Test for Air-Conditioned, Heated, and Ventilated Off-Road Self-Propelled Work Machines	SAE	SAE J1503
	Heating Value of Fuels	SAE	SAE J1498
	Selection and Application Guidelines for Diesel, Gasoline, and Propane Fire Liquid Cooled Engine Pre-Heaters	SAE	SAE J1350
	Electric Engine Preheaters and Battery Warmers for Diesel Engines	SAE	SAE J1310
	Heater, Aircraft, Internal Combustion Heat Exchanger Type	SAE	SAE AS8040
	Total Temperature Measuring Instruments (Turbine Powered Subsonic Aircraft)	SAE	SAE AS793
	Heaters, Aircraft, Internal Combustion Heat Exchanger Type	SAE	SAE AS143
	Heater, Airplane, Engine Exhaust Gas to Air Heat	SAE	SAE ARP86
	Installation, Heaters, Airplane, Internal Combustion Heater Exchange Type	SAE	SAE ARP266
	Performance of Low Pressure Ratio Ejectors for Engine Nacelle Cooling	SAE	SAE AIR1191
	HVAC Systems—Applications, 1st ed. (1986)	SMACNA	SMACNA
	Installation Standards for Residential Heating and Air Conditioning Systems (1988)	SMACNA	SMACNA
	Electric Baseboard Heating Equipment (1987)	UL	ANSI/UL 1042-1986
	Electric Central Air Heating Equipment (1986)	UL	ANSI/UL 1096-1985
	Heating and Cooling Equipment (1990)	CSA/UL	ANSI/UL 1995-1992
Humidifiers	Appliance Humidifiers	AHAM	ANSI/AHAM HU 1
	Central System Humidifiers	ARI	ANSI/ARI 610-89
	Self-Contained Humidifiers	ARI	ANSI/ARI 620-89
	Humidifiers and Evaporative Coolers	CSA	C22.2 No. 104-M1983
	Humidifiers (1987)	UL	ANSI/UL 998-1985

Table 1 Codes and Standards Published by Various Societies and Associations (*Continued*)

Subject	Title	Publisher	Reference
Ice Makers	Automatic Commercial Ice Makers	ARI	ARI 810-91
	Ice Storage Bins	ARI	ANSI/ARI 820-88
	Methods of Testing Automatic Ice Makers	ASHRAE	ANSI/ASHRAE 29-1988
	Refrigeration Equipment	CSA	CAN/CSA-C22.2 No. 120-M91
	Automatic Ice-Making Equipment	NSF	NSF-12
	Ice Makers (1992)	UL	ANSI/UL 563-1991
Incinerators	Incinerators, Waste and Linen Handling Systems and Equipment	NFPA	ANSI/NFPA 82-1990
	Residential Incinerators (1993)	UL	UL 791
Induction Units	Room Air-Induction Units	ARI	ANSI/ARI 445-87
	Frame Assignments for Alternating Current Integral-Horsepower Induction Motors	NEMA	NEMA MG 13-1984 (R 1990)
Industrial Duct	Rectangular Industrial Duct Construction (1980)	SMACNA	SMACNA
	Round Industrial Duct Construction (1977)	SMACNA	SMACNA
Insulation	Specification for Adhesives for Duct Thermal Insulation	ASTM	ASTM C916
	Specification for Thermal and Acoustical Insulation (Mineral Fiber, Duct Lining Material)	ASTM	ASTM C1071-86
	Test Method for Steady-State Heat Flux Measurements and Thermal Transmission Properties by Means of the Guarded Hot Plate Apparatus	ASTM	ASTM C177-85
	Test Method for Steady-State Heat Flux Measurements and Thermal Transmission Properties by Means of the Heat Flow Meter Apparatus	ASTM	ASTM C518-85
	Test Method for Steady-State Heat Transfer Properties of Horizontal Pipe Insulations	ASTM	ASTM C335-89
	Test Method for Steady-State and Thermal Performance of Building Assemblies by Means of a Guarded Hot Box	ASTM	ASTM C236-89
	Thermal Insulation, Mineral Fibre, for Buildings	CSA	A101-M1983
	National Commercial and Industrial Insulation Standards	MICA	MICA 1993
Louvers	Test Method for Louvers, Dampers, and Shutters	AMCA	AMCA 500-89
Lubricants	Method of Testing the Floc Point of Refrigeration Grade Oils	ASHRAE	ANSI/ASHRAE 86-1983
	Practice for Calculating Viscosity Index from Kinematic Viscosity at 40 and 100°C	ASTM	ASTM D2270-91
	Practice for Conversion of Kinematic Viscosity to Saybolt Universal Viscosity or to Saybolt Furol Viscosity	ASTM	ASTM D2161-87
	Method for Estimation of Molecular Weight of Petroleum Oils from Viscosity Measurements	ASTM	ASTM D2502-87
	Method for Separation of Representative Aromatics and Nonaromatics Fractions of High-Boiling Oils by Elution Chromatography	ASTM	ASTM D2549-91
	Classification for Viscosity System for Industrial Fluid Lubricants	ASTM	ASTM D2422-86
	Test Method for Carbon-Type Composition of Insulating Oils of Petroleum Origin	ASTM	ASTM D2140-86
	Test Method for Dielectric Breakdown Voltage of Insulating Liquids Using Disk Electrodes	ASTM	ASTM D877-87
	Test Method for Dielectric Breakdown Voltage of Insulating Oils of Petroleum Origin Using VDE Electrodes	ASTM	ASTM D1816-84a (90)
	Test Method for Mean Molecular Weight of Mineral Insulating Oils by the Cryoscopic Method	ASTM	ASTM D2224-78 (1983)
	Test Method for Molecular Weight of Hydrocarbons by Thermoelectric Measurement of Vapor Pressure	ASTM	ASTM D2503-82 (1987)
	Test Methods for Pour Point of Petroleum Oils	ASTM	ASTM D97-87
	Semiconductor Graphite	NEMA	NEMA CB 4-1989
Measurements	Procedure for Bench Calibration of Tank Level Gaging Tapes and Sounding Rules	ASME	ANSI MC88.2-1974 (R 1987)
	Standard Methods of Measuring and Expressing Building Energy Performance	ASHRAE	ANSI/ASHRAE 105-1984 (RA 90)
	Engineering Analysis of Experimental Data	ASHRAE	ASHRAE Guideline 2-1986 (RA 90)
	Standard Method for Temperature Measurement	ASHRAE	ANSI/ASHRAE 41.1-1986 (RA 91)
	Standard Method for Pressure Measurement	ASHRAE	ANSI/ASHRAE 41.3-1989
	Standard Method for Measurement of Proportion of Oil in Liquid Refrigerant	ASHRAE	ANSI/ASHRAE 41.4-1984
	Standard Method for Measurement of Moist Air Properties	ASHRAE	ANSI/ASHRAE 41.6-1982
	Standard Method for Measurement of Flow of Gas	ASHRAE	ANSI/ASHRAE 41.7-1984 (RA 91)
	Standard Methods of Measurement of Flow of Liquids in Pipes Using Orifice Flowmeters	ASHRAE	ANSI/ASHRAE 41.8-1989
	A Standard Calorimeter Test Method for Flow Measurement of a Volatile Refrigerant	ASHRAE	ANSI/ASHRAE 41.9-1988
	Glossary of Terms Used in the Measurement of Fluid Flow in Pipes	ASME	ANSI/ASME MFC-1M-1991
	Guide for Dynamic Calibration of Pressure Transducers	ASME	ANSI MC88-1-1972 (R 1987)

Table 1 Codes and Standards Published by Various Societies and Associations (*Continued*)

Subject	Title	Publisher	Reference
Measurements (continued)	Measurement of Fluid Flow in Pipes Using Orifice, Nozzle, and Venturi	ASME	ASME MFC-3M-1989
	Measurement of Fluid Flow in Pipes Using Vortex Flow Meters	ASME	ASME/ANSI MFC-6M-1987
	Measurement of Gas Flow by Means of Critical Flow Venturi Nozzles	ASME	ASME/ANSI MFC-7M-1987
	Measurement of Gas Flow by Turbine Meters	ASME	ANSI/ASME MFC-4M-1986 (R 1990)
	Measurement of Industrial Sound	ASME	ANSI/ASME PTC 36-1985
	Measurement of Liquid Flow in Closed Conduits Using Transit-Time Ultrasonic Flowmeters	ASME	ANSI/ASME MFC-5M-1985 (R 1989)
	Measurement of Rotary Speed	ASME	ANSI/ASME PTC 19.13-1961 (R 1986)
	Measurement Uncertainty	ASME	ANSI/ASME PTC 19.1-1985 (R 1990)
	Measurement Uncertainty for Fluid Flow in Closed Conduits	ASME	ANSI/ASME MFC-2M-1983 (R 1988)
	Pressure Measurement	ASME	ASME/ANSI PTC 19.2-1987
	Temperature Measurement	ANSI	ANSI/ASME PTC 19.3-1974 (R 1986)
Mobile Homes and Recreational Vehicles	Load Calculation for Residential Winter and Summer Air Conditioning	ACCA	Manual J
	Recreational Vehicle Cooking Gas Appliances (with 1989 addenda)	AGA	ANSI Z21.57-1990; Z21.57a-1991
	Mobile Homes	CSA	CAN/CSA-Z240 MH Series-92
	Mobile Home Parks	CSA	Z240.7.1-1972
	Oil-Fired Warm-Air Heating Appliances for Mobile Housing and Recreational Vehicles	CSA	B140.10-1974 (R 1991)
	Recreational Vehicle Parks	CSA	Z240.7.2-1972
	Recreational Vehicles	CSA	CAN/CSA-Z240 RV Series-M86 (R 1992)
	Gas Supply Connectors for Manufactured Homes	IAPMO	IAPMO TSC 9-1992
	Manufactured Home Installations	NCSBCS	ANSI A225.1-93
	Recreational Vehicles	NFPA	NFPA 501C-1993
	Plumbing System Components for Manufactured Homes and Recreational Vehicles	NSF	ANSI/NSF-24
	Gas Burning Heating Appliances for Mobile Homes and Recreational Vehicles (1965)	UL	UL 307B
	Gas-Fired Cooking Appliances for Recreational Vehicles (1993)	UL	UL 1075
	Liquid Fuel-Burning Heating Appliances for Manufactured Homes and Recreational Vehicles (1990)	UL	ANSI/UL 307A-1989
	Low Voltage Lighting Fixtures for Use in Recreational Vehicles (1993)	UL	UL 234
	Roof Jacks for Manufactured Homes and Recreational Vehicles (1990)	UL	UL 311
	Roof Trusses for Manufactured Homes (1990)	UL	UL 1298
	Shear Resistance Tests for Ceiling Boards for Manufactured Homes (1992)	UL	UL 1296
Motors and Generators	Steam Generating Units	ASME	ANSI/ASME PTC 4.1-1964 (R 1991)
	Testing of Nuclear Air-Treatment Systems	ASME	ANSI/ASME N510-1989
	Nuclear Power Plant Air Cleaning Units and Components	ASME	ANSI/ASME N509-1989
	Energy Efficiency Test Methods for Three-Phase Induction Motors (Efficiency Quoting Method and Permissible Efficiency Tolerance)	CSA	C390-93
	Motors and Generators	CSA	CAN/CSA-C22.2 No. 100-92
	Guide for the Development of Metric Standards for Motors and Generators	NEMA	NEMA 10407-1980
	Motion/Position Control Motors and Controls	NEMA	NEMA MG 7-1987
	Motors and Generators	NEMA	NEMA MG 1-1987 (R 1992)
	Renewal Parts for Motors and Generators—Performance, Selection, and Maintenance	NEMA	NEMA RPI-1981 (R 1987)
	Electric Motors (1989)	UL	ANSI/UL 1004-1988
	Electric Motors and Generators for Use in Hazardous (Classified) Locations (1989)	UL	UL 674
	Impedance-Protected Motors (1982)	UL	ANSI/UL 519-1984 (R 1989)
	Thermal Protectors for Motors (1991)	UL	ANSI/UL 547-1991
Operation and Maintenance	Preparation of Operating and Maintenance Documentaion for Building Systems	ASHRAE	ASHRAE Guideline 4P-1993
Outlets and Inlets	Method of Testing for Rating the Performance of Air Outlets and Inlets	ASHRAE	ANSI/ASHRAE 70-1991
Pipe, Tubing, and Fittings	Power Piping	ASME	ANSI/ASME B31.1-1989
	Refrigeration Piping	ASME	ASME/ANSI B31.5-1987
	Scheme for the Identification of Piping Systems	ASME	ANSI/ASME A13.1-1981 (R 1985)
	Specification for Acrylonitrile-Butadiene-Styrene (ABS) Plastic Pipe, Schedules 40 and 80	ASTM	ASTM D1527-89
	Specification for Polyethylene (PE) Plastic Pipe, Schedule 40	ASTM	ASTM D2104-90
	Specification for Polyvinyl Chloride (PVC) Plastic Pipe, Schedules 40, 80, and 120	ASTM	ASTM D1785-91
	Specification for Seamless Copper Pipe, Standard Sizes	ASTM	ASTM B42-89

Table 1 Codes and Standards Published by Various Societies and Associations (*Continued*)

Subject	Title	Publisher	Reference
Pipe, Tubing, and Fittings (continued)	Specification for Seamless Copper Tube for Air Conditioning and Refrigeration Field Service	ASTM	ASTM B280-88
	Specification for Welded Copper and Copper Alloy Tube for Air Conditioning and Refrigeration Service	ASTM	ASTM B640-90
	Standards of the Expansion Joint Manufacturers Association, Inc., 6th ed. (1993)	EJMA	EJMA
	Corrugated Polyolefin Coilable Plastic Utilities Duct	NEMA	NEMA TC 5-1990
	Corrugated Polyvinyl-Chloride (PVC) Coilable Plastic Utilities Duct	NEMA	NEMA TC 12-1991
	Electrical Nonmetallic Tubing (ENT)	NEMA	NEMA TC 13-1986
	Electrical Plastic Tubing (EPT) and Conduit Schedule EPC-40 and EPC-80	NEMA	NEMA TC 2-1990
	Extra-Strength PVC Plastic Utilities Duct for Underground Installation	NEMA	NEMA TC 8-1990
	Filament Wound Reinforced Thermosetting Resin Conduit and Fittings	NEMA	NEMA TC 14-1984 (R 1986)
	Fittings, Cast Metal Boxes, and Conduit Bodies for Conduit and Cable Assemblies	NEMA	NEMA FB 1-1988
	Fittings for ABS and PVC Plastic Utilities Duct for Underground Installation	NEMA	NEMA TC 9-1990
	Polyvinyl-Chloride (PVC) Externally Coated Galvanized Rigid Steel Conduit and Intermediate Metal Conduit	NEMA	NEMA RN 1-1989
	PVC and ABS Plastic Utilities Duct for Underground Installations	NEMA	NEMA TC 6-1990
	Smooth Wall Coilable Polyethylene Electrical Plastic Duct	NEMA	NEMA TC 7-1990
	National Fuel Gas Code	NFPA/ AGA/ANSI	ANSI/NFPA 54-1992/ ANSI Z223.1-1992
	Plastics Piping Components and Related Materials	NSF	ANSI/NSF-14
	Refrigeration Tube Fittings (1977)	SAE	ANSI/SAE J513 OCT77
	Seismic Restraint Manual Guidelines for Mechanical Systems (1991)	SMACNA	SMACNA
	Rubber Gasketed Fittings for Fire-Protection Service (1993)	UL	UL 213
	Tube Fittings for Flammable and Combustible Fluids, Refrigeration Service, and Marine Use (1993)	UL	UL 109
Plumbing	BOCA National Plumbing Code, 8th ed. (1990)	BOCA	BOCA
	Uniform Plumbing Code (1991) (with IAPMO Installation Standards)	IAPMO	IAPMO
	National Standard Plumbing Code (NSPC)	NAPHCC	NSPC 1993
	Standard Plumbing Code (1991 ed. with 1992/1993 revisions)	SBCCI	SBCCI
Pumps	Centrifugal Pumps	ASME	ASME PTC 8.2-1990
	Displacement Compressors, Vacuum Pumps and Blowers	ASME	ANSI/ASME PTC 9-1974 (R 1992)
	Liquid Pumps	CSA	CAN/CSA C.22.2 No.108-M89
	Performance Standard for Liquid Ring Vacuum Pumps, 1st ed. (1987)	HEI	HEI
	Centrifugal Pump	HI	HI 1.1-1.6 (1993)
	Vertical Pump	HI	HI 2.1-2.6 (1993)
	Rotary Pump	HI	HI 3.1-3.6 (1993)
	Sealless Rotary Pump	HI	HI 4.1-4.6 (1993)
	Sealless Centrifugal Pump	HI	HI 5.1-5.6 (1993)
	Reciprocating Power Pump	HI	HI 6.1-6.6 (1993)
	Controlled Volume Pump	HI	HI 7.1-7.5 (1993)
	Direct Acting (Steam) Pump	HI	HI 8.1-8.5 (1993)
	General Pump	HI	HI 9.1-9.6 (1993)
	Hydraulic Institute Engineering Data Book, 2nd Edition (1991)	HI	HI
	Circulation System Components and Related Materials for Swimming Pools, Spas/Hot Tubs	NSF	ANSI/NSF-50
	Swimming Pool Pumps, Filters and Chlorinators (1986)	UL	ANSI/UL 1081-1985
	Motor-Operated Water Pumps (1991)	UL	ANSI/UL 778-1991
	Pumps for Oil-Burning Appliances (1993)	UL	ANSI/UL 343-1992
Radiation	Ratings for Baseboard and Fin-Tube Radiation (1993)	HYDI	IBR
	Testing and Rating Code for Baseboard Radiation, 6th ed. (1990)	HYDI	IBR
	Testing and Rating Code for Finned-Tube Commercial Radiation (1990)	HYDI	IBR
Receivers	Refrigerant Liquid Receivers	ARI	ANSI/ARI 495-85
Refrigerant-Containing Components	Refrigerant-Containing Components for Use in Electrical Equipment	CSA	C22.2 No.140.3-M1987 (R 1993)
	Refrigerant-Containing Components and Accessories, Non-Electrical (1993)	UL	ANSI/UL 207-1993
Refrigerants	Performance of Refrigerant Recovery, Recycling, and/or Reclaim Equipment	ARI	ARI 740-91
	Specifications for Fluorocarbon Refrigerants	ARI	ANSI/ARI 700-88
	Methods of Testing Discharge Line Refrigerant-Oil Separators	ASHRAE	ANSI/ASHRAE 69-1990
	Number Designation and Safety Classification of Refrigerants	ASHRAE	ANSI/ASHRAE 34-1992
	Reducing Emission of Fully Halogenated Chlorofluorocarbon (CFC) Refrigerants in Refrigeration and Air-Conditioning Equipment and Applications	ASHRAE	ASHRAE Guideline 3-1990

Table 1 Codes and Standards Published by Various Societies and Associations (*Continued*)

Subject	Title	Publisher	Reference
Refrigerants (continued)	Refrigeration Oil Description	ASHRAE	ANSI/ASHRAE 99-1981 (RA 87)
	Sealed Glass Tube Method to Test the Chemical Stability of Material for Use Within Refrigerant Systems	ASHRAE	ANSI/ASHRAE 97-1983 (RA 89)
	Refrigerant Recovery/Recycling Equipment (1989)	UL	ANSI/UL 1963-1991
	Recommended Service Procedure for the Containment of HFC-134a	SAE	SAE J2211
	HFC-134a Recycling Equipment for Mobile Air Conditioning Systems	SAE	SAE J2210
	CFC-12 (R-12) Extraction Equipment for Mobile Automotive Air Conditioning Systems	SAE	SAE J2209
	HFC-134a (R-134a) Service Hose Fittings for Automotive Air Conditioning Service Equipment	SAE	SAE J2197
	Standard of Purity for Recycled HFC-134a for Use in Mobile Air Conditioning Systems	SAE	SAE J2099
	Recommended Service Procedure for the Containment of R-12	SAE	SAE J1989
Refrigeration	Safety Code for Mechanical Refrigeration	ASHRAE	ANSI/ASHRAE 15-1992
	Refrigeration Equipment	CSA	CAN/CSA-C22.2 No.120-M91
	Equipment, Design and Installation of Ammonia Mechanical Refrigeration Systems	IIAR	ANSI/IIAR 2-1984
	Refrigerated Medical Equipment (1978)	UL	UL 416
Refrigeration Systems			
Steam Jet	Ejectors	ASME	ASME PTC 24-1976 (R 1982)
	Standards for Steam Jet Vacuum Systems, 4th ed. (1988)	HEI	HEI
Transport	Mechanical Transport Refrigeration Units	ARI	ARI 1110-92
	Mechanical Refrigeration Installations on Shipboard	ASHRAE	ANSI/ASHRAE 26-1978 (RA 85)
	General Requirements for Application of Vapor Cycle Refrigeration Systems for Aircraft (1973) (reaffirmed 1983)	SAE	SAE ARP731A
	Safety Practices for Mechanical Vapor Compression Refrigeration Equipment or Systems Used to Cool Passenger Compartment of Motor Vehicles (1987)	SAE	SAE J639 JAN87
Refrigerators	Method of Testing Open Refrigerators for Food Stores	ASHRAE	ANSI/ASHRAE 72-1983
	Methods of Testing Closed Refrigerators	ASHRAE	ANSI/ASHRAE 117-1992
Commercial	Food Carts	ANSI/NSF	NSF 59
	Food Equipment	ANSI/NSF	NSF-2
	Food Service Refrigerators and Storage Freezers	ANSI/NSF	NSF 7
	Soda Fountain and Luncheonette Equipment	NSF	NSF 1
	Commercial Refrigerators and Freezers (1992)	UL	ANSI/UL 471-1991
	Refrigerating Units (1989)	UL	ANSI/UL 427-1989
	Refrigeration Unit Coolers (1980)	UL	ANSI/UL 412-1984
Household	Refrigerators Using Gas Fuel	AGA	ANSI Z21.19-1990; Z21.19a-1992
	Household Refrigerators and Household Freezers	AHAM	AHAM HRF 1
	Capacity Measurement and Energy Consumption Test Methods for Refrigerators, Combination Refrigerator-Freezers, and Freezers	CSA	CAN/CSA C300-M91
	Household Refrigerators and Freezers (1983)	UL	ANSI/UL 250-1984
Roof Ventilators	Commercial Low Pressure, Low Velocity Duct Systems	ACCA	Manual Q
	Power Ventilators (1984)	UL	ANSI/UL 705-1984
Solar Equipment	Method of Measuring Solar-Optical Properties of Materials	ASHRAE	ANSI/ASHRAE 74-1988
	Method of Testing to Determine the Thermal Performance of Flat-Plate Solar Collectors Containing a Boiling Liquid Materials	ASHRAE	ANSI/ASHRAE 109-1986 (RA 90)
	Method of Testing to Determine the Thermal Performance of Solar Collectors	ASHRAE	ANSI/ASHRAE 93-1986 (RA 91)
	Methods of Testing to Determine the Thermal Performance of Solar Domestic Water Heating Systems	ASHRAE	ASHRAE 95-1981 (RA 87)
	Methods of Testing to Determine the Thermal Performance of Unglazed Flat-Plate Liquid-Type Solar Collectors	ASHRAE	ANSI/ASHRAE 96-1980 (RA 90)
Solenoid Valves	Solenoid Valves for Use with Volatile Refrigerants	ARI	ANSI/ARI 760-87
Sound Measurement	Methods for Calculating Fan Sound Ratings from Laboratory Test Data	AMCA	AMCA Standard 301-90
	Laboratory Method of Testing—In-Duct Sound Power Measurement Procedure for Fans	AMCA	ANSI/AMCA 330-86
	Reverberant Room Method for Sound Testing of Fans	AMCA	AMCA 300-85
	Application of Sound Rated Outdoor Unitary Equipment	ARI	ARI 275-84
	Method of Measuring Machinery Sound within Equipment Rooms	ARI	ARI 575-87
	Method of Measuring Sound and Vibration of Refrigerant Compressors	ARI	ANSI/ARI 530-89
	Rating the Sound Levels and Sound Transmission Loss of Packaged Terminal Equipment	ARI	ANSI/ARI 300-88
	Sound Rating of Large Outdoor Refrigerating and Air-Conditioning Equipment	ARI	ARI 370-86
	Sound Rating of Non-Ducted Indoor Air-Conditioning Equipment	ARI	ARI 350-86
	Sound Rating of Outdoor Unitary Equipment	ARI	ARI 270-84

Table 1 Codes and Standards Published by Various Societies and Associations (*Continued*)

Subject	Title	Publisher	Reference
Sound Measurement (continued)	Guidelines for the Use of Sound Power Standards and for the Preparation of Noise Test Codes	ASA	ASA 94; ANSI S12.30-1990
	Method for the Calibration of Microphones (reaffirmed 1986)	ASA	ANSI S1.10-1966 (R 1986)
	Specification for Sound Level Meters (reaffirmed 1986)	ASA	ASA 47; ANSI S1.4-1983; ANSI S1.4A-1985
	Measurement of Industrial Sound	ASME	ASME/ANSI PTC 36-1985
	Procedural Standards for Measuring Sound and Vibration	NEBB	NEBB-1977
	Sound and Vibration in Environmental Systems	NEBB	NEBB-1977
	Sound Level Prediction for Installed Rotating Electrical Machines	NEMA	NEMA MG 3-1974 (R 1990)
Space Heaters	Electric Air Heaters	CSA	C22.2 No. 46-M1988
	Electric Air Heaters (1980)	UL	ANSI/UL 1025-1991
	Fixed and Location-Dedicated Electric Room Heaters (1992)	UL	UL 2021
	Movable and Wall- or Ceiling-Hung Electric Room Heaters (1992)	UL	UL 1278
	Gas-Fired Room Heaters, Vol. I, Vented Room Heaters	AGA	ANSI Z21.11.1-1991
	Gas-Fired Room Heaters, Vol. II, Unvented Room Heaters	AGA	ANSI Z21.11.2-1992
Symbols	Graphic Electrical Symbols for Air-Conditioning and Refrigeration Equipment	ARI	ARI 130-88
	Graphic Symbols for Electrical and Electronic Diagrams	IEEE	ANSI/IEEE 315-1975
	Graphic Symbols for Heating, Ventilating, and Air Conditioning	ASME	ANSI/ASME Y32.2.4-1949 (R 1984)
	Graphic Symbols for Pipe Fittings, Valves and Piping	ASME	ANSI/ASME Y32.2.3-1949 (R 1988)
	Graphic Symbols for Plumbing Fixtures for Diagrams used in Architecture and Building Construction	ASME	ANSI/ASME Y32.4-1977 (R 1987)
	Symbols for Mechanical and Acoustical Elements as used in Schematic Diagrams	ASME	ANSI/ASME Y32.18-1972 (R 1985)
Testing and Balancing	Site Performance Test Standard—Power Plant and Industrial Fans	AMCA	AMCA 803-87
	General Pump (Including "Measurement of Airborne Sound")	HI	HI 9.1-9.6 (1993)
	Procedural Standards for Certified Testing of Cleanrooms (1988)	NEBB	NEBB-1988
	Procedural Standards for Testing, Adjusting, Balancing of Environmental Systems, 5th ed. (1991)	NEBB	NEBB-1991
	HVAC Systems—Testing, Adjusting and Balancing (1993)	SMACNA	SMACNA
Terminals, Wiring	Quick Connect Terminals	NEMA	NEMA DC 2-1982 (R 1988)
	Electrical Quick-Connect Terminals (1991)	UL	UL 310
	Equipment Wiring Terminals for Use with Aluminum and/or Copper Conductors (1988)	UL	ANSI/UL 486E-1987
	Splicing Wire Connectors (1991)	UL	ANSI/UL 486C-1990
	Wire Connectors and Soldering Lugs for Use with Copper Conductors (1991)	UL	ANSI/UL 486A-1990
	Wire Connectors for Use with Aluminum Conductors (1991)	UL	ANSI/UL 486B-1990
Thermal Storage	Commissioning of HVAC Systems	ASHRAE	ASHRAE Guideline 1-1989
	Metering and Testing Active Sensible Thermal Energy Storage Devices Based on Thermal Performance	ASHRAE	ANSI/ASHRAE 94.3-1986 (RA 90)
	Method of Testing Active Latent Heat Storage Devices Based on Thermal Performance	ASHRAE	ANSI/ASHRAE 94.1-1985 (RA 91)
	Methods of Testing Thermal Storage Devices with Electrical Input and Thermal Output Based on Thermal Performance	ASHRAE	ANSI/ASHRAE 94.2-1981 (RA 89)
	Practices for Measurement, Testing and Balancing of Building Heating, Ventilation, Air-Conditioning, and Refrigeration Systems	ASHRAE	ANSI/ASHRAE 111-1988
Turbines	Land Based Steam Turbine Generator Sets	NEMA	NEMA SM 24-1991
	Steam Turbines for Mechanical Drive Service	NEMA	NEMA SM23-1991
Unit Heaters	Gas Unit Heaters	AGA	ANSI Z83.8-1990; Z83.8a-1990
	Oil-Fired Unit Heaters (1988)	UL	ANSI/UL 731-1987
Valves	Automatic Gas Valves for Gas Appliances	AGA	ANSI Z21.21-1987; Z21.21a-1989; Z21.21b-1992
	Manually Operated Gas Valves for Appliances, Appliance Connection Valves, and Hose End Valves	AGA	ANSI Z21.15-1992
	Relief Valves and Automatic Gas Shutoff Devices for Hot Water Supply Systems	AGA	ANSI Z21.22-1986; Z21.22a-1990
	Combination Gas Controls for Gas Appliances	AGA	ANSI Z21.78-1992
	Requirements for Manually Operated Gas Valves for Use in House Piping Systems	AGA	3-88
	Requirements for Automatic Non-Shutoff Modulating Gas Valves	AGA	1-92
	Requirements for Manually Operated Valves for High Pressure Natural Gas	AGA	2-93
	Requirements for Gas Operated Valves for High Pressure Natural Gas	AGA	3-93
	Refrigerant Access Valves and Hose Connectors	ARI	ANSI/ARI 720-88
	Refrigerant Pressure Regulating Valves	ARI	ARI 770-84

Table 1 Codes and Standards Published by Various Societies and Associations (*Continued*)

Subject	Title	Publisher	Reference
Valves (continued)	Solenoid Valves for Use with Volatile Refrigerants	ARI	ANSI/ARI 760-87
	Thermostatic Refrigerant Expansion Valves	ARI	ANSI/ARI 750-87
	Methods of Testing Nonelectric, Nonpneumatic Thermostatic Radiator Valves	ASHRAE	ANSI/ASHRAE 102-1983 (RA 89)
	Face-to-Face and End-to-End Dimensions of Valves	ASME	ASME/ANSI B16.10-1986
	Large Metallic Valves for Gas Distribution (Manually Operated, NPS-2 1/2 to 12, 125 psig Maximum)	ASME	ANSI/ASME B16.38-1985
	Manually Operated Metallic Gas Valves for Use in Gas Piping Systems up to 125 psig	ASME	ANSI B16.33-1990
	Manually Operated Thermoplastic Gas Shutoffs and Valves in Gas Distribution Systems	ASME	ANSI/ASME B16.40-1985
	Safety and Relief Valves	ASME	ANSI/ASME PTC25.3-1988
	Valves—Flanged Threaded, and Welding End	ASME	ANSI/ASME B16.34-1988
	Electrically Operated Valves (1982)	UL	ANSI-UL 429-1988
	Pressure Regulating Valves for LP-Gas (1985)	UL	ANSI/UL 144-1985
	Safety Relief Valves for Anhydrous Ammonia and LP-Gas (1993)	UL	UL 132
	Valves for Anhydrous Ammonia and LP-Gas (Other than Safety Relief) (1993)	UL	UL 125
	Valves for Flammable Fluids (1980)	UL	UL 842
Vending Machines	Methods of Testing Pre-Mix and Post-Mix Soft Drink Vending and Dispensing Equipment	ASHRAE	ANSI/ASHRAE 91-1976 (RA 91)
	Vending Machines	CSA	CAN/CSA-C22.2 No.128-M90
	Vending Machines for Food and Beverages	NSF	ANSI/NSF-25
	Refrigerated Vending Machines (1989)	UL	ANSI/UL 541-1988
Vent Dampers	Automatic Vent Damper Devices for Use with Gas-Fired Appliances	AGA	ANSI Z21.66-1988; Z21.66a-1991; Z21.66b-1991
	Vent or Chimney Connector Dampers for Oil-Fired Appliances (1988)	UL	ANSI/UL 17-1988
Venting	Draft Hoods	AGA	ANSI Z21.12-1990; Z21.12a-1993
	National Fuel Gas Code	AGA	ANSI Z223.1-1992
	Requirements for Electrically Operated Automatic Combustion and Ventilation Air Control Devices for Use with Gas-Fired Appliances	AGA	1-88
	Requirements for Mechanical Venting Systems	AGA	6-90
	Chimneys, Fireplaces, Vents and Solid Fuel Burning Appliances	NFPA	ANSI/NFPA 211-1992
	Explosion Prevention Systems	NFPA	ANSI/NFPA 69-1992
	Guide for Steel Stack Design and Construction (1983)	SMACNA	SMACNA
	Draft Equipment (1993)	UL	UL 378
	Gas Vents (1991)	UL	ANSI/UL 441-1991
	Low-Temperature Venting Systems, Type L (1986)	UL	ANSI/UL 641-1985
Ventilation	Commercial Low Pressure, Low Velocity Duct Systems	ACCA	Manual Q
	Guide for Testing Ventilation Systems	ACGIH	ACGIH
	Industrial Ventilation (1992)	ACGIH	ACGIH
	Method of Determining Air Change Rates in Detatched Dwellings	ASHRAE	ANSI/ASHRAE 136-1993
	Method of Testing for Room Air Diffusion	ASHRAE	ANSI/ASHRAE 113-1990
	Ventilation for Acceptable Indoor Air Quality	ASHRAE	ANSI/ASHRAE 62-1989
	Residential Mechanical Ventilation Systems	CSA	CAN/CSA F326-M91
	Ventilation Directory (1990)	NCSBCS	NCSBCS
	Parking Structures; Repair Garages	NFPA	ANSI/NFPA 88A-1991; 88B-1991
	Removal of Smoke and Grease-Laden Vapors from Commercial Cooking Equipment	NFPA	ANSI/NFPA 96-1991
	Food Equipment	NSF	ANSI/NSF-2
	Class II (Laminar Flow) Biohazard Cabinetry	NSF	NSF-49
	Performance Test for Air-Conditioned, Heated, and Ventilated Off-Road Self-Propelled Work Machines	SAE	SAE J1503
	Test Procedure for Battery Flame Retardant Venting Systems	SAE	SAE J1495
	Hose, Air Duct, Flexible Nonmetallic, Aircraft	SAE	SAE AS1501
	High Pressure Oxygen System Filler Valve	SAE	SAE AS1225
	Heater, Airplane, Engine Exhaust Gas to Air Heat	SAE	SAE ARP86
	Aerothermodynamic Systems Engineering and Design	SAE	SAE AIR1168/3
Water Heaters	Gas Water Heaters, Vol. I, Storage Water Heaters with Input Ratings of 75,000 Btu per Hour or Less	AGA	ANSI Z21.10.1-1990; Z21.10.1a-1991; Z21.10.1b-1992
	Gas Water Heaters, Vol. III, Storage, with Input Ratings Above 75,000 Btu per Hour, Circulating and Instantaneous Water Heaters	AGA	ANSI Z21.10.3-1990; Z21-10.3a-1990; Z21.10.3b-1992
	Requirements for Non-Metallic Dip Tubes for Use in Gas-Fired Water Heaters	AGA	1-89
	Requirements for Indirect Water Heaters for Use with External Heat Source	AGA	1-91

Table 1 Codes and Standards Published by Various Societies and Associations (*Continued*)

Subject	Title	Publisher	Reference
Water Heaters (continued)	Methods of Testing for Rating Commercial Gas, Electric and Oil Water Heaters	ASHRAE	ANSI/ASHRAE 118.1-1993
	Methods of Testing for Rating Residential Water Heaters	ASHRAE	ANSI/ASHRAE 118.2-1993
	Methods of Testing to Determine the Thermal Performance of Solar Domestic Water Heating Systems	ASHRAE	ANSI/ASHRAE 95-1981 (RA 87)
	Methods of Testing for Rating Combination Space-Heating and Water-Heating Appliances	ASHRAE	ANSI/ASHRAE 124-1991
	Construction and Test of Electric Storage-Tank Water Heaters	CSA	CAN/CSA-C22.2 No. 110-M90
	Performance of Electric Storage Tank Water Heaters	CSA	CAN/CSA-C191-SERIES-M90
	Oil Burning Stoves and Water Heaters	CSA	B140.3-1962 (R 1991)
	Oil-Fired Service Water Heaters and Swimming Pool Heaters	CSA	B140.12-1976 (R 1991)
	Water Heaters, Hot Water Supply Boilers, and Heat Recovery Equipment	NSF	NSF-5
	Commercial-Industrial Gas Heating Equipment (1973)	UL	UL 795
	Electric Booster and Commercial Storage Tank Water Heaters (1988)	UL	ANSI/UL 1453-1987
	Household Electric Storage Tank Water Heaters (1989)	UL	ANSI/UL 174-1989
	Oil-Fired Storage Tank Water Heaters (1988)	UL	ANSI/UL 732-1987
Woodburning Appliances	Method of Testing for Performance Rating of Woodburning Appliances	ASHRAE	ANSI/ASHRAE 106-1984
	Installation Code for Solid Fuel Burning Appliances and Equipment	CSA	CAN/CSA-B365-M91
	Solid-Fuel-Fired Central Heating Appliances	CSA	CAN/CSA-B366.1-M91
	Chimneys, Fireplaces, Vents and Solid Fuel Burning Appliances	NFPA	ANSI/NFPA 211-1992
	Commercial Cooking, Rethermalization and Powered Hot Food Holding and Transport Equipment	NSF	ANSI/NSF-4
	Room Heaters, Solid-Fuel Type (1988)	UL	ANSI/UL 1482-1988

ABBREVIATIONS AND ADDRESSES

ABMA	American Boiler Manufacturers Association, 950 N. Glebe Road, Suite 160, Arlington, VA 22203
ACCA	Air Conditioning Contractors of America, 1513 16th Street, NW, Washington, D.C. 20036
ACGIH	American Conference of Governmental Industrial Hygienists, 6500 Glenway Avenue, Building D-7, Cincinnati, OH 45211
ADC	Air Diffusion Council, Suite 200, 111 E. Wacker Drive, Chicago, IL 60601
AGA	American Gas Association, 1515 Wilson Boulevard, Arlington, VA 22209
AHAM	Association of Home Appliance Manufacturers, 20 N. Wacker Drive, Chicago, IL 60606
AIHA	American Industrial Hygiene Association, 2700 Prosperity Avenue, Suite 250, Fairfax, VA 22031
AMCA	Air Movement and Control Association, Inc., 30 W. University Drive, Arlington Heights, IL 60004-1893
ANSI	American National Standards Institute, 11 West 42nd Street, New York, NY 10036
ARI	Air-Conditioning and Refrigeration Institute, 4301 North Fairfax Drive, Suite 425, Arlington, VA 22203
ASA	Acoustical Society of America, Standards Secretariat, 120 Wall Street, New York, NY 10005
ASHRAE	American Society of Heating, Refrigerating and Air-Conditioning Engineers, Inc., 1791 Tullie Circle, NE, Atlanta, GA 30329
ASME	The American Society of Mechanical Engineers, 345 E. 47 Street, New York, NY 10017
	For ordering publications: ASME Marketing Department, Box 2350, Fairfield, NJ 07007-2350
ASTM	American Society for Testing and Materials, 1916 Race Street, Philadelphia, PA 19103
BOCA	Building Officials and Code Administrators International, Inc., 4051 W. Flossmoor Road, Country Club Hills, IL 60478-5795
BSI	British Standards Institution, 2 Park Street, London, W1A 2BS, England
CABO	Council of American Building Officials, 5203 Leesburg Pike, Suite 708, Falls Church, VA 22041
CAGI	Compressed Air and Gas Institute, 1300 Sumner Avenue, Cleveland, OH 44115
CSA	Canadian Standards Association, 178 Rexdale Boulevard, Rexdale (Toronto), Ontario M9W 1R3, Canada
CTI	Cooling Tower Institute, P.O. Box 73383, Houston, TX 77273
EJMA	Expansion Joint Manufacturers Association, Inc., 25 N. Broadway, Tarrytown, NY 10591
HEI	Heat Exchange Institute, 1300 Sumner Ave., Cleveland, OH 44115
HI	Hydraulic Institute, 9 Sylvan Way, Suite 180, Parsippany, NJ 07054-3802
HYDI	Hydronics Institute, 35 Russo Place, Berkeley Heights, NJ 07922
IAPMO	International Association of Plumbing and Mechanical Officials, 20001 Walnut Drive South, Walnut, CA 91789-2825
ICBO	International Conference of Building Officials, 5360 Workman Mill Road, Whittier, CA 90601
IFCI	International Fire Code Institute, 5360 Workman Mill Road, Whittier, CA 90601-2298
IIAR	International Institute of Ammonia Refrigeration, 111 East Wacker Drive, Chicago, IL 60601
MICA	Midwest Insulation Contractors Association, 2017 South 139th Circle, Omaha, NE 68144
NAPHCC	National Association of Plumbing-Heating-Cooling Contractors, P.O. Box 6808, Falls Church, VA 22040
NCSBCS	National Conference of States on Building Codes and Standards, 505 Huntmar Park Drive, Suite 210, Herndon, VA 22070
NEBB	National Environmental Balancing Bureau, 1385 Piccard Drive, Rockville, MD 20850
NEMA	National Electrical Manufacturers Association, 2101 L Street, NW, Suite 300, Washington, D.C. 20037-1526
NFPA	National Fire Protection Association, 1 Batterymarch Park, P.O. Box 9101, Quincy, MA 02269-9101
NSF International	National Sanitation Foundation, P.O. Box 130140, Ann Arbor, MI 48113-0140
SAE	Society of Automotive Engineers, 400 Commonwealth Drive, Warrendale, PA 15096
SBCCI	Southern Building Code Congress International, Inc., 900 Montclair Road, Birmingham, AL 35213-1206
SMACNA	Sheet Metal and Air Conditioning Contractors' National Association, 4201 Lafayette Center Drive, Chantilly, VA 22021
TEMA	Tubular Exchanger Manufacturers Association, Inc., 25 N. Broadway, Tarrytown, NY 10591
UL	Underwriters Laboratories Inc., 333 Pfingsten Road, Northbrook, IL 60062-2096

ADDITIONS AND CORRECTIONS

This section supplements the current handbooks and notes technical errors found in the series. Occasional typographical errors and nonstandard symbol labels will be corrected in future volumes. The authors and editor encourage you to notify them if you find other technical errors. Please send corrections to: Handbook Editor, ASHRAE, 1791 Tullie Circle NE, Atlanta, GA 30329.

1991 HVAC Applications

p. 47.5, Equation (5). Include units conversion multiplier with equation so it reads:

$$Q = 776\ CA\sqrt{2\ \Delta p/\rho}$$

1992 HVAC Systems and Equipment

p. 31.4, Equation (10). The term πd_i^2 is missing from the denominator of the last term in parenthesis in the equation. The equation should read:

$$\Delta p = \frac{k\rho_m V^2}{5.2(2g)} = \frac{k}{5.2(2g)}\left(\frac{T_m}{1.325B}\right)\left(\frac{144 \times 4IM}{3.6 \times 10^6\ \pi d_i^2}\right)^2 \quad (5)$$

1993 Fundamentals

p. 3.20, 2nd column. Revise unnumbered equation in the middle of the column to read:

Schmidt's empirical solution is given by:

$$\Phi = \tanh\ (mr_i\Phi)/(mr_i\Phi)$$

p. 3.20, 2nd column, 7th line up. Delete superscript 2 from the variable a under the square root sign so the line reads:

by Figure 24 as $a/2$ or b (whichever is less) and 0.5 $\sqrt{(a/2)^2 + b^2}$,

p. 3.20, Figure 23. The dimension L is half the horizontal distance between two tubes. The vertical line marking the right end of L should be moved to the right edge of the crosshatched area.

p. 6.5, Table 2. Enthalpy of moist air at saturation (h_s) at 143 °F should read 224.233, rather than 244.233.

p. 6.9, Table 3. Enthalpy of vaporization (h_{fg}) at 121 °F should read 1024.63, rather than 1023.62. Also, enthalpy of vaporization at 123 °F should read 1023.47 rather than 1024.47.

p. 6.20, 2d column. $h_{fg} = h_g - h_f$ = enthalpy of vaporization, Btu/lb

p. 8.7, Table 4. Change the heading in the second column to read Btu/(h · ft^2).

p. 16.3, Table 1. Correct Reference Number to read 152a, not T152a.

p. 20.15, 1st column. Delete last sentence in paragraph before Example 1. Chapter 19 no longer includes a vapor flow calculation example.

p. 20.19, Equation (21). A division symbol is missing in the last part of the equation. It should read:

$$q = (t_b - t_p)/R_s = (t_p - t_o)/R_i \quad (21)$$

p. 20.19, Figure 12. Change two variable to read t_b and t_p, instead of D_b and D_p.

p. 21.13, 2nd column. Change last sentence in the column to read:

Table 4 shows estimated flow rates needed to prevent freezing in insulated pipes.

p. 21.14, Equation (1). The variable H in Equation (1) should be θ. Also, b should be 62.4 lb/ft^3, which is the density of water near freezing.

p. 22.9, Table 4. Change C-value of 0.375 in. plywood to 1.69 instead of 1.59.

p. 22.9, Table 4. Under *Siding*. change Aluminum or Steel^m, over sheathing to read, Aluminum, vinyl, or steel^m, over sheathing.

p. 22.9, Table 4. Change C-value of hollow-backed aluminum, vinyl, or steel siding to 1.64 instead of 1.61. Also, add superscript n to specific heat value of 0.29.

p. 22.9. Table 4. Add a new footnote n to read, "Vinyl specific heat = 0.25 Btu/lb · °F." Then change current footnotes n to o and o to p.

p. 22.9, Table 4. Change first part of footnote m to read, "Values for metal or vinyl siding..."

p. 23.2, Equation (2). Add ρ = air density, lb$_m$/ft^3 to list of equation variables.

p. 23.16, Table 4. Change the area ratio for tight elevator shaft walls to read 0.18×10^3.

p. 24.20. Correct spelling for Amadabad, India.

p. 25.12, 2nd column. The variable F_2 that is listed for Equation (6) should refer to Table 16, not Table 5.

p. 26.5, Equation (4). Radiation difference term should be shown as ΔR in list of variables.

p. 26.6, Table 1. Change equation table heading to read:

$$t_e = t_o + \alpha I_t/h_o - \epsilon\Delta R/h_o$$

p. 26.8, Table 3. The latent heat gains listed in Table 3 are not correct. Please enter the correct values or replace the table with the following one.

Table 3 Rates of Heat Gain from Occupants of Conditioned Spaces

Degree of Activity		Total Heat, Btu/h Adult Male	Adjusted, M/F[a]	Sensible Heat, Btu/h	Latent Heat, Btu/h	% Sensible Heat that is Radiant[b] Low V	High V
Seated at theater	Theater, matinee	390	330	225	105		
Seated at theater, night	Theater, night	390	350	245	105	60	27
Seated, very light work	Offices, hotels, apartments	450	400	245	155		
Moderately active office work	Offices, hotels, apartments	475	450	250	200		
Standing, light work; walking	Department store; retail store	550	450	250	200	58	38
Walking, standing	Drug store, bank	550	500	250	250		
Sedentary work	Restaurant[c]	490	550	275	275		
Light bench work	Factory	800	750	275	475		
Moderate dancing	Dance hall	900	850	305	545	49	35
Walking 3 mph; light machine work	Factory	1000	1000	375	625		
Bowling[d]	Bowling alley	1500	1450	580	870		
Heavy work	Factory	1500	1450	580	870	54	19
Heavy machine work; lifting	Factory	1600	1600	635	965		
Athletics	Gymnasium	2000	1800	710	1090		

Notes:
1. Tabulated values are based on 75 °F room dry-bulb temperature. For 80 °F room dry bulb, the total heat remains the same, but the sensible heat values should be decreased by approximately 20%, and the latent heat values increased accordingly.
2. Also refer to Table 4, Chapter 8, for additional rates of metabolic heat generation.
3. All values are rounded to nearest 5 Btu/h.

[a]Adjusted heat gain is based on normal percentage of men, women, and children for the application listed, with the postulate that the gain from an adult female is 85% of that for an adult male, and that the gain from a child is 75% of that for an adult male.
[b]Values approximated from data in Table 6, Chapter 8, where V is air velocity with limits shown in that table.
[c]Adjusted heat gain includes 60 Btu/h for food per individual (30 Btu/h sensible and 30 Btu/h latent).
[d]Figure one person per alley actually bowling, and all others as sitting (400 Btu/h) or standing or walking slowly (550 Btu/h).

p. 26.10, 2nd column. In line above equation (20), change (FFL) to (F_{FL}).

p.26.11, Table 6. In last column, change the Load Factor for a fryer from 0.056 to 0.006.

p. 26.15, 2nd column. Line 2 above the heading Special conditions. should refer to Table 9 in Chapter 22, not Table 7.

p. 30.15, Equation (10). Change θ_∞ to read t_∞.

p. 31.6, Equation (1). In list of variables, change t_o to read t_x.

p. 31.7, Table 2. The heading for the last column should read **Range of T_{50}/L**, not **Range of $T_{0.25}/L$**. Also, delete the * from the values in the third column for Ceiling slot diffusers.

p. 31.11, Equation (14). The equation should read:

$$\frac{Q_x}{Q_0} = \sqrt{\frac{2X}{K'H_o}} \qquad (14)$$

p. 33.19, Table 21. The pipeline capacities listed are for pressure drops of 0.3 in. of water. Correct Note 1 to read a pressure drop of 0.3 in. of water instead of 0.5 in. of water.

p. 34.5, 1st column. In the Mathematical Symbols table, the symbol for "a raised to the power n" should read a^n, not a_n.

p. 35.1. To convert footcandle to lux, multiply by 10.76, not 1.076.

COMPOSITE INDEX

ASHRAE HANDBOOK SERIES

This index covers the current Handbook volumes published by ASHRAE. Listings from each volume are identified as follows:

R = 1994 Refrigeration

F = 1993 Fundamentals

S = 1992 Systems and Equipment

A = 1991 Applications

The index is alphabetized in a *word-by-word* format; for example, *Air diffusers* is listed before *Aircraft*, and *Heat flow* is listed before *Heaters.*

Note that the code for a volume includes the chapter number followed by a decimal point and the beginning page or page number range within the chapter. For example, R32.4 means the information may be found in the Refrigeration volume, chapter 32, starting on page 4.